D0985012

WORLD *of*

Robyn V. Young, *Editor*
Suzanne Sessine, *Assistant Editor*

Detroit
San Francisco
London
Boston
Woodbridge, CT

GALE GROUP STAFF

Chris Jeryan, Kristine Krapp, Jacqueline Longe, Kimberley McGrath, *Contributing Editors*

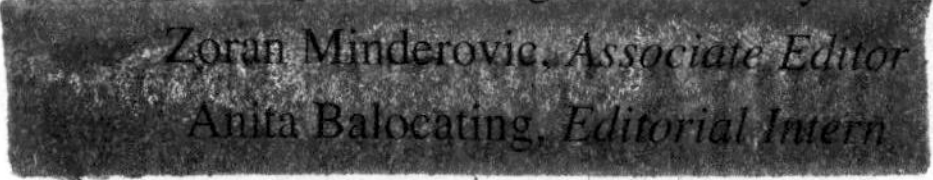

Zoran Minderovic, *Associate Editor*
Anita Balocating, *Editorial Intern*

Mary K. Fyke, *Editorial Technical Specialist*

Maria Franklin, *Permissions Manager*
Margaret A. Chamberlain, *Permissions Specialist*
Shalice Shah-Caldwell, *Permissions Associate*

Mary Beth Trimper, *Composition Manager*
Evi Seoud, *Assistant Production Manager*
Wendy Blurton, *Senior Buyer*

Cynthia D. Baldwin, *Product Design Manager*
Michelle DiMercurio, *Art Director*
Barbara Yarrow, *Graphic Services Manager*
Randy Bassett, *Image Database Supervisor*
Mike Logusz, *Imaging Specialist*
Pamela A. Reed, *Photography Coordinator*
Leitha Etheridge-Sims, *Junior Image Cataloguer*
Indexing provided by Synapse, the Knowledge Link Corporation
Illustrations created by Electronic Illustrators Group, Fountain Hills, Arizona

Gale Group
27500 Drake Road
Farmington Hills, Michigan 48331-3535

ISBN 0-7876-3650-9
Printed in the United States of America
10 9 8 7 6 5 4 3 2 1

Library of Congress Cataloging-in-Publication Data
World of chemistry / Robyn V. Young, editor, Suzanne Sessine, assistant editor.
p. cm.
Includes bibliographical references and indexes.
Summary: Articles on theories, discoveries, concepts, and notable people in chemistry.
ISBN 0-7876-3650-9 (alk. Paper)
1.Chemistry. [1. Chemistry-Encyclopedias.] I. Young, Robyn V., 1958- II. Sessine, Suzanne, 1976-
QD33.W873 2000
540—dc21 99-044325

Contents

INTRODUCTION

In *World of Chemistry* the reader will find over 1,000 entries that provide basic information about chemical terms and concepts, applications of chemistry encountered in everyday life, descriptions of the chemistry behind industrial and commercial products, natural phenomena, and biographical sketches of individuals who have made major contributions to the development of chemical ideas and inventions. *World of Chemistry* is designed for individuals who are interested in science and the application of science. Each article has been reviewed to ensure that it is accurate, clear, concise and accessible to the general public. Arcane jargon has been avoided and technical terms are defined using standard vocabulary. Analogies to common, familiar phenomena are also used to explain chemical concepts.

Many chemists refer to chemistry as the "central science." Modern biology, certainly from the time of James Watson and Francis Crick's elucidation of the structure of DNA in 1953, has borrowed heavily from chemical concepts and techniques. Biotechnology is, in large measure, applied chemistry. Great strides in the treatment of disease have been made by the chemical synthesis of new pharmaceuticals. Furthermore, the development of biologically-compatible materials for joint replacements and other implants requires an understanding of chemical interactions. Hybrid names such as biochemistry, cosmochemistry, and, geochemistry, and combinations such as atmospheric chemistry, marine chemistry and medicinal chemistry demonstrate the broad range of areas of knowledge that involve chemistry. *World of Chemistry* provides the user with information on many exciting aspects of chemistry's contributions to our understanding of our planet, its atmosphere, the biosphere, hydrosphere, and geosphere, as well as technological applications of chemistry in other fields.

The beginning of a new century provides an unusual vantage point for marking the progress of a field of knowledge. Chemistry has undergone tremendous change since 1900, when prominent scientists could openly doubt the existence of atoms. Ideas we now take for granted as the most basic foundations of chemical science were not accepted, or, in many cases, even developed until decades after the beginning of the last century. It was not known, for example, that atoms are made up of elementary particles such as electrons, protons, and neutrons. The participation of electrons in chemical bonding, covalent bonding, and the concept of valence had not been postulated. Acids and bases had not been defined in terms of the exchange of protons. In addition, there was no theoretical basis to understand why some reactions are faster than others, or how reactions occur on a molecular scale. In *World of Chemistry,* the reader will not only be introduced to these important ideas, but will learn who developed the idea and when. There are also over 400 biographical sketches to provide information on the individuals themselves.

Chemists are now able to determine the detailed structure of very large molecules such as DNA, RNA's, and enzymes, to synthesize in the laboratory natural biomolecules of staggering complexity, and to determine the presence of submicroscopic amounts of elements and compounds in a sample. Much of what the modern chemist can accomplish is possible only because of technological advances in instrumentation. Highly sophisticated NMR spectrometers and x-ray diffractometers have become indispensable tools for identifying the structure of molecules. Chemists have the ability to separate and purify substances by chromatographic methods and identify molecules by the use of a variety of types of molecular spectrometers, electrochemical devices, and mass spectrographs with speed and accuracy on a scale that could not have been imagined even a few decades ago. In *World of Chemistry* the user will learn about these and many other techniques that chemists employ, the information they provide, and why they work.

Entirely new types of matter have entered our vocabulary. Synthetic polymers, superconductors, nanostructures, and a new form of elemental carbon, the fullerenes, are the subjects of science articles in newspapers and magazines. There are also new twists on types of materials that have been known for many years, such as ceramics, alloys, paints, fibers, and adhesives. *World of Chemistry* provides an explanation of not only the chemical compositions and structures of such materials, but how they were developed as well.

One of the most exciting aspects of chemistry is the extent to which it impacts our daily lives. The clothes and cosmetics we wear, many of the foods we eat, the audio tapes and compact discs we listen to, and even the automobiles we drive all rely on industrial applications of chemistry. *World of Chemistry* will introduce the user the chemistry behind such products.

Although this book provides extensive coverage of chemistry and the individuals who have influenced its development, no single volume can comprehensively describe all of the discoveries, inventions and applications of so broad a field. We have not attempted to duplicate the degree of detail that is available in college-level textbooks or technical encyclopedias designed for professional engineers and scientists. Other resources the reader may wish to consult are the second editions of *World of Invention* and the *World of Scientific Discovery.*

David Lavallee
September 1999

ACKNOWLEDGMENTS

Advisory Board

In compiling ***World of Chemistry***, we have been fortunate in collaborating with our panel of subject advisors who contributed greatly to the accuracy of the information presented in this edition. We would like to express our sincere appreciation:

Jed F. Fisher
Scientist, Discovery Chemistry Research
Pharmacia & Upjohn, Inc.
Kalamazoo, Michigan

Sharlene Kissonergis
Chemistry Instructor
Regina High School
Harper Woods, Michigan

David Lavallee
Office of the Provost
State University of New York
New Paltz, New York

Richard J. L. Phillips, MD
Bloomfield Hills, Michigan

Richard Robinson
Medical Writer
Tucson, Arizona

Contributing Writers

William Acree, Jr., Hilda Bachrach, Julia Barrett, Steve R. Boone, Marlene Bradford, Ray Brogan, Amanda De la Garza, Marie-Ange Djieugoue, Ileana Flintoff, Donald Franceschetti, Randall Frost, Theresa Golden, Wang Haixiang, Bethany Halford, Shelby Hatch, Robert Havlik, David Helwig, George B. Kauffman, Susan M. Kutay, Ann Lenkiewicz, K. Lee Lerner, Jennifer McGrath, Peter Mitrasinovic, J. William Moncrief, David Newton, Tan Pham, Jeanne Robinson, Perry Romanowski, Tracy Lynn Ross, Gordon Rutter, Randy Schueller, Lydia Scratch, Laurel Sheppard, Alexandra Stenson, Stephen Summers, Jens Thomas, Ruthanne Thomas, Marie L. Thompson, Jennifer Truchan, Rashmi Venkateswaran, Terry A. Watkins, David Watson, Todd Whitcombe, David Wiedenfeld, Armen Zakarian, Jennifer A. Zarutskie.

Special thanks are due to William Acree, Jr., Randall Frost, K. Lee Lerner, Randy Schueller, Rashmi Venkateswaran, and David Wiedenfeld, who copyedited and revised many entries with competence and enthusiasm under an unyielding production schedule.

HOW TO USE THIS BOOK

The first edition of ***World of Chemistry*** has been designed with ready reference in mind:

- **Entries are arranged alphabetically**, rather than by chronology or subdiscipline.
- **Bold terms** direct the reader to related entries.
- **Cross-references** at the end of entries alert the reader to related entries not specifically mentioned in the body of the text.
- A **Sources Consulted** section lists the most worthwhile print and electronic material encountered during the compilation of this volume.
- The **Historical Chronology** includes important events in the chemical sciences from 50,000 B.C. to 1999.

A

ABEL, JOHN JACOB (1857-1938)

American pharmacologist

John Jacob Abel is called the father of modern American experimental pharmacology. He was the first to isolate a form of the **hormone** epinephrine and to crystallize insulin.

Abel was born and raised on a farm near Cleveland. He interrupted undergraduate studies at the University of Michigan to teach and serve as superintendent of schools in LaPorte, Indiana. He then spent seven years studying medicine in Europe, earning a doctorate in physiology from the University of Leipzig, Germany in 1886 and an M.D. from the University of Strasbourg (then in Germany) in 1888.

In 1891 he established the first modern department of pharmacology in the United States at the University of Michigan. Two years later he set up a similar department at Johns Hopkins University, where he remained for the rest of his life.

Abel began investigating adrenal gland extracts in 1895, shortly after their actions were noted by Edward Sharpey-Schäfer and others. Using a complicated chemical process involving a benzoyl derivative, in 1899 Abel isolated several salts of a substance secreted by the adrenal medulla. He named his product epinephrine.

A visit to Abel's laboratory inspired the chemist **Jokichi Takemine** to devise a simpler process that isolated the crystalline hormone itself.

In 1912 Abel developed vividiffusion, a method of dialysis, to remove waste products from the blood. In his dialysis machine, an animal's blood treated with the anti-clotting agent hirudin flowed through membrane tubing immersed in a salt solution. Osmotic pressure removed impurities but not cells and plasma, while oxygen went from the salt solution into the blood. The now-purified blood flowed back into the animal's body. Abel also discovered how to store large amounts of blood without damaging the blood cells. The method, called plasmapheresis, was a forerunner of blood banks.

Abel went on to study poisons, especially the mushroom *Amanita phalloides* and the venom of the tropical toad *Bufo agua*, which contains seven percent epinephrine. He isolated the crystalline form of insulin in 1926. Abel's drive and intellect did not diminish with age. At 75 he began studying how tetanus toxin affected the body's nervous system.

Among his many honors were memberships in the National Academy of Sciences of the United States of America and the Royal Society (London).

ABSORPTION SPECTROSCOPY

Absorption **spectroscopy** is a ''workhorse'' technology widely used in industry and in **chemistry**, biology, medicine, and other fields of scientific research. Numerous sciences, including chemistry and astronomy, have reached their current state of advancement only though use of absorption spectroscopic instruments.

Spectroscopy studies the way electromagnetic radiation interacts with **matter**. When radiation meets matter, the radiation is either scattered, emitted, or absorbed. This gives rise to three principal branches of spectroscopy. Emission spectroscopy observes light *emitted* by atoms excited by radiation/matter interactions. Raman spectroscopy monitors light *scattered* from molecules. Absorption spectroscopy studies radiation *absorbed* at various frequencies (wavelengths).

The roots of absorption spectroscopy can be traced to 1802, when the English chemist William Hyde Wollaston (1766-1828) observed what later became known as the Fraunhofer lines—the many dark lines seen in the spectrum of sunlight. In 1859, the German physicist Gustav Kirchhoff (1824-1887) determined that these lines resulted from absorption of specific frequencies of solar radiation by vapors in the Sun's atmosphere. Subsequent study of these lines taught scientists many things about the chemical composition of the Sun and other astronomical objects.

In its early years, spectroscopy was largely confined to the study of emitted radiation. By the mid-1950s, however,

John Jacob Abel.

several technical difficulties had been overcome, and manufacturers began producing absorption spectroscopy equipment for routine laboratory analysis.

In absorption spectroscopy, light illuminates a sample of material to be analyzed. The sample, which can be liquid, solid, or gas, is usually enclosed in an absorption cell, which in turn may be enclosed in a oven to vaporize and atomize the material. Each element or compound in the sample absorbs particular wavelengths of light, resulting in one or more dark lines on its spectrum. These lines are a ''fingerprint'' identifying what chemical substances are present in the sample and their quantities, as well as other information about detailed structure and activity.

The technology is capable of identifying both basic elements and highly complex molecules in concentrations as small as parts per billion. It is used in a wide range of applications, including analysis of body fluids and tissues, municipal **water** supplies, workplace atmospheres, metal alloys, and geological samples. In forensic science, absorption spectroscopy detects **arsenic** and other poisons in the bodies of crime victims, analyses gunshot residues, and even allows investigators to detect counterfeit whisky. In industrial settings it can monitor the amount of fat in peanut butter or **carbon** dioxide in soda, as well as the quality of **cosmetics**, detergents, paints, drugs, and thousands of other manufactured products.

One common use of absorption spectroscopy is called a *chem 20.* This is a widely-used series of 20 or 30 tests on a patient's blood, prescribed by doctors to help determine the cause of illness and prescribe appropriate treatment.

Another example of the usefulness of this technology is in determining potential environmental impacts of using methyl **alcohol** instead of **gasoline** in internal **combustion** engines. Samples of exhaust gas produced by this fuel were placed in a chamber about 13 ft (4 m) long, through which infrared light was passed. The resulting spectra indicated the presence of methyl nitrite—a potentially dangerous substance whose presence was not indicated by other analytical methods.

As the millennium drew to a close, Western scientists were expressing great interest in a technique known as intracavity laser absorption spectroscopy that was quietly developed in the Soviet Union during the 1970s. **Lasers** amplify light by bouncing small numbers of photons back and forth in a cavity with mirrors at either end. By placing the sample to be analyzed inside a laser's light amplification cavity, the Soviet scientists found they could amplify faint signals from just a few molecules of **gases** that would otherwise go undetected. Because of the Cold War, news of this development took some time to reach researchers in other parts of the world, and was largely ignored when it did arrive. However, by the late 1990s, Western scientists acknowledged that the laser technique was far more sensitive than traditional absorption spectroscopy, and were using it for many applications, including research aimed at understanding the rapid chemical changes that occur during combustion.

See also Infrared spectrometry

ACETIC ACID

Acetic acid is a clear, colorless liquid with the chemical formula $C_2H_4O_2$. It has a melting point of 62.06°F (16.7°C) and boils at 244.4°F (118°C). In high concentrations, it is a corrosive organic acid that has a pungent odor and can cause severe burns on skin.

Acetic acid has been known to humans for centuries. It is most likely that it was discovered accidently during the wine making process. When the process of fermenting fruit juices is allowed to go on too long, the wine spontaneously forms vinegar, a dilute form of acetic acid. Consequently, the name acetic acid is derived from the Latin word *acetum* which means vinegar.

While vinegar was known for centuries the corrosive component was not isolated immediately. The first known attempt to isolate the acid was done during the 700s. At this time, the Arab alchemist Jabir ibn Hayyan Geber produced concentrated acetic acid by distilling vinegar. However, it was not until a millenium later in 1700 that the pure form of acetic acid was isolated by chemist **Georg Ernst Stahl**. As **chemistry** and chemical theories became more sophisticated, scientists were able to better identify and produce various materials. In 1844, the German chemist **Adolf Kolbe** synthesized acetic acid from pure **carbon** and **water** using various catalysts.

Today, the production of acetic acid can be accomplished by various methods. It can be obtained by the destructive **distillation** of **wood**, and can also be produced from acetylene and water using an **oxidation** process with air. One of the first manufacturing methods was a separation process. A dilute **solution** of acetic acid was cooled below its freezing point. The acid would solidify and separate from the water. For this reason, pure acetic acid is known as glacial acetic acid. Vinegar is produced using a two-step **fermentation** process. Naturally occurring starches are first converted to sugars. These sugars are then allowed to ferment with yeast producing **alcohol**. The alcohol is then exposed to an acetobacterium which converts it to vinegar.

The biological role of acetic acid was discovered by biochemist Konrad Emil Bloch in the mid-1900s. He found that acetic acid is the primary precursor in the production of body **cholesterol**. The acetic acid is converted to cholesterol in the liver through a series of 36 chemical reactions. Bloch was able to use radioactive tagging methods to determine which carbons from acetic acid were incorporated into cholesterol. This research was important for our current understanding of cholesterol **metabolism** and its role in heart disease.

In small concentrations, acetic acid is suitable for ingestion. Vinegar is one of the primary flavor components of many types of salad dressings where it is present at about 5%. It provides a biting, sour or tangy **taste**. In other food products, acetic acid is used as a preservative. White vinegar is used as a household cleaner because it has certain antibacterial characteristics. Acetic acid is used in other industries. It is the chemical precursor for important materials like acetic annhydride, acetate esters, **cellulose** acetate, and acetate rayon. It can be used as a solvent for many other types of processes such as the production of **plastics**, rubber, gums, resins, and volatile oils, and is also an important acidifier in pharmaceutical products.

Acetone

Acetone is a colorless, flammable, and volatile liquid with a characteristic odor that can be detected at very low concentrations. It is used in consumer goods such as nail polish remover, model airplane glue, lacquers, and paints. Industrially, it is used mainly as a solvent and an ingredient to make other chemicals.

Acetone is the common name for the simplest of the **ketones**. The formula of acetone is $CH_3 \cdot CO \cdot CH_3$.

The International Union of Pure and Applied Chemistry's (IUPAC) systematic name for acetone is 2-propanone; it is also called dimethyl ketone. The **molecular weight** is 58.08. Its **boiling point** is 133°F (56°C) and the melting point is -139.63°F (-95.4°C). The specific gravity is 0.7899.

Acetone is the simplest and most important of the ketones. It is a polar organic solvent and therefore dissolves a wide variety of substances. It has low chemical **reactivity**. These traits, and its relatively low cost, make it the solvent of choice for many processes. About 25% of the acetone produced is used directly as a solvent.

About 20% is used in the manufacture of methyl methacrylate to make **plastics** such as acrylic plastic, which can be used in place of **glass**. Another 20% is used to manufacture methyl isobutyl ketone, which serves as a solvent in surface coatings. Acetone is important in the manufacture of artificial fibers, explosives, and polycarbonate resins.

Styrofoam dissolving in acetone. ***(Photograph by Charles D. Winters, Photo Researchers, Inc. Reproduced by permission.)***

Because of its importance as a solvent and as a starting material for so many chemical processes, acetone is produced in the United States in great quantities. In 1999, the worldwide acetone market reached 9.4 billion lb (4.27 billion kg) at a steady growth rate of 2-3% per year. Acetone is prepared by several routes, from petrochemical sources. The methods of its synthesis include **oxidation** of 2-propanol (isopropyl alcohol), the hydration of propene, and as a co-product (with phenol) of the O_2-oxidation of cumene.

Acetone is normally present in low concentrations in human blood and urine. Diabetic patients produce it in larger amounts. Sometimes ''acetone breath'' is detected on the breath of diabetics by others and wrongly attributed to the drinking of liquor. If acetone is splashed in the eyes, irritation or damage to the cornea will result. Excessive breathing of fumes causes headache, weariness, and irritation of the nose and throat. Drying results from contact with the skin.

Acetylsalicylic acid

Acetylsalicylic acid, commonly known as aspirin, is an analgesic (pain-killing), antipyretic (fever-reducing), and anti-inflammatory sold without a prescription as tablets, capsules, powders, or suppositories. The drug reduces pain and fever, is believed to decrease the risk of heart attacks and strokes, and may deter colon cancer and help prevent premature birth.Often called the wonder drug, aspirin can have serious side effects, such as irritating the stomach lining, causing Reye's syndrome in children between the ages of three and 15 years, adversely affecting breathing in people with sinusitis or asthma, and possibility delaying the onset of labor in full term pregnancies. More more accidental poisoning deaths in children under five years of age and 10% of all accidental or suicidal episodes reported by hospitals are related to aspirin.

In the mid-to-late 1700s, English clergyman Edward Stone chewed on a piece of willow bark and discovered its analgesic property. The bark's active ingredient was isolated in 1827 and named salicin for the Greek word *salix*, meaning willow. Salicylic acid, first produced from salicin in 1838 and synthetically from phenol in 1860, was effective in treating rheumatic fever and gout, but caused severe nausea and intestinal discomfort. In 1898, a chemist named Felix Hoffmann, working at Bayer Laboratories in Germany and whose father suffered from severe rheumatoid arthritis, synthesized acetylsalicylic acid in a successful attempt to eliminate the side effects of salicylic acid, which until then was the only drug that eased his father's pain. The process for making large quantities of acetylsalicylic acid was patented, and aspirin—named for its ingredients acetyl and spiralic (salicylic) acid—became available by prescription. Its popularity was immediate and worldwide. It became available without a prescription in the United in 1915.

Aspirin's recommended therapeutic adult dosage ranges from 600-1000 milligrams and works best against "tolerable" pain; extreme pain is virtually unaffected, as is pain in internal organs. Aspirin inhibits (blocks) production of **hormones** (chemical substances formed by the body) called **prostaglandins** that may be released by an injured cell, triggering release of two other hormones that sensitize nerves to pain. The blocking action prevents this response and is believed to work in a similar way to prevent tissue inflammation. Remarkably, aspirin only acts on cells producing prostaglandins—for instance, injured cells. Its effect lasts approximately four hours.

Aspirin also reduces fever by inhibiting the production of prostaglandins in the hypothalamus, a portion of the brain that regulates such functions as heart rate and body **temperature**. The body naturally reduces its **heat** through perspiration and the dilation (expansion) of blood vessels. Prostaglandins released in the hypothalamus inhibit the body's natural heat-reducing mechanism. As aspirin blocks these prostaglandins, the hypothalamus is free to regulate body temperature.

Another prostaglandin that is inhibited by aspirin is thromboxane A_2, which aids platelet aggregation (accumulation of blood cells). Because aspirin inhibits thromboxane production, thus "thinning the blood," it is frequently prescribed in low doses over long periods for at-risk patients to help prevent heart attacks and strokes.

ACID RAIN

Acid rain is the collective name given to any rain containing acidic impurities. These impurities are generally the product of industrial **pollution**.

Normal, unpolluted rain has a **pH** of 5.6 due to the presence of dissolved **carbon** dioxide. Acid rain is regarded as rain with a pH more acidic than this value (a pH value less than 5.6). The greater acidity of acid rain is due to different oxides of **nitrogen** and **sulfur** dissolved in the rain. These oxides are generally referred to as NO_x and SO_x to indicate their variable composition. These pollutants are released into the atmosphere by the **combustion** of fossil fuels such as **coal**, oil, and gas.

Acid rain has been implicated in a number of environmental problems. In the 1960s and 1970s, fish stocks were depleted in a large number of Scandinavian lakes and this occurred at exactly the same time as an increase in the acidity of the lake **water**. With lakes and other bodies of fresh water the acid pH itself is not the killer, but the problem is that the increased acidity allows toxic elements, such as **aluminum**, to exist in greater **concentration** in **solution**. It is the increased level of these poisons that causes the damage to wildlife.

A more obvious problem, particularly in the United States, is the damage caused to trees and forests. Where trees are exposed more directly to rain, such as at the edge of forests or on the sides of hills, there is an increased level of damage evident. This environmental damage takes the form of tree die-back where the outer branches are progressively denuded of their leaves. The damage caused by acid rain also can be seen by studying lichens. Lichens are intolerant to the presence of sulfur dioxide in acid rain and different species of lichen have different threshold levels before death occurs. This means it is possible to quantify the approximate level of pollution by studying which species of lichen are present and which have disappeared from an area.

Acid rain does not just affect plants and animals. It is also strong enough to damage stone. Many old buildings and statues show pitting caused by the acidic rain which dissolves the stone.

Many of the developed countries of the world have entered into agreements to reduce the levels of sulfur and nitrogen oxides released into the atmosphere. The measures taken will eventually reduce the amount of acid rain and the damage caused will consequently be reduced. One action that has been taken is the cleaning of **gases** released from factory and power station chimneys. The simplest technique for cleaning these emissions is known as scrubbing and consists of firing a fine spray of water through the waste gases. This dissolves the sulfur and nitrogen oxides before they reach the atmosphere. The acidic water produced is collected and treated before release from the factory.

Acid rain is a problem associated with industrialization and the burning of fossil fuels. Countries without high levels of industrialization generally do not show evidence of acid rain damage, unless they are neighboring an industrial country. The gases responsible for acid rain are blown around in the atmosphere. As a result, the damage is not always recorded near to the site of production. To assess where the acid rain and pollution originate factors such as the direction of the prevailing wind must be taken into account.

Acid rain is a problem that was first recognized in the mid-twentieth century. It was discovered by checking records of plants and looking at levels of damage to buildings that this form of pollution had existed since the Industrial Revolution. By the end of the twentieth century, a number of international protocols were successfully reducing the levels of pollutants in the atmosphere.

See also Acids and bases

Acids and bases

Acid and base are terms used by chemists to categorize chemicals according to their **pH**. An acid is generally considered to be any material that gives up a **hydrogen ion** in **solution**, while a base is any material that creates a hydroxide ion in solution. Many of these acids and bases are familiar in everyday life. The vinegar we use in salad dressings gets its tart flavor from **acetic acid**, and one of the most common household drugs, aspirin, is a type of acid. **Proteins**, butter and oils, fruits, and berries all contain a number of natural organic acids. Bases feel slippery when dissolved in **water** and are used to make soap and other household products. When people get heartburn, they might take baking soda or an **antacid** tablet, both of which are mild bases. Countless industrial processes use acids and bases as reactants or catalysts to make a variety of consumer goods.

People were probably aware of common acids and bases in prehistoric times, ever since they learned how to make wine. When wine turns sour, it changes to vinegar, or dilute acetic acid. In early times, people roasted limestone to obtain **lime** (calcium oxide), a base. Gradually, scientists learned to formulate new acidic and basic substances. In the 700s, an Arabian alchemist named Geber (c. 721-815) prepared **nitric acid** and acetic acid obtained by distilling vinegar. Some time before 1300, **sulfuric acid** was prepared, and alchemists created **aqua regia**, a mixture of sulfuric and nitric acids that is capable of dissolving **gold**, **platinum**, and many other materials. When strong acids became widely available during the Middle Ages, they launched an experimental revolution. For the first time, alchemists were able to decompose substances without high temperatures and long waiting periods.

During the 1600s, alchemical methods of preparing acids were improved. Johann Rudolf Glauber (1604-1670), a German chemist, set up a small factory for making acids and salts that are formed when acids and bases neutralize each other. Soon chemists became more interested in studying the properties of acids and bases and the **neutralization** reaction between the two substances. Dutch physician Franciscus Sylvius (1614-1672) diagnosed the human body in terms of its **balance** between acids and bases. Although Sylvius's ideas were simplistic, it is true that our health depends on a proper balance of acids and bases in our cells and in body fluids such as blood.

During the 1660s, **Robert Boyle** discovered that certain plant extracts, such as litmus, can be used to distinguish acids from bases. Litmus **paper** turns red when dipped in an acid, blue when exposed to a base. Since then several other indicator substances have been found that change **color** at different levels of acidity in a solution. Boyle went on to characterize acids, noting their sour or tart **taste** and their ability to corrode **metals**. Scientists speculated that acids were made of sharply pointed particles that literally pricked the tongue or scratched the metal. Neutralization, they theorized, occurs when an acid particle's spikes fit into a basic particle's pores.

A micrograph image of butanedioic acid. ***(Photograph by Astrid & Hanns Frieder-Michler/Science Photo Library, National Audubon Society Collection/Photo Researchers, Inc. Reproduced by permission.)***

During the 1700s, chemists attempted to describe the neutralization process in terms of the affinity, or degree of attraction, between acidic and basic particles. In 1791, German chemist Jeremias Benjamin Richter (1762-1807) demonstrated that a particular acid and base always neutralize each other in the same proportions. The idea that chemicals react in certain fixed proportions is called **stoichiometry**, upon which quantitative **chemistry** is based.

A new but erroneous definition of acids was developed by **Antoine-Laurent Lavoisier** in the 1770s. He mistakenly believed that all acids contain **oxygen** (named from the Greek words meaning acid-producing) because he observed that acids are formed when oxygen compounds are dissolved in water. Lavoisier's theory was disproved in the early 1800s, when chemists began using the new tool of **electricity** to break compounds into elements. First, **Humphry Davy** demonstrated that **hydrochloric acid** (HCl) contains no oxygen. His finding was supported by Joseph Gay-Lussac, who proved that oxygen is not a component of prussic acid (now known as hydrocyanic acid [HCN]).

Our modern understanding of acids began to take root in the 1830s, when German chemist **Justus von Liebig** defined an acid as a compound that contains hydrogen in a form that can be displaced by a metal. Bases, however, were understood only in terms of their ability to neutralize acids.

Then in the 1880s, **Svante August Arrhenius** proposed that when acids and bases dissolve in water, their molecules break up into electrically charged particles called ions (a term introduced by Michael Faraday). Acids produce positively charged hydrogen ions (H^+), while bases produce negatively charged hydroxyl ions (OH^-). Arrhenius's theory also explained, in very simple terms, what happens when an acid and base neutralize each other: the positive and negative ions unite to form water (H_2O), which is neutral.

The strength of a particular acid or base depends on its **concentration** of hydrogen ions that is measured by the pH system on a scale of one (strongest acid) to 14 (strongest base). Strong bases are sometimes called alkalis. Because acidic and basic, or alkaline, solutions both conduct electricity, their strength can also be quantified by measuring their **electrical conductivity**.

Although Arrhenius's theory represented a giant step in our understanding of acids and bases, it had its limitations. What about **solvents** other than water, for example? And what about **ammonia**, which contains no oxygen but produces hydroxyl ions when dissolved in water? Another complication was the fate of the hydrogen ion in water. Instead of floating free, hydrogen ions combine with water molecules to produce a positively charged **hydronium ion** (H_3O^+).

In 1923, Arrhenius's concept was refined by Danish chemist **Johannes Nicolaus Bronsted** (1879-1947), who broadened our definition of acids and bases. The hydrogen ion, he pointed out, is a proton—a hydrogen **atom** without its **electron**, or negatively charged particle. So Bronsted defined acids as **proton** donors (they release hydrogen ions) and bases as proton acceptors (any substance that will combine with a loose proton). The same idea was proposed simultaneously by British chemist Thomas M. Lowry. The Bronsted-Lowry definition holds up no matter what the solvent is, and it explains why pure acids and dissolved acids behave differently.

The same year that Bronsted's work was published, American chemist **Gilbert Newton Lewis** suggested a slightly different way of looking at the new definition. Instead of donating protons, acids accept unattached pairs of electrons; conversely, instead of accepting protons, bases supply pairs of electrons. Under Lewis's definition, even substances that do not produce hydroxyl ions can be considered bases.

Despite all of these refinements, most common acids and bases behave just as Arrhenius described. Today, these substances are used in refining oil and sugar and in manufacturing a great variety of products, including **fertilizers**, **explosives**, **plastics**, soap, paper, film, drugs, synthetic fabrics, dyes, solvents, and **pesticides**.

ACROLEIN

Acrolein (also called acrylaldehyde) is the generic or general name for the chemical substance 2-propenal. It is a very simple compound, consisting of three carbons, one **oxygen**, and four hydrogens, but its simplicity is deceiving. Acrolein has a number of interesting features from a chemical point of view. First, the oxygen is contained in a carbonyl, an organic **functional group** where the oxygen and **carbon** are double-bonded. But the other two carbon atoms are also double-bonded to each other, which results in **conjugation** of the **electron density** in the bonds. This means that the two bonds form a single system that covers the whole **molecule**. The result is that acrolein absorbs ultraviolet light, a property that industrial manufacturers take advantage of in making such things as sunglasses which absorb ultraviolet radiation.

Unfortunately, acrolein has a bad side as well. At high concentrations, it is thought to be both a mutagen (a substance capable of causing mutations) and a carcinogen (a substance capable of causing cancer). The use of acrolein is carefully monitored as it is a very useful compound in the manufacture of **plastics**, artificial resins, synthetic fibers, and **polyurethane** foams, but also is a potentially toxic substance for humans and the environment. The highly reactive nature of acrolein is the basis of its toxicity. Acrolein's vapors are very irritating to the eyes, skin, and mucous membranes, and this chemical contributes significantly to the irritating quality of cigarette smoke and photochemical smog.

ACTINIDES

The actinides are the elements with atomic numbers 89 (actinium) through 103 (Lawrencium). With each unit increase in **atomic number**, the **atom** of each succeeding **actinium** element has one additional 5f **electron**. Thus, the actinium series of fourteen elements is homologous with the lanthanide series for fourteen elements in which each successive element in the series has one additional 4f electron. Like the lanthanide elements, the actinides have very similar chemical properties. And like the lanthanides, the most common **oxidation** state in aqueous **solution** is 3+. In **oxide** and fluoride compounds, among others, the actinide elements are found in higher oxidation states, from IV to VII as well as III.

The first four actinide elements, actinium (atomic number 89), **thorium** (90), **protactinium** (91) and **uranium** (92) are naturally occurring and have been known for a considerable time. The dates of discovery of these elements as 1789 for uranium, 1829 for thorium, 1899 for actinium and 1913 for protactinium. The next two elements in the actinide series, **neptunium** (93) and **plutonium** (94) were discovered as products of **nuclear fission** experiments. Neptunium (93), the first synthetic element, was discovered in 1940 and plutonium (94) discovered shortly afterward in the same year. Many different isotopes of the remaining elements in the series were produced purposefully using **neutron** sources of particle accelerators of various types over the following two decades: **americium** (95) and **curium** (96) in 1944, **berkelium** (97) in 1949, **californium** (98) in 1950, **einsteinium** (99) and **fermium** (100) in 1952, mendelevium (101) in 1955, nobelium (102) in 1957 and lawrencium (103) in 1961. The privilege of naming a new element has traditionally been accorded to the first group of scientists that can convince their colleagues that they have successfully produced the element experimentally. As might be surmised from the names of the actinides, credit for their synthesis was given principally the group of scientists led by Glenn Seaborg at the University of California at Berkeley and later the Lawrence Livermore Laboratories.

Two of the actinides, thorium and uranium, are found in relatively high abundance naturally and can often be detected by standard chemical techniques. The other actinides have very low or no known natural abundance and must be analyzed by highly sensitive radiation detection techniques.

Two isotopes of actinides, uranium-235 and plutonium-239, are used extensively in nuclear fission power generators and are also used in nuclear weapons. Tons of plutonium-239 have been produced in fission reactors used for providing **electricity**. Plutonium-239 poses a great challenge for storage because it is highly toxic due to its intense **alpha particle** emission and it has a very long **half-life** of nearly 25,000 years. Because the technology for fabricating nuclear weapons is now so widespread and many nations would be capable of manufacturing these weapons if they have access to uranium-235 or plutonium-239, the quantities of these isotopes that are available also pose a security challenge.

Other isotopes of the actinide elements that have found application are Pu-238 which is used to power heart pacemakers and space instrumentation, americium-241 which is used in smoke detectors, and californium-252 which is used in analytical techniques that involve production of isotopes of other elements by neutron-capture.

Actinium

Actinium is the third element in Row 7 of the **periodic table**. Its **atomic number** is 89 and chemical symbol is Ac. Actinium is a radioactive element, whose most common **isotope** has a **mass** of 227.

Properties

Only a limited amount of information is known about the properties of actinium. It is a **silver** metal with a melting point of 1,922°F (1,050°C) and a **boiling point** of 5,792°F (3,200°C). Its chemical properties are similar to those of **lanthanum**, the element above it in the periodic table.

Occurrence and Extraction

Actinium occurs only rarely in the Earth's crust. It is usually found with ores of **uranium**, where it is produced by the radioactive decay of that element.

Discovery and Naming

Actinium was discovered by the French chemist André Debierne (1874-1949) in 1899. Debierne was a close friend of **Marie Curie** and her husband Pierre. The Curies were among the first scientists to study the new phenomenon of **radioactivity** in the 1890s. They discovered two new elements themselves, **polonium** and **radium**. Actinium was named after the Greek words *aktis* or *aktinos*, meaning "beam" or "ray."

Uses

There are no practical uses for actinium, although it is sometimes used in research.

Activated complex

A crucial conceptual element of the transition state theory, activated complex is an unstable, high-energy grouping of atoms. According to transition state theory, there is an intermediate point, in a chemical reaction, when the bonds of the reactants have not been quite broken, and the reaction product's new bonds have not yet been formed. The entity reflecting that point is the activated complex. For example, in a HI + Cl reaction, which yields HCl + I, the I-H-Cl group of atoms appears during the reaction: this is an activated complex. Since, as transition state theory posits, a chemical reaction requires a specific level of energy, known as the activation energy, every initiated reaction is not automatically completed. If a reaction is successful, the activated complex will yield the product; if the reaction process cannot be completed due to insufficient energy, the activated complex will revert to the original reactants. In our example, the I-H-Cl group would become HI and Cl. Michael Polanyi (1891-1976) and Henry Eyring (1901-1981), who developed the transition state theory in the 1920s and 1930s, postulated that every chemical reaction goes through the activated complex phase.

Activation energy

The term activation energy refers to the minimum amount of **energy** required for a chemical reaction to occur. Most reactions require that **atoms** and **molecules** crash into each other with a great deal of force. These violent collisions must occur with enough energy to cause the chemical bonds within the molecules to be weakened or broken. When this occurs, the reactant molecules form an **activated complex** from which the product molecules are formed. Depending on the strength of the bonds within the reactant molecules the activation energy can be quite large, making the activated complex difficult to form.

Fortunately most activation energies are large enough so that many combinations of substances can coexist at room temperature without reacting to a great extent, even if they are favored by **thermodynamics**. For example, it is thermodynamically favorable for organic materials and fuels such as **wood**, **coal**, oil, and gas to react with **oxygen** to form **carbon dioxide** and **water**. Yet fuels and other organic compounds, including living organic tissue, come into contact with oxygen all the time under normal conditions and are very stable. We know from experience that these reactions don't occur unless we provide a spark that supplies the needed activation energy. Only then do these reactions occur readily.

Thanks to activation energies many combinations of substances remain stable and non-reactive under normal conditions for a very long time. If this were not the case, many of the materials and fuels that we now take for granted would long ago have literally "go up in smoke."

ADAMS, ROGER (1889-1971)

American chemist

Roger Adams was a member of the **chemistry** department at the University of Illinois from 1916-1954 and headed the department for the last 28 years of that period. Some of his most important work involved the elucidation of complex organic structures such as tetrahydrocannabinol (from marijuana), chaulmoogric acid (used in the treatment of leprosy), and gossypol (a toxic agent found in cottonseed oil). He is probably best known, however, for inventing a method for the preparation of **platinum** in a form usable in catalysis, a substance now known as Adams's catalyst. Throughout his career, Adams was also very active in the administrative aspects of science. Many of his students assumed positions of leadership in the chemical industry, and Adams is credited with having contributed significantly to the rapid development of chemistry in the United States.

Roger Adams was born in Boston on January 2, 1889. He was the youngest of four children born to Austin Winslow Adams and the former Lydia Curtis. He attended the Boston and Cambridge Latin schools and then, at the age of sixteen, entered Harvard University. In spite of his solid precollege training, Adams apparently "did not take college seriously" until he was a junior at Harvard, according to biographer Nelson J. Leonard in a 1969 retrospective on Adams in the *Journal of the American Chemical Society.* From that point forward, he became intent on his chemical education, earning in quick order his A.B. in 1909, his A.M. in 1910, and his Ph.D. in 1912.

For his postdoctoral studies, Adams traveled to Europe where he studied with **Otto Diels** at the University of Berlin and with **Richard Willstätter** at the Kaiser Wilhelm Institute in Dahlem. He began a second postdoctoral year at Harvard in 1913 under C. L. Jackson and was also soon asked to teach an introductory course in **organic chemistry** at the college. Adams continued to teach and do research at Harvard until 1916.

In 1916, W. A. Noyes, chairman of the chemistry department at the University of Illinois, offered Adams a position as assistant professor of chemistry. Adams accepted and over the next four decades was promoted successively to professor in 1919, to head of the department in 1926, to research professor in 1954, and finally to professor emeritus in 1957. During his long academic career at the University of Illinois, Adams was awarded honorary doctorates from ten universities and 24 medals and awards from organizations around the world. Among these were the Willard Gibbs Medal of the Chicago Section of the American Chemical Society (ACS) in 1936, the Elliott Cresson Medal of the Franklin Institute in 1944, the Davy Medal of the Royal Society in 1945, the Priestly Medal of the ACS in 1946, the Medal for Merit of the U.S. government in 1948, the Gold Medal of the American Institute of Chemists in 1964, and the National Medal of Science in 1964.

Adams's tenure at Illinois coincided with a growing demand in the chemical industry for academically trained doctorate holders. Convinced of the interdependence of universities and industry, Adams set about strengthening links between the two spheres by training graduates specifically to go into the field. His department's ability to supply quality industrial chemists in bulk reinforced the demand for them, and Adams therefore garnered much of the credit for the growth of chemistry in the United States.

Shortly before moving from Harvard to Illinois, Adams was married on August 29, 1918, to Lucile Wheeler. The Adamses later had one daughter, Lucile, who later married William E. Ranz, a chemical engineer at the University of Minnesota.

Adams studied three major subjects during his career as a researcher: catalysis, **stereochemistry**, and structural analysis. Perhaps his best-known discovery, Adams's catalyst, came about as the result of an accident in a University of Illinois laboratory. During a student experiment, a mixture of chloroplatinic acid, **formaldehyde**, and alkali was spilled on a laboratory desk. Hoping to salvage the materials, Adams gave directions by which the mixture should be treated and the reactants should be recovered.

When these directions were carried out, the platinum was regenerated in a new form, brown platinum **oxide**. Treatment of the oxide with **hydrogen** gas resulted in the formation of a very finely divided platinum metal. Adams discovered that in this form the platinum was an extraordinarily efficient catalyst in the conversion of unsaturated organic compounds to their saturated (that is, consisting of molecules that have only single bonds) counterparts, a process with many important industrial applications. For nearly a half century thereafter, Adams's catalyst continued to be one of the most valuable intermediates for a number of industrial preparations.

During the 1920s and 1930s, Adams focused his research on stereochemistry, a study of the properties and behaviors of molecules determined by their three-dimensional structures. In particular, he investigated the ways in which free rotation around a single bond can be blocked and how this kind of restriction affects the properties of a **molecule**.

Adams was also successful in elucidating the structures of a number of complex organic molecules. In the 1930s, for example, he was asked by the U.S. Narcotics Bureau to study the components of marijuana. Adams was able to separate out and determine the molecular composition of cannabinol, cannabidol, tetrahydrocannabinol, and other related compounds found in marijuana. As an unexpected side effect of this research, Adams found that the bureau's standard test for marijuana actually identified a harmless substance that occurs in conjunction with the marijuana itself.

In later research, Adams found the structure of gossypol, a toxic compound that occurs in cottonseed oil and had long limited the uses of that material. This work allowed the development of a procedure for removing gossypol from cottonseed, vastly increasing the uses of that important substance. Adams also determined the structures of chaulmoogric and hydrocarpic acids, components of the chaulmoogra oil that had been used for centuries in the treatment of leprosy, and of necic acid and its derivatives, responsible for cirrhosis of the liver in cattle, sheep, and horses. Beyond his many research interests, Adams was actively involved in a wide variety of political and professional activities throughout his life. In the mid–1930s, for example, he served on President Franklin D. Roosevelt's short-lived Science Advisory Board. From 1941 to 1946 he was a member of the National Defense Research Committee. During the last year of that term, he served as science advisor to Lieutenant General Lucius D. Clay, U.S. deputy military governor in the American occupation zone of Germany. In 1947 Adams was the leader of a scientific advisory committee whose job it was to recommend to General Douglas MacArthur the most effective methods for democratizing Japanese science.

Adams's first major professional position was as chair of the organic division of the American Chemical Society in 1920. He later served on the board of directors of the ACS (from 1930 to 1935 and from 1940 to 1950), as president of the organization (1935) and as chair of the ACS board (from 1944 to 1950). He was also president of the American Association for the Advancement of Science in 1950, a member of its executive committee from 1941-1946 and from 1948-1952, and chair of the board of directors in 1951. Adams died at his home in Urbana, Illinois, on July 6, 1971.

Adelard of Bath (1075-1160)

English translator

Among the series of medieval English translators, mathematicians, and natural philosophers of England who traveled extensively in search of Arabic texts was Adelard of Bath. He is responsible for the conversion of Arabic-Greek learning into Latin.

Abelard was born in approximately 1075 in Bath, England. An extensive traveler, he went to study at Tours and later taught at Laon. Leaving Laon, for the next seven years he journeyed to various cities and countries, including Salerno, Syria, Cilicia, Spain and possibly Palestine before returning to Bath in 1130. During his travels, however, it was possible that he learned Arabic in Sicily and received Spanish-Arabic texts from other Arabists who had lived in or visited Spain.

Adelard made significant contributions to the field of philosophy with the writing of two treatises: *De eodem et diverso*, dedicated to William, bishop of Syracuse and written before 1116, and *Quaestiones naturales* written prior to 1137 and perhaps even earlier. Speaking as a quasi Platonist in *De eodem et diverso*, Adelard draws on the major themes of Platonism, but opposing the Platonic doctrine of realism in his theory of universals. His second treatise *Quaestiones naturales* consists of a dialogue with an unnamed nephew of 76 scientific discussions derived from Muslim science. In Adelard's first work, there is no trace of Arabic influence; however, there are descriptions of a pipette-like vessel mentioned in other Arabic works by the unnamed nephew in *Quaestiones naturales*.

While Adelard's contributions to the area of philosophy were significant, he is more renowned for his translations of Arabic scientific texts. With these germinal translations, Adelard was also able to culminate the use of algorithms and Arabic numerals, as opposed to the intractable Roman numerals.

Using *Ysagoga minor Iapharis matematici in astronomicam per Adhelardum bathoniensem ex arabico sumpta*, a translation of Abu Ma shar's *Shorter Introduction to Astronomy*, Adelard gave the first sampling to the Latin Schoolmen. This contained some astrological rules and axioms, and was abridged by Abu Ma shar for his longer *Introductorium maius*. Adelard's translation of this work proved pivotal, as the longer version was translated twice thereafter into Latin. Additionally, Adelard translated *Liber prestigiorum Thebidis (Elbidis) secundum Ptolomeum et Hermetem per Adelardum bathoniensem translatus*, an astrological work on images and horoscopes by Thabit ibn Qurra.

Adelard also produced a variety of works involving arithmetic, the earliest being *Regule abaci*. Another work, *Liber ysagogurm Alchorismi in artem astronomicam a magistro A. compositus*, is composed of books concerning arithmetic, geometry, music, and astronomy.

Before Adelard translated Euclid's *Elements* into Latin, there were only a few incomplete versions, such as that of Boethius, in existence from the Greek. While Adelard's name is associated with three different versions, each somewhat incomplete, each codex has been pieced together to present a full version.

Among the most important works he translated were the astronomical tables of Al-Majriti, and a treatise on Arabic arithmetic by the mathematician al-Khwarizmi, whose name later became synonymous to the mathematical system of algorithms.

Adelard's writings and translations remain an important bridge between Muslim science and Western learning.

Adhesives and glues

Adhesives and glues are substances that are capable of bonding two solid materials together at their surfaces. While they have been used by humans for centuries, it is only in the twentieth century that synthetic adhesives were developed. Archeological evidence suggests that ancient Egyptians used resinous adhesives at least 6,000 years ago. Other adhesives including starch, sugar, casein, bitumen, shellac, pitch and glues made from animals and fish were used around 1500 B.C. These first adhesives were of natural origin and are still used today in a form that is not substantially different.

The first synthetic adhesive was produced in 1869. This material was incorrectly termed nitrocellulose and was created by a reaction between **nitric acid**, **sulfuric acid**, and **cellulose**. Today, this product is known as cellulose nitrate. In 1912, **Leo Baekeland** produced phenol-**formaldehyde** resins, a basic material for many of today's adhesives. High strength, elastomeric adhesives were available in 1928 when a reaction that produced polychloroprene was developed. Later in the 1930s, pressure sensitive tapes were developed.

The first metal bonding adhesive was developed by Nicholas de Bruyne in 1941. This material was used in the construction of aircraft. Later in the decade, epoxy resin adhesives were introduced. During the 1960s the extremely strong cyanoacrylate adhesives were developed. These products, called super glues, became adhesive when exposed to moisture

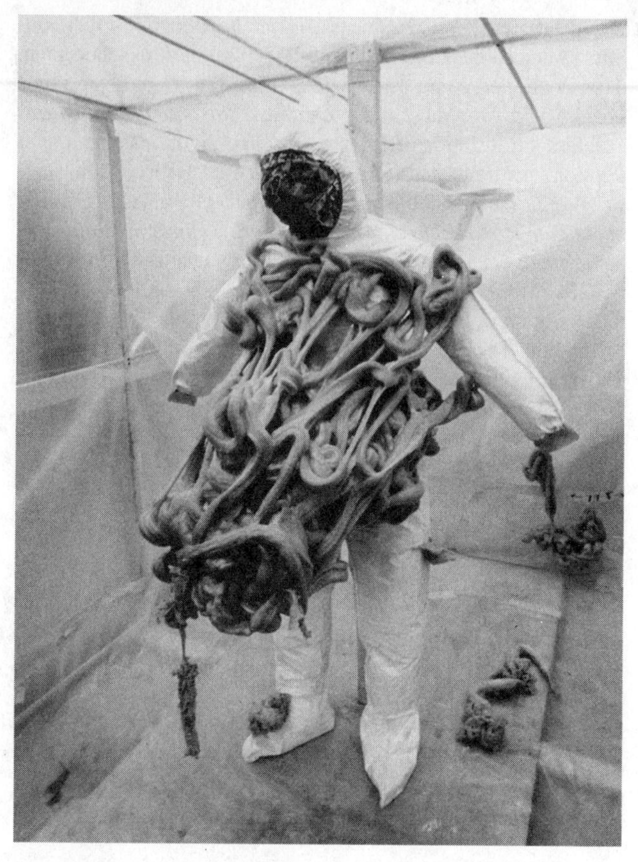

A mannequin covered in 'sticky foam' developed at the Sandia National Laboratories. The foam is made by a chemical fired by a gun-like dispenser from a range of up to 10 meters, and covers the target with an adhesive, entangling mass. Although this substance is harmless, it immobilizes the target long enough to allow other restraining measures to be applied, such as handcuffs. The sticky foam dispenser is part of the Less-Than-Evil program sponsored by the U.S. National Institute of Justice, and is intended to replace the use of firearms by police officers and prison guards. *(Photograph by Sandia National Laboratories/Science Photo Library, Photo Researchers, Inc. Reproduced by permission.)*

in the air. Other adhesives that were developed during this time include silicones and anaerobic adhesives. Since that time, most of the advances in adhesive technology have been the result of formulation modifications using varied polymers. In the late 1980s Post-It™ notes were introduced using a microstructured adhesive.

A variety of theories about how adhesives work have been proposed. While none of these satisfactorily describes all aspects of adhesion, they do attempt to explain observed phenomena. The leading theories of adhesion include the **diffusion**, electrostatic, surface energetics, and mechanical theories.

The diffusion theory states that adhesion is a result of the **solubility** of the adhesive to the substrate. When the adhesive is applied to the substrate, it is suggested that a **solution** of the substrate and adhesive is formed if the solubility characteristics are equal. This creates a stable phase that bonds the two surfaces together. While this theory provides some insights, it is mostly applicable to situations in which polymers adhere to each other. It is not applicable to systems where the substrate and adhesive are radically dissimilar.

The electrostatic theory suggests that adhesion is the result of differences in the **electronegativities** of adhering materials. According to the theory, when two materials are brought into contact there is an amount of electron transfer that occurs. This results in a charge layer being formed which causes the materials to stay together. An example of this theory can be seen when using static **electricity** to make a balloon stick to a wall.

The surface energetics and wettability theory describes adhesion in terms of intermolecular and interatomic forces. For these forces to have an effect, the adhesive must come in close contact with the substrate. This is only achieved through Awetting® of the surface. According to this theory adhesion is a result of **bonding** across the interface. An example of this theory is chemical adhesion that occurs when the chemical groups from the adhesive covalently bond with those of the substrate. Secondary adhesion occurs similarly through **hydrogen bond**ing.

The mechanical theory describes adhesion in terms of a physical interlocking of the adhesive with the substrate. Mechanical adhesion occurs when the adhesive material flows into and on the microscopically rough substrate surface. This creates a lock and key effect similar to velcro. The **viscosity** of the adhesive and the contact time with the substrate are important parameters for mechanical adhesion. Viscosity adhesion occurs by restricted movement due to the viscous nature of the substrates.

For a material to be a good adhesive it must has a variety of characteristics. It should have a liquid surface tension that is lower than the wetting tension of the substrate. It should be applied to a surface that is significantly rough to improve adhesion. For polymeric substrates, the adhesive should be somewhat mutually soluble allowing diffusion between the two to occur.

Adhesives can be classified by the type of delivery of the adhesive or by the polymer used in the adhesive. In general, there are five categories of adhesives including structural, natural, pressure-sensitive, hot melt, and **solvent**-based adhesives.

Structural adhesives are some of the strongest adhesive materials available. They are based on resin systems, typically thermosets, and are meant to serve as permanent bonds. They are supplied as low-molecular-weight polymers that solidify when polymerized. They are sold in a variety of forms including two-part systems, pastes, and films. The most common examples include epoxy resins, acrylic adhesives, phenolic resins, elastomeric adhesives, high-temperature-resistant adhesives, and urethane adhesives.

Epoxy resins are typically based on the polymerization reaction of bisphenol A with epichlorohydrin. The product is sold as a two-part system in which the user applies the epoxy resin, then an amine hardener that causes the resin to cure. Epoxy resins are used in the construction of aircraft and automobiles. They are also a component of plastic cement.

Phenolic resins have been used as adhesives since the early 1900s. They are produced by the polymerization reaction of phenol and formaldehyde. To cure, heat is typically required to drive off excess solvent. This type of adhesive is used in the production of plywood. They also are noted as the most adhesive material to aluminum.

A variety of acrylic adhesives are available. Anaerobic adhesives are unique reactive adhesives. They consist of a mixture of hydroperoxides and dimethacrylates that polymerize in the absence of **oxygen**. This type of material is useful in anchoring screws and bolts. One of the most famous and strongest acrylic adhesives is Super Glue™, or Krazy Glue™. This adhesive is based on cyanoacrylates which spontaneously polymerizes in moist air.

Many structural adhesives that are used for aerospace purposes are lightweight and high-**temperature** resistant. These are typically resins that have high glass-transition temperatures. An example of this type of adhesive is polyimide that is formed by the polymerization reaction of an aromatic amine with an aromatic anhydride. Since many of the structural adhesives become brittle after they cure, adhesives formulators developed elastomeric adhesives. These materials contain compounds that reduce brittleness without significantly reducing adhesion. This property is useful in applications where vibrational forces can stress adhesives. The final class of structural adhesives is the urethanes. These materials are based on the polymerization reaction of a diol with a diisocyanate. They are primarily sold as two-part structural adhesives. An important application of these kinds of adhesives is in the automotive industry. **Polyurethanes** are used to bond **polyester** cords in rubber tires.

Many natural-based adhesives are available. In general, these adhesives are not as strong as the synthetic adhesives. Natural rubber has been used as an adhesive for over 100 years. In 1825, the Macintosh raincoat, which consisted of two layers of cotton bonded by a layer of natural rubber, was introduced. **Protein** adhesives are used as structural adhesives. They are based on polyamino acids obtained from various animal and plant sources. Curing is typically dependent on heat. Starch-based adhesives are derived from plants. They are primarily used for binding paper and as envelope adhesives. Cellulose is another natural material that provides adequate adhesion. Cellulose nitrate was one of the first modified adhesives produced. It is a general all purpose adhesive which is waterproof and flexible. Modified methylcellulose is used to make wallpaper paste. Other natural adhesives are takifying resins that are derived from **coal**, petroleum or wood tar.

Pressure sensitive adhesives are polymeric-based adhesives that melt at room temperature. When pressure is applied to the adhesive, they become flowable thereby covering the substrate. As the pressure is removed adhesion takes place. Many tapes use this kind of adhesive material. Pressure-sensitive tape uses a blend of glycerol and abietic acid esters with natural rubber on cellophane. In addition to pressure-sensitive adhesives, hot melt adhesives have been developed. Since thermoplastics melt when heated and reform when cooled, they make good adhesives. This is the principle behind glue guns. **Nylon** polymers are often used for their formulation.

Solvent-based adhesives work through the action of the substrate or adhesive. The material is put on the substrate and when the solvent evaporates bonding occurs. Adhesion is aided if the solvent interacts with the substrate. An example is model airplane glue that tends to dissolve some of the **plastic**. This helps to create a solid weld. Latex adhesives are another type of solvent-based adhesive. They are polymeric materials that provide adhesion after their aqueous solvent evaporates. These are the same types of polymers that are used for latex paints. Latex adhesives are used for bonding pile to carpets.

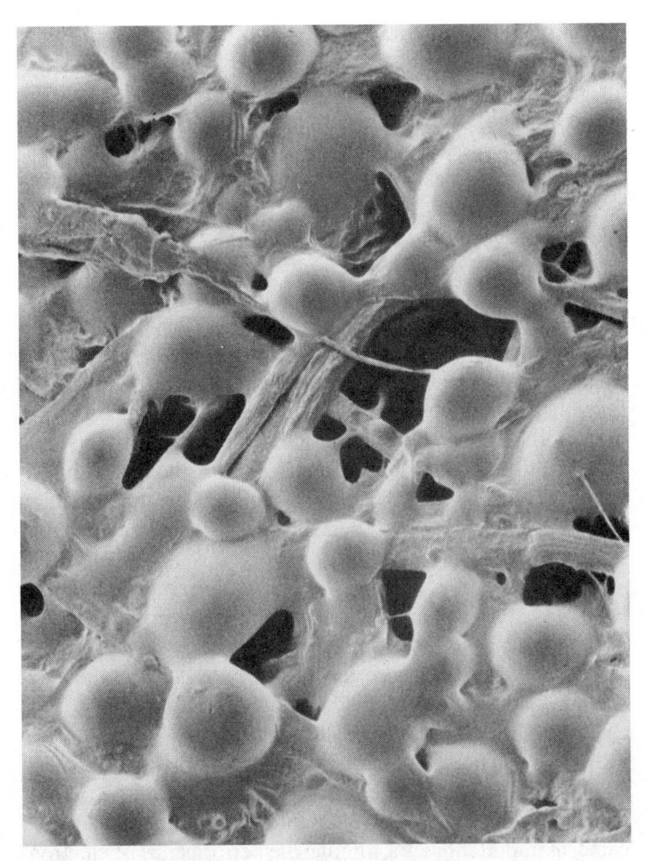

A colored scanning electron micrograph of the sticky surface of a Post-It note. The bumps are resin spheres containing glue embedded in the paper. The spheres, 15-40 micrometers in diameter, are made from urea formaldehyde resin. Every time the paper is pressed to a surface, a few spheres burst and release a film or glue. The paper can be used numerous times until the spheres of glue have been used up. *(Science Photo Library, Photo Researchers, Inc. Reproduced by permission.)*

While natural resins dominated the adhesive market years ago, in 1999 they account for less than 10% of the market. Instead **elastomers**, thermoplastics, and thermosets have the bulk of market share in this three billion-dollar plus market.

ADSORPTION

Adsorption is the accumulation of atoms, molecules, or ions at the surface of a solid or liquid as the result of physical or chemical forces. It differs from absorption, in that an adsorbed substance remains at the surface while an absorbed substance spreads throughout the absorbing material. An adsorbed substance is termed an adsorbate while the material on which adsorption occurs is the substrate. The release of an adsorbate is termed desorption.

Chemists make a distinction between chemical adsorption, or *chemisorption*, characterized by the formation of chemical bonds with the substrate, and physical adsorption or *physisorption*, which results from the van der Waals force. Chemisorption plays an essential role in **corrosion**, heterogeneous catalysis, and **electrochemistry**. Physisorption is a factor to be dealt with in the design of vacuum systems and is also used as a tool to study highly irregular surfaces. Further distinctions can be drawn between localized adsorption, in which the adsorbed molecules remain at the sites at which they were initially adsorbed, and delocalized adsorption, in which the adsorbed molecules can move along the surface. A distinction can also be made between monolayer adsorption and multilayer adsorption.

Adsorption is a very widespread phenomenon. In the open atmosphere many solid materials are covered by a layer of molecules adsorbed from the gas. Our understanding of the adsorption of **gases** has largely been a by-product of the early vacuum-based electronics technology. Beginning with the electric light, which involved heating thin metal filaments in an evacuated enclosure, through the development of electronic vacuum tubes and the television tube, the adsorption of gases and their subsequent desorption under reduced pressure became an important factor limiting the performance of electronic components. One of the first major researchers in the field, the American physical chemist **Irving Langmuir**, began his work on adsorption as an employee of the General Electric Company, seeking to extend the life of light bulbs with **tungsten** filaments. It was Langmuir who first realized that the surfaces of **metals** and other **solids** were generally characterized by unsatisfied valences and that adsorption could be the result of **covalent bond** formation with the substrate. In 1932, Langmuir received the Nobel Prize for his work in surface **chemistry**.

The adsorption of gases on solids is generally described by an equilibrium relation called an isotherm, which describes how the amount of adsorbed material varies with the partial pressure of the gas (with the **temperature** held constant). The amount of material adsorbed is sometimes expressed as the coverage, that is, as a fraction of the number of adsorbing sites that are occupied by an adsorbed **molecule**. One of the simplest isotherms, now generally known as the Langmuir isotherm, is obtained by assuming a rate of adsorption proportional to the product of the gas partial pressure and the amount of unoccupied surface balanced by a rate of desorption proportional to the amount of occupied surface. A slightly more complicated isotherm, the Brunauer, Emmett, Teller (BET) isotherm, allows for the possibility of multilayer adsorption. The latter isotherm is frequently used to determine the total surface area of a finely divided powder, by measuring the amount of **nitrogen** or another gas it absorbs as a function of pressure.

The adsorption of reactants is the first step in the heterogeneous catalysis. Many of the transition metals are adsorbers of **hydrogen**, **oxygen**, **hydrocarbon** molecules, and **carbon** monoxide. **Nickel** is used, for instance, to add hydrogen to oils with multiple bonds, referred to as unsaturated or polyunsaturated oils, to produce a solid form for cooking use. **Iron** is used in the synthesis of **ammonia** from nitrogen and hydrogen. Amorphous alumina-silica **composites** are used to crack or split hydrocarbon **chains** in petroleum refining. A dispersion of **platinum** on alumina particles is used in petroleum reforming, a combination of reactions that turn a mixture of hydrocarbons into a fuel better suited to use in internal **combustion** engines.

Adsorption is also a key factor in electrochemical reactions. When a metal electrode is immersed in an electrolyte **solution** it is generally covered by physically adsorbed **water** molecules. Anions like chloride and bromide also adsorb on electrode surfaces while smaller cations like **lithium**, **sodium** and **potassium** are generally too strongly bound to a small number of water molecules to bind directly to the electrode. The number and arrangement of adsorbed ions and molecules is a function of the electrode charge and may strongly affect the approach of ions in solution to the electrode, and thus the rates of electrochemical reactions.

An important case of adsorption on liquid surfaces is that of surface-active or surfactant molecules at the water-air or water-oil interface. These molecules typically involve a charged or polar hydrophillic group and a nonpolar hydrophobic group. By aggregating at the water-air interface, the molecules are able to obtain an energetically favorable configuration with their hydrophobic ends protruding from the water. The development of this type of adsorbed layer acts to lower the surface tension of the water, an effect predicted in the Gibbs adsorption isotherm, introduced by the American Physicist J. Willard Gibbs (1839-1903) in 1875. The ability of detergents to remove oily material from fabric is due to the presence of surfactant molecules which stabilize the oil-water interface, reducing its surface tension and enabling the oil to break up into small droplets which disperse through the water.

AEROGELS

The first aerogels were prepared in 1931 by Steven S. Kistler of the College of the Pacific in Stockton, California. Kistler was trying to prove that a **gel** contains a continuous solid network of the same size and shape as the wet gel. Kistler observed that a wet gel, if allowed to dry on its own, would shrink and become severely cracked. He correctly surmised that the solid component of the gel was microporous, and that the pore structure collapsed due to the liquid-vapor surface tension forces of the evaporating liquid. Kistler noted that in order to produce an **aerogel**, it would be necessary to replace the liquid of a wet gel with air by some means in which the surface of the liquid does not recede within the gel. The gels studied by Kistler were silica gels prepared by the acidic condensation of aqueous **sodium** silicate. Kistler succeeded in creating aerogels by washing the silica gels with **water**, exchanging the water for **alcohol**, and finally converting the alcohol to a supercritical fluid and allowing it to escape. Kistler's aerogels were very similar to those prepared today. They were transparent, of low **density**, and highly porous. Kistler later prepared aerogels from many other materials, including alumina, **tungsten oxide**, ferric oxide, **tin** oxide, **nickel** tartarate, **cellulose**, cellulose nitrate, **gelatin**, agar, egg albumen, and rubber. Upon leaving the College of the Pacific, Kistler joined Monsanto, which shortly thereafter began marketing an aerogel for use as a thixotropic agent in **cosmetics** and toothpaste. In the 1960s, the development of inexpensive fumed silica undercut the market for aerogel, and Monsanto ceased production of its aerogel.

In the late 1970s, the French government approached Professor Stanislaus Teichner at Universite Claud Bernard in Lyon, France, about developing a method for storing **oxygen** and rocket fuels in porous materials. Teichner's laboratory replaced the sodium silcate used by Kistler with tetramethylorthosilicate (TMOS). Hydrolyzing TMOS with **methanol** produced a gel in a single step. Drying these gels under supercritical alcohol conditions produced high quality silica aerogels.

In the 1980s, elementary particle physics researchers began using silica aerogels as a medium for the production and detection of Cherenkov radiation.

In 1983, researchers at Lawrence Berkeley Laboratory found that TMOS, which is highly toxic, could be replaced with tetraethylorthosilicate (TEOS), without sacrificing quality in the aerogel produced. They also found that they could replace the alcohol in the gel with liquid **carbon** dioxide, thereby easing the severity of the processing conditions.

In 1984 silica gel technologists in Sweden suffered a setback when the first pilot plant for the production of silica aerogel monoliths using TMOS exploded. The plant was subsequently rebuilt and is now operated by Airglass Corporation.

Also in the 1980s, workers at Lawrence Livermore National Laboratory prepared the world's least dense silica aerogel (and the world's least dense solid material). This aerogel had a density of 0.003 g/cm^3, about 3 times the density of air. Subsequent work at Lawrence Livermore led to the preparation of aerogels of organic polymers, including resorcinil-formaldehyde and melamine-formaldehyde aerogels. It was found that the resorcinol-formaldehyde gels could be pyrolyzed to give aerogels of pure carbon.

Silica aerogel prepared at the Jet Propulsion Laboratory in Pasadena, California, has flown on several Space shuttle missions, where it has been used to collect samples of high **velocity** cosmic dust.

The 1990s saw several new commercializations of aerogel technology, including operations by Nanopore and Aerojet Corporation in the United States, and Hoechst Corporation in Germany.

Aerosol

An aerosol is a **suspension** of very small particles of solid or liquid dispersed in a gas medium. Solid particles may be either inorganic or organic materials. Smoke, smog, fog and mists are all aerosols. Airborne pollutants and pollens that bother asthma sufferers are often contained in aerosols. Aerosols are also of concern with regard to contamination of sensitive electrical components. The term aerosol is also used to describe self-contained pressurized spray systems used to dispense various solid, liquid, and gas products. Aerosol products have four key components: a package which is capable of containing relatively high pressures; a valve which seals the can and controls dispensing; a gaseous propellant which forces the contents out of the can, and the active ingredients make the formula functional.

Aerosol spray can. ***(Photograph by Robert J. Huffman. Field Mark Publications. Reproduced by permission.)***

The first aerosol products were created during the early years of World War II to deliver **pesticides**. Over the last 40 years, aerosols have become tremendously popular for a variety of product types including personal care, **paints and coatings**, and foods. In 1996 aerosolized personal care products alone sold over a billion units in the U.S.

Containers used for aerosols are typically made of either tinplated steel or **aluminum** because of their superior strength. These metal cans are capped with valves that seal the pressurized propellant in the can and control dispensing of the contents. A valve consists of a diptube which feeds product from inside the can to the valve body where the propellant and liquid concentrate are mixed. The top of the valve is equipped with a button (or actuator) which is depressed to release the can's contents.

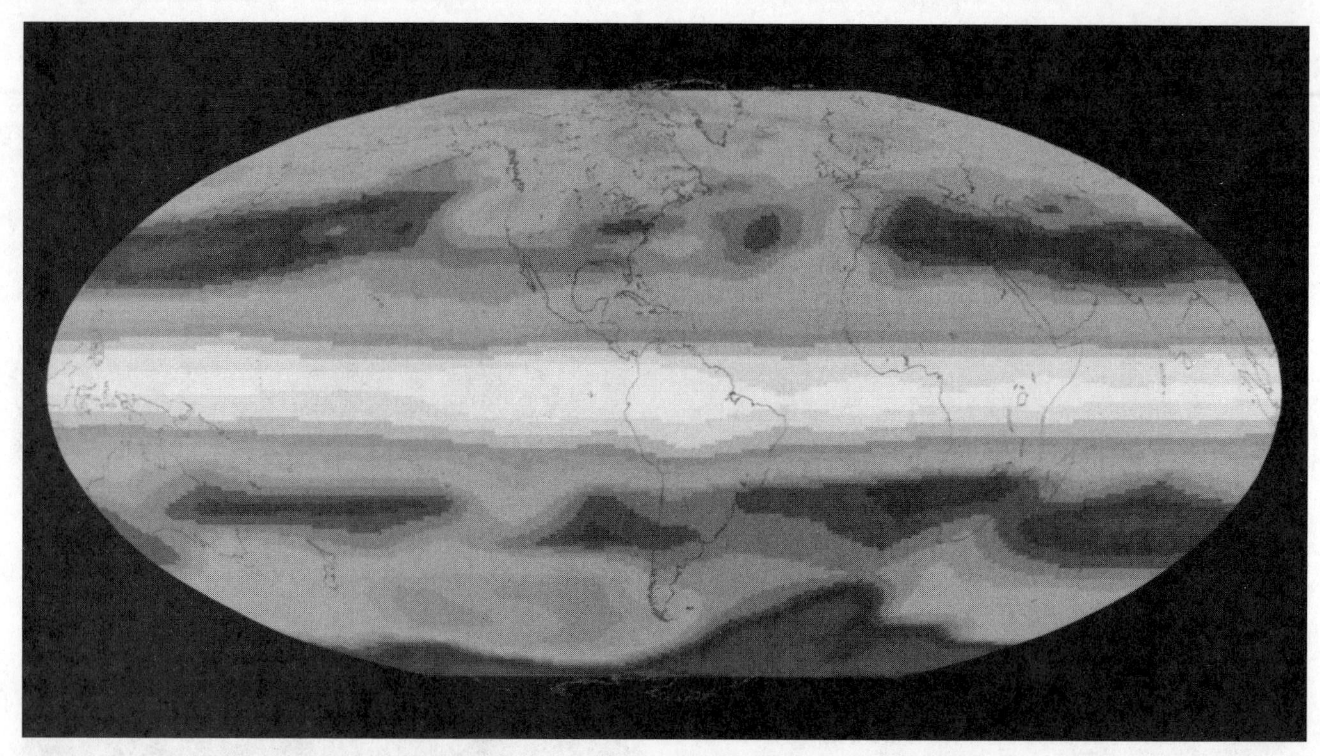

A false-colored satellite image of the Earth showing the aerosol optical depth of the atmosphere from region to region. Aerosol optical depth provides a measure of the concentration of aerosols in the atomosphere. The band along the equator correspond to the highest concentrations. This image was recorded by the SAGE II satellite between August 27 and October 6, 1986. *(Science Photo Library, Photo Researchers, Inc. Reproduced by permission.)*

Aerosol products contain a mixture of active ingredients, **solvents**, and propellants. Active ingredients are responsible for the product's primary function. For example, the active ingredients in a spray paint are the **dyes and pigments** that impart **color**. In a hair spray the actives are the resins that hold the hair in place. Solvents, such as solvents are **water, alcohol**, and **acetone**, are used to dilute active ingredients to the appropriate **concentration**. Propellants are liquified gasses which have a low **boiling point** and **vapor pressure** high enough to expel the concentrate from the container.

For over thirty years, chloroflurocarbons (CFCs) were the most common type of propellant used in aerosol products because of their low flammability and good solvency. But in the late 1970s, the discovery of CFCs' connection with destruction of the ozone layer led to the eventual ban of these materials, and alternatives, like hydrocarbons, became popular substitutes. However, in the late 1980s and early 1990s, hydrocarbons came under fire as possible contributors to air **pollution**. Some states, most notably California, have passed legislation which limits their use in certain products. Hydrofluorocarbons and dimethyl **ether** are two newer propellants that are being considered as **hydrocarbon** replacements. As of 1999, many states are still finalizing "clean air" legislation limiting key aerosol ingredients. While the aerosol industry maintains that these products have a minimal environmental effect, it is still anticipated that this regulatory trend will have serious impact on the future of aerosol formulations.

See also Ozone layer depletion

Agricola, Georgius (1494-1555)

German metallurgist and physician

Georgius Agricola, born Georg Bauer on March 24, 1494, in Glauchau, Saxony, was the son of a draper. He studied Greek and philosophy at the University of Leipzig and medicine and natural science in Bologna and Venice. In Venice, he edited Galen's works on medicine and became friends with the scholar Erasmus who encouraged him to write his own books.

In 1526 Agricola returned to Saxony as town physician in Joachimsthal, a mining center where he pursued his interest in minerals that could be used as drugs. He visited mines and **smelting** plants and acquired a substantial knowledge of all aspects of mining, including miners' diseases. Agricola moved to Chemnitz, a small mining town that had a **copper** smelter, to serve as town physician and later as mayor. He remained in Chemnitz the rest of his life.

Agricola's major work *De re metallica*, published a few months after his death, is an illustrated compendium of Saxon mining knowledge including the geology of ores, mine construction and operation, and smelting and refining processes.

Herbert and Lou Hoover translated this work into English in 1912. Agricola's other writings described a system of mineral classification based on geometrical form (no chemical analysis of ores existed in Agricola's time) and his theories of the creation of ore deposits. These works earned him the designation "father of **mineralogy**."

AGRICULTURAL CHEMISTRY

Modern agriculture depends quite heavily on the advances that have been made in science, and **chemistry** in particular, to maximize the yield of crops and animal products. **Fertilizers**, **pesticides**, and antibiotics play ever increasing roles in this field.

Fertilizers are perhaps the most widely used form of chemical in agriculture. Fertilizers are added to the soil in which crops are growing to provide nutrients required by the plants. Fertilizers can be divided into two categories: organic and inorganic. Organic fertilizers are derived from living systems and include animal manure, guano (bird or bat excrement), fish and bone meal, and compost. These organic fertilizers are decomposed by microorganisms in the soil to release their nutrients. These nutrients are then taken up by the plants. Inorganic or chemical fertilizers are less chemically complex and usually more highly concentrated. They can be formulated to provide the correct **balance** of nutrients for the specific crop that is being grown. Both organic and inorganic fertilizers supply the nutrients required for maximum growth of the crop. Inorganic fertilizers contain higher concentrations of chemicals that may be in short supply in the soil. The major or macro-nutrients in inorganic fertilizers are **nitrogen**, phosphorous, and **potassium**. These fertilizers also may provide other nutrients in much smaller quantities (micro-nutrients).

With the expansion of cities due to increases in population, there has been a loss of agricultural land. Appropriate use of fertilizers to increase crop yield has in part counterbalanced this loss of land. The use of fertilizers is not without controversy, however. There are concerns that adding supplements of nitrogen, particularly in the form of inorganic fertilizers, can be detrimental. It is thought that adding additional nitrogen to the soil can disrupt the action of nitrogen-fixing bacteria, an important part of the **nitrogen cycle**. If these nitrogen-fixing bacteria in the soil are killed, then less nitrogen is added naturally. As a consequence, more and more fertilizer must be applied. Inorganic nitrogen fertilizers are relatively cheap and they are often added to arable land in excessive amounts. The crop does not assimilate all of this extra nitrogen, but instead the nitrogen can run off the land and enter the **water** supply. High levels of nitrogen in water can lead to eutrophication which can trigger algal and bacterial blooms. These organisms remove **oxygen** from the water faster than it is replaced by **diffusion** and **photosynthesis**, causing some other aquatic animals to die from oxygen deprivation. High levels of nitrate in drinking water, which can be due to agricultural runoff, have been implicated in human health problems, such as blue baby syndrome (methemoglobinemia).

Pesticides are another important group of agricultural chemicals. They are used to kill any undesired organism interfering with agricultural production. Pesticides can be divided into fungicides, **herbicides**, and insecticides. Fungicides are used to control infestations of fungi, and they are generally made from **sulfur** compounds or heavy metal compounds. Fungicides are used primarily to control the growth of fungi on seeds. They are also used on mature crops, although fungal infestation are harder to control at this later stage.

Herbicides are weed killers that are used to destroy unwanted plants. Generally herbicides are very selective, since they would be useless for most applications if they were not. A general non-selective herbicide can be used to clear all plants from a particular area. However, appropriate treatment must be carried out to remove the herbicide or render it ineffective if that area is to be used for subsequent plant growth. Herbicides can be used to kill weeds that grow among crops and reduce the value of the harvest. They can also be used to kill plants that grow in fields used for grazing by animals, since some plants can be poisonous to livestock or can add unpleasant flavors to the meat or milk obtained from the livestock. Breeding and genetic manipulation are used to introduce herbicide resistance to crops, allowing the use of more broad-spectrum herbicides that can kill more weed species with a single application. Herbicides include a wide range of compounds, such as common **salt**, sulfates, and ammonium and potassium salts. In the 1940s 2,4-D (2,4 trichlorophenoxyacetic acid) was developed and this herbicide is still widely used today. The use of a related compound, 2,4,5-T (2,4,5 trichlorophenoxyacetic acid), is now controlled because of its potentially harmful effects. 2,4,5-T was a constituent of Agent Orange, a defoliant used during the Vietnam War.

Insecticides are chemicals that are used to kill insect pests. Insects can spread livestock diseases, can eat stored grain, and can feed on growing crops. Not all insects are harmful, and certain species of insects are needed to pollinate plants to ensure that they set seed. Many insecticides are non-selective and kill all insects, beneficial as well as harmful. Some insecticides, which are very effective at killing insects, have other problems associated with them. For example, DDT (dichlorodiphenyltrichloroethane) persists in the environment and is concentrated in the food chain. With high levels of exposure, DDT can directly kill fish and birds by paralyzing their nerve centers. In lower concentrations, it can weaken bird's egg shells and cause sharp declines in reproductive rates. Insecticides work in a number of ways. Some are direct poisons (chrysanthemic acids, contact poisons, systemic poisons), while others are attractants or repellents that move the insects to a different location (fumigation acrylonitrile). Some insecticides will only attack a particular stage of an insect's life cycle and this can make them more specific.

Antibiotics and growth **hormones** are routinely used as feed supplements for a number of animals. These additives are supplied to keep the animals free from disease and to help them grow to a marketable size as quickly as possible. However, the indiscriminate use of antibiotics can cause problems, since this can lead to the development of resistant strains of microorganisms or sensitization to the antibiotic among people who eat these animal products. The effects on humans of eating the meat of animals treated with growth hormones are poorly understood at the present time.

Agricultural chemistry has provided us with more and cheaper food than ever before. It has also allowed food to be produced in areas that previously were unsuitable for agriculture. The application of chemicals to farming has been one of the chemical success stories of the twentieth century. This is not to say that there have not been problems, the most famous being DDT. In the 1980s and 1990s there has been a backlash against the application of chemicals to foodstuffs in the western world. This has led to the production of organic and green products that are produced without artificial application of chemicals. These products are often more expensive to produce and this increased price is passed on to the consumer. However, there is much research to suggest that appropriate organic techniques can be competitive in cost with typical chemical agriculture and that Third World countries would benefit greatly from many organic practices.

In addition to the applications outlined above, chemicals also have other agricultural uses. For example, sulfur dioxide can be used to keep grain fresh and useable for a longer period of time than untreated grain. Other chemicals can be added to promote the ripening of fruits or the germination of seeds. It is difficult to estimate the monetary value of agricultural chemicals, but many multi-national corporations are involved in their manufacture and use. Agricultural chemistry has increased the diversity of the human diet and has led to a greater overall availability of food, both animal and plant.

See also Antibiotic drugs

ALBERT THE GREAT (1206-1280)

German philosopher, naturalist, and alchemist

Albert the Great (variously known as Albertus Magnus, Albert the German, Albert of Ratisborn, or Albert of Cologne), was the only scholar of his age to have received this title of ''the Great,'' which was also used before his death. He is also referred to as *doctor universalis* for his wide spectrum of knowledge and interests, as an advocate for Aristotelianism at the University of Paris, and as the teacher of Thomas Aquinas.

Albert was the eldest son of the Count of Böllstadt, a wealthy German lord, and was born in 1206 in the castle of Lauingen on the Danube River in the southern German province of Swabia. Albert's early schooling included instruction in the arts as well as acquiring detailed knowledge of natural phenomena, which later proved its significance in his botanical writings. After his early schooling, Albert attended the University of Padua in northern Italy to study the liberal arts. In the summer of 1223, when Albert was 16 years old, a group who called themselves ''Brothers'' (''Friars'') of the Dominican Order caught Albert's attention. This group's purpose was to counteract the heresies of the Roman Catholic Church, and to evangelize Christians. Jordan of Saxony, the master general of the Dominican Order, journeyed to Padua in hopes of attracting new members. Of the 10 students seeking admission, Albert became one of them, despite the strong opposition posed by his family. He continued his studies at the universities of Padua and Bologna, and in Germany, and then taught theology at several convents in Germany, lastly at Cologne.

At one point before 1245, Albert was sent to the Dominican convent of Saint-Jacques at the University of Paris. Here, he first was exposed to Aristotle's works, which were recently translated from Greek and Arabic, and the commentaries on Aristotle's works by Averroës, a twelfth century Spanish-Arabian philosopher. Albert began to teach and lecture on the Bible and theology. He obtained the Dominican chair ''for foreigners'' and was graduated master in the theological faculty.

While in Paris, Albert wrote commentaries and ''digressions'' on the Bible, all of Aristotle's known works, and Peter Lombard's *Sentences*, the theological textbook of the medieval universities. With this project, Albert was able to carry out his definition of the term ''experiment'' as a careful process of observing, describing, and classifying. He undertook this project so as to ''to make... intelligible to the Latins'' all the branches of natural science, logic, rhetoric, mathematics, astronomy, ethics, economics, politics, and metaphysics. Albert believed there was no ''double truth'' in opposition against each other; instead, faith and reason by way of philosophy and science are joined in harmony. Albert also defended the Averroist teachings, which held that only one intellect, which is common to all human beings, remains after death.

Albert began to gain a reputation as a renowned scholar due to his lectures and publications. He was soon ranked as an authority of natural history and natural sciences, and his stature was even equal to that of Aristotle's in some circles, who had so far produced any truly comprehensive treatises on the natural sciences. Furthermore, even Roger Bacon, a contemporary of Albert's, spoke of him as ''the most noted of Christian scholars.'' He continued to teach, preach and study in and organize the first *studium generale* (''general house of studies'') in Cologne in the summer of 1248. Albert became the Provincial of ''Teutonia,'' the German province of the Dominicans, and took on many burdensome administrative duties. Despite the extra work, Albert continued his writing and scientific observation and research.

In 1256, Albert became more involved in Church politics. Pope Alexander IV ordered Albert to the papal court, along with Thomas Aquinas and Bonaventure, a noted Franciscan scholar. They had to defend the mendicant orders against the attacks of the masters of the University of Paris at the Papal Curia at Anagni. The three men were successful in defending the right of the mendicants to teach at the university.

Although Albert resigned from his office of provincial in 1257 to resume teaching at the University of Cologne, he was appointed by the Pope as Bishop of Regensburg in 1259. Despite the objections from the Dominican Order and his own reluctance, he became bishop in 1260. When Pope Alexander IV died in 1261, Albert was able to resign, and again resumed reaching at Cologne.

Albert died in Cologne on November 15, 1280, exhausted by his work and austere life. He is buried in the Dominican Church St. Andreas in Cologne. In 1931, he was canonized a saint and heralded as a doctor of the church; in 1941 he became the patron of those studying the natural sciences.

ALCHEMY

Alchemists employed both physical procedures and magic to convert base metals, such as **lead,** into **gold,** and to create elixirs of eternal life. While alchemy is commonly defined as a field of knowledge developed in Europe from the Middle Ages until the eighteenth century, its roots reach into the distant past. The term *alchemy* is derived from the Arabic definite article *al* and *chymia,* a term which may have been used to indicate the practice of manipulating substances. Indeed, the practice of manipulating substances, exemplified by the art of cosmetics, was known to the ancient Egyptians, and predates alchemy itself.

Before the practice of alchemy was established in Alexandria, Egypt, at around 300 B.C., humankind had been changing substances and developing techniques that would be used by alchemists and later by chemists. Heat was used to cook and dry foods to make them more readily digestible and/or preserve them. Heat along with making specific combinations of materials was used in making pottery. Parts of specific plants, as well as seashells and some types of rock were treated with fermented liquids, slat and other materials to make pigments and dyes. Glass was first made from sand and **sodium carbonate**, found in dry lake beds around Alexandria, and later improved greatly by the addition of lime, calcium oxide. Techniques such as extraction, crystallization and fermentation were used to produce medicinals.

A number of techniques were also developed to obtain and purify metals. **Copper**, **silver**, gold and a combination of gold, and silver known as electrum can all be found as the metals in nature. **Iron** was probably also first known in its metallic state from meteors (in Arabic it was called the metal of heaven). Around 3000 B.C, the Sumerians found that they could obtain copper from certain earths, or ores, by heating the earths with a source of carbon such as straw. The processing of using heat and a reducing agent to obtain a metal is called smelting. Iron was produced at about the same time by smelting. Until it was discovered that carbon was necessary to produce a strong material, iron was not as useful as bronze, which is a combination of copper and **tin**. Artisans learned to distinguish ores by their characteristics such as color, texture, color produced in a flame, smell and so forth so that they could eventually produce a number of different metals. By 300 B.C., seven metals were widely known in the region at the eastern end of the Mediterranean Sea: gold, silver, electrum (thought to be a separate metal), copper, iron, tin, and lead. Iron and some high quality steel was also known in sub-Saharan Africa, India, where steel was used for surgical instruments, China, and Japan, where it was used for swords and knives, and in what is now Turkey, where it was used for weapons.

Artisans fabricated apparatus to carry out techniques that are still important today. ''Maria the Jewess'' is known as the inventor of several types of apparatus, including a distillation still, the hot-ash bath and water bath (the double boiler is referred to in France as the bain-marie for this reason). Others developed extraction apparatus much like a percolator, and numerous types of vessels for extraction, **distillation** and sublimation.

The period of the practice of alchemy is generally accepted to be from about 300 B.C. to the eighteenth century. The beginning is marked by the effort to determine the nature of matter in a fundamental way using physical processes to manipulate matter and observe the results. This period marked the initial attempts to understand transformations of matter rather than develop and repeat recipes. The beginning of alchemy is attributed to the spread of ideas of the Greek philosophers to what is now Egypt as a result of the conquests of Alexander the Great (356-323 B.C.). Several of the Greek philosophers developed concepts of the material world based on observations of natural phenomena rather than on myths. Among the most important of these concepts was the extension of the four-element idea of Empedocles (fifth century B.C.) by Aristotle (384-323 B.C.). Empedocles postulated that all matter is composed of various proportions of the elements (which he called roots): earth, water, air and fire. Aristotle associated qualities to each of these four elements: cold and dry were associated with earth, cold and moist with water, hot and moist with air and hot and dry with fire. Aristotle justified his idea with the observations that wood when burned produces smoke (air), a gummy residue (water), ash (earth) and fire, so all four elements are present in wood. The approach of the Greek philosophers such as Aristotle that led to alchemy and eventually chemistry was the use of observation to develop a theory that was then tested for logical consistency with other observations. Alchemists went beyond the philosophers by using experiments, transforming matter themselves, to test their ideas.

Alchemists struggled to find ways that qualities, such as inherent hotness or moisture or purity (gold representing the purity of metals) could be extracted from materials and recombined to make valuable products such as precious metals or medicines. The Arab world expanded greatly during the time of the founder of Islam, Mohammed (c. 570-632), and a few decades after his death. From the mid-seventh century to the mid-eighth century, Moslems conquered the region from Persia (now Iran, Iraq and Turkey) to Spain. Alchemy spread with the Moslem conquest, bringing into Europe the techniques developed by artisans over many centuries, as well as Greek ideas, absorbed by Moslem scholars, about the nature of matter.

Alchemists in Europe were especially intrigued, as many Arab alchemists were, with mercury and sulfur. They could quite readily convert mercury from brightly colored solids (actually oxides or sulfur compounds of mercury) to shiny, metallic liquid metal and reverse the process. They also readily converted sulfur from a bright yellow solid to solids of other colors and to a gas and to the very reactive acid, sulfuric acid (called vitriol or vitriolic acid for its destructive action). Attempts to transform matter into gold were often based on the idea that gold could be achieved by transforming a material so that its properties became more and more like those of gold; properties such as yellowness, heaviness and softness. Alchemists found many ways to make yellow solids, heavy alloys, and malleable alloys but, of course, never gold itself.

Another objective of many alchemists was an elixir that would prolong life. A Swiss alchemist who called himself Paracelsus, or ''better than Celsus,'' the influential Roman medical writer who lived in the first century A.D., assembled the most extensive compendium of chemical treatments. Paracelsus (1493-1541) undertook a systematic study of chemical reactions to develop medicines and recorded his results early in the sixteenth century. He found that solutions of a variety of materials in alcohol (tinctures) were effective in treating several skin disorders and that mercury was effective against syphilis. Several dramatic cures that he effected led others to adopt his approach.

Alchemy in other parts of the world, in particular India and China, has been associated mainly with the development of medicinals. As early as the third century, Chinese alchemists used formulations of mercury as elixirs and attempted to transmute other substances into gold to use the gold as an elixir to prolong life. Gold colored powders and liquids were a mainstay of the Chinese alchemist's practice. From the second century, Indian alchemists recorded their procedures for producing medicines as well as metals and other practical substances such as alcohol, acids, alkalis and useful salts of the common metals.

Alchemy is often associated with secrecy and unscrupulousness, and with good reason: some alchemists were charlatans. A typical trick was to coat a gold object with a metal that could be dissolved by an acid. The object would be dipped into the acid, most likely with appropriate fanfare and perhaps incantations and ceremony, to reveal the transmuted gold. Others purported to have transmuted metals or concocted elixirs and sold their recipes to others.

Several alchemists and others interested in the nature of matter wrote extensive accounts of what was known at the time that were not shrouded in secrecy. The thirteenth-century scholar, Bartholomew the Englishman, who taught at the University of Paris, wrote a nineteen-volume treatise entitled *Book of the Properties of Things* that included Aristotle's ideas as well as ideas of the Arab alchemists concerning the importance of mercury and sulfur. A French Dominican priest named Vincent de Beauvais compiled an 80-volume encyclopedia titled *Mirror of Nature*, meant to summarize all known science and natural history. **Vannoccio Biringuccio** wrote the first printed book on metallurgical chemistry, *Concerning Pyrotechnics* in 1540. **Georgius Agricola** wrote *On Metals*. His descriptions of assaying, smelting procedures, refining, production of glass and other processes in metallurgy and geological chemistry were used for over two centuries. Even those writings of alchemists who did not hide their procedures in secret codes are often hard for modern scientists to decipher, however. The materials alchemists used were typically impure mixtures whose composition varied according the site from which they originated. This also made it hard for the procedures developed by one alchemist to be repeated reproducibly by others even when the procedures were revealed.

The transition from alchemy to **chemistry** occurred in the eighthteenth century. The establishment of chemistry as the science we know it today required several critical changes in approach. The scientific method of testing hypotheses by experimentation, using only information that could be observed, not the pronouncement of an authority figure, was the first step. The idea of basing judgements about nature only on observation has been attributed to Roger Bacon (c.1214-1294), who espoused this basic concept of empirical science at a time when theology was the dominant intellectual discipline. It took another five hundred years, however, for Bacon's ideas to become relevant to changes in the nature of **matter**, because many of the most common chemical processes (such as combustion, oxidation of metals, and reduction of metal ores to metals) involve invisible gases, and early scientists, including alchemists, alchemists did not appreciate the importance of weighing materials before and after chemical changes occur.

It was not until the latter half of the seventeenth century that experimenters such as **Robert Boyle** determined properties of gases and even later that gases were identified as specific, distinct substance. The first gas to be so characterized was carbon dioxide, discovered by the physician **Joseph Black**, who termed it ''fixed air'' and determined that it was a product of combustion and respiration as well as a component on calcium carbonate. Nitrogen was discovered by **Daniel Rutherford** as the unreactive component of the atmosphere, the gas that remained after oxygen and carbon dioxide were removed. **Joseph Priestley** (1733-1804), who also prepared hydrogen chloride, ammonia, nitrous oxide, and other gases, discovered oxygen. Priestley also invented soda water (for which he received his greatest award, the Copley Medal of the Royal Society), by forcing carbon dioxide produced by fermentation into water.

Perhaps the most crucial step in the development of chemistry as a science was the careful experimentation of **Antoine Lavoisier** in the latter half of the eighteenth century, culminating in his postulation of the existence of a relatively small number of distinct and immutable chemical elements that could be combined in fixed ratios to produce chemical compounds with their own distinct properties. Tragically, Lavoisier's own work was terminated when he was beheaded, principally for his role as a tax collector, in the French revolution, but his ideas were widely accepted, leading to a fundamental change in the way that chemical changes were examined and explained.

ALCOHOL

What is commonly referred to as alcohol is just the most widely known example of a group of organic compounds which are collectively known as alcohols. The alcohol we drink is correctly called **ethanol** or ethyl alcohol and it has the chemical formula of C_2H_5OH. The alcohol functional group contains an -OH, or hydroxyl functional group, attached to **carbon**. Phenols are a subgroup of alcohol in which the hydroxyl groups are attached directly to the carbon atoms of a **ring** structure. Phenols have markedly different characteristics from the rest of the alcohols and the phenol group is discussed in a separate entry. Aliphatic dihydric alcohols are known as **glycols**.

If a compound possesses only one hydroxyl group then it is said to be a monohydric alcohol. Di-, tri-, and polyhydric alcohols also exist. Monohydric alcohols derive their names from the appropriate **hydrocarbon** with an ending of -ol. For example, the alcohol derivative of **methane** is **methanol** and of ethane is ethanol. If the hydroxyl group is bound to a carbon **atom** that is itself bound to three other carbon atoms then the resultant alcohol is a tertiary alcohol. A secondary alcohol has the hydroxyl group bound to a carbon atom bound to two other carbon atoms. A primary alcohol (such as ethanol) has the hydroxyl attached to a carbon atom which is itself attached to either one or no other carbon atoms.

The production of alcohols is carried out in different ways depending on the start products and the alcohol required. When unsaturated organic compounds are used as the starting point the alcohol is made by the addition of **water** to the double or triple bond. Methanol can be made by mixing **hydrogen** and **carbon monoxide** at high temperatures and pressure over a **zinc oxide**-zinc chromate catalyst.

One of the simplest ways to manufacture alcohol is by **fermentation**. Sugar is taken and in the presence of an **enzyme** (a biological catalyst) ethyl alcohol and **carbon dioxide** are produced. When alcohol is produced for human consumption, zymase—provided by yeast—is used as the catalyst.

The lower alcohols (those with only a small number of carbon atoms) are readily soluble in water. As the size of the **alkyl group** increases, i.e. as the number of carbon atoms increases, the water **solubility** decreases.

The **boiling point** of an alcohol is always higher than the hydrocarbon from which it is derived. This is due to the hydrogen **bonding** that occurs between the hydroxyl groups. Extra **energy** must be added to the system to overcome these extra bonds in order for boiling to occur. With alcohols containing several hydroxyl groups the level of hydrogen bonding is much higher and consequently the level of energy required to move from the liquid to the gaseous phase is yet higher. The hydrogen bonding can still occur in the gaseous phase, giving paired molecules of alcohol. This can then give a **molecule** with twice the expected **molecular weight** when analysis, such as gas **chromatography**, is undertaken.

When an alcohol is reacted with a strongly electropositive metal, such as **sodium**, the products are hydrogen and a crystalline, ionic salt of the metal, called an alkoxide. In this type of reaction primary monohydric alcohols are the most reactive.

Oxidation occurs with alcohols, and the products depend upon the alcohol reacted and the conditions under which the reaction occurs. Primary alcohols lose two hydrogen atoms to produce an aldehyde. Further oxidation, by the addition of **oxygen**, converts the aldehyde to a carboxylic acid. For example ethanol loses two hydrogen atoms to give ethanal and when an oxygen atom is added to this the product is ethanoic (acetic) acid. Secondary alcohols undergo the same process but **ketones** are produced. Tertiary alcohols are oxidized with difficulty, but the result of the first stage is a mixture of ketone and **carboxylic acid**. These oxidation reactions can be used as a test to distinguish between the different classes of alcohols. The most regularly employed oxidizing agent for this set of reactions is a mixture called chromic acid, a mixture of sodium or **potassium** dichromate and **sulfuric acid**. It is also possible to oxidize primary alcohols using atmospheric oxygen with **platinum** or **copper** as a catalyst. In the case of ethanol, oxidation can occur under the action of certain species of bacteria. This reaction is responsible for the souring of wine and it is commercially employed in the manufacture of vinegar.

Reduction of an alcohol produces the corresponding alkane, by loss of the hydroxyl group. This is carried out using an aqueous **solution** of hydrogen iodide reacting in the presence of red phosphorus. This mixture is one of the most powerful reduction agents for use on organic compounds. The red phosphorus is present to make more hydrogen iodide from the **iodine** gas that would be otherwise liberated.

Complete **combustion** of alcohols produces carbon dioxide and water. Alcohols can be dehydrated in two distinct manners. If the alcohol (with the exception of methanol) is heated with **aluminum** oxide, **phosphoric acid** or excess sulfuric acid, an alkene is formed along with a molecule of water. The second type of alcohol **dehydration** occurs when the alcohol is reacted, in excess, with sulfuric acid. This reaction produces an **ether** and water.

Esterification of alcohols can occur with either an organic or an inorganic acid. Both types of reaction are reversible and produce an ester and water. This reaction is catalyzed by hydrogen ions. The hydrogen is usually supplied by sulfuric acid which also acts as a dehydrating agent, thus giving the maximum yield of ester.

Alcohols are used for a number of commercial and industrial processes. Methanol is manufactured chiefly for conversion into **formaldehyde**. The formaldehyde is then reacted with **urea** or phenol to make a number of **plastics**, including bakelite. Methanol is also used as a paint solvent, antifreeze, and as an additive to ethanol to make it undrinkable. Ethanol is used to provide the alcohol content in alcoholic drinks. It is also used as an additive to **gasoline** to make a compound called gasohol. Gasohol has been extensively used in some countries in South America to decrease their reliance on imported oil.

See also Alkane functional group; Alkyl group; Phenol functional group

ALDEHYDES

Aldehydes are a class of highly reactive organic chemical compounds that contain a **carbonyl group** (in which a **carbon atom** is double-bonded to an **oxygen** atom) and at least one **hydrogen** atom bound to the alpha carbon (the central carbon atom in the carbonyl group). The aldehydes are similar to the **ketones**, which also contain a carbonyl group. In the aldehydes, however, the carbonyl group is attached to the end of a chain of carbon atoms, which is not the case with the ketones. The name aldehyde was given to this group of compounds in the nineteenth century by German chemist **Justus von Liebig** (1803-1873) and was taken from a mock Latin phrase, **alcohol** *dehydrogenatus*, describing their preparation. (The first aldehyde was prepared by removing two hydrogen atoms (dehydrogenation) from **ethanol**.) The general formula for an aldehyde is RCHO, with R representing an **alkyl group**.

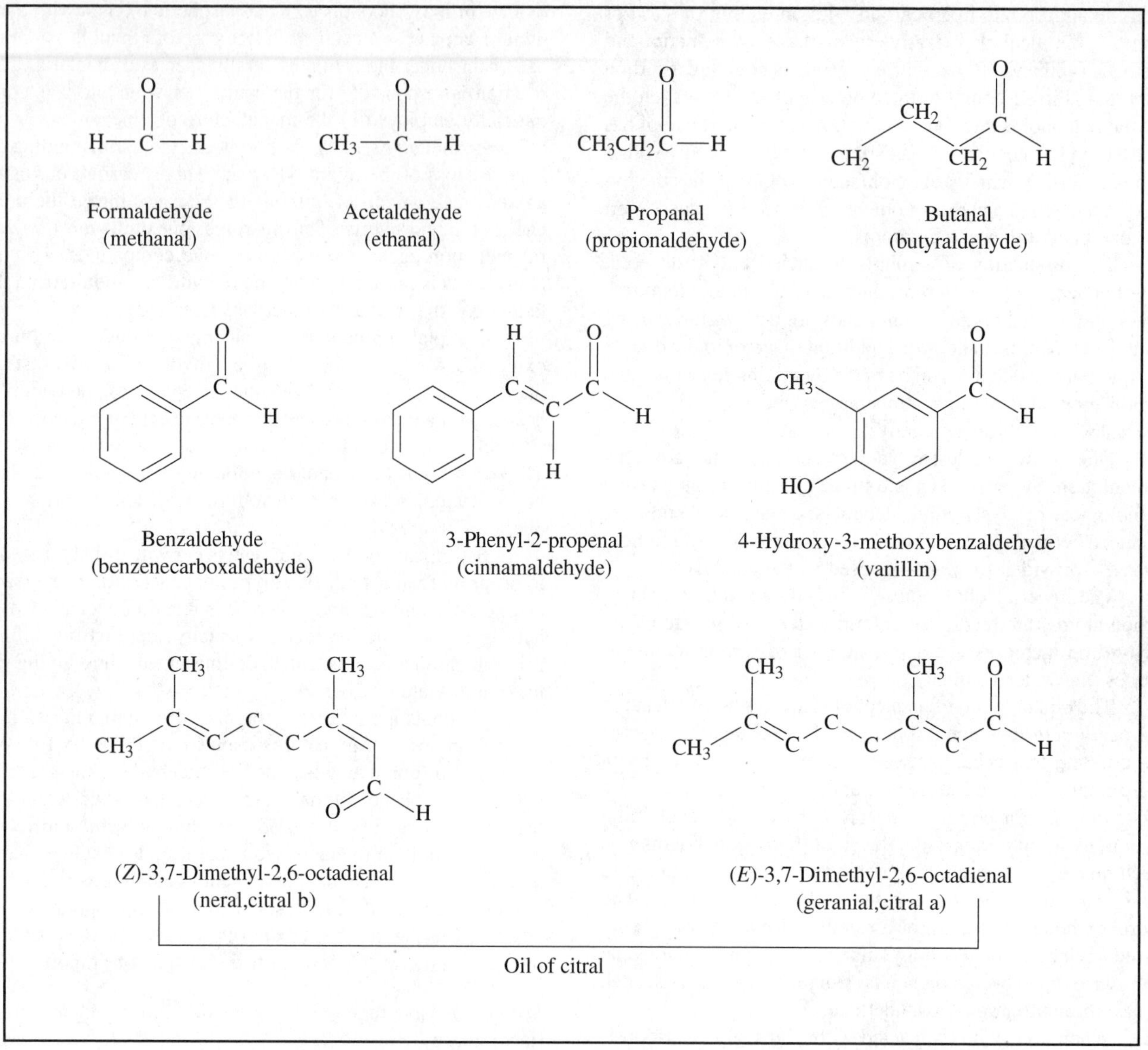

Formaldehyde (methanal)

Acetaldehyde (ethanal)

Propanal (propionaldehyde)

Butanal (butyraldehyde)

Benzaldehyde (benzenecarboxaldehyde)

3-Phenyl-2-propenal (cinnamaldehyde)

4-Hydroxy-3-methoxybenzaldehyde (vanillin)

(*Z*)-3,7-Dimethyl-2,6-octadienal (neral,citral b)

(*E*)-3,7-Dimethyl-2,6-octadienal (geranial,citral a)

Oil of citral

Aldehyde functional group. ***(Illustration by Electronic Illustrators Group.)***

The names of specific aldehydes are generally taken from the name of the alkyl group with the addition of an ''-al'' ending, e.g. methanal, ethanal. Aldehydes are generally colorless **liquids** or (for those with a **ring** structure) **solids.** They can be prepared from the **oxidation** of primary alcohols. The CHO group of the aldehydes does not form hydrogen bonds, unlike the OH group of the alcohols. As a result, the boiling points of aldehydes are considerably lower than the boiling points of corresponding alcohols. For example, menthanol boils at 150.8°F (66°C) and methanal (formaldehyde) boils at -5.8°F (- 21°C). Most aldehydes are soluble in **water**; their **solubility** decreases as their **molecular weight** increases. Aldehydes are very easy to detect by smell. Some are very fragrant, and others have a smell resembling that of rotten fruit.

Aldehydes have a strong tendency to join together to produce polymers. This makes the whole group very important in the manufacture of **plastics**, and they react quite readily with a large number of other molecules by simple addition. Aldehydes can be converted to alcohols by the addition of two hydrogen atoms to the central carbon oxygen double bond (reduction). Oxidation of an aldehyde breaks the double bond within the carbonyl group to give a carboxylic acid. Aldehydes also readily undergo substitution reactions with **halogens**. If **chlorine** gas is bubbled through ethanal, the hydrogen atoms of the **methyl group** are replaced by chlorine atoms to give chloral. Chloral is a strong hypnotic and is also the first step in the manufacture of DDT, carbon tetrachloride. Chloroform, and silicone rubbers.

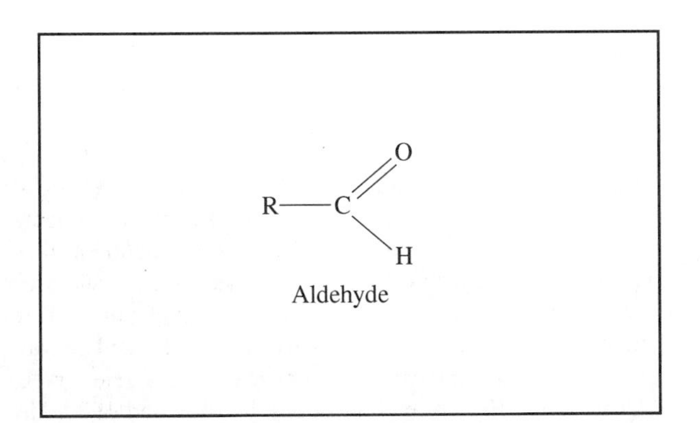

Chemical structure of aldehyde. ***(Illustration by Electronic Illustrators Group.)***

Methanal (formaldehyde) is the simplest aldehyde. The central carbon atom in the carbonyl group is bound to two hydrogen atoms. Its chemical formula is $H_2C=O$. Methanal, discovered by Russian chemist Aleksandr Butlerov in 1859, is a gas in its pure state. It is either mixed with water and sold as formalin solutions or as a solid polymer called paraformaldehyde. The rather small methanal **molecule** is very reactive and has found applications in the manufacture of many organic chemicals such as dyes and medical drugs. Methanal is also a good insecticide, and it is used to kill germs in warehouses and ships. It is probably most familiar to the general public in its application as a preservative. In biological laboratories, animals and organs are suspended in **formaldehyde** solutions, which are also used as embalming fluid to preserve dead bodies from decay.

Ethanal (acetaldehyde) is the name of the shortest carbon chain aldehyde. It has a central carbon atom that has a double bond to an oxygen atom (the carbonyl group), a single bond to a hydrogen atom, and a single bond to another carbon atom connected to three hydrogen atoms (methyl group). Its chemical formula is written as CH_3CHO. Ethanal is one of the oldest known aldehydes and was first made in 1774 by Swedish chemist Carl Wilhelm Scheele (1742-1786). Its structure was not completely understood until 60 years later, when Justus von Liebig determined the constitution of ethanal, described its preparation from ethanol, and gave the name of aldehydes to the chemical group.

The next larger aldehyde molecules have longer carbon atom **chains** with each carbon atom connected to two hydrogen atoms. This group of aldehydes is called aliphatic and has the general formula $CH_3(CH_2)_nC$ HO, where n=1–6. When n=1, the aldehyde formula is CH_3CH_2CHO and is named propanal (propionaldehyde); when n=2, $CH_3(CH_2)_2C$ HO or butanal (butyraldehyde). The aliphatic aldehydes have irritating smells. For example, the smell of butanal, in low concentrations, resembles that of rotten butter. These medium-length aldehyde molecules are used as intermediates in the manufacture of other chemicals such as **acetone** and ethyl acetate used in nail polish remover. They are also important in the production of plastics.

A silver mirror test for aldehydes. ***(Photograph by Jerry Mason, Photo Researchers, Inc. Reproduced by permission.)***

Fatty aldehydes contain long chains of carbon atoms connected to an aldehyde group. They have 8-13 carbon atoms in their **molecular formula**. The fatty aldehydes have a very pleasant odor, with a fruity or a floral scent, and can be detected in very low concentrations. Because of these characteristics, the fatty aldehydes are used in the formulation of many perfumes. They are also added to soaps and detergents to give them their "fresh lemon scent."

The aromatic aldehydes have a **benzene** or phenyl ring connected to the aldehyde group. The aromatic aldehyde molecules have very complex structures but are probably the easiest to identify. The odor of cinnamon found in various products is due to an aromatic aldehyde of complex structure named 3-phenyl-2-propenal (cinnamaldehyde). Another aldehyde, 4-hydroxy-3-methoxy-benzaldehyde (vanillin), is a constituent of many vanilla-scented perfumes.

ALDER, KURT (1902-1958)

German chemist

Kurt Alder was recognized for his contribution to synthetic organic **chemistry**, especially for the reaction which has been

Kurt Alder.

called the diene reaction, or the **Diels-Alder reaction**, after Alder and his mentor-colleague, **Otto Diels**. This reaction, discovered in 1928, was so useful in the synthesis of every type of organic compound that Diels and Alder were awarded the Nobel Prize in chemistry in 1950 for their development of the method. It was Alder who explored the reaction deeply, and wrote many papers on the nature of the reactants and products, including the **stereochemistry**, or geometric consequences, of the reaction.

Alder was born on July 10, 1902, in Königshütte, Germany. His father, Joseph Alder, was a schoolteacher in the nearby town of Kattowitz. Kurt attended the local schools, but at the end of World War I, the region in which the Alders lived became part of Poland, and Joseph Alder moved his family to Kiel in order to retain their German citizenship. Kurt Alder completed his secondary education in Berlin, and enrolled in the University of Berlin in 1922. He began to study chemistry at Berlin, but transferred to the Christian Albrecht University (now University of Kiel) and worked under Otto Diels, a professor of **organic chemistry** and the director of the chemical institute. Alder obtained his Ph.D. in 1926, and remained at the university as Diels's assistant. Diels and Alder collaborated in their research at Kiel for a decade. Alder was promoted to reader in organic chemistry in 1930 and extraordinary professor in 1934. The direct interaction between the two chemists ended in 1936 when Alder accepted an appointment as director of scientific research at the Bayer Werke laboratory at Leverkusen, a branch of I. G. Farbenindustrie. This position gave Alder direct experience with **industrial chemistry**, and his expertise was directed to the development of synthetic rubber, which was a goal of organic chemists in every nation at that time. Alder left Bayer in 1940 and returned to academic life. He was appointed to the chair of experimental chemistry and chemical technology at the University of Cologne, and he was also director of the university's Chemistry Institute. He served as Dean of the Faculty of Philosophy in 1949 and 1950. He remained at Cologne until his death on June 20, 1958.

The diene synthesis is one of a few methods which has proven to be so useful to organic chemists that their development led to Nobel Prizes for the chemists who discovered them. The diene synthesis requires two components, the diene and the dienophile. The diene contains four atoms, in which the first and second are joined by a double bond, as are the third and fourth: 1=2–3=4. The greatest number of dienes have **carbon** as the numbered atoms, but **nitrogen**, **oxygen**, or **sulfur** may be substituted for a carbon **atom**. The dienophile contains a double bond: 5=6. Again, the atoms are usually carbon, but may be other elements. When the diene and dienophile react, a **ring** containing the six atoms is formed, with new single bonds between all the atoms except 2 and 3, which is joined by a double bond. In general, all of the new bonds are formed in one step, in what is termed a concerted reaction. Because six-membered rings are found in many **natural products**, the diene synthesis has been used successfully in the syntheses of such complex molecules as reserpine, **morphine**, and many **steroids**. Commercial products have been synthesized with the reaction, including the insecticides dieldrin and aldrin (named for the chemists) and chlordane. Alder and his students carried out extensive investigations on the geometry of the ring formation, and discovered regularities which allowed the geometry to be predicted with certainty.

In the course of his research at Bayer Werke, Alder found that the reaction between butadiene, a diene, and styrene, a dienophile, gave a normal diene reaction under certain experimental conditions. However, by allowing the reaction to take place in the presence of peroxides, a polymer—a **molecule** consisting of similar or identical small molecules linked together—was formed, which was the synthetic rubber known as ''Buna S.'' This was developed commercially in Germany in the 1930s, and was ultimately important during World War II, when supplies of natural rubber were curtailed.

Alder discovered another reaction in 1943, which has been named the ene reaction. This is also a concerted reaction, and Alder recognized its essential similarity to the diene synthesis. In the ene reaction, one reactant contains three atoms, with a double bond between atoms 1 and 2, and a **hydrogen** atom attached to atom 3:1=2–3-H. The other reactant contains a double bond: 4=5. In the product, the hydrogen atom moves from atom 3 to atom 4, atom 1 joins atom 5, and a double bond forms between atoms 2 and 3: H–4–5–1–2=3. Alder called this type of reaction a substituting addition. The ene reaction also has found widespread use in synthesis. The mechanism of the diene reaction and the ene reaction was a **matter** of great interest for many years, for Alder and others. The theoretical foundation was not completely established until 1965, when **Robert B. Woodward** (Nobel Prize, 1965) and **Roald Hoffmann** (Nobel Prize, 1981) showed that these concerted reactions were governed by rules determined by quantum mechanics. The Diels-Alder reaction is one of many known ''cycloaddition reactions'' in which rings are formed from open-chain compounds, and the Woodward-Hoffmann rules accurately predict the course of these reactions.

Alder never married, and his greatest interests were his science, his colleagues, and students. He was committed to world peace and joined other Nobel laureates in 1955 in an appeal to governments to end war. His failing health limited his activities, and he died in Cologne at the early age of 55.

ALDOL REACTION

The aldol reaction, or aldol condensation, is a reaction between two open chain (aliphatic) aldehyde molecules to produce a compound called a beta hydroxyaldehyde (an aldol). This reaction is an example of a polymerization reaction and it occurs only with those compounds containing the CH_2CHO **functional group**.

An aldol is both an aldehyde and an **alcohol** at the same time, meaning that it possesses both -COH (aldehyde) and -OH (alcohol) functional groups. Aldols can be produced by reacting an aldehyde with a range of alkaline condensing reagents (for example, **sodium** ethanoate or dilute sodium hydroxide)in the presence of a catalyst (frequently **potassium** cyanide) at low temperatures. **Aldehydes** with a closed **ring** structure (aromatic aldehydes) undergo a similar reaction called the Cannizzaro reaction. [This reaction is named for Italian chemist **Stanislao Cannizzaro** (1826-1910) who discovered it in the nineteenth century.]

The product of this reaction is capable of losing **water** quite easily; this then returns to the starting point of unsaturated aldehydes. Aliphatic **ketones** (ketones with an open chain structure) react in a similar manner. This reaction can take place between aldehydes or between ketones and it can also take place between aldehydes and ketones. When a ketone is involved the resultant product is a beta hydroxyketone.

The simplest aldol—$CH_3CHOHCH_2CHO$—is produced from ethanal. This is the first stage in the manufacture of 1-butanol which is an important solvent for resins and lacquers. The United States produces some 40,000 metric tons of 1-butanol annually.

See also Polymer chemistry

ALKALI METALS

The alkali **metals** are **lithium** (Li), **sodium** (Na), **potassium** (K), **rubidium** (Rb), **cesium** (Cs), and **francium** (Fr). All of the alkali metals are found in group IA (1) of the **periodic table** of the elements.

Because chemical properties are determined by the outer **electron** configuration of elements, it is important to note that the **valence** (outer) shell of all alkali metals contains a lone s^1 electron. Lithium has a $2s^1$ outermost electron. Sodium has a $3s^1$ outermost electron. Potassium has a $4s^1$ outermost electron. Rubidium has a $5s^1$ outermost electron. Cesium has a $6s^1$ outermost electron. Francium has a $7s^1$ outermost electron.

All of the alkali metals seek to lose their outermost s^1 electron in to achieve the electron configuration of the nearest noble gas. Because they lose their valence electron easily, these elements are very reactive. In nature, alkali metals are found only as ions (charged particles formed when atoms gain or lose electrons) in compounds, never as free metals.

The atomic properties of the alkali metals vary according to well-established periodic trends regarding chemical and physical properties.

With regard to group IA (1) on the periodic table, the alkali metals increase in **atomic radius** as one moves down the group. As a result, lithium has the smallest atomic radius and francium has the largest.

Because the first **ionization energy** (energy required remove an element's outermost electron) decreases down a group on the periodic table, lithium has the highest first ionization energy (ionization potential) and francium the lowest. It requires far less energy to remove the $7s^1$ electron from francium than to remove the $2s^1$ electron from lithium.

All alkali metals have relatively low melting points. The melting point of alkali metals lowers as one moves down group IA (1) on the periodic table. The melting point of lithium is 393.062°F (200.59°C). Francium's melting point is only 80.69°F (27.05°C). Accordingly, the boiling points of alkali metals also decrease down the group.

The **boiling point** of lithium is 2456.33°F (1346.85°C). Cesium's boiling point is 1300.73°F (704.85°C). The boiling point for francium is undetermined.

The word alkali comes from the Arabic word for ''ashes,'' where the oxides of these metals were first found. Alkali is now used as another word for base; when the ashes containing the oxides of the alkali metals are dissolved in **water**, a base is formed. For example, sodium **oxide** dissolved in water yields **sodium hydroxide**, or lye.

The free element sodium was first produced in 1807 by the English chemist **Humphry Davy** by **electrolysis** of molten sodium hydroxide. He isolated potassium, another well known alkali metal, from potassium hydroxide in the same year. Lithium was discovered in 1817. Rubidium and cesium were discovered in 1860 by Robert Bunsen using the newly developed spectroscope which shows light-emission patterns that are characteristic for each element. Francium was discovered in 1939 as by the disintegration of an **isotope** of **actinium**.

The alkali metals are silver colored except for cesium, which is pale gold.

Alkali metals are so soft that they can be cut with a knife. Because the freshly exposed surface quickly oxidizes in air, these metals are stored in mineral oil or other non-aqueous **solvents**.

When an alkali metal is dropped into water, **hydrogen** is released, sometimes explosively. All of these metals dissolve in **ammonia** to form blue solutions that conduct **electricity** and function as strong reducing agents.

All alkali metals exhibit the **photoelectric effect** and are used in photoelectric cells. Cesium and rubidium lose their valence electrons especially easily when light strikes their polished surfaces and are photosensitive over the full visible spectrum.

Sodium compounds produce a distinctive yellow flame when burned. Potassium compounds burn with a lilac flame often masked by the stronger sodium yellow. A dark blue **glass** can be used to mask the yellow of sodium, allowing the potassium flame to be seen as red. Lithium compounds produce a crimson red flame. Rubidium and cesium flames are reddish violet or magenta. The compounds are used in fireworks, along with potassium perchlorate, chlorate, and/or nitrate as oxidizing agents.

Nearly all compounds of the alkali metals are soluble in water. **Sodium chloride**, containing ions of sodium and ions of **chlorine**, is well known as table **salt**. The global supply of sodium chloride is mostly mined as rock salt from man-made caves resulting from evaporation of ancient inland seas. Large deposits, consisting of up to 70% sodium chloride exist in a number of countries. The salt mine in Wielitzka, Poland, has been worked continuously for over 600 years. The deposit is 500 miles long, 20 miles wide, and 1200 feet thick, and includes a chapel carved into the salt. Other alkali salts are also mined. The Stassfurt mine in Germany yields potassium chloride and potassium **sulfate** as well as sodium chloride. Searles Lake in California, a brine-encrusted lake in the Mojave Desert, is a major source of lithium salts as well as sodium and potassium salts. Vast deposits of sodium nitrate are found in South America

Sodium and potassium compounds are leached from their parent rocks by weathering. Sodium concentrates in seas, but potassium is strongly absorbed by clays. Potassium compounds are usually derived from plants, although a deposit in Saskatchewan, Canada, formed from the crystallization of minerals in an ancient sea, is thought to contain ten billion tons of potassium chloride.

There are many industrial uses for alkali metals and their compounds.

Sodium is transported in railroad cars with heating units so that the metal can be easily loaded or unloaded as a liquid. Liquid sodium's low melting point and good **heat** transfer properties lead to use as a coolant in nuclear breeder reactors, especially on submarines. In sodium vapor lights, sodium vaporizes when electricity is passed through it. For most of the twentieth century, metallic sodium was heated with lead to form an **alloy** as the first step in the production of tetraethyl **lead**, an antiknock **gasoline** additive. Because of environmental concerns, lead in gasoline was gradually phased out in the United States and was no longer commercially available as of 1990. Sodium is also used as a chemical reducing agent in producing **titanium**, **zirconium**, **niobium**, and **tantalum** from their fused salts.

Sodium hydroxide, the most common base in teaching laboratories, is an important industrial chemical used making soap, dyes, and other chemical compounds such as sodium cyanide. Sodium hydroxide, also known as caustic soda, is used in the production of **aluminum** and by the pulp and **paper** industry. It is also a popular drain cleaner. Sodium hydroxide is produced from the electrolysis of brine, yielding chlorine as an important byproduct.

Sodium carbonate has been obtained from natural deposits since prehistoric times. Most of the sodium carbonate used in North America comes from large deposits of the mineral trona, a sodium carbonate/bicarbonate found in Wyoming. It can also be produced from sodium chloride. It is known as washing soda or soda ash, and is used for softening water, in the manufacture of borax, and in making glass, paper, detergents, and soap. Borax, or sodium borate, is used as a washing agent and as an insecticide; although effective against cockroaches and ants it is not highly toxic to mammals.

Because of the **solubility** of alkali metal salts, they find many uses in which it is the **anion** rather than the **cation** that is important. **Sodium bicarbonate** (baking soda) is used to treat excess stomach acidity. Sodiumfluoride, sodium fluorosilicate, sodium arsenate, sodium borate (borax), and sodium chlorate are all used as weed or pest-killers. Sodium chlorite, hypochlorite, perborate, and peroxide are used to bleach paper, cotton, and rayon. Sodium hyposulfite is used in the **reduction** of certain dyes. Sodium thiosulfate is used to dissolve unreduced silver salts in photographic processes. Sodium sulfide is used as a depilatory and in the manufacture of **sulfur** dyes. Sodium silicate is used to impregnate **wood**, to weight silk, as a mordant, and as an adhesive. **Sodium benzoate** is used as a food preservative. Sodium and potassium dichromate and potassium permanganate are powerful laboratory oxidizing agents. Potassium iodide is added to table salt because the iodide **ion** prevents goiter. Trisodium **phosphate**, or TSP, is a strong cleansing and disinfecting agent. Potassium chlorate is a powerful oxidizing agent and a source of **oxygen** in the laboratory.

Cream of tarter, potassium tartrate, is used in cooking. It is a hard, crystalline substance which forms as grapes are fermented into wine. Potassium perchlorate is the oxidizing agent used to set off fireworks.

Searles Lake supplies about one-half the world's supply of lithium, primarily as lithium chloride. Lithium has the lowest **density** of any metal. Lithium/magnesium alloys have an exceptionally high strength-to-weight ratio and so are important in aircraft and spacecraft design. Metallic lithium is used in lightweight **batteries**.

Lithium aluminum hydride is an important reducing agent in organic **chemistry**. One of the largest uses of metallic lithium, industrially and in the laboratory, is in the preparation of organolithium compounds such as methyl lithium as starting compounds for organic syntheses.

Alkali metals and their associated compounds are also important to biological systems.

Sodium and potassium ions are essential components of the human nervous system and brain and are present in all body fluids and essential for proper electrolyte **balance**. Potassium and **magnesium** ions are important cellular cations. Potassium chloride is used as a salt substitute by people who, because of high blood pressure, are on a restricted sodium diet.

Sodium and potassium phosphates and nitrates are used as **fertilizers**. Not only is the phosphate ion necessary, but potassium is necessary in maintaining water balance. Without potassium, plants are unable to absorb water and will not grow.

Lithium carbonate is used to treat manic depression, although the mechanisms for its effectiveness is not well understood it is known that the body cannot always distinguish between lithium and sodium. In order to maintain a constant lithium level, the body must maintain a relatively constant sodium level. If sodium levels fall, more lithium is retained, sometimes even reaching toxic levels.

ALKALINE EARTH METALS

Group 2 (IIA) consists of **beryllium**, **magnesium**, **calcium**, **strontium**, **barium**, and **radium**. This family of elements is known as the alkaline earth **metals**, or just the alkaline earths. Like those of the Group 1 metals, the oxides of the alkaline earths can be dissolved in **water** to form bases. Calcium **oxide** dissolves to form calcium hydroxide, magnesium oxide dissolves to form magnesium hydroxide, and so on.

Early chemists gave the name ''earths'' to a group of naturally occurring substances that were unaffected by **heat** and insoluble in water. The more soluble of the earths were called ''alkaline earths,'' a group of substances which we now know as the oxides of the elements of Group 2.

The compounds of Group 2 are usually found in the earth, in contrast to Group 1 compounds which concentrate in the ocean. **Calcium carbonate** is familiar as limestone, marble, coral, pearls, and chalk—all derived mainly from the shells of small marine animals. The weathering of calcium silicate rocks over millions of years converted the insoluble calcium silicate into soluble calcium salts, which were carried to the oceans. The dissolved calcium was used by marine organisms to form their shells. When the organisms died, the shells were deposited on the ocean floor where they were eventually compressed into sedimentary rock. Collisions of tectonic plates caused this rock to rise above the ocean floor to become the deposits that we know as limestone.

Caverns throughout the world are formed by the action of atmospheric carbonic acid (water plus **carbon** dioxide) on limestone to form the more soluble calcium bicarbonate. When the **solution** of calcium bicarbonate reaches the open cavern and the water evaporates, **carbon dioxide** is released and calcium carbonate remains. The calcium carbonate is deposited as stalagmites if the drops hit the ground before evaporating, or as stalactites if the water evaporates while the drop hangs from above.

Other minerals of alkaline earth metals are beryllium **aluminum** silicate (beryl), calcium magnesium silicate (asbestos), **potassium** magnesium chloride (carnallite), calcium magnesium carbonate (dolomite), magnesium **sulfate** (epsomite), magnesium carbonate (magnesite), **hydrogen** magnesium silicate (talc), calcium fluoride (fluorspar), calcium fluorophosphate (fluorapatite), calcium sulfate (gypsum), strontium sulfate (celestite), strontium carbonate (strontianite), barium sulfate (barite), and barium carbonate (witherite). Radium compounds occur in pitchblende, which is primarily **uranium** oxide, since radium is a product of the radioactive disintegration of U-238. Most pitchblende in the United States is found in Colorado.

Sir **Humphry Davy** first obtained magnesium, calcium, and strontium by **electrolysis** of their molten salts in 1808. Beryllium was isolated in 1828 by **Friedrich Wöhler**. Radium was not isolated until 1910, when **Marie Curie** succeeded in the electrolysis of radium chloride.

The alkaline earth metals, like the **alkali metals**, are too reactive to be found in nature except as their compounds; the two **valence** electrons completing an s-subshell are readily lost, and ions with +2 charges are formed. The alkaline earth metals all have a **silver** luster when their surfaces are freshly cut, but, except for beryllium, they tarnish rapidly. Like all metals, they are good conductors of **electricity**.

Only magnesium and calcium are abundant in the Earth's crust. Magnesium is found in seawater and as the mineral carnallite, a combination of potassium chloride and magnesium chloride, at the Stassfurt Mine in Germany. Calcium carbonate exists as whole mountain ranges of chalk, limestone, and marble. Its most abundant mineral is feldspar, which accounts for two-thirds of the Earth's crust. Beryllium is found as the mineral beryl, a beryllium aluminum silicate. With a chromium-ion impurity, beryl is known as emerald. If **iron** ions are present, the gemstone is blue-green and known as aquamarine.

Beryllium is lightweight and as strong as steel. It is hard enough to scratch **glass**. Beryllium is used for windows in x ray apparatus and in other nuclear applications, allowing the rays to pass through with minimum absorption.

Because beryllium is rather brittle, it is often combined with other metals in alloys. Beryllium-copper alloys have unusually high tensile strength and resilience, which makes them ideal for use in springs and in the delicate parts of many instruments. The **alloy** does not spark, and so finds use in tools employed in fire-hazard areas. Since beryllium-nickel alloys resist **corrosion** by **salt** water, they are used in marine engine parts.

Magnesium, alone or in alloys, replaces aluminum in many construction applications because the supply of this metal from seawater is virtually unlimited. Magnesium is fairly soft and can be machined, cast, and rolled. Magnesium-aluminum alloys (trade name Dowmetal) are used in airplane construction.

Because magnesium is very reactive, it is used as a sacrificial **anode** (a substance designed to be oxidized first, thus protecting other metals) for the hulls of ships and in underground pipelines. Magnesium can replace **titanium** from titanium(IV) chloride, which provides a method for obtaining the corrosion-resistant titanium from its ores. Magnesium is used to make Grignard reagents, alkyl- or aryl-magnesium compounds used as starting materials in organic syntheses.

Magnesium hydroxide is used as milk of magnesia for stomach upsets. Epsom salts is magnesium sulfate. Soapstone, a form of talc, is used for laboratory table tops and laundry tubs. Magnesium oxide is used for lining furnaces.

Calcium carbonate as limestone and marble are important in construction, not only as cut rock but also pulverized for use in concrete. Slaked **lime**, or calcium hydroxide, is the principal ingredient in plaster and mortar, in which the calcium hydroxide is gradually converted to calcium carbonate by reaction with the carbon dioxide in the air. Slaked lime is an important flux in the **reduction** of iron in blast furnaces. It is also used as a mild germ-killing agent in buildings that house poultry and farm animals, in the manufacture of cement and **sodium** carbonate, for neutralizing acid soil, and in the manufacture of glass.

Calcium carbide, made by reacting calcium oxide with carbon in the form of coke, is the starting material for the production of acetylene. Calcium propionate is added to foods to inhibit mold growth. Calcium carbonate and calcium pyrophosphate are ingredients in toothpaste.

Plaster of Paris is $2CaSO_4•H_20$, which forms $CaSO_4•2H_2O$ (gypsum), as it sets. Gypsum is used to make wallboard, or sheet rock. **Asbestos**, which is no longer used, is a naturally occurring mineral, a calcium magnesium silicate. Calcium and magnesium chlorides, byproducts of **sodium chloride** purification, are used in the de-icing of roads. Calcium chloride absorbs water from the air, so is used in the prevention of dust on roads, **coal**, and tennis courts and as a drying agent in the laboratory.

Florapatite, a calcium fluorophosphate, is an important starting material in the production of **phosphoric acid**, which, in turn, is used to manufacture **fertilizers** and detergents. The mines in Florida account for about one-third of the world's supply of this **phosphate** rock. Fluorspar, or calcium fluoride, is used as a flux in the manufacture of steel. It is also used to make hydrofluoric acid, which is then used to make fluorocarbons such as **Teflon**.

Calcium is involved in the function of nerves and in blood coagulation. Muscle contraction is regulated by the entry or release of calcium ions by the cell. Calcium phosphate is a component of bones and teeth. Hydroxyapatite, calcium hydroxyphosphate, is the main component of tooth enamel. Cavities are formed when acids decompose this apatite coating. Adding fluoride to the diet converts the hydroxyapatite to a more acid-resistant coating, fluorapatite or calcium fluorophosphate. Magnesium is the metal **ion** in **chlorophyll**, the substance in plants that initiates the **photosynthesis** process in which water and carbon dioxide are converted to sugars. Calcium ion is needed in plants for cell division and cell walls. Calcium pectinate is essential in holding plant cells together. Calcium and magnesium ions are required by living systems, but the other Group 2 elements are toxic.

The word barium comes from the Greek *barys*, meaning heavy. Barium salts are opaque to x rays, and so a slurry of barium sulfate is ingested in order to outline the stomach and intestines in x- ray diagnosis of those organs. Although barium ions are poisonous, the very low **solubility** of barium sulfate keeps the **concentration** low enough to avoid damage.

Both barium and strontium oxides are used to coat the filaments of vacuum tubes, which are still used in some applications. Because these elements act to remove traces of **oxygen** and **nitrogen**, a single layer of barium or strontium atoms on a filament may increase the efficiency more than a hundred million times.

Radium is a source of radioactive rays traditionally used in cancer treatment, though other radioactive isotopes are now more commonly used. A radioactive **isotope** of strontium, strontium-90, is a component of nuclear fallout. When it falls on grazing land and is incorporated into milk, it can accumulate in bone in place of calcium, killing cells and sometimes causing leukemia.

The alkaline earths and their compounds burn with distinctive colors. The green of barium, the red of strontium, and the bright white of magnesium are familiar in fireworks. Strontium is also used in arc lamps to produce a bright red light for highway flares.

ALKALOIDS

Alkaloids are a class of compounds that typically contain **nitrogen** and have complex, **ring** structures. They naturally occur in seed-bearing plants and are found in berries, bark, fruit, roots, and leaves. Often, they are bases that have some physiological effect.

Evidence suggests that alkaloids have been used by humanity for thousands of years. The first civilizations to use them were probably the ancient Sumarians and Egyptians. However, it was not until the early nineteenth century that these compounds were reproducibly isolated and analyzed. Advances in analytical separation techniques, such as **chromatography** and **mass spectroscopy**, led to the elucidation of the chemical structure of alkaloids. The term for these compounds is thought to have originated from the fact that the **alkaloid, morphine**, had similar properties to basic salts derived from the alkali ashes of plants; thus, it was called a vegetable alkali or alkaloid. Since the first alkaloids were isolated, thousands more have been identified and classified.

Although numerous alkaloids exist, they have similar properties when separated. In general, they are colorless, crystalline **solids** that are basic, have a ring structure, and have definite melting points. They are also derived from plants and have a bitter **taste**. However, some exceptions are known. For instance, some alkaloids are not basic and others are brightly colored or liquid. Other alkaloids are produced synthetically. Most alkaloids are also chiral molecules, meaning they have nonsuperimposable mirror images. This results in isomers that have different chemical properties. For example, one isomer may have a physiological function while the other does not.

It is mostly unknown why plants produce alkaloids. Various theories have been proposed to explain their existence. Some suggest that alkaloids are byproducts of normal plant **metabolism**. It is also thought that alkaloids may provide a means of defense against insects and animals. Alkaloids may also be a reservoir for molecules that plants often use. It is likely that all of these theories are correct to some extent.

Many of the thousands of alkaloids can be classified in certain families. **Nicotine** is an alkaloid found in the tobacco plant. It is concentrated in this plant's leaves and is derived from pyridine molecules. In addition to being a component of cigars and cigarettes, it is a poison that is used as an insecticide.

The opium poppy contains a variety of alkaloids. Morphine, which gets its name from the Greek god of dreams Morpheus, is a powerful painkiller. It is often given to terminally ill patients. Codeine, similar in structure to morphine, is also obtained from the poppy. It functions much like morphine but is less potent. Heroin is a synthetic derivative of morphine that is highly addictive. The muscle relaxer papavarine is also derived from the opium poppy. The majority of opiates are produced in India.

Some alkaloids are based on chemical structures called indole rings. **Strychnine** is an example of this type of compound. It is a powerful central nervous system stimulant. Lysergic acid, which is produced by a fungus that grows on rye, is another example. A synthetic variation of this compound called lysergic acid diethylamide is a powerful hallucinogen called LSD. Another class of alkaloids are based on structures called piperidine rings. These include compounds such as cocaine and atropine. Cocaine is a powerful stimulant that can be addictive. Atropine is an important medicine that is used to dilate the pupils of the eye, or act as a smooth muscle relaxer. Other important alkaloids are **caffeine**, ricinine, and **quinine**.

To produce commercial quantities of alkaloids, manufacturers begin by drying large quantities of the plants in which they occur. Since most alkaloids are basic, they can then be separated from their biomass sources by extraction with a dilute mineral acid. Using high pressure liquid chromatography, the alkaloids can be purified and crystalized. In this way, large amounts of physiologically active compounds can be obtained.

The physiological effects of alkaloids have made them important compounds in medicine. They have been used as painkillers, stimulants, muscle relaxers, tranquilizers, and **anesthetics**. The four types of alkaloids that have the most important economic impact include opiates, cocaine, caffeine, and nicotine.

ALKANE FUNCTIONAL GROUP

The alkanes are the hydrocarbons where all of the carbons are bonded to the other **carbon** atoms by single bonds. Alkanes may exist in a chain structure (either straight or branched) or in a **ring** structure. Alkanes that are acyclic (not in a ring) possess an **empirical formula** of $C_nH_{(2n+2)}$. Alkanes that are cyclic possess an empirical formula of C_nH_{2n}. In discussing the alkanes as a **functional group**, the terms aliphatic (for the acyclic alkane substituent) and alicyclic (for the cyclic alkane substituent) are used. The alkanes were originally called paraffins, which came from the Latin phrase *parum affinis*, which means little affinity. This phrase was used due to the low **reactivity** of these hydrocarbons. When the alkanes do react the reactions are usually quite vigorous.

Alkanes are denoted by the ending -ane in their name. The simplest example is **methane**, CH_4. The first four members of the straight-chained group—methane, ethane (CH_3CH_3), propane ($CH_3CH_2CH_3$), and butane ($CH_3CH_2CH_2CH_3$)—are all **gases**. The next members are all **liquids**, and from $C_{16}H_{34}$ on they are all **solids**. These states of **matter** are at standard **temperature** and pressure. The alkanes are very poorly soluble in **water** but are soluble with many of the other organic **solvents** such as chloroform and **benzene**.

Alkanes are produced by the **reduction** of alkenes, where **hydrogen** adds across to form a saturated double bond, or by reacting alkyl iodides with **sodium** in a **solution** of **ether** (known as the Wurtz reaction).

Alkanes burn readily in air to give **carbon dioxide** and water. If there is a restricted amount of **oxygen** only partial **combustion** takes place and **carbon monoxide** and carbon are produced along with the carbon dioxide and water. Some of the **halogens** will react with alkanes. **Chlorine** and **bromine** react quite readily, usually under free radical substitution conditions (in sunlight or in the presence of an **iron** catalyst). **Iodine** reacts to a much lesser extent with the alkanes due to the weak stability of the carbon-iodine bond. If methane is reacted with chlorine in strong sunlight the reaction is explosive and carbon and **hydrochloric acid** are the products. If the sunlight is less powerful a series of reactions takes place with the gradual replacement of the hydrogen molecules by chlorine. The products of this reaction are chloromethane (methyl chloride, CH_3Cl), dichloromethane (methylenedichloride, CH_2Cl_2), trichloromethane (chloroform, $CHCl_3$), and tetrachloromethane (carbon tetrachloride, CCl_4). This is an example of a **substitution reaction**. Alkanes can be reacted with **nitric acid** in the vapor phase at 932°F (500°C) to yield nitroalkane. By careful use of appropriate conditions and catalysts alkanes can be converted into other hydrocarbons such as the alkenes and alkynes.

The majority of alkanes are used as fuels, as solvents, and as starting products for the production of other hydrocarbons. Methane reacts with steam (gaseous water) at 1650°F (900°C) over a **nickel** catalyst to produce carbon monoxide and hydrogen. This mixture is known as synthesis gas. Synthesis gas is used to manufacture methanal (formaldehyde, $CH_2{=}O$), and also **methanol** (CH_3OH). The reverse process can be carried out where synthesis gas is reacted with a catalyst to give high **molecular weight** hydrocarbons. This is a process known as the Fischer Tropsch reaction. Ethane and propane, as well as being fuels, are important precursors in the manufacture of **ethene** (acetylene) which is then further converted to a range of products. Higher alkanes are chiefly used as fuels or precursors of other hydrocarbons.

Isomers of alkanes exist depending on the arrangement and joining of the carbon atoms. For example C_5H_{12} is pentane when all five carbon atoms are arranged in a straight line. If the arrangement of the **molecule** is such that there is a backbone of four carbon atoms then a **methyl group** can be added to one of these carbon atoms. If the methyl group were added to the second carbon **atom** in the chain then the resultant compound would be 2 methylbutane. These isomers differ in their physical properties, in the example given pentane has a melting point of 202°F (130°C) and a **boiling point** of 97°F (36°C), whereas the isomeric molecule 2 methylbutane has a melting point of 256°F (160°C) and a boiling point of 82°F (28°C). These differences are noticed in all of the other isomers within the series.

The alkanes are a simple group of hydrocarbons. They are found in petroleum and are most frequently used either as fuels or as precursors in the manufacture of other hydrocarbons.

See also Alkene functional groups; Alkyne functional groups

ALKENE FUNCTIONAL GROUP

The alkenes are a **functional group** characterized by the presence of a carbon-carbon double bond. Acyclic alkene hydrocarbons are represented by the **empirical formula** C_nH_{2n}. Alkenes were originally called olefins. This name came from olefiant gas, which is an old name for ethylene, the first member of the group. The modern names of the alkenes follow the standard numbering of the **carbon** atoms forming the chain (eth-, prop-, but-, etc.), with an ending of -ene. In addition, alkenes are also described as ''unsaturated,'' reflecting their ability to react with H_2 gas (in the presence of a metal catalyst) and so to undergo **reduction** of their carbon-carbon double bond. The product of this reduction is an alkane. The term ''polyunsaturated'' indicates the presence of more than one carbon-carbon double bond in the **molecule**.

The alkene carbon-carbon double bond is made from two sp^2 trigonally hybridized carbons, where one of the carbon-carbon bonds is a sigma bond (along the internuclear line) and the other is a pi bond (with side to side bonding). The result are bond angles of approximately 120° for the three atoms bonded to each of the alkene carbons. Rotation around the double bond is not possible, since it requires an **activation energy** equivalent to the complete breaking of the pi bond.

This has important consequences for the structure of alkenes. If an alkene is 1,2- disubstituted, there are two possible structures—one with the substituents on the same side and one with the substituents on the opposite sides. For example, butene ($CH_3CH=CHCH_3$) exists as two different isomers with different physical properties. In one isomer the two methyl groups (CH_3) are on the same side of the alkene (called a *cis*- or *Z*- arrangement), while in the other they are on opposite sides (called *trans*- or *E*- arrangement). This difference is important. For example, the primary fatty acid found in olive oil is the *cis* alkenoic acid, (*Z*)-octadec-9-enoic acid, commonly called oleic acid. At room **temperature** oleic acid is a liquid (melting point, 55.4°F/13°C), while its isomer [(*E*)-octadec-9-enoic acid, commonly called elaidic acid] is a solid (melting point 111.2°F/44°C).*Cis*-fatty acids are easily metabolized when eaten whereas *trans*-fatty acids are not. Thus, the nutritional and biochemical consequences of the formation of *trans*-fatty acids during the processing of polyunsaturated plant oils is an active area of research.

The principal reactivities of the alkenes are those involving the double bond present between the two carbon atoms. Addition reactions of electrophilic reagents to the alkene double bond (yielding substituted alkenes) are the most typical. A reduction reaction with H_2 produces the corresponding alkane by adding **hydrogen** across the double bond and changing the molecule from an unsaturated to a saturated state. This reaction is carried out using a metal catalyst. Among the metal catalysts useful for this reaction in the laboratory are

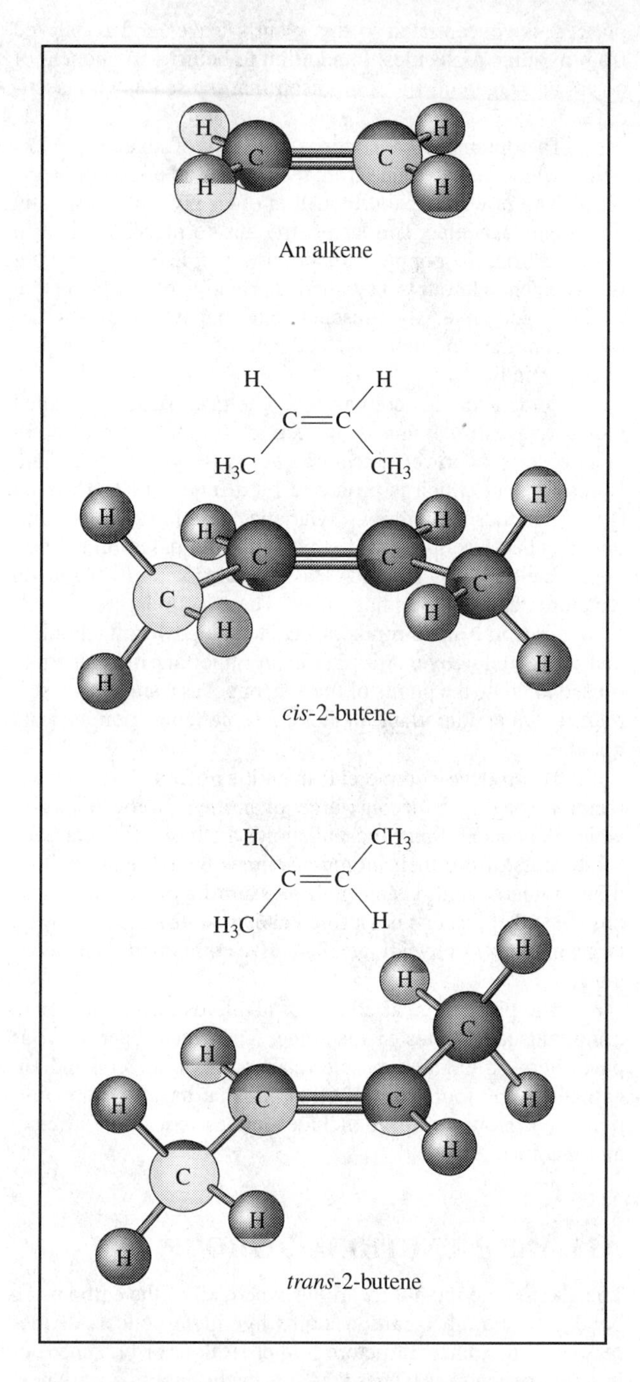

An alkene molecule and two butene isomers. ***(Illustration by Electronic Illustrators Group.)***

finely divided **nickel** (called Raney nickel after the chemist who developed it in 1927), **platinum**, **palladium**, and **rhodium**. At higher temperatures and pressures **iron**, **chromium**, **cobalt**, and **copper** may be used. This important reaction is called **hydrogenation** and it is used extensively to transform oils (unsaturated hydrocarbons) to saturated fats in the manufacture of ma rgarine.

Halogenation is a similar process to hydrogenation with **halogens** adding across double bonds. **Chlorine** and **bromine** undertake this reaction readily but **iodine** only reacts with difficulty. The discoloration of bromine **water** is a practical test for alkenes.

Alkyl halides are produced by adding hydrogen halides across the double bonds. Hydrogen bromide and iodide react readily, but hydrogen chloride requires moderate **heat**. When cold, concentrated **sulfuric acid** reacts with an alkene, alkyl hydrogen sulfates are formed, and these are soluble in the sulfuric acid. Alkanes do n ot undergo this reaction, and consequently this reaction is used to separate alkanes and alkenes. When oil is refined this reaction is used to remove unwanted alkenes. When alkyl hydrogen sulfates are added to water, sulfuric acid is recovered and the product obtained from the alkene is an **alcohol**. For example, ethyl hydrogen **sulfate** and water give sulfuric acid and **ethanol**. Water and alkenes can be made to react directly to produce alcohols, but high temperatures and pressures are required.

In common with other hydrocarbons, **combustion** of an alkene produces water and **carbon dioxide**. This is an example of an **oxidation** reaction. Another oxidation reaction of alkenes is the addition of **oxygen** to form a three-membered **ring** consisting of two carbons from the double bond and oxygen. This reaction is called epoxidation, and this ring is named an oxirane (or epoxide). For example, oxidation of **ethene** over a **silver** catalyst at 570°F (300°C) yields oxirane (ethylene oxide) that is used as a precursor in the manufacture of ethanal and several organic **solvents**. When lower temperature mixing with air occurs over a palladium catalyst **ketones** are formed directly from the alkene in a process called the Wacker-Hoescht process.

Alkenes can be polymerized. Polymerization of ethene gives polythene (polyethylene). The structure of this polymer is that of a very long chained alkane. Initial manufacturing processes involved very high pressures (some 1,500 atmospheres), but the Ziegler process from the 1950s uses a pressure of 2-6 atmospheres in conjunction with **titanium** tetrachloride and **aluminum** trialkyl catlysts. Polythene is a thermoplastic, meaning it can be repeatedly reheated and reshaped. It is also a useful electrical insulator. The Ziegler process can be used in the polymerization of many different alkenes. Many of the alkenes are used to make **plastics** and rubber substitutes of great economic value.

The alkenes are a large group of chemicals that can be used to produce a wide range of socially and economically important products. The principal reason for this is the unsaturated nature of the alkene that allows the double bond to polymerize and to participate in many other reactions that the saturated alkanes cannot.

See also Alkane functional group

ALKYL GROUP

The term ''alkyl'' refers to a **functional group** (or substituent) that is derived from the alkanes by the removal of a **hydrogen atom**. The name of the particular alkyl group is taken from the stem name of the alkane, with the -ane being replaced with -yl. For example, the derived alkyl group from **methane** is the **methyl group**. Methane has the chemical formula CH_4 and the methyl group is CH_3. The first four alkyl groups are methyl, ethyl, propyl, and butyl, with the number of **carbon** atoms being one, two, three, and four respectively. Isomers are possible with an alkyl just as with the alkane parent, so, for example, isopropyl exists as well as propyl. The isopropyl group $[(H_3C)_2HC\text{-}]$ has a different point (carbon atom) of attachment than the **propyl group** ($H_3CH_2CH_2C$-). The symbol often used to denote an alkyl group is the letter R. R is used when a reaction is being discussed and the functional group is specified, but the rest of the **molecule** is not given since it does not actively take part in the reaction. For example, if the molecule being discussed is identified as ROH that would show the reaction would work with any organic **alcohol**, regardless of the particular alkyl group attached to the alcohol group.

In discussing specific compounds, the location of the alkyl group is indicated in the name of the compound. For example, 2-methylhexane indicates that the methyl group is attached to the second carbon atom of a six carbon (hexane) chain. Alkyl groups may also correspond to **ring** structures. Cyclohexylbromide is the compound where **bromine** replaces one of the hydrogens of cyclohexane. In addition, alkyl groups may correspond to reagents and intermediates in organic **chemistry**. For example, methyl **lithium** is CH_3Li, where the carbon has a full negative charge (a carbanion). Methyl lithium is very reactive and acts as a nucleophilic methylating reagent to electrophilic functional groups, such as **ketones**. Alkyl **carbocations** and alkyl **free radicals** are also encountered frequently as reaction intermediates.

Alkylation reactions are reactions that introduce an alkyl group into a compound. These reactions are carried out using appropriate alkylating agents. For example, methylation of the cytosine nucleoside base of **DNA** is a method for the biochemical regulation of gene expression. Methylbromide (bromomethane), on the other hand, is cytotoxic as a result of its indiscriminant ability to methylate biomolecules, including the guanosine base of DNA. Yet it is this cytotoxic property that had made methylbromide uniquely valuable as a fumigant for the preservation of grain stored in silos.

An important property of alkyl groups that is observed in biochemical and laboratory mechanisms is their ability to stabilize, compared to hydrogen, electron-deficient carbon atoms, such as carbocations and carbon radicals. Thus, a trialkyl-substituted carbon **cation** (referred to as a tertiary carbocation) is much more stable than a primary carbocation (a carbon cation with one alkyl group and two hydrogen substituents). Tertiary carbocations are encountered as reaction intermediates, whereas primary carbocations are never encountered.

The term alkyl refers to the **hydrocarbon** functional group derived formally by the loss of a hydrogen atom from

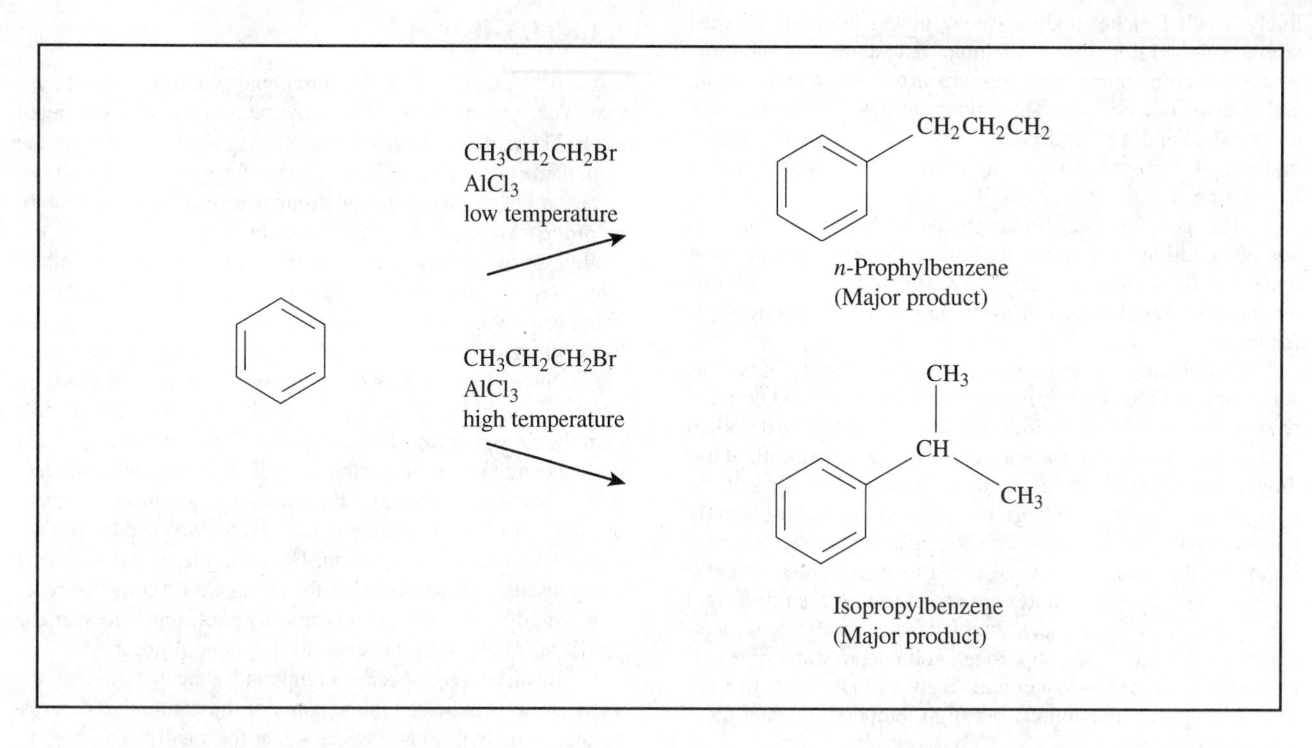

Figure 1. Alkylation reactions. ***(Illustration by Electronic Illustrators Group.)***

the alkane. Alkyl groups modify the **reactivity** of organic molecules, for example, by acting as electron-donating substituents.

See also Alkane functional group

ALKYLATION REACTIONS

Alkylation reactions are reactions where an **alkyl group** is introduced into a **molecule**. Many different types of alkylation reactions are known. For example, alkyl groups can be introduced into aliphatic hydrocarbons or **aromatic hydrocarbons** by alkylation reactions. Alkylation reactions can also be used to prepare ethers from alkyl alcohols or aromatic alcohols. Alkylated amines can be prepared by alkylation; however, special methods are normally required to control the number of alkyl groups introduced onto **nitrogen**. Important synthetic methodology also exists for the alkylation of enolate ions. Alkylation is often promoted through the use of **heat** or of an appropriate catalyst such as an acid catalyst.

One very important alkylation reaction is the Friedel Crafts Alkylation Reaction, used for the alkylation of aromatic species, which was discovered in 1877. Many alkyl halides, alcohols, or alkenes can be reacted with **benzene** in the presence of certain catalysts to give an alkyl benzene. Other aromatic hydrocarbons give similarly substituted alkyl aromatic products. A Lewis acid, such as **aluminum** chloride, is often the catalyst used for this reaction. This reaction is used industrially for the synthesis of hydrocarbons. If the reaction is carried out at high temperatures, rearrangement of the alkyl group may occur. For example reaction of n-propyl halides with benzene at low temperatures affords mainly n-propyl benzene but at higher temperatures the major product is isopropyl benzene (Figure 1). In Friedel-Crafts reactions, alkyl chlorides are more reactive than alkyl bromides and alkyl iodides react only with difficulty.

Alkylation is used industrially to produce basic building blocks for the synthesis of more elaborate materials. One commonly used application is in the production of anti-knock **gasoline**.

See also Acids and bases; Reactions (types of)

ALKYNE FUNCTIONAL GROUP

The alkynes are a **functional group** characterized by the presence of a carbon-carbon triple bond. Originally the alkynes were called the acetylenes. This is still the common name given to the simplest member of the group, although the IUPAC name is **ethyne** (Figure 1). The alkynes are all distinguished by the ending -yne.

Non-cyclic alkyne hydrocarbons, such as butyne ($CH_3C{=}CCH_3$), are characterized by an **empirical formula** of $C_nH(_{2n-2})$. The **bonding** relationship of the triple alkynes is most commonly described as bonding between two sp- p^2 hybridized **carbon** atoms. The alkyne carbon atoms bond together through one sigma bond, comprised of one of the two sp orbitals, and two pi bonds. The sigma bond occupies the internuclear space, and the two pi bonds are at 90° angles to each

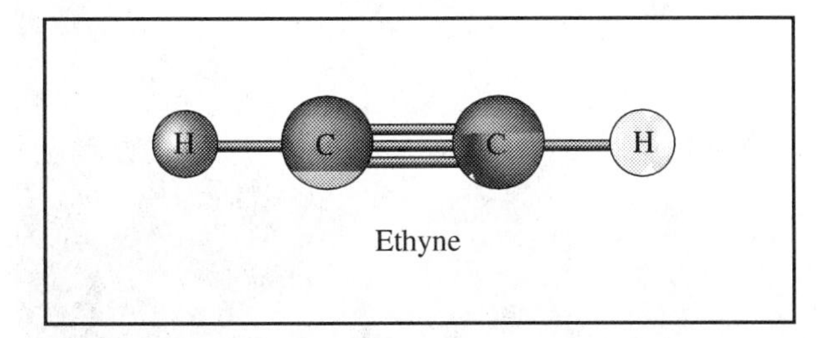

Chemical structure of ethyne. ***(Illustration by Electronic Illustrators Group.)***

other surrounding the carbon-carbon internuclear space. This results in a linear relationship of the four carbons of butyne (i.e., all four carbon nuclei of butyne are constrained by the bonding to the same line), and a substantial pi **electron density** surrounding all sides of the carbon-carbon triple bond. Due to this density of the pi electrons, and reflecting the lower thermodynamic stability of the pi bond compared to the sigma bond, the dominant **reactivity** of the alkyne is **electrophile** addition to the triple bonds. For example, molecular **hydrogen** (H_2) in the presence of metal catalysts will add to (hydrogenate) alkynes in an energetically favorable (i.e., exothermic) reaction. The addition of one **mole** of H_2 to an alkyne gives the alkene, whereas the addition of a second mole of H_2 gives the alkane.

Halogens, such as Cl_2 and Br_2, add as electrophiles to the alkyne to give the dihalo-substituted alkene, that continues to react with a second mole of the halogen to give the tetrahaloalkane. Hydrohalogenation, exemplified by the reaction of an alkyne with HCl, also occurs in the same manner. Acetylene (ethyne, HC=CH), for example, reacts with HCl at high pressure and in the presence of a metal catalyst to give chloroethene (vinyl chloride, H_2C=CHCl) as the product. Polymerization of this product gives polyvinyl chloride (PVC) which is extensively used in the **plastics** industry. **Oxidation** also is used to break down the triple bond and this can be an explosive process. Ethyne burned in **oxygen** produces **carbon dioxide** and **water** (with complete combustion) and a very hot flame. This is the principle behind oxy-acetylene welding.

Alkynes contain a carbon-carbon triple bond, and the reactivity of this triple bond dominates their chemical behavior.

See also Alkane functional group; Alkene functional group

ALLENE FUNCTIONAL GROUP

The allenes are a group of hydrocarbons with the general formula of C_nH_{2n-2}, but they differ from the alkynes in that there are two double bonds present, involving three **carbon** atoms, instead of the triple bond with two carbon atoms found in the alkynes. The **functional group** of the allenes is this twin double bond. All allenes are derivatives from 1,2 propadiene (allene) and they are isomers (compounds with the same chemical formulas but different structures and properties) of alkynes. Lower allenes are all **liquids** with a characteristic odor, and the higher members of the series are **solids**.

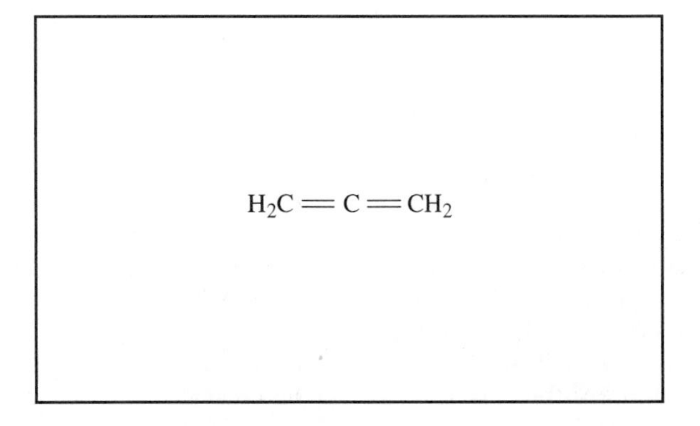

Allene functional group. ***(Illustration by Electronic Illustrators Group.)***

The allenes are not generally as stable as the alkenes which have only one double bond. Allenes can usually be converted to the appropriate alkyne by the simple addition of a base. Allenes are prepared by the removal of **bromine** and **hydrogen** bromide from 1,2,3 tribromopropane and its derivatives under the action of **potassium** hydroxide, **zinc**, and **ethanol**. The reactions which allenes partake in are typical of the alkenes, generally addition reactions involving the breaking down of the double bonds that are present.

Although allenes are far less common in nature than alkenes, a number of naturally occurring allenes have been identified and described, including pyrethrolone (an insecticide) and mycomycin (an antibiotic).

See also Alkene functional group; Alkyne functional group

ALLOSTERIC REGULATION

Allosteric regulation is the major mechanism by which enzymes are controlled in cells. Since enzymes perform virtually every function in a cell, their regulation is a vital part of cell **biochemistry**.

Enzymes are highly specific catalysts for cell reactions. They promote reactions by binding to reactants, called substrates, at a cleft in their surface called the active site. When substrate binds to the active site, it is transformed into the product, which then detaches to open the site for new substrates.

In order to control the rate of the catalyzed reaction, the cell must have a way to turn the **enzyme** on and off as needed. This can either be accomplished by blocking the active site directly, or by inducing a change in the enzyme's shape (called a **conformation** change) that alters the shape of the active site. Enzymes that undergo such shape changes are called allosteric **proteins**, and the substances that induce them are known as allosteric regulators. *Allosteric* means ''other site,'' and refers to the fact that the regulator is not binding at the active site, but at another site on the surface of the enzyme.

There are several common types of allosteric regulators. One is the **phosphate ion**, which can be linked to an exposed amino acid chain on the enzyme surface, usually a threonine, serine, or tyrosine. The bulky and negatively charged oxygens of the phosphate interact with the enzyme in such a way so as to distort the enzyme's shape, altering the conformation at the active site. Depending on the enzyme's initial conformation, this can either activate it or inactivate it.

Phosphorylation of an enzyme is itself an enzyme-catalyzed process, and the enzymes responsible are referred to as kinases. Phosphorylation by kinases does not occur randomly. Instead, the phosphorylating enzyme is itself activated by some signal. The most common intracellular signaling **molecule** is cyclic AMP (adenosine monophosphate), or cAMP. When cAMP binds to a kinase, it allosterically activates the kinase.

To understand how these multiple **chains** of regulation and activation serve their purpose, it is helpful to consider a specific example, the control of glucose **metabolism** in a muscle cell. The hormone epinephrine is released by the body in times of stress, and triggers a series of changes that increase the availability of glucose to the muscle to allow ''fight or flight.'' When epinephrine lands on the cell membrane, it triggers the allosteric activation of adenylate cyclase, the enzyme that creates cAMP. cAMP then allosterically activates a protein kinase. This protein kinase has two separate actions. First, it phosphorylates **glycogen** synthase, an enzyme that binds glucoses into the storage molecule, glycogen. This phosphorylation allosterically inactivates this enzyme, thereby keeping free glucose available. The same protein kinase also phosphorylates another enzyme, allosterically activating it, and this enzyme goes on to phosphorylate yet another enzyme, glycogen phosphorylase, allosterically activating it. Glycogen phosphorylase then acts on glycogen to release glucose molecules for breakdown, supplying **energy** to the muscle cell.

While this system is complicated, it has two advantages over a simpler system. First, it provides amplification: one molecule of epinephrine can trigger the activation of dozens of kinases, and ultimately hundreds of glycogen phosphorylases. Second, it allows a very fine level of control, since the system can be regulated at multiple levels.

Another common allosteric regulator is calmodulin, a protein that accounts for as much as 1% of total cell protein. Calmodulin binds **calcium**, triggering a conformation change. This activated form of calmodulin can then bind to regulatory sites on a wide variety of proteins, activating or deactivating them. Calcium, therefore, acts as a signal, and calmodulin transmits the signal throughout the cell. Calcium is stored inside the cell in the endoplasmic reticulum, and its release can be triggered by a variety of stimuli, depending on the cell type, including **hormones** and neurotransmitters.

Allotrope

An allotrope is one of several different physical forms of an element in the same state of **matter**. **Diamond** and **graphite** are both formed from **carbon** but, due to the different **bonding** present in the two substances, they have entirely different physical characteristics—they are both allotropes of carbon. The carbon atoms in graphite are arranged in flat sheets that slide easily over each other, while the atoms in diamond are bonded in a complex, honeycombed structure that makes the solid much harder. The difference between the allotropes may be in the bonding present as in the diamond carbon example or it may be due to the crystal form produced as is common with most **metals**. When compounds show different characteristic forms they are termed *polymorphic*.

Graphite (used in pencils) and diamonds are carbon allotropes. *(Photograph by Paul Silverman, Photo Researchers, Inc. Reproduced by permission.)*

In 1830, Swedish chemist Jöns Berzelius (1779-1848) questioned whether a given element could exist in two or more forms with different chemical and physical properties. A decade later he suggested the name *allotropy* for this phenomenon. Berzelius already knew of elements with this property—carbon, **phosphorus**, and **sulfur** in particular. But he did no further research on the subject.

Atmospheric **oxygen** (O_2) and ozone (O_3) are allotropes of oxygen. The properties of allotropes can be very different. As a gas **diatomic** oxygen is essential for life whereas **triatomic** oxygen is poisonous. Both forms occur naturally in the atmosphere, but diatomic oxygen is present throughout the atmosphere and ozone is normally present only in the upper atmosphere, except when it occurs as a pollutant at ground level or when it is manufactured by electrical discharge during storms. Diatomic oxygen is a more stable form than ozone. With the exception of **nitrogen**, all group V elements show allotropy. For example, phosphorous occurs in three forms—white, red, and black. White phosphorous is poisonous and very reactive, red phosphorous is not poisonous and it is only moderately reactive, and black phosphorous is nearly inert. Each of these allotropes of phosphorous also has its own subset of forms. Group VI elements also show allotropism. The classic example of this is rhombic and monoclinic sulfur. Both forms are **ring** structures containing eight sulfur molecules. Rhombic sulfur is the most stable form, but after heating and allowing the sulfur to melt and then cool down monoclinic sulfur is formed. Both are crystalline forms of sulfur, but their structures are slightly different. If monoclinic sulfur is left standing it will slowly revert back to the rhombic form over several days. These different forms of sulfur have slightly different characteristics, such as altered melting points. There are several other allotropes of sulfur that exist as well. Sulfur also shows allotropes in the liquid form. These are generally due to the breaking down of the ring structure and having different ratios of broken and unbroken rings. This effect is also enhanced by unpaired electrons at the broken ends of the ring structure. These broken ends can bond with other broken **chains** giving long chain molecules with a very high **viscosity** and decreased mobility. This latter effect is due to increased tangling of the molecules. Sulfur vapor also shows allotropic behavior.

See also Isomer

ALLOY

An alloy is a substance with metallic properties that is composed of a mixture of two or more elements. Alloys can be classified as interstitial or substitutional. In an interstitial alloy, smaller elements fill holes that are in the main metallic structure. The smaller element may be a nonmetallic element, such as **boron**, **carbon**, **nitrogen**, or **silicon**. For example, steel is an interstitial alloy in which carbon atoms fill the holes between the crystal structure of **iron**. In substitutional alloys, some of the atoms of the main metal are substituted with atoms of another metal. If the two metal atoms are about the same size and have the same crystallographic structure, then the two **metals** may form a solid **solution**. Hume-Rothery rules predict which metals will form solid-solutions based on the relative sizes and electronic properties of the metal atoms. Brass, an alloy composed of **copper** and **zinc**, is an example of a substitutional alloy.

Alloys have played an important role in human history. Early Mesopotamian civilizations first used bronze, an alloy of copper and **tin** as early as 3000 B.C. Bronze is more easily cast than pure copper because of its lower melting point. It is also stronger than copper after it solidifies. Because of the widespread use of weapons and other implements made of bronze, this period has become known as the Bronze Age. Around 1200 B.C. the raw materials to make bronze were becoming scarce in the Mesopotamian area, so local metal workers began to use the more readily available iron. Pure iron is soft, ductile, and malleable because it is made of small spherical atoms that can move easily past each other. It is also more difficult to work with than bronze and it rusts easily. Early metal workers improved the pure iron by a process called carburizing which added 0.5 - 1% carbon. This produced an interstitial alloy which is very similar to modern day steel. Adding carbon creates strong iron to carbon bonds which prevent atoms from sliding past each other as easily as they do in pure iron. The resulting alloy is stronger and harder than iron or bronze.

Metallurgy, or the study of metals and their alloys, remained relatively unchanged from antiquity until the end of the eighteenth century. The Industrial Revolution greatly in-

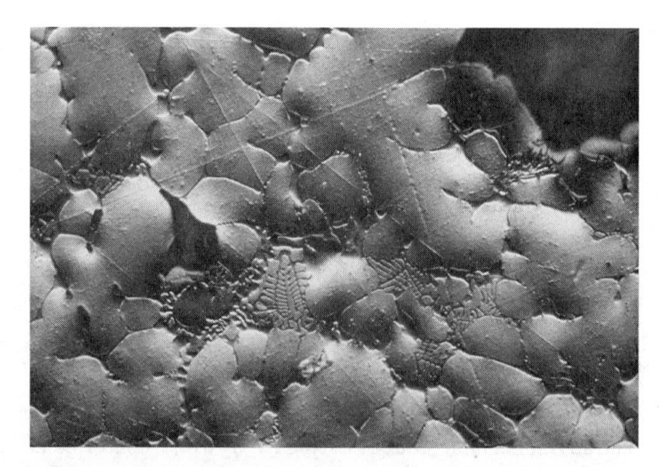

Alumimum alloys. ***(Photograph by Jan Hinsch/Photo Researchers, Inc. Reproduced by permission.)***

creased the need for steel so practical inventors and scientists developed new techniques for making alloys. For example, in 1850 the steel making industry was drastically changed by the Bessemer process which burned out impurities in iron through the use of a blast furnace. The study of alloys was also advanced by two developments in the latter half of the nineteenth century. In 1863 Henry Clifton Sorby of Sheffield (1826-1908) developed a technique for polishing and etching metals so that they could be observed under a microscope. This enabled scientists to correlate crystalline structures with the strength, **ductility**, and other properties of alloys. In 1887 Hendrik Willem Bakhuis Roozeboom (1854-1907) applied Josiah Willard Gibb's (1839-1903) phase rule to alloys. The phase rule applied thermodynamic principles to chemical equilibria and allowed Roozeboom to develop a phase diagram of the iron carbon system. A phase diagram shows the phases that can be present in an alloy at different temperatures, pressures, and compositions at thermodynamic equilibrium. Roozeboom's phase diagram enabled him and others to improve the quality of steel. Later, other procedures, such as **electron microscopy** and x-ray techniques, also contributed greatly to the study of alloys.

Alloys have many applications. Alloys based on tin, **cadmium**, copper, or **silver** are used to make bearings which reduce friction between two sliding surfaces. Dental fillings can be made from alloys of silver and **mercury** or alloys of **gold**, silver, and copper. Stainless steels, iron alloys with more than 12% **chromium**, are examples of corrosion-resistant alloys. Other alloys, such as Nichrome, a **nickel** based alloy with 12-15% chromium and 25% iron, are very strong at high temperatures. They are used in power-generating plants, jet engines, and gas turbines. Light weight alloys, like the **aluminum**, zinc, and **magnesium** system, are used in aircraft. Some alloys, like solder, which is usually 60% tin and 40% **lead**, are used to make electrical circuits. Prosthetic devices, like artificial knees and hips, are made from high-strength, corrosion-resistant alloys that are usually based on iron, **cobalt**, or **titanium**. Alloys can also be superconductors, which are materials that have zero resistance to the flow of electrical current at low temperatures. One alloy of **niobium** and titanium becomes superconducting at -442.3°F (263.5°C). Alloys of **precious metals**, like gold, silver, and **platinum**, are used as coins, catalysts for chemical reactions, electrical devices, **temperature** sensing devices, and jewelry. Yellow gold contains gold, silver, and copper in a 2:1:1 ratio. Some iron-based alloys like Alnico-4, which is 55% iron, 28% nickel, 12% aluminum, and 5% cobalt, are used as magnets. Many other applications exist for the over 10 thousand different types of alloys that have been developed.

Allyl Functional Group

The allyl **functional group** is CH_2=$CHCH_2$ with a double bond being present between the first and second **carbon** atoms. The International Union of Pure and Applied **Chemistry** (IUPAC) name for the allyl group is the propenyl group. The name allyl comes from the scientific name for garlic, *Allium*. Garlic contains a number of allyl-containing molecules, notably allyl **alcohol** (CH_2=$CHCH_2OH$) and diallyl disulfide (CH_2=$CHCH_2SSCH_2$ CH=$_2$) that gives garlic its characteristic odor.

One of the most important uses of allyls is in the manufacture of various **plastics**. After polymerization using **heat**, light, radiation, or catalysis a range of strong thermosetting plastics can be produced. Compounds that are made from polymerization of the allyl functional group are used as plastics, adhesives, furniture finishes and they can also have flame retardant properties. Some possess such good resistance to chemical and abrasive attack that they are used for the manufacture of plastic eyeglass lenses.

The most important **reactivity** of this group is its ability to impart reactivity to the carbon of the allyl immediately adjacent to the alkene double bond. The biosynthesis of many important primary metabolites (such as cholesterol) proceed through intermediates that contain the allyl alcohol functional group. This allyl alcohol is activated by phosphorylation to give a reactive allyl **cation** that is used to build up more complex structures. Allyl halides, such as allyl bromide (CH_2=$CHCH_2Br$), are used extensively in organic synthesis for their ability to transfer easily the allyl group to nucleophilic atoms.

Alpha Particle

The alpha particle is emitted by certain radioactive elements as they decay to a stable element. It consists of two protons and two neutrons; it is positively charged. The element that undergoes "alpha decay" changes into a new element whose **atomic number** is down two and atomic **mass** is down four from the original element. Alpha decay occurs when a **nucleus** has so many protons that the strong nuclear force is unable to counterbalance the strong repulsion of the electrical force between the protons. Because of its mass, the alpha particle travels relatively slowly (less than 10% the speed of light), and it can be stopped by a thin sheet of **aluminum** foil.

When **Henri Becquerel** first discovered the property of **radioactivity** in 1896, he did not know that the radiation consisted of particles as well as **energy**. **Ernest Rutherford** began experimenting to determine the nature of this radiation in 1898. One experiment demonstrated that the radiation actually consisted of 3 different types: a positive particle called "alpha," a negative particle, and a form of electromagnetic radiation that carried high energy. By 1902 Rutherford and his colleague **Frederick Soddy** were proposing that a different chemical element is formed whenever a radioactive element decays, a process known as **transmutation**. Rutherford was awarded the Nobel Prize in **Chemistry** in 1908 for discovering these basic principles of radioactivity. In Rutherford's classic "**gold** foil" experiment to determine the structure of an **atom**, his assistants Hans Geiger and E. Marsden used positively charged high speed alpha particles that were emitted from radioactive **polonium** to bombard a gold foil. The results of this experiment showed that an atom consisted of mostly empty space, with essentially all its mass concentrated in a very small, dense, positively charged center called the nucleus. In fact, the alpha particle itself was identified as the nucleus of the **helium** atom! Soddy, working with **William Ramsay**, who in 1895 had discovered that the element helium was a component of earth minerals, verified that helium is produced when **radium** is allowed to decay in a closed tube.

The first production in the laboratory of radioisotopes—as opposed to the naturally occurring radioactive isotopes—was achieved by bombarding stable isotopes with alpha particles. In 1919, Rutherford succeeded in producing oxygen-17 by bombarding ordinary nitrogen-14 with alpha particles; a **proton** was set free in the reaction. The 1935 Nobel Prize in Chemistry was awarded to Irène and Frédéric Joliet-Curie for their work in producing radioisotopes through the bombardment of stable elements with alpha particles from polonium decay. After bombarding aluminum with alpha particles, they found that when the nucleus absorbed an alpha particle, it changed into a previously unknown radioisotope of **phosphorus**; an unidentified neutral particle was set free in the reaction. James Chadwick repeated their experiment using a **beryllium** target, then captured the particle that was emitted and used principles of classical physics to identify its mass as the same as that of the proton. This particle was named the **neutron**, and Chadwick received the 1935 Nobel Prize in Physics for its discovery. Subsequently, scientists began accelerating alpha particles to very high energies in specially constructed devices like the cyclotron and linear accelerator before shooting them at the element they wanted to transmute.

A number of radioisotopes, both natural and man made, have been identified as alpha emitters. The decay chain of the most abundant **uranium isotope**, uranium-238, to stable lead-206 involves eight different alpha decay reactions. Uranium has a very long half life (4.5 billion years) and because it is present in the earth in easily measured amounts, the ratio of uranium to **lead** is used to estimate the age of our planet Earth. More recently, Guenther Lugmair at the University of California at San Diego introduced age-dating using samarium-147, which undergoes alpha decay to produce neodymium-143.

Most of the transuranic radioactive elements undergo alpha decay. During the 1960's when an instrument was needed to analyze the lunar surface, Anthony Turkevich developed an alpha scattering instrument, using for the alpha source curium-242, a transuranic element produced by the alpha decay of americium-241. This instrument was used on the moon's surface by Surveyors 5, 6, and 7. **Plutonium** has been used to power more than twenty spacecraft since 1972; it is also being used to power cardiac pacemakers.

Ionization smoke detectors detect the presence of smoke using an ionization chamber and a source of ionizing radiation (americium-241). The alpha particles emitted by the radioactive decay of the americium-241 ionize the **oxygen** and **nitrogen** atoms present in the chamber, giving rise to free electrons and ions. When smoke is present in the chamber, the electrical current produced by the free electrons and ions (as they move toward positively and negatively charged plates in the detector) is neutralized by smoke particles; this results in a drop in electrical current and sets off an alarm.

Because of their low penetrability, alpha particles do not usually pose a threat to living organisms, unless they are ingested. This is the problem presented by the radioactive gas **radon**, which is formed through the natural radioactive decay of the uranium (present in rock, soil, and **water** throughout the world). As radon gas seeps up through the ground, it can become trapped in buildings, where it may build up to toxic levels. Radon can enter the body through the lungs, where it undergoes alpha decay to form polonium, a radioactive solid that remains in the lungs and continues to emit cancer-causing radiation.

AL-RAZI, ABU BAKR MOHAMMAD IBN ZAKARIYYA (865-925)

Arab chemist

Abu Bakr Mohammad ibn Zakariyya al-Razi, also known as Rhazes in the Latin West, was born at Rayy, Persia (near present-day Tehran, Iran) in 865. Early on in al-Razi's life, he became interested in singing and music, but soon his interests took residence in the study of **alchemy** and **chemistry**, philosophy, logic, mathematics, physics, and medicine. Al-Razi studied from a student of Hunayn Ibn Ishaq, who was well versed in the ancient Greek, Persian, and Indian systems of medicine and other subjects; he also studied under Ali Ibn Rabban.

Al-Razi was first appointed as head of the first Royal Hospital at Rayy, due to his fame in medicine. Soon thereafter, al-Razi moved to the Muqtadari Hospital in Baghdad as the head during the reign of the Adhud-Daulah. Adhud-Daulah held al-Razi responsible for choosing the best building site for the Muqtadari Hospital. He is said to have chosen the position by hanging pieces of meat in various parts of Baghdad and noted which putrefaction of meat was the slowest. He chose the spot of the least rotten meat as the building site for the hospital.

Al-Razi practiced for over 35 years and wrote about 200 books, of which more than half are medical and 21 concern alchemy. His medical commentaries contributed mainly to the fields of ophthalmology, obstetrics, and gynecology. He also wrote on other subjects, such as physics, mathematics, astronomy, and optics. Al-Razi's medical works include monographs, one of which was the first treatise on smallpox, chicken-pox, and measles, distinguishing between smallpox and measles. Later this was treatise was twice translated into Latin in the eighteenth century when interest rose in inoculation or variolation around 1720 following the description of the procedure in Turkey by Lady Mary Wortley Montagu.

The most famous of al-Razi's works is the *al-Hawi* or *The Comprehensive Book*, which includes Greek, Syrian, and early Arabic medical knowledge and also virtually every topic of medical importance in their entirety. Al-Razi took meticulous notes from all the books available to him on medicine, combined with his own medical experience. It was originally comprised of 20 volumes, of which 10 have survived.

In addition to his work in the medical field, al-Razi also compounded medicines, and later in his life, turned to experimental and theoretical sciences. He took careful notes in great detail about several chemical reactions and also gave full descriptions of and designs for about 20 instruments used in chemical investigations. His book *Kitab-al-Asrar* concerns the preparation of chemical materials and their utilization. Another book called *Liber Experimentorum*, which was translated in Latin, al-Razi divides substances into plants, animals, and minerals, giving way for organic and **inorganic chemistry**. As a chemist, by mixing different acids, he was the first to produce **sulfuric acid**, and also prepared **alcohol** by fermenting sweet products.

Al-Razi made a variety of other contributions to the field of medicine. He wrote a few smaller treatises on colic, on stones in the kidney and bladder, on curing disease within an hour (such as headache, toothache, hemorrhoids, and dysentery in small children), on children's diseases, on diabetes, on food for the sick, on maladies of the joints, on medicine for those unable to see a physician, on medical aphorisms, and on medical diagnoses and treatments. He was the first to use music as therapy by arranging his students in concentric circles with the most advanced in the inner circle. During a specific period when roses bloomed, al-Razi noticed head swelling occurring, and he then apparently became the first to relate hay fever to the scent of roses.

In addition to his contributions to the field of medicine, al-Razi was a Hakim, which means he was also a philosopher besides an alchemist. The basic elements in his philosophical system are the creator, spirit, **matter**, space, and time. He discusses their characteristics in detail and his concepts of space and time as constituting a continuum are outstanding. These views, on the other hand, were criticized by other Muslim scholars of the time. Most of al-Razi's philosophical and anti-religious works are lost.

Al-Razi moved at various times to different cities, especially between Rayy and Baghdad. He ultimately returned to Rayy, where he died in 925.

Sidney Altman.

ALTMAN, SIDNEY (1939-)

American molecular biologist

In the early 1980s, Sidney Altman discovered that ribonucleic acid (RNA) molecules can act as enzymes. This disclosure, independently and concurrently made by **Thomas R. Cech** of the University of Colorado, broadened our understanding of the origins of life. Before this discovery, it was believed that all enzymes were made of protein and that primitive cells, therefore, used **proteins** to catalyze biochemical processes. Now, it appears that **RNA** may have acted as a catalyst. Altman and Cech's work has not only had a ''conceptual influence on basic natural sciences,'' according to the Royal Swedish Academy of Sciences, but in addition, ''the discovery of catalytic RNA will probably provide a new tool for gene technology, with potential to create a new defense against viral infections.'' As a result of their findings, Altman and Cech were jointly awarded the 1989 Nobel Prize for **chemistry**.

Altman was born in Montreal, Quebec, on May 8, 1939, the second son of Victor Altman, an immigrant grocer, and Ray Arlin, who before her marriage worked in a textile mill. He attended West Hill High School in Montreal and the Massachusetts Institute of Technology, from which he graduated with a bachelor of science in physics in 1960. Between 1960-1962, he was a teaching assistant in the Department of Physics at Columbia University, while he waited for a suitable position in a lab. Around this time, Altman switched from physics to the newly emerging interdisciplinary field of molecular biology. He moved to the University of Colorado in Boulder in late 1962 to work as a research assistant under Leonard S. Lerman. He was mainly preoccupied with studying the replication of the T4 bacteriophage, a substance that infects bacterial cells in much the same way as a virus infects human cells. Altman received his Ph.D. in biophysics in 1967.

After graduation, Altman briefly worked as a research assistant in molecular biology at Vanderbilt University before winning a grant from the Damon Runyon Memorial Foundation for Cancer Research. This permitted him to work as a research fellow in molecular biology at Harvard University. From 1967 to 1969, working under the biochemist **Matthew Meselson**, he continued his research into the genetic structure of the T4 bacteriophage. His receipt of the Anna Fuller Foundation Fellowship in 1969 enabled him to transfer to the Medical Research Council Laboratory of Molecular Biology in Cambridge, England, to work with molecular biologists **Sydney Brenner** and **Francis Crick**. The latter, in partnership with **James D. Watson**, discovered DNA's double-helix structure in 1954.

It was clear to scientists that genetic information is carried by **DNA** (deoxyribonucleic acid) into a cell's **nucleus**. In the cytoplasm (the substance inside the cell wall, surrounding the nucleus), the genetic code is copied into RNA. It is then converted into proteins, which are built of **chains** of **amino acids**. Altman originally intended to study the three-dimensional structure of transfer RNA (tRNA), which is a small component of RNA that transfers amino acids onto a growing polypeptide chain as proteins are made. Much of the breakthrough work in this area had already been accomplished, however, so Altman decided to switch his attention to the transcription of tRNA from DNA. He found that the DNA from which tRNA is produced is not directly copied into tRNA but first undergoes an intermediary stage when it becomes a long strand of what is called ''precursor RNA.'' This is composed of a strand of tRNA with additional genetic sequences at each end which are somehow later removed before it becomes tRNA.

While still working at the Medical Research Council in Cambridge, Altman studied the tRNA genes of the *Escherichia coli (E. coli)* bacterium, to which he added toxic chemicals. Subsequent mutations in the tRNA enabled him to isolate precursor tRNA from bacterial cells. Altman discovered that the additional sequences at each end of the strand of precursor tRNA were removed by something in the cells of the bacteria, probably an **enzyme**. The scientists found that the enzyme, named ribonuclease P (RNase P), would only cut off the extra sequences at a precise point.

When he returned the United States in 1971, Altman joined Yale University's biology department as an assistant professor. In 1972, he married Ann Korner. They have a son, Daniel, and a daughter, Leah. Altman was promoted to associate professor in 1975. In 1978, he published the results of an experiment carried out by one of his graduate students, Benjamin Stark. It demonstrated that RNase P was at least partially composed of RNA, which meant that RNA itself played an integral part in the activity of the enzyme. This finding was highly unorthodox, as it was then presumed that enzymes are made of protein, not **nucleic acids**.

In 1980, Altman attained a full professorship of biology at Yale. The following year, Cech at the University of Colorado published independent results similar to Altman's. Cech discovered that the precursor RNA from the protozoan *Tetrahymena* was reduced to its final size as tRNA without the assistance of protein, and suggested that the precursor RNA catalyzed this itself. His findings lent weight to Altman's. Cech's use of the word "catalyst" to describe the action of the RNA was questioned, however, because rather than just speeding up a reaction, it used itself up in the process.

Three years later, by which time Altman had become chairman of Yale's biology department, his colleague, Cecilia Guerrier-Takada, was testing the catalytic activity of RNase P. She discovered catalysis even in the control experiments that used the RNA subunit of RNase P (the M1 RNA) but which contained no protein. Altman was able to prove that the M1 RNA demonstrated all the classical properties of a catalyst, especially as, unlike that studied by Cech, it remained unchanged by the reaction. This removed the last shadow of a doubt that RNA could act as an enzyme.

In 1984, Altman became a naturalized American, but retained his Canadian citizenship. From 1985 until 1989, as the dean of Yale College, Altman established a greater role for scientific education in all of Yale's curriculums. In 1989 he and Cech jointly received the Nobel Prize for chemistry for their discovery of RNA's catalytic ability. Their work put an end to the conundrum regarding proteins and nucleic acids which had long mystified scientists. They had been unable to discover which came first in the development of life, proteins or nucleic acids. Proteins catalyze biological reactions, whereas nucleic acids, such as RNA, transport the genetic codes that create the proteins. Altman and Cech proved that nucleic acids were the building blocks of life, acting as both codes and enzymes.

High hopes exist for the practical applications of their discovery, which was described by the Nobel Academy as one of "the two most important and outstanding discoveries in the biological sciences in the past 40 years," the other being Crick and Watson's discovery of DNA's double helix structure. If RNA enzymes are able to cut additional sequences of tRNA from a strand of precursor tRNA, doctors could possibly use RNA enzymes to cut infectious RNA from the genetic system of a person with an infectious viral disease. Research into this field is ongoing and, if fruitful, could contain the key to curing viral infections such as cancer and AIDS.

In addition to receiving the Nobel Prize, Altman was honored with the Rosentiel Award for Basic Biomedical Research in 1989. He is a member of the National Academy of Sciences and the American Society of Biological Chemists, of the Genetics Society of America, and is a fellow of the American Association for the Advancement of Science. He holds honorary degrees from the Université de Montréal, York University in Toronto, Connecticut University, McGill University, the University of Colorado, and the University of British Columbia. In 1991, he was selected to present the DeVane lecture series at Yale on the topic "Understanding Life in the Laboratory."

Altman has held a number of other part-time positions in addition to his full-time academic positions, including associate editor of *Cell* from 1983 to 1987, member of the Board of Directors of the Damon Runyon-Walter Winchell Fund for Cancer Research, member of the Board of Governors of the Weizmann Institute of Science, member of the Scientific Advisory Board of Bio-Méga, Inc., and Special Consultant to the Pathogenesis Corporation in Seattle.

ALUMINUM

Aluminum is the second element in Group 13 of the **periodic table**. It has an **atomic number** of 13 and an atomic **mass** of 26.98. Its chemical symbol is Al.

Aluminum is a silver-like metal with a slightly bluish tint. It has a melting point of 1,220°F (660°C), a **boiling point** of about 4,440°F (2,450°C), and a **density** of 2.708 grams per cubic centimeter. Aluminum is both ductile and malleable.

Aluminum is a very good conductor of **electricity**, surpassed only by **silver** and **copper** in this regard. However, aluminum is much less expensive than either silver and copper. For that reason, engineers are currently trying to discover new ways in which aluminum can be used to replace silver and copper in electrical wires and equipment.

Aluminum has one interesting and very useful chemical property. In moist air, it combines readily with **oxygen** to form aluminum oxide: $4Al + 3O_2 \rightarrow 2Al_2O_3$.

The aluminum **oxide** forms a thin, whitish coating on the aluminum metal that prevents the metal from reacting further with oxygen. That is, it protects the metal from further **corrosion**.

Aluminum is the third most abundant element in the Earth's crust, ranking only behind oxygen and **silicon**. It makes up about 9% of the Earth's crust, making it the most abundant of all **metals**.

Aluminum occurs in nature as a compound, never as a pure metal. The primary commercial source for aluminum is the mineral bauxite, a complex compound consisting of aluminum, oxygen, and other elements. Bauxite is found in many parts of the world, including Australia, Brazil, Guinea, Jamaica, Russia, and the United States. In the United States, aluminum is produced in Montana, Oregon, Washington, Kentucky, North Carolina, South Carolina, and Tennessee.

Aluminum is extracted from bauxite in a two-step process. In the first step, aluminum oxide is separated from bauxite. Then aluminum metal is produced from aluminum oxide.

At one time, the extraction of pure aluminum metal from aluminum oxide was very difficult. That process requires that aluminum oxide first be melted, then electrolyzed. But aluminum oxide melts at only very high temperatures.

An inexpensive method for carrying out this operation was discovered in 1886 by **Charles Martin Hall**, at the time, a student at Oberlin College in Ohio. Hall found that aluminum oxide will melt at a much lower **temperature** if it is first mixed with a mineral known as cryolite. When an electric current is passed through a molten mixture of aluminum oxide and cryolite, aluminum metal is produced very easily.

At the time of Hall's discovery, aluminum was a very expensive metal. It sold for about $20 per kilogram ($10 per

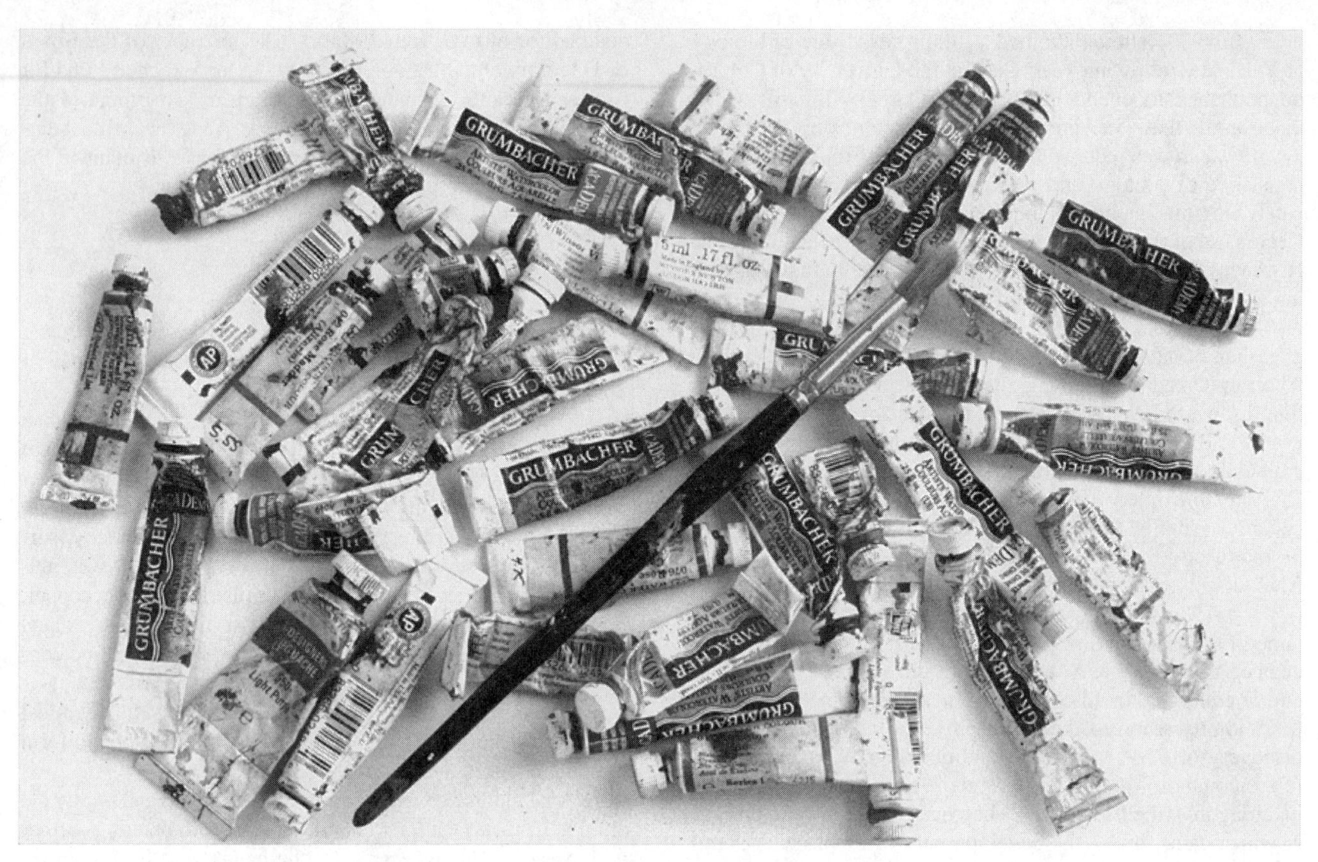

Paint tubes are made from aluminium. ***(Photograph by Robert J. Huffman. Reproduced by permission.)***

pound) and was displayed at the 1855 Paris Exposition next the the French crown jewels. As a result of Hall's research, the price of aluminum dropped to less than $1 per kilogram (about $.40 per pound).

Aluminum was named for one of its most important compounds, alum, a compound of **potassium**, aluminum, **sulfur** and oxygen. The chemical name for alum is potassium aluminum **sulfate**, $KAl(SO_4)_2$.

Alum has been widely used by humans for thousands of years. It was mined in ancient Greece and then sold to the Turks who used it to make a beautiful red dye known as Turkey red. Alum has also been long used as a mordant in dyeing. In addition, alum was used as an astringent to treat injuries.

Eventually, chemists began to realize that alum might contain a new element. The first person to actually produce aluminum from a mineral was the Danish chemist and physicist Hans Christian Oersted (1777-1851). Oersted was not very successful, however, in producing a very pure form of aluminum.

The first pure sample of aluminum metal was not made until 1827 when the German chemist **Friedrich Wöhler** heated a combination of aluminum chloride and potassium metal. Being more active, the potassium replaces the aluminum, as shown by the following equation:

$3K + AlCl_3 \xrightarrow{heat} 3KCl + Al$

The chemical symbol for aluminum, Al, is taken from the first two letters of the element's name. In some parts of the world, the element's name is spelled and pronounced somewhat differently, as aluminium (or al-yoo-MIN-ee-um).

Aluminum is usually used not as a pure metal, but as an **alloy**. The aluminum industry has developed a special system for labeling the alloys of aluminum. The 1,000 aluminum alloys, for example, are nearly pure aluminum metal. The 2000 series consists of alloys of aluminum and copper. The 3000 series includes alloys of aluminum and **manganese**. The 4000 series are alloys of aluminum and silicon; the 5000 series are alloys of aluminum and **magnesium**; the 6000 series are alloys of aluminum, magnesium, and silicon; the 7000 series are alloys of aluminum and **zinc**.

The largest single use of aluminum alloys is in the transportation industry. Car and truck manufacturers like aluminum alloys because they are strong, but lightweight. Another important use of aluminum alloys is in the packaging industry. Aluminum foil, beer and soft drink cans, paint tubes, and containers for home products are all made of aluminum alloys. Other uses of aluminum alloys include window and door frames, screens, roofing, siding, electrical wires and appliances, automobile engines, heating and cooling systems, kitchen utensils, garden furniture, and heavy machinery.

Aluminum is also made into a large variety of compounds with many industrial and practical uses. Aluminum ammonium sulfate, $Al(NH_4)(SO_4)_2$, is used as a mordant, in **water** purification and sewage treatment systems, in **paper** production and the tanning of leather, and as a food additive. Aluminum borate ($Al_2O_3 \cdot B_2O_3$) is used in the production of **glass** and **ceramics**.

One of the most widely used compounds is aluminum chloride ($AlCl_3$), employed in the manufacture of paints, antiperspirants, and synthetic rubber. It is also important in the process of converting crude petroleum into useful products, such as **gasoline**, diesel and heating oil, and **kerosene**.

Alum still has many important applications, such as in the manufacture of paper and in leather tanning, in fire extinguisher systems, as a mordant, and as a fireproofing and fire retardant material.

Americium

Americium is a transuranium element, one of the elements that lies beyond **uranium** in the **periodic table**. Its **atomic number** is 92, and the **mass** of its most stable **isotope** is 243. Its chemical symbol is Am.

Occurrence and Extraction

Americium does not occur in nature. It is produced artificially in **nuclear reactors**. The first step in the preparation of americium is the manufacture of **plutonium** by the bombardment of uranium-238 with neutrons. Plutonium then decays to produce americium.

Properties

Americium is a silvery white metal with a melting point of about 2,150°F (1,175°C) and a **density** of about 13.6 grams per cubic centimeter. All of its isotopes are radioactive. The half life of the most stable isotope, americium-243, is 7,380 years.

Discovery and Naming

Americium was discovered as a byproduct of military research during World War II. At that time, the U.S. government maintained a major research center at the University of Chicago for the purpose of developing the materials and technology needed to build the first fusion (''atomic'') bomb.

During this research, a team of scientists from the University of California discovered a new element. The team consisted of Glenn Seaborg, **Albert Ghiorso**, Ralph A. James, and Leon O. Morgan (1919-). The team decided to name the element americium. They chose the name for two reasons. First, it was intended to honor the United States of America, where it was discovered. Second, its location in the periodic table places it just below element 63, **europium**, which was named for the continent of Europe.

Uses

The only important commercial use of americium is in smoke detectors. In a smoke detector, americium gives off a weak form of radiation that travels across the smoke detector and connects with an electrical circuit. As long as the smoke detector is empty, there is no break in the connection. If smoke enters the detector, however, the radiation can no longer travel across the detector. The connection is broken and a signal is set off. The signal may be a flashing light, a buzzer, or some other sound.

Ames, Bruce N. (1928-)

American biochemist and molecular biologist

Bruce N. Ames is a professor of **biochemistry** and molecular biology at the University of California at Berkeley. He is best known for the development of a test that has been used as an indicator of the carcinogenicity (cancer-causing potential) of chemicals. Known as the Ames test, it measures the rate of **mutation** in bacteria after the introduction of a test substance. His research led to a greater appreciation of the role of genetic mutation in cancer and facilitated the testing of suspected cancer-causing chemicals. He also developed a data base of chemicals that cause cancer in animals, listing their degree of virulence. Ames has been involved in numerous controversies involving scientific and environmental policies relevant to cancer prevention. In the 1970s he vociferously advocated strict government control of synthetic chemicals. In the 1980s, however, the discovery that many natural substances were also mutagenic (causing gene mutation), and thus possibly cancer causing, led him to reverse his original position.

Bruce Nathan Ames was born on December 16, 1928, in New York City, the son of Dr. Maurice U. and Dorothy Andres Ames. His father taught high school science and then became assistant superintendent of schools. Ames himself graduated from the Bronx High School of Science in 1946. He received a B.A. in biochemistry from Cornell University in 1950 and a Ph.D. in the same field from the California Institute of Technology in 1953. He recalled in *Omni* that although he ''was never terribly good at getting A's,'' he ''was always good at problem solving.'' He worked at the National Institutes of Health, primarily in the National Institute of Arthritis and Metabolic Diseases, from 1953 to 1967. In 1968 he moved to the Department of Biochemistry and Molecular Biology at the University of California at Berkeley as a full professor. He was Chairman of the Department from 1984 to 1989. In addition, he became Director of the National Institute of Environmental Health Science at the University in 1979. In 1960 he married Dr. Giovanna Ferro-Luzzi, a biochemist who is also on the faculty at Berkeley. They have two children, Sofia and Matteo.

In the 1960s and early 1970s, Ames developed a test that measured the degree to which synthetic chemicals cause gene mutation (a change in the deoxyribonucleic acid, or **DNA**, the **molecule** that carries genetic information). He began by deliberately mutating a *Salmonella* bacterium. The changed bacterium could not produce an amino acid called histidine that normal bacteria produce and that they need to survive. The next step was to add just enough histidine to allow the bacteria to live, and to add, as well, the synthetic chemical being tested. If the added chemical caused genetic mutation, the abnormal gene of the *Salmonella* bacteria would mutate and again be able to produce histidine. When this happened, the added chemical was marked as a suspected carcinogen, since cancer is associated with somatic cell mutation, that is to say, mutation of any cells with the exception of germ cells.

Over eighty percent of organic chemicals known to cause cancer in humans tested positive as **mutagens** in the test developed by Ames and his colleagues. This result gave support to the theory that somatic mutation causes cancer and helped to validate the use of the test for initial identification of mutagens when considering synthetic chemicals for industrial and commercial use. In addition to these practical results, the research of Ames and a colleague, H. J. Whitfield, Jr., led to important advances in understanding the biochemistry of mutagenesis. Beyond his work in genetic **toxicology**, Ames made important discoveries in molecular biology including ground-breaking studies on the regulation of the histidine operon (the gene or locus of the gene that controls histidine) and the role of transfer ribonucleic acid (RNA) in that regulation.

In the 1980s Ames set up a database of animal cancer test results with colleague Lois Swirsky Gold of Lawrence Berkeley Laboratory. The database can be used to determine whether a chemical has tested positive as a carcinogen and gives the degree of its virulence. From these data, Ames developed a value measuring the carcinogenic danger of a chemical to humans. HERP (daily Human Exposure dose/Rodent Potency dose) is the value determined by comparing the daily dose of a chemical that will cause cancer in half a group of test animals with the estimated daily dose to which humans are normally exposed. The result is a percentage that suggests the degree of carcinogenicity of a chemical for humans.

In the 1970s Ames was a conspicuous advocate of particular regulatory and environmental public policies that relate to the cancer-causing potential of synthetic substances. In the 1970s he believed that even trace amounts of mutagenic chemicals could cause a mutation (and thus possibly cancer). He found that *tris* **phosphate** tris, the chemical that was used as a flame retardant on children's pajamas, was a mutagen in the Ames test; he was instrumental in getting it banned. Similarly, he found that some hair dyes contained mutagens. His advocacy led to governmental regulations that forced manufacturers to reformulate their products. In his position on the regulation of synthetic chemicals, he was a natural ally of environmentalists.

However, in the early 1980s he reversed his position, arguing that there is no scientific evidence that most synthetic chemicals cause human cancers in the small doses that humans absorb them and that, in the absence of such evidence, they should not be controlled. This about-face was partly a result of a growing body of knowledge concerning the mutagenic properties of numerous chemicals found in nature. Ames began arguing against the existing large public expenditures for **pollution** control and the regulation of synthetic chemicals, noting that cancer might just as plausibly be caused by the chemicals in plants. His arguments were based primarily on three factors: his view that more scientific evidence should be required before controls are implemented; his attitude toward the setting of priorities, which he believed should be centered on basic research rather than regulation; and finally his belief that the large public expenditures incurred by the regulatory process hurt American economic competitiveness.

Ames and his colleague Gold have also argued that the use of bioassays (animal tests) of chemicals to predict their carcinogenic potential in humans should be abandoned. In a typical bioassay, rats are given a ''maximum tolerated dosage'' (MTD) of a particular chemical daily for a period of time (such as a year). The maximum tolerated dosage is as much as the animal can be given without immediately becoming ill or dying. At the end of the time period, the number of animals that have developed cancers is tabulated as an indicator of the cancer causing potential of the chemical being tested. Ames suggested that it is often the large dosage itself, rather than the nature of the particular chemical, that induces the rat cancers. He argued that, since humans are not normally exposed to such large doses, the assays were not valid for predicting human cancers.

Ames's views have some support both within and outside scientific communities. However, he also has numerous critics among scientists and others. Critics note that pollution control involves issues that include but also go beyond cancer (such as acid rain). They suggest that Ames has not offered a substitute for animal assays (the Ames test has not proved to be such a substitute), and that neither he nor they have a good idea of what goes on at low dosages. They say he has an oversimplified view of the regulatory process, which is based on a consideration of animal assays but also on other factors. It has also been argued that the discovery that natural chemicals have a high mutagenic rate just as synthetic chemicals do should not lead to the assumption that synthetic chemicals pose less risk than was previously supposed. Such an assumption places too much emphasis on mutagenic rate as a sole indicator of carcinogenicity, ignoring the complex, multi-stage developmental process of the disease.

Yet the disagreements between Ames and his critics are based on several points of commonality—that cancer is a complex multi-stage process that is not fully understood; that there is no perfect test or group of tests that can fully predict the potential carcinogenicity of many substances in humans; and that public regulatory and environmental policies must be made and carried out in spite of this deficiency of knowledge. As for Ames, he has described his public policy activism as a hobby, and has noted that his recent scientific work includes studies in the biochemistry of aging.

Elected to the National Academy of Sciences in 1972, Ames has received many awards, including the Eli Lilly Award of the American Chemical Society (1964), the Mott Prize of the General Motors Cancer Research Foundation (1983), and the Gold Medal of the American Institute of Chemists (1991). He is the author or coauthor of more than 250 scientific articles.

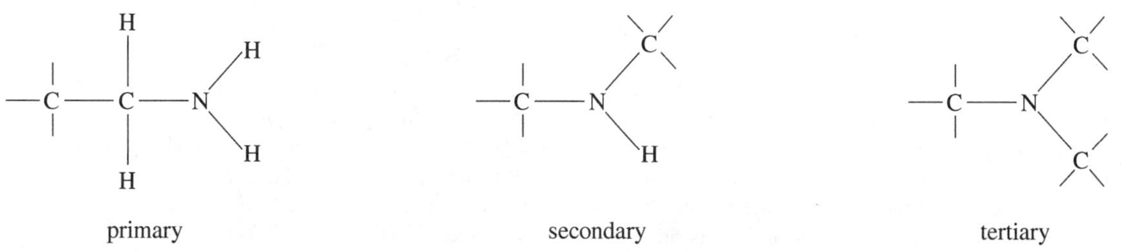

Primary, secondary, and tertiary amines. ***(Illustrations by Electronic Illustrators Group.)***

AMIDE GROUP

The amides are a group of organic compounds derived from **ammonia** (NH_3). One or more of the hydrogens of the ammonia is replaced with an organic acid group to produce a primary, secondary, or tertiary amide. The simplest form of amide—a primary amide—has the **functional group** -$CONH_2$ (a double bond exists between the **carbon** and the oxygen). A secondary amide is produced when two **hydrogen** atoms are replaced and has a general formula of $(RCO)_2NH$. A tertiary amide has the general formula of $(RCO)_3N$. All amides have the ending -amide as part of their name. There is no distinction made between the three types in their naming.

The amides are generally crystalline **solids** which can dissolve in **alcohol** and **ether**. Amides are hydrolyzed to ammonium salts with catalysis by acids or alkalis. This process is the starting point for the manufacture of a number of organic acids. Reaction with nitrous acid yields carboxylic acid directly as a product. **Dehydration** of amides is brought about by heating with phosphorous pentoxide and the product of this reaction is the alkyl cyanide. If reacted with **sodium** hydroxide **solution** and **bromine**, amides are transformed into the appropriate amines. This reaction is known as the Hofmann degradation after its discoverer, German chemist August von Hofmann (1818-1892). Another way to produce amines from amides is by **reduction**. This is carried out at a pressure of 250 atmospheres and a **temperature** of 490°F (250°C). This reduction reaction requires the use of **copper** oxides, **chromium** oxides, or **lithium aluminum** hydride as a catalyst. Halogenation of amides can occur to give the appropriate chloro- or bromoamides.

Amides can be distinguished from most other compounds by boiling with **sodium hydroxide** solution. After a short delay they give off ammonia. Ammonium salts treated in the same way give off ammonia immediately, and other **nitrogen** containing compounds yield no ammonia.

Primary amides are prepared by reacting ammonia or amines with acid chlorides, anhydrides, or esters. Secondary and tertiary amides are prepared by reacting primary amides or nitriles with organic acids. Primary amides are weakly basic and can form compounds with **metals** such as sodium, **potassium**, **mercury**, and **cadmium**, although heavy metal amides can be explosive. Alkylated amides are used in replacement reactions involving ionic reagents, since they are good **solvents** for this reaction.

The simplest amide is methanamide, $HCONH_2$. This is the only member of the group that is a liquid at standard temperature and pressure (STP). Methanamide is manufactured by reacting **carbon monoxide** and ammonia together under pressure. Like the other amides, it is a good solvent for a range of organic and inorganic compounds. The amides have elevated melting and boiling points due to the presence of hydrogen bonds.

AMINE GROUP

Amines are organic compounds formed from **ammonia** (NH_3 by the replacement of one or more **hydrogen** atoms by **hydrocarbon** groups (unlike the amides where the hydrogen in ammonia is replaced by organic acid groups). Primary amines have one hydrogen **atom** replaced by a hydrocarbon group, secondary amines have two hydrogen atoms replaced, and tertiary amines have all three hydrogen atoms replaced. In 1850 German chemist August von Hofmann (1818-1892) discovered that heating alkyl halides with an alcoholic **solution** of ammonia produces a mixture of the three types of amine. This process was modified in 1890 by Hinsberg to produce separate products, based on the different **reactivity** and solubilities of the three forms. This was a much easier method of obtaining the required chemicals than collecting them from herring brine, where they occur naturally.

Primary amines also are manufactured by a reaction known as the Hofmann degradation which takes amides as the starting point. The amide is heated with **bromine**, and the resulting mixture is then treated with excess **sodium** hydroxide to give the required primary amine. **Reduction** of amides will also yield amines, as will reduction of alkyl cyanides and nitroalkenes. Secondary amines are manufactured by heating a primary amine with an alkyl **halide**, although care must be used to ensure that excess alkyl halide is not used. Tertiary amines are manufactured by heating an alcoholic solution of ammonia with excess alkyl halide.

The simpler amines are generally **gases** or **liquids** at temperatures just above the freezing point of **water**. Weak hydrogen **bonding** elevates their boiling points. The amines are soluble in water and they are also weak bases when in solution. The reactions of amines are very similar to those of ammonia. **Combustion** yields water, **carbon** dioxide, and **nitrogen**. Reaction with nitrous acid is used to distinguish between primary, secondary, and tertiary amines. When a primary amines is reacted with nitrous acid, nitrogen is produced as well as a non-gaseous product which is soluble in water (the product varies depending on the amine under test but alcohols are common). Secondary amines react with nitrous acid to give nitroso compounds, pale yellow oils that are insoluble in water. Tertiary amines dissolve in nitrous acid without evolving any gas. In general, amines are recognized by their **solubility** in **hydrochloric acid** coupled with their ammonia aroma.

The simplest amine is monomethylamine (commonly referred to as methylamine). It has the formula CH_3NH_2 and is a colorless, flammable gas that is soluble in water. It is manufactured by heating methanal with ammonium chloride. Methylamine is used extensively in the manufacture of **herbicides** and fungicides.

The simplest secondary amine is dimethylamine. It has the chemical formula CH_3NHCH_3 and is a colorless, flammable liquid that is made by reacting a hot solution of **sodium hydroxide** with nitrosodimethylaniline. This compound also is used as a **fungicide** and herbicide, but it is chiefly used in the manufacture of other compounds such as **solvents** and rocket fuels.

The simplest tertiary amine is trimethylamine. It has a chemical formula of $(CH_3)_3N$ and is also a colorless, flammable liquid. It is used as a starting point in the manufacture of methanal.

See also Amide group

AMINO ACIDS

Amino acids are the building blocks of **proteins** and serve many other functions in living organisms. An amino acid is a **molecule** that contains a terminal acidic **carboxyl group** (COOH) and a terminal basic amino group (NH_2). The approximately 20 amino acids (plus a few derivatives) that have identified as protein constituents are alpha-amino acids in which the $-NH_2$ group is attached to the alpha-carbon next to the -COOH group. Thus, their basic structure is $NH_2CHRCOOH$, where R is a side chain. This side chain, which uniquely characterizes each alpha-amino acid, determines the molecules overall size, shape, chemical **reactivity**, and charge. There are hundreds of alpha-amino acids, both natural and synthetic.

The amino acids that receive the most attention are the alpha-amino acids that genes are codes for, and that are used to construct proteins.

These amino acids include: glycine NH_2CH_2COOH
alanine $CH_3CH(NH_2)COOH$
valine $(CH_3)2CHCH(NH_2)COOH$
leucine $(CH_3)_2CHCH_2CH(NH_2)COOH$
isoleucine $CH_3CH_2CH(CH_3)CH(NH_2)COOH$
methionine $CH_3SCH_2CH_2CH(NH_2)COOH$
phenylalanine $C_6H_5CH_2CH(CH_2)COOH$
proline C_4H_8NCOOH
serine $HOCH_2CH(NH_2)COOH$
threonine $CH_3CH(OH)CH(NH_2)COOH$
cysteine $HSCH_2CH(NH_2)COOH$
asparagine glutamine $H_2NC(O)(CH_2)2CH(NH_2)COOH$
tyrosine $C_6H_4OHCH_2CHNH_2COOH$
tryptophan $C_8H_6NCH_2CHNH_2COOH$
aspartate $COOHCH_2CH(NH_2)COOH$
glutamate $COOH(CH_2)2CH(NH_2)COOH$
histidine $HOOCCH(NH_2)CH_2C_3 H_3H_2$
lysine $NH_2(CH_2)_4CH(NH_2)COOH$
arginine $(NH_2)C(NH)HNCH_2CH_2CH_2CH(NH_2)COOH$

Proteins are one of the most common types of molecules in living **matter**. There are countless members of this class of molecules. They have many functions from composing cell structure to enabling cell-to-cell communication. One thing that all proteins have in common is that they are composed of amino acids.

Proteins consist of long **chains** of amino acids connected by peptide linkages (-CO•NH-). A protein's *primary* structure refers to the sequence of amino acids in the molecule. The protein's *secondary* structure is the fixed arrangement of amino acids that results from interactions of amide linkages that are close to each other in the protein chain. The secondary structure is strongly influenced by the nature of the side chains, which tend to force the protein molecule into specific twists and kinks. Side chains also contribute to the protein's *tertiary* structure, i.e., the way the protein chain is twisted and folded. The twists and folds in the protein chain result from the attractive forces between amino acid side chains that are widely separated from each other within the chain. Some proteins are composed of two of more chains of amino acids. In these cases, each chain is referred to as a subunit. The subunits can be structually the same, but in many cases differ. The protein's *quaternary* structure refers to the spatial arrangement of the subunits of the protein, and describes how the subunits pack together to create the overall structure of the protein.

Even small changes in the primary structure of a protein may have a large effect on that protein's properties. Even a single misplaced amino acid can alter the protein's function. This situation occurs in certain genetic diseases such as sickle cell anemia. In that disease, a single glutamic acid molecule has been replaced by a valine molecule in one of the chains of the hemoglobin molecule, the protein that carries **oxygen** in red blood cells and gives them their characteristic **color**. This seemingly small error causes the hemoglobin molecule to be misshapen and the red blood cells to be deformed. Such red blood cells cannot distribute oxygen properly, do not live as long as normal blood cells, and may cause blockages in small blood vessels.

Enzymes are large protein molecules that catalyze a broad spectrum of biochemical reactions. If even one amino acid in the enzyme is changed, the enzyme may lose its catalytic activity.

The amino acid sequence in a particular protein is determined by the protein's genetic code. The genetic code resides in specific lengths (called genes) of the polymer **deoxyribonucleic acid (DNA)**, which is made up of from 3,000 to several million nucleotide units, including the nitrogeneous bases: adenine, guanine, cytosine, and thymine. Although there are only four nitrogenous bases in DNA, the order in which they appear transmits a great deal of information. Starting at one end of the gene, the genetic code is read three nucleotides at a time. Each triplet set of nucleotides corresponds to a specific amino acid.

Occasionally an error, or **mutation**, may occur in the genetic code. This mutation may correspond to the substitution of one nucleotide for another or to the deletion of a nucleotide. In the case of a substitution, the result may be that the wrong amino acid is used to build the protein. Such a mistake, as demonstrated by sickle cell anemia, may have grave consequences. In the case of a deletion, the protein may be lose its functionality or may be completely missing.

Amino acids are also the core construction materials for neurotransmitters and **hormones**. Neurotransmitters are chemicals that allow nerve cells to communicate with one another and to convey information through the nervous system. Hormones also serve a communication purpose. These chemicals are produced by glands and trigger metabolic processes throughout the body. Plants also produce hormones.

Important neurotransmitters that are created from amino acids include serotonin and gamma-aminobutyric acid. Serotonin ($C_{10}H_{12}N_2O$) is manufactured from tryptophan, and gamma-aminobutyric acid ($H_2N(CH_2)_3COOH$) is made from glutamic acid. Hormones that require amino acids for starting materials include thyroxine (the hormone produced by the thyroid gland), and auxin (a hormone produced by plants). Thyroxine is made from tyrosine, and auxin is constructed from tryptophan.

A class of chemicals important for both neurotransmitter and hormone construction are the catecholamines. The amino acids tyrosine and phenylalanine are the building materials for catecholamines, which are used as source material for both neurotransmitters and for hormones.

Amino acids also play a central role in the immune system. Allergic reactions involve the release of histamine, a chemical that triggers inflammation and swelling. Histamine is a close chemical cousin to the amino acid histidine, from which it is manufactured.

Melatonin, the chemical that helps regulate sleep cycles, and melanin, the one that determines the color of the skin, are both based on amino acids. Although the names are similar, the activities and component parts of these compounds are quite different. Melatonin uses tryptophan as its main building block, and melanin is formed from tyrosine. An individual's melanin production depends both on genetic and environmental factors.

Proteins in the diet contain amino acids that are used within the body to construct new proteins. Although the body also has the ability to manufacture certain amino acids, other amino acids cannot be manufactured in the body and must be gained through diet. Such amino acids are called the essential

A computerized graphic of glycine amino acid. ***(Photograph by Scott Camazine, National Audubon Society Collection/Photo Researchers, Inc. Reproduced with permission.)***

dietary amino acids, and include arginine, histidine, isoleucine, leucine, lysine, methionine, phenylalanine, threonine, tryptophan, and valine.

Foods such as meat, fish, and poultry contain all of the essential dietary amino acids. Foods such as fruits, vegetables, grains, and beans contain protein, but they may lack one or more of the essential dietary amino acids. However, they do not all lack the same essential dietary amino acid. For example, corn lacks lysine and tryptophan, but these amino acids can be found in soy beans. Therefore, vegetarians can meet their dietary needs for amino acids as long by eating a variety of foods.

Amino acids are not stockpiled in the body, so it is necessary to obtain a constant supply through diet. A well-balanced diet delivers more protein than most people need. In fact, amino acid and protein supplements are unnecessary for most people, including athletes and other very active individuals. If more amino acids are consumed than the body needs, they will be converted to fat or metabolized and excreted in the urine.

However, it is vital that all essential amino acids be present in the diet if an organism is to remain healthy. Nearly all proteins in the body require all of the essential amino acids in their synthesis. If even one amino acid is missing, the protein cannot be constructed. In cases in which there is an ongoing deficiency of one or more essential amino acids, an individual may develop a condition known as kwashiorkor. Which is characterized by severe weight loss, stunted growth, and swelling in the body's tissues. The situation is made even more grave because the intestines lose their ability to extract nutrients from whatever food is consumed. Children are more strongly affected by kwashiorkor than adults because they are still growing and their protein requirements are higher. Kwashiorkor often accompanies conditions of famine and starvation.

See also Biochemistry; Metabolism; Neurochemistry; Nucleic acids; Nutrition; Protein synthesis

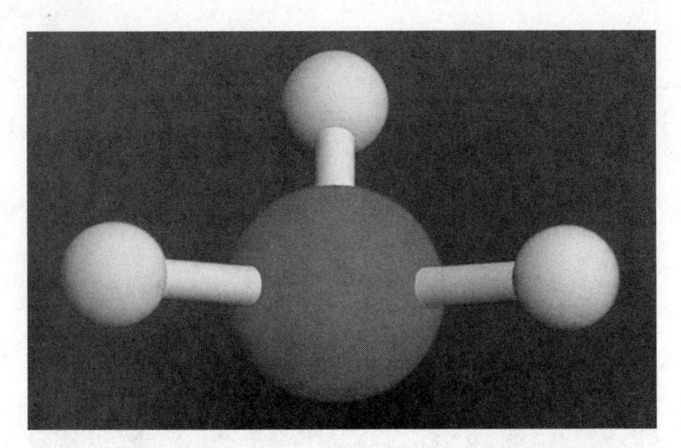

A computerized graphic of an ammonia molecule. ***(Photograph by Ken Eward, Photo Researchers, Inc. Reproduced by permission.)***

AMMONIA

Ammonia is a compound composed solely from the elements **hydrogen** and **nitrogen**. At normal room temperatures and pressures ammonia exists as a colorless gas that is lighter than air and has a characteristic pungent odor. Each single ammonia **molecule** contains one nitrogen **atom** bonded covalently to three hydrogen atoms; hence, the chemical formula of ammonia is NH_3. The structure of the ammonia molecule is best described as pyramidal. The nitrogen atom occupies the apex of the pyramid, while the three hydrogen atoms constitute its triangular base. As a consequence of the **electron** structure of the nitrogen atom, two of its electrons do not participate in the **bonding** to the hydrogen atoms, and constitute a **lone pair** that points directly up from the top of the pyramid. It is this lone pair of electrons that is responsible for the prolific chemical **reactivity** observed for ammonia. Most notably, the **chemistry** of ammonia is rich in acid/base reactions, whereby the alkaline ammonia reacts with acids to form ammonium salts.

The derivation of the name ammonia can be traced back to the Sun god of ancient Egypt, Ammon. At the temple of Ammon near Karnak, it is said that the priests collected camel dung from passing travelers to burn as fuel. Upon burning, the dung produced acrid vapors and would leave behind a white residue which they called *sal ammoniac*, meaning **salt** of Ammon. Later, in the Middle Ages, ammonia was formed by the **distillation** of either stale urine, or the hooves and horns of oxen and deer, the latter concoction giving rise to the now defunct name, spirits of hartshorn. In 1774, the first deliberate preparation and isolation of ammonia, or *alkaline air* as he called it, was performed by the English chemist **Joseph Priestley** (1733-1804). However, it was not until a few years later (c. 1780) that the composition of ammonia was determined by the French chemist Claude Louis Berthollet (1748-1822). Ammonia is not abundant on Earth, however, it is a major component in the atmospheres of some other planets in our solar system. The most notable among these planets is Jupiter, which ironically is the name the Romans gave to the Egyptian Sun God, Ammon, when they adopted this deity for their own religion.

During the nineteenth century ammonia was produced, rather inefficiently, by the dry distillation of **coal**. However, a major breakthrough in ammonia synthesis was made in the first decade of the twentieth century by the German chemist **Fritz Haber** (1868-1934). He discovered that by reacting nitrogen and hydrogen at high pressures (approximately 150-200 times atmospheric pressure) and elevated temperatures [typically around 900°F (480°C)] in the presence of a catalyst (osmium or uranium), an efficient conversion of these materials into ammonia could be achieved. Haber's work was the culmination of an in-depth study into the chemical equilibrium that exists between ammonia and its constituent elements, hydrogen and nitrogen, and was regarded as a triumph for the application of **thermodynamics** to practical chemistry. In 1918 Haber was awarded the Nobel Prize for Chemistry in recognition of the importance of his work. As the twentieth century has progressed, Haber's method has been optimized, i.e., new catalysts have been discovered, but the procedure remains essentially unchanged.

The initial impetus behind the search for an efficient ammonia synthesis lay in man's desire to *fix* atmospheric nitrogen. It was recognized in the 1800s that nitrogen-containing **fertilizers** were of great importance to the agricultural industry. At the turn of the nineteenth century, the main source of nitrogenous compounds necessary for the production of these fertilizers was to be found in a 220-mile (356-km) long guano (bird droppings) deposit along Chile's coastline. This natural resource was fast running out and consequently spurred the effort to take atmospheric nitrogen and convert or *fix* it into useful nitrogen-containing substances such as ammonia. Haber solved this problem, and ammonia now plays a significant role, not only in agriculture, but also in the chemical industry as a whole.

The major use of ammonia is as a fertilizer, and it is applied to the soil in either its liquefied state, or as an ammonium salt, such as ammonium nitrate or **phosphate**. Ammonium nitrate, made by the direct combination of ammonia and **nitric acid** (also produced from ammonia) is also a common ingredient in **explosives**. Ammonia is highly soluble in **water**, generating ammonium hydroxide, dilute solutions of which are used in household cleaning products. Ammonia is a very important feedstock compound for the chemical industry, and is the starting point for the syntheses of many different substances that ultimately find their way into pharmaceuticals, dyes, and polymers. The liquefaction of ammonia occurs readily under the application of pressure, or upon cooling down to its **boiling point** of -28.3°F (-33.4°C). It is this property, in conjunction with the relatively large amount of **heat** absorbed upon going from the liquid to gaseous state (a large heat of vaporization), that makes ammonia an ideal refrigerant. Indeed, the demand for ammonia-based refrigeration systems is increasing as a consequence of the concern over chlorofluorocarbons (CFCs) and the damage they cause to the ozone layer. Liquid ammonia is also utilized as a solvent in certain chemical reactions.

ANALGESICS

Analgesics are medicines used to relieve pain such as headaches, backaches, joint pain, sore muscles, menstrual cramps, and pain that results from surgery, injury, or illness. While these drugs do not treat whatever is causing the pain, they can provide enough relief to make people more comfortable and to allow them to carry out their daily routines.

Among the most common analgesics are aspirin (acetylsalicylic acid), ibuprofen, naproxen **sodium**, and ketoprofen—all in the general category known as nonsteroidal anti-inflammatory drugs (NSAIDs). NSAIDs relieve pain and also reduce inflammation. Another common analgesic, acetaminophen (Tylenol, Panadol), provides pain relief but does not reduce inflammation. For some types of serious, chronic pain **narcotic** analgesics (also called opioid analgesics) are used. Analgesics in this category include natural narcotics (derived from the opium poppy), such as **morphine** and codeine (morphine-3-methyl ether), and synthetic narcotics, such as propoxyphene (Darvon) and meperidine (Demerol).

Salicylic acid, the compound from which the active ingredient in aspirin was first derived, was found in the bark of the willow tree (*Salix alba*) in 1763 by Reverend Edmund Stone of Chipping Norton, England. Earlier accounts indicate that the ancient Greek physician Hippocrates (c.460-c.377 B.C.) used willow leaves to reduce fever and relieve pain. During the 1800s various scientists extracted salicylic acid from willow bark and produced the compound synthetically. In 1897, German chemist Felix Hoffmann working at the Bayer division of I.G. Farber developed an improved method for synthesizing aspirin and it was first marketed in 1899. Today, Americans alone consume about 16,000 tons of aspirin tablets annually. Ibuprofen, a type of carboxylic acid (carboxylic acid is the agent that gives aspirin its anti-inflammatory action), was developed by British drug manufacturer and retailer Boots and Company in the 1960s. It has the advantages of being more effective than aspirin for treating certain kinds of pain while causing fewer stomach problems.

Morphine (named for Morpheus, the Greek god of dreams) was first isolated from the juice of the opium poppy in 1805 by German pharmacist Friedrich Sertürner. It is an **alkaloid**, which means it is an organic compound that contains **carbon**, **hydrogen**, and **nitrogen** and which forms a water-soluble **salt**. Morphine is one of the most powerful pain relievers know-milligram for milligram it is about 50 times more powerful than aspirin. Codeine, another alkaloid derived from the juice of the opium poppy, was isolated by French chemist Pierre-Jean Robiquet (1780-1840) in 1832 in an attempt to produce an analgesic without some of morphine's undesirable side effects (drowsiness, confusion, constipation, etc.).

Pain is the body's signal that something is wrong. Pain can result from an injury, such as a broken bone, a burn or a sprain; from overuse of muscles (including muscle tension due to stress); from infections, such as sinus infections or meningitis; or from natural events, such as childbirth.

Pain begins at the level of the cells. In response to injury or inflammation, cells release chemical messengers. These

The analgesic drug Zomax. It is usually prescribed for arthritis, post-surgical pain, and muscle-skeletal injury. ***(Photo Researchers, Inc. Reproduced by permission.)***

chemical messengers alert other specialized cells called pain **receptors**. The pain receptors send signals to the brain. The brain interprets the signals, and we perceive pain. Analgesics work by either blocking the signals that go to the brain or by interfering with the brain's interpretation of the signals.

Determining the best pain reliever depends, in part, on the type of pain. The two main categories are acute pain and chronic pain. Acute pain is usually temporary and results from something specific, such as a surgery, injuries, or infections. Chronic pain is any pain that lasts more than three months and may disrupt daily life. Sometimes chronic pain is just a nagging discomfort, but it can flare up into severe pain. Although narcotic analgesics may be used, for most types of chronic pain a combination of non-narcotic medication and lifestyle changes is recommended.

Overuse of pain relievers can actually make some types of pain worse. To manage long-term pain, such as recurring headaches, chronic backache, or arthritis pain, many pain treatment specialists recommend an approach that helps people cope without depending on large or frequent doses of drugs. Relaxation techniques, biofeedback, massage, exercise, proper diet, and good sleep habits can all be helpful. Caution is also recommended when exercising after taking an analgesic to relieve the pain of an injury. Since the analgesic depresses the body's pain signals, the injury can actually be aggravated by exercising when pain signals are muted.

Side effects are one reason to be careful about frequent or long-term use of pain relievers. Narcotic analgesics are very effective, but can cause addiction. Aspirin and ibuprofen can irritate the stomach. Acetaminophen does not produce the side effects that aspirin and ibuprofen do, but high doses can cause liver damage, especially in people who drink **alcohol** regularly. Some pain relievers contain **caffeine**, which enhances their effectiveness. Taking these drugs near bedtime can interfere with sleep.

ANALYTICAL CHEMISTRY

Imagine that a small child or a loved one is rushed to a hospital's emergency room, comatose or delirious because they have just consumed an unknown white powder. The doctor takes immediate action, pumping the child's stomach and using activated **charcoal** as a treatment. But the most important question remains: what was it that they ate? What are the chemical compounds in the white powder and how much of each is present?

This is where analytical **chemistry** steps in. It is the job of the analytical chemist to determine the chemical composition of any unknown, whether it is a toxic white solid, **gold** ore from a mine, or the compounds in the discharge from a pulp mill. Analytical chemistry is about answering the questions ''What is it?'' and ''How much is there?'' (qualitative and **quantitative analysis**, respectively).

Analytical chemistry, like most of the sub-disciplines of chemistry, is neither simple nor homogeneous. There are many different forms that analytical chemistry can take. The simplest are the ''classical'' methods such as titrimetry and gravimetric analysis. ''Classical'' because these are the methods that were used by chemists hundreds of years ago. But they are still used today.

An example of a simple **titration** is an acid/base titration used to determine the **pH** or acidity of an unknown **solution**. In this case, the unknown quantity of acid or base is neutralized using a known quantity of the opposite species (e.g., an unknown acid is titrated with a known base). By knowing both the **concentration** and the **volume** added, it is easy to determine how much of the unknown compound was originally present. This also confirms that the substance is either an acid or a base, allowing for a better judgement as to the chemical composition of the unknown.

Gravimetric analysis is similar. A known chemical compound is added to an unknown with the intent of causing all of the unknown species to precipitate. The resulting solid is weighed, providing a measure of how much was in the original sample. This only really works where there is enough sample to actually weigh something and a suitable **precipitation** reaction is known.

What about minuscule amounts of an unknown substance? In this case, there are a number of more sophisticated analytical techniques that have been developed in the past 50 years. Perhaps the most useful and widespread is ''nuclear magnetic **resonance**'' **spectroscopy**. NMR spectroscopy exposes the **nucleus** to a large magnetic field. If the nucleus has a magnetic moment (and every element has at least one **isotope** that does), then the nuclear spins will align in the field. By focussing radio waves of different frequencies on the sample, the different nuclei will ''flip'' or absorb radiation. The technique relies on the fact that not only do different elements show up at different frequencies, but so do atoms of the same element if they have different chemical locations. That is, the way that all of the atoms are connected is information that can be gained from NMR spectroscopy.

This allows an analytical chemist to reconstruct a picture of the **molecule** or substance which, in turn, identifies the specific compound. NMR is only one of a battery of techniques, such as **infrared spectroscopy**, U.V./visible light spectroscopy, and **mass** spectroscopy, that allow a chemist to definitively state the chemical nature of an unknown substance. That is, the spectroscopic methods allow chemists to answer the question ''what is it?''

To answer the question of quantity, chemists use a number of other techniques such as **chromatography**. Literally, this means ''**color** writing,'' and the technique originates with the separation of the colored compounds found in leaves using filter **paper** or thick paper. A number of instruments employ chromatographic columns to separate and quantify the substances present.

Gas chromatography is probably the most widely used technique. In its simplest form, it involves injecting (using a syringe) a small amount of the unknown into a flowing gas, which is pushed through a long 100 ft (33 m) column containing silica or some other porous material. The different compounds in the sample have differing levels of ''stickiness'' towards the silica and ''affinities'' for the gas.

As an analogy, imagine a tourist on a busy New York City street lined with shops. The speed with which they move down the street is not the same as the rest of the pedestrian traffic. If there were several tourists that started together in a tour group, some would move slightly faster than others with the result that the tour group would get spread out. A tour guide waiting at the end of the block would have no trouble identify each individual as they reached the end. Gas chromatography does much the same thing, only on a molecular level, with the tourists being the unknown substances. And, using a detector, it is able to tell exactly how many have reached the end, under one condition. Different compounds have different detector sensitivities necessitating that there is a pure sample of the compound available so that the detector can be calibrated. That is, if a detectors response is say 1,000 counts for a 1 mg sample, then it will be 2,000 counts for a 2 mg sample of the same compound. But it could be quite a different count for 2 mg of something else.

This limits the usefulness of chromatographic techniques. While they are very good at answering the question of ''how much?'' they are not good at answering the question of ''what is it?'' and must rely on other analytical methods for making that determination. But once an instrument is calibrated for a particular compound, once both the time it takes for a compound to travel through the column and the relationship between counts and concentration are known, a chromatograph can provide answers to both the questions.

An alternative method is the utilization of complex instruments. One of the most popular is the G.C./Mass Spec., a gas chromatograph to determine quantity with a mass spectrometer as a detector to determine composition. This is the sort of instrument that is used in drug testing. It is a very powerful combination because it can answer both questions on unknown substances. Indeed, if a patient does arrive at an emergency room, having eaten a toxic substance, a G.C./Mass Spec. is one of the techniques used to determine the composition of the unknown. It tells them how many substances are present, what are their relative concentrations, and what are their chemical compositions.

One of the other important aspects of analytical chemistry is actually "mathematics" or, more specifically, the field of statistical analysis. It is one thing to know that a particular lake has 0.238 ppm of **lead** in it, but what does that mean? Do we really know that it is "0.238" and not "0.236" or even "0.220"? Is this value high, low, or perfectly normal?

To answer these questions, analytical chemists rely on such mathematical tools as the "standard deviation" and "mean, median, and mode" for a sample set. They also routinely analyze multiple samples, sometimes at completely different laboratories to ensure that their results are accurate. Questions about the significance of a reading must always be put in context. For example, if an analysis of lake **water** was to give a reading of 0.238 ppm for ten samples and the standard deviation was 0.008 ppm, this means that the chemist doing the analysis is very confident in the value giving. A large number of samples gave essentially the same results. If there is no systematic error in the technique, then the value of lead is likely to be about 0.238 ppm.

However, a result of 0.238 ppm from a single sample with a standard deviation of 0.189 ppm would mean that the results could be anywhere from 0.049-0.427 ppm, quite a wide range in uncertainty. Clearly, this would leave the question of the lead content uncertain. This may not seem so important, but consider that some environmental contaminants have very controlled limits. A high degree of uncertainty brings into question whether or not an industrial site is in compliance.

Further, the question of whether the results are significant or not must be framed in terms of the larger picture. If 0.238 ppm was the highest reading ever recorded for a lake, then there may be cause for some concern. On the other hand, a reading of 0.238 ppm may be an average reading for all of the lakes in a particular reading.

The challenge to the chemist is to determine both the concentration and the identity of the unknown substance, be it in a human body, from the environment, or synthesized by a researcher in the laboratory. Analytical chemistry is the discipline devoted to answering the questions of "what is it?" and "how much is there?"

Gloria L. Anderson. ***(AP/Wide World Photos. Reproduced by permission.)***

ANDERSON, GLORIA L. (1938-)

American chemist

Gloria L. Anderson is a distinguished chemist, educator, and college administrator. Her scientific research has involved industrial, medical and military applications of fluorine–19 **chemistry**. As an educator, she has served as the Callaway professor of chemistry, chair of the chemistry department, and dean of academic affairs at Morris Brown College in Atlanta. Anderson, in addition, has been a board member and vice-chair of the Corporation for Public Broadcasting, for which she has lectured nationally on issues related to minorities and women in mass media and public television.

Anderson was born in Altheimer, Arkansas, on November 5, 1938, the daughter of Charley Long and Elsie Lee Foggie. She enrolled at the Arkansas Agricultural, Mechanical and Normal College (now the University of Arkansas at Pine Bluff), where she was awarded a Rockefeller Scholarship from 1956 to 1958. Anderson received her B.S. degree summa cum laude in 1958. She married Leonard Sinclair Anderson on June 4, 1960; they have one son, Gerald. In 1961, Anderson was awarded her M.S. degree from Atlanta University. For the next year, she worked as a chemistry instructor at South Carolina State College in Orangeburg. From 1962 to 1964, she held an instructorship at Morehouse College in Atlanta, then went on to take a position as a teaching and research assistant at the University of Chicago, where she received her doctorate in **organic chemistry** in 1968.

Anderson's dissertation and aspects of her subsequent research have related to fluorine–19 chemistry. (The '19' following **fluorine** refers to a particular **isotope** of fluorine that, like other elements with odd numbered masses, has magnetic properties.) Fluorine–19 chemistry became an important field of research shortly before World War II when many commercial uses for fluorine compounds were discovered. Much of Anderson's research has involved **nuclear magnetic resonance** (NMR) **spectroscopy**, a method of investigating organic compounds by analyzing the nucleic responses of molecules subjected to radio-frequency radiation within a slowly changing magnetic field. NMR spectroscopy, which has been widely exploited for chemistry, **biochemistry**, biophysics, and solid-state physics research, enables extremely sophisticated analysis of the molecular structures and interactions of various materials. The small size, low **reactivity**, and high sensitivity of fluorine–19 make it particularly suited for NMR spectroscopy. Since the late 1960s, fluorine NMR spectroscopy has been applied to a range of biochemical problems, including the study of the human **metabolism** and the formulation of new pharmaceuticals.

Anderson joined the faculty of Morris Brown College in Atlanta in 1968 as associate professor and chair of the chemistry department. From 1973 to 1984, Anderson was the Fuller E. Callaway professor of chemistry at Morris Brown, and continued her service as the chemistry department chair. Anderson left the chemistry department to serve as dean of academic affairs at Morris Brown for the years 1984–89. In 1990, Anderson resumed her post as the Callaway professor of chemistry. In 1976, Anderson was recognized as an Outstanding Teacher at Morris Brown, and received a Scroll of Honor award from the National Association of Negro Business and Professional Women. In 1983, she received a Teacher of the Year award and was voted into the Faculty/Staff Hall of Fame at Morris Brown. In 1987, she received an Alumni All-Star Excellence Award in Education from the University of Arkansas at Pine Bluff.

In addition to her work at Morris Brown, Anderson has conducted research through a number of independent and government facilities. Beginning in 1971 she continued her investigations of fluorine–19 chemistry—first in association with the Atlanta University Center Research Committee, then under the National Institutes of Health, the National Science Foundation, and the Office of Naval Research. She also conducted research on amantadines (a drug used to prevent viral infection) under the Minority Biomedical Support Program of the National Institutes of Health. She held a faculty industrial research fellowship with the National Science Foundation in 1981, and with the Air Force Office of Scientific Research in 1984. In 1985, Anderson investigated the synthesis of potential **antiviral drugs** as a United Negro Fund Distinguished Scholar. In that same year, she conducted research on the synthesis of solid rocket propellants under the Air Force Office of Scientific Research. Since 1990, she has been affiliated with BioSPECS of The Hague, Netherlands, as a research consultant.

In 1972, Anderson was appointed to a six-year term on the board of the Corporation for Public Broadcasting (CPB). At the CPB, Anderson chaired committees on Minority Training, Minorities and Women, and Human Resources Development; she was vice-chair of the CPB board from 1977–79. She is a member of the American Institute of Chemists, the American Chemical Society, the National Institute of Science, the National Science Teachers Association, the Association of Computers in Mathematics and Science Teaching, the Georgia Academy of Science, and the Atlanta University Science Research Institute, among other scientific and professional bodies. She has served as a proposal review panel member, contract reviewer, or field reader for the Department of Health, Education and Welfare's Office of Education, the National Science Foundation's Women in Science Program, the Nation Cancer Institute, the Department of Education, and the National Institute of Drug Abuse.

ANDERSON, W. FRENCH (1936-)

American biochemist and geneticist

The age of gene therapy—the treatment of disease by genetic engineering—began on September 14, 1990, when a four-year-old girl suffering from a hereditary immune deficiency was transfused with her own white blood cells that had been mixed with the gene needed to cure her. The doctor in charge of the treatment, W. French Anderson, had labored long and hard not only in the laboratory to make this therapy a reality but in the halls of government as well, working his clinical approach or protocol through a myriad of bureaucratic and scientific review boards. Gene therapy, like all revolutionary procedures, had as many detractors as supporters: the specter of genetic designing hovers over the field. But Anderson had insisted to the committees that he was concerned with curing the sick, not creating test-tube super humans. The test—the first ever approved gene therapy—was allowed to proceed as scheduled. It was the pinnacle of achievement for a man who had spent his career researching human biochemical genetics, hematology, and the synthesis of hemoglobin and **proteins**.

William French Anderson was born in Tulsa, Oklahoma, on December 31, 1936. His father was an engineer and his mother a journalist, and Anderson grew up preferring books to sports. He was something of a child prodigy with grade-school and high-school test scores going off the charts. By the age of fourteen, he was already reading college-level medical texts and studying astronomy. Even as a child, Anderson showed a determination to overcome any obstacle. A stutterer, he decided that a good way to cure his stammer would be to join a debating team. This crude form of homeopathy worked. In 1953, he applied for early entrance to Harvard University, writing on his application that he wanted to gain knowledge of disease at the molecular level. That same year, **James Watson** and **Francis Crick** unlocked the structure of **DNA** (deoxyribonucleic acid —the carrier of genetic information).

Anderson was a natural mathematician, publishing a paper in 1956 that demonstrated how to perform higher arithmetic with Roman numerals, a concept thought unworkable because such numerals do not have place values like Arabic numerals do. But he soon gave up on a mathematical career, opting instead for one in medicine, doing original research even as an undergraduate. A course in DNA and genetics taught by James Watson set Anderson on the track of genetic therapy. Graduating magna cum laude from Harvard in 1958, he went to Cambridge University in England to study under Francis Crick, working on research that led to the discovery of the underlying structure of the genetic code: its organization in triplets. It was while at Cambridge that Anderson also met his future wife, Kathryn, who was studying medicine. He graduated with honors from Cambridge in 1960, earning his M.A., and returned to Harvard University's medical school where he received his M.D. magna cum laude in 1963. He then interned in pediatric medicine at Boston's Children's Hospital Medical Center.

In 1965 Anderson joined the staff of the National Institutes of Health at Bethesda, Maryland. There he initially worked with **Marshall Warren Nirenberg**, who broke the genetic code, and whose name in genetics ranks with those of Watson and Crick. Anderson soon proved his own worth at NIH in research that led to synthesizing hemoglobin, the oxygen-carrying pigment in red blood cells, as well as discovering how cells initiate **protein synthesis**. Blood research was his focus for many years, and he became well known for his work in thalassemia, a hereditary and fatal form of anemia, pioneering the use of **iron** chelators or organic reagents which absorb excess iron in the body, thus prolonging the lives of the victims of this disease. He also conducted important research in sickle-cell anemia. Meanwhile, he worked his way up the ladder of promotion at NIH. By 1977 he had become the director of the molecular hematology laboratory at the National Heart, Lung, and Blood Institute, as well as a research fellow at the American Cancer Society.

Once gene-splicing techniques had been developed in the 1970s, Anderson turned his attention back to his original goal: dealing with disease on the molecular level—gene therapy. (Gene splicing refers to the process of recombining, or engineering, specific fragments of DNA, usually from more than one species of organism.) In particular, Anderson sought ways in which engineered genes could be transferred into cells. One of his developments was a microscopic needle used for injection of genes directly into cells, but this technique was too limited. For gene therapy to be effective, a wide spectrum of cells had to be reached, Anderson knew. Other researchers around the world were looking at this same problem, and some—including Richard Mulligan of the Whitehead Institute for Biomedical Research in Cambridge, Massachusetts—thought they had come up with a solution. The vector, or agent to transport genetic material into the cell, they hit on was a form of virus. These simple organisms function by getting into cells and taking over the genetic works. Most viruses, like those of the cold or smallpox, simply kill the cells off; but one strain, the retrovirus, actually tricks the cell's DNA into reproducing the virus's genetic material over and over again. Mulligan and others saw how these retroviruses—acquired immune deficiency syndrome (AIDS) is caused by a retrovirus—could be used as a Trojan horse for genes they wanted to insert into cells. By the early 1980s, techniques had been perfected for stripping the viruses of their harmful genetic material and for inserting cloned genes in their stead. The retroviral vector had come into existence.

Anderson, meanwhile, was hoping to use these new technologies to tackle thalassemia, but because of its complexity—several genes are responsible for the disease—he finally had to give up on it as a pioneer case in gene therapy. Other researchers hit on another hereditary disease of the more than four thousand that plague humans: adenosine deaminase (ADA) deficiency. ADA is an **enzyme** vital to the functioning of the immune system, and in rare cases a single gene responsible for its creation is missing. Here was a more workable disease on which researchers could focus. By 1984 John Hutton of the University of Cincinnati had managed to clone the ADA gene and shared his research with Anderson at the NIH. Two parts of the puzzle were in place now: the vector and the gene. They needed to discover what type of cells in which to insert the gene.

Logically, the stem cells of the bone marrow—those that manufacture blood cells and have a long lifespan—would be the place to begin, using bone marrow transplant surgery to implant genetically altered retroviruses. But as these stem cells occur only once in 100,000 bone marrow cells, the probability of reaching them was considered too low. Instead, Anderson decided to use a white blood cell called the T cell. A crucial step in perfecting the procedure was a 1988 trial in which Anderson, in collaboration with Steven A. Rosenberg of the National Cancer Institute, managed successfully to put genetically tagged cells into terminal cancer patients. Tracing the progress of these earmarked cells, the scientists could see that in fact such cells would reproduce once in the body. Still, there were political hurdles to jump, and it took until September 14, 1990, to clear them all. On that day Anderson, in collaboration with R. Michael Blaese and Kenneth W. Culver, treated a young ADA patient with T cells withdrawn from her own body, which had been "infected" with a genetically engineered virus, and then cultured in the laboratory. The initial transfusion replaced more than one billion of these genetically altered cells—with the ADA manufacturing gene in place—and further treatments and tests indicated that the cells were reproducing in the patient's body, creating ADA. Gene therapy, in fact, appeared to work.

From this initial trial, the number of ongoing and planned gene therapy protocols has mushroomed worldwide, including experimental treatments for melanoma, cystic fibrosis, cardiovascular disease, high **cholesterol**, and AIDS. In 1991 Anderson also took research out of federal laboratories, helping to found Genetic Therapy, a Maryland-based company that works closely with NIH to provide genetically engineered vectors to hospitals and universities. In 1992 Anderson moved his laboratory to the University of Southern California's Kenneth J. Norris Jr. Comprehensive Cancer Center so his wife could accept the position of surgeon-in-chief at the Children's Hospital of Los Angeles.

Anderson has been widely acknowledged for his work, receiving the Scientific Achievement Award for Biological Sciences from the Washington Academy of Sciences in 1971; the Thomas B. Cooley Award for Scientific Achievement in 1977; and the Mary Ann Liebert Biotherapeutic Award in 1991, among others. He and his wife have no children: they

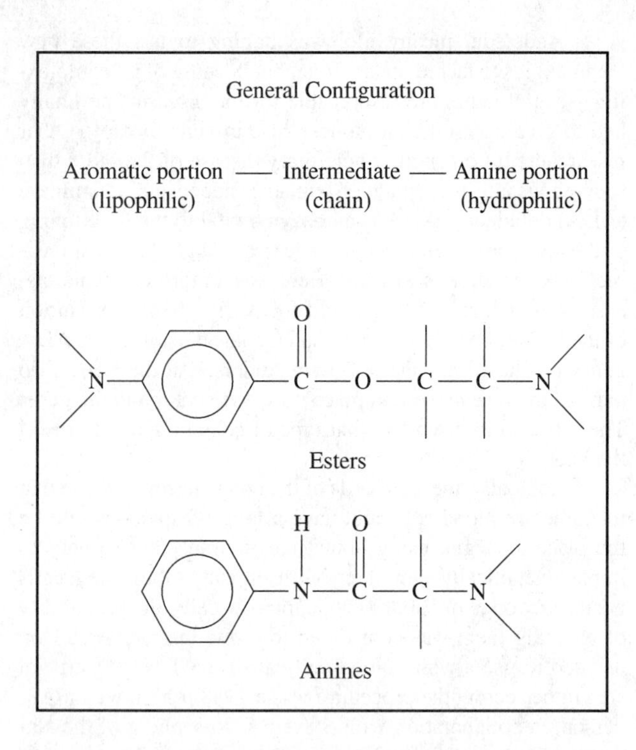

Chemical structures of various types of local anesthetics. ***(Illustration by Electronic Illustrators Group.)***

opted early in their marriage for career over family life. In his spare time, Anderson practices Korean martial arts—tae kwon do—in which he has a fourth-degree black belt.

Anesthetics

Anesthesia is the loss of sensation of painful stimuli either with or without loss of consciousness. Anesthetic agents produce a state of anesthesia. Anesthetics can be broken down into two basic categories: general anesthetics (those that cause loss of consciousness), and local or regional anesthetics (those that do not cause a loss of consciousness but rather "deaden" or "numb" the area).

The history of general anesthesia begins in ancient times with the inhalation of opium and other plant alkaloids. In the recent past dentists were instrumental in the introduction of diethyl **ether** and **nitrous oxide**. In 1846 William T. G. Morton (1819 - 1868), a Boston dentist, gave his famous demonstration of ether in the "ether dome" of the Massachusetts General Hospital. Morton was late for the procedure, and as he ran into the operating theater the surgeon, Doctor Warren, remarked, "Well, sir, your patient is ready." There surrounded by a somewhat hostile and silent audience Morton quietly worked. When the patient was unconscious and anesthetized Morton announced, "Doctor Warren, your patient is ready." The operation was begun, strong men surrounded the table prepared to restrain the patient, but they were not needed. When the procedure was completed Warren turned to the astonished audience and made the famous statement, "Gentlemen, this is no humbug." The age of general anesthesia had begun.

After 1846, ether was widely used in the United States. In 1847, chloroform was introduced by the Scottish obstetrician, James Simpson (1811- 1870). Although it became popular in England chloroform was never so in the United States due to its severe hepatotoxicity (damaging to the liver) and cardiovascular depressant effects. In 1929, chemists discovered cyclopropane. It became the most popular anesthetic for the next 30 years, however it was an explosive gas, and as more electrical devices came into use in the operating room the search for a new agent was intensely pursued. British chemists discovered halothane, a nonflammable anesthetic agent. In 1956, it was introduced into clinical practice and was a great success. Since that time most of the inhalation anesthetics which are halogenated hydrocarbons and ethers are modeled after halothane.

Unfortunately, in the late 1800s and early 1900s, to get adequate muscle relaxation for a sugical procedure, dangerously deep levels of anesthetic were needed. But in the 1940s chemists found that curare, a poison used by bow hunters in South America, acted by paralyzing certain muscle tissue. Curare began to be used as a muscle relaxant in surgery and allowed a much safer depth of anesthetic by inhalation.

Finally, intravenous compounds completed the spectrum of general anesthetic agents. In 1935, Lundy, introduced thiopental, a rapidly acting barbiturate for intravenous use. Short acting opioids have also been found effective. Today's anesthesia then consists of rapid inductions with a short acting barbiturate (or in young children inhaling halothane which, acts very rapidly due to their small body mass) and/or short acting intravenous opioids, enhanced by muscle relaxation provided by curare like drugs. Variation of this technique are common, but the basic pattern is fairly standard.

Local anesthetics act by blocking the transmission of electrical impulses through the axons of pain receptive nerve fibers. The first such drug was cocaine which grows in the Andean mountains between 3,000 and 9,000 ft. Used by the natives for centuries for its stimulatory effects to enhance hard work at altitude, it was first isolated by Albert Nieman (1834-1861) in 1860. When he tasted the newly isolated compound he noted a numbness of his tongue. William Halstead (1852-1922), a surgeon at the Johns Hopkins Hospital in Baltimore, demonstrated cocaine's local anesthetic use when injected. All the local anesthetics of 1999 stem from these early observations.

General anesthetics come from many types of chemical compounds. Figure 1 shows the structures of characteristic compounds from each major class. Nitrous oxide is discussed under its own section in this volume. Diethyl ether (commonly known as "ether") is the parent compound of the class of organic compounds known as ethers. At room **temperature** and standard **pressure** ethers are colorless, neutral **liquids** with pleasant odors. Ethers are easily soluble in organic liquids. Diethyl ether has a **boiling point** of 62.5°F (35°C): therefore it can be inhaled. It can be bottled or in the case of medical

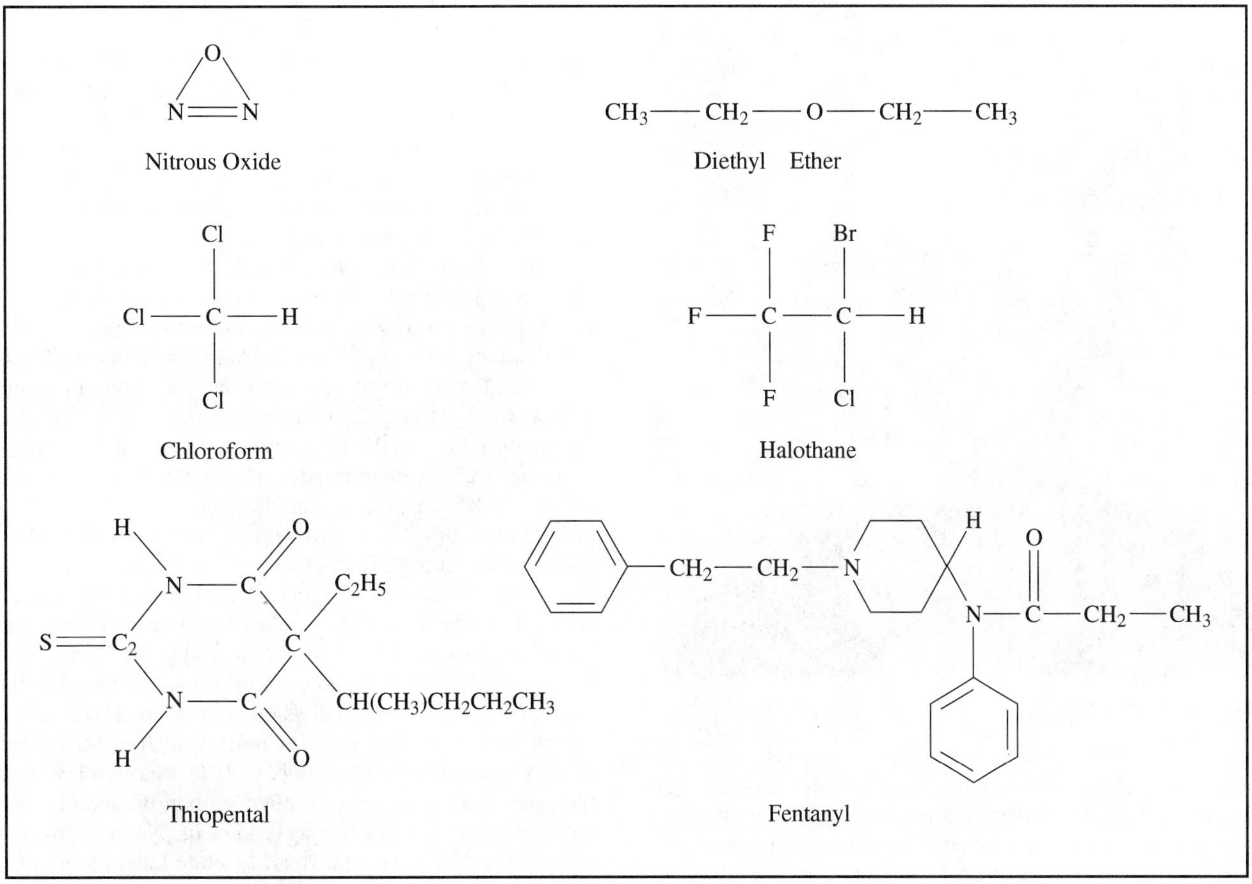

$CH_3—CH_2—O—CH_2—CH_3$

Nitrous Oxide

Diethyl Ether

Chloroform

Halothane

Thiopental

Fentanyl

Chemical structures of various types of general anesthetics. ***(Illustration by Electronic Illustrators Group.)***

uses contained in small cans with small cylindrical nozzles which can be capped. Diethyl ether is flammable and in the presence of an electric spark will ignite.

Chloroform (trichloromethane) is a clear, colorless, nonflammable, heavy, liquid. Its specific gravity is 1.476 at 68°F (20°C). The vapor which is approximately four times as dense as air will sink to the floor in an open operating room. Chlorform has a low **solubility** in **water** but is readily dissolved into animal fats and other organic **solvents**. It may be decomposed to toxic compounds when exposed to **heat**, flame, light, or **oxygen**. Halothane (2-bromo-2-chloro-1,1,1-triflouro-ethane) is a synthetic anesthetic liquid which is vaporized for inhalation as a general anesthetic. It has an asymmetric **carbon atom** and therefore has levo and dextro isomers but is supplied as a racemic mixture for anesthetic use.

Fentanyl (N-(1-phenethyl-4-piperidyl) propionanilide) is a commonly used synthetic opium **alkaloid**. It is very soluble in intravenous fluids and is rapid in onset and short—acting in duration, making it an ideal ''use as you go'' **narcotic** anesthetic. By balancing its effects and side-effects against other agents being used a more effective and safer state of anesthesia may be obtained.

The chemistry of local anesthetics is more straightforward; all are either esters or amides. They have in common a general configuration of an organic **ring**, an intermediate chain, and an amine portion of their molecules (Figure 2). The aromatic end is lipophilic and binds to fatty tissue, whereas the hydrophilic end binds to water. Chemical alterations of either end of the **molecule** will also alter its water-fat (water-lipid) coefficient, its protein binding, and its activity as an anesthetic agent. In a biological system degradation of the molecule occurs at the intermediate bond level and accounts for differences in allergic reactions and **metabolism**. Commonly used amino esters are procaine, (Novocaine), and cocaine, and commonly used amino amides are lidocaine (Xylocaine) and bupivicaine (Marcaine). In general, allergic reactions to amino esters are much more common than amino amides

In order to demonstrate the clinical use of anesthetic agents, a typical surgical procedure, using all the classes of anesthetics in use in 1999, will be described. Closure of a cleft lip (incomplete formation of the lip, or ''harelip'') in an infant is such a procedure. The child is brought to the operating room, usually in the arms of its parents. Then while the infant is being rocked and held, a mask with flowing halothane is held very near the face. Within a few minutes the baby is unconscious, the parents leave the room, and continuing to use inhaled halothane an intravenous line is inserted. Through this

Christian Boehmer Anfinsen.

line succinylcholine (a curare-type drug, but very short acting) is given. A tube is then placed through the mouth into the trachea and connected to a ventilator. With the airway thus controlled nitrous oxide, halothane, and oxygen are given in proper concentrations. Fentanyl or thiopental may be given intravenously if a deeper level of anesthesia is required during some parts of the procedure.

The surgeon then prepares the operative site for lip closure by cleansing the area, drawing proposed incision lines, and then injecting lidocaine (Xylocaine) with epinepherine to locally anesthetize the tissue which allows the anesthesiologist to lighten the level of general anesthesia. The procedure is performed, then the infant is placed on 100% oxygen until awakening, the tube is then removed, further oxygen is supplied by mask, and the child is taken to the recovery room. This scenario describes the usual case although many variations are used according to the needs of the patient and the training of the anesthesiologist.

One must admit that anesthetics have greatly improved since the days of the use of ''strong men'' to hold a reluctant and frightened patient down in order to achieve surgical intervention.

Anfinsen, Christian Boehmer (1916-1995)

American biochemist

Biochemist Christian Boehmer Anfinsen is known for establishing that the structure of an **enzyme** is intimately related to its function. This discovery was a major contribution to the scientific understanding of the nature of enzymes. For this achievement, Anfinsen shared the 1972 Nobel Prize for **Chemistry** with the research team of **Stanford Moore** and **William Howard Stein**.

Anfinsen was born on March 26, 1916, in Monessen, Pennsylvania, a town located just outside of Pittsburgh. He was the child of Christian Anfinsen, an engineer and emigrant from Norway, and Sophie Rasmussen, who was also of Norwegian heritage. Anfinsen earned his B.A. from Swarthmore College in 1937. Subsequently, he attended the University of Pennsylvania, earning an M.S. in **organic chemistry** in 1939. After earning his master's degree, Anfinsen received a fellowship from the American Scandinavian Foundation to spend a year at the Carlsberg Laboratory in Copenhagen, Denmark. Upon his return in 1940, he entered Harvard University's Ph.D. program in **biochemistry**. His doctoral dissertation involved work with enzymes; he described various methodologies for discerning the enzymes present in the retina of the eye, and he earned his Ph.D in 1943.

After receiving his Ph.D., Anfinsen began teaching at Harvard Medical School, in the department of biological chemistry. From 1944 to 1946 he worked in the United States Office of Scientific Research and Development. He then worked in the biochemical division of the Medical Nobel Institute in Sweden under Hugo Theorell, as an American Cancer Society senior fellow, from 1947 to 1948. Harvard University promoted him to associate professor upon his return, but in 1950 he accepted a position as head of the National Institutes of Health's (NIH) National Heart Institute Laboratory of Cellular Physiology. He served in this position until 1962. Anfinsen returned to teaching at Harvard Medical School in 1962, but he returned to NIH a year later. This time he was named director of the Laboratory of Chemical Biology at the National Institute of Arthritis, **Metabolism**, and Digestive Diseases. He held this position until 1981; he spent a year at the Weizmann Institute of science and then in 1982 accepted an appointment as professor of biology at Johns Hopkins University, where he remained until his death. A former editor of *Advances in Protein Chemistry*, he was also on the editorial boards of the *Journal of Biological Chemistry*, *Biopolymers*, and the *Proceedings of the National Academy of Sciences.*

Anfinsen began his research concerning the structure and function of enzymes in the mid–1940s. Enzymes are a type of protein; specifically, they are what drives the many chemical reactions in the human body. All **proteins** are made up of smaller components called peptide **chains**, which are **amino acids** linked together. Amino acids are, in turn, a certain class of organic compounds. The enzymes take on a globular, three-dimensional, form as the amino acid chain folds over. The unfolded chain form of an enzyme is called the primary structure. Once the chain folds over, it is said to be in the tertiary structure. From one set of amino acids for one particular enzyme there are 100 different possible ways in which these amino acids can link together. (Only certain amino acids can ''fit'' next to other amino acids.) However, only one configuration will result in an active enzyme. In general, Anfinsen's research concerned finding out how a particular set of amino acids knows to configure in a way that results in the active form of the enzyme.

Anfinsen chose to study the enzyme ribonuclease (RNase), which contains 124 amino acids and is responsible for breaking down the ribonucleic acid (RNA) found in food. This reaction enables the body to recycle the resultant smaller pieces. He felt that by determining how a particular enzyme assumes its particular active configuration, the structure and function of enzymes could be better understood. He reasoned that he could determine how an enzyme protein is built and when the enzyme becomes functional by observing it adding one amino acid at a time. He utilized techniques developed by Cambridge University's **Frederick Sanger** to conduct this research. Another research team headed by Stanford Moore and William Howard Stein was working simultaneously on the same enzyme as Anfinsen, ribonuclease; in 1960, using ribonuclease, Moore and Stein were the first to determine the exact amino acid sequence of an enzyme. However, Anfinsen remained more concerned with how the enzyme forms into its active configuration.

Anfinsen eventually changed his methodology of research during an opportunity to study abroad. While at the NIH, Anfinsen took yet another leave of absence when a Rockefeller Public Service Award allowed him to spend 1954 to 1955 at the Carlsberg Laboratory studying under the physical chemist Kai Linderstrøm-Lang. Anfinsen had been studying ribonuclease by building it up; Linderstrøm-Lang convinced him to start with the whole **molecule** and study it by stripping it down piece by piece. Anfinsen began with the whole ribonuclease molecule and then successively broke the various bonds of the molecule. The process is called denaturing the protein or, in other words, causing it to lose its functional capacity. By breaking certain key bonds, other bonds formed between the amino acids resulting in a random, inactive form of ribonuclease. By 1962, Anfinsen had confirmed that when this inactive form is placed into an environment that mimics the environment in which ribonuclease normally appears in the body, that inactive form would slowly revert to the active configuration on its own and thus regain its enzymatic activity. This discovery revealed the important fact that all the information for the assembly of the three-dimensional, active enzyme form was within the protein's own sequence of amino acids.

For uncovering the connection between the primary and tertiary structure of enzymes, Anfinsen received half of the 1972 Nobel Prize for Chemistry. Moore and Stein and were awarded the other half. In addition to his numerous journal articles on protein structure, enzyme function, and related matters, in 1959, Anfinsen published a book entitled *The Molecular Basis of Evolution.* After receiving the Nobel Prize, Anfinsen began focusing his research on the protein interferon, known for its key role as part of the body's immunity against both viruses and cancer. He succeeded in isolating and characterizing this important human protein.

Anfinsen's honors in addition to the Nobel Prize include The Rockefeller Foundation Public Service Award, a Guggenheim Fellowship, as well as honorary degrees from Georgetown University and New York Medical College and five other universities. Anfinsen was a member of the National Academy of Sciences, the American Society of Biological Chemists, and the Royal Danish Academy. As an opponent of biological weapons, he belonged to the Committee for Responsible Genetics. He married Florence Bernice Kenenger in 1941, and they had three children. Anfinsen and Kenenger divorced in 1978. In 1979, Anfinsen married Libby Esther Schulman Ely. He died on 14 May, 1995, at Northwest Hospital Center in Randallstown, MD.

ANILINE

Aniline is the common, and the systematic, name given to aminobenzene. It is, therefore, the phenyl-substituted derivative of **ammonia**. The **empirical formula** of aniline is C_6H_7N, although it is more commonly written as either $C_6H_5NH_2$ (aminobenzene) or $PhNH_2$ (phenylamine). It is an example of an aromatic primary amine. At standard **temperature** and pressure it is a colorless, oily liquid with an unpleasant odor that turns brown due to slow **oxidation** by air. It has a melting point of 22°F (-6°C) and a **boiling point** of 326°F (184°C). Aniline is manufactured by **reduction** of nitrobenzene, using **hydrogen** gas and a **copper** catalyst, or using **iron**, **water**, and **hydrochloric acid** for the reduction.

Aniline is an industrially important compound. It is used in the manufacture of **antioxidants**, pharmaceuticals, dyes, and in the vulcanization of rubber. Production of this chemical in the United States was at a level of 400,000 tons annually in the late 1990s.

Aniline is a weakly basic compound that gives water soluble salts with the mineral acids. When aniline is reacted with excess acetic (ethanoic) acid under **dehydration** conditions a white, crystalline material is formed, acetanilide. Acetanilide, $C_6H_5NHC(O)CH_3$, is formed by acylation of the aniline by the **acetic acid** with loss of water. Acetanilide is used in the manufacture of rubber and dyes. Dyes are prepared by a nitration addition reaction to aniline, followed by reduction of the nitro group using **sodium** sulfide. The products from these reactions are used in the production of an important class of dyes, the azo dyes. The azo dyes are used for wool, cotton, and **color** photography. They are an acid fast category of dyes.

Phenol can be produced from aniline by reacting aniline with a mixture of sodium nitrite and hydrochloric acid to give **benzene** diazonium chloride, that when heated gently, gives off **nitrogen** to leave phenol. The diazonium group can also be replaced by halogen, cyano, or pseudohalogen groups or atoms in a double decomposition reaction called Sandmeyer's reaction. The benzene diazonium chloride is treated with (e.g.) copper chloride to yield nitrogen gas and chlorobenzene. The diazonium **salt** can also be reacted and joined to phenolic groups and amines to give azo compounds.

Halogenation occurs readily with aniline by electrophilic aromatic substitution. For example, **bromine** water is rapidly discolored and a precipitate of 2,4,6-tribromoaniline is obtained.

Heating aniline with a variety of compounds yields a wide range of products. Heating with alkyl chlorides gives a mixture of the mono and the dialkyl aniline. Skraup's reaction (the addition of **sulfuric acid** and **glycerol** to aniline) gives quinoline which is used in the manufacture of dyes and pharmaceuticals. Quinaldine can be made by reacting aniline, hydrochloric acid, and paraldehyde. This compound can be used in the manufacture of photosensitive dyes.

Aniline is an arylamine that has a large industrial usage in the manufacture of various dyes for a wide range of applications as well as being a precursor for the manufacture of a number of pharmaceutical compounds.

See also Amine group; Azo, diazo, azine and azide groups; Dyes and pigments

Anion

An anion is a negatively charged **ion**. During **electrolysis** an anion is attracted to the **anode** (positive electrode). The name was coined in the 19th century and is a combination of *ana* (Greek for upwards) and ion. An anion can be a single, negatively charged **atom** or it can be a group of negatively charged ions. Anions are to be found in **solids** or **liquids**, for example in **sodium** chloride the chloride is the anion and it is represented by Cl^-. An anion will form an **ionic bond** with a **cation** (positive ion) to produce an ionic compound. An anion has a greater number of electrons than protons, giving it an overall negative charge.

Anions are made of atoms that will readily accept an **electron** when the conditions are right. All of the **halogens** will form anions as will all non-metals (except hydrogen) and a range of compounds will also readily form anions and accept electrons. Examples of compounds or groups that accept anions include the nitrate and hydroxide radicals. An anion can have an overall negative charge of -1 or -2.

Anions are negatively charged particles found in ionic compounds.

Anode

The term anode, from the Greek words *hodos*, meaning way, and *ana*, meaning up, was used by experimenters in the late ninteenth and early twentieth century to designate the conducting to which negatively charged particles, electrons, were drawn when they subjected partially evacuated tubes to high voltage. The tubes were made of **glass** and typically had a voltage source connected to two pieces of conducting materials sealed into the glass and separated by a few inches. The piece of conducting material from which the electrons emanated was called the cathode and the one to which they were attracted was called the anode.

In **chemistry** the most common use of the term anode occurs in **electrochemistry**. Electrochemical cells consist of one half-cell in which **oxidation** occurs and a second half-cell in which **reduction** occurs. Electrons flow though a conductor connecting the two half-cells. At the two ends of the conductor are the electrodes, which gather and disperse the electrons. The anode is the electrode at which oxidation takes place and the cathode is the electrode at which reduction takes place.

A voltaic cell, or a sources of **energy**, is an electrochemical cell is one in which an **oxidation-reduction reaction** takes place simultaneously. The everyday **batteries** we commonly use are voltaic cells. Because electrons originate at the anode in a voltaic cell, the anode has a (-) charge; electrons enter the cathode, which has a (+) charge. This polarity of the terminals and the role of the electrodes are accurate for any cell that is operating spontaneously and producing **electricity**.

In an electrolytic cell, energy must be supplied in the form of electric current from an external source. Electrolytic cells are used to electroplate tableware and jewelry and to produce useful chemicals from naturally occurring materials. In an electrolytic cell, the power source pumps electrons into the electrode to which it is attached; that electrode is the cathode. Therefore, in electrolytic cells, the cathode is the negative terminal and the anode is the positive terminal.

The polarities of the anode and cathode are reversed in comparison to their assignments in voltaic cells, but the function of the anode, being the site of oxidation, and of the cathode, being the site of reduction, is the same.

The materials that are useful for anodes must be good conductors and must not corrode too easily under oxidizing conditions. If the anode itself is the substance that one wishes to **oxide**, it should be the most easily oxidized material in the half-cell. If another substance is to be oxidized, the anode material should not be readily oxidized. Some materials commonly used as unreactive anodes are **platinum** and **graphite**.

One of the most common materials using an anode is **zinc** metal. The zinc containers of carbon-zinc and alkaline dry cell batteries and **mercury** batteries serve as the anode and are oxidized as the battery is used. In NiCad batteries, the anode is composed of **cadmium** metal and cadmium hydroxide. As the battery is used, cadmium metal is oxidized to cadmium hydroxide and as it is recharged the process is reversed.

Electrochemical techniques are also widely used in chemical analysis. In electrochemical analysis by such methods as anodic stripping voltammetry, cyclic voltammetry, differential pulse polarography, and potentiometry, the anode always functions as the site of oxidation. In general, the anode is a highly conductive and relatively inert material such as platinum, graphite or mercury.

Anodized Surfaces

Few metallic elements exist in nature in the metallic state—**gold** is often found in the metallic state, **silver** and **copper** are sometimes found as **metals** and **iron** and **nickel** are present in meteorites. Most metals are so readily oxidized by atmospheric **oxygen** that they are only found naturally as oxides, sulfides or salts. After ores are reduced to produce metals, means must be found to prevent them to be reoxidized by the atmosphere. One method for protecting metal surfaces is anodization.

When a metal surface is anodized, the metal is connected to a source of electrical potential such that it serves as the **anode**. Oxygen is provided by the **electrolysis** of **water**. Under carefully controlled conditions, the surface is oxidized to produce a tough **oxide** coating that is relatively impenetrable to atmospheric oxygen. Natural oxide layers are often too thin to provide sufficient protection: normal abrasion can remove the oxide allowing further **oxidation** of the metal or the oxide layer is permeable to additional oxygen. In addition to providing protection, the thickness of the anodized oxide layer, as well as added dyes or pigments, can determine the **color** of the surface, providing vivid, attractive and durable surfaces.

One of the most common metals that is anodized is **aluminum**. A wide variety of electrolytes are used to produce the desired surface properties: chromic acid, **sulfuric acid**, and **oxalic acid** are the most common but borates, citrates and oxoacids are also sometimes used. The surface that is produced by the electrochemical process is often quite irregular (amorphous) so that there may be regions that atmospheric oxygen can still penetrate. To form a more even, less penetrable surface the electrochemically treated metal may be subjected to a second treatment, such as submersion in boiling water to form a hydrated layer or immersion in hot dichromate **salt** solutions or **sodium** silicate to form oxoanion layers. The properties of the surface of the anodized metal are sensitive to the type of electrolyte that is used, its **concentration**, the **temperature** at which the process is carried out and the nature of the secondary treatment. When anodized surfaces are colored by the addition of dyes or pigments, a secondary treatment with acetates salts such as **cobalt** or nickel acetates is used to help seal the coloring agents to the surface. The surface may also be treated with non-polar, insulating substances such as oils or **waxes** to deter **corrosion** by aqueous solutions.

One of the more common electrolytes for the anodization of aluminum is sulfuric acid. At the concentrations typically used for this process, of about 1 to 2 molar, the sulfuric acid is able to dissolve any of the naturally produced oxide film on the surface, allowing good control for the formation of a new oxide surface. Because of the high conductivity of sulfuric acid solutions, the process can be carried out quite rapidly, typically less than an hour at about 20 V of DC current with a current **density** of 1 Ampere for a metal surface about 1 inch by a foot (130 Amperes per square meter). Under these conditions, the anodized surface is about 0.005-0.010 mm thick and is transparent.

Oxalic acid was first used in Japan, then Germany and now worldwide instead of sulfuric acid to produce surfaces that are not only hard and corrosion and abrasion resistant but attractively colored with added dyes or pigments. The colors depend on the conditions used for the electrolysis and, in particular, the thickness of the film. The drawback to the oxalic acid process is its expense relative to the sulfuric acid process.

Impurities such as iron and **silicon** in aluminum can lead to vulnerable spots for corrosion in the anodized surface and they also decrease the transparency of the oxide layer because compounds other than aluminum oxide are produced by the impurities during the anodization process. For this reason, very pure aluminum must be used to produce the highly reflective surfaces desired for reflectors and for appliance and automobile trim.

Among the metals that are often anodized, **magnesium** is notable because the untreated metal is so very readily corroded. Anodization of magnesium provides a hard, corrosion resistant surface that is also highly adhesive to paints. Magnesium metal has a low density and magnesium alloys are critically important in the aircraft industry, where integrity of the metal and prevention of corrosion are a must.

Anodization is also used to produce surfaces on metals that will adhere well to paints and other coatings. For these applications, secondary treatments are also used to bind species, typically oxoanions, that form strong linkages to pigments or organic species that bind to the oxide surface, allowing organic coatings to adhere. The feature of anodization that lends itself to this application is that the metal surface can be covered evenly in a predictable manner and the oxide surface is very strongly (covalently) bound to the bulk metal.

ANTACID

Roughly 95 million Americans suffer from heartburn, which can result from stress, eating too much or too fast, or eating spicy or fatty foods. Symptoms include a burning sensation, gas, nausea, and pain. Many people treat this condition by using antacids, an over-the-counter (OTC) drug taken orally that comes in liquid or tablet form. Typical well-known brands include Tums, Rolaids, and Alka-Seltzer. Mylanta is one brand that comes in liquid form. Brands like Alka-Seltzer are taken in **suspension** form. Tablets consist of fine antacid powder combined with flavorings and binders. Simethicone is often added to antacids as an anti-gas agent.

The antacid market covers a wide range of products. In the United States alone there are over 120 different formulations, composed of single ingredients or mixtures. In 1997, traditional antacids (acid neutralizers) generated sales of $770 million in the United States, out of the antacid-antigas market totaling $1.32 billion. The rest of the market goes to acid blockers, which prevent the action of histamine on the acid-secreting cells of the stomach, reducing acid secretion. Commercial OTC brands include Tagamet and Zantac, which use cimetidine and ranitidine, respectively, as the blocking agents.

Antacids based on **calcium** carbonate have been used for over 2,000 years. This material occurs naturally in coral and limestone and is still used in products such as Tums. However, it wasn't until the nineteenth century that scientists gained a better understanding of the digestive processes of the stomach. A bizarre incident led to the discovery that the stomach produces **hydrochloric acid**. In 1822, a man accidentally shot himself in the stomach while cleaning his shotgun. The wound never properly healed, leaving a hole in the stomach covered by a flap of skin. As a result, the contents of the stomach could be observed, including the excretion of gastric juices. These were removed and analyzed and were found to contain hydrochloric acid (HCl).

Over the last century, antacids were developed based on the hydroxides and carbonates of the group II and III **metals**, as well as the bicarbonates of the **alkali metals**. All currently marketed antacids contain at least one of the following metals: **aluminum**, calcium, **magnesium**, **sodium**, **potassium**, or **bismuth**. Antacids help neutralize excess acid produced in the stomach, i.e. the **hydrogen ion concentration** is reduced. The effectiveness of antacids is determined by its rate of reaction and residence time, which in turn are affected by various factors. Since metal-containing antacids can interfere with the absorption of many prescribed medications, especially antibiotics, non-metal antacids also have been developed.

Antacids alone or in combination with simethicone may also be used to treat the symptoms of stomach or duodenal ulcers. With larger doses than those used for the antacid effect, magnesium hydroxide (magnesia) and magnesium **oxide** antacids produce a laxative effect. Some antacids, like aluminum carbonate and aluminum hydroxide may be prescribed with a low-phosphate diet to treat hyperphosphatemia (too much **phosphate** in the blood). Aluminum carbonate and aluminum hydroxide may also be used with a low-phosphate diet to prevent the formation of some kinds of kidney stones.

The action of antacids is based on **neutralization**, the chemical process in which an acid and a base react to form a **salt** and **water**. For example, when an acid reacts with a hydroxide, a salt and water are produced as in the following equation: $HCl (aq) + NaOH (aq) \rightarrow NaCl (aq) + H_20$. (Hydrochloric acid reacts with **sodium hydroxide** to produce salt and water.)

In the case of **sodium bicarbonate**, or baking soda ($NaHCO_3$), a rapid reaction occurs with gastic acid to produce an increase in gastic **pH**. **Sodium chloride**, **carbon** dioxide gas, and water are formed during this reaction. One gram of sodium bicarbonate will neutralize 11.9 mEq of acid. However, too large a dose can **lead** to the production of alkaline urine, resulting in kidney problems. Since neutralization is due to the bicarbonate, other soluble bicarbonates (potassium bicarbonate) can be used as bicarbonates.

Calcium carbonate ($CaCO_3$) and other calcium compounds are used either alone or with magnesium compounds. With calcium carbonate, neutralization of acid involves the formation of calcium chloride, **carbon dioxide**, and water. One gram of this antacid will neutralize 20 mEq of acid. Negative side effects include constipation and acid rebound.

Magnesium compounds include magnesium oxide (MgO), magnesium hydroxide ($Mg(OH)_2$), and magnesium carbonate ($MgCO_3$-$Mg(OH)_2$- $3H_2O$. In the presence of water, magnesium oxide is converted to the hydroxide. The latter reacts with gastric acid to produce magnesium chloride and water. The commercial antacid made with this ingredient is commonly known as milk of magnesia, which contains 7-8.5% of the compound. One gram of magnesium hydroxide neutralizes 32.6 mEq of gastric acid and one ml of milk of magnesia will neutralize approximately 2.7 mEq of acid.

Magnesium compounds have the advantage of producing little absorption, having a prolonged action, and generating no carbon dioxide (except for magnesium carbonate). However, the magnesium chloride that is produced acts as a laxative; therefore, most commercial formulations also contain calcium carbonate or aluminum hydroxide to prevent this problem.

Aluminum compounds include aluminum hydroxide ($Al(OH)_3$), aluminum carbonate (Al_2O_3- CO_2), and aluminum glycinate, which contains aluminum oxide (Al_2O_3) and an acid called glycine. The commercial version of aluminum hydroxide usually consists of an aqueous suspension containing a small amount of aluminum oxide. During neutralization, aluminum hydroxide forms aluminum chloride and water; each milliliter of the suspension neutralizes from 0.4-1.8 mEq acid in 30 minutes. However, if the pH goes above 5, the neutralization reaction is likely to be incomplete. Though aluminum hydroxide has a long shelf life it can cause constipation so is often used in combination with magnesium antacids. Aluminum carbonate has similar antacid properties to aluminum hydroxide.

The demand for antacids is expected to grow as the number of people suffering from heartburn increases due to an aging population, more stressful lifestyles, and eating out more often. Acid blockers may continue to increase in popularity at the expense of acid neutralizers. Drug manufacturers of both types will continue to improve their antacid products regarding **taste**, cost, and effectiveness to make it even easier for the consumer to treat themselves.

ANTIBIOTIC DRUGS

Antibiotic drugs (frequently referred to as antibiotics) are substances, usually obtained from microorganisms, that inhibit the growth of or destroy other microorganisms, particularly bacteria. Synthetic antibiotics also exist but the majority are analogs of naturally occurring antibiotics or their derivatives.

Antibiotics were first discovered in the nineteenth century when French chemist and bacteriologist **Louis Pasteur** demonstrated that one species of microorganism is capable of killing another. This idea was further developed by German bacteriologist **Paul Ehrlich**, who realized that selective toxicity was possible. Selective toxicity implies that a chemical can be toxic to one set of organisms, but leave others entirely unaffected. Many antibiotics come from bacteria, actinomycetes, and fungi. The most widely known antibiotic is **penicillin**, and its derivatives. Penicillin is produced by the mold *Penicillium chrysogenum* and the fact that it produces a substance capable of killing bacterial cultures was first noted by Scottish bacteriologist Alexander Fleming (1881- 1955) in 1928 (the discovery was made using a related species, *Penicillium notatum*). Fleming was unable to purify the active ingredient and it was left to Australian biochemist Howard Florey (1898-1968) and German biochemist Ernst Chain to isolate, characterize, and mass produce the antibiotic some ten years later.

Antibiotics are generally produced by **fermentation** in large vats of nutritive media. Other techniques are also used, such as de novo chemical synthesis and modification of **natural products**.

There are several different classification systems for antibiotics, based on chemical structure, origin, range, or action. Some antibiotics are broad spectrum and can be used against a wide variety of microorganisms, while others are very specific in their action and kill only microorganisms of one species or group.

Within naturally occurring populations of bacteria there are generally antibiotic-resistant strains. Once an antibiotic is used, resistant strains of the bacteria develop, and new antibiotics must be formulated to combat them. For this reason, it is important that antibiotics are not overused, since overuse ultimately will make them less effective. Antibiotics are frequently given to livestock animals to enhance their growth. Regarded a poor practice by some, the use of antibiotics encourages the growth of resistant bacteria in the animals and may lead to disease epidemics for which there are no treatments. When people consume meat or other products from antibiotic-treated livestock, they are exposed to low levels of these antibiotics. This can lead to sensitization, making individuals who consumed these food products unable to take that type of antibiotic when it is needed to treat a disease. Some antibiotics are useful for human diseases, while others are used only on other animals, and others still are generally only used in laboratories because their toxicity is so high to living organisms.

Humans take antibiotics orally, by injection, or topically (applied to the skin). The method of administration is governed by where the antibiotic needs to be to work, its stability, and its mode of action. Some antibiotics have side effects, e.g. they may produce rashes or send a patient into shock. An individual can be allergic to specific antibiotics and, in such a case, administration of these antibiotics can be fatal. Some antibiotics, such as the tetracycline group, are broad spectrum and kill useful bacteria as well as harmful ones. This can lead to an opportunistic infection by fungi or other microorganisms.

Tetracycline is a broad spectrum antibiotic that kills bacteria from both the gram-negative and gram-positive groups. It is isolated from bacteria in the genus *Streptomyces* and the active compound is based on a napthacene skeleton—an aromatic structure with four benzene-type rings. Erythromycin is active only against gram-positive bacteria. It is used for individuals who are allergic to penicillin and tetracycline and has the chemical formula $C_{37}H_{67}NO_{13}$.

See also Antiviral drugs

ANTIBODIES

The immune system uses two main approaches to fight off infection. The first is a general approach known as cell-mediated immunity in which certain immune system cells attack and destroy infectious invaders. The second approach is more specialized and is known as humoral immunity. The cornerstone of humoral immunity is formed by antibodies. Antibodies latch onto infectious invaders such as viruses, bacteria, molds, and parasites and swiftly aid in their destruction.

Antibodies are proteins that belong to a class of protein molecules called immunoglobulins (Ig), also known as gamma-globulins. Immunoglobulins are the most important molecules in the immune system. They have been reported to consist of 19,996 atoms associated in 1,320 **amino acid** units. The three main groups of immunoglobulins are IgG (a **glycoprotein** and the immunoglobulin present in the body in the largest concentration), IgA, and IgM. Immunoglobulin G has a **molecular weight** of about 160,000. The larger chains each have a molecular weights of about 50,000, and the smaller chains molecular masses of about 25,000.

The cells that are involved in humoral (antibody) immunity are B cells and helper T cells. Both the B cells and the T cells belong to a class of immune system cells called lymphocytes. Only the B cells produce antibodies, but they can't release them without the assistance of helper T cells.

When infectious invaders enter the body, the T cells and B cells identify them as foreign owing to their antigens. Antigens are **macromolecules**, usually proteins having a molecular weight of at least 10,000, on the invader that are unlike molecules normally found in the body, and which trigger the formation of antibodies in the blood. An antibody attacks an invader by binding to an antigen. (In autoimmune diseases, such as rheumatoid arthritis, the immune system confuses a factor in the body with an antigen and launches an attack against the body itself.) The antibody-antigen complex alerts the immune system that there is an invader and other immune system components move quickly to destroy it and stop, or even prevent, disease. Antigens appear to ''mate'' with antibodies much like enzymes bond to substrates, i.e., in such a way as to match certain radicals (especially polar and quaternary ammonium groups) with complementary structures in the antibody **molecule**.

The relationship between an antibody and an antigen is very specific as each individual antibody can only recognize a specific antigen. Since the immune system may encounter countless different antigens over a lifetime, it has to be prepared to produce a wide variety of antibodies. This challenge is handled at the genetic level.

Each antibody molecule is composed of four subunits—two identical heavy chains and two identical light chains. Both chains are coded for in three or more sections of **DNA**, and each section of DNA contains many different genes. There can be more than 300 genes in a section, but only one is used for each antibody molecule. Because there is a large choice of which gene to use from each section, and because the choice is made randomly, the odds are very good that each individual B cell will produce a unique antibody.

Once all four chains are produced and matched up with one another, they form a large Y-shaped molecule. The end of each arm of the Y is called an antigen-binding site; therefore, each antibody has two antigen-binding sites. These antigen-binding sites are presented on the surface of every B cell. If a B cell encounters an antigen to which its unique antibody can bind, it becomes activated. However, it cannot produce and release copies of its antibody until it also encounters an activated helper T cell.

Helper T cells are activated by antigen-presenting cells. Antigen-presenting cells are a type of macrophage—an immune system cell that consumes an infectious invader, processes it, and displays the pieces. Once activated, the number of helper T cells increases greatly. When one of these helper T cells encounters a B cell that has been activated by the same antigen, it secretes cytokines, chemicals that cause the B cell to clone itself. The resulting clones, called **plasma** cells, are a mature version of the activated B cell. They are able to secrete the antibody which helps the immune system fight infection.

To protect the body against future attacks by the same infectious invader, some of the activated B cells become memory B cells. These memory B cells do not secrete antibody, but they remain in the body for years and provide a swift response to any further encounters with the infection. Because the infectious invader can be recognized so quickly, it does not have the opportunity to cause an infection and a person is said to be immune to the disease.

The antibody produced by the cloned B cells is called a monoclonal antibody. Monoclonal antibodies can also be produced by fusing a single plasma cell with an immortal myeloma cell line in the laboratory. The myeloma cell line contains cells that can be replicated repeatedly but cannot produce antibody. The combined cells are called a hybrid cell line or hybridoma. The hybridoma produces the exact antibody that was created by the plasma cell.

The cells in the hybridoma cell line can be stored and cultured (i.e., grown under suitable conditions), which allows medical researchers to generate monoclonal antibodies in the laboratory. Because these monoclonal antibodies are highly specific and it is known to which antigen they bind, they can be used to identify and possibly treat a specific diseases. Such monoclonal antibodies are capable of identifying the presence of various diseases in from 20-30 minutes (compared to the three to six days required by conventional techniques). Monoclonal antibodies have been used to combat viruses, bacteria, parasites, and some forms of cancer.

ANTI-INFLAMMATORY AGENTS

There exists in medical practice at the close of the 20th century two basic categories of anti-inflammatory drugs: steroidal (from steroid compounds) and non-steroidal. **Steroids** are given their own section in this volume. They are potent inhibitors of inflammation and the immune system, but also have a host of serious, even deadly, side-effects. The non-steroidal agents are in general less potent but also have fewer side-effects and will be the subject of this section.

In the mid 1700s, Reverend Edmund Stone wrote a letter to the English Royal Society noting the ability of the bark of the willow tree to cure fever. He was reiterating an observation known for ceturies by many cultures, but the actual ingredient remained unknown until 1829 when Leroux isolated salicin. Salicin indeed proved to decrease fever. Chemically salicin can be converted to salicilic acid, a very successful treatment of fever and gout. In 1875, Hoffman, a chemist working for the Bayer Company, synthesized acetylsalicilic acid which was made available for clinical use under the name aspirin. Many synthetic variations of aspirin were used in the early 1900s, but in 1999 only acetaminophen (Tylenol) remains. Other synthetic non-steroidal anti-inflammatory drugs (NSAIDs) have moved to the foreground. Aspirin, nonetheless, remains not only the parent compound but also is still widely used clinically throughout the world.

Inflammation is the response of the human body to numerous noxious stimuli (e.g.,infection; chemical injury such as impaired blood supply or a chemical burn; trauma including flame, electric burn, and frostbite injury; and antigen-antibody interactions). Clinically, inflammation is characterized by redness, swelling, **heat**, pain, loss of function, and tenderness. A classic example would be a scald burn or a wasp sting. The inflammatory response is used by the body to control the injury and is intrinsically the same for all stimuli with minor variations based on the nature of the offending agent. The body's ability to mount an inflammatory resonse is a survival mechanism, but on occasion can be overwhelming (e.g. dying from a bee sting) or remain active for a prolonged time (as in rheumatoid arthritis). The exact events at the cellular and biochemical level which make up inflammation are extremely complex and are not yet entirely known. Cells involved include white and red blood cells, ''helper'' cells such as tissue lymphocytes, and scar tissue forming cells. But it is at the biochemical level where the complexities become extreme. One group of chemicals known to be essential are called **prostaglandins** (first isolated from the prostate glands of laboratory rats), and the first **enzyme** to be triggered by inflammation in the production of prostaglandins is cyclooxygenase-2 (COX-2).

It is the COX-2 enzyme that NSAIDs inhibit and thus exert their effects. Unfortunately, cyclooxygenase-1 (COX-1) is also usually inhibited. COX-1 is a normal tissue enzyme required to prevent stomach ulceration and to allow blood to clot. So aspirin may well relieve pain and fever but may also cause stomach ulceration and a tendency to bleed. This lattter side-effect is useful in lowering the risk of heart attack due to blood clots forming in the arteries which supply the heart muscle but makes the situation worse in the case of a bleeding ulcer. Nonetheless, NSAIDs are a very useful group of compoounds when the appropriate patient, condition to be treated, dosage, and side-effects are all considered in the choice of drug. They are organic acids and thus able to be taken by mouth, bind well to **plasma proteins** so as to be widely distributed throughtout the body, and accumulate at the inflammed area.

Chemically there are at least eight classes of NSAIDs but only those in common use are described here. Salicylic acid derivatives include aspirin; Para-aminophenol derivatives are not used in 1999 except for acetaminophen (Tylenol, etc.) which is like aspirin in its fever and pain reducing properties but has no COX-2 enzyme anti-inflammatory effect (or COX-1 side-effects). For a time in the early 1990s one Heteroaryl **acetic acid**, ketorolac (Toradol) was used primarily for post-operative pain, however it has fallen into some disfavor due to the amount of side-effects. In 1999 the Propionic acid derivatives such as ibuprofen (Motrin, Advil, etc.), naproxen (Naprosyn, Alleve, etc.), and ketoprofen (Orudis, etc.) have become the mainstays of NSAIDs used in the United States. Many are available without prescription, and sometimes one may be much more effective, varying from patient to patient. Pyrazolon derivatives have been used in the past but were discontinued because of severe adverse effects. In 1999, however, celecoxib (Celebrex), a diaryl substituted pyrazole was introduced as the first selective COX-2 inhibitor available in the United States. It has good anti-inflammatory activity and much lower stomach and intestinal side-effects suggesting that it is much more selective for COX-2 than COX-1 inhibition.

The downside to any anti-inflammatory agent is that it is inhibiting a body protecting mechanism which has evolved over millions of years. Although in many situations alleviation of inflammation seems desirable, on occasion it may result in undesirable consequences such as allowing an infection to worsen. Chemists synthesize an amazing array of compounds, but only the clinical accumen of a health care provider skilled in their use can bring the chemists' brilliance to the patient. Both disciplines are called upon in the search for improved anti-inflammatory drugs.

See also Prostaglandins

Antimony

Antimony is the fourth element in Group 15 of the **periodic table**. Its **atomic number** is 51, its atomic **mass** is 121.75, and its chemical symbol is Sb.

Properties

Antimony is a **metalloid**. It exists in three allotropic forms: a silvery white metal; a yellow, crystalline solid; and an amorphous black powder. Its melting point is 1,166°F (630°C) and its **boiling point** is 2,975°F (1,635°C). Its most common allotropic form, the silver-white metal, is a relatively soft material that can be scratched by **glass**. Its **density** is 6.68 grams per cubic centimeter.

Occurrence and Extraction

Antimony is rarely found in nature as an element. Its most common ore is the mineral stibnite, a form of antimony sulfide (Sb_2S_3). Pure antimony can be obtained from antimony sulfide by heating the compound with hot iron: $2Fe + Sb_2S_3 \rightarrow Fe_2S_3 + 2Sb$.

The usual source of most antimony produced today is the **recycling** of metal alloys. About half of the antimony produced in the United States is recycled from old **lead** storage **batteries**, in which the antimony was originally alloyed with lead.

Discovery and Naming

Compounds of antimony have been known and used by humans for centuries. Probably the first person to describe the element in detail was the French chemist Nicolas Lemery (1645-1715). The origin of the element's name is uncertain, but probably comes from two Arabic words *anti* and *monos* that mean ''not alone.'' The name was chosen because antimony does not occur alone in nature, only in compounds.

Uses

Antimony is usually used in the form of an **alloy**. Lead-antimony alloys were once very widely used for solder, ammunition, fishing tackle, covering for electrical cables, low-melting alloys, and batteries. Such uses are now decreasing because of the serious health problems posed by lead. Antimony and its compounds are also used in transistors, the manufacture of **ceramics** and glass, and the production of plastic.

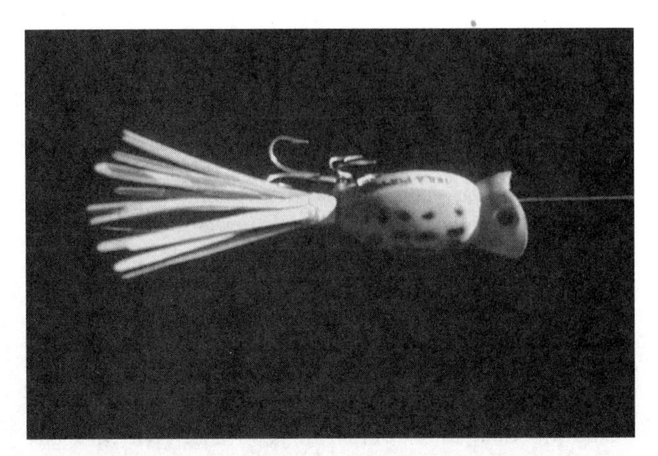

Lead-antimony alloys are often used in the manufacture of fishing tackle and lures, as shown above. ***(Photograph by Robert Huffman. Reproduced by permission.)***

Antioxidants

Abundant in vegetables, whole grains, and fruits, antioxidants are naturally occurring nutrients that neutralize harmful, unstable molecules in the body called **free radicals**. Research shows that free radicals may be responsible for cancer, heart disease, respiratory ailments, and the aging process itself. By protecting the immune system and repairing damage inflicted by free radicals, antioxidants (also known as ''free radical scavengers'') may help prevent disease and early aging.

Free radicals are constantly present in the body, formed by the myriad chemical reactions that occur daily and by exposure to various factors. These include ultraviolet radiation, cigarette smoke, rancid fat, polyunsaturated fats, chronic inflammation, **herbicides**, and **pollution**. Free radicals steal nearby stable molecules' electrons to complete their own molecular structures, thus destabilizing the target **molecule**. This causes a lightning-fast **chain reaction** that produces many more free radicals, resulting in what is clinically known as oxidative stress. Many scientists believe that this stress leads to disease and premature aging if an organism has insufficient antioxidants on hand to stop the reaction.

The medical community as a whole generally agrees that antioxidants are more effective when consumed in the form of 5-8 daily servings of fruits and vegetables than as supplements, although some studies indicate that higher intake can benefit people with known illnesses. Some of the substances can even be toxic when taken in excessive doses. Indeed, it may be that other chemicals found in antioxidants' natural sources are responsible for the beneficial effects presently associated with antioxidants—scientists are still investigating the matter. In the meantime, a heated debate continues over the most appropriate doses of these nutrients for humans.

The main recognized antioxidants are the carotenoids (precursors of vitamin A), vitamin C (ascorbic acid), and vitamin E (tocopherol). Beta-carotene, lutein, and lycopene comprise the carotenoid antioxidants. They occur naturally in such foods as dark-green leafy vegetables and orange, yellow, and red foods, such as yams and carrots. Lending credence to the widespread belief in this antioxidant's power to slow the aging process, some clinical studies have shown that the carotenoid level in a mammal's tissues is an effective predictor of its life span.

Vitamin C, the most plentiful water-soluble antioxidant, is present in all citrus fruits and such vegetables as broccoli, sweet peppers, and potatoes. This vitamin has often been associated with reduced rates of all kinds of cancer, but especially those affecting the larynx, esophagus, and mouth. Vitamin C is also important for its ability to recycle vitamin E after the latter neutralizes a free radical. Studies have indicated that it may be particularly effective in combating the free radicals generated by smoking and pollution.

Vitamin E, the most plentiful fat-soluble antioxidant, is a component of some oil-containing foods, such as peanut butter, seeds, vegetable oils, and whole grains. It may offer significant protection against cardiovascular diseases by fighting **oxidation** of LDL (low-density lipoprotein) **cholesterol** and preventing the buildup of arterial plaque. Some scientists believe vitamin E is the most powerful antioxidant. In the late 1990s, small studies suggested that the vitamin may provide some protection against Alzheimer's disease by preventing the brain's frontal and temporal lobes from falling prey to free radical damage.

Researchers have identified many other natural antioxidants, including the trace minerals **selenium**, **manganese**, and **copper** (which work in conjunction with enzymes); flavonoids (plant pigments such as those found in milk thistle and gingko biloba); and **amino acids** containing **sulfur** (methionine and cysteine, which are present in beans, eggs, nuts, brewer's yeast, liver, and fish). More recently, researchers have identified antioxidants in such substances as green tea (polyphenols), grapes (ellagic acid), seeds and nuts (coumarines), and blueberries and peaches (chlorogenic acid). In addition, the human body produces its own antioxidants that take the form of enzymes, including catalase, glutathione peroxidase, and superoxide dismutase.

Each of these neutralizes a specific kind of oxidation. Some antioxidants—for example, butylated hydroxytoluene (BHT), butylated hydroxyanisole (BHA), propyl gallate, ethylenediamine tetraacetic acid (EDTA), and tert butylhydroquinone (TBHQ)—are also commonly used as **food additives** to slow fat spoilage and prevent **color** changes. They work by serving as a decoy for a product's fat, since atmospheric **oxygen** prefers to react with the antioxidants instead of oxidizing the fat. However, several studies have suggested that these food-preserving antioxidants may in fact be carcinogenic or cause mild to life-threatening allergies.

See also Vitamins

ANTIPARTICLE

One of the seminal scientific discoveries of the twentieth century was Louis Victor de Broglie's theory of the wave nature of the **electron**. By integrating theories by **Albert Einstein** dealing with **mass** and **energy** and Max Planck's Quantum relationship between energy and wavelength, de Broglie posited that any material particle should also exhibit wave like properties, just as Einstein earlier had shown that light, which had been described as a wave phenomenon, also exhibited particle like behavior.

Building on de Broglie's theory, **Erwin Schrödinger** developed his famous wave equation for both free electrons and those bound to atoms. A limitation of **Schrodinger's equation** was that it did not include relativity. In the late 1920s, the English physicist Paul Adrien Maurice Dirac extended Schrodinger's theory by incorporating relativity into the wave equation for the electron. He showed that particles, such as the electron should always exist in two energy states: one positive and one negative. When applied to the electron, the theory suggested that, in addition to the negatively-charged electron already known, there should also exist a positively-charged "twin." The twin would be identical to the electron in every respect except for its electrical charge.

Dirac's prediction, announced in 1930, was confirmed experimentally within two years. In his 1932 studies of cosmic ray interactions, Carl David Anderson found the positive electron that Dirac had anticipated. Anderson suggested the name positron for the new particle. Anderson discovered that the collision of an electron and a positron resulted in the annihilation of both particles, with their mass being converted into energy. Similarly, an electron-positron pair can be created when a high-energy **photon** interacts with **matter**. The mathematical equivalence of mass-energy calculated for these conversions provided dramatic confirmation of Einstein's mass-energy equation ($e = mc^2$) of 1905.

Dirac's theory applied not only to electrons, but to all other particles. Just as he predicted an antielectron, he also suggested the existence of an antiproton, a particle identical to a **proton** in every respect except for its electrical charge.

The negatively-charged proton proved to be much more elusive than had been the positron. With a mass more than 1,800 times greater than that of the positron, the antiproton requires far greater energy for its production. Such energies are available in cosmic rays, but the rate of antiproton production from this source is too low to have resulted in their discovery.

It was not until the invention of particle accelerators that a systematic search for the antiproton could be launched. Then, in 1955, the particle was discovered by a research team led by Emilio Segrè and Owen Chamberlain at the University of California. The antiproton was produced when protons from a cyclotron were used to bombard a **copper** target. As predicted by Dirac, the antiproton was similar to the proton in every respect except for its electrical charge.

Is there also an antineutron? Lacking an electrical charge, an exact analogue to the electron and proton for the **neutron** is obviously impossible. However, Dirac's original research had anticipated and solved this problem. Along with the existence of antiparticles, Dirac had predicted that all particles possess an intrinsic property known as *spin*. The spin of a particle is designated as either "up" or "down." An antineutron would, Dirac suggested, be identical to a neutron except for its spin. In the presence of an external magnetic field, an antineutron and a neutron would have equal but opposite intrinsic spins. The first antineutron was discovered at the University of California at Berkeley only a year after the discovery of the first antiproton.

One can imagine atoms that are made of antiparticles: antiprotons, antineutrons, and positrons. Substances composed of such atoms are known as antimatter. Thus far, the antideuteron, made of one antiproton and one antineutron, and an **isotope** of antihelium-3 have been made in the laboratory.

Theory suggests that, at the moment the universe was created, equal amounts of matter and antimatter were formed. Today, however, we are able to detect only matter in the universe. One of the great questions of modern physics, therefore, is what happened to the original antimatter that was created at the time of the Big Bang.

Modern theory suggests that all particles have antiparticles. For example, the **neutrino**, hypothesized to explain energy changes that occur during some **nuclear reactions**, was found in 1963 to have its own antiparticle, the antineutrino. In some cases, the theory produces somewhat bizarre predictions. For example, since the photon has no mass and no electrical charge, it is, in fact, its own antiparticle.

Antiviral drugs

Antiviral drugs are those used to treat viral infections such as herpes simplex, herpes zoster, cytomegalovirus (CMV), influenza, AIDS, and certain AIDS related infections.

Viral diseases are harder to control than bacterial diseases due to their mode of operation. A virus is a small packet of infectious nucleic acid material surrounded by a protein coat. The nucleic acid is either **DNA** or **RNA**. This nucleic acid must use the machinery of the host's cells to reproduce and successfully complete the life cycle of the virus. With some viruses as soon as the RNA enters the cell it is translated and more viral particles are released; with others the nucleic material must first incorporate itself into the nucleic material of the host. This latter category of viruses is the retroviruses. An example of a retrovirus is Human Immunodeficiency Virus (HIV) which causes AIDS. An intimate knowledge of the life cycle of a virus is important, since there are only certain stages when the virus is susceptible to attack. Many viruses are very difficult to eradicate and all that can be achieved is control of the infection. In particular once a retrovirus has integrated itself into the genome the only way it can be removed is by physically cutting it out. Many antiviral drugs act on the machinery that the virus uses to reproduce itself, and this is a much easier method of control to employ.

One example of an antiviral compound is interferon which is a chemical produced by the body in response to viral infection. Interferon is used artificially and naturally in the treatment of viral infection. Interferon acts by stimulating uninfected cells to produce a protein which blocks entry of viral particles and also halts RNA production in those cells. Interferon is a non-specific antiviral chemical that is most effective when produced by the body. By the time an infection is observable, externally administered interferon is of little use.

Other antiviral drugs act in other ways, for example azidodideoxythymidine (AZT), which is a drug administered to HIV patients, acts on the genetic material of the virus, reducing its capacity to replicate. AZT may also decrease the frequency of infection by other diseases. However, it does not cure AIDS, and the best that AZT can do is hold the infection at a manageable level. In common with many antiviral drugs, AZT has many deleterious side effects. The side effects of antiviral drugs are caused by the way in which they operate, i.e. disrupting the machinery used by the virus to reproduce. This is the same machinery which normal healthy cells need to operate. When the viral cells are disrupted so too are the activities of non-infected cells. AZT side effects can include anemia, and a reduction in the presence of certain types of white blood cells.

Any antiviral drug which was capable of excising viral genetic material from the host genome would disrupt the genome so dramatically that cell death would occur. This is why retroviral infections, in particular, are so difficult to eradicate. Other viral infections are often held to a level where no disease is caused, but, when the body is infected by another agent, then the viral infection can spread while the body is weakened. Many diseases caused by viruses have no appropriate treatment. The body must be left to heal itself.

The development of antiviral drugs is a challenging undertaking. This is due to the mode of infection and the action of the virus. Many drugs that are effective on viral particles cause disruption of the body's own cells as well.

See also Antibiotic drugs

Aqua Regia

One of the major quests for the alchemists of northern Africa, China, India and Europe, the forerunners of modern chemists, was a process to transform the common, corruptible **metals** into **gold**. Over several centuries, they developed the ability to produce a number of important chemical reagents. Since they did not have our current concept of elements, they named the reagents they prepared by the way they react (such as vitriolic acid, for what we now call **sulfuric acid** because it is so very corrosive) or from where they were found (such as Epsom's **salt**, which is **magnesium sulfate** found in large quantities near Epsom, England) or what was used to produce them. Among the most powerful reagents was **nitric acid** (aqua fortis or strong water) which is both a strong oxidizing acid and **hydrochloric acid** (called muriatic acid from the Latin word muria, for brine, since it was made by treating sea salt with sulfuric acid). A mixture of three parts concentrated hydrochloric acid to one part nitric acid is called "aqua regia" or "royal **water**," because it dissolves even the "royal" (normally untouchable or incorruptible) metals **silver** and gold.

The ability of aqua regia to dissolve silver in gold is due both to the oxidizing power of the nitrate **ion** and the ability of chloride ion to form highly stable, partially covalent polyatomic ions with the metal ions once they are oxidized. By removing the free metal ions from **solution**, the formation of the chloride containing polyatomic ions allows the **oxidation** reaction to continue toward equilibrium. In the case of gold, the metal atoms are first oxidized by the nitrate ion to gold(III) ions. The gold(III)ions then react with chloride ion to form polyatomic ions containing one gold(III) **atom** and four chloride(-I) atoms, giving the overall reaction: $Au(s) + 3NO_3^- + 4Cl^- + 6H^+ = AuCl_4^- + 3NO_2(g) + 3H_2O$

When alchemists discovered that aqua regia dissolved gold, they were likely convinced that their ability to transform gold into a solution and, by evaporation, into a salt made it possible that other types of materials could transform back into gold if only they could find the right combination and the right conditions.

ARGON

Argon is a noble gas, one of the elements that make up Group 18 of the **periodic table**. Its **atomic number** is 18, its atomic **mass** is 39.948, and its chemical symbol is Ar.

Properties

Argon is a colorless, odorless, tasteless gas with a **density** of 1.784 grams per liter. Its melting point is -302.55°F (-185.86°C) and its freezing point, -308.7°F (- 189.3°C). Argon is chemically inert.

Occurrence and Extraction

Argon is the third most abundant gas in the atmosphere, after **nitrogen** and **oxygen**. It makes up about 0.93% of the atmosphere. It is also found in very small amounts in the Earth's crust.

Discovery and Naming

Argon was discovered in 1894 by the English chemist John William Strutt (Lord Rayleigh) and the Scottish chemist **William Ramsay**. The element's existence had been predicted more than a century earlier by the English chemist and physicist **Henry Cavendish**. Cavendish had removed the two most abundant **gases**, nitrogen and oxygen, from air, but found that a very small amount of an unidentified gas still remained from the air. Unfortunately, he was not able to identify what that remaining portion was.

The name chosen by Rayleigh and Ramsay for the new element comes from the Greek word *argos*, meaning "lazy." They thought the name was a good one because the new element seemed too "lazy" to react with any other elements or compounds.

Uses

Argon is used to provide an inert atmosphere that will protect materials from reacting with oxygen or other gases. For example, the inside of a light bulb is often filled with argon. No matter how hot the filament inside the bulb gets, it will not react with argon. Argon is also used in welding to prevent the two **metals** being heated from reacting with oxygen before they join to each other.

AROMATIC HYDROCARBONS

Aromatic hydrocarbons contain one or more **carbon** rings. They are typified by **benzene**, which contains six carbon atoms and three double bonds. Derived chiefly from petroleum and **coal** tar, aromatic hydrocarbons tend to be reactive. The designation *aromatic* comes from the strong and not unpleasant odor produced by these molecules.

Benzene was first identified in compressed oil gas by **Michael Faraday** in 1825. Although benzoin and benzoic acid had been called aromatics before 1820, it was not until the 1850s that benzene and its derivatives were placed in the chemical family of aromatics. The so-called aromatic spices and resins such as frankincense, camphor, turpentine, were shown in the 1880s not to be chemically similar to benzene, but rather to be derivatives of a family of ten-carbon compounds, called terpenes. Nevertheless, use of the term aromatic has persisted to this day as a descriptor of these fragrant, but chemically non-aromatic, substances.

In 1925, **Robert Robinson** attributed the properties of benzene and its analogs to the six extra electrons that produce the stable association responsible for the aromatic character. In 1931, Erich Huckel showed that benzene's stability arose from six sigma bonds in the plane of the **ring**, and six pi electrons in orbits at right angles above and below the plane. The pi orbitals give rise to benzene's diamagnetism.

Benzene has traditionally been written as a six-membered ring with alternating single and double bonds. But this representation suggests that the carbon-carbon bonds (alternating double and single bonds) should be of unequal lengths, whereas in fact they are all the same length. This is explained by **resonance** theory, according to which the bonds should share the characteristics of both double and single bonds.

See also Bonding; Magnetism

ARRHENIUS, SVANTE AUGUST (1859-1927)

Swedish chemist

Svante August Arrhenius was awarded the 1903 Nobel Prize in **chemistry** for his research on the theory of electrolytic **dissociation**, a theory that had won the lowest possible passing grade for his Ph.D. two decades earlier. Arrhenius's work with chemistry was often closely tied to the science of physics, so much so that the Nobel committee was not sure in which of the two fields to make the 1903 award. In fact, Arrhenius is regarded as one of the founders of physical chemistry—the field of science in which physical laws are used to explain chemical phenomena. In the last decades of his life, Arrhenius became interested in theories of the **origin of life** on Earth, arguing that life had arrived on our planet by means of spores blown through space from other inhabited worlds. He was also one of the first scientists to study the heat-trapping ability of **carbon** dioxide in the atmosphere in a phenomenon now known as the **greenhouse effect**.

Arrhenius was born on February 19, 1859, in Vik (also known as Wik or Wijk), in the district of Kalmar, Sweden. His mother was the former Carolina Thunberg, and his father was Svante Gustaf Arrhenius, a land surveyor and overseer at the castle of Vik on Lake Ma angstrom lar, near Uppsala. Arrhenius's uncle, Johan, was a well-known botanist and agricultural writer who had also served as secretary of the Swedish Agricultural Academy.

Young Svante gave evidence of his intellectual brilliance at an early age. He taught himself to read by the age of three and learned to do arithmetic by watching his father keep books for the estate of which he was in charge. Arrhenius began school at the age of eight, when he entered the fifth-grade class at the Cathedral School in Uppsala. After graduating in 1876, Arrhenius enrolled at the University of Uppsala.

At Uppsala, Arrhenius concentrated on mathematics, chemistry, and physics and passed the candidate's examination for the bachelor's degree in 1878. He then began a graduate program in physics at Uppsala, but left after three years of study. He was said to be dissatisfied with his physics advisor, Tobias Thalén, and felt no more enthusiasm for the only advisor available in chemistry, Per Theodor Cleve. As a result, he obtained permission to do his doctoral research in absentia with the physicist Eric Edlund at the Physical Institute of the Swedish Academy of Sciences in Stockholm.

The topic Arrhenius selected for his dissertation was the **electrical conductivity** of solutions. The problem was of some interest to scientists because of the fact that while neither pure **water** nor dry salts conduct an electrical current, a **solution** made by dissolving salts in water does. A number of other phenomena related to solutions were also puzzling. For example, a given amount of **sodium** chloride causes twice the lowering in freezing point of a solution as does a comparable amount of sugar, while other salts, such as **barium** chloride and **aluminum** chloride, cause a freezing point depression of three or four times as much as that of a sugar solution.

In 1884 Arrhenius submitted his thesis on this topic. In the thesis he hypothesized that when salts are added to water, they break apart into charged particles now known as ions. What was then thought of as a **molecule** of **sodium chloride**, for example, would dissociate into a charged sodium **atom** (a sodium ion) and a charged **chlorine** atom (a chloride ion). The doctoral committee that heard Arrhenius's presentation in Uppsala was totally unimpressed by his ideas. Among the objections raised was the question of how electrically charged particles could exist in water. In the end, the committee granted Arrhenius his Ph.D., but with a score so low that he did not qualify for a university teaching position.

Svante August Arrhernius.

Convinced that he was correct, Arrhenius had his thesis printed and sent it to a number of physical chemists on the continent, including Rudolf Clausius, Jacobus van't Hoff, and **Wilhelm Ostwald**. These men formed the nucleus of a group of researchers working on problems that overlapped chemistry and physics, developing a new discipline that would ultimately be known as **physical chemistry**. From this group, Arrhenius received a much more encouraging response than he had received from his doctoral committee. In fact, Ostwald came to Uppsala in August 1884 to meet Arrhenius and to offer him a job at Ostwald's Polytechnikum in Riga. Arrhenius was flattered by the offer and made plans to leave for Riga, but eventually declined for two reasons. First, his father was gravely ill (he died in 1885), and second, the University of Uppsala decided at the last moment to offer him a lectureship in physical chemistry.

Arrhenius remained at Uppsala only briefly, however, as he was offered a travel grant from the Swedish Academy of Sciences in 1886. The grant allowed him to spend the next two years visiting major scientific laboratories in Europe, working with Ostwald in Riga, Friedrich Kohlrausch in Würzburg, **Ludwig Boltzmann** in Graz, and van't Hoff in Amsterdam. After his return to Sweden, Arrhenius rejected an offer from the University of Giessen, Germany, in 1891 in order to take a teaching job at the Technical University in Stockholm. Four years later he was promoted to professor of physics there. In 1903, during his tenure at the Technical University, Arrhenius was awarded the Nobel Prize in chemistry for his work on the dissociation of electrolytes.

Arrhenius remained at the Technical University until 1905 when, declining an offer from the University of Berlin, he became director of the physical chemistry division of the Nobel Institute of the Swedish Academy of Sciences in Stockholm. He continued his association with the Nobel Institute until his death in Stockholm on October 2, 1927.

Although he was always be remembered best for his work on dissociation, Arrhenius was a man of diverse interests. In the first decade of the twentieth century, for example, he became especially interested in the application of physical and chemical laws to biological phenomena. In 1908 Arrhenius published a book entitled *Worlds in the Making* in which he theorized about the transmission of life forms from planet to planet in the universe by means of spores.

Arrhenius's name has also surfaced in recent years because of the work he did in the late 1890s on the greenhouse effect. He theorized that **carbon dioxide** in the atmosphere has the ability to trap **heat** radiated from the Earth's surface, causing a warming of the atmosphere. Changes over time in the **concentration** of carbon dioxide in the atmosphere would then, he suggested, explain major climatic variations such as the glacial periods. In its broadest outlines, the Arrhenius theory sounds similar to current speculations about climate changes resulting from global warming.

Arrhenius was married twice, the first time in 1894 to Sofia Rudbeck, his pupil and assistant at the Institute of Technology. That marriage ended in 1896 after the birth of one son, Olev Wilhelm. Nine years later Arrhenius was married to Maria Johansson, with whom he had two daughters, Ester and Anna-Lisa, and one son, Sven. Among the honors accorded Arrhenius in addition to the Nobel Prize were the Davy Medal of the Royal Society (1902), the first Willard Gibbs Medal of the Chicago section of the American Chemical Society (1911), and the Faraday Medal of the British Chemical Society (1914).

ARSENIC

Arsenic is the third member of the **nitrogen** family, which consists of elements in Group 15 of the **periodic table**. Its **atomic number** is 33, its atomic **mass** is 74.9216, and its chemical symbol is As.

Properties

Arsenic occurs in two allotropic forms, the more common of which is a shiny, gray, brittle, metallic-looking solid. The less common **allotrope** is a yellow crystalline solid produced when vapors of arsenic are cooled suddenly. When heated, arsenic does not melt, as do most **solids**. Instead, it sublimes. However, under high pressure, arsenic can be forced to melt at a **temperature** of about 1,500°F (814°C). Arsenic's **density** is 5.72 grams per cubic centimeter.

Chemically, arsenic is a **metalloid**. When heated in air, it reacts with **oxygen** to form arsenic **oxide** (As_2O_3). The reaction results in the formation of a characteristic blue flame and a distinctive garlic-like odor.

Occurrence and Extraction

Arsenic is a relatively uncommon element. Its abundance on Earth is thought to be about 5 parts per million. It rarely occurs in nature as a pure element. Its most common ores are the minerals arsenopyrite (FeAsS), orpiment (As_2S_3), and realgar (As_4S_4). The world's largest producers of arsenic are China, Chile, Mexico, Belgium, Namibia, and the Philippines. Arsenic is not mined in the United States.

Discovery and Naming

Arsenic can be produced from its ores very easily, so many early craftspeople may have seen the element without realizing what it was. Since arsenic is somewhat similar to **mercury**, early scholars probably confused the two elements with each other. Credit for the actual discovery of arsenic is usually given to the alchemist Albert the Great (Albertus Magnus). He heated a compound of arsenic—orpiment—with soap to produce nearly pure arsenic.

Uses

A small amount of arsenic is used to make alloys. The most common alloying element is **lead**. Lead-arsenic alloys were once used widely in lead storage battery and lead shot. The toxic properties of both lead and arsenic have, however, led to a dramatic reduction in the manufacture of these products. Small amounts of arsenic are also used to make transistors and light-emitting diodes (LEDs), used in hand-held calculators, clocks, watches, and other electronic devices.

Arsenic is much more widely used as a compound. Historically, one of the most notable uses of arsenic compounds has been as a poison for use with both rodents and humans. A number of famous murders in history have been accomplished by using compounds of arsenic. The use of arsenic compounds in rat poisons has been largely discontinued because of the risk such compounds pose to human health.

Today, the most important use of arsenic is in the manufacture of a compound known as chromated **copper** arsenate (CCA). CCA is used as a **wood** preservative in the construction of homes and other wooden structures. It prevents termites and other wood-eating organisms from destroying wood treated with this compound. CCA now accounts for about 90% of all arsenic produced in the United States.

Health Issues

The health effects of arsenic are well known. In low doses, it can produce nausea, vomiting, and diarrhea. In larger doses, it causes abnormal heart beat, damage to blood vessels, and a feeling of "pins and needles" in the hands and feet. In the largest doses, it can cause death. Direct contact with the skin can cause redness and swelling. Long term exposure to arsenic vapors can result in lung cancer. Digestion of the element or its compounds may lead to cancer of the bladder, kidneys, liver, and lungs.

ARTIFICIAL FATS

Artificial fats are **food additives** that are created to simulate the flavor of fats without providing the same level of calories. They also have the added benefit of reducing excessive fat consumption which has been linked to both coronary heart diseases and obesity. These fats have been produced from three different types of compounds including **proteins**, **carbohydrates**, and non-digestible fats. During the 1990s the popularity of artificial fat products increased significantly and they were consumed regularly by 172 million Americans.

While many consumers want to reduce the amount of fat in their diets, it has been difficult because fat containing products have a unique **taste**, texture, mouth-feel, and **viscosity**. Low fat foods have had to sacrifice many of these characteristics to reduce fat. Artificial fats have been developed to solve the problems associated with low fat foods. In the United States three primary artificial fats have been introduced including Olestra, Simplesse, and Stellar.

One of the first artificial fats that was discovered is olestra. Olestra was discovered in the late 1960s by researchers at the Proctor & Gamble company. In 1987, the first petition to the Food and Drug Administration (FDA) related to olestra was submitted. It was approved for use in food products during 1996.

Olestra is a material derived from sucrose and **fatty acids**. It can be described as a sucrose **polyester** with six to eight fatty acids bound to a sucrose **molecule**. The fatty acids are derived from either soybean or corn oil. The molecular composition of olestra is so large compared to a fat molecule, it can not be hydrolyzed by digestive enzymes in the body. This causes olestra to pass through the digestive system unchanged. Consequently, it does not contribute calories or fat to the diet. Since olestra is derived from fat molecules, it has similar chemical and physical properties. It reportedly has the same look, feel and flavor of fat. Olestra has been used as an ingredient in salty snack products. It may also be used as cooking oil for products such as french fries and onion rings. Additionally, it may be used for baking.

Even though olestra is not absorbed by the body, there have been some concerns related to its general use. Certain studies have suggested that olestra can cause abdominal cramping and loose stools. It has also been found that olestra may inhibit the absorption of some **vitamins** and nutrients. These studies resulted in two significant actions by the FDA related to products containing olestra. First, the products are required to have a warning label telling consumer of the possible problems with olestra. Second, foods that use olestra as an additive must also have extra vitamins A, D, E, and K added. Repeat studies of earlier findings have shown that there are no significant problems with the use of olestra. In 1999, after reviewing all the available data, the FDA declared that olestra does not significantly impact public health.

Another fat replacement is called simplesse. It is a material produced by the NutraSweet company which is derived from a protein. The first petition for use of simplesse was submitted to the FDA in 1988, and in 1990, they approved its use in frozen desserts. Simplesse is made by blending cooked egg whites with milk or whey protein. The result of this process is a system containing round protein particles that are extremely small. In fact, these microparticulated proteins are too small to be perceived individually by the mouth, but in bulk they create a creamy taste and texture similar to fat.

Unlike olestra, simplesse is absorbed and digested by the body. However, since it is protein it only contributes 1 to 2 calories per gram. This is significantly less than the 9 calories per gram obtained from fat. It can be used in a variety of food products, such as ice cream, yogurt, dips, and oil-based foods like salad dressing. It can not be used for cooked foods because the protein tends to lose its creamy quality when heated.

A third fat replacement was introduced in 1991. This material, known as stellar, is a carbohydrate-based artificial fat produced by the A. E. Staley Manufacturing Company. Since it meets FDA regulations regarding modified starches, it did have to go through the same approval process as olestra. Stellar is a fine white powder composed of an acid-hydrolyzed, corn derived starch. The powder is blended with **water** using high speeds and pressures to produce an opaque, smooth cream. This material is white and has properties similar to shortening. It has a particle **gel** structure that simulates the mouth-feel and texture of fat.

Since stellar is a carbohydrate it provides only 1 calorie per gram when used in a recipe. It can be used in a wide variety of foods, including baked goods, frostings, margarine, frozen deserts, meat products, soups, and cheese products. It does not have the same flavor as fat so a certain amount of fat or artificial flavoring is required. Stellar is stable over a large **temperature** range, however, it cannot be used for frying food because it can break down at extremely high temperatures.

See also Lipids

ARTIFICIAL SWEETENERS

Artificial sweeteners are chemicals designed to mimic the **taste** of sugar without providing the same amount of dietary calories. They can be derived from synthetic or natural sources. In the United States, four artificial sweeteners are approved for use: saccharin, aspartame, sucralose, and acesulfame-K.

The desire for sweet taste is thought to be an innate human trait. Cave paintings at Arana in Spain show a Neolithic man getting honey from a bee hive. Scientists have suggested that early humans used sweetness as a guide to foods that were safe to eat.

One drawback to foods that contain large amounts of sugar is that they are also high in calories. Artificial sweeteners were developed to provide sweetening with fewer or no calories. They also have the added benefits of improving the palatability of foul-tasting drugs, helping people who suffer from diabetes. The first artificial sweetener was saccharin. It was discovered serendipitously in 1879 during experiments with toluene derivatives. It was used worldwide for years until concerns that it could cause cancer arose. The possible link be-

A polarized light crystal (magnified 25 times) of aspartame, a synthetic sugar substitute. ***(Photograph by Leonard Lessin, FBPA, Photo Researchers, Inc. Reproduced by permission.)***

tween saccharin and cancer led Canada to ban the substance in 1977. In the United States the Food and Drug Administration (FDA) determined that the studies linking saccharin and cancer were inconclusive and did not ban it. They did, however, require all products that contain saccharin to have a warning label.

The sweet taste of aspartame was also discovered accidently. In 1965, James Schlatter was working on an anti-ulcer drug and inadvertently spilled some on his hand. When he licked his finger to pick up a piece of **paper** he noticed a sweet taste. After several years of intensive testing Schlatter and his parent company G.D. Searle & Co. were issued U.S. patent 3,492,131. The sweetener was approved for use in the United States in 1980. By 1993, annual worldwide sales of aspartame reached $705 million.

Various other artificial sweeteners have been submitted to the FDA for approval. In 1994, acesulfame-K was approved. Sucralose was approved in 1999. Other sweeteners that are waiting for FDA approval include alitame and neotame.

Saccharin was the first artificial sweetener to be discovered. It is also called 2,3-dihydro-3-oxobenzisosulfonazole and is represented by the chemical formula $C_6H_4CONHSO_2$. Saccharin is a white, crystalline powder that can be as much as 500 times sweeter than sucrose. In high concentrations it has a bitter taste, but it is sweet in more dilute solutions. Saccharin is not metabolized by the human digestive system so it does not provide any caloric content. It is used in a variety of products such as toothpaste, mouthwash, sugarless gum, table-top sweetener, yogurt, and fruit-based products.

Aspartame is an odorless, white crystalline powder derived from two **amino acids**, aspartic acid and phenylalanine. It is represented by the chemical formula $C_{14}H_{18}N_2O_5$ and is also known by its chemical name *N*-L-aspartyl-L-phenylalanine-1-methyl ester. It is approximately 200 times sweeter than sugar and can be easily dissolved in **water**. It has a sweet taste without a bitter aftertaste and contributes a relatively small number of calories when it is eaten. Aspartame is approved as a table-top sweetener and can be used in frozen desserts, gelatins, beverages, and chewing gum. Since aspartame tends to interact with other food flavors, its flavor may not be identical to sugar.

Acesulfame-K is a compound that is about 200 times sweeter than sucrose. It has good **heat** stability and can be incorporated into baking products. It was approved for use in the United States in 1994. Unlike saccharin and aspartame, acesulfame-K does not require health warnings. It is typically used in chewing gum, pudding desserts, instant coffee, and dairy products.

Sucralose is a modified sucrose **molecule** that is about 600 times sweeter than sugar. It was approved for use in foods and beverages in 1999 in the United States. It is produced from sucrose using a patented process that replaces the three hydroxyl groups (OH) with three **chlorine** atoms. This modified sugar is minimally absorbed by the body and passes out unchanged. It is stable in heat and can be used in food products that are baked or fried.

FDA approval is pending for two other compounds. Alitame, a compound that is 2,000 times sweeter than sucrose, has been used in Australia, New Zealand, and China for some years. When approved, it may be used in baked goods, beverages, frozen desserts, and chewing gum. Neotame is another artificial sweetener that is expected to be approved. It is reported to be 8,000 times sweeter than sucrose. At that potency it promises to be a cost-effective ingredient for the food industry.

See also Carbohydrates

ASBESTOS

The term asbestos is used for several hydrated silicate minerals that occur in fibrous forms. Since **silicates** are not flammable and only melt at very high temperatures, asbestos fibers have been used to make flame-retardant fabrics and composite materials. The asbestos minerals are divided into two major classifications based upon their crystal structures, serpentines and amphiboles. The most commonly used type of asbestos, accounting for 95% of all commercial usage, is chrysotile, also called white asbestos. It is a hydrated **magnesium** silicate found in large amounts in Canada and Russia. Fibers of chrysotile are easily spun, are strong and fuse readily both to **glass** and organic compounds, but they must be protected from acid and **water** to increase their durability. Another asbestos mineral, crocidolite, is used for more specialized applications, when, for example, better acid resistance is desirable. Crocidolite is obtained from Australia and South Africa.

Asbestos was widely used as a fireproofing material, flame-retardant reinforcing structural material and for such products as automobile brake linings until about 1975, when its adverse health effects became widely appreciated. One of the strong indications that asbestos is a health hazard came from the high rates of lung cancer, gastrointestinal cancer and blood disorders suffered by workers exposed to large amounts of asbestos over several year periods. Even though is has been known for decades that the length of time and amounts of asbestos to which workers were exposed lead to increased probability of such adverse effects, no accurate dose-response relationships for humans or animals have been achieved.

The most direct adverse health effect of asbestos exposure is occupational asbestosis characterized by diffuse fibrosis of the thorax and lung. It has been associated only with workers who were exposed to high levels of airborne asbestos particles on a daily basis over several years. Since asbestosis develops in the respiratory system, it seems reasonable to conclude that airborne particles are responsible for the disease. Elevated levels of lung cancer have also been reported for asbestos industry workers. The types of lung cancer are the same as seen in the general population and the extent to which smoking is responsible is not completely understood. Although asbestosis has not been reported in the general population, and the relationship of exposure to development of lung cancer is not clear, great precautions are now taken to reduce the possibility of respiration of asbestos particles.

Asbestos is now essentially banned as a construction material in the United States. Billions of dollars have been spent to remove or sequester asbestos (asbestos abatement) in previously constructed buildings. The target of most asbestos abatement efforts is easily crumbled or ''friable'' asbestos or asbestos that is likely to be disturbed during the renovation or demolition of a building. The Environmental Protection Agency does not recommend removing intact asbestos materials or non-friable asbestos that is unlikely to be disturbed. Discovery of asbestos in a building can greatly reduce is market value.

The molecular structures of the asbestos minerals is directly related to their fibrous nature. The silicate **ion** is tetrahedral, with four **oxygen** atoms covalently bound around a **silicon atom** with an overall charge of 4-. Silicic acid is the corresponding acid of the silicate ion, with each oxygen atom bound to a **hydrogen** atom and a net charge of 0. The structure of the fibrous silicates is a chain of silicate units that would result from removing two water molecules from two neighboring silicic acid molecules to form two Si-O-Si bridges, leaving two -OH groups on each silicon atom, then adding another silicic acid **molecule** and removing two water molecules to form two more O-Si-O bridges and so forth. As each silicic acid molecule is added, the chain lengthens and the only -OH groups are on the silicon atoms at the two ends of the chain. In the actual mineral, the **chains** are not perfect. Metal ions, such as magnesium in the case of chrysotile, bind to the exposed oxygen atoms wherever the chain has -OH groups exposed.

The fibrous silicates consist of one-dimensional chains of linked silicate units. Some silicates, such as quartz, have structures in which the O-Si-O bridges extend in three dimensions and others, such as mica, have sheet structures in which the O-Si-O bridges extend in two dimensions.

ASTATINE

Astatine is a member of the halogen family, the elements that make up Group 17 of the **periodic table**. Its **atomic number** is 85, its atomic **mass** is 209.9871, and its chemical symbol is At.

Properties

All isotopes of astatine are radioactive. They have such short half lives that it has been difficult to determine the element's properties. No melting point, **boiling point**, or **density** data are available for the element. Experiments that have been conducted on the element show that its chemical properties are similar to those of the other **halogens** in Group 17.

Occurrence and Extraction

Astatine is a very rare element. Scientists estimate that no more than about 25 grams of the element exist on the Earth's surface. Astatine can also be prepared artificially in particle accelerators, although no more than about a millionth of a gram has been produced by that method so far.

Discovery and Naming

Astatine was one of the last of the naturally occurring elements to be discovered. It was found in 1940 by three chemists working at the University of California, Dale R. Corson, Kenneth R. Mackenzie, and Emilio Segrè. The element was named after the Greek word for ''unstable,'' which is *astatos*.

Uses

Astatine is too rare to have any practical uses. Even its use in research is very limited because of its scarcity.

Health Issues

There is evidence that astatine behaves like **iodine** in the body. That is, it tends to concentrate in the thyroid gland. If that fact is true, one possible future use for astatine is the treatment of thyroid disorders.

ASTBURY, WILLIAM (1898-1961)

English chemist and molecular biologist

William Astbury's standing in the history of science lies primarily with his work in the structure of organic fibers (e.g., wool). He is also an early figure of importance in the race to discover the structure of deoxyribonucleic acid (DNA), the genetic material, and therefore a founder of molecular biology. His work as an applied scientist, slowly decoding the nature of **molecular structure** of virtually the largest organic materials, fibrous and globular **proteins**, was valuable to both science and industry.

Born on February 25, 1898, at Longton, England, later to be incorporated into Stoke-on-Trent, William Thomas Astbury was the son of a potter, though he, like his brother, the physicist N. F. Astbury, had the fortune of an excellent collegiate education. Having won a scholarship to Jesus College, Cambridge University in 1916, he began the study of mathematics and physics. Only two years into his work, Astbury was drafted into the Royal Army Medical Corps X-ray unit. That introduction in World War I to the use of X-ray methods in human war injuries later became the central tool in his lifelong study of organic structure.

Upon returning to Cambridge, Astbury added **chemistry** and **mineralogy** to his interests. It was under the mineralogist A. Hutchinson that Astbury discovered crystallography, and he spent the rest of his career studying how X rays could be applied to crystallographic problems. After graduation in 1921, he joined the crystallographic group at University College, London, headed by William Henry Bragg. Two years later, Bragg's laboratory moved to the Royal Institution and Astbury went with it. In 1928, Astbury was appointed a lecturer in textile physics at the University of Leeds. During his years with Bragg, he had begun his investigation into the structure of wool, and within two years of his arrival at Leeds, he had explained how wool can stretch and fold. The protein that is wool, keratin, was the subject of his first book, *Fundamentals of Fibre Structure.* Explaining the molecular structure of keratin secured his reputation in science and became a springboard for much of his later research.

Bragg had recommended him to Leeds and it turned out to be a happy relationship; Astbury remained at the University of Leeds moving up the professional ladder from lecturer to reader (1937) to honorary reader (1945) and, finally, to the first occupancy of the then new Chair of Biomolecular Structures (from 1945 until his death). Astbury had been successful in Bragg's laboratory even before his departure for Leeds, however. With **Kathleen Lonsdale**, Astbury produced the first table of space groups. These groups define the internal symmetry of crystals, whether inorganic (e.g., table salt) or organic (e.g., keratin). Astbury and Lonsdale's work remains central in crystallography.

Unfortunately, part of the detail of how Astbury believed keratin to be structured turned out to be wrong. The brilliant chemist and (later) two-time Nobelist, **Linus Pauling**, replaced much of the Astbury model for keratin's structure with his notion of the alpha helix and beta-pleated sheet configurations for large proteins. Pauling's explanation has endured, but the groundwork laid by Astbury had been crucial to the movement forward in large **molecule** crystallography. Another major feature of Astbury's career is his work in globular protein structure. Wool, hair, and fingernails are made up of a long, fibrous protein (keratin) that is mostly inert, but globular proteins include the active proteins such as enzymes and others, such as hemoglobin—carrier of blood-bound **oxygen**. Astbury showed that such globular proteins were three-dimensional, folded **chains** that could be denatured (unfolded) and, in some examples, renatured (refolded to their original shapes). This was a significant discovery.

In the 1930's, Astbury discovered what was a major feature of **DNA** in one of the X-ray pictures taken in his laboratory, the famous 3.4 Angstrom spot (an X-ray dense repeating shadow indicating a helical molecule of a certain size). The X-ray picture of the 3.4 Angstrom spot was published jointly by Astbury and doctoral student Florence Bell, the individual who actually made the picture. Astbury was the living expert on fibrous materials and as such was the most informed about their complex structure, and he grasped the importance of DNA as a molecule that might be able to transmit hereditary information. That was the more remarkable because **Oswald Theodore Avery**, **Colin Munro Mac Leod** and **Maclyn Mc Carty** had not yet published their seminal paper (of 1944) showing that DNA surely was the genetic material.

Astbury's Leeds laboratory had been, for a number of years, supported by generous funding from the Rockefeller Foundation of New York; no similar resources were at that time available in the United Kingdom. In this well-financed situation, expensive X-ray studies were possible. Even with that, X-ray pictures taken by **Rosalind Elsie Franklin** in 1952–53 at King's College, London, proved better and were then used by **James Watson** and **Francis Crick** as they defined the structure of DNA. The Leeds pictures made by Bell, Astbury, and others were inferior to those of Franklin due to a number of technical reasons. Even so, by 1951, excellent pictures emanated from Astbury's laboratory and Mansel Davies, a student of Astbury's who has written on this critical period, has argued that Astbury had all that was necessary in front of him to solve the structure of DNA. The only suggestion Davies could make for why Astbury did not move forward was due to his individualistic personality. Astbury had, in another situation said: "I am not prepared to be anybody's lackey." Davies argued that since Astbury realized how close Watson and Crick were in the early 1950s, he (Astbury) would not preempt their work, but would, rather, follow his own investigations on other topics. Nevertheless, the discovery of the 3.4 Angstrom spot provided essential grist for the Watson-Crick mill and what some have called the greatest biological discovery of the 20th century. Six years before Watson and Crick, Astbury even proposed a model for the structure of DNA. His model, however, was far from correct (not even helical). For his lifetime of creative research, Astbury was elected a Fellow of the Royal Society. Author of two books and over 100 articles, Astbury remains a figure of considerable stature in his several specialties.

ASTON, FRANCIS W. (1877-1945)

English chemist and physicist

Francis W. Aston was an English chemist and physicist whose motto—"Make more, more, and yet more measurements"—summed up the hard work and dedication he brought to a lifetime of achievement. Among his most important contributions were detailed observations of atomic phenomena with a **mass** spectrograph that he built himself. This device allowed him to articulate the theory that the atomic weight of each element is a whole number, but that most elements have isotopes (atoms of the same element with the same number of protons in their **nucleus**, but different numbers of neutrons). For these insights he received the 1922 Nobel Prize in **chemistry**. From evidence he gathered in the 1920s, Aston went on to note that the weights of atoms vary minutely from whole numbers in proportion to the **density** of their nuclei.

Francis William Aston was born on September 1, 1877, in Harborne, Birmingham, England, the third of seven children. His parents were William Aston, farmer and metal merchant, and Fanny Charlotte Hollis Aston, the daughter of a gunmaker. From an early age Aston showed great scientific curiosity, performing experiments in his makeshift laboratory on the family farm. In 1893 he graduated from high school at the top of his class in math and science. While at Mason College (which became the University of Birmingham), he worked with P. F. Frankland in **organic chemistry** and optics, issuing a paper in 1901 on his results. Lacking scholarship money for continued studies, he went to work for a brewing company as resident chemist in the early 1900s. He also performed experiments in **electricity** with sophisticated devices he built at home.

For this work he was awarded a scholarship in 1903 to the new University of Birmingham. There he discovered a phenomenon, now called the Aston space, that appears in electrical currents passed through **gases** at low pressures. In 1910, working with **Joseph John Thomson** both at Cambridge University and the Royal Institution in London, he began experiments on the gas **neon**. When his work was interrupted by World War I, Aston returned to Cambridge. In 1919 he became a fellow of Trinity College, Cambridge, where he stayed for the rest of his life. That same year he managed to perfect the mass spectrograph. In this instrument, a beam of neon atoms directed onto a photographic plate angled away from the flow of atoms created a distinctive pattern when the heavier atoms deflected farther down the plate than the lighter atoms. Taking the average of these deflections, Aston was able to calculate the proportion of heavier to lighter atoms in the element neon. From this information Aston deduced that most elements are mixtures of isotopes and that the weights of atoms are always whole numbers (the whole number rule). He posited that isotopic constituents accounted for the fractional weights observed for some atoms.

Not one to rest on his 1922 Nobel Prize laurels, Aston built larger and even more accurate spectrographs. From a new round of observations taken with one of these instruments in 1927, Aston measured fractional deviations from the whole number rule. He discovered that the tighter the packing of the particles in an atomic nucleus, the greater a fraction of its mass became converted to **energy** devoted to keeping the nucleus together. He incorporated these ''packing fractions'' into calculations from which physicists and chemists have derived essential information about the abundance and stability of the elements.

Politically conservative, Aston preferred working alone to collaborating with colleagues. He never married, preferring to keep busy with outdoor sports, traveling by sea and becoming an accomplished photographer and amateur musician. He also loved animals. Aston acquired some financial skills, leaving behind a large estate to Trinity College and various scientific enterprises. He held many honorary degrees and received, in addition to the Nobel Prize, awards such as the 1938 Royal Medal of the Royal Society, of which he was a member, and the 1941 Duddell Medal and Prize of the Institute of Physics. He died on November 20, 1945, in Cambridge.

Francis W. Aston.

ASTROCHEMISTRY

The movie ''Alien'' used the line ''In space, no one can hear you scream'' as its advertising campaign. The reason? Outer space is a vacuum. That is, it is empty of matter and sound does not travel through a vacuum.

Or, at least, space is a vacuum by Earth standards. We live at the bottom of a very thick and dense ocean of air (surface density is about 0.074 lb/cu.ft. or 1.184 g/liter). But this atmosphere thins out considerably with increasing altitude until outer space is reached (by definition, about 200 miles or 300 kilometers above the Earth's surface). Even though gas densities are several orders of magnitude lower, in the immediate outer space around our planet, there are perhaps 100 molecules per cubic meter. Between our solar system and the nearest star, this density drops to a few atoms per cubic meter which is a vacuum far better than any that we can create in the laboratory. There simply is not a lot of matter out there.

So why study astrochemistry? With such little material to work with, does chemistry actually occur in outer space? The answer is ''yes.'' Even with only a couple of atoms per cubic meter, atoms and chemical compounds still undergo collisions and react to give a variety of species. Indeed, evidence would suggest that mere contact is sufficient for a reaction as the amount of available energy in space is sufficient to overcome activation barriers.

But astrochemistry is not simply the study of chemistry between the stars. It is also the study of chemistry in nebulae where new stars are born. And it borrows from nuclear physics for an understanding of nucleosynthesis which is the process for the creation of the elements within the interior of stars.

Indeed, all of the matter in the universe was created in the first few moments after the big bang. However, assembling it into atoms of carbon, oxygen, nitrogen, iron, and such has taken many cycles of stellar birth and decay. It is only in the stars and, more importantly, in the death of stars that the heavy atoms are created through the fusion of simple nuclei into the more complex and massive elements. All of the chemical elements that we have discovered on Earth had their origins in the heart of a star. This realization prompted Carl Sagan to quip that "we are made from the stuff of stars."

Other than the synthesis of elements, stars are not the home for chemical compounds. The intense heat and gravitational pull prevent the assembly of molecules. But between the stars and within the slightly denser clouds of hydrogen called "nebulae" from stars are born, simple chemical compounds have been observed. These compounds are generated by photolysis of atoms and simpler chemical compounds. The harsh U.V. radiation in outer space is capable of providing sufficient activation energy for most (if not all) chemical reactions.

The method of observation for these compounds relies on spectroscopic methods. By looking at the light from distant stars, astronomers and spectroscopists are able to detect the line spectra associated with small molecules and compounds. This task is complicated by the multitude of chemical compounds in the interstellar medium and the elemental line spectra in the originating star light.

The spectra that are observed indicate that small molecules, such as cyanide, methane, and cyanogen, can assemble in outer space. Indeed, there are indications that pyrroles can form and this has led to the speculation that it is possible that porphyrins are synthesized in the nether region between worlds.

This speculation has become the basis for theories suggesting that life on Earth is extraterrestrial in origin. The basis of these theories is that small molecules assembled in outer space through photolysis, were captured in the interstellar dust by meteors and such, and were subsequently deposited on Earth. There, the molecules began the complex series of reactions that eventually lead to self-propagating chemical sequences that eventually became replicating life forms. Two important conclusions have been derived from these theories. The first is that this scenario is likely to have happened all over the galaxy suggesting that there is hope for finding other life forms out there. The second is that life is an extraterrestrial phenomenon.

The energy-rich vacuum of outer space thrives with chemical reactions. Astrochemistry is an attempt to explore and understand this chemistry.

See also Spectroscopy

ATMOSPHERIC CHEMISTRY

We live at the bottom of an ocean of air. We take the atmosphere pretty much for granted; we are generally much more concerned with the weather. But this ocean of air has profound consequences for life on Earth.

The surface **density** of air is about 0.074 lb/ft^3 (1.184 g/l) and surface pressure is about 14 lb/ft^2 (1 atm). This **mass** of air presses down upon us at all times. But at a higher altitude, both the **pressure** and the density of air decrease. This is why passenger jets, which fly at 40,000 ft (12,192 m) to take advantage of the thin or low density air, require pressurized cabins. Without them, passengers would not be able to breath properly.

The atmosphere is generally divided into four zones or layers. Starting at sea level and increasing in altitude, they are the *troposphere* (0-10 mi [0-16.1 km]), the "stratosphere"(10-30 mi [16.1-48.3 km]), the *mesosphere* (30-60 mi [48.3-96.6 km]), and the *thermosphere* (beyond 60 mi [96.6 km]). These altitudes are very approximate and depend upon a variety of conditions, and are clearly distinct in both their physical properties (e.g., temperature) and their **chemistry**.

The troposphere is the region of air closest to the ground. It is where our "weather" occurs or, at least, the clouds and storm systems are to be found. It is also the region that is in direct contact with effluent chemicals generated by living things. This can range from the **carbon** dioxide and **water** vapor we exhale to industrial or automotive pollutants. In the absence of such compounds, **atmospheric chemistry** is very simple. Since the splitting of both the **nitrogen** and **oxygen** molecules requires a great deal of **energy**, the atmospheric composition is pretty much constant. That is, at sea level, without interfering compounds, there is not much atmospheric chemistry.

"Smog" is the term applied to the mixture of nitrous oxides, spent hydrocarbons, **carbon monoxide**, and ozone that is generated by automobiles and industrial **combustion**. It is the thick brown haze that hovers over large populated areas. This combination of **gases** is reactive. The addition of water vapor or rain drops, for example, can result in the scrubbing of these compounds from the air but also the generation of nitrous, nitric, and carbonic acid. Ozone is a powerful oxidizing agent and results in the degradation of **plastics** and other materials. However, it is also capable of reacting with spent hydrocarbons to generate noxious chemicals.

Industrial pollutants, such as **sulfur** dioxide generated by coal-burning power plants, can generate acid rain as the sulfur dioxide is converted to sulfurous and **sulfuric acid**. Even forest fires contribute a large variety of chemical compounds into the atmosphere and induce chemical reactions. And, the largest of all natural disasters, a volcanic eruption, spews tons of chemical compounds into the troposphere where they react to produce acids and other compounds.

The stratosphere is the home of the ozone layer, which is misleading as it implies a distinct region in the atmosphere that has ozone as the major constituent. Ozone is never more than a minor constituent of the atmosphere, although it is a very important minor constituent. The **concentration** of ozone achieves its maximum in the stratosphere. It is here that the chemistry occurs that blocks incoming ultraviolet radiation.

The complete spectrum of radiation from the sun contains a significant amount of high energy ultraviolet light and the energy of these photons is sufficient to ionize atoms or molecules. If this light penetrated to the Earth's surface, life as we know it could not exist as the ionizing radiation would continually break down complex molecules.

Within the ozone layer, this ultraviolet energy is absorbed by a delicate **balance** of two chemical reactions. The first is the photolytic reaction of molecular oxygen to give atomic oxygen, which subsequently combines with another oxygen **molecule** to give ozone. The second reaction is the absorption of another **photon** of ultraviolet light by an ozone molecule to give molecular oxygen and a free oxygen **atom**.

$$O_2 \xrightarrow{h\nu} O + O$$
$$O + O_2 \rightarrow O_3$$
$$O_3 \xrightarrow{h\nu} O_2 + O$$
$$3O_2 \overset{h\nu}{\leftrightarrows} 2O_3$$

It is the combination of these two reactions that allows the ozone layer to protect the planet. These two reactions actually form an equilibrium with the forward reaction being the formation of ozone and the backwards reaction being the depletion. The ozone concentration is thus at a constant and relatively low level. It occurs in the stratosphere because this is where the concentration of gases is not so high that the excited molecules are deactivated by collision but not so low that the atomic oxygen generated can not find a molecular oxygen with which to react.

In the last half of the twentieth century, the manufacture of **chlorofluorocarbons (CFCs)** for use as propellants in aerosol sprays and **refrigerants** has resulted in a slow mixing of these compounds with the stratosphere. Upon exposure to high energy ultraviolet light, the CFCs break down to atomic **chlorine** which interferes with the natural balance between molecular oxygen and ozone. The result is a shift in the equilibrium and a depletion of the ozone level. The occurrence of ozone depletion was first noted over the Antarctic. Subsequent investigations have demonstrated that the depletion of ozone also occurs over the Arctic, resulting in higher than normal levels of ultraviolet radiation reaching many heavily populated regions of North America. This is, perhaps, one of the most important discoveries in atmospheric chemistry and has lead to major changes in legislation in all countries in an attempt to stop ozone depletion.

Beyond the stratosphere, the energy levels increase dramatically and the available radiation is capable of initiating a wide variety of poorly characterized chemical reactions. Understanding all of the complexities of atmospheric chemistry is subject for much ongoing research.

See also Equilibrium, chemical; Photochemistry

A diagram of the structure of the beryllium atom, which has 4 electrons, 4 protons, and 5 neutrons (only 3 of which are shown). ***(Image by Michael Gilbert/Science Photo Library, Photo Researchers, Inc. Reproduced by permission.)***

ATOM

Atoms are the elementary building blocks of material substances. Although the term atom, derived from the Greek word *atomos*, meaning indivisible, would seem inappropriate for an entity that, as science has established, is divisible, the word atom still makes sense, because, depending on the context, atoms can still be regarded as indivisible. Namely, once split, the atom loses its identity. For example, an atom of **gold** is the building block of gold. If we split a gold atom, there is no more gold. As Carl H. Snyder has written, an atom is "the smallest particle of an element that we can identify as that element."

Atoms share many characteristics with other material objects: they can be measured, and they also have mass and weight. For example, a gold atom, which is could be regarded as an atom of average size, measures approximately 3×10^{-8} in diameter, and its mass is about 3.3×10^{-22} g. Because traditional methods of measuring are difficult to use for atoms and subatomic particles, scientists have created a new unit, the atomic mass unit (amu), which is defined as the one-twelfth of the **mass** of the average carbon atom.

The principal subatomic particle are the **nucleus**, the **proton**, the neutron, and the **electron** (*elektron* is Greek for yellow amber; the ancient Greeks were among the first to observe static electricity, produced when a piece of amber is rubbed). A nucleus, the atom's core, consists of protons, which are positively charged particles, and neutrons, particles without any charge. Electrons are negatively charged particles with negligible mass which move around the nucleus. Compared to an atom's total mass, the subatomic particles are truly minuscule. For example, neutrons and protons have the mass of 1.67×10^{-24}, while electrons have the hardly detectable mass of 0.0009×10^{-24}. An electron's mass is so small that it is usually given a 0 value in atomic mass units, compared to the value of 1 assigned to neutrons and protons. In fact, as the nucleus represents more than 99% of an atom's mass, it is interesting to note that an atom is mostly space. For example, if a hydrogen atom's nucleus were enlarged to the size of a marble, the atom's diameter would be around 0.5 mi (800 m).

Scientists used to believe that electrons circled around the nucleus in planet-like orbits. Because all **subatomic particles**, including electrons, exhibit wave-like properties, it is makes no sense to conceptualize the movement of electrons as planetary rotation. Scientists therefore prefer terms like "electron cloud patterns," or "shells," indicating an electron's pattern of movement in relation to the nucleus. Thus, for example, **hydrogen** has one electron shell, containing one electron; **lithium** has two shells, with one electron in the inner shell, and two in the outer. Scientists have posited the existence of four shells, each containing one or more electrons, and each defined by a particular level of energy.

While subatomic particles may, in a certain way, be regarded as generic and interchangeable, they, in fact, determine an atom's identity. For example, we know that an atom with a nucleus consisting of one proton must be hydrogen (H). An atom with two protons is always a **helium** (He) atom. Thus, we see that the key to an atom's identity is to be found in the atom's inner structure. In addition, a chemical element is an instance of atomic equilibrium: for example, in a chemical element, the number of positively charged particles (protons) always equals the number of negatively charged particles (electrons). While elements are not always stable, sometimes appearing as different isotopes, the balance of forces within an atom is regarded as a fundamental feature.

Research into the atom's nucleus has uncovered a variety of subatomic particles, including quarks and gluons. Considered by some researchers the true building blocks of matter, quarks are the particles which form protons and neutrons. Gluons hold smaller clusters of quarks together.

Although the atom, particularly its mysterious inner world of quantum laws and wave-like particles, may seem far removed from the practical concerns of **chemistry**, and more relevant to theoretical physics than to anything else, there is a clear connection between atomic structure and the tangible, observable characteristics of substances, which is obviously the domain of chemistry. In particular, chemists study reactions, the behavior of elements in interaction, and reactions, such as those leading to the formation of chemical compounds, involve electrons. For example, the formation of sodium chloride, also known as table salt, would be impossible without specific changes at a subatomic level. The genesis of sodium chloride (NaCl) starts when a sodium (Na) atom, which has 11 electrons, loses an electron. With ten electrons, the atom now has one proton too many, thus becoming a positively charged ion, or cation. The electron lost by the **sodium** atom is gained by a **chlorine** atom, which has 17 electrons distributed in three shells. The newly acquired electron goes into the outer shell, also known as the valence shell, which contains seven electrons. The chlorine atom's behavior also exemplifies the general tendencies of atoms to enter reactions in order to attain eight electrons in the valence shell. But the chlorine atom not only reaches an octet balance; it also becomes a negatively charged ion, or **anion**, since it now has 18 electrons to its 17 protons. Crystals of table **salt** consist of equal numbers of sodium cations and chlorine anions, cation-anion pairs being held together by a force of attraction.

Interestingly, not long after scientists realized that an atom is divisible, transmutation, or the old alchemic dream of turning one substance into another, became a reality. Scientists even succeeded in creating gold by bombarding platinum-198 with neutrons to create platinum-199 which decays to gold-199. Although clearly demonstrating the reality of transmutation, this particular is by no means a cheap method of producing gold—quite the contrary, since platinum, particularly the platinum-199 isotope is more expensive than gold. However, the symbolic value of the experiment is immense, as it shows that the idea, developed by ancient alchemists and philosophers, of material transmutation does not essentially contradict our understanding of the atom.

ATOMIC FORCE MICROSCOPE

In recent years, tremendous advances have been made in the field of **microscopy**. In 1985, the atomic force microscope (AFM) was invented by Gerd Binnig (co-inventor of the STM), Christoph Gerber in Zurich, Switzerland, and Calvin Quate in California. The AFM represents the technological pinnacle of microscopy.

The AFM uses a tiny needle made of **diamond**, **tungsten**, or **silicon**, much like those used in the STM. While the STM relies upon a subject's ability to conduct **electricity** through its needle, the AFM scans its subjects by actually lightly touching them with the needle. Like that of a phonograph record, the AFM's needle reads the bumps on the subject's surface, rising as it hits the peaks and dipping as it traces the valleys. Of course, the topography read by the AFM varies by only a few molecules up or down, so a very sensitive device must be used to detect the needle's rising and falling. In the original model, Binnig and Gerber used an STM to sense these movements. Other AFM's use a fine-tuned laser.

The AFM has already been used to study the super-microscopic structures of living cells, objects that could not be viewed with the STM. American physicist Paul Hansma and his colleagues at the University of California, Santa Barbara, are quickly becoming experts in AFM research. In 1989, this team succeeded in observing the blood-clotting process within blood cells. Hansma's team presented their findings in a 33-minute movie, assembled from AFM pictures taken every ten seconds.

Other scientists are utilizing the AFM's ability to remove samples of cells without harming the cell structure. By adding a bit more force to the scanning needle the AFM can scrape cells, making it the world's most delicate dissecting tool. Scientists hope to apply this method to the study of living cells, particularly floppy protein cells, whose fragility makes them nearly impossible to view without distortion.

Atomic number

Atomic number is defined as the number of protons in the **nucleus** of an **atom**. This concept was historically important because it provided a theoretical basis for the periodic law. **Dmitri Mendeleev's** discovery of the periodic law in the late 1860s was a remarkable accomplishment. It provided a key organizing concept for the chemical sciences. One problem that remained in Mendeleev's final analysis was the inversion of certain elements in his **periodic table**. In three places, elements arranged according to their chemical properties, as dictated by Mendeleev's law, are out of sequence on the basis of their atomic weights.

The solution to this problem did not appear for nearly half a century. Then, it evolved out of research with x rays, discovered in 1895 by Wilhelm Röntgen. Roentgen's discovery of this new form of electromagnetic radiation had inspired a spate of new research projects aimed at learning more about x rays themselves and about their effects on **matter**. Charles Grover Barkla, a physicist at the Universities of London and Cambridge, initiated one line of x-ray research. Beginning in 1903, he analyzed the way in which x rays were scattered by gasses, in general, and by elements, in particular. He found that the higher an element was located in the periodic table, the more penetrating the rays it produced. He concluded that the x-ray pattern he observed for an element was associated with the number of electrons in the atoms of that element.

Barkla's work was brought to fruition only a few years later by the English physicist H. G. J. Moseley. In 1913, Moseley found that the x-ray spectra for the elements changed in a simple and regular way as one moved up the periodic table. Moseley, like Barkla, attributed this change to the number of electrons in the atoms of each element and, thus, to the total positive charge on the nucleus of each atom. (Since atoms are electrically neutral, the total number of positive charges on the nucleus must be equal to the total number of negatively charged electrons.)

Moseley invented the concept of *atomic number* to describe his findings. He assigned atomic numbers to the elements in such a way as to reflect the regular, integral, linear relationship of their x-ray spectra. It soon came to be understood that the atomic number of an atom is equal to the number of protons in the atom's nucleus.

Moseley's discovery was a very important contribution to the understanding of Mendeleev's periodic law. Mendeleev's law was a purely empirical discovery. It was based on properties that could be observed in a laboratory. Moseley's discovery provided a theoretical basis for the law. It showed that chemical properties were related to **atomic structure** (number of electrons and nuclear charge) in a regular and predictable way.

Arranging the periodic table by means of atomic number also resolved some of the problems remaining from Mendeleev's original work. For example, elements that appeared to be out of place when arranged according to their atomic weights appeared in their correct order when arranged according to their atomic numbers.

Atomic radius

The radius of an **atom** is defined as half the distance between the nuclei of two atoms in a pure sample of the substance. The measurements have been obtained through x-ray diffraction techniques. The **periodic table** displays two significant relationships about the radii of atoms: 1) as elements increase in **atomic number** in a period, going from Group IA to Group 7A, the atomic radius decreases; 2) as elements in the same group increase in atomic number, the atomic radius increases. The largest atom has a radius that is only twice the size of the smallest.

The reason that the radii of the atoms within a period decreases relates to the electric force between the positively charged **nucleus** and the negatively charged electrons orbiting it. The number of protons in the nucleus is increased from Group IA to Group 7A. Adding protons to the nucleus results in a greater electric force. When the force of attraction increases, the electrons are drawn closer to the nucleus.

The reason that the radius of the atom increases within a group as the period number increases is the location of outer electrons in higher **energy** levels. These electrons have a higher probability of being located farther from the nucleus, resulting in a larger atomic radius. There is also a shielding effect by the core electrons, decreasing the force of attraction between the nucleus and the **valence** electrons.

Atomic theory

"Atomic theory," wrote Bernard Pullman, "is the most important and enduring legacy bequeathed by antiquity." Indeed, the atomic theory provides a brilliant example of pure speculation which develops into a universally accepted scientific theory. First developed by the Greek philosophers Leucippus and his follower **Democritus,** atomic theory remained purely speculative until the time, in the early twentieth century, when scientists started probing atomic structure through experimental methods. Furthermore, Greek atomism is a striking departure from other cosmogonic theories, such as the idea, suggested by Thales (c. 624-545 B.C.), that everything is derived from water, since atomism did not in any way rely on empirical observation.

Leucippus, who probably lived in the fifth century B.C. is widely considered as the father of atomic theory. Because no writings by Leucippus have come down to us (in fact Epicurus, a prominent atomist, denied his existence), his ideas are difficult to separate from those of Democritus. According to Leucippus and Democritus, the material world consists of invisible and indivisible (in Greek: *atomos*) corpuscles. Atoms are hard, compact, incompressible, infinite in number, and in permanent motion. However, atoms are not all identical: the differ according to shape, arrangement, and position. Building a comprehensive theory of reality on the basis of atoms, Democritus used atoms to explain a wide variety of phenomena, including the structure of the inanimate world, life processes, as well as human memory and perception. No less important than the obvious idea of the atom is the concept of void, which Democritus introduced to explain atomic movement. Later Greek philosophers, such as Aristotle (384-322 B.C.), rejected atomism because they found the idea of a void unacceptable. Atomic theory was further developed by Epicurus (c.341-270 B.C.), who theorized that atoms had weight and moved at a steady speed. His best-known contribution to atomism, however, is the idea of the *clinamen*, or swerve. According to Epicurus, atoms had a tendency to suddenly, and randomly, deviate from a seemingly predictable downward trajectory. These deviations, Epicurus taught, account for the variety of material phenomena. The atomism of Epicurus inspired the Roman poet Lucretius Carus (c. 99-55 B.C.) to compose his *De rerum natura*, a poetic exposition of atomism which is regarded as one of the great works of Roman literature. Epicurus has been criticized, somewhat anachronistically for his atheism, mainly because of his materialism, according to which even the human soul, which consists of atoms, dies with the body. Epicurus never questioned the existence of a divine reality, manifested by immortal gods; however, he taught that the gods were indifferent to the human realm.

Christian thinkers condemned the atomic theory, because they equated materialism with atheism. This bias was strengthened in the Middle Ages, when Christian theology incorporated the teachings of Aristotle, who not only rejected the concept of void, an integral element of the atomic theory, but also accepted, in modified, the theory of four elements. However, atomic theory, although neglected, was not completely forgotten in the Middle Ages. For example, Adelard of Bath, who lived in the twelfth century, defined atomism as a rational explanation of the material world. In a more direct fashion, William of Ockham (1300-1350) stated that the physical reality consisted of elementary particles.

The most prominent representative of atomism in Renaissance was Giordano Bruno (1548-1600). Introducing elements of Platonic idealism into atomic theory, he expanded the traditional definition of the atom, adding specific spiritual attributes, such as the possession of a soul, in an effort to reconcile a corpuscular theory of matter with his grand vision of an essentially spiritual universe. It is ironic that this passionate believer in the divine origin of scientific knowledge was burned as a heretic in Rome.

In the seventeenth century, as scientists and philosophers started turning to empirical research in order test their theories, atomists approached atomic theory as a key to nature's secret. Thus **Robert Boyle**, who was a chemist, strove to establish the connection between particular types of atoms, which he imagined appeared in specific shapes, and certain perceived characteristics of material substances. Isaac Newton (1642-1727) accepted atomism, but raised the question of creation as a result of pure chance which traditional atomic theory implied. For Newton, creation could only the work of a Divine Intelligence, and he accordingly adapted atomism to fit his conception of God's role in creation; however, Newton's universal law of gravitation, which applies to both macroscopic and microscopic phenomena, opened new theoretical vistas which greatly enhanced the development of atomic theory. For example, Rudjer Bošković (1711-1787), building on Newton's theory of universal gravitation, created an atomic theory that was clearly ahead of its time. Namely, Bošković, who accepted Newton's gravitation theory as valid in macrocosmic realm, posited that, as the distance between physical objects diminishes, attraction is replaced by a repulsive force. Thus, while attraction provides the atomic cohesion needed for the construction of physical objects, a repulsive force keeps individual atoms at a certain distance from each other. It is Bošković's conception of an atom—lacking spatial extension, resembling a geometrical point—that conjures up the world of subatomic particles, discovered in the twentieth century. Yet Boškovuć's world view, like Newton's, did not allow the possibility of a universe without God.

In 1789, in his *Elementary Treatise of Chemistry*, **Antoine-Laurent Lavoisier**, widely considered the father of modern chemistry, defined an element as a substance that cannot be chemically analyzed. Lavoisier's fundamentally analytical approach to chemistry set the stage for the great breakthroughs in atomic theory in the nineteenth century. Thus, **John Dalton**, in *A New System of Chemical Philosophy* (1808), asserted, building on Lavoisier's observation that each element was a unique substance, that atoms forming a particular element were similar and had the same weight. In 1815, having observed that the densities of several gases were whole number multiples of the **density** of **hydrogen**, **William Prout** theorized that the atomic weights of different elements were whole number multiples of the **atomic weight** of hydrogen. Throughout the nineteenth century, scientists studied the behavior of gases to strengthen the case for atomic theory. For example, the kinetic molecular theory, proposed independently by **James Clerk Maxwell**, in 1859, and by **Ludwig Boltzmann**, in the 1870s, followed the hypothesis that gases consisted of atoms and molecules in constant motion. Furthermore, **Dmitri Mendeleev** developed his **periodic table** of elements by demonstrating the connection between, on the one hand, an element's atomic weight, and its chemical properties and physical characteristics, on the other.

The doctrine of an atom's indivisibility was shattered in 1897, when **J. J. Thomson**, published his discovery that cathode rays, which are emitted by the cathode, or negative electrode, in a gas-filled tube through which electricity is discharged negatively charged particles whose mass was a mere fraction of an atom's. When Thomson discovered that these particles, named *electrons* by George Johnstone Stoney (1826-1911), were identical regardless of the gas from which they emanated, he realized that what he had found were elements of an atom's inner structure. The existence of electrons was confirmed by numerous subsequent findings, including **Marie Curie**'s demonstration that the beta-particles, emitted by decaying uranium atoms, which **Henri Becquerel** had identified in 1896 as radioactivity, were in fact electrons. Thus, at the end of the nineteenth century, two axiomatic features of the atom were invalidated: indivisibility and stability.

In 1902, Lord Kelvin in an attempt to describe an atom's structure, proposed a model according to which electrons were evenly distributed throughout a positively charged space, effectively creating a balanced state. Two years later, J. J. Thomson modified Kelvin's model, having the electrons move in concentric circle within a positively charged spheric space. However, the Kelvin-Thomson model was challenged by **Ernest Rutherford**, who noted, on the basis of experiments in which thin metal foil was bombarded by alpha particles, or positively charged **ions**, that the characteristic scattering pattern of the alpha particles indicated the existence of a positively charged **nucleus** within the atom. In other words, the alpha particles usually encountered no resistance within the atom, as if the atom were empty, only occasionally ricocheting back. If the Kelvin-Thomson model had been accurate, the alpha particles would have been repelled by a positively charged space. In 1911, Rutherford explained the results of these experiments: most of the atom is, in fact, empty space, the nucleus representing a minuscule fraction of an atom's volume. Rutherford's model provided the foundation for several important discoveries, including the positively charged particle, or particles, constituting the nucleus, which Rutherford named *proton* in 1920. In 1932, James Chadwick demonstrated the existence of neutral particles, or neutrons in the nucleus. Also in 1932, Carl Anderson (1905-1990) discovered the positron, or positively charged **electron**, the first known particle of **antimatter**, anticipating future discoveries of other antiparticles. Building on the pioneering work of **Henry Moseley**, who was killed at Gallipoli, scientists also defined the fundamental concept of atomic number as the number of protons in a given atom's nucleus. In addition, concepts such as atomic number and atomic mass (number of protons plus number of neutrons) enabled scientists to explain the phenomenon of isotopes, or different manifestations of a particular chemical element. For example, with the atomic number remaining constant, different isotopes were defined as consisting of atoms with a varying atomic mass.

However, there were several problems with the Rutherford model, as Rutherford himself admitted. For example, while the planetary model worked for the macroscopic world, in which a planet's orbit is maintained by the equilibrium between gravitation and the centrifugal force, within the atom, where the principal force is the attraction between positively and negatively charged particle, an electron orbiting around the nucleus is bound to gradually lose energy, eventually spiraling into the nucleus. Furthermore, the planetary model fails to explain the particular spectral lines, or characteristic lines of the light spectrum, which particular atoms emit when exposed to heat. If the electron kept losing energy, the emitted radiation would not be constant, effectively excluding the possibility of consistent light patterns.

The atomic model which replaced Rutherford's was based on **Max Planck**'s theory of quanta. In 1900, while observing light emitted by a heated object at various temperature, Planck observed a certain regularity in the vibration of a heated object's atom. In fact, the frequency of a vibrating atom is always the value of a physical constant, later named Planck's constant, multiplied by a whole number. In other words, energy can only be emitted in discrete chunks.

In 1913, **Niels Bohr** used Planck's quantum theory to do develop a new theoretical model of the atom. Bohr's model was based on two fundamental postulates: an electron has specific energy values, or levels; and an electron can change its energy value only by jumping from one energy level to another. According to Bohr, transition between energy levels explained the line spectrum: when an electron dropped to a lower energy level, the lost energy was emitted in the form of a photon, or particle of light. Furthermore, Bohr explained the atom's stability by positing that an electron can rise to a higher energy level: in other words, a gradual loss of energy, eventually leading to an electron's collapse into the nucleus, simply does not occur. An electron's ability to ascend to a higher energy level also explain the phenomenon of light absorption: by absorbing a photon an electron jumps to a higher energy level.

Initially, each electron in an atom was assigned three quantum numbers: the principal quantum number (describing an electron's energy), the angular momentum quantum number (describing the shape of an electron's path), and the magnetic quantum number (describing an electron path's orientation in space). Wolfgang Pauli (1900-1958) introduced a fourth number, the spin quantum number (describing an electron's movement similar to a spin on its axis). Magnetism caused by electron spin was first observed by Otto Stern (1888-1969) and Walther Gerlach (1889-1979) in 1921. In 1924, Pauli formulated his famous exclusion principle, according to which no two electrons in an atom may have all four identical quantum numbers.

The application of quantum theory to the behavior of subatomic particles opened a filed known as quantum mechanics, also termed wave mechanics, because scientists expanded quantum theory to include wave phenomena. Thus, in 1923, Louis de Broglie (1982-1987), generalized **Albert Einstein**'s insight that light has both particle and wave properties by positing that particles of matter have wave properties. Four years later, Lester Germer (1896-1971), Clinton Davisson (1881-1958), and J. J. Thomson's son George Paget Thomson (1892-1975) demonstrated the wave property of electrons when they diffracted an electron bean by using a crystal. This discovery enabled Ernest Ruska (1906-1988) to develop the first electron microscope in 1933. De Broglie formulated a mathematical equation to define the wave behavior of particles; however, since his equation only applied to particles in a force-free environment, it could not work for electrons, which are affected by the attractive force of the nucleus. However, in 1926, **Erwin**

Schrödinger developed a theory which attempted to explain the wave property of electrons. According to Schrödinger, the movement of an electron in relation to the nucleus could not be described as an orbit: in order to conceptualize any such movement, one would need to know both the position and momentum of a particle. This, as **Werner Heisenberg** demonstrated in 1927, is impossible (Heisenberg's uncertainty principle). True, Schrödinger attempted to mathematically track an electron's movement by using a wave function; however, this mathematical function, in light of the uncertainty principle, only yielded the probability of encountering an electron at a given spot. Instead of the misleading term *orbit*, scientists started using the word *orbital* to describe an electron's movement. Scientists have used mathematical to describe orbitals, but in reality their exact shapes cannot be known. "Nevertheless," as Lothar Schäfer has written "we say that these wave forms exist, because *the results of their interference are apparent in observable phenomena.* For example, an important type of interference, and the basis of all chemistry, is that between different atoms when they are forming a chemical bond. The results are apparent in the observable properties of molecules; their structures, for example, depend directly on the atomic wave forms by whose interference they are created."

While the dogma of an atom's indivisibility was refuted in at the end of the nineteenth century, the dogma of invisibility endured almost a century longer, fueled by claims that the creation of an appropriate microscope was precluded by the laws of physics and human physiology. However, in 1981, Gerd Binnig (1947-) and Heinrich Rohrer (1933-) invented the scanning tunneling microscope, which enabled scientists to see individual atoms. In 1986, Binnig and Rohrer received a half of the Nobel Prize for physics; the other half went to Ruska, who invented the first electron microscope.

As scientists continued studying the subatomic world, they realized that, while electrons remained definable as elementary particles, neutrons and protons may be defined as constructs of smaller particles. For example, according to Murray Gell-Mann (1929-) particles such as neutrons or protons consist of smaller particles, which he named quarks. Since, for example, a proton has a charge of +1, a quark would have a fractional charge.

One of the great accomplishments of contemporary atomic theory, as Pullman, has written, is the "remarkable effort to synthesize the two major theoretical contributions of the twentieth century, namely, the theory of relativity, with its fundamental law of equivalence between mass and energy, and quantum mechanics." According to Pullman, the paradigm emerging as a result of this synthesis is the relativistic quantum theory of fields, according to which reality can be defined in terms of interacting fields. "Fields," Pullman asserted, "are the ultimate reality, and there are as many fundamental fields as there are elementary particles."

Among the significant corollaries of the theory of fields, which offers a unified theory of reality, is a profound revision of the concept of vacuum. Thus, if reality is defined as a system of interconnected fields, the dichotomy, postulated by Democritus, between atoms and a vacuum in which they exist becomes meaningless. According to Pullman, vacuum "is a latent state of reality, while matter, made of elementary particles, is its actualized state." Finally, while many scientists accept the field paradigm as intellectually satisfying, the consensus among researchers is that further work in atomic theory will lead to many surprising discoveries.

ATOMIC WEIGHT/MASS

The atomic **mass** of an element is the mass of the number of particles in its **nucleus** given in atomic mass units. Although it is sometimes referred to as atomic weight, this is inaccurate, since weight is a measurement that includes the effect of gravity. One atomic mass unit (amu) is 1.66×10^{-24} grams. This is equal to one-twelfth the mass of an **atom** of carbon-12. This standard was adopted internationally in 1961, replacing an arbitrarily assigned value of 16.000 amu for the atomic mass of an atom of **oxygen**.

The instrument most often used to measure atomic mass is the mass spectrometer (spectrograph). In this instrument, atoms are vaporized and then changed to positively charged particles by knocking off electrons. These charged particles are passed through a magnetic field that causes them to be deflected different amounts, depending on the size of the charge and mass. The particles are eventually deposited on a detector plate where the amount of deflection can be measured and compared with the charge. Very accurate relative masses are determined in this way.

In the early years of the study of the behavior of **matter**, it became generally accepted that mass was conserved in physical and chemical change. By the end of the eighteenth century, studies of compounds suggested that when elements combine, the compounds that form follow the law of constant (or definite) proportions. This law was formally stated by the French chemist Joseph Proust (1754-1826) in 1799: the ratio of the masses of the elements in a compound will always be the same, independent of the method of preparation or the proportions that are mixed in its preparation.

Early in the nineteenth century, **John Dalton**, the English chemist who is sometimes called the father of modern **chemistry**, introduced the concept of atomic mass when he formulated a model of the atom as a round, solid, indivisible particle that represents the elementary unit involved in chemical change. He postulated that all the atoms of the same element have the same atomic mass, while the atoms of a different element have a different atomic mass. He determined that compounds always have the same percentage composition by mass. His explanation was that compounds have a fixed ratio based on the number of atoms of each element in the compound. He expanded on the law of constant proportions, which deals with compounds made of any two single elements, adding to it the law of multiple proportions, to explain situations where more than one compound is formed from the same two elements. The law of multiple proportions states that the ratio of the element to itself in multiple compounds will be a small whole number, like **carbon** monoxide and **carbon dioxide**.

In 1815, the English chemist **William Prout** observed that the atomic weights of several **gases** were whole number multiples of the atomic weight of **hydrogen**. His conclusion based on this experimental result was that every atom is made up of a number of hydrogen atoms that are somehow held together.

In 1864, the English chemist J.A.R. Newlands (1837-1898) and the German chemist Lothar Meyer (1830-1895) recognized independently that there was a connection between the atomic masses of the elements and the periodic repetition of chemical and physical properties. In 1869, the Russian chemist **Dmitryi Mendeleev** tried to arrange the 63 known elements in a pattern that would reflect similar chemical behavior. He found that the arrangement was essentially one of increasing atomic mass with three exceptions where he had to reverse the order of atomic mass in order to maintain the pattern of similar chemical behavior.

After the turn of the century, changes in the model of the atom led to an awareness of a relationship between the atomic mass and the number of particles in the nucleus. At first, the nuclear particles were thought to be only those that carry the positive charges of the atom, but it was soon obvious that, except for hydrogen, the weight of an atom was around twice the weight of the number of protons it contained. **Ernest Rutherford** suggested in 1920 that the atom must have neutral particles (neutrons) whose mass is similar to that of the **proton**. This particle was identified by the English physicist James Chadwick (1891-1974) in 1932.

See also Mole; Molecular weight

AVERY, OSWALD THEODORE (1877-1955)

American bacteriologist

By the early 1940s, scientists knew that chromosomes existed and that they were composed of smaller units called genes. Chemical analysis had revealed that the eucaryotic **chromosome** consists of about 50% protein and 50% deoxyribonucleic acid (DNA). There was no particular interest in **DNA** for several previous decades because no role had been assigned to it. This changed when a Canadian-born American named Oswald Avery showed that DNA is responsible for the transmission of heritable characteristics.

Avery moved from Canada to New York City in 1887. He attended Colgate University, and in 1904, he received his medical degree from the College of Physicians and Surgeons at Columbia University. He practiced medicine for several months before he became more interested in the transmission of infectious diseases. In 1913, Avery arrived at Rockefeller Institute, where he worked as a bacteriologist for over 43 years.

Avery and his coworkers studied the life cycle and chemical make-up of *Diplococcus pneumoniae*, or pneumonocci, a species of bacteria that causes pneumonia. Avery's interest was sparked by the work of Frederick Griffith, and by 1932 Avery focused on transformation—a process by which heritable characteristics of one species are incorporated into another different species. In 1928, Griffith described an experiment in which he injected mice with a mixture of a harmless strain of living pneumonococci and the dead remains of a virulent strain of the bacteria. The mice died from infection by the *live* organisms of the virulent strain, though they had been dead when they were administered.

Oswald Theodore Avery.

In an attempt to duplicate Griffith's work, Avery and his colleagues began to grow large quantities of virulent type III capsulated pneumonococcus. They purified the live virulent encapsulated bacteria and then killed them by extreme **heat**. The bacteria's polysaccharide protein that makes up the capsule or outer envelope was then removed. The remaining portion of the dead bacteria, its polysaccharide gone but capsules intact, was added to living, unencapsulated bacteria. It was found that the offspring of these bacteria had capsules. Avery had determined that the active transforming principle, as Griffith had described earlier, still remained. Because the polysaccharide protein had been removed for the test, it could not be the transforming factor.

Avery wanted to be certain that the active agent was the DNA and not a small amount of protein contamination. To verify the result, a quantity of DNase, an **enzyme** that would destroy the DNA without affecting the protein, was prepared and added to the sample. When a portion of bacteria was tested, it could no longer transform the unencapsulated bacteria into encapsulated bacteria. Avery and his coworkers had conclu-

Amedeo Avogadro.

sively proven that DNA was the transforming principle responsible for the development of polysaccharide capsules in the unencapsulated bacteria.

This experiment, first published in 1944, was extremely important because, for the first time, scientists had proven that DNA controls the development of a cellular feature. It also implicated DNA as the basic genetic material of cells and stimulated James Watson and **Francis Crick** to later discover its structure and method of replication. Today, we understand that DNA is the fundamental **molecule** involved in heredity.

Dr. Avery also worked in the field of immunology, where he determined that polysaccharides play an important role in immunity and helped to develop diagnostic tests used to identify many disease-causing bacteria. He received many awards and honors throughout his distinguished career, including induction into the prestigious National Academy of Sciences and the Royal Society of London.

Avogadro, Amedeo (1776-1856)

Italian chemist

Avogadro was born in Turin, Italy, on August 9, 1776, the son of Count Filippo Avogadro and Anna Maria Vercellone. Count Avogadro was a lawyer, civil servant, and senator for the state of Piedmont. Amedeo followed his father into the law and received his doctorate in ecclesiastical law in 1796.

He practiced law for only a few years, however, before his interests shifted to the sciences. He studied mathematics and physics on his own and, in 1806, was appointed demonstrator in physics at the Academy of Turin. In 1809, he was appointed professor of natural philosophy at the College of Vercelli. A decade later he became the first professor of mathematics at Turin although he lost his chair in the revolution of 1822. He was reappointed in 1835 and remained at Turin until he retired in 1850. He died in Turin on July 9, 1856.

Avogadro's name is best known in connection with his hypothesis about the composition of **gases**. In 1809, the French chemist and physicist Joseph Gay-Lussac had published reports on his research on the combining volumes of gases. Gay-Lussac reported that, under conditions of equal **temperature** and pressure, gases combine with each other in simple whole number ratios. For example, when **hydrogen** and **oxygen** combine to form **water**, they do so in the proportions of two volumes of hydrogen to one **volume** of oxygen, producing two volumes of water vapor in the process.

The conclusion suggested by these results was obvious to Avogadro. Equal volumes of all gases, he said, must contain equal numbers of particles. This statement is known as Avogadro's hypothesis.

The same thought had occurred to **John Dalton**, originator of the **atomic theory**. But Dalton had decided to reject this hypothesis, as well as Gay-Lussac's experimental results. Dalton believed that gases were made of atoms that were in contact with each other. Since he thought that atoms of different elements were of different sizes, then equal volumes could not contain equal numbers of particles. He suggested that Gay-Lussac's research was faulty and carried out experiments of his own that produced results more consistent with his own ideas.

It was Dalton, however, who was in error this time. Gay-Lussac and others were able to demonstrate that the particles in a gas are not in contact with each other and actually make up only a very small fraction of the gas itself. Most of the gas is empty space. So the size and shape of atoms in the gas are irrelevant.

Avogadro made one further suggestion to clarify Gay-Lussac's results. Consider the case with water again. Gay-Lussac showed that: two volumes of hydrogen + one volume of oxygen = two volumes of water.

If we assume the simplest case, in which each volume consists of only one particle, the same result can be expressed as: two particles of hydrogen (H H) + one particle of oxygen (O) = two particles of water (H-O H-O). (At the time, water was thought to consist of one **atom** of hydrogen and one atom of oxygen.)

Obviously, this situation is impossible. One cannot begin with one particle of oxygen and end with two. The solution Avogadro suggested is that the smallest particle of an element can sometimes consist of *two* atoms joined together in a **molecule**, a term he invented for the new particle. Thus, if the smallest particle of oxygen is a *molecule* of oxygen, we can explain Gay-Lussac's results as follows: two particles of hydrodgen (H H) + one particle of oxygen (O-O) = two particles of water (H-O H-O).

We now know the smallest particle of hydrogen also consists of two atoms together. If we draw the reaction with modern forms of the components and products, Avogadro's solution works as follows: H-H H-H + O-O = H-O-H H-O-H.

Avogadro's hypothesis can also be applied to non-gasses. Equal atomic weights of any solid or liquid have the same number of particles. For example, an atom of **carbon** has an atomic weight of 12, while a molecule of water is composed of an atom of oxygen with an atomic weight of 16 plus two atoms of hydrogen with atomic weights of one each for a total of 18. Therefore, twelve grams of carbon will have the same number of particles as eighteen grams of water. This number of particles turns out to be 6.026×10^{23} and is called **Avogadro's number**.

Avogadro's suggestions expanded and improved on Dalton's atomic theory. For a number of reasons, however, his ideas were largely ignored for half a century. Not until Stanislao Cannizarro began to spread Avogadro's ideas in the 1850s did chemists finally understand and adopt them.

AVOGADRO'S LAW

Avogadro's law (sometimes called Avogadro's principle) states that equal volumes of different **gases** at the same **temperature** and pressure contain equal numbers of molecules. **Amedeo Avogadro** (1776-1856), an Italian physicist, formulated this law in 1811 as a direct consequence of Gay-Lussac's law which was articulated by the French chemist **Joseph-Louis Gay-Lussac** (1778-1850) in 1802.

Gay-Lussac's law states that gases combine in simple whole number ratios (all volumes measured at the same temperature and pressure). Avogadro's law explains Gay-Lussac's law without violating John Dalton's idea that atoms are indivisible. Avogadro's law also led to the concepts of **Avogadro's number** and the **mole**, and to the ideal gas law. Avogadro's ideas were largely ignored for half a century until Stanislao Cannizarro (1826-1910) began to promote them in the 1850s.

See also Gas laws; Gases, behavior and properties of

AVOGADRO'S NUMBER

Historically, Avogadro's number (or the Avogadro constant) is the number of particles, atoms, formula units, or molecules, in one **mole** of a given substance. The **metric system** now more precisely defines it as the number of atoms in exactly 0.024 lb (12g) of ^{12}C. In equations, Avogadro's number is given the symbol L, numerically it is equal to 6.023×10^{23}.

Avogadro's number is the number of particles present when the amount of material is the same as the atomic weight (or relative molecular **mass**, or relative atomic mass or weight) expressed in grams. This is one mole of the substance. For example, with **water**, H_2O, the relative molecular mass is 18 (16 for **oxygen** and 1 for each of the two hydrogens), so in 0.036 lb (18 g) of water there are 6.023×10^{23} HOH molecules. With a gas there is a slight difference because the gas may be encountered in the **diatomic** state in the atmosphere. For example, oxygen and **nitrogen** in the atmosphere are generally encountered as O_2 and N_2 respectively. With these diatomic molecules, there is an Avogadro's number of diatomic molecules in the amount of gas that is equivalent to the relative molecular mass. There may be an ambiguity if correct terminology is not applied. One mole of oxygen could refer to one mole of oxygen atoms (6.023×10^{23} atoms) or it could refer to one mole of diatomic, gaseous oxygen, one mole of oxygen molecules (12.046×10^{23} oxygen atoms). The first case is elemental oxygen and the latter is molecular oxygen. A mole of anything contains the Avogadro number of those objects, whether it be a mole of atoms, molecules, or ions.

Amedeo Avogadro was an Italian physicist who lived from 1776-1856, and his main contribution to **chemistry** is Avogadro's hypothesis which gave us Avogadro's Law, which in turn lead to Avogadro's number. These were worked out by looking at volumes of gas reacting together and Avogadro noticed that they always reacted in a constant ratio of whole numbers of volumes of gas. In other words, one **volume** of gas A would always react completely with one volume of gas B or two volumes of gas C. Avogadro realized there was a constant relationship, and by further observation and hypothesizing, he eventually came up with what we now know as **Avogadro's law** and number.

A number as large as Avogadro's number is difficult to comprehend, but the following example is often quoted to give an indication of exactly how large this number is. When standing on a beach with an uninterrupted view to the horizon both left and right, the number of sand grains that are present are still not enough to make one mole of sand grains. Another way of looking at this would be to consider an Avogadro's number of soft drink cans: if they were stacked together, they would cover the entire surface of the earth to a depth of 200 mi (322 km).

The numerical value of Avogadro's number has been confirmed by several different experimental techniques, including **Brownian motion**, electronic charge, and the counting of alpha particles.

Avogadro's hypothesis states that equal volumes of **gases** at the same **temperature** and pressure contain equal numbers of molecules. Avogadro's law states that a gas at constant temperature and pressure has a volume directly proportional to the number of moles of gas. From these two statements, and from what has been previously said, it can be seen that the volume of a gas is directly proportional to the number of molecules present. Also, one mole or an Avogadro's number of molecules of any gas would occupy the same volume as one mole or one Avogadro's number of any other gas, at constant temperature and pressure, irrespective of the size of the molecules considered. Avogadro's number can be used in working out the amount of substance required to completely take part in a chemical reaction. Obviously dealing routinely with numbers of the size of Avogadro's number is unwieldy. To overcome the problems associated with this, the mole is used.

Avogadro's number is not just true for gases, it holds true whatever the state of **matter** under consideration. When a substance is dissolved in a solvent, the strength of **solution** can be discussed in terms of Avogadro's number. When an Avogadro's number of particles is dissolved into 0.264 gal (1 l) of solvent the strength of the solution is one molar. This is the **molarity** of the solution. **Molality** is a similar concept to molarity. Instead of dealing with the volume of the solution or solvent, molality is concerned with the mass of solution or solvent. As such a one molal solution has one mole dissolved in 2.205 lb (1 kg) of solution. A one molal solution has an Avogadro's number of particles of the solute.

Avogadro's number is a conversion factor between the number of moles present and the actual number of physical particles. For example, the number of particles present in 0.1 lb (50 g) of **carbon** dioxide can easily be calculated. The **molecular weight** of **carbon dioxide** is 12 for the carbon plus 16 for each of the two oxygens, which totals a molecular weight of 12 + 16 + 16 = 44. One mole of carbon dioxide would weigh 0,088 lb (44 g). Moles are calculated by dividing the weight by the weight of one mole, this is 50 divided by 44, which equals 1.14. In this example, the result is 1.14 moles of carbon dioxide in 0.1 lb (50 g) of carbon dioxide. The number of molecules of carbon dioxide is the number of moles multiplied by Avogadro's number, or $1.14 \times 6.023 \times 10^{23}$ which gives us an answer of 6.866×10^{23}. In other words, in 50 g or 1.14 moles of carbon dioxide there are 6.866×10^{23} molecules of carbon dioxide. The calculations listed here will work in any direction relating moles to particles to weights of materials, simple rearrangement of the equations will provide the appropriate answer.

Avogadro's number is a powerful and useful concept. It illustrates how much of a material is actually participating in a reaction at a basic level. The numbers that it generates are too large to comprehend with any validity so the mole concept is used. One mole of substance contains an Avogadro's number of particles. It is easier to visualize one mole simply because the number one is a more comfortable number to work with than 6.023×10^{23}, even though, by definition, the quantities involved are exactly the same. Conversion is easy between Avogadro's number of particles, moles, weights and also the figures used in chemical equations. Such figures help to optimize a reaction so reactants are not overused.

Avogadro's number is a measure of the number of particles of any type that are present in one mole of a substance. It is numerically equal to 6.023×10^{23}.

See also Avogadro's law; Molality; Molarity; Mole

AXELROD, JULIUS (1912-)

American biochemist and pharmacologist

Julius Axelrod is a biochemist and pharmacologist whose discoveries relating to the role of neurotransmitters in the sympathetic nervous system earned him the Nobel Prize in physiology or medicine in 1970, together with **Ulf Euler** of Sweden and Sir **Bernard Katz** of Great Britain. As Axelrod himself has said, he was a late starter as a distinguished scientist, due to both the humble circumstances of his birth and his coming of age in the Great Depression of the 1930s. He only began real scientific research in 1946, and earned his Ph.D. in 1955. From then on he compensated for lost time and became the first chief of the pharmacology section of the National Institute of Mental Health, a branch of the prestigious National Institutes of Health.

Axelrod was born on May 30, 1912, in a tenement house in New York City, the son of Isadore Axelrod, a maker of flower baskets for merchants and grocers, and Molly Leichtling Axelrod. His parents had immigrated to the United States from Polish Galicia in the early years of the century, met and married in New York, and settled in the heavily Jewish area of the Lower East Side of Manhattan. Julius Axelrod attended public elementary and high schools near his home but later recalled that he got his real education in the neighborhood public library, reading voraciously through several books a week, everything from pulp novels to Upton Sinclair and Leo Tolstoy. He studied for a year at New York University, but when his money ran out he transferred to the tuition-free City College of New York, from which he graduated in 1933 with majors in biology and **chemistry**. He later claimed that he did most of his studying on the long subway rides between his home and the uptown Manhattan campus of City College.

Axelrod applied to several medical schools but was not admitted to any. It has been widely reported, in the *New York Times,* for example, that he failed to get into medical school because of quotas for Jewish applicants. It was difficult to find any work in New York in the depths of the Depression, and Axelrod was fortunate to find employment in 1933 as a laboratory assistant at the New York University Medical School at $25 per month. In 1935 he took a position as chemist at the Laboratory of Industrial Hygiene, a nonprofit organization set up by the New York City Department of Public Health to test vitamin supplements added to foods. He married Sally Taub on August 30, 1938, and they eventually had two sons, Paul Mark and Alfred Nathan. Axelrod took night courses and received an M.A. in chemistry from New York University in 1941. In the early 1940s he lost the sight of one eye in a laboratory accident.

Axelrod later speculated that he might have remained at the Laboratory of Industrial Hygiene for the rest of his working life. The work, he said, was moderately interesting, and the pay adequate. However, in 1946, quite by chance, he received the opportunity to do some real scientific research and found it exciting. The laboratory received a small grant to study the problem of why some persons taking large quantities of acetanilide, a non-aspirin pain-relieving drug, developed methemoglobinemia, the failure of hemoglobin to bind **oxygen** for delivery throughout the body. Axelrod, who had little experience in such work, consulted Dr. Bernard B. Brodie of Goldwater Memorial Hospital of New York. Brodie was intrigued with the problem and worked closely with Axelrod in finding its solution. He also found Axelrod a place among the research staff at New York University. The two men soon discovered that the body metabolizes acetanilide into a substance with an analgesic effect, and another substance that causes methemoglobinemia. They recommended that the beneficial metabolic product be administered directly, without the use of acetanilide. Related **analgesics** were investigated in the same manner.

In 1949, Axelrod, Brodie, and several other researchers at Goldwater Hospital were invited to join the National Heart Institute of the National Institutes of Health in Bethesda, Maryland. There Axelrod studied the physiology of **caffeine** absorption and then turned to the sympathomimetic amines, drugs which mimic the actions of the body's sympathetic nervous system in stimulating the body to prepare for strenuous activity. He studied such compounds as amphetamine, mescaline, and ephedrine and discovered a new group of enzymes which allowed these drugs to metabolize in the body. By the mid 1950s, Axelrod decided that he needed a doctorate to advance in his career at the National Institutes of Health. He took a year off to prepare for comprehensive examinations at George Washington University in the District of Columbia, submitted research work he had already done to satisfy the thesis requirements, and received a Ph.D. in pharmacology in 1955, at the age of forty-three. He was then offered the opportunity to create a section in pharmacology within the Laboratory of Clinical Sciences at the National Institute of Mental Health, another branch of the National Institutes of Health. He became chief of the section in pharmacology and held that position until his retirement in 1984.

In 1957 Axelrod began the research which eventually led to the Nobel Prize. He and his colleagues and students studied the manner in which neurotransmitters, the chemicals which transmit signals from one nerve ending to another across the very small spaces between them, operate in the human body. In the 1940s the Swedish scientist Ulf von Euler had discovered that noradrenaline, or norepinephrine, was the neurotransmitter of the sympathetic nervous system. Axelrod was concerned with the way in which noradrenaline was rapidly deactivated in order to make way for the transmission of later nerve signals. He discovered that this was accomplished in two basic ways. First, he found a new **enzyme**, which he named catechol-O-methyltransferase (COMT), which was essential to the **metabolism**, and hence the deactivation, of noradrenaline. Second, through a series of experiments on cats, he determined that noradrenaline was reabsorbed by the nerves and stored to be reused later. These seemingly esoteric discoveries in fact had enormous implications for medical science. Axelrod demonstrated that psychoactive drugs such as antidepressants, amphetamines, and cocaine achieved their effects by inhibiting the normal deactivation or reabsorption of noradrenaline and other neurotransmitters, thus prolonging their impact upon the nervous system or the brain. His experiments also pointed the way to many new discoveries in the rapidly growing field of neurobiological research and the chemical treatment of mental and neurological diseases. The 1978 Nobel Prize in physiology or medicine, shared with Ulf von Euler and Bernard Katz, crowned his achievements in this area.

In his later years, Axelrod has worked in many areas of biochemical and pharmacological research, notably in the study of **hormones**. Especially important to the advancement of medical science was his development of many new experimental techniques which could be widely applied in the work of other researchers. He also had a great impact through his training of and assistance to a long line of visiting researchers and postdoctoral students at the National Institutes of Health. He continued his own research at the National Institute of Mental Health following his formal retirement in 1984. Early in 1993 Axelrod had the unusual experience of having his own life saved through a scientific discovery he had made many years before. At the age of eighty, he suffered a massive heart attack. The cardiologists at Georgetown University Medical Center soon determined that several of his coronary arteries were almost completely blocked by blood clots and that he must have immediate triple coronary-artery bypass surgery. The complication was that his blood pressure had fallen so dangerously low that he might not survive the operation. The solution to this crisis was to inject a synthetic form of noradrenaline to stimulate the contractions of his heart and thus raise his blood pressure to a more acceptable level. Axelrod survived the operation and within two months was back at work and attending conferences in foreign countries.

Azeotrope

One of the oldest of all chemical techniques is **distillation**. An example of distillation in its simplest form is boiling **water**. Indeed, boiling water and then condensing it on a cold surface is one of the best methods for **water purification**.

With water as the solvent, the **boiling point** of a **solution** is relatively constant (slight variations can occur due to the presence of impurities at high concentrations). And while this is generally the case for any pure solvent, it is not the case for a mixture of two or more **solvents**. In a mixture, the boiling point is dependent upon the relative concentrations of all of the species or, more accurately, on their **mole** fractions.

In the case of mixed solvents, the ability of one component to vaporize can exceed the ability of any other component to vaporize. To understand this, it is necessary to consider what vaporization requires. A **molecule** must approach the surface of solvent with sufficient **velocity** of kinetic **energy** to overcome the inter-molecular interactions that hold it to the surrounding molecules. It is a bit like someone struggling to get clear of a crowd—their ability to escape depends upon their strength and determination.

In the case of molecules, the kinetic energy of the molecule is dependent upon the **temperature**. But the ability of a molecule to escape is also dependent upon its **mass** and the type of interactions it has with the other solvent molecules.

All of this is an explanation as to why the vapor composition above a mixture of solvents does not necessarily have the same chemical composition as the solvent itself. But a curious thing happens during a distillation. If the more volatile solvent evaporates more quickly—if the vapor is enriched in one of the components—then the remaining solution is depleted in that component. That is, if a solution consists of a two component mixture, A and B, and A is more volatile, then the mole fraction of A in the vapor will be higher than the mole fraction of A in the liquid, and vice versa for B.

The result is that the composition of the solution changes as the distillation occurs. At some point though, the relative rate of vaporization of the components becomes equal. That is, A may be more volatile but there is less of it in solution. When this occurs, the vapor is no longer enriched in A. Rather, the vapor has the exact same composition as the solution. This is the azeotrope—the point in a distillation curve where the chemical composition of the vapor and the solution are the same.

Azeotropes or azeotropic mixtures occur for a number of important chemical compounds but perhaps the most important is the ethanol-water mixture. Distillation of **ethanol** mixtures is the basis of the distillery industry which makes such alcoholic beverages as bourbon and gin. However, industrial ethanol is "95%" as this is the azeotropic mixture for the combination of pure ethanol and water. That is, if yeast are cultivated in a sugar water solution, the resulting **alcohol** that can be distilled is the azeotropic mixture consisting of 95.6% ethanol and 4.4% water.

One of the many industrial uses of ethanol involves blending it with **gasoline** to make gasohol. There are several reasons for doing this. It produces a cleaner **combustion** and reduces the consumption of non-renewable resources. However, ordinary ethanol—the 95% ethanol-water azeotrope—is unsuitable for this use as the water interferes with the gasoline combustion and prevents the ethanol from being miscible with the gasoline.

This problem is solved by taking advantage of another azeotrope. The benzene-ethanol-water ternary azeotrope boils at a lower temperature than the ethanol-water azeotrope. By introducing **benzene** to a water ethanol mixture, the water can be extracted through distillation leaving behind the pure ethanol which contains a minor amount of benzene as an impurity. The result is that almost pure ethanol can be obtained which is suitable for blending with gasoline. (In fact, the benzene in the ethanol is an octane booster. The disadvantage of this fuel additive is that the presence of any water in the gas tank will result in the separation of the ethanol as the water-ethanol mixture will be reestablished.)

The industrial use of azeotropes to distill or purify mixtures of **liquids** is one of the more important aspects in any distillation process.

Azo, diazo, azine, and azide groups

Each of these groups is characterized by the presence of nitrogen-nitrogen bonds. The azo group contains two **nitrogen** atoms joined by a triple bond (R—N N—R); the diazo group has two nitrogens with a double bond (R=N=N); the azine group has a **carbon** double bonded to a nitrogen which is single bonded to another nitrogen and this is double bonded to another carbon ($R_2C=N—N=CR_2$); and the azide group has three nitrogens and two double bonds (R—N=N=N). The names of these groups are derived from the Greek word *azote* meaning "without life," the name for nitrogen proposed by French chemist **Antoine-Laurent Lavoisier**. (Lavoisier proposed this name due to the inertness of elemental nitrogen in chemical reactions.)

Many of the compounds containing these groups are extremely good dyes. The are used to dye fabrics such as wool and cotton, and also as dyes in photographic processes. The alkali salts of these groups are generally stable, although there is a tendency for them to degrade on heating. Heavy metal salts of these compounds are generally very explosive and are used in the manufacture of such devices as percussion caps.

The **chemistry** of these groups is related to the multiple-bonded nitrogens. Splitting across this bond can occur to yield such products as amines. These compounds can be prepared from **reduction** of nitro compounds (azo), by the action of **nitric acid** on amines (diazo), from compounds containing the **carbonyl group** (azines), and from the salts of hydrazoic acid (azide).

B

BACHRACH, HOWARD L. (1920-)

American biochemist and molecular biologist

Howard L. Bachrach has been awarded more than two dozen honors, including the National Medal of Science in 1983, for his pioneering research in the molecular biology of viruses. After earning a bachelor's degree in **chemistry** from the University of Minnesota in 1942, he chose to specialize in **organic chemistry** and **biochemistry** and received his Ph.D. in these fields in 1949. Bachrach spent the war years doing research on the development of chemical **explosives** and then was sent by the U.S. Department of Agriculture (USDA) to Denmark to learn more about foot-and-mouth disease (FMD). In 1953, Bachrach accepted an appointment as head of biochemical research at the USDA's Plum Island Animal Disease Center, an affiliation he maintained for the next four decades. During the 1970s, he developed a method for producing FMD vaccine by means of recombinant **DNA** (genetic engineering) techniques.

Bachrach was born on May 21, 1920, in Faribault, Minnesota. His parents were Elizabeth P. and Harry Bachrach, the owner of a clothing store for men and boys. Bachrach attended Faribault High School, from which he graduated as salutatorian in 1938. That fall, he entered the University of Minnesota, to major in chemistry. He graduated with a B.S. *cum laude* in that field four years later, having also been inducted into the Phi Lambda Upsilon national honorary fraternity in chemistry and the Gamma Alpha honorary fraternity in science. After graduation, Bachrach worked briefly for the Joseph E. Seagram Company before joining the war effort with the Office of Scientific Research and Development at the Carnegie Institute of Technology. His first work there involved research on chemical explosives. He was later assigned to a problem closer to his own field, being asked to study the chemical changes that take place as bread becomes stale. This research had been requested by the Quartermaster Corp of the U.S. Army in its attempt to find ways of preserving food for longer periods of time.

At war's end, Bachrach returned to the University of Minnesota, where he undertook a doctoral program in biochemistry. He completed that program in 1949 and received his Ph.D. His doctoral thesis involved a study of the virus that causes cholera in hogs, a disease that cost the swine industry millions of dollars in losses each year. Bachrach was able to demonstrate that hog cholera is produced not only by the virus itself, but also by a soluble protein that it produces.

Bachrach's background in viral immunology made him a logical candidate for important research then going on in connection with FMD. Prior to the 1930s, this disease had caused enormous losses in the U.S. livestock industry. It had been brought under control, however, and no cases had been reported in this country since 1929. During the 1940s, however, the disease had reappeared in Mexico and was spreading rapidly. The USDA became concerned about the possible spread of the disease into the United States and had begun a crash program to find ways of protecting the U.S. livestock industry against its reappearance here.

As part of that program, the USDA invited Bachrach to spend a year in Denmark studying at the USDA's European Commission on Foot-and-Mouth Disease laboratories. That experience provided him with invaluable knowledge about the FMD virus and about viral immunology in general.

Having completed his year in Denmark, Bachrach accepted an appointment at the University of California's Virus Laboratory in Berkeley, where he worked with Nobel Laureate biochemist **Wendell Meredith Stanley**. At Berkeley, Bachrach's major accomplishment was the first purification of the poliomyelitis virus. He was also able to obtain the first pictures of the virus using the university's **electron** microscope.

At the conclusion of his three years at the Virus Laboratory, in 1953, Bachrach was offered an appointment as head of the Chemical and Physical Investigations Section of the USDA's Plum Island Animal Disease Center in Greenport, New York. There he continued his efforts to develop a vaccine for FMD. Among the many discoveries he made during his four decades at Plum Island was that certain portions of the virus known as capsid **proteins** are able to produce an immune response in an organism even though the proteins themselves are not infectious. Working with scientists at the Genentech Corporation, he was able to incorporate these proteins into carrier molecules by means of gene splicing techniques. The resulting product was the first effective vaccine for use in humans or other animals produced by **genetic engineering** techniques.

The techniques developed by Bachrach hold promise for the development of other types of viral vaccines. One line of research, for example, involves the search for a human immunodeficiency virus (HIV) vaccine, one that involves the incorporation by gene splicing of a capsid protein from the virus into a carrier **molecule**.

In 1961 Bachrach became chief scientist at Plum Island, besides continuing in his post as the head of biochemical research. Twenty years later, Bachrach ended his most intensive duties at Plum Island but maintained his relationship with the research center. Although officially retired, he continues his research on viral diseases there. From 1981 on he has also been particularly active as a consultant to a number of organizations, including the Walter Reed Army Institute for Research, the Office of Technology Assessment, the National Research Council, the National Cancer Institute, and the Texas A&M University Institute of Biosciences and Technology.

Bachrach was awarded the National Medal of Science in 1983 for his work in molecular virology and his role in developing gene-splicing techniques. He has also been awarded a USDA Certificate of Merit (1960), a U.S. Presidential Citation (1965), the AAAS-Newcomb Cleveland Prize (1982), the USDA Distinguished Service Award (1982), and the Alexander von Humboldt Award (1983). He was elected to the National Academy of Sciences in 1982 and to the USDA's Agricultural Research Service's Science Hall of Fame in 1987.

Bachrach was married to Shirley Faye Lichterman on June 13, 1943. They had two children, Eve Elizabeth, an attorney, and Harrison Jay, a physician. Bachrach lists his hobbies as golf, walking, gardening, photography, and his family.

BACKGROUND RADIATION

Background radiation is the low intensity radiation from the small amounts of radioisotopes in the environment to which we are all exposed.

Background radiation is a result of naturally occurring radioactive decay from a number of elements found in the air, rocks, soil, and in all living things, as well as a component of bombardment by cosmic rays. There is also an amount that has been added by humans. This latter radiation comes from all uses that **radioactivity** is put to, from nuclear bombs to power stations, x-ray machines, and irradiation of food. The greatest component of background radiation is naturally occurring **radon** gas. This accounts for 55% of all background radiation. The next largest component at 11% is medical x-rays, which is the same level of contribution as radioisotopes naturally occurring in the human body. The level of exposure to background radiation for a person for one year is approximately 360 mrem. A typical dental x-ray would be the equivalent of 0.5 mrem. There are no detectable clinical effects for a short-term exposure to ionizing radiation below 25 rem. Since the 1940s when nuclear material started to be used by humans the level of background radiation has increased by a tiny amount. This increase has to be taken into account when calculating dates using the carbon-14 dating technique, but, other than such techniques that rely on measuring minute quantities of radioactive material, the increase in background radiation has had no noticeable affect.

Background radiation is mostly natural levels of radiation which have been around throughout most of the history of the planet.

See also Carbon dating; Radiation chemistry; Radioactivity

BAEKELAND, LEO (1863-1944)

Belgian chemist

Leo Hendrik Baekeland was born November 14, 1863 in Ghent, Belgium. He graduated from high school first in his class when he was 16 and earned a scholarship to the University of Ghent. At 21 he received his doctorate degree with highest honors. He arrived in the United States by way of a traveling fellowship, settled in the state of New York and went to work for a photographic firm.

In 1891, Baekeland perfected the manufacturing process for ''Velox,'' a gelatine **silver** chloride **paper** invented by Josef Eder (1885-1944). The paper made it possible to develop photographic prints under artificial light. He sold his invention to George Eastman, owner of *Kodak*, for one million dollars in 1899.

Baekeland bought a home in Yonkers, New York, with his fortune and built a laboratory where he began experiments in **electrochemistry**. He was granted patents for his work with electrolytic cells. Baekeland then began searching for a substitute for shellac which at the time was an entirely natural product. Baekeland felt the market was ready for a cheaper substitute.

Baekeland centered his research on finding a solvent that would dissolve a resinous substance formed by a **condensation reaction** of **formaldehyde** with phenol. As noted by Baeyer in the 1800s, this tacky residue was nearly impossible to remove from laboratory glassware. Baekeland felt that a solvent capable of breaking down this residue would have to possess the shellac-like properties he was looking for.

After long research turned up no appropriate solvent, it occurred to Baekeland that a residue impervious to **solvents** might be more than a nuisance after all. He began attempting to *create* an impervious resin. He built a reaction vessel, which he called a Bakelizer, and began experimenting with the phenol-formaldehyde reaction. By controlling the chemical proportions, catalysts, pressure, and **temperature**, he eventually succeeded in forming a clear solid that was **heat**, **water**, and solvent resistant, and nonconductive. Baekeland patented the solid in 1907, naming it Bakelite. It could be easily machined and could be dyed any **color** with no adverse effects on its physical properties. Bakelite was first used in automotive applications. Soon it replaced hard rubber and amber for electrical uses and is still used in industrial arts where thermoplastics are unsuitable. Though Baekeland did not fully understand the chemical structure of his invention his careful record keeping and observation during his experiments made his search a success.

Bakelite was the first totally synthetic plastic and the first thermoset plastic. After Bakelite was successfully developed, the search was on for other artificial substitutes for natural materials such as rubber and silk. By the 1940s, this research began to pay dividends—and today, petrochemical **plastics** and fabrics are important in almost every aspect of our daily lives.

Baeyer, Johann Friedrich Wilhelm Adolf von (1835-1917)

German chemist

Johann Friedrich Wilhelm Adolf von Baeyer was a German organic chemist best known for synthesizing a wide variety of important compounds, including barbituric acid and indigo. Additionally, Baeyer also conducted research on phthalein dyes, concentrating his later research efforts to expanded knowledge of synthetic compounds and to develop a theory explaining the stability of five- and six-carbon rings. For his accomplishments in compound synthesis, Baeyer was awarded the 1905 Nobel Prize in **chemistry**.

Baeyer was born in Berlin on October 31, 1835. His father, Johann Jacob Baeyer, was an officer in the Prussian army who also conducted geodetic surveys for the Prussian government, and his mother, Eugenie Hitzig, was the daughter of a prominent authority on criminal law and historian of literature. Baeyer developed an interest in science at an early age, and chemistry was the subject that intrigued him most. In his autobiography, *Erinnerungen aus meinem Leben,* he reports that he carried out his first chemical experiments at the age of nine and, three years later, discovered a previously unknown double carbonate of **copper** and **sodium**. At age thirteen, Baeyer performed his first experiments with indigo, the compound that later made him famous.

Baeyer attended the Friedrich Wilhelm Gymnasium in Berlin, where he assisted his science teacher with chemistry lectures. Upon his graduation in 1853, Baeyer entered the University of Berlin, intending to major in mathematics and physics. Two years later he left the university for a year of military service and then decided to continue his college education. By now Baeyer was committed to a program in chemistry and elected to attend the University of Heidelberg as Berlin had no chemistry laboratories. Heidelberg's chemistry department was headed by **Robert Bunsen**, a brilliant scholar responsible for developing such things as the electric cell and the **Bunsen burner**. Bunsen, however, had little interest in **organic chemistry**, the field that had become Baeyer's passion. As a result,

Leo Baekeland. *(Corbis Corporation. Reproduced by permission.)*

Baeyer soon found himself gravitating toward the laboratories of **Friedrich Kekulé**, a German chemist known for his work on organic compounds. At this time Kekulé was a *privatdozent* at Heidelberg and one of the few organic chemists of the time. Although his collaboration with Kekulé was interesting and productive, Baeyer returned to Berlin in 1858 to complete his doctoral work on the compound known as cacodylic (arsenic methyl chloride).

After receiving his degree, Baeyer returned to work with Kekulé, who had by now accepted a call to the University of Ghent. Two years later, in 1860, Baeyer returned to Berlin and took a position at the Berlin Institute of Technology. Although his salary was very low, he had a large, relatively well-equipped laboratory in which to work and he remained at the institute for twelve years. It was during this time that he made most of his important discoveries, the first of which was developing a derivative of uric acid, in 1863. Barbituric acid is the parent compound of a group of drugs known as the **barbiturates**, which have a number of medical applications.

It was also at this time that Baeyer began his classic research on the dye indigo. For centuries the beautiful blue dye had been obtained from the plant of the same name, but its extraction was a costly process. By 1866 Baeyer had found a method for determining the approximate **structural formula** of indigo and then, four years later, first produced the compound

Johann Friedrich Wihelm Adolf von Baeyer.

synthetically in his laboratory. Baeyer continued to work on indigo for another two decades, finally announcing an even more precise formula for the compound in 1883.

Baeyer's years at the Berlin Institute of Technology were marked by a number of other important discoveries. In 1871, for example, he first reported on the structures of the dyes phenolphthalein and fluorescein. One of his students, Karl Graebe, determined the structure of another important dye, alizarin, by means of a technique outlined by Baeyer and after Baeyer had ordered him to do the experiment. In 1872 Baeyer was called to be professor of chemistry at the University of Strasbourg, the first significant academic appointment in his career. His most notable accomplishment at Strasbourg was the development of methods for preparing condensation products from the reaction between phenol and **formaldehyde**. Three decades later, **Leo Baekeland** would adapt Baeyer's method in the synthesis of a phenol-formaldehyde resin known as bakelite, one of the world's first commercial **plastics**.

Baeyer reached the pinnacle of his career in 1875 when he was appointed professor of organic chemistry at the University of Munich. In the forty years he held this post, Baeyer wielded enormous influence on his profession, largely through the students he trained but also as a result of his continued research. His most important achievement during this period was the theory of strain that he developed to explain the structure of **ring** compounds. Extending Kekulé's work on the tetrahedral **bonding** of the **carbon atom**, Baeyer concluded that five- and six-membered rings are dynamically more stable than are rings with greater or lesser numbers of atoms, thus accounting for the much larger number of compounds of the former type.

Baeyer continued to perform most of his academic duties into his eightieth year. He then retired to his country house near Lake Starnberg, where he died on August 20, 1917. He had been married in 1868 to Lida Bendemann, and the couple had three children: Eugenie, Hans, and Otto. In addition to the 1905 Nobel Prize in chemistry, the honors accorded Baeyer included the Liebig Medal of the Congress of Berlin Chemists and the Davy Medal of the Royal Society. In 1885, King Ludwig II of Bavaria made Baeyer a member of the nobility, allowing him to add the honorific ''von'' to his name.

BALANCING EQUATIONS

All **matter**, for practical purposes, is composed of atoms. They are discrete entities and indivisible under normal circumstances. This has a profound effect on all of **chemistry** in that any compound must be composed of a whole number of atoms. That is, **oxygen** may occur as a compound containing 2 oxygen atoms (which is commonly referred to a molecular oxygen) or a compound containing 3 oxygen atoms (called ''ozone'') but never as a compound containing 2.5 oxygen atoms. Fractional coefficients can be used in chemical formulas but they have very specific meanings. Discrete molecular entities must always contain a whole number of atoms.

Another important property of atoms is that they can not change into one another. That is, under normal circumstances, an **atom** of oxygen will always be an atom of oxygen (see **nuclear chemistry** for exceptions). Certainly within the context of ordinary chemical reactions this must be the case.

The result of these two properties is that balance must always be maintained in chemistry between the reacting species and the products of the reaction. If three oxygen molecules go into a reaction then six oxygen atoms must be found somewhere in the products. Matter is neither created nor destroyed, only transformed.

An example of a balanced equation is the **combustion** of **hydrogen** to produce water: $2H_2 + O_2 \rightarrow 2H_2O$. By convention, this equation is written with whole numbers as the coefficients for the species. This reaction is very simple but it illustrates ''balance.'' Each side of the reaction contains four hydrogen atoms and two oxygen atoms. The same atoms appear on both sides and the reaction balances.

The process of balancing equations is referred to as **stoichiometry** and is an important aspect of all chemical reactions. The number of atoms in the reactants must always correspond to the number of atoms in the products. However, sometimes balancing reactions can be confusing if there are polyatomic species present. For example, the **phosphate ion** is a discrete ionic entity, a polyatomic ion, in which the atoms are covalently bound. It is balanced as a single unit much in the same way as the other atoms are treated.

In this case, the number of atoms on each side is the same (i.e. three **sodium** ions are on each side) but the phosphate ion can be treated as a single species for the purposes of keeping track of the atoms. Thus, each side of the reaction has only a single phosphate group.

A slightly more complicated example is the following reaction: $3Cu + 8 HNO_3 \rightarrow 2NO + 3Cu(NO_3)_2 + 4H_2O$

Again, it is fairly straightforward to verify that there are the same number of each type of atom on each side of the reaction. There are 3Cu, 8N, 8H, and 24O. But the discrete grouping of atoms in the nitrate ion ($NO3^-$) has undergone a reaction to give the **nitric oxide** (NO) and **water**. Still, balance is maintained throughout the reaction. Matter is neither created nor destroyed.

Balance, in chemistry, is also found in such things as the total **mass** of the elements involved in a reaction. This is a simple extension of the principle that all of the atoms must be conserved. It was, in fact, the historical observation that preceded the notion of atoms and balanced equations. By careful measurement of the mass of reactants before a reaction and the mass of the products after, chemists were able to devise the law of definite proportions and the notion that matter is conserved.

See also Atom; Matter; Stoichiometry

Baltimore, David (1938-)

American molecular biologist

David Baltimore spent his career working with viruses. His research involved finding the relationship between cancer viruses and the **DNA** of the cells they infect. In 1975, Baltimore shared the Nobel Prize in Physiology or Medicine with Renato Dulbecco, an Italian-born virologist and Howard M. Temin, an American virologist. In 1965, Temin, an assistant professor of oncology at the University of Wisconsin, proposed for the first time a process called reverse transcription. During this process, viral **RNA** inserts its own genes into its host cell's DNA. At the Massachusetts Institute of Technology while studying Rauscher mouse-leukemia virus, Baltimore tested Temin's hypothesis and discovered an RNA viral **enzyme** that alters the host DNA. Temin also found a similar enzyme in the Rous-sarcoma virus. Viruses that alter host DNA in this manner are called retroviruses, and human immunodeficiency virus, the virus that causes AIDS is an example of a retrovirus. The enzyme involved in reverse transcription is known as reverse transcriptase. Baltimore and Temin's work on reverse transcriptase showed how a retrovirus disrupts a cell's replication mechanism and causes cancer. Their findings led to certain cancer treatments.

Born in New York City, Baltimore became interested in science as a child, particularly mathematics and biology. While in high school, he attended summer school at Jackson Memorial Laboratory, Bar Harbor, Maine, which sparked his interest in biological research. In 1956, Baltimore began as a biology major at Swarthmore College in Pennsylvania and later switched to **chemistry**. His interest in **biochemistry** developed after a summer at Cold Spring Harbor Laboratories working with George Streisinger. Baltimore graduated from Swarthmore in 1960 and received high honors in chemistry. He attended graduate school at the Massachusetts Institute of Technology in the field of biophysics, and then transferred to Rockefeller University in New York. There, he worked on his thesis and continued his research in animal virology. In 1964, he received his Ph.D. from Rockefeller University. From 1964-1965, he studied virus-specific enzymes with Dr. Jerard Hurwitz at the **Albert Einstein** College of Medicine in New York.

David Baltimore.

Baltimore worked as a research associate from 1965-1968 at the Salk Institute for Biological Studies in California. His colleague, Renato Dulbecco, was studying the differences between normal cells and viral-induced tumor cells. He then joined the faculty at the Massachusetts Institute of Technology in 1968, and was named full professor in 1972. In 1974, Baltimore conducted research at the MIT Center for Cancer Research where Salvador Luria was the director. There he focused on the relationship between viruses and cancer.

In 1990, he was appointed president of Rockefeller University in New York, but was forced to resign the following year as a result of a paper written in 1986 containing data that his associate, Thereza Imanishi-Kari, a Tufts University professor, allegedly falsified. The NIH Office of Scientific Integrity investigated the scientific behavior of three of the authors.

The case drew national attention and took on political overtones. Baltimore was never accused of falsifying data, but a committee from the House of Representatives found the fraud charges against Imanishi-Kari believable. Nevertheless, the prestigious name of David Baltimore and the questionable credibility of the Tufts graduate student who was the whistle blower in the case left the matter unresolved. Finally, in 1996, 10 years after the paper's publication, a federal appeals panel rejected the fraud charges against Imanishi-Kari. David Baltimore continues to maintain the position of Professor of Biology at MIT. In 1998, he was appointed President of Caltech in California. He is also serves as a Professor of Microbiology for the American Cancer Society and Chairman of the AIDS Vaccine Research Committee of the National Institutes of Health.

Barbiturates

Barbiturates are derived from barbituric acid, an **alkaloid** with nitrogen-containing rings. Belonging to a class of mood-altering drugs, barbiturates induce relaxation and sleep. By manipulating the central nervous system, barbiturates can effect degrees of behavioral depression from mild sedation to coma, and, if used improperly, can cause death. Barbiturates depress the activity of nerves, muscles, heart tissue, and the brain. They can impair a person's ability to engage in rational thought, thus diminishing their reasoning capacity.

The use of depressants likely began with **alcohol** consumption. Alcohol was once used as a remedy or anesthetic for practically all diseases and problems; in the Middle Ages, alcohol was viewed as a life-giving elixir. Many barbiturates have been prescribed to treat problems for which alcohol was once administered.

The first barbituric acid was prepared in 1864 by Adolf von Baeyer, but it was not until 1903 that his student, **Emil Fischer**, introduced the first barbiturate derivative for use as a sedative. Fischer produced 5.5-diethylbarbituric acid, a hypnotic and sedative known by the trade names Barbital, Veronal, and Dorminal. By 1912, a phenylethyl derivative was developed and commercially introduced as Phenobarbital and Luminal. Since then, more than 2,500 barbiturates have been synthesized, of which more than 50 have been marketed.

Doctors often prescribe barbiturates to help patients relax during times of great stress or to help patients suffering from any number of anxiety disorders; however, as barbiturates are highly addictive and may trigger severe depression, they are usually dispensed on a short-term basis and their use is closely monitored. Severe withdrawal effects follow prolonged periods of barbiturate-induced depression. Since barbiturate-induced sleep is characterized by slow brain wave activity, it prevents a person from entering the deepest and most restful stage of sleep, known as the REM (rapid-eye-movement) stage. Because of REM sleep deprivation, some doctors believe individuals using barbiturates will suffer marked drowsiness and hangover-like symptoms after they stop taking the drug. Furthermore, because dreaming takes place during REM sleep, individuals may experience a period of intense dreaming shortly after discontinuing their use of the drug.

Some barbiturates are used to treat symptoms of alcohol withdrawal, while others are commonly used as minor tranquilizers. The use of barbiturates must be carefully considered because excessive use of such sedatives may hinder the development of a person's coping skills and result in severe depression.

People are sternly warned not to drink alcohol or take other central nervous system depressants (medications that slow down the nervous system, like antihistamines) while taking barbiturates because mixing these substances can prove fatal. Alcohol itself is a depressant, therefore combining it with a barbiturate can depress the nervous system to such an extent that it ceases functioning altogether. Usually, whatever depressant effects substances possess separately are intensified when they are combined.

Barium

Barium is the fifth element in Group 2 of the **periodic table**, the alkaline earth elements. Its **atomic number** is 56, its atomic **mass** is 137.34, and its chemical symbol is Ba.

Properties

Pure barium is a pale yellow, somewhat shiny, somewhat malleable metal. It has a melting point of about 1,300°F (700°C) and a **boiling point** of about 2,980°F (1,640°C).

Barium is an active metal that combines easily with **oxygen**, the **halogens**, and other non-metals. It also reacts with **water** and with most acids. It is so reactive that it must be stored under **kerosene** or some other inert liquid to prevent it from reacting with oxygen and moisture of the air.

Occurrence and Extraction

Barium is the fourteenth most abundant element in the Earth's crust, with abundance estimated at about 0.05%. The most common ores of barium are barite (barium **sulfate**; $BaSO_4$) and witherite (barium carbonate; $BaCO_3$. The world's major sources of barium ores are China, India, Morocco, the United States, Turkey, and Kazakhstan.

Pure barium metal is produced by heating barium **oxide** (BaO) with **aluminum** or **silicon**, as, for example: $3BaO + 2Al \rightarrow 3Ba + Al_2O_3$.

Discovery and Naming

The first mention of barium compounds goes back to the early seventeenth century when records spoke of a ''Bologna stone,'' named for the city of Bologna, Italy. The Bologna stone attracted the attention of scholars because it glowed in the dark. For more than a century, researchers attempted to discover the composition of the Bologna stone. Finally, in 1774, the Swedish chemist **Carl Wilhelm Scheele** announced that he had found a new element in the Bologna stone. He suggested the name barium for the element based on the scientific name for the Bologna stone, barite.

Uses

Barium metal has relatively few uses. One of those uses is as a "getter" or "scavenger" for the removal of unwanted oxygen from sealed **glass** containers. In such cases, oxygen would react with other components within the glass tube and interfere with some operation that occurs within the tube. The barium reacts with oxygen to form barium oxide.

The most important industrial use of barium compounds is in the petroleum industry. Barium sulfate is used as a "weighting agent" during the process of drilling for petroleum. A weighting agent adds body to petroleum and prevents the formation of gushers. Barium sulfate is also used to add body to or as a coating for **paper** products; as a white coloring agent in paints, **inks**, **plastics**, and textiles; in the manufacture of rubber products; in the production of **lead** storage **batteries**; and in a variety of medical applications.

Perhaps the best known applications of barium sulfate are in the field of medicine. The compound is widely used in diagnostic imaging procedures of the gastrointestinal (GI) system. A patient is first asked to drink a **suspension** of barium sulfate in water. An x-ray photograph is then taken of his or her GI as the barium sulfate moves through it. Since barium sulfate is opaque to x rays, a clear image of the details of the GI system can be obtained, and most defects can be detected. Alternatively, the barium sulfate suspension can be injected through the patient's rectum, a procedure known as a barium enema. The principle is the same in that the barium sulfate lines the inside of the lower GI system, making it opaque to x rays.

Barium carbonate is used in the production of **chlorine** and **sodium** hydroxide; as a rat poison; and in the manufacture of certain special types of glass.

BARTHOLOMEW THE ENGLISHMAN (c. 13th century)

English encyclopedist

Bartholomew the Englishman (also known as Bartholomaeus Anglicus) was an English Franciscan monk during the thirteenth century, notable for his work as an encyclopedist. He was educated at Oxford, and later went to Paris at about 1220 as a lector, and then to Magdeburg. In 1250, he wrote a famous encyclopedia, *Liber de Propietatibus Rerum (On the Properties of Things)*, of which the scientific side is of particular interest, dealing with all the known branches of medical science, astronomy and geography. This encyclopedia, comprised of 19 books, quickly became an accepted work of reference for the natural sciences. In the fourteenth century, Bartholomew's encyclopedia was translated into several languages. However, only the French translation by Jean Corbechon in 1372 for King Charles V proved successful.

BARTLETT, NEIL (1932-)

American chemist

Neil Bartlett has been called "the foremost **fluorine** chemist in the world" by a colleague, as reported in *Chemical and Engineering News.* In 1962 he used his skill with that highly active reagent to produce the first-ever compound of a noble gas. Bartlett's success forced a reexamination of basic **valence** theory, which proposes that the number of free electrons in an **atom** is the prime factor in determining that atom's **bonding** behavior.

Bartlett was born September 15, 1932, in Newcastle-upon-Tyne, England, the middle sibling in a family of three children. His father, Norman Bartlett, was a shipwright, a trade plied by the Bartletts for over a century. His mother was Anne Vock Bartlett. After attending Heaton Grammar School from 1944 to 1951, Bartlett entered King's College of the University of Durham, where he received his bachelor of science degree in 1954 and his doctorate in 1958. In 1957 he married Christina Isabel Cross. They have four children: Jeremy, Jane, Christopher, and Robin.

Following graduation, Bartlett taught at the Duke's School, then emigrated to Canada when he was appointed a lecturer in **chemistry** at the University of British Columbia in Vancouver in 1958. By 1964 he had worked his way up to full professor of chemistry. In 1966 Bartlett was named professor of chemistry at Princeton University; simultaneously, he became a member of the research staff at Bell Telephone Laboratories. In 1969 he joined the faculty of the University of California at Berkeley as a professor of chemistry and faculty senior scientist at the Lawrence Berkeley Laboratory.

While at the University of British Columbia, Bartlett began studying the factors that limit the combining capacity, or **oxidation** states, of various elements. He concentrated on the noble **metals**, such as **gold** and **platinum**, because they offered a range of oxidation states. He was particularly interested in the relationship of the geometry of the molecules to their valence, or outer shell **electron** configurations.

As part of this work, Bartlett was using fluorines, the most powerful oxidizing agents (electron acceptors) of the known elements, and reacting them with the noble metals. Treating platinum or platinum compounds with fluorine, Bartlett produced a highly reactive red solid, which was thought to be platinum oxyfluoride. After devising special techniques to study the solid, Bartlett and D. H. Lohmann determined that it was actually a **salt**, dioxygenyl hexafluoroplatinate, and the first compound to contain both positively and negatively charged ions. This discovery paved the way for what *Chemical and Engineering News* called "one of the most important developments in **inorganic chemistry** in modern times": Bartlett's creation of a compound of a noble gas.

Since the discovery of **argon** and **helium** in 1894 by Sir **William Ramsay** and Lord John Rayleigh, the noble gases—which also included **neon**, **krypton**, and xenon—had proved remarkably inert. Valence theory offered an explanation. Atoms are brought together by the electrons orbiting their nuclei. These orbits, or shells, can hold only a certain number of electrons. For example, oxygen's outermost shell can hold eight electrons, but the atom itself has only six in the outer shell. **Oxygen** atoms seek to fill their outer **ring** by joining with atoms that can provide two electrons. **Hydrogen** atoms have only one electron each, so two hydrogen atoms are a perfect complement to an oxygen atom.

The outer shell of electrons in a noble gas atom, however, is already full. Helium, for instance, has two electrons—its maximum—orbiting its **nucleus**; the other noble **gases** have eight. Because the outer shell is complete, a noble gas atom does not need to share electrons with any other atoms. Thus, valence theory reasoned, the noble gases are completely inert. In 1933 **Linus Pauling** surmised that **xenon**, the heaviest stable noble gas, might react with a very active compound, perhaps a fluorine. A number of experiments to create a compound with xenon failed. Attempts continued for years without success.

But in 1962 Bartlett succeeded, using platinum hexafluoride to oxidize (remove electrons from) xenon. Since his discovery, scientists have become aware of the limitations of simple valence theory. Noble gas compounds have been the subject of a new field of study, and other researchers, building on Bartlett's work, have prepared new compounds of xenon and two other noble gases, **radon** and krypton. While he is best known for his work with noble gases, Bartlett's other research includes preparing new synthetic metals from **graphite** or graphite-like **boron** nitride; synthesizing salts containing perfluoroaromatic cations; preparing new binary fluorides, and discovering, with B. Žemva and his co-workers, a new method of synthesizing thermodynamically unstable high **oxidation state** fluorides.

The author of more than one hundred scientific papers, Bartlett has received numerous accolades from his peers in recognition of his work. Besides honorary degrees from universities in the United States, Canada, and Europe, he was awarded the Corday-Morgan Medal and Prize of the Chemical Society, London. In 1965 he received both the Research Corporation Prize and the Steacie Prize in Natural Sciences (with **John C. Polanyi**). Bartlett received the Dannie-Heineman Prize from Göttingen Academy in Germany in 1971, and in 1976, the Robert A. Welch Award. In 1988 he received the Prix Moissan (with George Cady) in Paris, and in 1992 was recognized with the Bonner Chemiepreis from Friedrich-Wilhelms University of Bonn, Germany.

BARTON, DEREK H. R. (1918-1998)

English chemist

Derek H. R. Barton had a long and distinguished career in several universities in different countries in the field of natural products chemistry. He was active in structure determination, synthesis, and biosynthesis of a number of complex molecules, and had a special interest in the invention of new and useful chemical reactions. However, he is primarily known for his brilliant understanding of the importance of geometry in the behavior of organic compounds. Barton shared the 1969 Nobel Prize for **Chemistry** with Odd Hassel, a Norwegian physical chemist, for developing and applying the principles of conformation in chemistry, that is, for showing how the shapes of molecules determine their physical and chemical properties.

Derek Harold Richard Barton was born in Gravesend, Kent, England, on September 8, 1918. His grandfather and father were carpenters, and his father, William Thomas Barton, owned a successful lumberyard. Derek was able attend a good private school, but was forced to leave at seventeen without a degree because of his father's sudden death. He helped his mother, Maude Lukes Barton, in lumber business for two years, then enrolled in Gillingham Technical College. After a year at the college, he entered Imperial College, University of London, a center of science in England. He could not afford to live in London, and commuted two hours each way. At Imperial College, he received his B.Sc. with first-class honors in 1940, and his Ph.D. is 1942. He did his graduate thesis work on the synthesis of vinyl chloride (the starting compound for vinyl plastics) under the supervision of two eminent organic chemists, I. M. Heilbron and E. R. H. Jones. After completing his Ph.D., Barton remained at Imperial College to work on the formulation of secret inks for military intelligence, and in 1944 left to work on the synthesis of organic phosphorous compound for a company in Birmingham. After a year in the chemical industry, Barton returned to Imperial College as a junior lecturer in inorganic chemistry. He taught inorganic and physical chemistry for four years until a position in organic chemistry became available.

When Barton returned to Imperial College, he began research on the structures of complex organic compounds, including triterpenoids and steroids. He correlated structures with a physical property of the molecules, molecular rotation, and was able to assigned structures based on a simple physical measurement. During his work with these complex molecules, he became aware of the work of Odd Hassel, who had determined the precise geometry of cyclohexane, a compound that is a **ring** of six **carbon** atoms, with each carbon bonded to two **hydrogen** atoms. Cyclohexane is a structural unit commonly found in sterpods and triterpenoids, and Barton extended Hassel's structure to the complex molecules. He designed a set of models that accurately represented the actual geometry of steroisa, and had them built in 1948. These models provided Barton with an understanding of the three-dimensional geometry (stereochemistry) of these molecules, which was unknown to other chemists at the time.

Barton's work on steroids had come to the attention of Louis Fieser, a professor at Harvard University, and an eminent authority on steroids. Fieser invited Barton to Harvard as a visiting lecturer, replacing Robert B. Woodward for his sabbatical year. Barton arrived at Harvard in 1949, at a time of intense interest in the chemistry of steroids because their spectacular use in medicine (cortisone therapy) had just been announced. During a seminar lecture by Fieser, in which he discussed unsolved problems in the chemistry of steroids, Barton realized that the precise shape of the molecules, which he knew from his models, could explain the results. He formulated a four- page paper and submitted it to the Swiss journal *Ex-*

perientia, which had a modest readership. This paper provided a stimulus for countless investigations in practical and theoretical chemistry, and was the basis for Barton's Nobel Prize in 1969. Barton's description of the influence of molecular geometry on chemistry is called conformational analysis; the principles are readily understood, and are introduced early in undergraduate organic chemistry textbooks. But conformational analysis is also a powerful tool in solving complex biochemical problems, such as **enzyme** catalysis and pharmacological studies. Although Barton's contribution to the field was seminal, he left its development to others. He used the principles of conformational analysis to understand the chemistry of the molecules in which he was interested, but his primary concern was always the molecules themselves.

Another of his major interests was one-electron **oxidation** of organic compounds, which he exploited to explain how complex molecules, such as morphine, are produced in the opium poppy (biosynthesis). Key intermediates in the electron oxidation are reactive species called **free radicals**, and Barton vigorously explored the use the use of free-radical chemistry in synthesis. Free radicals may be formed in chemical reactions, or by using energy to rupture chemical bonds. The latter method may involve energy from ultraviolet light, and the radicals are thereby generated photochemically. Barton was able to use photochemical reactions effectively in synthesis, and the Barton reaction was invented in 1958 to synthesize the steroid aldosterone, a hormone that regulates electrolyte balance in the body. At the time, the world supply of aldosterone was only several milligrams, and Barton's synthesis yielded 60 grams of the hormone by a simple procedure. Barton devoted many of his studies to the invention, rather than the discovery, of new reactions. He and his co-workers contributed many new reagents and procedures that accomplish otherwise difficult chemical transformations.

Barton's great command of all areas of chemistry enabled him to see connections between what appear to be unrelated facts; he called this ability gap-jumping. For example, his knowledge of chemical physics and steroid chemistry led to the development of conformational analysis. From an early age, he developed the habit of closely reading the literature, and was associated with many other outstanding chemists throughout his career who kept him informed of the latest discoveries. His routine brought him to the laboratory daily to check on the progress of his students and associates, and even in his seventies his work day lasted from three or four in the morning until seven in the evening.

Barton was married to Jeanne Kate Wilkins, and had a son, William Godfrey Lukes Barton. The marriage ended in divorce, and Barton married Christiane Cognet. Barton held several positions in academic institutions, and was also associated with companies in the chemical and pharmaceutical industries. After his return to England from Harvard in 1950, he accepted a position as reader in organic chemistry at Birkbeck College, University of London, where all classes were held in the evening. He was promoted to professor at Birkbeck College in 1953, and left in 1955 to become Regious Professor at the University of Glasgow. After two years, he returned to Imperial College in London as professor of organic chemistry, remaining there for twenty years.

In 1977, at the age of fifty-nine, Barton was appointed director of research of the Centre National de la Recherché Scientifique (CNRS) at the Institut de Chimie de Substances Naturells (ICSN) at Gif-sur-Yvette, France. A year later, he retired from Imperial College. He had an excellent command of the French language, and his French wife was delighted to return home. At the ICSN, Barton continued his work inventing reactions, producing a series of Gif reagents, named for the site of the ICSN. After eight years, Barton retired again, and this time accepted a Distinguished Professorship at Texas A&M University in College Station. He continued to pursue chemical research at his usual active pace in his newly adopted country. Barton had previously visited the United States many times to give lectures and courses, and he spent several summers at the Research Institute for Medicine and Chemistry (RIMAC) in Cambridge, Massachusetts. It was at RIMAC that the Barton reaction for the synthesis of aldosterone was invented. Recognized for his achievement by chemical societies and universities of many nations, he is regarded as one of the most prominent organic chemists of the twentieth century. Barton died in 1998.

Batteries

Batteries are used for the storage of electrical **energy**. They range in size from gigantic boxes as large as a house that are used by utility companies to paper-thin devices for the protection of memory in electronic devices. Storage batteries, and indeed all batteries, are simply electrochemical reactors. An oxidizer (an **electron** acceptor) and a reducer (an electron donor) react together by transferring electrons to form products and release chemical energy in the form of **electricity**. By arranging the components of a battery in a special way, it is possible to control the rate of the reaction and to release the electrical energy on demand.

The generation of electrical energy from chemical reactions in battery-like devices dates to the very early days of the investigation of electricity. At the end of the eighteenth century, Alessandro Volta (1745-1827) put together layers of substances in such a way that he could generate voltage differences and then obtain ever greater potentials by linking each set of materials—each cell—in series.

A common feature of typical batteries is that their components are **solids**. The reason for this relates directly to the need for a battery to provide a constant potential. The potential of a cell depends critically on **concentration**. If we measure the potential of a cell in which the electrochemical conversion of species dissolved in **solution** occurs, the potential of the cell changes as the concentration of the dissolved species in one half-cell increases and the concentration of the dissolved species in the other half-cell decreases. It is important that the voltage of a battery be constant or very nearly constant. A relatively constant voltage output can be maintained if the concentration of species involved does not change. The concentration of a pure solid does not change as some of it is used up. As long as a solid maintains its composition and its **density**, it has the same concentration. If half a solid electrode is used up, its **mass** and **volume** are both decreased by the same factor, so its concentration remains the same. Some components of a battery that are involved in the electrochemical reaction may not be solids, such as the **sulfuric acid** in the **lead** storage battery. If the concentration of such a species is high enough that it does not change much as the battery is discharged, then the voltage output remains nearly constant.

The Lead Storage Battery

The lead-acid storage battery used to start automobile engines is one of the most successful electrochemical systems and the most successful storage battery developed to date. Indeed, the production of these batteries accounts for the largest single use of lead and its compounds. Gaston Plante developed the first working lead-acid battery in 1859, but the storage of electrochemical energy was not practical then because no efficient way of recharging the battery was available. The principal early commercial application of the lead-acid battery was in the telegraphic industry.

Modern lead storage batteries consist of several cells within one container. The individual cells consist of positive electrodes or plates made of insoluble PbO_2; these plates are separated from negative electrodes containing a porous, spongy PbO by thin sheets of a microporous, nonconducting fiberglass or PVC. The electrodes are immersed in aqueous sulfuric acid, which acts as the electrolyte. Each individual cell generates approximately 2 V; thus, three elements are joined to make a 6 V battery, six for a 12 V battery, and so forth.

When electricity is generated by the battery, the metallic lead at the **anode** is oxidized: $PbO(s) + HSO_4^-(aq) \rightarrow PbSO_4(s) + H^+(aq) + 2e^-$

and PbO_2 at the cathode is reduced: $PbO_2(aq) + 3H^+(aq) + HSO_4^-(aq) + 2e^- \rightarrow PbSO_4(aq) + 2H_2O(l)$

The sum of these two reactions is called the double **sulfate** reaction: $PbO(s) + PbO_2(s) + 2H_2SO_4(aq) \rightarrow 2\ PbSO_4(s) + 2H_2O(l)$

As soon as the circuit of the lead storage battery cell is closed and the cathode and anode are connected, the reaction begins and electrons flow from the anode to the cathode. If a voltmeter (an instrument used for measuring voltage) is placed in the circuit, it will register the voltage difference between the two electrodes. A voltage difference is registered for any electrochemical reaction that proceeds spontaneously. This voltage difference is called the electromotive force, or emf. The emf of a voltaic cell can be thought of as the force that pushes or drives the electrons from the anode toward the cathode. It arises from the difference in energy between an electron at one site and at the other.

If an electrical device, such as a light bulb, is placed in the circuit, the voltage generated by the voltaic cell can be used to accomplish electrical work by heating the filament in the bulb and causing it to emit light. The redox reaction in the battery is the source of the electrical energy; batteries are voltaic cells.

The voltage of a cell depends upon the identity of its components, their concentration, the **temperature**, and a number of other variables. To make comparisons and predictions involving electrochemical reactions possible, a scale of relative potentials for half-reactions has been developed by determining the voltage for a number of half-cells connected to a single, standard half-cell. The standard or reference half-cell used for this purpose is composed of a solution of 1 molar acid under an atmosphere of **hydrogen** gas at a pressure of 1.00 bar and containing a **platinum** electrode. This is called the standard hydrogen electrode or SHE. A potential of 0.00 V has been assigned to this cell.

The electrical discharge from the lead storage battery provides the energy to start the internal **combustion** process in a car. When the battery is being charged by the alternator or generator of the automobile engine, the discharge process is reversed. Since hydronium ions are consumed during the discharge cycle, the acid concentration in the lead storage battery decreases. Sulfuric acid is more dense that **water**, so as the acid concentration decreases, the density of the solution decreases, too. Conversely, when the battery is being charged, the acid concentration increases and the density increases. Thus, a simple test for the degree to which the battery is charged is to measure the density of the solution in the battery.

Dry Cells

So-called primary batteries are designed to be discharged only once and then discarded. Probably the batteries most familiar to you are dry cells, a category that includes carbon-zinc cells, the most commonly found primary cells worldwide, and alkaline batteries, which use **sodium** hydroxide or **potassium** hydroxide as the electrolyte. Dry cell batteries have been used for many applications since their invention in 1866 by Georges Leclanché (1839-1882). Although the actual chemical reactions taking place within a battery are very complex, the major cathodic and anodic reactions can be idealized based on a knowledge of the parts of a battery and their electrochemical function.

Traditionally carbon-zinc batteries use a **carbon** rod as an electrode to collect current. The rod is immersed in a paste of **manganese** dioxide, carbon, and **zinc** chloride that serves as the cathode. The zinc container itself is the anode. The name "dry cell" reflects the absence of excess liquid electrolyte, although some water must be present for the cell to function; the zinc chloride in the cathode mixture serves as the electrolyte. The potential of this battery is about 1.5 V.

The cathodic reaction can best be described by two one-electron steps: 1) $MnO_2 + H_3O^+ + e^- \rightarrow MnOOH + H_2O$ 2) $MnOOH + H_3O^+ + e^- \rightarrow Mn(OH)_2(s) + H_2O$

The anode reaction is: $ZnO(s) \rightarrow Zn_2^+ + 2e^-$

Primary alkaline batteries use **sodium hydroxide** or potassium hydroxide as the electrolyte. In an alkaline battery, the components are slightly different, but the voltage (1.54 V) is nearly the same as for the standard dry cell. The steel can of the battery serves as the current collector for the cathode, which is a dense mass of MnO_2 and carbon packed within the can. This cathodic mass has a hollow center that is lined to separate it from the anode, which is another mixture consisting of zinc powder and the alkaline electrolyte. Contact between the anode and cathode is achieved by a metal pin or leaf inserted into the anode mix. The alkaline battery derives its power from the **reduction** of the MnO_2 cathode and the **oxidation** of the zinc anode. Although both anodic and cathodic reactions are very complicated, the following reactions provide a useful overview of the chemistry:

Anode reaction: $Zn + 2OH^- \rightarrow ZnO + H_2O + 2e^-$

Cathode reaction: $2e^- + 2MnO_2 + 2H_2O \rightarrow 2MnOOH + 2OH^-$

Overall reaction: $ZnO + 2MnO_2 + H_2O \rightarrow ZnO + 2MnOOH$

A third type of dry cell, the **mercury** cell, is used where compactness is essential, suc as in small devices like calculators, cameras, and watches. It produce about 1.35 V. Again, the anode is zinc, but, in this case, the cathode reaction involves mercuric **oxide** paste:

Anode reaction: $Zn(s) + 2OH^- \rightarrow ZnO(s) + H_2O + 2e^-$

Cathode reaction: $HgO(s) + H_2O + 2e^- \rightarrow Hg(l) + 2OH^-$

Overall reaction: $HgO(s) + Zn(s) \rightarrow ZnO(s) + Hg(l)$

NiCad Cells

The NiCad cell is a battery that can be used interchangeably with a dry cell, but it has the advantage of being rechargeable. This cell has one electrode that consists of **cadmium** and cadmium hydroxide and another consisting of nickel(IV) oxide, NiO_2, and **nickel** (II) hydroxide, $Ni(OH)_2$. The voltaic reaction consists of the oxidation of cadmium metal to cadmium(II) **ion**, which is precipitated as a hydroxide, and the reduction of Ni(IV) to Ni(II), which is also precipitated as a hydroxide. Aside from water, all the components are solid. All reactants and products readily adhere to the electrodes, and this battery can be sealed into a compact device. The reactions are:

Anode reaction: $Cd + 2OH^- \rightarrow Cd(OH)_2(s) + 2e^-$

Cathode reaction: $NiO_2(s) + 2\ H_2O + 2e^- \rightarrow Ni(OH)_2(s) + 2OH^-$

Overall reaction: $NiO_2(s) + Cd(s) + 2H_2O \rightarrow Ni(OH)_2(s) + Cd(OH)_2(s)$

Lithium-Polymer Cells

The continuing challenge to battery manufacturers is to find lightweight, cheap materials from which reliable batteries can be made. For many uses it is essential that the electrochemical reactions in the battery run at room temperature in both directions so that the units can be discharged and recharged repeatedly. The materials of construction must also be resilient and not break down during these processes.

Lithium-polymer batteries are already used in small devices like hearing aids. The electrolyte is a solid polymer roughly the thickness of plastic food wrap that cannot leak, corrode or emit dangerous **gases**. The polymer sheet is sandwiched between a solid anode composed of a **lithium alloy** and a solid cathode consisting of an oxide in carbon.

The advantages of a lithium-polymer battery are numerous. Lithium is the lightest metal on the **periodic table**, in contrast to lead in lead-acid batteries, which is one of the heaviest. In addition, the other components of the battery are also very light. The batteries can be assembled in thin sandwiches less than one-fiftieth of an inch thick. The solid electrolyte is environmentally safer than the **liquids** used in other batteries. Finally, the batteries can be formed into almost any shape and are not restricted to cylindrical canisters.

Zinc-silver oxide battery

Miniature zinc-silver oxide batteries are used in electronic watches. They function over a wide temperature range and have a good storage life. These batteries are designed for applications where usage is continuous and the cell is open to air. The cathode has two duties in these batteries—it is the site of reduction, and it catalyzes the reaction of the **oxygen** gas at its surface.

The reaction at the anode is: $Zn \rightarrow Zn_2^+$

The reaction at the cathode is: $Ag_2O(s) + H_2O + 2e^- \rightarrow 2Ag + 2OH^-$

Zinc-air battery

Among the more exotic batteries are the so-called zinc-air batteries. Zinc once again functions as the anode. The cathodic reaction is: $O_2 + 2H_2O + 4e^- \rightarrow 4\ OH^-$

Fuel Cells

Fuel cells are electrochemical cells that convert the chemical energy of a fuel directly into electrical and thermal energy. A combustion reaction is, after all, an **oxidation-reduction reaction** in which oxygen is normally the oxidizing agent. Unlike a typical battery that has all the reactants sealed in a container, in a hydrogen-oxygen fuel cell, hydrogen is fed continuously to the anode and oxygen to the cathode. The electrochemical reactions take place at the electrodes to produce an electric current. The fuel cell theoretically has the capability of producing electrical energy for as long as the fuel and oxidant are fed to the electrodes. In addition, the cells produce **heat** which is available for other applications.

The most successful application of the alkaline fuel cell technology was the U.S. Space program—the Apollo missions to the moon and the Space Shuttle. The fuel cell for Apollo used pure H_2 and O_2 gases and concentrated electrolyte (85% KOH).

In a hydrogen/oxygen fuel cell, a solution of potassium hydroxide is used as the electrolyte. The reactions are:

Anode reaction: $2H_2 + 4OH^- \rightarrow 4H_2O(l) + 4e^-$

Cathode reaction: $O_2 + 2H_2O\ (l) + 4e^- \rightarrow 4OH^-$

Overall reaction: $2H_2 + O_2 \rightarrow 2H_2O$

Fuel cells can operate which energy losses of only about 25%. This constitutes a major improvement over combustion reactions, which typically lose from one-half to three-quarters of the chemical energy they produce. The major impediment to the routine use of fuel cells is the problem of contamination of the electrodes.

Johann Joachim Becher.

Becher, Johann Joachim (1635-1682)

German chemist and physician

The death of Becher's father, a Protestant minister, left him at the age of eight with the need to help support his mother and brothers. The lack of money for a formal education forced Johann to educate himself, mainly by traveling through Sweden, Holland, Italy, and Germany. He developed interests in a wide- range of subjects including **alchemy**, medicine, politics, economics, theology, history, and mathematics. In 1661, Becher's first published book appeared, and the University of Mainz awarded him a M.D. A restless individual, Becher served for a short period of time as court physician to the Electors of Mainz and Munich before turning to economics and politics. Emperor Leopold I in Vienna appointed him alchemical advisor and imperial economic counselor in 1666. In this capacity, Becher instituted educational reforms, including the establishment of technical schools and proposed a Rhine-Danube canal and colonial settlements in South America. During this period he produced his most important work *Physica subterranea* in 1669 and two supplements in 1671 and 1675. After some of his policies provoked imperial disfavor and a short prison sentence, Becher moved to Holland in 1678 where he submitted to the Dutch parliament a plan for extracting **gold** from sea sand. A small-scale test of the process in 1679 proved successful, but Becher moved to England before it could be tried on a larger scale. In England he wrote the third supplement to *Physica* (1680) which included the gold extraction process. From that time until his death in London in 1682 he toured mines in Scotland and Cornwall and completed several books including *Chymischer Glückshafen,* which detailed fifteen hundred chemical processes. Although Becher was presented a book and was a candidate for membership, to his dismay he was never elected a fellow of the Royal Society.

Becher's theory of the elements was a melding of alchemical ideas with the growing chemical knowledge of the seventeenth century. He believed that the three basic substances were air, **water**, and earth and that all inorganic bodies were composed of water and earth, with air serving only as a mixing instrument. To explain the differences, Becher proposed the presence in varying amounts of three distinct types of earth: vitreous earth (*terra fusilis,*) which gave a body substance and made it virtually incapable of alteration, combustible earth (*terra pinguis*), a moist, oily substance which gave a body odor, **taste**, **color** and combustibility, and mercurial earth (*terra fluida*), which provided weight, **ductility**, and **volatility**. He insisted that every flammable body contained combustible earth, but he had no definite position on how this substance played a role in the burning process. This was not a new idea, and Becher did little to attempt to prove it by experimentation. His great contribution to the matter lies in the influence he had upon Georg Stahl, who half a century later expanded Becker's idea of combustible earth into the **phlogiston** theory of **combustion**.

Becher supported the theories of spontaneous generation, metallic **transmutation**, and the belief that **metals** grow in the earth. Among his more practical suggestions were that sugar and air were needed for **fermentation** and that **coal** could be distilled to produce tar.

Becquerel, Antoine-Henri (1852-1908)

French physicist

Antoine-Henri Becquerel's landmark research on X rays and his discovery of radiation laid the foundation for many scientific advances of the early twentieth century. X rays were discovered in 1895 by the German physicist Wilhelm Conrad Röntgen, and in one of the most serendipitous events in science history, Becquerel discovered that the **uranium** he was studying gave off radiation similar to X rays. Becquerel's student, **Marie Curie**, later named this phenomenon **radioactivity**. His later research on radioactive materials found that at least some of the radiation produced by unstable materials consisted of electrons. For these discoveries, Becquerel shared the 1903 Nobel Prize in physics with Marie and **Pierre Curie**. Becquerel's other notable research included the effects of **magnetism** on light and the properties of luminescence.

Becquerel was born in Paris on December 15, 1852. His grandfather, Antoine-César Becquerel, had fought at the Battle of Waterloo in 1815 and later earned a considerable reputation as a physicist. He made important contributions to the study of **electrochemistry**, meteorology, and agriculture. Antoine-

Henri's father was Alexandre-Edmond Becquerel, who also made a name for himself in science. His research included studies on photography, **heat**, the conductivity of hot **gases**, and luminescence.

Becquerel's early education took place at the Lycée Louis-le-Grand from which he graduated in 1872. He then enrolled at the Ecole Polytechnique, and two years later he moved on to the Ecole des Ponts et Chaussées. It appears that there was never any question about the direction of Becquerel's career, as he concentrated on scientific subjects throughout his schooling. In 1877 he was awarded his engineering degree and accepted an appointment as an *ingénieur* with the National Administration of Bridges and Highways.

During his years at the Ecole des Ponts et Chaussées, Becquerel became particularly interested in English physicist Michael Faraday's research on the effects of magnetism on light. Faraday had discovered in 1845 that a plane-polarized beam of light (one that contains light waves that vibrate to a specific pattern) experiences a rotation of planes when it passes through a magnetic field; this phenomenon was called the Faraday effect. Becquerel developed a formula to explain the relationship between this rotation and the refraction the beam of light undergoes when it passes through a substance. He published this result in his first scientific **paper** in 1875, although he later discovered that his initial results were incorrect in some respects.

Although the Faraday effect had been observed in **solids** and **liquids**, Becquerel attempted to replicate the Faraday effect in gases. He found that gases (except for oxygen) also have the same ability to rotate a beam of polarized light as do solids and liquids. Becquerel remained interested in problems of magneto-optics for years, and returned to the field with renewed enthusiasm in 1897 after Dutch physicist **Pieter Zeeman**'s discovery of the **Zeeman effect**, whereby spectral lines exposed to strong magnetic fields split, provided new impetus for research.

In 1874 Becquerel had married Lucie-Zoé-Marie Jamin, daughter of J.-C. Jamin, a professor of physics at the University of Paris. She died four years later in March of 1878, shortly after the birth of their only child, Jean. Jean later became a physicist himself, inheriting the chair of physics held by his father, grandfather, and great-grandfather before him. Two months prior to Lucie's death, Becquerel's grandfather died. At that point, his son and grandson each moved up one step, Alexandre-Edmond to professor of physics at the Musée d'Histoire Naturelle, and Antoine-Henri to his assistant. From that point on, Becquerel's professional life was associated with the Musée, the Polytechnique, and the Ponts et Chaussées.

In the period between receiving his engineering degree and discovering radioactivity, Becquerel pursued a variety of research interests. In following up his work on Faraday's magneto-optics, for example, he became interested in the effect of the Earth's magnetic field on the atmosphere. His research determined how the Earth's magnetic field affected **carbon** disulfide. He proposed to the International Congress on Electric Units that his results be used as the standard of electrical current strength. Becquerel also studied the magnetic properties

Antoine-Henri Becquerel.

of a number of materials and published detailed information on **nickel**, **cobalt**, and ozone in 1879. He also reported the surprising discovery that nickel-plated **iron** becomes magnetic when heated to redness.

In the early 1880s Becquerel began research on a topic his father had been working on for many years—luminescence, or the emission of light from unheated substances. In particular, he made a detailed study of the spectra produced by luminescent materials and examined the way in which light is absorbed by various crystals. Becquerel was especially interested in the effect that **polarization** had on luminescence. For this work Becquerel was awarded his doctoral degree by the University of Paris in 1888, and he was once again seen as an active researcher after years of increasing administrative responsibility.

When his father died in 1891, Becquerel was appointed to succeed him as professor of physics at the museum and at the conservatory. The same year he was asked to replace the ailing Alfred Potier at the Ecole Polytechnique. Finally, in 1894 he was appointed chief engineer at the Ecole des Ponts et Chaussées. Becquerel married his second wife, Louise-Désirée Lorieux, the daughter of a mine inspector, in 1890; the couple had no children.

The period of quiescence in Becquerel's research career came to an end in 1895 with the announcement of Röntgen's discovery of X rays. The aspect of the discovery that caught

Becquerel's attention was that X rays appeared to be associated with a luminescent spot on the side of the cathode-ray tube used in Röntgen's experiment. Given his own background and interest in luminescence, Becquerel wondered whether the production of X rays might always be associated with luminescence.

To test this hypothesis Becquerel wrapped photographic plates in thick layers of black paper and placed a known luminescent material, **potassium** uranyl **sulfate**, on top of them. When this assemblage was then placed in sunlight, Becquerel found that the photographic plates were exposed. He concluded that sunlight had caused the uranium **salt** to luminesce, thereby giving off X rays. The X rays then penetrated the black paper and exposed the photographic plate. He announced these results at meeting of the Academy of Sciences on February 24, 1896.

Through an unusual set of circumstances the following week, Becquerel discovered radioactivity. He began work on February 26th as usual by wrapping his photographic plates in black paper and taping a piece of potassium uranyl sulfate to the packet. Since it was not sunny enough to conduct his experiment, however, Becquerel set his materials aside in a dark drawer. He repeated the procedure the next day as well, and again a lack of sunshine prompted him to store his materials in the same drawer. On March 1st Becquerel decided to develop the photographic plates that he had been prepared and set aside. It is not clear why he did this since, according to his hypothesis, little or no exposure would be expected. Lack of sunlight had meant that no luminescence could have occurred; hence, no X rays could have been emitted.

Surprisingly, Becquerel found that the plates had been exposed as completely as if they had been set in the sun. Some form of radiation—but clearly not X rays—had been emitted from the uranium salt and exposed the plates. A day later, according to Oliver Lodge in the *Journal of the Chemical Society,* Becquerel reported his findings to the academy, pointing out: ''It thus appears that the phenomenon cannot be attributed to luminous radiation emitted by reason of phosphorescence, since, at the end of one-hundredth of a second, phosphorescence becomes so feeble as to become imperceptible.''

With the discovery of this new radiation Becquerel's research gained a new focus. His advances prompted his graduate student Marie Curie to undertake an intensive study of radiation for her own doctoral thesis. Curie later suggested the name radioactivity for Becquerel's discovery, a phenomenon that had until that time been referred to as Becquerel's rays.

Becquerel's own research continued to produce useful results. In May of 1896, for example, he found uranium metal to be many times more radioactive than the compounds of uranium he had been using and began to use it as a source of radioactivity. In 1900 he also found that at least part of the radiation emitted by uranium consists of electrons, particles that were discovered only three years earlier by **Joseph John Thomson**. For his part in the discovery of radioactivity, Becquerel shared the 1903 Nobel Prize in physics with Curie and her husband Pierre.

Honors continued to come to Becquerel in the last decade of his life. On December 31, 1906, he was elected vice president of the French Academy of Sciences, and two years later he become president of the organization. On June 19, 1908, he was elected one of the two permanent secretaries of the academy, a post he held for less than two months before his death on August 25, 1908, at Le Croisic, in Brittany. Among his other honors and awards were the Rumford Medal of the Royal Society in 1900, the Helmholtz Medal of the Royal Academy of Sciences of Berlin in 1901, and the Barnard Medal of the U.S. National Academy of Sciences in 1905.

BEER'S LAW

When light shines on **matter**, it may be transmitted (as with a glass of water), refracted and scattered (as with dust particles in the air at sunset), reflected (as with the chrome trim of an automobile), or absorbed (as with exposed skin at the beach).

The light which we can see, known as visible light, is made up of a continuum of different wavelengths of electromagnetic radiation. Each wavelength in this range of wavelengths, or spectrum, which makes up visible light, has its own associated color, ranging from red to violet. Also, the visible spectrum of light is only one part of a larger continuum of electromagnetic radiation. These wavelengths that are not visible to the human eye range from radio waves at one extreme to x rays at the other.

The nature of visible light and its properties, including how light interacts with matter, has remained the subject of study and speculation by scientists from the very early stages of the development of scientific thought into the twentieth century. The works of Plato (427-347 B.C.) and Aristotle (384-322 B.C.) formed the basis for much of the speculation which preceded the Scientific Revolution (c.1550-1700). Isaac Newton (1642-1727) laid the groundwork for the modern scientific consideration of the properties of light, and **Albert Einstein** applied quantum ideas to the properties of electromagnetic radiation.

Among the phenomena that scientists sought to explain was the fact that the intensity of light was diminished as it passed through substances, including chemical solutions. In 1729, P. Bouguer (1698-1758), was the first to state the law of absorption: the fraction of light absorbed by a particular material (i.e., the decrease in the intensity of the light beam as it passes through the material) is directly proportional to the thickness of the material. The proportionality constant is called the absorption coefficient or the extinction coefficient. For instance, if the intensity of light is 1/4 as strong after passing through a 5-inch thickness of an aqueous solution of a dye, it will be diminished to 1/2 its original intensity upon passing through a 10-inch sample. The absorption coefficient in this case is 0.05.

As sometimes happens in the history of science, Bouguer's discovery did not make a significant impact and was forgotten. Sometime later, a better known scientist, J. H. Lambert (1728-1777), independently rediscovered and published this law of absorption. Although Bouguer had priority in the discovery, confusion remains, and the law is known both as the Bouguer Law and the Lambert Law.

As additional observations and more accurate measurements were made, it was noticed that the amount of light absorbed by solutions also depends on other factors. In 1852, J. Beer announced a more complete law of absorption which is known variously as Beers law, the Lambert-Beer law, and the Bouguer-Beer law. Given the history of the discoveries and the rules of scientific priority, the law should carry the name Bouguer-Beer.

Beer observed that, in addition to the effect of the thickness of the sample, the amount of radiation absorbed by a solution is proportional to the concentration of the dissolved substance which is absorbing the radiation. Written mathematically, the Bouguer-Beer law (Beer's law) is: $\log (P_0/P)=ebc$, where P_0 is the power of the incoming radiation, P is the power of the radiation after passing through the sample, e is the extinction coefficient (or absorption coefficient), b is the length of the radiation path through the solution, and c is the concentration of the absorbing material in solution.

This law has formed the basis for the development of quantitative spectroscopy, particularly as applied to **analytical chemistry**, where it provides, for instance, a method of determining concentrations without having to destroy a portion of the sample.

In addition to thickness and concentration, the amount of radiation absorbed by a sample depends on the chemical identity of the sample and on the wavelength of the radiation. The determination of the wavelengths absorbed by a particular molecule and the relationship to the energy levels in atoms and molecules is the basis of the field of spectroscopy, on which much of the understanding of the nature of atoms and molecules and their interactions is based. Spectroscopy owes much of its basic quantitative framework to the pioneering work of Bouguer, Lambert, and Beer.

BENZENE

Benzene is the simplest of the **aromatic hydrocarbons**. It is a **ring** structure with the **empirical formula** C_6H_6. Benzene was first discovered by **Michael Faraday** in 1825 from the liquid condensed by compressing oil gas.

The chemical structure of the **molecule** was originally worked out by Kekulé in 1865. The structure of benzene is formally that of a regular hexagonal array of the **carbon** atoms, termed a ring structure, comprising three alternating carbon-carbon double bonds and three carbon-carbon single bonds. The electrons of the double bond are now known not to be static, however, but form a symmetrical "π-cloud" molecular orbital above and below the ring where these electrons are delocalized (or resonating) around the ring. This **resonance** strongly stabilizes benzene and profoundly influences its chemical properties.

Benzene is a clear, highly refractive liquid at standard **temperature** and pressure with a characteristic, sweet smell. It is highly flammable, and burns with a smoky yellow flame (indicating a high level of carbon in the structure). Benzene is a good solvent for fats and lower **molecular weight** aromatic compounds. It is freely miscible with **ethanol**, diethylether, **acetone** (propanone), and **acetic acid** (ethanoic acid). Excessive exposure to the vapor is toxic, and chronic exposure to benzene results in damage to the bone marrow, liver, and kidneys. Chronic exposure is also associated with an increased risk for leukemia and possibly other cancers.

A computerized graphic of a benzene molecule. ***(Photograph by Ken Eward/Science Source, National Audubon Society Collection/Photo Researchers, Inc. Reproduced by permission.)***

Like other hydrocarbons with double bonds, addition reactions can occur to change the double bonds to single bonds. For example, catalytic **hydrogenation** adds six **hydrogen** atoms to produce cyclohexane, a colorless liquid with the properties of an alkane. The requirements to affect this **reduction** of benzene to cyclohexane, in terms of catalyst, reaction temperature, and reaction time, are much more vigorous than for the reduction of simple alkenes. This increased difficulty reflects the resonance stabilization that benzene possesses. Benzene burns in air to produce **carbon dioxide**, **water**, and carbon but it is very resistant to chemical **oxidation**. Thus **potassium** permanganate ($KMnO_4$) will not decolorize in its presence. This resistance to chemical oxidation is likewise due to the resonance stability of the benzene. Under rigorous conditions benzene will undergo substitution for one of its hydrogen substituents, by a process called electrophilic aromatic substitution. For example, a mixture of concentrated nitric and sulfuric acids will nitrate benzene. A yellow oil separates after this mixture and benzene are added to cold water. This liquid is nitrobenzene, $C_6H_5NO_2$. Benzene will also form a range of compounds with the transition **metals**. Halogenation will occur in the presence of strong sunlight or ultraviolet light. Benzene will react with **chlorine** to give hexachlorobenzene, which once was used as an insecticide. A similar reaction occurs with **bromine** but not with **iodine** or **fluorine**. When atoms or functional groups are added to the benzene ring, their position is indicated by numbering. Each carbon **atom** is numbered from 1-6 and when the two substitutents are located at the 1,2 position it is ortho-; 1,3 is meta-; and 1,4 is para-.

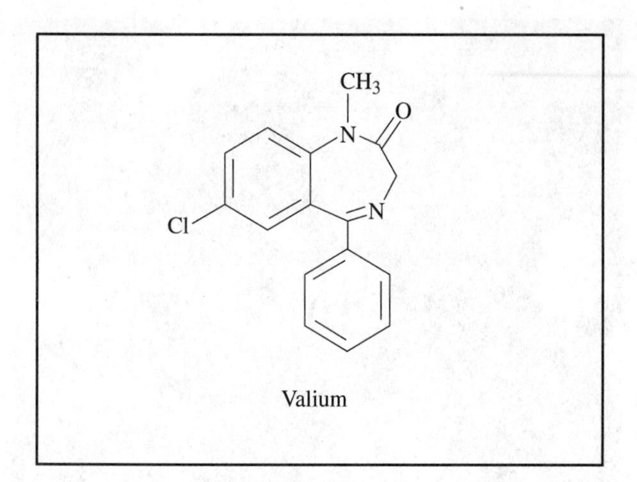

Figure 1. Valium with benzodiazepine ring bolded. ***(Illustration by Electronic Illustrators Group.)***

Benzene is manufactured industrially by **dehydrogenation** and dealkylation of appropriate fractions of petroleum. In the laboratory benzene can be made by the action of **heat** on a mixture of **sodium** benzoate and **sodium hydroxide**. The Friedel-Crafts reaction, the common name for electrophilic aromatic substitution, can be used to make **hydrocarbon** derivatives of benzene. For example, benzene and ethyl bromide can be use to make ethylbenzene. In the 1990s, United States production of benzene was in the order of 5 million megatons (1 megaton = 1 million metric tons). Industrially, benzene is used in the manufacture of **nylon**, phenol, styrene (and by polymerization polystyrene), and cyclohexane. It is a common constituent of automobile **gasoline**. Benzene is also used in the manufacture of the insecticide DDT.

Benzene is the simplest aromatic hydrocarbon and it has many commercial uses.

Benzodiazepines

Benzodiazepines are antianxiety drugs that help relieve nervousness, tension, and other symptoms by slowing the central nervous system. The group of drugs known as benzodiazepines includes alprazolam (Xanax), chlordiazepoxide (Librium), diazepam (Valium), and lorazepam (Ativan). These medicines take effect fairly quickly, usually within one hour after they are taken. They are available only with a doctor's prescription, and should not be used to relieve the nervousness and tension of normal everyday life. The recommended dosage depends on the type of benzodiazepine, its strength, and the condition for which it is being taken.

While anxiety is a normal response to stress, some people have unusually high levels of anxiety that can interfere with everyday life. For these people, benzodiazepines can help bring their feelings under control. The medicine can also relieve troubling symptoms of anxiety, such as pounding heartbeat, breathing problems, irritability, nausea, and faintness.

The molecular **nomenclature** of these compounds is taken from the core seven-member **ring**, containing two **nitrogen** atoms, that is annulated with a **benzene** ring (Figure 1). Two of the best known and most widely used benzodiazepines are Librium and Valium. Both of these drugs were synthesized by Leo Sternbach of the Roche Drug Company in Nutley, New Jersey and introduced to the pharmaceutical market in the early 1960s. Each of these compounds generates emotional tranquility with usually only mild accompanying sedation of the patient. As such, this class of drugs represented a major breakthrough in the treatment of stress and anxiety disorders.

The apparent selectivity of these drugs for the relief of anxiety has attracted the attention of brain researchers. Various theories for the action of this class of compounds have been proposed. In 1977, two groups discovered that there are specific Valium binding sites in the brain. These specific **receptors** are suited to the molecular shape of and are able to recognize all benzodiazepines. The strength of binding is correlated with the effectiveness of the drug as an antianxiety compound, suggesting that these receptors are the active site for the drug's actions.

Further studies, by John Tallman, have shown that the effectiveness of the benzodiazepines is enhanced by gamma-aminobutyric acid (GABA) and vice versa. This compound is a common inhibitory neurotransmitter. The calming effects of the benzodiazepines are therefore connected with the inhibitory effects of GABA on neurons in various parts of the brain. What remains to be discovered is the naturally occurring compounds for which the benzodiazepines are analogues.

Unfortunately, the picture for these compounds is not without its dark side. Over time, the benzodiazepines are slightly addictive. Tolerance develops with habitual use and withdrawal symptoms occur when administration of the drug is stopped abruptly. Still, for many people, the relief from anxiety and stress provided far outweighs any potential problems.

Berg, Paul (1926-)

American biochemist

Paul Berg made one of the most fundamental technical contributions to the field of genetics in the twentieth century: he developed a technique for splicing together deoxyribonucleic acid (DNA)—the substance that carries the genetic information in living cells and viruses from generation to generation—from different types of organisms. His achievement gave scientists a priceless tool for studying the structure of viral chromosomes and the biochemical basis of human genetic diseases. It also let researchers turn simple organisms into chemical factories that churn out valuable medical drugs. In 1980 he was awarded the Nobel Prize in **chemistry** for pioneering this procedure, now referred to as recombinant **DNA** technology.

Today, the commercial application of Berg's work underlies a large and growing industry dedicated to manufacturing drugs and other chemicals. Moreover, the ability to recombine pieces of DNA and transfer them into cells is the basis of an important new medical approach to treating diseases by a technique called gene therapy.

Berg was born in Brooklyn, New York, on June 30, 1926, one of three sons of Harry Berg, a clothing manufacturer, and Sarah Brodsky, a homemaker. He attended public schools, including Abraham Lincoln High School, from which he graduated in 1943. In a 1980 interview reported in the *New York Times,* Berg credited a "Mrs. Wolf," the woman who ran a science club after school, with inspiring him to become a researcher. He graduated from high school with a keen interest in microbiology and entered Pennsylvania State University, where he received a degree in **biochemistry** in 1948.

Before entering graduate school, Berg served in the United States Navy from 1943 to 1946. On September 13, 1947, he married Mildred Levy and they had one son, John Alexander. After completing his duty in the navy, Berg continued his study of biochemistry at Western Reserve University (now Case Western Reserve University) in Cleveland, Ohio, where he was a National Institutes of Health fellow from 1950 to 1952 and received his doctorate degree in 1952. He did postdoctoral training as an American Cancer Society research fellow, working with **Herman Kalckar** at the Institute of Cytophysiology in Copenhagen, Denmark, from 1952 to 1953. From 1953 to 1954 he worked with biochemist **Arthur Kornberg** at Washington University in St. Louis, Missouri, and held the position of scholar in cancer research from 1954 to 1957.

He became an assistant professor of microbiology at the University of Washington School of Medicine in 1956, where he taught and did research until 1959. Berg left St. Louis that year to accept the position of professor of biochemistry at Stanford University School of Medicine. Berg's background in biochemistry and microbiology shaped his research interests during graduate school and beyond, steering him first into studies of the molecular mechanisms underlying intracellular **protein synthesis**.

During the 1950s Berg tackled the problem of how **amino acids**, the building blocks of **proteins**, are linked together according to the template carried by a form of **RNA** (ribonucleic acid, the "decoded" form of DNA) called messenger RNA (mRNA). A current theory, unknown to Berg at the time, held that the amino acids did not directly interact with RNA but were linked together in a chain by special molecules called joiners, or adapters. In 1956 Berg demonstrated just such a **molecule**, which was specific to the amino acid methionine. Each amino acid has its own such joiners, which are now called transfer RNA (tRNA).

This discovery helped to stoke Berg's interest in the structure and function of genes, and fueled his ambition to combine genetic material from different species in order to study how these individual units of heredity worked. Berg reasoned that by recombining a gene from one species with the genes of another, he would be able to isolate and study the transferred gene in the absence of confounding interactions with its natural, neighboring genes in the original organism.

In the late 1960s, while at Stanford, he began studying genes of the monkey tumor virus SV40 as a model for understanding how mammalian genes work. By the 1970s, he had mapped out where on the DNA the various viral genes occurred, identified the specific sequences of nucleotides in the genes, and discovered how the SV40 genes affect the DNA of host organisms they infect. It was this work with SV40 genes that led directly to the development of recombinant DNA technology. While studying how genes controlled the production of specific proteins, Berg also was trying to understand how normal cells seemed spontaneously to become cancerous. He hypothesized that cells turned cancerous because of some unknown interaction between genes and cellular biochemistry.

Paul Berg.

In order to study these issues, he decided to combine the DNA of SV40, which was known to cause cancer in some animals, into the common intestinal bacterium *Escherichia coli.* He thought it might be possible to smuggle the SV40 DNA into the bacterium by inserting it into the DNA of a type of virus, called a bacteriophage, that naturally infects *E. coli.*

A DNA molecule is composed of subunits called nucleotides, each containing a sugar, a **phosphate** group, and one of four nitrogenous bases. Structurally, DNA resembles a twisted ladder, or helix. Two long **chains** of alternating sugar and phosphate groups twist about each other, forming the sides of the ladder. A base attaches to each sugar, and **hydrogen bonding** between the bases—the rungs of the ladder—connects the two strands. The order or sequence of the bases determines the genetic code; and because bases match up in a complementary way, the sequence on one strand determines the sequence on the other.

Berg began his experiment by cutting the SV40 DNA into pieces using so-called restriction enzymes, which had been discovered several years before by other researchers. These enzymes let him choose the exact sites to cut each strand of the double helix. Then, using another type of **enzyme** called terminal transferase, he added one base at a time to one side of the double-stranded molecule. Thus, he formed a chain that extended out from the double-stranded portion. Berg performed the same biochemical operation on the phage DNA, except he changed the sequence of bases in the reconstructed phage DNA so it would be complementary to—and therefore readily bind to—the reconstructed SV40 section of DNA extending from the double-stranded portion. Such complementary extended portions of DNA that bind to each other to make recombinant DNA molecules are called "sticky ends."

This new and powerful technique offered the means to put genes into rapidly multiplying cells, such as bacteria, which would then use the genes to make the corresponding protein. In effect, scientists would be able to make enormous amounts of particular genes they wanted to study, or use simple organisms like bacteria to grow large amounts of valuable substances like human growth hormone, antibiotics, and insulin. Researchers also recognized that **genetic engineering**, as the technique was quickly dubbed, could be used to alter soil bacteria to give them the ability to "fix" **nitrogen** from the air, thus reducing the need for artificial **fertilizers**.

Berg had planned to inject the monkey virus SV40-bacteriophage DNA hybrid molecule into *E. coli*. But he realized the potential danger of inserting a mammalian tumor gene into a bacterium that exists universally in the environment. Should the bacterium acquire and spread to other E. coli dangerous, pathogenic characteristics that threatened humans or other species, the results might be catastrophic. In his own case, he feared that adding the tumor-causing SV40 DNA into such a common bacterium would be equivalent to planting a ticking cancer time bomb in humans who might subsequently become infected by altered bacteria that escaped from the lab. Rather than continue his ground-breaking experiment, Berg voluntarily halted his work at this point, concerned that the tools of genetic engineering might be leading researchers to perform extremely dangerous experiments.

In addition to this unusual voluntary deferral of his own research, Berg led a group of ten of his colleagues from around the country in composing and signing a letter explaining their collective concerns. Published in the July 26, 1974, issue of the journal *Science,* the letter became known as the "Berg letter." It listed a series of recommendations supported by the Committee on Recombinant DNA Molecules Assembly of Life Sciences (of which Berg was chairman) of the National Academy of Sciences.

The Berg letter warned, "There is serious concern that some of these artificial recombinant DNA molecules could prove biologically hazardous." It cited as an example the fact that *E. coli* can exchange genetic material with other types of bacteria, some of which cause disease in humans. "Thus, new DNA elements introduced into *E. coli* might possibly become widely disseminated among human, bacterial, plant, or animal populations with unpredictable effects." The letter also noted certain recombinant DNA experiments that should not be conducted, such as recombining genes for antibiotic resistance or bacterial toxins into bacterial strains that did not at present carry them; linking all or segments of DNA from cancer-causing or other animal viruses into plasmids or other viral DNAs that could spread the DNA to other bacteria, animals or humans, "and thus possibly increase the incidence of cancer or other disease."

The letter also called for an international meeting of scientists from around the world "to further discuss appropriate ways to deal with the potential biohazards of recombinant DNA molecules." That meeting was held in Pacific Grove, California, on February 27, 1975, at Asilomar and brought together a hundred scientists from sixteen countries. For four days, Berg and his fellow scientists struggled to find a way to safely balance the potential hazards and inestimable benefits of the emerging field of genetic engineering. They agreed to collaborate on developing safeguards to prevent genetically engineered organisms designed only for laboratory study from being able to survive in humans. And they drew up professional standards to govern research in the new technology, which, though backed only by the force of moral persuasion, represented the convictions of many of the leading scientists in the field. These standards served as a blueprint for subsequent federal regulations, which were first published by the National Institutes of Health in June 1976. Today, many of the original regulations have been relaxed or eliminated, except in the cases of recombinant organisms that include extensive DNA regions from very pathogenic organisms. Berg continues to study genetic recombinants in mammalian cells and gene therapy. He is also doing research in molecular biology of HIV–1.

The Nobel Award announcement by the Royal Swedish Academy of Sciences cited Berg "for his fundamental studies of the biochemistry of **nucleic acids** with particular regard to recombinant DNA." But Berg's legacy also includes his principled actions in the name of responsible scientific inquiry.

Berg was named the Sam, Lula and Jack Willson Professor of Biochemistry at Stanford in 1970, and was chairman of the Department of Biochemistry there from 1969 to 1974. He was also director of the Beckman Center for Molecular and Genetic Medicine (1985), senior postdoctoral fellow of the National Science Foundation (1961–68), and nonresident fellow of the Salk Institute (1973–83). He was elected to the advisory board of the Jane Coffin Childs Foundation of Medical Research, serving from 1970–80. Other appointments include the chair of the scientific advisory committee of the Whitehead Institute (1984–90) and of the national advisory committee of the Human Genome Project (1990). He was editor of *Biochemistry and Biophysical Research Communications* (1959–68), and a trustee of Rockefeller University (1990–92). He is a member of the international advisory board, Basel Institute of Immunology.

Berg received many awards in addition to the Nobel Prize, among them the American Chemical Society's Eli Lilly Prize in biochemistry (1959); the V. D. Mattia Award of the Roche Institute of Molecular Biology (1972); the Albert

Lasker Basic Medical Research Award (1980); and the National Medal of Science (1983). He is a fellow of the American Academy of Arts and Sciences, and a foreign member of the Japanese Biochemistry Society and the Académie des Sciences, France.

BERGIUS, FRIEDRICH (1884-1949)

German chemist

Friedrich Bergius was an organic chemist who, as research director of the Goldschmidt Company in Essen, Germany, was able to develop two **hydrogenation** processes that were widely used in industry. These high-pressure methods enabled both Germany and England to have sufficient supplies of motor fuel during World War II. Bergius also developed high-pressure methods for breaking **wood** down into edible products, a process that was called "food from wood." For his work with these methods, Bergius was awarded the 1931 Nobel Prize in **chemistry**.

Friedrich Karl Rudolf Bergius was born October 11, 1884, in Goldschmieden, Germany (now part of Poland). His father, Heinrich Bergius, was the head of a local chemical factory and his mother, Marie Haase Bergius, was the daughter of a classics professor. Both of Bergius's parents valued education, and at a young age Bergius observed the chemical processes in his father's factory. In addition, his father sent him to study metallurgy at a factory in the Ruhr valley, an area known for its heavy industry.

Bergius began his formal training in chemistry at the University of Breslau in 1903 under Albert Ladenburg and Richard Abegg. He conducted some doctoral research under Arthur Hantzsch at the University of Leipzig and completed his research under Abegg on concentrated **sulfuric acid** as a solvent. He was awarded his doctorate from the University of Breslau in 1907.

For several years after receiving his doctorate, Bergius worked with **Walther Nernst** in Berlin and **Fritz Haber** in Karlsruhe to develop a way of making **ammonia** from **hydrogen** and atmospheric **nitrogen**. By 1909 Bergius was working with Ernest Bodenstein at his chemical laboratory in Hanover. Under Bodenstein, Bergius used pressure as high as 300 atmospheres (one atmosphere equals 14.7 pounds per square inch) to study the breakdown of **calcium** peroxides. It was during these apprentice years that Bergius developed leakproof high-pressure apparatus that enabled him to extend his research into other areas. In order to advance his research in more than one area, Bergius established his own laboratory in Hanover. He used high-pressure techniques to transform heavy oils and oil residues into lighter oils. This procedure boosted **gasoline** output, of great interest at the time since the automobile was becoming the preferred form of transportation. In 1913 he was granted a patent for the manufacture of liquid hydrocarbons (compounds containing **carbon** and hydrogen) from **coal**. During this time Bergius was also teaching physical and **industrial chemistry** at a university in Hanover.

In addition to his position as research director for the Goldschmidt Company in Essen from 1914 to 1945, Bergius

Friedrich Bergius.

was instrumental in the construction of a plant at Rheinau to facilitate the development of coal-hydrogenation processes on a large scale. When the demand for gasoline decreased after World War I, however, this project was neglected until 1921. After selling his patent rights to German and foreign companies, Bergius was able to develop new equipment to process the coal hydrogenation. Previously, equipment had only been able to use high-pressure methods with **gases**.

Although he was able to improve the process of coal hydrogenation in numerous ways during the years between 1922 and 1925, it never became economically feasible. In 1926 he sold his patents to Badische Anilin-und Sodafabrik (BASF), a large German chemical company, which later joined other German chemical companies to form I. G. Farben. Farben expanded hydrogenation research, improved Bergius's processes, and increased the yield of gasoline from coal. Two years later Farben built a plant to produce oil from coal.

Another of Bergius's research projects was to produce sugar from wood and convert it into **alcohol**, yeast, and dextrose. He hoped to do this by using concentrated **hydrochloric acid** and **water** to promote the breakdown of wood **cellulose**. This work proved valuable to Germany during World War II, because it furnished a great deal of the carbohydrate material that was needed during food shortages.

Both of Bergius's lifetime interests, the hydrogenation of coal into gasoline and the hydrogenation of wood into food products, became widely used commercially and industrially. In 1931 he shared the Nobel Prize in chemistry with **Karl Bosch**, who had continued Bergius's work after he sold his patents to BASF. During the presentation of the award at the Royal Swedish Academy of Sciences, the presenter called Bergius's high-pressure methods an extraordinary improvement in the field of chemical technology.

Through the 1930s and the 1940s, Bergius continued his research on the **hydrolysis** (the breakdown of a substance using water) of wood. The plant he established in Rheinau in 1943 provided basic products that Germany needed during World War II. After the war, Bergius left Germany and lived briefly in Austria. For a time he lived in Madrid, where he founded a company at the invitation of the Spanish government. From 1946 to 1949 he was a technical research adviser to the government of Argentina in its ministry of industries. He died in Buenos Aires on March 30, 1949.

Bergius was married to Ottilie Krazert, and the couple had two sons and one daughter. Besides the Nobel Prize, Bergius was awarded the Liebig Medal of the German Chemical Society and received honorary degrees from the University of Heidelberg and the University of Hanover. He also contributed many articles to newspapers and to scientific and technical magazines and was a member of the American Chemical Society and the Verein Deutscher Chemiker.

Bergström, Sune Karl (1916-)

Swedish biochemist

Sune Karl Bergström is best known for his research on **prostaglandins**. These substances, which were first discovered in the prostate gland and seminal vesicles, were found by Bergström and his colleagues to affect circulation, smooth muscle tissue, and general **metabolism** in ways that can be medically beneficial. Certain prostaglandins, for example, lower blood pressure, while others prevent the formation of ulcers on the stomach lining. For his research, Bergström shared the 1982 Nobel Prize in medicine or physiology with John R. Vane and **Bengt Samuelsson**.

Sune Bergström was born in Stockholm on January 10, 1916, to Sverker and Wera (Wistrand) Bergström. Upon completion of high school he went to work at the Karolinska Institute as an assistant to the biochemist Erik Jorpes. The young Bergström was assigned to do research on the **biochemistry** of fats and **steroids**. Jorpes was impressed enough with his assistant to sponsor a year-long research fellowship for Bergström in 1938 at the University of London. While there, Bergström focused his research on bile acid, a steroid produced by the liver which aids in the digestion of **cholesterol** and similar substances.

Bergström had planned to continue his research in Edinburgh the following year thanks to a British Council fellowship, but the fellowship was canceled after World War II broke out. He did, however, receive a Swedish-American Fellowship in 1940, which allowed him to study for two years at Columbia University and to conduct research at the Squibb Institute for Medical Research in New Jersey. At Squibb, Bergström researched the steroid cholesterol, particularly its reaction to chemical combination with **oxygen** at room **temperature**, a process called auto-oxidation.

Bergström returned to Sweden in 1942, receiving doctorates in medicine and biochemistry from the Karolinska Institute two years later. He was appointed assistant in the biochemistry department of Karolinska's Medical Nobel Institute. While there, he continued experiments with auto-oxidation, working with linoleic acid, which is found in some vegetable oils. He discovered a particular **enzyme** was responsible for the **oxidation** of linoleic acid, and helped attempt to purify the enzyme while working with biochemist **Hugo Theorell**.

While attending a meeting of Karolinska's Physiological Society in 1945, Bergström met the physiologist Ulf von Euler. Von Euler, who was better known as the discoverer of the hormone norepinephrine, had been doing research on prostaglandins. Scientists had observed in the 1930s that seminal fluid used in artificial insemination stimulated contraction and subsequent relaxation in the smooth muscles of the uterus. Von Euler isolated a substance from the seminal fluid of sheep and found it had the same effect in relaxing the smooth muscle of blood vessels. Impressed with Bergström's work on enzyme purification, von Euler gave him some of the extract for further purification.

Bergström began initial experiments but put his work on hold when in 1946 he was named a research fellow at the University of Basel. Returning from Switzerland in 1947, he was appointed professor of physiological **chemistry** at the University of Lund. His first task was to help revitalize the university's research facilities, which had fallen into disuse during the war. Afterwards, he resumed his research on prostaglandins, assisted by graduate students such as Bengt Samuelsson. Working with new large supplies of sheep seminal fluid, Bergström and his colleagues were able to isolate and purify two prostaglandins by 1957. Bergström was appointed professor of chemistry at Karolinska a year later, and brought his research on prostaglandins and his collaboration with Samuelsson with him. By 1962, six prostaglandins, identified as A through F, had been identified.

Bergström and Samuelsson then worked on determining how prostaglandins are formed. They discovered that prostaglandins are formed from common **fatty acids**, and further identified specific functions performed by each prostaglandin. Over the next few years, Bergström and Samuelsson surmised that certain prostaglandins could be used to treat high blood pressure, blocked arteries, and other circulatory problems by relaxing muscle tissue. These prostaglandins were also shown to prevent ulceration of the stomach lining and to protect against side effects of such drugs as aspirin, long known to irritate the stomach lining. Other prostaglandins could be used to raise blood pressure or stimulate uterine muscle by their contracting effect.

Bergström remained at Karolinska, serving as dean of its medical school from 1963 to 1966 and as rector of the institute from 1969 to 1977. He was chairman of the Nobel Foundation's Board of Directors from 1975 to 1987, and from 1977 to 1982 he served as chairman of the World Health Organization's Advisory Committee on Medical Research. He retired from teaching in 1981, choosing to devote his full time to research at Karolinska.

A modest, reserved man, Bergström's reaction upon learning of his Nobel award was gratitude—first, that his colleagues appreciated his efforts, and second, that his former student Samuelsson had also been named. The book *Nobel Prize Winners* reports him as saying that there is "no greater satisfaction than seeing your students successful." His connection with the Nobel Foundation had led some to wonder whether he might be passed over for a prize of his own. But the *New York Times,* reporting on the 1982 awards, noted that "it was only a matter of time, most scientists agree, before Dr. Bergström's research would be honored by the foundation he directs—for the work was too important to be ignored through any concern over apparent conflicts of interest."

The scientist married the former Maj Gernandt in Sweden in 1943; the couple has one son. Bergström's memberships include the Royal Swedish Academy of Science (he served as its president from 1983 to 1985), the American Philosophical Society, and the American Academy of Arts and Sciences. Other awards given to Bergström besides the Nobel include the Albert Lasker Award in 1977, Oslo University's Anders Jahre Prize in Medicine in 1970, and Columbia's Louisa Gross Horwitz Prize in 1975.

BERKELIUM

Berkelium is a transuranic element, with an **atomic number** of 97 and an atomic **mass** of 274.0703. Its chemical symbol is Bk. Berkelium is also classified as an actinide element, one of the elements that makes up Row 7 of the **periodic table**.

Properties

Berkelium exists in such small quantities that very little is known about its properties. Its melting point is estimated to be about 1,810°F (986°C). Two allotropes of the element are known with densities of 13.25 and 14.78 grams per cubic centimeter. All known isotopes of berkelium are radioactive, with the longest-lived being berkelium-247, with a half life of 1,380 years.

Occurrence and Extraction

Berkelium has not been found in the Earth's crust. It is produced artificially by bombarding other transuranic elements in a particle accelerator.

Discovery and Naming

Berkelium was discovered in 1949 by a team of researchers at the University of California at Berkeley (UCB). The UCB team was analyzing reactions that occur when very heavy atoms are bombarded with alpha particles. When **americium** was used as a target in this research, evidence for the formation of a new element was obtained. The UCB research team selected the name berkelium for the new element in honor of the city in which the research was done.

Uses

There are no commercial uses for berkelium, although it is sometimes used in research projects.

BERKOWITZ, JOAN B. (1931-)

American physical chemist

Joan B. Berkowitz is a physical chemist who has worked in a number of different research areas throughout her career, including the **thermodynamics** of inorganic systems, the **electrochemistry** of flames, **oxidation** studies of refractory **metals** and alloys, inorganic coating technology, and high-temperature vaporization. Berkowitz has specialized in the area of environmental management since 1972, when she contributed to the U.S. Environmental Protection Agency's first report to Congress on hazardous waste. Berkowitz was the first woman president of the Electrochemical Society, and in 1983 she received the Achievement Award of the Society of Women Engineers for her pioneering contributions in the field of hazardous-waste management. Berkowitz is currently managing director of Farkas, Berkowitz, and Company, an environmental consulting agency based in Washington, D.C., which focuses on the evaluation of **waste treatment** and disposal, as well as remediation technologies and market-potential assessment.

Berkowitz was born in Brooklyn, New York, on March 13, 1931. Her father, Morris Berkowitz, worked as a salesman for the Englander Mattress Company and struggled to support his family during the Great Depression. Although her mother, Rose Gerber Berkowitz, did not work, she had been influenced by the women's suffrage movement and the extension of the vote; her interest in women's rights played a major role in Berkowitz's own life decisions. At the age of twelve, Berkowitz knew she would always earn a living and she saw the best opportunities for herself in science. She attend Swarthmore College on a scholarship and received a B.A. degree in **chemistry** in 1952. She wanted to follow her high-school boyfriend Arthur Mattuck to Princeton, where he was studying mathematics, but the chemistry department there did not then admit women. She was accepted at the University of Illinois at Urbana, however, and she completed her Ph.D. in **physical chemistry** in 1955 with both theoretical and experimental research on electrolytes (nonmetallic substances which conduct electricity).

Berkowitz accepted a postdoctoral position at Yale University from 1955 to 1957, and on September 1, 1959, she married Mattuck, who had obtained a professorship at the Massachusetts Institute of Technology. Berkowitz then began her long consulting career with Arthur D. Little, Inc. Here, she worked on studies of high-temperature oxidation, which was part of the research involved in the space program; compounds that could withstand the reentry temperatures of the upper atmosphere were important for rocket design and manned flights into space. Berkowitz worked mainly with the transition metals (those elements lying roughly in the middle of the periodic table), and in particular **molybdenum**, **tungsten**, and **zirconi-**

um, and investigated the properties of alloys (which are mixtures of pure elements) for their strength and hardness at extremely high temperatures. With this work, Berkowitz developed a patent for manufacturing reusable molds for **iron** and steel castings from molybdenum and tungsten—a patent which was eventually used to make space vehicles.

Berkowitz served as adjunct professor in the department of chemistry at Boston University from 1963 to 1968, teaching undergraduate courses in physical chemistry. Her work on high-temperature oxidation led to her investigations in hazardous-waste disposal. In 1975 Berkowitz headed a team that evaluated various physical, chemical, and biological techniques in the treatment of hazardous wastes and produced a two-volume report which still serves as an important reference work in the field of hazardous-waste treatment.

Berkowitz participated in the senior executive program of the Sloan School of Management at the Massachusetts Institute of Technology in 1979, and the following year became vice president of A. D. Little. Divorced in 1977, she moved to Washington, D.C., in 1986, where she first served as chief executive officer at Risk Science International. In 1989, along with Allen Farkas, she formed the firm of Farkas, Berkowitz, and Company, which specializes in management consulting on environmental projects and hazardous-waste management.

BERNAL, JOHN DESMOND (1901-1971)

Irish physicist

Although John Desmond Bernal was highly instrumental in the pioneering stages of x-ray **crystallography** and microbiology, he is perhaps most well-known for his philosophical studies of the social aspects of science. Marxist in thinking and communistic in politics, Bernal wrote a classic book on the ''science of science'' entitled *The Social Function of Science*. This work reflected the ideas of a large school of intellectuals and scientists influenced by Bernal loosely called the ''Invisible College.'' The basic premise of the book, that science is for everyone and that, used appropriately, could greatly improve the fate of humanity, was opposed by the scientific school of the day. Bernal, noted for his **energy** and exuberance, was professor of physics at the University of London and became a fellow of the Royal Society at age 36. He held positions on literally hundreds of scientific and political committees, and played a highly influential role in many other similar organizations.

Bernal was born in County Tipperary, Ireland, to upper-class parents from whom he gained a passion for science. His father, a patriotic Irish nationalist and devout Catholic, died when Bernal was 18. His mother was a well-educated and refined woman formerly from the United States. He attended Stonyhurst Jesuit Public School that taught no science; therefore, with the help of his mother, he planned his own education from the age of 12, showing considerable aptitude in mathematics and physics. His exposure to Marxist philosophy as an undergraduate at Cambridge ultimately led him to forgo Catholicism, not because of its theology, but because of its social teaching. In the book *Society and Science* edited by Maurice Goldsmith and Alan Mackay, C.P. Snow, friend and associate of Bernal, writes ''this conversion was linked with his insight to what applied science could do for his fellow human being.''

At Cambridge, Bernal was reputed for his physical toughness, quick wit, and fearlessness, and was highly popular among his peers. After leaving Cambridge, he went to the Davy-Faraday Research Laboratory where he continued his developing interest in crystallography. Here, he worked under the leadership of Sir Lawrence Bragg with other researchers who were embarking on the exciting new field of the study of materials. Bernal was reputed for his wild imagination, passion, and high energy; however, in spite of his personality, he painstakingly, patiently, and excitedly applied himself to the somewhat boring task of measuring **graphite** structures and surfaces.

In 1927, Cambridge instituted a new lectureship in crystallography, to which Bernal was appointed. He taught at Cambridge for 10 years, becoming highly influential in science, politics, and social forecasting. Known as a ''dazzling thinker and talker'' and recognized as a major power in the field of science, he was nicknamed ''Sage'' by his intellectual peers because of his wisdom and the profound effect he had upon both scientific and social thought. He also became a central figure in a revolution that ultimately took physics and **chemistry** into the field of biology. Through his experience with crystallography, Bernal had developed a highly skilled technique to unravel the structure of materials. He turned this technique and his focus to important biological agents such as **amino acids**, **vitamins**, **water** (because water is the primary substance of most organisms), **proteins**, and viruses. This was literally the beginning of microbiology. He was not only highly instrumental in drawing other scientists into the developing field, but most microbiologists in England in the 1960s were former students of his.

During World War II, Bernal served on a committee to help improve air raid precautions in England. He ultimately became scientific advisor to Lord Mountbatten, traveling to countries such as Burma, India, and Africa, which exposed him to the dire needs of the poor. He described the atomic bomb, the only scientific development that ever shocked him, as ''that wretched discovery.'' Bernal firmly believed that the most important factor in the advancement of science and humanity was the absence of war. Therefore, preventing future military conflict became one of his priorities. After the war, in his role as chairman of the Presidential Committee of the World Council of Peace, he negotiated with the Russian leader, Nikita Khrushchev, and other world leaders, for an agreement against the use of atomic weapons.

While Bernal engulfed himself in a multitude of peace councils and global social concerns, he continued to be deeply engrossed in pure and creative scientific endeavors. In the introduction to their book, Goldsmith and Mackay quote Bragg's remark that, ''if one traces back almost any fruitful line of crystallographic work, it will be found that Bernal assisted at its conception but left the child to be brought up by foster-parents. This is particularly so in the case of molecular biology and in the analysis of protein crystals. Immediately on seeing the first x-ray differentiation pictures from protein crystals...he assumed that protein structures would sooner or later be solved, and handed out problems to his students and to anyone whom he could persuade to take them up.''

In the mid 1960s, at the age of 66, Bernal was still deeply immersed in solving the structure of **liquids**, studying continental drifts and conditions surrounding the life of meteorites, analyzing the exodus of scientists from Britain, and working on the third edition of another monumental volume, *Science in History*.

Of his colleague, Snow writes: ''[Bernal] had, all his life, along with his wild natural generosity, something of the carelessness about money...has lived simply...and has been a freer soul than most men. He had a passionate revulsion from cruelty or even unkindness, and...in the simplest words, was a good man.''

Berner, Robert A. (1935-)

American geochemist

Robert A. Berner's research in sedimentary **geochemistry** led to the application of mathematical models to describe the physical, chemical, and biological changes that occur in ocean sediment. Berner, a professor of geology and geophysics at Yale University, also developed a theoretical approach to explain larger geochemical cycles, which led to the creation of a model for assessing atmospheric **carbon** dioxide levels and the **greenhouse effect** over geological time. A prolific researcher, Berner has written many scientific journal articles and is one of the most frequently quoted earth scientists in the *Science Citation Index.*

Robert Arbuckle Berner was born in Erie, Pennsylvania, on November 25, 1935, to Paul Nau Berner and Priscilla (Arbuckle) Berner. As a young man Berner decided to become a scientist because of his propensity for logical thinking. ''Science forces you to seek the truth and see both sides of an argument,'' he told Patricia McAdams. Berner began his academic studies at the University of Michigan where he earned his B.S. in 1957 and his M.S. a year later. He then went to Harvard University and earned his Ph.D. in geology in 1962. He married fellow geology graduate student Elizabeth Marshall Kay on August 29, 1959; they have three children—John Marshall, Susan Elizabeth, and James Clark.

Berner began his professional career at the Scripps Institute of Oceanography in San Diego, where he won a fellowship in oceanography after graduating from Harvard. In 1963 he was appointed assistant professor at the University of Chicago, and two years later he became an associate professor of geology and geophysics at Yale University. Since 1968 Berner has also served as associate editor or editor of the *American Journal of Science.* He was promoted to full professor at Yale in 1971, and in 1987 he became the Alan M. Bateman Professor of geology and geophysics.

Principles of Chemical Sedimentology, which Berner published in 1971, reflects the interest that has fueled much of his research. Berner sees the application of chemical **thermodynamics** and **kinetics** as a valuable tool in unveiling the secrets of sediments and sedimentary rocks. Thus, Berner's is an unconventional approach to sedimentology (the chemical study of sediments rather than the study of chemical sediments). Berner identifies his goal in *Principles of Chemical Sedimentology* as illustrating ''how the basic principles of physical **chemistry** can be applied to the solution of sedimentological problems.'' Berner's *Early Diagenesis,* published in 1980, is a study of the processes over geological time whereby sedimentary materials are converted into rock through chemical reactions or compaction. Because of the frequency with which *Early Diagenesis* has been quoted, it was declared a Science Citation Classic by the Institute for Scientific Information.

Berner observes in *Scientific American* that ''the familiar biological carbon cycle—in which atmospheric carbon is taken up by plants, transformed through **photosynthesis** into organic material and then recovered form this material by respiration and bacterial decomposition—is only one component of a much larger cycle: the geochemical carbon cycle.'' Berner has studied an aspect of this geochemical **carbon cycle** that is analogous to the transfer of carbon between plants, animals, and their habitats—the ''transfer of carbon between sedimentary rocks at or near the earth's surface and the atmosphere, biosphere and oceans.'' **Carbon dioxide** is vital to both these aspects of the geochemical carbon cycle, as carbon is primarily stored as carbon dioxide in the atmosphere. Berner's research has contributed to the ''BLAG'' model (named after Berner and his associates Antonio L. Lasaga and Robert M. Garrels) for assessing the changes in atmospheric levels of carbon dioxide throughout the earth's geological eras. First published in 1983 and subsequently refined, the BLAG model quantifies factors such as degassing (whereby carbon dioxide is released from beneath the earth), carbonate and silicate rock weathering, carbonate formation in oceans, and the rate at which organic **matter** is deposited on and buried in the earth that enable scientists to assess the climactic conditions of the planet's previous geological eras.

Berner's research on atmospheric carbon dioxide levels includes the study of the greenhouse effect, whereby carbon dioxide and other **gases** trap excessive levels of radiated **heat** within the earth's atmosphere, leading to a gradual increase in global temperatures. Since the nineteenth-century industrial revolution, this phenomenon has increased primarily because of the burning of fossil fuels such as **coal**, oil, and **natural gas**, and also because of deforestation. Berner reports in *Scientific American* that ''slow natural fluctuations of atmospheric carbon dioxide over time scales of millions of years may rival or even exceed the much faster changes that are predicted to arise from human activities.'' Thus, the study of the carbon cycle is essential to an objective evaluation of the greenhouse effect within larger geological processes. In 1986 Berner published the textbook *The Global Water Cycle: Geochemistry and Environment* which he co-authored with his wife Elizabeth, who is also a geochemist. *The Global Water Cycle* reviews the properties of **water**, marine environments, and water/energy cycles, and includes a discussion of the greenhouse effect. Berner's research has since focused on Iceland where he is investigating how volcanic rock is broken down by weathering and by the plant-life that gradually takes root on it.

Berner is enthusiastic about every aspect of his work as a geochemist. He enjoys the travelling associated with his research and likes to help students learn to think creatively for themselves. "I'm very proud of the twenty or so graduate students that have received Ph.D.s working with me. I've learned as much from them as they have from me," he told McAdams. Berner served as president of the Geochemical Society in 1983, and he is also a member of the National Academy of Sciences, the American Academy of Arts and Sciences, the Geological Society of America, and the Mineral Society. He has chaired the Geochemical Cycles Panel for the National Research Council and served on the National Committee on Geochemistry, the National Science Foundation Advisory Committees on Earth Sciences and Ocean Sciences, and the National Research Council Committee on Oceanic Carbon. He has received numerous awards, including an honorary doctorate from the Université Aix-Marseille III in France in 1991 and Canada's Huntsman Medal in Oceanography in 1993. His hobbies include Latin American music, tennis, and swimming.

BERTHOLLET, CLAUDE LOUIS (1748-1822)

French chemist

Claude Berthollet was born into a French family living in the Savoy, a region of France that was then part of Italy. Although the family's finances had declined, his parents were able to send him to a college in Turin, Italy, where he earned his medical degree in 1768. A few years later, he moved to Paris, France, where he studied **chemistry** and continued his medical work.

In 1784 Berthollet became director of the Gobelins textile factory. There he began research on the bleaching properties ofchlorine, which had been discovered by **Carl Wilhelm Scheele** in 1774. Like Scheele, Berthollet originally believed that **chlorine** was a compound of **oxygen** rather than an element. Through other experiments he determined the composition of **ammonia** (NH_3) and that the compound **potassium** chlorate explodes when mixed with **carbon**. He thought that this compound might replace conventional substances used to make **gunpowder**. In 1788 Berthollet attempted a public demonstration of this new, more powerful gunpowder, but it ended in disaster. The mixture proved to be too explosive, and four people in the crowd were killed on the spot.

When he first came to Paris, Berthollet met **Antoine-Laurent Lavoisier**, whose reputation and influence were then at their peak. Berthollet joined Lavoisier's team of chemists in developing a new, logical system of chemical **nomenclature**, which was proposed in 1787 and quickly adopted by chemists worldwide. However, Berthollet disagreed with Lavoisier's theory that all acids contain oxygen; Berthollet's analysis of prussic acid (hydrocyanic acid, or HCN) eventually proved Lavoisier's theory wrong.

Claude Louis Berthollet. *(Archive Photo. Reproduced by permission.)*

Berthollet's career survived the French Revolution, which began in 1789, and in the 1790s, Berthollet became scientific advisor and chemistry teacher to Napoleon Bonaparte (1769-1821). On a trip to Egypt with Napoleon in 1798, Berthollet noticed deposits of soda (sodium carbonate) on the shores of a saltwater lake. He concluded that the soda was formed because of prevailing physical conditions in the area. This was the beginning of Berthollet's theory that the speed of chemical reactions depends on more than just the attraction of one reactant to another. He determined that other factors such as the **concentration** of each reactant must also be considered. This idea, which Berthollet formally proposed in 1803, foreshadowed the *fundamental law of* **mass** *action*, which states that reaction speed depends on the masses of the reactants.

Berthollet also believed that the composition of a **chemical compound** depends on the masses of reactants, but this theory was proved wrong. Beginning in 1799, Berthollet carried on a friendly argument with fellow chemist Joseph Louis Proust, who maintained, correctly, that pure reactants always combine in the same proportions to produce exactly the same compound. Berthollet died in 1822.

Beryllium

Beryllium is the first member of Group 2 of the **periodic table**. The members of this group are also called the alkaline earth elements. Beryllium's **atomic number** is 4, its atomic **mass** is 9.012182, and its chemical symbol is Be.

Properties

Beryllium is a hard, brittle metal with a grayish-white surface. It is the least dense metal that can be used in construction. It has a melting point of 2,336°F (1,280°C) and a **boiling point** estimated to be about 4,500°F (2,500°C). Its **density** is 1.8 grams per cubic centimeter, and it has a high **heat** capacity and heat conductivity.

One of its important physical properties is transparency to x rays. This property is utilized in the production of beryllium windows for x-ray machines.

Chemically, beryllium is moderately active. It reacts with acids and **water** to form **hydrogen** gas and with **oxygen** to form beryllium **oxide** (BeO). The beryllium oxide formed in this reaction acts as a protective skin that prevents the metal itself from further **oxidation**.

Occurrence and Extraction

Beryllium is relatively common in the Earth's crust, with an abundance estimated at 2-10 parts per million. It never occurs as a free element in nature. Its most common compound is a complex beryllium **aluminum** silicate known as beryl, with the chemical formula $Be_3(Al_2(SiO_3))_6$.

Beryllium is extracted from its ores by first converting the ore to beryllium oxide. The beryllium oxide is then converted to beryllium chloride ($BeCl_2$) or beryllium fluoride (BeF_2), and then electrolyzed: $BeCl_2$ —electric current→ Be + Cl_2.

Discovery and Naming

Beryl was well known to ancient peoples, at least as far back as the time of ancient Egypt. However, nothing was known about its chemical composition until the late eighteenth century. Then, the French chemist Louis-Nicolas Vauquelin analyzed beryl and found a new element. Vauquelin suggested the name glucinium, meaning ''sweet tasting'' for the element because some of its compounds have a sweet **taste**. That name continued to be used by at least some chemists until 1957, when the element was officially given the name beryllium. The name is based on the element's most important ore, beryl.

Uses

By far the greatest use of beryllium metal is in the manufacture of alloys. Beryllium alloys are popular because they are tough, stiff, and lighter than similar alloys. For example, a new **alloy** of beryllium and aluminum called Beralcast was released in 1996. Beralcast is three times as stiff and 25% lighter than pure aluminum. Its widest applications are likely to be in the manufacture of helicopters and satellite guidance systems.

The most popular alloys of beryllium at the present time are those made with **copper** metal. These alloys contain about

Jöns Jacob Berzelius *(Swedish Information Service. Reproduced by permission.)*

2% beryllium. They conduct heat and **electricity** almost as well as pure copper, but are stronger, harder, and more resistant to fatigue and **corrosion**. Beryllium-cooper alloys are used in circuit boards, radar systems, computers, home appliances, aerospace systems, automatic systems in factories, automobiles, oil and gas drilling equipment, and heavy machinery.

About 15% of the beryllium used in the United States is in the form of beryllium oxide. The compound is a white powder that can be made into many different shapes. It is used in high-speed computers, auto ignition systems, **lasers**, microwave ovens, and sophisticated military applications.

Health Issues

Beryllium is a very toxic metal, especially in powder form. Inhalation of powdered beryllium can produce acute disorders similar to pneumonia or chronic damage to the respiratory system that may lead to bronchitis and lung cancer.

Berzelius, Jöns Jacob (1779-1848)

Swedish chemist

The Swedish chemist Jöns Jacob Berzelius (1779-1848) was one of the first European scientists to accept John Dalton's

atomic theory and to recognize the need for a new system of chemical symbols. He was a dominant figure in chemical science.

Jöns Jacob Berzelius, the son of a clergyman-schoolmaster, was born on Aug. 20, 1779, at Väversunda, Sweden. He studied for six years at the medical school at Uppsala and then studied **chemistry** at the Stockholm School of Surgery. In 1808 he was elected to the Swedish Academy of Science and was appointed its secretary in 1818. He married Elisabeth Poppius in 1835 and on that occasion was made a baron by the Swedish king, Charles XIV.

During the first decade of the nineteenth century, chemists were becoming aware that chemicals combined in definite proportions. This concept, sometimes known as Proust's law after the French chemist Joseph Louis Proust, showed that no matter under what circumstances separate elements combined, their proportions would always be in whole-number ratios. Berzelius was the first to prove beyond a doubt the validity of Proust's law and having been impressed by Dalton's theory of atoms, he proceeded to determine atomic weights. By 1818, Berzelius had obtained, with a high degree of accuracy, the atomic weights of no fewer than 45 elements.

While engaged in this work, Berzelius came to the conclusion that the system of full names for the elements was a hindrance, and he also rejected Dalton's set of symbols for the elements. As a substitute (and this system became the international code for the elements), Berzelius suggested that the initial of the Latin name or the initial plus the second letter be used to designate the element. Now O could be written for **oxygen**, H for **hydrogen**, and CO for **carbon** monoxide. By adding subscriptive numbers, other compounds could be symbolized, such as CO_2 for **carbon dioxide** and H_2O for **water**. Thus, a new international language of chemistry came into use.

The numerous experiments on the effects of an electrical current on chemical solutions had caught the imagination of the scientific world quite early in the nineteenth century. The electrical current used was that obtained from one of Volta's "galvanic piles." Berzelius and Wilhelm Hisinger worked with the voltaic pile, and in 1803, they reported that, just as an electrical current could decompose water, it could separate solutions of salts so that the acids formed would go to one pole while the alkalies would be collected at the opposite one. In further experiments, with M. M. Pontin, Berzelius succeeded in producing amalgams of **potassium**, **calcium**, and **ammonia**, by using **mercury** as the negative electrode.

From these experiments in **electrochemistry**, Berzelius arrived at his own electrochemical theory, which stated that all compounds can be divided into their positive and negative parts. This so-called dualistic theory held that all compounds are divided into two groups: those that are electropositive and those that are electronegative. In any chemical reaction, there is a **neutralization** of opposite electricities, and depending on the strength of the components, this reaction may vary from a very feeble one to ignition and **combustion**. The opposite of chemical combination, in Berzelius's view, was **electrolysis**, in which electric charges are restored and the combined molecular groups are separated.

In 1807, when he was appointed professor of medicine at the Stockholm School of Surgery, Berzelius began his researches in **organic chemistry**. At this time very little was known about organic chemistry, especially its involvement in life processes. Berzelius realized that he himself knew nothing of physiological chemistry. He thought that there might be some chemical process associated with the functions of the brain but admitted that the understanding of this seemed impossible. He began analyzing animal substances such as blood, bile, milk, membranes, bones, fat, flesh and its fluids, and animal semen. He discovered that blood contains **iron** and that muscular tissue contains lactic acid, the same acid found in sour milk. Most of his work in this field was inconclusive, as he was the first to realize. He concluded that analyses of animal products needed to await the day of more sophisticated techniques and apparatus, and he gave up his studies.

At an early point in his career, Berzelius became interested in a rare mineral, Bastnäs **tungsten**, and undertook an analysis of it. He came to the conclusion that it contained an unknown metal, and he and Hisinger named it **cerium** after the recently discovered asteroid Ceres.

Some years later, Berzelius discovered the element **selenium**, which was isolated from the sediment in **lead** tanks used in the manufacture of **sulfuric acid**. He named his new discovery after the Greek word for the moon. His next discoveries, of the elements **vanadium** and **thorium**, were named after the Norse goddess Vanadium and the god Thor.

Berzelius's *Textbook of Chemistry* went through many editions and was translated into the principal European languages. To this work he added his tables of atomic weights. He devised new methods of analysis and obtained values for combining weights not very different from those found today. He started by using the atomic weight figure of 100 for oxygen and related all of the other elements to it.

Much of Berzelius's work involved studies of minerals. He found that previous systems of classification were unreliable, so he proceeded to devise his own system, based not on description of crystal forms but on chemical composition. In 1836 the Royal Society of London awarded him the Copley Medal for this work.

Of great importance to chemical knowledge in the nineteenth century were two concepts in theory, both of which are associated with Berzelius: isomerism and catalysis. He remembered that the lactic acid he had discovered in muscle tissue behaved differently toward polarized light than the lactic acid of **fermentation**. Other examples of such behavior could be found, and Berzelius suggested that compounds of the same chemical composition which possess different chemical properties be called isomers, from the Greek word meaning equal parts. The importance of understanding isomerism was that it demonstrates that there is more involved in chemical structure than the ratios of the elements and atomic weight. The manner in which atoms are distributed in a **molecular structure** is a determining factor in the chemical properties of a compound.

In 1835, Berzelius advanced the theoretical concept of catalysis, or chemical change in which one agent produces the reaction without itself being changed. Berzelius wrote about this process as it applied to plant chemistry. He believed that in inorganic chemical reactions **metals** can act as catalytic agents. In summing up catalysis, Berzelius wrote, "Thus it is certain that substances . . . have the property of exerting an effect . . . quite different from ordinary chemical affinity, in that they promote the conversion . . . without necessarily participating in the process with their own component parts. . .."

As secretary of the Swedish Academy of Science, and also for some years as librarian of the academy, Berzelius began in 1821 to publish the *Annual Surveys of Progress in the Sciences*. Publication was continued until his death, at which time 27 volumes had been issued. His massive correspondence with scientists has been published, and it is a comprehensive picture of the great chemical world which was unfolding in his day.

It was perhaps natural that Berzelius, who achieved such great eminence early in the century, should have insisted on dominating the chemical sciences as they progressed. He was cheerful as a youth, but as he grew older and developed more and more health problems, he became conservative, argumentative, and even dictatorial. It has been said that when Berzelius condemned a new idea, it might just as well be forgotten, and that his insistence on the acceptance of his own ideas in part blocked the progress of chemistry. In his last years, he was still denouncing some of his colleagues for what he termed their "Swedish laziness." He died on August 7, 1848, and was buried in Stockholm.

BETA RADIATION

Beta radiation is the emission of an **electron** from the **nucleus** of a radioactive **isotope**. This electron comes from one of the neutrons in an unstable nucleus. The weak nuclear force is involved, and the **neutron** is converted into a **proton** when the beta particle is emitted. This produces an isotope of the next element in the **periodic table**, a process known as **transmutation**. The emitted beta particle travels through air at close to the speed of light. However, it can be stopped by a sheet of **aluminum** foil greater than 0.12 in (3 mm) thick.

When French physicist **Henri Becquerel** (1852-1908) first discovered the property of **radioactivity** in 1896, he did not know that radiation consists of particles as well as **energy**. Beginning in 1898, **Ernest Rutherford** (1871-1937) conducted experiments to determine the nature of this radiation. One experiment demonstrated that the radiation actually consisted of three different types: a positive particle, a negative particle, and a form of electromagnetic radiation that carried high energy. Further studies on the mass/charge ratio of the particles supported the idea that the negative radiation, which he had labeled "beta", had the same charge and **mass** as the particle identified by J.J. Thomson (1856-1940) as the electron. By 1902 Rutherford and his colleague **Frederick Soddy** (1877-1956) proposed that a different chemical element is formed whenever a radioactive element decays. Rutherford was awarded the 1908 **Chemistry** Nobel prize for his work in explaining these processes.

Although they had discovered a new type of reaction, physicists strongly believed that the conservation laws of classical physics would still apply. This meant that the decay reaction should exhibit **conservation of energy**, conservation of linear momentum, conservation of angular momentum, conservation of electric charge, and conservation of the number of particles in the nucleus. Three of the quantities were conserved when the beta particle was emitted; however, energy and angular momentum appeared to be missing. To satisfy these last two conservation laws, Wolfgang Pauli (1900-1958) suggested in 1934 that beta decay must involve a third particle that is neutral, has negligible rest mass and a spin of one-half, and carries away from the reaction the energy that appears to be missing. Enrico Fermi (1901-1954) named this particle the **neutrino**, Italian for "little neutral one"—its existence was accepted without evidence. It was not until 1953 that Frederick Reines (1918-) devised an experiment that successfully detected the neutrino; he received the 1995 Nobel Prize in Physics for that work.

During the 1930s, when scientists were experimenting with reactions that were produced by neutrons, Enrico Fermi and his associates discovered that a heavier isotope of **uranium** than is found in nature, uranium-239, will spontaneously give off a beta particle and change into a new element of one higher **atomic number**, 93; this element will also undergo beta decay to change into an additional new element, 94. The new elements were named **neptunium** and **plutonium**, respectively. This is the first record of the production of synthetic elements, known as transuranic elements. Early in 1999 synthesis of the element with atomic number 114 was reported.

More detailed explanations of beta decay were developed in the late twentieth century, following theories of the existence of unique particles that transmit the weak nuclear force. These particles, identified as W and Z, were finally discovered in 1983 by Carlo Rubbia (1934-), who was awarded the Physics Nobel Prize the following year (1984).

The decay chain of the most abundant uranium isotope uranium-238 to stable lead-206 involves six different beta decay reactions. In addition, a number of radioisotopes, both natural and man-made, have been identified as beta emitters. Rubidium-87, one of the relatively abundant minerals in the Earth's crust, is a beta emitter; its **half-life** is 49 billion years. The ratio of the amount of the element produced (strontium-87) to **rubidium** remaining in the sample is one of the methods that is being used by geologists to determine the age of the rocks of the Earth. Some of these isotopes have been put to practical use in industry and medicine. Often their application depends on the half-life of the element. **Tritium**, the radioactive isotope of **hydrogen** that is produced in the atmosphere, is also a beta emitter. Its half-life of 12.33 years is the right range for age dating fine wines and other materials that contain a high percentage of **water**. Cobalt-60 is an example of a beta emitter of shorter half-life (5.27 years); it is being used in medical applications. Iodine-131 (half-life 8.04 days) is also used in medical applications.

BIOCHEMISTRY

Biochemistry is the study of the **chemistry** of living organisms. Researchers in this area of science study life on a molecular scale. Biochemistry has its roots in biology, chemistry, and physics. It is not clear when biochemistry was first considered a separate field of science, but the synthesis of **urea**, an organic compound, from non-organic reactants by Friedrick Wöhler in 1828 serves as a convenient benchmark. In 1953, James Watson and **Francis Crick** made history when they used methods from biology, chemistry, and physics to describe the double helix structure of **DNA**.

Although biochemistry is a relatively young field of science, its principles have been used by humans for thousands of years. People in ancient China, India and the Mediterranean region employed biochemistry for making bread with yeast, fermenting beer and wine, and treating diseases with plant and animal extracts. However, it wasn't until the early 19th century that people began to examine the chemical properties of life.

Discoveries in the biological sciences and in chemistry and physics from this point in history onward laid the groundwork for the development of biochemistry. In the field of biology, important advances included discoveries about cell structure and function, the laws of genetics, and DNA. From the fields of chemistry and physics, a foundation for biochemistry could be found in the syntheses of organic chemicals, advances in the understanding of metabolic pathways, and the uncovering of the secrets of **molecular structure**.

Researchers in the biological sciences were interested in describing life from the perspective of cell organization and function. Other researchers sought to apply physical and chemical laws to uncovering how living cells functioned. Both approaches were equally valid, but their combination provided an even more powerful approach.

Biochemistry has evolved rapidly in the fifty years since 1950. The field can now be divided into three general research areas: cellular structure and function, **metabolism** and **energy** flow, and storage and transfer of genetic information.

The chemical basis of life rests on the 28 elements that naturally occur in living organisms. Some of these elements are present in large quantities; others exist in very small amounts. The most abundant elements are **carbon**, **hydrogen**, **oxygen**, **nitrogen**, **sulfur**, and **phosphorus**. Less abundant elements, which are equally necessary for life, include **calcium**, **manganese**, **iron**, and **iodine**. The elements found in living organisms serve as building blocks for molecules.

Cells are the smallest organized units in living **matter** that possess all the atributes of life. The molecules that make up living cells associate with one another to create very large molecules, called macromolecules. The four major components from which cells are constructed are **proteins**, **carbohydrates**, **nucleic acids**, and **lipids**. Proteins are formed when alpha **amino acids** are linked by peptide bonds. There are twenty common amino acids which combine to form proteins and they have different side groups with certain chemical properties. The sequence in which the amino acids combine determines the properties of the protein and its three-dimensional shape. This causes the surface of each protein to have certain unique features, which are responsible for the activity and function of the protein. The term carbohydrate is used to describe three types of molecules: monosaccharides, oligosaccharides, and polysaccharides. Monosaccharides are simple sugars, while oligo- and polysaccharides are polymers of monosaccharides which vary in length. Carbohydrates have either structural functions (such as **cellulose** which forms the walls of plant matter) or energy-storage functions. Nucleic acids are polymers of nucleotides, which themselves are base-sugar-phosphate compounds. The shape of nucleic acids is due to the presence of hydrogen bonds between complementary nucleic acid strands. DNA, which is the genetic material of chromosomes, consists of two complementary polynucleotides coiled into a double helix. The relationship between the nucleotide sequence and the protein amino-acid sequence determines the genetic code. The different types of **RNA**, which are also nucleic acids, have varied functions in the duplication of the genetic code. Lipids, or fats and oils, have low **solubility** in **water**. Phospholipids are structural lipids and are the main component of membrane walls. Phospholipids are molecules which contain a portion that is hydrophobic (dislikes water) and a portion which is hydrophilic (likes water). The hydrophobic parts of two layer of molecules (phospholipid bilayer) form the center of the membrane wall. Water-soluble matter cannot pass through the phospholipid bilayer unless transported by a protein. Proteins present on or in the phospholipid bilayer also catalyze many other important reactions. Exploring the chemical properties of the cell membrane's components can reveal how the membrane accomplishes its tasks. For example, in 1943, **Hans Adolf Krebs** was largely responsible for the formulation of the **citric acid** or **Krebs cycle**. This cycle describes a principal way in which the **oxidation** of carbohydrates to water and **carbon dioxide** is carried out with the production of a number of high-energy molecules for each **molecule** of carbon dioxide that is formed.

There are many other structures associated with cells in addition to the cell membrane. The interior of a cell contains structures called organelles that can be compared to the organs in a body. The organelles handle processes such as metabolism and energy flow. When the cell absorbs nutrients, chemical processes come into play that allow the cell to use these nutrients to support life.

Explorations in the three areas of biochemistry have lead to advances in health, medicine, and **nutrition**. For example, our need for **vitamins** is due to the fact that they function as coenzymes in many metabolic processes. As a result, diseases such as diabetes, sickle cell disease, and cystic fibrosis are now known to be caused by problems at the molecular level. This knowledge has lead to therapies that can help people with these diseases. Further explorations may help fight diseases such as AIDS, cancer, and Alzheimer's disease.

Another advancement that has been fueled by biochemical studies is recombinant DNA technology. This technology forms the foundation for the development of plants that can resist disease and contain more nutrients, and for creating new drugs that will aid in the fight against disease. Biochemistry also helps advance the field of biotechnology in which cells, organelles, and biological processes can be used in a variety of ways from creating medicines to cleaning up oil spills and toxic wastes.

Researchers today continue to explore life on the molecular level. Their discoveries pave the way for understanding diseases and finding cures, improving nutrition, and advancing the frontiers of other fields such as biotechnology and molecular biology.

BIOENERGETICS

Bioenergetics is the study of **energy** transformations in a living organism. It encompasses the fundamental energy-harvesting processes of **photosynthesis** and respiration, as well as other metabolic reactions that require or release energy. Energy in a cell is used for reproduction, synthesis of vital molecules, movement, and other basic functions. Virtually every reaction in the cell either requires or releases energy.

According to the Second Law of **Thermodynamics**, the **entropy** of the universe increases with every chemical change. And yet, in the face of this increasing disorder, living organisms carry out exceedingly complex reactions and remain highly organized. How is this possible? This maintenance of order within the living system is accomplished by constantly taking in high-energy, well-ordered materials from the environment, and excreting low-energy, less-ordered waste products. Glucose, for instance, is a high-energy compound used as a food source. Much of its energy is extracted during glycolysis and cell respiration. Its atoms are excreted in the form of **carbon** dioxide and **water**. The entropy of the organism therefore remains low, while the entropy of the universe becomes correspondingly higher.

The overall flow of energy through biological systems involves both catabolic and anabolic processes. Together, these two processes constitute **metabolism**, the sum of the biochemical processes in an organism. Catabolism is the breakdown of complex molecules into simpler constituents, usually with the release of energy. Anabolism is the reverse, involving the synthesis of complex molecules from simpler ones, with the input of energy. Photosynthesis is an anabolic reaction, requiring energy from the sun to build up sugars from **carbon dioxide** and water. Cell respiration is a catabolic process, returning glucose to carbon dioxide and water, and releasing stored energy. **Protein synthesis** is an anabolic process, using cellular energy and simple **amino acids** to build a complex protein.

Another central question addressed by bioenergetics is how reactions are regulated to allow them to occur at rates fast enough to support life. Although glucose and **oxygen** react spontaneously to liberate energy, they do so exceedingly slowly at room **temperature** outside of a cell. This is why a bowl of sugar remains essentially unchanged for months or even years, although it is exposed to copious amounts of oxygen during that time. Inside a cell, however, sugar is rapidly reacted.

The key to this difference in reaction rates is the presence of enzymes, **proteins** that catalyze reactions. A catalyst is a **molecule** that speeds the rate of a reaction without being used up, and without affecting the net energy change of the reaction.

Like all catalysts, enzymes speed reactions by lowering the **activation energy** required to begin the reaction. Most reactions initially require that bonds in the reactants be weakened, allowing the reactant atoms to pass into a less stable transition state, and finally to rearrange into their product state. Activation energy is the energy needed to achieve the transition state. Enzymes lower the activation energy by forming temporary weak bonds to the reactants. This stabilizes the transition state, making it less energetic and therefore easier to attain. Since more sets of reactants can achieve the transition state in the presence of the **enzyme**, the reaction can proceed faster. Regulation of enzymes is a key part of the cellular control of metabolism.

However, not all reactions will proceed faster simply because an enzyme is present. Reactions that release energy (exergonic reactions) will, but those that require energy (endergonic reactions) will not. To speed an endergonic reaction, the cell not only needs an enzyme, but a source of energy to feed into the reaction.

The fundamental way that an endergonic reaction is driven in the cell is by coupling it with an exergonic reaction. In this way, the energy released by the exergonic reaction can be partially harnessed to drive the endergonic one. The energy transfer is not 100% efficient, and the excess energy is released as **heat**, or thermal motion of the surroundings. In fact, 100% efficiency would generally be counterproductive, since more efficient transfers require longer times, and are more easily reversible. By wasting some of the energy as heat, the cell can insure the reaction happens quickly and irreversibly.

The most common exergonic reaction used for coupling is the **hydrolysis** of adenosine triphosphate (ATP). ATP reacts with water to cleave off the last of its three **phosphate** groups, forming adenosine diphosphate (ADP) and inorganic phosphate **ion** (Pi). This reaction releases approximately 12 kilocalories per **mole** of ATP consumed, enough to drive most endergonic reactions in the cell. A second phosphate can be hydrolyzed as well, releasing an additional 12 kcal/mole, for reactions requiring a stronger driving force.

Formation of ATP is also accomplished by coupling, in this case to reactions that are even more exergonic, such as the **oxidation** of glucose. The oxidation of one molecule of glucose via glycolysis and cell respiration releases enough energy to form 38 ATP molecules. Fats and other food molecules can also be oxidized to form ATP. In this way, ATP serves as the "energy currency" of the cell, acting as an intermediate between the energy-releasing reactions of catabolism, and the energy-requiring reactions of anabolism.

Hydrolysis and **dehydration** are so ubiquitous in cell metabolism that it is worth looking more closely at them. Dehydration—the removal of a water molecule—serves to link small molecules into longer **chains**. Amino acids are dehydrated to form proteins, nucleotides are dehydrated to form ribonucleic acid (RNA) and deoxyribonucleic acid (DNA) molecules, sugars are dehydrated to form complex **carbohydrates**, and **fatty acids** are dehydrated with **glycerol** to form fats. Dehydration is endergonic not only because the products are less strongly bonded, but because entropy is decreased in the synthesis of the more complex **macromolecule**. Hydrolysis—the addition of a water to split up a large molecule—

serves to return these complex macromolecules to their simpler constituents. The exergonic character of hydrolysis is due to both greater bond strength and greater entropy of the products.

ATP is also used to drive other energy-requiring processes, including the creation of **concentration** gradients across the **plasma** membrane of a cell. The transport protein Na^+/K^+ ATPase uses the energy from ATP hydrolysis to transport **sodium** ions out of the cell, and bring **potassium** ions in, creating large differences in the concentration of each ion across the membrane. Such gradients can be used for a variety of purposes. In cells of the intestinal lining, glucose is scavenged from the gut by a cotransporter for glucose and sodium. The high external concentration of sodium creates a strong drive for glucose transport, even though glucose is being moved ''uphill'' against its concentration gradient. In neurons, the potassium and sodium gradients are used for intracellular transmission of the nerve signal.

The ultimate source of energy for living organisms is the sun. Photosynthesis is the process by which solar energy is captured for the creation of high-energy compounds. Within the chloroplast, light is captured by the molecule **chlorophyll**, whose **electron** structure allows the absorption of light in the visible range. Energy is transferred to an electron removed from a water molecule, creating oxygen gas in the process. The energized electron is passed down a series of electron acceptors of increasing **electronegativity**. At several points along this electron transport chain, the energy released is used to pump H^+ ions across the chloroplast membrane, creating a gradient. The energy of this gradient is used to drive the production of ATP by the enzyme ATP synthase, as the H^+ ions flow through the enzyme. After being reenergized by another chlorophyll, the electron is combined with other H^+ ions and the energy carrier nicotinamide adenine dinucleotide phosphate ($NADP^+$) to form NADPH, a carrier of high-energy hydrogens used in anabolic reactions. The NADPH and ATP are used in the Calvin cycle to drive the creation of a three-carbon sugar, using carbon dioxide as the carbon source.

Energy is harvested from sugars by a three-step process. In glycolysis, the six-carbon glucose is oxidized to form two molecules of the three-carbon organic acid, pyruvate. This forms two ATPs, and two molecules of nicotinamide adenine dinucleotide (NADH), a carrier of high-energy hydrogens used in catabolic reactions. Pyruvate is transported to the mitochondrion, where it undergoes further catabolism, resulting in the production of carbon dioxide and more NADH, as well as another **hydrogen** carrier, flavin adenine dinucleotide ($FADH_2$), and guanosine triphosphate, similar to ATP. All NADHs and $FADH_2$ are processed in the mitochondria as well. Hydrogens, with their energy-rich electrons, are stripped off. The electrons pass down an electron transport chain similar to the one in the chloroplast, creating an H^+ gradient which is used to drive ATP synthesis. The electrons and hydrogens link up with oxygen to form water.

Bioinorganic chemistry

Chemistry is far from a homogeneous subject. It is composed of many sub-disciplines (e.g., **inorganic chemistry**, **biochemistry**, etc.) and these are, in turn, composed of still more sub-disciplines (e.g., organometallic chemistry, coordination chemistry, etc.). A relatively new sub-discipline in bioinorganic chemistry—a borderline field between biochemistry and inorganic chemistry. Bioinorganic chemistry studies the interactions and chemistry of inorganic elements (usually metal **ions**) within the biosphere.

But as young as this discipline is, bioinorganic chemistry is as old as life itself. Substantial arguments suggest that life evolved through bioinorganic-type reactions and compounds. This antiquity is observed in the prevalence of certain types of compounds in biological organisms of all shapes and sizes. One of the more commonly accepted notions in evolutionary biology is that the more prevalent a compound is, the older it is. Bioinorganic compounds, such as the iron heme complex, are literally everywhere—in every living organism. They are very old and very successful compounds in evolutionary terms.

What sorts of molecules qualify as bioinorganic? Bioinorganic compounds run the gamut from **zinc** fingers to hemoglobin to **chlorophyll** to bone material. All of these compounds rely upon an ''inorganic ion'' for their functionality and utility to biological organisms. From the smallest microbe to the blue whale, all organisms require inorganic elements to survive.

The organism, though, that we are most familiar with is the human being. To examine human requirements for inorganic compounds, the inorganic elements present in the human body are divided into three broad categories. The first is the ''bulk'' elements, which include **calcium**, **phosphorus**, **sulfur**, **potassium**, **chlorine**, **sodium**, and **magnesium**. All of these are present in quantities greater than one ounce (30 g) and, for the most part, participate in ''big'' processes in the body. For example, calcium is present in bones and teeth as the compound hydroxyapatite—a calcium-hydroxide-phosphate complex. But calcium also plays a role in the contraction of muscles and as a neurotransmitter.

The second category is the ''trace elements.'' **Iron**, zinc, **copper**, **fluorine**, and even **silicon** fall into this group. These are essential elements that are required for very specific enzymatic or structural roles within metalloproteins (**proteins** that require a metal ion to function). Usually, they occur as single atoms. An example of a trace element metalloprotein is zinc fingers. Zinc fingers have an amino acid complex with zinc that holds the protein backbone in a very specific configuration. Typically, in each finger, there are two zinc-binding sites. There are three or four fingers in the whole molecule. The functional role of this protein is to aid **DNA** replication. The protein grabs one end of a section of DNA and untwists it—much the same way that we can grab a piece of twisted rope and open it up. The result allows a portion of the DNA to replicate or do its job, while leaving the rest of the helix undisturbed. Zinc is, therefore, a critical factor in gene activation, which is why zinc deficiencies are linked to growth defects.

The final category of elements could be called the ''nonessentials.'' These are elements that are present in the human body, but they do not appear to have a biological role. Elements such as **rubidium**, **zirconium**, and **aluminum** are mostly incorporated into human biochemistry as a consequence of their association with essential elements present in the human diet. For instance, in digesting potassium and sodium, humans accidentally pick up a bit of rubidium, since it has similar chemical properties. Most of these elements are present in "ultra-trace" levels—levels well below one gram, and this sometimes makes them difficult to detect. Because they are difficult to detect, it is also sometimes difficult to determine whether or not they are essential. An example of this is **chromium**. Chromium was originally thought to be a contaminant of iron and non-essential for life, but it has now been recognized that it plays a vital role in regulating blood sugar levels.

Probably the most important of all the metal containing biological species is one that doesn't occur in humans but in plants. Chlorophyll-a is a magnesium complex that, in combination with iron-containing species, is capable of harvesting light **energy** from the Sun and converting it into an electrical potential. It is this electrical potential that drives the **enzymes** that convert **carbon dioxide** and **water** into sugar and **oxygen**. Without this vital bioinorganic molecule, life as we know it could not exist on Earth. We are oxygen-dependent creatures, tied to the energy from the Sun through a simple bioinorganic compound.

See also Blood chemistry

Biological Membranes

Biological membranes are those membranes enclosing the cell or any of the organelles within it. Membranes serve not only to enclose and define the cell or organelle, but to regulate the flow of materials passing in and out of it. The membrane surrounding a cell is known as the **plasma** membrane, while those inside it are called internal membranes.

Biological membranes contain approximately equal weights of two different types of molecules: phospholipids and **proteins**. Phospholipids give the membrane its basic structure, while proteins perform most of the regulatory functions.

The structure of the membrane is best understood by looking at the structure of the phospholipid **molecule**. In a phospholipid, a **glycerol** molecule is bonded to two fatty acid **chains** and one **phosphate** group. The **fatty acids** are long and nonpolar, while the phosphate group is relatively short and polar. This gives the molecule as a whole a polar (hydrophilic) end that attracts **water** and a nonpolar (hydrophobic) end that repels it. Molecules with both hydrophilic and hydrophobic parts are known as amphipathic.

When placed in water, phospholipid molecules tend to spontaneously orient themselves in a layer at the surface one molecule thick (a monolayer) with the polar end touching the water's surface and the nonpolar end facing away from it. The long fatty acid tails bond weakly with van der Waals attractions. Under the right conditions, two monolayers can join together in water to form a spherical bilayer, with the nonpolar ends of each layer touching on the interior of the bilayer, and the polar ends facing the water both outside and within the sphere. This phospholipid bilayer is the fundamental structure of all biological membranes.

The weak attractions between phospholipid molecules allow individual molecules in the membrane to move in two dimensions, and biological membranes are sometimes called two-dimensional fluids. With the large number of proteins embedded in membranes, they are also called fluid mosaics. The fluidity of the membrane is stabilized by **cholesterol**, a steroidal lipid found in all animal cell membranes. Cholesterol helps prevent drastic changes of fluidity that would otherwise occur with large **temperature** changes.

Phosphate groups of phospholipids each have other small groups covalently attached to them. These groups differ between the internal and external lipid layers, with more positively charged groups to the outside and negatively charged groups on the inside. This provides each side with a unique chemical signature, which may be important for orienting and embedding membrane proteins, for instance.

Membrane proteins may span the entire membrane, or be exposed to one surface or the other. Proteins of either type rely on hydrophobic **amino acids** on their surfaces to embed them in the membrane. Some proteins act as open channels for the movement of ions through the membrane, while others act as carriers for specific molecules, often ones in very low **concentration** that must be pumped into the cell using **energy**. **Glycoproteins** on the external surface act as cellular identity tags, allowing different cells to recognize one another for the purposes of forming tissues, or for protection by the immune system.

Small, uncharged molecules can cross the membrane without any transporter. These molecules include water, **oxygen**, and **carbon** dioxide. The concentration of each within the cell tends to equalize with the concentration outside the cell, because of **diffusion**. Diffusion of water across a semipermeable membrane such as the cell's plasma membrane is known as diffusion.

Channel proteins form pores to allow the diffusion of certain ions, such as **potassium**. Although it would seem the potassium channel would also allow the flow of other, smaller ions such as **sodium**, in fact the channel can be selective for potassium, due to its structure. All ions in **solution** are hydrated, surrounded by water molecules that bond to it through ionic attractions. Ions are unlikely to shed this cage of water unless they are stabilized by attraction to other negatively charged groups. The potassium channel is lined with such groups, spaced to accommodate the large potassium **ion**, but too far apart to effectively stabilize smaller ions. As a result, potassiums easily shed their hydrating waters upon entering the channel and pass through rapidly, while sodium does not. Other channels exist for sodium, chloride, and **calcium**, with the distribution and concentration of each depending on cell type.

Sodiums and potassiums are also actively transported by one of the most important membrane proteins, the K^+/Na^+ ATPase. This protein pumps three sodiums out of the cell, and two potassiums in, using the energy of one adenosine triphosphate (ATP). This creates transmembrane gradients for both ions. The potassium gradient dissipates by diffusion of potassium through its channel, leaving the cell with a net negative charge inside. The combination of the charge gradient and the sodium gradient, known as an electrochemical gradient, is a powerful force promoting the movement of sodium back into the cell. In the intestine, this force is used to drive the transport of glucose out of the gut and into the intestinal cell, by linking its transport to that of sodium. This is accomplished by another membrane protein, the sodium/glucose cotransporter. Binding of both triggers a **conformation** change that transports both into the cell.

The Na^+/K^+ ATPase is also used to regulate cell **volume**. The concentration of water inside the cell is usually lower than outside, due to the high concentration of large molecules such as proteins and **carbohydrates**, as well as ions. Therefore, water tends to flow into the cell by **osmosis**, down its concentration gradient. To prevent this, the cell runs the Na^+/K^+ ATPase, whose net effect is to decrease the number of ions in the cell, and increase it outside. In addition, since it creates a positive charge outside the cell, chloride ions tend flow out of the cell as well, further decreasing the osmotic potential for water. The Na^+/K^+ ATPase is so important to the function of cells that its operation consumes approximately 30% of the entire energy output of the cell.

Larger molecules such as proteins cannot be transported across the membrane by carriers. Instead, the movement of these molecules is accomplished by endocytosis (taking within the cell) and exocytosis (releasing outside the cell). In endocytosis, the substance to be transported (the ligand) first attaches to receptor proteins in the plasma membrane. These receptor-ligand complexes move across the outer surface of the membrane until they become engaged in special regions known as coated pits. Proteins on the inner surface of the membrane then cause this region of the membrane to pull in, forming a pocket that eventually seals and pinches off from the plasma membrane. This small sphere, known as a vesicle, is transported to processing organelles within the cell, where its contents are directed to their final destination. Cholesterol enters cells from the blood stream via this mechanism. It is transported through the bloodstream attached to low-density lipoprotein, or LDL. LDL attaches to LDL **receptors** on the cell surface.

Exocytosis is the reverse of endocytosis, in that an internal vesicle fuses with the membrane, releasing the contents to the exterior surface. Many protozoans, such as the amoeba, use endocytosis for capturing and ingesting food, and exocytosis for releasing waste products.

As noted previously, all internal organelles have membranes as well. These include lysosomes, the Golgi complex, the endoplasmic reticulum, and vacuoles (not found in human cells). Other organelles have double membranes, consisting of two phospholipid bilayers. These organelles are the **nucleus** and the mitochondria. In plants and algae, chloroplasts have three membranes—two forming the outer surface, and a third forming the grana.

Internal membranes serve to divide the cell into functional compartments, allowing highly specialized reactions to occur in different parts of the cell, free from interference with other parts.

The endoplasmic reticulum is the largest single membrane system in the cell. Its functions include formation of phospholipids, modification of proteins, and creation of vesicles. Vesicles join with the Golgi complex for further processing, and then bud off for exocytosis. The Golgi also creates lysosomes, organelles responsible for digestion of worn out cell components and endocytosed substances.

The multiple membranes of the mitochondrion and chloroplasts are thought to reflect their origins as free-living bacteria that developed a symbiotic relationship with host cells hundreds of millions of years ago. It is speculated that these bacteria were endocytosed, but rather than being consumed by the host cell, they became permanent residents, maintaining their own membranes while remaining surrounded by a host membrane as well.

See also Bioenergetics; Enzymes; Lipids; Transport proteins

BIOLUMINESCENCE

Bioluminescence refers to the light-emitting characteristics of certain living organisms. It is a special type of the more general process, **chemiluminescence** that relates to the production of light from a chemical reaction. The reaction that creates bioluminescence is generally the same in all organisms. It is a complex one that involves the **oxidation** of luciferin (a protein) catalyzed by luciferase **enzyme**.

A wide variety of organisms throughout nature produce light. The most familiar are fireflies and glow worms. However, bioluminescence is more commonly found in marine animals such as jellyfish, dinoflagellates, mollusks, shrimp, octopuses, fish, and sponges. Bacteria and fungi also have species that exhibit bioluminescence. These organisms create bioluminescence in many ways and for different reasons. Fireflies have glowing abdomens, which function in mating. Marine fireworms use bioluminescence for a similar reason. Deep sea fish have organs called photophores which create a center for the bioluminescent reactions. These organs not only help attract mates, they also attract prey and help illuminate the search for food. The angler fish has a photophore located on a long fin which it dangles as a lure in front of its mouth. Other uses of bioluminescence include camouflage and schooling. While it is evident why certain organisms display bioluminescence, some organisms such as mushrooms glow for unknown reasons.

As varied as the types of organisms are that produce bioluminescence, so too are the ways that they produce it. Some bacteria and fungi emit light constantly. Other organisms must be stimulated to produce light. Still others produce it only when required. There are various bioluminescent organisms that do not actually produce the light themselves, but have a symbiotic relationship with a luminescent organism. The most common colors are yellow, blue, red, and green. Certain shrimp have developed lenses and **color** filters which makes their bodies light up in a multicolor fashion.

Bioluminescence is a unique chemical reaction. It is a form of chemiluminescence that involves the generation of light without the production of **heat**. In these types of reactions the reactants form intermediates which have electrons in an excited state. When the electrons return to a relaxed state, they emit photons and produce light. The wavelength at which the **photon** is emitted is responsible for the color that is observed. The basic chemical reactions that produce bioluminescence are the same in all organisms. The principal chemicals involved are luciferin, luciferase, **oxygen**, and adenosine triphosphate (ATP). Most of the research performed to understand these reactions has been done using dried firefly tails.

The firefly luciferin has a chemical formula $C_{13}H_{12}N_2O_3S_2$. It is an organic compound with a complex **ring** system. In the presence of ATP and the luciferase enzyme, a complex containing all three is formed. It reacts with oxygen, releasing adenosine monophosphate (AMP), inorganic **phosphate**, the enzyme, **carbon** dioxide, and a photon of light. In the firefly, this reaction occurs in an instant flash in the abdomen. Other organisms like bacteria or fungi have a sustained bioluminescence. The sustained effect is a result of the reaction of riboflavin phosphate with oxygen and luciferase.

While the use of bioluminescent materials is limited by their expense, they have been used for certain applications. For example, in many countries, lanterns are made using ground up fireflies. Luciferin and luciferase have been used in biological research, space exploration, and oceanography to detect traces of ATP.

Bismuth crystals. ***(Photograph by Earl Scott, National Audubon Society Collection/Photo Researchers, Inc. Reproduced by permission.)***

BIRINGUCCIO, VANUCCIO (1480-1539)

Italian metallurgist

Vanuccio Biringuccio was a sixteenth century pioneer in the field of metallurgy. He deviated from and discounted the centuries-old idea of **alchemy** and **transmutation**, that a particular (and as yet undiscovered substance) would transform common **metals** into **silver** or **gold**. Biringuccio's theories and experiments laid the groundwork for what we know today as **chemistry**.

Biringuccio recorded his knowledge in the famous book, *Pirotechnia*, published in 1540. This ten-volume set was the first printed work to cover the entire field of contemporary metallurgy and touch on applied chemistry. In it, he addresses a multitude of subjects including the liquidation of silver from **copper**; the manufacture of steel; the description of semi-minerals such as alum, **arsenic**, common **salt**, saltpeter, and **manganese**; how **mercury** is obtained; and how to separate and purify **sulfur**. He distinguishes saltpeter from soda; formulates a method for casting bronze and other metals; and gives several compositions for **gunpowder**. He also describes a method for manufacturing weapons, discusses military arts, and explains the use of machinery driven by waterpower.

Little is know of Biringuccio's private life. He was born in Siena, Italy in 1480 and died in Rome in 1538 or 1539. In his youth, he worked for Pandolfo Petrucci, known as the Tyrant of Siena, falling in and out of popularity with the ruler. His already comprehensive knowledge of technology and metallurgy expanded greatly following a trip to Germany. Later in his career, he entered the service of the Duke of Parama, Duke Alphonso I of Ferrara, and the republic of Venice. He became known for his adroit military skills, which earned him the role of director of the arsenal for the Pope, the position he held until he died.

BISMUTH

Bismuth is the fifth and last element in Group 15 of the **periodic table**. Elements in this group are often referred to as the **nitrogen** family. Bismuth's **atomic number** is 83, its atomic **mass** is 208.9804, and its chemical symbol is Bi.

Properties

Bismuth is a soft, silvery metal with a bright, shiny surface and a yellowish or pinkish tinge. The metal breaks easily and can not be fabricated at room **temperature**. Its melting point is 520°F (271°C) and its **boiling point** is 2,840°F (1,560°C). Its **density** is 9.78 grams per cubic centimeter.

A somewhat unusual and valuable property of bismuth is its tendency to expand as it cools. Most **liquids** contract as they cool, so bismuth is an exception to the general rule. This property leads to some important applications of the metal. For example, when used as an **alloy** in type metal, it expands to fill all the corners of a mold, forming clean, crisp letters, numbers, and symbols.

Chemically, bismuth is a relatively active metal that combines with **oxygen** at room temperature. The bismuth **oxide** (Bi_2O_3) thus formed is responsible for the yellowish or pinkish tinge of the metal's surface.

Occurrence and Extraction

Bismuth is a relatively rare element in the Earth's crust with an abundance estimated at about 0.2 parts per million. It is rarely found in its elemental state. Its compounds tend to be found with ores of other **metals**, such as **lead**, **silver**, **gold**, and **cobalt**. The most important ore of bismuth is bismuthinite, also known as bismuth glance (Bi_2S_3). The largest producers of bismuth in the world are Mexico, Peru, China, Belgium, and Japan. Bismuth is produced in the United States only as a byproduct of lead refining at a plant in Nebraska.

Discovery and Naming

Bismuth and its compounds have been known for hundreds of years. During the fifteenth century, some of the first printing presses used type made of bismuth alloys. Scholars had a great deal of trouble distinguishing **arsenic**, **antimony**, and bismuth from each other. In fact, the first clear description of bismuth as an element was probably a book written by the French scholar Claude-Françoise Geoffrey in 1753.

The origin of bismuth's name is also uncertain. Historians of science believe that the name may have come from two German words, *weisse masse*, meaning "white mass." This phrase describes the appearance of the element in nature. Later the name was shortened to *wismuth*, and then to *bisemutum* before bismuth came into common use.

Uses

The primary use of bismuth metal is in the manufacture of alloys. A number of these alloys have low melting points, some as low as 158°F (70°C). Since this temperature is less than the boiling point of **water**, such alloys can be used in fire sprinkler systems, fuel tank safety plugs, solder, and other applications.

Recently there has been an increased interest in the use of bismuth as a substitute for lead in alloys. Bismuth adds many of the same properties to an alloy as does lead, but it is much less toxic. As an example, an alloy containing 97% bismuth and 3% **tin** is popular as shot used in waterfowl hunting.

Two-thirds of the bismuth produced in the United States is made into drugs, pharmaceuticals, and other chemicals. The most widely used compound is bismuth subsalicylate [$Bi(C_7H_5O_3)_3$], the active ingredient in many over-the-counter stomach remedies. Other compounds of bismuth are used for other medical purposes, such as the treatment of burns, stomach ulcers, and intestinal disorders. They also find wide application in **cosmetics**. Bismuth oxychloride (BiOCl), for example, is a lustrous white powder that is used in making face powder.

Joseph Black.

Black, Joseph (1728-1799)

Scottish chemist

The Scottish chemist Joseph Black is famous for his discovery of "fixed air" (**carbon dioxide**). He also discovered latent **heat** and was the first to recognize clearly the difference between intensity and quantity of heat.

Joseph Black was born on April 16, 1728, in Bordeaux, France, the son of a Scottish merchant settled in that city. Educated first at the University of Glasgow, he proceeded to the University of Edinburgh to complete his medical studies and presented his thesis there in 1754. This thesis, submitted, as was then customary, in Latin, was published in English in an expanded form in 1756 under the title *Experiments upon Magnesia Alba, Quicklime, and Some Other Alcaline Substances.*

The work described in this thesis sounded the death knell of the **phlogiston** theory and led in due course to the development of the modern system of **chemistry** through the work of Lavoisier and others. In his thesis Black showed by careful quantitative experiments that magnesia alba, a mild alkali, lost weight on heating; that this loss in weight was due to the release of an air, different from ordinary atmospheric air, which he named "fixed air" (now known as **carbon** dioxide); and that the ignited magnesia no longer effervesced with acids. Mild alkalies were thus shown to differ from caustic alkalies by containing "fixed air" in combination, and the same "fixed air" was later found by him to be produced in respiration, in **fermentation**, and in the **combustion** of **charcoal**. To appreciate the full significance of these results, it should be re-

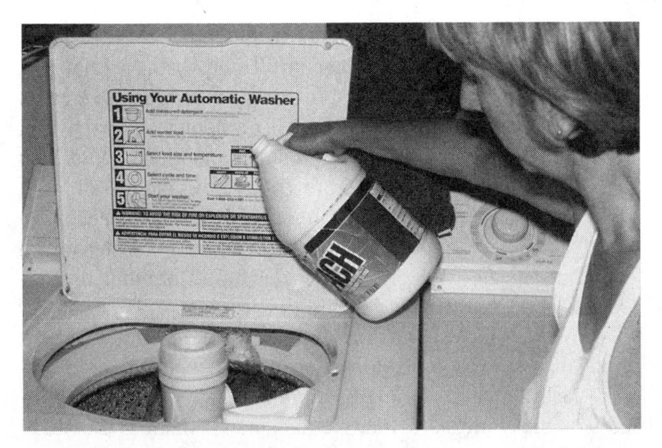

Woman pouring chlorine bleach into washing machine. ***(Photograph by Robert J. Huffman.)***

membered that prior to Black's work it was believed that limestone (a mild alkali) on heating absorbed fiery particles (phlogiston) and was thereby converted to quicklime (a caustic alkali). Black's application of the chemical **balance** to the study of such chemical reactions demonstrated the falsity of this view and in the broader sense was perhaps his greatest contribution to science.

When Black moved to Glasgow in 1756 as professor of anatomy and chemistry, he turned his attention to the study of heat, applying to it the same quantitative approach he had used in his chemical work. He showed that different substances have different capacities for heat. Further studies led him to the discovery of latent heat and to the first reasonably accurate measurements of the latent **heat of vaporization** and freezing of **water**. James Watt later applied these discoveries in his development of the steam engine. Black returned to the University of Edinburgh in 1766 as professor of chemistry and medicine, a position which he occupied until his death on December 6, 1799.

Bleaches

Bleaches are used to eliminate **color** imperfections such as grayness from natural fibers, including wool and cotton, and to impart a pleasing white, laundered effect. They are also used to lighten the color of flour and other foodstuffs; to treat **wood** pulp prior to its manufacture into **paper**; as disinfectants; to sanitize swimming pools; to remove mold; as deodorants; and as cleaning agents. Although bleaches are commonly used to remove color from clothing, there are other ways, such as exposure to sunlight, to achieve the same effect. The bleaching action is accomplished by **oxidation** of the colored impurities by the bleach.

There are two basic types of bleaches: **chlorine** (hypochlorite) bleaches, and peroxy type bleaches. The hypochlorite bleaches tend to be more powerful oxidizers than the peroxy compounds. **Sodium** bromide is sometimes used in combination with the hypochlorites to bleach **cellulose** materials.

Sodium hypochlorite, which is now sold commercially as a liquid bleach in household (5.25 wt % NaClO) and industrial (13 wt % NaClO) strengths, has been available since 1787. **Calcium** hypochlorites, which are produced by passing chlorine gas over slaked **lime**, have been used since 1799 as bleaching powders. Other commonly used chlorine compounds include **potassium** dichloroisocyanurate and chlorinated trisodium **phosphate**.

The chlorine bleaches deliver the whitening action of chlorine without requiring the direct handling of liquid or gaseous chlorine. When properly used, chlorine bleach is a simple and effective disinfectant. However, this type of bleach and its vapors are irritating to the skin, eyes, nose, and throat. Dermatitis may result from direct skin contact. Ingestion can cause esophageal injury, stomach irritation, and prolonged nausea and vomiting. And chlorine bleaches, when mixed with acidic substances such as **ammonia**, toilet bowl cleaners, drain cleaner, or vinegar, form toxic **gases** that may prove fatal if inhaled.

Chlorine bleach is a highly reactive chemical that rapidly converts, through oxidative reactions, chiefly into NaCl (table salt), **oxygen**, and **water**. Chlorine bleach manufacturers, therefore, claim that problems with biodegradability will not arise when this bleach is used in the correct manner and in recommended quantities. But critics are quick to point out that, legally, a product can be said to be biodegradable if at least 80% of it biodegrades in 90-180 days, and this definition can leave as much as another 20% in the environment for longer periods of time as toxic waste.

Of the peroxy type bleaches, sodium perborate ($NaBO_3 \cdot 4H_2O$) is the least expensive. Organic peroxides such as dibenzoyl peroxide, are used in the bleaching of flour. **Hydrogen** peroxide is generally made from nonrenewable resources, but decomposes into hydrogen and oxygen. It is safer than chlorine (hydrogen peroxide is often used as a topical antiseptic and mouthwash). It is used to bleach cotton, wool, and groundwood pulp, and is a component in hair bleaching preparations.

Optical bleaches (or fluorescent whitening agents) are organic **chromophores** that absorb ultraviolet light and then emit part of this **energy** as visible light. They thus enhance the appearance of laundered fabrics by causing the fabric to appear brighter and whiter.

Bloch, Konrad (1912-)

American biochemist

Konrad Bloch's investigations of the complex processes by which animal cells produce **cholesterol** have helped to increase our understanding of the **biochemistry** of living organisms. His research established the vital importance of cholesterol in animal cells and helped lay the groundwork for further research into treatment of various common diseases. For his contributions to the study of the **metabolism** of cholesterol, he was awarded the 1964 Nobel prize for Physiology or Medicine.

Konrad Emil Bloch was born on January 21, 1912 in the German town of Neisse (now Nysa, Poland) to Frederich (Fritz) D. Bloch and Hedwig Bloch. Sources list his mother's

Konrad Bloch.

maiden name variously as Steiner, Steimer, or Striemer. After receiving his early education in local schools, Bloch attended the Technische Hochschule (technical university) in Munich from 1930 to 1934, studying **chemistry** and **chemical engineering**. He earned the equivalent of a B.S. in chemical engineering in 1934, the year after Adolf Hitler became chancellor of Germany. As Bloch was Jewish, he moved to Switzerland after graduating and lived there until 1936.

While in Switzerland, he conducted his first published biochemical research. He worked at the Swiss Research Institute in Davos, where he performed experiments involving the biochemistry of phospholipids in tubercle bacilli, the bacteria that causes tuberculosis.

In 1936, Bloch emigrated from Switzerland to the United States; he would become a naturalized citizen in 1944. With financial help provided by the Wallerstein Foundation, he earned his Ph.D. in biochemistry in 1938 at the College of Physicians and Surgeons at Columbia University, and then joined the Columbia faculty. Bloch also accepted a position at Columbia on a research team led by Rudolf Schoenheimer. With his associate David Rittenberg, Schoenheimer had developed a method of using radioisotopes (radioactive forms of atoms) as tracers to chart the path of particular molecules in cells and living organisms. This method was especially useful in studying the biochemistry of cholesterol.

Cholesterol, which is found in all animal cells, contains 27 **carbon** atoms in each **molecule**. It plays an essential role in the cell's functioning; it stabilizes cell membrane structures and is the biochemical ''parent'' of **cortisone** and some sex **hormones**. It is both ingested in the diet and manufactured by liver and intestinal cells. Before Bloch's research, scientists knew little about cholesterol, although there was speculation about a connection between the amount of cholesterol and other fats in the diet and arteriosclerosis (a buildup of cholesterol and lipid deposits inside the arteries).

While on Schoenheimer's research team, Bloch learned about the use of radioisotopes. He also developed, as he put it, a ''lasting interest in intermediary metabolism and the problems of biosynthesis.'' Intermediary metabolism is the study of the biochemical breakdown of glucose and fat molecules and the creation of **energy** within the cell, which in turn fuels other biochemical processes within the cell.

After Schoenheimer died in 1941, Rittenberg and Bloch continued to conduct research on the biosynthesis of cholesterol. In experiments with rats, they ''tagged'' **acetic acid**, a 2-carbon compound, with radioactive carbon and **hydrogen** isotopes. From their research, they learned that acetate is a major component of cholesterol. This was the beginning of Bloch's work in an area that was to occupy him for many years—the investigation of the complex pattern of steps in the biosynthesis of cholesterol.

Bloch stayed at Columbia until 1946, when he moved to the University of Chicago to take a position as assistant professor of biochemistry. He stayed at Chicago until 1953, becoming an associate professor in 1948 and a full professor in 1950. After a year as a Guggenheim Fellow at the Institute of Organic Chemistry in Zurich, Switzerland, he returned to the United States in 1954 to take a position as Higgins Professor of Biochemistry in the Department of Chemistry at Harvard University. Throughout this period he continued his research into the origin of all 27 carbon atoms in the cholesterol molecule. Using a mutated form of bread mold fungus, Bloch and his associates grew the fungus on a culture that contained acetate marked with radioisotopes. They eventually discovered that the two-carbon molecule of acetate is the origin of all carbon atoms in cholesterol. Bloch's research explained the significance of acetic acid as a building block of cholesterol, and showed that cholesterol is an essential component of all body cells. In fact, Bloch discovered that all steroid-related substances in the human body are derived from cholesterol.

The transformation of acetate into cholesterol takes 36 separate steps. One of those steps involves the conversion of acetate molecules into squalene, a **hydrocarbon** found plentifully in the livers of sharks. Bloch's research plans involved injecting radioactive acetic acid into dogfish, a type of shark, removing squalene from their livers, and determining if squalene played an intermediate role in the biosynthesis of cholesterol. Accordingly, Bloch traveled to Bermuda to obtain live dogfish from marine biologists. Unfortunately, the dogfish died in captivity, so Bloch returned to Chicago empty-handed. Undaunted, he injected radioactive acetate into rats' livers, and was able to obtain squalene from this source instead. Working with Robert G. Langdon, Bloch succeeded in showing that squalene is one of the steps in the biosynthetic conversion of acetate into cholesterol.

Bloch and his colleagues discovered many of the other steps in the process of converting acetate into cholesterol. **Feodor Lynen**, a scientist at the University of Munich with whom he shared the Nobel Prize, had discovered that the chemically active form of acetate is acetyl **coenzyme** A. Other researchers, including Bloch, found that acetyl coenzyme A is converted to mevalonic acid. Both Lynen and Bloch, while conducting research separately, discovered that mevalonic acid is converted into chemically active isoprene, a type of hydrocarbon. This in turn is transformed into squalene, squalene is converted into anosterol, and then, eventually, cholesterol is produced.

In 1964, Bloch and his colleague Feodor Lynen, who had independently performed related research, were awarded the Nobel Prize for Physiology or Medicine "for their discoveries concerning the mechanisms and regulation of cholesterol and fatty acid metabolism." In presenting the award, Swedish biochemist **Sune Bergström** commented, "The importance of the work of Bloch and Lynen lies in the fact that we now know the reactions that have to be studied in relation to inherited and other factors. We can now predict that through further research in this field... we can expect to be able to do individual specific therapy against the diseases that in the developed countries are the most common cause of death." The same year, Block was honored with the Fritzsche Award from the American Chemical Society and the Distinguished Service Award from the University of Chicago School of Medicine. He also received the Centennial Science Award from the University of Notre Dame in Indiana and the Cardano Medal from the Lombardy Academy of Sciences the following year.

Bloch continued to conduct research into the biosynthesis of cholesterol and other substances, including glutathione, a substance used in protein metabolism. He also studied the metabolism of olefinic **fatty acids**. His research determined that these compounds are synthesized in two different ways: one comes into play only in aerobic organisms and requires molecular **oxygen**, while the other method is used only by anaerobic organisms. Bloch's findings from this research directed him toward the area of comparative and evolutionary biochemistry.

Bloch's work is significant because it contributed to creating "an outline for the chemistry of life," as E.P. Kennedy and F.M. Westheimer of Harvard wrote in *Science.* Moreover, his contributions to an understanding of the biosynthesis of cholesterol have contributed to efforts to comprehend the human body's regulation of cholesterol levels in blood and tissue. His work was recognized by several awards other than those mentioned above, including a medal from the Societe de Chimie Biologique in 1958 and the William Lloyd Evans Award from Ohio State University in 1968.

Bloch served as an editor of the *Journal of Biological Chemistry,* chaired the section on metabolism and research of the National Research Council's Committee on Growth, and was a member of the biochemistry study section of the United States Public Health Service. Bloch has also been a member of several scientific societies, including the National Academy of Sciences, to which he was elected in 1956, the American Academy of Arts and Sciences, and the American Society of Biological Chemists, in addition to the American Philosophical Society.

Katharine Burr Blodgett. ***(AP/Wide World. Reproduced by permission.)***

Bloch and his wife, the former Lore Teutsch, met in Munich and married in the United States in 1941. They have two children, Peter and Susan. Bloch is known for his extreme modesty; when he was awarded the Nobel Prize, the *New York Times* reported that he refused to have his picture taken in front of a sign that read, "Hooray for Dr. Bloch!" He enjoys skiing and tennis, as well as music.

BLODGETT, KATHARINE BURR (1898-1979)

American chemist

Katharine Burr Blodgett, the first woman to become a General Electric (GE) scientist, made several significant contributions to the field of industrial **chemistry**. The inventor of invisible, or non-reflecting, **glass**, Blodgett spent nearly all of her professional life working in the Schenectady, New York, GE plant. Although Blodgett's name has little household recognition, some of the techniques in surface chemistry that she and her supervisor and mentor **Irving Langmuir** developed are still used in laboratories; in addition, Blodgett's invisible glass is used extensively in camera and optical equipment today.

Blodgett was born on January 10, 1898 in Schenectady, New York, the town in which she spent most of her life. Her parents had moved to Schenectady earlier in the decade from their native New England when Blodgett's father, George Bedington Blodgett, became the head of the patent department at the GE plant opening up in town. Blodgett never knew her father, who died a few weeks before she was born. Left widowed with two small children, Blodgett's mother, Katharine Buchanan Burr, decided to move back east to New York City; three years later, she moved the family to France so that her children would be bilingual. After a few years of French schooling, Blodgett spent a year at an American school in Saranac Lake, New York, followed by travel in Germany. While in her mid-teens, Blodgett returned with her family to New York City where she attended the now-defunct Rayson School. Blodgett later won a scholarship to the all-women's Bryn Mawr College, where she excelled at mathematics and physics.

After college, Blodgett decided that a career in scientific research would allow her to further pursue both of these academic interests. During Christmas vacation of her senior year, she traveled to upstate New York to explore employment opportunities at the Schenectady GE plant. Some of George Blodgett's former colleagues in Schenectady introduced his daughter to research chemist Irving Langmuir. After conducting to a tour of his laboratory, Langmuir told eighteen-year-old Blodgett that she would need to broaden her scientific education before coming to work for him.

Taking Langmuir's advice, Blodgett enrolled in the University of Chicago in 1918 to pursue master's degree in science. Since she knew that a job awaited her in industrial research, Blodgett picked a related thesis subject: the chemical structure of gas masks. Upon graduating, Blodgett returned to GE, where Langmuir hired her as his assistant (the first female research scientist the company had ever employed). At the time, Langmuir—who had worked on vacuum pumps and light bulbs early in his GE career—had turned his attention to studying current flow under restricted conditions. Blodgett soon started working with Langmuir on these studies; between 1918 and 1924, the two scientists wrote several papers about their work. Blodgett's collaboration with the 1932 Nobel winner lasted until Langmuir's death in 1957.

Blodgett soon realized that she would need a doctoral degree if she wanted to further her career at GE. Six years after Blodgett started working for him, Langmuir arranged for his associate to pursue doctoral studies in physics at the Cavendish Laboratory at England's Cambridge University. Blodgett needed her mentor's help to gain admission to Cavendish because laboratory administrators hesitated to give one of their few open spots to a woman. With Langmuir's endorsement, however, Blodgett was able to persuade the Cambridge physicists—including Nobel winner **Ernest Rutherford**—to allow her entrance. In 1926, Blodgett became the first woman to receive a doctorate in physics from Cambridge University.

When Blodgett returned to Schenectady, Langmuir encouraged her to embellish some of his earlier discoveries. First, he set her to work on perfecting **tungsten** filaments in electric lamps (the work for which he had received a patent in 1916). Langmuir later asked his protege to concentrate her studies on surface chemistry. In his own long-standing research on the subject, Langmuir had discovered that oily substances formed a one-molecule thin film when spread on **water**. By floating a waxed thread in front of **stearic acid** molecules, the scientist showed that this layer was created by the molecules' active ends resting on the water's surface. Blodgett decided to see what would happen if she dipped a metal plate into the molecules; attracted to the metal, a layer of molecules formed similar to that on the water. As she inserted the plate into the **solution** again and again, Blodgett noticed that additional layers—all one molecule—formed on top of one another. As the layers formed, different colors appeared on the surface, colors which could be used to gauge how many layers thick the coating was. Because this measurement was always constant, Blodgett realized she could use the plate as a primitive gauge for measuring the thickness of film within one micro-inch.

Not long after Blodgett's discovery, GE started marketing a more sophisticated version of her **color** gauge for use in scientific laboratories. The gauge was comprised of a sealed glass tube that contained a six-inch strip on which successive layers of molecules had formed. To measure the thickness of film few millionths of an inch thick, the user need only compare the color of film with the molecular grades. The gauge could measure the thickness of a transparent or semitransparent substances within one to twenty millionths of an inch as effectively as much more expensive optical instruments, a very effective device for physicists, chemists, and metallurgists.

Blodgett continued working in the field of surface chemistry. Within five years, she had found another practical application that stemmed from Langmuir's original studies: nonreflecting, or invisible, glass. Blodgett discovered that coating sheets of ordinary glass with exactly forty-four layers of one-molecule thick transparent liquid soap rendered the glass invisible. This overall layer of soap—four-millionths of an inch thick and one quarter the wave length of white light—neutralized the light rays coming from the bottom of the glass with those coming from the top so that no light was reflected. Since the transparent soap coating blocked only about one percent of the light coming in, invisible glass was perfect for use in optical equipment—such as cameras and telescopes—in which multiple reflecting lenses could affect performance.

Blodgett did not hold sole credit for creating invisible glass. Two days after she announced her discovery, two physicists at the Massachusetts Institute of Technology (MIT) publicized that they had found another method of manufacturing non-reflecting glass using **calcium** fluoride condensed in a vacuum. Both groups of scientists, however, were concerned that their coatings were not hard and permanent enough for industrial use. Using some Blodgett's insights, the MIT scientists eventually found a more appropriate method of producing invisible glass. Today, the fruits of Blodgett's discovery can be found in almost all lenses used in cameras and other optical equipment, as well as automobile windows, showcases, eyeglasses, picture frames, and submarine periscopes.

During World War II, GE moved away from studies such as the one that lead to invisible glass in favor of tackling problems with more direct military applications. Following suit, Blodgett temporarily shelved her glass research, but did not move far from the field of surface chemistry. Her wartime experiments lead to breakthroughs involving plane wing deicing; she also designed a smoke screen that saved numerous lives during various military campaigns.

When the war ended, Blodgett continued doing research that had military ramifications. In 1947, for example, she worked with the Army Signal Corps, putting her thin film knowledge to use by developing an instrument that could be placed in weather balloons to measure humidity in the upper atmosphere. As Blodgett worked, plaudits for her research continued to pour in. Along with receiving numerous honorary degrees, Blodgett won the 1945 Annual Achievement Award from the American Association of University Women for her research in surface chemistry. In 1951, she accepted the Francis P. Garvan Medal from the American Chemical Society; that same year, Blodgett also had the distinction of being the only scientist honored in Boston's First Assembly of American Women in Achievement. To top off the year, Schenectady decided to honor its own by celebrating Katharine Blodgett Day.

Blodgett spent all of her adult life in the home she bought overlooking her birthplace. She was active in civil affairs in her beloved Schenectady, serving as treasurer of the Travelers Aid Society. Blodgett summered in a camp at Lake George in upstate New York, where she could pursue her love of gardening. She also enjoyed amateur astronomy, collecting antiques, and playing bridge with her friends. Blodgett died at her home on October 12, 1979, at the age of eighty-one.

BLOOD CHEMISTRY

Blood in the human body is a complex fluid. It is responsible for transport of dissolved **gases**, nutrients and waste products. It regulates the **pH** of not only itself, but of all the intercellular fluid in the body. In addition, blood carries **hormones** and critical parts of the immune system throughout the body, and aids in thermoregulation by redistributing **heat**. Blood also contains the factors needed for clotting, thereby preventing its own loss in the event of injury.

An adult human has 4-6 liters of blood. Approximately half of this is **water**, and slightly less than half is red blood cells. The rest is made up of **proteins**, sugars, salts, and other small molecules, plus white blood cells and platelets. The noncellular portion is termed **plasma**, while the cellular parts are collectively referred to as the formed elements.

While a small amount of **oxygen** can be dissolved directly in plasma, the quantity is far too low to support human life. Instead, transport of oxygen and **carbon** dioxide is accomplished by the red blood cells, of which there are approximately 25 trillion in circulation at any one time. Red blood cells are packed full of the protein hemoglobin, whose iron-containing heme group binds oxygen at high concentrations, and releases it at low concentrations. This allows hemoglobin to pick up oxygen in the lungs where the **concentration** is high, and deposit it in the tissues where it is low. Similarly, hemoglobin binds **carbon dioxide** (not with the heme group), and carries it to the lungs. **Carbon monoxide**, a product of incomplete **combustion**, binds to heme directly, and much more tightly than oxygen. For this reason, carbon monoxide is a poison. Genetic changes in the hemoglobin **molecule** that affect the shape of the red blood cell and the oxygen carrying capacity of the cell result in the inheritable diseases of sickle-cell anemia and beta-thalassemia.

Nutrients from the gut are dissolved directly in the plasma for transport. Several mechanisms prevent wide oscillations in plasma nutrient levels, despite irregular supply from the intestine. All materials absorbed into the bloodstream first pass directly to the liver, where excess nutrients are stored and then released as needed. (In contrast to all other nutrients, fats are not absorbed into the bloodstream, but instead are picked up by the lymphatic system.) The concentration of sugar in the blood is regulated by the hormones insulin and glucagon, which, respectively, lower or raise the blood sugar level in response to changes in supply and demand. Excesses that cannot be absorbed by the liver are excreted in the kidney, whose active transport and reclamation mechanisms make it one of the most finely tuned regulators of blood composition. The kidney is principally responsible for regulating the levels of ions, water, and wasted products in the blood.

The pH of the blood and fluid surrounding cells is normally 7.4. The most common threat to the maintenance of this pH is the production of organic acids by metabolic processes in the cell. Of most significance is lactic acid, produced by **fermentation** in the absence of oxygen, especially in muscle cells during strenuous exercise. To prevent lactic acid and other acids from changing the blood's pH, the blood uses a buffering system. While there are several such systems, the most important is the bicarbonate **buffer** system. In this system, the carbon dioxide released during cell **metabolism** is acted upon by the **enzyme** carbonic anhydrase to create carbonic acid (H_2CO_3), which immediately dissociates to form H+ and HCO_3^-, the bicarbonate **ion**. H+ ions released by lactic acid react with the bicarbonate ion to form carbonic acid again, preventing a change in pH. The kidney also participates in pH regulation by secreting excess H+ as necessary.

While carbon dioxide is central to pH control, it is itself an acid, and too much CO_2 accumulation can cause acidosis, or a blood pH below normal. Respiratory acidosis most commonly occurs from hypoventilation, or too little gas exchange in the lungs. As blood pH falls, the respiratory centers of the brain increase the breathing rate, to try to remove the excess carbon dioxide. Conversely, hyperventilation, or too-rapid breathing, can deplete the blood of carbon dioxide, leading to respiratory alkalosis.

The process of preventing blood loss from a damaged blood vessel is called hemostasis. While immediate hemostasis involves contraction of the muscles surrounding the blood vessel, long-term hemostasis requires the formation of a blood clot. This process involves a complex cascade of more than a dozen different plasma proteins. Clot formation begins when cell fragments known as platelets adhere to the wound, due to chemical changes undergone by wounded tissue. Once the platelets stick, they release "platelet thromboplastic factor," which, in conjunction with **calcium** ions in the plasma, triggers the activation of a series of other factors. This cascade acts to amplify the original signal from the platelets, rapidly causing more and more factors to become activated at each level. The next to last step in this chain is the activation of the circulating proenzyme prothrombin into its active form, thrombin. This enzyme then activates the circulating fibrous protein fibrinogen, into its active form, fibrin. Fibrin binds to the platelets at the wound site, forming a meshwork in which red blood cells and other elements become trapped, forming a clot. Clotting usually begins within 15 seconds after the wound occurs.

Hemophilia is an inherited disorder of clotting, due to a defect in the gene for one of the clotting factors. Hemophilia can be treated by regular administration of the factor, isolated from donated blood. More recently, this factor has been made by **genetic engineering** in a bacterium.

Blood is routinely analyzed during the diagnosis or treatment of disease, because it provides a window on the metabolic processes occurring in the body. For instance, immediately after a heart attack, the level of a certain muscle enzyme is elevated in the blood, due to the damaged heart muscle. Within several days, the level of this enzyme falls, but another rises, and then it falls in turn. By measuring the concentrations of each after a heart attack, the exact time of the attack can be estimated, providing important information for treatment planning.

See also Buffers; Enzymes

Hermann Boerhaave.

BODENSTEIN, ERNST (1871-1942)

German chemist

Ernst Max Bodenstein was born in Germany in 1871. During his life, he gained fame through his study of kinetic reactions in the gas phase. Although he found **chemistry** boring as a student, he gradually recognized the value of gas phase studies until he was recognized as the foremost researcher on the topic.

Bodenstein, like **James Prescott Joule**, was born into a wealthy family in the brewery business. He began chemistry research as a young boy working in a crude laboratory in the basement of his father's shop. He also liked to build things and to climb mountains. While his mountain climbing competed with chemistry for his interest, his technical skill enhanced his research since he built his own equipment. After college, Bodenstein continued in graduate school working toward a Ph.D in chemistry. His mentor was Victor Meyer, who was famous for organizing **stereochemistry** as a field of study. Bodenstein was remembered for exactness in setting up his experiments.

After earning his doctorate, Bodenstein continued in his research in gas phase reactions, sometimes funding the projects by himself. When he was offered a position as director of the Institute for Physical Chemistry in Berlin, he accepted. A very strict director, Bodenstein imposed a regulation that all students should construct their own equipment as he had done so in his youth.

Bodenstein began studying **kinetics** at a time when the field was undergoing substantial improvements in approach and methodology. An important development was a monograph on chemical kinetics published by Jacobus van't Hoff in the 1880s. One of the points that van't Hoff clarified in his work was that reaction rates were a function of **temperature**. The main reason that Bodenstein chose to study reactions in the gas phase was that molecules are not as densely packed as in solutions. This means that there are fewer interactions, besides the target reaction, needing an explanation. In his studies of gas phase reactions, he discovered that certain chemical reactions might create radicals. These are molecules with an odd number of electrons. These radicals quickly (and with little hindrance) react with other molecules. Radical-based reactions create more radicals, which quickly and easily react with other molecules creating even more radicals. This process means that the chain reactions where radicals are present have a increase rate of speed. It took many years between the first observation and the complete explanation of what happened. During this time, Bodenstein offered some conclusions that were not well constructed.

Bodenstein's research advanced the study of polymolecular reactions in kinetic field, but it did not offer any insight into unimolecular reactions. Unimolecular reactions were the quagmire of his time. However, many researchers built on his work. One of these was William C. M. Lewis, who was the first to apply quantum methodology to the study of reaction rates. Although he went beyond what Bodenstein did, his early work confirmed the phenomenon that Bodenstein had discovered.

Boerhaave, Hermann (1668-1738)

Dutch physician and chemist

The Dutch physician and chemist Hermann Boerhaave (1668-1738) was the leading medical teacher of the early eighteenth century. His works on medicine and **chemistry** had widespread use as basic textbooks.

Hermann Boerhaave was born on December 31, 1668, at Voorhout, Holland, the son of a minister in the Dutch Reformed Church. A painful leg ulcer which affected him for five years during his youth excited his interest in medicine. He aimed first to combine a career as a pastor and physician. After entering the University of Leiden in 1684, he took courses in mathematics, natural philosophy, botany, and languages, as well as in theology.

In 1690, Boerhaave obtained the degree of doctor of philosophy and began medical studies. As a physician, he was almost entirely self-taught, medical instruction at Leiden being at a low ebb. He obtained his medical degree in 1693 from the University of Harderwijk.

Having come under suspicion of being sympathetic to the doctrines of Spinoza, Boerhaave abandoned the idea of an ecclesiastical career and began to devote himself exclusively to medicine and science. His private practice in Leiden was not lucrative but left him time to continue his studies and begin extensive experiments in chemistry.

His highly successful teaching career began in 1701. He taught medicine at the University of Leiden and gave private courses in chemistry. During the next eight years, he published in Latin his two major medical works, *The Institutes of Medicine* and *The Aphorisms concerning the Knowledge and Cure of Diseases*. Numerous editions were produced and the works were widely translated, even into Japanese. They continued to be used as textbooks for at least 50 years after his death.

Boerhaave was appointed professor of medicine and botany in 1709. In this post he greatly improved the collection of the celebrated botanical garden of the University of Leiden and carried out an extensive correspondence with the world's leading botanists. In 1714 he became professor of medicine and a physician to St. Cecilia Hospital in Leiden. There in his small clinic he established the value of bedside teaching for medical training.

He obtained the chair of chemistry in 1718 and for 11 years held three chairs simultaneously. His definitive *Elements of Chemistry* (1732) became very famous and was the source of his influence on eighteenth-century chemistry.

A tall and robust man of immense erudition, Boerhaave was a superb teacher. He was patient, unaffected, and readily approachable by his students. They flocked from all parts of

Niels Bohr.

Europe to hear his lectures, thereby increasing the renown of the University of Leiden. Boerhaave died, universally esteemed, in 1738 of heart disease.

Bohr, Niels (1885-1962)

Danish physicist

Niels Bohr received the Nobel Prize in physics in 1922 for the quantum mechanical model of the **atom** that he had developed a decade earlier, the most significant step forward in scientific understanding of **atomic structure** since English physicist **John Dalton** first proposed the modern **atomic theory** in 1803. Bohr founded the Institute for Theoretical Physics at the University of Copenhagen in 1920, an Institute later renamed for him. For well over half a century, the Institute was a powerful force in the shaping of atomic theory. It was an essential stopover for all young physicists who made the tour of Europe's center of theoretical physics in the mid-twentieth century. Also during the 1920s, Bohr thought and wrote about some of the fundamental issues raised by modern quantum theory. He developed two basic concepts, the principles of complementarity and correspondence, that he said must direct all future work in physics. In the 1930s, Bohr became interested in problems of the atomic **nucleus** and contributed to the development of the liquid-drop model of the nucleus, a model used in the explanation of **nuclear fission**.

Niels Henrik David Bohr was born on October 7, 1885, in Copenhagen, Denmark. He was the second of three children born to Christian and Ellen Adler Bohr. Bohr's early upbringing was enriched by a nurturing and supportive home atmosphere. His mother had come from a wealthy Jewish family involved in banking, government, and public service. Her father, D. B. Adler, had founded the Commercial Bank of Copenhagen and the Jutland Provincial Credit Association. Bohr's father was a professor of physiology at the University of Copenhagen. His closest friends met every Friday night to discuss events, so that, as a young boy, Bohr "learned much from listening to these conversations," according to J. Rud Nielsen in *Physics Today.*

Bohr became interested in science at an early age. His biographer, Ruth Moore, has written in her book *Niels Bohr: The Man, His Science, and the World They Changed* that as a child he "was already fixing the family clocks and anything else that needed repair." Bohr received his primary and secondary education at the Gammelholm School in Copenhagen. He did well in his studies, although he was apparently overshadowed by the work of his younger brother Harald, who later became a mathematician. Both brothers were also excellent soccer players.

On his graduation from high school in 1903, Bohr entered the University of Copenhagen, where he majored in physics. He soon distinguished himself with a brilliant research project on the surface tension of **water** as evidenced in a vibrating jet stream. For this work he was awarded a **gold** medal by the Royal Danish Academy of Science in 1907. In the same year, he was awarded his bachelor of science degree, to be followed two years later by a master of science degree. Bohr then stayed on at Copenhagen to work on his doctorate, which he gained in 1911. He thesis dealt with the **electron** theory of **metals** and confirmed the fact that classical physical principles were sufficiently accurate to describe the qualitative properties of metals but failed when applied to quantitative properties. Probably the main result of this research was to convince Bohr that classical electromagnetism could not satisfactorily describe atomic phenomena. The stage had been set for Bohr's attack on the most fundamental questions of atomic theory.

Bohr decided that the logical place to continue his research was at the Cavendish Laboratory at Cambridge University. The director of the laboratory at the time was English physicist **J. J. Thomson**, discoverer of the electron. Only a few months after arriving in England in 1911, however, Bohr discovered that Thomson had moved on to other topics and was not especially interested in Bohr's thesis or ideas. Fortunately, however, Bohr met English physicist **Ernest Rutherford**, then at the University of Manchester, and received a much more enthusiastic response. As a result, he moved to Manchester in 1912 and spent the remaining three months of his time in England working on Rutherford's nuclear model of the atom.

On July 24, 1912, Bohr boarded ship for his return to Copenhagen and a job as assistant professor of physics at the University of Copenhagen. Also waiting for him was his bride-to-be Margrethe Nørland, whom he married on August 1. The couple later had four sons: Hans, Erik, Aage, and Ernest. Two other sons, one named Christian, died young. Aage earned a share of the 1975 Nobel Prize in physics for his work on the structure of the atomic nucleus.

The field of atomic physics was going through a difficult phase in 1912. Rutherford had only recently discovered the atomic nucleus, which had created a profound problem for theorists. The existence of the nucleus meant that electrons must be circling it in orbits somewhat similar to those traveled by planets in their motion around the sun. According to classical laws of electrodynamics, however, an electrically charged particle would continuously radiate **energy** as it traveled in such an orbit around the nucleus. Over time, the electron would spiral ever closer to the nucleus and eventually collide with it. Although electrons clearly *must* be orbiting the nucleus, they could *not* be doing so according to classical laws.

Bohr arrived at a solution to this dilemma in a somewhat roundabout fashion. He began by considering the question of atomic spectra. For more than a century, scientists had known that the heating of an element produces a characteristic line spectrum; that is, the specific pattern of lines produced is unique for each specific element. Although a great deal of research had been done on spectral lines, no one had thought very deeply about what their relationship might be with atoms, the building blocks of elements.

When Bohr began to attack this question, he decided to pursue a line of research begun by the German physicist Johann Balmer in the 1880s. Balmer had found that the lines in the **hydrogen** spectrum could be represented by a relatively simple mathematical formula relating the frequency of a particular line to two integers whose significance Balmer could not explain. It was clear that the formula gave very precise values for line frequencies that corresponded well with those observed in experiments.

When Bohr's attention was first attracted to this formula, he realized at once that he had the solution to the problem of electron orbits. The solution that Bohr worked out was both simple and elegant. In a brash display of hypothesizing, Bohr declared that certain orbits existed within an atom in which an electron could travel *without* radiating energy; that is, classical laws of physics were suspended within these orbits. The two integers in the Balmer formula, Bohr said, referred to orbit numbers of the "permitted" orbits, and the frequency of spectral lines corresponded to the energy released when an electron moved from one orbit to another.

Bohr's hypothesis was brash because he had essentially no theoretical basis for predicting the existence of "allowed" orbits. To be sure, German physicist **Max Planck**'s quantum hypothesis of a decade earlier had provided some hint that Bohr's "quantification of space" might make sense, but the fundamental argument for accepting the hypothesis was simply that it worked. When his model was used to calculate a variety of atomic characteristics, it did so correctly. Although the hypothesis failed when applied to detailed features of atomic spectra, it worked well enough to earn the praise of many colleagues.

Bohr published his theory of the ''planetary atom'' in 1913. That paper included a section that provided an interesting and decisive addendum to his basic hypothesis. One of the apparent failures of the Bohr hypothesis was its seeming inability to predict a set of spectral lines known as the Pickering series, lines for which the two integers in the Balmer formula required half-integral values. According to Bohr, of course, no ''half-orbits'' could exist that would explain these values. Bohr's solution to this problem was to suggest that the Pickering series did not apply to hydrogen at all, but to **helium** atoms that had lost an electron. He rewrote the Balmer formula to reflect this condition.

Within a short period of time spectroscopists in England had studied samples of helium carefully purged of hydrogen and found Bohr's hypothesis to be correct. Although a number of physicists were still debating Bohr's theory, at least one—Rutherford—was convinced that the young Danish physicist was a highly promising researcher. He offered Bohr a post as lecturer in physics at Manchester, a job that Bohr eagerly accepted and held from 1914 to 1916. He then returned to the University of Copenhagen, where a chair of theoretical physics had been created specifically for him. Within a few years he was to become involved in the planning for and construction of the University of Copenhagen's new Institute for Theoretical Physics, of which he was to serve as director for the next four decades.

In many ways, Bohr's atomic theory marked a sharp break between classical physics and a revolutionary new approach to natural phenomena made necessary by quantum theory and relativity. He was very much concerned about how scientists could and should now view the physical world, particularly in view of the conflicts that arose between classical and modern laws and principles. During the 1920s and 1930s, Bohr wrote extensively about this issue, proposing along the way two concepts that he considered to be fundamental to the ''new physics.'' The first was the principle of complementarity that says, in effect, that there may be more than one true and accurate way to view natural phenomena. The best example of this situation is the wave-particle duality discovered in the 1930s, when particles were found to have wavelike characteristics and waves to have particle-like properties. Bohr argued that the two parts of a duality may appear to be inconsistent or even in conflict and that one can use only one viewpoint at a time, but he pointed out that both are necessary to obtain a complete view of particles and waves.

The second principle, the correspondence principle, was intended to show how the laws of classical physics could be preserved in light of the new quantum physics. We may know that quantum mechanics and relativity are essential to an understanding of phenomena on the atomic scale, Bohr said, but any conclusion drawn from these principles must not conflict with observations of the real world that can be made on a macroscopic scale. That is, the conclusions drawn from theoretical studies must correspond to the world described by the laws of classical physics.

In the decade following the publication of his atomic theory, Bohr continued to work on the application of that theory to atoms with more than one electron. The original theory had dealt only with the simplest of all atoms, hydrogen, but it was clearly of some interest to see how that theory could be extended to higher elements. In March, 1922, Bohr published a summary of his conclusions in a paper entitled ''The Structure of the Atoms and the Physical and Chemical Properties of the Elements.'' Eight months later, Bohr learned that he had been awarded the Nobel Prize in physics for his theory of atomic structure, by that time universally accepted among physicists.

During the 1930s, Bohr turned to a new but related, topic: the composition of the atomic nucleus. By 1934, scientists had found that the nucleus consists of two kinds of particles, protons and neutrons, but they had relatively little idea how those particles are arranged within the nucleus and what its general shape was. Bohr theorized that the nucleus could be compared to a liquid drop. The forces that operate between protons and neutrons could be compared in some ways, he said, to the forces that operate between the molecules that make up a drop of liquid. In this respect, the nucleus is no more static than a droplet of water. Instead, Bohr suggested, the nucleus should be considered to be constantly oscillating and changing shape in response to its internal forces. The greatest success of the Bohr liquid-drop model was its later ability to explain the process of nuclear fission discovered by German chemist **Otto Hahn**, German chemist **Fritz Strassmann**, and Austrian physicist **Lise Meitner** in 1938. It is somewhat ironic that the Nobel Prize won by Bohr's son Aage in 1975 was given for the latter's elucidation of his father's nuclear model (to which many other scientists had also contributed).

Bohr continued to work at his Institute during the early years of World War II, devoting considerable effort to helping his colleagues escape from the dangers of Nazi Germany. When he received word in September, 1943, that his own life was in danger, Bohr decided that he and his family would have to leave Denmark. The Bohrs were smuggled out of the country to Sweden aboard a fishing boat and then, a month later, flown to England in the empty bomb bay of a Mosquito bomber. The Bohrs then made their way to the United States, where both Bohr and his son became engaged in work on the Manhattan Project to build the world's first atomic bombs.

After the War, Bohr, like many other Manhattan Project researchers, became active in efforts to keep control of atomic weapons out of the hands of the military and under close civilian supervision. For his long-term efforts on behalf of the peaceful uses of atomic energy, Bohr received the first Atoms for Peace Award given by the Ford Foundation in 1957. Meanwhile, Bohr had returned to his Institute for Theoretical Physics and become involved in the creation of the European Center for Nuclear Research (CERN). He also took part in the founding of the Nordic Institute for Theoretical Atomic Physics (Nordita) in Copenhagen. Nordita was formed to further cooperation among and provide support for physicists from Norway, Sweden, Finland, Denmark, and Iceland.

Bohr reached the mandatory retirement age of seventy in 1955 and was required to leave his position as professor of physics at the University of Copenhagen. He continued to serve as director of the Institute for Theoretical Physics until his death in Copenhagen on November 18, 1962.

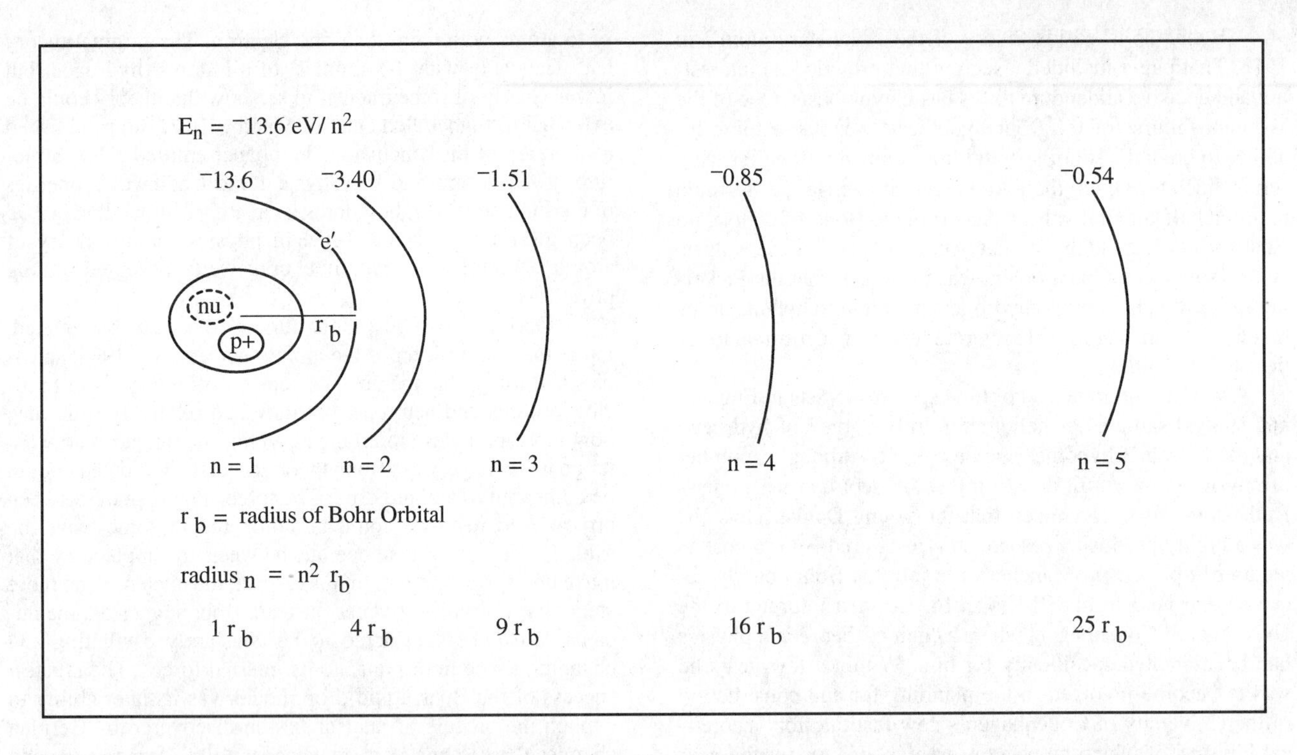

The Bohr model accounts for much of physical phenomena encountered in chemistry and classical physics. According to the Bohr model, electrons occupy only a selected number of concentric orbits or energy shells. The maximum number of electrons allowed in each shell is given by the formula $2n^2$ (where n=the principal quantum number). Potential energy increases with orbital distance from nucleus. The radii of the energy shells is given by $r^n=n^2r_b$, where r_b is radius of Bohr Orbital $_{n=1}$ =5.3 X 10^{-11}m. *(Illustration by Electronic Illustrators Group.)*

Bohr was held in enormous respect and esteem by his colleagues in the scientific community. American physicist **Albert Einstein**, for example, credited him with having a ''rare blend of boldness and caution; seldom has anyone possessed such an intuitive grasp of hidden things combined with such a strong critical sense.'' Among the many awards Bohr received were the Max Planck Medal of the German Physical Society in 1930, the Hughes (1921) and Copley (1938) medals of the Royal Society, the Franklin Medal of the Franklin Institute in 1926, and the Faraday Medal of the Chemical Society of London in 1930. He was elected to more than twenty scientific academies around the world and was awarded honorary doctorates by a dozen universities, including Cambridge, Oxford, Manchester, Edinburgh, the Sorbonne, Harvard, and Princeton.

Bohr theory

The Bohr theory or, more properly, the Bohr model of **atomic structure** was developed by Danish physicist and Nobel laureate **Niels Bohr**. Published in 1913, Bohr's model improved the classical atomic models of physicists **Sir Joseph John Thomson** and **Ernest Rutherford** by incorporating quantum theory. While working on his doctoral dissertation at Copenhagen University, Bohr studied German physicist **Max Planck**'s quantum theory of radiation. After graduation Bohr worked in England with Thomson and subsequently with Rutherford. During this time Bohr developed his model of atomic structure.

Before Bohr, the classical model of the **atom** was similar to the Copernican model of the solar system, where, just as the planets orbit the Sun, electrically-negative electrons moved in orbits around a relatively massive, positively-charged **nucleus**.

The classical model of the atom allowed electrons to orbit at any distance from the nucleus. This predicted that when, for example, a **hydrogen** atom was heated it should produce a continuous spectrum of colors as it cooled because its **electron**, moved away from the nucleus by the **heat energy**, would gradually give up that energy as it spiraled back closer to the nucleus. Spectroscopic experiments, however, showed that hydrogen atoms produced only narrow bands of **color** when heated. In addition, Scottish physicist **James Clerk Maxwell**'s influential studies on electromagetic radiation (light) predicted that an electron orbiting around the nucleus according to Newton's laws would continuously lose energy and eventually fall into the nucleus. To account for the observed properties of hydrogen, Bohr proposed that electrons existed only in certain orbits and that, instead of traveling between orbits, electrons made instantaneous quantum leaps or jumps between allowed orbits.

In the Bohr model the most stable, lowest **energy level** is found in the innermost orbit. This first orbital forms a shell around the nucleus and is assigned a principal quantum number (n) of n=1. Additional orbital shells are assigned values n=2, n=3, n=4, etc. The orbital shells are not spaced at equal distances from the nucleus and the radius of each shell increases rapidly as the square of n. Increasing numbers of electrons can fit into these orbital shells according to the formula $2n^2$. The first shell can hold up to two electrons, the second shell (n=2) up to eight electrons, the third shell (n=3) up to 18 electrons. Subshells or suborbitals (designated *s, p, d,* and *f*) with differing shapes and orientations allow each element a unique electron configuration.

As electrons move farther away from the nucleus they gain **potential energy** and become less stable. Atoms with electrons in their lowest energy orbits are in a "ground" state, those with electrons jumped to higher energy orbits are in an "excited" state. Atoms may acquire energy that excites electrons by random thermal collisions, collisions with **subatomic particles**, or by absorbing a **photon**. Of all the photons (quantum packets of light energy) that an atom can absorb, only those having an energy equal to the energy difference between allowed electron orbits will be absorbed. Atoms give up excess internal energy by giving off photons as electrons return to lower energy (inner) orbits.

The electron quantum leaps between orbits proposed by the Bohr model accounted for Plank's observations that atoms emit or absorb electromagnetic radiation only in certain units called quanta. Bohr's model also explained many important properties of the **photoelectric effect** described by physicist **Albert Einstein.**

According to the Bohr model, when an electron is excited by energy it jumps from its ground state to an excited state (i.e., a higher energy orbital). The excited atom can then emit energy only in certain (quantized) amounts as its electrons jump back to lower energy orbits located closer to the nucleus. This excess energy is emitted in quanta of electromagnetic radiation (photons of light) that have exactly same energy as the difference in energy between the orbits jumped by the electron. For hydrogen, when an electron returns to the second orbital (n=2) it emits a photon with energy that corresponds to a particular color or spectral line found in the Balmer series of lines located in the visible portion of the electromagnetic (light) spectrum. The particular color in the series depends on the higher orbital from which the electron jumped. When the electron returns all the way to the innermost orbital (n=1) the photon emitted has more energy and forms a line in the Lyman series found in the higher energy, ultraviolet portion of the spectrum. When the electron returns to the third quantum shell (n=3), it retains more energy and therefore the photon emitted is correspondingly lower in energy and forms a line in the Paschen series found in the lower energy, infrared portion of the spectrum.

Because electrons are moving charged particles, they also generate a magnetic field. Just as an ampere is a unit of electric current, a magneton is a unit of magnetic **dipole** moment. The orbital **magnetic moment** for the hydrogen atom is called the Bohr magneton.

Bohr's work earned a Nobel Prize in 1922. Subsequently, more mathematically complex models based on the work of physicists **Louis Victor de Broglie** and **Erwin Schrödinger** that depicted the particle and wave nature of electrons proved more useful to describe atoms with more than one electron. The standard model incorporating quark particles further refines the Bohr model. Regardless, Bohr's model remains fundamental to the study of **chemistry**, especially the **valence** shell concept used to predict an element's reactive properties.

The Bohr model remains a landmark in scientific thought that poses profound questions for scientists and philosophers. The concept that electrons make quantum leaps from one orbit to another, as opposed to simply moving between orbits, seems counter-intuitive, that is, outside the human experience with nature. Bohr said, "Anyone who is not shocked by quantum theory has not understood it."

See also Atomic theory

Boiling point

The boiling point of a substance is the **temperature** at which the liquid boils. This is the temperature at which the **vapor pressure** of the liquid is the same as that of the atmosphere. When the boiling point is reached the chemical changes from the liquid to the gas or vapor phase. This change does not occur instantly as all molecules have to gain sufficient **energy** to change from one state of **matter** to another.

Different **liquids** have different boiling points due to their chemical make up. Any liquid that has a high number of **hydrogen** bonds will have an elevated boiling point because extra energy must be added into the system to break the hydrogen bonds. For example, **water** (H_2O) has a boiling point of 212°F (100°C). Boiling points can also be altered by changes in pressure, the lower the pressure, the lower the boiling point, this is because the vapor pressure of the liquid at a lower atmospheric pressure does not have to be as high to become equal to that atmospheric pressure. Boiling points also can be altered by adding other chemicals to the liquid. The boiling point of water can be increased by several degrees by adding **sodium** chloride (common salt).

See also Phase changes

Boisbaudran, Paul Émile Lecoq de (1836-1912)

French chemist

Boisbaudran was born into a well-to-do family of wine merchants in Cognac, France, on April 18, 1836. Although he had no formal education, he was personally instructed by his mother, a woman learned in the classics, history, and foreign languages. Boisbaudran became interested in science early in life and taught himself out of textbooks used in the Ècole Polytechnique. He carried out many experiments in a home-built laboratory.

Boisbaudran's special area of interest was **spectroscopy**, a science developed by Gustav Kirchhoff in 1859. Over more than a decade, Boisbaudran carefully examined the spectral lines produced by 35 different elements. In the process, he discovered, in 1875, a new element that he named **gallium**. Most people believe that the name was given in honor of his homeland, since the Latin name for France was *Gaul*. Some writers, however, have suggested that he may have named the element after himself since the French word *lecoq* means rooster, which, in Latin, is *gallus*.

The discovery of gallium has added significance. Research soon proved that it was the element predicted by Dmitri Mendeleev four years earlier. Mendeleev referred to the proposed element as eka-aluminum, since it was presumed to be the next element in the **aluminum** series. The close match between the predicted properties of eka-aluminum and the actual properties of gallium provided a valuable confirmation of Mendeleev's periodic theory.

By 1879, Boisbaudran had shifted his interests to the field of the **rare earth elements**. Over the next three decades, he examined the spectral lines of ores and compounds of these elements. One of the earliest results of this research was the discovery in 1879 that the didymium discovered by Carl Gustav Mosander a few years earlier actually consisted of two parts, one of which was a new element that Boisbaudran named **samarium**.

Finally, in 1886, Boisbaudran discovered yet a third element, **dysprosium**. Boisbaudran recognized the new element from the properties of its compounds and its spectral lines, but never obtained the element itself in pure form.

Boisbaudran's health began to fail after 1895, and he produced relatively few discoveries in the last two decades of his life. He died in Paris, France, on May 28, 1912.

Boltwood, Bertram Borden (1870-1927)

American chemist

At the turn of the twentieth century, when the science of **radioactivity** was still young, Bertram Boltwood was considered to be the United States' foremost authority. This was a reputation he had earned, along with Great Britain's **Ernest Rutherford**, by advancing the experiments of **Henri Becquerel**, **Marie Curie**, and **Pierre Curie**. Among his accomplishments was the proof of a radioactive series (the transformation of one radioactive element into another) that led eventually to a reliable method for determining the age of Earth.

Boltwood came from an academic family: his grandfather had helped found Amherst College in Massachusetts, and his cousin was the poet Ralph Waldo Emerson (1803-1882). Though his father died when Bertram was only two, his mother continued her husband's legacy of education. He entered the **chemistry** department of Yale University in 1889, graduating at the top of his class three years later. After spending two years in Germany at the Ludwig-Maximilian University in Munich, Boltwood enrolled in graduate classes at Yale, receiving his Ph.D. in 1897.

Boltwood spent the next few years as an academic consultant to Yale's departments of analytical and **physical chemistry**. He also spent a good deal of time and effort designing new laboratory apparatus and teaching materials.

In 1900 Joseph Hyde Pratt, another Yale graduate, teamed up with Boltwood to form a partnership of "mining engineers and chemists," in which Pratt collected ore samples that Boltwood would later examine in the laboratory. While this enterprise earned the two young scientists a moderate living and, perhaps more importantly, gave Boltwood valuable experience in the handling of radioactive materials.

About this time Rutherford and his colleague **Frederick Soddy** proclaimed that radioactive elements decayed, and in that decaying process, transmuted into other elements, both active and inactive. The evidence they presented to support their claim was substantial; however, Boltwood felt that he could express the relationship between radioactive elements more clearly and irrefutably. Thus, in 1904, he turned his research efforts toward radioactivity, a direction from which he would never again stray.

Boltwood began by showing that, in a given amount of **uranium**, a constant amount of **radium** was present; if the amount of uranium were increased, the amount of radium would increase proportionally. It seemed clear, then, that radium must be a "daughter" element of uranium. Still, there seemed to be a missing link—an element that existed between uranium and radium in the radioactive chain. Boltwood never found this intermediate element, although it was later discovered to be a form of **thorium**.

Though other scientists had discovered the "parent of radium," Boltwood continued his experiments with radioactive isotopes. He began to notice that traces of **lead** could always be found in uranium-containing minerals. Using this as a springboard he theorized that lead was the ultimate final step in the **transmutation** of uranium; this was supported by the fact that the older a mineral sample was, the greater was the amount of lead.

The most important application of this research was the development of the uranium-lead method of geologic dating. Until this time, the best estimate for the age of Earth was that by **William Thomson** (Lord Kelvin). Using the **temperature** of the Sun as a starting point, Thomson determined the amount of time necessary for our planet to cool to its present temperature. The age he arrived at was 40 million years—a number considered by most geologists to be far too low. Rutherford, too, attempted to date Earth, this time by determining the amount of **helium** (also a radioactive decay product) in the crust; however, this method was found to be unreliable as well.

Boltwood first determined the amount of time it took for lead to be produced from uranium—about one billion years for one gram of uranium to produce 0.15 grams of lead. Using this formula he tested rock samples of many different ages, ranging from millions to billions of years old. However, none older than 3.7 billion years could be found; this number agreed with geologists' theories, and it became the accepted age of the Earth. Later experiments on meteorite and lunar rock samples showed them to be older than the Earth by 0.9 billion years. The explanation for this is that the solar system was formed nearly 5 billion years ago; the Moon and meteorites cooled 4.7

billion years ago, but the Earth remained a liquid ball for 0.9 billion years, with the first rocks forming about 3.7 billion years ago.

In 1910 Boltwood was given a full professorship at Yale; along with this position went the responsibility for the construction of a new physics laboratory. As his reputation grew he was also requested to speak to many different organizations, as well as act as a consultant to engineers and other laboratories. In 1918, he was made the director of the Yale chemical laboratory and was required to design its new Sterling Chemistry Laboratory. Though he completed this task, it exacted a heavy mental and physical toll on him. Boltwood suffered a nervous breakdown, and throughout his remaining years, he experienced periods of intense depression. It was during such a period in 1927 that he committed suicide.

BOLTZMANN, LUDWIG EDUARD (1844-1906)

Austrian physicist

Ludwig Eduard Boltzmann, a brilliant physicist, was called ''The Father of Statistical Mechanics,'' the science which became the foundation of quantum mechanics. Boltzmann applied the laws of mechanics to describe the random motion of molecules (atoms) and its effects on the physical properties of **matter**. His statistical analysis included the kinetic-molecular theory of **gases** and **thermodynamics**, the nature of **heat** and its conversion into other forms of **energy**. He also described mathematically the principle of blackbody, an ideal substance that absorbs all radiant energy that strikes it and reflects none. So radical and complicated were his hypotheses that many other scientists violently disputed them. Suffering serious illness, a depressive disorder, and perhaps feeling hopeless that his scientific endeavors would never be recognized or understood, he committed suicide just before his theories were proven.

Boltzmann, whose father was a tax official, was born on February 20, 1844 in Vienna, Austria. He received his doctorate from the University of Vienna in 1866, choosing for his thesis the kinetic theory of gases. This theory explains the physical action of gases by assuming that the extremely minute particles (molecules) of which any gas is composed, are also extremely far apart in relation to their size. The only time they affect each other (exert a force) is during a ''rare, perfectly elastic collision.'' The application of the laws of mechanics applied to the behavior of gas in this ''ideal state,'' establishes ''**gas laws**.'' Pressure is then the collision of large numbers of molecules against the walls of the container in which the gas is held.

After graduating, Boltzmann first became assistant to his instructor, Josef Stefan (1869-1893), then taught at the University of Graz before moving first to Heidelberg and then Berlin. In 1869 he became chair of theoretical physics at Graz, retaining the post for four years before becoming chair of mathematics at Vienna. Once again, he returned to Graz as chair of experimental physics only to return in 1894 to Vienna as chair of theoretical physics.

A year later, Ernest Mach (1838-1916) became chair of history and philosophy of science at Vienna. The two clashed badly, not just differing in scientific philosophy, but personally, as well. Boltzmann moved to Leipzig in 1900 where he began working with Wilhelm Oswald (1853-1932), one of his most adamant opponents. On good terms personally, their scientific arguments apparently led Boltzmann to an unsuccessful suicide attempt. When Mach retired from Vienna because of ill health in 1901, Boltzmann returned to Vienna in 1902 to assume the still vacant position of chair of theoretical physics. Here, he taught mathematical physics, and was also given the philosophy class of his old opponent, Mach. His lectures became so popular that his audience grew too large for the lecture hall.

Boltzmann's work continued to receive severe opposition from many contemporaries, however. The difference in their analytical processes has been likened to a ''scientific war between the atomists (those who based their calculations on the movement of atoms) and the energists (who declared all physical science was based on energy only).'' Boltzmann was one of the first European physicists to understand the importance of a theory of electromagnetics proposed by England's James Maxwell. He also derived a hypothesis known as the Maxwell-Boltzmann kinetic theory of gases, which simplistically states that the average amount of energy used for each different direction in which an **atom** moves is exactly the same. It involved the statistical probability theory, that **temperature** and heat involved only molecular movement and that molecules at high temperature have only a high *probability* of moving toward those at low temperature. This flew in the face of the firmly-held concept of certainty, heat flowing from hot to cold.

Boltzmann published a series of papers in 1870 in which he declared that the laws of mechanics and the probability theory, which explains and predicts how the properties of atoms (mass, charge, and structure), could be successfully applied to determine the visible properties of matter (viscosity, thermal conductivity, and diffusion).

In 1904, Boltzmann travelled to the United States to attend the World's Fair in St. Louis. He lectured on applied mathematics and visited Berkeley and Stanford where he saw new discoveries being made in relation to radiation. Tragically, he did not understand these discoveries would ultimately prove his theories to a doubting scientific community. He continued on, however, attempting to explain in *Populäre Schiriften* in 1905 how the physical world could be adequately described by equations yet fail to describe the underlying **atomic structure**. He wrote ''May I be excused for saying with banality that the forest hides the trees for those who think that they disengage themselves from atomistics by the consideration of differential equations.''

Suffering throughout life from severe mood swings, Boltzmann's depression had a negative impact upon his professional and personal relationships. Continuing to undergo attacks upon his work, and suffering from depression and failing health, Boltzmann hanged himself on October 5, 1906 while on vacation with his wife and daughter at the Bay of Duino near Trieste. Shortly thereafter, his mathematical interpretation of the atomic world would be confirmed and the foundation laid for quantum mechanics.

BONDING

A molecule is the smallest unit of a substance capable of independent existence. As early as the 17th century, there were proposals that molecules were made up of atoms of elements held together either by the mutual attraction of the atoms or by the atoms fitting together in a manner similar to a jigsaw puzzle. Well into the 19th century, however, there were continuing disagreements on the molecular nature of chemical compounds. After **Edward Frankland** proposed in 1866 that molecules are made up of atoms held together by localized directional chemical bonds, the supporting experimental evidence led to general acceptance of the idea. In 1874, **August Kekulé** and the team of **J. H. van't Hoff** and **J. A. Le Bel** independently proposed a tetrahedral structure for carbon in polyatomic molecules. In the methane molecule CH_4, for instance, the carbon atom is located at the center of a tetrahedron with the four hydrogen atoms located at the four apexes of the tetrahedron. The geometry of molecules involving the atoms of a number of other elements was soon proposed.

This new way of thinking of molecules as stable groupings of atoms, with these atoms having a definite arrangement in space relative to one another, revolutionized chemistry. A chemical or molecular bond was now regarded as the combination of forces that holds two atoms in fixed relationship within a molecule. Chemical reactions could now be thought of as the formation and the breaking of bonds, with chemical reactivity dependant on the strength of bonds and on their spatial arrangement. Indeed, the concept of chemical bonds lies at the heart of much of the work with which chemistry is involved.

The exact nature of chemical bonds and the explanation of the source of the attraction between atoms which holds them in such definite spatial relationships remained a mystery. Any significantly useful bonding theory had to explain why bonds form, why some bonds are stronger and some more reactive than others, and why a particular spatial arrangement is the most stable.

In 1902, soon after **J. J. Thomson** proved, in 1897, that negatively charged electrical units known as electrons exist and are a basic constituent of matter, **G. N. Lewis** proposed that atoms form ionic bonds by the transfer of one or more electrons from one atom to the other. In 1916, Lewis proposed that non-ionic, or covalent, bonds are formed by the sharing of electrons between the atoms which are held together by such a bond. It remained for quantum mechanics to provide the methods for understanding the chemical bond and the chemical phenomena associated with it on the basis of this shared electron concept.

The two extremes of molecular bonding are ionic bonds and covalent bonds. When one atom comes close to another, the electrons of the two atoms are attracted electrostatically to the nuclei of both atoms. Ionic bonds form between atoms with widely different electronegativities, i.e., abilities to hold onto their electrons. When two atoms with large differences in electronegativity come into close proximity, the atom with the greater attraction for electrons not only holds onto its own electrons but may pull one or more electrons away from the other, less electronegative atom. One atom thus becomes a negative ion and the other a positive ion. The two ions of opposite charge attract each other and an ionic bond is formed. The energy of attraction between oppositely charged particles is inversely proportional to the distance between them. At great distances of separation, the attraction between the two ions is essentially zero; as the distance decreases, the attraction increases. The maximum energy of attraction would come when the distance of separation is zero. This does not happen, however, because the electron clouds of the two ions repel each other at very close distance, as do the nuclei of the two ions. Therefore, when an ionic bond forms, the bond length is the equilibrium distance at which the forces of attraction exactly balance the forces of repulsion.

A covalent bond forms when the participant atoms do not differ greatly in electronegativity and, consequently, both have essentially the same hold on their electrons. When two such atoms come close together, the electrons of the two atoms are attracted, as in the case of an ionic bond, to the nuclei of both atoms. Now, however, neither atom is strong enough to pull electrons away from the other, and ions do not form. One or more of the electrons from each of the atoms may, however, be able to move around both nuclei, spending part of the time more closely associated first with one, then the other nucleus. These electrons are shared by the two nuclei, and a covalent bond forms. The bond length is determined by the equilibrium position at which the attraction resulting from sharing an electron is exactly balanced by the repulsion of positive nuclei and negative electrons.

Bonds between atoms of the same element, such as hydrogen, H_2, are purely covalent: the shared bonding electrons are equally attracted to both nuclei. The covalent bonds of atoms with unequal electronegativities, however, are polar covalent bonds. In such a bond the electrons that the two atoms share are attracted more to the atom with the greater electronegativity, and they spend more time near it. Consequently, the more electronegative atom has a partial negative charge. The less electronegative atom has a partial positive charge since the shared electrons spend less time with it. Because there is a separation of charges, with a partial positive charge at one end of the bond and an equal negative partial charge at the other, a polar bond results.

This simplified qualitative discussion of molecular bonding is placed on a much firmer theoretical and quantitative basis through the utilization of quantum mechanics. A number of scientists have contributed to the development of quantum mechanical applications to molecular systems. The work of **Linus Pauling** was central to this effort. His book *The Nature of the Chemical Bond* (1960) is a classic and provides an excellent introduction to the field of quantum chemistry.

Attention has been focused on two principal methods for the quantum mechanical treatment of molecular bonding: valence bonding theory and molecular orbital theory. Only a very general introduction, incorporating ideas from both approaches, can be given here.

The solutions of the quantum mechanical wave equation for an atom are a set of wave functions, known as atomic orbitals. Each electron of the atom may be assigned to a wave function, and the region of space where the electron is located can be determined from its wave function. The region occupied by some atomic orbitals is spherical with the nucleus of the atom at the center of the sphere. Other orbitals have directional character with their associated electrons located in a particular direction relative to the nucleus. Since only the electrons that are more distant from the nucleus are shared in bond formation, only the atomic orbitals associated with these electrons, known as valence electrons, need be considered in the discussion of bond formation between atoms.

As atoms approach each other, their orbitals overlap, i.e., they occupy the same region in space. Mathematically, the two atomic orbitals add together, resulting in a molecular orbital associated with both atoms. An electron that was initially associated with only an atomic orbital on its own atom, is now associated with a molecular orbital and may move about in the space corresponding with this new orbital, space that includes both atoms. If the quantum mechanical solution for the energy associated with an orbital indicates that the electrons in the molecular orbital are more stable than in the two isolated atomic orbitals, a bond is formed, and the molecular orbital is a bonding orbital.

The bonds that have been discussed up to this point involve the overlapping of a single atomic orbital on each of the two bonded atoms to form a molecular orbital that contains two shared electrons, one from each of the bonded atoms. This is called a single bond. It is also possible for two orbitals on each atom to overlap, forming two molecular orbitals between the two atoms. The result is a double bond. If three atomic orbitals on each of the two atoms overlap to form three molecular orbitals, a triple bond is formed.

The approaches used above to explain the bonding between two atoms may be generalized to polyatomic molecules. The bonds in many polyatomic molecules can be considered to be localized covalent bonds: the bond between each pair of atoms in the molecule is thought of as being made up of only the overlapping atomic orbitals of those two atoms, and the electrons shared by each pair of atoms are confined to the space between the two atoms. The molecular orbital description for such a polyatomic molecule is simply the collection of the localized molecular orbitals of all its individual bonds. The properties of the polyatomic molecule, such as reactivity and geometry, result from the properties of all the localized molecular orbitals associated with the bonds between the individual pairs of atoms.

In some cases, however, the properties of a polyatomic molecule are quite different from what might be expected from the sum of individual localized bonds making up the molecule. In these cases, nonlocalized molecular orbitals must be considered. Nonlocalized molecular orbitals are made up of the atomic orbitals of more than two atoms; they stretch over more than two atoms and the shared electrons are associated with all of the atoms whose atomic orbitals make up the molecular orbital. A good example of nonlocalized bonds is the benzene molecule. One way of picturing the bonding in the benzene ring is with alternating localized single and double bonds. Bond strength is the energy needed to break apart two bonded atoms. Bond length is the distance between two bonded atoms. The energy needed to break a single bond is significantly different from that needed to break a double bond. A double bond is significantly shorter than a single bond. Experimentally, however, the distance between adjacent carbon atoms in benzene is found to be the same for each pair of carbons. Also, the energy needed to separate any pair of carbon atoms is found to be the same. Molecular orbital theory explains this by using atomic orbitals on all six carbons of the benzene ring to form six nonlocalized molecular orbitals associated with all six atoms. The shared electrons in these molecular orbitals are free to move throughout the ring. This model of benzene predicts that the properties of all six carbon-carbon bonds (e.g., bond distances and bond energies) should be the same, in agreement with the experimental observations. Also, the values predicted for the bond angles (the angle formed by adjacent bonds in a polyatomic molecule) agrees with experiment.

The use of atomic orbitals on each of the six carbon atoms in the benzene ring to form molecular orbitals is an example of one of the most powerful computational methods used in the study of bonding in molecules. In this approach, the wave function of molecular orbitals for a polyatomic molecule are assumed to be made up of contributions from all atoms in the molecule. The molecular orbital is said to be a linear combination of atomic orbitals (LCAO), with the contribution of each atomic orbital to the resulting molecular orbital given an appropriate weight. Sophisticated computer programs are then used to determine the values for the contribution of each atomic orbital which gives the best fit to experimental data. For benzene, the contribution of each of the six atomic orbitals will be the same. In molecules in which there is little interaction between bonds, a set of molecular orbitals will be obtained which correspond to localized bonds involving only two adjacent atoms in each molecular orbital.

The use of quantum mechanical approaches has revolutionized the understanding of chemical bonds and has proved highly successful in explaining the phenomena related to bonds, including chemical reactivity and molecular geometry. The field of quantum chemistry continues to apply, with great success, the methods of quantum mechanics to virtually all areas of concern to chemists.

BORLAUG, NORMAN ERNEST (1914-)

American plant pathologist and geneticist

Norman Borlaug began his career as a plant pathologist and became a force in international politics through a stint as a consultant in agronomy (the science of raising crops) to the Mexican government. Through the work he performed in Mexico, he created a system of plant breeding and crop man-

Norman Ernest Borlaug. *(AP/Wide World. Reproduced by permission.)*

agement which was then exported to countries throughout the world to create what was dubbed the Green Revolution. Borlaug's unique combination of technical innovation, idealism, **energy**, and impatience with bureaucratic inefficiency took entire countries from starvation to self-sufficiency in the space of a few years. Some countries even became net exporters. Borlaug was aware, however, that his innovations were no final answer to the world's population explosion, but had only bought humankind time to deal with this essential ecological problem. In 1970, Borlaug became the first agricultural scientist, and the fifteenth American, following 1964 laureate Martin Luther King, Jr., to win the Nobel Peace Prize for his service to humanity.

Norman Ernest Borlaug was born on March 25, 1914, to Henry O. and Clara Vaala Borlaug, Norwegian immigrant who owned a fifty-six-acre farm near Cresco, Iowa. With his two sisters, Borlaug grew up on his family's farm. Cresco and its surroundings had a large, hard-working Norwegian population whose lifestyles reflected experiences with hunger and privation that led them to migrate to the New World. The importance of careful planning and hard work in order to survive came home to Borlaug at an early age. In Cresco, he attended the grade school and high school, where he was the captain if the football team. Harry Schroeder, a teacher of agriculture at Cresco High School, recalled that Borlaug showed considerable interest in crop and soil management. Schroeder rewarded his interest by supplying Borlaug with supplementary teaching on different aspects of agriculture.

In 1932, Borlaug graduated from high school, and, instead of becoming a farmer, entered the University of Minnesota to satisfy his grandfather's wish that he get a college education. Working his way through college, Borlaug earned a B. S. Degree in forestry in 1937. During his freshman year, he attended a lecture by the head of the plant pathology department, Elvin Charles Stakman, an authority on crop research. Stakman's lecture had such an impact on Borlaug that he decided to study plant pathology under the professor's direction. At Stakman's urging, Borlaug remained at the University for post-graduate study. Working part-time as a forester, completed his M.S. in 1939, earning his Ph.D. in 1942 in the field of plant pathology. In his doctoral thesis, he discussed a fungal rot endemic to the flax plant. After graduating from college, Borlaug married Margaret G. Gibson on September 24, 1937. They had two children, a daughter, Norma Jean (Borlaug) Rhoda, and a son, William Gibson Borlaug.

Before Borlaug left graduate school, widespread use of chemical **pesticides** had begun. Paul Müller had already discovered, in 1939, that dichlorodiphenyltrichloroethane (DDT) was a powerful insecticide. During World War II, the United States government made extensive use if DDT, especially in the military, and pushed hard to place agriculture on an industrial footing, with the development of new chemicals to control plant diseases and insects. In 1942, Borlaug went to E. I. du Pont de Nemours and Company in Wilmington, Delaware, to apply his expertise in plant pathology to determine the effects these new chemicals had on plants and their diseases. He stayed with du Pont for two years, researching ways to chemically counteract the fungus and bacteria that attack plants.

In 1944, worried by a succession of crop failures in wheat, the Mexican Ministry of Agriculture asked the Rockefeller Foundation to send a team of agricultural scientists to share the technological advances the United States had made in agronomy with the Mexican people. The Rockefeller Foundation named George Harrar, a plant pathologist and future head of foundations, as a leader of this group, and left it up to him to gather the scientists he wanted. Harrar chose Edward Wellhouse, a corn breeder noted for his expertise in Mexico's major grain crop, William Colwell, and agronomist, and Borlaug, who was appointed director of the Cooperative Wheat Research and Production Program in Mexico.

Apart from some modernization in Sonora in the northwest, most wheat production in Mexico had not changed since the conquering Spaniards had established the crop in the sixteenth century. The fields were prepared with wooden ploughs pulled by draft animals, and the harvesting and winnowing were done by hand. Borlaug recalled that there was only one scientist in Mexico available to conduct wheat-breeding experiments, and he was so burdened with other duties that he could contribute only a fraction of his time. There were no government programs for soil management or disease and insect control.

At first, Borlaug sought to improve the variety of wheat commonly grown. This bread was tall and thin-stemmed, an adaptation of centuries of competition with weeds for sunlight. Together with Mexican agricultural scientists, Borlaug's team of researchers began breeding this species of wheat to develop a high-yield, disease-resistant strain that would thrive in the wide range of growing conditions that Mexico offered. Borlaug aimed for results and showed that it was possible to speed up crop productions by harvesting two generations of this new wheat every year, one in Sonora, close to the sea level, and the other in the mountains of Mexico City.

Wheat production improved so dramatically that by 1948, instead of importing half its wheat to feed its population, Mexico had become self-sufficient. However, problems surfaced with the strain of wheat that Borlaug had developed. In the 1950s, wheat yields stagnated. Considerable losses were incurred as the heads on the plants, enlarged by the use of fertilizer and irrigation, grew too heavy from the thin stalks to support, causing the plants to fall over or lodge. Borlaug realized that he would have to breed a wheat plant with a shorter, thicker stem to support the larger heads. In 1954, he and his assistants created a hybrid strain using the improved Mexican grain and a Japanese dwarf variety called Gaines that was perfected by Orville A. Vogel, of Washington State University.

Not only did Borlaug's Mexican-Gaines hybrid prove effective at preventing lodging, but it actually used fertilizer more efficiently by concentrating the growth in the head rather than the stalk. This latest strain was twice as productive as Borlaug's improvement and ten times as productive than the original Mexican strain. In 1961, Mexican farmers began growing the new dwarf hybrid. By the late 1950s, the International Center for Maize and Wheat Improvement, Borlaug's research team, began expanding its activities to other countries, He visited Pakistan in 1959 and India in 1963. His experiences in both countries prompted him to point out that even the best grain varieties would be of no use if bureaucratic inertia stood in the way. In other words, the way a country fed its population was as much a political issue as an agricultural one.

Borlaug was not shy in articulating what resources a nation had to supply for his agrarian reforms to succeed: a stable governing body with the political will to enact his proposals; the ability to provide the chemicals and machinery necessary to modern architecture; and, of greatest importance, a commitment to training young scientists in agronomy. Without this brain trust, there would be no ongoing improvement of crop strains to counteract new diseases or pests.

As noted in *Nobel Prize Winners*, Borlaug became so weary of red tape that he remarked on the subject, "One of the greatest threats to mankind today is that the world may be choked by an explosively pervading but well-camouflaged bureaucracy." His desire to see results quickly led him to inaugurate his Green Revolution programs in the mid-1960s. His aim was to double wheat yields in the host country in the first year of his agricultural improvements. The purpose was twofold: one, to greatly reduce the country's reliance on food imports for a quick economic boost, and, two, to break through the skepticism of the country's officials. Borlaug's programs ultimately benefited countries in Latin America, the Middle East, and Asia.

In the 1960s, scientists at the International Rice Research Institute in the Philippines succeeded in breeding strains of dwarf rice using techniques similar to those Borlaug used to develop dwarf wheat in Mexico. These new rice strains allowed the Green Revolution to spread to Southeast Asia, a region of the world where the staple grain is not wheat but rice.

In 1970, Borlaug received the Nobel Peace Prize for founding the Green Revolution, which allowed a greater measure of peace and prosperity throughout the less-developed world. In his Nobel Prize speech, he made it very clear that an adequate food supply, although essential to a stable world order, was only a first step and that there was much left to be done. He identified population growth as the biggest problem facing humanity, saying, "Since man is potentially a rational being, I am confident that within the two decades he will recognize the self-destructive course he steers along the road of irresponsible population growth and will adjust the growth rate to levels which will permit a decent standard of living for all mankind."

Environmentalists in the 1970s began to criticize Borlaug's techniques on the grounds that it relied on polluting industrial products such as **fertilizers** and insecticides. Borlaug remarked that these accusations missed the point he had made in his Nobel Prize speech, which was that the greatest danger to the environment came not from industrialization but from the population explosion. Having an ample food supply actually helped countries to break the cycle of poverty and starvation that led to overpopulation. While admitting that even after the Green Revolution problems of distribution remained, he asserted that these difficulties were nevertheless an improvement over the ones caused by famine.

In 1979, Borlaug retired from the International Center for Maize and Wheat Improvement in Mexico City, but kept his position as associate director of the Rockefeller Foundation. In this capacity he continued cooperating with the Mexican Ministry of Agriculture in research projects. He became Distinguished Professor of International Agriculture at Texas A & M University in 1984.

After receiving his Nobel Peace Prize, Borlaug gave the benefit of his expertise to the Renewable Resources Foundation, the United States Citizens' Commission of Science, Law, and Food Supply, the Commission on Critical Choices for America, and the Foundation for Population Studies in Mexico. He co-authored several books and wrote over seventy articles. Throughout his career, he received awards and honors from the governments of Mexico, Pakistan, and the United States, among others, as well as from scientific and agricultural societies the world over.

Borodin, Aleksandr Profirevich (1833-1887)

Russian composer, physician, and chemist

The Russian composer Aleksandr Porfirevich Borodin was also a physician and research chemist. He epitomized the group of composers known as the "Mighty Five" and used folk music in conscious pursuit of a "national style."

Aleksandr Borodin was born in St. Petersburg. The name Borodin was that of a retainer to Prince Gedeanov; the

Aleksandr Profirevich Borodin.

prince acknowledged paternity and provided the mother and the boy with a name. Borodin was raised with many of the privileges of the nobility, and his education was broad in the tradition of the European gentleman. This included musical training and preparation for a profession: medicine.

While still a young medical intern, Borodin gained entry to the Mighty Five, partly on the strength of his keyboard ability—a defining factor of the nineteenth-century romantic Russian composer. His training had been that of the gifted dilettante; he now came under the influence of the taskmaster of the group, Mili Balakirev, and subsequently under the influence of the other members of the Mighty Five: Modest Mussorgsky, César Cui, and Nicolai Rimsky-Korsakov. Of them, Borodin alone stuck by his original and primary profession, although he gave up actual medical practice for research.

Although his works are relatively few, Borodin ranks a close second to Mussorgsky as a creative artist among the Mighty Five. His gift is marked neither by the uncertainty nor the verbosity of some of his colleagues and most of his musical heirs. Moreover, his confidence is not marred by the self-righteous certainty that led the next generation of Russian composers into relatively insignificant utterance.

Borodin's Second Symphony (the *Bogatyr* or *Heroic*) and his opera *Prince Igor* (finished posthumously by Rimsky-Korsakov and Aleksandr Glazunov) are his principal works of large proportions. In both he uses a developed folk style effectively, and in the opera he makes a major contribution to the subgenre of "Russian music about the East." Borodin's happy gift for beguiling melody is attested to by the adaptation of his *Prince Igor* music for the American musical *Kismet*. Other than the symphony and the opera, his most-played works are, perhaps, the two String Quartets, some of whose themes are also heard in *Kismet*. A few other chamber works and some 18 art songs nearly round out Borodin's complete list of works.

Some elements of Borodin's personal life and his creative procedures remain obscure. A significant store of Borodiniana has been, since the composer's death, in the hands of the Dianin family. Although the family has tried to present the composer to the world (the first Dianin was Borodin's laboratory assistant), they are too closely involved and Soviet puritanism is far too strong to allow for frankness about personal things; and the Dianins, none of them professional musicians, misjudge what is significant about the creative procedure. Sergei Dianin, a mathematician, in his Borodin biography (1963) supposed the composer to have combined musical elements as a chemist combines chemicals.

Borodin did not teach. He died in 1887, and his legacy was preserved by his friends and reappears in some of the work of Sergei Prokofiev. Borodin's few works, like those of Mussorgsky, are disproportionately important.

Boron

Boron is the first element in Group 13 of the **periodic table**. It is the only element in this chemical family that is not a metal. Its **atomic number** if 5, its atomic **mass** is 10.811, and its chemical symbol is B.

Properties

Boron exists in at least three distinct allotropic forms. One consists of clear red crystals with a **density** of 2.46 grams per cubic centimeter. A second **allotrope** consists of black crystals with a metallic appearance and a density of 2.31 grams per cubic centimeter. A third allotrope occurs as an amorphous brown powder with a density of 2.350 grams per cubic centimeter. All allotropes of boron have very high melting points, from 4,000-4,200°F (2,200-2,300°C).

One property of special interest is boron's ability to absorb neutrons in abundant amounts. This property is utilized in control rods of **nuclear reactors**, which regulate the number of free neutrons passing through the reactor.

Boron reacts with **oxygen** at room **temperature** to form boron trioxide (B_2O_3), which forms a thin film on the surface to prevent further reaction with oxygen. Boron does not react with most acids although it will dissolve in very hot **nitric acid** (HNO_3) or **sulfuric acid** (H_2SO_4).

Occurrence and Extraction

The abundance of boron in the Earth's crust is estimated to be about 10 parts per million, placing about in the middle among the elements in terms of their abundance in the Earth. Boron never occurs as a free element, but always as a compound in minerals such as borax (sodium borate; $Na_2B_4O_7$), colemanite (calcium borate; $Ca_2B_6O_{11}$), and ulexite (sodium **calcium** borate; $NaCaB_5O_9$). These minerals usually occur as white crystalline deposits in desert areas. The two largest producers of boron compounds in the world are Turkey and the United States.

Discovery and Naming

The first mention of boron compounds is found in a book by the Persian alchemist Rhazes (Abu Bakr Mohammad ibn Zakariya al-Razi). Rhazes classified minerals into six classes, one of which was the *boraces*. This class included the compound we now know as borax. Borax was widely used by ancient craftspeople because it reduces the melting point of materials from which **glass** is made. It was also used in the production of **metals** from their ores.

Pure boron was first produced in 1808 by the English chemist **Humphry Davy** by a method that he had earlier used to isolate active metals such as **sodium**, **potassium**, and calcium. The names borax and boron probably originated as far back as the time of Rhazes as *buraq* (in Arabic) or *burah* (in Persian).

Uses

As an element, boron is used to make alloys. Among the most important of these alloys are those containing **iron** and **neodymium** along with boron. They are used to make some of the strongest magnets ever developed. These magnets are used for microphones, magnetic switches, loudspeakers, headphones, particle accelerators, and a variety of technical applications.

The most important boron compound commercially is sodium borate ($Na_2B_4O_7$), used in the manufacture of borosilicate glass, glass fiber insulation, and textile glass fiber. Glasses made with sodium borate have a high resistance to thermal shock and are used in kitchenware (such as Pyrex dishes) and laboratory equipment. Boron also forms important compounds with two other elements, **carbon** and **nitrogen**. Boron carbide (B_4C) and boron nitride (BN) are among the hardest substances known. Both compounds have high melting points: 4,262°F (2,350°C) for boron carbide and more than 5,400°F (3,000°C) for boron nitride. These compounds are used to make high-speed tools, military aircraft and spacecraft, **heat** shields, and specialized heat-resistant fibers.

Bosch, Karl (1874-1940)

German chemist

Karl Bosch, a German chemist, engineer, and industry leader, developed a commercial process for converting gaseous **hydrogen** and **nitrogen** into **ammonia**, an important component in the production of **fertilizers** and **explosives**. In addition, methods that he helped develop to synthesize **gasoline**, **methanol**, and hydrogen had a profound influence on the chemical industry. In 1931, Bosch and **Friedrich Bergius** shared the Nobel Prize in **chemistry** for their pioneering work in chemical high-pressure methods.

Bosch was born in Cologne, Germany, on August 27, 1874, the eldest child of Karl and Paula Bosch. The elder Bosch sold gas and plumbing supplies and prospered as a businessman and entrepreneur. Young Bosch, who showed an interest in and talent for science and technology, was prompted by his father to study metallurgy. In 1894, Bosch enrolled at the Technical University in Charlottenburg, Germany. He studied metallurgy and mechanical engineering, and acquired practical machine-shop experience. But he was disappointed with the semi-empirical methods used in his technical classes, and in 1896 he decided to enter the University of Leipzig to study chemistry. Two short years later, after submitting a dissertation on the study of **carbon** compounds, he received his doctorate.

In 1899, Bosch went to work as a chemist for the Badische Anilin-und Sodafabrik (BASF) at Ludwigshafen am Rhein. BASF was a company specializing in making **coal** tar dyes. Bosch's first assignment was to find an inexpensive method of producing indigo, a dark blue dye important in dyeing cotton.

At the beginning of the twentieth century, Germany was importing a half million tons of Chilean **sodium** nitrate every year for use in producing fertilizers and explosives. Because ammonia, a scarce natural resource, can be used to produce sodium nitrate, many investigators had sought a simple way of using **electricity** to produce ammonia from hydrogen and nitrogen. The existing method was costly, however, because Germany lacked a plentiful and inexpensive source of hydroelectric power. In 1904, **Fritz Haber** found that large quantities of ammonia could be produced by combining hydrogen and nitrogen at high pressures and temperatures and using **osmium** and **uranium** as catalysts.

In 1909, under Bosch's leadership, BASF acquired from Haber the patent rights for the ammonia process. Bosch realized that for the Haber process to be commercially feasible he would need huge quantities of hydrogen and nitrogen, an effective cheap catalyst, and equipment able to withstand extreme pressures and temperatures. He devised a method of separating large quantities of hydrogen from a mixture of hydrogen and **carbon monoxide**. Nitrogen was obtained in pure form by collecting fractions of liquid air. More than 20,000 experiments were necessary to find a suitable catalyst to replace Haber's expensive uranium and osmium. But Bosch's greatest challenge was to construct a reaction chamber that could withstand temperatures of 500 degrees Celsius and pressures that would easily rupture most vessels. Haber had used a steel chamber that had become brittle and dangerously unstable because the hydrogen used in the process caused the steel to lose its carbon content. Bosch cleverly substituted a double-walled chamber. The inside chamber was made of soft steel and could leak hydrogen. The outer chamber was fortified with heavy-duty carbon steel. By forcing a cold mixture of hydrogen and nitrogen gas at 200 atmospheres into the space between the inner and outer chambers, Bosch was able to equalize the pressure on the inner chamber while keeping the outer chamber cool. In 1911, only two years after acquiring the Haber process, BASF began producing commercial quantities of ammonia at a plant near Oppau. During World War I, BASF expanded its production facilities, and by 1918 Germany was producing more than 200,000 tons of synthetic ammonia annually.

In 1919 Bosch became the managing director of BASF. Nonetheless, he remained active in the laboratory, and four years later he succeeded in developing a commercial method for preparing methyl **alcohol** by combining carbon monoxide and hydrogen at high pressures. In 1925, faced with increasing competition from dye industries in Britain and America, BASF merged with six other German chemical firms to become I. G. Farben, and Bosch was appointed its president. The same year, Friedrich Bergius sold BASF his rights to a method of making gasoline from coal dust and hydrogen. Although Bosch succeeded in applying his technical skills to the Bergius method, the process never became profitable. However, for their work in large scale chemical synthesis, Bosch and Bergius shared the 1931 Nobel Prize in chemistry. Four years later Bosch became chairman of the I.G. Farben board of directors.

In addition to his work on ammonia and gasoline, Bosch studied catalytic methods, phase relationships, **photochemistry**, and polymers. He influenced the design of large-scale chemical reactors, compressors, and monitoring devices. He published articles on chemical reactions and received many awards, including the Liebig Medal of the German Chemical Society and the Carl Lueg Memorial Medal of the Association of German Metallurgists. As a tribute to his intellect and leadership abilities he was elected in 1937 to Germany's highest scientific position, president of the Kaiser Wilhelm Institute (later renamed the **Max Planck** Society). Ironically, while Bosch had helped Germany become independent from Chilean sodium nitrate and assured his country a steady supply of ammonia for the weapons industry, he openly opposed Hitler and the Nazi regime.

Bosch married Else Schilbach in 1902; they had two children, a son and a daughter. Committed to education and life-long learning, Bosch relaxed by collecting butterflies, beetles, plants, and minerals. He also enjoyed stargazing from his private observatory near Heidelberg. He died in 1940 in Heidelberg after a long illness.

Boussingault, Jean-Baptiste-Joseph-Dieudonné (1802-1887)

French agricultural chemist

After studying **mineralogy** in Paris, Boussingault found employment with a mining company that sent him to South America in 1821. There, he became involved in the Spanish colonies' wars of independence and, at one point, served in Simon Bolívar's (1783-1830) army. While supervising mines in Columbia, which had just won its freedom from Spain, Boussingault conducted geological and meteorological research.

Jean-Baptiste-Joseph Dieudonné Boussingault.

Boussingault's work in South America boosted his scientific reputation. Following his return to France in 1832, he devoted himself to the study of agricultural science at his farm in northeastern France. Elected to the French Academy of Sciences in 1839, Boussingault is considered to be the founder of experimental **agricultural chemistry**.

One of the problems that Boussingault researched was the source of plant **nitrogen**. Although scientists knew that plants contain nitrogen compounds, no one was sure where the nitrogen came from. Most people believed that it was restored to the soil by humus, decaying organic. Boussingault showed that plants could flourish without humus, as long as other sources of nitrogen, such as nitrates or ammonium salts, were supplied.

Air, which consists of nearly 80% nitrogen, is another obvious source. In experiments with beans, peas, clover, and other legumes, Boussingault demonstrated that these plants restore nitrogen to soil that has been exhausted of the element. When Boussingault grew legumes in soil and **water** that did not contain nitrogen, the plants were still able to produce nitrate compounds. He concluded that legumes must be extracting, or assimilating, nitrogen from the air—a basic step in the **nitrogen cycle**. Boussingault also showed that other plants depend entirely on fertile soil for their nitrogen, because they cannot obtain it from the air as legumes do.

Boussingault found that animals, like most plants, are unable to absorb atmospheric nitrogen. His tests proved that animals receive all of their nitrogen by eating food, not from the air. The first scientist to study the weight loss in animals when their diet is deficient in certain substances, Boussingault measured the nutritive value of specific foods by determining the exact quantity needed to prevent weight loss. This early research on **nutrition** was put to use later when scientists began to study the role of **vitamins** and trace minerals in animal health. Boussingault was also the first scientist to suggest that **iodine** compounds might be able to cure goiter, an enlargement of the thyroid gland in humans. Although he found iodine in certain salts used by South American Indians to treat goiter, it was not until over half a century later that his suggested cure was confirmed. As the culmination of his career, Boussingault published an eight-volume work on agricultural chemistry between 1860-1874, which was also published in English and German.

Boyer, Herbert W. (1936-)

American biochemist

Herbert W. Boyer has long been one of the leaders in both the science and the business of biotechnology, the engineering of genetic material. It was Boyer, in collaboration with the Stanford biochemist **Stanley Cohen**, who first cloned, or artificially constructed, new and functional deoxyribonucleic acid (DNA) from two separate gene sources. Following the work of **Paul Berg**, who had developed the technique of gene splicing, in 1973 Boyer and Cohen managed to take genes from two bacteria, recombine them, and insert them in another cell which divided itself and reproduced the new genetic material. It was the dawn of a new biological age, one full of potential and fraught with ethical problems.

Herbert Wayne Boyer was born on July 10, 1936, in Pittsburgh, Pennsylvania; his father was a **coal** miner and railroad worker. He grew up in the nearby town of Derry and attended both grammar and high school there, where he showed more of an aptitude for football than for science. But his football coach was also his science teacher, and Boyer soon became interested in that subject. In 1954 he entered St. Vincent College in Latrobe, Pennsylvania, a Benedictine liberal arts school, where he embarked on premedical studies. He soon abandoned medicine, however, when he discovered the field of **DNA** research. **James Watson** and **Francis Crick** had made their ground-breaking discovery of the double-helix structure of DNA in 1953, and scientists were racing to map human genes. Boyer was fascinated; he soon dubbed his pet cats Watson and Crick. Boyer graduated from St. Vincent in 1958.

In 1958, Boyer married a young woman from Latrobe, Grace Boyer, and entered the University of Pittsburgh as a graduate student in bacterial genetics. He earned his Ph.D. in bacteriology in 1963, and he went on to Yale for three years as a postdoctoral fellow, studying enzymology and protein **chemistry**. It was here that he began to focus more and more on restriction enzymes, **proteins** which are designed to cut up and destroy foreign DNA invading the cell. In 1966 he was appointed assistant professor at the University of California at San Francisco.

At the University of California, Boyer was given a laboratory where he could continue his research on restriction enzymes, specifically those of the bacterium *Escherichia coli* (*E. coli*). This one-celled bacterium has long been a favorite for laboratory study because of its simple structure and because it rapidly reproduces and is comparatively easy to culture. Within ten hours of inception, a single cell of *E. coli* will multiply into over a billion daughter cells. Along with another young biochemist, Howard Goodman, Boyer isolated certain of the restriction enzymes in *E. coli,* and he began to see that they could be used as chemical scissors to cut the DNA **molecule** at certain points. One such **enzyme**, *Eco* RI Eco, always cut the same genetic phrase along the DNA molecule. Not only did it cut the DNA, but it did so in such a way that it left two flaps or "sticky ends"; these ends made it possible to bind other cut pieces of DNA if they were inserted into the incised section. *Eco* RI Eco was, Boyer began to see, a revolutionary biochemical tool, which could be used not only for cutting DNA but for pasting it as well.

At this time, during the early 1970s, scientists at both Stanford and Harvard were researching gene splicing and recombining. At Stanford, Paul Berg had managed to splice genes or segments of the DNA molecule of one virus—SV40, a simian virus—into another virus to be used as a transporter or "vector." He had stopped short, however, at then inserting such recombined or recombinant DNA into the *E. coli*, for fear that his viral vector might somehow escape his laboratory and then become established in the human intestine. But Berg's work had pointed the way for others. Two floors below Berg's lab at Stanford, Stanley Cohen was at work isolating genetic material from *E. coli.* But Cohen was not concerned with the one round **chromosome** of the bacterium which contains most of *E. coli*'s genetic material—nearly 4,000 genes. Rather he was isolating small rings of genetic material separate from the chromosome known as plasmids. Using detergents, Cohen could isolate these plasmids and their DNA molecule more easily than it was possible to do with the chromosome itself.

Cohen heard Boyer describing the action of his *Eco* RI Eco enzyme at a conference in Hawaii, and, in a meeting that is now legendary, the two talked over a sandwich at a delicatessen and agreed to work together. Combining their technologies, they saw how they could splice genes from two bacteria and insert them in a third which could then be replicated billions of times. In San Francisco, Boyer and Cohen worked out the basics of **genetic engineering**. Boyer's restrictive enzyme would cut segments of DNA along Cohen's plasmids. The segments could then be inserted or spliced in the cut segments of different plasmid rings using the enzyme ligase—whose function is the opposite of a restriction enzyme—to secure the splice. This hybrid plasmid or vector could then be inserted into *E. coli* where it would replicate itself, creating a new, genetically engineered organism. Cohen called the resulting organism a chimera after the mythical Greek monster—part goat, part serpent, and part lion—but they are more commonly referred to as clones.

Boyer and Cohen first published their results in 1973, and they followed it up with work involving gene splicing and cloning of the African clawed toad, again employing *E. coli* as the replicating organism. Boyer soon saw the commercial potential of the process he helped pioneer. By splicing in segments of higher organisms—the genetic material responsible for the creation of proteins or **hormones**, for instance—one could make the bacteria into factories, producing these various materials as they replicated themselves. It was this sort of thinking that created the biotechnology business; Boyer invested 500 hundred dollars with the young venture capitalist Robert Swanson and became one of the first in the field, cofounding the company Genentech in 1976.

That same year, Boyer became a full professor of **biochemistry** at the University of California in San Francisco and an investigator at the Howard Hughes Medical Institute. He continued his research as well as his job as vice president of Genentech, contracting his own university laboratory for results in many instances. Genentech began producing somatostatin, a brain hormone, in 1977; they were manufacturing insulin by the next year and a growth hormone the year after, and by 1980 they were manufacturing interferon, a protein which acts as an antiviral within the cell. Following the initial, difficult years of research and a 1980 Supreme Court ruling which made it legal to patent life forms, Boyer's company went public. With the public offering of Genentech, Boyer became a millionaire many times over, and he began to draw criticism from the academic community for possible conflicts of interest. He was also stung by criticism of the new gene technology and its ethics. Many laypersons and scientists alike argued that the new technology was like playing god, and some believed that the manufacture and sale of body parts was a new form of slavery. Boyer withdrew from the public eye as a result of such criticism; he resigned from the vice presidency of Genentech and began to work for them as a consultant, though retaining his stock shares. He returned to his laboratory work and began new research in methylation patterns of DNA —places that show how vital proteins of the cell such as restriction enzymes actually work with the double helix. In collaboration with a team from the University of Pittsburgh, Boyer helped discover how the restriction enzyme, *Eco* RI, actually interacts on the atomic level with DNA.

Boyer has been widely honored for his research. Among other awards, he has won the V. D. Mattia Award in 1977, the Albert and Mary Lasker Medical Research Award in 1980, the Moet Hennessy-Louis Vuitton Prize in 1988, and the National Medal of Science in 1990. He was elected a member of the National Academy of Sciences in 1985 and is a fellow of American Academy of Arts and Sciences and the American Society of Microbiology. Boyer and his wife have two children.

BOYER, PAUL D. (1918-)

American biochemist

Paul Boyer, professor emeritus of **chemistry** at the University of California at Los Angeles (UCLA), shared 1997's Nobel Prize in chemistry with **John E. Walker** of England, for their discovery of how enzymes synthesize the compound adenosine triphosphate (ATP), an essential part of how cells store and release **energy**.

Paul D. Boyer was born July 31, 1918, in Provo, Utah, son of a physician, Daryl D. Boyer, and Grace (Guymon) Boyer. Paul graduated from Provo High School's college preparatory course in 1935, and went on to study chemistry at Brigham Young University (BYU); he did not follow in his father's footsteps, he says, "because I didn't want to have to worry about people." While a student at BYU, he met Lyda Whicker; they were married in 1939, the same year Paul earned his bachelor of science degree. They have three children.

While at BYU, Boyer noticed a flyer for a scholarship to the University of Wisconsin. He applied, and won the scholarship. "Sometimes one has to make a choice, and just hope that it's the right one," he said in a 1997 interview with contributor Fran Hodgkins. "I've been lucky, that all my choices seem to have been right choices."

The atmosphere at the University of Wisconsin was "superb," stimulating Boyer's interest in enzymes. After college graduation, Boyer decided to study **biochemistry** at the University of Wisconsin at Madison. He received his master's in 1941, and his Ph.D. in 1943.

He then became a research associate at Stanford University, working there from 1943 to 1945. From there, he went to the University of Minnesota. He remained with the University of Minnesota from 1945 to 1963, rising from assistant professor in the department of biochemistry to full professor (in 1953); from 1956-1963, he was the Hill Professor of biochemistry.

From 1963 to 1990, he was a professor in the department of chemistry and biochemistry at the University of California at Los Angeles. During his time at UCLA, he was a member of the UCLA Molecular Biology Institute, of which he served as director from 1965 to 1983. He was also director of the University of California Program for Research and Training in Biotechnology from 1985 to 1989. He is now professor emeritus at UCLA.

Since the 1950s, Boyer has been studying enzymes, **proteins** that cause most of the chemical reactions that take place in a cell. He was especially interested in the enzymes at work in the process of converting food nutrients to ATP. ATP serves as the energy source for everything that cells do—from building bones to transmitting nerve impulses.

"My interest in how ATP is made likely arose from my graduate student studies that included the first demonstration of a requirement of K^+ by an **enzyme**, in this instance the transphosphorylation from phosphoenolpyruvate to ADP to form ATP," Boyer wrote in "From Human Serum Albumin to Rotational Catalysis by ATP Synthase." "The intellectual milieu at the University of Wisconsin, where I was enrolled over 50 years ago, was superb. My studies were stimulated by a symposium on Respiratory Enzymes at which Cori and others mentioned the exciting reports of Ochoa and of Beitzer and Tsibakova showing that more than one ATP was made for each **oxygen atom** consumed. Although much of my subsequent career has concerned studies with other enzymes, that later portion increasingly focused on the mechanism of the ATP synthase.''

ATP was discovered in 1929 by Karl Lohmann, a German chemist. **Fritz Lippman**, who studied ATP extensively during 1939 to 1941, discovered that the compound carries chemical energy in the cell (he received the Nobel Prize in medicine in 1953). It has been called the cell's ''energy currency.'' In 1948 **Alexander Todd** synthesized ATP chemically (he received the Nobel Prize in chemistry in 1957). Researchers during these decades learned that the mitochondria of animal cells and the chloroplasts of plant cells form most ATP, through cell respiration and **photosynthesis**, respectively. In 1960 researcher Efraim Racker discovered ATP synthase, the enzyme that creates ATP, and described its structure.

Further research showed that the **concentration** of **hydrogen** ions (also known as pH) inside and outside the mitchondria's membrane changes during cell respiration. A stream of hydrogen ions drives the formation of ATP, suggested **Peter Mitchell** in 1961—a suggestion that Boyer was, at first, ''reluctant to accept.'' Yet despite all this vital information about the compound's structure and function, no one knew how it was created from **phosphate** and adenosine diphosphate (ADP).

Racker and his co-workers discovered that ATP synthase consists of two major parts: one that anchors the enzyme to the cell membrane (which he called the F_0 part) and the part containing the ''engine'' or catalytic center (the F_1 part). Racker called ATP synthase ''F_0 F_1 ATPase.''

The F_0 part is disc-shaped and consists of smaller units (called ''c'' subunits). Jutting from the center of this disk is the asymmetrical main shaft of the F_1 part, called the ''gamma'' subunit. Surrounding the gamma subunit are three alpha and beta subunits, forming a cylinder around the gamma subunit. Over decades, Boyer created a model of how the parts of ATP synthase work together to generate energy.

ATP synthase's structure is like a mixer's. The F_0 part is like the part of the mixer that holds the beaters firmly attached to the mixer's body. The shaftlike, asymmetrical ''gamma'' subunit is attached to the F_0 like a beater, and projects into the main part of the F_1 as a beater sticks into a bowl. Just like a beater moves with the movement of its base, so does the gamma section move with the F_0 part when a stream of H ions hits it.

However, the alpha and beta parts of F_1 do not move. The gamma section ''beater'' strikes them as it moves and makes them change shape—like a beater changes the shape of a hard stick of butter in the bowl (unlike the butter, the alpha and beta sections bounce back to their original shapes once the gamma subunit has spun by). This shape changing creates energy, which binds ADP and phosphate into the ATP **molecule**. The ATP molecule stores extra energy from the reaction. (Walker, who shared the 1997 Nobel with Boyer, verified the model.)

In addition to the Nobel Prize, Paul Boyer and his work have been recognized many times over the years. He received the American Chemical Society Award in Enzyme Chemistry in 1955, the same year he was named a Guggenheim Fellow. He received UCLA's McCoy Award in 1976, the Tolman Medal from the Southern California chapter of the American Chemical Society in 1981, and the Rose Award from the American Society of Biochemistry and Molecular Biology in 1989. He has received honorary doctorates from the University of Stockholm (1974) and the University of Minnesota (1996). He had also been a member of the National Academy of Sciences and a fellow of the American Academy of Arts and Sciences.

Boyer has also served on the editorial board of the journals *Archives of Biochemistry and Biophysics, Biochemistry*, and the *Journal of Biological Chemistry*. He also served as coeditor of *The Enzymes* (as well as editor of its third edition), coeditor of *Biochemical and Biophysical Research Communications*, and associate editor and editor of the *Annual Review of Biochemistry*.

From 1957 to 1961, Boyer was a member of the biochemistry study section of the National Institutes of Health and chairman of that group from 1962-1967. He was chairman of the American Chemistry Society's biological chemistry division (1959-60), and a member of the U.S. National Committee for Biochemistry (1965-1971). With the American Society of Biological Chemists, he was a council member from 1965-1971, president 1969-1970, and chairman of the public affairs advisory committee, 1982-1987. He also was a councilor of the American Academy of Arts and Sciences (1981-1985), and vice president, biological sciences, 1985-1987.

BOYLE, ROBERT (1627-1691)

English physicist and chemist

Boyle is considered by many as the father of modern chemisty. He offered the first accurate definitions of elements and chemical reactions, is credited with pioneering modern scientific method, and also formulated the law that bears his name which describes the relationship between pressure and **volume** in **gases**.

For centuries people believed that everything was made of just three or four substances, which were mistakenly called elements. It was Boyle who first set science on the right track and asserted the true nature of elements and compounds.

Boyle's father was an Englishman who made his fortune in Ireland and became a successful landowner there. Boyle was his seventh son and the youngest of fourteen children. By the time Boyle was born, his father had become an earl and was one of the wealthiest men in the country. Like his father, Boyle was an industrious worker; before he entered the prestigious Eton school at the age of eight, he was already speaking Greek and Latin. His passion for reading and learning continued to grow, and Boyle proved to be a gifted student with an excellent memory.

At an early age, Boyle and his brother went to Europe with a tutor to study French, mathematics, and many other subjects. For six years, they lived in Switzerland and traveled ex-

Robert Boyle. ***Corbis Corporation. Reproduced by permission.)***

tensively through France and Italy, where Boyle learned of Galileo's experiments on the effects of gravity and other physical laws. When Boyle was 13, he witnessed a sudden, violent thunderstorm that changed his whole outlook on life. But in Boyle's view, his religious faith did not contradict scientific beliefs but reinforced his admiration for the creator of such a complex universe. From then on, he was a devout Christian, and he learned several ancient languages, including Hebrew and Aramaic, so that he could read the Bible in its original texts.

A civil war between England and Ireland demanded Boyle's return from continental Europe, and he reached home in 1645. By this time, Boyle had become interested in performing experiments in order to understand the way things work. Previously, people believed in making up a theory and then judging how well the facts fit the theory. Boyle, however, agreed with the philosopher Francis Bacon (1561-1626) that facts should be observed first, and then a theory should be developed to explain them. Boyle and other scholars interested in experimentation began meeting regularly in London, England, to discuss these new ideas. At first, they called themselves the ''Invisible College,'' but when King Charles II (1630-1685) was restored to the throne in 1663, he granted a charter to the group of scientists; the group thus became known as the Royal Society.

When Boyle moved to Oxford in 1654, he met many more scientists and became interested in **chemistry**, because of its relation to medicine. During his 14 years at Oxford, Boyle contributed greatly to scientific philosophy in the fields of physics and chemistry. He set up an elaborate research laboratory and hired skilled assistants to conduct experiments. Unlike most scientists of his day, Boyle believed in meticulously recording his experiments and publishing them so that others could repeat his tests and confirm the data. He is credited with pioneering modern scientific method and today this practice is universal in the research world.

In 1661 Boyle published his most famous work, *The Sceptical Chymist*, which revolutionized scientific thought and formed the basis of modern chemistry. In this work, Boyle defined an element as the simplest form of **matter**, one that cannot be broken down into any simpler form or changed into a different substance. Boyle's ideas contradicted beliefs held ever since the ancient Greeks proposed that all things are made of only four elements—air, earth, fire, and water—which could be changed, or *transmuted*, into other substances. In another version of this idea, only three substances existed in nature (salt, **sulfur**, and mercury). But according to Boyle, none of these substances were true elements. Boyle argued that elements could be identified only by scientific experimentation. He also pointed out that a compound will usually have chemical properties that are very different from its parent elements.

Boyle's concept of an element arose from his experiments with gases, and he was the first scientist to succeed in collecting **hydrogen** in a device now called a *pneumatic trough*. In 1660 Boyle discovered a fundamental law of physics that helps explain the behavior of gases. When a gas is pressurized, Boyle found that the amount of space it takes up is related to the amount of pressure being exerted on it, as long as the gas's **temperature** doesn't change. For example, if the pressure on a given quantity of gas is doubled, the gas's volume is cut in half; if pressure is tripled, volume is reduced to one-third. (This relationship is called *inversely proportional.*) **Boyle's law**, along with a similar law that explains the effects of temperature, allows chemists today to calculate the volume of gases under any pressure or temperature conditions. Boyle also realized that if air could be compressed, it must be composed of tiny particles separated by space. It was this conclusion that led Boyle to envision a universe composed of numerous tiny particles, and, in doing so, he anticipated the modern concept of **atomic theory**.

Vacuums were poorly understood but of much scientific interest to Boyle and his assistant, Robert Hook. Airpumps were used in early laboratories to create a vacuum inside cylinders. Robert Hook built an improved airpump based on German engineer, Otto von Guerricke's airpump design. Together, Boyle and Hook developed a better vaccum. This new vacuum had better placement of the pumps valves and a preferable method of cranking its piston and supporting the air pump's cylinder. Boyle also proved for the first time that all objects, no matter how light or heavy, fall through a vacuum at the same speed. This showed, as Galileo had predicted, that the force of gravity is uniform. In another experiment, Boyle demonstrated that the sound of a clock ticking could not be heard in a vacuum, proving that sound waves depend on air for their transmission. Boyle showed, however, that electrical attraction could be felt through a vacuum.

Boyle was also interested in the nature of **color**, and he accurately described how the absorption and reflection of light produces the appearance of black and white, studying the changes in color that occur in certain plant extracts, such as *litmus*. He discovered that these substances, now called *indicators*, can be used to distinguish acids from bases. Boyle went on to develop tests for identifying other substances, such as **copper**, **silver**, and sulfur, via chemical reaction. He not only coined the term *analysis* in its modern sense, but also encouraged generations of chemists to determine the composition of substances through meticulous experimentation. In the late 1660s Boyle became the first scientist to study the phenomenon of **bioluminescence**, showing that certain bacteria and other organisms will glow in the dark if supplied with air. Boyle also found that **water** begins to expand just before it freezes. Throughout this period, Boyle and his staff published immense amounts of information for use by other scholars and scientists.

In 1668 Boyle returned to London to live with his favorite sister. In 1680 he invented the first match by coating a piece of course **paper** with **phosphorus**. He produced a flame by drawing a sulfur-tipped wooden splint through a fold in the paper. Also in 1680, he was elected president of the Royal Society, but he declined the honor, believing that the oath of office would conflictwith his strict religious beliefs, and he continued to refuse all titles and other honorary positions. In his later years, Boyle wrote about medicine and diseases and devoted greater effort to promoting Christian ideals; in his will he left money for a series of lectures to defend Christianity from atheists and other ''infidels.''

Boyle's law

Boyle's law is one of the **gas laws**. It states that at constant **temperature**, the **volume** of a fixed **mass** of gas is inversely proportional to the pressure. The other way of expressing this law is pV = constant, where p is the pressure and V is the volume of the gas. It is named after British physicist and chemist **Robert Boyle** (1627-1691).

Boyle's law is sometimes known as the constant temperature law. It can be combined with Charles' law and the pressure law to give the ideal gas law (also known as the universal gas law), pV = nRT, where p is pressure, V is volume, n is the number of moles of gas, R is the universal gas constant, and T is temperature.

Boyle's law is an approximation and works perfectly only for an ideal(theoretical) gas. In practice it works best at low pressures but once the pressures become high the predictions become less accurate. This inaccuracy is due to the size of the gas molecules and weak **intermolecular forces**, such as **van der Waals forces**. At high pressures the molecules are forced together whereas at low pressures the molecules are free to move with very little interaction from neighboring molecules.

Boyle's law is a good indicator of how a gas will react if the temperature is kept constant and the pressure and volume are altered.

See also Gases, behavior and properties of

Brackett series

The middle of the nineteenth century saw a surge in the spectroscopic investigation and identification of the elements. German chemist **Robert Bunsen** and German physicist **Gustav Robert Kirchhoff** discovered the process of chemical analysis called **spectroscopy**. Swedish physicist Anders Ångström (1814-1874) measured the wavelengths of the **hydrogen** spectrum. Swiss mathematician and physicist Johann Balmer (1825-1898) calculated the mathematical relationship between the wavelengths and showed that they form a simple series. And, at the turn of the century, German physicist **Max Planck** made sense of it all with his ''quantization'' of light.

In 1906, the American physicist Theodore Lyman (1874-1954) observed additional lines in the spectrum of hydrogen in the far ultraviolet region with the same mathematical relation calculated previously by Balmer except that these lines required an n-value of 1. This led to further investigations and the discovery of other spectroscopic series, all of which have been named after their discoverers. The **Brackett series** (discovered by American physicist F.S. Brackett) are electronic transitions between the n = 4 state (an excited **energy** state) and higher energy levels. The energy gap is so small that these absorptions occur in the infrared portion of the spectrum and are only visible using specialized instruments. However, they provide further evidence for the discrete nature of the electronic states in the **atom** and confirmation of the principles expressed in **Niels Bohr**'s theory of the **atom** and subsequently, quantum mechanics.

See also Quantum chemistry

Brady, St. Elmo (1884-1966)

American chemist

St. Elmo Brady was the first African American to receive the Ph.D. degree in **chemistry**. He taught general and **organic chemistry** to a great number of scientists and health professionals at four historically black colleges in a long and distinguished career. Primarily a teacher, he followed the example of **George Washington Carver** and carried out research on plants native to the southern United States, searching for useful chemical products.

Brady was born in Louisville, Kentucky, in 1884. He attended elementary and high school there and graduated from high school with honors. He began Fisk University in Nashville, Tennessee in 1904, where he studied under Thomas W. Talley, one of the early teachers of modern chemistry in black colleges. When he graduated from Fisk in 1908, he accepted a position at Tuskegee Institute (now Tuskegee University) in Alabama. At Tuskegee he became the friend of educator and Tuskegee founder Booker T. Washington and agricultural researcher George Washington Carver and learned from them the value of working for the advancement of others.

He took leave from Tuskegee in 1913 to attend the graduate program in chemistry at the University of Illinois. At the time, no African American had earned a doctorate in chemistry, and few had advanced degrees in any academic field. He received the M.A. degree in 1914 and was given a graduate fellowship which enabled him to continue his study towards the Ph.D. His research director was Clarence G. Derick, who was exploring the effect of structure on the strength of organic acids. Brady added to the knowledge of this subject, studying the effect of the divalent **oxygen atom**, which played an important role in the development of theoretical organic chemistry later in the century. At Illinois, Brady was the first African American admitted to Phi Lambda Upsilon, the chemistry honor society (1914), and was one of the first inducted into Sigma Xi, the science honor society (1915). Derick left Illinois in 1916 for the rapidly expanding chemical industry, and Brady, who received the Ph. D. in 1916, returned to Tuskegee. Brady realized, as quoted by **Samuel P. Massie** in the *Boule Journal,* that his work in a modern research laboratory could not be reproduced in a ''school in the **heat** of Alabama, where I wouldn't even have a **Bunsen burner**,'' but his spirit of service led him back to Tuskegee where he was appointed head of the division of science.

In 1920, Brady accepted an offer to be professor and head of the chemistry department at Howard University in Washington, DC. He remained at Howard for seven years, building the undergraduate program in chemistry, but left when he had the opportunity to return to his alma mater, Fisk University, to head the chemistry department there. Brady spent the major part of his career at Fisk, retiring in 1952 after 25 years in the chemistry department. He taught general and organic chemistry to hundreds of students, and published his research on the chemical constituents of magnolia seeds and castor beans. He supervised the construction of the first modern chemistry building at a black college, which was eventually named after him and his teacher, Thomas Talley. He was also able to begin a graduate program in chemistry, which was the first in a black college. In conjunction with the graduate program, Brady established the Thomas W. Talley lecture series, which brought outstanding chemists to Fisk. He established, in conjunction with faculty from the University of Illinois, a summer program in **infrared spectroscopy** at Fisk which was open to faculty members of all colleges and universities. Because of limited opportunities for employment for educated black women in Nashville, Brady's wife, Myrtle, had lived and worked in Washington while he traveled often between Washington and Nashville. They had one son, a physician, who died very young, and two granddaughters. When Brady retired from Fisk, he moved to Washington. In his retirement, Brady was asked to assist the development of the chemistry department at Tougaloo College, a small college in Mississippi, and he was eager to help design their new science building and obtain new faculty members. St. Elmo Brady died in Washington on Christmas Day, 1966, at the age of 82.

Bragg, William Lawrence (1890-1971)

English physicist

William Lawrence Bragg shared a remarkable two-year collaboration with his father and fellow physicist, William Henry Bragg, during which they founded the new science of X-ray **crystallography**. The methods developed by this father-son team made it possible to explore the **atomic structure** of **matter** very precisely and in great detail. The Braggs shared the 1915 Nobel Prize in physics for their work.

Bragg was born in Adelaide, Australia. His father was professor of physics and mathematics at the University of Adelaide; his mother, Gwendoline Todd Bragg, was the daughter of Sir Charles Todd, South Australia's postmaster general and government astronomer. Bragg had a brother one year younger than he, Robert, who was killed at Gallipoli in Turkey during World War I, and a sister, Gwendolen, seventeen years his junior. The children's parents were a contrast; Gwendolen later wrote in the biography *William Henry Bragg* that their father wanted his children to be absolutely free and avoided advising them on what they should do, whereas their mother ''always knew exactly what one ought to do, and said so.'' Foreshadowing the field of the Braggs' future work, Bragg's father built a primitive X-ray machine within weeks after Wilhelm Conrad Röntgen's 1896 announcement of his discovery of the rays. Soon thereafter, five-year-old W. L. fell from his tricycle and broke his elbow. Professor Bragg used his device to reveal the location and extent of the injury. This was the first recorded use of X rays for medical diagnosis in Australia.

As a child, Bragg was a loner, being academically ahead of boys his age and poor at sports because, as he explained it, he wasn't aggressive or self-confident enough. His sister quoted him as telling her, ''You and I find *things* easier than people.'' He expressed his interest in natural sciences by collecting shells; a new cuttlefish species he discovered, *Sepia braggii,* was named after him. With his brother, he enjoyed creating mechanical devices out of discarded scraps, and a **chemistry** master aroused his interest in scientific experimentation.

A gifted scholar, Bragg attended St. Peter's College (a secondary school) in Adelaide and entered the University of Adelaide in 1905 at the age of fifteen. His father was beginning to experiment with X rays at the university at that time, and W. H. often talked to his son about his results. ''I lived in an inspiring scientific atmosphere,'' Bragg wrote of those years in a chapter he contributed to *Fifty Years of X-Ray Diffraction* entitled ''Personal Reminiscences.'' He graduated in 1908, after just three years, with first-class honors in mathematics. In January 1909 the family left Australia for England, where W. H. Bragg had accepted a professorship at the University of Leeds. W. L. enrolled at Trinity College, Cambridge, where he began by studying mathematics. At the end of his first year, taking an exam while sick in bed with pneumonia, he won a major scholarship. Then, following his father's suggestion, he switched his concentration to physics. Again he graduated with first-class honors, this time in natural sciences, in 1912.

He stayed at Cambridge, doing research under **J. J. Thomson** at the Cavendish Laboratory, and began his very fruitful collaboration with his father in the fall of 1912.

Earlier that year, **Max Laue** had discovered the diffraction of X rays in crystals: X rays passing through crystals are bent, producing distinct patterns. At the time, physicists were engaged in a lively debate about the nature of X rays: Were they particles or waves? Bragg's father had a deep interest in X rays and favored the particle theory. He was excited by Laue's discovery, even though the diffraction patterns could be explained only if X rays were waves, and he thought further investigation of the new phenomenon might provide the missing evidence to support either the wave or the particle theory, or even a new theory incorporating both wave and particle properties. W. L. Bragg, too, was excited by Laue's discovery, especially as it related to the structure of crystals.

During the summer holiday of 1912, father and son talked often about Laue's discovery, and when W. L. returned to Cambridge that fall, he launched into a series of experiments using **X-ray diffraction**. He concluded that Laue's wave interpretation was correct, but that his explanation of diffraction was unnecessarily complex. W. L. suggested that since atoms are arranged in a regular way in crystals, the diffraction patterns might be caused by X rays reflecting off the planes of atoms within the crystals. He developed an equation relating angles of rays, wavelength, and perpendicular distance between atomic planes, which became known as Bragg's law, with the glancing angle of the X rays called Bragg's angle. Both Braggs saw the great implications of W. L.'s idea: the reflected patterns of the X rays would reveal the previously hidden arrangement of the atoms within the crystals. The Braggs plunged into investigations of crystals, with W. L. first analyzing **sodium** chloride (table salt) and **potassium** chloride (a crystalline **salt** used as a fertilizer), though they were hampered by inadequate equipment. Bragg's father solved this problem in 1913 by his invention of the X-ray spectrometer, a device that measured X-ray angles and intensities precisely and allowed for analysis of complex crystals. The elder Bragg used the spectrometer to continue his studies of radiation, while W. L. used it to pursue his interest in analyzing the structure of crystals.

W. L. Bragg published his first suggestion about the wave nature of X rays as shown by Laue's discovery in November 1912. W. L. and his father published their first joint paper early in 1913 in the *Proceedings of the Royal Society,* which established the basic principles of the new science of crystal analysis by X-ray diffraction. W. L. followed this a few months later with his own paper on the structure of **sodium chloride**, which he showed was not made up of molecules, as had been thought, but rather of ions (charged atoms) of sodium and chloride. This finding became very important in the analysis of solutions, as it was the first distinction made between compounds that consist of ions and those made up of molecules. In July 1913 the Braggs published a joint paper on the structure of **diamond**, and W. L. published another paper in November describing more crystal structure discoveries. By 1914 the Braggs had set the standards for **X-ray crystallography**. They had also transformed crystallography from its role as a secondary, if interesting, branch of science to its new position as a fundamental branch of modern physics with applications to many other sciences. In "Personal Reminiscences" W. L. said of this period of collaboration with his father, "We had a thrilling time together in an intense exploitation of the new fields of research."

The Braggs were recognized for their ground-breaking achievements in exploring the arrangements of atoms by the 1915 Nobel Prize in physics, awarded—in the Nobel Committee's words—"for their contributions to the study of crystal structure by means of X rays." W. L. was only twenty-five years old, the youngest Nobel laureate ever. But the "thrilling time" of collaboration had ended abruptly with the outbreak of World War I in August 1914. Bragg volunteered for service and found himself assigned to a horse artillery battery among "hunting men." After a year, he was sent to France to adapt the new method of locating enemy guns by sound, called sound ranging, for use by the British forces. He later recalled that it was while setting up a sound-ranging base in Belgium that he learned of his Nobel Prize. The parish priest in whose house he was lodged broke out a bottle of wine to celebrate. After the war, Bragg returned to Cambridge as a lecturer.

In 1919 Bragg was named to succeed **Ernest Rutherford** as professor of physics at the University of Manchester. The first years were difficult, as Bragg was inexperienced and he had to handle classes full of war veterans. Gradually, however, he built up a fine research facility, concentrating on improving methods of using X rays to determine crystal structure. He published a list of atomic sizes and measured absolute intensities of X-ray reflections. He then turned to the silicate family of minerals, whose complex structure had proved elusive. This work, completed around 1930, was of fundamental importance to the science of **mineralogy**. In 1927 Bragg spent four months at the Massachusetts Institute of Technology as a visiting professor, lecturing on X-ray diffraction and crystal structure. During the 1930s Bragg oversaw and encouraged investigations into **metals** and metal alloys, which provided new basic knowledge about the chemistry of metals, and he encouraged the application of these X-ray diffraction techniques to industrial firms in northern England, where the university was located.

In 1938 Bragg left Manchester to became director of the National Physical Laboratory. But a year later Rutherford died, and Bragg was invited to take his place as professor of physics at the Cavendish Laboratory at Cambridge. During World War II, Bragg remained at Cambridge (except for eight months in Canada as scientific liaison), advising the British navy on methods of underwater detection of submarines and assisting with further development of sound ranging. After the war had ended, Bragg organized an international meeting of crystallographers in London, which resulted in the formation of the International Union of Crystallography; he was named its first president in 1948. He also secured funds from British industries to help establish *Acta Crystallographica,* a scientific journal first published in 1948 and devoted to the new crystallography.

Although in his later years Bragg was more involved with administration than with hands-on, day-to-day scientific research, he continued to have a powerful influence on the development of X-ray analysis. He organized and found resources to support the very challenging work of **Max Perutz**, later joined by **John C. Kendrew**, in investigating globular **proteins**, molecular structures that contain many thousands of atoms. This work dramatically culminated in the structural analysis of hemoglobin by Perutz and Kendrew and of **DNA** (deoxyribonucleic acid, the **molecule** that carries the genetic formula) by **Francis Crick**, **James Watson**, and **Maurice Wilkins**.

In 1954 Bragg became director of the Davy-Faraday Research Laboratory as well as professor of chemistry at London's Royal Institution. He revitalized the laboratory with much-needed infusions of research funds from sources such as the Rockefeller Foundation, the National Institutes of Health, and industrial firms, and by pooling resources with the Cavendish Laboratory. He was actively interested in the Royal Institution's long-standing tradition of sponsoring public lectures on science to nonscientific audiences, and he took particular pleasure in developing a program to bring the excitement of science to schoolchildren. Bragg himself was a popular lecturer owing to his ability to explain complex scientific concepts in clear, simple terms, and he gave a series of scientific talks on television. In the 1960s, at the government's request, he created a series of elementary science lectures to provide civil servants with basic scientific knowledge.

After his retirement in 1966, Bragg continued lecturing, especially to young people, and writing. He completed a definitive book on the history of his field, *The Development of X-Ray Analysis,* only a week before he died on July 1, 1971. In 1921 Bragg had married Alice Hopkinson, the daughter of a doctor. Mrs. Bragg was mayor of Cambridge in 1945, and served on a number of public bodies. The couple had four children, two sons and two daughters. Bragg was knighted in 1941 and was named a Companion of Honour in 1967. He received many awards and honorary doctorates and was a member of numerous scientific academies around the world. His contributions to science were immense, touching as they did on a number of fields in addition to physics, including chemistry, metallurgy, mineralogy, and molecular biology. These contributions came both from Bragg's own direct research and from his notably energetic, effective, and committed organization and leadership of the research efforts of others. In *William Henry Bragg,* Bragg's sister summed up the lives of both W. H. and W. L. Bragg in these words: "To each, science was an art, research an adventure, and life an experimental journey which they lived with enthusiasm."

BRENNER, SYDNEY (1927-)

English geneticist and molecular biologist

Sydney Brenner is a geneticist and molecular biologist who has worked in the laboratories of Cambridge University since 1957. Recognized as one of the founders of molecular biology, Brenner played an integral part in the discovery and understanding of the triplet genetic code of **DNA**. He was also a member of the first scientific team to introduce messenger **RNA**, helping to explain the mechanism by which genetic information is transferred from DNA to the production of **proteins** and enzymes. In later years, Brenner conducted a massive, award-winning research project, diagramming the nervous system of a particular species of worm and attempting to map its entire genome.

Brenner was born in Germiston, South Africa, on January 13, 1927. His parents were neither British nor South African—Morris Brenner was a Lithuanian exile who worked as a cobbler, and Lena Blacher Brenner was a Russian immigrant. Sydney Brenner grew up in his native town, attending Germiston High School. At the age of fifteen, he won an academic scholarship to the University of the Witwatersrand in Johannesburg, where he earned a master's degree in medical biology in 1947. In 1951 Brenner received his bachelor's degree in medicine, the qualifying degree for practicing physicians in Britain and many of its colonies. The South African university system offered no Ph.D. degrees and could offer him no further education, so he embarked on independent research. He studied chromosomes, cell structure, and staining techniques, built his own **centrifuge**, and laid the foundation for his interest in molecular biology.

Frustrated by his lack of resources and eager to pursue his interest in molecular biology, Brenner decided to seek education elsewhere, and was encouraged by colleagues to contact **Cyril Hinshelwood**, professor of physical **chemistry** at Oxford University. In 1952 Hinshelwood accepted Brenner as a doctoral candidate and put him to work studying a bacteriophage, a virus that had become the organism of choice for studying molecular biology in living systems. Brenner's change of location was an important boost to his career; while at Oxford he met Seymour Benzer, with whom Brenner collaborated on important research into gene mapping, sequencing, mutations and colinearity. He also met and exchanged ideas with **James Watson** and **Francis Crick**, the Cambridge duo who published the first paper elucidating the structure of DNA, or deoxyribonucleic acid, the basic genetic **molecule**. Brenner and Crick were to become the two most important figures in determining the general nature of the genetic code.

Brenner earned his Ph.D. from Oxford in 1954, while still involved in breakthrough research in molecular biology. His colleagues tried to find a job for him in England, but he accepted a position as lecturer in physiology at the University of the Witwatersrand and returned to South Africa in 1955. Brenner immediately set up a laboratory in Johannesburg to continue his phage research, but missed the resources he had enjoyed while in England. Enduring almost three years of isolation, Brenner maintained contact with his colleagues by mail. Distracted by the tense political situation in South Africa, Brenner, as quoted in Horace Judson's *The Eighth Day of Creation,* wrote, "It is worse here than I ever imagined in my most terrible nightmares."

In January 1957 Brenner was appointed to the staff of the Medical Research Council's Laboratory of Molecular Biology at Cambridge, and he and his family were able to settle in England permanently. Brenner immediately attended to theoretical research on the characteristics of the genetic code that he had begun in Johannesburg, despite the chaotic atmosphere. At the time, the world's foremost geneticists and molecular biologists were debating about the manner in which the sequences of DNA's four nucleotide bases were interpreted by an organism. The structure of a DNA molecule is a long, two-stranded chain that resembles a twisted ladder. The sides of the ladder are formed by alternating **phosphate** and sugar groups. The nucleotide bases adenine, guanine, thymine, and cytosine—or A, G, T, and C—form the rungs, a single base anchored to a sugar on one side of the ladder and linked by **hydrogen** bonds to a base similarly anchored on the other side. Adenine bonds only with thymine and guanine only with cytosine, and this complementarity is what makes it possible to replicate DNA. Most believed that the bases down the rungs of the ladder were read three at a time, in triplets such as ACG, CAA, and so forth. These triplets were also called codons, a term coined by Brenner. Each codon represented an amino acid, and the **amino acids** were strung together to construct a protein. The problem was understanding how the body knew where to start reading; for example, the sequence AACCGGTT could be read in several sets of three-letter sequences. If the code were overlapping, it could be read AAC, ACC, CCG, and so forth.

Brenner's contribution was his simple theoretical proof that the base triplets must be read one after another and could not overlap. He demonstrated that an overlapping code would put serious restrictions on the possible sequences of amino acids. For example, in an overlapping code the triplet AAA, coding for a particular amino acid, could only be followed by an amino acid coded by a triplet beginning with AA—AAT, AAA, AAG, or AAC. After exploring the amino acid sequences present in naturally occurring proteins, Brenner concluded that the sequences were not subject to these restrictions, eliminating the possibility of an overlapping code. In 1961 Brenner, in collaboration with Francis Crick and others, confirmed his theory with bacteriophage research, demonstrating that the construction of a bacteriophage's protein coat could be halted by a single "nonsense" **mutation** in the organism's genetic code, and the length of the coat when the transcription stopped corresponded to the location of the mutation. Interestingly, Brenner's original proof was written before scientists had even determined the universal genetic code, although it opened the door for sequencing research.

Also in 1961, working with Crick, **François Jacob**, and **Matthew Meselson**, Brenner made his best-known contribution to molecular biology, the discovery of the messenger RNA (mRNA). Biologists knew that the DNA, which is located in the **nucleus** of the cell, contains a code that controlled the production of protein. They also knew that protein is produced in structures called ribosomes in the cell cytoplasm, but did not know how the DNA's message is transmitted to, or received by, the ribosomes. RNA had been found within the ribosomes, but did not seem to relate to the DNA in an interesting way. Brenner's team, through original research and also by clever interpretation of the work of others, discovered a different type of RNA, mRNA, which was constructed in the nucleus as a template for a specific gene, and was then transported to the ribosomes for transcription. The RNA found within the ribosomes, rRNA, was only involved in the construction of proteins, not the coding of them. The ribosomes were like protein factories, following the instructions delivered to them by the messenger RNA. This was a landmark discovery in genetics and cell biology for which Brenner earned several honors, including the Albert Lasker Medical Research Award in 1971, one of America's most prestigious scientific awards.

In 1963 Brenner set out to expand the scope of his research. For most of his career, he had concentrated on the most fundamental chemical processes of life, and now he wanted to explore how those processes governed development and regulation within a living organism. He chose the nematode *Caenorhabditis elegans,* a worm no more than a millimeter long. As reported in *Science,* Brenner had initially told colleagues, "I would like to tame a small metazoan," expecting that the simple worm would be understood after a small bit of research. As it turned out, the nematode project was to span three decades, involve almost one hundred laboratories and countless researchers, make *C. elegans* one of the world's most studied and best understood organisms, and become one of the most important research projects in the history of genetics.

Brenner's nematode was an ideal subject because it was transparent, allowing scientists to observe every cell in its body, and had a life cycle of only three days. Brenner and his assistants observed thousands of *C. elegans* through every stage of development, gathering enough data to actually trace the lineage of each of its 959 somatic cells from a single zygote. Brenner's team also mapped the worm's entire nervous system by examining **electron** micrographs and producing a wiring diagram that showed all the connections among all of the 309 neurons. This breakthrough research led Brenner to new discoveries concerning sex determination, brain chemistry, and programmed cell death. Brenner also investigated the genome of the nematode, a project that eventually led to another milestone, a physical map of virtually the entire genetic content of *C. elegans.* This physical map enabled researchers to find a specific gene not by initiating hundreds of painstaking experiments, but by reaching into the freezer and pulling out the part of the DNA that they desired. In fact, Brenner's team was able to distribute copies of the physical map, handing out the worm's entire genome on a postcard-size piece of filter paper.

Brenner's ultimate objective was to understand development and behavior in genetic terms. He originally sought a chemical relationship that would explain how the simple molecular mechanisms he had previously studied might control the process of development. As his research progressed, however, he discovered that development was not a logical, program-driven process—it involved a complex network of organizational principles. For example, Brenner would say that the simple mechanics of the genetic code can explain how to make insulin, but cannot explain how to make a hand or a foot. The instructions for making complex body parts are not explicitly coded; they are embedded in higher organizational principles. Brenner's worm project was his attempt to understand the next level in the hierarchy of development. What he and his assistants have learned from *C. elegans* may have broad implications about the limits and difficulties of understanding behavior through gene sequencing. The Human Genome Project, for instance, is an effort to sequence the entire human DNA. James Watson has pointed to Brenner's worm experiments as a model for the project.

Brenner's research has earned him worldwide admiration. He has received numerous international awards, including the 1970 Gregor Mendel Medal from the German Academy of Sciences, the prestigious Kyoto Prize from Japan, as well as honors from France, Switzerland, Israel, and the United States. He has been awarded honorary degrees from several institutions, including Oxford and the University of Chicago, and has taught at Princeton, Harvard, and Glasgow Universities. Brenner is known for his aggressiveness, intelligence, flamboyance, and wit. His tendency to engage in remarkably ambitious projects such as the nematode project, as well as his ability to derive landmark discoveries from them, led *Nature* to claim that Brenner is "alternatively molecular biology's favorite son and *enfant terrible*."

While still in Johannesburg in 1952, Brenner married May Woolf Balkind. He has two daughters, one son, and one stepson. In 1986 the Medical Research Council at Cambridge set up a new molecular genetics unit, and appointed Brenner to a lifelong term as its head. Research at the new unit is centered on Brenner's previous work on *C. elegans* and the mapping and evolution of genes.

Bressani, Ricardo (1926-)

Guatemalan biochemist

Ricardo Bressani is a prominent Central American food scientist who has contributed significantly to the knowledge of human **nutrition** and food production. Long associated with the Institute of Nutrition of Central America and Panama (INCAP), Bressani has focused his chief efforts toward increasing the availability of high quality foods for humans. His contributions include improving production of high nutrition foods; investigating the composition and nutritional value of basic foods such as maize, sorghum rice, beans, and amaranth; studying the effects of food processing on nutritional value; evaluating food storage techniques; and analyzing the efficient biological utilization of foods.

Ricardo Bressani Castignoli was born on September 28, 1926, in Guatemala City, Guatemala, to Primina (Castignoli), a homemaker, and César Bressani, a farmer. He obtained a B.S. degree from the University of Dayton (Ohio) in 1948. In 1951 he was awarded a master's degree from Iowa State University and began directing the food analysis laboratories at INCAP. Then two years old, INCAP had assessed the nutritional status of Central America, finding widespread malnutrition and a heavy reliance by Central Americans on cereals and legume grains. This first, brief association with INCAP stimulated Bressani's interest in the serious nutrition problems of populations in Central America.

In 1952 Bressani enrolled in Purdue University in Indiana, where he was a graduate research assistant at the Biochemical Research Institute and where he obtained his Ph.D. in 1956. He then returned to Guatemala to head INCAP's agricultural and food sciences division, a position that he held for 32 years. During this time, from 1963 to 1964, he was a visiting professor in the Department of Food Science at Rutgers State University in New Jersey; and in 1967 he was a visiting lecturer in the food science and nutrition department at the Massachusetts Institute of Technology. Other positions he has held for INCAP include that of research coordinator from 1983 to 1988; research advisor in food science and agriculture from 1988 to 1992; and consultant in food science, agriculture and nutrition beginning in 1993.

Bressani conducted important studies regarding the nutritional value of resources already in abundance in Central America, such as Brazil nuts, rubber tree seeds and jicara seeds, caulote, jack beans, African palm, corozo, and buckwheat. These studies resulted from his concern that these resources held food value yet would vanish from the area because populations were ignorant of their nutritional value. The grain and vegetable amaranth, for instance, cultivated by Aztec, Mayan, and Inca civilizations, was rediscovered and converted into highly nutritional flour. In addition, Bressani's research on legumes not indigenous to Central America nor normally consumed by its populations—such as the jack bean and the cowpea—spurred agricultural production of some high-yield as well as highly nutritional beans. For instance, he obtained a protein isolate from the jack bean, uncovering it as a valuable food resource. The Central American Cooperative Program for the Improvement of Food Crops (PCCMCA) honored Bressani and his staff for their outstanding work in cereal and legume research.

As early as 1956, in Bressani's Ph.D. dissertation, he expressed concern about the nutritional problems of people whose diets consisted mainly of corn and beans. He conducted research into the effects of soil fertilization with minor elements on the yield and protein value of cereals and legumes. According to his dissertation, he was convinced that "in order to produce a corn of a high nutritive value, the ratio of germ to whole grain must be increased, or else the relative quantities of **proteins** other than zein should be increased in the endosperm." Eight years later at Purdue, scientist Edwin T. Mertz successfully isolated the Opaque–2 gene, the chemical-nutritive qualities of which follow the postulations in Bressani's dissertation. Subsequently, such organizations as INCAP, ICTA (Guatemala), and CIMMYT (Mexico) developed a superior maize called NUTRICTA. It was ready for agricultural production in October 1983.

Bressani has made numerous other contributions to nutrition and agriculture in Central America and has published over 450 articles in scientific journals. He is married to Alicia Herman, and they have seven children. He enjoys farming, horseback riding, reading, and photography.

Bromine

Bromine is the third member of the halogen family, the elements that make up Group 17 in the **periodic table**. Its **atomic**

number is 35, its atomic **mass** is 79.904, and its chemical symbol is Br.

Properties

Bromine is one of only two elements that is liquid at room **temperature**, the other being **mercury**. It is a beautiful deep reddish brown liquid that evaporates easily, giving off strong fumes that irritate the throat and lungs. Bromine boils at a temperature of 137.8°F (58.8°C), and its **density** is 3.1023 grams per cubic centimeter. Bromine's freezing point is 18.9°F (-7.3°C).

Like all **halogens**, bromine is a very reactive element, less reactive than **fluorine** or **chlorine**, but more reactive than **iodine**. It reacts vigorously with many **metals**, sometimes explosively. Bromine even reacts with relatively inert elements such as **platinum** and **palladium**.

Occurrence and Extraction

Bromine is a moderately abundant element with an estimated abundance of about 1.6-2.4 parts per million in the Earth's crust. It is far more abundant in seawater, where its abundance is estimated to be about 65 parts per million.

The element is far too reactive to occur as an element in the Earth's crust, and is found primarily in the form of **sodium** bromide (NaBr) and **potassium** bromide (KBr). Since both compounds are highly soluble in **water**, they tend to dissolve when water washes over the Earth's surface. They are then carried into the oceans. Very large underground deposits of sodium and potassium bromide were formed when ancient seas dried up and were buried by earth movements. Those deposits, known as **salt** domes are a major dry-land source of bromine today.

Discovery and Naming

Compounds of bromine have been known for many hundreds of years. One of the most famous of these compounds was a dye known as Tyrian purple, or royal purple. The dye was obtained from a particular mollusk found on the shores of the Mediterranean Sea. The cost of extracting the dye from the mollusks was so great that only very rich people could afford it, thus accounting for its alternative name.

Bromine was discovered at almost the same time in 1826 by the German chemist Carl Löwig (1803-1890) and the French chemist Antoine-Jérôme Balard (1802-1876). Löwig, a student at the University of Heidelberg at the time, probably made his discovery first. However, Balard was the first to publish the results of his research. As a consequence, credit for the discovery of bromine is usually given to both men. Bromine was named after the Greek word *bromos*, which means "stench."

Uses

The most important use of bromine today is in making flame retardant materials. Many of the materials used in making clothing, carpets, curtains, and drapes are highly flammable. To reduce the risk that such materials will catch fire, they may be soaked in a bromine compound, or the bromine compound may be applied after the product has been manufactured. In either case, the bromine compound prevents the material from catching fire, although the material may still smolder or char.

A vessel filled with bromine. ***(Photograph by Charles N. Winters, National Audubon Society Collection/Photo Researchers, Inc. Reproduced by permission.)***

A highly controversial use of one bromine compound, methyl bromide (CH_3Br), is as a pesticide. Methyl bromide is sprayed on the surface or injected directly into the ground on a farm. Some of the methyl bromide escapes into the air where it damages the ozone layer. Concern about this problem has led world leaders to recommend an end to the use of methyl bromide as a pesticide shortly after the beginning of the new millennium.

Bromine and its compounds are becoming increasingly popular in the purification of public water supplies and swimming pools. Bromine kills bacteria in the same way that chlorine does, but it appears to be even more effective than chlorine.

BROOKS, RONALD E. (1935-1989)

American chemist

Ronald E. Brooks led a research unit at General Electric which developed oil-eating microorganisms to clean up environmentally-damaging oil spills in the ocean. His management of the

General Electric Research and Development Center's environmental unit in the 1970s and 1980s crossed the scientific and technological boundaries of **chemistry**, electronics, research physics, engineering and environmental science.

Born in New York City on May 28, 1935, Ronald Elmer Brooks was the son of Elmer, a minister, and Oretha (Beverley) Brooks. Both parents were known for their exhaustive religious and civic activities in the New York City region. During the Korean War era, Brooks served as a second lieutenant in the United States Army. His academic career began shortly afterward, and he graduated with a bachelor's degree in chemistry from the City College of New York in 1958. At the time of his graduation, the field of chemistry had opened new frontiers in the development of research instruments that were utilized in electrochemical analyses. From 1958 to 1960 Brooks worked as an assistant chemist at Burroughs Wellcome Company Research Laboratories in Tuckahoe, New York. He left to complete graduate studies at Brown University, and in 1965 received his Ph.D. in **organic chemistry**. Brooks' dissertation topic required detailed research in solvolytic rearrangement, the adding of a liquid substance such as **water** or **hydrogen** peroxide to a **solution** to form a new compound. This artificial production of a compound was historically used to make **alcohol**, **ammonia** and rubber.

The Research and Development Center of General Electric (GE) in Schenectady, New York, employed about 600 nationally-recognized scientists. In 1964 Brooks began his career with GE in the Information Physics Branch as a research chemist. His talents and leadership skills very quickly earned him a promotion as a project manager of the Photocharge Recording Materials Program, a unit studying the composition of organic materials with the aid of advanced measuring instruments. Photocharge recording was, at the time, superior to early twentieth-century inventions such as the spectroscope and to the mid-century developments of vacuum tubes and photoelectric cells.

Brooks also led his research team through studies of photoresist, impactless printing, and photoconduction. These processes required the use of apparati such as chemical etching, photographic and thermoplastic film and **electron** beam recording to ''make pictures'' of patterns found in organic compounds. Each pattern of each compound would show different **color** spectra, diffraction patterns or wavelengths once they were analyzed. Analyses and comparisons could then be made of the content of each compound. As a result of his efforts, Brooks was eventually placed at the helm of the Environmental Technology Program of the Materials Engineering Laboratory.

In 1969 the first report of the President's Panel for Oil Spills stated that the United States did not have an adequate oil spill technology. A year later, the Marine Science Affairs Committee of the U.S. government reported to Congress that oil **pollution** was a major source of concern in the marine environment and, in 1971, world oil pollution was estimated to be one million metric tons a year. The problem of water pollution was also discussed within the context of the residuals causing the pollution, the high costs of controlling the pollutants and the difficulty in tracing pollution sources. With these conditions in mind, Brooks was appointed manager of the General Electric Research and Development Center's Environmental Unit. According to GE, Brooks' position involved the development of ''innovative technical solutions to... national environmental problems.''

In 1972, Dr. Ananda M. Chakrabarty, a member of Brooks' research unit, with the consensus of General Electric, applied for a patent on an oil-digesting microorganism. The microbe, within laboratory conditions, was found to metabolize several hydrocarbons that were found in petroleum, and was effective in converting hydrocarbons into **carbon** dioxide and water which lessened the toxicity of the petroleum. With decreased toxicity, microorganisms that appeared naturally in water systems could eat the petroleum. The advantage of this microbe was that it could degrade crude oil at increased rates and thus quickly counteract the damage caused by an oil spill.

The microbe gained international attention. By 1976, it had been patented in England and the following year it was patented in France, but the U.S. Patent and Trademark Office did not grant rights to General Electric for the discovery until 1981. Originally denied patent rights, in a landmark case the Supreme Court overturned the earlier decision and found that Chakrabarty's microbe could not be denied patent protection merely because it was a living organism. It was the first time a genetically-engineered microorganism was awarded a U.S. patent. Brooks began personal research on microbial PCB degradation in 1982. PCBs (polychlorinated biphenyls) are industrial chemicals used primarily as dielectric fluid in transformers and capacitors. However, Brooks died without seeing the formulation and implementation of his plan to neutralize PCBs.

Brooks was killed at the age of 54 on August 13, 1989, in a motor vehicle accident near his home in Guilderland, New York. His wife Elsa survived the accident and went on to raise their three sons, Hodari, Dahari and Bakari. Along with his scientific accomplishments, Brooks' was known for his association with the Baptist church and for his work with inner city youth. He served as chairman of the deacon board, treasurer, trustee and a Sunday school teacher at his church and founded Project Mercury, a federally funded program that trained minority and disadvantaged students as chemical technicians in Albany, New York. His professional memberships included the American Chemical Society, the Chemical Society of London, the Society of Photographic Scientists and Engineers, and the American Society of Microbiology. Brooks was also a founding member of the National Organization for the Professional Advancement of Black Chemists and Chemical Engineers and served as president of the Capital District Region of the American Chemical Society. John Brown, in the *General Electric R&D Center Post,* said: ''Ron was a very kind and caring man.... who clearly appreciated his fellow man and whose devotion to his friends and family was obvious to all who knew him.''

BROOM, ROBERT (1866-1951)

South African paleontologist

As a child, Robert Broom suffered from respiratory illnesses, and spent a year at the seashore to recuperate. It was there that a marine officer sparked his interest in marine biology. His father, a fabric designer, was an amateur botanist, and he encouraged the boy's interest in the natural sciences. Like his father, Broom had a talent for drawing. Broom attended Hutcheson's Grammar School, and entered the University of Glasgow in 1883. He received his medical degree in 1889 and went to Australia in 1892. There he married Mary Baillie who had followed him from Scotland, and in 1897, they made South Africa their permanent home. For a good part of his life, Broom practiced medicine in rural communities. In addition, he worked extensively in paleontology, a branch of geology that deals with prehistoric life.

Broom's investigations into the comparative anatomy and embryology of mammals and reptiles led him to believe that mammals originated from reptiles and not from amphibians as advocated by T.H. Huxley. He made a series of reconstructions that showed the way that the mammalian ear evolved from the reptilian ear. He identified bones in the skull and axial skeleton that were homologous, of the same origin, in mammals and reptiles. He described and sketched large numbers of mammal-like reptile fossils.

After World War I, Broom turned his attention to prehistoric humans in Africa, and the study of anthropology. In 1925, Raymond Dart announced his discovery of an infantile Australopithecus skull at Taungs, South Africa. Australopithecines are a group of primates whose fossils date back 1.5-4 million years ago. Dart considered them a link in human evolution. Broom studied the skull, and agreed with Dart about its relationship to humans and its evolutionary significance. He retired from medical practice in 1928, and wrote three books. In his book entitled, *The Origin of the Human Skeleton*, he surveyed vertebrate evolution and discussed the transition from reptiles to mammals. In 1934, Broom joined the Transvaal Museum in Pretoria. In the years that followed, he made some impressive finds. Amazingly, he discovered an adult australopithecine at Sterkfontien one week after he arrived. He discovered other significant fossils at Kromdraai and Swartkrans. His most famous discovery occurred in 1937 when he unearthed Australopithecus Robustus. Broom described the excitement of these discoveries in a book he wrote in 1950 entitled *Finding the Missing Link*. For the rest of his career, Broom explored these sites and interpreted the early hominid remains that were discovered there.

Broom wrote in great detail on a variety of subjects such as evolution, zoology, anthropology and medicine. He was elected to the Royal Society of London in 1920. In his later years, he received many honorary degrees, fellowships in distinguished societies, and medals. Broom helped unravel problems about evolution that had been raised by nineteenth century scientists. His ability to synthesize paleontology and embryology enabled him to contribute valuable information about the origin of mammals.

BRØNSTED, JOHANNES NICOLAUS (1879-1947)

Danish physical chemist

Johannes Nicolaus Brønsted carried on the tradition of earlier distinguished Danish scientists who made major contributions to the field of **chemistry** and left a legacy that included a widely used model of how acids and bases work and the establishment of a major research institute. Known among friends for his intellectual curiosity beyond the sciences, Brønsted believed that the lack of logic and clear thinking in politics was the cause of many social problems.

Brønsted was born on February 22, 1879, in the small West Jutland town of Varde. His mother died shortly after he was born, and his father, an engineer, remarried several years later. Young Brønsted was greatly influenced by his father's work with Hedeselskabet, a corporation that reclaimed moors by draining, irrigating, and planting. Until 1891 the family lived at one of the farms of the Society for Cultivation of Heaths. It was here that his love of unspoiled nature flowered, and he developed a lifelong interest in birds. For the rest of his life he worked to protect them from the encroachment of civilization.

When Brønsted was twelve, the family moved to Aarhus, the second largest city in Denmark, where he attended school, excelling in mathematics, and enjoyed the surrounding countryside. He also developed an interest in chemistry after finding an agricultural dictionary in the attic of his house and performed primitive experiments at home. His life in Aarhus came to an end after his father died in 1893, leaving the family in a precarious financial situation. Against the advice of family friends who thought young Johannes should start earning a living, his stepmother insisted on ensuring that he and his sister received a good education; she moved her family to Copenhagen, where Brønsted attended the Metropolitan School. At the school, he met Niels Bjerrum, a student who was to become a lifelong friend and chief rival among Danish chemists whose work brought fame to their country.

In 1897 Brønsted entered the Polytechnic Institute in Copenhagen, where he studied engineering, intending to follow his father's career. Brønsted apparently did not find his studies to be a burden, and he was able to cultivate interests such as philosophy, art, and poetry. He also enjoyed singing and music, becoming a performer in the circle of his family and friends. While at the Polytechnic Institute, he met Charlotte Louise Warberg, a chemical engineer, whom he married in 1903. The couple settled down in the small town of Birkero slash d. Through his sister-in-law's husband, the painter Johannes Larsen, Brønsted's life became filled with art as well as science; it may have been this influence that led Brønsted to take up painting during World War I.

After earning his engineering degree in 1899, Brønsted switched his studies to chemistry and was awarded his magister scientiarum (M.Sc.) degree in chemistry in 1902. The degree was so rare at the time that Brønsted's friends nicknamed him Magister. He worked for a time in an electrical engineering business until 1905, when he accepted a position as assistant at the University of Copenhagen's chemical laboratory. He earned his doctorate in 1908, presenting as his doctoral thesis the third in a series of papers on measurements of chemical affinities in reactions, in this case between **water** and **sulfuric acid**. In all, Brønsted published thirteen monographs on chemical affinity.

The year he obtained his doctorate, Brønsted was appointed professor of chemistry at the University of Copenhagen, a position in which he taught elementary **inorganic chemistry** to students at the Polytechnic Institute and **physical chemistry** to chemical engineers and the students of the university. In 1912 he wrote a small textbook on elementary physical chemistry called *Outlines of Physical Chemistry,* which by the 1930s would be outdated and require the addition of his work in **thermodynamics**. In 1919 he was exempted from teaching inorganic chemistry, which gave him more time for his work. That work extended Danish chemist Julius Thomsen's studies on chemical processes, an important problem in the field of physical chemistry. A major area of research at this time was the measure of affinities: the strengths of acids and the tendency of substances to react with each other.

Formerly, such phenomena were described in vague terms, but in the early twentieth century, chemists were learning to express affinities mathematically and precisely by using the principles of thermodynamics. These principles describe changes in the **heat** of substances that react with each other, the direction of chemical processes, and the **energy** absorbed or released in the reactions. Thermodynamics allowed chemists to make predictions of how substances would behave and to describe accurately what happened in observed reactions. Julius Thomsen originally believed that the amount of heat released by chemical processes could be used to measure affinity. Yet by 1900 it was clear that the true measure of affinity was the maximum work the processes produced, not the heat of the reaction. However, there were very few accurate or systematic measurements of affinity before Brønsted set to work.

Brønsted's series of thirteen papers on affinity, which appeared from 1906 to 1918, was a major step in consolidating and explaining thermodynamic principles and a significant contribution to the field of chemical thermodynamics. After 1913 Brønsted's work on thermodynamics included the measurement of specific heat (the number of calories absorbed in raising the **temperature** of a gram of pure material by one degree centigrade) and the **solubility** of substances.

Brønsted's continuing interest in thermodynamics also led him to study electrolytes, substances that conduct **electricity** in **solution**, including both electromotive force and solubility measurements. At the time, the term electrolyte activity was new and reflected the newly developed concept that electrostatic forces between ions (charged atoms) might be chemically important. Brønsted embarked on several years of study of the properties of ionic solutions, publishing a series of papers on the topic beginning in 1919. In 1921 he published "The Principle of the Specific Interaction of Ions," which states that individual properties of an **ion** depend mainly on the presence of oppositely charged ions in the same solution.

Brønsted incorporated the use of electrochemical cells into his work. Because the voltage of the electrodes or conductors of these cells is proportional to the **concentration** of ions involved in reactions at a particular electrode, these cells can measure concentrations of ions. When current flow is produced by the electrochemical reactions, the cell is called a galvanic cell. Brønsted built his own galvanic cells and measured their electromotive force. During World War I, he even built **batteries** from such cells and used them to light his house in Birkerød.

During the 1920s Brønsted's new interests tumbled one onto another, his solubility studies leading into the field of reaction **kinetics**, the study of the rate at which chemical reactions occur and the intermediate steps that take place during the reactions. Brønsted's first studies in kinetics measured the effect of salts on acid-base equilibria, the point at which the conversion of an acid to a base equals the reverse reaction. For example, he demonstrated that the concentration of **hydrogen** ions released by **acetic acid** (vinegar) increases with the addition of **salt**, while salt has little effect on hydrogen ions released by ammonium ions.

Brønsted's extensive work with the effect of salt on acid-base equilibria led him to redefine the terms acid and base. According to the original definition, proposed by the Swedish chemist **Svante Arrhenius** in 1887, acids are compounds that dissociate or break up in water to yield hydrogen ions, and bases ionize in water to yield hydroxide ions. The extent to which these reactions occur determines the strength or weakness of the acid or base. Whereas a strong acid produces many hydrogen ions in an aqueous solution, a strong base yields many hydroxide ions, suggesting that certain substances act only as acids, others only as bases.

In 1923 Brønsted published simultaneously with Thomas Lowry in Britain a new theory of **acids and bases** that distinctly changed the concept of acids and bases. The theory had the advantage of applying to reactions that take place in all **solvents**, not just water. But more important, it explained that an acid is a substance that tends to release a **proton**, while a base tends to take up a proton. Thus any acid in releasing a proton becomes a base, which can take up a proton to become an acid again. And any base can accept a proton, becoming an acid in the process. This concept that all acids and bases can be arranged in conjugate, or corresponding, pairs, broadened the range of substances that were recognized either as acids or bases.

By 1921 Brønsted's work was drawing visiting scientists from the United States and England who crowded into his modest laboratory at the Polytechnic Institute to study under his supervision. The fame these collaborations brought him bore impressive fruit during his 1926–27 visit to the United States, where he was visiting professor at Yale University. While in the United States, he sought funding to build a new physicochemical laboratory in Copenhagen. His efforts were rewarded when the International Education Board agreed to finance a new University Physicochemical Institute, which he helped to design. When the institute opened in 1930, Brønsted not only had ideal working conditions but also a comfortable residence, noted for the charm and hospitality of the Brønsteds, who entertained often.

While his close friends enjoyed his eclectic interests and charming personality, his students and younger colleagues, who were generally more in awe of him, found Brønsted distant. This may have been due in part to the intellectual rigor he demanded during discussion and debate, whether on topics of science, politics, or some other issue. Brønsted was a fervent Danish patriot but traveled widely in Europe. While his early contacts were with German scientists, after World War I his sympathies turned to England, which he visited often, particular for meetings of the Faraday Society. He spoke fluent English, enjoyed the English people and their literature, and the quaint and beautiful English countryside. The German occupation of Denmark during World War II made these visits impossible; but Brønsted contented himself with studying Anglo-European politics of the nineteenth and twentieth centuries and listening to news broadcasts from London.

After the war Brønsted continued to be preoccupied with politics, particularly the question of whether Germany or Denmark should control Schleswig, an area on the southern part of the Denmark's Jutland peninsula. The question had troubled both countries since the nineteenth century, and in 1947, vexed by the old debate, Brønsted took action. He accepted nomination as a candidate for the Danish Parliament, and on October 28 he was elected to that body. Brønsted threw himself into his new endeavor and began to study parliamentary procedure. However, before he could begin his government service, he died on December 17, 1947.

Brønsted was an honorary fellow of the British Chemical Society (1935), and in 1949, the organization hosted a memorial lecture honoring the Danish chemist and patriot. He had also been honored by his British colleagues by being made an honorary member of the Academy of Arts and Sciences in 1929 and an honorary doctor of London University in 1947; in addition he was awarded an honorary degree of doctor of science from the University of London in 1947. He was a member of the Royal Danish Academy of Sciences and Letters beginning in 1914 and belonged to the Danish Academy of Technical Sciences since its establishment in 1937. In 1928 he was awarded the Ørsted Medal.

BROWN, HERBERT C. (1912-)

American chemist

Herbert C. Brown opened an entirely new field of **chemistry** for study with his discovery of the **organoboranes** (an important chemical compound). Much of Brown's career has focused on the investigation of **boron** reagents, molecules that can temporarily link larger molecules together during a reaction. Because these boron-based molecules are highly active, they can foster chemical reactions that had previously been unachievable; they have become valuable in **organic chemistry** and in the manufacture of synthetics and pharmaceuticals. A professor at Purdue University, Brown received the 1979 Nobel Prize in chemistry.

Herbert Charles Brown was born in London on May 22, 1912. His parents, Charles Brovarnik and Pearl Gorinstein Brovarnik, were Ukranian Jews who had emigrated to London in 1908. Brown's paternal grandparents had already settled in Chicago and anglicized their surname to Brown; when his family arrived in Chicago in 1914 they followed suit. The only son, Herbert was the second of four children. Charles Brown had been a cabinetmaker in England and worked in the United States as a carpenter. He opened a hardware store in 1920, and the family lived upstairs. After his father died in 1926 of an infection, Herbert dropped out of high school to run the store and support the family. Returning to high school in 1929, he completed two years of work in one year to graduate in 1930.

After the hardware store failed, Brown worked as a shoe salesman and packer of notebook paper and belts, which strengthened his resolve to return to school. In February 1933 he entered Crane Junior College in Chicago. There he met Sarah Baylen, his future wife. The Depression forced Crane, like many other city colleges, to close. Brown worked odd jobs and took classes at Lewis Institute. Both Sarah and Brown attended Wright Junior College in 1934, graduating in 1935. Brown then won a partial scholarship to the University of Chicago. Because the tuition (and the scholarship) amounted to one hundred dollars per semester with no course limit, Brown loaded himself down with course work. Once again, he completed two years of work in one year, working with the noted chemist H. I. Schlesinger to investigate diborane—a rare but excellent reagent that could at the time be produced only in small quantities. In 1936 Brown's graduation gift was a copy of Alfred Stock's *The Hydrides of Boron and Silicon*, prophetically inscribed by his wife "to the future Nobel Prize winner."

Brown intended to find a job in industry after graduation, but his mentor Julius Stieglitz convinced him to go on to graduate school at the university. Again working with Schlesinger, Brown investigated the reactions of diborane with carbonyl compounds (molecules containing a carbon-oxygen double bond), and received his Ph.D. in 1938. Brown and Sarah married while he was in graduate school. Their son, Charles, became a chemist.

Brown remained at the University of Chicago on a one-year postdoctoral fellowship with M. S. Kharasch, one of his mentors; he then became assistant to Schlesinger in his study of the borohydrides. In 1940 the National Defense Research Committee approached the researchers and asked them to search for volatile **uranium** compounds that could be used to produce the pure uranium needed for the atomic bomb. They used diborane to create uranium borohydride, but the process of preparing diborane was so slow and difficult that they sought a new method. They succeeded in preparing diborane by reacting **lithium** or sodium hydride with boron trihalides, but by then the war department had obtained what it needed from other sources. Their efforts had not, however, been in vain. Not only did Brown's group discover a new way to create what had been a rare and expensive compound, but they also discovered a new reagent, sodium borohydride, which was to become an important reducing agent, or **electron** donor, in organic chemistry.

Brown left the University of Chicago in 1943 for a position as assistant professor of chemistry at Wayne (later Wayne State) University in Detroit. Much of his research during his four years at Wayne focused on steric strains, the deviations of the bond angles of a **molecule** from their norms. He showed that these angles could have as much effect on how a compound reacts as could the constituent atoms' electrical charges.

In 1947 he became a full professor at Purdue University in Indiana. There he experimented with adding diboranes to carbon-carbon double bonds, and accidentally discovered that a process called hydroboration permitted rapid and easy conversion to organoboranes. Subsequent work with these compounds showed them to be of substantial utility in synthetic chemistry, particularly as intermediaries, or temporary links, in the creation of new carbon-carbon bonds. Brown has often compared the investigation of the organoboranes to the discovery of a new continent. As he told Malcolm Brown in the *New York Times,* "I feel that we have uncovered a new continent, just beginning to explore its mountain ranges and valleys. But it will take another generation of chemists to fully explore and apply this new chemistry of boron hydrides and organoboranes."

For his work with boron reagents Brown shared the 1979 Nobel Prize in chemistry with **Georg Wittig** of West Germany. In addition, he has received the **Linus Pauling** Medal in 1968, the National Medal of Science in 1969, the American Chemical Society's Award in Synthetic Organic Chemistry in 1960, the Chemistry Pioneer Award in 1974, the Priestly Medal in 1981, and the Perkin Medal in 1982.

In the course of his career, Brown has written over seven hundred scientific papers and four books and influenced countless students and colleagues with his thoroughness. In the January 1980 issue of *Science,* James Brewster and Ei-ichi Negishi noted, "Brown has been almost religious in resisting facile conclusions and has thereby succeeded in avoiding erroneous ones. Some of his most significant discoveries and developments have resulted from a dogged, logical pursuit of the kind of everyday chemical puzzle that many would dismiss with a glib rationalization."

BROWN, RACHEL FULLER (1898-1980)

American biochemist

Rachel Fuller Brown, with her associate **Elizabeth Hazen**, developed the first effective antibiotic against fungal disease in humans—the most important biomedical breakthrough since the discovery of **penicillin** two decades earlier. The antibiotic, called nystatin, has cured sufferers of life-threatening fungal infections, vaginal yeast infections, and athlete's foot. Nystatin earned more than $13 million in royalties during Brown's lifetime, which she and Hazen dedicated to scientific research.

Brown was born in Springfield, Massachusetts, on November 23, 1898, to Annie Fuller and George Hamilton Brown. Her father, a real estate and insurance agent, moved the family to Webster Groves, Missouri, where she attended grammar school. In 1912, her father left the family. Brown and her younger brother returned to Springfield with their mother, who worked to support them. When Brown graduated from high school, a wealthy friend of the family financed her attendance at Mount Holyoke College in Massachusetts.

Rachel Fuller Brown.

At Mount Holyoke, Brown was initially a history major, but she discovered **chemistry** when fulfilling a science requirement. She decided to double-major in history and chemistry, earning her A.B. degree in 1920. She subsequently went to the University of Chicago to complete her M.A. in **organic chemistry**. For three years, she taught chemistry and physics at the Francis Shimer School near Chicago. With her savings, she returned to the University to complete her Ph.D. in organic chemistry, with a minor in bacteriology. She submitted her thesis in 1926, but there was a delay in arranging her oral examinations. As her funds ran low, Brown took a job as an assistant chemist at the Division of Laboratories and Research of the New York State Department of Health in Albany, New York. Seven years later, when she returned to Chicago for a scientific meeting, Brown arranged to take her oral examinations and was awarded her Ph.D.

Brown's early work at the Department of Health focused on identifying the types of bacteria that caused pneumonia, and in this capacity she helped to develop a pneumonia vaccine still in use today. In 1948, she embarked on the project with Hazen, a leading authority on fungus, that would bring them their greatest acclaim: the discovery of an antibiotic to fight fungal infections. Penicillin had been discovered in 1928, and in the ensuing years antibiotics were increasingly used to fight bacterial illnesses. One side effect, however, was the rapid growth of fungus that could lead to sore mouths or upset stomachs. Other fungal diseases without cures included infections attacking the central nervous system, athlete's foot, and ringworm. Microorganisms called actinomycetes that lived in soil were known to produce antibiotics. Although some killed fungus, they also proved fatal to test mice. Hazen ultimately narrowed the search down to a microorganism taken from soil near a barn on a friend's dairy farm in Virginia, later named streptomyces norsei. Brown's chemical analyses revealed that the microorganism produced two antifungal substances, one of which proved too toxic with test animals to pursue for human medical use. The other, however, seemed to have promise; it wasn't toxic to test animals and attacked both a fungus that invaded the lungs and central nervous system and candidiasis, an infection of the mouth, lungs, and vagina.

Brown purified this second antibiotic into small white crystals, and in 1950 Brown and Hazen announced at a meeting of the National Academy of Sciences that they had found a new antifungal agent. They patented it through the nonprofit Research Corporation, naming it "nystatin" in honor of the New York State Division of Laboratories and Research. The license for the patent was issued to E. R. Squibb and Sons, which developed a safe and effective method of mass production. The product—called Mycostatin—became available in tablet form in 1954 to patients suffering from candidiasis. Nystatin has also proved valuable in agricultural and livestock applications, and has even been used to restore valuable works of art.

In 1951, the Department of Health laboratories promoted Brown to associate biochemist. Brown and Hazen, in continuing their research, discovered two additional antibiotics, phalamycin and capacidin. Brown and Hazen were awarded the 1955 Squibb Award in **Chemotherapy**. Brown won the Distinguished Service Award of the New York State Department of Health when she retired in 1968, and the Rhoda Benham Award of the Medical Mycological Society of the Americas in 1972. In 1975, Brown and Hazen became the first women to receive the Chemical Pioneer Award from the American Institute of Chemists. In a statement published in the *Chemist* the month of her death, Brown hoped for a future of "equal opportunities and accomplishments for all scientists regardless of sex."

On retirement, Brown maintained an active community life, and became the first female vestry member of her Episcopalian church. By her death on January 14, 1980, she had paid back the wealthy woman who had made it possible for her to attend college. Using the royalties from nystatin, more importantly, she helped designate new funds for scientific research and scholarships.

Brown, Robert (1773-1858)

Scottish naturalist

Robert Brown got his intellectual honesty and solid character from his father, an Episcopalian clergyman. Unlike his father, he did not develop a calling for religion. Instead, Brown completed college in Aberdeen, Scotland, then joined the army. Brown's journals during his five years as an assistant surgeon in the army show his determination to master details and his far-reaching curiosity.

In 1798, Brown was introduced to English botanist Joseph Banks (1743-1820), who was clearly impressed with Brown's talents. It may have been Banks' influence that got Brown the position of naturalist aboard the *Investigator* in 1801 to survey the plant life along the unfamiliar coasts of Australia.

When Brown returned to England in 1805, he brought with him over 4,000 species of plants and set about the arduous task of reporting on each species. The two systems of plant classification then in use were not suitable for the unusual Australian plant varieties, so Brown developed a modified system, making use of the *natural system* for the first time in England.

Brown concentrated on the classification of plants until 1827 when he made an unusual discovery. Using a microscope, he began observing grains of orchid pollen suspended in fluid. To his astonishment he saw that particles within the grain were moving. After ruling out the fluid and its gradual evaporation as possible causes for the movement, he proposed that the particle itself might be "alive." He tested fresh pollen from a variety of other plants—all with the same result. When he expanded his experiment to test powdered **glass**, **coal**, rocks and **metals**, he found the same movement. All minuscule particles that could be suspended in **water** exhibited what is now called **Brownian motion**.

Despite his experimentation, Brown could not explain the movement, and ultimately he left it to others for interpretation. (Eventually, Brownian motion was linked to important theories in kinetic **energy**. This phenomenon was a visible manifestation of the idea that water was composed of particles.)

In 1831, Brown published a pamphlet on his observations of orchid pollen. The pamphlet contains a remarkably casual passage pinpointing a major revelation. Brown reported that the cells in orchid leaves contain a "single circular areola"—only one in each cell—generally more opaque than the cell membrane. He also noticed that this areola appeared in the plant tissue as well. He dubbed it the cell **nucleus** (Latin for little nut). It had been observed earlier, but Brown was the first to name it and to recognize it as a general feature in all living cells.

Brownian Motion

Brownian motion is the incessant random motion exhibited by microscopic particles immersed in a fluid. The effect is named after its discoverer, **Robert Brown**, a Scottish botanist who first noticed the effect in a **suspension** of pollen grains in 1827. It was determined quickly by Brown and others that the effect is exhibited by small particles of any material when they are suspended in a liquid or gas, and is not unique to pollen grains or other once-living **matter**.

We now understand Brownian motion as the result of the many millions of collisions with molecules of the fluid experienced by the particle being observed each second. While the average **velocity** of the molecules in a stationary fluid is zero, the average velocity of the small number of molecules which will collide with the particle at any instant will fluctuate about zero, producing a net force continuously changing in magnitude and direction. As a result, each particle executes a highly random motion. Since the path of each particle executing Brownian motion will be different, a theory of Brownian motion must deal with the statistical properties of the paths of a large number of particles. It is found by observation that the average square of the displacement of Brownian particles over any time interval is directly proportional to the elapsed time. The proportionality constant is a characteristic of the fluid, the particle size, and the **temperature**.

Albert Einstein, who chose the topic for his doctoral dissertation at the University of Zurich, provided the first detailed explanation of the effect. He also submitted a paper on the subject for publication in 1905, the same year as the publication of his famous theory of relativity. An important result of his theory was a method of measuring the number of molecules in a **volume** of liquid, by observing the motion of the much larger Brownian particles.

Einstein's paper, and its subsequent validation by the French physicist Jean-Baptiste Perrin (1870-1942), for which the latter was awarded the Nobel prize in 1926, is regarded by some as the final proof of the existence of atoms and molecules. By the end of the Nineteenth Century, as a result of Maxwell's kinetic theory of **gases** and Avogadro's molecular interpretation of Gay-Lussac's law of combining volumes for reacting gases, it had been generally accepted that gases consisted of molecules moving rapidly through space. Scientists were less certain that the structure of **solids** and **liquids** was also atomic in character, since it was conceivable that atoms only formed as a liquid or solid evaporated. Some eminent scientists, including the Austrian physicist and philosopher, Ernst Mach (1838-1916) opposed basing physical theory on the existence of unobservable entities. Einstein's work showed that, not only was it possible to demonstrate the continuing existence of molecules in the liquid, but that the Brownian particles could be viewed themselves as gigantic molecules in thermal equilibrium with the other molecules in **solution**. In accordance with the fundamental principles of statistical mechanics, this means that the average kinetic **energy**, equal to one-half the **mass** times the square of the velocity, for each Brownian particle is the same as the average kinetic energy for the molecules of the solution. Because the mass of the Brownian particles is so much greater than that of the liquid molecules, the velocity is smaller, but the difference is only quantitative, not qualitative. The diffusive motion of Brownian particles is of the same character as that of the molecules of the liquid.

The paths of particles exhibiting Brownian motion is of some mathematical interest in its own right. A motion picture of a Brownian particle's path will look pretty much the same if it is viewed at double speed but magnified four times or at triple speed and magnified nine times. The motion is thus *self-similar* and has no single characteristic length scale. Further the path taken by the Brownian particle, on all but the fastest time scales is highly irregular. The trajectory of the particle is constantly changing directions. It is, in fact, one of the objects characterized by mathematician Benoit B. Mandelbrot as *fractal*, that is an object of fractional dimension. This is most easily seen if one thinks of drawing line segments connecting the locations of the particle at equal time intervals. For a particle moving along a smooth curve, the total length of the line segments over a given total time will begin to approach a clear limit as the time interval becomes smaller and smaller. For a Brownian particle, this distance appears to increase without limit, as long as the time intervals are long enough for several atomic collisions to occur. In this sense fractal geometry assigns to the Brownian paths a dimensionality greater than one but less than two.

Brownian motion is very important from the standpoint of theory but it also has practical application in the study of colloidal materials. One of the simplest ways of demonstrating that an apparent solution is actually a suspension of microscopic particles is through light scattering. A true solution will scatter light only weakly while a colloidal suspension will display a pronounced Tyndall effect, scattering a concentrated beam of light to the side with an efficiency that is dependent on both the number and volume of the colloidal particles. Because the colloidal particles will also be undergoing Brownian motion, the scattered light will also exhibit a Doppler shift towards higher or lower frequency depending on the distribution of particle velocities. Measuring the amount of light scattering and the range of Doppler shifts together permits accurate determination of particle size. Such information is important in developing better, more stabilized colloidal suspensions for use in the paint, ink, and dairy industries. Noncolloidal suspensions having larger particle sizes, such as fine sand grains dispersed in **water**, tend to settle out with time.

Buchner, Eduard (1860-1917)

German chemist

Eduard Buchner is credited with introducing the field of modern **enzyme chemistry**. His research put an end to the widely accepted theory that **fermentation** of sugar to **alcohol** required the action of living (vital) yeast, which was promoted by such leading scientists as the French chemist **Louis Pasteur**. His work also discredited the mechanists' view that decomposing yeast cells acted as the catalyst for such change. For his pioneering efforts to scientifically explain the ancient process of fermentation, and for initiating the systematic study of enzymes, Buchner received the Nobel Prize in chemistry in 1907. In addition to being an outstanding chemist, he was a German patriot and soldier.

Eduard Buchner was born in Munich, Germany, May 20, 1860. He was descended from an old and scholarly Bavarian family. His father, Ernst Buchner, was a professor of obstetrics and forensic medicine, and the editor of a medical publication. Buchner's mother was Friederike Martin Buchner. After graduating from the Realgymnasium (high school) in Munich, he served in the field artillery, and later enrolled at Technische Hochschule (Technical College) in Munich, where he studied chemistry. Financial troubles forced him to temporarily leave his studies and work in canneries around Munich. With the help of his older brother, Buchner returned to school in 1884, attending the Bavarian Academy of Sciences in Munich to study under the eminent chemist Adolf von Baeyer.

While still a chemistry student at the Academy of Sciences, Buchner began work on the problems of fermentation of sugar to alcohol at the Institute for Plant Physiology. By 1886, he had published his first paper on the subject, disagreeing with Pasteur's opinion that fermentation had to be carried out in an oxygen-free environment. In 1888, Buchner received his doctorate in chemistry from the Academy of Sciences, where he was appointed Baeyer's teaching assistant in 1890, advancing to *Privatdozent,* or lecturer, a year later. Baeyer also acquired the funds for a laboratory that would allow Buchner to continue his work on fermentation.

In 1893, Buchner accepted a position at the University of Kiel, where he was in charge of the **analytical chemistry** section for three years, while continuing his research on fermentation. Buchner's brother, Hans, was conducting similar research on extracts from bacteria, trying to find medically useful products. His assistant, Martin Hahn, showed Buchner a technique to break down yeast cell walls to extract the cell juices that Buchner required for his research.

In his Nobel address of 1907, translated in part in *Nobel Prize Winners in Chemistry,* Buchner described the extraction process, in which a mixture of *kiesulguhr* (commonly known as diatomite) and black sand is used to reduce the difficulty of grinding the yeast. Buchner attributed a similar technique, in which quartz powder was used, to Marie von Manassein in Vienna, who, he believed, introduced the process in 1872. Buchner further explained that "the initially dust-dry **mass**... becomes dark gray and plastic, like dough. When this thick dough is wrapped in a strong cloth and put into a hydraulic press, a liquid juice seeps out under pressure which is gradually increased to 90 kg./sq. cm.... Within a few hours, 500 cu. cm. of liquid can be obtained from 1000 grams of yeast."

Because the clear, yellow fluid pressed from the ground yeast cells readily decomposed, Buchner, his brother, and Hahn added sugar to the syrup as a preservative. To their amazement, gas bubbles soon formed. Fermentation was in progress, yet there were no live, or decomposing, yeast cells in the mixture. Alcohol was being produced by cell-free fermentation; they had made a revolutionary, and controversial, discovery.

Buchner was appointed professor of analytical **pharmaceutical chemistry** at University of Tübingen in 1896, and published the discovery of cell-free fermentation in the paper, "Alkoholische Gärung ohne Hefezellen" ("Alcoholic Fermentation without Yeast Cells"), the following year. Buchner left Tübingen for the Agricultural College in Berlin, where he was appointed professor of chemistry in 1898.

Buchner had a greater opportunity for his research in Berlin. Along with his appointment to the College, he was made director of the Institute for the Fermentation Industry. In 1900, he married Lotte Stahl, the daughter of a mathematician at the University of Tübingen. They had two sons and a daughter. Within four years of his arrival in Berlin, Buchner had published fifteen papers on cell-free fermentation. In 1903, the three researchers published a detailed account of their discoveries titled "Die Zymase: Gärung" ("The Zymase: Fermentation"). Buchner named the specific active agent in yeast cell extract "zymase" from the Greek word "zyme," which means yeast or ferment. As his research progressed, Buchner recognized that zymase was one example of an important class of natural substances called enzymes. He described his results in his Nobel address: "The cells of plants and animals appear, with increasing distinctness, as factories where in separate workshops all kinds of products are produced. The foremen in this work are the enzymes."

In 1909, Buchner accepted the position of head of physiological chemistry at the University of Breslau. Two years later, he was pleased to be invited to the University of Würzburg, in a region of Germany where he could also enjoy his hobbies of hunting and climbing. At the outset of World War I in 1914, Buchner volunteered for the army. He saw action as a captain of an ammunition supply unit, and was promoted to major in 1916, just before he returned to the University of Würzburg. Buchner again volunteered for active duty the following year and was sent to Focsani, Rumania, where he died from a shrapnel wound on August 13, 1917.

BUFFER

A buffer is a **solution** that resists changes in **pH** upon the addition of acid or base. Buffers typically contain a species in solution that reacts with added acid and another that reacts with added base. Although acids or bases themselves could accomplish this task in theory, in practice if they are in solution together they interact and neutralize each other. Buffers are most often prepared one of two ways from conjugate acid-base pairs: from a weak acid and its conjugate base, or from a weak base and its conjugate acid.

If 10 mL of 0.10 M HCl is added to a liter of **water** at pH 7, the pH drops four units to pH 3. If 10 mL of 0.10 M NaOH is added to a liter of water at pH 7, the pH rises about four units to pH 11. But if the 10 mL of 0.10 M strong acid or strong base is added to a liter of blood, the normal pH of the blood (7.4) changes only about 0.1 pH units. Blood and many other bodily fluids are naturally buffered to resist changes in pH.

To explain the properties of a buffer, it is useful to consider a specific example, the acetic acid/acetate buffer system. In one liter of a solution that is 1.0 M in **sodium** acetate and 1.0 M in **acetic acid**, the pH drops less than one-tenth of a pH unit when 10 mL of 0.10 M HCl is added. This solution is a buffer; acetate **ion** from the sodium acetate is the conjugate base of the weak acid acetic acid. Acetic acid by itself in water exists in equilibrium with a small amount of acetate ion, its conjugate base:

$$CH_3COOH(aq) + H_2O(l) \Leftrightarrow H_3O^+ + CH_3COO^-(aq)$$

In the buffer solution, a much larger amount of the conjugate base is present than would arise from the **dissociation** of the weak acid. The relationship among the species present in solution, however, is still governed by the dissociation constant (K_a) for the weak acid, acetic acid.

When acid is added to this buffer, the added **hydronium ion** reacts with the strongest base in the medium, namely the acetate ion, and forms more acetic acid. This reaction consumes added hydronium ion, preventing the pH from rising drastically, and is responsible for the buffering effect. As a result of adding acid to the buffer, the **concentration** of acetate decreases and the concentration of acetic acid increases. The solution acts as a buffer because nearly all of the added hydronium ion is consumed by reaction with acetate. As the **hydrogen** ion concentration increases, the acetate concentration and acetic acid concentration must adjust for the value of the equilibrium constant to be maintained. The pH changes slightly to reflect the shift in the equilibrium concentrations, but the change is much smaller than in the absence of the buffer because most of the added acid is consumed by its reaction with the acetate ion. This example of an acetic acid/acetate ion buffer is typical of other buffer systems.

When the concentrations of species involved in equilibria change, the equilibrium constant does not change. The equilibrium is temporarily out of **balance**, and all concentrations shift to restore the equilibrium. Any change in conditions that disrupts the balance within a system at equilibrium is referred to as a stress on that equilibrium. The system is stressed by being out of balance, and concentrations change to relieve that stress. The general principle that governs such changes is called Le Chatelier's Principle: if a system at equilibrium is subjected to a stress, the system will react in a way that partially relieves that stress. In buffered systems, the addition of an acid or base to a buffer stresses the equilibrium established in that system. By adding acid to the acetic acid-acetate buffer, the concentration of hydronium ion on the right in the equation is increased and the system responds by consuming the hydronium ion, forming more acetic acid. The reaction proceeds to the left until equilibrium is reestablished. When base is added, it reacts with hydronium ion to decrease its concentration, thereby stressing the system. In response, more acetic acid dissociates to partially restore the hydronium ion concentration and balance the equilibrium. The reaction proceeds to the right until equilibrium is reestablished.

Buffers are of great importance in living systems. Both the rates of biochemical reactions and their equilibrium constants are very sensitive to the availability of hydronium ions. Many biochemical reactions involved in vital processes like **metabolism**, respiration, the transmission of nerve impulses and muscle contraction and relaxation take place within a narrow pH range. **Le Chatelier's Principle** and the same types of chemical reactions that apply for the acetic acid-acetate buffer system govern the behavior of the physiological buffers.

An important buffer in the blood consists of bicarbonate ion and dissolved **carbon** dioxide in the form of carbonic acid. These two species constitute the conjugate acid-base pair of the buffering system. The pH of the blood can be altered by the ingestion of acidic or basic substances, and the carbonate/bicarbonate buffer system compensates for such additions and maintains the pH within the required range. This buffering system is intimately tied to our respiration, and an exceptional feature of pH control by this system is the role of ordinary breathing in maintaining the pH.

Carbon dioxide is a normal product of metabolism. It is transported to the lungs, where it is eliminated from the body every time we exhale. However, carbon dioxide in water—which is essentially what carbon dioxide in blood is—hydrolyzes, forming carbonic acid, which dissociates to produce the hydrogen carbonate ion and the hydronium ion:

$$O_2(aq) + H_2O(l) \Leftrightarrow H_2CO_3(aq) \Leftrightarrow H_2O^+(aq) + HCO_2^-(aq)$$

If a chemical reaction or the ingestion of an acidic material increases the hydronium ion concentration in the blood, bicarbonate ion reacts with the added hydronium ion and is transformed into carbonic acid, which means that the concentration of dissolved carbon dioxide in the blood increases. Respiration increases, and more carbon dioxide is expelled from the lungs. In the terminology of Le Chatelier's Principle, the conjugate acid-base equilibrium is stressed by the addition of acid. In response to that stress, the conjugate base reacts with the hydronium ion, producing more carbon dioxide which is removed from the system by increasing respiration.

If a base is ingested, the hydronium ion reacts with it, causing a decrease in the concentration of hydronium ion. The equilibrium compensates for this change by shifting to the right, so that more carbonic acid dissociates to restore the hydronium ion consumed by the base. This requires more carbon dioxide to be dissolved in the blood, so respiration is decreased and more gas retained. In the language of Le Chatelier, the base reacts with the hydronium ion, stressing the equilibrium for its production to the left to restore balance. More carbonic acid dissociates to compensate for the hydronium ion consumed; in turn, respiration decreases so more CO_2 is retained in the lungs to restore the carbonic acid concentration.

Because the pH of the blood is in large part under respiratory control, simple alterations in normal breathing can change the pH of the blood. The state of respiratory acidosis arises as a result of hypoventilation. Slow, shallow breathing causes more CO_2 to be retained in the lungs, which in turn causes more CO_2 to be dissolved in water in lungs and blood, more carbonic acid to be formed, and the concentration of bicarbonate and hydronium ion in the blood to rise. All equilibria involved are stressed toward the production of more hydronium ion; hence, the pH drops. Some drugs are alkaline, and one of the clinical signs of drug overdose is hypoventilation.

Hyperventilation induces the opposite situation. Increased expulsion of CO_2 from the lungs causes the equilibrium to be stressed to the left, consuming hydronium ion and resulting in a pH more basic than the normal 7.4.

In the kidneys, the buffer system is more complex and involves the dihydrogen phosphate/monohydrogen **phosphate** system in addition to the bicarbonate buffer of blood. If acidity increases, the hydrogen phosphate ion acts as a base and accepts the added protons, thereby forming more dihydrogen phosphate. If a base consumes the hydronium ions, the dihydrogen phosphate dissociates to restore the hydronium ion concentration, thereby forming more hydrogen phosphate. In either case, the added acid or base is consumed by its reaction with the appropriate component of the buffer and the pH does not change much.

To act as a good buffer, a solution must maintain a nearly constant pH when either acid or base is added. Two considerations must be made when a buffer is prepared: (1) What pH do we wish to maintain? The desired pH defines the range of the buffer. (2) How much acid or base does the solution need to consume without a significant change in pH? This defines the capacity of the buffer. The desired pH also determines the conjugate acid-base pair used in making up the buffer. The quantity of acid or base the buffer must be able to consume determines the concentrations of the buffers components that must be used.

The meanings of the terms ''nearly constant pH'' or ''significant change in pH'' depend entirely on the situation, and quantitative and specific definitions require some indication of the requirements of the system in which the buffer will be used. Many biochemical reactions are sensitive to a pH change to within 0.1 pH units, whereas some synthetic reactions carried out industrially may proceed satisfactorily over a much larger pH range, such as ± 1 pH unit. Clearly, these two situations may require different buffer capacities. To prepare an appropriate buffer or to determine the limits of a given buffer, the pH limits for the system must be known and the amount of acid or base that might be added must be known.

In terms of range, a simple rule of thumb is useful in selecting the most appropriate buffer. The best buffering system for laboratory applications is typically one whose conjugate acid-base pair has a pK_a value within one unit of the desired pH of the solution. The range of such a buffer is defined as $pH = pK_a +/- 1$. A buffer made with nearly equal concentrations of a weak acid and its conjugate base, will have a concentration of hydronium ion near the value of the K_a of the acid used. This is the same as saying that the pH of the buffer is at or near the pK_a of the acid. Of course, one can change the pH of the buffer by selecting other concentrations of acid and conjugate base, but the range of pH values over which a given buffer functions most effectively are close to the pK_a of the acid.

The second consideration in preparing a buffer—its capacity—is just as important as selecting the right pair of components, because the buffer capacity insures that the solution will be buffered well at the desired pH. To maintain a nearly-constant pH, the ratio of the concentrations of weak acid to conjugate base (or weak base to conjugate acid) must remain nearly constant. The capacity of the buffer defines the concentration of components needed in solution to consume added acid or base without significant change in pH. This will only occur in solutions in which the initial concentrations of weak acid and conjugate base (or weak base and conjugate acid) are comparable to each other and greater than the concentration of acid or base added.

One of the most important applications of the acid-base properties of salts is the formation of buffer solutions. The pH of a buffer solution changes to a relatively small extent when acid or base is added to it. A solution containing a weak acid and the **anion** that is its conjugate base or a solution containing a weak base and the **cation** that is its conjugate acid acts as a buffer. Two crucial properties of a buffer are its range, the portion of the pH scale over which a particular buffer is effective, and its capacity, the amount of acid or base that can be added without causing a large change in pH. Buffers in living systems maintain the pH in a range that prevents denaturation of **proteins** and degradation of other pH-sensitive biomolecules and which allows biological reactions to take place consistently.

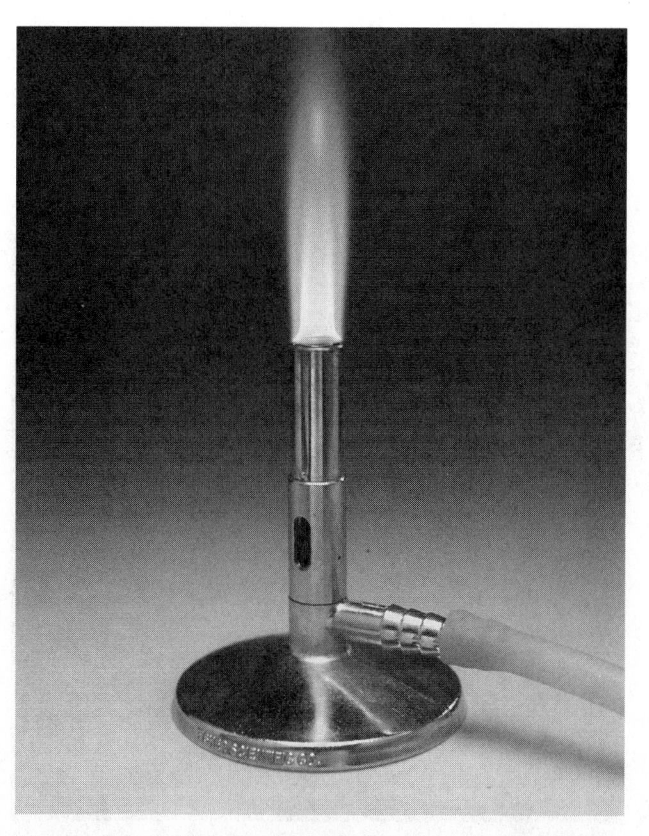

A bunsen burner with flame. Bunsen burners are a staple instrument in chemical laboratories. ***(Photograph by Charles D. Winters, Photo Researchers, Inc. Reproduced by permisision.)***

BUNSEN BURNER

The Bunsen burner was invented in Germany in 1855 by German chemist **Robert Wilhelm Bunsen**. One of the most commonly used pieces of equipment for heating in a laboratory, a Bunsen burner uses **natural gas** as a fuel supply. Due to its design the flame is very easy to control.

The entire device stands on a base that has a chimney emerging from it. The gas enters at the base of the chimney through a small, upward-facing hole that produces a jet of gas. Around the base of the chimney, level with the jet is a movable collar. This collar has a hole in it that can be lined up with a similar hole in the chimney. When both holes are aligned the

Robert Wilhelm Eberhard Bunsen.

maximum amount of air is allowed to enter. By altering the ratio of air and gas different types of flame can be produced. When the hole is closed and air is excluded from mixing with the jet of gas, a yellow, luminous flame is produced. This occurs because with the reduced air supply not all of the fuel is burned and any unburned particles of **carbon** become hot and glow. By increasing the air flow at the base more of the fuel can be burned. When the two holes are exactly aligned a blue flame is produced with a roaring sound. This is the hottest flame that can be produced by a Bunsen burner and it will not leave carbon deposits on whatever is being heated. The blue flame has an inner cone of unburned gas and a paler outer cone of complete **combustion**. The hottest part of the flame (about 2,700°F/1,500°C) is immediately above the inner cone.

A Bunsen burner is a gas burner which can have variable heating properties by adjusting the mixture of gas and air.

BUNSEN, ROBERT WILHELM (1811-1899)

German chemist

Many people associate the word *Bunsen* with a gas burner used in school laboratories. Although generally known for his work on perfecting the **Bunsen burner**, which was actually an invention of **Michael Faraday**, Robert Bunsen's true claim to fame is **spectroscopy**. Along with Gustav Kirchhoff, Bunsen discovered the process of chemical analysis called spectroscopy.

The son of a university professor, Bunsen graduated from the Gymnasium at Holzminden in 1828. He began his graduate studies in **chemistry**, physics, and mathematics at the University of Gottingen in his hometown (the same school at which his father taught), earning his doctorate in 1830. Using grant money awarded by the Hanoverian government, Bunsen spent several years traveling in Europe, visiting Berlin, Bonn, Paris, and Vienna, and meeting with such prominent scientists as **Joseph Louis Gay-Lussac**, Henri-Victor Regnault (1810-1878), and Cesar Despretz (1798-1863). Upon his return Bunsen served on the academic staff of several universities before accepting a position at the University of Heidelberg in 1852, a post he held until his retirement in 1889.

Bunsen conducted scientific research throughout his career. He began in his late 20s, studying the properties of **arsenic** and the compounds that contained it, most of which were inorganic. During his research he found that hydrated ferric **oxide** made an effective antidote for arsenic poisoning. Some of Bunsen's experiments were quite dangerous: fires and explosions occurred frequently, and twice he nearly lost his life due to the inhalation of arsenic fumes. The last of these experiments was in 1843, when an explosion of cacodyl cyanide cost him the use of his right eye.

For the next several years Bunsen focused on understanding heated **gases**. He visited geysers and volcanoes as well as refineries and industrial furnaces. He published his only book, *Gasometrische Methoden* as a compilation of his gas research findings. Bunsen had become interested in the chemical properties of **alkali metals**, such as **barium** and **sodium**. In order to isolate these elements, he invented new types of galvanic and carbon-zinc **batteries** (many of which are still called Bunsen batteries). To properly analyze the elements, he constructed a very sensitive ice calorimeter, measuring the **volume** of melted ice, rather than its **mass**. Along the way, he invented scientific instruments such as an ice calorimeter and the Bunsen battery. Still, the behavior of inorganic compounds and their components remained his primary interest. He became curious about the effects of incandescence—heating an element until it glows or and studying the **color** of its light or flame. It was at this time that Bunsen met **Gustav Kirchhoff**, an instructor at Breslau. Kirchhoff was also interested in the information that could be obtained by analyzing the light emitted from certain elements. Recognizing a kindred spirit, Bunsen persuaded Kirchhoff to come to Gottingen so that they might work together.

Bunsen and Kirchhoff knew that the key to analyzing incandescent light was to view it through a spectroscope, a device that split the light into its component colors. (A prism is a very simple spectroscope.) In order to better see the spectral patterns, they first shone the light through a slit, reducing the light to a thin beam. When passed through a spectroscope, this beam revealed a series of colored lines called an emission spectrum. However, the spectra produced from Bunsen and Kirchhoff's initial experiments were tainted: light from the gas burner used to **heat** the specimens to incandescence would creep into the beam, skewing the spectrum. To maintain a pure incandescent light, Bunsen modified a gas burner which had as very hot yet nearly invisible flame. Using the Bunsen burner, the scientists' research yielded spectral patterns of remarkable clarity.

After only a few weeks, it became clear that each element was characterized by a specific emission spectrum, as individual as a fingerprint. It did not take long for Bunsen and Kirchhoff to catalog the spectral patterns of all the known elements. With knowledge of these spectra, a chemist could deduce the components of any unknown substance. Using this method, Bunsen and Kirchhoff discovered two new elements, **cesium** and **rubidium**.

Over the course of his career, Bunsen also invented a grease-spot photometer (used for measuring light), a process for mass-producing **magnesium**, a laboratory filter pump for washing precipitate samples, and a steam calorimeter.

Buret

A **buret** (also spelled burette) is a long **glass** tube open at both ends, that is used to measure out precise volumes of **liquids** or **gases**. Most burets are about 0.394 in (1 mm) in diameter and 29.55 in (75 cm) long. The bottom of a buret is tapered so that its diameter is only about 0.1 mm in diameter. Burets are most commonly designed to hold volumes of 1 mL or less.

Fluid is dispensed from a buret through a glass stopcock at the lower end of the glass tube. The stopcock consists of an inner piece of ground glass that fits tightly into the glass tube and that can be rotated in a tightly fitting casing. The stopcock allows a fluid to be released in very small, precise amounts. Commercially available burets can usually be read with an accuracy of +/-0.01 mL.

Probably the most familiar application of a buret is in the process known as **titration**. In this process, accurately measured amounts of two solutions are allowed to react with each other in order to determine the **concentration** of one. Burets have far more uses, however. A single buret can be used, for example, to release a known **volume** of a **solution** of known concentration in order to determine the **mass** of an unknown solid. Oxidation-reduction reactions can also be studied quantitatively using burets.

Gas-dispensing burets consist of arrangements in which some gas is contained by and forced out of a graduated cylindrical tube by means of some liquid, such as **mercury**. In their appearance, gas burets look something like an upside-down version of their liquid counterparts.

Burets became necessary in chemical research only with the development of relatively precise analytical techniques in the eighteenth century. Credit for their invention is usually given to the French chemist Joseph Louis Gay-Lussac, who first developed them for the purpose of assaying **silver**.

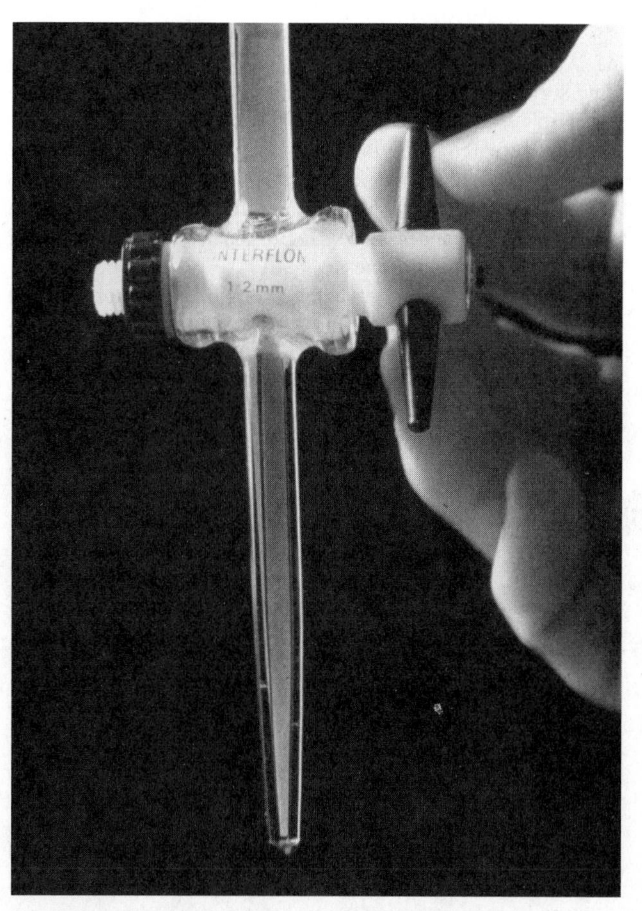

A close-up view of the valve and tip of a laboratory buret. This instrument is typically used in volumetric analysis. ***(Andrew McClenagha/Science Photo Library, Photo Researchers, Inc. Reproduced by permission.)***

Burger, Alfred (1905-)

American chemist

Alfred has spent close to seven decades researching drugs, their effects on the human body, their chemical design, and their abuse. By methodically analyzing their chemical makeup, Burger considerably expanded the knowledge of medicinal **chemistry** and of its far-reaching effects on the human population. Humans generally live longer than 50 years ago owing to a greater understanding of the medicinal qualities of certain drugs. Effective treatments for addiction have been developed thanks to a greater knowledge of some chemicals' addictive qualities. In fact, drugs can either help or harm human beings through their reactions with other chemicals. As Burger explained in an article for the *Charlottesville Daily Progress*, "Drugs are just chemicals. They can't do anything else but react with other chemicals. They can't subdue your pain or raise your expectations by themselves. They can only react with other substances." Through his studies, Burger has helped to broaden our understanding of medicinal chemicals, an understanding that has led to new cures, treatments, and therapies.

Alfred Burger was born on September 6, 1905, in Vienna, Austria-Hungary (now Austria), the son of S. L. Burger (a civil servant) and Clasriss Burger. After receiving his Ph.D. in chemistry from the University of Vienna and conductind post-doctoral research in Switzerland in 1928, Burger immigrated to the United States the following year. On August 1, 1936, he married Frances Page Morrison, who eventually bore him one daughter, Frances. In 1937, Burger became a naturalized citizen of the United States.

Upon arrival in the United States in 1929, Burger went to work as a research associate at the newly created Drug Addiction Laboratory at the University of Virginia. Throughout his long research career, he remained linked to the University. Burger's work, which focused on laboratory research from 1929-1938, later included teaching. An assistant professor from 1938-1939, Burger was an associate professor from 1939-1952, when he was promoted to full profession, a position he held until 1970, acting as department chair from 1962-1963. In 1970, he was named professor emeritus.

Burger's entire career has been devoted to studying chemicals and their medicinal properties. As an expert in organic chemistry he pinpointed the syntheses of **morphine** substitutes, and also researched and designed numerous drugs, including antimalarials, antituberculous drugs, organic **phosphorous** compounds, antimetabolites, and psychopharmacological drugs. He also did much to further **chemotherapy**, acting as a member of the chemistry panel and then as a medical chairman of the Cancer Chemotherapy National Service Center from 1956 through 1964. His most recent work focused on monoamine oxidase (MAO) inhibitors and antidepressant drugs.

In 1958, Burger founded the *Journal of Medicinal Chemistry*, providing a formal communication venue for his discipline. He was editor of the journal from its inception through 1971. Today, the journal still remains a voice for medicinal chemist from around the globe.

Burger's scientific accomplishments did not go unnoticed by his peers. In 1953, he was awared the Pasteur Medal by the Pasteur Institute in Paris. In 1971, he received an honorary degree from the Philadelphia College of Pharmacy and Science, as well as an award from the American Pharmacological Society Foundation. In 1977, The American Chemical Society recognized his contribution to medicinal chemistry by giving him the Smissman Award. Burger received his greatest accolade when the American Chemical Society created the Alfred Burger Award in Medicinal Chemistry in his honor.

Burger is a member of the American Chemical Society and the American Pharmacological Society. A prolific writer, he has published nine books, as well as about 200 articles.

BUTENANDT, ADOLF (1903-1995)

German biochemist

Adolf Butenandt's groundbreaking research into sex **hormones** led to the formulation of the compounds estrone and androsterone, hormones involved in the regulation of sexual processes in the body. He has worked on both male and female **sex hormones** using microanalytical methods developed by the Austrian chemist **Fritz Pregl**. By uncovering the underlying structure of sex hormones, Butenandt opened biochemical study to the relationship of the chemical structure of sex hormones and carcinogenic substances. For his work, Butenandt was awarded the Nobel Prize in **chemistry** in 1939, an award that he shared with **Leopold Ružička** but was unable to receive until 1949 because the Nazi government did not allow him to accept it.

Adolf Butenandt.

Adolf Friedrich Johann Butenandt was born in Bremerhaven-Lehe (now Wesermünde), Germany, on 24 March 1903, to Otto Louis Max Butenandt, a businessman, and Wilhelmina Thomfohrde Butenandt. He received his basic education in Bremerhaven at the Oberrealschule, after which he went to the University of Marburg in 1921 to study chemistry and biology. When he continued his studies at the University of Göttingen in 1924, he was inspired to study **biochemistry** by his professor, **Adolf Windaus**.

Upon completion of his dissertation on a compound used in insecticides, Butenandt was granted a doctorate by the University of Göttingen in 1927. He was also made an assistant at the Institute of Chemistry in Göttingen in 1931. He remained in Göttingen until 1933, when he was appointed professor of **organic chemistry** at the Danzig Institute of Technology. He remained in Danzig until 1936, having been coerced by the Nazi government to reject an appointment to Harvard University in 1935. By 1929 Butenandt had isolated a female sex hormone in pure crystalline form. This research was made possible by his association with Walter Schoeller, the director of research for a pharmaceutical firm, Schering

Corporation. Schoeller had asked Windaus for help to investigate the female sex hormones and their chemical structures. Windaus recommended his student, Butenandt, for this research. Schoeller provided Butenandt with the necessary hormonal substances needed to carry out the study. Butenandt first called the hormone he isolated folliculin, because it is secreted in the lining of the follicles of the ovary. It was later renamed estrone, however, because it is an estrogen hormone that controls a number of female processes.

In 1931 Butenandt married Erika von Ziegner, his assistant in his early research. They had two sons, Otfrid and Eckart, and five daughters, Ina, Heide, Anke, Imme, and Maike. Also in 1931, Butenandt was able to confirm the existence of another female sex hormone, estriol, which had been discovered in London by G. F. Merrian. (Another biochemist, the American **Edward A. Doisy**, had also isolated estrone at about this time.) He also isolated and purified in crystalline form the male sex hormone androsterone, which is secreted from the testes. This hormone is related to testosterone, the main male sex hormone. He continued his research with sex hormones, and by 1934 he and his associates had isolated the hormone progesterone. In five years he was able to synthesize progesterone from its **cholesterol** precursor.

An important aspect of Butenandt's research with sex hormones was the discovery that the exact location of male sex hormone activity is in the **nucleus** of the **carbon** atoms. This was a major contribution to the study of human biochemistry; it enabled scientists to produce various medical products that alleviate the symptoms of major diseases. **Cortisone**, a synthetic product closely related to some of the hormones Butenandt researched, has been used in the treatment of arthritis and is one example of the medical applications of hormone research. By 1935 Butenandt completed some significant research on testosterone (the main male sex hormone) that led to his discovery of the chemical sites of biological activities. He found that male and female sex hormones were chemically related by a common sterol nucleus.

Butenandt was asked to become director of the Kaiser Wilhelm Society, which oversaw all scientific research in Germany, in 1936. He accepted this position from the physicist **Max Planck**, and the institution now bears Planck's name. The award of the Nobel Prize in chemistry in 1939 was made to Butenandt and Leopold Ružička for their contributions to the study of sex hormones. Because of the outbreak of World War II and the intervention of the Nazi government, Butenandt was not able to receive his award until 1949.

During the war Butenandt worked on genetic problems relating to eye pigmentation in insects. This research led Butenandt to the one-gene-one **enzyme** theory that was shared by other researchers. After the war the Kaiser Wilhelm Institute moved to Tübingen; Butenandt became professor of physiological chemistry there and continued his research with insects. By 1953 he had isolated the first insect hormone, ecdysone, which stimulates the transformation of a caterpillar into a butterfly. His associate, Peter Karlson, later showed that ecdysone is derived from cholesterol and is also related to sex hormones in mammals.

In 1956 the Kaiser Wilhelm Institute moved again, this time to Munich, and Butenandt became professor of physiological chemistry at the University of Munich. There, he studied a substance that is synthesized by female silkworms to attract males. Butenandt continued his association with the Max Planck Society for the Advancement of Science, serving as its president from 1960 until 1972. He retired from his position at the University of Munich in 1971. Butenandt died in Munich on 18 January 1995.

Butenandt received many awards, including the Grand Cross for Federal Services of West Germany and the Adolf von Harnack Medal of the Max Planck Society. He was made a commander of the Legion of Honor of France in 1969. He received honorary degrees from many universities throughout Europe and held honorary memberships in scientific societies all over the world. He published numerous articles in scientific journals and wrote a number of books.

BUTYL GROUP

The butyl group corresponds to the group C_4H_9. The name comes from butyric acid, an acid that has the smell of rancid butter. This fact gives rise to its name, i.e., butyl comes from the Latin word *butyrum*, meaning butter.

There are two possible arrangements of the atoms in the butyl group, which means there are two isomers: n-butyl and isobutyl. With n-butyl, all four **carbon** atoms lie in a straight chain, i.e., $CH_3CH_2CH_2CH_2$-; in the case of isobutyl, there are three carbon atoms in a straight chain with the remaining one being joined to the central carbon **atom**, i.e., $(CH_3)_2CHCH_2$.

The butyl group is found in a number of organic compounds. The characteristics of these groups are determined mainly by the **functional group**. The presence and size of the butyl group modifies such physical properties as the melting and boiling points of chemical compounds.

Butyl rubber is a synthetic rubber made by copolymerizing 2-methylpropene and methyl-1,3-diene. This rubber can be vulcanized (treated under **heat** and pressure with **sulfur** to improve elasticity and strength). It has been extensively used in the manufacture of inner tubes for cars and bicycles.

See also Alcohols; Aldehydes; Amine group

C

CADMIUM

Cadmium is a transition metal, one of the elements found in Rows 4 through 7 between Groups 2 and 13 in the **periodic table**. Cadmium has an **atomic number** of 48, an atomic **mass** of 112.41, and a chemical symbol of Cd.

Properties

Cadmium is a shiny metal with a bluish cast. It is very soft and can almost be scratched with a fingernail. Its melting point is 610°F (321°C) and its **boiling point** is 1,410°F (765°C). The **density** of cadmium is 8.65 grams per cubic centimeter. When cadmium is used in alloys, it tends to lower the melting point of the **alloy**. Cadmium reacts slowly with **oxygen** in moist air at room **temperature** forming cadmium **oxide** (CdO). It does not react with **water**, but it does react with most acids.

Occurrence and Extraction

Cadmium is relatively rare in the Earth's crust with an estimated abundance of about 0.1-0.2 parts per million. The only important ore of cadmium is greenockite, or cadmium sulfide (CdS). The largest producers of cadmium are Canada, Japan, Belgium, the United States, China, Kazakhstan, and Germany. Cadmium commonly occurs in **zinc** ores and is produced commercially as a by-product of the production of zinc from its ores.

Discovery and Naming

Cadmium was discovered in 1817 by the German chemist Friedrich Stromeyer (1776-1835). Stromeyer was studying an interesting problem involving the pharmaceutical compound known as zinc carbonate ($ZnCO_3$). Under some circumstances, the zinc carbonate supplied by manufacturers to pharmacies turned yellow. Pharmacists refused to accept shipment of this product because they thought it was impure. In an effort to determine the nature of this impurity, Stromeyer discovered the presence of an as-yet-unidentified element, cadmium. The close connection between cadmium and zinc is reflected in the choice for the new element's name. *Cadmia* is the ancient name for zinc oxide.

Uses

At one time, the most important use for cadmium was in the **electroplating** of steel. However, medical scientists have discovered that cadmium is very toxic to humans and other animals. As a result, the use of cadmium for this purpose has been reduced significantly. Today, the vast majority of cadmium is used in the production of nickel-cadmium (nicad) **batteries** that can be re-charged and re-used many times. Such batteries are used in a large variety of appliances, including compact disc players, cellular telephones, pocket recorders, handheld power tools, cordless telephones, laptop computers, camcorders, and scanner radios.

Health Issues

In low levels, cadmium causes nausea, vomiting, and diarrhea. If inhaled, cadmium dust causes dryness of the throat, choking, headache, and pneumonia-like symptoms. The long-term effects of cadmium are not fully understood, but are thought to include heart and kidney disease, high blood pressure, and cancer. One of the most famous environmental diseases, itai-itai (Japanese for ''ouch-ouch'') has been attributed to cadmium poisoning.

CAFFEINE

Caffeine is a naturally occurring drug belonging to a group of compounds called alkaloids. It has a **molecular weight** of 194.19 and is composed of **carbon**, **hydrogen**, **nitrogen**, and **oxygen**. It is found in tea leaves, coffee beans, guarana paste, and cola nuts. Caffeine is the most widely used, non-medicinal

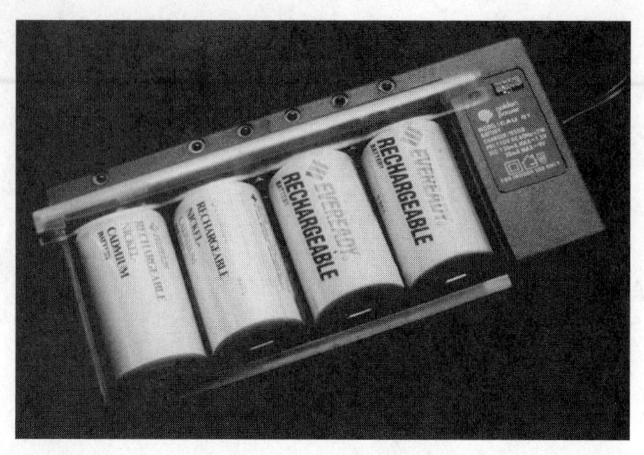

Nicad (nickel-cadmium) batteries can be recharged in a plug-in unit such as this one.

A microscopic image of caffeine. *(Photograph by Phillip A. Harrington, The Stock Market. Reproduced by permission.)*

stimulant worldwide and has been a staple of the human diet for centuries. While some studies have indicated that caffeine may have deleterious health effects, no definitive conclusions have suggested moderate amounts are harmful.

The first evidence of caffeine use is from Aztec records during the time of the leader Montezuma. This civilization consumed caffeine in a hot drink made from cacao leaves. Later cultures drank caffeine in the form of coffee. Coffee was first introduced as a medicine in England, but it became a fashionable beverage between 1670 and 1730. While humans have consumed caffeine for thousands of years, it was not until the early 1800s that the chemical was isolated and characterized. During the 1820s, the stimulating agents in coffee, tea, and chocolate were identified. At the time scientists did not realize that they had found the same ingredient. In 1840, two researchers, T. Martins and D. Berthemot, working independently showed that all of these active agents were chemically identical to caffeine isolated from coffee beans.

During the rest of the century, the chemical characterization of caffeine was worked out. In 1848, Edward Nicholson published important research on the constitution of caffeine. The German chemist **Emil Fischer** (1852-1919) worked out the **molecular structure** of caffeine during the 1890s. He first synthesized caffeine from basic raw materials in 1895. Two years later he ascertained its **structural formula**.

Caffeine is classified as an **alkaloid**. Alkaloids are nitrogen-containing molecules that have a slightly bitter **taste** and are physiologically active. The specific effect on the body varies greatly depending on the alkaloid. Caffeine is a mild stimulant and is found in food products like coffee, tea, soft drinks, and chocolate.

The chemical name for caffeine is 1,3,7- trimethylxanthine,1,3,7, trimethyl-2,6-dioxopurine or methyltheobromine. It is represented by the **molecular formula** $C_8H_{10}N_4O_2 \cdot H_2O$. The composite elements are arranged in a bicyclic fashion which is a derivative of a purine **ring** system.

When isolated and purified, caffeine is a white, crystalline powder. It is composed of long hexagonal prisms. It ultimately melts at 458.2 °F (236.8°C), losing **water** at 176°F (80°C) and subliming at 352.4°F (178°C). It is odorless and has a bitter taste. It is slightly soluble in water and **alcohol** and is typically heated when incorporated into beverages. Aqueous solutions of caffeine have a neutral **pH**.

Guarana paste which is made from the Paullinia tree has the highest **concentration** of caffeine (about 4%). Tea leaves contain about 3.5% caffeine. Coffee beans 1-2.2%, and kola nuts have about 1.5%. Another source of caffeine is maté leaves. When added to food products, caffeine is isolated from these sources and refined. It is then incorporated into the product at a specific dose. For example, the average cup of coffee contains about 100 mg of caffeine. A can of soda contains about 55 mg, while a cup of tea has about 60 mg of the compound. A chocolate candy bar contains about 20 mg. During the late 1990s it became fashionable to add caffeine to other types of products such as bottled water and chewing gum. These products are designed to help keep the users alert.

There are a variety of physiological effects that caffeine has on the human body. It is a mild stimulant to the central nervous system. It is thought to work by binding to select neurotransmitter receptor sites on nerve cells, causing them produce a continual signal. These signals are responsible for the effects of caffeine consumption such as alertness, excitability, increased mental awareness, and restlessness. Caffeine is also used to treat migraine headaches because it constricts dilated blood vessels. It is also put into aspirin products. Additionally, it is thought to relieve asthma attacks because it can widen bronchial airways.

In large amounts, caffeine can produce undesirable side effects such as nervousness, insomnia, rapid and irregular heartbeats, excess stomach acid, heartburn, and elevated blood sugar levels. It is possible to overdose on caffeine; the fatal dose estimated at 10 g. This would be the equivalent of the rapid consumption of 100 cups of coffee.

While the health risks of too much caffeine are well established, the effects of long term exposure to lower dosages are unknown. Caffeine is known to be a teratogen which means it can produce birth defects. When a pregnant woman ingests caffeine the fetus is exposed and more effected by the drug because of its smaller size. Some studies have also suggested that caffeine has a mutagenic effect in large doses.

However, no study has ever indicated that the average coffee or tea drinker is in danger from caffeine.

Chemical structure of caffeine

Chemical structure of caffeine. ***(Illustration by Electronic Illustrators Group.)***

CALCINATION

In calcination, a substance is heated to a high **temperature** below its melting point to bring about thermal decomposition or a phase transition other than melting. Calcination often has the effect of making a substance friable. The process usually takes place in long cylindrical kilns. Calcined materials may lose **water** (as in the conversion of ferric hydroxide to ferric oxide) or some other volatile constituent (such as **carbon** dioxide upon heating limestone), or may undergo **oxidation** or **reduction**. In oxidative calcination (also known as roasting), a substance is heated in the presence of air or **oxygen**. Oxidative calcination is commonly used to convert metal sulfide ores to oxides in the first step of recovering such **metals** as **zinc**, **lead**, and **copper**. Reductive calcination (smelting) involves the heating and reduction of metals from their ores, followed by separation out of the valueless material.

Calcination reactions may include thermal **dissociation**, including the destructive **distillation** of organic compounds, i.e., heating a highly carbonaceous material in the absence of air or oxygen, to produce **solids**, **liquids**, and gases. Examples of other calcination reactions include the **concentration** of **aluminum** by heating bauxite; polymorphic phase transitions such as the conversion of anatase to the rutile; and thermal recrystallizations such as the devitrification of **glass**. Materials that are commonly calcined include **phosphate**, aluminum **oxide**, **manganese** carbonate, petrol coke, and sea water magnesite.

See also Oxidation-reduction reaction; Phase changes

CALCIUM

Calcium is the third element in Group 2 of the **periodic table**. The members of this group are commonly described as the alkaline earth elements. Calcium's **atomic number** is 20, its atomic **mass** is 40.08, and its chemical symbol is Ca.

Properties

Calcium is a fairly soft metal with a shiny **silver** surface when first cut. The surface quickly becomes dull, however, as calcium reacts with **oxygen** to form a coating of white or gray calcium **oxide**. Calcium's melting point is 1,560°F (850°C) and its **boiling point** is 2,620°F (1,440°C). The element has a **density** of 1.54 grams per cubic centimeter.

Calcium is a moderately active element that combines readily with oxygen to form calcium oxide (CaO): $2Ca + O_2 \rightarrow 2CaO$. Calcium also reacts with the **halogens**, with cold **water**, with most acids, and with most nonmetals, such as **sulfur** and **phosphorus**.

Occurrence and Extraction

Calcium is the fifth most common element in the Earth's crust with an abundance estimated at about 3.64%. It is also the fifth most abundant element in the human body. Calcium does not occur free in nature and always exists as a compound. The most common calcium compound is **calcium carbonate** ($CaCO_3$), the major component of rocks and minerals such as aragonite, calcite, chalk, limestone, marble, and travertine. Calcium carbonate is also found in oyster shells and coral.

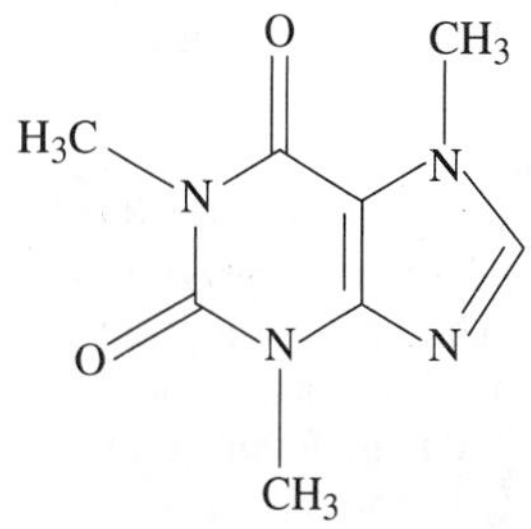

Pure calcium metal can be made by electrolyzing molten calcium chloride in a method similar to that by which the element was first prepared in 1807: $CaCl_2$ —electric current→ Ca + Cl_2.

Discovery and Naming

The abundance of calcium compounds in nature and their value as building materials assured that they would be found and used by humans very early in history. Limestone and marble have long been popular building materials, while mortar and plaster of paris have also found a variety of applications in the construction business. All of these materials contain calcium.

Calcium compounds are very stable, however, and finding a way to extract the pure metal from its compounds proved to be a serious challenge to chemists. Finally, in 1807, the great English chemist and physicist **Humphry Davy** found a method for breaking down calcium compounds. First he melted one of those compounds, calcium chloride, and then he passed an electrical current through the molten material. Pure calcium metal was released at one electrode of the apparatus. The procedure invented by Davy is essentially the same as the one used today for the manufacture of the element. The name selected for the element comes from a Latin term *calx*, meaning limestone.

Uses

Elemental calcium has relatively few uses, most of those involving an **alloy** of the element. For example, an alloy of calcium and **cerium** is used in flints in cigarette and other types of lighters.

By contrast, the compounds of calcium have a myriad number of uses. By far the most important of these compounds is **lime** (calcium oxide; CaO). Lime usually ranks in the top five chemicals produced in the United States. Each year, close to 20 billion kilograms (45 billion pounds) of the compound are produced in the United States.

The most important use of lime is in the production of **metals**. It is used during the manufacture of steel to remove unwanted sand (silicon dioxide; SiO_2) present in **iron** ore: $CaO + SiO_2 \rightarrow CaSiO_3$. The calcium silicate ($CaSiO_3$) formed in this reaction is removed as slag.

Lime is also used in **pollution** control. Waste **gases** from manufacturing and industrial processes often contain noxious gases which must be removed as they leave the plant. One way to remove unwanted gases is to pass them through a **solution** of calcium oxide in the smokestack. The calcium oxide reacts with and removes certain harmful gases, such as sulfur dioxide (SO_2): $CaO + SO_2 \rightarrow CaSo_3$.

Another use of lime is in **water purification** and water treatment plants. Lime combines with water to form calcium hydroxide [$Ca(OH)_2$], also known as slaked lime. Slaked lime is a sticky precipitate that sinks to the bottom of a tank, carrying with it impurities such as suspended particles and disease-causing microorganisms.

Lime is also the starting point from which more than 150 other industrial chemicals are produced. Some examples include: 1) calcium alginate: a thickening agent used in food products, such as ice cream and cheese products; 2) calcium arsenate [$Ca_3(AsO_4)_2$]: an insecticide; 3) calcium carbide (CaC_2): used to make acetylene gas for acetylene torches and in the manufacture of **plastics**; 4) calcium chloride ($CaCl_2$): used for ice removal and dust control on dirt roads, as a conditioner for concrete, and as an additive for canned tomatoes; 5) calcium gluconate [$Ca(C_6H_{11}O_7)_2$]: used as a food additive and in vitamin pills; 6) calcium hypochlorite [$Ca(OCl)_2$]: used as a swimming pool disinfectant, bleaching agent, algicide, and **fungicide**; 7) calcium permanganate [$Ca)MnO_4)_2$]: used as a rocket propellant in textile production, as a water sterilizing agent, and in dental procedures; 8) calcium **phosphate** [$Ca(PO_4)_2$]: used as a supplement in animal feed, as a fertilizer, in the commercial production of dough and yeast products, and in the manufacture of **glass**; 9) calcium phosphide (Ca_3P_3): used in fireworks, rodenticide, torpedoes, and flares; 10) calcium stearate [$Ca(C_{18}H_{35}O_2)_2$]: used in the manufacture of wax crayons, cements, plastics, **cosmetics**, water-resistant materials, paints, and as a food additive; 11) calcium tungstate ($CaWO_4$): used in luminous paints, fluorescent lights, and x-ray studies in medicine.

Health Issues

Calcium is essential to both plant and animal life. In humans, it makes up about 2% of a person's body weight, and nearly all of it is found in bones and teeth. The body uses calcium in a compound known as hydroxyapatite [$Ca_{10}(PO_4)_6(OH)_2$] that makes teeth and bones hard and resistant to wear.

Calcium also has other important functions in the human body, such as controlling the way the heart beats. An excess of calcium can increase the rhythm of the heart beat, and a deficiency can decrease heart beat. Serious health problems result in either case.

Calcium Carbonate

Calcium carbonate ($CaCO_3$) is one of the most common compounds on Earth, making up about 7% of Earth's crust. It occurs in a wide variety of mineral forms, including limestone, marble, travertine, and chalk. Calcium carbonate also occurs combined with **magnesium** as the mineral dolomite, $CaMg(CO_3)_2$. Stalactites and stalagmites in caves are made of calcium carbonate. A variety of animal products are also made primarily of calcium carbonate, notably coral, sea shells, egg shells, and pearls.

Calcium carbonate has two major crystalline forms—two different geometric arrangements of the calcium ions and carbonate ions that make up the compound. These two forms are called aragonite and calcite. All calcium carbonate minerals are conglomerations of various-sized crystals of these two forms, packed together in different ways and containing various impurities. The large, transparent crystals known as Iceland spar, however, are pure calcite.

In its pure form, calcium carbonate is a white powder with a specific gravity of 2.71 in the calcite form or 2.93 in the aragonite form. When heated, it decomposes into calcium **oxide** (CaO) and **carbon** dioxide gas (CO_2). It also reacts vigorously with acids to release a froth of **carbon dioxide** bubbles. It is said that Cleopatra, to show her extravagance, dissolved pearls in vinegar (acetic acid).

Every year in the United States alone, tens of millions of tons of limestone are dug, cut, or blasted out of huge deposits in Indiana and elsewhere. It is used mostly for buildings and highways and in the manufacture of steel, where it is used to remove silica (silicon dioxide) and other impurities in the **iron** ore; the calcium carbonate decomposes to calcium oxide in the **heat** of the furnace, and the calcium oxide reacts with the silica to form calcium **silicates** (slag), which float on the molten iron and can be skimmed off.

Deposits of calcium carbonate can be formed in the oceans when calcium ions dissolved from other minerals react with dissolved carbon dioxide (carbonic acid [H_2CO_3]). The resulting calcium carbonate is quite insoluble in **water** and sinks to the bottom.

However, most of the calcium carbonate deposits found today were formed by sea creatures millions of years ago when oceans covered much of what is now land. From the calcium ions and carbon dioxide in the oceans, they manufactured shells and skeletons of calcium carbonate, just as clams, oysters, and corals still do today. When these animals die, their shells settle on the sea floor where, long after the seas have gone, we now find them compressed into thick deposits of limestone. The White Cliffs of Dover in England are chalk, a soft, white porous form of limestone made from the shells of microscopic sea creatures called Foraminifera that lived about 136 million years ago. Blackboard ''chalk'' isn't made of chalk; it is mostly gypsum ($CaSO_4$).

In pearls—which mollusks make out of their shell-building material when they are irritated by a foreign body in their flesh—and in sea shells, the individual $CaCO_3$ crystals are invisibly small, even under a microscope. But they are laid down in such a perfect order that the result is smooth, hard, shiny, and sometimes even iridescent, as in the rainbow colors of abalone shells. In many cases, the mollusk makes its shell by laying down alternating layers of calcite and aragonite. This gives the shell great strength, as in a sheet of plywood where the grain of the alternating **wood** layers runs in crossed directions.

CALIFORNIUM

Californium is a transuranium elements, one of the elements found beyond **uranium** in Row 7 of the **periodic table**. It has an **atomic number** of 98, and atomic **mass** of 251.0796, and a chemical symbol of Cf.

Properties

All isotopes of californium are radioactive. The most stable **isotope**, californium-251, has a half life of 898 years. The element exists in such small amounts that very little is known about its chemical and physical properties. Its melting point has been estimated to be about 1,650°F (900°C).

Occurrence and Extraction

Californium does not occur naturally in the Earth's crust. It is prepared artificially by bombarding **curium** (atomic number 96) with alpha particles in a particle accelerator. By means of this reaction, microgram-size amounts of the element can be prepared.

Discovery and Naming

Californium was discovered in 1950 by a research team at the University of California at Berkeley (UCB). The team was made up of **Glenn T. Seaborg**, **Albert Ghiorso**, Kenneth Street, Jr., and Stanley G. Thompson (1912-). The discoverers chose to name the new element after the state of California in which they had done their research.

Uses

Californium has very few important applications. One of its uses has been in non-destructive testing, such as inspections of airline baggage. The isotope californium-252 has also found some use in determining the amount of moisture in soil, information that is very important to road builders and construction companies. All isotopes of californium destined for commercial use are now produced at the High Flux Isotope Radiator at the Oak Ridge National Laboratory in Tennessee.

CALORIMETRY

Calorimetry is the measurement of the **heat** absorbed or given off in a chemical or physical change. A device which measures the heat involved in such a change is called a calorimeter.

Historically, among the earliest observations made of chemical changes was that heat was generally involved. Over the years, as investigators attempted to give a rational interpretation for this observation, a number of theories were proposed. Among them were theories which assumed that heat was a material substance, called caloric. Caloric was thought to flow in or out of materials as they changed during chemical transformations. It was necessary to develop methods that were capable of accurately measuring the amount of heat transferred during chemical processes in order for an adequate explanation of the heat phenomenon to be developed.

In 1783, **Antoine Lavoisier** and P. S. de Laplace (1749-1827) published a description of an ice calorimeter which they used to measure the amount of heat given off in the burning of substances (oxidation). Ice was packed around the container in which a chemical reaction occurred, and the amount of ice that melted during the process was measured. The relative amount of ice which melted during the burning of various substances was a quantitative indication of the amount of heat emitted in these oxidation processes and provided a comparative measure of the readiness with which substances reacted with oxygen. Lavoisier and Laplace regarded respiration by animals to be a form of **combustion** (**oxidation**) as well. They used the ice calorimeter to measure the heat given off by a guinea pig as it breathed oxygen, relating this heat to that of a chemical reaction occurring inside the animal.

Subsequently, the design of calorimeters was improved and more accurate measurements of the heat involved in chemical reactions were made. This facilitated the steady advances being made in the field of thermodynamics and in **thermochemistry**, which is the application of **thermodynamics** to chemical processes. These fields of study have contributed dramatically to our present understanding of chemical compounds and their reactions.

In order to measure heat quantitatively and to compare the amounts of heat involved in different chemical processes, it is necessary to define a standard unit of heat. A calorie is defined as the amount of heat required to raise the temperature of one gram of liquid water by one degree celsius. The specific heat capacity of a substance is the amount of heat necessary to raise the **temperature** of one gram by one degree celsius. The specific heat capacity of water is, therefore, by definition, equal to 1. The heat capacity of other substances is determined relative this standard. For instance, the specific heat capacity of sucrose is 0.30. This means that 0.30 calories of heat are required to raise the temperature of one gram of sucrose by one degree celsius; in other words, only 30% as much heat is required to increase the temperature of a given weight of sucrose by one degree as it would to raise the same amount of water by one degree.

The calorimeter consists of a container in which the chemical reaction that releases (or absorbs) heat is caused to occur. The heat evolved (or absorbed) during the reaction flows into (or out of) a known amount of water or other material which surrounds the reaction vessel. If the heat capacity of the material in this surrounding jacket is known, and the rise (or fall) of its temperature is measured accurately, we may calculate the quantity of heat released (or absorbed) in the reaction. For instance, suppose that the calorimeter contains 100 grams of water in the jacket and that a chemical reaction occurs in the enclosed vessel; and suppose that the temperature of the water rises from 30°-75°C, an increase of 45°. Since it takes one calorie of heat to raise the temperature of one gram of water by one degree, there must have been 45/100=0.45 calories of heat transferred from the reaction to the water. The amount of heat produced in the chemical reaction was, therefore, 0.45 calories.

In the preceding example, by calculating the heat released in the reaction using only the heat capacity of the water, we assumed that the amount of heat absorbed in this process by the calorimeter and by the reaction mixture itself may be ignored. In real experimental situations, where accuracy is important, this assumption is not made, and the heat capacities of all parts of the system are carefully determined. This will include, in addition to the heat capacity of the water, the amount of heat needed to raise the temperature of the reaction vessel, the calorimeter container, and the reaction mixture by one degree celsius. The change in the temperature observed in the chemical reaction is then multiplied by the sum of these heat capacities to determine the total amount of heat absorbed by all parts of the system during the process and thus the amount of heat released in the chemical reaction.

Another type of calorimeter that is commonly used is the bomb calorimeter. Whereas the reactions that occur in a simple calorimeter take place under conditions of constant atmospheric pressure, the reactions in a bomb calorimeter occur under constant volume conditions. The reactants are sealed into a vessel that is then immersed in water in an insulated container. The reaction is triggered, usually by an electrical discharge, and heat is absorbed from or released to the surrounding water as the reaction proceeds. The amount of heat involved in the reaction is determined as above by measuring the temperature change of the surrounding medium and multiplying by the heat capacity of the medium. The bomb calorimeter is particularly useful for combustion reactions and for reactions involving **gases**.

Manufacturers are required to list the number of Calories contained in a serving of a food product on its container. This nutritional Calorie is actually a kilocalorie which is equal to 1,000 of the calories as defined above: a kilocalorie is the amount of heat necessary to raise the temperature of 1,000 grams of water by one degree celsius. The number of Calories listed for a given serving size of a particular product is the amount of heat released by its combustion in a calorimeter. It is correlated with the amount of heat energy that would be released as the body uses this food product. For instance, the label of the bottle of a certain chocolate syrup contains the information that in a normal serving of 2 Tbsp (39 grams) of chocolate there are 100 Calories. The combustion of 39 grams of the syrup in a calorimeter or in your body should produce 100 kilocalories or 100,000 calories of heat energy.

In addition to heats of reaction, calorimeters are used to measure heats of solution, heat capacities, and the heat involved in other chemical and physical changes. An early example of use of calorimeters was the successful work of **James Joule** in the period 1843-1878 to determine the equivalence of work and heat. He showed that when 4.184 joules of work is done, 1.0 calorie of heat is produced. This is known as the mechanical equivalent of work and is of basic importance in the study of thermodynamics.

Calvin, Melvin (1911-1997)

American chemist

Melvin Calvin began his academic career with an interest in the practical and physical aspects of **chemistry**. His greatest accomplishments, however, have been in the interactions between chemistry and the life sciences. In 1961, Calvin was honored with the Nobel Prize in chemistry for his elucidation of the mechanism by which **carbon** dioxide is incorporated into green plants. In the years that followed, he pursued his interest in some unusual applications of chemistry, such as researching oil-bearing plants for their possible development as alternative **energy** sources and in the search for other forms of life that may exist in the universe.

Calvin was born in St. Paul, Minnesota, on April 8, 1911, to Elias and Rose Irene (Hervitz) Calvin. Both Calvin's parents had emigrated from Russia in the 1880s—his father from an urban area in northern Russia and his mother from a rural region in southern Russia. Calvin's father had apparently been well-educated before coming to the United States and, in spite of the fact that he ended up as a factory worker, always put a high value on developing intellectual skills.

The Calvins moved to Detroit, Michigan, when Melvin was a young boy so that his father could take a job at the Cadillac factory there. Calvin received both his grade school and high school education in the Detroit Public Schools, but his studies made little lasting impact on him. Calvin's only recollection from his high school science classes was a physics teacher telling him that he would never become a scientist because he was too impulsive, because he didn't wait to collect *all* the data needed to solve a problem. Calvin observed that the physics teacher didn't really understand the process of scientific advancement: if one were really to know *all* the data, a computer alone would be all that would be necessary to derive a conclusion.

Calvin's interests in science developed as a result of internal forces. He describes walking home from school with a friend and, in a sudden flash of insight, suddenly understanding the role of atoms as the building blocks of all **matter** in the universe. For Calvin, that moment was a great thrill because it was his own "personal discovery." By the time he reached high school, Calvin knew that he wanted to become a chemist or, more precisely, a chemical engineer.

Calvin pursued his dream of becoming a chemist after he graduated from high school in 1927. He enrolled at the Michigan College of Mining and Technology (now Michigan Technological University) in Houghton, but had to leave at the end of two years. The first rumblings of the Great Depression were being felt, and Calvin could not afford to stay in school. Instead, he got a job at a brass factory in Detroit where he rapidly became familiar with a number of chemical procedures.

The experience convinced him to continue with his plans to major in **chemical engineering** because "I figured I would always be in demand." He looked closely at the world in which he lived and saw chemical applications everywhere. In a time of economic depression, with his father out of work, it was the possibility of making a living rather than "grand questions about the universe," that, he told Swift, determined his career choice.

In any case, Calvin soon returned to the Michigan College of Mining and completed his bachelor of science degree in 1931. He then entered the doctoral program in chemistry at the University of Minnesota. At Minnesota, he gave evidence of the wide-ranging chemical interests that later characterized his professional career. After pursuing problems in both physical and **organic chemistry**, he finally settled on a problem involving the **electron** affinity of **iodine** and **bromine** for his doctoral thesis. Successful completion of that research earned him a Ph.D. degree in 1935.

In the same year, Calvin was awarded a Rockefeller Foundation fellowship allowing him to spend two years of postgraduate study at the University of Manchester, England. At Manchester, Calvin worked under Michael Polanyi, professor of **physical chemistry**. One of Calvin's research assignments involved studying the role of metalloporphyrins—organic molecules from which are derived **chlorophyll** and hemoglobin—in various catalytic reactions. Such assignments were a modest preview of the research he would undertake three decades later when studying chlorophyll and that compound's role in **photosynthesis**. According to *Nobel Prize Winners,* Calvin's interest in coordination catalysis remained "paramount" for many years after his work at Manchester, eventually resulting "both in theoretical (the chemistry of metal **chelate** compounds) and practical (oxygen-carrying synthetic chelate compounds) applications."

At the conclusion of his two years in Manchester, Calvin accepted an appointment as instructor of chemistry at the University of California at Berkeley. Two important influences at Berkeley were Gilbert N. Lewis and G. E. K. Branch, fellow chemists who spurred Calvin's interests in the structure and behavior of organic molecules.

Calvin's first promotion at Berkeley—to assistant professor—came in 1941, only months before the United States' entry into World War II. Although he continued to teach during the war, Calvin became actively involved in the national war effort, first as an investigator for the National Defense Research Council, and later as a researcher in the Manhattan Project. His most important wartime contribution was the development of a process for obtaining pure **oxygen** directly from the atmosphere. Variations of that process now have a number of applications, as in machines that provide a continuous supply of oxygen for patients with breathing problems.

After the war, Calvin remained active in national and military organizations. He served as a member of the chemistry advisory committee of the Air Force Office of Scientific Research from 1951 to 1955 and as a delegate to the International Conference on Peaceful Uses of Atomic Energy in Geneva in 1955. In 1942, Calvin was married to Marie Genevieve Jemtegaard, a social worker whose parents were Norwegian immigrants. The Calvins had two daughters, Elin Bjorna and Karole Rowena, and one son, Noel Morgen.

Melvin Calvin.

Calvin was promoted to the position of associate professor in 1945 and to full professor in 1947. In the intervening years, he was also appointed director of the Bio-organic Chemistry Group at the University of California's Lawrence Radiation Laboratory. Calvin pointed out to Swift that this appointment was one of the very few administrative positions he had ever held because "You can't do both jobs [research and administration]." He only agreed to the Lawrence post, he said, in order to insure that he would have "an infrastructure on which I could do my job."

By 1948, Calvin had begun the research for which he is most famous, the elucidation of the process of photosynthesis. Scientists had known the general outlines of that process since the late eighteenth century, a process with which all beginning science students are familiar. In that process, **carbon dioxide** and **water** combine with each other in the presence of sunlight to form complex organic compounds known as **carbohydrates**. Scientists had also long known that photosynthesis is a far more complex process than is suggested by this simple summarizing statement. They knew that the conversion of carbon dioxide to carbohydrates involves many discrete chemical reactions, some of which were then vaguely known, but most of which were not.

Calvin's foray into the photosynthesis question was not without its problems. The only instruction in biology he had ever received came by way of a course in paleontology at Michigan Tech. Thus, when colleagues in the biology department at Berkeley learned that Calvin was about to take on one of the fundamental problems in biology, they could have been forgiven for some doubts about the successful conclusion of that work. Still, Calvin applied himself to mastering the study of biology for more than a decade, from about 1945 to the late 1950s. Eventually, he was able to convince biologists that he knew what he was talking about when he spoke to them about photosynthesis.

As with most scientific discoveries, unraveling the process of photosynthesis was possible only after the development of certain essential research tools and techniques. In this case, the most important of those tools and techniques were radioactive tracer isotopes and **chromatography**, the ability to separate the compounds within a **solution**. The radioactive tracer **isotope** that Calvin needed—carbon-14—had been available only since 1945. Carbon-14 is an extremely valuable research tool in biological research since, while it behaves in plants and animals in exactly the same way as non-radioactive carbon does, its emission of beta and **gamma radiation** make it continuously detectable to a researcher.

The design of Calvin's research on photosynthesis was elegantly simple. He maintained a water **suspension** of the green alga called chlorella in a thin **glass** flask that could be exposed to light. He then introduced to the flask, under controlled conditions, a certain amount of carbon dioxide consisting of carbon-14. As it carried out its normal life processes, the chlorella incorporated the radioactive carbon-14, converting it to carbohydrate. All Calvin had to in order to study the photosynthesis taking place was to stop the reaction at various points and analyze the compounds present in the chlorella.

The analysis required the use of the second new research tool, chromatography. In **paper** chromatography—one of many methods available—a mixture of compounds such as that obtained from the chlorella is allowed to diffuse along a strip of paper. Each compound diffuses at its own characteristic rate and can be identified by its position on the strip after a given period of time. The presence of a tracer isotope such as carbon-14 makes the process even simpler. By placing a photographic film in contact with the paper strip, the radioactive isotope "takes its own picture" as a result of the radiation it releases. The film offers a distinct record of the isotope's position on the paper strip.

Probably the greatest technical problem Calvin faced was deciding what compound was represented by each spot on the chromatogram. A decade after beginning his research, however, he had the answer he was seeking. The first set of reactions and compounds he proposed were not entirely accurate, but he re-worked the series of reactions until correct. That set of reactions is now known to all biochemists as the Calvin cycle. It was in recognition of his determination of the cycle of carbon in photosynthesis that Calvin was awarded the 1961 Nobel Prize in chemistry, as well as a number of other honors, including his 1959 election to the Royal Society and receipt of its prestigious Davy Medal five years later.

But receiving the Nobel Prize did not end Calvin's career in chemistry. Shortly after he traced the path of carbon through the photosynthetic process, Calvin did the same for oxygen, this time using a radioactive isotope of that element.

In 1960, Calvin assumed the directorship of the Laboratory of Chemical Biodynamics at Berkeley, a place where many new and exciting types of research were taking place, including studies on brain chemistry, **radiation chemistry**, solar energy conversion, and the origins of life on Earth. The last of these topics was one in which Calvin had been particularly interested for some time. During the 1950s, a vigorous debate had been going on among scientists as to whether the earth's primitive atmosphere had consisted exclusively of reducing **gases**, such as **hydrogen**, **methane**, and **ammonia**, or whether it was an oxidizing atmosphere that also included gases such as carbon dioxide. During this time, Calvin carried out a series of experiments in which a hypothetical primitive atmosphere consisting of hydrogen, carbon dioxide, and water was exposed to intense radiation provided by Berkeley's 60-inch cyclotron. The experiment resulted in the formation of a number of simple organic molecules, such as **formaldehyde**, **formic acid**, and glycolic acid. When a similar experiment was later repeated by Calvin's student **Cyril Ponnamperuma**, with **nitrogen** included this time, simple amino acids—the building blocks of life—were also found among the products.

Calvin's interest in the origins of life on Earth led him in another direction also: the possibility of life elsewhere in the universe. His own feeling has been that the conditions that led to the formation of life on Earth could hardly have been unique in the universe. Instead, he has argued, "we can assert with some degree of scientific confidence that cellular life as we know it on the surface of the Earth does exist in some millions of other sites in the universe."

During the 1970s, Calvin began research on yet another somewhat unusual application of chemistry, the development of alternative fuels. He discovered that certain members of the rubber tree family produce a sap-like material that can be burned in much the same way as petroleum. He suggested the possibility that such trees could be grown on huge plantations in order to provide an alternative source of energy as our supply of crude oil continues to diminish. With his wife, Calvin eventually established an experimental farm in Northern California to test out this idea.

As Calvin's academic career came to a close, he continued to receive the recognition of his peers in the field of science. In 1978, he was awarded the Priestley Medal of the American Chemical Society (ACS) and the Gold Medal of the American Institute of Chemists. In 1981, Calvin was awarded the Oesper Prize by the ACS. To balance a public life dedicated to science, Calvin maintained a number of personal interests, including photography, gardening, politics, and sports. Although he retired as University Professor of Chemistry in 1980, he continued his scientific work. Calvin died on January 8, 1997 at Alta Bates Hospital in Berkeley, California.

CANNABINOIDS

The leaves and flowering parts of the (*cannabis indica*) and (*cannabis sativa*) plants, commonly known as marijuana, have been used for thousands of years for a variety of purposes ranging from recreational drug to medicinal analgesic to religious sacrament. The term cannabinoid refers to chemicals found exclusively in these cannabis plants. Although there are 70 known cannabinoids found in marijuana, the chemical delta-9-tetrahydrocannabinol (THC) has been identified as the principal psychoactive and pharmacologically significant component.

There is a rich history concerning the medicinal use of marijuana and THC as pain killers. Archeological findings at an ancient site in Jerusalem suggest that marijuana was consumed to alleviate labor pains 1,600 years ago. Upon its introduction to the United Kingdom and United States in the 1800s, marijuana was reportedly used to treat a number of ailments including tetanus, migraine, depression, gonorrhea, and opium addiction with varying results. It is rumored that marijuana was used to relieve Queen Victoria's menstrual cramps. Its medicinal use in the United States fell out of favor during the prohibition of the 1920s when it was classified as a **narcotic**. Its status as an illegal substance remains to this day. Little research was done on the use of cannabinoids as **analgesics** for most of the twentieth century as most animal studies proved to be inconclusive. Scientists largely attributed the pain relieving properties of marijuana to its psychoactive effects such as euphoria, changes in attention, movement deficits, and cognitive impairment. Any medicinal value of cannabinoids lingered in anecdotal obscurity until the 1980s.

Research done in the 1980s and early 1990s led to the discovery of cannabinoid **receptors** in the brain where THC is known to bind and produce its psychoactive effects. This type of receptor is known as the CB1 receptor. The known locations of the CB1 receptors are in the basal ganglia, which controls unconscious muscle movements; the limbic system, which is associated with memory integration and strong emotions such as rage and fear; and in the cerebellum, which is responsible for balance and the planning of motion. Since its discovery, many chemicals not found in the cannabis plants have been identified as being able to bind to the CB1 receptor. These chemicals are also termed cannabinoids.

The discovery of a cannabinoid receptor suggested that there must be a cannabinoid that occurs naturally in the brain. In 1992, researchers identified arachidonyl ethanolamine amide, or anandamide (from the Hindu word ananda meaning eternal bliss) as a chemical in the brain that is a naturally occurring cannabinoid or "endogenous ligand of CB1." Anandamide is thought to play a crucial role in the inhibition of movement and may block the neurotransmitter dopamine. This has led to the study of cannabinoids as potential therapies in the treatment of diseases such as schizophrenia and Tourette's syndrome associated with an overabundance of or hypersensitivity to dopamine. In a similar vein, research directed towards the inhibition of the normal function of anandamide potentially provides therapeutic routes for diseases associated with dopamine deficiency such as drug addiction and Parkinson's disease. Interestingly, anandamide, along with two other chemicals thought to prevent its destruction, have been found in chocolate. It has been postulated that the presence of cannabinoids accounts for the unique cravings associated with chocolate. Subsequently, other endogenous CB1 ligands have been identified such as the chemical sn-2 arachidonylglycercerol, or 2-AG, which researchers believe is involved in the memory process.

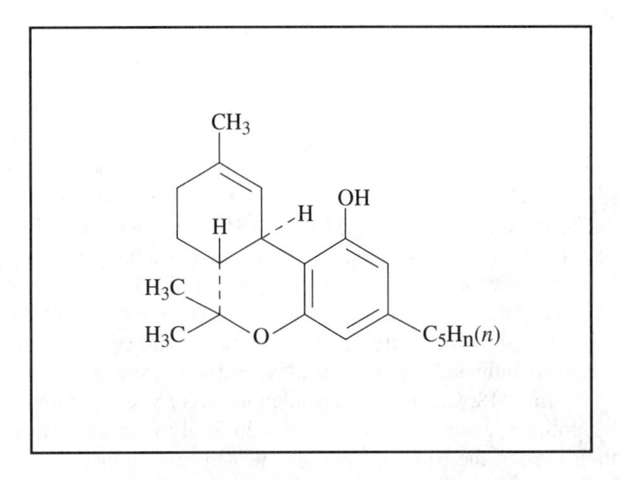

Chemical structure of δ-9-tetrahydrocannabinol. ***(Illustration by Electronic Illustrators Group.)***

In 1993, a second type of cannabinoid receptor, known as CB2, was discovered. While CB1 is located exclusively in the brain, CB2 is found principally on cells of the immune system. This has stimulated pharmaceutical research towards developing synthetic cannabinoids that could act at CB2 as anti-inflammatory agents or immunosupressants. Furthermore, current research is focused upon developing cannabinoids that act upon only one of the two receptors, enabling researchers to focus on either neurological or immunological effects of cannabinoids.

Current research is also directed at exploiting the previously mentioned analgesic effects of cannabinoids. They have been shown to be particularly useful pain relievers for patients suffering from cancer, AIDS, and multiple sclerosis. Known to increase appetite, cannabinoids are especially attractive pain relievers for AIDS patients who usually suffer weight loss exacerbated by traditional analgesics such as **morphine** that cause nausea. Non-analgesic applications of cannabinoids include their use as anticonvulsants, muscle relaxants, vasodilators for bronchial asthma, and to decrease the intraocular pressure associated with glaucoma. Much of the research in the creation of THC derivatives and other synthetic cannabinoids is directed towards drugs that are selective in their ability to enhance the positive therapeutic effects of marijuana while eliminating the negative side effects such as short-term memory loss and paranoia.

CANNIZZARO, STANISLAO (1826-1910)

Italian chemist

From time to time in science, someone comes along whose original contributions are less significant than his or her ability to explain another's ideas. Such a person was Stanislao Cannizzaro. Cannizzaro was born in Palermo, Italy, on July 13, 1826, and died in Rome on May 10, 1910. He entered the University of Palermo as a medical student in 1841, but soon realized that the university could not provide the instruction in **chemistry** that he wanted and needed to learn. As a result, he moved to the University of Pisa, where he studied under the foremost Italian chemist of the day, Raffaele Piria.

In 1848, Cannizzaro's academic career was interrupted by political events. Revolutionaries in Sicily rose up against their master, the King of Naples and Cannizzaro joined the rebellion as an artillery officer and a member of the revolutionary government. When the rebellion failed in April 1849, he fled to France. Eventually, he took a position in the laboratory of **Michel-Eugène Chevreul** in Paris, France.

In 1851, Cannizzaro returned to Italy where he held successive posts at the Collegio Nazionale in Alessandria, the University of Genoa (1855), the University of Palermo (1861), and the University of Rome (1871). Throughout his academic career, Cannizzaro remained politically active. He took part in Garibaldi's successful rebellion to liberate Sicily in 1860 and, in Rome, became a senator of the new nation of Italy.

Cannizzaro devoted his research efforts entirely to the study of organic compounds, especially **natural products**. The discovery for which he is probably best known is a method for converting benzaldehyde into benzyl **alcohol** and benzoic acid. Students of **organic chemistry** today still know this reaction as *Cannizzaro's reaction.*

By far his greatest contribution to chemistry did not involve his own original research. Chemistry in the mid-1800s was in a terrible state of disarray. **John Dalton**'s **atomic theory** had seemed to provide a promising theoretical basis for the science and had been widely and quickly adopted by most chemists. But problems of interpreting and applying Dalton's ideas began to appear almost immediately.

Chemists found it difficult to agree on exactly what the term **atom** really meant, how an atom was different from what Dalton called a *compound atom* and what we now know as a **molecule**, what were the true atomic and molecular weights for known elements and compounds, how to accurately represent the formulas of compounds, and so on. As a symptom of this confusion, **Friedrich Kekulé** was, at one point, able to list 19 different formulas for **acetic acid**.

The key concept needed to resolve much of this confusion had actually been proposed as early as 1811. In that year, **Amedeo Avogadro** had proposed and explained the concept of a molecule and had outlined his hypothesis that equal volumes of **gases** contained equal numbers of molecules. The problem was that Avogadro's ideas were not well known or well understood among chemists throughout Europe. As a result, his ideas disappeared into oblivion for over four decades.

Then, in 1858, Cannizzaro rediscovered his countryman's hypothesis and immediately saw the key it offered to resolving some current controversies in chemistry. He wrote a letter to a friend at the University of Pisa, Sebastiano de Luca, outlining Avogadro's ideas and showing their relevance to current debates in chemistry. The letter was later published in the Italian journal, *Nuovo cimento* (*New Test*) and reprinted as a pamphlet.

The main impact of Cannizzaro's ideas occurred at the First International Chemical Congress, held at Karlsruhe, Germany, in 1860. The Congress was organized at the request of a number of young chemists who were eager to bring some logic and organization to their profession. Cannizzaro read his pamphlet at the Congress and handed out copies to the delegates. He reiterated Avogadro's hypothesis, explained how atoms and molecules were different from each other, described a new method for determining atomic and molecular weights, and showed that the same chemical principles apply to both inorganic and organic chemistry.

Not all chemists understood or adopted Cannizzaro's presentation immediately. But, gradually they were convinced by the logic of his arguments. **Julius Lothar Meyer**, for example, claimed that "the scales fell from his eyes" when he read Cannizzaro's pamphlet. In the end, the pamphlet was probably the single most important factor in forging a new consensus in chemistry during the second half of the nineteenth century.

CARBANIONS

Carbanions are **carbon** anions, having the general formula R_3C^-. This negatively charged **ion** is highly reactive. Carbanions can be produced by the cleavage of certain single bonds such as those found between carbon and **hydrogen**, carbon and **halogens**, carbon and **metals**, and carbon and carbon. It is quite difficult to recognize the presence of these carbanions directly, they have to be inferred from their actions.

With rare exception carbanions are not isolated, but are intermediate in reactions. These reactions are both numerous and important. They include the biological biosynthesis of **fatty acids**, where the important event in the elongation of the acetate starting material is cyclation of the acetate carbanion.

Carbanions are commonly prepared by either deprotonation (cleavage of a carbon-**hydrogen bond**, invariably adjacent to a **functional group** that is capable of **resonance** or inductive stabilization of the negative charge on the carbon) or metalation of a carbon-halogen bond. In the process of metalation a metal (usually an alkali metal such as Li, Na, or K) inserts into the carbon-halogen bond (for example: $2Li\bullet + R_3C\text{–}Br \rightarrow R_3C^-Li^+ + LiBr$).

Two important (and uncommon) examples of stable carbanions are the cyanide **anion** (NC^-) and the metalocenes. The metalocenes, an extremely important class of catalysts for modern polymer synthesis, are made from the reaction of two cyclopentadienylide anions (obtained by deprotonation of cyclopentadiene by an organometallic reagent, such as a Grignard reagent) with a metal. The result is the metalocene **salt**, where the cationic metal is located between the two cyclopentadienylide rings in what is commonly called a "sandwich" arrangement.

Carbanions are negatively charged carbon atoms of extraordinary importance to the synthesis of both biologically and man-made chemicals.

See also Carbocations

CARBENE

A carbene has a general formula of R_2C:. Carbenes are one of the principle groups of reactive organic intermediaries and they are highly reactive due to the two unpaired electrons. Photolysis of a diazoalkane or a simple **elimination reaction** will produce a carbene. Dihalocarbenes such as Cl_2C: have the greatest synthetic value as they frequently act as intermediates in organic syntheses.

Carbenes can act as **electron** donors in chemical reactions. They are intermediates in such reactions as the change of a carboxylic acid to its next highest homologue or to a derivative of a homologous acid such as an ester or an amide. (A homologous series is a group of organic compounds that have the same **functional group** and a regular structural pattern so that a member of the series differs from the next member by a fixed number of atoms.) The reaction of a base with a trihalogen derivative of **methane** is an example of a commonly encountered reaction which proceeds via a carbene intermediate.

Carbenes are so reactive that they are unlikely to be frequently encountered. In any reaction of which they are a part, they rapidly move to the next stage of the synthesis. It is only in reactions with transition **metals** that carbenes are stable, due to the distribution of electrons within the structure formed.

CARBOCATIONS

Carbocations (also called carbenium ions or carbonium ions) are positively charged ions of the general formula R_3C^+. Like the negatively charged **carbanions**, they are found as intermediates in a large number of reactions, for example the dissolution of alkenes in strong acids and the dehydrohalogenation of alkyl halides.

Carbocations exhibit a broad range of stability and **reactivity**. As intermediate compounds they are unstable and rapidly converted to the product of the reaction. These intermediates can only be detected by such techniques as **nuclear magnetic resonance**. There are several stable compounds that can be formed such as triphenylmethylcarbenium and 1,3,5-cycloheptatriene carbenium (tropylium carbenium). The stability of carbocations increases depending upon whether they have one, two, or three R groups attached to the **carbon atom**.

Alkyl carbocations can be prepared by reacting the alkyl fluoride and **antimony** fluoride at low **temperature** in the presence of **sulfur** dioxide.

All carbocations are strong alkylating agents with a high affinity for groups which donate or share their electrons (nucleophiles), such as **halide** ions, hydroxyl anions and amines.

Carbocations are a group of strongly reacting molecules which are rarely encountered in their native form due to their reactivity.

See also Carbanions

CARBOHYDRATES

Carbohydrates are one of the most widely occurring types of organic compounds. The carbohydrates include sugars, starches, and celluloses. They all contain **carbon**, **hydrogen**, and **oxygen** following a general formula of $C_x(H_2O)_y$, the ratio of hydrogen to oxygen is always 2:1. It must be stressed that while all carbohydrates fit this formula not all organic compounds which fit this formula are carbohydrates, for example ethanoic acid $C_2H_4O_2$, is classified as an acid, not a carbohydrate. On **combustion** carbohydrates produce **carbon dioxide**, **water**, and **energy**. Plants manufacture simple carbohydrates from carbon dioxide and water by **photosynthesis**. It is estimated that some 2×10^{11} ton of carbohydrate are manufactured by plants each year. The most abundant form of carbohydrate is glucose, a simple sugar with the chemical formula $C_6H_{12}O_6$. The relationship between the carbon atoms and the groups of atoms attached to them control the nature of the sugar and its properties. One form of glucose is alpha glucose. If the H and OH groups are interchanged at the first carbon **atom** position, beta glucose is produced. Strictly speaking carbohydrates are not hydrated carbon, in fact they are polyhydroxy **aldehydes** and **ketones**. Glucose is a six member **ring** aldehyde sugar. Fructose is the six member ketone form (fructose is twice as sweet as sucrose and occurs commonly in fruit). Glucose and fructose are isomers of each other, that is, they have the same **molecular formula** but different structures. Glucose can react with itself to produce a six membered ring structure with the reaction occurring between the aldehyde and **alcohol** functional groups. A similar reaction can occur with fructose but it can produce either a six or five membered ring. The latter form is the one in which it is found in **natural products**, but when it crystallizes it does so in the six membered form.

Both glucose and fructose are monosaccharides, sugars that cannot be broken into smaller subunits by **hydrolysis** with a weak acid. Monosaccharides are all white, crystalline **solids** that are soluble in water and have a sweet **taste**.

A **condensation reaction** can link together two monosaccharide sugars to form a disaccharide with the elimination of water. For example a glucose **molecule** and a fructose molecule together form the disaccharide sugar sucrose (common table sugar).

Disaccharides can be hydrolyzed to form their constituent monosaccharides. When sucrose is hydrolyzed the mixture of glucose and fructose is called invert sugar. This is actually sweeter than the original sucrose. It is the thick, sweet syrup often found in cans of fruit. Sucrose is found in many plants and is refined from sugar cane and sugar beet for commercial usage. Sucrose can be dehydrated by heating to give a brown caramel, or it can be dehydrated by the action of concentrated **sulfuric acid**. This reaction yields a porous **mass** of carbon.

Chemical structure of glucose. *(Illustration by Electronic Illustrators Group.)*

Chemical structure of alpha glucose. *(Illustration by Electronic Illustrators Group.)*

Chemical structure of beta glucose. *(Illustration by Electronic Illustrators Group.)*

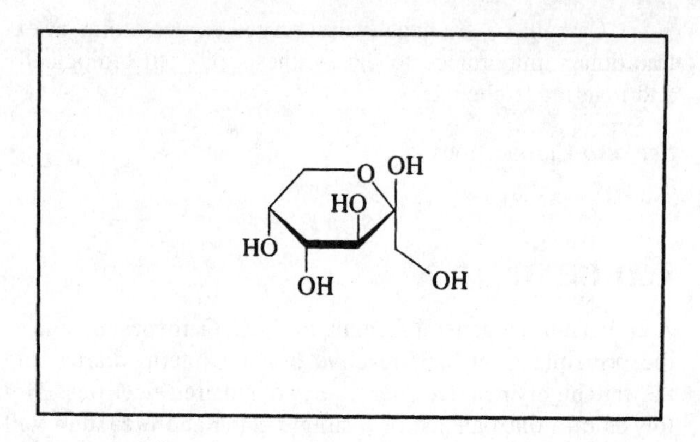

Chemical structure of fructose. *(Illustration by Electronic Illustrators Group.)*

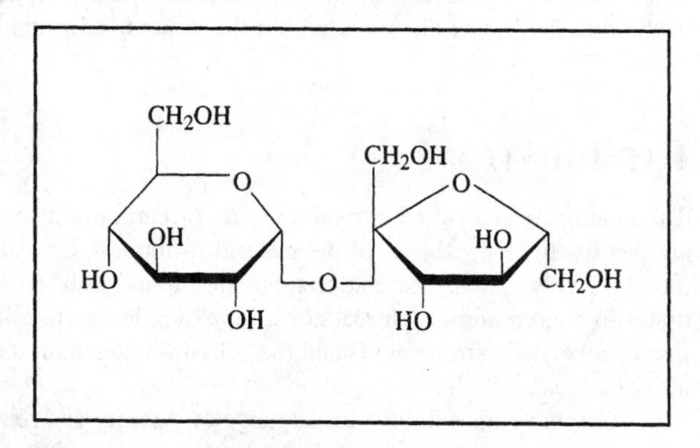

Chemical structure of table sugar (saccharose) and sucrose. *(Illustration by Electronic Illustrators Group.)*

All mono- and disaccharides are commonly referred to as sugars. Some sugars are known as reducing sugars (for example glucose and fructose) and they can be recognized by their ability to reduce hot Fehling's or Benedict's **solution**, producing a brick red precipitate of **copper** (I) **oxide**.

Polysaccharides are produced when several monosaccharides are joined together by repeated condensation reactions. Polysaccharides are generally tasteless and do not dissolve in water. The most important polysaccharides are starch, **cellulose** and **glycogen**.

Starch is a storage product for plants and it is the second most abundant plant product (after cellulose), it is made of repeating units of alpha glucose with occasional branching. Starch is found in seeds and tubers of plants. Rice and potatoes

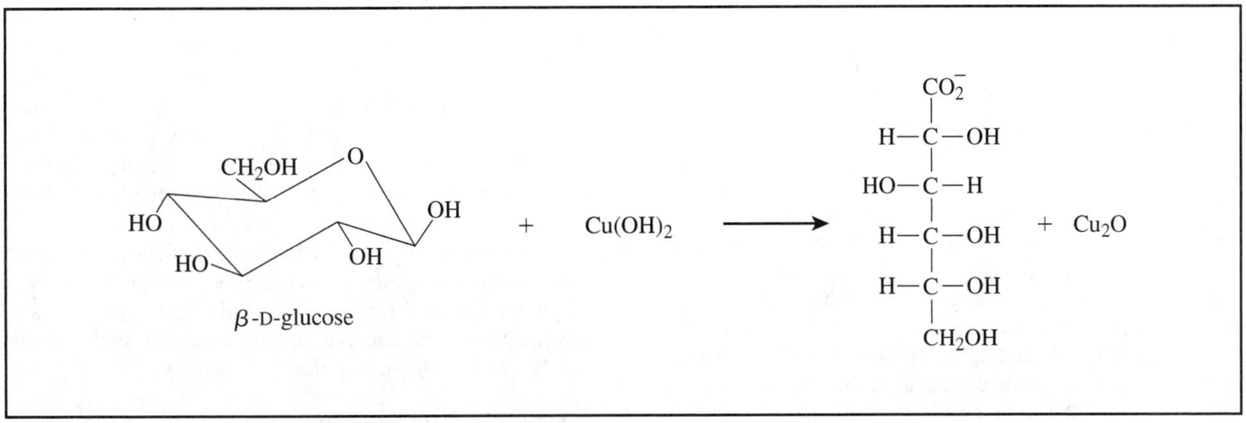

Chemical structure of a reducing sugar. *(Illustration by Electronic Illustrators Group.)*

are all major sources of starch. Storage products such as this are one of the main food supplies for animals, enzymes promote digestion of starch molecule to the disaccharide maltose and other enzymes hydrolyze this to glucose that is eventually oxidized to release energy. Hydrolysis of starch with dilute mineral acids directly yields glucose. Starch can also be used in **fermentation** to produce alcohol. Starch can be recognized by the fact that it produces a dark blue black coloration when it is mixed with **iodine** solution. Starch does not melt, but instead decomposes on heating.

Cellulose, a long chain molecule of beta glucose molecules, is the main structural molecule of plants and it is the most abundant plant product. Cellulose is a long chain molecule of beta glucose molecules. The structure of starch is such that the surface of the molecule is essentially covered by OH groups, this allows for the production of hydrogen bonds between neighboring molecules. This cross **bonding** produces a three dimensional structure which is very strong. Starch does not do this because most of the OH groups are directed toward the inside of the molecule and are not available for hydrogen bonding. Cotton fibers are almost pure cellulose and **wood** is nearly 50% cellulose. Cellulose is made entirely of glucose molecules, up to several hundred in a chain. Because the three dimensional structure of cellulose is different from that of starch, enzymes which are capable of breaking down starch will not necessarily have the same effect on cellulose. Cellulose is not digested by any enzymes found naturally occurring in the human body. A range of species of bacteria can break down cellulose. These bacteria are found in the digestive systems of animals that eat grass, such as cattle and sheep. The cellulose is digested by the bacteria and they supply the animals with short chain **fatty acids**.

Cellulose has no melting point and it decomposes on strong heating. It is insoluble in water and it has a relative chemical inertness. Acid hydrolysis will convert cellulose completely to glucose. **Methane** and carbon dioxide are released when cellulose is enzymatically broken down. Cellulose is used in the production of **paper**, textiles, clothing materials, various **plastics** and **explosives**.

Artificial textiles can be made by dissolving cellulose in a suitable solvent and then squirting the solution through a narrow nozzle into a drying chamber. Rayon can be manufactured in this manner and this is a substitute for silk (silk is actually a protein fiber). Cellulose reacts with nitric and sulfuric acids to produce cellulose nitrate. This material is used in the manufacture of plastics, and is inflammable. Cellulose trinitrate on the other hand is an explosive (guncotton). **Cordite** is made from guncotton and nitroglycerine.

Glycogen is a starch like compound found in the human body. Glycogen is also made of long **chains** of branching glucose molecules although there is more branching than in starch. Glycogen can be ten times the length of cellulose. Like starch, glycogen can be hydrolyzed to glucose. It is used as an energy reserve in the body. Glycogen is concentrated in the liver and muscles. In the liver it serves as storage for glucose and is used to regulate the blood sugar level. In the muscles it can be rapidly broken down to provide a quick burst of energy to allow the muscles to respond quickly. Glycogen is sometimes referred to as animal starch.

Carbohydrates can be converted between the three principal forms relatively easily. Interconversion in the body is carried out by the action of specific enzymes. In the laboratory various chemical treatments are required, such as treatment with hot acid.

Carbohydrates can form more complex molecules when linked with other molecules. Sugar and **phosphoric acid** form the backbone chains of nucleotides in **DNA**. Mucopolysaccharides are sugars containing **nitrogen**, they occur most commonly in chitin which is the compound in the cell walls of fungi and in the exoskeleton of arthropods. A polymer of various sugars and **amino acids** combine to give a structure called lignin. Lignin acts as a rigid support in plants.

Carbohydrates are an important group of organic compounds. They provide food and raw materials for animals, after they have been made by plants from carbon dioxide and water.

See also Lipids; Nucleic acids; Proteins

CARBON

Carbon is the first element in Group 14 of the **periodic table**. The elements in this group are sometimes called the carbon family. Carbon has an **atomic number** of 6, an atomic **mass** of 12.01115, and a chemical symbol of C.

Properties

Carbon exists in a number of allotropic forms. Two that have crystalline structure are **diamond** and **graphite**. These two allotropes have very different physical properties. Diamond is the hardest known naturally-occurring substance, while graphite is one of the softest. The melting point of diamond is about 6,700°F (3,700°C), its **boiling point** about 7,600°F (4,200°C), and its **density** 3.50 grams per cubic centimeter. By contrast, graphite does not melt when heated, but sublimes at a **temperature** of about 6,600°F (3,650°C). Its density ranges from about 1.5-1.8 grams per cubic centimeter. The exact value depends on the source from which the graphite was obtained.

Among the non-crystalline allotropes of carbon are **coal**, lampblack, **charcoal**, carbon black, and coke. Like other amorphous materials, the non-crystalline allotropes of carbon do not have clear-cut melting and boiling points. Their densities also vary depending on the sources from which they come.

Elemental carbon is a relatively inert element, although it does combine with **oxygen** at high temperatures to form both **carbon dioxide** (CO_2) and **carbon monoxide** (CO). One of carbon's most striking chemical properties is its ability to form long **chains**, rings, and other structures containing dozens, hundreds, or even thousands of carbon atoms. Carbon is unique among the elements in its ability to form macromolecules of this kind. These macromolecules form the basis of nearly all chemical compounds found in living organisms as well as a very large number of synthetic products.

In 1985, an interesting new **allotrope** of carbon was discovered. The molecules that make up this allotrope consist of 60 atoms joined together in a soccer-ball-like structure. Each face of the "soccer ball" consists of either a six- or five-carbon **ring**. The allotrope was given the name fullerene in honor of the American architect Buckminster Fuller (1895-1983). Fuller had become famous in part because of a number of structures he designed using a pattern similar to the one that occurs in fullerene. The architectural structure is known as a geodesic dome. Molecules of fullerene are more commonly referred to as "buckyballs."

Occurrence and Extraction

Carbon is the sixth most common element in the universe and the fourth most common element in the solar system. It is the 17th most common element in the Earth's crust and the second most common element in the human body. About 18% of a person's body weight is due to carbon.

Carbon occurs in both native and combined form in the Earth's crust. The most common forms of elemental carbon are coal and diamonds. Both substances are formed in the Earth's crust over millions of years as plants and animals die, decay, and decompose. The first stage of this decomposition results in the formation of coal, while increased pressure can eventually convert coal to diamond.

Carbon also occurs in a number of minerals. Among the most common of these minerals are the carbonates of **calcium** ($CaCO_3$) and **magnesium** ($MgCO_3$). Carbon also occurs in the form of carbon dioxide in the Earth's atmosphere. It makes up only a small fraction of the atmosphere (about 400 parts per million), but it is crucial to all life processes that occur on the Earth. Plants use carbon dioxide in the atmosphere in the process of **photosynthesis**, by which carbon dioxide and **water** are converted into starches and sugars.

Diamonds can now be made synthetically in large quantities. These diamonds are usually not very large, but they are in great demand for a variety of industrial applications. One advantage of synthetic over natural diamonds is that the latter can be made without any flaws in them.

Discovery and Naming

Humans have known about various forms of carbon for thousands of years. When cave people made a fire, they observed the formation of soot (finely divided carbon particles), although they had, of course, no detailed understanding of the **chemistry** involved. The production of charcoal by heating **wood** in the absence of air and of coke by heating coal in the absence of air were both well known to ancient peoples. Lampblack was also mixed with olive oil or balsam gum to make ink by early peoples, and Egyptians are known to have used lampblack as eyeliner.

The first recognition that carbon might be a chemical element is often attributed to the French physicist René Antoine Ferchault Réaumur (1683-1757), who attributed the difference among wrought **iron**, cast iron, and steel to the presence of some "black combustible material" that he knew was present in charcoal. Carbon was finally classified officially as an element near the end of the eighteenth century. The name chosen for the element was based on an earlier Latin name for charcoal, *chabon*.

Uses

Diamond and graphite both have a variety of important commercial and industrial uses. Most people may think of jewelry first when thinking of diamonds, although their industrial applications are of at least as much importance. Because they are so hard, diamonds are used to polish, grind, and cut **glass**, **metals**, and other materials. As an example, the bit on an oil-drilling machine may be made of diamonds.

Perhaps the best known use of graphite is in lead pencils. The material is used because it is so soft that it comes off easily when rubbed on a piece of **paper**. Graphite is also widely used as a **lubricant** because its molecules slide back and forth across each other so smoothly. Graphite is also used as a refractory in high-temperature furnaces, to make black paint, in **explosives** and matches, and in certain kinds of cathode ray tubes.

Amorphous forms of carbon also have many uses. For example, they are used to provide the black **color** in **inks**, pigments, rubber tires, stove polish, typewriter ribbons, and phonograph records.

One commercially important form of carbon is activated charcoal. The term *activated* means that the charcoal has been ground to a very fine powder. In this form, the charcoal can adsorb impurities from **liquids** that pass over it, and it is commonly used to remove color and odor from oils and water solutions, such as the water being purified in a water treatment plant.

Carbon dioxide is used to make carbonated beverages, in fire extinguishers, and as a propellant in aerosol products. In the solid form known as dry ice, it is widely used as a refrigerant. Carbon monoxide has somewhat fewer uses than carbon dioxide, the most important being in the extraction of pure metals from their oxides. When heated together, carbon monoxide and the **oxide** react to produce the pure metal: $3CO + Fe_2O_3 \rightarrow 3CO_2 + 2Fe$.

The vast majority of carbon compounds are classified as organic compounds. At one time, the term organic was used for any compound found in a living organism. Nearly all such compounds do contain carbon. However, over time, the meaning of the word organic evolved to include all carbon-containing compounds, whether or not they originated in a living organism. The few exceptions to this definition are the oxides of carbon (CO_2 and CO), the carbonates, the cyanides (compounds of carbon, **nitrogen**, and at least one more element), and a few other families of compounds. Well over ten million organic compounds have now been identified. The number increases by the tens of thousands every year. Included among these compounds are some of the best known of all chemical families, including the hydrocarbons, alcohols, ethers, **aldehydes** and **ketones**, and organic acids.

It would take a book many times the size of this one to discuss even a fraction of the organic compounds for which important uses have been found. As an example, hydrocarbons, compounds that contain carbon and **hydrogen** only, are widely used as fuels. Some familiar examples include **methane**, ethane, propane, and **kerosene**. Methyl **alcohol** (wood alcohol) and ethyl alcohol (grain alcohol) are the most familiar members of the alcohol family, compounds that contain carbon and one or more hydroxyl (-OH) groups. Methyl alcohol is used as a raw material in the manufacture of many organic compounds and as a solvent. Ethyl alcohol is used for the same purposes as well as being an important component of alcoholic drinks such as beer, wine, and hard liquor.

The list of everyday products made from organic compounds is very long and includes drugs, artificial fibers, dyes, artificial colors and flavors, **food additives**, **cosmetics**, **plastics** of all kinds, detergents, synthetic rubber, adhesives, antifreeze, **pesticides** and **herbicides**, synthetic fuels, and **refrigerants**. In addition, organic compounds make up nearly all of the important chemical families found in living bodies, such as **carbohydrates**, **proteins**, fats, oils, and **nucleic acids**.

CARBON CYCLE

The series of chemical, physical, geological, and biological changes by which **carbon** moves through the Earth's air, land, **water**, and living organisms is called the carbon cycle.

Carbon makes up no more than 0.27% of the **mass** of all elements in the universe and only 0.0018% by weight of the elements in the Earth's crust. Carbon occurs in many different chemical combinations, including **calcium** carbonate ($CaCO_3$), **carbon dioxide** (CO_2), **methane** (CH_4), and a huge diversity of organic compounds (including hydrocarbons and biochemicals). In contrast to carbon's relative scarcity in the environment, it makes up 19.4% by weight of the human body. Along with **hydrogen**, carbon is the only element to appear in every organic **molecule** in every living organism on Earth.

The most abundant mineral forms of carbon in the rocks and soil of the Earth's crust are limestone ($CaCO_3$) and dolomite [$CaMg(CO_3)_2$]. These mostly occur in sedimentary rocks, which were formed in ancient marine environments through biological influences that resulted in the **precipitation** of limestone and dolomite from ions of calcium (Ca^{2+}), **magnesium** (Mg^{2+}), and bicarbonate (HCO_3^-) dissolved in water. The amount of carbon stored in sedimentary rocks has not yet been accurately estimated, but is thought to be much larger than that occurring in any other compartment of the carbon cycle.

Carbon also occurs in spaces within sedimentary crustal rocks in the form of hydrocarbons (i.e., compounds only containing carbon and hydrogen), such as **coal**, petroleum, and **natural gas** (collectively, these are known as fossil fuels). These hydrocarbons are extremely important, but non-renewable, natural resources used as sources of **energy** and for the manufacturing of **plastics** and other materials. Wherever mining and drilling technology can access fossil fuel deposits, these deposits are being rapidly used up. Other organic compounds of carbon also occur within the rocks of the Earth's crust (these may contain additional elements, such as **oxygen**, **sulfur**, and nitrogen), although in much smaller amounts than the hydrocarbons. All of these various kinds of organic carbon are derived from the partially decomposed biomass of ancient plants and other organisms, which became buried deep beneath marine sediment and were transformed very slowly (under intense pressure and **heat** in the absence of oxygen) into their present forms.

In the atmosphere, carbon exists almost entirely as gaseous carbon dioxide (CO_2). Its global **concentration** is about 360 parts per million (ppm), or 0.036% by **volume**. This makes carbon dioxide the fourth most abundant gas in the atmosphere after **nitrogen**, oxygen, and **argon**. Some carbon is also released as methane (CH_4) and **carbon monoxide** (CO) to the atmosphere by natural and human mechanisms. Carbon monoxide reacts readily with oxygen in the atmosphere, however, converting it to carbon dioxide.

Carbon returns to the hydrosphere when carbon dioxide dissolves in the oceans, as well as in lakes and other bodies of water. The **solubility** of carbon dioxide in water is not especially high, 88 milliliters of gas in 100 milliliters of water. Still, the Earth's oceans are such a vast reservoir that experts estimate that approximately 36,000 billion tons of carbon are stored there. They also estimate that about 93 billion tons of carbon flow from the atmosphere into the hydrosphere each year.

Carbon moves out of the oceans in two ways. Some escapes as carbon dioxide from water solutions and returns to the atmosphere. That amount is estimated to be very nearly equal

(90 billion tons) to the amount entering the oceans each year. A smaller quantity of carbon dioxide (about 40 billion tons) is incorporated into aquatic plants.

On land, green plants remove carbon dioxide from the air through the process of photosynthesis—a complex series of chemical reactions in which carbon dioxide is eventually converted to starch, **cellulose**, and other **carbohydrates**. About 100 billion tons of carbon are transferred to green plants each year, and a total of 560 billion tons of the element is thought to be stored in land plants alone.

The carbon in green plants is eventually converted into a large variety of organic (carbon-containing) compounds. When animals eat green plants, they use the carbohydrates and other organic compounds as raw materials for the manufacture of thousands of new organic substances. The total collection of complex organic compounds stored in all kinds of living organisms represents the reservoir of carbon in the Earth's biosphere.

The cycling of carbon through the biosphere involves three major kinds of organisms. Producers are organisms with the ability to manufacture organic compounds such as sugars and starches from inorganic raw materials such as carbon dioxide and water. Green plants are the primary example of producing organisms. Consumers are organisms that obtain their carbon (i.e., their food) from producers. All animals are consumers. Finally, decomposers are organisms such as bacteria and fungi that feed on the remains of dead plants and animals. They convert carbon compounds in these organisms to carbon dioxide and other products. The carbon dioxide is then returned to the atmosphere to continue its path through the carbon cycle.

Land plants return carbon dioxide to the atmosphere during the process of respiration. In addition, animals that eat green plants exhale carbon dioxide, contributing to the 50 billion tons of carbon released to the atmosphere by all forms of living organisms each year. Respiration and decomposition both represent, in the most general sense, a reverse of the process of **photosynthesis**. Complex organic compounds are oxidized with the release of carbon dioxide and water—the raw materials from which they were originally produced.

At some point, land and aquatic plants and animals die and decompose. When they do so, some carbon (about 50 billion tons) returns to the atmosphere as carbon dioxide. The rest remains buried in the Earth (up to 1,500 billion tons) or on the ocean bottoms (about 3,000 billion tons). Several hundred million years ago, conditions of burial were such that organisms decayed to form products consisting almost entirely of carbon and hydrocarbons. Those materials exist today as pockets of the fossil fuels. Estimates of the carbon stored in fossil fuels range from 5,000 to 10,000 billion tons.

The processes that make up the carbon cycle have been occurring for millions of years, and for most of this time, the systems involved have been in equilibrium. The total amount of carbon dioxide entering the atmosphere from all sources has been approximately equal to the total amount dissolved in the oceans and removed by photosynthesis. However, a hundred years ago changes in human society began to unbalance the carbon cycle. The Industrial Revolution initiated an era in which the burning of fossil fuels became widespread. In a short period of time, large amounts of carbon previously stored in the Earth as coal, oil, and natural gas were burned, releasing vast quantities of carbon dioxide into the atmosphere.

Between 1850 and 1998, measured concentrations of carbon dioxide in the atmosphere increased from about 280 ppm to about 360 ppm, an increase of 29%. Scientists estimate that fossil fuel **combustion** now releases about five billion tons of carbon dioxide into the atmosphere each year. In an equilibrium situation, that additional five billion tons would be absorbed by the oceans or used by green plants in photosynthesis. Yet this appears not to be happening. Measurements indicate that about 60% of the carbon dioxide generated by fossil fuel combustion remains in the atmosphere.

The problem is made even more complex because of deforestation. As large tracts of forest are cut down and burned, carbon dioxide from forest fires is added to that from other sources, and the loss of trees decreases the worldwide rate of photosynthesis. Overall, it appears that these two factors have resulted in an additional one to two billion tons of carbon dioxide in the atmosphere each year.

No one can be certain about the environmental effects of this disruption of equilibria in the carbon cycle. Some scientists believe that the additional carbon dioxide will augment the Earth's natural **greenhouse effect**, resulting in long-term global warming and climate change. Others argue that we still do not know enough about the way oceans, clouds, and other factors affect climate to allow such predictions.

CARBON DATING

Carbon dating is a technique used to determine the approximate age of once-living materials. It is based on the decay rate of the radioactive carbon **isotope** C-14, a form of carbon taken in by all living organisms while they are alive.

Before the twentieth century, determining the age of ancient artifacts was considered the job of archaeologists, not nuclear physicists. By comparing the placement of objects with the age of the rock and silt layers in which they were found, archeologists could usually make a general estimate of their age. However, many objects were found in caves, frozen in ice, or in other areas whose ages were not known; in these cases, it was clear that a method for dating the actual object was necessary.

In 1907, the American chemist Bertram Boltwood (1870- 1927) proposed that rocks containing radioactive **uranium** could be dated by measuring the amount of **lead** in the sample. This was because uranium, as it underwent radioactive decay, would transmute into lead over a long span of time. Thus, the greater the amount of lead, the older the rock. Boltwood used this method, called radioactive dating, to obtain a very accurate measurement of the age of the Earth. While the uranium-lead dating method was limited (being only applicable to samples containing uranium), it was proved to scientists that radioactive dating was both possible and reliable.

The first method for dating organic objects (such as the remains of plants and animals) was developed by another

American chemist, Willard Libby (1908-1980). He became intrigued by carbon-14, a radioactive isotope of carbon. Carbon has isotopes with atomic weights between 9 and 15. The most abundant isotope in nature is carbon-12, followed in abundance by carbon-13. Together carbon-12 and carbon-13 make up 99% of all naturally occurring carbon. Among the less abundant isotopes is carbon-14, which is produced in small quantities in the Earth's atmosphere through interactions involving cosmic rays. In any living organism, the relative **concentration** of carbon-14 is the same as it is in the atmosphere because of the interchange of this isotope between the organism and the air. This carbon-14 cycles through an organism while it is alive, but once it dies, the organism accumulates no additional carbon-14. Whatever carbon-14 was present at the time of the organism's death begins to decay to nitrogen-14 by emitting radiation in a process known as beta decay. The difference between the concentration of carbon-14 in the material to be dated and the concentration in the atmosphere provides a basis for estimating the age of a specimen, given that the rate of decay of carbon-14 is well known. The length of time required for one-half of the unstable carbon-14 nuclei to decay (i.e., the half-life) is 5,730 years.

Libby began testing his carbon-14 dating procedure by dating objects whose ages were already known, such as samples from Egyptian tombs. He found that his methods, while not as accurate as he had hoped, were fairly reliable. He continued his research and, through improvements in his equipment and procedures, was eventually able to determine the age of an object up to 50,000 years old with a precision of plus-or-minus 10%. Libby's method, called radiocarbon or carbon-14 dating, gave new impetus to the science of radioactive dating. Using the carbon-14 method, scientists determined the ages of artifacts from many ancient civilizations. Still, even with the help of laboratories worldwide, radiocarbon dating was only accurate up to 70,000 years old, since objects older than this contained far too little carbon-14 for the equipment to detect.

Starting where Boltwood and Libby left off, scientists began to search for other long-lived isotopes. They developed the uranium-thorium method, the potassium-argon method, and the rubidium-strontium method, all of which are based on the transformation of one element into another. They also improved the equipment used to detect these elements, and in 1939 scientists first used a cyclotron particle accelerator as a **mass** spectrometer. Using the cyclotron, carbon-14 dating could be used for objects as old as 100,000 years, while samples containing radioactive **beryllium** could be dated as far back as 10-30 million years. A new method of radioactive tracing involves the use of a new clock, based on the radioactive decay of uranium-235 to protactinium-231. A mass spectrometer was used for one of the most famous radioactive dating experiments, the dating of the Shroud of Turin, the supposed burial cloth of Jesus. Tests in 1988 dated the linen cloth of the shroud to the 1300s and not to the time of Christ, although some experts believe these results may still be open to question.

See also Atmospheric chemistry; Mass spectroscopy; Radiation chemistry

Pure carbon dioxide gas can be poured because it is heavier than air. *(Photograph by R. Folwell. National Audubon Society Collection/ Photo Researchers, Inc. Reproduced by permission.)*

CARBON DIOXIDE

Carbon dioxide was the first gas to be distinguished from ordinary air, perhaps because it is so intimately connected with the cycles of plant and animal life. It is released during respiration and **combustion**. When plants store **energy** in the form of food, they use up carbon dioxide. Early scientists were able to observe the effects of carbon dioxide long before they knew exactly what it was.

Around 1630, Flemish scientist **Jan van Helmont** discovered that certain vapors differed from air that was then thought to be a single substance or element. Van Helmont coined the term gas to describe these vapors and collected the gas given off by burning **wood**, calling it gas sylvestre. Today, we know this gas to be carbon dioxide, and van Helmont is credited with its discovery. He also recognized that carbon dioxide was produced by the **fermentation** of wine and from other natural processes. Before long, other scientists began to notice similarities between the processes of respiration and combustion, both of which use up and give off carbon dioxide. For example, a candle flame will eventually be extinguished when enclosed in a jar with a limited supply of air, as will the life of a bird or small animal.

Then in 1756, **Joseph Black** proved that carbon dioxide, which he called fixed air, is present in the atmosphere and that it combines with other chemicals to form new compounds. Black also identified carbon dioxide in exhaled breath, determined that the gas is heavier than air, and characterized its chemical behavior as that of a weak acid. The pioneering work of van Helmont and Black soon led to the discovery of other **gases** by **Henry Cavendish**, **Antoine-Laurent Lavoisier**, **Carl Wilhelm Scheele**, and other chemists. As a result, scientists began to realize that gases must be weighed and accounted for in the analysis of chemical compounds, just like **solids** and **liquids**.

The first practical use for carbon dioxide was invented by **Joseph Priestley**, an English chemist, in the mid 1700s. Priestley had duplicated Black's experiments using a gas produced by fermenting grain and showed that it had the same properties as Black's fixed air, or carbon dioxide. When he dissolved the gas in **water**, he found that it created a refreshing drink with a slightly tart flavor. This was the first artificially carbonated water, known as soda water or seltzer. Carbon dioxide is still used today to make colas and other soft drinks. In addition to supplying bubbles and zest, the gas acts as a preservative.

The early study of carbon dioxide also gave rise to the expression to be a guinea pig, meaning to subject oneself to an experiment. In 1783, French physicist Pierre Laplace (1749-1827) used a guinea pig to demonstrate quantitatively that **oxygen** from the air is used to burn carbon stored in the body and produce carbon dioxide in exhaled breath. Around the same time, chemists began drawing the connection between carbon dioxide and plant life. Like animals, plants breathe, using up oxygen and releasing carbon dioxide. But plants also have the unique ability to store energy in the form of **carbohydrates**, our primary source of food. This energy-storing process, called **photosynthesis**, is essentially the reverse of respiration. It uses up carbon dioxide and releases oxygen in a complex series of reactions that also require sunlight and **chlorophyll** (the green substance that gives plants their color). In the 1770s, Dutch physiologist Jan Ingen Housz established the principles of photosynthesis that helped explain the age-old superstition that plants purify air during the day and poison it at night.

Since these early discoveries, chemists have learned much more about carbon dioxide. English chemist **John Dalton** guessed in 1803 that the **molecule** contains one carbon **atom** and two oxygen atoms (CO_2); this was later proved to be true. The decay of all organic materials produces carbon dioxide very slowly, and Earth's atmosphere contains a small amount of the gas (about 0.033%). Spectroscopic analysis has shown that in our solar system, the planets of Venus and Mars have atmospheres very rich in carbon dioxide. The gas also exists in ocean water, where it plays a vital role in marine plant photosynthesis.

In modern life, carbon dioxide has many practical applications. For example, fire extinguishers use CO_2 to control electrical and oil fires that cannot be put out with water. Because carbon dioxide is heavier than air, it spreads into a blanket and smothers the flames. Carbon dioxide is also a very effective refrigerant. In its solid form, known as dry ice, it is used to chill perishable food during transport. Many industrial processes are also cooled by carbon dioxide, which allows faster production rates. For these commercial purposes, carbon dioxide can be obtained from either **natural gas** wells, fermentation of organic material, or combustion of fossil fuels.

Recently, carbon dioxide has received negative attention as a greenhouse gas. When it accumulates in the upper atmosphere, it traps the earth's **heat**, eventually causing global warming. Since the beginning of the industrial revolution in the mid 1800s, factories and power plants have significantly increased the amount of carbon dioxide in the atmosphere by burning **coal** and other fossil fuels. This effect was first predicted by **Svante August Arrhenius**, a Swedish physicist, in the 1880s. Then in 1938, British physicist G. S. Callendar suggested that higher CO_2 levels had caused the warmer temperatures observed in America and Europe since Arrhenius's day. Modern scientists have confirmed these views and identified other causes of increasing carbon dioxide levels, such as the clearing of the world's forests. Because trees extract CO_2 from the air, their depletion has contributed to upsetting the delicate balance of gases in the atmosphere.

In very rare circumstances, carbon dioxide can endanger life. In 1986, a huge cloud of the gas exploded from Lake Nyos, a volcanic lake in northwestern Cameroon, and quickly suffocated more than 1,700 people and 8,000 animals. Scientists have attempted to control this phenomenon by slowly pumping the gas up from the bottom of the lake.

Carbon fixation

Carbon fixation refers to the chemical transformation of simple, inorganic compounds of carbon into more complex forms of organic **matter**. Examples of the simple compounds include **carbon dioxide** (CO_2) and bicarbonate (HCO_3^-), while the more complex forms include calcite ($CaCO_3$), which is inorganic, and the organic matter of organisms.

In ecosystems, carbon is fixed by autotrophs—organisms that utilize an external **energy** source to drive the synthesis of CO_2 and **water** (H_2O) into simple sugars. Usually, sunlight is the energy source for the fixation reactions, which is referred to as **photosynthesis**, and the organisms as photoautotrophs. Examples of photoautotrophs include plants, algae, and blue-green algae. Expressed simply, the photosynthetic reaction is:

(1) $CO_2 + 2H_2O$ - light $\rightarrow CH_2O + 2O_2 + H_2O$

In reaction 1, the term CH_2O refers to a carbohydrate, which is a primary product of photosynthesis. The molecular **oxygen** (O_2) is a ''waste'' product of the photosynthetic reaction and is usually released by the autotroph. Through the many complex reactions of **metabolism**, the carbohydrate can then be used to synthesize the additional biochemicals needed by autotrophs for survival, such as complex **carbohydrates**, **proteins**, and **lipids**.

Some autotrophs are non-photosynthetic, meaning they utilize energy sources other than sunlight to drive their carbon-fixation reactions. These so-called chemoautotrophs use the stored energy of certain chemicals [usually sulfides such as **hydrogen** sulfide (H_2S) or **iron** sulfide (FeS_2)] to drive chemosynthesis. Chemoautotrophs are the basis of ecosystems that are independent of solar radiation.

In addition to photosynthetic fixations of CO_2, inorganic forms of carbon can be fixed by other biotic reactions occurring in aquatic ecosystems. One of these involves the fixation of dissolved carbon dioxide by a series of simple reactions, as follows:

(2) $CO_2 + H_2O \rightarrow H_2CO_3$

(3) $H_2CO_3 \rightarrow H^+ + HCO_3^-$

(4) $HCO_3^- \rightarrow H^+ + CO_3^{-2}$

(5) $Ca^{+2} + 2HCO_3^- \rightarrow CaCO_3 + CO_2 + H_2O$

In reaction 2, dissolved carbon dioxide combines with water to form carbonic acid (a weak acid). In reaction 3, the carbonic acid dissociates to form hydrogen **ion** and bicarbonate. In reaction 4, the bicarbonate dissociates to hydrogen ion and carbonate. In reaction 5, **calcium** ion combines with bicarbonate to form calcite plus carbon dioxide and water. Reaction 5 commonly occurs within the bodies of certain aquatic organisms, which utilize calcite to construct their shells (invertebrate animals) or bones (vertebrates). The calcite is an insoluble mineral, which upon death of the organisms sinks to the floor of the body of water and accumulates in the sediment. Through extremely slow geological processes occurring in deep sediment, the calcite may eventually metamorphose into rocks, such as limestone or chalk.

CARBON MONOXIDE

Carbon monoxide is a compound of carbon and **oxygen** with the chemical formula CO. It is a colorless, odorless, tasteless, toxic gas. It has a **density** of 1.250 g/L at 32°F (0°C) and 760 mm Hg pressure. Carbon can be converted into a liquid at its **boiling point** of -312.7°F (-191.5°C) and then to a solid at its freezing point of -337°F (-205°C).

The discovery of carbon monoxide is often credited to the work of the English chemist and theologian **Joseph Priestley**. From 1772-1799, Priestley gradually recognized the nature of this compound and showed how it was different from **carbon dioxide**, with which it often appeared. Nonetheless, carbon monoxide had been well known and extensively studied in the centuries prior to Priestley's work. As early as the late 1200s, the Spanish alchemist Arnold of Villanova described a poisonous gas produced by the incomplete **combustion** of **wood** that was almost certainly carbon monoxide.

In the five centuries between the work of Arnold and that of Priestley, carbon monoxide was studied and described by a number of prominent alchemists and chemists. Many made special mention of the toxicity of the gas. In 1644, Johann (or Jan) Baptista van Helmont wrote that he nearly died from inhaling *gas carbonum*, apparently a mixture of carbon monoxide and carbon dioxide.

An important milestone in the history of carbon monoxide came in 1877 when the French physicist Louis Paul Cailletet found a method for liquefying the gas. Two decades later, a particularly interesting group of compounds made from carbon monoxide—the carbonyls—were discovered by the French chemist **Paul Sabatier**.

Carbon monoxide is the twelfth most abundant gas in the atmosphere. It makes up about 1.2×10^{-5} percent of a sample of dry air in the lower atmosphere. The major natural source of carbon monoxide is the combustion of wood, **coal** and other naturally occurring substances on Earth's surface. Huge quantities of carbon monoxide are produced, for example, during a forest fire or a volcanic eruption. The amount of carbon monoxide produced in such reactions depends on the availability of oxygen and the combustion **temperature**. High levels of oxygen and high temperatures tend to produce complete **oxidation** of carbon, with carbon dioxide as the final product. Lower levels of oxygen and lower temperatures result in the formation of higher percentages of carbon monoxide in the combustion mixture.

Commercial methods for producing carbon monoxide often depend on the direct oxidation of carbon under controlled conditions. For example, producer gas is made by blowing air across very hot coke (nearly pure carbon). The final product consists of three gases: carbon monoxide, carbon dioxide, and **nitrogen** in the ratio of 6:1:18. **Water** gas is made by a similar process, by passing steam over hot coke. The products in this case are **hydrogen** (50%), carbon monoxide (40%), carbon dioxide (5%) and other **gases** (5%). Other methods of preparation are also available. One of the most commonly used involves the partial oxidation of hydrocarbons obtained from **natural gas**.

The toxic character of carbon monoxide has been well known for many centuries. At low concentrations, carbon monoxide may cause nausea, vomiting, restlessness, and euphoria. As exposure increases, a person may lose consciousness and go into convulsions. Death is a common final result. The U.S. Occupational Safety and Health Administration has established a limit of 35 ppm (parts per million) of carbon monoxide in workplaces where a person may be continually exposed to the gas.

The earliest explanation for the toxic effects of carbon monoxide was offered by the French physiologist Claude Bernard in the late 1850s. Bernard pointed out that carbon monoxide has a strong tendency to replace oxygen in the respiratory system. Someone exposed to high concentrations of carbon monoxide may actually begin to suffocate as his or her body is deprived of oxygen.

Today, there is a fairly sophisticated understanding of the mechanism by which carbon monoxide poisoning occurs. Normally, oxygen is transported from the lungs to cells in red blood cells. This process occurs when oxygen atoms bond to an **iron atom** at the center of a complex protein **molecule** known as oxyhemoglobin. Oxyhemoglobin is a fairly unstable molecule that decomposes in the intercellular spaces to release free oxygen and hemoglobin. The oxygen is then available to carry out metabolic reactions in cells, reactions from which the body obtains **energy**.

If carbon monoxide is present in the lungs, the sequence is disrupted. Carbon monoxide bonds with iron in hemoglobin to form carbonmonoxyhemoglobin, a complex somewhat similar to oxyhemoglobin. Carbonmonoxyhemoglobin is, however, more stable compound than oxyhemoglobin. When it reaches cells, it has much less tendency to break down, but continues to circulate in the bloodstream in its bound form. As a result, cells are unable to obtain the oxygen they need for **metabolism** and energy production dramatically decreases. The clinical symptoms of carbon monoxide poisoning are manifestations of these changes.

At moderate levels, carbon monoxide poisoning is common in everyday life. Poorly vented **charcoal** fires, improperly installed gas appliances, and the exhaust from internal combustion vehicles are among the most common sources of the gas. In fact, levels of carbon monoxide in the air can become dangerously high in busy urban areas where automotive transportation is extensive. Cigarette smokers may also be exposed to dangerous levels of the gas. Studies have shown that the one to two pack-a-day smoker may have up to 7% of the hemoglobin in her or his body tied up in the form of carbonmonoxyhemoglobin.

Carbon monoxide is a very important industrial compound. In the form of producer gas or water gas, it is widely used as a fuel in industrial operations. The gas is also an effective reducing agent. For example, when carbon monoxide is passed over hot iron oxides, the oxides are reduced to metallic iron, while the carbon monoxide is oxidized to carbon dioxide.

In another application, a mixture of metallic ores is heated to 122-176°F (50-80°C) in the presence of producer gas. All oxides except those of **nickel** are reduced to their metallic state. This process, known as the Mond process, is a way of separating nickel from other **metals** with which it commonly occurs.

Yet another use of the gas is in the Fischer-Tropsch process for the manufacture of hydrocarbons and their oxygen derivatives from a combination of hydrogen and carbon monoxide. Carbon monoxide also reacts with certain metals, especially iron, **cobalt**, and nickel, to form compounds known as carbonyls. Some of the carbonyls have unusual physical and chemical properties that make them useful in industry. The highly toxic nickel tetracarbonyl, for example, is used to produce very pure nickel coatings and powders.

CARBONYL GROUP

The carbonyl group is an organic **functional group** that consists of a **carbon atom** double bonded to an **oxygen** atom (Figure 1). It is one of the most common reactive components in organic structures, and any compound that contains the carbonyl group is generally categorized as a carbonyl compound. It is important to distinguish these organic compounds from the organometallic group known as the metal carbonyls. Metal carbonyls are organometallic compounds where one or more of the metal ligands are **carbon monoxide** (CO). The organic carbonyl functional group is found in a variety of consumer products including **plastics**, adhesives, foods (proteins and fats), and drugs such as aspirin and **penicillin**. The structural features of the carbonyl group give these compounds their characteristic properties.

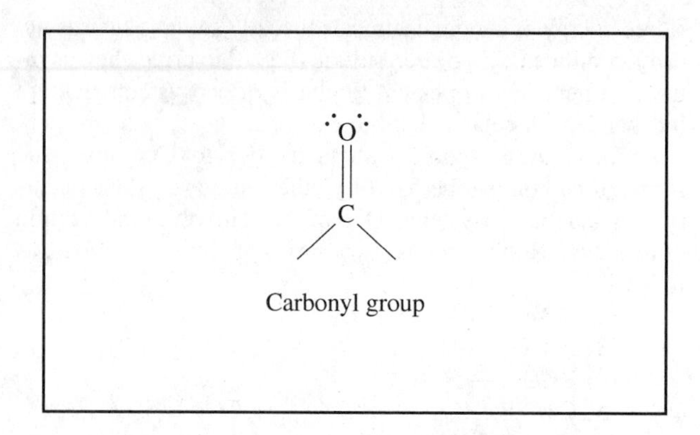

Figure 1 (Carbonyl group). ***(Illustration by Electronic Illustrators Group.)***

The carbonyl group possesses a trigonal planar arrangement of groups around the carbon atom which is sp^2 hybridized. As a result, the carbonyl carbon, and the other three attached atoms lie in the same plane with the bond angles being approximately 120°.

The carbonyl group is a polar group because the carbonyl carbon atom bears a partial positive charge and the carbonyl oxygen atom bears a partial negative charge, creating a **dipole** moment. This charge distribution arises because the electronegative oxygen atom makes the **resonance** contribution of the second structure significant, emphasizing the Lewis acid character of the carbon of the carbonyl.

The carbonyl group is further subdivided into many different functional groups, all of which contain the carbon-oxygen double bond but differ in the other atoms bonded to carbon. These other atoms profoundly influence the relative amount of partial positive charge at the carbon and, thus, the relative **reactivity** pf the carbonyl. Among the common carbonyl-containing functional groups, listed in approximate order of decreasing reactivity, are acyl chlorides [R-C(O)-Cl], acid anhydrides [R-C(O)-O-C(O)-R'], **aldehydes** [R-C(O)-H] (Figure 2), thioesters [R'-C(O)-SR'], **ketones** [R- C(O)-R'] (Figure 3), esters [R-C(O)-OR'], **carboxylic acids** [R-C(O)-OH], carbamates [RO-C(O)-NR'_2], amides [R-C(O)-NR_2], and ureas [R_2N- C(O)-NR'_2]. Of these, the two important carbonyl functional groups that illustrate the behavior of the carbonyl are aldehydes and ketones.

Aldehydes contain the carbonyl group which is bonded to at least one **hydrogen**, and are generally produced by the **dehydrogenation**, or **oxidation**, of a primary **alcohol**. Aldehydes have been valued since ancient times for their characteristic strong fragrance and **taste**. The appealing smell of vanilla is given off by the aldehyde vanillin, and the enticing scent of cinnamon is due to the presence of cinnamaldehyde. A widely used aldehyde in manufacturing is **formaldehyde**. Formaldehyde is readily soluble in **water**, and is used to preserve biological specimens. Aldehydes are important intermediates in the manufacturing of many industrial compounds such as dyes, **solvents**, and pharmaceuticals because of their high chemical reactivity.

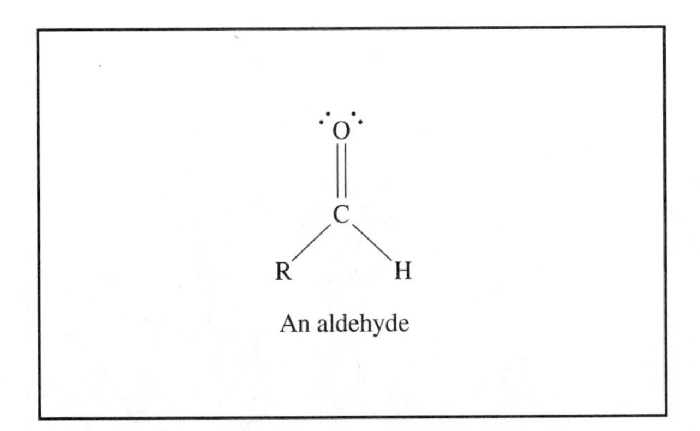

Figure 2 (Carbonyl group). ***(Illustration by Electronic Illustrators Group.)***

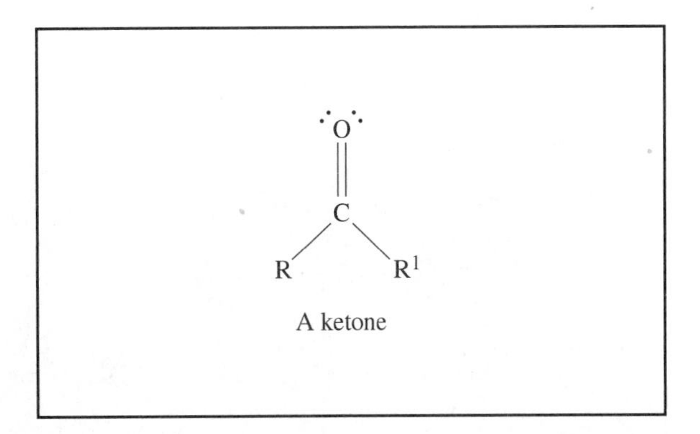

Figure 3 (Carbonyl group). ***(Illustration by Electronic Illustrators Group.)***

Ketones contain the carbonyl group, singly bonded to two carbon atoms. Industrially, they are mainly manufactured as the products of the dehydrogenation, or oxidation, of secondary alcohols. A commonly used ketone is **acetone**. Acetone is a widely used solvent, for example, it is used in nail polish, and it is also the starting material for the synthesis of many polymers. Ketones are also used in the flavoring and perfumery industry. Camphor is a frequent component of medicinal remedies such as cough drops. Although ketones are not as reactive as aldehydes, they are also important intermediates in the synthesis of many compounds.

Aldehydes and ketones have higher boiling points than other hydrocarbons of the same **molecular weight** because the carbonyl group is polar. However, since aldehydes and ketones do not form strong hydrogen bonds between their molecules, they have lower boiling points than the corresponding alcohols. Low molecular weight aldehydes and ketones are soluble in water because the carbonyl oxygen atom allows the molecules to form strong hydrogen bonds with the molecules of water.

One highly characteristic reaction of aldehydes and ketones is nucleophilic addition to the carbon-oxygen double bond. The trigonal planar arrangement of the groups around the carbonyl carbon atom enables the carbon atom to be relatively open to attack from above, or below. Its partial positive charge means that is susceptible to attack by a **nucleophile**. The partial negative charge on the oxygen atom makes it susceptible to acid catalysis. Many nucleophilic additions to carbon-oxygen double bonds are reversible, and the equilibrium position of the reaction determines the overall results. This is in contrast with most electrophilic additions to carboncarbon double bonds, and with nucleophilic substitutions at saturated carbon atoms. The latter reactions are essentially irreversible. In simple additions, the carbonyl compounds can behave as both Lewis acids and Lewis bases depending on what reagents are present.

The stability of the product of a nucleophilic addition to the carbon-oxygen double bond will depend on the reaction conditions. By deliberately manipulating the reaction conditions, the formerly stable product can be used as a stepping stone for a subsequent reaction; this is why these compounds are frequently used in many synthesis reactions. Aldehydes are much more easily oxidized than ketones because the carbon-hydrogen bond of the aldehyde is realtively easily broken by oxidizing agents. Since ketones lack this carbon-hydrogen bond and are instead flanked by two carbon-carbon bonds, they are more resistant to oxidation.

From the synthetic chemist's point of view, a very useful nucleophilic addition to the carbon-oxygen double bond that produces an alkene as the reaction product is the Wittig reaction. This reaction occurs between an ylide and a carbonyl group (usually an aldehyde or ketone) which ultimately yields the alkene. An ylide is a compound containing opposite charges on adjacent atoms, which can act as a nucleophile. German chemist **Georg Wittig** discovered it in his laboratories, and shared the 1979 Nobel Prize in **chemistry** largely for his work on this reaction.

A second important characteristic of carbonyl compounds is an acidity of the hydrogen atoms on the carbon atoms adjacent to the carbonyl group. These hydrogens are usually called the α-hydrogens, and the carbon to which they are attached to is known as the α-carbon. The α-hydrogens are reactive to strong bases, as **proton** donors, with the pK_{α}'s of the carbonyl α-hydrogens on the order of 10^{-19} to 10^{-20}. As a result, carbonyl compounds can form carbon anions (carbanions) and act as nucleophiles rather than electrophiles. The removal of an α-hydrogen forms a resonance stabilized carbanion known as an enolate. The negative charge of this **anion** is delocalized, giving the oxygen and the α-carbon a partial negative charge. The enolate can reversibly accept a proton in two ways: it may accept the proton at the carbon to revert to the original carbonyl compound (the keto form), or it may accept the proton at the oxygen atom to form an enol (Figure 4). The enolate is important for many synthetic reactions. For example, the **aldol reaction** involves the addition of enolate ions to aldehydes or ketones. This is an important reaction in organic synthesis as it provides a method to join two smaller molecules through a carbon-carbon bond.

The presence of the carbonyl group can easily be detected using various spectroscopic techniques. Carbonyl groups of aldehydes and ketones give rise to very strong carbon-oxygen

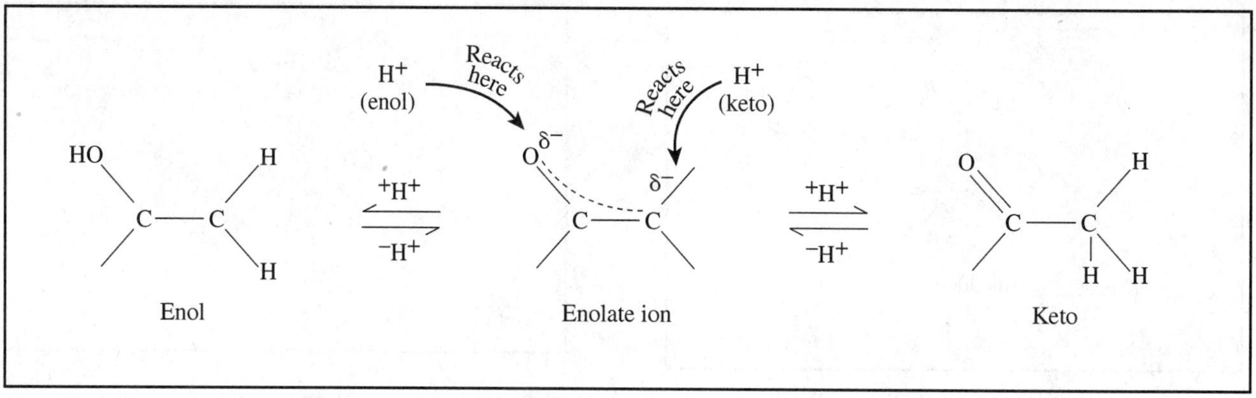

Figure 4 (Carbonyl group). ***(Illustration by Electronic Illustrators Group.)***

double bond stretching bands in the 1665-1780 cm^{-1} region of the infrared (IR) spectrum. The aldehydic proton gives a strongly deshielded resonance signal (δ = 9-10) in the H-1 NMR spectrum. In addition, the carbonyl groups of saturated aldehydes and ketones give a weak absorption band in the ultraviolet (UV) region between 270 and 300 mm. The exact location of the signals depends on the structure of the aldehyde or ketone.

Carborane

Next to **carbon**, **boron** probably has the richest **chemistry** with regard to catenation, which is the formation of **chains** of atoms of the same element. The number of ''boranes'' (hydrides of boron) that have been synthesized illustrates this rich chemistry. In general, these compounds have a formula of $B_n H_n^{2-}$. Replacement of two BH^- units with the isoelectronic CH moiety leads to the formation of carboranes. This results in neutral compounds with the general formula $B_{n-2}C_2H_n$, where n = 5 to 12.

Structurally, the boranes and carboranes fall into three classes of polyhedra. The *closo* compounds follow the regular polyhedra. The *nido* compounds are the regular polyhedra with one vertex missing. The *arachno* compounds have two missing vertices. Extensive rules have been developed for determining the structure of the resulting carborane from the **electron** count of the framework atoms.

The most common and important compounds are the two isomers of $B_{10}C_2H_{12}$. The carbon atoms are found either adjacent (called ''1,2'') or separated by a single boron (called ''1,7''). These carboranes exhibit a rich chemistry as the **hydrogen** atoms on the carbons react readily with metallic **lithium**. The lithium atoms transfer two electrons to the carborane and the resulting dianion reacts with a number of electrophiles. The number of compounds that have been synthesized by this method is significant. The principle driving force for this research is the industrial application of carboranes in the synthesis of high molecular-weight polymers. Dual functionalization allows their incorporation into, for example, silicones where the carborane enhances the thermal stability of the polymer.

Carboranes have also been utilized in the synthesis of discrete metal **ion** compounds. Although not strictly speaking ''organometallic,'' the resulting species bear a strong resemblance to the **metallocenes** in their chemistry and are isolobal (have the same number, symmetry properties, shape and approximate energies of their frontier orbitals) with the cyclopentadienyl radical. The chemistry of these compounds is still an area of active investigation.

Carboxyl Group

The carboxyl group, R-COOH, is one of the most common and important organic functional groups. Chemicals that contain the carboxyl group belong to the large family of organic compounds that contain the **carbonyl group**, C=O, a **carbon atom** doubly bonded to an **oxygen** atom. The carboxyl group consists of a carbonyl group whose carbon atom is singly bonded to a hydroxyl group, -OH. The carboxyl group carbon atom also makes a bond to a carbon atom within the main chain of the parent compound. Some familiar compounds which contain a carboxyl group are **acetic acid** (the main component of vinegar), **formic acid** (the compound responsible for the ''sting'' of various ants and plant nettles), **amino acids** (the ''building blocks'' of every protein), and butyric acid (the chemical responsible for the stench of rancid butter).

The carboxyl group is sp^2 hybridized and as a result has planar geometry with bond angles of 120°. The length of the C=O carbonyl bond is shorter than that of the C-OH bond due to its double-bonded character, making the carboxyl group slightly asymmetric. The **hydrogen** atom of the -OH group can participate in a **hydrogen bond** with the carbonyl oxygen atom of a neighboring carboxyl group. In a pure **solution** of a compound containing at least one carboxyl group, a network of hydrogen bonds can form among the groups, causing the molecules to become strongly associated with each other. This network of hydrogen bonds causes the characteristic boiling points of compounds containing a carboxyl group to be very high (212-392°F [100- 200°C]), since much **energy** is needed to break the hydrogen **bonding** interactions. The hydrogen, or

proton, of the -OH group can also dissociate to form a **hydronium ion** with **water** if the compound containing a carboxyl group is water-soluble. A carboxyl group that has lost its -OH hydrogen due to **dissociation** is called a carboxylate ion. The carboxylate ion has a negative charge, which is delocalized over both oxygen atoms to form a stable **resonance** hybrid. Because a carboxylate ion has low energy due to resonance stabilization, its formation via dissociation in aqueous solution is somewhat favorable. For this reason, water-soluble compounds containing a carboxyl group are generally called **carboxylic acids**, since the carboxyl group acts as a proton donor when it dissociates in water. However, carboxylic acids are much weaker than mineral acids such as HCl since they have a lower ionization constant and do not dissociate completely in aqueous solution.

Compounds containing carboxyl groups are synthesized through both natural processes and man-made methods such as the Grignard reaction, in which an organometallic reagent reacts with **carbon dioxide** to form a carboxylated derviative of the reagent. The carboxyl group is fairly reactive, and carboxyl group-containing compounds can undergo nucleophilic substitution to form many useful derivatives such as acid chlorides, anhydrides, esters, and amides. In all cases, the -OH of the carboxyl group is replaced with another chemical group. These derivatives are the starting material for many other organic reactions. The carboxyl group can also be converted into an **alcohol** by **reduction** of the carbonyl carbon. Carboxyl groups can also react with alkali metal bases such as **sodium** hydroxide to form water-soluble carboxylic salts. The salts of long-chain carboxylic, or "fatty" acids, are one of the main ingredients of soap, and are formed in a process known as saponification.

Carboxylic acids were among the first organic compounds to be purified and identified in the late 18th and early 19th centuries. At the time, it was known only what the chemical composition and gross chemical properties of the compounds were. In 1845, **Adolph Wilhelm Hermann Kolbe** was the first person to synthesize a naturally occurring carboxyl compound, acetic acid. His success at synthesizing this biologically relevant compound, along with Friedrich Woehler's 1828 synthesis of **urea**, another chemical found in living organisms, disproved the widely held "**vitalism**" theory. Vitalists believed that living organisms were composed of special **matter** containing a "vital force." By changing "nonliving" precursors into "living matter" through simple chemical conversions, Woehler and Kolbe made it clear that the matter comprising living and nonliving objects were one and the same. It was not until the early to mid 20th century that spectroscopic and **x-ray diffraction** studies on many different carboxylated compounds allowed the detailed physical and structural properties of the carboxyl group to be identified and defined.

See also Carbonyl group; Carboxylic acids; Organic functional group

Carboxylic acids

Carboxylic acids are organic chemical compounds that contain one or more **carboxyl group** (-COOH).The **oxygen atom** is attached to the **carbon** by a double bond, while the OH is attached to the carbon atom with a single bond. The carboxyl group is joined to another **hydrogen** atom or to one end of an organic compound. Because the bond between the carboxyl group and the hydrogen atom are highly polar (having distinct positive and negative regions that do not overlap), many carboxylic acids are **water** soluble and have higher boiling points than alcohols. Solutions of many carboxylic acids have a sour **taste** to them, a characteristic of many acids. Carboxylic acids also react with alkalis, or bases. Generally, however, carboxylic acids are not as chemically active as the nonorganic mineral acids, such as **hydrochloric acid** or **sulfuric acid**.

Carboxylic acids that have very long, unbranched **chains** of 12-24 carbon atoms attached to them are called **fatty acids**. As their name suggests, they are important in the formation of fat in the body. Many carboxylic acids are present in the foods and drinks we ingest, like malic acid (found in apples), **tartaric acid** (grape juice), and **oxalic acid** (spinach and some parts of the rhubarb plant). Two other simple carboxylic acids are propionic acid and butyric acid. Propionic acid is partly responsible for the flavor and odor of Swiss cheese. Butyric acid is responsible not only for the smell of rancid butter, but also contributes to the odor of sweat. A form of Vitamin C is called ascorbic acid and is a carboxylic acid.

Lactic acid is generated in muscles of the body as the individual cells metabolize sugar and do work. A buildup of lactic acid, caused by overexertion, is responsible for the fatigue one feels in the muscles by such short-term use. When one rests, the lactic acid is gradually converted to water and **carbon dioxide**, and the feeling of fatigue passes. It is manufactured for use in pharmaceuticals, leather tanning, textile dyeing, **plastics**, lacquers, and **solvents**.

A special form of carboxylic acids are the **amino acids**, which are carboxylic acids that also have a nitrogen-containing group called an **amine group** in the **molecule**. Amino acids are very important because combinations of amino acids make up the **proteins**. Proteins are one of the three major components of the diet, the other two being fats and **carbohydrates**. Much of the human body, like skin, hair, and muscle, is composed of protein.

Carboxylic acids are also very important industrially. Perhaps one of the most important industrial applications of compounds with carboxyl groups is the use of fatty acids—carboxyl groups attached to long carbon chains—making soaps, detergents, and shampoos. In some such compounds, the hydrogen atom in the carboxyl group is neutralized by reaction with a base, to form the metal salt of the fatty acid. The modified carboxyl group is soluble in water, while the long chain of carbons remains soluble in fats, oils, and greases. This double **solubility** allows water to wash out the fat- and oil-based dirt. Many shampoos are based on lauric, palmitic, and stearic acids, which have long chains of 12, 16, and 18 carbon atoms, respectively. This reaction makes a soap molecule that has one end soluble in water and the other soluble in fat or grease or oil. Various fatty acids are used to make soaps and detergents that have different applications in society. Carboxylic acids are also important in the manufacture of greases, crayons, and plastics.

Compounds with carboxyl groups are relatively easily converted to compounds called esters, which have the hydrogen atom of the carboxyl group replaced with a group containing carbon and hydrogen atoms. Such esters are considered derivatives of carboxylic acids. Esters are important because many of them have characteristic tastes and odors. For example, methyl butyrate, a derivative of butyric acid, smells like apples. Benzyl acetate, from **acetic acid**, has a jasmine odor. Carboxylic acids are thus used commercially as raw materials for the production of synthetic odors and flavors. Other esters, derived from carboxylic acids, have different uses. For example, the ester ethyl acetate is a very good solvent and is a major component in nail polish remover.

See also Soap and detergents

Carcinogens

Carcinogens are substances, usually chemicals, that can cause cancer. Although chemical carcinogens are often linked with industry and technology, it is important to recognize that many carcinogens are naturally occurring substances.

Researchers have identified four factors, each of which is able to trigger cancer. These factors are chemicals, radiation (including ultraviolet radiation in sunlight), certain viruses, and inherited genes. In some cancer cases, more than one cause is responsible for generating the disease. For example, it is possible that certain chemicals might only cause cancer if a person has a genetic susceptibility to the disease or because there is a genetic error which prevents **DNA** repair.

Carcinogens are generally divided into two classes: those that are organic compounds and those that are inorganic compounds. Organic compounds always contain the element **carbon**. Chemicals in this class of carcinogens include **solvents**, certain **pesticides**, and naturally occurring chemicals produced by some plants and molds. Some organic compounds are not carcinogenic until they have been biochemically altered, or metabolized, in the body. Inorganic compounds that are carcinogenic include **arsenic**, **asbestos**, certain forms of **chromium**, and **nickel**. There is less evidence for **beryllium**, **cadmium**, **lead**, and silica, but these chemicals have been shown to cause cancer in animals.

The term cancer was first used by Hippocrates in the fifth century B.C. to describe the abnormal growth of tissues in the body. Cancer can occur in almost every tissue in the body. Cancers are classified according to the type of tissue that is affected. The three classifications are: carcinoma, sarcoma, and lymphomas and leukemias.

Carcinomas are the most common types of cancers. These cancers affect epithelial cells—the cells that cover internal and external surfaces in the body. Common carcinomas include lung, breast, and colon cancers. Sarcomas are less common and affect tissues that serve as cell and tissue framework. These cancers include those that affect bone, cartilage, connective tissue, and muscles. Lymphomas and leukemias affect blood and lymph cells.

Normally, the cells in a tissue have a balanced cycle of cell division, cell maturation, and cell death. This balanced cycle allows new cells to be created at the same rate that old cells need to be replaced. Under some circumstances, certain cells develop an unbalanced cycle, and more new cells are produced than are necessary to replace the old ones. As a result, an abnormal **mass** of tissue, called a tumor, forms.

Not all tumors are cancerous, however. Benign tumors are confined to a specific area and are rarely life-threatening. Malignant tumors, however, can be dangerous. Malignant tumor cells have abnormal properties that allow them to spread from the site of the original tumor. The cancer cells can spread into nearby tissues or migrate to other areas of the body and establish tumors elsewhere.

The link between chemicals and cancer was first noted in the early eighteenth century. In 1761, John Hill (1707?-1775), a doctor living in London, noted that people who used snuff (a form of tobacco that is snorted) had a high incidence of nasal cancer. Later, in 1775, an English physician and surgeon named Percival Pott (1714-1788) wrote about the high incidence of cancer of the scrotum among chimney sweeps. Due to the nature of their work, chimney sweeps had a lot of contact with soot, and Pott suggested that some element in the soot caused the chimney sweeps to develop cancer.

Throughout the nineteenth century, the industrial use of chemicals sky-rocketed and reports linking certain types of cancer with specific industries accumulated. For example, many scientific reports detailed increased incidence of skin and bladder cancer in connection to the dye industry. However, factories continued to use the chemicals, because the link between chemicals and cancer was not clear. Cancer itself was not a well-understood disease.

By the early twentieth century, researchers were beginning to study the link between chemicals and cancer in a systematic way. The first detailed studies involved applying **coal** tar to the skin of animals. The animals developed malignant skin tumors, which proved that there was something in the coal tar that caused cancer. It wasn't until the early 1930s that the substance was shown to be dibenz(*a, h*)anthracene. Following this discovery, further animal studies were completed in which other substances were tested and shown to be carcinogens.

For ethical reasons, such experimental studies would never be conducted on humans. But researchers can observe the incidence of cancer among groups of people and try to establish if a common factor exists. This kind of study is called an epidemiological study. Epidemiologists, the scientists who conduct epidemiological studies, must have a very good understanding of statistics. They must consider many different factors that can affect people's exposure to the suspected carcinogen and they must also consider alternate explanations for their observations.

Through epidemiological studies, scientists showed that cigarette smokers developed lung cancer at a much higher rate than people who did not smoke. This finding alone was not enough to prove the connection between cigarettes and cancer, but it served as very good evidence. Scientists determined which chemicals were in cigarette smoke and did experimental studies in animals to test which were carcinogens. They found that tobacco smoke contains a complex mixture of more than 30 carcinogens, each of which is capable of causing cancer on its own.

Carcinogens trigger cancer by interacting with cellular molecules. There are two ways in which carcinogens can act: genetic and non-genetic. Genetic effects occur with a two-stage attack on the DNA. The first stage is called initiation. During initiation, the carcinogen causes an irreversible change in the DNA. The second stage is called promotion. During promotion, the cell with the abnormal DNA is stimulated to multiply. Because the DNA is damaged, cell multiplication is uncontrolled and a tumor can form. Many years can pass between initiation and promotion, and the effects of being exposed to a carcinogen may not be apparent for decades. Carcinogens can also cause cancer by non-genetic means. They act by altering cell signaling pathways or by encouraging the growth of abnormal cells.

Although epidemiological studies and animal experiments have identified many carcinogens, they tend to be time-consuming and expensive. Further, they are not helpful in predicting which substances may be carcinogens. By the time a substance is discovered to be a carcinogen, it may have already affected many people.

To address these problems, a test was developed by an American molecular geneticist, Bruce Ames (1928-), in the early 1970s. The so-called Ames test is based on the fact that most carcinogens are **mutagens** (substances that damage DNA).

The Ames test uses a strain of bacteria, *Salmonella typhimurium*, that has been genetically altered so that it cannot produce histidine. Bacteria that cannot produce histidine will not grow. If the bacteria grow after being exposed to the suspected carcinogen, it is very likely that the substance has altered the DNA and may be considered a carcinogen.

Once a carcinogen has been identified, scientists and others investigate how much threat it poses to people and what can be done to lessen the risks. In workplaces, carcinogens are replaced with other chemicals where possible. If substitution is not possible, workers wear protective gear and use the substances under strictly regulated conditions.

Environmental carcinogens are a source of much concern because it can be difficult to cleanse the environment of the compound. Carcinogens may be present in the environment through both human activities and natural causes. Many countries now have laws that restrict the release of carcinogens into the environment; however, other countries either do not have such laws or do not enforce them rigorously.

Exposure to naturally occurring carcinogens can be difficult to control. Humans and other organisms have evolved mechanisms to prevent or repair damage due to many naturally occurring carcinogens; however, these mechanisms cannot handle all naturally occurring carcinogens. For example, arsenic in drinking **water** is a continuing problem in several countries including areas in China, India, and Bangladesh. People who rely on wells for their drinking water receive continued exposure to arsenic, resulting in an increased incidence of cancer in those areas.

Nicholas Léonard Sadi Carnot.

Individuals can decrease their cancer risk from certain carcinogens by making lifestyle changes. Cancer researchers estimate that at least 30% of the cancer deaths in the United States can be linked to smoking. The longer a person smokes, the greater his or her risk of developing cancer. Other cancer deaths seem associated with diets heavy in fat and low in fruits and vegetables. Fruits and vegetables contain compounds called **antioxidants**. Antioxidants may help the body's natural cancer-fighting mechanisms.

Carnot, Nicolas Léonard Sadi (1796-1832)

French physicist and engineer

Nicholas "Sadi" Carnot studied the process by which **heat** converts into other forms of **energy**. His unique mathematical theory pertaining to the conversion relationship between heat and work (energy) preceded and predicted the first and second laws of **thermodynamics**. A major interest to Carnot was industrial development and, in particular, the steam engine. He focused much of his research attempting to improve its effectiveness and efficiency, studying the relationship between pressure, **temperature**, and energy output.

Carnot was born in Palais du Petit-Luxembourg, the eldest son of Lazare Carnot, a French revolutionary and Napoleon's minister of war. In 1807, Lazare resigned his post to devote his time to educating his two sons, Sadi (named after a medieval poet and philosopher Sa'di of Shiraz) and Hippolyte. Carnot was an exceptional student and entered the Ecole Polytechnique at the age of 16, the youngest age for entry into the elite institute. Carnot interrupted his studies when he and several other students volunteered to join Napoleon in the ultimately unsuccessful battle to defend Vincennes. He graduated in 1814, ranked sixth in his class, and embarked on a two-year course at the Ecole du Génie at Metz in military engineering. When Napoleon was again defeated after returning from exile to assume what became known as his ''Hundred Days Rule,'' Carnot's father was exiled and lived in Germany for the remainder of his life.

Remaining in France as a second lieutenant in the Metz engineering regiment, Carnot was transferred by the military from place to place on different assignments, which seriously interfered with his intellectual scientific activities. Tiring of the humdrum assignments, he sought relief when, in 1819 he, sat for and passed a difficult exam that enabled him to join the General Staff Corps in Paris. Immediately, he took leave with half pay, moved into his father's empty apartment in Paris, and began taking courses at the Sorbonne, the College de France, and other prestigious institutions. He developed a keen interest in industry and, in particular, the steam engine, which ultimately led him to the study of **gases**.

In 1821, Carnot took a trip to visit his brother and exiled father in Magdeburg where, three years earlier, the town's first steam engine became operational. Upon his return to Paris, his excitement for steam power ultimately led to a manuscript, *Réflexions*, the only work published in his lifetime. This 118-page essay gave a succinct report of the ''industrial, political, and economic importance of the steam engine.'' In it, he also introduced his concept of the ideal engine, which he called the ''Carnot engine,'' the idea of reversibility, ''that motive power can be used to produce the temperature difference within the engine.'' He also addressed two significant areas which remained elusive in the advancement of the steam engine: 1) Is there a limit to the motive power of heat, and 2) Do agents exist which are more capable of producing motive power than steam? He perceived his entire essay as a ''deliberate examination of these questions.''

To answer these questions, researchers had been using (unsuccessfully) two methods: 1) Empirical observation of the relationship between fuel input and energy output of individual engines; and 2) Application of mathematical analysis of gases to abstract engine functions. Carnot's methodology was radically different and became the focal point of what would ultimately become his important contribution to the development of thermodynamics. In *Réflexions*, he set out three premises: 1) The impossibility of perpetual motion in relationship to mechanics; 2) His ''caloric'' theory of heat, which proposed that heat (*calorique*) must be viewed as a ''weightless fluid that could neither be created nor destroyed in any process...and that the quantity of heat absorbed or released by a body depends only on the initial and final states of the body;'' and 3) Whenever a temperature difference exists, motive power can be produced...by the movement of caloric from a warm body to a cold body.

Réflexions was released by the foremost scientific publishing company in Paris in June 1824 and received positive acclaim upon its presentation at the Académie des Sciences in July. However, a critique published sometime later in the *Revue encyclopédique* focused only the application of Carnot's methodology in relation to steam engines; it would be years before the originality of his mathematical reasonings in heat-to-energy transference would be recognized and appreciated.

In 1831, Carnot turned his primary focus to studying physical properties of gases and vapors in relation to temperature and pressure. His work was tragically cut short when, in June 1832 he developed scarlet fever. Because of his weakened state, he died on August 24, 1832, just 24 hours after contracting cholera. Due to the practice of burning all personal effects of those suffering with the disease, most of Carnot's research notes were lost. His only existing writings include *Réflexions*, some other manuscript notes (which years later became noted as a major work), fragments of mathematics and physics lecture notes, and ''Recherche.'' This latter piece was a 21-page manuscript compiled around 1823, the earliest of his written works, in which he was seeking a mathematical expression for motive power as well as a general **solution** to encompass the heat/energy dynamics for every type of steam engine. Although never published, its careful preparation in comparison to his notes indicated he intended it to be so. This manuscript remained undiscovered until 1966.

Following his death, Carnot's work received brief attention in 1834 then virtually none until **William Thomson** who, in 1850, working directly from *Réflexions*, confirmed some of Carnot's theories. Extensions of these theories by Thomson and others ultimately led to the development of the second law of thermodynamics.

Twenty-three sheets of rough and disjointed manuscript notes, not discovered nor published until 1878, contained details of a new theory for kinetic energy. Carnot was obviously aware that this theory refuted some theories developed in his *Réflexions*; however, undaunted, he calculated a ''conversion coefficient for heat and work and went on to assert that the total quantity of motive power in the universe was constant.'' These hypotheses hinted at almost all the basic theories that would ultimately become the first law of thermodynamics.

CARO, HEINRICH (1834-1910)

German chemist

Caro was born in Posen, Prussia (now Poznan, Poland) in 1834. He moved to Berlin in 1834 and eventually studied dyeing at the *Gewerbeinstitut* and **chemistry** at the University of Berlin. In 1855 a calico factory hired Caro as a colorist. There he was able to apply his chemical training to solve production problems, and he was soon sent to England for training in more advanced dyeing processes. He studied there for two years, returning to Germany to work with synthetic dyes, which were growing in popularity, for Roberts, Dale and Company.

Caro was able to build on the work of others to discover improved processes. He enhanced **William Henry Perkin**'s technique for deriving mauve dye from **aniline** and was made a partner in the firm. His work on induline was built on the Johann Peter Griess's (1829-1888) work with aniline dye. Caro also worked with C. A. Martius to produce Bismarck brown and Martius yellow. Caro's research helped to define the structure of triphenylmethane, the parent compound of rosaniline dyes.

In 1866 Caro moved back to Berlin, where he soon became the director of the Badische Anilin und Soda Fabrik (BASF) laboratory. Caro improved the technique of Carl Graebe (1841-1927) and Liebermann to produce alizarin. He added **bromine** to **Adolf von Baeyer**'s flourescein and produced flourescent red. In collaboration with Baeyer, Caro developed the dye known as methylene blue. In the mid 1870s numerous azo dyes were developed and discovered. Caro, with others, developed chrysoidine, orange and fast red. He went on to discover naphthol yellow, persulfuric acid, also called Caro's acid. Caro resigned from BASF in 1889. Heinrich Caro was responsible for more of Germany's successes in the dye industry than any other individual. He died in Dresden in 1910.

Wallace Hume Carothers.

Carothers, Wallace Hume (1896-1937)

American chemist

By the time he was forty, Wallace Hume Carothers had made significant contributions to the field of organic **chemistry**. Heading a research team at the Du Pont Company in Wilmington, Delaware, Carothers, in the late 1920s and 1930s, framed a general theory for the behavior and synthesis of polymers, huge chainlike molecules that indefinitely repeat the same structure. Two years after Carothers's death, Du Pont began mass production of a polymer he invented—nylon. This and Carothers's other well-known creation, synthetic rubber, laid the foundation for two industries. In recognition of his services, Du Pont in 1946 named its new laboratory for synthetic fibers research the Carothers Research Laboratory.

Carothers was born on April 27, 1896, in Burlington, Iowa. His father, Ira Hume Carothers, was descended from Scottish farmers and artisans who had settled in Pennsylvania before the American Revolution. Born on an Illinois farm in 1869, Ira Carothers began teaching country school at nineteen. He then went to teach at the Capital City Commercial College in Des Moines, Iowa, where he stayed for forty-five years, eventually becoming vice-president. Hume's mother, Mary Evalina McMullin Carothers of Burlington, Iowa, descended from Scotch-Irish forebears who were mostly farmers and artisans as well. Mary Carothers's family had a strong musical background, which might explain her son's deep appreciation for music. Wallace Hume Carothers, the only scientist in his family, was the eldest of four siblings. One sister, Isobel, was a particular favorite of his. She became a star on the radio as Lu in the musical trio of Clara, Lu, and Em. Her death in 1936 was a loss from which Carothers never recovered.

Carothers began his education in the public schools of Des Moines, Iowa. As a boy he was recognized for his thoroughness in his schoolwork and exhibited an aptitude for experimenting with machines and working with his hands. Graduating from the North High School in 1914, he entered his father's school, the Capital City Commercial College, that fall. He completed the accountancy curriculum at an accelerated pace, graduating in 1915. In September of the same year, he enrolled at Tarkio College in Missouri to pursue courses in science as well as to assist teaching in the accounting department. From the outset of his college years, Carothers excelled in chemistry and physics.

During World War I, the head of the institution's chemistry department, Arthur M. Pardee, left Tarkio, and the college could not find anyone qualified to fill the position. Carothers, who had taken all the chemistry courses the college had to offer, took over Pardee's duties. Ineligible for military service for health reasons, he served as head of the department all throughout the war. All of the four chemistry majors at Tarkio who studied under Carothers's supervision subsequently obtained doctorates in that field and later testified to his leadership capabilities.

In 1920, after graduating from Tarkio with a bachelor of science degree, Carothers entered the University of Illinois chemistry department, completing his master's degree by the summer of 1921. Arthur Pardee, the former head of the Tarkio chemistry department, had become the head of the chemistry department at the University of South Dakota. He needed someone to teach courses in chemistry and enticed Carothers to take the position for the fall and spring semesters. Carothers accepted Pardee's offer, intending to work just long enough to earn the money to pay for his graduate work.

While teaching, Carothers began to work on some of **Irving Langmuir**'s ideas as they related to **organic chemistry**. In one of the first papers Carothers published, "The Double Bond" (1924), he described the double bond between atoms, borrowing the notions from physics that Langmuir had developed. Although Carothers enjoyed his contact with the students at South Dakota, it became clear to him that teaching would always remain a distant second to research, which he pursued with prodigious **energy** and determination. To devote more time to it, he returned to the University of Illinois in 1922 to pursue his Ph.D., which he completed in 1924.

At the University of Illinois, Carothers mastered both organic and **inorganic chemistry**, and during the years 1923 and 1924 held the Carr fellowship, the chemistry department's highest award. His thesis paper, prepared under the supervision of **Roger Adams**, addressed the reactions undergone by organic compounds known as **aldehydes** with **platinum** as a catalyst, or enhancer, of the reactions. After Carothers received his doctorate, the university offered him a position in the chemistry department, which Carothers accepted in the fall of 1924.

Harvard University began a search for an instructor of organic chemistry in 1926 and selected Carothers out of a wide field of applicants. During his first year at Harvard, Carothers taught courses in experimental organic chemistry and structural chemistry, switching his second year to lecturer and lab leader for a course on elementary organic chemistry. He developed a reputation as a researcher with highly original ideas. While at Harvard, Carothers began his experimental investigations into chemical structures of high **molecular weight** polymers.

In 1928 the Du Pont Company, having learned of Carothers's work with polymers, invited the scientist to become the director of a new research program at their flagship laboratory, the Experimental Station, at Wilmington, Delaware. The program's goal was fundamental research into industrial applications for what was then an emerging field of artificial materials, such as **vinyl**. Carothers would head an entire team of trained chemists whose duties would be to assist in his research aims. The position was an immense responsibility, but one that offered opportunities on a scale that Harvard could not match. Carothers accepted Du Pont's offer, and in the next nine years he and his team made significant advancements in the field of organic chemistry.

Carothers's creativity, enthusiasm, and capacity to bring out the best in his workers fostered an atmosphere that produced dozens of major contributions to both the theory of polymers and its applications. His initial investigations focused on molecules in the acetylene family, using methods based on discoveries of chemist and botanist Julius Arthur Nieuwland. This line of inquiry produced over twenty papers and patents. Carothers and his colleagues managed to combine vinylacetylene with a **chlorine** compound to create a polymer with properties like rubber. In 1931 Du Pont started to manufacture this product, which was marketed under the name neoprene.

In what was to prove even more fruitful, a separate area of research dealt with the creation of fibers whose structures resembled natural polymers such as silk and cotton. Carothers began experimenting with reactions among so-called **fatty acids** that were well understood by organic chemists at that time. What Carothers needed as the basis for his polymer chain was a reaction that would produce long molecules oriented in the same direction. His intuitions on the subject, which, before coming to Du Pont, he had already expressed in a letter to John R. Johnson of Cornell University on February 14, 1928, eventually paid off with the publication of thirty-one papers. With these writings he codified the terminology in the new field of polymers and proposed a general theory to explain the factors that govern their creation. His theories resulted not only in the production of the first synthetic fiber, **nylon**, in 1939, but also illuminated how polymers in the natural world operate. New techniques in polymerization proliferated rapidly after Carothers's findings spread through the scientific community.

As Carothers's reputation grew, chemists throughout the world looked to him for advice and guidance. He wrote for and edited scientific journals, among them the *Journal of the American Chemical Society* and *Organic Synthesis.* A frequent speaker at chemists' groups, he made a famous address in 1935 entitled "Polymers and the Theory of Polymerization" to a summer colloquium at John Hopkins. A year later he became the first industrial organic chemist elected to the National Academy of Sciences.

Carothers married Helen Everett Sweetman, an employee of the patent division in Du Pont's chemical division, on February 21, 1936. They had one daughter, Jane, born on November 27, 1937. In spite of success and burgeoning fame, Carothers suffered from periodic bouts of depression that worsened over time, despite the efforts of his friends and colleagues. The death of his favorite sister, Isobel, in January of 1937 plunged Carothers into an emotional slide that proved irreversible. On April 29, 1937, he committed suicide in Philadelphia. His family buried his ashes in Glendale Cemetery in Des Moines, Iowa. Carothers was remembered as a shy man with a generous streak. He disliked publicity but was always ready to help anyone who came to him with a problem or a question. Quiet in a group, he became lively and humorous when talking one-on-one. He took great interest in the world about him, staying well informed on politics, business, and economics. A voracious reader since boyhood, he amassed a great knowledge of literature and philosophy. He enjoyed tennis, although his duties as head researcher at Du Pont eventually forced him to give up playing. Above all he loved music and possessed a talent for singing, a trait he shared with his mother and siblings.

Carver, George Washington (1864-1943)

American professor, agricultural/food scientist, farmer

George Washington Carver devoted his life to research projects connected primarily with southern agriculture. The products he derived from the peanut and the soybean revolutionized the economy of the South by liberating it from an excessive dependence on cotton.

George Washington Carver was an agricultural chemist and botanist whose colorful life story and eccentric personality transformed him into a popular American folk hero to people of all races. Born into slavery, he spent his first 30 years wandering through three states and working at odd jobs to obtain a basic education. His lifelong effort thereafter to better the lives of poor Southern black farmers by finding commercial uses for the region's agricultural products and natural resources—in particular the peanut, sweet potato, cowpea, soybean, and native clays from the soil—brought him international recognition as a humanitarian and chemical wizard. An accomplished artist and pianist as well, Carver was among the most famous black men in the United States during the early twentieth century.

Carver was born a slave on the plantation of Moses Carver near Diamond Grove, Missouri, sometime during the American Civil War of 1861 to 1865. His father appears to have died in a log-rolling accident shortly after George's birth. The Carver farm was raided several times throughout the war and on one occasion, according to legend, bandits kidnapped George, who was then an infant, and his mother, Mary, and took them to Arkansas. Mary was never found, but a neighbor rescued young George and returned him to the Carver farm, accepting as payment a horse valued at $300.

Now orphaned, George and his older brother, Jim, were raised by Moses and Susan Carver. George was frail and sickly and his frequent bouts with croup and whooping cough temporarily stunted his growth and permanently injured his vocal chords, leaving him with a high-pitched voice throughout his life. While his healthy brother grew up working on the Carver farm, George spent much of his childhood wandering in the nearby woods and studying the plants. Here he formed the interests and values that determined his later life—love and understanding of nature, long morning walks in the woods spent thinking and observing, strong religious training, and a taste of racial prejudice.

The Carvers realized that George was an extremely intelligent and gifted child eager for an education. But since he was black, he was not allowed to attend the local school. In 1877 he left home to study in a school for blacks in nearby Neosho, getting his first exposure to a predominantly black envi-

George Washington Carver.

ronment. He roomed with a local black couple, paying his way by helping with the chores. Soon exhausting his teacher's limited knowledge, he hitched a ride to Fort Scott, Kansas, in the late 1870s with another black family, becoming part of the mass exodus of Southern blacks to the Great Plains during that decade in search of a better life.

Carver worked as a cook, launderer, and grocery clerk while continuing to pursue his education. Witnessing a brutal lynching in March of 1879, he was terrified. As quoted by Linda O. McMurry in *George Washington Carver: Scientist and Symbol,* more than sixty years after the incident he wrote: "As young as I was the horror haunted me and does even now." He immediately left Fort Scott and moved to Olathe, Kansas, again working odd jobs while attending school. There he lived with another local black couple, Ben and Lucy Seymour, following them to Minneapolis, Kansas, the next year. Obtaining a bank loan, Carver opened a laundry business, joined the Seymours' local Presbyterian Church, and entered a school with whites, finally completing his secondary education.

In 1884 he moved to Kansas City, working as a clerk in the Union Depot. Accepted by mail at a Presbyterian college in Highland, Kansas, he was refused admission when he arrived because of his race. Though humiliated, he stayed in Highland to work for the Beelers, a cordial and supportive white family. Carver followed one of their sons to western Kansas in 1886 and tried homesteading, building a 14-square-foot sod house. But at that time he seemed more interested in playing the piano and organ and in painting than farming.

Carver moved again in 1888 to Winterset, Iowa, where he worked at a hotel before opening another laundry. A local white couple he met at church, Dr. and Mrs. Milholland, persuaded him to enter Simpson College, a small Methodist school open to all, in nearby Indianola, Iowa. He enrolled in September of 1890 as a select preparatory student, one allowed to enter without an official high school degree. Carver was unique in more ways than one: besides being the only black student on campus, he was the only male studying art.

By all accounts his Simpson experience was enjoyable. Carver took in laundry to support himself, was accepted by his fellow students, and had many friends. But his art teacher, impressed by his talent with plants, strongly encouraged his transfer to the Iowa State College of Agriculture in Ames, which housed an agricultural experiment station considered one of the country's leading centers of farming research. Three future U.S. secretaries of agriculture came from this university, including Professor James Wilson, who took Carver under his wing.

Again Carver was the only black on campus. He lived in an old office, ate in the basement, supported himself with menial jobs, and was active in the campus branch of the Young Men's Christian Association (YMCA). Soon he stood out for his talent as well. One of his paintings was among those chosen to represent Iowa at the 1893 World's Columbian Exposition in Chicago. The faculty, equally impressed by his ability to raise, cross-fertilize, and graft plants, persuaded him to stay on as a post-graduate after he graduated in 1894.

Carver was appointed to the faculty as an assistant botanist in charge of the college greenhouse. He continued his studies under Louis Pammel, an authority in mycology (fungi and other plant diseases), receiving a master's degree in science in 1896.

The new graduate was in great demand. Iowa State wanted him to continue working there. Alcorn Agriculture and Mechanical College, a black school in Mississippi, was interested in his services. But when school principal Booker T. Washington, the most respected black educator in the country, asked Carver to establish an agricultural school and experiment station at Tuskegee Institute in Alabama, he accepted. According to Barry Mackintosh in *American Heritage,* Carver responded: ''Of course it has always been the one great ideal of my life to be of the greatest good to the greatest number of my people possible, and to this end I have been preparing my life for these many years, feeling as I do that this line of education is the key to unlock the golden door of freedom for our people.''

Tuskegee was an entirely new world for Carver—an all-black, industrial trade school located in the segregated deep South. He went because he agreed with Washington's efforts to improve the lives of the country's black citizens through education, economic development, and conciliation rather than political agitation. He would devote the rest of his life to the institution and its goals.

Carver arrived at Tuskegee in the fall of 1896 and immediately ran into problems. Many of the faculty members resented him because he was a dark-skinned black from the North who was educated in white schools and earned a higher salary than they did. Carver was a trained research scientist, not a teacher, at a primarily industrial trade school. He had few pupils, for the simple reason that most black students viewed a college education as a way to escape from the farm. In addition, he proved to be a poor administrator and financial manager of the school's two farms, barns, livestock, poultry, dairy, orchards, and beehives.

Washington and Carver often clashed. The realistic and pragmatic school principal expected practical results, while his idealistic, research-oriented professor preferred working at the school's 10-acre experimental farm. In 1910 Carver was removed as head of the agriculture department and put in charge of a newly-formed department of research. He gradually gave up teaching except for his Sunday evening Bible classes.

Carver found his true calling as head of the Tuskegee Experiment Station, working on research projects designed to help Southern agriculture in general and the poor black farmer, ''the man farthest down,'' in particular. Alabama agriculture was in a sorry state when he arrived. Many farmers were impoverished, and much of the state's soil had been exhausted and eroded by extensive single-crop cotton cultivation. Carver set out to find a better way and to make Tuskegee a leading voice in Southern agricultural reform, as well as an important research, information, and educational center.

He encouraged local farmers to visit the school and to send in soil, **water**, crops, feed, **fertilizers**, and insects to his laboratory for analysis. Most of his findings and advice stressed hard work and the wise use of natural resources rather than expensive machinery or fertilizer that the area's poor farmers could not afford. Realizing that his discoveries and those of other agricultural researchers nationwide would have little effect unless publicized, Carver brought Tuskegee to the countryside by creating the Agriculture Movable School, a wagon that traveled to local farms with exhibits and demonstrations.

He also attempted to reach a wide audience with the experiment station's bulletins and brochures that he wrote and published from 1898 until his death. Rarely containing new ideas, Carver's bulletins instead publicized findings by agricultural researchers throughout the country in simple, nontechnical language aimed at farmers and their wives. His early bulletins stressed the need for planting crops other than cotton to restore the soil, the importance of crop rotation, strategies for managing an efficient and profitable farm, and ways to cure and keep meat during the hot southern summers. They also offered instructions on pickling, canning and preserving foods and lessons on preparing balanced meals.

To replace cotton, the longtime staple of Southern agriculture, Carver experimented with sweet potatoes and cowpeas (also known as black-eyed peas), along with crops new to Alabama like soybeans and alfalfa, the soil-building qualities of which would revitalize cotton-exhausted soil. He publicized his results in several bulletins from 1903 to 1911, providing growing tips and listing uses ranging from livestock feed to recipes for human consumption. But none of these crops became as popular with farmers or caught the public's fancy as his work with the ordinary peanut.

When Carver arrived in Tuskegee in 1896, the peanut was not even recognized as a crop. A few years later, Carver grew some Spanish peanuts at the experiment station. Recognizing its value in restoring **nitrogen** to depleted Southern soil, he mentioned the peanut in his 1905 bulletin, *How to Build Up Worn Out Soils.* Eleven years later another bulletin, *How to Grow the Peanut and 105 Ways of Preparing It for Human Consumption,* focused on the peanut's high protein and nutritional value, using ideas and recipes published previously in other U.S. Department of Agriculture bulletins.

A revolution was underway in Southern agriculture, and Carver was right in the middle of it. Peanut production increased from 3.5 million bushels in 1889 to more than 40 million bushels in 1917. Following a post-World War I decline in production, peanuts became the South's second cash crop after cotton by 1940.

After publicizing the peanut and encouraging Southern farmers to grow it, Carver turned his attention to finding new uses for the once-lowly goober. Learning of his work, the United Peanut Associations of America asked Carver to speak at their 1920 convention in Montgomery, Alabama. His address, "The Possibilities of the Peanut," was noteworthy for two reasons: a black addressing a white organization in the segregated South and Carver's knowledge and enthusiasm about the product.

The following year he testified before the U.S. House Ways and Means Committee, captivating congressional representatives with his showmanship and ideas for multiple derivatives from the crop including candy, ink, and ice cream flavoring. Mackintosh noted that Carver established his new celebrity nationwide, telling the lawmakers, "I have just begun with the peanut." From then on he was known as the "Peanut Man." After his death, the Carver Museum at Tuskegee credited him with developing 287 peanut by-products, including food and beverages, paints or dyes, livestock feed, **cosmetics**, and medicinal preparations. Peanut butter, however, was not among his discoveries. His similar laboratory work with the sweet potato totaled 159 commodities like flour, molasses, vinegar, various dyes, and synthetic rubber.

But in reality, most of these by-products were more fanciful than practical and could be mass-produced more easily from other substances. Peanuts continued to be used almost entirely for peanut butter, peanut oil, and for baked goods instead of the plethora of products Carver concocted. For all his discoveries, he only held three patents: two for paint products and one for a cosmetic. None was commercially successful.

Carver's laboratory methods were equally unorthodox and not in accord with standard scientific procedures. He usually worked alone, was uncommunicative with other researchers, and rarely wrote down his many formulas or left detailed records of his experiments. Instead, he claimed to work by divine revelation, receiving instructions from "Mr. Creator" in his laboratory.

Carver's prestige began to rise after Booker T. Washington's death in 1915. Given his growing celebrity status, he became Tuskegee's unofficial spokesman and a popular speaker nationwide at black and white civic groups, colleges, churches, and state fairs. He often played the piano at fund-raising events for the school. Carver was named a fellow of the British Royal Society for the Arts in 1916 and received the NAACP's Spingarn Medal in 1923 for advancing the black cause.

A fanciful 1932 article in *American Magazine* solely credited Carver with increasing peanut production and developing important new peanut products that transformed Southern agriculture. Reprinted in the *Reader's Digest* in 1937, it boosted his soaring popularity as a scientific wizard. Backed by automobile manufacturer Henry Ford and inventor Thomas Edison, Carver became the unofficial spokesman of the chemurgy movement of the 1930s that combined **chemistry** and related sciences for the benefit of farmers. Continuing his work with peanuts, he encouraged the use of peanut oil as a massage to help in the recovery of polio victims.

With his soft-spoken manner, strong Christian beliefs, scientific reputation, seeming disregard for money, and accomodationist viewpoint toward the nation's racial question, Carver became a national symbol for both races. Southern whites approved of his seeming acceptance of segregation and used his accomplishments as an example of how a talented black individual could excel in their separate but equal society. Blacks and liberal whites saw Carver as a positive role model and much-needed symbol of black success and intellectual achievement, a man who visited U.S. presidents Franklin Roosevelt and Calvin Coolidge and dined with Henry Ford.

Carver left his life savings of $60,000 to found the George Washington Carver Foundation—to provide opportunities for advanced study by blacks in botany, chemistry, and agronomy—and the Carver Museum, to preserve his scientific work and paintings at Tuskegee. The site of Moses Carver's farm is now the George Washington Carver National Monument. A U.S. postage stamp was issued in the agricultural pioneer's honor, and Congress has designated January 5, the day of his death, to pay tribute to him each year.

At his death from complications of anemia in 1943, Carver remained the most famous African-American of his era, world renowned as a scientific wizard. However, none of his hundreds of formulas for peanut, sweet potato, and other by-products became successful commercial products. Nor was he solely instrumental in diversifying Southern agriculture from cotton to peanuts and other crops. The great boom in Southern peanut production occurred prior to World War I and Carver's bulletins promoting the crop.

Carver's true importance in history lay elsewhere. For nearly 50 years he remained in the South, working to improve the lives of the region's many poor farmers, black and white. Through his talents as an interpreter and promoter, he put the agricultural discoveries and technical writings of leading scientists in everyday language that ill-educated farmers could understand and use. And in an age of strict racial segregation, his importance as a role model and national symbol of black ability, education, and achievement cannot be undervalued.

George Castro. *(Arte Publico Press. Reproduced by permission.)*

CASTRO, GEORGE (1939-)
American physical chemist

George Castro's work in photoconductors and superconductors has led the way to new and improved electrophotographic copying machines as well as digital information storage systems. Additionally, he has worked for twenty-five years in civic activities on behalf of the Hispanic American community to ensure adequate education and employment opportunities.

Castro was born on February 23, 1939, in Los Angeles, California, the second of five children. His parents are both of Mexican descent. Castro grew up in Los Angeles, where he attended Roosevelt High School. Graduating in 1956, Castro won the Los Romanos Scholarship. Although it provided him with only a small amount of money, the scholarship was an important vote of confidence from the local Hispanic community. Castro was the first of his family ever to attend college. He went to UCLA, where he earned a B.S. in **chemistry**. He then became a research fellow in the chemistry department at Dartmouth College. Castro returned to Los Angeles in 1963 and married Beatrice Melendez, with whom he had attended both junior high and high schools. He finished his graduate studies at the University of California at Riverside, earning his Ph.D. in **physical chemistry** in 1965. For the next three years, Castro served as a postdoctoral fellow, first at the University of Pennsylvania, and then at Caltech.

In 1968, Castro joined the staff of IBM at its San Jose Research Lab. By 1971, he had become a project manager, and by 1973, a department manager. His early work was in photoconduction—the increased **electrical conductivity** of a substance when subjected to light waves. In particular, Castro discovered how organic photoconductors were generated by intrinsic charge carriers, an essential element in the technology of both electrophotographic copying machines and laser printers. His early work contributed to the understanding of the principles of the photogeneration process, which is essential to both copying and printing technology. Research teams he managed also discovered a mechanism called photochemical hole burning. By this mechanism, very high densities of digital information can be stored in **solids** at greater numbers than accomplished by other optical processes.

In 1975, Castro was made manager of the entire physical sciences division of the IBM San Jose Research Lab. Under his directorship the lab has made breakthrough discoveries in superconductors, high-resolution laser techniques, and new methods for investigating magnetic materials. From 1986 to 1992, Castro worked jointly with a team from Stanford University to develop a synchrotron X-ray facility—one which utilizes the mechanism of synchronized acceleration produced in atomic particles passing through an electrically energized magnetic field. Other research he has supervised includes the construction of a photoelectron microscope for high resolution spectroscopic studies.

Castro has consistently taken time from his busy work schedule to become involved in the education of Hispanic youth. He has taught math and science classes on a volunteer basis, served on the boards of numerous local associations and schools, and developed on-the-job training programs at IBM. Castro won the Outstanding Innovation Award from IBM in 1978, the Outstanding Hispanic Professional award from the San Jose Mexican-American Chamber of Commerce in 1984, and the Hispanic in Technology National Award from the Society of Hispanic Professional Engineers in 1986; he was elected a member of the American Physical Society in 1990. He and his wife have four children.

CATALYST AND CATALYSIS

A catalyst increases the rate of a particular reaction without itself being used up. A catalyst can be added to a reaction and then be recovered and reused after the reaction occurs. The process or action by which a catalyst increases the **reaction rate** is called catalysis. The study of reaction rates and how they change when manipulated experimentally is called **kinetics**.

The term catalysis was proposed in 1835 by the Swedish chemist **Jöns Berzelius** (1779-1848). The term comes from the Greek words *kata* meaning down and *lyein* meaning loosen. Berzelius explained that by the term catalysis he meant "the property of exerting on other bodies an action which is very different from chemical affinity. By means of this action, they produce decomposition in bodies, and form new compounds into the composition of which they do not enter."

Most chemical reactions occur as a series of steps. This series of steps is called a pathway or mechanism. Each individual step is called an elementary step. The slowest elementary step in a pathway determines the reaction rate. The reaction rate is the rate at which reactants disappear and products appear in a chemical reaction, or, more specifically, the change in **concentration** of reactants and products in a certain amount of time.

While going through a reaction pathway, reactants enter a transitional state where they are no longer reactants, but are not yet products. During this transitional state they form what is called an **activated complex**. The activated complex is short-lived and has partial **bonding** characteristics of both reactants and products. The **energy** required to reach this transitional state and form the activated complex in a reaction is called the **activation energy**. In order for a reaction to occur, the activation energy must be reached. A catalyst increases the rate of reaction by lowering the activation energy required for the reaction to take place. The catalyst forms an activated complex with a lower energy than the complex formed without catalysis. This provides the reactants a new pathway which requires less energy. Although the catalyst lowers the activation energy required, it does not affect reaction equilibrium or **thermodynamics**. The catalyst does not appear in the overall chemical equation for a pathway because the mechanism involves an elementary step in which the catalyst is consumed and another in which it is regenerated.

Catalysts exist for all types of chemical reactions. A specific catalyst can be classified into one of two main groups; homogeneous and heterogeneous. A catalyst that is in the same phase as the reactants and products involved in a reaction pathway is called a homogeneous catalyst. When a catalyst exists in a different phase than that of the reactants, it is called a heterogeneous catalyst. For example, **nickel** is a catalyst in the **hydrogenation** of vegetable oils. Nickel is a solid, while the oil is a liquid, therefore nickel is a heterogeneous catalyst. An advantage of using heterogeneous catalysts is their ease of separation from the reactants and products involved in a pathway. **Metals** are often used as heterogeneous catalysts because many reactants adsorb to the metal surface, increasing the concentration of the reactants and therefore the rate of the reaction. Ionic interactions between metals and other molecules can be used to orient the reactants involved so that they react better with each other, or to stabilize charged reaction transition states. Metals also can increase the rate of oxidation-reduction reactions through changes in the metal ion's **oxidation** state.

Another group of catalysts are called enzymes. Enzymes are catalysts that are found in biological systems. The role of catalysts in living systems was first recognized in 1833. French chemists **Anselme Payen** (1795-1871) and Jean François Persoz isolated a material from malt that accelerated the conversion of starch to sugar. Payen called the substance diatase. A half century later German physiologist Willy Kühne suggested the name **enzyme** for biological catalysts.

Enzymes are **proteins** and therefore have a highly folded three-dimensional configuration. This configuration makes an enzyme particularly specific for a certain reaction or type of reaction. Synthetic catalysts, on the other hand, are not nearly as specific. They will catalyze similar reactions that involve a wide variety of reactants. Enzymes, in general, will lose activity more easily than synthetic catalysts. Very slight disturbances in the protein structure of enzymes will change the three-dimensional configuration of the **molecule** and, as a result, its **reactivity**. Enzymes tend to be more active, i.e., they catalyze reactions faster, than synthetic catalysts at ambient temperatures. Catalytic activity for a reaction is expressed as the turnover number. This is simply the number of reactant molecules changed to product per catalyst site in a given unit of time. When **temperature** is increased, synthetic catalysts can become just as active as enzymes. With an increase in temperature, many enzymes will become inactive because of changes to the protein structure.

There are endless reactions that can undergo catalysis. One example is the decomposition of **hydrogen** peroxide (H_2O_2). Without catalysis, **hydrogen peroxide** decomposes slowly over time to form **water** and **oxygen** gas. A 30% **solution** of hydrogen peroxide at room temperature will decompose at a rate of 0.5% per year. The activation energy for this reaction is 75 kJ/mol. This activation energy can be lowered to 58 kJ/mol with the addition of iodide ions (I^-). These ions form an intermediate, HIO^-, which reacts with the hydrogen peroxide to regenerate the iodide ions. When the enzyme catalase is added to the hydrogen peroxide solution, the activation energy is lowered even further to 4 kJ/mol. The catalase is also regenerated in the reaction and can be separated from the solution for reuse. This example shows how a catalyst can lower the activation energy of a reaction without itself being used up in the reaction pathway.

Another example of catalysis is the catalytic converter of an automobile. Exhaust from the automobile can contain **carbon** monoxide and **nitrogen** oxides, which are poisonous **gases**. Before the exhaust can leave the exhaust system these toxins must be removed. The catalytic converter mixes these gases with air and then passes them over a catalyst made of **rhodium** and **platinum** metals. This catalyst accelerates the reaction of **carbon monoxide** with oxygen and converts it to **carbon dioxide**, which is not toxic. The catalyst also increases the rate of reactions for which the nitrogen oxides are broken down into their elements.

A well-known example of catalysis is the destruction of the ozone layer. Ozone (O_3) in the upper atmosphere serves as a shield for the harmful ultraviolet rays from the Sun. Ozone is formed when an oxygen molecule (O_2) is split into two oxygen atoms (O) by the radiation from the Sun. The free oxygen atoms then attach to oxygen molecules to form ozone. When another free oxygen **atom** reacts with the ozone molecule, two oxygen molecules are formed. This is the natural destruction of ozone. Under normal circumstances, the rate of destruction of ozone is the same as the rate of ozone formation, so no net ozone depletion occurs. When **chlorine** (Cl) atoms are present in the atmosphere, they act as catalysts for the destruction of ozone. Chlorine atoms in the atmosphere come from compounds containing chlorofluorocarbons, or CFCs. CFCs are compounds containing chlorine, **fluorine**, and carbon. CFCs are very stable and can drift into the upper atmosphere without first being broken down. Once in the upper atmosphere, the energy from the Sun causes the chlorine to be released. The chlorine atom reacts with ozone to form chlorine monoxide (ClO) and an oxygen molecule. The chlorine monoxide then reacts with another oxygen atom to form an oxygen molecule and the regenerated chlorine atom. With the help of the chlorine catalyst, the degeneration of ozone occurs at a faster rate than its formation, which has caused a net depletion of ozone in the atmosphere.

The previous examples illustrate some of the many practical applications of catalysis. Almost all of the chemicals produced by the chemical industry are made using catalysis. Catalytic processes used in the chemical industry decrease production costs as well as create products with higher purity and less environmental hazards. A wide variety of products are made using catalytic processes. Catalysis is used in industrial **chemistry**, **pharmaceutical chemistry**, and **agricultural chemistry**, as well as in the specialty chemical industry. Useful chemicals such as **sulfuric acid**, **penicillin**, and fructose are made more efficiently using catalytic processes. Research and development efforts in the chemical industry are significantly more productive with the use of catalysis in fields such as fuel refining, petrochemical manufacturing, and environmental management.

The majority of manufacturing processes in use today by the chemical industry employ catalytic reactions. These reactions are highly efficient, but research is continuing to increase the efficiency even more. The focus of this research is on separation and regeneration of the catalysts in order to decrease costs of production while increasing the purity of the product. The field of catalysis research is rapidly growing and will continue to do so as new catalysts and catalytic processes are discovered.

See also Ozone layer depletion

CATHODE

The term cathode, from the Greek words *hodos* meaning way and *kata* meaning down, was used by experimenters in the late 19th and early 20th century to designate the conducting material from which negatively charged particles, electrons, emanated when they subjected partially evacuated tubes to high voltage. The tubes were made of **glass** and typically had a voltage source connected to two pieces of conducting material sealed into the glass and separated by a few inches. The piece of conducting material from which the electrons emanated was called the cathode and the one to which they were attracted was called the **anode**.

Electrochemical cells consist of one half-cell in which **oxidation** occurs and a second half-cell in which **reduction** occurs. Electrons flow though a conductor connecting the two half-cells. At the two ends of the conductor are the electrodes which gather and disperse the electrons. These electrodes are called the *anode* (from a Greek word meaning ''up and out'') and the cathode. The cathode is the electrode at which reduction takes place and the anode is the electrode at which oxidation takes place.

A voltaic cell is an electrochemical cell is one in which an **oxidation-reduction reaction** takes place simultaneously. Voltaic cells are sources of **energy**. The **batteries** we use everyday are voltaic cells. In voltaic cells, electrons enter at the cathode, which has a (+) charge. Electrons originate at the anode in a voltaic cell, so the anode has a (-) charge. This polarity of the terminals and the role of the electrodes are accurate for any cell that is operating spontaneously and producing **electricity**.

In an electrolytic cell, energy must be supplied in the form of electric current from an external source. Electrolytic cells are used to electroplate tableware and jewelry and to produce useful chemicals from naturally occurring materials. In an electrolytic cell, the power source pumps electrons into the electrode to which it is attached; that electrode is the cathode. Therefore, in electrolytic cells, the cathode is the negative terminal and the anode is the positive terminal.

The polarities of the anode and cathode are reversed in comparison to their assignments in voltaic cells, but the function of the cathode, being the site of reduction and the anode, being the site of reduction, are the same.

The materials that are useful for cathodes must be good conductors and must not corrode too easily under reducing conditions. If the cathode itself is the substance that one wishes to reduce, it should be the most easily reduced material in the half-cell. If another substance is to be reduced, the cathode material should not be readily reduced. Some materials commonly used as unreactive cathodes are **platinum** and **graphite**.

Some examples of common cathodes are: (1) a **carbon** rod in typical non-alkaline dry cells batteries, (2) the steel container of alkaline dry-cell batteries, and 3) a solid paste consisting of **nickel** (IV) **oxide**, NiO_2, and nickel (II) hydroxide, $Ni(OH)_2$ in rechargeable NiCad batteries.

Electrochemical techniques are also widely used in chemical analysis. In electrochemical analysis by such methods as anodic stripping voltammetry, cyclic voltammetry, differential pulse polarography, and potentiometry the cathode always functions as the site of reduction. In general, the cathode is a highly conductive and relatively inert material such as platinum, graphite or **mercury**.

See also Electrochemistry; Oxidation and Reduction

CATION

A cation is a positively charged **ion**. The name comes from a 19th century combination of cata (from the Greek for down, kata) and ion. During **electrolysis**, cations are reduced at the cathode.

Cations have all lost one or more electrons, giving them a net positive charge. Thus, cations have more protons than electrons. Monatomic (single atom) cations are typically either metallic elements or **hydrogen**. This is because these are the elements with a tendency to lose electrons in a reaction, making them positively charged. Cations can, however, also be positively charged groups of atoms (polyatomic cations) such as ammonium NH_4. In nature, cations cannot exist by themselves. They are always in the presence of anions (negative ions). Cations and anions can only exist in ionic compounds, nearly all of which are **solids** at room **temperature**, or in **solution**. A cation will form an **ionic bond** with an **anion**. In an ionic compound, the sum of the positive charge(s) of the cation(s) and the negative charge(s) of the anion(s)must be zero. The actual formula of the ionic compound (and the number of cations and anions) will thus depend on the relative charges of the anion and cation. The charge on a cation is typically +1, +2, or +3. The majority of **metals** have a cation with a +2 charge, although some can exist in different forms. For example, **iron** exists as both a +2 and a +3 cation. The characteristics of the +2 and +3 form are slightly different. Iron nitrate can be found as iron (II) nitrate or iron (III) nitrate, the former being green in **color** while the latter is pale violet.

See also Anion; Electrolysis; Ion

Henry Cavendish.

Cavendish, Henry (1731-1810)

English physicist and chemist

Today, Henry Cavendish is recognized as one of the most brilliant scientists of his day, but his contemporaries knew him to be an extremely unusual person. Throughout his life, Cavendish avoided women entirely; he even went so far as to order his female servants to stay out of his way, or they would be fired. He was so shy he almost never left his house, except to attend the weekly dinners and meetings of the scientific Royal Society. Sometimes, he would hang around outside the meeting room, waiting to slip in when no one would notice him. If a stranger came near him, he quickly walked away. The only picture we now have of Cavendish is a watercolor sketch that was painted without his knowledge. He never changed his style of dress; he wore old-fashioned clothes, which he replaced with an identical outfit according to a regular schedule.

Whatever he lacked in social graces, Cavendish, a Cambridge scholar, made up for in scientific excellence. His laboratories and experiments were all that he cared about, and he managed to take part in the scientific activities of his day even though he preferred to work alone. After leaving Cambridge without getting a degree, which was not unusual then, Cavendish moved in with his father in London, England, where he set up a laboratory and workshop. Cavendish also maintained a large library in a separate building in London, and he had a country house equipped with another laboratory.

In 1766 Cavendish presented a paper to the Royal Society telling how he discovered **hydrogen** gas, which he called *inflammable air*. Cavendish produced hydrogen by combining **metals** with acids. For example, he found that hydrogen is produced when **zinc**, **iron**, or **tin** is dropped into **hydrochloric acid** or a diluted **sulfuric acid**. Cavendish's inflammable air was given its modern name of hydrogen in 1791 by **Antoine-Laurent Lavoisier**. Although a few chemists had collected hydrogen previously, Cavendish was the first to distinguish it from ordinary air and to investigate its specific properties. He found that hydrogen was extremely light in comparison to other **gases**. Eventually, this discovery was used to develop hydrogen-filled balloons, which could travel higher and farther than balloons filled with hot air. In some of his most important experiments, done during 1784 and 1785, Cavendish discovered that when hydrogen is combined with air, the mixture explodes and creates **water**. Cavendish even figured out approximately how much hydrogen and air combine to produce water, which anticipated the development of water's modern chemical formula (H_2O). At that time, many people still believed that water itself was a basic chemical element, along with fire, air, and earth.

Cavendish also discovered the inert gas **argon** while experimenting with **nitrogen**. Cavendish isolated nitrogen, although he is not credited with its discovery because his work remained unpublished, and then combined it electrically with **oxygen**. When finishing the experiment, he found that about one percent of the original gas was unaccounted for in the reaction. In 1894 this mysterious gas was identified as argon by Sir William Ramsey. Cavendish also discovered the composition of **nitric acid**. By dissolving the nitrogen-oxygen gas in water, he created nitric acid, proving that it is composed of nitrogen, oxygen, and hydrogen.

In a series of about 400 experiments, Cavendish established that the composition of air is the same regardless of the air's geographic origin. He determined that air contains about twenty-one percent oxygen, and he studied the properties of **carbon** dioxide, comparing its specific properties with those of air. During the same period Cavendish studied the freezing points of **mercury** and various acids.

Cavendish also performed many important physical and electrical experiments. He was the first scientist to calculate the Earth's mass—and he was right, to within about 10% of the best modern estimates. In this spectacular experiment, Cavendish made the most important advance in the understanding of gravity since Isaac Newton proposed the *law of gravity* in 1665. Newton's law lacked one key ingredient—a number called the gravitational constant that represents gravity's force of attraction between objects. Without this number scientists could not calculate gravity's effect on the orbits of the moon and planets, the ebb and flow of the ocean's tides, or the path of falling objects and projectiles. Cavendish measured gravity's force of attraction, using a method suggested earlier by John Michell (1724-1793). The gravitational attraction between objects that are small enough to work with in the lab is very weak, so Cavendish's equipment had to be very sensitive to detect the force. His equipment consisted of a lightweight rod suspended by a wire; small balls were placed at both ends of the rod. When Cavendish placed larger balls near the small ones, the gravitational attraction between each pair of balls caused the rod to twist. From this he calculated the strength of attraction, which enabled him to solve Newton's equation for the gravitational constant. Once that number was known, Cavendish was then able to calculate the **mass** of the Earth as well as its average **density**. Cavendish had made the law of gravity complete for the first time.

In Cavendish's day, electrical experimentation had become quite fashionable in scientific and popular circles throughout England and continental Europe. Cavendish anticipated many basic concepts that were later developed by **Michael Faraday**, Charles-Augustin Coulomb, and others. He explained how charged objects attract or repel each other, and he introduced the idea of voltage, or electrical potential. He also showed that different materials conduct different amounts of electric current and studied the capacity of those materials to store an electrical charge.

Although Cavendish published two papers on **electricity** during the 1770s, his work remained unknown until his notebooks and manuscripts were discovered and published by **James Clerk Maxwell** in 1879. Much of Cavendish's other scientific research in optics, geology, **magnetism**, and pure mathematics was not revealed during his lifetime, but Cavendish did not care whether his research was appreciated or whether he got credit for his work. He studied science to satisfy his own curiosity. When Cavendish died, his relatives left much of his fortune to Cambridge University, which established its Cavendish Professorship and the world-renowned Cavendish Laboratory.

CECH, THOMAS R. (1947-)

American biochemist

The work of Thomas R. Cech has revolutionized the way in which scientists look at **RNA** and at **proteins**. Up to the time of Cech's discoveries in 1981 and 1982, it had been thought that genetic coding, stored in the **DNA** of the **nucleus**, was imprinted or transcribed onto RNA molecules. These RNA molecules, it was believed, helped transfer the coding onto proteins produced in the ribosomes. The DNA/RNA nexus was thus the information center of the cell, while protein molecules in the form of enzymes were the workhorses, catalyzing the thousands of vital chemical reactions that occur in the cell. Conventional wisdom held that the two functions were separate—that there was a delicate division of labor. Cech and his colleagues at the University of Colorado established, however, that this picture of how RNA functions was incorrect; they proved that in the absence of other enzymes RNA acts as its own catalyst. It was a discovery that reverberated throughout the scientific community, leading not only to new technologies in RNA engineering but also to a revised view of the evolution of life. Cech shared the 1989 Nobel Prize for **Chemistry** with **Sidney Altman** at Yale University for their work regarding the role of RNA in cell reactions.

Cech was born in Chicago on December 8, 1947, to Robert Franklin Cech, a physician, and Annette Marie Cerveny Cech. As he recalls in an autobiographical sketch for *Les Prix Nobel,* he grew up in "the safe streets and good schools" of Iowa City, Iowa. His father had a deep and abiding interest in physics as well as medicine, and from an early age Cech took an avid interest in science, collecting rocks and minerals and speculating about how they had been formed. In junior high school he was already conferring with geology professors from the nearby university. He went to Grinnell College in 1966; at first attracted to **physical chemistry**, he soon concentrated on biological chemistry, graduating with a chemistry degree in 1970.

It was at Grinnell that he met Carol Lynn Martinson, who was a fellow chemistry student. They married in 1970 and went together to the University of California at Berkeley for graduate studies. His thesis advisor there was John Hearst who, Cech recalled in *Les Prix Nobel,* "had an enthusiasm for **chromosome** structure and function that proved infectious." Both Cech and his wife were awarded their Ph.D. degrees in 1975, and they moved to the east coast for postdoctoral positions—Cech at the Massachusetts Institute of Technology (MIT) under Lou Pardue, and his wife at Harvard. At MIT Cech focussed on the DNA structures of the mouse genome, strengthening his knowledge of biology at the same time.

In 1978, both Cech and his wife were offered positions at the University of Colorado in Boulder; he was appointed assistant professor in chemistry. By this time, Cech had decided that he would like to investigate more specific genetic material. He was particularly interested in what enables the DNA **molecule** to instruct the body to produce the various parts of itself—a process known as gene expression. Cech set out to discover the proteins that govern the DNA transcription process onto RNA, and in order to do this he decided to use **nucleic acids** from a single-cell protozoa, *Tetrahymena thermophila.* He chose *Tetrahymena* because it rapidly reproduced genetic material and because it had a structure which allowed for the easy extraction of DNA.

By the late 1970s much research had already been done on DNA and its transcription partner, RNA. It had been determined that there were three types of RNA: messenger RNA, which relays the transcription of the DNA structure by attaching itself to the ribosome where **protein synthesis** occurs; ribosomal RNA, which imparts the messenger's structure within the ribosome; and transfer RNA, which helps to establish **amino acids** in the proper order in the protein chain as it is being built. Just prior to the time Cech began his work, it was discovered that DNA and final-product RNA (after copying or transcription) actually differed. In 1977 Phillip A. Sharp and others discovered that portions of seemingly noncoded DNA were snipped out of the RNA and the chain was spliced back together where these intervening segments had been removed. These noncoded sections of DNA were called introns.

Cech and his coworkers were not initially interested in such introns, but they soon became fascinated with their function and the splicing mechanism itself. In an effort to understand how these so-called nonsense sequences, or introns, were removed from the transcribed RNA, Cech and his colleague Arthur Zaug decided to investigate the pre-ribosomal RNA of the *Tetrahymena,* just as it underwent transcription. In order to do this, they first isolated unspliced RNA and then added some *Tetrahymena* nuclei extract. Their assumption was that the catalytic agent or **enzyme** would be present in such an extract. They also added small molecules of salts and nucleotides for **energy**, varying the amounts of each in subsequent experiments, even excluding one or more of the additives. But the experiment took a different turn than they expected.

What Cech and Zaug discovered was that RNA splicing would occur even without the nucleic material being present. This was a development they did not understand at first; it was a long-held scientific belief that proteins in the form of enzymes had to be present for catalysis to occur. But here was a situation in which RNA appeared to be its own catalytic motivator. At first they suspected that their experiment had been contaminated. Cech did further experiments involving recombinant DNA in which there could be no possibility of the presence of splicing enzymes, and these had the same result: the RNA spliced out its own intron. Further discoveries in Cech's laboratory into the nature of the intron led to his belief that the intron itself was the catalytic agent of RNA splicing, and he decided that this was a sort of RNA enzyme which they called the ribozyme.

Cech's findings of 1982 met with heated debate in the scientific community, for it upset many beliefs about the nature of enzymes. Cech's ribozyme was in fact not a true enzyme, for thus far he had shown it only to work upon itself and to be changed in the reaction; true enzymes catalyze repeatedly and come out of the reaction unchanged. Other critics argued that this was a freak bit of RNA on a strange microorganism and that it would not be found in other organisms. They were soon proved wrong, however, when scientists around the world began discovering other RNA enzymes. In 1984, **Sidney Altman** proved that RNA carries out enzyme-like activities on substances other than itself.

A micrograph image of adenosine triphosphate (ATP). ***(Photograph by M. W. Davidson, National Audubon Society Collection/Photo Researchers, Inc. Reproduced by permission.)***

The discovery of catalytic RNA has had profound results. In the medical field alone RNA enzymology may lead to cures of viral infections. By using these rybozymes as gene scissors, the RNA molecule can be cut at certain points, destroying the RNA molecules that cause infections or genetic disorders. In life sciences, the discovery of catalytic RNA has also changed conventional wisdom. The old debate about whether proteins or nucleic acids were the first bit of life form seems to have been solved. If RNA can act as a catalyst and a genetic template to create proteins as well as itself, then it is rather certain that RNA was first in the chain of life.

Cech and Altman won the Nobel Prize for Chemistry in 1989 for their independent discoveries of catalytic RNA. Cech has also been awarded the Passano Foundation Young Scientist Award and the Harrison Howe Award in 1984; the Pfizer Award in Enzyme Chemistry in 1985; the U. S. Steel Award in Molecular Biology; and the V. D. Mattia Award in 1987. In 1988, he won the Newcombe-Cleveland Award, the Heineken Prize, the Gairdner Foundation International Award, the Louisa Gross Horwitz Prize, and the Albert Lasker Basic Medical Research Award; he was presented with the Bonfils-Stanton Award for Science in 1990.

Cech was made full professor in the department of chemistry at the University of Colorado in 1983. He and his wife have two daughters. In the midst of his busy career in research, Cech still finds time for skiing and backpacking.

CELLULAR RESPIRATION

Cellular respiration is the process by which cells obtain **energy** from food through chemical reaction with an inorganic **electron** acceptor, usually **oxygen**. The principal product is adenosine triphosphate (ATP), a high energy compound used for a wide variety of energy-requiring processes in the cell. Cellular respiration occurs in three main stages: glycolysis, **Krebs cycle**, and oxidative phosphorylation. (In some textbooks, cellular respiration refers only to the Krebs cycle and oxidative phosphorylation, while in others, it refers only to oxidative phosphorylation). Cellular respiration is contrasted with physiologic respiration, which refers to the mechanisms of gas exchange at the lungs.

Sugars and **fatty acids** are the primary food sources for cells. Each contains large numbers of C-C and C-H bonds, which are relatively weak compared to C-O and H-O bonds. During cellular respiration, these weaker bonds are broken, while the stronger bonds with oxygen are formed, thus releasing energy. This energy is used to form the weakly bonded ATP **molecule** from its constituents, adenosine diphosphate (ADP) and inorganic **phosphate** (Pi). The formation of ATP absorbs energy, which is thus stored and available for driving reactions elsewhere in the cell.

Glycolysis, the first stage of cellular respiration, occurs in the cytosol of the cell. Only sugars undergo glycolysis. During glycolysis, a glucose molecule ($C_6H_{12}O_6$) is split to form two molecules of pyruvic acid ($C_3H_8O_3$). Hydrogens from glucose are removed by the carrier molecule nicotinamide adenine dinucleotide (NAD^+), forming NADH. The bond joining the H to the NAD^+ is weak, meaning the electrons of the bond are still high-energy electrons. In this way, NADH serves as a transporter of high-energy electrons from the cytosol into the mitochondria, where the rest of cellular respiration takes place. Two NADH are formed for each glucose reacted, along with two molecules of ATP. Most of the energy of the glucose remains in the pyruvates, however.

Pyruvic acid passes into the inner compartment of the mitochondrion, the cell organelle chiefly responsible for ATP production. In this compartment, called the matrix, the pyruvic acid is decarboxylated to a two-carbon acetyl group, releasing a CO_2 molecule, and is enzymatically joined to **Coenzyme** A, a large carrier molecule. This reaction creates another NADH molecule. Fatty acids are also linked with coenzyme A, two carbons at a time. The product in all cases is acetyl-coA.

Acetyl-coA then enters the Krebs cycle. In this series of reactions, the two-**carbon** acetyl group is linked to a four-carbon compound, forming **citric acid**. (The Krebs cycle is also called the citric acid cycle, and, because of the three carboxyl groups in citric acid, it is also known as the tricarboxylic acid cycle.) In a series of transformations, the four-carbon compound is regenerated, **carbon dioxide** is released, and ATP, NADH, and $FADH_2$ are formed. $FADH_2$ is another high-energy electron carrier.

The final stage of cellular respiration occurs in two steps. In the first step, NADH and $FADH_2$ are stripped of their electrons, regenerating the original carrier molecules, which are recycled to their original locations. The electrons are attracted away from their carriers by NADH-Q reductase, the first in a series of increasingly electronegative **proteins** that form an electron transport chain in the inner membrane of the mitochondria. Each protein in turn is first reduced when it accepts the electrons, then oxidized as they are removed by the next protein in the chain. In succession, these carriers are ubiquinone, cytochrome reductase, cytochrome c, cytochrome oxidase. The electrons are finally accepted by molecular oxygen, which together join with H^+ ions to form **water**. The energy released during this series of redox reactions is used to transport other H^+ ions across the inner mitochondrial membrane, creating an electrochemical gradient.

The second, final step of this stage uses the energy stored in the electrochemical gradient to produce ATP. H^+ ions flow through a membrane protein called ATP synthase. The energy released by this flow drives the synthesis of ATP from ADP and Pi. This process is known as **chemiosmosis**. The combination of electron transport and chemiosmostic ATP synthesis is known as oxidative phosphorylation, sometimes abbreviated as OXPHOS.

The overall ATP harvest from one glucose molecule is either 36 or 38 ATP, depending on the cell type involved. Of these, all but four are formed by ATP synthase.

In some instances, electron transport can be ''uncoupled'' from ATP synthesis, so that food molecules are consumed, but no ATP is created. Instead, the mitochondria releases the energy harvested as **heat**. This occurs naturally in brown fat, a mitochondria-rich tissue found in human babies and in adults of some other species that hibernate. This fat tissue is brown due to the high numbers of mitochondria present. In these mitochondria, the inner membrane is porous, so that the H^+ ions transported during the electron transport chain flow back through, without ATP production. In this way, heat is generated directly, without requiring creation and consumption of ATP. Fat-soluble **proton** carriers such as dinitrophenol act as chemical uncouplers. Such compounds have been tried as diet drugs, but the effective dose is too close to the lethal dose to allow them to be used safely.

See also Bioenergetics; Carbohydrates; Fermentation

CELLULOSE

Cellulose is a polymer of glucose, and is formed by plants, fungi, and algae. Cellulose is the principal component of the plant cell wall, and provides the raw material for a wide variety of plant-based industries.

Cellulose forms ribbon-like **chains**, each composed of thousands of glucose molecules bonded together end to end. These chains are linked together by a network of **hydrogen** bonds, allowing cellulose to form microfibrils, which are themselves linked by non-cellulose molecules to form the cell wall. The plant cell membrane is enclosed within this meshwork of cellulose.

Cellulose is a principal component of **wood**, flax, hemp, jute, and other plant products used for fuel or fiber. Cotton is virtually pure cellulose. During **paper** manufacture, wood is processed to remove much of the non-cellulose material, leaving the pulp, whose fibers are then pressed and dried to make the paper.

Although both starch and cellulose are pure glucose, a slight chemical difference between the two prevents humans from digesting cellulose, but allows us to obtain **energy** from starch. As in starch, glucose molecules in cellulose are linked by **dehydration** between the 1 position of one glucose **ring** and the 4 position of the next. However, in cellulose, the geometric orientation of the H and OH at the 1 position differ from those in starch, and as a result, the linkage takes on a different shape. This so-called beta-1,4 bond cannot be broken by starch-digesting enzymes, and in fact cannot be broken down at all by animals. Fungi, certain protozoans, and some bacteria do contain cellulose-digesting enzymes, called cellulases. These organisms form the important class of decomposers within the food web, without which cellulose would continue to accumulate in the environment. Termites host some species of cellulose-digesting protozoans in their gut, allowing them to use wood as a food source. Cows and other ruminants employ the same strategy. Humans have no such organism in their digestive systems, and as a result, cellulose passes through the intestines unchanged.

Celsius, Anders (1701-1744)

Swedish astronomer

Celsius is a familiar name to much of the world since it represents the most widely accepted scale of **temperature**. It is ironic that its inventor, Anders Celsius, the inventor of the Celsius scale, was primarily an astronomer and did not conceive of his temperature scale until shortly before his death.

The son of an astronomy professor and grandson of a mathematician, Celsius chose a life within academia. He studied at the University of Uppsala where his father taught, and in 1730 he, too, was given a professorship there. His earliest research concerned the aurora borealis (northern lights), and he was the first to suggest a connection between these lights and changes in the earth's magnetic field.

Celsius traveled for several years, including an expedition into Lapland with French astronomer Pierre-Louis Maupertuis (1698-1759) to measure a degree of longitude. Upon his return he was appointed steward to Uppsala's new observatory. He began a series of observations using colored **glass** plates to record the magnitude of certain stars. This constituted the first attempt to measure the intensity of starlight with a tool other than the human eye.

The work for which Celsius is best known is his creation of a hundred-point scale for temperature, although he was not the first to have done so since several hundred-point scales existed at that time. Celsius' unique and lasting contribution was the modification of assigning the freezing and boiling points of **water** as the constant temperatures at either end of the scale. When the Celsius scale debuted in 1747 it was the reverse of today's scale, with zero degrees being the **boiling point** of water and one hundred degrees being the freezing point. A year later the two constants were exchanged, creating the temperature scale we use today. Celsius originally called his scale centigrade (from the Latin for "hundred steps"), and for years it was simply referred to as the Swedish thermometer. In 1948 most of the world adopted the hundred-point scale, calling it the Celsius scale.

Anders Celsius.

Cement and concrete

Concrete, from the Latin word *concretus*, meaning having grown together, almost always consists of a relatively unreactive filler, or aggregate; Portland cement (from the Latin *caementum*, meaning quarry stone; **water**; and voids. The filler is usually a conglomerate of rocks and sand, and more specifically, of gravel, pebbles, sand, broken stone, and blast-furnace stony **matter** known as slag.

Portland cement is made up of finely pulverized stony matter produced by heating mixtures of **lime**, silica, alumina, and **iron oxide** in air to about 2,642°F (1,450°C). The properties of the cement may be varied by changing the relative proportions of the ingredients, and by grinding to different degrees of fineness. Chemically, Portland cement is a mixture of **calcium aluminum silicates**, typically including tricalcium silicate ($3CaO \cdot SiO_2$), dicalcium silicate ($2CaO \cdot SiO_2$), and tricalcium aluminate ($3CaO \cdot Al_2O_3$); it may also contain tetracalcium aluminoferrate ($4CaO \cdot Al_2O_2 \cdot Fe_2O_2$). (The component CaO, commonly referred to as quicklime, is prepared industrially by heating **calcium carbonate** ($CaCO_2$) to 1,562°F

[850°C].) Small amounts of **sulfur**, **potassium**, **sodium**, and **magnesium** ions may also be present in the cement.

When Portland cement is mixed with water, the cement and water form a rigid **gel** with the approximate formula $Ca_2Si_2O_7{\cdot}3H_2O$. (A gel is a mixture of a solid and usually a liquid that behaves mechanically as a solid. But the water in this gel is bonded to the solid phase only by non-covalent interactions.) For a short time, the resultant mix remains malleable, but as chemical reactions take place, the mix begins to stiffen, i.e., set. Even after the mix has finished setting, it continues to combine chemically with water, lending rigidity and strength to the composite. This process is called hardening. The wet concrete mix is characterized by its **viscosity**.

Ordinarily, an excess of water is added to the concrete mix to decrease the viscosity and allow the concrete to form. When the chemical reactions have more or less reached completion, the excess water, which is only held by non-covalent interactions, either escapes, leaving behind voids, or remains trapped in tiny capillaries. Hardening and setting result from chemical reactions; they do not occur as the result of the mixture drying out. In the absence of water, the reactions stop. But as long as moisture is present, these reactions continue, sometimes for many years. Hardening does not have to take place in air; it can even take place under water.

The strength of concrete is determined by the extent to which chemical reactions have taken place between the cement and water, by the filler size and distribution, by the void **volume**, and by the amount of water used.

Concretes have been modified with plastic (polymeric) materials to improve their properties, these materials sometimes being referred to as polymer-reinforced hydrate ceramic **composites**. One such composition consists of a super-durable **alloy** of poly(ethylene terephthalate), poly(butylene terephthalate), and concrete. More traditional reinforced concretes very often contain steel.

See also Iron; Oxide; Polymer chemistry; Silicon

CENTRIFUGE

A centrifuge is a device used to separate **liquids** from **solids** by spinning. Long ago, people saw that gravity could eventually separate a sediment from a liquid or separate two liquids which do not mix. The heavier element within a container would descend, while the lighter element would rise to the surface. This process was extremely slow if left up to nature alone and was also wasteful, as evidenced by the way farmers used to separate cream from milk. They would let whole milk stand for several hours until the lighter cream rose to the top. They then skimmed off the cream with a wooden spoon, but as much as 40 percent of the cream was left in the milk. Later, small strainer dishes were used to extract the cream, yet this too was a slow and tedious process.

In 1877, Swedish inventor Carl Gustaf Patrik de Laval introduced a high-speed centrifugal cream separator. Milk was placed in a chamber where it was heated and then sent through tubes to a container that was spun at 4,000 revolutions per minute by a steam engine. The centrifugal force separated the lighter cream, causing it to settle in the center of the container. The heavier milk was pushed to the outer part and forced up to a discharge pipe. Thus, only the cream was left in the container. Several years later an improved cream separator was introduced with the capability for self-skimming and self-emptying. Other separators can extract impurities from lubricating oils, beer and wine, and numerous other substances. Other types of centrifuges were created in which spin dryers were used for filtering solids: a perforated drum was spun, driving any separated liquids to the outside where they were collected. These spin dryers can now develop accelerations of up to 2,000 times the force of gravity. They are used in the food, chemical, and mineral industries to separate **water** from all sorts of solids. Other centrifuges remove blood serum (plasma) from the heavier blood cells.

However, some scientists needed faster rotations for separating smaller particles. Such particles, like **DNA** (deoxyribonucleic acid), **proteins**, and viruses, are too small to settle out with normal gravity; the banging of water molecules is enough to keep the particles from separating. The key was to build an ultracentrifuge that could spin fast enough to cause these small particles to settle out. In 1923 a Swedish chemist, **Theodor Svedberg**, developed a device that could spin fast enough to create gravity over 100,000 times normal. It could take small samples in **glass** containers, balance them on a cushion of air, and send jets of compressed air that touched the outer surface. By 1936 Svedberg had produced an ultracentrifuge that spun at 120,000 times per minute and created a centrifugal force equal to 525,000 times that of normal gravity. Newer models can accelerate samples to 2,000,000 times the force of gravity. This machine enabled biologists, biochemists, physicians, and other life scientists to examine viruses; cell nuclei; small parts within cells; and individual protein and nucleic acid molecules. Thus, **genetic engineering** became a field ripe with possibility. Centrifuges are also commonly used in the petroleum industry to separate oil components. In 1997, state of the art centrifuges, such as those built by U.S. Centrifuge, use a finite-element-analysis software made by Algor Inc. to design advanced centrifuge systems.

CERAMICS

The term ceramic refers to an inorganic material that is solid and nonmetallic. Ceramic materials are generally made from compounds of metallic and nonmetallic elements. Typical examples would include metal oxides (*i.e.*, Al_2O_3, BeO and ZrO_2, $BaTi_3$, Ti_2), metal and semimetal nitrides (*i.e.*, Si_3N_4 and metal and semimetal carbides (*i.e.*, B_4C and SiC). These compounds exhibit either covalent or ionic **bonding**, depending upon the location of the metallic and nonmetallic elements in the **periodic table**. Ionic bonds result whenever a metallic and a nonmetallic element from opposite sides of the periodic table are combined. Combinations closer together in the periodic table, like **silicon** with **nitrogen** and silicon with **carbon**, tend to be covalent in nature.

The type of bonding involved determines to a large extent the material's mechanical, chemical and electrical properties. For example, silicon nitride and silicon carbide are stable to air **oxidation** up to 2,552-2732°F (1,400-1,500°C). Metal **oxide** ceramic materials also do not react with **oxygen**. Ceramic materials are generally less dense than steel, and generally possess fairly high melting point temperatures. Many ceramic materials are unfortunately brittle, which does limit their use in a number of applications.

From a historical perspective, the discovery of ceramic materials can be traced back to prehistoric man. At some point in our past, and no one really knows when, our ancestors realized that a large leaf or an animal hide could be used for transporting items such as grain or fruits. The increase in carrying capacity over just the arms probably enhanced the survivability of the group. But hides are heavy and leaves are brittle. Again, an unknown ancestor began to weave together leaves and bark, generating a tough, light basket that was excellent for carrying dry material. Unfortunately, such baskets tended to leak like the proverbial sieve. The problem could be solved by using mud to line the basket. It would hold **water** for a short period of time although it did tend to make the basket a little more fragile as the mud was subject to cracking. Drying the basket in the sun increased the toughness of the material and the durability of the basket. Surely, then, drying the basket using the increased **heat** of a fire would further enhance the toughness and make a better basket.

Such a basket may have been the foundation for the first clay vessel. The interior would be lined with clay and then placed in a fire. As the exterior burnt away and the clay was fired, a serviceable, waterproof, and relatively lightweight pot would remain. Is this how it happened? Perhaps. Perhaps not. Crude earthenware pots, made from sun dried clay have been dated back to about 9000 B.C. and fired clay pots back to about 7000 B.C. Such pots are an early example of the use of fire for something other than cooking. That is, the first attempts by civilized man to change the world around him to fit his needs rather than accepting what nature had to offer.

A small community with access to high grade kaolinite, and consequently producing high grade pottery, could trade for food or other resources. The advent of glazes and ''slips'' (a fine water **suspension** of clay used to provide a non-porous exterior) lead to the decoration of clay pots which, in turn, further stimulated trade between adjacent groups. In addition, glazing, which essentially coats the clay in **glass**, lead to the development of glass as a separate and identifiable material.

Arguably, ceramics is one of the pillars upon which modern civilization was built, both literally and figuratively. We still use ceramics. They find their way into the kitchen cupboard as bowls, plates, and cups and on our shelves as flower vases. Indeed, a ceramic cup is still the most environmentally friendly method for holding one's morning coffee. Bricks have always been used for years by the construction industry, and for lining of ore **smelting** furnaces. Ceramic blades are being tested in jet engine turbines and automobile engines. Ceramics stovetops are common, perhaps the most famous use of ceramics is the ceramic tiles that coat the space shuttles, preventing re-entry burn-up.

Sand, clay and feldspar (aluminum silicates) were the basic raw materials for many of the very early ceramic materials. High temperatures were needed to process ceramics. The conversion of feldspar to cristobalite: $3\,(Al_2O_3 \cdot 2\,SiO_2 \cdot 2\,H_2O) \rightarrow 3\,Al_2O_3 \cdot 2\,SiO_2 + 4\,SiO_2 + 6\,H_2O$ involves the removal of water. Cristobalite is a crystalline form of silica that has a diamondlike structure. The silicon atoms are placed as are the carbon atoms in **diamond**, with the oxygen atoms being situated midway between each Si-Si pair.

Ceramic materials can also be produced by a sol-gel method, which involves the synthesis of the metal oxide from its respective metal alkoxide. In the commercial synthesis of titania, TiO_2, pure $Ti(OCH_2CH_3)_4$ reacts with water in an appropriate organic solvent $Ti(OCH_2CH_3)_4 + 4\,H_2O \rightarrow Ti(OH)_4 + 4\,CH_3CH_2OH$ to form a colloidal dispersion of extremely small solid particles of $Ti(OH)_4$. The **pH** of the **solution** is then judiciously adjusted to cause a water **molecule** to be expelled from between two adjacent Ti-OH bonds. When the material is heated to a **temperature** of between 392-932°F (200-500°C) all of the liquid is removed, leaving behind a finely divided **titanium** dioxide powder. Fine powders of **zirconium** dioxide, **aluminum** oxide and silicon dioxide can be prepared in similar fashion from their respective metal alkoxides.

In 1987, a Swiss physicist, Karl Alex Mueller, and a German colleague, Johannes Georg Bednorz, studying ceramics of metallic oxides, found that they were able to obtain **superconductivity** at 30K. This broke the previous record high by over ten degrees and suddenly focussed the efforts and energies of physicists everywhere on this ''new'' material. Ceramics were suddenly the ''hot topic'' for physicists around the world. Shortly thereafter, superconductivity slightly above the liquid nitrogen temperature range (77K) was demonstrated using Barium-Yttrium-Copper Oxides, Mercury-Calcium-Barium-Copper Oxides and Thallium-Calcium-Barium-**Copper** Oxides, bringing about predictions of room temperature ceramic superconductors within a few years. Unfortunately, the predictions were overly optimistic and there are many inherent problems that must be overcome before the full potential of ceramic superconductors is realized. The present processes are ''cookbook'' **chemistry**, with little reproducibility between successive batches. The ceramic materials generated are not easily incorporated into the electrical wires or films necessary for practical applications. The brittleness of the substance leads to stress failures.

CERIUM

Cerium is the fourth element in Row 6 of the **periodic table**. It belongs to the family known as the lanthanides, named after the first member of that series, **lanthanum** (atomic number 47). The members of this family are also known as the **rare earth elements**. The name arose originally not because of their scarcity in the Earth's crust, but because of the difficulty of separating the lanthanide elements from each other. Cerium's **atomic number** is 58, its atomic **mass** is 140.12, and its chemical symbol is Ce.

Properties

Cerium is an iron-gray metal with a melting point of 1,463°F (795°C), and a **boiling point** of 5,895°F (3,257°C). Its

density is 6.78 grams per cubic centimeter. It is ductile and malleable and exists in four different allotropic forms.

Cerium is a relatively active metal. It can be set on fire simply by scratching the surface with a knife. It also reacts slowly with cold **water**, rapidly with hot water, and with most acids.

Occurrence and Extraction

Cerium is thought to rank about 26th in abundance among elements found in the Earth's crust with an estimated abundance of 40-66 parts per million. It occurs most commonly in the minerals known as cerite, monazite, and bastnasite.

Discovery and Naming

Cerium was discovered in 1839 by the Swedish chemist Carl Gustav Mosander. Mosander was studying a new rock that had been found outside the town of Bastnas, Sweden, a rock from which six other elements were also to be discovered eventually. Mosander suggested the name of cerium for the element in honor of the asteroid Ceres that had been discovered in 1801.

Uses

Cerium and its compounds have a great variety of uses, many of them in the field of **glass** and **ceramics**. Cerium and its compounds are added to these materials to add **color** (yellow), remove unwanted color, make glass sensitive to certain forms of radiation, add special optical qualities to glass, and strengthen certain kinds of dental materials. Cerium is also used in the manufacture of **lasers** and phosphors used in cathode ray tubes.

CESIUM

Cesium is the fifth member of the alkali family, the elements that make up Group 1 of the **periodic table**. Its **atomic number** is 55, its atomic **mass** is 132.9054, and its chemical symbol is Cs.

Properties

Cesium is a silvery white, shiny metal that is very soft and ductile. Its melting point is so low (83.3°F or 28.5°C) that it melts easily in the **heat** of a person's hand. Its **boiling point** is 1,301°F (705°C) and its **density** is 1.90 grams per cubic centimeter.

Cesium is a very active metal, one of the most active of all elements. It reacts violent with **oxygen** in the air and with **water**. So much **energy** is released in the latter reaction that the **hydrogen** formed as a product ignites. Cesium is normally stored under **kerosene** as a safety precaution to prevent it from reacting with oxygen or water vapor in the air. Cesium also reacts vigorously with acids, the **halogens**, **sulfur**, and **phosphorus**.

Occurrence and Extraction

Cesium is moderately abundant in the Earth's crust with an estimated abundance of about 1-3 parts per million. It is found in an ore of **lithium** called lepidolite but, most abundantly, in the ore known as pollucite ($Cs_4Al_4Si_9O_{26}$).

Discovery and Naming

Cesium was discovered in 1861 by two German chemists, **Robert Bunsen** and **Gustav Kirchoff**. The two had been instrumental in developing the new analytical technique of **spectroscopy**, and they found spectral lines for a previously unidentified element while studying a sample of spring water. They suggested the name cesium, from the Latin word *caesius* meaning ''sky blue'' because blue is the **color** of the element's spectral lines.

Uses

Cesium and its compounds have relatively few commercial uses. It is used in photovoltaic cells that convert the energy of sunlight into **electricity**. It is also used in the cesium clock, a device that measures time by means of the wavelength of light given off by one of the elements isotopes, cesium- 133.

CHAIN, ERNST BORIS (1906-1979)

English biochemist

Ernst Boris Chain was instrumental in the creation of **penicillin**, the first antibiotic drug. Although the Scottish bacteriologist Alexander Fleming discovered the *Penicillium notatum* mold in 1928, it was Chain who, together with Howard Florey, isolated the breakthrough substance that has saved countless victims of infections. For their work, Chain, Florey, and Fleming were awarded the Nobel Prize in physiology or medicine in 1945.

Chain was born in Berlin on June 19, 1906 to Michael Chain and Margarete Eisner Chain. His father was a Russian immigrant who became a chemical engineer and built a successful chemical plant. The death of Michael Chain in 1919, coupled with the collapse of the post-World War I German economy, depleted the family's income so much that Margarete Chain had to open up her home as a guesthouse.

One of Chain's primary interests during his youth was music, and for a while it seemed that he would embark on a career as a concert pianist. He gave a number of recitals and for a while served as music critic for a Berlin newspaper. A cousin, whose brother-in-law had been a failed conductor, gradually convinced Chain that a career in science would be more rewarding than one in music. Although he took lessons in conducting, Chain graduated from Friedrich-Wilhelm University in 1930 with a degree in **chemistry** and physiology.

Chain began work at the Charite Hospital in Berlin while also conducting research at the Kaiser Wilhelm Institute for Physical Chemistry and **Electrochemistry**. But the increasing pressures of life in Germany, including the growing strength of the Nazi party, convinced Chain that, as a Jew, he could not expect a notable professional future in Germany. Therefore, when Hitler came to power in January 1933, Chain decided to leave. Like many others, he mistakenly believed the Nazis would soon be ousted. His mother and sister chose not to leave, and both died in concentration camps.

Chain arrived in England in April 1933, and soon acquired a position at University College Hospital Medical School. He stayed there briefly and then went to Cambridge to work under the biochemist **Frederick Gowland Hopkins**. Chain spent much of his time at Cambridge conducting research on enzymes. In 1935, Howard Florey became head of the Sir William Dunn School of Pathology at Oxford. Florey, an Australian-born pathologist, wanted a top-notch biochemist to help him with his research, and asked Hopkins for advice. Without hesitation, Hopkins suggested Chain.

Florey was actively engaged in research on the bacteriolytic substance lysozyme, which had been identified by Fleming in his quest to eradicate infection. Chain came across Fleming's reports on the penicillin mold and was immediately intrigued. He and Florey both saw great potential in the further investigation of penicillin. With the help of a Rockefeller Foundation grant, the two scientists assembled a research team and set to work on isolating the active ingredient in *Penicillium notatum.*

Fleming, who had been unable to identify the antibacterial agent in the mold, had used the mold broth itself in his experiments to kill infections. Assisted in their research by fellow scientist Norman Heatley, Chain and Florey began their work by growing large quantities of the mold in the Oxford laboratory. Once there were adequate supplies of the mold, Chain began the tedious process of isolating the "miracle" substance. Succeeding after several months in isolating small amounts of a powder which he obtained by freeze-drying the mold broth, Chain was ready for the first practical test. His experiments with laboratory mice were successful, and it was decided that more of the substance should be produced to try on humans. To do this, the scientists needed to ferment massive quantities of mold broth; it took 125 gallons of the broth to make enough penicillin powder for one tablet. By 1941 Chain and his colleagues had finally gathered enough penicillin to conduct experiments with patients. The first two of eight patients died from complications unrelated to their infections, but the remaining six, who had been on the verge of death, were completely cured.

One potential use for penicillin was the treatment of wounded soldiers, an increasingly significant issue during the Second World War. However, for penicillin to be widely effective, the researchers needed to devise a way to mass-produce the substance. Florey and Heatley went to the United States in 1941 to enlist the aid of the government and of pharmaceutical houses. New ways were found to yield more and stronger penicillin from mold broth, and by 1943 the drug went into regular medical use for Allied troops. After the war, penicillin was made available for civilian use. The ethics of whether to make penicillin research universally available posed a particularly difficult problem for the scientific community during the war years. While some believed that the research should not be shared with the enemy, others felt that no one should be denied the benefits of penicillin. This added layers of political intrigue to the scientific pursuits of Chain and his colleagues. Even after the war, Chain experienced firsthand the results of this dilemma. As chairman of the World Health Organization in the late 1940s, Chain had gone to Czechoslovakia to supervise the operation of penicillin plants established there by the United Nations. He remained there until his work was done, even though the Communist coup occurred shortly after his arrival. When Chain applied for a visa to visit the United States in 1951, his request was denied by the State Department. Though no reason was given, many believed his stay in Czechoslovakia, however apolitical, was a major factor.

Ernst Boris Chain.

After the war, Chain tried to convince his colleagues that penicillin and other antibiotic research should be expanded, and he pushed for more state-of-the-art facilities at Oxford. Little came of his efforts, however, and when the Italian State Institute of Public Health in Rome offered him the opportunity to organize a biochemical and microbiological department along with a pilot plant, Chain decided to leave Oxford.

Under Chain's direction, the facilities at the State Institute became known internationally as a center for advanced research. While in Rome, Chain worked to develop new strains of penicillin and to find more efficient ways to produce the drug. Work done by a number of scientists, with Chain's guidance, yielded isolation of the basic penicillin **molecule** in 1958, and hundreds of new penicillin strains were soon synthesized.

In 1963 Chain was persuaded to return to England. The University of London had just established the Wolfson Laboratories at the Imperial College of Science and Technology, and Chain was asked to direct them. Through his hard work the Wolfson Laboratories earned a reputation as a first-rate research center.

In 1948, Chain had married Anne Beloff, a fellow biochemist, and in the following years she assisted him with his research. She had received her Ph.D. from Oxford and had worked at Harvard in the 1940s. The couple had three children.

Chain retired from Imperial College in 1973 but continued to lecture. He cautioned against allowing the then-new field of molecular biology to downplay the importance of **biochemistry** to medical research. He still played the piano, for which he had always found time even during his busiest research years. Over the years, Chain also became increasingly active in Jewish affairs. He served on the Board of Governors of the Weizmann Institute in Israel, and was an outspoken supporter of the importance of providing Jewish education for young Jewish children in England and abroad—all three of his children received part of their education in Israel.

In addition to the Nobel Prize, Chain received the Berzelius Medal in 1946 and was made a commander of the Legion d'Honneur in 1947. In 1954 he was awarded the Paul Ehrlich Centenary Prize. Chain was knighted by Queen Elizabeth II in 1969. Increasing ill health did not slow Chain down initially, but he finally died of heart failure on August 14, 1979.

CHAIN REACTION

A chain reaction is a situation in which one action causes or initiates a similar action. In a nuclear chain reaction, for example, a **neutron** strikes a uranium-235 **nucleus**, causing the nucleus to undergo fission, which in turn produces a variety of products. Among these products is one or more neutrons. Thus, the particle needed to initiate this reaction (the neutron) is itself produced as a result of the reaction. Once begun, the reaction continues as long as uranium-235 nuclei are available. Nuclear chain reactions are important sources of fission and fusion **energy**.

CHAINS

One of the more important properties of **carbon**, from the point of view of organic **chemistry**, is its ability to ''catenate'' or form long chains with itself. Carbon is by no means unique in this regard as most of the non-metallic elements are capable of forming chains. However, carbon is unique in being able to form stable chains. That is, the structurally analogous silanes—mixtures of **silicon** and **hydrogen**—are spontaneously flammable in air.

Examples of simple chains are the **hydrocarbon**s. In essence, these are built from the successive addition of -CH_2 units. **Methane** is therefore one -CH_2 inserted into a H-H bond. Ethane is two -CH_2 units. Propane is three and so on. Extension of these chains to the millionth -CH_2 unit provides polymers and **plastics**. In essence, this is the basic structure of **polyethylene**.

Simple chains are not the only possibility. A variety of polymeric substances, built from repeating units, are also possible. But more important and interesting are the chains of life. These are the protein molecules, made from chains of **amino acids** strung together by amide bonds, and DNA/RNA, which are chains of **nucleic acids** strung together by **phosphate** linkages between the deoxyribose (for **DNA**) and ribose (for **RNA**) saccharides, respectively. Peptides, such as many human messengers and hormones, contain between about four to 100 amino acids. **Proteins** are longer chains, containing between 100-1,000 amino acids. DNA, on the other hand, is a chain of millions of base pairs. But in either case, it is the ability of carbon to form chains with itself that make these molecules possible.

CHANCE, BRITTON (1913-)

American biochemist and biophysicist

Combining an interest in electronics with his specialties of **chemistry** and biology, Britton Chance developed new equipment and techniques for research in **biochemistry** and biophysics, including invention of the double-beam spectrophotometer and the reflectance fluorometer and application of computer methods to the study of **enzyme** action and metabolic control. Furthermore, the experimental results he has obtained using his own innovative procedures are of major importance in such areas as determining the actions of narcotics and poisons on living cells.

Born in Wilkes-Barre, Pennsylvania, on July 24, 1913, Chance was the son of Edwin M. Chance, a chemist who was honored for research in mine **gases** and who worked with the Chemical Warfare Service during World War I. His mother was Eleanor Kent Chance. Both parents were from Philadelphia, Pennsylvania, and the family returned there after Britton's birth, with the senior Chance becoming an engineer. Summers were spent at Barnegat Bay, New Jersey, where the family kept a 100-foot cruiser that they sailed as far as the Caribbean and Europe.

After graduating from the Haverford School, Chance studied at the University of Pennsylvania, where he earned a B.S. degree in chemistry in 1935, and an M.S. the following year. His experience with sailing led him to an interest in navigation and automatic steering, which he thought could be improved using electronics; he found room in his curriculum to study enough physics and electrical engineering to pursue that interest. Under the guidance of Martin Kilpatrick, his chemistry mentor at Pennsylvania, Chance investigated rapid chemical reaction techniques. Using his knowledge of electronics, he developed better instrumentation for observing the small, quick changes in optical **density** that accompany reactions involving enzymes—the protein produced by living cells that stimulates such important functions as food digestion and the release of **energy**.

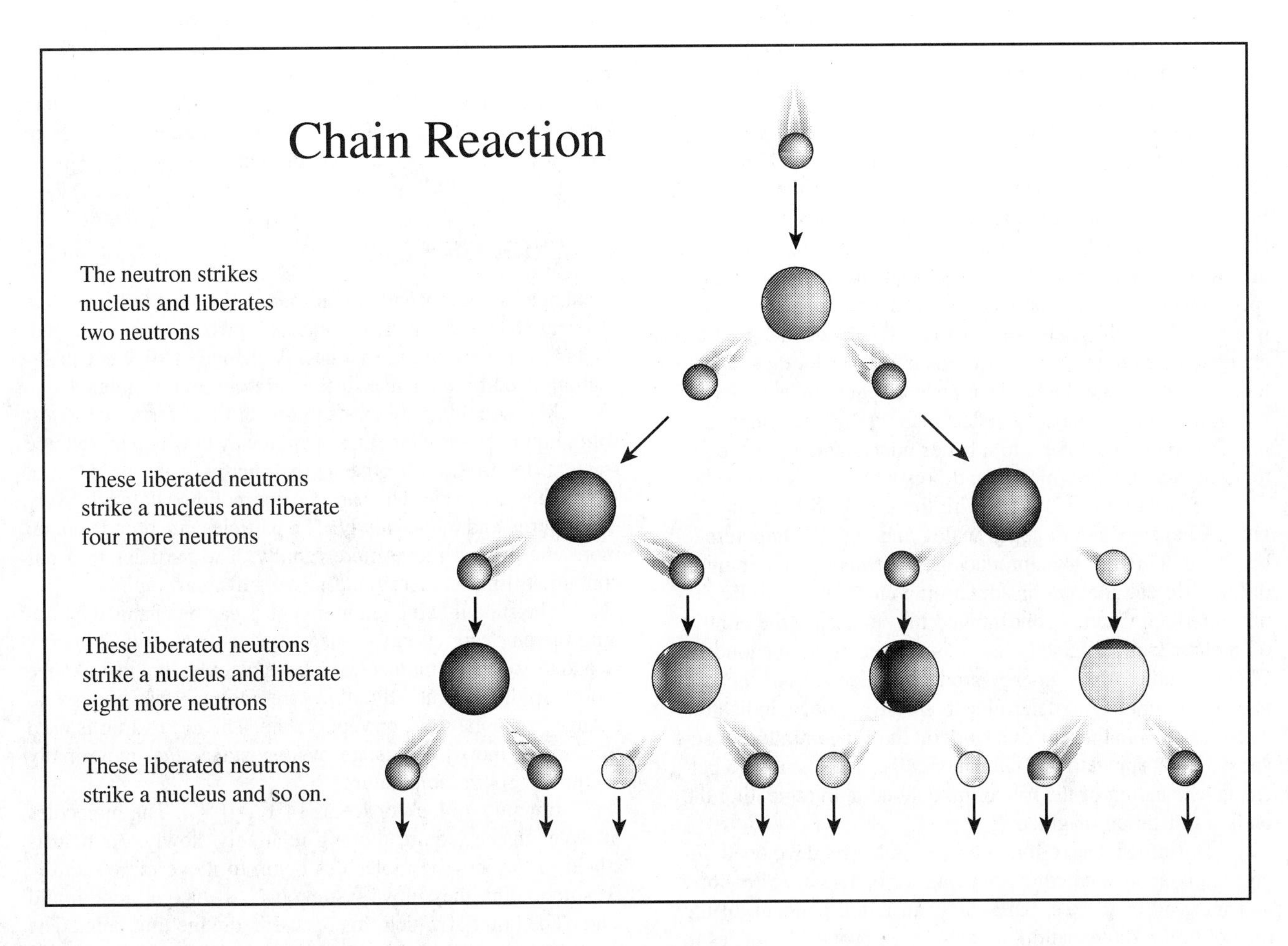

(Illustration by Electronic Illustrators Group.)

After receiving his master's degree, Chance traveled to England to supervise the installation of the marine electronic steering system he had invented, which was first used on a 12,000-ton ship. In 1938 he married his childhood sweetheart, Jane Earle, and they used the ship's trial cruise to Australia as their honeymoon. Upon returning to England, Chance enrolled as a research student at Cambridge University. He studied under F. J. W. Roughton, who had developed a fundamental procedure for observing rapid reactions, and G. A. Millikan, a physiologist. Millikan was using Roughton's technique to study muscle pigments, and Chance adapted the same procedure to reactions between small quantities of substances. Altering both the experimental technique and the equipment, Chance devised the "accelerated flow modification" procedure: reactants were simultaneously injected into a fine tube, where the changes in light absorption of the reacting material were continuously monitored with a highly sensitive oscilloscope. This important modification allowed experiments to be conducted more quickly, was more sensitive to reactions, and required less materials to complete experiments.

At the outbreak of World War II in 1939, Chance returned to the United States and joined the faculty at the University of Pennsylvania's Eldridge Reeves Johnson Foundation of Medical Physics. There, he finished developing his technique for measuring the reaction between **hydrogen** peroxide and the enzyme peroxidase. He was awarded a Ph.D. in **physical chemistry** in 1940 and became acting director of the Johnson Foundation. Two years later, he was awarded a Ph.D. in biology and physiology by Cambridge.

During World War II, Chance was invited to work on radar at the Massachusetts Institute of Technology's Radiation Laboratory. Between 1941 and 1946, he headed the Precision Components Group, served as associate head of the Receiver Components Division, and was one of the younger members of the laboratory's Steering Committee. Among the devices he helped develop were precision timing and computing circuits for bombing and navigation. In 1942, he applied advanced electronic circuits to measuring small changes in light absorption, and in 1943, he experimentally validated the Michaelis-Menton theory of enzyme action, which had first been proposed in 1913.

When the war ended, Chance devoted himself to the study of the nature of enzymes and won a Guggenheim Fellowship to study at the Nobel Institute in Sweden and the Molteno Institute in Great Britain from 1946 to 1948. His interest in boating continued; while in Europe, he and his wife (a skilled helmsman) won numerous races for Class E scows that were held in the Baltic Sea.

In 1949, Chance returned to the University of Pennsylvania as a professor, chairperson of the department of biophysics and physical biochemistry, director of the Johnson Foundation, and faculty adviser for the university yacht club. His research on enzymatic reactions in living systems directed him to the creation of the double-beam or dual-wavelength spectrophotometer, an optical device that was used to study mitochondria—those elements of the cell **nucleus** that produce energy—as well as cell suspensions and other biological systems. Adapting the double-beam spectrophotometer to living materials, he developed the reflectance fluorometer, an instrument that led, eventually, to a better understanding of the actions of narcotics, **hormones** and poisons on human cells.

In addition to his work with these ground-breaking devices, Chance began to apply analog and digital computers to his research in the 1940s, producing the first computer **solution** of the differential equations describing enzyme action. Beginning in 1955, Chance concentrated his studies on the control of **metabolism**, especially as it is related to mitochondria. Chance studied these energy-producing elements in the cell nucleus, attempting to determine their role as optical indicators of change within the cell. His work on the **concentration** of adenosinediphosphate (ADP) in tumor cells gave scientists a better understanding of the role of mitochondria in regulating the body's utilization of glucose.

During the next four decades, Chance developed increasingly sophisticated instruments for optical **spectroscopy** and imaging of tissues, particularly in living patients. In the area of tumor oxygenation, he found significant differences in hemoglobin oxygenation between the surface and deeper regions of solid tumors, which a 1992 article in *Proceedings of the IEEE* described as being "of great importance clinically, for example, in relation to the oxygen-dependent response to radiation therapy." In another study, he made time-resolved measurements of **photon** migration on the forehead of a patient with Alzheimer's disease, finding the scattering coefficient of the brain tissue in some regions to be one-half to one-third of normal values.

Recognition of Chance's fundamental contributions to biomedical electronics and other fields came from all over the world. His pioneering work with rapid enzyme reactions was described in a 1949 issue of *Nature:* "The value and scope of this experimental technique in the field of reaction **kinetics** can scarcely be overestimated." His achievements were further acknowledged in 1954 with his election to the National Academy of Sciences, and he became a life fellow in the Institute of Electrical and Electronics Engineers (IEEE) in 1961. He holds several honorary medical degrees, including an honorary M.D. from the Karolinska Institute of the University of Stockholm. Chance has also been honored with the Dutch Chemical Society's Genootschaps-Medaille and the Franklin Medal of the Franklin Institute. President Gerald Ford presented him with the National Medal of Science in 1974, and in 1976, the University of Pennsylvania named him university professor, its highest academic appointment. He retired from the university in 1983.

In 1956, Chance married for the second time, to Lilian Streeter Lucas. He has twelve children, including the four from his first marriage and four stepchildren.

CHANGE OF STATE

A change of state occurs when **matter** is converted from one physical state to another. For example, when **water** is heated, it changes from a liquid to a gas. A change of state is usually accompanied by a change in **temperature** and/or **pressure**.

Matter commonly exists in one of three forms, or states: solid, liquid, or gas. One fundamental way in which these three states differ from each other is the **energy** of the particles of which they are made. The particles in a solid contain relatively little energy and move slowly. The particles in a liquid contain more energy and move more rapidly. The particles in a gas contain still more energy and move still more rapidly.

The state in which matter occurs can be changed by adding or removing energy. When water is heated, heat energy is added to water molecules. The molecules begin to move more rapidly. Eventually, they are moving fast enough to change to the gaseous, or vapor, state. The term vapor is used to describe the gaseous state of a substance that is normally a liquid at room temperature.

Imagine a block of ice at 14°F (-10°C). The molecules of water in the ice are moving relatively slowly. As **heat** is added to the ice, the molecules begin to move more rapidly. At some point, they move fast enough to change to the liquid state. The point at which this occurs is the melting point. The melting point is the temperature at which a solid changes to a liquid. The melting point of ice is 32°F (0°C).

If additional heat is added to the liquid water, water molecules move even faster. The increase in speed with which they move is measured as an increase in temperature. The temperature of the liquid water increases from 32°F (0°C) to 212°F (100°C). At a temperature of 212°F (100°C), the water molecules are moving fast enough to change to a vapor, called steam. The temperature at which a liquid changes to a gas is called its **boiling point**.

If the steam formed in this process is heated further, its temperature continues to increase. But, under normal circumstances, it undergoes no further changes of state.

Changes of state occur also when a material is cooled. Suppose the steam in this example is cooled below 212°F (100°C). When that happens, the temperature changes back to a liquid. The steam is said to condense to a liquid. The condensation point is the temperature at which a gas or vapor changes to a liquid. It is the same as the boiling point of the liquid.

Further cooling of a liquid eventually causes one more change of state. At some point, the liquid cools sufficiently to change to a solid. At this point, the liquid is said to have become frozen. The freezing point of a liquid is the temperature at which the liquid changes to a solid. The freezing point is the same as the melting point of the solid.

Some materials behave differently from water when they are heated. They may pass directly from the solid state to the gaseous state. **Iodine** is an example. When solid iodine is heated, it does not melt. Instead, it changes directly into a vapor. Substances that behave in this way are said to sublime. The **sublimation** point of a substance is the temperature at which it changes directly from a solid to a vapor.

Dry ice (solid CO_2), which rapidly undergoes sublimation from solid to vapor at room temperatures, is often used to create fog on stage and movie sets. Dry ice, a white opaque solid, is also widely used as a cryogenic agent in industry to reduce bacteria growth and maintain low temperatures. At atmospheric pressures found on Earth, dry ice undergoes sublimation at -109.3°F (-78.5°C). Special care must be taken when working with dry ice to avoid frostbite. In addition, dry ice must be used in a well-ventilated environment because, as it undergoes sublimation, dry ice will reduce the percentage of available **oxygen**.

Each kind of state change requires a specific amount of energy. For example, it takes about 80 calories of heat to melt one gram of ice. By contrast, it takes only about 6 calories of heat to melt one gram of **lead**. And it takes about 50 calories of heat to melt one gram of **copper**. The amount of heat needed to melt one gram of a substance is called its **heat of fusion**.

Heat of crystallization is the quantity of heat given up at a constant temperature when one gram of liquid is changed to a solid. For any given substance the heat of crystallization is equal to the heat of fusion.

Similarly, it takes a specific amount of energy to convert a given amount of liquid into a gas. That amount of energy is defined as that substance's **heat of vaporization**. The heat of vaporization of liquid water is about 540 calories per gram. In comparison, the heat of vaporization of lead is about 226 calories per gram and for copper, 1270 calories per gram.

The heat of vaporization for any given substance is equal to the heat ofcondensation.

By analogy, the heat of sublimation is the amount of heat energy needed to convert a given amount of solid directly to the vapor state. The heat of sublimation of solid iodine is about 15 calories per gram.

See also Gases; Liquids; Solids

CHAOS THEORY

Chaos theory is used to model the overall behavior of complex systems. Despite its name, chaos theory is used to identify order in complex and seemingly unpredictable systems.

Chaos theory is used to understand explosions, complex chemical reactions (e.g., the Belousov-Zhabotinsky oscillating reaction that yields a red **solution** that turns blue at varying intervals of time), and many biological and biochemical systems. Chaos theory is now an important tool in the study of population trends and in helping to model the spread of disease. Epidemiologists use chaos theory to help predict the spread of epidemics.

Deterministic dynamical systems are those systems that are predictable based on accurate knowledge of the conditions of the system at any given time. When systems are, however, sensitive to their initial conditions they eventually become unpredictable. In particular, chaos theory deals with complex nonlinear dynamic (i.e., nonconstant, nonperiodic, etc.) systems. Nonlinear systems are those described by mathematical recursion and higher algorithms. *Deterministic chaos*, is mainly devoted to the study of systems the behavior of which can, in principle, be calculated exactly from equations of motion.

The roots of chaos theory are to be found in the classical mechanics introduced in the venerated English physicist Sir Isaac Newton's (1642-1727) *Philosophy Naturalis Principia Mathematica (Mathematical Principles of Natural Philosophy)*, first published in 1686. It was Newton, one of the inventors of the calculus, who revolutionized astronomy and physics by showing that the behavior of all bodies, celestial and terrestrial, was governed by the same laws of motion, which could be expressed as differential equations. These equations relate the rates of change of physical quantities to the values of those quantities themselves.

Differential equations make it possible to predict the future behavior of a system of bodies from knowledge of the positions and the velocities of the bodies at any one time (the initial conditions), provided the forces acting between the bodies are known. Newton also showed that every body in the universe attracts every other with a gravitational force proportional to the product of the masses and inversely proportional to the square of the distance between them. This meant that the orbits of the planets could be calculated. Considering only the sun's attraction yields an elliptical orbit, an excellent first approximation. The effects of the other planets can be incorporated by approximation methods, such as lead to the discovery of the planet Neptune in 1846.

A problem arises, however, in that the approximation methods, which involve adding up very large number of very small quantities, can not produce a unique prediction for behavior in the distant future. The question arose as to whether the solar system was stable in the long term, that is, whether the planets would continue to orbit the sun as opposed to some of them falling into the sun or being ejected into space. The definitive and unexpected answer was provided in 1890 when the great French mathematician, Jules Henri Poincaré (1854-1912), proved that if the solar system were in a stable state, an infinitesimal change could result in an unstable state and vice versa. Since the state of the solar system cannot be specified with infinite precision, the question cannot be answered at all.

The study of such mathematical irregularities remained a relatively unnoticed corner of advanced mathematics until the advent of the digital computer. In 1956, Edward Lorenz, a professor of meteorology at the Massachusetts Institute of Technology was studying the numerical solution to a set of three differential equations in three unknowns, a highly simplified version of the type of equations meteorologists were then using to describe atmospheric phenomena. Lorenz came to the conclusion that his set of differential equations displayed a sensitive dependence on initial conditions, a sensitivity of the same type that Poincaré had discovered for the Newtonian equations applied to the solar system. Lorenz, however, gave this phenomenon a new and highly appealing name, the butterfly effect, suggesting that the flapping of a butterfly's wings in Kansas might be responsible for a monsoon in India a month later.

Lorenz made a further discovery about the behavior of his three variables with regard to time. For any system described by a finite number of variables that depend upon time, one can think of the system as described by a point that traces out a continuous curve in an abstract space. For simple systems, as time goes on, the curve will either approach a limit point or a closed curve, a limit cycle. Lorenz found that in his case, he obtained neither of these, but rather an infinitely long curve that repeatedly passed through a definite range of values for his three variables without ever closing back on itself. This curve was named a strange attractor, strange because of its infinite length and because the system variables tended toward values found on the curve as time went on.

The term chaos as used to describe unpredictable behavior of this type was introduced in 1975 by Tien-Yien Li and James A. Yorke in an article in the *American Mathematical Monthly*.

It is now understood that chaotic behavior may be characterized by sensitive dependence on initial conditions and attractors (including, but not limited to strange attractors). A particular attractor represents the behavior of the system at any given time. The actual state of any system (i.e., measured characteristics) depends upon earlier conditions. If initial conditions are changed even to a small degree the actual results for the original and changed systems become different (sometimes drastically different) over time even though the plot of the attractor for both the original and changed systems remains the same. In other words, although both systems yield different values as measured at any given time the plots of their respective attractors (i.e., the overall behavior of the system) look the same.

One chaotic chemical system that has been well studied is a mixture of equal numbers of moles of **carbon** monoxide and **oxygen** with a small amount of molecular **hydrogen**. Depending on the precise **temperature** this system will exhibit slow formation of **carbon dioxide**, periodic ignition or chaotic reaction. New examples of chemical chaos are discovered each year.

See also Thermodynamics

CHARCOAL

Anyone who has sat around a campfire or witnessed a forest fire knows that **wood** burns. But they are also aware that wood generates a great deal of smoke as it burns. This is a result of the incomplete **combustion** of the oils and resin in the wood, along with the high moisture content. Unfortunately, this limits the practicality of wood as a fuel for a variety of purposes including indoor cooking and the **smelting** of **metals**. In the latter case, the impurities in the wood smoke result in a lower grade of ore being obtained.

It was for the purposes of smelting **copper** and subsequently **iron** that charcoal was first used. Its origin is unknown but it has been used in Europe for over 5,500 years. It was the smelting fuel for the bronze and iron ages of civilization, being used in smelters and the blacksmith's shop.

In essence, charcoal is distilled wood. It is simply the **carbon** content of the wood with little or no impurities. This almost pure carbon has two distinct advantages. It burns much hotter than wood itself, allowing access to the temperatures necessary to melt ores and it burns without a flame and little smoke. Consider the backyard barbecue. Until some fat or barbecue sauce is spilt on the coals, they do not smoke but simply produce a great deal of **heat**.

The combustion of wood converts it back into the **carbon dioxide** and **water** vapor from which it is made with little ash remaining. Making charcoal, on the other hand, is an art as it requires that wood be burnt to the point of carbon without actually being consumed in flames. The burn removes the impurities from the wood and leaves behind the carbon content. This requires careful control, with limited **oxygen**, just enough to sustain combustion but not enough to ignite the wood proper. Charcoal making was one of the more time consuming, and expensive, components of the smelting process.

The collier worked from late spring through to early fall, requiring the hot, dry heat of summer to sustain his efforts. During his off season, he traded in his tools for an ax to restock his wood supply for the following burn season. The supply of wood was rapidly depleted in many areas and, in Great Britain, laws were passed to protect the dwindling forests. This resulted in the conversion of English iron works to coke—a partially combusted form of **coal** generated in much the same way as charcoal. Ironically, this alternative carbon source resulted in an increase in iron production and **lead** to the demise of the charcoal industry itself.

Charcoal has also been used historically for the production of **glass**, as a purifier of food and water, in the making of **gunpowder** and in the hearths of smithies. Its byproducts included a oil that was used as part of the Egyptian embalming process. In more recent times, charcoal finds usage in the backyard barbecue, providing the right combination of heat with little flame.

Another important use for charcoal or carbon is in the ''activated'' form. The porous structure of charcoal as it is formed fills with the **hydrocarbon** residue from the combustion of the wood. Removal of this hydrocarbon content, by heating in an inert atmosphere, results in a form of charcoal that is extremely porous. Five pounds of activated charcoal is estimated to have a surface area of approximately one square mile. In addition, the surface atoms are capable of both **adsorption** and absorption of chemical compounds. Activated charcoal is capable of absorbing between 25% and 100% of its own **mass** in contaminants.

Activated charcoal is used in the purification of liquid sugar during the production of white sugar, in the purification of water, in the clean up of other solutions including industrial process streams, for air purification in large airport concourses and small personal gas masks, and in the control and recovery of vapors during manufacturing processes. Its large surface area ensures that maximum absorption occurs and it has the advantage that it can be reused. Simple heating results in the desorption of absorbed material. Still, despite its usefulness, activated charcoal is produced on a smaller scale than either charcoal or coke, which are still used in the smelting of ores and other processes.

Chardonnet, Hilaire Bernigaud, Comte de (1839-1924)

French chemist

Hilaire Bernigaud, Comte de Chardonnet's career in science began with engineering studies at the École Polytechnique; he also assisted **Louis Pasteur** (1822-1845) in his efforts to save the French silk industry from a devastating silkworm epidemic. Realizing there was a market for an artificial silk, Chardonnet built upon the work of the Swiss chemist George Audemars and Sir Joseph Swan of England to develop cellulose-based fibers. Audemars had received a patent in 1855 for the manufacture of synthetic fibers; by 1880 Swan had developed threads from nitrocellulose.

Chardonnet first treated cotton with nitric and sulfuric acids and then dissolved the mixture in **alcohol** and **ether**. He then passed the **solution** through **glass** tubes, forming fibers, and allowed them to dry. These fibers, called rayon (the term used in referring to any fiber developed from cellulose) were highly flammable until they were denitrated. Reportedly, some garments made of early rayon burst into flames when lit cigarettes were nearby, Unfortunately the techniques that existed at that time to denitrate the material weakened it and made it unsuitable for the textile industry. Chardonnet used ammonium sulfide to denitrate these fibers thus reducing the flammability and retaining fiber strength comparable to that of silk. He received the first patent for his work in 1884 and began manufacturing rayon in 1891. The material was displayed at the 1891 Paris Exposition where it won the grand prize. He was also awarded the Perkin medal in 1914 for his development of rayon.

After his work with fibers, Chardonnet went on to study a number of other subjects including ultraviolet light, telephony, and the movements of birds' eyes. He died in 1924 in Paris, France.

Chargaff, Erwin (1905-)

American biochemist

In his own eyes he was a natural philosopher, part of an extinct species of researcher. But in the world of science, Erwin Chargaff was a pioneer in **biochemistry** whose demonstration that genes were comprised of deoxyribonucleic acid (DNA), was one of the most fundamental discoveries in the study of heredity. His accomplishments ranged over much of the field of biochemistry, including the study of **lipids** of microorganisms, blood coagulation, and the use of radioactive isotopes in the study of **metabolism**.

Erwin Chargaff. *(Archive Photo. Reproduced by permission.)*

Chargaff was born on August 11, 1905 in Czernowiz, Austria, which at the time was the provincial capital of the Austrian monarchy. Although he considered his education at Vienna's Maximilians Gymnasium excellent in quality, he found it limited in scope. While his parents were not wealthy, having lost their money to an inflated economy, Chargaff successfully earned his doctoral degree in **chemistry** at the University of Vienna's Spath's Institute in 1928, working under the direction of Fritz Feigl.

His family's dire financial problems made a strong impression on Chargaff, who chose the field of chemistry because he believed it would be the most likely to offer employment. In fact, because students had to pay for their own chemicals and equipment, he chose to do his research with a particular chemistry professor whose work did not require much time or money.

Chargaff began his long, productive career in biochemistry at Yale University, where he worked under Rudolph J. Anderson from 1928 to 1930. From 1930 to 1933, he was an assistant at the University of Berlin before moving to Paris to work for nearly two years at the Pasteur Institute. In 1935 he settled into a permanent home at Columbia University, assuming the position of assistant professor of biochemistry. He became full professor in 1952 and was chairman of the department from 1970 until 1974, when be became professor emeritus. Chargaff was presented Columbia's Distinguished Service Award in 1982.

Chargaff's early work included studies of the complex lipids, the fats or **fatty acids** that occur in microorganisms. He helped to discover the unusual fatty acids and **waxes** in so-called acid-fast mycobacteria. This initial work led him to study the metabolism and biological role of lipids in the body, especially the lipoproteins, which are lipids attached to **proteins**.

Chargaff was a pioneer in the use of radioactive isotopes of **phosphorus** as a tool to study the synthesis and breakdown of phosphorus-containing lipid molecules (phospholipids) in living cells. As a result, he published the first paper on the synthesis of a radioactive organic compound called alpha-glycerophosphoric acid. Blood coagulation also caught his interest, and Chargaff studied the biochemistry of this important phenomenon, including the control of blood clotting by enzymes.

In 1944, while at Columbia, Chargaff's path veered sharply away from lipid metabolism and blood coagulation to the study of **nucleic acids**. Until then, most scientists believed that amino acids—the building blocks of proteins—carried genetic information. This seemed reasonable, since it was thought that the twenty **amino acids** that occur in cells could create enough combinations to form the complex code needed to make many thousands of different genes. **DNA** was also believed to be an unspecific aggregate of "tetranucleotides" made up of adenine, guanine, cytosine and thymine, that served as an attachment site for the amino acids that made up genes. But in 1944, **Oswald T. Avery** and his collaborators determined that DNA was a key element in property transfer between certain bacteria, and recognized that DNA could be the principle constituent of genes.

It was already known that a cell's **nucleus** is comprised in part by DNA, which is itself composed of sugar, **phosphate** and two types of complex molecules called purine bases (adenine and guanine) and pyrimidine bases (cytosine and thymine). Using two newly developed experimental techniques, Chargaff isolated DNA from cell nuclei and broke the giant, parent molecules down into constituent purines and pyrimidines. He first separated the bases from each other using a technique called paper **chromatography**. By then exposing these to ultraviolet light he identified the individual bases. Since different bases absorb ultraviolet light of specific wavelengths, he was able to determine how much of which bases were present by measuring the amount of light each quantity of base absorbed.

The results of this work represented a major contribution to the understanding of the structure of DNA. Chargaff and his colleagues showed that adenine and thymine occur in DNA in equal proportions in all organisms; likewise, cytosine and guanine are also found in equal quantities. However, the proportions of each pair of these nucleic acids differs among organisms. In other words, the presence of each pair of nucleotides is linked to the presence of the other.

Chargaff concluded that DNA, rather than protein, carries genetic information, and while there are only four different nucleic acids, the number of different combinations in which they can appear in DNA provides enough complexity to form the basis of heredity. In addition, he concluded that the identity of combinations differs from species to species; thus DNA strands differ from species to species.

These conclusions helped trigger a rush of new insights into DNA, the most important of which led **James Watson** and **Francis Crick** of the Cavendish Laboratory in Cambridge, England, to determine the exact structure of DNA. The Cavendish team showed that DNA consisted of two strands of sugar and phosphate connected by crosslinks of purines and pyrimidines. They concluded that the nucleic acids that make up each pair always occur in the same proportion because they always bond together, adenine to thymine, cytosine to guanine—known as the base-pairing rules. The entire **molecule** is stabilized by being twisted into a double helix structure. For their success in clarifying the structure of DNA, the Cavendish team won the Nobel Prize in 1962.

Chargaff continued to contribute to the understanding of how DNA and **RNA** (ribonucleic acid) work during the years after the double helix model was proposed. In 1962, Chargaff attended a symposium at Columbia University at which the nature of the genetic code was debated. He challenged a current concept of how the DNA code is translated into a precise sequence of amino acids. The theory held that each amino acid within a protein was coded by a specific series of three nucleic acids in the gene. But Chargaff's investigation of the genes for the protein bovine ribonuclease showed that the code, as then understood, could not be responsible for the specific amino acid sequences in this protein. Either the code was wrong, he suggested, or some amino acids may be coded by more than one sequence of nucleic acids. This theory, supported by the Spanish-American biochemist **Severo Ochoa**, turned out to be correct.

Despite his important contributions to the understanding of the biochemistry of nucleic acids, Chargaff was also very concerned with the state of the world at large. In his book, *Heraclitean Fire: Sketches From a Life Before Nature,* he wrote that society in general—manners, literature, music, and even science—had declined during his lifetime. He was particularly disillusioned with both his adopted country and the scientific community after the **atom** bomb was unleashed upon Hiroshima and Nagasaki at the close of World War II.

Chargaff also wrote longingly of the simpler days of science when research was not as influenced by money and politics. He summed up his disappointment with the way the world had turned out by saying that he had been "searching for a destiny that did not exist." His destiny, however, lead to important contributions in the field of biochemistry, including the addition of a key piece in the puzzle of the structure of DNA. In turn, this knowledge led to major developments in the field of medical genetics, and, ultimately helped pave the way for gene therapy and the birth of the biotechnology industry.

A visiting professor at Cornell University in 1967, Chargaff served that function in many countries around the world, including Sweden (1949), Japan (1958), and Brazil (1959). He also held the Einstein Chair at the College de France, Paris, in 1965. Among the many awards conferred upon him throughout his career, Chargaff received the Charles Leopold

Mayer Prize in 1963, one of the highest awards granted by the French Academy of Sciences. The following year, he became the first recipient of the Heineken Prize in biochemistry from the Royal Netherlands Academy of Sciences. He received honorary degrees from both Columbia University and the University of Basel in 1976.

Charge, electrical

The ancient Greeks first conceived of the notion of an **atom**, which was their word for indivisible. Modern theory divides the atom into the **proton**, **electron**, and **neutron**. These particles all differ in weight and charge. The proton has a positive charge, the electron a negative charge, and the neutron a zero or negligible charge. Because the atom is ordinarily electrically neutral, the net positive charge due to the protons in the atom's **nucleus** must cancel out the negative charge of the atom's electrons, which is to say that the number of protons and electrons must be the same. Electric charge always comes as multiples of the elementary charge of the electron, i.e., 1.6×10^{-19} coulombs.

Electrons are about one six-million-millionth of an inch in diameter, and have a **mass** of about 0.9 billionth of a billionth of a billionth of a gram. (Actually, relativistic theory tells us that the electron's mass depends on its **velocity**; since mass increases with speed, the faster the electron moves, the more it weighs.) Electrons possess inertia, so remain at rest or in uniform motion in the same direction unless acted upon by some external force. All electrons are alike: they all carry the same negative charge, and they repel each other. Oppositely charged particles, such as a proton and an electron, attract.

To say that **matter** is positively charged or negatively charged is strictly meaningful only in relative terms. If one removes electrons from a body, that body becomes positively charged. Since, ordinarily, it is not possible to remove all the electrons from an object, an object's net positive charge is just an indication of how many electrons were removed. Simply stated, an object that has an excess of electrons is negatively charged; one that has a deficiency of electrons is positively charged.

The science that concerns itself with electric charges at rest is called electrostatics. When a **glass** rod and a piece of fur are rubbed together, a process referred to as charging takes place. Ordinarily, the electrons in the glass rod are distributed fairly uniformly, and the rod will not have any affect when brought into contact with small pieces of **paper**. But when the glass rod is rubbed with the fur, electrons are redistributed to the portion of the rod that was brought into contact with the fur, and that part of the rod acquires the ability to attract the pieces of paper. This is a simple example of contact electrification involving a redistribution of charge, without net charge transfer, so the glass rod as a whole remains electrically neutral.

When charges in a conductor are in motion, they produce an electric current, defined as the rate at which charge is transported past a given point in the conductor. Here the term conductor includes **metals**, alloys, **semiconductors**, electrolytes, ionized **gases**, imperfect dielectrics, and even vacuum (under certain conditions). In some conductors such as metallic **solids**, one or more of the electrons bound to each atom may be liberated and become free to wander about the material. In electrolytes, ions having atomic mass are free to move about, producing an electric current as they do so. In gases, atoms may become ionized, so that the resultant free electrons and ions are free to conduct **electricity**. In still other materials, i.e., semiconductors, some of the bound charges can be liberated by thermal vibrations, light, or the application of externally produced electric fields. That the current flowing in metals and the liberation of charges from matter can be measured in the laboratory provides a basis for the macroscopic description of electromagnetism. Macroscopic theory, however, provides us with no insight into why the basic unit of charge is fixed and why it is the same for all particles.

See also Electrical conductivity

Charles, Jacques-Alexandre-César (1746-1823)

French experimental physicist

Jacques-Alexandre-César Charles is famous for his contribution to ballooning, (the science of aerostation). He was the first to use **hydrogen** (''inflammable air'') instead of hot air in an aeronautical balloon, and designed and developed almost all the essential components of a balloon so perfectly that few changes or additions have been made in the ensuing 200 years of ballooning.

Charles was born in Beaugency on November 12, 1746. The only information surviving about his childhood is that he received a liberal education with no scientific focus. When he moved to Paris, he worked at the bureau of finances and in 1799 became interested in nonmathematical, experimental physics. By 1781, he began giving public lectures and experimental demonstrations, attracting a large number of attendees, among them people who had already achieved considerable notoriety. He was ultimately appointed a resident member of the Académie de Sciences in 1795. He also became professor of experimental physics and librarian at the Conservatoire de Arts et Métiers and, in 1816, president of the Class of Experimental Physics at the academy. In 1804, he married Julie-Françoise Bouchard des Hérettes, who died in 1817 after suffering from a long illness.

Charles' association with ballooning began when, in 1782, Joseph Montgolfier (1740-1810) discovered that hot air entering a **paper** bag held over an open fire made the bag rise to the ceiling. He and his brother, Étienne (1745-1799), experimented with larger ''envelopes'' and, on April 25, 1783, launched a balloon large enough that it was considered capable of carrying a person. On June 4 they successfully launched, from Annonay in France, a manned balloon made of linen covered with paper. Their creation, called *montgolfières*, consisted of an open fire hanging in a cage under the basket. The fire had to be fed by the pilot for the balloon to remain afloat.

Jacques Charles.

News of the Montgolfier's experiment quickly reached Paris and the French Academy of Science assigned Charles the task of studying the invention. He had already learned of a recent scientific discovery by British scientist **Henry Cavendish** of a gas 14 times lighter than air and which was being called "flammable air." (It would ultimately be named hydrogen.)

Hydrogen was probably first produced by alchemists in the fifteenth century. Although the element itself was unknown, alchemists were aware of metal/acid reactions that, according to Phillippus Aureolus Paracelsus caused "air to arise and break forth like the wind." French chemist **Antoine-Laurent Lavoisier**, continuing Cavendish's experiments, became the first to produced both hydrogen and **oxygen** in a laboratory by dissolving **metals** in acid. He named them "flammable air" and "life-sustaining air" respectively.)

Charles either assumed Montgolfier's balloon utilized this new gas, or simply decided to use it in his own ballooning experiments. Regardless, on August 27, 1783, he launched his first balloon from the Champs de Mars in Paris (the site on which the Eiffel Tower was ultimately constructed), creating the gas by pouring **sulfuric acid** over scrap **iron**. Among the spectators at this launch was the American ambassador to France, Benjamin Franklin. The balloon descended after about 45 minutes, landing in a field where terrified farmers attacked the "monster" with pick axes and shovels, "inspired by the beast's behavior to sigh and groan and emit a horrible smell."

Charles' creation included almost all the essential components still used in ballooning. He constructed the balloon sack of silk, coating it with a **solution** of rubber dissolved in turpentine to prevent the gas from escaping. At the top of the sack was an apex, open tube through which expanded gas could escape to prevent the balloon from exploding. A network of ropes covering the sack attached to a wooden hoop holding a wicker basket suspended below the sack. This basket would ultimately hold pilot and passengers. Charles also invented a valve line, allowing the pilot to release gas to initiate descent.

In December 1783, Charles, along with Nicholas Marie-Noel Robert (1761-1826), one of two brothers who helped in the design of the balloon sack, entered the basket and launched his creation called *Charlière*. This was the first manned gas balloon flight. Leaving Tuileries, the two aeronauts travelled for about two hours, landed in the village of Nesles-la-Vallée, and Charles then ascended alone to soar for another half hour. For some reason, this was the only flight Charles made, even though he lived until 1823.

Because of Charles' success with hydrogen, hot air balloons quickly disappeared, and it was not until recent times have they made a resurgence in popularity.

CHARLES' LAW

One of the **gas laws**, Charles' law states that at a constant pressure the **volume** of a fixed **mass** of gas is directly proportional to the absolute **temperature**. This law also can be expressed by the equation V/T = constant where V is the volume of the gas and T is the temperature in Kelvin.

Charles' law also is known as Gay-Lussac's law and the constant pressure law. This law was discovered independently by French physicist **Jacques Charles** in 1787 (who did not publish his findings) and French chemist **Joseph Gay-Lussac** in 1802. Although Gay-Lussac was the first scientist to announce this law, it is commonly known as Charles' law for its earlier discoverer. Charles' law can be combined with **Boyle's law** and the pressure law to give the ideal gas law (also known as the universal gas law), $pV = nRT$, where p is pressure, V is volume, n is the number of moles of gas, R is the universal gas constant, and T is temperature.

Charles' law is one of the ways in which absolute zero is calculated. This is done by extrapolating a graph of volume against temperature. At the point on the graph where the volume of the gas is zero (gas cannot have a volume less than zero), the theoretical temperature is absolute zero, the coldest temperature possible. This point has never actually been achieved, since **gases** liquefy before they reach this temperature and they are no longer subject to the gas laws.

Charles' law is a good indicator of what happens to a gas, since the volume and temperature are altered while the pressure remains constant.

See also Gases, behavior and properties of; Gay-Lussac's laws

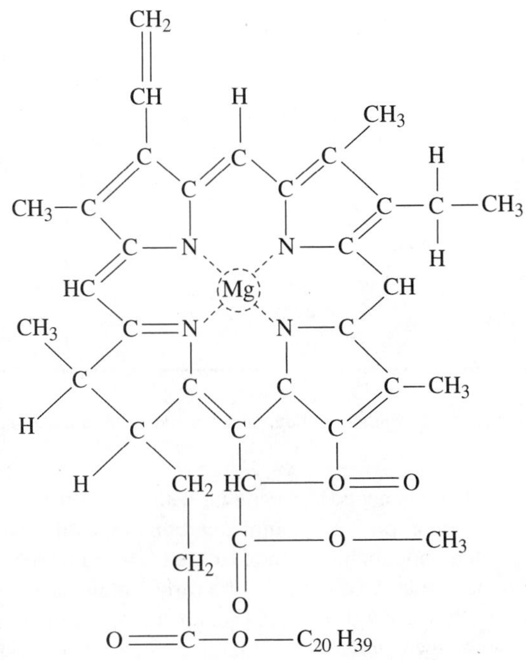

Chlorophyll

Figure 1 (Chelate). *(Illustration by Electronic Illustrators Group.)*

Heme

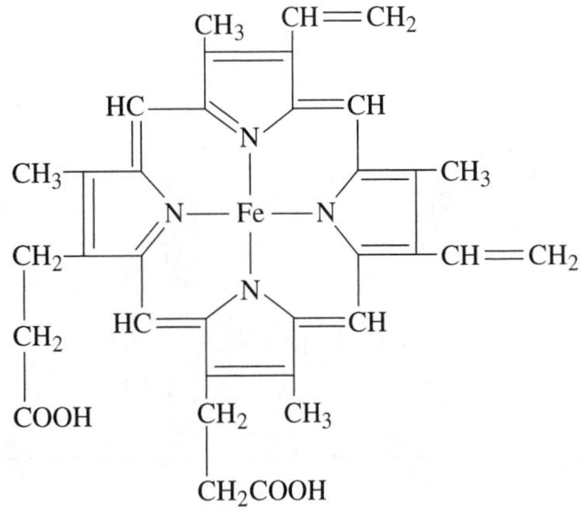

Figure 2 (Chelate). *(Illustration by Electronic Illustrators Group.)*

CHELATE

The word chelate comes from the Greek word *chele*, meaning crab's claw. A chelate is a **chemical compound** in which one **atom** is enclosed within a larger cluster of atoms that surround it like an envelope. A chelate is formed from a chelating agent plus a metal **ion**. The chelating agent wraps around a metal ion and attaches in several places. Compounds that attach to metal ions by sharing their electrons are known as ligands. Ligands that can attach in two places are called bidentate (two-bite) ligands. Those that attach in three places are tridentate, or three-bite ligands. There are even hexadentate ligands such as ethylene diammine tetraacetic acid, commonly known as EDTA. Chelating agents are bidentate or higher ligands. When you picture a ligand ''biting'' the metal ion from two sides, you can picture the pincers of a crab. Chelate compounds are very stable.

Chelates occur naturally. Chemicals in humus form chelation complexes with **calcium** and other ions. Plants are able to release chelated ions and incorporate them. The humus thus acts as a nutrient reservoir, so that essential ions are not leached into the soil.

A **magnesium** ion is the center of the chelate compound known as **chlorophyll** (Figure 1), a substance that allows plants to absorb light **energy** and synthesize **water** and **carbon** dioxide into glucose and **oxygen**. Four pyrrole nitrogens link to a central iron(II) ion in the heme **molecule** (Figure 2). Myoglobin and hemoglobin are macromolecules containing one and four heme groups, respectively. Vitamin B_{12} is a cobalt-centered chelate. **Platinum** chelates show antitumor activity.

Synthetic chelates are often brightly colored. The compound **nickel** dimethylglyoxime looks like strawberry-pink lipstick, insoluble in water. Its discovery by L. Tschugaeff in 1905 spurred chelation chemistry's rapid development. Because the size of the metal ion must be exactly right for the ligand to wrap around it, chelating agents are highly specific. Dimethylglyoxime (Figure 3) reacts with only two of the many known metal ions—nickel in basic **solution**, and **palladium** in acid solution. 8-quinolinol (Figure 4) is nearly specific for **aluminum**. Cupron is highly selective for **copper** ion in ammoniacal solution and for **molybdenum** in acid.

Oxine, or 8-hydroxyquinoline, can be made to form insoluble chelates with almost all metal ions except for those in Group 1. By regulating **pH**, some selectivity can be obtained. Among the least soluble are the hydroxyquinolates of palladium, copper(II), and iron(III) which can be precipitated at pH 3. At a pH just over 4, aluminum precipitates. In ammoniacal solution, magnesium is precipitated. Substituted hydroxyquinolines are more specific: 2-methyl-8-hydroxyquinoline will precipitate **zinc** but not aluminum, and 5,7-dibromo-8-hydroxyquinoline precipitates only **titanium**, copper, and **iron** from acid solution.

Ethylene diamine tetraacetic acid (EDTA) reacts with magnesium and calcium, providing the basis for hard water analysis and treatment and prevention of boiler scale. It completely complexes iron, preventing its **precipitation** as the rust-colored iron hydroxide. EDTA is added to foods to protect various **vitamins** from **oxidation**, to guard against air-induced discoloration of canned potatoes, fish, and many other foods, and to protect unsaturated side **chains** of triglycerides from the

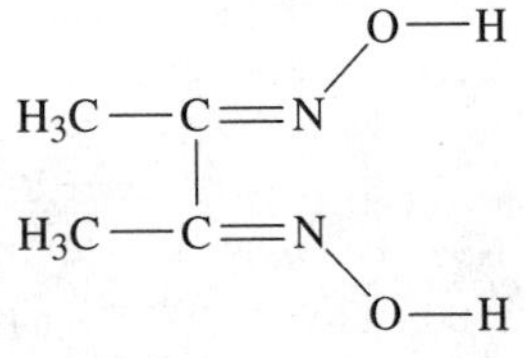

Dimethylglyoxime

Figure 3 (Chelate). ***(Illustration by Electronic Illustrators Group.)***

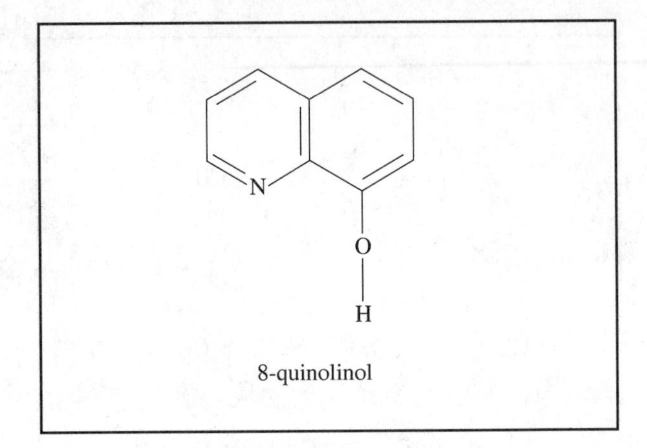

Figure 4 (Chelate). ***(Illustration by Electronic Illustrators Group.)***

oxidation that causes rancidity. It protects food by acting as a sequestrant, tying up metal ions such as aluminum, iron, and zinc as soluble chelates so that they cannot act as catalysts for oxidation. The soluble chelates are eventually flushed from the body.

In a similar manner, EDTA has found a use in treatment of heavy-metal poisoning. Chelation therapy consists of treating the patient with EDTA intravenously so that the chelate that is formed can be excreted from the body before heavy **metals** such as **lead** or **mercury** have an opportunity to denature the body's **proteins**. This is the preferred method for treatment of lead poisoning. It has also been used to attack calcium deposits such as those formed in the joints from arthritis, and can also be used to tie up **plutonium** in cases of radioactive poisoning. These treatments are not without side effects, since EDTA also removes calcium ions and other essential trace ions from body tissue, which must then be replenished. Often the EDTA is administered as the calcium complex, so that calcium ions are introduced into the body as they are replaced by the heavy metal.

BAL, or British Anti-Lewisite, a derivative of **glycerol** developed to protect against the gas Lewisite, is also an effective chelating agent for mercury. Unfortunately, as in lead poisoning, the symptoms may not appear until the neurological damage is irreversible.

CHEMICAL COMPOUND

The majority of chemical substances are compounds. A compound is a pure substance consisting of two (or more) elements that are chemically bonded in a fixed proportion. Unlike a mixture, in which components retain their characteristics, a compound is a new entity. Thus, for example, atoms of **oxygen** and **hydrogen** can be mixed, randomly, in any proportion. The resulting mixture will contain hydrogen molecules and oxygen **molecules**. **Water**, however, which is a compound of oxygen and hydrogen, also contains oxygen and hydrogen, but its molecules combine both elements, in a fixed proportion, as expressed by the formula H_2O.

Certain elements can combine with each other in more than one proportion. For example, **carbon monoxide** and **carbon dioxide** both contain carbon oxygen. Thus a molecule of carbon monoxide (CO) consists of a carbon **atom** and an oxygen atom. In carbon dioxide (CO_2), a carbon atom is bonded with two oxygen atoms to form a molecule of the compound. It is important to note that these two compounds are quite different. Unlike carbon dioxide, carbon monoxide is a lethal gas.

According to the type of bond with which its atoms are held, a compound can be either covalent or ionic. A covalent compound is held together by covalent bonds (a **covalent bond** holds two atoms together to create diatomic molecules, such as H_2 or N_2. Covalent compounds include water, carbon monoxide, and carbon dioxide. An ionic compound contains cations (positively charged ions) and anions (negatively charged ions). In sodium chloride, a typical ionic compound, there is an equal number of sodium ions and chlorine ions. A sodium **cation** (Na^+) is created when a **sodium** atom loses an **electron**; a chlorine anion (Cl! is created when a chlorine atom gains an electron. In an ionic compound, the ions are held together by the ionic bond—the attraction between the positively charged ion and its negatively charged counterpart.

Compounds can also be divided into organic and inorganic. As a rule, organic compound molecules contain two or more carbon atoms. Examples of organic compounds are **methane** (CH_4), a small **hydrocarbon** compound, and large polymers, which include **proteins**, **carbohydrates**, and **nucleic acids**. Inorganic compounds, which include water and many minerals, usually contain elements other than carbon. Examples of inorganic compounds are water and minerals (with some exceptions). There is a third group of compounds, however, which combines an organic and an inorganic substance are combine. Because the inorganic component is usually a metal, these compounds are called organometallics. Examples of natural **organometalics** are **chlorophyll**, hemoglobin, and vitamin B_{12}.

By the late eighteenth century, chemists knew that an individual compound's components bonded in a definite proportion by mass. While **Antoine Lavoisier** and his contemporaries were aware of the phenomenon of definite proportionality of a compound's components, it was Joseph-Louis Proust (1754-1826) who formulated the law of definite proportions. Proust validated this law by experimentally examining a considerable number of compounds. For example, a gram of **sodium chloride** contains 0.3934 g of sodium and 0.6066 g of **chlorine**. This proportion is constant. However, chemists later discovered exceptions to this law. Compounds exhibiting slight deviations from the law of definite proportions are called nonstoichiometric compounds, and they may occur as a result of lattice defects during **crystallization**. A crystal is a solid substance in which atoms, molecules, or ions are arranged in a definite geometrical manner. Sodium chloride, for instance, is a crystal in which each Na^+ ions is surrounded by six Cl^- ions, while each Cl^- ions is surrounded by six Na^+ ions. This particular geometric regularity, which is called a lattice, may fail in some compounds, leading to compounds which violate the law of definite proportions. For example, certain spots in a crystal lattice may remain vacant, or a lattice may have an unequal number of **anion** and cation sites. Thus, when iron oxide (FeO) crystallizes, iron sites often remain vacant. Because there are fewer iron atoms, each must assume a +3 charge in order to create a crystal in which the electrical forces are balanced. As a result, however, there are more oxygen atoms than iron atoms in the crystal. While iron oxide in its nonstoichometric form is still represented by the FeO formula, the proportion of its components will deviate from that of the compound in its ideal form. Other examples of nonstoichometric compounds include ceramic superconductors.

CHEMICAL CONTRACEPTIVES

Throughout history, men and women have looked for safe, reliable, convenient ways to prevent pregnancy. Some have been based on medical knowledge, but many have arisen from superstition and old wives' tales. However, one category of contraceptives, those that use chemicals, have proved themselves valuable and trustworthy. Employed at least since ancient Egyptian and Greek times, chemical contraceptives are still the most successful and popular agents for birth control today.

Most modern chemical birth control methods are used by women and rely on synthetic versions of the **hormones** estrogen and progesterone. These simulate the biochemical effects of pregnancy. Thus, the user's body will not release an egg to be fertilized, since all signs indicate that a pregnancy is already under way. For this reason, the synthetic hormones are technically known as ovulation inhibitors.

Depending on the formulation, ovulation inhibitors can be as much as 99% effective at preventing pregnancy. The hormones used in these products are usually synthetic because the stomach and intestinal absorption of natural ones is poor. Some of the most common synthetic estrogens used in birth control products are chlormadinone acetate, ethynylestradiol, and mestranol, while synthetic progestins include norgestrel, norethindone, ethynodiol diacetate, norethynodrel, and dimethisterone.

Ovulation inhibitors, which most people know collectively as "the pill," were first introduced to the American public in the 1960s. Because this form of oral birth control meant that a woman using it was (theoretically, at least) able to have intercourse at any time at a moment's notice without fear of pregnancy, the pill had a dramatic effect on society and sexual mores. It also empowered women to take control of their own reproductive fates because of the pill's effectiveness, relative safety (although early high- estrogen-dose products could have significant side effects), and ease of use.

Ovulation inhibitors generally contain a combination of synthetic estrogen, which prevents bleeding in between periods, and synthetic progestin, which actually prevents ovulation. By the 1990s, most products no longer contained 100-150 micrograms of estrogen, since researchers had found that doses of only 30-35 micrograms of estrogen were safer and actually more effective as contraception. However, these low-dose (or second-generation) pills also caused more women to experience "breakthrough" (between-period) bleeding. The solution turned out to be packaging the product so that it contained multiple phases of the chemicals, usually represented by different-colored pills, that varied the ratio of synthetic estrogen to progesterone throughout a woman's cycle. Thus, a woman received the lowest dose of estrogen during the menstrual period. The first of these multiphasic pills went on the market in 1984.

Other chemical contraceptives use different delivery methods, but similar ingredients. For instance, Depo-Provera (medroxyprogesterone acetate) is injected four times a year, while Norplant (levonorgestrel) is surgically implanted under the skin of the inner upper arm and effectively prevents pregnancy for up to five years. NET EN (norethindrone enanthate) is another injectable contraceptive, although it must be injected at two-month intervals. All of these are termed "long-acting progestins."

Researchers at the turn of the millenium were investigating the practicality of another group of chemical contraceptives based on an even more carefully balanced estrogen-progesterone ratio. These included gestodene, desogestrel, and norgestimate, which were already being called the third generation of oral contraceptives. Another new group of chemical contraceptives are the antiprogestins, which include such products as RU486 (mifepristone). These prevent or even abort an in-progress pregnancy by blocking the **receptors** for progesterone.

Nonhormone-based chemicals can also be effective in pregnancy prevention, although these are far fewer in number. One of these is centchroman. This chemical works by using powerful antiestrogenic factors to keep a fertilized egg (zygote) from attaching itself to the lining of the uterus, where the zygote's cells would normally begin dividing and eventually form the fetus.

Nonhormone chemical contraceptives also include spermicides, which, as their name indicates, deactivate and kill sperm before they can fertilize the egg. The main chemical used in spermicides is nonoxynol-9, which along with octoxynol in the 1940s became the first spermicide introduced to the public. Nonoxynol-9 is most often found in vaginal suppositories, where they form a lethal barrier (whether in a sponge, foam, cream, or jelly) against sperm trying to make their way to the cervix to fertilize an egg. However, the chemical can also coat condoms or be dispersed by intrauterine devices to enhance contraception. In general, nonoxynol-9 is about 70% effective on its own.

CHEMICAL ENGINEERING

Engineering, in general, is a business which deals with the design, construction, and operation of the systems and equipment for both industrial and public use. Chemical engineering, in particular, is devoted to the development of systems and equipment for the manufacture of products such as acids, dyes, drugs, **plastics**, and synthetics for the chemical industry. Chemical engineers are responsible for designing practical applications for basic chemical research in order to transform raw materials into useful products. They must use not only **chemistry**, but also mathematics, physics, and engineering in order to solve problems. Chemical engineering utilizes many aspects of chemistry, including **energy** transfer, **thermodynamics**, **mass**, momentum, and chemical **kinetics**.

Because the field of chemical engineering utilizes many different aspects of chemistry, there are numerous applications in which a member of the profession may specialize. A chemical engineer may specialize in **pharmaceutical chemistry**, **petrochemicals**, food additives, **ceramics**, environmental clean-up, safety engineering, or nuclear chemistry. Other important specializations include biotechnology, chemical production, **electrochemistry**, **paints and coatings**, and **water** technology. Two people who both call themselves chemical engineers may actually do two very different things.

The chemical engineering profession was not always so widely accepted and diverse. In the late 1800s there were people who called themselves "chemical engineers," but there was no unity in their education or specializations. The earliest chemical engineers were mechanical engineers who dabbled in chemistry or worked in a chemical plant. Chemical engineering, as we know it today, started in 1888, when the Massachusetts Institute of Technology first offered a formal degree in the field. In 1892 the University of Pennsylvania and later, in 1894, Tulane University both offered formal chemical engineering programs as well.

The field of chemistry existed long before that of chemical engineering. It became necessary to educate chemists differently as the Industrial Revolution began. Many so-called "industrial chemicals" were becoming necessary in great quantities in order to continue the expansion of industry. One such chemical was **sulfuric acid**. If a company could produce sulfuric acid in large quantities quickly and with low cost, it would enjoy large profits because of the great demand for the chemical. The method which was used to produce this acid in the early 1800s was the Lead-Chamber Method. This method simply required air, water, **sulfur** dioxide, a nitrate, and a large **lead** container. The nitrate was quite expensive and not very efficient to work with. The chemical plants which produced sulfuric acid could barely keep up with the demand. In 1859 a new method was introduced, called the Glover Tower. This utilized a mass transfer tower which recovered nitrate lost to the atmosphere in the Lead-Chamber Method. Engineers were becoming necessary in the chemical industry because of the economic demand for more modern and more productive chemical plants.

Another important industrial chemical at this time was soda ash (Na_2CO_3), which was used in the production of **glass**, soap, and textiles. This alkali compound was originally harvested from natural sources, such as trees or kelp. As these natural sources became depleted, a new source was needed. The Le Blanc Process, designed in the early 1800s, converted **salt** into soda ash. One problem with this process was the high levels of **pollution** and the potential health hazards to anyone living near a soda ash plant. Another process, called the Solvay Process, could produce soda ash in a more direct way which created much less pollution. Because it was a more direct process, complex engineering had to be employed in order to use it in a large-scale chemical plant.

The need for a new chemist who had an understanding of engineering processes was quickly becoming apparent. In England in 1880 a "Society of Chemical Engineers" was unsuccessfully attempted. Chemical engineering was still not considered a separate profession, and there was no clear- cut definition of what a chemical engineer actually did. It was only a few years later when formal chemical engineering programs were started in colleges and universities. The main focus of these programs was to prepare chemists to fulfill the demands of the chemical industry. The course work involved an emphasis on mechanical engineering in combination with **industrial chemistry**. Competition between chemical plants to manufacture the most product at the lowest cost increased the demand for the chemical engineer. Early chemical engineers focused on optimizing the chemical plants of the industrial revolution, utilizing such processes as continuously operating reactors, purification of products, and **recycling** reactants.

Despite the need for chemical engineers in industry, chemical engineering as a separate profession was not immediately recognized. Many members of industry believed that chemists could solve just as many problems. In 1908 the American Institute of Chemical Engineers (AIChE) was formed in order to validate and unite the profession. Despite early conflicts with the American Chemical Society, AIChE survived and is still in existence today. The formation of this organization not only gave the field of chemical engineering formal recognition, but also helped convince the chemical industry that chemical engineers should be used in plant design and operation instead of mechanical engineers.

Despite the recognition of chemical engineering as a profession and the introduction of several formal chemical engineering programs, there were still inconsistencies in the education and training of chemical engineers. To solve this problem, AIChE started an accreditation program for schools offering chemical engineering degrees. In 1925 a list of 14 schools was published which had earned accreditation. Chemical engineering was the first profession which utilized accreditation in order to gain consistency and ensure the appropriate education of its members. Eventually, other branches of engineering followed suit and in 1932 the Accreditation Board for Engineering and Technology was formed.

The focus of the chemical engineering profession changed with the onset of World War I. Instead of being concerned with industrialization, the chemical engineer was enlisted to create materials which could be used in the war. Chemical industries in America were now working toward a common goal instead of competing with each other. As a result, **ammonia** plants were built which produced not only **fertilizers**, but the necessary **explosives** to help win the war.

During World War II, new applications of chemical engineering were introduced. In the beginning of the war, Japan captured rubber producing lands, including 90% of America's natural rubber sources. Rubber was very important during the war and was used by the military for tires, gaskets, hoses, and boots. Chemical engineers had to design factories to produce synthetic rubber and actually increased synthetic rubber production by over a hundred times. Efficient high-octane **gasoline** was also important for war efforts and in 1940 the Standard Oil Company developed a catalytic reforming process which produced not only high octane fuel from less expensive petroleum, but also toluene for **trinitrotoluene** (TNT). In 1942, the chemical engineers at Du Pont began the design and eventually the operation of a **plutonium** production plant to use for the new atomic bomb. This plant was called Hanford Engineering Works and was a major contribution of chemical engineering to the war efforts.

After the war, the chemical engineering profession began to focus on the petroleum industry, which is still a major branch of chemical engineering today. With the continuous introduction of new technologies, the field of chemical engineering is constantly evolving. Chemical engineers today need to respond to industrial as well as technological demands. Chemical engineering education is also changing, with a much stronger mathematical and technical background now than was found in the original chemical engineering programs. New fields of specialization are constantly being introduced, especially in the areas of biotechnology, electronics, food processing, pharmaceuticals, and environmental clean-up. Chemical engineering is an important profession and as long as technological advances continue to be made, the demand for chemical engineers will continue to rise.

CHEMICAL EVOLUTION

Charles Darwin introduced his theory of biological evolution in *The Origin of Species*, published in 1859. This development made people wonder what else in Nature had evolved; could, for example, chemical elements evolve? This question was answered in the mid-twentieth century.

In 1950s, several researchers from the United States concluded that all chemical elements originate from **hydrogen**. The notion that all elements derive from hydrogen was not a completely new hypothesis. In the nineteenth century, a couple of chemists proposed that chemical elements originated from hydrogen. The mechanism for the evolution of elements from hydrog en though was not developed until 1957 by Margaret Burbidge, Geoffrey Burbidge, William Fowler, and Fred Hoyle, who presented their mechanism in a paper entitled ''Synthesis of the Elements in Stars.'' They proposed that the stars are the seat of origin of the elements, a process called nucleosynthesis. They concluded that they could expl ain the abundances of practically all the elements from hydrogen through synthesis in stars.

In 1812, **Humphry Davy** thought that hydrogen might be the primordial **matter** from which all other chemical elements derived, because he obtained hydrogen from most **metals** and from other unlikely materials. In 1815, **William Prout** concluded that the atoms of all the chemical elements must have been formed by multiple combinations of hydrogen atoms. This conclusion was based on his finding that atomic weights were close to whole numbers. Prout calculated the specific gravities of elements and then compared the results with the specific gravity of hydrogen, which is one. All of his calculations produced whole numbers, and this led h im to wonder whether hydrogen was the basis of all matter, whether all the chemical elements somehow evolved from hydrogen. An example of Prout's thinking is if the atomic weight of **oxygen** is 16, then 16 volumes of hydrogen had condensed to form this element.

Over the rest of the nineteenth century developments were made in calculating the atomic weights of elements. Prout's hypothesis supposed that the atomic weights of elements should be whole numbers, but as the experiments and calculations improved, it was found that some elements did not have whole number atomic weights. **Neon**, for example, has an **atomic weight** of 20.18. In 1913, **F. W. Aston** and **J. J. Thomson** found that neon is composed of two main species, or isotopes, with atomic weights of 20 and 22 in the approximate ration of 10:1. Finding that the isotopes of neon have atomic weights that are whole numbers vindicated Prout's hypothesis that hydr ogen was the basis for all the elements. The evidence was building in favour of the hypothesis that hydrogen was the source for the evolution of elements.

Henry Russel (1877-1957) in 1929 demonstrated that hydrogen is the major elementary constituent of the Sun. This led to the hypothesis that the origin of the elements was in the stars, a process known as stellar nucleosynthesis. There were two mechanisms put forward to account for how the hydrogen in stars, like the Sun, could evolve into all of the elements we know of today. The first and earliest mechanism was based on a high-**temperature** equilibrium between atomic nuclei, giving the relative distribution of elements upon cooling. The second mechanism was a kinetic mechanism whereby the elements evolve sequentially with the addition of protons and neutrons from the hydrogen **atom**.

The mechanism put forward by the Burbidges, Fowler, and Hoyle combined the two earlier proposed mechanisms. They proposed eight main nucleosynthetic processes in star interiors, all starting from hydrogen. The evolution of chemicals follows these eight processes: 1) the conversion of hydrogen to **helium**; 2) the combustion of helium to **carbon**, oxygen, and neon; 3) the capture of successive alpha particles by oxygen and neon producing **magnesium**, **silicon**, **argon**, and **calcium**; 4) an equilibrium process to allow for the high abundance of **iron** group elements; 5) a slow process of **neutron** capture by iron grou p elements (iron, **cobalt**, and nickel); 6) a rapid process of neutron capture in supernovae explosions; 7) a **proton** capture process producing the rare light isotopes of the heavy elements; and 8) a light-atom process where high- **energy** collisions break up heavier nuclei.

Davy and Prout then, early in the nineteenth century, were not wrong in thinking that hydrogen was the primary matter that all elements evolved from. Darwin's theory of biological evolution equipped scientists with a mechanism to show how all chemical elements originate from hydrogen.

CHEMICAL INSTRUMENTS

Chemical instruments are pieces of laboratory equipment that are used to make quantitative measurements. They are distinguishable from chemical apparatus in that they directly measure quantities, or enable a quantity to be measured, whereas chemical apparatus are used solely in the conduction of experiments. For example, burettes and spectro photometers are chemical instruments—one measures **volume** and the other measures absorbance. Fractional **distillation** columns, Bunsen burners, and beakers on the other hand are chemical apparatus because they are used in conducting experiments, but do not measure any quantity.

Historically, the development of analytical techniques has closely followed the introduction of new chemical instruments. The first quantitative analyses were gravimetric analyses (weighing things) that were made possible by the invention of a precise **balance**. English physicist Francis Hauksbee (1666- 1713) constructed the first chemical precise balance in 1710. In the 1770s, French chemist **Antoine-Laurent Lavoisier** dismantled the **phlogiston** theory of **combustion** by accurately and precisely measuring the weight of the chemicals both before and after his experiments. This soon became the standard method of performing gravimetric a nalyses.

Chemical instruments were not solely introduced for increasingly ambitious experiments. Governments and other regulatory agencies also used them. For example, in 1789, English instrument maker Jesse Ramsden (1735-1800) made a balance for the purpose of measuring the specific gravity of water-alcohol mixtures to facilitate government tax legislation.

Chemical instruments have become far more complex than their predecessor, the simple chemical balance. In modern laboratories, many chemical instruments are electrical. This began in the 1930s when a rapid development in electronics resulted in a major revolution in analytical chemical instrumentation. The basic task of electrical instruments is to translate chemical information about a substance into a form that a chemist can directly observe. Electrical chemical instruments allow for the naturally unobservable, such as molecules, to be observed. Realizing the steps between the observer and the observable is important in order to understand the role of chemical instruments in modern laboratories. There are four main components to electrical instruments. 1) Signal generators allow for direct or indirect interaction of the substance being analyzed with some form of **energy**, such as **electricity** or light. 2) Input transducers, also known as detectors, are devices that transform the chemical or physical property of the substance being analyzed into an electronic signal. 3) Signal modifiers are electronic components that perform operations, such as amplification and filtering, on the generated signal. 4) Output transducers convert the modified electrical signal into information that can be read, recorded, and interpreted by a chemist.

Consider the basic spectrophotometer as an example. This instrument is used widely in laboratories for analyzing a substance and its components based on its absorbance. The signal generator is the light that is shone on the substance. The input transducer is the photo detector that detects the radiation emitted from the radiated substance. The signal modifier is the computing device that calculates the absorbance of the substance. And finally, the output transducer is the dial on the spectrophotometer itself that shows the absorbance of the substance. This process of finding the absorbance of a substance would be much more cumbersome without the use of electronically enhanced chemical instruments.

Chemical instruments are useful in expanding the scope of our senses. Precise experiments cannot be conducted without knowing quantities. Instruments allow chemical and physical qualities that we cannot see, hear, smell, **taste**, or touch to be observed. With increased use of chemical instruments deeper probing of the properties and behavior of chemical substances is possible.

See also Quantative analysis; Thermometers

CHEMICAL NOTATION

Chemical notation is the specialized system of signs and symbols used in **chemistry**. Chemical notation is a short hand system to represent what is happening without having to write everything in long hand.

The most basic part of chemical notation is the abbreviations for elements used in the **periodic table**. Each individual element has a unique one or two letter code to identify it. The first letter of this code is upper case and the second, if present, is lower case. For example the element **hydrogen** is represented by the symbol H, and the element **calcium** is represented by Ca. When an element is encountered it is given the appropriate symbol. The majority of times when materials are encountered they are compounds, consisting of two or more of elements. Compounds can be represented in the chemical notation system. Compounds are represented by the symbols for the elements involved. This is known as the chemical formula. The name of the compound is part of the chemical **nomenclature** system. There are rules that are laid down by the **International Union of Pure and Applied Chemistry (IUPAC)** for how a chemical formula is formed. There are two types of material inorganic compounds and organic compounds. With inorganic compounds there are compounds and molecular compounds. To consider the ionic compound first. In an ionic compound

An assortment of empty laboratory glassware. From the right of image: conical flask and dropping pipette, beaker and test tubes, a round-bottomed flask. At the far left are a stack of petri dishes, used to culture microorganisms. ***(Photograph by Oscar Burriel/LatinStock/Science Photo Library, Photo Researchers, Inc. Reproduced by permmission.)***

there are two different types of ions present, the positively charged cations cations and the negatively charged anions. The cations are either **metals** or hydrogen. When the chemical formula is produced the symbol for the metal or hydrogen appears first. The next symbol to appear is the **anion**. For example this would give us the formula NaCl for **sodium** chloride. This is the situation for two elements combining together.

This rule of **cation** followed by anion is universal. Some compounds have polyatomic ions, where the **ion** is composed of two or more atoms. The ammonium ion is a polyatomic cation (NH_4^+) and the nitrate ion is an example of a polyatomic anion (NO_3^-). When polyatomic ions are considered the symbol of the element farthest to the left in the periodic table comes first. If both elements are in the same group, then the lower one comes first. These polyatomic ions are also called molecular compounds.

To produce the chemical formula for an organic compound the same general rules are followed with some modification. Because of the potential size and complexity of organic compounds there are slight variations. All organic compounds are covalently bonded. The simplest version of the formula is the **empirical formula**. Here, **carbon** is listed first, then hydrogen, and then the other elements that are present. The other elements after hydrogen are listed alphabetically. This is known as the Hill System. When the **molecular formula** is considered the, situation is more complex. To provide the full molecular formula, the **structural formula** must also be considered. Here, the **molecule** is drawn out with the longest chain of carbon atoms as the central part. The various aryl and alkyl groups are then added onto the molecule in the appropriate positions. For the full chemical formula, the modifier groups are listed first, then the backbone chain and finally the **functional group**. For example, **ethanol** is CH_3CH_2OH. If the functional group has two bonds such as a ketone, then it is included in the middle of the formula. An example of this situation is methyl ethyl ketone, or butanone, $CH_3COCH_2CH_3$.

If the rules of producing chemical formulae are not followed, less information is conveyed. Other symbols use in chemistry include the Greek letter delta to denote a change in **energy** levels and an arrow to indicate the direction a reaction proceeds in. A plus symbol has the normal meaning, and when structural formulas are used. a line is used to represent a chemical bond, multiple bonds being represented by multiple lines. **Ring** structures are represented by a ring showing the bonds, the carbon and hydrogen atoms are not drawn in they are implied. Superscripts are used to denote charge and its size and subscripts are used to show how many molecules are present in a compound.

Chemical notation is relatively easy. Once the basics are understood chemical formulas can be rapidly interpreted and produced. Chemical equations can also be understood and the maximum amount of information can be seen. Chemical notation is merely a system use to present as much information as rapidly as possible.

See also Formula, chemical; Nomenclature

Chemical oxygen demand

Chemical oxygen demand (COD) is a measure of the ability of chemical reactions to oxidize matter in an aqueous system. The results are expressed in terms of oxygen so that they can be compared directly to the results of biochemical **oxygen** demand (BOD) testing. The test is performed by adding the oxidizing **solution** of a dichromate **salt** (e.g. **potassium** dichromate, $K_2Cr_2O_7$) to a sample, boiling the mixture on a refluxing apparatus for two hours, and then titrating the amount of dichromate remaining after the refluxing period. The titration procedure involves adding ferrous ammonium **sulfate** (FAS), at a known normality, to reduce the remaining dichromate. The amount of dichromate reduced during the test—the initial amount minus the amount remaining at the end—is then expressed in terms of oxygen. The test has nothing to do with oxygen initially present or used. It is a measure of the demand of a solution or suspension for a strong oxidant. The oxidant will react with most organic materials and certain inorganic materials under the conditions of the test. For example, Fe(II) and Mn(II) will be oxidized to Fe(III) and Mn(IV), respectively, during the test.

Generally, the COD is larger than the BOD exerted over a five-day period (BOD_5), but there are exceptions in which microbes of the BOD test can oxidize materials that the COD reagents cannot. For a raw, domestic wastewater, the COD/BOD_5 ratio is in the area of 1.5-3.0/1.0. Higher ratios would indicate the presence of toxic, non- biodegradable or less readily biodegradable materials.

The COD test is commonly used because it is a relatively short-term, precise test with few interferences. However, the spent solutions generated by the test are hazardous. The **liquids** are acidic, and contain **chromium, silver, mercury**, and perhaps other toxic materials in the sample tested. For this reason laboratories are doing fewer or smaller COD tests in which smaller amounts of the same reagents are used.

Chemical Revolution

Practical **chemistry** is as old, if not older, than civilization. It can be seen in ancient wall paintings, pottery shards, **glass** work, and metal processing. Its origins are lost in past. However, the development of the philosophy or theoretical aspects of chemistry are reasonably well documented. The Greeks and, in particular, Aristotle believed that all things in the world around us could be broken down into combinations of earth, **water**, air, and fire. These were the four elements and any compound could be made by adjusting the relative proportions of each. This is basis of **alchemy** and the belief in the **transmutation** of the elements.

Centuries later, chemistry began to emerge out of the dark realms of alchemy. Scientists adopted a systematic approach, trying to find common patterns and explanations for the observed chemical diversity in nature. The decomposition of minerals and the discovery of the different components of air lead to new explanations for the nature of **matter**.

Historically, the modern science of chemistry stems from the early 1800s and two remarkable men. The first was **Antoine Laurent Lavoisier**, a French aristocrat, and the second was **John Dalton**, an English school teacher. Both men helped to systematize chemistry, to define the law of definite proportions and the underlying atomic and elemental principles. They transformed the foundations of chemical thought and opened the way for those around them to create a chemical revolution that was as powerful and influential as any political revolution of the time. Arguably, it was their genius that transformed chemistry from an art to a science. The repercussions of their work lead to the discovery of the remaining elements, the foundation of the **periodic table**, an understanding of **stoichiometry** and structure, and ultimately to the chemical processes that presently enrich our world.

Chemical Standards

Analytical chemistry pertains to determining the chemical composition of a substance and the quantity of each substance present. From the nanogram to kilogram scale, there is a method for determining what is there and how much. But the important question asked many times is''how do analytical chemists know that they are right?'' The answer to this question is chemical standards.

In its simplest terms, a chemical standard is a substance for which the exact composition is known. Or, at least, known as far as possible. Detection limits for most analytical techniques are in the parts per billion (ppb) or parts per trillion level (ppt) and although this may seem a very small amount, it still means a billion molecules of an impurity would not be detected in a drop of **water**. This is the nature of chemistry. Chemists deal in incredibly large numbers of molecules.

This inherent limit in detection of substance is a difficulty for analytical chemists and why the use of chemical standards has been adopted. The essential idea of a chemical standard is that all chemists agree on its composition and then build from there. By comparing solutions or other compounds to the standard, their composition is determined and so on.

To illustrate this process, consider the procedure for standardizing a **solution** that is approximately 1 molar **hydrochloric acid**. The exact **concentration** of the hydrochloric acid is unknown, but because it will be used for titration, knowing the concentration is very helpful. To determine the exact concentration, titrating the hydrochloric acid against a solution of base, for example sodium hydroxide, would be appropriate. The titrated **volume** and **molarity** of the base can be used to determine the acid concentration. But how is the concentration of sodium hydroxide in its solution obtained? That is determined by titrating against a primary standard, potassium hydrogen phthalate.

In other words, the **potassium hydrogen** phthalate (a primary standard) is used to determine the concentration of the **sodium** hydroxide (a secondary standard) which is then used to determine the concentration of hydrochloric acid. Why go through this complicated process? To ensure that the results are accurate and valid. By reference back to aprimary standard, providing that the analytical technique is valid and carried out properly, analytical chemists can provide some degree of assurance for their results.

Potassium hydrogen phthalate is used as a primary standard because it is a solid, organic acid **salt** of moderate **molecular weight** that is of known purity. The fact that it is a solid meansthat it can be weighed on an analytical **balance** quite accurately — to seven decimal places if necessary. The fact that it is the salt of an organic acid—a diacid, actually means that it is readily soluble in water or aqueous solution. The fact that it has a moderate molecular weight means not a lot is required for a **titration** but enough that a good weight can be obtained. And the purity of the substance has been checked and verified many times over. In addition, it is easy to obtain and reacts in a rapid, quantitative reaction with any of the simple bases, such as sodium hydroxide. All of these attributes allow it to be used as a primary standard. There are surprisingly few compounds that qualify. For acid-base reactions, **sodium carbonate**, sodium tetraborate, benzoic acid, potassium hydrogen phthalate, and potassium hydrogen iodate make up the complete list. Consequently, analytical chemistry invariably requires the use of secondary standards.

Secondary standards are substances that can be referred back to primary standards, such asthe **sodium hydroxide** in the above illustration. These are generally stable substances that will not change composition in the short term. However, long-term storage is not possible as the secondary standards will eventually "go off." For example, sodium hydroxide stored in a **glass** bottle or volumetric will slowly dissolve the glass, leading to a change in composition. Even storage in a nalgene bottle results in the slow absorption of water and impurities from the nalgene. Although storing sodium hydroxide this way extends its usefulness as a secondary standard, fresh solutions need to be made and calibrated against a primary chemical standard on a regular basis.

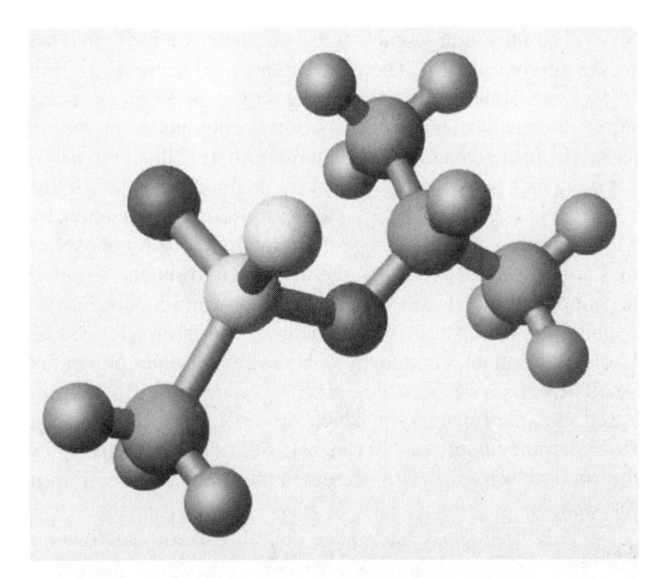

A computer-generated molecular model of the poisonous nerve agent SARIN. ***(Photo Researchers, Inc. Reproduced by permission.)***

A further set of standards for chemical analysis are those provide by the National Instituteof Standards and Technology (NIST). These are chemical compounds or mixtures that have beentested at multiple laboratories, using a variety of methods, and provide the basis for the validation of new tests and techniques, along with providing quality assurance for the laboratories themselves. For example, an analytical chemist might be interested in determining the composition of a sample of dirt. Having devised a technique for doing so, he or she would request a sample of "standardized dirt" from NIST intending to test their technique against the known chemical composition of the sample. If their technique provides the same results as the analysis of the standard sample, within the accepted margins of error, then they have a valid method for measuring the composition of dirt. This is important if the analytical chemist plans to use this technique for research or for testing soil samples for commercial clients. The quality of the results is assured by their accuracy.

In the end, chemical standards are the only method that chemists have of assuring that analytical results are accurate and valid. Testing against standards is a necessary part of chemical analysis.

CHEMICAL WEAPONS

Chemical weapons are weapons that achieve their deadly effects not by explosion, **heat**, or penetration, but by biochemical interaction with the target. Chemical weapons include agents that act on the nervous system, such as sarin and VX, and those that affect all exposed tissues, such as mustard gas. Other chemical weapons include botulinum toxin, **hydrogen** cyanide, psychoactive compounds such as phencyclidine, and tear gas. The use of chemical weapons in warfare has been regulated by a series of international treaties, although some nations have not signed these agreements and continue to use them.

The most common chemical weapons are those that act on the nervous system. These are often called nerve **gases**, but in fact very few chemical weapons are actually gases. Most nerve agents are organophosphorus compounds, including sarin (isopropyl methylphosphonofluoridate), the first nerve agent, which was widely produced beginning in the 1930s. Since then, it has been supplanted in most modern arsenals by VX (O-ethyl S-diisopropylaminomethyl methylphosphonothiolate), which persists longer in the environment. Another organophosphate, tabun (O-ethyl dimethylamidophosphorylcyanide), is the easiest to manufacture, and is therefore more likely to be manufactured by less developed countries or by small terrorist organizations. These compounds can be prepared as binary agents, in which the final product is formed from two precursors just before release, or during delivery of the shell or warhead. This increases the safety of handling of the weapon.

All organophosphates act by binding to and inactivating the **enzyme** acetylcholinesterase. This enzyme normally breaks down the neurotransmitter acetylcholine, which stimulates muscle contraction, as well as performing other functions within the body. Without acetylcholinesterase, muscle cells and other organs become overactive. Weakness, tremors, and convulsions follow exposure. Respiratory paralysis is the direct cause of death. Organophosphates are also used as insecticides, and rely on the same mechanisms of action for this application.

Mustard gas (bis-[2-chloroethyl]sulphide), more properly called mustard agent, was first used as a chemical weapon in World War I, where it was responsible for the deaths of large numbers of soldiers on both sides. It was also used by Iraq during the Iran-Iraq war of 1979-1988, where it is though to be responsible for up to 1,000 deaths, and many more injuries. Mustard agent acts by binding covalently to many different types of molecules, disrupting their structures and destroying their normal functions. Damage is delayed for a number of hours after exposure, making it a less effective weapon in this respect than nerve agents.

Chemiluminescence

Luminescence is a process by which a **molecule** loses **energy** through the emission of visible or invisible radiation, unaccompanied by high **temperature**. The emission may result from the absorption of exciting energy in the form of photons, charged particles, or chemical change.

Chemiluminescence is a special case of luminescence in which the excitation source is a chemical reaction. One important chemiluminescent processes discovered during the middle of the last century involves the **oxidation** of the organic molecule luminol by **hydrogen** peroxide (H_2O_2) in the presence of $Fe(CN)_6^{3-}$. A green-blue light is emitted. Other well known chemiluminescent systems are based on the decomposition of esters of **oxalic acid**. **Hydrogen peroxide** acts as a catalyst to form an energy-rich decomposition product of the ester. This energy can be transferred to a fluorescer, which then emits light. In the case of green lightsticks, an oxalate ester and fluorescer are present in an outer plastic tube, with H_2O_2 held in a breakable inner tube.

Because the emission of radiation involves a loss of energy, chemiluminescent reactions must be exothermic. A general expression for these reactions shows that the **photon** of light given off has an energy equal to **Planck's constant** times the frequency of the light emitted. This energy needs to be on the order of 40-70 kcal per **mole** if the emitted light is to be visble to the human eye. This corresponds to light with wavelengths between 400 and 700 nanometers.

Chemiluminescent systems often involve electrochemical reactions. In electrochemiluminescent reactions, the emitting species, the luminophore, is excited to a higher electronic state by a charge-transfer reaction occurring at an electrode. One possible reaction involves one luminophore molecule being oxidized (losing an electron), and another being reduced (gaining an electron). A charge transfer reaction occurs in which the reduced and oxidized species combine to form two neutral molecules, one of which is in an excited state. The excited luminophore emits light as it move to its ground state.

Electrogenerated chemiluminescence can be employed as an analytical tool. The light intensity given off by the chemiluminescent reaction is proportional to the luminophore **concentration**. Certain luminophores can be bound to biological materials, such as **DNA**, **proteins**, or **antibodies**, and in this way act as labels. The concentration of the biological species can be ascertained by measuring the chemiluminescent light intensity. Chemiluminescence has also been used as a detection method in **chromatography**.

Besides chemiluminescence, other types of luminescence include photoluminescence (with photon excitation), electroluminescence (with electric field excitation), **triboluminescence** (with mechanical excitation) and **bioluminescence** (with biochemical excitation). The glow of the firefly, which is due to the oxidation of the molecule luciferin, with the **enzyme** luciferase acting as a catalyst in the oxidation process and **oxygen** acting as the **electron** acceptor, is an example of bioluminescence.

The terms **fluorescence and phosphorescence** refer to specific characteristics of luminescent transitions. In fluorescence, the glow accompanying the emission of energy is very short (10^{-8} to 10^{-3} seconds), while in phosphorescence, the glow may last for several hours. This is because fluorescence involves an electronic transition from a higher to a lower electronic state, e.g., a triplet to singlet emission; whereas the electronic transition in phosphoresence is a same-state transition, e.g., a singlet to singlet emission. The afore-mentioned light from the firefly is an example of phosphorescence.

See also Spectroscopy

Chemiosmosis

All life from the simplest bacteria to the most complex life forms rely on cellular chemical activity. The number of biochemical pathways and associated compounds has been estimated at well in excess of 100,000. To run all of these processes and make all of these compounds, cells need **energy**.

The energy coinage for most organisms is adenosine triphosphate (ATP). The cleavage of the high energy **phosphate** linkage provides about 7.5 kcal or 30 kJ per **mole** of ATP. The adult human body makes and uses about 77 lb (35 kg) of ATP per day. This is slightly over 40% of the average male adult body **mass**. And, all of this ATP is synthesized in a simple organelle within the cells called the mitochondria.

The process for the synthesis of ATP is respiration. The end products are the consumption of glucose and **oxygen** to produce **water** and **carbon** dioxide. However, it is by no means as simple as **combustion**. Both glucose and oxygen are involved in their own enzymatic cycle, located within the mitochondria.

In particular, oxygen is consumed via a series of transmembrane protein complexes in the inner wall or membrane of the mitochondria. Roughly speaking, the pathway involves the consumption of reducing **equivalents** at one end and the **reduction** of oxygen to water at the other. Along the way, this results in the shuttling of protons from the inner parts of the mitochondria to the inter-membrane fluid. This has two effects. It creates a chemical potential or **concentration** gradient between the opposite sides of the membrane while also creating an electrical potential. It is this concentration gradient that is used by two other trans-membrane **proteins**, labeled the F_1 and F_0 complexes, that is used to generate ATP from adenosine diphosphate and inorganic phosphate ions.

At one point, there were competing theories as to how this chemical and electrical potential was created. The chemical hypothesis postulated that the chemical components were coupled at all stages of the process but the absence of any high energy intermediates linking **oxidation** with phosphorylation has been seen as telling evidence against this theory. The other theory and the one presently accepted by most biochemists, is the chemiosmotic theory, in which the oxidation and phosphorylation are uncoupled pathways. That is, the oxidation of chemical species in the respiratory chain results in the evolution and ejection of protons from the inner mitochondrial matrix separately from the inner rush of protons that drives the ATP synthesis.

The evidence for chemiosmosis is extensive and the experimental findings are consistent with this pathway. For example, the outer membrane of the mitochondria can be stripped away leaving the working inner membrane intact. Addition of protons to a treated mitochondria results in the increased synthesis of ATP without a concurrent increase in the oxidative pathways. The use of uncouplers (compounds that increase the permeability of the mitochondrial membranes to proton) effectively shut down the synthesis of ATP despite continued operation of the oxidative pathway. The experimental evidence demonstrates the absence of a link between the two biochemical pathways, consisting with the chemiosmotic theory.

Ion pumping seems to be the active mechanism in many biochemical paths including neuronal transmission and muscular movement. Pumping against a gradient is like pushing a cart up hill. It requires work but does allow for a ride back down.

CHEMISTRY

Chemistry is the science that studies why materials have their characteristic properties, how these particular qualities relate to their simplest structure, and how these properties can be modified or changed. The term chemistry is derived from the word alchemist which finds its roots in the Arabic name for Egypt or the ''black country,'' *al-Kimia.* The Egyptians are credited with being the first to study chemistry. They developed an understanding of the materials around them and became very skillful at making different types of **metals**, manufacturing colored **glass**, dying cloth, and extracting oils from plants. Today, chemistry is divided into four traditional areas: organic, inorganic, analytical, and physical. Each discipline investigates a different aspect of the properties and reactions of the substances in our universe. The different areas of chemistry have the common goal of understanding and manipulating **matter**.

Organic chemistry is the study of the chemistry of materials and compounds that contain **carbon** atoms. Carbon atoms are one of the few elements that bond to each other. This allows vast variation in the length of carbon **atom chains** and an immense number of different combinations of carbon atoms from which to form the basic structural framework of millions of molecules. The word organic is used because most natural compounds contain carbon atoms and are isolated from either plants or animals. Rubber, **vitamins**, cloth, and **paper** represent organic materials we come in contact with on a daily basis. Organic chemistry explores how to change and connect compounds based on carbon atoms in order to synthesize new substances with new properties. Organic chemistry is the backbone in the development and manufacture of many products produced commercially, such as drugs, food preservatives, perfumes, food flavorings, dyes, etc. For example, recently scientists discovered that **chlorofluorocarbons** (CFCs), are depleting the **ozone** layer around the earth. One of these CFCs is used in refrigerators to keep food cold. Organic chemistry was used to make new carbon atom containing compounds that offer the same physical capabilities as the chlorofluorocarbons in maintaining a cold environment, but do not deplete the ozone layer. These compounds are called hydrofluorocarbons (HFCs) and are not as destructive to the earth's protective layer.

Inorganic chemistry is the study of the chemistry of all the elements in the **periodic table** except for carbon. Inorganic chemistry is a very diverse field because it investigates the properties of many different elements. Some materials are **solids** and must be heated to extremely high temperatures to react with other substances. For example, the powder that is responsible for the light and **color** of fluorescent light bulbs is manufactured by heating a mixture of various solids to thousands of degrees of **temperature** in a poisonous atmosphere. An inorganic compound may alternatively be very unreactive and require special techniques to change its chemical composition. Inorganic chemistry is used to construct electronic components such as transistors, diodes, **computer chips**, and various metal compounds. In order to make a new gas for refrigerators that does not deplete the ozone layer, inorganic chemistry was used to make a metal catalyst that facilitated the large scale production of HFCs for use throughout the world.

Physical chemistry is the branch of chemistry that investigates the physical properties of materials and relates these properties to the structure of the substance. Physical chemistry studies both organic and inorganic compounds and measures such variables as the temperature needed to liquefy a solid, the **energy** of the light absorbed by a substance, and the **heat** required to accomplish a chemical transformation. The computer is used to calculate the properties of a material and compare these assumptions to laboratory measurements. Physical chemistry is responsible for the theories and understanding of the physical phenomenon utilized in organic and inorganic chemistry. In the development of the new refrigerator gas, physical chemistry was used to measure the physical properties of the new compounds and determine which one would best serve its purpose.

Analytical chemistry is that area of chemistry that develops methods to identify substances by analyzing and quantitating the exact composition of a mixture. A material is identified by a measurement of its physical properties, such as the **boiling point** (the temperature where the physical **change of state** from a liquid to a gas occurs) and the refractive index (the angle which light is bent as it shines though a sample), and the **reactivity** of the material with various known substances. These characteristics that distinguish one compound from another are also used to separate a mixture of materials into their component parts. If a **solution** contains two materials with different boiling points, then they can be separated by heating the liquid until one of the materials boils out and the other remains. By measuring the amount of the remaining liquid, the component parts of the original mixture can be calculated. Analytical chemistry develops instruments and chemical methods to characterize, separate, and measure materials. In the development of HFCs for refrigerators, analytical chemistry was used to determine the structure and purity of the new compounds tested.

Chemists are scientists who work in the university, the government, or the industrial laboratories investigating the properties and reactions of materials. These people research new theories and chemical reactions as well as synthesize or manufacture drugs, **plastics**, and chemicals. The chemist of today may have many ''non-traditional'' occupations such as a pharmaceutical salesperson, a technical writer, a science librarian, an investment broker, or a patent lawyer, since discoveries by a ''traditional'' chemist may expand and diversify into a variety of related fields which encompass our whole society.

CHEMOMETRICS

Chemometrics is the application of mathematical, statistical, and other logic- based methods to the field of **chemistry**. The term was coined in the early 1970s to describe the growing use of these approaches in chemistry.

Individual researchers were performing what later became known as chemometrics for several decades before it became a formal field of study. One of the first was William Gossett, a brewmaster at the Guinness brewery in Dublin, Ireland. In 1908, he published the statistical methods (the t- test and later the F-test) that he developed to improve the quality and taste of Guinness beer. In the 1940s and 1950s, Jack Youdan and Grant Wernimont performed data analysis and wrote articles and books on quality control. George Box, who later became a famous statistician, began his career performing biochemical determinations on the effect of poisonous **gases** on small animals during World War II. His tests produced varied results that he was unable to interpret. Since there were no statisticians available to help him analyze the uncertainties in his experiments, he learned and applied statistical techniques on his own.

However, it was not until the late 1960s and early 1970s that tools such as automated data collection and the growing use of computers led to both the necessity and the ability of chemists to use mathematics and statistics more extensively. Pioneers Bruce Kowalski, D.L. Massart, and Svante Wold thus began developing chemometrics into a formal field of study. As such, chemometrics can be applied to all aspects of a chemical experiment. This includes experimental design, analytical measurement, data analysis, and data evaluation.

Randomization is one of the most widely used chemometric methods for optimizing experimental design. Randomization eliminates the effects of systematic errors in many experiments. For example, if the time of day is thought to influence the experiment, the experiment can be repeated at different times during the day. Factorial design is another method chemometricians apply to experimental design. It tests combinations of many different variables at the same time, so that effects and interactions between those variables can be measured.

Measurement errors can occur in accuracy, precision, and measurement characterization. Thus, one application of chemometrics to analytical measurement involves calibration of the instrument. The calibration should be such that accurate and precise values can be obtained for the particular chemical variable being studied. An instrumental response must be transformed into a chemical variable such as **concentration**. Also, a chemist needs sensitive and accurate means by which to detect changes in the variable being measured as function of controllable variables. For example, if the amount of heat released from a reaction is being measured, variations in other factors such as amount of reactants and pressure of the vessel must be taken into account. Chemometrics employs calculations that determine the number of samples needed to ensure that the average value reported equals the true value within a given confidence interval. Chemometrics can also be used to validate a model of analytical measurement. This can help determine whether or not (or to what degree) the measurement being made can be transformed into the desired chemical variable.

Data analysis is probably the area in which chemometrics is used most extensively. A large quantity of overlapping data can be disentangled using multivariate analysis. This is known as data resolution. A closely related situation exists in that very complex data may have underlying, non-measurable factors that can be extracted from the data using chemometrics. Once they have been extracted, the influences of these factors on a system can be understood.

Lastly, chemometrics can assist in data evaluation. For example, theory may suggest that data for a particular experiment will lie on a straight line. Chemometrics can be used to calculate how closely the actual data lie to the straight line predicted by theory. Furthermore, it can indicate the level of confidence in data collected. If data are being used validate a theory, it is important to know how good the data are.

The future place of chemometrics in chemistry as a whole is not yet clear. Chemometrics is often thought of as a subset of **analytical chemistry** and those who do chemometrics as chemometricians. However, mathematical and statistical methods are beginning to permeate all areas of chemistry, and, as such, chemometrics may become more of a tool for all chemists rather than a subset of one branch of chemistry.

See also Quantitative analysis

CHEMORECEPTION

Chemoreception is a sensory system used by organisms to detect chemicals in the environment. There are two primary forms of chemoreceptors: gustatory and olfactory, which are responsible for the senses of **taste** and smell. In this process, molecules interact directly with receptor sites to initiate reactions which the brain interprets as sensory data.

For example, olfactory receptor cells in vertebrates and marine invertebrates are arranged in various places where they are in contact with air or **water** which carries the molecules to be detected. These **receptors** are embedded in a layer of supporting cells below the surface. One end of the receptor cell extends to the epithelium where it forms a knob-like projection with cilia protruding from it; the other end of the receptor cell connects to a neuron. The supporting cells secrete a layer of mucous that covers the cilia of the receptor cell. The molecules being detected must diffuse through the layer of mucous to reach the receptor cells. For odors, the receptors are transmembrane **proteins** located in the cilia membrane. A variety of proteins are found in the cilia epithelium which indicates there are many different receptors for distinct **odors**.

When the odor molecules bind to the receptor, a G protein on the intracellular surface of the cilia is activated. The G protein consists of alpha, beta, and gamma subunits. Activation of the this protein causes the G alpha subunit to activate an **enzyme**, adenylate cyclase which catalyzes the formation of a cyclic nucleic acid, cAMP. The strength of the signal resulting from this detection can be quite high because a single odor **molecule** can cause the release of many molecules of cAMP. The cAMP molecule changes the **ion** channels in the membrane of the olfactory receptor cell and causes **sodium** ions to flow into the cell. This ionic flow depolarizes the receptor cell and results in a nerve impulse which travels via the central nervous system to olfactory centers in the brain where they are decoded. The organism then recognizes the stimulation as a distinct smell.

Chemoreception is used by the organisms for a variety of functions. For example, single cell organisms use chemoreception for ''chemotaxis'' which is the ability of a microorganism to direct its motion in response to chemical gradients. From an evolutionary standpoint this is a valuable talent because chemical clues can indicate the direction of food, the presence of an enemy, or the approach of a potential mate. It could indicate regions of toxicity and regions of safety.

Higher organisms like insects and animal chemically communicate using compounds called **pheromones**. For example, Bombykol is the female sex attractant for the Gypsy Moth. The male is thought to be able to detect as little as a single molecule and will follow the scent for miles to find an eligible female. Queen bees emit pheromones that prevent other females from maturing sexually to ensure that the queen's genes remain dominant. Certain female fish release scent markers cause the sperm of male fish to quintuple overnight. Some amphibians, when injured, emit a compound that warns others of their species to keep out of harm's way. Many mammals, from wolves to musk oxen, mark their territory by urinating around their borders. And male voles excrete a powerful aphrodisiac chemical in their urine that causes females to ovulate within 48 hours.

The ability to communicate using chemicals is not limited to just animals. Plants communicate with animals through chemicals. The smell of flowers, no matter how lovely or putrid, is meant as an attractant to the insects necessary for pollination. The smell of terpenes in the forest, which provides that wonderful pine scent, is a method of communication between plants. The trees emit different terpenes depending upon their state of health. Ironically, when a tree is attacked by insects, it releases certain compounds which some insects actually use to their advantage. They detect the compound and swarm the injured tree. Such is the world of chemical ecology, where communication is by a nonspecific method and everyone can have access.

Chemoreception is not just about detecting the external world, though. Complex organisms, such as human beings, have countless billions of receptors within our body that provide the clues for daily living. Receptors for **hormones** regulate metabolic processes. Neurotransmitters regulate thought. Of course, from another point of view, this is external detection if the body is viewed as a collection of individually acting cells operating under a very special type of chemical ecology. Each cell monitors the environment external to it. Just because it happens to be inside your brain or arteries doesn't mean that it is not external to the cell.

Chemoreception is one of the oldest methods for detecting the environment. Whether it is smelling a rose for pleasure, tracking down food, enemies, or a potential mate, or communicating amongst the myriad of neurons that compose our brains, the ability to send and detect chemical compounds is fundamental.

See also Biological membranes; Neurochemistry; Proteins

Chemotherapy

Chemotherapy is the treatment of disease with chemicals or drugs and is most commonly associated with treating cancer. The term was first coined in the early 1900s by the German bacteriologist, **Paul Ehrlich**, after he discovered a dye called Trypan red that combined with and killed trypanosomes, a type of protozoa that caused sleeping sickness. This development earned Ehrlich a share of the Nobel Prize in physiology and medicine in 1908. He also discovered a treatment for syphilis using arsphenamine in 1910. Also during this time, the bone marrow suppressive effect of **nitrogen** mustard was discovered. Previous attempts to use chemicals to cure disease several centuries earlier failed since there were no methods at that time to determine the causes of disease.

Since Ehrlich's discovery, much progress has been made in developing drugs and chemicals for treating or curing disease. Some of the more famous ones have been antibiotics like **penicillin**, antibacterials like the sulfa drugs, and anti-inflammatory drugs like **cortisone**. For treating infectious diseases, drugs may block a biochemical reaction, such as penicillin which blocks synthesis of bacterial cell walls. The effectiveness of such drugs depends on their rate of absorption, how fast they act, how much is stored in the body, how critical to the pathogen is the pathway that they block, and other factors. Drugs have also been developed for treatment of mental illnesses such as depression, cardiovascular diseases such as hypertension, and metabolic diseases such as growth hormone deficiency.

More recently, since 1950, chemical agents have been developed for treating cancer and malignant tumors. The majority of these chemicals kill cancer cells by affecting **DNA** synthesis. The major categories are alkylating agents, antimetabolites, plant alkaloids, anti-tumor antibiotics, and steroid **hormones**. Alkylating agents, which kill cells by directly attacking DNA, include nitrogen mustards, alkyl sulfonates, triazenes, and platinum. These chemicals are used in the treatment of leukemia, Hodgkin's disease, lymphomas, and lung, breast, prostate, testicular, and ovarian cancers.

Antimetabolites block cell growth by interfering with certain activities, usually DNA synthesis. These drugs are used to treat leukemia, tumors of the gastrointestinal tract, and breast and ovary cancers. Mercaptopurine and fluorouacil are common types. Anti-tumor antibiotics bind with DNA and prevent **RNA** synthesis and are used for a wide range of cancers. Bleomycin falls into this category. Plant or vinca alkaloids are derived from plants and block cell division during mitosis. These are used for treating leukemia, Hodgkin's disease, and lung cancer, among others.

Steroid hormones—including estrogens, progesterones, and androgens—modify the growth of certain hormone-dependent cancers. These include breast and prostate cancers. Since **platinum** coordination complexes can be used in many types of cancer, they are one of the largest classes of chemotherapeutic agents, at an estimated $800 million in annual sales.

Each category affects the cell cycle and cell **chemistry** in different ways. The cell cycle involves a period of cell division, preceded and followed by RNA and protein synthesis, a resting period when cells are inactive, and a period of DNA synthesis. When the resting cells are stimulated, they begin multiplying again, synthesizing RNA and protein. With cancer, genes called oncogenes that produce cell growth become overactive, while tumor-suppressing genes become disabled. Growth-factor **receptors**—molecules embedded in the membranes of cells that receive growth signals—are produced in greater than normal amounts or have mutations that cause them to be overactive. These receptors can be shut off by using **antibodies** that bind to them.

Research continues in the fight against cancer, and **genetic engineering** may provide a **solution**. A new class of DNA **molecule**, similar to the naturally occurring RNA molecule ribozyme, has been developed. This molecule recognizes specific chemical sequences of messenger RNA and destroys them, preventing the production of of harmful **proteins**. This discovery could lead to cheaper chemotherapy drugs that produce far fewer side effects.

Another type of drug that may reduce side effects combines radiation and chemotherapy by binding radioactive atoms to antibodies. These drugs deliver the radiation directly to the cancerous areas (50,000 times more radiation than to noncancerous tissues), reducing healthy tissue's exposure. Drug companies are also looking at small molecules that can block growth-factor receptors that would be less expensive than conventional antibodies. Modified viruses that kill tumor cells lacking tumor-suppressing genes is another approach under investigation.

Chevreul, Michel-Eugène (1786-1889)

French chemist

Although Michel Chevreul's career was distinguished by his skills in many different fields, his research on organic substances stands as his greatest achievement. It led to development of new candles and better soap-making processes and even gave science its first clue as to the cause of diabetes.

When Chevreul began studying animal fats in 1809, organic **chemistry** was in its infancy. In effect, he established the entire branch of chemistry that covers the nature of fats, and he proved that organic substances obey the same laws as inorganic chemicals. With few rules to go by, Chevreul was able to isolate and name several **fatty acids**, including stearic, palmitic, and oleic—the three most important components of fats and oils. In doing so, he introduced new analytical techniques such as using the melting point of a substance to determine its identity and purity. By 1816, he had found that all animal fats yielded fatty acids and glycerin which could be combined with an alkali to produce soap. Chevreul's research on these raw materials transformed soap-making from a primitive art into a quantitative, scientific process.

At the time, animal fat in the form of tallow was widely used in candles, which were still a primary source of lighting. Chevreul's work revealed that the unpleasant smell of burning tallow was caused by glycerin in the fat. With fellow chemist **Joseph Gay-Lussac**, Chevreul patented a new type of candle made of purified fatty acids. These candles smelled better, gave more light, and burned more easily. Since they were also harder than paraffin wax candles, they softened less quickly during hot weather.

Among other organic substances, Chevreul also studied sugar, which he isolated from the urine of a diabetic in 1815. By showing that this sugar was identical to glucose (grape sugar), Chevreul took the first step toward the realization that diabetes occurs when the human body cannot burn sugar effectively. However, it took medical science another hundred years to fully exploit his discovery.

The properties of **color** also fascinated Chevreul. In addition to developing natural dyes that intensified and preserved color in wool fabric, Chevreul investigated the basic question of why one color contrasts with another. Precise as always, he created a chromatic circle and defined almost fifteen thousand tones which could be faithfully reproduced by following Chevreul's standards. This work was valuable not only to dyers and designers, but also to an emerging school of oil painters, the French Impressionists. From Chevreul's principles of color perception, neo-impressionists derived the technique of using dots of pure pigments, which the human eye blends into shades when the painting is viewed.

Chevreul also worked to expose spiritualist quacks and, in his nineties, pioneered the science of gerontology (aging)—a subject on which he was by then an expert. His one hundredth birthday was celebrated with great festivity, and scientists from around the world gathered to pay their respects. He continued to publish his work until the age of 102, and when he died, thousands of people attended his funeral at the Notre Dame Cathedral in Paris.

Michel-Eugène Chevreul.

CHLORINE

Chlorine is the second element in Group 17 of the **periodic table**, a group of elements known as the **halogens**. Chlorine's **atomic number** is 17, its atomic **mass** is 35.453, and its chemical symbol is Cl.

Properties

Chlorine is a greenish yellow poisonous gas with a **density** of 3.21 grams per liter. Its specific gravity compared to air is 2.49. Chlorine's **boiling point** is -29.29°F (-34.05°C) and its melting point is -149.8°F (-101°C). The gas is soluble in **water** and reacts with water as it dissolves.

Chlorine is one of the most active of all elements. It combines with all elements except the noble **gases** of Group 18 of the periodic table. Chlorine does not undergo **combustion**, although it does support combustion in much the same way as does **oxygen**.

Occurrence and Extraction

Chlorine occurs abundantly in the Earth's crust and in the Earth's hydrosphere. Its abundance in the earth is about 100-300 parts per million, making it about the 20th most abundant element in the crust. Its abundance in seawater is about 2%, where it occurs primarily as **sodium** chloride ($NaCl$) and **potassium** chloride (KCl). The most common minerals of chlorine are halite, or rock **salt** ($NaCl$), sylvite (KCl), and carnallite ($KCl \cdot MgCl_2$). Large amounts of these minerals are mined from underground salt beds that were formed when ancient oceans dried up.

Discovery and Naming

Chlorine was discovered by the Swedish chemist **Carl Wilhelm Scheele**. Scheele produced chlorine by treating the mineral pyrolusite (primarily **manganese** dioxide, MnO_2) with **hydrochloric acid** (HCl). Scheele described the chlorine gas formed as having a greenish yellow **color** and a suffocating odor "most oppressive to the lungs."

Scheele was somewhat confused about his discovery, believing that the gas was a compound of oxygen with a new element. The true nature of the gas as an element was described in 1807 by the English chemist and physicist **Humphry Davy**. Davy suggested the name chlorine for the gas based on the Greek word *chloros* meaning "greenish yellow."

Uses

Chlorine traditionally ranks among the top 10 chemicals produced in the United States. More than 10 billion kilograms

(23 billion pounds) of the element are produced in the United States alone every year. One of the best known uses of chlorine is in the purification of water. Many cities and towns treat their public water supplies and their public swimming pools with chlorine to kill disease-causing organisms.

Water purification actually accounts for only a small fraction of the chlorine produced each year, however. About three times as much of the element is used as a bleach in the **paper** and pulp industry. The most important use of chlorine is to make other chemicals. For example, chlorine can be reacted with **ethene** (ethylene; C_2H_2) gas to make ethylene dichloride ($C_2H_2Cl_2$). About a third of all the chlorine produced in the United States is used for this purpose. Ethylene dichloride is the starting material for the production of polyvinyl chloride, used to make piping, tubing, flooring, siding, film, coatings, and many other products. Chlorine is also used extensively in the manufacture of propylene **oxide**, from which the class of **plastics** known as polyesters is made. Polyesters are found in a wide range of materials, including car and boat bodies, bowling balls, fabrics for clothing, and rugs.

At one time, the manufacture of chlorofluorocarbons (CFCs) was one of the major uses of chlorine. Those compounds were extremely popular for many years for a variety of applications. They are chemically very stable, nonflammable, non-toxic, and easily liquefied. They found use in applications such as air conditioning and refrigeration systems, aerosol spray products, and cleaning materials. At the peak of their popularity, more than 700 million kilograms (1.5 billion pounds) of these compounds were being made every year.

Over the past two decades, scientists have learned that CFCs can cause serious harm to the Earth's atmosphere. CFC molecules react with and destroy ozone in the Earth's stratosphere. This reaction has serious consequences for life on Earth since ozone blocks out ultraviolet radiation from the Sun that may have harmful effects on both plants and animals. Today, the production and use of CFCs is largely banned in most parts of the world.

Health Issues

Chlorine gas is extremely toxic. In small doses, it irritates the nose and throat, causing sneezing, runny nose, and red eyes. Larger doses of chlorine can be fatal. In fact, chlorine gas has been used in a number of wars throughout history, most notably in World War I. German armies used the gas as a weapon, causing many deaths and permanent injuries to Allied soldiers.

On the other hand, compounds of chlorine are essential in maintaining good health in humans and other animals. The average human body contains about about 3.5 ounces (95 grams) of chlorine, primarily in the form of hydrochloric acid (HCl; ''stomach acid''), **sodium chloride** (NaCl), and potassium chloride (KCl).

CHLOROFLUOROCARBON

A chlorofluorocarbon (CFC) is an organic compound typically consisting of **chlorine**, **fluorine**, **carbon**, and **hydrogen**. Freon, a trade name, is often used to refer to CFCs, which were invented in the 1930s and have been used widely as aerosol propellants, **refrigerants**, and **solvents**. Odorless, colorless, nontoxic, and nonflammable, CFCs are considered valuable industrial products and have proven an especially safe and reliable aid in food preservation. However, the accumulation of CFCs in the stratosphere that may be linked to **ozone layer depletion** has generated considerable public debate and has led to legislation and international agreements banning the production of CFCs by the year 2000.

In the late 1920s, researchers had been trying to develop a coolant that was both nontoxic and nonflammable. At that time, methyl chloride was used, but if it leaked from the refrigerator, it could explode. This danger was demonstrated in one case when methyl chloride gas escaped, causing a disastrous explosion in a Cleveland hospital. **Sulfur** dioxide was sometimes used as an alternative coolant because its unpleasant odor could be easily noticed in the event of a leak. The problem was brought to the attention of **Thomas Midgley** Jr., a mechanical engineer at the research laboratory of General Motors. He was asked by his superiors to try to manufacture a safe, workable coolant. (At that time, General Motors was the parent company to Frigidaire.) Midgley and his associate chemists thought that fluorine might work because they had read that carbon tetrafluoride had a **boiling point** of 5° F (-15° C). The compound, as it turns out, had accidentally been referenced. Its actual boiling point is 198° F (92.2° C), not nearly the level necessary to produce refrigeration. Nevertheless, the incident proved useful because it prompted Midgley to look at other carbon compounds containing both fluorine and chlorine. Within three days, Midgley's team discovered the right mix: dichlorodifluoromethane, a compound whose molecules contain one carbon, two chlorine, and two fluorine atoms. It is now referred to as CFC-12 or F-12 and marketed as Freon—as are a number of other compounds, including trichlorofluoromethane, dichlorotetrafluoroethane, and chlorodifluoromethane.

Midgley and his colleagues had been correct in guessing that CFCs would have the desired thermal properties and boiling points to serve as refrigerant **gases**. Because they remained unreactive, and therefore safe, CFCs were seen as ideal for many applications. Through the 1960s, the widespread manufacture of CFCs allowed for accelerated production of refrigerators and air conditioners. Other applications for CFCs were discovered as well, including their use as blowing agents in **polystyrene** foam. Despite their popularity, CFCs became the target of growing environmental concern by certain groups of researchers. In 1972, two scientists from the University of California, **F. Sherwood Rowland** and **Mario Jose Molina**, conducted tests to determine if the persistent characteristics of CFCs could pose a problem by remaining indefinitely in the atmosphere. Soon after, their tests confirmed that CFCs do indeed persist, until they gradually ascend into the stratosphere, break down due to ultraviolet radiation, and release chlorine, which in turn affects ozone production. Their discovery set the stage for vehement public debate about the continued use of CFCs. By the mid-1970s, the United States government

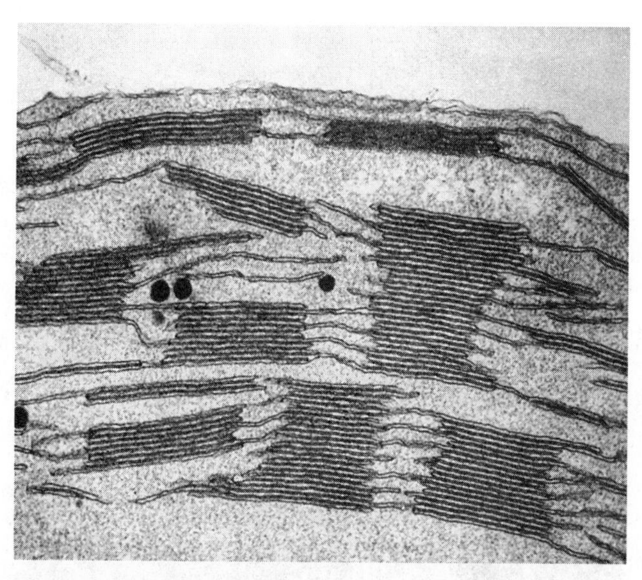

A false-colored transmission electron micrograph of stacks of grana in a chloroplast from a leaf of maize, *Zea mays*. The grana in chloroplasts are the sites photosynthesis in higher plants, bearing the light-receptive green pigment chlorophyll which contains the enzymes that convert carbon dioxide and water into useful chemical energy in the prescence of sunlight. *(Photograph by Dr. Kenneth R. Miller/Science Photo Library, Photo Researchers, Inc. Reproduced by permission.)*

banned the use of CFCs as aerosol propellants but it resisted a total ban for all industries. Instead, countries and industries began negotiating the process of phasing out CFCs. As CFC use is allowed in fewer and fewer applications, a black market has been growing for the chemical. In 1997, the U.S. Environmental Protection Agency, Customs Service, State Dept., Justice Dept., IRS and the FBI initiated enforcement actions to prevent (CFC) smuggling in the United States.

Many CFCs have been replaced with related materials known as hydrochlorofluorocarbons (HCFCs) which reportedly do not interfere with the ozone layer. However, these materials are not free from controversy either. A 1997 Belgian health officials report indicate hydrochlorofluorocarbon-123 may be responsible for a rise in acute hepatitis among workers of a **smelting** plant. Liver damage found in nine workers is believed to be caused by a leaking refrigerant pipe in the workers' cabin. The U.S. Environmental Protection Agency, disputes these findings.

CHLOROPHYLL

Plants derive **energy** from the sun through a process called **photosynthesis**. At the center of this process is the pigment chlorophyll. Chlorophyll is also found in several one-celled organisms. Algae, diatoms, dinoflagellates, and photosynthetic bacteria all belong to this group of organisms. The chlorophyll in these organisms may be slightly different from that found in multi-celled plants, but the mechanism for deriving energy from sunlight is the same.

Chemical structure of chlorophyll. *(Illustration by Electronic Illustrators Group.)*

The chlorophyll **molecule** is composed of a central porphyrin **ring** and side **chains**. A **magnesium atom** is found at the center of the porphyrin ring—a complex multi-ring structure composed of **carbon**, **hydrogen**, **nitrogen**, and **oxygen** atoms. The conjugated double bonds of this structure allow the chlorophyll molecule to absorb energy from sunlight.

There are several types of chlorophylls that all have the central porphyrin ring and magnesium atom in common. The differences are found in their side chains and allow absorption of slightly different light wavelengths. Plants usually contain at least two types of chlorophyll: chlorophyll *a* and chlorophyll *b*. Chlorophyll *a* occurs in nearly every photosynthetic organism, but chlorophyll *b* is replaced by chlorophyll *c* in brown algae, diatoms, and dinoflagellates. Further forms of chlorophyll, called bacteriochlorophylls, are found in certain strains of photosynthetic bacteria.

In plants, the most complex of the photosynthetic organisms, chlorophyll is found in the chloroplasts. Chloroplasts are small energy-producing organelles within plant cells.

Visible light from the sun is a type of electromagnetic radiation and is transmitted in wavelengths from 400-700 nm. When this light passes through a crystal prism, the light bends and the different wavelengths are seen as the colors of the rainbow. Sunlight also has properties of particles, because the energy contained in light is transmitted in specific amounts. These particles are called photons. Photons with the short wavelengths have the greatest amount of energy and correspond with the violet portion of the spectrum. Photons at the higher end of the wavelength range have less energy and correspond with the red portion of the spectrum.

When sunlight hits a chlorophyll molecule, the energy of the photons is transmitted to chlorophyll molecules. Chlorophyll absorbs energy from the red and blue portions of the spectrum and reflects the green portion. For that reason, the leaves and stems of plants appear green. More colors are apparent in some plants and in flowers owing to the presence of other pigments.

With the absorption of a **photon**, a chlorophyll molecule enters an energy-rich, excited state. In its excited state, an **electron** in the molecule is pushed into a higher atomic orbit. The energy contained in this molecule is funneled into an electron transport chain, made up of a series of chemical compounds. The energy from the electron is passed from one compound to the next.

Chlorophyll is involved in two separate electron transport chains—photosystem I and photosystem II. In photosystem I, a chlorophyll-protein complex absorbs light with the wavelength of 700 nm. In photosystem II, a different chlorophyll-protein complex absorbs light at the wavelength of 680 nm. All plants have both photosystems, as do algae and other photosynthetic organisms that emit oxygen. Photosynthetic bacteria, which do not emit oxygen, have photosystem I only.

Both photosystems feature chemicals that are naturally found in plants. As the energy of the excited electron is passed down the electron transport chain, the energy from the sunlight is transformed into chemical energy stored as ATP and NADPH. This energy can be later used to fuel processes necessary for life. One of the key uses of this energy is to produce **carbohydrates** from **carbon dioxide**.

See also Photosynthesis

CHOLESTEROL

Cholesterol (sometimes also known as cholesterin) is the principle sterol (an ester of a fatty acid and an aromatic (ring structure) alcohol) found in all animal tissues. The name comes from the Greek *chole* and *stereos* and it means ''solid bile.'' High levels of cholesterol have been implicated in some forms of atherosclerosis (a disease caused by the thickening of arterial walls).

Cholesterol is found in humans in both free and esterified forms. It was first isolated from humans in gallstones. Cholesterol is now commercially obtained from cattle spinal cords or from lanosterol—the fatty coating of sheep wool.

Cholesterol is manufactured by all animals from ethanoate (a **salt** of ethanoic acid) molecules via the intermediates mevalonic acid, squalene, and lanosterol. Cholesterol is biologically important as a constituent of the **plasma** membranes of many animal cells. It is also present in smaller amounts in the membranes of the mitochondria and endoplasmic reticulum within the cells. Cholesterol is a precursor of other **steroids** required in **metabolism**, including bile acids, fecal sterols, sex **hormones**, and adrenocortical hormones.

The chemical formula of cholesterol is $C_{27}H_{46}O$ and it has a melting point of 300°F (149°C). Cholesterol is a white, crystalline solid which is optically active. The scientific or IUPAC name for cholesterol, reflecting its structure, is 5 cholesten-3 beta-ol. Cholesterol is a biochemically important **alcohol**. It is a physically large **molecule** and the hydroxyl radical is only a small part. Since cholesterol has only one hydroxyl group per molecule, it has only a limited amount of **solubility** in **water**. In fact, only 0.26 g of cholesterol will dissolve per 100 ml of water. Cholesterol is soluble in chloroform, **ether**, **benzene**, or hot alcohol and other organic **solvents**. These solvents are used to extract cholesterol from cells.

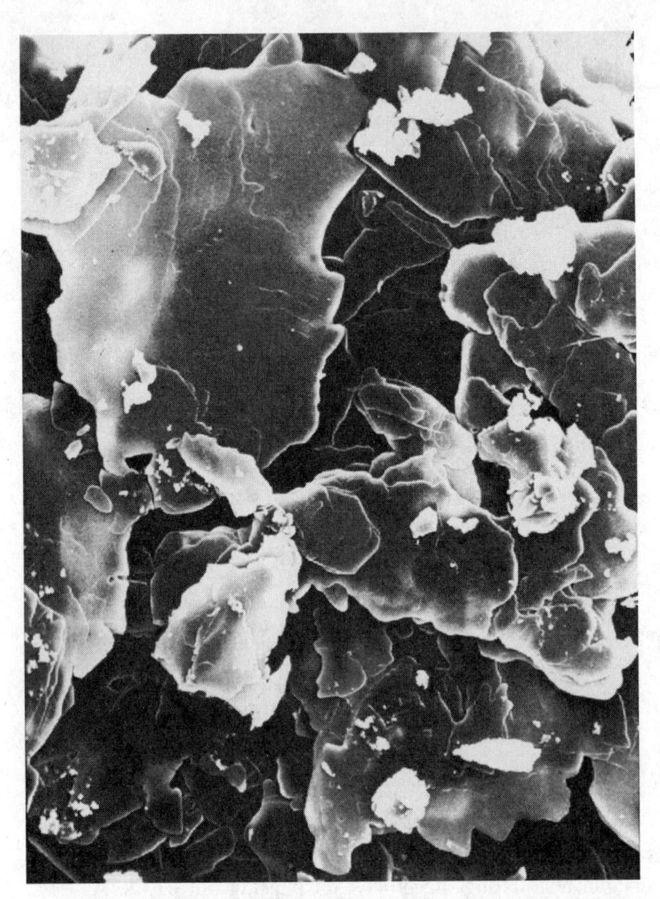

A micrograph image of monohydrate cholesterol percipitated crystals. ***(Photograph by Dr. Jeremy Burgess/Science Photo Library, National Audubon Society Collection/Photo Researchers, Inc. Reproduced by permission.)***

When cholesterol is present in the body in large amounts it may precipitate from **solution**. When it precipitates in the gall bladder it forms crystalline **solids** called gallstones. When it precipitates in arteries and veins it can obstruct them, narrowing the diameter of the vessels and reducing blood flow. This can in turn lead to high blood pressure and other cardiovascular problems.

The amount of cholesterol in a person's body is determined by the amount of cholesterol the body manufactures combined with the individual's total dietary intake of cholesterol. (Dietary cholesterol comes from the cholesterol manufactured by other animals.) There is evidence to suggest that a large intake of calories signals the body to synthesize large amounts of cholesterol. Cholesterol can mix with glycerides and phospholipids and it can apparently endow lipid mixtures with the ability to absorb water.

Cholesterol is not present in the tissues of plants which instead contain phytosterols. Fungi do not contain cholesterol either—they have mycosterols.

The pathway for the enzymatic synthesis of cholesterol was first worked out by **Konrad Bloch**, **Feodor Lynen**, George Popjak, and John Cornforth in the 1940s. This work was carried out by the systematic degradation of biologically labeled cholesterol and this showed the origin of each **carbon atom** in the structure. This synthesis pathway was originally studied by feeding rats radioactively labeled acetate. The **radioactivity** in the acetate was later found to be incorporated into cholesterol and the liver. It was eventually shown that the process proceeds from acetate to mevalonic acid to squalene to cholesterol. This is an example of a metabolic pathway.

An adult on a low cholesterol diet can manufacture some 800 mg of cholesterol per day. The feedback mechanism that controls cholesterol manufacture in humans is mediated by changes in the activity of a reductase **enzyme**. (This enzyme catalyzes the formation of mevalonate.) The naturally occurring biosynthesis of cholesterol in the body is suppressed by dietary cholesterol. It is believed that a cholesterol-containing lipoprotein, a bile acid, or a protein, is the inhibitor. Fasting also inhibits cholesterol biosynthesis. Degradation of cholesterol in the liver yields bile acids.

The entire cholesterol molecule is hydrophobic (water repellent) with the exception of the OH group at the position of carbon atom number three. This hydroxyl group is hydrophilic (water loving), due to **hydrogen bonding**. In eucaryotes, cholesterol is a key regulator of cell membrane fluidity. It prevents the crystallization of fatty acyl **chains** by fitting between them. If these chains solidify they can cause blockages in blood vessels and reduce the permeability of cell walls. Cholesterol can also sterically block large molecules of fatty acyl chains, making the cell membrane less fluid, thus controlling the membrane fluidity. All of the carbon atoms in cholesterol come from the biologically active molecule acetyl-coenzyme A (normally found as part of the **Krebs cycle**).

Bile salts (the major one is glycocholate) are polar derivatives of cholesterol, they are highly effective detergents due to the presence of both polar and nonpolar regions. Bile salts, which are synthesized in the liver, are stored and concentrated in the gallbladder and then released into the small intestine. They solubilize dietary **lipids** facilitating their **hydrolysis** by lipases and their absorption into the bloodstream.

Cholesterol is transported in the body by a series of lipoproteins that are classified according to their **density**. They are called chylomicrons, very low density lipoproteins (VLDL), low density lipoproteins (LDL), and high density lipoproteins (HDL). These lipoproteins consist of a core of hydrophobic lipids surrounded by polar lipids and then a shell of protein. This type of molecule, a lipid and a protein joined together, is more correctly termed an apolipoprotein. Eight types of apolipoproteins have been categorized—AI, AII, B, CI, CII, CIII, D, and E. These complexes solubilize highly hydrophobic lipids such as cholesterol. Also their protein components contain signals that regulate the entry and exit of particular lipids at specific targets. The cholesterols that are secreted into the blood plasma are manufactured and secreted by the liver. Chylomicrons transport dietary cholesterol to the liver and to adipose tissue in the body. It is in the liver that cholesterol is transformed into other products required by the body. If the cholesterol is in excess of what the body requires, it is excreted via the gall bladder. This cholesterol can precipitate out to form hard gall stones, which are almost pure cholesterol. When gall stones are present in the ducts of the gall bladder they can cause blockages and subsequent pain.

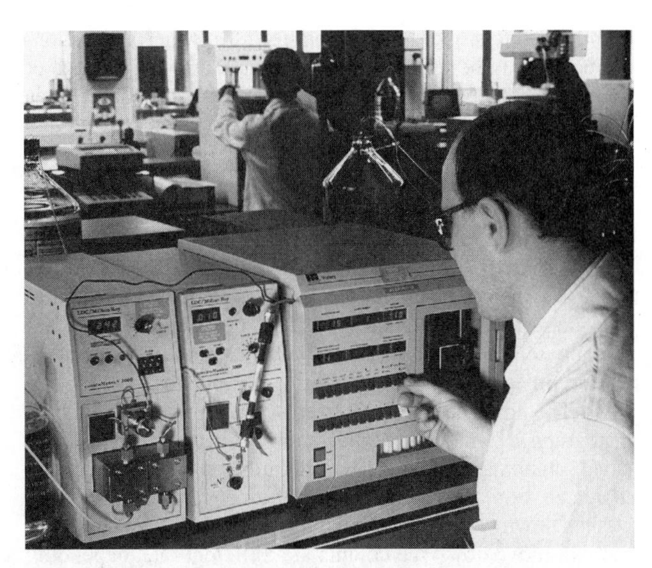

High-Performance Liquid Chromatography (HPLC) machine. ***(Photograph by Simon Fraser/Searle Pharmaceuticals, National Audobon Society Collection/Photo Researchers, Inc. Reproduced by permission.)***

See also Nutrition

Chromatography

The term chromatography was originally used about 100 years ago by a Russian botanist, Mikhail S. Tswett (1872-1919) to describe the separation of bands of plant pigments (chlorophylls) extracted from green leaves. The process used with petroleum **ether** on **calcium** carbonate packed in a vertical **glass** column. Though chromatography (from the Greek word for **color** writing) was descriptive of colored bands, most modern chromatographic methods do not involve separation of colored compounds. Chromatography now describes the process of separating compounds and ions by a variety of matrices on large numbers and types of columns. The International Union of Pure and Applied Chemistry has defined chromatography as follows: "A method, used primarily for separation of the compounds of a sample, in which the components are distributed between two phases, one of which is stationary while the other moves. The stationary phase may be a solid, or a liquid supported on a solid, or a **gel**. The stationary phase may be packed in a column, spread as a layer, or distributed as a film, etc.; in this definition chromatographic bed is used as a general term to denote any of the different forms in which the stationary phase may be used. The mobile phase may be gaseous or liquid. Therefore, a chromatographic system consists of three components described as solute, solvent, and sorbant or more appropriately as the sample, mobile phase, and stationary phase.''

The goal of chromatography is to separate mixtures of compounds into separate bands or peaks of individual compounds within a reasonable length of time. Separation is achieved when the solutes (or sample compounds) in the mobile phase demonstrate different affinities for the stationary solid phase, the mobile phase, or both, resulting in different retention times for the various sample compounds.

The simplest type of chromatography is column chromatography, in which a vertical tube is filled with a finely divided stationary phase. The mixture of materials to be separated is placed at the top of the column and is slowly washed down with a suitable mobile phase. Each type of material will move down the column at a different rate, depending on the its **solubility** and its tendency to be adsorbed. If the stationary phase used is a liquid adsorbed on a solid carrier, the process is called partition chromatography, since the mixture to be analyzed will be partitioned, or distributed, between the stationary liquid and a separate liquid mobile phase. If the stationary phase is solid, the separation process is called **adsorption** chromatography.

In thin-layer chromatography (TLC), the stationary phase is a thin layer on a glass plate or plastic film. The thin layer may be an adsorbent such as silica gel or alumina, which is made into a slurry, placed in a layer on the glass plate, and then dried. The sample mixture is dissolved in a volatile solvent. A small portion of this **solution** is placed on the thin layer. The solvent evaporates, leaving the mixture to be separated on the plate in the form of a small spot. The plate is placed upright in a jar, and a suitable developing solvent is added to the bottom. The jar is closed so that the atmosphere of the jar becomes completely saturated with the vapor of the solvent. The solvent rises up the plate by capillary action. When it has risen 4-6 in (10-15 cm) in about 10 to 20 minutes, the plate is removed and dried. Separation of compounds is determined by examination under ultraviolet light or by spraying with a reagent that colors the various compounds. **Paper** chromatography, in which **water** adsorbed on paper acts as the stationary phase and an organic liquid is used as the mobile phase, is similar to TLC.

Gas chromatography (GC) usually uses a liquid on a solid support as the stationary phase and an inert gas such as **nitrogen**, **hydrogen**, **helium**, or **argon**, as the mobile phase. The stationary phase is contained in a narrow, coiled column from 4-15 ft (1.5-5 m) in length. The mixture to be separated is injected with the mobile gas phase into the column, which is heated so that the mixture is vaporized. The different substances in the mixture pass through the column at different rates. Upon leaving the column, the substances pass through a detector, which gives a signal to a recording device. The resulting gas chromatogram shows a series of peaks, each of which is characteristic of a particular substance.

High-performance liquid chromatography (sometimes called high-pressure liquid chromatography [HPLC]) is a more developed type of column chromatography. The particles that hold the stationary phase in the column are very small (0.004 in [0.01 mm]) and uniform in size, provide a large surface area for the sample substances in the mobile liquid phase. The large pressure drop created in a column filled with such small particles is overcome by using a high-pressure pump to push the mobile liquid phase through the column and to the detector. The advantages of HPLC are high resolution and sensitivity. HPLC and GC are the two most commonly used separation techniques in analytical laboratories.

Granules of chromium, a hard, silvery transition metal used in the manufacture of alloy steels and as a decorative, corrosion-resistant electroplated coating. ***(Photograph by Klaus Guldbrandsen/Science Photo Library, Photo Researchers, Inc. Reproduced by permission.)***

Gel permeation chromatography is based on the filtering or sieving action of the stationary phase, which has pores of uniform size in the range of 20-30 nm. A substance dissolved in a mobile liquid phase while moving down a column will be excluded from the stationary phase if its size is greater than that of the pores. If its molecular size is smaller, it will become trapped. Intermediate-sized molecules will be trapped by some pores but not others. Separation is based on molecular size, with larger molecules separating out first and smaller molecules last. This type of chromatography is used to measure the **molecular weight** of polymers, **proteins**, and other biological substances of high molecular weight.

Chromatography is used for the separation of pure substances from complex mixtures and is important in the analysis of environmental contamination, foods, drugs, blood, petroleum products, and radioactive fission products.

CHROMIUM

Chromium is a transition metal, one of the elements found between Groups 2 and 13 in Rows 4 through 6 of the **periodic table**. Its **atomic number** is 24, its atomic **mass** is 51.996, and its chemical symbol is Cr.

Properties

Chromium is a hard, steel-gray, shiny metal that breaks easily. It has a melting point of 3,452°F (1,900°C), a **boiling point** of 4,788°F (2,642°C), and a **density** of 7.1 grams per cubic centimeter. A physical property that greatly adds to chromium's commercial importance is that it can be polished to a high shine.

Chromium is a relatively active metal that does not react with **water**, but does react with most **metals**. It combines slowly with **oxygen** at room **temperature** to form chromium **oxide** (Cr_2O_3). The chromium oxide formed in this reaction then acts as a protective layer, preventing the metal from reacting further with oxygen.

Occurrence and Extraction

Chromium ranks about 20th among the elements present in the Earth's crust with an abundance of about 100-300 parts per million. It never occurs as a free element, but is found primarily in the form of chromite, or chrome **iron** ore ($FeCr_2O_4$). The element is extracted from this ore by converting it first to chromium oxide (Cr_2O_3) and then heating it in combination with **charcoal** or **aluminum**: $2Cr_2O_3 + 3C \xrightarrow{\text{heat}} 3CO_2 + 4Cr$ or by converting the ore to chromium chloride ($CrCl_3$) and then electrolyzing this compound: $2CrCl_3 \xrightarrow{\text{electric current}} 2Cr + 3Cl_2$. In some cases, chromite is converted directly to an **alloy** known as ferrochromium (or ferrochrome), which has a number of important commercial uses: $FeCr_2O_4 + C \xrightarrow{\text{heat}} Co_2$ + ferrochromium.

Discovery and Naming

Chromium was discovered in 1797 by French chemist Louis- Nicolas Vaquelin (1763-1829) in a mineral known as Siberian red **lead**. The element was named after the Greek word *chroma*, meaning "**color**" because many chromium compounds have a distinctive color, ranging from purple to black to green to orange to yellow.

Uses

Nearly 90% of the chromium used in the United States goes to the production of alloys. The addition of chromium to an alloy makes the final product harder and more resistant to **corrosion**. About 70% of all chromium is used in the production of stainless steel employed in a great variety of products, including automobile and truck bodies, plating for boats and ships, construction parts for buildings and bridges, parts for chemical and petroleum equipment, electric cables, machines parts, eating and cooking utensils, and reinforcing materials in tires.

Some compounds of chromium also have relatively limited commercial applications. Because of their distinctive colors, they are often used in printing, dyeing, and coloring cloth, plastic, and other materials.

Health Issues

Plants and animals require very small amounts of chromium in order to maintain their health. But larger doses of chromium can be harmful. Some compounds are especially dangerous and can cause a rash or sore if spilled on the skin. If swallowed, some chromium compounds can seriously damage the throat, stomach, intestines, kidneys, and circulatory system. Chromium compounds are also thought to be carcinogenic.

Chromophores

Chromophores (from the Greek chroma, or "**color**," and phoros, or "bearer") are groups of atoms in an organic compound that absorb light at certain wavelengths. A particular chromophore gives the compound its distinctive color by causing it to absorb light selectively. An example of a chromophoric compound is **chlorophyll**, the plant pigment that gives vegetation its green color.

Photosynthetic organisms such as trees, plants, algae, and some bacteria have evolved a set of chromophoric pigments that efficiently capture sunlight, which they convert to **energy** that they use to sustain themselves. Other examples include beta-carotene which is responsible for the orange color of carrots. It is also added to most butter and margarine products to provide a yellow color. Lycopene is another type of plant pigment that gives tomatoes their red color. An example of a synthetically derived chromaphore is malachite green, a material that is used as a textile dye.

Scientists once thought chromophores such as these were the source of colored light, but now they know that color comes from the reflection of particular wavelengths of light that the chromophore does not absorb. Chromophores interact with light in a unique way because they contain conjugated bonds, which consist of a network of alternating double and single bonds. The electrons in these bond arrangements are spread throughout the **molecule** in such a way that they absorb light in a specific fashion. The unabsorbed light (either reflected or transmitted) is what we see as the color of the material. While researchers have discovered all colored compounds do indeed contain one or more chromophores, not all chromophore containing compounds are colored.

One of the major practical applications of chromophore **chemistry** is in the manufacture of synthetic dyes, or dyestuffs, for textiles. Each dye molecule has one or more chromophores that give the textile its characteristic color. In fact, the dyes are named after their chromophores, e.g., the nitroso group, the azo group, the **carbonyl group**, the thio group, and the nitro group.

Chromophore chemistry varies widely in industrial dyeing applications. In acid dyes, which are good for dyeing wool, silk, and acrylics, the chromophores are part of a negative **ion**. Chromophores in metallized dyes contain a chelated (tightly bound) metal **atom**. In basic dyes, used mainly for acrylics, the chromophores are part of a positive ion. Some of the dyes to which chromophores give a vibrant color are naphthol yellow, vat blue, congo red, and methylene blue. Chromophores are also important in direct, vat, sulphur, disperse, and reactive dye chemistry.

The color produced by chromophores may be intensified or shifted by other groups of atoms on the molecule. These

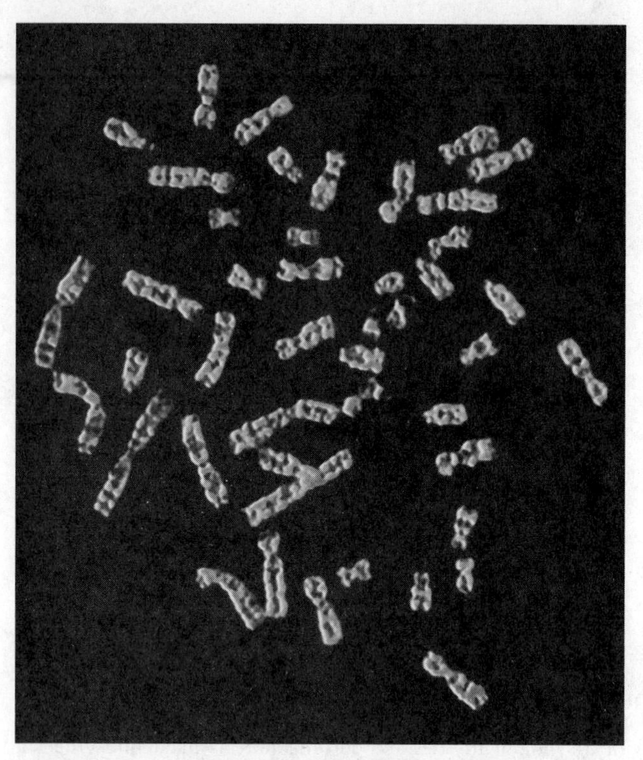

A false-colored light micrograph image of normal human chromosomes, obtained by amniocentesis. *(CNRI/Science Photo Library, Photo Researchers, Inc. Reproduced by permission.)*

auxiliary groups, known as auxochromes, can change the absorption characteristics of chromophores. Examples of auxochromes include amino groups, halogen atoms, hydroxyl groups, and alkoxyl groups.

See also Conjugation

CHROMOSOME

Chromosomes are thread-like bodies in the cell **nucleus** of all plants and animals that hold the genes—the blueprints of heredity. Each **chromosome** carries a single strand of deoxyribonucleic acid (DNA) that threads together about 1,000 genes.

Not much was known about chromosomes prior to the 1880s, due to the lack of adequate cell staining techniques and poor microscopes. In 1879, however, Walther Flemming, using new synthetic dyes, was able to discern bodies in cells that had previously gone undetected. He noticed that some material scattered throughout the nucleus heavily absorbed the dye and coined the word *chromatin* to describe this dark, stainable substance. Upon further observation, he noted that when a cell divided into two daughter cells, the chromatin first doubled, then split lengthwise, leaving each daughter cell with the same amount of chromatin as the parent cell. By 1882, Flemming had identified all the stages of this process, a fundamental operation of cell division now termed mitosis. In 1887, Edouard van Beneden observed a different form of division in sex cells. Rather than doubling, the chromatin split in half, with each half being distributed to the daughter cells. This phenomena was later named meiosis.

In 1888, the German anatomist Wilhelm von Waldeyer-Hartz renamed Flemming's chromatin *chromosomes*, meaning colored bodies. However, the connection between chromosomes and heredity was not made until 1902, when Walter S. Sutton published a short scientific article concerning the newly discovered work of Gregor Mendel. Sutton proposed that the ''factors'' which Mendel could not identify, but believed controlled heredity, were indeed contained in the chromosomes. Theodor Boveri independently came to a very similar conclusion, and in 1903, their work became known as the *Chromosomal Theory of Inheritance.*

For many scientists, the theories of Mendel and Sutton provided a sufficient explanation for heredity and evolution. However, the American geneticist Thomas Hunt Morgan remained skeptical of their work because the conclusions were speculative, based on nothing more than observation, inference, and analogy. Morgan wanted to draw firm conclusions based on quantitative and analytical data and so set out to test their theories using the fruit fly as his subject. The results of his experiments and those of his assistant Hermann Muller contributed greatly to the understanding of chromosomes and their role in heredity. Morgan found that genes—the term Wilhelm Johannsen coined for Mendel's ''factors''—were located on chromosomes. For the first time, the association of one or more hereditary traits with specific chromosomes was clear. He also discovered that genes on the same chromosome were often inherited together—an occurrence known as *autosomal linkage.* (Autosome is the name given to all the chromosomes that are not sex chromosomes.) However, chromosome pairs would sometimes break apart and exchange pieces—a process known as crossing over. The findings of Morgan and his colleagues focused the attention of the scientific world on chromosomes and prompted further research in the area of genetics.

In 1997, artificial human chromosomes were created for the first time by Huntington Willard and his colleagues at Case Western Reserve University in Cleveland, Ohio. Such artificial chromosomes may help scientists understand better how natural chromosomes work. They may also one day be used as a vehicle for carrying **DNA** into patients receiving gene therapy.

CITRIC ACID

Citric acid is an organic (**carbon** based) acid found in nearly all citrus fruits, particularly lemons, limes, and grapefruits. It is widely used as a flavoring agent, preservative, and cleaning agent. The COOH group is a carboxylic acid group, so citric acid is a tricarboxylic acid, possessing three of these groups.

Citric acid is produced commercially by **fermentation** of sugar by several species of mold. As a flavoring agent, it can help produce both a tartness [caused by the production of **hydrogen** ions (H^+)] and sweetness (the result of the manner in

A microscopic image of citric acid. ***(Photograph by Alfred/Science Photo Library, National Audubon Society Collection/Photo Researchers, Inc. Reproduced by permission.)***

which citric acid molecules ''fit'' into ''sweet'' **receptors** on our tongues). Receptors are protein molecules that recognize specific other molecules.

Citric acid helps to provide the ''fizz'' in remedies such as Alka-Seltzer trademark. The fizz comes from the production of **carbon dioxide** gas that is created when **sodium** bicarbonate (baking soda) reacts with acids. The source of the acid in this case is citric acid, also helps to provide a more pleasant **taste**.

Citric acid is also used in the production of hair rinses and low **pH** (highly acidic) or slightly acidic shampoos and toothpaste. As a preservative, citric acid helps to bind (or sequester) metal ions that may get into food via machinery used in processing. Many **metals** ions speed up the degradation of fats. Citric acid prevents the metal ions from being involved in a reaction with fats in foods and allows other preservatives to function much more effectively. Citric acid is also an intermediate in metabolic processes in all mammalian cells. One of the most important of these metabolic pathways is called the citric acid cycle (it is also called the **Krebs cycle**, after the man who first determined the role of this series of reactions). Some variants of citric acid containing **fluorine** have been used as rodent poisons.

CLATHRATE

A clathrate is any compound in which one type of **molecule** is physically enclosed in the crystal structure of a second compound, e.g., a gas is trapped inside a crystal. The properties of the aggregate are essentially those of the enclosing compound.

The molecules SO_2, CO_2, and CO, as well as the noble **gases**, form clathrate compounds with cages of open-structured ice or hydroquinone crystals. In ice clathrates, the ice forms a hydrogen- bonded cage enclosing the small guest atoms or molecules. Another example of a clathrate is the benzene-nickel cyanide compound. The petroleum industry has frequently been cursed by hydrocarbon- clathrates blocking gas pipelines in the Arctic regions.

Methane hydrate is an ice-like compound in which methane molecules are enclosed in cavities formed by **water** molecules. The compound, which looks like ice, forms deposits on the deep ocean floor and in permafrost regions. Most of it is of biological origin. By some estimates, the **energy** locked up in methane hydrate deposits is more than twice the global reserves of all conventional gas, oil, and **coal** deposits combined. There has been speculation that methane released from these clathrates could alter climatic conditions in the event of global warming, but no one is really sure just how much methane hydrate exists on Earth.

Liquid clathrates are two-phase systems consisting of an upper layer (typically pure solvent) with a clathrate species residing in a lower layer in a non-stoichiometric, but fixed, proportion of solvent. The solvent is frequently an aromatic species such as toluene, although chloroform has also been used. Liquid clathrates have been of interest in solvent-extraction and coal-liquefaction studies.

See also Aromatic hydrocarbons; Crystallography; Oil recovery; Petrochemicals; Stoichiometry

CLAUDE, GEORGES (1870-1960)

French chemist and inventor

Trained as a chemist, Georges Claude was a prolific inventor who was responsible for technological developments ranging from innovations in the use of acetylene to the invention of **neon** lights. His inventions made him wealthy, but he spent much of this money during the 1920s and 1930s on experiments that his contemporaries considered unconventional. Claude was an eccentric man, with strong, right-wing political views. Having collaborated with the Nazis during the occupation of France in World War II, he was imprisoned after the war's end. Claude was born in Paris, France, on September 24, 1870, and was educated at the École de Physique et Chimie, a municipal school for physics and **chemistry**. Graduating in 1886, he was employed in the engineering department of an **electricity** company, and then as an engineer in several different industrial plants. He married in 1893 and subsequently had three children.

It was while working in industry that Claude started upon his long career of inventions and discoveries. He determined in 1897 that acetylene, a highly combustible gas, could be safely transported if dissolved in **acetone**. Acetylene was—and still is—highly valued in industry because of its use in the cutting and welding of **metals**, and Claude's innovation greatly increased its demand in industry. By 1927, the production of acetylene was valued at ten million dollars per year in the United States alone.

In 1902, Claude developed a method for liquefying air, which was used by other scientists to identify the inert **gases**. Once these gases had been identified, Claude found a way to separate **nitrogen** and **oxygen** from liquefied air and then developed methods to produce these gases in large quantities. He proposed the use of liquid oxygen in **iron smelting** in 1910, and although he proved the technique could be successful, it was not widely adopted until after World War II. During World War I, however, Claude adapted the procedures he had developed during these experiments to manufacture liquid **chlorine**, which was employed for poisonous gas attacks.

Claude's work with inert gases also led to more peaceful inventions. In 1910, he discovered that the electrification of neon (a mostly inert gas) produced a bright, red light. He then produced specially coated tubes—which could be bent and twisted into any shape—to keep the light active. Such was the beginning of neon lights, a major transformation in the lighting and advertising industries. He introduced his invention to the United States at a conference in 1913, but World War I put off the general use of neon lights until close to 1920. The invention eventually brought Claude fame and a substantial sum of money. In the 1930s, the potential for neon lighting increased with yet another innovation: the coating of the interior of the tubes with fluorescent material, which created the white lights still in use today. Fluorescent bulbs soon replaced American inventor **Thomas Edison** 's incandescent lighting in businesses and factories, earning Claude the nickname ''Edison of France.''

In 1917, Claude invented a method to produce synthetic **ammonia**, a procedure which was similar to the Haber Process developed in Germany at around the same time. Claude used the Haber principles, but he applied them with four times greater pressure, thus producing an ammonia with greater efficiency (among the most important practical applications of ammonia is as an element in fertilizer). Claude's thinking was scorned by many contemporary scientists, but his idea was successful, producing ammonia and earning him election to the French Academy of Sciences in 1924.

After 1926, Claude concentrated on developing new sources of **energy**. He was particularly interested in the **temperature** differences in ocean waters—which in some regions can be quite extreme—and he theorized that there was a way to exploit them. **Water** under lower than normal pressure boils at temperatures below one hundred degrees centigrade, and Claude believed that he could create a turbine-type system using the upper ocean water for steam production and the lower, cooler water for condensation. The steam, caught in transit between the boiler and the condenser, could easily be harnessed to power a turbine. With the scientific knowledge of the day indicating that Claude was at least theoretically correct, he set out to build a small model. Although the model was successful, prominent scientists still gave Claude a great deal of opposition. Determined to prove his theory, Claude used his own money to build a plant in Belgium in which a turbine was, indeed, successfully operated by harnessing differences in ocean temperature.

After building a successful plant in Belgium, Claude received financial backing and moved his operations to Cuba, where he believed his ideas would be better received. His project, however, suffered large-scale problems. To reach water that was deep enough—and therefore cold enough—special tubing had to be developed. A tube six thousand feet long and over six feet in diameter was designed to reach over a third of a mile into the ocean. The first of these tubes was set out in 1930 but was lost when they attempted to put it in place. A second tube was also lost; Claude reported that it had been sabotaged. A third tube was finally put in place, but it was soon ruined because of waves and currents, though not before sucking up a previously unknown species of fish subsequently named after Claude by the Havana Academy of Science. After the destruction of the third tube, Claude permanently abandoned the project. Later, however, Claude's plans were reviewed by a French bureaucrat, improvements in the tubing were made, and experimentation with the theory continued. By this time Claude himself was involved in the activities which would eventually lead to his prison sentence.

Claude had joined the political organization *Action Francaise* in 1919. During World War II, he lectured at his own expense on the benefits of National Socialism and encouraged the Vichy government—which ruled as a Nazi puppet state following the fall of France—to take stronger action against the resistance movement. Following the Allied army landing in North Africa, Claude is reported to have attempted suicide with an overdose of **strychnine**. At the end of the war, he was put on trial for his lecturing activities, found to be a Vichy sympathizer, and sentenced to life imprisonment. Claude also lost his membership in the French Academy of Sciences.

Imprisoned at the age of seventy-four, Claude still conducted experiments during his confinement. He attempted to develop a deep-water tube for fishing, which would suck fish directly from the water into boats; the fish would then be immediately frozen with liquid air. Through the efforts of friends, Claude was released from prison in 1950 after serving four and a half years of his sentence. He passed away in Saint-Cloude, France, on May 23, 1960, actively working and writing his memoirs up until his death.

Claude's innovative methods often met with strong criticism, and if he was at all successful in having his inventions adopted, it was because of years of salesmanship. Claude realized that timing played a key role in invention. In an interview with *Scientific American* in 1929, he said: ''The difficulty of an inventor is not to make an invention, but to choose from among the multitude of inventive ideas which strike his mind the one which is really worth while.'' In a different interview with *Scientific American* the following year, he observed that his strict rule concerning inventing was that ''only simple solutions are worth considering.''

CLUSTERS

Molecular clusters are complexes of atoms. The term is most often applied to metal **atom** complexes involving between 2 and 12 metal atoms.

Atomic and molecular clusters provide means to study large scale molecules on a smaller scale. In some cases they represent small analogues for the large scale reactions (e.g., heterogeneous catalysis at metal surfaces). In addition, complex phenomena such as crystal formation and protein folding can be better analyzed in systems with smaller numbers of particles.

In metallic clusters, the metal atoms are either directly bonded through metal-metal interactions or are bridged by appropriate ligands. Cluster properties are influenced by the delocalization of electrons (electrons not associated with a particular atom) within the complex. In delocalized clusters, electrons have the ability to respond collectively.

A variety of geometries are observed for clusters. Simple binuclear and trinuclear complexes form linear and trigonal arrangements but a tetranuclear complex can occur as a tetrahedron (i.e., white **phosphorus** or P_4) or a cubane structure (i.e., Fe_4S_4). Cluster compounds with five to eight vertices are also possible and exhibit a variety of geometries. In most clusters (e.g., metal and semiconductor clusters) the geometric and electronic shell structure interact to determine the overall patterns of molecular shape and stability.

Perhaps the richest and certainly the most extensively studied cluster compounds are the carbonyl complexes. A large variety of structures, particularly with second and third row transition metal complexes, have been identified. These include clusters with more than one type of metal atom present. In addition, substitution of the neutral carbonyl ligand for phosphines, amines, or other neutral species significantly extends the number of possible compounds.

COAL

Coal is a naturally occurring combustible material consisting primarily of the element **carbon**, but with low percentages of solid, liquid, and gaseous hydrocarbons and other materials, such as compounds of **nitrogen** and **sulfur**. Coal is usually classified into the sub-groups known as anthracite, bituminous, lignite, and peat. The physical, chemical, and other properties of coal vary considerably from sample to sample.

Coal forms primarily from ancient plant material that accumulated in surface environments where the complete decay of organic **matter** was prevented. For example, a plant that died in a swampy area would quickly be covered with **water**, silt, sand, and other sediments. These materials prevented the plant debris from reacting with **oxygen** and decomposing to **carbon dioxide** and water, as would occur under normal circumstances. Instead, anaerobic bacteria (bacteria that do not require oxygen to live) attacked the plant debris and converted it to simpler forms: primarily pure carbon and simple compounds of carbon and **hydrogen** (hydrocarbons). Because of the way it is formed, coal (along with petroleum and **natural gas**) is often referred to as a fossil fuel.

The initial stage of the decay of a dead plant is a soft, woody material known as peat. In some parts of the world, peat is still collected from boggy areas and used as a fuel. It is not a good fuel, however, as it burns poorly and with a great deal of smoke.

If peat is allowed to remain in the ground for long periods of time, it eventually becomes compacted as layers of sediment, known as overburden, collect above it. The additional pressure and **heat** of the overburden gradually converts peat into another form of coal known as lignite or brown coal. Continued compaction by overburden then converts lignite into bituminous (or soft) coal and finally, anthracite (or hard) coal. Coal has been formed at many times in the past, but most abundantly during the Carboniferous Age (about 300 million years ago) and again during the Upper Cretaceous Age (about 100 million years ago).

High-grade, low-sulfur coal. Coal is formed from fossilized plants from the Carboniferous era (approximately 300 million years ago). High-grade coal, or anthracite, contains 85-88% carbon. ***(Kaj R. Svensson/Science Photo Library, Photo Rsearchers, Inc. Reproduced by permission.)***

Today, coal formed by these processes is often found in layers between layers of sedimentary rock. In some cases, the coal layers may lie at or very near the earth's surface. In other cases, they may be buried thousands of feet or meters under ground. Coal seams range from no more than 3-197 ft (1-60 m) or more in thickness. The location and configuration of a coal seam determines the method by which the coal will be mined.

Coal is classified according to its heating value and according to its relative content of elemental carbon. For example, anthracite contains the highest proportion of pure carbon—about 86%-98%—and has the highest heat value—13,500-15,600 Btu/lb (British thermal units per pound)—of all forms of coal. Bituminous coal generally has lower concentrations of pure carbon (from 46%-86%) and lower heat values (8,300-15,600 Btu/lb). Bituminous coals are often subdivided on the basis of their heat value, being classified as low, medium, and high volatile bituminous and sub-bituminous. Lignite, the poorest of the true coals in terms of heat value (5,500-8,300 Btu/lb) generally contains about 46%-60% pure carbon. All forms of coal also contain other elements present in living organisms, such as sulfur and nitrogen, that are very low in absolute numbers, but that have important environmental consequences when coals are used as fuels.

By far the most important property of coal is that it combusts. When the pure carbon and hydrocarbons found in coal burn completely, only two products are formed:carbon dioxide and water. During this chemical reaction, a relatively large amount of **energy** is released. The release of heat when coal is burned explains the fact that the material has long been used by humans as a source of energy, for the heating of homes and other buildings, to run ships and trains, and in many industrial processes.

The complete **combustion** of carbon and hydrocarbons described above rarely occurs in nature. If the **temperature** is not high enough or sufficient oxygen is not provided to the fuel, combustion of these materials is usually incomplete. During the incomplete combustion of carbon and hydrocarbons, other products besides carbon dioxide and water are formed, primarily **carbon monoxide**, hydrogen, and other forms of pure carbon, such as soot.

During the combustion of coal, minor constituents are also oxidized. Sulfur is converted to sulfur dioxide and sulfur trioxide, and nitrogen compounds are converted to nitrogen oxides. The incomplete combustion of coal and the combustion of these minor constituents results in a number of environmental problems.Carbon monoxide formed during incomplete combustion is a toxic gas and may cause illness or death in humans and other animals. Oxides of sulfur and nitrogen react with water vapor in the atmosphere and then are precipitated out as **acid rain**. Acid rain is thought to be responsible for the destruction of certain forms of plant and animal (especially fish) life.

In addition to these compounds, coal often contains a few percent of mineral matter: quartz, calcite, or perhaps clay minerals. These do not readily combust and so become part of the ash. The ash either escapes into the atmosphere or is left in the combustion vessel and must be discarded. Sometimes coal ash also contains significant amounts of **lead**, **barium**, **arsenic**, or other compounds. Whether airborne or in bulk, coal ash can therefore be a serious environmental hazard.

Coal is extracted from the earth using one of two major techniques, sub-surface or surface (strip) mining. The former method is used when seams of coal are located at significant depths below the earth's surface. The first step in sub-surface mining is to dig vertical tunnels into the earth until the coal seam is reached. Horizontal tunnels are then constructed laterally off the vertical tunnel. In many cases, the preferred method of mining coal by this method is called room-and-pillar mining. In this method, vertical columns of coal (the pillars) are left in place as coal around them is removed. The pillars hold up the ceiling of the seam preventing it from collapsing on miners working around them.

Surface mining can be used when a coal seam is close enough to the earth's surface to allow the overburden to be removed economically. In such a case, the first step is to strip off all of the overburden in order to reach the coal itself. The coal is then scraped out by huge power shovels, some capable of removing up to 100 cubic meters at a time. Strip mining is a far safer form of coal mining, but it presents a number of environmental problems. In most instances, an area that has been strip mined is terribly scarred, and restoring the area to its original state is a long and expensive procedure. In addition, any water that comes in contact with the exposed coal or overburden may become polluted and require treatment.

Coal is regarded as a non-renewable resource, meaning that it was formed at times during the Earth's history, but significant amounts are no longer forming. Large supplies of coal are known to exist (proven reserves) or thought to be available (estimated resources) in North America, the former Soviet Union, and parts of Asia, especially China and India. According to the most recent data available, China produces the largest amount of coal each year, about 22% of the world's total, with the United States 19%, the former members of the Soviet Union 16%, Germany 10% and Poland 5% following. China is also thought to have the world's largest estimated resources of coal, as much as 46% of all that exists. In the United States, the largest coal-producing states are Montana, North Dakota, Wyoming, Alaska, Illinois, and Colorado.

Coal is still used in industries such as **paper** production, cement and ceramic manufacture, **iron** and steel production, and chemical manufacture for heating and for steam generation.

Another use for coal is in the manufacture of coke. Coke is nearly pure carbon produced when soft coal is heated in the absence of air. In most cases, one ton of coal will produce 0.7 ton of coke in this process. Coke is of value in industry because it has a heat value higher than any form of natural coal. It is widely used in steel making and in certain chemical processes.

A number of processes have been developed by which solid coal can be converted to a liquid or gaseous form for use as a fuel. Conversion has a number of advantages. In a liquid or gaseous form, the fuel may be easier to transport, and the conversion process removes a number of impurities from the original coal (such as sulfur) that have environmental disadvantages.

One of the conversion methods is known as gasification. In gasification, crushed coal is reacted with steam and either air or pure oxygen. The coal is converted into a complex mixture of gaseous hydrocarbons with heat values ranging from 100 Btu to 1000 Btu. One suggestion has been to construct gasification systems within a coal mine, making it much easier to remove the coal (in a gaseous form) from its original seam.

In the process of liquefaction, solid coal is converted to a petroleum-like liquid that can be used as a fuel for motor vehicles and other applications. On the one hand, both liquefaction and gasification are attractive technologies in the United States because of our very large coal resources. On the other hand, the wide availability of raw coal means that new technologies have been unable to compete economically with the natural product.

During the last century, coal oil and coal gas were important sources of fuel for heating and lighting homes. However, with the advent of natural gas, coal distillates quickly became unpopular, since they were somewhat smoky and foul smelling.

Cobalt

Cobalt is a transition metal, one of the elements that occurs in the middle of the **periodic table**, between Groups 2 and 13. It is located between **iron** and **nickel**, with whom it shares many common properties. Cobalt's **atomic number** is 27, its atomic **mass** is 58.9332, and its chemical symbol is Co.

Properties

Cobalt is a hard, gray metal that looks much like iron and nickel. It is ductile, but only moderately malleable. Cobalt is one of three naturally occurring magnetic elements, the other two being iron and nickel. The melting point of cobalt is 2,719°F (1,493°C), its **boiling point** is about 5,250°F (2,900°C), and its **density** is 8.9 grams per cubic centimeter. Cobalt is a moderately reactive element that combines slowly with **oxygen** in the air. It does not react with **water** at room **temperature**, but it does react with most acids to produce **hydrogen**.

Occurrence and Extraction

Cobalt is a relatively abundant element with an abundance of about 10-30 parts per million in the Earth's crust. It usually occurs in the form of a compound with its most common minerals being cobaltite, smaltite, chloranthite, and linnaeite. The major supplies of cobalt in the world are in Zambia, Canada, Russia, Australia, Zaire, and Cuba. No cobalt is mined in the United States. Cobalt is obtained from its ores by converting the naturally-occurring minerals to cobalt **oxide** (Co_2O_3) and then reducing the oxide with aluminum: $2Al + Co_2O_3 \rightarrow Al_2O_3 + 2Co$.

Discovery and Naming

Cobalt was discovered in 1735 by the Swedish chemist Georg Brandt (1694-1768) who was analyzing a dark blue pigment found in **copper** ore. Chemists later found that cobalt often appears in conjunction with copper ores.

The element was given a name that had been associated with cobalt minerals for many years, *Kobold.* That word is German for "goblin" or "evil spirit." The name arose because certain cobalt minerals are very difficult to mine and, when refined, give off an offensive gas that can cause illness. The gas was later identified as **arsenic** trioxide (As_4O_6), which often occurs with cobalt minerals in the earth.

Cobalt samples. ***(Photograph by Russ Lappa/Science Source, National Audubon Society Collection/Photo Researchers, Inc. Reproduced by permission.)***

Uses

About 65% of the cobalt used in the United States is used to make alloys, primarily superalloys. Superalloys consist primarily of iron, cobalt, or nickel with smaller amounts of other **metals**, such as **chromium**, **tungsten**, **aluminum**, and **titanium**. Superalloys are so-called because they are very resistant to **corrosion** and retain their properties at high temperatures. Superalloys are used in the manufacture of jet engines and gas turbines, where temperatures in the thousands of degrees are produced routinely. Cobalt alloys are also widely used in the manufacture of strong electromagnets.

The most important use of cobalt compounds is as coloring agents in **glass**, glazes, **cosmetics**, paints, rubber, **inks**, and pottery. Some of the compounds frequently used are cobalt oxide, or cobalt black (Co_2O_3); cobalt **potassium** nitrite, or cobalt yellow [$CoK_3(NO_2)_6$]; cobalt aluminate, or cobalt blue ($Co(AlO_2)_2$); and cobalt ammonium **phosphate**, or cobalt violet ($CoNH_4PO_4$).

Health Issues

Cobalt is needed in very small amounts to maintain good health in animals. It is used in the synthesis of certain enzymes in the body. Rare examples have been observed of cobalt deficiency disorders, such as a condition known as Coast disease, which appears among sheep in Australia in areas where the soil is deficient in cobalt. An excess of cobalt may cause health problems. For example, people who work with the metal may inhale its dust or get dust on their skin, producing vomiting, diarrhea, or skin rashes.

Coenzyme

While all enzymes belong to the protein family, many of them are unable to participate in a catalytic reaction until they link with a nonprotein component, or coenzyme. This can be a metal ion—**copper**, **iron**, or **manganese**, for instance—or a moderately-sized **molecule** called a prosthetic group. Quite often, though, coenzymes are composed wholly or partially of **vitamins**. Although some enzymes are attached very tightly to their coenzymes, others can be easily parted. In either case, the parting almost always deactivates both partners.

The first coenzyme was discovered by English biochemist Sir **Arthur Harden**. Toward the end of the nineteenth century, Harden began an intense study of the **fermentation** process, particularly alcoholic fermentation. Inspired by Eduard Buchner—who, in 1897, had discovered an active **enzyme** in yeast juice that he had named zymase—Harden used an extract of yeast in most of his studies.

While working with William J. Young in 1904, Harden made a surprising discovery. He'd already learned that boiling yeast juice appeared to destroy all its enzyme activity. However, he found that, when he added some of the boiled and presumably useless yeast juice to an active batch, the active yeast juice suddenly showed an increased capacity to ferment glucose. Some active principle, he reasoned, must have survived the boiling. To solve the chemical mystery, Harden used a **filtration** process called **dialysis**. He placed another batch of yeast juice in a semipermeable bag, then left the bag in a container of pure **water**. Before long, the juice's smaller molecules filtered through the bag's membrane and into the water, leaving the larger molecules behind in the bag. After further testing, Harden discovered that the yeast enzyme apparently consisted of two parts: a large-molecular part that could not survive boiling and was almost certainly a protein; and a small-molecular part that could survive boiling and was probably not a protein. (Harden called the nonprotein a coferment, but others soon began calling it a coenzyme.)

Several researchers quickly began studying the newly discovered component's chemical nature and, roughly 20 years later, **Hans von Euler-Chelpin,** a German-Swedish chemist, was able to define its structure as that of niacin, a component of the NADH and NADPH redox cofactors. (Harden and Euler-Chelpin shared the 1929 Nobel Prize in **chemistry** for their work.)

It soon became clear that virtually all coenzymes were composed of vitamins, particularly those in the water-soluble B family. For the most part, they functioned primarily in **energy** transfers and in the **metabolism** of fats, **carbohydrates**, and **proteins**. During the 1930s, Swedish biochemist **Axel Theorell** and American biochemist Conrad Arnold Elvehjem (1901-1962) greatly furthered the understanding of vitamins through independently conducted research on **oxidation** enzymes and pellagra, respectively.

In 1947, a coenzyme particularly important in the metabolic process, Coenzyme A, or CoA, was discovered by **Fritz Lipmann**, a German-born American biochemist who received the 1953 Nobel Prize in medicine and physiology for the discovery. **Feodor Lynen** is also remembered for his research on CoA; because he succeeded in isolating acetylcoenzyme A (CoA combined with the two-carbon fragment first theorized by Lipmann), he, along with **Konrad Emil Bloch**, received the 1964 Nobel Prize for medicine and physiology.

Because most vitamins are inactive when first taken into the body, a two-step process must take place. The vitamin must be activated to its coenzyme form (with vitamins B_1, B_2, and B_6, for instance, this means the addition of a **phosphate** group.) After that, the coenzyme must combine with its proper enzyme partner. Only then can the catalytic activity, for which both are programmed, be set in motion.

COHEN, STANLEY (1922-)

American biochemist

A pioneer in the study of growth factors —the nutrients that differentiate the development of cells—Stanley Cohen is best known for isolating nerve growth factor (NGF), the first known growth factor, and for subsequently discovering and fully identifying the epidermal growth factor (EGF). Cohen shared the 1986 Nobel Prize for physiology or medicine with his colleague, Italian American neurobiologist **Rita Levi-Montalcini**, who first discovered NGF. Research on NGF has led to better understanding of such degenerative disorders as cancer and Alzheimer's disease, while studies concerning EGF have proved useful in exploring alternative burn treatments and skin transplants.

Cohen was born in Brooklyn, New York, in 1922 to Russian immigrant parents. Though his father earned only a modest living as a tailor, both parents, Louis and Fannie (Feitel) Cohen, ensured that their four children received quality educations. As a child, Cohen was stricken with polio, imparting him with a permanent limp. His illness, however, influenced him to pursue intellectual interests. While a student at James Madison High School he earnestly studied science as well as classical music, learning to play the clarinet. Cohen entered Brooklyn College to study **chemistry** and zoology, graduating in 1943 with a B.A. Following his undergraduate studies, Cohen received a scholarship to Oberlin college in Ohio, where he earned an M.A. in zoology in 1945. He then attended the University of Michigan on a teaching fellowship in **biochemistry**, earning his Ph.D. in 1948.

From 1948 until 1952, Cohen worked at the University of Colorado School of Medicine in Denver, holding a research and teaching position in the Department of Biochemistry and Pediatrics. There Cohen earned the respect of his peers for his collaborative studies with pediatrician Harry H. Gordon on the metabolic functions of creatinine (a chemical found in blood, muscle tissue, and urine) in newborn infants. Cohen moved to St. Louis, Missouri, in 1952 to work as a postdoctoral fellow in the radiology department at Washington University. The following year, he was asked to become a research associate in the laboratory of renowned zoologist Viktor Hamburger, who was conducting studies on growth processes. Levi-Montalcini, who had been researching nerve cell growth in chicken embryos that had been injected with the tumor cells of male mice, had just returned from Rio de Janeiro, where she had conducted successful tissue culture experiments that definitively proved the existence of NGF. Working at the lab in St. Louis, Levi-Montalcini relied on Cohen's expertise in biochemistry to isolate and analyze NGF.

The collaboration between Levi-Montalcini and Cohen combined two similar personalities. Both scientists have been characterized by their unassuming manners despite their obvious intellectual abilities and perceptive intuitions. Describing her early recollections of Cohen, Levi-Montalcini wrote in her autobiography *In Praise of Imperfection,* "I had been immediately struck by Stan's absorbed expression, total disregard for appearances—as evidenced by his motley attire—and modesty.... He never mentioned his competence and extraordinary intuition which always guided him with infallible precision in the right direction." Between the years 1953 and 1959, Cohen and Levi-Montalcini conducted intense research, both enthusiastically pursuing thier findings concerning NGF.

By 1956 Cohen had succeeded in extracting NGF from a mouse tumor; however, this proved to be a difficult substance to work with. Upon the suggestion of biochemist **Arthur Kornberg**, Cohen added snake venom to the extract, hoping to break down the **nucleic acids** that made the extract too gelatinous. Fortuitously, the snake venom produced more nerve growth activity than the tumor extract itself, and Cohen was able to proceed more rapidly with his studies. In 1958 he discovered that an abundant source of NGF could be found in the salivary glands of male mice—glands not unlike the venom sacs of snakes. Cohen's biochemical advances enabled Levi-Montalcini to study the neurological effects of NGF in rodents.

At a time when Levi-Montalcini and Cohen were advancing rapidly in their collaborative research, funding for Hamburger's laboratory could no longer support Cohen. Before leaving Washington University, Cohen was able to purify NGF as well as produce an antibody for it; however, its complete chemical structure was not fully determined until 1970 when researchers at Washington University completed analysis of NGF's two identical **chains** of **amino acids**. Before departing St. Louis, Cohen also observed an unusual occurrence in newborn rodents that had been injected with unpurified salivary NGF. Unlike control mice, whose eyes opened on the thirteenth or fourteenth day, those injected with the unpurified NGF opened their eyes on the seventh day; they also sprouted teeth earlier than did the control group.

Cohen left Washington University in 1959 to join a research group at Vanderbilt University in Nashville, Tennessee; there he continued his work with growth factors, focusing on identifying the unknown factor in unpurified NGF that had caused the mice to open their eyes earlier than normal. By 1962, Cohen had extracted the contaminant in these samples of NGF and was able to purify a second substance, a protein that promoted skin cell and cornea growth which he called epidermal growth factor, or EGF. This protein has found widespread use in treating severe burns; a **solution** rich in EGF can promote the speedy healing of burned skin, while a skin graft soaked in EGF will quickly bond with damaged tissue. Cohen also isolated the protein which acted as a receptor for EGF—an important step toward understanding the transmission of signals that stimulate normal and abnormal cell growth—that has been particularly crucial in studying cancer development. Cohen was successful in fully identifying the amino acid sequence of EGF by 1972.

Despite his significant contributions, Cohen has never managed a large laboratory, and for many years his work went unacknowledged. He remarked in *Science* that while the scientific community took little notice of his early studies on growth factors, this anonymity proved beneficial. "People left you alone and you weren't competing with the world," he recalled. "The disadvantage was that you had to convince people that what you were working with was real." Cohen's work has subsequently gained wide recognition, and he has received numerous awards in addition to the Nobel, including the Alfred P. Sloan Award in 1982, as well as both the National Medal of Science and the Albert Lasker Award in 1986.

Stanley Cohen (left). ***(Archive Photo. Reproduced by permission.)***

Cohn, Mildred (1913-)

American biochemist and biophysicist

Mildred Cohn overcame both gender and religious prejudice to have a profound impact on **biochemistry** and biophysics. Her research contributed to the scientific understanding of the mechanisms of enzymatic reactions and the methods of studying them. Cohn authored numerous papers that are considered classics and received many honors, including the 1982 National Medal of Science presented by President Ronald Reagan. She was a member of the National Academy of Sciences, the American Academy of Arts and Sciences, and the American Philosophical Society.

Cohn was born on July 12, 1913, to Isidore M. and Bertha (Klein) Cohn, the second of their two children. Her parents were both immigrants from Sharshiv, a small town in Russia. Her father was a businessman who did linotype work for the printing trade and published a journal on printing. Cohn attended public schools in New York City, demonstrating an interest in mathematics and **chemistry** by the time she reached high school. In 1928, at age fourteen, she enrolled at Hunter College in New York to study chemistry and physics. She received her B.A. *cum laude* little more than three years later, at age seventeen. Cohn was determined to pursue a graduate education in the physical sciences in spite of the many barriers raised against her.

Cohn entered the doctoral program at Columbia University but was not accepted as a teaching assistant, because the positions were awarded only to men. She worked as a babysitter to help pay for her first year of education in **thermodynamics**, classical mechanics, molecular **spectroscopy**, and **physical chemistry**. In 1932, after being awarded an M.A., she accepted a job in the laboratory at the National Advisory Committee for Aeronautics at Langley Field, Virginia. She was initially assigned computational work and later a research position in the engine division. The project was to develop a fuel-injection, spark-ignition airplane engine that operated on the diesel cycle. Cohn believed that her positions at Langley impressed upon her the importance of attacking problems on many levels, including the practical and theoretical.

In 1934, Cohn decided that her opportunities for scientific advancement had declined at Langley and she returned to Columbia to seek a Ph.D. She worked under **Harold Clayton Urey**, separating stable isotopes. She was not successful in trying to separate the isotopes but learned experimental and theoretical methods from which she benefited throughout her career. Cohn wrote her Ph.D. dissertation in 1937 and published it with Urey under the title "**Oxygen** Exchange Reactions of Organic Compounds and **Water**." Upon graduation, she considered applying for an industrial position, as many of the other graduates did. Due to her sex and religion, however, she was not even granted interviews with large corporations, including Du Pont and Standard Oil.

Cohn was eventually offered and accepted a postdoctoral position with **Vincent Vigneaud**, professor of biochemistry at George Washington University Medical School. He wanted to introduce isotopic tracers into his research on sulfur-amino acid **metabolism**. Isotopic tracers are forms of chemical elements; because of their difference in nuclear structure, either **mass** or **radioactivity**, they can be observed as they progress through a metabolic pathway. By following an **isotope**, Cohn was able to understand more clearly the mechanisms of chemical reactions in animals. For example, one study, which Cohn told contributor John Henry Dreyfuss was "the most elegant tracer experiment done in du Vigneaud's lab," involved what she called "doubly labeled" methionine. The researchers used the labeled methionine—a large, complicated **molecule** to which two isotopes had been added—to observe the mechanism by which methionine was converted to the amino acid cystine in a rat.

Cohn married Henry Primakoff, a physicist, on May 31, 1938. During World War II, Cohn and the draft-exempt men in du Vigneaud's lab continued their research, while du Vigneaud and the others supported the war effort. In 1946, her husband accepted a position in the physics department at Washington University in St. Louis. Cohn took a position in the biochemistry department, where she worked with the Nobel Prize-winning husband and wife team of **Gerty T. Cori** and **Carl Ferdinand Cori**. One of her major objectives at Washington University was to study the mechanisms of enzyme-catalyzed reactions. Cohn used an isotope of oxygen to gain insight into the enzyme-catalyzed reactions of organic phosphates and in 1958 initiated work with **nuclear magnetic resonance** (NMR) toward the same goal.

Two years later, her husband was named Donner Professor of Physics at the University of Pennsylvania, and Cohn joined the biophysics department there. At Pennsylvania, Cohn pursued her research on **energy** transduction within cells and cellular reactions in which adenosinetriphosphate (ATP) is utilized using NMR. Cohn told Dreyfuss that she began to look more deeply into the structure and function of enzymes by studying manganese-enzyme-substrate complexes, utilizing "every technically feasible aspect of magnetic resonance." Cohn performed other important collaborative studies, including NMR of transfer ribonucleic acid (RNA), a key chemical in cellular **protein synthesis**. Cohn's work with nuclear magnetic resonance of various types of molecules, structures, and reactions was probably her most important contribution to science and medicine.

Cohn and Primakoff had three children. Besides her scientific work, she enjoyed the theater, hiking, writing, and reading. In 1958, after nearly twenty-one years as a research associate, she was promoted to associate professor; she was a Career Investigator of the American Heart Association for fourteen years after 1964. In 1982, she was named Benjamin Bush Professor in Biochemistry and Biophysics at the University of Pennsylvania. From 1982 to 1985, she was a senior scientist at the Fox Chase Cancer Center. In 1982, she was awarded the National Medal of Science and in 1987 the Distinguished Award of the College of Physicians. From 1978 to 1979, she was president of the American Society of Biological Chemistry. Cohn was also elected to the American Academy of Arts and Sciences and American Philosophical Society.

Cold Fusion

Cold fusion is the term proposed to describe controlled **nuclear fusion** reactions occurring at or near room **temperature**. The attainment of cold fusion, with a net release of **energy**, was announced at a press conference in Salt Lake City, Utah, on March 23, 1989, by B. Stanley Pons of the University of Utah and Martin Fleishman of Southampton University in England. The following day, a paper claiming the achievement of cold fusion at a much lower rate was mailed to the British journal *Nature* by Steven E. Jones, a professor of physics at Brigham Young University, and a group of his colleagues. The announcement by Pons and Fleishman received worldwide media coverage, but also a great deal of criticism as it did not provide adequate technical details for other scientists to independently verify what was claimed. As the details of the experiment became known, attempts to reproduce it in a number of other laboratories gave inconsistent results. In addition, Pons and Fleishman found it necessary to retract some of their initial claims. To deal with the controversy, the Energy Research Advisory Board of the United States Department of Energy appointed a fact-finding panel. By the end of the year, this panel concluded that there was "no convincing evidence that useful sources of energy would result from" the cold fusion work. Within a few years, the majority of chemists and physicists reached consensus that the Fleishman and Pons experiment had failed to demonstrate the release of fusion energy on a significant scale. While some scientists believe that Jones and others following his approach had demonstrated an interesting new effect, there is currently only limited interest in funding research pertaining to cold fusion as a **potential energy** source.

Nuclear fusion is the process responsible for the production of **heat** energy in the sun and other stars. It can be understood using a few basic ideas from nuclear physics. The structure of the **nucleus** is the result of two different fundamental forces: the long range electrical repulsion which exists between protons, and the short range nuclear strong force which is attractive and of roughly equal strength between protons and protons, protons and **neutron** and neutrons and neutrons. Just as electrons outside the **atom**, the protons and neutrons separately occupy set of energy levels within the nucleus, with the number in each level limited by the **Pauli exclusion principle**. A third fundamental force, the nuclear weak force, allows for neutrons to decompose into a **proton** plus an **electron** plus a massless particle called an antineutrino, or a proton, if it has enough energy available, to split into a neutron and a positron plus a massless **neutrino**. These weak-force induced properties account for the roughly equal number of protons and neutrons in the common elements, since an excess of protons or neutrons requires that the excess particles be in higher energy levels and can spontaneously change into the other type of particle.

In the sun and other stars, **hydrogen** is fused into **helium** in a cyclic reaction in which two protons first combine to form a deuteron (heavy hydrogen nucleus), a third proton is added to form a nucleus of helium-3, and then two helium-3 nuclei combine to form helium-4 and release two protons. This reaction occurs only in the core of the sun at a temperature of about ten million degrees Kelvin. The high kinetic energies of the particles in this case are sufficient to allow the particles to approach closely enough to allow the short-range strong force to act. Each day approximately 1.5×10^{19} kJ of energy reaches the earth's surface as a result of the fusion reactions taking place in the interior of the sun. The direct fusion of deuterons is also possible, with three possible outcomes: the formation of a helium-three nucleus and release of a neutron, the formation of a **tritium** (hydrogen-3) nucleus with the release of a proton, and the direct formation of a helium-4 nucleus with the emission of a gamma ray. Both theoretical and experimental studies indicate that the third possibility occurs with the least frequency and the first two with nearly equal frequency.

The harnessing of fusion energy to replace fossil and nuclear (fission) fuels has been a goal of energy researchers since the 1960's. The main obstacle to controlled thermonuclear fusion had been being able to confine **plasma**, a mixture of ions and electrons, at the extremely high temperature required for thermonuclear fusion—temperatures hot enough to melt the walls of any container. The cold fusion researchers had in mind another possibility. The tunnel effect of quantum mechanics allows the spontaneous fusion of the two nuclei in a **diatomic deuterium molecule**, although at an astronomically slow rate. If one of the **bonding** electrons is replaced by a muon, an elementary particle with the same charge as the electron and 200 times the **mass**, the bond is much shorter and tunneling becomes far more probable. Indeed, Steven E. Jones, was a recognized researcher in muon- catalyzed fusion.

The premise behind the cold fusion experiments was that one could achieve a comparable increase the rate of deuterium fusion by electrochemically driving the nuclei into one of the transition **metals** known to have a large capacity to absorb hydrogen gas. **Palladium**, for example, is able to absorb six hydrogen atoms for every ten palladium atoms. The experiment would require energy input to establish the required current in an electrochemical cell, but the energy output could be determined by **calorimetry**. The initial report by Pons and Fleishman claimed a fourfold return on the energy input and a net production of helium. The experiment did not however, show a significant production of neutrons, as would be expected from the helium-3 producing process, or the level of gamma ray production to be expected with helium-4, and the original claims of heat and helium production were also brought into doubt by subsequent investigation. Many other investigators have elaborated the more careful experiment of Jones, with generally lower estimates obtained for the possible fusion rate. These studies, however, leave open the possibility that some real but as of yet unexplained phenomenon may be at the root of the reported results. It may be several years, perhaps even decades, before all of the experimental research pertaining to cold fusion is explained to everyone's satisfaction. In the meantime small pockets of research are still being conducted at several prominent research laboratories both in the United States and abroad.

COLLIGATIVE PROPERTIES

A colligative property is a physical property that is independent of the size, **mass**, or characteristics of the solute particles present in a **solution**. Instead, it depends on the **concentration** of the solution, in other words, the number of solute particles present in the solution. The word colligative means "depending on the collection." There are four different types of colligative properties. These are **vapor pressure reduction**, **boiling point** elevation, freezing point depression, and **osmotic pressure**. The latter three properties are directly proportional to the **molality** of the dissolved solute, which is defined as the number of moles of solute per kilogram of solvent.

Vapor pressure is due to the presence of gas molecules over a liquid surface. Molecules constantly pass from the liquid to the vapor phase, and then in the reverse direction. Vaporization and condensation occur simultaneously. When these two opposing processes are in equilibrium, they occur at exactly the same rate. At equilibrium, the number of molecules per unit **volume** in the vapor state remains constant, and the equilibrium vapor pressure is then reached.

When a nonvolatile solute is added to a solvent, the vapor pressure is lowered. This phenomenon is called vapor pressure reduction, the first of the colligative properties. A nonvolatile substance is one that does not easily become a gas under existing conditions. When a nonvolatile solute is added to a solution, the dissolved solute molecules take up some of the space in the solution, blocking the way for the solvent molecules to escape from the liquid surface. The gas molecules that are returning to the liquid can still do so at the same rate. This creates a nonequilibrium system where more molecules are leaving the gaseous state than are entering. Equilibrium is re-established as more gas molecules join the liquid, hence lowering the vapor pressure.

Francois-Marie Raoult (1830-1901) studied the vapor pressures of solutions and discovered that the partial pressure of each component above a liquid mixture is directly proportional to the **mole** fraction composition of the component in the liquid mixture. A nonvolatile solute does not contribute to the total pressure, hence the observed pressure above a mixture is simply that of the solvent molecules. The presence of the solute in the solvent means that $x_{solvent}$ is less than unity. Doubling the concentration of solute doubles the effect of vapor pressure reduction. This fact is known as Raoult's Law, which is strictly obeyed only in the case of ideal solutions. The mathematical relationship does provide a very good approximation for nonideal solutions that are sufficiently dilute. A solution in which x solvent > 0.95 generally meets this requirement. Vapor pressure reduction only depends on the solute concentration, and not on the identity of the nonvolatile solute, therefore it is a colligative property.

Another colligative property is called boiling point elevation. The boiling point of a solution will increase with the concentration of solute. At the temperature where the vapor pressure of the liquid is equal to the atmospheric pressure, the liquid is said to be at its normal boiling point. As discussed above, adding a nonvolatile solute lowers the vapor pressure. As a result, more **heat energy** must be added to raise the vapor pressure of the solvent to that of atmospheric pressure, which means the boiling point of the solvent has been raised. Boiling point elevation, ΔT_{bp} is proportional to the molality of the dissolved solute $\Delta T_{bp} = k_{bp}\ m_{solute}$ where k_{bp} is the solvent-dependent proportionality constant. Representative numerical values of the boiling point depression constant are: $k_{bp} = 0.512$ °C molal^{-1} for **water**; $k_{bp} = 2.53$°C molal^{-1} for **benzene**; and k_{bp} = 3.07°C molal^{-1} for **acetic acid**. Because the boiling point elevation depends on the solute concentration, not the identity of the solute (although electrolytes have a greater effect than do nonelectrolytes), it is considered a colligative property.

The third colligative property is called freezing point depression. The freezing point of a solution is the temperature at which the vapor pressure of the liquid is equal to that of the solid. Adding a nonvolatile solute to the solvent will lower the vapor pressure and, as a result, the temperature where the vapor pressures of the liquid and the solid can be equal. Freezing point depression, ΔT_{fp} is directly proportional to the molality of the dissolved solute $\Delta T_{fp} = k_{fp}m_{solute}$ The proportionality constant has a negative numerical value and is again solvent dependent. For water the freezing point depression constant is $k_{fp} = -1.86$°C molal^{-1}. Freezing point depression, like boiling point elevation, does not depend on the identity of the solute (again, electrolytes have a greater effect than do nonelectrolytes) and is therefore considered a colligative property.

The last colligative property is osmotic pressure. Osmotic pressure results when **osmosis** occurs. Osmosis is the net flow of solvent molecules from a less concentrated solution to a more concentrated solution through a semipermeable membrane. A semipermeable membrane allows small solvent molecules to pass through, but not the larger solute molecules. As a result, the liquid levels between two solutions separated between a semipermeable membrane will become uneven as the solvent molecules move from one side to the other. Eventually, a pressure difference between the two heights of the solutions occurs which is so large that osmosis cannot continue. The amount of pressure that prevents osmosis from occurring is called the osmotic pressure. For dilute nonelectrolyte solutions, the mathematical relationship for osmotic pressure (π)

$$\pi = (n_{solute}/V)\ RT$$

$$\pi = m_{solute}\ d_{solvent}\ RT$$

looks very similar to the ideal gas law equation. The osmotic pressure is given in atmospheres, n_{solute} is the number of moles of solute in V liters of solution, T refers to the absolute Kelvin temperature, $d_{solvent}$ denotes the **density** of the solvent in kg L^{-1}, and R is the universal gas law constant. As indicated by the mathematical relationship a solution of greater concentration will have greater osmotic pressure. Because osmotic pressure depends only on the concentration of the solution, it is a colligative property.

There are many practical examples of the colligative properties. For example, freezing point depression and boiling point elevation are both involved in the use of antifreeze in the cooling system of automobiles. If water alone were used as the engine coolant, then one might experience the engine overheating in the hot summer months, and/or radiator freezing during the cold winter months when the outside temperature dropped below 32°F. Addition of antifreeze (made of ethylene glycol) lowers the freezing point of the engine coolant and also raises its boiling point temperature. When ice and snow make driving hazardous in the winter, rock **salt** is often placed on the roads. The rock salt dissolves in the water and lowers the freezing point of water several degrees. This prevents ice from forming on the streets. The original zero point of the Fahrenheit temperature scale was set at the lowest temperature that could be reached in a mixture of salt and ice.

Any solute will affect the colligative properties of a solution. By definition, the colligative properties are dependent only on the amount of solute present, not the identity. The effect the solute has depends on the number of solute particles that are dissolved in a given amount of the solvent. The more solute particles that are present, the greater the effect on the colligative properties. The number of solute particles which form in a solution depends on the chemical nature of the solute. For example, table sugar ($C_{12}H_{22}O_{11}$) does not dissociate into ions when placed in a solution. Therefore, for every sugar **molecule**, there is only one solute particle in the solution. Ionic molecules will form more than one particle when placed in solution. For example, NaCl will form two particles; Na^+ and Cl^-. $CaCl_2$ will form three particles in solution; Ca^{2+} and two Cl^- ions. If 100 molecules of NaCl were added to a solution, they would affect the colligative properties of the solution twice as much as if 100 molecules of $C_{12}H_{22}O_{11}$ were added to the solution, because there would be twice as many NaCl particles.

Freezing point depression and osmotic pressure provide a convenient experimental method for determining the molar mass of unknown substances, particularly biological and polymeric compounds. To illustrate the method, assume that 1.500 grams of an organic compound is dissolved in 20.000 grams of benzene (k_{fp} = - 5.12°C molal^{-1}). The measured freezing point temperature is measured to be T_{fp} = 3.38°C, which is 2.15°C lower than benzene's normal freezing point temperature of T_{fp} = 5.53°C. The concentration of the unknown organic compound is calculated by substituting ΔT_{fp} = -2.15°C and k_{fp} = -5.12°C molal^{-1} into the freezing point depression equation: -2.15 = - 5.12 m_{solute} and then solving for molality: m_{solute} = 2.15/5.12 = 0.420. The molar mass of the organic compound is calculated from the definition of molality: Molality = (grams of solute/molar mass of solute)/mass of solvent in kilograms. Molar mass of solute = (1.500)/(0.420 x 0.0200) = 178.57 grams mole^{-1}. Molar masses can be similarly determined from osmotic pressure measurements using $\pi = m_{solute}\, d_{solvent}\, RT$. Molar masses of biological compounds are rarely determined using the boiling point elevation method because the compounds often decompose at the higher boiling point temperatures.

COLLISION THEORY

Collision theory is a special type of kinetic theory that states that for a chemical reaction to take place there must be a collision between reactants. In addition, to cause a reaction the colliding reactant molecules must surpass a minimum energy termed the **activation energy** and, in many cases, the reactant molecules must have a certain spatial orientation with respect to each other.

The collision theory provides an explanation for the experimental observation that, in a chemical reaction, if two types of molecules in the gas phase are combined, not all of the molecules instantly enter the reaction.

Reactions between molecules can occur in the solid, liquid, or gaseous state. Reactions also occur in solution. The simplest reactions are those which occur between molecules in the gas phase. The reason for this is that gas phase molecules do not have bonds between molecules or bonds to solvents that must be broken before a reaction can occur. Even so, there are other obstacles to overcome in the gas phase.

By definition, the molecules of a gas are very far apart from one another. Also, gas molecules travel at very high speeds. For example, at room **temperature**, 77°F (25°C), the average speed of an **oxygen** molecule is almost 1,000 miles per hour (444 meters per second), while the average speed of a **hydrogen** molecule is almost 4,000 miles per hour (about 1,788 meters per second).

The motion of gas molecules is linear but random. That means that the molecules travel in straight lines until they hit the wall of the container (at which point they bounce off the wall and travel in a direction determined by the angle at which they hit the wall) or another **molecule**. If a molecule collides with another molecule of the same type, the collision will not result in a reaction. The two molecules will simply bounce off one another and continue their random linear motion. It is only possible to have a reaction when two different molecules come into contact with one another. This explains why the presence of more molecules increases the chance that a reaction will occur. There are more likely to be collisions between different molecule types if there is a higher concentration of molecules, just as it is more likely to bump into someone in a crowded room than in an empty room. Even so, a reaction will only occur if the two molecules are chemically capable of reacting with one another. For example, air consists predominantly of oxygen and **nitrogen**, but under atmospheric conditions, they will not react with one another. This does not mean that oxygen and nitrogen will never react with one another, just that the conditions have to be specific for reaction to occur.

Under normal laboratory conditions, molecules collide about ten billion times per second. However, when two reactive materials are placed in a container, not all of these collisions between the two types of molecules will produce a reaction. The energy of the reactant molecules is critical. If the collision is too gentle the energy of the collision will be insufficient to produce a reaction. Thus, even if two molecules that could react collide, if the energy of the collision is too low the molecules simply bounce off one another.

In order to produce a chemical reaction, the electrons of one molecule must come into close contact with the electrons of the other molecule with enough energy to permit the transfer of electrons. Because electrons are attracted to each of the molecules, it requires energy to detach them. This minimum energy required for a reaction to occur is called the activation energy. When the activation energy is low, the reaction occurs more easily as not much energy is needed for the two colliding molecules. When the activation energy is high, a much greater energy is needed on collision of the two molecules. The higher the activation energy, the smaller the proportion of molecules there are at room temperature that possess the energy required to cause a chemical reaction. Thus, reactions with a high activation energy usually proceed more slowly. This theory explains the experimental observation that when two types of molecules in the gas phase are combined, not all of the molecules instantly enter the reaction.

When the temperature of the reaction is increased, the molecules move faster because they have increased energy as measured in the form of heat. This increased speed allows more molecules to react. Because they are moving faster, a larger proportion of molecules of both types achieve enough kinetic energy to overcome the activation energy barrier. Therefore, at higher temperatures more molecules have the required kinetic energy to supply the activation energy. Accordingly, more molecules are now capable of reacting.

A small increase in temperature does not, however, usually result in a small increase in the rate of reaction. The effect of a temperature increase is not linear and a small increase in temperature can have a large effect on the rate of reaction. A useful rule of thumb is that a 10% increase in temperature will approximately double the rate of reaction.

Collision theory also states that the orientation of the molecules is an important criteria in determining whether or not molecules will enter into a reaction. Even if the molecules that are colliding have enough energy to overcome the activation energy barrier, sometimes a reaction will not occur because of the way in which the molecules collide. If the reactive part of one molecule collides with the unreactive part of another molecule, it is possible that they will bounce off one another rather than react. For some reactions, it is necessary for the two reacting molecules to approach each other with the right geometry or orientation, as well as enough energy, for the molecules to react.

See also Gases; Kinetics; Reactions (types of); Reactivity

COLLOID

A **colloid** is a mixture of substances in which the particles of one substance are of greater than molecular size but are stabilized, at least for a time, with respect to forming larger particles or settling under gravity. The particles are referred to as the disperse phase while the other phase is termed the dispersion medium or continuous phase. Smoke is a colloidal **suspension** of solid particles in air. Fog is a colloidal suspension of **water** droplets. Milk is a colloidal suspension of oil droplets in water. Chemists are most often concerned with colloids in which solid or liquid particles are suspended in a liquid. The special properties of colloids result from the large contact area between particle and solvent, which may reach hundreds of square meters per gram.

An ingestible colloidal suspension of **gold** particles in oil was used as ''potable gold'' by medical alchemists. The first systematic studies of inorganic colloids were reported between 1845 and 1850 by Francesco Selmi (1817-1881), an Italian chemist. The term colloid itself was introduced in 1861 by the British physical chemist, **Thomas Graham** to distinguish ''gluelike'' materials which would not pass through a parchment filter from the majority of substances which in **solution** pass through filters with ease. The latter he termed ''crystalloids'' as they could be crystallized from solution. Graham also coined the term **dialysis**, to describe the use of membranes to remove dissolved substances from a colloidal suspension, and the terms sol and **gel** to indicate the fluid and more rigid phase of a colloidal suspension

Colloids are generally classified as either *lyophyllic* (solvent-loving), either *lyophobic* (solvent hating), or as either *association* colloids. When the continuous phase is an aqueous solution, the older terms ''hydrophillic'' and ''hydrophobic'' are also used. Lyophillic colloids are those in which the interaction of the particle surfaces and the solvent is energetically favorable. Aqueous solutions of **proteins** and other macromolecules are colloids of this type. They will form spontaneously when the solvent is added to the dry particles. When water is the solvent, the particles of a lyophillic colloid typically carry a significant surface charge, compensated by ions of the opposite sign in true solution. In lyophobic colloids the particle-solvent interaction is energetically unfavorable and the suspension will sooner or later separate. Lyophobic colloids are often prepared by vigorous agitation. The homogenization of milk is a process of this type. Association colloids are formed in solutions of molecules that include both lyophilic and lyophobic regions. The most important examples are the micelles formed by surfactant molecules in water in which the nonpolar regions of the molecules aggregate together in the center so that only the polar groups are exposed to the surface.The ability of surfactant micelles to accommodate additional oily material is the basis of their detergent action.

Physically, colloids are characterized by light scattering and **microscopy**, and by their electrial properties. Colloidal particles have typical dimensions comparable to or less than the wavelengths of visible light, roughly 500 nm and below. As a result colloids scatter light strongly, a phenomenon known as the *Tyndall effect*, but can not be clearly imaged under the light microscope. The invention of the ultramicroscope by H. F. W. Siedentopf and Richard **Zsigmondy** in 1903 made it possible to image the scattered light well enough to allow the particles to be counted and their **Brownian motion** to st udied. The subsequent invention of **electron** microscopy has allowed the imaging of colloidal particles with the solvent evaporated. Charged colloidal particles will migrate in an applied electric field, a process called **electrophoresis**. The ratio of their **velocity** to the strength of the driving field is called the electrical mobility of the particles, from which one can calculate the zeta potential, the electrostatic potential of the particles at their radius of contact with the continuous phase.

Understanding the interaction between colloidal particles is a **matter** of great theoretical and practical importance. The particles are constantly being brought into contact with each other through Brownian motion and should they adhere to each other would rapidly coagulate into a single **mass**. An ''atmosphere'' of compensating charges ionic will surround each charged colloidal particle. Beyond a distance known the Debye length, the particle and its atmosphere will appear to be electrically neutral. Particles separated by more than a Debye length will attract each other weakly due to the van der Waals force. As the particles move closer, the atmospheres will overlap and the electrostatic repulsion will begin to result in repulsion. If the Debye length is decreased, however, through the addition of additional ions, particularly ions of a higher charge, the colloidal particles may approach closely enough for a bond to form. The **concentration** at which this occurs is known as the critical coagulation concentration (c.c.c.) for the particular **salt** being added and the process, as a whole is sometimes called the ''salting-out'' of the colloid.

The flow characteristics of colloids are also of great interest. A very desirable property in many applications is *thix-*

otropy, in which the material behaves as a gel or very viscous liquid at rest or subject to mild shear, but flows freely when subjected to a larger shear. This property is highly desirable in paints which must be transported on a brush but then flow freely as the brush is moved against a stationary surface.

There are numerous practical applications of colloid science. The **wood** pulp and clays used in papermaking are both colloidal, as are the **inks** used in writing and printing. Colloidal phenomena are key in the separation of minerals from their ores by particle flotation. The colloidal natures of detergents, paint, and milk products were mentioned above. Colloidal phenomena are also the basis for numerous adhesive, cosmetic and other products in widespread use today.

COLOR

When we look out on the world we see a rich tapestry of colors. Colors are not just adornment or luxurious sensory experiences, they also provide important information about the world around us. The region of the **electromagnetic spectrum** that contains light at frequencies and wavelengths that stimulate the eye is termed the visible region of the electromagnetic spectrum. Color is the association the eye makes with selected portions of that visible region.

This distinction is not trivial. Indeed, the colors we ordinarily assign to objects are not directly a part of the object. The oceans and sky are not blue—we perceive them as blue. The blue light that we see in the sky has been selectively filtered from sunlight by the interaction of the sunlight with gaseous and **water** molecules in the atmosphere that differentially (i.e., to a different degree) absorb, transmit, and reflect the wavelengths of light that comprise the visible spectrum.

Colors are associated with specific wavelengths of light. Because the wavelength of light is inversely proportional to the frequency—and hence the energy—of light, colors are also characteristic of a certain frequency and **energy** of the visible spectrum. Although one could relate frequencies to color just as easily as one could relate wavelengths to color, wavelengths are traditionally used to divide the visible region of the electromagnetic spectrum into bands of color. A nanometer (10^{-9}m) is the most common unit used for characterizing the wavelength of light.

The visible portion of the electromagnetic spectrum is located between 380 nm and 750 nm. The component color regions of the visible spectrum are Red (670-770 nm), Orange (592-620 nm), Yellow (578-592 nm), Green (500-578 nm), Blue (464-500 nm), Indigo (444-464 nm), and Violet (400-446 nm). The color region can easily be remembered by use of the acronym ROYGBIV. Red light is longest in wavelength and lowest in energy. Violet light is shortest in wavelength, highest in frequency and highest in energy.

Electromagnetic fields exert forces that excite electrons. Colors of light are related to the effects exerted on the **valence electron** of an excited **atom**. Electrons vibrate (jump between orbitals) only at certain frequencies characteristic for every atom and **molecule**. The energy of light of a particular color

Light shining through a solution of sodium hydroxide (left) and a colloidal mixture. The size of colloidal particles makes the mixture, which is neither a solution or a suspension, appear cloudy. *(Photograph by Charles D. Winters, National Audubon Society Collection/Photo Researchers, Inc. Reproduced with permission.)*

is, therefore, proportional to its frequency. As the frequency of light increases, so does the energy of light associated with that frequency. As a result blue light is more energetic than red light and hot flames from well adjusted Bunsen burners emit blue rather than yellow light.

Atoms and molecules absorb and emit light only at very specific frequencies characteristic of the atom or molecule. Correspondingly, because an atom or molecule absorbs only a limited range of frequencies and wavelengths of the spectrum, it follows that the atom or molecule must reflect all the other frequencies and wavelengths of light. These reflected frequencies and wavelengths, being only a portion of the white-light spectrum, determine the color of the substance made up of the atoms or molecules reflecting the light.

There is a fundamental difference between the color of light that **matter** emits or reflects and the color of light that matter absorbs. In white light, objects that appear blue actually absorb all the other wavelengths of the white light and reflect the wavelengths that the human eye interprets as the color blue. In other words, after reflecting off the object, only the blue light reaches the human eye. Objects in white light (that is, light that contains all the wavelengths of the visible spectrum) that appear white are reflecting all wavelengths of light.

Objects that appear black are absorbing all of the wavelengths of light in the visible portion of the electromagnetic spectrum. A truly black object that did not reflect any light would, therefore, reflect no light to the human eye and thus could only be discerned by the contrast it makes with its background.

Accordingly, if we describe paint as green in color that is because the properties of pigments in the paint allow them to absorb all the other wavelengths of visible light except those that, when reflected to the human eye, are interpreted as a neural signals the mind associates and labels with the word green.

In reality, light that is comprised of one wavelength or hue of light is rare. What is perceived as color is actually an interpretation of a combination or mixture of reflected wavelengths. In addition, the mind labels certain combinations of wavelengths perceived by the eye as ''green'' light. Because of this, colors can be created by mixing other colors—or the pigments in paints that reflect them. A green leaf on a tree, is perceived as green because the atoms of the **chlorophyll** molecules in the leaf are absorbing most of the frequencies of visible light except those the eye eventually labels as green.

Visible light can also be separated into its component colors through other means than atomic absorption and emission. Sir Isaac Newton described a rainbow of colors displayed when sunlight passes through prism or other crystalline substances. Water droplets in the atmosphere produce rainbows of color by refracting sunlight.

Some scientists describe colors as the light-related properties of matter than can be emitted, absorbed, reflected, and refracted (bent and scattered). Other scientists, particularly in the behavioral sciences emphasize the perception-dependent nature of color. Regardless, the properties of color have had a profound influence on the development of art, culture and on human physiology.

See also Planck's constant

COMBUSTION

Combustion is the chemical term for a process known more commonly as burning. It is certainly one of the earliest chemical changes noted by humans, at least partly because of the dramatic effects it has on materials. Today, the mechanism by which combustion takes place is well understood and is more correctly defined as a form of **oxidation** that occurs so rapidly that noticeable **heat** and light are produced.

Probably the earliest reasonably scientific attempt to explain combustion was that of **Johannes Baptista van Helmont**, a Flemish physician and alchemist. Van Helmont observed the relationship among a burning material, smoke, and flame and said that combustion involved the escape of a *spiritus silvestre* (wild spirit) from the burning material. This explanation was later incorporated into a theory of combustion—the **phlogiston** theory—that dominated alchemical thinking for the better part of two centuries.

According to the phlogiston theory, combustible materials contain a substance—phlogiston—that is emitted by the material as it burns. A non-combustible material, such as ashes, will not burn, according to this theory, because all phlogiston contained in the original material (such as wood) had been driven out. The phlogiston theory was developed primarily by the German alchemist **Johann Becher** and his student **Georg Ernst Stahl** at the end of the seventeenth century.

Although scoffed at today, the phlogiston theory satisfactorily explained most combustion phenomena known at the time of Becher and Stahl. One serious problem was a quantitative issue. Many objects weigh more after being burned than before. How this could happen when phlogiston escaped from the burning material? One possible explanation was that phlogiston had negative weight, an idea that many early chemists thought absurd, while others were willing to consider. In any case, precise measurements had not yet become an important feature of chemical studies, so loss of weight was not an insurmountable barrier to the phlogiston concept.

As with so many other instances in science, the phlogiston theory fell into disrepute only when someone appeared on the scene who could reject traditional thinking almost entirely and propose a radically new view of the phenomenon. That person was the great French chemist **Antoine Laurent Lavoisier**. Having knowledge of some recent critical discoveries in **chemistry**, especially the discovery of **oxygen** by **Carl Wilhelm Scheele** in 1771 and **Joseph Priestley** in 1774, Lavoisier framed a new definition of combustion. Combustion, he said, is the process by which some material combines with oxygen. By making the best use of precise quantitative experiments, Lavoisier provided such a sound basis for his new theory that it was widely accepted in a relatively short period of time.

Lavoisier initiated another important line of research related to combustion, one involving the amount of heat generated during oxidation. His earliest experiments involved the study of heat lost by a guinea pig during respiration, which Lavoisier called ''a combustion.'' In this work, he was assisted by a second famous French scientist, Pierre Simon Laplace (1749-1827). As a result of their research, Lavoisier and Laplace laid down one of the fundamental principles of **thermochemistry**, namely that the amount of heat needed to decompose a compound is the same as the amount of heat liberated during its formation from its elements. This line of research was later developed by the Swiss-Russian chemist Henri Hess (1802-1850) in the 1830s. Hess' development and extension of the work of Lavoisier and Laplace has earned him the title of father of thermochemistry.

From a chemical standpoint, combustion is a process in which chemical bonds are broken and new chemical bonds formed. The net result of these changes is a release of **energy**, the **heat of combustion**. For example, suppose that a gram of **coal** is burned in pure oxygen with the formation of **carbon dioxide** as the only product. In this reaction, the first step is the destruction of bonds between carbon atoms and between oxygen atoms. In order for this step to occur, energy must be added to the coal/oxygen mixture. For example, a lighted match must be touched to the coal.

Once the carbon-carbon and oxygen-oxygen bonds have been broken, new bonds between carbon atoms and oxygen atoms can be formed. These bonds contain less energy than did the original carbon-carbon and oxygen-oxygen bonds. That energy is released in the form of heat, the heat of combustion. The heat of combustion of one **mole** of **carbon**, for example, is about 94 kcal.

Humans have been making practical use of combustion for millennia. Cooking food and heating homes have long been two major applications of the combustion reaction. With the development of the steam engine by Denis Papin, Thomas Savery, Thomas Newcomen, and others at the beginning of the eighteenth century, however, a new use for combustion was found: performing work. Those first engines employed the combustion of some material, usually coal, to produce heat that was used to boil **water**. The steam produced was then able to move pistons and drive machinery. That concept is essentially the same one used today to operate fossil-fueled electrical power plants.

Before long, inventors found ways to use steam engines in transportation, especially in railroad engines and steam ships. However, it was not until the discovery of a new type of fuel-gasoline and its chemical relatives and a new type of engine—the internal combustion engine—that the modern face of transportation was achieved. Today, most forms of transportation depend on the combustion of a **hydrocarbon** fuel such as **gasoline**, **kerosene**, or diesel oil to produce the energy that drives pistons and moves the vehicles on which modern society depends.

When considering how fuels are burned during the combustion process, "stationary" and "explosive" flames are treated as two distinct types of combustion. In stationary combustion, as generally seen in gas or oil burners, the mixture of fuel and oxidizer flows toward the flame at a proper speed to maintain the position of the flame. The fuel can be either premixed with air or introduced separately into the combustion region. An explosive flame, on the other hand, occurs in a homogeneous mixture of fuel and air in which the flame moves rapidly through the combustible mixture. Burning in the cylinder of a gasoline engine belongs to this category. Overall, both chemical and physical processes are combined in combustion, and the dominant process depends on very diverse burning conditions.

The use of combustion as a power source has had such a dramatic influence on human society that the period after 1750 has sometimes been called the Fossil Fuel Age. Still, the widespread use of combustion for human applications has always had its disadvantages. Pictorial representations of England during the Industrial Revolution, for example, usually include huge clouds of smoke emitted by the combustion of **wood** and coal in steam engines.

Today, modern societies continue to face environmental problems created by the prodigious combustion of carbon-based fuels. For example, one product of any combustion reaction in the real world is **carbon monoxide**, a toxic gas that is often detected at dangerous levels in urban areas around the world. Oxides of **sulfur**, produced by the combustion of impurities in fuels, and oxides of **nitrogen**, produced at high **temperature**, also have deleterious effects, often in the form of **acid rain** and smog. Even **carbon dioxide** itself, the primary product of combustion, is suspected of causing global climate changes because of the enormous concentrations it has reached in the atmosphere.

Composites

Composite materials consist of two or more distinct materials with a recognizable interface between them. The term composite is usually reserved for materials in which the distinct phases are separated on a scale larger than atomic, and in which the composite's mechanical properties have been significantly altered from those of the constituent materials.

Nature has been creating composites for millions of years. Natural composites include **wood** and bone. Wood is a composite of **cellulose** and lignin. Cellulose fibers are strong in tension and are flexible. Lignin cements these fibers together to make them stiff. The toughness and strength of bone arises from the dispersal of hard plate-like crystals of hydroxyapatite (mineral) in a soft matrix of collagen fibers (protein). As a composite, the resulting microstructure yields physical properties that even synthetic materials have been unable to match.

For the past several decades, scientists, taking their cue from nature, have been designing new materials with composite structures whenever there is a need to rectify the weakness in one material by the strength in another. Composites tend to be developed when no single, homogeneous material can be found that had all of the desired characteristics for a given application. Specific applications for composites now include structural materials such as concrete, high performance coatings, catalysts, electronics, photonics, magnetic materials, and biomedical materials, grinding wheels, underground electrical cables, superconducting ribbons, and ceramic fiber composites.

The discontinuous filler phase in a composite is usually stiffer or stronger than the binder phase. There must be a substantial **volume** fraction of the reinforcing phase present to provide reinforcement. Examples do exist, however, of composites where the discontinuous phase is more compliant and ductile than the matrix.

Particle-reinforced composites consist of particles of one material dispersed in a matrix of a second material. The particles may have any shape or size, but they are generally spherical, ellipsoidal, polyhedral, or irregular in shape.

Fiber-reinforced materials are typified by fiberglass in which there are three components: **glass** filaments (for mechanical strength), a polymer matrix (to encapsulate the filaments), and a **bonding** agent (to bind the glass to the polymer).

The mechanical properties of composite materials usually depend on the composite's structure. Thus these properties typically depend on the shape of inhomogeneities, the volume fraction occupied by the inhomogeneities, and the interfaces between the components. The strength of composites depends on such factors as the brittleness or **ductility** of the inclusions and matrix.

High-performance composites often exhibit better performance than conventional structural materials such as steel and **aluminum** alloys. Such composites are almost all continuous fiber-reinforced composites, with organic (resin) matrices. Fibers used in high performance composites include glass fibers, **carbon** fibers, aromatic polyamide fibers, **boron** fibers, silicon-carbide fibers, and aluminum **oxide** fibers. Matrices include epoxy resins, bismaleimide resins, polyimide resins, and thermoplastic resins.

See also Aromatic hydrocarbons; Fibers, natural; Fibers, synthetic; Intermolecular forces; Magnetism; Metals; Minerology; Oxirane functional group; Paints and coatings; Polymer chemistry

Compound microscope

Microscopes have been in use in various forms for more than 3,000 years. The first types were extremely simple magnifiers made of globes of water-filled **glass** or chips of transparent crystal. Ancient Romans were known to use solid, bead-like glass magnifiers; Emperor Nero (A.D. 37-68) often used a bit of cut emerald to augment his poor vision.

The first lenses, which were used in primitive eyeglasses, were manufactured in Europe and China in the late thirteenth century. By this time, lenscrafters realized that most clear glass or crystal could be ground into a certain shape (generally with the edges thinner than the center) to produce a magnifying effect. All of these single-lens magnifiers are called simple microscopes.

Until the turn of the seventeenth century, most simple microscopes could provide a magnification of 10 power (magnifying a specimen to ten times its diameter). About this time, the Dutch draper and amateur optician Antoni van Leeuwenhoek (1632-1723) began constructing magnifying lenses of his own. Though still relying upon single lenses, Leeuwenhoek's unparalleled grinding skill produced microscopes of very high power, with magnifications ranging to 500 power.

In order to achieve such results, Leeuwenhoek manufactured extremely small lenses, some as tiny as the head of a pin. Because of the very short *focal length* of these lenses, the microscope had to be held a fraction of an inch away from both the observed specimen and the observer's eye. Through his minute lenses Leeuwenhoek observed tiny ''animalcules''—what we now know as bacteria and protozoa—for the first time. His findings earned him international acclaim, and the simple microscopes he designed are still among the best-crafted. However, the limitations of the single-lens magnifier were apparent to scientists, who labored to develop a practical system to increase microscope magnification.

The next breakthrough in **microscopy** was the invention of the **compound microscope**; however, the origin of this device, as well as the identity of its inventor, is the subject of some debate. Generally, credit for the invention of the compound microscope has been given to another Dutchman, the optician Zacharias Janssen (1580-c. 1638). Around 1590, Janssen reportedly stumbled upon an idea for a multiple-lens microscope design, which he then constructed. Though he affirmed its ability, no record exists of Janssen actually using his invention. Currently, it is believed that Janssen's son fabricated the story. Meanwhile, yet another Dutch-born scientist, Cornelius Drebbel, claimed that he had constructed the first compound microscope in 1619. Galileo also reported using a two-lens microscope to examine and describe the eye of an insect.

Regardless of its inventor, the design of the original compound microscope is very similar to those used today. Two or more lenses are housed in a long tube. Individually, none of the lenses are particularly powerful; however, the image produced by the first lens is further magnified by the second (and the third and fourth, in a multiple-lens system), producing a greatly enlarged image. In addition, multiple lenses allow for a much longer focal length, permitting both the specimen and the eye a greater distance from the lenses.

The first scientist to further improve the compound design was the Englishman Robert Hooke in the 1670s. Hooke was the first to use a microscope to observe the structure of plants, consisting of tiny walled ''chambers'' that he called cells. After Hooke, little advancement occurred in microscopy until the work of Carl Zeiss (1816-1888) and Ernst Abbe in the mid-1800s. Abbe, generally recognized as the first optical engineer, took over the design duties at the Zeiss Optical Works in 1876. The scientific instruments that resulted from Zeiss and Abbe's collaboration set new standards for optical equipment. Among their inventions were lenses that corrected blurring and **color** aberrations.

By the twentieth century, the essential design and shape of the compound microscope had evolved into those we know today. Microscopes used in schools and small laboratories can achieve magnification of up to 400 power. More advanced microscopes used in research laboratories can magnify a specimen to almost 1000 power. These research microscopes often have binocular eyepieces, relying upon a series of prisms to split the image so that it may be viewed with both eyes. Even trinocular microscopes—creating a third image for a camera to view—have been designed.

At its most powerful, the practical limit for any compound microscope is 2,500 power. This limited magnification capability frustrated scientists, who, in the early twentieth century, were anxious to view the world on submicroscopic and subatomic levels. In 1931, Ernst Ruska constructed the **electron** microscope to permit such investigations.

Designed much like a compound microscope, the **electron microscope** uses a beam of electrons focused through magnetic lenses. Since electrons have much smaller wavelengths than does visible light, the electron microscope can provide much higher magnification than light-based instruments. Through electron microscopes, scientists first viewed strands of **DNA**. Since Ruska's invention, instruments like the **scanning tunneling microscope** and the **field ion microscope** have been developed with the ability to observe the activities and structures of individual atoms.

Compressibility

The *isothermal compressibility* of a material is defined as the fractional decrease of **volume** per unit increase of pressure, at constant **temperature**. This is the compressibility usually employed in thermodynamic calculations.

Sometimes, as when considering the propagation of sound waves, the *adiabatic compressibility* is required. This quantity is defined as the fractional decrease of volume per unit increase of pressure, when no **heat** flows in or out of the system. Because there is no heat flow, the **entropy** remains unchanged according to the second law of **thermodynamics** (in a reversible process), so the adiabatic process is also isentropic, i.e., a process that takes place at constant entropy. In adiabatic compression, the temperature rises, so the pressure increases more sharply than in isothermal compression. Therefore, the compressibility at constant entropy is always smaller than that at constant temperature.

It follows from thermodynamic relationships that the ratio of the isentropic (adiabatic) compressibility to the isothermal compressibility is equal to the ratio of the heat capacity at constant volume to the heat capacity at constant pressure.

In the case of an ideal gas, the pressure-volume-temperature relations are governed by the kinetic **energy** of the gas particles. For denser **gases** and **liquids**, it is necessary to take **potential energy** effects arising from atomic interactions into account, and the pressure-volume-temperature relations tend to be very complex. For crystals, the pressure-volume-temperature relationships depend on the binding forces in the crystal. In ionic crystals such as **sodium** chloride, the compressibility typically depends on the distance of closest approach of the ions in the crystal under equilibrium conditions, and on the repulsive interactions between ions. In the general case, the calculation of the compressibility of **solids** reduces to a quantum mechanical problem.

Dilute gases obey the ideal gas law very closely, so the isothermal compressibility is equal to the gas volume divided by the gas constant times the temperature (V/RT); the isentropic compressibility is equal to the isothermal compressibility times the heat capacity at constant volume divided by the heat capacity at constant pressure. However, in the case of *compressed* gases, the compressibility at high densities falls to a small fraction of the value predicted for the ideal gas.

According to kinetic molecular theory, pressure arises from the impact of gas molecules on container walls. If a gas is compressed at constant temperature, the speed and force of impact of the molecules on the wall remain the same, but the number of collisions per unit area with the container increases. If the gas is compressed adiabatically, however, the heat of compression is not lost, and average molecular speed and force at impact on the walls of the vessel increase as well, so the compressibility is smaller than in the case of isothermal compression.

A related quantity, the bulk modulus, is defined as the reciprocal of the compressibility.

See also Crystallography; Intermolecular forces; Quantum chemistry; Thermochemistry

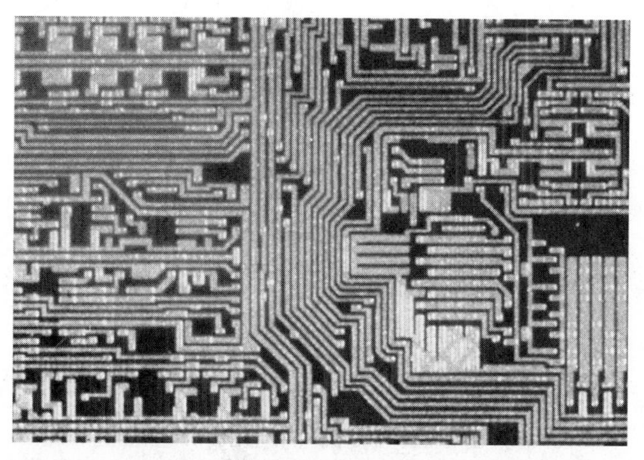

A light micrographic image of a portion of a silicon wafer microchip. The rows of electronic components that make up the integrated circuit are imprinted into the silicon wafer by various deposition and etching processes. ***(Photograph by Astrid and Hanns-Frieder Michler/ Science Photo Library, Photo Researchers, Inc. Reproduced by permission.)***

COMPUTER CHIPS

The building block of the computer revolution in the 1970s and the basis of today's multibillion-dollar computer industry is the computer chip. Usually about one-third of an inch square, these chips are the ''brains'' critical to many sophisticated and not-so-sophisticated devices such as calculators, appliances, televisions and radios. They also have great importance in areas of environmental monitoring, medicine, automobiles and other vehicles, and traffic control. They have greatly reduced the size and cost of most electronic products, while at the same time increasing their power and versatility.

Computer chips are integrated circuits called microprocessors built up from transistors and other components. Transistors are small binary electrical switches. Transistors have no moving parts and are switched between their two allowed states, ''on'' and ''off'', electronically. Microprocessors are fabricated on a single crystal of a semiconducting material such as **silicon** or **germanium**. A semiconductor is a substance whose ability to conduct **electricity** is between that of an insulator like rubber and a full conductor like **copper**. Often the conducting properties of a semiconductor can be varied by adding an impurity known as a dopant so that a semiconductor can be made to act like either an insulator or a conductor. A single chip's microscopically small components, including transistors, capacitors, and resistors, can number more than 1 million for an ''ultra large-scale integration,'' or just 100 for a ''small-scale integration.''

There are two main kinds of chips. The first, called central processing unit (CPU) chips, give computers their information-processing power. An example of a CPU chip is Microsoft's Pentium III. Most personal computers (PCs) and small networks only require a single chip for their purposes, but larger systems link many chips together to support their greater demands. The second kind of chip, memory chips, come in many different configurations, ranging from those with only a few kilobytes of memory to those with 10 megabytes. The power of a computer's memory chip affects how fast the machine runs, because as the CPU processes instructions for the machine to carry out, it needs to store that information somewhere. The memory chip is that storage area, so the larger this space is, the more rapidly and efficiently the CPU can work.

Computer chips are manufactured in an industrial process that depends upon precision and extreme cleanliness, since trace contamination can ruin a complex chip's design. The process has analogy to children's stencils; masks are used to cover portions of a chip and desired patterns are formed on it.

The first step is to prepare a single crystal of an ultrapure material such as a silicon wafer that has a thin silicon dioxide surface. The wafer is then covered with a substance called a photoresist. A photoresist is a substance, which becomes soluble when exposed to ultraviolet (UV) light. The photoresist is in turn covered with a mask, or stencil, and then strong UV light is projected onto the assembly. Many different masks are used to control the end properties of the computer chip. Portions of the chip's photoresist that are not covered by the mask become soluble upon irradiation and those that are do not. The soluble photoresist is removed by washing with solvent and the resulting pattern can be etched onto the chip permanently by removal of the revealed silicon dioxide surface. Either chemicals (''wet etching'') or corrosive **plasma gases** perform the etching in a vacuum chamber. The process of producing the patterns forming the basis for the chip's circuitry through the use of masks, chemicals, and UV light is known as photolithography.

Removal of the remaining photoresist (that which was not exposed to the UV light since part of the mask covered it) gives a wafer with a silicon dioxide surface with a controlled pattern etched out of it to reveal a polysilicon pattern. Another layer of polysilicon can then be applied and the etching procedure repeated to make a more elaborate assembly.

A ''doping'' procedure is used to controllably introduce impurities into the exposed polysilicon and change its crystalline structure. Ionized forms of the impurities are showered onto the chip and ions become implanted into the silicon. The portions of the chip still covered by the silicon dioxide surface are not affected in the doping process. Thus, the doping happens only where the silicon dioxide surface has been removed during the etching process. These impurities modulate the silicon's ability to conduct electricity (conductivity). Introduction of **boron** impurities results in electron-deficient or ''positive'' silicon; introduction of phosphorous impurities results in electron-rich or ''negative'' silicon.

The microprocessor is completed after multiple layering and further processing; ''windows'' are left in some layers so that connections can be made between surfaces as necessary. Metal is then deposited into the windows and more layers are formed. The resulting metal strips are the electrical connections between layers. A twenty-layered microprocessor is not uncommon.

The individual components of the microprocessor perform the functions of switching, amplifying, and detecting electricity running through the chip. Since microprocessors are rather fragile, they must be inserted into protective packaging. The completed chip then goes on a small plastic mount with **gold** wires linking its components to metal pins. These pins plug into the circuit board of the product for which the chip is intended.

CONCENTRATION

The proportional amount of a component present in a mixture of substances is given by its concentration. In a liquid **solution**, one of the components is designated the solvent, while all other components are called solutes. The solvent is either the only liquid component of the solution or the liquid in the highest concentration. In aqueous solutions, **water** is the solvent and all dissolved substances are solutes. The concentration of a solute in a liquid solution may be given in the following ways:

1. Percentage composition on weight basis: the weight of solute divided by the total weight of all substances in the solution, including the solvent, multiplied by 100.
2. Molarity: the number of moles of the solute present in 1 liter of solution. The **molarity** is usually designated by a capitol M. After percentage composition, molarity is the most common way to express concentration.
3. **Mole** fraction: the number of moles of the solute present divided by the total number of moles of all substances in the solution, including the solvent.
4. Molality: the number of moles of the solute present in 1000 grams of solvent. The **molality** is usually designated by a lower case m.
5. Formality: the number of formula weights of the solute in a liter of solution. Formality is usually designated by F.
6. Normality: the weight of a solute that would react with 8 grams of **oxygen** (or one equivalent of another oxidant) or with 1 gram of **hydrogen** (or one equivalent of another reductant) present in one liter of solution. It is usually designated by N.

The concentration in other mixtures, such as a mixture of **gases**, may be expressed in a similar manner to that in solutions.

CONDENSATION REACTION

A condensation reaction is a reaction where two molecules join together and in doing so create a complex **molecule** and a much simpler molecule that is a byproduct of the reaction. The simple molecule eliminated can be **water**, **ammonia**, or an **alcohol**. The two combining molecules each contribute a single moiety (a part of portion of a molecule having a characteristic chemical property) to the eliminated molecule.

The formation of a peptide bond between two **amino acids** is an example of a condensation reaction, as is the manufacture of an ester (esterification).

Certain types of condensation reactions have their own name, for example the Claisen condensation. In this reaction, an ester joins with another ester, a ketone, or a nitrile. This must be carried out in the presence of **sodium** ethoxide that acts as a catalyst for the reaction. This type of reaction gives as its product a beta ketonic ester, ketone, or nitrile (depending upon the starting compounds) and an alcohol. The Claisen reaction is important in cell biology as it is this reaction that starts the acetyl group of acetyl CoA into the **citric acid** cycle (the Krebs or tricarboxylic acid cycle). The **Krebs cycle** is part of the **energy** pathway in all living cells.

A catalyst is usually, but not always, required. Catalysts commonly encountered in condensation reactions include acids, bases, cyanide, and complex metal ions.

If the molecules reacting together are identical then the reaction is known as self condensation. A condensation reaction can also occur between two separate parts of the same molecule.

Aldehydes, **ketones**, esters, alkynes, amino acids, **nucleic acids**, and amines are examples of several organic compounds that are capable of combining with each other and, with the exception of amines, themselves. Many of these compounds are important intermediate steps in organic synthesis reactions and in the manufacture of biologically important molecules.

If the condensation reaction process is repeated many times a condensation polymerization reaction is said to be taking place. This is the type of reaction used to form polyesters such as terylene and **nylon**. Both of these molecules contain the results of approximately 100 separate condensation reactions. This multiple condensation reaction is sometimes called a polycondensation reaction. This is the type of reaction that is used to build large organic polymers.

The first synthetic polymer produced by a condensation reaction was Bakelite. Bakelite is produced from phenol and **methanol** and the eliminated product is water. Bakelite has been used extensively as an electrical insulator.

Condensation reactions occur with the general formula of: $XOH + HOY = XOY + H_2O$, where X and Y may be the same or different. This example gives water as an elimination product, but as has been stated other small, simple molecules may be manufactured instead.

The opposite of a condensation reaction is **hydrolysis**. In hydrolysis water is added to a compound to break it down.

Condensation reactions are an important class of reactions. They are used extensively in the manufacture of many commercially important products as well as being the driving force behind many of the processes necessary for life itself.

CONFORMATION

Many of a macromolecule's fundamental characteristics are determined by that molecule's degree of polymerization, but the molecule's physical structure also makes a significant contribution to its macroscopic properties.

The terms configuration and conformation are frequently, and mistakenly, used interchangeably to describe the geometric structure of a **macromolecule**. More correctly, a macromolecule's configuration is that part of its structure that is determined by its chemical bonds. A macromolecule's configuration cannot be altered without breaking and reforming chemical bonds. Conformation, on the other hand, refers to structural effects that arise from the rotation of molecular segments about single bonds. But the two terms are related, as shown below.

There are two types of configurations: cis and trans. The cis configuration arises when substituent groups are located on the same side of a carbon-carbon double bond. The trans configuration describes substituents on opposite sides of the double bond. These structures cannot be changed by any physical means, e.g., rotation, short of breaking and reforming chemical bonds.

Stereoregularity is used to describe the configuration of macromolecular **chains**, in which three distinct structures are possible: In an isotactic configuration, all substituent groups are located on the same side of the macromolecular chain; the substituent groups of a syndiotactic polymer are located on alternating sides of the chain; and an atactic polymer has a random placement of substituent groups.

If two atoms are joined by a single bond, rotation may occur about the bond because there is no need to break the bond for it to occur, which is different from the case of a double bond. When an **atom** rotates about a single bond relative to the atom with which it is joined, there is an adjustment of torsional angle. If the two atoms are bonded to other atoms or groups, those configurations that vary in torsional angle are referred to as *conformations*. Because, in general, each conformation corresponds to a different set of interactions between neighboring atoms or molecular groups, different conformations usually correspond to different potential **energy** states for the **molecule**.

In the case of carbon-carbon single bonds in a macromolecule such as **polyethylene**, there are three possible conformations. The energy barriers separating these three conformational states are much greater than thermal energy fluctuations, so the time spent in each conformational state will be longer than the time of a thermal vibration.

Unlike polyethylene, a protein molecule contains many molecular groups that interact with each other, either on the same or on adjacent molecules. Interactions such as those between **hydrogen** bonds and amido-hydrogen atoms and carbonyl-**oxygen** atoms, forces between charged groups, and solvent-polymer effects may all contribute to the conformation of a protein molecule. Hydrogen **bonding** gives rise to two familiar protein conformations in biological systems: planar sheets and the alpha helix. **X-ray diffraction** has proven an important technique for determining the conformations of protein molecules.

See also Amide group; Carbonyl group; Crystallography; Polymer chemistry

CONJUGATION

Conjugation is the term used to describe an arrangement of chemical bonds in which two double bonds are separated by one single bond. Chemical structures with this configuration are very stable and can show unusual behavior during the course of a chemical reactions. This effect results from the ability of the electrons in conjugated bonds to delocalize, or spread their effect throughout the **molecule**. Two important examples of a conjugated system are isoprene and butadiene, both of which are important components used in the manufacture of synthetic rubber.

In molecules with conjugated bonds, saturated bonds (carbon atoms with four attached atoms) alternate with unsaturated bonds (carbon atoms with less then four attached atoms.) This bond arrangement is possible between atoms whose p-orbital electons are capable of interacting. (The term ''p-orbital'' describes how the electrons orbit the **nucleus** of the **atom**.) For example in 1,3-butadiene, the unsaturated bonds can interact or overlap with one another. This interaction is maximized when the electron's bonds overlap in a single plane. This overlap results in stabilization of the system and lowers the **reactivity** of the molecule. For this reason, conjugated double bonds in 1,3-butadiene are less reactive than the isolated one in 1-butene. Not only is the conjugated system less reactive, but the chemical transformations it undergoes are somewhat different than nonconjugated systems.

Conjugated systems also interact with light waves in a characteristic manner. The overlap of the p orbitals results in a narrowing of the **energy** gap between the highest energy filled electronic orbital and lowest unfilled electronic orbital. Consequently, light absorption occurs at lower energy in conjugated systems than in nonconjugated systems. This absorption corresponds to electronic excitation from the highest filled molecular orbital to lowest unfilled orbital. The more conjugated p-bonds present in a molecule, the lower the energy of the lowest energy transition. When eight double bonds are in conjugation, the molecule absorbs visible light and is colored.

Conjugation can also occur between other p-systems such as triple bonds between **carbon** atoms as well as bonds between carbon and other atoms. It is important to note that overlap can occur between like or unlike p-systems and it is not restricted to organic systems (i.e., molecules containing carbon atoms.) When conjugation occurs in planar monocyclic carbon rings, the total number of p-electrons fits the formula 4n+2 (where n = a nonnegative integer), and conjugated orbitals form an uninterrupted cyclic array, the system has a special stability called aromaticity. Aromatic systems are an important topic in their own right and have a dramatically different chemical reactivity than nonaromatic systems. A representative aromatic species is **benzene**. Benzene is planar and has an uninterrupted cyclic array of three conjugated double bonds arranged in a monocyclic six membered **ring**. The stability of benzene is even greater than nonaromatic conjugated species, which have 6 p-electrons. An example of a similar nonaromatic system is 1,3,5-hexatriene. The chemical reactivity of benzene is also rather different than that of 1,3,5-hexatriene.

While conjugation is most common for systems with alternating single bonds and unsaturated bonds, it can occur whenever the p-bonded systems have a spatial arrangement that allows the orbitals to overlap. Thus conjugation can occur in molecules in which the unsaturated sites are close in space but are separated by more than a single **covalent bond**.

Unsaturated systems are also able to overlap or conjugate with unfilled or partially filled orbitals. For example, the allyl **cation** has a cationic (i.e., positively charged) center conjugated with a double bond and the species is more stable than nonconjugated cations. Likewise, a radical center can be conjugated with a p-system resulting in a specially stabilized radical such as the allyl radical.

While conjugation is most often associated with p systems, other bond orbitals can also exhibit this phenomenon. For instance, electrons in the s orbital can be conjugated with p- orbitals when in the proper spatial relationship. For carbon systems, this is known as hyperconjugation. A typical form is the overlap or conjugation of the electrons of a s-bond with an adjacent radical or cationic center. The more of these interactions there are, the more stable the radical or cation. Hyperconjugation also accounts for why the more alkyl substituents an alkene bears, the more stable it is upon comparison to similar less substituted alkenes. The stability associated with hyperconjugation is generally not as great as that associated with normal conjugation.

Another form of conjugation is exhibited by heavy group 14 elements such as **silicon**, **germanium**, and **lead**. When linked together in a chain the atoms in these atoms can form delocalized systems through conjugation of neighboring s-bonds. This arrangement is known as s-conjugation. One consequence of s-conjugation is that certain species of silicon and germanium strongly absorb energy in the near ultraviolet range. These absorption bands correspond to electronic transitions from s-bonding to s-antibonding orbitals. Bicyclic **tin** compounds are known that also exhibit s-conjugation; as a consequence, some of these tin compounds are even colored.

CONSERVATION OF ENERGY

Energy is defined as the capacity to do work. The English physicist and physician Thomas Young (1773-1829) was the first person to use the term energy in this sense. The concept of energy unites almost all branches of science through its various manifestations (light, **heat**, atomic and subatomic behavior, etc.). Energy can be converted from one form to another and the total energy in any closed system remains constant. In classical physics, this principle was known as conservation of energy; in modern physics, it is termed the **conservation of mass** and energy.

In classical physics, there are two types of energy: kinetic and potential. (Modern physics recognizes a third type, rest-mass energy.) When two moving objects collide, the first object may bounce off the second in a new direction but with no change in speed, or it may slow down, or it may be blown to pieces. The work performed by a moving object in coming to rest is known as kinetic energy. The larger an object is and the faster it moves, the greater will be its kinetic energy. Mathematically, the kinetic energy of a moving object is equal to one-half the product of its mass times the square of its **velocity**.

An object at rest near the edge of a cliff has **potential energy**. If the object falls toward the bottom of the cliff, the object's potential energy is converted into kinetic energy. The magnitude of the object's kinetic energy on impact at the bottom of the cliff will be equal to whatever work was required to raise it to its position at the top of the cliff from the ground.

The action of a pendulum provides an excellent example of the interconversion of potential and kinetic energy. The potential energy achieves maximum value when the pendulum is at either extreme of its swinging motion; it decreases to zero when the pendulum is vertical to the ground. The kinetic energy is zero when the pendulum is at its extreme position, and maximum when the pendulum is in the vertical position.

Another example is provided by the toss of a ball into the air. The ball's kinetic energy gradually decreases as it rises, and eventually reaches zero (having been completely transformed to potential energy) at the point that the ball stops ascending and begins descending. The fact that the ball regains its kinetic energy while descending can be taken as evidence that potential energy can be converted back to kinetic energy.

In the Sun, the process of thermonuclear fusion converts atoms of **hydrogen** into **helium** atoms, producing radiant energy. Some of this radiant energy reaches the Earth. Some of this energy reaches plants, which may eventually form **coal** or be used by other organisms dependent on plants for food. Part of the solar energy contributes to the evaporation of ocean **water**, which returns to Earth in the form of rain. And this rainwater may be converted into hydroelectric energy by generating plants.

In mechanical systems, there is always a certain amount of energy lost as heat due to frictional processes and inelastic collisions between moving parts. The notion that heat is a form of energy derives in part from observations in an eighteenth century arsenal that the mechanical energy expended in boring a cannon is roughly equivalent to the heat produced in the process. The discovery that heat is a form of energy led to the formulation of the first law of **thermodynamics**, i.e., that energy can neither be created nor destroyed, only converted into another form, which is a restatement of the law of conservation of energy.

As science evolved, other forms of energy were recognized. When **James Clerk Maxwell** succeeded in formulating the laws of electromagnetism, it became possible to recognize electrical energy as yet another manifestation of energy.

The discovery of the **neutrino** was predicted on the basis of the principle of conservation of energy. When the energy changes accompanying the decay of certain radioactive nuclei were not found to properly **balance**, physicists speculated that an unseen particle must be responsible for the discrepancy. It was 30 years before scientists were able to verify the existence of this particle, but eventually it was found.

CONSERVATION OF MASS

Although Flemish chemist Jan van Helmont (1579-1644) and British chemist **Robert Boyle** stated versions of a law of indestructibility of **matter**, it is French chemist **Antoine Lavoisier** who is usually given the credit for first clearly expressing and proving the law of conservation of **mass**. He performed a series of careful experiments and quantitative measurements that showed clearly that there is no change in total mass in a chemical reaction. English chemist **John Dalton** explained this observation by proposing (in 1803) that chemical compounds are made up of combinations of elements and that these elements consist of indivisible, indestructible atoms. In a chemical reaction, atoms are not destroyed nor are they changed into atoms of other elements. Atoms are simply rearranged in the product compounds. Since the products contain the same atoms as the reactants, mass is neither destroyed nor created in a chemical process.

With the discovery of the **electron** by **J. J. Thomson** in 1897 and of the **nucleus** by **Ernest Rutherford** in 1911, it became evident that atoms are, in fact, not indivisible. Atoms are made up of smaller particles: electrons, protons, and neutrons. In 1905 **Albert Einstein** proposed that in very high-energy processes the mass of these **subatomic particles** may be converted into **energy**. Subsequent experimental evidence proved his proposal to be correct. In these processes, the law of conservation of mass is not obeyed. Since, however, these high-energy transformations do not occur in ordinary chemical reactions, chemists may assume that the law of conservation of mass is followed in the processes with which they are normally involved.

COOLIDGE, WILLIAM D. (1873-1975)

American physical chemist

William D. Coolidge was an American physical chemist and inventor who made important contributions to three areas of twentieth-century technical development. He developed a pliable filament that improved the electric light, making it possible to mass produce the electric light bulb for commercial, industrial, and personal use. He developed X-ray technology so that it was possible to adapt X-ray techniques to many areas of medicine and technology. He also played a role in the development of the atomic bomb used by the United States at the end of World War II.

William David Coolidge was born on a farm in Hudson, Massachusetts, on October 23, 1873, to Albert Edward Coolidge, a shoe factory worker, and Martha Alice Shattuck. As a young boy, Coolidge showed an ability for electrical and mechanical projects. This ability was evident from the fact that he spent much of his spare time in a local machine shop. To help ease his family's economic difficulties, he temporarily left high school during his junior year to take a job in a rubber factory. He graduated from Hudson High School in 1891 after a six month delay.

A state scholarship and a loan from a friend's father made it possible for Coolidge to enter the Massachusetts Institute of Technology in 1891 to study electrical engineering. On account of illness, his graduation was delayed until 1896, when he also was appointed to the position of assistant in physics at M.I.T. His debts at the time amounted to $4,000, a large sum at that time, and he was not able to pay off the debt until he was in his early thirties.

In 1897, Coolidge received a graduate fellowship from the University of Leipzig, where he earned his Ph.D. in 1899. On his return to the United States, Coolidge did research in physical **chemistry** at M.I.T. for five years. He was an instructor from 1901 to 1903, and from 1904 to 1905 he was an assistant professor. In 1905, his former professor of chemistry at M.I.T., Willis R. Whitney, invited Coolidge to become a researcher in physico-chemistry at the General Electric Company in Schenectady, New York, where Whitney had become director of the General Electric laboratory in 1900.

Although the General Electric laboratory was expanding with the company's growth after the 1892 merger of the Edison General Electric Company with the Thomson-Houston Electric Company, Coolidge hesitated to accept the position offered to him because of doubts he had about the nature of the problems on which he would be working in industry. A promise was made to him that he could devote as much time as he wanted to conductivity measurement research (which had engaged his interests at M.I.T.) and Coolidge accepted the job. His early association with General Electric proved to be fruitful both in terms of the research he was able to do and in terms of his career status. He was appointed assistant director in 1908, associate director in 1928, director in 1932, and became vice president and director of research in 1940 before retiring as director emeritus in 1944.

During his long association with the General Electric Company, Coolidge developed a pliable form of **tungsten** called ductile tungsten, which could be drawn into fine wires only a sixth of the diameter of a human hair. Earlier, Thomas Edison had used **carbon** fibers, but they had proved too brittle to handle. Coolidge patented his technique in 1909. This technique is still used today in the production of incandescent light bulbs, and it has also proven useful in the development of radio tubes, **electron** tubes, contacts, and high-power water-cooled transmitting tubes.

Coolidge's work with the electric light was successful by 1911, and by 1913, he began work on the X-ray tube. He was able to utilize the ductile tungsten he had developed for light bulbs to produce X-ray tubes that were accurate and stable. His version of the X-ray tube became known as the ''Coolidge tube ,'' and these tubes found wide use in the industrial and medical communities. By 1917, during World War I, he produced a portable X-ray generating outfit and a C-Tube that could be used as a listening device for submarines and for underwater signaling purposes.

In 1924, Coolidge developed an oil-immersed self-contained X ray that could be handled safely by technicians, dentists, and others. Significant applications of Coolidge's X-ray tube are found in cancer treatments and in the industrial application of gauging the thickness of metal as it is being processed. In 1928, Coolidge added to his contributions by explaining the cold-cathode effect that limits how much voltage can be used on a given tube. The Coolidge cathode-ray tube beam is also able to cause some chemicals to undergo changes and to destroy germs and insects rapidly.

From 1932 until his retirement in 1944, Coolidge served mainly in an advisory capacity at General Electric. During World War II, he was enlisted to help study the value of **uranium** for military use and was involved in the development of the atomic bomb. Shortly after the war, he was engaged in nuclear research at the Hanford Engineer Works in Richland, Washington. His work was considered important in so far as he could bring expert an engineering sense to the project.

During his lifetime, Coolidge held eighty-three patents and wrote numerous articles, beginning with ''Ductile Tungsten'' and ''A Powerful X-Ray Tube with a Pure Electron Discharge,'' and going on to such titles as, ''Measurement of the Dielectric Constant of **Liquids**,'' ''Development of Modern X-Ray Generating Apparatus,'' and ''The Production of X-Rays of Very Short Wave-Length.'' His many awards included the Rumford Medal, the Edison Medal, the John Scott Award, the Faraday Medal of the Institution of Electrical Engineers of England, and the Duddell Medal of the Physical Society of England.

Coolidge was married to Ethel Westcott Woodard in 1908. They had two children: a daughter, Elizabeth, and a son, Lawrence David. Ethel Coolidge died in 1915, and Coolidge married Dorothy Elizabeth MacHaffie in 1916. He died in Schenectady on February 3, 1975, at the age of 101.

Cooper, Thomas (1759-1839)

American scientist and educator

English-born American scientist and educator Thomas Cooper was also a controversial political pamphleteer.

Thomas Cooper was born in Westminster, England, on October 22, 1759. He studied at Oxford but failed to take a degree. He then heard anatomical lectures in London, took a clinical course at Middlesex Hospital, and attended patients briefly in Manchester. Having also qualified for the law, he traveled as a barrister, engaged briefly in business, and dabbled in philosophy and **chemistry**.

Being a materialist in philosophy and a revolutionist by temperament, Cooper believed that the English reaction against the French Revolution proved that freedom of thought and speech was no longer possible in England; in 1794 he emigrated to the United States with the scientist **Joseph Priestley**. He settled near Priestley at Northumberland, Pa., where he practiced law and medicine and began writing political pamphlets on behalf of the Jeffersonian party. In 1800, Cooper was jailed and fined under the new Alien and Sedition Acts.

After Thomas Jefferson's election to the U.S. presidency, Cooper served as a commissioner and then as a state judge, until in 1811 he was removed on a charge of arbitrary conduct by the Pennsylvania Legislature. Driven from politics, Cooper was elected to the chair of chemistry in Carlisle (now Dickinson) College and then served as professor of applied chemistry and **mineralogy** at the University of Pennsylvania until 1819. The following year (when clerical opposition denied him the chair Jefferson had created for him at the University of Virginia) Cooper became professor of chemistry in South Carolina College (now University of South Carolina). Elected president of the college, he maintained his connection with it until 1834.

Cooper served mainly as a disseminator of scientific information and as a defender of science against religious encroachments. He edited the *Emporium of Arts and Sciences;* published practical treatises on dyeing and calico printing, gas lights, and tests for **arsenic**; and edited several European chemistry textbooks for American use. In *Discourse on the Connexion between Chemistry and Medicine* (1818) he upheld the materialist position. In *On the Connection between Geology and the Pentateuch* (1836) Cooper attacked those who sought to correlate geological findings with the biblical account of creation.

A member of the American Philosophical Society, Cooper received an honorary medical degree from the University of New York in 1817. He was twice married: to Alice Greenwood, with whom he had three children; and in 1811 to Elizabeth Hemming, with whom he had three children. He died on May 11, 1839.

Thomas Cooper.

Coordination chemistry

Coordination compounds are formed by the union of a metal **ion** (usually a transition metal) with a nonmetallic ion or **molecule**. For example, when white, anhydrous **copper sulfate**, $CuSO_4$, is exposed to **ammonia** gas, a deep blue crystalline product is formed. This product consists of four moles of ammonia per **mole** of copper sulfate; the positive copper ion Cu^{2+} bonds to four ammonia molecules. The **nitrogen atom** of each NH_3 molecule contributes a pair of shared electrons to form a **covalent bond** with the Cu^{2+} ion. This type of bond, where both electrons are contributed by the same atom, is referred to as a coordinate covalent bond.

The reaction of the Cu^{2+} ion with four ammonia molecules r esembles the bond between a **proton** and an ammonia molecule to form the ammonium ion. In both cases, the ammonia molecule is acting as a Lewis base, donating the pair of electrons required to form the coordinate covalent bond. The Cu^{2+} ion acts as a Lewis acid in accepting a pair of electrons.

Complex ions are ions that have a **molecular structure** consisting of a central atom bonded to other atoms by coordinate covalent bonds. Compounds containing a complex ion are known as coordination compounds, or complex compounds. In the broadest sense, complex ions are charged species consisting of more than one atom. However, in a more restricted sense, complex ions are charged species in which a metal atom is joined by coordinate covalent bonds to neutral molecules and/or negative ions.

The metal atom in the complex ion is referred to as the central atom. The molecules or anions attached to the central atom are called coordinating groups or ligands. The ligand may be either positively or negatively charged, or may be a molecule of **water** or ammonia. (When ammonia is the ligand, the compounds are called ammines.) Ligands have **electron** pairs on the coordination atom that can be either donated or shared with the metal ions. The **metals** that most commonly form stable coordination compounds are **cobalt**, **platinum**, **iron**, copper, and **nickel**. The number of bonds formed by the central atom is referred t o as its coordination number. The coordination number is usually 2, 4, or 6, often depending on the type of ligand involved. The **bonding** between the ligand and the metal ion is intermediate between covalent and electrostatic. The charge on the complex ion is the sum of the charges on the metal ion and the ligands.

Besides being important in the chemical industry, coordination compounds play a primary role in sustaining life on Earth. Two well known coordination compounds in biological systems are **chlorophyll** and hemoglobin. Chlorophyll, the chief molecule in plant **photosynthesis**, is a **magnesium** complex; hemoglobin, a major component of animal blood that carries **oxygen** to cells in the body, is an iron complex.

In the chemical industry, coordination compounds are used in **qualitative analysis** to separate metal ions, and to identify unknown ions in **solution**. One test for the presence of **silver** ions in solution is to add chloride ions to the solution. Any silver ions present form a white precipitate (silver chloride). Addition of excess ammonia dissolves the silver chloride, causing it to forms the stable metal complex, $[Ag(NH_3)_2]^+$.

The formation of a coordination compound is often accompanied by a **color** change. Invisible ink, for example, is an aqueous solution of $CoCl_2$ having a very pale pink color, or being essentially colorless. If one writes with this ink on a sheet of **paper**, and then heats the sheet, the ink turns blue. When the paper cools, the ink once again becomes colorless (invisible). The reason for this is changes in the metal complex. In dilute solution, the cobalt forms the complex

$[Co(H_2O)_6]Cl_2$. With the application of **heat**, some of the water molecules evaporate, and a new complex, $[CoCl_2(H_2O)_2]$ is formed. The new complex is blue color instead of pale pink like the first. Upon cooling, water is again taken up by the complex and the original pink complex reforms.

Coordination compounds has been of interest to scientists for over 200 years. The first metal complex, discovered in 1798 by Tassaert, was hexaamminecobalt(III) chloride ($CoCl_3 \cdot 6NH_3$). Eighteenth century chemists were unable to explain how two stable compounds, $CoCl_3$ and NH_3, could combine to form another stable complex.

Metal complexes are generally prepared by reacting a **salt** with another molecule or ion. The early coordination compounds prepared using ammonia were metal amines. Other early coordination compounds used the anions CN^-, NO_2^-, and Cl^-, an example being Reinecke's salt ($NH_4 \cdot [Cr(NSC)_4(NH_3)_2]$.

As coordination **chemistry** evolved, other important discoveries followed. By comparing the electrical conductivities of solutions having the same concentrations, it was shown that the number of ions in solution could be estimated. Also, the existence of isomers of coordination compounds was verified.

Nearly 100 years after the discovery of the first coordination compound, **Alfred Werner** established the foundations for the modern science of coordination chemistry, with the elucidation of such concepts as primary **valence** and coordination number.

Although Werner's theories explained the existence and structure of coordination compounds, they did not describe the coordinate bond, or secondary valence. At least three theories have been proposed to describe the coordinate bond: valence bond theory, crystal field theory, and **molecular orbital theory**.

COPPER

Copper is a transition metal, one of the 30 elements that lie between Groups 2 and 13 in the **periodic table**. Its **atomic number** is 29, its atomic **mass** is 63.546, and its chemical symbol is Cu

Properties

Copper is a fairly soft, reddish brown metal that is quite ductile. It is one of the best conductors of **heat** and **electricity** of all elements. Its melting point is 1,982°F (1,083°C), its **boiling point** is 4,703°F (2,595°C), and its **density** is 8.96 grams per cubic centimeter.

Copper is a moderately reactive metal that dissolves in most acids and alkalis. One of its most interesting properties is its tendency to react in moist air with **water** and **carbon** dioxide to form hydrated copper carbonate $[Cu_2(OH)_2CO_3]$. Hydrated copper carbonate has a beautiful greenish **color** that accounts for the patina that develops over time on buildings and other structures coated with copper.

Occurrence and Extraction

Copper is a relatively common element in the Earth's crust, with an abundance of about 70 parts per million. Very small amounts (about one part per billion) also occur in seawater.

At one time, it was not unusual to find copper nuggets lying on the ground. However, such obvious sources of the element were collected long ago. Today, copper is obtained almost entirely from minerals rich in the element, minerals such as azurite, or basic copper carbonate $[Cu_2(OH)_2CO_3]$; chalcocite, or copper glance (copper sulfide; Cu_2S); chalcopyrite, or copper pyrites, or copper **iron** sulfide ($CuFeS_2$); cuprite, or copper **oxide** (Cu_2O); and malachite, or basic copper carbonate $[Cu_2(OH)_2CO_3]$.

Copper is mined in more than 50 nations, from Albania and Argentina to Zambia and Zimbabwe. The world's leading producers are Chile and the United States. Copper is extracted from copper ores in a series of steps that involves converting the copper first to copper **sulfate** ($CuSO_4$) and then reacting the copper sulfate with iron metal to produce elementary copper: $CuSO_4 + Fe \rightarrow FeSO_4 + Cu$ Since very pure copper is often needed for many applications, the copper thus obtained is then further refined electrolytically.

Discovery and Naming

Objects made from copper metal have been found that date back as far as 9000 B.C.. By 5000 B.C., copper was being used routinely to make tools, weapons, and jewelry. In the New World, Native Americans were using copper objects as early as 2000 B.C..

Humans had learned how to make one of the first alloys, bronze, from copper and **tin** by about 4000 B.C.. Over the next thousand years, the role of bronze tools, weapons, jewelry, and other objects had such a profound impact on human civilization that the period is generally known as the Bronze Age. The Bronze Age came to an end around 3000 B.C. as iron came to replace the **alloy** for the manufacture of objects.

The symbol for copper comes from the Latin word *cuprum*, the ancient name of the island of Cyprus. The Romans obtained much of their copper from mines on that island.

Uses

The most important application of copper metal is electrical wiring. Nearly every electrical device relies on copper wiring because copper metal is highly conductive and relatively inexpensive. Some of the everyday appliances in which copper wiring can be found include electric clocks, stoves, portable CD players, and high-tension electrical transmission wires. Two of the oldest and most widely used of all alloys, bronze and brass, also contain copper. These alloys find their way into roofs, heating and plumbing systems, and many parts of office buildings and factories.

A number of copper compounds are used as **pesticides**, including basic copper acetate $[Cu_2O(C_2H_3O_2)_2]$, an insecticide and **fungicide**; copper chromate ($CuCrO_4 \cdot 2CuO$), a fungicide for the treatment of seeds; copper fluorosilicate ($CuSiF_6$), a fungicide used in grape vineyards; copper **methane** arsenate ($CuCH_3AsO_3$), an algicide; and copper-8-quinolinolate $[Cu(C_9H_6ON)_2]$, used to protect fabric from mildew.

Other compounds of copper are found in battery fluid; fabric dye; fire retardants; **food additives** for farm animals;

fireworks; manufacture of **ceramics** and enamels; photographic film; pigments in paints, metal preservatives, and marine paints; **water purification**; and **wood** preservatives.

Health Issues

Copper is an essential micronutrient for both plants and animals. It is used in the manufacture of certain enzymes, such as those that occur in blood vessels, tendons, bones, and nerves.

Cordite

Patented by Sir Frederick Abel and Sir **James Dewar** in 1889, **cordite** is a smokeless powder derived from nitroglycerine and nitrocellulose (guncotton). Although not the first of the explosive mixtures to supersede **gunpowder**, cordite nonetheless represented an important advance due to its plasticity, its ability to be molded into cord-like shapes and then be divided precisely into various sizes. Like all smokeless powders, cordite enabled armies to conceal their battle positions more effectively and also brought about improvements in gun and bullet technology. Variants of cordite and Paul Vieille's (1833-1896) *Poudre B*, invented in 1884, formed the basic ammunition during World War I and are still in use today. Alfred Nobel's development of ballistite in 1888 made possible Abel and Dewar's invention, which was essentially ballistite with the addition of **acetone** and petroleum jelly for increased stability.

Cluster of copper pennies. From its introduction in 1909 to 1982, the penny was 95% percent copper. In 1982, the U.S. Mint switched the penny's core to an inexpensive zinc coated with copper. *(Photograph by Robert J. Huffman. Reproduced by permission.)*

Corey, Elias James (1928-)

American chemist

Elias James Corey, a specialist in the synthesis of organic chemicals, developed many of the theories and methods that now define his field. Since his research career began in the 1950s, one of his goals has been to make the synthesis of chemicals more systematic, and he is best known for his logical approach to the creation of new substances, which he has named "retrosynthetic analysis." Corey's achievements in research, which resulted in far-reaching benefits for medicine and human health, were recognized in 1990 when he was awarded the Nobel Prize for **Chemistry**.

Corey was born July 12, 1928, to Fatina (Hasham) and Elias Corey in Methuen, Massachusetts. His father died eighteen months later. Corey had been named William at birth, but after his father's death his mother renamed him in his memory. Corey had three siblings, a brother and two sisters, and they were all raised by their mother, with the help of her sister and brother-in-law. His aunt and uncle actually moved in with them, and Corey still credits his mother's sister with being an important influence on his life. Corey enjoyed football and baseball as a child; he was also a good student, graduating from high school in 1945 and entering the Massachusetts Institute of Technology at the age of sixteen. He intended to study engineering but quickly developed an interest in chemistry. It was here that Corey began his research on organic synthesis, the manual formation of chemicals. He worked under John Sheehan on the organic synthesis of **penicillin** and received his Ph.D. in 1951.

After earning his doctorate, Corey accepted a position as instructor in chemistry at the University of Illinois at Champaign-Urbana. He continued his research on organic synthesis; in 1954 he became an assistant professor, and in 1955 he was named full professor of chemistry. In 1957, Corey received a Guggenheim fellowship. He left Illinois on sabbatical, and during this time he did research that laid the foundation for the rest of his career. He spent a portion of his sabbatical working on chemical synthesis with **Robert B. Woodward** at Harvard. The rest of the time he devoted solely to research in Europe, examining the problems of synthesizing **prostaglandins** (hormone-like substances with many different effects, found in the tissue of the body).

During the time Corey was a Guggenheim fellow, chemical synthesis was widely considered an intuitive process and often called an "art form." In a 1990 article in *Science,* Corey recalled his field as he found it in the 1950s: "Chemists approached each problem in an ad hoc way. Synthesis was taught by the presentation of a series of illustrative—and generally unrelated—examples of actual synthesis." It was while working with Woodward that Corey began his effort to systematize

Elias James Corey. *(Corbis Corporation. Reproduced by permission.)*

this intuitive process. Success in organic synthesis was often personal and difficult for the individual scientist to explain in full; Corey wanted the methods, as well as the results, to be both reproducible and teachable.

In 1959, Corey was offered a professorship of chemistry at Harvard, which he accepted. In his new position, he continued to search for what he has called the ''deep logic'' of chemical synthesis. His central innovation was to reverse the usual order of procedure by adding a planning process that began with the desired result, instead of the initial chemicals. Corey planned the process backwards from the **molecule** he wanted to synthesize, creating a chart or ''tree'' that included many possible compounds and reactions. This was retrosynthetic analysis, a formal system which eliminated much of the guesswork, as well as making it easier to use chemicals that were readily available or easy to synthesize. The system also made it possible to use computers for chemical synthesis, and Corey has been a pioneer in this application of artificial intelligence.

The actual results of Corey's work in chemical synthesis have been almost as important as his methodological innovations. In 1968, Corey and his colleagues were able to synthesize five different prostaglandins, which are involved in regulating many functions in the body including blood pressure, blood coagulation, and reproduction. Before this time only a small quantity of these substances was available, as they had to be extracted from the testes of Icelandic sheep. With scientists able to synthetically produce prostaglandins, their applications in medicine have increased profoundly. Eventually Corey's work on prostaglandins led to the development of what is now commonly known as the Corey lactone aldehyde. From this, prostaglandins of all three familial types can be derived.

Another result of Corey's work was the 1988 synthesis of ginkgolide B, a chemical naturally extracted from the ginkgo tree. This chemical is used for the treatment of asthma and circulatory problems in the elderly, and has grown to a market of over 500 million dollars a year. Besides these accomplishments, Corey also has improved or started over fifty new reactions. This has broadened the application of organic synthesis by increasing the tools available to the scientist.

Corey was awarded the 1990 Nobel Prize not for any specific scientific achievement but for his career as a whole. In conferring the award, the Nobel Prize committee said of him: ''No other chemist has developed such a comprehensive and varied assortment of methods, often showing the simplicity of genius, which have become commonplace in organic synthesis laboratories.'' Corey continues his association with Harvard as head of an organic synthesis laboratory which operates with the help of graduate students. The ''Corey research family'' has contributed to the training of over one hundred-fifty university professors and an even greater number of scientists working in industry.

Corey has received eleven honorary degrees including doctorates from the University of Chicago in 1968 and Oxford University in 1982. He has received more than three dozen other awards from universities and scientific societies, including a 1971 Award for Creative Work in Synthetic Organic Chemistry, the Linus Pauling Award in 1973, and in 1988 the Robert Robinson Medal from the Royal Society of Chemistry. Corey was a member of the American Academy of Arts and Sciences from 1960 to 1968. He has been a member of the American Association for the Advancement of Science since 1966. He has served on the editorial board of several scientific journals and contributed over seven hundred articles for publication. He is co-author of a 1989 book, *The Logic of Chemical Synthesis.*

Corey was married in September 1961 to Claire Higham. He and his wife have three children (two sons and a daughter), and they reside in Cambridge, Massachusetts.

CORI, CARL FERDINAND (1896-1984)

American biochemist

Carl Ferdinand Cori and his wife, biochemist **Gerty T. Cori**, were prominent researchers in physiology, pharmacology, and biology. Their most important work involved carbohydrate **metabolism** (especially in tumors), **phosphate** processes in the muscles, the process of glucose-glycogen interconversion, and the action of insulin. The Coris shared the 1947 Nobel Prize for physiology or medicine (along with the Argentine physiologist Bernardo Houssay) ''for their discovery of the course of the catalytic conversion of **glycogen**.''

Cori was born on December 5, 1896, in Prague, which was then part of the Austro-Hungarian empire. His parents were Carl Isidor Cori, a professor of zoology at the German University of Prague, and Maria Lippich Cori. When Cori was still young, the family moved to Trieste, Italy, where his father had been appointed director of the Marine Biology Station. Cori studied at the Gymnasium in Trieste from 1906 to 1914, and then returned to Prague and began medical studies at the German University. His studies, however, were interrupted by World War I; serving in the Austrian army, he worked in hospitals for infectious diseases on the Italian front.

It was during his first term at the University of Prague that Cori met his wife, who was also a medical student. Described as redheaded and vivacious, Gerty Theresa Radnitz was the daughter of a Prague businessman and a lifelong resident of that city. As medical students, the Coris coauthored their first scientific publication; ultimately, they would publish over two hundred research articles together. They were married on August 5, 1920, shortly after receiving their medical doctorates.

From 1920 to 1922, Cori served first as a researcher at the First Medical Clinic in Vienna, Austria, and then in the same capacity at Austria's University of Graz. During this time, his wife worked as an assistant at a children's hospital in Vienna. In 1922, Cori accepted a position at the New York State Institute for the Study of Malignant Diseases in Buffalo. Gerty Cori joined him soon thereafter, and they continued their research together. During this period, the Coris were studying carbohydrate metabolism, particularly in tumor cells. They also researched the effects of the surgical removal of the ovaries on the incidence of tumors.

The Coris became American citizens in 1928, and in 1931 they accepted positions at the medical school of Washington University in St. Louis, Missouri, where Cori was to remain until 1966. Their research on carbohydrate metabolism now centered on glucose, or "blood sugar," the **energy** source for animal life. They developed methods to analyze the relationship of glucose to glycogen, the starchlike form in which glucose is stored in the liver and muscles. In the 1930s, the Coris performed groundbreaking research on the biochemical processes involved in the interconversion of glucose to glycogen, a process now called the Cori cycle. This interconversion is responsible for maintaining the blood sugar at a constant level.

In 1936, the Coris isolated glucose–1-phosphate, now known as the Cori ester, which is involved in the formation and breakdown of glycogen. The Coris also analyzed the function of insulin, a hormone in the pancreas that is vital to the body's processing of glucose. In 1938, the Coris analyzed the conversion of glucose–1-phosphate to glucose–6-phosphate. Then, in 1943, they isolated phosphorylase, an **enzyme** important to the glucose-glycogen interconversion, in crystalline form. The Coris were able in 1944 to synthesize glycogen in a test tube, the first such synthesis of a high molecular substance.

Cori was appointed professor of **biochemistry** at Washington University in 1944, and two years later he became chairman of the department. In 1947, he and his wife were awarded the Nobel Prize for physiology or medicine for their research on the relationship between glucose and glycogen. This led to many comparisons in the press between the Coris and the first husband-and-wife team to win the Nobel Prize, **Pierre** and **Marie Curie**.

In addition to sharing the Nobel Prize in 1947, Cori received numerous awards and honors, including the Isaac Adler Prize from Harvard University in 1944, the Midwest Award of the American Chemical Society in 1945, and the Harry M. Lasker Award of the American Society for the Control of Cancer in 1946. He also received the Squibb Award of the Society of Endocrinologists, which was bestowed on him along with his wife in 1947, and the Willard Gibbs Medal of the American Chemical Society in 1948. Cori received honorary degrees from Cambridge, Yale, and other universities, and was a member of various scientific societies.

In the same year they won the Nobel Prize, his wife was diagnosed with myelosclerosis, a disease of the blood. She died ten years later, on October 26, 1957, of complications from the disease, and Cori suffered the loss of both wife and scientific partner. They had one child, a son, Carl Thomas Cori. Cori subsequently remarried, wedding Anne Fitzgerald Jones on March 23, 1960. In 1966, after retiring from Washington University, Cori served as a visiting professor at the Harvard University School of Medicine. Cori died on October 20, 1984, in Cambridge, Massachusetts.

CORI, GERTY T. (1896-1957)

American biochemist

Gerty T. Cori made significant contributions in two major areas of **biochemistry**, which increased understanding of how the body stores and uses sugars and other **carbohydrates**. For much of her early scientific career, Cori performed pioneering work on sugar **metabolism** (how sugars supply **energy** to the body), in collaboration with her husband, **Carl Ferdinand Cori**. For this work they shared the 1947 Nobel Prize in physiology or medicine with Bernardo A. Houssay, who had also carried out fundamental studies in the same field. Cori's later work focused on a class of diseases called **glycogen** storage disorders. She demonstrated that these illnesses are caused by disruptions in sugar metabolism. Both phases of Gerty Cori's work illustrated for other scientists the importance of studying enzymes (special **proteins** that permit specific biochemical reactions to take place) for understanding normal metabolism and disease processes.

Gerty Theresa Radnitz was the first of three girls born to Otto and Martha Neustadt Radnitz. She was born in Prague, then part of the Austro-Hungarian Empire, on August 15, 1896. Otto was a manager of sugar refineries. It is not known if his work helped shape his eldest daughter's early interest in **chemistry** and later choice of scientific focus. However, her maternal uncle, a professor of pediatrics, did encourage her to pursue her interests in science. Gerty was first taught by tutors at home, then enrolled in a private girls' school. At that time,

Gerty T. Cori.

girls were not expected to attend a university. In order to follow her dream of becoming a chemist, Gerty first studied at the Tetschen *Realgymnasium.* She then had to pass a special entrance exam (*matura*) that tested her knowledge of Latin, literature, history, mathematics, physics, and chemistry.

In 1914 Gerty Radnitz entered the medical school of the German University of Prague (Ferdinand University). There she met a fellow classmate, Carl Ferdinand Cori, who shared her interest in doing scientific research. Together they studied human complement, a substance in blood that plays a key role in immune responses by combining with **antibodies**. This was the first of a lifelong series of collaborations. In 1920 they both graduated and received their M.D. degrees.

Shortly after graduating, they moved to Vienna and married. Carl worked at the University of Vienna's clinic and the University of Graz's pharmacology department, while Gerty took a position as an assistant at the Karolinen Children's Hospital. Some of her young patients suffered from a disease called congenital myxedema, in which deposits form under the skin and cause swelling, thickening, and paleness in the face. The disease is associated with severe dysfunction of the thyroid gland, located at the base of the neck, which helps to control many body processes, including growth. Gerty's particular research interest was in how the thyroid influenced body **temperature** regulation.

In the early 1920s, Europe was in the midst of great social and economic unrest in the wake of World War I, and in some regions, food was scarce; Gerty suffered briefly from malnourishment while working in Vienna. Faced with these conditions, the Coris saw little hope there for advancing their scientific careers. In 1922 Carl moved to the United States to take a position as biochemist at the New York State Institute for the Study of Malignant Diseases (later the Roswell Park Memorial Institute). Gerty joined him in Buffalo a few months later, becoming an assistant pathologist at the institute.

Life continued to be difficult for Gerty Cori. She was pressured to investigate malignant diseases, specifically cancers, which were the focus of the institute. Both she and Carl did publish studies related to malignancies, but studying cancer was not to be the focus of either Gerty's or Carl's work. During these early years in the United States, the Coris' publications covered topics from the biological effects of X rays to the effects of restricted diets on metabolism. Following up on her earlier work on the thyroid, Gerty published a report on the influence of thyroid extract on paramecium population growth, her first publication in English.

Colleagues cautioned Gerty and Carl against working together, arguing that collaboration would hurt Carl's career. However, Gerty's duties as an assistant pathologist allowed her some free time, which she used to begin studies of carbohydrate metabolism jointly with her husband. This work, studying how the body burns and stores sugars, was to become the mainstream of their collaborative research. During their years in Buffalo, the Coris jointly published a number of papers on sugar metabolism that reshaped the thinking of other scientists about this topic. In 1928 Gerty and Carl Cori became naturalized citizens of the United States.

In 1931 the Coris moved to St. Louis, Missouri, where Gerty took a position as research associate at Washington University School of Medicine; Carl was a professor there, first of pharmacology and later of biochemistry. The Coris' son, Carl Thomas, was born in 1936. Gerty become a research associate professor of biochemistry in 1943 and in 1947 a full professor of biochemistry. During the 1930s and 1940s the Coris continued their work on sugar metabolism. Their laboratory gained an international reputation as an important center of biochemical breakthroughs. No less than five Nobel laureates spent parts of their careers in the Coris' lab working with them on various problems.

For their pivotal studies in elucidating the nature of sugar metabolism, the Cori's were awarded the Nobel Prize for physiology or medicine in 1947. They shared this honor with Argentine physiologist Bernardo A. Houssay, who discovered how the pituitary gland functions in carbohydrate metabolism. Gerty Cori was only the third woman to receive a Nobel Prize in science. Previously, only **Marie Curie** and Iréne Joloit-Curie had been awarded such an honor. As with the previous two women winners, Cori was a co-recipient of the prize with her husband.

In the 1920s, when the Coris began to study carbohydrate metabolism, it was generally believed that the sugar called glucose (a type of carbohydrate) was formed from another carbohydrate, glycogen, by the addition of **water** molecules (a process known as hydrolysis). Glucose circulates in the blood and is used by the body's cells in virtually all cellular processes that require energy. Glycogen is a natural polymer (a large **molecule** made up of many similar smaller molecules) formed by joining together large numbers of individual sugar molecules for storage in the body. Glycogen allows the body to function normally on a continual basis, by providing a store from which glucose can be broken down and released as needed.

Hydrolysis is a chemical process that does not require enzymes. If, as was believed to be the case in the 1920s, glycogen were broken down to glucose by simple hydrolysis, carbohydrate metabolism would be a very simple, straightforward process. However, in the course of their work, the Coris discovered a **chemical compound**, glucose–1-phosphate, made up of glucose and a **phosphate** group (one **phosphorus atom** combined with three **oxygen** atoms—sometimes known as the Cori ester) that is derived from glycogen by the action of an **enzyme**, phosphorylase. Their finding of this intermediate compound, and of the enzymatic conversion of glycogen to glucose, was the basis for the later understanding of sugar metabolism and storage in the body. The Coris' studies opened up research on how carbohydrates are used, stored, and converted in the body.

Cori had been interested in **hormones** (chemicals released by one tissue or organ and acting on another) since her early thyroid research in Vienna. The discovery of the hormone insulin in 1921 stimulated her to examine its role on sugar metabolism. Insulin's capacity to control diabetes lent great clinical importance to these investigations. In 1924 Gerty and Carl wrote about their comparison of sugar levels in the blood of both arteries and veins under the influence of insulin. At the same time, inspired by earlier work by other scientists (and in an attempt to appease their employer), the Coris examined why tumors used large amounts of glucose.

Their studies on glucose use in tumors convinced the Coris that much basic research on carbohydrate metabolism remained to be done. They began this task by examining the rate of absorption of various sugars from the intestine. They also measured levels of several products of sugar metabolism, particularly lactic acid and glycogen. The former compound results when sugar combines with oxygen in the body.

The Coris measured how insulin affects the conversion of sugar into lactic acid and glycogen in both the muscles and liver. From these studies, they proposed a cycle (called the Cori cycle in their honor) that linked glucose with glycogen and lactic acid. Their proposed cycle had four major steps: (1) blood glucose becomes muscle glycogen, (2) muscle glycogen becomes blood lactic acid, (3) blood lactic acid becomes liver glycogen, and (4) liver glycogen becomes blood glucose. Their original proposed cycle has had to be modified in the face of subsequent research, a good deal of which was carried out by the Coris themselves. For example, scientists learned that glucose and lactic acid can be directly inter-converted, without having to be made into glycogen. Nonetheless, the Coris' suggestion generated much excitement among carbohydrate metabolism researchers. As the Coris' work continued, they unraveled more steps of the complex process of carbohydrate metabolism. They found a second intermediate compound, glucose–6-phosphate, that is formed from glucose–1-phosphate. (The two compounds differ in where the phosphate group is attached to the sugar.) They also found the enzyme that accomplishes this conversion, phosphoglucomutase.

By the early 1940s the Coris had a fairly complete picture of carbohydrate metabolism. They knew how glycogen became glucose. Rather than the simple non-enzymatic hydrolysis reaction that, twenty years earlier, had been believed to be responsible, the Coris' studies painted a more elegant, if more complicated picture. Glycogen becomes glucose–1-phosphate through the action of one enzyme (phosphorylase). Glucose–1-phosphate becomes glucose–6-phosphate through the action of another enzyme (phosphoglucomutase). Glucose–6-phosphate becomes glucose, and glucose becomes lactic acid, each step in turn mediated by one specific enzyme. The Coris' work changed the way scientists thought about reactions in the human body, and it suggested that there existed specific, enzyme-driven reactions for many of the biochemical conversions that constitute life.

In her later years, Cori turned her attention to a group of inherited childhood diseases known collectively as glycogen storage disorders. She determined the structure of the highly branched glycogen molecule in 1952. Building on her earlier work on glycogen and its biological conversions via enzymes, she found that diseases of glycogen storage fell into two general groups, one involving too much glycogen, the other, abnormal glycogen. She showed that both types of diseases originated in the enzymes that control glycogen metabolism. This work alerted other workers in biomedicine that understanding the structure and roles of enzymes could be critical to understanding diseases. Here again, Cori's studies opened up new fields of study to other scientists. In the course of her later studies, Cori was instrumental in the discovery of a number of other chemical intermediate compounds and enzymes that play key roles in biological processes.

At the time of her death, on October 26, 1957, Cori's influence on the field of biochemistry was enormous. She had made important discoveries and prompted a wealth of new research, receiving for her contributions, in addition to the Nobel Prize, the prestigious Garvan Medal for women chemists of the American Chemical Society as well as membership in the National Academy of Sciences. As the approaches and methods that she helped pioneer continue to result in increased scientific understanding, the importance of her work only grows greater.

CORNFORTH, JOHN WARCUP (1917-)

Australian chemist

Sir John Warcup Cornforth was awarded the 1975 Nobel Prize in **Chemistry** "for his work on the **stereochemistry** of enzyme-catalyzed reactions" and the **molecular structure** of **cholesterol**. He shared this prize with **Vladimir Prelog**, whose research focused on the stereochemistry of organic molecules and reactions. Although profoundly deaf by the age of 20, Cornforth's research into **steroids** led him to discover what ultimately would prove to be a key reaction in steroid synthesis.

The press release by the Royal Swedish Academy of Sciences on October 17, 1975 described Cornforth's Nobel Prize winning research: ''This subject is difficult to explain to the layman as it is a question of geometry in three dimensions; it is concerned with the delicate mechanism of important reactions of biological systems, where a group of atoms takes the place of a certain **hydrogen atom** among two or three, which may appear to be equivalent. The problem is to decide which of the hydrogen atoms is replaced and if nearby groups retain their positions or if they are rearranged in some way. The **enzyme** leads the process in a quite uniform way. Without this guidance, chaos would break out in the biological system.''

Cornforth was born on September 7, 1917 in Sydney, Australia, the second of four children. His father, the son of an English-born Oxford graduate, married his mother, the daughter of a German minister of religion who immigrated to New South Wales in 1832. Growing up in New South Wales, he lived in both the big city of Sydney and the rural community of Armidale. By the age of 10, the first signs of deafness were becoming obvious. By 16, he could no longer hear lectures, and by 20, he was completely deaf. This ''handicap'' may have been the catalyst for his tremendous contribution to chemistry, for he writes in an autobiographical sketch, ''...I was attracted by laboratory work in **organic chemistry** (which I had done in an impoverished laboratory at home since the age of 14) and by the availability of the original chemical literature.''

His hearing was sufficient to allow him to attend Sydney Boys' High School where he was highly influenced by a young instructor, Leonard Basser, toward the study of chemistry. By the time he completed university in 1937, Cornforth had earned first-class honors and a university medal. A year of postgraduate research won him an 1851 Exhibition scholarship—one of two such scholarships awarded each year-to Oxford University in London, England, under the supervision of **Robert Robinson**, the 1947 Nobel laureate in chemistry. The second scholarship was awarded to Rita Harradence, also from Sydney and also an organic chemist. The two researchers worked closely together, married in 1941, and ultimately had three children and two grandchildren. Of his alliance with his wife, Cornforth writes, ''Throughout my scientific career, my wife has been my most constant collaborator. Her experimental skill made major contributions to the work; she has eased for me beyond measure the difficulties of communication that accompany deafness; her encouragement and fortitude have been my strongest supports.''

As the pair left Sydney for Oxford, World War II broke out and, after completing their research into steroid synthesis for their Ph.D.s, the joined Robinson in his work on **penicillin**, a major chemical project in his laboratory during the war. Cornforth contributed to the compilation of *The Chemistry of Penicillin*, a written record of the major international focus on developing the drug, which was published in 1949 by Princeton University Press. However, after the war, he returned to his work on the synthesis of steroids, which ultimately uncovered the principle reaction in the synthesis process. Collaborating again with Robinson on the scientific staff of the Medical Research Council at its National Institute at Hampstead and then Mill Hill, he completed the first total synthesis of the non-aromatic steroids simultaneously.

It was at the institute that he began collaboration with biological scientists, in particular, a George Popják. This alliance initiated a combined research effort through both chemistry and **biochemistry** into the structure of cholesterol, ultimately leading to the carbon-by-carbon degradation of its 19-ring structure. They also identified the arrangement of **acetic acid** molecules, the building blocks of the system, through the use of radioactive tracers.

In 1962, Cornforth and Popják left the Medical Research Council and became co-directors of the Milstead Laboratory of Chemical Enzymology where they developed the study of stereochemistry of enzymic reactions stimulated by isotopic substitution. Collaboration with a German researcher in 1967 led to the understanding of the ''asymmetric **methyl group**,'' opening the way for similar research in many other biological arenas.

Cornforth left Milstead in 1975 to become Royal Society Research Professor the University of Sussex. His outstanding achievements have earned him voluminous awards and honors. These include election to the Royal Society in 1953 and being awarded the Society's Davy Medal in 1968; the Corday-Morgan Medal and Prize, Chemical Society (1953); a joint award with Popják of the Biochemical Society's Ciba Medal (1965); the American Chemical Society's Ernest Guenther Award (1968); Australian Man of the Year (1975); Foreign Member, Royal Netherlands Academy of Sciences (1978); and the Copley Medal from the Royal Society (1982).

Corrosion

Corrosion is the deterioration of a material, or of its properties, as a consequence of reaction with its environment. The direct and hidden costs of corrosion amount to billions of dollars per year throughout the world.

There are several schools of thought regarding the inclusion of the term ''corrosion.'' Some would limit the term to the corrosion of **metals**, while others include the effects of soils, atmospheres, chemicals and **temperature** upon all types of materials. The need to understand and control corrosion has given rise to the new sciences of corrosion technology and corrosion control, both of which are solidly based upon **chemistry**.

The corrosion of metals is caused by the electrochemical transfer of electrons from one substance (oxygen for example) to another. This may occur from the surfaces of metals in contact, or between a metal and another substance when a moist conductor or electrolyte is present. Depending upon the conditions, various types of corrosion may occur. These include, general, intergranular and pitting corrosion, stress corrosion

A shipwreck on an American Samoa sea coast. The rusting occurred due to the corrosion of iron, the main component of the steel alloys used to manufacture ships. The iron takes part in a chemical reaction to combine with both water and oxygen to form rust (hydrated iron oxide). ***(Photograph by John Mead/Science Photo Library, Photo Researchers, Inc. Reproduced by permission.)***

cracking, corrosion fatigue, galvanic and cavitation corrosion, impingement attack, fretting corrosion, **hydrogen** embrittlement, graphic corrosion, dezincification and parting, and biological corrosion.

The mechanism of corrosion in plastic materials is different than in metallic materials. Corrosion of plastic materials involves modifications to the molecular **chains** that form the material. Corrosion in plastic materials involves both physical ageing and chemical degradation. The primary causes of corrosion through physical ageing involve the action of **solvents**, loss of material, or cracking due to stress. The main causes of chemical degradation are **oxidation**, photodegradation (the effect of ultraviolet rays on the material), and **heat**. The chemical attack of plastic materials is known as solvation. In this case chemical agents penetrate the plastic, causing swelling, softening, charring, crazing, delamination, blistering, embrittlement, discoloration, dissolving, and ultimate failure.

Perhaps the earliest recognition of corrosion was the effect of sea **water** and sea atmospheres on ships. **Salt** water, continual dampness, and the growth of marine life such as marine borers, led to the decay of wooden hulls. Because of its toxicity, **copper** cladding of the hulls was widely used to discourage marine growth. In 1824, to protect the copper from deterioration, the team of Sir **Humphry Davy** and **Michael Faraday** applied **zinc** protector plates to the copper sheathing. This was the first successful application of cathodic protection, in which a more readily oxidized metal is attached to the metal to be protected. This procedure was widely used until after World War II when most hulls were replaced by steel or newer materials.

With the development of the industrial age, and the increased use of **iron**, the oxidation of iron, or rust, forced the development of steels and the search for new metals and metal coatings to protect surfaces. This gave birth to the science of corrosion control that involves measures of material selection, inhibition, painting, and novel design. In turn, the growth of chemical industries also led to attempts to solve the problem of corrosive effects encountered in the manufacture of hazardous inorganic and organic compounds.

The advent of the nuclear and space age has brought on new environmental corrosion challenges. Both new and old materials can now be stressed, in a **matter** of minutes, due to high launch temperatures or the super cold of outer space. Materials may also suffer long-term exposure to various types of radiation or **ionization**. Material failure under these conditions can have disastrous results.

Not all research on corrosion, however, is oriented toward its prevention. In today's modern world, we are overrun with plastic waste and refuse and the high cost of its disposal. By applying knowledge gained on the processes of corrosion, engineers can now plan intentional degradation in advance. The most common and welcome example of this is the development of ''biodegradable'' **plastics**, a product that will grow in importance in the future.

Cortisone

Cortisone is one of several steroid **hormones** secreted by the cortex (outer covering) of the adrenal gland. These hormones, called corticoids, are classified according to their functions, glucocorticoids controlling sugar **metabolism** and mineralocorticoids controlling the metabolism of minerals and **water**. The principal glucocorticoids are corticosterone and hydrocortisone (cortisol) and the principal mineralocorticoid is aldosterone. Cortisone is in both categories, because it quickly converts protein to the carbohydrate glucose and it helps regulate **salt** metabolism. Cortisone also helps the body withstand stress. It is used medically to reduce inflammation.

The adrenal cortex's production of cortisone is controlled by the hormone ACTH (adrenocorticotropic hormone), which is secreted by the pituitary gland. The pituitary, in turn, responds to corticotropin-releasing factor, a hormone-like substance produced by the portion of the brain called the hypothalamus.

Knowledge of cortisone is due primarily to three scientists, the Swiss chemist **Tadeus Reichstein** and the Americans Edward Kendall, a biochemist, and **Philip Showalter Hench**, a medical researcher. Kendall first began work on adrenal cortex hormones because an extract had been used successfully against Addison's disease, which is caused by adrenal gland dysfunction. The original hormone theory, developed by the British physiologists William Bayliss and Ernest Starling, held that each type of gland secreted only one hormone. But by the mid-1930s, Kendall and others believed that the adrenal gland produced many hormones. In 1936 Reichstein was the first to isolate what later was named cortisone. Kendall isolated a series of adrenal substances and converted the one he called Compound E into an active substance. He deduced that it was a steroid.

Hench and Kendall studied Compound E for possible use in treating arthritis. In 1948 and 1949, Hench and Kendall gave the name cortisone to Compound E, and the next year Hench and another colleague were the first to use it to successfully treat arthritis. For their work with cortisone and other adrenal hormones, Reichstein, Hench, and Kendall shared the 1950 Nobel prize in physiology or medicine.

Soon after cortisone's first successful medical use, treatments were discontinued. Researchers found that rheumatoid arthritis is not caused by hormone deficiency, and cortisone treatments have some serious side effects. These included edema (fluid retention), high stomach acidity, damage to bone, and abnormal metabolism of **sodium**, **potassium**, and **nitrogen**. Further experiments, however, yielded a refined product that reduced the side effects.

Cortisone (17-hydroxy-11-dehydrocorticosterone) has been synthesized by several different methods. It was originally derived from the bovine bile constituent deoxycholic acid in a costly 37-step process. A less expensive mass-production method was developed in 1948 by the American chemist **Percy Julian**, who was widely known for synthesizing chemicals from soybeans. Julian had already synthesized a substance that he called cortexolone, which was very similar to cortisone, except that it had one less **oxygen atom**. Julian added oxygen to his cortexolone, turning it into cortisone at a greatly reduced production cost.

The American chemist **Carl Djerassi** in 1951 was the first to synthesize cortisone from raw plant materials—yams and sisal (diosgenin and hecogenin). That same year, the American chemist Robert Burns Woodward synthesized cortisone from orthotoluidine, a coal-tar derivative whose structure was one **carbon ring** with an attached methyl (CH_3) group. By adding various groups of atoms, Woodward transformed it into the basic steroid pattern of four rings and two methyl groups. Though twenty steps were involved, Woodward's process was considerably more streamlined than that for converting bovine bile.

Today, cortisone is prescribed to reduce inflammation in allergy and in arthritis and other connective tissue diseases. It is also prescribed as a replacement hormone in Addison's disease, and for people whose adrenal glands have been removed. Other uses include cancer therapy, asthma treatment, and **reduction** of the body's immune response to prevent rejection of transplanted organs.

Cosmetics

Cosmetics are products designed to improve the appearance and condition of the body. They accomplish this by cleaning, coloring, and moisturizing specific areas of the body, including the skin, nails, hair, lips, and eyes. First used by humans as early as 4000 B.C., cosmetic products have steadily improved over the years. Today, a wide variety of cosmetics are sold representing a multibillion dollar industry.

Cosmetics have been used by humans for centuries. Archeological discoveries indicate that people in prehistoric cultures around the globe colored their faces with pigmented greases and **waxes**. This was done to improve attractiveness, ward off evil spirits, and during times of war. Ancient people in the Middle east used cosmetics for religious and aesthetic reasons. They also used eye cosmetics to protect the eyes from the sun. The oldest cosmetic product found is a lipstick dating from about 4000 B.C. These early materials were made from naturally occurring compounds such as animal excretions, pastes made from burnt **wood**, and natural deposits of **antimony** and **lead** ore.

The most sophisticated of the first cosmetics were developed in Ancient Egypt. In this culture, men and women used red pigments on the face to give a more god-like appearance. They used antimony to color their eyebrows and malachite for coloring eyelids. Hair coloring with henna and indigo was also done. Egyptians used bath oils, white powders, and abrasives to clean their teeth. Cosmetics were also important to the ancient Greek and Roman societies. In fact, the word cosmetics is derived from the Greek word *kosmetikos*, meaning skilled in decorating. Ancient Greek women adorned themselves with highlighted eyebrows and painted lips. They used lead carbonate to make skin lighter. To the Romans, cosmetics were status symbols, representing wealth and power. Pliny the Elder wrote an encyclopedia that included a section on cosmetic formulas and perfume components. In 200 A.D., Galen recorded methods for producing cosmetics that continued to be used for centuries.

During the time of the Crusades, Europeans were introduced to various cosmetics that were brought back from the Middle East. By the 1500s, the French were considered the most skilled at applying cosmetics. They used flower pollen, lemon juice, and other natural materials. They also began the scientific study of ingredients used in cosmetics. An unfortunate fact of many of these early cosmetics was that they could severely injure the user. For example, toxic materials such as lead, **sulfuric acid**, and mercuric sulfide were used.

The use of cosmetics was at its height during the eighteenth century. However, with the French Revolution and the Victorian Age cosmetics fell out of favor, and more natural looks were encouraged. By the late 1800s, the cosmetic industry was slowly revived. When motion pictures were developed in the 1920s, cosmetics experienced a tremendous level of renewed popularity. Two early entrepreneurs who contributed to the growth of the cosmetics industry were Elizabeth Arden and Helena Rubenstein. In 1938, the cosmetics industry was large enough to warrant regulation by the United States government. Since this time, one of the primary changes in cosmetics is the improved safety of the materials.

Modern cosmetics are designed to modify or improve the appearance of various areas of the body. The primary targets include the skin, hair, and mouth. Products that treat the skin have been around for centuries. They are designed to clean, color, and protect skin. The basic skin cosmetics include cleansers, creams, and colorants.

Skin cleansing products help remove dirt and germs from the skin surface and pores. The first products for this purpose were paste and bar soaps. **Soap** is a **salt** formed by a process called saponification—a chemical reaction between an alkali metal, like **sodium**, and fatty **carboxylic acids**. It is composed of surfactants that can surround dirt particles on a surface and lift them off, allowing it to be rinsed away. In the 1930s, synthetic detergents were developed that have the same function as soap. Today, there are a wide variety of skin soaps available including luxury soaps, deodarant soaps, moisturizing soaps, baby soaps, abrasives, and protective soaps. While bar soap has been the traditional skin cleanser of choice, body washes composed of synthetic detergents have become preferred by many consumers.

Some skin products are designed to improve the physical appearance and condition of skin. These products include skin creams and lotions. Skin creams are mixtures of ingredients including fats, oils, and **water** soluble ingredients. They also contain special ingredients called emulsifiers that allow water and oil to combine together in a stable manner. This system, called an **emulsion**, is one of the primary delivery systems used for cosmetic products.

When a cream or lotion is applied to the skin surface it can improve the skin in various ways. The oils and emollients in the formula help reduce roughness and improve feel. Materials like petrolatum help alleviate dryness and increase moisture. This makes skin look smoother and feel softer. Other ingredients can reduce redness and itching and increase shine.

Another important skin cosmetic is make-up. Make-up has been used by humans since before 1000 B.C. They contain ingredients that change the hue of the face, lips, nails, or eye region. Numerous coloring agents are used. To produce a white **color**, talcum, **zinc** oxide, and **titanium** dioxide are used. Colored products use **iron** oxides, chromic oxide, and various organic pigments. Of key importance to these products is safety. Particularly products that will be used around the eyes.

Make-up include vanishing cream, face powder, compressed powders, and cream makeup. Products that are used around the eye include eye shadow, eyeliner, eyebrow pencils, and eye makeup remover. The most common products used for the lip area include lipsticks and balms. These products have the added benefit of improving lip feel and moisturizing. Nail polishes are designed to color and protect the nails. These products are applied with a brush and require special solvents for removal.

Certain cosmetics are available that have specific benefits such as protecting from the damaging effects of sun, warding off sweat and body **odors**, preventing acne, and aiding in the removal of unwanted hair.

In the last 20 years, tanned skin has become a desirable beauty characteristic. Unfortunately, exposure to UV light generated by the sun can cause burns and even skin cancer. For this reason, cosmetics such as sunscreens and tan lotions are produced that block this damage. These products utilize special chemicals that can absorb or scatter UV light and reduce damage. Compounds that protect against sun damage include such things as titanium dioxide, para amino benzoic acid (PABA), and benzyl cinnamate. Depending on the concentration of sunscreen in a product, sunscreens are given a Sun Protection Factor (SPF) rating. While creams are the primary type of sunblock other forms include gels, lotions, and sprays. Additionally, products such as tanning oils and tan accelerators are made to promote tanning. Artificial tanning products use chemicals such as dihydroxyacetone (DHA) that react with proteins in the skin to create a brown color.

Reducing and eliminating body odor and sweat is function of deodorants and antiperspirants. Most deodorants are odor maskers that cover the unpleasant odor with a more appealing one. Odor absorbing deodorants are also available. Certain deodorant products contain materials, such as triclosan, that eliminate the bacteria responsible for body odor. Antiperspirants are similar to deodorants, however, they are designed to reduce body odor by reducing sweat production. The primary active ingredient are aluminum salts such as alkaline aluminum chloride. Deodorants and antiperspirants are sold as aerosols, roll-ons, and sticks.

To improve the appearance of skin by reducing blemishes and acne is the function of anti- acne cosmetics. Skin blemishes are caused by bacteria that attacks excess sebum on the face or body. Anti-acne creams contain compounds such as salicylic acid, sulphur, and triclosan that are effective at killing bacteria and reducing the signs of acne.

Since body hair is often visually undesired, a range of cosmetics are made to aid in its removal. In general, there are three ways to eliminate body hair. First the hair can be physically removed. This typically involves shaving or plucking. Shaving creams are made up of oils and emollients that help glide the blade across the skin reducing pain and incidents of cutting. To remove hair from the root, epilatories are used. These products are composed of waxes. When put on the skin they harden and allow hair to be physically pulled out. Depilatories are put on skin to chemically degrade body hair. They contain ingredients such as inorganic sulfides, stannites, and mercaptan salts that react with hair keratin to break it down.

While skin care represents an important area for cosmetic improvements, the hair also provides an opportunity to be improved. One of the most common products for hair is shampoo. Shampoos are designed to remove dirt and oils from the hair. Additionally, they help remove dandruff, unwanted odors, and residues from other hair care products. Prior to the 1930s, hair- cleansing products were based on soaps. These products tended to be irritating to the eyes and were negatively effected by hard water. Synthetic detergent shampoos were introduced in 1933. These products foamed well and did not have the problems associated with soaps. While the detergent is the primary active ingredient in a shampoo, other ingredients are also added to moisturize and condition hair. Dandruff shampoos are a special type of shampoo that solves a specific problem. They contain salicylic acid that combats the production of dandruff.

While shampoos clean hair, they can also leave it in a poorly manageable condition. For this reason, cream rinse conditioners were developed. They are applied to hair after washing and help improve combing, shine, and static control. The primary conditioning ingredients are cationic surfactants, silicones, and **polymers**.

To hold hair in place a variety of hair styling products are available. The most important of these are hairsprays, gels, and mousses. These products contain resins that form a film on hair and become solid when dry. They can be modified to improve feel and appearance.

Hair can also be chemically treated to make it more or less curly or change its color. Hair perm products use thioglycolic acid—a chemical that reacts with protein bonds in the hair. When the hair is curled, it is treated with the reducing treatment. It is then treated with an oxidizing solution that reforms protein bonds and causes the hair to maintain a curl. Hair relaxers work in a similar manner to perms however, they are used to straighten curly hair. Permanent hair-coloring products are based on oxidative dyes. These are small molecules that can penetrate the hair. They are then treated with a reacting compound that causes the molecules to polymerize and change the color of hair. Typically, a material such as **hydrogen peroxide** is first applied to remove the hair's natural color. Semi-permanent hair colorings based on **dyes** are also available.

The oral cavity represents a third important area for treatment with cosmetics. Toothpastes are designed to help clean teeth and remove plaque. Mouthwashes are used to reduce unpleasant odors in the mouth. These products are typically fortified with **fluoride**, which has been proven to reduce the incidence of cavities.

Cosmetic products are put into a wide variety of packages. The most common are bottles and jars. These can be made from plastic or glass. Other types of packaging includes stick packages as used for products like lip sticks or deodorants. Hair styling products are often put into tubes that can be squeezed into the hand. Aerosols are another important form of cosmetic packaging. A variety of plastics are used to produce cosmetic packaging. Typical kinds are **polypropylene**, **polyethylene**, and **polyvinyl chloride**.

COUPER, ARCHIBALD SCOTT (1831-1892)

Scottish chemist

The British chemist Archibald Scott Couper (1831-1892) shares with **Friedrich Kekulé** the distinction of recognizing the tetravalency of **carbon** and the capacity of carbon atoms to combine to form **chains**, thereby providing the basis for structural **organic chemistry**.

Archibald Scott Couper was born on March 31, 1831, at Kirkintilloch in Dumbartonshire, Scotland, the son of a prosperous cotton weaver. He commenced his university studies at Glasgow mainly in classics, spent the summer semester of 1852 in Berlin, and returned to Scotland to complete his university course in logic and metaphysics at Edinburgh. He spent the period 1854-1856 in Berlin and during this time decided to study chemistry.

Couper entered the laboratory of Charles Wurtz in Paris in the autumn of 1856 and remained there until his return to Scotland in 1858; during these two years he made all his contributions to chemistry: two papers containing experimental contributions and his now famous memoir "On a New Chemical Theory.'' A few months after his return to Edinburgh to be assistant to Lyon Playfair, in the autumn of 1858, he suffered a severe nervous breakdown, followed by a general breakdown in health. He retired to Kirkintilloch and lived there incapable of intellectual work and completely lost to chemistry until his death 34 years later.

The story of Couper's work, its subsequent disappearance from view, and its later recognition, largely through the efforts of Richard Anschütz, as a major piece of chemical history is one of the most remarkable in science. Early in 1858 Couper, then 27 and after only some three years' contact with chemistry, asked Wurtz to present Couper's manuscript ''On a New Chemical Theory'' to the French Academy. Wurtz, however, delayed taking any steps, and in the interim August Kekulé's paper ''On the Constitution and Metamorphoses of Chemical Compounds and on the Chemical Nature of Carbon'' appeared, containing essentially similar proposals. Couper protested to Wurtz about his procrastination but was, it is said, shown out of the laboratory.

Couper's paper was, however, finally presented by Jean Baptiste Dumas to the academy on June 14, 1858, and published in the *Comptes rendus;* fuller versions were subsequently published in English and French. After pointing out the inadequacy of current theories, Couper wrote in his paper: ''I propose to consider the single element carbon. This body is found to have two highly distinguished characteristics: (1) It combines with equal numbers of **equivalents** of **hydrogen**, **chlorine**, **oxygen**, sulphur, etc. (2) It enters into chemical combination with itself. These two properties, in my opinion, explain all that is characteristic of **organic chemistry**. This will be rendered apparent as I advance. This second property is, so far as I am aware, here signalized for the first time.''

Couper also introduced the use of a line to indicate the **valence** linkage between two atoms and, had he used 16 rather than eight for the atomic weight of oxygen, his chemical formulas would have been almost identical with those used today. It is also remarkable that in his paper he represents cyanuric acid by a formula containing a **ring** of three carbon and three **nitrogen** atoms joined by valence lines—the first ring formula ever published. The introduction of ring formulas is often ascribed to Kekulé, who in 1865 used this concept to develop his formula for **benzene**. It is interesting to speculate whether Couper might have anticipated Kekulé's formulation of aromatic compounds had he been able to continue his chemical work. But Couper's paper ''On a New Chemical Theory'' remains a landmark in the history of organic chemistry.

COURTOIS, BERNARD (1777-1838)

French chemist, manufacturer, and pharmacist

Bernard Courtois was born in Dijon, France. Exposed to **chemistry** at an early age, he divided his time between the saltpeter (potassium nitrate) works of his father, Jean- Baptiste Courtois, and the laboratory at the Dijon Academy. In 1791, he became an apprentice for three years to a pharmacist at Auxerre. Courtois was admitted to the famous engineering school in Paris, the École Polytechnique, where he studied under the chemist Antoine François, Comte de Fourcroy.

In 1799, Courtois was drafted in the French military and served as a pharmacist. He then became assistant to chemist Louis Jacques Thénard (1801) and assistant to chemist Armand Seguin (1804). Courtois returned to Dijon to take over his father's business. In 1808, he married a young, nearly-illiterate peasant girl, Madeleine Eulalie Morand.

During the Napoleonic Wars, the British naval blockade cut off foreign imports of saltpeter needed to prepare **gunpowder**. Courtois used seaweed ash from algae found in Normandy and Britanny as a source of valuable **potassium** and **sodium** salts. Because the mother liquor obtained by leaching the ash with **water** contained undesirable impurities, he removed these by treatment with sulfuric acid. In late 1811 Courtois apparently used too much acid and observed clouds of violet vapor, which had an irritating odor similar to that of **chlorine** and condensed on cold surfaces, to form dark, almost black, lustrous crystals with a metallic sheen. He had accidentally produced **iodine** by oxidizing the sodium and potassium iodides in the mother liquor. Suspecting that the new substance was a previously unknown element, Courtois spent approximately six months investigating it but was unable to conduct a thorough study in his primitive laboratory. Lacking the financial resources to continue the research and barely able to support his family by the saltpeter business, he abandoned his work on the substance.

In July 1812, Courtois informed two chemist friends, Charles Bernard Desormes and Nicolas Clément, of his discovery and allowed them to report it to the Institut de France, which they did on November 9, 1813. Upon his request, Desormes and Clémenty continued his research at the laboratory of the Conservatoire des Arts et Métiers. Clément extensively studied the new substance and concluded that it was an element similar to chlorine. That same year, French chemist Joseph Louis Gay- Lussac called the substance ''iode'' from the Greek word for violet, and English chemist Sir **Humphry Davy** confirmed its elementary nature. Their results convinced skeptics that chlorine was an element, for if iodine was an element, the chemically-similar chlorine probably was as well. In 1826, Antoine Jérôme Balard confirmed the elementary nature of both iodine and chlorine by discovering a third halogen element, **bromine**.

During the 1820s Courtois abandoned his saltpeter business and began manufacturing iodine compounds and other chemicals, but this business likewise failed to prosper, and he abandoned it in 1835. Courtois was acknowledged as the discoverer of iodine, which was widely used in the treatment of goiter, and in 1831 the Dijon Academy awarded him the Montyon prize of 6,000 francs ''for having improved the art of healing.''

Courtois died in Paris on September 27, 1838, without ever having published a scientific paper during his lifetime. On November 9, 1913, the centenary of Desormes and Clément's announcement of his discovery of iodine, a ceremony was held in honor of Courtois' achievements at the Dijon Academy, and a commemorative plaque was placed on the house in which he was born. In 1914 a street in Dijon was named after him.

COVALENT BOND

A covalent bond is a bond formed when two atoms share a pair of electrons. A neutral group of atoms bonded together covalently is called a **molecule**, and a substance which is made up of molecules is called a molecular substance. Covalent molecules make up many common substances, including **plastics**, **paper**, and human tissue.

Another type of bond that occurs when electrons are transferred between atoms is called an ionic bond(or electrovalent bond). Compounds held together by ionic bonds are called ionic compounds. Molecular compounds have lower melting and boiling points than do ionic compounds. Molecular compounds often occur as individual molecules, whereas ionic compounds occur as a crystal lattice. This crystal lattice structure gives the ionic compounds their higher melting and boiling points.

One characteristic that both ionic and covalent compounds share is that they both adhere to the **octet rule**. The octet rule is the principle that describes the **bonding** in atoms. Individual atoms are unstable unless they have an octet of electrons in their highest **energy** level. The electrons in this level are called **valence** electrons. When atoms gain, lose, or share electrons with other atoms, they satisfy the octet rule and form chemical compounds. An example of this is the formation of molecular **fluorine** (F_2) from two individual fluorine atoms. A fluorine **atom** has seven valence electrons. According to the octet rule, eight valence electrons is the most stable configuration, so each fluorine atom needs one more **electron**. If each fluorine atom shares one electron with the other, the octet rule will be satisfied and a covalent bond is formed. Another example of a molecule formed by covalent bonds is **ammonia** (NH_3). A **nitrogen** atom has five valence electrons, so it needs three more to satisfy the octet rule. **Hydrogen** has one valence electron, so it needs one more to fill its outer **energy level** (hydrogen has electrons in only the first energy level, which fills when two electrons are present, unlike the other energy levels that require eight). If one nitrogen atom shares an electron with each of three hydrogen atoms, and the three hydrogen atoms each share an electron with the nitrogen atom, three covalent bonds are formed and the octet rule is satisfied for all four atoms.

Covalent bonds are formed due to the forces of electric attraction between atoms, for example, the covalent bond that forms between two fluorine atoms. An attractive force occurs between the negatively charged electrons on the first fluorine atom and the positively charged protons on the second fluorine atom. A repulsive force also occurs between the electrons in each atom and between the protons in each atom. It would seem that this repulsive force would cause the atoms to move away from each other. The distance between the protons in one fluorine atom and the protons in the other fluorine atom (and, similarly, between the electrons between each atom as well) is greater than the distance between the electrons of one atom and the protons of the other. Therefore, the attractions between the two atoms are greater than the repulsions, and the two fluorine atoms are held together. Making the individual flourine atoms more stable (satisfying the octet rule) is also important. The stabilization of the atoms outweighs any repulsive forces there may be between the two atoms.

The fact that there are repulsive as well as attractive forces between two atoms means that the covalent bond must be flexible. If two atoms start to move apart, the attractive forces will draw them back toward each other. If they then move too close to each other, the repulsive forces will push them apart. The atoms in a covalent bond are actually vibrating back and forth around an average distance where the two forces are balanced. This average distance is called the bond length. The bond length is related to bond energy, which is the energy required to break a covalent bond. A general rule relating bond length and bond energy is the shorter the bond length, the greater the bond energy; the two are inversely related.

Covalent bonds can occur as single bonds or multiple bonds. Molecular fluorine and ammonia are both examples of single covalent bonds—in each case, a single pair of electrons is shared between atoms. In a multiple bond, two atoms share more than one pair of electrons. For example, a double covalent bond occurs when two atoms share two pairs of electrons. This occurs in a **formaldehyde** (H_2CO) molecule. **Carbon** has four valence electrons, so it needs four more to satisfy the octet rule. It can only share one pair of electrons from each of the hydrogen atoms for a total of two valence electrons gained. In order to satisfy the octet rule and gain two more valence electrons, it must share two pairs with the **oxygen** atom. Oxygen has six valence electrons, and when it shares two pairs with the carbon atom it gains the two it needs to satisfy the octet rule. The two shared pairs of electrons between carbon and oxygen create a double covalent bond.

Two atoms can share three pairs of electrons as well, forming a triple covalent bond. This occurs in an **ethyne** (C_2H_2) molecule. Each carbon atom needs to gain four valence electrons. Each can gain one by sharing a pair of electrons with a hydrogen atom, and then gain the other three by sharing three pairs of electrons with each other. The three shared pairs of electrons between the two carbon atoms create a triple covalent bond.

When electrons are shared between atoms, they are not always shared equally. One atom may attract the electrons more than the other. This has to do with an atom's **electronegativity**. Electronegativity is a way to measure how much an atom attracts electrons in a chemical bond. The greater the electronegativity, the greater the attraction. When two atoms with two different electronegativities share electrons, the atom with the greater electronegativity will have a stronger attraction towards the electrons. When one atom has a much greater electronegativity than the other, the covalent bond between them is said to be polar. A **polar bond** results in an uneven distribution of charge. The atom with the greater electronegativity becomes slightly negative because it has a greater electron **density**, while the atom with the lower electronegativity becomes slightly positive. A molecule that has a polar covalent bond is called a polar molecule. An example of a polar molecule is **water** (H_2O). Oxygen is more electronegative than hydrogen. The electrons that are shared between the oxygen and the two hydrogen atoms are pulled closer to the oxygen atom, giving the oxygen a slight negative charge. The hydrogen atoms, as a result, have a slight positive charge.

When two atoms with similar electronegativities share electrons, the covalent bond that results is said to be nonpolar. In a nonpolar covalent bond, both atoms attract the electrons equally. Nonpolar covalent bonds occur in covalent bonds between atoms of the same element, for example, molecular fluorine. Both atoms have the same electronegativity, so they exert an equal pull on the electrons shared between them. Electronegativity can be used to predict whether a bond will be a nonpolar covalent bond, a polar covalent bond, or an **ionic bond**. If the difference between the electronegativities of two atoms is 2.0 or greater, the bond is an ionic bond. If the difference is 0.4 or less, the bond is nonpolar. If the electronegativity difference between two atoms is between 0.4 and 2.0, the bond is considered to be a polar covalent bond. The greater the electronegativity difference, the greater the polarity.

Covalent bonds, or bonds formed by the sharing of electrons between two atoms, can occur in several forms. Single, double, and triple covalent bonds can occur. A covalent bond can also be polar or nonpolar. Covalent bonds between atoms are quite common and result in many different kinds of molecules. The phenomenon of two atoms sharing electrons is what builds many of the structures, both living and nonliving, in the world around us and understanding covalent bonds is an important part of both the physical and the life sciences.

CRAM, DONALD J. (1919-)

American organic chemist

Organic **chemistry** underwent profound changes in the second half of the twentieth century, and one of the scientists responsible for these advances is Donald J. Cram. When he entered the profession in the 1940s, **organic chemistry** was primarily concerned with elucidating **molecular structure** and with synthesizing new molecules by mixing reagents with organic compounds by a method that was more or less ad hoc. The mechanisms of reactions were infrequently exploited in directing reactions towards a desired product. In the years after World War II reaction mechanisms attracted new attention; the exact three-dimensional details of how molecules combine to form products became known, and chemists realized that compounds of very specific shapes could be constructed. This was called **stereochemistry**, and it had valuable applications for the discipline of making molecules that make other molecules—that is, building compounds that can hold other compounds in a specific configuration, which in turn can lead to a specific reaction that would not otherwise have taken place. It was for his studies in this area, specifically his work in host-guest molecules, that Cram shared the Nobel Prize in 1987.

Donald James Cram was born April 22, 1919, in Chester, Vermont, the fourth child and only son of William and Joanna Shelley Cram, who had recently come from Ontario. The family moved to Brattleboro, Vermont, in 1921, and Cram's father died of pneumonia in 1923. Many years later,

Donald J. Cram.

Cram recalled that this loss ''forced me to construct a model for my own character that was composed of pieces taken from many different individuals; some being people I studied and others I lifted from books.'' He spent his childhood in Brattleboro, a curious, mischievous, bookish teenager who read through most of the standard classics but also played varsity sports. He supported himself and the family with a succession of odd jobs paid by barter, including dental work in exchange for lawn mowing; these taught him self-discipline, but convinced him that he did not want to spend his life in a job that was repetitive and uninspiring. In 1935, when he was sixteen, his family dispersed and he entered Winwood, a small, private school on Long Island, where he finished his high school studies in 1937.

Cram received a scholarship to Rollins College in Florida, where he earned his B.S. in 1941. The chemistry department at Rollins was small and underfunded, but it was here that Cram realized research could provide the ever new experience he had hoped to find in a career. He went on to receive an M.S. in chemistry at the University of Nebraska in 1942, and then spent the war years with Merck and Company, a pharmaceutical firm, in their **penicillin** program. Three years later, with a research fellowship and a strong recommendation from Merck's **Max Tishler**, he moved to Harvard University, where he received his Ph.D. under Louis Fieser in 1947. After a three-month postdoctoral stint with John D. Roberts at the Massachusetts Institute of Technology, Cram accepted an assistant professorship at the University of California at Los Angeles. He would remain here for the rest of his career, becoming a full professor in 1956 and Saul Winstein Professor of Chemistry in 1985.

Cram's research divides chronologically into two sections. In the first phase, from 1948 to about 1970, he concentrated on reaction mechanisms. He conducted his first mechanistic study on the **substitution reaction** of a compound with two adjacent asymmetric **carbon** atoms (carbons with four different groups attached, arranged in a specific order in space). As the asymmetry was preserved during the reaction, it was clear that something prevented rotation of the carbon atoms on their common bond in the transition state. Cram proposed that they were held in place by what he named a phenonium **ion**, formed by a phenyl (benzene-ring) group on one of the carbons; he believed this acted as a bridge between them in the transition state. Cram adduced other evidence to support the existence of this new ion, and he carried this kind of study to other organic molecules. The implications of such studies were particularly important in biological systems, where the greater number of large molecules contain asymmetric carbons.

Cram then turned to elimination reactions with the same sort of compound, containing two adjacent asymmetric carbon atoms. In an **elimination reaction**, an **atom** or group is removed from each carbon, creating a double bond between them, and the adjacent asymmetric carbon atoms show how the remaining groups will be arranged on the resulting double-bonded compound. He formulated his findings in what came to be called "Cram's rule." He went on to study many more molecules that formed a negative carbon atom in the transition state, and he showed that the associated positive ion could do many previously unsuspected things—including migrating to an adjacent carbon atom, or skating around a double-bond system and ending up on the other side of the carbon to which it was originally attached. At the same time that he performed his work on reaction mechanisms, Cram created and studied a new class of compounds, called the cyclophanes, in which two **benzene** rings are fastened together at each end by bridges of two or more carbon atoms. This brings the rings into close juxtaposition and also creates considerable angular strain.

Eventually, Cram decided that his research was becoming repetitive—precisely the situation he had resolved to avoid many years before. At age fifty he turned to a new field, the investigation of host-guest molecules. For his first host molecules he chose the "crown ethers "that had been synthesized by **Charles John Pedersen** of DuPont Chemical's research laboratories. Crown ethers are cyclic compounds in which **oxygen** atoms recur regularly around the **ring**, spaced apart by two or more carbons. In some conformations the oxygen atoms stick up like the points of a crown; hence the name. These atoms, which are polar and possess unbonded **electron** pairs, can form complexes with a variety of positive or incipiently positive atoms or molecules.

The simplest of these crown-ether structures was already known to form complexes with **potassium** ions by turning its oxygen atoms into the center to form what Cram called a corand. He discovered that a corand can be used to separate potassium from other ions. By constructing other corands and basket-shaped molecules he called cavitands, with interiors of carefully controlled size, Cram was able to select out each of the alkali metal ions (**lithium**, **sodium**, potassium, **rubidium**, **cesium**) from **solution** with a high degree of specificity. This had important applications in analytical chemistry, particularly in medical and biological systems. Other cavitands were synthesized that looked less and less like crown ethers, except that they had oxygen or **nitrogen** atoms in their interiors for complexation. Cram extended these studies to organic molecules. A special asymmetric compound was devised that could form complexes with either right- or left-handed **amino acids**; this was worked into a continuous mechanical separator for these asymmetric molecules.

The ultimate goal of those working with artificial enzymes has long been to produce large molecules. This has not yet been attained, but Cram's work has made great strides in this direction. For his research on host-guest molecules, he shared the 1987 Nobel Prize in chemistry with Pedersen and **Jean-Marie Lehn** of Strasbourg University. Cram delivered a lecture at the awards ceremony entitled "The Design of Molecular Hosts, Guests, and Their Complexes." Newspaper accounts emphasized the ramifications of his discoveries for both medical and industrial research, and it was observed that Cram had taught many of the chemists working on molecular recognition around the world.

Cram has co-authored an undergraduate textbook on organic chemistry, organized not by types of compounds (like nearly all other such works) but by types of **reaction mechanism**; it has gone through four editions and has been translated into thirteen languages. He also wrote another lower-level text, *Essence of Organic Chemistry,* with his second wife, Jane Maxwell Cram. These publications attest to his ongoing interest in undergraduate teaching. Cram's *Fundamentals of Carbanion Chemistry,* published in 1965, summarizes work in the field. In 1990, he produced the autobiographical *From Design to Discovery,* which contains relatively little personal information but is of great interest to chemists who want a review, with bibliography, of his research over four decades.

Other awards and honorary degrees have been presented to Cram in addition to the Nobel Prize. In 1974, he received the California Scientist of the Year award and the American Chemical Society's Arthur C. Cope Award. He was presented with the Richard Tolman Medal, the Willard Gibbs Award, and the Roger Adams Award, all in 1985. He also has received honorary doctorates from six institutions, including his undergraduate alma mater.

Cram has been married twice, first to Jean Turner from 1940 to 1968, and then in 1969 to Jane Maxwell, who is also a chemist. Both marriages have been childless. A man of abundant drive and **energy**, Cram spends his leisure time surfing and downhill skiing; he also sings folksongs and plays the guitar.

Crick, Francis (1916-)

English molecular biologist

Francis Crick is one half of the famous pair of molecular biologists who unraveled the mystery of the structure of deoxyribonucleic acid (**DNA**), carrier of genetic information, thus ushering in the modern era of molecular biology. Since this fundamental discovery, Crick has made significant contributions to the understanding of the genetic code and gene action, as well as of molecular neurobiology. In Horace Judson's book *The Eighth Day of Creation,* the Nobel laureate Jacques Lucien Monod is quoted as saying, "No one man created molecular biology. But Francis Crick dominates intellectually the whole field. He knows the most and understands the most." Crick shared the Nobel Prize in medicine in 1962 with **James Watson** and Maurice Wilkins for the elucidation of the structure of **DNA**.

The eldest of two sons, Francis Harry Compton Crick was born to Harry Crick and Anne Elizabeth Wilkins on June 8, 1916, in Northampton, England. His father and uncle ran a shoe and boot factory. He attended grammar school in Northampton, and was an enthusiastic experimental scientist at an early age, producing the customary number of youthful chemical explosions. As a schoolboy, he won a prize for collecting wildflowers. In his autobiography, *What Mad Pursuit,* Crick describes how, along with his brother, he "was mad about tennis," but not much interested in other sports and games. At the age of fourteen, he obtained a scholarship to Mill Hill School in North London. Four years later, at eighteen, he entered University College, London. At the time of his matriculation, his parents had moved from Northampton to Mill Hill, and this allowed Crick to live at home while attending university. He obtained a second-class honours degree in physics, with additional work in mathematics, in three years. In his autobiography, Crick writes of his education in a rather light-hearted way. He feels that his background in physics and mathematics was sound, but quite classical, while he says that he learned and understood very little in the field of **chemistry**. Like many of the physicists who became the first molecular biologists and who began their careers around the end of World War II, Crick read and was impressed by **Erwin Schrödinger**'s book *What Is Life?,* but later recognized its limitations in its neglect of chemistry. Nonetheless, it is clear that Crick read widely and grasped the essence of the argument and logic of what he read.

Following his undergraduate studies, Crick conducted research on the **viscosity** of **water** under pressure at high temperatures, under the direction of Edward Neville da Costa Andrade, at University College. It was during this period that he was helped financially by his uncle, Arthur Crick. In 1940, Crick was given a civilian job at the Admiralty, eventually working on the design of mines used to destroy shipping. Early in the year, Crick married Ruth Doreen Dodd. Their son Michael was born during an air raid on London on November 25, 1940. By the end of the war, Crick was assigned to scientific intelligence at the British Admiralty Headquarters in Whitehall to design weapons.

Francis Crick.

Realizing that he would need additional education to satisfy his desire to do fundamental research, Crick decided to work toward an advanced degree. Surprisingly, he found himself fascinated with two areas of biology, particularly, as he describes it in his autobiography, by "the borderline between the living and the nonliving, and the workings of the brain." He chose the former area as his field of study, despite the fact that he knew little about either subject. After preliminary inquiries at University College, Crick settled on a program at the Strangeways Laboratory in Cambridge under the direction of Arthur Hughes in 1947, to work on the physical properties of cytoplasm in cultured chick fibroblast cells. Two years later, he joined the Medical Research Council Unit at the Cavendish Laboratory, ostensibly to work on protein structure with British chemists **Max Perutz** and **John Kendrew** (both future Nobel Prize laureates), but eventually to work on the structure of DNA with Watson.

In 1947, Crick divorced Doreen, and in 1949 married Odile Speed, an art student whom he had met during the war, when she was a naval officer and Crick was working for the admiralty. Their marriage coincided with the start of Crick's Ph.D. thesis work on the **x-ray diffraction** of **proteins**. X-ray diffraction is a technique for studying the crystalline structure of molecules, permitting investigators to determine elements of three-dimensional structure. In this technique, x rays are directed at a compound, and the subsequent scattering of the x-ray beam reflects the molecule's configuration on a photographic plate.

In 1941 the Cavendish Laboratory where Crick worked was under the direction of physicist Sir **William Lawrence Bragg**, who had originated the x-ray diffraction technique forty years before. Perutz had come to the Cavendish to apply Bragg's methods to large molecules, particularly proteins. In 1951, Crick was joined at the Cavendish by James Watson, a visiting American who had been trained by Italian physician Salvador Edward Luria and was a member of the Phage Group, a group of physicists who studied bacterial viruses (known as bacteriophages, or simply phages). Like his phage colleagues, Watson was interested in discovering the fundamental substance of genes and thought that unraveling the structure of DNA was the most promising **solution**. The informal partnership between Crick and Watson developed, according to Crick, because of their similar "youthful arrogance" and similar thought processes. It was also clear that their experiences complemented one another. By the time of their first meeting, Crick had taught himself a great deal about x-ray diffraction and protein structure, while Watson had become well informed about phage and bacterial genetics.

Both Crick and Watson were aware of the work of biochemists Maurice Wilkins and **Rosalind Franklin** at King's College, London, who were using x-ray diffraction to study the structure of DNA. Crick, in particular, urged the London group to build models, much as American chemist **Linus Pauling** had done to solve the problem of the alpha helix of proteins. Pauling, the father of the concept of the chemical bond, had demonstrated that proteins had a three-dimensional structure and were not simply linear strings of **amino acids**. Wilkins and Franklin, working independently, preferred a more deliberate experimental approach over the theoretical, model-building scheme used by Pauling and advocated by Crick. Thus, finding the King's College group unresponsive to their suggestions, Crick and Watson devoted portions of a two-year period discussing and arguing about the problem. In early 1953, they began to build models of DNA.

Using Franklin's x-ray diffraction data and a great deal of trial and error, they produced a model of the DNA **molecule** that conformed both to the London group's findings and to the data of Austrian-born American biochemist **Erwin Chargaff**. In 1950 Chargaff had demonstrated that the relative amounts of the four nucleotides (or "bases") that make up DNA conformed to certain rules, one of which was that the amount of adenine (A) was always equal to the amount of thymine (T), and the amount of guanine (G) was always equal to the amount of cytosine (C). Such a relationship suggests pairings of A and T, and G and C, and refutes the idea that DNA is nothing more than a "tetranucleotide," that is, a simple molecule consisting of all four bases.

During the spring and summer of 1953, Crick and Watson wrote four papers about the structure and the supposed function of DNA, the first of which appeared in the journal *Nature* on April 25. This paper was accompanied by papers by Wilkins, Franklin, and their colleagues, presenting experimental evidence that supported the Watson-Crick model. Watson won the coin toss that placed his name first in the authorship, thus forever institutionalizing this fundamental scientific accomplishment as "Watson-Crick."

The first paper contains one of the most remarkable sentences in scientific writing: "It has not escaped our notice that the specific pairing we have postulated immediately suggests a possible copying mechanism for the genetic material." This conservative statement (it has been described as "coy" by some observers) was followed by a more speculative paper in *Nature* about a month later that more clearly argued for the fundamental biological importance of DNA. Both papers were discussed at the 1953 Cold Spring Harbor Symposium, and the reaction of the developing community of molecular biologists was enthusiastic. Within a year, the Watson-Crick model began to generate a broad spectrum of important research in genetics.

Over the next several years, Crick began to examine the relationship between DNA and the genetic code. One of his first efforts was a collaboration with Vernon Ingram, which led to Ingram's 1956 demonstration that sickle cell hemoglobin differed from normal hemoglobin by a single amino acid. Ingram's research presented evidence that a "molecular genetic disease," caused by a Mendelian **mutation**, could be connected to a DNA-protein relationship. The importance of this work to Crick's thinking about the function of DNA cannot be underestimated. It established the first function of "the genetic substance" in determining the specificity of proteins.

About this time, South African-born English geneticist and molecular biologist **Sydney Brenner** joined Crick at the Cavendish Laboratory. They began to work on "the coding problem," that is, how the sequence of DNA bases would specify the amino acid sequence in a protein. This work was first presented in 1957, in a paper given by Crick to the Symposium of the Society for Experimental Biology and entitled "On Protein Synthesis." Judson states in *The Eighth Day of Creation* that "the paper permanently altered the logic of biology." While the events of the transcription of DNA and the synthesis of protein were not clearly understood, this paper succinctly states "The Sequence Hypothesis... assumes that the specificity of a piece of nucleic acid is expressed solely by the sequence of its bases, and that this sequence is a (simple) code for the amino acid sequence of a particular protein." Further, Crick articulated what he termed "The Central Dogma" of molecular biology, "that once 'information' has passed into protein, it cannot get out again. In more detail, the transfer of information from nucleic acid to nucleic acid, or from nucleic acid to protein may be possible, but transfer from protein to protein, or from protein to nucleic acid is impossible." In this important theoretical paper, Crick establishes not only the basis of the genetic code but predicts the mechanism for **protein synthesis**. The first step, transcription, would be the transfer of information in DNA to ribonucleic acid (RNA), and the second step, translation, would be the transfer of information from **RNA** to protein. Hence, to use the language of the molecular biologists, the genetic message is "transcribed" to a messenger, and that message is eventually "translated" into action in the synthesis of a protein.

A few years later, American geneticist **Marshall Warren Nirenberg** and others discovered that the nucleic acid sequence U-U-U (polyuracil) encodes for the amino acid phenylalanine, and thus began the construction of the DNA/RNA dictionary. By 1966, the DNA triplet code for twenty amino acids had been worked out by Nirenberg and others, along with details of protein synthesis and an elegant example of the control of protein synthesis by French geneticist François Jacob, Arthur Pardée, and French biochemist Jacques Lucien Monod. Brenner and Crick themselves turned to problems in developmental biology in the 1960s, eventually studying the structure and possible function of histones, the class of proteins associated with chromosomes.

In 1976, while on sabbatical from the Cavendish, Crick was offered a permanent position at the Salk Institute for Biological Studies in La Jolla, California. He accepted an endowed chair as Kieckhefer Professor and has been at the Salk Institute ever since. At the Salk Institute, Crick began to study the workings of the brain, a subject that he had been interested in from the beginning of his scientific career. While his primary interest was consciousness, he attempted to approach this subject through the study of vision. He published several speculative papers on the mechanisms of dreams and of attention, but, as he stated in his autobiography, "I have yet to produce any theory that is both novel and also explains many disconnected experimental facts in a convincing way."

An interesting footnote to Crick's career at the Salk Institute was his proposal of the idea of "directed panspermia." Along with Leslie Orgel, he published a book, *Life Itself,* which suggested that microbes drifted through space, eventually reaching Earth and "seeding" it, and that this dispersal event has been caused by the action of "someone." Crick himself was ambivalent about the theory, but he and Orgel proposed it as an example of how a speculative theory might be presented.

During his career as an energetic theorist of modern biology, Francis Crick has accumulated, refined, and synthesized the experimental work of others, and has brought his unusual insights to fundamental problems in science. There is probably no better description of Crick's intellectual gifts than that of François Jacob, who, in his book *The Statue Within,* describes Crick's famous paper "On Protein Synthesis" by noting, "On this difficult subject, Crick was dazzling. He had the gift of going straight to the crux of the matter and ignoring the rest. Of extracting from the hodge-podge of the literature, the solid and the relevant, while rejecting the soft and the vague."

Critical phenomena

Everything in the universe that occupies space is considered **matter**. All matter is classified as being a solid, liquid, gas or **plasma** (very hot ionized gases) depending on its physical state. Each of these **states of matter** is characterized by different physical properties. Critical phenomena relates to the physical properties of matter that occur at the phase transition points and specifically at the critical points.

While the first theories of matter were proposed by the ancient Greeks as early as 500 B.C., it was not until the nineteenth century that the modern understanding of matter was proposed. At this time, English chemist **John Dalton** introduced the **atomic theory** that postulated that all matter is composed of individual particles called atoms. He also suggested that various atoms exist with different masses and sizes to make up all of the known elements. Today, it is known that these atoms can bond with each other to produce molecules.

Building on Dalton's theory, scientists deduced that the different states of matter are a result of the physical arrangement of atoms in space. For example, a solid substance like ice is composed of **water** molecules that are bound relatively close together and neatly ordered. When the ice is exposed to **heat** under atmospheric pressure, the molecules move farther apart and become less ordered. When the **temperature** is raised above the first transition temperature (the melting point) the solid ice abruptly becomes liquid water. Similarly, when the temperature is increased above the second transition temperature (**boiling point**) the molecules move so far apart that the liquid water quickly becomes water vapor, or steam. When the temperature is raised so high that the electrons are separated from their atoms (the atoms are 100% ionized), then plasma is formed. The transitions from solid to liquid and liquid to gas represent first-order phase transitions.

Critical phenomena relate to the changes that occur at the second-order or continuous phase transition. This is a physical transition that occurs above the critical point. The critical temperature is the temperature above which vapor cannot be liquefied no matter how much pressure is applied to the system. The critical pressure is the amount of pressure required to liquefy water at the critical temperature. Together, these values define the critical point. For water, the critical point occurs at a temperature of 705.2°F (374°C) and a pressure of 2.21 x 10^7 pascals.

Different materials have different critical points. Unlike first-order phase transitions, second-order transitions are gradual. In a pressurized system of water vapor, as the temperature is decreased toward the critical point, the vapor goes through an intermediate "fluid" critical phase transition that is neither liquid nor a gas. It is milky and turbid and represents a phenomenon known as critical opalescence. Another example of a critical phenomenon occurs in solid, ferromagnetic materials such as **iron** and **nickel**. At the critical point, also called the **Curie point**, these materials lose their natural magnetic properties and can only be magnetized by applying a magnetic field. **Superconductivity** is another example of a critical phenomenon.

One of the primary goals of scientists who study critical phenomena is to develop a common theory to explain and predict the behavior of matter above the critical point. A common feature of all systems experiencing critical phase transitions is order parameters. These are quantities that are zero on one side of the critical point and nonzero on the other. Net **magnetism** is an order parameter in ferromagnetic systems because above the critical point magnetism is zero. Below this point, it gradually increases. **Density** and **concentration** are other properties that can be order parameters. In addition to order parameters, other thermodynamic quantities, such as heat capacity, fluctuate significantly at the critical point.

The mean field theories represent the earliest attempts at describing the behavior of systems experiencing critical transitions. These include the van der Waals model for fluids, proposed in 1873, and the Weiss model of ferromagnets, suggested in 1907. Characteristic of these theories is the assumption that every particle in a system can be described by the average properties of the whole system. While these theories were successful in predicting numerous aspects of critical phenomena, they did not provide adequate quantitative results. This desire for quantitative results fueled the development of the modern theory of critical phenomena.

The modern theory began with the scaling hypothesis suggested by Ben Widom in 1965 and the universality hypothesis proposed by Leo Kadanoff in 1967. These theories were based on the assumption that the values of critical exponents of a system depend only on the general features such as dimensionality and symmetry and not on particle- particle interactions. Using a mathematical procedure originally developed for quantum mechanics, Ken Wilson translated the ideas of the scaling hypothesis and the universality hypothesis into the renormalization group method. This method resulted in more solvable equations.

One of the difficulties in verifying these critical phenomena hypotheses is encountered when measuring the critical parameters experimentally. This is because the experiments must be extremely precise and carefully done. In 1996, Robert Gammon and his colleagues overcame some of these problems by performing a critical phenomena experiment on the space shuttle *Columbia*. High above Earth, they were able to maintain the critical temperature to millionths of a degree without the interference of gravity. This allowed them to take readings of various properties at the critical point. It is thought that through more experiments like these researchers will gain a fundamental understanding for what actually occurs during critical phase transitions.

See also Phase changes; States of matter

William Crookes.

Crookes, William (1832-1919)

English chemist and physicist

The English chemist and physicist Sir William Crookes discovered the element **thallium** and invented the radiometer, the spinthariscope, and the Crookes tube.

William Crookes was born in London on June 17, 1832. His education was limited, and despite his father's wish that he become an architect, he chose industrial **chemistry** as a career. He entered the Royal College of Chemistry in London, where he began his researches in chemistry. In 1859, he founded the *Chemical News,* which made him widely known, and remained its editor and owner all his life.

Most notable among Crookes's chemical studies is that one which led to his 1861 discovery of thallium. Using spectrographic methods, he had observed a green line in the spectrum of **selenium**, and he was thus led to announce the existence of a new element, thallium. While determining the atomic weight of thallium, using a delicate vacuum **balance**, he noticed several irregularities in weighing, which he attributed to the method. His investigation of this phenomenon led to the construction in 1875 of an instrument that he named the radiometer.

In 1869, J. W. Hittorf first studied the phenomena associated with electrical discharges in vacuum tubes. Not knowing of this, Crookes, 10 years later, made a parallel but more extensive investigation. In his 1878 report, he pointed out the significant properties of electrons in a vacuum, including the fact that a magnetic field causes a deflection of the emission. He suggested that the tube was filled with **matter** in what he called the ''fourth state;'' that is, the mean free path of the molecules is so large that collisions between them can be ignored. Tubes such as this are still called ''Crookes tubes,'' and his work was honored by naming the space near the cathode in low **pressure** ''Crookes dark space.''

Crookes also made useful contributions to the study of **radioactivity** in 1903 by developing the spinthariscope, a device for studying alpha particles. He foresaw the urgent need for nitrogenous **fertilizers**, which would be used to cultivate crops to meet the demands of a rapidly expanding population. Crookes did much to popularize phenol (carbolic acid) as an antiseptic; in fact, he became an expert on sanitation. Mention should also be made of the serious and active interest he took in psychic phenomena, to which he devoted most of four years.

Crookes was knighted in 1897. His marriage lasted from 1856 until the death of his wife in 1917; they had 10 children. He died in London on April 4, 1919.

Cross, Charles Frederick (1855-1935)

English chemist

Charles Cross was born in 1855 in Brentford, England. Upon his graduation from King's College of London in 1878, he set up an analytical and consulting chemical firm with Edward John Bevan in London.

In 1892 Cross, Bevan, and Beadle developed viscous rayon which is formed by first dissolving **cellulose** from **wood** fibers in a **sodium** hydroxide **solution** and then treating it with **carbon** disulfide after drying. The resulting solution—called viscose—is extruded through a small hole into an acid bath to form. After drying these fibers can be spun into fabric.

Cross wrote several technical books, including *Cellulose*, before his death in Sussex in 1935.

Crutzen, Paul J. (1933-)

Dutch meteorologist

Paul Crutzen is one of the world's leading researchers in mapping the chemical mechanisms that affect the ozone layer. He has pioneered research on the formation and depletion of the ozone layer and threats placed upon it by industrial society. Crutzen has discovered, for example, that **nitrogen** oxides accelerate the rate of ozone depletion. He has also found that chemicals released by bacteria in the soil affect the thickness of the ozone layer. For these discoveries he has received the 1995 Nobel Prize in **Chemistry**, along with **Mario Molina** and **Sherwood Rowland** for their separate discoveries related to the ozone and how **chlorofluorocarbons** (CFCs) deplete the **ozone** layer. According to Royal Swedish Academy of Science, ''by

Paul J. Crutzen. *(Corbis Corporation. Reproduced by permission.)*

explaining the chemical mechanisms that affect the thickness of the ozone layer, the three researchers have contributed to our salvation from a global environmental problem that could have catastrophic consequences.''

Paul Josef Crutzen was born December 3, 1933, to Josef C. Crutzen and Anna Gurek in Amsterdam. Despite growing up in a poor family in Nazi-occupied Holland during 1940-1945, he was nominated to attend high school at a time when not all children were accepted into high school. He liked to play soccer in the warm months and ice skate 50-60 miles (80-97 km) a day in the winter. Because he was unable to afford an education at a university, he attended a two-year college in Amsterdam. After graduating with a civil engineering degree in 1954, he designed bridges and homes.

Crutzen met his wife, Tertu Soininen, while on vacation in Switzerland in 1954. They later moved to Sweden where he got a job as a computer programmer for the Institute of Meteorology and the University of Stockholm. He started to focus on **atmospheric chemistry** rather than mathematics because he had lost interest in math and did not want to spend long hours in a lab, especially after the birth of his two daughters, Illona and Sylvia. Despite his busy schedule, Crutzen obtained his doctoral degree in Meteorology at Stockholm University at the age of 35.

Crutzen's main research focused on the ozone, a bluish, irritating gas with a strong odor. The ozone is a **molecule** made up of three **oxygen** atoms (O_3) and is formed naturally in the atmosphere by a photochemical reaction. The ozone begins approximately 10 miles (16 km) above Earth's surface, reaching between 20-30 miles (32-48 km) in height, and acts as a protective layer that absorbs high-energy ultraviolet radiation given off by the sun.

In 1970 Crutzen found that soil microbes were excreting **nitrous oxide** gas, which rises to the stratosphere and is converted by sunlight to **nitric oxide** and nitrogen dioxide. He determined that these two **gases** were part of what caused the depletion of the ozone. This discovery revolutionized the study of the ozone and encouraged a surge of research on global biogeochemical cycles.

In 1977, while he was the director of the National Center for Atmospheric Research (NCAR) in Boulder, Colorado, Crutzen studied the effects of burning trees and brush in the fields of Brazil. Every year farmers cleared the forests by burning everything in sight. The theory at the time was that this burning caused more **carbon** compounds or trace gases and **carbon monoxide** to enter the atmosphere. These gases were believed to cause the **greenhouse effect**, or a warming of the atmosphere. Crutzen collected and examined this smoke in Brazil and discovered that the complete opposite was occurring. He stated in *Discover* magazine: ''Before the industry got started the tropical burning was actually decreasing the amount of **carbon dioxide** in the atmosphere.'' The study of smoke in Brazil led Crutzen to further examine what effects larger amounts of different kinds of smoke might have on the environment, such as smoke from a nuclear war.

The journal *Ambio* commissioned Crutzen and John Birks, his colleague from the University of Colorado, to investigate what effects nuclear war might have on the planet. Crutzen and Birks studied a simulated worldwide nuclear war. They theorized that the black carbon soot from the raging fires would absorb as much as 99% of the sunlight. This lack of sunlight, coined ''nuclear winter,'' would be devastating to all forms of life. For this theory Crutzen was named ''Scientist of the Year'' by *Discover* magazine in 1984 and awarded the prestigious Tyler Award four years later.

As a result of the discoveries by Crutzen and other environmental scientists, a very crucial international treaty was established in 1987. The Montreal Protocol was negotiated under the auspices of the United Nations and signed by 70 countries to slowly phase out the production of chlorofluorocarbons and other ozone-damaging chemicals by the year 2000. However, the United States had ended the production of CFCs five years earlier, in 1995. According to the *New York Times,* ''the National Oceanic and Atmospheric Administration reported in 1994, while ozone over the South Pole is still decreasing, the depletion appears to be leveling off.'' Even though the ban has been established, existing CFCs will continue to reach the ozone, so the depletion will continue for some years. The full recovery of the ozone is not expected for at least 100 years.

From 1977-80, Crutzen was director of the Air Quality Division, National Center for Atmospheric Research (NCAR), located in Boulder, Colorado. While at NCAR, he located he taught classes at Colorado State University in the department of Atmospheric Sciences. Since 1980 he has been a member of the Max Planck Society for the Advancement of Science, and he is the director of the Atmospheric Chemistry division at Max Planck Institute for Chemistry. In addition to Crutzen's position at the institute, he is a part-time professor at Scripps Institution of Oceanography at the University of California. In 1995 he was the recipient of the United Nations Environmental Ozone Award for outstanding contribution to the protection of the ozone layer. Crutzen has co-authored and edited several books, as well as having published several hundred articles in specialized publications.

CRYSTALLIZATION AND CRYSTAL GROWTH

Crystal lattices consist of repeating units of a **molecule**, **ion**, or **atom** arranged in a symmetrical array. One can imagine a wallpaper pattern with a repeating design that extends in three dimensions. Similarly, the only difference between each repeating molecule in a crystal is a translation and perhaps a rotation extended in three dimensions to comprise a beautiful, well-ordered crystalline solid with smooth edges. Crystals occur in several forms, from those we encounter daily—ice, **salt**, table sugar, metals—to the more precious gems—quartz and diamond—to finally the ones artificially created—protein, and even **DNA**.

Crystallographers describe the repeating unit of a crystal lattice, the unit cell, by the length of its edges and angles between them. This aids in classifying the 14 different kinds of lattices that are categorized into seven crystal systems: Cubic, Tetragonal, Orthorhombic, Monoclinic, Triclinic, Rhombohedral, and Hexagonal. Each of the 14 lattices differs by its unit-cell dimensions. The type of lattice and **molecular structure** can be deciphered by subjecting the crystal to a beam of x rays. This generates a pattern of spots on a film that translates into a molecular structure using sophisticated calculations done by computer software. Structures of several **proteins** and organic molecules were determined with this very useful method, as well as the helical structure of DNA.

Although molecules, ions, and atoms are limited to only 14 lattices similar crystals can have different physical properties. The differences lie in the packing of the molecules, their attractive and repulsive forces, and the type of **bonding** that supports lattice stability. The weakest crystals, molecular crystals, consist of weak attractive forces between the molecules, such as **hydrogen** bonds, hydrophobic interactions, and **dipole** interactions. For example, ice crystals are composed of **water** molecules connected by hydrogen bonds, such that each water molecule can coordinate with four other water molecules. Other examples include dry ice (CO_2) and artificial crystals as protein and DNA. Ionic crystals are composed of ions at the lattice points sustained by stronger, electrostatic forces. Examples include limestone ($CaCO_3$) and table salt (NaCl), which is composed of a cubic lattice with Na+ and Cl- ions occupying the lattice points. These types of crystals tend to be harder and have higher melting points than the weaker molecular crystals. However, these crystals are still not as sturdy as covalent crystals. Held together by a network of covalent bonds, covalent crystals have very high melting temperatures and are tough substances. **Diamond**, the hardest natural substance, contains elemental **carbon** that is covalently bonded to four neighboring carbon atoms. Quartz (SiO_2) is an integral component of many sands. Covalent crystals are often used to produce cutting tools since they are so hard. None of the types of crystals mentioned so far are electrical conductors. Metallic crystals have this property because of the way in which the atoms are arranged. Solid metallic crystals, such as **copper** (Cu) and **iron** (Fe) usually form cubic or hexagonal lattice systems. The nuclei and inner electrons are localized at the lattice points while the outer electrons are free moving and shared by the entire l attice, enabling **metals** to conduct **electricity**. Metals are flexible and usually have high melting temperatures.

Crystals grow in supersaturated solutions. These solutions contain an amount of pure substance that exceeds the **solubility** of that substance. That is, the substance is barely dissolved in water and the **solution** is unstable. The strategy lies in finding the point of supersaturation that coerces the substance to leave the aqueous phase in an organized fashion to form a crystal. The very first molecules or atoms join (known as nucleation) and as more water evaporates, more molecules join the lattice and the crystal grows. This is the premise of crystal growth. Eventually, crystals cease growing when the substance depletes, or when crystal defects arise. Depending on the chemical substance, crystals grow immediately, within hours, or sometimes within months.

There are several ways of reaching supersaturation. Heating a saturated solution followed by slow cooling is the simplest laboratory method. This also occurs naturally in the earth; hot **gases** evaporate and leave behind saturated minerals which then crystallize. A more controlled version of this technique is vapor **diffusion**. A concentrated solution containing a pure is placed in a closed chamber separating it from an outer reservoir of **alcohol** or any substance that readily absorbs water. As the water evaporates, nucleation begins. Another common technique, **dialysis**, employs a semi-permeable membrane that allows only small molecules to pass. The substance to be crystallized is trapped inside the membrane and as new molecules enter the membrane, the solution becomes more saturated and crystals form. Ionic and molecular crystals are commonly grown using these laboratory techniques.

These procedures are not always so simple and frequently result in noncrystalline precipitates. Further manipulation by the crystallographer is then required. Sometimes a single pre-existing crystal is placed in the concentrated solution as a ''seed'' to stimulate growth. Other parameters affecting crystal growth—**pH**, **temperature**, time, and pressure—can also be manipulated. Artificial diamonds require high temperature and pressure for growth. Solutions concentrated too rapidly or too slowly could result in defective crystals or unwanted salt crystals. Solution impurities retard crystal growth since they impede nucleation and assembly. Researchers are also investigating the gravitational affect of crystal growth by attempting to grow crystals in space.

Crystals exemplify the beauty, order, and symmetry of nature and are often visually attractive. One can engage in this wondrous process by making sucrose crystals, or rock candy. Make a supersaturated sugar solution by dissolving 2.5 cups

Pyrite crystal. *(Photograph by Joe Bator; The Stock Market. Reproduced by permission.)*

of sugar into 1 cup of water over low **heat**. Then cook without stirring over medium heat until the solution reaches 250°F. Pour the solution into a pan or **glass** jar. Drop in a string or stick not allowing it to reach bottom, or instead add a few sugar-grains as seeds. Cover the solution and do not disturb. Observe crystal for mation in a few days.

Crystallography

Crystallography is the study of materials in which the atoms stack in a three-dimensionally ordered geometric arrangement. In a single crystal a single pattern extends throughout the entire material. Polycrystalline materials have discontinuities in the **periodicity** of the material. In amorphous materials, there is still less periodicity within the material; the amount of non-ordered atoms in amorphous materials is at least comparable with that exhibiting periodicity.

Detailed analyses of crystal structures are carried out by **x-ray diffraction**. In 1912, Max von Laue predicted that the spacing of crystal layers is small enough to diffract light of the appropriate wavelength. William Henry Bragg and his son, **William Lawrence Bragg**, were awarded the Nobel Prize in **chemistry** (1915) for their development of crystal structure analysis using x-ray diffraction. The Braggs' found that when x-ray radiation was scattered by a crystalline material both constructive and destructive interference occurred; this interference occurs because the wavelength of the x rays is of similar magnitude to the spaces between the atoms of the crystal. Analysis of the resulting diffraction pattern, the position of the lines of the scattered radiation along with their relative intensities, is the basis of the x-ray diffraction technique; this analysis allows the determination of the precise location of the atoms in the crystal.

The x-ray diffraction technique has been one of the most important structural methods throughout the twentieth century. It has expanded our knowledge by providing detailed structures of **vitamins**, **proteins** (enzymes, bacterial membranes, **liquid crystals**, polymers, organic compounds and inorganic compounds.

The idea that a crystal is composed of identical structural subunits was first proposed in 1784 based on observations of the cleavage of calcite. Subsequent investigations have shown that the structures of these subunits can be inferred from a crystal's symmetry. Even casual observation suggests that the symmetry of a crystal as a whole is related to some smaller subunit within it. The subunit is called a unit cell and it contains all of the essential information, such as the symmetry and elemental composition, of the crystal. Repeated translation along the edges of the unit cell can be used to derive the entire crystal lattice; in other words, the crystal lattice is the unit cell repeated many times in a periodic fashion.

The unit cells are often categorized in terms of space lattices called Bravais lattices; in such lattices, imaginary points called lattice points replace all of the atoms of the crystals. There are only fourteen distinct types of Bravais lattices, and these are associated with seven crystal systems. The crystal systems are all parallelepipeds whose shapes are completely defined by the lengths of the three sides and by the three angles characterizing the parallelepiped.

The most important symmetry elements in the consideration of crystal structures are axes of rotation and mirror planes. An n-fold rotation axis brings the crystal into self-coincidence after rotation by 360° a mirror plane occurs when the crystal can be bisected in such a way that one half is the reflection of the other. The seven crystal systems, distinguished by their axes of symmetry are as follows (see Figure 1 for examples): 1) triclinic no symmetry elements; all unit cell lengths different, no 90° unit cell angles. 2) monoclinic one 2-fold axis or one plane; all unit cell lengths different, two 900 unit cell angles and one non-900 unit cell angle. 3) orthorhombic three mutually perpendicular axes, or two planes intersecting in a 2-fold axis; all unit cell lengths different, all unit cell angles = 900. 4) tetragonal one 4-fold axis or a 4-fold inversion axis; two unit cell lengths are identical and one is different, all unit cell angles = 900. 5) trigonal one 3-fold axis; all unit cell lengths are identical, all unit cell angles are identical but are not = 900. 6) hexagonal one 6-fold axis; two unit cell lengths identical and one different, two unit cell angles = 900 and the third = 1200. 7) cubic four 3-fold axes; all bond lengths identical, all bond angles = 900. The cubic form is the most symmetric, and the triclinic form the least. Generally speaking, crystals of lower symmetry are more common in nature than those of higher symmetry.

Another issue in crystallography is the physical basis for the crystal packing. How and why do crystals form? In what patterns do different types of materials crystallize? Crystals form since the ordered states of most **solids** are energetically more favorable (lower in energy) than disordered or irregularly packed ones. Hardness is one indication of the packing efficiency of atoms in a material. Materials with small atoms, packed closely together with strong covalent bonds throughout tend to be the hardest materials. The softest materials may contain metallic bonds or weaker van der Waals interactions.

The placement of atoms in a crystal can be described in terms of their layering. Within each layer, the most efficient packing occurs when the particles are staggered with respect to one another, leaving small triangular spaces between the particles. The second layer is placed on top of the first, in the depressions between the particles of the first layer. In one packing arrangement, the third layer lies in the depressions of the second and directly over depressions of the first layer. These so-called *face-centered cubic* close-packed structures are common in **metals** such as **aluminum**, **rhodium**, **iridium**, **copper**, **silver**, and **gold**. If the third layer particles are directly over atoms of the first, one has *hexagonal close packing*. This packing arrangement also is observed for many metals, including **rubidium**, **osmium**, **cobalt**, **zinc**, and **cadmium**.

Although ionic solids follow similar patterns as described above for metals, the detailed arrangements are more complicated, because the positioning of two different types of ions, cations and anions, must be considered. In general, it is the larger **ion** (usually, the anion) that determines the overall packing and layering, while the smaller ion fits in the holes (spaces) that occur throughout the layers.

The **energy** associated with crystal formation, the lattice energy, can be calculated by consideration of the different types of bonds within the solid: van der Waals bonds, ionic bonds, **hydrogen** bonds, covalent bonds, and/or metallic bonds. In the case of van der Waals solids (e.g., Ne, CO_2), the lattice energy can be calculated by summing up the pair potentials of interacting atoms using a secondary bond potential for atomic interactions. For ionic solids (e.g., NaCl, ZnS), the Coulomb interaction, supplemented with a strongly repulsive force, is used in place of the secondary bond potential. In covalent (e.g., **diamond**, graphite) and hydrogen-bonded (e.g., H_2O) materials, the calculation of the lattice energy is much more complicated, and the lattice energy cannot simply be calculated as a sum over pair potentials acting between atoms.

The growth and size of any crystal depends on the conditions of its formation. **Temperature**, pressure, the presence of impurities, etc., will affect the size and perfection of the crystal. As a crystal grows, different imperfections may occur which can be classified as either point defects, line defects (or dislocations), or plane defects. Point defects include missing atoms or substituted atoms; line defects are defects that extend along straight or curved lines in a crystal; plane defects extend along true planes or curved surfaces within crystals.

Sometimes, imperfections are introduced to crystals intentionally. For example, the conductivity of **semiconductors** such as **silicon** and **germanium** can be modulated by the intentional addition of **arsenic** or **antimony** impurities. This procedure is called "doping." In this case, the additional electrons provided by arsenic or antimony impurities result in increased conductivity of the semiconductors.

Experiments in decreased gravity conditions aboard the space shuttles and in *Spacelab I* demonstrated that proteins formed crystals rapidly, and with fewer imperfections, than is possible under normal gravitational conditions. This is important because macromolecules are difficult to crystallize, and usually will form only crystallites whose structures are difficult to analyze. Protein analysis is important because many diseases (including Acquired Immune Deficiency Syndrome, AIDS) involve enzymes, which are the highly specialized protein catalysts of chemical reactions in living organism. The analysis of other biomolecules may also benefit from these experiments. It is interesting that similar advantages in crystal growth and degree of perfection have also been noted with crystals grown under high gravity conditions.

There is continued scientific interest in learning new ways to grow crystals due to the promise of new or improved crystalline materials with valuable properties. Although methods of synthesizing larger diamonds are expensive, polycrystalline diamond films can be made cheaply by a method called chemical vapor deposition (CVD). The technique involves **methane** and hydrogen **gases**, a surface on which the film can deposit, and a microwave oven. Energy from microwaves breaks the bonds in the gases, and, after a series of reactions, **carbon** films in the form of diamond are produced. The method holds much promise for a) the tool and cutting industry since diamond is the hardest known substance; b) electronics applications since diamond is a conductor of **heat**, but not **electricity**; and c) medical applications since it is tissue- compatible and tough. Thus it is suitable for use in items such as joint replacements and heart valves.

See also Bonding; Electrical conductivity; X-ray diffraction

Curie, Marie (1867-1934)

French physicist and radiation chemist

Marie Curie was the first woman to win a Nobel Prize, and one of very few scientists ever to win that award twice. In collaboration with her physicist-husband **Pierre Curie**, Marie Curie developed and introduced the concept of **radioactivity** to the world. Working in very primitive laboratory conditions, Curie investigated the nature of high **energy** rays spontaneously produced by certain elements, and isolated two new radioactive elements, **polonium** and **radium**. Her scientific efforts also included the application of X rays and radioactivity to medical treatments.

Curie was born to two schoolteachers on November 7, 1867, in Warsaw, Poland. Christened Maria Sklodowska, she was the fourth daughter and fifth child in the family. By the age of five, she had already begun to suffer deprivation. Her mother Bronislawa had contracted tuberculosis and assiduously avoided kissing or even touching her children. By the time

Marie Curie.

Curie was eleven, both her mother and her eldest sister Zosia had passed away, leaving her an avowed atheist. Curie was also an avowed nationalist (like the other members in her family), and when she completed her elementary schooling, she entered Warsaw's "Floating University," an underground, revolutionary Polish school that prepared young Polish students to become teachers.

Curie left Warsaw at the age of seventeen, not for her own sake but for that of her older sister Bronya. Both sisters desired to acquire additional education abroad, but the family could not afford to send either of them, so Marie took a job as a governess to fund her sister's medical education in Paris. At first, she accepted a post near her home in Warsaw, then signed on with the Zorawskis, a family who lived some distance from Warsaw. Curie supplemented her formal teaching duties there with the organization of a free school for the local peasant children. Casimir Zorawski, the family's eldest son, eventually fell in love with Curie and she agreed to marry him, but his parents objected vehemently. Marie was a fine governess, they argued, but Casimir could marry much richer. Stunned by her employers' rejection, Curie finished her term with the Zorawskis and sought another position. She spent a year in a third governess job before her sister Bronya finished medical school and summoned her to Paris.

In 1891, at the age of twenty-four, Curie enrolled at the Sorbonne and became one of the few women in attendance at the university. Although Bronya and her family back home were helping Curie pay for her studies, living in Paris was quite expensive. Too proud to ask for additional assistance, she subsisted on a diet of buttered bread and tea, which she augmented sometimes with fruit or an egg. Because she often went without **heat**, she would study at a nearby library until it closed. Not surprisingly, on this regimen she became anemic and on at least one occasion fainted during class.

In 1893, Curie received a degree in physics, finishing first in her class. The following year, she received a master's degree, this time graduating second in her class. Shortly thereafter, she discovered she had received the Alexandrovitch Scholarship, which enabled her to continue her education free of monetary worries. Many years later, Curie became the first recipient ever to pay back the prize. She reasoned that with that money, yet another student might be given the same opportunities she had.

Friends introduced Marie to Pierre Curie in 1894. The son and grandson of doctors, Pierre had studied physics at the Sorbonne; at the time he met Marie, he was the director of the École Municipale de Physique et Chimie Industrielles. The two became friends, and eventually she accepted Pierre's proposal of marriage. Their Paris home was scantily furnished, as neither had much interest in housekeeping. Rather, they concentrated on their work. Pierre Curie accepted a job at the School of Industrial Physics and **Chemistry** of the City of Paris, known as the EPCI. Given lab space there, Marie Curie spent eight hours a day on her investigations into the magnetic qualities of steel until she became pregnant with her first child, Irene, who was born in 1897.

Curie then began work in earnest on her doctorate. Like many scientists, she was fascinated by French physicist **Antoine-Henri Becquerel**'s discovery that the element **uranium** emitted rays that contained vast amounts of energy. Unlike Wilhelm Röntgen's X rays, which resulted from the excitation of atoms from an outside energy source, the ''Becquerel rays'' seemed to be a naturally occurring part of the uranium ore. Using the piezoelectric quartz electrometer developed by Pierre and his brother Jacques, Marie tested all the elements then known to see if any of them, like uranium, caused the nearby air to conduct **electricity**. In the first year of her research, Curie coined the term ''radioactivity'' to describe this mysterious force. She later concluded that only **thorium** and uranium and their compounds were radioactive.

While other scientists had also investigated the radioactive properties of uranium and thorium, Curie noted that the minerals pitchblende and chalcolite emitted more rays than could be accounted for by either element. Curie concluded that some other radioactive element must be causing the greater radioactivity. To separate this element, however, would require a great deal of effort, progressively separating pitchblende by chemical analysis and then measuring the radioactivity of the separate components. In July, 1898, she and Pierre successfully extracted an element from this ore that was even more radioactive than uranium; they called it polonium in honor of Marie's homeland. Six months later, the pair discovered another radioactive substance—radium—embedded in the pitchblende.

Although the Curies had speculated that these elements existed, to prove their existence they still needed to describe them fully and calculate their atomic weight. In order to do so, Curie needed an abundant supply of pitchblende and a better laboratory. She arranged to get hundreds of kilograms of waste scraps from a pitchblende mining firm in her native Poland, and Pierre Curie's EPCI supervisor offered the couple the use of a laboratory space. The couple worked together, with Marie performing the physically arduous job of chemically separating the pitchblende and Pierre analyzing the physical properties of the substances that Marie's separations produced. In 1902 the Curies announced that they had succeeded in preparing a decigram of pure radium chloride and had made an initial determination of radium's atomic weight. They had proven the chemical individuality of radium.

Pierre Curie's father had moved in with the family and assumed the care of their daughter, Irene, so the couple could devote more than eight hours a day to their beloved work. Pierre Curie's salary, however, was not enough to support the family, so Marie took a position as a lecturer in physics at the École Normal Supérieure; she was the first woman to teach there. In the years between 1900 and 1903, Curie published more than she had or would in any other three-year period, with much of this work being coauthored by Pierre Curie. In 1903 Curie became the first woman to complete her doctorate in France, summa cum laude.

The year Curie received her doctorate was also the year she and her husband began to achieve international recognition for their research. In November the couple received England's prestigious Humphry Davy Medal, and the following month Marie and Pierre Curie—along with Becquerel—received the Nobel Prize in physics for their efforts in expanding scientific knowledge about radioactivity. Although Curie was the first woman ever to receive the prize, she and Pierre declined to attend the award ceremonies, pleading they were too tired to travel to Stockholm. The prize money from the Nobel, combined with that of the Daniel Osiris Prize—which she received soon after—allowed the couple to expand their research efforts. In addition, the Nobel bestowed upon the couple an international reputation that furthered their academic success. The year after he received the Nobel, Pierre Curie was named professor of physics of the Faculty of Sciences at the Sorbonne. Along with his post came funds for three paid workers, two laboratory assistants and a laboratory chief, stipulated to be Marie. This was Marie's first paid research job.

In December, 1904, Marie gave birth to another daughter, Eve Denise, having miscarried a few years earlier. Despite the fact that both Pierre and Marie frequently suffered adverse effects from the radioactive materials with which they were in constant contact, Eve Denise was born healthy. The Curies continued their work regimen, taking sporadic vacations in the French countryside with their two children. They had just returned from one such vacation when on April 19, 1906, tragedy struck; while walking in the congested street traffic of Paris, Pierre was run over by a heavy wagon and killed.

A month after the accident, the University of Paris invited Curie to take over her husband's teaching position. Upon acceptance she became the first woman to ever receive a post in higher education in France, although she was not named to a full professorship for two more years. During this time, Curie came to accept the theory of English physicists **Ernest Rutherford** and **Frederick Soddy** that radioactivity was caused by atomic nuclei losing particles, and that these disintegrations caused the **transmutation** of an atomic **nucleus** into a different element. It was Curie, in fact, who coined the terms disintegration and transmutation.

In 1909, Curie received an academic reward that she had greatly desired: the University of Paris drew up plans for an Institut du Radium that would consist of two branches, a laboratory to study radioactivity—which Curie would run—and a laboratory for biological research on radium therapy, to be overseen by a physician. It took five years for the plans to come to fruition. In 1910, however, with her assistant André Debierne, Curie finally achieved the isolation of pure radium metal, and later prepared the first international standard of that element.

Curie was awarded the Nobel Prize again in 1911, this time ''for her services to the advancement of chemistry by the discovery of the elements radium and polonium,'' according to the award committee. The first scientist to win the Nobel twice, Curie devoted most of the money to her scientific studies. During World War I, Curie volunteered at the National Aid Society, then brought her technology to the war front and instructed army medical personnel in the practical applications of radiology. With the installation of radiological equipment in ambulances, for instance, wounded soldiers would not have to be transported far to be x-rayed. When the war ended, Curie returned to research and devoted much of her time to her work.

By the 1920s, Curie was an international figure; the Curie Foundation had been established in 1920 to accept private donations for research, and two years later the scientist was invited to participate on the League of Nations International Commission for Intellectual Cooperation. Her health was failing, however, and she was troubled by fatigue and cataracts. Despite her discomfort, Curie made a highly publicized tour of the United States in 1921. The previous year, she had met Missy Meloney, editor of the *Delineator,* a woman's magazine. Horrified at the conditions in which Curie lived and worked (the Curies had made no money from their process for producing radium, having refused to patent it), Meloney proposed that a national subscription be held to finance a gram of radium for the institute to use in research. The tour proved grueling for Curie; by the end of her stay in New York, she had her right arm in a sling, the result of too many too strong handshakes. However, with Meloney's assistance, Curie left America with a valuable gram of radium.

Curie continued her work in the laboratory throughout the decade, joined by her daughter, **Irene Joliot-Curie**, who was pursuing a doctoral degree just as her mother had done. In 1925, Irene successfully defended her doctoral thesis on alpha rays of polonium, although Curie did not attend the defense lest her presence detract from her daughter's performance. Meanwhile, Curie's health still continued to fail and she was forced to spend more time away from her work in the laboratory. The result of prolonged exposure to radium, Curie contracted leukemia and died on July 4, 1934, in a nursing home in the French Alps. She was buried next to Pierre Curie in Sceaux, France.

CURIE, PIERRE (1859-1906)

French physicist

Pierre Curie was a noted physicist who became famous for his collaboration with his wife **Marie Curie** in the study of **radioactivity**. Before joining his wife in her research, Pierre Curie was already widely known and respected in the world of physics. He discovered (with his brother Jacques) the phenomenon of piezoelectricity—in which a crystal can become electrically polarized—and invented the quartz **balance**. His papers on crystal symmetry, and his findings on the relation between **magnetism** and **temperature** also earned praise in the scientific community. Curie died in a street accident in 1906, a physicist acclaimed the world over but who had never had a decent laboratory in which to work.

Pierre Curie was born in Paris on May 15, 1859, the son of Sophie-Claire Depouilly, daughter of a formerly prominent manufacturer, and Eugène Curie, a free-thinking physician who was also a physician's son. Dr. Curie supported the family with his modest medical practice while pursuing his love for the natural sciences on the side. He was also an idealist and an ardent republican who set up a hospital for the wounded during the Commune of 1871. Pierre was a dreamer whose style of learning was not well adapted to formal schooling. He received his pre-university education entirely at home, taught first by his mother and then by his father as well as his older brother, Jacques. He especially enjoyed excursions into the countryside to observe and study plants and animals, developing a love of nature that endured throughout his life and that provided his only recreation and relief from work during his later scientific career. At the age of 14, Curie studied with a mathematics professor who helped him develop his gift in the subject, especially spatial concepts. Curie's knowledge of physics and mathematics earned him his bachelor of science degree in 1875 at the age of sixteen. He then enrolled in the Faculty of Sciences at the Sorbonne in Paris and earned his *licence* (the equivalent of a master's degree) in physical sciences in 1877.

Curie became a laboratory assistant to Paul Desains at the Sorbonne in 1878, in charge of the physics students' lab work. His brother Jacques was working in the **mineralogy** laboratory at the Sorbonne at that time, and the two began a productive five-year scientific collaboration. They investigated pyroelectricity, the acquisition of electric charges by different faces of certain types of crystals when heated. Led by their knowledge of symmetry in crystals, the brothers experimentally discovered the previously unknown phenomenon of piezoelectricity, an electric **polarization** caused by force applied to the crystal. In 1880 the Curies published the first in a series of papers about their discovery. They then studied the opposite effect—the compression of a piezoelectric crystal by an electric field. In order to measure the very small amounts of **electricity** involved, the brothers invented a new laboratory instrument: a piezoelectric quartz electrometer, or balance. This device became very useful for electrical researchers and would prove highly valuable to Marie Curie in her studies of radioactivity. Much later, piezoelectricity had important practical applications. Paul Langevin, a student of Pierre Curie's, found that inverse piezoelectricity causes piezoelectric quartz in alternating fields to emit high-frequency sound waves, which were used to detect submarines and explore the ocean's floor. Piezoelectric crystals were also used in radio broadcasting and stereo equipment.

In 1882 Pierre Curie was appointed head of the laboratory at Paris' new Municipal School of Industrial Physics and **Chemistry**, a poorly paid position; he remained at the school for 22 years, until 1904. In 1883 Jacques Curie left Paris to become a lecturer in mineralogy at the University of Montpelier, and the brothers' collaboration ended. After Jacques's departure, Pierre delved into theoretical and experimental research on crystal symmetry, although the time available to him for such work was limited by the demands of organizing the school's laboratory from scratch and directing the laboratory work of up to 30 students, with only one assistant. He began publishing works on crystal symmetry in 1884, including in 1885 a theory on the formation of crystals and in 1894 an enunciation of the general principle of symmetry. Curie's writings on symmetry were of fundamental importance to later crystallographers, and, as Marie Curie later wrote in *Pierre Curie,* "he always retained a passionate interest in the physics of crystals" even though he turned his attention to other areas.

From 1890 to 1895 Pierre Curie performed a series of investigations that formed the basis of his doctoral thesis: a study of the magnetic properties of substances at different temperatures. He was, as always, hampered in his work by his obligations to his students, by the lack of funds to support his experiments, and by the lack of a laboratory or even a room for his own personal use. His magnetism research was conducted mostly in a corridor. In spite of these limitations, Curie's work on magnetism, like his papers on symmetry, was of fundamental importance. His expression of the results of his findings about the relation between temperature and magnetization became known as Curie's law, and the temperature above which magnetic properties disappear is called the **Curie point**. Curie successfully defended his thesis before the Faculty of Sciences at the University of Paris (the Sorbonne) in March 1895, thus earning his doctorate. Also during this period, he constructed a periodic precision balance, with direct reading, that was a great advance over older balance systems and was especially valuable for chemical analysis. Curie was now becoming well-known among physicists; he attracted the attention and esteem of, among others, the noted Scottish mathematician and physicist **William Thomson** (Lord Kelvin). It was partly due to Kelvin's influence that Curie was named to a newly created chair of physics at the School of Physics and Chemistry, which improved his status somewhat but still did not bring him a laboratory.

In the spring of 1894, at the age of 35, Curie met Maria (later Marie) Sklodowska, a poor young Polish student who had received her *licence* in physics from the Sorbonne and was then studying for her *licence* in mathematics. They immediately formed a rapport, and Curie soon proposed marriage. Sklodowska returned to Poland that summer, not certain that she would be willing to separate herself permanently from her family and her country. Curie's persuasive correspondence convinced her to return to Paris that autumn, and the couple married in July, 1895, in a simple civil ceremony. Marie used a cash wedding gift to purchase two bicycles, which took the newlyweds on their honeymoon in the French countryside and provided their main source of recreation for years to come. Their daughter Irene was born in 1897, and a few days later Pierre's mother died; Dr. Curie then came to live with the young couple and helped care for his granddaughter.

The Curies' attention was caught by **Henri Becquerel**'s discovery in 1896 that **uranium** compounds emit rays. Marie decided to make a study of this phenomenon the subject of her doctor's thesis, and Pierre secured the use of a ground-floor storeroom/machine shop at the School for her laboratory work. Using the Curie brothers' piezoelectric quartz electrometer, Marie tested all the elements then known to see if any of them, like uranium, emitted "Becquerel rays," which she christened "radioactivity." Only **thorium** and uranium and their compounds, she found, were radioactive. She was startled to discover that the ores pitchblende and chalcolite had much greater levels of radioactivity than the amounts of uranium and thorium they contained could account for. She guessed that a new, highly radioactive element must be responsible and, as she wrote in *Pierre Curie,* was seized with "a passionate desire to verify this hypothesis as rapidly as possible."

Pierre Curie too saw the significance of his wife's findings and set aside his much-loved work on crystals (only for the time being, he thought) to join Marie in the search for the new element. They devised a new method of chemical research, progressively separating pitchblende by chemical analysis and then measuring the radioactivity of the separate constituents. In July 1898, in a joint paper, they announced their discovery of a new element they named **polonium**, in honor of Marie Curie's native country. In December 1898, they announced, in a paper issued with their collaborator G. Bémont, the discovery of another new element, **radium**. Both elements were much more radioactive than uranium or thorium.

The Curies had discovered radium and polonium, but in order to prove the existence of these new substances chemically, they had to isolate the elements so the atomic weight of each could be determined. This was a daunting task, as they would have to process two tons of pitchblende ore to obtain a few centigrams of pure radium. Their laboratory facilities were woefully inadequate: an abandoned wooden shed in the School's yard, with no hoods to carry off the poisonous **gases** their work produced. They found the pitchblende at a reasonable price in the form of waste from a uranium mine run by the Austrian government. The Curies now divided their labor. Marie acted as the chemist, performing the physically arduous job of chemically separating the pitchblende; the bulkiest part of this work she did in the yard adjoining the shed/laboratory. Pierre was the physicist, analyzing the physical properties of the substances that Marie's separations produced. In 1902 the Curies announced that they had succeeded in preparing a decigram of pure radium chloride and had made an initial determination of radium's atomic weight. They had proven the chemical individuality of radium.

The Curies' research also yielded a wealth of information about radioactivity, which they shared with the world in a series of papers published between 1898 and 1904. They announced their discovery of induced radioactivity in 1899. They wrote about the luminous and chemical effects of radioactive rays and their electric charge. Pierre studied the action of a magnetic field on radium rays, he investigated the persistence of induced radioactivity, and he developed a standard for measuring time on the basis of radioactivity, an important basis for geologic and archaeological dating techniques. Pierre Curie also used himself as a human guinea pig, deliberately exposing his arm to radium for several hours and recording the progressive, slowly healing burn that resulted. He collaborated with physicians in animal experiments that led to the use of radium therapy—often called ''Curie-therapie'' then—to treat cancer and lupus. In 1904 he published a paper on the liberation of **heat** by radium salts.

Through all this intensive research, the Curies struggled to keep up with their teaching, household, and financial obligations. Pierre Curie was a kind, gentle, and reserved man, entirely devoted to his work—science conducted purely for the sake of science. He rejected honorary distinctions; in 1903 he declined the prestigious decoration of the Legion of Honor. He also, with his wife's agreement, refused to patent their radium-preparation process, which formed the basis of the lucrative radium industry; instead, they shared all their information about the process with whoever asked for it. Curie found it almost impossible to advance professionally within the French university system; seeking a position was an ''ugly necessity'' and ''demoralizing'' for him (*Pierre Curie*), so posts he might have been considered for went instead to others. He was turned down for the Chair of Physical Chemistry at the Sorbonne in 1898; instead, he was appointed assistant professor at the Polytechnic School in March 1900, a much inferior position.

Appreciated outside France, Curie received an excellent offer of a professorship at the University of Geneva in the spring of 1900, but he turned it down so as not to interrupt his research on radium. Shortly afterward, Curie was appointed to a physics chair at the Sorbonne, thanks to the efforts of Jules Henri Poincaré. Still, he did not have a laboratory, and his teaching load was now doubled, as he still held his post at the School of Physics and Chemistry. He began to suffer from extreme fatigue and sharp pains through his body, which he and his wife attributed to overwork, although the symptoms were almost certainly a sign of radiation poisoning, an unrecognized illness at that time. In 1902, Curie's candidacy for election to the French Academy of Sciences failed, and in 1903 his application for the chair of mineralogy at the Sorbonne was rejected, both of which added to his bitterness toward the French academic establishment.

Recognition at home finally came for Curie because of international awards. In 1903 London's Royal Society conferred the Davy medal on the Curies, and shortly thereafter they were awarded the 1903 Nobel Prize in physics—along with Becquerel—for their work on radioactivity. Curie presciently concluded his Nobel lecture (delivered in 1905 because the Curies had been too ill to attend the 1903 award ceremony) by wondering whether the knowledge of radium and radioactivity would be harmful for humanity; he added that he himself felt that more good than harm would result from the new discoveries. The Nobel award shattered the Curies' reclusive work-absorbed life. They were inundated by journalists, photographers, curiosity-seekers, eminent and little-known visitors, correspondence, and requests for articles and lectures. Still, the cash from the award was a godsend, and the award's prestige finally prompted the French parliament to create a new professorship for Curie at the Sorbonne in 1904. Curie declared he would remain at the School of Physics unless the new chair included a fully funded laboratory, complete with assistants. His demand was met, and Marie was named his laboratory chief. Late in 1904 the Curies' second daughter, Eve, was born. By early 1906, Pierre Curie was poised to begin work—at last and for the first time—in an adequate laboratory, although he was increasingly ill and tired. On April 19, 1906, leaving a lunchtime meeting in Paris with colleagues from the Sorbonne, Curie slipped in front of a horse-drawn cart while crossing a rain-slicked rue Dauphine. He was killed instantly when the rear wheel of the cart crushed his skull. The world mourned the untimely loss of this great physicist. True to the way he had conducted his life, he was interred in a small suburban cemetery in a simple, private ceremony attended only by his family and a few close friends. In his memory, the Faculty of Sciences at the Sorbonne appointed Curie's widow Marie to his chair.

CURIE POINT

Substances with positive magnetic susceptibilites are said to be paramagnetic, which implies that they will have induced magnetic moments in the direction of an applied magnetic field. A ferromagnet, on the other hand, has a spontaneous **magnetic moment**, i.e., a magnetic moment even in the absence of an applied magnetic field.

In ferromagnetic materials, a transition occurs between the magnetized state (for example, of **iron** or nickel) at low temperatures and an unmagnetized state at high temperatures. The magnetization exhibits decreases gradually to zero rather than changing abruptly, but there does exist a **temperature** (called the Curie point after **Pierre Curie** who investigated it) at which the drop in magnetization becomes especially pronounced. The Curie point is thus considered to be the temperature where the spontaneous magnetization vanishes. It separates the disordered paramagnetic phase (at higher temperatures) from the ordered ferromagnetic phase (at lower temperatures).

The magnetic susceptibility of a paramagnetic material is inversely proportional to absolute temperature. The modified law proposed by Curie and French physicist Pierre Weiss (1865-1940) for ferromagnetic materials above their Curie points, which states the susceptibility is proportional to the reciprocal of the difference between the absolute temperature and the Curie temperature, is often obeyed.

See also Magnetism

Curium

Curium is a transuranium element, one of the elements in Row 7 of the **periodic table** that follows **uranium** (atomic number 92). Curium's **atomic number** is 96, its atomic **mass** is 247.0703, and its chemical symbol is Cm.

Properties

All isotopes of curium are radioactive with curium-247 having the longest half life, about 16 million years. Relatively little is known about its physical and chemical properties, although its melting point has been measured to be 2,444°F (1,340°C) and its **density**, 13.5 grams per cubic centimeter.

Occurrence and Extraction

Very small amounts of curium are thought to exist in the Earth's crust as the result of the decay of uranium and other naturally occurring radioactive elements. The element has never actually been found in the earth, however, and it has only been seen as the product of **nuclear reactions** that occur in particle accelerators and **nuclear reactors**.

Discovery and Naming

Curium was discovered in 1944 by **Glenn T. Seaborg**, Ralph A. James, and **Albert Ghiorso**, researchers from the University of California at Berkeley. The discovery was made at the Metallurgical Research Laboratory at the University of Chicago, where work on the first atomic bomb was being conducted. The element was named in honor of the Polish-French physicist **Marie Curie** and her husband, **Pierre Curie**, who carried out pioneering research in the field of **radioactivity**.

Uses

One application of curium has been as a portable source of electrical power in certain highly specialized situations. One of those situations was the Mars *Pathfinder* spacecraft that was sent to that planet's surface in 1997. Curium is particularly useful for such applications since it generates a very large amount of **heat** per unit of mass.

Curl, Robert Floyd (1933-)

American physical chemist

American scientist Robert F. Curl, Jr., a professor of physical **chemistry** at Rice University, won the 1996 Nobel Prize for Chemistry, along with fellow Rice professor **Richard E. Smalley** and Briton **Harold W. Kroto** from the University of Sussex, for the discovery of a new form of the element **carbon**, called Carbon 60. The third molecular form of carbon (the other two forms are diamonds and graphite), C60 consists of 60 atoms of carbon arranged in hexagons and pentagons and is called a "buckminsterfullerene," "fullerene," or by its nickname "Buckyball" in honor of Buckminster Fuller, whose geodesic domes it resembles.

Fullerenes, which consist of 60 atoms of carbon arranged in hexagons and pentagons in a structure resembling a soccer ball, have practical applications. Extraordinarily stable and resistant to radiation and chemical destruction, fullerenes promise to be the basis for remarkably strong but lightweight materials, new drug delivery systems, computer **semiconductors**, solar cells, and superconductors.

Curl was born in Alice, Texas, on August 23, 1933, to Robert and Lessie (Merritt) Floyd. His father was a Methodist minister and moved the family all around southern and southwest Texas while Curl was young. Curl credits receiving a chemistry set for Christmas when he was nine years old for sparking his interest in chemistry. According to a biography of Curl on Rice University's web site, Curl decided to get his B.A. from Rice University because of their famous football team. He graduated in 1954 and a year later, on December 21, he married Jonel Whipple. They would have two sons, Michael and David.

Scientific interest, rather than football, led him to work with Kenneth Pitzer at the University of California at Berkeley, whom he had read about in an undergraduate course. Pitzer became Curl's research advisor and remembered him as a quiet student with exceptional skills. Curl and Pitzer shared the Clayton Prize of the Institution of Mechanical Engineers, and this began a fruitful collaboration that would continue throughout Curl's career. Curl excelled in graduate school, publishing five research papers and obtaining a National Science Foundation fellowship. After completing his Ph.D. in 1957, he decided to go to Harvard to study with E.B. Wilson, who was working on microwave **spectroscopy**. His talents then landed him an assistant professorship at his alma mater. In 1967 Curl became a full professor at Rice and again joined Pitzer, who was president of the university. They would collaborate on papers again, although not on the research that led to the Nobel Prize.

At the same time Curl was researching and publishing papers on microwave and **infrared spectroscopy** at Rice, his acquaintance Harold Kroto, a professor in Sussex, England, was researching **chains** of carbons in space. Kroto thought these chains might be the products of red-giant stars, but wasn't sure how the chains actually formed. In 1984, Kroto contacted Curl, who told him about fellow Rice professor Richard Smalley. Smalley had designed and built a special laser beam apparatus that could break materials down into a **plasma** of atoms that could then be controlled and studied. Curl often worked with Smalley on experiments using Smalley's microwave and infrared spectroscopy expertise. Kroto thought that he could use Smalley's apparatus to simulate the temperatures in space needed to form the carbon chains he was interested in, so Curl introduced the two and began what would later make science history.

Smalley and Curl had been looking at semiconductors like **silicon** and **germanium** in Smalley's laser apparatus, and had no reason to look at simple carbon. It was something of a favor as well as a break in their research when Kroto asked them to look at carbon in order to verify his research. So that September, the scientists turned the laser beam on a piece of **graphite** and found something they were not looking for, a **molecule** that had 60 carbon atoms. Carbon had previously been known to have only two molecular forms, **diamond** and graphite. They surmised correctly that the molecule had a cage-like structure that looked like a soccer ball. They named the structure buckminsterfullerene, which later became fullerenes, and then by the nickname "buckyballs." It was Curl who attained the right degree of equilibrium in the carbon vapor that permitted the scientists to look at how the carbon molecule differed from all the rest. Curl examined the spectroscopic graph and saw the Carbon 60 had the largest peak on the graph. Evidence for the existence of large carbon **clusters** had existed before, but Curl, Smalley, and Kroto were the first scientists to fully identify and stabilize Carbon 60. In October of 1996, all three were recognized for this remarkable discovery with the Nobel Prize in Chemistry.

Fullerene research took off quickly and today scientists are manufacturing pounds of buckyballs every day. Extraordinarily stable because of their **molecular structure** and resistant to radiation and chemical destruction, fullerenes promise to be the basis for remarkably strong but lightweight materials, new drug delivery systems, computer semiconductors, solar cells, and superconductors.

In addition to the Nobel Prize, Curl has also received the Alexander von Humboldt Senior Scientist Award and the APS International Prize for New Materials in 1992. He has also been a NATO fellow, an Alfred P. Sloan fellow, and an Optical Society of America fellow. Curl was the chairman of the chemistry department at Rice from 1992 to 1996. He continues to study how molecules react with each other in **combustion** processes. This research promises to offer solutions to monitoring emissions monitoring from cars, forest fires, and chemical plants.

Robert Floyd Curl. ***(Archive Photo. Reproduced by permission.)***

CYANIDE AND NITRILE

A nitrile is an organic cyanide. Both cyanides and nitriles have the general formula RCN with a triple bond between the **carbon** and **nitrogen** (R-CN). The compound commonly called cyanide is actually **hydrogen** cyanide, HCN. When this compound is dissolved in **water**, an acidic **solution** of hydrogen cyanide, also known as prussic acid, is produced. Cyanides are the salts of hydrogen cyanide. All of the inorganic cyanides have inorganic groups such as a metal attached to the CN group. The nitriles have an organic group attached to the CN group. This gives rise to compounds such as methyl cyanide. The nitriles were formerly known as the alkyl cyanides.

The industrial preparation of most cyanides starts with the addition of **potassium** carbonate to carbon and **ammonia** gas (Beilby's process) to produce potassium cyanide. The potassium cyanide is then reacted with **silver** nitrate to give silver cyanide as a precipitate. The silver cyanide is then reacted with the chloride of the desired end product. This yields the appropriate cyanide and silver chloride.

Nitriles are formed by heating amides with phosphorous pentoxide or by treating organic halogen compounds with **sodium** cyanide. Nitriles can be decomposed by acids or alkalis to give the corresponding carboxylic acid or they can be reduced to give primary amines. Nitriles are **solvents** for a large range of metal complexes and salts. They are also used as intermediates in a number of processes.

All nitriles and cyanides are poisonous, with the cyanides being the more potent form. They act by paralyzing the respiratory center of the brain. Cyanides and nitriles are sweet smelling (the smell is very reminiscent of almonds), colorless **liquids** that dissolve in water. Water **solubility** decreases with increasing **molecular weight**. Nitriles are also soluble in organic solvents.

As well as the -CN structure of cyanides/nitriles a form also exists which can be expressed as -NC. This form is an isocyanide/isonitrile. The properties of the isocyanides/isonitriles are very similar to the properties of the cyanides/nitriles, but there is a slight rearrangement of the distribution of the electrons within the compound and this has a minor effect on some reactions. **Hydrolysis** of a cyanide gives an amide (or a carboxylic acid if the reaction is carried to completion), and this reaction can be carried out by boiling the cyanide under reflux with acids or alkalis. Hydrolysis of an isocyanide is catalyzed by acid but not by alkali and the result is an amine. The differences are caused by the fact that the R group is attached to the nitrogen (RNC) in isocyanides but to the carbon (RCN) in cyanides. Some chemicals, such as hydrogen cyanide, exist in both forms in equilibrium. The rate of conversion is very rapid and it is not possible to isolate the two forms. This is an example of isomerism.

Cyanides and nitriles are commercially used in the synthesis of artificial fibers and as powerful fumigants.

Cyanides and nitriles have the same **functional group**, -CN; cyanides are the inorganic form and nitriles are the organic form. All are poisonous with the inorganic form being the more toxic.

CYTOCHROMES

The cytochromes are highly colored components of all aerobic cells (the word cytochrome comes from words meaning cell and color). They are moderate size **proteins**, typically consisting of about one hundred to a few hundred amino acid units. The active site or prosthetic group of all the cytochromes is an **iron** containing porphyrin, or heme. In a number of cytochromes, the heme unit is the same as the one in the **oxygen** carrying proteins hemoglobin and myoglobin, called iron-protoporphyrin IX. In other cytochromes there are different organic groups around the periphery of the porphyrin **ring**. The biological origin of the heme unit in all cytochromes is the same, porphobiligen, a five membered ring with one **nitrogen atom**, that undergoes a series of reactions to form the porphyrin ring which has four of the nitrogen atom containing five membered rings connected to one another by an intervening **carbon** atom.

Most cytochromes play an active part in the metabolic reactions known as oxidative phosphorylation, by which molecular oxygen is reduced to form **water** and foodstuffs are oxidized to form **carbon dioxide**, water, and other oxidized products, resulting in a release of **energy**. Some cytochromes carry out other crucial biological functions. For example, a group of cytochromes known as the cytochromes P-450 catalyze the introduction of oxygen into -CH bonds of organic molecules, producing –C–OH groups that solubilize the organic molecules in water. By this means, organic molecules that would ordinarily be toxic are altered in such a way that they can be excreted. High concentrations of the cytochromes P-450 are found in liver tissue.

Because the cytochromes are so intensely colored, their visible absorption spectra were readily detectable even with relatively unsophisticated apparatus. The cytochromes were named and classified by the wavelength of the peak in the visible absorption spectrum that occurs closest to the infrared end of the spectrum. Cytochrome a had the longest wavelength band, cytochromes b the next longest, and so forth. As more cytochromes were identified, this classification system became cumbersome because new cytochromes were found with bands between the ones of those that had already been named. Cytochromes are now named by the actual wavelength in the center of the longest wavelength visible band, hence the designation cytochrome P-450, for example.

Some cytochromes, including a_1, a_{31}, b, and c_1 are tightly associated with membranes. This keeps the active iron-porphyrin units in a relatively fixed orientation to make **electron** transport from one to another, and finally to oxygen very efficient. The **amino acids** on the exterior of these cytochromes are hydrophobic so that they are more stable within the nonpolar interior of the membrane than in the aqueous environment of the cytoplasm.

Another important cytochrome, cytochrome c, has very different characteristics. Its exterior is polar and cytochrome c is quite soluble in the cytoplasm. Cytochrome c is also relatively small, so that its amino acid sequence can be determined relatively easily. Because it is easy to isolate and analyze and it is found in aerobic cells of all animals, cytochrome c has been used extensively to study the evolutionary trees of many species. These studies are based on two ideas: (1) the closer the amino acids sequences of two species are the more likely they are related in evolutionary development and (2) species that vary more in their amino acid composition are likely to have evolved further apart in time than those with fewer differences. These ideas follow from the assumption that mutations are rare and statistically independent events.

D

DALTON, JOHN (1766-1844)

English chemist

Although he dabbled in meteorology and studied color-blindness, the contribution for which Dalton is most famous, without question, is his **atomic theory** and the list of atomic weights for the known elements that he created.

Dalton was born on approximately September 5, 1766 in Eaglesfield, England. Since his Quaker parents did not register his birth officially, no one knows his exact birth date.

Strongly influenced by his Quaker faith, Dalton grew up in modest surroundings and continued to live simply throughout his life. He attended the local school at Eaglesfield until the age of 11. A year later, he returned as a teacher at the same school, confronted with the task of instructing a number of students his own age or older. Then, as throughout his life, Dalton continued studying a number of topics on his own.

At the age of 15, Dalton became an assistant at the Quaker school in Kendal. About four years later, he was appointed principal of the school, a post he held for eight more years. While at Kendal, Dalton also offered a series of public lectures on natural philosophy. The series was unsuccessful, however, as Dalton was shy and a less than fascinating speaker.

In 1793, Dalton accepted an assignment as tutor in mathematics and natural philosophy at the New College in Manchester. Before long, however, he decided that the post took far more time than he cared to give and, in 1799, he resigned his post at New College. For the rest of his life, he supported himself as a private tutor in mathematics and natural philosophy.

Dalton's earliest scientific interests were in the field of meteorology. For 57 years he kept a daily record of the **temperature**, barometric pressure, rainfall, dew point, and other weather conditions. In his lifetime, he accumulated more than 200,000 individual observations.

Another topic of interest was color-blindness, a condition that afflicted both Dalton and his brother. His first scientific paper, read to the Literary and Philosophical Society of Manchester on October 31, 1794, dealt with this topic. As a consequence of his interests, the condition of color- blindness was given the name (and is still sometimes referred to as) Daltonism.

Without question, the contribution for which Dalton is most famous is his atomic theory. Dalton's interest in weather caused him to think about the nature and composition of air. He eventually concluded—as had a few scholars before him—that air consists of tiny, individual particles. But Dalton went beyond the musings of his predecessors and hypothesized that *all* forms of matter—solid, liquid, and gas—consisted of these tiny particles.

He first developed a formal theory about these particles between 1803 and 1805. The ideas he presented were by no means new ones. The Greek philosopher **Democritus** of Abdera had proposed such a theory of **matter** in the fourth century B.C. He had called the tiny particles *atomos*, Greek for "indivisible."

Democritus' ideas were quite different from those of Dalton. He used the term *atomos* to describe an abstract, theoretical concept, not the concrete, marble-like particles Dalton had in mind. The Greek's ideas were not widely accepted by his contemporaries and, over the centuries, they became less fashionable than competing theories of matter proposed by Aristotle.

Still, Democritus' ideas never really died out. They filtered in and out of the writings of natural philosophers and scientists for 22 centuries until Dalton expressed them in modern form.

The *atoms* that Dalton had in mind (for he retained the ancient Greek term) were tiny, indivisible particles that constituted all chemical elements. If you could imagine looking at a piece of **gold**, for example, with the very highest magnification imaginable, what you would see is a collection of gold atoms. Similarly, an examination of a piece of **copper** under maximum magnification would reveal a collection of copper atoms.

John Dalton.

The atoms of any one element, Dalton said, were all exactly alike. A single gold **atom** would look like any other gold atom anywhere in the world. But atoms of different elements were different from each other. A gold atom looked different from a copper atom.

Dalton's theory also dealt with the composition of compounds. The smallest particle of any compound, he said, was a compound atom. Thus, he taught that **water** was composed of *compound water atoms*. Two decades later, **Amedeo Avogadro** was to clarify the difference between atoms and compound atoms, which he proposed calling *molecules*.

Dalton is called the father of modern atomic theory partly because of his clear statement of that theory and partly because of his emphasis on atomic weights. No proponent of the concept of atoms had previously made clear the fact that atoms must have weights that can be determined experimentally. Dalton did. He said that finding the weights of atoms was a relatively straight-forward task that any chemist could accomplish.

For example, he said, suppose that we assume that the elements **oxygen** and **hydrogen** are both composed of atoms. When oxygen gas combines with hydrogen gas to form water, then, what happens on the simplest level is that one atom of oxygen combines with one atom of hydrogen to form one compound atom of water, or: $H + O \rightarrow HO$ (Dalton had no way of knowing the actual combining ratio of hydrogen and oxygen in water [2:1], so he assumed the simplest possible ratio [1:1]).

Next, Dalton said, it is easy to determine in the laboratory what the *weight ratio* of hydrogen to oxygen is. A fixed weight of hydrogen will consume eight times its weight in oxygen to form water. It followed that each atom of oxygen must weigh eight times as much as each atom of hydrogen.

Dalton's discussion of atomic weights gave chemists a concrete plan for exploring at least this aspect of the atomic theory. Within a few years, Dalton had prepared a list of atomic weights for the known elements.

Dalton's theory was quickly accepted by the vast majority of chemists. One reason for its rapid success was that it explained certain experimental results that had recently been announced, most notably Joseph Proust's law of constant composition which stated that a **chemical compound** always contains the same constituents in the same proportions.

In his later years, Dalton was flooded with honors and awards. He died quietly at his home in Manchester on July 27, 1844.

Dalton's law of partial pressures

Dalton's law of partial pressures states that the total pressure of a mixture of **gases** equals the sum of the pressures that each gas would exert if it was present alone. The pressure exerted by one gas in a mixture of gases is the partial pressure of that individual gas. This law assumes that the gases do not react with each other, but that each gas is a separate component of the system.

Dalton's law of partial pressures can be mathematically expressed by the following. If P_t is the total pressure of a mixture of gases and P_1, P_2, etc. are the pressures that each gaseous component would exert by itself, then the total pressure is given by $P_t = P_1 + P_2 +$ etc. Dalton's law of partial pressures tells us that each gas behaves independently of the other gases in the mixture, i.e., each exerts its own pressure (the partial pressure) providing the gases do not react.

The total pressure of a mixture of gases is a function of the number of moles of gas present, irrespective of what type of gas is considered. One **mole** of any gas has the same **volume** as one mole of any other gas, providing the **temperature** and pressure are the same. This is the basic theory behind **Avogadro's law**. One mole of gas under the same conditions as one mole of a second gas will has the same number of particles as the second gas. The same volume (number of moles) of two gases exerts the same pressure in a container because each gas has the same number of particles and hence the same number of collisions between the particles and the container walls. Similarly one mole of gas in a fixed volume exerts the same pressure as one mole of any other gas in an identical volume. If both gases are in the same container then the total pressure is twice the pressure exerted by either of the gases, providing the same number of moles of each gas was present. The pressure exerted by any one of these gases is the partial pressure.

The pressure of a gas is a measure of how frequently molecules of the gas hit the sides of the container the gas is held in. Gases have molecules that are widely separated due to their high **energy** levels. As a result, the molecules within a volume of gas have plenty of space between them and they move in an entirely random manner. Consequently if another gas is introduced into the same volume the molecules of this second gas are equally free to move in a random manner. There are no space limitations and the number of collisions between gas molecules is very small compared to the number of collisions between the gas molecules and the much larger walls of the container (which gives us the pressure). The molecules of each gas strike the walls of the container independently of each other. The total pressure exerted is the pressure exerted by all molecules hitting the container walls, irrespective of which gas they belonged to. Each gas behaves as if it is the only gas present in the system, and the partial pressure that each gas exerts is the same as if no other gases were present. Because the pressure exerted is a function of the speed of the molecules, kinetic theory states that the pressure will increase if the energy of the molecules increases. One way of increasing the kinetic energy of a gas is to increase its temperature. If both gases are heated, the overall pressure increases and the partial pressure of each component of the system also increases.

Atmospheric pressure is the sum of the partial pressures of all of the gases present in the atmosphere. The atmosphere is a mixture of several main gases and many other gases present in much smaller quantities, with a large amount of local variation. The gas with the greatest **concentration** in the atmosphere is **nitrogen** with 78%. Next comes **oxygen** with a concentration of 21%. **Argon**, **carbon** dioxide, **neon**, **krypton**, **xenon**, and **radon** are present in the atmosphere at lower concentrations. Each of these gases exerts a pressure as if the other gases were not present, i.e., its partial pressure. The total pressure of the air is the sum of the partial pressures of all of the gases present. Atmospheric pressure varies depending on the height above sea level. When standard temperature and pressure are discussed the pressure used is the pressure at sea level—this is known as one atmosphere.

John Dalton was an English schoolteacher who lived from 1766 to 1844. He was interested in **chemistry** and meteorology. His first contribution to chemistry was an early version of **atomic theory**. It was in 1801 while studying the properties of air that Dalton formulated his law of partial pressures.

Dalton's law of partial pressures works for low, normal, and moderate pressures, but it starts to fail at very high pressures. This failure at high pressure is due to the fact that the molecules become more densely packed. Because the molecules of gas are close together, they are no longer able to move freely in the system and collisions between molecules become more frequent. The other factor adding to the failure of the law of partial pressures at high pressure is the presence of weak, **intermolecular forces**. When the molecules are free to move widely separated from each other the effect of these attractive forces is negligible. As the molecules are pushed closer together, the effects of small forces, such as **van der Waals forces**, become apparent. This leads to a clumping of the molecules present. If the molecules are clumped together in an aggregate, they move more slowly because they are larger. The immediate effect of this is a slight **reduction** in pressure because the number of collisions of gas molecules with the walls of the container is reduced.

Dalton's law of partial pressures states that in a mixture of gases each component gas exerts a pressure that is totally independent of the other gases present in the mixture. The total pressure of the mixture is the sum of all of the pressures of the gases present. Dalton's law of partial pressures only works for gases that do not react with each other.

See also Gas laws; Gases, behavior and properties

DALY, MARIE M. (1921-)

American biochemist

Marie M. Daly was the first African American woman to earn a Ph.D. in **chemistry**. Throughout her career, her research interests focused on areas of health, particularly the effects on the heart and arteries of such factors as aging, cigarette smoking, hypertension, and **cholesterol**. In addition to research, she taught for fifteen years at Yeshiva University's **Albert Einstein** College of Medicine.

Marie Maynard Daly was born in Corona, Queens, a neighborhood of New York City, on April 16, 1921. Her parents, Ivan C. Daly and Helen (Page) Daly, both valued learning and education and steadily encouraged her. Her father had wanted to become a chemist and had attended Cornell University, but was unable to complete his education for financial reasons and became a postal clerk. Daly attended the local public schools in Queens and graduated from Hunter College High School in Manhattan. She credits her interest in science to both her father's scientific background and to influential books such as Paul DeKruif's *The Microbe Hunters.*

Daly enrolled in Queens College as a chemistry major, graduating with a B.S. degree in 1942. The following year she received her M.S. from New York University and then went to Columbia University where she entered the doctoral program in **biochemistry**. In 1948 she made history at that university, becoming the first African American woman to earn a Ph.D. in chemistry.

Daly began teaching during her college days as a tutor at Queens College. She began her professional career a year before receiving her doctorate, when she accepted a position at Howard University in Washington, D.C., as an instructor in physical sciences. In 1951 she returned to New York first as a visiting investigator and then as an assistant in general physiology at the Rockefeller Institute. By 1955 she had become an associate in biochemistry at the Columbia University Research Service at the Goldwater Memorial Hospital. She taught there until 1971 when she left Columbia as an assistant professor of biochemistry to become associate professor of biochemistry and medicine at the Albert Einstein College of Medicine at Yeshiva University in New York.

Daly conducted most of her research in areas related to the biochemical aspects of human **metabolism** (how the body processes the **energy** it takes in) and the role of the kidneys in that process. She also focused on hypertension (high blood pressure) and atherosclerosis (accumulation of **lipids** or fats in the arteries). Her later work focused on the study of aortic (heart) smooth muscle cells in culture.

During her career, she held several positions concurrently with her teaching obligations, such as investigator for the American Heart Association from 1958 to 1963 and career scientist for the Health Research Council of New York from 1962 to 1972. She was also a fellow of the Council on Arteriosclerosis and the American Association for the Advancement of Science, a member of the American Chemical Society, a member of the board of governors of the New York Academy of Science from 1974 to 1976, and a member of the Harvey Society, the American Society of Biological Chemists, the National Association for the Advancement of Colored People, the National Association of Negro Business and Professional Women, and Phi Beta Kappa and Sigma Xi. In 1988 Daly contributed to a scholarship fund set up at Queens College to aid African American students interested in the sciences. Daly, who married Vincent Clark in 1961, retired from teaching in 1986.

DAM, HENRIK (1895-1976)

Danish biochemist

Henrik Dam is best known for his discovery of vitamin K, which gives blood the ability to clot, or coagulate. The discovery of vitamin K dramatically reduced the number of deaths by bleeding during surgery, and for the discovery Dam received the 1943 Nobel Prize in medicine and physiology. (**Edward A. Doisy**, the American biochemist who isolated and synthetically produced vitamin K, shared this prize with Dam.)

Carl Peter Henrik Dam was born in Copenhagen, Denmark, on February 21, 1895. His interest in science was shaped at least in part by his background. His father, Emil Dam, was a pharmaceutical chemist who wrote a history of pharmacies in Denmark. His mother, Emilie Peterson Dam, was a schoolteacher. He attended the Polytechnic Institute in Copenhagen, from which he received his master of science degree in 1920. He was associated with the Royal School of Agriculture and Veterinary Medicine in Copenhagen for the next three years, after which he spent five years as an assistant at the University of Copenhagen's physiological laboratory. He became assistant professor of **biochemistry** in 1928 and associate professor in 1929 (a post he held until 1941).

During these years Dam studied microchemistry under **Fritz Pregl** in Austria (1925) at the University of Graz, and collaborated with biochemist Rudolf Schoenheimer in Freiburg, Germany (on a Rockefeller Fellowship) from 1932 to 1933. He was awarded a doctorate in biochemistry by the University of Copenhagen in 1934. Afterwards, he worked with the Swiss chemist Paul Xarrer at the University of Zurich in 1935. Dam specialized in **nutrition**, which became his area of expertise.

It was while Dam was studying in Copenhagen that he became interested in what would become the vitamin K factor. In the late 1920s he began experimenting with hens in an attempt to discover how the animals synthesized **cholesterol**. Providing them with a synthetic diet, Dam discovered that they developed internal bleeding in the form of hemorrhages under the skin—lesions similar to those found in the disease scurvy. He added lemon juice to the diet (citrus fruits, high in vitamin C, had been found by the eighteenth century Scottish physician James Lind to cure scurvy in sailors), but the supplement did little to reverse the hens' condition.

After experimenting with a variety of **food additives**, Dam came to the conclusion that some vitamin must exist to give blood the ability to clot—and that this vitamin was what was missing from his synthetic hen diet. He made his findings known in 1934, naming the vitamin "K" from the German word *Koagulation.* Dam's continued research, along with the work of Doisy and other biochemists, led to the isolation of vitamin K and its synthetic production.

Dam's discovery proved vitally important in two areas: in surgical procedures and in treatment of newborn babies. Prior to surgery, patients are given vitamin K to assist in clotting the blood and reduce the risk of death by hemorrhage. Newborns are born deficient in vitamin K. Normally, beneficial bacteria that exist in the environment enter the intestinal tracts of infants and induce production of vitamin K. Modern hospitals are disinfected to such an extreme, however, that they kill these good bacteria along with the harmful ones. Mothers are injected with vitamin K shortly before giving birth to ensure that adequate amounts of the vitamin will be in the newborn's system.

Dam's discovery led not only to the Nobel Prize but also the Christian Bohr Award in Denmark in 1939. Dam came to the United States in 1940 for a series of lectures in the U.S. and Canada under the auspices of the American-Scandinavian Foundation. During his visit Nazi Germany invaded Denmark. Dam chose not to return to his native country and accepted a position as senior research associate at the University of Rochester's Strong Memorial Hospital. Because of the war, the Nobel Prize Committee decided to present the awards in New York in 1943. The Nobel recipients of that year, including Dam, were the first to be awarded their prize in the United States. In 1945, Dam became an associate member of the Rockefeller Institute for Medical Research.

After Denmark was liberated, Dam returned in 1946 to accept the position of head of the biology department at the Polytechnic Institute (the position had been awarded to him in absentia in 1941). He returned to the U.S. in 1949 for a three-month lecture tour, this time to discuss vitamin E. In 1956, he was named head of the Danish Public Research Institute. He was a member of numerous organizations including the American Institute of Nutrition, the Society for Experimental Biology and Medicine, the Royal Danish Academy of Science, the Société Chimique of Zurich, and the American Botanical Society. During his career he published more than one hundred articles in scientific journals on vitamin K, vitamin E, cholesterol, and a variety of other topics. Dam married Inger Olsen in 1925. His primary form of recreation was travel. After he returned to Denmark, he pointedly criticized the American hospital system, saying it was hurt by too much emphasis on the business of running hospitals. He died in Copenhagen at the age of eighty-one on April 17, 1976. At his request, news of his death was delayed by one week to allow for private services.

DAVIS, JR., RAYMOND (1914-)

American astrochemist

Raymond Davis, Jr. has devoted much of his scientific career to pursuing one of the universe's great phantoms, the **neutrino** (low-mass grains of **matter** from the sun that travel at the speed of light). Davis created the first working neutrino detector while at Brookhaven National Laboratory, and the results he and his collaborator John Bahcall obtained have led physicists to reevaluate the assumptions made concerning the relationship between the internal fires of stars to the internal workings of the **atom**.

Davis was born October 14, 1914, in Washington, D.C., the older of two brothers. His father, Raymond, worked in the photographic division of the National Bureau of Standards. His mother was Ida Rogers Younger Davis. He attended Washington, D.C. public schools. Although an indifferent student, Davis loved learning and early on became enamored of **chemistry**; he would frequently visit the library and read the *Smithsonian Technical Reports.* He received his bachelor's degree in chemistry from the University of Maryland in 1937, and then worked for a year for Dow Chemical Company in Midland, Michigan. In 1939, he earned his master's degree in **physical chemistry** from the University of Maryland. He then went on to Yale University, receiving his Ph.D. in 1942.

Davis served in the U.S. Army Air Corps from 1942 to 1946. After the war, he worked as a research chemist for Monsanto Chemical Company's Mound Laboratory in Dayton, OH, from 1946 to 1948. In that year he moved to the Brookhaven National Laboratory on Long Island, NY, where he worked as a research chemist from 1948 to 1984. In 1984 he was named a research professor in the astronomy department at the University of Pennsylvania, concurrently serving as a consultant to the Los Alamos National Laboratory in New Mexico.

Neutrinos, named by Italian physicist Enrico Fermi, who believed they could never be detected, are particles of extremely low **mass**. During the 1950s astrophysicists theorized that these particles formed during fusion reactions within the sun's internal furnace, and as a result, the sun threw off neutrinos by the trillion.

Davis was fascinated by the new theory that depicted the sun as a gigantic nuclear furnace. If neutrinos could be found, he thought, they could provide a clue to the sun's internal workings. But the question was, How to detect something smaller than an **electron** with almost no mass at all? Neutrinos pass unnoticed—and unimpeded—through the sun's matter, through space, through stone, and through our bodies. Besides having little mass, some neutrinos (called pp neutrinos) have very little **energy** and interact weakly with atomic nuclei and with each other.

In 1955, Davis built his first neutrino detector. Located 20 ft (6 m) underground at the Brookhaven National Laboratory (the underground location would keep cosmic rays from interfering with the results), it was filled with 1,000 gal (3,780 l) of **chlorine** compound. ''I didn't really expect to see anything,'' Davis told *Discover* magazine. ''There was still too much **background radiation**, and the pp neutrino energies were too low.''

The Brookhaven detector failed to find even a single pp neutrino. But in 1958, other researchers discovered there were more energetic neutrinos—thirty times more energetic than the pp neutrinos Davis had been seeking—shooting out of the sun. These energetic neutrinos resulted from a different reaction, and were called boron–8 neutrinos. In 1962, Davis contacted astrophysicist John Bahcall, who calculated that about eight of these neutrinos might be detected each day. Davis built another neutrino detector in an abandoned **gold** mine a mile beneath the Black Hills of South Dakota. Completed in 1967, the 48-ft (14.6 m) long metal tank is filled with 100,000 gal (378,000 l) of perchlorothylene, a kind of dry cleaning fluid.

The detector worked when neutrinos, passing through rock and metal, would travel through the fluid in the tank. Occasionally, one of them would strike a chlorine atom, splitting a **neutron** into a **proton** and an electron, and changing the chlorine into radioactive **argon**. Every couple of months, the argon would be drawn off and its atoms counted by measuring the rate of radioactive decay; the result would indicate how many neutrinos had passed through the tank and reacted with the chlorine. The results were surprising. The tank detected not eight neutrinos a day as calculations had predicted, but one neutrino every two days. These figures remained consistent for over a quarter of a century, and have been confirmed by another neutrino detector deep within the Kamioka Mine in Japan.

Davis's work has created a worldwide effort by researchers to probe the secret of the neutrino. Astrophysicists and particle physicists, due in part to Davis's research, have developed a closer working relationship as they try to learn the neutrino's secrets. Davis, for his part, is planning a new detector using **iodine**, thought to be even more sensitive to neutrino bombardment than chlorine.

In addition to his work with neutrinos, Davis also took part in the analysis of lunar samples gathered by the Apollo moon missions of the 1970s, and measured the **radioactivity** of material from the moon's surface. Davis has also written more than sixty scientific papers, primarily in **nuclear chemistry**, **geochemistry**, and **astrochemistry**, and contributed articles to a number of books. He is a member of the National Academy of Sciences, the American Academy of Science, the Geochemical Society, Meteorite Society, the American Geophysical Union, the American Physical Society and the American Astronomical Society.

Among the honors he has received are the W.H.K. Panofsky Prize from the American Physical Society, 1992; the Bonner Prize for Nuclear Physics from the American Physical Society, 1988; the American Chemical Society Award for Nuclear Chemistry, 1979; the Comstock Award from the National Academy of Sciences, 1978; and the Boris Pregel Prize from the New York Academy of Sciences, 1957. In 1948, Davis married Anna Marsh Torrey, whom he met while working at

Brookhaven, where she was a biology technician. They have five children: Andrew M. Davis, Martha S. Kumler, Nancy E. Klemm, Roger W. Davis, and Alan P. Davis.

In 1958 Racine's Women's Civic Council recognized Davis for her contributions as a civic leader. Davis died in Racine in September 19, 1967, three days after her eightieth birthday.

DAVIS, MARGUERITE (1887-1967)

American chemist

Marguerite Davis is best known as co-discoverer of **vitamins** A and B. Her research at the University of Wisconsin in Madison with biochemist **Elmer Verner Mc Collum** led to definitive identification of both vitamins and paved the way for later research in **nutrition**.

Davis was born on September 16, 1887, in Racine, Wisconsin. Her father, Jefferson J. Davis, was a physician and botanist who taught at the University of Wisconsin. Her grandmother, Amy Davis Winship, was a social worker and an early champion of women's rights. Her background, coupled with her own interest in science, led her to enroll at the University of Wisconsin in 1906. She transferred to the University of California at Berkeley in 1908 and received her bachelor of science degree there in 1910. Upon graduation, she returned to the University of Wisconsin and pursued graduate studies, although she never completed the master's program. She worked briefly for the Squibbs Pharmaceutical Company in New Brunswick, New Jersey, but returned to Wisconsin.

It was during her time at the University of Wisconsin that she began her work with McCollum, who had been studying nutrition for several years. The Dutch physician **Christiaan Eijkman** and the British biochemist Sir **Frederick Gowland Hopkins** had determined that traces of as-yet unidentified elements in foods were essential for adequate nutrition. The Polish-American biochemist Casimir Funk, believing the substances were amines, proposed the name "vitamine"—literally, "life-giving amine" (when it later became clear that not all the substances were amines the "e" was dropped). McCollum was trying to create simple mixtures that could replace natural food in animal diets. Although his efforts were unsuccessful, he wanted to find out whether natural food contained some special substance like that proposed by Eijkman and Hopkins.

Davis and McCollum worked with various food components and in 1913 discovered a factor in some fats that apparently was essential to life. Because the substance differed chemically from one described earlier by Eijkman, Davis and McCollum named theirs fat-soluble A and Eijkman's water-soluble B. These were later called vitamins A and B. The identification of A and B led later to the discovery of the other vitamins and their specific roles in nutrition, as well as which foods contain them.

Davis joined the University of Wisconsin's chemical research staff and founded its nutrition laboratory. She later went on to Rutgers University in New Jersey and organized a similar lab for its school of pharmacy. She retired and moved back to Racine in 1940 but continued to serve as a **chemistry** consultant for many years. She became active in Racine civic affairs and pursued other interests, including history and gardening.

DAVY, HUMPHRY (1778-1829)

English chemist

Humphry Davy grew up poor helping his mother pay off debts left by his father, a woodcarver who had lost his earnings in speculative investments. As a result, Davy's education was haphazard, and he disliked being a student. The schools in his part of the country (Cornwall, the southwest tip of England) were far from outstanding at that time. Still, Davy managed to absorb knowledge of classic literature and science. In later life, he said he was happy he didn't have to study too hard in school so that he had more time to think on his own.

Without money for further education, Davy began at age 17 to serve as an apprentice to a pharmacist/surgeon. During this time, he took it upon himself to learn more about whatever interested him, such as geography, languages, philosophy, and science. He also wrote poems that later earned him the respect and friendship of William Wordsworth, Samuel Coleridge, and other leading poets of his time. When he was 19 years old, Davy read a book on **chemistry** by the famous French scientist **Antoine-Laurent Lavoisier** that convinced him to concentrate on that subject. For the rest of his life, Davy's career was marked by brilliant, if impetuous, scientific explorations in chemistry and **electrochemistry** that led to the numerous discoveries for which he is known today.

One of Davy's scientific trademarks was his willingness, even eagerness, to use himself as a guinea pig. In early tests with **hydrogen**, for example, he breathed four quarts of the gas and nearly suffocated. He also tried to breathe pure **carbon** dioxide, which is actually a product of human respiration. These experiments were conducted while Davy was working at an institute for studying the therapeutic properties of various **gases**.

In one instance, Davy's fondness for risk paid off. While studying **nitrous oxide** gas, he discovered that its unusual properties made him feel giddy and intoxicated. When he encouraged his friends to inhale the gas with him, he found that their inhibitions were lowered and their feelings of happiness or sadness intensified. Davy's poet friend Robert Southey (1774-1843) referred to his experience as being "turned on," and nitrous oxide became known as laughing gas. Beyond Davy's circle, nitrous oxide parties became a fad among the wealthy people. Davy recognized, however, that the gas could be used to dull physical pain during minor surgery. Although the medical profession ignored Davy's discovery for nearly half a century, nitrous oxide eventually became the first chemical anesthetic. In an 1844 experiment, a dentist had a tooth successfully extracted while under the influence of nitrous oxide (having first taken the precaution of writing his will). Some dentists still use the gas today for apprehensive patients.

Davy was also known for his skill as a public speaker. In the early 1800s, he was hired to lecture for the Royal Institution, a new scientific institution that was having financial problems. Davy's poise and the polish of his lectures and demonstrations drew enthusiastic crowds from London's high society and soon reversed the institution's fortunes. (Some historians have also referred to Davy's good looks, which probably contributed to his popularity with the fashionable women in the audience.)

After a brief foray into **agricultural chemistry**, Davy entered his most prolific period of discovery. His style in the laboratory was to work quickly and intensely, pursuing one new idea after another. He aimed at originality and creativity, rather than tediously repeating tests and confirming results. Stimulated by the Italian physicist Alessandro Volta's invention of the electric battery, Davy rushed into the new field of electrochemistry and, in 1805, produced an electric arc by making a strong current leap from one electrode to another. This discovery led to arc lighting and arc welding, which are still in use today.

Greatly excited with this new tool of **electricity**, Davy went on to build his own large battery—the strongest one at the time—and used it to decompose substances most scientists thought were pure elements. According to many sources, Davy danced around the room exuberantly when he discovered the element **potassium**, which he created by electrolyzing potash. Just a week later, he isolated **sodium** from soda in a similar way. Then, using a slightly modified method, he isolated the elements **calcium**, **magnesium**, **barium**, and **strontium**.

By this time other scientists, spurred by Davy's triumphs, had begun to compete with him. Although Davy discovered **boron**, French chemists Joseph Gay-Lussac and Louis Thénard also received credit for their boron experiments. Indeed, Gay-Lussac and Davy mined the same fields for a time. And while Davy proved that **hydrochloric acid** did not contain **oxygen** (which contradicted the prevailing theory), Gay-Lussac did the same for prussic acid. Both of them showed that **iodine**, discovered by French chemist **Bernard Courtois** (1777-1838), was an element. In his analysis of hydrochloric acid, Davy identified a greenish-colored gas and named it **chlorine**, after the Greek word for green. This work led to his proof that chlorine was also an element. Davy was also the first to notice that **platinum** could catalyze, or speed up, chemical reactions—a discovery that would later be exploited to a much greater extent.

While in his early thirties, after being knighted in 1812, Davy married a wealthy Scottish widow and began to travel extensively, enjoying his fame wherever he went. In Italy, using a great microscopic lens available in Florence, he studied diamonds and concluded that they are a form of carbon. Davy was accompanied on some of these trips by his assistant/valet **Michael Faraday**, who was destined to eclipse his mentor's reputation in the realm of science.

After returning to England, Davy was summoned to study coal-mine explosions, which were killing hundreds of miners yearly. He invented a miner's safety lamp, also called the Davey lamp, in less than three months. While testing samples of the "fire-damp" gas that caused the explosions, he verified that it was mostly **methane** and that it would ignite only at high temperatures. In Davey's lamp, wire gauze surrounds the flame to dissipate **heat** and prohibit the natural gases from igniting. The invention of this lamp marks the first major attempt at safety in the **coal** mining industry.

Humphry Davy.

Davy's work was rewarded by many honors and medals. In addition to his knighthood, he was made a baronet in 1818 for his service to the mining industry and was elected president of the prestigious Royal Society in 1820. In his conflicts with other scientists, however, Davy made some enemies who thought he was arrogant. He even tried to prevent his associate Faraday from being elected to the Royal Society.

Health problems began to plague Davy while he was still in his thirties. The same curiosity that drove him to discover and invent with such success had also taken its toll on his body. By sniffing and tasting unknown chemicals, he had poisoned his system, and his eyes had been damaged in a laboratory explosion. Although Davy continued to pursue scientific interests, he suffered a stroke when he was only 49 and died abroad just two years later.

DDT

DDT (*dichloro-diphenyl-trichloroethane*) is perhaps the most recognized of all insecticides because it revealed the many

DDT sperulites magnified 50 times. *(Photograph by David Malin, Photo Researchers, Inc. Reproduced by permission.)*

hazards associated with using synthetic **pesticides**. This colorless, odorless, insoluble toxic pesticide contains up to 14 chemical compounds and is known for its ability to eradicate pesky insects such as flies, lice, mosquitoes, and agricultural pests. Although first synthesized in 1874 by German chemist Othmar Zeidler, DDT was not used as an insecticide until 1939, when Swiss scientist Paul Hermann Müller (1899-1965) discovered its insect-killing properties. The benefits of DDT were demonstrated in the 1940s, however, when it was used in World War II to clear out mosquito-infested areas prior to invasion. Even after the war, the use of DDT in the United States almost completely wiped out malaria and yellow fever. In tropical areas, the use of DDT has helped save millions of lives that would otherwise have been lost to disease. DDT was also routinely applied as a crop dust or **water** spray on orchards, gardens, fields, and forests. At one point, it was registered for use on 334 agricultural crops.

Unfortunately, DDT is too durable; in some applications, it is effective for up to 12 years. Water cannot wash it away and it resists breakdown by light and air. This strength and persistence has resulted in DDT's transfer to non-target living organisms. Once in an ecosystem, it can pass on from crops to birds and from water to fish, eventually affecting the whole food chain. When ingested by humans, DDT is stored in body fats and can be passed on to nursing babies. Low levels of DDT in humans are harmless but large concentrations can cause severe health problems such as liver cancer. When applied to an insect, DDT is easily absorbed through the body surface and after attacking the nervous system, causes paralysis. Some insects, however, have a resistance to DDT, thereby making the insecticide ineffective. These resistant insects are able to reproduce and pass this trait on to their offspring. Many problems arise when larger animals are exposed to DDT or eat smaller animals that have ingested the toxin. For example, while DDT is more toxic to fish than birds, it still causes widespread bird deaths. With high levels of exposure, DDT causes convulsions and paralyzes the birds' nerve centers. In smaller concentrations, it can weaken their egg shells and can cause sharp declines in the species' reproductive rate. DDT ingestion by peregrine falcons is thought to have caused their almost complete extinction in most regions of the United States. Evidence documenting DDT's adverse impact on the environment and human health was detailed in Rachel Carson's landmark book, *Silent Spring* (1962) which exposed the dangers of unregulated pesticide use. Spurred by public pressure, state and federal governments turned their attention to the regulation of pesticides, and, in 1972, the United States Environmental Protection Agency banned the use of DDT. Today, DDT is restricted in the United States, Europe, and Japan. However, many other countries still use DDT widely for malaria control, delousing, and the eradication of other disease-spreading insects.

The debate on the adverse health effects of DDT continues today. A 1997 report in the *New England Journal of Medicine* indicates that, despite all previous studies to the contrary, DDT and its derivative DDE can not be conclusively linked to breast cancer.

DE BROGLIE, LOUIS VICTOR (1892-1987)

French theoretical physicist

Louis Victor de Broglie, a theoretical physicist and member of the French nobility, is best known as the father of wave mechanics, a far-reaching achievement that significantly changed modern physics. Wave mechanics describes the behavior of **matter**, including **subatomic particles** such as electrons, with respect to their wave characteristics. For this groundbreaking work, de Broglie was awarded the 1929 Nobel Prize for physics.

Louis Victor Pierre Raymond de Broglie was born on August 15, 1892, in Dieppe, France, to Duc Victor and Pauline d'Armaille Broglie. His father's family was of noble Piedmontese origin and had served French monarchs for centuries, for which it was awarded the hereditary title *Duc* from King Louis XIV in 1740, a title that could be held only by the head of the family. A later de Broglie assisted the Austrian side during the Seven Years War and was awarded the title *Prinz* for his contribution. This title was subsequently borne by all members of the family. Another of de Broglie's famous ancestors was his great-great-grandmother, the writer Madame de Stael.

The youngest of five children, de Broglie inherited a familial distinction for formidable scholarship. His early education was obtained at home, as befitted a great French family of the time. After the death of his father when de Broglie was fourteen, his eldest brother Maurice arranged for him to obtain his secondary education at the Lycée Janson de Sailly in Paris.

After graduating from the Sorbonne in 1909 with baccalaureates in philosophy and mathematics, de Broglie entered the University of Paris. He studied ancient history, paleography, and law before finding his niche in science, influenced by the writings of French theoretical physicist **Jules Henri Poincaré**. The work of his brother Maurice, who was then engaged in important, independent experimental research in X rays and **radioactivity**, also helped to spark de Broglie's interest in theoretical physics, particularly in basic **atomic theory**. In 1913, he obtained his Licencié ès Sciences from the University of Paris's Faculté des Sciences.

De Broglie's studies were interrupted by the outbreak of World War I, during which he served in the French army. Yet even the war did not take the young scientist away from the country where he would spend his entire life; for its duration, de Broglie served with the French Engineers at the wireless station under the Eiffel Tower. In 1919, after what he considered to be six wasted years in uniform, de Broglie returned to his scientific studies at his brother's laboratory. Here he began his investigations into the nature of matter, inspired by a conundrum that had long been troubling the scientific community: the apparent physical irreconcilability of the experimentally proven dual nature of light. Radiant **energy** or light had been demonstrated to exhibit properties associated with particles as well as their well-documented wave-like characteristics. De Broglie was inspired to consider whether matter might not also exhibit dual properties. In his brother's laboratory, where the study of very high frequency radiation using spectroscopes was underway, de Broglie was able to bring the problem into sharper focus. In 1924, de Broglie, with over two dozen research papers on electrons, **atomic structure**, and X rays already to his credit, presented his conclusions in his doctoral thesis at the Sorbonne. Entitled "Investigations into the Quantum Theory," it consolidated three shorter papers he had published the previous year.

In his thesis, de Broglie postulated that all matter—including electrons, the negatively charged particles that orbit an atom's nucleus—behaves as both a particle and a wave. Wave characteristics, however, are detectable only at the atomic level, whereas the classical, ballistic properties of matter are apparent at larger scales. Therefore, rather than the wave and particle characteristics of light and matter being at odds with one another, de Broglie postulated that they were essentially the same behavior observed from different perspectives. Wave mechanics could then explain the behavior of all matter, even at the atomic scale, whereas classical Newtonian mechanics, which continued to accurately account for the behavior of observable matter, merely described a special, general case. Although, according to de Broglie, all objects have "matter waves," these waves are so small in relation to large objects that their effects are not observable and no departure from classical physics is detected. At the atomic level, however, matter waves are relatively larger and their effects become more obvious. De Broglie devised a mathematical formula, the matter wave relation, to summarize his findings.

American physicist **Albert Einstein** appreciated the significant of de Broglie's theory; de Broglie sent Einstein a copy of his thesis on the advice of his professors at the Sorbonne, who believed themselves not fully qualified to judge it. Einstein immediately pronounced that de Broglie had illuminated one of the secrets of the universe. Austrian physicist **Erwin Schrödinger** also grasped the implications of de Broglie's work and used it to develop his own theory of wave mechanics, which has since become the foundation of modern physics. Still, many physicists could not make the intellectual leap required to understand what de Broglie was describing.

De Broglie's wave matter theory remained unproven until two separate experiments conclusively demonstrated the wave properties of electrons—their ability to diffract or bend, for example. American physicists **Clinton Davisson** and Lester Germer and English physicist **George Paget Thomson** all proved that de Broglie had been correct. Later experiments would demonstrate that de Broglie's theory also explained the behavior of protons, atoms, and even molecules. These properties later found practical applications in the development of magnetic lenses, the basis for the **electron** microscope.

De Broglie devoted the rest of his career to teaching and to developing his theory of wave mechanics. In 1927, he attended the seventh Solvay Conference, a gathering of the most eminent minds in physics, where wave mechanics was further debated. Theorists such as German physicist **Werner Karl Heisenberg**, Danish physicist **Niels Bohr**, and English physicist **Max Born** favored the uncertainty or probabilistic interpretation, which proposed that the wave associated with a particle of matter provides merely statistical information on the position of that particle and does not describe its exact position. This interpretation was too radical for Schrödinger, Einstein, and de Broglie; the latter postulated the "double solution," claiming that particles of matter are transported and guided by continuous "pilot waves" and that their movement is essentially deterministic. De Broglie could not reconcile his pilot wave theory with some basic objections raised at the conference, however, and he abandoned it.

The disagreement about the manner in which matter behaves described two profoundly different ways of looking at the world. Part of the reason that de Broglie, Einstein, and others did not concur with the probabilistic view was that they could not philosophically accept that matter, and thus the world, behaves in a random way. De Broglie wished to believe in a deterministic atomic physics, where matter behaves according to certain identifiable patterns. Nonetheless, he reluctantly accepted that his pilot wave theory was flawed and throughout his teaching career instructed his students in probabilistic theory, though he never quite abandoned his belief that "God does not play dice," as Einstein had suggested.

In 1928, de Broglie was appointed professor of theoretical physics at the University of Paris's Faculty of Science. De Broglie was a thorough lecturer who addressed all aspects of wave mechanics. Perhaps because he was not inclined to encourage an interactive atmosphere in his lectures, he had no noted record of guiding young research students.

In 1929, at the age of thirty-seven, de Broglie was awarded the Nobel Prize for physics in recognition of his contribution to wave mechanics. In 1933, he accepted the specially created chair of theoretical physics at the Henri Poincaré Institute—a position he would hold for the next twenty-nine years—where he established a center for the study of modern physical theories. That same year, he was elected to the Académie des Sciences, becoming its Life Secretary in 1942; he used his influence to urge the Académie to consider the harmful effects of nuclear explosions as well as to explore the philosophical implications of his and other modern theories.

In 1943, anxious to forge stronger links between industry and science and to put modern physics, especially quantum mechanics, to practical use, de Broglie established a center within the Henri Poincaré Institute dedicated to applied mechanics. He was elected to the prestigious Académie Française in 1944 and, in the following year, was appointed a counsellor to the French High Commission of Atomic Energy with his brother Maurice in recognition of their work promoting the peaceful development of **nuclear energy** and their efforts to bridge the gap between science and industry. Three years later, de Broglie was elected to the National Academy of the United States as a foreign member.

During his long career, de Broglie published over twenty books and numerous research papers. His preoccupation with the practical side of physics is demonstrated in his works dealing with cybernetics, atomic energy, particle accelerators, and wave-guides. His writings also include works on x rays, gamma rays, atomic particles, optics, and a history of the development of contemporary physics. He served as honorary president of the French Association of Science Writers and, in 1952, was awarded first prize for excellence in science writing by the Kalinga Foundation. In 1953, Broglie was elected to London's Royal Society as a foreign member and, in 1958, to the French Academy of Arts and Sciences in recognition of his formidable output. With the death of his older brother Maurice two years later, de Broglie inherited the joint titles of French duke and German prince. De Broglie died of natural causes on March 19, 1987, at the age of ninety-five, having never fully resolved the controversy surrounding his theories of wave mechanics.

DE BROGLIE RELATION

Our current understanding of atomic and **molecular structure** is based on the theory of quantum mechanics (also known as wave mechanics). The first major break with the classical mechanics of English physicist Isaac Newton (1642-1727) came with the proposal in 1900 by German physicist **Max Planck** that **energy** is absorbed or given off by **matter** in small packets called quanta. Subsequent work by **Albert Einstein** supported and extended Planck's quantum ideas. It was not until the revolutionary proposal by French physicist **Louis de Broglie** in 1923, however, that the pieces came together which resulted in the development of wave mechanics. De Broglie proposed that minute moving particles such as electrons, atoms, and molecules might possess certain properties commonly associated with waves, including a wavelength. The de Broglie relation that predicted the wave length (l) of a particle of **mass** (m) traveling at a **velocity** (v) is $l = \hbar mv$, where $\hbar$ is a universal constant known as **Planck's constant** (6.63×10^{-27} erg-sec).

De Broglie's proposal was confirmed by Clinton Davisson (1881-1958), Lester H. Germer (1896-1971) and George P. Thomson (1892-1975) who observed that a stream of electrons is diffracted as it passes through a crystal which acts as a diffraction grating. This would be expected behavior for a wave but not for a particle unless it possesses wave properties.

In 1926 Austrian physicist **Erwin Schrödinger** (1887-1961) combined the developing ideas of quantized systems and the wave nature of particles into a set of postulates which form the basis of wave mechanics, a theory that has led to a new understanding of the world of atoms and molecules.

DE DUVÉ, CHRISTIAN RENÉ (1917-)

English biochemist and cell biologist

Christian René de Duvé's ground-breaking studies of cellular structure and function earned him the 1974 Nobel Prize in physiology or medicine (shared with Albert Claude and George Palade). However, he did much more than discover the two key cellular organelles—lysosomes and peroxisomes—for which the Swedish Academy honored him. His work, along with that of his fellow recipients, established an entirely new field, cell biology. De Duvé introduced techniques that have enabled other scientists to better study cellular anatomy and physiology De Duvé's research has also been of great value in helping clarify the causes of and treatments for a number of diseases.

De Duvé's parents, Alphonse and Madeleine (Pungs) de Duvé, had fled Belgium after its invasion by the German army in World War I, escaping to safety in England. There, in Thames-Ditton, Christian René de Duvé was born on October 2, 1917. De Duvé returned with his parents to Belgium in 1920, where they settled in Antwerp. (DeDuvé later became a Belgian citizen.) As a child, de Duvé journeyed throughout Europe, picking up three foreign languages in the process, and in 1934 enrolled in the Catholic University of Louvain, where he received an education in the ''ancient humanities.'' Deciding to become a physician, he entered the medical school of the university.

Finding the pace of medical training relaxed, and realizing that the better students gravitated to research labs, de Duvé joined J. P. Bouckaert's group. Here, he studied physiology,concentrating on the hormone insulin and its effects on uptake of the sugar glucose. De Duvé's experiences in Bouckaert's laboratory convinced him to pursue a research career when he graduated with an M.D. in 1941. World War II disrupted his plans, and de Duvé ended up in a prison camp. He managed to escape and subsequently returned to Louvain to resume his investigations of insulin. Although his access to experimental supplies and equipment was limited, he was able to read extensively from the early literature on the subject. On September 30, 1943, he married Janine Herman, and eventually had four children with her: Thierry, Anne, Françoise, and Alain. Even before obtaining his Ph.D. from the Catholic University of Louvain in 1945, de Duvé published several works, including a four-hundred page book on glucose, insulin, and diabetes. The dissertation topic for his *Agrégé de l'Enseignement Supérieur* was also insulin. De Duvé then obtained an M.Sc. degree in **chemistry** in 1946.

After graduation, de Duvé decided that he needed a thorough grounding in biochemical approaches to pursue his research interests. He studied with **Axel Theorell** at the Medical Nobel Institute in Stockholm for eighteen months, then spent six months with **Carl Ferdinand Cori**, Gerty Cori, and Earl

Sutherland at Washington University School of Medicine in St. Louis. Thus, in his early postdoctoral years he worked closely with no less than four future Nobel Prize winners. It is not surprising that, after this hectic period, de Duvé was happy to return to Louvain in 1947 to take up a faculty post at his alma mater teaching physiological chemistry at the medical school. In 1951, de Duvé was appointed full professor of **biochemistry**. As he began his faculty career, de Duvé's research was still targeted at unraveling the mechanism of action of the anti-diabetic hormone, insulin. While he was not successful at his primary effort (indeed the answer to de Duvé's first research question was to elude investigators for more than thirty years), his early experiments opened new avenues of research.

As a consequence of investigating how insulin works in the human body, de Duvé and his students also studied the enzymes involved in carbohydrate **metabolism** in the liver. It was these studies that proved pivotal for de Duvé's eventual rise to scientific fame. In his first efforts, he had tried to purify a particular liver **enzyme**, glucose–6-phosphatase, that he believed blocked the effect of insulin on liver cells. Many enzymes would solidify and precipitate out of **solution** when exposed to an electric field. Most could then be redissolved in a relatively pure form given the right set of conditions, but glucose–6-phosphate stubbornly remained a solid precipitate. The failure of this electrical separation method led de Duvé to try a different technique, separating components of the cell by spinning them in a **centrifuge**, a machine that rotates at high speed. De Duvé assumed that particular enzymes are associated with particular parts of the cell. These parts, called cellular organelles (little organs) can be seen in the microscope as variously shaped and sized grains and particles within the body of cells. It had long been recognized that there existed several discrete types of these organelles, though little was known about their structures or functions at the time.

The basic principles of centrifugation for separating cell parts had been known for many years. First cells are ground up (homogenized) and the resultant slurry placed in a narrow tube. The tube is placed in a centrifuge, and the artificial gravity that is set up by rotation will separate material by weight. Heavier fragments and particles will be driven to the bottom of the tube while lighter materials will layer out on top. At the time de Duvé began his work, centrifugation could be used to gather roughly four different fractions of cellular debris. This division proved to be too crude for his research, because he needed to separate out various cellular organelles more selectively.

For this reason, de Duvé turned to a technique developed some years earlier by fellow-Belgian Albert Claude while working at the Rockefeller Institute for Medical Research. In the more common centrifugation technique, the cells of interest were first vigorously homogenized in a blender before being centrifuged. In Claude's technique of differential centrifugation, however, cells were treated much more gently, being merely ground up slightly by hand prior to being spun to separate various components.

When de Duvé used this differential centrifugal fractionation technique on liver cells, he did indeed get better separation of cell organelles, and was able to isolate certain enzymes to certain cell fractions. One of his first findings was that his target enzyme, glucose–6-phosphatase, was associated with microsomes, cellular organelles which had been, until that time, considered by cell biologists to be quite uninteresting. De Duvé's work showed that they were the site of key cellular metabolic events. Further, this was the first time a particular enzyme had been clearly associated with a particular organelle.

De Duvé was also studying an enzyme called acid phosphatase that acts in cells to remove **phosphate** groups (chemical **clusters** made up of one **phosphorus** and three **oxygen** atoms) from sugar molecules under acidic conditions. The differential centrifugation technique isolated acid phosphatase to a particular cellular fraction, but measurements of enzyme activity showed much lower levels than expected. De Duvé was puzzled. What had happened to the enzyme? He and his students observed that if the cell fraction that initially showed this low level of enzyme were allowed to sit in the refrigerator for several days, the enzyme activity increased to expected levels. This phenomenon became known as enzyme latency.

De Duvé believed he had a solution to the latency mystery. He reasoned that perhaps the early, gentle hand-grinding of differential centrifugation did not damage the cellular organelles as much as did the more traditional mechanical grinding. What if, he wondered, some enzymes were not freely exposed in the cells' interiors, but instead were enclosed *within* protective membranes of organelles. If these organelles were not then broken apart by the gentle grinding, the enzyme might still lie trapped within the organelles in the particular cell fraction after centrifugation. If so, it would be isolated from the chemicals used to measure enzyme activity. This would explain the low initial enzyme activity, and why over time, as the organelles' membranes gradually deteriorated, enzyme activity would increase.

De Duvé realized that his ideas had powerful implications for cellular research. By carefully observing what enzymes were expressed in what fractions and under what conditions, de Duvé's students were able to separate various enzymes and associate them with particular cellular organelles. By performing successive grinding and fractionations, and by using compounds such as detergents to break up membranes, de Duvé's group began making sense out of the complex world that exists within cells.

De Duvé's research built on the work of other scientists. Previous research had clarified some of the roles of various enzymes. But de Duvé came to realize that there existed a group of several enzymes, in addition to acid phosphatase, whose primary functions all related to breaking down certain classes of molecules. These enzymes were always expressed in the same cellular fraction, and showed the same latency. Putting this information together, de Duvé realized that he had found an organelle devoted to cellular digestion. It made sense, he reasoned, that these enzymes should be sequestered away from other cell components. They functioned best in a different environment, expressing their activity fully only under acidic conditions (the main cell interior is neutral). Moreover, these enzymes could damage many other cellular components if set loose in the cells' interiors. With this research, de Duvé identified lysosomes and elucidated their pivotal role in cellular digestive and metabolic processes. Later research in de Duvé's laboratory showed that lysosomes play critical roles in a number of disease processes as well.

De Duvé eventually uncovered more associations between enzymes and organelles. The enzyme monoamine oxidase, for example, behaved very similarly to the enzymes of the lysosome, but de Duvé's careful and meticulous investigations revealed minor differences in when and where it appeared. He eventually showed that monoamine oxidase was associated with a separate cellular organelle, the peroxisome. Further investigation led to more discoveries about this previously unknown organelle. It was discovered that peroxisomes contain enzymes that use oxygen to break up certain types of molecules. They are vital to neutralizing many toxic substances, such as **alcohol**, and play key roles in sugar metabolism.

Recognizing the power of the technique that he had used in these early experiments, de Duvé pioneered its use to answer questions of both basic biological interest and immense medical application. His group discovered that certain diseases result from cells' inability to properly digest their own waste products. For example, a group of illnesses known collectively as disorders of **glycogen** storage result from malfunctioning lysosomal enzymes. Tay Sachs disease, a congenital neurological disorder that kills its victims by age five, results from the accumulation of a component of the cell membrane that is not adequately metabolized due to a defective lysosomal enzyme.

In 1962, de Duvé joined the Rockefeller Institute (now Rockefeller University) while keeping his appointment at Louvain. In subsequent years, working with numerous research groups at both institutions, he has studied inflammatory diseases such as arthritis and arteriosclerosis, genetic diseases, immune dysfunctions, tropical maladies, and cancers. This work has led, in some cases, to the creation of new drugs used in combatting some of these conditions. In 1971, de Duvé formed the International Institute of Cellular and Molecular Pathology, affiliated with the University at Louvain. Research at the institute focuses on incorporating the findings from basic cellular research into practical applications.

De Duvé's work has won him the respect of his colleagues. Workers throughout the broad field of cellular biology recognize their debt to his pioneering studies. He helped found the American Society for Cell Biology. He has received awards and honors from many countries, including more than a dozen honorary degrees. In 1974, de Duvé, along with Albert Claude and George Palade, both also of the Rockefeller Institute, received the Nobel Prize for physiology or medicine, and were credited with creating the discipline of scientific investigation that became known as cell biology. De Duvé was elected a foreign associate of the United States National Academy of Sciences in 1975, and has been acclaimed by Belgian, French, and British biochemical societies. He has also served as a member of numerous prestigious biomedical and health-related organizations around the globe. De Duvé became professor emeritus at the University of Louvain in 1985, receiving the same honor at Rockefeller University three years later.

DEBYE, PETER (1884-1966)

American chemical physicist

Most of Peter Debye's professional work involved the application of physical laws to the structure and behavior of molecules. In the 1910s, for example, he determined the **dipole** moments of many molecules, obtaining results that allowed him to calculate the polarity of such molecules. In recognition of this work, the unit of dipole moment, the debye, was named in his honor. Debye was also awarded the 1936 Nobel Prize in **chemistry** for this research, although he is perhaps best known for his contribution to the theory of electrolytic **dissociation**, the Debye-Hückel theory, announced in 1923. Driven out of Germany and the Netherlands during World War II, Debye immigrated to the United States and became professor of chemistry at Cornell University.

Debye was born Petrus Josephus Wilhelmus Debije on March 28, 1884, in Maastrict, the Netherlands. He is better known by the Anglicized form of his name, Peter Joseph William (or Wilhelm) Debye. His father, Joannes Wilhelmus Debije, was a foreman at a metalware manufacturer, while his mother was the former Maria Anna Barbara Ruemkens, a theater cashier prior to her marriage. The Debije's had one other child, a daughter four years younger than Peter.

Debye attended the Hoogere Burger School in Maastricht from 1896 to 1901. He then enrolled at the Technische Hochschule in Aachen, thirty kilometers from his home across the Dutch-German border. The cost of an advanced education placed a severe strain on the modest budget of the Debije family. But, as Mansel Davies writes in the *Biographical Memoirs of Fellows of the Royal Society,* Peter's father vowed that he "would work night and day" to keep his son in school. As a result, Debye eventually completed his studies at Aachen and received his degree in electrical engineering in 1905.

By a stroke of good fortune, one of Debye's teachers at Aachen had been the great German physicist Arnold Sommerfeld. When Sommerfeld was called to the chair of theoretical physics at the University of Münich in 1906, he asked Debye to join him as an assistant. Debye remained at Münich for five years, earning his doctorate in physics in 1908. The subject of his thesis was the effect of radiation on spherical particles with a variety of refractive properties. After earning his degree, Debye continued his research at Münich, serving as lecturer in physics during his last year there.

In 1911, Debye was offered the prestigious chair of theoretical physics at the University of Zürich, a post most recently held by **Albert Einstein**. Debye remained only a year at Zürich, but it appears to have been an important one. During this time he seems to have made his first serious attack on the question of the physical properties of molecules.

By the early 1900s, a fair amount of information was known about the chemical properties of molecules, but relatively little was known about their physical structure and behavior. Debye chose to deal with one aspect of this topic, the dipole moment of molecules, during his year at Zürich. The dipole moment of a **molecule** is its tendency to rotate in an external magnetic field, a property that is a function of the distribution of electric charge in (the polarity of) the molecule.

Hoping to do more experimental work than was expected at Zürich, Debye moved to the University of Utrecht in 1912, but stayed in this post only two years. He completed some exploratory research on the dipole of molecules there, but published relatively little on the subject. It was not until nearly a decade later that this research would be brought to fruition, with spectacular success.

In the meanwhile, Debye moved on again, this time to the University of Göttingen in 1914 as professor of theoretical and experimental physics. During his stay at Utrecht, Debye had married Matilde Alberer on April 10, 1913. Matilde was one of three daughters at the boarding house where Debye lived. The Debyes eventually had two children, Peter Paul Ruprecht and Mathilde Maria Gabiele. The younger **Peter Debye** became a physicist and collaborated with his father on a number of occasions.

Debye's most important work at Göttingen involved **X-ray diffraction** studies. The use of X rays to determine the structure of materials had been developed only a few years earlier by Max Laue and by **William Henry** Bragg and **William Lawrence Bragg**. The most serious problem with the Laue-Bragg discoveries was that they required the preparation of relatively large crystals. Debye, working with a colleague named Paul Scherrer, found that X-ray diffraction could also be used with powders. They developed this technique and eventually reported on the structure of a number of materials examined by this method.

In 1920, Debye returned to the University of Zürich where he served as professor of experimental physics and director of the physics laboratory at the Federal Institute of Technology. It was during his stay at Zürich that Debye developed the concept for which he is probably best known, the Debye-Hückel theory of electrolytic dissociation. Pioneering work on the behavior of electrolytes, substances that conduct **electricity** through the movement of ions, had been conducted by the Swedish chemist Svante Arrhenius in the late 1880s. Arrhenius had argued that molecules break up spontaneously in **solution**, with the liberated ions becoming electrolytic agents.

Debye took a different approach to the problem by way of redefining the mathematical application to physicochemical data, instead of to each possible configuration of ions. Electrolytes *must* dissociate almost completely in solution, he proposed, because they are already completely ionic in the solid state. The reason they do not behave that way in solution, he said, is that each **ion** has become surrounded by other ions of opposite charge. The movement of ions through a solution, then, is disturbed by the dragging effect of the surrounding ions. Working with a colleague, **Erich Hückel**, Debye generated a mathematical theory that precisely described the behavior of electrolytes in solution.

After his success with solution theory, Debye returned to his research on X-ray diffraction. In 1923, he also developed a theory that mathematically explained the Compton effect —the way the wavelengths of X rays change when they collide with electrons—and provided additional support for the wave-particle theory of electromagnetic radiation. He continued

Peter Debye.

those studies when he moved to the University of Leipzig from 1927 to 1934 and then to the University of Berlin in 1934. The Nobel Prize in chemistry he received in 1936 was given in recognition of his contributions in many fields, including "his contributions to our knowledge of **molecular structure** through the investigations of dipole moments and on the diffraction of X rays and electrons in **gases**."

The rise of National Socialism resulted in an increase of political issues in German research, and Debye was soon required to become a German citizen in order to retain his post in Berlin. He chose instead to accept an appointment as professor of chemistry and head of the department at Cornell University in Ithaca, New York, in 1940. He became a U.S. citizen in 1946. Debye officially retired from his Cornell post in 1952, but continued his research in the field of **polymer chemistry** for another decade.

During this time, Debye was much in demand as a lecturer, both in the United Sates and Europe and remained active in the scientific community until the age of 81. He suffered a heart attack at Kennedy International Airport in April 1966 while awaiting a flight to Europe and then a fatal attack, on November 2, 1966, at his home in Ithaca. During his lifetime, Debye collected a host of honors and awards, including the

Acid dehydration of sugar. ***(Photograph by David Taylor. Science Photo Library, National Audubon Society Collection/Photo Researchers, Inc. Reproduced by permission.)***

Rumford Medal of the Royal Society (1930), the Lorentz Medal of the Royal Netherlands Academy of Sciences (1935), the Franklin Medal of the Franklin Institute (1937), the Faraday Medal (1949), as well as the Gibbs Medal (1949), the Kendall Award (1957), the Nichols Medal (1963), and the Priestley Medal (1963) of the American Chemical Society.

Dehydration

Dehydration is the removal of **water** from a compound, set of compounds, or other material. Dehydration refers not only to the removal of preexisting water molecules from a **solution**, but also to the chemical formation of water by its removal from a compound to form a new compound.

The simplest form of dehydration is evaporation, in which water spontaneously leaves the liquid phase. The evaporation rate can be increased with **heat**. As water leaves a solution, the remaining solution becomes increasingly concentrated. At sufficiently high concentrations, solute molecules or ions may exceed their **solubility**, and begin to crystallize. This process is one of the most common means of obtaining crystals, both in the laboratory and in nature.

Dehydration for the purpose of crystallization may be aided by vacuum, which lowers the **vapor pressure** above the solution to increase the rate of evaporation. Freeze-drying uses evacuation and cold to prevent deterioration of solutes during dehydration. Freeze-drying is an especially effective means of food preservation, since the water content can be brought down so low that most microorganisms cannot grow on the food.

For substances with lower boiling points than water, heating can be used to drive off the more volatile substance, which can then be collected by condensing it. This process is known as **distillation**, and is used to purify many low-boiling organic compounds, such as **ethanol**. Ethanol produced by **fermentation** is at most a 20% solution in water. Distillation can raise its percentage up to 95%.

Simple distillation cannot completely dehydrate ethanol, because it forms an **azeotrope** with water. An azeotrope is a mixture with a constant **boiling point** that cannot be separated by distillation. By adding a small amount of other organic **liquids**, the ethanol can be distilled to 100%.

Gases and nonaqueous liquids can be dehydrated by passing them over or through a hygroscopic substance, one that readily absorbs water. Solid **calcium** chloride is used for this purpose both in the laboratory and in the home, where cans of anhydrous (''without water'') calcium chloride may be placed in damp closets, for instance, to prevent mildew. Anhydrous **sodium** hydroxide is another common laboratory desiccant, or substance that removes water. ''Molecular sieves'' made from zeolite clays are even more effective in many applications. The high internal surface area of these clays allow them to bind tightly with very large amounts of water.

Dehydration, in the sense of formation of a water **molecule** through chemical reaction, is an important part of organic **chemistry** and **biochemistry**. Alcohols such as ethanol (H_3CCH_2OH) can be dehydrated in this sense by reaction with heat in the presence of **sulfuric acid**, which acts not only as a catalyst, but as a hygroscopic medium to remove water as it is formed. The product, **ethene** ($H_2C{=}CH_2$) forms by removal of the OH from one **carbon** and an H from the other, leaving a double bond uniting the two carbons. This is one of the most common ways of synthesizing alkenes, double-bonded hydrocarbons.

Dehydration is perhaps the most common and vital reaction in the synthesis of biological macromolecules, such as **proteins**, deoxyribonucleic acid (DNA) and ribonucleic acid (RNA), **carbohydrates**, and fats. These four types of molecules are polymers, each formed from smaller building blocks, or monomers, which become linked by dehydration.

The basic reaction is A OH + HO B $\rightarrow$ A O B $+H_2O$. (For proteins, the basic reaction is A OH + H B $\rightarrow$ A B $+H_2O$.) For proteins, A and B represent **amino acids**; for **DNA** and **RNA** they represent nucleotides; for carbohydrates they represent single sugars; and for fats, they represent a fatty acid and a **glycerol**.

DEHYDROGENATION

Dehydrogenation is the removal of two **hydrogen** atoms from an organic **molecule**. The removal may result in the formation of hydrogen gas, or alternatively, the two hydrogen atoms may be transferred sequentially to an oxidizing agent. In the latter case, the process usually takes several steps, which might involve: (a) the stepwise transfer of a hydride **anion** (H^-) followed by a hydrogen **cation** (H^+); (b) the stepwise transfer of two hydrogen atoms; (c) the stepwise transfer of a hydrogen **atom**, followed by a hydrogen cation and an **electron**; and so on. Irrespective of the **reaction mechanism** the overall net effect is the same, namely the removal of two hydrogen atoms or two protons plus two negatively charged electrons. Dehydrogenation is thus a two-electron process, and is one of the two most important processes (the other being oxygenation) whereby organic molecules are oxidized. Most oxidations in organic **chemistry** involve a gain of **oxygen** and/or loss of hydrogen. The reverse is true for reductions. Coupling reactions of the type: 2 $CH_3CH_2CH_2CH_3 \rightarrow CH_3(CH_2)_6CH_3 + H_2$ are not generally classified as dehydrogenation reactions. Here, the molecular hydrogen release requires two molecules of reactant, rather than one.

Thermal dehydrogenation of alkanes of fewer than six **carbon** atoms can be accomplished with metal **oxide** catalysts (*i.e.*, **chromium** oxides, **molybdenum** oxides and **vanadium** oxides) at temperatures between 700-850 K. Provision must be made for the continuous removal of the hydrogen gas that is generated. Hydrogen removal from the alkane produces carbon-carbon double bonds. Location of the double bond(s) in open chain molecules is not easily controlled, and the product is usually a mixture of alkenes. Little fragmentation is observed in the case of the smaller alkanes.

Larger alkanes undergo thermal dehydrogenation with degradation. The reaction is often accompanied by cyclization: $CH_3CH_2\ CH_2CH_2CH\ _2CH_2CH\ _3 \rightarrow C_6H_5CH_3 + 4H_2$ which can lead to the formaton of **aromatic hydrocarbons**. Where structural features permit, bicyclic and polycyclic compounds may be formed. Many heterocyclic aromatic compounds have also been made in this fashion. The majority of rings formed as a result of dehydrogenation are either six-membered or five-membered. Dehydrogenation is the commercial process for preparing toluene from heptane as given above. Toluene is an aviation fuel, and is an important organic solvent found in paints, gums, resins, liquid **paper** and most oils, and is both a diluent and thinner for nitrocellulose lacquers. It is also a starting material in the manufacture of numerous dyes, medicines and detergents. Styrene is prepared in similar fashion, by heating ethylbenzene to about 900 K in the presence of a chromium oxide-aluminum oxide catalyst.

Other methods for achieving aromatization involve the use of elemental **palladium**, **platinum**, **nickel**, **sulfur** (S_8) or **selenium**, activated **charcoal** or quinones (*i.e.*, 2,3- dichloro5,6dicyanobenzene). Sulfur and selenium combine with the hydrogen evolved to give H_2S and H_2Se, respectively. In the case of quinones, the reaction mechanism involves the transfer of a hydride anion to the quinone oxygen, followed by the transfer of a **proton** to the resulting phenolate **ion**. Hydrogen gas is not released. Rather, hydrogen is transferred to the quinone, which functions as an external oxidizing agent. Dehydrogenation reactions are also used to analyze the number of fused rings in the carbon atom skeleton of complex biological molecules. This helps in the identification of biological molecules. For example, the reaction of **cholesterol** with elemental sulfur yields the aromatic **hydrocarbon** 3'-methyl-1,2-cyclopentenophenanthrene, which has the same number of fused rings. Cholesterol and 3'-methyl-1,2- cyclopentenophenanthrene both have a three sixmembered plus one five- membered **ring** system.

Methane, CH_4, can be converted into **carbon dioxide** through a series of alternating oxygenation and dehydrogenation steps. The first step involves the insertion of oxygen into one of the four carbonhydrocarbon bonds to give **methanol**, CH_3OH, which is a racing car fuel. The dehydrogenation step removes both the hydrogen from the hydroxyl functional group and one of the hydrogen atoms attached to carbon. The product formed, $H_2C{=}O$, contains an aldehyde group. This particular aldehyde (called formaldehyde) is used as an embalming fluid by funeral homes, and is a starting material in the manufacture of **plastics** and polyesters. The second oxygenation again involves insertion of oxygen, this time into one of the two remaining carbon- hydrogen bonds to yield **formic acid**, HCOOH. Carbon dioxide is formed when formic acid is dehydrogenated. In both dehydrogenation steps the components of molecular hydrogen are removed from adjacent carbon and oxygen atoms. Hydrogen removal leads to the formation of the C=O chemical bond.

Finally, catalytic dehydrogenation provides a convenient synthetic means for converting both primary alcohols into aldehydes: $RCH_2OH \rightarrow R\ CHO$ and second alcohols into ketones: RCHOHR→RCOR where R denotes an alkyl chain having an **molecular formula** of C_nH_{2n+1}. **Copper** chromite and **zinc** oxide are the catalytic agents most often used; however, it is possible to employ metallic catalysts such as **silver** and copper. Catalytic dehydrogenation is the preferred method for the industrial preparation of several important **aldehydes** and **ketones**. Dehydrogenation avoids the problem of further **oxidation** of the aldehyde to the carboxylic acid that is encountered when using strong oxidizing agents like acid dichromate and **potassium** permanganate.

DEISENHOFER, JOHANN (1943-)

German biochemist and biophysicist

Johann Deisenhofer is a biochemist and biophysicist whose career has been devoted to analyzing the composition of molecular structures. An expert in the use of X-ray technology to analyze the structure of crystals, he became part of a team of scientists in the 1980s who were studying photosynthesis—the process by which plants convert sunlight into chemical **energy**. In 1988, he shared the Nobel Prize for **Chemistry** with **Robert Huber** and **Hartmut Michel**, awarded for their work in mapping the chemical reaction at the center of **photosynthesis**.

Deisenhofer was born September 30, 1943, in Zusamaltheim, Bavaria, approximately fifty miles from Munich, Germany. He was the only son of Johann and Thekla Magg Deisenhofer; his parents were both farmers and they expected him to take over the family farm, as was the tradition. It was clear from an early age, however, that Deisenhofer was not interested in agriculture, and his parents sent him away to school in 1956. Over the next seven years, Deisenhofer attended three different schools, graduating from the Holbein Gymnasium in 1963. He then took the *Abitur,* an examination German students must take in order to qualify for university. He passed the exam and was awarded a scholarship. He then spent eighteen months in the military, as was required for young German men, before enrolling at the Technical University of Munich to study physics. His interest in physics had been developed through reading popular works on the subject, and he had an early passion for astronomy. Deisenhofer soon found himself doing an increasing amount of work in solid-state physics, which concerns the structures of condensed **matter** or **solids**. He secured a position in the laboratory of Klaus Dransfeld, and there he narrowed his interests further to biophysics, the application of the principles of the physical sciences to the study of biological occurrences. In 1971, Deisenhofer published his first scientific paper and received his diploma, roughly equal to a master's degree. He then began work on his Ph.D. in **biochemistry** at the **Max Planck** Institute in Munich under the direction of Robert Huber. Here, Deisenhofer began using a technique known as X-ray **crystallography**, which had first been demonstrated by **Max Laue** in 1912.

A crystal is a solid characterized by a very ordered internal **atomic structure**. The structural base of any crystal is called a lattice, which is defined by M.F.C. Ladd and R.A. Palmer in *Structure Determination by X-ray Crystallography* as "a regular, infinite arrangement of points in which every point has the same environment as any other point." Crystallography, the study of crystals, is considered a field of the physical sciences, and **X-ray crystallography** is the study of crystals using radiation of known length. When X rays hit crystals, they are scattered by electrons. Knowing the wavelength of the X rays used, and measuring the intensities of the scattered X rays, the crystallographer is able to determine first the specific **electron** structure of the crystal and then its atomic structure.

Deisenhofer finished work for his Ph.D. in 1974. He chose to remain in Huber's laboratory and continue his work with X-ray crystallography, first on a postdoctoral basis, and later as a staff scientist. At the same time, he was developing computer software to be used in the mapping of crystals. While working on his doctorate, Deisenhofer had embarked on a collaborative effort with Wolfgang Steigemann; they studied crystallographic refinement of the structure of Bovine Pancreatic Trypsin Inhibitor, and their findings were published in *Acta Crystallographica* in 1975.

In 1979, Hartmut Michel joined Huber's laboratory. He had been studying photosynthesis for several years and was trying to develop a method for a detailed analysis of the molecules essential to this reaction. Photosynthesis is a very complicated process, about which much is still not known. The photosynthetic reaction center, which is a membrane protein, is considered a key to understanding the process, since it is here the electron receives the energy which drives the reaction. In 1981, Michel discovered a way to crystallize the photosynthetic reaction center from the purple bacterium *Rhodopseudomonas viridis.* Once Michel had developed this technique, he turned to Huber for help in analyzing it. Huber directed Michel to Deisenhofer, and a four-year collaboration began.

Deisenhofer, with Kunio Miki and Otto Epp, used his X-ray crystallography techniques to determine the position of over 10,000 atoms in the **molecule**. They produced the first three-dimensional analysis of a membrane protein. *New Scientist* magazine, as quoted in *Nobel Prize Winners Supplement 1987–1991,* called the combined efforts "the most important advance in the understanding of photosynthesis for twenty years." The Royal Swedish Academy of Sciences awarded the 1988 Nobel Prize for Chemistry jointly to Huber, Michel, and Deisenhofer for this work. Their findings opened the possibility of creating artificial reaction centers, but the scientists were credited with more than an increase in knowledge of photosynthesis. Their findings will aid efforts to increase the scientific understanding of other functions, such as respiration, nerve impulses, hormone action, and the introduction of nutrients to cells. Deisenhofer and Michel were also recipients of the 1986 Biological Physics Prize of the American Physical Society and the 1988 Otto-Bayer Prize.

In 1987, Deisenhofer accepted the Virginia and Edward Linthicum Distinguished Chair in Biomolecular Science at the University of Texas Southwestern Medical Center at Dallas; his goal there is to establish a major center for X-ray crystallography. He has continued his research interests in the areas of protein crystallography, macromolecules, and crystallographic software. Deisenhofer has been awarded the Knight Commander's Cross of the Order of Merit of the Federal Republic of Germany, as well as the Bavarian Order of Merit. He is a fellow of the American Association for the Advancement of Science and a member of the American Crystallographic Association, the German Biophysical Society, and Academia Europa. In 1993, Deisenhofer, with James R. Norris of the Argonne National Laboratory, published a two-volume book called *The Photosynthetic Reaction Center,* based on work that grew out of Diesenhofer's collaboration with Michel.

Deisenhofer was married in 1989 to a fellow scientist, Kirsten Fischer Lindahl. He enjoys music, history, skiing, swimming, and chess in his free time. After Diesenhofer won the Nobel Prize, Dr. Kern Wildenthal, president of the Southwestern Medical School, described him to the *New York Times* as "very shy" and a man whose "life was his work." Wildenthal further observed that the scientist is "quiet, peaceful and calm. But beneath that exterior, he is scientifically fearless."

DEMOCRITUS (460-370 B.C.)

Greek philosopher and naturalist

Democritus's background is uncertain. Many stories of his life are tradition or later inventions. Among the few undisputed facts are his birthplace (Abdera, Thrace) and his teacher (Leucippus). Only fragments of his writings survived, and many of them concerned ethics. Most of the knowledge of Democritus's scientific ideas came from his students, especially Nausiphanes, the teacher of Epicurus.

According to Democritus, the physical universe was composed of atoms and void. An infinite number of atoms were in perpetual motion throughout the void. An **atom** was an eternal, unchangeable, solid body without pores or voids, so small it could not be seen or divided (the word *atomon* means indivisible). Atoms only differed from each other in shape, size, arrangement and position, which explained the properties of different substances. For example, **water** atoms were smooth and round so water flowed easily and had no permanent shape. Contrarily, atoms of fire, which were thorny caused pain, and atoms of earth which were jagged and rough held together in a definite form. A substance's feel and **taste** depended upon the observer's sense organs and convention while **color** was a result of the position of the atoms of compounds. A substance's change of nature, such as water turning into steam, was the result of atoms separating and rejoining in a new pattern.

Democritus was one of the earliest mechanists who believed that the eternal, unbreakable laws of nature, not the whims of gods or demons, determined the creation and operation of the universe, and even the human mind and soul.

Democritus.

Dendrimers

Traditionally, polymers are long **chains** of monomers, repeated over and over until a specific size range is achieved. Essentially, this process results in a "one dimensional" **molecule** having length but little breadth or width. In addition, each chain is individual grown and the reactions are concurrent, which results in limited control of the exact **molecular weight** and dimensions of the product.

An alternative approach is to grow the polymer in three dimensions to produce tree like structures and this is exactly what chemist Donald A. Tomalia and co-workers accomplished in the late 1970s. They called these compounds "dendrimers" which comes for "dendron," Greek for tree, and polymers. Retaining the allusion to trees, some chemists have adopted the name "arborols" although dendrimer is the preferred name.

In essence, these molecules start with a three dimensional core and, by adding successive layers, rapidly blossom into high molecular weight molecules. They are a molecular example of the doubling process. For example, starting with 1,4-diaminobutane as the core, four acrylonitriles can be added by a conjugate addition of the amine alkene of the acrylonitrile. After **hydrogenation**, each of the four new amines on the ends of the molecule can react with acrylonitrile, resulting in eight new ends. The next generation involves sixteen acrylonitriles and the next 32. The molecule quickly blossoms in size and, because of steric crowding, it adopts a three dimensional or spherical shape. With each generation a separate reaction, control of the domains in the polymer is significantly and exquisitely enhanced.

Probably one of the most exciting aspects of dendrimers is their demonstrated capacity to encapsulate small molecules for controlled release. This property will eventually allow these polymers to act as drug delivery agents at the molecular level. The specificity of the release mechanism will ultimately may allow site specific delivery which will benefit patients undergoing **chemotherapy** for cancer.

Density

Density is a measure of how much **mass** a substance has in comparison to its size. Density is defined as the ratio of mass to the **volume** of a particular substance. The mass of a material is directly proportional to the volume. If mass versus volume for a particular substance were plotted on a graph the result would be a straight line. The slope of this line would be the density of the substance. When calculating the density of a sample, the mass and the volume of the sample are determined and the mass of the sample is then divided by the volume. The *Système International d'Unités* (S.I.) unit for density measurements is kg/m^3, although it is often expressed in g/cm^3 or g/mL.

Conceptually, density can be described as how close together the molecules in a substance are packed. Every substance has a specific density. The density of a particular substance is a property which can be used to identify the substance. It is an intensive physical property of a particular material and does not depend on the amount of material present. For example, a cube of rubber with a mass of 100 g and a volume of 350 mL would have a density of 0.29 g/mL (100 g divided by 350 mL). If this piece of rubber were cut in half so the mass was now 50 g and the volume was now 175 mL, the density would still be 0.29 g/mL (50 g divided by 175 mL). The density of this particular sample of rubber remains the same, regardless of how much rubber is present.

Because the density of a particular substance is constant, the density of an unknown substance can be used to determine its identity. The mass and volume of a sample of the material could be used to calculate the density, and this value can then be compared to the known densities of various materials. The density of a particular substance can also be used to set up a conversion factor for use in **stoichiometry** problems.

Another use for the density value of a particular substance is in determining whether this substance will sink or float. A material will float in a particular medium if the medium has a greater density than the material. The material will sink if its density is greater than that of the medium. For example, **water** has a density of 0.998 g/mL and cork has a density of 0.24 g/mL. Cork is less dense than water so it will float when placed in water. **Lead** has a density of 11.35 g/mL, therefore it would sink when placed in water. **Helium** is less dense than air, so a balloon filled with helium gas will ''float'' on the air. ''Heavy'' cream, which floats on top of milk, is actually less dense than the milk, despite its thicker consistency.

Even though the density of a particular material is a constant, physical property of that material, it does vary with **temperature**. Most materials expand with increasing temperature. When a material expands, it occupies a greater volume. The mass of the material remains the same, so when mass is divided by the greater volume, the density is actually lower with the increased temperature. It is because of this phenomenon that sidewalks are built with cracks or spaces between concrete squares and bridges are designed with spacers at various intervals. When the weather is hot, the concrete expands, and without the spaces would buckle and crack. Another example is the operation of hot air balloons. The fire underneath the opening of the balloon heats the air inside the balloon. As the temperature of the air increases, so does its volume, and the balloon inflates. As the volume of the air increases, its density decreases, so that the density of the hot air balloon is less than the air surrounding it. Because it is less dense, it ''floats'' on the air, rising off of the ground. When the air inside of the balloon is allowed to cool, the air contracts and becomes more dense, and the balloon begins to descend.

Because density varies with temperature, it follows that density also varies with the states of **matter**. A material in the liquid state has a different density than the same material in the solid or gaseous state. Density is generally greatest in the solid state, and continues to decrease as the material becomes liquid and then gaseous. A solid material is made up of particles which are closer together than either a liquid or a gas. As a solid moves to a liquid, the particles become farther apart, and as the liquid changes to a gas, the particles are even more widespread. The density of a material in the gaseous state is only approximately 0.1% as dense as it would be in the liquid or solid state. The density of a gas follows the Ideal Gas Law and varies inversely with temperature (measured in Kelvin) and directly with molar mass and pressure.

The difference in density between a material in the gaseous state and the same material in the liquid state is far greater than that between the material in the liquid state and the solid state. Most materials are only slightly denser as **solids** than as **liquids**. One exception to this rule is liquid versus solid water (ice). The density of ice is actually less than that of liquid water. This means that water expands when it freezes. It is because of this phenomenon that cracks in windshields seem to ''grow'' faster in the winter. Moisture gets inside of a crack, and as the temperature decreases, the water in the crack freezes and expands, causing the crack to expand as well.

Determining a material's density has many practical applications. When designing a boat, for example, the structure of the boat needs to be such that it is less dense than water, even when loaded with passengers and fuel. This is generally accomplished by including many large air pockets in the frame (if a boat were solid steel, it would sink rather quickly). ''Lava lamps'' utilize the changing density of paraffin wax with temperature to create the illusion of lava bubbling through the lamp. The temperature of the wax is increased by the **heat** from a light bulb at the base of the lamp. With the increase in temperature comes a decrease in density, and the paraffin floats up to the top of the lamp. Once it reaches the top, it is allowed to cool again and the density increases, causing the paraffin wax to sink back to the bottom of the lamp. Ships which travel through icy water need to be aware of the density difference between ice and water when approaching an iceberg. The density of ice is about 89% that of liquid water, which means 89% of a body of ice will remain below the surface of the water. When a seaman spots an iceberg, it is only 11% of the actual block of ice. What may appear to be a small iceberg is actually much larger.

Besides these practical applications, chemists utilize density values in many ways. They can be used to determine the identity of an unknown substance. They can also be used to create conversion factors for stoichiometry problems. The density of different fractions of petroleum is the basis behind petroleum refining. Density is a simple calculation that can be used in many aspects of **chemistry**.

Desalination

Desalination is the process of removing **salt** from sea **water**. This process is also known as desalinization, desalting or saline water reclamation. Approximately 97% of the earth's water is either sea water or brackish (salt water contained in inland bodies), both of which are undrinkable by humans. Sea water contains 35,000 parts per million (ppm) (3.5% by weight) of dissolved **solids**, mostly **sodium** chloride and **calci-**

um and **magnesium** salts. Brackish water typically contains 5,000-10,000 ppm dissolved solids. To be consumable, or potable, water must contain less than 500 ppm dissolved solids. The method used to reach this level depends on the local water supply, the water needs of the community, and economics. Growing populations in arid or desert lands, contaminated groundwater, and sailors at sea all created the need for desalting techniques.

In the fourth century B.C., Aristotle told of Greek sailors desalting water using evaporation techniques. Sand filters were also used. Another technique used a wool wick to siphon the water. The salts were trapped in the wool. During the first century A.D., the Romans employed clay filters to trap salt. **Distillation** was widely used from the fourth century on—salt water was boiled and the steam collected in sponges. The first scientific paper on desalting was published by Arab chemists in the eighth century. By the 1500s, methods included filtering water through sand, distillation, and the use of white wax bowls to absorb the salt. The techniques have become more sophisticated, but distillation and filtering are still the primary methods of desalination for most of the world. The first desalination patent was granted in 1869, and in that same year, the first land-based steam distillation plant was established in Britain, to replenish the fresh water supplies of the ships at anchor in the harbor. A constant problem in such a process is scaling. When the water is heated over 160°F (71°C), the dissolved solids in water will precipitate as a crusty residue known as scale. The scale interferes with the transfer of **heat** in desalting machinery, greatly reducing the effectiveness. Today, the majority of desalting plants use a procedure known as multistage flash distillation to avoid scale. Lowering the pressure on the sea water allows it to boil at temperatures below 160°F (71°C), avoiding scaling. Some of the water evaporates, or flashes, during this low pressure boiling. The remaining water is now at a lower **temperature**, having lost some **energy** during the flashing. It is passed to the next stage at a lower temperature and pressure, where it flashes again. The condensate of the previous stage is piped through the water at the following stage to heat the water. The process is repeated many times. The water vapor is filtered to remove any remaining brine, then condensed and stored. It is the reuse of heat that makes these plants economical. Over eighty percent of land-based desalting plants are multistage flash distillation facilities.

A host of other desalinization processes have been developed. An increasingly popular process, reverse **osmosis**, essentially filters water at the molecular level, by forcing it through a membrane. The pressures required for brackish water range from 250-400 pounds per square inch (psi), while those for sea water are between 800-1,200 psi. The pressure required depends on the type of membrane used. Membranes have been steadily improving with the introduction of polymers. Membranes were formerly made of **cellulose** acetate, but today they are made from polyamide **plastics**. The polyamide membranes are more durable than those of cellulose acetate and require about half the pressure. Solar distillation is used in the subtropical regions of the world. Sea water is placed in a black tray and covered by a sloping sheet of **glass** or plastic. Sunlight passes through the cover. Water evaporates and then condenses on the cover. It runs down the cover and is collected. The salts are left behind in the trays. This method has been used successfully in the Greek islands.

In the 1990s, desalination technology is advancing in several directions. One increasingly popular approach is to use desalinated water to supplement regular drinking water. The state of Florida, for example, is using dozens of reverse-osmotic plants to treat undrinkable brackish water and then mixing the treated water with the regular water supply. The intent is to extend the local water supply. Another approach is to make traditional methods, like distillation, more economically feasible. In 1998, engineers at California's Metropolitan Water District reported they have designed a new, improved distillation plant that uses cheaper materials (e.g., aluminium instead of **titanium**, concrete instead of steel) to keep construction costs down.

DEUTERIUM

Deuterium is an **isotope** of **hydrogen** with atomic **mass** of 2. It is represented by the symbols 2H or D. Deuterium is also known as heavy hydrogen. The **nucleus** of the deuterium **atom**, consisting of a **proton** and a **neutron**, is also known as a deuteron and is represented by the symbol d.

The possible existence of an isotope of hydrogen with atomic mass or two was suspected as early as the late 1910s after **Frederick Soddy** had developed the concept of isotopes. Such an isotope was of particular interest to chemists. Since the hydrogen atom is the simplest of all atoms —consisting of a single proton and a single electron—it is the model for most atomic theories. An atom just slightly more complex-one that contains a single neutron also—could potentially contribute valuable information to existing atomic theories.

Among those who sought for the heavy isotope of hydrogen was **Harold Urey**, at that time professor of **chemistry** at Columbia University. Urey began his work with the realization that any isotope of hydrogen other than hydogen-1 (also known as protium) must exist in only minute quantities. The evidence for that fact is that the atomic weight of hydrogen is only slightly more than 1.000. The fraction of any isotopes with mass greater than that value must, therefore, be very small. Urey designed an experiment, therefore, that would allow him to detect the presence of heavy hydrogen in very small concentrations.

Urey's approach was to collect a large **volume** of liquid hydrogen and then to allow that liquid to evaporate very slowly. His hypothesis was that the lighter and more abundant protium isotope would evaporate more quickly than the heavier hydrogen-2 isotope. The volume of liquid hydrogen remaining after evaporation was nearly complete, then, would be relatively rich in the heavier isotope.

In the actual experiment, Urey allowed 4.2 qt (4 l) of liquid hydrogen to evaporate until only 0.034 oz. (1 ml) remained. He then submitted that sample to analysis by **spectroscopy**. In spectroscopic analysis, **energy** is added to a sample. Atoms in the sample are excited and their electrons are raised to higher energy levels. After a moment at these higher energy levels, the electrons return to their ground state, giving off their excess energy in the form of light. The bands of light emitted in this process are characteristic for each specific kind of atom.

By analyzing the spectral pattern obtained from his 0.034 oz. (1 ml) sample of liquid hydrogen, Urey was able to identify a type of atom that had never before been detected, the heavy isotope of hydrogen. The new isotope was soon assigned the name deuterium. For his discovery of the isotope, Urey was awarded the 1934 Nobel Prize in chemistry.

Deuterium is a stable isotope of hydrogen with a relative atomic mass of 2.014102 compared to the atomic mass of protium, 1.007825. Deuterium occurs to the extent of about 0.0156% in a sample of naturally occurring hydrogen. Its melting point is 18.73 K (compared to 13.957 K or protium) and its **boiling point** is 23.67 K (compared to 20.39 K for protium).

Deuterium is most commonly found as part of the compound deuterium **oxide** (D_2O), usually referred to as heavy **water** (an appropriate name considering that deuterium has twice the weight of the normal hydrogen atom). Given that the deuterium isotope occurs naturally in the ratio 1:4500; one would expect to find D_2O at a level of about 1 in 20 million water molecules. Heavy water was first separated from ordinary water in 1932 by G N Lewis, a chemist at the University of California. Compounds containing deuterium have slightly different properties from those containing protium, and the melting and boiling points of heavy water (see below) are, respectively, 38.86°F (3.81°C) and 214.56°F (101.42°C).

Heavy water can be prepared by the prolonged **electrolysis** of water. During the electrolysis, molecular **oxygen** is generated at the **anode**, and molecular hydrogen at the cathode. The deuterium tends to remain behind, so the **concentration** of heavy water increases as electrolysis progresses. When electrolysis has been carried to completion, all that remains is pure heavy water. It takes approximately 100,000 gal (378,000 l) of water to produce a single gallon of pure heavy water by electrolysis.

Heavy water is a suitable and convenient moderator of neutrons in nuclear reactors. In heavy water reactors, **plutonium** can be bred from natural **uranium**. Therefore, the production of heavy water is closely monitored, and the material is export controlled. Nations seeking large quantities of heavy water immediately become suspect of wanting to use the material to moderate a reactor, with possible intentions of producing plutonium. However, Canada's CANDU reactors, which use heavy water, are designed and used for commercial electric power production.

Heavy water is not radioactive, nor is it especially dangerous to humans or other lifeforms. But seeds will not germinate in heavy water, and some animals, including tadpoles, cannot live in it. This is because replacing hydrogen with deuterium slows down the rate of any chemical reaction in which the chemical bond to the hydrogen atom is broken, including many chemical reactions occurring in biological systems. Hydrogen atoms from water end up in a large number of biomolecules, so any process involving these hydr ogen atoms will also be slowed down if heavy water is substituted for water. Thus, heavy water slows down many metabolic processes.

Deuterium has primarily two uses, as a tracer in research and in thermonuclear fusion reactions. A tracer is any atom or group of atoms whose participation in a physical, chemical, or biological reaction can be easily observed. Radioactive isotopes are perhaps the most familiar kind of tracer. They can be tracked in various types of changes because of the radiation they emit.

Deuterium is an effective tracer because of its mass. When it replaces protium in a compound, its presence can easily be detected because it weighs twice as much as the protium atom. Also, as mentioned above, the bonds formed by deuterium with other atoms are slightly different from those formed by protium with other atoms. Thus, it is often possible to figure out what detailed changes take place at various stages of a chemical reaction using deuterium as a tracer.

Deuterium plays a critical role in most thermonuclear fusion reactions. In the solar process, for example, the fusion sequence appears to begin when two protium nuclei fuse to form a single deuteron. The deuteron is used up in later stages of the cycle by which four protium nuclei are converted to a single **helium** nucleus.

In the late 1940s and early 1950s, scientists found a way of duplicating the process by which the sun's energy is produced in the form of thermonuclear fusion weapons, the socalled hydrogen bomb. The detonating device in this type of weapon was **lithium** deuteride, a compound of lithium metal and deuterium. The detonator was placed on the casing of an ordinary fission (''atomic'') bomb. When the fission bomb detonated, it set off further nuclear reactions in the lithium deuteride that, in turn, set off fusion reactions in the larger hydrogen bomb.

For more than four decades, scientists have been trying to develop a method for bringing under control the awesome fusion power of a hydrogen bomb for use in commercial power plants. So far, the technical details for making such a process commercially viable have not been completely worked out.

DEWAR, JAMES (1842-1923)

Scottish chemist and physicist

James Dewar was born in Kincardine, Fife, Scotland, on September 20, 1842. He attended the University of Edinburgh and developed a wide range of interests, including **electricity**, **chemistry**, **spectroscopy**, and the measurement of high **temperature**. He was served as professor at both Cambridge University and the Royal Institution in London, England.

In 1874, Dewar published two papers: *The Latent Heat of Liquid Gases* and *A New Method of Obtaining a Very Perfect Vacua*. He had discovered that when **charcoal** was cooled it became extremely efficient at absorbing molecules. That meant that after a vacuum had been created with an air pump, charcoal could be used to absorb any remaining molecules, creating a better vacuum. This improved vacuum would come in very handy for Dewar in the future.

With George D. Liveing, Cambridge professor of chemistry, Dewar published 78 papers on spectroscopy between 1877-1904. They discovered the absorption spectrum of many elements and compounds. By comparing the spectra of known objects with that of the stars, it was possible to identify the elements in the stars.

Dewar became interested in the nature of materials at extremely low temperatures after Louis Paul Cailletet and Raoul Pictet had created small amounts of **oxygen** and **nitrogen** in nearly liquid form in 1877. They had been able to attain a temperature of eighty degrees above absolute zero.

Dewar, however, was more concerned with studying the characteristics of liquified gasses. He built a cooling device and, in 1885, became the first person to produce a quantity of liquid oxygen which, he later discovered, was influenced by a magnet. He eventually reached a temperature of 14° above absolute zero, becoming the first person to create solid oxygen and **hydrogen**.

The biggest impediment to Dewar's work was in keeping the gases cold enough to remain liquid. To better insulate the gases, he created a double-walled flask in 1872 that kept **liquids** cold by keeping out the warm surrounding air. Unfortunately, Dewar did not patent the idea and the Dewar flask was eventually adapted for use as an insulated beverage container better known as a Thermos ™ brand bottle.

Dewar and his colleague Frederick Abel (1827-1902) did patent their discovery of **cordite**, a form of smokeless **gunpowder**. This discovery had come about following long discussions with **Alfred Nobel**, who sued them and lost.

Intending to explore the entire field of cryogenics, Dewar joined forces with John A. Fleming, professor of electrical engineering at London's University College, who would later invent the vacuum tube. They studied the electrical and magnetic properties of **metals** and alloys at low temperatures. They discovered that as temperature dropped, electrical resistance dropped as well. At absolute zero resistance would vanish completely. This paved the way for the science of **superconductivity**. Creating a calorimeter to measure specific and latent heat at low temperatures in 1913, Dewar determined the atomic heats of elements and molecular heats of compounds.

Dewar was knighted in 1904. While many of his discoveries came about through the collaboration of colleagues, Dewar did not work well in a team situation; he could be brusque and individualistic at times. He was indifferent as a teacher, but excellent as a public lecturer. At the time of his death, on March 27, 1923, the 80-year-old was still experimenting, using a charcoal-gas thermoscope to measure infrared radiation from the sky.

James Dewar.

Dialysis

Dialysis is a process by which small molecules in a **solution** are separated from large molecules. The principle behind the process was discovered by the Scottish chemist **Thomas Graham** in about 1861. Graham found that the rate at which some substances, such as inorganic salts, pass through a semipermeable membrane is up to 50 times as great as the rate at which other substances, such as **proteins**, do so. We now know that such rate differences depend on the fact that the openings in semipermeable membranes are very nearly the size of atoms, ions, and small molecules. That makes possible the passage of such small particles while greatly restricting the passage of large particles.

In a typical dialysis experiment, a bag made of a semipermeable membrane is filled with a solution to be dialyzed. The bag is then suspended in a stream of running **water**. Small particles in solution within the bag gradually diffuse across the semipermeable membrane and are carried away by the running water. Larger molecules are essentially retained within the bag. By this process, a highly efficient separation of substances can be achieved.

Electrodialysis is a form of dialysis in which the separation of ions from larger molecules is accelerated by the presence of an electrical field. In one arrangement, the solution to be dialyzed is placed between two other solutions, each of which contains an electrode. Cations within the middle solution are attracted to one electrode and anions to the other. Any large molecules in the middle solution remain where they are.

One possible application of electrodialysis is the **desalination** of water. In this procedure, **sodium** ions from seawater migrate to the cathode and chloride ions to the **anode** of an electrodialysis apparatus. Relatively pure water is left behind in the central compartment.

A mixture of natural and synthetic industrial diamonds. Diamonds are naturally occurring forms of carbon which had crystallized under great pressure and is the hardest known mineral on Earth. *(Photograph by Sinclair Stammers/Science Photo Library, Photo Researchers, Inc. Reproduced by permission.)*

DIAMOND

Substances that exist in two or more forms that are significantly different in physical or chemical properties are called allotropes. Diamond is one of the three allotropic forms of **carbon**; the other two are amorphous carbon and **graphite** (Buckminsterfullerene may be recognized by some as a fourth allotropic form). Unlike many other elements, carbon does not readily convert to different allotropic forms in the absence of extreme conditions, e.g., high pressure, high **temperature**, and/or long times. Besides diamonds, other forms of carbon found in nature include **charcoal**, **coal**, and soot. Synthetic diamonds are produced by forcing an allotropic transition from graphite to diamond under conditions of extremely high temperature and mechanical pressure over a period of several days or weeks.

The French chemist **Antoine Lavoisier** showed in the eighteenth century that, when air is present, diamonds are combustible, producing **carbon dioxide**. Sir **Humphry Davy** demonstrated in 1814 that the sole product of the **combustion** of diamonds in **oxygen** is carbon dioxide. He also proved that diamond and charcoal both consist of carbon atoms, so are chemically identical. This was the first demonstration that two materials with the same chemical composition need not have the same physical properties.

The beauty of diamonds are due to their high refractive indices and their ability to disperse the colors of ordinary light. Diamonds have cleavage planes in four directions, making them highly susceptible to shattering when struck by a hard blow. The hardness of diamond is due to its symmetrical structure. Each carbon **atom** is surrounded by four others in a tetrahedral arrangement; as a result, each diamond is a single **molecule**.

```
    H   H   H   H            H   H   H
    |   |   |   |            |   |   |
H—C— C—C—C—H         H—C—C—C—H
    |   |   |   |            |   |   |
    H   H   H   H            H   |   H
                                 |
                               H-C-H
                                 |
                                 H
      Butane                Isobutane
```

Figure 1. ***(Illustration by Electronic Illustrators Group.)***

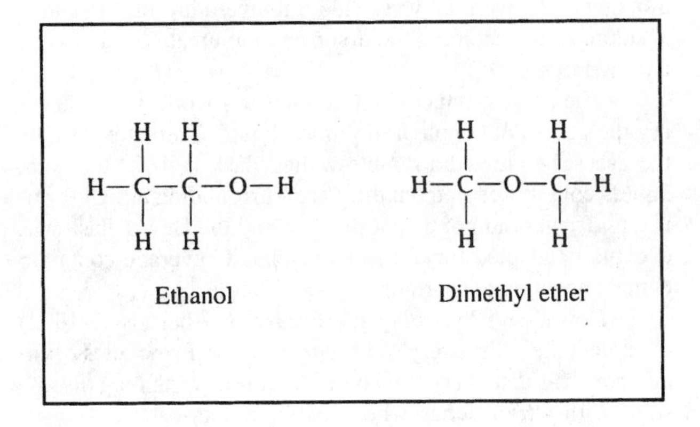

Figure 2. ***(Illustration by Electronic Illustrators Group.)***

The word diamond comes from the Greek word *adamas*, meaning invincible. Diamonds were first found in the sands of India. Alexander the Great (356- 323 B.C.) introduced them to Europe in 327 B.C.

DIASTEREOMERS

Isomers are compounds that have the same molecular formulas but different structural arrangements of atoms. They fall into two categories: constitutional isomers and stereoisomers. Constitutional isomers are isomers that have different atomic connectivities. Examples of constitutional isomers include butane and isobutane (both have the **molecular formula** C_4H_{10}, but different structures; Figure 1) and **ethanol** and dimethyl **ether** (both have the formula C_2H_6O, but again the two differ structurally; Figure 2). Stereoisomers are isomers whose constituent atoms are connected in the same sequence, but in different spatial patterns. Examples of stereoisomers include the cis and trans isomers of the alkenes (Figure 3).

Stereoisomers can be further subdivided into: **enantiomers** and diastereomers. Enantiomers are stereoisomers whose molecules are nonsuperposable mirror-images of each other. Diastereomers are stereoisomers whose molecules are

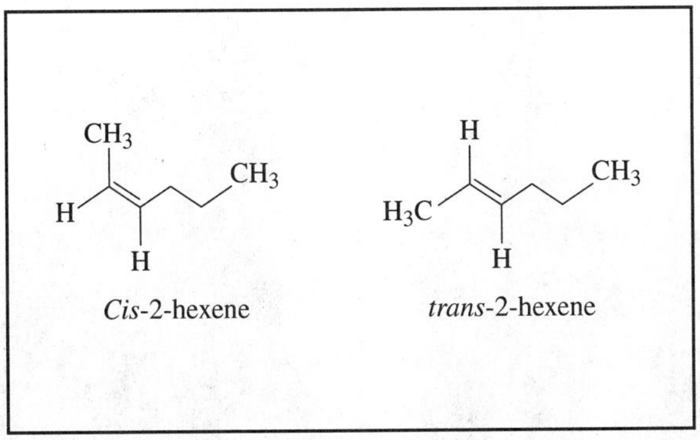

CH3 H CH3 H H3C CH3 H H
Cis-2-hexene trans-2-hexene

Figure 3. ***(Illustration by Electronic Illustrators Group.)***

H CH3 H H H3C H H3C CH3
trans-2-butene cis-2-butene

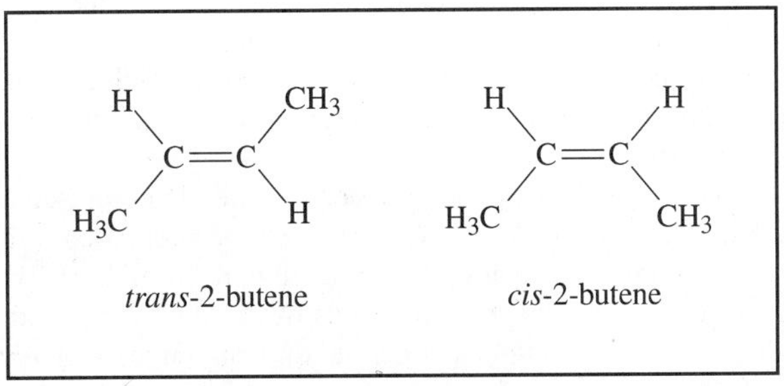

Figure 4. ***(Illustration by Electronic Illustrators Group.)***

not mirror images of each other. The atoms of diastereomers are joined together in the same order but differ in the way the atoms are arranged in space (Figure 4). Pairs of diastereomers are not related as are enantiomers, i.e., they are not mirror images and they have different physical properties.

One way to conceptually distinguish enantiomers from diastereomers is to hold a model of one of the stereoisomers in question in front of a mirror, and check whether its mirror image corresponds to the other stereoisomer. If it does not, the stereoisomers are diastereomers; if it does, the stereoisomers are enantiomers.

Enantiomers can be distinguished from each other by the way each rotates polarized light. An ordinary light beam consists of a group of electromagnetic waves that oscillate in all directions. When such a beam passes through a polarizer, only those waves oscillating in a specific plane are transmitted. If a polarized beam is passed through a **solution** consisting of only one enantiomeric type, the beam emerges with its plane of **polarization** rotated. Enantiomers rotate a beam's plane of polarization in opposite directions, and, in general, by equal amounts.

Enantiomers have identical physical properties. A pair of diastereomers, on the other hand, may have entirely different physical properties. They may have different melting points, boiling points, densities, refractive indices, etc.

The most useful method for resolving enantiomers involves converting them into diastereomers. If an enantiomeric pair can be converted into a pair of diastereomers, the physically separated diastereomers can then be converted back to the enantiomers. This works in the following way: If group X on each enantiomer reacts with group Y of an enantiomer from another source, a pair of diasteromers is formed. The diastereomers are then separated by physical means, e.g., **distillation** or fractionation. The separated diastereomers are then individually converted back to enantiomers. Such a separation of enantiomers is referred to as a resolution.

The study of isomeric molecules belongs to the field of **stereochemistry**, which is concerned with the three- dimensional structures of molecules. The history of stereochemistry began in 1852 with English chemist Edward Frankland's (1825-1899) suggestion that an element has a definite capacity for combining with other elements. In 1861, Russian chemist Aleksandr Butlerov (1828-1886) used the term "chemical structure" for the first time, stating that the properties of a compound are determined by its **molecular structure** and reflect the way that the atoms in the **molecule** are bound together. In 1874, Dutch chemist **Jacobus van't Hoff** pointed out that the **optical activity** of **carbon** atoms arose from the way the carbon bonds are directed in space; van't Hoff's observation became the basis for subsequent work in organic stereochemistry.

See also Isomer and isomerism; Molecular structure

DIATOMIC

In nature, chemical elements are often occur as individual atoms, which are also known as monatomic molecules. However, many elements appear as diatomic molecules, as a result of attraction between individual atoms. For example, many **gases**, such as **hydrogen**, **oxygen**, and **nitrogen** appear as diatomic molecules (H_2, O_2, N_2).

The creation of a diatomic **molecule** is a process: two atoms first approach each other; the atoms' outer orbital(s) then converge to create molecular orbitals. Thus, in order to create a H_2, the simplest molecule (because hydrogen is the simplest atom), two hydrogen atoms combine there single orbitals into a molecular orbital. Diatomic molecules have two basic types of orbitals. For example, when, in the hydrogen molecule, the values of the two atomic orbitals are added, the resulting molecular orbital, also known as a **bonding** orbital, occurs in the area between the nuclei. When, however, the value of one atomic orbital is subtracted from the other, the resulting molecular orbital, with a value of almost zero, occurs in other areas than the space between two nuclei. This particular orbital is characterized as anti-bonding.

The formation process is more complex when an **atom** has more than one orbital. For example, when two **lithium** atoms (atomic number 3, two orbitals) start forming a L_2 molecule, only the outer orbitals of each atom connect, creating two molecular orbitals (bonding and anti-bonding). It is important to note that an outer orbitals of one atom will not interact with the other atom's inner orbital. Particular orbitals differ greatly in **energy**, and only similar orbitals will interact.

Diatomic molecules such as H_2 or Li_2 are known as homonuclear: they consist of two identical nuclei. Heteronuclear diatomic molecules, exemplified by **carbon** monoxide (CO) contain two different nuclei. As in the simplest homonuclear diatomic molecules, orbital interaction leads to molecule formation, except that the process is more intricate, given the variety of orbitals.

Heteronuclear diatomic molecules are also represented by a number of salts, including table **salt**, or **sodium** chloride (NaCl), **potassium** iodide (KI), and lithium bromide (LiBr). These particular salts are compounds of an alkali metal and a halogen (the term *halogen*, a derivation from Greek, conveys the idea of creating salt).

In the early days of modern **atomic theory**, the idea of diatomic molecules seemed counter-intuitive. Thus, for example, **John Dalton** held that all elements appear as single atoms, or monatomic molecules. Because he rejected the idea of a hydrogen molecule (H_2 unnatural, Dalton insisted on representing **water** as HO. However, the diatomic nature of a number of common molecules, including hydrogen, was clearly understood by **Auguste Laurent**, who used more accurate notation.

Diels, Otto (1876-1954)

German chemist

A skillful organic chemist, Otto Diels is known primarily for the **Diels-Alder reaction**, which he and his student Kurt Alder developed in 1928, and for which they received the Nobel Prize for **chemistry** in 1950. This reaction involves a synthesis between a **molecule** containing a double bond and a second molecule containing two adjacent double bonds. Diels's work in determining the structure of compounds led to the discovery of **carbon** suboxide, and helped other scientists determine the structure of **cholesterol**. As a professor, he was a masterful lecturer and produced a popular **organic chemistry** textbook.

Otto Paul Hermann Diels was born to Hermann and Bertha (née Dübell) Diels on January 23, 1876, in Hamburg, Germany. The second of three sons, he grew up in an atmosphere of learning and culture. His father was a noted classical philologist and professor at the Friedrich Wilhelm University in Berlin. In later life, Diels commented on the amount of time that his father was able to spend with his sons, despite a heavy commitment to teaching and research. His mother, the daughter of a district judge, was an unpretentious and forthright woman, though nervous and rather melancholy. Diels believed he had inherited his tendencies toward pessimism and plain speaking from her. In addition to their regular schooling, the boys were taught book binding, fret work, chess, cards, and sketching, at which Otto was especially proficient. Diels's older brother, Ludwig, became a professor of botany, a subject in which Diels himself held a lifelong interest. His younger brother, Paul, became a professor of Slavic philology. Diels also had an abiding love of theater and music, favoring the composers Wagner and Verdi.

Diels was fascinated by the chemical demonstrations performed by one of his teachers in the sixth grade. However, it was not until several years later, when he was thirteen or fourteen years old, that he made the decision to become a chemist as a result of carrying out chemical experiments at home. Initially acting only as his older brother's assistant, Diels soon became the leader in the activity as Ludwig's attention turned more and more to botany. The experiments were halted by his mother after he spilled **sulfuric acid** on his father's desk, and turned some window curtains red by spraying them with a powdered dye.

Otto Diels. *(Corbis Corporation. Reproduced by permission.)*

Diels received his early education in a classical gymnasium. He only went to school, he said, because his father insisted, and told him a diploma was essential for a worthwhile career. In 1895, still determined to study chemistry, he entered Friedrich Wilhelm University, where his father was a professor. He had wished to go somewhere else, but his parents decreed otherwise. With time out for his year of military service in 1896, he received a doctorate in 1899 for his work on cyanuric compounds under the direction of the eminent German chemist **Emil Fischer**. Upon graduation, he accepted an offer to become Fischer's assistant at the university's Institute of Chemistry. In 1904, he became a lecturer in organic chemistry and began overseeing a display of chemical apparatus that was part of Germany's chemistry exhibit at the Louisiana Purchase Exposition in St. Louis, Missouri. An international jury awarded it a gold medal.

At Fischer's suggestion, Diels's first independent research was an investigation of certain compounds that could be extracted in large quantities from **coal** tar, but for which there was no commercial demand. From one of these, fluorene, he was able to synthesize a number of dyes. Unfortunately, their cost was too high for commercial success. Disappointed by this failure, he sought a more fruitful area of research and turned his attention to a class of compounds called diacetyls. In 1906, in connection with this work, he dehydrated diethyl malonate with **phosphorus** pentoxide, producing an evil-smelling gas instead of the products he'd expected. On examination, the gaseous material turned out to be a new **oxide** of carbon made up of three carbon atoms and two **oxygen** atoms. The existence of this compound, which Diels named carbon suboxide, had never been suspected. Diels is said to have prized this discovery above any of the others that he had made.

On Fischer's recommendation, Diels was appointed assistant professor at Friedrich Wilhelm University in 1906, and assumed the post of division head in 1913. In 1916, he was called to Christian Albrecht University in Kiel as full professor and director of its Chemical Institute. Like Fischer, Diels constantly tried to improve his lectures. Simplicity, exactness, and clarity were the hallmarks of his numerous lectures and of his textbook, *Einführung in die Organische Chemie.* First published in 1907, it proved very popular and remained in print through 1966, having gone through twenty-two editions. He also wrote a lab manual for elementary **inorganic chemistry**, which appeared in 1922. His graduate students numbered more than one hundred, and went on to play important roles in the academic and industrial worlds.

Cholesterol was first isolated from gall stones in 1775, and its composition established in 1894. Found to occur in a large variety of animal tissues, its **molecular structure** was not known when, in 1903, Diels began a study of the compound at the suggestion of Swiss chemist and physiologist Emil Abderhalden, who had obtained a quantity of gall stones from a Berlin hospital. Diels isolated pure cholesterol from these in an attempt to establish the structure of the compound. Although unsuccessful, his work did supply an essential step in the final determination of the structure by fellow German chemists **Heinrich Wieland** and **Adolf Windaus**. The demands of supervising a growing number of doctoral students caused Diels to abandon work on cholesterol for a time, but he returned to it in the early 1920s to make his second major discovery, a new method for dehydrogenating compounds.

In attempting to work out the structure of a compound, chemists would dehydrogenate it, i.e. carry out a reaction to remove **hydrogen** atoms from the molecule in order to produce a simpler molecule from which, perhaps, some features of the unknown structure could be inferred. Although none of the standard **dehydrogenation** procedures worked satisfactorily with cholesterol, Diels thought that one of them, a reaction commonly produced by **sulfur**, might be suitable if the sulfur were replaced with **selenium**, an element just below sulfur in the **periodic table**, and less active. The process of dehydrogenation with selenium, which he and his co-workers developed, was published in 1927. When Diels applied it to cholesterol, he obtained a **hydrocarbon**, 3'methyl–1,2 cyclopentenophenanthrene, sometimes called the "Diels hydrocarbon." This compound is the basic structural unit not only of cholesterol, but also of many other important **natural products**, such as the **steroids** (including the sex and adrenocortical hormones), bile acids, and the active components of digitalis, which are used medically as a heart stimulant. It was this work that Diels chose to talk about in his Nobel Prize address, perhaps to avoid duplicating what Alder might say, but also because of the importance he attached to it.

The work for which Diels is best known is the Diels-Alder reaction, the union of a dienophile, a molecule containing a double bond, with a second molecule containing two adjacent double bonds, called a conjugated diene. This versatile synthesis, which produces a six-membered **ring** compound, is one of the most useful processes in organic chemistry. It takes place under quite mild conditions, room **temperature** often sufficing, and without the need of catalysts or condensing agents. As a result, the formation of by-products is minimized, and yields of the desired product are maximized. Furthermore, the two reactants join together in a very specific way, so that the structure of the product is known with certainty. The structure of camphor, for example, was confirmed when it was synthesized by means of a Diels-Alder reaction. During World War II, both the Allies and Germany used the reaction of butadiene (the diene) with styrene (the dienophile) to make a synthetic rubber. Since suitable dienes occur widely in nature and the reaction conditions are so mild, Diels thought that many natural products were synthesized in nature through a Diels-Alder reaction. The resultant paper appeared in 1928, and Diels continued to work on diene reactions for the next sixteen years, publishing thirty-three papers on the subject. The last experimental work he published, in 1944, was on this topic. Because of the theoretical and practical developments that the reaction made possible, the Nobel Foundation recognized Diels and Alder with the 1950 Nobel Prize in chemistry.

In the turmoil that accompanied the end of World War I in Germany, Diels played an important role in reviving the university. There, he carried out his own laboratory research, lectured, and oversaw the research of his graduate students until 1944, when the bombing raids on Kiel destroyed the institute (as well as Diels's home). This brought a halt to his research activity, and he was unable to resume it after the war's end. His request for retirement was granted early in 1945, but he remained on staff from 1946 to 1948, when his successor was named. He continued to lecture until 1950, when, at the age of seventy-four, his deteriorating physical and mental condition forced him to stop. He was able, nevertheless, to summon enough **energy** to go to Stockholm in December of that year to receive the Nobel Prize in person.

In spite of his professional success, Diels's personal life was marked by loss. In 1909, after a six-year acquaintance, Diels married Paula Geyer, the daughter of a government official, with whom he had five children, Volker, Hans Otto, Klaus, Joachim, Renate, and Marianne. As to so many others, World War II brought a great deal of sorrow to the Diels family. In the winter of 1943–44, Hans and Klaus were killed within three months of each other on the Russian front. The death of Diels's wife in 1945 was another severe blow. As if this marked the end of his own life, he ceased making diary entries. Invited by the Norwegian Chemical Society in 1950 to speak at the University of Oslo in connection with his Nobel Prize, Diels withdrew at the last minute so that he could return to Germany to visit his wife's grave on her seventieth birthday. In addition to the emotional pain occasioned by these losses, his last years were burdened with arthritis, which limited his beloved nature walks and eventually left him housebound. Diels regretted that ten years of war and its aftermath had prevented him from contributing as much to chemistry as he was capable. Still, a bibliography of his research papers appended to Sigurd Olsen's biographical sketch of Diels in *Chemische Berichte* lists 180 papers over forty-six years. On March 7, 1954, at the age of seventy-eight, Diels died of a heart attack.

Diels-Alder reaction

Two German chemists, **Otto Diels** and **Kurt Alder**, in 1928 reported an organic reaction that would eventually bear their names and win them the Nobel Prize for **chemistry**. They had reacted a conjugated diene with a simple alkene to generate a cycloalkane. Their study of the chemistry of this reaction **lead** to significant advances in the understanding of organic reactions. It provided a useful synthetic route to a wide variety of cyclic organic compounds and, in particular, **natural products** such as the terpenoids.

The Diels-Alder reaction involves the interaction of a dienophile (literally, a ''diene loving'' compound) with a diene, under certain conditions. The simplest example of the reaction, the combination of butadiene with ethylene, is actually a very slow reaction requiring a **temperature** of 392°F (200°C) and pressure to react. However, the addition of an **electron** withdrawing group to the dienophile (i.e., a conjugated carbonyl) and electron releasing groups to the diene (i.e., a methyl group) results in significant improvement in the rate of reaction. For example, 2,3-dimethyl-1,3-butadiene and 3-buten-2-one react readily at room temperature and pressure.

The utility of the Diels-Alder reaction is extensive but it does have limits. The conjugated diene must be able to adopt a *cis* geometry with respect to the single bond. Bulky substituents can interfere with the reaction yield. But, because the mechanism for the reaction is pericyclic, there is retention of geometry such that quite specific products can be generated. A *cis* dienophile will generate a *cis* product and a *trans* will generate a *trans*.

Overall, the Diels-Alder reaction is one of the more useful ''named'' reactions in **organic chemistry** and certainly one of the most studied.

Diffusion

Diffusion commonly refers to the spontaneous movement of a substance (gas, liquid, or solid) into its surrounding area. The molecules, or particles, that make up the substance distribute over time from an area of higher **concentration** to an area of lower concentration in order to create, at equilibrium, a uniform distribution of particles throughout the system. Diffusion is a natural process that requires no added **energy** to occur. It increases the **entropy** of the system and hence is an energetically favorable and irreversible process.

An example of diffusion is the release of a drop of ink into a beaker of **water**. The ink will be visibly distinguishable from the water for some amount of time, but it will diffuse eventually to all areas of the beaker. The collision of the ink molecules with the water molecules keeps diffusion from happening quickly. This is an example of a liquid diffusing into another liquid. In comparison to a liquid, a gas has a much lower **density**. With fewer molecules with which to collide, diffusion of **gases** occurs much more quickly than diffusion of **liquids** (or solids). Comparing two gases, the lighter gas (i.e., the gas with the smaller molar mass) diffuses faster that the heavier gas (i.e., the gas with the larger molar mass). **Solids** diffuse into one another as well, but at an even slower rate that liquids. The molecules in a solid cannot move very much at all. Not only does a solid have greater density compared to a liquid or gas, resulting in the very short movement of one **molecule** before its collision with another, but energetically favorable intermolecular interactions of a solid also slow diffusion. As a result, diffusion of a solid into another solid takes place over a very long period of time.

Temperature also influences the rate of diffusion. Temperature is a measure of the average thermal energy of the system. As the thermal energy increases, the molecules move faster. Therefore, diffusion occurs faster at higher temperatures.

Scottish chemist **Thomas Graham** (1805-1869) measured the relative ability of substances to pass through a membrane. Graham's law of diffusion states that the rate at which a gas diffuses is directly proportional to temperature, but inversely proportional to the square root of the molar **mass** of the gas. In other words, the smaller the molar mass of the gas, the faster it diffuses.

Common examples of diffusion include the diffusion of solids into liquids, such as **salt** into water. The formation of a **solution** of the salt in the water by this process is called dissolution. A liquid may diffuse into a gas (such as water into air) by the process of evaporation, or a solid into a gas (such as camphor into air) by the process of **sublimation. Osmosis** is the diffusion of a liquid (the solvent) across a semipermeable membrane. The term semipermeable refers to the ability of the membrane to prevent passage of the solids that are dissolved in the liquid (the solutes) through the membrane. Osmosis is important in the extraction by plants of water from the soil.

Dialysis, also discovered by Graham, is an important application of diffusion. In dialysis a solution is passed over a semipermeable membrane, allowing solutes up to a certain size (but not larger molecules) to diffuse across the membrane to a second solution. Artificial kidney machines use dialysis to remove metabolic waste products, such as **urea** and creatinine, from the blood. In these machines, blood is circulated on one side of a semipermeable membrane (made from cellophane), while a dialysis fluid, which closely matches the chemical composition of blood, is circulated on the other side of the membrane. The waste products diffuse from the blood into the dialysis fluid and are then discarded. Important blood components, such as the oxygen- carrying protein hemoglobin, are too large to enter the pores of the membrane and hence are retained in the blood.

A final example of the biological importance of diffusion is the exchange of gases to and from the blood that occurs in the alveolar membrane of the lungs. This membrane separates the flowing blood from the gases within the lung. **Carbon** dioxide (CO_2), a chemical end product of biological **metabolism**, is plentiful in venous blood that enters the lung. Release of the CO_2 from this blood occurs by its diffusion across the alveolar membrane, and this CO_2 is expelled upon exhalation. Inhalation brings air into the lung, and air contains 20.95% by **volume** of **oxygen** (O_2). Diffusion of O_2 across the alveolar membrane, in the other direction, allows its dissolution in the blood. Oxygen is carried, bound to hemoglobin, by the arterial blood to the cells where it is released—again by diffusion—for its use by the cells as the terminal oxidant of aerobic respiration.

DIOXIN

Dioxin is the short name for the chemical 2, 3, 7, 8-tetrachlorodibenzo-*p*-dioxin (TCDD). The chemical is a by-product of several paper- and chemical-manufacturing processes. It can be detected in emissions from incinerators and hazardous waste sites. It has also been found as a contaminant in certain **pesticides**.

In laboratory studies with animals, dioxin has been shown to have biological effects at very low levels. It has been identified as a carcinogen (cancer-causing chemical) in animals, and it is strongly suspected to cause cancer in humans. Another potential health problem that may be linked to dioxin is suppression of the immune system and decreased resistance to infectious diseases and cancer. Dioxin has been shown to cause birth defects and impaired development in animals, but there is no evidence for such effects in humans. Dioxin may also affect the reproductive system. In males, there may be decreased synthesis of testosterone (the male hormone); effects in females are unclear.

The amounts of dioxin in the environment are very low. It is estimated that 80 lb (36 kg) of dioxin are released into the environment each year. The health effects of dioxin in the environment are not as clear-cut as they appear in the laboratory. Dioxin breaks down very slowly in the body. Owing to very sensitive measuring techniques, it can be detected at extremely low levels in blood and breast milk (e.g., parts per trillion). It is unknown whether such levels have definite health effects in the general population.

Among industrial workers and others who experience higher exposure to dioxin than the general population, medical research has shown a small, but measurable increase in certain cancers and other diseases. However, these people also have greater exposure to other chemicals which may be **carcinogens** as well.

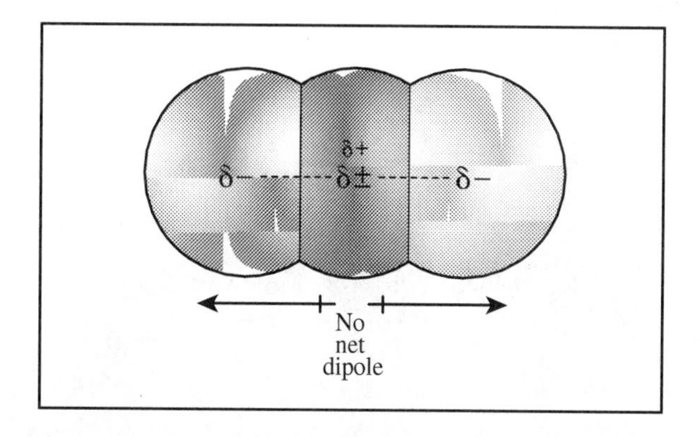

Figure 1. Carbon dioxide molecule. *(Illustration by Electronic Illustrators Group.)*

DIPOLE

An understanding of molecular dipoles provides insight into how molecules interact and helps to explain many of their properties.

A **molecule** possesses a dipole if it is electrically asymmetric, i.e., if the centers of the distributions of positive and negative charge within the molecule are different. When a molecule possesses a dipole, it is said to be polar. The dipole moment is equal to the magnitude of the equal and opposite charges multiplied by the distance between the centers of the charge distributions.

In a **covalent bond** between atoms with unequal electronegativities (i.e., having unequal attraction for electrons), the **atom** with the larger **electronegativity** will attract the electrons shared in the **bonding** molecular orbital more, and the shared electrons will spend more time near that atom. It will, consequently, have a partial negative charge. The less electronegative atom will have a partial positive charge since the shared electrons spend less time near it. Because there is a separation of charges, with a partial positive charge at one end of the bond and an equal negative partial charge at the other, a dipole results. Such a bond is called a polar covalent bond.

In a covalent **diatomic** molecule, a difference in the electronegativity of the atoms will result in a dipole moment. The larger the difference in the electronegativities of the atoms, the larger the dipole moments. For instance, electronegativities in the **halide** sequence **fluorine**, **chlorine**,bromine, and **iodine** decrease from fluorine through iodine. The dipole moments of the hydrogen-halide sequence HF, HCl, HBr, and HI are 1.98, 1.03, 0.79, and 0.38 respectively.

A polyatomic molecule with bonds which are polar is not necessarily polar itself. In **carbon** dioxide (CO_2), for instance, the bonds between each of the **oxygen** atoms and the central carbon atom are polar: oxygen is more electronegative than carbon and, therefore, has a partial negative charge while the carbon atom has a partial positive charge. CO_2 is however, as shown in Figure 1, a linear molecule so that, in effect, its two ends (oxygen atoms) have partial negative charges with the center of negative charge distribution in the center of the

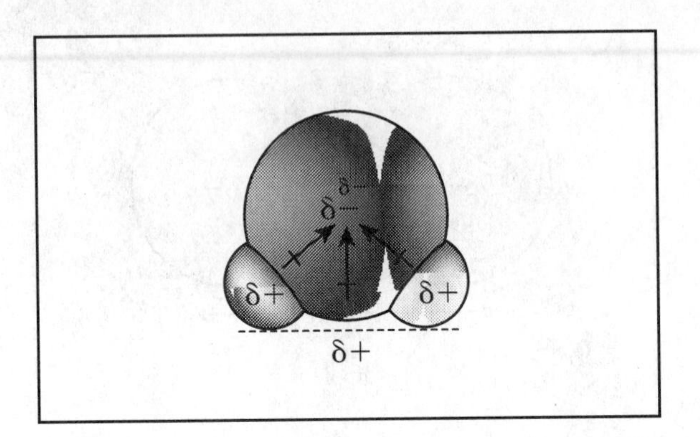

Figure 2. Water molecule. ***(Illustration by Electronic Illustrators Group.)***

molecule, precisely at the point where the center of positive charge lies on the carbon atom. There is, as a result, no separation of charges and the CO_2 molecule is not polar. In **water** (H_2O) each of the bonds between a **hydrogen** atom and the central oxygen atom is polar with the oxygen atom having a partial negative charge and the hydrogen atoms having partial positive charges. The H_2O molecule is, however, non-linear so that the effective center of positive charge, halfway between the two hydrogen atoms, does not lie at the oxygen atom, which is the center of negative charge (Figure 2). Since there is a separation between the effective centers of charge, the $_2O$ molecule is polar and possesses a dipole.

The unit assigned to dipole moments is the debye. It is named to honor **Peter Debye** who did extensive work with dipoles, developing methods for measuring their magnitudes and publishing, in 1912, a theoretical treatment of dipoles and their properties.

The measurement of the capacitances of various substances is a quantitative indication of their relative dipole moments. When polar molecules are placed between the parallel plates of a condenser, the molecules attempt to orient themselves so that their positive ends will be nearest the negatively charged plate (the cathode) with their negative ends nearest the positively charge plate (the anode). Even though thermal motion tends to disrupt this preferred orientation of the dipoles, a sufficient number of molecules, on the average, are oriented in this manner to cause the capacitance of the system to be increased. The molecules with higher dipole moments have higher capacitances. They are called dielectrics. Substances with low capacitance are insulators. The dielectric constant of a substance is the ratio of the capacitance of a pair of condenser plates with the substance between it and the capacitance of the plates in a vacuum.

When dipolar molecules are adjacent, they are attracted to each other through dipole-dipole forces. These forces result from the fact that the negative end of one dipole is attracted to the positive end of another dipole. This attraction leads to a more stable arrangement of molecules. Consequently, upon solidification, molecules align themselves to take maximum advantage of the dipole-dipole forces of attraction. In **liquids**, the molecules are also oriented in this way but they do not maintain the specific arrangement because the molecules are in constant motion. Additional **energy** is necessary to break apart molecules that are held in stable relationships with their neighbors because of the mutual attractive force between dipoles, and therefore, the boiling and melting points of materials whose molecules have dipolės are generally higher than they would be if there were no dipoles. For example, the **boiling point** of **hydrochloric acid**, HCl, a polar diatomic molecule is about -64°F (-53°C, 220 K). In contrast, the boiling point of **argon** (a non-polar fluid of atoms with nearly the same molar mass) is about -280°F (-173°C, 100 K). HCl has a higher boiling point because of its polar nature.

Molecules may change their rotational energy levels by absorbing energy from electromagnetic radiation in the microwave region of the spectrum. In order for a molecule to interact with and absorb radiation it must possess a dipole. As a result, such symmetrical non-polar molecules as CO_2, CH_3Cl, and N_2 cannot change their rotational energy by absorbing electromagnetic radiation. On the other hand, asymmetrical molecules such as HCl, HCN, and H_20, that have dipole moments, absorb radiation and exhibit absorption spectra in the microwave region.

Disproportionation

Literally, disproportionation is a chemical reaction in which a single metal **ion** species undergoes an internal redox reaction and forms two different products in stoichiometric quantities. It is to be distinguished from such processes as partial **combustion**, which leads to a significant variety of chemical species. A disproportionation leads to very specific compounds, of varying **oxidation** state and chemical composition.

Disproportionation is a reaction that is more characteristic of the heavier **transition elements** and the f-block elements. Thus, while the **chemistry** of platinum(II) and platinum(IV) is well developed, attempts at the synthesis of the platinum(III) species were foiled by the disproportionation reaction yielding the +II and +IV oxidation states. The synthesis of Pt(III) complexes requires extraodinary ligands and special conditions. Note that the same is not true of **palladium**, where a well-developed +III **oxidation state** has been observed, although this oxidation state is often achieved through the disproportionation of the Pd(II) species giving Pd(O) as the other product. Also, disproportionation is sometimes masked by the chemical composition of the product. TlBr3 undergoes a **reduction** to give a compound that has the formula ''$TlBr_2$''. This species is actually the compound $Tl^+[TlBr_4]^-$ in which both Tl(I) and Tl(III) are present.

With the actinides, disproportionation reactions are quite common. For example, plutonium(IV) in **water** spontaneously reacts via disproportionation to yield the plutonium(VI) **oxide** and plutonium(III):

$$3Pu^{+4} + 2H_2O \rightarrow PuO_2^{+2} + 2Pu^{+3} + 4H^+$$

Indeed, in 1M $HClO_4$ at room **temperature**, all four oxidation states of **plutonium**, Pu(III) to Pu(VI) are observed in equilibrium.

Dissociation

Dissociation is the process by which a chemical combination breaks up into simpler constituents. Dissociation may be accomplished by the addition of **energy**, as in the case of gaseous molecules dissociated by **heat**; or by the action of a solvent on a polar compound (i.e., electrolytic decomposition). All electrolytes dissociate to some extent in polar **solvents**. The degree of dissociation can be used to calculate the equilibrium constant for the dissociation reaction.

In the case of **diatomic** molecules, the dissociation energy refers to the energy required to break the gaseous molecules into their constituent atoms. Generally, for families of molecules such as HF, HCl, HBr, and HI; or C_2, Si_2, and Pb_2, the heavier molecules tend to have smaller dissociation energies. In the case of N_2, O_2, and F_2, the dissociation energy increases with increasing net **bonding**. In general, there is an inverse relationship between dissociation energy and bond length for related molecules. In the case of polyatomic molecules, resulting in the formation of molecular fragments, the energy required to break a single bond is referred to as a bond dissociation energy.

When the acid **hydrogen** chloride (HCl) dissolves in **water**, a dipole-dipole interaction occurs between the negative **dipole** of one **molecule** and the positive side the other. The water molecules pull on the HCl molecules, tending to dissociate them into H^+ ions and Cl^- ions. Opposing this tendency is the **covalent bond** holding the HCl molecule together. In the case of the water-HCl interaction, the dissociation reaction predominates, and H^+ ions are formed in the **solution**. The production of H^+ ions in aqueous solutions is a property common to all acids. Bases, on the other hand, produce OH^- ions in water, forming basic (or alkaline) solutions.

Acids and bases, for all practical purposes, exist, because water dissociates into H^+ and OH^- ions. In equilibrium, only a few water molecules are dissociated. The general law of equilibrium predicts an equilibrium constant for this dissociation of $[H^+][OH^-]/H_2O]$, but since the **concentration** of water in aqueous solutions is very large, this constant is usually put into the form $[H^+][OH^-] = 1.0 \times 10^{-14}$ at 25 deg C. In plain language, this expression says that any water solution at 25 deg C will contain H^+ and OH^- ions in such concentrations that their product will be 1.0×10^{-14}.

Ordinarily the concentrations of H^+ and OH^- ions in a solution are not equal. A water solution in which the H^+ **ion** concentration exceed the OH^- concentration is called acidic; solutions in which the OH^- concentration is larger are basic.

Weak acids and bases form H^+ and OH^- ions, respectively, in solution only to a small extent. In the case of the hydrogen fluoride (HF), the *acid dissociation constant* is defined as $[H^+][F^-]/HF]$. For **ammonia** (NH_3; a base) in water, the *base dissociation constant* is $[H^+][OH^-]/NH_3]$. When a compound containing a particular complex ion is added to water, it is customary to treat the ensuing reaction as if it were a simple dissociation. The dissociation constant of the complex ion is written analogously.

Distillation

Distillation is one of the most important processes for separating the components of a **solution**. The solution is heated to form a vapor of the more volatile components in the system, and the vapor is then cooled, condensed, and collected as drops of liquid. By repeating vaporization and condensation, individual components in the solution can be recovered in a pure state. Whiskey, essences, and many products from the oil refinery industry are processed via distillation.

Distillation has been used widely to separate volatile components from non-volatile compounds. The underlying mechanism of distillation is the differences in **volatility** between individual components. With sufficient **heat** applied, a gas phase is formed from the liquid solution. The liquid product is subsequently condensed from the gas phase by transferring heat from the vapor. Therefore, heat is used as the separating agent during distillation. Feed material to the distillation apparatus can be liquid and/or vapor, and the final product may consist of liquid and vapor. A typical apparatus for simple distillation used in **chemistry** laboratories is one in which the still pot can be heated with a **water**, steam, or oil bath. When **liquids** tend to decompose or react with **oxygen** during the course of distillation, the working pressure can be reduced to lower the boiling points of the substances and hence the **temperature** of the distillation process.

In general, distillation can be carried out either with or without reflux. For the case of single-stage differential distillation, the liquid mixture is heated to form a vapor that is in equilibrium with the residual liquid. The vapor is then condensed and removed from the system without any liquid allowed to return to the still pot. This vapor is richer in the more volatile component than the liquid that is removed as the bottom product at the end of the process. However, when products of much higher purity are desired, part of the condensate has to be brought into contact with the vapor on its way to the condenser and recycled to the still pot. This procedure can be repeated for many times to increase the degree of separation in the original mixture. Such a process is normally called ''rectification.''

Distillation has long been used as the separation process in the chemical and petroleum industries because of its reliability, simplicity, and low-capital cost. Recorded applications date back nearly 2,000 years. It is employed to separate **benzene** from toluene, **methanol** or **ethanol** from water, **acetone** from **acetic acid**, and many multicomponent mixtures. Fractionation of crude oil and the production of **deuterium** also rely on distillation.

Today, with 40,000 distillation towers in operation, distillation accounts for about 95% of all current industrial separation processes; even so, distillation systems also have relatively high **energy** consumption. Significant effort, therefore, has been made to reduce the energy consumption and to improve efficiency in distillation systems. This includes incorporating new analytical sensors and reliable hardware into the system to achieve advanced process control, using heat rejected from a condenser of one column to reboil other columns, and coupling other advanced process such as **adsorption** and crystallization with distillation to form energy-saving hybrid operation systems.

DJERASSI, CARL (1923-)

American chemist

Carl Djerassi has been called the "father of the birth-control pill." As a youth in his twenties Djerassi headed a research team of young scientists in a small, obscure Mexican laboratory where, using a locally grown yam, he first was able to synthesize **cortisone**, a drug used to treat rheumatoid arthritis. Within twelve months, he also created the first steroid that effectively blocked fertillization. The resulting use of the contraceptive, commonly called the Pill, has raised controversy around the world.

Djerassi was born in Vienna on October 29, 1923. His parents, Samuel and Anna Friedmann, both physicians, were Austrian Jews. Djerassi had his early schooling in Vienna, but when the German army invaded Poland in 1938, Djerassi's mother—separated from Djerassi's father in 1929—moved with her son to Bulgaria. After a year in Sofia, mother and son left for the United States, arriving penniless in America in the winter of 1939.

Djerassi attended Newark Junior College before enrolling in Kenyon College in Ohio in 1941 with a major in **chemistry**. He received his A.B. the next year. Following graduation, Djerassi returned to his home in New Jersey and accepted a position as junior chemist with the Swiss pharmaceutical firm CIBA, where he worked as an assistant to another Austrian emigre, Charles Huttrer. Within a year, the pair discovered one of the first antihistamines, pyribenzamine, used to combat allergy and cold symptoms. They received their first patent for the discovery. After a year at CIBA, Djerassi took leave to study for his doctorate in chemistry at the University of Wisconsin, Madison. Within three years, he received his degree, became a U.S. citizen, and married Norma Lundholm. (He divorced in 1976, and in 1985 married Diane Middlebrook.)

Djerassi returned to CIBA in 1945 where he remained for another four years before being invited to head a research group working in the field of steroid chemistry at Syntex, Inc., a small drug company in Mexico City. While at Syntex, Djerassi was involved in synthesizing cortisone on a commercial basis from chemicals found in yams. In the process of developing a method to derive **steroids** from plants, Djerassi also produced the hormone progesterone, the first synthesis of a steroid oral contraceptive. To test the effectiveness of the ovulation-inhibiting properties of the steroid on humans, Djerassi worked with Harvard Medical School gynecologist **John Rock** to undertake the long period of clinical studies.

Over the next several decades, Djerassi was involved in trying to influence public policy regarding birth control. He has advocated, for example, that the government develop a global contraceptive policy for developing countries. Djerassi also believes that too often governmental agencies, such as the Food and Drug Administration, are more concerned with the possible side effects of the Pill, rather than in promoting the Pill's benefits.

In 1952, Djerassi left Mexico to accept the position of associate professor of chemistry at Detroit's Wayne University (now Wayne State University). He continued his work on steroids at Wayne State, developing the spectropolarimeter (a device to conduct measurements on steroids), as well as various projects in the field of antibiotics. In 1959, he left Michigan with most of his research staff to become professor of chemistry at Stanford University—a position he currently holds. His research at Stanford has ranged from studying chemical structures obtained by satellite spectrography to developing methods to detect **lead** poisoning in urine.

In 1989, Djerassi was named to the board of Monoclonal Antibodies, a biotechnology company in California. He is working with the company trying to develop a home test for progesterone. His goal is to provide a "red light, green-light" test that will tell a woman exactly when she is fertile, and when she is not.

For his synthesis of an oral contraceptive, Djerassi was awarded the National Science Medal in 1973. In 1978, he was named to the Inventor's Hall of Fame. Djerassi is the recipient of many other honors, including the Perkin Medal, 1975; the Wolf Prize in chemistry, 1978; the John and Samuel Bard Award in science and medicine, 1983; the Esselen Award, 1989; the National Medal of Technology, 1991; and the Nevada Medal, 1992. In 1992, he was also awarded the American Chemical Society's prestigious Priestley Medal for his life's work. In addition to his scientific research and professional writing, Djerassi has written more than a dozen works of fiction, non-fiction, and poetry, including his autobiography, *The Pill, Pygmy Chimps, and Degas' Horse,* published in 1992.

DNA

The letters DNA stand for deoxyribonucleic acid, the cellular material that stores and transmits genetic information. The information contained in DNA directs the synthesis of countless **proteins**. These proteins provide the structural and functional foundations necessary for life.

DNA was discovered in 1869 by German scientist Johann Friedrich Miescher (1844-1895). He called this newly discovered **molecule** nuclein, otherwise known as nucleic acid, because it was found in the **nucleus** of cells. (The nucleus is a structure inside plant and animal cells.) Just a few years previous to Miescher's discovery, an Austrian monk named Gregor Johann Mendel (1822-1884) had made his first reports concerning the laws of genetics. However, it wasn't until the middle of the twentieth century that the two discoveries were joined and DNA was definitively accepted as the basis for genetic inheritance.

By the late nineteenth century, most scientists accepted the idea that inheritable characteristics were somehow linked to cellular **chemistry**. Nucleic acid was considered a potential candidate, but proteins were the favored possibility. Proteins were preferred over nucleic acid because they are more abundant in cells and their chemical structures are more diverse.

This diversity seemed to imply a more likely ability to transmit detailed information from generation to generation. In 1910, a scientist named Phoebus Levene demonstrated that nucleic acid was composed of only four building blocks. Owing to inadequate analysis, he mistakenly estimated that all four building blocks, or nucleotides, were present in equal amounts. Compared with proteins, which use 20 **amino acids** in varying amounts for construction, nucleic acid just seemed too simple for storing genetic information.

In the 1940s, scientists began to reconsider the possibility that nucleic acid might be the correct candidate after all. The first breakthrough came as a result of experiments conducted by **Erwin Chargaff**, an Austrian-American biochemist. His findings confirmed Levene's discovery that nucleic acid was composed of four nucleotides each containing a **phosphate** group, a sugar molecule, and one of four nitrogen-containing bases—adenine, cytosine, guanine, or thymine. However, he disproved Levene's hypothesis that the nucleotides were present in equal amounts. From his analysis of numerous samples of different DNA's, Chargaff instead proposed that the number of adenine bases equaled those of thymine and the amount of cytosine bases equaled those of guanine.

The next advance was made in 1944 by three scientists: **Oswald Avery**, Colin MacLeod (1909-1972), and Maclyn McCarty (1911-). Building on research conducted by Frederick Griffith (c.1879- 1941) in 1927, Avery, MacLeod, and McCarty conducted experiments using two strains of the bacteria *Diplococcus pneumoniae*. One strain, III*S*, can cause pneumonia. The other strain, II*R* does not cause pneumonia.

Griffith's experiments had shown that neither heat-killed III*S* nor living II*R* could cause pneumonia in mice. But if the two were combined, the resulting mixture caused pneumonia. From this result, he concluded that heat-killed III*S* contained a factor that transformed the usually harmless II*R* into an infectious bacteria. Avery, MacLeod, and McCarty set out to identify the transforming factor.

They met with success in a set of experiments that started by isolating the transforming factor from a large amount of III*S* bacteria. They demonstrated that this factor could transform II*R* except when it was treated with deoxyribonuclease, an **enzyme** that destroys DNA. (Enzymes are proteins that catalyze chemical reactions.) Further testing of the transforming factor confirmed that it was DNA and established that genetic information is stored and transmitted using this nucleic acid. Avery, MacLeod, and McCarty's conclusions were confirmed by several scientists, including Alfred Hershey (1908-1997) and Martha Chase. In 1952, Hershey and Chase showed that phages, a type of virus that attacks bacteria, use DNA rather than protein to transmit infection.

The next step in DNA research was uncovering the molecule's structure. Between 1940 and 1953, many scientists worked on this puzzle, including Chargaff, Maurice Wilkens, **Rosalind Franklin**, **Linus Pauling**, **Francis Crick**, and **James Watson**. All of these scientists supplied pieces of information, but Watson and Crick were the first in putting the pieces together to solve the puzzle.

Working at Cavendish Laboratory in Cambridge, England, Watson and Crick published their proposed structure of DNA in 1953. The so-called Watson-Crick model depicted a double-helix (spiral) structure containing two long **chains** of nucleotides. Each nucleotide of the chains contained a phosphate base, a sugar (called deoxyribose), and one of four nitrogenous bases.

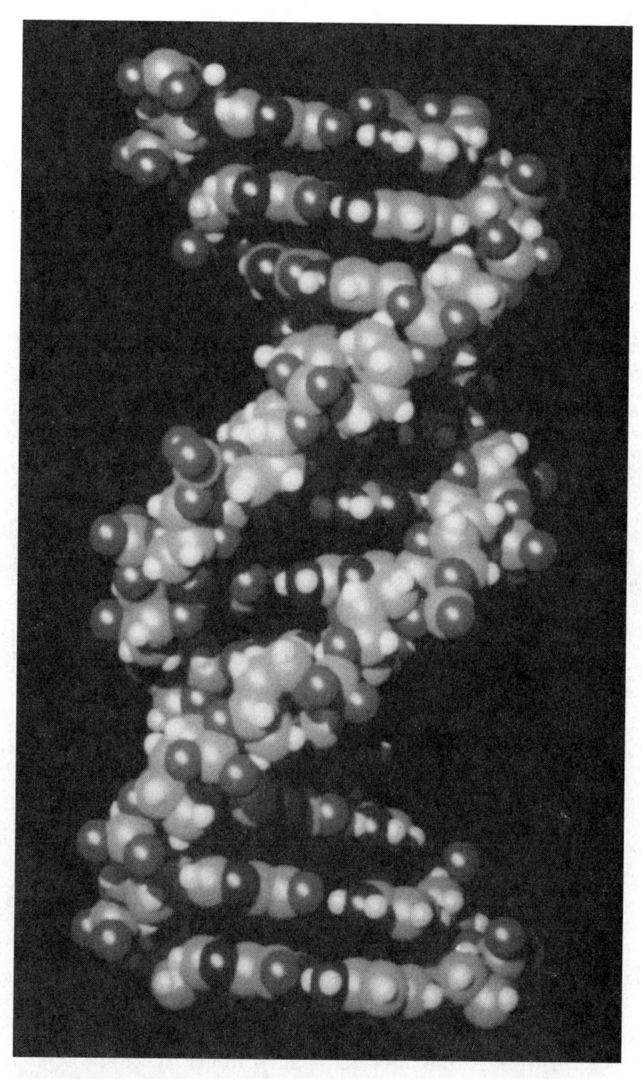

A computerized image of a section of a DNA molecule, the genetic material of most living organisms. ***(Photograph by Clive Freeman, The Royal Institution/Science Photo Library, Photo Researchers, Inc. Reproduced by permission.)***

The nitrogenous bases belong to two classes of molecules—purines and pyrimidines. Both types have molecular rings composed of **carbon** and **nitrogen**. Pyrimidine molecules are single- **ring** molecules. Cytosine and thymine are included in this group. Purines are larger two-ring molecules. Guanine and adenine are included in this group.

The backbone of both strands in the double helix is a chain of linked phosphate and deoxyribose molecules. When the two strands are lined up, the nitrogenous bases face one another like the teeth of a zipper. The strands are joined by **hydrogen** bonds between bases. Adenine always matches with thymine, and guanine is always paired with cytosine. This pattern is called Chargaff's rule, in honor of Chargaff's discoveries about nucleotide ratios. Because of the tetrahedral arrangement of substituents on the carbon **atom** to which each nitrogenous base is attached and the tetrahedral shape of substituents around the phosphate phosphorous atom along the backbone, the double-stranded molecule is forced into a spiral shape.

Virtually all living organisms contain DNA as a double-stranded molecule. Living organisms are divided into two categories: eukaryotes and prokaryotes. Eukaryotes store DNA in the nucleus of their cells. They tend to be multicellular organisms such as plants and animals. (The mitochondria of eukaryotes also contain DNA. Mitochondrial DNA directs the construction and maintenance of mitochondria only.) Prokaryotes, which are generally single-celled organisms such as bacteria, lack a nucleus. In these organisms the DNA is simply stored in the cell's cytoplasm.

Because DNA is such a large molecule, a great deal of information can be contained. Storage of DNA is related to its ability to twist and coil into a compact form. In eukaryotes, a protein called histone forms multi-part complexes known as nucleosomes. The double helix of DNA winds about individual nucleosomes, forming a superhelix. The superhelix is coiled further until the DNA-protein complex, or chromatin, is very compact. In its most condensed form, DNA is stored in bundles called chromosomes. DNA storage in prokaryotes is very similar to that in eukaryotes, other than the fact that prokaryotes do not have nuclei.

Regardless of the storage method, DNA functions similarly in all species. The genetic code contained in DNA carries the blueprint for the proteins that build and maintain living cells. The information is conveyed by means of triplet nucleotide sequences. A total of 64 triplets are possible; 61 correspond to amino acids and the remaining three serve as stop codes.

However, the information flow from DNA to protein is not direct. One strand of the double helix—the coding strand or the template—serves as a master copy. The cell uses a related nucleic acid called **RNA** (ribonucleic acid) to construct a working copy of that strand. RNA is chemically similar to DNA except it is single-stranded, the DNA base thymine is replaced by uracil (a very similar pyrimidine), and the sugar portion of the nucleotide has an -OH group in place of a hydrogen atom.

The process of transferring the genetic code to RNA is called transcription. Once the DNA has been transcribed, the RNA is modified and then released into the cell's interior where specialized structures called ribosomes translate the information and help construct the new protein.

When a cell divides and forms two new cells, it must pass on exact copies of its DNA to both new cells. DNA replication begins with a partial separation of the two strands, each of which serves as a template for a new matching strand. The separation between the template strands grows as replication continues until they are completely dissociated from one another and each has a new matching strand.

The task of constructing the new strands falls to a set of enzymes called DNA polymerases. The polymerases add nucleotides to the growing strands one at a time, always pairing the bases properly to those on the template strand. As replication continues, a special polymerase proofreads the newly forming DNA to ensure an accurate copy. If a mistake is found, the error is snipped and corrected.

When replication is complete, the cell holds two double-stranded molecules of DNA. Both molecules have one old strand and one newly constructed strand. A short time after DNA replication is accomplished, the cell divides and passes an equal amount of DNA to both new cells.

With the discoveries of DNA structure, function, and replication, new areas for investigation opened up. Two of the most exciting advances have been DNA sequencing and gene cloning.

DNA sequencing is a process through which it is possible to determine the exact nucleotide order of a DNA sample. The Sanger dideoxy sequencing method is one of the more common techniques, although there are others that work as well. In the Sanger method, a sample of single-stranded DNA is divided into four test tubes. Added to each test tube are all four nucleotides, DNA polymerase, and a very small amount of one of four dideoxynucleotides. Each dideoxynucleotide has a different base—adenine, thymine, guanine, or cytosine—and chemically is slightly different from the regular nucleotides. The dideoxynucleotides are also weakly radioactive.

In each tube, the polymerase works to make new matching strands for the single-stranded DNA molecules. The polymerase uses single strands as templates, but whenever a dideoxynucleotide is added, further construction is stopped on that particular molecule. Since there is only a small amount of dideoxynucleotide in each tube, construction on other molecules continues. After a set amount of time, the four tubes contain double-stranded DNA at various stages of completion. A small amount is withdrawn from each tube and loaded into separate wells of a polyacrylamide **gel**. A polyacrylamide gel is simply a thin slab of jello-like material with small divots, or wells, along its top edge.

This gel is in a tank of fluid through which an electrical current runs. The electrical current causes the DNA molecules to move through the gel. The molecules move according to their size. With each nucleotide that was added to a matching strand, individual molecules are larger and move a smaller distance from the wells. After several hours, the samples in each well have formed a column of short horizontal bands through the gel. Only the bands that contain a dideoxynucleotide can be detected on a sheet of photographic film owing to their **radioactivity**.

The bands form a ladder-like pattern with each band representing a particular dideoxynucleotide addition to the matching strand. Each step down the ladder corresponds to the next nucleotide in the sequence. The step can be easily identified by knowing the test tube from which it was drawn.

To clone a gene—the DNA that codes for a specific product—a technique called reverse DNA transcription can be used. In this method, the RNA that corresponds to a particular gene is treated with an enzyme called reverse transcriptase. Reverse transcriptase directs the manufacture of DNA, using the RNA strand as a template. With one strand of DNA, DNA polymerases can be used to construct the second strand. The end result is a double-stranded DNA molecule called cDNA. (A similar sequence of events is carried out naturally by retroviruses. Retroviruses are viruses that use RNA as their genetic material rather than DNA. HIV, the virus that causes AIDS, is a retrovirus.) Once the cDNA is complete, it can be inserted into another cell which then begins to produce the cloned product.

See also Catalyst and catalysis; Protein synthesis

DNA Fingerprinting

DNA fingerprinting is the overall term applied to a range of techniques that are used to show dissimilarites between the DNA present in different individuals of the same or different species.

DNA fingerprinting requires DNA to be broken down, or digested by, enzymes. This digested DNA is then placed into an agarose **gel** along with other samples of DNA. These other samples may be test samples or they may be controlled samples, because it is vital to always include a standard piece of DNA to calibrate the results. The loaded gel is then placed in a liquid bath and an electric current is passed through the system. The fragments of DNA are of different sizes and different electrical charges. As a result, the fragments migrate down the gel in various distances. The DNA can be seen by the application of dye, producing a gel which has a series of lines showing where the DNA has migrated. The enzymes used for the digestion cut at specific locations and different-sized fragments are produced depending on the bases (that is, the particular sequence of nucleotides) present in the DNA. Fragments of the same size in different lanes indicate the DNA has been broken into segments the same size. This indicates homology between the sequences under test. The greater the number of enzymes used in the digestion, the finer the resultant resolution.

DNA fingerprinting is used in forensics to examine DNA samples taken from a crime scence and compare them to those of a suspect. The statistical chance of two samples of DNA producing identical digestion patterns different individuals is very small.

Döbereiner, Johann Wolfgang (1780-1849)

German chemist

The earliest years of Döbereiner's life did not suggest a brilliant future. He was born in Hof, Bavaria, on December 13, 1780, to poor parents. His father was a coachman who could provide Johann with only the most basic education. Although Döbereiner did attend a few lectures in **chemistry**, botany, mineralogy, philosophy, and languages, he was largely self-taught. Yet, he developed unusual skill in chemical research and caught the eye of Duke Carl August in 1810. Duke Carl appointed Döbereiner to the position of professor extraordinary in chemistry at Jena, a position he held throughout the rest of his academic life.

Döbereiner was a man of far-ranging interests and accomplishments. He conducted research on the manufacture of vinegar, the abundance of elements in the earth's crust, the use of mineral waters for medical purposes, and many other topics in general, pharmaceutical, and **analytical chemistry**. In 1831, he discovered the **chemical compound furfural**, obtained from corn cobs, oat and rice hulls, and other cellulose-containing materials. He was one of the first chemists to offer laboratory instruction in chemistry.

In the early 1820s, Döbereiner studied the role of **platinum** metal as a catalyst. Somewhat earlier, Sir **Humphry Davy** had observed that heated platinum wire greatly increased the rate at which organic compounds oxidize. Döbereiner's contribution was to show that finely divided platinum (''platinum sponge'') was even more effective than was solid platinum metal. He even invented a lighter that would generate a flame when gaseous **hydrogen** came into contact with a spongy platinum catalyst.

Most students recognize Döbereiner's name, however, for his contribution to the development of the periodic law. One consequence of Jöns Berzelius' work on atomic weights was the realization by chemists that the properties of elements might be related to these atomic weights. Around 1817, Döbereiner noticed a pattern among three elements with similar chemical properties, **chlorine**, **bromine**, and **iodine**. Specifically, he noted that the atomic weight of bromine (80.970) was the arithmetic mean of the atomic weights of chlorine (35.470) and iodine (126.470). The currently accepted atomic weight for bromine was 80.470. Furthermore, the properties of the three elements varied in an orderly manner, from chlorine to bromine to iodine. Döbereiner spoke of this group of elements as a triad. He found two other triads among the known elements. One triad consisted of **calcium**, **strontium**, and **barium**; the other of **sulfur**, **selenium**, and **tellurium**.

Other chemists attempted to find other triads among the elements, but, overall, Döbereiner's discovery seemed to be a dead end. Thus, chemists largely ignored the Law of Triads. Not until Dmitri Mendeleev's discovery of the periodic law four decades later did the significance of Döbereiner's discovery finally become apparent.

Döbereiner died at Jena on March 24, 1849, 20 years before Mendeleev published his periodic law.

Doisy, Edward A. (1893-1986)

American biochemist

Edward Adelbert Doisy was an acclaimed biochemist whose contributions to research involved studying how chemical substances affected the body. In addition to research on antibiotics, insulin, and female **hormones**, he is remembered for his successful isolation of vitamin K, a substance that encourages blood clotting. Because he was able to synthesize this sub-

Edward A. Doisy.

stance, many thousands of lives are saved each year. For this research, Doisy shared the 1943 Nobel Prize in medicine or physiology with Danish scientist **Henrik Dam**.

Doisy, one of two children, was born November 13, 1893, in Hume, Illinois, to Edward Perez Doisy, a traveling salesman, and Ada (Alley) Doisy. His parents, while themselves having little in the way of higher education, encouraged him to attend college. Doisy received his baccalaureate degree in 1914 from the University of Illinois at Champaign and then obtained his master's in 1916. The advent of World War I interrupted his schooling for two years, during which time he served in the Army. After the war, Doisy received his Ph.D. from Harvard University Medical School in 1920. Beginning in 1919 he rapidly rose through the academic ranks, achieving the position of associate professor of **biochemistry** in the Washington University School of Medicine, St. Louis. He left this position in 1923 to go to the St. Louis University School of Medicine, and a year later he was appointed to the chair of biochemistry, where he engaged in research and teaching. He also was named the biochemist for St. Mary's Hospital. Doisy held these positions until his retirement in 1965.

For 12 years—from 1922 until 1934—Doisy worked with biologist Edgar Allen to study the ovarian systems of rats and mice. During this time he participated in research that isolated the first crystalline of a female steroidal hormone, now called oestrone. He later isolated two other related products, oestriol and oestradiol–17β. When Doisy administered these in tiny quantities to female mice or rats whose ovaries had been removed, the creatures acted as if they still had ovaries. Many women have benefitted from this research, as these compounds and their derivatives have been used to treat several hormonally-related problems, including menopausal symptoms.

Doisy, in 1936, turned from this line of research to trying to isolate an antihemorrhagic factor that had been identified by Danish researcher **Henrik Dam**. Dam had discovered a chemical in the blood of chicks that decreased hemorrhaging; he called this substance *Koagulations Vitamine,* or vitamin K. Using Dam's work as a springboard, Doisy and his co-workers spent three years researching this new vitamin. They discovered that the vitamin had two distinct forms, called K1 and K2, and successfully isolated each—K1 from alfalfa, K2 (which differs in a side chain) from rotten fish. Alter Doisy had isolated these two compounds he successfully determined their structures, and was able to synthesize the extremely delicate vitamin K1.

Synthesizing vitamin K enabled large quantities of it to be produced relatively inexpensively. It has since been used to treat hemorrhages that would previously have been fatal, especially in newborns and other individuals who lack natural defenses; it is estimated that the use of vitamin K saves almost five thousand lives each year in the United States alone. For these research advances, Doisy shared the 1943 Nobel Prize for medicine or physiology with Dam. Some of this research was funded by the University of St. Louis and some of the funds were contributed by the pharmaceutical manufacturer Parke-Davis and Co.—a financial arrangement that Doisy saw as a model for future industry-university research relations.

Over the course of his career, most of Doisy's research focused on how various chemical substances worked in the human body. In addition to vitamin K, his team studying the effects of certain antibiotics, **sodium**, **potassium**, chloride, and **phosphorus**. He also developed a high-potency form of insulin, for use in treating diabetes.

Doisy was made St. Louis University's distinguished service professor in 1951, and later was named emeritus professor of biochemistry. As a sign of his contributions, the university's department of biochemistry was named in his honor in 1965, and he was made its emeritus director. Because of his prominence and his loyalties to the University, there are numerous plaques and buildings bearing his name.

Doisy's contributions to the field of biochemistry are recognized by the numerous honorary awards he held and the scientific societies to which he belonged. He was member of the League of Nations Committee on the Standardization of Sex Hormones from 1932 to 1935, and in 1938 was elected to the National Academy of Sciences. In 1941 he was honored with the Willard Gibbs Medal of the American Chemical Society, which is perhaps the highest distinction in chemical science. He served as both the vice president and then president, from 1943 to 1945, of the American Society of Biological Chemists, and was the 29th president of the Endocrine Society in 1949.

Doisy married Alice Ackert on July 20, 1918, and they had four children: Edward Adelbert, Robert, Philip, and Richard. His second marriage, after his first wife died, was to Margaret McCormick, on April 19, 1965. He died October 23, 1986.

DOMAGK, GERHARD (1895-1964)

German biochemist

Gerhard Domagk was a biochemist who discovered sulfonamide therapy for bacterial infections. Prior to his work, only a few chemical compounds had been found effective against these infections, and most of these had serious side effects. Domagk was awarded the Nobel Prize in physiology or medicine in 1939 for this discovery, but the German government forced him to decline it. In 1947, he was awarded the Nobel Prize Medal. In presenting this award, Nanna Svartz of the Royal Caroline Institute said that Domagk's discovery ''meant nothing less than a revolution in medicine.'' The introduction of sulfonamide therapy prior to World War II undoubtedly saved many thousands of lives.

Domagk was born October 30, 1895, in Lagow, Brandenburg, Germany, to Paul and Martha Reiner Domagk. His father was assistant headmaster of a school, and he sent his son to a grade school that specialized in the sciences. Domagk enrolled in the University of Kiel as a medical student in 1914. His studies, however, were almost immediately interrupted by World War I. He enlisted in the German Army, fought at Flanders, and was transferred to the eastern front in December of 1914, where he was wounded. He was then transferred to the medical corps. He served in several hospitals, and his experience attempting to treat wounds and infectious diseases with the inadequate tools of the time undoubtedly influenced the direction of his later research.

Domagk resumed his studies at the University of Kiel following the war and earned his medical degree in 1921. In 1924 he took up the post of lecturer of pathological anatomy at the University at Greifswald, and in 1925 he moved on to a similar post at the University at Münster. In 1927, Domagk took a leave of absence from the university, which reshaped his career. He left to work in the laboratories of a company called I. G. Farbenindustrie, where he would remain for the rest of his professional life.

Domagk's career was profoundly influenced by the work of **Paul Ehrlich**. In 1907, Ehrlich had discovered arsphenamine, a compound specifically developed to be toxic to trypanosomes, and in 1909 this drug had been found to be quite effective against the bacterium that causes syphilis. Ehrlich's work had stimulated a number of searches for other antibacterials, and Domagk systematically continued this work at I. G. Farbenindustrie.

Domagk investigated thousands of chemicals for their potential as antibacterials. He would first test them against bacterial cultures in the test tube, then find the doses tolerated by animals such as mice, and lastly determine if compounds that worked in the test tube also worked against bacteria in living animals. For five years Domagk searched in vain for a ''magic bullet'' that would be toxic to bacteria and not to animals. His success illustrates Pasteur's dictum that chance favors the prepared mind. Methodically checking thousands of compounds for antibacterial activity, Domagk found in 1932 that a red leather dye showed a small effect on bacteria in the test tube. Developed by others at the company, the compound was called Prontosil Rubrum, and it proved quite non-toxic to mice.

Gerhard Domagk.

Domagk's original experiment to determine the effectiveness of Prontosil Rubrum was straightforward. He injected twenty-six mice with a culture of hemolytic streptococcal bacteria. Fourteen mice served as controls, receiving no therapy, and all died within four days, as expected from previous experiments with untreated animals. The remaining twelve mice were injected with a single dose of Prontosil Rubrum an hour and a half after being infected with the bacteria. All twelve survived in good condition. In 1932, I. G. Farbenindustries began clinical testing of Prontosil Rubrum. For reasons that are unknown, however, Domagk delayed publishing the results of his experiment for three years. But it is clear that he understood its implications. During this time his daughter contracted a streptococcal infection from a needle prick and failed to respond to traditional therapies. As she lay near death, Domagk injected her with Prontosil Rubrum, and she subsequently recovered.

There was some initial skepticism when Domagk first published his experimental results, but rapid replication of his findings led to widespread acceptance of the value of Prontosil Rubrum therapy. Throughout Europe, hospitals treated a variety of illnesses—including pneumonia, meningitis, blood poisoning, and gonorrhea—with Prontosil Rubrum and closely allied compounds. Subsequent laboratory studies have shown that it is only a part of the Prontosil Rubrum **molecule**, the sulfonamide group itself, that is responsible for its effect on bacteria. Moreover, the compound does not kill bacteria but interferes with their **metabolism** and therefore with their ability to reproduce.

Although the importance of his work was widely recognized by physicians and fellow scientists, the world of politics obstructed formal acknowledgement of his discovery. Carl von Ossietzky, a German pacifist incarcerated in a prison camp, had been awarded the Nobel Peace Prize in 1936, and Hitler had declared that no German citizen could accept a Nobel Prize. When he was awarded the prize in 1939, Domagk notified the German government and was promptly arrested. He was soon released but was forced to decline the prize. He was awarded the Nobel Medal after the war, but the prize money had reverted to the foundation.

During the late 1930s and throughout World War II, Domagk continued to investigate other compounds for their antibacterial effects. He concentrated considerable effort on antitubercular drugs, recognizing the problem of increasing resistance to streptomycin. His work resulted in some drugs of limited use against tuberculosis, though the class of compounds he studied proved to be somewhat toxic. Domagk retired in 1958 but remained active in research. He spent the last few years of his career attempting without success to find an anti-cancer drug.

In addition to the Nobel Prize, Domagk received numerous other accolades. In 1959, he was elected to the Royal Society of London. He was awarded medals by both Spain and Japan, and several German universities conferred honorary doctorates upon him.

Domagk married Gertrude Strube in 1925. They had four children, three sons and a daughter. Domagk died of a heart attack on April 24, 1964.

DORN, FREIDRICH (1848-1916)

German Chemist

Friedrich Ernst Dorn was the German physicist who discovered **radon** in 1900, although his discovery was nearly ignored. Dorn's announcement of his findings came upon the same time as English physicist **William Ramsay** revealed his the discoveries of the five other noble **gases**. Ramsay had already predicted the existence of radon and even specified some of its characteristics. Ramsay and his associate Whytlaw-Gray later established radon's atomic weight in a very intricate experiment. The discovery of radon was Dorn's only contribution to **chemistry**. Although Ramsay was awarded the Nobel Prize in Chemistry in 1904 for his work leading to the discovery of the noble gases, Dorn's achievement should not be diminished. It was unique in the history of chemistry because Dorn noted that radon develops as an emission of **radium**. This is the first time that one element has been shown to transmutate from another. This finding is not on the same scale as turning common **metals** into **gold**, but it brings chemistry back to the founding concept of **alchemy** from which chemistry evolved.

Friedrich Dorn was born in what is now Dobre Miastro in Poland. Like **Gustav Robert Kirchhoff**, he attended college at Königsberg but Dorn was there a generation (around twenty years) after Kirchhoff. In 1873, Dorn was offered a position as professor of physics at the University of Breslau and he accepted. He stayed at Breslau for thirteen years until he was offered a professorship at University at Halle. University officials promised him the chance to do research on x-rays. He conducted his x-ray research for over thirteen years before he observed the phenomenon that led to the discovery of radon. At first, he believed that he was observing the ordinarily expected radiation from radium. On closer study, he reported that radium actually emanates a radioactive gas. Once this gas was analyzed discretely, it was found to be the sixth noble gas. Dorn called his discovery ''radium **emanation**.'' Ramsay eight years later gave it the name ''niton.'' It was not given the name ''radon'' until 1923. This was seven years after Friedrich Dorn had died at the age of 68.

Perhaps the significance of Dorn's discovery would have remained dormant for many years if Ramsay had not isolated radon. The importance of Dorn's discovery also relied on Ramsay's work to establish the characteristics and atomic weight of the element. Still, Dorn's work was an important contribution to furthering the knowledge of what elements exist. His discovery also allowed other researchers to determine if other elements also transmutate one from another. The unique characteristic of the radium to radon **transmutation** affords much further research.

DOW, HERBERT HENRY (1866-1930)

American chemist

Dow was born in Belleville, Ontario, Canada, on February 26, 1866. His parents were Americans who had moved to Canada when his father could no longer find work in their native New England. Only a few months after Herbert's birth, the Dow family moved back to the United States. They settled first in Derby, Connecticut, then moved on to Cleveland, where Herbert's father got a job at the Chisholm Steel Shovel works. In 1884, Dow entered the Case School of Applied Science (now Case Western Reserve University). He majored in **chemistry** and became especially interested in the study of brines, a form of **salt water**. During his senior year at Case, he presented a **paper** on brines before the American Association for the Advancement of Science.

After graduation in 1888, Dow began to work on methods for extracting **bromine** from brine by electrolytic methods.

He received his first patent on the topic in 1889 and decided to open a plant based on the process. The plant, located in Canton, Ohio, had an impressive name—The Canton Chemical Company—but consisted of little more than an ill-equipped shed. Less than a year after it opened the business failed and Dow looked for a new location where he could begin again.

The site he selected was Midland, Michigan at the base of the state's ''thumb'' region. One reason for this choice was the presence of enormous underground reserves of brine beneath the eastern part of the state. The brine, a remnant of an ancient sea, constituted an essentially limitless supply of raw material for Dow's new process.

In August 1890, Dow moved to Midland and opened his new company, The Midland Chemical Company. At first his neighbors were puzzled by the newcomer's business, and they referred to him behind his back as ''Crazy Dow.'' However, less than six months later Dow's plant was in operation producing bromine from brine by using electric current.

Dow rapidly proved that he was a chemical genius, soon developing methods for producing **chlorine**, **magnesium**, and other commercially valuable chemicals from brine.

He was not as successful at first as a businessman, however. At one point, he lost control of his own patents and of the company he had founded. By 1897, however, his business skills had improved and he founded another new company, the Dow Chemical Company. Dow Chemical has since grown to become one of the half dozen largest chemical companies in the United States. The business is now a multimillion dollar operation that produces hundreds of different chemicals, including drugs, agricultural chemicals, **plastics**, dyes, caustic soda, **hydrochloric acid**, rubber, industrial **solvents**, and the chlorine-and bromine-based products with which the company started.

Dow became ill in September 1930 and entered the Mayo Brothers Clinic at Rochester, Minnesota. He was diagnosed with cirrhosis of the liver and failed rapidly. He died at the Clinic on October 15, 1930.

DRICKAMER, HARRY G. (1918-)

American chemist

Harry George Drickamer is recognized as a world leader in the use of high pressure to investigate the electronic structure of **matter**. His fundamental discovery was that pressure affects different kinds of electronic orbitals to different degrees. An electronic orbital is the wave function of an **electron** moving in a **molecule** or **atom**, corresponding to the orbit of the electron in earlier theory. The finding of this pressure effect, known as the ''pressure tuning'' of electronic **energy** levels, had two main consequences. One, it explained how the electronic transition to a new ground state, the state of least possible energy, can lead to very different and often unanticipated physical and chemical properties in a variety of materials. Two, it provided new ways to test theories about physical and chemical phenomena using pressure as a variable.

Drickamer was born on November 19, 1918, in Cleveland, Ohio. His father, George Henry Drickamer, who died when his son was two years old, had a degree in mechanical engineering. Drickamer's mother, Louise Strempel Drickamer, had only an eighth-grade education, but she taught herself to be a secretary after her husband's death. As Drickamer later recalled in a 1997 letter to author Linda Wasmer Smith, ''She had no knowledge or interest in science but was determined that my sister and I would go to college—and we had better get good grades or else.''

Harry G. Drickamer.

Drickamer attended the University of Michigan, where he received a bachelor's degree in 1941 and a master's degree in 1942. Up until this point, he had shown little interest for science himself, and he had completed math through calculus plus one year each of physics and **chemistry**. However, on a bet, Drickamer took the written **chemical engineering** qualifying exams. A few months later, he learned that he had passed. Meanwhile, he had taken a job at the Pan American Refining Corporation in Texas City, Texas, where he was introduced to more advanced mathematical and science concepts. Drickamer's appetite for science was whetted.

While at Pan American, Drickamer began working toward a doctoral degree from the University of Michigan. He was granted permission to use Pan American's lab equipment on nights and Sundays for his thesis. However, he still needed 22 hours of course work. He completed the necessary courses in a single semester, receiving his Ph.D. degree in 1946. That same year, Drickamer accepted a position as assistant professor of chemical engineering at the University of Illinois. He spent the rest of his career at this institution, rising to the position of professor of chemical engineering, chemistry, and physics. Although Drickamer formally retired in 1989, he continues to play an active role in research at the college as a professor emeritus.

When Drickamer first arrived at the University of Illinois, high pressure was still a very specialized tool used in relatively few labs. As Drickamer noted in a 1990 paper, published in *Annual Reviews of Materials Science*, ''Now [high pressure] is used in hundreds of laboratories around the world for studies in such diverse fields as physics, chemistry, geology, and **biochemistry**. The central feature in this development has been the realization that pressure (really compression) is a powerful, indeed essential, tool for investigating the molecular and electronic properties of matter.''

Drickamer has been a key figure in this burgeoning field. Techniques were developed in his lab to study phenomena such as optical absorption, luminescence, electrical resistance, Mössbauer **resonance**, and **x-ray diffraction** under high pressure. Summing up his achievements in his letter to author Linda Wasmer Smith, Drickamer wrote: ''(1) We were the first to show that pressure could be used to study atomic, molecular and electronic phenomena—the stuff of modern chemistry and physics... (2) We have been fortunate to pick the right techniques and the right systems to study so that one could demonstrate the wide applicability of high **pressure**... (3) We have been diligent enough to study a wide variety of materials by a variety of techniques so that the power and versatility of pressure tuning **spectroscopy** has become apparent.''

Drickamer has received many honors over the years. He is a member of the National Academy of Sciences, the National Academy of Engineering, and the American Philosophical Society, and a fellow of the American Academy of Arts and Sciences. His numerous awards include the Colburn, Alpha Chi Sigma, William H. Walker, and Warren K. Lewis Awards of the American Institute of Chemical Engineers; the Ipatieff Prize and the **Irving Langmuir** and **Peter Debye** Awards of the American Chemical Society; the Chemical Pioneers Award and **Gold** Medal of the American Institute of Chemists; and the Buckley Solid State Physics Award of the American Physical Society. In 1989, Drickamer received the National Medal of Science. The citation recognized ''his discovery of the 'pressure tuning' of electronic energy levels as a way to obtain new and unique information on the electronic structure of **solids**.''

Drickamer married Mae Elizabeth McFillen on October 28, 1942. She later became a family planning nurse practitioner. The couple have five grown children: Lee, a zoologist; Lynn, a library employee; Kurt, a biochemist; Margaret, a physician; and Priscilla, a reference librarian. In his leisure time, Drickamer enjoys walking several miles a day and reading history books and mysteries. However, a quiet retirement is not his style. He maintains an active research program at the University of Illinois, where he is currently involved in studies of nonlinear optical phenomena.

DU VIGNEAUD, VINCENT (1901-1978)

American biochemist

Vincent du Vigneaud, an American biochemist, received the 1955 Nobel Prize for **chemistry** for his breakthrough achievement of synthesizing oxytocin—a hormone released by the posterior pituitary gland used to induce labor and lactation in pregnant women, and for his work with **sulfur**. Throughout his career, du Vigneaud was recognized for isolating and synthesizing **penicillin** and the hormone vasopressin, which is used to suppress urine flow, identifying the chemical composition of insulin, discovering the structure of vitamin H, otherwise known as biotin, and his pioneering work with methyl groups.

Du Vigneaud was born in Chicago, Illinois, on May 18, 1901, to Alfred, an inventor and designer of machines, and Mary Theresa (O'Leary) du Vigneaud. Early in his high school education in Chicago's public school system, du Vigneaud demonstrated an aptitude for chemistry and physiology. He constructed a laboratory in his parents' basement, where he carried out his first experiments. Du Vigneaud enrolled in the University of Illinois as an **organic chemistry** major and graduated in 1923. He stayed on to earn a masters degree in 1924, studying under C. S. Marvel. Also in 1924, du Vigneaud married Zella Zon Ford. Both of their children went on to become doctors. Their son, Vincent du Vigneaud, Jr., became an obstetrician and gynecologist, and their daughter, Marilyn Renee Brown, became a pediatric gastroenterologist.

From 1924 to 1925, du Vigneaud was an assistant biochemist at the University of Pennsylvania's Graduate School of Medicine and also worked in Philadelphia General Hospital's clinical chemistry laboratory. He then moved to the University of Rochester in New York to study for his Ph.D. under John R. Murlin at the School of Medicine. For his doctoral research, he undertook an examination of the chemical makeup of insulin, the protein hormone and sulfur compound that is secreted by the islets of Langerhans located in the pancreas. Du Vigneaud's investigations, which were inspired by a lecture given by renowned biochemist W. C. Rose at the University of Illinois, sparked a lifelong interest in the range of sulfur compounds, but especially the sulfur-containing **amino acids** methionine, homocystine, cystine, cysteine, and cystathionine.

After receiving his doctorate in 1927, du Vigneaud became a National Research Council fellow. He worked first at Johns Hopkins Medical School's Department of Pharmacology under **John J. Abel**, where he continued his research into the structure of insulin. His suspicion that insulin was a derivative of the amino acid cystine was justified when he succeeded in isolating cystine from insulin crystals. He was thereby able to prove that insulin consists only of amino acids and an **ammonia** by-product.

Du Vigneaud left the United States for Germany in 1928 on a brief overseas tour. He first stopped at the Kaiser Wilhelm Institute in Dresden, where he worked under Max Bergman, an expert on the chemistry of amino acids and **peptides** (chains of amino acids). Du Vigneaud later turned down an assistantship position with Bergman to proceed to the University of Edinburgh's Medical School, where he worked with biologist George Barger. He also spent time at the University College Hospital Medical School at the University of London, where he worked with Charles Harrington.

Upon returning to the United States, du Vigneaud joined the University of Illinois's physiological chemistry staff under his mentor, W. C. Rose. In 1932, he left his alma mater to take up a position as head of the Department of **Biochemistry** at the George Washington University School of Medicine in Washington, DC. One of his innovations there was to add a course in biochemistry to the medical school curriculum. His own research lead him to investigate his hypothesis that insulin's blood sugar-lowering effects were related to disulfide bonds of cystine.

In 1936 he and his staff succeeded in artificially creating glutathione, a tripeptide containing the amino acids cysteine, glycine, and glutamic acid, that is widely occurring in plant and animal tissues and which plays a vital part in biological oxidation-reduction processes and the activation of some enzymes. He also continued to pursue his research into insulin. By the following year, he was in a position to prove that the amino acid cystine comprises insulin's entire complement of sulfur, and that insulin can be deactivated by the **reduction** of its bonds of insulin by cystine or glutathione. Also in the late 1930s, du Vigneaud's work with methionine revealed how the body shifts a **methyl group** (CH_3) from one compound to another.

In 1938 du Vigneaud was appointed head of Cornell University Medical College's biochemistry department. Within two years, he had succeeded in isolating biotin (vitamin H). He spent the next few years carefully studying the substance and by 1942, had figured out its structure. He next turned to the human posterior pituitary gland, especially the study of the **hormones** oxytocin and vasopressin that it produces. Oxytocin is known to stimulate the contraction of uterine muscles and the secretion of milk in women during labor. Vasopressin, also known as the antidiuretic hormone, is a polypeptide hormone responsible for causing increased blood pressure and decreased urine flow. Du Vigneaud and his colleagues managed to isolate a highly purified form of these hormones from the pituitary gland and set about discovering their chemical nature.

To his surprise, du Vigneaud discovered that oxytocin is made up of only eight amino acids. Most **proteins** are comprised of several hundred amino acids. It took du Vigneaud another ten years to determine their sequence in an oxytocin **molecule**. Once he had cracked this puzzle, he was finally able to synthesize oxytocin. The importance of du Vigneaud's achievement lay not only in its making available an unlimited supply of the protein, but also in the light it shed on the relationship between **molecular structure** and biological function. The synthetic protein was tested on pregnant women at the Lying-in Hospital of the New York Hospital-Cornell Medical Center, where it was found to be as effective in inducing labor and milk flow as pure oxytocin. In 1946 the journal *Science* announced another du Vigneaud breakthrough: his synthesis of penicillin. Although du Vigneaud carried out the decisive experiments at Cornell University, it was one of the greatest international efforts of its kind, said *Science,* the culmination of five years of concerted effort by thirty-eight teams of scientists in the U.S. and Britain.

Du Vigneaud's illustrious scientific career was widely recognized. In 1955, he was awarded the Nobel Prize for chemistry for "his work on biochemically important sulfur compounds and especially for the first synthesis of a polypeptide hormone." Du Vigneaud's other awards include the Nichols Medal of the American Chemical Society in 1945, the Association of Medical Colleges' Borden Award in the Medical Sciences in 1947, the Public Health Association's Lasker Award in 1948, the Osborne and Mendal Award of the American Institute of Nutrition in 1953, Columbia University's Charles Frederick Chandler Medal, and the Willard Gibbs Medal of the American Chemical Society. In addition, he was a member of the American Philosophical Society, the National Academy of Sciences, and the New York Academy of the Arts and Sciences. Du Vigneaud's leisure interests included bridge and horse riding.

From 1967 to 1975, du Vigneaud served as Cornell University's professor of chemistry. In 1975 he advanced to the level of emeritus professor of biochemistry at Cornell. Du Vigneaud died at St. Agnes Hospital in White Plains, New York, on December 11, 1978.

DUCTILITY

When a crystalline body is subjected to an applied force, it will, under most circumstances, undergo a change in shape. If the forces are small, the deformation will be elastic, i.e., the material will revert to its former shape as soon as the external forces are released. If, however, the forces are sufficiently large, the material will only partly revert to its original shape when they are released. Such deformation is referred to as plastic.

At high enough stresses, irreversible processes accompany deformation. Brittle materials break into separate pieces (i.e., fracture or mechanically fail) at the point at which the applied stress exceeds the value at which there is no further plastic deformation. *Ductile* materials, on the other hand, exhibit a time-dependent extension (i.e., one that depends on how fast the material is stressed) that is not recovered when the stress is removed. The ductility of a material is characterized by the strain (or elongation) at fracture; the more ductile a material is, the easier it is to draw that material into a wire. The **metals tungsten** and **copper** both exhibit very high ductilities. Here strain is defined as the relative change in dimensions or shape in a body as the result of an applied stress (it is a dimensionless quantity); stress is the magnitude of the applied force per unit area (usually measured in units of pounds per square inch or pascals).

Certain types of metals undergo a transition from ductile to brittle fracture when the **temperature** is decreased sufficiently, the strain rate is increased sufficiently, and/or the surface of the metal has been notched. The ductile to brittle transition temperature is typically measured by the change in **energy** absorbed, the change in ductility, or the change in the fracture appearance.

See also Strain energy

Duhem, Pierre Maurice Marie (1861-1916)

French scientist, physicist, chemist, and historian

The French physicist, chemist, and historian of science Pierre Maurice Marie Duhem published work in **thermodynamics**, physical **chemistry**, hydrodynamics, elasticity, **electricity** and **magnetism**, and the history and philosophy of science.

Pierre Duhem was born on June 9, 1861, in Paris. He entered the École Normale in 1882 and qualified for a teaching certificate in 1885. His first published **paper** on **physical chemistry** appeared in 1884. That year he also presented a doctoral dissertation in physics, which attacked the ''maximum-work principle'' of Marcelin Berthelot, a powerful figure in the French academic world. Berthelot succeeded in having the thesis rejected and is reported to have said that Duhem would never teach in Paris. The prediction came true.

Duhem stayed at the École Normale for another two years and in 1888 presented a doctorate in mathematics on the theory of magnetism. Meanwhile he published his first thesis and 30 articles on physics and chemistry. In 1887, he was named lecturer at Lille, but in 1893, after a fight with the dean of the faculty, he was transferred to Rennes and in 1894 to Bordeaux. There he remained for the rest of his life, deprived of the position at the Science Faculty in Paris to which his work would seem to have entitled him. He died at Cabrespine on September 14, 1916.

Duhem believed that physical theories describe, condense, and classify experimental results rather than explain them. He also believed that physical theories evolve by successive changes to conform to experiment and thus gradually approach a ''natural classification'' that somehow reflects underlying reality. These philosophical ideas led him after 1895 to investigate the history of science, especially in the Middle Ages and Renaissance. His *Studies on Leonardo da Vinci* (three volumes, 1906-1909) revealed the works of medieval scholastics in physics and astronomy that Leonardo had used. He explored these works in *The System of the World* (10 vols., 1913-1959). Although Duhem approached his subject almost exclusively from the point of view of the ancient and medieval contribution to modern science, this history ranks him as the rediscoverer of medieval science.

As a chemist, Duhem contributed to the Gibbs-Duhem equation, which describes the relation between variations of chemical potentials. From 1884 until 1900 and after 1913, his work was predominantly concerned with thermodynamics and electromagnetism; from 1900-1906 he concentrated on hydrodynamics and elasticity. Trained before the discovery of **radioactivity**, Duhem opposed those scientists who sought a mechanical explanation of the universe through the use of atomic and molecular models. He believed that classical mechanics was a special case of a more general continuum theory and spent much of his career working on a generalized thermodynamics that would serve as a descriptive theory for all of physics and chemistry. He expressed his views most fully in his *Treatise on Energetics* (two volumes, 1911).

Dulong and Petit's law

In 1819, two French scientists, Pierre Dulong and Alexis Petit, proposed the first direct approach to calculating atomic weights. They proposed that the amount of **heat** required to raise the **temperature** of a single **atom** of a solid by a given amount should be independent of the type of atom. According to Dulong and Petit, then, the amount of heat required to raise the temperature of 1 gram atomic weight of a solid element by 1°C should be the constant (because 1 gram atomic weight of every element contains the same number of atoms).

Dulong and Petit 's law is usually expressed in terms of specific heat, which is the amount of heat required to raise the temperature of one gram of a substance by 1°C. In its modern form, the law says that the product of the specific heat of a solid element multiplied by its gram atomic weight should be approximately 6 cal/degree C.

In practice, Dulong and Petit's law is far from exact, with many elements showing deviations by 10% or more. Nevertheless, it does yield approximate values for the **metals** that can be refined by data from chemical analysis. It is however only employed for **solids** at high temperatures (on the order room temperature). At very low temperatures, the specific heat is proportional to the temperature raised to the third power.

See also Analytical chemistry; Molecular weight; Thermochemistry

Dumas, Jean Baptiste André (1800-1884)

French chemist, scientist, and politician

The French chemist Jean Baptiste André Dumas worked in the field of organic **chemistry** and developed the ''type'' theory of organic structure.

On July 14, 1800, Jean Baptiste Dumas was born at Alais. In his youth he was apprenticed to an apothecary. In 1816 he moved to Geneva and studied physiological chemistry in the laboratory of A. Le Royer. In Geneva, Dumas met the famous scientist Alexander von Humboldt, who persuaded Dumas to move to Paris, where he would find greater scientific opportunities. This he did in 1823, and he was engaged as a lecture assistant in chemistry at the École Polytechnique; he became professor of chemistry in 1835. During this period Dumas began to work on his major book, *Treatise on Chemistry,* and he also participated in the founding of the Central School for Arts and Manufactures.

In 1830 Dumas challenged the so-called dualistic theory of the great Swedish chemist Jöns Jacob Berzelius. The dualistic theory stated that all compounds could be divided into positive and negative parts. Dumas presented instead a unitary theory which held that atoms of opposite charges could be substituted in compounds without causing much alteration in the basic properties of the compound. This theory was related to his belief in families of organic compounds, in which substitutions could be made with the fundamental characteristics of the

Jean Baptiste André Dumas.

family remaining unchanged. At this time Berzelius was at the height of his eminence and would accept no affront to his authority; such was the strength of his attack on Dumas that the latter did not continue the dispute. Later researches proved Dumas to have been more correct in his theories than was the Swedish master.

Dumas isolated various essences and oils from **coal** tar; developed a method for measuring the amount of **nitrogen** in organic compounds, which made quantitative organic analysis possible; and developed a new method of determining vapor densities. He also concerned himself with determining the atomic weights of such elements as **carbon** and **oxygen** and published a new list of the weights of some 30 elements in 1858-1860.

In addition to his scientific achievements, Dumas led an active public life during the reign of Napoleon III. He was minister of agriculture and commerce and then minister of education. He was also a senator, master of the French mint, and president of the municipal council of Paris. His public life ended with the downfall of the Second Empire in 1871. Dumas died in 1884 in Paris.

Dumas Method

The Dumas method is a protocol with methodology allowing the **molecular weight** of an unknown substance to be determined. The molecular weight of a compound is the sum of the atomic weights of the atoms which comprise the **molecule**. For instance, the molecular weight of **water** (H_2O) is equal to the atomic weight of **oxygen** plus the atomic weights of two **hydrogen** atoms. Often, the Dumas method is used to determine the molecular weights of volatile organic substances that are **liquids** at room **temperature**.

Using the Dumas method, molecular weight is calculated by measuring the **mass** of a known **volume** of a vaporized liquid. The mass of the vapor produced is measured by condensing it into liquid. The Ideal Gas Law, $PV=nRT$, is then used to determine the volume of the gas, and from that, its **density**. In the Ideal Gas Law, P is pressure, V is the volume of gas in liters, n is the number of moles of the gas, R is the Ideal Gas Constant, and T is the temperature of the gas (on Kelvin scale). From the density of the substance, the molecular weight can be estimated. Once the molecular weight of a substance is known, estimates of its atomic composition, or which elements it is made from and in what proportions, can be made. The method was designed by a french chemist named **Jean Baptiste André Dumas**, after whom the procedure is now named. Even though, during the nineteenth century the science of **chemistry** was at a relatively early state, Dumas was able to show that the vapor densities of some organic compounds are directly proportional to their molecular weights.

Dupont, E. Irèneè (1771-1834)

French-American industrialist

Eleuthére Irèneè Dupont survived the French Revolution and the Reign of Terror to immigrate to America in 1800. As an industrialist with chemical experience, he applied to his business enterprise the exacting standards found in science and created a company that is still successful almost 200 years later. The company today is diversified with many products ranging from industrial anticorrosion solutions to synthetic fibers. However, it all began with Irèneè Dupont and his one product, **gunpowder**.

Dupont came from a distinguished family in France. His grandfather was a royal financier for Louis XVI. His brother was a diplomat and counted Thomas Jefferson among his friends. His father was a political economist and had many distinguished academics as friends. One of these friends was the French **chemistry** researcher **Antoine-Laurent Lavoisier**.

Irèneè was born in 1771 and grew up on his father's estate. He was tutored at home, but did not show much interest in his academic topics. As a young teen, he showed a tremendous interest in **explosives** and willingly read and wrote about gunpowder. He entered the Royal College in Paris at age fourteen. Two years later, his father arranged for him to work with Lavoisier as an apprentice at the government agency for manufacturing gunpowder. Soon, Irèneè left the agency to work in

Eleuthére Irèneè duPont de Nemours.

the factory. After four years, Lavoisier was no longer associated with the gunpowder industry; so, Irèneè took a position managing his father's publishing house. About the time that he started in the publishing business, Irèneè met Sophie Madeleine Dalmas and wanted to marry. Her father refused to allow the wedding. Irèneè would not give up and was willing to fight two duels for the woman he loved. Her father eventually conceded and in 1791, they were married.

The politics of the Dupont family contained many contradictions. They had connections through the grandfather with the court of Louis XVI. However, the father and sons were in favor of reform in the government. The elder Dupont was a pacifist while Irèneè supported the revolutionaries. Still, the Duponts were part of a group that rescued the king and were nearly executed by the revolutionaries. However, they ran a newspaper perceived to be anti-government. The contrary assumptions of their politics by the mobs did not assure their safety in the turbulent times of France in the late eighteenth century. In 1797, after Irèneè and his father were arrested by the revolutionaries, they decided to leave France to live in America. It took two years to make arrangements and in January 1800, they settled in Bergen Point, New Jersey.

Once in America, Irèneè's father and brother set about to establish a business. However, they took a long time to decide what business. Irèneè was considered an introvert and was not usually included in the discussions. On a hunting trip with a fellow French expatriate, Irèneè noticed that American gunpowder was ineffective, in addition to being costly. He investigated the American gunpowder industry and realized there was a strong need for the quality gunpowder that he had had experience with in France. With the help of such notables as Thomas Jefferson and Alexander Hamilton, Irèneè spurred his family into starting the company that they named after him. Even the French government wanted to help. They perceived that the Dupont company would undercut the dominance of the British in the American gunpowder business. When Irèneè suddenly died in 1834, his company was one of the largest companies in America.

Before the establishment of chemical companies, such as the Dupont Company, the major source of funding for research was the university and other academic-based institutions. The realization of the need for more research in academia coincides in modern times with the realization of the need for the products of that research in industry. When in World War II, the Army needed unavailable silk for parachutes, they founded a substitute in a synthetic fiber called **nylon**, invented by **Wallace Hume Carothers**, who was funded by a company called Dupont.

DYES AND PIGMENTS

Dyes and pigments are substances that are used to impart **color** to liquid solutions and solid materials. Dyes are generally soluble (or partly soluble) organic compounds used in the textile industry to color fabrics, textile fibers and clothing. In additional to their application in textiles, dyes are used on **paper** and leather. Dyes are also added, under government regulation, to foods, drugs and **cosmetics**. Pigments, on the other hand, are insoluble inorganic and organic chemicals that are added to color paints, glazes and printing **inks**.

Natural plant materials have been used to dye wool since ancient times. Tyrian purple, from the shellfish murex, was a dye reserved for the emperor; this dye was worth several times its weight in **gold**. Woad, used since the time of the ancient Celts, produces a blue dye. Dyer's broom, or greenweed, yields a yellow dye, often added to woad-dyed cloth to turn it green. The roots of lady's bedstraw, a roadside weed in the northeastern United States, produce a red dye on wool yarn, as does the root of the madder plant, a perennial originating in the Mediterranean regions. Dyer's chamomile makes a yellow textile dye or hair coloring, and dyer's coreopsis provides a bright gold dye. Alkanet root or dyer's bugloss, yields a red dye. Indigo, the parent compound of the indigoid class of dyes, has been in use as a vat due since before recorded history. Natural indigo is obtained from the plant *indigofera*. It is no longer commercially important since synthetic indigo is produced inexpensively.

In order to get the color to stick well on the fiber, the wool was usually pretreated with a metal compound called a mordant; common mordants, still used in craft dying, are compounds of **aluminum**, **iron**, **copper**, **tin**, and **chromium**. The mordant reacts with the dye to form an insoluble complex called a ''lake.'' An example of a mordant is aluminum hydroxide precipitated into cotton fiber. This mordant reacts with the dye alizarin to form a read lake, exactly as it does in a test tube in the typical analytical test for aluminum. Mordant dyes are no longer used commercially, since equal or better results can be obtained less expensively. They are still popular with home dyers; the vegetable dye madder yields a substance that decomposes into alizarin, requiring an aluminum hydroxide mordant to result in the color known as turkey red. A tin mordant produces an orange color with madder. Carmine red can be formed from cochineal, an insect dye, with an aluminum **ion** mordant.

The first synthetic dye, mauve, was prepared in 1856 by the English chemist **William Henry Perkin**. In 1876, Otto Witt described the structural features of organic molecules that cause them to be colored. The color of a dye is due to its ability to absorb light in the visible region of the spectrum, an ability enhanced by unsaturation (presence of double bonds in the molecule) and **resonance**. The main structural unit of an organic dye, which is always unsaturated, is called the chromophore. Dyes are classified according to their **chromophores**. Today we recognize azo dyes, one of the most important classes of dyes because they can be used on synthetic fibers, which contain one or more azo (-N=N-) groups. Polyazo dyes can have four or more azo linkages. Some other dye systems are based on the chromophores triphenylmethane, xanthene, anthroquinone, indigo, and phthalocyanine.

The raw materials for today's dyes are mainly aromatic hydrocarbons: **benzene**, toluene, **naphthalene**, anthracene, pyrene, and others. These compounds once came from the **distillation** of **coal** tar, and the dyes are still known as coal-tar dyes. The most commonly used, benzene and toluene, are now produced from petroleum and **natural gas**. Indigo is now derived from naphthalene, and anthracene yields alizarin, the dye formerly obtained from madder root.

The process of dyeing can be carried out in batches or on a continuous basis. Either the yarn or the finished fabric can be dyed. A single dye or a mixture of up to a dozen dyes can be used. Vat dyes are insoluble, but a soluble form with an affinity for cotton is formed when the dye is reduced. This **reduction** was originally carried out in wooden vats, giving rise to the name ''vat dye.''

Pigments are also used to impart desired color to solid substances. Important pigments include inorganic oxides and insoluble salts, which are mechanically mixed in a coating material. The pigment is deposited on the substrate as the coating dries. Several hundred pigments have been used commercially. Iron oxides of varying composition (*i.e.*, Spanish red, Indian red and Venetian red), **cadmium** selenide and select organic toluidines are the most popular red pigments. Basic **lead** chromate and molybdate orange (lead chromate-molybdate) give a yellow color. Blue pigments include Prussian blue (ferric ferrocyanide) and phthalocyanine blue. **Titanium** dioxide and **zinc oxide** are added to coating materials to produce a white color. The vast majority of black pigments consist of finely divided particles of carbon-carbon black or lampblack. Finely, flakes of bronze, copper, lead, **nickel**, aluminum or **silver** can be used to give a metallic appearance.

See also Oxidation-reduction reaction

Dysprosium

Dysprosium is one of 15 **rare earth elements**, a family also known as the lanthanides. The lanthanides make up the elements between **barium** and **hafnium** in Row 6 of the **periodic table**. Dysprosium's **atomic number** is 66, its atomic **mass** is 162.50, and its chemical symbol is Dy.

Properties

Dysprosium is a soft metal with a shiny **silver** luster. It is so soft that it can easily be cut with a knife. The element has a melting point of 2,565°F (1,407°C), a **boiling point** of about 4,200°F (about 2,300°C), and a **density** of 8.54 grams per cubic centimeter. Dysprosium is relatively unreactive at room temperatures, but does react with both dilute and concentrated acids.

Occurrence and Extraction

The term rare earth elements is a misnomer, since the elements that belong to this family are not particularly rare. The name was originally used for these elements because they were so difficult to separate from each other. Dysprosium, as an example, is actually more common in the Earth's crust than some better known elements, such as **bromine**, **tin**, and **arsenic**. It has an abundance of about 8.5 parts per million. It is extracted from ores by being converted first to dysprosium trifluoride (DyF_3) and then reacted with **calcium** metal: $2DyF_3 + 3Ca \rightarrow 3CaF_2 + 2Dy$.

Discovery and Naming

Dysprosium was proved to be an element in 1886 by the French chemist **Paul-Émile Lecoq de Boisbaudran**. The research that led to Boisbaudran's discovery involves a long and complex story of efforts to analyze a rock known as ytterite first discovered in 1787 by a Swedish army office, Carl Axel Arrhenius (1757-1824). Ytterite was eventually to yield eight other new elements in addition to dysprosium. The element's name is taken from the Greek word *dysprositos*, meaning ''difficult to obtain.''

Uses

Dysprosium has a relatively limited number of uses, most of them involving specialized alloys. For example, some dysprosium alloys have very good magnetic properties that make them suitable for use in CD players. Dysprosium alloys are also used in control rods used to moderate the flow of neutrons through a nuclear reactor.

E

EDELMAN, GERALD MAURICE (1929-)

American biochemist

For his "discoveries concerning the chemical structure of **antibodies**," Gerald M. Edelman and his associate **Rodney Porter** received the 1972 Nobel Prize for physiology or medicine. During the lecture Edelman gave upon acceptance of the prize, he stated that immunology "provokes unusual ideas, some of which are not easily come upon through other fields of study.... For this reason, immunology will have a great impact on other branches of biology and medicine." He was to prove his own prediction correct by using his discoveries to draw conclusions not only about the immune system but about the nature of consciousness as well.

Born in New York City on July 1, 1929, to Edward Edelman, a physician, and Anna Freedman Edelman, Gerald Maurice Edelman attended New York City public schools through high school. After graduating, he entered Ursinus College, in Collegeville, Pennsylvania, where he received his B.S. in **chemistry** in 1950. Four years later, he earned a M.D. degree from the University of Pennsylvania's Medical School, spending a year as medical house officer at Massachusetts General Hospital.

In 1955, Edelman joined the United States Army Medical Corps, practicing general medicine while stationed at a hospital in Paris. There, Edelman benefited from the heady atmosphere surrounding the Sorbonne, where future Nobel laureates Jacques Monod and François Jacob were originating a new scientific discipline, molecular biology. Following his 1957 discharge from the Army, Edelman returned to New York City to take a position at Rockefeller University, studying under Henry Kunkel. Kunkel, with whom Edelman would conduct his Ph.D. research, was examining the unique flexibility of antibodies at the time.

Antibodies are produced in response to infection in order to work against diseases in diverse ways. They form a class of large blood **proteins** called globulins—more specifically, immunoglobulins—made in the body's lymph tissues. Each immunoglobulin is specifically directed to recognize and incapacitate one antigen, the chemical signal of an infection. Yet they all share a very similar structure.

Through the 1960s and 1970s, a debate raged between two schools of scientists to explain the situation whereby antibodies share so many characteristics yet are able to perform many different functions. In one camp, George Wells Beadle and **Edward Lawrie Tatum** argued that despite the remarkable diversity displayed by each antibody, each immunoglobulin must be coded for by a single gene. This has been referred to as the "one gene, one protein" theory. But, argued the opposing camp, led by the Australian physician, Sir Frank Macfarlane Burnet, if each antibody required its own code within the deoxyribonucleic acid (DNA), the body's master plan of-protein structure, the immune system alone would take up all the possible codes offered by the human **DNA**.

Both camps generated theories, but Edelman eventually disagreed with both sides of the debate, offering a third possibility for antibody synthesis in 1967. Though not recognized at the time because of its radical nature, the theory he and his associate, Joseph Gally, proposed would later be confirmed as essentially correct. It depended on the vast diversity that can come from chance in a system as complex as the living organism. Each time a cell divided, they theorized, tiny errors in the transcription—or reading of the code—could occur, yielding slightly different proteins upon each misreading. Edelman and Gally proposed that the human body turns the advantage of this variability in immunoglobulins to its own ends. Many strains of antigens when introduced into the body modify the shape of the various immunoglobulins in order to prevent the recurrence of disease. This is why many illnesses provide for their own cure—why humans can only get chicken pox once, for instance.

But the proof of their theory would require advances in the state of biochemical techniques. Research in the 1950s and 1960s was hampered by the difficulty in isolating immunoglobulins. The molecules themselves are comparatively large, too

Gerald Maurice Edelman.

large to be investigated by the chemical means then available. Edelman and **Rodney Porter**, with whom Edelman was to be honored with the Nobel Prize, sought methods of breaking immunoglobulins into smaller units that could more profitably be studied. Their hope was that these fragments would retain enough of their properties to provide insight into the functioning of the whole.

Porter became the first to split an immunoglobulin, obtaining an ''active fragment'' from rabbit blood as early as 1950. Porter believed the immunoglobulin to be one long continuous **molecule** made up of 1,300 amino acids—the building blocks of proteins. But Edelman could not accept this conclusion, noting that even insulin, with its 51 **amino acids**, was made up of two shorter strings of amino acid **chains** working as a unit. His doctoral thesis investigated several methods of splitting immunoglobulin molecules, and, after receiving his Ph.D. in 1960, he remained at Rockefeller as a faculty member, continuing his research.

Porter's method of splitting the molecules used enzymes that acted as chemical knives, breaking apart amino acids. In 1961, Edelman and his colleague, M. D. Poulik succeeded in splitting IgG—one of the most studied varieties of immunoglobulin in the blood—into two components by using a method known as ''reductive cleavage.'' The technique allowed them to divide IgG into what are known as light and heavy chains. Data from their experiments and from those of the Czech researcher, František Franek, established the intricate nature of the antibody's ''active sight. The sight occurs at the folding of the two chains which forms a unique pocket to trap the antigen. Porter combined these findings with his, and, in 1962, announced that the basic structure of IgG had been determined. Their experiments set off a flurry of research into the nature of antibodies in the 1960s. Information was shared throughout the scientific community in a series of informal meetings referred to as ''Antibody Workshops,'' taking place across the globe. Edelman and Porter dominated the discussions, and their work led the way to a wave of discoveries.

Still, a key drawback to research remained. In any naturally obtained immunoglobulin sample a mixture of ever so slightly different molecules would reduce the overall purity. Based on a crucial finding by Kunkel in the 1950s, Porter and Edelman concentrated their study on myelomas, cancers of the immunoglobulin-producing cells, exploiting the unique nature of these cancers. Kunkel had determined that since all the cells produced by these cancerous myelomas were descended from a common ancestor they would produce a homogeneous series of antibodies. A pure sample could be isolated for experimentation. Porter and Edelman studied the amino acid sequence in subsections of different myelomas, and in 1965, as Edelman would later describe it: ''Mad as we were, [we] started on the whole molecule.'' The project, completed in 1969, determined the order of all 1,300 amino acids present in the protein, the longest sequence determined at that time.

Throughout the 1970s, Edelman continued his research, expanding it to include other substances that stimulate the immune system, but by the end of the decade the principle he and Poulik uncovered led him to conceive a radical theory of how the brain works. Just as the structurally limited immune system must deal with myriad invading organisms, the brain must process vastly complex sensory data with a theoretically limited number of switches, or neurons.

Rather than an incoming sensory signal triggering a predetermined pathway through the nervous system, Edelman theorized that it leads to a selection from among several choices. That is, rather than seeing the nervous system as a relatively fixed biological structure, Edelman envisioned it as a fluid system based on three interrelated stages of functioning.

In the formation of the nervous system, cells receiving signals from others surrounding them fan out like spreading ivy—not to predetermined locations, but rather to regions determined by the concert of these local signals. The signals regulate the ultimate position of each cell by controlling the production of a cellular glue in the form of cell-adhesion molecules. They anchor neighboring groups of cells together. Once established, these cellular connections are fixed, but the exact pattern is different for each individual.

The second feature of Edelman's theory allows for an individual response to any incoming signal. A specific pattern of neurons must be made to recognize the face of one's grandmother, for instance, but the pattern is different in every brain. While the vast complexity of these connections allows for some of the variability in the brain, it is in the third feature of the theory that Edelman made the connection to immunology. The neural networks are linked to each other in layers. An incoming signal passes through and between these sheets in a specific pathway. The pathway, in this theory, ultimately determines what the brain experiences, but just as the immune system modifies itself with each new incoming virus, Edelman theorized that the brain modifies itself in response to each new incoming signal. In this way, Edelman sees all the systems of the body being guided in one unified process, a process that depends on organization but that accommodates the world's natural randomness.

Dr. Edelman has received honorary degrees from a number of universities, including the University of Pennsylvania, Ursinus College, Williams College, and others. Besides his Nobel Prize, his other academic awards include the Spenser Morris Award, the Eli Lilly Prize of the American Chemical Society, **Albert Einstein** Commemorative Award, California Institute of Technology's Buchman Memorial Award, and the Rabbi Shai Schaknai Memorial Prize.

A member of many academic organizations, including New York and National Academy of Sciences, American Society of Cell Biologists, Genetics Society, American Academy of Arts and Sciences, and the American Philosophical Society, Dr. Edelman is also one of the few international members of the Academy of Sciences and the Institut de France. In 1974 he became a Vincent Astor Distinguished Professor, serving on the board of governors of the Weizmann Institute of Science and is also a trustee of the Salk Institute for Biological Studies. Dr. Edelman married Maxine Morrison on June 11, 1950. They have two sons, Eric and David, and a daughter, Judith.

EFFUSION

Effusion is the movement of gas molecules through a small opening into an area of lower pressure. **Thomas Graham**, a Scottish chemist studied effusion and **diffusion** of **gases** and found that lighter gases move faster than denser gases. Based on his experiments with **density** he formulated Graham's law of effusion or diffusion which states that the rate of effusion or diffusion of gases at the same **temperature** and pressure are inversely proportional to their densities. Since density of a gas is proportional to its molar **mass** the law can also be written as the rate of effusion of gases at the same temperature are inversely proportional to their molar masses.

To mathematically explain his experiments the following facts can be used, one, the average kinetic energies of the molecules of two different gases at the same temperature and pressure are equal, two, the **velocity** of a gas varies inversely with its mass, and three, the kinetic **energy** is equal to one-half the mass times the velocity squared. The kinetic energies of two different gases can therefore, be set equal to each other. By substituting the rate of effusion for velocity the final formula supports Graham's law.

William Aston, a physicist used Graham's law to separate isotopes of **neon**. **Uranium** and other radioactive substances are also separated and used in medicine, **chemistry**, and nuclear energy. Another application of Graham's law is the ability to calculate the molar mass of an unknown gas if the mass of one gas is known and the rate of effusion has been determined. For example, if **hydrogen** effuses 5.65 times faster than an unknown gas the molar mass of the unknown gas can be determined. The square root of the molar mass of the unknown gas divided by the square root of the molar mass of hydrogen is equal to 5.65. The molar mass of the unknown gas is 64 grams per **mole**.

To compare the rate of effusion of two different gases such as **methane** and **sulfur** dioxide you must first calculate the molar mass of each gas. The molar mass of methane is 16 grams and the molar mass of sulfur dioxide is 64 grams. The square root of the heavier gas divided the square root of the lighter gas is two. Therefore, methane travels twice as fast as sulfur dioxide.

See also Diffusion; Kinetics

EHRLICH, PAUL (1854-1915)

German bacteriologist and immunologist

Paul Ehrlich's pioneering experiments with cells and body tissue revealed the fundamental principles of the immune system and established the legitimacy of chemotherapy—the use of chemical drugs to treat disease. His discovery of a drug which cured syphilis saved many lives and demonstrated the potential of systematic drug research. His studies of dye reactions in blood cells helped establish hematology—the scientific field concerned with blood and blood-forming organs—as a recognized discipline. Many of the new terms he coined as a way to describe his innovative research, including "**chemotherapy**," are still in use. From 1877 to 1914, Ehrlich published 232 papers and books, won numerous awards, and received five honorary degrees. In 1908, Ehrlich received the Nobel Prize in medicine or physiology. Along with **Robert Koch**, Nobel Prize winner in 1905, and **Emil Behring**, Nobel Prize winner in 1901, he is considered one of the "Big Three" in medicine. Ehrlich, Koch, and von Behring were contemporaries and frequent collaborators.

Ehrlich was born on March 14, 1854, in Strehlen, Silesia, once a part of Germany, but now a part of Poland known as Strzelin. He was the fourth child after three sisters in a Jewish family. His father, Ismar Ehrlich, and mother, Rosa Weigert, were both innkeepers. As a boy, Ehrlich was influenced by several relatives who studied science. His paternal grandfather, Heimann Ehrlich, made a living as a liquor merchant but kept a private laboratory and gave lectures on science to the citizens of Strehlen. Karl Weigert, cousin of Ehrlich's mother, became a well-known pathologist. Ehrlich, who was close friends with Weigert, often joined his cousin in his lab, where he learned how to stain cells with dye in order to see them better under the microscope. Ehrlich's research into the dye reactions of cells continued during his time as a university student. He studied science and medicine at the universities of Breslau,

Paul Ehrlich.

Strasbourg, Freiburg, and Leipzig. Although Ehrlich conducted most of his course work at Breslau, he submitted his final dissertation to the University of Leipzig, which awarded him a medical degree in 1878.

Ehrlich's 1878 doctoral thesis, "Contributions to the Theory and Practice of Histological Staining," suggests that even at this early stage in his career he recognized the depth of possibility and discovery in his chosen research field. In his experiments with many dyes, Ehrlich had learned how to manipulate chemicals in order to obtain specific effects: Methylene blue dye, for example, stained nerve cells without discoloring the tissue around them. These experiments with dye reactions formed the backbone of Ehrlich's career and led to two important contributions to science. First, improvements in staining permitted scientists to examine cells—healthy or unhealthy—and microorganisms, including those that caused disease. Ehrlich's work ushered in a new era of medical diagnosis and histology (the study of cells), which alone would have guaranteed Ehrlich a place in scientific history. Secondly, and more significantly from a scientific standpoint, Ehrlich's early experiments revealed that certain cells have an affinity to certain dyes. To Ehrlich, it was clear that chemical and physical reactions were taking place in the stained tissue. He theorized that chemical reactions governed all biological life processes. If this were true, Ehrlich reasoned, then chemicals could perhaps be used to heal diseased cells and to attack harmful microorganisms. Ehrlich began studying the chemical structure of the dyes he used and postulated theories for what chemical reactions might be taking place in the body in the presence of dyes and other chemical agents. These efforts would eventually lead Ehrlich to study the immune system.

Upon Ehrlich's graduation, medical clinic director Friedrich von Frerichs immediately offered the young scientist a position as head physician at the Charite Hospital in Berlin. Von Frerichs recognized that Ehrlich, with his penchant for strong cigars and mineral **water**, was a unique talent, one that should be excused from clinical work and be allowed to pursue his research uninterrupted. The late nineteenth century was a time when infectious diseases like cholera and typhoid fever were incurable and fatal. Syphilis, a sexually transmitted disease caused by a then unidentified microorganism, was an epidemic, as was tuberculosis, another disease whose cause had yet to be named. To treat human disease, medical scientists knew they needed a better understanding of harmful microorganisms.

At the Charite Hospital, Ehrlich studied blood cells under the microscope. Although blood cells can be found in a perplexing multiplicity of forms, Ehrlich was with his dyes able to begin identifying them. His systematic cataloging of the cells laid the groundwork for what would become the field of hematology. Ehrlich also furthered his understanding of **chemistry** by meeting with professionals from the chemical industry. These contacts gave him information about the structure and preparation of new chemicals and kept him supplied with new dyes and chemicals.

Ehrlich's slow and steady work with stains resulted in a sudden and spectacular achievement. On March 24, 1882, Ehrlich had heard Robert Koch announce to the Berlin Physiological Society that he had identified the bacillus causing tuberculosis under the microscope. Koch's method of staining the bacillus for study, however, was less than ideal. Ehrlich immediately began experimenting and was soon able to show Koch an improved method of staining the tubercle bacillus. The technique has since remained in use.

On April 14, 1883, Ehrlich married 19-year-old Hedwig Pinkus in the Neustadt Synagogue. Ehrlich had met Pinkus, the daughter of an affluent textile manufacturer of Neustadt, while visiting relatives in Berlin. The marriage brought two daughters, Steffa and Marianne. In March, 1885, von Frerichs committed suicide and Ehrlich suddenly found himself without a mentor. Von Frerichs's successor as director of Charite Hospital, Karl Gerhardt, was far less impressed with Ehrlich and forced him to focus on clinical work rather than research. Though complying, Ehrlich was highly dissatisfied with the change. Two years later, Ehrlich resigned from the Charite Hospital, ostensibly because he wished to relocate to a dry climate to cure himself of tuberculosis. The mild case of the disease, which Ehrlich had diagnosed using his staining techniques, was almost certainly contracted from cultures in his lab. In September of 1888, Ehrlich and his wife embarked on an extended journey to southern Europe and Egypt and returned to Berlin in the spring of 1889 with Ehrlich cured.

In Berlin, Ehrlich set up a small private laboratory with financial help from his father-in-law, and in 1890 he was honored with an appointment as Extraordinary Professor at the University of Berlin. In 1891, Ehrlich accepted Robert Koch's invitation to join him at the Institute for Infectious Diseases, newly created for Koch by the Prussian government. At the institute, Koch began his immunological research by demonstrating that mice fed or injected with the toxins ricin and abrin developed antitoxins. He also proved that **antibodies** were passed from mother to offspring through breast milk. Ehrlich joined forces with Koch and von Behring to find a cure for diphtheria, a deadly childhood disease. Although von Behring had identified the antibodies to diphtheria, he still faced great difficulties transforming the discovery into a potent yet safe cure for humans. Using blood drawn from horses and goats infected with the disease, the scientists worked together to concentrate and purify an effective antitoxin. Ehrlich's particular contribution to the cure was his method of measuring an effective dose.

The commercialization of a diphtheria antitoxin began in 1892 and was manufactured by Höchst Chemical Works. Royalties from the drug profits promised to make Ehrlich and von Behring wealthy men. But Ehrlich—possibly at von Behring's urging—accepted a government position in 1885 to monitor the production of the diphtheria serum. Conflict-of-interest clauses obligated Ehrlich to withdraw from his profit-sharing agreement. Forced to stand by as the diphtheria antitoxin made von Behring a wealthy man, he and von Behring quarreled and eventually parted. Although it is unclear whether or not bitterness over the royalty agreement sparked the quarrel, it certainly couldn't have helped a relationship that was often tumultuous. Although the two scientists continued to exchange news in letters, both scientific and personal, the two scientists never met again.

In June of 1896, the Prussian government invited Ehrlich to direct its newly created Royal Institute for Serum Research and Testing in Steglitz, a suburb of Berlin. For the first time, Ehrlich had his own institute. In 1896 Ehrlich was invited by Franz Adickes, the mayor of Frankfurt, and by Friedrich Althoff, the Prussian Minister of Educational and Medical Affairs, to move his research to Frankfurt. Ehrlich accepted and the Royal Institute for Experimental Therapy opened on November 8, 1899. Ehrlich was to remain as its director until his death sixteen years later. The years in Frankfurt would prove to be among Ehrlich's most productive.

In his speech at the opening of the Institute for Experimental Therapy, Ehrlich seized the opportunity to describe in detail his ''side-chain theory'' of how antibodies worked. ''Side-chain'' is the name given to the appendages on **benzene** molecules that allow it to react with other chemicals. Ehrlich believed all molecules had similar side-chains that allowed them to link with molecules, nutrients, infectious toxins and other substances. Although Ehrlich's theory is false, his efforts to prove it led to a host of new discoveries and guided much of his future research.

The move to Frankfurt marked the dawn of chemotherapy as Ehrlich erected various chemical agents against a host of dangerous microorganisms. In 1903, scientists had discovered that the cause of sleeping sickness, a deadly disease prevalent in Africa, was a species of trypanosomes (parasitic protozoans). With help from Japanese scientist Kiyoshi Shiga, Ehrlich worked to find a dye that destroyed trypanosomes in infected mice. In 1904, he discovered such a dye, which was dubbed ''trypan red.''

Success with trypan red spurred Ehrlich to begin testing other chemicals against disease. To conduct his methodical and painstaking experiments with an enormous range of chemicals, Ehrlich relied heavily on his assistants. To direct their work, he made up a series of instructions on colored cards in the evening and handed them out each morning. Although such a management strategy did not endear him to his lab associates—and did not allow them opportunity for their own research—Ehrlich's approach was often successful. In one famous instance, Ehrlich ordered his staff to disregard the accepted notion of the chemical structure of atoxyl and to instead proceed in their work based on his specifications of the chemical. Two of the three medical scientists working with Ehrlich were appalled at his scientific heresy and ended their employment at the laboratory. Ehrlich's hypothesis concerning atoxyl turned out to have been correct and would eventually lead to the discovery of a chemical cure for syphilis.

In September of 1906, Ehrlich's laboratory became a division of the new Georg Speyer Haus for Chemotherapeutical Research. The research institute, endowed by the wealthy widow of Georg Speyer for the exclusive purpose of continuing Ehrlich's work in chemotherapy, was built next to Ehrlich's existing laboratory. In a speech at the opening of the new institute, Ehrlich used the phrase ''magic bullets'' to illustrate his hope of finding chemical compounds that would enter the body, attack only the offending microorganisms or malignant cells, and leave healthy tissue untouched. In 1908, Ehrlich's work on immunity, particularly his contribution to the diphtheria antitoxin, was honored with the Nobel Prize in medicine or physiology. He shared the prize with Russian bacteriologist **Élie Metchnikoff**.

By the time Ehrlich's lab formally joined the Speyer Haus, he had already tested over 300 chemical compounds against trypanosomes and the syphilis spirochete (distinguished as slender and spirally undulating bacteria). With each test given a laboratory number, Ehrlich was testing compounds numbering in the nine hundreds before realizing that ''compound 606'' was a highly potent drug effective against relapsing fever and syphilis. Due to an assistant's error, the potential of compound 606 had been overlooked for nearly two years until Ehrlich's associate, Sahashiro Hata, experimented with it again. On June 10, 1909, Ehrlich and Hata filed a patent for 606 for its use against relapsing fever.

The first favorable results of 606 against syphilis were announced at the Congress for Internal Medicine held at Wiesbaden in April 1910. Although Ehrlich emphasized he was reporting only preliminary results, news of a cure for the devastating and widespread disease swept through the European and American medical communities and Ehrlich was besieged with requests for the drug. Physicians and victims of the disease clamored at his doors. Ehrlich, painfully aware that mishandled dosages could blind or even kill patients, begged physicians to wait until he could test 606 on ten or twenty thousand more patients. But there was no halting the demand and the Georg Speyer Haus ultimately manufactured and distributed 65,000 units of 606 to physicians all over the globe free of charge. Eventually, the large-scale production of 606, under the commercial name "Salvarsan," was taken over by Höchst Chemical Works. The next four years, although largely triumphant, were also filled with reports of patients' deaths and maiming at the hands of doctors who failed to administer Salvarsan properly.

In 1913, in an address to the International Medical Congress in London, Ehrlich cited trypan red and Salvarsan as examples of the power of chemotherapy and described his vision of chemotherapy's future. The City of Frankfurt honored Ehrlich by renaming the street in front of the Georg Speyer Haus "Paul Ehrlichstrasse." Yet in 1914, Ehrlich was forced to defend himself against claims made by a Frankfurt newspaper, *Die Wahrheit* (The Truth), that Ehrlich was testing Salvarsan on prostitutes against their will, that the drug was a fraud, and that Ehrlich's motivation for promoting it was personal monetary gain. In June 1914, Frankfurt city authorities took action against the newspaper and Ehrlich testified in court as an expert witness. Ehrlich's name was finally cleared and the newspaper's publisher sentenced to a year in jail, but the trial left Ehrlich deeply depressed. In December, 1914, he suffered a mild stroke.

Ehrlich's health failed to improve and the start of World War I had further discouraged him. Afflicted with arteriosclerosis, a disease of the arteries, his health deteriorated rapidly. He died in Bad Homburg, Prussia (now Germany), on August 20, 1915, after a second stroke. Ehrlich was buried in Frankfurt. Following the German Nazi era—during which time Ehrlich's widow and daughters were persecuted as Jews before fleeing the country and the sign marking Paul Ehrlichstrasse was torn down—Frankfurt once again honored its famous resident. The Institute for Experimental Therapy changed its name to the Paul Ehrlich Institute and began offering the biennial Paul Ehrlich Prize in one of Ehrlich's fields of research as a memorial to its founder.

EIGEN, MANFRED (1927-)

German physical chemist

Manfred Eigen shared the Nobel Prize for chemistry in 1967 with **George Porter** and **Ronald G. W. Norrish** for their combined work on fast chemical reactions. Whereas previously scientists had no means of calculating the rates of these reactions, Eigen discovered that high-frequency sound waves could be used to create pulses of **energy** in a chemical system. Observing the change as the system returned to a state of equilibrium enabled him to measure rates of reactions that lasted only a billionth of a second. Most of his long career has been spent at the **Max Planck** Institute for Physical Chemistry in Göttingen. Eigen was born in the town of Bochum in the Ruhr region of Germany on May 9, 1927, to Ernst Eigen and Hedwig Feld Eigen. His father was an accomplished chamber musician, and Eigen was a talented pianist. He served briefly with an army anti-aircraft artillery unit at the end of World War II. He then returned to the University of Göttingen (where he had begun his education), earning his doctorate in 1951.

For several years Eigen worked as a research assistant at the University of Göttingen, and then he joined the staff at the Max Planck Institute for Physical Chemistry. In 1958 he was appointed a research fellow at the institute, and in 1962 he became head of the department of biochemical **kinetics**. In 1964 he was made the director.

Eigen discovered in his early research that the reason sound waves are absorbed by seawater more quickly than by **water** is that they disrupt the charged particles of **magnesium sulfate**, small amounts of which are dissolved in seawater. The sound wave causes the loss of a small amount of energy. This discovery led Eigen to use high-frequency sound waves to produce disturbances in a chemical system in order to measure rates of chemical processes that had not been measured before. He was able to study chemical reactions measuring from a thousandth to a billionth of a second. This technique is called a relaxation technique because it measures a new state of equilibrium in a chemical system.

Eigen's interest in fast chemical reactions is related to his interest in biology. He concentrated his research on extremely rapid biochemical body reactions in an effort to discover how molecules formed and evolved into the first forms of life. Basically, his theory is built on the premise that the first organisms evolved from a chance set of circumstances coming together. In *Laws of the Game,* a book that he coauthored with his associate Ruthild Winkler, Eigen explores how the principles of nature govern chance. The authors believe that chance and necessity cause all events. They develop their models from general science, philosophy, sociology, and aesthetics, as well as from biology. The models are used to explore complex scientific concepts. The authors see the play of games as being basic to both the organization of the physical world and to human behavior.

Eigen's relaxation techniques have been used to study enzyme-catalyzed reactions and the coding of biological information. During the 1970s he worked on hypercycles (the self-organization of **nucleic acids** into complex structures and their interaction with proteins). Eigen's work has been extremely valuable in many other areas of scientific investigation; it has been used in areas such as **radiation chemistry** and **enzyme** kinetics, where the sequence of processes becomes converted into products.

Much of Eigen's research is considered by his colleagues to be groundbreaking, since he opened new areas of application for relaxation techniques. He has received wide recognition for his work, winning a number of important awards, including the **Otto Hahn** Prize from the German Physical Society in 1962, the **Linus Pauling** Medal of the American Chemical Society in 1967, and the Faraday Medal of the British Chemical Society in 1977, and receiving honorary doctorates from universities in Europe and the U.S. He has written more than 100 papers and several books, and has traveled and lectured widely.

Eigen married Elfriede Mueller in 1952. The Eigens have two children, Gerald and Angela. Hobbies that Eigen enjoys are hiking and wild mushroom collecting (in which he is considered an expert).

EINSTEIN, ALBERT (1879-1955)

American physicist

Albert Einstein ranks as one of the most remarkable theoreticians in the history of science. During a single year, 1905, he produced three papers that are among the most important in twentieth-century physics, and perhaps in all of the recorded history of science, for they revolutionized the way scientists looked at the nature of space, time, and matter. These papers dealt with the nature of particle movement known as **Brownian motion**, the quantum nature of electromagnetic radiation as demonstrated by the **photoelectric effect**, and the special theory of relativity. Although Einstein is probably best known for the last of these works, it was for his quantum explanation of the photoelectric effect that he was awarded the 1921 Nobel Prize in physics. In 1915, Einstein extended his special theory of relativity to include certain cases of accelerated motion, resulting in the more general theory of relativity.

Einstein was born in Ulm, Germany, on March 14, 1879, the only son of Hermann and Pauline Koch Einstein. Both sides of his family had long-established roots in southern Germany, and, at the time of Einstein's birth, his father and uncle Jakob owned a small electrical equipment plant. When that business failed around 1880, Hermann Einstein moved his family to Munich to make a new beginning. A year after their arrival in Munich, Einstein's only sister, Maja, was born.

Although his family was Jewish, Einstein was sent to a Catholic elementary school from 1884 to 1889. He was then enrolled at the Luitpold Gymnasium in Munich. During these years, Einstein began to develop some of his earliest interests in science and mathematics, but he gave little outward indication of any special aptitude in these fields. Indeed, he did not begin to talk until the age of three and, by the age of nine, was still not fluent in his native language. His parents were actually concerned that he might be somewhat mentally retarded.

In 1894, Hermann Einstein's business failed again, and the family moved once more, this time to Pavia, near Milan, Italy. Einstein was left behind in Munich to allow him to finish school. Such was not to be the case, however, since he left the *gymnasium* after only six more months. Einstein's biographer, Philipp Frank, explains that Einstein so thoroughly despised formal schooling that he devised a scheme by which he received a medical excuse from school on the basis of a potential nervous breakdown. He then convinced a mathematics teacher to certify that he was adequately prepared to begin his college studies without a high school diploma. Other biographies, however, say that Einstein was expelled from the *gymnasium* on the grounds that he was a disruptive influence at the school.

In any case, Einstein then rejoined his family in Italy. One of his first acts upon reaching Pavia was to give up his German citizenship. He was so unhappy with his native land

Albert Einstein.

that he wanted to sever all formal connections with it; in addition, by renouncing his citizenship, he could later return to Germany without being arrested as a draft dodger. As a result, Einstein remained without an official citizenship until he became a Swiss citizen at the age of 21. For most of his first year in Italy, Einstein spent his time traveling, relaxing, and teaching himself calculus and higher mathematics. In 1895, he thought himself ready to take the entrance examination for the Eidgenössiche Technische Hochschule (the ETH, Swiss Federal Polytechnic School, or Swiss Federal Institute of Technology), where he planned to major in electrical engineering. When he failed that examination, Einstein enrolled at a Swiss cantonal high school in Aarau. He found the more democratic style of instruction at Aarau much more enjoyable than his experience in Munich and soon began to make rapid progress. He took the entrance examination for the ETH a second time in 1896, passed, and was admitted to the school. (In *Einstein,* however, Jeremy Bernstein writes that Einstein was admitted without examination on the basis of his diploma from Aarau.)

The program at ETH had nearly as little appeal for Einstein as had his schooling in Munich, however. He apparently hated studying for examinations and was not especially interested in attending classes on a regular basis. He devoted much of this time to reading on his own, specializing in the works of **Gustav Kirchhoff**, Heinrich Hertz, **James Clerk Maxwell**, Ernst Mach, and other classical physicists. When Einstein graduated with a teaching degree in 1900, he was unable to find a regular teaching job. Instead, he supported himself as a tutor in a private school in Schaffhausen. In 1901, Einstein also published his first scientific paper, "Consequences of Capillary Phenomena."

In February, 1902, Einstein moved to Bern and applied for a job with the Swiss Patent Office. He was given a probationary appointment to begin in June of that year and was promoted to the position of technical expert, third class, a few months later. The seven years Einstein spent at the Patent Office were the most productive years of his life. The demands of his work were relatively modest and he was able to devote a great deal of time to his own research.

The promise of a steady income at the Patent Office also made it possible for Einstein to marry. Mileva Marić (also given as Maritsch) was a fellow student in physics at ETH, and Einstein had fallen in love with her even though his parents strongly objected to the match. Marić had originally come from Hungary and was of Serbian and Greek Orthodox heritage. The couple married on January 6, 1903, and later had two sons, Hans Albert and Edward. A previous child, Liserl, was born in 1902 at the home of Marić's parents in Hungary, but there is no further mention or trace of her after 1903 since she was given up for adoption.

In 1905, Einstein published a series of papers, any one of which would have assured his fame in history. One, "On the Movement of Small Particles Suspended in a Stationary Liquid Demanded by the Molecular-Kinetic Theory of Heat," dealt with a phenomenon first observed by the Scottish botanist Robert Brown in 1827. Brown had reported that tiny particles, such as dust particles, move about with a rapid and random zigzag motion when suspended in a liquid.

Einstein hypothesized that the visible motion of particles was caused by the random movement of molecules that make up the liquid. He derived a mathematical formula that predicted the distance traveled by particles and their relative speed. This formula was confirmed experimentally by the French physicist Jean Baptiste Perrin in 1908. Einstein's work on the Brownian movement is generally regarded as the first direct experimental evidence of the existence of **molecule**s.

A second paper, "On a Heuristic Viewpoint concerning the Production and Transformation of Light," dealt with another puzzle in physics, the photoelectric effect. First observed by Heinrich Hertz in 1888, the photoelectric effect involves the release of electrons from a metal that occurs when light is shined on the metal. The puzzling aspect of the photoelectric effect was that the number of electrons released is not a function of the light's intensity, but of the color (that is, the wavelength) of the light.

To solve this problem, Einstein made use of a concept developed only a few years before, in 1900, by the German physicist **Max Planck**, the quantum hypothesis. Einstein assumed that light travels in tiny discrete bundles, or "quanta," of energy. The energy of any given light quantum (later renamed the **photon**), Einstein said, is a function of its wavelength. Thus, when light falls on a metal, electrons in the metal absorb specific quanta of energy, giving them enough energy to escape from the surface of the metal. But the number of electrons released will be determined not by the number of quanta (that is, the intensity) of the light, but by its energy (that is, its wavelength). Einstein's hypothesis was confirmed by several experiments and laid the foundation for the fields of quantitative photoelectric chemistry and quantum mechanics. As recognition for this work, Einstein was awarded the 1921 Nobel Prize in physics.

A third 1905 paper by Einstein, almost certainly the one for which he became best known, details his special theory of relativity. In essence, "On the Electrodynamics of Moving Bodies" discusses the relationship between measurements made by observers in two separate systems moving at constant velocity with respect to each other.

Einstein's work on relativity was by no means the first in the field. The French physicist Jules Henri Poincaré, the Irish physicist George Francis FitzGerald, and the Dutch physicist Hendrik Lorentz had already analyzed in some detail the problem attacked by Einstein in his 1905 paper. Each had developed mathematical formulas that described the effect of motion on various types of measurement. Indeed, the record of pre-Einsteinian thought on relativity is so extensive that one historian of science once wrote a two-volume work on the subject that devoted only a single sentence to Einstein's work. Still, there is little question that Einstein provided the most complete analysis of this subject. He began by making two assumptions. First, he said that the laws of physics are the same in all frames of reference. Second, he declared that the velocity of light is always the same, regardless of the conditions under which it is measured.

Using only these two assumptions, Einstein proceeded to uncover an unexpectedly extensive description of the properties of bodies that are in uniform motion. For example, he showed that the length and mass of an object are dependent upon their movement relative to an observer. He derived a mathematical relationship between the length of an object and its velocity that had previously been suggested by both FitzGerald and Lorentz. Einstein's theory was revolutionary, for previously scientists had believed that basic quantities of measurement such as time, mass, and length were absolute and unchanging. Einstein's work established the opposite—that these measurements could change, depending on the relative motion of the observer.

In addition to his masterpieces on the photoelectric effect, Brownian movement, and relativity, Einstein wrote two more papers in 1905. One, "Does the Inertia of a Body Depend on Its Energy Content?," dealt with an extension of his earlier work on relativity. He came to the conclusion in this paper that the energy and mass of a body are closely interrelated. Two years later he specifically stated that relationship in a formula, $E=mc^2$ (energy equals mass times the speed of light squared), that is now familiar to both scientists and non-scientists alike. His final paper, the most modest of the five, was "A New Determination of Molecular Dimensions." It was this paper that Einstein submitted as his doctoral dissertation, for which the University of Zurich awarded him a Ph.D. in 1905.

Fame did not come to Einstein immediately as a result of his five 1905 papers. Indeed, he submitted his paper on relativity to the University of Bern in support of his application to become a *privatdozent,* or unsalaried instructor, but the paper and application were rejected. His work was too important to be long ignored, however, and a second application three years later was accepted. Einstein spent only a year at Bern, however, before taking a job as professor of physics at the University of Zurich in 1909. He then went on to the German University of Prague for a year and a half before returning to Zurich and a position at ETH in 1912. A year later Einstein was made director of scientific research at the Kaiser Wilhelm Institute for Physics in Berlin, a post he held from 1914 to 1933.

In recent years, the role of Mileva Einstein-Marić in her husband's early work has been the subject of some controversy. The more traditional view among Einstein's biographers is that of A. P. French in his "Condensed Biography" in *Einstein: A Centenary Volume.* French argues that although "little is recorded about his [Einstein's] domestic life, it certainly did not inhibit his scientific activity." In perhaps the most substantial of all Einstein biographies, Philipp Frank writes that "For Einstein life with her was not always a source of peace and happiness. When he wanted to discuss with her his ideas, which came to him in great abundance, her response was so slight that he was often unable to decide whether or not she was interested."

A quite different view of the relationship between Einstein and Marić is presented in a 1990 paper by Senta Troemel-Ploetz in *Women's Studies International Forum.* Based on a biography of Marić originally published in Yugoslavia, Troemel-Ploetz argues that Marić gave to her husband "her companionship, her diligence, her endurance, her mathematical genius, and her mathematical devotion." Indeed, Troemel-Ploetz builds a case that it was Marić who did a significant portion of the mathematical calculations involved in much of Einstein's early work. She begins by repeating a famous remark by Einstein himself to the effect that "My wife solves all my mathematical problems." In addition, Troemel-Ploetz cites many of Einstein's own letters of 1900 and 1901 (available in *Collected Papers*) that allude to Marić's role in the development of "our papers," including one letter to Marić in which Einstein noted: "How happy and proud I will be when both of us together will have brought our work on relative motion to a successful end." The author also points out the somewhat unexpected fact that Einstein gave the money he received from the 1921 Nobel Prize to Marić, although the two had been divorced two years earlier. Nevertheless, Einstein never publicly acknowledged any contributions by his wife to his work.

Any mathematical efforts Mileva Einstein-Marić may have contributed to Einstein's work greatly decreased after the birth of their second son in 1910. Einstein was increasingly occupied with his career and his wife with managing their household; upon moving to Berlin in 1914, the couple grew even more distant. With the outbreak of World War I, Einstein's wife and two children returned to Zurich. The two were never reconciled; in 1919, they were formally divorced. Towards the end of the war, Einstein became very ill and was nursed back to health by his cousin Elsa. Not long after Einstein's divorce from Marić, he was married to Elsa, a widow. The two had no children of their own, although Elsa brought two daughters, Ilse and Margot, to the marriage.

The war years also marked the culmination of Einstein's attempt to extend his 1905 theory of relativity to a broader context, specifically to systems with non-zero acceleration. Under the general theory of relativity, motions no longer had to be uniform and relative velocities no longer constant. Einstein was able to write mathematical expressions that describe the relationships between measurements made in *any* two systems in motion relative to each other, even if the motion is accelerated in one or both. One of the fundamental features of the general theory is the concept of a space-time continuum in which space is curved. That concept means that a body affects the shape of the space that surrounds it so that a second body moving near the first body will travel in a curved path.

Einstein's new theory was too radical to be immediately accepted, for not only were the mathematics behind it extremely complex, it replaced Newton's theory of gravitation that had been accepted for two centuries. So, Einstein offered three proofs for his theory that could be tested: first, that relativity would cause Mercury's perihelion, or point of orbit closest to the sun, to advance slightly more than was predicted by Newton's laws. Second, Einstein predicted that light from a star will be bent as it passes close to a massive body, such as the sun. Last, the physicist suggested that relativity would also affect light by changing its wavelength, a phenomenon known as the redshift effect. Observations of the planet Mercury bore out Einstein's hypothesis and calculations, but astronomers and physicists had yet to test the other two proofs.

Einstein had calculated that the amount of light bent by the sun would amount to 1.7 seconds of an arc, a small but detectable effect. In 1919, during an eclipse of the sun, English astronomer Arthur Eddington measured the deflection of starlight and found it to be 1.61 seconds of an arc, well within experimental error. The publication of this proof made Einstein an instant celebrity and made "relativity" a household word, although it was not until 1924 that Eddington proved the final hypothesis concerning redshift with a spectral analysis of the star Sirius B. This phenomenon, that light would be shifted to a longer wavelength in the presence of a strong gravitational field, became known as the "Einstein."

Einstein's publication of his general theory in 1916, the *Foundation of the General Theory of Relativity,* essentially brought to a close the revolutionary period of his scientific career. In many ways, Einstein had begun to fall out of phase with the rapid changes taking place in physics during the 1920s. Even though Einstein's own work on the photoelectric effect helped set the stage for the development of quantum theory, he was never able to accept some of its concepts, particularly the uncertainty principle. In one of the most-quoted comments in the history of science, he claimed that quantum mechanics, which could only calculate the probabilities of physical events, could not be correct because "God does not play dice." Instead, Einstein devoted his efforts for the remaining years of his life to the search for a unified field theory, a single theory that would encompass all physical fields, particularly gravitation and electromagnetism.

Since the outbreak of World War I, Einstein had been opposed to war, and used his notoriety to lecture against it during the 1920s and 1930s. With the rise of National Socialism in Germany in the early 1930s, Einstein's position became difficult. Although he had renewed his German citizenship, he was suspect as both a Jew and a pacifist. In addition, his writings about relativity were in conflict with the absolutist teachings of German leader Adolf Hitler's party. Fortunately, by 1930, Einstein had become internationally famous and had traveled widely throughout the world. A number of institutions were eager to add his name to their faculties.

In early 1933, Einstein made a decision. He was out of Germany when Hitler rose to power, and he decided not to return. Instead he accepted an appointment at the Institute for Advanced Studies in Princeton, New Jersey, where he spent the rest of his life. In addition to his continued work on unified field theory, Einstein was in demand as a speaker and wrote extensively on many topics, especially peace. The growing fascism and anti-Semitism of Hitler's regime, however, convinced him in 1939 to sign his name to a letter written by American physicist informing President Franklin D. Roosevelt of the possibility of an atomic bomb. This letter led to the formation of the **Manhattan Project** for the construction of the world's first nuclear weapons. Although Einstein's work on relativity, particularly his formulation of the equation $E=mc^2$, was essential to the development of the atomic bomb, Einstein himself did not participate in the project. He was considered a security risk, although he had renounced his German citizenship and become a U.S. citizen in 1940.

After World War II and the bombing of Japan, Einstein became an ardent supporter of nuclear disarmament. He also lent his support to the efforts to establish a world government and to the Zionist movement to establish a Jewish state. In 1952, after the death of Israel's first president, Chaim Weizmann, Einstein was invited to succeed him as president; he declined the offer. Among the many other honors given to Einstein were the Barnard Medal of Columbia University in 1920, the Copley Medal of the Royal Society in 1925, the Gold Medal of the Royal Astronomical Society in 1926, the Max Planck Medal of the German Physical Society in 1929, and the Franklin Medal of the Franklin Institute in 1935. Einstein died at his home in Princeton on April 18, 1955, after suffering an aortic aneurysm. At the time of his death, he was the world's most widely admired scientist and his name was synonymous with genius. Yet Einstein declined to become enamored of the admiration of others. He wrote in his book *The World as I See It:* "Let every man be respected as an individual and no man idolized. It is an irony of fate that I myself have been the recipient of excessive admiration and respect from my fellows through no fault, and no merit, of my own. The cause of this may well be the desire, unattainable for many, to understand the one or two ideas to which I have with my feeble powers attained through ceaseless struggle."

EINSTEINIUM

Einsteinium is a transuranium element, located in Row 7 of the **periodic table**. The elements that make up this family are also known as the actinides, after the first member of the family. Einsteinium's **atomic number** is 99, its atomic **mass** is 252.0828, and its chemical symbol is Es.

Properties

All isotopes of einsteinium are radioactive, the most stable being einsteinium-252 with a half life of 20.47 days. Too little of the element has been prepared thus far for scientists to have made determinations of its properties.

Occurrence and Extraction

Einsteinium does not occur naturally in the Earth's crust. Instead, it is prepared in particle accelerators by bombarding isotopes of heavy transuranium elements with alpha particles.

Discovery and Naming

Einsteinium was discovered in November 1952 at Eniwetok Atoll, Marshall Islands, in the Pacific Ocean as a byproduct of the first **hydrogen** bomb test. The element was produced in trace amounts during the explosion of that bomb. The research team that discovered the element was led by **Albert Ghiorso**, who has also been involved in the discovery of many other transuranium elements. The team chose to name the new element after the German-American physicist **Albert Einstein**, sometimes regarded as the greatest scientist who has ever lived.

Uses

Einsteinium is sometimes used for research purposes, but it has no practical applications.

ELASTOMERS

An elastomer is a cross-linked, non-crystalline polymer above its **glass** transition **temperature** (a characteristic transition point at which a material's physical behavior changes from rigid to rubbery). Rubber bands probably provide the most familiar examples of elastomers. When an elastomer is deformed and released (provided the temperature exceeds the material's glass transition temperature), it snaps back with typical rubber-like behavior. The stress required to deform an elastomer depends on the number of linkages (called crosslinks) among the material's constituent macromolecules, the extent of deformation, and the temperature.

Elastomers may fall into any one of four categories: diene elastomers, saturated elastomers, thermoplastic elastomers, or inorganic elastomers. Diene elastomers have structures based on the molecules butadiene, isoprene, and/or their derivatives or copolymers. Natural rubber (polyisoprene), the first known elastomer, is a member of the diene elasomer family. So are polybutadiene, polychloroprene, and styrene-butadiene rubber. Diene elastomers can be recognized by the presence of double bonds in the main **chains** of the macromolecular molecules.

Saturated elastomers, which include the polyacrylates, have no double bonds in their main chains, which is to say that the chains are saturated. Saturated elastomers are valued for their resistance to **oxygen**, **water**, and ultraviolet light.

Unlike diene and saturated elastomers, thermoplastic elastomers are physically, rather than chemically, cross-linked. They behave just like chemically cross-linked polymers at room temperature, but like uncross-linked polymers at higher temperatures. Thus, the elastomeric properties come and go with changes in temperature. This is a relatively new class of elastomer; examples include block copolymers containing at least three blocks (or segments).

Inorganic elastomers, such as silicone rubber, have good thermal properties and typically find use in high temperature applications.

See also Electromagnetic spectrum; Silicon

ELECTRICAL CONDUCTIVITY

Electrical conductivity is the ability of different types of **matter** to conduct an electric current. The electrical conductivity of a material is defined as the ratio of the current per unit cross-sectional area to the electric field producing the current. Electrical conductivity is an intrinsic property of a substance, dependent on the **temperature** and chemical composition, but not on the amount or shape. The unit of electrical conductivity in the SI system is the *siemens per meter*, where the siemens is the reciprocal of the ohm, the unit of electrical resistance, represented by the Greek capital letter omega. An older name for the siemens is the mho, which, of course, is ohm spelled backwards (which was written as an inverted Greek omega). Electrical conductivity is due to the presence of electrons or ions able to move through the material of interest.

Electrical conductivity is the inverse quantity to electrical resistivity. For any object conducting **electricity**, one can define the resistance in ohms as the ratio of the electrical potential difference applied to the object to current passing through it in amperes. For a cylindrical sample of known length and cross sectional area, the resistivity is obtained by dividing the measured resistance by the length and then multiplying by the area. The conductivity is then found by taking the reciprocal of the resistivity.

Metals generally have very high electrical conductivity. The electrical conductivity of **copper** at room temperature, for instance, is over 70 million siemens per meter. On an atomic level this high conductivity reflects the unique character of the **metallic bond** in which pairs of electrons are shared not between pairs of atoms, but among all the atoms in the metal, and are thus free to move over large distances. Many metals undergo a transition at low temperatures to a superconducting state, in which the resistance disappears entirely and the conductivity becomes infinite. The superconduction process involves a coupling of **electron** motion with the vibration of the atomic nuclei and inner-shell electrons, to allow net current flow without **energy** loss.

Electrical conductivity in the liquid state is generally due to the presence of ions. Substances that give rise to ionic conduction when dissolved are called electrolytes. The conductivity of a one molar electrolyte is of the order of 0.01 siemens per meter, far less than that of a metal, but still very much larger than that of typical insulators. **Sodium** chloride (common table salt), composed of sodium ions and chloride ions, is a very poor conductor in the solid state. If it is dissolved in **water**, however, it becomes a good ionic conductor. Likewise, if it is melted, it becomes a good conductor. Substances such as **hydrogen** chloride or **acetic acid** are nonconductors in the pure state but give rise to ions and thus electrical conductivity when dissolved in water. In modern **electrochemistry**, substances of the **sodium chloride** type, which are actually composed of ions, are termed true electrolytes, while those which require a solvent for **ion** formation, like hydrogen chloride, are termed potential electrolytes.

The electrical conductivity of a solution. ***(Photograph by Charles D. Winters, Photo Researchers, Inc. Reproduced by permission.)***

As might be expected, the conductivity of a dissolved electrolyte, depends on its **concentration** and so physical chemists find it useful to define molar conductivity of an electrolyte as the electrical conductivity of the **solution** divided by the concentration of the electrolyte in moles per liter. Because different electrolytes can give rise to ions of different charge, for example, Mg^{2+} ions instead of Na^{+} ions, an even more useful measure of conductivity is given by the equivalent conductivity, obtained by dividing the molar conductivity by the number of moles of positive charge released by one **mole** of the electrolyte. Because the ions+ of an electrolyte solution interact with each other, the equivalent conductivity is found to decrease gradually as the concentration increases. For dilute solutions, the equivalent conductivity can be expressed as a constant, the equivalent conductivity at infinite dilution minus a term proportional to the square root of the concentration. This important relation was discovered by the German physicist Friedrich Wilhelm Georg Kohlrausch (1840-1910). It is the limiting conductivities at infinite dilution that are additive for the ions involved. Thus, there are tabulations in handbooks of these conductivities for different ions (Na^+, K^+, Mg^+, Cl^-, Br^-, etc.) from which the corresponding conductivities of different electrolytes (NaCl, KCl, KBr, etc) can be calculated. The equivalent conductivities of the individual ions are proportional to their electrical mobilities (ratio of particle **velocity** to electrical field strength).

Ionic crystals do exhibit a limited electrical conductivity even in the crystalline state. This arises from point defects present at any temperature above absolute zero. For compounds like sodium chloride, composed of ions of roughly equal size, there will be a certain number of **cation** and **anion** vacancies, which allows a jumping motion of the corresponding cations and anions. For compounds like **silver** chloride, in which one of the ions, the cation, is much larger than the other ion, the anion, vacancies and interstitials (out of place ions) of the smaller ion will allow an ionic conductivity. In addition, compounds like silver chloride tolerate a range of **stoichiometry**, that is a deviation of the Ag to Cl ratio of one to one. An excess of silver is accommodated by a delocalization of the extra electrons to allow an electronic conductivity in addition to the conductivity of silver ions. A deficit of silver introduces mobile electron vacancies, or holes, in to the electronic structure of the material adding a component to the conductivity which behaves as if positively charged.

Semiconductors are covalently bonded materials for which **heat** energy is sufficient to promote electrons from **bonding** energy levels to delocalized energy levels. The conductivity of semiconductors is a sensitive function of impurity content (the substitution of semiconductor atoms by atoms with one fewer or more **valence** electrons) introducing positive (hole) or negative charge carriers into the material.

See also Superconductivity

ELECTRICITY

The word electricity comes from the Greek word, *electron*, meaning amber. Amber is a translucent, yellowish mineral made of fossilized resin. When rubbed with a cloth, amber becomes charged with static electricity. We now know that such electrostatic effects arise whenever a body has an excess or deficiency of **electrons** or protons. These effects usually accompany the transfer of electrons, such as occurs when two dissimilar materials are rubbed together. A fundamental rule of electrostatics is that bodies of like charge repel each other, and bodies of opposite charge attract each other. The Greeks described these attractive and repulsive electrostatic forces as electric force.

At the time of the American Revolution, little more was known about electricity than at the time of the Greeks. Benjamin Franklin was one of the first persons to experiment with electricity. In 1821, Hans Oersted demonstrated that electricity and **magnetism** were interrelated. Later, **James Clerk Maxwell** synthesized all that was known about electricity and magnetism into a set of equations that explain why a moving electric charge gives rise to a magnetic field, and why a changing magnetic field produces an electric field.

In the classical picture of the **atom**, an electron cloud surrounds a small **nucleus** (composed of neutrons and protons). The **proton** and electron are considered the fundamental particles of electricity, the two particles having equal and opposite charges (convention assigns a positive charge to the proton, and a negative one to the electron). Under normal conditions, the number of electrons equals the number of protons, so the atom is electrically neutral.

Since its discovery at the end of the nineteenth century, the electron has been considered a fundamental building block of nature that cannot be decomposed into more basic particles. Electrons are intrinsically stable, and possess **mass**, charge, and spin (i.e., intrinsic angular momentum). In recent years, despite extensive experimental and theoretical research in elementary particle physics, many physicists have concluded that the electron is indeed an ultimate, indivisible constituent of **matter** that will not be found to have more fundamental parts. The electron is thus considered a basic unit of electricity and matter.

When electrons in a conducting material drift under the influence of an electric field, they produce an electric current. Electric currents can also be produced by the motion of positive nuclei, as well as by ions moving in a liquid or gaseous electrical discharge.

In order for electric current to flow, there must exist an excess of electrons and a conductor to carry the current. As electrons enter the conductor, they displace the electrons in the conductor's atoms. In turn, these displaced electrons displace other electrons. Thus, the electron that exits the conductor will not be the same one that entered it. When a current flows in a conductor, the charges are impeded by collisions with the atoms that make up the material. The charges give up **energy** to the atoms, often as **heat**.

The electrons in the atoms of the conductor have fixed orbits. The maximum number of electrons in any orbit is fixed. For all orbits except the first one, this maximum number is eight. The number of **valence** electrons in an atom is the number of electrons that would be required to fill the outer orbit. If an electron orbit is full and unable to accept electrons, there must be eight electrons in that orbit. When an outside electron enters the conductor, it joins the partially empty orbit of one of the conductor's atoms, giving the atom a net negative charge. In the presence of an electric field, the displaced electron moves on to another atom, and the process repeats itself.

It is the electrons in the outer orbit that are available for electron flow. These electrons determine whether a particular element is classified as a conductor, semiconductor, or insulator. Freeing one of the electrons in the inner orbits of the atom requires considerably more energy than required to displace one in the outer orbit, and usually does not occur. **Metals** are excellent conductors, especially **silver**, **copper**, and **aluminum**. Metals are sometimes represented as a gas of free electrons. In the **alkali metals**, the outer electrons are only loosely bound to the nucleus, so are free to move from **ion** to ion in the solid state. To describe the conducting properties of more complex metals, the interactions of the various atoms and the periodic potential in which the electrons move must be taken into account. In pure **semiconductors**, there are no free electrons at absolute zero **temperature**. As the temperature is increased, some of the electrons are excited to current-carrying states (the conducting band); the valence states that are left unoccupied (called holes) are also available to carry electric current. Insulators may be thought of as semiconductors with very large gaps between the valence and conduction bands. In some insulating **solids** such as the alkali halides, ions carry current by hopping between vacant atomic sites in the crystal.

Electricity as a natural phenomenon shows up in many forms, including piezoelectricity, static electricity, atmospheric effects, and cosmic rays. When a piezoelectric material such as Rochelle **salt**, quartz, or tourmaline is squeezed, mechanical energy is converted into electrical energy. The Earth resembles a large battery, with electrons continuously being lost to the atmosphere. Thunderstorms return the electrons that have been conducted away from the Earth's surface back to Earth. When cosmic rays strike the upper atmosphere, they release electrons, giving rise to such phenomena as the northern lights.

Although electricity has been known to man since ancient times, and scientists have identified many of its consistent and predictable characteristics, it is still not completely understood.

See also Bonding; Electrical conductivity

ELECTROCHEMISTRY

Electrochemistry is the study of reactions caused by electric current and reactions which are capable of generating electric current. It may also be viewed as the study of the interconversion of chemical and electrical **energy**. Electrochemistry involves oxidation-reduction reactions. These reactions involve the transfer of electrons and consequent changes in the **oxidation** states of atoms during chemical reactions. **Oxidation-reduction** (or redox) **reactions** occur without passage of electric current when they take place between reagents that were in direct contact: ions or molecules in **solution**, **gases** mixed in the vapor phase, or **liquids** contacting **solids**. In electrochemistry, however, the reagents undergoing oxidation and **reduction** are actually physically separated from one another and contained in the compartments of a device called a cell. The reagents that are coupled in these oxidation-reduction reactions use some conduit to transfer charge from one compartment to another—in many cases a wire, in others a **gel** or solid that allows the passage of electrons from one species to another.

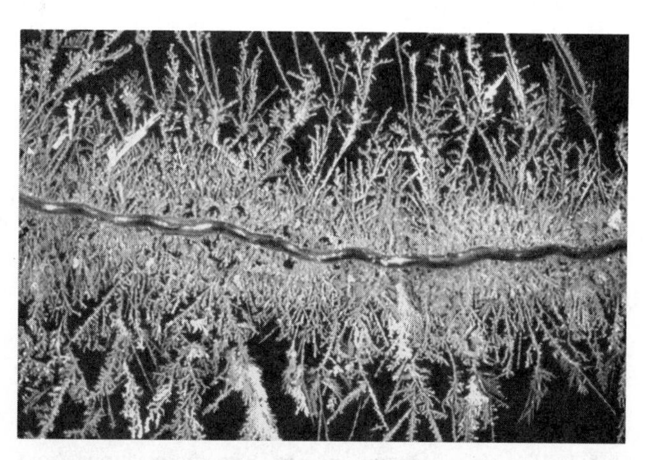

An electrochemical exchange. Copper wire is suspended in a solution of silver nitrate, and dissolves into a solution displacing the silver, which is deposited as crystals on the wire. In time, all of the silver is displaced, leaving a solution of copper nitrate. *(Photograph by Andrew Syred/Science Photo Library, Photo Researchers, Inc. Reproduced by permission.)*

Redox reaction are central to electrochemistry because they are chemical processes that involve the flow of electrons from one chemical entity to another. They can be either a source of electric current, as they are in a battery, or can be driven by electric current, as they are in **electroplating** processes.

Electrons flowing through a wire or ions flowing through an aqueous solution are both examples of electric current. If the electrons gained and lost by the components in a spontaneous redox reaction can be channeled through a wire or some other conducting material, then the conversion of chemical energy to electrical energy can be measured and the energy released by the spontaneous reaction can be used to do work. By the same token, if electrical current can be added to a system, a nonspontaneous redox reaction can be made to occur by converting electrical energy back into chemical energy to force a reaction to take place.

A voltaic cell generates electrical energy from a spontaneous redox reaction. Common **batteries** in watches, calculators and other devices are voltaic cells. In the cell, two electrodes, conductors by which electrons enter or leave a conducting medium, are connected by a material that carries the electrons between separate chambers containing the species involved in the electrochemical reaction. Both electrodes are immersed in solutions or other conducting materials, such as a moist paste, containing ions called electrolytic solutions or electrolytes. As the electrochemical reaction proceeds in a cell, ions are produced on one side and consumed on the other. To maintain a charge **balance** in an electrochemical reaction taking place in two different containers, a **salt** bridge can be used. The salt bridge contains a high **concentration** of an electrochemically unreactive salt, such as KCl, supported in an aqueous gel. Ions flow out of the salt bridge into the cell to maintain electrical neutrality in the solutions.

When the circuit is complete, electrons spontaneously flow from the electrode in contact with the more easily oxidized substance (in some case, the electrode material itself) to the electrode in contact with the more readily reduced material (again, this can be the electrode material). The reactions resulting from this flow of electric current occur at the electrode surfaces within each chamber in the cell. Oxidation occurs at the electrode from which electrons flow and reduction takes place at the electrode that receives the electrons. Oxidation and reduction occur simultaneously and to the same extent.

Each electrode and its surrounding electrolyte make up a half-cell. By definition, reduction takes place in one half-cell at the electrode called the cathode, and oxidation takes place in the other at the electrode called the **anode**. The sum of the half-reactions taking place in both cells is the overall redox reaction in the electrochemical cell. The term anode comes from the Greek *anodes*, or "up and out," for the electrode from which electrons leave a cell and oxidation occurs. Cathode comes from the Greek *kathodos*, or "down and in," for the electrode by which the electrons enter a cell and reduction occurs.

Because electrons originate at the anode in a voltaic cell, the anode has a (-) charge; electrons enter the cathode, which has a (+) charge. This polarity of the terminals and the role of the electrodes are accurate for any cell that is operating spontaneously and producing **electricity**. The electric current produced by this type of cell can be used to perform work.

Voltaic cells are often called galvanic cells in honor of Luigi Galvani, who is credited with discovering the phenomenon of electricity. The volt and, hence, voltaic cell are named in honor of Alessandro Volta, who developed some of the first cells that could generate a voltage.

A short-hand notation has been developed to depict the physical arrangement and define the chemical reactions taking place in an electrochemical cell:

anodic half-cell cathodic half-cell

(electrode)|(ionn+)aq||(ionn+)aq|(electrode)

For example, the reaction: Zno + Cu2+ ⇒Zn2+ + Cuo, would be represented by:

Anodic half-cell Cathodic half-cell

Zno|Zn2+(aq)||Cu2+(aq)|Cuo

The single vertical line "|" indicates the boundary between a solid electrode and the solution in which it is immersed. The double line "||" symbolizes the salt bridge.

Half-cells are characterized by their standard reduction potential, which is a measure of how readily a particular species is reduced relative to the **hydronium ion**. The more readily reduced a species is, the more positive its standard reduction potential. The better an oxidizing agent a species is the more readily it will itself be reduced. Therefore, the standard reduction potential is also a measure of the strength of an oxidizing agent.

For example, the **silver** ion is a stronger oxidizing agent than the hydronium ion, and so the silver ion is spontaneously reduced and the **hydrogen** spontaneously oxidized when a silver metal/silver ion half cell is connected to a hydrogen gas/ hydronium ion standard half-cell and the standard reduction potential for the silver ion is positive. in this cell. The Ag+(aq) ion is also a stronger oxidizing agent than the Cu2+(aq) ion and has a more positive standard reduction potential (+0.80V vs. +0.34V). If the standard half-cell for **copper** is connected to the standard half-cell for silver, a spontaneous reaction occurs in which the silver half-cell with its higher standard reduction potential undergoes reduction. Therefore, it functions as the cathode and the copper half-cell functions as the anode.

Electrical current is also used to cause non-spontaneous reactions to occur. An important industrial use of electrochemistry is the production of **metals** and halogen gases from salts. When a salt such as **sodium** chloride is melted and electrodes are immersed in it, application of a voltage difference across electrodes leads to a reduction of the sodium **cation** to metallic sodium and oxidation of the chloride ion to **chlorine** gas. This decomposition of an electrolyte caused by passing an electric current through it is called **electrolysis**. The apparatus in which an oxidation-reduction reaction occurs as a result of an applied voltage is called an electrolytic cell.

In an electrolytic cell, energy must be supplied in the form of electric current from an external source. The negative terminal of a power source pumps electrons into the electrode to which it is attached; that electrode is the cathode at which the reduction of sodium occurs. The positive terminal of the power source is connected to the electrode that functions as the anode; the oxidation of chloride ion to chlorine gas occurs at this site. In an electrolytic cell the reduction takes place at the cathode, just as it does in the voltaic cell. But because electrons must be supplied to the cathode by an external power source to drive this process, the cathode in an electrolytic cell has a negative charge. Its polarity is reversed in comparison to cathode in the voltaic cell, but its function—being the site of reduction—is the same.

Because electrolytic reactions require the input of electrical energy to cause a non-spontaneous reaction to take place, the products of electrolytic reactions typically must be physically separated from each other to prevent them from reacting with each other in the spontaneous (reverse) reaction direction. Electrolysis experiments established the composition of **water**—the fact that it is made up of hydrogen and **oxygen** and that the ratio of these components in the water **molecule** is 2:1. To electrolyze water or other non- ionic substances, an ionic substance, an electrolyte, is added to the water so that current will flow and the reaction will proceed at a relatively low voltage. The ions are said to "carry" the current by "migration" to the electrodes.

Despite the explanation used above, however, individual ions do not "migrate" toward electrodes. They are not "attracted" at long distances through water across cells. The insulating power of water is so great that the effect of a positively charged ion on a negatively charged ion falls rapidly to insignificance in a distance equivalent to only a few molecular diameters. The attraction of an ion toward an electrode is certainly not important over distances of millimeters or centimeters within cells. Ions change position over time as an electrochemical reaction proceeds, but the change is caused by normal processes of mixing and charge **neutralization**, not by the attraction of individual ions by the charge at or near an electrode.

As an electrochemical reaction such as the electrolysis of water proceeds, charge is reduced on species at the cathode and charge is increased on species in the vicinity of the anode. If there were no way to neutralize this change in charge, the electrochemical reaction would soon cease because too much energy would be required to maintain the charge separation. The region of excess positive charge would no longer be able to give up electrons and the region of excess negative charge would no longer accept additional electrons. However, other ions in the vicinity of the electrodes are buffeted about by the motion of the solvent molecules and, once near an ion of the opposite charge, tend to stay nearby. Eventually, the concentrations of ions throughout the solution provide for a uniform neutrality.

The Chlor-alkali Process for the production of **sodium hydroxide** chlorine gas is carried out on a very large scale; it consumes 1% of the total electricity produced in the United States annually in 1988. The applications of chlorine and sodium hydroxide are so varied that hardly a consumer product exists that does not depend on one or both of these compounds. The chlor-alkali process involves the electrolysis of brines, which are saturated aqueous solutions of **sodium chloride**. The decorative electroplating of jewelry using **gold** and silver was the first electrochemical process to be patented and is one of the oldest uses of electrochemistry. Electroplating consists of applying a metallic coating to an article by passing an electric current through a solution of electrolyte in contact with the article. Items ranging from jewelry to tableware (''silverware'') to automobile bumpers (''chrome'') are covered with **precious metals** or protective coatings by electroplating.

In this process, the less expensive metal is pre-formed into the desired shape and immersed in a solution of a salt of the plating metal. In the cell, the object to be plated is the cathode, the electrolyte contains the ionic form of the metal to be deposited on the cathodic surface, and an anode is immersed in the solution to complete the circuit. A power source is used to drive the electrochemical reaction. Its positive terminal is connected to the anode and its negative terminal to the cathode where the reduction of the metal ion takes place. When the circuit is complete, the cationic form of the metal to be coated on the object is reduced to the metal on the cathode. The reduced metal typically forms a metallic **alloy** on the **cathode**.

The quantitative relationships that govern the amount of material that is produced in electrochemical processes are called **Faraday's Laws**, after **Michael Faraday**, a British scientist of the nineteenth century.

The amount of energy that a voltaic cell can generate or the energy that is required for an electrolytic cell to produce a certain amount of material can be calculated from the total number of mols of electrons that flow during the process and the potential (also called the electromotive force) of the cell. The calculated energy can, in turn, be used to determine the equilibrium constant for the overall reaction that takes place in the cell.

Electrochemical potentials also arise when half-cells containing the same materials but at different concentrations are connected to one another. The potential difference between the two half-cells is greatest when the concentration difference is large. If there is a means for ions to migrate to bring the concentrations of particular ions closer to the same concentration in each half-cell the potential difference decreases. The Nernst equation defines the relationship of the potential of a cell as a function of the concentration of its components and the **temperature**. Many processes that result in concentration changes in electrochemical cells can be analyzed using the Nernst equation. The concept of concentration cells and the dependence of cell potential on ion concentration is central to our understanding of nerve impulse transmission.

Another important area in which the application of the concepts of electrochemistry is useful is metallic **corrosion**. Many metals are readily oxidized by oxygen in the atmosphere. Typically the way in which atmospheric corrosion occurs is through the action of oxygen dissolved in water in contact with the metal. The corrosive action is best understood as involving neutral metal atoms, oxidized metal ions, hydronium ions and hydroxide ions in the water and dissolved oxygen gas.

A very common application of electrochemistry is the use of batteries to store electrical energy for use on demand. Although the actual reactions taking place within most batteries are very complex, they can be understood using simple considerations of oxidation and reduction reactions. Dry cells, NiCad batteries and fuel cells are common devices used to supply electrical energy on demand. Research continues on ways to harness electrochemical energy that use lighter, cheaper materials to meet the personal and industrial demands of our increasingly energy intensive world.

ELECTROLYSIS

Electrolysis is the process of causing a chemical reaction to occur by passing an electric current through a substance or mixture of substances, most often in liquid form. Electrolysis frequently results in the decomposition of a compound into its elements. To carry out an electrolysis, two electrodes, a positive electrode (anode) and a negative electrode (cathode), are immersed into the material to be electrolyzed and connected to a source of direct (DC) electric current.

The apparatus in which electrolysis is carried out is called an electrolytic cell. The roots *-lys* and *-lyt* come from the Greek *lysis* and *lytos*, meaning to cut or decompose; electrolysis in an electrolytic cell is a process that can decompose a substance.

The substance being electrolyzed must be an electrolyte, a liquid that contains positive and negative ions and therefore is able to conduct **electricity**. There are two kinds of electrolytes. One kind is a **ion** compound **solution** of any compound that produces ions when it dissolves in **water**, such as an inorganic acid, base, or **salt**. The other kind is a liquefied ionic compound such as a molten salt.

In either kind of electrolyte, the liquid conducts electricity because its positive and negative ions are free to move toward the electrodes of opposite charge—the positive ions toward the cathode and the negative ions toward the **anode**. This transfer of positive charge in one direction and negative

Hoffman's apparatus for the electrolysis of water. ***(Photograph by Charles D. Winters, Photo Researchers, Inc. Reproduced by permission.)***

charge in the opposite direction constitutes an electric current, because an electric current is, after all, only a flow of charge, and it doesn't matter whether the carriers of the charge are ions or electrons. In an ionic solid such as **sodium** chloride, for example, the normally fixed-in-place ions become free to move as soon as the solid is dissolved in water or as soon as it is melted.

During electrolysis, the ions move toward the electrodes of opposite charge. When they reach their respective electrodes, they undergo chemical oxidation-reduction reactions. At the cathode, which is pumping electrons into the electrolyte, chemical **reduction** takes place—a taking-on of electrons by the positive ions. At the anode, which is removing electrons from the electrolyte, chemical **oxidation** takes place—a loss of electrons by the negative ions.

In electrolysis, there is a direct relationship between the amount of electricity that flows through the cell and the amount of chemical reaction that takes place. The more electrons are pumped through the electrolyte by the battery, the more ions will be forced to give up or take on electrons, thereby being oxidized or reduced. To produce one mole's worth of chemical reaction, at least one **mole** of electrons must pass through the cell. A mole of electrons, that is, 6.02×10^{23} of them, is called a faraday. The unit is named after **Michael Faraday**, the English chemist and physicist who discovered this relationship between electricity and chemical change. He is also credited with inventing the words anode, **cathode**, electrode, electrolyte, and electrolysis.

Various kinds of electrolytic cells can be devised to accomplish specific chemical objectives.

Perhaps the best known example of electrolysis is the electrolytic decomposition of water to produce **hydrogen** and oxygen:

$2H_2O$ + **energy** $\rightarrow 2H_2 + O_2$

Because water is such a stable compound, we can only make this reaction go by pumping energy into it—in this case, in the form of an electric current. Pure water, which doesn't conduct electricity very well, must first be made into an electrolyte by dissolving an acid, base, or salt in it. Then an anode and a cathode, usually made of **graphite** or some non-reacting metal such as **platinum**, can be inserted and connected to a battery or other source of direct current.

At the cathode, where electrons are being pumped into the water by the battery, they are taken up by water molecules to form hydrogen gas:

$4H_2O + 4e^- \rightarrow 2H_2 + 4OH^-$

At the anode, electrons are being removed from water molecules:

$2H_2O - 4e^- \rightarrow O_2 + 4H^+$

The net result of these two electrode reactions added together is

$2H_2O \rightarrow 2H_2 + O_2$

(Note that when these two equations are added together, the four H^+ ions and four OH^- ions on the right-hand side are combined to form four H_2O molecules, which then cancel four of the H_2O molecules on the left-hand side.) Thus, two molecules of water have been decomposed into two molecules of hydrogen and one **molecule** of **oxygen.**

The acid, base, or salt that makes the water into an electrolyte is chosen so that its particular ions cannot be oxidized or reduced (at least at the voltage of the battery), so they don't react chemically and serve only to conduct the current through the water. **Sulfuric acid**, H_2SO_4, is commonly used.

By electrolysis, common salt, **sodium chloride**, NaCl, can be broken down into its elements, sodium and chlorine. This is an important method for the production of sodium; it is used also for producing other **alkali metals** and **alkaline earth metals** from their salts.

To obtain sodium by electrolysis, we'll first melt some sodium chloride by heating it above its melting point of 1,474°F (801°C). Then we'll insert two inert (non-reacting) electrodes into the melted salt. The sodium chloride must be molten in order to permit the Na^+ and Cl^- ions to move freely between the electrodes; in solid sodium chloride, the ions are frozen in place. Finally, we'll pass a direct electric current through the molten salt.

The negative electrode (the cathode) will attract Na^+ ions and the positive electrode (the anode) will attract Cl^- ions, whereupon the following chemical reactions take place.

At the cathode, where electrons are being pumped in, they are being grabbed by the positive sodium ions:

$Na^+ + e^- \rightarrow Na$

At the anode, where electrons are being pumped out, they are being ripped off the chloride ions:

$Cl^- - e^- \rightarrow Cl$

(The chlorine atoms immediately combine into **diatomic** molecules, Cl_2.) The result is that common salt has been broken down into its elements by electricity.

Another important use of electrolysis is in the production of magnesium from sea water. Sea water is a major source of that metal, since it contains more ions of magnesium than of any other metal except sodium. First, magnesium chloride, $MgCl_2$, is obtained by precipitating magnesium hydroxide from seawater and dissolving it in **hydrochloric acid**. The magnesium chloride is then melted and electrolyzed. Similar to the production of sodium from molten sodium chloride, above, the molten magnesium is deposited at the cathode, while the chlorine gas is released at the anode. The overall reaction is $MgCl_2 \rightarrow Mg+Cl_2$.

Sodium hydroxide, NaOH, also known as lye and caustic soda, is one of the most important of all industrial chemicals. It is produced at the rate of 25 billion pounds a year in the U.S. alone. The major method for producing it is the electrolysis of brine or ''salt water,'' a solution of common salt, sodium chloride in water. Chlorine and hydrogen **gases** are produced as valuable byproducts.

When an electric current is passed through salt water, the negative chloride ions, Cl^-, migrate to the positive anode and lose their electrons to become chlorine gas.

$Cl^- + e^- \rightarrow Cl$

(The chlorine atoms then pair up to form Cl_2molecules.) Meanwhile, sodium ions, Na^+, are drawn to the negative cathode. But they don't pick up electrons to become sodium metal atoms as they do in molten salt, because in a water solution the water molecules themselves pick up electrons more easily than sodium ions do. What happens at the cathode, then, is

$2H_2O + 2e^- \rightarrow H_2 + OH^-$

The hydroxide ions, together with the sodium ions that are already in the solution, constitute sodium hydroxide, which can be recovered by evaporation.

This so-called *chloralkali* process is the basis of an industry that has existed for well over a hundred years. By electricity, it converts cheap salt into valuable chlorine, hydrogen and sodium hydroxide. Among other uses, the chlorine is used in the purification of water, the hydrogen is used in the **hydrogenation** of oils, and the lye is used in making soap and **paper**.

The production of aluminum by the Hall process was one of the earliest applications of electrolysis on a large scale, and is still the major method for obtaining that very useful metal. The process was discovered in 1886 by **Charles M. Hall**, a 21-year-old student at Oberlin College in Ohio, who had been searching for a way to reduce aluminum **oxide** to the metal. Aluminum was a rare and expensive luxury at that time, because the metal is very reactive and therefore difficult to reduce from its compounds by chemical means. On the other hand, electrolysis of a molten aluminum salt or oxide is difficult because the salts are hard to obtain in anhydrous (dry) form and the oxide, Al_2O_3, doesn't melt until 3,762°F (2,072°C).

Hall discovered that Al_2O_3, in the form of the mineral bauxite, dissolves in another aluminum mineral called cryolite, Na_3AlF_6, and that the resulting mixture could be melted fairly easily. When an electric current is passed through this molten mixture, the aluminum ions migrate to the cathode, where they are reduced to metal:

$Al^{3+} + 3e^- \rightarrow Al$

At the anode, oxide ions are oxidized to oxygen gas:

$2O^{2-} - 2e^- \rightarrow O_2$

The molten aluminum metal sinks to the bottom of the cell and can be drawn off.

The production of aluminum by the Hall process consumes huge amounts of fuel. The **recycling** of beverage cans and other aluminum objects has become an important energy conservation measure.

Unlike aluminum, copper metal is fairly easy to obtain chemically from its ores. But by electrolysis, it can be refined and made very pure-up to 99.999%. Pure copper is important in making electrical wire, because copper's **electrical conductivity** is reduced by impurities. These impurities include such valuable metals as **silver**, **gold** and platinum; when they are removed by electrolysis and recovered, they go a long way toward paying the electricity bill.

In the electrolytic refining of copper, the impure copper is made from the anode in an electrolyte bath of copper **sulfate**, $CuSO_4$, and sulfuric acid H_2SO_4. The cathode is a sheet of very pure copper. As current is passed through the solution, positive copper ions, Cu^{2+}, in the solution are attracted to the negative cathode, where they take on electrons and deposit themselves as neutral copper atoms, thereby building up more and more pure copper on the cathode. Meanwhile, copper atoms in the positive anode give up electrons and dissolve into the electrolyte solution as copper ions. But the impurities in the anode do not go into solution because silver, gold and platinum atoms are not as easily oxidized (converted into positive ions) as copper is. So the silver, gold and platinum simply fall from the anode to the bottom of the tank, where they can be scraped up.

Another important use of electrolytic cells is in the electroplating of silver, gold, **chromium** and **nickel**. Electroplating produces a very thin coating of these expensive metals on the surfaces of cheaper metals, to give them the appearance and the chemical resistance of the expensive ones.

In silver plating, the object to be plated (let's say a spoon) serves as the cathode of an electrolytic cell. The anode is a bar of silver metal, and the electrolyte (the liquid in between the electrodes) is a solution of silver cyanide, AgCN, in water. When a direct current is passed through the cell, positive silver ions (Ag^+) from the silver cyanide migrate to the negative anode (the spoon), where they are neutralized by electrons and stick to the spoon as silver metal:

$2H_2O + \text{energy} \rightarrow 2H_2 + O_2$

Meanwhile, the silver anode bar gives up electrons to become silver ions:

$Ag - e^- \rightarrow Ag^+$

Thus, the anode bar gradually dissolves to replenish the silver ions in the solution. The net result is that silver metal has been transferred from the anode to the cathode, in this case the spoon. This process continues until the desired coating thickness is built up on the spoon-usually only a few thousandths of an inch-or until the silver bar has completely dissolved.

In electroplating with silver, silver cyanide is used in the electrolyte rather than other compounds of silver such as silver nitrate, $AgNO_3$, because the cyanide ion, CN^-, reacts with silver ion, Ag^+, to form the complex ion $Ag(CN)_2^-$. This limits the supply of free Ag^+ions in the solution, so they can deposit themselves only very gradually onto the cathode. This produces a shinier and more adherent silver plating. Gold plating is done in much the same way, using a gold anode and an electrolyte containing gold cyanide, AuCN.

ELECTROMAGNETIC SPECTRUM

Visible light is just a small part of a continuum of electromagnetic radiation that extends from radio waves to gamma rays. Phenomena as diverse as radio waves used to transmit information, microwaves used to cook food, and x rays used for medical purposes are all part of the electromagnetic spectrum. Although each type of radiation has distinct properties, they can all be described in the same simple terms.

The characteristic of electromagnetic radiation that causes its properties to vary is its wavelength. The wave-like properties of electromagnetic radiation are similar to the waves created when an object is dropped into **water**. The wavelength of the wave is the distance between two successive peaks. The wavelength of radiation is sometimes given in units with which we are familiar, such as inches or centimeters, but for very small wavelengths, they are often given in angstroms (Å). (10,000,000,000 Å equals 1 m.)

An alternative way of describing a wave is by its frequency—the number of peaks that pass a particular point in one second. Frequencies are normally given in cycles per second (hertz [Hz]), named after Heinrich Hertz who was the first to artificially generate radio waves. Other common units are kilohertz (kHz, or thousands of cycles per second), megahertz (MHz, millions of cycles per second), and gigahertz (GHz, billions of cycles per second). The frequency and wavelength, when multiplied together, give the speed of the wave. For electromagnetic waves in empty space, that speed is the speed of light, which is approximately 186,000 miles per second (300,000 km per sec).

In addition to the wave-like properties of electromagnetic radiation, it also can behave as a particle. The **energy** of a light particle or **photon**, can be calculated from its frequency by multiplying by **Planck's constant**. Thus, higher frequencies (and shorter wavelengths) have higher energy. A common unit used to describe the energy of a photon is the **electron** volt (eV). Multiples of this unit, such as keV (1,000 electron volts) and MeV (1,000,000 eV), are also used.

Properties of waves in different regions of the spectrum are commonly described by different notation. Visible radiation is usually described by its wavelength, for example, while x rays are described by their energy. All of these schemes are equivalent, yet they are just different ways of describing the same properties.

The electromagnetic spectrum is typically divided into wavelength or energy regions, based on the characteristics of the waves in each region. Because the properties vary on a continuum, the boundaries are not sharp, but are rather loosely defined.

Radio waves are familiar to us due to their use in communications. The standard AM radio band is at 540-1650 kHz, and the FM band is 88-108 MHz. This region also includes shortwave radio transmissions and television broadcasts.

We are most familiar with microwaves because of microwave ovens, which heat food by causing water molecules to rotate at a frequency of 2.45 GHz. In astronomy, emission of radiation at a wavelength of 8.2 in (21 cm) has been used to map neutral **hydrogen** throughout the galaxy. Radar is also included in this region.

The infrared region of the spectrum lies just beyond the visible wavelengths. It was discovered by William Herschel in 1800 by dispersing sunlight with a prism, and measuring the **temperature** increase just beyond the red end of the spectrum.

The visible wavelength range is the range of frequencies with which we are most familiar. These are the wavelengths to which the human eye is sensitive, and which most easily pass through the Earth's atmosphere. This region is further broken down into the familiar colors of the rainbow, which fall into the wavelength intervals.

A common way to remember the order of colors is through the name of the fictitious person ROY G. BIV (the I stands for indigo).

The ultraviolet range lies at wavelengths just shortward of the visible. Although we do not use UV to see, it has many other important effects on Earth. The ozone layer high in Earth's atmosphere absorbs much of the UV radiation from the sun, but that which reaches the surface can cause suntans and sunburns.

We are most familiar with x rays due to their uses in medicine. X radiation can pass through the body, allowing doctors to examine bones and teeth. Surprisingly, x rays do not penetrate Earth's atmosphere, so astronomers must place x-ray telescopes in space.

Gamma rays are the most energetic of all electromagnetic radiation, and we have little experience with them in everyday life. They are produced by nuclear processes, for example, during radioactive decay or in **nuclear reactions** in stars or in space.

ELECTRON

An electron is one of the three **subatomic particles** which make up an **atom**. The other two subatomic particles are protons and neutrons, which are found in the **nucleus** (center) of the atom. Electrons are found in the area outside the nucleus of an atom. They have a negative electrical charge (as opposed to protons which are positively charged and neutrons which have no charge) and are quite small. An electron is only about 1/2,000 the size of a **proton** or a **neutron**. Protons and neutrons have masses of approximately 1 atomic **mass** unit (amu) each, whereas electrons only have a mass of.0006 amu (9.11×10^{-28} g). The **atomic number** of an element is equal to the number of protons in one atom of the element. In a neutral atom, the number of electrons is equal to the number of protons, therefore, the atomic number also indicates the number of electrons in an atom. An atom can gain or lose electrons at which point it gains an electric charge and is called an **ion**.

Exactly where the electrons in an atom are located has been the topic of much research and debate. Until the early

1800s the idea that elements were made up of smaller particles called atoms was unknown. English chemist **John Dalton** performed various experiments that led him to develop his **atomic theory**. This theory stated that all elements are made up of atoms. According to Dalton, atoms were the smallest particles possible and could not be divided into smaller particles. Atoms of the same element were exactly alike, and atoms of different atoms were different. Atoms of different elements combined to form compounds. Protons, neutrons, and electrons were still undiscovered.

In 1897, English physicist **J.J. Thomson** was experimenting with passing an electric current through a gas. The gas was made of uncharged atoms, but when an electric current passed through it, negatively charged particles in the form of rays were given off. Thomson was faced with the problem of explaining where these negatively charged particles originated. He reasoned that the only place these particles could have come from was the individual atoms in the gas. Therefore, the atom must be made up of even smaller particles. He used the term *corpuscles* to describe the negatively charged particles that we now call electrons.

Now that Thomson had identified electrons, the problem he faced was determining the placement of these particles inside the atom. Thomson developed a model of electrons being scattered throughout a positively charged material, much like plums would be scattered throughout plum pudding (this model of the atom is often referred to as the ''plum pudding'' model). In 1911, British physicist **Ernest Rutherford** proposed that there are positively charged particles in an atom called protons. The protons are located in the nucleus of an atom, and the electrons are scattered outside of the nucleus around the edge of the atom. The negatively charged electrons are held in place by the force that results from the attraction toward the positively charged protons. This force is called the electromagnetic force. This model explained further the idea of subatomic particles, but still could not describe the exact location of the electrons in an atom.

A few years later, in 1913, Danish physicist **Niels Bohr** improved upon the Rutherford model. He developed the idea of **energy** levels in an electron. According to this new atomic model, the electrons were placed in definite orbits—called energy levels—around the nucleus. Each **energy level** is a certain distance away from the nucleus. This model likens electrons orbiting the nucleus to planets orbiting the Sun. Today, scientists realize that this is not exactly correct. Electrons do not move in a definite orbit, always a certain distance from the nucleus. In fact, the exact location of the electrons in an atom cannot be determined. Only the approximate location where an electron is likely to be found can be predicted.

The electrons in an atom take up a particular amount of space as they move around the nucleus at a rate of billions of time per second. This space is referred to as the electron cloud. The electron cloud is not a definite shape or size, rather it is simply the space where electrons are likely to be located. Each electron appears to stay in a certain area within the cloud. The area of the cloud in which any particular electron is likely to be found depends upon how much energy the electron has. This area is an energy level within the electron cloud. Electrons with low energies are in the lowest energy level, closest to the nucleus. Electrons with higher energies are in higher energy levels, increasing in distance from the nucleus as energy increases. Each energy level can hold a certain number of electrons. The lowest energy level can hold only two electrons; the second level, eight; and the third level, 18.

How the energy levels are filled in an atom determines the properties of the different elements. This electron arrangement is important in predicting the chemical properties of an element. One of the most important properties determined by electron arrangement is the **bonding** characteristics of an element. Some elements bond readily to other elements and others hardly bond at all. This is determined by the number of electrons—called **valence** electrons—present in the outermost energy shell. An atom is most stable when its valence electrons satisfy what is called the **octet rule**. This means that an atom with eight electrons in its outermost energy level is very stable. An atom will bond with another in order to satisfy the octet rule. When two atoms combine, they can gain or lose electrons. When one atom transfers electrons to another, it is called ionic bonding. When electrons are shared between two atoms, it is called covalent bonding.

The number of valence electrons of an atom can also be used to group elements together into families on the **Periodic Table**. Elements within a family share many properties. **Metals** have one, two, three, or four valence electrons. These electrons are in the outer energy level quite weakly, so they are lost easily. This is the reason metals react readily with **water** or other atmospheric elements in a reaction known as **corrosion**. Elements that are nonmetals have five, six, seven, or eight valence electrons. Nonmetals tend to gain electrons during chemical reactions. Nonmetals with eight valence electrons are chemically unreactive.

The first family on the periodic table is the **alkali metals**. Atoms of these elements have only one valence electron, so they tend to bond easily. In fact, they are so reactive that they are rarely found in nature not combined with other elements. Family 2 on the periodic table is the **alkaline earth metals** with two valence electrons. They lose these electrons easily. Since they have two electrons that they must lose, they are not quite as reactive as the alkali metals. The **transition elements** have one or two valence electrons, which they lose when they react. They can also lose an electron from the energy level that is next to the highest. Atoms of these elements can also share electrons when they react with other elements. Family 13, the **boron** family, has three valence electrons. Family 14 is known as **carbon** family, and atoms of these elements have four valence electrons. Family 15, the **nitrogen** family, has five valence electrons. Atoms of these elements tend to gain or acquire electrons when they react with other atoms. Family 16 is called the **oxygen** family. Atoms of these elements have six valence electrons and tend to gain or acquire electrons in chemical reactions. Members of the halogen family, family 17, have seven valence electrons. A halogen atom only needs to gain one electron to fill its outer electron shell. This property makes elements in family 17 the most active nonmetals. These elements are also rarely found uncombined in nature. **Halogens** tend to react with the alkali metals rather easily. The last family, family 18, contains the noble **gases**. These are the elements with eight valence electrons, or complete energy levels. These elements are unreactive.

Electrons not only determine an element's **reactivity**, they also are the cause of various phenomena, for example, **beta radiation**. Some electrons can be formed inside of the nucleus of an atom when a neutron breaks apart. When this occurs, the electron shoots out of the atom and is called a beta particle, a type of **radioactivity**. Another example of an electron-caused phenomenon is static **electricity**. Sometimes the electrons in the outermost energy levels are easily lost from an atom. When an atom loses an electron, it becomes positively charged. When it gains an electron, it becomes negatively charged. An object can also become charged if the atoms of which it is composed gain or lose electrons. An example of an entire object becoming charged is when a balloon is rubbed with a piece of cloth. The cloth loses electrons that are transferred to the balloon, and the balloon acquires a negative charge. When the balloon is held up to a wall, the negative charge causes the electrons in the wall to move away from the area. That area of the wall then becomes positively charged. The negative balloon is attracted to the positive wall, and will be held on the wall by the electromagnetic force.

This example illustrates the phenomenon of induction—an electrical charge built up due to the rearrangement of atoms. The wall became charged because of induction. When electrons flow from one object to another it is called conduction. Certain materials—called conductors—allow for this electron flow better than others. Metals are good conductors because the electrons in these atoms are loosely held and free to move. Materials that do not allow for electrons to flow are called insulators. Insulators do not conduct electricity well because the electrons are tightly held and are not free to move.

Another well known phenomenon that occurs because of electrons is lightning. Clouds can build up a negative charge, and if they pass near the surface of the Earth, the Earth can become positively charged because of induction. Electrons then are attracted to the positively charged Earth and jump from the cloud. As they move through the air, they produce a great quantity of light and **heat**. The light is the lightning bolt we see. The heat causes the air around it to expand quickly, which is the thunder we hear.

Many electronic devices and equipment are made possible because of electrons. Televisions, computer monitors, and video games all use a device called a cathode ray tube. The cathode ray tube is a vacuum tube that emits a beam of electrons onto a screen. As the electrons hit the screen, which is coated with a fluorescent material, they cause the screen to glow, producing an image. Photocells in cameras are based on a phenomenon called the **photoelectric effect**. When light shines on a metal, the loosely held electrons are shot off of the surface of the metal. For any particular metal, a particular frequency of light must be used for the photoelectric effect to take place.

See also Covalent bond; Ionic bond

ELECTRON DIFFRACTION

When a beam of electrons passes through a material, an electron-sensitive film placed beyond the sample will show, when developed, a pattern of concentric circles around the central intense **electron** beam. This phenomenon is due to the diffraction of the electron beam by the material through which it passes. C. Davisson (1881-1958), L. H. Germer (1896-1971), and George P. Thomson (1892-1975) used this method to prove the hypothesis of **Louis de Broglie** that electrons have wave characteristics. Further research on electron diffraction showed that the angles of the diffracted electrons and the intensity of the diffracted electron beams can be used to determine the structure of the molecules of the gas.

Diffraction is a phenomenon that is characteristic of all types of waves. Perhaps the simplest case to understand is the diffraction of monochromatic light. Whenever an advancing wavefront of light encounters a barrier in which there is a hole, this hole acts as if it were a point source of light, with radiation of the same wavelength moving beyond the barrier away from the hole. When the barrier contains a series of holes, then each hole acts as a point source. The radiation observed beyond the barrier will be the sum of the radiation from all of these individual point sources. If the holes are uniformly placed at distances similar to the wavelength of the light, the waves from all these individual point sources will add together to produce beams of light only in certain definite directions. This is caused by constructive interference and results from the circumstance that the waves from all the point sources are in phase only in these directions. In other directions, destructive interference occurs among the waves, and they tend to cancel each other out. This phenomenon is called diffraction.

If, instead of holes, a series of parallel narrow slits are cut in the barrier at uniform distances similar in magnitude to the wavelength of the impending wave, the result is a diffraction grating. A photographic film placed a short distance beyond the grating will show a series of lines when exposed. The direction of each line on the film from the diffraction grating is given by an equation which relates the sine of the angle of the diffracted line with an integral multiple of the wavelength divided by the distance between the slits. Joseph von Fraunhofer (1787-1826) invented the diffraction grating. The results of his experiments on the diffraction of light played a large part in convincing the scientific community that light was made up of waves.

In 1859, Julius Plücker (1801-1868) discovered that when a high voltage difference is established between two electrodes in a container of dilute gas or in a vacuum, an electrical discharge occurs between the electrodes and cathode rays are produced. In 1897, **Joseph John Thomson** showed that cathode rays are made up of electrons. In subsequent experiments, the **mass** of the electron was shown to be 9.1×10^{-28} gram with an electrical charge of 1.6×10^{-19} coulomb.

Max von Laue (1879-1960) recognized that the regularly spaced planes of atoms and molecules in crystals resemble a diffraction grating. In 1912, he demonstrated that x rays, with wavelengths in the range of 10^{-8} cm, comparable to the magnitude of crystal spacings, was diffracted when passed through crystalline substances.

In 1923, de Broglie proposed that moving particles possess some of the properties commonly associated with waves. He predicted that the wavelength of such a particle wave would be equal to **Planck's constant** (6.6×10^{-27} erg-second) divided by the product of the particle's mass and the **velocity** with which the particle is moving (the particle's momentum). For a "particle" with a mass as large as a baseball, the wavelength would be very small indeed, and could not be detected. For a particle whose mass is of similar magnitude to Planck's constant, such as an electron, the wavelength should be large enough for the effects of its wave nature to be observed.

The wave length that de Broglie predicted for electrons is determined by their velocity and can be adjusted to the same order of magnitude as atomic separations in molecules and crystals by changing the voltage of the charged electrodes between which the electrons move. Davisson, Germer, and Thomson employed this approach when they proved the wave nature of electrons. They used von Laue's observation that the regular spaced atoms in crystals act as a diffraction grating for wavelengths of the order of magnitude of 10^{-8} cm. Diffraction of a beam of electrons by a crystal of **nickel** was observed when the voltage of the electrodes was adjusted to produce electron beams with the appropriate wavelength.

Von Laue showed that x rays are diffracted in specific directions determined by the distances of separation between the parallel planes of molecules in crystals. X-ray crystallographers subsequently demonstrated that the intensities of these diffracted beams were determined by the arrangement of the atoms within the molecules themselves.

A similar phenomenon is observed when electron beams pass through **gases**. A diffraction pattern of concentric circles is obtained from an exposed electron-sensitive film placed beyond the diffracting gas. A detailed analysis of this scattering, begun by R. Wierl in his pioneering research in and expanded by other researchers, led to an equation (known as the Wierl equation) that relates the intensity of the scattered beam at a particular angle to the inter-atomic distances within the molecules. Refinements developed by **Linus Pauling**, L.O. Brockway, and others have resulted in techniques which have been useful in determining molecular structures with accurate values of interatomic distances and angles.

Further advances in theory, technique, and instrumentation have led to the development of electron microscopes that are used to examine the fine structure of biological samples.

ELECTRON MICROSCOPE

Described by the Nobel Society as "one of the most important inventions of the century," the **electron** microscope is a valuable and versatile research tool. The first working models were constructed by German engineers Ernst Ruska and Max Knoll in 1932, and since that time, the **electron microscope** has found numerous applications in **chemistry**, engineering, and medicine.

At the turn of the twentieth century, the science of **microscopy** had reached an impasse: because all microscopes relied upon visible light, even the most powerful could not detect an image smaller than one wavelength of light. This was tremendously frustrating for physicists, who were anxious to study the structure of **matter** on an atomic level. Around this time, French scientist, **Louis de Broglie** theorized that **subatomic particles** sometimes act like waves, but with much shorter wavelengths. Ruska, then a student at the University of Berlin, wondered why a microscope couldn't be designed that was similar in function to a normal microscope but used a beam of electrons instead of a beam of light. Such a microscope could resolve images thousands of times smaller than a wavelength.

There was one major obstacle to Ruska's plan, however. In a **compound microscope**, a series of lenses are used to focus, magnify, and refocus the image. In order for an electron-based instrument to perform as a microscope, some device was required to focus the electron beam. Ruska knew that electrons could be manipulated within a magnetic field, and in the late 1920s, he designed a magnetic coil that acted as an electron lens. With this breakthrough, Ruska and Knoll constructed their first electron microscope. Though the prototype model was capable of magnification of only a few hundred power (about that of an average laboratory microscope), it proved that electrons could indeed be used in microscopy.

The microscope built by Ruska and Knoll is very similar in principle to a compound microscope. A beam of electrons is directed at a specimen sliced thin enough to allow the beam to pass through. As they travel through, the electrons are deflected according to the **atomic structure** of the specimen. The beam is then focused by the magnetic coil onto a photographic plate; when developed, the image on the plate shows the specimen at very high magnification.

Scientists worldwide immediately embraced Ruska's invention as the breakthrough in optical research, and they directed their own efforts toward improving upon its precision and flexibility. A Canadian-American physicist, James Hillier, constructed a microscope from Ruska's design that was nearly 20 times more powerful. In 1939, modifications made by Vladimir Kosma Zworykin enabled the electron microscope to be used for studying viruses and protein molecules. Eventually, electron microscopy was greatly improved, with microscopes able to magnify an image 2,000,000 times. One particularly interesting outcome of such research was the invention of holography and the hologram by Hungarian-born engineer Dennis Gabor in 1947. Gabor's work with this three-dimensional photography found numerous applications upon development of the laser in 1960.

There are now two distinct types of electron microscopes: the transmission variety (such as Ruska's), and the scanning variety. Scanning electron microscopes, instead of being focused by the scanner to peer through the specimen, are used to observe electrons that are scattered from the surface of the specimen as the beam contacts it. The beam is moved along the surface, scanning for any irregularities. The scanning electron microscope yields an extremely detailed three-dimensional image of a specimen but can only be used at low resolution; used in tandem, the scanning and transmission electron microscopes are powerful research tools.

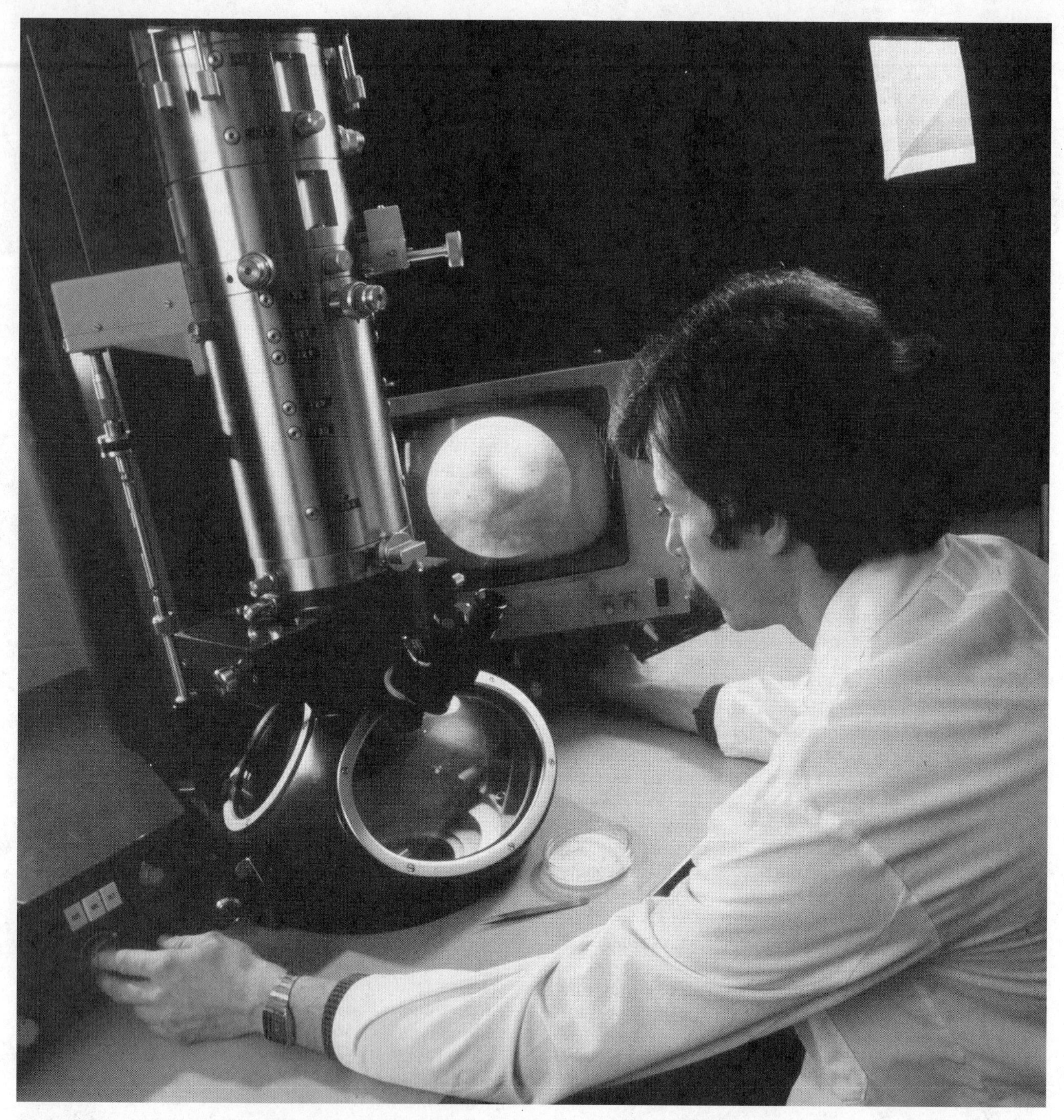

A biochemistry researcher working at a transmission electron microscope (TEM). This particular TEM is used to micro-engrave an aluminum fluoride target. ***(Photograph by James King/Science Photo Library, Photo Researchers, Inc. Reproduce by permission.)***

Today, electron microscopes can be found in most hospital and medical research laboratories. One of the more interesting applications for the invention is that by petroleum companies, who use electron microscopy to study the molecular links in **petrochemicals**. The results of their studies help in the search for petroleum substitutes and synthetic fuels, especially those derived from **coal**.

The advances made by Ruska, Knoll, and Hillier have contributed directly to the development of the **field ion microscope** (invented by Erwin Wilhelm Muller) and the **scanning tunneling microscope** (invented by Heinrich Rohrer and Gerd Binnig), now considered the most powerful optical tools in the world. For his work, Ruska shared the 1986 Nobel Prize for physics with Binnig and Rohrer.

Electronegativity

Electronegativity is a measure of the ability of an **atom** to attract and hold electrons. **Robert Mulliken** suggested that electronegativity could be expressed essentially as an average of the **ionization** potential and the **electron** affinity of the atom. The ionization potential is the minimum amount of **energy** necessary to remove an electron completely from the outer electron shell of an atom. The electron affinity is the energy given off when a neutral atom acquires an electron. **Linus Pauling** proposed a somewhat different approach for determining electronegativies, but the values calculated by the two methods are in good agreement.

Atoms with widely different electronegativities will form ionic bonds, since the atom with the highest eletronegativity is able to pull an electron away from the atom with the lower electronegativity. For atoms whose electronegativity difference is not large, the atom with the higher electonegativity is not able to pull an electron completely away from the other atom, but it will pull the electrons toward itself and they will spend more time in its vicinity. The result is a partial negative charge on the atom with the larger electronegativity and an equal partial positive charge on the atom with the lower electronegativity. The resulting bond is called a polar **covalent bond**, the electrons of the bond are shared by the two atoms, but the atom with greater electronegativity will have greater possession of them.

The electronegativities of the atoms **chlorine** (Cl), **carbon** (C), and **sodium** (Na) are 3.0, 2.5, and 0.9 respectively. The electonegativity difference between C and Cl is 0.5; between Na and Cl the difference is 2.1; between C and C, it is 0.0. The C-Cl bond is polar covalent, Na-Cl is ionic, and the C-C bond is pure covalent with each atom sharing the **bonding** electrons equally.

If there is a separation of charge in an bond, it possesses a **dipole**. The relative magnitudes of the molecular dipoles can be explained by the relative electronegativity differences of the atoms forming the bonds.

The relative magnitude of an atom's electronegativity can be understood by considering its **atomic structure**. The electrons of an atom are associated with atomic orbitals. In the atom's lowest energy state, the inner orbitals are filled with electrons, and only the highest energy orbitals are partially filled. Only the electrons in the outer orbitals are shared in molecular bonds; it is these electrons that may be pulled away by other atoms to form ions (ionization potential), and it is into these outer orbitals that electrons may be attracted from other atoms (electron affinity). An atom's electronegativity is its ability to pull electrons into these outer orbitals and hold onto them. Its ability to do so depends on the net electrostatic attraction that electrons in these outer orbitals feel from the **nucleus**.

The energy of attraction between opposite charges is reciprocally related to the distance between the charges. The further electrons are away from the positively charged nucleus, the less attraction they feel. Therefore, the electrons in the outer orbitals feel less of an attraction than electrons in the inner orbitals. Also the electrons in inner shells shield outer electrons from the full attractive force of the nucleus.

Only two electrons may be located in each orbital. The atomic orbital of lowest energy is called the 1s orbital. It contains two electrons. The orbitals of successively higher energy and greater distance from the nucleus are: the 2s orbital containing two electrons, three 2p orbitals containing a total of six electrons, one 3s orbital with two electrons, three 3p orbitals with six electrons, five 3d orbitals containing a total of 10 electrons, and so forth to higher and higher energy levels further and further from the nucleus.

With these ideas in mind, consider the nuclear attraction felt by electrons in the outer shell of carbon and sodium atoms. Carbon has six electrons and a relative nuclear charge of 6. Two of its electrons occupy the lowest 1s orbital, and two will be in the 2s orbital. The partially filled outer **energy level** containing the two other electrons is 2p. Sodium has eleven electrons: 2 in 1s, 2 in 2s, 6 in 2p, and the outer electron is in the 3s orbital. In comparing the relative ability of carbon and sodium to attract and hold electrons in their outer shells, first notice that the outer shell of sodium is further from the nucleus. Next, notice that in sodium there are ten electrons in orbitals of lower energy, closer to the nucleus, shielding the attractive force of the nucleus from the outer shell electrons. In carbon, there are only four inner shell electrons shielding the attractive force of the nucleus from the outer electrons. The net effect is, therefore: in carbon the electrons attracted and held in the outer shell will feel the attraction of a nuclear charge of 6 shielded by four inner electrons (a net unshielded nuclear charge of 2), whereas electrons in the outer shell of sodium will feel the attraction of a nuclear charge of 11 shielded by 10 inner electrons (a net charge of 1). The outer electrons of sodium are also at a greater distance from the nucleus than those in carbon and, therefore, feel less electrostatic attraction for that reason, as well. The result is that the outer shell electrons of sodium feel less of the attraction of their nucleus. This somewhat simplified discussion predicts that the electronegativity of carbon should be greater than that of sodium. The observed values are 2.5 and 0.9 respectively.

Electrophile

An electrophile is any compound or reaction intermediate that is electron-deficient and thus accepts electrons from another source. The word electrophile is derived from the Greek words *electros* meaning **electron**, and *philos* meaning loving. Coined by Sir **Christopher Ingold**, the common notation used by chemists to represent an electrophile is E+, where E stand for the electrophile and the positive sign emphasizes its electron deficiency.

However, electrophiles can be neutral, partially positively charged, or have a full positive charge. Some examples of positively charged electrophiles are the **hydronium ion**, nitronium ions, metal ions, and **carbocations**. Some examples of neutral electrophiles are ozone (O_3), Lewis acids (such as BF_3), sulfonating agents, halogenating agents (such as Cl_2), and peracids. Electrophiles can also be reaction intermediates that do not have an octet of **valence** electrons, such as **car-**

benes, nitrenium ions, and halogen atoms. Ingold was among the earliest chemists to recognize that many chemical mechanisms may be understood in terms of electrophiles reacting with nucleophiles.

Electrophiles are encountered most often in organic **chemistry**, and they are involved in many general reaction groups. Among the reactions involving electrophiles are electrophilic addition to multiple bonds (alkenes and alkynes), reactions with alkanes (as rearrangements, free radical halogenation, and reactions with peracids), and electrophilic aromatic reactions (such as Friedel-Crafts alkylation and acylation).

The most fundamental electrophile in **organic chemistry** is the **proton** written as H^+ or H_3O^+ (the hydronium ion). The addition of the proton to a double bond creates a carbocation that, in turn, is subject to nucleophilic attack. Electrophilic addition to a multiple bond is one example of the use of the proton as an electrophile. The first step is the addition of the proton to the pi bond of the alkene or alkyne, acting as the electron source. This leads to an electron-deficient carbocationic intermediate. In the second step, this carbocation acts as the electrophile and bonds with a **nucleophile**.

Rearrangements of hydrocarbons are often found in the presence of Lewis acids. One example is the complete skeletal rearrangement of *endo*-tetrahydrodicyclopentadiene to adamantane in the presence of $AlCl_3$. Although the mechanism is uncertain, it has been found that this reaction is initiated by electrophiles that generate carbocationic intermediates. Hydride abstraction processes initiated by these electrophiles are the origin of these carbocation intermediates.

Carbocations make excellent electrophiles, and in the presence of Lewis acids they are able to substitute onto an aromatic **ring**. Two men discovered that in the presence of Lewis acids (such as **aluminum** chloride or ferric chloride), alkyl halides will alkylate benzenes. French chemist Charles Friedel (1832-1899) and American chemist James Crafts (1839-1917) were the first scientists to study this aromatic addition in 1877. In this electrophilic aromatic substitution, the first electrophile is the Lewis acid that accepts the halogen from an alkyl **halide** (the nucleophile). This produces the second electrophile, the positively charged **alkyl group**. This carbocation attacks the pi system of the **benzene** ring leading to an electron-deficient arenium ion. The arenium intermediate loses a proton to reestablish the **resonance** of the $6e^-$ pi system.

There are many important industrial processes that use electrophilic reactions. One important reaction is the production of isooctane (2,2,4-trimethylpentane) for **gasoline**. This reaction is initiated by the electrophilic protonation of isobutylene. Electrophilic addition of the t-butyl **cation** to isobutylene, followed by hydride transfer from isobutane completes the process and regenerates the t-butyl cation. The end result is the addition of isobutane to isobutylene to form isooctane, the **molecule** that is the basis of octane ratings seen at the gas pumps.

In the mining industry, the production of TNT (trinitrotoluene) is made possible by the electrophilic nitration of toluene (methyl benzene). A nitronium ion, made by treating **nitric acid** with **sulfuric acid**, is the electrophile. The nitronium ion attacks the pi electrons of the benzene ring to give—by electrophilic aromatic substitution—nitrotoluene. This nitration is repeated an additional two times (with each nitration requiring increasingly rigorous reaction conditions) to yield the 1-methyl-2,4,6-trinitrobenzene as the final product. The presence of the three stongly electrophilic nitro substituents in the TNT molecule makes the molecule reactive (in this case, as an explosive).

The concept of an electrophile is very general one. There are literally hundreds of reagents not discussed here that can be categorized as electrophiles. In each case, electrophiles are simply molecules or ions in search of electrons.

ELECTROPHORESIS

Electrophoresis is a sensitive analytical form of **chromatography** that allows the separation of charged molecules in a **solution** medium under the influence of an electric field.

The degree of separation and rate of molecular migration of mixtures of molecules depends upon the size and shape of the molecules, the respective molecular charges, the strength of the electric field, the type of medium used (e.g., **cellulose** acetate, starch gels, **paper**, agarose, polyacrylamide **gel**, etc.) and the conditions of the medium (e.g., electrolyte **concentration**, **pH**, **ionic strength**, **viscosity**, **temperature**, etc.)

Some mediums (also known as support matrices) are porous gels that can also act as a physical sieve for macromolecules.

A wide range of molecules may be separated by electrophoresis, including, but not limited to **DNA**, **RNA**, and protein molecules.

In general, the medium is mixed with buffers needed to carry the electric charge applied to the system. The medium/buffer matrix is placed in a tray. Samples of molecules to be separated are loaded into wells at one end of the matrix. As electrical current is applied to the tray, the matrix takes on this charge and develops positively and negatively charged ends. As a result, molecules such as DNA and RNA that are negatively charged, are pulled toward the positive end of the gel.

Because molecules have differing shapes, sizes, and charges, they are pulled through the matrix at different rates and this, in turn, causes a separation of the molecules. Generally, the smaller and more charged a **molecule**, the faster the molecule moves through the matrix.

When DNA is subjected to electrophoresis, the DNA is first broken by restriction enzymes that act to cut the DNA in selected places. After being subjected to restriction enzymes, DNA molecules appear as bands (composed of similar length DNA molecules) in the electrophoresis matrix. Because **nucleic acids** always carry a negative charge, separation of nucleic acids occurs strictly by molecular size.

Proteins have net charges determined by charged groups of **amino acids** from which they are constructed. Proteins can also be amphoteric compounds—a compound that can take on a negative or positive charge depending on the surrounding conditions. A protein in one solution might carry a positive charge in a particular medium and migrate toward the negative end of the matrix. In another solution, the same protein might carry a negative charge and migrate toward the positive end of the matrix. For each protein there is an isoelectric point related to a pH characteristic for that protein where the protein molecule has no net charge. By varying pH in the matrix, additional refinements in separation are possible.

The advent of electrophoresis revolutionized the methods of protein analysis. Swedish biochemist **Arne Tiselius** was awarded the 1948 Nobel Prize in **chemistry** for his pioneering research in electrophoretic analysis. Tiselius studied the separation of serum proteins in a tube (subsequently named a Tiselius tube) that contained a solution subjected to an electric field.

Sodium dodecyl **sulfate** (SDS) polyacrylamide gel electrophoresis techniques pioneered in the 1960s provided a powerful means of protein fractionation (separation). Because the protein bands did not always clearly separate (i.e., there was often a great deal of overlap in the protein bands) only small numbers of molecules could be separated. The subsequent development in the 1970s of a two-dimensional electrophoresis technique allowed greater numbers of molecules to be separated.

Two-dimensional electrophoresis is actually the fusion of two separate separation procedures. The first separation (dimension) is achieved by isoelectric focusing (IEF) that separates protein polypeptide **chains** according to amino acid composition. IEF is based on the fact that proteins will, when subjected to a pH gradient, move to their isoelectric point. The second separation is achieved via SDS slab gel electrophoresis that separates the molecule by molecular size. Instead of broad, overlapping bands, the result of this two-step process is the formation of a two-dimensional pattern of spots, each comprised of a unique protein or protein fragment. These spots are subsequently subjected to staining and further analysis.

Some techniques involve the application of radioactive labels to the proteins. Protein fragments subsequently obtained from radioactively labels proteins may be studied my radiographic measures.

There are many variations on gel electrophoresis with wide-ranging applications. These specialized techniques include Southern, Northern, and Western Blotting. Blots are named according to the molecule under study. In Southern blots, DNA is cut with restriction enzymes then separated. In Northern blotting, RNA is separated. Western blots target proteins with radioactive or enzymatically-tagged **antibodies**.

Modern electrophoresis techniques now allow the identification of homologous DNA sequences and have become an integral part of research into gene structure, gene expression, and the diagnosis of heritable diseases. Electrophoretic analysis also allows the identification of bacterial and viral strains and is finding increasing acceptance as a powerful forensic tool.

See also Buffer; Chromatography; DNA fingerprinting

Electroplating

Electroplating is the process of coating an electrically conducting surface with a thin layer (seldom more than 0.001 in [0.025 mm] thick) of metal by electrolytic deposition. In electroplating, the object to be plated is made the cathode in an electrolytic bath of salts of the metal to be plated. The **anode** may be an unaffected metal, or, more commonly, the metal to be plated. An electric current is passed through the **solution**, which results in the deposition of a thin metal plating in the desired thickness on the cathode. Traces of organic materials are usually added to the plating solution to give a more adherent coating, though the reasons for this effect are not well understood.

Electroplating may be used to increase the value or improve the appearance of an object. For example, the technique is used to silver-plate table utensils and to weatherproof objects with **cadmium** or **chromium** plating. Coatings such as **zinc** and **tin** provide protection against **corrosion**. Other plated **metals** include **nickel**, **copper**, and **gold**.

While the electroplating is in progress, additional salts of the metal to be plated must continually be added to the plating solution, or else the anode must be renewed, if it consists of the plating metal. If the coating metal does not form a strong **alloy** with the metal to be plated, it may be necessary to first coat with an intermediate metal; for example, when plating **silver** on steel, it is customary to first place a coating of copper over the steel.

When plating gold or silver, it is customary to use a solution containing a double cyanide of the coating metal and **potassium** (the cyanide **ion** lowers the **concentration** of free metal ions, and prevents the plating from taking place too rapidly). In nickel plating, an electrolytic solution containing nickel ammonium **sulfate** may be used.

See also Electrical conductivity; Electrochemistry; Paints and coatings

Electrorefining and electrowinning

In the field of electrometallurgy, electrorefining and electrowinning (different parts of the same procedure) are important techniques that allow purification of nonferrous (noniron-based) **metals** to an extreme degree. Economical and straightforward, they use electrical current to separate metals from their accompanying impurities. The overall process is known as **electrolysis**, which literally means to break down using **electricity**. Electrorefining is the least expensive way to purify metals because it is so selective in terms of what it produces. In other words, it produces a pure metal in just one step, saving time and money. There are commonly used alternate names for both techniques: electrorefining is also called electrolytic refining, while electrowinning is also called electroextraction.

Electrorefining works by running a current through an electrolyte **solution** via an electrolytic cell. An **anode** is the positive electrode (conductor of electricity) through which electrons leave an electrolytic cell, and a cathode is the negative electrode through which electrons enter the cell. An electrolyte solution is a conductive medium containing the particular metal's salts; when current moves through the medium, ions (electrically charged atoms) move with it.

To purify a metal by electrorefining, it must be made the anode in the electrolytic cell. For instance, to purify a big chunk of **copper**, the chunk would be put into the electrolytic cell as the anode. There would also be a small piece of very pure copper in the cell to act as the cathode. As the electricity passed through the cell, electrons would be removed from the impure copper anode. Copper ions would form and dissolve into the electrolyte. Those same ions would then flow to the cathode, and after electrons were added, would end up as pure copper.

Some metals, including **arsenic** and **nickel**, do not collect at the cathode after leaving the anode, but remain in the electrolyte solution. They require a different method of collection involving **reduction** and purification of the electrolyte solution.

The collection process itself is called electrowinning. The amount of metal that is electrowon varies greatly according to the size of the original piece of ore at the anode, but large electrolytic cells can electrowin hundreds or thousands of pounds of a metal in one day.

Another big benefit of electrorefining is that after the process for the original target metal is complete, there might be valuable ''leftovers.'' For example, after electrorefining copper, there are usually tiny amounts of **silver** and **gold** that are released from the anode as the copper is oxidized. but which are not deposited at the cathode. These metals sink to the bottom of the electrolytic cell, where they form a substance called ''anode sludge'' or ''anode slime.'' Just like the copper, the gold and silver can be collected and purified, although they need to go through more steps to reach the same state. Otherwise, the ore's original impurities remain at the anode.

The most common electrowon metals are **lead**, copper, gold, silver, **zinc**, **aluminum**, **chromium**, **cobalt**, **manganese**, and the rare-earth, alkali, and alkaline-earth metals. All of these metals are essential ingredients in many industrial manufacturing processes, including those in the communications, transportation, and electronic fields. In the mid-1990s, the U.S. Department of Energy even began experimenting with using electrorefining to help the nation deal with its huge amounts of nuclear waste. Electrorefining could separate such dangerous heavy metals as **plutonium**, **cesium**, and **strontium** from the less-toxic bulk of the nuclear waste, not only leaving less to dispose of, but removing many materials that could harm the environment or potentially be made into weapons.

ELEMENT, CHEMICAL

A chemical element is a fundamental substance of the material world. Each element has an identity; for example, **gold** consists of only gold atom, and a gold atom is unlike any other **atom**. Indeed, a gold atom can be split, but the particles (electrons, protons, etc.) that constitute a gold atom are not gold. It could be said that subatomic particles are generic, interchangeable. Atoms, on the other hand, have an identity, and constitute the identity of an element.

Chemists use the concept of **atomic number** to identify particular elements. Atomic number is the number of protons in an atom's **nucleus**, which means that atomic number constitutes, so to speak, a particular atom's identity tag. Thus, when we ascertain that a particular atom has 11 protons, we can immediately identify it as a **sodium** atom. No other atom has 11 protons. Unfortunately, despite their clearly defined identity, chemical elements sometime exhibit puzzling behaviors that have led scientists to question the concept of a distinct chemical element. Namely, a particular element can occur as two or more **isotopes**. If sodium (atomic number 11) has 11 protons, we assume that the number of neutrons will also be 11, giving us a total of 22, which chemists call atomic **mass**. Ideally, atomic mass should be the atomic number multiplied by two, but that is not always the case. In fact, **beryllium** (atomic number 4) usually occurs with a **mass number** of nine. Isotopes can thus be defined as atoms of the same element showing a discrepancy between atomic number (which never changes) and mass number.

A **hydrogen** (atomic number 1) atom does not have a **neutron**, which means that its atomic number and atomic mass are the same. When a neutron is added, hydrogen's atomic mass rises to two, yielding **deuterium**, an isotope which is necessary for the manufacture of the hydrogen bomb. However, if a **proton** is added to the hydrogen atom, a new element, **helium** is created. An additional proton will create **lithium** (atomic number 3), a metal widely used in industry, and so on.

There are 92 elements occurring in nature, with **uranium** (atomic number 92) at the end of the sequence which starts with hydrogen. However, since 1940, when Philip Abelson (b. 1913) and **Edwin McMillan** created a new element, **neptunium** (atomic number 93) by bombarding uranium with neutrons, scientist have created a number of artificial elements, some of which are highly unstable, lasting only a fraction of a second. Transuranic (beyond uranium) elements now go up to the atomic number of 118. However, elements 113, 155, and 117 have yet to be created.

At room temperature, most elements appear in solid form; only two are **liquids** (including **mercury**, which is a metal), and 11 are **gases**. In **chemistry**, each element is known by its symbols: these symbols, usually a letter or two letter, usually represent an element's archaic name (in languages including Greek and Latin). Thus the chemical symbol for **iron** is Fe, from ferrum, the Latin word for iron. Thus, a chemical symbol often does not represent an element's everyday name, and that can cause some confusion. For example, while Al (aluminum, or aluminium —in Europe) seems straightforward, K (potassium) and Na (sodium) seem somewhat misleading.

Early Greek science postulated the existence of a primordial element as the foundation of the material universe. Thus, for example, Thales (c.624-545 B.C. regarded **water** as the fundamental element. Empedocles (died c. 430 B.C.) identi-

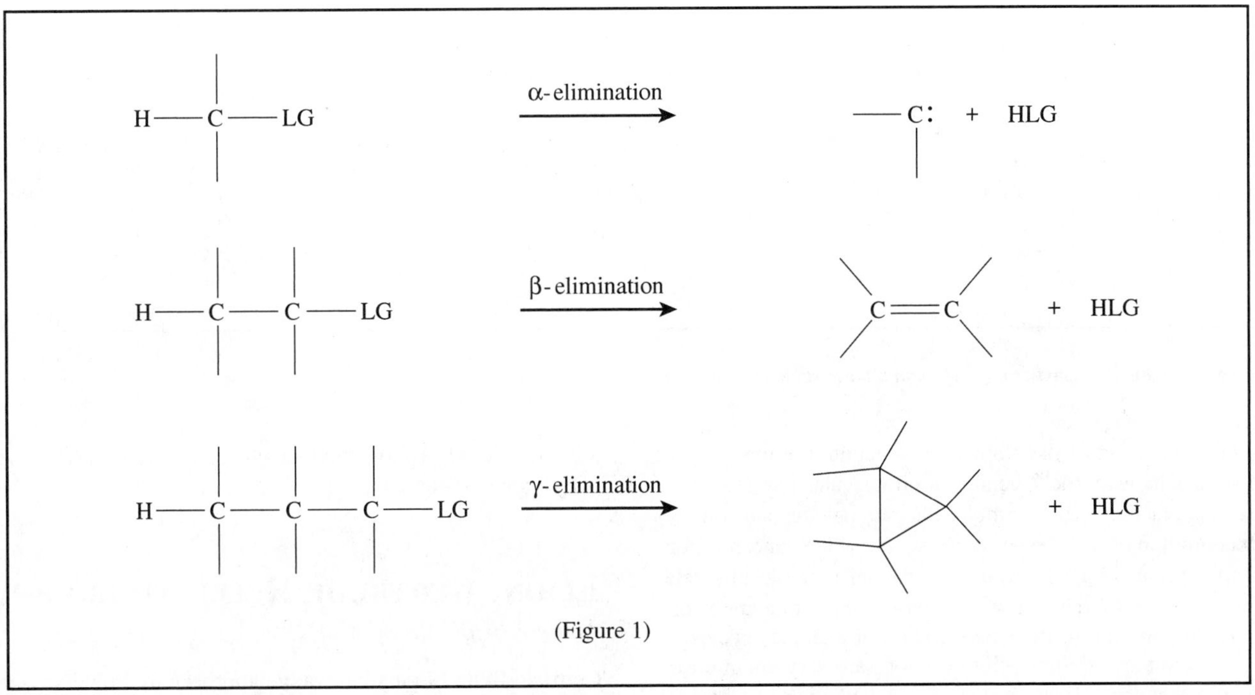

(Figure 1)

Figure 1 (Elimination Reaction). ***(Illustration by Electronic Illustrators Group.)***

fied four basic elements: earth, water, air, and fire. While the quaternary scheme of Empedocles eventually became part of official science thanks to the enormous influence of Aristotle (384-322 B.C.), who incorporated the idea of four element into his theory of nature, early alchemists identified particular substances as elements. For example, they were familiar with certain **metals**, including gold, which inspired mystical reverence as the perfect element. Other elements discussed by both Western and Arabic alchemists are sulphur and mercury, regarded as important spiritual principles. During the Renaissance, Philippus Paracelsus (1491-1541) introduced the idea of three primal elements. Expanding on the sulphur-mercury duality in **alchemy**, Paracelsus created a trinity consisting of **salt**, representing the body, sulphur, representing the soul, and mercury, representing the spirit. In 1661, **Robert Boyle** published *The Sceptical Chymist*, criticized both the tradition conceptions of elements, including the Aristotelian four element s and the alchemical triad of salt, sulphur, and mercury. In his view, elements were pure substances.

The modern scientific concept of element was not formulated until the late eighteenth century. In 1789, in his influential *Elementary Treatise on Chemistry*, **Antoine-Laurent Lavoisier** defined an element as a substance that cannot be chemically decomposed into a more basic substance. Essentially, Lavoisier's definition is in agreement with contemporary **atomic theory**. Lavoisier's list of 33 elements include light, **heat**, certain compounds, but also around 20 substances which he correctly identified as elements: among his elements are **oxygen**, **nitrogen**, hydrogen, sulphur, **antimony**, **arsenic**, **bismuth**, **cobalt**, **copper**, gold, iron, **lead**, **manganese**, mercury, **molybdenum**, **nickel**, **platinum**, **silver**, **tin**, **tungsten**, and **zinc**. Coincidentally, Lavoisier's book was published in year the French Revolution started. In 1794, Lavoisier was executed as an enemy of the Revolution. Those who pleaded for his life were told that "the Republic has no need for scientists."

ELIMINATION REACTION

An elimination reaction is a reaction in which atoms or groups of atoms are removed from a **molecule**. There are many known elimination reactions among the various classes of molecules including but not limited to organic, inorganic and polymeric materials. Elimination reactions are the reverse process of the corresponding addition reactions. There are three common spatial arrangements for the components being eliminated among organic substrates (LG = leaving group): 1) α-elimination of a leaving group and a **proton** from the same **carbon** to give a reactive carbene species, 2) β-elimination of a leaving group and a proton from adjacent carbons to afford an alkene, and 3) γ-elimination of a leaving group and a proton from carbons separated by a third carbon to give a cyclopropane (see Figure 1). The exact mechanism of each of these processes has been the focus of much research and is influenced greatly by the exact substrate, identity of the leaving group and reaction conditions. The leaving groups are typically weak bases such as halides (e.g., Br^-, or I^-) or sulfonate groups (e.g., $MeSO_2O^-$ or $ArSO_2O^-$).

In abbreviated form elimination reactions are described as E reactions with added notation (e.g., E2, E1, etc.) to speci-

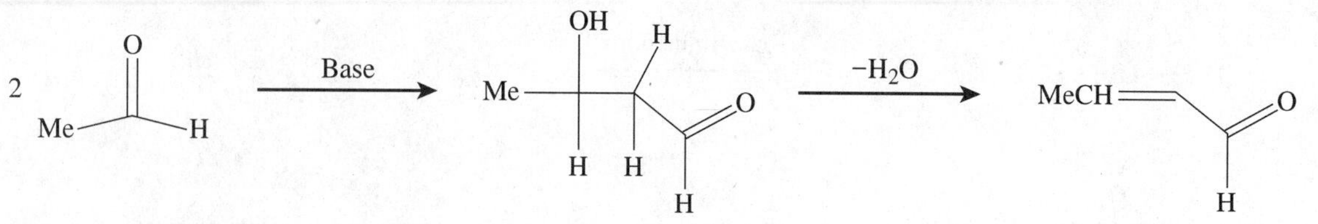

(Figure 2)

Figure 2 (Elimination Reaction). ***(Illustration by Electronic Illustrators Group.)***

fy the mechanism of the elimination reaction. For example in an E2 mechanism, the 2 denotes a bimolecular rate law. The rate of a reaction with a bimolecular rate law depends on the **concentration** of two species or the square of the concentration of one species. An E1 mechanism has an unimolecular rate law. The rate of a reaction with a unimolecular rate law is directly proportional to the concentration of a single species.

Examples of elimination reactions are very common in organic chemistry. Examples are the **dehydration** of alcohols and the dehydrohalogenation of alkyl halides. The dehydration of alcohols is the method most frequently used in the laboratory to produce alkenes. If the starting material is **ethanol** (C_2H_5OH), then **ethene** ($CH_2{=}CH_2$) is the product. As the **water** is eliminated, a double bond is formed between the two carbon atoms. These elimination reactions usually require specific conditions to occur. The production of ethene can be carried out by passing the **alcohol** vapor over **aluminum oxide** heated to 570-660°F (300-350°C). This catalytic reaction is the one most frequently used in the industrial manufacture of alkenes not produced by the refining of oil.

Condensation reactions are a common synthetic method in which two molecules are linked together through an addition/elimination sequence. Generally, the molecule that is eliminated during a condensation sequence is a simple one such as water. An example of this type of **condensation reaction** is the aldol condensation of **aldehydes** (see Figure 2). The starting aldehyde shown is known as acetaldehyde); in the initial addition step an enolate **ion** adds to a aldehyde; in the second step, the elimination step, water is eliminated to condense the two molecules together and form an $\alpha\beta$-unsaturated system.

Repeated condensation reactions find utility in polymerization or chainlengthening sequences. A polymer forming condensation sequence is the production of **nylon** (see Figure 3), the given example is known as nylon-6,6.

In the production of nylon an amine reacts with a carboxylic acid. A bond is formed between the N and the C, along with the formation, or elimination, of water. When this happens many times over nylon is formed. The identity of the nylon product depends on the amine and carboxylic acid used as starting materials.

See also Alkyl group; Polymer chemistry; Reactions (types of); Synthesis, chemical

ELION, GERTRUDE BELLE (1918-1999)

American biochemist

Gertrude Belle Elion's innovative approach to drug discovery advanced the understanding of cellular **metabolism** and led to the development of medications for leukemia, gout, herpes, malaria, and the rejection of transplanted organs. Azidothymidine (AZT), the first drug approved for the treatment of AIDS, came out of her laboratory shortly after her retirement in 1983. One of the few women who has held a top post at a major pharmaceutical company, Elion worked at Wellcome Research Laboratories for nearly five decades. Her work, with colleague George H. Hitchings, was recognized with the Nobel Prize for physiology or medicine in 1988. Her Nobel Prize was notable for several reasons: few winners have been women, few have lacked the Ph.D., and few have been industrial researchers.

Elion was born on January 23, 1918, in New York City, the first of two children, to Robert Elion and Bertha Cohen. Robert, a dentist, immigrated to the United States from Lithuania as a small boy. Bertha came to the United States from Russia at the age of fourteen. Elion, an excellent student who was accelerated two years by her teachers, graduated from high school at the height of the Great Depression. As a senior in high school, she had witnessed the painful death of her grandfather from stomach cancer and vowed to become a cancer researcher. She was able to attend college only because several New York City schools, including Hunter College, offered free tuition to students with good grades. In college, she majored in **chemistry** because that seemed the best route to her goal.

In 1937, Elion graduated Phi Beta Kappa from Hunter College with a B.A. at the age of nineteen. Despite her outstanding academic record, Elion's early efforts to find a job as a chemist failed. One laboratory after another told her that they had never employed a woman chemist. Her self-confidence shaken, Elion began secretarial school. That lasted only six weeks, until she landed a one-semester stint teaching **biochemistry** to nurses, and then took a position in a friend's laboratory. With the money she earned from these jobs, Elion began

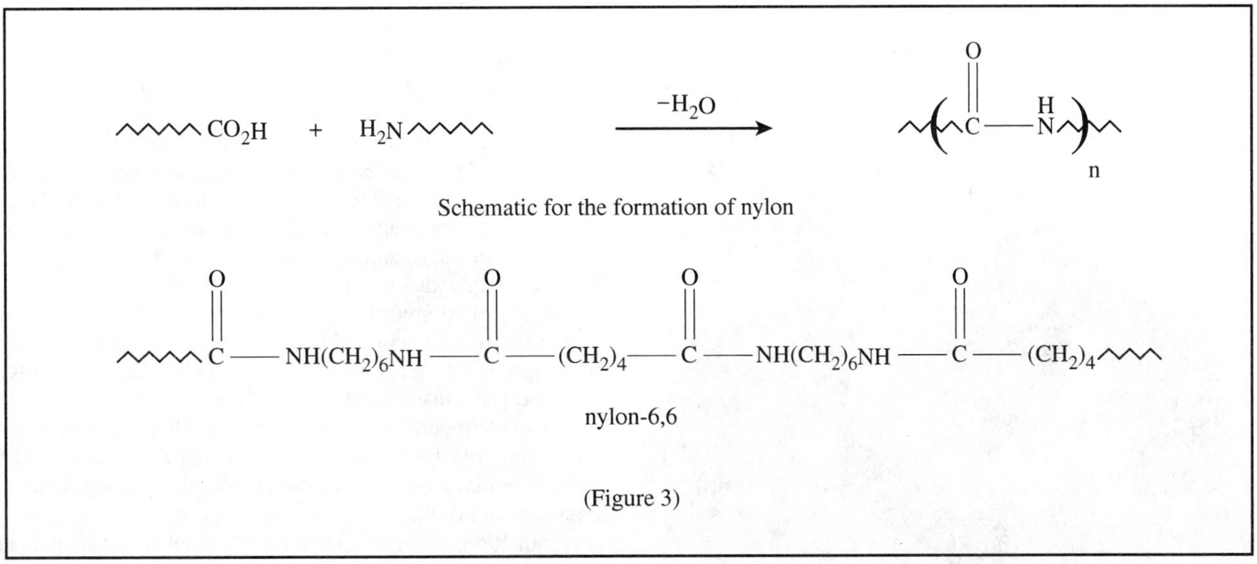

$$\sim\!CO_2H + H_2N\!\sim \xrightarrow{-H_2O} \sim\!\left(\!\sim\!\overset{O}{\overset{\|}{C}}-\overset{H}{N}\!\sim\!\right)_n\!\sim$$

Schematic for the formation of nylon

$$\sim\!\overset{O}{\overset{\|}{C}}-NH(CH_2)_6NH-\overset{O}{\overset{\|}{C}}-(CH_2)_4-\overset{O}{\overset{\|}{C}}-NH(CH_2)_6NH-\overset{O}{\overset{\|}{C}}-(CH_2)_4\!\sim$$

nylon-6,6

(Figure 3)

Figure 3 (Elimination Reaction). ***(Illustration by Electronic Illustrators Group.)***

graduate school. To pay for her tuition, she continued to live with her parents and to work as a substitute science teacher in the New York public schools system. In 1941, she graduated summa cum laude from New York University with a M.S. degree in chemistry.

Upon her graduation, Elion again faced difficulties finding work appropriate to her experience and abilities. The only job available to her was as a quality control chemist in a food laboratory, checking the color of mayonnaise and the acidity of pickles for the Quaker Maid Company. After a year and a half, she was finally offered a job as a research chemist at Johnson & Johnson. Unfortunately, her division closed six months after she arrived. The company offered Elion a new job testing the tensile strength of sutures, but she declined.

As it did for many women of her generation, the start of World War II ushered in a new era of opportunity for Elion. As men left their jobs to fight the war, women were encouraged to join the workforce. ''It was only when men weren't available that women were invited into the lab,'' Elion told the *Washington Post.*

For Elion, the war created an opening in the research lab of biochemist George Herbert Hitchings at Wellcome Research Laboratories in Tuckahoe, New York, a subsidiary of Burroughs Wellcome Company, a British firm. When they met, Elion was 26 years old and Hitchings was 39. Their working relationship began on June 14, 1944, and lasted for the rest of their careers. Each time Hitchings was promoted, Elion filled the spot he had just vacated, until she became head of the Department of Experimental Therapy in 1967, where she was to remain until her retirement 16 years later. Hitchings became vice president for research. During that period, they wrote many scientific papers together.

Settled in her job and thrilled by the breakthroughs occurring in the field of biochemistry, Elion took steps to earn a Ph.D., the so-called ''union card'' that all serious scientists are expected to have as evidence that they are capable of doing independent research. Only one school offered night classes in chemistry, the Brooklyn Polytechnic Institute (now Polytechnic University), so that's where Elion enrolled. Attending classes meant taking the train from Tuckahoe into Grand Central Station and transferring to the subway to Brooklyn. Although the hour-and-a-half commute each way was exhausting, Elion persevered for two years, until the school accused her of not being a serious student and pressed her to attend full-time. Forced to choose between school and her job, Elion had no choice but to continue working. Her relinquishment of the Ph.D. haunted her, until her lab developed its first successful drug, 6-mercaptopurine (6MP).

In the 1940s, Elion and Hitchings employed a novel approach in fighting the agents of disease. By studying the biochemistry of cancer cells, and of harmful bacteria and viruses, they hoped to understand the differences between the metabolism of those cells and normal cells. In particular, they wondered whether there were differences in how the disease-causing cells used **nucleic acids**, the chemicals involved in the replication of **DNA**, to stay alive and to grow. Any dissimilarities discovered might serve as a target point for a drug that could destroy the abnormal cells without harming healthy, normal cells. By disrupting one crucial link in a cell's biochemistry, the cell itself would be damaged. In this manner, cancers and harmful bacteria might be eradicated.

Elion's work focused on purines, one of two main categories of nucleic acids. Their strategy, for which Elion and Hitchings would be honored by the Nobel Prize forty years later, steered a radical middle course between chemists who randomly screened compounds to find effective drugs and scientists who engaged in basic cellular research without a thought of drug therapy. The difficulties of such an approach

Gertrude Belle Elion, with unidentified male. ***(AP/Wide World. Reproduced by permission.)***

were immense. Very little was known about nucleic acid biosynthesis. Discovery of the double helical structure of DNA still lay ahead, and many of the instruments and methods that make molecular biology possible had not yet been invented. But Elion and her colleagues persisted with the tools at hand and their own ingenuity. By observing the microbiological results of various experiments, they could make knowledgeable deductions about the biochemistry involved. To the same ends, they worked with various species of lab animals and examined varying responses. Still, the lack of advanced instrumentation and computerization made for slow and tedious work. Elion told *Scientific American,* "if we were starting now, we would probably do what we did in ten years."

By 1951, as a senior research chemist, Elion discovered the first effective compound against childhood leukemia. The compound, 6-mercaptopurine (6MP; trade name Purinethol), interfered with the synthesis of leukemia cells. In clinical trials run by the Sloan-Kettering Institute (now the Memorial Sloan-Kettering Cancer Center), it increased life expectancy from a few months to a year. The compound was approved by the Food and Drug Administration (F.D.A.) in 1953. Eventually 6MP, used in combination with other drugs and radiation treatment, made leukemia one of the most curable of cancers.

In the following two decades, the potency of 6MP prompted Elion and other scientists to look for more uses for the drug. Robert Schwartz, at Tufts Medical School in Boston, and Roy Calne, at Harvard Medical School, successfully used 6MP to suppress the immune systems in dogs with transplanted kidneys. Motivated by Schwartz and Calne's work, Elion and Hitchings began searching for other immunosuppressants. They carefully studied the drug's course of action in the body, an endeavor known as pharmacokinetics. This additional work with 6MP led to the discovery of the derivative azathioprine (Imuran), that prevents rejection of transplanted human kidneys and treats rheumatoid arthritis. Other experiments in Elion's lab intended to improve 6MP's effectiveness led to the discovery of allopurinol (Zyloprim) for gout, a disease in which excess uric acid builds up in the joints. Allopurinol was approved by the F.D.A. in 1966. In the 1950s, Elion and Hitchings's lab also discovered pyrimethamine (Daraprim and Fansidar) a treatment for malaria, and trimethoprim, for urinary and respiratory tract infections. Trimethoprim is also used to treat Pneumocystis carinii pneumonia, the leading killer of people with AIDS.

In 1968, Elion heard that a compound called adenine arabinoside appeared to have an effect against DNA viruses. This compound was similar in structure to a chemical in her own lab, 2,6-diaminopurine. Although her own lab was not equipped to screen antiviral compounds, she immediately began synthesizing new compounds to send to a Wellcome Research lab in Britain for testing. In 1969, she received notice by telegram that one of the compounds was effective against herpes simplex viruses. Further derivatives of that compound yielded acyclovir (Zovirax), an effective drug against herpes, shingles, and chickenpox. An exhibit of the success of acyclovir, presented in 1978 at the Interscience Conference on Microbial Agents and **Chemotherapy**, demonstrated to other scientists that it was possible to find drugs that exploited the differences between viral and cellular enzymes. Acyclovir (Zovirax), approved by the F.D.A. in 1982, became one of Burroughs Wellcome's most profitable drugs. In 1984, at Wellcome Research Laboratories, researchers trained by Elion and Hitchings developed azidothymidine (AZT), the first drug used to treat AIDS.

Although Elion retired in 1983, she continued at Wellcome Research Laboratories as scientist emeritus and kept an office there as a consultant. She also accepted a position as a research professor of medicine and pharmacology at Duke University. Following her retirement, Elion has served as president of the American Association for Cancer Research and as a member of the National Cancer Advisory Board, among other positions.

In 1988, Elion and Hitchings shared the Nobel Prize for physiology or medicine with Sir James Black, a British biochemist. Although Elion had been honored for her work before, beginning with the prestigious Garvan Medal of the American Chemical Society in 1968, a host of tributes followed the Nobel Prize. She received a number of honorary doctorates and was elected to the National Inventors' Hall of Fame, the National Academy of Sciences, and the National Women's Hall of Fame. Elion maintained that it was important to keep such awards in perspective. "The Nobel Prize is fine, but the drugs I've developed are rewards in themselves," she told the *New York Times Magazine.*

Elion never married although she was engaged once. Sadly, her fiance died of an illness. After that, Elion dismissed thoughts of marriage. She was close to her brother's children and grandchildren, however, and on the trip to Stockholm to receive the Nobel Prize, she brought with her 11 family members. Elion once said that she never found it necessary to have women role models. "I never considered that I was a woman and then a scientist," Elion told the *Washington Post.* "My role models didn't have to be women—they could be scientists." Her other interests were photography, travel, and music, especially opera. Elion, whose name appears on 45 patents, died on February 21, 1999.

El-Sayed, Mostafa Amr (1933-)

American physical chemist

Mostafa Amr El-Sayed is a physical chemist who has used **lasers** to study changing **energy** states in molecules; he is the Julius Brown Professor of **Chemistry** at the Georgia Institute of Technology. El-Sayed was one of the first to employ two-color lasers and **ionization** direction, and he has used laser technology to study the mechanism of photosynthetic systems. El-Sayed was born on May 8, 1933, in Zifta, a small town on the western branch of the Nile River in Egypt. He was the youngest of seven children born to Amr and Zakia (Ahmed) El-Sayed. His father was a high school mathematics teacher, and two of El-Sayed's strongest dreams were to become a university professor and to travel to the United States. In 1953, when El-Sayed completed his undergraduate studies at Ein Shams University in Cairo, he was offered his first teaching job. He began work as a chemistry instructor in the fall, but the job was only to last a few months. He had applied for a doctoral fellowship at Florida State University, and late in December of that year he received word that he had been accepted. By February he was headed for Florida; he recalls Tallahassee as "a delightful town which I consider my birthplace in this country."

While in graduate school, El-Sayed developed an interest in molecular **spectroscopy**, the study of a molecule's response to light waves. This was to become his primary field of research. He received his Ph.D. in **physical chemistry** from Florida State in 1958. El-Sayed did postdoctoral work at Harvard in 1959 in the area of molecular spectroscopy and continued his research on another postdoctorate degree at the California Institute of Technology from 1960 to 1961. In 1961 he was offered a position in the chemistry and **biochemistry** department at the University of California, Los Angeles (UCLA). He became full professor there in 1967 and remained affiliated with that institution until 1994.

Throughout his career El-Sayed's research has been aimed at understanding how atoms within molecules respond to light. In some molecules light can be absorbed and transformed into useful chemical energy. When this transformation occurs, it is because an **atom** absorbs a **photon** and lifts some of its electrons into a higher energy state. This change of the energy state of electrons is what makes possible such molecular functions as **photosynthesis** in plants and vision in animals. Using sophisticated laser techniques that he developed, El-Sayed has been able to explore the minute details of this energy transfer. His molecular subjects have varied from isolated **gases** to the photosynthetic system of a microbe called bacteriorhodopsin.

El-Sayed received the Fresenius National Award in Pure and Applied Chemistry in 1967, the McCoy Award in 1969, and the Gold Medal Award of the American Chemical Society in 1971. He was elected to the National Academy of Science in 1980. In 1988 he received the Egyptian American Outstanding Achievement Award, and in 1990 he won both the King Faisal International Prize in the Sciences and the Tolman Award. He has published over 300 professional articles about spectroscopy and molecular dynamics, and since 1980 he has served as the chief editor of the *Journal of Physical Chemistry* and the United States editor of the *International Reviews in Physical Chemistry.* El-Sayed and his wife, Janice, have five children.

Emanation

Emanation refers to the most stable radioactive **isotope** of **radon**, which emanates naturally from rocks as a gas. Radon is a chemical element with the symbol Rn. The most dense gas known, radon is colorless and chemically unreactive. Along with **helium**, **neon**, **argon**, **krypton**, and **xenon**, it is classed as an inert (or noble) gas within group 0 of the **periodic table**. Twenty isotopes of radon are known, but only three occur naturally. **Ernest Rutherford** discovered thoron (radium-220) in 1899. By 1900, Friedrich Dorn discovered the isotope radon-222 and called it **radium** *emanation* (Em). Radon-222 (**half-life** 3.82 days) is produced by the decay of radium-226. For a period of time, each isotope of radium had its own name. In 1920, however, the name radon was adopted to refer to all the isotopes of the element, although the name emanation and symbol Em still are sometimes used. Radon is highly radioactive and has a very short half-life. The chief use of radon is in the treatment of cancer by radiotherapy because it decays by the emission of alpha particles. Principally found in rock and soil, the element is also found in some spring **water**, streams, and to a very limited extent in air. Radon is created by the radioactive decay of its precursors in minerals, most often uranium-238, from which it diffuses in tiny quantities into air as a gas. Its presence in homes and other buildings is important because radon produced from surrounding soil and rock can reach levels regarded as being dangerous to people due to its suspected role in causing lung cancer. Radon emissions are more prevalent in some areas of the U.S. than others, but the seriousness of the problem is under debate.

Emerson, Gladys Anderson (1903-1984)

American biochemist and nutritionist

Gladys Anderson Emerson was an eminent biochemist who conducted valuable research on vitamin E, **amino acids**, and the B vitamin complexes. She later studied the biochemical bases of **nutrition** and the relation between disease and nutrition. An author and lecturer, Emerson helped establish dietary allowance for the United States Department of Agriculture.

Emerson was born in Caldwell, Kansas, on July 1, 1903, the only child of Otis and Louise (Williams) Anderson. When the family moved to Texas, Emerson attended elementary school in Fort Worth. She later graduated from high school at El Reno, Oklahoma, where she excelled in debate, music, languages, and mathematics. In 1925, she received a B.S. degree in **chemistry** and an A.B. in English from the Oklahoma College for Women.

Following graduation, Emerson was offered assistantships in both chemistry and history at Stanford University, and earned an M.A. degree in history in 1926. She eventually became head of the history, geography, and citizenship department at an Oklahoma junior high school, and, a short time later, accepted a fellowship in **biochemistry** and nutrition at the University of California at Berkeley, where she received her Ph.D. in animal nutrition and biochemistry in 1932. That year, Emerson was accepted as a postdoctoral fellow at the University in Göttingen, Germany, where she studied chemistry with the **Adolf Windaus**, Nobel laureate, and **Adolf Butenandt**, a prominent researcher who specialized in the study of **hormones**.

In 1933, Emerson returned to the United States and began work as a research associate in the Institute of Experimental Biology at the University of California at Berkeley. She remained there until 1942, conducting pioneering research on vitamin E. Using wheat germ oil as a source, Emerson was first to isolate vitamin E. In 1942, she joined the staff of Merck and Company, a pharmaceutical firm, eventually leading its department of animal nutrition. Staying with Merck for fourteen years, Emerson directed research in nutrition and pharmaceuticals. In particular, she studied the structure of the B vitamin complex and the effects of vitamin B deprivation on lab animals. She found that when vitamin B_6 was withheld from rhesus monkeys, they developed arteriosclerosis, or hardening of the arteries.

During World War II, Emerson served in the Office of Scientific Research and Development. From 1950 to 1953, she worked at the Sloan-Kettering Institute, researching the link between diet and cancer. From 1962 to her retirement in 1970, she was professor of nutrition and vice-chairman of the department of public health at the University of California at Los Angeles.

In 1969, President Richard M. Nixon appointed Emerson vice president of the Panel on the Provision of Food as It Affects the Consumer (The White House Conference on Food, Nutrition, and Health). In 1970, she served as an expert witness before the Food and Drug Administration's hearing on **vitamins** and mineral supplements and additives to food. A photography enthusiast who won numerous awards for her work, Emerson was also a distinguished board member of the Southern California Committee of the World Health Organization and was active on the California State Nutrition Council.

EMPIRICAL FORMULA

Empirical is derived from a Greek word meaning experienced, and empirical formulas are determined from experimental values. The empirical or simplest formula uses element symbols and subscripts that represent the smallest whole number ratio of atoms in a compound not the actual number of atoms. Empirical formula represents both ionic and molecular compounds. For an ionic compound the empirical formula is also called the formula unit and represents the ratio of ions. For example, KCl has an equal number of **potassium** and **chlorine** ions. **Barium** chloride, $BaCl_2$ has twice as many chlorine ions as barium. Molecular formulas represent compounds that exist as molecules. The formula $C_2 H_4$ has two atoms of **carbon** and four atoms of **hydrogen**. The empirical formula is the smallest whole number ratio and would be CH_2. The empirical formula can be determined through analysis and the **molecular formula** can be calculated from the empirical formula. The subscripts of the molecular formula are a multiple of the empirical formula. In some cases, the empirical formula is the only possible formula.

The empirical formula of a compound can be determined from the masses of the individual elements in the compound. The **mass** amount of each element must be converted to **mole** values to determine **atom** ratio. A mole is a unit that contains **Avogadro's number** of particles (6.022×10^{23}). This is the reason ratios are determined by comparing mole amounts of each element, since one mole of any element has the same number of particles. Masses cannot be compared because no two elements have the same mass. One mole of **sodium** and one mole of chlorine have the same numbers of particles but their masses are different.

The empirical formula is calculated by dividing the mass of each element in the compound by its atomic mass on the **periodic table**. This value is the mole value of the element. The following is an example. It is determined that a compound contains 0.5 grams of **magnesium** and 1.46 grams of chlorine. To find the empirical formula for this compound, divide each gram amount by the element's atomic mass. For magnesium, 0.5 divided by 24.3 equals 0.02 moles of magnesium. For chlorine, 1.46 is divided by 35.5, equaling 0.04 moles of chlorine. To obtain the smallest whole number ratio between these two elements divide both mole value by the smallest, 0.02. The ratio is one mole of magnesium to two moles of chlorine. The formula is $MgCl_2$.

Empiricial formulas may also be calculated from percentages of the elements in a compound. The size of the sample is not important. The percentage of each of the elements in the compound is assumed to be the same. To convert or change percentages to grams you can assume you have a 100 gram sample, then each percentage would also represent the gram amount of the substance. For example, to calculate the empirical formula for a compound containing 36% **aluminum** and 64% **sulfur**, drop the percentage sign and replace it with grams, then follow the same steps as previously mentioned. Thirty six grams of aluminum divided by 27 grams per mole equals 1.33 (repeating) moles. Sixty four grams of sulfur divided by 32 grams per mole equals two moles. Divide by the lowest value, 1.33. The results indicate the mole ratio of one mole of aluminum to 1.5 moles of sulfur. Since atoms in a compound can occur in only small whole number ratios, the decimals can be eliminated by multiplying both values by two. The smallest whole number ratio would then be two moles of aluminum to three moles of sulfur or $Al_2 \backslash S_3$. Any fractional or decimal value must be multiplied by a whole number to eliminate fractional subscripts. For example, if the mole ratios of a compound containing potassium, **chromium**, and **oxygen** were determined to be one mole of potassium to 3.5 moles of chromium and one mole of oxygen then the smallest whole number ration would be 2:7:2. The resulting empirical formula is $K_2Cr_7O_2$. If the mole ratio of carbon to hydrogen in a compound is one mole of carbon to 2.667 or 2-2/3 moles of hydrogen multiplying by three would eliminate the decimal. The empirical formula would be C_3H_8.

Empiricial formulas for **gases** can also be calculated from liter or dm^3 amounts. One mole of any gas occupies 22.4 liters at STP (standard **temperature** and pressure), This information allows conversion of liters to moles and ratios can then be calculated. Divide the liter amount of the gas by 22.4 liters per mole and continue with the previous procedure.

Compounds that have the same empirical formula do not necessarily have the same molecular or **structural formula**. Butene and **ethene** have the same empirical formula CH_2 but different molecular and structural formulas. Butene, C_4H_8 and ethene C_2H_4 have different properties because of these differences.

See also Molecular geometry

EMULSION

An emulsion is a system that has at least one immiscible liquid distributed in another liquid, or solid, in the form of tiny droplets. It is an unstable system that is made more stable by the presence of a surfactant. Emulsions have been used by humans for centuries and continue to find application in a variety of industries.

Emulsions are typically made up of a hydrophilic, aqueous portion and a hydrophobic, oil portion. These two portions constitute the internal and external phases of the emulsion. The internal phase is composed of the tiny dispersed droplets, while the external phase is made up of the rest of the materials. The more abundant phase is called the external or continuous phase. When **water** is the external phase and oil is dispersed, it is known as an oil-in-water emulsion (o/w). If water is the internal phase it is called a water in oil emulsion (w/o). Many factors are responsible for determining whether an emulsion is of the o/w or w/o type, including the concentration of the components, the type of emulsifier, and processing steps.

More complex emulsion systems are possible when there are numerous internal phases. These systems are called multiple emulsions and can be described as emulsion in emulsions. For instance, if water is dispersed in an oil and then further dispersed in another water phase, the system is described as a water in oil in water (w/o/w) emulsion. Other multiple emulsions can have many times this number of internal phases.

The particles that make up the emulsion internal phase are polydisperse, meaning they have variable sizes. Emulsions can be classified by the size of these particles. When the average diameter is less than 100å the system is called a micellar emulsion. A microemulsion has a particle diameter of 100-2,000å. Systems with larger sized particles are known as macroemulsions.

Typical emulsions have an oil phase and an aqueous phase. The oil phase is composed of non-polar materials that are not soluble in water. These include materials such as fats, oils and waxes. The aqueous phase of an emulsion is made up water and all of the other hydrophillic materials in the system.

Compounds that stabilize emulsions are called emulsifiers. Also known as surfactants, these materials help stabilize the system by reducing the interfacial tension between the two phases. Most emulsifiers have a molecular structure that includes a lipophilic portion and a hydrophilic portion. These materials are classified by the nature of their hydrophilic portion. They can be either anionic, cationic, nonionic, or amphoteric emulsifiers. An example of an anionic emulsifier is stearic acid. It is made up of a number of carbon molecules that are lipophilic attached to a hydrophilic **carboxylic acid** head group.

When an emulsifier is put into water the molecules have a tendency to align themselves in a manner that reduces the interaction between its hydrophilic and lipophilic ends. If enough emulsifier is present spherical structures called micelles form. These micelles are particles that have the lipophilic tails oriented toward the center and hydrophilic heads on the outer surface.

By themselves, oil and water will not mix. If they are combined in a container and shaken, the oil breaks up into smaller particles and may be dispersed momentarily. However, after the agitation is stopped, the dispersed oil particles quickly separate from the water. When an emulsifier is present, the oil particles are stabilized because they become incorporated into the interior part of the micelles. The oil particles are shielded from the water and each other. This prevents them from combining and separating with the water. The result is a more stable system.

Although emulsifiers help stabilize the interaction between the oil phase and the water phase, emulsions are still inherently unstable according to the second law of thermodynamics because the entropy of the system would be greater if all molecules were dispersed randomly. One of the basic facts about emulsions is that given enough time or energy they will separate into the original phases. The speed and efficiency at which this occurs depends on the composition of the emulsion. For example, a system of mineral oil and water forms an emulsion when agitated that immediately separates completely upon standing. If a small amount of an emulsifier is added, the system may remain stable for a few days. Other emulsions can remain stable for years.

Examples of emulsions are found throughout nature. For instance, homogenized milk is an emulsion made up of butterfat droplets dispersed in water. The emulsifier is a protein called casein. Yogurt, ice cream, and whipped cream are also emulsions. Emulsions are also used for making various products. They are the most common type of delivery system used for cosmetic products, enabling a wide variety of ingredients to be quickly and conveniently put on hair and skin. Additionally, emulsions are used for things such as cleaning products, paints, coatings, adhesives, **inks**, and dyes.

Enantiomers

Certain pairs of molecules differ from one another only in the fact that one is the mirror image of the other. These molecules are known as enantiomers (from the Greek word for ''opposite''). The relationship between two enantiomers is the same as the left hand to the right; in fact, you could think of your two hands as enantiomers. By analogy, one enantiomer is called the R-isomer (from the Latin *rectus*, meaning right), and the other, the S-isomer (from the Latin *sinister*, or left). Although enantiomers are identical in most of their chemical and physical properties, they can exhibit profound differences under certain conditions, particularly in biological systems.

Molecules that differ from their mirror images are said to be chiral (from the Greek word *cheir* for ''hand''); those that are identical to their mirror images are said to be achiral. The easiest way to tell whether a **molecule** is chiral or not is to try to superpose it on its mirror image. Chiral molecules are not superposable; achiral molecules are. Your two hands are not superposable; when your palms face in the same direction, for instance, one thumb will be up, while the other will be down.

Enantiomers exhibit different behavior when they interact with other chiral substances, e.g., they undergo different rates of reaction with other chiral molecules, and they show different solubilities in **solvents** that contain a single enantiomer or an excess of one enantiomer. Chiral molecular interactions can show up in many ways. For example, one enantiomeric form of the compound limonene is primarily responsible for our perception of the odor of oranges, while the other enantiomer supplies the odor of lemons.

Another way in which enantiomers can be distinguished is by the way they rotate polarized light. This is, in fact, the most important method for telling them apart. An ordinary light beam consists of a group of electromagnetic waves that vibrate in all directions. When such a beam passes through a polarizer, only those waves vibrating in a specific plane are transmitted. If a polarized beam is passed through a **solution** consisting of only one enantiomeric type, the beam emerges with its plane of **polarization** rotated. Enantiomers rotate a beam's plane of polarization in opposite directions, and, in general, by equal amounts.

If a sample of a compound contains equal amounts of two enantiomers, it is said to be a racemic mixture. Racemic mixtures do not rotate the plane of polarized light because a beam of polarized light passing through this type of sample encounters equal numbers of right- and left-handed enantiomers.

The easiest way to obtain a pure source of one enantiomer is to isolate it from some natural source, because nearly all chiral molecules found in living systems occur as single enantiomers. The enantiomers of dietary **carbohydrates**, such as glucose, do not exist to any great extent in nature, for example. Similarly, the **proteins** that are responsible for catalysis, nutrient transport, and bone structure in the human body consist of only one enantiomeric type of amino acid; their enantiomers only appear in a few life forms such as bacteria. (Over time, the principal type of **amino acid** partially converts into its enantiomer; this fact provides a basis for estimating the ages of fossilized proteins.) No one knows why biological systems exhibit preferences for one enantiomer over another, although some scientists have speculated that when life first appeared on Earth, the circularly polarized light from stars may have destroyed one amino acid enantiomer while leaving the other one untouched.

In the absence of a natural source, however, pure enantiomers can be obtained by resolving a racemic mixture. **Louis Pasteur** performed the first enantiomeric resolution in 1848 while studying crystals of a **salt** of racemic **tartaric acid**. With the aid of a magnifying **glass**, he noticed that some of the crystals were mirror images of each other. Using tweezers, he was then able to separate the two types of crystals. When dissolved in **water**, one crystal type rotated the plane of polarization in one direction, and the other crystal type in the opposite direction.

As it turned out, however, Pasteur's mechanical resolution of two isomers was of more historical than practical significance because very few enantomers crystallize as separate mirror image crystals. The most useful method for resolving enantiomers involves converting them into **diastereomers**. Diastereomers have their atoms joined in the same order but differ in the way the atoms are arranged in space. Pairs of diastereomers are not related as enantiomers, e.g., they are not mirror images and they have different physical properties. If an enantiomeric pair can be converted into a pair of diastereomers, the physically separated diastereomers can then be converted back to the enantiomers. The way that this works is as follows: if group X on each enantiomer reacts with group O of an enantomer from another source, a pair of diastereomers is formed. The separated diastereomers are then individually converted back to enantiomers.

Not all examples of enantiomeric interactions are benign. At one time, the drug thaidomide was prescribed to alleviate the symptoms of morning sickness in pregnant women. It was later discovered that this drug can cause severe birth defects in children whose mothers used it. Evidence now suggests that one enantiomer does indeeed cure morning sickness, but the other one causes birth defects.

Endothermic reactions

Chemical changes alter the composition and structure of a substance. These kinds of changes are always accompanied by **energy** changes. If there is net energy released in forming the new structure, the reaction is said to be *exothermic*. If more **heat** than is available from the reactants must be absorbed from the surroundings to create the new structure, the reaction is said to be *endothermic*.

The change in heat content of the products relative to the reactants in a chemical reaction is known as the change in **enthalpy**. If the heat content of the products is greater than the heat content of the reactants, the change in enthalpy is positive, and the reaction is endothermic. If the change in enthalpy is negative the reaction is exothermic.

Because the change in enthalpy that accompanies any given reaction varies with **temperature**, scientific convention has adopted the standard of reporting heat data at 13°F (25°C) and 1 bar pressure. The standard enthalpy of formation is the change in enthalpy that accompanies the formation of a compound from its elements with all substances in their standard states at 25°C.

An example of an endothermic reaction is the decomposition of **mercury** (II) **oxide**. To dissociate two moles of mercury (II) oxide into mercury and molecular **oxygen**, one must supply 43,400 calories of energy to the system.

The energy changes that accompany chemical reactions are not always limited to heat. In voltaic cells, for example, energy is produced in the form of **electricity**. In **photosynthesis**, the conversion of **water** and **carbon** dioxide into sugar and oxygen, energy is absorbed in the form of light.

See also Energy; Heat

ENERGY

Energy is defined as the capacity to perform work or to produce **heat**. The *Système International d'Unités* (S.I.) unit for energy is the joule (J). Lifting a medium-sized potato a distance of one meter against gravity would require approximately one joule of energy. Energy is often expressed as the calorie (cal), which is the amount of heat needed to raise the **temperature** of one gram of **water** by one degree Celsius. One calorie is equal to 4.184 joules. The calorie (Cal), which is how the energy in food is expressed, is 1,000 calories. Energy is not only involved in chemical reactions and technology, but also in our basic bodily functions. It is because of the energy of the reactions inside of our body that we maintain a constant body temperature, our heart and lungs can function, and our muscles can move. Energy can take many forms, all of which can be classified into three categories. These categories are radiant energy, kinetic energy, and **potential energy**.

An example of radiant energy is sunlight. Kinetic energy is the energy of an object in motion. Examples of kinetic energy are mechanical energy (caused by motion of parts) and thermal energy (caused by the random motion of particles of matter). Potential energy is the energy of position of objects. It is also known as stored energy. An example of potential energy is water in an elevated tank. If the water is allowed to fall on a wheel and the wheel turns, the wheel can be used to produce **electricity**. The water in the tank has the potential to perform work once it falls, therefore it has gravitational potential energy (gravity is what converted the potential energy of the water as it fell into kinetic energy). Other forms of potential energy include chemical potential energy, the energy stored in fuels such as **kerosene** or in the food you eat, and electrical potential energy, the energy when objects with different electrical charges are separated such as in **batteries**.

Two forms of energy that are most familiar are light and heat. Light is a form of radiant energy. Scientists used to believe that light was a beam of energy which took the form of waves. In the 1900s, however, experiments were performed which showed that light behaved like very small, fast moving particles. Scientists today realize that light has properties of both waves and particles and describe it accordingly.

Heat is a form of thermal energy that is transferred from one object to another when there is a temperature difference between them. More specifically, it is the transfer of kinetic energy from something of greater temperature to something of lesser temperature. Heat energy can cause physical or chemical changes. A physical change that heat can cause is the melting of ice into water, or the evaporation of water into vapor. This water vapor can undergo a chemical change if enough additional heat is added, when it decomposes into **oxygen** gas and **hydrogen** gas.

An English scientist named **James Joule** performed an experiment in 1845 which demonstrated a relationship between energy and heat. The experiment was not complicated—he placed a paddle wheel in a tank of water, and measured the temperature of the water. He then cranked the wheel in the water for a period of time, and read the temperature of the water again. He found that the temperature of the water rose as he cranked the paddle wheel. He quantified this observation and discovered that an equal amount of energy was always required to raise the temperature of water by one degree. He also discovered that it did not have to be mechanical energy, it could be energy in any form. He obtained the same results with electrical or magnetic energy as he did with mechanical energy. Joule's experiments showed that different forms of energy are equivalent and can be converted from one form to another. In the process of converting energy, no energy is lost and no energy is created.

These observations led to what is now called the "Law of **conservation of energy**." The law of conservation of energy states that any time energy is transferred between two objects, or transformed from one form to another, no energy is created, and none is destroyed. The total amount of energy involved in the process remains the same. This law applies to any process except those that occur in the sun and in **nuclear reactors**, which involve nuclear energy, for a combination of **mass** and energy are conserved but mass and energy are interconverted

Energy is not only involved in macroscopic processes, but in microscopic processes as well. Most chemical reactions involve changes in energy. A chemical reaction is simply breaking bonds between atoms and making new ones. When bonds are broken in a chemical reaction, the reactants involved are colliding with enough kinetic energy to break the atoms apart. During the collision, the kinetic energy is transformed into potential energy as the atoms are rearranged and new bonds are formed. In 1888, **Svante Arrhenius**, a Swedish chemist, theorized that the particles involved in a chemical reaction must have a certain minimum amount of kinetic energy in order to react. This minimum amount of energy is called the

activation energy of a reaction. Bond breaking requires energy, while bond making releases energy. Most chemical reactions, therefore, involve the overall absorption or release of energy. The energy flow that results is heat.

Energy is involved in physical as well as chemical changes. As the temperature of a substance rises, the individual particles of the substance tend to gain kinetic energy. As the particles gain kinetic energy, they obtain different physical properties, and this results in a phase change. A phase change is the conversion of a substance from one of the three states of **matter** (solid, liquid, or gas) to another. A phase change always involves a change of energy. The particles which make up a solid have very low kinetic energies, so they stay fixed together and immobile. As the temperature increases, they gain kinetic energy, and they can move around more easily. When the particles in a substance have enough kinetic energy so that the substance can flow, it is in the liquid phase. As the temperature increases, and the kinetic energy of the particles increase, the particles that make up the liquid begin to move in a speedy, random fashion. At this point, the liquid has become a gas, the phase with the highest kinetic energy. As the temperature of a gas increases, the particles begin to move faster and faster, and their collisions with themselves and their container become more forceful. As this occurs, gas pressure increases.

There are many practical applications to the study of energy. One is the development of fuels. A fuel is a **chemical compound** that burns easily and releases a large amount of heat. When **gasoline** is burned in an automobile engine, for example, it releases 48 kilojoules of heat per gram. This heat then undergoes several **energy transformations** to power the automobile. Two other familiar fuels are oil and **coal**. Both of these fuels release a large amount of energy when burned. Oil releases 20,900 British Thermal Units (BTU) per pound, and coal releases 9,600 BTU per pound (one BTU is the amount of heat needed to raise the temperature of one pound of water one degree Fahrenheit). Plastic products could also release similar quantities of energy. The plastic in **polyethylene** terephthalate (PET) soda bottles can release 10,900 BTU per pound, and the plastic used in high **density** polyethylene (HDPE) milk jugs releases 18,700 BTU per pound. The potential of re-using these plastic containers by burning them in a "waste-to-energy incinerator" is currently under investigation.

The use of alternative forms of fuel, such as the plastic containers described above, is being investigated in order to conserve energy. The law of conservation of energy states that energy can neither be created nor destroyed. Why, then, is there concern about "running out" of energy? The energy scientists are concerned about is useable energy. The majority of our energy comes from coal and petroleum, two of the three fossil fuels (the third fossil fuel is natural gas). Approximately 90% of our energy comes from fossil fuels. Coal is burned in power plants to generate electricity and petroleum is converted into the fuel which is used for transportation. When the potential energy stored in these fossil fuels is released, it is converted into other forms of energy such as heat and sound which cannot be recaptured and reused. As Earth's coal and petroleum are used up, we will need to find different sources of energy. Alternatives which are currently under investigation are wind, water, geothermal energy, and solar energy.

The many forms of energy are an integral part of daily life. The amount of energy in the universe remains constant, but the amount of useable energy is decreasing. The study of energy and its different forms will become increasingly important as the need for alternative forms of energy rises.

Energy level

Energy level is a particular value of the total energy that an atomic particle or group of particles can acquire, according to quantum theory. In this theory the energy levels for **matter** and radiation must fall into a discrete, quantized set.

Based on observations of absorption and emission spectra, a number of scientists pointed out in the latter half of the 19th century that the elements each have a set of characteristic wavelengths of electromagnetic radiation which they absorb when cool and emit when hot. An element only absorbs or emits radiation of its characteristic wavelengths. Any useful model of the **atom** must account for this observation.

J. J. Thomson proved the existence of the **electron** in 1897. In 1911 **Ernest Rutherford** proposed the nuclear model of the atom with a heavy **nucleus** at the center and much lighter electrons moving in orbits around the nucleus, analogous to the solar system. In 1913, **Niels Bohr** combined Rutherford's nuclear model of the atom with the quantum ideas of **Max Planck** to propose a model of the **hydrogen** atom that introduced the concept of quantized energy levels and explained the characteristic absorption and emission spectra of hydrogen.

Bohr assumed that the energy which an electron in an atom may have is quantized, i.e. it can possess only certain amounts of energy. If an electron is to change energy, it can do so only by moving from one of the allowed quantized energy levels to another allowed level. It absorbs energy to move from a permitted level of lower energy to one of higher energy. It emits energy when it moves from a higher energy level to a lower energy level. Each element would be expected to have its own characteristic set of allowed quantized energy levels, based on the attractive force between its nucleus and electrons. This explained the absorption and emission spectra of an element; only frequencies of radiation whose energies correspond exactly with the energy gaps between allowed electron energy levels could be absorbed or emitted. Bohr's model predicted the frequencies which the quantized hydrogen atom could absorb or emit; the predicted values matched the experimentally observed values. Unfortunately, efforts to extend Bohr's theory to atoms with more than one electron failed

In 1923, **Louis de Broglie** proposed that particles possess wave properties, and this was confirmed in 1927 by C. Davisson (1881-1958), L. H. Germer (1896-1971), and George Thomson (1892-1975). In 1926, based on de Broglie's proposal, **Erwin Schrödinger** proposed a theory, known as wave or quantum mechanics, that is based on a quantized wave equation and could potentially explain the activity and properties of particles such as electrons in atoms. The solutions of this equation for hydrogen match exactly the predictions of the **Bohr theory**. Although Schrödinger's wave equation is not soluble exactly for systems of more electrons than hydrogen, appropriate approximations lead to solutions for the allowed energies of electrons in other atoms which match observed absorption and emission spectra. Predictions made by wave mechanics for other properties also agree with experiment. Thus the working model of the atom has been revolutionized, and the concept of quantized energy levels has become firmly established.

Each **solution** of Schrödinger's equation for atomic systems is called an orbital. Physically, the orbital describes the space in which the electron can move. Each orbital has a set of 4 quantum numbers associated with it. The principal quantum number n may have any positive, non-zero integer value (1,2,3...). It is the primary determinant of the allowed energy levels of electrons. The angular momentum quantum number l may have only integer values from 0 to (n-1). The energy of an orbital depends slightly on the value of l; orbitals with higher values of l are higher in energy. Different values of l correspond to different orbital shapes which designated s,p,d, and f. (The letters are derived from spectroscopic observations and refer to "sharp," "principal," "diffuse," and "fine.") The s orbital is spherical while the p orbital is dumbbell shaped.

The magnetic quantum number m_l may have any of the integer values -l to +l including 0 (...-1, 0, +1...). The spin quantum number s may have only the values +1/2 and -1/2. The energy is not affected by the value of the magnetic and spin quantum numbers. Each orbital has a different set of these four quantum numbers.

In 1925, Wolfgang Pauli (1900-1958) showed that no two electrons in an atom may occupy the same orbital. Since the spin quantum number may have only the values +1/2 and -1/2 and since the value of spin does not affect the energy nor any other part of the wave function, often we speak of two electrons being in each orbital, one with spin of +1/2 and the other with spin of -1/2.

The quantum numbers corresponding to the possible energy levels of an atom are as follows (with energy increasing as the value of n and l increase):

There is only one orbital with n="1", because when n="1", l=0 and m=0. This is called the 1s orbital, and two electrons may occupy it (one with spin +1 and one with spin -1).

There are 5 orbitals with n=2: a 2s orbital with n=2, l=0, m=0 containing 2 electrons; three orbitals, called 2p orbitals, of slightly higher energy for which n=2, l=1, and m has the values -1, 0, and +1; each of these p orbitals may contain 2 electrons, and 6 electrons may thus be accommodated in these p orbitals.

There are 9 orbitals with n=3: a 3s orbital which may contain 2 electrons; 3 3p orbitals of slightly higher energy, which may contain as many as 6 electrons; and 5 orbitals with m values of -2, -1, 0, +1, and +2 that are called 3d orbitals and that have slightly higher energy than the 3p orbitals. The 3d orbitals may contain a total of 10 electrons. Similar orbital schemes may be constructed for larger values of the principal quantum number.

The ground state of an atom is the state of lowest energy. It exists when the electrons of the atom are occupying the orbitals of lowest energy. An excited state is one in which one or more of the electrons occupy orbitals of higher energy than in the ground state.

The lone electron of the hydrogen atom will occupy the 1s orbital in the ground state. It could absorb energy and move into a 2s, 2p, 3s, or higher energy level orbital. It would then be in an excited state. The frequency of radiation absorbed in the transition from the ground state to one of the other orbitals corresponds the difference in energy between the ground state and the excited state.

A **chlorine** atom has 17 electrons. In the ground state, its 17 electrons will be located in the orbitals of lowest possible energy: 2 electrons in the 1s orbital, 2 in 2s, 6 in 2p, 2 in 3s, and 5 in 3p. The following notation is used to described chlorine's electron configuration: $1s^2\ 2s^2\ 2p^6\ 3s^2\ 3p^5$. A **bromine** atom has 35 electrons, distributed in atomic orbitals as follows: 2 in 1s, 2 in 2s, 6 in 2p, 2 in 3s, 6 in 3p, 10 in 3d, 2 in 4s, and 5 in 4p. ($1s^2\ 2s^2\ 2p^6\ 3s^2\ 3p^6\ 3d^{10}\ 4s^2\ 4p^5$) The absorption spectra of these atoms can be correlated with the change of energy of electrons in the transition between the ground state configuration to an excited state configuration: e.g. an electron in a 5p orbital of bromine excited to a 6s orbital.

The energy levels of electrons in a **molecule** can be understood in a similar way. Generally only the occupied orbital of the highest energy, called the **valence** shell, in the bonded atoms need to be considered. The electrons in orbitals of energy below the valence shell remain associated with their own atom and their energy is negligibly affected by the other atoms of the molecule. The wave theory treatment of the shared electrons of a molecule results in a set of molecular orbitals that spread over two or more atomic centers. A molecular orbital is a **bonding** orbital if its energy is lower than that of the isolated atomic orbitals. Each of the electrons in the valence shell that are shared by the bonding atoms belongs to a molecular orbital. The ground state of the molecule is that in which the bonding electrons are in the molecular orbitals of lowest energy. An excited state occurs when an electron in a low energy molecular orbital moves to an orbital of higher energy. The absorption and emission spectra of molecules can be correlated with such transitions of electrons between allowed quantized energy levels.

The wave lengths of electromagnetic radiation that correspond to transitions between energy levels in atomic and molecular systems lie in the visible and ultraviolet regions of the **electromagnetic spectrum**.

The energy of the chemical bond is the energy required to break the atoms apart with the shared electrons in the molecular orbitals reverting to atomic orbitals on the individual atoms. The bond energy is less for a molecule in an excited state than for one in the ground state.

Chemical species possess energy levels other than those of their electrons. There are also translational, rotational, vibrational, and nuclear energies. The energy states of the nucleus of an atom are of little consequence in **chemistry** and may be ignored in this discussion.

The quantum mechanical treatment of molecular rotational energy yields a set of quantized energy levels with allowed energies determined by the product J(J+1), where J is a quantum number with integer values, and reciprocally related to the moment of inertia of the molecule. The moment of inertia is related to the **mass** of the molecule's atoms and to the bond distance. The result is a set of quantized energy levels corresponding to the various integer values of J, the lower values of J having lower energy. The molecule will be in its ground state when J=0 which corresponds to no rotation at all. An electromagnetic absorbtion can occur when the energy of the radiation corresponds to the energy gap between a occupied wave function and an excited state wave function. An absorption spectrum results. The wavelengths of electromagnetic radiation that rotating molecules are capable of absorbing lies in the microwave region of the spectrum.

In addition to translation and rotation, a molecule vibrates about an equilibrium position. The quantum mechanical treatment of the vibration of molecules yields a set of wave functions with associated energies. The energies of the allowed quantized vibrational states are related to **Planck's constant**, a quantum number v, the strength of the bond and reciprocally to the masses of the atoms involved. The stronger the bond, the higher the energy required to make it vibrate. The molecule absorbs radiation whose frequency corresponds to the difference in energy between vibrational energy levels. Molecular absorption spectra are observed in the infrared and microwave portion of the electromagnetic radiation spectrum. The strength of the molecular bond can be determined from data obtainable from the vibrational absorption spectrum.

See also Molecular orbital theory; Schrödinger equation

ENERGY TRANSFORMATIONS

Energy is the capacity to do work or to produce **heat**. There are many forms of energy, each of which can be classified into three categories: radiant energy, kinetic energy, and **potential energy**. Energy can be changed, or transformed, from one form into another. Energy transformation is also called energy conversion. The *Système International d'Unités* (SI) unit for energy is the joule (J), after **James Joule**, who demonstrated that work can be converted into heat. Lifting a medium-sized potato a distance of 1 m (3.28 ft) would require approximately one joule of energy. Energy is often expressed as the calorie (cal), which is the amount of heat needed to raise the **temperature** of one gram of **water** by one degree Celsius. One calorie is equal to 4.184 joules. The calorie (Cal), which is used to express the energy in food, is 1,000 calories.

Kinetic energy is the energy of an object in motion. An object in motion can cause another object to do work by colliding with it, causing it to move a particular distance. The colliding objects can be a hammer swinging down on a nail, or two atoms colliding in a chemical reaction. Example of kinetic energy are mechanical energy (caused by motion of parts) and thermal energy (caused by the random motion of particles of matter). An object that has potential energy is not moving or doing work. At some point, that object had work performed on it which resulted in energy storage. One example of performing work on an object to give it potential energy is the action of winding a watch. As it is being tightened, the spring is gaining potential energy. Winding a watch, therefore, transforms mechanical into potential energy. Another example of potential enegy is **water** in an elevated tank. If water is allowed to fall on a wheel and the wheel turns, the wheel can be used to produce **electricity**. The water in the tank has the potential work once it falls; therefore, it has gravitational potential energy (gravity is what converted the potential energy of the falling water into kinetic energy). The potential energy of the water was transformed into mechanical energy of the wheel, which was further transformed into electrical energy. Conversion between potential and kinetic energies is quite common. As the potential energy of an object increases, its kinetic energy tends to decrease. Likewise, if the kinetic energy of an object increases, its potential energy decreases. For example, a rubber ball held out of a third-story window has potential energy. If it is released, gravity causes it to fall. As it falls, it loses potential energy becease its height decreases. It gains kinetic energy beceause its velocity is increasing. Eventually, when the ball reaches the ground, it has lost all of its potential energy (because it cannot fall any further) and at the same time reached a maximum kinetic energy.

In 1845, Joule performed an experiment which demonstrated energy transformation both qualitatively and quantitatively. The experiment was not complicated—he placed a paddle wheel in a tank of water, and measured the temperature of the water. He then cranked the wheel in the water for a period of time, and read the temperatutre again. He found that the temperature of the water rose as he cranked the paddle wheel. He quantified this observation and discovered that an equal amount of energy was always required to raise the temperature of the water by one degree. He also discovered that it did not have to be mechanical energy; it could be energy in any form. He obtained the same results with electrical or magnetic energy as he did with mechanical energy. Joule's experiments showed that different forms of energy are equivalent and can be converted from one form to another.

These observations led to what is now called the "Law of **conservation of energy**." This law states that any time energy is transferred between two objects, or converted from one form into another, no energy is created and none is destroyed. The total amount of energy involved in the process remains the same. It should be noted, however, that this does not apply to processes, such solar explosions and the generation of energy in **nuclear reactors**, which involve nuclear energy.

Most chemical reactions involve transformations in energy. A chemical reaction is simply the process whereby bonds are broken between atoms and new ones are made. When bonds are broken in a chemical reaction, the reactants, the reactants involved are colliding with sufficient kinetic energy to break the atomic bonds apart. During the collision, the kinetic energy is transformed into potential energy as the atoms are rearranged and new bonds are formed. Most chemical reactions involve the overall absorption or release of energy. The energy flow resulting from transforming kinetic energy into potential energy is heat.

There are many everyday examples of energy transformation which can be cited. One simple example is the act of lighting a match, which transforms chemical energy stored in the match head into heat and light, two other forms of energy. Another example of energy transformation is the stored chemical energy in flashlight batteries. When the flashlight is turned on, the chemical energy is first transformed into electrical energy and ultimately into light energy. Compressing a spring gives it potential energy; releasing the spring converts the potential energy into kinetic energy. Similarly, stretching a rubber band between your fingers gives the rubber band potential energy, and when it is released, the rubber band is given the kinetic energy to fly away. Electromagnetic energy in an electric motor is converted into mechanical energy. In a heat engine, the chemical energy in the fule is converted into heat energy as it is burned. The heat energy is then converted into mechanical energy. Some energy transformations seem to go "full circle." In a microphone-loudspeaker system, for example, sound (which is a form of mechanical energy) is converted by the microphone into electrmagnetic energy (electricity). The electricity reaches the loudspeaker, which converts it back into sound.

Quite often, a whole series of energy transformations is needed to complete a particular job. For example, many transformations occur in the operation of an electric hair dryer. The dryer needs to use electromagnetic energy (electricity), which does not simply originate in the outlet on the wall. A fuel source, such as **coal**, is burned in a power plant, releasing the chemical energy of fuel. The chemical energy is converted to heat energy as it is burned. This heat energy is used to change water into steam, and the rising steam is used to turn a turbine. In this process, heat energy has been transformed into mechanical energy. The mechanical energy performs work on a generator, which converts the mechanical energy into electromagnetic energy. This energy reaches the hair dryer in the form of electricity. The electricity is changed to mechanical energy inside of the dryer, where it is transformed again into heat and sound.

The ultimate example of energy transformation is that of the radiant energy of the sun. All of the energy on Earth originated from the sun. The sun's radiant energy is converted by plants into chemical energy tnrough the process of **photosynthesis**. This chemical energy is stored in the form of sugars and starches. When these plants are eaten by animals, this chemical energy is either transformed into another form of chemical energy (fats or muscle) or used for mechanical or thermal energy. Whenever an animal eats another animal, a similar process occurs. Petroleum originated from ancient plants and animals buried under many layers of sediment. Ultimately, then, even the fuels we use comes from the sun's energy. The energy we use every day is the result of the transformation of solar energy over millions of years.

Energy transformations occur every time a chemical reaction takes place, a light switch is turned on, or a person takes a breath. Joule's work enabled scientists to understand energy transformations which in turn made many technological advancements possible. Understanding the nature of energy transformation is one step toward understanding how the world works.

See also Calorimetry; Kinetics

Enols and enolate group

The enol **functional group** is -CCOH, with a double bond between the two **carbon** atoms. Compounds containing this group are enols, and their conjugate bases—the -C=COH anion—are enolates. This arrangement is also present in phenols and some of the characteristics of the enols are the same as those of the phenols. The **hydrogen atom** in the **alcohol** group (OH) is acidic in nature and is replaceable by **metals**. It can also be replaced by a carboxylic acid.

The enol form of a **molecule** is often found in equilibrium with the appropriate ketone form. This is tautomerism, where two isomers exist in dynamic equilibrium with each other. This tautomerism is brought about by the reversible move of one hydrogen atom. Most often, the ketone is the much more stable of the two tautomers. **Nucleic acids** also contain heterocyclic structures capable of this tautomerism, and can exist in the enol form, although they are normally found as the ketone form. The enol form can have different characteristics due to the presence of the alcohol group and an extra double bond in the main structure. A phosphorylated enol functional group is found in the compound phosphoenol pyruvate, an important intermediate in the glycolysis biochemical pathway. Phosphoenol pyruvate is made by the catalytic action of the enzyme enolase, and is used in turn to transfer its phosphate group to the nucleotide adenosine diphosphate (ADP), synthesizing the triphosphate-containing nucleotide adenosine triphosphate (ATP).

Enols are the tautomeric isomers of certain **ketones** enolates are the conjugate bases of enols and are important intermediates in both synthetic and biosynthetic chemistry, where the act as nucleophiles in many different substitution reactions..

See also Isomer and isomerization; Ketones; Tautomerization

Enthalpy

Most chemical reactions involve changes in **energy**. A chemical reaction is simply breaking bonds between atoms and making new ones. When bonds are broken in a chemical reaction, the reactants involved are colliding with enough kinetic energy to break the atoms apart. During the collision, the kinetic energy is transformed into **potential energy** as the atoms are rearranged and new bonds are formed. Bond breaking requires energy while bond making releases energy. Most chemical reactions, therefore, involve the overall absorption or release of energy. The energy flow that results is **heat**. The heat absorbed or released in a reaction depends on a quantity called enthalpy, represented by the capital letter H.

The enthalpy of a system can be defined as the amount of heat in the system at constant pressure. The enthalpy of a substance is the energy of the substance (the sum of the kinetic and potential energies of all of its particles) plus a small term that accounts for the pressure and **volume** of the substance. The enthalpy of a substance, like energy, depends upon its **temperature**, composition, and physical state. The numerical value of the enthalpy of a substance is very close to its energy. Enthalpy is more useful to chemists than energy when measuring the heat involved in chemical reactions. The heat absorbed or released in a reaction at constant pressure is the difference in enthalpy between the reactants and the products, not the difference in energy.

There is no way to directly measure the enthalpy of a system. The only enthalpy measurement that can be made is the change in enthalpy. An enthalpy change is the amount of heat absorbed or lost during a chemical reaction at constant pressure. The enthalpy change for a reaction is symbolized as δH (read as delta H). The Greek letter δ is used to represent a change or difference. The enthalpy change for a reaction is written as the enthalpy of the products minus the enthalpy of the reactants, and is expressed in the units of kilocalories per **mole** (kcal/mol) or kilojoules per mole (kJ/mol).

The standard enthalpy change, symbolized δH^o, is the enthalpy change measured when reactants in their standard states react to form products in their standard states. The standard state of a substance is its pure form at one atmosphere (atm) pressure (defined as the standard pressure). If the substance is a pure element, it must be in its most stable form. The standard temperature is defined as 77° F (25° C), and if a reaction occurs at a different temperature (for example, a **combustion** reaction which generates a lot of heat), the standard enthalpy change must take into account the heat required in restoring the products to standard conditions when the reaction is complete. Once δH is calculated, it can be used to determine if a reaction is endothermic or exothermic. A reaction is endothermic if heat is absorbed. In this case, δH would have a positive value. A reaction is exothermic if heat is released. An exothermic reaction would have a negative value for δH. The enthalpy change for a given reaction can increase or decrease, depending on the amount of reactants and products present. When finding the enthalpy change in a **chemistry** problem, it can be treated much like a **stoichiometry** problem. The amount of enthalpy change which occurs depends on the number of moles of the reactants. If there are twice as many moles of reactants, the enthalpy for the reactions will also be doubled.

Sometimes, a reaction cannot be performed directly. When this is the case, it is still possible to determine the change in enthalpy, thanks to the work of Swiss chemist G. H. Hess (1802-1850). In 1840, Hess showed that the heat absorbed or released in a given reaction is the same whether the reaction occurs in one or several steps. The net enthalpy change of a series of steps will be the same as if the reaction occurred all at once. This method of adding the enthalpy changes of each step is called Hess's law of heat summation. This method is quite useful if a reaction can be written as a series of reactions for which the enthalpy changes are known.

Special terms are given to enthalpy changes that occur in specific situations. The enthalpy change which occurs when a solid is melted is called the **heat of fusion**. The enthalpy change which occurs when a liquid is vaporized is called the **heat of vaporization**. The enthalpy change that occurs during the complete combustion of one mole of a substance is called the molar **heat of combustion**, symbolized δH_c^o. The heat which is absorbed or released when one mole of a substance is formed is called the molar **heat of formation**, symbolized by δH_f^o. The value of δH_f^o for a particular substance can give valuable information about the substance. If the heat of formation is a large negative number, then a large amount of energy is released as the compound is formed. Compounds that release a lot of energy during formation are usually quite stable. Once the reactions that form them start, they usually tend to continue vigorously and without any catalysis. If the heat of formation of a particular substance is a positive (or small negative) number, the substance is quite unstable and will eventually spontaneously decompose back into its elements. A compound that has a high positive number for its heat of formation is very unstable and may decompose in a violent reaction.

Scientists measure the enthalpy change in a reaction using a process called **calorimetry**. Calorimetry is the study of heat flow and heat measurement in chemical reactions. Calorimetry experiments are used to determine the enthalpy changes of reactions by accurately measuring the temperature of the reaction from initiation through completion. The reaction takes place in an apparatus called a calorimeter, which is specially designed to completely insulate the reaction system so no heat is gained from or lost to the surrounding environment.

Whether a reaction will occur on its own or will require energy input into the system, in other words, the spontaneity of the reaction, can be determined using the enthalpy changes of the reaction. A reaction which has a negative value for δH is an exothermic reaction and releases heat. In this type of reaction, the products contain less enthalpy than the reactants. This situation can be pictured as the reactants sitting on top of a hill while the products are at the bottom. The reactants contain potential energy to "roll down" the hill to become products at the bottom. It can easily be seen from this model that exothermic reactions would tend to be spontaneous. A reaction which has a positive value for δH is an endothermic reaction and requires heat. In this type of reaction, the reactants contain less enthalpy than the products. In this situation, the reactants would be at the bottom of the hill, and would need to "roll up" the hill to become products. This model illustrates the nonspontaneous tendency of **endothermic reactions**.

Most spontaneous processes are exothermic, although there are exceptions to this rule. One such exception is the melting of ice at a temperature above 32° F (0° C), a spontaneous, endothermic process. The spontaneity of a reaction, therefore, does not entirely depend on the enthalpy change involved. It also depends on a quantity called **entropy**, symbolized S. Entropy is a quantitative measurement of the randomness, or disorder, of the substances involved in a reaction. J. Willard Gibbs (1839-1903), an American mathematician, proposed a concept that would incorporate the concepts of enthalpy and entropy. This concept is called Gibbs **free energy** and is symbolized G. The change in Gibbs free energy, δG, is given by $\delta G = \delta H - T\ \delta S$. The change in Gibbs free energy equals the enthalpy change for the reaction minus the product of the absolute temperature and the change in entropy for the reaction. If δ G is negative, the reaction is spontaneous. If δG is positive, the reaction is not spontaneous. If δG is zero, the reaction is in equilibrium.

The enthalpy of a reaction describes the heat involved in the process. It has a numerical value similar, but not equal to, the energy involved in the reaction. The enthalpy change in a reaction can be used to determine if a reaction is endothermic or exothermic. It can also be used with the change in entropy to determine the spontaneity of a reaction.

Entropy

Entropy is a measure of the disorder present in a system. Entropy is disorder or randomness.

Entropy is a thermodynamic quantity, its value is equal to the amount of **heat** absorbed or emitted divided by the thermodynamic **temperature** (the absolute temperature, in Kelvin). The units of entropy are joules per Kelvin per **mole**. In chemical equations entropy is normally given the symbol S. The higher the entropy value the more disordered a system is, so for example of the three states of **matter** a gas, with the greatest degree of disorder, has the highest entropy value. Next are **liquids** and finally of the three **states of matter** the most ordered form are the **solids** and they have the lowest degree of entropy. Any change that increases entropy is a positive entropy change. Most spontaneous thermodynamic processes are accompanied by an increase in entropy.

Entropy is a measure of disorder, it is **energy** in a system that is unavailable for work. All compounds have an absolute entropy level that is an indication of the level of randomness of the molecules and the amount of energy which cannot be freed for use. For example, at 68°F (20°C), liquid **water** has an absolute entropy level of 70 $J^{-}\ mol^{-1}$ and water in the vapor phase has an absolute entropy value of 189 $J^{-}\ mol^{-1}$. The values of absolute entropy for some other compounds are **oxygen** gas 205, **carbon** monoxide gas 198, **carbon dioxide** gas 214, solid **magnesium** 33, and solid magnesium **oxide** 27. All values are in $J^{-}\ mol^{1}$ at 68°F (20°C). Absolute entropy values are different at different temperatures.

The third law of **thermodynamics** states that for a perfect crystal at a temperature of absolute zero on the Kelvin scale the entropy value is zero. This agrees with the work of **Albert Einstein** who predicted that the specific heat of a substance would approach zero at 0K. Subsequent to Einstein, **Max Planck** said that the entropy value of pure solids and liquids approaches zero as they approach 0K. Obviously if a substance is not pure then the entropy value will be different.

When a reaction occurs spontaneously the entropy of the system and surroundings increases as the change takes place. As time progresses the reaction starts to slow down and it eventually stops. The system is in equilibrium and the entropy no longer changes. A **diamond** is made of carbon in a very ordered manner. As this ordered form becomes more random the diamond reverts back to carbon. In other words, as the entropy of the system increases the diamond reverts to a more disordered form, an incredibly slow process.

A gas will expand spontaneously to fill a given **volume**, ice will melt spontaneously at a temperature above its melting point, and other similar reactions occur. With these examples the products are in a more random state than the reactants, meaning an increase in entropy. In ice the water molecules are held rigidly in place in a crystal lattice. As the water melts the water molecules are now free to move about. In solid ice the water molecules are rigidly held in place, it can be reasonably shown where an individual **molecule** is at any given time. Once that ice has melted the water molecules are in constant motion with respect to each other, it is now much harder to predict where the molecules are, the randomness and entropy of the system has increased. Many other systems show this change, for example changes from solid to liquid to gas and many substances dissolving in a solvent.

The total change of entropy in a system is dependant only upon the start and final points of the system, the pathway taken from start to finish is irrelevant. A measure of something which does not depend upon the pathway taken is known as a state function, consequently entropy is a state function. A positive entropy value indicates an increase in disorder, a negative entropy value shows that there is a move towards a more ordered state.

The second law of thermodynamics is normally stated as being that heat cannot of itself pass from a colder to a warmer body. The second law can be rewritten in the following way (taking into account entropy): any system of its own accord will always undergo change in such a way as to increase entropy. When applying the second law of thermodynamics it is important to take into account the surroundings as well as the system under consideration. In a reversible process the total change in entropy in the system and the total change in entropy in the surroundings is zero. In an irreversible process the total change in entropy in the system and in the surroundings is greater then zero, it is a positive value. The effect of this is that in the universe as a whole entropy increases. Unlike energy, entropy is not conserved and the universal level of entropy is constantly increasing.

Cleaning a house can be used as an example of increasing entropy. The house starts in a random, untidy form. A person comes along and cleans so the house is now neat. The house (the system) is now more ordered, it has had a decrease in entropy. The person doing the work however has had to metabolize food, this has given out heat (energy) into the surroundings. The second law of thermodynamics tells us that the increase in entropy in the surroundings must be greater than the decrease in entropy in the system, so the overall entropy level in the universe has been increased by the act of tidying up a house, even though at the level of a human observation it might appear that the entropy has decreased as the house is now more ordered.

Humans add order to the environment around them. They extract **metals**, build things, and generally produce ap-

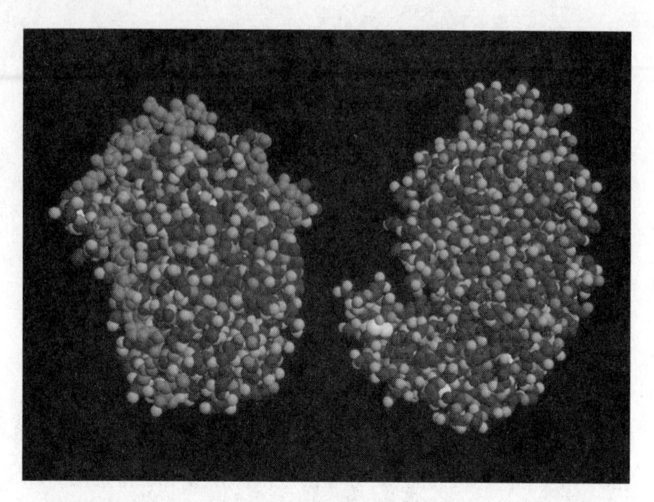

Computer-generated models of pepsinogen (left), a pre-enzyme found in the stomach, and pepsin, a digestive enzyme. In the presence of incresed acidity, pepsinogen cleaves a segment, transforming it into pepsin. ***(Photograph by Ken Eward/Science Source/National Audubon Society Collection/Photo Researchers, Inc. Reproduced by permission.)***

parent order out of chaos. However, the overall level of disorder is increasing. For each of the events listed energy is required, which will add heat to the surroundings and cause entropy to increase in the universe. As solid food is ingested some of the waste and by-products are released as liquids and **gases**, states with higher levels of entropy. In a chemical reaction when two substances react together chemical bonds are formed. These chemicals now have the constituent components joined together reducing their freedom to move independently, so the molecules concerned now have less entropy than before. The overall entropy of the universe will still increase in some other way, for example heat may be liberated by the reaction.

Ludwig Boltzmann realized that the disorder of a system is related to the number of possible arrangements of the molecules. Boltzmann devised the equation $S = k \ln W$, where S is entropy, K is the Boltzmann constant (1.38×10^{-23}J K^{1}) and ln W is the natural log of the number of possible arrangements in the system. For example a pure crystal at absolute zero has only one possible arrangement.

Entropy is a measure of randomness in a system. The universe tends towards disorder as a result of constantly increasing entropy.

See also Energy level; Energy transformations

ENZYMES

Enzymes are **proteins** used by cells to catalyze reactions. A cell carries on thousands of reactions, virtually all of them regulated by enzymes, with a unique enzyme for each reaction. The central importance of enzymes is illustrated by the fact that of the estimated 100,000 genes in the human genome, almost all code for enzymes.

The unique feature of enzyme catalysts is their specificity, meaning that most enzymes can catalyze reactions involving only one set of reactants, referred to as the enzyme's substrates. In contrast, inorganic catalysts, such as **platinum**, tend to accelerate many different reactions. While this makes them useful in the **chemistry** laboratory, where the experimenter can add a select set of reactants and create a desired product, it would be disastrous in a cell, where there are many thousands of reactions possible, only a small handful of which should be running at once. Enzyme specificity gives the cell great control over the reactions occurring within it.

Like all catalysts, enzymes speed reactions by lowering the **activation energy** required to begin the reaction. Most reactions initially require that bonds in the reactants be weakened, allowing the reactant atoms to pass into a less stable transition state, and finally to rearrange into their product state. Activation energy is the energy needed to achieve the transition state. Even if a reaction is exergonic (energy-releasing), it may not proceed if its activation energy is too high. The **combustion** of sugar is a good example. Although this reaction gives off a large amount of energy, its rate at room **temperature** is essentially zero, because of the high activation energy needed. To get the reaction to proceed, either the activation energy must be supplied in the form of **heat**, or the activation energy must be lowered, using a catalyst.

Enzymes lower the activation energy by forming temporary weak bonds to the substrates. This stabilizes the transition state, making it less energetic and therefore easier to attain. Since more sets of substrates can achieve the transition state in the presence of the enzyme, the reaction can proceed faster. Once in the transition state, the atoms can rearrange into the products, and separate from the enzyme.

Like other proteins, enzymes are composed of **amino acids** linked together to form a chain. The sequence of amino acids, called the primary structure of the enzyme, is determined by the gene coding for the enzyme. In virtually all enzymes, the amino acid chain folds back on itself in one of several regular substructures. These include the alpha-helix, in which the chain winds into a spiral, and the beta- pleated sheet, in which portions of the chain line up to form a flattened sheet. These form the secondary structure of the protein, and are responsible for creating its inner framework. The rest of the chain drapes around this framework, forming the final, tertiary, structure. Separate amino acid **chains** may then link together by **hydrogen bonding** to form a multichain quaternary structure. Enzymes are called globular proteins because their tertiary or quaternary structures are generally blob-like in shape.

The catalytic region of the enzyme, called the active site, is a very small pocket on its outer surface. Active sites may be highly exposed, or may be deeply buried within a cleft in the outer surface. Some enzymes maintain their structure rigidly, and the substrate fits in like a lock into a key. Others are more flexible, and undergo a **conformation** change to accommodate the substrate as it binds. This is known as induced fit. As discussed below, one method of enzyme regulation is to block the active site.

The specificity and catalytic activity of the enzyme depend on the particular amino acids lining the active site. The side chains of these amino acids take part directly in forming the weak bonds with substrates, and therefore the proper position, orientation and charge on these amino acids are vital to enzyme function.

Gene mutations that affect the active site can have significant effects on cell **metabolism**. While most mutations are harmful, some may allow the enzyme to react with new substrates, allowing adaptation to new food sources, for instance. For instance, the set of enzymes known as serine proteases are protein-digesting enzymes in the intestine. This group, including trypsin and chymotrypsin, evolved from a common ancestral protease, and all use the same mechanism of action at their active site. However, they differ in the particular amino acid linkages they will cleave. Trypsin cleaves only when the linkage involves lysine or arginine, which are positively charged, while chymotrypsin cleaves only linkages involving large nonpolar amino acids, such as tyrosine and phenylalanine. These differences are due to small changes in the amino acids of the enzyme at the active site.

That protein enzymes can digest proteins raises the important question of how enzymes are regulated. In the case of the serine proteases, they are synthesized as *proenzymes*, larger molecules that require modification before they become active. All are secreted together by the pancreas, in response to **hormones** signaling the arrival of food in the intestine. Trypsin is activated by another enzyme, enteropeptidase, that cleaves a unique linkage in it that allows it to take on its proper shape. Trypsin then cleaves bonds in the other proteases, activating them. Proenzymes are also important in the blood-clotting cascade, with activation by platelets.

There is a wide variety of other enzyme-regulating schemes. One of the most common is end-product inhibition, in which the product of the reaction binds to the enzyme in such a way as to change its shape slightly. This conformation change reduces the activity at the active site until the product is taken away or used up by other cell reactions, allowing the enzyme to resume working. End-product inhibition keeps the level of products fairly constant in the cell, and is the most common type of regulation in metabolic pathways such as glucose breakdown or amino acid synthesis. The inhibitor may not be the direct product of the enzyme's reaction, but instead be a **molecule** produced by another enzyme further downstream. For instance, high levels of adenosine triphosphate (ATP) inhibit the activity of phosphofructokinase, a key enzyme in glycolysis. ATP is not the direct product of this enzyme (in fact, it is a substrate), but the enzyme does promote eventual production of ATP. When the cell already has enough, the enzyme shuts down until levels fall.

Inhibition may be either competitive or noncompetitive, depending on where on the enzyme the interaction occurs. In competitive inhibition, the inhibitor directly competes for the active site, preventing the substrate from entering. **Alcohol** dehydrogenase, for instance, is the catalyst for the first step in metabolism of the **ethanol** in alcoholic drinks. It will also act on other small molecules, including **methanol** and ethylene glycol. When it does so, however, it leads to the formation of highly toxic products—formic acid and **oxalic acid**, respectively. When a person ingests either methanol or ethylene glycol, medical treatment is to give large amounts of ethanol, which competes for the active site, preventing creation of the toxic products, until the precursors are excreted.

Some competitive inhibitors bind irreversibly to the active site, preventing reaction until the cell creates new enzyme molecules. Many such inhibitors are poisons. For instance, **penicillin** kills bacteria because it permanently inhibits the final enzyme used in creating the bacterial cell wall. Such inhibitors often provide important information about the active site, because they allow the active site to be studied with a substrate-like molecule in place, but without the enzyme undergoing changes due to reaction.

In noncompetitive inhibition, the inhibitor binds to the enzyme away from the active site, but induces a conformation change that distorts the active site to prevent catalytic activity. This type of regulation is known as allosteric (other site) regulation. Unlike in competitive inhibition, noncompetitive inhibition does not require regulator molecules to be structurally similar to substrates. This allows a much larger range of regulatory possibilities. One of the most common mechanisms of **allosteric regulation** is phosphorylation, or addition of a **phosphate** group to an exposed amino acid side chain. This reversibly induces the conformation change needed to open or close the active site.

The activity of enzymes can be described by several quantities. The turnover number is the number of reactions catalyzed per second. Maximum turnover numbers range from less than one for lysozyme, to 100 for chymotrypsin, to an astonishing 600,000 for carbonic anhydrase, which catalyzes the transformation of **water** and **carbon** dioxide to carbonic acid in the bloodstream. Actual turnover numbers are dependent on **concentration** of substrates and speed of removal of products.

A related measure is the maximum **velocity**, or V_{max}, expressed in moles of product formed per second. K_M is a constant used to convey the binding strength of an enzyme to its substrates. Derived from the rate constants of the forward and reverse reactions, K_M is equal to the substrate concentration at which the **reaction rate** is half of its maximal value. Thus, a low K_M value means the enzyme is working quickly at low substrate concentration, because of its tight binding to substrate. These two quantities can be combined with the substrate concentration [S] to give the velocity at a particular concentration:

$$V=V_{max}([S]/([S] + K_M))$$

This is known as the Michaelis-Menten equation, after the two scientists who developed it.

A type of non-protein enzyme has recently been discovered made from **RNA**, or ribonucleic acid. Such "ribozymes" are often used to catalyze reactions involving deoxyribonucleic acid (DNA) or RNA. In human cells, for instance, messenger RNA is edited by RNA-containing enzymes before it is shipped out of the **nucleus** to the ribosome, which is itself a very large RNA-protein whose catalytic activity is probably dependent on the RNA it contains. Ribozymes are thought to be a remnant of the "RNA world" that preceded the DNA- and protein-based life forms we are familiar with.

EQUATION, CHEMICAL

Chemical equations represent, and are a summary of, chemical **reactions**.

Chemical equations are powerful and useful notations that can be used to quickly and concisely convey large amounts of information regarding chemical reactions.

Chemical reactions are composed of reactants (the initial substances that enter into the reaction) and products (the final substances that are present at the end of the reaction). Chemical equations are written with the reactants on the left side of the equation and with the reaction products on the right side of the equation. An arrow separates the reactants and products.

The arrow in a chemical equation represents the process of the chemical reaction. Because the chemical reaction can be simple or complex, the arrow can stand for a simple process or a complex reaction that involves many steps. Chemists often use other symbols in association with the arrow (above or below the arrow) to specify the conditions under which a chemical reaction takes place or to identify chemicals that are required for a reaction to take place but that do not directly enter into the reaction to become reactants or products (e.g., catalysts).

In chemical equations, the arrow always points in the direction of the products, that is, from the reactants to the products. The arrow is usually read as the word ''yield'' or ''produce''. In some cases the arrow points in both directions. A two-headed (bi-directional) arrow ($\leftrightarrows$) pointing both from reactants to products and from products to reactants signifies that a reaction can also run in reverse (i.e., with products of the original reaction becoming new reactants for a reverse reaction that reforms the original reactants).

In theory, all chemical reactions are reversible. In some reactions there are simultaneous reactions that form products and that revert products back to reactants. In these reactions, the formation of products represents the net chemical reaction. The direction of a reaction can usually be controlled by altering the conditions of the reaction. In theory then, all chemical reactions should contain bi-directional arrows. In practice, however, conditions are usually selected that ensure the maximum conversion of reactants into products. Although it is not always feasible to control conditions that ensure complete irreversibility, reactions that take place under conditions that minimize the reversibility of the reaction are appropriately designated as chemical equations using a single headed (directional) arrow pointing from reactants to products.

Reactants and products are separated by addition symbols (+). The addition signs represent the interaction of the reactants and are used to separate and list the products formed. The chemical equations for some reactions may have a lone reactant or a single product.

The subscript numbers associated with the chemical formula designating individual reactants and products represent the number of atoms of each element that are in in each molecule (for covalently bonded substances) or formula unit (for ionicly associated substances) of reactants or products.

In a balanced reaction, all of the **matter** (i.e., atoms or molecules) that enters into a reaction must be accounted for in the products of a reaction. Accordingly, associated with the symbols for the reactants and products are numbers (**stoichiometry** coefficients) to the left of the reactants and products that represent the number of molecules, formula units, or moles of a particular reactant involved in the reaction, or the number of molecules, formula units, or moles that result from a balanced chemical reaction.

Balanced chemical equations reflect the law of **conservation of mass**. For a chemical reaction to be balanced, all of the atoms present in molecules or formula units or moles of reactants to the left of the equation arrow must be present in the molecules, formula units and moles of product to the right of the equation arrow. The combinations of the atoms may change (indeed, this is what chemical reactions do) but the number of atoms present in reactants must equal the number of atoms present in products.

Charge is also conserved in balanced chemical reactions. Accordingly, chemical equations must reflect the conservation of electrical charge between reactants and products.

Chemical equations are usually concerned only with reactants and products. As a result chemical equations often do not show intermediate steps in a reaction. Some reactions, however, are written as multi-step reactions where the products of one reaction become the reactants for the next step in the reaction sequence. When equations are written with more than one step, each step is represented by an arrow. In multistep equations, products not written to the right of the last arrow in the multi-step equation are termed intermediate products (or intermediary products). It is these intermediate products that become new reactants for the next step in the reaction sequence.

Chemical equations that illustrate intermediate steps in a reaction may be listed as a continuous equation or may be broken up into intermediate steps with each step represented on a new line where the products of the previous step become the reactants for the next step in the reaction.

Chemical equations also convey information regarding the **states of matter** of the reactants and products of a chemical reaction. Solids, liquids and gases are designated by the respective subscripts (*s*), (*l*), and (*g*). Products and/or reactants dissolved in water are designated by the symbol for aqueous (*aq*). Subscripts designating states of matter are always enclosed in parentheses. When chemical equations lack phase notations the aqueous phase is understood to be assumed.

Reaction catalysts are usually written above or below the reaction arrow. In some cases, the energy changes associated with a particular reaction (e.g., **heat** given off by an exothermic reaction) are also designated.

Chemical equations are sometimes written in a shorthand type form where chemicals that are present both as reactants and products are not included in what is termed the *overall chemical equation*). These reaction schemes replace the formal chemical equation. The use of reaction schemes is widespread in scientific literature to summarize chemical reactions.

To be an accurate representation of a chemical reaction, a chemical equation must be entirely consistent with experimental data regarding the chemical reaction it describes. It must specifically and accurately state the reactants used up and the products formed in the chemical reaction it describes. Chemical equations accurately reflect various types of chemical reactions.

Combustion reactions are those where **oxygen** combines with another compound to form **water** and **carbon dioxide**. The equations for these reactions usually designate that the reaction is exothermic (heat producing). Equations for synthesis reactions demonstrate that two or more simple compounds combine to form a more complicated compound. Chemical equations for decomposition reactions reflect the reversal of synthesis reactions (e.g., reactions where complex molecules are broken down into simpler molecules). The **electrolysis** of water to make oxygen and **hydrogen** is an excellent example of a decomposition reaction.

Equations for single displacement reactions, double displacement, and acid-base reactions reflect the appropriate reallocation of atoms in the products.

Regardless of the type of chemical reaction, the chemical equation represents, it is important to first identify the reactants and the products. When attempting to balance chemical equations, it is critical that the chemical formulas for the reactants and the products remain unchanged. Balancing should only be accomplished through manipulation of the stoichiometry coefficients associated with reactants and products in accord with experimental observation.

See also Balance; Chemical formula

EQUATION OF STATE

An equation of state is a quantitative relationship among the variables that determine the state of a given system, that is, whether it is solid, liquid, or vapor, its **enthalpy** or **heat** content, etc. The state of a system is its condition or situation, at a given instant, and is determined by its properties. Such an equation can be used to calculate the value of one state variable, for example, pressure, if the values of the other variable are known.

Scientific studies generally involve the examination of changes in the states of systems. A system is the portion of the physical world which we set apart for study, e.g., a chemical reaction or a container of gas. The values of a specific number of the properties or variables of the system must be specified in order to completely define its state.

Variables of state, also known as state functions or properties of state, are those properties of a system which can be determined without reference to the history of the system. They depend only on the present state of the system and can be measured or calculated directly from the system in its present condition. Regardless of the previous states that the system has passed through, regardless of the values its properties have had previously, and regardless of the means used to bring it to its present condition, the present values of its state variables (functions or properties) will be the same. The change in a function of state depends only on the beginning state and the ending state, not on the path taken between the two states.

Pressure (P), **temperature** (T), **volume** (V) and the amount of material (n) are especially important state properties for many systems. Knowledge of the values of these variables determines the state of the system and facilitates the calculation of the values of other properties.

The term state is also used to designate the physical phase of the system, i.e., solid, liquid, or gas. In some cases it is necessary to give the phase of a substance, in addition to values of such properties as P, T, V, and n, in order to completely and unequivocally designate its state. For instance, **water** can exist as a supercooled liquid below the normal melting point, and we must stipulate whether we are dealing with ice or liquid water at that temperature. There are also cases for which a particular solid form must be designated in order to establish a material's state completely. For instance, solid **carbon** exists as both **graphite** and **diamond** at a wide range of pressure, volume, and temperature.

As stated above, an equation which gives the relationship among functions of state for a given system is called its equation of state. Such an equation can be used to calculate the values of a state variable using the known values of other variables.

Of particular interest are the equations of state for **gases**. Based on experimental measurements, **Robert Boyle** developed an equation in which there is a reciprocal relationship between the pressure and the volume of gases, when the temperature and the amount of gas are kept constant. The work of **Jacques Charles** in 1787 and of **Joseph-Louis Gay-Lussac** in 1808 demonstrated that the volume of a given amount of gas is directly proportional to the temperature, when its pressure remains constant.

Amedeo Avogadro proposed in 1811 that equal volumes of all gases, at the same pressure and temperature, contain the same number of molecules. In other words, an equal number of moles of all gases occupy the same volume at constant temperature and pressure. This proposal, combined with the relationships discovered by Boyle, Charles, and Gay-Lussac led to an equation of state for gases: $PV=nRT$. T is the absolute temperature in degree Kelvin (equal to the temperature in C plus 273°) and R is the universal gas constant (equal to 0.08, when V is is measured in liters, P in atmospheres, and n in moles; or equivalently, 8.3 joules per degree **mole**.)

This equation is now known as the equation of state for **ideal gases** or the ideal gas law. Although the behavior of real gases closely follows the ideal gas equation under circumstances of low pressure and reasonably high temperature, significant deviations from the ideal expectations do occur. Studies of the departures from ideal behavior by gases has led to a better understanding of atoms and molecules and their interactions in the gaseous state.

The most widely used equation of state for real gases is that of **Johannes van der Waals**. Instead of using the measured value of pressure in the equation of state for ideal gases, he substituted a term which takes into consideration the reduced pressure on the walls of a container due to the attractive forces. This magnitude of the pressure term increases with the **density** of molecules: when they are closer together, the molecules are more constrained by their mutual attraction. The van der Waals equation also corrects the volume term in the ideal gas equation by subtracting the space that is taken up by the molecules themselves. This molecular volume is not available for other molecules to move about in. The study of real gases in relationship to the van der Waals equation yields valuable information about **intermolecular forces** and the volume of molecules.

Although the van der Waals equation of state provides helpful information, it does not fit the behavior of gases at all values of pressure and temperature. A number of other theoretical and empirical equations of state have been proposed. One of these is the virial equation of state suggested by Heike Kamerlingh-Onnes (1853-1926) in 1901. It adds correction terms to the ideal gas equation by multiplying the right side of the equation (nRT) by the mathematical progression $(1 + B/V + C/V^2 + D/V^3 + E/V^4 + ...)$ where the constants A, B, C, etc. are temperature dependent. Computers are used to fit this equation to the experimental data, and values for the constants B, C, D, E, etc. are determined for the particular gas. Although this is a purely empirical procedure, there has been success in using the values of the constants to understand the behavior of gases.

EQUILIBRIUM

When chemicals that are capable of reacting with each other are mixed together, a reaction ensues and product is produced. In some cases, the reaction continues to completion—the reactants are used up and only products are present at the completion of the reaction. In many cases, however, the reaction proceeds rapidly at first, slowing down as products accumulate, until reaction ceases, even though there may still be a significant amount of the reactants present. This state of the system is called chemical equilibrium.

In most chemical reactions, if a set of reactants A and B come together and react to form products C and D, then these products are capable of accomplishing the reverse, i.e., product molecules come together and react to form the original reactants. In other words, if the reaction $A + B \rightarrow C + D$ occurs, then the reverse reaction, $C + D \rightarrow A + B$, is usually possible also. Such a reaction is said to be reversible and is represented by $A + B \leftrightarrow C + D$. The double arrow $\leftrightarrow$ indicates the reaction may proceed in either direction and will reach equilibrium. A reaction for which the reverse reaction is not possible is called an irreversible reaction.

If the forward reaction occurs when a **molecule** of A collides with a molecule of B, it should be expected that, as the reaction proceeds and molecules of A and B are used up, collisions between A and B become fewer and the reaction to products occurs less often, e.g., the rate of the reaction slows as it proceeds. Conversely, the reverse reaction occurs when a molecule of C collides with a molecule of D. Early in the reaction, when few molecules of C and D have been formed, collisions between them do not occur very often, and consequently the rate of the reaction $C + D \rightarrow A + B$ is relatively slow. As the forward reaction produces more and more C and D molecules, however, the rate of this reverse reaction speeds up. With more molecules of C and D present, they collide more often and produce A and B more rapidly.

Thus, as the reaction $A + B \rightarrow C + D$ proceeds, A and B molecules continue to be used up at a decreasing rate. But additional A and B molecules are being produced in the reverse reaction at an increasing rate. At some point, A and B are being produced in the reverse reaction as quickly as they are being used up in the forward reaction. At this point the amounts of reactant and product remain constant, and the reaction appears to cease. The reaction is now at equilibrium.

The reversibility of chemical reactions, which the concept of equilibrium requires, was first suggested by French chemist **Claude-Louis Berthollet** in 1799. In 1864, Norwegian chemists Cato M. Guldberg and **Peter Waage** published the law of **mass** action. This law states that when the reaction $A + B \leftrightarrow C + D$ comes to equilibrium, the relationship among the concentrations of the various constituents of the reaction is given by the equation: $K = [C] \times [D] / [A] \times [B]$, where the brackets [] indicate the **concentration** of a substance. K is a constant for this particular system and is called the equilibrium constant. Every chemical reaction has its own unique equilibrium constant. Regardless of the initial concentrations of A, B, C, and, C, D, the value of K will always have the same value when the reaction reaches equilibrium. Even if we begin with only C and D, and with no A or B, the value of K will be the same at equilibrium.

Dutch chemist **Jacobus H. van't Hoff** generalized the law of mass action in 1884. For the stoichiometrically balanced, reversible, general reaction $aA + bB \leftrightarrow cC + dD$, the equilibrium constant equation is written: $K = [C]^c \times [D]^d / [A]^a \times [B]^b$.

For instance, for the reaction in which ozone decomposes to oxygen: $2\ O_3 \leftrightarrow 3\ O_2$, the equilibrium constant equation is written: $K = [O_2]^3 / [O_3]^2$. The concentration of a gas may be expressed as its partial pressure, P. The equilibrium constant equation for this decomposition may, therefore, be written: $K = (P_{O2})^3 / (P_{O3})^2$.

If a reactant or product is a solid or a pure liquid, its concentration may be taken as 1 in the equilibrium constant equation. For instance for the heterogeneous reaction of solid (s) **carbon** and gaseous (g) oxygen: $2C(s) + O_2(g) \leftrightarrow 2CO(g)$, the equilibrium constant equation is: $K = (P_{CO})^2 / (PO_2)$, and the amount of **carbon monoxide** (CO) formed at equilibrium is independent of the amount of carbon initially present.

Another use of this equilibrium relationship is in the case of **solubility**. When the **salt** NaCl dissolves in **water**, it dissociates into the ions Na^+ and Cl^- in **solution**. This reversible process may be written as: $NaCl(s) \leftrightarrow Na^+(aq) + Cl\ (aq)$. The designation (aq) indicates that the **ion** is in aqueous solution (dissolved in water). The equilibrium constant equation is written: $K_{sp} = [Na^+] \times [C^1]$. K_{sp} is called the solubility product constant. The equilibrium constant for such processes may also be called the **dissociation** constant.

The relative magnitude of the equilibrium constant gives a clear indication of whether the reaction is likely to proceed to completion. For instance, the gaseous reaction: $I_2(g) + H_2(g)$

$\leftrightarrow$ 2HI(g), has an equilibrium constant with a value of 54. The very similar reaction, $Cl_2(g) + H_2(g) \leftrightarrow 2HCl(g)$, has an equilibrium constant of 2 x 10^7. The equilibrium constant for another similar reaction, involving **nitrogen** and **hydrogen**, $N_2(g) + 3H_2(g) \leftrightarrow 2NH_3(g)$, is 10 x 10^4. Since the product formed in these reactions appears in the numerator of the equilibrium constant equation, an equilibrium constant with such a high value as 2 x 10^7 indicates that the product is favored to such an extent that, at equilibrium, virtually all of the hydrogen and **chlorine** gas have reacted to form gaseous hydrogen chloride. On the other hand, a reversible reaction with an equilibrium constant whose value is very low, such as 10 x 10^{-4}, favors the reactants and very little product will be present at equilibrium. When nitrogen and hydrogen are mixed and their reaction is triggered, very little **ammonia** is formed. An equilibrium constant with a value of 54, however, indicates that, although product is favored, there will still be a significant amount of reactant present at equilibrium.

The following values are given for the solubility product constants of various substances in aqueous solution: **barium** carbonate, 7 x 10^{-9}; **cadmium** sulfide, 4 x 10^{-29}; **aluminum** hydroxide, 4 x 10^{-13}; **silver** bromate, 4 x 10^{-5}. Since K_{sp} is small for all of the compounds, none of them will be very soluble. However, if an aqueous solution was formed with each, silver bromate would be the most soluble, while the solubility of the others would be in the order: barium carbonate, aluminum hydroxide, and cadmium sulfide.

Another important use of the equilibrium constant concept is with weak acids. In general, when an acid dissolves completely in solution, it releases a hydrogen ion (H^+) which gives the solution its acidic property. For the generic acid dissociation: $HA(aq) \leftrightarrow H^+(aq) + A(aq)$, the dissociation constant may be written: $K_{sp} = [H^+] \times [A^-] / [HA]$. The larger the K_{sp} for an acid, the greater will be the concentration of hydrogen ion in solution and the greater the acidity (and the smaller the **pH**.)

If an acid, such as **hydrochloric acid** (HCl), breaks down completely, when dissolved in water, into H^+ and Cl^- ions, there will be a negligible concentration of HCl left in solution. The resulting zero in the denominator of the K_{sp} equation means that its solubility constant is undefined. HCl and similar acids are known as strong acids.

If, however, an acid dissolves completely in aqueous solution, but does not dissociate completely, it is known as a weak acid. For instance, the dissociation constant of **acetic acid** is 1.8 x 10^{-5} and for boric acid is 6.4 x 10^{-10}. The relative size of K_{sp} for these two acids indicates that while acetic acid is considerably more acidic than boric acid, both are weak acids, and both are much weaker than a strong acid such as HCl.

EQUIPARTITION OF ENERGY

In an ideal gas, each **atom** can move in three directions. The atoms's ability to move in each of these directions is counted as one degree of freedom; hence each atom is said to have three degrees of freedom. According to the principle of equipartition of **energy**, the energy per degree of freedom for this kind of motion (called translational motion) is equal to 0.5RT, where R is the molar gas constant and T is the absolute **temperature**. The equipartition principle further predicts a measurable molar **heat** capacity for such an atom of 0.5R (about 1 cal/°C-mole) per degree of freedom. Thus, for the case of an ideal gas, since each atom has three degrees of freedom, and the equipartition principle predicts a molar heat capacity of 1.5R.

In the case of a rigid **diatomic molecule**, the molecule can rotate and change the position of its center of **mass**; in this case there are three translational and two rotational degrees of freedom. Thus the equipartition theorem predicts the molar heat capacity of this system to be 2.5R. If the molecule is not rigid, it will have two extra vibrational degrees of freedom, and the molar heat capacity will be 3.5R.

Actual experimental values for many **gases** bear out the predictions of this theory. Data for monatomic gases like **argon** and **helium** give exact agreement, and many diatomic gases (e.g., **carbon** dioxide, **hydrogen** chloride, **nitrogen**, nitric **oxide**, oxygen) show experimental results in close accord with the theory.

Dulong and Petit's law is easily derived from the equipartition theorem by assuming that the internal energy of a solid consists of the vibrational energy of the molecules. This model predicts that there are six degrees of freedom for such systems, for which the molar heat capacity should be 3R.

See also Ideal gases; Thermochemistry

EQUIVALENTS

The equivalent of a substance is the **mass** which supplies or consumes one **mole** of another substance in a reaction. Equivalents can be used to simplify balancing chemical equations for many reactions. When using equivalents it is often not necessary to write balanced equations to find the amount of substances participating in the reaction. As a result, although to determine the equivalent weight of a substance you must specify the reactants and the products involved in a reaction, the equation does not have to be balanced. Because one equivalent always reacts with one equivalent, the use of equivalents can be of great help when trying to work with reactions where the balanced equation is not known.

In acid-base reactions the equivalent weight is the gram **molecular weight** of acid that yields an **Avogadro number** (6.022 x 10^{23}) of **hydrogen** ions (or, depending on the acid-base scheme, the mass required to form one mole of hydronium, H_3O+, ions). For acids such as **hydrochloric acid** (HCl) that contribute one H+ **proton** to an acid-base reaction, the equivalent weight is the gram molecular weight of the acid. According to this definition, 35.5g of HCl can contribute 6.022 x 10^{23} hydrogen protons to an acid-base reaction.

For diprotic acids—those acids such as **sulfuric acid** (H_2SO_4) that contribute two H+ protons to an acid-base reaction— the equivalent weight is one-half the gram molecular weight of the acid. For triprotic acids—those acids such as phosphoric acid (H_3PO_4) that contribute three H+ protons to an acid-base reaction—the equivalent weight is one-third the acid's gram molecular weight.

The equivalent weight of a base is the gram molecular weight of base that contributes an Avogadro number of hydroxyl ions (OH-). Bases such as **sodium** hydroxide (NaOH) that contribute one hydroxyl **ion** to an acid-base reaction, have an equivalent weight equal to their gram molecular weight.

Bases such as **calcium** hydroxide, $Ca(OH)_2$, that contribute two hydroxyl ions to an acid-base reaction, have an equivalent weight one-half their gram molecular weight. Bases such as **aluminum** hydroxide, $Al(OH)_3$, that contribute three hydroxyl ions to an acid-base reaction have an equivalent weight equal to one-third their gram molecular weight.

With regard to oxidation-reduction reactions, one equivalent of a reducing substance is the mass of the substance that gives up or releases an Avogadro number (6.022×10^{23}) of electrons in a reaction. Correspondingly, one equivalent of an oxidizing substance is the amount of the substance that collects an Avogadro number or one mole of electrons. Balancing oxidation-reduction reactions is thus made easier because one equivalent of an oxidizing agent reacts with one equivalent of reducing agent.

The equivalent weight of any compound entering into an oxidation-reduction reaction is determined by dividing the weight of one mole by the number of electrons gained or lost per formula unit. The products of the reaction must be known to determine equivalent weights in **oxidation reduction reactions** because it is necessary to know the changes in oxidation numbers between the products and the reactants. When V_2O_5 is reduced to VO_2, for example,its equivalent weight is 90.9g. When V_2O_5 is reduced to VO, however, its equivalent weight is 30.3g.

Equivalent weight is also called combining weight because elements, compounds, and ions enter into combinations proportional to their equivalent weights.

Normality is based on the concept of equivalent weight. Normality is defined as the number of equivalents per liter of **solution**. When one mole of sulfuric acid (H_2SO_4) is put into solution there are two equivalents present in the solution and therefore the solution is a two normal (2 N) solution.

See also Acid-base reactions; Normality; Valence

ERBIUM

Erbium is one of the **rare earth elements** found in Row 6 of the **periodic table**. The elements in this family are also known as the lanthanides, after the first member of the family. Erbium's **atomic number** is 68, its atomic **mass** is 167.26, and its chemical symbol is Er.

Properties

Erbium is a typical metal with a bright, shiny surface. It is soft and malleable with a melting point of 2,772°F (1,522°C), a **boiling point** of about 4,500°F (about 2,500°C), and a **density** of 9.16 grams per cubic centimeter. Erbium tends to be moderately active, and does not react with **oxygen** as quickly as do most other lanthanides. Erbium compounds tend to be pink or red, making them useful for coloring **glass** and **ceramics**.

Occurrence and Extraction

The term rare earth elements is somewhat inappropriate as the elements in this family are not really very rare. The term actually refers to the difficulty in separating the elements from each other because of their chemical similarity. Erbium ranks about number 42 in abundance in the Earth's crust, making it more common than **bromine**, **uranium**, **tin**, **silver**, **mercury**, and other better-known elements. It is extracted from its ores by first being converted to erbium fluoride (EbF_3), which is then electrolyzed to obtain the pure metal: $2ErF_3 \rightarrow 2Er + 3F_2$.

Discovery and Naming

Erbium was discovered in 1843 by the Swedish chemist **Carl Gustav Mosander**. Mosander spent many years analyzing an unusual new mineral that had been discovered in 1787 by the Swedish army office Carl Axel Arrhenius (1757-1824). That mineral, called yttria, was later to yield a total of nine new elements. Mosander suggested the name erbium for the element in honor of the town near which the mineral was discovered, Ytterby. Three other elements were also named after this small town: **terbium**, **yttrium**, and **ytterbium**.

Uses

The most important uses of erbium are in **lasers** and optical fibers. Erbium lasers are now being used to treat skin problems, such as removing wrinkles and scars. They work well for this purpose because the light they produce has relatively little **energy** and does not penetrate the skin very deeply. They also produce relatively little **heat** and cause few side effects. Erbium optical fibers are being used in long distance communication systems and in military applications. There are no commercially important erbium compounds.

ERLENMEYER, RICHARD (1825-1909)

German chemist

Richard August Carl Emil Erlenmeyer made many contributions to the field of **chemistry**. His career demonstrates that being engrossed in a field leads to the invention of improvements in methods and organization of knowledge. Erlenmeyer improved the methods of chemistry by creating a beaker that now bears his name. The beaker is conical shaped which makes it easier for mixing substances. It is an invention that grew out of his recognizing a necessity. This recognition of the need is testimony to his many hours in the laboratory. In addition, Erlenmeyer endorsed and improved a new representational system for molecules. However, he is most remembered for his research in the synthesis and constitution of aliphatic compounds. From this work, he discovered previously unknown compounds and the ''Erlenmeyer rule.''

Erlenmeyer was born in Germany in early 1825. When he was twenty years old, he entered the University of Giessen as a medical student. As part of his study, he attended a lecture by the researcher Leibig. That lecture led Erlenmeyer to a career in chemistry. He studied with Leibig in Giessen and later with Kekulé at Heidelberg. When Erlenmeyer was 43, he accepted a position at the Munich Polytechnic School as professor of chemistry. He stayed in that position for fifteen years. During his career, he was the editor of a journal started by Kekulé and another one founded by Leibig. He helped co- author a book on the organization of chemistry. In addition, he also wrote many research articles on experimental and theoretical **organic chemistry**.

Erlenmeyer is credited with being the first to synthesize several compounds. One of these was isobutyric acid, which he discovered in 1865. He was the first to synthesize guanide because he was the first to identify correctly its **structural formula**. He also correctly identified the structural formulas for creatine and creatinine, paving the way for their successful synthesis. Another compound that he successfully synthesized was tyrosine.

In 1880, Erlenmeyer illustrated the formation and structure of lactones derived from hydroxy acids. In this same year, he also reported his failed endeavors to produce **alcohol** from hydroxyls. Because of the presence of a double bond **carbon atom**, he ended up with isomeric carbonyl compounds. After many attempts, he concluded that all such alcohols would change immediately to **aldehydes** or **ketones**. This maxim became known as the Erlenmeyer rule.

Because he was so involved with research and the issues involved, he could see the needs and the direction of the field better than most. His explanations of the concepts of **valence** and structure **lead** the way to their greater acceptance. He actually coined the terms "mononvalent" and "divalent" to replace the theoretically awkward "monatomic" and "**diatomic**" (respectively). Another term he coined was "unsaturated," as applied to compounds that formed addition compounds. His use, and subsequent modifications, of the Crum Brown structural chart also lead to its faster acceptance by the field. He promoted Kekulé's **ring** structure for **benzene** and from this he created a more clear formula for benzene.

Erlenmeyer spent much time clarifying concepts. He did this by producing new terms and demonstrating why those new terms better explain a concept than what went before. He demonstrated with his sincerity that to organize an issue is to prepare for its resolution. His ideas could be readily accepted because all in the field knew his hard work was behind his recommendations.

ERNST, RICHARD R. (1933-)

Swiss chemist

Richard R. Ernst was congratulated by the president of Switzerland and friends in Zurich, who organized a party to celebrate his award of the 1991 Nobel Prize in **chemistry**, while he was in flight to receive another award from Columbia University. The pilot of the aircraft called Ernst into the cockpit to give him the news. The Royal Swedish Academy of Sciences noted in its citation of the award that Ernst's development of the methodology of high-resolution **nuclear magnetic resonance** (NMR) **spectroscopy** is the most important instrumental measuring technique within chemistry. Ernst's contributions in this field led to the development of magnetic resonance imaging (MRI), the biomedical instrument widely used today to perform noninvasive diagnosis of the human body.

Born on August 14, 1933, to Robert Ernst and Irma Brunner in Winterthur, Switzerland, Richard Robert Ernst was educated at the Eidgenössiche Technische Hochschule (ETH) in Zurich, where he received his doctorate in 1962. In 1976 he became a full professor at the ETH. After receiving his Ph.D., Ernst moved to Palo Alto, California, to become a research scientist at Varian Associates. There he worked with Weston A. Anderson on efforts to make NMR spectroscopy more sensitive.

Ernst's contribution built on the NMR experiments reported in 1945 by **Felix Bloch** at Stanford University and **Edward Mills Purcell** at Harvard. Bloch and Purcell shared the Nobel Prize in physics for this work. They demonstrated that some atomic nuclei can be knocked out of alignment in a strong magnetic field when exposed to a slow sweep of radio frequencies. The nuclei realign in response to "resonant" frequencies, emitting signals that are like a chemical signature. When he received his prize, Purcell predicted NMR would become a tool for chemical analysis. For that to happen, however, it was necessary to overcome limited sensitivity of the early NMR method to the chemical signature of the substance being analyzed. Only a few substances—**hydrogen**, **fluorine**, and **phosphorus**—had spectra strong enough to identify reliably.

In 1966 Ernst and Anderson were able to enhance NMR spectra by replacing the slow sweep of radio frequencies with short, intense pulses. Spectra too weak to identify previously became discernable. The spectra resulting from exposure to the pulse of radio frequencies were complex, and to analyze them, Ernst made use of a Fourier transformation, which computers of the mid–1960s could use to interpret the small fluctuations in brightness of the NMR spectra. NMR equipment became widely available to the chemical research community by the early 1970s.

Ernst later developed an even more sophisticated variant, two-dimensional NMR spectroscopy, which replaced single pulses of radio frequencies with a sequence of pulses, just as the pulse had replaced a sweep of radio frequencies. Building further on the concept underlying this advance, subsequent multidimensional techniques (COSY, for example, which stands for correlation spectroscopy; others are SECSY, NOESY, and ROESY) made it possible to analyze the three-dimensional structures of **proteins** and other large biological molecules. They were the basis for the development of magnetic resonance imaging or MRI, which is used in medical diagnosis. The methods also enable researchers to gather information about the chemical environment of the molecules under study. Ernst's contribution to this multidimensional study of molecules is admired by his colleagues for its elegance.

The award that Ernst was on his way to accept when he heard the airborne news about the Nobel Prize was the 1991 Louisa Gross Horwitz Prize at Columbia University. This was a joint award to Ernst and his colleague Kurt Wüthrich for their work in developing NMR methods that could show both the behavior and structure of complex biological molecules. The scientific community was not surprised with Ernst's awards since his MRI method is so widely used in clinical studies. He also received the Wolf Prize in Chemistry for 1991.

Commenting in an interview with *Physics Today* on his work with NMR spectroscopy and MRI, Ernst said, "I did not expect that it would become as useful and practical as it has." He and his group are engaged in studying how molecules interact with one another over time and how they change shape. His work is considered cross-disciplinary since it falls between chemistry and physics and also involves problems in quantum mechanics (a theory that assumes that **energy** exists in discrete units). His book on NMR is considered a classic in the field, invaluable for its cross-references to related literature.

On October 9, 1963, Ernst married Magdalena Kielholz; they had two daughters, Anna Magdalena and Katharina Elisabeth, and a son, Hans-Martin Walter. Ernst holds a number of patents for his inventions and has received recognition from many scientific societies for his work. He serves on the editorial boards of several journals dealing with magnetic resonance. He enjoys music for relaxation and pursues an interest in Tibetan art.

ESTER FUNCTIONAL GROUP

Esters are organic compounds that are formed by the joining of an acid and an **alcohol** with the resultant elimination of a **water molecule**. This is a **condensation reaction**. The **functional group** of esters is a **carbonyl group** (COO) in which the **carbon** is double bonded to one of the **oxygen** atoms and the other oxygen and the carbon both have an alkyl radical attached. The general formula of an ester is R^1COOR^2 where R^1 and R^2 may be the same or different. The majority of the reactions involving this functional group involve a break between the single bonded oxygen and carbon to yield R^1CO^+ and R^2OH. The reaction to produce an ester is reversible and to obtain the maximum yield of ester the water must be removed as it is formed. This can be carried out using **sulfuric acid** (this is the Fischer esterification reaction). The ester functional group is composed of two different groups. From the organic acid there is the acyl portion, RCO, and from the alcohol is the alkyl or aryl group, OR. In the naming of esters the alkyl or aryl group is quoted first followed by the acyl group as a separate word (the -ic ending of the acid is replaced by an -ate ending), for example butanol and propanoic acid react to produce the ester butyl propanoate and water.

Esters can be hydrolyzed back to the original acid and alcohol by heating the ester with **sodium** hydroxide. Reaction with **ammonia** yields the appropriate amide. Esters can be used to manufacture tertiary alcohols by using a Grignard reagent.

Because of their properties, esters are used as **solvents**, flavorings, perfumes, and in chemical processes. Esters also are used as organic solvents particularly for paints, varnishes, and nitrocellulose. One ester is used to extract **penicillin** from the fermenter in which it is manufactured.

Esters are colorless **liquids** with generally pleasant, fruity smells. **Solubility** in water decreases as **molecular weight** increases, although all are soluble in organic solvents. Boiling an ester with an alkali, such as **sodium hydroxide**, reclaims the alcohol and produces the sodium **salt** of the acid in a process called saponification. This process is sometimes also called base catalyzed **hydrolysis**. This is how soap is made.

Esters can be reduced to a mixture of alcohols by breaking the double bond on the oxygen. This is carried out by the addition of **hydrogen** either nascent hydrogen produced by the action of sodium on **ethanol** or the reaction can be carried out at high pressure and moderate **temperature** in the presence of a **copper** and **chromium oxide** catalyst. This is an easy way to produce an alcohol from an acid using the ester stage as an intermediary. The ester functional group can also be split by ammonia to give amides and alcohols. This reaction is sometimes referred to as ammonolysis. In the bodies of animals enzymes can break the ester bond as well to produce the corresponding acid and alcohol.

The polymerization of an ester can produce a **polyester**, an artificial fiber that is used as a cotton substitute. These polyesters also form the base material of audio tapes and video tapes as well as being used in the manufacture of clothing. Many naturally occurring fats and oils consist of esters.

See also Soap and detergents

ETHANOL

Ethanol, sometimes called methylated spirits, ethyl **alcohol**, or just alcohol, is a member of a group of organic chemicals called alcohols. Ethanol has the chemical formula of CH_3CH_2OH, with the OH (hydroxyl ion) being the **functional group**. Because of the presence of the -O H **hydrogen bonding** is quite common. Due to the fact that the hydroxyl group is attached to a **carbon atom** which is itself attached to only one other carbon atom, ethanol is referred to as a primary alcohol. Ethanol is a colorless liquid with a **boiling point** of 172°F (78°C), and a characteristic sweet smell. Ethanol can be found in nature as a natural product of the **fermentation** of **carbohydrates** such as starch and, most commonly, sugar by certain yeasts, bacteria, and molds. For human consumption, ethanol is still manufactured by fermentation. Industrially ethanol is now mostly frequently manufactured by the catalytic hydration of **ethene** or by the **hydrolysis** of ethyl sulfates.

Ethanol burns with a clear, hot, soot free flame and it is often used as a fuel for portable stoves because of this. Complete **combustion** yields **carbon dioxide** and **water**. When reacted with **sodium**, ethanol produces hydrogen gas and sodium ethoxide (CH_3CH_2ONa). Ethanol can be chemically oxidized using a mixture of concentrated **sulfuric acid** and sodium dichromate. The aldehyde, ethanal (also called acetaldehyde), is the product of this reaction. This is a pungent smelling compound. Further **oxidation** yields ethanoic acid (also called **ace-**

tic acid), a carboxylic acid. **Reduction** can occur to produce the alkane ethane. This reaction is mediated by the action of concentrated hydrogen iodide and red phosphorus. The alkene ethene can be produced by heating ethanol with (for example) **aluminum** oxide. The production of diethyl **ether** is an **elimination reaction** where one **molecule** of water is produced from two molecules of ethanol. This is carried out using concentrated sulfuric acid. The ester ethyl acetate can be produced using either an organic or inorganic acid. Water is produced in this reaction and as the reaction is reversible, the water must be removed to maximize the yield of ethyl acetate. This removal is normally carried out by adding sulfuric acid to the mixture. Ethyl halides are produced by the reaction of alkyl halides with ethanol. Chloroform can be made from ethanol by reacting the ethanol with **calcium** hypochlorite.

Ethanol is completely miscible with water and in the manufacture of ethanol it is found that a constant boiling mixture of 95.6% ethanol and 4.4% water is produced. This mixture (an **azeotrope**) is produced by fractional **distillation** and it is known as rectified spirit. To obtain a higher **concentration** of ethanol, other means must be used. In the laboratory higher concentration ethanol, with less water, can be produced by refluxing the rectified spirit with quicklime and then distilling the alcohol mixture. The product of this is absolute alcohol. Industrially the same process can be carried out by azeotropic distillation with another liquid. **Benzene** is added to the rectified spirit and the mixture is fractionally distilled, the third fraction produced from this is the absolute alcohol. Once absolute alcohol has been produced it must be stored in such a manner as to exclude water, since the high hygroscopic nature of the ethanol will readily reabsorb the water.

Different grades of ethanol have different uses. Absolute alcohol is used as a solvent and a chemical intermediate or starting point in the manufacture of other chemicals, chiefly ethanal. Mineralized methylated spirit is rectified spirit that has been made unfit for drinking by the addition of **methanol** and a blue dye, and it is this form that is often used as a fuel.

In the 1990s annual production of non-fermented alcohol was in the order of 240,000 metric tons.

Ethanol acts on humans as a depressant. In low doses ethanol can appear to have the opposite effect because it removes some inhibitions.

Ethanol is an organic alcohol with a wide range of uses, both industrially and recreationally. It has a relatively simple manufacturing process making it readily available and cheap to manufacture. Many countries charge high levels of duty on the manufacturers of ethanol which is not intended for industrial use.

Ethene

Ethene, or ethylene, is the simplest member of the alkene group. It has a chemical formula of CH_2CH_2, with a double bond between the two **carbon** atoms. Ethene is a colorless gas that can be collected from **natural gas** and crude oil. It is the most commonly manufactured organic chemical in the world.

A computerized graphic of crown ether. ***(Photograph by Ken Eward/ BioGrafx/Science Source, National Audubon Society Collection/Photo Researchers, Inc. Reproduced with permission.)***

In the 1990s production of ethene in the United States totaled approximately 15 million metric tons. The majority of the manufacturing is from the catalytic cracking of ethane, petroleum fractions, and crude oil. Ethene also is obtained from **ethanol** and it can be readily converted back to ethanol by the addition of **water** in the presence of a catalyst or at high temperatures.

Ethene has a large number of commercial and industrial applications. When ethene is absorbed by a **solution** of **potassium** permanganate the anti-freeze ethylene glycol is produced. Polymerization at high pressures or in the presence of a catalyst can produce polythene. The United States produces over six million metric tons of this one plastic annually. Ethene also is used in the manufacture of polyvinyl chloride (PVC) and **polystyrene**, and ethene gas promotes the ripening of many fruits, such as apples and bananas.

In common with all unsaturated hydrocarbons **hydrogen** can be added across the double bond to produce ethane. When ethene is burned it does so with a luminous, sooty flame and when mixed with **oxygen** it can produce an explosive mixture.

Ethene is the most commonly manufactured organic chemical and it is then processed in various ways to make many of the **plastics** on which modern society depends.

See also Alkene functional group

Ether

Ether, which is also called diethyl ether, diethyl **oxide**, ethyl ether, and ethoxyethane, has the chemical formula C_2H_5O C_2H_5. At standard **temperature** and pressure ether is a colorless liquid, with a sweet smell. Ether is very volatile and in air its vapor can form an explosive mixture. The **boiling point** of ether is 94°F (34.5°C).

Ether can be formed by the **dehydration** of excess **ethanol** by **sulfuric acid** at a temperature of 284°F (140°C). Ether does not have a **hydrogen atom** attached to the **oxygen** atom, this reduces the number of reactions it can take part in (for example it cannot function like an alcohol). Ether does not occur naturally so all ether used has to be manufactured. Ether is most commonly manufactured using the process previously described, this manufacturing process was first devised in the early nineteenth century by Alexander Williamson (1824-1904).

Ether has very few chemical properties as it is relatively inactive. It is only partly miscible with **water** but it is an excellent organic solvent. When burnt in air complete **combustion** occurs yielding **carbon** dioxide and water, as has already been stated a mixture of ether vapor and air is explosive. A strong mineral acid, such as concentrated hydrochloric or sulfuric, will dissolve ether to produce an oxonium **salt**. Ether can be oxidized by **nitric acid** to produce ethanoic acid. On standing in air ether will slowly oxidize to produce small quantities of ether peroxide, this is an unstable compound with a high boiling point and its presence can lead to explosions when old samples of ether are distilled. Storage of ether in tightly closed bottles made of dark **glass** can prevent the formation of ether peroxide.

One of the most common uses of ether is as an organic solvent. In this manner it is used in the preparation of alkanes, in reactions involving **sodium**, and as a solvent for Grignard reagents. Ether is also particularly useful in solvent extraction. In this process the ether is shaken with an organic solute in aqueous **solution**. The organic solute dissolves into the ether, which will not itself dissolve in the water. The ether and solute can then be removed for the mixture and the ether can be boiled or evaporated off, leaving a pure sample of the organic compound. Ether is commonly used as a solvent for oils, fats, **waxes**, and alkaloids.

One of the minor uses of ether but perhaps the most well known is as an anesthetic. Ether was first used as an anesthetic in 1842 by Crawford Long, in Georgia. As an anesthetic ether is reversible, predictable, and controllable. Ether also has the benefit of acting as a muscle relaxant. Application is by inhalation.

Ether is a relatively unreactive compound which has found an extensive use as an organic solvent.

See also Ether functional group

Ether functional group

The **ether** functional group is R^1OR^2, where R^1 and R^2 are alkyl or aryl groups which may or may not be the same. The central part of the ether **molecule** is C-O-C, and because the **oxygen** atom is attached only to carbons and not to a **hydrogen** atom the ethers are relatively unreactive. If a hydrogen **atom** was attached to the oxygen many of the properties would be similar to those of the alcohols. Instead the relative unreactivity is more closely related to the alkanes.

One of the main uses of ethers are as organic **solvents**. The oxygen atom of an ether **functional group** has two lone pairs of electrons. This is important in solutions of high hydrogen **ion** concentration where one of these pairs of electrons can form a dative **covalent bond** (where one of the two atoms involved contributes both electrons) with a hydrogen ion. This gives a positively charged oxonium ion. In this reaction the oxygen atom is behaving the same way as it does in a **water** molecule. In aqueous solutions free hydrogen atoms are not encountered because they join with water molecules to give hydroxonium ions, H_3O^+. Some ether compounds have the oxygen atom from the functional group attached to **carbon** atoms which are cyclic in nature, that is they are part of a **ring** structure. This is an example of a heterocyclic compound. Due to the greater number and stronger bonds that are present the **boiling point** of these cyclic ethers is higher than might be anticipated from their chemical formula. For example ether is $C_2H_5OC_2H_5$ and it has a boiling point of 94°F (34.5°C). The cyclic ether tetrahydrofuran, which has a chemical formula of $C_2H_4OC_2H_4$, has a boiling point of 149°F (65°C). This difference in the boiling point of these two very similar molecules is due to the cyclic nature of the latter and the greater **energy** needed to break some of the hydrogen bonds which are formed.

One class of compounds with the ether functional group that are an exception to the rule of unreactivity of ethers are the epoxides. Originally epoxides were named as oxides of alkenes. Ethylene **oxide** is an example of an epoxide ether. It has the chemical formula CH_2OCH_2 with the C-O-C forming a ring structure. Ethylene oxide has the **International Union of Pure and Applied Chemistry (IUPAC)** name of epoxyethane or oxirane. The C-O-C functional group when present in an epoxide has a far greater **reactivity** towards nucleophiles. The basic mode of action is to make the cyclic structure a straight chain molecule. For example ethylene oxide is added to the strong **nucleophile** water in the presence of the catalyst **sulfuric acid** at 140°F (60°C) to produce ethylene glycol (antifreeze).

The ether functional group is generally very unreactive except when found in the form of an epoxide. The oxygen atom has two lone pairs of electrons which make ethers very powerful organic solvents.

See also Alcohol; Alkane functional group; Alkene functional group

Ethyl Group

The ethyl group is a modifier of chemical behavior. The ethyl group has the chemical formula CH_3CH_2—which is often written as C_2H_5. It is frequently abbreviated to Et—.

The ethyl group is a modified alkane, it is an ethane molecule with a hydrogen atom removed. The ethyl group is an example of an **alkyl group**, or an alkyl radical.

One of the effects of the ethyl group is that as the molecular weight of the entire molecule increases, the characteristics of the molecule change. The ethyl group has a greater molecular weight than the smallest alkyl radical which is the methyl group. If we consider the **alcohol** containing each of these two groups we can see several important differences. The two alcohols are **methanol** (CH_3OH) and **ethanol** (C_2H_5OH). Ethanol has a boiling point of 172°F (78°C) whereas methanol will boil at 148°F (64.5°C). The increase in boiling point is not due totally to the increase in molecular weight, with this larger alkyl group there is an increased tendency in the alcohol to form **hydrogen bonds**. As the alkyl group increases in size the alkane properties of the alcohol become more marked. Some reactions that ethanol can undergo are impossible for methanol. For example reduction (**dehydrogenation**) of ethanol gives the alkene ethene. Methanol cannot be reduced to the corresponding alkene, because there is only the single carbon atom. For humans methanol is poisonous in small quantities whereas ethanol at low concentrations is harmless, although it can be poisonous at higher concentrations.

The free radical of the ethyl group is known to exist as an entity in its own right. This is as an intermediary a number of reactions. The ethyl group is found for example during the thermal degradation of alkanes as is encountered during various petrochemical processes.

When an ethyl group is added to a compound the process is called ethylation. If an aliphatic (straight chain) molecule is being ethylated it occurs by the substitution of the hydrogen atom found in a hydroxyl or **amine group**. In aromatic compounds (ring molecules) the substitution may be of one of the hydrogens on the ring structure itself. This substitution can be carried out by the Friedel Crafts reaction, which in the case of **benzene** (C_6H_6) is represented as:

$$C_6H_6 + CH_3CH_2Cl \rightarrow C_6H_5CH_2CH_3 + HCl$$

Ethylbenzene is produced, and the ethylation is facilitated by the addition of a small amount of anhydrous aluminum chloride.

Functional groups attached to a benzene ring affects the reactivity of the **ring** and determines to a large extent where the substitution will take place. A group that makes the ring more reactive than benzene is called an activating group, whereas a deactivating group makes the ring less reactive. The ethyl group is considered to be weakly activating in the case of electrophilic aromatic substitution reactions. For example, if equimolar amounts of benzene and ethylbenzene are treated with a small amount of **nitric acid** (a nitrating reagent) about 25 times more nitroethylbenzene is produced than nitrobenzene. The added ethyl group makes ethylbenzene about 25 times more reactive to nitration than benzene. Nitration occurs almost exclusively at the ortho- and para-ring positions in ethylbenzene. Very little of the meta-nitroethylbenzene isomer is produced.

There are several important compounds that contain an ethyl group as part of the molecule. Diethyl ether, $CH_3CH_2OCH_2CH_3$, was used for many years as a surgical anesthetic agent until being replaced by safer nonflammable alternatives. Ethyl formate, $HCOOCH_2CH_3$, is used to flavor lemonade. Ethyl chloride, CH_3CH_2Cl, is a refrigerant and is sometimes applied to skin as a topical anesthetic. Ethylbenzene is a raw material in the commercial manufacturing of styrene.

See also Alkane functional group; Alkene functional group

Ethylenediaminetetraacetic Acid

Ethylenediaminetetraacetic acid [$(HOOC\text{-}CH_2)_2N\text{-}CH_2\text{-}CH_2\text{-}N(CH_2\text{-}COOH)_2$], typically shortened to EDTA, is a **chemical compound** with the ability to form multiple bonds with metal ions, making it an important chemical to analytical scientists and industry alike.

The compounds used to create EDTA include ethylenediamine, **formaldehyde** and **sodium** cyanide. When these compounds are mixed in an appropriate fashion, a series of chemical reactions take place. Formaldehyde reacts with the sodium cyanide to form formaldehyde cyanohydrin. In the presence of **sulfuric acid**, this compound then reacts with ethylenediamine forming an intermediate compound that eventually reacts with **water** to form EDTA.

Solid EDTA is readily dissolved in water where it can form multiple chemical bonds with many metal ions in a **solution**, in effect tying up the metal ions. A **molecule** such as EDTA that has at least one free pair of unbonded electrons and therefore can form chemical bonds with metal ions. These compounds are known as ligands. Ligands are typically classified by the number of available free **electron** pairs that they have for creating bonds. **Ammonia** (NH_3), for example, which has one pair of unbonded electrons, is known as a monodentate ligand (from Latin root words meaning one tooth). EDTA has six pairs of unbonded electrons and is called a hexadentate ligand. Ligands such as EDTA, that can form multiple bonds with a single metal **ion**, in effect surrounding the ion and caging it in, are known as chelating agents (from the Greek *chele,* meaning crab's claw). The EDTA molecule seizes the metal ion as if with a claw, and keeps it from reacting normally with other substances.

Chelating agents play an important roll in many products, such as food, soda, shampoo, and cleaners. All of these products contain unwanted metal ions which affect **color**, odor, and appearance. In addition to affecting the physical characteristics of these products, some metal ions also promote the growth of bacteria and other microorganisms. EDTA ties up metal ions in these products and prevents them from doing their damage.

EDTA is also used extensively by analytical chemists for titrations. A **titration** is one method for finding the amounts of certain **metals** in various samples based on a known chemical reaction. Since EDTA bonds with metal ions, the amount of metal in a sample can be calculated based on the amount of EDTA needed to react with it. In this way, chemists can determine the amount of such things as **lead** in drinking water or **iron** in soil.

Ethyne

Ethyne, or acetylene, has the chemical formula CH_2CH_2, with a triple bond between the two carbons. It is the simplest member of the alkynes. At standard **temperature** and pressure, ethyne is a colorless gas with a pleasant smell. It is virtually insoluble in **water**. Both the gas and liquid forms of ethyne are very explosive. Ethyne burns with a very sooty flame, showing its high percentage of **carbon** content.

Hans von Euler-Chelpin.

Traditionally, the manufacture of ethyne has been by the controlled action of water on **calcium** carbide, and this reaction has been used to provide the acetylene that is burned by gas miners' and cavers' lamps. This method of production is still used although **oxidation** of **methane**, pyrolysis of alkanes, and cracking of hydrocarbons are more commonly encountered. Ethyne essentially does not naturally occur.

Ethyne is mixed with **oxygen** to produce a hot flame in oxy acetylene welding. It is also used as a starting point in the manufacture of many organic compounds. Polyvinyl chloride (PVC), various other **plastics**, and ethanal (acetaldehyde) are the most important.

In common with other unsaturated hydrocarbons, the triple bond can be broken down with the addition of **hydrogen**, or a number of **metals**, to produce the appropriate metal **salt**. The properties of ethyne are typical of the alkynes, and very similar to those of the alkenes due to the unsaturated nature of the bond between the carbons.

See also Alkene functional group, Alkyne functional group

EULER-CHELPIN, HANS VON (1873-1964)

Swedish biochemist

Hans von Euler-Chelpin described the role of enzymes in the process of **fermentation** and also researched **vitamins**, tumors, enzymes, and coenzymes. He was an important contributor in the discovery of the structure of certain vitamins. In 1929, he shared the Nobel Prize in **chemistry** with **Arthur Harden** for their research on the fermentation of sugar and enzymes. Euler-Chelpin's research has far-reaching implications in the fields of **nutrition** and medicine.

Hans Karl Simon August von Euler-Chelpin was born in Augsburg in the Bavarian region of Germany on February 15, 1873, to Rigas, a captain in the Royal Bavarian Regiment, and Gabrielle (Furtner) von Euler-Chelpin. His mother was related to the Swiss mathematician Leonhard Euler. Shortly after his birth, Euler-Chelpin's father was transferred to Munich and Euler-Chelpin lived with his grandmother in Wasserburg for a time. After his early education in Munich, Würzburg, and Ulm, he entered the Munich Academy of Painting in 1891 intending to become a artist. The problems of pigmentation led him to change his professional interest to science.

In 1893, Euler-Chelpin enrolled at the University of Munich to study physics with **Max Planck** and Emil Warburg. He also studied **organic chemistry** with **Emil Fischer** and A. Rosenheim, after which he worked with **Walther Nernst** at the University of Göttingen on problems in **physical chemistry**. This post-doctoral work in the years 1896 to 1897 was undertaken after Euler-Chelpin received his doctorate in 1895 from the University of Berlin.

The summer of 1897 was the first of several that Euler-Chelpin spent in apprentice roles in Stockholm and in Berlin. He served as an assistant to **Svante Arrhenius** in his laboratory at the University of Stockholm, becoming a privatdocent (unpaid tutor) there in 1899. Returning to Germany that summer, he studied with **Eduard Buchner** and **Jacobus Van't Hoff** in Berlin until 1900. His studies during this period centered on physical chemistry, which was receiving a great deal of attention at that time in both Germany and Sweden. Recognition came early to Euler-Chelpin for his work. He received the Lindblom Prize from Germany in 1898.

It was evident that there were new opportunities in organic chemistry. The new equipment used to measure properties could be applied to the complexities of chemical changes that took place in organisms. Euler-Chelpin's interests, therefore, shifted to organic chemistry. He visited the laboratories of others working in the field, such as Arthur Hantzsch and Johannes Thiele in Germany and G. Bertrand in Paris. These contacts contributed to his developing interest in fermentation.

In 1902, Euler-Chelpin became a Swedish citizen and in 1906 he was appointed professor of general and organic chemistry at the University of Stockholm, where he remained until his retirement in 1941. By 1910, Euler-Chelpin was able to present the fermentation process and **enzyme** chemistry into a systematic relationship with existing chemical knowledge. His book, *The Chemistry of Enzymes,* was first published in 1910 and again in several later editions.

In spite of being a Swedish citizen, Euler-Chelpin served in the German army during World War I, fulfilling his teaching obligations for six months of the year and military service for the remaining six. In the winter of 1916–1917 he took part in a mission to Turkey, a German ally during World War I, to accelerate the production of munitions and **alcohol**. He also commanded a bomber squadron at the end of the war.

After the war, Euler-Chelpin began his research into the chemistry of enzymes, particularly in the role they played in the fermentation process. This study was important because enzymes are the catalysts for biochemical reactions in plant and animal organisms. An integral aspect of Euler-Chelpin's work with enzymes was to identify each substrate (the **molecule** upon which an enzyme acted) in the reaction. He succeeded in demonstrating that two fragments (hexose) that split from the sugar molecule were disparate in **energy**. He further illustrated that the less energetic fragment, which is attached to the **phosphate**, is destroyed in the process. Apart from tracing the phosphate through the fermentation sequence, Euler-Chelpin detailed the chemical makeup of cozymase, a non-protein constituent involved in **cellular respiration**.

In 1929, Euler-Chelpin was awarded the Nobel Prize in chemistry, which he shared with Arthur Harden "for their investigations on the fermentation of sugar and of fermentative enzymes." The presenter of the award noted that fermentation was "one of the most complicated and difficult problems of chemical research." The **solution** to the problem made it possible, the presenter continued, "to draw important conclusions concerning carbohydrate **metabolism** in general in both the vegetable and the animal organism."

In 1929, Euler-Chelpin became the director of the Vitamin Institute and Institute of **Biochemistry** at the University of Stockholm, which was founded jointly by the Wallenburg Foundation and the Rockefeller Foundation. Although he retired from teaching in 1941, he continued research for the remainder of his life. In 1935 he had turned his attention to the biochemistry of tumors and developed, through his collaboration with Georg von Hevesy, a technique for labeling the **nucleic acids** present in tumors which subsequently made it possible to trace their behavior. He also helped elucidate the function of nicotinamide and thiamine in compounds which are metabolically active.

Euler-Chelpin was twice married, each time to a woman who assisted him in his research. His first wife, Astrid Cleve, was the daughter of P. T. Cleve, a professor of chemistry at the University of Uppsala. She helped him in his early research in fermentation. They married in 1902, had five children, and divorced in 1912. Euler-Chelpin married Elisabeth, Baroness Ugglas in 1913, with whom he had four children. This marriage lasted for fifty-one years. A son by his first wife, **Ulf Euler**, later also won a Nobel Prize. His award was made in 1970 in the field of medicine or physiology for his work on neurotransmitters and the nervous system.

Euler-Chelpin was awarded the Grand Cross for Federal Services with Star from Germany in 1959. He also received numerous honorary degrees from universities in Europe and America. He held memberships in Swedish science associations, as well as many foreign professional societies. He is the author of more than eleven hundred research papers and over half a dozen books. Euler-Chelpin died on November 6, 1964, in Stockholm, Sweden.

EUROPIUM

Europium is a member of the rare earth family of elements, found in Row 6 of the **periodic table**. The element has an **atomic number** of 63, an atomic **mass** of 151.96, and a chemical symbol of Eu.

Properties

Europium is a steel-gray solid with a bright, shiny surface. Its melting point is 1,519°F (826°C), its **boiling point** is about 2,712°F (1,489°), and its **density**, 5.24 grams per cubic centimeter. It is a very active metal that reacts vigorously with **oxygen** in the air, catching fire spontaneously. It also reacts easily with **water** to produce **hydrogen** gas.

Occurrence and Extraction

Europium is not very abundant in the Earth's crust, with a **concentration** of no more than about one part per million. It is one of the least abundant of the **rare earth elements**. Europium is extracted from its ores by heating with **lanthanum** metal: $Eu_2O_3 + 2La \rightarrow La_2O_3 + 2Eu$.

Discovery and Naming

Europium was discovered in 1901 by French chemist Eugène-Anatole Demarçay (1852-1904) while he was analyzing the mineral cerite, found more than a century earlier near the town of Bastnas, Sweden. Cerite was eventually found to contain six new elements in addition to europium. Demarçay chose to name the new element after the continent of Europe.

Uses

Although there are no commercially important uses for europium metal, some of its compounds do have useful applications. Some europium compounds, for example, are used as phosphors in television picture tubes, producing a red **color** when bombarded with electrons. Europium **oxide** phosphors are also used in printing postage stamps because they allow machines to read the stamp and determine its monetary value.

EVAPORATION

Evaporation is the change from a liquid state to a vapor phase. This change, unlike boiling, can occur at any **temperature**. Evaporation occurs when a liquid is left with its surface exposed to the air. A commonly encountered example is when puddles of **water**, left on the grounds after rain, eventually disappear due to evaporation.

A number of factors control the rate of evaporation of a liquid. Surface area is one, where the greater the surface area, the faster evaporation takes place. The temperature is also important; the higher the temperature, the more rapidly evaporation proceeds.

Evaporation can be used to separate a dissolved solid from its solvent. A good example of this is in the separation of **salt** from seawater. If a bowl of seawater is left standing then eventually the water will evaporate leaving behind a white residue of salt. If this is carried out slowly (by having a small surface area or by ensuring the whole system is at a cool temperature) then the resultant salt is deposited as evenly-shaped crystals. This method of producing crystals by evaporation is called crystallization.

Evaporation is an important process for collecting salt from seawater in hot countries, drying clothes, and obtaining sugar from sugar cane. In industry evaporation can mean the removal of water from a **solution**. The opposite of evaporation is condensation where vapor changes to a liquid. When the two processes are combined so a liquid is evaporated and then condensed the process is called **distillation**. Distillation is a useful method for concentrating **liquids** from a solution.

Evaporation works because of the **energy** of the molecules of the liquid. Not all of the molecules have the same energy, some have considerably more energy than the molecules around them. If these high energy molecules are near the surface then they may be able to escape into the surrounding atmosphere. Molecules can gain extra energy by having other molecules collide with them. As the temperature of the liquid increases the number of molecules with energy sufficient to escape into the vapor phase increases. Eventually a temperature is reached where all molecules have sufficient energy to change into the vapor phase. This temperature is the **boiling point** of the liquid. During boiling the temperature and energy of all of the molecules of the liquid rises, resulting in the formation of vapor bubbles within the body of the liquid. Evaporation occurs only at the surface of the liquid.

As evaporation takes place the overall energy content decreases and the net movement of molecules in the liquid decreases. The practical effect of this is that an evaporating liquid feels cold. As the evaporating liquid cools it can take energy from its surroundings making them cool down in turn. This is the mechanism by which sweating cools down the body. Evaporation also cools down animals by transferring energy to water during breathing and panting. This process also occurs to plants during transpiration.

Evaporation is part of the water cycle. Water evaporates from seas, lakes, and rivers and then condenses to form clouds and eventually rain, which returns the water to the seas, lakes and rivers.

If a liquid is kept in a sealed container it will reach an **equilibrium** over time where the number of molecules evaporating is the same as the number of molecules condensing back into the liquid. When equilibrium is reached the pressure exerted by the vapor is known as the **vapor pressure**. If the system is not closed then the vapor can be constantly removed and so with time all of the liquid will evaporate.

Evaporation is the change from liquid to vapor phase at a temperature below the boiling point of the liquid.

See also Gases; States of matter

EXOTHERMIC REACTIONS

An exothermic reaction is a chemical reaction which produces **heat**. During an exothermic reaction heat flows away from the reactants and into the surrounding environment. If an exothermic reaction is taking place inside a vessel the walls of the vessel take up the heat and feel warm to the touch. Many spontaneously occurring reactions are exothermic.

Methane reacting with **oxygen** is an example of an exothermic reaction. In this example more **energy** is produced from the formation of bonds than is required to break the existing bonds, creating an overall surplus of energy. This excess energy is released to the environment surrounding the system as heat. The rusting of **iron** is an exothermic reaction and this process is used in hand warmers which are utilized in cold regions.

The opposite of an exothermic reaction is an endothermic reaction, in which energy is taken in. Exothermic reactions are more common than endothermic ones.

The **enthalpy** change in an exothermic reaction is negative. Any burning is an example of an exothermic reaction, so fuels being burned are all exothermic reactions.

Exothermic reactions involve the production of energy which is given to the surrounding environment as heat.

See also Endothermic reactions

EXPLOSIVES

Explosives are substances that produce violent chemical or **nuclear reactions**. These reactions generate large amounts of **heat** and gas in a fraction of a second. Shock waves produced by rapidly expanded **gases** are responsible for much of the destruction seen following an explosion.

The power of most chemical explosives comes from the reaction of **oxygen** with other atoms such as **nitrogen** and **carbon** or in the decomposition of solid or liquid compounds to form gases. This split-second chemical reaction results in a small **volume** of material being transformed into a large amount heat and rapidly expanding gas. The heat released in an explosion can incinerate nearby objects. The expanding gas can smash large objects like boulders and buildings to pieces. Chemical explosives can be set off, or detonated, by heat, **electricity**, physical shock, or another explosive.

The power of nuclear explosives comes from **energy** released when the nuclei of particular heavy atoms are split apart, or when the nuclei of certain light elements are forced together. These nuclear processes, called fission and fusion, release thousands or even millions of times more energy than chemical explosions. A single nuclear explosive can destroy an entire city and rapidly kill thousands of its inhabitants with lethal radiation, intense heat, and blast effects.

Chemical explosives are used in peacetime and in wartime. In peacetime they are used to blast rock and stone for mining and quarrying, project rockets into space, and fireworks into the sky. In wartime, they project missiles carrying warheads toward enemy targets, propel bullets from guns, artillery shells from cannon, and provide the destructive force in warheads, mines, artillery shells, torpedoes, bombs, and hand grenades. So far, nuclear explosives have been used only for tests to demonstrate their force and at the end of World War II in Japan.

The first chemical explosive was **gunpowder**, or black powder, a mixture of **charcoal**, **sulfur**, and **potassium** nitrate (or saltpeter). The Chinese invented it approximately 1,000 years ago. For hundreds of years, gunpowder was used mainly to create fireworks. Remarkably, the Chinese did not use gunpowder as a weapon of war until long after Europeans began using it to shoot stones and spear-like projectiles from tubes and, later, metal balls from cannon and guns.

Europeans probably learned about gunpowder from travelers from the Middle East. Clearly by the beginning in the thirteenth century gunpowder was used more often in the West to make war than to make fireworks. The English and the Germans manufactured gunpowder in the early 1300s. It remained the only explosive for 300 hundred years, until 1628, when another explosive called fulminating **gold** was discovered.

Gunpowder changed the lives of both civilians and soldiers in every Western country that experienced its use. (Asian countries like China and Japan rejected the widespread use of gunpowder in warfare until the nineteenth century.) Armies and navies who learned to use it first-the rebellious Czech Taborites fighting the Germans in 1420 and the English Navy fighting the Spanish in 1587, for example-scored influential early victories. These victories quickly forced their opponents to learn to use gunpowder as effectively. This changed the way wars were fought, and won, and so changed the relationship between peoples and their rulers. Royalty could no longer hide behind stone walls in castles. Gunpowder blasted the walls away and helped, in part, to end the loyalty and servitude of peasants to local lords and masters. Countries with national armies became more important than local rulers as war became more deadly due, in large part, to the use of gunpowder. It was not until the seventeenth century that Europeans began using explosives in peacetime to loosen rocks in mines and clear fields of boulders and trees.

Other chemical explosives have been discovered since the invention of gunpowder and fulminating gold. The most common of these are chemical compounds that contain nitrogen such as azides, nitrates, and other nitrocompounds.

In 1846, Italian chemist Ascanio Sobrero invented the first modern explosive, **nitroglycerin**, by treating glycerin with nitric and sulfuric acids. Sobrero's discovery was, unfortunately for many early users, too unstable to be used safely. Nitroglycerin readily explodes if bumped or shocked. This inspired Swedish inventor **Alfred Nobel** in 1862 to seek a safe way to package nitroglycerin. In the mid-1860s, he succeeded in mixing it with an inert absorbent material. His invention was called dynamite.

Dynamite replaced gunpowder as the most widely used explosive (aside from military uses of gunpowder). But Nobel continued experimenting with explosives and in 1875, invented a gelatinous dynamite, an explosive jelly. It was more powerful and even a little safer than the dynamite he had invented nine years earlier. The addition of ammonium nitrate to dynamite further decreased the chances of accidental explosions. It also made it cheaper to manufacture.

These and other inventions made Nobel very wealthy. Although the explosives he developed and manufactured were used for peaceful purposes, they also greatly increased the destructiveness of warfare. When he died, Nobel used the fortune he made from dynamite and other inventions to establish the Nobel Prizes, which were originally awarded for significant accomplishment in the areas of medicine, physics, and peace.

Picric acid crystals. Picric acid is an explosive commonly used in homemade bombs. ***(Photograph by Andrew Syred/Science Photo Library, Photo Researchers, Inc. Reproduced by permission.)***

Continued research has produced many more types of chemical explosives than those known in Nobel's time: percholates, chlorates, ammonium nitrate-fuel oil mixtures (ANFO) and liquid oxygen explosives are examples.

Explosives are not only useless but dangerous unless the exact time and place they explode can be precisely controlled. Explosives would not have had the influence they have had on world history if two other devices had not been invented. The first device was invented in 1831 by William Bickford, an Englishman. He enclosed gunpowder in a tight fabric wrapping to create the first safety fuse. Lit at one end, the small amount of gunpowder in the core of the fuse burned slowly along the length of the cord that surrounded it. When the thin, burning core of gunpowder reached the end of the cord, it detonated whatever stockpile of explosive was attached. Only when the burning gunpowder in the fuse reached the stockpile did an explosion happen. This enabled users of explosives to set off explosions from a safe distance at a fairly predictable time.

In 1865, Nobel invented the blasting cap, a device that increased the ease and safety of handling nitroglycerin. Blasting caps, or detonators, send a shock wave into high explosives causing them to explode. It is itself a low explosive that is easily ignited. Detonators are ignited by primers. Primers burst into flame when heated by a burning fuse or electrical wire, or when mechanically shocked. A blasting cap may contain both a primer and a detonator, or just a primer. Another technique for setting off explosives is to send an electric charge into them, a technique first used before 1900. All these control devices helped increase the use of explosives for peaceful purposes.

In 1905, nine years after Nobel died, the military found a favorite explosive in TNT (**trinitrotoluene**). Like nitroglycerin, TNT is highly explosive but unlike nitroglycerin, it does not explode when it is bumped or shocked under normal conditions. It requires a detonator to explode. Many of the wars in this century were fought with TNT as the main explosive and with gunpowder as the main propellant of bullets and artillery shells. Explosives based on ammonium picrate and picric acid were also used by the military.

A completely different type of explosive, a nuclear explosive, was first tested on July 16, 1945, in New Mexico. Instead of generating an explosion from rapid chemical reactions, like nitroglycerin or TNT, the atomic bomb releases extraordinary amounts of energy when nuclei of **plutonium** or **uranium** are split apart in a process called **nuclear fission**. This new type of explosive was so powerful that the first atomic-bomb exploded with the force of 20,000 tons of TNT.

Beginning in the early 1950s, atomic bombs were used as detonators for the most powerful explosives of all, thermonuclear **hydrogen** bombs, or H-bombs. Instead of tapping the energy released when atoms are split apart, hydrogen bombs deliver the energy released when types of hydrogen atoms are forced together in a process called **nuclear fusion**. Hydrogen bombs have exploded with as much force as 15 million tons of TNT.

All chemical explosives, whether solid, liquid, or gas, consist of a fuel, a substance that burns, and an oxidizer, a substance that provides oxygen for the fuel. The burning and the resulting release and expansion of gases during explosions can occur in a few thousandths or a few millionths of a second. The rapid expansion of gases produces a destructive shockwave. The greater the pressure of the shockwave, the more powerful the blast.

Fire or **combustion** results when a substance combines with oxygen gas. Many substances that are not explosive by themselves can explode if oxygen is nearby. Turpentine, **gasoline**, hydrogen, and **alcohol** are not explosives. In the presence of oxygen in the air, however, they can explode if ignited by a flame or spark. This is why drivers are asked to turn off their automobile engines, and not smoke, when filling fuel tanks with gasoline. In the automobile engine, the gasoline fuel is mixed with oxygen in the cylinders and ignited by spark plugs. The result is a controlled explosion. The force of the expanding gases drives the piston down and provides power to the wheels.

This type of explosion is not useful for most military and industrial purposes. The amount of oxygen in the air deep in a cannon barrel or a mine shaft may not be enough ensure a dependably powerful blast. For this reason, demolition experts prefer to use explosive chemicals that contain molecules that readily decompose to form gases.

Many chemical explosives contain nitrogen because it readily combines to form nitrogen gas (N_2). Nitrogen is usually introduced through the action of **nitric acid**, which is often mixed with **sulfuric acid**. Nitrogen is an important component of common chemical explosives like TNT, nitroglycerin, gunpowder, guncotton, nitrocellulose, picric acid, and ammonium nitrate.

Another type of explosion can happen when very fine powders or dust mixes with air in an enclosed space. Anytime a room or building is filled with dust of flammable substances such as **wood**, **coal**, or even flour, a spark can start a fire that will spread so fast through the dust cloud that an explosion will result. Dust explosions such as these have occurred in silos where grain is stored.

There are four general categories of chemical explosives: blasting agents, primary, low, and high explosives. Blasting agents such as dynamite are relatively safe and inexpensive. Construction workers and miners use them to clear rock and other unwanted objects from work sites. Another blasting agent, a mixture of ammonium nitrate and fuel oil, ANFO, has been used by terrorists around the world because the components are readily available and unregulated. Ammonium nitrate, for instance, is found in **fertilizers**. One thousand pounds of it, packed into a truck or van, can devastate a large building.

Primary explosives are used in detonators, small explosive devices used to set off larger amounts of explosives. **Mercury** fulminate and **lead** azide are used as primary explosives. They are very sensitive to heat and electricity.

Low, or deflagrating, explosives such as gunpowder do not produce as much pressure as high explosives but they do burn very rapidly. The burning starts at one end of the explosive and burns all the way to the other end in just a few thousandths of a second. This is rapid enough, however, that when it takes place in a sealed cylinder like a rifle cartridge or an artillery shell, the gases released are still powerful enough to propel a bullet or cannon shell from its casing, though the barrel of the rifle or cannon toward a target hundreds or thousands of feet away. In fact this relatively slow burning explosive is preferred in guns and artillery because too rapid an explosion could blow up the weapon itself. The slower explosive has the effects of building up pressure to smoothly force the bullet or shell out of the weapon. Fireworks are also low explosives.

High, or detonating, explosives are much more powerful than primary explosives. When they are detonated, all parts of the explosive explode within a few millionths of a second. Some are also less likely than primary explosives to explode by accident. TNT, PETN (pentaerythritol tetranitrate), and nitroglycerin are all high explosives. They provide the explosive force delivered by hand grenades, bombs, and artillery shells. High explosives that are set off by heat are called primary explosives. High explosives that can only be set off by a detonator are called secondary explosives. When mixed with oil or wax, high explosives become like clay. These plastic explosives can be molded into various shapes to hide them or to direct explosions. In the 1970s and 1980s, plastic explosives became a favorite weapon of terrorists. Plastic explosive can even be pressed flat to fit into an ordinary mailing envelope for use as a "letter bomb."

The power of chemical explosives comes from the rapid release of heat and the formation of gases when atoms in the chemicals break their bonds to other atoms. The power of nuclear explosives comes not from breaking chemical bonds but from the core of the **atom** itself. When unstable nuclei of heavy elements, such are uranium or plutonium, are split apart, or when the nuclei of light elements, such as the isotopes of hydrogen, **deuterium** or **tritium**, are forced together, in nuclear explosives they release tremendous amounts of uncontrolled energy. These nuclear reactions are called fission and fusion. Fission creates the explosive power of the atomic bomb. Fusion creates the power of the thermonuclear or hydrogen bomb. Like chemical explosives, nuclear weapons create heat and a shock wave generated by expanding gases. The power of nuclear explosive, however, is far greater than any chemical explosive. A ball of uranium-239 small enough to fit into your hand can explode with the force equal to 20,000 tons of TNT. The heat or thermal radiation released during the explosion travels with the speed of light and the shock wave destroys objects in its path with hurricane-like winds. Nuclear explosives are so much more powerful than chemical explosives that their force is measured in terms of thousands of tons (kilotons) of TNT. Unlike chemical explosives, nuclear explosives also generate radioactive fallout.

Explosives continue to have many important peacetime uses in fields like engineering, construction, mining, and quarrying. They propel rockets and space shuttles into orbit. Explosives are also used to bond different **metals**, like those in U.S. coins, together in a tight sandwich. Explosives carefully applied to carbon produce industrial diamonds for as cutting, grinding and polishing tools.

Today, dynamite is not used as often as it once was. Since 1955 different chemical explosives have been developed. A relatively new type of explosive, "slurry explosives," are liquid and can be poured into place. One popular explosive for industrial use is made from fertilizers like ammonium nitrate or **urea**, fuel oil, and nitric or sulfuric acid. This "ammonium nitrate-fuel oil" or ANFO explosive has replaced dynamite as the explosive of choice for many peacetime uses. An ANFO explosion, although potentially powerful and even devastating, detonates more slowly than an explosion of nitroglycerin or TNT. This creates more of an explosive "push" than a high **velocity** TNT blast. ANFO ingredients are less expensive than other explosives and approximately 25% more powerful than TNT. As of 1995, sale of ANFO components were not regulated as TNT and dynamite were. Unfortunately, terrorists also began using bombs made from fertilizer and fuel oil. Two hundred and forty-one marines were killed when a truck loaded with such an ANFO mixture exploded in their barracks in Beirut Lebanon in 1983. Six people were killed and more than 1,000 injured by a similar bombing in the World Trade Center in New York in 1993. In 1995, terrorists used the same type of explosive to kill more than 167 people in Oklahoma City.

Other popular explosives in use today include PETN (pentaerythrite tetranitrate), Cyclonite or RDX, a component of plastic explosives and Amatol, a mixture of TNT and ammonium nitrite.

Nuclear explosives have evolved too. They are more compact than they were in the mid-part of the century. Today, they fit into artillery shells and missiles launched from land vehicles. Weapons designers also have created "clean" bombs that generate little radioactive fallout and "dirty" bombs that generate more radioactive fallout than older versions. Explosions of "**neutron**" bombs have been designed to kill humans with neutron radiation but cause little damage to buildings compared to other nuclear explosives.

EYRING, HENRY (1901-1981)

American physical chemist and educator

Henry Eyring was born on February 20, 1901, in the Mormon colony of Colonia Juárez, Chihuahua, México, to Edward Christian Eyring, a wealthy cattle rancher, and Caroline Cottam Eyring (Romney). He called himself a "little Mexican" and a "cowpoke" and remained a Mexican citizen until he became a naturalized United States citizen in 1935.

Because of the revolution against dictator Porfirio Díaz, in 1912, the Eyring family fled to El Paso, Texas, leaving virtually all their possessions behind. In 1914, they moved to Pima, Arizona, where Henry's father bought a farm. Henry attended the Gila Academy at Thatcher, near Pima, and although he worked on the family farm, he earned high grades. In 1919, he entered the University of Arizona at Tucson, from which he received his B.S. degree (1923).

Work in the Inspiration Copper Mining Company of Miami, Arizona, and in the mines of the Butte-Anaconda area of Montana convinced Eyring that mining was uninteresting and dangerous. He therefore obtained a United States Bureau of Mines fellowship to pursue graduate work in metallurgy at the University of Arizona, where he took his first course in **physical chemistry**. After receiving his M.S. degree in 1924, he worked during the summer at the United Verde smelter at Clarksdale, Arizona, where he decided against metallurgy as a career. He returned to the University of Arizona as Instructor in Chemistry, where he served from 1924 to 1925.

In 1925, Eyring enrolled as a graduate student and teaching fellow at the University of California at Berkeley, where he earned his Ph.D. in chemistry in two years. He was the first student to work at Berkeley on nuclear chemistry, a field for which the UC School of Chemistry later became internationally renowned.

Eyring served as Instructor at the University of Wisconsin during the academic year 1927-28. In the following year, he became a Research Associate and worked on the rate of decomposition of nitrogen(V) **oxide** in various **solvents**, an experience that led him from radiation studies to chemical **kinetics**, the field in which he would become world-famous. On August 25, 1928, he married Mildred Bennion, with whom he had three sons. His wife died in 1969, and on August 13, 1971, he married Winifred Brennan Clark, who had four daughters by a previous marriage.

In 1929, Eyring obtained a National Research Foundation fellowship to spend a year working at the Kaiser Wilhelm Institute for Physical Chemistry and **Electrochemistry** in Berlin-Dahlem with Michael Polanyi. The two used semiempirical procedures to apply the relatively new quantum theory to explain reaction rates. In 1931, they were the first to develop a three-dimensional potential **energy** surface showing how **potential energy** varied with the distances within and between the reacting molecules for a chemical reaction. Their paper was one of the most influential and important contributions that either of the two scientists ever made because it led to a deeper understanding of various chemical and physical reactions.

Eyring returned to Berkeley as a Lecturer in 1930 and extended his Berlin work to hydrogen-halogen reactions, which garnered him an invitation to participate in an American Chemical Society (ACS) symposium. There, Hugh Stott Taylor, Chairman of the Princeton University Chemistry Department, invited Eyring to present two lectures on quantum mechanics at Princeton. This developed into a one-year appointment at the univesiry, which resulted in a fifteen-year tenure.

Eyring's first work at Princeton presented an equation for the addition of dipoles, which is fundamental to high polymer theory. Other work involved reactions caused by high- energy radiation, the first work to demonstrate the mechanistic similarity between radiation-induced chemical reactions and photochemical reactions; the separation of isotopes; the theory of the electrolytic separation of heavy **water**; **optical activity**; a theory of **liquids** in which he introduced the controversial idea of treating a liquid in terms of "holes," and overvoltage.

Eyring's most important work of the 1930s was his formulation of a general treatment of reaction rates, which involved what he called "activated complexes" in a state of quasi-equilibrium with the reactants. This theory of absolute reaction rates, now called transition-state theory, appeared in one of the most significant papers ever written on chemical kinetics. In 1942, Eyring began his three- decade collaboration with Princeton biologist Frank H. Johnston on the interpretation of **bioluminescence** according to his **transition state theory**. During World War II, he worked for the United States Army and Navy on the theory of smoke and high **explosives**.

Because Eyring and his wife believed that their family's religious and social needs could better be satisfied in a Mormon community, he accepted the position of first Dean of the Graduate School and Professor of Chemistry (1946-47) and later Distinguished Professor of Chemistry and Metallurgy (1967-81) at the University of Utah at Salt Lake City. Here, in addition to establishing a program of graduate study and research, he continued an active research program until shortly before his death on December 26, 1981 of a cancer-related illness.

The author or co-author of eleven books and more than 600 scientific papers, Eyring held fifteen honorary doctorates and received numerous honors. A friendly, gregarious person, he was affectionately known on the University of Utah campus as "Uncle Henry."

Although Eyring's work on kinetics and other areas of physical chemistry was primarily theoretical, he always possessed an abiding interest in experimental matters, and he designed his theoretical work so as to interpret experimental results. His transition-state theory is one of the most important developments in twentieth-century chemistry, and his failure to win a Nobel prize has been a matter of surprise to many physical chemists.

F

Fahrenheit, Daniel Gabriel (1686-1736)

Dutch physicist

Fahrenheit invented the first truly accurate thermometer using **mercury** instead of **alcohol** and **water** mixtures. In the laboratory, he used his invention to develop the first **temperature** scale precise enough to become a worldwide standard.

The eldest of five children born to a wealthy merchant, Fahrenheit was in Danzig (Gdansk), Poland. When he was fifteen his parents died suddenly, and he was sent to Amsterdam to study business. Instead of pursuing this trade, Fahrenheit became interested in the growing field of scientific instruments and their construction. Sometime around 1707 he began to wander the European countryside, visiting instrument makers in Germany, Denmark, and elsewhere, learning their skills. He began constructing his own **thermometers** in 1714, and it was in these that he used mercury for the first time.

Previous thermometers, such as those constructed by Galileo and Guillaume Amontons, used combinations of alcohol and water; as the temperature rose, the alcohol would expand and the level within the thermometer would increase. These thermometers were not particularly accurate, however, since they were too easily thrown off by changing air pressure. The key to Fahrenheit's thermometer was a new method for cleaning mercury that enabled it to rise and fall within the tube without sticking to the sides. Mercury was an ideal substance for reading temperatures since it expanded at a more constant rate than alcohol and is able to be read at much higher and lower temperatures.

The next important step in the development of a standard temperature scale was the choosing of fixed high and low points. It was common in the early eighteenth century to choose as the high point the temperature of the body, and as the low point the freezing temperature of an ice-and-salt mixture—then believed to be the coldest temperature achievable in the laboratory. These were the points chosen by Claus Roemer, a German scientist whom Fahrenheit visited in 1701. Roemer's scale placed blood temperature at 22.5°F (-5.2°C) and the freezing point of pure water at 7.5 °F (-13.5°C). When Fahrenheit graduated his own scale he emulated Roemer's fixed points; however, with the improved accuracy of a mercury thermometer, he was able to split each degree into four, making the freezing point of water 30°F (-1.1°C) and the temperature of the human body 90°F (31.9°C). In 1717 he moved his points to 32°F (0°C) and 96°F (35.2°C) in order to eliminate fractions.

These points remained fixed for several years, during which time Fahrenheit performed extensive research on the freezing and boiling points of water. He found that the **boiling point** was constant, but that it could be changed as atmospheric pressure was decreased (such as by increasing elevation to many thousand feet above sea level). He placed the boiling point of water at 212°F (99°C), a figure that was actually several degrees too low. After Fahrenheit's death scientists chose to adopt this temperature as the boiling point of water and to shift the scale slightly to accommodate the change. With 212°F as the boiling point of water and 32°F as the freezing point, the new normal temperature for the human body became 98.6°F (36.6°C).

In 1742 Fahrenheit was admitted to the British Royal Society despite having had no formal scientific training and having published just one collection of research papers.

Faraday, Michael (1791-1867)

English physicist and chemist

Michael Faraday's early life had a remarkable resemblance to that of Benjamin Franklin. Faraday was born on September 22, 1791, in Newington, Surrey, England. Like Franklin, Michael Faraday was part of a large family. His father was a blacksmith who lacked the resources to obtain a formal education for Michael. Franklin had been apprenticed to a printer; young Fara-

Michael Faraday.

day went a similar route, becoming apprenticed to a bookbinder. In each case, this led to a voracious love of books. Michael was especially interested in **chemistry** and **electricity**. He studied the articles about electricity in the *Encyclopaedia Britannica*. His employer not only allowed him to read all that he wanted, he encouraged the boy to attend scientific lectures.

A turning point in Faraday's life occurred in 1812 when he obtained tickets to hear the lectures of **Humphry Davy** at the Royal Institution. Faraday took careful notes extending to 386 pages, which he had bound in leather. He sent a copy to Sir Joseph Banks (1743-1820), president of the Royal Society of London, who wielded great influence over European scientific investigation. Faraday hoped to make a favorable impression, but if Banks ever looked through the book with its carefully drawn and colored diagrams, Faraday never knew it. Determined not to be ignored, Faraday sent a copy of his notes to Davy and included an application for a job as Davy's assistant. Davy was extremely impressed, but did not offer Faraday work since he already had an assistant. However, after firing the assistant in 1813, Davy contacted Faraday. The job description was not quite what Faraday had in mind. A trustee of the Royal Institution had said, ''Let him wash bottles. If he is any good, he will accept the work; if he refuses, he is not good for anything.'' Faraday accepted, even though it meant he would be paid less than what he was making as a bookbinder.

Shortly thereafter, Davy resigned his post at the Royal Institution, married a wealthy widow, and decided to travel through Europe. Faraday accompanied the couple and met such illustrious men as Italian physicist Alessandro Volta and French chemist **Louis Nicholas Vauqueline**.

The next major event in Faraday's life occurred in 1820 when Danish physicist Hans Christian Oersted amazed scientists with the discovery that electric current produced a magnetic field. Faraday had a greater goal in sight: Oersted had converted electric current into a magnetic force; Faraday intended to reverse the process and create electricity from **magnetism**. Within a year Faraday, now back in England, constructed a device which essentially consisted of a hinged wire, a magnet and a chemical battery. When the current was turned on, a magnetic field was set up in the wire, and it began to spin around the magnet. Faraday had just invented the electric motor. Faraday's motor was certainly an interesting device, but it was treated as a toy. At this point, Davy, realizing that Faraday had the potential to eclipse him, jealously claimed that Faraday had taken his own idea for the experiment.

Faraday's first major contribution to chemistry came a few years later. In 1823 he unknowingly became the ''father'' of cryogenics by producing laboratory temperatures that were below freezing. He discovered how to liquify **carbon** dioxide, **chlorine**, **hydrogen** sulfide, and hydrogen bromide **gases** by placing them under pressure. But, once again Davy took credit for Faraday's work. Two years later Faraday discovered the compound **benzene**, which became his greatest contribution in chemistry. In 1865 German chemist **Friedrich Kekulé** added to Faraday's discovery by determining the structure of benzene which lead to a greater understanding of molecular structures in general.

Electricity and magnetism, were still main interests of Faraday, so he elaborated on Davy's pioneering work in **electrochemistry**. Davy had passed an electric current through a variety of molten **metals** and created new metals in the process. Faraday named this process **electrolysis**. He also bestowed names that were suggested by the British scholar Whewell and are still in use today: electrolyte, for the compound or **solution** that conducts electricity; electrode, for the metal rod inserted in the object; and **anode** and cathode for the positively and negatively charged electrodes, respectively.

In 1832 he devised what became known as **Faraday's laws** of electrolysis, which hold that the **mass** of the substance liberated at an electrode during electrolysis is proportional to the amount of electricity going through the solution; and the mass liberated by a given amount of electricity is proportional to the atomic weight of the element liberated and inversely proportional to the ''combining power'' of the element liberated. The two laws showed there was a connection between electricity and chemistry. They also supported the suggestion that Franklin had made nearly 100 years earlier when he claimed electricity was composed of particles, a theory that would be another 50 years in the making.

In another experiment Faraday sprinkled **iron** filings on a **paper** which was held over a magnet and noticed the filings had arranged themselves along what he called "lines of force." The connections along the lines showed where the strength of the field was equal. With the magnetic field now "visible," scientists began to wonder if space itself was filled with interacting fields of various types and this helped establish a new way of thinking about the universe. Up to this point most scientists had believed in the mechanical nature of the universe as established by Galileo and Isaac Newton. Taking the concept of his lines of force one step farther, Faraday realized that when an electric current began to flow it caused lines of force to expand outward. When the current stopped, the lines collapsed. If the lines expanded and collapsed across an intervening wire, an electric current would be induced to flow through it, first forward then in reverse.

By this time Faraday was giving public lectures, which were very popular, at the Royal Institute, just as Davy had. Faraday reasoned that if electricity could induce a magnetic field, then it should be possible for the reverse to be true. Taking an iron ring during one demonstration, Faraday wrapped half of it with a coil of wire that was attached to a battery and switch. André Ampère (1775-1836) had shown that electricity would set up a magnetic field in the coil. The other half of the ring was wrapped with a wire that led to a galvanometer. In theory, the first coil would set up a magnetic field that the second coil would intercept and convert back to electric current which the galvanometer would register. Faraday threw the switch: the experiment worked. He had just invented the transformer. However, the result was not exactly what he expected. Instead of registering a continuous current, the galvanometer moved only when the circuit was opened or closed. Ampère had observed the same effect a decade earlier but ignored it because it did not fit his theories. Deciding to make the theory fit the observation, instead of the other way around, Faraday concluded that when the current was turned on or off, it caused magnetic "lines of force" from the first coil to expand or contract across the second coil, inducing a momentary flow of current in the second coil. With this observation Faraday had defined electrical induction. Meanwhile, in the United States, physicist Joseph Henry had independently made the same discovery.

Faraday's affiliation with Davy had been suffering because Davy was extremely jealous of his former assistant, who was now overshadowing him. The situation escalated following Faraday's invention of the transformer; Davy claimed the idea for the experiment had been his. When Faraday was nominated to become a member of the Royal Society in 1824, Davy cast the only negative vote.

In 1839, at the age of 48, Faraday had a nervous breakdown. Like Newton, who had suffered a similar fate, Faraday never completely recovered. It is also possible that he was afflicted with a low-grade chemical poisoning. This was a common ailment affecting chemists at the time; Davy had suffered from it as well. In any event, Faraday's failing memory forced him to leave the laboratory. He published a book describing his lines of force in 1844, but because he lacked a formal education it was written without mathematical equations and it was therefore not taken seriously. When **James Clerk Maxwell** investigated the subject, he essentially came to the same conclusion as Faraday had, but used mathematics to prove his theory.

Although out of the laboratory, Faraday was by no means inactive. He investigated the effect of weak magnetic fields on nonmetallic substances and coined the terms *paramagnetic* and *diamagnetic* to differentiate between the force of attraction and repulsion. Development of a theory to explain the two opposing forces, however, had to wait more than 50 years for the work of Paul Langevin. In the 1850s, during the Crimean War, the British government sought his opinion on the feasibility of using poisonous **chemical weapons** and asked him to oversee their development. Faraday immediately said the project was very feasible, but he refused to have anything to do with its initiation.

Faraday died on August 25, 1867, at Hampton Court, Middlesex, England. The word "farad," which is a unit of capacitance, was named in his honor.

FARADAY'S LAWS

English chemist and physicist **Michael Faraday** developed the first scientific understanding of fundamental relations in electrochemical processes. He discovered that the amount of a substance produced or consumed in an electrochemical process depends quantitatively on the amount of **electricity** that flows as the reaction takes place. For example, the thickness of the **silver** plate that is deposited electrochemically on a spoon is greater if the electrical current is allowed to flow longer.

Just as the amount of **water** flowing through a pipe is equal to the rate of flow, or the current, multiplied by the length of time that the water flows, the amount of electricity that passes through a wire is equal to the electrical current multiplied by the time. The unit for current flow is the ampere. One ampere or amp is equivalent to one coulomb per second; a coulomb is the measure of charge in the SI system and is equal to 6.6242 x 10^{18} electrons.

Faraday discovered several relationships that apply to the quantitative measure of electricity. His laws can be summarized using the modern term for quantities of electricity named after him, the faraday. One faraday (1 F) of electricity is equal to one **mole** of electrons, which is equal to 96,487 coulombs of electricity. If one faraday of electricity flows through an external circuit, 6.022 x 10^{23} electrons have passed through the wire. Faraday's laws may be condensed into one statement: during an electrochemical process, the passage of one faraday through a circuit results in the transfer of one mole of electrons at the cathode and the transfer of one mole of electrons at the **anode**. Because we know the relationship between numbers of moles and the **mass** of chemical substances involved in a chemical reaction, we can determine the mass of a material which is oxidized at the anode or reduced at the cathode in an electrochemical cell if we know how many electrons are transferred for each **atom** of the substance that undergoes the chemical reaction.

Faraday's laws allow us to determine the amount of electricity (the number of coulombs) required to obtain the electrochemical conversion of a known quantity of a substance. Given a particular current flow in coulombs per second, we can also determine the time required to produce a certain amount of a substance electrochemically. The important quantities we must use to carry out these calculations are the molecular weights of the substances, the number of electrons required to oxidize or reduce each **molecule** of the substance, and, in the case of determining the time required for an electrochemical conversion, the rate of current flow.

FARR, WANDA K. (1895-1983)

American biochemist

Wanda K. Farr solved a major scientific mystery in botany by showing that the substance an important compound found in all plants, is made by tiny, cellular structures called plastids. The discovery was all the more notable because the process of **cellulose** synthesis had been obscured by the very techniques that previous researchers were using to study the phenomenon under the microscope.

Farr was born Wanda Margarite Kirkbride in New Matamoras, Ohio, on January 9, 1895, the daughter of C. Fred and Clara M. Kirkbride. Farr's father died of tuberculosis when she was a child, and her budding interest in living things was nurtured by her great-grandfather, Samuel Richardson, who was a prominent local physician. Farr had initially decided to attend medical school, but her family insisted she not become a doctor, fearing she too would be exposed to tuberculosis. Farr enrolled at Ohio University in Athens, where she received her B.S. in botany in 1915. She did graduate work in botany at Columbia University, earning her M.S. in 1918. It was at Columbia that she met her future husband, Clifford Harrison Farr, who was completing work for a Ph.D. in botany.

When Clifford went to the Agricultural and Mechanical College of Texas to teach plant physiology, Farr accepted a position as instructor of botany at Kansas State College in 1917 to be closer to him. After they were married, the young couple moved to Washington, where Clifford Farr worked at the Department of Agriculture during World War I. Here, their son Robert was born, and Farr completed her research for her master's degree from Columbia.

After the war, they moved back to Texas A&M. In 1919, the couple moved to Iowa, where Clifford became a faculty member of the University of Iowa's botany department. There, the two botanists continued to pursue their research on root hair cells—the tiny, fine hairs on plant roots that absorb **water** and nutrients from the soil. In 1925, Clifford accepted a position at Washington University in St. Louis, and the couple again moved. It was there, in February, 1928, that her husband died from a long-standing heart condition. Farr later married E. C. Faulwetter.

After her husband's death, Farr abandoned her plan to return to Columbia to finish her doctorate, remaining at Washington University to teach her late husband's classes. She also became a research assistant at the Barnard Skin and Cancer Clinic in St. Louis from 1926 to 1927, where she learned new techniques for growing animal cells in culture dishes. After the school year ended, the Bache Fund of the National Academy of Sciences awarded her a grant to continue her studies of root hairs. Her work led in 1928 to a position with the U.S. Department of Agriculture working as a cotton technologist at the Boyce Thompson Institute, which was at that time located in Yonkers, New York.

The cotton industry was eager to learn more about cotton so it could train their employees to be better judges of the quality of this crop. Farr's work centered on the origin of cellulose, which makes up most of the cell walls of cotton fibrils and provides its form and stiffness. Though she and her co-workers studied the chemical content and other characteristics of the fibrils, they were confounded—as were scientists before them—by the origin of cellulose. Researchers knew that sugar was made in tiny structures called plastids, which capture the sun's **energy** during the process of **photosynthesis**. Molecules of cellulose, however, seemed to spring into existence fully formed within the cytoplasm of cells.

The problem, Farr discovered, was that plastids seemed to disappear into the cytoplasm when mounted in water for study under the microscope. Normally, light waves refract or bend when traveling from one medium, such as air, into another medium, such as water. (This refraction of light waves is what makes a spoon placed into a glass of water seem to bend.) The cellulose plastids, however, do not do this; rather, light passes directly through them, and they are rendered indistinguishable from the liquid medium. When these plastids fill with cellulose, the pressure within the structures makes them explode, spilling the cellulose molecules into the cytoplasm, where they are then visible. Thus, when viewed through a microscope, these molecules appear to arise suddenly, fully formed.

In 1936, during her studies of cotton cellulose, Farr was appointed Director of the Cellulose Laboratory of the Chemical Foundation at the Boyce Thompson Institute, and she later worked for the American Cyanamide Company (from 1940 to 1943) and the Celanese Corporation of America (from 1943 to 1954). In 1954, she was the twelfth annual **Marie Curie** lecturer at Pennsylvania State University. In 1956, Farr established her own laboratory, the Farr Cytochemistry Lab, in Nyack, New York. She also taught botany and cytochemistry at the University of Maine from 1957 to 1960, later serving as an occasional lecturer. At the time, she was one of the few women to become director of a research laboratory.

In a 1940 *New York Times* article about Farr, she was described as "versatile, chic, vivacious, [and] as modern as tomorrow." Her versatility included using her knowledge of cellulose to analyze the sheets from a 3,500-year-old Egyptian tomb for New York's Metropolitan Museum of Art to determine if the fabric was made of cotton or linen (she determined they were linen). Among the organizations Farr held a membership were the American Association for the Advancement of Science; the American Chemical Society; the New York Academy of Science; the Royal Microscope Club (London); the American Institute of Chemists; Phi Beta Kappa; Sigma Xi; and Sigma Delta Epsilon. Farr died in 1983.

FATTY ACIDS

Fatty acids are a type of lipid composed of a long **carbon** skeleton with a **carboxyl group** on one end. They are organic acids that have the general formula RCOOH, where R represents a straight chain **hydrocarbon**. They are produced by the **hydrolysis** of fats and oils. First investigated over 200 years ago, fatty acids have become useful in producing various materials such as detergents, lubricants, and thickeners.

The **chemistry** of fatty acids was first studied by the French chemist **Michel Eugène Chevreul** in 1811. During this time, he performed experiments on a **potassium** soap produced from pig fat. One of these experiments yielded a crystalline material that had acidic properties. This was the first fatty acid discovered. During the next few years, Chevreul studied other types of soaps and produced numerous other fatty acids. He named these materials and some of these names, like **stearic acid** and butyric acid, are still used today.

Fatty acids can be thought of as having two ends. One end is the "head," which contains a carboxylic acid group (-COOH). The other end is the "tail," which is composed of a long carbon skeleton that can have anywhere from 4-24 or more atoms in length. The most common fatty acids contain 14, 16, or 18 carbon atoms. The carbon atoms are covalently bonded and surrounded by **hydrogen** atoms. Nearly all naturally occurring fatty acids have an even number of carbon molecules.

The carboxyl group on the fatty acid contributes some **water solubility** to the **molecule**. In an aqueous **solution**, the head groups tend to orient themselves toward the water while the tails are oriented towards each other. This produces structures known as micelles, giving fatty acids the ability to form emulsions. The water solubility of fatty acids is reduced as the chain length increases. When the number of carbons is greater than 12, the materials are nearly insoluble in water.

Fatty acids are classified by their carbon chain lengths and the number and location of their double bonds. Saturated fatty acids are those that contain no double bonds in their carbon **chains**. The term is used because all of the bonds that could possibly be occupied by hydrogen atoms are filled. Fatty acids that contain double bonds between carbon atoms in their chain are termed unsaturated fatty acids. For this reason, the shape of an unsaturated fatty acids is slightly kinked wherever a double bond occurs. Some common saturated fatty acids include capric, lauric, palmitic, myristic, and stearic acid. Unsaturated fatty acids include palmitoleic, oleic, linoleic, and linolenic acid. Linoleic and linolenic acid are essential fatty acids needed by the body to make important. Since the body can not produce these saturated fatty acids itself, they must be ingested.

Plants and animals provide the primary source of fatty acids. The fats or oils from these sources are isolated and hydrolyzed to produce **glycerol** and fatty acids. In general, animal fats contain saturated fatty acids. Saturated fatty acids have higher melting points than unsaturated materials and are typically solid at room **temperature**. The fatty acids from plant and fish sources are unsaturated and are liquid at room temperature. The most abundant sources of industrial fatty acids are coconut oil, palm oil, and soybean oil.

Lloyd N. Ferguson.

The simplest fatty acids include **formic acid** (HCOOH) and **acetic acid** (CH_3COOH). These materials both have a sharp smell, tangy **taste**, and are irritating to skin in large concentrations. Slightly more complex fatty acids include butyric, caprylic, and capric acid. These materials are characterized by unpleasant odors. Longer chain fatty acids such as palmitic, stearic, and myristic acid are greasy.

Fatty acids and their derivatives are important industrial chemicals. In the leather industry, they are used to soften and polish leather after tanning. Fatty acids are incorporated into rubber during the vulcanization process. They are also used as plasticizers to improve the characteristics of polymers. In the food industry, fatty acids are used as emulsifiers, stabilizers, lubricants, and defoamers. Additionally, fatty acids are added to cosmetic products to build **viscosity** and produce soaps.

FERGUSON, LLOYD N. (1918-)

American chemist

After a long and distinguished career as a chemist and educator at the California State University, Los Angeles, Lloyd N. Ferguson achieved emeritus status in 1986. In addition to teaching, Ferguson conducted important research on the relationship between the **chemistry** of organic compounds and properties such as odor and **taste**, alicycles (organic compounds with unusual molecular structures), and cancer **chemotherapy** during his years at California State. Despite all these accomplishments, Ferguson considers his efforts to encourage minority youth to pursue careers in science as one of his more significant contributions through the years. An active educator and writer, Ferguson has published six textbooks and numerous pedagogical articles, and he is the recipient of several awards, including the Distinguished Teaching Award of the Manufacturing Chemists Association in 1974, the American Chemical Society Award in chemical education in 1978, and the Outstanding Teaching Award from the National Organization of Black Chemists in 1979.

Lloyd Noel Ferguson was born in Oakland, California, on February 9, 1918, the son of Noel Swithin and Gwendolyn Louise (Johnson) Ferguson. He studied chemistry at the University of California at Berkeley, receiving his Bachelor of Science degree, with honors, in 1940 and his doctorate in 1943. Additionally, at intervals between 1941 and 1944, Ferguson worked on National Defense research projects, and he was assistant professor at the Agricultural and Technical College at Greensboro, North Carolina, during 1944–45. He married Charlotte Olivia Welch on January 2, 1944; they have two sons and a daughter.

In 1945, two years after receiving his doctorate, Ferguson joined the faculty of Howard University in Washington, D.C., becoming a full professor in 1955 and chairing the chemistry department from 1958 to 1965. Ferguson's research during this period included studies of the chemical properties of aromatic molecules; in particular, he investigated halogenation, the complex mechanisms by which aromatic molecules combine with a halogen. Ferguson also studied the molecular components and biochemical processes of taste—research that is valuable, as Ferguson argued in his 1958 article titled "The Physicochemical Aspects of the Sense of Taste," in gaining a fuller understanding "about the ways chemicals stimulate biological activity." Exploring one aspect of such research, Ferguson investigated whether a chemical compound's structural configuration has an effect on its taste by measuring the absorption of sweet and nonsweet compounds by various surfaces. Ferguson wrote three of his textbooks as a professor at Howard University: *Electron Structures of Organic Molecules, Textbook of Organic Chemistry,* and *The Modern Structural Theory of Organic Chemistry.* In 1953, Ferguson was awarded a Guggenheim fellowship, which took him to the Carlsberg Laboratory in Copenhagen, Denmark. Between 1961 and 1962 he was a National Science Foundation fellow at the Swiss Federal Institute of Technology in Zurich, Switzerland.

In 1965 Ferguson joined the faculty at California State University in Los Angeles as professor of chemistry; he then chaired the chemistry department from 1968–71. During this period, Ferguson's areas of research included the chemistry of alicycles. In his 1969 article "Alicyclic Chemistry: The Playground for Organic Chemists," Ferguson describes alicycles as providing "ideal systems for measuring electrical and magnetic interaction between nonbonded atoms and for studying the [structural] and mechanistic aspects of organic reactions," and as supplying "models for elucidating the chemistry of **natural products** such as **steroids**, alkalids, **vitamins**, **carbohydrates**, [and] antibiotics."

In 1970, Ferguson received an honorary doctorate of science degree from Howard University. He published three additional textbooks during the following decade: *Organic Chemistry: A Science and an Art, Highlights of Alicyclic Chemistry,* both volume 1 and 2, and *Organic Molecular Structure.* Along with his national teaching and educational awards, Ferguson received Outstanding Professor awards from California State University in 1974 and 1981. Ferguson's interest in cancer chemotherapy is reflected by his service on the chemotherapy advisory committee of the National Cancer Institute from 1972–75 and by articles such as "Cancer: How Can Chemists Help?" In 1973 Ferguson was appointed to the United States national committee to the **International Union of Pure and Applied Chemistry (IUPAC)** for three years. He also served on the National Sea Grant Review Panel from 1978–81 and was affiliated with the National Institute of Environmental Health Sciences from 1979–83. In 1986, Ferguson retired as emeritus professor of chemistry at California State University.

Ferguson is a member of the American Chemical Society, the National Cancer Institute, the American Association for the Advancement of Science, and the Royal Chemical Society, among other professional and scientific bodies. Ferguson accepted a post as visiting professor at the University of Nairobi in Kenya during 1971–1972. In 1976, he was awarded the Distinguished American Medallion from the American Foundation for Negro Affairs. In 1984–85, Ferguson taught at Bennett College in Greensboro, North Carolina, as a United Negro College Fund scholar-at-large. He has helped establish both the National Organization of Black Chemists and Engineers (1989) and the American Chemical Society's SEED (Support of the Educationally and Economically Disadvantaged) program.

FERMENTATION

Fermentation is the biochemical process in which **energy** is extracted from sugar without the use of **oxygen**. Fermentation can be performed by virtually all organisms, and some rely on it exclusively for their energy needs. Fermentation by yeast is the basis of the alcoholic beverage industry.

The **chemistry** of fermentation were first investigated by **Louis Pasteur** in 1860, who called the process *la vie sans air*, or life without air. Fermentation is indeed carried out without oxygen, and is therefore called an anaerobic process (*an-* meaning without; *aero* meaning air). Organisms that can either use air or not are called facultative aerobes, while those that must live without air are called obligate anaerobes. When oxygen is low, facultative aerobes, such as yeast, switch from **cellular respiration**, which requires oxygen, to fermentation.

In 1897, Hans and Eduard Beuchner discovered that fermentation could occur in a cell-free extract of yeast. This work led to the elucidation of the enzymes involved, and also dealt a blow to **vitalism**, the belief that life possessed a special force that distinguished it from non-living chemicals.

In all organisms, breakdown of the six-carbon sugar glucose begins with a multi-step pathway known as glycolysis. During glycolysis, glucose is split to form two molecules of the three-carbon compound pyruvate. This reaction occurs in the cytosol of the cell. In aerobes with sufficient oxygen, pyruvate is transported to the mitochondrion to undergo further transformations in the **Krebs cycle**. Without oxygen, the pyruvate instead remains in the cytosol, and undergoes anaerobic transformation. Depending on the organism, the product will be either lactic acid, **ethanol** plus **carbon** dioxide, or another compound. Some scientists use the term fermentation to mean only this last transformation, while others include glycolysis within the scope of fermentation. Here we shall use this more inclusive definition.

The purpose of glucose breakdown is to harvest energy from it, to allow the cell to perform various energy-requiring processes such as **protein synthesis** or locomotion. These processes invariably use adenosine triphosphate (ATP) as the direct supplier of **free energy**. When ATP is hydrolyzed by **water** to adenosine diphosphate (ADP) plus inorganic **phosphate** (Pi), energy is released. When this reaction is coupled to an energy-requiring reaction, that reaction is driven forward toward the product state. This coupling is the key to cell function, and indeed to life itself.

ATP can be made by dehydrating ADP and Pi, but of course this requires a supply of energy. This energy is supplied by the glucose reacting in glycolysis. By the end of glycolysis, enough energy has been extracted from the conversion of glucose to pyruvate to allow the net formation of two ATP molecules. To perform this conversion, though, some of the hydrogens from glucose have to be removed, and placed on **hydrogen** carrier molecules called nicotinamide adenine dinucleotide (NAD^+) to form NADH.

NAD^+ plays a critical role in glycolysis, and further rounds of glycolysis cannot occur unless it is regenerated from NADH. If oxygen is present, the NADH is shuttled to the mitochondria, where it is stripped of its hydrogen before returning to the cytosol. Without oxygen, however, it must be regenerated by the final stage of fermentation. During this stage, the hydrogen is donated back to pyruvate, forming lactic acid (in animals) or ethanol and **carbon dioxide** (in yeast).

By 1940, the details of the complete glycolytic pathway had been elucidated, by a number of biochemists, most notably Gustav Embden (1872-1933) and Otto Myerhoff. Glycolysis is sometimes called the Emden-Myerhoff pathway. The details are as follows:

1. Glucose is first destabilized by the addition of a phosphate group, to form glucose-6-phosphate. The phosphate comes from reaction of 1 ATP supplied from the cytosol. This reaction is catalyzed by the **enzyme** hexokinase.

2. Glucose-6-phosphate is rearranged to form fructose-6-phosphate, by the enzyme phosphoglucoisomerase. This rearrangement exposes an OH group needed in the next step.

3. Another phosphate is added from another ATP, to form fructose-1,6- bisphosphate. This step is catalyzed by phosphofructokinase. This is the final- energy requiring step. Both of the phosphates will later be taken back to form ATPs again.

4. The **molecule** is split by aldolase to form glyceraldehyde phosphate and dihydroxyacetone phosphate. The latter compound is converted into another molecule of glyceraldehyde phosphate by isomerase. Each of the following steps, therefore, should be thought of as occurring twice, once for each glyceraldehyde phosphate.

5. Glyceraldehyde phosphate is oxidized by removal of a hydrogen **atom** and its two **bonding** electrons, which join NAD^+ to form NADH. At the same time, a negatively charged phosphate (Pi) from the cytosol bonds with its electrons to the carbon that lost the hydrogen, forming 1,3-bisphosphoglycerate. This phosphate bond is weak, and therefore can be harvested later to form ATP. Since there are two of these molecules at this stage, this provides the two ATPs which are the energy profit of glycolysis. The whole reaction is catalyzed by triose phosphate dehydrogenase.

6. The phosphate just added is transferred to ADP to form ATP, and forming 3- phosphoglyceric acid. It is structurally identical to glyceraldehyde phosphate from step 4, except that the terminal group is a carboxylic acid instead of an aldehyde. It is that **oxidation** from aldehyde to acid that supplied the energy to make the ATP and the NADH.

7. 3-phosphoglyceric acid is isomerized to 2-phosphoglyceric acid, by phosphoglyceromutase. This places the phosphate in a more advantageous position for the next step.

8. 2-phosphoglyceric acid is dehydrated and rearranged to form phosphoenolpyruvic acid, catalyzed by enolase.

9. The final phosphate is removed by pyruvate kinase, forming an ATP and pyruvate.

Overall, then, glycolysis has converted on glucose molecule to 2 pyruvates, invested 2 ATP and harvested 4, for a net of 2 ATP formed, and converted 2 NAD^+ into 2 NADH.

As noted above, regeneration of NAD^+ is the goal of the next steps. In animals, this is accomplished by donating the hydrogen on NADH to pyruvate, forming lactic acid. This step does not produce more ATP, and is only done when no oxygen is available in the mitochondrion to accept the hydrogen there. Lactic acid build-up in muscle triggers pain **receptors**, signaling the animal to stop exerting itself until more oxygen can arrive from the bloodstream. The muscle ache you feel after a rapid bout of heavy exercise is due to lactic acid. Lactic acid formed in muscle is transported to the liver, where it is converted back into pyruvate.

In yeast and some bacteria, NAD^+ is regenerated in a two-step process. Pyruvate is first decarboxylated to produce acetaldehyde and carbon dioxide, catalyzed by pyruvate decarboxylase. Acetaldehyde then accepts the hydrogen of NADH, forming ethanol (ethyl alcohol) and NAD^+.

This reaction is the basis of the brewing industries. Sugars from grapes (for wine) or grain (for beer) is fermented by strains of yeast chosen for their tolerance of **alcohol** and other characteristics. These yeast continue to ferment until either the sugar is depleted, or their fermenting enzymes become inhibited by the build-up of products. This occurs at 10-20% alcohol for most wines, and 3-6% for most beers. Stronger alcoholic beverages, such as whiskey or vodka, are made by **distillation** of alcohol.

In both wine- and beer-making, initial fermentation occurs in large vats, allowing the carbon dioxide to bubble off. For most wines, bottling is done only after fermentation has ceased. Champagne and other ''sparkling'' wines are bottled before the end of fermentation, causing some carbon dioxide to be trapped in the bottle, where it becomes dissolved in the liquid. Most beers are also completely fermented before bottling, with carbon dioxide added back in just before capping.

Fermentation is also important in baking. Yeast are mixed in with the dough, where they consume sugar. The carbon dioxide gas causes the dough to expand during the rising phase. Once in the oven, fermentation ceases as the yeast die from the **heat**. Ethanol produced during fermentation is mostly driven off by the heat, but some may remain in the bread.

Different types of bacteria have evolved different fermentation strategies. Some bacteria convert pyruvate and NADH to **acetic acid** and carbon dioxide. The word *vinegar* derives from the French *vin aigre*, meaning sour wine, in reference to the sour **taste** acetic acid imparts to wine contaminated by such bacteria. A variety of other organic acids are produced by other bacteria, including butyric acid, responsible for the taste of rancid butter. Other bacteria, such as Clostridium, create hydrogen gas from the extra hydrogens on NADH. This bacterium is an obligate anaerobe, and is responsible for gaseous gangrene.

See also Bioenergetics; Carbohydrates

Fermium

Fermium is a transuranium element with **atomic number** 100. Its atomic **mass** is 257.0951 and its chemical symbol is Fm.

Properties

All isotopes of fermium are radioactive, with fermium-257 having the longest half life, 20.1 hours. Thus far, too little fermium has been prepared for scientists to determine its physical properties. Research suggests that its chemical properties are similar to those of **erbium**, above it in the **periodic table**.

Occurrence and Extraction

Fermium has never been discovered in the Earth's crust and is not thought to be present there. It is prepared artificially in particle accelerators by bombarding heavy transuranium elements, such as **californium**, with alpha particles.

Discovery and Naming

Fermium was discovered in 1952 as one of the byproducts of the first **hydrogen** bomb test at Eniwetok Atoll, Marshall Islands, in the Pacific Ocean. **Albert Ghiorso** led the team of researchers from the University of California that made the discovery. The team decided to name the element in honor of the Italian physicist Enrico Fermi (1901-54) who made many important scientific discoveries and was a leader of the U.S. effort to build the world's first atomic bomb during World War II.

Uses

Fermium exists in such small amounts that it has no practical uses.

Fersman, Aleksandr Evgenievich (1883-1945)

Russian geochemist

Aleksandr Evgenievich Fersman was a Russian geochemist and mineralogist. He made major contributions to Russian geology, both in theory and exploration, advancing scientific understanding of **crystallography** and the distribution of elements in the Earth's crust, as well as founding a popular scientific journal and writing biographical sketches of eminent scientists. He was known as a synthesist of ideas from different subdisciplines.

Fersman was born on November 8, 1883 in St. Petersburg to a family that valued both art and science. His father, Evgeny Aleksandrovich Fersman, was an architect; his mother, Maria Eduardovna Kessler, a pianist and painter. Fersman's maternal uncle, A. E. Kessler, had studied **chemistry** under Russian chemist Aleksandr Mikhailovich Butlerov.

At the family's summer estate in the Crimea, Fersman first discovered minerals and began to collect them. When his mother became ill, the family traveled to Karlovy Vary (Carlsbad) in Czechoslovakia. There the young Fersman explored abandoned mines and added to his collection of crystals and druses (crystal-lined rocks).

Fersman graduated from the Odessa Classical Gymnasium in 1901 with a gold medal and entered Novorossisk University. He found the **mineralogy** course so dull that he decided to study art history instead. He was dissuaded by family friends (the chemist A. I. Gorbov and others) who encouraged him to delve into molecular chemistry. He subsequently studied **physical chemistry** with B. P. Veynberg, who had been a student of Russian chemist **Dmitri Ivanovich Mendeleev**. Veynberg taught Fersman about the properties of crystals.

The Fersman family moved to Moscow in 1903 because Aleksandr's father became commander of the First Moscow Cadet Corps. Fersman transferred to Moscow University, where his interest in the structure of crystals continued. Studying with mineralogist **V. I. Vernadsky**, he became an expert in goniometry (calculation of angles in crystal) and published seven scientific papers on crystallography and mineralogy as a student. When Fersman graduated in 1907, Vernadsky encouraged him to become a professor.

By 1908 Fersman was doing postgraduate work with **Victor Goldschmidt** at Heidelberg University in Germany. Goldschmidt sent him on a tour of western Europe to examine the most interesting examples of natural **diamond** crystals in the hands of the region's jewelers. This work formed the basis of an important monograph on diamond crystallography Fersman and Goldschmidt published in 1911.

While a student in Heidelberg, Fersman also visited French mineralogist François Lacroix's laboratory in Paris and encountered pegmatites for the first time during a trip to some islands in the Elbe River that were strewn with the rocks. Pegmatites are granitic rocks that often contain rare elements such as **uranium**, **tungsten**, and **tantalum**. Fersman was to devote years to their study later in his career.

In 1912 Fersman returned to Russia, where he began his administrative and teaching career. He became curator of mineralogy at the Russian Academy of Science's Geological Museum. He would be elected to the Academy and become the museum's director in 1919. During this period Fersman also taught **geochemistry** at Shanyavsky University and helped found *Priroda,* a popular scientific journal to which he contributed throughout his life.

Fersman participated in an Academy of Science project to catalogue Russia's natural resources starting in 1915, traveling to all of Russia's far-flung regions to assess mineral deposits. After the Russian Revolution, Lenin consulted Fersman for advice on exploiting the country's mineral resources. During World War I, Fersman consulted with the military, advising on strategic matters involving geology, as he was also to do in World War II.

In the early 1920s, Fersman devoted himself to one of geochemistry's major theoretical questions: the distribution of the chemical elements in the Earth's crust. Fersman worked out the percentages for most of the elements and proposed that these quantities be called "clarkes" in honor of Frank W. Clarke, an American chemist who had pioneered their study. Clarkes had traditionally been expressed in terms of weight percentages; Fersman calculated them in terms of atomic percentages. His work showed different reasons for the terrestrial and cosmic distribution of the elements. He was very interested in the ways in which elements are combined and redistributed in the earth's crust. He coined the term "technogenesis" for the role of humans in this process, concentrating some elements and dispersing others through extraction and industrial activities.

Over the next twenty years, Fersman was responsible for a reassessment of the U.S.S.R.'s mineral resources. There were many areas, such as Soviet Central Asia and Siberia, which were thought to be resource-poor. Fersman showed otherwise, traveling from the Khibiny Mountains north of the Arctic Circle near Finland to the Karakum Desert north of Iran. He found rich deposits of apatite (a **phosphorus**-bearing mineral useful in fertilizers) in the former and a lode of elemental **sulfur** in the latter.

Fersman was acutely aware of the history of his profession and of science in general, passing on to his students his respect for his predecessors, especially Mendeleev and Vernadsky. He wrote many biographical sketches of distinguished scientists and published a number of popular works on mineral collecting. He was active in the Academy of Science of the U.S.S.R., serving in five different administrative posts, and received a number of honors, including the Lenin Prize. He died in the Soviet Georgian city of Sochi on May 20, 1945.

Fertilizers

Fertilizers are chemicals added to soil for the purpose of improving plant growth quality and yield. First discovered by ancient farmers, fertilizer technology has significantly improved as the chemical needs of growing plants were discovered. Synthetic fertilizers are composed primarily of **nitrogen**, phosphorous, and **potassium** compounds. Secondary nutrients that are important for plant growth are also included. The use of fertilizers has had an important impact on civilization, significantly improving the quality and quantity of the food available today.

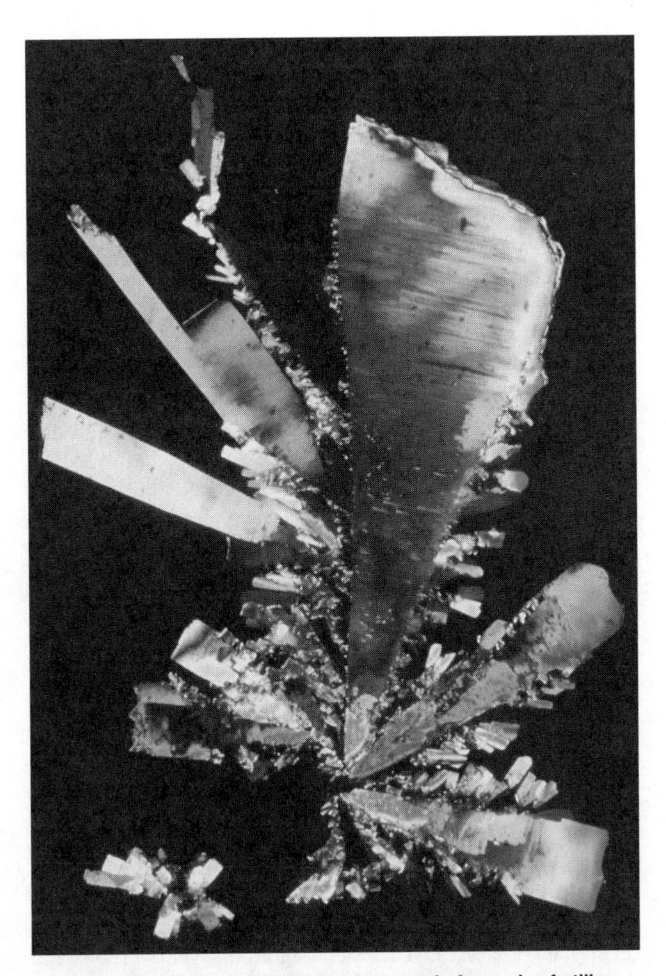

A polarized light micrograph image of a crystal of a garden fertilizer. This water-soluble product is used by amateur gardeners as a general plant fertilizer. The product contains 15% nitrogen, 30% phosphorous, and 15% potassium. *(Photograph by Sidney Moulds/ Science Photo Library, Photo Researchers, Inc. Reproduced by permission.)*

The technique of increasing growing capacity by adding materials to soil has been used since the early days of agriculture. Ancient farmers discovered that plant yield could be increased on a plot of land by spreading animal manure throughout. As humans became more experienced, fertilizer technology improved. Other materials were added to soil. For example, the Egyptians added ashes from burnt weeds. The ancient Greeks and Romans added various animal excrements depending on soil or plant variety. Seashells, clay, and vegetable waste were also early fertilizers.

The development of modern fertilizers began in the early seventeenth century with discoveries made by scientists such as Johann Glauber (1604-1668). Glauber created a mixture of saltpeter, **lime, phosphoric acid**, nitrogen and potash which was the first completely mineral fertilizer. As the chemical needs of plants were discovered, improved fertilizer formulas were made. The first patent for a synthetic fertilizer was granted to Sir John Lawes (1814-1900). It outlined a method for making **phosphate** in a form that was an effective fertilizer. The fertilizer industry experienced significant growth when **explosives** manufacturing facilities were converted to fertilizer factories after World War I.

Since plants do not consume food like animals, they depend on the soil to provide the basic chemicals needed to power biological processes. However, the supply of these compounds is limited and depleted each time the plants are harvested. This causes a **reduction** in the quality and yield of plants. Fertilizers help replace these chemical components. But fertilizers also improve the growing potential of soil and make it a better growing environment than it is naturally. Since different plants have different requirements, fertilizers are modified depending on the type of crop being grown. The most common components of fertilizers are nitrogen, **phosphorus**, and potassium compounds. Additionally, they contain trace elements needed for plant growth.

The primary chemical components in fertilizers are nutrients vital for plant growth. Nitrogen is an important material used by the plant for processes such as the synthesis of **proteins**, **nucleic acids**, and **hormones**. A plant that is nitrogen deficient is smaller and has more yellow leaves. Phosphorus is another important compound and it is used by plants for the production of nucleic acids, phospholipids, and several proteins. It is also part of the **energy** process that drives metabolic reactions. Phosphorus deficient plants are much smaller than non-deficient plants. Potassium is another key substance plants obtain from soil. It is used for **protein synthesis** and various other plant processes. Weak stems and roots, yellow spots, and dead tissue are all symptoms of a lack of potassium.

Other materials are vital for healthy plant growth, but needed in smaller amounts. **Calcium**, **magnesium**, and **sulfur** are examples of these. They are included in small amounts because most soils naturally contain adequate amounts. Micronutrients, such as **iron**, **chlorine**, **copper**, **manganese**, **zinc**, **molybdenum**, and **boron**, are primarily cofactors in enzymatic reactions.

The raw materials used in producing fertilizers are obtained from a variety of sources. These materials can be isolated from naturally occurring sources or mined. Examples include seaweed, bones, guano, **sodium** nitrate, potash, and phosphate rock. Components can also be organically synthesized from more basic chemicals. These include things such as **urea**, **ammonia**, **nitric acid**, and ammonium phosphate. When ammonia is used as the nitrogen source in a fertilizer it may be produced from **natural gas** and air. The phosphorus component is made using the raw materials sulphur, **coal**, and phosphate rock. Potassium chloride derived from potash is the primary potassium source. Calcium is obtained from limestone. Magnesium is derived from dolomite. Sulphur is another material that is mined and added to fertilizers. Other mined materials include iron from ferrous sulphate, copper, and molybdenum.

Fertilizers are produced rapidly in fully integrated factories. Here the primary components including nitrogen, potassium, and phosphorous are prepared. They may be synthesized or isolated from feedstock. They are then mixed with secondary components and granulated. This material is then packaged and supplied to farmers.

The environmental impact of fertilizers has been a concern since a relatively small amount of the nitrogen that is applied is actually assimilated into plants. When it rains, the fertilizer is washed away adding significant amounts of nitrates to drinking **water**. Certain medical studies have indicated that health disorders of the urinary and kidney systems may be the result of excessive nitrates in drinking water. The nitrates themselves are not thought to be harmful but certain bacteria in the soil convert them to nitrite ions. Research has shown that ingested nitrite ions can bond with hemoglobin in the bloodstream causing the hemoglobin to lose its ability to store **oxygen**. Excessive amounts can cause serious health problems. Nitrosamines are another potential byproduct derived from nitrates in fertilizers. These have been shown to cause tumors in laboratory animals raising fears that the same may be happening in humans. However, no study has shown a definitive link between fertilizer use and human tumors.

FEYNMAN, RICHARD PHILLIPS (1918-1988)

American physicist

Feynman was born in New York City on May 11, 1918. He was educated in the public schools of New York City and rapidly impressed his teachers with his remarkable grasp of difficult concepts. It is reported that his high school physics teacher allowed Feynman to sit in the back of the room and solve assigned problems using calculus while other students struggled along with simple algebra

Feynman attended the Massachusetts Institute of Technology, from which he received a bachelor's degree in 1939. His doctoral work was completed at Princeton University in 1942. Along with nearly all physicists of the time, Feynman spent World War II working in the **Manhattan Project** on the development of the first nuclear weapons. After the war, he accepted a teaching position at Cornell University. In 1950, Feynman became professor of theoretical physics at the California Institute of Technology, where he remained until his death.

Feynman's special field of interest was quantum electrodynamics. That term refers to the study of the interactions among electrons, positrons, and photons. During the 1920s, a great deal of effort was expended on the application of modern quantum mechanics to classical electromagnetic theory. A number of leading physicists, including Paul Adrien Maurice Dirac, Werner Heisenberg, Wolfgang Pauli, and Enrico Fermi, all looked for ways to salvage **James Clerk Maxwell**'s theory of electromagnetism by incorporating quantum principles.

Many of these efforts were at least partially successful. However, they too often failed to account for detailed phenomena, such as the specific **energy** levels observed for **electrons** in the **hydrogen atom**. There was some concern that either

Richard Phillips Feynman.

classical theory or its modern modifications might have to be totally abandoned.

While at Princeton University, Feynman attacked this problem. He developed a program for ''renormalizing'' some of the fundamental terms in equations that had been giving troublesome results. This program eliminated many of the nonsensical answers obtained from the equations. It became apparent that existing theories could be modified and successfully applied. For his work on quantum electrodynamics, Feynman shared the 1965 Nobel Prize for physics with Julian Schwinger and Sin-itiro Tomonaga. Schwinger and Tomonaga had independently developed methods of analysis similar to those of Feynman's.

Feynman was widely known and admired outside the field of science. His autobiography, *Surely You're Joking, Mr. Feynman*, became a best-seller. On another level, his *Feynman Lectures on Physics* is widely regarded as a lucid explanation of some fundamental ideas in modern physics.

Feynman died in 1988 after a long battle with cancer.

FIBERS, NATURAL

Natural fibers may be of animal, vegetable, or mineral origin. Although the annual production of vegetable fibers outweighs that of animal or mineral fibers, all have long been useful to humans.

A colored scanning electron micrograph image of the weave of a piece of linen. Linen is made from the fibers of the stem of the flax plant, *Linum usitatissimum.* *(Photograph by Dr. Jeremy Burgess/ Science Photo Library, Photo Researchers, Inc. Reproduced by permission.)*

Animal fibers

Animal hair fibers consist of a protein known as *keratin.* It has a composition similar to human hair. Keratin **proteins** are actually quite similar to the synthetic fiber **nylon**, where the repeating units are **amino acids**. The fibrous proteins form crystals. They also crosslink through disulfide bonds present in the cystine amino acid.

Wool forms the protective covering of sheep, screening them from **heat** and cold, and allowing them to maintain even body temperatures. The following are important characteristics of wool fibers. (1) They are 1-14 in (2.54-35.56 cm) or more in length, with diameters of 0.0016-0.0003 in (0.04 - 0.008 mm). (2) Their average chemical compositions are: **carbon**, 50%; **hydrogen**, 7%; **oxygen**, 22-25%; **nitrogen**, 16-17%; and sulphur, 3-4%. (3) They are extremely flexible and can be bent 20,000 times without breaking. (4) They are naturally resilient. (5) They are capable of trapping air and providing insulation. (6) They absorb up to 30% of their weight in moisture. (7) They are thermally stable, and begin to decompose only at 212°F (100°C).

Silk is a continuous protein filament spun by the silkworm to form its cocoon. The principle species used in commercial production is the mulberry silkworm, which is the larva of the silk moth, *Bombyx mori.*

Silk and sericulture (the culturing of silk) probably began in China more than 4,000 years ago. The Chinese used silk for clothing, wall hangings, paintings, religious ornamentation, interior decoration, and to maintain religious records.

A false-color scanning electron micrigraph image of cotton fibers magnified 550 times. Many of these cellulose fibers make up a single thread of cotton. *(Photograph by Andrew Syred/Science Photo Library, Photo Researchers, Inc. Reproduced by permission.)*

Knowledge of the silkworm passed from China to Japan through Korea. The production of silk transformed the tiny, technologically backward Japanese islands into a world power.

Silk was also passed to Persia and Central Asia where it was encountered by the Greeks. Aristotle was the first Western writer to describe the silkworm. In 550 A.D., the Emperor Justinian acquired silkworm eggs and mulberry seeds, beginning the varieties of silkworms that supplied the Western world with silk for 1,400 years.

After World War II, the women's silk hosiery market, silk's single largest market, was mostly overtaken by nylon.

Silk fibers are smooth, translucent, rod-like filaments with occasional swellings along their length. The raw silk fiber actually consists of two filaments called fibroin bound by a soluble silk gum called *sericin.* Fibroin and sericin are made up of carbon, hydrogen, nitrogen, and oxygen.

Silk is a partially crystalline protein fiber, and has several important qualities. (1) It is lower in **density** than wool, cotton, or rayon. (2) It is a poor conductor of heat and **electricity**. (3) It is capable of soaking up to 30% of its weight in moisture. (4) It is extremely strong, with a breaking strength as high as 65,000 psi. (5) It will stretch to as much as 20% of its length without breaking. (6) It is thermally stable; it is able to withstand temperatures as high as 284°F (140°C). (7) It becomes smooth, lustrous, and luxurious when processed. (8) It is remarkably resilient, and shows excellent wrinkle recovery.

In addition to wool and silk, a number of specialty fibers are also obtained from animals. In most cases, animal fibers are similar to each other. They grow in two principal coats: the shiny and stiff outer coat or hair; and the undergrowth or fur. Hair forms a protective shield around the animal's body against the elements; fur is closer to the skin and consists of shorter fibers than the hair that acts as insulation against heat or cold.

Fabrics containing specialty fibers are expensive because of the difficulties in obtaining the fibers, and the amount of processing required to prepare the fibers for use. Unlimited combinations of specialty fibers with wool are possible. Specialty fibers may be used to add softness or luster to fabrics. They also enhance the insulating properties of blended fabrics.

Vegetable fibers were used by ancient man for fishing and trapping. Evidence exists that man made ropes and cords as early as 20,000 B.C.. The Egyptians probably produced ropes and cords from reeds, grasses, and flax around 4000 B.C.. They later produced matting from vegetable fibers, rushes, reeds, and papyrus grasses bound with flax string.

Vegetable fibers consist of **cellulose**, i.e., polysaccharides made up of anhydroglucose units joined by an oxygen linkage to form long molecular **chains** that are essentially linear, bound to lignin, and associated with various amounts of other materials.

Vegetable fibers are classified according to the part of the plant that they come from. The four groupings are: seed-hair fibers, leaf fibers, bast fibers, and miscellaneous fibers.

Mineral fibers

The three principal fibers derived from minerals are **asbestos**, **glass**, and **aluminum** silicate. But of the three fibers only asbestos is a true natural fiber. Glass and aluminum silicate fibers require human intervention in their processing, and might be better considered man-made fibers.

Asbestos is a fibrous mineral mined from rock deposits. There are approximately 30 types of minerals in the asbestos group. Of the six that have commercial importance, only oneotile, a hydrated silicate of **magnesium** that contains small amounts of **iron** and aluminum, oxides, is used in fiber processing.

Asbestos probably formed prehistorically when hot waters containing **carbon dioxide** and dissolved salts under high pressure acted upon rock deposits of iron, magnesia, and silica.

The ancient Greeks knew of asbestos as early as the first century A.D.; the name comes from the Greek word for inconsumable. But asbestos did not find commercial use until it was used for packing and insulation when the steam engine was invented. Currently the world's leading suppliers of asbestos are Canada and the former Soviet Union.

Useful for heat protection, the most notable characteristic of asbestos is that it will not burn. It can be spun and woven into textiles. However, asbestos is a known carcinogen. It is highly toxic when inhaled as dust particles. The American Conference of Governmental Industrial Hygienists has established a maximum exposure level to chrysotile fibers. A worker may be exposed, without adverse effects, to 2 fibers/cc more than 5 microns long.

FIBERS, SYNTHETIC

Uses for synthetic fibers range from **nylon** stockings and clothing to cables and tire reinforcement. Synthetic fibers are made from polymers that are either melted into a **solution** by **heat** or dissolved by a solvent. The solution is then passed through a metal plate with fine holes, called spinnerets. This process forms the polymers into strands. The fibers are then either cooled or passed through a jet of air to allow the solvent to evaporate. Most fibers at this point are subjected to cold drawing, a strengthening technique developed by **Wallace Carothers**'s team in its search for artificial silk. After drawing, the fibers are washed, dried, dyed, and woven.

The first patent for synthetic fiber was granted to George Audemars in 1855. A related patent was granted to Sir Joseph Swan (1828-1914) in 1880. Both of these men produced fibers from **cellulose**, which, unfortunately, were not very strong. **Hilaire Comte de Chardonnet** later found that by denitrating the fibers he could strengthen them until they were as durable as silk. Edward John Bevan (1856-1921) and **Charles F. Cross** developed the industrial production process for this material, which he called rayon. In the modern production process, rayon is prepared by subjecting a solution of cellulose to chemical reaction, aging, or solution ripening; followed by **filtration** and removal of air, spinning the fiber, combining the filaments into yarn, and finishing (bleaching, washing, oiling, and drying). Strength is achieved by orienting rayon molecules when they are made. Hydroxyl groups in the cellulose **molecule** cause rayon to absorb **water**; in the dry state the fibers are **hydrogen** bonded, and dry rayon retains its strength even at high temperatures.

Acrylonitrile, which was produced by Charles Moureau (1863-1929) in 1893, is used in the production of nitrile rubber, acrylic fibers, insecticides, and **plastics**. The fibers formed by acrylonitrile are unstable at their melting point, so melt extrusion is impractical. Solution spinning was not possible for many years because no appropriate solvent had been found. Today cold drawing is used to strengthen the fibers by orienting the molecules and allowing hydrogen **bonding** to occur. The fibers are then dried and woven into a fabric that resembles soft wool, which is used for sails, cords, blankets, and clothing.

Nylon 66 is a polyamide based on hexamethylene diamine and adipic acid. Nylon 6 is based on caprolactam. Nylon was developed by DuPont Company researchers as a substitute for silk. Carothers and his team of assistants had been researching long chain polyesters and polyamides. In 1938 they had almost given up hope on finding a suitable fiber when two members of the team discovered cold drawing. Nylon is used in clothing, laces, tooth brushes, sails, fish nets, and carpets. It is resistant to alkalis, molds, **solvents**, and moths; but damaged by strong acids, phenol, **bleaches**, and heat above 338°F (170°C).

John R. Whinfield and J. Dickson continued Carothers's research and made a **polyester** with terephthalic acid in 1941. The fiber was christened Terylene and marketed as Dacron in the U.S. by the DuPont Company. Acrilan, produced by the Chemstrand Corporation, is an acrylic fiber used in fabrics and may be blended with wool or cotton to form clothing, carpeting, linens, draperies, and upholstery. Fabrics made from Acrilan resist mildew, moths, and wrinkling. They also tend to dry quickly. Dacron polyester is resistant to weak acids and alkalis, solvents, oils, mildew, and moths; it is damaged by phenol, and heat above 338°F (170°C).

Orlon, which is an acrylic fiber containing at least 85% acrylonitrile units, is a class of synthetic fibers first produced commercially by the DuPont Company in 1950. The fibers vary in size, texture, and ability to hold dyes. They can be woven or knitted, usually into bulky garments. Orlon is made by dissolving acrylonitrile in an organic solvent, then filtering the solvent and dry-spinning the fibers, and finally drawing the fibers. Orlon is used in upholstery and carpets. It is resistant to dilute acids and alkalis, solvents, insects, mildew, and weather; it is damaged by alkalis and acids, heat (above 356°F [180°C]), **acetone**, and **ketones**.

Vinyon filaments and fibers were developed by the Carbide and Carbon Chemicals Corporation, which licensed American Viscose Corporation to produce them in 1939. The fiber is a copolymer of 88 % **vinyl** chloride and 12 % vinyl acetate. It was the first plastic fiber produced on a large scale in the United States. The fibers are stretched in a process similar to cold drawing. The stretching increases the strength of the fibers but lowers its elasticity. The fiber does not take dyes and becomes sticky if heated to over 149°F (65°C). At 167°F (75°C), garments made of the fiber will shrink. Further research led to the development of a fiber Vinyon N, which is a copolymer of vinyl chloride with acrylonitrile. For fibers, the copolymer ranges from 56-60% copolymerized vinyl chloride. It was patented in 1947.

Kevlar is a polyamide fiber developed by **Stephanie Kwolek** of DuPont in 1965. Kevlar is a polyamide in which at least 85 wt% of the amide linkages are directly attached to two aromatic rings. Its molecular structure of alternating aromatic rings and amide groups crosslinked by hydrogen bonding makes it incredibly strong. Kevlar's light weight and high strength make it very marketable. It is commonly used to make bulletproof vests, heavy duty conveyor belts, and composite structures; and to reinforce radial tires.

Other important synthetic fibers based on high polymers (i.e., macromolecules) include acetate (consisting of esters of cellulose, but not regenerated cellulose); spandex (segmented polyurethanes); vinyls and vinylidenes (copolymers of vinyl chloride, and vinylidene chloride or vinyl acetate); fluorocarbons (long chain molecules with bonds saturated by fluorine); olefins (at least 85 wt% ethylene, propylene or other olefin other than amorphous rubber olefins); vinal (at least 50 wt% of vinyl alcohol units in a long polymer chain); and azlon (any regenerated naturally occurring protein).

Besides these organic synthetic fibers, **glass** fibers consisting primarily of silica (used for thermal insulation), metal fibers (used in composite materials), and ceramic

See also Acids and bases; Fibers, natural; Polymer chemistry

FIELD ION MICROSCOPE

The field **ion** microscope is a remarkably powerful optical device. It represents several generations of scientific evolution of the **electron** microscope and is probably the most powerful magnifying instrument yet invented.

Before German electrical engineer Ernst Ruska (1906-1988) and Max Knoll invented the **electron microscope** in 1932, the science of **microscopy** had reached a standstill.

Though scientists were theorizing about the structure and composition of the **atom**, no microscope yet invented had the ability to resolve an image smaller than a wavelength of light; thus, there was no way to actually observe atomic and **subatomic particles**. Ruska and Knoll overcame this barrier by substituting a beam of electrons for the microscope's light source. With wavelengths far shorter than those of visible light, electrons could be used to view specimens at magnification levels thousands of times higher than those possible with simple microscopes. The electron microscope provided the first views of viruses and bacteria, as well as large molecules.

Though the power of the electron microscope was unsurpassed at the time, many scientists were unsatisfied with its results. Ruska's invention had revealed a new world of submicroscopic particles, but it was not powerful enough to view individual atoms. Though this seems like a purely academic goal, it was of crucial importance to a scientific community struggling to understand the properties of these universal building blocks.

Erwin Wilhelm Müller began work on a new type of particle microscope while still attending the Technical University at Berlin. In 1936—just a year after he graduated—he developed the field emission microscope. Based on J. Robert Oppenheimer's theories of quantum mechanics, as well as on Ruska's groundbreaking research, the field emission microscope allowed magnification up to two million times.

In the field emission microscope, a very thin metal needle is placed within a vacuum tube. When a powerful electric current (up to 30,000 volts) is applied, electrons are emitted from the sharpened tip of the negatively charged needle. The electrons, spreading as they move away from the needle, strike the surface of a (positively charged) fluorescent screen. The image produced is an enlarged view of the atomic lattice structure of the tip of the needle.

The field emission microscope provided unparalleled images of **atomic structure**. However, there remained several shortcomings: first, the process could only observe the atomic structure of **metals**, and sturdy ones at that, since the intense electric current necessary to liberate the electrons would destroy any other substance (organic and biological **matter** could not be observed); second, the random nature of electron movement created images of high magnification that often were too blurry to be useful. Most frustrating to Müller was his invention's inability to achieve its primary goal: the imaging of individual atoms.

Müller remained in Germany until 1952, when he moved to the United States to work at the Pennsylvania State University Field Emission Laboratory. Here, he tried a new design, utilizing ion emission rather than electrons. This involved switching the polarity of the system, so that the needle was positively charged and the screen negatively charged. The random travel of ions was much easier to contain by supercooling the needle tip (usually with liquid **nitrogen** or hydrogen). Lastly, he added a small amount of inert gas to the vacuum tube. Müller's new design, called the field ion microscope, proved to be even more powerful than the emission microscope—and without the latter's resolution problems. Using this device, Müller finally was able to resolve individual atoms.

The field ion microscope is used primarily in atomic research, particularly in the study of how metals crystallize. It was still impossible to examine biological specimens until the development of field evaporation technology; this allows scientists to evaporate metal surrounding a needle of organic material. While this technique has provided some insights into the structure of biological molecules, the field ion microscope is still not flexible enough to be used on living cells; this was later made possible by the invention of the **scanning tunneling microscope**.

FIESER, MARY PETERS (1909-1997)

American organic chemist

Mary Peters Fieser's substantial contributions to the field of **organic chemistry** include her work on the Harvard research team headed by her husband, **Louis Fieser**, and her authorship of numerous key texts and reference books in the field. She was involved in numerous important areas, including the synthesis of vitamin K, the development of an antimalarial drug, and the synthesis of **cortisone** and carcinogenic chemicals for medical research. For her research, publications, and skill in teaching chemistry students how to write, she was awarded the prestigious Garvan Medal in 1971.

Fieser was born in 1909 in Atchison, Kansas, to Robert Peters, an English professor, and Julia (Clutz) Peters, a bookstore owner and manager. Her father accepted a position at what is now Carnegie-Mellon University, and Fieser grew up in Harrisburg, Pennsylvania. Her family believed strongly in educational and professional achievement for women: Fieser's mother did graduate work in English, and her sister, Ruth, became a mathematics professor. Her grandmother, who had educated her seven children herself at home until they were college age, impressed upon Fieser the importance of using her education constructively.

After attending a private girls' school, Fieser went to Bryn Mawr College, where she earned a B.A. in chemistry in 1930. There, she met her future husband, Louis Fieser, who was a chemistry instructor at the college. She enjoyed his courses, finding his emphasis on experimental rather than **theoretical chemistry** to be especially interesting. When Louis left Bryn Mawr in 1930 to teach at Harvard, she went with him. There, she performed chemistry research in his laboratory while earning a master's degree in organic chemistry, which she received in 1936.

When the couple married in 1932, Fieser continued her professional association with her husband on his research team. This arrangement benefited Fieser enormously in her professional career, because bias against women in the field of chemistry very strong at that time. For instance, her **analytical chemistry** professor at Harvard, Gregory Paul Baxter, refused to allow her to perform her experiments in the laboratory with the rest of the class. Instead, Fieser had to perform experiments in the deserted basement of another building, with little or no supervision. Once married, however, she was free to conduct research on her husband's team unhampered. As she commented during an interview with the *Journal of Chemical Education,* there were too many obstacles to an academic career in chemistry as a single woman, but after she was married, "I could do as much chemistry as I wanted, and it didn't matter what Professor Baxter thought of me."

As part of Louis Fieser's research team, Mary Fieser helped develop a practical method of obtaining substantial amounts of vitamin K. The antihemorrhagic properties of vitamin K had been discovered during the 1930s by **Henrik Dam** in Copenhagen. Researchers had discovered this vitamin in green plants and especially in dried alfalfa, but the amount available from these sources was too small to be of practical use in medical therapy. The Fieser research group developed a method of synthesizing large amounts of vitamin K in a short period of time. The vitamin's blood-clotting characteristic has proved useful in prenatal therapy and other therapeutic purposes as well.

The Fiesers also focused on the use of naphthoquinones as antimalarial drugs. **Quinine** was one of the standard drugs used to treat malaria. When Japan invaded the East Indies during World War II, most of the world's supply of quinine became inaccessible to the Allies. The Fieser research team undertook a study of naphthoquinones as an alternative treatment. The Fiesers' research focused on isolating and identifying intermediate compounds along the reaction pathway. Their work ultimately contributed to the synthesis of the antimalarial drug lapinone.

Fieser worked on numerous other projects, including studies of the chemical causes of cancer. She helped develop methods of synthesizing various carcinogenic chemicals for use in medical research. She also played an important role in one of the Fiesers' more well-known projects: their contribution to the synthesis of cortisone, a steroid hormone used in the treatment of rheumatoid arthritis.

Fieser was highly regarded by her colleagues for her skill as a research chemist. Harvard chemist William von Doering is quoted in the *Journal of Chemical Education* as saying that she was "a very gifted experimentalist" and an "active, influential part of the team." In addition to her research, Fieser wrote or co-wrote with her husband a dozen chemistry texts and reference books, beginning in 1944 with the best-selling textbook *Organic Chemistry.* One of their most widely used publications, *Reagents for Organic Synthesis,* was the first reference work of its kind for researchers in organic chemistry. It was the result of a comprehensive, international review of organic chemistry literature from which Mary Fieser culled the results of studies in chemical synthesis.

Fieser's books were especially noteworthy because of her expert writing skills—an unusual ability for a chemist at that time. Fieser attempted to improve the quality of writing in her field by publishing *Style Guide for Chemists.* She and her husband, also a skilled writer, often argued at length over minor stylistic issues, such as the placement of a comma. These differences over writing style prompted Fieser's sister, Ruth, to suggest that their by-line "Fieser and Fieser" be changed to "Fieser versus Fieser."

In 1971, Fieser was awarded the Garvan Medal for her research contributions, her writing, and her skill in teaching chemistry students how to write. The Garvan Medal was established to "honor an American woman for distinguished service in chemistry." Her colleagues also noted that the awards her husband received were due at least in part to her efforts in the laboratory. In her leisure time, Fieser enjoyed indulging her strong competitive streak by organizing games for her husband's research group after work hours and setting up contests in ping-pong, badminton, and horseshoes for the graduate students. She and her husband owned many cats, including one named in honor of their work on synthesizing vitamin K. Their cats' photographs were used in their published work and came to be their trademark. Mary Fieser died on March 22, 1997 in her home in Belmont, Massachusetts.

FILTRATION

Filtration is the process of separating material, usually a solid, from a substrate (liquid or gas) by passage of the substrate through a septum or membrane which retains most of the materials on or within itself. The septum is called a filter medium, and the equipment assembly that holds the medium and provides space for the accumulated materials is called a filter. The filtration is a physical process; any chemical reaction is inadvertent and normally unwanted. The object of filtration may be to purify the fluid by clarification or to recover clean, fluid-free particles, or both.

There are two kinds of filters: *surface filters* and *depth filters*. With the surface filter, filtration is essentially an exclusion process. Particles larger than the filter's pore or mesh dimensions are retained on the surface of the filter; all other matter passes through. Examples are filter papers, membranes, mesh sieves, and the like. These are frequently used when the solid is to be collected and the filtrate is to be discarded. Depth filters, however, retain particles both on their surface and throughout their thickness; they are more likely to be used in industrial processes to clarify fluid for purification.

Filtration most commonly is used in one of the following ways:

Solid-liquid filtration: the separation of solid particulate matter from a carrier liquid. The solid-liquid separations are important in the manufacture of chemicals, polymer products, medicinals, beverages, and foods; in mineral processing; in **water purification**; in sewage disposal; in the chemistry laboratory; and in the operation of machines such as internal combustion engines: the oil filter, and the fuel filter.

Solid-gas filtration: the separation of solid particulate matter from a carrier gas, which is often but not always to remove the solids (called dust) from a gas-solid mixture. Normally, three kinds of gas filters are in common use. Bag filters are bags of woven fabric, felt, or paper through which the gas is forced; the solids are deposited on the inter wall of the bag. A very simple example is the dust collector of a household vacuum cleaner. Air filters are light webs of fibers through which air containing a low concentration of dust can be passed to cause entrapment of the dust particles. A simple example is the air filter installed in the air-intake lines of the car. Granular-bed separators consist of beds of sand, **carbon**, or other particles, which will trap the **solids** in a gas suspension that, is passed through the bed. The bed depth can range from less than an inch or several feet.

Liquid-liquid separation: a special class of filtration resulting in the separation of two immiscible **liquids**. One of them is **water**, by means of a hydrophobic medium.

Gas-liquid filtration: the separation of gaseous matter from a liquid in which it is usually, but not always, dissolved.

There are two general methods of filtration: *gravity*, and *vacuum*. During the gravity filtration the filtrate passes through the filter medium under the combined forces of gravity and capillary attraction between the liquid and the funnel stem. In vacuum filtration a **pressure** differential is maintained across the filter medium by evacuating the space below the filter medium. Vacuum filtration adds the force of atmospheric pressure on the **solution** to that of gravity, with a resultant increase in the rate of filtration.

Normally in a chemistry laboratory, filtration is generally used to separate solid impurities from a liquid or a solution or to collect a solid substance from the liquid or solution from which it was precipitated or recrystallized. This process can be accomplished with the help of gravity alone or speeded up by using vacuum techniques. Vacuum filtration provides the force of atmospheric pressure on the solution in addition to that of gravity, and thus increases the rate of filtration.

See also Gases

Fischer, Edmond H. (1920-)

American biochemist

Edmond H. Fischer was the joint recipient with his longtime associate, **Edwin Krebs**, of the Nobel Prize in physiology or medicine in 1992 for discoveries dealing with reversible protein phosphorylation as a biological regulatory mechanism. Responsible for a wide range of basic processes, including cell growth and differentiation, regulation of genes, and muscle contraction, protein phosphorylation is now the subject of one in every twenty papers published in biology journals. Application of Fischer and Krebs's work to medicine has elucidated mechanisms of diseases such as cancer and diabetes, and has yielded drugs that inhibit the body's rejection of transplanted organs. The recognition accorded Fischer and Krebs, who began a collaboration at the University of Washington in Seattle in the early 1950s, was hailed within the scientific community as long overdue.

Edmond H. Fischer was born on April 6, 1920, in Shanghai, China. His father, Oscar Fischer, had come to China from Vienna, Austria, after earning degrees in business and law. Fischer's mother, Renée Tapernoux Fischer, was born in France. She had come to Shanghai with her family after first arriving in Hanoi, where her father was a journalist for a Swiss publication. In Shanghai, Fischer's grandfather founded the first French newspaper published in China and helped to establish a French language school that Fischer attended.

At the age of seven, Fischer was sent, along with two older brothers, to a Swiss boarding school near Lake Geneva. One of his brothers studied engineering at the Swiss Federal Polytechnical Institute in Zurich and the other went to Oxford to study law. While he was in high school, Fischer developed a lifelong friendship with Wilfried Haudenschild, whose inventiveness and unusual ideas impressed him. The two decided that one would be a scientist and the other a physician, so that together they would cure all the ills of the world. Fischer was also influenced in his youth by classical music and for a time entertained the idea of becoming a professional musician. Instead he decided to become a scientist.

After entering the School of Chemistry at the University of Geneva at the start of World War II, Fischer was able to earn a degree in biology and another in chemistry. He received his doctorate at Geneva in 1947 and worked at the university on research until 1953. American universities at the time afforded more opportunities in the new field of **biochemistry**, and Fischer soon found himself in the United States. His first position was at the California Institute of Technology, where he was given a postdoctoral fellowship. Fischer was amazed that wherever he went in the United States he was offered a job. In Europe, research positions in his field were next to impossible to obtain.

Hans Neurath, chair of the department of biochemistry at the University of Washington, invited Fischer to Seattle. On his first visit he found the scenery reminiscent of Switzerland and accepted Neurath's offer of an assistant professorship at the new medical school at the university. Thus began a long association with Edwin Krebs. Krebs had worked in the laboratory of **Carl Ferdinand Cori** and **Gerty T. Cori** in St. Louis on the **enzyme** phosphorylase in the late 1940s (enzymes are **proteins** that encourage or inhibit chemical reactions). The Coris won the Nobel Prize in 1947 for their isolation of phosphorylase, showing its existence in active and inactive form. Fischer had worked on a plant version of the same enzyme while he was in Switzerland.

In the mid–1950s Fischer and Krebs set out to determine what controlled the protein's activity. Their experiments centered on muscle contraction. A resting muscle needs **energy** (stored as **glycogen** in the body) in order to contract, and phosphorylase frees glucose from glycogen for use by the muscle. Fischer and Krebs discovered that an enzyme they called protein kinase was responsible for adding a **phosphate** group from the compound ATP (adenosine triphosphate, the cell's energy store) to phosphorylase, which activated the enzyme. In a reverse reaction, an enzyme called protein phosphatase removed the phosphate, turning phosphorylase off. Protein kinases are present in all cells and are critical for many phases of cell activity, including **metabolism**, respiration, **protein synthesis**, and response to stress.

By the 1970s biochemical research in the area that Fischer and Krebs opened up was so extensive that 5% of papers in biology journals dealt with protein phosphorylation. Between 1-5% of the genetic code may be concerned with protein kinases and phosphatases. Science has made connections that show the role of protein kinases in diseases, including cancer, diabetes, and muscular dystrophy. Fischer and Krebs have also been able to demonstrate in their research how the immune system is activated. They showed how a surface protein starts a **chain reaction** that recruits lymphocytes to fight infection.

In the field of organ transplants, drugs that influence phosphorylation prevent rejection of the transplants by the body's immune system. The drug cyclosporin has been developed and is widely used to prevent the rejection of liver, kidney, or pancreatic transplants in human beings. Cyclosporin and another drug, FK–506, inhibit protein phosphatase, thereby preventing the rejection of tissues in organ transplant operations. Irregular protein kinase activity can cause abnormal cell growth leading to tumors and cancer. Philip Cohen, a colleague from the University of Dundee, comments in a *New Scientist* interview that protein kinases and phosphatases "will be the major drug targets of the 21st century."

Edmond H. Fischer has received many honors for his scientific research over the course of his long career. Notable, besides the Nobel Prize, are his election to the American Academy of Arts and Sciences in 1972 and his winning of the Werner Medal from the Swiss Chemical Society as early as 1952. Fischer has been married twice. His first wife, Nelly Gagnaux, died in 1961. He had two sons, François and Henri, with her. In 1963, he married Beverly Bullock, a native Californian. Besides his accomplishment in classical piano, Fischer also enjoys painting, piloting a plane, and mountaineering. Along with his colleague, Krebs, Fischer annually joins research groups in retreats at the University of Washington Park Forest Conference Center in the foothills of the Cascade Mountains. There they review their latest findings and lay plans for their future research, which they continue in the role of emeritus professors at the University of Washington. Philip Cohen, commenting on the nature of their collaboration in *New Scientist,* said, "They complement each other. Fischer has lots of brilliant ideas; Krebs has the judgment to know which of the ideas are worth following."

Emil Fischer.

Fischer, Emil (1852-1919)

German chemist

Emil Fischer was a professor of **chemistry** for forty years who also served as director of the German chemical industries during World War I. Fischer's research on important organic substances such as sugars, **enzymes**, and **proteins**, built the foundation for modern **biochemistry**. He was the scientist who initially described the action of enzymes as a lock and key mechanism where the structure of an enzyme fits exactly into the **molecule** with which it reacts to "unlock" a biochemical reaction. In 1902 he received the Nobel Prize in chemistry for his laboratory synthesis of sugars and purine, a substance found naturally in all deoxyribonucleic acid (**DNA**). Fischer was dedicated to academic research and was among the first scientists in the world to promote substantial industrial as well as governmental support for university laboratories.

Emil Hermann Fischer was born on October 9, 1852 in Euskirchen, Germany, near Bonn and Cologne. With five older sisters, he was the only son of Laurenz Fischer and Julie Poensgen Fischer. His father was a successful businessman who started as a grocer, then added a wool spinning mill and a brewery as he prospered. Fischer described his youth as happy in his unfinished autobiography, *Aus meinem Leben* (*Out of my Life*). Fischer was a brilliant student, graduating in 1869 at the top of his class from the Gymnasium (high school) of Bonn. After graduation, Fischer tried working in business with an uncle, but he was much more interested in building a laboratory. He entered the University of Bonn in the spring of 1871.

After less than a year at the University of Bonn, Fischer transferred to Strasbourg where he studied under the noted chemist, **Adolf Baeyer**. Fischer's creativity flourished in the academic atmosphere of Strasbourg; he especially noted in his autobiography the accessibility of his professors, and the opportunities to travel and visit other chemical laboratories. For his doctorate Fischer did research on fluorescein, a **coal** tar dye that shows a fine yellow-green fluorescence in **solution**, and is used to trace **water** through systems. Fischer's researches into coal, coal tar, and the synthesis of organic chemicals, did much to build the German dye industry. Dyes manufactured in Germany soon captured the world market.

Fischer received his doctoral degree in 1874 from Strasbourg, but he continued his research on coal tar dyes with a cousin, Otto Philipp Fischer, until 1878. Ultimately he acquired a number of patents for industrially useful chemicals.

In 1875 Fischer was invited to follow Baeyer to the University of Munich where Fischer became associate professor of analytical chemistry in 1879. His researches included the discovery of a new compound, phenylhydrazine, a chemical he later used extensively in research on sugars. By 1878 he figured out the chemical formula for phenylhydrazine, and this discovery stimulated other researches leading to the development of such synthetic drugs as novocaine. In 1881 Fischer began investigations into a new field, purine chemistry (part of a group of **nucleic acids**), identifying three amino acids and synthesizing many more. This research resulted in many more advances in the German drug industry.

Fischer left in 1882 to accept the position of professor of chemistry at the University of Erlangen, near Nuremberg. At Erlangen, Fischer continued his work on purines and began to study **carbohydrates** in 1884. His subsequent work with phenylhydrazine in an unventilated laboratory caused him to suffer the effects of phenylhydrazine poisoning which attacks the kidney, liver, and respiratory system. Fischer had, from an early age, periodically suffered from stomach disorders; the added contamination to phenylhydrazine made him extremely ill. Upon his recovery in 1885 he accepted a chair in Würzburg, where, he wrote in his autobiography, ''gaiety and humor flourished.'' In 1888 Fischer married Agnes Gerlach. They had three sons before she died in 1895. While Fischer was at Würzburg he was honored with a Bavarian medal.

In 1892 Fischer accepted the position of professor in charge of the chemistry department at the University of Berlin, the most prestigious position for an academic chemist in Germany at that time. He was offered full freedom in the construction of a new building at the chemical institute of Berlin, and his subsequent design of a well-ventilated laboratory became a model for university laboratories all over the world. In addition, his teaching methods led to the formation of small groups of students involved in basic scientific research. With the help of his cooperative teams of students, and fellow researchers from many countries, he designed a careful plan for each research project. As the work progressed he always looked for deviations from the expected results. Each unusual occurrence was researched systematically to its conclusion. This strategy permanently influenced both graduate education in chemistry and the expectations of universities for research and publication from their professors worldwide.

Fischer's researches into sugar and purines had proven especially successful. He synthesized about one hundred and thirty purines, which included **caffeine**, theophylline (used in the preparation of the motion sickness drug Dramamine), and uric acid. In addition, after studying the three-dimensional shapes of sugar molecules, Fischer synthesized glucose as well as about thirty other sugars. By 1899 Fischer finished most of his work on sugars and purines and began research on proteins and enzymes in an effort to identify their chemical nature. Fischer was elected to membership in the Academy of Sciences, and, in 1902, he received the Nobel Prize ''for his synthesis in the groups of sugars and purine,'' as quoted by Eduard Farber in *Nobel Prize Winners in Chemistry.* In 1909 he received the Helmholtz Medal for his work on sugar and protein chemistry.

Fischer believed in basic research. Determined to keep the preeminent position of world leader in chemical research for Germany, a position he did much to create, he gathered support from industry, government, and other scientists to establish a number of research institutes—the Kaiser Wilhelm Society for the Advancement of Sciences, the Kaiser Wilhelm Institute for Chemistry in Berlin-Dahlem, and the Kaiser Wilhelm Institute for Coal Research in Mulheim-Ruhr. Fischer was interested in research in every branch of chemistry. As director of the University of Berlin laboratories he started a radiochemistry laboratory where, years after his death, scientists **Otto Hahn** and **Lise Meitner** worked on research that led to the fission of **uranium** and the ultimate development of the atomic bomb.

World War I took Fischer away from most of his experimental investigations as he redirected his research concentrations toward the war effort. Besides being the leading chemist in Germany, he had long worked closely with industry and government. The British blockade would have brought the defeat of Germany by 1915 had Fischer and his colleagues not succeeded in using the resources they had to synthesize much of what they could no longer get on the world market. He led the development of synthetic saltpeter (**potassium** nitrate) and **nitric acid** both used in the manufacture of **explosives**. As food became in short supply he coordinated research and production of synthetic **fertilizers**. Fischer directed research to replace diminishing supplies of camphor (used to stabilize gunpowder) and pyrites which supplied **sulfur** for explosives.

Before World War I scientists had enjoyed the freedom to travel and communicate with other scientists regardless of political differences and skirmishes between their respective countries. However, World War I brought a change. Scientists became national resources. Fischer ended his long friendship with British chemist, Sir **William Ramsay**, also a Nobel laureate. But research alone could not win the war, and not all of Fischer's projects were successful. It was obvious to Fischer that Germany would be defeated. In an effort to organize the rebuilding of chemical research and industry in Germany to gain back the leadership it had before the war, Fischer and a friend made plans to form the German Society for the Advancement of Chemical Instruction.

The war years were personally tragic for Fischer. He lost his two younger sons, which left him depressed, and he was suffering from cancer. Emil Fischer died in Berlin, July 15, 1919. Some reports say he died of cancer, most say it was suicide. His remaining son, Hermann Otto Laurenz Fischer (1888–1960) went on to become a Professor of Biochemistry at the University of California in Berkeley. On October 9, 1952, Fischer's son dedicated the Emil Fischer Library at the University of California which is the repository of the collected works of Fischer, including the manuscript for his autobiography, research files, and Fischer's correspondences in World War I.

FISCHER, ERNST OTTO (1918-)

German inorganic chemist

The field of organometallic chemistry—the study of compounds of metal and carbon—is tremendously important not only for the understanding of such basic structures as the B **vitamins**, but also of the chemical industry as a whole. The growth of **plastics** as well as the refining of petroleum hydrocarbons all involve at some stage the metal-to-carbon bond which is at the heart of organometallic **chemistry**, and Ernst Otto Fischer has played a crucial role in furthering this science. A co-recipient of the 1973 Nobel Prize for chemistry for his X-ray analysis of the structure of a particular iron-to-**carbon** bond in so-called ''sandwich compounds,'' Fischer, working with members of his research laboratory in Munich, was also on the cutting edge of transition-metal research, synthesizing totally new classes of compounds.

Fischer was born on November 10, 1918, in the Munich suburb of Solln. The third child of Valentine Danzer Fischer and Karl Tobias Fischer, a physics professor at Munich's Technische Hochschule, Fischer attended the Theresien Gymnasium (high school), graduating in 1937. He subsequently did his compulsory service in the German army, the compulsory two-year period being extended due to the outbreak of World War II in 1939. In between serving in Poland, France, and Russia, Fischer managed, in the winter of 1941–42, to begin his studies in chemistry at the Technische Hochschule in Munich. Captured by the Americans, he was held in a prisoner of war camp until repatriation in the fall of 1945. He resumed his chemistry studies in Munich in 1946, working under Walter Hieber, well known for his early work on combining **metals** with molecules of carbon and **oxygen**, or metal-carbonyl chemistry. Fischer earned his Ph.D. degree in 1952 for research on carbon-to-nickel bonds; his course was well set by this time for a career in the new field of organometallic chemistry.

After earning his doctorate, Fischer remained at the Technische Hochschule, working as an assistant researcher. He and his first research students were drawn to a puzzling compound reported by the chemists T. Kealy and P. Pauson. In an attempt to link two cyclopentadiene—five-carbon—rings together, these scientists discovered an unknown compound which they believed involved an **iron atom** linked between two consecutive longitudinal rings of carbon. The intervening iron atom seemed to join with a carbon atom on each of the rings. That such metal-to-carbon bonds exist was not the surprising thing. In fact, such unstable bonds are necessary for catalytic processing of such compounds. What was interesting about this compound (initially called dicyclopentadienyl iron) was that it was not unstable at all. It was in fact highly stable both thermally and chemically. Such stability made no sense to Fischer given the nature of the proposed structure of the compound, and he theorized that it was in fact an entirely new sort of molecular complex. An English chemist, **Geoffrey Wilkinson**, soon proposed an alternate structure to the compound (now renamed ferrocene). He described ferrocene as made up of an atom of iron sandwiched between two parallel rings, one on top of the other rather than in a line on the same plane. Thus the iron formed bonds not just with a single atom on each **ring**, but with all of the atoms and also with the electrons *within* the rings, accounting for its stability. From this description came the term ''sandwich compounds.'' Meanwhile, Fischer and his research team, including W. Pfab, carried out meticulous X-ray **crystallography** on ferrocene, elucidating the compound's structure, and proving Wilkinson's theory correct. The examination and discovery of the structure of ferrocene was a watershed event in the field of organometallic chemistry, providing work for a new generation of inorganic chemists.

From ferrocene, Fischer and his team went on to determine the structure of, as well as synthesize, other transition metals—those substances at a stage in between metal and organic—particularly dibenzenechromium, an aromatic **hydrocarbon**. Such substances are termed aromatic not because of smell, but because of structure: they are hydrocarbons in closed rings which are capable of uniting with other atom groups. Fischer showed dibenzenechromium to be another sandwich compound with two rings of **benzene** joined by an atom of **chromium** in between. This bit of research earned him world-wide renown in scientific circles, as the neutral chromium **molecule** and neutral benzene molecules had been thought to be uncombinable. Fischer's rise in academia paralleled the swift advance of his research: by 1954 he was an assistant professor at the Technische Hochschule; by 1957, a full professor at the University of Munich; and in 1964 he returned to the Technische Hochschule—by now called the Technische Universität or Technical University—as director of the Institute for Inorganic Chemistry, replacing the retiring director and his former mentor, Professor Hieber. Fischer's laboratory, equipped with all the latest equipment for spectrographic and structural analysis, soon became a center for worldwide organometallic research, and Fischer, who excelled both as a lecturer and as researcher, soon became the leading spokesperson for the new study. He also began lecturing around the world, and spent two visiting professorships in the United States in 1971 and 1973.

In 1973 Fischer, shared the Nobel Prize for chemistry with the English chemist Sir Geoffrey Wilkinson. The two scientists were cited for their ''pioneering work, performed independently, on the chemistry of the organometallic, so-called sandwich compounds.'' Around this time, Fischer and his team at Munich's Technical University were successfully synthesizing both the first carbene complexes and carbyne complexes—carbon atoms triply joined to metal atoms—which heralded an entirely new class of metal complexes of a transitional sort and spurred research in the field.

In addition to the Nobel, Fischer—a life-long bachelor—has also won the Göttingen Academy Prize in 1957 and the **Alfred Stock** Memorial Prize of the Society of German Chemists in 1959, as well as honorary membership in the American Academy of Arts and Sciences and full membership in the German Academy of Scientists. Among the many commercial and industrial spin-offs of his work is the creation of catalysts employed in the drug industry and also in **oil refining**, leading to the manufacture of fuels with low **lead** content.

Hans Fischer.

FISCHER, HANS (1881-1945)

German chemist

Hans Fischer was a medically-minded chemist who won the Nobel Prize for **chemistry** for his pioneering investigations into the chemical structure of pyrroles, molecular compounds which give the specific **color** to many important biological substances, including blood, bile, and the leaves of plants. Building on the foundations laid by his predecessors and colleagues, many of them from Fischer's homeland of Germany, he spearheaded a series of investigations lasting more than two decades that led to the synthesis of hemoglobin, bilirubin, and (more than 25 years after his death) **chlorophyll**. During the course of his investigations, Fischer developed and oversaw an extremely productive microanalytical approach to studying chemical compounds, especially the pigments that occur in nature. By overseeing specific laboratory procedures conducted simultaneously by several labs, Fischer was able to conduct more than 60,000 microanalyses of chemical substances. In 1930, he won the Nobel Prize, primarily for his work in elucidating the structure of and synthesizing the blood pigment hemin.

Fischer was born at Höchst am Main in Germany on July 27, 1881, to Eugen Fischer, a dye chemist, and Anna Herdegen Fischer. Through his father's work as laboratory director at the Kalle Dye works, Fischer developed an early interest in the chemical nature of pigments, or coloring **matter**. Interested in both chemistry and medicine, Fischer received his doctorate in chemistry in 1904 from the University of Marburgh and his M.D. in 1908 from the University of Munich. After working on chemical structure of **peptides** and sugars with **Emil Fischer** (no relation) in Berlin, Fischer went to the Physiological Institute in Munich, where he first began his study (under Freidrich von Müller) of pigments, an area that was to become the overriding focus of his scientific pursuits. Fischer's dual expertise in chemistry and medicine led him to become chair of medical chemistry at the University of Innsbruch in 1916. Although he published his first notable scientific paper (on the subject of bilirubin, or bile pigment) in 1915, his research efforts soon came to a standstill due to World War I and the following years of reconstruction after Germany's defeat. Fischer's ill health also impeded his research efforts. He contracted tuberculosis when he was 20 years of age and had a kidney removed in 1917 due to complications from the disease.

In 1921, Fischer's investigation of pigments began in earnest as he accepted an appointment as director of the Institute für Anoreganische Chemise at the Technische Hochschule, or Technical University, in Munich. It was there that Fischer would conduct his groundbreaking research into pyrrole chemistry for nearly a quarter of a century. Fischer immediately reinitiated his studies of bile pigments and organized a number of specialized laboratories to simultaneously conduct the specific tasks needed to determine their chemical structures. Using a process known as Gattermann aldehyde synthesis to systematically prepare the numerous compounds needed for pyrrole derivatives, Fischer organized teams of microanalysts, sometimes referred to as "Gattermann cooks." He also set up specific laboratories to work on individual segments of a chemical problem, such as making calorimetric determinations and developing X-ray diagrams. By segmenting the work, Fischer's laboratory turned out more new chemical compounds than any laboratory that had preceded it.

Fischer's first major advance was the discovery of porphyrin synthesis in 1926. **Porphyrins** are made up of pyrroles joined in a chemical **ring** and are the pigments that appear throughout nature. The accepted view in chemistry prior to Fischer's work was that a single basic porphyrin structure was the primary component for all pigments occurring in nature. Fischer began to unravel the fundamental chemical structure of the porphin (the nucleic core of porphyrins), which had been proposed by W. Küster in 1912. This accomplishment led to the discovery of specific molecular structures of individual porphyrin groups that make up certain pigments. Specifically, Fischer had found that porphyrins are made up of four pyrrole nuclei bound by **methane** groups into a ring structure. This led to the creation of porphyrin in a laboratory setting. With the ability to synthesize porphyrin, Fischer and his colleagues were able to further determine thousands of specific porphyrin structures. In *Great Chemists,* **Heinrich Wieland**, an organic chemist, describes Fischer's attempt to synthesize porphyrins. "Fischer began to put the pyrrole segments together in mosaiclike arrangements and then to weld together, by brilliant synthetic procedures, the semimolecules of the pyrromethenes produced in this manner.'' Fischer soon recognized that porphyrins differed primarily through the components that made up the rings. He also discovered that bilirubin was derived from hemin and identified it as a porphyrin.

In 1929, Fischer successfully synthesized hemin, showing that its ring had a center **atom** of **iron**. Fischer received the Nobel Prize in chemistry in 1930 for his synthesis of hemin, which is one of two components of hemoglobin, the red respiratory protein of erythrocytes (red blood cells or corpuscles). During the Nobel Prize ceremonies, Fischer was also noted for his demonstration that hemin is related to chlorophyll, the light absorbing, green plant pigment. In 1944, Fischer finally worked out the chemical structure of and synthesized the pigment bilirubin, which he had first begun investigating during World War I. Over the years, Fischer's laboratory had synthesized approximately 130 porphyrins. He also conducted in-depth studies of the specific structure of chlorophyll and published 129 papers on the topic. He successfully identified chlorophyll's pyrrole rings, which had a center of **magnesium** rather than iron like hemin's pyrrole rings. The synthesis of chlorophyll, while based largely on Fischer's work, was not accomplished until 1960, 15 years after his death.

Fischer was a dedicated scientist who had few outside interests. He was also secretive and seldom discussed his work with other scientists outside of his laboratory. Fischer's lack of outside interests extended to politics as well. Although he privately expressed concern over the rise of dictator Adolf Hitler and Nazi Germany, he chose not to speak out publicly. In 1935, Fischer married Wiltrud Haufe. Despite being three decades older than his bride, Fischer was a happily married man and once confided to Wieland, who was a personal friend, that his wife had greatly enriched his life.

Despite Fischer's dedication to his work, which some colleagues called obsessive, he did enjoy taking long motoring vacations in his car. His other love was the outdoors. Although he constantly battled the debilitating effects of tuberculosis, Fischer was an expert skier, accomplished hiker, and an avid mountain climber, a passion he shared with his father until an accident claimed the elder Fischer's life. Germany's involvement in World War II added to Fischer's woes. Because of supply restrictions and frequent bombing raids made by Allied forces, his work was seriously restricted. When a bombing run destroyed Fischer's institute, the scientist gave in to despair. In 1945, Fischer committed suicide, despondent over what he viewed as the destruction of his life's work.

Although he was able to organize large scientific efforts and had an intuitive feel for the chemical structures involved in the field of pyrrole chemistry, Fischer was not noted for his ability to clearly write or lecture on such topics. Despite this fact, he published the definitive work on pyrrole chemistry in three volumes, *Die Chemie des Pyrrols,* which has remained a standard text on the subject. In addition to the Nobel Prize, Fischer received the Leibig Memorial Medal in 1929 and the Davy Medal in 1936. He also received an honorary degree from Harvard University in 1935.

FLAME ANALYSIS

When German chemist **Robert Wilhelm Eberhard Bunsen** invented the **Bunsen burner**—a device used in almost every chemistry laboratory—he also opened the door to the analysis of **matter** via flame analysis, a technique now grouped with other procedures more commonly known as atomic emission **spectroscopy** (AES).

Working with **Gustav Kirchhoff**, Bunsen helped to establish the principles and techniques of spectroscopy. A distinguished scientist, Bunsen discovered the elements **cesium** and **rubidium** during a long and productive career. Using the techniques he pioneered, scientists have been able to determine the chemical composition of stars.

Bunsen examined the spectra—the colors of light—emitted when a substance was subjected to intense flame. When air is admitted at the base of a Bunsen burner, it mixes with the gas to produce a very hot flame at approximately 3,272°F (1,800°C). This **temperature** is sufficient to cause the emission of light from certain elements. The **color** of the flame and its spectrum (component colors) is unique for each element.

To examine the spectra of elements, Bunsen used a simple apparatus that consisted of a prism, slits, and a magnifying **glass** or photo-sensitive film. Bunsen determined that the spectrum of elements that emitted light were different, that is, each did not contain all the colors of the spectrum. Instead, Bunsen determined that only portions of the spectrum were present.

Bunsen's fundamental observation that flamed elements that emit light only at specific wavelengths and that every element produced a characteristic spectra paved the way for the subsequent development of quantum theory by **Max Planck**, **Niels Bohr** and others.

Flame analysis or atomic emission spectroscopy is based on the physical and chemical principle that atoms—after being heated by flame—return to their normal **energy** state by giving off the excess energy in the form of light. The frequencies of the light given off are characteristic for each element.

Flame analysis is a qualitative test and not a quantitative test. A qualitative chemical analysis is designed to identify the components of a substance or mixture. Quantitative tests measure the amounts or proportions of the components in a reaction or substance.

The unknown to be subjected to flame analysis is either sprayed into the flame or placed on a thin wire that is then put into the flame. Very volatile elements (chlorides) produce intense colors. The yellow color of **sodium**, for example, can be so intense that it overwhelms other colors. To prevent this the wire to be coated with the unknown sample is usually dipped in **hydrochloric acid** and subjected to flame to remove the volatile impurities and sodium.

The flame test does not work on all elements. Those that produce a measurable spectrum when subjected to flame include, but are not limited to, **lithium**, sodium, **potassium**, rubidium, cesium, **magnesium**, **calcium**, **strontium**, **barium**, **zinc**, and **cadmium**. Other elements may need hotter flames to produce measurable spectra.

It takes some special techniques to properly interpret the results of flame analysis. The colors produced by a potassium flame (pale violet) can usually be observed only with the assistance of glass that can filter out interfering colors. Some colors are similar enough that line spectrum must be examined to make a complete and accurate identification of the unknown substance, or the presence of an identifiable substance in the unknown.

Flame analysis can also be used to determine the presence of metal elements in **water** by measuring the spectrum produced by the **metals** exposed to flame. The water is vaporized and then the emissions of the vaporized metals can be analyzed.

Flame tests are useful means of determining the composition of substances. The colors produced by the flame test are compared to known standards. In this way the presence of certain elements in the sample can be confirmed.

See also Bohr theory; Electromagnetic spectrum; Spectroscopy

FLAVONOID

Flavonoids are a large family of complex, phenolic compounds found universally in vascular plants. Extracted from plant material or synthesized artificially, flavonoids are colorless, crystalline, tricyclic (consisting of three **carbon** rings) subtances that form the base molecules of certain yellow pigments. Consisting of approximately 2,200 compounds, the flavonoid family of chemicals consists also of the closely related flavones, flavonones, and flavonols. Flavonoids are found naturally as flower and other plant structure pigments, as intermediates in general plant **metabolism**, and as specially produced plant *secondary compounds*. Secondary compounds are bioactive compounds which many plants produce as defense mechanisms against herbivory, or to reduce competition with rival plants for optimal habitat. Many commonly used spices are secondary plant compounds, including cinnamon and anise. Flavonoid secondary compounds that act as insect and mammalian repellants make plant material unpalletable or toxic. Flavonoids have diverse toxic effects including disruption of **cellular respiration**, inhibition of **enzyme** function, and interference with reproduction. Some flavonoids, such as proanthocyanidins and anthocyanins, are protective pigments which absorb harmful ultraviolet radiation. Other, closely related compounds like tannins reduce the digestibility of plant material, making it undesirable to animals. Flavonoids have gained notoriety as so-called phytonutrients or phytochemical extracts which are sold as nutritional supplements. Cranberries, citrus fruits, and grapes are reported to have high levels of flavonoids, presumably useful in the prevention of such ailments as cancer and circulatory diseases. Because of their powerful antioxidant properties, flavonoids are also touted as anti-aging supplements.

FLORY, PAUL J. (1910-1985)

American chemist

Paul Flory is widely recognized as the founder of the science of polymers. The Nobel Prize in **chemistry** he received in 1974 was awarded not for any single specific discovery, but, more generally, ''for his fundamental achievements, both theoretical and experimental, in the **physical chemistry** of macromolecules.'' That statement accurately reflects the wide-ranging character of Flory's career. He worked in both industrial and academic institutions and was interested equally in the theory of macromolecules and in the practical applications of that theory.

Paul John Flory was born in Sterling, Illinois, on June 19, 1910. His parents were Ezra Flory, a clergyman and educator, and Martha (Brumbaugh) Flory, a former school teacher. Ezra and Martha's ancestors were German, but they had resided in the United States for six generations. Both the Flory and the Brumbaugh families had always been farmers, and Paul's parents were the first in their line ever to have attended college.

After graduation from Elgin High School, Flory enrolled at his mother's alma mater, Manchester College, in North Manchester, Indiana. The college was small, with an enrollment of only 600. He earned his bachelor's degree in only three years, at least partly because the college ''hadn't much more than three years to offer at the time,'' as he was quoted as having said by Richard J. Seltzer in *Chemical and Engineering News.* An important influence on Flory at Manchester was chemistry professor Carl W. Holl. Holl apparently convinced Flory to pursue a graduate program in chemistry. In June of 1931, therefore, Flory entered Ohio State University and, in spite of an inadequate background in mathematics and chemistry, earned his master's degree in **organic chemistry** in less than three months. He then began work immediately on a doctorate, but switched to the field of physical chemistry. He completed his research on the **photochemistry** of **nitric oxide** and was granted his Ph.D. in 1934.

Flory's doctoral advisor, Herrick L. Johnston, tried to convince him to stay on at Ohio State after graduation. Instead, however, he accepted a job at the chemical giant, Du Pont, as a research chemist. There he was assigned to a research team headed by **Wallace H. Carothers**, who was later to invent the process for making **nylon** and neoprene. Flory's opportunity to study polymers was ironic in that, prior to this job, he knew next to nothing about the subject. Having almost *any* job during the depths of the Great Depression was fortunate, and Flory was the envy of many classmates at Ohio State for having received the Du Pont offer.

Flory's work on the Carothers team placed him at the leading edge of chemical research. Chemists had only recently begun to unravel the structure of macromolecules, very large molecules with hundreds or thousands of atoms, and then to understand their relationship to polymers, molecules that have chemically combined to become a single, larger **molecule**. The study of polymers was even more difficult than that of macromolecules because, while the latter are very large in size, they have definite chemical compositions that are always the same for any one substance. Polymers, on the other hand, have variable size and composition. For example, **polyethylene**, a common polymer, can consist of anywhere from a few hundred to many thousands of the same basic unit (monomer), arranged always in a straight chain or with cross links between **chains**.

With his background in both organic and physical chemistry, Flory was the logical person to be assigned the responsibility of learning more about the physical structure of polymer molecules. That task was made more difficult by the variability of size and shape from one polymer molecule to another—even among those of the same substance. Flory's solution to this problem was to make use of statistical mechanics to average out the properties of different molecules. That technique had already been applied to polymers by the Swiss chemical physicist Werner Kuhn and two Austrian scientists, **Herman Mark** and Eugene Guth. But Flory really developed the method to its highest point in his research at Du Pont.

During his four years at Du Pont, Flory made a number of advances in the understanding of polymer structure and reactions. He made the rather surprising discovery, for example, that the rate at which polymers react chemically is not affected by the size of the molecules of which they are made. In 1937, he discovered that a growing polymeric chain is able to terminate its own growth and start a new chain by reacting with other molecules that are present in the reaction, such as those of the solvent. While working at Du Pont, Flory met and, on March 7, 1936, married Emily Catherine Tabor. The Florys had two daughters, Susan and Melinda, and a son, Paul John, Jr. Flory's work at Du Pont came to an unexpected halt when, during one of his periodic bouts of depression, Carothers committed suicide in 1937. Although deeply affected by the tragedy, Flory stayed on for another year before resigning to accept a job as research associate with the Basic Science Research Laboratory at the University of Cincinnati. His most important achievement there was the development of a theory that explains the process of gelation, which involves cross-linking in polymers to form a gel-like substance.

Flory's stay at the University of Cincinnati was relatively brief. Shortly after World War II began, he accepted an offer from the Esso (now Exxon) Laboratories of the Standard Oil Development Company to do research on rubber. It was apparent to many American chemists and government officials that the spread of war to the Pacific would imperil, if not totally cut off, the United States' supply of natural rubber. A massive crash program was initiated, therefore, to develop synthetic substitutes for natural rubber. Flory's approach was to learn enough information about the nature of rubber molecules to be able to predict in advance which synthetic products were likely to be good candidates as synthetic substitutes (''**elastomers**''). One result of this research was the discovery of a method by which the structure of polymers can be studied. Flory found that when polymers are immersed in a solvent, they tend to expand in such a way that, at some point, their **molecular structure** is relatively easy to observe.

In 1943, Flory was offered an opportunity to become the leader of a small team doing basic research on rubber at the Goodyear Tire and Rubber Company in Akron, Ohio. He accepted that offer and remained at Goodyear until 1948. One of his discoveries there was that irregularities in the molecular structure of rubber can significantly affect the tensile strength of the material.

Paul J. Flory. *(AP/Wide World. Reproduced by permission.)*

In 1948, Flory was invited by **Peter Debye**, the chair of Cornell University's department of chemistry, to give the prestigious George Fisher Baker Lectures in Chemistry. Cornell and Flory were obviously well pleased with each other as a result of this experience, and when Debye offered him a regular appointment in the chemistry department beginning in the fall of 1948, Flory accepted—according to Maurice Morton in *Rubber Chemistry and Technology*—''without hesitation.'' The Baker Lectures he presented were compiled and published by Cornell University Press in 1953 as *Principles of Polymer Chemistry*. Flory continued his studies of polymers at Cornell and made two useful discoveries. One was that for each polymer solution there is some **temperature** at which the molecular structure of the polymer is most easily studied. Flory called that temperature the theta point, although it is now more widely known as the Flory temperature. Flory also refined a method developed earlier by the German chemist **Hermann Staudinger** to discover the configuration of polymer molecules using **viscosity**. Finally, in 1956, he published one of the first papers ever written on the subject of **liquid crystals**, a material ubiquitous in today's world, but one that was not to be developed in practice until more than a decade after Flory's paper was published.

In 1957, Flory became executive director of research at the Mellon Institute of Industrial Research in Pittsburgh. His charge at Mellon was to create and develop a program of basic research, a focus that had been absent from that institution, where applied research and development had always been of primary importance. The job was a demanding one involving the supervision of more than a hundred fellowships. Eventually, Flory realized that he disliked administrative work and was making little progress in refocusing Mellon on basic research. Thus, when offered the opportunity in 1961, he resigned from Mellon to accept a post at the department of chemistry at Stanford University. Five years later, he was appointed Stanford's first J. G. Jackson-C. J. Wood Professor of Chemistry. When he retired from Stanford in 1975, he was named J. G. Jackson-C. J. Wood Professor Emeritus. In 1974, a year before his official retirement, Flory won three of the highest awards given for chemistry—the National Medal of Science, the American Chemical Society's Priestley Medal, and the Nobel Prize in chemistry. These awards capped a career in which, as Seltzer pointed out, Flory had "won almost every major award in science and chemistry."

Flory's influence on the chemical profession extended far beyond his own research work. He was widely respected as an outstanding teacher who thoroughly enjoyed working with his graduate students. A number of his students later went on to take important positions in academic institutions and industrial organizations around the nation. His influence was also felt as a result of his two books, *Principles of Polymer Chemistry,* published in 1953, and *Statistical Mechanics of Chain Molecules,* published in 1969. Leo Mandelkern, a professor of chemistry at Florida State University, is quoted by Seltzer as referring to the former work as "the bible" in its field, while the latter has been translated into both Russian and Japanese.

Flory was also active in the political arena, especially after his retirement in 1975. He and his wife decided to use the prestige of the Nobel Prize to work in support of human rights, especially in the former Soviet Union and throughout Eastern Europe. He served on the Committee on Human Rights of the National Academy of Sciences from 1979 to 1984 and was a delegate to the 1980 Scientific Forum in Hamburg, at which the topic of human rights was discussed. As quoted by Seltzer, Morris Pripstein, chair of Scientists for Sakharov, Orlov, and Scharansky, described Flory as "very passionate on human rights.... You could always count on him." At one point, Flory offered himself to the Soviet government as a hostage if it would allow Soviet scientist Andrei Sakharov's wife, Yelena Bonner, to come to the West for medical treatment. The Soviets declined the offer, but eventually did allow Bonner to receive the necessary treatment in Italy and the United States.

Flory led an active life with a special interest in swimming and golf. In the words of Ken A. Dill, professor of chemistry at the University of California, San Francisco, as quoted by Seltzer, Flory was "a warm and compassionate human being. He had a sense of life, a sense of humor, and a playful spirit. He was interested in, and cared deeply about, those around him. He did everything with a passion; he didn't do anything half way." Flory died on September 8, 1985, while working at his weekend home in Big Sur, California. According to Seltzer, at Flory's memorial service in Stanford, James Economy, chair of the American Chemical Society's division of **polymer chemistry**, expressed the view that Flory was "fortunate to depart from us while still at his peak, not having to suffer the vicissitudes of old age, and leaving us with a sharply etched memory of one of the major scientific contributors of the twentieth century."

Fluorescence and phosphorescence

Fluorescence and phosphorescence are both luminescence phenomena that occur due to the de-excitation of electrons. When a source of **energy** such as thermal or radiant energies is applied to a fluorescent material, high energy particles or photons collide with the electrons in the material, causing them to become excited and jump to a higher **energy level**. The electrons in the excited state have the same spin as they did in the ground state. The electrons in this excited state are not stable and almost spontaneously become de-excited and move back down to the lower energy levels, releasing energy in the form of light. When the source of energy is removed, the light emission ceases. A fluorescent state can exist for only 10^{-5} to 10^{-8} seconds. One familiar example of this phenomenon is a "black light" poster. "Black light" posters are painted with fluorescent ink. When the ink is exposed to "black light" (a light bulb which emits mostly blue and ultraviolet light), it absorbs the blue light, which excites the electrons in the ink. As the electrons move back to their ground state, they emit light in the green and yellow wavelengths.

Phosphorescence is similar to fluorescence except the light emission continues even after the source of energy is removed. This event was first observed in the seventeenth century but was not studied until the nineteenth century. The current model was proposed by Philipp Lenard (1862-1947) of Kiel University, Germany. He had been interested in luminescent the phenomenon since he was a boy in Hungary and would apply **heat** to crystals of **fluorine** to make them glow. In 1888, working at the University of Heidelberg, he began work with cathode rays and found that they were not equivalent to fluorescent light, as was the current thinking. He worked for many years on the **photoelectric effect**, trying to find an explanation for the phenomenon. In 1902, he published his theory of **electron** excitation and luminescence. His theory could not be explained, however, until 1905 when **Albert Einstein** developed the theory of photons. Lenard won the 1905 Nobel Prize in physics for his work in cathode rays.

A phosphorescent material emits light after it absorbs energy, for instance, radiant energy. The absorbed energy excites electrons in the phosphorescent material and causes them to be caught in potential energy troughs. While excited, the electrons experience a change in spin called intersystem crossing. This change in spin places the electrons in a metastable position from which they cannot move back into the ground state. They are freed from these troughs by thermal energy within the crystal structure of the material, which causes them to be raised to a higher, less stable energy level where they can eventually fall back to lower energy levels. As they do so, they emit light. Because of the slow decay due to the intersystem crossing, the light continues after the exciting source is removed.

Phosphorescence is **temperature** dependent since thermal energy is required for the electrons to become de-excited. Phosphorescent states can exist from 10^{-4} seconds to several hours. Phosphorescent materials are used in "glow-in-the-dark" toys that will emit light after exposure to some form of radiant energy. The hands on certain watches and alarm clocks are coated with a phosphorescent material which will emit light for many hours after a light source is removed. The most common form of phosphorous exhibits this phenomenon, hence the term "phosphorescence." Some fluorescent materials, under the appropriate conditions, will also phosphoresce.

Fluorescence has been utilized for many purposes, the most obvious being fluorescent lamps. A fluorescent lamp is an evacuated **glass** tube that contains **mercury** vapor and is coated with an apatite analog ($Sr_5(PO_4)_3F$), small amounts of rare earth elements, and **antimony**, which is used as an indicator. When a voltage is applied across the tube (approximately 600 volts), the mercury ionizes and conducts. An electron is stripped from a mercury **atom**. When the stripped electron collides with an electron in another mercury atom, it causes the second electron to become excited and jump to a higher energy level. When that electron returns to its ground state, it releases energy as ultraviolet light, which is invisible. The coating on the inside of the tube is referred to as phosphors. Phosphors absorb ultraviolet energy, causing the electrons in the material to become excited. When these electrons return to their ground state, they release light in the visible region. Phosphors are also used in laundry detergents to make white clothes appear whiter. When the ultraviolet radiation from sunlight is absorbed the phosphors in the clothes glow. Sometimes a mixture of **krypton** and **argon** are used instead of the mercury.

There are many naturally occurring fluorescent materials. Many minerals, for example, will emit light when activated by ultraviolet light, x rays, or electron beams. Electrons in the mineral absorb the energy from the activator and become excited. As they return to their ground state, they emit light. The light emitted after activation is sometimes a very different **color** from the mineral itself.

Many mineral collectors use fluorescence to identify minerals. A special fluorescent lamp that emits ultraviolet light while filtering out white light is required to view the fluorescence in many minerals. Some minerals are considered fluorescent even though not all samples of the mineral will fluoresce. This is often the result of the different samples of the mineral being obtained from different locations. Many times impurities in the sample will inhibit, or change the color of, the mineral's fluorescence. Sometimes the color of a mineral's fluorescence can be used to determine its place of origin. For example, calcite fluoresces yellow, white, green, red, or just about any of the colors of the visible spectrum; but if it is from Terlingua, Texas, it is always bluish-white and if it is from Franklin, New Jersey, it is always red.

Two kinds of ultraviolet light can be used to detect fluorescence in minerals, long wave and short wave. Some minerals fluoresce the same under both wavelengths, some fluoresce different colors under each wavelength, and some only fluoresce under one of the two types of ultraviolet light. Some minerals which fluoresce under long wave ultraviolet light include agate, calcite, fluorite, and halite, which produce a white light; amber, which produces an orange light; and quartz and opal, which produce green light. Minerals which fluoresce under short wave ultraviolet light include gypsum, which produces white light; ruby, which produces red light; talc, which produces yellow light; and **diamond**, which produces green light. Fluorescent diamonds were once considered more precious than non-fluorescent stones, although recently the fluorescent property is considered an impurity.

Fluorescence and phosphorescence are two similar luminescent phenomena. Fluorescence is the emission of light from a material while being exposed to an energy source, while phosphorescence is the emission of light from a material during, as well as after, exposures to an energy source. The difference between the two is the result of the spin of the excited electrons. There are many familiar examples of fluorescence and phosphorescence in our daily lives. Fluorescent materials can be synthetic, such as phosphors, or occur in nature, for example, fluorescent minerals.

FLUORIDATION

Fluoridation consists of adding fluoride to a substance (often drinking water) to reduce tooth decay. Fluoridation was first introduced into the United States in the 1940s in an attempt to study its effect on the **reduction** of tooth decay. Since then many cities have added fluoride to their **water** supply systems. Proponents of fluoridation have claimed that it dramatically reduced tooth decay, which was a serious and widespread problem in the early twentieth century. Opponents of fluoridation have not been entirely convinced of its effectiveness, are concerned by possible side effects, and are disturbed by the moral issues of personal rights that are raised by the addition of a chemical substance to an entire city's water supply. The decision to fluoridate drinking water has generally rested with local governments and communities and has always been a controversial issue.

Tooth decay occurs when food acids dissolve the protective enamel surrounding each tooth and create a hole, or cavity, in the tooth. These acids are present in food, and can also be formed by acid-producing bacteria that convert sugars into acids. There is overwhelming evidence that fluoride can substantially reduce tooth decay. When ingested into the body, fluoride concentrates in bones and in dental enamel which makes the tooth enamel more resistant to decay. It is also believed that fluoride may inhibit the bacteria that convert sugars into acidic substances that attack the enamel.

Fluoride is the water soluble, ionic form of the element **fluorine**. It is present in most water supplies at low levels and nearly all food contains traces of fluoride. When water is fluoridated, chemicals that release fluoride are added to the water. In addition to fluoridation of water supplies, toothpaste and mouthwash also contain added fluoride.

In 1901 Frederick McKay (1874-1959), a dentist in Colorado Springs, Colorado, noticed that many of his patients' teeth were badly stained. Curious about the cause of this staining, or dental fluorosis as it is also known, McKay concluded after three decades of study that the discolorations were caused by some substance in the city's water supply. Analysis of the water indicated high levels of fluoride, and it was concluded that the fluoride was responsible for the stained teeth. McKay also observed that although unsightly, the stained teeth of his patients seemed to be more resistant to decay. The apparent connection between fluoride and reduced decay eventually convinced H. Trendley Dean (1893-1962) of the U.S. Public Health Service (USPHS) to examine the issue more closely.

In the 1930s, Dean studied the water supplies of some 345 U.S. communities and found a low incidence of tooth decay where the fluoride levels in community water systems were high. He also found that staining was very minor at fluoride concentrations less than or equal to one part fluoride per million parts of water (or one ppm). The prospect of reducing tooth decay on a large scale by adding fluoride to community water systems became extremely appealing to many public health officials and dentists. By 1939, a proposal to elevate the fluoride levels to about one ppm by adding it artificially to water supplies was given serious consideration, and eventually several areas were selected to begin fluoridation trials. By 1950, USPHS administrators endorsed fluoridation throughout the country.

The early fluoridation studies apparently demonstrated that fluoridation was an economical and convenient method to produce a 50-60% reduction in the tooth decay of an entire community and that there were no health risks associated with the increased fluoride consumption. Consequently, many communities quickly moved to fluoridate their water supplies in the 1950s. However strong opposition to fluoridation soon emerged as opponents claimed that the possible side effects of fluoride had been inadequately investigated. It was not surprising that some people were concerned by the addition of fluoride to water since high levels of fluoride ingestion can be lethal. However, it is not unusual for a substance that is lethal at high **concentration** to be safe at lower levels, as is the case with most **vitamins** and trace elements.

Opponents of fluoridation were also very concerned on moral grounds because fluoridation represented compulsory **mass** medication. Individuals had a right to make their own choice in health matters, fluoridation opponents argued, and a community violated these rights when fluoride was added to its water supply. Fluoridation proponents countered such criticism by saying that it was morally wrong not to fluoridate water supplies because this would result in many more people suffering from tooth decay which could have easily been avoided through fluoridation.

The issue of fluoridation had become very much polarized by the 1960s since there was no middle ground: water was either fluoridated or not. Controversy and heated debate surrounded the issue across the country. Critics pointed to the known harmful effects of large doses of fluoride that led to bone damage and to the special risks for people with kidney disease or those who were particularly sensitive to toxic substances. Between the 1950s and 1980s, some scientists suggested that fluoride may have a mutagenic effect (that is, it may be capable of causing human birth defects). Controversial claims that fluoride can cause cancer were also raised. Today, some scientists still argue that the benefits of fluoridation are not without health risks.

The development of the fluoridation issue in the United States was closely observed by other countries. Dental and medical authorities in Australia, Canada, New Zealand, and Ireland endorsed fluoridation, although not without considerable opposition from various groups. Fluoridation in Western Europe was greeted less enthusiastically and scientific opinion in some countries, such as France, Germany, and Denmark, concluded that it was unsafe. Widespread fluoridation in Europe is therefore uncommon.

Up until the 1980s the majority of research into the benefits of fluoridation reported substantial reductions (50-60% on average) in the incidence of tooth decay where water supplies had fluoride levels of about one ppm. By the end of the decade however, the extent of this reduction was being viewed more critically. By the 1990s, even some fluoridation proponents suggested that observed tooth decay reduction, directly as a result of water fluoridation, may only have been at levels of around 25%. Other factors, such as education and better dental hygiene, could also be contributing to the overall reduction in tooth decay levels. Fluoride in food, **salt**, toothpastes, rinses, and tablets, have undoubtedly contributed to the drastic declines in tooth decay during the twentieth century. It also remains unclear as to what, if any, are the side effects of one ppm levels of fluoride in water ingested over many years. Although it has been argued that any risks associated with fluoridation are small, these risks may not necessarily be acceptable to everyone. The fact that only about 50% of U.S. communities have elected to adopt fluoridation is indicative of people's cautious approach to the issue. In 1993, the National Research Council published a report on the health effects of ingested fluoride and attempted to determine if the maximum recommended level of four ppm for fluoride in drinking water should be modified. The report concluded that this level was appropriate but that further research may indicate a need for revision. The report also found inconsistencies in the scientific studies of fluoride toxicity and recommended further research in this area.

FLUORINE

Fluorine is the first element in Group 17 of the **periodic table** and the first member of the halogen family. Its **atomic number** is 9, its atomic **mass** is 18.998404, and its chemical symbol is F.

Properties

Fluorine is a pale yellow gas with a **density** of 1.695 grams per cubic centimeter, about 1.3 times that of air. Fluorine's **boiling point** is -306.6°F (-188.13°C) and its freezing point is -363.3°F (-219.61°C). Fluorine has a strong and characteristic odor that can be detected in amounts as low as 20 parts per billion.

Fluorine is the most reactive of all elements. It combines easily and even explosively with every other element except **helium**, **neon**, and **argon**. It also reacts violently with most compounds and, for this reason, must be handled with great care.

Occurrence and Extraction

Fluorine never occurs as an element in nature, but is found in minerals such as fluorspar, fluorapatite, and cryolite. The element is thought to have an abundance of about 0.06%, making it the 13th most abundant element in the Earth's crust. It is obtained commercially by electrolyzing a liquid mixture of **hydrogen** fluoride and **potassium** hydrogen fluoride.

Discovery and Naming

Fluorine was discovered in 1886 by the French chemist **Henri Moissan**. Moissan's discovery was a remarkable accomplishment because a number of his predecessors had been unable to find a way to tame the dangerous element and extract it from its compounds. The method Moissan used is essentially the same one by which fluorine is produced today. Moissan suggested that name fluorine for the element after one of its most common minerals, fluorspar. Fluorspar, in turn, had been given its name centuries before after a Latin word *fluere*, meaning "to flow." The name comes from the fact that fluorspar is frequently used as a flux to reduce the **temperature** at which ores and other **natural products** will melt when heated.

Uses

Fluorine has relatively few applications as a free element, largely because it is so active. Compounds of fluorine find more uses, however. For example, **sodium** fluoride (NaF), **calcium** fluoride (CaF_2), and stannous fluoride (SnF_2) are all used to help reduce tooth decay. They are included in "fluoridated" toothpaste and are added to the public **water** supply by some cities and other municipalities. Fluorides are incorporated into the structure of growing teeth and make those teeth stronger and more resistant to decay.

Fluorine was once widely used also in the production of chlorofluorocarbons (CFCs). These compounds were discovered in the late 1920s by the American chemical engineer **Thomas Midgley, Jr.** They have properties that make them highly desirable for a number of commercial and industrial applications, such as cooling and refrigeration systems, cleaning agents, aerosol sprays, and specialized polymers. The production of CFCs grew from about 1 million kilograms (2 million pounds) in 1935 to more than 700 million kilograms (1.5 billion pounds) in 1985.

Beginning in the early 1980s, however, research suggested that CFCs contribute to the destruction of the ozone layer, a thin layer of the stratosphere rich in ozone (O_3). The ozone layer is critical to the survival of life on Earth because it absorbs ultraviolet radiation that would otherwise cause injuries and deaths. For this reason, the production and use of CFCs have been essentially banned in most parts of the world.

Another well-known use of fluorine is in the manufacture of **Teflon**, a widely popular non-stick plastic. Teflon is the trade name for polytetrafluorethylene, or PTFE. The material is now used in a host of products ranging from kitchen cookware to baking sprays to stain repellents used for fabrics and textiles.

Health Issues

Fluorine gas is one of the most toxic **gases** known to humans. Inhaled in even small amounts, it causes severe irritation to the respiratory system and, in larger doses, it can cause death. The highest recommended exposure to fluorine is one part per million of air over an eight-hour period. Fortunately, few people are likely to encounter elementary fluorine in their daily lives.

Food additives

Food additives help preserve the freshness and appeal of food between the time it is manufactured and when it finally reaches the market. Additives may also improve nutritional value of foods and improve their **taste**, texture, consistency or **color**. All food additives approved for use in the United States are carefully regulated by federal authorities to ensure that foods are safe to eat and are accurately labeled.

Food additives have been used by man since earliest times. Early peoples, for example, used **salt** to preserve meats and fish; they also added herbs and spices to improve the flavor of their foods. It has only been in recent times, however, that governments have sought to ensure the safety of additives through regulation. Today, food and color additives are more strictly regulated than at any time in history.

Additives may be incorporated in foods to maintain product consistency, improve or maintain nutritional value, maintain palatability and wholesomeness, provide leavening or control acidity/alkalinity, and/or enhance flavor or impart desired color.

Loosely, a food additive is any substance added to food. The Food and Drug Administration (FDA), however, recognizes additives as any substance whose intended use will affect, or may reasonably be expected to affect, the characteristics of any food. This definition includes any substance used in the production, processing, treatment, packaging, transportation or storage of food. A color additive, when applied to food additives, is any dye, pigment, or substance that can impart color when added or applied to a food. Federal law prohibits the use of any additive that has been found to cause cancer in humans or animals.

Substances that are added to a food for a specific purpose are referred to as direct additives. Indirect food additives are those introduced into a food in trace amounts in the course of packaging, storage, or other handling.

To market a new food or color additive, a manufacturer must first petition the FDA for its approval. Federal regulations require evidence that each substance is safe at its intended levels of use before it may be added to foods. In deciding

A false-colored scanning electron micrograph (SEM) image of monosodium glutamate crystals. Monosodium glutamate (MSG) is a well known food flavoring prepared from wheat gluten and other proteins.) ***Science Photo Library, Photo Researchers, Inc. Reproduced by Permission.)***

whether an additive should be approved, the agency considers the composition and properties of the substance, the amount likely to be consumed, its probable long-term effects and various safety factors. All additives are subject to ongoing safety review as scientific understanding and methods of testing continue to improve.

New techniques for producing food additives are under investigation. Biotechnologists are attempting to use simple organisms to produce additives that are the same food components found in nature. In 1990, the FDA approved the first bioengineered **enzyme**, rennin, which traditionally has been extracted from calves' stomachs for use in making cheese.

Alginates, lecithin, mono- and diglycerides, methyl **cellulose**, carrageenan, glyceride, pectin, guar gum, and **sodium** aluminosilicate are added to foods to impart and/or maintain desired consistency. **Vitamins** A and D, thiamine, niacin, riboflavin, pyridoxine, folic acid, ascorbic acid, **calcium** carbonate, **zinc oxide**, and **iron** are used as additives to improve and/or maintain nutritive value. Propionic acid and its salts, ascorbic acid, butylated hydroxy anisole (BHA), butylated hydroxytoluene (BHT), benzoates, sodium nitrite, and **citric acid** are added to foods to maintain their palatability and wholesomeness. To give food a light texture, and/or control its acidity/alkalinity, yeast, **sodium bicarbonate**, citric acid, fumaric acid, **phosphoric acid**, lactic acid, and tartrates may be added. To enhance flavor or impart a desired color, cloves, ginger, fructose, aspartame, saccharin, FD&C Red no. 40, monosodium glutamate, caramel, annatto, limonene, and turmeric can be added.

See also Artificial sweeteners; Acids and bases; Carcinogens; Lipids; Nitrates and nitrites; Silicates

FORENSIC CHEMISTRY

Forensic chemistry is the application of chemistry in the pursuit of the law, particularly in solving crimes.

Forensic chemistry is a rapidly advancing science with many new tests being constantly added. Until these tests become established, there is often much controversy over their usage. One well known example of forensic chemistry is **DNA fingerprinting**. This is a technique that was invented by Alec Jeffreys in 1985. By looking at small samples of bodily tissues, a pattern can be produced by digesting the DNA using a series of enzymes known as restriction enzymes, or restriction endonucleases. This digested material is then placed on a sheet of agarose **gel**, and the fragments are separated by passing an electric current across the gel. Different enzymes produce different banding patterns and normally several enzymes are used in conjunction to produce a high definition digestion. Because these digestions of DNA are particular to an individual they are known as DNA fingerprints. These tests can be carried out on small pieces of material due to DNA amplification techniques. Different combinations of enzymes produce different restriction patterns and the likelihood of a match between the test DNA and an individual differs. Presently, the chance of a match is given as a statistical likelihood. Insufficient DNA fingerprint tests have been carried out to give the chances of a false match with certainty. Certain patterns are specific to ethnic groups. With most DNA fingerprint tests, the chance of a random match is several million to one against. This is usually taken as a strong indicator that the person with DNA matching that from a crime scene is the culprit.

One of the first examples of chemistry used in forensic studies was in the application of the blood type test to crime scenes and evidence. Due to the **proteins** present in blood, there are several different categories into which blood can be classified. The simplest and most widely known is the blood group system where two different proteins can be present in the blood, type A and type B. Because of the way blood is made up an individual may be type A, B, AB, or O (when neither protein is present). If type A blood is present at a crime scene, then an individual who is type B cannot be the same individual. There are various other proteins that are used in blood typing that can only rule out individuals as described: the next most common after the ABO system is the rhesus protein (rh), where an individual can be either rhesus positive (protein present) or rhesus negative (protein absent). Blood group proteins are routinely tested for by an agglutination test.

Blood is added to known samples and the blood will not stick together when added to its own type. There are also a number of blood specific stains that can be added to a compound to see if it is blood or not. The majority of these tests involve a protein binding to the **iron** in hemoglobin. Specific proteins can now be added to blood samples so that even minute quantities of blood can be typed.

There are very specific forensic tests that exist. For example, some can detect the presence of minute traces of **gunpowder** on clothes or skin. This group of tests can show if an individual has recently fired a gun. This test first came into use in the 1970s, and it has been much refined since to make it more reliable. In the 1970s, a number of false positives were achieved using the earliest versions of this test. These false positives were created by common household chemicals reacting with the test. This showed that the individual had recently used a firearm when in reality all that had occurred had been the handling of a common material such as domestic bleach.

Certain chemicals can also be detected at a crime scene using forensic chemistry. Depending on the type of chemicals, their presence can then be tested for on the clothing of a suspect. Even apparently identical chemicals can be dissimilar, and forensic chemistry can distinguish the differences. For example even though it may apparently be of the same **color**, paint can have an individual fingerprint—a recognizable and testable chemical makeup. Consequently, paint samples from a crime scene can be taken and compared with those found in association with a suspect. These tests can be carried out by looking at the absorption spectra of paint samples or by looking at their composition in flurometers.

Many other products that appear identical to the naked eye are not. Many fibers can be distinguished on their chemical basis. Again, this is due to differences in composition at a molecular level and analysis is carried out by looking at absorption spectra and fluorescent characteristics.

In a case where poisoning is suspected, chemical tests can be carried out to detect poisons. These tests are basically a form of **qualitative analysis**, determining the individual molecules that are present. These sorts of tests are useless in isolation. Over time, data have been produced showing the composition of different poisons, drugs, and false positives. Although specific chemical tests are available, the chemical analysis is usually carried out by some form of photometry. In many fictionalized detective stories, the investigating police are shown tasting a white powder and instantly identifying it as a **narcotic**. This simply does not work. Many narcotics do not have a distinctive **taste** and a whole battery of tests must be carried out to show the nature of the substance. This is even more of a problem with some **natural products**. ''Magic mushrooms,'' *Psilocybe* species, are notoriously difficult to identify. Even when the species is correctly identified, there are variations in the amount of active substance present. Again, chemical tests examining the molecules present have to be carried out to confirm the presence of narcotic substances.

Forensic chemistry is a constantly advancing science. New detection methods are constantly being tried to make identification more certain. With forensic chemistry, an unreliable test is worse than useless. An unreliable test can convict the wrong person. In court, both defense and prosecution argue over the validity of forensic tests and use expert witnesses. It is important that these sorts of arguments are applied to forensic evidence. It is this sort of questioning that drives the advancement of forensic chemistry, allowing new and more accurate tests to be formulated. Forensic chemistry is only one way of gathering evidence for court cases and it should be considered in conjunction with all other evidence that is offered.

Forensic chemistry is not a quick process. Many tests are carried out, the majority using analytical machines that look at the composition of the molecules under test. This whole range of techniques is covered under the heading **spectroscopy**. It is a very advanced science in terms of the technology that is employed. Advances in forensic chemistry are constantly being made, allowing the identification of substances with a greater certainty.

See also Analytical chemistry

Formaldehyde

Formaldehyde (HCHO) is the simplest member of the class of organic compounds known as aldehydes. Aldehydes are distinguished by the presence and end position of a **carbonyl group** (-OH) attached to only one carbon **atom**. At room temperature, formaldehyde is an extremely reactive colorless gas with a suffocating odor. It is commonly sold as an aqueous solution (formalin) or in solid polymeric forms (paraformaldehyde and trioxane). Formaldehyde is used in the manufacture of dyes, in the production of synthetic resins, and in embalming and as a preservative for biological specimens.

Formaldehyde was first intentionally produced by **August Hofmann**, a German chemist, in 1867. Justus von Liebig had researched aldehydes earlier but had never succeeded in producing formaldehyde. Alexander Mikhailovich Butlerov (1828-1886) had hydrolyzed methylene acetate while trying to form methyl glycol and produced formaldehyde gas. He had noted that the product of the reaction behaved as an aldehyde. Hofmann, in 1867, exposed a mix of methyl alcohol vapors and air over a hot platinum spiral to form stable formaldehyde. Formaldehyde today is produced either from methanol, as Hofmann did, or by oxidizing hydrocarbons.

Due to health concerns, in 1992 the Occupational Safety and Health Administration (OSHA) lowered the legal exposure limits for formaldehyde from 1.0 parts per million to 0.75 parts per million. Nonetheless, the U.S. formaldehyde market remains strong with production anticipated to reach 12 billion lb (5.4 billion kg) annually by the end of the century.

Formic acid

Formic acid belongs to the family of **carboxylic acids** and has the chemical formula HCOOH or HCO_2H. The **carbon atom** has a double bond with one of the **oxygen** atoms. It is the simplest carboxylic acid and has the lowest **molecular weight** (46.03). A colorless, corrosive liquid with a sharp odor, formic acid boils at 213.3°F (100.7°C) and solidifies at 47.1°F (8.4°C). Like other acids, it reacts with most alcohols to form esters and decomposes when heated; it is also easily oxidized.

Formic acid decomposes slowly at room **temperature** into **carbon monoxide** (CO) and **water** (H_2O). Decomposition is dependent on temperature and **concentration**. Though it is neither explosive nor spontaneously flammable in air, formic acid is flammable under certain conditions and is corrosive. It is also considered an environmental contaminant of air and water and has been identified as the toxic intermediate (formate) in **methanol** poisoning.

Formic acid occurs naturally in a variety of plants and fruits (the fruit of the soaptree), mammalian tissues, and insect venoms, including the bodies of red ants and the stingers of bees. It occurs synthetically as a byproduct in the manufacture of acetaldehyde and **formaldehyde** and during the atmospheric **oxidation** of turpentine. Formic acid is prepared commercially by heating carbon monoxide and **sodium** hydroxide to form sodium formate. The sodium formate is then carefully treated with **sulfuric acid** at low temperatures and distilled in a vacuum to yield formic acid. Another preparation method involves acid **hydrolysis** of methyl formate.

A single-step low-cost method has also been developed at the Boreskov Institute of Catalysis to produce formic acid with a yield of 90%. This method, claimed to be the first single stage method to produce formic acid from formaldehyde-containing **gases**, involves direct oxidation of formaldehyde by oxygen over an original **oxide** catalyst in a tube fixed-bed reactor followed by the product condensation. The final product contains 55-85% of formic acid depending on the raw material type and no more than 0.1% of organic admixtures. **Distillation** can increase the content of formic acid up to 95% and higher.

Formic acid has a wide range of industrial uses and over 300,000 tons is produced on a global basis every year. As an intermediate, it is used as a solvent for polyamides and **cellulose** acetate and in the manufacture of various chemicals and pharmaceuticals, such as **caffeine**, enzymes, antibiotics, artificial sweeteners, polyvinyl chloride plasticizers, and rubber **antioxidants**. In the dyeing of natural and synthetic fibers, formic acid regulates the **pH** and is also used to help impregnate waterproof textiles. Formic acid is also used in the preservation of green feed/fodder and in the coagulation of natural rubber (latex).

In addition, formic acid is an active ingredient in commercial cleaning products, such as descalers, rust removers, and degreasers. Due to its bactericidal properties, formic acid is used in the disinfection of the **wood** barrels used for storing wine and beer. A more recent application of formic acid is for pH regulation of flue gas desulfurization. The flue gas is passed through an aqueous limestone slurry containing formic acid. The **sulfur** dioxide in the gas reacts with the limestone to form gypsum (calcium sulfate). Formic acid is also used in fumigants, **refrigerants**, **solvents** for perfumes and lacquers, brewing, and silvering **glass**.

In 1998, a pesticide based on formic acid was developed by the U.S. Department of Agriculture as an alternative way to save beehives from varroa mites, pests that are becoming resistant to the conventional pesticide, fluvalinate. Since the formic acid is applied as a **gel**, it is safer to use than the formic acid spray employed in Europe. In field tests, the formic acid gel killed up to 84% of varroa mites and 100% of tracheal mites, another bee pest.

FORMULA, CHEMICAL

Each chemical element has a chemical symbol and a combination of chemical symbols gives a chemical formula.

The chemical formula shows what elements are present in a substance. If a pure element is present the chemical formula is merely the chemical symbol for the element. This situation is encountered most often with the noble **gases**. In some cases, representing an element by its chemical symbol is technically inaccurate but the symbol is used as a shorthand convention. For example, **sodium** is normally represented by the symbol Na, although in natural situations sodium is encountered as a giant metallic lattice with many sodium molecules bonded together. Each sodium **molecule** is held in place adjacent to the next molecule by weak **intermolecular forces**. This produces a large sheet of identical repeating units of sodium. Many layers are joined in a similar manner. Some chemicals are generally encountered as a **diatomic** bonded pair. For example, **oxygen** when encountered as a gas has the chemical formula of O_2. The **halogens** are also generally encountered as diatomic molecules.

The chemical formula of an ionic compound is simply the whole number ratio of the positive to negative ions encountered in the compound. Consequently table **salt**, **sodium chloride**, has the chemical formula NaCl. This is because for every positively charged sodium **ion** there is one negatively charged **chlorine** ion. The overall electrical charge on the compound must be neutral and the positive and negative charges must **balance** each other. An ionic compound such as sodium chloride exists in a similar manner to the sodium previously described. Each ion surrounded by several of the opposite ions and this situation is repeated throughout the structure. Any one sodium ion is surrounded by chlorine ions and each chlorine ion is surrounded by sodium ions.

The same convention is used for the production of all chemical formulas. A balanced molecule must be produced for a **chemical compound.** Al^{3+} has to combine with three Cl^- ions to give the balanced molecule with the formula $AlCl_3$. This and the previous examples represent the empirical formulas. An **empirical formula** is the lowest ratio of all of the atoms that are present. It does not necessarily represent the true number of atoms that are in association with each other. The formula that does represent the true numbers of atoms is the **molecular formula**. The molecular formula shows the number of atoms of each component is a system. In some cases the molecular formula may be the same as the empirical formula, but in others it is impossible to define the molecular formula. Such is the case with ionic compounds and **metals** which form a crystal lattice.

The chemical formula of an ionic compound is determined by knowing what elements are present, the valency of the elements, and the charge on the ions. The valency of an element is its combining power. In the preceding example the **aluminum** has a valency of three and the chlorine has a valency of one. These two elements must combine in a ratio of 1:3 to give an electrically balanced compound. The charge of an aluminum ion is +3, and the change of a chlorine ion is -1. So by combining the correct number of the appropriate ions an electrically neutral compound is produced. By knowing the valency and the charges of the elements involved the correct chemical formula can be calculated quite easily.

To calculate the chemical formula of an ionic compound, the correct chemical symbol for each element present is first written down. Next the valency of each ion is specified (this is both the charge and the number associated with that charge). Finally the quantities of the ions present are balanced so that a neutral compound is obtained. With some compounds this is relatively easy. For example, sodium chloride is NaCl, a simple 1:1 mixture. $AlCl_3$ is a 1:3 mixture. It becomes slightly more complicated when the valancies do not readily match as with the ionic compound aluminum **oxide**. Aluminum has a +3 charge and oxygen has a -2 charge. To balance these requires a ratio of 1:1.5, but by convention only whole numbers are used. To get around this the whole compound is multiplied by two to eliminate the fraction, giving a ratio of 2:3. Thus the formula is Al_2O_3.

The same rules outlined above also apply when polyatomic ions are considered, i.e. those ions that consist of more than one element. **Calcium** carbonate consists of Ca which has a +2 charge and carbonate (CO_3) which has a -2 charge. So the correct, balanced chemical formula for **calcium carbonate** is $CaCO_3$. When working out the chemical formula it is convenient to put parentheses around the polyatomic ion to ensure that only the number of polyatomic ions is altered and not the number of the constituent atoms as this would give an entirely different compound. When the final formula is produced, the parentheses can be removed if there is only one example of the polyatomic ion present in the compound. If, however, there are several of the polyatomic ions present, then the parentheses are retained. This logic produces such chemical formulas as NaOH, $CuSO_4$, and $Mg(NO_3)_2$.

In covalent compounds there are no ions. Instead the atoms involved share electrons. Covalent compounds lack the crystal lattices common in ionic compounds. This means that the chemical formula is a lot less complex. The true situation, i.e. the molecular formula, can be represented more easily. Organic compounds are covalently bonded. Thus, the chemical formula of an organic compound represents exactly what is present and in the correct ratios.

Once the chemical formulas for molecules are determined, then chemical equations can be worked out. By knowing the correct chemical formula for each component in a reaction, the amount of each reactant and the formula of the reactions products can be determined.

Chemical formulas can also note the appropriate state of **matter** for each participant in a reaction. A subscript is given after the formula to indicate if it is present as a solid, liquid, gas, or if it is in **solution**. The subscripts used for this are s, l, g, and aq, respectively. The state of matter may affect the way in which the molecule takes part in a chemical reaction.

The **structural formula** of a compound is based on the chemical formula. At its most basic the structural formula is the chemical formula showing the bonds that are present. More advanced structural formulas will indicate bond angles and relative placements of the different atoms present.

The chemical formula of a substance is a way of showing what elements are present in a compound and what relationships they have to each other. The chemical formula provides a great deal of information about a substance in shorthand form. In addition, the chemical formula of a substance makes it easier to determine what is happening in chemical reactions and chemical equations.

See also Chemical notation; Symbol, chemical

FOX, SIDNEY W. (1912-)

American biochemist

The biochemist Sidney W. Fox is best known for his research on thermal polymerization. Polymerization is a chemical process by which molecules of different elements are combined into more complex structures. Fox has studied such polymerization processes under conditions similar to those thought to have reigned on earth before the advent of life. His work, though highly controversial, may shed light on the **chemistry** of the origins of life.

Sidney Walter Fox was born in Los Angeles, California, on March 24, 1912, to Jacob and Louise (Burmon) Fox. His lifetime of interest in the principles of life began at the age of twelve, reading biology texts in San Diego's Balboa Park, where he spent his summers with his father. It was this early exposure, he would later recall, that inspired him to take up the study of chemistry. He enrolled at the University of Southern California at Los Angeles and received a bachelor's degree in 1933. Following his graduation, Fox traveled to New York to work in the laboratory of Max Bergmann—one of the world's leading authorities in the new field of biochemistry—at the Rockefeller Institute.

In 1935, Fox returned to California to continue his education with the **thermodynamics** expert Hugh M. Huffman, then at the California Institute of Technology. Fox presented his Ph.D. thesis on thermal data in the **urea** cycle in 1940. He then worked briefly at Cutter Labs and as a researcher at the University of Michigan in 1941, before again returning to California to take a position at F. E. Booth & Co., to work on a process of extracting vitamin A from shark's oil. When that project ended in 1943, Fox returned to pure research while teaching protein chemistry at Iowa State College, where he remained until 1955. First at Iowa State's Institute for Atomic Research and later at the university's Agricultural Experimental Station, studying the genetics of corn, Fox pursued research that would gradually **lead** him into the study of life's origins. He became head of the Agricultural Station in 1949.

By this time, the theories of Russian chemist **Aleksandr Ivanovich Oparin** on the chemistry of the origins of life were beginning to replace the theory of spontaneous generation.

Oparin pictured an inhospitable early Earth. Its air and young seas composed primarily of **ammonia** and such **carbon**-containing (organic) compounds as **methane**, it would nevertheless be the perfect crucible for the **origin of life**. Still, while all the separate components abounded, there would be little tendency for them to combine, forming the more complex molecules of living **matter**. The difficulty Oparin faced was to find the source of **energy** needed to drive the reactions that would turn this primordial soup into the building blocks of life.

In 1953, **Stanley Lloyd Miller**, then a graduate student at the University of Chicago, tested Oparin's theory. In a sealed flask, he combined the elements Oparin had suggested and then supplied what he considered a reasonable form of energy, **electricity**, as might have come from lightning striking the ancient seas. What formed was a variety of the most critical organic particles in life today in a thick, organic ooze coating the vessel's walls. Although the experiment was a success of sorts, the organic compounds thus created were highly unstable. How could these fragile structures survive in such a harsh climate long enough to evolve into life's vastly complex organic systems?

In the early 1950s, Fox began his research struggling with these issues. He found that when nearly dry mixtures of pre-organic chemicals were heated, long protein-like molecules called polymers were readily formed. He also noted that if these polymers were then exposed to water—as could happen in a case as simple as a rain shower washing a small volcanic pool—they themselves would break down into the very **amino acids** that are the building blocks of **proteins**.

In testing his theory, Fox, almost by accident, hit upon an astounding finding, providing not only a possible **solution** to the problem of stability but also perhaps offering an explanation of how cells could have arisen. While washing out test tubes in which amino acids had been cooked, a milky residue formed. At first Fox couldn't account for the layer, but, while on a lecture tour in 1959, an idea came to him. Telephoning his lab assistant, Jean Kendrick, he suggested she look for something resembling bacteria. She did find something similar. A large number of small bubbles, uniform in size, that Fox would come to call proteinoid microspheres, filled the samples. Oparin had earlier proposed that cell-like structures might have been the mechanism early life employed to concentrate organic materials. He had noted that certain chemicals form clumps when surrounded by **water** and showed that these structures, which he called coacervate droplets, mimicked some of the functions of living cells.

The crucial difference that separated Fox's work from Oparin's was its starting point. While Oparin began with assumptions he could not prove, including the presence of an already organic **molecule**, methane, in the original matrix, Fox needed to assume little. Fox stressed that he chose conditions for his experiments that were similar to what would actually have existed at the time of primordial synthesis. His experiments had revealed fundamental, empirical, and repeatable evidence that early precursors to cells may have formed as a natural by-product of thermal polymerization. The next phase was to discover what properties these proto-cells exhibited.

Assuming the directorship of Florida State University's Oceanographic Institute in 1955, Fox began the process of identifying the properties of the microspheres. Avoiding all presuppositions as to what life must have been like when it began, he could not easily narrow the search. The work, laborious and time consuming, eventually uncovered more than twenty unique properties that were essential aspects of life.

Early on Fox noted that the microspheres had a semipermeable double membrane that allowed only certain materials to enter while holding others outside. This may have allowed the cells to accumulate just the right mixture of chemicals to develop life's increasingly complex systems. And they proved to be remarkably stable, some remaining intact for up to six years. The stability combined with semi-permeability led to another interesting possibility. If a microsphere survived long enough, using its membrane to grow in size, and, instead of bursting like a soap bubble, divided into two new microspheres both with a similar mix inside, these two new microspheres might in turn grow and divide. In other words, they may have developed an early form of reproduction, one of life's universal attributes. Fox did indeed note that his microspheres were capable of this behavior, budding like yeast. And, while this form of reproduction was primitive in relation to today's cellular reproduction, which makes use of **DNA**, it may have been just such a pathway that led to the development of DNA itself. Working with Allen Vegotsky at Florida State and Joseph Johnson of the Massachusetts Institute of Technology, Fox was able to provide some experimental evidence of this. The team noted that a key intermediate of modern nucleic acid, ureidosuccinic acid, was formed under conditions no more remarkable than those found at hot springs in America's national parks.

Speaking to the American Association for the Advancement of Science during a 1959 plenary session in Chicago, Fox discussed his theory of the origins of life. Publicity from the talk caught the eye of one of the government's newest and best funded research programs, the National Aeronautics and Space Administration (NASA). The public's fascination with NASA helped change the attitude then prevalent toward research into the origin of life, bringing studies once considered too radical to even be considered into the mainstream. In 1961, NASA created the Institute for Space Biosciences with Fox as director. With ample resources, Fox's investigation of the properties of the microspheres widened.

In 1963 Fox moved to the University of Miami, where he would remain until 1989, continuing to investigate the microspheres. While there he took up an entirely new direction in his research, studying a property of the microspheres that he daringly speculated might be a precursor of mental functioning. Having long known the microspheres to be composed of a double membrane, it was not until Fox exposed them to sunlight that he discovered their electrical properties. They exhibited an electrical gradient remarkably similar to that of neurons, or brain cells. Working with Aristotel Pappelis at SIUC, he discovered that the bubbles could even react to the external environment by adapting their charges. This behavior was so similar to the function of the neuron that it led him to announce in 1991 that the "mind is there from the very beginning—not the mind in the sense we take it but rather the fundamental principle that separates organic matter—and brain tissue in particular—from all other matter, the ability to discriminate."

In August 1993, Fox joined the faculty at the University of South Alabama as distinguished research scientist, where he continues his research at the Coastal Research and Development Institute. A member of many national and international organizations, Fox has three sons: Jack Lawrence, Ronald Forrest, and Thomas Oren. He has been the recipient of many honors and awards, including the Honors Medal and a citation as Outstanding Scientist of Florida, given by the Florida Academy of Science in 1968; the Texas Christian University's Distinguished Scientist of the Year Award (1968); and the Iddles Award (1973).

FRAENKEL-CONRAT, HEINZ (1910-1999)

American biochemist

Heinz Fraenkel-Conrat is an internationally known German-born biochemist who became a naturalized citizen of the United States. The majority of his research, and the studies for which he is best known, was conducted at the University of Berkeley, California. Fraenkel-Conrat's research helped advance the study of viruses. He determined that under certain conditions, a virus could be separated into its component parts. These studies revealed both the virus's infective agent and its method of replication. Fraenkel-Conrat's research inspired numerous studies of viruses, which proved useful in the explanation of molecular biological processes such as replication and **mutation**.

Heinz Ludwig Fraenkel-Conrat was born July 29, 1910, in Breslau, Germany (now Wrocław, Poland) to Ludwig Fraenkel, a gynecologist who was famous for his discoveries concerning mammalian ovulation, and Lili Conrat Fraenkel. Fraenkel-Conrat was educated in Munich, Vienna, Geneva, and at the University of Breslau, where he received his M.D. in 1933. He left Germany when Adolf Hitler and his Nazi party came into power. In 1936, Fraenkel-Conrat obtained his Ph.D. in **biochemistry** from the University of Edinburgh, Scotland, for studies on ergot alkaloids and thiamine. He subsequently came to the United States and studied a type of **enzyme** at the Rockefeller Institute for Medical Research in New York. Fraenkel-Conrat unexpectedly discovered that enzymes formed peptide bonds, which, in turn, form the building blocks of **proteins**. Fraenkel-Conrat next joined his brother-in-law, K. H. Slotta, as a research associate at the Instituto Butantan at Sao Paulo, Brazil, where he began to study the components of snake venoms. The work resulted in the isolation of a protein from rattlesnake venom that acted as a neurotoxic and also destroyed red blood cells.

Fraenkel-Conrat left Brazil and returned to the United States, becoming a naturalized citizen in 1941. He became a member of the H. M. Evans Institute of Experimental Biology at the Berkeley Campus of the University of California in 1938. For more than ten years, his research involved purifying **hormones**, particularly follicle-stimulating hormones, and studying how structural changes effected hormonal activity. Some of this work was carried out at the Western Regional Research Laboratory of the U.S. Department of Agriculture, where he worked from 1938 to 1942, first as an associate chemist and then later as a chemist. His work at this time also focused on modifying protein groups. Fraenkel- Conrat and his co-workers documented how modifying a protein's structure changed its function. Several of their techniques were later used by others studying proteins.

Fraenkel-Conrat joined the virus laboratory of the University of California, Berkeley, in 1952. In 1960, using techniques similar to those in his protein work, he and his collaborators were able to determine the complete amino acid sequence—consisting of 158 amino acid residues—of the tobacco mosaic virus (TMV), making it the biggest protein of known structure at the time. Several years before Fraenkel-Conrat's virus research, scientists had determined that viruses contained a protein shell and nucleic acid the latter was believed to carry the virus's genetic information. From his studies of protein structure, Fraenkel-Conrat began further studies with the tobacco mosaic virus. He developed techniques that enabled him to gently separate the protein material from the nucleic acid, in the form of ribonucleic acid (RNA), without seriously damaging either part. He then recombined the protein and nucleic acid. If both molecules were intact, the particles rejoined and were once more infective.

Fraenkel-Conrat's subsequent research proved to be his most distinguished work. Continuing his experiments, he showed that when the two substances were separated, the protein coat had no infective properties but the ribonucleic acid still was somewhat infective. Subsequent studies showed that the protein shell was needed to get the nucleic acid, which carried the virus's genetic material, into a host cell. Once inside the cell, the nucleic acid took over the host cell's own genetic material, deoxyribonucleic acid (DNA), and began reproducing itself, making not only more infective nucleic acid but compatible protein coatings as well. This study provided definitive proof that **RNA** can act like **DNA** as the genetic blueprint for cell reproduction. Fraenkel-Conrat continued to study RNA and, along with B. Singer and other colleagues, developed new methods for stabilizing the acid for better structural studies.

From 1952 until 1958, Fraenkel-Conrat was a professor of virology at the University of California, Berkeley; he later became a professor of molecular biology. In 1968, his research emphasis concentrated on how RNA was translated in viruses and how viruses replicated this material. In 1982, when he retired, he became emeritus professor of molecular biology. For his contributions to the field of molecular biology, he was honored by a Lasker Award, also receiving the first California Scientist of the Year Award in 1958. He was also a member of the National Academy of Sciences. After retiring, his interests remained with the field of virology and he wrote a number of virology texts. For close to ten years, Fraenkel-Conrat was a contributing editor of the journal *Comprehensive Virology,* starting in 1973.

Fraenkel-Conrat married Jane Operman on July 15, 1938. They had two sons, Richard and Charles. His second marriage, on June 1, 1964, was to Beatrice Brandon Singer, who worked with him in his laboratory. He made his retirement home in California, where he occasionally involved himself in research. Fraenkel-Conrat died on April 10, 1999.

FRANCIUM

Francium is the last member of the alkali metal family, the elements that make up Group 1 of the **periodic table**. Its **atomic number** is 87, its atomic **mass** is 223.0197, and its chemical symbol is Fr.

Properties

All isotopes of francium are radioactive, with francium-223 being the most stable with a half life of 22 minutes. The element is so rare and its isotopes so short-lived that there is little information available about its chemical and physical properties. In 1991, scientists were able to confirm that the element is chemically similar to other **alkali metals** above it in the periodic table.

Occurrence and Extraction

Francium is probably the rarest element found in the Earth's crust. Some experts believe that no more than 15 grams (less than an ounce) of the element exists there. Francium is now prepared for research purposes in a particle accelerator. The most extensive research on the element is being done by a special team of researchers at the State University of New York at Stony Brook who have found a way to trap a collection of francium atoms in the middle of a magnetic field. This process allows a more careful study of the element's properties.

Discovery and Naming

Francium was discovered in 1939 by the French physicist **Marguerite Perey** while she was analyzing the products formed during the radioactive decay of **actinium** (atomic number 89). Perey suggested the name francium in honor of her homeland, France.

Uses

There are no commercial uses for francium or any of its compounds.

FRANKLAND, EDWARD (1825-1899)

English chemist

As with many famous chemists, Edward Frankland entered science by way of pharmacy. He was born in Churchtown, England, on January 18, 1825. After completing grammar school, he was apprenticed at a local apothecary (pharmacy) shop, in preparation for a medical career. Instead, he became interested in **chemistry** and studied under Lyon Playfair (1818-1898) at the University of London, **Robert Bunsen** at the University of Marburg, and **Justus von Liebig** at the University of Giessen

Edward Frankland.

In 1851, Frankland was appointed the first professor of chemistry at Owens College in Manchester, England. He moved to St. Bartholomew's Hospital in London, England in 1857, to the Royal Institution in 1863, and to the Royal School of Mines in 1865.

Much of Frankland's research was related to the concept of organic radicals. In the early 1800s, organic compounds were thought to be composed of radicals, just as inorganic compounds were composed of elements. In attempting to isolate organic radicals, Frankland synthesized the first organometallic compounds which consist of metal atoms joined to organic groups. He ultimately prepared organometallic compounds of **zinc**, **mercury**, **tin**, **boron**, **sodium**, and **potassium**. He found these compounds to be highly reactive. These compounds, which today are referred to as Grignard reagents, are still used widely in the synthesis of organic compounds.

While working with the organometallic compounds, Frankland noted that each metal **atom** would combine with only a definite number of organic groups. Zinc, for example, always combined with two groups, while **phosphorus** could combine with either three or five. (Phosphorus was among the semimetals or metalloids Frankland also studied in his research on radicals.) Frankland concluded from this evidence that every atom had a certain specific "combining power" to which he gave the name atomicity. The term was later changed to **valence**. Frankland's concept of valence was crucial in the later development of structural theory and **structural formula** s, the attempts to show how atoms are physically connected to each other in molecules.

Late in his life, Frankland concentrated on a topic in which he had long been interested, **water** supply and purification. He became very active in the study of the causes of water **pollution** and its possible solutions.

In the late 1860s, Frankland, working with the astronomer Joseph Norman Lockyer (1836-1920), discovered spectral lines in sunlight that were eventually agreed to be those of a new element, **helium**.

Frankland was knighted in 1897 and died while on vacation in Norway on August 9, 1899.

FRANKLIN, ROSALIND ELSIE (1920-1958)

English molecular biologist

The story of a great scientific discovery usually involves a combination of inspiration, hard work, and serendipity. While all these ingredients play a part in the discovery of **DNA**, the relationships between the four individuals who pieced together the double-helix model of the master **molecule** provides a subplot tainted by controversy. At the center of this quartet stands British geneticist Rosalind Franklin, who made key contributions to studies of the structures of coals and viruses, in addition to providing the scientific evidence upon which **James Watson** and **Francis Crick** based their double-helix model. Compounding the irony that Franklin died four years before Watson, Crick and **Maurice Wilkins** shared the Nobel Prize for this discovery (the Nobel Committee honors only living scientists), is James Watson's characterization of Franklin in his personal chronicle of the search for the double-helix as a competitive, stubborn, unfeminine scientist. Despite his account, Franklin has been depicted elsewhere as a devoted, hard-working scientist who suffered from her colleagues' reluctance to treat her with respect.

Franklin was born in London on July 25, 1920, to a family with long-standing Jewish roots. Her parents, who were both under the age of twenty-five when she was born, were avowed socialists. Ellis Franklin devoted his life to fulfilling his socialist ideals by teaching at the Working Men's College, while his wife, Muriel Waley Franklin, cared for their family, in which Rosalind was the second of five children and the first daughter. From an early age, Franklin excelled at science. She attended St. Paul's Girls' School, one of the few educational institutions that offered physics and **chemistry** to female students. A foundation scholar at the school, Rosalind decided at the age of fifteen to pursue a career in science, despite her father's exhortations to consider social work. In 1938, Franklin enrolled at Newnham College, Cambridge, the second youngest student in her class.

She graduated from Cambridge in 1941 with a high second degree and accepted a research scholarship at Newnham to study gas-phase **chromatography** with future Nobel Prize winner **Ronald G. W. Norrish**. Finding Norrish difficult to work with, she quit graduate school the following year to accept a job as assistant research officer with the British Coal Utilization Research Association (CURA). At CURA, she applied the **physical chemistry** experience she had garnered at Cambridge to studies concerning the microstructures of coals, using **helium** as a measurement unit. From 1942 to 1946, she authored five papers, three of them as sole author, and submitted her thesis to Cambridge. Franklin moved to Paris in 1947 to take a job with the Laboratoire Central des Services Chimiques de l'Etat. There she became fluent in French and, under the tutelage of Jacques Mering, learned the technique known as **X-ray diffraction**. Using this technique, Franklin was able to describe in exacting detail the structure of **carbon** and the changes that occur when carbon is heated to form **graphite**. In 1951, she left Paris for an opportunity to try her new skills on biological substances. As a member of Sir John T. Randall's Medical Research Council at King's College, London, Franklin was charged with the task of setting up an X-ray diffraction unit in the laboratory to produce diffraction pictures of DNA.

Eager to apply Franklin's X-ray diffraction skills to the problem of DNA structure, Randall had lured her to his lab with a Turner Newall Research Fellowship and the promise that she would be working on one of the more pressing research problems of the era—puzzling out the structure and function of DNA. When she arrived in Randall's research unit, she started working with a student, Raymond Gosling, who had been attempting to capture pictures of the elusive DNA. No stranger to the sexism rife in science at that time, Franklin made no apologies for the fact she was a woman. Maurice Wilkins, already well ensconced in the lab and working on the same problem as Franklin, took a disliking to her the first time they met. Franklin's biographers have difficulty ascertaining exactly why Wilkins and Franklin did not find common ground. Anne Sayre has suggested that the discomfort might have stemmed from the fact that Wilkins, only four years older than Franklin, may have misinterpreted her presence in his lab as a subordinate, whereas she considered herself an equal. Their mutual dislike of one another was not helped by the fact that the staff dining room was open only to the male faculty. In addition, she was the only Jew on staff. But the animosity between the two did not detract Franklin from her work, and shortly after arriving at King's, she started x-raying DNA fibres that Wilkins had obtained from a Swiss investigator.

Within a few months of joining Randall's team, Franklin gave a talk describing preliminary pictures she had obtained of the DNA as it transformed from a crystalline form, or A pattern, to a wet form, or B pattern, through an increase in relative humidity. The pictures showed, she suggested, that **phosphate** groups might lie outside the molecule. In the audience that November day sat James Watson, a twenty-four-year-old American who was also working on unraveling the **molecular structure** of DNA. Working with Francis Crick at Cambridge, Watson was even more disinclined than Wilkins to like and respect Franklin. Compounding his dislike for her was Franklin's refusal to set aside hard crystallographic data in favor of model building. Perhaps for that reason, Franklin remained publicly scornful of the notion gradually gaining adherents that perhaps the DNA molecule had a helical structure. In her unpublished reports, however, she suggested the probability that the B form of DNA exhibited such a structure, as did, perhaps, the A form. Throughout the spring of 1952, she continued studying the A form, which seemed to produce more readable X-ray photographs. This presumed legibility proved deceptive, however, because the A form does not show the double-helical structure as clearly as the B form.

In the late spring of 1952, Franklin travelled to Yugoslavia for a month, where she visited coal research labs. When she returned, she and Gosling continued to investigate the A form, to no avail. In January 1953, she started model building, but could think of no structure that would accommodate all of the evidence she had gleaned from her diffraction pictures. She ruled out single and multi-stranded helices in favor of a figure-eight configuration. Meanwhile, Watson and Crick were engaged in their own model building, hastened by the fear that the American scientist **Linus Pauling** was nearing a discovery of his own. Although Watson had not befriended Franklin in the past two years, he had grown quite close to Wilkins. In *The Double-helix,* Watson recalls how Wilkins showed him the DNA diffraction pictures Franklin had amassed (without her permission), and immediately he saw the evidence he needed to prove the helical structure of DNA. Watson returned to Crick in Cambridge and the two began writing what would become one of the best-known scientific papers of the century: ''A Structure for Deoxyribose Nucleic Acid.'' Franklin and Gosling, who had been working on a paper of their own, quickly revised it so that it could appear along with the Watson and Crick paper. Although it is unclear how close Franklin was to a similar discovery—in part because of the misleading A form—unpublished drafts of her paper reveal that she had deduced the sugar-phosphate backbone of the helix before Watson and Crick's model was made public.

On April 25, 1953, Watson and Crick published their article in the British science journal *Nature,* along with a corroborative article by Franklin and Gosling providing essential evidence for the double-helix theory. In July of 1953, she and Gosling published another paper in *Nature* that offered ''evidence for a 2-chain helix in the crystalline structure of **sodium** deoxyribonucleate.'' But Franklin's interest in the world of DNA research had already begun to wane by the spring of 1953. Despite all the excitement surrounding the double-helical structure, she had decided to move on to a lab that she hoped would offer a more congenial working environment. When she informed Randall of her intention to leave King's College for J. D. Bernal's unit at Birkbeck College, he made it clear that the DNA project was to stay in his lab. Although Gosling had been warned against further associating with Franklin, they continued to meet in private and finish their DNA work. She also continued to work on coal, but devoted the bulk of her efforts to applying crystallographic techniques to uncover the structure of the tobacco mosaic virus (TMV).

Franklin did, in fact, find the Birkbeck lab more to her liking, even though she complained to some of her friends that Bernal, a strong Marxist, attempted to foist his political views on anyone who would listen. In comparison to the situation at King's College, however, she found this bearable. She did not even complain about her lab situation. At Birkbeck she worked in a small lab on the fifth floor while her x-ray equipment sat in the basement. Because there was no elevator in the building, she made frequent treks up and down the stairs. The roof leaked, and she had to set up pots and pans to catch the **water**. But Franklin didn't mind adversity. In fact, she told friends she preferred the challenge it presented, whether at work or even while travelling. She loved to travel and once journeyed to Israel in the steerage of a slow boat sheerly for the adventure of it. She said she preferred to travel with little money ''because then you need your wits,'' an attitude that stood her in good stead in 1955 when the Birkbeck lost its backing from the Agricultural Research Council, in part, Franklin thought, because they did not approve of a project headed by a woman. Franklin successfully sought funding from another government source—the U.S. Public Health Service. The year after Franklin began at Birkbeck, the South African scientist **Aaron Klug** joined the laboratory. By 1956, Franklin had obtained some of the best pictures of the crystallographic structure of the TMV and, along with her colleagues, disproved the then-standard notion that TMV was a solid cylinder with **RNA** in the middle and protein subunits on the outside. While Franklin confirmed that the protein units did lay on the outside, she also showed that the cylinder was hollow, and that the RNA lay embedded among the protein units. Later, she initiated work that would support her hypothesis that the RNA in the TMV was single-stranded.

Franklin spent the summer of 1956 in California with two American scientists, learning from them techniques by which to grow viruses. Upon returning to England, Franklin fell ill, and friends began to suspect she was in pain a great deal of the time. That fall, she was operated on for cancer and, the following year, she had a second operation, neither of which stopped either her work or the disease. Franklin knew she was dying, but did not let that impede her progress. She began working on the polio virus, even though people warned her it was dangerous and highly contagious. She died of cancer at the age of 37 on April 16, 1958. Four years later, Watson, Crick, and Wilkins won the Nobel Prize in medicine or physiology, and Watson penned his potboiler account of the discovery of DNA. Although he vilifies her throughout his account, he tones down his earlier depiction of her as the mad, feminist scientist in an epilogue: ''Since my initial impressions of her, both scientific and personal (as recorded in the early pages of this book), were often wrong, I want to say something here about her achievements.'' He continues that he and Crick ''both came to appreciate greatly her personal honesty and generosity, realizing years too late the struggles that the intelligent woman faces to be accepted by a scientific world which often regards women as mere diversions from serious thinking.''

Frasch, Herman (1851-1914)

American chemist

Frasch, the son of a prosperous apothecary, was born in Gaildorf, Württemberg (now part of Germany) on Christmas Day 1851. He studied at the gymnasium in Halle but rather than attend the university, he decided to migrate to the United States in 1868. Frasch taught at the Philadelphia College of Pharmacy and continued to study **chemistry** with an eye to becoming an expert in a newly-emerging field, petroleum.

The oil industry in the United States began with the opening of the Titusville, Pennsylvania, oil field in 1859. In 1870, John D. Rockefeller formed Standard Oil, which refined a majority of the oil in the country, in Cleveland, Ohio. Frasch sold his patent for an improved process for refining paraffin wax to a subsidiary of Standard Oil in 1877 and moved to Cleveland to open a laboratory and consulting office. Soon he became the city's outstanding chemical consultant. In 1882, he sold to the Imperial Oil Company in Ontario, Canada, a process for reducing the high **sulfur** content of petroleum which gave it an disagreeable odor and caused the **kerosene** refined from it to burn poorly. When Standard Oil discovered a field of "sour oil" in Indiana and Ohio, the company hired Frasch as a full time consultant, bought his process and the Empire Oil Company he had recently purchased in Ontario, and gave him charge of the American petroleum industry's first experimental research program. Frasch's process for removing sulfur, patented in 1887, was to treat the petroleum with a variety of metallic oxides to precipitate the sulfur and recover the oxides for further use. He continued with Standard Oil as special consultant for the development of new petroleum by- products and became wealthy. He refused to join Standard Oil as an executive, choosing instead to be a lifetime consultant.

Frasch turned his attention to sulfur, the substance his process removed from petroleum. The island of Sicily held a virtual monopoly on this valuable mineral from which **sulfuric acid**, industry's most vital chemical, was made. While Sicilian sulfur deposits were near the Earth's surface and more easily mined, sulfur deposits in Texas and Louisiana were deeper, and American laborers were unwilling to go into sulfur mines. Frasch believed that sulfur could be melted and pumped from the ground in much the same manner petroleum was, but boiling **water** was not hot enough to liquify the sulfur. He organized the Union Sulfur Company in 1892, and two years later began employing the method he had patented a year earlier. His process required three concentric pipes to be sunk into the sulfur deposit. Water, superheated under pressure to above 241°F (116°C), was pumped into the sulfur deposit through the outside pipe. Compressed air was forced down the center pipe, and through the center pipe the melted sulfur flowed to the surface where it was pumped into bins to solidify. The major problem with this method was the cost of heating the water, but the discovery of the East Texas oil fields in the early twentieth century provided an inexpensive, readily available fuel supply. Frasch expanded his research into the use of sulfur as an insecticide and a **fungicide**. Other companies infringed on his patent rights, and his company disappeared, but the use of the Frasch process enabled the United States to become self-sufficient in the production of sulfur needed to supply its growing chemical industry.

Frasch died in Paris on May 1, 1914. Among his honors was the Perkin Medal in 1912. His greatest honor was the distinction of having two chemical processes, one for producing sulfur and the other for removing sulfur from petroleum, carry his name.

Fraser-Reid, Bertram Oliver (1934-)

Canadian chemist

Bert Fraser-Reid is a distinguished researcher in organic synthesis and sugar **chemistry**. In 1966, at the University of Waterloo in Ontario, he developed a process to make synthetic **pheromones**, the chemical attractants produced naturally by insects and other species. For this discovery, he received the Merck, Sharpe and Dohme Award for outstanding contribution to **organic chemistry** in Canada in 1977. At Duke University in Durham, North Carolina, Fraser-Reid has conducted groundbreaking research on the synthesis of organic compounds from simple sugars. In addition, he has led a research team that developed a unique process to combine simple sugars into oligosaccharides, complex sugars composed of two or more monosaccharides. This process, as Fraser-Reid indicated in *Black Enterprise,* "may have some potential to facilitate the development of a cure for AIDS." A *Black Enterprise* writer speculates that Fraser-Reid's research may ultimately earn him a Nobel Prize in chemistry, for which he has already been nominated.

Bertram Oliver Fraser-Reid, who has dual Jamaican and Canadian citizenship, was born on February 23, 1934, in Coleyville, Jamaica, to William Benjamin Reid and Laura Maria Fraser. Fraser-Reid was working as a teacher when, at the age of twenty-two, he became interested in chemistry. This led him to Queen's University in Kingston, Ontario, Canada, where he earned his B.S. degree in 1959 and his M.S. degree in 1961. Fraser-Reid married Lillian Lawryniuk on December 21, 1963; they have two children, Andrea and Terry. Fraser-Reid received his doctorate in chemistry from the University of Alberta in 1964. From 1964 to 1966, he held a postdoctoral fellowship at the Imperial College of the University of London.

In 1966, Fraser-Reid joined the faculty of the University of Waterloo, where he remained until 1980. During this period, he was able to effect the organic synthesis (the preparation of complex organic structures from simpler compounds) of pheromones, which social insects emit to transmit messages about food sources, the presence of predators, and reproductive behavior. The Canadian Forestry Service was subsequently able to control insect populations that were damaging timber by using synthetic pheromones to disrupt the insect's mating cycles; relying on synthetic pheromones allowed the Forestry Service to discontinue their use of DDT, the controversial insecticide now banned throughout much of the world. Since glucose (a form of sugar) was the basis for Fraser-Reid's synthetic pheromones, the larger implication of his achievement is that sugars can be used in place of petroleum for industrial applications of organic synthesis—a discovery with potentially global economic ramifications in terms of the manufacture of **plastics** and pharmaceuticals.

Fraser-Reid was a professor of chemistry at the University of Maryland from 1980 to 1982. He joined the faculty of Duke University in 1982, becoming the James B. Duke professor of chemistry in 1985. Commenting on his experiences as an educator, Fraser-Reid told *Black Enterprise,* ''It never ceases to amaze me how young impressionable minds can mature into competent scientists.'' He encourages all students, and especially black students at Duke, to consider careers in science and engineering. Nonetheless, he is quoted in *Science* as regretting that ''The black students in my classes are first and foremost Americans, and over the past 15 to 20 years Americans have not been going into science and engineering.''

During his tenure at Duke, Fraser-Reid's research turned to the **biochemistry** of oligosaccharides, which his team of researchers have been able to synthesize from simple sugars. As Fraser-Reid noted in *Black Enterprise,* oligosaccharides are ''among the most important biological regulators in nature... They regulate the whole immune system.'' Oligosaccharides are involved in the biological functioning of the antigenic substances found in the blood, which are complex **carbohydrates** capable of stimulating an immune response. Thus, advances in the understanding of oligosaccharides and in oligosaccharide synthesis promise to play an important role in the medical battle against AIDS.

Fraser-Reid was named Senior Distinguished U.S. Scientist by the Alexander von Humboldt Foundation in 1989. In 1990, he received both a Jamaican National Foundation Award and the American Chemical Society Claude S. Hudson Award in Carbohydrate Chemistry. In 1991, he received the National Organization of Black Chemists and Chemical Engineers' Percy Julian Award. In 1993, he served as chair of the Gordon Conference on Carbohydrates. Fraser-Reid was also awarded the 1995 Haworth Memorial Medal and Lectureship of the Royal Society of Chemistry. His professional memberships include the Chemical Institute of Canada, the American Chemical Society, the American Institute of Chemistry, and the British Chemical Society. He also serves as consultant to Blackside Films on minorities in science. An accomplished organist, Fraser-Reid has performed at recitals internationally.

Fraunhofer lines

When a source (such as the sun) gives off light, that light can be dispersed into a rainbow spectrum by a prism or diffraction grating. If the light is pure white, the spectrum will contain all the colors of the spectrum. If, however, the light is not perfectly white, there will be areas within the spectrum where no **color** is present. These areas are distinguished by dark lines called Fraunhofer lines, after the German optician Joseph von Fraunhofer who first discovered them.

Fraunhofer earned a reputation as the premier glassmaker of his time. He was particularly famous for his optical instruments, such as the lenses used in telescopes, that were highly valued among European physicists. In order to test the quality of his lenses, Fraunhofer would calibrate them by passing through each lens the light from a candle and observing the spectrum produced. On one occasion, however, he chose to use the light from the sun to calibrate his lens. Upon examining the solar spectrum, Fraunhofer was surprised to find hundreds of dark lines crossing the colored band.

Though he did not know the cause of the dark lines, Fraunhofer knew he had made an important discovery. He was able to detect 574 different lines, and he diagramed these in detailed illustrations. Later, through painstaking calculation, he succeeded in estimating the wavelengths at which each line occurred.

It is now known that Fraunhofer's lines are absorption lines and are caused by elements absorbing certain wavelengths of light. In the case of sunlight, white light is emitted by the sun's core. However, the outer areas of the solar atmosphere contain many different elements, all of which will partially block the passage of light. Thus, when the sun's light reaches Earth, it is missing a number of wavelengths. These missing wavelengths are characterized by dark lines in the spectrum.

The absorption lines were further studied in 1859 by the German physicist **Gustav Kirchhoff**. Through experimentation, Kirchhoff and his partner Robert Bunsen were able to distinguish which elements corresponded to each line. Using this information, the two scientists identified what elements in the sun's atmosphere were causing Fraunhofer lines.

Bunsen and Kirchhoff soon discovered that elements, when heated, would emit light at the same wavelengths they absorbed, producing bright lines in the spectrum. Thus, by heating an unknown compound and examining its spectrum, scientists could identify the elements that made up the compound. This formed the basis of the modern science of **spectroscopy**.

Free energy

A spontaneous process is one that proceeds by itself. Such a process may not occur without the addition of a certain amount of energy, such as the combustion of a fuel. However, once it has begun, the reaction will proceed until the reactants are exhausted or the system reaches a state of equilibrium. A nonspontaneous process is the opposite of a spontaneous process. It will only proceed if energy is supplied to the system. However, the moment the energy supply is removed, the reaction will cease. The amount of useful work that can be obtained from a spontaneous reaction is limited by the difference in free energy between the products and reactants. If the free energy of a system is negative, the reaction at constant **temperature** and **pressure** can produce useful work, and is said to be spontaneous. If the free energy change of a system is positive, work must be performed to carry out the reaction at constant temperature and pressure; the reaction in this direction is nonspontaneous but the reverse reaction is spontaneous. If the free energy change of the system is zero, the reaction is in **equilibrium**, meaning that the forward reaction and the reverse reaction proceed at the same rate. Standard free energy changes are those that accompany changes from reactants in their standard states to products in their standard states. For solids and liquids, the standard state is the pure solid or liquid under standard pressure. For gases, the standard state is the ideal gas at standard pressure. Thermodynamic relations make it possible to obtain information about the reactants and products at equilibrium from the standard free energy change.

A reaction that has a positive free energy change at 13°F (25°C) and atmospheric pressure might be made to proceed by raising the temperature or lowering the pressure. The dissociation of calcium carbonate to calcium oxide and carbon dioxide at 25°C and atmospheric pressure provides an example; raising the temperature or lowering the pressure drives the reaction in the forward direction.

The standard free energy of formation of a pure substance is defined as the free energy change when one mole of the substance is formed from its constituent elements in their standard states.

The free energy change is proportional to the amount of substance that reacts or is produced in a reaction. It is equal in magnitude but opposite in sign to the free energy change for the reverse reaction. If a reaction can be regarded as the sum of two or more reactions, the free energy change for the overall reaction must equal the sum of the free energy changes for the other reactions.

For a reaction occurring under constant pressure, the free energy change is equal to the **enthalpy** change of the system minus the absolute temperature times the **entropy** change of the system; if the reaction is spontaneous, this quantity, known as the Gibbs free energy, must be less than zero. For spontaneous reactions occurring at constant volume, the equivalent quantity is the Helmholtz free energy, defined as the internal energy change minus the absolute temperature times the entropy change.

Because, experimentally, it is easier to control the pressure than the volume during a reaction, the Gibbs free energy has more practical value than the Helmholtz free energy. By studying the variation of the Gibbs function with pressure at constant temperature, and with temperature under constant pressure, it is possible to gather useful volume and entropy data, which are indispensible to understanding the **kinetics** of reactions.

The standard free energy change is also related to the equilibrium constant of a reaction. Thus, the free energy change can be used to find the value of the equilibrium constant and vice versa.

Thermodynamics can be used to determine whether a change can take place spontaneously and how far a change may proceed before the reaction reaches equilibrium. It does not, however, provide any information about how fast a change will take place or about how changes take place.

See also Thermochemistry

FREE RADICALS

A free radical is an **atom** or group of atoms with a single unpaired **electron**. It is usually produced by breaking a **covalent bond**.

The concept of the covalent (or electron-pair) bond was first formulated by the American chemist **Gilbert Newton Lewis** in 1916. Lewis showed that, by sharing electrons, atoms can achieve the stability of the noble **gases**. For example, when two **hydrogen** atoms, each containing one electron, combine to form an H_2 **molecule**, each atom shares the two electrons and the hydrogen molecule takes on the electronic configuration of the noble gas **helium** with **atomic number** 2.

The only electrons that take part in a covalent bond are the electrons in the outer orbitals, i.e., the **valence** electrons. The rule that atoms in covalently bonded species prefer noble gas electron configurations is sometimes referred to as the **octet rule** because nonmetals, with the exception of hydrogen, achieve a noble gas structure by having eight electrons in their outer electron shells. Although many molecules obey this rule, some, such as **nitric oxide** (with 11 valence electrons) and **nitrogen** dioxide (with 17 valence electrons), have odd numbers of valence electrons, and clearly do not. Other molecules that fail to conform to the octet rule include the fluorides of **beryllium** (BeF_2) and **boron** (BF_3), in which the central atom is surrounded by four and six electrons, respectively. Species containing unpaired, i.e., an odd number of, valence electrons are called free radicals.

Free radicals such as nitric oxide and nitrogen dioxide show weak attractions toward magnetic fields—they are called paramagnetic. Superoxides, i.e., compounds characterized by the presence in their structures of the O_2^- **ion** (containing an odd number of electrons), are also paramagnetic. Stable superoxides include **sodium** superoxide, **potassium** superoxide, and **rubidium** superoxide.

Energy must be supplied to start reactions in which there is a splitting of a covalent bond and the production of free radical (unpaired electron) intermediates. This energy is usually supplied as **heat** or by irradiation with light. Peroxides, i.e., compounds with oxygen-oxygen single bonds, readily form free radicals when heated because the **oxygen** oxygen bond is weak. Halogen molecules also contain relatively weak bonds, so they too readily form free radicals when heated or irradiated with light.

Most small free radicals are highly reactive. When they react with other molecules, they tend to do so in such a way that they pair their unpaired electrons. They can do this in many cases by appropriating an electron from the other molecule. When a halogen free radical, for example, encounters a hydrogen atom in an alkane molecule, it takes an electron from the hydrogen atom and pairs it with its own unpaired electron, converting the alkane into an alkyl radical. Another way free radicals can react is by combining with a compound to form a new, larger radical. An example of this process is the formation of free radicals from chlorofluorocarbons in the upper atmosphere (leading to the depletion of the ozone layer).

In the *free radical polymerization* of ethylene to form **polyethylene**, a small amount of a peroxide is added to a large

volume of ethylene. When the peroxide is heated, it forms a free radical. The free radicals from the peroxide immediately seek other electrons to complete their octet electron shells, which they find in the *pi* bond of the ethylene molecule. But whenever a peroxide free radical appropriates an electron from an ethylene molecule, a new ethylene free radical is formed. This free radical in turn attacks yet another ethylene molecule, with the result that another ethylene free radical is added to the growing polyethylene chain. This polymerization continues until two growing **chains** meet end-to-end and form a complete polymer molecule.

Free radicals are important in medicine and biology, since they are produced in the normal course of **metabolism**. Because free radicals are highly reactive, they are capable of damaging the body. Thus, they have been implicated in the aging process, e.g., in life-limiting, chronic diseases. In particular, free radicals are believed to be important in the development of cancers and atherosclerosis. There is also evidence that free radicals in cigarette smoke inactivate antiprotease in the lungs, leading to emphysema.

See also Magnetism; Polymer chemistry

FREEZING AND MELTING

Freezing is the change that occurs when a liquid changes into a solid as the **temperature** decreases. Melting is the opposite change, from a solid to a liquid as the temperature increases. These are both examples of changes in the states of **matter** of substances.

Substances freeze at exactly the same temperature as they melt. As a consequence of this the temperature at which, under a specified pressure, liquid and solid exist in equilibrium is defined as the melting or freezing point. When the pressure is one atmosphere this temperature is known as the normal freezing (or melting) point. A change in pressure will change the temperature at which the change in the state of matter occurs. A decrease in pressure will decrease the temperature at which this occurs and an increase in pressure will increase the temperature required.

At a fundamental level freezing and melting represent changes in the **energy** levels of the molecules of the substance under consideration. Freezing is a change from a high energy state to one of lower energy, the molecules are moving less as their temperature falls. They become more ordered and fixed in shape. When a substance melts the average **energy level** of the constituent molecules increases. The molecules are moving more rapidly and in a less ordered manner in a liquid than in a solid. It is this greater freedom of movement that allows a liquid to flow to touch the walls of its container whereas a solid is fixed in a rigid shape. This consideration of the energy of the molecules is known as the **kinetic molecular theory**.

The temperature at which substances freeze and melt is different for different chemicals. The chemical formula of a substance is not necessarily a true indicator of what the freezing or melting point may be. Isomers of substances can have different physical properties including freezing and melting points. Similarly the presence of **hydrogen** bonds and other attractive forces such as **van der Waals forces** can influence the **bonding** within the substance and hence the freezing and melting points. If any **intermolecular forces** are present more energy must be added to the system to change from a solid to a liquid. This is because the intermolecular bonds have to be overcome to allow the molecules to move more freely. This is less of a change than occurs from the change from liquid to gas, because the molecules are still touching each other in both **liquids** and **solids**.

The purity of the compound can influence the temperature at which the solid liquid change takes place. For example adding **sodium** chloride (common salt) to **water** depresses the freezing point, this is why **salt** is put on roads to stop them icing over. A pure substance has a definite melting or freezing point, the addition of an impurity as well as lowering this temperature also spreads it so that there is a less definite, more diffuse melting/freezing point. This means that we can use the freezing/melting point as an indicator of the purity of a substance. When a solid is melted or a liquid frozen while being heated or cooled the temperature stays constant as it occurs. Thus, if a graph of temperature is plotted against **heat** added a shoulder or plateau will be seen which represents the freezing/melting point. With an impure substance this shoulder will not be so precise. A graph of this nature is known as a heating curve. The conversion between solid and liquid occurs at a constant temperature.

With most substances the solid is denser than the liquid phase. As a result of this when freezing the solid will sink to the bottom of the liquid. Water does not behave in this manner. Ice is less dense than water and consequently ice will float on water. Water has its maximum **density** at 39°F (4°C). This is caused by hydrogen bonding, which in the liquid phase is unordered. When the water freezes to form ice, the molecules assume an open ordered pattern which allows the maximum amount of hydrogen bonding. This characteristic has had a profound effect on life on Earth as well as causing many burst pipes.

Normally, when we talk about a substance being a solid or a liquid we are referring to its appearance at standard temperature and pressure, this is a pressure of one atmosphere and a temperature of 68°F (20°C). If the melting point is below this temperature and the **boiling point** is above it then the chemical is a liquid at standard temperature and pressure.

It is possible to cool a liquid below its freezing point and still have it remain as a liquid. This is known as a supercooled liquid. This represents an unstable equilibrium and in time the liquid freezes. It is very easy to supercool water and it is possible to cool it down to 12°F (20°C) and still have it remain as a liquid. The supercooled liquid will not start to freeze until there is a point for the ice to start to form. This may be a single piece of dust, which acts as a nucleation point for the ice to start forming. Supercooled water is not encountered in nature because there is too much particulate material in the atmosphere. If any of these particles lands in a super cooled liquid it will instantly turn into the solid form.

Some chemicals do not have a point at which they turn from solid to liquid they can change directly from solid to gas, this is a property called **sublimation**. Dry ice, solid **carbon** dioxide, exhibits this. Like melting and freezing this also happens at one specific temperature.

Solids and liquids are both densely packed at a molecular level. One difference in terms of the molecules is that with a liquid the molecules are more readily capable of slipping over each other. It is this property that makes it easier to pour a liquid. The molecules in a liquid are still touching each adjacent **molecule** (as they do in a solid), although they are less freely held.

Ionic compounds generally have a higher melting point than covalent compounds. This is because the intermolecular forces in an ionic compound are much stronger. If the pressure is increased the molecules are forced closer together and this means that the intermolecular forces are holding the particles closer together and more tightly, so a higher temperature is required to make the material melt.

Melting is also called fusion, and the energy required to bring about this **change of state** is called the **heat of fusion** or the **enthalpy** of fusion. For ice to turn into liquid water the heat of fusion is 6.01 kJ/mol. Melting and sublimation are both endothermic processes and freezing is an exothermic process. Whenever a material changes from one state to another there is an energy change within the system. For melting the order of the system is decreasing so energy must be supplied to increase the randomness of the molecules. For freezing the molecules are becoming more ordered, so energy is lost from the system.

Freezing and melting are the change of state from liquid to solid and from solid to liquid. For any given pure chemical they happen at a specific temperature, which is the same for freezing and melting.

See also States of matter

FUELS AND FUEL CHEMISTRY

That modern society is the consumer of a significant amount of **energy** is readily apparent on the streets of any large city. Automobiles and buses are everywhere. Electric lighting and signs light the way. Everything from air conditioners to computers to airplanes requires energy. And, invariably, this energy comes from some form of chemical fuel. Indeed, better than 90% of the energy we use originates with a chemical fuel.

In essence, a fuel is any compound that has stored energy. This energy is captured in chemical bonds through process such as **photosynthesis** and respiration. It is released during **oxidation**. The most common form of oxidation is the direct reaction of a fuel with **oxygen** through **combustion**. **Wood**, **gasoline**, **coal**, and any number of other fuels have energy rich chemical bonds created using the energy from the sun which is released when the fuel is burnt. Indeed, this is what burning is—the release of chemical energy. Chemical fuels or the "fossil fuels" just happen to be a particularly useful reserve of energy and are therefore used extensively to satisfy the demands of an energy-hungry civilization.

Fossil fuels are principally **hydrocarbons** with minor impurities. They are so named because they originate from the decayed and fossilized remains of plants and animals that lived millions of years ago. Yes, the ultimate fate of the dinosaurs is probably in a tank of unleaded from your local service station.

Fossil fuels can be separated into three categories. The first is petroleum or oil. This is a mixture of light, simple hydrocarbons dominated by the fractions with six to 12 carbons but also containing some light hydrocarbons (i.e. **methane** and ethane) and heavier fractions. Fully half of the energy consumed in the United States is from petroleum, be this in automobiles, recreational vehicles, home heating, or industrial production.

The principal use of petroleum is the production of gasoline. Over 40% all of all production ends up consumed in automobiles and such. Smaller fractions are turned into fuel oil (27%), jet fuel (7.4%), and other miscellaneous fuels, while the small fraction (about 10%) is used for the synthesis of the thousands of **petrochemicals** used in our daily lives. Indeed, many food compounds and pharmaceuticals owe their synthesis to a petrochemical precursor.

The second most prominent and naturally most abundant fossil fuel is coal. Coal also originates from decayed vegetative material buried eons ago, but the process is slightly different, being less oxidizing. The resulting material still has some of the original lignin-like structure exhibiting many fused rings and a large fraction of aromatic compounds. Consequently, coal is more of a polymeric substance than petroleum and is found as a solid not a liquid. The **carbon** to **hydrogen** ratio in coal is close to 1:1 (depending upon the type of coal), whereas the carbon to hydrogen ration in petroleum is closer to the 1:2 value expected for a **hydrocarbon** chain.

Minable coal is defined as 50% of the coal in a seam of at least 12 in thickness. The proven reserves of minable coal are sufficient to supply the industrial needs of modern society for the next four to five hundred years. Unfortunately, as a fuel source, coal has many disadvantages. It is a very "dirty" fuel producing a large amount of unburnt hydrocarbon, particulate, and most damaging of all significant quantities of **sulfur** dioxide. Indeed, it is the coal burning power plants of the eastern United States that are responsible for much of the **acid rain** and environmental damage observed in upstate New York and eastern Canada. The other significant disadvantage of coal is that it is not liquid, making it awkward to transport and store and limiting its use in applications like automobiles. A great deal of research has been done on the liquefaction of coal but with little success so far.

The third major fossil fuel is **natural gas**. This is a generic term for the light hydrocarbon fractions found associated with most oil deposits. Natural gas is mostly methane with small quantities of ethane and other **gases** mixed in. It is hydrogen-rich, since methane has a carbon to hydrogen ratio of 1:4. It is also an excellent fuel, burning with a high **heat** output and little in the way of unwanted **pollution**. It does produce **carbon dioxide** which is a greenhouse gas, but any and all organic compounds generate carbon dioxide on combustion. Natural gas is also easy to transport through pressurized pipelines.

All of this would appear to make natural gas the perfect fuel. But it is not without its drawbacks. This includes the presence of hydrogen sulfide in some gas fields, leading to the term "sour gas." Hydrogen sulfide is the smell of rotten eggs, but if smell were the only problem, this would be of little concern. However, hydrogen sulfide is extremely corrosive to the pipes used to transport natural gas and is a very toxic compound being lethal at levels around 1500 ppm.

In addition, natural gas is potentially explosive as the gas must be maintained under pressure, and any hydrocarbon, in a gaseous state, can explode. This is in contrast to the use of gasoline, which is a much safer fuel, despite Hollywood movies where a simple fender bender causes a car to explode! Nevertheless, both the gaseous and liquid forms of hydrocarbons are much more volatile and represent a hazard compared to coal.

Of course, the most obvious question that people ask when discussing organic fuels is "how long will they last?". Unfortunately, no one knows the real answer to this question. It has been suggested that since all of the oxygen in the atmosphere came from the splitting of carbon dioxide via photosynthesis, the total oxygen content may lead to an estimate of the total carbon reserves. This calculation would suggest that we can keep using fossil fuels at our present rate of consumption for the next 50,000 years.

However, there are serious flaws in this argument, as a significant amount of the world's carbon reserves are tied up in **calcium** carbonate rock formations. Loosely speaking, **calcium carbonate** is limestone and there is an awful lot of limestone present throughout the world. Industry analysts place our fossil fuel reserves at much lower levels. It is estimated by a wide variety of sources that we will reach maximum oil production in the next twenty years. After that, production will decline world wide and we will be forced to wean ourselves from an oil- based society. The estimates for coal provide a slightly better prognosis giving a window of about 500 years for consumption of all known reserves. Natural gas production already appears to be in decline although this can be somewhat deceiving as significant reserves may be available in the arctic.

Natural gas also has one significant advantage over the other fossil fuels and that is that it is "renewable." This is not to say that if we tap and drain a deposit, it will somehow mysteriously fill up again. Once natural gas has been taken from the ground, it is no longer available. But natural gas is found as a side product of any decaying material. Methanogenic bacteria—literally, methane-making bacteria—exist in the garbage dumps and waste disposal sites of the industrial world, busily producing methane from all of the garbage. Enough that the Fresh Kills garbage dump on Staten Island, New York is capable of heating 16,000 homes. Tapping into garbage as a method of generating natural gas holds out some hope for fossil fuels in the future.

Of course, why stay with organic compounds? Why not switch to some other chemical species such as hydrogen? Indeed, this approach holds out tremendous promise for the future. Various companies have been exploring the use of hydrogen as a fuel. When used in simple combustion, hydrogen has some of the problems associated with natural gas. It must be stored under pressure and is extremely explosive upon ignition in air. To witness how explosive hydrogen can be, just recall the Hindenburg disaster, in which a hydrogen-filled airship burnt up in a matter of minutes. The type of explosion—the shape of the detonation—also makes hydrogen unsuitable as an alternative fuel for the conventional automobile. The sharpness of the explosion would quickly rattle pistons to pieces.

However, hydrogen does not need to be burnt directly with oxygen to provide energy. Fuel cells combine hydrogen and oxygen at electrodes to produce **electricity**, which can then be used to run an electric motor or a spacecraft. NASA has been employing hydrogen/oxygen fuel cells for years to provide the electricity for both manned and unmanned spacecraft. And fuel cells have the added bonus of providing crew members with fresh drinking **water** as the only product is pure water, eliminating any form of pollution. Through the use of fuel cells, hydrogen could potentially be used for conventional automobiles. Questions about storing hydrogen and long term viability of the fuel cells need to be addressed, but the future of this technology looks promising. There is only one drawback. At present, our supply of hydrogen is obtained from fossil fuels. Hydrogen is released from hydrocarbons during the refining process. Using hydrogen to run automobiles does not solve the problem of a limited fossil fuel reserve.

At present, the petroleum industry produces hydrogen but that is not our only potential source. Both the use of sunlight and solar panels to create sufficient electricity to electrolyze water and bacteria capable of splitting water to generate hydrogen and oxygen may make hydrogen the chemical fuel of the future. In addition, research is underway into the use of **methanol** as a potential partner for a fuel cell, eliminating the need for hydrogen and greatly reducing the difficulties with storage and filling the tank.

With an annual energy consumption of over 75 quadrillion Btu annually in the United States alone and 90% of this energy originating with fossil fuels, it is easy to see that the rapid and irreversible consumption of these chemical fuels will lead to a significant change in society in the future. However, a switch to a methane and hydrogen based fuel economy will reduce the impact and, in the end, provide for a less polluting source of energy.

FUKUI, KENICHI (1918-1998)

Japanese theoretical chemist and engineer

Kenichi Fukui is a theoretical chemist whose career was devoted to explaining the nature of chemical reactions. His work is distinguished from that of other chemists by its mathematical structure, and he has made a major contribution to bridging the gap between quantum theory, a mathematical theory of the behavior of molecules and atoms, and practical **chemistry**. He made it easier both to understand and predict the course of chemical reactions, and he shared the 1981 Nobel Prize in chemistry with **Roald Hoffmann** for his achievements.

Fukui was born October 4, 1918, in Nara, on the island of Honshu, Japan. He was the eldest of three sons born to Chie and Ryokichi Fukui. His father was a merchant and factory manager who played a major role in shaping his son's career; he persuaded Fukui to study chemistry. Fukui had no interest in chemistry during high school, and he later described his father's persuasiveness as the "most decisive occurrence in my educational career." He enrolled at the Department of Industrial Chemistry at Kyoto Imperial University, and he remained associated with that university throughout his career. Fukui graduated from the university in 1941, and he spent most of World War II at a fuel laboratory, performing research on the chemistry of synthetic fuel.

Fukui returned to Kyoto University in 1945, when he was named assistant professor. He received his Ph.D. in engineering in 1948 and was promoted to a full professorship in **physical chemistry** in 1951. At the beginning of his career, his research interests ranged broadly through the areas of chemical reaction theory, **quantum chemistry**, and physical chemistry. But during the 1950s, Fukui began theorizing about the role of **electron** orbitals in molecular reactions. Molecules are groups of atoms held together by electron bonds. Electrons circle the nuclei in what are called orbitals, similar to the orbit of planets around the sun in our solar system. Whenever molecules react with one another, at least one of these electron bonds is broken and altered, forming a new bond and thus changing the **molecular structure**. At the time Fukui began his work, scientists understood this process only when one bond was changed; the more complex reactions, however, were not understood at all.

During the 1950s, Fukui theorized that the significant elements of this interaction occurred in the highest occupied molecular orbital of one **molecule** (HOMO) and the lowest unoccupied molecular orbital of another (LUMO). Fukui named these "frontier orbitals." The HOMO has high **energy** and is willing to lose an electron, and the LUMO has low energy and is thus willing to accept an electron. The resulting bond, according to Fukui, is at an **energy level** between the two starting points. Over the next decade, Fukui developed and tested his theory using complex mathematical formulas, and he attempted to use it to predict the process of molecular interaction and **bonding**.

Fukui continued to break new ground in **theoretical chemistry** through the 1960s. Other chemists began research on these same problems during this period, but Fukui's work was largely neglected. His use of advanced mathematics made his theories difficult for most chemists to understand, and his articles were published in journals that were not widely read in the United States and Europe. In an interview quoted in the *New York Times,* Fukui also attributed some of his obscurity to resistance from Japanese colleagues: "The Japanese are very conservative when it comes to new theory. But once you get appreciated in the United States or Europe, then after that the appreciation spreads back to Japan."

Two of the chemists who had been working independently of Fukui were **Roald Hoffmann**, of Cornell University, and **Robert B. Woodward**, of Harvard, and in 1965 they came to conclusions that were similar to his, arriving, however, along a different path. Staying away from complex math, these two developed a formula almost as simple as a pictorial representation. Taken together, the work of Fukui and the American team enabled research scientists to predict how reactions would occur and to understand many complexities never before explained. These formulae answered questions about why some reactions between molecules occurred quickly and others slowly, as well as why certain molecules reacted better with some molecules than with others. They removed much of guesswork from this area of chemistry research.

For the advancements in knowledge their work had brought, Fukui and Hoffmann shared the 1981 Nobel Prize in chemistry. Woodward, who might have shared in the prize, had died two years before. Fukui was one of the first Japanese to receive the Nobel Prize in any field, and the very first in the area of chemistry. After receiving the Nobel Prize, Fukui remained at Kyoto University. He continued his research on chemical reactions, expanding his formula to predict the interaction of three or more molecules.

Fukui was elected senior foreign scientist of the American National Science Foundation in 1970. In 1973, he participated in the United States-Japan Eminent Scientist Exchange Program. In 1978 and 1979, he was vice-president of the Chemical Society in Japan, and he served as their president from 1983 to 1984. In 1980, he was made a foreign member of the National Academy of Sciences, and in 1982 he was named President of the Kyoto University of Industrial Arts and Textile Fibers. He was a member of the International Academy of Quantum Molecular Science, the European Academy of Arts, Sciences, and Humanities, and the American Academy of Arts and Sciences.

Fukui was married in 1947 to Tomoe Horie. They had one son and one daughter. In his spare time Fukui enjoyed walking, fishing, and golf. He died in 1998.

FULLERENE

As recently as 1984, **carbon** was thought to exist in only two solid forms (allotropes). There was **graphite**, in which the **carbon** atoms arranged themselves as layered sheets of hexagonally bonded atoms, and there was **diamond**, in which the carbon atoms formed octahedral structures in which each carbon **atom** had four nearest neighbors.

Then, in 1985, chemists **Richard E. Smalley**, Robert F. Curl Jr., J. R. Heath, and S. O'Brien at Rice University, and H. W. Kroto of the University of Sussex in England observed that a hollow truncated icosahedron, similar in shape to a soccer ball, and consisting of 60 carbon atoms, tends to form spontaneously when carbon vapor condenses. In 1990, physicists D. R. Huffman and L. Lamb of the University of Arizona, working with W. Kratschmer and K. Fostiropoulos of the Max Planck Institute in Germany, discovered a way to make bulk quantities of this C-60 **molecule**, which investigations using high resolution **electron** microscopes have shown to have sizes of about one billionth of a meter.

As C-60 has the same structure as the geodesic dome developed by American engineer and philosopher R. Buckmin-

A computerized illustration of a fullerene, or buckyball. The buckyball's constituent spheres represent carbon atoms while the lines represent the bonds between them. The bonds link the atoms in pentagons and hexagons to form a football-shaped sphere. ***(Image by Laguna Design/Science Photo Library, Photo Researchers, Inc. Reproduced by permission.)***

ster Fuller, these molecules were christened buckminsterfullerenes by the group at Rice University. The Swiss mathematician Leonhard Euler had proved that a geodesic structure must contain 12 pentagons to close into a spheroid, although the number of hexagons may vary. Later research by Smalley and his colleagues showed that there should exist an entire family of these geodesic-dome-shaped carbon **clusters**. Thus, C-60 has 20 hexagons; whereas its rugby-ball shaped cousin C-70 has 25. Research has since shown that laser vaporization of graphite produces clusters of carbon atoms whose sizes range from two to thousands of atoms. These molecules are now known as fullerenes. All the even numbered species between C-3 and C-600 are hollow fullerenes, but below C-32, the fullerene cage is too brittle to remain stable. Helical microtubules of graphitic carbon have also been found.

Although many examples are known of five-membered carbon rings attached to six-membered rings in stable organic compounds (for example, the **nucleic acids** adenine and guanine), only a few occur whose two five-membered rings share an edge. The smallest fullerene in which the pentagons need not share an edge is C-60; the next is C-70. C-72 and all larger fullerenes adopt structures in which the five-membered carbon rings are well separated, but the pentagons in these larger fullerenes occupy strained positions. This makes the carbon atoms at such sites particularly vulnerable to chemical attack. Thus, it turns out that the truncated icosahedral structure of C-60 distributes the strain of closure equally, producing a molecule of great strength and stability. This molecule will, however, react with certain **free radicals**.

When compressed to 70% of its initial **volume**, the fullerene is expected to become harder than diamond. After the pressure has been released, the molecule would be expected to take up its original volume. Experiments in which these molecules were thrown against steel surfaces at about 17,000 m/h (25,744 km/h) showed them to just bounce back.

Fullerenes are, in fact, the only pure, finite form of carbon. Diamond and graphite both form infinite networks of carbon atoms. Under normal circumstances, when a diamond is cut, the surfaces are instantly covered with **hydrogen**, which tie up the unattached surface bonds. The same is true of graphite. Because of their symmetry, fullerenes need no other atoms to satisfy their surface chemical **bonding** requirements.

The fullerene seems to have an incredible range of electrical properties. It is currently thought that it may alternately exist in insulating, conducting, semiconducting, or superconducting forms.

Fullerenes with metal atoms trapped inside the carbon cage have also been studied. These are referred to as endohedral metallofullerenes. Reports of **uranium**, **lanthanum**, and **yttrium** metallofullerenes have appeared in the literature. It has been exceptionally difficult to isolate pure samples of these shrink-wrapped metal atoms, however.

For reasons that are not yet fully understood, C-60 seems to be the inevitable result of condensing carbon slowly at high temperatures. At high temperatures when carbon is vaporized, most of the atoms initially coalesce into clusters of 2-15 atoms. Small clusters from **chains**, but clusters containing at least 10 atoms commonly form monocyclic rings. Although these rings are favored at low temperatures, at very high temperatures they break open to form linear chains of up to 25 carbon atoms. These carbon chains may then link together at high temperatures to form graphite sheets, which somehow manage to form the geodesic fullerenes. One theory has it that the carbon sheets, when heated sufficiently, close in on themselves to form fullerenes.

Kratschmer and his coworkers in Germany managed to prepare the first concentrated **solution** of fullerenes in 1990 by mixing a few drops of **benzene** with specially prepared carbon soot.

Scientists later demonstrated that fullerenes can be conveniently generated by setting up an electric arc between two graphite electrodes. In their method, the tips of the electrodes are screwed toward each other as fast as the graphite is evaporated to maintain a constant gap. The process has been found to work best in a **helium** atmosphere in which other **gases** such as hydrogen and **water** vapor have been eliminated.

Fullerenes have been reported to occur naturally in certain coals, as well as in structures produced by lightening known as fulgurites, and in the soot of many flames.

Fullerenes have so far failed to realize their commercial potential. This is partly for reasons of cost and partly because it has proven difficult to isolate large quantities of sought-after types. Fullerenes are actively being studied for the following applications: optical devices, hardening agents for carbides, chemical sensors, gas separation devices, thermal insulation, diamonds, **batteries**, catalysts, hydrogen storage media, polymers and polymer additives, and medical applications.

FUNCTIONAL GROUP

A molecule's functional group is the chemically reactive portion, or segment, of an organic chemical **molecule**. In this sense, the functional group determines most of a compound's chemical (and some of its physical) properties. The functional group of an alkene is its carbon-carbon double bond. The **alkyne functional group** is the carbon-carbon triple bond. Alkanes, which contain carbon-carbon single bonds and carbon-hydrogen bonds that are much less reactive than the other common functional groups, are sometimes not considered to contain functional groups.

Functional groups characterize the various families of organic compounds, and the naming conventions used in **organic chemistry**. For example, the functional group common to the alcohols is the -OH, or hydroxyl group. The three molecules CH_3-OH, CH_3-CH_2-OH, and CH_3-CH_2-CH_2-OH (known as **methanol**, **ethanol**, and 1-propanol, respectively) share certain properties in common because they all contain this functional group, despite there being a different number of **carbon** atoms in each molecule. Other important functional groups are R-F, R-Cl, R-Br, and R-I (for the alkyl halides); R-O-R' (for ethers); R-SH (for thiols, also known as mercaptans); R-NH_2, R_2NH, and R_3N (for amines); R-(C=O)-H (for aldehydes); R-(C=O)-R' (for ketones); R-(C=O)-OH (for carboxylic acids); R-(C=O)-OR' (for carboxylic esters); and R-(C=O)-NH_2, R-(C=O)-NHR', and R-(C=O)-NR'R'' (for amides). In these examples, R represents any **alkyl group**.

More complex molecules may contain two or more functional groups. The **amino acids**, for example, contain both amino and carboxyl groups. These biologically important molecules form the building blocks from which large protein molecules are formed.

See also Alkane functional group; Alkene functional group; Amide group; Amine group; Thiol functional group

FUNGICIDE

Fungicides are chemicals that inhibit the growth or attack of fungi. Fungi can attack agricultural crops, garden plants, **wood** and wood products (dry rot in particular is a major problem), and many other items of use to humans. Fungicides usually kill the fungus that is causing the damage. **Sulfur**, sulfur containing compounds, organic salts of **iron**, and heavy **metals** are all used as fungicides. Other fungicide types include carbamates or thiocarbamates such as benomyl and ziram, thiozoles such as etridiazole, triazines such as anilazine, and substituted organics such as chlorothalonil. Fungicides generally have low mammalian toxicity. Many have been shown to cause cancer or reproductive toxicity in experimental animal studies.

Fungicides operate in different ways depending upon the species that they are designed to combat. Many are poisons and their application must be undertaken carefully or over application may kill other plants in the area. Some fungicides disrupt some of the metabolic pathways of fungi by inhibiting **energy** production or biosynthesis, and others disrupt the fungal cell wall, which is made of chitin, as opposed to the **cellulose** of plant cell walls. Chitin is a structural polysaccharide and is composed of **chains** of N-acetyl-D-glucosamine units. Each **mole** of chitin is made of approximately 500 N-acetyl-D-glucosamine units. Fungal pathogens come from two main groups of fungi, the ascomycetes (rusts and smuts) and the basidiomycetes (the higher fungi—mushrooms, toadstools, and bracket fungi).

Human fungal infections, such as athlete's foot, can be treated by fungicides normally referred to as antifungal agents or antimycotics. Compounds such as clotrimazole and nystatin are used to treat human fungal infections.

See also Agricultural chemistry

FURFURAL

Furfural—also called furfuraldehyde, fural, 2-furaldehyde, pyromucic aldehyde, or 2-furancarboxaldehyde—is a viscous, colorless liquid with a freezing point of-37.6°F (-3.11°C) and a **boiling point** of 323°F (161°C). Exposed to air, it turns dark brown. A highly reactive aldehyde, **furfural** is derived from corn cobs, hulls of rice, cottonseeds, or oats that have been treated with hot **hydrochloric acid**. It is useful as an intermediate in the manufacture of many polymers and to dissolve impurities in petroleum compounds and vegetable and lubricating oils. Furfural is a selective solvent meaning that it dissolves only some materials.

Johann Doebereiner, (1780-1849), a German chemist accidentally discovered furfural in 1832 when he treated sugar with **sulfuric acid** and **manganese** dioxide. Large-scale manufacturing techniques were developed in the early 1920s. Furfural consists of a string of **carbon** atoms connected in a ring-shaped structure which contains both an aldehyde group (R-COH) and an **ether** bond (C-O-C). Because the bonds in furfural alternate between double and single bonds, there are several **bonding** sites which makes furfural highly reactive. The **ring** is usually opened at the ether linkage. Furfural reacts with phenols, **ketones**, and esters as an aldehyde; removal of the aldehyde group yields furan, which is converted to tetrahydrofuran (THF), used in the manufacture of **nylon**. Butadiene is a derivative of THF, as well. Furfural is also used in the manufacture of several synthetic resins contained in plastic products. Additionally, it is used in fungicides, germicides, **herbicides**, and insecticides, and as a catalyst in the vulcanization process. Today furfural is used to produce furfuryl alcohols which are used used in foundry sand binders.

Under the Clean Water Act, furfural is listed as a hazardous substance. Because it is a skin irritant, protective measures such as gloves, goggles, protective clothing, or engineering controls must be used to prevent contact.

G

GADOLINIUM

Gadolinium is a rare earth element, one of the elements that occurs in Row 6 of the **periodic table** between **barium** (atomic number 56) and **hafnium** (atomic number 72). Its **atomic number** is 64, its atomic **mass** is 157.25, and its chemical symbol is Gd.

Properties

Gadolinium is a typical metal with a shiny metallic luster and a slightly yellowish tint. It is both ductile and malleable. It has a melting point of 2,394°F (1,312°C), a **boiling point** of about 5,400°F (3,000°C), and a **density** of 7.87 grams per cubic centimeter. Gadolinium is strongly magnetic and has the highest neutron-absorbing ability of any element. The element is not particularly reactive, although it does react slowly with cold **water**, it dissolves in most acids, and it reacts with **oxygen** at high temperatures.

Occurrence and Extraction

Gadolinium is a relatively common element with an abundance of about 4.5-6.4 parts per million, belying its classification as a ''rare earth element.'' The term ''rare earth'' actually refers more to the difficulty of separating the elements in this family from each other rather than their relative abundance in the Earth's crust. Gadolinium is extracted from its ores by being converted first to gadolinium chloride ($GdCl_3$) or gadolinium fluoride (GdF_3) and then being electrolyzed: $2GdCl_3$ —electric current→ $2Gd + 3Cl_2$.

Discovery and Naming

Gadolinium was discovered in 1880 by the French chemist Jean-Charles Galissard de Marignac (1817-1894). The element was found in a complex mineral called cerite that had first been discovered in 1803. Over nearly a century, chemists continued to analyze cerite, eventually discovering a total of seven new elements in it. The element was named for Finnish chemist Johan Gadolin (1760-1852) who conducted some of the earliest research on cerite.

Uses

Because of its neutron-absorption capacity, gadolinium is used in the manufacture of control rods for nuclear power plants. Compounds of the element are also used to make phosphors that coat the screen of television tubes and CRTs (cathode ray tubes). Gadolinium is also used in the manufacture of synthetic minerals known as gadolinium **yttrium** garnets, used in microwave ovens.

GAIA THEORY AND CHEMISTRY

Since the earliest recorded history, philosophers, scientists, and others have described an interplay between life and Earth's environment. The Gaia hypothesis draws on this concept but incorporates modern scientific disciplines such as microbiology, geology, and **atmospheric chemistry**. The name Gaia is drawn from Greek mythology and means Mother Earth.

James Lovelock, an atmospheric chemist, and Lynn Margulis, a microbiologist, published their first paper on the Gaia hypothesis in the early 1970s. Their hypothesis stated that the biota, which consists of all living organisms, maintain an environment on Earth that is optimal for life. This environment features an atmosphere that has a stable **temperature** and chemical composition. Because these factors stay within certain limits, other factors such as the climate and the amount of sunlight that reaches Earth's surface are also relatively stable. The Gaia hypothesis was later refined to emphasize that the environment is regulated through feedback between living and nonliving systems. The refined hypothesis, often called the Gaia theory, includes the idea that as biota evolved over millions of years, their ability to regulate their environment also evolved.

The Gaia hypothesis has its roots in Lovelock's work with National Aeronautics and Space Administration (NASA) during the 1960s. Lovelock's team was investigating whether the identity and molecular composition of a planet's atmosphere could be used to predict whether life existed on it. In contrast to Venus and Mars, Earth, the only planet known with life, has an atmosphere that is not at chemical equilibrium. It has too much **oxygen** and **nitrogen** and too little **carbon** dioxide. Because of the high amount of oxygen, other **gases** such as **methane** should burn off rapidly. However, this non-equilibrated atmosphere has endured for billions of years, and it is ideal for life. Lovelock concluded that biology must be considered alongside chemistry and physics to answer questions about the formation and regulation of a planet's atmosphere.

The Gaia theory goes a step further and proposes that the biota have an active role in controlling Earth's atmosphere, therefore Earth's environment. The scientific community generally agrees that the biota help form the gases in the atmosphere; however, they do not necessarily agree that the biota help regulate their amounts. One of the first objections to the Gaia hypothesis was that it seemed to imply a purpose behind atmospheric regulation—as though the biota had planned to create a hospitable environment. Another objection was based on Charles Darwin's theory of evolution through natural selection. Natural selection means only individual organisms that can survive in a particular environment will reproduce and pass that ability to the next generation. In stark contrast to organisms evolving in response to the environment, the Gaia hypothesis asserts evolution occurred to affect their environment.

These criticisms prompted Lovelock and Margulis to refine their hypothesis. One refinement was Lovelock's creation of the Daisy World concept as a model for Gaia theory. Daisy World is a simplified Earth system with a plant community of white and black daisies. It incorporates natural selection and shows that foresight is not necessary for regulating an atmospheric quality such as temperature.

Initially, Daisy World is too cold for any daisies to grow. But, as with Earth's sun, solar output increases with time. At a certain point, Daisy World becomes warm enough for the daisies to germinate. The white daisies are at a disadvantage because sunlight is reflected from their petals and the surrounding temperature remains cool. However, the black daisy petals absorb the sun's warmth and increase the surrounding temperature. Since the black daisies are warmer, they grow better and produce more seeds. Eventually, there are so many black daisies that the temperature on Daisy World increases and white daisies also grow well. Because the white daisies are doing well, there is a **balance** between absorbed and reflected solar radiation and the temperature remains steady.

Daisy World indicates that the phenomena described by Gaia theory could be possible, but it is not proof. A key problem in proving Gaia theory is that Earth is the only planet that has been shown to have life. One positive outgrowth of Gaia theory is increased cooperation between scientists from different fields in the study of Earth. Through their work, they may be able to determine how much humans affect Earth. Of particular interest is the possibility that human activities may be counterbalanced by other parts of the biota. In recent decades, human activities such as deforestation and so-called greenhouse gas emissions (e.g., **carbon dioxide** and methane) have been blamed for global warming and related climate change. Strikingly, Gaia theory predicts that other parts of the biota would have a counterbalancing effect to prevent dramatic climatic changes such as runaway warming.

Margulis cites several systems to demonstrate the selfregulation proposed by Gaia theory. Two proposed feedback systems are:

Carbon dioxide. Over the last four billion years the luminosity of the sun has increased by about 25% and yet the temperature on earth has remained roughly constant. Decline in atmospheric carbon dioxide levels has been linked to the decline in the ability of the atmosphere to absorb solar radiation. Proponents of the Gaia theory say the decline in atmospheric carbon dioxide was caused by the biota's ability to fix carbon dioxide and form limestone.

Ocean salinity. Normal weathering releases so much **salt** into the ocean that it would have long ago become too salty for most life if the salt was simply dissolved. However over the course of millions of years, the salt content of seawater has remained below 10% of saturation. Proponents of Gaia theory say the excess salt is sequestered by bacteria to form protective sheaths within which the bacterial colonies can safely live. Thus the bacteria help to regulate the salinity of seawater.

These and other proposed feedback mechanisms have detractors. For example, critics of the Gaia theory believe the drop in carbon dioxide levels can be explained by rain; they propose that rainwater simply reacts with carbon dioxide to form carbonic acid which is in turn neutralized by salt.

GALLIUM

Gallium is the third element in Group 13 of the **periodic table**, a group of elements sometimes known as the **aluminum** family. It has an **atomic number** of 31, an atomic **mass** of 69.72, and a chemical symbol of Ga.

Properties

Gallium is a soft, silvery metal with a shiny surface. It is quite soft and can be cut with a knife. It has a very low melting point of only 85.5°F (29.7°C), such that a sample of gallium will melt if held in a person's hand (body **temperature** is about 98.6°F or 37°C). Gallium's **boiling point** is about 4,350°F (2,400°), and its **density** is 5.9037 grams per cubic centimeter.

Chemically, gallium is a fairly reactive element. It combines with most non-metals at high temperatures and reacts with both acids and alkalis.

Occurrence and Extraction

Gallium is a moderately abundant element in the Earth's crust, occurring in a **concentration** of about five parts per million. The world's largest producers of the element are Australia, Russia, France, and Germany. The United States produces no gallium. Pure gallium metal can be extracted from its ores by converting the ore first to gallium **oxide** (Ga_2O_3) and then electrolyzing the product: $2Ga_2O_3$ — electric current → 4Ga + $3O_2$.

Discovery and Naming

The existence of an element with atomic number 31 was predicted by the Russian chemist **Dmitry Mendeleev**, creator of the periodic table. Mendeleev noted an empty place in the periodic table he constructed in position #31 and listed the properties of the element he expected to be present in that position. In 1875, the French chemist Paul-Émile Lecoq de Boisbaudran discovered the element and found that it had essentially the properties predicted by Mendeleev. Lecoq de Boisbaudran suggested the name gallium for the new element in honor of the ancient Latin name for France, Gallia.

Uses

About 95% of all gallium produced is used to make a single compound, gallium arsenide (GaAs). This compound is used in devices that convert electric current to light. For example, the lighted numbers on hand-held calculators are produced by a device known as a light-emitting diode (LED), made from gallium arsenide. Gallium arsenide is also used in **lasers**, such as those present in a compact disc player, in transistors, and in photovoltaic cells.

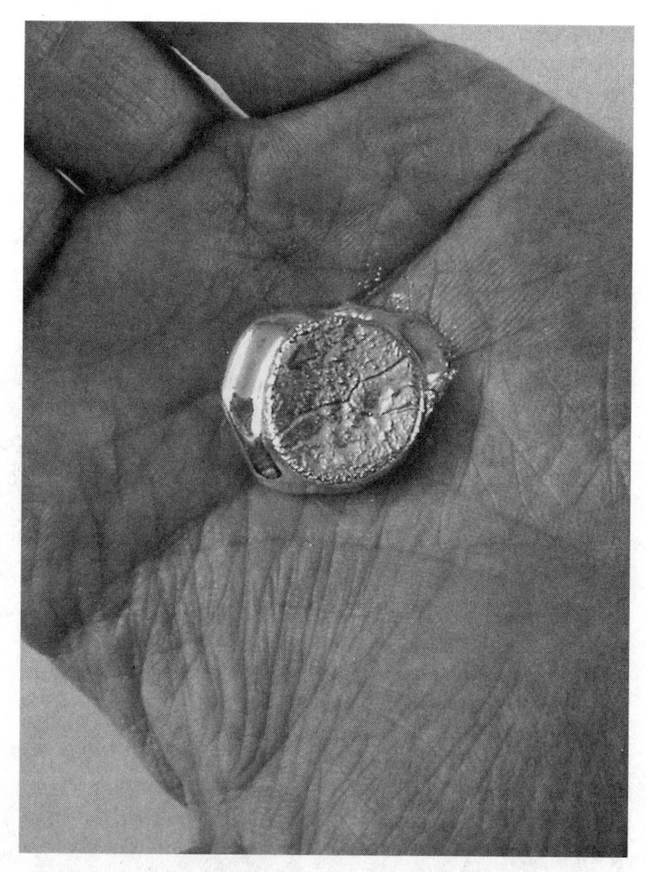

Gallium melts when held in the hand. ***(Photograph by Russ Lappa/Science Source, National Audubon Society Collection/Photo Researchers, Inc. Reproduced by permission.)***

GAMMA RADIATION

Gamma rays are a form of electromagnetic radiation, just like visible light or x rays, but with a much higher **energy**. The gamma ray spectrum is usually defined as light having a frequency between 10^{18} and 10^{21} Hertz. Gamma rays are more energetic than x rays, but are less energetic than cosmic rays. Compared to ultraviolet light (light that corresponds in energy to electronic transitions within molecules and having a frequency of approximately 8 x 10^{14} Hertz), a gamma ray **photon** with of frequency of 10^{19} Hertz is approximately 10,000-fold more energetic. Gamma rays are sufficiently energetic to cause nuclear transitions within atoms and, thus, also correspond to the energy release that accompanies the transition of an excited **nucleus** to its ground (i.e. most stable) state.

The discovery of gamma rays generally is attributed to the French physicist Paul Villard (1860-1934). Following the discover of x rays by the German physicist Wilhelm Röntgen (1845-1923) in 1895, and natural **radioactivity** for the French physicist **Antoine Becquerel** in 1896, Villard worked in this new field of nuclear physics. In studying the rays emitted during the radioactive decay of **radium**, he observed a new radiation that was unaffected by a magnetic field (was not electrically charged). Moreover, the radiation passed directly through many materials and was capable of penetrating several centimeters into **lead**. This new radiation, that of a highly energetic photon emitted from the decayed radium nucleus, was called a gamma ray.

Gamma rays are a product of the radioactive decay of an **atom**. Most commonly, the unstable nucleus of an atom decays to a more stable nucleus of another atom by emitting an **alpha particle** (the positively charged **helium** ^4He nucleus) or beta particle (a negatively charged **electron** or positively charged positron). When a nucleus decays, through the emission of either an alpha or beta particle, the resulting new nucleus is often left in an excited state. This excited nucleus can discharge its excess energy, and so decay to a lower (more stable) state by emitting a photon of gamma radiation. The energy of this photon corresponds to the difference in energy between the two nuclear states and is generally in the range of 0.1 x 10^6 to 10 x 10^6 eV.

The encounter of the highly energetic gamma ray with another **molecule** can result in the ejection of electrons from that molecule. This process is called **ionization**, and this capability had led to gamma radiation being termed ionizing radiation. Ionization provides a mechanism for the detection of the gamma ray (for example, by the Geiger counter) and also represents the mechanism by which biological molecules are damaged.

The Geiger counter is named after its inventor, Hans Geiger (1882-1945). In collaboration with his colleague Walther Müller, Geiger devised the Geiger-Müller tube in 1928. The invention of this tube enabled the development and use of the Geiger-Müller counter as a practical method for the detection of radioactivity. This tube consists of two electrical plates, maintained at a potential difference, and surrounded by gas. The passage of radiation through this tube results in **ion** formation, and these ions then accelerate towards one of the two plates. This results in a measurable voltage change. Unfortunately, gamma radiation is so energetic that most rays pass through the tube with no discernible effect. In this circumstance, scintillation counters are used. In scintillation counting, a fluorescent material absorbs the energy of the photon and emits a new photon as a flash of light (the scintillation) in response. This flash can then be amplified into an electrical pulse, indicating the presence of the radiation.

When the molecules of a living organism are exposed to radiation that results in the loss of an electron from the molecule, reactive intermediates (called free radicals) are formed. While all cells have mechanisms to minimize free radical damage, if the quantity of **free radicals** that are formed exceeds the cell's detoxification capacity or if the molecular damage is difficult to repair (such as certain **DNA** lesions), the cells of the organism are damaged. The damage is categorized as either somatic or genetic. Somatic effects are those that directly affect the individual who received the original dose. The short-term effects can include burns, nausea, and hair loss, although some damage is not apparent until years later when it can surface in the form of cancers. It was their exposure to this radiation that led to the premature deaths of several of the first scientists to study radioactivity, including Paul Villard and **Marie Curie**. Genetic effects of radiation are caused by damage to the DNA of reproductive cells that causes harmful mutations in the children and grandchildren of the exposed individual.

Because of the harmful effects of ionizing radiation, it is important to be able to measure exposure to it. The unit of absorbed dose is call the *gray*, which is the amount of energy in joules from the radiation absorbed per kilogram of the absorbing material. For biological organisms, however, the effect not only depends on the amount of energy absorbed but also on how it was absorbed. If all the energy was absorbed over a very small distance, on the scale of a cell, the effect is more significant than if the energy is spread over a larger distance. The *dose equivalent* of radiation is, therefore, the amount of radiation absorbed multiplied by a quality factor (sometimes called the linear energy transfer), which depends on the energy deposited per unit path length. Fast moving gamma photons have a quality factor of about one, whereas alpha particles (which lose all of their energy over a very small path length) have quality factors nearer 20.

Gamma radiation has several uses. Since it created by the decay of a nucleus, it can be used as a nuclear probe, the nature of the emitted radiation giving an insight into the nature of the nuclear state. Gamma rays also are emitted as a result of high-energy processes throughout the universe and can travel large distances before they reach the Earth. They, therefore, provide a useful information source for astronomers on everything from black holes to processes occurring within our own Sun. Gamma rays, such as those emitted from the unstable nuclei of cobalt-60 and cesium-137, are used in industry to assess the metallurgical integrity of, for example, pipes, girders, and turbine blades.

Gamma radiation is also important in the medical sciences. Ionizing radiation is more harmful to cells when they are dividing (replicating their DNA), and this has led to the use of gamma radiation for the treatment of cancer. The rapidly dividing cancer cells are more susceptible to the radiation than the healthy cells and are destroyed selectively without the need for surgery. Molecules that contain very short-lived isotopes are used as tracers to study the distribution of the molecules within the body. Due to the sensitivity with which radiation is detected only minute quantities of the chemical are required. This information is used for both diagnostic and research purposes. For example, it has enabled the three dimensional imaging of the brain as a function of particular stimuli.

See also Radiation chemistry

GARROD, ARCHIBALD (1857-1936)

English physician and chemist

Archibald Garrod was a physician whose innovative work in clinical medicine and **chemistry** led him to discover a new class of human disease based on hereditary factors. A pioneer in **biochemistry**, Garrod stressed the chemical uniqueness of each person. For his work on inborn errors of **metabolism**, Garrod was elected to the Royal Society and received a knighthood.

Archibald Edward Garrod was born in London on November 25, 1857, the fourth and youngest son of Sir Alfred Baring Garrod and Elisabeth Ann Colchester. Garrod's father, a distinguished professor of medicine at University College in London, was the first physician to note the presence of uric acid in patients suffering from gout. In later years, Garrod would cite his father's discovery as the first quantitative biochemical investigation performed on living humans.

As a child, Garrod demonstrated an early talent for illustration and a lasting interest in **color**. He studied physical geography at Marlborough and astronomy at Oxford, where he graduated with first-class honors in natural science. Deciding to follow in his father's footsteps, Garrod began the study of medicine at St. Bartholomew's Hospital in London. He received a number of scholarships and spent a year attending medical clinics in Vienna, resulting in the publication of a book on the laryngoscope, a device used to examine the interior of the throat. A tall, handsome man, Garrod became a skilled clinician whose reassuring manner enabled him to gather detailed medical histories from his patients. In 1884, Garrod joined the staff of St. Bartholomew's Hospital, but promotion was slow and for nearly three decades he had ample time to pursue his interest in chemistry and disease. He wrote a number of papers on joint disorders, his father's specialty, pointing out the difference between rheumatism and rheumatoid arthritis as diseases.

Garrod's interest in joint disease led him to study the chemistry of pigments in urine. While working as a visiting physician at the Great Osmond Street Hospital for Sick Children, he examined a three-month old boy, Thomas P., whose urine was stained a deep reddish-brown. Garrod's diagnosis was alkaptonuria, which is caused by an abnormal build-up of homogentisic acid, or alkapton. In a normal person, the acid is broken down through a series of chemical reactions into **carbon** dioxide and **water**. But in rare cases, the metabolic process is interrupted and the acid is excreted in the urine, where it turns black on contact with the air. According to the germ theory of disease, which had transformed the study of medicine in Garrod's time, alkaptonuria was thought to be a bacterial infection of the intestine. The disorder was almost always diagnosed in infancy, lasted throughout life and was thought to be contagious. Garrod's training in physical science, however, led him to investigate the disease as a series of chemical reactions. He reviewed 31 cases of alkaptonuria from his own practice and from the medical literature, and presented his findings to the Royal Medical and Chirurgical (Surgical) Society of London in 1899. Alkaptonuria, he noted, although rare, tended to appear among children of healthy parents. It was not contagious and seemed to be a harmless error in metabolism.

When a third child with alkaptonuria was born to the parents of Thomas P., Garrod suspected that something more than mere chance was involved. When he learned that Thomas P.'s parents were blood relations—their mothers were sisters—he inquired into the backgrounds of other families with one or more children with alkaptonuria. In every instance, their parents were also first cousins. It was while walking home from the hospital one afternoon that Garrod conceived of the possibility that alkaptonuria might be a disease caused by heredity (genetics). Gregor Mendel's work on the principles of heredity, newly discovered in England, offered a simple explanation. The mating of first cousins apparently created conditions under which a rare, recessive Mendelian factor (or gene) appeared in the offspring. Garrod's classic paper on alkaptonuria was published in *Lancet* in 1902.

Garrod went on to study other metabolic disorders, including the pigment disorders porphyria, the cause of George III's madness, and albinism. Like alkaptonurics, albinos tended to be children of parents who were first cousins. In a series of lectures delivered before the Royal College of Physicians in 1908, Garrod described such disorders as "inborn errors of **metabolism**." In each instance, he claimed, a genetic factor caused a deficiency in a certain **enzyme** which led to a premature block in the chemistry of normal metabolism. In his book, *Inborn Errors of Metabolism* (1909), Garrod described an important new class of diseases which were genetic, not bacteriological in origin.

In recognition of his contributions to science, Garrod was made a fellow of the Royal Society in 1910, and was knighted in 1918. He spent World War I in the Army Medical Service on the island of Malta as consulting physician to the British forces in the Mediterranean. Two of his sons were killed in combat, a third died of the Spanish influenza following the armistice.

After the War, Garrod returned briefly to St. Bartholomew's, but was soon summoned to Oxford to become Regius Professor of Medicine. In his lectures, Garrod urged students to think of disease in terms of biochemistry. Clinicians, he argued, were uniquely placed to observe anomalies of nature which they could then investigate in the laboratory. In his later writings, Garrod hypothesized that there might be a molecular (genetic) basis for all variations in life functions, including physical appearance, susceptibility to disease, even behavior.

Garrod retired in 1927. He and his wife, Laura Elizabeth, whom he married in 1886, moved to Cambridge to be near their daughter, Dorothy, a noted archaeologist and teacher at Newnham College. Archibald Edward Garrod died at home on March 28, 1936. He was 78.

The significance of Garrod's contribution to the science of genetics was not appreciated in his lifetime; he was an elderly physician when most young geneticists were botanists and zoologists. It was not until the 1940s that Garrod's path-breaking work in human genetics was rediscovered and applied to gene theory.

Gas laws

A substance in the gaseous state of **matter** is one which has no definite shape or **volume**. The particles in a gas are moving with high energies and speeds. There are many familiar properties of **gases**. For example, a gas will take on the shape of any container in which it is placed. Likewise, a gas will occupy any volume which is made available to it. A gas can also be easily compressed when pressure is exerted on it. All gases show the same physical properties. They all have **mass**, they can move through each other, they exert pressure, and the pressure of all gases depends on their **temperature**.

The physical properties of gases are explained by a model called the **kinetic molecular theory** of gases. This theory is composed of several postulates. First of all, a gas is made up of tiny particles. These tiny particles each have a particular mass. Second, the distance between the particles in gases is quite large in comparison with the distance between particles in **liquids** and **solids**. Third, the particles of a gas are in a constant motion. They move rapidly and randomly. Fourth, as the gas particles collide with each other or with the wall of a container, their collisions are perfectly elastic. That means no **energy** is lost by a particle as it makes a collision. Fifth, the only factor which can affect the kinetic energy of the gas particles is the temperature of the gas. As the temperature of the gas increases, so does the kinetic energy of the particles. Finally, a gas particle does not exert a force on another gas particle.

The postulates of the kinetic-molecular model built the foundation for several theories which are now called the gas laws. The gas laws were formed by several scientists and collectively summarize the current understanding of the gaseous state of matter. Each of the gas laws assumes that the gas is under specific conditions, called standard temperature and pressure (STP). The standard temperature is zero degree Celsius or 273 Kelvin (K). The standard pressure is one atmosphere (atm), which is equivalent to 760 mm **mercury** (Hg) or 101.325 Pascal (Pa). These laws give a mathematical model for the observed properties of pressure, volume, temperature, and quantity of a gas.

The first of the gas laws is known as **Boyle's law**. Boyle's law describes the relationship between the pressure and the volume of a gas. Gases, as stated above, can be easily compressed. This is because the particles of a gas have a great distance between them. It is this property of gases that make them useful as cushioning devices such as the air bags in an automobile. An English chemist and physicist named **Robert Boyle** was the first scientist to describe this property of gases. In fact, he wrote about "the spring of air" in more than one publication.

Boyle performed many experiments which explored the "spring of air." In one such experiment, he trapped a certain volume of air and measured the new volume after changing the pressure. To do this, he used a J-shaped tube. The short end of the "J" was closed off, and the long end was left open. Boyle trapped some air in the end of the J-tube by adding mercury in the longer end. By changing the height of the mercury in the long end of the tube, Boyle could alter the pressure on the air in the short end. He measured the volume of the air in the short end of the tube at several different pressures, keeping the temperature constant throughout the experiment. He found that as he increased the pressure on the air, the volume of the air would decrease.

Boyle used these results to formulate his law. Boyle's law maintains that if the temperature of a gas is constant, the product of the pressure times the volume is constant. The mathematical formula which describes this law is $PV = k_1$. P is the pressure of the gas, V is the volume, and k_1 is the constant. The value of k_1 depends on the units of pressure and volume used, the amount of gas present, and the temperature of the gas. Using this equation, it can be stated that the volume of a gas and its pressure are inversely proportional to each other. This law can be useful if you measure the pressure and volume of a gas in two trials at constant temperature. If you rearrange the above equation accounting for the two trials, you will come up with $P_1V_1=P_2V_2$, where P_1 and V_1 are the pressure and volume of the gas in the first trial, and P_2 and V_2 are the pressure and volume of the gas in the second trial.

The next of the gas laws is **Charles's law**. **Jacques Charles** formed this law in order to describe the relationship between the temperature of a gas and its volume. Charles also conducted several experiments in order to develop his theory. In one experiment, he trapped a gas sample in a cylinder with a moveable piston. The temperature of the gas sample was changed by placing the cylinder in **water** baths at different temperatures. As the gas' temperature changes, so does its volume, which is observed by the piston moving up or down. The volume of the gas can be measured by the position of the piston. Charles found, through these experiments, that the volume of a gas is directly proportional to its temperature. As the temperature rises, so does the volume of a gas. Anyone who has bought a **helium** balloon during cold weather has observed this phenomenon. When the balloon is removed from the heated store and walked through the cold parking lot, it collapses. As the temperature of the helium decreases, so does its volume. If the temperature is measured using the Kelvin temperature scale and the pressure is kept constant, the relationship can be described mathematically by the equation $V = k_2T$. V is the volume of the sample, T is the temperature in Kelvin, and k_2 is called the Charles's law constant of proportionality. Again, this equation can be rearranged for an experiment with two trials to give the relationship $V_1T_1 = V_2T_2$.

The next of the gas laws is known as **Avogadro's law**. This law describes the relationship between the amount of a gas and its volume. It was proposed by the Italian chemist **Amedeo Avogadro** and states that two samples of gases of equal volume at the same temperature and pressure have the same number of particles in each. This relationship can be expressed by the equation $V = k_3n$. V is the volume of the sample, n is the number of moles of the gas and k_3 is called Avogadro's law constant.

The fourth gas law is called **Dalton's law of partial pressures**. **John Dalton**, an English chemist, explored the nature of gases in mixtures. Dalton determined that each gas present in a mixture exerts the same pressure as it would if it were not in a mixture at the same temperature. In a mixture, each gas exerts a pressure which is called the partial pressure of that gas. The total pressure of the mixture of gases is equal to the sum of all of the partial pressures of the individual gases present in the mixture.

Each of the above laws can be combined to give an equation called the ideal gas equation, $PV = nRT$. P is the pressure of the gas, V is the volume, n is the number of moles of the gas, T is the temperature, and R is called the gas constant (or, sometimes, the universal gas constant). At STP, R = 0.0821 atm-L/mol-K. This equation summarizes the physical properties of an "ideal" gas. An ideal gas is one which follows each of the postulates given by the kinetic-molecular theory of gases. In reality, no gas completely adheres to the kinetic-molecular theory, so there is no such thing as an ideal gas. Scientists still use the ideal gas equation to describe "real" gases. Real gases generally adhere to the kinetic-molecular postulates, it is only at high pressures and low temperatures that they deviate significantly.

The gas laws described above are a culmination of the works of many scientists. These laws are very useful in describing and predicting the behavior of gases. Even though no **ideal gases** exist, real gases, for the most part, adhere to the principles outlined above.

Gases

A gas is a state of **matter**, the others being **solids** and **liquids**. Gases are distinguished from the other states by the amount of **energy** the constituent molecules have. With gases the molecules making up the gas have virtually completely free movement within their container. If the **volume** is increased then the gas expands to fill the new volume. As this happens the pressure that the gas is exerting on the walls of the chamber decreases as there are fewer molecules colliding with a given area of the containers walls. A gas is distinguished from a vapor in that a gas is above the critical point at which the liquid boils.

There are several laws and equations that describe the behavior of gases under specified conditions, collectively they are known as the **gas laws**. **Boyle's law** describes the behavior of a gas when the **temperature** is kept constant. At constant temperature the volume of a given **mass** of gas is inversely proportional to the pressure. This law works very well at low pressures but at high pressures the predictions become less accurate. This failure of Boyle's law is due to the fact that the molecules of which the gas are composed are being pressed so close together that they are interacting. These interactions are attractions and repulsions due to **intermolecular forces** and the fact that the molecules have a definite size.

Charles's law describes the behavior of a gas at a constant pressure. At constant pressure the volume of a given mass of gas is directly proportional to the absolute temperature, that is the temperature expressed in Kelvin. A third law, the constant volume law, states that when the volume is kept constant the temperature is inversely proportional to the pressure of the gas.

These laws are combined together to give the ideal gas equation. The ideal gas equation can be written as $PV = nRT$ where P is the pressure, V is the volume, n is the number of moles of gas, T is the temperature of the gas (in Kelvin), and R is the universal gas constant. This equation can be used to calculate the behavior of a gas as conditions are altered.

There are several other laws pertaining to gases including **Dalton's law of partial pressures.** In a mixture of gases each gas (assuming there is no reaction between the gases) exerts the same pressure on the container walls as if it were the only gas present. As a consequence of this the total pressure in a container of a mixture of gases is the sum of the individual pressure that each gas would exert. Graham's Law concerns itself with the **effusion** of two or more gases through a hole in a container. This law states that the rate of effusion of a gas is inversely proportional to the square root of the molar mass of the gas. **Avogadro's law** is an explanation of the relationship between the quantity of a gas and its volume. At a constant temperature and pressure the volume of a gas will increase if the quantity of gas is increased. Avogadro also worked out that identical volumes of gas under the same conditions contained the same number of molecules.

Gases are a high energy state, compared to liquids and solids. The molecules are more free to move and expand to fill the space into which they are placed. The change from liquid to gas is boiling and some chemicals can change directly from solid to gas in a process known as **sublimation**.

See also Gases, Behavior and properties of; States of matter

GASES, BEHAVIOR AND PROPERTIES OF

Gases are the most diffuse form of **matter**. In a gas the molecules (the smallest particles of which the gas is made—for some gases these equate to atoms) are highly energetic and they can be widely separated from each other. The behavior of gases and their properties derive from these facts.

Sometimes the term vapor is used to describe a gas. Strictly speaking a gas is a substance at a **temperature** above its **boiling point**. A vapor is the gaseous phase of a substance that, under ordinary conditions, exists as a liquid or solid.

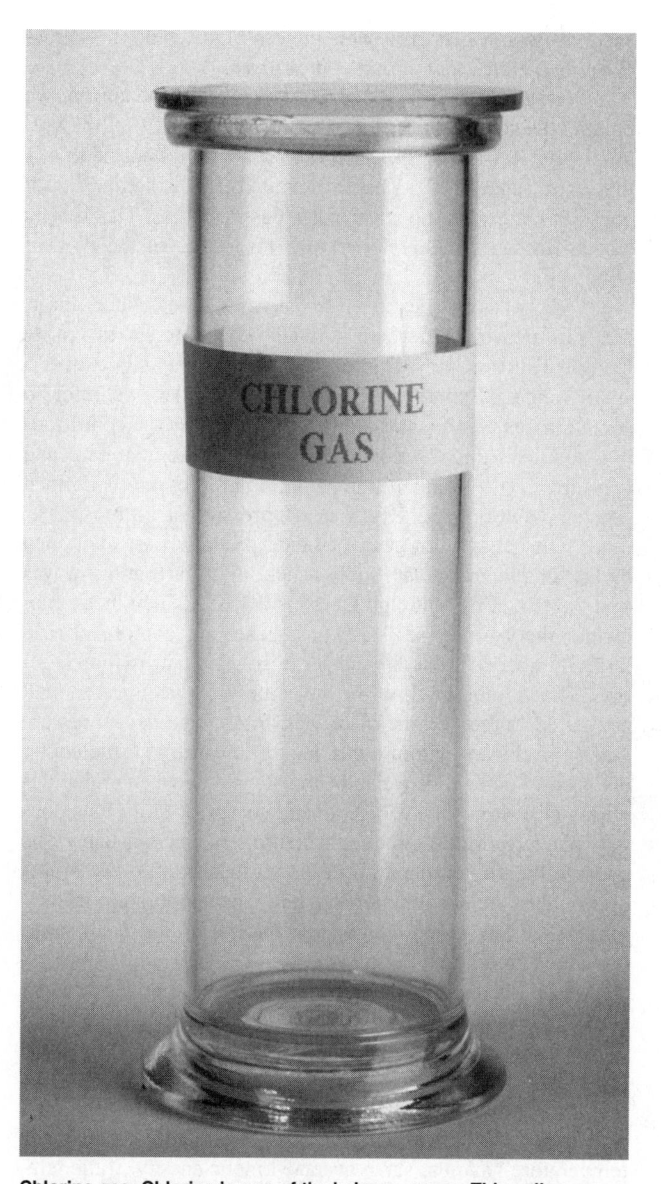

Chlorine gas. Chlorine is one of the halogen gases. This yellow-green gas is active, reacting with water, organic compounds, and many metals. The gas will not burn in air, but will support combustion itself; for example, a paraffin candle will burn in a pure chlorine atmosphere. ***(Photograph by Martyn F. Chillmaid/Science Photo Library, Photo Researchers, Inc. Reproduced by permission.)***

One of the more obvious characteristics of gases are their ability to expand and fill any **volume** they are placed in. This contrasts with the behavior exhibited by **solids** (fixed shape and volume) and **liquids** (fixed volume but indeterminate shape). Gases have an indeterminate volume and shape. This is due to the freedom of movement exhibited by the molecules comprising the gas. A gas can be compressed to a smaller volume or it can expand to fill a larger volume. Gases will form homogenous mixtures with each other. Providing no reaction occurs between the gases, the gases will disperse throughout the volume of the container they are kept in. The pressure inside the container is the sum of the pressures that each gas would exert if it were present alone. This is summarized in **Dalton's law of partial pressures**.

Gases are mostly space. For example, in the air that we breathe the actual molecules only account for 0.1% of the volume—the rest is empty space. The practical effect of this is that a **molecule** in a gas behaves as if it is in isolation, i.e., it does not react with the other molecules in the gas. This is why all gases have similar properties, irrespective of the type of gas.

Many substances are referred to as gases, since this is the state in which they are normally encountered. In reality many substances may only exist in the gaseous phase over a small range of conditions. Consequently when we refer to something as being a gas and we do not specify the conditions it is assumed that the condition is standard temperature and pressure (STP). Standard temperature and pressure is defined as a temperature of 32°F (0°C) and a pressure of 1 atmosphere.

The behavior of gases and their properties are explained by the **kinetic molecular theory** which, in its current form, was first published by Rudolph Clausius in 1857. The kinetic molecular theory of gases states that gases are comprised of a large number of molecules that are in constant, random motion. The volume of all of the molecules of a gas is very small compared to the volume of the gas. **Intermolecular forces** are largely irrelevant within a gas due to the large intermolecular distances. While **energy** can be transferred between molecules during collisions the average kinetic energy of molecules does not change (providing the temperature remains constant). The average kinetic temperature of the molecules of a gas is proportional to the absolute temperature (the temperature in Kelvin) and at any given temperature the molecules of all gases have the same kinetic energy.

The kinetic molecular theory allows us to understand the behavior of gases. The pressure exerted by a gas is a measure of how frequently (and how hard) the molecules of a gas strike the walls of the container in which it is being stored. The absolute temperature of a gas is a measure of the average kinetic energy of the molecules of the gas. Different gases at the same temperature have the same average kinetic energy. If the temperature of a gas is altered then the kinetic energy of the molecules is altered accordingly. If the absolute temperature of a gas is halved then the average kinetic energy of the molecules is also halved. The average kinetic energy is the same between all molecules of a gas, but some individual molecules in a gas are moving faster and some slower than the other molecules. The distribution of energies within a body of gas follows a normal distribution, if the number of molecules is plotted against the energy of the molecules.

Effusion (the movement of gas molecules through a pinhole in a container) and **diffusion** (the uniform spread of gas molecules throughout a mixture of gases) are both related to the energy levels of the gas. The higher the energy levels the more rapidly the molecules are moving and the more rapidly the processes of effusion and diffusion take place. This is due to the fact that the molecules are more likely to hit the pinhole the more rapidly they are moving, and the more rapidly they are moving the quicker they will spread into an unoccupied space. The law governing the rates of effusion of two gases is called **Graham's law**. Both effusion and diffusion are faster for lighter gases (diffusion will also occur with solids but the rate of diffusion is infinitesimal when compared to gases). Related to the rate of diffusion is the mean free path of a molecule. This is the distance traveled between collisions. The greater the distance traveled between collisions the more rapid will be diffusion, since collisions make gas molecules move from one place to another in a staggered manner. The length of the mean free path is governed by the **density** of the gas—the more the molecules are packed together the greater is the likelihood of a collision between molecules. At sea level the mean free path distance for a gas molecule in air is about 100 nm. At an altitude of 62 mi (100 km) the air density is much lower and the mean free path is some million times longer than at sea level. The mean free path in air is about 6 in (16 cm) at this altitude.

The universal gas law can be used to describe the behavior of a gas. Derived from **Boyle's law**, **Charles's law**, and the constant volume law, it can be written as pV/T = a constant, where p is the pressure, V is the volume, and T is the absolute temperature. Another way of writing this law is $p_1V_1/T_1 = p_2V_2/T_2$ where the subscript 1 represents the start conditions and the subscript 2 represents the end conditions. A third way of writing the universal gas law is as $pV = nRT$ where n represents the number of moles of gas being dealt with and R represents the universal gas constant. The universal gas constant is 0.08204 liter atmosphere per degree, or 8.314 $Jmol^{1}K^{-1}$.

Avogadro's hypothesis states that at equal pressure and temperature equal volumes of gases contain the same number of molecules. This hypothesis lead to **Avogadro's law**, which states that the volume of a gas maintained at constant temperature and pressure, is directly proportional to the number of moles of the gas. So if the temperature and pressure of the gas remain constant and the number of moles of gas present is doubled then the volume will also double.

See also Gas laws

GASOLINE

Gasoline is a mixture of liquid hydrocarbons distilled from crude oil that is used as a fuel for internal-combustion engines. Gasoline was invented after discovery of crude oil in the late 1850s when various refining processes were developed to make the oil useable. Early types of gasoline were produced as a byproduct of the process used to make **kerosene** fuel for oil lamps. Since this was before the development of the internal **combustion** engine, much of this early gasoline was discarded because no one had any use for it.

When the automobile was invented, a new market was created for gasoline. At first, automobile engines used ''straight-run'' gasoline—the natural gasoline fraction produced by distilling crude oil—a process that removes gasoline from other compounds in crude oil. But this process alone yielded less than 15 barrels of gasoline from each barrel of oil. After the mass production of cars began in 1908, oil refiners could not keep up with the growing demand for gasoline. In

1913, just in time for World War I, a process was invented to increase the amount of gasoline produced from crude oil. William Burton, who worked for Standard Oil of Indiana, developed *thermal cracking*, a process by which heavy hydrocarbons are broken down by **heat** and pressure into the lighter compounds used in gasoline. Since Burton's discovery, this basic process has been greatly improved. During the 1930s, catalysts were introduced to promote chemical reactions during cracking. Catalysts such as **aluminum**, **platinum**, processed clay, and acids are added to petroleum to break down larger molecules so that it will possess the desired compounds of gasoline. Besides increasing gasoline yields, *catalytic cracking* produces a higher quality gasoline than does thermal cracking.

During World War II catalytic cracking and other new refining processes greatly increased the United States' output of gasoline. More than 80% of the aviation fuel used by the Allies during the war was supplied by the United States. In Europe, gasoline became extremely scarce, and the German army had to rely on cruder types of gasoline that were produced from **coal** and heavy oil. The **hydrogenation** process for making this fuel had been developed in the 1920s by **Friedrich Bergius**, a German chemist who later fled his native country. A similar process developed in 1923 is called Fischer-Tropsch synthesis, which produces gasoline and other **liquids** from coal-derived synthesis gas (hydrogen and **carbon** monoxide). Gasoline can be made from just about any substance containing **hydrogen** and carbon.

Today's gasolines are blended from hundreds of hydrocarbons, and different combinations are produced to meet the needs of different engines. For example, engines vary in how hard they compress the fuel mixture of gasoline vapor and air. Although higher compression improves the engine's performance, it can also cause the gasoline to ignite too soon, creating a metallic ''knocking'' or pinging sound in the engine. This means that the engine is not burning fuel efficiently, and severe knocking can actually damage the engine. A gasoline's resistance to knocking is measured by its octane rating; if the gasoline's performance is 90% as good as that of a reference fuel (pure iso-octane), it gets an octane number of 90. In the early 1900s, engine knock was recognized as a problem, and the auto industry began searching for a fuel that could withstand high pressures without knocking. While engineers experimented with different engine designs, chemists explored ''additives''—substances that could be added to gasoline to prevent knocking. In 1921, a team of American chemists led by **Thomas Midgley**, Jr. and T. A. Boyd made a spectacular breakthrough at General Motors. After much trial and error, Midgley began a systematic study of promising compounds, based on the position of each compound's elements on the **periodic table**. ''What had seemed at times a hopeless quest,'' he recalled, ''rapidly turned into a fox hunt. Predictions began fulfilling themselves instead of fizzling.'' The essential compound turned out to be tetraethyl **lead**. When added to gasoline in minute amounts, tetraethyl lead prevents engine knock and increases the gasoline's octane rating. From 1920-1950, gasoline octane numbers increased from 55 to 85, allowing automakers to nearly double engine performance by using higher internal pressures.

Unfortunately, tetraethyl lead pollutes the air with poisonous lead compounds when the gasoline is burned, and today leaded gasoline is being phased out. Instead, new engines have been designed to run on lower octane gasoline, which is made of hydrocarbons that are resistant to knock. New additives have also been formulated to increase the octane numbers of unleaded gasoline. Other modern additives preserve fuel quality and prevent rust, ice, and deposits of burned **solids** in the engine and fueling system. **Antioxidants** are added to prevent the formation of gum in the engine. Gum is a resin formed in gasoline that can coat the internal parts of the engine and increase wear. During the 1970s, leaded gasoline became associated with another problem. When lead is present in exhaust fumes, it ruins the car's anti-pollution equipment. In 1970, the government required automakers to sharply reduce emissions of **carbon monoxide** and ''unburned'' hydrocarbons that are produced when the engine is out of tune. To meet these standards, automakers introduced catalytic converters—devices that are attached to the exhaust system just behind the manifold. Most converters use platinum or **palladium** metal catalysts that convert carbon monoxide and hydrocarbons to **carbon dioxide** and **water** vapor. The catalysts are easily poisoned by lead, however, which clogs their reactive surfaces. That is why most cars must now use unleaded gasoline.

The abundance of gasoline in the United States through most of the twentieth century has made many aspects of life convenient and pleasurable. People are able to travel greater distances to get to their jobs or to go on vacation. Farmers are able to produce more food by using gasoline-fueled machinery. But by the early 1970s, consumption of gasoline had grown so enormously that oil refiners began depending on imported oil. When foreign oil supplies were disrupted, gasoline supplies suddenly became limited. People waited for hours to fill up their tanks at service stations and gasoline prices skyrocketed from less than 40 cents a gallon to more than a dollar. Since then, automakers have introduced smaller cars that use less fuel. It was the gasoline shortages of the 1970s that made compact Japanese cars popular in the United States for the first time. For a time, the government also encouraged people to conserve gasoline by using public transportation, and states reduced highway speed limits. During the 1980s, these conservation measures were neglected, and higher speeds are now allowed on some stretches of highway. However, the government still requires automakers to continue increasing fuel economy and reducing pollutant emissions.

In the 1990s, concern about **pollution** is even greater and cleaner-burning gasoline formulas have been developed to meet increasingly stringent clean-air standards. These reformulated gasolines contain either corn-derived **ethanol** or natural-gas derivatives such as MTBE (methyl tertiary butyl ether). Both additives increase the **oxygen** content of the gasoline which causes it to burn more cleanly and evaporate more slowly. Another recent antipollution measure requires service stations to control the gasoline vapors released into the atmosphere during fueling operations. This involves anti-evaporation equipment which is built into the gas pumps.

Joseph-Louis Gay-Lussac.

GAY-LUSSAC, JOSEPH-LOUIS (1778-1850)

French chemist

Born about a decade before the French Revolution, Joseph Gay-Lussac was only 14 when his comfortable family life was disrupted by his father's arrest for suspected royalist sympathies. Although Gay-Lussac's tutor fled the country, the boy's education resumed at a private school in Paris, and he successfully competed for entrance to the new Ecole Polytechnique. The hallmark of his scientific career was the analysis of **gases**, which he studied not only in the laboratory, using ingenious new techniques, but also on daring balloon flights that reached record altitudes.

Gay-Lussac's career began with a major revelation. In 1802, when he was just 24, he published a new fundamental law of physics that helps explain the behavior of gases. Although everyone knew that **heat** would make gases expand in **volume**, Gay-Lussac's careful experiments proved that different gases would all expand by the same amount with the same rise in **temperature**. Now known as Gay-Lussac's law (sometimes called **Charles's law**), this idea was crucial to **Amedeo Avogadro**'s discovery a few years later that equal volumes of gases at the same temperature and pressure contain the same number of molecules.

Not content to remain in the laboratory, Gay-Lussac took to the sky in a balloon, armed with various instruments to test the Earth's magnetic field, the atmosphere's **electricity**, and the composition of high-altitude gases. On his second trip, in 1804, Gay-Lussac flew higher than the tallest mountain in the Alps. This record of more than 4 m (6.4 km) above sea level was not reached again for another 50 years. The balloon tests showed that the composition of the air and the magnetic force of the Earth were the same at high altitudes as they were on the ground. After these flights, Gay-Lussac spent two years studying the Earth's magnetic intensity at ground level with his collaborator Alexander von Humboldt.

By then, however, word had spread of Sir **Humphry Davy**'s astonishing results in discovering new elements by using electricity to split compounds. In the name of French nationalism, Napoleon Bonaparte (1769-1821) funded construction of a bigger, more powerful battery for Gay-Lussac and a co-worker, Louis Thénard, to perform similar experiments. But after all, the battery wasn't needed; the two scientists were able to make **potassium** and **sodium** by purely chemical means, without electricity. What's more, they produced these elements in greater quantities than Davy had, making the **metals** available to others at lower cost. Pursuing this technique further, Gay-Lussac and Thenard discovered the element **boron** in 1808, just nine days before Davy did.

Around the same time, Gay-Lussac was also trying to determine the proportions in which various gases combined with each other. Collaborating with Humboldt, Gay-Lussac found that when gases form compounds, they always combine in simple proportions—for example, two parts of **hydrogen** by volume would react with one part **oxygen** to form **water**, or two parts **carbon** monoxide with one part oxygen to form **carbon dioxide**. Some scientists believe that this discovery, called the law of combining volumes, was even more important than Gay-Lussac's earlier gas law.

After some dispute with Davy over who had discovered **chlorine**, Gay-Lussac went on to study **iodine**, which had first been isolated by **Bernard Courtois**. Again, Gay-Lussac's research overlapped Davy's, but it was Gay-Lussac who gave the element its name, after the Greek word for "violet-colored." Gay-Lussac then performed a series of experiments on cyanides, proving conclusively that prussic acid contains no oxygen. At the time, chemists supposed that oxygen was present in all acids (in fact, hydrogen is). During these analyses, Gay-Lussac produced a carbon-nitrogen gas that he named cyanogen, from the Greek for "dark blue." He carefully examined several cyanide compounds, as well as a popular dye called Prussian blue.

Another basic idea in **chemistry** was formulated by Gay-Lussac in 1814—the concept of isomers, which are different compounds composed of the same elements in identical quantities, but differently arranged. Gay-Lussac, an ingenious technician, also created or perfected many of the techniques still practiced in **analytical chemistry**. In introducing a volumetric method for estimating the purity of silver—which was much more accurate than the primitive method then in use—Gay-Lussac popularized the use of **titration** (which involves the careful addition of tiny, exact volumes). He was also the first to use the terms **pipette** and burette to describe **glass** apparatus that is found in every chemistry lab today.

Although Gay-Lussac is better known for his achievements in **inorganic chemistry**, he and Thenard also studied organic compounds. They divided vegetable substances into three classes, one of which is now called **carbohydrates**, and they were the first to determine the elementary composition of sugar.

With another French chemist, **Michel Eugène-Chevreul**, Gay-Lussac also developed a new type of candle that greatly improved on the candles of his day. Made of certain **fatty acids** isolated by Chevreul, the new candle smelled better, gave more light, and burned more easily than tallow candles. It was also harder, meaning that it would soften less quickly during hot weather.

In a way, Gay-Lussac may also have been one of the first environmental scientists. At the time, noxious gases (nitrogen oxides) were released into the atmosphere during the manufacture of **sulfuric acid**. Gay-Lussac devised a way to capture these polluting gases in an absorption tower. His method was later adopted in a process to recover gases.

Most of Gay-Lussac's important work was done while he was relatively young. In his later years, he had to support a large family and perhaps became too overworked to produce the brilliant results of his youth.

Gay-Lussac's law

By the late 1600s, scientists had learned that **heat** would make **gases** expand in **volume**, and British chemist **Robert Boyle** had begun to explain the relationship between the volume, pressure, and **temperature** of gases. He summarized his ideas in an expression, known as which states that the volume of a given amount of a gas varies inversely with pressure. This principle is important since chemistry often involves the study of gases that are hard to measure because they expand to fill whatever container they are in. The value of **Boyle's law** lies in its ability to calculate a gas's volume at a standard pressure without actually having to measure the volume at that pressure.

It was more than a century later, however, until a French chemist **Joseph Gay-Lussac** discovered, with a series of careful experiments, that different gases all expand in volume by the same amount with the same rise in temperature. Although he was the first to announce this principle in 1802, today it is more commonly known as **Charles's law** after French chemist **Jacques Charles**, who actually had the same idea earlier but did not publish his findings before Gay-Lussac.

Further, Gay-Lussac found that gases combine with each other in simple proportions—for example, two parts of **hydrogen** by volume reacts with one part **oxygen** to form **water**. This important elaboration came to be known as *Gay-Lussac's law of combining volumes*. These fundamental concepts were crucial to **Amedeo Avogadro**'s discovery a few years later that equal volumes of gases at the same temperature and pressure contain the same number of molecules. Although

Silica gel, used to absorb moisture in electronic component chemistry. ***(Photograph by Mark C. Burnett, Photo Researchers, Inc. Reproduced by permission.)***

Gay-Lussac's laws strictly apply only to certain theoretical gases (called ''ideal''), many simple substances that are gases at room temperature and normal air pressure do behave as if they were **ideal gases**. Thus, the laws have proved extremely useful in explaining the behavior of gases, and they are considered fundamental concepts in chemistry and physics.

Gel

Colloidal systems are intimate mixtures of two or more substances, in which a dispersed phase is uniformly distributed though a dispersion medium. It is conventional to refer to a **colloid** system that resembles a liquid as a sol, and one that resembles a solid, jelly-like substance, a *gel*. When **water** is one component of the colloid, the system may be referred to as a hydrosol or hydrogel. The reversible transformations of a sol into a gel, and a gel into a sol, are called solation and gelation, respectively. It is difficult to make clear distinctions between gels, sols, and colloids.

The materials ordinarily called gels include silica gels (usually prepared from a **sodium** metalsilicate solution), agar (a carbohydrate polymer derived from seaweed), **gelatin** (closely related to proteins), soft soaps (potassium salts of higher fatty acids), oleates and stearates, polyvinyl **alcohol**, and various hydroxides in water.

The formation of gels from suspensions or solutions is accompanied by the formation of three-dimensional cross-linkages between molecules of one component. The second component, meanwhile, permeates the first as a continuous phase. Thus the gel resembles a loosely interlinked polymer.

Gelation can be brought about in a number of ways; these include cooling a sol, initiaing a chemical reaction, or by adding a precipitating agent or incompatible solvent. Gelatin is an example of a material that is readily soluble in hot water, but that gels upon cooling. Gelation may require from minutes to days to complete, depending on the material, its history, and the **temperature**.

See also Aerogel; Macromolecule; Polymer chemistry

Gelatin

Gelatin is a mixture of water-soluble **proteins** derived from the partial **hydrolysis** of collagen obtained from the skin, connective tissue, and bones of animals. It is a colorless to slightly yellow material that is produced in sheets, flakes, or powder. It is practically odorless and tasteless. In an aqueous **solution** it absorbs 5-10 times its weight of **water** and creates a stable **gel**. Since gelatin is derived from collagen, it is composed of various amino acids. The most prevalent **amino acids** are glycine, proline, and hydroxyproline.

The origin of gelatin is unknown but it was likely to have been accidently discovered after boiling the remains of animals that were killed for food. This process would produce a substance that remained liquid when hot and solid when cooled. The first known commercial production of gelatin was in Holland in 1685. By 1700, England was producing gelatin. It was first produced in the United States in 1808. In 1993, about 75 million pounds of gelatin were being produced in America annually.

Gelatin has numerous uses. Since it is generally recognized as safe for ingestion, it is used extensively in food products as a stabilizer, thickener, binding agent, emulsifier, adhesive agent, film former, and texturizer. Products that contain gelatin include gummi candies, confections, marshmallows, meats, beer, jellies, and cream soups. In the pharmaceutical industry gelatin is an important suspending or encapsulating compound. Hard or soft gelatin capsules are produced and filled with various pharmaceutical doses. Gelatin is also used as a coating and binder for tablets. Other important applications for gelatin include film processing, the manufacture of cement, adhesives, rubber substitutes, and artificial silk. In addition, it has been used as a **plasma** expander and for the preparation of bacterial cultures.

Gene

Genes are the physical units of heredity and are located along each **chromosome** in the cells of the human body. For each physical trait—eye **color**, height, hair color—a person inherits two genes or two groups of genes, one from each parent. Because both of these genes cannot be expressed together, one usually overpowers the other. The more powerful gene is called the dominant gene and the weaker, the recessive gene.

All genes on the same chromosome are called linked genes because they are usually inherited together. Genes on the X and Y chromosomes are called sex-linked because the X and Y chromosomes are the ones that determine sex.

Sometimes genes on the same chromosome are not inherited together. When sex cells divide to form an egg or sperm cell (a process called meiosis), each chromosome pairs off with a partner. As the chromosomes lie side by side, groups of genes from one chromosome may change places with groups of genes from the partner chromosome. This is called crossing over and thus explains how families inherit different combinations of linked traits.

An Austrian monk, Gregor Mendel, introduced the world to hereditary factors—genes—that determine all hereditary traits. During his experiments with pea plants, Mendel noticed that the plants inherited traits in a predictable way. It was as if the pea plants had a pair of factors responsible for each trait. Even though he never actually saw them, Mendel was convinced that tiny independent units determined how an individual would develop. Before Mendel's findings, traits were thought to be passed on through a blending of the mother and father's characteristics, much like a blending of two **liquids**.

When Mendel's laws of heredity were rediscovered in 1900, they became vitally important to biologists. Among other things, Mendel's laws established heredity as a combining of independent units, not a blending of two liquids. Wilhelm Johannsen, a strong supporter of Mendel's theories, coined the term ''gene'' to replace the variety of terms used to describe hereditary factors. His definition of the gene led him to distinguish between genotype (an organism's genetic makeup) and phenotype (an organism's appearance).

In 1910, Thomas Hunt Morgan began to uncover the interesting relationship between genes and chromosomes. Morgan discovered that genes are located on chromosomes. He also found that the appearance of the characteristics associated with linked genes occur together in the offspring. In other words, genes located on the same chromosome are usually inherited together, or ''linked.'' But as Morgan and his colleagues worked with more and more characteristics simultaneously, they discovered that genes on the same chromosome were not always linked. Sometimes crossing over occurred; paired chromosomes would break apart and rejoin during meiosis, causing many different gene combinations. Morgan and his colleagues went on to develop and perfect these and other gene concepts, which laid the groundwork for all genetically based medical research.

Currently, almost 4,000 genetic diseases are known to affect humans. In recent years, scientists have been identifying genes associated with various diseases and conditions at a furious pace. For example, in 1995, scientists found genes linked to at least some cases of male infertility, epilepsy, schizophrenia, Alzheimer's disease, and breast cancer. In 1996, they found genes associated with some cases of retinitis pigmentosa (a progressive form of blindness) and basal cell carcinoma (the most common form of skin cancer), and in 1997, with heroin addiction, Parkinson's disease, fetal **alcohol** syndrome, glaucoma, and obsessive-compulsive disorder. Given enough time and effort, scientists may one day learn how to prevent or treat many such diseases.

Gene and genetics

Genetics is the branch of biology concerned with the science of heredity, or the transfer of specific characteristics from one generation to the next. Genetics focuses primarily on genes, coded units found along the **DNA** molecules of the chromosomes, housed by the cell nucleus. Together, genes make up the blueprints that determine the entire development of the species of organisms down to the specific traits, such as the color of eyes and hair. Geneticists are concerned with two primary areas of **gene** study: how genes are expressed and regulated in the cell and how genes are copied and passed on to successive generations. Although the science of genetics dates back at least to the nineteenth century, little was known about the exact biological makeup of genes until the 1940s. Since that time genetics has moved to the forefront of biological research. Scientists are now on the verge of identifying the location and function of every gene in the entire human genome. The result will not only be a greater understanding of the human body, but new insights into the origins of disease and the formulation of possible treatments and cures.

Although humans have known about inheritance for thousands of years, the first scientific evidence for the existence of genes came in 1866, when the Austrian monk and scientist Gregor Mendel published the results of a study of hybridization of plants-the combining of two individual species with different genetic make ups to produce a new species. Working with pea plants with specific characteristics such as height (tall and short) and color (green and yellow), Mendel bred one type of plant for several successive generations. He found that certain characteristics appeared in the next generation in a regular pattern. From these observations, he deduced that the plants inherited a specific biological unit (which he called factors and which are now called genes) from each parent. Mendel also noted that when factors or genes pair up, one is dominant (which means that it determines the trait, like tallness) while the other is recessive (which means that it has no bearing on the trait).

The period of classical genetics, in which researchers had no knowledge of the chemical constituents in cells that determine heredity, lasted well into the twentieth century. However, several advances made during that time contributed to the growth of genetics. In the eighteenth century, scientists used the relatively new technology of the microscope to discover the existence of cells, the basic structures in all living organisms. By the middle of the nineteenth century, they had discovered that cells reproduce by dividing.

Although Mendel laid the foundation of genetics, his work began to take on true significance in 1903 when William Sutton discovered chromosomes, threadlike structures that split and then pair off as the cell divides. Sutton found that this pairing followed the pattern of gene transmission in successive generations. In 1910, Thomas Hunt Morgan (1866-1946) confirmed the existence of chromosomes through experiments with fruit flies. He also discovered a unique pair of chromosomes called the sex chromosomes, which determined the sex of offspring. From his observation that an X-shaped **chromosome** was always present in the flies that had white eyes, Morgan deduced that specific genes reside on chromosomes. A later discovery showed that chromosomes could mutate or change structurally, resulting in a change in characteristics which could be passed on to the next generation.

But more than three decades would pass before scientists began to delve into the specific molecular and chemical structures that make up chromosomes. In the 1940s, a research team led by **Oswald Avery** took a small nucleic acid called DNA (deoxyribonucleic acid) from one type of bacteria and inserted it into another. They found that the bacteria that received this DNA took on certain traits of the first bacteria. The final proof that DNA was the specific molecule that carries genetic information was made by Alfred Hershey and Martha Chase in 1952. By transferring DNA from a virus to an animal organ, they found that an infection resulted just as if an entire virus had been inserted.

The watershed discovery in genetics occurred in 1953, when **James Watson** and **Francis Crick** solved the mystery of the exact structure of DNA. The two scientists used chemical analyses and x-ray diffraction studies performed by other scientists to uncover the specific structure and chemical arrangement of DNA. X-ray diffraction is a procedure in which parallel x-ray beams are diffracted by atoms in patterns that reveal the atoms' **atomic weight** and spatial arrangement. A month after their double-helix model of DNA appeared in scientific journals, the two scientists showed how DNA replicated. Armed with these new discoveries, geneticists embarked on the modern era of genetics, including efforts like **genetic engineering**, gene therapy, and a massive project to determine the exact location and function of all of the more than 100,000 genes that make up the human genome.

The basis for heredity through the transfer of genes to the next generation lies in the cell, the smallest basic unit of all life. Organs and tissues that make up complex organisms are, in turn, made up of specific types of cells. Although there are several components of cells, the information for heredity is carried within the cell nucleus, a structure located approximately in the middle of a cell. Within the nucleus lie the chromosomes, threadlike structures which carry genetic information organized in DNA molecules.

Multicellular organisms contain two types of cells-body cells and germ cells. Inheritance begins with the germ or reproductive cell. These cells duplicate through a process called meiosis, which ensures that the germ cells have a single set of chromosomes, a condition called haploid or n. The body cells of humans have 23 pairs of chromosomes or 46 chromosomes overall, a condition known as diploid or 2n. Through the process of meiosis, a new cell (called a haploid gamete) is created with only 23 chromosomes: this is either the sperm cell of the father and 23 chromosomes from the egg cell of the mother. The fusion of egg and sperm restores the diploid chromosome number in the zygote. This cell carries all the genetic information needed to grow into an embryo and eventually a full grown human with the specific traits and attributes passed on by the parents. Offspring of the same parents differ because the sperm cells and egg cells vary in their genetic codes, which can recombine in various, or random, arrangements.

The somatic or body cells are the primary components of functioning organisms. The genetic information in these cells is passed on through a process of cell division called mitosis. Unlike meiosis, mitosis is designed to transfer the identical number of chromosomes during cell regeneration or renewal. This is how cells grow and are replaced in exact replicas to form specific tissues and organs, such as muscles and nerves. Without mitosis, an organism's cells would not regenerate, resulting not only in cell death but possible death of the entire organism. (It is important to note that some organisms reproduce asexually by cell division alone.)

To understand genes and their biological function in heredity, it is necessary to understand the chemical makeup and structure of DNA. Although some viruses carry their genetic information in the form of **RNA**, most higher life forms carry genetic information in the form of DNA, the **molecule** that make up chromosomes.

The complete DNA molecule is often referred to as the blueprint for life because it carries all the instructions in the form of genes for the growth and functioning of most organisms. This fundamental molecule is similar in appearance to a spiral staircase, which is also called a double helix.

The sides of the DNA ladder are made up of alternate sugar and phosphate molecules, like links in a chain. The rungs, or steps, of DNA are made from a combination of four nitrogen-containing bases-two purines (adenine [A] and guanine [G]) and two pyrimidines (cytosine [C] and thymine [T]). The four letters designating these bases (A, G, C, and T) are the alphabet of the genetic code. Each rung of the DNA molecule is made up of a combination of two of these letters, one jutting out from each side. In this genetic code A always combines with T, and C with G to make what is called a base pair. Specific sequences of these base pairs, which are bonded together by atoms of hydrogen, make up the genes.

While a four-letter alphabet may seem rather small for constructing the comprehensive vocabulary that describes and determines the myriad life forms on Earth, the sequences or order of these base pairs are nearly limitless. For example, various sequences or rungs that make up a gene could be ATCG-GC, or TAATCG, or AGCGTA, or ATTACG, and so on. Each one of these combinations has a different meaning, providing the code not only for the type of organism but for specific traits like brown hair and blue eyes. The more complex an organism, from bacteria to humans, the more rungs or genetic sequences appear on the ladder. The entire genetic makeup of a human, for example, may contain 120 million base pairs, with the average gene unit being 2,000 to 200,000 base pairs long. Except for identical twins, no two humans have exactly the same genetic information.

Genetic information is duplicated during the process of DNA replication, which is initiated by chemical substances in the cells like **proteins** and **enzymes**. An extremely stable structure, DNA rarely mutates during replication. To produce identical genetic information during cell mitosis, the DNA hydrogen bonds unzip, cleaving the DNA in half. This process begins a few hours before the initiation of cell mitosis. Once completed, each half of the DNA ladder can form a new DNA molecule with the identical genetic code because of specific chemical catalysts that help synthesize the complementary strand.

An identical replication occurs because each strand has one half of the code for the same genetic pattern. Since T always combines with A and C with G, no matter which of these letters is on the remaining half of the ladder, the corresponding chemical (or letter) is synthesized to create the appropriate bond with the other letter. The result is the exact same base pair sequence. When cell mitosis is completed, each new cell contains an exact replica of the DNA.

The functioning of specific cells and organisms is conducted by proteins (enzymes) synthesized by the cells. Cells contain hundreds of different proteins, complex molecules that comprise over half off all solid body tissues and control most biological processes within and among these tissues. A cell functions according to which of the thousand of types of different proteins it contains. Proteins are made up of chains of amino acids. It is the genetic base-pair sequence in DNA that determines, or codes for, the arrangement of the amino acids to build specific proteins.

Since the sites of protein production lie outside the cell nucleus, coded messages, in the form of the genes for protein production, pass out of the nucleus into the cytoplasm. The messenger that carries these instructions is messenger RNA or mRNA (a single stranded version of DNA that has a mirror image of the base pairs on the DNA), which is itself manufactured by specific genes to produce one type of protein. Other types of RNA molecules complete the amino acid sequencing process, which is called translation. Once a protein had been coded for a specific function, it cannot be changed, which is why acquired characteristics during an organisms life, such as larger muscles or the ability to play the piano, cannot be inherited. However, people may have genes that make it easier for them to acquire these traits through exercise or practice.

The expression of the products of genes is not equal, and some genes will override others in expressing themselves as an inherited characteristic. The offspring of organisms that reproduce sexually contain a set of chromosome pairs, half from the father and half from the mother. The genes residing in the chromosome's DNA are also present in alternative forms, known as alleles. These alleles are coded to produce different characteristics, such as blue or brown eyes, or blond or brown hair. Normally, people do not have one blue eye and one brown eye, or half brown hair and half blond hair because most genetic traits are the result of the expression of either the dominant or the recessive genes or alleles. (It is important to note, however, that geneticists are uncovering more and more evidence that most traits are the result of several genes working together.) If a dominant and recessive allele appear together (the heterozygous condition), the dominant will always win out, producing the trait it is coded for. The only time a recessive trait or allele (such as the one for blond hair) expresses itself is when two recessive alleles are present (the homozygous condition). As a result, parents with heterozygous alleles for brown hair could produce a child with blond hair if the child inherits two recessive blond-hair alleles from the parents.

This hereditary law also holds true for genetic diseases. Neither parent may show signs of a genetic disease, caused by a defective gene, but they can pass the double-recessive combination on to their children. Some geneticdiseases are dominant and others are recessive. Dominant genetic defects are more common because it only takes one parent to pass on a defective allele. A recessive genetic defect requires both parents to pass on the recessive allele that causes the disease. A few inherited diseases (such as Down syndrome) are caused by abnormalities in the number of chromosomes, where the offspring has 47 chromosomes instead of the normal 46.

The molecule is extremely stable, ensuring that the offspring have the same traits and attributes that will enable them to survive as well as their parents. However, a certain amount of genetic variation is necessary if species are to adapt by natural selection to a changing environment. Often this change in genetic material occurs when chromosome segments from the parents physically exchange segments with each other during the process of meiosis. This is known as cross over or genetic recombination.

Genes can also change by mutations on the DNA molecule, which occur when something alters the chemical or physical makeup of DNA. Genetic mutations in somatic (body) cells result in malfunctioning cells or a mutant organism. These mutations result from a change in the base pairs on the DNA, which can alter cell functions and even give rise to different traits. Somatic cell mutations can result in disfigurement, disease, and other biological problems within an organisms. These mutations occur solely within the affected individual.

When mutations occur in the DNA of germ or reproductive cells, these altered genes can be passed on to the next generation. A germ cell mutation can be harmful or even result in an improvement, such as a change in body coloring that acts as camouflage. If the trait improves an individual organism's chances for survival within a particular environment, it is more likely to become a permanent trait of the species because the offspring with this gene would have a greater chance to survive and pass on the trait to succeeding generations.

Mutations are generally classified into two groups: spontaneous and induced. Spontaneous mutations occur naturally from errors in coding during DNA replication. Induced mutations come from outside influences called environmental factors. For example, certain forms of radiation can damage DNA and cause mutations. A common example of this type of mutating agent is the ultraviolet rays of the sun, which can cause skin cancer in some people who are exposed to intense sunlight over long periods of time. Other mutations can occur due to exposure to man-made chemicals. These types of mutations modify or change the chemical structure of base pairs.

Population genetics is the branch of genetics that focuses on the occurrence and interactions of genes in specific populations of organisms. One of its primary concerns is evolution, or how genes change from one generation to the next. By using mathematical calculations that involve an interbreeding population's gene pool (the total genetic information present in the individuals within the species), population geneticists delve into why similar species vary among different populations that may, for example, be separated by physical boundaries such as bodies of water or mountains.

As outlined in the previous section, genetic mutations may cause changes in a population if the mutation occurs in the germ cells. Many scientists consider mutation to be the primary cause of genetic change in successive generations. However, population geneticists also study three other factors involved in genetic change or evolution: migration, genetic drift, and natural selection.

Migration occurs when individuals within a species move from one population to another, carrying their genetic makeup with them. Genetic drift is a natural mechanism for genetic change in which specific genetic traits coded in alleles (alternate states of functioning for the same gene) may change by chance often in a situation where organisms are isolated as on an island.

Natural selection, a theory first proposed by Charles Darwin in 1858, is an extremely slow process that occurs over successive generations. The theory states that genetic changes that enhance survival for a species will gradually come to the forefront over successive generations because the gene carriers are better fit to survive and are more likely reproduce, thus establishing a new gene pool, and eventually, perhaps, an entirely new species.

More than any other biological discipline, genetics is responsible for the most dramatic breakthroughs in biology and medicine today. Scientists are rapidly advancing in their ability to reengineer genetic material to achieve specific characteristics in plants and animals. The primary way to genetically engineer DNA is called gene splicing, in which a segment of one DNA molecule is removed and spliced, or inserted, into another DNA molecule. This approach has applications in agriculture and medicine, as well as in forensic science, where it has been used to help identify criminals or to determine parentage of a child.

In agriculture, **genetic engineering** is used to produce transgenic animals or plants, in which genes are transplanted from one organism to another. This approach has been used to reduce the amount of fat in cattle raised for meat, or to increase proteins in the milk of dairy cattle that favor cheese making. Vaccines have also been genetically re-engineered to trigger an immune response that will protect against specific diseases. One approach is to remove genetic material from a diseased organism, thus making the material weaker. As a result, it initiates an immune response without causing the disease. Fruits and vegetables have also been genetically engineered so that they do not bruise as easily, or so that they have a longer shelf life.

In the realm of medicine, genetic engineering is aimed at understanding the cause of disease and developing treatments for them. For example, recombinant DNA (a slice of DNA containing DNA fragments from another cell's DNA) is being used to developed antibiotics, hormones and other disease-preventing agents.

Researchers are also using restriction enzymes to cut DNA into fragments of different lengths. The ends of these fragments are ''sticky'' in that they have an affinity for complimentary ends of other DNA fragments. This technique has been used to analyze genetic structures in fetal cells and to diagnose certain blood disorders such as sickle cell anemia. This works because the enzyme will seek out the corresponding sticky ends on the target DNA, in essence cleaving or cutting out a certain section of the DNA. By looking at the size of this identified fragment, the investigators can determine whether the gene has the proper genetic code-whether or not it is normal.

Gene therapy is another outgrowth of genetics. The idea behind gene therapy is to introduce specific genes into the body to either correct a genetic defect or to enhance the body's capabilities to fight off disease and repair itself. Since many inherited or genetic diseases are caused by the lack of an enzyme or protein, scientists hope to one day treat the unborn by inserting genes to provide the missing enzyme.

In the realm of law, genetic fingerprinting or DNA typing is based on each individual's unique genetic code. To identify parentage or the presence of someone at a crime, scientists use genetic engineering and restriction enzymes to seek out specific DNA segments (usually microsatellites), which can be acquired from hair, semen, blood, or a fingernail fragments. These unique segments or sequences are called restriction fragment length polymorphisms (RFLPs).

One of the most exciting recent development in genetics is the initiation of the Human Genome Project. This project is designed to provide a complete genetic road map outlining the location and function of the 100,000 or so genes that are found in human cells. As a result, genetic researchers will have easy access to specific genes to study how the human body works and to develop therapies for diseases. Gene maps for other species of animals are also being developed.

Despite the promise of genetics research, many ethical and philosophical questions arise. Many of the concerns about this area of research focus on the increasing ability to manipulate genes. There is a fear that the results will not always be beneficial. For example, some fear that a genetically reengineered virus could turn out to be extremely virulent, or deadly, and may spread if there is no way to stop it.

Another area of concern is the genetic engineering of human traits and qualities. The goal is to produce people with specific traits such as better health, improved looks, or even high intelligence. While these traits may seem to be desirable on the surface, the concern arises about who will decide exactly what traits are to be engineered into human offspring, and whether everyone will have equal access to an expensive technology. Some fear that the result could be domination by a particular socio-economic group.

Despite these fears and concerns, genetic research continues. In an effort to ensure that the science is not abused in ways harmful to society, governments, in the United States and abroad, have created panels and organizations to oversee genetic research. For the most part, the benefits for medicine and agriculture seem to far outweigh the possible abuses.

Genetic engineering

Genetic engineering is the artificial altering of genes. This usually takes the form of the insertion of alien genes into the genome of a recipient organism. Genetic engineering is also called recombinant **DNA** technology. This name is slightly more descriptive of the process, which is actually recombining the DNA of existing organisms into new organisms.

Before scientists had the ability to directly alter the genes of an organism, many attempts were made to change living organisms. At its simplest, these attempts involved selective breeding, whereby plants or animals with the desired characteristics were bred together to enhance those characteristics. The offspring not showing the required combinations were not used to breed again, whereas those showing them were bred again. Successive generations gave new combinations. Examples of this technique are the many different breeds of dogs that exist today. Selective breeding is a slow process, the exact time being dependent upon the length of the breeding cycle and the age at which sexual maturity of the organisms concerned is reached.

Cell fusion is a quicker way of altering the DNA. This technique, however, is suitable only for individual cells. Two selected cells under the mediation of viral attack will stick together and fuse. The nuclei of the two cells remain intact, resulting in a binucleate cell. The cell then starts to undergo mitosis. Not all of the chromosomes are capable of lining up properly, so the cell is usually incapable of developing into a living organism. However, it is still often capable of producing **proteins** and polypeptides.

Genetic engineering is a more direct approach, resulting in quicker, more reliable, and more usable products. With this process, a desired piece of DNA, which may have been manufactured artificially or removed from another organism, is relocated into another cell. The cell which has had the DNA inserted now becomes able to manufacture the appropriate protein. The cells into which DNA is most frequently inserted are generally bacterial cells. Their rapid reproductive cycle (sometimes as fast as every 20 minutes) allows for a quick build of modified cells so that a large quantity of product can be effectively gathered.

Another scientific stride is the cloning of the **gene**. This procedure has been undertaken to allow bacteria to manufacture useful proteins. It is also used to mark bacteria so they can be rapidly screened. This is carried out most frequently with the addition of a gene for resistance to antibiotics in conjunction with the desired gene. By screening the resultant bacteria in antibiotic containing media, only the recombinants will survive. Fluorescent genes from jelly fish have also been added to organisms in this manner to show which are successful recombinants.

It is also possible to splice genes into higher organisms. This procedure has been undertaken with a number of plants and animals. Genes have been successfully inserted into cattle to produce human blood clotting agents, and into various plants to make the fruit to last longer. Due to the contentious nature of such work in humans, very little research has been carried out in this area.

Genetic engineering in the form of recombinant DNA technology is a very powerful tool. It can produce great benefits, however, it poses great moral issues as well. Ethical committees from many disciplines are in place to oversee this sort of work and to consider the future of this technology.

Geochemistry

Geochemistry is the study of the chemical processes that form and shape the Earth. The Earth is essentially as large **mass** of crystalline **solids** that are constantly subject to physical and chemical interaction with a variety of solutions (e.g., **water**) and substances. These interactions allow a multitude of chemical reactions.

It is through geochemical analysis that estimates of the age of the Earth are formed. Because radioactive isotopes decay at measurable and constant rates (e.g., **half-life**) that are proportional to the number of radioactive atoms remaining in the sample, analysis of rocks and minerals can also provide reasonably accurate determinations of the age of the formations in which they are found. The best measurements obtained via radiometric dating (based on the principles of nuclear reactions) estimate the age of the Earth to be four and one half billion years old.

Dating techniques combined with spectroscopic analysis provide clues to unravel Earth's history. Using **neutron** activation analysis, Nobel Laureate, Luis Alvarez, discovered the presence of the element **iridium** when studying samples from the K-T boundary layer (i.e., the layer of sediment laid down at the end of the Cretaceous and beginning of the Tertiary periods). Fossil evidence shows a mass extinction at the end of the Cretaceous period, including the extinction of the dinosaurs. The uniform iridium layer—and presence of quartz crystals with shock damage usually associated only with large asteroid impacts or nuclear explosions—advanced the hypothesis that a large asteroid impact caused catastrophic climatic changes that spelled doom for the dinosaurs.

Although **hydrogen** and **helium** comprise 99.9% of the atoms in the universe, Earth's gravity is such that these elements readily escape Earth's atmosphere. As a result, the hydrogen found on Earth is found bound to other atoms in molecules.

Geochemistry generally concerns the study of the distribution and cycling of elements in the crust of the earth. Just as the **biochemistry** of life is centered on the properties and reaction of **carbon**, the geochemistry of the Earth's crust is centered upon **silicon**. (Si). Also important to geochemistry is **oxygen**. Oxygen is the most abundant element on earth. Together, oxygen and silicon account for 74% of the Earth's crust.

The type of magma (basaltic, andesitic or ryolytic) extruded by volcanoes and fissures (magma is termed lava when at the Earth's surface) depends on the percentage of silicon and oxygen present. As the percentage increases, the magma becomes thicker, traps more gas, and is associated with more explosive eruptions.

The eight most common elements found on Earth, by weight, are oxygen (O), silicon (Si), aluminum(Al), **iron** (Fe), **calcium** (Ca), **sodium** (Na), **potassium** (K) and **magnesium** (Mg).

Unlike carbon and biochemical processes where the **covalent bond** is most common, however, the **ionic bond** is the most common bond in geology. Accordingly, silicon generally becomes a **cation** and will donate four electrons to achieve a noble gas configuration. In quartz, each silicon **atom** is coordinated to four oxygen atoms. Quartz crystals are silicon atoms surrounded by a tetrahedron of oxygen atoms linked at shared corners.

Rocks are aggregates of minerals and minerals are composed of elements. A mineral has a definite (not unique) formula or composition. All minerals have a definite structure and composition; Diamonds and **graphite** are minerals that are polymorphs (many forms) of carbon. Although they are both composed only of carbon, diamonds and graphite have very different structures and properties. The types of bonds in minerals can affect the properties and characteristics of minerals.

Pressure and **temperature** affect the structure of minerals. Temperature can determine which ions can form or remain stable enough to enter into chemical reactions. Olivine$(Fe, Mg)_2 SiO_4$), for example is the only solid that will form at 1800°C. According to Olivine's formula, it must be composed of two atoms of either iron or magnesium. Olivine is built by the ionic substitution of iron and magnesium—the atoms are interchangeable because they carry the same electrical charge and are of similar size—and thus Olivine exists as a range of compositions termed a solid **solution** series. Olivine can thus be said to be ''rich'' in iron or rich in magnesium. As magma cools larger atoms such as potassium ions to enter into reactions and additional minerals form.

The determination of the chemical composition of rocks involves the crushing and breakdown of rocks until they are in small enough pieces that decomposition by hot acids (hydrofluoric, nitric, hydrochloric, and perchloric acids) allows the elements present to enter into solution for analysis. Other techniques involve the high temperature fusion of powdered inorganic reagent (flux) and the rock. After melting the sample, techniques such as X-ray **fluorescence** spectrometry may be used to determine which elements are present.

Chemical and mechanical weathering break down rock through natural processes. Chemical weathering of rock requires **water** and air. The basic chemical reactions in the weathering process include solution (disrupted ionic bonds), hydration, **hydrolysis**, and **oxidation**.

The geochemistry involved in many environmental issues has become an increasing important aspect of scientific and political debate. The effects of **acid rain** are of great concern to geologists not only for the potential damage to the biosphere but also because acid rain accelerates the weathering process. Rain water is made acidic as it passes through the atmosphere. Although rain becomes naturally acidic as it contacts **nitrogen**, oxygen and **carbon dioxide** in the atmosphere, many industrial pollutants bring about reactions that bring the acidity of rainwater to dangerous levels. Increased levels of carbon dioxide from industrial **pollution** can increase the formation of carbonic acid. The rain also becomes more acidic. **Precipitation** of this ''acid rain'' adversely affects both geological and biological systems.

According to Plate Tectonic theory, the crust (lithosphere) of the Earth is divided into shifting plates. Geochemical analysis of the Earth's tectonic plates reveals a continental crust that is older, thicker and more granite-like than the younger, thinner oceanic crusts made of basaltic (iron, magnesium) materials.

See also Crystallography; Mineralogy

Gerhardt, Charles (1816-1856)

French chemist

Charles Gerhardt was born in 1816. When Charles was fifteen, he was sent to study science at the Karlsruhe Polytechnikum for two years, and later studied **chemistry** at the University of Leipzig. His father, a Swiss-Alsatian businessman, had wanted him to take over the family's white **iron** business, but Gerhardt left school to join the Army. After several years, a family friend arranged for his discharge and he began his studies more earnestly in 1836. After studying for two years with Leibig at Giessen, Leibig suggested he continue his studies in Paris with Dumas. Gerhardt became Dumas' lecture assistant and worked for his doctorate, which he received in 1841. For ten years, he worked as a professor of chemistry at Montpelier, but his unsociable behavior toward his co-workers lead to his dismissal. He died five years later at age forty.

As a researcher, he worked closely with **August Laurent** and the two built their reputations together. They were both interested in the same phenomena and issues even before they met. Their research was remarkably similar in organization and findings. There had been no interaction between them although they knew each other and were both former students of Dumas. After they started collaborating, it is difficult to determine who offered what in the collaborations. Most likely the brilliant insights were from Laurent, while the expression and subsequent experiments were from Gerhardt. Because they were both Republicans in an authoritarian state, it was difficult to have their ideas accepted. Furthermore, each had an abrasive personality and unnecessarily made enemies. Many French scientists argued against their ideas simply because it was Gerhardt or Laurent presenting them. However, those ideas were received favorably in England and outside of France. Politics became less of a problem after the Republicans won the Revolt of 1848.

Gerhardt developed a theory of residues, which was closely related to the controversial theory of types. His theory stated that in organic reactions, some organic material is left with the **water** and **carbon** dioxide as a by-product of the reaction. These residues will combine to form new organic material. His residues were noted to be similar to the types in Dumas's theory. However, unlike types, residues were not indivisible nor contained a charge.

One of Gerhardt's important discoveries was the Principle of Homology. This principle states that the properties of a series are similar and almost identical regardless of **molecular weight**. Although there were several flaws in his original expression, the discussion of homology lead the field closer to a more unified classification system between **physical chemistry** and **organic chemistry**.

Another step toward unifying the field was to resolve the differences in the calculations of molecular weights between physical chemistry and organic chemistry. He demonstrated how the formulas according to organic chemistry would produce a molecular weight twice as high as that of physical chemistry. Gerhardt clearly laid out the problems and the issues involved. However, Laurent established the definitions of molecular weight and atomic weights, vis a vis equivalent weight. With this explanation from Laurent, it was easier for Gerhardt to endorse the formula according to physical chemistry.

Germanium

Germanium is the third element in Group 14 of the **periodic table**, a group of elements sometimes known as the **carbon** family. Germanium has an **atomic number** of 32, an atomic **mass** of 72.59, and a chemical symbol of Ge.

Properties

Germanium looks like a metal, with a bright, shiny, silvery luster, but it is brittle and breaks apart rather easily. Since it has properties of both a metal and a **nonmetal**, it is generally classified as a **metalloid**. Germanium's melting point is 1,719°F (937.4°C), a **boiling point** of 5,126°F (2,830°C), and a **density** of 5.323 grams per cubic centimeter. Germanium is a relatively inactive element that does not react with **oxygen** at room **temperature**, but will react with hot acids and oxygen at high temperatures.

Occurrence and Extraction

The abundance of germanium in the Earth's crust is estimated to be about 7 parts per million, making it a relatively uncommon element. The element usually occurs in conjunction with ores of **zinc**, although the mineral germanite contains about 8% of the element. Important producers of the element are China, the United Kingdom, Ukraine, Russia, and Belgium.

Discovery and Naming

The existence of an element with atomic number 32 was predicted by the Russian chemist **Dmitry Mendeleev** in the 1860s. When Mendeleev drew the first periodic table of the elements, he found an empty place below **silicon** and above **tin** in the chart. He not only predicted that an element would be found to fill that spot, but also predicted the properties of that element. In 1885, the German chemist Clemens Alexander Winkler (1838-1904) discovered an element whose atomic number is 32 and whose properties match those predicted by Mendeleev. Winkler proposed the name germanium for the element in honor of his homeland, Germany

Uses

Nearly half of the germanium produced in the United States is used in the manufacture of optical fibers for communication systems. Some germanium is also used to make **semiconductors**, catalysts, and specialized **glass** for military weapon-sighting systems.

GHIORSO, ALBERT (1915-)

American nuclear chemist

Trained as an electrical engineer, Albert Ghiorso was drawn into nuclear physics through his work with the **Manhattan Project**, which built the first atomic bomb. This led to his participation over a period of thirty years in the discovery of twelve new elements—organizing and leading the effort for the last six—and in the development of the Heavy **Ion** Linear Accelerator and its successor, the Super HILAC.

Ghiorso was born July 15, 1915, in Vallejo, California, one of the seven children of John and Mary Ghiorso. His father was a riveter. Ghiorso attended the University of California at Berkeley, where he received his B.S. degree in electrical engineering in 1937. He then went to work for a small electronics firm called Cyclotron Specialties Company in Moraga, California, where he designed and constructed various kinds of radio equipment. While with Cyclotron Specialties, Ghiorso engineered and built the first commercial Geiger-Mueller counters for measuring radiation and sold them to the radiation laboratory at Berkeley.

When the United States entered World War II, Ghiorso decided to seek a commission in the U.S. Navy and contacted **Glenn T. Seaborg**, whom he knew from the radiation laboratory at Berkeley, for a reference. Instead, Seaborg invited Ghiorso to join him at the wartime metallurgical laboratory at the University of Chicago. Only after Seaborg assembled his team could he reveal what their project was: to perform nuclear and chemical research on **plutonium**, as part of the Manhattan Project. Seaborg had discovered the new element in 1940, but the discovery had been kept secret. Furthermore, although plutonium had been detected by highly sensitive tracers, it had not yet actually been seen, and nothing was yet known about its properties. Drawing on his electrical engineering background, Ghiorso helped develop the methods and intricate instrumentation needed to separate plutonium from **uranium** and fission products. In the process, he learned nuclear physics and **nuclear chemistry**.

In 1944, Seaborg decided to extend his work to a search for elements with a higher **atomic number** than plutonium. (Uranium, with an atomic number of 92, is the last naturally occurring element; elements from atomic number 93 and higher are synthetically produced and are called transuranium elements.) To do this, Seaborg chose two chemists, Ralph A. James and Leon O. Morgan, to master the difficult chemical separation techniques, and he asked Ghiorso to develop techniques to measure the alpha-particle **energy** needed to identify the elements. First, outside labs bombarded plutonium in cyclotrons and **neutron** reactors; the samples were then flown to Chicago, where the group examined the fractions for unusual alpha-particle **radioactivity**, using the measuring counter Ghiorso had developed called a mica-absorber. After a series of unsuccessful experiments, in July, 1944, they found a new **alpha particle** in a sample of irradiated plutonium. They had discovered the first transplutonium element —curium, element 96. They followed this with the discovery of element 95, **americium**, in October, 1944.

In 1946, Seaborg returned to the University of California at Berkeley, and Ghiorso—along with other members of the Chicago team—came with him. This team established and equipped a very strong nuclear chemical division at the radiation laboratory. Meanwhile, **nuclear reactors** were producing large enough quantities of americium and **curium** to serve as target materials for the production of other transuranic elements. These target elements were extremely radioactive; any new elements had to be rapidly and very efficiently separated from them. This was a complicated technical problem, and to overcome it Ghiorso developed a forty-eight-channel pulse-height analyzer, along with a gridded ion chamber to detect alpha recoils. In December, 1949, he participated in the discovery of element 97, **berkelium**, and in February, 1950, he was part of the group that discovered element 98, **californium**; the separations for this last element had to be made even more quickly, as it had a **half-life** of only forty-five minutes.

In a speech at the Robert A. Welch Foundation Conference in October, 1990, Ghiorso recalled the discovery of the next two transuranic elements as the most dramatic in the series: "It was so unexpected, absolutely unexpected and out of the blue; within four days... we had discovered element 99... and a few months later element 100.... [T]o leapfrog from uranium to **fermium** was pretty spectacular." The drama began in November, 1952, with the first large-scale thermonuclear explosion, which was set off in the South Pacific on Eniwetok Atoll. Fallout from the explosion revealed that the intense neutron flux had instantaneously created previously unknown heavy isotopes. Ghiorso calculated that it might be possible to find traces of neutron-heavy elements up to atomic number 100 in this nuclear debris. He convinced his colleagues at Berkeley to acquire and examine a filter **paper** from airplanes that had flown through the radiation cloud. Within a few days, he and Stanley G. Thompson had found element 99, **einsteinium**, using Ghiorso's forty-eight-channel analyzer. In January, 1953, Ghiorso and his colleagues chemically isolated element 100, fermium, from larger samples of fallout collected from the bombed island's coral. Scientists at the Argonne National Laboratory and the Los Alamos Scientific Laboratory also contributed to the discovery of elements 99 and 100.

New techniques were needed for the discovery of the next transuranic elements, because the half-lives of these were so short that the standard method of chemically separating transmuted atoms from the target material was ineffective. Ghiorso solved the problem by harnessing the recoil he had previously observed in cyclotrons and applying it to a very thin layer of target material. As the target material (in this case, einsteinium) was struck in the cyclotron by an ion beam, the newly transmuted atoms recoiled onto a catcher foil. These recoil atoms could then be recovered from the catcher foil rather than from the entire highly radioactive target. The mechanics of the process were unusual: the bombardment was done at the Berkeley cyclotron, and the extraction from the foil was performed in Ghiorso's Volkswagen as he sped up the hill to the laboratory, where the final chemistry and pulse analysis were performed. This method produced single atoms of element 101, mendelevium, discovered by Ghiorso and his colleagues in February, 1955.

Discovery of the next elements required the acceleration of heavy ions, for which improvements to the existing linear accelerator were necessary. With his colleagues, Ghiorso designed the Heavy Ion Linear Accelerator (HILAC), which was in operation by October, 1957. The HILAC was the first accelerator ever to use magnetic strong focusing. Ghiorso and his fellow scientists combined the use of the HILAC with the devices they had developed to measure the alpha energies they were discovering. Although an international group working at the Nobel Institute in Sweden and a Soviet group both reported finding what they thought to be element 102 (nobelium), Ghiorso and his colleagues made the definitive discovery in 1958 using a double-recoil technique. They followed this with the identification of element 103 (lawrencium) in 1961, element 104 (rutherfordium) in 1969, element 105 (hahnium) in 1970, and element 106 in 1974. (A dispute by Russian experimenters about the discovery of element 106 delayed its naming until March, 1994.) In the course of the search for the transuranic elements, Ghiorso and his coworkers in 1964 invented a highly advanced double-ring synchrotron called the Omnitron that could accelerate all elements up to uranium to high energies. The Vietnam War blocked funding for the Omnitron, so Ghiorso instead improved the HILAC into the Super HILAC, capable of accelerating heavy ions to high energies. This was completed in 1971. Ghiorso then invented the Bevelac, which was able to accelerate heavy ions up to velocities close to the speed of light. The Bevelac became a valuable tool for research in high-energy physics, nuclear chemistry, biology, and medicine.

Ghiorso also was a direct participant and, in many instances, research leader in the discovery of approximately 150 isotopes, including many of the known isotopes in the heavy element range. He has also been deeply involved in attempts to find the hypothetical "island of stability" of superheavy elements, also called super transuranics; these are elements, as yet undiscovered, with atomic numbers around 114, and scientists believe they may have greater nuclear stability because of longer half-lives. To aid in this search, Ghiorso built a recoil separator, the Small Angle Separating System (SASSY) and its second-generation version, SASSY2, which provided a very effective means of looking for superheavy elements with extremely short half-lives—the half-lives of some of these elements are as short as a microsecond. After the discovery in the 1990s of two new isotopes of element 106 by a Russian team working with American counterparts, Ghiorso and his Berkeley colleagues began working on the possibility of creating superheavy elements of atomic numbers 110 through 114 from these isotopes.

Ghiorso was a senior scientist at the Lawrence Berkeley Laboratory from 1946 until 1982, and he was named director of Berkeley's HILAC in 1969. Although he officially retired in 1982, he has continued to work at the Berkeley lab as senior scientist emeritus. He received an honorary doctor of science degree from Gustavus Adolphus College in 1966, and in 1973 he received the American Chemical Society's Award for Nuclear Applications in Chemistry. He is a fellow of the American Physical Society and a member of the American Academy of Arts and Sciences, and has written numerous articles about his findings for scientific journals. He married Wilma Belt in 1942; they have a daughter and a son, who has worked with Ghiorso at the Lawrence Berkeley Laboratory as a technician.

William F. Giauque.

Giauque, William F. (1895-1982)

American chemist

William F. Giauque is best known for his research in two areas, thermodynamic studies at very low temperatures and the isotopic composition of **oxygen**. In order to carry out the first of these investigations, Giauque found it necessary to develop a new technique of cooling **gases** to temperatures close to absolute zero, a method known as adiabatic demagnetization. By using this system, Giauque was able to obtain temperatures within a few thousandths of a degree from absolute zero. For his research, Giauque was awarded the 1949 Nobel Prize in chemistry. His findings led to improvements in the production of **gasoline**, steel, rubber, and **glass**. His discovery of the previously unknown oxygen isotopes 17 and 18 resulted in the recalibration of atomic weight scales.

William Francis Giauque was born in Niagara Falls, Ontario, Canada, on May 12, 1895. His mother, Isabella Jane (Duncan) Giauque, and his father, William Tecumseh Sherman Giauque, were United States citizens. When Giauque was young, his family moved back to the United States where his father took a job with the Michigan Central Railroad as a station master. When the elder Giauque died in 1908, his family returned to Canada, where William enrolled in the Niagara Falls Collegiate and Vocational School.

At first, it seemed that Giauque would have to assume the responsibility of supporting his family. After completing his two-year course at the vocational school, he found a job at the Hooker Electro-Chemical Company in Niagara Falls, New York. But his mother was determined to convince him to pursue a college education, and her employer, chemist J. W. Beckman, influenced his decision to enroll at the University of California at Berkeley in 1916. Giauque subsequently received his bachelor of science in chemistry in 1920 and then entered graduate school at Berkeley. There he was exposed to some of the nation's greatest chemists, including **Gilbert Newton Lewis**, George E. Gibson (his thesis advisor), Joel Hildebrand, and G. E. K. Branch. Giauque's dissertation concerned the properties and behavior of materials at very low temperatures, a subject which remained the focus of his academic career.

The focus of Giauque's earliest research was the Third Law of **Thermodynamics**, first proposed by German physical chemist **Walther Nernst** in 1906. Nernst hypothesized that the **entropy** (the measure of randomness in a closed system) of a pure crystalline substance at absolute zero is zero. In order to test this hypothesis, it was necessary to find ways of cooling materials to temperatures as close to absolute zero as possible.

The "father" of low-temperature research of this kind had been the Dutch physicist **Heike Kamerlingh-Onnes**. During the first decade of the twentieth century, Kamerlingh-Onnes had developed methods for cooling gases by causing them to evaporate under reduced pressure. Using this method, he had obtained temperatures of about one degree kelvin (1 K). While other researchers had produced even lower temperatures to within a few tenths of a degree kelvin over the next two decades, Kamerlingh-Onnes' method seemed to reach its limit by the early 1920s.

In 1924 Giauque suggested a new technique for producing cooling; **Peter Debye** proposed the same procedure independently at about the same time. In this method, a weakly magnetic material is magnetized in such a way that all of its molecules line up in exactly the same direction. This material is then surrounded with liquid **helium** and demagnetized. During demagnetization, the material's molecules go from a highly ordered to a highly disordered state. The **energy** needed to bring about this change is removed from the liquid helium, causing its temperature to fall even further.

This method was beset with a number of practical problems, including the difficulty of measuring the very low temperatures produced. Giauque devoted nearly twenty years to refining the technique and obtaining results with it. Finally, in 1938, he was able to report having reduced the temperature of a sample of liquid helium to 0.004 K, by far the lowest temperature yet observed.

During the 1920s, Giauque was also working on the study of entropy in gases. As part of this work, he regularly examined the band spectra produced by such gases. In 1928, while working on this problem, Giauque made an unexpected find. He observed that the spectrum of normal atmospheric oxygen consists of one set of strong lines and two sets of very weak, barely visible lines. The conclusion he drew was that normal oxygen must consist of three isotopes, one (oxygen–16) that is abundant and two (oxygen–17 and oxygen–18) that are relatively rare.

Walter Gilbert. *(AP/Wide World. Reproduced by permission.)*

Giauque's professional reputation continued to grow at Berkeley and throughout the world. He was promoted to assistant professor in 1927, associate professor in 1930, and full professor in 1934. At his retirement in 1962, he was named emeritus professor of chemistry. In addition to his 1949 Nobel Prize in chemistry, he was also recognized by a number of other awards, including the Chandler Foundation Medal in 1936, the Elliot Cresson Medal in 1937, and the American Chemical Society's Gibbs and Lewis Medals in 1951 and 1955, respectively.

Giauque was married to the former Muriel Frances Ashley, a physicist, on July 19, 1932. They had two sons, William Francis Ashley and Robert David Ashley. Giauque died at his home in Berkeley on March 28, 1982 as a result of a fall.

GILBERT, WALTER (1932-)

American molecular biologist

Walter Gilbert is a molecular biologist who shared the 1980 Nobel Prize in **chemistry** for his discovery of how to sequence, or chemically describe, deoxyribonucleic acid (DNA) molecules. Gilbert also identified repressor molecules, which modify or repress the activity of certain genes, and collaborated with Noble laureate biologist **James Watson**'s in his efforts to isolate messenger ribonucleic acid (RNA). Later in his career, Gilbert helped form and was chief executive officer of the biotechnology firm Biogen, and became a moving force in the medical research project known as the **human genome project**.

Gilbert was born in Cambridge, Massachusetts on March 21, 1932. His father, Richard V. Gilbert, was an economist at Harvard University, and his mother, Emma Cohen Gilbert, was a child psychologist who provided her children's early education at home. In 1939 the family moved to Washington, D.C., where Gilbert initially performed poorly in school. He did, however, show a great deal of interest in science. He was fascinated by astronomy, and at age twelve he ground his own **glass** for telescopes; he also "nearly blew himself up brewing **hydrogen** in the pantry," writes Anthony Liversidge in *Omni*. As a senior at Sidwell Friends High School, he would go to the Library of Congress to expand his knowledge of nuclear physics.

In 1949, Gilbert entered Harvard University, where he majored in chemistry and physics, earning his B.A. *summa cum laude* in 1953. Gilbert remained at Harvard for a master's degree in physics, which he received in 1954. He went to Cambridge University for doctoral work in theoretical physics, studying under the physicist **Abdus Salam**. His doctoral dissertation focused on mathematical formulae that could predict the behavior of elementary particles in so-called "scattering" experiments. He received his Ph.D. in mathematics in 1957. Gilbert returned to Harvard as a National Science postdoctoral fellow in physics, and he gained an appointment as assistant professor in 1959.

While at Cambridge, Gilbert had met biologists **James Watson** and **Francis Crick**; just a few years before, these two men had established the structure of **DNA** and constructed a three-dimensional model of it. Their work had launched a new field of science called molecular biology, and when Watson moved to Harvard in 1960 he and Gilbert met again. Watson discussed with Gilbert his interest in isolating messenger **RNA**. This is the substance believed to be responsible for transmitting information from DNA to ribosomes, which are the cellular structures in which **protein synthesis** takes place. At Watson's invitation, Gilbert joined him and his colleagues to work on this project. This collaboration with Watson began Gilbert's move into molecular biology. He became convinced that his future lay in this field, and he made up for a lack of formal training in **biochemistry** through hard work. Within a few years he was publishing articles on molecular biology. He was made a tenured associate professor of biophysics at Harvard in 1964, and he became a full professor of biochemistry in 1968. In 1972, Harvard named him the American Cancer Society Professor of Molecular Biology.

In the middle of the 1960s, Gilbert began research on how genes are activated within cells. This was a problem that had been introduced by the French geneticists **François Jacob** and **Jacques Lucien Monod**; DNA should, in theory, encode **proteins** continually, and they wondered what kept the genes from being activated. If cells contain some sort of element that in effect represses some genes, this would explain in part how cells perform different functions even though each cell contains the same complement of genetic instructions. Gilbert set out to determine whether actual "repressor" substances exist within each cell.

Working with the *Escherichia coli* bacterium, Gilbert attempted to find what he called the lac repressor. *E. coli* manufactures the **enzyme** betagalactosidase when the milk sugar lactose is present, and he hypothesized that the gene responsible for producing the enzyme was repressed by a substance that would only detach itself from the DNA **molecule** in the presence of lactose. If this hypothesis could be proven, it would confirm the existence of repressors. In 1966, Gilbert and his colleagues added radioactive lactose-like molecules to a **concentration** of the bacteria as a means of tracing any potential lac repressor activity. As he had hoped, the lac repressor bonded with the radioactive material. By 1970, he was able to determine the precise region of DNA (called the lac operator) to which the repressor bonds in the absence of lactose.

The next phase of Gilbert's research focused on sequencing DNA. His aim was to identify and describe chemically any strand of DNA. Working with graduate student Allan Maxam, he began to sequence parts of the DNA strand. It was known that the molecules could be "broken" at specific chemical junctures by using certain chemical substances, and a colleague introduced Gilbert and Maxam to a procedure that broke DNA molecules into fragments that were easier to describe. After breaking radioactively labeled DNA into fragments, Gilbert and Maxam worked to separate them. They used a technique known as **gel electrophoresis**, in which an electric current causes the fragments to pass through a gel substance. Upon exposure to X-ray film, the fragments can be read and the chemical code of DNA can be identified. Working independently, the British scientist **Frederick Sanger** developed a similar sequencing technique. In recognition of both their contributions, Gilbert and Sanger were awarded half the 1980 Nobel Prize in chemistry. The other half went to the American biochemist **Paul Berg** for his work with recombinant DNA, more commonly referred to as "gene splicing."

With the breakthroughs that were being made by Gilbert, Sanger, Berg, and others, the concept of applying technological principles to biology came of age in the 1970s. The idea of being able to alter the genetic composition of a cell, for example, opened up such possibilities as curing or even eradicating many diseases. The possibilities in biotechnology intrigued not only scientists but the business community. Here was a concept that could be both revolutionary and lucrative—a company that held the patent to a definitive cure for cancer, for example, could become quite wealthy and powerful. Business leaders began to approach scientists, Gilbert among them. Most scientists were skeptical at first, but in 1978 Gilbert met with a group of venture capitalists who wanted to start a biotechnology firm. After receiving assurances that they would have considerable control over research and development, he and other scientists formed Biogen N.V., with Gilbert as the chairman of the scientific board of directors. Gilbert was so convinced of Biogen's potential that he left Harvard in 1981 to become the company's chief executive officer.

Despite widespread belief in the company's potential, Biogen had some difficult years in the beginning. After four years, it was still unprofitable and Gilbert had become increasingly disillusioned with business. He found the differences between the business world and the scientific community difficult to reconcile. The science of creating new products is vastly different from the business of bringing them to market. Scientists need to be patient because their breakthroughs might take years to obtain, but sound business practice dictates cutting one's losses when a project fails to produce after a reasonable time, and these differences led to conflicts with others at Biogen. Gilbert also found it time-consuming and expensive to run a company (although he personally profited from the venture), and in late 1984 he resigned his position at Biogen, while maintaining some involvement with the firm. In 1985, he was named H. H. Timken Professor of Science in Harvard's cellular and developmental biology department. In 1987 he became chairman of the department and Carl M. Loeb University Professor.

Gilbert resumed his research, unhindered by the responsibilities of running a business. But he soon became interested in an undertaking that was bigger than Biogen: the human genome project. This project, wrote Robert Kanigel in the *New York Times Magazine,* "would reveal the precise biochemical makeup of the entire genetic material, or genome, of a human being.... It would grant insight into human biology previously held only by God." The plan was to create a map or library of human DNA. Such information would help researchers not only find cures for diseases but also identify potentially harmful gene mutations. Gilbert spoke out enthusiastically in favor of the genome project, and along with many others he encouraged Congress to support it with federal funds. Frustrated by the political process and believing that the project would be damaged by the bureaucracy federal participation would impose, he tried a different approach. In 1987 he announced plans to create his own company, which would sequence DNA, copyright the information, and sell it. Although he failed to get adequate backing for that project, he did win a two million dollar annual grant from the U.S. government to conduct his work at Harvard under the auspices of the National Institutes of Health.

In an interview with *Omni,* Gilbert has explained what he sees as some of the benefits of a complete genetic map of the human being: "The differences between people are what the genetic map [is] about. That knowledge will yield medicine tailored to the individual. One will first identify obvious genetic defects like cystic fibrosis. The next round of genetic mapping will show us clusters of genes for common diseases from arthritis to schizophrenia. We will be able to predict the side effects of those drugs and tailor the right dose to each person." Gilbert has estimated that the project would take at least ten to twenty years to complete.

Gilbert has also expressed concern as a researcher, arguing in favor of what he calls a "paradigm shift in biology." As scientific techniques are perfected, he contends, new scientists should be able to concentrate on new research, not repeating old research. Writing in *Nature,* he has noted that "in 1970, each of my graduate students had to make restriction enzymes in order to work with DNA molecules; by 1976 the enzymes were all purchased and today no graduate student knows how to make them." While it is important for scientists to understand what they are doing and why, he continues, "this is not the meaning of their education. Their doctorates should be testimonials that they [have] solved a novel problem, and in so doing [have] learned the general ability to find whatever new or old techniques were needed." As more and more of the technological problems of molecular biology are solved, he believes that biological research will begin with theoretical rather than experimental work.

In addition to the Nobel Prize, Gilbert shared with Sanger the Albert Lasker Basic Medical Research Award in 1979. He also won the Louisa Horwitz Gross Prize from Columbia University in 1979, and the Herbert A. Sober Memorial Award of the American Society of Biological Chemists in 1980. His memberships include the American Academy of Arts and Sciences, the National Academy of Sciences, and the British Royal Society.

Gilbert has been married to Celia Stone since 1953; the couple have a son and a daughter. Although his career has sometimes seemed almost as complex as the substances he studies, he is widely respected as a scientist. Younger scientists have called him "intimidating," but despite his successes he remains a committed teacher and researcher.

Glass

A glass is a substance that is non-crystalline yet almost completely undeformable. This state is called the vitreous state. Vitreous substances, when heated, will transform slowly through stages of decreasing **viscosity**. As a sample of glass is heated, it becomes more and more deformable, eventually reaching a point where it resembles a very viscous liquid. Ice, on the other hand, does not go through these changes as it is heated. It changes directly from a solid to a liquid. Ice, therefore, is not a vitreous substance. Glasses are only very slightly deformable. Glasses tend to bend and elongate under their own weight, especially when formed into rods, plates, or sheets. Glasses can be either organic or inorganic materials.

Many definitions of glass include the idea that they do not crystallize as they solidify. Since the definition of solidification is the act of **crystallization**, this idea of glass as a non-crystalline solid may not be entirely correct. Crystallization is when the molecules of a substance arrange themselves in a systematic, periodic fashion. The atoms or molecules of glass do not exhibit this **periodicity**. This has led to the idea of glass as an extremely viscous, or "supercooled"' liquid.

Glass is often referred to as an amorphous solid. An amorphous solid has a definite shape without the geometric regularity of crystalline **solids**. Glass can be molded into any shape. If glass is shattered, the resulting pieces are irregularly shaped. A crystalline solid would exhibit regular geometrical shapes when shattered. Amorphous solids tend to hold their shape, but they also tend to flow very slowly. If left undisturbed for a long period of time, a glass will very slowly crystallize. Once it crystallizes, it is no longer considered to be glass. At this point, it has devitrified. This crystallization process is extremely slow and in many cases may never occur.

The chemical make-up of standard window glass, which will be described in greater detail below, is quite similar to the mineral quartz. An x-ray crystallographic picture of quartz would show atoms arranged in an orderly, periodic sequence. X-ray **crystallography** studies of glass show no such arrangement. The atoms in glass are disordered and show no periodic structure whatsoever. This irregular arrangement of atoms not only defines a substance as glass, but also determines several of its properties.

The bonds between the molecules or ions in a glass are of varying length, which is why they show no symmetry or periodic structure. Because the bonds are not symmetrical, glass is isotropic and has no definite melting point. The melting of glass instead takes place over a wide **temperature** range. Changing the state of a substance with asymmetric bonds requires more **energy** than a crystalline structure would. The tendency of glass to devitrify is a result of the atoms moving from a higher to a lower energy state.

The most common glasses are **silicon** based. Most glasses are 75% silicate. These glasses are based on the SiO_2 **molecule**. This molecule creates an asymmetric, aperiodic structure. Some of the **oxygen** atoms are not bridged together, creating ions which need to be neutralized by metal cations. These metal cations are randomly scattered throughout the glass structure, adding to the asymmetry. The oxides of elements other than silicon can also form glasses. These other oxides include Al_2O_3, B_2O_3, P_2O_5, and As_2O_5.

The production of glasses is a complicated process. In general, certain molten materials are cooled in a specific manner so that no crystallization occurs, i.e., they remain amorphous. There are four basic materials that are used in glass production. These materials are the glass-forming substances, fluxes, stabilizers, and secondary components.

A glass-forming substance is any mineral which remains vitreous when cooled. Glass-forming substances are usually silica, boric **oxide**, phosphorous pentoxide, or feldspars. Sometimes Al_2O_3 is used. As mentioned earlier, silica is the most commonly used material to make glass. This silica is obtained from sand, which is 99.1-99.7% SiO_2. Occasionally, natural silica deposits are discovered that are pure enough to use in glass manufacturing, but these deposits are rare and the silica found in them is usually expensive to obtain. Even the lowest quality sands can be purified rather economically. Impurities in the natural silica are important because they can dramatically alter the quality of the glass produced. The most common impurities found in natural silica are **iron** sesquioxide (Fe_2O_3), alumina (Al_2O_3), and **calcium** compounds. Ferric oxide is sometimes found as an impurity. Even if the amount of ferric oxide in a natural silica sample is only 0.1% of the sample, the glass produced would have a deep yellow-green **color** and the impurity would have detrimental effects on the thermal and mechanical properties of the glass.

Occasionally impurities are added to the glass-forming substances to give the glass certain qualities. For example, phosphorous pentoxide, which can be added as **phosphoric acid**, calcium **phosphate**, or bone ash, will increase the transparency of the glass produced. One drawback of using this compound is that it reduces the mechanical properties of the glass. Feldspars are added as orthoclase in order to reduce the devitrification of the glass. When alumina is added to the glass-forming substances SiO_2 or B_2O_3 it can stabilize the glass by neutralizing the natural alkalinity of glass.

Fluxes are used in the production of glass to increase the fusibility of the glass as well as to increase the workable temperature range of the glass. These can be feldspars or alkali metal oxides, in particular, **sodium** oxide and **potassium** oxide. Stabilizers are used to protect the glass. They reduce the **solubility** of glass in **water** and in other chemical agents, especially those found in the atmosphere. Stabilizers also are used to give the finished product particular characteristics. For example, **calcium carbonate** can be added as a stabilizer which will make the glass produced insoluble in water. **Lead** oxide added as a stabilizer gives the glass extreme transparency, brightness, and a high refractive index. Lead oxide also makes glass easier to cut. **Zinc** oxide can be added to glass to make it more resistant to changes in temperature as well as to increase its refractive index. **Aluminum** oxide can also be added as a stabilizer to increase the physical strength of the glass. Secondary components are added to determine some of the final properties of the glass and also to correct any defects in the glass. The secondary components can be classified as decolorants, opacifiers, colorants, or refiners.

The production of glass includes many steps that can be generalized as follows. First, the fluxes, glass-forming substances, and stabilizers are crushed and milled, then blended and mixed together. They are then re-milled and granulated. At this point, the secondary components are added, if needed. The granules are then fused, refined, homogenized, and corrected, using more secondary components if necessary. Finally, the glass is formed and finished.

The final product is one of many different types of glass. There are hundreds of different kinds of glass. The most common glasses are composed of SiO_2, CaO, and Na_2O. Sometimes Al_2O_3 or Fe_2O_3 is added, depending on the type of glass being produced. These glasses are used for green bottles, colorless bottles and jars, mirror glasses, lamp glasses, semicrystal glasses, and sheet glasses.

One popular type of glass, especially in laboratory settings or for use in the kitchen, is borosilicate glass. Some well-known borosilicate glasses are Jena, Pyrex, Durax, and Thermoglass. These glasses contain 12% or more B_2O_3. The addition of the **boron** oxide increases the softening temperature of the glass, making it more resistant to high temperatures such as those experienced while cooking or while performing laboratory experiments. Borosilicate glasses are also used in the production of **thermometers**, television tubes, and other objects that need to have constant dimensions or a high softening point.

GLYCEROL

Glycerol is the common name of the organic compound whose chemical structure is $HOCH_2$-CHOH- CH_2OH. Propane-1,2,3-

triol or glycerin (USP), as it is also called, consists of a chain of three **carbon** atoms with each of the end carbon atoms bonded to two **hydrogen** atoms (C-H) and a hydroxyl group (-OH) and the central carbon **atom** is bonded to a hydrogen atom (C-H) and a hydroxyl group (-OH). Glycerol is a trihydric **alcohol** because it contains three hydroxyl or alcohol groups. Glycerin is a thick liquid with a sweet **taste** that is found in fats and oils and is the primary triglyceride found in coconut and olive oil. It was discovered in 1779, when the Swedish chemist **Carl Wilhelm Scheele** washed glycerol out of a heated a mixture of **lead oxide** (PbO) and olive oil. Today, it is obtained as a byproduct from the manufacture of soaps.

One important property of glycerol or glycerin is that is not poisonous to humans. Therefore it is used in foods, syrups, ointments, medicines, and **cosmetics**. Glycerol also has special chemical properties that allow it to be used where oil would fail. Glycerol is a thick syrup that is used as the "body" to many syrups, for example, cough medicines and lotions used to treat ear infections. It is also an additive in vanilla extracts and other food flavorings. Glycerin is added to ice cream to improve the texture, and its sweet taste decreases the amount of sugar needed. The base used in making toothpaste contains glycerin to maintain smoothness and shine. The cosmetic industry employs glycerin in skin conditioning lotions to replace lost skin moisture, relieve chapping, and keep skin soft. It is also added in hair shampoos to make them flow easily when poured from the bottle. The raisins found in cereals remain soft because they have been soaked in glycerol. Meat casings and food wrapping papers use glycerin to give them flexibility without brittleness. Similarly, tobacco is treated with this thick chemical to prevent the leaves from becoming brittle and crumbling during drying. It also adds sweetness to chewing tobacco. Glycerol is added during the manufacture of soaps in order to prepare shiny transparent bars. The trihydric alcohol structure of glycerin makes it a useful chemical in the manufacture of various hard foams, like those that are placed under siding in buildings and around dish washers and refrigerators for insulation and sound proofing. Analogously, the chemical structure of glycerol makes it an excellent catalyst in the microbiological production of vinegar from alcohol.

In the manufacture of foods, drugs, and cosmetics, oil cannot be employed as a **lubricant** because it might come in contact with the products and contaminate them. Therefore, the nontoxic glycerol is used to reduce friction in pumps and bearings. **Gasoline** and other **hydrocarbon** chemicals dissolve oil-based greases, so glycerin is used in pumps for transferring these fluids. Glycerol is also applied to cork gaskets to keep them flexible and tough when exposed to oils and greases as in automobile engines. Glycerin is used as a lubricant in various operations in the textile industry, and can be mixed with sugar to make a nondrying oil. Glycerol does not turn into a solid until it is cooled to very cold **temperature**. This property is utilized to increase the storage life of blood. When small amounts of glycerin are added to red blood cells, they can be frozen and maintained for up to three years.

Chemical derivatives of glycerol or propane-1,2,3-triol are important in a wide range of applications. **Nitroglycerin** is the trinitrate derivative of glycerol. One application of this chemical is as the key ingredient in the manufacture of dynamite **explosives**. Nitroglycerin can also be used in conjunction with gun cotton or nitrocellulose as a propellant in military applications. In the pharmaceutical industry, nitroglycerin is considered a drug to relieve chest pains and in the treatment of various heart ailments. Another derivative, guaiacol glyceryl **ether**, is an ingredient in cough medicines, and glycerol methacrylate is used in the manufacture of soft contact lenses to make them permeable to air. Glycerol esters are utilized in cakes, breads and other bakery products as lubricants and softening agents. They also have similar applications in the making of candies, butter and whipped toppings. A specially designed glycerol ester called *caprenin* can be used as a low calorie replacement for cocoa butter.

The acetins are derivatives of glycerol that are prepared by heating glycerol with **acetic acid**. Monoacetin is used in the manufacture of dynamite, in tanning leather, and as a solvent for various dyes. Diacetin, another derivative of glycerol, is used as a solvent and a softening agent. Triacetin, the most useful of the acetins, is used in the manufacture of cigarette filters and as a component in solid rocket fuels. It is also used as a solvent in the production of photographic films, and has some utility in the perfume industry. Triacetin is added to dried egg whites so that they can be whipped into meringues.

GLYCOGEN

Like fat, glycogen is a source of stored **energy** in the body. However, glycogen is composed of glucose rather that fat. Glucose, a simple sugar **molecule**, is the main fuel for many of the body's energy needs. Like other large molecules built from smaller sugar molecules, glycogen is a carbohydrate.

Glycogen occurs mostly in the liver and in muscles. The glycogen in the liver serves energy needs throughout the body. It is particularly important for the brain that requires glucose for proper function but cannot store the fuel itself.

Whenever there is more glucose than is required for the body's immediate needs—such as after a meal—glycogen synthesis is increased. The synthesis is aided by the **enzyme**, glycogen synthase. (Enzymes are protein molecules that trigger speedy chemical reactions.)

Glycogen synthase exists in inactive and active forms. The inactive form becomes active when it is switched on by insulin. Insulin is a hormone that is secreted by the pancreas in response to high glucose levels in the blood. Blood glucose levels normally rise following meals.

Excess glucose molecules are first tagged with another molecule, UDP or uridine diphosphate, to form UDP-glucose. The UDP serves as a signal to the activated synthase that the glucose may be stored for future use. One by one, the glucose molecules are added to a growing glycogen molecule like links in a chain.

As the body uses up available glucose, it draws on glycogen for more. Glycogen is broken down during glycogenolysis, the process which releases glucose. Glycogenolysis

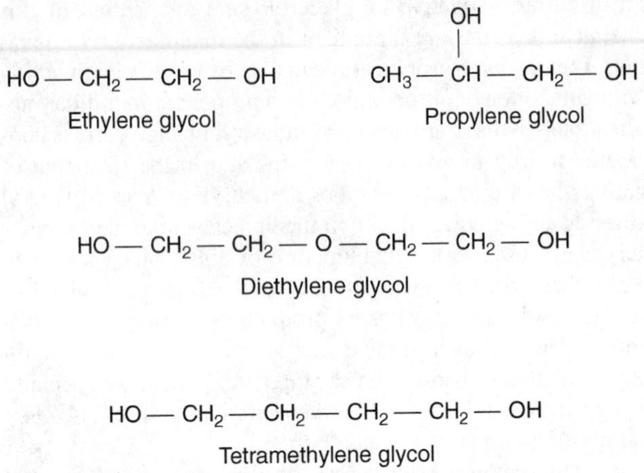

HO— CH_2— CH_2— OH
Ethylene glycol

CH_3— CH(OH)— CH_2— OH
Propylene glycol

HO— CH_2— CH_2— O— CH_2— CH_2— OH
Diethylene glycol

HO— CH_2— CH_2— CH_2— CH_2— OH
Tetramethylene glycol

Figure 1. Structures of common glycols. ***(Illustration by Electronic Illustrators Group.)***

$$>C{=}C< \xrightarrow[H_2O]{KMnO_4} -\underset{OH}{C}-\underset{OH}{C}-$$

An alkene A glycol

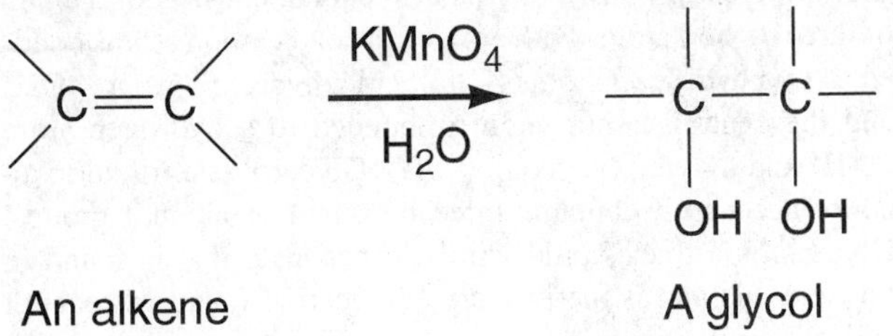

Figure 2. Laboratory preparation of a glycol. ***(Illustration by Electronic Illustrators Group.)***

requires the enzyme, glycogen phosphorylase. Like the synthase, the phosphorylase exists in active and inactive forms and requires a hormone to be activated. The necessary hormone is glucagon, which is secreted by the pancreas when blood glucose levels get too low.

GLYCOLS

A glycol is an aliphatic organic compound in which two hydroxyl (OH) groups are present. The most important glycols are those in which the hydroxyl groups are attached to adjacent **carbon** atoms, and the term glycol is often interpreted as applying only to such compounds. The latter are also called vicinal diols, or 1,2-diols. Compounds in which two hydroxyl groups are attached to the same carbon **atom** (geminal diols) normally cannot be isolated.

The most useful glycol is ethylene glycol (IUPAC name: 1,2-ethanediol). Other industrially important glycols include propylene glycol (IUPAC name: 1,2-propanediol), diethylene glycol (IUPAC name: 3-oxa-1,5-pentanediol) and tetramethylene glycol (IUPAC name: 1,4-butanediol).

The common glycols are colorless **liquids** with specific gravities greater than that of **water**. The presence of two hydroxyl groups permits the formation of **hydrogen** with water, thereby favoring **miscibility** with the latter. Each of the glycols shown above is completely miscible with water. Intermolecular hydrogen **bonding** between glycol molecules gives these compounds boiling points which are higher than might otherwise have been expected; for example, ethylene glycol has a **boiling point** of 388.5°F (198°C).

The most convenient and inexpensive method of preparing a glycol in the laboratory is to react an alkene with cold dilute **potassium** permanganate, $KMnO_4$.

Yields from this reaction are often poor and better yields are obtained using **osmium** tetroxide, OsO_4. While O_sO_4 is both expensive and toxic, powerful synthetic methods have been developed (notably by Barry Sharpless) that use catalytic quantities of osmium oxidants, often with excellent stereochemical control with respect to the 1,2-diol product (the Sharpless asymmetric dihydroxylation reaction). For this reason, osmatebased oxidants have nearly completely displaced permanganate oxidants, such as KM_nO_4, for the laboratory preparation of glycols.

In the industrial preparation of ethylene glycol, ethylene (IUPAC name: ethene) is oxidized to ethylene **oxide** (IUPAC name: oxirane) using **oxygen** and a **silver** catalyst. Ethylene oxide is then reacted with water at high **temperature** or in the presence of an acid catalyst to produce ethylene glycol. Diethylene glycol is a useful by-product of this process.

Alternative methods of preparing ethylene glycol that avoid the use of toxic ethylene oxide are currently being investigated.

In 1995, the production of ethylene glycol totaled 6.2 billion lb (2.8×10^9 kg), with a steady annual growth rate of 5-6%. Much of this ethylene glycol is used as antifreeze in automobile radiators. The addition of ethylene glycol to water causes the freezing point of the latter to decrease, thus the damage that would be caused by the water freezing in a radiator can be avoided by using a mixture of water and ethylene glycol as the coolant. An added advantage of using such a mixture is that its boiling point is higher than that of water, which reduces the possibility of boil-over during summer driving. In addition to ethylene glycol, commercial antifreeze contains several additives, including a dye to reduce the likelihood of the highly toxic ethylene glycol being accidentally ingested. Concern over the toxicity of ethylene glycol—the lethal dose of ethylene glycol for humans is 1.4 ml/kg—resulted in the introduction, in 1993, of antifreeze based on the less toxic propylene glycol.

The second major use of ethylene glycol is in the production of poly(ethylene terephthalate), or PET. This polymer, a **polyester**, is obtained by reacting ethylene glycol with terephthalic acid (IUPAC name: 1,4-benzenedicarboxylic acid) or its dimethyl ester.

Poly(ethylene terephthalate) is used to produce textiles, large soft-drink containers, photographic film, and overhead transparencies. It is marketed under various trademarks including DACRON®, Terylene®, Fortrel®, and Mylar®. Textiles containing this polyester are resistant to wrinkling, and can withstand frequent laundering. Poly(ethylene terephthalate) has been utilized in the manufacture of clothing, bed linen, carpeting, and drapes.

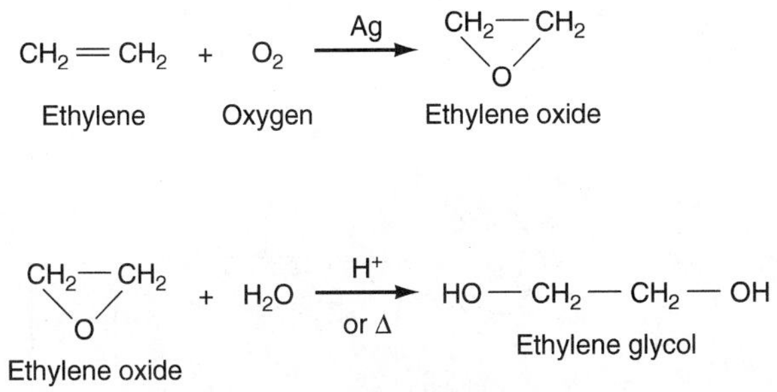

$$CH_2{=}CH_2 + O_2 \xrightarrow{Ag} \text{(cyclic) } CH_2{-}CH_2{-}O$$

Ethylene Oxygen Ethylene oxide

$$\text{(cyclic) } CH_2{-}CH_2{-}O + H_2O \xrightarrow[\text{or } \Delta]{H^+} HO{-}CH_2{-}CH_2{-}OH$$

Ethylene oxide Ethylene glycol

Figure 3. Industrial preparation of ethylene glycol. *(Illustration by Electronic Illustrators Group.)*

Other glycols are also used in polymer production; for example, tetramethylene glycol is used to produce polyesters, and diethylene glycol is used in the manufacture of **polyurethane** and unsaturated polyester resins. Propylene glycol is used in the manufacture of the polyurethane foam used in car seats and furniture. It is also one of the raw materials required to produce the unsaturated polyester resins used to make car bodies and playground equipment.

GLYCOPROTEINS

Glycoproteins are **proteins** to which one or more carbohydrate **chains** are attached. Glycoproteins are most commonly found embedded in the **plasma** membranes of cells, with the carbohydrate exposed to the cell interior. Glycoproteins are thought to function in cell-cell recognition.

The carbohydrate portion of a glycoprotein is commonly a branched chain of fifteen or fewer monosaccharides. Glucose, galactose, fructose, and other common monosaccharides may be combined in the chain, often with linkages not normally found in non-protein linked **carbohydrates**. A single glycoprotein may have multiple carbohydrate chains. Carbohydrate chains are added after **protein synthesis**, by special enzymes in the endoplasmic reticulum.

The exact function of most glycoproteins is not known, although many are thought to aid in the interactions between cells. One of the most important groups of glycoproteins are those of the major histocompatibility complex, or MHC. MHC glycoproteins display antigens, molecular fragments formed within the cell. The antigen-MHC complex is detected by cells of the immune system to determine whether the cell is infected. Other glycoproteins are thought to act during development to aid cells as they migrate to their proper location in the body. Viruses employ glycoproteins as well, where they may function to promote entry into the host cell. The human immunodeficiency virus (HIV) virus, responsible for Acquired Immune Deficiency Syndrome (AIDS), uses a surface glycoprotein known as gp120 to attach itself to its cell surface receptor, called CD4. Once this occurs, the virus can enter the cell. Because of its critical role in infection, gp120 is a target for development of an HIV vaccine.

GOLD

Gold is one of the transition **metals**, a group of elements found in the center of the **periodic table** in Rows 4, 5, and 6 between Groups 2 and 13. Its **atomic number** is 79, its atomic **mass** is 196.9665, and its chemical symbol is Au.

Properties

Gold is a fairly soft metal with a beautiful golden **color**. It is unusually ductile and malleable and can be hammered into very thin sheets with a thickness of only 0.00001 in (0.00025 cm). Gold foil of this thickness is used to make lettering on window signs. Gold has a melting point of 1,947°F (1,064°C), a **boiling point** of about 5,070°F (2,800°C), and a **density** of 19.3 grams per cubic centimeter. Gold is an excellent conductor of **electricity** and has a high optical reflectivity.

Generally speaking, gold is not very reactive. It does not react with **oxygen**, the **halogens**, or most acids at room **temperature**. This property accounts for the metal's use in jewelry, artwork, and coins.

Occurrence and Extraction

Gold is thought to be one of the ten rarest elements in the Earth's crust, with an abundance of about 0.005 parts per million. It is thought to occur more commonly in the oceans with an estimated 70 million metric tons of gold dissolved in seawater and as much as 10 billion metric tons deposited on the ocean floors. So far, no method for retrieving this gold has been developed.

Gold occurs in both elemental and compound states. At one time, large nuggets of gold could be found lying on the Earth's surface. Pieces of such size have long since been collected, although very finely divided gold dust still exists in a number of locations. The most common minerals of gold are the tellurides, such as gold telluride ($AuTe_2$), present in the mineral calavarite. About a quarter of the world's gold comes from South Africa. Other leading producers of the metal are the United States, Australia, Canada, China, and Russia.

Discovery and Naming

Gold has been known to humans for nearly 5,000 years. Objects made of the metal dating to 2600 B.C. have been found in royal tombs of ancient civilizations. It has been used routinely by nearly all cultures for coins, jewelry, and artwork. The element's chemical symbol, Au, is derived from the ancient Latin word for the element, *aurum*.

Uses

In a 1986 study, experts estimated that 121,000 metric tons of gold have been mined throughout history. Of that amount, about 18,000 metric tons were used for industrial, research, health, and other uses in which the metal was discarded after it was used. Of the remaining 103,000 metric tons of gold, about a third has been made into gold bars and held by national banks as security for national money systems. The remaining 68,000 tons of gold are owned by private individuals in the form of jewelry, coin, artwork, or bullion.

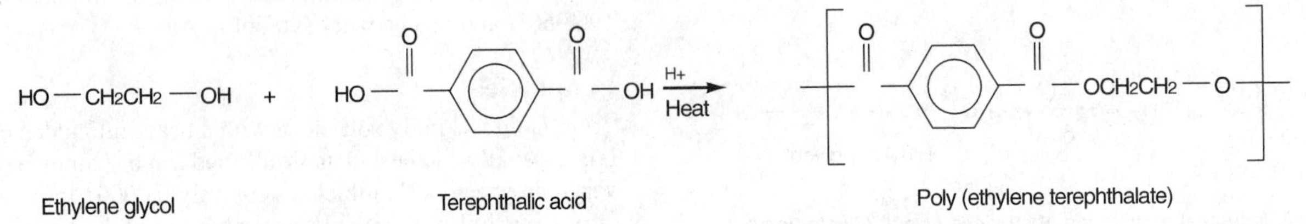

HO—CH2CH2—OH + HO—C(=O)—C6H4—C(=O)—OH →(H+, Heat) [—C(=O)—C6H4—C(=O)—OCH2CH2—O—]

Ethylene glycol Terephthalic acid Poly (ethylene terephthalate)

Figure 4. Synthesis of poly(ethylene terephtalate), a glycol. ***(Illustration by Electronic Illustrators Group.)***

A radioactive **isotope** of gold, gold-198, is sometimes used in the detection and treatment of diseases. In one application, the gold-198 is prepared in a colloidal form and then injected into the patient's body, where it travels to the liver. The radiation released by the isotope allows a study of the liver. Needles made of gold-198 are also sometimes used to treat certain forms of cancer. The isotope is used because it can provide radiation needed for treatment without producing reactions from the body to the metal itself.

See also Precious metals

GOLDSCHMIDT, VICTOR (1888-1947)

Norwegian geochemist, petrologist, and mineralogist

Victor Goldschmidt, called the founding father of modern **geochemistry**, helped lay the foundations for the field of crystal **chemistry**. He was a highly esteemed mineralogist, petrologist, and geochemist who devoted the bulk of his research to the study of the composition of the earth. During his many years as a professor and director of a mineralogical institute in Norway, he also investigated solutions to practical geochemical problems at the request of the Norwegian government.

Victor Moritz Goldschmidt was born on January 27, 1888, in Zurich, Switzerland, to Heinrich Jacob Goldschmidt, a distinguished professor of **physical chemistry**, and Amelie Kohne. His family left Switzerland in 1900 and moved to Norway, where his father took a post as professor of physical chemists at the University of Christiania (now Oslo). Goldschmidt's family obtained Norwegian citizenship in 1905, the same year he entered the university to study chemistry, geology, and **mineralogy**. There he studied under the noted geologist and petrologist Waldemar Brogger, becoming a lecturer in mineralogy and **crystallography** at the university in 1909.

Goldschmidt obtained his Ph.D. in 1911. His doctoral dissertation on contact metamorphic rocks, which was based on rock samples from southern Norway, is considered a classic in the field of geochemistry. It served as the starting point for an investigation of the chemical elements that Goldschmidt pursued for three decades. In 1914, he became a full professor and director of the University of Christiana's mineralogical institute. In 1917, the Norwegian government asked Goldschmidt to conduct an investigation of the country's mineral resources, as it needed alternatives to chemicals that had been imported prior to World War I and were now in short supply. The government appointed him Chair of the Government Commission for Raw Materials and head of the Raw Materials Laboratory.

This led Goldschmidt into a new area of research—the study of the proportions of chemical elements in the earth's crust. His work was facilitated by the newly developed science of **X-ray crystallography**, which allowed Goldschmidt and his colleagues to determine the crystal structures of 200 compounds made up of seventy-five elements. He also developed the first tables of atomic and ionic radii for many of the elements, and showed how the hardness of crystals is based on their structures, ionic charges, and the proximity of their atomic particles.

In 1929, Goldschmidt moved to Gottingen, Germany, to assume the position of full professor at the Faculty of Natural Sciences and director of its mineralogical institute. As part of his investigation of the apportionment of elements outside the earth and its atmosphere, he began studying meteorites to ascertain the amounts of elements they contained. He researched numerous substances, including **germanium**, **gallium**, **scandium**, **beryllium**, **selenium**, **arsenic**, **chromium**, **nickel**, and **zinc**, using materials from both the earth and meteorites to devise a model of the earth. In this model, elements were distributed in different parts of the earth based on their charges and sizes. Goldschmidt stayed at Gottingen until 1935, when Nazi anti-Semitism made it impossible for him to continue his work. Returning to Oslo, he resumed work at the university there and assembled data he had collected at Gottingen on the distribution of chemical elements in the earth and the cosmos. He also began studying ways to use Norwegian olivine rock for use in industry.

When World War II began, Goldschmidt had confrontations with the Nazis that resulted in his imprisonment on several occasions. He narrowly escaped internment in a concentration camp in 1943 when, after the Nazis arrested him, he was rescued by the Norwegian underground. They managed to secretly get him onto a boat to Sweden, where fellow scientists arranged for a flight to Scotland.

In Scotland, Goldschmidt worked at the Macaulay Institute for Soil Research in Aberdeen. Later during the war, he worked as a consultant to the Rothamsted Agricultural Experiment Station in England. As reported in *Chemists,* Goldschmidt carried with him a cyanide suicide pill for use in the event the Nazis invaded England. When a colleague asked him for one, he responded, ''Cyanide is for chemists; you, being a professor of mechanical engineering, will have to use the rope.''

After the war, Goldschmidt returned to Oslo and his job as professor and director of the geological museum. There he worked on a newly-equipped raw materials laboratory supplied by the Norwegian Department of Commerce. He continued his work until his death on March 20, 1947.

Goldschmidt was a member of the Royal Society and the Geological Society of London, the latter of which awarded him the Wollaston Medal in 1944. He was also an honorary member of the British Mineralogical Society, the Geological Society of Edinburgh, and the Chemical Society of London. He wrote over 200 papers as well as a treatise, *Geochemistry,* which was published posthumously in 1954. Although he never married, he had many friends among his colleagues and students, many of whom became notable geochemists and heads of university departments.

GOOD, MARY L. (1931-)

American chemist

Mary L. Good is a highly regarded chemist whose multifaceted career has ranged from academia, to the industrial sector, to the national government, where she serves as undersecretary of technology in the Department of Commerce. Good is described by Jeffrey Trewhitt in *Chemical Week* as an unabashed proponent of **industrial chemistry**. She told Trewhitt that ''we've... gotten ourselves into a trap.... The word 'chemical' has become a bad name, and yet, without chemicals the world doesn't move. We've got to develop a better perspective.''

Mary Lowe Good was born on June 20, 1931, in Grapevine, Texas, the daughter of John W. and Winnie (Mercer) Lowe. In 1950, she received a bachelor of science degree from the University of Central Arkansas, with a major in chemistry and with minors in both physics and mathematics. She received her master of science degree from the University of Arkansas in 1953, with majors in **inorganic chemistry** and radiochemistry. She completed her doctorate at Arkansas in 1955.

In 1954, Good accepted a post as instructor of chemistry at Louisiana State University in Baton Rouge, where she became assistant professor in 1956. She moved to Louisiana State University in New Orleans as associate professor in 1958, and became full professor three years later. At various intervals over the next two decades, she also was affiliated with the medicinal chemistry commission at the National Institutes of Health, the Lawrence-Berkeley Laboratory at the University of California, the research office of the United States Air Force, and the Brookhaven and Oak Ridge National Laboratories. She received the Agnes Faye Morgan Research Award in 1969, the Garvan Medal in 1973, and the Herty Medal in 1975. From 1972 to 1980, she served on the American Chemical Society's board of trustees, chairing the board from 1978 to 1980.

Good was named Boyd Professor of Chemistry within the LSU system in 1974, and returned to Baton Rouge as Boyd Professor of Materials Science in 1979. Among her areas of expertise is catalysis, a process whereby a substance—which remains unchanged chemically at the end of the reaction—induces a modification (an especially an increase) in the rate of the chemical reaction. An important facet of her research is the chemistry of **ruthenium**, a rare **platinum** metal that is used as a catalyst in the synthesis of hydrocarbons.

In 1980, Good left academia to serve as vice-president and director of research at Universal Oil Products (UOP)—an affiliate of the Allied Signal Corporation—in Des Plaines, Illinois. Two years later she was named Scientist of the Year by *Industrial Research & Development* magazine for ''her work as a chemist, educator, lecturer, author, research administrator, and industry spokesperson.'' Good told Barbara H. Brown in *Industrial Research & Development* that ''the greatest challenge in applied science today is to devise more efficient and effective ways to reduce the transition time from laboratory 'proof of principle' to the commercial marketplace.'' In 1985, Good was promoted to president of the engineered materials research division of Allied Signal in Des Plaines. Here, she supervised research in the industrial application of catalysis and polymerization (the chemical process whereby molecules combine into larger molecules characterized by repeating structural units). In 1988, she became senior vice-president at Allied Signal in Morristown, New Jersey, supervising the corporation's entire research and development department, whose 1992 budget amounted to $821 million.

Good was appointed to the National Science Board in 1980. From 1980 to 1985, she directed the inorganic division of the International Union of Pure and Applied Chemistry. In 1983, she was awarded the American Institute of Chemistry's **Gold** Medal. She became president of the American Chemical Society, the second woman elected to this honor, in 1987. In a 1987 *Chemical & Engineering News* article entitled ''ACS in a Changing Environment,'' Good observed: ''The challenge for ACS and individual chemists is to gain the recognition that chemistry, the molecular science, is at the heart of this new technology thrust. The chemical databases built over many years by chemists are the heart of the molecular design programs being utilized so aggressively in the development of new drugs, high-performance materials, specialty chemicals, and biotechnology products.''

Her stance on technological advances through chemistry has led Good to promote research such as the ''materials-by-design'' studies of the Energy Conversion and Utilization (ECUT) project within the Department of Energy. Trewhitt in *Chemical Week* quoted Good's testimony before a congressional panel in support of these studies: ''[Materials-by-design] will replace the current trial-and-error approach, in which one material after another is tested until the one having the desired properties is found. The new approach allows scientists to design the materials they want on a computer, and then use the computer-generated recipes to make materials in the lab.'' Good emphasized that the ''materials-by-design'' strategy would be particularly beneficial in the development of catalysts, thereby enhancing the global competitiveness of the American petrochemical industry and expanding its usage of non-petroleum fuel sources. As head of the ACS, Good also focused on attracting students—particularly women—to careers in the applied sciences.

In 1986, Good was reappointed to the National Science Board, which she chaired from 1988 to 1991. She received the Delmer S. Fahrney Medal of the Franklin Institute in 1988. Three years later she served on the President's Council of Advisers for Science and Technology, and was awarded an Industrial Research Institute Medal. In the same year, the American Chemical Society recognized her ''outstanding public service'' with their prestigious Charles Lathrop Parsons Award. In 1992, she received the National Science Foundation Distinguished Public Service Award, the American Association for the Advancement of Science Award, and the Albert Fox Demers Medal Award from the Rensselaer Polytechnic Institute. Good, whose publications include several monographs and more than one hundred articles in professional journals, has received many honorary degrees.

In May of 1993, President Bill Clinton nominated Good as undersecretary of technology in the Department of Commerce; she was confirmed for this post on August 5, 1993. As quoted in *Chemical & Engineering News,* Charles F. Larson, executive director of the Industrial Research Institute, endorsed Good's nomination: ''She has the perfect combination of background and experience to be an effective leader in that position and to interact with industry on important issues, rather than on those that are important just to the government.'' As undersecretary of technology, Good heads the Technology Administration, the National Institute of Standards and Technology, and the National Technology Information Service. In *Physics Today,* Irwin Goodwin declared that Good's task is to ''strengthen the nation's technology base through government-industry-academic partnerships of many kinds, encourage the introduction of advanced technology into small and medium-sized firms through extension services, and reduce the risks of private investments in new or more sophisticated technology.''

Mary Lowe married Billy Jewel Good on May 17, 1952. A former physics professor and college dean, Billy Good is now a full-time artist. The Goods have two sons, Billy John and James Patrick, and four grandsons. *Chemical Week* notes that Mary Good, an admirer of Edinburgh and the novels of Sir Walter Scott, is ''a Scottish history buff''; her other hobbies include fly-fishing.

GRAHAM, THOMAS (1805-1869)

Scottish chemist

Graham was born in Glasgow, Scotland. At age 14 he attended Glasgow University and studied **chemistry** against his father's wishes, which were that he would become a clergyman rather than a scientist. His father cut off Graham's funding and Graham supported himself by teaching and writing. Upon his graduation from Edinburgh University at age 19, Graham took a teaching position at Mechanics Institute, where he became a professor in 1830. Graham's first scientific paper was on the movement of gas, a subject of great interest to him.

Graham spent the greater part of the 1820s studying the dissolution and **diffusion** of various **gases** through different mediums. He was intrigued by **Johann Wolfgang Döbereiner**'s observation that when a bottle of **hydrogen** with a crack in it was partially submerged in **water**, the water level rose. The bottle was losing hydrogen more quickly than it was gaining air. In his studies Graham discovered that some gases do not behave as **William Henry** had postulated they would in Henry's law, which states that the **solubility** of a gas is dependent on the pressure on that gas. In 1831 Graham established the law that bears his name. It states that a gas diffuses at a rate inversely proportional to its **molecular weight**. Based on this, mixtures of gases could be separated based on their rates of diffusion, which indicates their specific gravity. For this law Graham is widely held to be one of the founders of **physical chemistry**.

In 1833 Graham presented a paper on arsenates, phosphates, and phosphoric acids. The paper is remarkable not only for its content but also for the way in which Graham conducted his research. Graham did not form theories and then conduct experiments to prove those theories. He investigated first and then formed his theories. He worked from facts rather than from the pre-established convictions common to many other chemists of his day. Graham also studied phosphoric acids and it was his paper that first delineated the differences between meta-, ortho-, and pyrophosphates. Until Graham's research there was no explanation for the anomalies of **phosphoric acid** and phosphates. He introduced the idea of polybasic compounds. These were hydrated acids that could replace more than one water **molecule** with a metallic **oxide** to form several different salts. The different degrees of hydration accounted for the differences in properties.

In 1837 Graham became the professor of chemistry at University of London. At this time he began publishing texts for students called *Elements of Chemistry*. He helped found the Chemical Society in 1841 and was its first president. Graham continued to lecture and conduct research. In 1849 he presented a paper to the Royal Society on the diffusion of **liquids**.

Although Graham left the University College in 1854 he continued his experiments. He began to study the diffusion rates of cupric sulphate and other colored chemicals through **solution**. Again Graham noticed that some chemicals diffused more slowly than expected. For instance, while some substances diffused easily through a piece of parchment placed in a solution others seemed to be blocked by the parchment. The process of diffusion through a membrane is called **osmosis**. A membrane is either permeable, allowing solvent and solute molecules to pass through, or semipermeable, allowing only solvent molecules through. Those substances that passed through easily could also be crystallized, so Graham called those substances crystalloids. He called the others colloids.

He found that colloids could be purified by placing them in a porous container and running water over them and through the container or membrane. Any crystalloid impurities were dissolved and passed through the barrier, leaving behind the pure **colloid**. This process was named **dialysis**. Crystalloids pass through the barrier because their particles are relatively small. In the body, **proteins**, **nucleic acids**, and protoplasm are all colloids. Because of his work with colloids, Graham is called the founder of colloid chemistry. Indeed, most of the **nomenclature** of the field comes from his work.

In 1866 Graham began studying **metals**. He found that heated metal foils are permeable to hydrogen. These same metals tended to absorb hydrogen at lower temperatures. Interestingly, Graham is also known as the first to advocate the use of denatured **alcohol**. He suggested that **ethanol** intended for purposes other than drinking be poisoned to prevent its consumption.

Graham is remarkable for the zest with which he pursued his research. His earliest research detailed the movement of one compound through another. He continued investigating the phenomenon in all its variations for the rest of his life. Where other chemists engaged in arguments of opinion, Graham simply did his research, carefully and conscientiously, and published his results and conclusions. Graham died in London in 1869.

Thomas Graham.

GRAHAM'S LAW

Graham's law states that the rate of **diffusion** of a gas is inversely proportional to the square root of its **molecular weight**. Thus, if the molecular weight of one gas is four times that of another, it would diffuse through a porous plug or escape through a small pinhole in a vessel at half the rate of the other. A complete theoretical explanation of Graham's law was provided years later by the kinetic theory of **gases**. Graham's law provides a basis for separating isotopes by diffusion—a method that came to play a crucial role in the development of the atomic bomb.

Graham's law was first formulated by Scottish physical chemist **Thomas Graham**, who became professor of **chemistry** at the University of Glasgow. Graham's research on the diffusion of gases was triggered by his reading about the observation of German chemist **Johann Döbereiner** that **hydrogen** gas diffused out of a small crack in a **glass** bottle faster than the surrounding air diffused in to replace it. Graham measured the rate of diffusion of gases through plaster plugs, through very fine tubes, and through small orifices. In this way her slowed down the process so that it could be studied quantitatively. He first stated the law as we know it today in 1831. Graham went on to study the diffusion of substances in **solution** and in the process made the discovery that some apparent solutions actually are suspensions of particles too large to pass through a parchment filter. He termed these materials colloids, a term that has come to denote an important class of finely divided materials.

At the time Graham did his work the concept of molecular weight was being established, in large part through measurements of gases. Italian physicist Amadeo Avogadro had suggested in 1811 that equal volumes of different gases contain equal numbers of molecules. Thus, the relative molecular weights of two gases are equal to the ratio of weights of equal volumes of the gases. Avogadro's insight together with other studies of gas behavior provided a basis for theoretical work by Scottish physicist **James Clerk Maxwell** who was trying to explain the properties of gases as collections of small particles moving through largely empty space.

Perhaps the greatest success of the kinetic theory of gases, as it came to be called, was the discovery that for gases, the **temperature** as measured on the Kelvin (absolute) temperature scale is directly proportional to the average kinetic **energy** of the gas molecules, where the kinetic energy of any object is equal to one-half its **mass** times the square of its **velocity**. Thus, to have equal kinetic energies, the velocities of two different molecules would have to be in inverse proportion to the square roots of their masses. Since the rate of diffusion is determined by the average molecular velocity, Graham's law for diffusion could be understood in terms of the molecular kinetic energies being equal at the same temperature.

After the outbreak of World War II, the United States mounted a massive project to develop an atomic bomb, the first weapon to use the recently discovered process of **nuclear fission** as an energy source. It was clear that a **chain reaction** could be sustained among nuclei of uranium-235, but this **isotope** constitutes less than 1% of the **uranium** present in uranium ore, the vast majority of which is in the form of the more stable isotope, uranium-238. To separate the two, the government built a gaseous diffusion plant at the then phenomenal cost of $100 million in Clinton, Tennessee. In this plant, uranium from uranium ore was first converted to uranium hexafluoride and then forced repeatedly to diffuse through porous barriers, each time becoming a little more enriched in the slightly lighter uranium-235 isotope.

See also Avogardo's law; Gas laws; Kinetics

GRAPHITE

Carbon, in the forms of **charcoal**, graphite, and **diamond**, was one of the earliest elements known to man. Although discovered later than charcoal, graphite was prized early on for its ability to produce a blue-gray streak when rubbed against a rough surface. The term graphite is derived from the Greek word, *grapho*, meaning "I write."

Substances that exist in two or more forms that are significantly different in physical or chemical properties are called allotropes. Graphite is one of the four allotropic forms of carbon; the other three are amorphous carbon, diamond, and fullerene. Unlike many other elements, carbon does not readily convert to different allotropic forms in the absence of extreme conditions, e.g., high pressure, high **temperature**, and/or long spans of time.

Structurally, graphite consists of sheets of carbon atoms linked hexagonally like chicken wire. Each layer is one, gigantic two-dimensional **molecule**. The bonds holding the layers together are similar to those in **metals**, with the result that graphite exhibits metallic conduction in the direction of the planes. Because the forces between the layers readily allow the layers to slide over each other, graphite is a good **lubricant**.

The natural supply of graphite is not sufficient to meet demand for it as a lubricant, so it is produced synthetically on a large scale. Most graphite is obtained from petroleum coke, i.e., the black tar that remains after all of the useful fuels and lubricants have been removed from crude oil. The coke is heated in an oven that has no access to outside air to burn off most of the impurities, leaving a fairly pure form of graphite.

In addition to its use as a lubricant, graphite is also used in the manufacture of "**lead**" pencils, pigments, electrodes for carbon-arc lamps and electrochemical cells, refractory crucibles, matches and **explosives**, and moderators for **nuclear reactors**.

See also Electrical conductivity; Petrochemicals

GRASSELLI, CAESAR AUGUSTIN (1850-1927)

American businessman and entrepreneur

Caesar Augustin Grasselli was the third generation to head the Grasselli Chemical Company. Following his death, one of America's oldest chemical businesses was merged into the E.I. du Pont Company.

Caesar Grasselli, son of Eugene Ramiro Grasselli, was born in Cincinnati, Ohio. He inherited his father's interest in and aptitude for **chemistry** and the family enterprise. Starting at age 15 he received on-the-job training, including stints as bricklayer and pipefitter, from his father. In addition, a chemistry professor from Karlsruhe University (in Germany) tutored him and he attended Mount St. Mary's College in Emmitsburg, Maryland. "C.A.," as he was known, married in 1871 and went on an extended honeymoon, in part a business holiday, visiting chemical plants in France, Germany, and Britain as well as the ancestral home at Torno, Italy; he made a similar chemical tour in 1899. He became a partner in Eugene Grasselli & Son in 1873, senior partner in 1882 after his father's death, and followed closely in his father's footsteps. Credited with a broad array of product and process innovations, he had a lifetime of entrepreneurial achievement.

President of the Grasselli Chemical Company (1885-1916) and chairman of the board of directors (1916-1927), C.A. led his company through decades of rapid growth; by 1916 it operated 14 factories. His notable friends included Mark Hanna, William McKinley, and John D. Rockefeller. He was honored in 1910 by King Victor Emmanuel III for his humanitarian philanthropy after natural disasters in Italy and in 1923 by both King Victor Emmanuel III and Pope Pius XI for helping Italy's World War I victims.

Under C.A. the Grasselli Chemical Company had been incorporated in 1885 with capital of $600,000; by 1913 its worth had risen to $15 million. Three years later Grasselli Chemical was operating 14 factories and had property worth over $30 million. Its capital jumped to $35 million in 1918, undoubtedly a consequence of World War I, giving it a rank of 192 among the 500 largest American industrials.

Having first specialized in heavy chemicals such as **sulfuric acid**, in 1904 Grasselli Chemical had introduced the production of **zinc** as spelter. By 1928 Grasselli Chemical, a vertically integrated enterprise, manufactured heavy chemicals, fertilizer, black scrap, zinc, and **explosives** with plants in 22 cities; owned zinc ore and pyrite deposits as well as **coal** deposits; and employed between 3,700 and 4,500 workers. Despite its size, Grasselli remained a family enterprise with the Grassellis and their in-laws serving as officers and dominating the board of directors.

Chemical manufacturers shared certain patterns in the decades before World War I. Owing to limited economies of scale, Grasselli Chemical expanded by adding more plants rather than by increasing the size of plants. Also, it integrated forward, replacing its sales agents with direct selling, and grew more by internal expansion than by merger.

Grasselli Chemical had gradually drifted into manufacturing explosives. It sold mixed acids for the production of nitroglycerine in the late 1860s and early 1870s. Since Grasselli supplied superior acids for the manufacture of dynamite, an explosives firm built a dynamite factory in Cleveland to take advantage of Grasselli Chemical as a supplier. The Grassellis improved the product by suggesting the substitution of nitrate of soda for the saltpeter. In 1900 E.I. du Pont de Nemours & Co. and the Grasselli Chemical Company jointly organized a company to supply nitroglycerine. Finally, in 1917 a Grasselli subsidiary acquired and consolidated three explosives companies.

World War I affected Grasselli Chemical in an unanticipated way; the United States government confiscated German patents, selling them to American interests through the Alien Property Custodian. This gave Grasselli Chemical the chance to become a producer of drugs and intermediates, broadening its product line. The Alien Property Custodian sold Bayer dye and drug patents to Grasselli Chemical, which transferred the medicinals to Sterling Products, and sold patents to Sterling Products, which resold them to Grasselli Chemical. However, Grasselli Chemical later sold out its coal-tar chemicals facility.

From the 15th century onward the Grassellis were chemists, druggists, perfumers, and the like at Torno on Lake Como in northern Italy. When Napoleon Bonaparte made this Po valley region a part of France, Giovanni Angelo Grasselli saw an opportunity. With another Italian emigrant he became a manufacturing chemist in Strasbourg in northeastern France. He built his first chemical plant there about 1800 and a second in 1810 at Mannheim in western Germany.

His son, Eugene Ramiro Grasselli (1810-1882), was born in Strasbourg, studied chemistry at the Universities of Strasbourg and Heidelberg, and worked with his father. Emigrating in 1836, he apprenticed in Philadelphia, Pennsylvania. In 1839 he started a chemical business, selling sulfuric acid, a good proxy for economic development, and other mineral acids to the meatpacking industry in Cincinnati, Ohio, known then as "Porkopolis." In response to his competitors, and to be nearer **oil refining** customers who wanted to integrate backwards by producing their own sulfuric acid, he opened a second factory in Cleveland, Ohio, in 1867 and also relocated the main office as well as his family there.

The new chemical works, on the Cuyahoga River, virtually adjoined the Rockefeller refinery; as Standard Oil prospered, so did the Grasselli firm. In that year Caesar Grasselli participated in Cleveland **sulfur** price-fixing; in 1872 the Cleveland oil refiners tried to depress the sulfuric acid price but the parties compromised. His partnership, doing business as Eugene Grasselli & Son (1873-1883), diversified in the late 1870s, manufacturing higher value-added products.

Under C.A.'s leadership the company continued to prosper as a private concern. In the year following his death, Grasselli Chemical embarked on a new direction; it obtained a listing on the New York Stock Exchange, becoming a publicly held corporation. Then came a bigger change. Later that year (1929) Grasselli Chemical was acquired by du Pont in the largest merger in which one of the top 200 nonfinancial companies acquired a company not on the list. Du Pont exchanged stock valued at $73 million for the Grasselli stock, 26 plants, and assets of $56 million. Thus the third-generation heirs of an Italian immigrant entrepreneur ended the independent family enterprise; such family enterprises typically do not endure beyond the third generation.

GREENHOUSE EFFECT

The **greenhouse effect**, the cause of global warming, is an unprecedented, and possibly irreversible, environmental condition in which damaging human-produced **gases** build up and trap **heat** within the earth's protective atmospheric shield, called the **ozone** layer.

The threat of global warming was first recognized in 1896 by Swedish chemist **Svante August Arrhenius**, who suggested that the burning of fossil fuels might have a serious impact on the earth's **temperature**. However, scientists at that time could not have predicted that it would become one of the world's most pressing environmental issues less than one hundred years later.

Arrhenius's early warning of the dangers of **carbon dioxide** was not taken seriously until 1938 when G. S. Callendar, an English physicist, pointed to meteorological records showing the gradual warming since 1880 of North America and Northern Europe. Callender was the first scientist to gather data from several sources on the danger of increasing **carbon dioxide** levels. Even with this data, Callender's study did not garner significant support because many scientists believed that the excess carbon dioxide would be absorbed in the oceans, not in the atmosphere. Also, the warming trend was inexplicably replaced by a temporary temperature decline around 1940. Callender remained adamant, however, and noted in 1958 that the warming had resumed in 1942. Finally, in the 1960s, concern about atmospheric **pollution** was taken seriously as the link between air pollution and the Earth's temperature became apparent.

Earth's fragile atmosphere is comprised primarily of **nitrogen**, **oxygen**, and carbon dioxide. These three ingredients, along with **methane**, provide the Earth with a protective blanket that regulates how much of the sun's enormous heat reaches the earth, as well as how much heat leaves our atmosphere. For this reason, these gases are called greenhouse gases, and they appeared to be balanced until the advent of the industrial revolution.

Through this delicate system, the Earth's atmosphere allows the visible and infrared wavelengths of radiation from the sun to reach the ground. Once it has hit the surface, this visible light is absorbed and reflected by the earth as infrared radiation that cannot be seen but which can be felt as heat. If not for the presence of atmospheric gases such as carbon dioxide, **water** vapor, and other greenhouse gases, the heat would escape out beyond Earth's atmosphere. Instead, it is absorbed by the greenhouse gases, and much of it is re-emitted down towards the surface, resulting in extra heat.

A great deal of the carbon dioxide that is released by industrialized societies is absorbed by forests, oceans, and the process of limestone deposition. However, these resources are either limited or depleting, and extra output of carbon dioxide collects in the atmosphere. Global warming from the build-up of greenhouse gases is also exacerbated by ozone depletion, which is allowing harmful radiation to reach the Earth's surface.

Recently, computers have allowed scientists to develop models that estimate the consequences of global warming while suggesting ways to slow the process. Measurements now show that between 1957–1975, the amount of carbon dioxide in the air has increased from 312 to 326 parts per million, a jump of approximately 5%. These measurements were collected all around the world from the top of the highest mountain, Mauna Loa in Hawaii, to the South Pole, where air was gathered through airplane air-intake systems. Overall, carbon dioxide concentrations are up about 30% since pre-industrial times. Methane is up 145%, and **nitrous oxide** is up 15%. A car emits about 20 lb (9.1 kg) of carbon dioxide for every gallon of gas it burns.

Most climatologists—scientists who study the Earth's climate—now believe it is clear that this human-produced greenhouse effect is changing the earth's climate. One recent comprehensive study used tree rings (which show temperature variations as well as **precipitation** variations), ice core samples (where trapped air can give clues about past environments), and coral records to trace climate patterns over the last 600 years. Researchers combined this evidence with temperature readings that have been available for only about the last 150 years and with historical records, and concluded that the twentieth century has been the warmest century in the last 600 years. Moreover, they concluded that the warmest years in all of that period were 1990, 1995, and 1997.

Although it is difficult to precisely predict exactly how greenhouse effect processes will affect the Earth's temperature in the future, one computer model estimates that by the year 2050 the temperature level could rise by as much as 4°F (2°C). This average increase would be unevenly distributed, ranging from as little as less than a degree at the equator to up to six degrees at higher latitudes. Although this may not sound like a large increase, it may be enough to affect the rate of glacial melting, raising sea levels, and to otherwise have an impact on the earth's climate. Scientists believe the global sea level has already risen by about 4-10 in (10-25 cm) over the past 100 years. By the year 2100, it is expected to rise an additional 20 in (50 cm), doubling the number of people in the way of storm surges. Warmer climates would affect growing seasons, shift crop zones, and could increase the risk of certain tropical diseases like malaria.

Valuable research continues as scientists try to compare today's atmospheric and air quality with much older historical data.

As research continues, industries and governments are trying to establish programs that would help protect the environment. At the 1997 climate-change conference in Kyoto, Japan, delegates from 159 nations agreed to a pact that takes the first steps toward legal regulations that will reduce industrial gases—Europe by 8%, Japan by 6%, and the United States by 7%. As these same gases are often behind the engines and industries of every country's economies, conflicts will need to be resolved between industrial users and environmental advocates.

GRIGNARD, FRANÇOIS AUGUSTE VICTOR (1871-1935)

French organic chemist

Organic synthesis owes a major debt to François Auguste Victor Grignard, whose pioneering studies were a critical early step in the advancement of the field. When he entered the profession at the turn of the century, the task of combining different organic chemical species was a major hurdle in the synthesis of new compounds. Grignard developed a process that enabled chemists to join many types of organic chemical compounds. It was a method that could be used with a broad array of organic reactants; it was also inexpensive and simple to perform, and it resulted in high yields. These features resulted in its rapid adoption by organic chemists around the world. Grignard shared the 1912 Nobel Prize in **chemistry** with **Paul Sabatier**. Grignard was born in Cherbourg, France, on May 6, 1871. His mother was Marie Hébert Grignard; his father, Théophile Henri Grignard, was a sailmaker and foreman at the local marine arsenal. During his early education, Grignard's parents steered him towards a career as a teacher, and at age eighteen he won a scholarship to the École Normale Spéciale, whose graduates routinely became secondary school teachers. But the school closed after he had completed only two years there, and then, despite further work at the University of Lyons, Grignard failed the mathematics exams that were then required for licensing as a teacher.

In 1892, Grignard did a year of compulsory military service, and upon his return to Lyons a friend urged him to study chemistry. Grignard had long considered chemistry scientifically uninteresting, believing it consisted mainly of unrelated facts and observations, but he soon found himself interested and began to pursue a graduate degree. Philippe Barbier became his supervisor and convinced him to do his doctoral thesis on chemical synthesis.

Barbier had been investigating the general problem of using a metal to attach together two organic radicals—chemical groups that remain unchanged by a reaction. Species made by joining an organic radical with a metal **atom** are known as organometallic compounds, and while the creation of these compounds was well known at the time, there were serious limitations on their uses. Some were extremely reactive (such as organosodium and organopotassium), while others were very unreactive (such as organomercury). **Zinc** had been used, but only a few compounds could be prepared with it, and organozinc compounds had the unfortunate tendency of spontaneously igniting at room **temperature**. Barbier had experimented with using **magnesium** to join organic radicals, but he had obtained inconsistent results.

This was the problem Grignard pursued for his thesis. He knew that one of his first challenges would be to solve the same problem that plagued the organozinc researchers; he had to avoid having his organomagnesium compounds spontaneously erupt in flame. His research showed him that workers had overcome this hazard when dealing with organozinc species by performing the reaction in anhydrous **ether**.

Thus Grignard added magnesium shavings in ether to methyl iodide. This **solution**, later to be known as the Grignard reagent, would react with an aldehyde or ketone, which were both readily obtained types of organic compounds. With the addition of dilute acid, the resultant compound broke apart with the formation of **water** to produce an **alcohol**. The particular alcohol produced depended solely on the particular aldehyde or ketone used in the reaction. The yield—the amount of alcohol produced relative to the aldehyde or ketone reactant—was high, the process simple, and the reaction consistent. Moreover, as Grignard and others rapidly discovered, the reaction was extremely flexible. Many different reactants besides **aldehydes** and **ketones** could be used, and the types of compounds that could be produced were virtually limitless.

Grignard's complete thesis was published by the University of Lyons in 1901, and several articles describing the process and its applications rapidly appeared in major chemistry journals. In 1909 Grignard left Lyons to take up a post at the University of Nancy, where he was soon named professor. In 1910, he married Augustine Marie Boulant. The first of their two children, Roger, was born the following year.

In 1912, Grignard's colleagues formally recognized the importance of his early studies by awarding him the Nobel Prize. He shared the 1912 Nobel Prize in chemistry with Paul Sabatier, also a Frenchman, who had developed another important methodology in organic synthesis, catalytic **hydrogenation**. The joint prize was the cause of great national pride, and one French newspaper announced that this proved chemistry was not "a predominately German science."

Grignard remained at Nancy for nearly a decade, and his research continued to focus on organomagnesium compounds and their reactions with a wide array of other chemical species. During World War I, Grignard worked on methods for synthesizing toluene, an important solvent, and on the analysis of chemical warfare **gases**. After the war, he returned to the University at Lyons, where he had been chosen to succeed Barbier, who was retiring. He remained at Lyons for the rest of his life, continuing his studies of organic synthesis processes and expanding his work to include other fields.

Grignard wrote two volumes of what he planned to be a fifteen-volume work on **organic chemistry**, but he died unexpectedly on December 13, 1935. Two more volumes of this work were published posthumously, with editorial assistance from his son Roger, who had followed his father and chosen a career in organic chemistry. Grignard's legacy includes 170 publications and a host of honors and awards for his work, including honorary doctoral degrees from the universities of Louvain and Brussels, and the title of Commander of the Legion of Honor.

GROUP THEORY

Group theory began as a branch of pure mathematics. Mathematicians, through their study of the way numbers behave and how they may be treated, often discover theorems and relationships that may appear to have no immediate application to the physical world. On the other hand, physical scientists—chemists and physicists—attempt to understand how objects in the physical world behave, for instance how chemical molecules interact with each other in chemical reactions. They seek ways of visualizing these reactions and methods of predicting and calculating whether or how fast a reaction may be expected to occur. Since the days of Gallileo (1564-1642) and Isaac Newton (1642-1727), scientists have found that mathematics provides useful methods for scientists to accomplish this task. Group theory is one of a number of branches of mathematics that have proven useful to chemists and physicists in their work.

In our everyday use of the term, we commonly employ the word "group" to mean a collection of things which have something in common: a group of children, a musical group, a group of ducks. In a mathematical group, however, these "things" which have something in common are called elements, and there are rules that define what a set of elements must have in common in order to be regarded as a mathematical group. In their pioneering work with groups, Evariste Galois (1811-1832), Camile Jordan (1838-1921), and Felix Klein (1849-1925) developed the rules for mathematical groups and the methods for working with them. The result is the branch of mathematics known as group theory.

There are a number of groups that are important in **chemistry**. The chemical elements are grouped into sets in the **periodic table** (such as halides and alkali metals) based on the chemical behavior they have in common. Functional groups (such as alcohols, acids, and amines) are identified in **organic chemistry**. The commonalties shared by these groups are not of a mathematical nature. These groups do not follow the rules for a mathematical group, and group theory is, therefore, of no help in understanding them.

Many chemical molecules are symmetrical. Their structure is such that the **molecule** may be moved in certain ways, and the molecule in its new position appears to be identical to that before the movement occurred. Let us suppose that a molecule is made up of one **atom** of element A and three atoms of element B (B', B'', B'''). Let us further stipulate that the atoms of the molecule are all located in the same plane. This molecule is shown as Figure I. Now rotate the molecule 120° about an axis which is perpendicular to the molecular plane and which passes through the central A atom. The configuration of the molecule after this rotation is shown in Figure II. But if all of the B atoms are chemically and physically the same, configurations shown in Figures I and II are equivalent as far as their chemical and physical properties are concerned. If we now rotate the molecule by another 120° about the same axis, the molecule shown in Figure III will appear. Again we see that the configuration shown in Figure III is equivalent to Figures I and II. Rotation of Figure III by 120° results in Figure IV, which is clearly not only equivalent to Figures I, II and III, but is identical to Figure I. The AB_3 molecule is thus said to possess a three-fold axis of symmetry.

The planar AB_3 molecule also possesses other symmetry elements. For instance, if you interchange the positions of

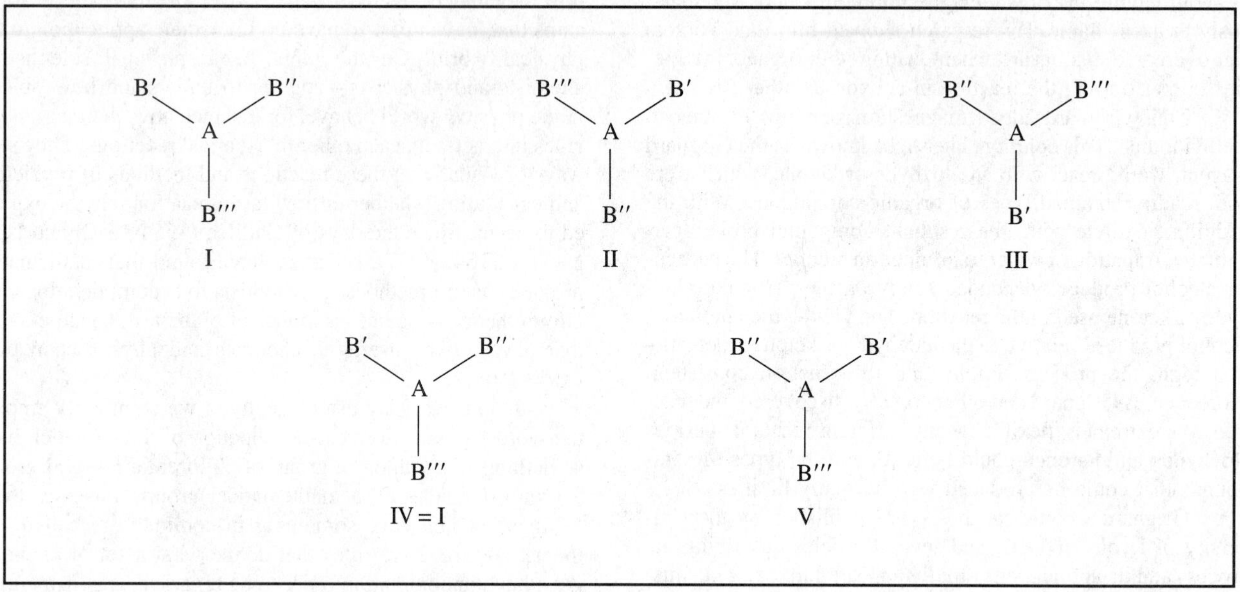

Group theory. ***(Illustration by Electronic Illustrators Group.)***

atoms B' and B'' leaving atoms A and B''' in their original places, the structure shown in Figure V is obtained. Again chemically and physically the molecule seen in Figure V is equivalent to the molecule in Figure I. The molecule is said to possess mirror image symmetry relative to a hypothetical mirror perpendicular to the plane of the molecule and passing through atoms A and B'''.

The collection of all of the symmetry elements that a molecule possesses forms a group. Interestingly, this group is found to follow the rules required for a mathematical group. The methods of group theory should, therefore, be applicable in the study of the behavior of symmetrical molecules such as AB_3.

The application of group theory to chemical molecules is called chemical group theory. It has been particularly useful in providing simpler methods for constructing and visualizing hybrid molecular orbitals and in simplifying the quantum mechanical calculations in **molecular orbital theory**. Group theory has also provided the basis for the development of **ligand field theory** and in understanding and predicting the vibrations of symmetrical molecules. The usefulness of group theory in these applications results from the fact that, as we have seen, some configurations of symmetrical molecules, as they move about, are equivalent. Because of this, restrictions are placed on the solutions of quantum mechanical equations that apply to the orbital structure and behavior of these molecules. As a result, solutions of these equations are greatly simplified.

GULDBERG, CATO (1836-1902)

Norwegian physical chemist and mathematician

Cato Maximilian Guldberg was born on August 11, 1836, in Christiania (now Oslo), Norway, the eldest of nine children of Carl August Guldberg, a minister and owner of a bookstore and printing office, and Hanna Sophie Teresia Bull Guldberg, who came from a highly esteemed and wealthy merchant family of Fredrikstad. When young Cato was 11 years old, his father was appointed minister at Nannestad, about 50 mi (80.5 km) north of Christiania. The foundation for Cato's later interest in the outdoors, hunting, and fishing was laid in his father's remote parish, but he could not receive a satisfactory education there. Therefore, at the age of 13, Cato was sent to live with his maternal grandmother, Catharina Ohlson, in Fredrikstad, where he attended secondary school and excelled in mathematics. This school could not grant admission certificates to the university so Cato returned to Christiania and spent his last year of secondary education at a private Latin school.

Guldberg entered the University of Christiania in 1854, the same year as his close friend **Peter Waage**, with whom he began a close lifetime friendship. Together with a number of other fellow students, the two founded a small, informal club whose members met on Saturday afternoons to discuss **chemistry** and physics problems. Guldberg majored in mathematics and prepared for secondary examinations in physics and chemistry. He worked independently on advanced mathematical problems, and his first published scientific article, ''On the Contact of Circles'' (published in 1861), won the Crown Prince's Gold Medal in 1859.

Guldberg graduated in 1859 and became a teacher at Nissen's secondary school in Christiania. He was appointed a mathematics teacher at the Royal Military Academy in 1860. In 1861 he made a one-year study tour of France, Switzerland, and Germany by means of a scholarship. Upon his return to Norway Cato became Professor of Applied Mechanics in 1862 and Professor of Advanced Mechanics the following year at the Royal Military College, two positions he held until his death in Christiania on January 14, 1902. In 1867, Cato was awarded a scholarship at the University of Christiania, where he became Professor of Applied Mathematics in 1869.

Guldberg's name is intimately linked with that of his friend Waage primarily for their joint discovery of the law of mass action. This fundamental law of chemistry, which today is known to every beginning chemistry student, had several forerunners, but the combined efforts of the theorist Guldberg and the empiricist Waage were ne eded to produce the first general, exact, mathematical formulation of the role of the amounts of reactants in chemical equilibrium systems.

But Guldberg and Waage were also related through two marriages; Guldberg married his cousin Bodil Mathea Riddervold, daughter of cabinet minister Hans Riddervold, and the couple had three daughters. Waage married Bodil's sister, Johanne Christiane Tandberg Riddervold by whom he had five children, and after her death in 1869, he became Guldberg's brother-in-law a second time, in 1870, by marrying one of Guldberg's sisters, Mathilde Sofie Guldberg, by whom he had six children.

Guldberg and Waage's collaboration on the studies of chemical affinity that led to the law of mass action began immediately after Guldberg's return from abroad in 1862. Waage presented their findings to the Division of Science of the Norwegian Academy of Science and Letters on March 14, 1864, where it elicited little response. Even after it s publication the following year in Norwegian in the academy's journal—which was not accessible to many scientists—it failed to attract attention. Moreover, their work remained almost completely unknown to scientists as did the more detailed description of their theory published in 1867 in French. The theory did not become generally known until 1877, when German chemist **Wilhelm Ostwald** published an article that adopted the law of mass action and proved its validity by experiments of his own. The following year Dutch chemist **Jacobus Henricus van't Hoff** derived the law from reaction **kinetics**, apparently without any awareness of Guldberg and Waage's previous work. Because their work was not become universally known and van't Hoff had not recognized their priority, Guldberg and Waage published their previous work for a third time, this time in the German journal *Annalen der Chemie* and in German, the *lingua franca* of 19th-century chemistry. In 1884, in his *Études de Dynamique Chimique*, van't Hoff finally mentioned their work, thus assuring their priority.

The law of mass action is Guldberg's most significant contribution to **physical chemistry**, but not his only one, for much of his early work, published in Norwegian, did not receive the attention that it deserved. He spent much time on a search for a general **equation of state** for **gases**, **liquids**, and **solids** from a kinetic-molecular approach. In 1867, 19 years before van't Hoff, he introduced the ideal gas equation in the form: $pV = 2T$. In 1869 he developed the concept of "corresponding temperatures" and derived an equation of state valid for all liquids of certain types. In 1890, he formulated the rule (Guldberg's rule) that the re duced boiling temperatures of most liquids are close to 2/3, a relationship discovered independently by the French-Swiss chemist Philippe Auguste Guye that same year. Guldberg also made a number of contributions to the **thermodynamics** of **solution** and of **dissociation**, and he discovered and explained cryohydrates. He wrote many articles on various practical problems as well as several textbooks on mechanics and mathematics.

Guldberg served as co-editor (1860-61) and Editor-in-Chief (1863-73) of *Polyteknisk Tidsskrift*, a journal devoted mainly to applications of science. The topics of some of his articles in this journal show the breadth of his interests: melting points of alloys (1860), fuels and their **heat** values (1860), specific weights (1860), a new system of weights and measures (1862), the cheapest form of lighting (1863), the mechanical theory of heat (1864), the steam engine (1870, 1871, 1872), the movement of **water** in tubes (1871), the theory of the movement of water and air on the Earth's surface (1872), superheated steam (1872), and new theories of the flow of gases and vapors (1873).

Guldberg served three terms as Chairman of the Polytechnical Society (1866-68, 1869-72, 1874-75), was an active member and officer of many scientific societies and commisions, and received several domestic and foreign honors. His work with the Norwegian Royal Commision of 1873 resulted in Norway's becoming the first Scandinavian country legally to adopt the **metric system** (1875).

GUNPOWDER

Gunpowder, the oldest known explosive, was probably discovered in China during the tenth century, though some sources cite a much earlier date. It consists of a mixture of saltpeter (potassium nitrate), sulphur, and **charcoal**, in variable proportions. Saltpeter is formed from the **oxidation** of organic **matter** containing **nitrogen**, and the process of making and purifying it was probably discovered in either India or China. A Chinese recipe for explosive powder dates to about 1000 A.D., and this was probably used to make incendiary bombs, fireworks, or terrifying explosions. It was probably not used to launch a projectile out of a gun or cannon until about 1304, the year in which the Arabs are said to have constructed the first gun, a crude device made of hollowed bamboo and reinforced with **iron** that was used to fire arrows. English and Italian manuscripts dating from around 1326 show the existence of guns in Europe at that time, so the technology of gunpowder seems to have spread very quickly.

By the mid-14th century, gunpowder as fuel for guns, cannons, and other destructive devices was commonplace throughout Europe and much of Asia. Prior to its first military uses, gunpowder was termed black powder. The recipe for black powder was first recorded by English philosopher Roger Bacon (c. 1214-1292) in 1242. The original formula, consisting of 7 parts saltpeter, 5 parts hazelwood (charcoal or carbon), and 5 parts sulphur, has been modified in its proportions to suit various uses, but the essential ingredients have remained the same. The mixture, though prone to rapid **combustion**, does not exhibit the commonly observed properties of thrust or loud explosion unless it is somehow confined. Consequently, early

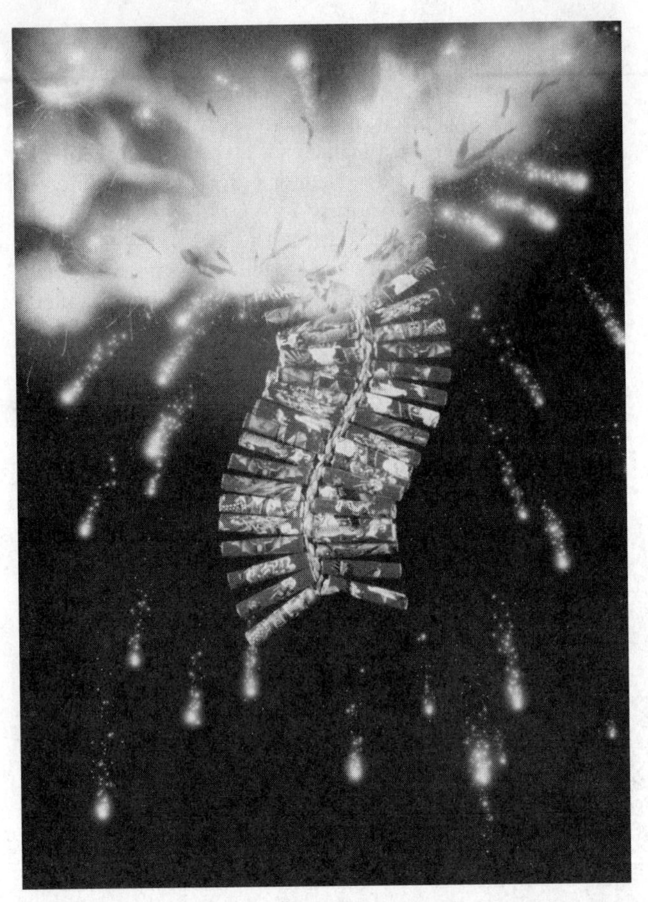

Gunpowder is still employed today for a number of specialized uses but is perhaps most commonly found in firecrackers. ***(Photograph by Erich Schrempp, Photo Researchers, Inc. Reproduced by permission.)***

gun designs revolved around barrel size and strength and the manner in which gunpowder was compacted. The proportions of essential ingredients used in gunpowder varied widely from place to place, and did become standardized until the eighteenth century.

The perfecting of gunpowder was an ongoing process. Eventually scientists realized that the powder was simply a mixture, not a new compound. It was limited by such military drawbacks as flashing and smoking. Moreover there was the additional disadvantage of delayed reaction time. These obvious flaws, coupled with the complaints of mining and engineering companies that gunpowder was incapable of pulverizing large pieces of rock, led independent researchers in several promising directions. Both nitroglycerine and TNT, two of the most powerful **explosives** in existence, can trace their origins to this race in research during the nineteenth century.

The discovery that directly supplanted gunpowder for use in firearms was guncotton, a forerunner of smokeless powder. Invented by **Christian Schönbein** in 1845, guncotton is formed by a nitrating process involving the dipping of cotton (cellulose) in nitric and sulfuric acids. This nitrocellulose explosive underwent improvements until it obtained final form as a reasonably safe and clean-burning powder in 1884. This powder was ground from a hardened **gelatin** created by the evaporation of **ether** and **alcohol**. Its inventor was French chemist Paul Vieille (1833-1896).

Gunpowder is still employed today for a number of specialized uses but is perhaps most commonly found in fireworks. Its production involves a number of important steps. After securing ingredients of the highest quality, the manufacturer must grind and mix the material while damp, then compress it into cakes, break it into grains of varying sizes, and, finally, ensure that the grains are glazed, dried, graded, and properly packaged. These exacting requirements were arrived at after centuries of experimentation and thousands of accidental deaths. Despite vast improvements in safety guidelines and preparedness, factory and operational accidents still occur due to the **volatility** of the substances involved.

H

Haagen-Smit, A. J. (1900-1977)

American chemist

In the mid–1940s, A. J. Haagen-Smit led investigations into the origins of smog. Through his research he discovered that smog is created by the **oxidation** of organic material in the air. Haagen-Smit spent a major part of his life challenging industry in an attempt to curtail air **pollution** caused by burning fuels. As a result of his constant battle, factories now use smoke stacks that filter **carbon** fumes, and the auto industry has incorporated components to reduce pollution-causing vapors from car and truck exhaust.

Arie J. Haagen-Smit was born in Utrecht, The Netherlands, on December 22, 1900, the son of Jan Willem Adrianus Haagen-Smit and Maria Geertruida van Maanen. His father was a chemist who maintained a large science library in the home. Haagen-Smit credited his interest in science and technology to the early enthusiasm of his father's discussions about his work, and the accessibility of his family's library. He attended the Rijks Hoogere Burger School and in 1922 graduated from the University of Utrecht with a major in organic **chemistry**. He acquired practical experience in **organic chemistry** while working summers in a local munitions plant and serving in the Dutch army chemical corps. By 1926 he had earned his M.A. degree and in 1929 he completed his Ph.D. While studying for his doctorate, he identified the structure of a group of **hydrocarbons**, called terpenes, found in volatile oils, or those vaporizing rapidly.

Haagen-Smit remained at the university for another five years, first as a chief assistant in organic chemistry, then as a lecturer. His primary interest was the chemical composition of **natural products**. In particular, he investigated and published his findings on plant **hormones** called "auxins." He also isolated and synthesized substances obtained from plant species related to poison ivy. Haagen-Smit rapidly gained a reputation as a specialist in the chemistry of plant hormones, as well as distinction as a highly talented researcher.

In 1936, he was invited to lecture at Harvard University. The following year, he was named associate professor of chemistry at the California Institute of Technology, and in 1940 he was named professor of bio-organic chemistry in the division of biology. His work during the 1940s focused on an extensive examination of essential oils—the volatile **liquids** extracted from plants by various means. His work on these essential oils was of great importance to the food industry. As Haagen-Smit explained in the May, 1961 issue of *Engineering and Science,* "an exact knowledge of these lost flavors becomes of decisive importance in the reconstitution of foods so that they regain most of their original quality." Other industries also were enriched by Haagen-Smit's work. The paint industry, for example, benefited from his research on the oil turpentine.

Following World War II, the residents of southern California experienced difficulty in breathing and a burning in the eyes when weather conditions caused fog and haze to blanket the region. Haagen-Smit was asked by several government agencies to try and find a solution. By using techniques originally developed in his work on essential oils, he determined that the phenomena called smog was the result of a chemical change in the atmosphere induced by an oxidation process. Specifically, he showed that the two principal ingredients of smog were the by-product of petroleum **combustion** in automobiles, and **nitrogen oxide** fumes being spewed into the air when fuels were burned by local industries. He deduced that since a by-product of oxidation is ozone, he could create smog in a laboratory by subjecting the gaseous ozone to **gasoline** fumes wafting out of a test tube. His experiment worked, and—according to *Current Biography*—Haagen-Smit told members of the National Municipal League, "It was luck. We hit the jackpot with the first nickel." Haagen-Smit developed an ozone test to measure smog intensity marking a major breakthrough in air pollution control.

Haagen-Smit became an outspoken advocate for establishing air pollution standards to promote clean air. He campaigned diligently to get industry to filter smoke stack fumes and was among the earliest to urge the automobile industry to develop hardware to filter exhaust vapors. He served as a consultant to the Los Angeles Air Pollution Control District, the United States Public Health Service, and the California Department of Public Health. From 1965 until his retirement in 1971, Haagen-Smit was the director of the Plant Environmental Laboratory at the California Institute of Technology. For his work in several fields of biochemistry he received many awards, including two by the American Chemical Society, of which he was a trustee: the Fritizche award in 1950, and the Tolman Award in 1964. He was also a recipient of the Smithsonian Medal, the $50,000 Alice Tyler Ecology Prize, and the Rhineland Award. In 1947, the Netherlands government conferred on him the Knight Order of Orange-Nassau. Haagen-Smit was elected a Fellow of the New York Academy of Sciences and the Royal Academy of Sciences of the Netherlands.

He married Petronella Francina Pennings in 1930, and before her death in 1933, they had a son, Jan Willem Adrianus. Maria Wilhelmina Bloemers became his second wife on June 10, 1935. Haagen-Smit died of cancer on March 17, 1977, in Pasadena, California. He was survived by his wife Maria and three daughters: Maria Van Pelt, Margaret Scott, and Joan Demers.

Haber-bosch process

The Haber-Bosch Process is a catalytic reaction in which **ammonia** is produced from **hydrogen** and **nitrogen gases**. The ammonia produced is used for production of artificial **fertilizers**. Prior to the invention of this process the United States and European countries purchased **sodium** and **potassium** nitrates from Chile and used them as natural fertilizers. During World War I, Germany was using nitrogen for **explosives**. The allies set up a naval blockade preventing them from receiving imports and another source of nitrogen was needed. In 1913, **Fritz Haber** developed the process for which he received the 1918 Noble prize in **Chemistry**. Although Fritz Haber invented the process, an engineer working for his company, **Karl Bosch** solved the difficult problems associated with the reaction conditions.

The reaction is a reversible reaction and requires a set of optimum conditions that favor the forward reaction, the formation of ammonia. It is a good example of **Le Chatelier's principle.** Optimum conditions include sufficient amounts of nitrogen and hydrogen, a **temperature** between 400°C and 500°C, a pressure of 100-350 atmospheres, a catalyst of **iron** mixed with oxides of other **metals**, and the removal of ammonia as it is formed. This process will give a 40%-60% yield of ammonia.

The United States produces about 22 billion lbs (10 billion kgs) of ammonia per year. This process is the first industrial process based on chemistry knowledge. Artificial fertilizers have allowed the agricultural industry to provide a greater food supply to meet the needs of a growing population.

See also Fertilizers, agricultural

Haber, Fritz (1868-1934)

German chemist

One the foremost chemists of his generation, Fritz Haber's legacy did not end with his considerable achievements of both theoretical and practical value in the fields of physical **chemistry**, **organic chemistry**, physics, and engineering. The Kaiser Wilhelm Institute for Chemistry, under his direction, became famous in the years after World War I as a leading center of research whose seminars attracted scientists from all nations. In his most outstanding contribution to chemistry—for which he won the 1918 Nobel Prize in Chemistry—Haber found an inexpensive method for synthesizing large quantities of **ammonia** from its constituent elements **nitrogen** and **hydrogen**. A steady supply of ammonia made possible the industrial production of fertilizer and **explosives**.

Haber was born on December 9, 1868, in Breslau (now Wroclaw, Poland), the only child of first cousins Siegfried Haber and Paula Haber. Haber's mother died in childbirth. In 1877, his father, a prosperous importer of **dyes and pigments**, married Hedwig Hamburger, who bore him three daughters. Haber and his father had a distant relationship, but his stepmother treated him kindly. From a local grade school, Haber went to the St. Elizabeth Gymnasium (high school) in Breslau. There he developed an abiding love of literature, particularly the voluminous writings of Goethe, which inspired him to write verse. Haber also enjoyed acting, considering it as a profession early on before settling on chemistry.

After entering the University of Berlin in 1886 to study chemistry, Haber transferred after a semester to the University of Heidelberg. There, under the supervision of Robert Bunsen (who gave his name to the burner used in laboratories everywhere), Haber delved into **physical chemistry**, physics, and mathematics. Getting his Ph.D. in 1891, Haber tried working as an industrial laboratory chemist but found its rigid routines too intellectually confining. He decided instead to enter the Federal Institute of Technology in Zurich, Switzerland, in order to learn about the most advanced **chemical engineering** techniques of his time, studying under Georg Lunge.

Haber then tried, without success, to work in his father's business, opting after six months to return to academia. In 1894, after a brief stint at the University of Jena, he took an assistant teaching position with Hans Bunte, professor of chemical technology at the Karlsruhe Technische Hochschule in Baden. Haber enjoyed Karlsruhe's emphasis on preparing its students for technical positions, stressing the connections between science and industry. His studies led him to investigate the breakdown at high temperatures of organic compounds known as hydrocarbons, an area pioneered by the French chemist Marcelin Berthelot. After correcting and systematizing Berthelot's findings, Haber's results, published in 1896 as a book entitled *Experimental Studies on the Decomposition and Combustion of Hydrocarbons,* led to his appointment that year as lecturer, a step below associate professor.

Haber married another chemist, Clara Immerwahr, in 1901. They had a son, Hermann, born in June, 1902. While a lecturer, Haber moved his experimental focus from organic chemistry to physical chemistry. Although he lacked a formal education in this area, with the help of a colleague, Hans Luggin, he began to research the effect of electrical currents on fuel cells and the loss of efficiency in steam engines through **heat**. Haber also devised electrical instruments to measure the loss of **oxygen** in burning organic compounds, outlining this subject in a book published in 1898, *Outline of Technical Electrochemistry on a Theoretical Basis,* which earned him a promotion to associate professor. Haber's exceptional abilities as a researcher, which included his precision as a mathematician and writer, induced a leading German science group to send him in 1902 to survey America's approach to chemistry in industry and education.

Haber published a third book, *Thermodynamics of Technical Gas Reactions,* in 1905. In the **volume** he applied thermodynamic theory on the behavior of **gases** to establish industrial requirements for creating reactions. His clear exposition gave him an international reputation as an expert in adapting science to technology. That same year, Haber began his groundbreaking work on the synthesis of ammonia. Europe's growing population had created a demand for an increase in agricultural production. Nitrates, used in industrial fertilizer, required ammonia for their manufacture. Thus, Haber's goal to find new ways to fabricate ammonia grew out of a very pressing need. Other scientists had been synthesizing ammonia from nitrogen and hydrogen but at temperatures of 1000°C, which were not practical for industrial production. Haber was able to get the same reaction but at manageable temperatures of three hundred degrees centigrade.

The chemist **Walther Nernst** had obtained the synthesis of ammonia with gases at very high pressures. He also had disputed Haber's results for his high-temperature reaction. Goaded by Nernst's skepticism, Haber executed high-pressure experiments and confirmed his earlier calculations. He then combined Nernst's technique with his own to greatly increase the efficiency of the process. To augment the yield even further, Haber found a superior catalyst for the reaction and redirected the heat it produced back into the system to save on the expenditure of **energy**.

The final step of bringing Haber's work into the factory fell to the engineer **Karl Bosch**, whose company, Badische Anilin- und Sodafabrik (BASF), had supported Haber's research. After Bosch solved some key problems such as designing containers that could withstand a corrosive process over a period of time, full-scale industrial output began in 1910. Today the **Haber-Bosch process** is an industry standard for the mass production of ammonia.

In 1912 Haber was appointed director of the newly formed Kaiser Wilhelm Institute for Physical Chemistry and Electrochemistry at Dahlheim, just outside of Berlin; **Richard Willstätter** and Ernst Beckmann joined as codirectors. With the outbreak of World War I in 1914, Haber volunteered his laboratory and his expertise to help Germany. At first, he developed alternate sources of antifreeze. Then, the German War Office consulted both Nernst and Haber about developing a chemical weapon that would drive the enemy out of their trenches in order to resume open warfare. In January, 1915, the

Fritz Haber. ***(AP/Wide World. Reproduced by permission.)***

German Army began production of a **chlorine** gas that Haber's team had invented. On April 11, 1915, in the first chemical offensive ever, five thousand cylinders of chlorine gas blanketed 3.5 miles of enemy territory near Ypres, Belgium, resulting in 150,000 deaths.

Haber claimed that he hated the war but hoped that in developing the gases he would help to bring it to a speedy end by breaking the deadlock of trench warfare. His wife, however, denounced his work as a perversion of science. After a violent argument with Haber in 1915, she committed suicide. Haber was married again in 1917 to Charlotte Nathan, who bore him a son and a daughter. Their marriage ended in divorce in 1927.

In 1916 Haber was appointed chief of the Chemical Warfare Service, overseeing every detail in that department. His process for developing nitrates from ammonia became incorporated into Germany's manufacture of explosives. Because of his duties as supervisor of chemical warfare, American, French, and British scientists vehemently contested his 1918 Nobel Prize in Chemistry. Although many of the Allied scientists had also contributed to the war effort, they charged that Haber was a war criminal for developing **chemical weapons**.

Since the 1918 prize had been reserved for until after the war ended, Haber accepted his Nobel Prize in November, 1919. Unquestionably, Haber had invented, in the words of the prize's presenter, A. G. Ekstrand of the Royal Swedish Academy of Sciences, "an exceedingly important means of improving the standards of agriculture and the well-being of mankind." Yet the controversy over his award, on top of Germany's defeat, his first wife's suicide, and his developing diabetes, depressed Haber greatly.

Nevertheless, Haber continued to turn his technical acumen to patriotic ends. In 1920, to help Germany pay off the onerous war reparations that the Versailles Treaty had imposed, Haber headed a doomed attempt to recover **gold** from seawater. Unfortunately, he had based his project on unverified nineteenth-century mineral analyses that had grossly overestimated the quantities for gold. It turned out after several abortive sea voyages that there was simply not enough gold present in seawater to make refining profitable. However, Haber did perfect a very precise method for measuring concentrations of gold.

Haber had much greater success as continuing director of the Kaiser Wilhelm Institute. His proven leadership ability attracted some of the best talent in the world to his laboratory in Karlsruhe and to the Institute, where in 1929 fully half of the members were foreigners from a dozen countries. In 1919 he began the Haber Colloquium, an ongoing seminar that during the postwar years brought together the best minds in chemistry and physics, among them **Niels Bohr**, **Peter Debye**, **Otto Meyerhof**, and **Otto Warburg**. Haber's sharp wit, critical intelligence and broad knowledge of science were greatly appreciated at the seminars. When he ceased attending regularly, they became markedly less popular. Haber traveled widely to foster greater cooperation between nations. As an example, he helped establish the Japan Institute in that nation to foster shared cultural interests with other countries. From 1929 to 1933 he occupied Germany's seat on the Union Internationale de Chimie.

When the Nazis came to power in 1933, the Kaiser Institute fell on hard times. After receiving a demand from the minister of art, science, and popular education to dismiss all Jewish workers at the institute, Haber—a Jew himself—resigned on April 30, 1933. He wrote in his letter of resignation that having always selected his collaborators on the basis of their intelligence and character, he could not conceive of having to change so successful a method.

Haber fled Germany for England, accepting the invitation of his colleague William J. Pope to work in Cambridge, where he stayed for four months. **Chaim Weizmann**, a chemist who would become the first president of Israel, offered Haber the position of director in the physical chemistry department of the Daniel Sieff Research Institute at Rehovot, in what is now Israel. Despite ill health, Haber accepted and in January, 1934, after recovering from a heart attack, began the trip. Resting on the way in Basel, Switzerland, he died on January 29, 1934. His friend and colleague Willstätter gave the memorial speech at his funeral. On the first anniversary of his death, over five hundred men and women from cultural societies across Germany converged on the institute—despite Nazi attempts at intimidation—to pay homage to Haber.

HACKERMAN, NORMAN (1912-)

American chemist

Norman Hackerman is internationally recognized as an expert in metal **corrosion**. He has devoted most of his scientific study to the **electrochemistry** of **oxidation**, and his work has resulted in the development of a number of processes—invaluable to the **metals** industry—which slow or prevent corrosion.

Hackerman was born March 2, 1912, in Baltimore, Maryland. He earned his undergraduate (1932) and doctoral degrees (1935) in **chemistry** at Johns Hopkins University and immediately embarked on a career as an educator by accepting an assistant professorship in the **physical chemistry** department at Loyola College in Maryland. Hackerman also worked concurrently as a research chemist for the Colloid Corporation until 1940, when he accepted a position as an assistant chemist with the United States Coast Guard.

Hackerman briefly returned to academic life in 1941 as an assistant professor of chemistry at Virginia Polytechnic Institute. Two years later, Hackerman accepted a post as a research chemist for the Kellex Corporation. In 1945, he accepted a post as assistant professor of chemistry with the University of Texas at Austin. During his 25-year tenure at the university, Hackerman served in numerous capacities, including chairman of the department of chemistry (1952-1962), director of the corrosion research laboratory and dean of research and sponsored programs (1961-1962), vice president and provost (1962), vice chancellor of academic affairs (1963-1967), and university president from 1967 to 1970. In 1970, he joined the faculty of Rice University as a professor of chemistry and served as president of the university, holding both appointments until his retirement in 1985. Rice University named him emeritus president and distinguished emeritus professor of chemistry in 1985, and the University of Texas at Austin honored him as an emeritus professor of chemistry in the same year.

Committed to educating the next generation of scientists—many of his students have earned the recognition of the international science community—Hackerman continued to pursue his own research despite the demands of teaching and administrative duties.

Hackerman's research has focused on the electrochemistry of corrosion. His goal has been to probe how solid metals react at the point of contact with solutions and the process of oxidation. By studying how the structures of molecules affect their function, the action of electrons, and how absorption strains molecular bonds, Hackerman has contributed to the development of products that retard corrosion and metal manufacturing techniques that produce corrosion-resistant metals. His research made it clear that simply preventing physical absorption does not prevent corrosion. Such understanding was crucial to manufacturing concerns that had emphasized protective coatings, for example, paint on automobiles, rather than the electrochemical nature of metals and the complex process of corrosion.

While continuing his work at Rice University, Hackerman broadened his research arena by accepting an appointment with the Robert A. Welch Foundation in 1982. As chairman of the foundation's Scientific Advisory Board, Hackerman again juggled administrative responsibilities with scientific investigation.

Although he retired from his academic post more than a decade ago, Hackerman continues his research with the Welch Foundation. His current investigations focus on thermally grown surface oxides, the electronic structure of surface films, and suppressing corrosion through passive and inhibitory techniques.

Since the beginning of his career, Hackerman has accepted the obligation to serve his profession and his nation as an integral part of his work. During World War II, he put his considerable talents at the disposal of the U.S. Coast Guard. A member of the national science board of the National Science Foundation since 1968, he served as board chairman from 1974 to 1980. Hackerman has also been a member of the National Board of Graduate Education and participated in Texas Governor's Task Force on Higher Education. He was the chairman of the board of **energy** studies for the National Academy of Science/NRC Commission on Natural Resources, a member of the Energy Research Advisory Board, and a member of the environmental **pollution** panel of the President's Scientific Advancement Committee.

Active in a number of academic and scientific organizations, Hackerman's numerous professional affiliations ranged from the chairmanship of the Gordon Research Conference on Corrosion in 1952 to serving as editor of the *Journal of Electrochemistry* since 1969. He is a member of the American Chemical Society, a member and former president of the Electrochemical Society and a member of the National Academy of Sciences.

Hackerman's work has been recognized with numerous awards and honors, including the Whitney Award from the National Association of Corrosion Engineers in 1956, the Palladium Medal from the Electrochemical Society in 1965, the Gold Medal of the American Institute of Chemists in 1978, the Charles Lathrop Parsons Award from the American Chemists Society in 1987, and the National Medal of Science in 1993. He is the author of more than 200 articles.

Hafnium

Hafnium is a transition metal, one of the elements in the middle of the **periodic table**. It is found in Group 4 and Row 6 of the table. Hafnium's **atomic number** is 72, its atomic **mass** is 178.49, and its chemical symbol is Hf.

Properties

Hafnium is a bright, silvery gray metal that is very ductile. Its melting point is 3,900°F (2,150°C), its **boiling point** is about 9,750°F (5,400°C), and its **density** is 13.1 grams per cubic centimeter. A very useful property of hafnium is its ability to absorb large quantities of neutrons per square centimeter of surface area. Chemically, hafnium is not very active. It reacts with **oxygen**, acids, and other reagents only at higher temperatures.

Hafnium. ***(Photograph by Russ Lappa/Science Source/National Audubon Society Collection/Photo Researchers, Inc. Reproduced by permission.)***

Occurrence and Extraction

Hafnium is a moderately common element in the Earth's crust with an abundance estimated at about 5 parts per million. The element is almost invariably found in conjunction with ores of **zirconium**, especially zircon and baddeleyite. Hafnium is separated from zirconium with considerable difficulty, usually by using the differential **solubility** of the two elements in various **solvents**.

Discovery and Naming

For most of history, hafnium and zirconium were not clearly distinguished from each other. The two elements usually occur together and they have similar chemical properties. Hafnium was finally identified as a distinct chemical element in 1923 by the Dutch physicist Dirk Coster (1889-1950) and the Hungarian chemist **Georg von Hevesy** by x-ray analysis of the mineral Norwegian zircon. The element's name is derived from the ancient name of the city of Copenhagen, Hafnia.

Uses

Hafnium and its compounds have relatively few commercial uses. The metal is sometimes used in the manufacture of control rods for nuclear power plants because of its strong attraction for neutrons. Some compounds of hafnium are also used as refractory substances to line the inside of high-temperature ovens.

Otto Hahn.

HAHN, OTTO (1879-1968)

German chemist

Otto Hahn is noted for his work on radioactive materials, which in 1938 led to his discovery, with physicist **Lise Meitner** and chemist **Fritz Strassmann**, of the process of **nuclear fission**. In recognition of their work, Hahn and Strassmann received the 1944 Nobel Prize in **chemistry**, and Hahn, Strassmann, and Meitner received the Fermi Award in 1966. Hahn was born in Frankfurt-am-Main on March 8, 1879, to Heinrich Hahn, a glazier, and Charlotte Giese Stutzmann Hahn. The Hahns' early years in Frankfurt were marked by poverty: according to R. Spence, writing in the *Biographical Memoirs of Fellows of the Royal Society,* the four Hahn boys—Otto, his two brothers and his half-brother from Charlotte's first marriage—"slept in an unheated attic bedroom and took their weekly bath in a tub on the landing." Gradually, Heinrich's business became more successful, and the family attained "middle-class respectability." Otto, who attended the Klinger Realschule, demonstrated some early interest in science, carrying out simple chemical experiments in the family laundry house. But other subjects seemed more important to him, and the honors he received upon graduation were for gymnastics, religious studies, and music. Hahn's parents had hoped that he would pursue a career in architecture, but when he entered the University of Marburg in 1897, it was a course in chemistry that he decided to pursue. He interrupted his studies at Marburg to spend one year at the University of Munich, but then returned to Marburg, where he was awarded his doctorate in **organic chemistry** in 1901.

Biographers—and Hahn himself—mention Hahn's devotion to non-academic pursuits in college, especially cigar smoking and beer drinking. He felt obligated to join one of the student societies that were ubiquitous in German universities and on one occasion even challenged another student to a duel. Hahn's membership in the Nibelungia Society apparently brought him considerable happiness until he resigned in the 1930s in opposition to Nazi policies adopted by the group.

After graduation, Hahn enlisted in the infantry for one year and then returned as an instructor at Marburg. Soon thereafter, hoping to better his chances at a job in the German chemical industry, Hahn decided to spend a year in an English-speaking institution where he could polish his language skills while advancing his knowledge in chemistry. Through the efforts of a former teacher, Hahn was offered a research post with Sir **William Ramsay** in his laboratory at University College, London. Hahn left for England in September, 1904.

Ramsay's fame rests primarily on his discovery of the inert **gases argon**, **neon**, **krypton**, and **xenon**. In the early 1900s, however, he had become interested in a new topic, **radioactivity**. When Hahn arrived in London, it was a problem in radioactivity, therefore, to which Ramsay assigned him. In some ways that decision was a peculiar one, since nothing in Hahn's background in organic chemistry had prepared him for research on radioactive materials. Ramsay seemed to believe that Hahn's lack of background in radioactivity might be an advantage, since he could proceed with no preconceived notions as to what to expect. The problem he assigned the young German chemist was to extract **radium** from a 100-gram sample of **barium** carbonate. After familiarizing himself with this new field of research, Hahn completed Ramsay's assignment, obtaining a few milligrams of radium. He discovered, however, that the radioactivity of the sample was greater than expected and eventually isolated a second radioactive material from the impure radium. He called the new substance "radiothorium," later identified as an **isotope** of **thorium** that decays into thorium-x.

At the conclusion of this research, both Ramsay and Hahn were convinced that the latter's future lay in radioactivity, not organic chemistry. Thus, Ramsay obtained for Hahn an appointment at Emil Fischer's Chemical Institute at the University of Berlin, where he could pursue this new interest. In preparation for the Berlin post, however, Hahn decided to spend another year of study, this time with the world's foremost authority on radioactivity, **Ernest Rutherford**, at McGill. During Hahn's year at McGill (1905–1906) he discovered a second radioactive substance, radioactinium, now known to be an isotope of thorium that decays into actinium-x.

Hahn began work at the Chemical Institute in Berlin in the fall of 1907, was appointed a Privat dozent (university lecturer) a year later, and was promoted to professor of chemistry in 1910. One of the most significant events of this period was the arrival of Lise Meitner as a student at the Chemical Institute. Prejudice against women in the sciences was very strong at the time, and Meitner was not allowed to work in the same laboratories as male students. Fischer did, however, allow her to share a tiny makeshift laboratory with Hahn in a converted workshop. Hahn and Meitner worked well together, with the former's chemical approach to problems complementing Meitner's outlook as a physicist. Thus a collaboration began that was to last for three decades. During their work together at the Chemical Institute, Hahn and Meitner concentrated on a study of beta emitters and, in the process, discovered more new radioactive isotopes.

In 1912, Kaiser Wilhelm authorized the establishment of a new research institute to consist of several separate departments. Invited to head up the section on radioactivity in the new Institute for Chemistry at Berlin-Dahlem, Hahn asked Meitner to join him there. One of the advantages of the new setting—in addition to much more space—was that radioactive materials had never been used in the rooms before, so that Hahn and Meitner were able to detect far lower levels of radiation than they could in their former laboratories. This allowed them to discover weakly radioactive isotopes of **potassium** and **rubidium** that had not yet been observed.

During this period, Hahn met Edith Junghans, whose father was a member of the Stettin City Council. The two became engaged in November of 1912. They were married in Stettin in March, 1913, and eventually had one child, a son, Hanno, born in April of 1922.

Hahn's personal and professional life were soon to be disrupted by the onset of World War I. In July, 1914, he was called into the army. After serving with distinction in the infantry at the battlefront, he was ordered back to Berlin, where he was assigned to work with a poison gas research unit headed by **Fritz Haber**. For the next three years, Hahn shuttled back and forth between Berlin and the battlefronts, developing and testing new gases. He was horrified by the results he saw when gases were used in battle, but he had become convinced that such atrocities might bring the war to an early end and save lives, the same argument eventually made for the use of nuclear bombs—built on the principle of nuclear fission discovered by Hahn—at the end of World War II.

During his stays in Berlin, Hahn was able to spend some time in his own laboratory at the Institute for Chemistry. One of the projects that he and Meitner pursued during this period was a study of a new radioactive element that had previously been announced by Kasimir Fajans and Oswald Göhring in 1913, an element they had named "brevium." Hahn and Meitner found the element in the residues of pitchblende ore and showed its relationship to parent and daughter isotopes. The name they suggested for the element, "protoactinium" (now protactinium), eventually became preferred to that recommended by Fajans and Göhring.

By the 1930s, Hahn's fame had begun to spread. In 1933, he was invited to spend a year as visiting professor at Cornell University and to deliver the prestigious Baker Lectures there. His visit to the United States was cut short, however, by news of the Nazi uprisings taking place in Germany. He decided that it was best for him to return to Berlin, which he did in the summer of 1933.

One of the great ironies of the 1930s was that while political turmoil was sweeping through the world's greatest scientific nation, Germany, momentous scientific discoveries with profound historical significance were also being made there. Hahn returned to a Germany where many of the world's finest scientists were either being expelled from their university posts or were fleeing Hitler's wrath for the United Kingdom, the United States, or other destinations.

In the midst of the political chaos around him, Hahn turned his attention to an exciting new field of research: the neutron bombardment of uranium and thorium. Two important recent discoveries convinced Hahn that he needed to go beyond the now nearly exhausted field of radioactivity. The first of these was the discovery by **Irène Joliot-Curie** and her husband **Frédéric Joliot-Curie** that an element can be made radioactive by bombarding it with alpha particles. The second was the discovery of the neutron by James Chadwick in 1932. These two discoveries had been utilized in the early 1930s by Enrico Fermi, who saw that neutrons would be far more effective "bullets" for bombarding atomic nuclei than were alpha or beta particles. In a short period of time, Fermi had used this technique to convert dozens of stable elements to radioactive forms. He was especially interested in trying this technique on uranium, since the predicted result of that reaction would be an element with an **atomic number** one greater than uranium. Since that element had never been observed in nature, a successful result of this experiment would be the formation of the world's first synthetic element. When Fermi carried out this experiment, the results he obtained were ambiguous, and he could draw no firm conclusion as to what had happened as a result of bombarding uranium with neutrons.

It was this problem to which Hahn turned in 1934. To work with him and Meitner on the problem, Hahn selected another radiochemist who had joined the Institute in 1929, Fritz Strassmann. Over the next four years, Hahn, Meitner, and Strassmann analyzed the complex mixture of isotopes formed when uranium is bombarded with neutrons. At first, they worked on the assumption that Fermi's hypothesis was correct and that isotopes with atomic numbers from about 90 to about 94 would predominate in the mixture. After all, the only **nuclear reactions** with which scientists were then familiar were those in which the atomic numbers of products differed from those of the original material by only one or two. By 1938, however, Hahn was convinced that something very different was taking place in this reaction. The evidence from his chemical analyses had become overwhelming. The product that had originally seemed to be an isotope of radium was actually an isotope of barium, whose atomic number is 36 less than that of uranium. Hahn and Strassmann were not completely willing to accept the apparent meaning of these results, however. In their January 6, 1939, paper announcing their results, they noted that, as chemists, they felt certain that barium was one of the products of the reaction, but they admitted that as nuclear physicists they were not yet willing to accept the "big step" that this conclusion suggested.

The paper carried the names of Hahn and Strassmann because their colleague Meitner had been forced to flee Germany as a result of the Nazi purge of Jewish scientists. From her new home in Sweden, however, Meitner stayed in contact with her colleagues in Berlin. When details of the forthcoming Hahn-Strassmann paper reached her, she continued to think about the problem. The solution came during a Christmas outing with her nephew **Otto Robert Frisch**. Meitner and Frisch finally realized that the Hahn-Strassmann results could only be explained by accepting that the uranium **nucleus** had been split into two large, roughly equal parts, a reaction to which Meitner gave the name *nuclear fission*. For his part in the discovery of nuclear fission, Hahn was awarded a share of the 1944 Nobel Prize in chemistry.

During World War II, Hahn remained in Germany. Although he had no love for the Nazi party, he felt a loyalty to his homeland. He was able to avoid working on the German atomic bomb project, however, and instead carried out research on fission fragments during the war. That research came to an end in March, 1944, after the Institute for Chemistry was destroyed in a series of bombing raids. Hahn and his wife soon moved to the southern German town of Tailfinger, where they were captured by an advance team of U.S. intelligence officers in April, 1945. The Hahns were sent to England, where they were held for almost a year.

On January 3, 1946, Hahn was permitted to return to Germany, where he became president of the Kaiser Wilhelm Society, then re-named the **Max Planck** Society. He devoted the next 15 years to the effort to rebuild the scientific community of his native land. In his later years, he was showered with honors and awards, including election to scientific societies in Berlin, Göttingen, Munich, Halle, Stockholm, Vienna, Madrid, Helsinki, Lisbon, Mainz, Rome, Copenhagen, and Boston. He also became active in the international movement to control nuclear weapons and helped draft the 1955 Mainau Declaration by Nobel laureates, warning of the dangers of misusing nuclear **energy**.

The last years of Hahn's life were filled with personal tragedy. He was shot in the back in 1951 by a disgruntled inventor and had scarcely recovered before his wife suffered a nervous breakdown. In 1960, his son Hanno and Hanno's wife were killed in an automobile accident in France, leaving their young son to Hahn's care. Distraught at the accident, Hahn's wife never recovered her health. In the spring of 1968, Hahn was seriously injured while getting out of a car. His health slowly deteriorated until he died on July 28. His wife also died two weeks later. He was buried in Stadtfriedhof, Göttingen, Germany.

HALF-LIFE

The half-life (or half-value period) of a substance is the time required for that substance to reduce to a size half of its initial value.

The half-life is most commonly encountered when talking about radioactive decay. Radioactive elements have different isotopes that decay at different rates. As a result the half-life has to be given in terms of the particular **isotope** under discussion. Some isotopes have very short half-lives, for example oxygen-14 has a half-life of only 71 seconds, some are even shorter with values measured in millionths of a second not being uncommon. Other elements' isotopes can have a much longer half-life, thallium-232 has a half-life of 1.4×10^{10} years and carbon-14 has a half-life of 5,730 years. This latter figure is used as the basis of radiocarbon dating of once-living organisms. While living, an organism takes in an amount of carbon-14 at a relatively constant rate. Once the organism dies no more carbon-14 is taken in and the amount of carbon-14 present overall starts to decrease, decreasing by half every 5,730 years. By measuring the ratio of carbon-12 to carbon-14 an estimate of the date when carbon-14 stopped being assimilated can be calculated. This figure can also be obtained by comparing the levels of **radioactivity** of the test material to that of a piece of identical material that is fresh. Other radioactive elements can be used to date older materials such as rocks.

Strontium-90 has a half-life of 29 years. If starting with a 2.2 lb (1 kg) **mass** of strontium-90, then after 29 years there will only be 1.11 lb (0.5 kg) of strontium-90 remaining. After a further 29 years there will only be 0.55 lb (0.25 kg). Strontium-90 decays to give yttrium-90 and one free **electron**. It can be seen from this example that the half life is independent of the mass of material present.

The half-life ($t_{1/2}$) of a material can be calculated by dividing 0.693 by the decay constant (which is different for different radionucleotides). The decay constant can be calculated by dividing the number of observed disintegrations per unit time by the number of radioactive nuclei in the sample. The decay constant is usually given the symbol k or λ.

When studying chemical reactions the presence of a half-life can also be seen. The half-life in this context is the time taken for a substance to change its **concentration** to half of its initial value, it is an indication of the rate of the reaction. As with radioactive decay this half life is independent of the concentration. Reactions that proceed very rapidly have a short half-life and slow reactions have a long half-life. The half-life in a chemical reaction must have all participants named otherwise the figure is meaningless.

The half-life of a material is a measure of how reactive it is either in terms of radioactive decay or in participation in a named reaction.

HALIDE

Halides are defined as binary compounds containing a halogen. Because **halogens** are highly electronegative, their atoms normally retain negative charge in these compounds. The magnitude of this negative charge depends on the difference in the two constituent's electronegativities.

The nature of the bonds between molecules determines the physical properties of the halides, and the distinction between ionic and covalent halides provides a basis for understanding these molecules. In the *ionic* halides, such as **sodium** chloride, all of the **bonding** electrons are transferred to the halogen, producing a halide **ion**. Interionic forces are large, so these halides have high melting and boiling points. In *covalent* halides, such as **carbon** tetrachloride, the bonding electrons are localized in the carbon-chlorine bonds of the individual molecules. The **intermolecular forces** in covalent halides are so weak that these molecules usually exist as **gases**, **liquids**, or low melting-point **solids**. Many elements form fluorides, however, that cannot be labeled covalent or ionic; these halides exhibit the characteristics of polymeric compounds.

The group IA elements (**lithium**, sodium, **potassium**, **rubidium**, **cesium**) readily lose their single **valence electron** to form singly charged cations. All of the halides of these elements are ionic compounds, and they all have high melting points. Examples include LiF, NaCl, and NaI.

The group IIA and IIB elements (**beryllium**, **magnesium**, **calcium**, **strontium**, **barium**, **zinc**, **cadmium**, **mercury**) also tend to form ionic halides. Examples include $MgCl_2$, and $CaBr_2$.

The group III and IV elements (**boron**, **aluminum**, **gallium**, **indium**, carbon, **silicon**, **germanium**, **tin**), on the other hand, tend to form covalent halides. Examples include BF_3 and $SiCl_4$.

The group V and VI elements (**nitrogen**, **phosphorus**, **arsenic**, **antimony**, **oxygen**, **sulfur**, **selenium**, **tellurium**) form covalent halides exclusively. Examples include NF_3 and SCl_2.

The group VII elements (**fluorine**, **chlorine**, **bromine**, and **iodine**) can form simple halides like I_2, but also more complex molecules like ClF_3.

See also Bonding; Elements; Polymer chemistry

Hall, Charles Martin (1863-1914)

American chemist

Charles Martin Hall was born in Thompson, Ohio, in 1863, the son of a minister. He attended Oberlin College in Oberlin, Ohio. There Hall studied **chemistry** and was influenced by one of his professors, F. F. Jewett, himself a former student of **Friedrich Wöhler** who had developed a process of purifying small quantities of **aluminum** in 1845. Wöhler's extraction method was so difficult that aluminum was considered a semiprecious metal. A remark by Professor Jewett that whoever could develop an efficient method of producing aluminum would become rich was taken seriously by Hall. Even before his graduation in 1885, Hall was experimenting with the process.

Only one year later, Hall succeeded in extracting a sample of pure aluminum from cryolite using an electrolytic process, instead of the thermal process then in use. In a short time, Hall had succeeded where others had been stymied for 60 years. Hall spent the next two years seeking financial support for his discovery. In 1889 he patented his extraction process. With the assistance of financier Andrew Mellon (1855-1937), Hall and a group of businessmen founded the Pittsburgh Reduction Company in 1888, which had its factory at Kensington, Pennsylvania.

In 1890 Hall became vice-president of the company, which was renamed Aluminum Company of America (Alcoa) in 1907. It produced commercially salable units of aluminum with the use of 2,000 ampere cells and required no external heating. The price of aluminum—once \$100 per pound (453g)—was reduced to \$0.70 per pound. By 1914 it was further reduced to \$0.18 per pound.

Halides in hexane solution. ***(Photograph by Jerry Mason/Science Photo Library, Photo Researchers, Inc. Reproduced by permission.)***

In the same year that Hall had made his discovery (1886), Frenchman Paul-Louis-Toussaint Héroult developed the same process. Hall was able to get his process into commercial production faster than Héroult and won a patent dispute with Héroult in 1893. Despite this, Hall and Heroult became best friends. Hall remembered Oberlin College with a \$3 million donation. Alcoa still preserves the original aluminum nodules that Hall produced in 1886.

Aluminum has become, along with steel, one of the two most important **metals** in the world. It is in demand for its strength, durability, and beauty. Aluminum is currently used in a variety of industries, including construction, airplane production, wiring, cookware, and packaging.

Hall, Lloyd Augustus (1894-1971)

American chemist

Chemist Lloyd Augustus Hall is best known for his work in the field of food technology, where he developed processes to cure and preserve meat, prevent rancidity in fats, and sterilize spices. In 1939, he cofounded the Institute of Food Technologists, establishing a new branch of **industrial chemistry**. Hall was born in Elgin, Illinois, on June 20, 1894. His father, Augustus Hall, was a Baptist minister and son of the first pastor of the Quinn Chapel A.M.E. Church, the first African American church in Chicago. Hall's mother, Isabel, was a high-school graduate whose mother had fled to Illinois via the Underground Railroad at the age of sixteen.

Hall became interested in chemistry while attending the East Side High School in Aurora, Illinois, where he was active in extracurricular activities such as debate, track, football, and baseball. He was one of five African Americans attending the school during his four years there. By the time he graduated among the top ten in his class, he'd been offered scholarships to four Illinois universities.

Hall chose to attend Northwestern University, working his way through school while he studied chemistry. During this time, he met Carroll L. Griffith, a fellow chemistry student, who would later play a part in his career. Hall graduated in 1916 with a bachelor of science degree in chemistry and continued his studies in graduate classes at the University of Chicago.

During World War I, Hall served as a lieutenant in the Ordnance Department, inspecting **explosives** at a plant in Wisconsin. However, he was subjected to such prejudice and discriminatory behavior that he asked to be transferred. The discrimination was also apparent in the civilian world: at one point, he was hired over the telephone by the Western Electric Company. When he arrived for work, he was told there was none.

In 1916, however, he was able to find a position in the Chicago Department of Health Laboratories. Within a year, he was made senior chemist. For the next six years, he worked at several industrial laboratories. In 1921, he was made chief chemist at Boyer Chemical Laboratory in Chicago. By then, he'd become interested in the developing field of food chemistry, and in 1922, he became president and chemical director of the Chemical Products Corporation, a consulting laboratory in Chicago.

In 1924, one of Hall's clients, Griffith Laboratories (his old lab partner at Northwestern had been Carroll L. Griffith) offered him a space where he could work for them while continuing his consulting practice. By 1925, Hall had become chief chemist and director of research at Griffith; in 1929, he gave up his consulting practice and devoted himself full-time to Griffith until 1959.

When Hall started at Griffith, current meat curing and preservation methods were highly unsatisfactory. It was known that **sodium** chloride preserved meat, while chemicals containing **nitrogen**—nitrates and nitrites—were used for curing. However, not much was known about how these chemicals worked, and food could not be preserved for an extended period of time.

In experiments, Hall discovered that nitrite and nitrate penetrated the meat more quickly than the **sodium chloride**, causing it to disintegrate before it had a chance to be preserved. The problem was to get the **salt** to penetrate the meat first, thereby preserving it before it was cured. Hall solved this through "flash-drying"—a quick method of evaporating a **solution** of all three salts, so that crystals of sodium chloride enclosing the nitrite and nitrate were formed. Thus, when the crystals dissolved, the sodium chloride would penetrate the meat first.

Hall's next accomplishment was in the area of spices. Although meat could now be preserved and cured effectively, the natural spices that were used to enhance and preserve it often contained contaminants. Spices such as allspice, cloves, cinnamon, and paprika as well as dried vegetable products like onion powder contained yeasts, molds, and bacteria. Hall's task was to find a way to sterilize the spices and dried vegetables without destroying their original flavor and appearance. Heating the foods above 240°F would sterilize them, but it would also destroy their **taste** and **color**. Hall discovered that ethylene **oxide**, a gas used to kill insects, would also kill the germs in the spices. He used a vacuum chamber to remove the moisture from the spices so that the gas could permeate and sterilize them when introduced into the chamber. The times and temperatures varied according to the type of bacteria, mold, or yeast to be destroyed.

The ability to sterilize spices had a major impact on the meat industry. The process also became popular in the hospital supplies industry and was used to sterilize bandages, dressings, and sutures. In fact, a number of industries in the United States benefited enormously from Hall's invention.

In his work at Griffith, Hall also discovered the use of **antioxidants** in preventing rancidity in foods containing fats and oils. Rancidity is caused by **oxidation** when constituents in the fats react with **oxygen**. By experimenting with various antioxidants, Hall found that certain chemicals in crude vegetable oil worked as antioxidants. Using some of these combined with salt, he produced an antioxidant salt mixture that protected foods containing fats and oils from spoiling.

During his thirty-five years at Griffith, Hall worked in several areas of food chemistry including seasoning, spice extracts, and enzymes. In 1951, Hall and an associate developed a way to reduce the time for curing bacon from between six and fifteen days to a few hours. The quality of the bacon was also improved, both in appearance and stability. He was also very interested in **vitamins** and the development of yeast foods. By 1959, Hall held over 105 patents in the United States and abroad and had published numerous papers on food technology. Hall served on various committees during his time at Griffith. During World War II, he was a member of the Committee on Food Research of the Scientific Advisory Board of the War Department's Quartermaster Corps; in that position he advised the military on the preservation of food supplies from 1943 to 1948. In 1944, he joined the Illinois State Food Commission of the State Department of Agriculture, serving until 1949.

As a further sign of the establishment of food chemistry as a field of science, the Institute of Food Technologists was founded in 1939. Hall was a charter member; he edited its magazine, *The Vitalizer,* and served on its executive board for four years. In 1954, Hall became chairman of the Chicago chapter of the American Institute of Chemists. The following year, he was elected a member of its national board of directors, becoming the first African American man to hold that position in the Institute's thirty-two-year history.

Upon his retirement from Griffith in 1959, Hall continued to serve as a consultant to various state and federal organizations. He also continued to work, and in 1961, he spent six months in Indonesia, advising the Food and Agricultural Organization of the United Nations. From 1962 to 1964, he was a member of the American Food for Peace Council, an appointment made by President John F. Kennedy. After retiring, Hall and his wife, Myrrhene, moved to California to benefit her health. Hall lived in Altadena, where he remained active in community affairs, until his death on January 2, 1971.

Hallucinogens

Hallucinogens are chemicals that are capable of producing hallucinations, or altered states of perception. Hallucinogens are both naturally-occurring and artificial in nature. They include material from plants and mushrooms (mescaline and psilocybin) as well as artificial compounds such as LSD, PCP, and MDMA.

Of the naturally-occurring hallucinogens, some are plant alkaloids such as hyosine (sometimes called scopolamine). Hyosine is a tropine **alkaloid** which can be found in *Atropa belladonna* (deadly nightshade) and *Datura stramonium* (Thorn apple, Jimsonweed). The symptoms of hyosine usage include dryness of the mouth and throat, dilated pupils, uncoordination, nausea, delirium, coma, and may cause death. Ingestion of this material has been carried out in the form of tea.

Historically, the alkaloids have been used for medicinal purposes. They mimic various nerve transmitters such as acetylcholine and serotonin. Certain fungi has also been used among native peoples of North and South America for their hallucinogenic properties during religious rituals. The most common hallucinogenic fungi of the mushroom and toadstool type are the *Amanitas* and the *Psilocybes*. These genera have been used by the Hindus (Soma, a plant deity, has been tentatively identified with hallucinogenic fungi), and in Mexico and South America. *Amanita muscaria* is a common hallucinogenic fungi of the *Amanita* group. Eugster and Waser first isolated muscimol from this species and it is this substance which is the main hallucinogen. Ibotenic acid is also found in this plant, and it naturally converts to muscimol during **metabolism**. The hallucinogenic ingredient of Psilocybe is the chemical psilocybin. This mushroom is perhaps the most widely used hallucinogenic fungus, commonly known as the magic mushroom. Ergot *(Claviceps purpurea)* has a hallucinogenic component called ergotamine. This is an acid amide derivative of lysergic acid. The **carboxyl group** of lysergic acid has been reduced to a **methyl group**.

One of the most widely known artificial hallucinogens is LSD (lysergic acid diethylamide). Many naturally-occurring organisms, including the fungus ergot, contain LSD derivatives. Synthetic LSD was first made by Hoffman in the 1950s. MDMA (ecstasy) is a synthetic drug with mind altering properties. LSD binds to neurotransmitter sites normally employed by serotonin. This produces an abnormally high level of stimulation of the receptor, and produces mood swings, hallucinations, and delusions. MDMA causes an excessive release of serotonin which overloads the **receptors**. PCP (angel dust) is, pharmacologically speaking, not a true hallucinogen but it may cause many of the same effects by releasing dopamine from neurons. PCP can cause schizophrenia and long-term usage can lead to memory loss and speech disorders.

Hallucinogens have powerful mind-altering properties. They can alter how the brain perceives time, reality, and the environment. There is little conclusive evidence that the use of hallucinogens causes permanent brain damage, but there is a high correlation between high usage and chronic mental disorders. Some hallucinogens, such as PCP and MDMA are addictive, while others, such as LSD, psilocybin, and mescaline are not.

Halogens

The halogens are a group of chemical elements that includes **fluorine**, **chlorine**, **bromine**, **iodine**, and **astatine**. Halogen comes from Greek terms meaning to produce sea **salt**. None of the halogens occur naturally in the form of elements, but, except for astatine, they are very widespread and abundant in chemical compounds where they are combined with other elements. **Sodium** chloride, common table salt, is the most widely known.

All of the halogens exist as **diatomic** molecules when pure elements. Fluorine and chlorine are **gases**. Bromine is one of only two liquid elements, and iodine is a solid. Astatine atoms exist only for a short time and then decay radioactively. Fluorine is the most reactive of all known elements. Chemical activity, the tendency to form chemical compounds, decreases with **atomic number**, from fluorine through iodine. Simple compounds of these elements are called halides. When one of the elements becomes part of a compound its name is changed to an -ide ending, e. g. chloride.

Chlorine was the first halogen to be separated and recognized as an element. It was named in 1811 by **Humphry Davy** from a Greek term for its greenish yellow **color**. Huge deposits of solid salt, mostly **sodium chloride**, and salts dissolved in the oceans are vast reservoirs of chloride compounds. Salt is a general term for a metal and **nonmetal** combination; there are many different salts. To obtain chlorine, an electrical current is applied to brine, a **water solution** of sodium chloride. Chlorine gas is produced at one electrode. The chlorine must be separated by a membrane from the other electrode, which produces **sodium hydroxide**.

Chlorine gas itself is toxic. It attacks the respiratory tract and can be fatal. For this reason, it was used as a weapon during World War I. Chlorine in solutions has been used as a disinfectant since 1801. It was very effective in hospitals in the 1800s, particularly in an 1831 cholera epidemic in Europe. Chlorine **bleaches** are employed in most water treatment systems in the United States, as well as over much of the rest of the world and in swimming pools.

Chlorine will combine directly with almost all other elements. Large amounts are used yearly for making chlorinated organic compounds, bleaches, and inorganic compounds. Organic compounds, ones which have a skeleton of **carbon** atoms bonded to each other, can contain halogen atoms connected to the carbon atoms. Low **molecular weight** organic chlorine compounds are **liquids** and are good **solvents** for many purposes. They dissolve starting materials for chemical reactions, and are effective for cleaning such different items as computer parts and clothing (dry cleaning). These uses are now being phased out because of problems that the compounds cause in Earth's atmosphere.

Chlorine-containing organic polymers are also widely employed. Polymers are large molecules made of many small units that hook together. One is polyvinyl chloride (PVC), from which plastic pipe and many other plastic products are made. Neoprene is a synthetic rubber made with another chlorine-containing polymer. Neoprene is resistant to the effects of **heat**, **oxidation**, and oils, and so is widely used in automobile parts.

Many medicines are organic molecules containing chlorine, and additional chlorine compounds are intermediate steps in the synthesis of a variety of others. Most crop protection chemicals, **herbicides**, **pesticides**, and **fungicides** have chlorine in them. Freon **refrigerants** are chlorofluorocarbons (CFCs). These perform well because they are volatile, that is they evaporate easily, but they are not flammable. Freon 12, one of the most common, is CCl_2F_2—two chlorine atoms and two fluorine atoms bonded to a carbon **atom**.

Chlorine is part of several compounds, such as the insecticide DDT, that are soluble in fats and oils rather than in water. These compounds tend to accumulate in the fatty tissues of biological organisms. Some of these compounds are **carcinogens**, substances that cause cancer. DDT and other pesticides, polychlorinated biphenyls (PCBs), and dioxins are substances that are no longer manufactured. However, they are still present in the environment, and disposal of materials containing these compounds is a problem.

The next heaviest element in the halogen family is bromine, named from a Greek word for stink, because of its strong and disagreeable odor. It was first isolated as an element in 1826. Bromine is a reddish-brown liquid that vaporizes easily. The vapors are irritating to the eyes and throat. Elemental bromine is made by oxidation, removal of electrons from bromide ions in brine. Brines in Arkansas and Michigan are fairly rich in bromide. Other worldwide sources are the Dead Sea and ocean water.

There are a variety of applications for bromine compounds. The major use at one time was in ethylene dibromide, an additive in leaded gasolines. This need has declined with the phase-out of leaded fuel. Several brominated organic compounds have wide utilization as pesticides or disinfectants. Currently, the largest **volume** organic bromine product is methyl bromide, a fumigant. Some medicines contain bromine, as do some dyes.

Halons, or halogenated carbon compounds, have been utilized as flame retardants. The most effective contain bromine, for example, halon 1301 is $CBrF_3$. Inorganic bromine compounds function in water sanitation, and **silver** bromide is used in photographic film. Bromine also appears in quartz-halide light bulbs.

The heaviest stable halogen is iodine. Iodine forms dark purple crystals, confirming its name, Greek for violet colored. It was first obtained in 1811 from the ashes of seaweed. Iodine is purified by heating the solid, which sublimes, or goes directly to the gas state. The pure solid is obtained by cooling the vapors. The vapors are irritating to eyes and mucous membranes.

Iodine was obtained commercially from mines in Chile in the 1800s. In the twentieth century, brine from wells has been a better source. Especially important are brine wells in Japan, and, in the United States, in Oklahoma and Michigan.

Iodine is necessary in the diet because the thyroid gland produces a growth-regulating hormone that contains iodine. Lack of iodine causes goiter. Table salt usually has about 0.01% of sodium iodide added to supply the needed iodine. Other compounds function in chemical analysis and in synthesis in a **chemistry** laboratory of organic compounds. Iodine was useful in the development of photography. In the daguerreotype process, an early type of photography, a silver plate was sensitized by exposure to iodine vapors.

Astatine could be described as the most rare element on Earth. All isotopes, atoms with the same number of protons in the **nucleus** and different numbers of neutrons, are radioactive; even its name is Greek for unstable. When an atom decays, its nucleus breaks into smaller atoms, **subatomic particles**, and **energy**. Astatine occurs naturally as one of the atoms produced when the **uranium** 235 **isotope** undergoes radioactive decay. However, astatine doesn't stay around long. Most of its identified isotopes have half-lives of less than one minute. That is, half of the unstable atoms will radioactively decay in that time.

Astatine was first synthesized in 1940 in cyclotron reactions by bombarding **bismuth** with alpha particles. The longest-lived isotope has a **half-life** of 8.3 hours. Therefore, weighable amounts of astatine have never been isolated, and little is known about its chemical or physical properties. In a **mass** spectrometer, an instrument that observes the masses of very small samples, astatine behaves much like the other halogens, especially iodine. There is evidence of compounds formed by its combining with other halogens, such as AtI, AtBr, and AtCl.

Fluorine was the most difficult halogen to isolate because it is so chemically reactive. **Henri Moissan** first isolated elemental fluorine in 1886, more than 70 years after the first attempts. Moissan received the 1906 Nobel Prize for Chemistry for this work. The technique that he developed, **electrolysis** of **potassium** fluoride in anhydrous liquid **hydrogen** fluoride, is still used today, with some modifications. The name fluorine comes from the mineral fluorspar, or **calcium** fluoride, in which it was found. Fluorspar also provided the term fluorescence, because the mineral gave off light when it was heated. Hydrofluoric acid has been used since the 1600s to etch **glass**. However, it, as well as fluorine, must be handled with care because it causes painful skin burns that heal very slowly. Fluorine and fluoride compounds are toxic.

Fluorine is so reactive that it forms compounds with the noble gases, which were thought to be chemically inert. Fluorine compounds have been extremely important in the twentieth century. Uranium for the first atomic bomb and for **nuclear reactors** was enriched in the 235 isotope, as compared to the more abundant 238 isotope, by gaseous **diffusion**. Molecules of a uranium atom with six fluorine atoms exist as a gas. Less massive gases will pass through a porous barrier faster than more massive ones. After passage through thousands of barriers, the uranium hexafluoride gas was substantially enriched in the 235 isotope.

Fluoride ions in low concentrations have been shown to prevent cavities in teeth. Toothpastes may contain "stannous fluoride," and municipal water supplies are often fluoridated. However, too high a **concentration** of fluoride will cause new permanent teeth to have enamel that is mottled. Chlorofluorocarbons were developed and used as refrigerants, blowing agents for **polyurethane** foam, and propellants in spray cans. Their use became widespread because they are chemically inert. Once the active fluorine is chemically bound the resulting **molecule** is generally stable and unreactive. The polymer polytetrafluoroethylene is made into **Teflon**, a non-stick coating.

Most of the organic halogen compounds mentioned are made synthetically. However, there are also natural sources. In 1968, there were 30 known naturally occurring compounds. By 1994, around 2,000 had been discovered, and many biological organisms, especially marine species, those in the oceans, had not been looked at as yet. Halogenated compounds were found in ocean water, in marine algae, in corals, jellyfish, sponges, terrestrial plants, soil microbes, grasshoppers, and ticks. Volcanoes are another natural source of halogens, and they release significant amounts into the air during eruptions. Chlorine and fluorine are present in largest quantities, mostly as hydrogen chloride and hydrogen fluoride.

In the 1980s, depletion of the layer of ozone (O_3) high in Earth's atmosphere was observed. Ozone absorbs much of the high energy ultraviolet radiation from the sun that is harmful to biological organisms. During September and October, in the atmosphere over the Antarctic, ozone concentration in a roughly circular area, the "ozone hole," drops dramatically.

Chlorine-containing compounds, especially CFCs, undergo reactions releasing chlorine atoms, which can catalyze the conversion of ozone to ordinary **oxygen**, O_2. Bromine and iodine-containing carbon compounds may also contribute to ozone depletion. Countries signing the Montreal Protocol on Substances That Deplete the Ozone Layer have pledged to eliminate manufacture and use of halocarbons. However, natural sources, such as volcanic eruptions and fires, continue to add halogen compounds to the atmosphere. Finding substitutes that work as well as the banned compounds and do not also cause problems is a current chemical challenge.

HAMMETT, LOUIS (1894-1987)

American chemist

Louis Plack Hammett offers a modern reminder of the classical challenge in **organic chemistry** to adapt methods from **physical chemistry**. The main method that Hammett offered the field is the use of quantitative statistics in measuring the relationships of substances in organic chemical reactions. The equation named after him describes specific effects that can be expected in reactions of a wide range of substances. The acidity function that he calculated provided a model for describing unseen reactions of **acids and bases**. This was particularly appreciated in areas that would have been restricted by difficult laboratory conditions.

Hammett was born in 1894 and spent most of his childhood in Portland, Maine. After graduating from Harvard University, he traveled to Switzerland for graduate work. He returned home in less than a year when America entered the First World War. After the war, he attended Columbia University for studies leading to his Ph.D. While a graduate st udent, he taught chemistry classes and continued as professor after earning his doctorate in 1923. He wrote three notable textbooks on chemistry. In these books, he made the effort to convince the student that the methods of studying physical chemistry were equally appropriate for the study of organic chemistry. He died in 1987.

HAMMOND, GEORGE S. (1921-)

American chemist

George S. Hammond is noted for creating and developing the field of organic **photochemistry**, the study of the interaction between light and various organic materials. He is also credited with training many of the important American organic photochemists over a period of three decades. Some of the products that resulted from work in this field include catalysts used in the production of **vinyl plastics**, chemicals used to form the intricate circuit patterns on **computer chips**, and materials used in solar cells to convert the sun's **energy** to electrical power.

George Simms Hammond was born in Auburn, Maine, on May 22, 1921. His father, Oswald Kenric Hammond, was a farmer. His mother's maiden name was Marjorie Thomas. George attended Bates College in nearby Lewiston, Maine, where he graduated magna cum laude with a B.S. degree in chemistry in 1943. After graduation he was employed as a chemist with Rohm and Haas Corporation for two years, followed by a position with the Office of Scientific Research and Development at Harvard University, where he worked on the development of insect repellents. In 1945 he married Marian Reese. During their marriage they had five children, Kenric, Janet, Steven, Barbara, and Jeremy.

In 1947 Hammond received both his M.S. and Ph.D. degrees in chemistry from Harvard University. After postgraduate work with the Office of Naval Research at the University of California, Los Angeles, he joined the faculty of Iowa State University in 1948. He stayed there until 1958, rising to a full professor in the chemistry department.

While working at Iowa State, he began his studies of photochemical reactions. One area that especially interested him was the concept of sensitization. In a sensitized reaction, light is absorbed by one chemical in a **solution**, called a photoinitiator, and the resulting increase in energy is then passed on to another chemical to start the reaction. The use of a photoinitiator "sensitizes" a solution and allows chemical reactions to be triggered by certain frequencies of light. This concept is widely used in the production of integrated circuits on computer chips. In this process, a photosensitive chemical on the surface of the chip reacts with ultraviolet light passing through a mask, or pattern, to form the image of the desired circuit.

In 1958 Hammond accepted a position as professor of **organic chemistry** at the California Institute of Technology, where he continued his work on photochemical reactions. Many of his papers published in the *Journal of the American Chemical Society* during the period 1959-1962 established the foundation for modern photochemistry.

During the 1960s, Hammond and his students made many important observations on various photochemical reactions. One of the areas of study was the chemistry of azo-bis-isobutrynitrile (AIBN). They observed that many of the components of this chemical were held in place by the surrounding solvent molecules and ended up reacting with each other, rather than with neighboring chemicals. They found that this ''cage effect'' could be altered to control the amount of reaction between the AIBN components and other chemicals. Based on this discovery, AIBN became one of the most frequently used catalysts to start and control the chemical reactions required for the production of vinyl plastics.

In 1964 Hammond was named the Arthur Amos Noyes Professor of Chemistry at Caltech, and in 1968 he became Chairman of the Division of Chemistry and Chemical Engineering. He was elected to the National Academy of Sciences in 1963 and to the American Academy of Arts and Sciences in 1965. The American Chemical Society honored Hammond three times while he was at Caltech: first with the award in petroleum chemistry in 1961, then with the James Flack Norris Award in Physical Organic Chemistry in 1967, and finally with the award in chemical education in 1972.

Hammond left Caltech in 1972 to become a professor of chemistry and vice-chancellor for natural sciences at the University of California, Santa Cruz. He gave up his position as vice-chancellor in 1974 in order to spend half his time as foreign secretary of the National Academy of Sciences, while continuing his research and teaching duties at the university. In 1976 the American Chemical Society honored him with their highest award, the Priestley Medal, for his distinguished service to the profession of chemistry. In 1977 Hammond's first marriage ended in divorce. He married Eva L. Menger shortly thereafter. Eva had two children by a previous marriage, Kirsten Menger-Anderson and Lenore Menger-Anderson.

In 1978 Hammond left the academic world to join Allied Signal Corporation in Morristown, New Jersey, as Associate Director of Corporate Research. Allied Signal was engaged in the manufacture of aerospace and automotive products, chemicals, fibers, plastics, and advanced materials. Hammond became Director of Integrated Chemical Systems in 1979, and Executive Director of Bioscience, Metals, and Ceramics in 1984.

Hammond retired from Allied Signal in 1988 to become a consultant. In 1991 he joined the faculty of Bowling Green State University in Bowling Green, Ohio, as Director of Materials Science, and in 1992 he was named a Distinguished Visiting Research Professor. In 1994, while holding the position of Senior McMaster Fellow at the Center for Photochemical Sciences at Bowling Green, Hammond was awarded the prestigious National Medal of Science for his work in organic photochemistry. He was also awarded the Seaborg Medal the same year.

Today, Hammond continues his research work at Bowling Green and is active in several scientific groups.

HARDEN, ARTHUR (1865-1940)

English chemist

Arthur Harden's groundbreaking work in the field of alcoholic **fermentation** has led to a greater understanding of metabolic processes, including the formation of lactic acid in muscles and the ossification of tissue. Apart from his discoveries in **biochemistry**, he distinguished himself as a Nobel laureate, professor, and contributor of scholarship to the field.

Born on October 12, 1865, Harden was the third of nine children—and the only son—of Manchester businessman Albert Tyas Harden and Eliza MacAlister Harden. His family's puritanical leanings and nonconformist attitude toward social conventions, such as the celebration of Christmas and interest in the theatre, remained an influence throughout Harden's life. Despite his austere upbringing, however, he was an accomplished skater and avid gardener, as well as a great fan of Charles Dickens and Victorian literature in general. He attended private school beginning at age seven, and then studied at Tattersall College between 1876 and 1881 in Staffordshire, reaching the age of sixteen at the time that he left there. His undergraduate studies at Owen College of the University of Manchester under the instruction of Henry Roscoe culminated in a degree with first class honors in **chemistry** in 1885. A year later he was awarded the Dalton scholarship and started graduate studies at the University of Erlangen under the tutelage of **Ernst Otto Fischer**. After completing his Ph.D. in 1888 by writing his dissertation on properties and purification of β-nitroso-a-naphthylamine, he returned to Owens where he served as junior and then senior lecturer under Roscoe's successor, H. B. Dixon.

Harden was more interested in teaching and writing than research. He was intrigued with the history of chemistry, and taught an honors class in that subject. Among the most prodigious writings he had done up until 1896 were several papers that had resulted from studying **John Dalton**'s notebooks with Roscoe in a joint research project. Their 1896 book, *A New View of the Origin of Dalton's Atomic Theory,* represented an area of interest which fascinated Harden for many years. (Dalton theorized his atomic principles by observing the **diffusion** of **gases**.) Harden then collaborated with F. C. Garrett on *Practical Organic Chemistry,* published in 1897, and revised and edited Roscoe's *Treatise on Inorganic Chemistry,* admirably supplementing his teaching salary of 200 or less a year.

After unsuccessfully applying for jobs as a private school principal and as an inspector, Harden was appointed professor of biochemistry in 1897 at the British Institute of Preventative Medicine (renamed a year later the Jenner Institute and, in 1903, the Lister Institute), where Roscoe was treasurer. While the institute staff taught students pursuing careers as public health officers, testing **water** for city officials and otherwise doing little scientific research, Harden was hired at Roscoe's suggestion to teach chemistry and bacteriology. As medical schools began to offer the same subjects, Harden's classes were phased out and the head of the bacteriology department recommended he consider research. Fermentation, or the breakdown of sugar by bacteria, became Harden's objective as he set upon discovering a chemical means of distinguishing various fermentation patterns in the bacteria *Escherichia coli*. Harden was able to show that the ratio of **alcohol** to **acetic acid**, two of the compounds formed during bacterial fermentation, was a useful guide in determining which variety of the bacterium was involved in a given fermentation process.

The year Harden arrived at the institute, **Eduard Buchner** had released his revolutionary research on fermentation. Buchner had discovered that fermentation could take place in the absence of living yeast cells, yielding an **enzyme** he named zymase. Although the reaction took longer than it would have had live cells been present, it produced the familiar end products of fermentation, **carbon** dioxide and alcohol. Buchner's experiment was the first evidence of the existence of enzymes. Like many others in the scientific community, Harden began to build on Buchner's work. He showed that fermentation could occur because zymase acted on **glycogen** (a sugar) that had been within the cells themselves. Assisted by William Young, Harden made significant discoveries about the role of **phosphate** in fermentation over the next decade.

In 1904, Harden put a semipermeable membrane bag full of yeast extract into pure water. Harden knew that the molecules, densely packed inside the bag, would move through the membrane into the water because of the lower **density** of yeast outside of the bag. He also knew that the membrane would allow only molecules of a given size to pass through—a process called **dialysis**. Zymase stopped breaking down the sugar inside the bag while reintroducing water from outside the bag which contained the small molecules that had diffused out through the membrane. When Harden boiled the yeast extract, it failed to cause fermentation at all, indicating that zymase actually consisted of two parts. Because zymase lost its activity after dialysis, he decided that the larger part remained trapped inside the bag; this, together with its loss of effectiveness led Harden to conclude that the larger part was probably a protein, the smaller part (having not perished during boiling), a nonprotein.

Pursuing the matter further, Harden and Young added boiled yeast juice to an ongoing fermentation and measured the amount of **carbon dioxide** released. Although all active agents should have been destroyed by the boiling, the addition sped up the process. They discovered the boiled juice contained a phosphate substance called cozymase. Harden's work showed that three factors are necessary for fermentation to occur: the ferment, the enzyme, and a **coenzyme**. By attending to the entire fermentation process—not only the end products as had been the previous practice—he laid the important groundwork for further understanding of metabolic processes, such as ossification.

In 1911, Harden was among the founders of the Biochemical Society, and the following year named coeditor of the *Biochemical Journal.* Although Harden left the institute in 1912 for a professorship in biochemistry at the University of London, his association with the institute was not over. He became acting director when the head of the institute went off to war in 1914. For the duration, his research focused on **nutrition**, particularly on the diseases beriberi and scurvy, while much of his free time was spent digging trenches with the Volunteer Reserve.

Arthur Harden.

Harden married Georgina Sydney Bridge, of Christchurch, New Zealand, in 1900; they had no children. She died in 1928, a year before Harden and **Hans Euler-Chelpin** received the Nobel Prize in chemistry for work in fermentation. In 1909 he was named a fellow of the Royal Society, being awarded its Davy Medal in 1935, and the following year he was knighted. He received honorary degrees from the universities of Manchester, Liverpool, and Athens, and was named an honorary member of the Institute of Brewing.

A year after winning the Nobel Prize Harden retired from teaching, although he stayed on as editor of the *Biochemical Journal* until 1937. His garden at Bourne End remained one of his great joys until he died at home on June 17, 1940, of a progressive neurological disorder.

HARE, ROBERT (1781-1858)

American chemist and inventor

Robert Hare, considered the leading American chemist of his time, was a productive inventor and writer.

Robert Hare was born in Philadelphia on January 17, 1781, the son of a prominent businessman and state senator.

Robert Hare.

He was educated at home, then studied **chemistry** under James Woodhouse. While managing his father's brewery, he found time for chemical research and gained international fame in 1801 with his invention of the oxyhydrogen blowpipe, which provided the highest degree of **heat** then known. (Its application led to the founding of new industries such as production of **platinum** and limelight illuminants.)

After teaching briefly at the College of William and Mary in Virginia, Hare was appointed professor of chemistry at the University of Pennsylvania in 1818, where he remained until 1847. Hare's classes were noted for his spectacular experiments. His inventions included a calorimeter, a deflagrator for producing high electric currents, and an improved electric furnace for producing artificial **graphite** and other substances.

Although primarily noted as an experimental chemist and inventor of experimental apparatus, Hare was keenly interested in theoretical speculations about both chemistry and meteorology. He published articles in the *American Journal of Science,* edited by his close friend and collaborator **Benjamin Silliman**. His famous controversies were with **Jöns J. Berzelius** over chemical **nomenclature**, **Michael Faraday** over **electricity**, and William C. Redfield over the nature of storms. Hare was especially committed to the theory of the materiality of heat.

In 1850 Hare published a historical novel, *Standish the Puritan*. In 1854, near the end of his career, he became a convert to spiritualism, much to the dismay of his rationally minded colleagues. He produced a book on the subject and went so far as to claim that Benjamin Franklin's spirit had validated his electrical theories. But he was unsuccessful in getting the American Association for the Advancement of Science to listen to his views.

Hare was a member of the American Philosophical Society and the American Academy of Arts and Sciences. Though his only degrees were honorary, he represented the newly emerging professional university scientist in contrast to the traditional gentleman-amateur.

Hare had married Harriet Clark in September 1811; one son, John James Clark Hare, became a distinguished lawyer. Hare died on May 15, 1858.

HASSEL, ODD (1897-1981)

Norwegian physical chemist

Through twenty-five years of painstaking work, Norwegian physical chemist Odd Hassel confirmed the long-suspected three-dimensional nature of organic molecules, and his work in this field, called conformational analysis, altered the perception of **chemistry**. He received the Nobel Prize for chemistry in 1969, which he shared with the English chemist **Derek H. R. Barton**. Although other Norwegians had won the prize before him, Hassel's win was a special source of pride for his countrymen, for he was the first winner whose work had been carried out almost entirely in Norway.

One of a set of twins, Odd Hassel was born May 17, 1897, in Kristiana (now Oslo), Norway. His father Ernst was a gynecologist. His mother, Mathilde Klaveness Hassel, raised her four sons and one daughter alone after her husband died when Odd was eight years old. While his brothers, including his twin Lars, entered law and civil engineering, Hassel chose a different route. He had disliked school except for mathematics and science. The interest he developed in chemistry during high school evolved into his major area of study at the University of Oslo, which he entered in 1915.

Hassel toured France and Italy for a year after his graduation in 1920, a common practice at the time. In 1922 he worked at K. Fajans's laboratory in Munich where he discovered **adsorption** indicators, organic dyes used in the analysis of **silver** and **halide** ions for greater accuracy. He returned to school to study at the University of Berlin, a center for chemistry and physics, where he was recommended for and received a Rockefeller scholarship. He earned his doctorate in 1924.

While in Berlin, Hassel worked at the Kaiser Wilhelm Institute, and learned the technique known as **X-ray crystallography**. In this method of analysis the **atomic structure** of a substance can be determined by striking a pure crystal of the substance with X rays. After passing through the crystal, the rays are bent, or diffracted; the pattern of this diffraction is captured on photographic film and, when analyzed, reveals the arrangement of the atoms within the substance.

In 1925 Hassel returned to the University of Oslo as an instructor, and a year later was named associate professor of

physical chemistry. In 1930 he began to investigate the three-dimensional structure of molecules, particularly ring-shaped **carbon** molecules. Many important organic molecules, including several **carbohydrates** and **steroids**, are built on a ring-shaped base. Although it was widely believed that all the carbon atoms in these molecules were arranged in one plane (rather like a doughnut lying on a plate), the possibility that they were actually three-dimensional had been proposed in 1885. Molecules having six or more carbon atoms, reasoned German chemist **Johann Friedrich Wilhelm Adolf Baeyer**, would be under too much strain to lie flat; in 1890 chemist Ulrich Sachse suggested two configurations of cyclohexane (a six **atom** carbon ring). One, the boat form, was represented as four atoms framing the "sides" laying in the same plane with the remaining atoms in the plane above them, like the bow and stern of a canoe. The second, or chair configuration, resembled a reclining shape having four atoms in the central plane, with one end atom above, and one below. In the absence of more conclusive experiments, however, most scientists maintained that cyclohexane resembled a doughnut on a plate.

Hassel's work was to correct that view. His primary investigations used the X-ray crystallography technique he had learned in Berlin; the drawback however, was that the technique could be used only with **solids**. A technique called **dipole** measurement, the analysis of positive and negative charges in a **molecule**, was also used. But **electron** diffraction proved to be the best method to investigate the structure of molecules because it could be used with **gases** and free molecules. By 1938 Hassel's laboratory was able to afford an **electron diffraction** unit, and he devoted the next five years to studying cyclohexane. Not only did he confirm that the boat and chair forms did indeed exist as predicted nearly fifty years before, Hassel also discovered that the molecules oscillated between the boat and chair forms at an enormous rate, with the latter form occurring predominantly. His investigations made it possible to predict the chemical properties of many organic substances whose base was cyclohexane. He also determined that the **hydrogen** atoms bonded to the carbon atoms either perpendicular to the four-atom plane (axial) or parallel (equatorial). These observations further deepened the behavioral chemistry of cyclohexanes and their related compounds—substituted cyclohexanes.

Hassel continued his work on cyclohexane even after Germany invaded Norway in 1940. He refused to publish his papers in German scientific journals, which limited the dissemination of his ideas. Some of his most important research was first reported in small Norwegian-language journals not circulated outside of Norway. In 1943, the Germans shut down the University of Oslo. Hassel, along with the other faculty members and scholars, was sent to a concentration camp at Grini, near Oslo. During his two years of imprisonment Hassel carried on his work, teaching physical chemistry without the consent of his captors. Despite his shy, reticent nature, he enlisted other scientists to work with him, including Per Andersen, and Ragnar Frisch, who remained a good friend and later received the Nobel Prize for economics the same year Hassel received one for chemistry. They were freed from Grini in November of 1944.

During the 1950s Hassel turned his attention to the physical structure of charge-transfer compounds. In such a compound, one part "donates" an electron to the other part, which "accepts" it. Because many of these compounds were too unstable to study in gaseous form, Hassel studied the solid forms with X-ray crystallography. He concluded that many of the theories about how these molecules worked were incorrect, and devised a new, simple set of rules that would inform the arrangement and size of the molecular bonds.

Hassel retired from the University of Oslo in 1964, but continued to research and publish until 1971. In the course of his career he published over 250 scientific papers, as well as *Kristallchemie* (1934), the first modern review of work in crystal chemistry. It was quickly translated into English, as *Crystal Chemistry,* and Russian. The book became a standard reference work for crystallographers and chemists throughout the field. From 1947 to 1957 Hassel was also the Norwegian editor of *Acta Chimica Scandinavica.* During his long career Hassel received numerous honors for his contributions to science. Apart from being honored with the Fridtjof Nansen Award in 1946, the Gunnerus Medal from the Royal Norwegian Academy of Sciences was awarded him in 1964, as well as the Guldber and Waage's Law of Mass Action Memorial from the Norwegian Chemical Society, of which he was an honorary fellow. In addition, he was a fellow of both the Royal Norwegian, and Royal Swedish Academies of Sciences, and Royal Danish Academy of Science. An honorary fellow of the British Chemical Society besides, Hassel received honorary degrees from the University of Copenhagen (1950) and the University of Stockholm (1960). He was made a knight of the Order of Saint Olav. In 1969, he shared the Nobel Prize for Chemistry with Derek Barton "for their contributions to the development of the concept of **conformation** and its application in chemistry." Speaking of the award to the *New York Times,* Hassel commented, "I had been among the chemistry candidates before, but did not expect to get the prize now. It was indeed very pleasing." He had doubts about going to Stockholm to accept the prize, however, saying, "I detest public appearances and have to think it over thoroughly." Hassel rarely attended international conferences and never married. "He prefers molecules," noted one of his students. After his twin brother died in 1980, Hassel reportedly lost his "zest for life." On May 15, 1981, Hassel died in Oslo, just two days before his eighty-fourth birthday.

Hatch, Marshall (1932-)

Australian biochemist

Marshall Davidson Hatch is a biochemist living in Australia. He gained international attention in 1968 when he discovered a previously unknown means by which a plant takes in **carbon** dioxide to merge with **carbohydrates**. Technically referred to as a "pathway," this new discovery was named after Hatch and his research colleagues: Slack and Kortschak. The description of this pathway provided a logical alternative to the Calvin photosynthetic pathway. Hatch later discovered an **enzyme** that aids in re-synthesis. The description of his findings, published in 1969, provided evidence of a cycle of synthesis and re-synthesis that had not been recognized before by any other researcher. Further study found that this cycle is important in the nutritional systems of plants in extreme climates.

Hatch was born in Perth, Australia. He attended college in Australia and the United States. After graduating from college, he began his research career with a position at the Commonwealth Scientific Industrial Research Organisation (CSIRO). He worked there from 1955 until 1959, during which time he published a comprehensive paper on the processes of plants changing starch and sugars into **ethanol** and **carbon dioxide**. After a two-year stay in the United States, he returned to Australia, in 1961, to accept a position with a sugar refining company. After the publication of his discoveries, he accepted the position of chief research scientist at a division of CSIRO.

HAUPTMAN, HERBERT A. (1917-)

American mathematician and biophysicist

Herbert A. Hauptman has spent most of his adult life in and around the laboratory. In the early 1950s, he and his fellow New Yorker and former classmate, **Jerome Karle**, developed a mathematical system, usually referred to as the "direct method," for the interpretation of data on **atomic structure** collected through X-ray **crystallography**. The system, however, did not come into general use until the 1960s, and it was only in 1985 that Hauptman and Karle were jointly awarded the Nobel Prize in **chemistry** for their accomplishment.

Herbert Aaron Hauptman was born in New York City on February 14, 1917, the son of Israel Hauptman, an Austrian immigrant who worked as a printer, and Leah (Rosenfeld) Hauptman. He grew up in the Bronx and graduated from Townsend Harris High School. At the City College of New York, he majored in mathematics and received a Bachelor of Science degree in 1937. Karle, his later collaborator, also graduated from City College the same year. Hauptman went on to complete a master's degree in mathematics at Columbia University in 1939. He married Edith Citrynell, a schoolteacher, on November 10, 1940; they eventually had two daughters, Barbara and Carol. Hauptman worked for two years as a statistician in the United States Bureau of the Census before serving in the United States Army Air Force from 1942 to 1947. After his period of service ended, Hauptman went to work as a physicist and mathematician at the Naval Research Laboratory in Washington, remaining there until 1970. While working at the laboratory, he enrolled in the doctoral program in mathematics at the University of Maryland and received his Ph.D. in 1955.

At the Naval Research Laboratory, Hauptman renewed his acquaintance with Karle, who had come to the laboratory in 1946. The two men soon began to work together on the problem of determining molecular structures through the methodology of **X-ray crystallography**. Most of the work which later led to their joint Nobel Prize was done between 1950 and 1956. A brief monograph, *Solution of the Phase Problem, 1. The Centrosymmetric Crystal,* was published in 1953 that revealed many of the results of their studies.

The German physicist Max Laue had discovered as far back as 1912 that it was possible to determine the arrangement of atoms within a crystal by studying the patterns formed on a photographic plate by X rays passed through a crystal. Since that time X-ray crystallography had become a standard tool for chemists, physicists, biologists, and other scientists concerned with determining the atomic structure of substances. X-ray crystallography, for example, had made possible the discovery of the double-helical structure of deoxyribonucleic acid (DNA) by molecular biologists **Francis Crick**, **James Watson**, and others in the 1950s. The problem with the technique was that interpreting the patterns on the photographic plates was a very difficult, laborious, and time-consuming task. The accurate determination of the atomic structure of a single substance could require one or more years of work based upon indirect inferences which often amounted to educated guesswork. The greatest difficulty arose from the fact that while photographic film could record the intensity of the X-ray dots that formed the patterns, it could not record the phases (the minute deviations from straight lines) of the X rays themselves.

The great achievement of Hauptman and Karle was to develop a very complex series of mathematical formulas, relying heavily on probability theory, which made it possible to correctly infer the phases from the data which was recorded on the photographic film. Their new mathematical system came to be known as the determination of **molecular structure** by "direct method." They demonstrated the workability of their new technique in 1954 by calculating by hand, in collaboration with researchers at the United States Geological Survey, the atomic structure of the mineral colemanite.

Hauptman and Karle's system met with a good deal of skepticism and resistance from the specialists in X-ray crystallography in the 1950s and was largely ignored for about ten years. This was partly due to the fact that most crystallographers of the time lacked the mathematical knowledge and sophistication to make use of the new technique. It also stemmed from the fact that the necessary mathematical calculations themselves were a laborious process. It was the introduction of computers and the development of special programs to deal with the Hauptman-Karle method in the 1960s that finally led to its widespread acceptance and use. The work that originally required months or years to complete could now be done in a matter of hours or, at most, days. By the mid 1980s the atomic structures of approximately 40,000 substances had been determined through use of the direct method, as compared to only some 4,000 determined by other methods in all the years prior to about 1970, and some 4,000 to 5,000 new structures were being determined each year.

Hauptman left the Naval Research Laboratory in 1970 to become head of the biophysics laboratory at the Medical Foundation of Buffalo, a small but highly regarded organization specializing in research on endocrinology. He also became professor of biophysical science at the State University of New York at Buffalo. Hauptman served as executive vice president and research director of the Medical Foundation from 1972 to 1985 and president from 1985 onwards. There he worked to perfect the direct method and to extend its use to the study of very large atomic structures. On June 2, 1993, the weekly scientific periodical *Inside R & D* announced that Hauptman, then aged seventy-six, had developed a new computer software package, called ''Shake-and-Bake,'' which could ''routinely solve structures of up to 300 or 400 atoms'' in a few hours or days. In recognition of his skill and knowledge, Hauptman has received numerous awards, including the 1985 Nobel Prize in chemistry shared with Karle, the Award in Pure Sciences from the Research Society of America in 1959, and also with Karle, the A. L. Patterson Memorial Award of the American Crystallography Association in 1984.

HAWORTH, WALTER (1883-1950)

English chemist

Walter Haworth's earliest research was influenced by his contact with William Perkin at the University of Manchester and involved a study of terpenes, a class of hydrocarbons often found in plants. The work for which he is best known, however, involves his studies of various **carbohydrates**, including a number of important monosaccharides, disaccharides, and polysaccharides. Among his finest achievements was the determination of the **molecular structure** for glucose, perhaps the most important of all monosaccharides. The method he used for designating the formula of glucose and those of other carbohydrates is well known today to any student of organic **chemistry** as the Haworth formula. The 1937 Nobel Prize in chemistry was awarded to Haworth in recognition not only of his work on carbohydrates but also for his elucidation of the structure of vitamin C and the first artificial synthesis of this important compound.

Walter Norman Haworth was born in Chorley, Lancashire, England, on March 19, 1883. He was the fourth child and second son of Thomas and Hannah Haworth. Thomas Haworth, whose family enjoyed a distinguished reputation in business, was the manager of a linoleum factory and took it for granted that his son would follow him into that business. And, indeed, after completing school at the age of fourteen, young Haworth did take a job at the linoleum factory. He soon decided, however, that he had no interest in making his career in that kind of work. Instead, he had become fascinated with the chemical applications he saw all around him and had decided that he wanted a career in that field.

That road was made all the more difficult, however, when Haworth's parents withheld their approval and support for any additional education for their son. It was only through great personal effort and the aid of a private tutor that he was finally able in 1903 to pass the entrance examination at Manchester University. There he studied chemistry under the department chairperson, **William Perkin, Jr.**, and became particularly interested in Perkin's own specialty, the chemistry of terpenes. Haworth received his degree in 1906, earning first-class honors in chemistry, and then stayed on at Manchester to work as Perkin's assistant.

In 1909 Haworth left Manchester to spend a year at the University of Göttingen studying with future (1910) Nobel Prize winner **Otto Wallach**, an expert on terpenes. In only one year, Haworth had earned his Ph.D. and was on his way back to Manchester. One year later, he had qualified for his second doctorate, a D.Sc. in **organic chemistry**. Over the next fifteen years, Haworth held posts at three institutions. He was senior demonstrator at the Imperial College of Science and Technology in London from 1911 to 1912, lecturer at United College in the University of St. Andrews from 1912 to 1920, and professor of organic chemistry at Armstrong (later King's) College in the University of Durham from 1920 to 1925. In the latter year he was appointed Mason Professor of Chemistry at the University of Birmingham, a post he held until his retirement in 1948.

Walter Haworth.

The most important period for Haworth in his pre-Birmingham days was his tenure at St. Andrews. It was there that he was introduced to the new field of carbohydrate chemistry by Thomas Purdie and James Colquhoun Irvine, two of England's foremost authorities in the field. In the early 1910s, scientists knew a fair amount about the chemical composition of the carbohydrates, but relatively little about their molecular structure. It was to the question of molecular structure that Haworth soon turned his attention at St. Andrews, and before long, he had abandoned his work on terpenes.

World War I interrupted Haworth's new line of research, however. For the duration of the war, the chemical laboratories at St. Andrews (like other such facilities) were completely given over to the manufacture of chemicals with military importance. At the war's conclusion, however, Haworth returned to his work on carbohydrates. The first stages of that research were devoted to the monosaccharides, the simplest of the carbohydrates. Haworth developed a method by which he could determine the sequence of linkages within a **molecule** and was able to elucidate the detailed formulas for many compounds. Among the most important of these was glucose, which Haworth showed in 1926 to exist as a six-membered **ring** consisting of five **carbon** atoms and one **oxygen atom**. The convention he used to represent the glucose structure, showing the three-dimensional orientation of its components, has since become known as a Haworth formula or Haworth projection.

In his later work at Birmingham, Haworth took on more and more complex structures, eventually finding formulas for lactose and sucrose, two important disaccharides. He also took on yet another challenge—the determination of the structure for hexuronic acid. Hexuronic acid had been discovered in 1932 by **Albert Szent-Györgyi** in extracts taken from the adrenal gland, in cabbages, and in oranges. Szent-Györgyi suspected that his hexuronic acid might be identical to vitamin C, the antiscurvy agent, that had also been discovered recently.

In his own research, Haworth was able to elucidate the structure of this compound and then to synthesize it in his laboratory. That accomplishment was historic since it was the first time that a vitamin had been produced synthetically. Because of the compound's antiscurvy properties, Haworth suggested that it be renamed ascorbic acid (''not-scurvy'' acid), a name by which it is now universally known. For his work both with carbohydrates and with vitamin C, Haworth was awarded a share of the 1937 Nobel Prize in chemistry with **Paul Karrer**.

Haworth's health failed him in 1938, but three years later he had recovered sufficiently to return to his research and other commitments. Included among those other commitments were a number of political and professional responsibilities. He served as chairperson of the British Chemical Panel for Atomic Energy during World War II. He also became dean of the faculty at Birmingham from 1943 to 1946 and served as president of the British Chemical Society from 1944 to 1946. At the same time, he continued an active program of research, concentrating on the most complex of all carbohydrates, the polysaccharides.

Haworth was married to Violet Chilton Dobbie in 1922. The couple had two sons. Haworth died at his home in Birmingham of a heart attack on March 19, 1950, his birthday. In addition to the Nobel Prize, he had been awarded the Longstaff Medal of the British Chemical Society in 1933, the Davy Medal in 1934, and the Royal Medal of the Royal Society in 1942. He was made a fellow of the Royal Society in 1928 and was knighted in 1948.

HAZEN, ELIZABETH (1885-1975)

American microbiologist

Hazen was born in Rich, Mississippi, on August 24, 1885, and raised by relatives in Lula, Mississippi, after the death of her parents. Hazen attended the public schools of Coahoma County, Mississippi, and earned a B.S. from the State College for Women, now Mississippi University for Women. She began teaching high school science and continued her own education during summers at the University of Tennessee and University of Virginia.

In 1916, Hazen began studying bacteriology at Columbia University, where she earned an M.A. the following year. World War I provided some opportunities for women scientists, and Hazen served in the Army diagnostic laboratories and subsequently in the facilities of a West Virginia hospital. Following the war, she returned to Columbia University to pursue a doctorate in microbiology, which she earned in 1927 at age 42.

After a four-year stint at Columbia University as an instructor, Hazen joined the Division of Laboratories and Research of the New York State Department of Health. She was assigned to special problems of bacterial diagnosis and spent the next few years researching bacterial diseases. She investigated an outbreak of anthrax, tracing it to a brush factory in Westchester County. Hazen discovered unknown sources of tularemia in New York and was the first to identify imported canned seafood that had spoiled as the cause of type E toxin deaths.

Her discoveries led Hazen to try to better understand mycotic (fungal) diseases. In 1944, she was given the responsibility of investigating such diseases, and she acquired cultures of fungi from local laboratories and specialized collections. Although Hazen was learning more about mycotic diseases, fungal infections continued to spread in epidemic proportions among school children in New York City. In addition to pneumonia, many other fungal diseases caused widespread ailments, such as moniliasis (thrush), a sore mouth condition that makes eating excruciatingly painful. Despite personal health problems and stressful working conditions, Hazen persevered.

In the mid-1940s, she teamed up with Rachel Brown (1898-1980), a chemist at the Albany laboratory who prepared extracts from the cultures sent by Hazen. In the fall of 1950, Hazen and Brown announced at a National Academy of Sciences meeting that they had successfully produced two antifungal agents from an antibiotic. This led to their development of Nystatin, the first **fungicide** safe for treating humans. Nystatin was immediately used nationwide, earning $135,000 in its first year.

Nystatin, which is still sold as a medication today under various trade names, turned out to be an extremely versatile substance. In addition to curing serious fungal infections of the skin and digestive system, it can also combat Dutch Elm disease in trees and even restore artwork damaged by **water** and mold. Remarkably, Hazen and Brown chose not to accept any royalties from the patent rights for Nystatin. Instead, they established a foundation to support advances in science. The donated royalties totaled more than $13 million by the time the patent expired. Hazen died on June 24, 1975.

HEAT

The concept of heat has always been an important consideration in scientific thought. Some things are warmer than others. When warm objects are placed in contact with cold ones, the warmer object cools down as the colder one heats up. Apparently something causes an object to be warm and flows from warm objects to cold ones. This "something" is called heat.

The controversy over whether heat is a material substance or is due to the kinetic motion of particles continued into the early twentieth century. Isaac Newton (1642-1727), Christiaan Huygens (1629-1695), Robert Hooke (1635-1695), **Henry Cavendish**, **Humphry Davy**, and **Benjamin Thompson** supported the kinetic interpretation. **Hermann Boerhaave**, **Antoine-Laurent Lavoisier** and Pierre-Simon Laplace (1749-1824) favored the material interpretation.

Joseph Black and Johann Wilcke (1732-1779) developed the concept of specific heat capacity, defined as the amount of heat which raises the **temperature** of a substance by 1 degree. The empirical definition of heat is based on this concept: the amount of heat involved in a process is equal to the product of the specific heat of the substance multiplied by the change in temperature which occurs when heat is added to or taken away from the substance. The unit given to the amount of heat is the calorie, defined as the amount of heat that raises the temperature of 1 gram of **water** by 1°Celsius.

James Joule showed that, unlike material substances, heat can be created and destroyed by changing it into work. One calorie of heat is equivalent to 4.2 joules of work. Subsequently, the kinetic theory proposed by A. K. Krönig (1822-1879) and Rudolf Clausius (1822-1888) explained heat as the result of translational, rotational, and vibrational motions of molecules: the faster they move, the higher the temperature. The work of **Albert Einstein** finally clinched the argument in favor of the kinetic interpretation of heat.

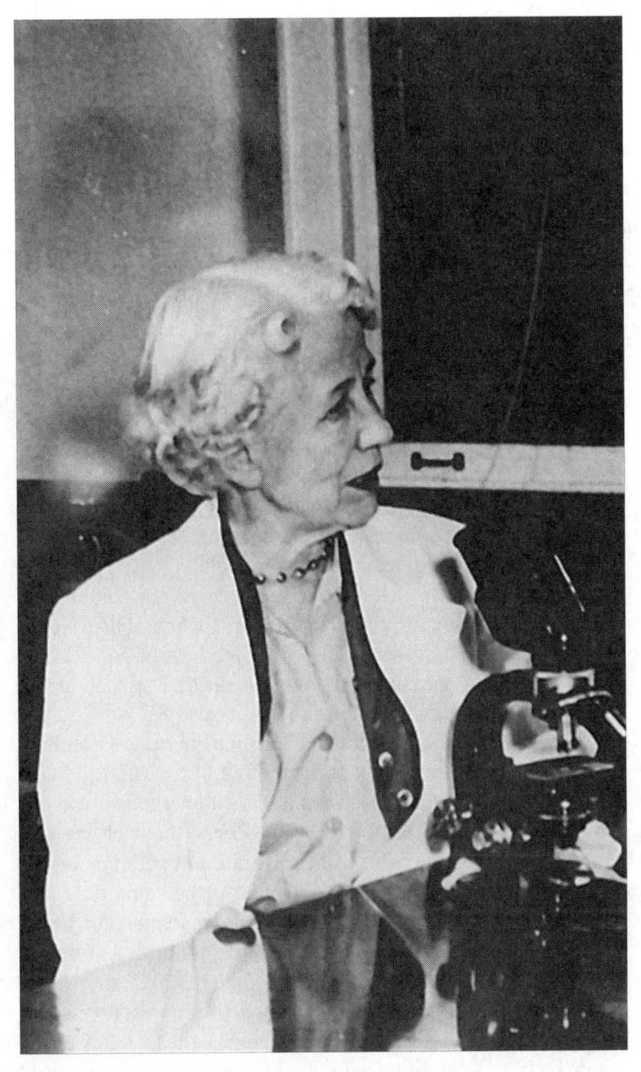

Elizabeth Hazen.

HEAT AND HEAT CHANGES

Heat is the **energy** exchanged when a difference in **temperature** exists between two regions. That is to say, changes in temperatures are produced by the addition or subtraction of heat from a substance.

In general, the more heat that is added to a particular body of **matter**, the more the temperature of that body rises. When heat is added to or removed from a physical system, the accompanying thermal changes may include a change in length and **volume**, and/or changes in physical states such as melting, evaporation, etc. The addition or removal of heat from a material may destroy the magnetic, superconductive, ferroelectric, shear, and/or cohesive properties of that material, and it may profoundly affect the mechanical properties of **solids**, fluidity of **liquids**, the **electrical conductivity** of **metals**, and the thermal conductivity of solids.

During the eighteenth century, heat was believed to be a fluid, called the caloric, that filled the space between the fundamental particles of matter. Once it was recognized that friction produces heat and that heat can be dissipated when mechanical work is performed, heat was viewed as another form of energy.

When physical changes in a body accompany the addition or removal of heat from that body, finite quantities of heat may be absorbed over an infinitesimal temperature range (at constant pressure), resulting in a change of phase. The heat absorbed at constant pressure per **mole** when a solid melts is referred to as the heat of melting. The **heat of vaporization** is the heat absorbed at constant pressure per mole when a liquid vaporizes. And the **heat of sublimation** is the heat absorbed at constant pressure per mole when the solid vaporizes.

See also Calorimetry; Enthalpy; Heat transfer; Phase changes; Thermochemistry

HEAT OF COMBUSTION

The **heat** of **combustion** is more precisely termed the **enthalpy** of combustion and is given the symbol ΔH_{comb}. The term heat of combustion is typically used interchangeably with the more precise term the enthalpy of combustion. The enthalpy of combustion of the **heat of reaction** for the combination of one **mole** of a substance with **oxygen** to form oxidized forms of the elements of the substance, such as **carbon** dioxide from carbon atoms, **water** from **hydrogen** atoms and **sulfur** dioxide from sulfur atoms. As is the case for other types of enthalpy, the heat of combustion is the heat absorbed for one mole of the particular substance that undergoes the reaction (combustion) under three specific conditions: (1) the pressure remains constant, (2) the only possible work that occurs is expansion against the atmosphere (so-called $P\Delta V$ work) and (3) the **temperature** remains constant during the process.

In practice, combustion reactions cause an increase in temperature. In addition, it is experimentally easier to use a sealed container of constant **volume** to carry out combustion reactions than to attempt to construct an apparatus that would maintain constant pressure. Therefore, the heat of combustion is not generally measured directly. Instead, it is calculated from other quantities that are directly measured.

In a typical experiment to determine the heat of combustion, a known amount of substance would be combusted in a sealed container (a bomb calorimeter) submerged in a well-insulated water bath. As the reaction occurs, the heat released from the combustion would increase the temperature of the water and the bomb calorimeter itself. At the same time, the pressure within the bomb calorimeter would change due to the temperature change and a change in the number of moles of gas contained within it, according to the **stoichiometry** of the combustion reaction. The **energy** change for the reaction is calculated from the temperature change and heat capacities of the water bath and calorimeter minus whatever energy was used to start the combustion reaction. The heat of combustion is then calculated by adding the equivalent of the $P\Delta V$ work, in this process actually ΔnRT from the Ideal Gas Law.

The heat of combustion experimentally determined for an actual combustion reaction is the same as the enthalpy for the reaction of the substance with oxygen under any conditions for which the same chemical equation pertains. This means that the enthalpy for the **metabolism** of foods, such as fats and sugars, is the same as the heat of combustion measured experimentally.

Many combustion reactions are very difficult or impossible to directly carry out under conditions that allow careful measurements to be taken. Often, however, the heats of combustion can be obtained indirectly from measurements of other reactions. Because the heat of combustion is an enthalpy of reaction, Hess's Law applies: when the equation for an reaction can be obtained by algebraically combining a series of other reactions, the enthalpy for the reaction can be obtained by algebraically combining the other equations in the same manner. For example, if reaction 1 and reaction 2 can be added to obtain the equation for the desired reaction, a combustion reaction that cannot be measured directly, the enthalpy of the desired reaction is the sum of the enthalpies of reactions 1 and 2.

Heats of combustion are often specifically designated to show the conditions under which they apply. The temperature for which the heat of combustion applies should always be designated. A subscript indicating the temperature on the Kelvin or absolute scale is generally used, for example, $\Delta H_{comb, 400}$ for a reaction at 400 K. If no temperature is indicated, it is assumed that the temperature is 298 K or 25 C. If the heat of combustion corresponds to a reaction in which all of the reactants and products are in their standard physical states, solid, liquid, or gas, at the stated temperature and they are present in standard concentrations, one molar for solutions, pure **solids** and **gases** at one bar pressure, it is called a standard heat of combustion, designated by a superscript ''o'', for example $\Delta H^{o}_{comb, 400}$. The units typically used for heats of **solution** are kilojoules per mole or kilocalories per mole.

See also Ideal gases

HEAT OF FORMATION

The **heat** of formation is the heat that is absorbed when a substance is formed from its elements. In common usage, the heat of formation is used in place of the more precise term the **enthalpy** of formation, which has the symbol $(\Delta)H_f$. The enthalpy of formation of a substance is the **heat of reaction** for the combination of elements in their standard (normal) physical states to form one **mole** of the substance, but only under three specific conditions: (1) the pressure remains constant (2) the only possible work that occurs is expansion against the atmosphere (so- called $P(\Delta)V$ work) and (3) the **temperature** remains constant during the process.

In practice, a formation reaction is likely to result in an increase or decrease in temperature (reactions that occur with an evolution of heat are called exothermic and reactions that require heat are termed endothermic). In addition, it is generally easier experimentally to use a sealed container of constant **volume** to carry out formation reactions than to attempt to construct an apparatus that would maintain constant pressure. Therefore, the heat of formation is not generally measured directly. Instead, it is calculated from other quantities that are directly measured.

Most reactions by which compounds are formed directly from their constituent elements are very difficult or impossible to carry out directly under conditions that allow careful measurements to be taken. In most cases, however, the heats of formation can be obtained indirectly from measurements of other reactions. Because the heat of formation is an enthalpy of reaction, Hess's Law applies: when the equation for an reaction can be obtained by algebraically combining a series of other reactions, the enthalpy for the reaction can be obtained by algebraically combining the other equations in the same manner. For example, if reaction 1 and reaction 2 can be added to obtain the equation for the desired reaction, a formation reaction, that cannot be measured directly, the enthalpy of the desired reaction is the sum of the enthalpies of reactions 1 and 2.

Heats of formation are often specifically designated to show the conditions under which they apply. The temperature for which the heat of formation applies should always be designated. A subscript indicating the temperature on the Kelvin or absolute scale is generally used, for example, $(\Delta)H_{f,\,300}$ for a reaction at 300 K. If no temperature is indicated, it is assumed that the temperature is 298 K or 25°C. If the heat of formation corresponds to a reaction in which all of the reactants and products are in their standard physical states, solid, liquid or gas, at the stated temperature and they are present in standard concentrations, one molar for solutions, pure **solids** and **gases** at one atmosphere pressure, it is called a standard heat of formation, designated by a superscript "o", for example $(\Delta)H^{o}_{f,\,300}$. The units typically used for heats of **solution** are kilojoules per mole or kilocalories per mole.

Heat of fusion

The **heat** of fusion is the heat that is absorbed to transform a substance from its solid state to its liquid state at constant, that is, to melt the solid substance. In common usage, the heat of fusion is used in place of the more precise term the **enthalpy** of fusion, which has the symbol $(\Delta)H_{fus}$. The enthalpy of fusion is the heat of fusion for melting of one **mole** of the substance under three specific conditions: (1) the pressure remains constant, (2) the only possible work that occurs is expansion against the atmosphere (so-called $P(\Delta)V$ work) and (3) the **temperature** remains constant during the process. A heat of fusion for a substance is only valid for conversion of the pure solid to the pure liquid state of the substance.

Unlike many other types of enthalpy changes, enthalpies of fusion for a large number of substances can be measured directly. Both the solid and liquid states are condensed phases and relatively small changes in **volume** occur as a substance is melted. In addition, if heat is applied slowly, a very nearly constant temperature can be maintained. To help maintain a constant temperature, it is useful to finely divide the solid substance to produce a large surface area to volume ratio. In this way, heat is readily distributed within the small particles.

For those cases in which it is not feasible to carry out a solid to liquid transformation without increasing the temperature of the system, the enthalpy of fusion can be determined by subtracting the heat absorbed to increase the temperature of the resulting liquid from the total heat absorbed in the process. The heat absorbed by the liquid is equal to the change in temperature multiplied by the molar heat capacity and the number of moles of solid melted.

Enthalpies of fusion are often specifically designated to show the conditions under which they apply. The temperature at which a substance melts depends upon the external pressure on the solid. If the pressure is other than one atmosphere, a subscript indicating the pressure should be used, for example, $(\Delta)H_{fus,\,2\,atm}$ for a reaction at twice atmospheric pressure. One atmosphere is the pressure of the atmosphere at sea level, previously defined by the atmospheric pressure that would support a 760 mm column of **mercury** in a tube closed at the top (a barometer) and now defined in the **metric system** or Systeme International, SI as 101,325 Pascals where one Pascal is one kilogram meter per seconds squared. Other pressure units that are commonly used are the torr (equal to 1.760 of a standard atmosphere), the Pascal, the bar (equal to 0.9869 atmospheres or 100,000 Pascals) and pounds per square inch (used principally by American engineers). The units typically used for heats of **solution** are kilojoules per mole or kilocalories per mole.

Enthalpies of fusion are positive because the solid state affords the molecules or ions that compose a pure substance the opportunity to maximize their attractive interactions with each other. When a substance is melted, the substance not only becomes fluid on the macroscopic scale that we can see, but within the material the molecules or ions are more mobile. In the liquid state they do not maintain the best geometric arrangement for maximizing their attraction for each other, as they do within the solid, but continually shift their orientations to neighboring molecules or ions.

Heat of reaction

The **heat** of reaction is the heat that is absorbed when a substance is formed from its elements. In common usage, the heat of reaction is used in place of the more precise term the **enthalpy** of reaction, which has the symbol $(\Delta)H_{rxn}$. The enthalpy of reaction is the heat absorbed when one **mole** of a specified substance completely reacts according to a specified chemical equation under three specific conditions: (1) the pressure remains constant, (2) the only possible work that occurs is expansion against the atmosphere (so-called $P(\Delta)V$ work) and (3) the **temperature** remains constant during the process.

In practice, a reaction is likely to result in an increase or decrease in temperature (reactions that occur with an evolution of heat are called exothermic and reactions that require heat are termed endothermic). In addition, it is generally easier experimentally to use a sealed container of constant **volume** to carry out reaction reactions than to attempt to construct an apparatus that would maintain constant pressure. Therefore, the heat of reaction is not generally measured directly. Instead, it is calculated from other quantities that are directly measured.

Many reactions are very difficult or impossible to carry out directly under conditions that allow careful measurements to be taken. In most cases, however, the heats of reaction can be obtained indirectly from measurements of other reactions. Because the heat of reaction is an enthalpy of reaction, Hess's Law applies: when the equation for an reaction can be obtained by algebraically combining a series of other reactions, the enthalpy for the reaction can be obtained by algebraically combining the other equations in the same manner. For example, if reaction 1 and reaction 2 can be added to obtain the equation for the desired reaction, that cannot be measured directly, the enthalpy of the desired reaction is the sum of the enthalpies of reactions 1 and 2.

Heats of reaction are often specifically designated to show the conditions under which they apply. The temperature for which the heat of reaction applies should always be designated. A subscript indicating the temperature on the Kelvin or absolute scale is generally used, for example, $(\Delta)H_{rxn, 300}$ for a reaction at 300 K. If no temperature is indicated, it is assumed that the temperature is 298 K or 25°C. If the heat of reaction corresponds to a reaction in which all of the reactants and products are in their standard physical states, solid, liquid or gas, at the stated temperature and they are present in standard concentrations, one molar for solutions, pure **solids** and **gases** at one bar pressure, it is called a standard heat of reaction, designated by a superscript "o", for example $(\Delta)H^{o}_{rxn, 300}$. The units typically used for heats of **solution** are kilojoules per mole or kilocalories per mole.

Heat of solution

The **heat** of **solution** is the heat that is absorbed when one substance dissolves in another to formation a homogeneous mixture. A homogeneous mixture is one which a sample of the mixture that contains a substantial number of molecules has the same composition as any other sample: the components of the mixture are distributed uniformly throughout. In common usage, the heat of solution is used in place of the more precise term the **enthalpy** of solution, which has the symbol $(\Delta)H_{soln}$. The enthalpy of solution is the heat of solution for dissolution of one **mole** of a substance in another substance such that the final solution is one molar and three conditions are fulfilled: (1) the pressure remains constant, (2) the only possible work that occurs is expansion against the atmosphere (so-called $P(\Delta)V$ work) and (3) the **temperature** remains constant during the process. The enthalpy of solution of a substance is only valid for dissolution of the substance in its pure state into another pure substance.

Enthalpies of solution for a large number of substances can be measured directly when the resulting solution is liquid, the substance that is dissolved (the solute) is either a solid or liquid and the substance into which the solute is dissolved (the solvent) is liquid. If the solute is dissolved slowly, a very nearly constant temperature can be maintained. The heat absorbed in gas-liquid dissolution processes is often readily measured, but if the measurements are made using a constant **volume** apparatus, the $P(\Delta)$ work must be accounted for. Some gas-solid solution processes, such as the dissolution of **hydrogen** in certain **metals**, occur under conditions that allow measurements to be made readily, but many do not. As in the case of gas-liquid solutions, the effect of the decreased volume as the gas is absorbed must be accounted for in determining the enthalpy of solution.

Gases that do not react with each other readily mix without substantial absorption or evolution of heat, so gas-gas enthalpies of solution are generally small. Enthalpies of solution for solid-solid solutions are difficult to measure accurately because the time required for dissolution is typically very long.

For those cases in which it is not feasible to carry out a dissolution process without increasing the temperature of the system, the enthalpy of solution can be determined by subtracting the heat absorbed to increase the temperature of the resulting solution from the total heat absorbed in the process. The heat absorbed by the solution is equal to the change in temperature multiplied by the specific heat capacity of the solution and the **mass** of the solution in grams.

Enthalpies of solution are often specifically designated to show the conditions under which they apply. The temperature is designated by a subscript, for example, $(\Delta)H_{soln, 350\,K}$ for a process in which one substance at 350 K is dissolved in another also at 350 K and no temperature change occurs. If no temperature is specified, the temperature is typically 298 K or 25°C. For some types of solutions, particularly gas-liquid and gas-solid solutions, the heat of solution as well as the **solubility** can depend on pressure. If there is a substantial effect of pressure, it should also be included in the subscript. The units typically used for heats of solution are kilojoules per mole or kilocalories per mole.

One of the most common types of heat of solution is that for the dissolution of a substance, most commonly salts, in **water**. The enthalpy of solution for aqueous solutions is called the enthalpy of hydration or, simply the heat of hydration, and has the symbol $(\Delta)H_{hyd}$, where *hyd* is the abbreviation for hydration. Enthalpies of hydration for individual ions have been calculated from the enthalpies of hydration of a large number of salts containing the ions. The solubilities of salts are related to their enthalpies of hydration, as the solubilities of other solutes in a particular solvent are related to their enthalpies of solution in those **solvents**. A more negative enthalpy increases the tendency for a solute to dissolve.

Enthalpies of solution can be positive or negative: the dissolution process may be endothermic, with heat being absorbed, or exothermic, with heat being evolved. If the **energy** of attraction of the solute molecules or ions with each other and the energies of attraction of the solvent molecules or ions with each other are greater than the energy of attraction of solute and solvent particles, heat will be required to make the solute disperse in the solvent. Heat will then be absorbed in the process and the heat of solution will be positive. In this case, the solubility of the material increases with temperature. The heat of solution for formation of aqueous solutions of most salts is positive. The absorption of heat as a result of a mixing process is the basis of commercial "cold-packs" used by athletic trainers to treat minor injuries. If the energy of attraction of solute molecules or ions and solvent molecules or ions is greater than the energies of attraction of solute particles with each other and solvent particles with each other, heat will be released as the solute dissolves. For these exothermic processes, the solubility decreases with an increase in temperature.

Heat of sublimation

The **heat** of **sublimation** is the heat that is absorbed to transform a substance from its solid state to its vapor. In common usage, the heat of sublimation is used in place of the more precise term the **enthalpy** of sublimation, which has the symbol $(\Delta)H_{sub}$. The enthalpy of sublimation is the heat of sublimation for vaporizing one **mole** of the solid substance under three specific conditions: (1) the pressure remains constant, (2) the only possible work that occurs is expansion against the atmosphere (so-called P(Δ)V work) and (3) the **temperature** remains constant during the process. A heat of sublimation for a substance is only valid for conversion of the pure solid to the pure gaseous state of the substance.

Relatively few substances sublime at or near atmospheric pressure. Some examples of those that do sublime readily are **carbon** dioxide and **iodine**. Most substances are either transformed from the solid to the liquid state as heat is added or they decompose into other substances. At pressures substantially below normal atmospheric pressure, more substances can be made to sublime. Enthalpies of sublimation can be measured directly for those substances that sublime at experimentally reasonable pressures and temperatures. The process is carried out sufficiently slowly to avoid temperature changes. To permit uniform transfer and heat and minimize the time required for the process, measurements are made on finely divided **solids** that provide a high surface area to **volume** ratio. Because there is a large change in volume, measurements made in a constant-volume apparatus are not enthalpies and the change in P(Δ)V work that is done must be accounted for to obtain the enthalpy for the process.

Enthalpies of sublimation are often specifically designated to show the conditions under which they apply. Unless the sublimation takes place at 298 K (25°C) or one atmosphere pressure, the temperature and pressure are indicated by subscripts: $(\Delta)H_{sub,\ 0.5\ atm}$ for a reaction at one-half atmospheric pressure. The temperature at which a substance sublimes can be strongly dependent upon the external pressure on the solid. One atmosphere is the pressure of the atmosphere at sea level, previously defined by the atmospheric pressure that would support a 760 mm column of **mercury** in a tube closed at the top (a barometer) and now defined in the **metric system** or Systeme International, SI as 101,325 pascals where one pascal is one kilogram meter per seconds squared. Other pressure units that are commonly used are the torr (equal to 1.760 of a standard atmosphere), the pascal, the bar (equal to 0.9869 atmospheres or 100,000 pascals) and pounds per square inch (used principally by American engineers). The units typically used for heats of sublimation are kilojoule per mole or kilocalories per mole.

The enthalpy of sublimation depends on the difference in the energies of attraction of molecules or ions in the solid state for each other and the energies of attraction for each other in the gaseous state. Since particles in the gaseous state have very small energies of attraction for each other and all molecules or combinations of negative and positive ions have strong attractions for each other when they are in close contact, as they are in the solid state, enthalpies of sublimation are always positive.

Heat of vaporization

The **heat** of vaporization is the heat that is absorbed to transform a substance from its liquid state to its vapor, that is, to boil or evaporate the liquid substance completely. In common usage, the heat of vaporization is used in place of the more precise term the **enthalpy** of vaporization, which has the symbol $(\Delta)H_{vap}$. The enthalpy of vaporization is the heat of vaporization for vaporizing one **mole** of the substance under three specific conditions: (1) the pressure remains constant, (2) the only possible work that occurs is expansion against the atmosphere (so-called P(Δ)V work) and (3) the **temperature** remains constant during the process. A heat of vaporization for a substance is only valid for conversion of the pure liquid to the pure gaseous state of the substance.

Enthalpies of vaporization can be measured directly for a large number of substances whose natural physical state at experimentally reasonable temperatures is liquid, if the process is carried out sufficiently slowly to avoid temperature changes. Because there is a large change in **volume**, measurements made in a constant-volume apparatus are not enthalpies and the change in P(Δ)V work that is done must be accounted for to obtain the enthalpy for the process.

Enthalpies of vaporization are often specifically designated to show the conditions under which they apply. Unless the vaporization takes place at 298 K (25°C) or one atmosphere pressure, the temperature and pressure are indicated by subscripts: $(\Delta)H_{vap,\ 0.5\ atm}$ for a reaction at one-half atmospheric pressure. The temperature at which a substance vaporizes can be strongly dependent upon the external pressure on the liquid. One atmosphere is the pressure of the atmosphere at sea level, previously defined by the atmospheric pressure that would support a 760 mm column of **mercury** in a tube closed at the top (a barometer) and now defined in the **metric system** or Systeme International, SI as 101,325 pascals where one pascal is one kilogram meter per seconds squared. Other pressure units that are commonly used are the torr (equal to 1.760 of a standard atmosphere), the pascal, the bar (equal to 0.9869 atmospheres or 100,000 pascals) and pounds per square inch (used principally by American engineers). The units typically used for heats of **solution** are kilojoules per mole or kilocalories per mole.

The enthalpy of vaporization depends on the difference in the energies of attraction of molecules or ions in the liquid state for each other and the energies of attraction for each other in the gaseous state. Since particles in the gaseous state have very small energies of attraction for each other and all molecules or combinations of negative and positive ions have strong attractions for each other, enthalpies of vaporization are always positive. The heat of vaporization of **water** is unusually large because the strong **hydrogen bonding** between water molecules in the liquid state. Because a great deal of **energy** is required to force water molecules apart from each other, water evaporation is an especially effective cooling process.

Heat transfer

Heat can be defined as the **energy** exchanged when a difference in **temperature** exist between two regions. The science of heat transfer attempts to describe the exchange of heat when the appropriate physical conditions characterizing that process have been specified. According to classical theory, there are three ways that heat can be transferred; these are by conduction, convection, and radiation.

In conductive heat transfer, heat is transported through solid materials or stagnant fluids from one body at a given temperature to another body at a lower temperature by the transfer of kinetic energy through molecular impact. In **gases** and **liquids**, where molecules are farther apart than in **solids**, the energy transferred by molecular collisions is smaller than in solids, where heat conduction is facilitated by atomic vibrations in the crystal. In solids that conduct **electricity**, heat conduction is further enhanced by the drift of free electrons.

In convective heat transfer, heat is carried from one place to another by the actual movement of hot **matter** from a hot region to a cooler one. While convection could involve solid transport, the most important cases of convection involve fluid transport. If a liquid is heated from below, the molecules in it begin to move more rapidly, until they are eventually pushed upward by the denser cooler fluid that remains with the result that convection currents arise in the fluid. The heating, cooling, and resultant movement of air molecules in the earth's atmosphere also give rise to convection currents, which we recognize as wind.

In radiant heat transfer, heat is transported via radiation. This is a completely different mechanism of heat transfer than those involved in convection and conduction. The thermal radiation emitted from a surface depends on the surface temperature, the type of surface, the wavelength of the emitted radiation, and the direction in which the heat is radiated. All objects radiate energy; objects at low temperatures emit long wavelength radiation, those at high temperatures emit short wavelength radiation. The heat that reaches the earth from the sun is transferred by radiation.

In convective and radiant heat transfer, the amount of heat transferred is usually a complex function of the temperatures of the two regions involved. In conduction, however, the amount of heat transferred is proportional to the temperature difference over a region of space.

When two objects, one hot and one cold, are brought into contact, heat is transferred from the hotter body to the colder one until their temperatures are equal. Among the best conductors of heat are **metals** (with loose outer electrons), which feel cool to the touch because they conduct heat away from our hands. **Wood**, wool, straw, and **paper** are poor conductors of heat because the outer electrons of these materials are firmly attached.

The most complex heat transfer problems involve the simultaneous occurrence of a phase change, i.e., condensation, boiling, solidification, or **sublimation**.

See also Electrical conductivity

HEAVY METALS AND HEAVY METALS POISONING

Heavy **metals** are metallic elements that have a high **atomic number** and are poisonous to living organisms. Approximately 30 metals have been shown to be poisonous to humans. Examples of heavy metals that are poisonous include **mercury**, **chromium**, **cadmium**, **arsenic**, and **lead**. Because they are poisonous, heavy metals are sometimes referred to as toxic metals. Heavy metals may be poisonous on their own or as part of chemical compounds.

It has been known for centuries that certain metals are toxic. For example, Theophrastus of Erebus (370-287 B.C.) and Pliny the Elder (23-79) both described poisonings that resulted from arsenic and mercury. Other heavy metals, such as cadmium, were not recognized as poisonous until the early nineteenth century.

Heavy metals occur naturally in the environment in rocks and ores. They cycle through the environment by geological and biological means. The geological cycle begins when **water** slowly wears away rocks and dissolves the heavy metals. The heavy metals are carried into streams, rivers, lakes, and oceans. The heavy metals may be deposited in sediments at the bottom of the water body, or they may evaporate and be carried elsewhere as rainwater. The biological cycle includes accumulation in plants and animals and entry into the food web.

Sometimes these natural cycles can pose a hazard to human health, because the levels of heavy metals exceed the body's ability to cope with them. A further complication is the addition of heavy metals to the environment as a result of human activity. Activities such as mining and manufacturing greatly increase the release of heavy metals from rocks and ores. The activities also create situations in which the heavy metals are incorporated into new compounds and may be spread worldwide.

The health hazards presented by heavy metals depend on the level of exposure and the length of exposure. In general, exposures are divided into two classes: acute exposure and chronic exposure. Acute exposure refers to contact with a large amount of the heavy metal in a short period of time. In some cases, the health effects are immediately apparent; in others, the effects are delayed. Chronic exposure refers to contact with low levels of the heavy metal over a long period of time. A person may not even be aware that the exposure is occurring. Heavy metal poisoning can be insidious because health problems may develop very slowly.

Depending on their form, heavy metals and their compounds can be taken up through the skin, respiratory system, or the digestive system. Contaminated air, water, soil, food, and drugs yield potential heavy metal exposures. Once in the body, heavy metals may be distributed to all parts by the blood. In both acute and chronic exposures, the symptoms of poisoning depend on the identity of the heavy metal or the compound in which it occurs.

Usually, immediate symptoms are linked to an acute exposure; however, not all acute exposures are immediately followed by symptoms. For example, ingesting a large dose of arsenic can cause fever, loss of appetite, enlarged liver, and heart failure soon after ingestion. However, the nerve damage resulting from brief skin contact with dimethyl mercury, a mercury compound used in some laboratories, may not be apparent for months.

Health problems linked to chronic exposure may include respiratory problems, altered blood cells, cancer, nerve damage, and liver or kidney damage. Chronic exposure to heavy metals may also have reproductive side effects. Some heavy metals and their compounds can accumulate in the body's tissues, such as in the bones or in nerves. In pregnant women, heavy metals and their compounds may cross the placenta and harm an unborn child.

One of the most famous cases of mercury poisoning resulting from chronic exposure occurred in Minamata, Japan. The city's harbor was contaminated with industrial wastes that contained methylmercury. Fish in the harbor were likewise contaminated and people who ate these fish were gradually poisoned. The methylmercury accumulated in their brains and caused severe, irreversible damage. More than 120 people were poisoned; 22 of them prior to birth because their mothers had unknowingly eaten contaminated fish.

Children are especially susceptible to health problems caused by heavy metals, because their bodies are smaller and still developing. Exposure to lead before birth or during infancy and childhood may lead to birth defects, developmental delays, and decreased intelligence.

Environmental controls are in place in many countries to limit exposures to heavy metals. However, these controls may be inadequate where the technology to limit heavy metal emissions is not available or the rules are not enforced. Also, accidents have occasionally occurred in which many people are exposed. Naturally occurring heavy metal contamination is difficult to control.

Other exposures occur because current rules do not apply to past actions. For example, lead was once commonly used in house paints before it was apparent that low levels of lead could present a health problem. Lead is not used in modern house paints, but many houses were built and painted before the rules went into effect. Therefore, older houses often pose a potential lead poisoning threat, especially for young children.

Heavy metal poisoning is treated by a class of drugs called chelating agents. Although the details of how each chelating agent works differ, they all share the same basic function. They work by binding to heavy metals in the body and forming nontoxic compounds with them. These nontoxic compounds can be removed much more easily than the heavy metals alone. Common chelating agents include British anti-lewisite (BAL) and related compounds, ethylenediaminetetraacetic acid (EDTA), and penicillamine.

See also Carcinogens; Chelate

Heisenberg uncertainty principle

Classical physicists were convinced that the only limitations on the knowledge would derive from limitations inherent in our ability to make measurements of the phenomena in question. However, in 1927, **Werner Heisenberg**, while studying the motion of the **electron**, discovered that a fundamental uncertainty exists in any attempt to accurately measure both the momentum and position of an electron. Heisenberg's discovery became the basis for his quantum mechanical uncertainty principle. The simplest statement of this principle is that it is not possible to determine at the same time the position and momentum of a particle. The more precise the measurement of position, the less precise the knowledge of the particle's momentum. (Mathematically, the principle states the product in the uncertainty in position and uncertainty in momentum cannot be less than **Planck's constant** divided by 2 π.) Trying to simultaneously determine both a particle's position and momentum becomes a little like trying to determine whether a tree falling in the forest makes any sound if there is nothing around to hear it. It simply cannot be done.

Of course Heisenberg's principle is completely contrary to what our everyday experience tells us about the nature of reality. Like the classical physicist, we find it difficult to accept a world in which relationships are fuzzy. When something is not clear, we tend to conclude that there must be something wrong with the way we are seeing it, not that the world just happens to be that way. (No less a scientist than **Albert Einstein** was sufficiently troubled by Heisenberg's results that he found it necessary to remark that he could not believe that God played dice with the universe.)

But, as Heisenberg showed us, when we venture further into the subatomic world, our perception of what is occurring there becomes less clear because of fundamental limitations on what we can measure. Nevertheless, a great deal of semi-quantitative information about the behavior of atomic systems can be obtained from the uncertainty principle. First, this principle tells us that a particle confined to a very small space cannot have zero kinetic **energy**. Instead, for an electron in an **atom**, Heisenberg's result predicts a finite kinetic energy (about 4 eV). Second, the uncertainty principle tells us that there is a minimum energy for molecular vibrations. Third, the size of the **hydrogen** atom's first electron orbit is accurately predicted. Finally, the uncertainty principle tells us that the energy of a system cannot be measured exactly unless an infinite time is available for the measurement. Thus, if an electron is in an energetically excited state, it will emit light having a variety of energies when it makes transitions to lower energy states.

Heisenberg was born in Würzburg, Germany, in 1901. The son of a professor at the University of Munich, Heisenberg's studies with Arnold Sommerfeld in Munich, Max Born in Göttingen, and **Niels Bohr** in Copenhagen thrust him at the forefront of the nascent discipline of quantum mechanics in the 1920s. At Bohr's suggestion, Heisenberg employed matrix mathematics to represent an electron's energy, momentum, position, and angular momentum. Heisenberg's intention was to use this mathematical model to explain relationships between the lines in atomic spectra.

Heisenberg was awarded the Nobel Prize in 1932. He was appointed professor at the University of Berlin in 1941 and director of the Kaiser Wilhelm Institute for Physics. At the end of World War II, he was taken prisoner by American troops, and held briefly in England. He returned to Germany to help found the Max Planck Institute for Physics in Göttingen. From 1957 until his death in 1976, Heisenberg worked on problems in **plasma** physics and thermonuclear processes.

Werner Karl Heisenberg.

See also Atomic structure; Bohr theory; Quantum chemistry; Spectroscopy

Heisenberg, Werner Karl (1901-1976)

German physicist

Werner Karl Heisenberg is best known for his discovery of the uncertainty principle, for which he was awarded the 1932 Nobel Prize in physics. The uncertainty principle states that it is impossible to specify precisely both the position and momentum of a particle at the same time. The enunciation of that principle contributed to the understanding of a number of problems in atomic physics. It was also to have a profound effect on a far more general level, in that it essentially invalidated the long-held and fundamental scientific thesis of cause and effect. One of Heisenberg's first great accomplishments was the development of a new approach for solving problems of **atomic structure**, matrix mechanics.

Heisenberg was born in Würzburg, Germany, on December 5, 1901. His father, August Heisenberg, taught Greek philology at the University of Würzburg and ancient languages at the Altes Gymnasium (secondary school) of the same city. His mother, Annie Wecklein, was the daughter of the headmaster of Munich's Maximilian Gymnasium. Werner entered elementary school at Würzburg, but at the age of nine he moved with his parents and his elder brother to Munich, where his father had accepted an appointment as professor of Greek philology at the University.

In Munich, Heisenberg enrolled in the Maximilian Gymnasium, where his grandfather was headmaster. His education was interrupted, however, in the summer of 1918, when he was called to assist the faltering wartime effort by harvesting crops on a Bavarian farm. Shortly thereafter, he returned to Munich and served as a messenger for the democratic socialist forces that managed after bitter street fighting to oust a communist-oriented government that had briefly taken control of the Bavarian state. He also became involved with a number of youth groups that had organized in an effort to build a new society out of the wreckage left in the wake of World War I. Eventually he was elected leader of a group associated with the Bund Deutscher Neupfadfinder (New Boy Scouts), which, according to David C. Cassidy writing in the *Dictionary of Scientific Biography,* "strove for a renewal of supposedly decadent German personal and social life through the direct experience of nature and the uplifting beauties of Romantic poetry, music, and thought." Already a gifted pianist, Heisenberg would maintain his interest in music throughout his life; later, after his marriage, the whole Heisenberg family often assembled for an evening of chamber music.

After the war, Heisenberg completed his studies at the Gymnasium and, in 1920, decided to enter the University of Munich. He hoped to pursue a degree in mathematics, a subject in which he had long been interested and for which he had shown great talent; while still a student at the gymnasium he had taught himself calculus and had written a paper on number theory. Heisenberg changed his mind, however, when Munich's professor of mathematics, Ferdinand von Lindemann, refused to accept him as a student in an advanced seminar. On his father's advice, Heisenberg then applied for his second choice, physics. He requested an interview with Arnold Sommerfeld, then a professor of theoretical physics at Munich, who agreed to admit Heisenberg to his seminar.

As Sommerfeld's student, Heisenberg not only became rapidly involved in the most fundamental problems in theoretical physics, but also became acquainted with a group of scientists—including the physicists **Max Born, Niels Bohr,** Wolfgang Pauli and Enrico Fermi—whose work would dominate the field of atomic physics for years to come. At Munich, Heisenberg immediately began investigating some of the problems related to refinements of the Bohr model of the **atom**. In 1913, Bohr had employed the new concept of quantum physics to develop a model of the atom that explained much empirical data with amazing accuracy. However, a number of problems remained with the Bohr model. For one thing, no theoretical basis existed for the concept of quantized orbitals professed by Bohr. Also, inadequacies in the model's predictive ability suggested that some essential points were still missing from the **Bohr theory**. One of the earliest improvements on the Bohr model had been suggested by Sommerfeld himself when, in 1916, he hypothesized the existence of elliptical orbitals for electrons.

The topic to which Heisenberg first addressed his attention was the **Zeeman effect**, the splitting of spectral lines, a phenomenon for which no explanation had yet been suggested. In a series of papers, two written with Sommerfeld, Heisenberg developed an explanation for the Zeeman effect. His approach required an abandonment of some fundamental principles in both classical physics and quantum mechanics. But the hypothesis was attractive, nonetheless, and, as Cassidy points out, was "the first indication of the radical changes required for solving the quantum riddle." Heisenberg obtained his Ph.D. in 1923, although not without some difficulty. As he was primarily interested in theoretical physics, he had spent little time on laboratory experimentation. During his oral examinations for the doctoral degree, he was unable to answer questions put to him on instrumental matters by Wilhelm Wien, a professor of experimental physics, who consequently recommended that Heisenberg not be passed. Largely as a result of Sommerfeld's negotiation, however, Wien was overruled and Heisenberg received his degree.

The Zeeman papers brought Heisenberg's name to the attention of physicists throughout Europe. Thus, when Sommerfeld spent a year as visiting professor at the University of Wisconsin in 1922–1923, he made arrangements for Heisenberg to spend that year studying under Born at Göttingen. Born was so impressed with his new student's work that he recommended Heisenberg for a teaching position. This decision was quite remarkable since Heisenberg was only 22 at the time and had just taken his Ph.D. Heisenberg remained in Göttingen until 1926, with the exception of a semester back in Munich and another spent working under Bohr in Copenhagen in the fall of 1924.

One of the life-long medical problems with which Heisenberg had to deal was hay fever. He eventually became accustomed to leaving the inhospitable climate of Bavaria whenever a serious attack occurred and traveling to the less pollen-dense island of Heligoland. One such trip in June, 1925, was to have profound significance for theoretical physics. With weeks of free time on his hands, Heisenberg devised a new approach to the problem of the Zeeman effect in particular and of **atomic theory** in general. Physicists had devoted too much effort, he said, to devising pictorial models that might have some physical reality. A better approach, he suggested, would be to work strictly with experimental data and to determine the mathematical implications of that data.

In order to pursue this approach, Heisenberg devised a system that came to be known as matrix mechanics. Although, as he later learned, the mathematics of this technique was already familiar to professional mathematicians, its application to the problems of atomic physics was entirely new. After his return from Heligoland, he collaborated with Born and Born's assistant, Pascual Jordan, to publish a refined version of the theory of matrix mechanics. The paper marked a turning point in physics. As biographers Nevil Mott and Rudolf Peierls wrote in the *Biographical Memoirs of Fellows of the Royal Society,* "It is fair to regard this paper as the start of a new era in atomic physics, since any look at the physics literature of the next two or three years clearly shows the intensity and success of the work stimulated by this paper."

Matrix mechanics did prove to be very fruitful in the development of atomic theory. Still, many physicists were somewhat uneasy with the mathematical abstraction of the approach because it didn't provide them with physical models to which to relate. When Austrian physicist **Erwin Schrödinger** devised his wave mechanics about a year later, many physicists switched their allegiance to his more physically-based approach. The conflict between the two theories was resolved fairly soon, however, when Schrödinger showed that matrix mechanics and wave mechanics are mathematically identical.

The Heisenberg-Born-Jordan paper was published in November, 1925, and six months later Heisenberg left Göttingen to become Bohr's assistant in Copenhagen. The almost daily contact between the two great physicists was to lead to one of the most productive periods in Heisenberg's life, including his enunciation of the principle for which he is best known, the uncertainty principle.

Heisenberg turned his attention to uncertainty in February, 1927. He was involved in a study of the fundamental quantum properties of an **electron** in its orbit around the **nucleus**. At one point in that analysis, he imagined using a gamma ray microscope to study an electron's motion. It occurred to him that the very act of measuring an electron's properties by shining gamma rays on it would disturb the electron in its orbit; the act of observing the electron, therefore, altered its behavior, and objectivity was lost. Out of this realization, he developed a fundamental principle of physics that is now familiar to most high school students of the subject. According to the uncertainty principle, one can measure the position of a particle *or* its momentum with as much precision as desired. However, the more precise one of these measurements becomes, the less precise the other will be, such that the product of their inaccuracies (to use Heisenberg's original term) must always be less than **Planck's constant** (the unvarying ratio of the frequency of radiation to its quanta of energy). The principle was quickly and widely accepted by most, though not all, physicists, who soon acknowledged the concept as a fundamental law of nature. For the development of the uncertainty principle, as well as his other contributions to theoretical physics, Heisenberg was awarded the Nobel Prize for physics in 1932.

In 1927, Heisenberg was offered chairs at the universities of Leipzig and Zürich. He chose the former and, upon assuming the post of professor theoretical physics in October of that year, became the youngest full professor in Germany. His duties at Leipzig, unlike those of modern professors, were extensive, including a full schedule of teaching and administrative responsibilities. Not surprisingly, his scientific output diminished somewhat as a result of these competing demands for his time and energy. Nonetheless, he continued to pursue research interests in a number of directions, including ferromagnetism, quantum electrodynamics, cosmic radiation, and nuclear physics. Perhaps his best-known contribution in this period occurred after English physicist **James Chadwick** discovered the **neutron** in 1932. Heisenberg proposed a model of the nucleus that consisted of protons and neutrons rather than protons and electrons, as favored by some physicists. This theory did not originate with Heisenberg, however.

The last half of Heisenberg's life was complicated by the political turmoil taking place in Germany and around the world in the years surrounding World War II. With the rise of National Socialism in the early 1930s, many German scientists fled the country to take up residence in the United Kingdom, the United States, and other free nations. Heisenberg, like his colleague **Otto Hahn**, was one of the few world-class scientists who determined to remain behind and protect, as well as he could, Germany's scientific traditions and institutions.

At first, Heisenberg and his colleagues resisted German leader Adolf Hitler's efforts to ''purify'' and nationalize German science, but the forces they faced were too great. Soon Nazi students were in command of universities, including that of Leipzig, ensuring that non-Aryan and ''politically unacceptable'' faculty members were weeded out. Heisenberg's position was especially tenuous since the Nazis regarded theoretical physics (as opposed to experimental physics) as a suspect, ''Jewish'' science. In 1935, an invitation from München for Heisenberg to succeed his former teacher Sommerfeld was met with violent opposition from German political leaders and professional colleagues, one of whom branded Heisenberg a ''white Jew.'' Any hope that Heisenberg may have had of accepting the Munich offer was dashed and, for a time, even his personal safety was at risk.

At the beginning of World War II, the Nazi government recognized Heisenberg's value and made him director of the German atomic bomb project. For the next five and a half years, all of his efforts were devoted to this project. Questions have since been raised as to the morality of Heisenberg's decision. The fact remains that he chose to stay in his native country and devote his abilities to its war effort.

When the war came to an end in 1945, Heisenberg and his bomb research team were captured in a remote region of southern Germany, where they had been sent to be safe from Allied bomb attacks. After a six-month internment in England, Heisenberg was allowed to return to Göttingen, where he reestablished the Kaiser Wilhelm Institute for Physics and renamed it the Max Planck Institute in honor of his colleague and friend, the originator of quantum theory.

In the postwar years, Heisenberg worked diligently to restore German science to its former high standing. He was largely responsible for the creation of the Deutscher Forschungsrat (German Research Council), of which he became president. In this role, he came to represent West Germany in many international organizations and at many international meetings. His continuing administrative and political responsibilities did not deter Heisenberg from his research interests, however. In the postwar years, he became particularly interested in the search for a unified field theory, which would link all physical fields, such as gravitation and electromagnetism. He also pursued his interest in more general topics, such as the relationship between physical theory and philosophy.

Heisenberg retired in 1970, although he continued to write on a variety of topics. His health began to fail in 1973, and two years later he became seriously ill. He died on February 1, 1976, in Munich. He was survived by his wife, Elisabeth Schumacher, whom he had married in 1937. He also left seven children, four daughters and three sons.

Helium

Helium is the first member of the noble gas family, the elements that make up Group 18 of the **periodic table**. Its **atomic number** is 2, its atomic **mass** is 4.002602, and its chemical symbol is He.

Properties

Helium is a colorless, odorless, tasteless gas with a number of unusual properties. For example, it has the lowest **boiling point** of any element, -452°F (-268.9°C). Its freezing point is -458°F (-272.2°C). At a **temperature** of about -456°F (-271°C), helium undergoes a striking change. It remains a liquid, but a liquid with unusual properties. One of the most interesting of those properties is superfluidity, a condition in which a liquid flows upwards out of a container against the force of gravity and may pass through very small holes that would normally be impermeable to the gas. Chemically, helium is completely inert. It does not form compounds or react with any other element.

Occurrence and Extraction

Helium is the second most abundant element after **hydrogen** in the universe and in the solar system. About 11.3% of all atoms in the universe are helium atoms. By contrast, about 88.6% of all atoms in the universe are hydrogen. Helium is much less abundant on Earth. It is the sixth most abundant gas in the atmosphere after **nitrogen**, **oxygen**, **argon**, **carbon** dioxide, and **neon**. It makes up about 0.000524% of the air. Helium is also formed in the Earth's crust during the radioactive decay of **uranium** and other radioactive elements, but it usually escapes into the atmosphere soon after being formed.

Discovery and Naming

Helium was first discovered not on Earth, but in the Sun. In 1868, the French astronomer Pierre Janssen studied light from the Sun during a solar eclipse using a spectroscope. He found spectral lines that had never been observed before and decided that they were produced by a new element. He suggested the name helium for the element after the Greek word for the Sun, *helios*. Helium was first discovered on Earth in 1895 by the English physicist Sir **William Ramsay** who found the element in a mineral containing the element uranium.

Uses

The most important single use for helium in the United States is in low-temperature cooling systems. Since helium can be cooled to a lower temperature than any other material, it can be used to cool any other material. Helium is also used to provide inert atmospheres for industrial operations. For example, it is used in welding systems to prevent the **metals** being heated from reacting with oxygen in the air before they join to each other. Helium is also used to purge and pressurize systems. Any gas could, in theory, be used for these purposes, but helium has the advantage of not reacting with any of the components of the system being purged or pressurized. One of helium's better-known, but less commercially important, applications is in lighter-than-air craft, such as dirigibles, and in weather and research balloons. For most purposes, helium has largely replaced hydrogen, which has a lower **density** but is more dangerous to work with.

HELMONT, JOHANNES BAPTISTA VAN (1579-1644)

Flemish physician, chemist, and physicist

Johannes van Helmont lived and worked during a unique period of history—the dawn of the scientific revolution. As Europe began to question the dominant ideas of the Middle Ages, scientists learned how to conduct experiments and make rational observations, rather than relying on religious dogma to explain the world around them. However, the clash between science and faith eventually caused difficulty for van Helmont. He was interrogated by the notorious Spanish Inquisition for giving scientific explanations for supernatural events.

Born into a noble family, van Helmont became a medical doctor in 1599. He often treated people for free, refusing to profit from the misery of his fellow human beings. In 1609, after extensive travels and medical experience (including treating victims of bubonic plague), he turned down several attractive job offers and devoted himself to pure research on the principles of nature.

In many respects, van Helmont followed the teachings of Paracelsus (1493-1541), a Swiss physician of the early 1500s who pioneered the use of **alchemy** to prepare medicine instead of trying to make **gold**. Like Paracelsus, van Helmont was interested in alchemy and mysticism. He also believed in the fallacious theory of spontaneous generation—that organisms can spring to life from decaying materials such as dusty grain or old rags.

But van Helmont disagreed with Paracelsus' belief in the ancient Greek theory that all **matter** is composed of four elements (earth, air, fire, and water). Instead, van Helmont asserted that the basic element of the universe was **water**, an idea that went even further back to the Greek philosopher Thales, who lived around 600 B.C. Van Helmont, to prove his theory in a scientific manner, grew a willow tree in a tub for five years, giving it nothing but pure water and even protecting it from dust in the air. He weighed the tree and the soil carefully before starting the experiment and then again at the end. Although the soil lost practically no weight, the tree had gained more than 160 pounds (72 kg), which van Helmont attributed, mistakenly, to the water he had added. Because this experiment represented the first application of quantitative methods to a biological question, van Helmont is sometimes called the father of **biochemistry**.

Van Helmont also discovered **gases** as a class of substances and first coined the term *gas* (possibly based on the Dutch pronunciation of the Greek word *chaos*) to describe vapors that differed from ordinary air. He applied scientific techniques to study several gases, most notably *gas sylvestre*, which he produced from burning **wood**. This gas is now known

Johannes Baptista van Helmont.

to be **carbon** dioxide. Van Helmont realized that **carbon dioxide** was also produced by other chemical processes, such as the **fermentation** of wine and the reaction of acids with limestone and other carbonates. He also described **carbon monoxide**, **chlorine** gas, digestive gases, **sulfur** dioxide, and a "vital gas" in the blood, and he showed that a burning candle would use up air in an enclosed space. In these studies, van Helmont paved the way for more definitive chemical analysis of gases by **Joseph Priestley**, **Antoine-Laurent Lavoisier**, **Joseph Black**, and other scientists who had better apparatus to work with.

Another popular belief of van Helmont's day was **transmutation**, which held that one metal could be changed into another or destroyed altogether. Relying upon scientific methods, van Helmont refuted this idea by showing that dissolved **metals** could be recovered in their original quantity. His insight was a forerunner of physical laws regarding the indestructibility of matter. Van Helmont also studied the behavior of pendulums and recommended that they be used to measure time.

Much of van Helmont's work, however, still focused on medicine and health. He demonstrated that acid is the stomach's digestive agent, and he suspected that the substance was **hydrochloric acid**. Through ingenious observations, van Helmont identified many causes of asthma. He also studied the symptoms of bronchitis, tuberculosis, epilepsy, and hysteria, and he diagnosed illness by analyzing the specific gravity of urine. Van Helmont realized that fever is part of the body's natural healing process. Instead of the traditional treatment of bloodletting or purging, he prescribed remedies according to the specific disease, its cause, and the bodily organ being affected. In fact, van Helmont's medical research introduced the concept that diseases are caused by specific harmful agents, rather than by a general imbalance of the body's "humors," or fluids.

Throughout his scientific career, van Helmont clung to his religious and mystical beliefs which, unfortunately, differed from those of the Catholic authorities. Spain had occupied the Flemish territories, and the Church still had the force of the law behind it. During the early 1600s, van Helmont was embroiled in a controversy over a pamphlet he had published on curative ointments. He was condemned for heresy by the General Inquisition of Spain, a special committee set up to punish those who contradicted the Catholic Church. Van Helmont was also denounced by the University of Louvain's medical and theological faculties. Eventually, he was detained and interrogated, then kept under house arrest for years. Perhaps because of this experience, van Helmont published little of his work. On his deathbed, he gave his writings to his son to edit and publish.

HENCH, PHILIP SHOWALTER (1896-1965)

American pathologist

Philip Showalter Hench, an American clinical pathologist, performed groundbreaking research in rheumatoid arthritis. His clinical tests of adrenal compound E, which Hench named **cortisone**, and of ACTH, which produces cortisone naturally by stimulating the adrenal cortex, offered the first hope for patients suffering from rheumatoid arthritis. Hench and his colleague, biochemist **Edward C. Kendall**, gained immediate worldwide attention when they filmed the miraculous recovery of arthritis patients—some of whom could barely walk—as they climbed stairs and even jogged in place. Although prolonged clinical trials showed that neither cortisone or ACTH was a viable long-term therapy for arthritis due to side effects such as high blood pressure and high glucose levels, Hench's efforts opened new vistas in medical research, particularly in the study of both **hormones** and rheumatoid arthritis. A meticulous researcher who methodically collected his clinical data before publishing his results, Hench shared the 1950 Nobel Prize in physiology or medicine with Kendall "for their discoveries relating to the hormones of the adrenal cortex, their structure and biological effects." (Chemist **Tadeus Reichstein** also received a share of the prize for his independent work with the adrenal cortex and its hormones.)

Hench was born in Pittsburgh, Pennsylvania, on February 28, 1896. The son of Jacob Bixler Hench, a classical scholar and school administrator, and Clara John Showalter, Hench attended a private high school, Shadyside Academy, and then enrolled at the University of Pittsburgh in 1916. His education looked as though it would be interrupted when he enlisted in the U.S. Army Medical Corps. But Hench was transferred to the reserves so he could return to his studies, and he enrolled in Lafayette College in Easton, Pennsylvania. He received his B.A. from Lafayette in 1916 and enrolled at the University of Pittsburgh School of Medicine, where he received his M.D. in 1920.

After completing his internship at St. Francis Hospital in Pittsburgh, Hench became a fellow in medicine at the Mayo Foundation of the University of Minnesota. The bright young physician and scientist would spend his entire career at the Mayo Clinic, where, in 1926, he cofounded the Department of Rheumatic Disease, which was the first training program in rheumatology in the United States. Hench spent the 1927–28 academic year on sabbatical studying research medicine with Ludwig Aschoff, a leading rheumatic fever investigator, at Freiburg University. He also studied with clinician Freidrich von Müller in Munich. Hench completed his formal education in 1931 when he received a master of science degree in internal medicine from the University of Minnesota.

A physician first, Hench's research was clinically based. He began studying rheumatoid arthritis in 1923. Unlike osteoarthritis, a degenerative joint disease common in later life, rheumatoid arthritis is a chronic inflammatory disease of the joints often contracted at the relatively young age of 30 to 35. In advanced stages, rheumatoid arthritis could cause deformity due to bone and surrounding muscle atrophy. In 1929, Hench took note of a patient who had suffered from severe arthritis for more than four years. The patient had entered the Mayo Clinic suffering from jaundice, a disease caused by excess bilirubin, a liver product, in the bloodstream. Amazingly, the man's arthritis had abated and remained dormant for several weeks after his recovery from jaundice. Carefully collecting data, Hench waited until he had authenticated nine similar cases, among them patients who experienced remissions from painful fibrositis and sciatica, two other inflammatory conditions, before publishing his data in 1933.

Hench was convinced that these cases held a vital clue to a therapy for arthritis and set out to induce jaundice artificially. Hench's initial experiments used infusion or ingestion of bile to emulate jaundice's production of excess bile in the blood or the liver. Although these experiments failed, Hench's attention was soon drawn to another group of patients, women whose arthritis vanished during pregnancy. He also observed that some arthritic patients went into less complete remission after surgical operations, anesthesia, or severe fasting. Looking for a common physiological denominator, Hench, who enjoyed reading Sir Arthur Conan Doyle's novels of Sherlock Holmes, had a prime suspect—glandular hormones. Furthermore, the fact that both jaundice and pregnancy caused remission in almost the exact same manner led Hench to believe that his missing compound was not bilirubin or a female-only sex hormone.

Fortunately, the Mayo Foundation's own Edward C. Kendall was a world renowned chemist in the field of **steroids**, a specific group of hormones. Kendall had isolated six steroids from the adrenal cortex, the outer part of the endocrine glands located atop the kidneys, which he alphabetized compound A though F. Hench's first try with compound A was a failure. Both Hench and Kendall then decided to try compound E. But at that time, in 1941, compound E was extremely difficult to synthesize and, as a result, costly. With both high (300° Fahrenheit) and low (–100° Fahrenheit) temperatures needed to produce compound E, the delicate work took time and attention and the slightest mistake could result in a useless compound. It wasn't until more than two years after World War II that scientists from the pharmaceutical firm Merck & Co. had developed a process that could produce enough compound E for Hench to attempt his experiment. Still, the compound was expensive to produce. Hench recalled in an interview for an article in the *Saturday Evening Post* that he and his colleagues "almost went into shock" when a $1,000 bottle of compound E was dropped on a marble floor.

Hench's results with compound E were miraculous. The first patient, a 29-year-old woman, experienced total remission of symptoms after three injections over three days. Hench's results were quickly confirmed by five other researchers across the country. Hench and his colleagues received instant public notoriety as a result of their studies both with compound E, which Hench named cortisone, and with adrenocorticotropic hormone (ACTH), a hormone produced by the pituitary gland which spurs the body's natural production of cortisone through the adrenal cortex.

Unfortunately, Hench's miraculous "cure" for arthritis turned out to be short lived. Without the use of cortisone or ACTH, rheumatic symptoms returned; and long-term use of cortisone or ACTH causes several side effects, including high blood glucose and high blood pressure, as well as obesity associated with adrenal or pituitary gland tumors. Much to Hench's credit, he maintained his scientific cautiousness throughout the heady early days of the discovery, quickly recognizing the harmful side effects and outlining future directions in research of these hormones. Nevertheless, the studies of Hench and Kendall, along with those of Tadeus Reichstein, opened entirely new avenues of medical research; as a result, the three scientists were awarded the Nobel Prize for medicine or physiology in 1950.

Hench retired from the Mayo Foundation in 1957. In addition to the Nobel Prize, he was a recipient of the numerous awards, including the prestigious Lasker Award, which he also shared with Kendall. Hench married Mary Genevieve Kahler in 1927, and the couple had two sons, Philip Kahler and John Bixler, and two daughters, Mary Showalter and Susan Kahler. His hobbies included photography, tennis, opera, and Sherlock Holmes novels. He died from pneumonia on March 30, 1965, while vacationing with his wife in Ocho Rios, Jamaica. To honor him, Hench's alma mater, the University of Pittsburgh, presents the annual Hench Award to a distinguished university alumnus.

HENRY, WILLIAM (1774-1836)

English chemist

William Henry's research of the **solubility** of **gases** in **liquids** led to what is now known as **Henry's law**. Henry was also the first to experiment with **electrochemistry**. Born in Manchester, England, on December 12, 1774, William Henry was the third son of Thomas Henry, a physician and industrial chemist who suggested the use of **chlorine** as a bleaching agent for textiles. Henry began his schooling at a private institution run by a local Unitarian minister, but later transferred to Manchester Academy, which offered a wide curriculum composed of scientific, practical, and mathematical subjects. At the age of ten, he was injured by a falling beam, and never fully recovered, remaining in poor health for the rest of his life. After leaving the Academy in 1790, Henry became secretary-companion to Dr. Thomas Percival, the founder of the influential Literary and Philosophical Society of Manchester. Here, Henry began his preliminary studies in medicine. He entered Edinburgh University's medical school five years later, but left a year later to help with his father's practice and to direct the family manufacturing business.

During this period in Henry's life, he began to conduct original research in **chemistry**. In 1796, he became a member of the Manchester Literary and Philosophical Society. Returning to Edinburgh in 1805, he received his M.D. in 1807 with his submission of a dissertation on uric acid. Henry later contributed to this field of study, submitting papers to medical journals on these and other related subjects. In 1808, he was elected a fellow of the Royal Society and was awarded the Royal Society's premier Copley Medal for papers already submitted.

In 1797, Henry published his first paper, which was a refutation of William Austen's claim in 1789 to have shown that **carbon** was not an element. Other chemical experiments that followed involved the composition of **hydrogen** chloride and **hydrochloric acid**. In the winter of 1798-99 at Manchester, Henry espoused the new chemical doctrines and **nomenclature** of **Antoine-Laurent Lavoisier**, grounding his first lecture demonstrations in Lavoisier's ideas. Based on these lectures and demonstrations, his textbook *Elements of Experimental Chemistry* was first published in 1801. *Elements of Experimental Chemistry* went to eleven editions by 1829, and expanded with each new edition.

In 1801, Henry read his paper rebutting Humphry Davy's arguments against the materiality of **heat** to the Manchester Literary and Philosophical Society. The following year, he presented to the Royal Society his paper describing the phenomenon that was to become "Henry's Law."

Henry published his paper that established Henry's law in 1803. He began to speculate why an atmosphere composed of gases of different densities did not separate into layers, but mixed together, which was a precursor of the theory of mixed gases. The results of his research on the solubility of gases formed the basis of Henry's law, which in its modern form states that the **mass** of gas dissolved by a given **volume** of liquid is directly proportional to the pressure of the gas with which it is in equilibrium, provided the **temperature** is constant. There are many deviations from the law and it is closely obeyed only by very dilute solutions.

At the beginning of the nineteenth century, **coal** gas was suggested for lighting purposes, and Henry began to analyze the different inflammable gaseous mixtures contained in coal. He also sought out to analyze other organic materials found in coal, with the purpose to determine their relative properties of illumination and to explain their compositional variations. In his investigative research, which covered a span of over 20 years, Henry confirmed **John Dalton's** results of the compositions of **methane** and ethane, and also substantiated Dalton's work showing that hydrogen and carbon combine in definite proportions to form a limited number of compounds. Henry also made use of the catalytic properties of **platinum** discovered by Johann Döbereiner in 1824.

In order to determine the composition of **ammonia**, Henry applied his analytical techniques to confirm and improve the earlier work of **Claude Berthollet**, A.B. Berthollet, and **Humphry Davy**. He confirmed that ammonia did contain hydrogen and **nitrogen** in the proportions suggested by them, but found that Davy was inaccurate in think ing it contained **oxygen**.

Henry's papers published in 1824 were the last of any kind of pertinence to experimental chemistry. He suffered from a disorder that affected his hands and after unsuccessful surgical operations in 1824, he was forced to give up manipulative experiments. Reverting back to the study of medicine, Henry turned to the study of contagious diseases, which was his belief that they were spread by chemical substances. While others with similar beliefs attempted to destroy the contagion by chemical reactions, Henry believed heat would inactivate the contagion. With the onset of Asiatic cholera in 1831, Henry was able to apply this concept.

Because of Henry's poor chronic ill health in addition to the neuralgic pains from his injury, he committed suicide in Manchester on September 2, 1836.

Henry's law

Henry's law states that the **mass** of gas which is dissolved by a **volume** of liquid, at constant **temperature**, is directly proportional to the pressure of the gas. The law is not followed exactly if there is any interaction between the gas and the liquid other than London dispersion forces.

The equation which describes Henry's law is $C_g = kP_g$ where C_g is the **solubility** of the gas in the **solution** phase, usually expressed as **molarity**, k is the Henry's law constant, and P_g is the partial pressure of the gas over the solution. The Henry's law constant is different for each solute and solvent pair, it is measured in moles per liter atmosphere.

Henry's law is used to indicate how much of a gas will dissolve in a liquid. By increasing the partial pressure of the gas more gas can be dissolved in the liquid. One practical effect of this law is shown in the production of carbonated drinks. These are bottled or canned under a pressure of **carbon** dioxide greater than one atmosphere. When the drinks are opened to the air the partial pressure of the **carbon dioxide** is decreased above the drink, the solubility of carbon dioxide decreases and carbon dioxide bubbles are produced in the drink.

See also Gas laws; Gases; Gases, Behavior and properties of

Herbicides

Herbicides are chemical **pesticides** that are used to manage vegetation. Usually, herbicides are used to reduce the abundance of weedy plants, so as to release desired crop plants from competition. This is the context of most herbicide use in agriculture, forestry, and for lawn management. Sometimes herbicides are not used to protect crops, but to reduce the quantity or height of vegetation, for example along highways and transmission corridors. The reliance on herbicides to achieve these ends has increased greatly in recent decades, and the practice of chemical weed control appears to have become an entrenched component of the modern technological culture of humans, especially in agroecosystems.

The total use of pesticides in the United States in the mid-1980s was 957 million lb per year (434 million kg/year), used over 592,000 square miles (148 million hectares). Herbicides were most widely used, accounting for 68% of the total quantity [646 million lb per year (293 million kg/year)], and applied to 82% of the treated land [484,000 square miles per year (121 million hectares/year)]. Note that especially in agriculture, the same land area can be treated numerous times each year with various pesticides.

A wide range of chemicals is used as herbicides, including: (1) chlorophenoxy acids, especially 2,4-D and 2,4,5-T, which have an auxin-like growth-regulating property and are selective against broadleaved angiosperm plants; (2) triazines such as atrazine, simazine, and hexazinone; (3) chloroaliphatics such as dalapon and trichloroacetate; (4) the phosphonoalkyl chemical, glyphosate, and (5) inorganics such as various arsenicals, cyanates, and chlorates.

The intended ecological effect of any pesticide application is to control a pest species, usually by reducing its abundance to below some economically acceptable threshold. In a few situations, this objective can be attained without important nontarget damage. For example, a judicious spot-application of a herbicide can allow a selective kill of large lawn weeds in a way that minimizes exposure to nontarget plants and animals.

Of course, most situations where herbicides are used are more complex and less well-controlled than this. Whenever a herbicide is broadcast-sprayed over a field or forest, a wide variety of on-site, nontarget organisms is affected, and sprayed herbicide also drifts from the target area. These cause ecotoxicological effects directly, through toxicity to nontarget organisms and ecosystems, and indirectly, by changing habitat or the abundance of food species of wildlife.

An important controversy related to herbicides focused on the military use of herbicides during the Vietnam war. During this conflict, the United States Air Force broadcast-sprayed herbicides to deprive their enemy of food production and forest cover. More than 5,600 square miles (1.4 million hectares) were sprayed at least once, about 1/7 the area of South Vietnam. More than 55 million lb (25 million kg) of 2,4-D, 43 million lb (21 million kg) of 2,4,5-T, and 3.3 million lb (1.5 million kg) of picloram were used in this military program. The most frequently used herbicide was a 50:50 formulation of 2,4,5-T and 2,4-D known as Agent Orange. The rate of application was relatively large, averaging about 10 times the application rate for silvicultural purposes. About 86% of spray missions were targeted against forests, and the remainder against cropland.

In addition, the Agent Orange used in Vietnam was contaminated by the **dioxin** isomer known as TCDD, an incidental by-product of the manufacturing process of 2,4,5-T. Using post-Vietnam manufacturing technology, the contamination by TCDD in 2,4,5-T solutions can be kept to a **concentration** well below the maximum of 0.1 parts per million (ppm) set by the United States Environmental Protection Agency (EPA). However, the 2,4,5-T used in Vietnam was grossly contaminated with TCDD, with a concentration as large as 45 ppm occurring in Agent Orange, and an average of about 2.0 ppm. Perhaps 243-375 lb (110-170 kg) of TCDD was sprayed with herbicides onto Vietnam. TCDD is well known as being extremely toxic, and it can cause birth defects and miscarriage in laboratory mammals, although as is often the case, toxicity to humans is less well understood. There has been great controversy about the effects on soldiers and civilians exposed to TCDD in Vietnam, but epidemiological studies have been equivocal about the damages. It seems likely that the effects of TCDD added little to human mortality or to the direct ecological effects of the herbicides that were sprayed in Vietnam.

HERSCHBACH, DUDLEY R. (1932-)

American physical chemist

Dudley R. Herschbach was a student from a non-scholarly background who, with the aid of fine teachers and university scholarships, developed an aptitude for scientific subjects into a distinguished career as a researcher and theorist. He was a pioneer in the field of molecular reaction dynamics —the study of the motions of atoms during chemical reactions. He was responsible for developing an experimental method called the molecular beam technique, which is used to examine the collision of molecules during the course of a reaction. Herschbach has also advanced the theoretical understanding of basic chemical processes, and for his contributions to this field, he shared with **Yuan T. Lee** and **John C. Polanyi** the 1986 Nobel Prize in **chemistry**.

The first of six children, Dudley Robert Herschbach was born in San Jose, California, on June 18, 1932, to Robert and Dorothy Beer Herschbach. His father was at that time a building contractor and later a rabbit breeder. Herschbach grew up in a rural area a few miles outside San Jose and engaged in farm activities such as feeding livestock, milking cows, and picking plums, apricots, and walnuts. Although Herschbach was often involved in outdoor activities, he was also an enthusiastic reader and he developed an early interest in science, especially astronomy. He took all the science and mathematics courses available at Campbell High School; he also became a star football player there and was offered athletic scholarships at several institutions. However, he chose an academic scholarship from Stanford University. He did play football during his freshman year but abandoned the sport when the coach tried to forbid him from taking lab courses during the season. He concentrated on his rapidly developing scientific interests, earning a B.S. in mathematics in 1954 and a M.S. in chemistry in 1955.

Herschbach continued his graduate study at Harvard University. Working under E. Bright Wilson, he decided to specialize in the dynamics of chemical reactions. He did his doctoral research on the microwave spectra of molecules in **methanol**, and he received his Ph.D. in chemical physics in 1958. While at Harvard, he was a member of the institution's prestigious Society of Fellows. Herschbach began his teaching and research career at the University of California at Berkeley; he became an assistant professor of chemistry in 1959 and an associate professor in 1961. He returned to Harvard as professor of chemistry in 1963. In 1976 he was named the Frank B. Baird, Jr. Professor of Science.

In the 1950s, the knowledge chemists had about the basic process of chemical reactions had not much advanced since the beginning of the century. Chemists mixed together specified quantities of substances under controlled conditions of **temperature** and pressure, and then they examined and measured the resulting products and the **energy** released or consumed during the reaction. In their measurements and examinations, what they observed was the combined products and energies of millions of individual molecules interacting in the substances. What actually took place among individual molecules during a reaction was not really clear. Herschbach's experiments provided the first real evidence of the details of a chemical reaction at the level of individual molecules.

It was his idea to borrow a technique from nuclear and particle physics; in this technique, directed beams of molecules were fired crosswise at each other and the resulting altered molecular products could be collected and studied as individual molecules or atomic particles. Herschbach had constructed a fairly simple apparatus at Berkeley, which heated substances into gaseous form and then directed beams of the molecules at each other at a ninety-degree angle in a vacuum. The molecules reacted at the crossing point and the altered molecules landed on a **tungsten** (later a platinum) filament through which an electric current was flowing. The dynamics of the reaction could then be studied by measuring the variation in the electric current caused by the altered molecules.

After returning to Harvard in 1963, Herschbach continued his work with a similar apparatus. However, his progress was hampered by the fact that these experiments were limited to reactions involving **alkali metals** and alkali **halides**, since these were the only kinds of molecules which the tungsten or **platinum** filaments could detect. The critical breakthrough in the study of the reactions of more complex molecules came in 1967 with the arrival of Taiwanese-born Yuan T. Lee. A postdoctoral fellow at Harvard, Lee designed and supervised the building of a very sophisticated machine which greatly improved the components of the earlier apparatus. He replaced the filaments with a **mass** spectrometer, which could detect and measure the reaction products of almost any kind of **molecule**. After Lee left for the University of Chicago in 1968, Herschbach continued to study increasingly complex chemical reactions at Harvard. He was particularly noted for his ability to devise theories which explained very complex molecular events in clear and relatively simple terms. His work and that of Lee and others transformed the entire field of chemistry, and for their achievements they shared with John C. Polanyi the Nobel Prize in chemistry in 1986.

Herschbach's success has been primarily due to his scientific talent. However, a portion of it has certainly stemmed from his warm and generous relationships with many colleagues and students. Progress in modern science is usually the result of a team effort, a fact of which Herschbach was well aware. In accepting the Nobel Prize, he referred to the speech he gave as a 'tour through a family album,'' and in an appendix to that address he listed the names of fifty-one graduate students and thirty-five postdoctoral fellows who had worked with him from 1962 to 1986. In the body of the address, he discussed the specific contributions of many of them, most notably Lee, and he also mentioned the assistance of many others, including the name of one staff member of the machine shop of the Harvard chemistry department. Unlike many famous scientists, he is also a popular teacher of both undergraduate and graduate students.

Herschbach married Georgene Botyos in 1964. She was then a graduate student in chemistry at Harvard; they had two daughters before she finished her Ph.D. in 1968. He and his wife served for five years as co-masters of Currier House, one of the Harvard residential colleges for undergraduates. In an autobiographical piece for *Les Prix Nobel 1986,* Herschbach recounted with relish how they were once summoned to a student's room to meet a seal in a bathtub. He now lives with his family in Lincoln, Massachusetts.

HERZBERG, GERHARD (1904-1999)

Canadian physical chemist

Gerhard Herzberg is known as the founding father of molecular **spectroscopy**, the science that observes the interaction of **energy** with **matter** to obtain information about the identity and structure of molecules. For his contributions to the knowledge of the elctronic structure and geometry of molecules, especially **free radicals**, Herzberg became the first Canadian to receive the Nobel Prize for **chemistry**, in 1971. Herzberg also did pioneering work in other scientific fields, including astrophysics, and, in association with the National Research Council of Canada, he founded the nation's premier molecular spectroscopy laboratory in 1948.

Herzberg was born on December 25, 1904, to Albin and Ella (Biber) Herzberg. Raised and schooled in Hamburg, Germany, Herzberg graduated in 1927 from the Darmstadt Institute of Technology, with a B.S. in engineering. In 1928, he completed his Ph.D. with a thesis on the interaction of electromagnetic radiation with matter. Herzberg also studied at the Univeristy of Göttingen and the University of Bristol, England. In 1930, he returned to Darmstadt as an instuctor.

In 1935, Herzberg was forced to flee Germany because of Hitler's anti-Jewish policies. He subsequently obtained a position as Carnegie guest professor at the University of Saskatchewan, Canada. Although the school lacked the resources and equipment Herzberg needed for his studies, the atmosphere was welcoming and supportive, and he accomplished a substantial amount of research while also publishing two books. While at Saskatchewan, Herzberg helped establish a graduate laboratory specializing in spectroscopy studies and obtained funding to improve the school's research facilities. Beginning in 1945, Herzberg spent three years as a professor at the Yerkes Observatory of the University of Chicago, later returning to Canada to set up a spectroscopic research laboratory in Ottawa for the National Research Council. This laboratory was commended by the Swedish Royal Academy of Science, the Nobel Prize awarding institution, as the foremost center for molecular spectroscopy in the world. Herzberg remained with the Canadian National Research Council for the remainder of his career, becoming the first Distinguished Research Scientist of that organization. He became a naturalized Canadian citizen in 1945.

Although Herzberg considered himself a physicist, his research in molecular spectroscopy had special significance for chemistry. His research was of great importance, as it developed a method of analyzing **molecular structure** by measuring light transmitted and absorbed by molecules. To do this, Herzberg used the spectroscope, a tool that enabled him to separate a molecule's radiant energy into different parts in the same way light is separated when passed through a prism. When molecules or atoms were passed through a spectroscope, the energy radiating from them separated into distinct lines, or spectra, allowing an accurate analysis of their structure. At the time Herzberg began his experiments, spectroscopy, or the study of molecular structure, was a primitive science. Very little was known about atomic spectra, and Herzberg made major contributions to the fields by analyzing complicated spectra obtained from molecules, particularly **diatomic** molecules, such as **nitrogen**, **oxygen**, and **hydrogen**. While still in Germany, for example, he and a colleague, Werner Heitler, showed that the nitrogen molecule's complex spectrum was much more significant than what contemporary scientific belief acknowledged. The particles then believed to comprise nitrogen's **nucleus** could not account for the intensity of some of the spectral bands. The neutrons responsible for the inconsistency were later discovered by the English physicist, James Chadwick.

Herzberg also discovered several molecular species in outer space and the upper atmosphere while at Yerkes Observatory. He found elemental hydrogen in some planetary atmospheres. He also discovered new bands, now called Herzberg Bands, in the oxygen spectrum. Spectroscopy, however, allowed him to do much more throughout his career than just identify molecules. He accurately measured the **electron energy level** in several species, using a combination of quantum mechanical theory and spectroscopy. By developing new experimental procedures, he was able to study short-lived molecules, or free radicals, that appear only briefly during chemical reactions. These chemical intermediates are difficult to study because they do not last long enough to respond to other types of tests. Two important examples of his success in this area are the spectra of the methylene radical, CH_2, and the methyl radical, CH_3.

Herzberg was also a voluminous writer throughout his career, publishing several hundred papers and many definitive books in the area of molecular and atomic spectra. He wrote in the introduction to his 1971 book, *The Spectra and Structure of Simple Free Radicals: An Introduction to Molecular Spectroscopy*, "My original plan, forty years ago, was to write a small book on [molecular spectroscopy] of no more than 200 pages. I was unable to prevent this original plan from leading to a three-volume work of over 2000 pages." Herzberg made a clear distinction between scientific research and technological research. He believed that the former concentrated on investigating nature, while the latter focused on the applications of science to society. In Herzberg's opinion, scientific research is the scientist's true vocation, while technological research is the concern of politicians. He also championed the scientist's right to work freely, without political or bureaucratic restrictions.

Herzberg received many honors in addition to his 1971 Nobel Prize for chemistry, including the Willard Gibbs Medal and the Linus Pauling Medal from the American Chemical Society, the Gold Medal of the Canadian Association of Physicists, and the Royal Medal of the Royal Society of London.

Herzberg married Luise H. Oettinger in 1929; the couple had two children, Paul and Agnes, both of whom became teachers. Luise died in 1971, and Herzberg married Monika Tenthoff a year later. Herzberg's hobbies included music and mountain climbing. He died in 1999.

Gerhard Herzberg (right). ***(AP/Wide World. Reproduced by permission.)***

HETEROCYCLIC AND HOMOCYCLIC COMPOUNDS

Complex molecules, consisting of many atoms chemically bonded together, may have atoms arranged linearly or in rings. *Heterocyclic* and *homocyclic* compounds are examples of molecules in which the constituent atoms form rings. The cyclic portion of the name indicates their **ring** nature. The *hetero-* (from Greek meaning different) or *homo-* (from Greek meaning same) prefix refers to the composition of atoms in the **molecule**.

Homocyclic compounds are molecules that are, or contain, ring structures that consist only of **carbon** atoms within the ring. An example is **benzene**. Benzene is a homocyclic compound of six carbon atoms bonded together in a hexagonal ring, with one **hydrogen atom** bonded to each of the six carbons. Benzene is a highly toxic and volatile compound sometimes used in cleaning solutions. Phenol is another common homocyclic compound. Phenol, sometimes used as an antiseptic, is a benzene ring with a hydroxide group substituted for one hydrogen in the ring. Cyclohexane is another homocyclic molecule. In contrast, heterocyclic compounds, or groups, are rings containing at least one non-carbon atom in the ring. An example is heterocyclic amines which are six-membered rings of five carbons and one **nitrogen** atom. Heterocyclic compounds have an immense range of properties and uses and even constitute their own branch of chemistry (heterocyclic chemistry). A small set of examples in which heterocyclic compounds are used includes dyes, photochromes of film, **pesticides**, antiviral medications, and **food additives**. Many biologically important molecules are heterocyclic. **DNA** nucleotide bases, for example, are heterocyclic molecules. Also, the development of heterocyclic antidepressants, namely Prozac, have contributed greatly to the treatment of disease.

HEVESY, GEORG VON (1885-1966)

Swedish radiochemist

Georg von Hevesy developed radioactive tracer analysis, a method widely used in **chemistry** and medicine. For this accomplishment, which had far-reaching consequences in physiology, **biochemistry**, and **mineralogy**, he was awarded the Nobel Prize in chemistry in 1943. Hevesy was also the codiscoverer of the element **hafnium**.

Georg Charles von Hevesy was born in Budapest, Hungary, on August 1, 1885, to Louis Bisicz and his wife, the former Baroness Eugenie Schosberger. The family, who was given a title by Emperor Franz Joseph I in 1895, first changed their name to Hevesy-Bisicz and then simplified it to Hevesy; Hevesy always used the "von" in German correspondence and publications. (Many sources refer to him as "de Hevesy," while his first name often appears as "George" or "György.") Both sides of the family were well-to-do; facing no financial obstacles, Hevesy moved smoothly through the Piarist Gymnasium in Budapest, then studied physics and chemistry at the University of Budapest, and finally earned his doctorate at the University of Freiburg in 1908 with a thesis on the chemical behavior of **sodium** hydroxide in fused sodium metal.

Hevesy received his degree and became an assistant at the Eidgenössische Technische Hochschule in Zürich, com-

Georg von Hevesy. *(Corbis Corporation. Reproduced by permission.)*

mencing a career in which he knew or worked with nearly every major scientist of the first half of the twentieth century. He was acquainted with **Albert Einstein**, for example, in Zürich, where he continued his work with fused salts. After two years he moved on to work with **Fritz Haber** at the Technische Hochschule in Karlsruhe, but he realized that he lacked the research techniques for the electron-emission studies that Haber had set for him. Hevesy suggested that he join **Ernest Rutherford**'s group at the University of Manchester in England, and Haber agreed. He received an honorary research fellowship and left in 1911.

At Manchester Hevesy worked with, among others, **Niels Bohr**, **Frederick Soddy**, **Henry Moseley**, and Hans Geiger. His first project was the chemistry of the radioactive decay products of **actinium**, and the effort yielded the finding that successive alpha decay products differ in chemical **valence** in steps of two. This provided support for the proposal Soddy had recently put forward concerning the existence of alpha-decay—the ejection of an **alpha particle**, or **helium nucleus**, from the radioactive nucleus. It was here, however, that Hevesy began the research which was to occupy him for the rest of his life, and it was Rutherford who set him on the road to it. The Austrian government had given Rutherford 100 kg (220 lb) of radioactive **lead** whose activity was known to be that of ''radium-D,'' a decay product of **radium**. Hevesy was assigned the task of separating the radium-D from ''all that lead.'' Over many months, he tried every chemical separation he knew, with a uniform lack of success. We now know that radium-D is the radioactive **isotope** lead–210; it is, in other words, a form of lead with the same number of protons and electrons, and hence the same chemistry, as any other lead, but with a different number of neutrons in its nucleus. Separation of isotopes can be done only by painstaking physical methods, and chemical separation is impossible.

Having failed in the separation study, Hevesy then acted on the principle of the popular saying, ''If life hands you lemons, make lemonade.'' Since he could not separate it, he decided to use radium-D to trace the course of lead in chemical processes. Working with Friedrich Paneth at the Vienna Institute of Radium Research in 1913, he was able to conduct precise **solubility** studies of lead salts by mixing an insignificant **mass** of radium-D with a regular lead **salt**, then determining the amount dissolved not by the usual tedious gravimetric methods, but by simple measurement of the proportion of **radioactivity** found in **solution**. He was also able to demonstrate lead exchange between solid and solution, and the migration of lead atoms in the metal. He and Paneth showed that the electrochemical properties of radium-D were identical with those of lead, thereby adding to the growing evidence of the existence of isotopes.

In 1913 Hevesy returned to Hungary, where he served for a time as a lecturer at the University of Budapest before joining the Austro-Hungarian army during World War I. His post during the war was as technical supervisor at the state electrochemical **copper** plant, and he was able to continue his research on a limited basis. After the war he became a full professor at Budapest, continuing his lead tracer work, but the political situation in Hungary was deteriorating rapidly, and in 1920 he accepted an invitation to join Bohr's Institute for Theoretical Physics at the University of Copenhagen.

His first project there, carried out with Johannes Bro, was an attempt at isotopic separation by fractional **distillation**. They had limited success with metallic **mercury**, but obtained fairly pure isotopic samples of **chlorine**, whose two stable isotopes differ by about six percent in mass. Hevesy wished to learn X-ray **spectroscopy**, and in 1923 he turned for help to physicist Dirk Coster. The two of them set about finding element seventy-two, which Bohr's recent revision of the **periodic table** suggested should be a transition metal corresponding to **zirconium**. They found the anticipated spectral lines in extracts of zirconium ores, and were able to isolate and characterize the new element as its fluoride. They named it ''hafnium'' after the Latin name for Copenhagen.

In 1923 Hevesy also returned to his work with radioactive lead tracers, and for the first time he ventured into biology to study the uptake of lead in bean seedlings. This work was published in 1924, the year in which Hevesy married Pia Riis, who would bear him three daughters and a son. Two years later, he moved his new family to the University of Freiburg, where he developed X-ray fluorescence as an analytical tool, while expanding the university's X-ray spectroscopy program. As an administrator, however, Hevesy came into increasing contact with the new Nazi regime, and this caused him to return to Copenhagen in 1934.

Heavy **water** (water in which some of the **hydrogen** atoms are the heavier isotope hydrogen–2, or deuterium) had just become available, and in Copenhagen Hevesy was pleased to have this first non-toxic isotope available to study animal and human physiology. He and his colleagues quickly demonstrated the rapid exchange of internal and external water in goldfish, and measured the average turnover time of a water **molecule** in the human body (about 13 days) and the approximate number of water molecules in the body (10^{27}).

In 1934, **Irène** and **Frédéric Joliot-Curie** succeeded in producing artificial radioisotopes by alpha particle bombardment. Hevesy seized the possibilities of this development by making radioactive phosphorus–32 from sulfur–32, a very large advance for the study of physiology. Here was an element central to all animal physiology, and a means of following its intake, circulation, exchange, and excretion. A number of discoveries followed from the use of this tracer, including the dynamic exchange of serum and bone **phosphate** and the synthesis and distribution of **DNA** and **RNA**. Today, these discoveries form the foundation of our understanding of body chemistry.

Phosphorus was only the first element that Hevesy used or introduced into use as a radioactive tracer. Others included calcium–45, potassium–42, sodium–24, chlorine–38, and carbon–14. There is little of physiological importance that lies outside this list except **nitrogen**, **oxygen**, and **sulfur**, and it is clear just how fundamental his contributions to science have been. It was in recognition of these accomplishments that Hevesy was awarded the Nobel Prize in chemistry in 1943. Announced in 1944 and overshadowed by the closing battles of World War II, the award received little public notice.

At the time he received the Nobel Prize, Hevesy had moved again. He had left Copenhagen in 1943, moving away from the Nazis for a second time, and settled in Stockholm. He worked at the Institute for Organic Chemistry there for the remainder of his life, becoming a Swedish citizen in 1945. Much of his later research focused on physiology and medicine, particularly the study of cancer. He published over four hundred books and papers in the course of his career and received many awards and honors, including honorary doctorates from nearly a dozen universities and honorary memberships in many scientific societies. He was also a foreign member of the Royal Society. In addition to the Nobel Prize, he received the Cannizzaro Prize in 1929 from the Academy of Sciences in Rome, the Copley Medal of the Royal Society in 1950, and the Faraday Medal in 1959. He also received the Atoms for Peace Award in 1959. Hevesy died at a clinic in Freiburg on July 5, 1966, after a long illness.

HEYROVSKÝ, JAROSLAV (1890-1967)

Czech physical chemist

Jaroslav Heyrovský dedicated his life to the study and improvement of a technique to analyze chemical solutions. His work led to his invention of the polarograph, a piece of scientific equipment used to quickly and efficiently determine the composition of a **solution**. For this, he was awarded the Nobel Prize in **chemistry** in 1959. Throughout his long career, Heyrovský collaborated with many other scientists to develop his methods. He also made valuable contributions to scientific publications in his home country, gaining international recognition for the work being done by Czech scientists.

Jaroslav Heyrovský.

Heyrovský was born in Prague, Austria-Hungary (now the Czech Republic) on December 20, 1890, to Leopold, a professor of Roman law at Charles University in Prague, and Klára (Hanl) Heyrovský. It was during his years at the gymnasium (high school) in Prague that Heyrovský developed a deep interest in mathematics and physics. By the time he entered the Charles University in 1909, his father had become rector of the college. Heyrovský studied physics, chemistry, and mathematics at Charles, becoming influenced by physicists František Záviška and Bohumil Kučera. He also studied with chemist Bohumil Brauner, an association that led him in 1910 to University College in London, where he received his bachelor of science degree in 1913.

Frederick G. Donnan had succeeded **William Ramsay** at University College and stimulated Heyrovský's interest in **electrochemistry**, which became the subject of his doctoral studies. He was detained in Prague, however, during a visit home at the onset of World War I. Although poor health had exempted him for military service, he was assigned to a military hospital as a chemist and radiologist between 1914 and 1918. Continuing his research despite the war, he presented his doctoral thesis on the electroaffinity of **aluminum** in the fall of

1918 to Charles University. In 1919 he became an assistant professor of chemistry there, and in 1920, was appointed lecturer.

Heyrovský published articles from his thesis and these earned him a second doctorate from University College in 1921. By 1922, he was appointed associate professor and head of the chemistry department at Charles University, and in 1924, he was named extraordinary professor and director of a new establishment, the Institute of Physical Chemistry at Charles. He held the position of full professor from 1926 until 1950, when he was named director of the Polarographic Institute of Czechoslovak Academy of Sciences.

Among his many associates and collaborators, Masuzo Shikata was instrumental in helping Heyrovský develop the polarograph. They published a description of the instrument they had designed, detailing how it automatically records the analysis of chemical solutions without altering them. This was one of the earliest laboratory instruments to help automate research. The two researchers had devised a method that reduced the process from over an hour to fifteen or twenty minutes. His work in this area is considered to have greatly aided the study of electrode processes, applications of which are useful in medicine and industry as well as in the research of biochemical reactions and the study of electrochemical processes of organic and inorganic compounds. In later work, Heyrovský was able to reduce the recording time of the process to fractions of seconds with great accuracy.

A Rockefeller Fellowship made it possible for Heyrovský to lecture at the University of Paris in 1926. The successful presentation of his work led to lectures on the polarographic process in the United States, notably at Berkeley in 1933 under a Carnegie visiting professorship. In 1934 he addressed the Mendeleev centenary in Leningrad. This worldwide recognition was aided, according to Heyrovský, by the help of Wilhelm Böttger in 1936. Böttger was the editor of an important chemical journal and invited Heyrovský to contribute an account on polarography for the second volume of his compendium on analytical methods in **physical chemistry**. During the 1940s, accounts of his discoveries were published in English and in German.

During World War II, Heyrovský's research continued in the midst of Nazi-occupied Czechoslovakia. While Czechs were removed from their teaching positions and replaced by Germans, Heyrovský's replacement was of mixed German-Czech parentage, and unsympathetic to Nazism. Heyrovský was permitted to carry on his work, and by the end of the war, he had completed important writing and begun new investigations.

Heyrovský believed that it was vital to Czech scientific culture that a journal be published in his native language. He was helped by the Royal Bohemian Society of Sciences to found a chemical journal for papers that had been written in French or English. Heyrovský accepted responsibility for translations of the English papers, while his associate, Emil Votoček, handled the French papers. The publication, *Collection of Czechoslovak Chemical Communications,* became internationally recognized. Heyrovsky also prepared a number of bibliographies on polarography with the help of a number of notable associates, including his wife, Marie Heyrovská.

Czech society was grateful for the work that Heyrovský did to reflect how the country's scientific community had matured. Specifically, he had created many Czech scientific terms in the belief that language was a critical tool in the development of research. In recognition, the Polarographic Institute of the Czechoslovak Academy of Sciences (founded as the Polarographic Institute at Charles University in 1950), was named the J. Heyrovský Institute of Polarography in 1964. Additionally, he was awarded the Czech State Prize in 1951 and the Order of the Czechoslovak Republic in 1955.

When Heyrovský received the Nobel Prize in chemistry in 1959, he revealed the intensity of his work. He had spent the better part of forty years working on polarography, sacrificing his spare time to work evenings and weekends on his projects. His dedication to his field and long hours of work did not totally preclude his pursuit of his interests in music, literature, and sports. He played the piano and he was a lover of opera. He has been noted for his good sense of humor and has been described as a generous host and an aficionado of good food and wine.

Heyrovsky married Marie Kořánova in 1926. They had a daughter, Jitka, who became a biochemist and a son, Michael, who followed his father to the Institute of Polarography. Heyrovsky died in Prague on March 27, 1967, at the age of seventy-six.

HILL, HENRY A. (1915-1979)

American chemist

Henry A. Hill was an expert on polymers, with a particular interest in resins, rubber, and **plastics**. Conscious of the limited opportunities for African Americans in the sciences in the 1940s and 1950s, Hill turned adversity to advantage and held a number of management positions in the chemical industry before starting his own company, Riverside Research Laboratory. Hill was frequently sought out by his colleagues for a range of consulting and advisory positions. He was responsible for developing guidelines for employers in the chemical industry, and was appointed by President Lyndon Johnson to the National Commission on Product Safety. In 1977 Hill served as president of the American Chemical Society.

Henry Aaron Hill was born May 30, 1915, in the small river town of St. Joseph, Missouri. His undergraduate education was completed at Johnson C. Smith University, a liberal arts school in North Carolina. Hill received a B.S. in chemistry in 1936. He then spent a year in graduate school at the University of Chicago, but went on to earn a Ph.D. in **organic chemistry** from the Massachusetts Institute of Technology (MIT) in 1942. At MIT Hill came briefly but memorably under the influence of James Flack Norris, who impressed Hill by being more interested in Hill's abilities as a chemist than in his heritage. Hill was later instrumental in establishing the American Chemical Society's Norris award.

Following his formal schooling, Hill held jobs involving several different research concerns, beginning as head of chemistry research at Atlantic Research Associates in Massachusetts from 1942 to 1943. In 1943 he was made a research director. In 1945 Hill was promoted to vice president in charge of research at what was now the National Atlantic Research Company. While moving quickly up the ranks, Hill spent his research time developing water-based paints, rubber adhesives, and synthetic rubber, among other projects. It was also there that Hill began to conceive of operating his own research laboratory.

Hill then spent six years, from 1946 to 1952, as group leader at the Dewey & Almy Chemical Company, working on polymer research. (Polymers are large molecules consisting of similar or identical small molecules or monomers linked together. Examples of naturally occurring polymers are **proteins** and silk; polymers synthesized in the laboratory include plastics and synthetic fibers.) Hill's experience led him to the collaborative development of National Polychemicals Inc., in Wilmington, Massachusetts, where he spent the next nine years beginning in 1952, the first four as assistant manager, and the last five as a vice president. This corporation was a manufacturer of chemical intermediaries used for polymers, and grew to have annual sales in 1971 of over ten million dollars. The company's success was largely credited to Hill's personal research contributions.

In 1961 Hill realized his ambition of operating his own research facility, establishing Riverside Research Laboratory. The mission of the corporation would be to provide research and development, as well as consulting, in the area of organic chemistry. Hill had a particular interest in resin, rubber, and plastics. By 1964 the company had moved to more spacious accommodations, where it would remain for the remainder of Hill's life. Hill eventually became known as an authority in **polymer chemistry** on fabric flammability.

Hill was active in the professional aspects of his field. In 1968 he served as chair of the committee on professional relations of the American Chemical Society. This committee produced widely used personnel guidelines for employers of chemists and chemical engineers. In 1968 Hill was appointed by President Lyndon Johnson to the National Commission on Product Safety, a position that galvanized Hill's interest in product liability and product safety. Hill was a fellow of both the American Association for the Advancement of Science and the American Institute of Chemists. He was a member of the American Chemical Society for thirty-eight years, served on its board of directors from 1971 to 1978, and was elected president in 1977. He was chair of the compliance committee of the National Motor Vehicle Safety Advisory Council, and a member of the Information Council on Fabric Flammability. He was married in 1943, and had one child.

Whatever obstacles Hill may have faced in his career owing to racial discrimination, his talent and persistence served him well in a highly competitive industry. In 1971 he was quoted in *Chemistry* as saying, ''My successes have hinged upon a scratch below the surface, a little extra persistence.'' Hill died of a heart attack on March 17, 1979.

Henry A. Hill. *(AP/Wide World. Reproduced by permission.)*

HINSHELWOOD, CYRIL N. (1897-1967)

English chemist and biochemist

Cyril N. Hinshelwood was not only a scientific thinker of the highest order but a great teacher who influenced a generation of chemists emerging from Oxford University in the decades before, during, and after World War II. His interests led him into the fields of physics, **chemistry**, biology, and even the philosophy of science, where he speculated on the nature of the scientific process. Hinshelwood's most notable achievement in a wide-ranging career took place in chemical **kinetics**, the study of the conditions under which chemical reactions occur. Specifically, he unraveled the daunting complexities of the reaction that produces **water**. In recognition of this work, Hinshelwood shared the 1956 Nobel Prize in chemistry with **Nikolay N. Semenov**, whose ideas he had used in his explanation.

An only child, Cyril Norman Hinshelwood was born on June 19, 1897, in London, England, to Norman MacMillan Hinshelwood, an accountant, and Ethel Smith Hinshelwood. The family moved to Canada, but because of Hinshelwood's health, he and his mother moved back to England in 1904. His father died soon afterward. Although he received a scholarship

Cyril N. Hinshelwood.

to Balliol College at Oxford in 1916, Hinshelwood delayed accepting it to work at the Queensferry Explosive Supply Factory during World War I. Promoted to assistant chief chemist in 1918, his work on solid **explosives** sparked a lifelong interest in chemical kinetics.

Hinshelwood entered Oxford in 1919, becoming a fellow of Balliol in 1920 and a fellow and tutor of Trinity College, Cambridge, in 1921. During the 1920s he began to apply kinetic theory, the study of bodies in motion, to the chemical reactions that occurred in **gases**. In 1927 he began to investigate the interaction of **hydrogen** and **oxygen**. At certain pressure thresholds, the reaction became explosive. Following Semenov's conclusions about his experiments with **phosphorus** and oxygen, Hinshelwood applied the theory of the branching **chain reaction**, which posits that the products of the reaction assist in spreading the reaction so rapidly that an explosion results. The reaction of hydrogen and oxygen is so basic to chemistry that Hinshelwood's findings opened up several avenues of research in both organic and **inorganic chemistry**.

By the late 1930s Hinshelwood had shifted the focus of his research to decipher the mechanisms of bacterial growth with the tools of chemical kinetics. For the rest of his career, he elucidated key processes such as environmental adaptation and cell regulation by breaking them down into discrete chemical reactions. Although his views were met with initial skepticism among biologists, they are now widely accepted principles in **biochemistry**. After decades of work as a professor, scientist, and college administrator, Hinshelwood retired from his chair at Oxford in 1964 and moved to London. However, as a senior research fellow at London's Imperial College, he continued his study of bacterial growth, while also serving as a trustee of the British Museum and chair of the Queen Elizabeth College in London.

Hinshelwood died in London on October 9, 1967. Devoted to his mother until she died in 1959, he was a lifelong bachelor. Well read in the classics and literature, he was a member of the Dante Society and president of the Classical Association at Oxford. He was also president of the Modern Language Association, and knew eight foreign languages. As recreational pursuits, Hinshelwood dabbled in oil painting, appreciated classical music, and collected Chinese porcelain and Persian rugs.

A member of the Royal Academy, Hinshelwood was knighted in 1948. Apart from the Nobel Prize, he won the 1942 Davy Medal, the 1947 Royal Medal, and the 1962 Copley Medal of the Royal Society. He also held numerous honorary degrees from various universities.

HOAGLAND, MAHLON BUSH (1921-)

American biochemist

Mahlon Hoagland is best known for discovering, in the 1950s, that ribonucleic acid (RNA) molecules in the cytoplasm retrieve specific **amino acids** and take them to the ribosomes for assembly into **proteins**.

Hoagland was born in Boston, the son of Hudson Hoagland (1899-1983), the American physiologist who co-founded the Worcester foundation for Experimental Biology with Gregory Pincus. Mahlon Hoagland received his M.D. in 1948 from Harvard Medical School, and served on the bacteriology department faculty from 1953 through 1967.

In the 1950s scientists had already determined that messenger **RNA** carried instructions for protein production from *codons*, triplets of base pairs on deoxyribonucleic acid (DNA), from a cell's **nucleus** into the cytoplasm. There, RNA-rich ribosomes somehow used the information to assemble amino acids into proteins. American biochemist **Paul Berg** had already determined that the amino acids were activated by combining with adenosine triphosphate (ATP), but it was not known how the amino acids recognized the *anticodons*, or the complimentary RNA triplets that carried their code.

Hoagland and his associates accidentally discovered that the amino acids first attached themselves not to the RNA in the ribosomes, but to specific locations on small molecules of what had been called *soluble RNA*. Hoagland theorized that these molecules, whose name soon became transfer RNA, were complementary to the ribosomal RNA, where the amino acids were then attached. Although he was unaware of it at the time, Hoagland's discovery fit a theory of **Francis Crick**, the co-discoverer of the **DNA** double helix, that the amino acids attached to an adaptor RNA **molecule** that is complementary to the ribosomal RNA. Both theories were proven to be correct.

Transfer RNA was also discovered independently by Paul Berg and by Robert Holley.

Hoagland's other work has included determining the cancer-causing properties of **beryllium**, biosynthesizing **coenzyme** A (which is required for cell metabolism), and studies of liver regeneration, growth control, and amino acid activation of **protein synthesis**.

In 1967 he joined the **biochemistry** faculty at Dartmouth Medical School, and also became president of the Worcester Foundation for Experimental Biology. He is a member of the U.S. National Academy of Sciences.

Dorothy Crowfoot Hodgkin. ***(Archive Photo. Reproduced by permission.)***

HODGKIN, DOROTHY CROWFOOT (1910-1994)

English chemist and crystallographer

Dorothy Crowfoot Hodgkin employed the technique of x-ray **crystallography** to determine the molecular structures of several large biochemical molecules. When she received the 1964 Nobel Prize in chemistry for her accomplishments, the committee cited her contribution to the determination of the structure of both **penicillin** and vitamin B_{12}.

Hodgkin was born in Egypt on May 12, 1910 to John and Grace (Hood) Crowfoot. She was the first of four daughters. Her mother, although not formally educated beyond finishing school, was an expert on Coptic textiles, and an excellent amateur botanist and nature artist. Hodgkin's father, a British archaeologist and scholar, worked for the Ministry of Education in Cairo at the time of her birth, and her family life was always characterized by world travel. When World War I broke out, Hodgkin and two younger sisters were sent to England for safety, where they were raised for a few years by a nanny and their paternal grandmother. Because of the war, their mother was unable to return to them until 1918, and at that time brought their new baby sister with her. Hodgkin's parents moved around the globe as her father's government career unfolded, and she saw them when they returned to Britain for only a few months every year. Occasionally during her youth she travelled to visit them in such far-flung places as Khartoum in the Sudan and Palestine.

Hodgkin's interest in chemistry and crystals began early in her youth, and she was encouraged both by her parents as well as by their scientific acquaintances. While still a child, Hodgkin was influenced by a book that described how to grow crystals of alum and **copper sulfate** and on x rays and crystals. Her parents then introduced her to the soil chemist A. F. Joseph and his colleagues, who gave her a tour of their laboratory and showed her how to pan for **gold**. Joseph later gave her a box of reagents and minerals which allowed her to set up a home laboratory. Hodgkin was initially educated at home and in a succession of small private schools, but at age eleven began attending the Sir John Leman School in Beccles, England, from which she graduated in 1928. After a period of intensive tutoring to prepare her for the entrance examinations, Hodgkin entered Somerville College for women at Oxford University. Her aunt, Dorothy Hood, paid the tuition to Oxford, and helped to support her financially. For a time, Hodgkin considered specializing in archaeology, but eventually settled on chemistry and crystallography.

Crystallography was a fledgling science at the time Hodgkin began, a combination of mathematics, physics, and chemistry. Max Laue, **William Henry** Bragg and **William Lawrence Bragg** had essentially invented it in the early decades of the century (they had won Nobel Prizes in 1914 and 1915, respectively) when they discovered that the atoms in a crystal deflected x rays. The deflected x rays interacted or interfered with each other. If they *constructively* interfered with each other, a bright spot could be captured on photographic film. If they *destructively* interfered with each other, the brightness was cancelled. The pattern of the x ray spots—*diffraction pattern*—bore a mathematical relationship to the positions of individual atoms in the crystal. Thus, by shining x rays through a crystal, capturing the pattern on film, and doing mathematical calculations on the distances and relative positions of the spots, the **molecular structure** of almost any crystalline material could theoretically be worked out. The more complicated the structure, however, the more elaborate and arduous the calculations. Techniques for the practical application of crystallography were few, and organic chemists accustomed to chemical methods of determining structure regarded it as a black art.

After she graduated from Oxford in 1932, Hodgkin's old friend A. F. Joseph steered her toward Cambridge University and the crystallographic work of J. D. Bernal. Bernal already had a reputation in the field, and researchers from many countries sent him crystals for analysis. Hodgkin's first job was as Bernal's assistant. Under his guidance, with the wealth of materials in his laboratory, the young student began demonstrating her particular talent for x-ray studies of large molecules such as sterols and **vitamins**. In 1934, Bernal took the first x-ray photograph of a protein crystal, pepsin, and Hodgkin did the subsequent analysis to obtain information about its **molecular weight** and structure. **Proteins** are much larger and more complicated than other biological molecules because they are polymers—long **chains** of repeating units—and they exercise their biochemical functions by folding over on themselves and assuming specific three-dimensional shapes. This was not well understood at the time, however, so Hodgkin's results began a new era; crystallography could establish not only the structural layout of atoms in a **molecule**, even a huge one, but also the overall molecular shape which contributed to biological activity.

In 1934, Hodgkin returned to Oxford as a teacher at Somerville College, continuing her doctoral work on sterols at the same time. (She obtained her doctorate in 1937.) It was a difficult decision to move from Cambridge, but she needed the income and jobs were scarce. Somerville's crystallography and laboratory facilities were extremely primitive; one of the features of her lab at Oxford was a rickety circular staircase that she needed to climb several times a day to reach the only window with sufficient light for her polarizing microscope. This was made all the more difficult because Hodgkin suffered most of her adult life from a severe case of rheumatoid arthritis, which didn't respond well to treatment and badly crippled her hands and feet. Additionally, Oxford officially barred her from research meetings of the faculty chemistry club because she was a woman, a far cry from the intellectual comradery and support she had encountered in Bernal's laboratory. Fortunately, her talent and quiet perseverance quickly won over first the students and then the faculty members at Oxford. Sir **Robert Robinson** helped her get the money to buy better equipment, and the Rockefeller Foundation awarded her a series of small grants. She was asked to speak at the students' chemistry club meetings, which faculty members also began to attend. Graduate students began to sign on to do research with her as their advisor.

An early success for Hodgkin at Oxford was the elucidation of **cholesterol** iodide's molecular structure, which no less a luminary than W.H. Bragg singled out for praise. During World War II, Hodgkin and her graduate student Barbara Low worked out the structure of penicillin, from some of the first crystals ever made of the vital new drug. Penicillin is not a particularly large molecule, but it has an unusual **ring** structure, at least four different forms, and crystallizes in different ways, making it a difficult crystallographic problem. Fortunately they were able to use one of the first IBM analog computers to help with the calculations.

In 1948, Hodgkin began work on the structure of vitamin B_{12}, the deficiency of which causes pernicious anemia. She obtained crystals of the material from Dr. Lester Smith of the Glaxo drug company, and worked with a graduate student, Jenny Glusker, an American team of crystallographers led by Kenneth Trueblood, and later with John White of Princeton University. Trueblood had access to state of the art computer equipment at the University of California at Los Angeles, and they sent results back and forth by mail and telegraph. Hodgkin and White were theoretically affiliated with competing pharmaceutical firms, but they ended up jointly publishing the structure of B_{12} in 1957; it turned out to be a porphyrin, a type of molecule related to **chlorophyll**, but with a single **atom** of **cobalt** at the center.

After the war, Hodgkin helped form the International Union of Crystallography, causing Western governments some consternation in the process because she insisted on including crystallographers from behind the Iron Curtain. Always interested in the cause of world peace, Hodgkin signed on with several organizations that admitted Communist party members. Recognition of Hodgkin's work began to increase markedly, however, and whenever she had trouble getting an entry visa to the U.S. because of her affiliation with peace organizations, plenty of scientist friends were available to write letters on her behalf. A restriction on her U.S. visa was finally lifted in 1990, shortly before the Soviet Union collapsed.

In 1947, she was inducted into the Royal Society, Britain's premier scientific organization. Professor Hinshelwood assisted her efforts to get a dual university/college appointment with a better salary, and her chronic money problems were alleviated. Hodgkin still had to wait until 1957 for a full professorship, however, and it was not until 1958 that she was assigned an actual chemistry laboratory at Oxford. In 1960, she obtained the Wolfson Research Professorship, an endowed chair financed by the Royal Society, and in 1964 received the Nobel Prize in chemistry. A year later, she was awarded Britain's Order of Merit, only the second woman since Florence Nightingale to achieve that honor.

Hodgkin still wasn't done with her research, however. In 1969, after decades of work and waiting for computer technology to catch up with the complexity of the problem, she solved the structure of insulin. She employed some sophisticated techniques in the process, such as substituting atoms in the insulin molecule, and then comparing the altered crystal structure to the original. Protein crystallography was still an evolving field; in 1977 she said, in an interview with Peter Farago in the *Journal of Chemical Education,* "In the larger molecular structure, such as that of insulin, the way the peptide chains are folded within the molecule and interact with one another in the crystal is very suggestive in relation to the reactions of the molecules. We can often see that individual side chains have more than one **conformation** in the crystal, interacting with different positions of solvent molecules around them. We can begin to trace the movements of the atoms within the crystals."

In addition to her scientific work, Hodgkin served as chancellor of Bristol University, a position she held until her retirement in 1988. She died on July 30, 1994 at her home in Shipston-on-Stour, Warwickshire, England.

In 1937, Dorothy Crowfoot married Thomas Hodgkin, the cousin of an old friend and teacher, Margery Fry, at Somerville College. He was an African Studies scholar and teacher, and, because of his travels and jobs in different parts of the world, they maintained separate residences until 1945, when he finally obtained a position teaching at Oxford. Despite this unusual arrangement, their marriage was a happy and successful one. Although initially worried that her work with x-rays might jeopardize their ability to have children, the Hodgkins produced three: Luke, born in 1938, Elizabeth, born in 1941, and Toby, born in 1946. The children all took up their parents scholarly, nomadic habits, and at the time of the Nobel Ceremony travelled to Stockholm from as far away as New Delhi and Zambia. After her retirement, she continued to travel widely and expanded her lifelong activities on behalf of world peace, working with the Pugwash Conferences on Science and World Affairs.

HOFFMANN, ROALD (1937-)

American chemist

Roald Hoffmann is a theoretical chemist who has straddled the traditional boundary between organic and inorganic **chemistry**. He has emphasized the role of aesthetics in science and the inherent beauty of chemical systems, and he values clarity and simplicity in the formulation of theories about chemistry. He is best known for constructing a method of predicting the course of chemical reactions that is based on the symmetry of **electron** orbitals. Called the Woodward-Hoffmann rules, he developed them in collaboration with **Robert B. Woodward** at Harvard, and these rules have enabled chemists to predict reactions without using complicated mathematical equations. The achievement has been widely recognized as the most important conceptual advance in **organic chemistry** since World War II, and for this Hoffmann shared the 1981 Nobel Prize in chemistry with **Kenichi Fukui**.

Hoffmann was born Roald Safran on July 18, 1937, in Zloczów, Poland on the eve of World War II. His father was Hillel Safran, a civil engineer; his mother, Clara Rosen, was a schoolteacher. In 1941 German troops occupied Zloczów, and the family was sent first to a Jewish ghetto and then interred at a labor camp. Safran managed to arrange for his wife and son to escape the camp, and the two were hidden by a Ukrainian teacher in the dark, cramped attic of a schoolhouse, where Hoffmann began his education under the tutelage of his mother. Hoffmann's father made plans to follow them, but his escape was discovered by the Nazis and he was executed. Hoffmann and his mother were able to remain undetected until 1944, when the Red Army liberated Zloczów, which later became part of the Soviet Ukraine. The two moved to Kraków, Poland, where his mother met and married Paul Hoffmann, whose spouse had also been killed in the war. They lived in several camps for displaced persons in Austria and Germany. In an autobiographical passage in *Chemistry Imagined,* Hoffmann remembers learning German, which was by then his fourth language, and being fascinated by the biographies of **George Washington Carver** and **Marie Curie**. The Hoffmanns were able to emigrate to the United States in 1949, and they settled in New York City.

Hoffmann learned English and attended public schools in Brooklyn, including Stuyvesant High School, which specialized in science. He told *Scientific American* that as a child he ''showed neither precocity nor early interest in chemistry.'' His mother wanted him to become a doctor, and he enrolled at Columbia University with this in mind. Taking an extra-heavy course load, Hoffmann exhibited a wide range of interests, including mathematics, French, and even art history, which nearly lured him away from science. He spent most of his summers studying the chemistry of cement and hydrocarbons at the National Bureau of Standards, and it was here his interest in chemistry really began. In 1958, after only three years, he graduated summa cum laude in chemistry, and then entered the doctoral program at Harvard University.

In 1959 Hoffmann was awarded a summer fellowship to attend a program in **quantum chemistry** at the University of Uppsala in Sweden. That same year he attended a summer symposium on quantum physics in Sweden and met Eva Börjesson. He married her in 1960, and the couple now has two children. He studied at Moscow University in the Soviet Union. After returning from Moscow in 1960, he began his doctoral work, studying under William Nunn Lipscomb, Jr. Hoffmann researched **theoretical chemistry** for his Ph.D. He examined questions relating to the electronic structure of certain organic molecules. He used computer programs to determine the electronic structure of **boron** hydrides and other polyhedral molecules and also to predict what shape these molecules would assume after a reaction. Hoffmann's work advanced the application of what is called the Hückel method, which is used to calculate the number of electrons in orbit around a **molecule**.

Hoffmann received his Ph.D. in chemical physics in 1962, and he remained at Harvard on a three-year fellowship from the Society of Fellows. This fellowship offered Hoffmann the time to shift the focus of his research away from purely theoretical to applied theoretical chemistry. In 1964 he began working with organic chemist Robert B. Woodward, who had observed an unusual and unexpected reaction during an attempt to synthesize vitamin B_{12}. Hoffmann left Harvard in 1965 to accept a position as associate professor at Cornell University, but he continued his collaboration with Woodward.

The reaction Woodward had observed was one of a class which is now called **pericyclic reactions**, whose course was very difficult to predict. Hoffmann and Woodward initially began to formulate their rules in an effort to identify the conditions that would produce certain results, and one of the difficulties they faced in this effort was the complexity of predicting how **energy** was released during these reactions. The release of energy during a reaction is determined by changes in the motion of electrons known as orbitals. What Hoffmann discovered was that the course of the reaction depended on the symmetry of these orbitals. Hoffmann and Woodward examined how orbital symmetry determined different reactions, and using quantum mechanics they were able to develop a mathematical procedure to predict these symmetries and thus whether certain combinations of chemicals would result in reactions.

The result of their work was a clear and relatively simple method of prediction that was based on diagrams; it was now possible to calculate the course of a reaction, as the *New York Times* observed, by ''jotting pictures on the back of an envelope.'' Andrew Streitwieser, Jr. wrote in *Science:* ''The results of orbital symmetry correlation diagrams lend themselves to alternative formulations that are frequently easier to apply.'' The Woodward-Hoffmann rules are now widely used, and they have had important practical applications in medical and industrial research. When Hoffmann was awarded the Nobel Prize in chemistry in 1981, Woodward had already died, but Hoffmann mentioned him frequently in his Nobel lecture and believed they would have shared the prize had he lived.

During the course of his work with Woodward, Hoffmann became convinced that similarities in the structure and function of electrons bridged many of the traditional divisions in chemistry, particularly the distinction between organic and **inorganic chemistry**. Following the formulation and publication of their rules, Hoffmann began conducting research to show that their method of predicting orbital symmetry could be applied to inorganic as well as organic compounds. Hoffmann and others working in his laboratory made detailed examinations of both inorganic and organometallic molecules—organic compounds which include metal. They were able to establish the unity of structure and function which he originally suspected, and his work has increased the ability of chemists to predict the course of inorganic reactions. The American Chemical Society presented Hoffmann with the Arthur C. Cope Award in Organic Chemistry in 1973, and the society named him as the recipient of their Inorganic Chemistry Award in 1982. He is the first American chemist to be honored in both disciplines.

Hoffmann has long been interested in the similarities between art and the creative process he believes is required in science. Inspired by memories of his undergraduate work with the literary critic Mark Van Doren at Columbia, and moved by the experience of reading the poems of Wallace Stevens, Hoffmann began writing poetry at the age of forty. He continues to work with a group of poets at Cornell, and he has published two volumes of his poems. He writes about science and the beauty of nature, as well as his childhood experiences during and after World War II. He has also written a book with artist Vivian Torrence on the relationship between chemistry and art.

Hoffmann was made a full professor at Cornell in 1968, and in 1974 he was named the John A. Newman Professor of Physical Science at that university. In addition to the Nobel Prize and his awards in organic and inorganic chemistry, Hoffmann received the Pauling Award in 1974, the Nichols Medal in 1981, and the Priestly Medal in 1990. He is a member of the National Academy of Sciences and a foreign member of the Royal Society in London.

August Wilhelm von Hofmann.

HOFMANN, AUGUST WILHELM VON (1818-1892)

German chemist

Hofmann was born in Giessen. He originally planned to study law at college, but after attending **chemistry** lectures by **Justus von Liebig**, he became interested in chemistry. In 1841 he received his doctorate. His dissertation was on **coal** tar, the precursor to **aniline**, which in turn could be derived from phenal and **ammonia**. Likewise, his many hundreds of papers were on coal tars and their derivatives. In 1843, after his father's death, Hofmann joined Liebig as an assistant. He taught for a few months at the University of Bonn in 1845 and then moved on to England, where he was appointed a professor of the Royal College of Chemistry. He remained there until 1863, when he returned to Berlin, publishing a textbook entitled *An Introduction to Modern Chemistry* in 1865.

Hofmann was more of an analyst than a bench chemist—that is, he worked more comfortably analyzing and interpreting data than physically manipulating substances in experimentation. For that reason, he searched out talented assistants to carry out the laboratory techniques. Many of his assistants and students, among them Frederick Abel, **William Crookes**, **William Henry Perkin**, William Nicholson (1753-1815), John Newlands (1837-1898), Peter Griess (1829-1888),

C. A. Martius, and Jacob Volhard (1834-1910), went on to develop synthetic dyes and establish the synthetic dye industries in England and Germany.

Hofmann spent many years studying **nitrogen** compounds like ammonia, which led to his development of polyammonias, now called triamines and diamines. Following the path started by his students, he also began to research synthetic dyes. In 1858 he reacted **carbon** tetrachloride with aniline which contained the impurities orthotoluidine and paratoluidine and formed a dye called rosaniline. Hofmann also succeeded in producing aniline blue, or diphenylaniline, by replacing **hydrogen** in rosaniline with aniline. He patented his Hofmann's violet in 1863 and successfully marketed it. He also developed new production techniques for synthetic dyes; the ''Hofmann degradation'' was widely used to produce amines. Hofmann also investigated methyl aldehyde, now called **formaldehyde**.

Hofmann was a founder of the German Chemical Society and was instrumental in standardizing the **nomenclature** for alkanes and alkane derivatives. He died in Berlin in 1892.

HOLLEY, ROBERT WILLIAM (1922-1993)

American biochemist

Robert William Holley.

Robert Holley is best known for isolating and characterizing transfer ribonucleic acid (tRNA). Essentially tRNA translates the genetic instructions within cells by first reading genes, the fundamental units of heredity, and then creating proteins, the building blocks of the body, from **amino acids**. Holley, along with **Har Gobind Khorana** and **Marshall Warren Nirenberg**, received the 1968 Nobel Prize for medicine or physiology for determining the sequence of tRNA. However, Holley's work on tRNA was only the beginning of a distinguished scientific career. Subsequently, he investigated the molecular factors that control growth and multiplication of cells. His work in this area has greatly contributed to scientific understanding of the processes that lead to cancer.

Robert William Holley was born in Urbana, Illinois, on January 28, 1922. His parents, Charles Elmer Holley and Viola Esther (Wolfe) Holley, were both teachers. Robert was one of four sons. He grew up in Illinois, California, and Idaho, developing, early in life, a life-long love of the outdoors and fascination with living things. The latter years of his childhood were spent in Urbana, where he attended high school. In 1938, he enrolled at the University of Illinois, majoring in **chemistry**.

After obtaining his B.A. in 1942, Holley took up graduate studies in **organic chemistry** at Cornell University. He served in various position at both the university and medical college for the next several years. In 1945, he married Ann Lenore Dworkin, a chemist and high school mathematics teacher. They had one son, Frederick.

During the mid-1940s, Holley participated as a civilian in war research for the United States Office of Research and Development. He was a member of the team of researchers that first succeeded in making synthetic **penicillin**. Supported by a fellowship from the National Research Council, he completed his doctorate in organic chemistry at Cornell University in 1947, and did a year of postdoctoral work at Washington State College (now University) in Pullman before returning east. In 1948, he became assistant professor at the New York State Agricultural Experiment Station, a branch of Cornell, in Geneva. Promoted to the position of associate professor in 1950, he was named full professor in 1964.

During a sabbatical on a Guggenheim Memorial Fellowship at the California Institute of Technology in 1955-1956, Holey started investigating **protein synthesis**. In the wake of **James Watson**'s and **Francis Crick**'s discovery that **DNA** contained the information of heredity, Holley targeted the chemistry of **nucleic acids**, which carry and transmit genetic information. His course may have been inspired, at least in part, by Crick's suggestion that adaptor molecules of some sort must be involved in the transition of genetic information into **proteins**. Toward the end of his year away from Cornell, Holley began to look specifically at the structure of transfer **RNA**, starting a non-year effort to unlock its secrets.

Back at Cornell in 1957, Holley was appointed research chemist at the United States Plant, Soil, and **Nutrition** Laboratory, where he continued his studies on tRNA. Heading up a research team, he meticulously planned and carried out a painstaking series of experiments. He and his colleagues spent three years developing a technique to isolate and partially purify different classes of tRNA from yeast. Finally, they succeeded in isolating a pure sample of alanine transfer RNA. The next five years were devoted to elucidating the sequence and structure of this particular transfer RNA.

To appreciate the profound impact of Holley's work on research into the **biochemistry** of life, it is useful to review some fundamental concepts. Cells carry the instructions for all of their necessary tasks in their chromosomes. Chromosomes within a cell are made up of very long molecules called deoxyribonucleic acid, or DNA. Genes, the basic units of heredity for all living things, are small sections of the long strands of DNA. Genes themselves are made up of a series of units called nucleotides. Nucleotides are molecules composed of a particular sugar (either ribose or deoxyribose), a **phosphate** group (one phosphorous **atom** combined with three **oxygen** atoms), and one of five specific bases. These bases—guanine, adenine, cytosine, thymine, and uracil—thus distinguish the nucleotides from one another. They are, in essence, the alphabet from which all of our genetic instructions are composed.

Cells and bodies, however, are built not of genes but of proteins. Proteins are the structural elements of cells, providing form and stability. Equally important, certain proteins, called enzymes, mediate critical biochemical reactions, allowing the formation and breakdown of innumerable chemicals that cells use during growth, functioning, and division. Proteins are sequences of amino acids. The amino acids are a group of about twenty different molecules that share certain chemical characteristics (e.g. the presence of an amine group).

Holley's work centered on the question of how sequences of nucleotides in DNA specify sequences of amino acids in proteins. It had been known that DNA did not directly create protein, but copied itself instead (in a complementary, or negative, sense) into strands of RNA. But it was not known how these long strands of RNA, called messenger RNA, or mRNA, functioned in the creation of proteins. Holley believed that the much smaller tRNA molecules played a key role. He knew that a triplet of bases, or codon, specifies each of the twenty amino acids. Examining the sequence of bases within alanine tRNA (which specifies creation of the amino acid alanine), he found an anti-codon for alanine. This anti-codon would be able to bond chemically with an alanine codon on an mRNA strand.

By studying the molecular sequence of alanine tRNA, Holley and his students were able to determine its structure and ten to deduce how it functioned. A tRNA anti-codon would bind to its matching codon along a strand of mRNA. The corresponding amino acid, held at the opposite end of the tRNA, would then be positioned to link up in series with the amino acid specified by the adjacent codon on the mRNA. In this manner, the series of nucleotides in a **molecule** of DNA would be translated into a series of amino acids that would make up a protein.

For his illumination of this vital process, Holley won a share of the 1968 Nobel Prize for physiology or medicine. He also received the prestigious Albert Lasker Award for Basic Medical Research in 1965.

From 1966 to 1967 Holley was on sabbatical at the Salk Institute for Biological Studies and the Scripps Clinic and Research Foundation in La Jolla, California. The following year, he joined the Salk Institute as a resident fellow. Like his earlier sabbatical, Holley's move proved pivotal for his research, as he launched an investigation of the molecular factors which regulate growth and multiplication of cells. Rooted somewhat in his previous work on how the protein molecules undelying cell growth are formed, the new investigations had quite differenr interpretations and implications.

The control of cell growth and division is critical for normal functional. Cancerous growths are characterized by uncontrolled cell division. Normally, a balanace of stimulating and inhibiting factors keeps cellular multiplication at its proper rate; the number of new cells produced roughly equals the number of cells that wear out and die. Rapid cell proliferation can be caused by over-production of the stimulating factors, a lack of inhibiting factors, or by some combination of these causes.

Holley examined the roles of **hormones**, blood-borne chemicals, usually proteins, which are released by various tissues and organs and which interact with one another. Hormones can either stimulate or inhibit cell proliferation, or even, as Holley later showed, do both.

Holley discovered that the **concentration** of two types of hormones, known as peptide and steroid hormones, in a **solution** with dividing cells would determine the rate of cell division, and, ultimately, cell **density**. Further, he found that types of cells prone to develop into tumors responded dramatically to these growth factors, dividing rapidly in response to very low hormone levels. Subsequent experiments demonstrated that peptide and steroid hormones could act synergistically: several of these growth factors together in a solution would produce a greater growth rate than the sum effect of each, individually. Holley also found that different types of cells responded differently to particular hormones, and that their responses could change with the cells' population density. At low densities, cells take up and utilize growth factors more efficiently than they do under conditions of high density. Cellular **receptors** for certain growth-promoting hormones increased under conditions of low cell density, whereas receptors for certain other hormones increased as cell density increased.

Holley also studied the effects of non-hormonal factors, such as certain sugars and amino acids, on cell proliferation. He found that while cell growth patterns were quite insensitive to the levels of many amino acids, they were strongly regulated by others, notably glutamine.

Looking at the other side of the coin, Holley and his collaborators also identified growth inhibitors. Some of these compounds suppress cell growth by blocking DNA replication. Holley discovered that, in addition to blocking DNA activity, growth inhibitors stimulated production of specific proteins whose functions were unknown. Growth factors, too, were found to have an associated protein synthesis in addition to stimulating DNA activity. Interestingly, Holley noted that while growth and inhibiting factors canceled each other's effects on DNA replication, they had no effect on each other's secondary production of hormones. With particular factors that increase both cell size and rate of cell division, Holley had noted similar effects. Adding a growth inhibitor would stop the cells from dividing, but would not stop the individual cells from growing.

As he and his co-workers had done with tRNA, Holley's team eventually sequenced some of the growth factors. These are considerably larger molecules than tRNA, but the techniques of molecular biology had improved so much over the years that these sequences were obtained much more readily. Holley identified the sequence of amino acids of a growth-inhibiting factor for a specific type of monkey cell. The sequence turned out to be identical to that for the human growth factor TGF-beta 2.

Holley's work during the later phase of his career shed new light on the factors that control how cells grow, differentiate, and divide. His research also had striking implications for the development of drugs to suppress tumor growth and for understanding the fundamental causes of cancer. Holley died in 1993.

Holmium

Holmium is a rare earth element, one of the elements found in Row 6 of the **periodic table** between Groups 2 and 3. Its **atomic number** is 67, its atomic **mass** is 164.9303, and its chemical symbol is Ho.

Properties

Holmium is a silvery metal that is soft, ductile, and malleable. It has a melting point of 2,680°F (1,470°C), a **boiling point** of 4,930°F (2,720°C), and a **density** of 8.803 grams per cubic centimeter. It tends to be chemically stable at room **temperature**, but becomes more active at higher temperatures when, for example, it reacts with **oxygen** to form holmium **oxide** (Ho_2O_3).

Occurrence and Extraction

The abundance of holmium in the Earth's crust is estimated to be about 0.7-1.2 parts per million, making it less common than most other **rare earth elements**, but more common than some familiar elements, such as **iodine**, **silver**, **mercury**, and **gold**. The most common ores of holmium are monazite and gadolinite. Holmium is extracted from its ores by heating **calcium** metal with holmium fluoride: $3Ca + 2HoF_3 \rightarrow 3CaF_2 + 2\ Ho$.

Discovery and Naming

Holmium was discovered in 1879 by the Swedish chemist Per Teodor Cleve (1840-1905). Cleve found the new element in a mineral called ytterite that had been discovered nearly a century earlier by a Swedish army officer, Carl Axel Arrhenius (1757- 1824). Ytterite was the subject of considerable research on the part of chemists and mineralogists and was eventually to yield nine new elements. Cleve named the new element after his birthplace, Stockholm.

Uses

Holmium has very few practical purposes, one exception being in the manufacture of specialized types of laser. Holmium **lasers** are now used to some extent in eye surgery where they are used to treat glaucoma and reduce abnormal eye pressure.

Frederick Gowland Hopkins.

Hopkins, Frederick Gowland (1861-1947)

English biochemist

Frederick Gowland Hopkins is considered the founder of British **biochemistry**. A pioneer in the study and application of what he called accessory food factors and what we call **vitamins**, he made important contributions to the study of uric acid, isolated tryptophan (a necessary component in nutrition), and developed the concept of essential **amino acids**. He also did pioneering work on cell **metabolism**, elucidating the role of enzymatic activity in **oxidation** processes. With the assistance of physiologist Walter M. Fletcher, he explained the relationship between lactic acid and muscle contraction. He became a member of the Royal Society in 1905, serving as its president in 1931. He was knighted in 1925 and received the Copley Medal of the Royal Society in 1926. For his contributions in the field of **nutrition**, he was awarded the 1929 Nobel Prize in physiology or medicine, sharing the award with the Dutch chemist Christiaan Eijkman. He was presented with the highest distinction of civil service, the Order of Merit, in 1935. In addition, many honorary degrees were bestowed on him by universities worldwide.

Hopkins was born in Eastbourne, Sussex, England, on June 20, 1861, to Frederick Hopkins and Elizabeth Gowland

Hopkins. The poet Gerard Manley Hopkins was his second cousin. His father died soon after his birth and he was taken by his mother to live with her family in London. He was a solitary and scholarly boy, given to reading Charles Dickens and writing poetry, although he showed no particular aptitude for any subject in school except **chemistry**. According to the *Dictionary of Scientific Biography,* he was captivated by what he saw through the microscope, saying that "the powers of the microscope thus revealed to me were something very *important*—the most important thing I had as yet come up against." But his scientific education was delayed when his uncle secured a position for him in the insurance business. The seventeen-year-old Hopkins stuck it out for six months before leaving. He did publish his first paper while working for the insurance company, on the vapor ejected by the bombardier beetle. Although entomology remained an interest throughout his life, Hopkins said that his work on the beetle had made him realize that his true vocation was biochemistry. A small inheritance enabled him to study chemistry at the Royal School of Mines and London's University College, where he received a B.Sc. in 1890. An exemplary performance on his final chemistry examination led to a position as an assistant to Thomas Stevenson, an expert in forensic medicine at Guy's Hospital, who also served as medical jurist to the Home Office. As assistant to Stevenson, Hopkins used his analytical skills to help secure the convictions of several notorious murderers.

Despite these achievements, Hopkins was aware that he needed more education. He entered Guy's Hospital to study medicine in 1890, earning his M.B. in 1894. He remained at Guy's to teach for four years after receiving his degree. In 1898, at the age of thirty-seven, he was invited to Cambridge to undertake the triple duties of teacher, tutor, and developer of the chemical physiology department. The demands of that heavy workload caused Hopkins to suffer a temporary breakdown in 1910. Even with that workload and his subsequent ill health, Hopkins managed to publish more than thirty papers. Trinity College made him a fellow and elected him to a praelectorship in biochemistry. His appointment at Trinity, when he was nearly fifty years old, allowed him to give full attention to research and the advancement of biochemistry within the university.

Hopkins became the first professor of biochemistry at Cambridge in 1914. He established an open admissions policy for his department and attracted biochemists from many nations, including some who had escaped from dire political situations. A great teacher as well as a brilliant and unassuming researcher, Hopkins was known for encouraging his students to pursue their own line of work and often handed over to them promising new research in which he had made a breakthrough. Any credit or distinction resulting from pursuit of that new work went to his students, not Hopkins. This generosity, coupled with his faith that biochemistry could provide important answers to biological questions, was largely responsible for the widespread development of biochemical thought and experimentation. At the time of his death, approximately seventy-five of his former students held professorial positions in universities throughout the world.

One of his earliest contributions to biochemistry, made while he was still at Guy's Hospital and used for many years, was the method he developed to detect the presence of uric acid in urine. His work on uric acid led him to the study of **proteins**, the presence of which in the diet affects uric acid excretion. Hopkins first developed methods to isolate and crystallize proteins. He isolated the amino acid tryptophan and determined its structure. He also studied the effect of bacteria on tryptophan, laying the foundation of bacterial biochemistry. Hopkins showed that tryptophan is essential in the diet, since proteins lacking the substance are nutritionally inadequate. He also studied the role of the amino acids arginine and histidine in nutrition, which led to the theory that the presence of different amino acids determines the nutritional quality of proteins.

Hopkins, however, was not satisfied that an adequate diet depended upon the presence of essential amino acids alone. He suspected that additional substances, which he called accessory food factors, also played an essential role in nutrition. Experiments with rats that were fed a synthetic diet of milk proteins, **carbohydrates**, fats and salts showed that such a diet caused a decline in growth in the animals. When milk was added to the diet, even in small quantities, the rats once again underwent normal growth. This preliminary work enabled Hopkins to isolate what are now called vitamins A and D. His observations led him to conclude that such diseases as rachitis (popularly called rickets, and caused by a vitamin D deficiency) and scurvy occurred when food lacked certain vitamins. Although it had been known for a long time that scurvy, a common shipboard ailment, could be prevented or cured by supplementing one's diet with lemons, Hopkins' work on isolating vitamin C was the first to explain this experiential finding scientifically.

In 1912 Hopkins published the work for which he is probably best known, "Feeding Experiments Illustrating the Importance of Accessory Food Factors in Normal Dietaries." During World War I, Hopkins continued his work on the nutritional value of vitamins. His efforts were especially valuable in a time of food shortages and rationing. He agreed to study the nutritional value of margarine and found that it was, as suspected, inferior to butter because it lacked the vitamins A and D. As a result of his work, vitamin-enriched margarine was introduced in 1926. Hopkins' nutritional theories were contested by colleagues until about 1920 but have been considered indisputable since then.

Although Hopkins won the 1929 Nobel Prize in medicine and physiology (shared with Christiaan Eijkman) for his work in nutrition, he was primarily interested in the biochemistry of the cell. The originality and vision of his research in this area set the standard for those who followed him. His study with Fletcher of the connection between lactic acid and muscle contraction was one of the central achievements of his work on the biochemistry of the cell. Hopkins had long studied how cells obtain **energy** from a complex metabolic process of oxidation and **reduction** reactions. He showed that **oxygen** depletion causes an accumulation of lactic acid in the muscle. The research techniques developed by Hopkins and Fletcher to study muscle stimulation were later used by others to study other aspects of muscle metabolism. Their work paved the way for the later discovery by Archibald Vivian Hill and **Otto Mey-**

erhof that a carbohydrate metabolic cycle supplies the energy used for muscle contraction. The discovery that **alcohol fermentation** under the influence of yeast is a process analogous to the formation of lactic acid in the muscle is also due to Hopkins's ground-breaking work. Hopkins' work on muscle metabolism led to an understanding of the importance of enzymes as catalysts to oxidation. He isolated glutathione, which plays an important role as an oxygen carrier in cells, and several oxidizing enzymes. Hopkins and his assistant, E. J. Morgan, made further contributions to this field in the late thirties.

Hopkins married Jessie Ann Stevens in 1898. They had three children. A gentle eccentric who struggled with feelings of insecurity about his worth as a scientist throughout his life, he considered himself an ''intellectual amateur,'' despite his achievements and the honors bestowed on him by his peers. Hopkins died at Cambridge on May 16, 1947.

HOPPE-SELYER, ERNST FELIX (1825-1895)

German biochemist

Ernst Hoppe-Selyer was one of the leaders in making **biochemistry** (or physiological **chemistry**, as it was called then) a scientific field distinct from medical physiology. He performed the first study of the **nucleic acids**, gave the name hemoglobin to the red blood cells, and discovered the **enzyme** invertase.

Ernst Hoppe was born in Freiburg-an-der-Unstrut, Germany. His father, a minister, and his mother died when he was a child. After he was adopted by his brother-in-law, he added Selyer to his name. He received his medical degree from the University of Berlin in 1851, then combined a medical practice with scientific research. Hoppe-Selyer's interest shifted gradually from physiology to chemistry. After serving on the faculties of the Universities of Berlin and Tübingen, in 1872 he became professor of physiological chemistry at the University of Strasbourg (then part of Germany). He established the first independent biochemistry laboratory in 1877 and the first biochemical journal.

Hoppe-Selyer's first important discovery came in 1862, when he used the newly invented spectrograph to determine the structure of the red blood cells, which he called hemoglobin. He later showed how hemoglobin binds and releases **oxygen** and how **carbon** monoxide can take oxygen's place in the blood cell. He also demonstrated some of the chemical similarities of hemoglobin and **chlorophyll**.

Hoppe-Selyer began studying the nucleic acids after they were discovered in 1869 by one of his students, the Swiss biochemist Johann Friedrich Miescher. Hoppe-Selyer showed that nucleic acids were present in yeast, and his work was extended by his one-time assistant, **Albrecht Kossel**.

Hoppe-Selyer's other research included the discovery in 1871 of invertase, the enzyme that converts sucrose (table sugar) into the simpler sugars glucose and fructose. He helped determine that lecithin is composed of **nitrogen**, **phosphorus**, fat, and choline (one of the B vitamins). And he demonstrated that lecithin and the steroid **cholesterol** are found in every cell.

HORMONES

Hormones are chemicals that naturally occur in the body. These chemicals direct many biochemical events, including **energy** storage and utilization, maintenance, and growth. Hormones are released mainly by the endocrine glands. Specific endocrine glands include the pituitary, thyroid, adrenals, parathyroids, gonads (e.g., testes and ovaries), and the islets of Langerhans which are located in the pancreas. The endocrine glands release very small quantities of hormones into the blood, but these small quantities have a large impact throughout the body.

From a chemical standpoint, hormones can be divided into three categories. The first category is made up of hormones that are derived from tyrosine, a type of amino acid. Specific hormones that belong to this category are epinephrine and norepinephrine. The second category of hormones are those that are derived from **steroids**, a class of chemicals based on **cholesterol**. Steroidal hormones include testosterone, estradiol, progesterone, cortisol, and aldosterone. The third category comprises hormones that are constructed from **peptides** and **proteins**. This category is by far the largest one and includes such hormones as insulin, glucagon, adrenocorticotropic hormone, and human growth hormone.

Most hormones are stored in the endocrine gland responsible for its construction. Steroidal hormones are the main exception to this pattern. These hormones are manufactured and released as needed. Regardless of whether a hormone is stored or not, endocrine glands rely on chemical signals from the body to start and stop hormone release, otherwise called secretion. Hormones are secreted in extremely small amounts. Their **concentration** in the blood is in the billionths or even the trillionths. Such small quantities are effective, however, because the hormones' targets are extremely specific. These targets are called hormone **receptors**. Each hormone in the body has a set of specific hormone receptors. Only these receptors will react to the presence of the hormone. The hormone receptors are located on the surface of target cells. If the hormone is present, it latches onto its receptor and triggers a biological response in the target cell.

The biological response depends on the hormone's identity and the target cell's purpose. As an example, glucagon, which is released by the islets of Langerhans, triggers a biochemical reaction in the liver. This reaction releases glucose, one of the main sugars serving the body's energy needs. The specific biochemical reaction to glucagon involves activating a particular **enzyme**. An enzyme is a type of protein that triggers speedy chemical reactions. The enzyme activated by glucagon is **glycogen** phosphorylase. In its active form, the phosphorylase breaks down glycogen, a substance composed of **chains** of glucose molecules.

In order for hormones to be effective, their activity must be tightly regulated. The body can only function when its **chemistry** is finely balanced. One of the main functions of hormones is to maintain this **balance**, often referred to as homeostasis. Once a hormone has done its job, there needs to be a system to stop the endocrine gland from secreting any more hormone. One such system is negative feedback. In the case of glucagon, a negative feedback system is based on the blood glucose level. Rising blood glucose levels indicate that glucagon is doing its job. When blood glucose levels have reached a certain point, the body does not need any more glucagon. The glucose itself signals the islets of Langerhans to stop the secretion of the hormone.

Since blood glucose levels move in both directions, however, there needs to be some means of lowering blood glucose when it exceeds the proper balance. Blood glucose levels normally rise following a meal. The need is met by insulin, which forms a closed negative feedback loop in partnership with glucagon. One of the actions of insulin, which is also secreted by the islets of Langerhans, is to activate a specific enzyme in the liver. This enzyme is glycogen synthase, which helps remove glucose from the blood and store it as glycogen. When blood glucose levels fall to the proper level, insulin secretion is switched off. Between glucagon and insulin, blood glucose levels remain stable, allowing the body to use its energy as efficiently as possible.

Certain hormones are not controlled by negative feedback, but through an opposite system called positive feedback. Positive feedback is based on hormones themselves triggering further secretion. This system is unusual and only comes into play with certain hormones. One such hormone is the steroidal hormone, oxytocin. Oxytocin has several uses in the body, one of which is to stimulate contractions of the uterus during childbirth. When the cervix, or opening of the uterus, begins to dilate (widen) in preparation for delivery, the brain signals the pituitary gland to secrete oxytocin. As the oxytocin circulates, the uterus contracts and cervix dilates further. The dilation triggers additional brain signals to the pituitary and even more oxytocin is released. This cycle continues until the contractions of the uterus are strong enough to force the baby through the birth canal to delivery.

Hormones remain in the blood for varying lengths of time. Their duration depends on their activity and how long it needs to last. Some hormones, such as epinephrine—the so-called fight-or-flight hormone—last only seconds in the bloodstream before breaking down. Other hormones, such as hormones secreted by the thyroid, endure for several days. The time delay between hormone secretion and biological response also varies.

Hormone secretion is a delicately balanced activity. Some systems require the coordination of more than one hormone to generate particular biological responses. The adrenal hormones are a notable example of this type of coordination. The adrenal glands, located near the kidneys, secrete many different hormones that fall into two categories: mineralcorticoids and glucocorticoids. Mineralcorticoids are important in maintaining a proper mineral balance in the body, while glucocorticoids are vital for managing **carbohydrates**.

Cortisol, also called hydrocortisone, is one of the most potent of the glucocorticoids. It is released in response to stress. Stress can arise from any situation in which a person feels nervous or afraid. One of the main functions of cortisol is to mobilize the body's energy sources to deal with the stressful situation. If cortisol levels remain high due to continuous stress, a person may suffer health effects such as lowered resistance to disease and high blood pressure. The secretion of cortisol is preceded by several events. First, there is a nerve impulse to the hypothalamus, an endocrine gland located near the brain. This nerve impulse causes the hypothalamus to secrete a ACTH-releasing hormone. This hormone stimulates the pituitary, another endocrine gland near the brain, to secrete adrenocorticotropic hormone (ACTH). The final step in the process occurs when the ACTH prompts the adrenal glands to manufacture and release cortisol.

ACTH is not the only hormone produced by the pituitary. Like other endocrine glands, the pituitary gland is able to produce several hormones, each of which has its own purpose. However, the amounts of each hormone are not equal. More than a third of the pituitary gland's cells are dedicated to manufacturing and secreting human growth hormone. Human growth hormone production increases sharply as a person enters adolescence. The hormone causes bones to become longer and organs to grow larger. It also alters the body composition, allowing for an increase in muscle and a relative decrease in fat. Once full growth is achieved, secretion of growth hormone drops to a lower level.

In the 1990s, scientists began studying a possible exception to the idea that certain hormone receptors only react to specific hormones. The exception is based on the molecular similarity between environmental contaminants and certain hormones. These environmental contaminants are referred to as endocrine disruptors, because they may have the potential to interfere with regular endocrine activity. The interaction of hormones and their associated receptors is highly specific. However, due to their resemblance to hormones, suspected endocrine disruptors may be able to interact with the receptors to some degree. The fit is not perfect, but the disruptors might be able to cause a partial response or possibly block the receptor from the proper hormone.

Whether there is a link between endocrine disruptors and health effects is controversial idea. Some studies suggest that endocrine disruptors in the environment may pose a serious health threat to both animals and people. In other studies, animals that have been exposed to endocrine disruptors have suffered reproductive effects such as lowered fertility or sterility. By the late 1990s, researchers were continuing to investigate suspected endocrine disruptors in the environment and the effects that these chemicals might have on human and animal health.

HOUDRY, EUGENE (1892-1962)

American chemist

Eugene Houdry was trained as a mechanical engineer in France and spent the first thirty-eight years of his life in that country. There he worked in his father's steel plant and ultimately became interested in the development of catalysts for use in a variety of industrial processes. In 1930 Houdry came to the United States and refined a process for the conversion of crude oil products to high-quality **gasoline** that he had first begun in France. Some observers credit the Houdry process of catalytic cracking with providing the Allies with a decisive technological edge that allowed them to win World War II. Later in life Houdry turned his attention to the development of catalysts that would reduce the amount of **pollution** produced by internal **combustion** engines.

Eugene Jules Houdry was born in Domont, France, near Paris, on April 18, 1892. His parents were the former Emilie Thais Julie Lemaire and Jules Houdry, owner of a successful steel manufacturing plant. As a young man, Houdry decided to continue his father's business and so entered the Ecole des Arts et Métiers in Paris to major in mechanical engineering. He graduated in 1911, earning a gold medal from the French government for receiving the best grades in his class.

After graduation Houdry entered his father's company as an engineer and became a junior partner. With the onset of World War I, he was drafted into the French Army, where he eventually became a lieutenant in the tank corps. During the battle of Juvincourt in 1917, he was severely wounded while directing the repair of damaged tanks. For his courageous actions, he was later awarded the Croix de Guerre.

Houdry's service in the tank corps was to have unanticipated long-term consequences for his own career and on the petroleum industry. His experience working with tanks encouraged a fascination with auto racing that in turn led to an interest in improving the quality of motor fuels. Although he returned to the steel manufacturing business at the end of World War I, he spent more and more time on his hobby—auto racing—and on research into automotive fuels.

In 1922 that research received an impetus from the French government. Faced with one of its recurring fuel shortages, the government asked Houdry to work on methods for converting fossil fuels such as bituminous **coal** and lignite, which France had in abundance, to a synthetic form of petroleum. By 1925 Houdry had accomplished that goal. His success was diminished, however, by the fact that the method he developed could not compete economically with products obtained from natural petroleum. Although his work was a failure in one respect, Houdry had become convinced while working on the project that research on fuels was his real passion. He gave up his job at the steel factory to devote his full time to research. His first objective was to find a way of converting crude oil to a high-quality gasoline.

Originally the term *gasoline* referred to one of the products obtained during the fractional **distillation** (purifying process) of petroleum. With the development of automobiles, this petroleum fraction became increasingly in demand. As internal combustion engines were improved, however, the "straight-run" gasoline obtained directly from fractional distillation proved to be a less and less satisfactory fuel. Chemists realized that modifications would have to be found that would increase the octane number (a measure of the efficiency of a fuel) of the gasoline obtained from petroleum.

The first widely successful method developed was called cracking. During the cracking process, crude petroleum is heated to high temperatures, causing large, saturated hydrocarbons (hydrogen-and-carbon-containing compounds containing single bonds) with low octane numbers to break apart into smaller, unsaturated hydrocarbons with higher octane numbers. Gasoline obtained by means of this "thermal" (for "**heat**") cracking process has octane numbers of about 72, about twenty points higher than the octane number of straight-run gasoline.

Eugene Houdry.

By 1927 Houdry had found that the efficiency of the thermal cracking process could be improved by the use of a catalyst. A catalyst is any material that changes the rate of a chemical reaction without undergoing any permanent change itself. The first catalysts used by Houdry were claylike materials made of silica and alumina. The "catalytically cracked" gasoline produced by the Houdry process had octane numbers of about 88. Later developments in the process raised that value to 100 or more.

Houdry tried to find an oil company in France that would finance the construction of a pilot plant utilizing the new catalytic process. He was without success, however, until 1930, when a U.S. firm, the Vacuum Oil Company, offered to finance continuation of his research. Houdry moved to the United States and formed the Houdry Process Corporation, one-third of which was owned by Vacuum Oil, one-third by the Sun Oil Company, and one-third by Houdry. By 1936 the first plant using Houdry's process had opened at Paulsboro, New Jersey, producing two thousand barrels a day of high-octane gasoline. Although other methods of improving the octane number in gasoline have since been developed, catalytic cracking has become a fundamentally important process in the petroleum industry. During World War II, for example, an estimated 90 percent of the high-quality fuel needed by airplanes was produced by the Houdry process.

Houdry became an American citizen in 1942, but never abandoned his native land. During World War II, he formed France Forever, an organization whose purpose it was to obtain support in the United States government for the Free French forces fighting under General Charles de Gaulle. Toward the end of World War II, Houdry turned his attention to a new problem, the release of carcinogenic materials in the exhaust **gases** from automobiles and industrial processes. He had become convinced that these pollutants were a major cause of lung cancer. Just a month before his death, Houdry was awarded a patent for a catalytic convertor for automobile exhaust systems of the type that is now standard on all U.S. automobiles.

Mr. Catalysis, as Houdry was often called, was married to Genevieve Marie Quilleret on July 1, 1922. The couple had two sons. Houdry died in Upper Darby, Pennsylvania, on July 18, 1962. Among the honors he received during his career were the Potts Medal of the Franklin Institute in 1948, the Perkin Medal of the Society of the Chemical Industry in 1959, and the Award in Industrial and Engineering Chemistry of the American Chemical Society in 1962.

HUBER, ROBERT (1937-)

German biochemist

The study of **photosynthesis**—the ability of plants, algae, and bacteria to translate sunlight into **energy** to build various chemical compounds—has long intrigued scientists, yet it is only since the 1950s that this process has begun to be understood in any detail. The analytic work of Robert Huber has played a significant role in the development of this understanding, and his most important achievement was matching the structure of a photosynthesizing protein complex to its function. Huber's work in **X-ray diffraction** enabled him and a co-worker to map the **atomic structure** of a bacterial photosynthetic reaction center—the basic unit or heart of the photosynthetic process. Such a description has helped advance not only photosynthesis research, but also various medical investigations. For his work in ''unraveling the full details of how [such a] protein is built up, revealing the structure of the **molecule atom** by atom,'' the Nobel committee awarded Huber and two other German co-researchers the 1988 Nobel Prize in **chemistry**.

Robert Huber was born on February 20, 1937 in Munich, Germany to Sebastian and Helen Kebinger Huber. His father was a bank clerk and the family had a hard time during World War II and the years following. In 1947 Huber entered the Humanistisches Karls-Gymnasium in Munich, a school with an emphasis on humanistic studies. In an autobiographical piece for *Les Prix Nobel,* Huber remembers the teaching of Latin and Greek as being ''intense,'' but it was here he developed an interest in chemistry. Few chemistry classes were available, so he taught himself ''by reading all the textbooks I could get.'' In 1956 he graduated from the gymnasium and entered the Technische Hochschule of Munich—later renamed the Technical University—to study chemistry. He graduated in 1960, and he married Christa Essig that same year. Various stipends and grants helped Huber and his growing family—they would have four children—through the years he spent as a graduate student in the **crystallography** laboratory of W. Hoppe.

As a graduate student, Huber worked with a number of prominent chemists, including Walter Hieber in the field of **inorganic chemistry**, **Ernst Fischer** who studied organometallic chemistry, and F. Weygand in **organic chemistry**. But it was crystallography that won Huber's interest. Though his thesis work for his 1963 doctorate was done on the crystal structure of a diazo compound, it was crystallographic studies on the insect metamorphosis hormone ecdysone that set him on the path of X-ray crystallography. Working with Hoppe at both the Technical University and at the Physiological-Chemical Institute of the University of Munich, Huber was able to determine the **molecular weight** and steroid nature of ecdysone. He employed X-ray diffraction techniques (where an X-ray beam is shot at a crystallized substance) to determine the atomic structure of ecdysone by analyzing how the beam was dispersed by the crystal. Huber was so impressed by the results he attained that he decided to concentrate on crystallographic research.

After determining the structure of several organic compounds and developing some improvements in existing **X-ray crystallography** methods, in 1967 Huber and H. Formanek set out to elucidate the structure of erythrocruorin, an insect protein. Their results showed a marked similarity between erythrocruorin and mammalian **proteins**. Their work also suggested for the first time that there might be a universal globin fold—the globin fold being the manner in which the chain of **amino acids** constituting the protein folds upon itself, endowing the protein with a shape specific to its function. In 1968, Huber became a lecturer at the Technical University, and three years later he accepted a position as a director at the prestigious **Max Planck** Institute for **Biochemistry** at Martinsried near Munich. He maintained his affiliation with the Technical University as well, becoming a full professor there in 1976.

Throughout the 1970s, Huber and his co-workers refined and perfected the techniques of X-ray crystallography, elucidating the structures of various proteins in collaboration with both foreign and domestic laboratories. His work in **enzyme** inhibitors and immunoglobulins has been of particular interest to researchers developing technologies for drug and protein design. Huber's laboratory at Martinsried became internationally recognized for the high quality of its work, and for Huber's delight in undertaking projects others thought impossible.

In 1982 a fellow researcher at Martinsried named **Hartmut Michel** came to Huber with a monumental task: to elucidate the atomic structure of the protein complex that powers photosynthesis in the purple bacteria, *Rhodopseudomonas viridis.* Michel had managed to isolate and crystallize a protein complex known as a membrane-bound protein, which is situated on the outer membrane of the bacterium. These proteins, made up of a tangle of four protein subunits and molecules of **chlorophyll**, help transport energy across the walls of cells. Yet they had been extremely difficult to isolate, because of their intermediary position on the membrane wall. Many believed these proteins were impossible to isolate, but by 1982 Michel had grown crystals of this protein complex, which functions as a photosynthesis reaction center. The reaction center is the place where electrons—released by a photon-excited chlorophyll molecule—create an electrical charge difference that produces the energy to power the synthesis of chemical compounds such as sugar, **carbohydrates**, and other nutrients.

Huber agreed to take on the task of developing a structural analysis of the proteins Michel had crystallized. Working with German biochemist **Johann Deisenhofer** at Martinsried and several other biochemists, his team used their improved X-ray crystallographic techniques to determine the exact atomic structure of the reaction center. By 1985 they had mapped over 10,000 separate atoms, and their structural analysis confirmed predictions as to the path that electrons follow in the reaction center. Though there are significant differences in the process of photosynthesis in green plants and in bacteria, the three-dimensional atomic model that Deisenhofer and Huber developed has proved to be of immense importance in further photosynthesis research in general. It has also been vital in research into the part that membrane-bound proteins may play in diseases such as cancer and diabetes. The work of the three main researchers in this project—Huber, Michel and Deisenhofer—was recognized by a joint award of the Nobel Prize for chemistry in 1988.

Huber has also been instrumental in developing computer models, such as FILME, PROTEIN, FRODO, and MADNESS to help in determining atomic structures through X-ray crystallography. Besides the Nobel Prize, Huber's work has been recognized by the E. K. Frey Medal from the German Society for Surgery in 1972, and the **Otto Warburg** Medal from the German Society for Biological Chemistry in 1977. He has received the Emil von Behring Medal from the University of Marburg in 1982, and the Keilin Medal from the London Biochemical Society and the **Richard Kuhn** Medal from the Society of German Chemists, both in 1987, as well as the Sir Hans Krebs Medal in 1992. He has also received numerous honorary doctorates and memberships in foreign chemical and biochemical societies.

HÜCKEL, ERICH (1896-1980)

German physicist

Erich Hückel was born in Charlottenburg, a suburb of Berlin, Germany, on August 8, 1896. He was the second of the three sons of Armand Hückel, a physician who was especially interested in **chemistry**. The father supervised the education of all his sons, and, in his private laboratory, he introduced them to chemistry, physics, and astronomy. Erich's critical, rational, and anti-metaphysical thinking probably originated from this influence of his father. Walter, the oldest son, became professor of **pharmaceutical chemistry** at the University of Tübingen and the author of a famous book entitled *Theoretische Grundlagen der organischen Chemie*. Rudi, the youngest son, became a physician.

Erich Hückel graduated from the Göttingen *gymnasium* (high school) and began to study physics and mathematics at the Göttingen University until the outbreak of World War I in 1914. During his military service Hückel worked in Ludwig Prandtl's aerodynamics laboratory in Göttingen. In 1919, he resumed his university studies, receiving his D. phil. degree in 1921 working under **Peter Debye** on the application of the Debye-Scherrer **X-ray diffraction** method to **liquid crystals**. According to Hückel, three Göttingen professors, who were or later became Nobel laureates, exerted the greatest influence on his career: Peter Debye, **Otto Wallach**, and **Richard Zsigmondy**, who was to become his father-in-law. At Göttingen, Hückel served as assistant to the mathematician David Hilbert in 1921 and physicist Max Born in 1922.

In 1922, Hückel began working as an assistant to Debye at the Eidgenössische Technische Hochschule (Swiss Federal Institute of Technology, ETH) in Zürich, Switzerland, where he remained until 1928. It was here that the two wrote their famous two joint articles in 1923 and 1924 on strong electrolytes, which could not be explained by the electrolytic **dissociation** theory of Swedish chemist **Svante August Arrhenius**. The second of these two articles served as Hückel's *Habilitationsschrift* (1925), a paper embodying the results of original and independent research which had to be accepted by the institute's faculty before he was granted the *venia legendi* or *venia docendi*, the privilege of lecturing as a *Privat-Dozent* (unsalaried lecturer) at a university. What became known as the Debye-Hückel theory was based on the simplifying assumption of an "ionic atmosphere" surrounding the ions that reduced their conductance and transport properties. In 1925, Hückel married Richard Zsigmondy's older daughter, Annemarie, by whom he had three sons and one daughter.

Between 1928 and 1929 Hückel worked as a Rockefeller Foundation fellow with Frederick G. Donnan in London, P. A. M. Dirac in Cambridge, and **Niels Bohr** in Copenhagen. In 1930, he worked with **Werner Heisenberg** and Friedrich Hund at Leipzig, where he began his most important achievement: the theory of the double bond and of unsaturated and aromatic organic compounds (those containing **benzene** rings). Far ahead of his time, Hückel may be considered the first modern quantum chemist. The Hückel molecular orbital (HMO) theory was first opposed by most chemists, but it is now the most widely used quantum model of π-electron delocalization.

Hückel's 1931 paper on predicting the stability of aromatic compounds, which is now known as "Hückel's (4n +2)p-electron rule" in **organic chemistry** textbooks, served as his second *Habilitationsschrift*, allowing him to become

Privat-Dozent in chemical physics at the Technical University of Stuttgart, but this irregular position provided him with little income. Hückel then became associate professor of theoretical physics in 1937, and finally full professor in 1961 at the University of Marburg, a year before his retirement. Although he was recognized early in his career, honors and awards came to him relatively late in his life of 83 years.

HUMAN GENOME PROJECT

The Human Genome Project began in October 1990. The goal of the project is to sequence the entire human genome (all the **DNA** in a cell). The potential benefits of this information are numerous. In the area of medicine, nearly 3,000 human diseases have been identified as having a genetic link. By sequencing the human genome, it may be possible to improve disease diagnosis, detect genetic diseases earlier, and develop new drugs and gene therapies. Other potential benefits include: accurately assessing risks due to chemicals and other environmental factors; increasing knowledge about human evolution and migration; identifying individuals who are crime or disaster victims; and establishing the guilt or innocence of suspected criminals.

The term genome refers to all of the genetic material contained in an organism. Genetic material is deoxyribonucleic acid, abbreviated DNA. DNA exists as a double-stranded, spiral-shaped **molecule**. The molecule is often called the double helix. The molecular building blocks of DNA are nucleotides. Each nucleotide is composed of deoxyribose (a type of sugar), a **phosphate** group, and one of four nitrogen-containing bases. The nitrogen-containing bases are adenine, thymine, cytosine, and guanine; these names are used to identify the nucleotides.

Each strand of DNA is built from a chain of the four nucleotides. The strands are complementary, which means there are set pairings between nucleotides at each position along the double helix. Adenine only pairs with thymine, and vice versa; cytosine only pairs with guanine, and vice versa.

To meet the goals of the Human Genome Project, it will be necessary to identify which nucleotide pair occupies each position along the double helix. The entire human genome contains approximately three billion base pairs. At the outset of the Human Genome Project, scientists estimated that it would take 15 years to sequence such a vast amount of information.

Scientists first began to consider the possibility of pursuing the project in 1986. In the United States, the National Institutes of Health and the U.S. Department of Energy drafted a plan for the Human Genome Project by 1988, and work began in 1990. Similar efforts were underway in other countries, such as France, Britain, and Japan. In order to coordinate these efforts, an international research group was created. This group is called the Human Genome Organization (HUGO) and its primary purpose is to coordinate the worldwide collaboration. As of the mid-1999, HUGO included more than 1,000 laboratories in 50 countries.

At the outset of the project, researchers concentrated on building a detailed map of each human **chromosome**. The human genome is divided into a total of 46 chromosomes, including the sex chromosomes, X and Y. Chromosomes are distinct bundles of genetic material which are organized into 23 matching pairs. Each parent contributes 23 chromosomes to a child. The end result is that the child has two copies of each chromosome, for a total of 46.

On each chromosome, the DNA is divided into units called genes. In general, each gene codes for a specific product, such as a protein. The human genome contains approximately 100,000 genes. All chromosomes carry a characteristic set of genes. For example, the gene that codes for the **enzyme**, glucose dehydrogenase, is always found on chromosome 1. It doesn't matter whose genetic material is examined; the gene for that enzyme will always be found on the same location on chromosome 1.

Genes and other well-characterized DNA segments are used as genetic markers. Genetic markers serve as sign posts over the entire genome. As of 1995, researchers had created a map of the genome that contained about 15,000 markers. However, this accomplishment is a small part of the overall sequencing effort; the average distance between markers is two million base pairs.

Since 1995, researchers have made great progress in identifying those base pairs, aided in large part by advances in laboratory techniques. Because of this success, it is estimated that a first draft of the human genome will be completed by the spring of 2000, five years ahead of schedule.

Alongside the scientific advances, ethical, legal, and social issues of genetic research have also been incorporated as a distinct research area within the Human Genome Project. Several areas of concern have been identified such as privacy and confidentiality, the potential consequences of genetic testing, and who should have access to individuals' genetic information. Other concerns include potential misuse, ownership and control, and fair use of the information.

The genetic techniques that were developed for the Human Genome Project have been applied to the Microbial Genome Project, started in 1994. This project focuses on the genomes of bacteria and opens possibilities in the areas of energy production and environmental clean-ups. In the area of agriculture, advances in genetic research may lead to crops with better resistance to disease, insects, and drought. Crops may also contain plants with more nutrients. Research is being done to genetically alter plants so that they contain vaccines that would protect people against certain diseases. Animal research is geared toward developing healthier and more productive farm animals.

The Human Genome Project is not the only group working on genetic sequencing, both of humans and of other organisms. In May 1998, J. Craig Venter, a scientist from the nonprofit Institute for Genomic Research (TIGR), described a new project on the human genome. It was a joint venture with the instrument company, Perkin-Elmer. He claimed he would complete the sequencing of the human genome by 2001 for a cost of $200-250 million. Although Venter plans to make information available to scientists, he will also patent 100-300 genes and charge for the use of databases. His technique is called whole-genome shotgun sequencing. First, the DNA is broken into thousands of fragments. Then, automated DNA sequencing machines read the read the base pairs at the ends of the various fragments. Finally, the information is fed into a computer that matches overlapping fragments. Although this technique leaves gaps, Venter maintains they are in repetitive areas of DNA that are not of interest. TIGR has already succeeded in sequencing the genome of a number of microorganisms, including the bacteria that cause syphilis, ulcers, and Lyme disease.

HUND'S RULE

Developed by the German scientist, Friedrich Hund (1896-1997), Hund's rule allows scientists to predict the order in which electrons fill an atom's suborbital shells. Hund's rule is based on the Aufbau principle that electrons are added to the lowest available **energy** level (shell) of an **atom**.

Around each atomic **nucleus**, electrons occupy energy levels termed shells. Each shell is identified with quantum number, n, that defines the main **energy level**. Each main level is made up of a number of sublevels. These sublevels are identified by their shapes: s sublevels have 1 orbital, p sublevels have 3 orbitals, d sublevels have 5 orbitals; and f sublevels have 7 orbitals. Each orbital can contain only 3 electrons spinning in opposite directions.

Although each suborbital can hold two electrons, the electrons all carry negative charges and, because like charges repel, electrons repel each other. In accord with Hund's rule, electrons space themselves as far apart as possible by occupying all available vacant suborbitals before pairing up with another **electron**. The unpaired electrons all have the same spin quantum number (represented in electron configuration diagrams with arrows all pointing either upward or downward).

The **Pauli exclusion principle** states that each electron must have its own unique set of quantum numbers that specify its energy. Accordingly, because all electrons have a spin of 1/2, each suborbital can hold up to two electrons only if their spins are paired +1/2 with -1/2. In electron configuration diagrams, paired electrons with opposite spins are represented by paired arrows pointing up and down.

For example, if there are three available p orbitals (p^x, p^y, p^z) the first three electons will fill these one at a time, each with the same spin. When the fourth electron is added, it will enter the (p^x orbital and will adopt the opposite spin since this is a lower energy configuration.

Although Hund's rule accurately predicts the electron configuration of most elements, exceptions exist, especially when atoms and ions have the opportunity to gain additional stability by having filled *s* orbitals or half- filled or filled *d* or *f* orbitals.

See also Atomic Structure; Bohr theory; Energy level; Quantum chemistry

HYBRIDIZATION

Hybridization is a model which is used to explain the behavior of atomic orbitals during the formation of covalent bonds. When an **atom** forms a **covalent bond** with another atom, the orbitals of the atom become rearranged. This rearrangement results in the "mixing" of orbitals. Hybridization occurs when two or more atomic orbitals of similar energies on the same atom mix together to form new orbitals. The new orbitals that are formed are all of equal energies. The number of new hybrid orbitals that are produced is the same as the number of original orbitals present in the atom before the covalent bond was formed. This hybridization of atomic orbitals occurs due to the perturbation of the orbitals as two atoms approach each other. The resulting hybrid orbitals have a combination of the properties of the original atomic orbitals present in the atom before the covalent bond was formed.

Electrons are located outside the **nucleus** of an atom in different **energy** levels, or orbitals. An atomic orbital is the region around the nucleus of an atom where an **electron** of a particular energy is most likely to be found. Atomic orbitals are sometimes referred to as "electron clouds." Atomic orbitals explain the electrons surrounding the nucleus of an unbonded atom, but do not explain those electrons involved in the bonds of a **molecule**. There are different kinds of atomic orbitals, each which their own characteristic shape, size, and energy. An atomic orbital is classified according to its shape and given a corresponding letter, either s, p, d, or f. An s orbital has a spherical shape, a p orbital has the shape of a dumbbell, while the d and f orbitals have extremely complicated shapes (too complex, in fact, to describe visually).

Atomic orbitals are not only distinguished by their shape. They are also distinguished by the particular **energy level** they occupy. An atom has a certain number of principal energy levels, designated by the quantum number "n." This is called the principal quantum number. This quantum number refers to the distance of the orbital from the nucleus of the atom.

Each principal energy level in the atom contains one or more sublevels. These sublevels divide the principal energy level into various parts. The sublevels exist in a very distinct pattern. The pattern of the sublevels is very important in understanding the atomic orbitals. The number of sublevels in each principal energy level differs from atom to atom. The quantum number, n, for each principal energy level designates the number of sublevels present. For example, if n=1, there is only one sublevel. If n=2, there are two sublevels. If n=3, there are three sublevels present in that principal energy level, and so on.

The sublevels are distinguished from one another by the quantum number and the letter (s, p, d, or f) which designates the shape of the orbital. For example, 2s means the s sublevel in the principal energy level two. 3d would mean the d sublevel in the principal energy level three. The sublevels are always present in the order s, then p, then d, then f. For the first energy level, n=1, there is only one sublevel (recall that the principal energy level designates the number of sublevels present). Therefore, the atomic orbital would be 1s. For the second energy level, there are two sublevels present. The atomic orbitals for this energy level would be designated 2s and 2p. The third energy level contains three sublevels. The atomic orbitals for the third energy level would therefore be designated 3s, 3p, and 3d. The fourth energy level has four sublevels: 4s, 4p, 4d, and 4f.

Before hybridization can be understood, the electron configurations of atoms need to be discussed. The electron configurations of atoms are how the electrons are distributed among the principal energy levels, sublevels, and orbitals. The electrons are distributed according to a set of standard principals that hold true for any atom. To determine the electron configuration of an atom, the electrons in the atom are added one at a time to the lowest energy orbitals available until all of the electrons of the atom have been accounted for. The idea of filling the energy levels from the lowest to the highest is called the Aufbau principle. Each orbital can only hold two electrons. These two electrons must be spinning in opposite directions in order to occupy the same orbital. When this occurs, and two electrons spinning in opposite directions are located in the same orbital, the electrons are "paired" with each other. If there is a single electron in an orbital it is "unpaired." The idea of paired and unpaired electrons is called the **Pauli exclusion principle**. The electrons occupy the energy orbitals in such a fashion that the number of unpaired electrons is at a maximum. This is called **Hund's rule**. If there are three orbitals available, and four electrons in the atom, the first three electrons would be placed one in each of the three orbitals. The fourth electron would then be placed in the lowest orbital, creating one orbital with paired electrons and two with unpaired electrons.

With this brief background of atomic orbitals, examples of hybridization can be illustrated. A good example of hybridization and its use to explain the geometry of molecular orbitals is the molecule **methane** (CH_4). Determining the electron configuration for a **carbon** atom shows that it has four **valence** electrons, or electrons occupying the outermost energy level. Carbon has one 2s orbital and three 2p orbitals. These four electrons are distributed such that there are two located in the 2s orbital and two located in the 2p orbitals. Experimental analysis has shown that methane exhibits a tetrahedral geometry, in other words, the carbon atom forms four equivalent covalent bonds with four **hydrogen** atoms. These covalent bonds are arranged tetrahedrally and are the result of orbital overlap between the carbon atom and the four hydrogen atoms. The idea of four equivalent covalent bonds seems impossible given the 2s orbital and the 2p orbitals have different shapes. In order to create four equivalent bonds, the 2s and 2p orbitals must hybridize with each other, in other words, mix equally with each other. The result of this hybridization is the formation of four new, identical orbitals. These orbitals are called sp^3 orbitals. The superscript 3 after the p indicates that there were three p orbitals involved in the hybridization. The idea that there was only one s orbital involved is understood by the lack of a superscripted number. The four sp^3 molecules all have the same shape and the same energy. The energy of the new hybrid orbitals is greater than that of the 2s orbital but less than that of the 2p orbitals.

Many different hybrid orbitals are possible. The result of the hybridization depends on the atom or atoms involved. The **molecular geometry** of covalently bonded substances can be predicted based on the type of hybrid orbitals that are formed. Hybrid orbitals that result in an sp designation are linear. An example of an sp hybrid is BeF_2. Orbitals that are designated sp^2 are trigonal-planar, for example, BCl_3. Hybrid orbitals that result in an sp^3 configuration are tetrahedral, given by the above example using CH_4.

The hybridization of orbitals is simply the mixing together of orbitals involved in covalent bonds to form new, equivalent orbitals. The hybrid orbitals are a mix of the shape and the energies of the original orbitals. Hybridization explains the molecular geometry of the resulting covalently bonded molecules.

See also Molecular orbital theory

HYDRATES

In the loosest sense, the term hydrate refers to the presence of **water** in compounds, where the water molecules may be bonded to positive or negative ions. Cases exist, however, in which water molecules occupy definite positions inside crystals without being coordinated with either positive or negative ions. And in noncrystalline materials such as colloidal gels, water may be present even in the absence of any chemical **bonding**. To add to the confusion, organic chemists sometimes refer to a compound formed by adding water to a carbon-carbon double bond as a hydrate; such compounds contain a hydroxyl **functional group** and usually cannot be dehydrated. And metallurgists frequently refer to a metal hydroxide as a hydrate; e.g., **calcium** hydrate is the same compound as calcium hydroxide.

By a more rigorous convention, however, a hydrate is defined as a **chemical compound** that contains water in a definite proportions, and in which the water **molecule** is not split. For example, in ionic crystals, the attraction between ions and water molecules can may lead to the incorporation of water molecules in the ionic crystal in the solid state. Such water molecules are usually referred to as water of hydration, and the crystals in which they are found are called hydrates. The crystal without water molecules is called anhydrous. Some anhydrous molecules became hydrated just by standing in air; such molecules are called hygroscopic.

The names of these hydrates are formed by adding a Greek prefix (corresponding to the number of water molecules in the compound's chemical formula) to the word "hydrate" and placing this designator behind the compound name. For example, $CuSO_4 \cdot 5H_2O$ is referred to as copper(II) **sulfate** pentahydrate, and $ZnSO_4 \cdot 7H_2O$ as **zinc** sulfate heptahydrate. By the same convention, gypsum (calcium sulfate dihydrate) and plaster of Paris (calcium sulfate monohydrate) can be recognized as hydrates of calcium sulfate.

Many binary ionic compounds such as **sodium** chloride and **magnesium oxide** melt without decomposing into pure **liquids**. Some of the more complex ionic **solids**, however, decompose when heated to give more than one pure substance. As a crystal, **barium** chloride assumes its hydrated form, $BaCl_2 \cdot H_2O$; at about 212°F (100°C), the hydrate loses its water molecule, forming anhydrous barium chloride. Another compound that forms more than one hydrate is sodium sulfate, which can combine with water to give sodium sulfate decahydrate, sodium sulfate heptahydrate, or sodium sulfate monohydrate. Upon exposure to dry air at room **temperature**, **copper** sulfate pentahydrate loses water to form the lower hydrates blue copper sulfate trihydrate and white copper sulfate monohydrate. At higher temperatures it forms anhydrous copper sulfate. (Anhydrous copper sulfate and copper sulfate pentahydrate have entirely different colors, densities, and crystal structures.) The loss of water at room temperature in dry air is known as efflorescence.

See also Crystallography; Metals; Organic chemistry

HYDROCARBONS

Hydrocarbons are compounds composed solely of **carbon** and **hydrogen**. Despite their simple composition, hydrocarbons include a large number of different compounds with a variety of chemical properties. Hydrocarbons are derived from oil deposits, and are the source of **gasoline**, heating oil, and other "fossil fuels." They also provide the carbon skeletons required for the thousands of chemicals produced by the chemical industry.

Hydrocarbons are classified on the basis of their structure and **bonding**. The three major classes are aliphatics, alicyclics, and aromatics. Aliphatics have carbon backbones that form straight or branched **chains**, with no rings. Alicyclics are **ring** compounds that, while they may have one or more double bonds, do not form conjugated sets of double bonds around the ring like **benzene**. Aromatics are compounds with at least one benzene ring. Aliphatic means fatty, and aromatic refers to odor, but these terms no longer have significance for the compounds they describe. All hydrocarbons are nonpolar. They are generally insoluble in **water**, and dissolve in nonpolar **solvents**.

The aliphatic hydrocarbons are further divided into alkanes, alkenes, and alkynes. Alkanes (sometimes called paraffins) have only single bonds, while alkenes (sometimes called olefins) have a carbon-carbon double bond, and alkynes have a carbon-carbon triple bond. Compounds with two double bonds are known as dienes. Compounds with double or triple bonds are referred to as "unsaturated," while those without are "saturated," meaning all of their carbons are bonded to the maximum number of hydrogens.

All alkanes have the general formula C_nH_{2n+2}, where "n" is the number of carbon atoms in the compound. Alkenes have the formula C_nH_{2n}, while alkynes are C_nH_{2n-2}. Because of these regularities, the members of each group are known as a homologous series.

The simplest alkane (and indeed, the simplest hydrocarbon) is **methane**, CH_4. The four C-H bonds are directed towards the four corners of a tetrahedron, with carbon at its center and hydrogens at each vertex. The C-H bonds are slightly polar, and equivalent in length and strength. The angle formed by any pair of bonds is 109.5 degrees, the "tetrahedral angle."

Despite the bond polarity, there is no net **dipole** because of the symmetry of the **molecule**, and methane has very weak intermolecular attractions, consisting only of van der Waals attractions. These attractions are proportional to surface area, which is small for the compact methane molecule, and as a result, methane has a very low melting point and **boiling point**, -297.4° F (-183° C) and -258.7° F (-161.5° C), respectively.

As are all bonds, the C-H bond in methane is formed by overlap of atomic orbitals from the two atoms. Carbon's outer-shell **electron** configuration is $1s^22p^2$. For carbon to form four equivalent bonds, it must combine these four atomic orbitals to form four hybrid orbitals, called sp^3 hybrids. Each hybrid orbital then overlaps with a hydrogen orbital to form a "sigma" bond. Sigma bonds are cylindrically symmetrical around the long axis connecting the two atoms.

Methane is found in oil deposits, forming the majority of the "**natural gas**" fraction. Methane is also formed by certain anaerobic bacteria, especially in swamp bottoms, where it bubbles to the surface as "marsh gas." It used in large quantities as fuel for heating and cooking, as it burns with **oxygen** to produce **carbon dioxide** and water, with very little soot or smoke. Methane can also be halogenated by free radical substitution.

The next alkane is ethane, C_2H_6. Its higher melting and boiling points (-277.6 [-172° C] and -127.3° F [- 88.5° C], respectively) reflect its larger surface area and consequently greater van der Waals attraction.

Unlike methane, ethane can exist in more than one **conformation**, or overall shape. The single bond joining the two carbons allows the two ends of the molecule to independently rotate about the carbon-carbon axis. The trio of hydrogens on each end can either line up with their opposite members to face them directly across the C-C bond, or, by rotating slightly, they can become staggered. Because it allows the hydrogen electron clouds to be slightly further apart, this staggered conformation is slightly more stable, by 3 kcal/mole, and is therefore the more prevalent conformation, especially at lower temperatures.

Propane, C_3H_8, is an important fuel because it can be liquified at pressures low enough to easily maintain in commercial and consumer apparatus, allowing easy transport and storage, but vaporizes to burn almost as cleanly as methane.

Higher alkanes continue the trend of increased surface area and higher melting and boiling points. However, as the number of carbons increase, structural isomerism becomes possible, allowing the same **molecular formula** to describe two or more compounds with different structures and different physical properties.

Butane, C_4H_{10}, for instance, can be either a straight-chain molecule, or a branched one, with three carbons in a straight chain and the fourth branching off from the middle carbon. The straight chain isomer is called n-butane (for "normal"), while the branched one is called either iso-butane, or 2-methyl propane. This latter name indicates that the longest straight-chain backbone within the molecule has three carbons (propane), and there is a single-carbon branch (methyl) at the second position in the main chain.

The extended structure of n-butane gives it a boiling point of 32° F (0° C), while the more spherical iso-butane boils twelve degrees lower, at 10.4° F (-12° C), due to its smaller surface area.

As the number of carbons increase, so too does the number of possible isomers. There are three isomers of pentane (C_5H_{12}), five of hexane (C_6H_{14}) nine of heptane (C_7H_{16}), eighteen of octane (C_8H_{18}), and by the time there are 20 carbons in the chain (eicosane), there are 366,319 possible isomeric structures. Stereochemical isomers are also possible with the alkanes, beginning with heptane.

The simplest alkene is **ethene**, C_2H_4, also called ethylene. Ethylene is a planar molecule, with the four hydrogens splayed out in a plane and angles between bonds of 120 degrees. Ethene melts at -272.2°F (-169°C, and boils at -151,6°F (-102°C).

The two bonds joining the carbons are not identical. One is a sigma bond, formed from sp^2 hybrid orbitals, while the other is formed from non-hybridized p orbitals, one from each carbon. The electron clouds from the two p orbitals overlap to form a pi bond, whose electron **density** is greatest above and below the plane formed by the atoms of the molecule. This arrangement prevents the molecule from spinning about the C=C axis, and as a result there is only one conformation of ethene.

Higher alkenes take their names from the corresponding alkanes: propene, butene, etc. However, beginning with butene, isomerism is possible based on the position of the double bond. If the bond is between C1 and C2, the compound is 1-butene, while if it lies between C2 and C3, it is 2-butene (the compound with a double bond between C3 and C4 is equivalent to 1-butene). In addition, the non-rotation of the double bond means there are two structural isomers of 2-butene, one with the two terminal carbons on the same side of the C=C long axis, termed *cis*-2-butene, and one with them on opposite sides, *trans*-2-butene. As might be expected, the number of possible isomers rises with an increase in number of carbons.

The **chemistry** of the alkenes is much richer than that of the alkanes, due to the presence of the double bond. The most common reaction is of addition across the double bond. For instance, addition of water to ethene creates ethyl **alcohol**. Alkenes are often prepared by the reverse of this reaction, **dehydration** across the double bond.

Dienes have two double bonds. Trienes, tetraenes, and so on are also possible. The most important class of these compounds are those in which the double bonds alternate with C-C single bonds. This creates a line of p orbitals down the length of the molecule, allowing **resonance** stabilization of the pi bonds. These systems are called conjugated dienes.

Alkynes have a triple bond. The simplest alkyne is C_2H_2, athyne, also called acetylene. Acetylene is an important high-temperature fuel used especially for metal cutting. The triple bond is composed of one sigma bond from overlap of sp hybrid orbitals, and two pi bonds, from overlap of two sets of p orbitals. The acetylene molecule is linear, with all four atoms in a straight line. The electron clouds of the pi bonds form a barrel shape along the carbon-carbon axis. Like alkenes, alkynes characteristically undergo addition.

The alicyclic compounds have one or more rings. The simplest compound is cyclopropane, C_3H_6, but its non-tetrahedral bond angles within the ring severely strain it, and it is very unstable. Cyclobutane (C_4H_8) is slightly more stable, and cyclopentane (C_5H_{10}) is a common industrial solvent.

The bond angles in cyclohexane (C_6H_{12}) are 109.5 degrees, the regular unstrained tetrahedral angle. To achieve these angles, the ring is not entirely planar, but instead bent into one of two configurations. In the ''boat'' configuration, two opposing carbons sit above the plane formed by the other four, while in the ''chair'' configuration, one is up while the other is down. The chair is slightly more stable, as it allows the two groups to be slightly further apart.

Polycyclic compounds also exist. The simplest is norborane (C_7H_{12}), formed from cyclohexane by bridging two opposing carbons with one more carbon that sits out of the main ring.

Cyclic alkenes, such as cyclohexene (C_6H_{10}), are common solvents, and also serve as starting points for a number of organic syntheses.

Benzene (C_6H_6) might be initially thought of as a cyclohexane with three double bonds alternating around the ring, but its properties are so importantly different from other alicyclic compounds that it defines its own class, the aromatics. Benzene is a flat molecule, with melting point of 41.9°F (5.5°C) and a boiling point of 176.2°F (80.1°C).

The characteristic that distinguishes benzene from the aliphatics and alicyclics, and that defines the aromatics, is the arrangement of electrons in the ring. The electrons in benzene's three pi bonds are highly delocalized, to the point that it is no longer accurate to describe them as between individual carbons. Instead, they form a cloud above and below the plane of the carbons, and as a result, all the carbon-carbon bonds are of equivalent length and strength.

Also unlike aliphatics, benzene and other aromatics do not generally undergo addition reactions. Instead, they undergo electrophilic substitution as their characteristic reaction, with hydrogens being replaced by electrophiles such as **halogens**.

HYDROCHLORIC ACID

Hydrochloric acid has the chemical formula HCl. It is an aqueous **solution** of **hydrogen** chloride, which when in a saturated form contains about 43% hydrogen chloride. Hydrogen chloride in aqueous solution is a strong electrolyte because it disassociates to form ions, H^+ and Cl^-. This disassociation is complete so no molecules of HCl are found. Hydrochloric acid is a **proton** donor, this is the hydrogen **ion**. When **ionization** occurs only one hydrogen ion is produced per **molecule** of HCl, which is an example of a monoprotic acid. Because ionization is so complete with hydrochloric acid it is regarded as a strong acid and the maximum number of hydrogen ions are freed up per molecule. Hydrochloric acid is used in a wide range of chemical reactions and conversions.

Hydrochloric acid can be used to clean metal surfaces and can be found naturally occurring in the stomachs of many animals as a relatively weak solution. Excess hydrochloric acid in the stomach can cause dyspepsia. It is also commonly encountered in the food industry. It is an extremely corrosive liquid and must be stored and handled appropriately. Because hydrochloric acid does not contain any organic molecules it is regarded as a mineral acid.

Hydrogen chloride is prepared by reacting **sodium** chloride (common table salt) with concentrated **sulfuric acid**. The products of this reaction are solid sodium hydrogen sulfide and gaseous hydrogen chloride. This gaseous hydrogen chloride can then be dissolved in **water**, by bubbling it through the water, to give hydrochloric acid. In the 1990s the annual United States production of hydrochloric acid was around 3 megatons.

The properties of hydrochloric acid are typical of any strong mineral acid. Hydrochloric acid is a corrosive substance, as such it can be used to clean metal surfaces. Any dirt or **oxide** is eaten away by the acid exposing pure metal. Hydrochloric acid when manufactured as explained earlier can have a **pH** value of 0.0, when encountered in the stomach the pH value is closer to 1.0, one tenth the strength. Hydrogen chloride as a gas is not acidic itself. It is a covalent compound and for acidic properties to be shown the molecules have to disassociate to the separate ions. For example if hydrogen chloride is dissolved in an organic compound such as methylbenzene no acidic properties are shown, unlike when it is dissolved in water.

When hydrochloric acid reacts with a base the **salt** of the base is produced. An example of this is potassium hydroxide and hydrochloric acid reacting together to give **potassium** chloride. Hydrochloric acid will neutralize a base in a chemical reaction. Hydrochloric acid reacted with a metal will dissolve the outer layer of the metal, cleaning it and evolving hydrogen gas in the process as well as producing the chloride salt of the metal. Some unreactive **metals** react only very slowly, such as **copper** and **lead**. More reactive metals such as sodium or **calcium** react very violently and it is not recommended to try these reactions. Hydrochloric acid will react with a chloride to produce **carbon** dioxide, water, and the salt of the chloride. For example **calcium carbonate** (chalk) will give water, **carbon dioxide**, and calcium chloride.

Hydrochloric acid is a typical mineral acid, undergoing all standard acid reactions. Because disassociation in water is so complete hydrochloric acid is a strong electrolyte, proton donor, and acid.

See also Acids and bases

HYDROGEN

Hydrogen is the first element in the **periodic table** and the first element in Group 1 of the table. It has an **atomic number** of 1, an atomic **mass** of 1.00794, and a chemical symbol of H.

Properties

Hydrogen is a colorless, odorless, tasteless gas with a **density** of 0.08999 grams per liter. By comparison, a liter of air weighs 1.29 grams, 14 times as much as a liter of hydrogen. Hydrogen's **boiling point** is -422.99°F (-252.77°C) and its freezing point is -434.6°F (-259.2°C). It is slightly soluble in **water**, **alcohol**, and a few other common **liquids**.

Hydrogen is relatively inactive at room **temperature**, but at elevated temperatures it reacts vigorously with many elements, include **oxygen**, **sulfur**, **phosphorus**, and the **halogens**. The reaction between hydrogen and oxygen results in the formation of water: $2H_2 + O_2 \rightarrow 2H_2O$.

Occurrence and Extraction

Hydrogen is the most abundant element in the universe. Nearly nine out of every ten atoms in the universe are hydrogen atoms. In space, hydrogen occurs in two forms, in stars and in the "empty" space between stars. In stars, hydrogen is the raw material used for the **nuclear reactions** by which stars make their **energy**. In the interstellar medium, hydrogen atoms occur at a very low density and very low temperature.

Hydrogen is also common on Earth. It is present in the atmosphere, hydrosphere, and lithosphere in the form of water and other compounds and is the third most abundant element on the Earth after oxygen and **silicon**. About 15% of all the atoms found on Earth are hydrogen atoms.

Three isotopes of hydrogen exist, hydrogen-1, hydrogen-2, and hydrogen-3. The three isotopes are sometimes known by the names of protium, **deuterium**, and **tritium**. Protium is by far the most common of the isotopes, making up about 99.9488% of all hydrogen in nature. Deuterium makes up an additional 0.0156%, while tritium accounts for the final small fraction of the hydrogen found naturally. Tritium is a radioactive element that decays by the loss of a beta particle to become a helium-3 **isotope**. The isotopes of hydrogen are of considerable research interest because their mass differences are so large. That is, deuterium has twice the atomic mass of protium, and tritium has three times the atomic mass of protium. Such mass differences are most unusual among other sets of isotopes.

The simplest way to prepare hydrogen is by electrolyzing water: $2H_2O$ —electric current:→ $2H_2 + O_2$. The cost of **electricity** is too high to make this process commercially feasible, however, and the element is usually prepared by some other method, such as the reaction between hot **charcoal** and steam: $H_2O + C \rightarrow CO + H_2$, or by the reaction between steam and hot methane: $H_2O + CH_4 \rightarrow CO + 3H_2$, or by the reaction between **carbon** monoxide and steam: $CO + H_2O$ —heat→ $CO_2 + H_2$.

Discovery and Naming

Hydrogen was probably "discovered" many times by early chemists, although none recognized the significance of his discovery. In 1671, for example, the English chemist **Robert Boyle** described experiments in which he added **iron** to **hydrochloric acid** (HCl) and to **sulfuric acid** (H_2SO_4) and got a gas that burned easily with a pale blue flame. The gas was obviously hydrogen.

The first person to appreciate the meaning of such experiments was the English chemist and physicist **Henry Cavendish**. Cavendish, like his predecessors, obtained a flammable gas when he added iron to acids, but he recognized the gas as a new element. The element was named not by Cavendish but by **Antoine-Laurent Lavoisier**, sometimes called the father of modern **chemistry**. Lavoisier suggested the name hydrogen after the Greek words meaning ''water former.''

Uses

The most important use for elemental hydrogen is in the production of **ammonia** (NH_3). Ammonia is made by combining hydrogen and **nitrogen** under high pressure and high temperature in the presence of a suitable catalyst. The reaction is known as the **Haber-Bosch process** after the two German scientists who developed it, **Fritz Haber** and **Karl Bosch**. The primary use of ammonia, in turn, is in the manufacture of synthetic **fertilizers**.

Hydrogen is also the starting point in the synthesis of other essential compounds. For example, it can be combined with **carbon monoxide** to produce **methanol** (methyl alcohol; CH_3OH) which, like ammonia, is used as a raw material in the production of many other important industrial compounds. Another use of hydrogen is in the manufacture of pure **metals**. Hydrogen gas is passed over hot metal oxides, reducing the **oxide** to the pure metal. For example: $MoO_2 + 2H_2$ —heat→ $2H_2O$ + Mo.

Hydrogen is also used by the food products industry to convert liquid oils to solid fats in a process known as **hydrogenation**. The purpose of the reaction is to convert oils into a form that makes them easier to pack and transport. Hydrogen is also used in oxyhydrogen and atomic hydrogen torches that produce flames with temperatures of a few thousand degrees. Such torches are used to cut through metals and to weld two metals to each other.

A once-important use of hydrogen is in lighter-than-air craft, such as dirigibles. This use was largely abandoned because of the dangers in handling gaseous hydrogen. The gas catches fire or explodes relatively easy, a property that led to some horrible airship disasters in the past. Improved methods of handling hydrogen may make it possible for this use to become important once more in the future.

Some people think that hydrogen may be the ''fuel of the future.'' As the world's supplies of fossil fuels diminish, some other fuel will have to be found to take their place. Hydrogen could be that fuel. It burns with a very hot flame, producing only water as a waste product. Imaginative inventors foresee buildings being heated by hydrogen stoves, cars being driven by hydrogen engines, and industrial operations being powered by hydrogen furnaces. The primary factor preventing the ''hydrogen economy'' from becoming a reality is the high cost of producing pure hydrogen gas in the first place. If and when that problem is solved, service stations of the future may be dispensing hydrogen gas rather than **gasoline** to keep the nation's economy on the move.

Millions of hydrogen compounds are known. In fact, hydrogen forms more different compounds than any other element. For example, nearly all organic and biochemical compounds containing hydrogen. One of the most important groups of hydrogen compounds is the acids. Each year, acids rank near the top of the list of the top 10 chemicals produced in the United States. Among the most important industrial acids are hydrochloric acid (HCl), sulfuric acid (H_2SO_4), **nitric acid** (HNO_3), **phosphoric acid** (H_3PO_4), **acetic acid** ($HC_2H_3O_4$), and hydrofluoric acid (HF). These acids are used directly for cleaning metals and other purposes as well as being used as the first step in many different synthetic chemical reactions.

$$H_2C{=}CH_2 \xrightarrow[\text{Catalyst}]{H_2} H_3C{-}CH_3$$

Hydrogen (simple double bond). ***(Illustration by Electronic Illustrators Group.)***

See also Acids and bases

HYDROGEN BOND

A **hydrogen** bond results from an interaction between two electronegative atoms (usually **oxygen**, **nitrogen**, or **fluorine**) and a hydrogen **atom** which is covalently bonded to one of them. **Electronegativity** is a measure of the ability of an atom to attract and hold electrons. It can be expressed as an average of the **ionization** potential (the minimum amount of **energy** necessary to remove an **electron** completely from the outer electron shell of an atom) and the electron affinity of the atom (the energy given off when a neutral atom acquires an electron).

A **covalent bond** between two atoms is formed by each atom contributing an electron to a molecular orbital that includes both atoms. The two atoms forming the bond share these two electrons. The shared electrons in a covalent bond between a highly electronegative atom (such as oxygen, nitrogen, or fluorine) and a hydrogen atom are located much closer to the more electronegative atom because it can attract these electrons more readily than the hydrogen. There is a resultant partial negative charge on the more electronegative atom and an equal and opposite positive charge on the hydrogen. A **polar bond** results, with the hydrogen at the positive end of the **dipole**.

When two polar molecules are in proximity, the positive end of one will be drawn to the negative end of the other, resulting in an attractive dipole-dipole interaction. When two molecules containing a polar bond between a hydrogen atom and a highly electronegative atom come close to each other, the hydrogen atom at the positive end of the dipole in one **mol-**

$$HC \equiv CH \xrightarrow[\text{Catalyst}]{H_2} H_2C = CH_2 \xrightarrow[\text{Catalyst}]{H_2} H_3C - CH_3$$

Hydrogen (triple bond). ***(Illustration by Electronic Illustration Group.)***

ecule is attracted to the electronegative atom at the negative end of the dipole in the other molecule. The hydrogen is, therefore, located between two electronegative atoms and is attracted to both of them. One of these interactions is in the form of a polar covalent bond; the other is electrostatic in nature. This arrangement is called a hydrogen bond.

The grouping of atoms in the hydrogen bond is more stable than the usual dipole-dipole interaction. This results from the fact that both hydrogen and the electronegative atoms that form strong hydrogen bonds (oxygen, nitrogen, and fluorine) are small. Thus, they can approach each other more closely than larger atoms before repulsion among the atoms becomes important. Since the electrostatic attraction between two equal and opposite charges is inversely proportional to the square of the distance between them, the attraction is greater when the distance is less. The hydrogen bond is, therefore, more stable than dipole-dipole interactions involving larger atoms that cannot approach each other as closely.

Molecular orbital theory also offers an explanation for the observed increase in the stability of hydrogen bonds compared with other dipole-dipole interactions. The two electronegative atoms involved in a hydrogen bond are not bonded covalently; they are non-covalent, electrostatic partners. Atoms of oxygen, nitrogen, and **fluorine** all have lone pairs of electrons, e.g., they all have exposed outer shell atomic orbitals that contain two electrons and are not participating in a molecular orbital. The overlap of a hydrogen 1s orbital that is almost empty (due to the shift of its electron toward the electronegative atom to which it is bonded) with the **lone pair** orbital of the non-bonded electronegative atom could result in a molecular orbital. As expected, the bond formed in this way is weaker than the usual covalent bond. While approximately 100 kcal **mole**$^{-1}$ of energy is required to break a normal covalent bond, a hydrogen bond can be disrupted with about 5 kcal mole^{-1}. In any case, although the hydrogen bond is primarily a result of an electrostatic dipole-dipole interaction, some added stability may be gained through this orbital overlap.

Low **molecular weight** molecules generally have low boiling points and most of them are **gases** at room **temperature**, even when they are polar. This is not true of molecules such as **water** (H_2O), **ammonia** (NH_3), hydrogen fluoride (HF), and **methanol** (CH_3OH) that contain hydrogen and a relatively small electronegative atom that are capable of forming hydrogen bonds. It is instructive to compare boiling points for several series of molecules made up of hydrogen and one other element.

The boiling points of molecules formed between the elements of group VIA and hydrogen are as follows: H_2Te, 24.8°F (-4°C); H_2Se, -43.6° F (-42°C); H_2S, -79.6°F (-62°C); and H_2O, 212°F (100°C). The boiling points decrease as the size of the electronegative atom decreases from hydrogen telluride to hydrogen sulfide. Then a major exception occurs: rather than decreasing further, the **boiling point** of the next member of the sequence, water, takes a large jump upward. The added attraction of the hydrogen bonds formed in liquid water makes it more difficult to break the molecules apart to release them into the gas phase.

A similar result is observed for molecules formed between hydrogen and elements of group VIIA. The boiling points are: HI, 248°F (120°C); HBr, -88.6°F (- 67°C); HCl, -119.2°F (-84°C); and HF, 66.2°F (19°C). Again, there is a decrease in the boiling points as the atomic size of the electronegative atom decreases from hydrogen iodide to hydrogen chloride. But again the boiling point increases abruptly for the molecule containing the smallest atom of the series, hydrogen fluoride. Like oxygen, fluorine forms hydrogen bonds that add stability to the liquid phase and require additional energy to disrupt.

The surface tension of a liquid is due to the **intermolecular forces** between molecules on the surface and molecules on the interior of the liquid. When there are stronger intermolecular forces between the molecules of a liquid, the surface tension should be greater. Indeed the surface tension of hydrogen-bonded **liquids** is greater than would be expected if such a phenomenon did not occur.

The **viscosity** of a liquid is the resistance it shows to the passage of a foreign body through it. The object passing through must break the intermolecular forces that hold the molecules together in the liquid state. Liquids with stronger intermolecular forces will show greater resistance. Consequently, liquids that form strong hydrogen bonds have significantly higher viscosities.

It is also found that when a hydrogen atom forms a hydrogen bond with an ancillary atom, the covalent bond with its principal partner is weakened. This should be expected since the hydrogen is pulled toward the ancillary electronegative atom, resulting in less participation by the hydrogen atom in the principal covalent **bonding** orbital. This effect can be observed in infrared spectra. Transitions between molecular vibrational energy levels occur in this region of the spectrum, and the energy values of these transitions are directly related to the strength of the vibrating bond (e.g., it takes more energy to stretch a stronger bond). It is observed that the radiation frequency (and, therefore, the energy) that is necessary to increase the level of vibrational energy of the O-H bond is greater if the molecule containing this bond is dissolved in a nonpolar solvent which is incapable of forming a hydrogen bond with O-H than it is if the molecule is dissolved in a solvent capable of forming hydrogen bonds, such as water. The hydrogen bond between OH and water weakens the O-H bond, and therefore, less energy is required to stretch it in a vibration.

Hydrogen bonds may be formed between two different molecules (intermolecular) or between two electronegative atoms of the same molecule (intramolecular). The examples that have been discussed are examples of intermolecular hydrogen bonds. The **conformation**, (e.g., the three dimensional structure) of **proteins** is, to a large extent, the result of intramolecular hydrogen bonding. Hydrogen bonds between different parts of the same protein molecule hold it in a stable spatial arrangement so that reactant molecules can approach the protein's active sites and fit onto them. Hydrogen bonds can be broken by **heat**. In proteins, this process results in denaturation. Cooking an egg is a common example of the denaturation of a protein by heat.

One of the most interesting examples of hydrogen bonding occurs in the deoxyribonucleic acid (DNA) molecule. The double helix nature of this molecule is the result of hydrogen bonds established between its two component **chains**. The fact that these bonds are strong enough to provide stability but weak enough so that the strands of the **DNA** molecule can separate relatively easily is essential for the activity of this most important genetic agent. The molecule must literally unzip itself, breaking all of its hydrogen bonds before it can replicate itself.

Hydrogen bonding allows molecules to display exceptional behavior. Because of this, water is a liquid at normal temperatures on Earth and in the bodies of animals, including humans. Because of this, the basic mode of regeneration of plants and animals, including humans, through nucleic acid replication is possible. Because of this, the chemical processes in plants and animals, including humans, can proceed through catalytic facilitation by proteins. Life as we know it would not be possible without hydrogen bonds.

See also Bonding; Intermolecular forces

Hydrogen peroxide

Like **water** (H_2O), **hydrogen** peroxide is a compound of hydrogen and **oxygen**. However, as the its formula (H_2O_2) indicates, water are quite different. At room **temperature**, hydrogen peroxide is a pale-blue, syrupy, weakly acidic liquid. Pure, it is a stable compound, but any impurities, such as dust, may cause an explosive decomposition to oxygen and water. Since this explosive reaction is always a possibility, hydrogen peroxide is stored in bottles that allow the release of oxygen.

Typically, hydrogen peroxide is an oxidizing and reducing agent. As an oxidizing agent, it is used to oxidize particular water pollutants, such as hydrogen sulfide (H_2S). For example, when hydrogen sulfide is exposed to hydrogen peroxide, it oxidizes to **sulfate ion** (SO_4^{2-}), which is not toxic.

In everyday life, hydrogen peroxide (a 3% solution) is used as an antiseptic. When hydrogen peroxide is applied to a superficial wound, it is reduced to oxygen and water by the action of a particular **enzyme**, catalase, which is found in blood and tissue.

Used as a hair bleach, hydrogen peroxide is also effective for bleaching **paper** pulp and textiles. It is also used in the manufacture of rocket fuels.

Hydrogenation

Hydrogenation is a chemical reaction in which **hydrogen** atoms add to carbon-carbon multiple bonds. In order for the reaction to proceed at a practical rate, a catalyst is almost always needed. Hydrogenation reactions are used in many industrial processes as well as in the research laboratory, and occur also in living systems.

Hydrogen gas, H_2, can react with a **molecule** containing carbon-carbon double or triple bonds known as an unsaturated compound. In its simplest form, a molecule with one double bond would react with one molecule of hydrogen gas. An example is shown in figure 1.

Some **carbon** compounds have triple bonds and need two molecules of hydrogen to completely saturate the carbon compound with hydrogen (figure 2).

Hydrogenation of a double or triple carbon-carbon bond will not occur unless the catalyst is present. Scientists have developed many catalysts for this kind of reaction. Most of them include a heavy metal, such as **platinum** or **palladium**, in finely divided form. The catalyst adsorbs both the carbon compound and the hydrogen gas on its surface, in such a way that the molecules are arranged in just the right position for addition to occur. This allows the reaction to proceed at a fast enough rate to be useful.

Because at least one of the reagents (hydrogen) is a gas, often the reaction will occur at an even faster rate if it occurs in a pressurized container, at a pressure several times higher than atmospheric pressure.

The hydrogenation reaction is a useful tool for a scientist trying to determine the structure of a new molecule. The **molecular formula**, showing the exact number of each kind of **atom**, can be determined in several ways, but discovering the arrangement of these atoms requires a large amount of detective work.

Sometimes, for example, a new substance is isolated from a plant, and a chemist needs to determine its structure. One method of attack is to find out how many molecules of hydrogen gas will react with one molecule of the unknown substance. If the ratio is, for example, two molecules of hydrogen to one of the unknown, the scientist can deduce that there are two carbon-carbon double bonds, or else one carbon-carbon triple bond in each molecule. Other kinds of chemical clues lead to the rest of the structure, and help the scientist to decide where in the unknown molecule the multiple bonds are located.

One of the simplest uses of hydrogenation in the research laboratory is to make new compounds. Almost any organic molecule that contains multiple bonds can undergo hydrogenation, and this sometimes leads to compounds that were unknown before. In this way, scientists have synthesized and examined many molecules not found in nature, or not found in sufficient quantity.

Many of the carbon compounds found in crude petroleum may contain multiple bonds, but can be converted to saturated compounds by catalytic hydrogenation. This is one source for much of the **gasoline** that is used today. Other chemicals besides gasoline are made from petroleum, and for these, too, the first step from crude oil may be hydrogenation.

Another commercial use of the hydrogenation reaction is the production of fats and oils in more useful forms. Fats and oils are not hydrocarbons, like the simple molecules previously mentioned, since they also contain **oxygen** atoms. However, they do contain long **chains** of carbon and hydrogen, joined together in part by carbon-carbon double bonds. Partial hydrogenation of these molecules, so that some, but not all of the double bonds react, creates compounds with different cooking characteristics. Partially hydrogenated vegetable oil is an example of such a product.

Many chemical reactions within the body require the addition of two atoms of hydrogen to a molecule in order to maintain life. These reactions are more complex because hydrogen gas is not found in the body. These kinds of reactions require ''carrier'' molecules that relinquish hydrogen atoms to the one undergoing hydrogenation. The catalyst in biological hydrogenation is an **enzyme**, a complex protein that allows the reaction to take place in the blood, at a moderate **temperature**, and at a rate fast enough for **metabolism** to continue.

Hydrogenation reactions can happen to many other types of molecules as well. However, the general features for all of the reactions are the same. Hydrogen atoms add to multiple bonds in the presence of a catalyst, to product a new compound, with new characteristics. This new compound has different properties than the original molecule.

HYDROLYSIS

Hydrolysis is a chemical reaction in which **water** reacts with another substance to split it into two or more new substances. Examples include the conversion of starch to glucose in water under the action of a suitable catalyst; conversion of sucrose to fructose and glucose in water in the presence of an acidic or **enzyme** catalyst; conversion of fats into **fatty acids** and **glycerol** in water in soap manufacture; and reactions of dissolved salts in water to form various products. Hydrolytic reactions are important in many biochemical processes, particularly in enzyme catalyzed reactions such as those used in digestion.

In each reaction, a water **molecule** separates into H^+ and OH^-, and the compound being hydrolyzed splits into two fragments, A^+ and B^-,where AB represents the original compound. The products of hydrolysis are AOH and HB. If, for example, ordinary **sodium** chloride (table salt) is placed in water, some of the **salt** will form sodium and **chlorine** ions, and these ions will combine with the **hydrogen** and hydroxide ions in water to form ionized **sodium hydroxide** (Na^+ + OH^-) and ionized hydrogen chloride (H^+ + Cl^-). Sodium hydroxide and hydrogen chloride essentially exist as ions in **solution** because they are a strong base and strong acid, respectively. Therefore, they are neutralized by the hydroxide and hydrogen ions, and the resultant solution is neutral.

The solution formed by hydrolysis may be acidic, basic, or neutral. In the case of **sodium carbonate** dissolved in water, the sodium ions react with the hydroxide ions of water to form ionized sodium hydroxide, but the hydrogen ions and carbonate ions form a weak acid that is only slightly ionized. The resultant solution is basic. When **zinc** chloride is placed in water, the hydrogen ions and chlorine ions form a strong acid, while the zinc and hydroxide ions form a weak base. In this case, an acidic solution is produced. Finally, if ammonium carbonate is placed in water, the weak base ammonium carbonate and weak acid carbonic acid are produced. Because there is neither an excess of hydrogen nor hydroxide ions, the resulting solution is neutral.

In the hydrolysis of carboxylic esters of the form R-C(=O)- OR', the carboxylic ester reacts with water under strong acid catalysis to produce the carboxylic acid R-C(=O)-OH and **alcohol** or phenol R'OH from which the ester was originally formed. Esters can also be hydrolyzed with a basic catalyst, in which case the carboxylic ester converts into a carboxylate **ion** R-C(=O)-O- and an alcohol. (Following the hydrolysis, the carboxylate ion can be converted into a free acid by adding hydrogen chloride.) The advantage of the basic catalysis over the acidic catalysis is that the former is a non-reversible. This is because the carboxylic acid and alcohol produced under acid catalysis can recombine to give the ester, but no recombination is possible when the carboxylate ion is formed under basic conditions. The base- catalyzed hydrolysis of carboxylic esters is referred to as *saponification*. Although, the human body contains many carboxylic esters, which are constantly being formed and hydrolyzed, these reactions are neither acid- nor base- catalyzed, being rather catalyzed by enzymes.

The case of the hydrolysis of amides (having the form R- C(=O)-NR_2') is similar to ester hydrolysis in that the hydrolysis requires a strong base or acid, but different in that neither the acid-catalyzed nor base-catalyzed reaction is reversible.

See also Amide group; Carbon dioxide; Carbohydrates; Ester functional group; Neutralization; Phenol functional group

HYDRONIUM ION

The Hydronium **ion** is an important participant in the chemical reactions that take place in aqueous (water, H_2O) solutions.

Through a process termed self-ionization, a small number of **water** molecules in pure water dissociate (separate) in a reversible reaction to form a positively charged H^+ ion and a negatively charged OH^- ion. In aqueous **solution**, as one water **molecule** dissociates, another is nearby to pick up the loose, positively charged, **hydrogen proton** to form a positively charged Hydronium ion (H_3O^+).

Water molecules have the ability to attract protons and form Hydronium ions because water is a polar molecule. **Oxygen** is more electronegative than hydrogen. As a result, the electrons in each of water's two oxygen-hydrogen bonds to spend more time near the oxygen **atom**. Because the electrons are not shared equally—and because the bond angles of the water molecule do not cancel out this imbalance—the oxygen atom carries a partial negative charge that can attract positively charged protons donated by other molecules.

In a sample of pure water, the **concentration** of hydronium ions is equal to $1x10^{-7}$ moles per liter (0.0000001 M). The water molecule that lost the hydrogen proton—but that kept the hydrogen **electron**— becomes a negatively charged hydroxide ion (OH^-).

The equilibrium (balance) between hydronium and hydroxide ions that results from self-ionization of water can be disturbed if other substances that can donate protons are put into solution with water. Hydrochloric Acid (HCl), for example, is an important acid found in the human stomach transfers a proton to water to form hydronium ions.

The potential hydronium ion concentration (pH) is equal to the negative logarithm of the hydronium ion concentration: **pH**=-log [H_3O^+].

See also Acid-base reactions; Acids and Bases.

HYDROPHOBIC EFFECT

In aqueous solutions, globular **proteins** usually turn their polar groups outward toward the aqueous solvent, and their nonpolar groups inward, away from the polar molecules. This is because the nonpolar groups prefer to interact with each other, and exclude **water** from these regions. These interactions are called hydrophobic, meaning water-hating, interactions. Hydrophobic interactions are usually weaker than **hydrogen** bonding, and act over large surface areas.

Both alcohols and ethers with up to three or four carbons are soluble in water because the OH groups in these molecules form hydrogen bonds with the water molecules. Alcohols and ethers having higher molecular weights do not dissolve well in water, however, because the water molecules can not completely surround those molecules. The molecule 1-heptanol, for example, consists of an alkyl chain of seven carbons and an OH group. The OH group forms hydrogen bonds with water molecules, but the alkyl portion of the molecule exerts no attraction on the water molecules. Thus, the alkyl portion of the molecule is called hydrophobic. Because this part of the molecule cannot be surrounded by water in an energetically favorable interaction, 1-heptanol in insoluble in water.

The **chemistry** of washing clothes provides an illustrative example of hydrophobicity. With water being a polar molecule, and dirt (such as fats and oils) being nonpolar and water-insoluble, one might expect the world to be a lot dirtier place than it is. The solution to this dilemma is provided by soaps, which are salts of long-chain **fatty acids**. The alkyl component of these molecules contains 12-18 carbon atoms, is completely nonpolar, and is therefore soluble in fats and oils but not in water. The other end of the molecule is a carboxylic acid **salt** and very polar, and thus is water soluble. Therefore soap contains a hydrophilic (water-loving) end and a hydrophobic end. By simultaneously dissolving in oils and water, soaps are able to remove oil and fats from dirty clothes and retains them in the water.

Considered microscopically, as the soap dissolves in water, the molecules orient themselves on the water's surface with the polar end submerged and the nonpolar **alkyl group** left bobbing above the surface. In this manner, the hydrophilic end is in contact with water, and the hydrophobic end sits out of the water. If more soap is added to the water than can conveniently orient in this manner, the soap molecules form three-dimensional clusters (called micelles), with the nonsoluble hydrophobic portions of the molecules filling the insides of the clusters, and the water-soluble (or hydrophilic) groups on the outside. If soiled clothing is dropped into the soapy water, the oils and fats are dissolved in the nonpolar centers of the micelles that remain suspended in the water (due to the repulsion between the charged outer surfaces of these clusters). The fats and oils are finally washed away as finely dispersed oil droplets.

Synthetic detergents act in much the same way as soaps. They consist of long nonpolar alkane chains with polar groups (usually sodium sulfonates or sodium sulfates) at their ends. The advantage of detergents over soaps is that the former function well in hard water containing **calcium**, **iron**, and **magnesium** ions. Because the salts of these ions and the alkane sulfonates and alkyl hydrogen sulfates are largely water soluble, the detergent remains dispersed in the water. In soaps, however, these ions would form precipitates.

See also Lipids; Polar bond; Soap and detergents

I

IDEAL GASES

An ideal or perfect gas is one which obeys the **equation of state**, PV = nRT, where T is the absolute **temperature** in Kelvin degrees (equal to the temperature in Celsius degrees plus 273 degrees) and R is the universal gas constant (equal to 0.08 when the **volume** V is in liters, the pressure P is in atmospheres, and n is the number of moles of the gas present; or equivalently, 8.3 joules per degree **mole**.)

This equation resulted from the examination of the behavior of air by **Robert Boyle**, **Jacques Charles**, and **Joseph-Louis Gay-Lussac**. Based on experimental measurements, Boyle showed that the product of the pressure and the volume of a sample of air, at constant temperature, is always equal to the same value. In other words, when the pressure placed on the sample is doubled, the volume is reduced to half its initial value. The work of Charles in 1787 and of Gay-Lussac in 1808 demonstrated that the volume of a given amount of air is directly proportional to the temperature, if its pressure remains constant; i.e. when the temperature of the sample is doubled the volume will double, if the pressure on the sample is kept constant.

Amadeo Avogadro proposed in 1811 that equal volumes of all **gases**, at the same pressure and temperature, contain the same number of molecules. In other words, an equal number of molecules (and, therefore an equal number of moles) of all gases occupy the same volume at constant temperature and pressure. This proposal, combined with the relationships discovered by Boyle, Charles, and Gay Lussac led to the equation of state for gases: PV=nRT.

In their work which led to this equation, Boyle, Charles, and Gay-Lussac certainly thought they were deriving quantitative relationships for actual, real gases. However, as subsequent studies were performed on a variety of gases over wide ranges of pressure, volume, and temperature and more precise measurements were made, it was discovered that, although the behavior of real gases generally closely follows the ideal gas equation under circumstances of low pressure and reasonably high temperature, significant deviations from the ideal expectations do occur. Rather than abandon the PV = nRT equation, scientists chose to keep the equation, refer to it as the equation of state for ideal or perfect gases (or, simply, as the ideal gas law), and use it as a first approximation when dealing with gases. They also proceeded to explore the reasons why gases do not always behave ideally and obey this equation. Studies of the departures from ideal behavior by gases have led to a better understanding of atoms and molecules and their interactions in the gaseous state.

This is not an unusual practice in science. In fact, the progress of science depends on the discovery of exceptions to quantitative relationships, such as the ideal gas law, and experimental determination of equations and explanations that fit the observed behavior more consistently. Examples include Newton's laws, which were discovered not to be applicable to atomic and molecular systems. For such systems, quantum mechanics fits the observed phenomena better and is a greater aid to our understanding. The classical mechanics of Newton has not been abandoned, however. Its elegant, simple equations work quite well for macroscopic systems, and the added complexity of quantum mechanics is not needed.

The kinetic-molecular theory of gases was developed by **Ludwig Boltzmann**, **James Clerk Maxwell**, and Rudolf Clausius. This theory explains the behavior of ideal gases by assuming that gases are made up of a collection of molecules moving about randomly in the container, colliding randomly and elastically with each other and with the walls of the container. If the molecules are assumed to be point particles with zero volume, the ideal gas equation is derived from this model. The pressure is due to the collision of gas molecules with the container walls, and the temperature is related to the kinetic **energy** of the molecules. An examination of this ideal kinetic model reveals reasons why real gases should be expected to depart from ideal gas behavior. Gases are not point particles: they occupy space. Collisions between molecules (and with the container wall) are not elastic: they involve the exchange of energy as molecular kinetic translational energy is changed into internal molecular energy such as the vibration of bonds, the rotation of the **molecule**, and even electronic energy. Although this model provides basic insight into the behavior of gases, a more accurate consideration of the behavior of real gases is needed.

The most widely used modification of the kinetic- molecular theory of gases is that proposed by **Johannes van der Waals** in 1873. He took into consideration both the attractive forces among the molecules and the space they occupy. The result is an equation of state for gases which is similar in form to the ideal gas equation but in which there is an additional internal pressure term due to the molecular attractions and the space occupied by the molecules is subtracted from the volume term.

Although the van der Waals equation of state was an advance in the understanding of real gases, and it can be used to determine helpful information, it also does not fit the behavior of gases at all values of pressure and temperature. A number of other theoretical and empirical equations of state have been proposed. One of these is the virial equation of state suggested by Heike Kammerlingh-Onnes (1853-1926) in 1901. This equation adds correction terms to the ideal gas equation by multiplying the right side of the equation (nRT) by the mathematical progression (1 + B/V + C/V2 + D/V3 + E/V4 +...), in which the constants A, B, C, etc. are temperature dependent. Computers are used to fit this equation to the experimental data, and the resulting values for the constants are determined for the particular gas. Although this is a purely empirical procedure, there has been some success in using the values of the constants to understand the behavior of gases.

IDENTIFICATION OF ELEMENTS AND COMPOUNDS

One of the most common activities of chemists is to determine the identity of elements and compounds present in a sample of material. If a chemist wishes to analyze a sample to find what elements or compounds are present, that is, to identify the elements and compounds present, she or he will choose a **qualitative analysis** method. Once the species that are present in a sample is known, the analytical chemist may choose to determine how much of each species is present using a quantitative analytical method. Some methods perform both qualitative and quantitative results, but most are used for one purpose or the other.

Literally hundreds of techniques for determining the identity of elements and compounds are available to chemists. The decision on which technique to use depends on many factors, including the physical state of the sample, the size of the sample, the level of accuracy required, and the time, money and expertise of the staff that are available. An important objective of analytical chemists is to develop reliable, inexpensive, rapid tests for the presence of specific compounds using the results from experiments with very sophisticated instruments.

If very little is known about a sample and a complete qualitative analysis is required, a great deal of effort using several sophisticated techniques would be requires. Much more commonly, however, there is a specific question about whether a particular element or compound is present. Is there **silver** present in an ore sample, **lead** in a paint sample, an **enzyme** present in a blood sample indicating that a heart attack has occurred and so forth? Most of the techniques to be described below are aimed at identifying a particular element or compound or a few closely related elements or compounds.

Qualitative elemental analysis methods comprise both instrumental and chemical or ''wet'' techniques. Most instrumental techniques for elemental analysis are destructive. A solid or liquid sample is subjected to harsh conditions such as a high **temperature** flame or a very high voltage electric arc or spark that converts the sample to gaseous atoms, ions, or a **plasma**. Under such conditions, the resulting species may emit light, the basis for the technique of atomic emission **spectroscopy**, which is very useful for identifying the presence of a wide variety of metallic elements. Modern atomic emission spectroscopy is a sophisticated extension of the flame tests that have been used for centuries to detect the presence of **metals** such as **sodium** and **potassium**. Elements of atomic **mass** greater than sodium can be made to emit x rays. By determining the wavelengths of the emitted x rays, many elements can be identified in a single sample simultaneously.

So-called wet techniques have been developed for many elements. They are often rapid and require little material. However, they may only provide a reliable result if the element is present in the sample in a particular form. For example, lead can be identified in a sample by the formation of a yellow precipitate when a **solution** of **sodium chloride** is added. But for this to happen the lead must be present in the sample as soluble Pb (II) ions. If the lead is present as insoluble PbO_2, it would go undetected. If the analyst knows enough about the nature of the sample and that it is likely to contain PbO_2, she might first add sufficiently strong acid and an appropriate reducing agent to convert it to soluble Pb (II). Typically, the choice of a wet method requires specific knowledge about the sample as well as the level of accuracy required.

Identification of compounds takes two forms: determining the nature of a compound whose identity is unknown and identifying the presence of a known compound in a sample. To accurately determine the nature of an unknown compound, it must first be carefully purified by techniques such as chromatographic separation, **distillation**, or crystallization. The strategy of identifying an unknown inorganic compound typically begins with qualitative and then **quantitative analysis** to establish its elemental formula. Analysis of an organic compound typically begins with spectroscopic techniques such as **infrared spectroscopy** (which reveals the presence of many types of functional groups), mass spectroscopy (which can provide the molecular mass as well as indicate the presence of certain functional groups from the molecule's fragmentation pattern), and **nuclear magnetic resonance** (which can be used to deduce presence and the connectivity of functional groups). Elemental analysis is often used to confirm the **molecular formula** deduced from spectroscopic experiments.

Once the structure of an organic species has been identified, the presence of the **molecule** in other samples can be detected by using spectroscopic patterns or chromatographic characteristics of the compound. Infrared spectra and mass spectrographs of organic compounds are typically very detailed, offering a unique ''fingerprint'' that can be used to identify the presence of the compound. Chromatographic methods are very useful for determining the presence of several, usually related, organic compounds in the same sample. In the most common chromatographic analysis techniques, gas **chromatography** and high performance liquid chromatography (HPLC), a sample is injected at one end of a column with a specific coating or containing a finely divided support material, respectively, and any organic compound is detected at the exit point of the column. A particular compound is identified in the sample if the detector indicates a compound at the same time it takes for the known compound to exit from the same column under the same conditions (for example, the same flow rate for the gas used in gas chromatography or solvent(s) used in liquid chromatography. Gas chromatography-mass spectroscopy (GC mass spec) is a widely used identification technique that combines a gas chromatograph, which separates and purifies compounds in a sample, with mass spectroscopy, which provides a fingerprint to allow rapid identification of each species. This combination is very powerful because the mass spectrogram of a complex mixture would consist of many overlapping ''fingerprints'' that could be very hard to decipher.

Many instrumental and wet chemical methods for the identification of elements and compounds have been developed for specialized applications best suited to the nature of the samples being analyzed, the accuracy required, and the speed and ease of performing the analysis. Before choosing a method, the chemist must understand the nature of the sample to be analyzed and the purpose of the analysis. The identification of elements and compounds is critical to all areas to which **chemistry** is applied, from **astrochemistry** and **biochemistry** to **geochemistry** and **industrial chemistry**.

See also Analytical chemistry

IMMUNOCHEMISTRY

Immunochemistry is the study of the **chemistry** of immune responses.

An immune response is a reaction caused by the invasion of the body by an antigen. An antigen is a foreign substance that enters the body and stimulates various defensive responses. The cells mainly involved in this response are macrophages and T and B lymphocytes. A macrophage is a large, modified white blood cell. Before an antigen can stimulate an immune response, it must first interact with a macrophage. The macrophage engulfs the antigen and transports it to the surface of the lymphocytes. The macrophage (or neutrophil) is attracted to the antigen by chemicals that the antigen releases. The macrophage recognizes these chemicals as alien to the host body. The local cells around the infection will also release chemicals to attract the macrophages; this is a process known as chemotaxis. These chemicals are a response to the infection. This process of engulfing the foreign body is called phagocytosis, and it leads directly to painful swelling and inflammation of the infected area. Lymphocytes are also cells that have been derived from white blood cells (leucocytes). Lymphocytes are found in lymph nodes, the spleen, the thymus, bone marrow, and circulating in the blood **plasma**. Those lymphocytes that mature inside mammalian bone marrow are called B cells. Once B cells have come into contact with an antigen they proliferate and differentiate into antibody secreting cells. An antibody is any protein that is released in the body in direct response to infection by an antigen. Those lymphocytes that are formed inside the thymus are called T lymphocytes or T cells. After contact with an antigen, T cells secrete lymphokines—a group of **proteins** that do not interact with the antigens themselves, instead they stimulate the activity of other cells. Lymphokines are able to gather uncommitted T cells to the site of infection. They are also responsible for keeping T cells and macrophages at the site of infection. Lymphokines also amplify the number of activated T cells, stimulate the production of more lymphokines, and kill infected cells. There are several types of T cells. These other types include T helper cells that help B cells mature into antibody-secreting cells, T suppresser cells that halt the action of B and T cells, T cytotoxic cells that attack infected or abnormal cells, and T delayed hypersensitivity cells that react to any problems caused by the initial infection once it has disappeared. This latter group of cells are long lived and will rapidly attack any remaining antigens that have not been destroyed in the major first stages of infection.

Once the **antibodies** are released by the B and T cells, they interact with the antigen to attempt to neutralize it. Some antibodies act by causing the antigens to stick together; this is a process known as agglutination. Antibodies may also cause the antigens to fall apart, a process known as cell lysis. Lysis is caused by enzymes known as lytic enzymes that are secreted by the antibodies. Once an antigen has been lysed, the remains of the antigen are removed by phagocytosis. Some antigens are still able to elicit a response even if only a small part of the antigen remains intact. Sometimes the same antibody will cause agglutination and then lysis. Some antibodies are antitoxins, which directly neutralize any toxins secreted by the antigens. There are several different forms of antibody that carry out this process depending upon the type of toxin that is produced.

Once antibodies have been produced for a particular antigen they tend to remain in the body. This provides immunity. Sometimes immunity is long term and once exposed to a disease we will never catch the disease again. At other times, immunity may only be short lived. Where the body produces its own antibodies to confer immunity that is a process called active immunity. Active immunity occurs after an initial exposure to the antigen. Passive immunity is where antibodies are passed form mother to child through the placenta. This form of immunity is short lived. Artificial immunity can be conferred by the action of immunization. With immunization, a vaccine is injected into the body. The vaccine may be a small quantity of antigen, or it may be a related antigen that causes a less serious disease, or it may be a fragment of the antigen, or it may be the whole antigen after it has been inactivated. If a fragment of antigen is used as a vaccine, it must be sufficient to elicit an appropriate response from the body. Quite often viral coat proteins are used for this. The first vaccine was developed by Edward Jenner (1749-1823) in 1796 to innoculate against smallpox. He used the mild disease cowpox to confer immunity for the potentially fatal but biochemically similar smallpox.

Within the blood there are a group of blood serum proteins called complement. These proteins become activated by antigen antibody reactions. Immunoglobulin is an antibody secreted by lymphoid cells called plasma cells. Immunoglobulins are made of two long polypeptide **chains** and two short polypeptide chains. These chains are bound together in a Y-shaped arrangement, with the short chains forming the inner parts of the Y. Each arm of the Y has specific antigen binding properties. There are five different classes of immunoglobulin that are based on their antigen-binding properties. Different classes of immunoglobulins come into play at different stages of infection. Immunoglobulins have specific binding sites with antigens.

One class of compounds in animals has antigens that can be problematical. This is the group called the histocompatibility complex. This is the group of usually surface proteins that are responsible for rejections and incompatibilities in organ transplants. These antigens are genetically encoded and they are present on the surface of cells. If the cells or tissues are transferred from one organism to another or the body does not recognize the antigens, it will elicit a response to try to rid the body of the foreign tissue. A body is not interested where foreign proteins come from. It is interested in the fact that they are there when they should not be. Even if an organ is human in origin, it must be genetically similar to the host body or it will be rejected. Because an organ is much larger than a small infection of an antigen when it elicits an immune response, it can be a greater problem. With an organ transplant, there can be a massive cascade reaction of antibody production. This will include all of the immune responses that the body is capable of. Such a massive response can overload the system and it can cause death. This is why tissue matching in organ transplants is so important. Often a large range of immunosuppressor drugs are employed until the body integrates a particular organ. In some cases, this may necessitate a course of drugs for the rest of the individuals life. Histocompatibility problems also exist with blood. Fortunately, the proteins in blood are less specific and blood transfusions are a lot easier to perform than organ transplants. The blood-typing systems that are in use are indications of the proteins that are present. If blood is mixed from the wrong types, it can cause clotting, which will lead to death. The main blood types are A, B, O, and AB. Group O individuals are universal donors, they can give blood to anyone. Group AB are universal recipients because they can accept blood from anyone. Type A blood has A antigens on the blood cells and B antibodies in the plasma. The combination of B antibodies and B antigens will cause agglutination. There are also subsidiary blood proteins such as the rhesus factor (rh) that can be positive (present) or negative (absent). If only small amounts of blood are transfused, it is not a problem due to the dilution factor.

Ph paper changing color. ***(Photograph by Yoav Levy/Phototake NYC. Reproduced by permission.)***

Immunochemistry is the chemistry of the immune system. Most of the chemicals involved in immune responses are proteins. Some chemicals inactivate invading proteins, others facilitate this response. The histocompatibility complex is a series of surface proteins on organs and tissues that elicit an immune response when placed in a genetically different individual.

Indicator, acid-base

An acid-base indicator is often a complex organic dye that undergoes a change in **color** when the **pH** of a **solution** changes over a specific pH range. Many plant pigments and other **natural products** are good indicators, and synthetic ones like phenolphthalein and methyl red are also available and widely used. **Paper** dipped in a mixture of several indicators and then dried is called pH paper, useful for obtaining the approximate pH of a solution. Blue litmus paper turns red in acidic solution, and red litmus paper turns blue in basic solution. The litmus test for acidity or basicity has been taken into the language to mean a definitive yes-or-no test for almost anything.

The pH at which the color of an indicator changes is called the transition interval. Chemists use appropriate indicators to signal the end of an acid-base **neutralization** reaction. Such a reaction is usually accomplished by **titration** : slowly adding a measured quantity of the base to a measured quantity of the acid (or vice versa) from a **buret**. (A buret is a long tube with **volume** markings for precise measurement and a stopcock at the bottom to control the flow of liquid.) When the reaction is complete, that is, when there is no excess of acid or base but only the reaction products, that is called the endpoint of the ti-

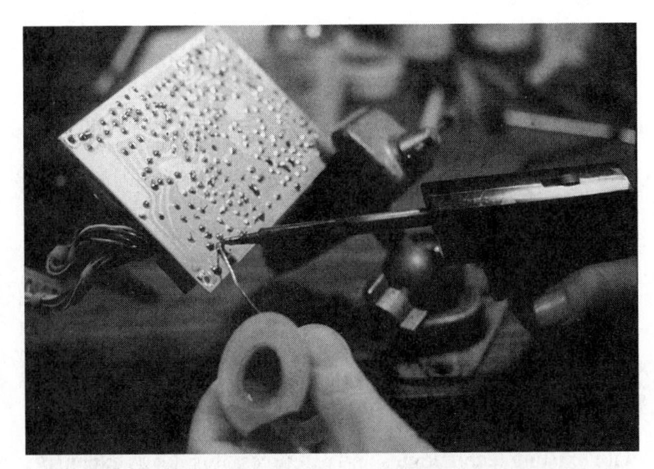

When indium is in its pure form, it sticks very tightly to itself or to other metals, making it useful as a solder. Here, a circuit board is being soldered. ***(Photograph by Robert J. Huffman. Reproduced by permission.)***

tration. The indicator must change color at the pH which corresponds to that endpoint.

The indicator changes color because of its own neutralization in the solution. Different indicators have different transition intervals, so the choice of indicator depends on matching the transition interval to the expected pH of the solution just as the reaction reaches the point of complete neutralization. Phenolphthalein changes from colorless to pink across a range of pH 8.2 to pH 10. Methyl red changes from red to yellow across a range of pH 4.4 to pH 6.2. Those are the two most common indicators, but others are available for much higher and lower pH values. Methyl violet, for example, changes from yellow to blue at a transition interval of pH 0.0 to pH 1.6. Alizarin yellow R changes from yellow to red at a transition interval of pH 10.0 to pH 12.1. Other indicators are available through most of the pH range, and can be used in the titration of a wide range of weak **acids and bases**.

Indium

Indium is the fourth element in Group 13 of the **periodic table**, a member of the **aluminum** family. It has an **atomic number** of 49, an atomic **mass** of 114.82, and a chemical symbol of In.

Properties

Indium is a silvery white, shiny metal with a **density** of 7.31 grams per cubic centimeter. It is one of the softest **metals**, even softer than **lead**. One unusual property of the element is that it produces a strange screeching sound when bent, a sound known as ''**tin** cry.'' Indium's melting point is 313.9°F (156.6°C) and its **boiling point** is 3,767°F (2,075°C). Chemically, indium is moderately active, dissolving in most acids, but not reacting with **oxygen** at room **temperature**.

Occurrence and Extraction

Indium is a relatively rare element in the Earth's crust with an abundance thought to be about 0.1 part per million. It is usually found in conjunction with ores of **zinc**. No indium is produced in the United States. Canada, China, and Russia are the world's largest producers of the element. Indium is obtained by extracting it from zinc ores while they are being processed to produce zinc metal.

Discovery and Naming

Indium was discovered in 1863 by German chemists Ferdinand Reich (1799-1882) and Hieronymus Theodor Richter (1824-98). Reich and Richter discovered the element while conducting a spectroscopic analysis of zinc ores. During this research, they found a set of spectral lines that had never been seen before and concluded that the lines were produced by a previously unknown element. They gave the element the name indium after the **color** of the most prominent of the element's spectral lines, a brilliant indigo blue.

Uses

The primary use of indium is in the production of alloys. Indium has sometimes been called a ''metal vitamin'' because the addition of very small amounts of the element can dramatically change the properties of an **alloy**. For example, very small amounts of indium are sometimes added to **gold** and **platinum** alloys to make them much harder and stronger. Such alloys are used in electronic devices and dental materials. The single most important application of indium alloys, however, is as coatings for materials that are subjected to high temperatures and stress, such as the outer surface or working parts of airplanes and space craft. Indium is also used in making optical devices, such as photovoltaic cells. One of the most advanced photovoltaic systems now available uses **copper** indium diselenide ($CuInSe_2$) in its solar cells.

Industrial Chemistry

Industrial **chemistry** is the branch of chemistry which deals with raw materials. More specifically, industrial chemists analyze and solve the problems associated with obtaining raw materials for industrial use, and are concerned with how to create new products from these raw materials, or how to improve upon existing production techniques. Industrial chemistry takes natural resources and improves upon them for practical use. This field of chemistry is a blend of science and technology and allows for society to exploit the benefits of the Earth's natural resources. Industrial chemists are concerned with optimizing the process involved with this science and technology to make the most product efficiently and cost-effectively.

Historically, there have been two separate and distinct branches of industrial chemistry, organic and inorganic. Both branches have been concerned with the transformations of raw materials into useful products. Recently, the distinction between the organic and inorganic industrial chemists has been fading and the disciplines seem to be merging as one. Traditionally, inorganic industrial chemistry has been concerned about using natural inorganic materials such as **water**, phosphates, or **uranium** to produce products that are useful technologically. Examples of products made by inorganic industrial chemists are glasses, **ceramics**, and cements. Inorganic industrial chemists work beyond the production of such materials. They are also concerned with manufacturing useful components from the materials, for example, **metals**, coke, and alloys. They have also focused on producing inorganic materials which help industry, such as **fertilizers** and pigments.

Organic industrial chemists have a similar focus, they simply base their research and development on organic products. They use resources such as rubber, natural fibers, and **cellulose** to make products such as soaps, fats, and sugars. They also derive compounds or manufacture other products from them, such as **sulfur**, **hydrogen**, or diesel fuels. They produce goods which are completely organic, such as alcohols, detergents, and pharmaceuticals, or they produce goods which are mixed in nature, such as **explosives**, **pesticides**, or varnishes.

The materials that are used in industrial chemistry are of a great variety. Many of the materials are natural resources obtained from the natural environment. These raw materials are converted into ''intermediates'' that are further processed to make the products that are of great industrial interest.

Most of the chemical elements, both metals and nonmetals, that are used in industrial chemistry are taken directly from Earth's crust. Many elements are chemically combined with other elements, and must undergo extensive processing to extract the ''pure'' element. Industrial chemists also use the fossil fuels—coal, **natural gas**, and petroleum—from Earth's crust, as well as salts that are used for chemical products such as **barium** and **calcium** sulphates. Fertilizers, for example, nitrates and phosphates, are also taken from Earth's crust, along with limestones, silica, and loams that are useful for making cement, **glass**, and ceramics.

Industrial chemistry utilizes materials from the oceans as well, such as sea **salt** and water. **Nitrogen**, **oxygen**, and noble **gases** are taken directly from the atmosphere for industrial use. Even the biosphere is a wealth of resources for the industrial chemist. Essences, cellulose, and saccharides can be taken from plants. Fats can be processed from both plants and animals. Enzymes for biochemical products are taken from animals. It is clear to see that the industrial chemist uses all of nature's resources for the benefit of technology.

Obtaining raw materials for industrial use is pointless if the chemists do not know what to do with them. This is where research comes in. Research in industrial chemistry is concerned with the development of new products, and the improvement of the processes currently in use. There are three main branches of research in industrial chemistry—innovative research, improvement and optimization, and application research. The research that is carried out in industry is of a completely different nature than that carried out in academia. Industrial chemistry research is concerned primarily with economics, making more product at less cost. Academic research is concerned with increasing knowledge. Industrial chemistry, therefore, focuses on applied research whereas academia focuses on theoretical or pure research.

The innovative research that takes place in industrial chemistry is usually carried out by specific chemical companies. A company performing innovative research is trying to make something new and interesting, something that will make the company stand out among its competition. This may be a new product or a new process for creating an old product. Innovative research involves more than just background study. This type of research involves laboratory experiments, pilot studies, and full-scale studies to determine if the product is useful and economically feasible. Innovative research is very costly, and the projected benefits must be great for a company to make this kind of investment.

Research into product development serves the purpose of developing new products from the old ones or to optimizing current production processes. This type of research is often initiated by members of the company who are interested in sales, in response to requests from their customers. Because it deals with existing technologies and resources, it is not as expensive and time consuming as innovative research.

Application research in industrial chemistry aims to increase the effectiveness of products and to determine if the products are safe to use. This type of research takes place in many different kinds of labs, from chemical to biological to technical. This branch of research is extremely important before a new product is introduced to the public. Teams of product specialists are usually involved in application research at a chemical company.

As profitable as industrial chemistry can be, there are many concerns about potential problems stemming from the industry. Many of the chemicals produced are toxic to plants and animals. Industrial chemical companies must be concerned about waste disposal and the potential for spills into the environment. Plant workers also must be concerned about exposure to toxic doses of chemicals. The chemical industry is also the industry that uses the greatest amount of **energy**. The cost of energy use in manufacturing processes constitutes about 35% of the final price of the products. Most of the energy used in the chemical industry is thermal energy, which is produced mainly by **electricity**. Most electricity ultimately comes from fossil fuels, of which there is a limited supply. As the fossil fuel supply decreases and energy costs increase, the costs of the products made in the chemical industry will also increase. Conserving energy while making affordable products is a tough problem for industrial chemists. In the future, alternative forms of **heat** energy will need to be considered.

INFRARED SPECTROSCOPY

Infrared **spectroscopy** is the study of the interaction between **matter** and electromagnetic radiation in the infrared region. This type of spectroscopy is used to study the vibrations of molecules; typically, the **energy** of molecular vibrations is the same as that of infrared light. Because of this, infrared spectroscopy can be used to determine molecular characteristics, such as bond length, bond angles, vibrational frequencies, and, indirectly, bond strengths. In addition, infrared spectroscopy is tremendously useful in determining the presence of certain characteristic groups in large molecules.

The **electromagnetic spectrum** consists of radiation having very high energy and frequency and short wavelength (such as gamma and x rays) to radiation having very low energy and frequency and long wavelength (such as radio waves). Visible light falls approximately in the middle of this spectrum. Infrared radiation has an energy that is lower than that of red light in the visible spectrum. Thus, infrared light lies in the energy range of about 0.5 kcal (2 kJ)-9.5 kcal (40 kJ). The energy of most molecular vibrations also lies in this range.

Molecular spectroscopy is very complicated because molecules contain two or more atoms. In fact, the more atoms there are in the **molecule**, the more complicated the interpretation of the spectrum can be. However, it is possible to simplify the interpretation of a molecular spectrum based on the types of movement of molecules and the energies required for these movements.

All molecules in the gas phase are capable of three basic types of movement. The first is translation, or movement of the entire molecule in a random path in space. The second is rotation, or movement of the molecule around an axis (usually a bond) of the molecule. Rotational energies are usually in the same range as microwave or far infrared (longer wavelength) energy. The third type of motion is vibration, or movement between the atoms of a molecule, and this applies to molecules that are solid, liquid, or gaseous. In this type of movement, the atoms can be compared to balls attached to springs. If a force is applied to one of the balls (or atoms) in a particular direction, the other balls (or atoms) are affected. For example, in a bent molecule containing three atoms (such as water), there are three types of vibrations possible. The first is a symmetric stretch, in which the two end atoms are pulled down along their axes, and the center **atom** moves up. The second is an asymmetric stretch in which one end atom is pulled away from the center atom and the other end atom is pushed towards the center atom. The third is a bending motion in which the two end atoms move towards one another, causing the central atom to move upwards, much like the motion caused by waving ones arms up and down if the head is considered as the central atom. As molecules become bigger, more types of vibrations are possible and this complicates the interpretation of data.

Classically, the movement of two spheres connected by a spring can be described by a harmonic oscillator (the two spheres bounce back and forth with smaller and smaller regular oscillations until they finally come to rest). In classical mechanics, the vibrations of a harmonic oscillator can assume any energy value (just as it is possible to stop anywhere on a ramp). The movement of atoms in a molecule, however, is described by quantum mechanics. The equation that describes the energy of the vibrations limits the values that the energy can assume (just as it is possible to stand on one step or another, but not in between steps when climbing stairs). In the quantum mechanical model, the energy of the ground state (or the lowest **energy level** possible in the molecule) is not zero, but a finite quantity which is a function of **Planck's constant** and the vibrational quantum number. This minimum energy of the ground state is called the zero-point energy. Most molecules normally exist in the ground state. The energy required to excite a molecule from the ground state to the first excited state corresponds to the energy of radiation in the infrared region of the electromagnetic spectrum. Thus, if a molecule is exposed to infrared radiation and a certain wavelength of light is absorbed by the molecule, the molecule can move from the ground vibrational state to the first excited vibrational state, which in turn yields information about the molecule itself. This is how infrared absorption spectra are obtained.

Experimentally, infrared absorption spectra are obtained using infrared spectrometers. In general, a spectrometer consists of a source of radiation, a means of separating radiation into its distinct frequencies, and a method to detect the radiation that passes through a sample. Simply put, if all frequencies of infrared radiation are passed through a sample, and some of these frequencies are absorbed by the sample in going from the ground state to the first excited vibrational state, then the ''spectrum'' will show a plot of the frequencies that are absorbed by the sample. That is, an infrared spectrum is a plot of the intensity of light absorbed as a function of the frequency of vibration. Reading the frequencies where light is absorbed gives the vibrational frequencies of the molecules in the sample. This method works for solid, liquid (or solution), and gaseous samples.

As mentioned earlier, the infrared spectra of small molecules can provide information about particular stretches and bends in the molecule. As molecules increase in size, their spectra become more complex. One advantage, however, is that even in a large molecule, certain stretches and/or bends occur at characteristic frequencies. These characteristic frequencies remain unchanged in spite of the differences in the geometry, the size, the **mass**, or other features of the molecule. For example, a bond between **oxygen** and **hydrogen** (O-H) in a molecule shows a vibration at 3600 wave numbers (a wave number is the reciprocal of the wavelength in the unit of centimeters), while a carbon-oxygen double bond (carbonyl bond) has a distinctive band at 1700 wave numbers. These characteristic vibrations serve as molecular fingerprints. If the spectrum of an unknown compound is analyzed, it is possible to determine if there are certain groups in the compound based on the presence of certain absorption bands.

Infrared radiation is associated with the production of **heat**. The element on a stove is ''hot'' long before one can see the red **color** that is only produced when the element becomes very hot; this heat is due to infrared radiation that the eye cannot detect. A typical infrared spectrometer consists of a source of radiation, usually a ceramic element that is heated to relatively high temperatures. This radiation is passed through the sample and the frequencies that are absorbed by the sample are measured by a relatively complex apparatus called a Fourier Transform Infrared Spectrometer (FTIR), details of which are beyond the scope of this article. The apparatus is relatively inexpensive, costing about $15,000, and it is completely computer controlled. In 1960, it took about 30 minutes to obtain an infrared spectrum. Currently, approximately 30 infrared spectra can be obtained in one second.

One unusual aspect of infrared spectroscopy is that common materials do not transmit infrared radiation in the regions of interest where molecular vibrations occur. Simple materials such as **glass** are completely opaque in infrared regions of interest. Glass is mainly composed of **silicon** dioxide. This molecule strongly absorbs almost all infrared radiation and is thus inappropriate for studies of this type. Therefore, in order to obtain a spectrum, the sample must be contained in a material that is transparent to this radiation—usually a substance which itself does not have vibrations in the infrared region. Interestingly, table **salt** (sodium chloride) does not absorb infrared radiation until the frequency is very low, and this is one of the most common materials now used to contain samples that are used in infrared spectroscopy. Today, there is a host of materials that can be used, and some of them are quite expensive and exotic (for example, **cadmium** telluride).

Currently, infrared spectroscopy is one of the standard work-horses of the manufacturing world, whether in semiconductor production, process control in **chemical engineering**, or environmental monitoring. It is relatively inexpensive, rapid, very sensitive, and can provide quick and efficient information on chemical reaction conditions through its ability to detect characteristic vibrations of molecules of interest in the process under investigation. Few techniques are so versatile at such a low cost, and it can easily be used as a remote sensor under extremely corrosive or dangerous reaction conditions, conditions that make many other potential *in situ* methods unsuitable.

INGENHOUSZ, JAN (1730-1799)

Dutch physician, chemist, and engineer

The Dutch physician, chemist, and engineer Jan Ingenhousz is noted for his demonstration of the process of **photosynthesis** in plants.

Jan Ingenhousz was born on Dec. 8, 1730, in Breda. He studied medicine at the University of Louvain and graduated in 1752. After spending some years in several European capitals in the typical 18th-century tradition, he settled in London in 1779 and worked with the celebrated naturalist John Hunter.

In that year Ingenhousz published his important book, *Experiments upon Vegetables—Discovering Their Great Power of Purifying the Common Air in the Sunshine and of Injuring It in the Shade and at Night.* In this work he anticipated by two years **Joseph Priestley**'s discovery of the principles of what is now called photosynthesis, that is, the process by which plants exude **oxygen** and absorb **carbon** dioxide, thus purifying the air for animals and man.

Unlike Priestley and other chemists who were working on the characteristics of oxygen from the point of view of chemical philosophy, Ingenhousz was preoccupied by the problem of the fundamental **balance** in the animal and vegetable kingdoms, and this led him to investigate the mutual interdependence of plants and animals. He introduced the concept that the leaves of plants are great laboratories for cleansing and purifying the air. He also noted that oxygen is emitted by the underside of the leaves and that this is a daylight process, whereas in darkness even plants emit small quantities of **carbon dioxide** instead of absorbing it. In his reflections at the end of the book, Ingenhousz said, "If these conjectures were well grounded, it would throw a great deal of new light upon the arrangement of the different parts of the globe and the harmony between all its parts would become more conspicuous."

The book was soon translated into many languages and became the foundation of that kind of research which in modern times led to a more basic understanding of the process of photosynthesis; however, his search for the concept of economy or balance in nature was not well understood by his contemporaries. As to the nature and origin of the oxygen which the plant emits, a controversy developed in the 1780s between Ingenhousz and Priestley. Ingenhousz thought that **water** which plants absorb changes into vegetation and that part of this water is then released as oxygen.

Ingenhousz built electrical machines and invented the plate electric machine. He also wrote a two-volume treatise dealing with problems in medicine which are relevant to the physicist and the medical man, and in a sense it could be said that his basic interest was in what is now called biophysics. Ingenhousz also opposed the theory of subtle electrical fluids and repeated some of the experiments on plant **electricity** to disprove the accepted view that positive electricity was good for the growth of plants and that negative electricity retarded it.

All of Ingenhousz's scientific work was motivated by a deeply religious attitude and the belief that balance in nature is the best expression of the harmony created by its Author. Ingenhousz died in Wiltshire, England, on Sept. 7, 1799.

INGOLD, CHRISTOPHER KELK (1893-1970)

English chemist

Christopher Kelk Ingold lived a life that reminds us that not every historical figure received prominent signs of destiny from childhood. Indeed, his first sign of brilliance was not recognized until he was ready to graduate from college. Later, when he was making major discoveries, acceptance of his authority was not naturally offered as it would have been to a recognized genius. Still, he had great influence in the field of organic **chemistry**, offering the landmark explanation of organic reactions. Furthermore, he addressed issues of methodology and, thus, had a crossover effect in **physical chemistry** in spite of the tendency for scientists to resist influence from outside their specialty. He is credited with systematizing modern **organic chemistry** because he clarified many issues through his prolific, comprehensible writings.

As quiet and unassuming as his research career was, it still contained some elements of a classic story of the struggle for success. (Science historian William Brock even suggests that a 1930 novel was partially based on Ingold's life.) Many classical novels include characters such as the loyalty-inspiring teacher and the ever-faithful wife. For Ingold, his teacher, J. F. Thorpe, inspired him long after their association. Ingold's wife Edith, an accomplished researcher herself, often helped in her husband's work. Ingold's drama also included the long-term friend (collaborator Edward David Hughes) and an unrelenting rival (Robert Robinson). The main point of this drama was that success was achieved through perseverance and self- confidence in the face of adversity.

Christopher Ingold was born in London in 1893. He lived an obscure childhood on the Isle of Wight. There was nothing remarkable about his youth and education. He entered Hartley University College of Southhampton in the year 1910 at age seventeen. He was very much interested in physics, but was drawn to chemistry by the enthusiasm shown for the subject by one of his teachers, David R. Boyd. Ingold wanted so much to impress Boyd that, when he achieved second honors in the class, he felt it was not good enough. However poorly he felt he did in that class, he made chemistry his life long obsession. Friends report that, until he turned sixty, he appeared to have no other interest.

After graduating from Southhampton, Ingold began graduate studies toward a Ph.D. at Imperial College. There, he was student and research assistant of Jocelyn Field Thorpe. Thorpe was on the breakthrough of important research dealing with tautomerism and the structure of compounds. Ingold was thrown immediately into the thick of this study. This collaboration was a mutually beneficial relationship. Ingold trusted Thorpe as mentor so much that when Thorpe arranged for him to work at a chemical company for over two years, Ingold considered the job as part of his studies. After the two years was over, Ingold returned to Imperial College and to collaborating with Thorpe. Only this time, Ingold was at college, not as student, but as lecturer. He had earned a Doctorate of Science (rather than a Ph.D.) based on his writings.

After three years as lecturer, Ingold took at turn in his life and in his career. He married his skillful research assistant, Edith Hilda Usherwood. The next year, 1924, Ingold was elected Fellow of the Royal Society and received a Chair of the Chemistry department at the University of Leeds. Ingold's wedding may have influenced his research interests. His wife Edith had already begun her own career in organic chemistry research. Around the time of his marriage, Ingold was turning from classical theoretical research, influenced by Thorpe, toward the physical studies from which he gained his international reputation. At this time, he also became serious (from a classical point of view) about overturning Robinson's (electron-based) theories of organic reactions. It was just the beginning of a long-term, sometimes uncivil, rivalry between the two.

In 1930, Ingold was offered second Chair of Chemistry at the University College of London. The position was vacant because of the promotion of Robinson who, incidentally, opposed Ingold's nomination. One of the lecturers in the department was Edward David Hughes who was to collaborate on Ingold's research for over thirty years. When F. G. Donnan retired from UCL in 1937, Ingold was made head of the Chemistry department where he stayed until his retirement in 1961. After his retirement, he accepted a fellowship at Vanderbilt University where he updated his well-known textbook on organic chemistry. He died in 1970 in Middlesex, England.

The debate between Robinson and Ingold centered on the applicability of the principles of physical chemistry to organic chemistry. When the personalities were extracted from the debate, it became clear that there were two major issues with Ingold or Robinson on the wrong side of either. Both were growing into an electron-based theory but from different paradigms. Robinson, before Ingold, recognized the value of attributing organic reactions to exchanges at the particle level. However, it was Ingold who stressed that the laws of organic reactions (when and if found) would hold true generally for all compounds and any differences seen within individual compounds are due to alternative influences. Ingold delivered 14 long experimental papers in 18 months on this issue. With almost all, Robinson challenged his findings in person or with critical papers. Both made embarrassing mistakes until they began to modify their stances through accepting points from each other.

In studying tautomerism, Ingold accepted that certain compounds must exist in two different forms. The question arose: "Which form was most descriptive of the properties of the compound?" Ingold studied the developing reactions of these compounds and recognized that there was an intermediate state before these compounds divided into their various forms. This intermediate state he called the mesomeric state. The most organized and useful description of the compound, he argued, was this mesomeric state. Through further study of the mesomeric state, Ingold reached one of his most important contributions to organic chemistry. This was the concept that there were four different types of reactions possible in elimination and substitution based on bimolecular or unimolecular reactions. Bimolecular elimination occurs through the influence of a base. Unimolecular elimination does not involve interaction with the base. Bimolecular substitution occurs when two reactants join to form a mesomeric **ion**. Unimolecular substitution is a two-step process involving the forming of an ion to react immediately with an **anion**. Ingold stressed that it was important to identify which type of reaction was occurring to best define and predict the effects of the reaction.

While others were making similar observations, Ingold was the first to elucidate all observations in a practical model. He thereby explained many different phenomena in a comprehensive manner. Ingold found the means to develop the models because he held to his belief that the mechanisms in organic chemistry were generally applied across all compounds. (Others experimented based on the concept that each compound followed principles of individual mechanisms.) Ingold was also one of the first to use the apparatus and methods of physical chemistry in the study of organic chemistry. These included the use of **isotope** effects and molecular **spectroscopy**.

It should be noted that Ingold's loyalty to Thorpe delayed his acceptance of an electron-based theory. Ingold's delay, in turn, influenced the delayed acceptance of the theory. When he finally accepted, he gave a most comprehensive explanation of the implications of the theory outshining proponents such as Robinson. One wonders how much organic chemistry could have been advanced if only these two giants of theory could have worked together. Ingold's refusing to acknowledge Robinson's influence on his work led to Robinson's accusations of plagiarism. This accusation probably influenced Ingold's not winning the Nobel Prize when he was nominated. However, Ingold did receive many honors, including a knighthood, for his work.

INGRAM, VERNON (1924-)

American biochemist

Vernon Martin Ingram escaped Nazi Germany as a young boy to grow up to become a noted **biochemistry** professor and researcher. Ingram is credited with the discovery of the exact molecular distinction between the hemoglobin carrying sickle cell anemia and normal hemoglobin. He is currently a professor at Massachusetts Institute of Technology and has been since 1958.

Ingram was born in 1924 in Breslau, Germany, (present-day Wroclaw, Poland). While he was still very young, his parents brought him to England because of the dangers they faced as the Nazis assumed power in Germany. As a young man, Ingram attended college at the University of London, receiving a B.S. degree in 1943. He did graduate studies at Birkbeck College and earned his Ph.D. in 1949. While working on his Ph.D., Ingram worked as a **chemistry** instructor at Birkbeck and at the University of London. He lived in America for two years (1950 -51) after winning fellowships from the Rockefeller Institute and Yale University. Ingram returned to England and for six years worked as part of the Medical Research Council at the Cavendish Laboratory at Cambridge. (**James Clerk Maxwell** and **J. J. Thomson** were associated with this laboratory.) It was here that he began his work on the causes of gene mutations. Ingram moved again to America, after he accepted a position at MIT where he is today. In 1961, Ingram received a Doctorate of Science based on his research writings. The American Society of Human Genetics awarded him the William Allen award in 1967. In 1988, MIT granted him the prestigious position as the John and Dorothy Wilson Professor of Biology.

Ingram's work has revolved around research on hemoglobins and their synthesis. He is trying to solve problems associated with genetically-based disorders, including sickle cell anemia, Alzheimer's disease, and Down's Syndrome through the study of biochemical mechanisms and molecular influences. While at the Cavendish Laboratory in the 1950s, Ingram worked with **Francis Crick**. Their research involved the painstaking effort to discover if **proteins** are coded by genes. This had been an underlying assumption in molecular biology but had yet to be substantiated. Even after joining MIT's faculty, Ingram did not lose his interest in this project. When he had the opportunity to contribute to the research on sickle cell anemia, he applied what he had learned with Crick. Through the research methods of the advancing field of molecular biology, Ingram was able to analyze the sickle cell hemoglobin down to a very minute level. This was the level necessary to illustrate that the sickle cell hemoglobin differed from normal hemoglobin at only one point. That is one amino acid out of five hundred. The years of frustration paid off in a way that promised hope of treatment for a debilitating disease. It was through his work at Cavendish that Ingram concluded that he would find genetic influence on proteins. This expectation kept him focused on the end as he went into increasingly distinct levels of the analysis. His methodology also had a substantial influence on the establishment of biochemistry as a different discipline separate from **physical chemistry**. This methodology included the effective use of **electrophoresis** and **paper chromatography**.

Ingram is a welcomed member of many notable research organizations in **organic chemistry** such as the American Chemical Society, British Biochemistry Society, and the Chemical Society. He is a fellow of the Royal Society in England and a member of the Academy of Arts and Sciences in America. Ingram makes his home in Massachusetts.

INKS

According to local tradition, a Chinese inventor by the name of Wei Tang who lived about 1,500 years ago developed the first ink suitable for both brush writing and block printing. His ink was prepared from the soot of lamp black mixed with **water**.

Modern printing inks contain four basic components: pigments, **solvents**, resinous binders, and performance additives. In most printing inks, the solvents are oil-based, with petroleum-based oils most commonly used for this purpose. Vegetable-based inks, in which vegetable and petroleum oils are mixed, frequently consist in part of soybean, corn, cottonseed, or linseed oil. Printing inks made with vegetable oils release fewer volatile organic compounds in the printing plants, thereby enhancing occupational safety. (Although most inks are oil-based, some printing inks used in corrugated packaging, magazines, and newspapers are water-based.)

Most newspapers use soy ink for **color** printing. Because the price of color soy ink for newsprint is competitive to that of petroleum-based color ink, newspapers favor soy ink because of its superior performance, environmental friendliness, and vibrant colors. Soy ink can be used on virtually any lithographic press with no modifications or special cleaning agents.

Soy ink is very similar to petroleum-based printing ink, except that it contains varying amounts of soybean oil instead of petroleum oil. Soybean oil is non-toxic, as evidenced by its use in cooking oils, margarine, and salad dressings.

Research aimed at making soy-based inks more technologically competitive with petroleum-based inks has targeted the following objectives: increase the soy oil's **molecular weight**; improve the oil's viscous properties; increase the degree of **conjugation** in the oil; modify the oil by placing hydroxyl groups along the fatty acid **chains**; and chemically modifying the oil to add more acid groups to the **molecule**.

INORGANIC CHEMISTRY

Inorganic **chemistry** is the branch of chemistry involving the reactions, structures, and properties of all *noncarbon-based* compounds. This includes all the chemical elements in their pure and semipure states (except carbon). Some simple **carbon** compounds, including **carbon monoxide**, **carbon dioxide**, and **hydrogen** cyanide, are nevertheless considered to be in the realm of inorganic chemistry.

The lines of definition are becoming increasingly blurry between inorganic chemistry and such other major areas of the science as **physical chemistry**, **biochemistry**, and even **organic chemistry**, due to the fields' many common interests. However, inorganic chemistry is unique in that it is the only branch of chemistry that specifically excludes the element carbon.

The modern concept of inorganic chemistry is virtually unrelated to the original sense of the word, which was strictly taken to mean ''not living.'' Until the beginning of the twentieth century, scientists believed that all living things (animals and plants) came within the confines of organic chemistry and that all nonliving, inanimate things (rocks, **water**, etc.) should be examined and analyzed using inorganic chemistry. In about 1910, researchers began to understand that all things, both living and inanimate, share the same basic chemistry, whether carbon or noncarbon based.

The distinction between organic and inorganic chemistry became highly relative, and it has continued to evolve in that direction. Take, for example, the protein hemoglobin. Hemoglobin falls within the discipline of inorganic chemistry for the sole reason that it contains an **iron molecule** at its center. The fact that the iron is surrounded by organic molecules makes no difference in hemoglobin's designation as an inorganic protein. This phenomenon works both ways—an organic chemist might find him- or herself studying a substance composed almost entirely of inorganic molecules, just because it contains one molecule of an organic element such as carbon.

The recognized subdisciplines within the area of inorganic chemistry include (1) **coordination chemistry**, which studies the bonds among the atoms of inorganic compounds; (2) inorganic technology, in which scientists manufacture and manipulate inorganic chemicals for industrial purposes; (3) **bioinorganic chemistry**, a science that specializes in the biological functions of metal complexes in living organisms; (4) **geochemistry**, which concentrates on the chemistry of all geological processes and which at the new millenium was especially pertinent in environment-related fields; (5) synthetic inorganic chemistry, which deals with creating, purifying, and studying inorganic materials; (6) reaction **kinetics** and mechanisms, which concerns the ways and rates at which chemicals react with each other; (7) **solid-state chemistry**, which involves the study of the crystals used in solid-state (vacuumless) electronics; and (8) nuclear science and **energy**, which concentrates on the chemistry of **nuclear reactions**.

INTERMOLECULAR FORCES

The attractions of molecules for each other are called intermolecular interactions to distinguish them from covalent and ionic **bonding**, forms of intramolecular interactions. For very large molecules, such as **proteins**, **nucleic acids**, and synthetic polymers, non-covalent interactions of groups that make up the molecules (functional groups) have the same characteristics as intermolecular interactions even though they may actually take place between groups within the same **molecule**. The interactions that lead to specific shapes of enzymes, referred to as protein-folding interactions, is an example. Intermolecular interactions are most significant in liquid and solid phases where molecules are very close together. In fact, even in **liquids** and **solids** intermolecular interactions are only strong for molecules that are next to each other. The interactions of molecules in the liquid and solid states have significant consequences that are readily observable. The strength of intermolecular interactions affects numerous properties, including boiling points, **miscibility**, and **solubility**.

All neighboring molecules in liquids and solids attract each other. The nature and strength of these interactions depend on the types of groups of atoms or functional groups that comprise the molecules. All molecules interact with other molecules through London dispersion forces, also called van der Waals interactions. London dispersion forces are the attractive forces of one transient **dipole** for another. A transient dipole is a temporary imbalance of positive and negative charge. At particular instants, even atoms that are spherical on average, such as those of the noble **gases**, will have greater **electron density** on one side of the **atom** than another. At that instant, the atom will possess a temporary dipole with a negative charge **concentration** on the side of the atom with greater electron density. If this happens in the case of an **argon** atom in liquid argon, for example, the argon atoms next to the one with temporary dipole would feel the effect of the dipole. An atom near the negative end of the dipole would have its own electrons slightly repelled from the negative concentration of charge, developing a dipole with its positive end near the negative charge of the original atom. An argon atom on the other side of the original temporary dipole would feel its electrons attracted to the positive end of the dipole, developing a dipole with the opposite orientation. In this way, temporary dipoles are propagated through a liquid or solid. The motion of the molecules in the liquid or solid soon disrupts the pattern, but similar events take place continually. The larger the size of atoms and the more electrons they possess, the greater the probability of forming substantial transient dipole interactions. Molecules which are non-polar and non-polar functional groups of molecules only experience London dispersion or van der Waals interactions with other molecules or functional groups.

Some molecules and functional groups have permanent dipoles that result from a non-symmetric geometric arrangement of atoms of different **electronegativity**. The portion of a molecule with a permanent dipole will be attracted to the portion of a neighboring molecule with its own permanent dipole. A polar-polar interaction is stronger than a van der Waals interaction.

The **hydrogen** bond is a specific and very important type of intermolecular interaction. A **hydrogen bond** occurs when the electrons forming the **covalent bond** between the hydrogen atom and the **oxygen**, **nitrogen**, or **fluorine** atom are pulled toward the electronegative atom. Only the first shell is occupied in the hydrogen atom, so when the electrons in the bond are pulled away, the **nucleus** is exposed. The extent of the exposure of positive charge on the hydrogen is the fundamental difference between it and other atoms which have an inner shell or several inner shells that continue to shield the nucleus even though the bonding electrons are pulled away. The **energy** of the hydrogen-bonding interaction can be as much as 100 kJ mol^{-1} for HF, but it is typically in the range of 20-40 kJ mol^{-1}. As such, it is the strongest of the intermolecular interactive forces. The specific geometric requirements for hydrogen bonding are profoundly important in **biochemistry**, notably in the operation of the genetic code.

The interactions of permanent dipoles with each other are stronger than London dispersion forces, and hydrogen-bonding interactions are stronger yet. Intermolecular interactions are additive, so that both the number of interactions between two molecules and the strength of each interaction determines the energy needed to separate molecules from each other. Molecules with many atoms and molecules that are highly polar and/or contain hydrogen bonds are especially difficult to melt or vaporize. Many phenomena familiar in everyday life arise because of forces among molecules.

The effects of the various types of intermolecular interactions can be seen from the boiling points of the compounds of elements formed with hydrogen (the hydrides) in the periodic groups containing **carbon**, nitrogen, oxygen, and fluorine. For all the hydrides except **water**, **ammonia**, and hydrogen fluoride, the boiling points of the hydrides of each periodic group (for example, CH_4, SiH_4, GeH_4, and SnH_4) increase as the **atomic number** of the element increases. This is the normal effect due to London forces—the greater the polarizability of the electron cloud, the more the condensed phase is stabilized by transient dipoles.

The hydrides of the **halogens** (HCl, HBr, and HI) all have permanent dipoles. They boil somewhat higher than the hydrides of the elements in the corresponding row of the preceding group of the **periodic table** and also show the trend of increasing **boiling point** as the polarizability of their electron clouds increases. The hydrides of oxygen, nitrogen, and fluorine, however, require the input of more energy to boil than would be predicted from either London forces or dipole-dipole interactions. Water, ammonia, and hydrogen fluoride all form hydrogen bonds. Each water molecule can form four hydrogen bonds with other water molecules—two involving its two hydrogen atoms, and two involving the unshared pairs of electrons on its oxygen—so the effect of hydrogen bonding on its boiling point is especially great.

Molecules forming hydrogen bonds are stabilized when the molecules align properly. The kinetic energy of molecules always competes against attractive forces to determine physical properties because molecular motion opposes proper alignment. More **heat** is required to disrupt the stable associations in a solid containing hydrogen- bonded molecules to form a liquid, in which the molecules are more mobile and not as well aligned. Similarly, compared with non-hydrogen-bonding substances, liquids in which the molecules can hydrogen bond require more heat to form a gas, in which there is very little interaction between molecules. The stronger the intermolecular attractive forces, the more heat energy is required to cause changes of phase. Each hydrogen bond between molecules is much weaker than the covalent bonds holding the atoms together within the molecule. It may require a large amount of heat to convert liquid water into steam, but it requires significantly more energy to break a molecule of water apart into hydrogen and oxygen atoms.

To understand the differences in properties of larger molecules, the additivity of intermolecular interactions becomes important. In effect, the interaction of each group of atoms of a molecule with a group of atoms of a neighboring molecule can be considered to be independent of the interactions of other groups of atoms of the molecules. The total energy required to move two molecules apart is then the sum of all the energies of the individual interactions. The more groups and the stronger each individual interaction, the greater the sum of energy of interactions. Among the non-polar linear alkanes, the boiling point for a molecule with many $-CH_2$ groups such as octane, $CH_3(CH_2)_6CH_3$ is higher than that of propane $CH_3\ CH_2CH_3$, with octane being liquid at room **temperature** and propane a gas, because of the greater number of London dispersion interactions between the octane molecules. If there are ten hydrogen bonding interactions per molecule, as there would be for a large molecule such as a sugar with many -OH groups, the total energy of interaction because of hydrogen bonds is about 300 kJ mol^{-1}, which is on the order of a strong chemical bond. Such molecules are typically solids at room temperature, while molecules containing only non-polar groups of the same molecular **mass** may be liquid.

Life on Earth may exist because of the hydrogen bond. The physical properties of water are in large part due to its extensive network of hydrogen bonds. Our external environment is aqueous—2/3 of Earth's surface is water—and 2/3 of our mass is water as well. Hydrogen bonding and intermolecular interactions are the very basis of the genetic code and the unique structures and shapes of the nonaqueous components of life: **DNA**, **RNA**, proteins, and other biomolecules making up living systems all owe their form and function to hydrogen bonds. The double helix of DNA is held together by hydrogen bonds. These are strong but still at least four times weaker than covalent bonds. Hydrogen bonds are the perfect strength to hold DNA together under most situations, but are weak enough to form and break readily to enable DNA to untwine for replication.

Stabilization through hydrogen bonding can also determine the ways in which molecules arrange themselves. The hydrogen bond is a directional bond, which means there is a specific architectural relationship among molecules hydrogen bonding with each other. An important illustration of the effect of the specific geometry of hydrogen-bonded molecules is the decrease in the density of water when it freezes to form ice. Most liquids increase in density as they solidify and, therefore, solid pieces of a substance typically sink when added to the liquid substance, whereas ice floats on water. For aquatic life and indeed for all life on Earth, this anomaly of water is important. As ice forms on the surface of a body of water, it eventually forms a sheet that tends to insulate the body of water below, helping to prevent the entire body of water from freezing solid. Aquatic life that would otherwise be killed by a freeze survives.

Many molecules are very complex, they have some polar parts, some non-polar parts and some functional groups that can hydrogen bond. As is true for the London forces in alkanes and as we have already indicated in the discussion of hydrogen bonding, polar interactions are also additive. The boiling points of propane, CH_3 CH_2CH_3 is -43.6°F (-42°C), while that of butane, $CH_3(CH_2)$ $_2CH_3$ is 32°F (0°C). Adding one CH_2 group increases the boiling point by 75.6°F (42°C). When the nonpolar -CH_2- unit of propane is replaced with a polar -C=O group (this gives us **acetone** $CH_3C{=}OCH_3$), the boiling point increases by 180°F (100°C), from 43.6°F (-42°C) to 136.4° (58°C). If additional groups are added the boiling point increases accordingly.

Intermolecular interactions also affect the properties of molecular solids and some types of covalent network solids. Molecular solids are composed of small molecules held together by London forces, dipole-dipole, and hydrogen-bonding interactions. Sugar is a molecular solid, composed of glucose molecules held in crystalline array by dipole-dipole interactions. Covalent network solids are composed of very large covalently bound molecules. **Graphite** is a covalent network. It consists of sheets of covalently bound carbon atoms held together by **van der Waals forces**. A **diamond** is a special kind of covalent network solid, because it consists entirely of covalently bonded carbon atoms. Each diamond is one molecule.

Some important observable consequences of intermolecular interactions are **miscibility**, solubility, the heat of mixing that occurs when different liquids are combined, and capillary action.

Some liquids mix and some do not. Familiar examples are the miscibility of **alcohol** and water and the immiscibility of oil and water, which can be explained by intermolecular interactions. A portion of the alcohol molecule (-OH) hydrogen bonds with water molecules and the interaction of the two molecules is energetically favorable. The pentane molecule, however, has neither hydrogen bonding nor polar groups to be attracted to water molecules. The water molecules tend to stick together and remain separate from the pentane. Because pentane is less dense than water, it floats above it. Substances that interact strongly with water, mixing well with it or dissolving in it, are called hydrophilic (water loving). Substances that interact poorly with water and tend to remain separate are called hydrophobic (water-fearing).

The effect of polarity (charge asymmetry) is also evident in the solubility of compounds. For example, a bent, polar molecule like **sulfur** dioxide, SO_2, is much more soluble in polar **solvents** than is linear, nonpolar **carbon dioxide**, CO SO_2. This effect is also additive when molecules are complex. Alcohols with very few nonpolar groups are much more soluble in water than are alcohols with many nonpolar groups. Also, compounds with many polar groups, such as the sugars and alcohols like **glycerol** with three -OH groups, dissolve better in water and low **molecular weight** alcohols than alcohols with a lower ratio of polar to nonpolar groups. This tendency is frequently referred to as "like dissolves like"; e.g., polar solvents like water dissolve polar solutes like sugar, whereas nonpolar solutes like carbon tetrachloride dissolve nonpolar solutes like oils and other hydrocarbons.

A readily observed phenomenon associated with hydrogen bonding is the release of heat when two hydrogen-bonding solvents are mixed. If the average intermolecular interaction in the combined system is stronger than in the two separate liquids, heat is released. There can be a noticeable temperature rise upon mixing whenever intermolecular interactions between different molecules are stronger than the interactions between like molecules, whether these associations are due to London forces, dipole-dipole interactions or hydrogen bonding. It is typically easiest to observe a heat of mixing when hydrogen bonding occurs, because this is the strongest class of intermolecular interaction.

When a capillary, a narrow tube, touches the surface of a liquid, fluid rises into the tube. The extent to which a liquid rises is different for different liquids. When a narrow tube is inserted into water, the water rises in the tube. This occurs because the surface of **glass** is quite polar. As water molecules rise along the inside surface of the capillary, they pull up other water molecules to which they have formed hydrogen bonds. The **balance** of gravity and the attraction of the water for the glass surface determine the height to which the water rises. Other polar liquids besides water also rise in capillaries, but some nonpolar liquids show the opposite effect, the height of the liquid inside the capillary is less than outside. The molecules of these liquids are attracted to each other more than they are to the surface of the glass. Liquid **mercury** shows an especially large effect because the mercury atoms are attracted to each other much more strongly than they are attracted to the glass surface. Capillary action is also responsible for absorption of liquids into **paper**, such as paper towels.

See also Van der Walls forces

INTERNATIONAL SYSTEM OF UNITS

Measurement is one of the hallmarks of civilization. How far is it? How much does it weigh? What time is it? These are all questions that we hear every day. Indeed, asking the time is probably the number one question that we ask, especially when we are doing something we don't like!

Simple units, such as the cubit which is the distance between a man's elbow and the tip of his middle finger, or the span, the distance from the tip of the thumb to the tip of the little finger of an outstretched hand, were used for much of recorded history. After all, the human body is always available for measuring distance. But it does suffer from one inherent problem. Not all men are created equal; some are taller than others so that not all forearms are the same length.

Of course, this led to disputes and difficulties. Solving such problems was probably the basis for the origin of civil law. How does one stop people from cheating by using a favorable measurement? The answer is to standardize the system of weights and measures. Records indicate that the Egyptians started this practice as early as 3000 B.C., probably as a result of the annual flooding of all of the arable land along the Nile. This necessitated accurate surveying every year and resulted in the early Egyptians developing an excellent understanding of geometry and surveying.

By the late eighteenth century, scientists and engineers were able to construct very precise measuring instruments but

Prefix	Symbol	Meaning
exa-	E	10^{18}
peta-	P	10^{15}
tera-	T	10^{12}
giga-	G	10^{9}
mega-	M	10^{6}
kilo-	k	10^{3}
hecto-	h	10^{2}
deka-	da	10^{1}
deci-	d	10^{1}
centi-	c	10^{2}
milli-	m	10^{3}
micro-	µ	10^{6}
nano-	n	10^{9}
pico-	p	10^{12}
femto-	f	10^{15}

Figure 1. Common prefixes. ***(Illustration by Electronic Illustrators Group.)***

the standard units for measurement were poorly defined. To develop a systematic and precise set of measurement standards, a commission, including the scientists Pierre-Simon Laplace (1749-1827), Joseph-Louis Lagrange (1736-1813), and **Antoine-Laurent Lavoisier**, was established. They developed a system based on objects or phenomena that could be measured reproducibly; these natural measurements were independent of man. For instance, the meter (from the Greek *metrein*, ''to measure''), the basic unit of length, was calculated to be 1/10,000,000 of the distance between the North Pole and the equator on a line running through Paris. Other units were worked out to interconnect with the meter. For example, the liter (a derived unit of volume) is the **volume** occupied by one cubic decimeter, while the kilogram (the basic unit of mass) is the **mass** of a liter of **water** at 4°C. Smaller and larger units were available by multiplying and dividing by powers of ten.

This so-called ''**metric system**'' was by far the most logical and scientifically based system of measurement devised. Metric units are related decimally (i.e. by powers of ten). There have been many meetings and conferences to establish exact interpretations for the metric system. The original meter was defined as the distance between two marks on a platinum/iridium bar kept in a locked, air-conditioned vault maintained by the International Bureau of Weights and Measures in Paris. In 1960, the General Conference of Weights and Measurements redefined the meter as 1,650,763.73 wavelengths of one of the spectroscopic lines of an **isotope** of the noble gas, **krypton**. In 1983, it was further refined to be 1/299,792,458 of the distance that light travels in a second.

The General Conference of Weights and Measurements in 1960 established a revised metric system as the ''International System of Units''. The French name is ''Le Système International d''Unités'' or the ''Système International'', for short, and it is this latter term that leads to the abbreviation ''SI''. Scientists all over the world have agreed to SI units as the basis for measurement. Indeed, with the exception of the United States, all of the countries in the world use SI units for commerce as well. The SI system is also called MKSA (meter, kilogram, second, ampere). The seven base units of the SI system, each of which is used for a particular physical quantity, are shown in figure 1.

Modification of the units is accomplished decimally using the prefixes shown in figure 2. For example, 5 teraseconds = 5 Ts = 5 x 10^{12}s, while 5 milliseconds = 5 ms = 5 x 10^{-3}s, and 5 femtoseconds = 5 fs = 5 x 10^{-15}s. While the base SI unit for weight is the kilogram (kg), for many purposes in **chemistry** the smaller weight unit, gram (g), is more convenient; 1 kg = 1000 g.

All other physical quantities can be derived from appropriate combinations of the seven basic units. Examples of these derived units include: 1) The SI unit of force is defined as the Newton (N) which is a kilogram times the acceleration in meters per second squared or N = kg m/ s^2. 2) The SI unit of **energy** is the Joule (J); this unit is defined as the amount of energy required to apply a Newton of force over the distance of one meter or J = N m. 3) The unit of charge, the Coulomb (C), is the amount of **electricity** in 1 Ampere during 1 second of time (C = A s). 4) The SI unit of area is the square meter

Physical Quantity	Name of Unit	Symbol
length	meter	m
mass	kilogram	kg
time	second	s
electric current	ampere	A
temperature	kelvin	K
luminous intensity	candela	cd
amount of a substance	mole	mol

Figure 2. Basic units of measurement. ***(Illustration by Electronic Illustrators Group.)***

or m^2. 5) The SI unit of volume is the cubic meter or m^3. 6) The unit of **density** (*d*) is the kilogram per cubic meter or $d = kg/m^3$.

The United States has been reluctant to completely adopt the SI units and still favors older units in some cases. Traditional units for distance such as the mile (= 1.609 km) and the yard (= 0.9144 m) are commonly employed. Other commonly used non-SI units are the pound (=0.45359 kg) for weight and the quart (=0.9463 L = 0.9463 dm^3 for volume. Moreover, there has not been complete adoption of the SI units in the scientific community and in some instances traditional metric units, called CGS units, continue to find use. This reluctance to change completely to SI units is likely mostly due to comfort and tradition. Nevertheless, the older units are slowly being supplanted by the SI system.

Constructing derived units is relatively straightforward in the SI system which is, in part, the reason that scientists have adopted it. However, it also has one other distinct advantage over some older systems of measurements and that is scalability. A meter is the basic unit of length but shorter and longer units are obtained by multiplication by factors of ten. For example, a centimeter is 1/100 of a meter, while a kilometer is 1,000 meters in length. Similarly, a centigram is 1/100 of a gram, while a kilogram is 1,000 grams. This commonality of prefixes for shifting powers of ten makes scientific measurement very much easier, particularly where scientists are investigating the very small and the very large. For example, the **nucleus** of an **atom** is about 1 fm or 1×10^{-15} m in diameter and our galaxy is about 10 Exameters or 1×10^{19} m in diameter.

INTERNATIONAL UNION OF PURE AND APPLIED CHEMISTRY (IUPAC)

The International Union of Pure and Applied Chemistry (IUPAC) is an international scholarly scientific organization dedicated to the advancement of chemistry. Formed in 1919, IUPAC is regarded by chemists as the preeminent authority on chemical **nomenclature**, measurements (e.g., atomic weights), and chemical physical constants.

In order to improve global communication among chemists, one of the principle aims of IUPAC is to standardize the terminology of chemistry. International standardization of chemical terms and standardization of names, symbols, and measures is important to foster cooperation and understanding among chemists working in many native languages.

IUPAC's role in standardizing chemical terminology has reduced the chaos of having different names for the same elements, molecules, or compounds. When new elements, for example, are discovered, they are reported to IUPAC so that they can be named according to the governing rules of IUPAC. Researchers who discover new elements are allowed to suggest and make recommendations regarding terminology as long as the proposed name fits within the guidelines for naming established by IUPAC. When, for example, a new organic compound is synthesized, IUPAC guidelines state that the name must based on the longest chain of **carbon** atoms—with sides **chains** also numbered in a particular manner.

Standardized chemical nomenclature allows scientists more than just an ability to identify an element or chemical by a name that everyone uses. According to IUPAC guidelines, chemical names must be based upon the structure of the compound, so that scientists can more easily and systematically understand the structure and consequently the various properties of a chemical. IUPAC maintains a number of historical names called trivial names that identify well-known elements and compounds.

IUPAC is also responsible for publishing up-to-date data on atomic standards (e.g., weights) and other data, constants, and standards of interest to chemists. In addition, IUPAC also helps chemists establish and standardize methods used in chemistry laboratories.

IUPAC members and committees participate in numerous geochemical and environmental studies. IUPAC has published the results of studies on trace elements in the environment, **pesticides** in **water**, and a glossary of atmospheric weather terms. IUPAC's publications include scholarly books, a bimonthly news magazine, *Chemistry International*, and a monthly scholarly periodical, *Pure and Applied Chemistry*.

IUPAC is actually an association of national bodies composed of more than 1,000 leading chemists representing more than 40 countries. In addition, more than a dozen countries informally participate in IUPAC as observers. IUPAC is organized into seven primary divisions: Physical Chemistry; Inorganic Chemistry; Organic Chemistry; Macromolecular; Analytical Chemistry; Chemistry and the Environment; and Chemistry and Human Health. IUPAC provides scientific advice to United Nations agencies [including the United Nations Education, Scientific and Cultural Organization (UNESCO)], the World Health Organization, the International Organization for Standardization (ISO), and other governmental and scholarly bodies.

In addition to being an authoritative scientific organization, IUPAC sponsors international meetings on many topics of concern to scientists. As a service, IUPAC sponsors conferences focused on social and political impacts of chemical research and applications. Topics have included the availability of raw materials and nutritional chemistry.

IODINE

Iodine is the fourth element in Group 17 of the **periodic table**, a group of elements known as the **halogens**. Iodine has an **atomic number** of 53, and atomic **mass** of 126.9045, and a chemical symbol of I.

Properties

Iodine is a heavy, grayish black, metallic-looking solid which, when heated, does not melt but, instead, sublimes to give a beautiful violet vapor. If a cold object, such as an **iron** bar, is placed in the violet vapors, attractive, delicate, metallic crystals of iodine condense on the bar. If heated under pressure, iodine can be made to melt at a **temperature** of 236.3°F (113.5°C) and to boil at a temperature of 363°F (184°C). The element has a **density** of 4.98 grams per cubic centimeter. It is only slightly soluble in **water**, but it dissolves in many organic **liquids**, such as **alcohol** and **carbon** disulfide, to which it imparts a distinctive purple **color**.

Like the other halogens, iodine is an active element. As its position in the periodic table would suggest, it is less active than the three halogens above—fluorine, **chlorine**, and **bromine**. Iodine's most common compounds are those formed with the **alkali metals sodium** and **potassium**. But it also forms compounds with many other elements, even with other halogens. Examples of the latter compounds are iodine monobromide (IBr), iodine monochloride (ICl), and iodine pentafluoride (IF_5).

Occurrence and Extraction

Iodine is not very abundant in Earth's crust. Its abundance is estimated to be about 0.3-0.5 parts per million, placing it in the bottom third of the elements in terms of abundance. Iodine never occurs as the free element, but usually as sodium iodide (NaI) or potassium iodide (KI). These compounds were present in ancient oceans and were deposited in massive **salt** beds when those oceans dried up. They were long ago buried underground by earth movements and exist now in the form of salt domes. Sodium iodide and potassium iodide mined from these salt domes is the primary source of the element today. Compounds of iodine are also present in sea water and in sea kelp, a form of seaweed.

Iodine can be extracted from its compounds by replacement with a more active halogen, as, for example: $2NaI + Cl_2 \rightarrow 2NaCl + I_2$.

Discovery and Naming

Iodine was discovered in 1811 by the French chemist **Bernard Courtois**. Courtois' family lived on the coast of Brittany, where they extracted chemicals from seaweed by treating the plants with **sulfuric acid**. On one occasion, Courtois accidentally added too much sulfuric acid to the seaweed and noticed that a beautiful violet vapor formed. Upon condensing the vapor, he obtained the first sample of pure iodine. He suggested the name iodine for the element after the Greek word *iodes*, which means "violet."

Uses

About two-thirds of all iodine and iodides produced are used to kill disease-causing organisms, either in sanitation systems or in the manufacture of various antiseptics and drugs. The element is also used to make dyes, photographic film, specialized soaps, and catalysts.

Health Issues

Iodine has both favorable and unfavorable effects on living organisms. It tends to kill bacteria, and other disease-causing organisms, accounting for its once widespread use in sanitation systems and as an antiseptic. Not so long ago, tincture of iodine was one of the most popular of antiseptics. The element and its compounds are now much less used for these purposes, however, since higher doses can irritate or burn the skin and can be toxic if taken internally.

ION

An ion is an **atom** or group of atoms that has lost or gained one or more electrons. This produces a charged particle, which can be either positive or negative.

Normally, an atom is neutrally charged. When negatively charged electrons are lost from an atom the resultant ion is positively charged and it is called a **cation**. An ion that has gained electrons is negatively charged and called an **anion**. The name ion comes from the Greek *ienai*, meaning to go. This is related to their behavior during **electrolysis** where the ions move to the two opposite electrodes. Properties of ions are generally very different from the atoms from which they are derived.

The net charge on an ion is denoted by a superscript showing both the size and charge. A cation can have +1, +2, or +3, these cases are brought about by the loss of one, two, or three electrons respectively. An anion can be -1, - 2, or -3, and these are brought about by the addition of one, two, or three electrons respectively. In general, metal atoms and **hydrogen** tend to lose electrons and non- metal atoms (except hydrogen) tend to gain electrons. The commonest charge on a cation is +2 and the commonest charge on an anion is -1. Some cations can have different charges due to different numbers of electrons lost. For example, **iron** can be +2 or +3. This is usually written as iron (II) and iron (III).

When the number of electrons of an atom are altered, it alters the physical shape of the atom. A cation has a smaller size than the atom from which it is created. Normally, the electrons surround the **nucleus** of an atom and contained within the nucleus are the positively charged protons. In a neutral atom, the charge of the electrons balances the charge of the protons, but when a cation is produced, the charges are in an imbalance. There is a net positive charge—there are more protons in the nucleus than electrons surrounding it. The protons are able to attract the smaller number of electrons towards the nucleus more strongly because of this imbalance in charge. This physically reduces the size of the atom. With an anion, the reverse is true. The anion is physically larger than the atom from which it was created because the electrons are being held less tightly—the attractive forces of the protons are spread over a greater number of electrons.

When atoms lose or gain electrons, they are attempting to become like the noble gas closest to their position in the **periodic table**. Noble **gases** are very unreactive and they have a stable **electron** structure. By losing or gaining electrons, the atoms are trying to achieve the same level of stability.

A single atom that loses or gains electrons produces a simple ion. There are more complex ions possible where several atoms are involved. Polyatomic ions are groups of atoms that are positively or negatively charged. The hydroxyl ion (OH^-) is an example of a polyatomic anion.

When two atoms swap electrons to produce a cation and an anion, the two ions are attracted to each other. This is due to the fact that opposite charges attract each other and the anion and the cation are oppositely charged. The bond that is formed between the two ions is an **ionic bond** (sometimes also called an electrovalent bond), and the compound produced is an ionic compound. Since **metals** tend to produce cations and nonmetals tend to produce anions, ionic compounds are usually combinations between a metal and a **nonmetal**. This is actually an oversimplification of what occurs as there is no discrete ionic compound containing only one cation and one anion. What is actually produced is a three dimensional matrix of cations and anions bound together. These regular patterns have each ion surrounded by examples of the other ion. They are packed together in a regular arrangement called a giant structure or a lattice. This usually leads to a physical structure that is crystalline in nature. These strong interactive forces between the ions give certain characteristics to ionic compounds. Ionic compounds tend to be **solids** at room **temperature**; they have high melting and boiling points; they are hard substances; and they cannot conduct **electricity** as a solid, but they can conduct electricity as a liquid or in aqueous phase. This last characteristic is because as solids the ions are not free to move but as a liquid they are. In **solution**, the cations move towards the cathode and the anions move to the **anode**. This is the principal behind electrolysis. Once at their respective electrodes electrons are gained or lost to return the ion to the neutrally charged state. The practical result of this is that the anions are given off at the anode, generally in the form of a gas and the cations are often deposited at the cathode as a metal layer.

When an ionic bond is formed, **energy** is released to give a stable **molecule**. This is due to the **ionization** energy (the energy required to remove an electron from an atom or ion) of the cation being small and the electron affinity (the energy released when an electron is added to a neutral atom) of the anion being large. As a general rule if a cation is from the left-hand side of the periodic table and the anion is from the right-hand side then an ionic bond will be formed. When this occurs, the larger the charge on the ions, or the smaller the size of the ions, the stronger will be the ionic attraction, and hence, the stronger will be the ionic bond produced.

The ionic bond is the name of the bond formed between two oppositely charged ions. The cation and anion bonded together are known as an ion pair. The compound produced when such a bond is formed is an ionic compound. A process called **ion exchange** exists whereby ions in solution are exchanged for ions that are bound on the surface of a resin. The resin is an ion exchange resin. This technique can be used to change hard **water** into soft water. An ion can be formed by ionizing radiation. This is an energetic form of radiation which when it strikes an atom can cause the atom to lose an electron, producing an ion. The energy required to remove an electron form an atom is called the ionization energy. Ionic compounds are electrically neutral because the number of cations balances the number of anions and the charges cancel each other out. When the chemical formula of an ionic compound is written down, it is the **empirical formula** that is used. This is because in reality in an ionic compound the ions are surrounded by ions of the other type, producing a giant structure.

See also Anion; Cation; Ionic bond

ION EXCHANGE

Ion exchange is a reversible exchange of ions between a solid ion exchanger and a liquid **solution**. The solid undergoes no structural change in the process. An ion exchanger, or ion exchange resin, can be a natural or synthetic substance which can exchange its own ions with the ions present in a solution which is passed through it. The ion exchanger is porous and allows for a solution to permeate through the material. As the solution passes through the ion exchanger and leaves as effluent, it contains ions different from those with which it began.

Ion exchange is usually associated with **water** treatment. The first industrial application of ion exchange technology was in 1905 when a sodium-aluminosilicate **cation** exchanger was used to soften water. As water passes through an ion exchange resin, harmful ions found in the water become attached to the resin, which releases its harmless ions into the water. The resin is usually composed of synthetic beads. Each bead is a polymer matrix containing ion exchange sites on the surface and within the matrix itself.

Anions can only be exchanged for other anions, and cations for other cations. The ion exchange resin that is used is therefore specific for the type of ion to be removed from the solution. An **anion** exchange resin has positively charged ion exchange sites with anions attached, and cation exchange resins have negatively charged ion exchange site with attached cations. The ion exchange resin usually originates with attached ions that have low affinities for the exchange sites. As the solution containing ions flows through the resin, the ions with the most affinity for the exchange sites will replace those with the lowest affinities. It is important, therefore, that the ion exchange resin contain ions with a lower affinity than those which need to be exchanged. For example, cation exchange resins usually come with **sodium** (Na) or **hydrogen** (H) ions attached to the exchange sites. Both of these ions have low affinities to the sites. Almost any cation which comes in contact with the resin will have a greater affinity and replace the hydrogen or sodium ions at the exchange sites. Anion exchange resins often use chloride (Cl) or hydroxyl (OH) ions because of their low affinities for the exchange sites as well.

Different ions can have different charges as well as affinities. A monovalent ion has a single charge, a divalent ion has two charges, and a trivalent ion has three charges. The charge on a resin bead must always remain neutral. This means that for every divalent ion of greater affinity which attaches to a bead, two of the ions must be released from the bead. For every trivalent ion which attaches to a bead, three of the ions must be released from the bead. In water treatment, the ion exchange resin which is used is usually more selective for divalent and trivalent ions, which is practical since the problematic ions are usually one of the two.

Ion exchangers are reversible, that is, they can be regenerated. Once a solution has passed through an ion exchange resin, the resin beads now contain the undesirable ion and the original ion which was attached to the bead can be found in the effluent. The unwanted ions can be removed and replaced by the original ions by passing a solution containing these ions in a much greater **concentration** than would be found in an untreated solution. For example, an ion exchange resin which originally contained Na^+ ions could be regenerated after use by washing it with a concentrated solution of **sodium chloride**, NaCl. When a solution of such great concentration is passed through an ion exchange resin, the resin is more selective for the monovalent ions simply because of the great number of ions bombarding the resin beads. When so many collisions occur, it is very likely that the unwanted ions will be removed and replaced. The early ion exchange resins could not be regenerated. The original resins were "**gel**" type materials which were not very porous at all. Because they were not porous, there were not as many opportunities for the collisions to occur during regeneration, and the higher affinity ions could not be released. In the 1950s, macroporous ion exchange resins were developed with the ability to be regenerated, which eventually evolved into the ion exchange beads and membranes in use today.

There are many different ion exchange resins available for use. Which resin used depends on the ion that needs to be removed from solution. Water softeners, for example, are designed to remove the **calcium** (Ca^{2+}) and **magnesium** (Mg^{2+}) cations from hard water. The ion exchange resin in water softeners contain cations such as Na^+ or H^+. When the hard water passes through the water softener, the resin exchanges its Na^+ or H^+ ions for the Ca^{2+} and Mg^{2+} ions. Ion exchange resins containing $OH/^-$ or Cl^- anions will exchange with anions such as HCO_3^-, CrO_4^{2-}, NO_3^-, or SO_4^{2-} in a solution. The first water softeners developed in the early 1900s only operated over a narrow **pH** range, in fact, only slightly basic water could be treated. The development of both cationic and anionic exchange materials did not occur until 1935.

Ion exchange resins have different abilities to exchange ions, called the ion exchange capacity of the resin. This is a numerical value which represents the quantity of ions that an ion exchange resin is able to release coupled with the quantity of ions it can acquire. This exchange capacity can be expressed as an absolute number (the total exchange capacity) or as an operational value (the practical exchange capacity). The total exchange capacity is measured from the time the solution begins to permeate through the resin until the resin cannot exchange any more ions. The practical exchange capacity is measured from the time the solution begins to permeate through the resin until the effluent reaches a set tolerable concentration of exchanged ions.

Certain conditions must be kept constant during the ion exchange process, especially if determining the ion exchange capacity of a resin, because the exchanging ability of a resin is a function of the affinity of the ions towards the exchanger. The contact time needs to be kept constant, for one. The contact time is the amount of water which passes through the exchanger with respect to the **volume** of the exchanger. The **temperature** needs to be kept constant, because temperature changes can affect the size of the pores in the ion exchange resin. The pH value of the water must be held constant as well, because the **dissociation** of the ions attached to the resin can be effected by changes in pH.

There are many applications of ion exchange in use today. Water treatment has already been mentioned as one of the leading uses of ion exchange resins. Other applications include the separation of **antibiotic drugs** from **fermentation** broths, recovery separation of products during laboratory procedures, and **pollution** control. Manufacturers utilize ion exchange to remove toxic substances from their waste waters before discharging the water into the environment. The use of ion exchange resins is widespread and growing quickly as more and more industries realize the ease with which it can be used.

IONIC BOND

A chemical bond is a mutual attraction between the nuclei and **valence** electrons of two different atoms. This attraction results in the two atoms binding together. An ionic bond, also called an electron-transfer bond, is a type of chemical bond that is a result of the electromagnetic attraction between ions of opposite charges, i.e., a **cation** (a positively charged ion) and an **anion** (a negatively charged ion). An **ion** is an **atom** or group of atoms that has acquired an electrical charge due to the loss or gain of electrons. In an ionic bond, an atom gives or receives electrons from another atom. This is in contrast to covalent **bonding**, where two atoms share **electron** pairs between them. An ionic compound consists of anions and cations combined such that the total charge of the **molecule** is zero. All salts are ionic compounds.

One characteristic that both ionic and covalent compounds share is that they adhere to the **octet rule**. The octet rule is the principle that describes the bonding in atoms. Individual atoms are unstable unless they have an octet of electrons in their highest **energy** level. The electrons in this level are called valence electrons. When atoms gain, lose, or share electrons with other atoms, they satisfy the octet rule and form chemical compounds.

Ionic bonding occurs when one atom transfers electrons to another atom. In doing so, the atoms may achieve a complete outer **energy level**, satisfying the octet rule. During the formation of an ionic bond, one atom gains electrons and the other atom loses electrons. As a result, the atoms gain an electric charge. The atom that gains electrons gains a negative charge, becoming an anion. The atom that loses electrons gains a positive charge, becoming a cation. An example of this is the ionic bond that is formed between **sodium** (Na) and **fluorine** (F) to make sodium fluoride, NaF. A fluorine atom has seven valence electrons. It needs one more to have a complete outer energy level and satisfy the octet rule. If it gains this electron, it will become a negatively charged fluorine anion (F^-). A sodium atom has only one valence electron. If it loses this electron, it is left with a complete outermost energy level that satisfies the octet rule. At the same time, it becomes a positively charged sodium cation (Na^+). As the sodium atom loses its electron, the fluorine atom picks it up. The two ions now have opposite charges and attract each other. This electromagnetic attraction is quite strong and holds the ions together, forming sodium fluoride. This binding of the two ions is called an ionic bond.

Energy is required for an atom to lose an electron. The process of losing an electron and forming an ion is called **ionization**. The energy that is needed for ionization to occur is called the ionization energy. The ionization energy needs to be sufficient to overcome the attraction between the positively charged **nucleus** and the negatively charged electron. If an atom has only a few valence electrons, the ionization energy is low. The removal of a small number of electrons does not require much energy. Atoms with few valence electrons tend to lose these electrons easily and become cations. Sodium is an example of this phenomenon. A sodium atom only has one valence electron, which it loses quite easily. Thus, a sodium atom has a low ionization energy. In contrast, atoms with many valence electrons have high ionization energies. The removal of several electrons requires more energy. These atoms do not lose electrons easily, instead, they tend to gain electrons and become anions. An atom that gains electrons easily is said to have electron affinity. Fluorine is an example of an atom with electron affinity. A fluorine atom has seven valence electrons, so it does not lose electrons easily. Instead, it tends to gain an electron in order to complete its outermost energy level.

Most bonds are not completely ionic nor are they completely covalent. How ionic or covalent a chemical bond is depends on how strongly the atoms of each element attract electrons. **Electronegativity** (a measure of an atom's tendency to attract electrons) can be used to predict whether a bond will be a nonpolar **covalent bond**, a polar covalent bond, or an ionic bond. If the difference between the electronegativities of two atoms is 2.0 or greater, the bond is an ionic bond. If the difference is 0.4 or less, the bond is nonpolar. If the electronegativity difference between two atoms is between 0.4 and 2.0, the bond is considered to be a polar covalent bond. The greater the electronegativity difference, the greater the polarity, and the greater the ionic character of the bond.

Many substances are formed through ionic bonding. As mentioned above, all salts are formed with ionic bonds. A familiar example of an ionic compound is table **salt**, found in nature as rock salt. Table salt is **sodium chloride** (NaCl). A sodium ion, Na^+, has a charge of 1+. A chloride ion, Cl^-, has a charge of 1-. When a sodium atom gives up an electron to become Na^+, and a **chlorine** atom gains this electron to become Cl^-, these atoms combine as NaCl, forming an ionic compound with no electrical charge. The majority of the rocks and minerals found on the Earth are formed using ionic bonding.

An ionic compound has a specific structure. Most ionic compounds are crystalline **solids**. An ionic compound is composed of a network of ions that results in a three-dimensional matrix of cations and anions. This crystalline structure is an orderly arrangement of ions known as a crystal lattice. A crystal lattice structure minimizes an ion's **potential energy**. Therefore, this structure is energetically favorable. The arrangement of ions in a crystal lattice represent the optimum **balance** between the forces of attraction among oppositely charged ions and the forces of repulsion among ions of the same charge.

The physical arrangement of ions in a crystal lattice depends upon the number of ions as well as the sizes and charges of the ions. An ionic compound cannot be isolated into individual, neutral units, like a molecular compound can. The chemical formula of an ionic compound, therefore, does not represent the formula for one molecule of the substance. Instead, it represents the simplest ratio of the ions that gives the compound no net electrical charge. The chemical formula of an ionic compound represents one formula unit of the compound, i.e., the simplest combination of atoms that gives the compound electrical neutrality. The ratio of ions in a formula unit depends on the charges of the ions in the compound. For example, the ionic compound **calcium** fluoride is composed of calcium cations (Ca^{2+}) and fluorine anions (F^-). In order to have a net charge of zero, two fluorine atoms must combine for every calcium atom. This relationship is represented in the compound's chemical formula, CaF_2.

The forces of attraction between ions in an ionic compound are very strong. The forces involved in ionic bonding are much stronger than the forces in a covalent bond. This difference in strength gives ionic compounds different properties than covalent compounds. The strong forces that hold ions together cause ionic compounds to have higher melting and boiling points than covalent compounds. Ionic compounds also do not vaporize as easily at room **temperature** as covalent compounds. Ionic compounds are very hard and also quite brittle, due to the crystal lattice structure. The ions in the crystal lattice cannot move very much without disturbing the overall balance between negative and positive charges. If one layer of atoms is moved, the result is a buildup of repulsive forces within the crystal structure, and the entire structure falls apart. Ionic compounds are good conductors of **electricity** when melted or dissolved in **water** because the ions dissociate in these states and are free to move and carry electrical current. In the solid state, the ions are not free to move, and the solid ionic compound does not conduct electricity.

Ionic compounds are found extensively in Earth's crust. They are very strong structures with unique properties. The properties of ionic compounds, for example, **electrical conductivity**, can be exploited for use in science and technology. Research into the nature of ionic bonding continues in order to find new uses for these compounds.

IONIC STRENGTH

Ionic strength is a characteristic of an electrolyte **solution** (a liquid with positive and negatively charged ions dissolved in it). It is typically expressed as the average electrostatic interactions among an electrolyte's ions. An electrolyte's ionic strength is half of the total obtained by multiplying the **molality** (the amount of substance per unit **mass** of solvent) of each **ion** by its **valence** squared.

Ionic strength is closely related to the **concentration** of electrolytes and indicates how effectively the charge on a particular ion is shielded or stabilized by other ions (the so-called ionic atmosphere) in an electrolyte. The main difference between ionic strength and electrolyte concentration is that the former is higher if some of the ions are more highly charged. For instance, a solution of fully dissociated (broken down) **magnesium sulfate** (Mg^{+2} SO_4^{-2}) has 4 times higher ionic strength than a solution of **sodium** chloride (Na $^+$Cl $^-$) of the same concentration. Another difference between the two is that ionic strength reflects the concentration of free ions, and not just of how much **salt** was added to a solution. Sometimes a salt may be dissolved but the respective ions still bound together pairwise, resembling uncharged molecules in solution. In this case the ionic strength is much lower than the salt concentration.

Ionic strength is an important factor in biochemical reactions and plays an essential role in the function of all living things. In order to survive, every organism must undergo a continuous series of chemical reactions that govern everything from **metabolism** to respiration. Ionic strength is a key factor in these reactions because it affects the rates at which ions react with each other and, thus, the extent to which the reaction occurs. Enzymes, protein molecules that catalyze and regulate reactions important to life, can also be extremely sensitive to ionic strength and may become insoluble or inactive if the organism's ionic strength is too high or too low, much in the same way that they are extremely sensitive to **temperature**.

Take for example, the case of a person running. When he or she begins to perspire he or she will lose moisture as well electrolytes, or ions. This loss of electrolytes is a practical example of how ionic strength works. If the runner does not replace those lost electrolytes, he or she will become thirsty, sluggish, and overheated. In chemical terms, the ionic strength of the runner's remaining electrolytes would be very high. Because the essential chemical reactions in the runner's body are affected by ionic strength, just as they are by temperature, for instance, the **balance** of chemical reactions of life would be disrupted and the runner would get sick. Ionic strength is one of the basic characteristics of an organism's chemical makeup that determines whether that organism can exist in a state of homeostatis, or internal stability. In higher animals, the kidneys regulate the body's ionic strength by maintaining electrolyte and **water** balances.

A more specific example of ionic strength's importance in the **chemistry** of life can be found in the substance acetylcholine, a positively charged ion that organisms release at the ends of certain neurons. Acetylcholine's jobs are to serve as a bridge between neurons, passing along nerve impulses from one to the next, and to start muscular contractions. Ionic strength determines the rate at which acetylcholine reacts with other chemicals in an organism, so if the ionic strength of the organism's electrolytes was too high, the acetylcholine would react at a rate too slow or too fast, or may bind too strongly or too weakly to its receptor for the organism to function normally.

Ionic strength is a also a useful parameter in the laboratory. If a researcher knows the ionic strength of an electrolyte, it can tell him or her a great deal about the dynamics of specific chemical reactions. In this respect, a known ionic strength value can be used as an experiment's control as a way to explore an unknown reaction. Scientists can also change an electrolyte's ionic strength to alter the outcome of a reaction in a measured, deliberate way.

High ionic strength chloride solutions are useful in many geological, environmental and industrial processes. For example, oil companies are researching the use of high ionic strength solutions in reducing the costs of controlling scale formation; the metallurgical industry uses them to improve the efficiency of **copper** extraction from chalcopyrite ores, and the nuclear waste industry uses them to understand the behavior of radionuclides in underground storage deposits where there are salt deposits. Because scientists do yet not fully understand the behavior of high ionic strength chloride solutions systems, most applications are based on observed models. But in the late 1990s, researches began focusing more powerful computers on the problem and have developed successful use of molecular dynamics in studying complex aqueous solutions. Such studies in molecular dynamics techniques may provide more practical applications of high ionic strength chloride solutions across a variety of industries.

IONIZATION

Ionization is the process in which an **electron** is pulled completely away from an **atom**, leaving an **ion** with positive charge. The **energy** necessary to accomplish this process is called the ionization energy or the ionization potential of the atom.

The first ionization potential of an element is the amount of energy required to remove one electron from the ground state orbital of highest energy, e.g., the **valence** shell, of the neutral atom. An ion with unit positive charge results. The second ionization potential is the energy needed to completely remove an electron from the valence shell of the ion that results from the first ionization. Similar definitions may be given for the third, fourth, and higher ionization potentials.

Ionization is closely related to **electronegativity**. The electronegativity of an atom is a measure of its ability to attract an electron and hold on to it. Its value is an average of the ionization potential and the electron affinity of the atom. The latter is a measure of the ability of a neutral atom to attract an extra electron.

In 1859, **Julius Plücker** discovered that when a high voltage difference is established between two electrodes in a container of dilute gas, an electrical discharge occurs between the electrodes and cathode rays are produced. **J. J. Thompson** showed that cathode rays are made up of electrons. The high voltage difference pulls electrons from the gas, and thus the gas is ionized. The cathode ray tube, a **glass** container in which this process occurs, has found a number of commercial applications, e.g., the television tube. A similar apparatus may be used to measure the ionization potentials of **gases** by measuring the voltage difference required to produce a flow of electrons. The values of ionization potentials are generally given in volts.

The variation of ionization potentials among the various elements is primarily due to differences in the attraction between the positively charged **nucleus** and the valence electron. This electrostatic attractive force is determined by the effective charge on the nucleus felt by the electron and by the distance between the nucleus and the valence shell electron. The effective nuclear charge is less than the actual nuclear charge since shielding by the other electrons of the atom diminishes the nuclear charge felt by the valence shell electron. A comparison of the ionization potentials of **neon** and **argon** will illustrate these principles. The first ionization potential of neon is 21.5 volts and that of argon is 15.7 volts.

The valence shell of neon (atomic number 10) is 2p; for argon (atomic number 18), the valence shell is 3p. The valence electron that is ionized in neon is shielded from the nuclear charge (10) by the 1s, 2s, and the other 2p electrons. In argon, the valence electron is shielded by 1s, 2s, 2p, 3s, and the other 3p electrons. Although argon has a larger nuclear charge than neon, much of the difference is cancelled due to additional shielding of the valence electron, and the effective nuclear charge of neon and argon are virtually the same. The distance between the 3p electron of argon and its nucleus is, however, greater than the distance between the 2p electron of neon and its nucleus. Since the electrostatic attraction between two charges is inversely proportional to the distance between them, the 2p electron of neon feels a greater attractive force than the 3p electron of argon. Therefore, since the effective nuclear charges felt by the valence electrons in neon and argon are essentially equal due to shielding, neon would be expected to have a larger ionization potential than argon.

Now compare the ionization potentials of **sodium** and **magnesium**. Their first ionization potentials are 5.1 and 7.6 volts respectively. The electron removed from both sodium and magnesium is a 3s electron. This electron is shielded, in both sodium and magnesium, by 1s, 2s, and 2p electrons. In addition, the 3s electron of magnesium is shielded by another 3s electron. However, the shielding due to an electron in the same shell is much less than that of electrons in inner electron shells and is not sufficient to offset the larger nuclear charge in magnesium. The ionization potential of magnesium should, therefore, be greater than that of sodium. Similar reasoning may be applied to other ionization potentials.

An element's second ionization potential is the energy required to remove a second electron. The second ionization potential of magnesium (whose first ionization potential is 7.6 volts) is 15.0 volts, and its third ionization potential is 80.0 volts. Magnesium has 12 electrons. Two of them occupy the 1s orbital; two electrons are in 2s; 6 in 2p; and two valence shell electrons are in the 3s orbital. An electron in a 3s electron orbital is shielded less by another 3s electron than it is by electrons located in orbitals for which n is less. It should not be surprising that it is easier to remove the first electron that is shielded by another 3s electron than to remove the second electron that is not so shielded. The difference is, however, not large since the shielding is by an electron in the same shell. The third electron to be removed, however, is a 2p electron and is, therefore, closer to the nucleus than the first two electrons removed. Since the attractive force between opposite charges increases as the distance between them decreases, the third ionization potential of magnesium should be significantly larger than the first and second. These predictions agree with the empirical measurements given above.

The ionizations discussed to this point occur in a gaseous elemental state. Ionizations also occur in **solution**. When an acid, base, or **salt** is placed in aqueous solution, some or all of the molecules break apart into ions. The solvent **water** molecules are polar and stabilize the resulting ions by orienting the negative ends of their dipoles around positive ions (anions) and their positive ends around negative ions (cations). This type of ionization in solution is also called **dissociation**.

In some cases, the substance dissolved in the solvent exists both in its ionized and un- ionized state: $AC(aq) \leftrightarrow A^{+}(aq)$

+ C^-(aq) indicates that the substance is dissolved in water, and the double arrow ↔ indicates that the process can occur in both the forward and reverse directions. For this situation, it is found that, when equilibrium is reached, the product of the concentrations of the **anion** and **cation** divided by the **concentration** of the unionized material is always equal to a constant. This relationship may be written as $K = [A^+] \times [C^-] / [AC]$. The brackets [] indicate concentration, often in moles $liter^{-1}$. The constant is called the ionization constant. For example, when **acetic acid** HOAc is dissolved in water, its ionization constant, $[H^+] \times [OAc^-] / [HOAc]$, is equal to 2×10^5. This very small number indicates that the denominator is much larger than the numerator, and acetic acid exists in solution (e.g., in vinegar) primarily in the un-ionized form, with small amounts of the two ions present. The acidic nature of the solution is due to the presence of the H^+ anion in solution.

When some substances are placed in a solvent, the un-ionized form remains a solid, while some of it ionizes into cations and anions in the solution: CD(solid) ↔C^+(aq) + D^-(aq). When the process reaches equilibrium, it is found that the product of the concentrations of the two ions in solution is always equal to a constant. In this case, the constant is called the **solubility** product constant. The relationship is written: $K_{SP} = [C^+] \times [D^-]$.

An example is dissociation of the the salt **silver** chloride AgCl in aqueous solution. When it dissolves in water, AgCl breaks down into the ions Ag^+ and Cl^-, and the un-ionized AgCl remains a solid: AgCl(s)↔ Ag^+(aq) + Cl^-(aq). The solubility product constant for AgCl is 1.6×10^{-10}. Since the K_{SP} is very small, AgCl is not very soluble, and very little of it ionizes in solution.

Another important ionization phenomenon occurs as a result of ionizing radiation. High-energy particles are emitted from radioactive substances. When these particles pass through a gas they collide with molecules of the gas and cause them to ionize. The Geiger counter, a device used to detect the presence of **radioactivity**, works on this principle. Charged plates within an enclosed container of gas attract the charged particles produced by the ionizing radiation to them, and a measurable current results. When high energy ionizing radiation strikes a biological sample, e.g., the human body, it will ionize molecules within the sample. The most common ill effects of exposure to radioactivity result from the ionization of water molecules by the radiation. In this process water looses an electron, becoming positively charged. The positively charged water **molecule** reacts further to produce highly reactive OH radicals which in turn react with almost any molecule they encounter. The net result is interference with biological processes and can lead to the formation of cancerous cells.

The **photoelectric effect**, discovered in 1890 by Wilhelm Hallwachs (1859-1922), is the result of a similar ionization process. When light shines on certain negative metallic electodes (cathodes) in a vacuum, the metal is ionized and cathode rays, i.e. electrons, are emitted. This phenomenon has found many technological applications.

Iridium

Iridium is one of the transition **metals**, the elements that occur in the middle of the **periodic table**. It is also classified as a member of the **platinum** group of metals, along with **ruthenium**, **rhodium**, **palladium**, **osmium**, and platinum. Iridium's **atomic number** is 77, its atomic **mass** is 192.217, and its chemical symbol is Ir.

Properties

Iridium is a silvery white metal with a melting point of 4,429°F (2,443°C) and a **boiling point** of about 8,130°F (4,500°C). It has a **density** of 22.65 grams per cubic centimeter, making it the most dense of all elements. Iridium is neither very ductile nor malleable at room **temperature**, although it becomes more ductile at higher temperatures. Iridium is a relatively inert element at room temperature, although it becomes more active at elevated temperatures and reacts with **oxygen**, the **halogens**, and some other elements and compounds.

Occurrence and Extraction

Iridium is one of the rarest elements in Earth's crust with an abundance estimated at about two parts per billion. The element usually occurs in combination with one or more of the other platinum family elements. It is extracted by a complex series of reactions by which it is separated from other platinum metals with which it occurs.

Discovery and Naming

Iridium was discovered in 1803 by the English chemist **Smithson Tennant**. Tennant was studying a new mineral that contained a number of platinum metals and discovered a black powder that other chemists had also seen, but had not analyzed. Tennant was able to demonstrate that the black powder was a new element. He chose the name iridium for the element after the Greek goddess Iris, whose symbol is a rainbow. Tennant chose the name because the compounds of iridium have a wide variety of colors ranging from dark red to olive-green to dark green to blue-black.

Uses

Very little iridium is produced each year, probably no more than a few metric tons. Most of the metal is used to make alloys with other members of the platinum family, especially platinum itself. Iridium adds strength to an **alloy**, but such alloys are quite expensive. They are used only for rather specialized purposes, such as sparkplugs in helicopters, electrical contacts, and special types of electrical wires.

Iron

Iron is a transition metal, one of the elements in the middle of the **periodic table** between Groups 2 and 13 in Rows 4, 5, and 6. Iron's **atomic number** is 26, its atomic **mass** is 55.847, and its chemical symbol is Fe.

Properties

Iron is a silvery white or grayish metal that is ductile and malleable. It is one of only three naturally occurring magnetic elements, the other two being its neighbors in the periodic table, **cobalt** and **nickel**. Iron has a very high tensile strength and is very workable, capable of being bent, rolled, hammered, cut, shaped, formed, and otherwise worked into some desirable shape or thickness. Iron's melting point is 2,797°F (1,536°C) and its **boiling point** is about 5,400°F (3,000°C). Its **density** is 7.87 grams per cubic centimeter.

Iron is an active metal that combines readily with **oxygen** in moist air to form iron **oxide** (Fe_2O_3), commonly known as rust. Iron also reacts with very hot **water** and steam to produce **hydrogen** gas and with most acids and a number of other elements.

Occurrence and Extraction

Iron is the fourth most common element in Earth's crust, and the second most common metal after **aluminum**. Its abundance is estimated to be about 5%. Scientists believe that Earth's core consists largely of iron, and the element is found commonly in the Sun, asteroids, and stars.

The most common ores of iron are hematite and limonite (both primarily ferric oxide; Fe_2O_3) and siderite (iron carbonate; $FeCO_3$). An increasingly important source of iron for commercial uses is taconite, a mixture of hematite and silica. Taconite contains about 25% iron. The largest iron resources in the world are found in China, Russia, Brazil, Canada, Australia, and India.

The traditional method for extracting pure iron from its ore is to **heat** the ore in a blast furnace with limestone and coke. The coke reacts with iron oxide to produce pure iron, while the limestone combines with impurities in the ore to form a slag that can then be removed from the furnace: $3C + 2Fe_2O_3 \xrightarrow{heat} 3CO_2 + 4Fe$.

Iron produced by this method is about 90% pure and is known as pig iron. Pig iron is generally too brittle to be used for most products and is further treated to convert it to wrought iron, cast iron, or steel. Wrought iron is an **alloy** of iron and any one of many different elements, while cast iron is an alloy of iron, **carbon**, and **silicon**. Steel is a generic term that applies to a very wide variety of alloys.

Discovery and Naming

Iron is one of a handful of elements that have been known and used since the earliest periods of human history. In the period beginning in about 1200 B.C. iron was so widely used for tools, ornaments, weapons, and other objects that the period was given the name the Iron Age. In some ways, one could refer to the current period in human civilization as the New Iron Age because of the metal's importance in such a wide variety of applications today. The chemical symbol for iron, Fe, comes from the Latin name for the element, *ferrum*.

Uses

The number of commercial products made of iron and steel is very large indeed. The uses of these two materials can generally be classified into about eight large groups, including (1) automotive; (2) construction; (3) containers, packaging, and shipping; (4) machinery and industrial equipment; (5) rail transportation; (6) oil and gas industries; (7) electrical equipment; and (8) appliances and utensils.

A block of iron. ***(Photograph by Charles D. Winters, Photo Researchers, Inc. Reproduced by permission.)***

A relatively small amount of iron is used to make compounds that have a large variety of applications, including dyeing of cloth, blueprinting, insecticides, **water purification** and sewage treatment, photography, additive for animal feed, fertilizer, manufacture of **glass** and **ceramics**, and **wood** preservative.

Health Issues

Iron is of critical important to plants, humans, and other animals. It occurs in hemoglobin, the **molecule** that carries oxygen in the blood. The U.S. Recommended Daily Allowance (USRDA) for iron is 18 mg (with some differences depending on age and sex) and it can be obtained from meats, eggs, raisins, and many other foods. Iron deficiency disorders, known as anemias, are not uncommon and can result in fatigue, reduced resistance to disease, an increase in respiratory and circulatory problems, and even death.

ISOELECTRONIC PRINCIPLE

The isoelectronic principle states that molecules with the same number of electrons and atoms will have similar structures and chemical properties. The word "isoelectronic" itself means "the same number of electrons," and should not be confused with the term "isoelectric," which concerns electrical potential.

Most generally, the **valence** electrons of molecules (those in the outer shell that form chemical bonds), are isoelectronic. However, the state can also apply only to the inner electrons of molecules (those closer to the nucleus), or both. If two molecules are isoelectronic for both the inner and outer electrons, they are likely to be even more similar in most ways. **Carbon dioxide** and **nitrous oxide** (laughing gas) are isoelectronic molecules (22 electrons for both), so their properties are remarkably similar. They currently share the dubious distinction of being two of the most potent greenhouse **gases** in the atmosphere.

Just as the **Periodic Table** is used to predict the behavior of isoelectronic elements, the number of electrons within a **molecule** ("**functional group**") can also be used to predict the behavior of these molecules or functional groups. Their isoelectronic molecules or functional groups will behave similarly. Also, if two unknown molecules are behaving similarly, at times it may be hypothesized that they are isoelectronic or contain an isoelectronic functional group, aiding in their characterization.

Chemists use isoelectronic molecules in a number of ways, but in general this property almost always provides a convenient shortcut for experiments and analysis. For instance, consider a chemist who is doing an experiment involving **carbon dioxide**. Because molecules of carbon dioxide are isoelectronic with nitrous oxide molecules, he or she can assume that nitrous oxide will react similarly to the way in which carbon dioxide reacts in the experiment. Working with isoelectronic molecules allows chemists to predict a whole range of properties and characteristics for multiple substances while only having to deal directly with a single material.

Another way chemists use the common denominators of isoelectronic molecules to advantage is to avoid dealing with a dangerous or inconvenient substance. Perhaps the subject of an experiment is a material that tends to explode when exposed to the air. It would be easier and safer to work with a material that is isoelectronic to the dangerous substance, collecting virtually the same information but avoiding the potential hazards of exposure to the target substance.

One way to determine whether molecules of different chemicals are isoelectronic is to examine them using **infrared spectroscopy**, or IR spectroscopy. This technology allows scientists to look at a chemical's spectrum (characteristic range of absorbed radiation) to determine its **molecular structure** according to how its **electron** bonds vibrate in the infrared range of radiation. One of the features that isoelectronic molecules (or functional groups) have in common is relative bond length or strength between atoms in the molecule (group). The substances absorb infrared light according to these bond lengths and strengths. Thus, if the molecules of two chemicals do not vibrate in the same way (i.e., absorb the same wavelength of light), they are unlikely to be isoelectronic, since the IR spectrography showed their bonds to behave differently when exposed to the same stimulus.

ISOMER AND ISOMERISM

Isomers are compounds having the same chemical formula but having different arrangements of atoms. There are many types of isomers that are defined based on the way in which the atomic arrangements differ.

Structural isomerism is a form of isomerism in which the atoms are arranged in a different order. For example, the ions SNC^- and SCN^- have the same atoms but they are connected to one another in a different order. Stereoisomers have the same bonds but in different spatial arrangements. For example, the **molecule** CHClBrI, in which the **carbon** (C) **atom** is the central atom with the other four atoms attached in a tetrahedral arrangement around the carbon, can be prepared in two distinct ways such that the two molecules are non-superimposable.

One of the subcategories within structural isomerism is chain isomerism which is due to different arrangements of atoms in a chain molecule. Chain isomerism is commonly encountered in organic chemicals. For example one molecule may be a straight chain, whereas an isomer of this molecule would be a chain with a branching of molecules at the end, such as n-butane, CH_3CH_2 CH_2 CH_3 and 2- methylpropane, $CH_3CH(CH_3)$ CH_3. Positional isomerism is an example of side chain groups being found on different atoms of the backbone or central chain molecule. An example of this would be the difference between the molecule 2-chloropentane ($CH_3CHClCH_2CH_2$ CH_3) and 3-chloropentane (CH_3CH_2 $CHClCH_2$ CH_3). **Functional group** isomerism (metamerism) is due to a rearrangement of atoms to produce a different functional group, for example many organic alcohols are isomers of **aldehydes**. A special case of this type of isomerism is dynamic isomerism or **tautomerization**. As the name suggests the structure can move from one form to another and back; it is reversible between the two isomers. This is caused by the movement of an atom or group of atoms from one position to another. This is often due to the presence of double bonds in the molecule, which can rearrange. These isomers usually exist in equilibrium. Structural isomers often exhibit significant differences in physical properties such as boiling points and refractive index.

The second major type of isomerism is stereoisomerism. Stereoisomers have the same atoms connected to each other in the same order but in different spatial arrangements. Optical isomerism is due to different arrangements of asymmetric molecules. Optical isomers are mirror images that cannot be superimposed on each other. Such isomers are called **enantiomers** and molecules that have enantiomers are said to be chiral or to show chirality. The most common examples of chiral molecules are enantiomers. These isomers have the ability to rotate the plane of **polarization** of polarized light. Isomers that rotate the plane of polarization to the right are dextrorotory isomers, those that rotate to the left are laevorotory isomers. A mixture of both types of isomer is known as a racemic mixture.

Geometrical isomerism is unique to molecules in which there is a double bond. The presence of the double bond restricts the motion of the two carbon atoms that contribute to the double bond. The nature of the double bond is such that the two carbon atoms that form the double bond and the two atoms attached to each of those carbon atoms lie in the same plane. For example, the accompanying figure shows two molecules which have the same chemical formula, CHClCHCl. However, in one molecule, the two H atoms are on the same side of the double bond (as are the two Cl atoms), while in the other molecule, there is one H and one Cl atom on each side of the double bond. These two molecules cannot be easily interconverted because of the high **energy** required to rotate a double bond. The naming of geometrical isomers reflects the position of the atoms or groups on each of the carbon atoms in the double bond. If both groups are on the same side of the double bond, that is known as the *cis* arrangement. Where they are on opposite sides of the double bond, the arrangement is the *trans* arrangement.

Different properties are exhibited by groups of isomers. With functional isomerism the reason is obvious, a different functional group is present. With other types of isomerism the change in properties may be more subtle. Steric hindrance may occur where the new arrangement of atoms makes it harder for other chemicals to get close to the part of the molecule where the reaction would take place.

Isomers are compounds that have the same **molecular formula**, but different structural formulas which may **lead** to different physical and chemical properties.

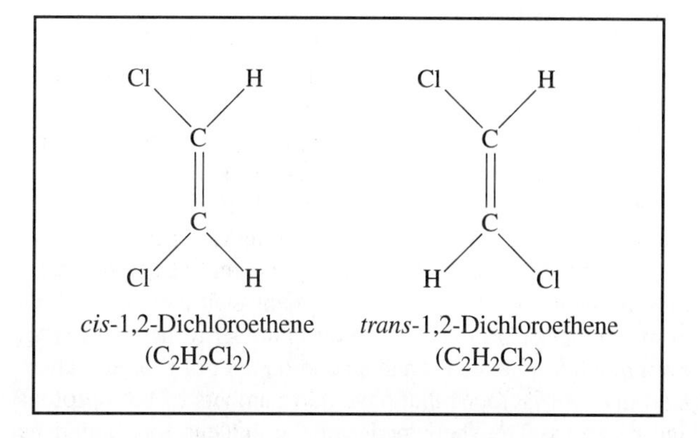

Although the isomers *cis*-1,2-Dichloroethene and *trans*-1,2-Dichloroethene have the same molecular formula, $C_2H_2Cl_2$, they are different compounds because they cannot be interconverted due to the large barrier to rotation of the carbon-carbon double bond. *Illustration by Electronic Illustrators Group.)*

Isomorphism

The term *isomorphism* of minerals refers to a similarity in crystal structure between two or more distinct substances. Many **solids** have a tendency to crystallize in definite geometric frameworks. Table **salt**, for example, has a characteristically cubic crystalline shape that can be observed with the naked eye. Quartz and gypsum are other familiar examples of crystalline structures. Crystalline solids such as these have a three-dimensional pattern that can be represented as a coordinate system or lattice. Like a network of interconnecting cubes, the crystalline lattice is composed of regularly arranged subunits. Lattices of ionic crystals, like table salt, consist of alternating ions, or charged atoms. The attraction between alternate cations (positive charge) and anions (negative charge) stabilizes the crystalline structure. Likewise, metallic crystals consist of lattices of positively charged metal ions regularly arranged in three dimensions among a virtual sea of electrons. The lattice structures of isomorphic substances are comparable. Therefore, they form crystals that appear to be nearly identical. Two or more isomorphic substances sometimes crystallize together to form a solid **solution** with a singular geometric configuration. Isomorphous substances usually have similar chemical formulas, and the relative distances between anions and cations are generally alike. **Sodium** nitrate and **calcium sulfate** are isomorphous, as are the sulfates of **barium**, **strontium**, and **lead**. Isomorphism was discovered by Eilhard Mitscherlich (1794-1863) in the early nineteenth century.

Isotope

Isotope is the term used for an **atom** of an element that differs in atomic **mass** from other atoms of the same element. The word is derived from the Greek words *isos*, meaning the same, and *topos*, meaning place, which refers to the fact that all the isotopes of an element will be in the same place in the **periodic table** of the elements. In the periodic table the number given for the atomic mass of an element represents the average mass of all the isotopes of that element, weighted by the relative abundance of each isotope. While all atoms of the same element have the same number of protons and electrons, different atoms of the same element often contain different numbers of neutrons. An atom that has a specific number of neutrons and protons is referred to as a nuclide.

Isotopes of the same element exhibit no difference in their chemical behavior, but they do differ slightly in certain physical properties, such as **density**, **boiling point**, and freezing point. Isotopes can be separated using these differences in physical properties as well as the effect of electric and magnetic fields on charged particles of different mass, which is the principle of operation of a mass spectrometer. Developing methods of separating isotopes became a very important practical problem as uses for specific isotopes were found. For example, since only uranium-235 will fission when it absorbs a **neutron**, it was important to be able to separate it from the more abundant uranium-238.

Frederick Soddy discovered and named isotopes in 1911 while he was working with **Ernest Rutherford**, although the neutron had not been identified yet. Soddy considered the difference to be one of radioactive properties, not atomic mass. (He was awarded the 1921 Nobel Prize in **chemistry** for his work on the nature and occurrence of isotopes.) In 1913, **J. J. Thomson** found the first stable isotopes due to atomic mass difference while working with **neon**. **Francis Aston** developed a mass spectrograph that he used to identify isotopes. He found 212 of the 281 naturally occurring stable isotopes and, thereafter, received the 1922 Nobel Prize in chemistry for this work.

Most elements have two or more isotopes, although there are 20 elements that have only one. **Tin** is the element with the largest number of isotopes; it occurs in nature as a mixture of 10 different isotopes ranging in mass from 112-124. A survey of the periodic table reveals that elements of even atomic numbers have more isotopes than do those of odd **atomic number**. Whether the **nucleus** has an odd or even number of neutrons seems to also enter into an atom's forming isotopes. The greatest number of stable nuclei have an even number of both protons and neutrons; there are not as many stable nuclei with an even number of protons and an odd number of neutrons or an even number of neutrons and an odd number of protons. There are only four isotopes that have odd numbers of both protons and neutrons. The shell model of the nucleus formulated by Maria Goeppert-Mayer (1906-1972), for which she received the 1963 Nobel Prize in physics, was an attempt to explain the high number of isotopes with 2, 8, 20, 28, 50, and 82 protons or neutrons. Nuclei with these "magic numbers" of protons or neutrons have many stable isotopes.

Isotopes of all elements, with the exception of **hydrogen**, are identified by their mass, which is the same as the number of nucleons in the nucleus of the atom; for example, carbon-12 or carbon-14. The isotopes of hydrogen have been given unique names: ordinary hydrogen, sometimes called protium, has a **mass number** of one and no neutrons; **deuterium**, which was discovered by **Harold Urey** in 1931, has a mass number of two and one neutron; and **tritium**, a radioactive isotope, has a mass number of three and two neutrons. In nature, the distribution of deuterium to protium is one part to 5,000 parts. **Water** made from deuterium, a naturally occurring substance known as heavy water, boils at 214.56°F (101.42°C) and freezes at 38.88°F (3.82°C) at standard pressure. Because heavy water is very efficient as a coolant and moderator in nuclear power plants, greater quantities were wanted than occur naturally. An enriching process can produce water that is 99.75% D_2O. Attempts to replicate thermonuclear fusion on Earth—the process by which the Sun keeps producing energy—involve forcing together the deuterium and tritium isotopes of hydrogen to produce **helium**. In the production of one **mole** of helium, 1.7 x 10^9 kilojoules (4.1 x 10^{11} calories) of **energy** are released.

Following James Chadwick's identification of the neutron in 1932, scientists experimented with reactions produced by neutrons. When the nucleus of most elements absorbed an additional neutron, the element changed into a heavier isotope of the same element. Although some were stable, this method led to the production of radioactive isotopes that do not occur in nature. It is believed that many of the chemical elements in the solar system were formed when neutrons were captured in the interior of the earlier stars, forming radioactive isotopes of elements that then changed to different elements through radioactive decay, transforming them into atoms of heavier elements. Proof of this was obtained with data from the Hubble Space Telescope during the 1990s, when eight elements heavier than **zinc** were discovered in interstellar gas: **arsenic**, **gallium**, **germanium**, **krypton**, **lead**, **selenium**, **thallium**, and tin.

Radioisotopes have become increasingly more important in many fields in the past 50 years. Some radioisotopes are produced by natural processes. Chlorine-36 is produced from chlorine-35 by thermal neutron activation; beryllium-10 is formed from atmospheric **oxygen** by cosmic ray action; carbon-14 is produced in the atmosphere when nitrogen-14 nuclei collide with high energy neutrons that are in cosmic radiation. The ratio of carbon-14 to carbon-12 taken in by living cells depends on the amount of cosmic radiation coming through the atmosphere. Scientists consider this to have remained fairly constant over geologic time, although it has been shown that it varies with the solar cycle as well as man-made activities that put high energy particles into the atmosphere, such as the detonation of atomic weapons or the explosion of the Chernobyl nuclear power plant. Once living organisms die, they no longer take in carbon-14, while the carbon-14 already part of the organism will decay according to the normal **half-life** decay period of the element. This lets one approximate the age of organic materials, a method known as "radiocarbon dating" for which **Willard Libby** received the 1960 Nobel Prize in chemistry. Ratios of naturally occurring radioactive minerals to their decay daughters can be used in determining the age of geological materials. To determine the age of the oldest materials requires isotopes with long half-lives like rubidium-87 that decays to strontium-87, samarium-147 that decays to neodymium-143, potassium-40 that decays to argon-40, rhenium-187 that decays to osmium-187, and the **uranium** isotopes that decay to stable isotopes of lead.

Many radioisotopes are produced in **nuclear fission** reactors either as fission fragments or decay daughters of fission fragments.Rubidium-86, cesium-136, and molybdenum-99, whose decay product technetium-99 is widely used in medical applications, are uranium-235 fission fragments. Other radioisotopes are produced in fission reactors through bombardment by the neutrons that were released in the fission **chain reaction**. An example is the production from cobalt-59 to cobalt-60, a good source of **gamma radiation** for many applications. Other radioisotopes are produced in machines where particles like protons, deuterons, and alpha particles, are accelerated to very high energies by an electric current as they move within a magnetic field and then are released to crash into a non-radioactive target, changing it into a radioactive isotope. In 1930, Sir John Cockcroft (1897-1967) and Ernest Walton (1903-1995) developed a machine that accelerated protons; they received the 1951 Nobel Prize in physics for **transmutation** of atomic nuclei by artificially accelerated atomic particles. In 1931, Ernest Lawrence (1901-1958) designed the first cyclotron, for which he received the 1939 Nobel Prize in physics. Lawrence used his cyclotron to accelerate deuterons that were then crashed into **sodium**, producing a therapeutically useful radioisotope of sodium. Newer radioisotopes produced this way for medical applications include thallium-201 and iodine-123.

Neutron activation analysis is a method of producing radioisotopes in very small samples of a material. Based on the properties of the radioisotopes produced, the elements in the original material can be verified. This technique was developed by Edward Sayre and Heather Lechtman at the Brookhaven National Laboratory in 1969, and, has been successfully used to authenticate art works.

Radionuclides are now being used in many fields—in research, industry, agriculture, and medicine, to name a few—because they can be identified in different locations by instruments that are sensitive to radiation. As early as 1911, **Georg von Hevesy** suggested that **radioactivity** could be used to label substances to make them easily recognizable. In 1934 he used a radioisotope of **phosphorus** to study how human tissue absorbs **phosphate**. He was awarded the 1943 Nobel Prize in chemistry for his work on tracers, which let an investigator follow the movement of a particular element. In medicine, tracers are used to diagnose and even treat some medical problems. A radioisotope is substituted for the stable form of a chemical that is normally used by a specific organ. Examples of widely used tracers in medicine are: calcium-47 for studies of bone formation, chromium-51 for red blood cell studies, cobalt-58 for diagnosing pernicious anemia, iodine-131 for the thyroid gland, chromium-51 for the spleen, gallium-67 for the lymph glands, phosphorus-32 for the liver, strontium-87 for bones, technetium-99 for the brain and liver, thallium7-201 for the heart, xenon-133 for blood flow studies. Dr. Rosalyn Yalow (1921-) received the 1977 Nobel Prize in medicine and physiology for her work in developing radio-immunoassay procedures that use radionuclides in laboratory tests to diagnose medical problems.

See also Carbon dating

J

JABIR IBN HAYYAN (c. 8th Century A.D.)
Alchemist

Jabir ibn Hayyan (active latter 8th century), called Geber by Europeans, was reputedly the father of Moslem **alchemy** and **chemistry**.

It seems clear that there was a real person called Jabir ibn Hayyan about whom we know little except that he lived in al-Kufa, an important city of Abbasid Iraq, and that he had the reputation for skill in alchemy. There exists a vast body of Arabic writings attributed to this Jabir which could not possibly have been written by someone living in the late 8th century because the bulk of Greek scientific and alchemical works had not been translated at that time; Arabic scientific terminology had not even been coined. The earliest biography of Jabir is contained in Ibn al-Nadim's *Fihrist,* a monumental bibliography compiled in 988; the author of the *Fihrist* is partially aware of such discrepancies but insists that Jabir was a historical personage.

Scholarship has shown that there was a sizable corpus of alchemical works attributed impossibly to the historical Jabir. The first references to these works by Jabir appear in the latter half of the 10th century. The corpus contains a great percentage of what medieval Islam knew of the scientific knowledge of the ancients, viewed through Islamic spectacles, and it appears impossible that the 9th- or 10th-century Jabir could have been a single man, however industrious. The Islamic point of view from which this encyclopedic collection of late Hellenistic science is viewed in the works of Jabir is an extremely heterodox one, and this is doubtless the reason for assigning its authorship to a long-deceased but actual Jabir of the 8th century.

Jabir's science of *al-kimiya,* from which Arabic word both "alchemy" and "chemistry" stem, was based upon the Hellenistic idea that all **metals** are fundamentally the same substance, but with varying impurities. The main object of alchemy was to discover a method which would transmute the base metals into the purest form of metal, **gold**; this could be done by means of a supposed substance called "red **sulfur**" by the Moslems and "the philosophers' stone" by Europeans. In the process of searching for red sulfur, Jabir and other Moslem alchemists developed a great many sound facts and processes which formed some of the basic building blocks for the science of chemistry.

In terms of practical methods evolved by Jabir and set forth in the almost 100 works ascribed to him, we are indebted to Moslem alchemy for methods of **distillation**, evaporation, crystallization, **filtration**, and **sublimation**. Methods of producing a considerable number of chemical substances are described: **nitric acid**, **sulfuric acid**, **mercury oxide**, **lead** acetate, and others.

JOHANSON, DONALD CARL (1943-)
American anthropologist

Johanson came from modest beginnings. His parents, Carl Torstgen Johanson and Salia Eugenia Johanson, were immigrants from Sweden. His father, a barber, died when Johanson was two years of age. Some time later, mother and son moved to Hartford, Connecticut, where she worked as a servant. A neighbor in Hartford taught anthropology, and he sparked the boy's interest in the subject. In high school, Johanson became interested in paleoanthropology, the study of fossils of human ancestors when he read about Louis and Mary Leakey's finds in Tanzania. In 1959, the Leakeys used the potassium-argon dating method to determine that their fossil skull of a hominid, Australopithecus boisei, was 1.8 million years old. Hominids are members of the human family. In 1962, they discovered Homo habilis, a true human fossil of about the same age. Johanson entered the University of Illinois as a **chemistry** major, but he changed to anthropology. In 1966, he earned his BA degree with honors. Johanson was mainly interested in paleontology, and he proceeded to do graduate work at the University of Chicago under the paleoanthropologist, F. Clark Howell. In 1974, Johanson earned his Ph.D. degree in the field of anthropology. As a graduate student he studied at museums in Europe and Africa and participated in field expeditions in Ethiopia.

In 1972, Johanson joined the faculty of the department of anthropology at Case Western Reserve University in Cleveland, Ohio. In addition, he assumed a position in anthropology at the Cleveland Museum of Natural History. He worked as researcher and curator at these institutions for nine years. In 1973, Johanson joined a team of scientists and students to search for early human fossils. They set up camp near the Awash River in the Hadar Valley of northeastern Ethiopia, an area rich in fossils. That year Johanson discovered a three million year old knee joint, fossil evidence of bipedal locomotion. In 1974, the team discovered fragments of four hominid skeletons. His greatest discovery came on November 30, 1974, when, along with graduate student Tom Gray, he unearthed a three million year old, three foot six inch (one meter five centimeter) tall, female hominid skeleton. The fossil was forty percent complete. The team of paleontologists called her Lucy after a song by the musical group, The Beatles, entitled, "Lucy in the Sky with Diamonds." Johanson decided that Lucy was too primitive a hominid to be classified as human, and classified her as Australopithicus (southern ape) afarensis (the region she was found).

Discovering Lucy made Johanson famous. In the fall of 1975, he returned to Hadar with an expanded team and better financing. This time the team discovered a cluster of 200 fossil bones and teeth from a group of about thirteen members of Australopithicus afarensis. The cluster of hominids supported Johanson's belief that human intelligence has its roots in cooperative behavior. In 1976, Johanson established a major anthropology laboratory at the Cleveland Museum of Natural History. This research center draws scholars from all over the world.

JOHNSTON, HAROLD S. (1920-)

American physical and atmospheric chemist

Harold S. Johnston has been recognized as one of the world's leading authorities in **atmospheric chemistry**. He was among the first to suggest that **nitrogen** oxides might damage the Earth's ozone layer. His research interests have been in the field of gas-phase chemical **kinetics** and **photochemistry**, and his expertise has been employed by many state and federal scientific advisory committees on air **pollution**, motor vehicle emissions, and stratospheric pollution.

Harold Sledge Johnston was born on October 11, 1920, in Woodstock, Georgia, to Smith L. and Florine Dial Johnston. He graduated with a chemistry degree from Emory University in 1941 and, later that year, entered the California Institute of Technology as a graduate student. During the early 1940s, he was a civilian meteorologist attached to a United States Army unit in California and Florida, after which time he returned to graduate studies and earned his Ph.D. in chemistry and physics in 1948, the same year he married Mary Ella Stay. The couple has four children: Shirley Louise, Linda Marie, David Finley, and Barbara Dial. Johnston was on the faculty of the chemistry department of Stanford University from 1947 to 1956 and of the California Institute of Technology from 1956 to 1957. He then became a professor of chemistry at the University of California, Berkeley, serving as dean of the College of Chemistry from 1966 to 1970.

Johnston's introduction to meteorology occurred when he was a civilian scientist working on a defense project in World War II. In 1941, Roscoe Dickinson, Johnston's research director at the California Institute of Technology, was overseeing a National Defense Research Council project, with which Johnston became involved. Dickinson's group tested the effects of poisonous volatile chemicals on charcoals that were to be used in gas masks. Later, they studied how gas clouds moved and dispersed under different conditions in order to appraise coastal areas that might be vulnerable to chemical attacks.

In 1943, Johnston moved with the Chemical Warfare Service to Bushnell, Florida, where he worked with, and eventually headed, the Dugway Proving Ground Mobile Field Unit of the U.S. Chemical Warfare Service. This unit carried out test explosions to assess how the dispersion of gas was affected by meterological changes. While he was there, Johnston and John Otvos developed an instrument to measure the **concentration** of various **gases** in the air.

Johnston applied his meteorology work to his Ph.D. studies, which he resumed in 1945. He wrote his thesis on the reaction between ozone, a naturally-occurring form of **oxygen**, and nitrogen dioxide, a pollutant formed during **combustion**. Later, during his tenure at Stanford, Johnston worked on a series of fast gas-phase chemical reactions. Using photo-electron multiplier tubes left over from the war, he pioneered a method of studying gas phase reactions that was a thousand times faster than existing techniques. Johnston then spent the years 1950 to 1956 researching high and low pressure limits of unimolecular reactions, and for the subsequent ten years, expanded his research to apply **activated complex** theory to elementary bimolecular reactions.

One of Johnston's most significant research efforts has been on the destruction of the ozone layer. This layer in the Earth's upper atmosphere protects people from the sun's ultraviolet rays. **Chlorofluorocarbons (CFCs)**, gaseous compounds often used in aerosol cans, **refrigerants**, and air conditioning systems, deplete this ozone layer, resulting in increased amounts of harmful sun rays reaching the Earth's surface. The Environmental Protection Agency has imposed production cutbacks on these harmful chemicals. Much like CFCs, nitrogen oxides also damage the ozone layer. During the late 1960s, the federal government financed the design and construction of two prototype supersonic transport (SST) aircraft. An intense political debate over whether the program should be expanded to construct five hundred SSTs was waged. Although Congress was split almost evenly, both houses voted to terminate the SST program in March, 1971. Johnston's articles and testimony suggesting the negative effects SSTs could produce on the atmosphere led two senators to introduce the Stratosphere Protection Act of 1971, which established a research program concerned with the stratosphere. The resulting program, with which Johnston was affiliated, was called the Climatic Impact Assessment Program (CIAP) and began its work in the fall of 1971. Among other things, CIAP concluded that nitrogen oxides from stratospheric aircraft would further reduce ozone. CIAP recommended that aircraft engines be redesigned to reduced nitrogen **oxide** emissions.

Throughout his career, Johnston has served on many state and federal scientific advisory committees. In the 1960s, he was a panel member of the President's Science Advisory Board on Atmospheric Sciences and was on the National Academy of Sciences (NAS) Panel to the National Bureau of Standards. Johnston served on the California Statewide Air Pollution Research Center committee and the NAS Committee on Motor Vehicle Emissions during the early 1970s. He also served on the Federal Aviation Administration's High Altitude Pollution Program from 1978 to 1982 and the NAS Committee on Atmospheric Chemistry from 1989 to 1992. He has been an advisor to High Speed Civil Transport Studies for the National Aeronautics and Space Administration (NASA) since 1988.

Johnston is the author of the book *Gas Phase Reaction Rate Theory* and the author or coauthor of more than 160 technical articles. He is a member of the NAS, the American Academy of Arts and Sciences, the American Chemical Society, the American Physical Society, the American Geophysical Union, and the American Association for the Advancement of Science. Among Johnston's numerous awards are the 1983 Tyler Prize for Environmental Achievement, the 1993 NAS Award for Chemistry in Service to Society, and an honorary doctor of science degree from Emory University.

JOLIOT-CURIE, FRÉDÉRIC (1900-1958)

French nuclear physicist

Frédéric Joliot-Curie was a French nuclear physicist who, together with his wife, **Irène Joliot-Curie**, discovered artificial **radioactivity**, for which they received the 1935 Nobel Prize in **chemistry**. Their efforts made **nuclear fission** and the subsequent development of both nuclear **energy** and the atomic bomb feasible. Joliot met his wife while working as an assistant at the **Radium** Institute at the University of Paris. Irène Curie was the daughter of **Marie** and **Pierre Curie**, the Nobel Prize laureates who discovered radium and founded the Radium Institute. Irène became Frédéric's lifelong research collaborator and they usually published their findings under the combined form of their last names, Joliot-Curie. After World War II, Frédéric Joliot-Curie brought France into the atomic age as director of France's atomic energy commission.

Jean-Frédéric Joliot was born on March 19, 1900, in Paris to Henri Joliot and Emilie Roederer. According to tradition, all the Joliot men were named Jean in honor of Jan Hus, a champion for spiritual freedom who had been burned at the stake in the fifteenth century. Henri Joliot came from a long

Frédéric Joliot-Curie.

line of liberal thinkers and had been part of the French Communard movement at the end of the Franco-Prussian War. He became a Parisian dry goods merchant and the family was settled into a middle-class life, yet Henri remained passionate about his leftist political concerns. He also had a great love of the outdoors and of music, combining the two by composing a number of calls for the hunting horn. His son Frédéric would someday be an avid outdoorsman despite his busy scientific career. Frédéric's mother Emilie was also from a liberal family and Frédéric was exposed to progressive social ideas at a young age. The social and political leanings of his parents had a profound influence on young Frédéric and he was an atheist and political leftist his entire life.

Joliot was educated at the Lycée Lakanal in a suburb of Paris, then at the École Primaire Supérieure Lavoisier in Paris. In 1920 he was admitted to the École Supérieure de Physique et de Chimie Industrielle of Paris, a preparatory school that turned out most of France's industrial engineers at the time. The director of the school, a brilliant physicist named Paul Langevin, recognized Joliot's interest and aptitude for scientific research and became Joliot's lifelong mentor.

After graduating at the head of his class, Joliot worked in industry for a short time. Following Langevin's advice, however, he took a position as research assistant at the Radium Institute in Paris under the guidance of **Marie Curie**, a Nobel laureate in nuclear physics. Joliot held Marie and Pierre Curie in high esteem, going so far as to have pictures of them hanging on the wall of his makeshift home laboratory. Working at the Institute, he was instructed by Marie Curie to be initiated into the rudiments of studying radioactivity by Irène Curie, the couple's elder daughter. For as affable and charming as Frédéric was, Irène was serious and aloof. Yet the two shared their love of research as well as political and social leanings and were married in 1926. They both adopted their combined last names to preserve the scientific Curie lineage.

Joliot-Curie began his research improving the Wilson chamber, a cloud chamber in which the charged particles of an **atom** can be detected as they leave a trail of **water** droplets in their wake. His engineering background made him a master at instrumentation and he redesigned the chamber so that the pressure could be lowered, making the tracks longer and more easily discernible. He also supervised the making of a camera that photographed what went on in the chamber. Further improvements allowed the energy of emissions to be measured using the amount of curvature shown by the tracks when placed in a magnetic field. Conducting experiments within the chamber had an elegance that Frédéric considered to be the most beautiful experience in the world.

In 1930, Joliot-Curie completed his doctoral thesis entitled *A Study of the Electrochemistry of Radioactive Elements,* thus beginning intensive research in the area of radioactivity. Later that same year, he and Irène became interested in the experiments of German physicists Walther Bothe and Hans Becker. The German pair had discovered that very strong radiation was emitted from some of the lighter elements when they were bombarded with alpha rays. An alpha ray is an energetic particle that resembles the **nucleus** of a **helium** atom and contains two positive charges. Such rays had only been discovered at the turn of the century as had the basic particles of the atom. Because of the readily available source of alpha rays—radium—stockpiled at the Radium Institute, Frédéric and Irène began doing research in which they bombarded nuclei of various elements with alpha particles.

In the course of experimentation, the Joliot-Curies had come very close to discovering two other **subatomic particles**, the **neutron** and the positron. But instead of investigating these anomalies, they focused their efforts on their alpha-particle research. As Joliot-Curie and his wife worked with radioactive **polonium** as a source of alpha rays they found that when an element such as **aluminum** is bombarded, some alpha particles are absorbed by the nuclei, transmuting the atoms to **phosphorus**. Aluminum usually has 13 protons in its nucleus, but when bombarded with alpha particles, which contain two positive charges each, the protons were added to the nucleus, forming a nucleus of phosphorus, the element that normally contains 15 protons. The phosphorus produced was different from naturally-occurring phosphorus because it gave off a strong radiation, as predicted by Bothe and Becker. It was a radioactive **isotope** produced artificially.

The Joliot-Curies' discovery of artificial radioactivity was announced to the Academy of Science in 1934 and won them the Nobel Prize for chemistry in 1935. (The prize was awarded in chemistry rather than physics because of their synthesis of new radioactive elements.) Sadly, though she had predicted the success and recognition of her daughter and son-in-law, Marie Curie died of leukemia the year before the award, a casualty of lifelong exposure to radiation.

In the latter part of the 1930s, experiments by Joliot-Curie showed that a nucleus of a radioactive element such as **uranium** could be divided into two smaller nuclei of comparable size with a release of energy emissions. It was predicted that this phenomenon, known as nuclear fission, could occur very quickly in a **chain reaction** and therefore needed to be controlled or moderated. This conclusion led other researchers such as **Otto Hahn** and Enrico Fermi to work on nuclear fission experiments which eventually led to the creation of the atomic bomb as well as atomic energy.

The Joliot-Curies enjoyed these productive years in the company of their two children, Hélène (born in 1927), and Pierre (born in 1932). Because of Irène's often fragile health, the family spent as much time away from work in Paris as possible, swimming or sailing at the beach or hiking or skiing in the mountains. It was said that rather than carry around pictures of his family in his wallet, Frédéric Joliot-Curie carried instead a picture of a giant pike he had once caught on a fishing trip. He was a very sociable man, telling not only fish stories, but entertaining both colleagues and family with his wit and charm.

By the time the Germans occupied Paris during World War II, Joliot-Curie, who was staunchly part of the French Resistance, continued to do research but kept a low profile. He wanted to prevent the Germans from gaining information about nuclear fission. In fact, he had a stockpile of heavy water, which was used as a moderator in nuclear fission experiments, sent out of the country so that the Germans could not gain possession of it. Inspired in part by the arrest of his friend and mentor, Paul Langevin, Joliot-Curie became more politically active. In 1942, when Langevin's son-in-law, Jacques Solomon, was tortured and shot by the Nazis for turning out a resistance newspaper, Joliot-Curie joined the Communist Party. Arrested twice and fearing for his safety and that of his family, Joliot-Curie went underground in Paris and sent Irène and their two children to Switzerland.

After World War II, Joliot-Curie was instrumental in convincing the government of Charles de Gaulle to establish an Atomic Energy Commission in France, modeled after the similar agency in the United States. After being appointed its Commissioner, Joliot-Curie oversaw the installation of a major nuclear research center. With the advent of the Cold War and tensions between the Soviet Union and the West, Joliot-Curie's membership in the Communist Party and his radical activism drew governmental concern over his reliability. He was relieved of his position as high commissioner of Atomic Energy.

Losing his position was a blow to Joliot-Curie both professionally and personally. He and his wife also began to suffer ill health from their lifelong exposure to radiation. Still dedicated to his leftist politics, he became president of the World Organization of the Partisans of Peace. In March of 1956, Irène finally succumbed to leukemia. Frédéric succeeded her as head of the Radium Institute and continued her efforts to build a new physics laboratory south of Paris. At 58, his liver badly damaged by exposure to radiation, Frédéric Joliot-Curie died following an operation made necessary by an internal hemorrhage.

Joliot-Curie, Irène (1897-1956)

French chemist and physicist

Irène Joliot-Curie, elder daughter of famed scientists **Marie** and **Pierre Curie**, won a Nobel Prize in **chemistry** in 1935 for the discovery, with her husband **Frédéric Joliot-Curie**, of artificial **radioactivity**. She began her scientific career as a research assistant at the **Radium** Institute in Paris, an institute founded by her parents, and soon succeeded her mother as its research director. It was at the Institute where she met her husband and lifelong collaborator, Frédéric Joliot. They usually published their findings under the combined form of their last names, Joliot-Curie.

Born on September 12, 1897, in Paris to Nobel laureates Marie and Pierre Curie, Irène Curie had a rather extraordinary childhood, growing up in the company of brilliant scientists. Her mother, the former Marie Sklodowska and her father, Pierre Curie, had been married in 1895 and had become dedicated physicists, experimenting with radioactivity in their laboratory. **Marie Curie** was on the threshold of discovering radium when little Irène, or "my little Queen" as her mother called her, was only a few months old. As Irène grew into a precocious, yet shy, child, she was very possessive of her mother who was often preoccupied with her experiments. If, after a long day at the laboratory, the little Queen greeted her exhausted mother with demands for fruit, Marie Curie would turn right around and walk to the market to get her daughter fruit. Upon her father Pierre Curie's untimely accidental death in 1908, Irène was then more influenced by her paternal grandfather, Eugène Curie. It was her grandfather who taught young Irène botany and natural history as they spent summers in the country. The elder Curie was also somewhat of a political radical and atheist, and it was he who helped shape Irène's leftist sentiment and disdain for organized religion.

Curie's education was quite remarkable. Marie Curie made sure Irène and her younger sister, Eve Denise (born in 1904), did their physical as well as mental exercises each day. The girls had a governess for a time, but because Marie Curie was not satisfied with the available schools, she organized a teaching cooperative in which children of the professors from Paris' famed Sorbonne came to the laboratory for their lessons. Marie Curie taught physics, and other of her famous colleagues taught math, chemistry, language and sculpture. Soon Irène became the star pupil as she excelled in physics and chemistry. After only two years, however, when Irène was 14, the cooperative folded and Irène enrolled in a private school, the Collège Sevigné, and soon earned her degree. Summers were spent at the beach or in the mountains, sometimes in the company of such notables as **Albert Einstein** and his son. Irène then enrolled at the Sorbonne to study for a diploma in nursing.

During World War I, Marie Curie went to the front where she used new X-ray equipment to treat soldiers. Irène soon trained to use the same equipment and worked with her mother and later on her own. Irène, who was shy and rather antisocial in nature, grew to be calm and steadfast in the face of danger. At age 21, she became her mother's assistant at the Radium Institute. She also became quite adept at using the Wilson cloud chamber, a device which makes otherwise invisible atomic particles visible by the trails of **water** droplets left in their wake.

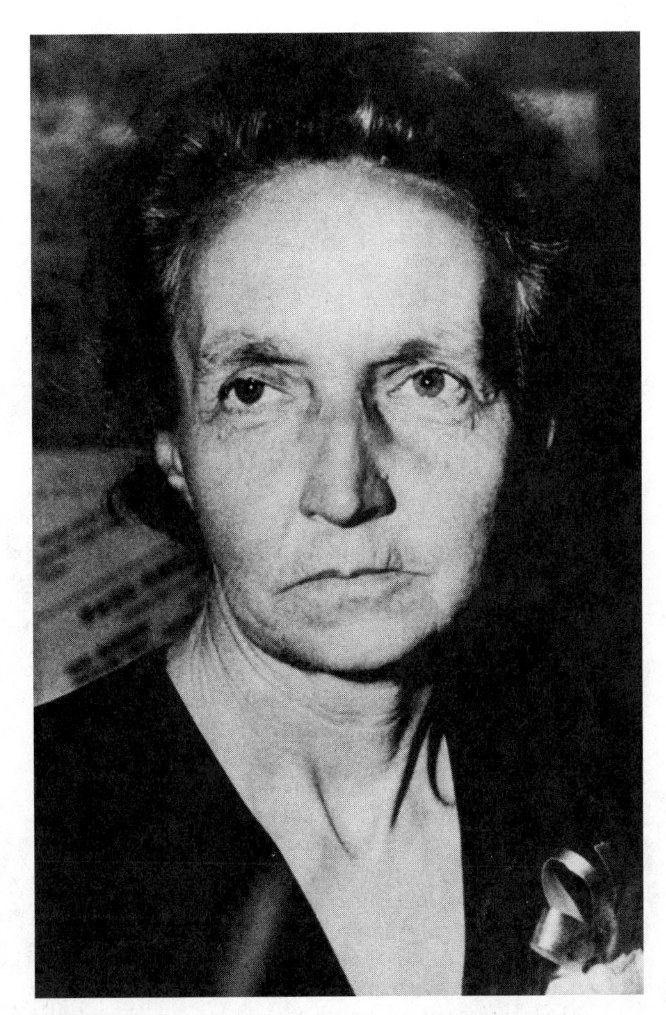

Irène Joliot-Curie.

In the early 1920s, after a jubilant tour of the United States with her mother and sister, Irène Curie began to make her mark in the laboratory. Working with Fernand Holweck, chief of staff at the Institute, she performed several experiments on radium resulting in her first paper in 1921. By 1925 she completed her doctoral thesis on the emission of alpha rays from **polonium**, an element that her parents had discovered. Many colleagues in the lab, including her future husband, thought her to be much like her father in her almost instinctive ability to use laboratory instruments. Frédéric was several years younger than Irène and untrained in the use of the equipment. When she was called upon to teach him about radioactivity, Irène started out in a rather brusque manner, but soon the two began taking long country walks. They married in 1926 and decided to use the combined name Joliot-Curie to honor her notable scientific heritage.

After their marriage, Irène and Frédéric Joliot-Curie began doing their research together, signing all their scientific papers jointly even after Irène was named chief of the laboratory in 1932. After reading about the experiments of German scientists Walther Bothe and Hans Becker, their attention focused on nuclear physics, a field yet in its infancy. Only at the turn of the century had scientists discovered that atoms contain a central core or **nucleus** made up of positively charged particles called protons. Outside the nucleus are negatively charged particles called electrons. Irène's parents had done their work on radioactivity, a phenomenon which occurs when the nuclei of certain elements release particles or emit **energy**. Some emissions are called alpha particles which are relatively large particles resembling the nucleus of a **helium atom** and thus contain two positive charges. In their Nobel Prize-winning work, the elder Curies had discovered that some elements, the radioactive elements, emit particles on a regular, predictable basis.

Irène Joliot-Curie had in her laboratory one of the largest supplies of radioactive materials in the world, namely polonium, a radioactive element discovered by her parents. The polonium emitted alpha particles which Irène and Frédéric used to bombard different elements. In 1933 they used alpha particles to bombard **aluminum** nuclei. What they produced was radioactive **phosphorus**. Aluminum usually has 13 protons in its nuclei, but when bombarded with alpha particles which contain two positive charges each, the protons were added to the nucleus, forming a nucleus of phosphorus, the element with 15 protons. The phosphorus produced is different from naturally-occurring phosphorus because it is radioactive and is known as a radioactive **isotope**.

The two researchers used their alpha bombardment technique on other elements, finding that when a nucleus of a particular element combined with an **alpha particle**, it would transform that element into another, radioactive element with a higher number of protons in its nucleus. What Irène and Frédéric Joliot-Curie had done was to create artificial radioactivity. They announced this breakthrough to the Academy of Sciences in January of 1934.

The Joliot-Curies' discovery was of great significance not only for its pure science, but for its many applications. Since the 1930s many more radioactive isotopes have been produced and used as radioactive trace elements in medical diagnoses as well as in countless experiments. The success of the technique encouraged other scientists to experiment with the releasing the power of the nucleus.

It was a bittersweet time for Irène Joliot-Curie. An overjoyed but ailing Marie Curie knew that her daughter was headed for great recognition but died in July of that year from leukemia caused by the many years of radiation exposure. Several months later the Joliot-Curies were informed of the Nobel Prize. Although they were nuclear physicists, the pair received an award in chemistry because of their discovery's impact in that area.

After winning the Nobel Prize, Irène and Frédéric were the recipients of many honorary degrees and named officers of the Legion of Honor. But all these accolades made little impact on Irène who preferred spending her free time reading poetry or swimming, sailing, skiing or hiking. As her children Hélène and Pierre grew, she became more interested in social movements and politics. An atheist and political leftist, Irène also took up the cause of woman's suffrage. She served as undersecretary of state in Leon Blum's Popular Front government in 1936 and then was elected professor at the Sorbonne in 1937.

Continuing her work in physics during the late 1930s, Irène Joliot-Curie experimented with bombarding **uranium** nuclei with neutrons. With her collaborator Pavle Savić, she showed that uranium could be broken down into other radioactive elements. Her seminal experiment paved the way for another physicist, **Otto Hahn**, to prove that uranium bombarded with neutrons can be made to split into two atoms of comparable **mass**. This phenomenon, named fission, is the foundation for the practical applications of nuclear energy—the generation of nuclear power and the atom bomb.

During the early part of World War II, Irène continued her research in Paris although her husband Frédéric had gone underground. They were both part of the French Resistance movement and by 1944, Irène and her children fled France for Switzerland. After the war she was appointed director of the Radium Institute and was also a commissioner for the French atomic energy project. She put in long days in the laboratory and continued to lecture and present papers on radioactivity although her health was slowly deteriorating. Her husband Frédéric, a member of the Communist Party since 1942, was removed from his post as head of the French Atomic Energy Commission in 1950. After that time, the two became outspoken on the use of **nuclear energy** for the cause of peace. Irène was a member of the World Peace Council and made several trips to the Soviet Union. It was the height of the Cold War and because of her politics, Irène was shunned by the American Chemical Society when she applied for membership in 1954. Her final contribution to physics came as she helped plan a large particle accelerator and laboratory at Orsay, south of Paris in 1955. Her health worsened and on March 17, 1956, Irène Joliot-Curie died as her mother had before her, of leukemia resulting from a lifetime of exposure to radiation.

JONES, MARY ELLEN (1922-1996)

American biochemist

Mary Ellen Jones, a prominent biochemist and enzymologist, is known for isolating carbamyl phosphate, one of a number of molecules that are the building blocks of biosynthesis. By synthesizing this substance, Jones contributed to laying the groundwork for major advances in biochemistry, particularly advances in research on deoxyribonucleic acid (**DNA**) and ribonucleic acid (**RNA**). She explored enzyme action, the process whereby the products of metabolism (metabolites) control enzyme activity, and metabolic pathways. The metabolic pathway is essential for cell division and differentiation, and studies of it are crucial for the understanding of a number of processes, including fetal development, cancer, and some mutations in humans. Jones was recognized for her work by being named the first woman Kennan Professor at the University of North Carolina at Chapel Hill, in 1980.

Mary Ellen Jones was born on December 25, 1922, in La Grange, Illinois, to Elmer Enold and Laura Ann (Klein) Jones. She earned her B.Sc. Degree from the University of Chicago in 1944. Jones later received a Ph.D. in biochemistry at Yale University, where she was a U.S. Public Health Service Fellow in the department of physiological chemistry from 1950 to 1951.

Jones solidly established herself as an enzymologist during her post-doctoral studies with Fritz Lipmann, a 1953 Nobel Prize laureate for physiology or medicine, who was then director of the Chemical Research Laboratory at Massachusetts General Hospital. In the 1950s, he and a team of researchers discovered a group of molecules that were considered the building blocks of biosynthesis. It was during this time that Jones isolated carbamyl phosphate, one of the most important of these essential molecules. The synthesis of this molecule facilitated important advances in biochemistry. Carbamyl phosphate is present in all life. Knowledge of carbamyl phosphate led to scientific understanding of two essential and universal pathways of biosynthesis, the production of a chemical compound by a living organism.

Jones and Lipmann noticed that, during certain biosynthetic reactions, the energy-releasing reaction was a splitting of adenosine triphosphate (ATP) that yielded a mononucleotide and inorganic pyrophosphate. The discovery suggested that DNA amd RNA synthesis might occur with the liberation of inorganic pyrophosphate from ATP and other trinucleotidesBa suggestion that was later corroborated by the biochemist **Arthur Kornberg**. Jones remained in the Biochemical Research Laboratory at Massachusetts General Hospital until 1957, and she served as a faculty member in the Department of Biochemistry at Brandeis University until 1966.

In 1966, Jones joined the University of North Carolina as an associate professor. Two years later, she was promoted to the rank of professor and appointed professor in the department of zoology. She left the University of North Carolina in 1971 for the University of Southern California, where she was professor of biochemistry until 1978. Jones returned to the University of North Carolina as a professor and chair of the biochemistry department, and was named Kenan Professor in 1980.

An author of over ninety papers related to biochemistry, Jones received international recognition for her creative scientific research. Elected a member of the Institute of Medicine in 1980, she was inducted into the National Academy of Sciences in 1984. In 1986, she served as president of the American Society of Biological Chemists. Jones was also awarded the Wilbur Lucius Cross Medal in 1982 by the graduate school at Yale University for her work as a Agifted investigator of the chemistry of life. She died in 1996.

James Prescott Joule.

Joule, James Prescott (1818-1889)

English physicist

James Prescott Joule is credited with the fundamental discovery that opened a new paradigm of the nature of **heat** which allowed many new discoveries and inventions. A rather unassuming man in appearance, he had difficulty for a long time in convincing his contemporaries of the importance of his discovery. However, once he received the endorsement of an accepted scientist **William Thomson** (Lord Kelvin), he received nearly immediate acclaim. Though he continued his work throughout his life, all of the discoveries for which he is now famous were accomplished before he was 34 years old. One truly noteworthy condition of Joule's career was that he was self-reliant in his education and his research. He was self-educated and, coming from a wealthy family, he funded his own research. This condition as respectable as it was, proved to be a great hindrance. The scientific community would not believe that Joule could discover something so important without the support, academically and financially, of an academic institution. Still he knew that he had the answer to a long puzzling scientific question and he persevered.

Joule is noted in the history of science as one of the most determined researchers. He even produced research while on his honeymoon. Joule and his bride were visiting Switzerland and Joule took the occasion to measure the **temperature** of a large waterfall as part of his research. By chance, Kelvin met him on the mountain that day and long after admired his determination. Joule was a great influence on Kelvin and many others for his breakthrough, his methods and his persistence.

Joule was born in 1818 in Salford, England. He grew up in a well-established wealthy family, which had owned a brewery for generations. As a child, Joule was not very healthy and was educated at home through tutors and his own readings. When he was sixteen, he went to study for two years with **John Dalton**. Dalton first stimulated Joule's interested in chemistry and experimentation. After Dalton could no longer teach because of his poor health, Joule built his own laboratory at his family's home. He conducted experiments and began writing research papers when he was twenty. In the autumn before he turned 23, Joule presented a paper at the meeting of the Manchester Literary and Philosophical Society. He was well received and was soon elected as a member of the society. He was an active member of the society for many years, culminating in his acceptance of the society presidency in 1860.

Joule never attended college. Instead, he read incessantly on whatever interested him and what did not interest him he avoided. He abhorred some of the topics usually considered part of a liberal college education. Still, he was fascinated with the study of physics in particular electromagnetic fields. In 1838, he described in a paper what he called an electromagnetic engine. He believed that these engines could replace steam but later had to admit their inefficiency. In 1840, he wrote a paper describing what is now called the Joule effect. His premise was that the amount of heat produced by electrical current is a function of the amount of resistance and the square of the current.

From 1843 through 1878, he regularly attended the meetings of the British Association for Advancement of Science. At his first meeting, he presented his seminal theory on the nature of heat. This theory stated that heat is derived from work and that heat and work are interchangeable. He worked to improve this theory and every year returned to the BAAS to present a paper reporting his progress. At the 1845 meeting, he brought a working model demonstrating the work-to-heat principle. This model consisted of **water** in a container that was agitated very fast by a paddle. The main problem with the conference attendees accepting his theory was that the experiments supporting the theory required highly precise measurements of heat. They did not appreciate that Joule, an amateur, had taken this preciseness into consideration.

After tolerating Joule for five years, the conference administrators requested that he give a summary of his findings and not a full presentation. This was the historic 1847 meeting where Kelvin rose to the defense of Joule's theory. After the short presentation, Kelvin asked so many questions that Joule had to go into his full presentation. Kelvin's fascination with the theory stirred up new interest among the scientists. Kelvin announced to the conference that he found Joule's methods and findings to be well developed and accurate. This was a turning point in Joule's career.

Only days after that turning point, Joule took a turning point in his life. He married Amelia Grimes who apparently was good-natured about his constantly measuring heat. When he saw the high waterfall on his honeymoon, he wanted to test his theory and she willingly helped him. He deduced that the water at the bottom of the fall should be hotter than at the top due to the work of the water running down the fall. As noted above, he met Kelvin again that day. Joule and Kelvin began collaboration from 1853 through 1862. During that time, they established the Joule-Thomson effect.

Joule's heat theory was the cornerstone of his research career. One of the embellishments involved the calculation of the **velocity** of a gas **molecule** in 1848. He developed relatively little more in his life but it was enough to eventually lead Kelvin and others to discover the second principle of **thermodynamics**. After 1862 when he was no longer doing research with Kelvin, Joule revised his theory as an expression of the **conservation of energy**. He died in 1889 at age 71.

Joule's heat-to-work theory was well founded in his experiments. He had done an exhaustive amount of work to determine that the relationship between heat and work was a constant. He even calculated the amount of work necessary to produce heat. Once his views were ''approved'' by Kelvin's endorsement, many researchers excitedly investigated his theory. They found that the number Joule used as his work-to-heat constant was too large. To find a more practical unit of measurement, they divided Joule's number by 10,000,000 and named the new unit after him. The measure of heat equivalent to work is now called the joule.

Working with Kelvin, Joule investigated different aspects of his theory. One of these was the difference in heat during expansion of a gas. They found that when an amount of gas was allowed to expand in a vacuum it tended to lose heat. They explained it in terms of heat-to-work, but it is now recognized as a property of the molecules separating one from another. By investigating this principle, industrialists founded the principles necessary to begin the refrigeration industry by the turn of the century.

Though it did not receive much recognition at the time, Joule's study of the velocity of gas molecules did much to encourage the kinetic theory of **gases**. His interest had been developed by the tutor from his teens, John Dalton. By this time, Joule was well recognized for his detailed thoughts and precise measurements. To have such a distinguished scientist present findings supporting the theory encouraged other researchers.

Joule received many awards in his lifetime. After presenting his research to the Royal Society in 1849, he was elected a fellow. As a member of the society, he earned the Royal medal and the honored Copley medal for the same research. It was well noted that this was a very rare achievement, to win both for the same research. Another honored event occurred when his family's industry went broke. The queen herself offered him a pension. However, the most treasured award he received was perhaps the first, when he was reunited with Dalton at his first meeting of the Literary and Philosophical society. Dalton recognized how impressed the audience was and uncharacteristically asked the audience to applaud the young man. This applause, though faded for many years, eventually included the nation.

Percy Lavon Julian.

Julian, Percy Lavon (1899-1975)

American chemist

Percy Julian is best known for formulating a drug to treat glaucoma and for synthesizing sex **hormones** and **cortisone**. A native of Montgomery, Alabama and the grandson of former slaves, Julian was born first of six children on April 11, 1899. His mother, Elizabeth Lena Adams Julian, taught school and his father, James Sumner Julian, also a teacher, worked as a railroad mail clerk.

Julian graduated with the standard eight years of education from the State Normal School for Negroes in 1916 and then took additional courses to prepare for a white college. In 1920, he graduated Phi Beta Kappa and valedictorian of his class at DePauw University, where he lived in a fraternity house attic and earned tuition by waiting tables. Because doctoral programs passed over black students, he was unable to obtain a fellowship. For two years he taught **chemistry** at Fisk University, then received an Austin Fellowship to study at Harvard University, where he earned a master's degree in chemistry in 1923. He then worked at Harvard an additional three years as a research fellow. Returning to the classroom, Julian taught one year at West Virginia State College, then two years at Howard University. With a grant from the General Education Board, he obtained a Ph.D.in **organic chemistry** in 1931 from the University of Vienna, where he specialized in alkaloids.

It was under the influence of Ernst Spath that Julian first began to realize the medical applicability of soya derivatives to human illness. On his return to the United States, he headed DePauw's chemistry department from 1932 to 1936. During this period, he married sociologist Anna John and fathered two children. With the support of the Rosenwald Fund, Julian made his first great laboratory breakthrough in 1935 with the artificial creation of physostigmine, a drug that forms naturally in the adrenal glands and which lowers eye pressure in victims of glaucoma. The drug is also used in the treatment of swelling of the brain, skin and kidney disease, bronchial asthma, and leukemia.

Denied a job at the Institute of Paper Chemistry because of his race, he headed Glidden's soya research center in Chicago and later, Julian Laboratories, where he extracted sterols from soybeans. With these substances, in 1950 he contained the cost of synthetic sterols at 20 cents per gram, thereby helping arthritics obtain a derivative, cortisone, at a low price. Other aspects of his research aided the formulation of paint, printing, paper-making, and waterproofing as well as the creation of low-cost milk substitutes, livestock and poultry feed, food emulsifier, and an oxygen-impermeable fire-fighting foam used by the military in World War II.

Julian also synthesized progesterone and testosterone, both of which are essential to the body's endocrine system. His discoveries helped treat cancer and relieved problem pregnancies and menstrual disorders. He founded his own institute in Franklin Park, Illinois, in 1953, and added a satellite plant in Mexico City, where he tapped local supplies of *diosorea* (a wild yam) as a rich *source of diosgenin*, which he developed into cortexolone.

In 1964, he sold his facilities to Smith, Kline, and French and retired to Oak Park, Illinois, but continued publishing technical articles and maintained the directorship of laboratory projects until his death from cancer on April 19, 1975. Fame and utility brought a mixed reception for Julian's contributions. A devoted family man and prominent citizen of Chicago, he was named man of the year by the Jaycees and the *Chicago Sun-Times*. In spite of his acceptance by the scientific community, however, he was barred from attending some professional functions held in segregated halls, and for four consecutive years he was turned down for admittance to the Inventors Hall of Fame before his induction in 1990. When he moved to a new home, he suffered harassment from local racists but refused to be bullied. His alma mater honored his work by naming the DePauw chemistry and mathematics building for him.

K

KARLE, ISABELLA (1921-)

American chemist, crystallographer, and physicist

Isabella Karle is a renowned chemist and physicist who has worked at the Naval Research Laboratory in Washington, D.C., since 1946 and heads the X Ray Diffraction Group of that facility. In her research, she applied **electron** and **X-ray diffraction** to **molecular structure** problems in **chemistry** and biology. Along with her husband **Jerome Karle**, she developed procedures for gathering information about the structure of molecules from diffraction data. For her work, she has received numerous awards such as the Annual Achievement Award of the Society of Women Engineers in 1968, the Federal Woman's Award in 1973, and the Lifetime Achievement Award from Women in Science and Engineering in 1986. Her work has been described in the book *Women and Success.*

Isabella Lugoski Karle was born on December 2, 1921, in Detroit, Michigan. Her parents were Zygmunt A. Lugoski, a housepainter, and Elizabeth Graczyk, who was a seamstress. Both her parents were immigrants from Poland, and Karle spoke no English until she went to school. While still in high school, she decided upon a career in chemistry, even though her mother wanted her to be a lawyer or a teacher. She received her B.S. and M.S. degrees in **physical chemistry** from the University of Michigan in 1941 and 1942. Determined to continue her studies, Karle ran into serious financial problems since teaching assistant positions at the University of Michigan were reserved exclusively for male doctoral students. She managed to stay in school on an American Association of University Women fellowship, however, and in 1943 also became a Rackham fellow. She received her Ph.D. in physical chemistry from the University of Michigan in 1944, at the age of twenty-two.

After receiving her doctorate, Karle worked at the University of Chicago on the Manhattan Project (the code name for the construction of the atomic bomb), synthesizing **plutonium** compounds. She then returned to the University of Michigan as a chemistry instructor for two years. In 1942 she had married Jerome Karle, then a chemistry student. In 1946 she and her husband joined the Naval Research Laboratory, where she worked as a physicist from 1946 to 1959. In 1959 she became head of the X-ray analysis section, a position she maintained through the 1990s.

When Karle began her work at the Naval Research Laboratory, information about the structure of crystals was limited. Scientists had determined that crystals were solid units, in which atoms, ions, or molecules are arranged sometimes in repeating, sometimes in random patterns. These patterns or networks of fixed points in space have measurable distances between them. Although chemists had been able to investigate the structure of gas molecules by studying the diffraction of electron or X-ray beams by the gas molecules, it was believed that information about the occurrences of the patterns—or phases—was lost when crystalline substances scattered an X-ray beam. The Karles, working as a team, gathered phase information using a heavy-atom or **salt** derivative. The position of a heavy **atom** in the crystal could be located by scattered X-ray reflections, even though light atoms posed more serious problems. When a heavy atom could not be introduced into a crystal, its structure remained a mystery. In 1950 Jerome Karle, in collaboration with the chemist **Herbert A. Hauptman**, formulated a set of mathematical equations that would theoretically solve the problem of phases in light-atom crystals. It was Isabella Karle who solved the practical problems and designed and built the diffraction machine that photographed the diffracted images of crystalline structures.

While investigating structural formulas and the make-up of crystal structures using electron and X-ray diffraction, Karle made an important discovery. She found that only a few of the phase values—no more than three to five—are sufficient to evaluate the remaining values. She could then use symbols to represent these initial values and also numerical evaluations. This process for determining the location of atoms in a crystal was amenable to processing in high-speed computers. Eventually, it became possible to analyze complex biological molecules in a day or two that previously would have taken years to analyze. The rapid and direct method for solving crystal structure resounded through chemistry, **biochemistry**, biology, and medicine, and Karle herself has been active in resolving applications in a range of fields.

In addition to describing the structure of crystals and molecules, Karle also investigated the **conformation** of **natural products** and biologically active materials. After a crystallographer determines the chemical composition of rare and expensive chemicals, scientists can synthesize inexpensive substitutes that serve the same purpose. Karle headed a team that determined the structure of a chemical that repels worms, termites, and other pests and occurs naturally in a rare Panamanian wood. The team was then able to produce a synthetic chemical that mimics the natural chemical and is equally effective as a pest repellent. In another application, Karle studied frog venom. Using extremely minute quantities of purified potent toxins from tropical American frogs, the team headed by Karle established three-dimensional models, called stereoconfigurations, of many of the toxins and showed the chemical linkages of each of these poisons. The inexpensive substitutes of the toxins were of great importance in medicine. The venom has the effect of blocking nerve impulses and is useful to medical scientists studying nerve transmissions. Karle has also researched the effect of radiation on deoxyribonucleic acid (DNA), the carrier of genetic information. She demonstrated how the structural formulas of the configurations of **amino acids** and **nucleic acids** in **DNA** may be changed when exposed to radiation. Her research into structural analysis also established the arrangement of peptide bonds, or combinations of amino acids.

Karle has held several concurrent positions, such as member of the National Committee on **Crystallography** of the National Academy of Science and the National Research Council (1974–1977). She has long been a member of the American Crystallographic Society and served as its president in 1976. She was elected to the National Academy of Sciences in 1978. From 1982 to 1990 she worked with the Massachusetts Institute of Technology, and she has been a civilian consultant to the Atomic Energy Commission.

Karle has received numerous awards including the Superior Civilian Service Award of the Navy Department in 1965, the Hildebrand Award in 1970, and the Garvan Award of the American Chemical Society in 1976. She has received several honorary doctorates. Her most recent awards have been the Gregori Aminoff Prize from the Swedish Academy of Sciences in 1988 and the Bijvoet Medal from the University of Utrecht, the Netherlands, in 1990. She has written over 250 scientific articles.

Isabella and Jerome Karle have three daughters, Louise Isabella, Jean Marianne, and Madeline Diane. All three have become scientists like their parents. Jerome Karle, who is chief scientist at the Laboratory for Structure and Matter of the U.S. Naval Laboratory, received the Nobel Prize in chemistry in 1985 for developing a mathematical method for determining the three-dimensional structure of molecules.

KARLE, JEROME (1918-)

American physical chemist

Jerome Karle is a chemist whose research into the structure of atoms, molecules, glasses, crystals, and solid surfaces has greatly advanced the understanding of chemical composition. He is best known for his contributions to the use of X-ray **crystallography** in determining the structure of crystal molecules, which earned him the 1985 Nobel Prize in **chemistry**. It was awarded to him jointly with his colleague **Herbert A. Hauptman** for "their outstanding achievements in the development of direct methods for the determination of crystals." The direct method has greatly increased the accuracy of **X-ray crystallography** and been of great importance in ascertaining the structure of large molecules. Using the direct method, scientists can determine which parts of molecules are biologically active; this has, among other advances, enabled pharmaceutical companies to engineer artificial compounds with identical properties which can be used as medicinal drugs.

Karle was born in Brooklyn, New York, on June 18, 1918, to Louis Karfunkle, a businessman, and Sadie Kun, an amateur pianist. His mother hoped that her son would inherit some of the family's artistic talents and perhaps become a professional pianist, but Karle decided at an early age that his career would be in science. He attended Abraham Lincoln High School in Brooklyn from 1929 to 1933. Upon graduating, he enrolled in chemistry and biology at the City College of New York, taking extra classes in mathematics and physics. In 1937, he was awarded his B.S. degree and the First Caduceus Award for Excellence in the Natural Sciences. He then studied biology for a year at Harvard and received his M.A. in 1938. Karle worked as a laboratory assistant at the New York State Health Department in Albany from 1939 to 1940, with the intention of saving money to return to graduate school. One of his discoveries, a method of measuring the amount of **fluorine** in **water**, became a standard in the field.

In 1940, Karle enrolled as a Ph.D student in the chemistry department of the University of Michigan. Sitting at an adjoining desk on the first day of class was Isabella Lugoski, a fellow chemist whom he would marry in 1942; they would have three children and collaborate closely with each other in their research. At Michigan, Karle and Isabella, both of whom were interested in **physical chemistry**, studied under Lawrence O. Brockway. By the end of the summer of 1943, Karle had completed work on his doctoral thesis and went to work on the Manhattan Project at the University of Chicago.

In 1944, having received his M.S. and Ph.D. for his work on gas **electron** diffraction, Karle and **Isabella Karle** returned to the University of Michigan. Karle joined the Naval Research Laboratory where he carried out some experiments on the structure of monolayers of long-chain **hydrocarbon** films; he also formulated a theory about **electron diffraction** patterns. In 1946, he and his wife joined the U.S. Naval Research Laboratory in Washington, D.C., where they worked together on developing a **quantitative analysis** of gas-electron diffraction.

The **solution** of a problem in gas-electron diffraction had, as Karle recalls in an autobiographical statement, "evident implications for crystal structure analysis.'' When Herbert A. Hauptman became affiliated with the laboratory, they moved to find ways to understand three-dimensional **molecular structure** using X-ray crystallography. But Karle has often insisted that it was his early collaboration with his wife that laid the groundwork for their Nobel Prize-winning work. With Hauptman, Karle developed the basic mathematics necessary to establishing a general method for solving all types of crystal structures. Later, after Hauptman left the laboratory, Karle and Isabella developed the general procedure of interpreting three-dimensional structures by means of X-ray crystallography.

X-ray crystallography uses the diffraction of X-ray beams to record the structure of a crystal; the crystal is exposed to the beams and a photographic plate captures the pattern of the X rays deflected by the atoms. When Karle, Hauptman, and Isabella Karle began their work, there was no accurate method for translating the pattern of dots the X rays left on the photographic plate into knowledge about the exact **atomic structure** of the crystal. Their pioneering direct method of interpreting the internal structure of crystal molecules relied on a mathematical formula for determining the phase of an X-ray beam as it traveled through a crystal—that is, the degree of displacement of the beam. From these mathematical calculations it was possible to construct an electron-density map depicting the precise position of each **atom** and hence, the molecule's structure. Ingenious though the direct method was, it was some time before it gained general acceptance. When it was introduced in a paper in 1953, many chemists were initially skeptical, partly because it relied on complex mathematics that few chemists understood. In time, however, its usefulness came to be widely recognized. In 1951, as their work on X-ray crystallography continued, Karle was made professor of chemistry at the University of Maryland, a position he held until 1970. Between 1954 and 1956, Karle was a member of the National Research Council. He would later serve on it again, from 1967 to 1987. In 1968, he was named to a specially created chair of science at the Naval Research Laboratory and became chief scientist of the Laboratory for the Structure of Matter. During the late 1960s, his collaborative work on the direct method was accorded wider acceptance in the scientific community as a result of practical demonstrations of its usefulness in analyzing large molecules given by Isabella Karle.

In the 1960s, Karle and his wife developed a method of crystal structure determination that encompassed the range of crystals, centrosymmetric as well as noncentrosymmetric. The ''symbolic addition procedure,'' as it was known, built on their work of the 1950s. Karle also dedicated himself to theoretical and experimental analysis of gas-electron diffraction. Since the 1970s, Karle has been investigating macromolecular structure and the evaluation of triplet phase invariants to discover their potential usefulness.

While working at the Naval Research Laboratory, Karle also spent some time as a visiting lecturer of mathematics and physics in England, Italy, Canada, Poland, Brazil, Japan, and Germany. He served as president of the American Crystallographic Association in 1972, president of the International Union of Crystallography between 1981 and 1984, and chair of the Chemistry Section of the National Academy of Sciences between 1988 and 1991. He is a fellow of the American Physical Society and the American Mathematical Society, and a member of the American Chemical Society, the American Philosophical Society, the American Crystallographic Association, and the American Association for the Advancement of Science. He has been a member of the National Science Foundation and the Committee on Human Rights of the National Academy of Sciences. Karle is also a member of the Union of Concerned Scientists and a National Sponsor of the Committee of Concerned Scientists.

Karle has won the Distinguished Public Service Award of the United States Navy in 1968, the Hillebrand Award of the American Chemical Society in 1970, the A. L. Patterson Memorial Award of the American Crystallographic Association in 1984, and the first National Research Laboratory Lifetime Achievement Award in 1993. He has received honorary degrees from Georgetown University, the University of Maryland, City College of New York, and the University of Michigan. His greatest accolade came in 1985, when, more than forty years after he and his wife created the basic mathematical formulas upon which the direct formula is based, he and Herbert A. Hauptman were jointly awarded the 1985 Nobel Prize for chemistry. Of his wife's involvement with the project, he has said: ''It is unfortunate that she was not included in the Nobel Prize.''

KARRER, PAUL (1889-1971)

Swiss organic chemist

Paul Karrer's long and distinguished career in **chemistry** included the study of sugars and plant pigments, subjects that led him to the description and synthesis of vitamin A as well as several other **vitamins**. Karrer's work with vitamins helped to solve their chemical riddle, enabling physiologists to define the way in which the body utilizes them. In 1937 Karrer shared the Nobel Prize for Chemistry for research that incorporated the vitamins A and B.

Paul Karrer was born in Moscow, Russia, on April 21, 1889, the son of Julie Lerch Karrer and Paul Karrer, a Swiss dentist who was practicing in Russia. When he was three, Karrer and his family returned to Switzerland, initially to Zürich, but later settling in the canton, of Aargau, a region in the north of the country. Karrer was educated in schools in this canton, and it was while in secondary school that he began showing a passion for science. In 1908 he entered the University of Zürich, ultimately studying chemistry under **Alfred Werner**, whose work on the linkage of atoms in molecules won him the Nobel Prize in 1913. Karrer, after completing his doctoral dissertation on **cobalt** complexes in 1911 and earning his Ph.D., became Werner's lecture assistant. His attention soon turned to organic arsenical compounds, and Karrer's first paper on the subject, published in 1912, caught the eye of **Paul Ehrlich**, a renowned chemist in Germany whose work at the turn of the twentieth century had helped to explain the action of poisons

Paul Karrer.

and how to neutralize their effects by antitoxins. Ehrlich subsequently invited Karrer to join him as a research assistant in Frankfurt-am-Main at the Georg Speyer-Haus, a research institute.

Karrer remained in Germany until the beginning of World War I when he was called back to Switzerland for national service. While serving in an artillery unit, he met and married Helene Frölich, the daughter of the director of a psychiatric clinic. They would have three sons, but only two of them, Jurg and Heinz, survived infancy. In 1915 Ehrlich died in Frankfurt, and Karrer accepted the position of researcher and director of the chemical research division of the Georg Speyer-Haus, returning to war-time Germany. Karrer stayed in Germany for the next three years, and during this time he focused more closely on plant product chemistry. Then in 1918 he returned to the University of Zürich, first taking a position as associate professor of **organic chemistry**, and with Werner's death in 1919, becoming a full professor of chemistry as well as director of the Chemical Institute. Karrer would remain at the University of Zürich for the rest of his career, acting as rector from 1950 to 1953.

Karrer, of necessity, had to split his time between administration, teaching, and research. With the latter, he turned his attention to the spatial or steric configuration of atoms in molecules of **amino acids**, **proteins**, and **peptides**. But by the late–1920s he had shifted his focus to the pigmentation of plants, and more specifically, to the anthocyanins, the blue and red colors of berries and flowers. Though these substances had been isolated by another researcher, Karrer—by splitting their macromolecules with enzymes—helped to clearly describe their chemical make-up. More importantly, it was these researches in plant pigments that eventually led him to his work on carotene.

From anthocyanin, Karrer moved on to crocin, the yellow pigment of flowers such as the crocus. In connection with yellow pigments, Karrer tackled the structure of carotenoids, the orange-to-yellow pigments found in foods such as carrots and sweet potatoes. **Richard Kuhn**, a German chemist, had managed to isolate beta-carotene at about this time, and he and Karrer became something of rivals in explaining beta-carotene's chemical constitution. By 1930 Karrer had solved the structure of the carotene **molecule**. It was a logical progression from the study of plant pigments to that of vitamins, for Karrer learned that the body actually synthesizes vitamin A from carotene. Thus, he was soon on the track of the chemical make-up of vitamins themselves. By 1931 Karrer had become the first scientist to describe the structure of a vitamin, successfully demonstrating that vitamin A is very similar to one half of the symmetrical carotene molecule. Up until the time of his discovery, scientists had thought vitamins to be some peculiar state of **matter**, perhaps a colloid—a dispersed **solution** in **suspension**. But Karrer managed to show that vitamin A, in specific, is made up of atoms of **hydrogen**, **carbon**, and **oxygen** in a regular ring-like formation.

Karrer carried on his vitamin research over the next decade, ultimately synthesizing vitamin A in the laboratory. He then went on to research the chemical structure of several B vitamins, riboflavin being the first that he actually synthesized. In 1937, for his work on carotenoids and flavins, he shared the Nobel Prize for Chemistry with **Walter Haworth**, an Englishman who researched the make-up of **carbohydrates** and vitamin C. Karrer, however, was not one to rest on his laurels, and the very next year he synthesized vitamin E, and soon after, vitamin K.

From vitamin research, Karrer turned to an investigation of the **enzyme** nicotinamide adenine dinucleotide (NAD) which is involved with the **energy** system of cells. By 1942 he had contributed greatly to the understanding of both the NAD structure and function in cellular **electron** transfer. In his sixties, Karrer went back to earlier work, both in carotenes and in poisons—this time alkaloids. In the former, he successfully synthesized all the carotenoids, some forty different compounds. His work in alkaloids helped to determine the structure of curare, a resinous extract from certain South American trees used by indigenous peoples for poison arrows. Its medicinal uses include general anesthesia and **reduction** of muscle spasms.

Apart from his research, Karrer was a tireless administrator and teacher, directing over 200 dissertations in his academic career. He was also a prolific writer, with over 1,000 publications to his credit, including the 1928 organic chemistry textbook, *Lehrbuch der organischen Chemie.* Aside from the Nobel, Karrer won numerous awards and prizes, among them the Marcel Benoist Prize from Switzerland in 1923, the Cannizzaro Prize from Italy in 1935, and the Officier de la Legion d'Honneur from France in 1954. Despite fame, wealth, and offers from universities around the world, Karrer stayed on in Zürich until his retirement in 1959, eschewing luxuries such as a car. He died on June 18, 1971, after a short illness.

Kekulé von Stradonitz, (Friedrich) August (1829-1896)

German chemist

Kekulé was born in Darmstadt, Germany, on September 7, 1829. After graduating from high school, he enrolled at the University of Giessen, intending to major in architecture. Instead, he became enraptured by the lectures of **Justus von Liebig** and changed his major to **chemistry**. After leaving Giessen in 1851, he studied for a year in Paris, France, and briefly worked at a private laboratory in Switzerland. He completed his education with a two-year stint as an assistant to John Stenhouse (1809-1880) at St. Bartholomew's Hospital in London, England.

When he returned to Germany, he became a lecturer at the University of Heidelberg and then professor of chemistry first at Heidelberg (1855-1858), then at the University of Ghent (1858-1867), and finally at the University of Bonn (1867-1896).

Kekulé was not a skilled experimentalist and only a mediocre teacher. His major contributions to chemistry involved theoretical ideas, specifically on the structure of compounds.

Until the mid-1800s, chemists had serious doubts about the value of using chemical formulas to describe the **molecular structure** of organic compounds. They acknowledged that molecules did consist of certain arrangements of atoms. But they were not convinced that those arrangements would survive chemical analysis and, therefore, thought that structures might be essentially unknowable. A leading proponent of this view, **Charles Gerhardt** (1816-1856), wrote that "chemical formulas are not destined to represent the arrangement of atoms."

As a result, organic chemists seldom wrote more than the symbols of the elements and the ratio of atoms in a compound. Chloroacetic acid, for example, might be represented simply as $C_2H_3ClO_2$.

In 1858, however, Kekulé presented a different view. In a paper entitled "On the constitution and metamorphoses of chemical compounds and the chemical nature of carbon," Kekulé stated two fundamental premises. First, he argued that **carbon** was always tetravalent. That is, any one carbon **atom** will always form four bonds—no more and no fewer—in a compound. Second, he suggested that carbon atoms had the ability to bond to each other in endless **chains**. A single carbon atom, for example, might well bond to four other carbon atoms. He also suggested that Gerhardt's position was wrong and that chemical analysis *would* yield reliable information about a compound's structure.

Kekulé developed a system of symbols for representing the structures of molecules. The system made use of ovals and circles that represented atoms of various elements. When combined to represent molecules, these symbols looked a bit like bunches of fruit. A colleague, **Adolph Kolbe**, referred to them as "Kekulé's sausages."

A much better system of notation was proposed by the Scottish chemist **Archibald Scott Couper** (1831-1892). Couper independently developed an analysis of molecular structure very similar to that of Kekulé. However, publication of his paper was delayed by two months and Kekulé received priority for the ideas both had developed. In place of Kekulé's "sausages," Couper suggested using dotted lines or short dashes to show the **bonding** between atoms in an organic **molecule**. The system Couper developed is essentially the one used by organic chemists today.

Kekulé's second great achievement was the determination of the structure of the **benzene** molecule. Analysis had shown that the **molecular formula** of benzene was C_6H_6. But chemists were mystified as to how these atoms could be arranged in a molecule so as to satisfy every atom's **valence** requirements.

The story of Kekulé's discovery and the debate over that story have occupied historians of science for more than a century. According to Kekulé's own account, he solved the problem when he fell asleep in front of the fireplace. Still half-thinking about the benzene problem, he wrote that "the atoms gamboled before my eyes. Smaller groups this time kept modestly to the background. My mind's eyes, trained by repeated visions of a similar kind, now distinguished larger formations of various shapes. Long rows, in many ways more densely joined; everything in movement, winding and turning like snakes. And look, what was that? One snake grabbed its own tail and mockingly the shape whirled before my eyes. As if struck by lightning I awoke; this time again I spent the rest of the night to work out the consequences."

The structure that Kekulé drew consisted of six carbon atoms arranged in a circle, joined by alternating single and double bonds. To each carbon atom was joined a single **hydrogen** atom. The benzene problem had been solved.

In 1876, Kekulé suffered a serious attack of the measles. He never fully recovered his health and produced little new research in the remaining two decades of his life. He died in Bonn, Germany, on July 13, 1896.

Kendall, Edward C. (1886-1972)

American biochemist

Edward C. Kendall is best remembered as a pioneer in the discovery and isolation of several important **hormones**. As a young scientist he isolated the hormone thyroxine from the thyroid glands of cattle; today, thyroxine is produced synthetically and used in the treatment of thyroid disorders. Later, he isolated six hormones produced by the adrenal cortex. One of

Edward C. Kendall.

these was **cortisone**, which proved to be a breakthrough in the treatment of rheumatoid arthritis. Kendall's work led to the 1950 Nobel Prize in medicine and physiology, which he shared with colleagues **Philip S. Hench** and **Tadeus Reichstein**.

Edward Calvin Kendall was born on March 8, 1886, in South Norwalk, Connecticut, the youngest of three children. His father, George Stanley Kendall, was a dentist, and his mother, Eva Frances (Abbott), was active with the Congregational Church. Kendall showed a curious nature early on, and when he entered Columbia University in 1904 he chose **chemistry** as his primary area of study. He earned his bachelor of science degree in 1908, his master's degree in 1909, and his Ph.D. in chemistry in 1910—all from Columbia.

Upon his graduation, he accepted a position with the pharmaceutical firm Parke, Davis, and Company in Detroit. He found the atmosphere stifling, however, because he had no opportunities for the kind of research he wanted to do. In his memoirs, *Cortisone: Memoirs of a Hormone Hunter,* he expressed his disdain for the rigid, controlled environment at Parke, Davis. As an example, he noted the company policy that all lab employees punch in and out on a time clock. "After working 18 hours a day for some weeks to finish my thesis," he recalled in his biography, "I could not accept the thought that my value to the company could be determined by the hours spent in the building."

After four months at Parke, Davis, Kendall left and returned to New York. He soon found a position at St. Luke's Hospital. It was at St. Luke's where Kendall began his work on isolating thyroxine. Nearly twenty years earlier, the German chemist Eugen Baumann had discovered high concentrations of **iodine** in the thyroid gland. Scientists were later able to obtain a protein called thyroglobulin; Kendall's aim was to isolate the active compound in this protein.

He was able to purify the protein, and early experiments with patients at St. Luke's proved successful. But while the physicians at St. Luke's were eager to find new ways to treat patients, their emphasis on actual research was not as strong as Kendall thought it should be. He left St. Luke's at the end of 1913 and headed west to the Mayo Clinic, where over the next four decades he did his most important work.

By the end of 1914, Kendall had isolated thyroxine. This breakthrough discovery eventually led to synthetic production of the substance, which in turn led to more effective treatment of thyroid disorders. For his work he was awarded the Chandler Prize by his alma mater in 1925. Kendall also isolated the peptide glutathione from yeast and determined its structure.

Now Kendall was ready to tackle the challenge of isolating hormones from the adrenal gland. During the 1930s he managed to isolate more than two dozen hormones, or corticoids (so called because they came from the cortex, or outer section, of the gland). The six most important hormones were each assigned a letter A through F. Compound E—cortisone—turned out to be the most significant of these.

Compound E was not easy to synthesize. Kendall worked for several years with a substance obtained from cattle bile and was finally successful in producing a small amount of the compound late in 1946. Kendall's research got a boost from the U.S. Government, which gave top medical priority to the investigation of cortisone during the Second World War. This was prompted in part by rumors (later proven untrue) that German scientists had been extracting adrenal gland extract from Argentine cattle and giving it to Nazi pilots to boost their strength (they were supposed to be able to fly planes at heights up to 40,000 feet). The U.S. Office of Scientific Research and Development (with which Kendall served as a civilian during the war) gave him support, and the pharmaceutical firm Merck and Company sent a scientist to help him complete the synthesis.

Research by Kendall's colleague Hench showed that cortisone might be useful in the treatment of rheumatoid arthritis. Actual experimentation with patients began in 1948, and the results were dramatic. Rheumatoid arthritis is a painful condition that causes severe pain and swelling in the joints; cortisone, though not a cure, was able to control the symptoms. It also controlled symptoms in some skin diseases and eye disorders. Reichstein, working independently of Kendall and Hench, also synthesized cortisone in Switzerland. It was for their work and research with cortisone that the three men were awarded the Nobel Prize. Kendall was also awarded several honorary degrees, including one from Columbia.

In 1951, Kendall accepted a position as visiting professor at Princeton University, where he remained for the rest of his life. Among his other awards were the American Public Health Association's Lasker Award and the American Medical Association's Scientific Achievement Award. He was a member of several organizations, and in addition to his book *Thyroxine* and his memoirs, he wrote articles for numerous scientific publications. He served as president of the American Society of Biological Chemists from 1925 to 1926 and the Endocrine Society from 1930 to 1931.

Kendall married Rebecca Kennedy in 1915 and had three sons and a daughter. All three of Kendall's sons died before their father passed away. Kendall himself died in Rahway, New Jersey, on May 4, 1972 and was buried in Rochester, Minnesota, home of the Mayo Clinic.

KENDREW, JOHN COWDERY (1917-1997)

English physicist and biochemist

The decades following World War II saw an increase in using physical methods to solve biological problems, which led to a greatly increased knowledge of living systems. John Kendrew's contribution to this trend was formulating the first three-dimensional structure of the protein myoglobin by using x-ray **crystallography**. For this achievement he shared the 1962 Nobel Prize for **chemistry** with his colleague **Max Perutz**, who had done similar work with hemoglobin.

John Cowdery Kendrew was born in Oxford, England, on March 24, 1917, the only child of Wilfrid and Evelyn Sandberg Kendrew. His father was a lecturer in climatology, and his mother was an art historian; thus he was nurtured in an academic atmosphere. He attended the Dragon School in Oxford and finished his secondary schooling at Clifton College in Bristol. He matriculated at Trinity College, Cambridge, where he studied chemistry and graduated in 1939.

His academic career was interrupted by World War II, when he served with the Ministry of Aircraft Production. He worked for a time on radar, and then acted as scientific advisor to the Allied air commander in chief in Southeast Asia. While serving in Asia, Kendrew met physicist J. D. Bernal. Bernal was convinced that the intersection between biology and the physical sciences would soon be an important area of research, and this greatly influenced Kendrew's career path. On a side trip to California, Kendrew also met the American physical chemist **Linus Pauling**, whose protein work would later serve as a foundation for his own.

After the war, Kendrew became a doctoral candidate at the Cavendish Laboratory at Cambridge, where he met Perutz, who had once been a student of Bernal's. Together they worked under physicist **William Lawrence Bragg**, who had helped establish the science of **X-ray crystallography** (a process that outlines a substance's atomic structure) in the early twentieth century. By 1947, Bragg had convinced the secretary of the medical research council that the government should finance the kind of work that Kendrew and Perutz were doing. A special unit for the study of the **molecular structure** of biological systems was founded for the two scientists, housed in an empty shed at Cambridge.

John Cowdery Kendrew.

Kendrew's doctoral thesis dealt with the differences between fetal and adult sheep hemoglobin, a component of red blood cells that contains **iron** and assists with **oxygen** transport. The project gave him valuable experience in x-ray methods and a chance to work with **proteins**, which he came to regard as the most important class of biological molecules. After receiving his Ph.D. in 1949, Kendrew commenced research on the protein myoglobin. Myoglobin is the **molecule** that binds and transports oxygen in the muscles, and was readily available from whales. Perutz was already working on determining the structure of hemoglobin. At the time Kendrew started his research, little was known about myoglobin, except that it was a protein.

The first problem Kendrew faced was to produce crystals of pure myoglobin. To do this, he planned to use x-ray crystallography, which required samples with regular crystal structures. The technique involved shining x rays through the crystal and observing on photographic **paper** the pattern of spots that leaves the crystal. The atoms in the crystal diffract the X rays through definite angles, concentrating them in patterns of bands or spots from which the scientist must calculate the kind of atomic arrangement that produced such a pattern.

Kendrew and his colleagues obtained their crystals and good X-ray pictures of them, but the pictures proved to be too complex to interpret. Fortunately, Perutz recalled a technique he had learned from Bernal in which a heavy metal **atom** bonded with a protein to serve as a marker for the diffracted x rays to be sorted. Kendrew, however, had to try a number of heavy **metals** before he could make interpretable pictures; in all he made well over 10,000 images. When he finally had the laboratory data he wanted, he made his mathematical calculations of **electron** densities with a computer.

Even after Kendrew obtained his densities, a formidable problem remained. The densities had to be plotted for planar slices through the crystal at intervals of a few angstrom units (one ten-billionth of a meter), and the contours of electron **density** (like elevations on a topographical map) had to be determined. High density indicated an atom, and certain atoms (like **nitrogen** and oxygen) could be distinguished by the magnitude of their density. Since computers were not yet capable of these calculations, Kendrick's group had to do all the plotting by hand. They announced their findings in 1960, in the same issue of *Nature* in which Perutz published his preliminary findings on hemoglobin.

The myoglobin molecule proved to be a dense, lumpy structure. It had none of the beauty and regularity that molecular biologists **Francis Crick** and **James Watson** had found in their x-ray work on deoxyribonucleic acid (**DNA**), for which they received the Nobel Prize for medicine or physiology in 1962—the same year that Kendrew and Perutz were awarded the Nobel Prize in chemistry. One writer commented that the significance of Kendrew's work lay not in the new insights it provided, but in the fact that it could be done at all. Kendrew had risked analyzing a complex structure when many simpler ones had not yet been attempted, and he had succeeded.

Although Kendrew continued his work on the structure of myoglobin after receiving the Nobel Prize, he was increasingly drawn into administration and government advisory positions. The department Kendrew and Perutz created at Cambridge was now known as the Laboratory for Molecular Biology, and Kendrew acted as deputy chairman of the organization from its inception until 1974. The following year, he established the European Molecular Biology Laboratory in Heidelberg, Germany, and served as its director until 1982. Earlier in his career he founded the *Journal of Molecular Biology* and acted as its editor-in-chief until 1987. From 1954-1968, he was a reader at the Davy-Faraday Laboratory of the Royal Institution, London, and from 1981-1987 he was president of St. John's College, Oxford. In addition, he served as president of the International Organization of Pure and Applied Biophysics, and as both secretary general and president of the International Council of Scientific Unions.

Kendrew was recognized for his achievements with many awards in addition to his Nobel Prize. He was an honorary fellow of Peterhouse, Cambridge, and St. John's, Oxford. He was also a fellow of the Royal Society, a foreign associate of the National Academy of Sciences in the United States, and a foreign honorary member of the American Academy of Arts and Sciences. He was knighted and given the Order of the British Empire in 1963. He was unmarried. Kendrew died in 1997.

KEROSENE

The invention of new oil lamps in the late 1700s and early 1800s greatly improved the quality of indoor lighting, but the supply of fuel for these lamps was limited. Whale oil and other fuels were too expensive to compete with gas lighting, which was gaining popularity at the time. Then, Abraham Gesner (1797-1864) produced kerosene, a pale liquid fuel that he distilled from thick crude oil. Gesner, a medical doctor from Canada who had become interested in geology, named the fuel after the word *keros* (Greek for wax) and obtained United States patents for preparing it. When kerosene was successfully introduced in America in the 1850s, Gesner remarked hopefully that it might save whales from being hunted for their oil.

Around the same time, the Scottish scientist James Young (1811-1883) also began promoting the use of kerosene, which he produced by **distillation** from **coal** and oil shale. Other competitors also introduced kerosene lamp oils made from various substances, but in 1859 the discovery of huge quantities of crude oil in America made all other sources of kerosene obsolete. By the end of the century, kerosene had become the chief product of American oil refineries.

Kerosene (or kerosine, as the oil industry spells it) is a close relative of **gasoline**. Both are produced by refining crude oil, but the kerosene fraction of the oil is a little heavier. Kerosene contains a mixture of **hydrocarbon** compounds, most of which have 10-14 atoms of **carbon** per **molecule**. Because kerosene belongs to the family of hydrocarbons called alkanes or paraffins, it is sometimes referred to as paraffin oil, in addition to the nicknames coal oil, lamp oil, and illuminating oil. At room **temperature**, kerosene is a thin liquid that evaporates easily and smells slightly sweet. Kerosene fuel, however, is poisonous. Drinking it causes vomiting and diarrhea, and breathing it causes headaches and drowsiness.

When kerosene was first produced, the lighter gasoline fraction of the oil and the heavier fuel-oil fraction were considered useless or even dangerous waste products. American refiners were only interested in making kerosene, most of which was exported to Europe for lighting purposes. Kerosene was also widely marketed as fuel for oil stoves, which were introduced in the late 1800s. But around the same time, incandescent electric lights began to compete with oil lamps. Refiners were forced to market other oil products, such as fuel oil for powering ships.

Most of the kerosene produced today is used as an engine fuel for jet aircraft and rockets. On a smaller scale, kerosene is still burned in special lamps, stoves, and space heaters, especially in rural areas that have no **electricity**. Kerosene appliances, however, can be dangerous; if the fuel does not burn properly, it can produce deadly **carbon monoxide**. Kerosene is also used as a fuel for tractors and power generators and as a solvent for garden chemicals such as weedkillers and insecticides. In 1997, President Clinton's budget proposed a tax on kerosene that was vehemently opposed by the Petroleum Marketers Association of America (PMAA) because they said it would hurt millions of U.S. homeowners, farmers, and others that purchase tax-free kerosine.

KETONE FUNCTIONAL GROUP

The **functional group** of the **ketones** is -CO-. The **carbon atom** is double bonded to the **oxygen**. The groups connected to the remaining bonds are alkyl or aryl groups (not **hydrogen**, which would result in the formation of an aldehyde). The alkyl or aryl groups modify the behaviour of the ketone in the normal way; for example as the molecular **mass** increases the **boiling point** of the resultant ketone increases. One of the most common ketones is **acetone**.

The functional group is prepared in one of two ways, either by partial chemical **oxidation** of a secondary **alcohol** or by **dehydrogenation**, with **heat** and a catalyst, of a secondary alcohol. Both of these processes remove the hydrogen from the hydroxyl group and allow a double bond to form between the connecting carbon and the oxygen.

The majority of reactions in which the ketone functional group takes place are addition reactions. Secondary alcohols can be readily recovered from ketones by breaking the double bond between the oxygen and carbon and adding hydrogen. This process can be carried out by adding the ketone to hydrogen in the presence of a catalyst or by using nascent hydrogen produced by the action of a dilute acid on a metal. A slightly different reaction occurs across the functional group if amalgamated **zinc** and concentrated **hydrochloric acid** are used. The **carbonyl group** is completely reduced to a -CH_2 group. This process produces an alkane and is named the Clemmensen reaction after its discoverer (1913). The carbonyl group can be broken down by the action of hydrogen cyanide; a hydroxyl group and a nitrile group are formed in this reaction. The resultant chemical is a ketone cyanhydrin. A similar reaction occurs with **sodium** hydrogen sulfite to give ketone hydrogen sulfite compounds. The ketone functional group can also be involved in condensation reactions. In this type of reaction hydrazine (H_2NNH_2) removes the oxygen from the carbonyl. This oxygen reacts with hydrogen from the hydrazine to give **water** as an elimination product. The result is a hydrazone, which has the new functional group -$CNNH_2$, with a double bond between the carbon and the first **nitrogen**. Derivatives of hydrazine react in a similar manner as do hydroxylamine (H_2NOH) which gives oximes and semicarbazide ($H_2NNHCONH_2$) which gives semicarbazones.

The ketone functional group can also take part in autocondensation reactions which eliminate water. These reactions will produce an aromatic **molecule** from a starting point of an aliphatic, straight chain molecule. Polymerization across the carbonyl group is a very uncommon reaction, particularly compared to the similar group in **aldehydes**. This reaction is very similar to the **aldol reaction**.

To identify the presence of a ketone functional group several tests must be carried out. First, the compound under test is dissolved in **methanol** to which is then added a chemical called Brady's reagent. A yellow or orange precipitate shows the presence of the carbonyl group. The precipitate is then collected and dried and further tests can then be carried out to determine if the starting material was an aldehyde or ketone. Aldehydes will show a reaction with the following substances whereas ketones will not: Fehling's solution, Schiff's reagent, and Tollen's reagent.

The ketone functional group is the carbonyl group with its properties modified by the presence of two alkyl or aryl groups. The properties of the ketone group are very similar to those of the aldehyde group which consist of a carbonyl group, one alkyl or aryl group and a hydrogen atom.

See also Aldehydes, Ketones

KETONES

Ketones have the general formula RCCOCR' where R and R' may be the same or different groups (they can be alkyl or aryl groups but not hydrogen). The central **carbon atom** has a double bond with the **oxygen**. Ketones differ from aldehydes in that the latter has a **hydrogen** atom attached as one of the R groups. The name of a ketone is derived from the longest chain that contains the **carbonyl group**, with the -e ending of the alkane being replaced by -one. The location of the carbonyl group is given by numbering the chain in the direction that will give the lowest number. It is also correct to name ketones by listing alphabetically the groups attached to the carbonyl group and following this with the word ketone. Some ketones are also known by common names. The ketone CH_3COCH_3 is known as propanone, dimethyl ketone, and **acetone**.

Ketones may be prepared by the partial **oxidation** of a secondary **alcohol** (complete oxidation would give **carbon dioxide** and water). This can be carried out using chemical oxidation agents such as **hydrogen peroxide**, ozone, or **potassium** dichromate. Manufacture of ketones can also be carried out by the **dehydrogenation** of a secondary alcohol at high **temperature**.

Typical reactions of ketones are addition reactions although as a group they are less reactive than the **aldehydes**. Secondary alcohols can be formed by reduction of the ketone by hydrogen gas in the presence of a catalyst. Ketones are extensively used as **solvents**, with acetone (propanone) being the most well known example. The carbonyl **functional group** gives polarity to the acetone, and acetone is completely soluble in **water** and dissolves a wide range of organic substances. Ethyl methyl ketone is also used as an industrial solvent. Aliphatic ketones are volatile **liquids** with a pleasant smell and aromatic ketones are usually **solids**. Aromatic ketones can be prepared by the Friedel-Crafts reaction. The **solubility** of ketones in water decreases as their molecular **mass** increases. Ketones are readily inflammable and will burn in air with a clear blue flame to produce carbon dioxide and water. Ketones can undergo a process known as autocondensation, where several ketone molecules react together to eliminate water. When this occurs with acetone in the presence of **sulfuric acid** (to remove the water) the product is 1,3,5- trimethylbenzene (mesitylene). This reaction is of interest because it is one of the few examples of a synthetic reaction where an aliphatic (straight chain) compound is transformed into an aromatic (ring structure) compound. This type of reaction is very common in biosynthesis, where many structures are made (especially by plants) by the "polyketide" pathway that uses large numbers of ketone condensation reactions. The reaction of ketones with strong bases gives enolate anions, which are the intermediates used in these condensation reactions.

Acetone is one of the oldest known ketones. It was known by the alchemists who called it spirit of Saturn. Antoine Bussy (1794-1882) gave acetone its current common name in 1833. Historically acetone was prepared by the destructive **distillation** of **wood**, followed by fractional distillation of the product. Ketones can display isomerism, for example methyl n-propyl ketone and isopropyl methyl ketone have the same formula but different arrangements of the atoms leading to different properties.

Ketones are a homologous series of organic chemicals containing the functional group -CCOC-. The first two members of the group are extensively used as organic solvents.

See also Ketone functional group

KHORANA, HAR GOBIND (1922-)

American biochemist

Har Gobind Khorana is considered a major contributor to the science of genetics. In addition to developing a relatively inexpensive method of synthesizing acetyl **coenzyme** A, a complex **molecule** used in biochemical research, he succeeded in cracking the genetic code of yeast by synthesizing parts of a nucleic acid molecule—an achievement for which he shared the Nobel Prize for physiology or medicine. Khorana went on to do other significant work, including the synthesis of the first completely artificial gene.

Khorana was born around January 9, 1922 in the small village of Raipur, India (now Pakistan), the youngest of five children of Ganpat Rai Khorana, a tax collector for the British colonial government, and Krishna (Devi) Khorana. His family, although poor, was one of the few literate families in his village. He received his early education under a tree in outdoor classes conducted by the village teacher, and went on to attend high school in Multan, Punjab (India). He later studied chemistry on a government scholarship at the Punjab University in Lahore, graduating with honors in 1943 and receiving a Masters of Science with honors in 1945.

After obtaining his M.S., Khorana went to the University of Liverpool on a Government of India Fellowship to study **organic chemistry**; there he earned a Ph.D. in 1948 for his research on the structure of the bacterial pigment violacein. From England, Khorana went to Zurich, Switzerland, as a postdoctoral fellow to study certain alkaloids (organic bases) under the tutelage of **Vladimir Prelog**, and after a brief visit to India in 1949, returned to England. From 1950 to 1952, Khorana worked at Cambridge University under Sir **Alexander Todd**, who later received the Nobel Prize for his work with **nucleic acids** (large molecules in the **nucleus** of the cell). While working with Todd, Khorana, too, became interested in the **biochemistry** of nucleic acids.

In 1952, Khorana took a position as director of the British Columbia Research Council's Organic Chemistry Section, located at the University of British Columbia in Vancouver. There he made his first contribution to the field of biochemistry when he and a colleague, John G. Moffat, announced in 1959 that they had developed a process for synthesizing acetyl coenzyme A, an essential molecule in the biochemical processing of **proteins**, fats and **carbohydrates** within the human body. A complex structure, this coenzyme had previously been available only by an astronomically expensive method of isolating the compound from yeast. Moffat and Khorana were able to synthesize acetyl coenzyme A relatively cheaply, thereby making it widely available for research. This work gave Khorana international recognition within the scientific community.

In 1960, Khorana moved to the University of Wisconsin in Madison to accept a position as co-director of the Institute for **Enzyme** Research. He became a professor of biochemistry in 1962, and in 1964 was named to the Conrad A. Elvelijem Professorship of the life sciences. Khorana then began focusing his research on genetics—specifically, on the biochemistry of nucleic acids, the biosynthesis of enzymes, and on deciphering the genetic code.

At the time Khorana began his research in genetics, scientists already knew much about genes and how they determine heredity. They had discovered that genes are located on chromosomes in the cell nucleus, and that genes are made of deoxyribonucleic acid (DNA), a nucleic acid which controls the biochemical processes of the cell and governs an organism's inherited traits. DNA's double-helix shape resembles a spiral staircase with regularly spaced steps, with each step consisting of a pair of chemical compounds called nucleotides. The four different types of nucleotides are arranged on the staircase in a pattern of heredity-carrying code "words."

To decipher this code, scientists needed to learn how those words were translated into a second "alphabet" consisting of 20 types of **amino acids**, the building blocks of protein. Part of this translation had been accomplished prior to Khorana's work. The **DNA** in the nucleus of a cell causes another nucleic acid called messenger ribonucleic acid (mRNA) to be produced; the messenger **RNA** then attaches itself to ribosomes, where the cell's proteins are produced. Another type of RNA, called transfer RNA, transports loose amino acids to the ribosomes, where messenger RNA uses them to construct proteins.

Scientists knew that the code word on each transfer RNA molecule indicated the kind of amino acid it would deliver, and instructed it to take it only to a complimentary messenger RNA. They had also figured out that there were 64 possible combinations of nucleotides, each with its own signal. What they did not know was which nucleotide word called for which amino acid.

In 1961, Dr. **Marshall Warren Nirenberg** of the National Institutes of Health successfully decoded most of the messages in the nucleotides. Khorana carried Nirenberg's work even further, adding significant details. In 1964 he synthesized parts of the nucleic acid molecule, and later was able to duplicate each of the 64 possible genetic words in the DNA staircase. He was able to map out the exact order of the nucleotides, and showed that the code is always transmitted by three-letter words. He also learned that certain nucleotides order the cell to start or stop making proteins.

Khorana's research was based in part on work done separately by both Nirenberg and **Robert W. Holley** of Cornell University, who completed the first delineation of a complete nucleic acid molecule in 1966. For their contributions to deciphering the genetic code, these three scientists were awarded the 1968 Nobel Prize for Physiology or Medicine. At the presentation ceremony, the three were commended for having "written the most exciting chapter in modern biology."

Two years later, Khorana made another contribution to the field of genetics when he and his colleagues succeeded in synthesizing the first artificial DNA gene of yeast. Khorana announced his achievement in a characteristically modest way, by informing a small symposium of biochemists at the University of Wisconsin in June 1970. He also announced that he and most of his research team would move to the Massachusetts Institute of Technology in the fall. As he explained to a friend, "You stay intellectually alive longer if you change your environment every so often." Khorana joined MIT's faculty as the Alfred P. Sloan Professor of Biology and Chemistry.

Khorana's accomplishments in the laboratory include the artificial synthesis of another gene found in *Escherichia coli,* an intestinal bacteria known commonly as E. coli. Outside the laboratory, his professional activities include membership in several scientific societies, including the National Academy of Sciences and the American Academy of Arts and Sciences. He also served on the editorial board of the *Journal of the American Chemical Society* for many years, and published more than 200 articles on technical subjects in that journal and other professional publications.

Khorana has received numerous accolades in addition to the Nobel Prize, including the Lasker Award and the American Chemical Society award. He also was awarded an honorary doctorate by the University of Chicago, and he was named a fellow by Churchill College in Cambridge, England. Khorana also has been a visiting professor at Rockefeller University and Stanford University, and has given lectures at numerous scientific meetings.

Khorana married Esther Elizabeth Sibler in 1952. The couple has two daughters, Julia Elizabeth and Emily Anne, and one son, Dave Roy. Khorana became an American citizen in 1966. Extremely committed to his work, he seldom takes time off, and once went 12 years without taking a vacation. He takes daily long walks, carrying with him an index card to record any ideas that might come to him. He also enjoys going on hikes and listening to music.

Har Gobind Khorana. ***(AP/Wide World. Reproduced by permission.)***

KINETIC ISOTOPE EFFECTS

Kinetic **isotope** effects (KIE) refer to the impact on **reaction rate** when one of the atoms in a **molecule** is replaced by its isotope. For example, replacing **hydrogen** (H) in a molecule with its isotope **deuterium** (D), which is heavier by one **neutron**, can slow down a reaction by up to 20 times. From rate information scientists learn about what happens to molecules during a reaction. They gain information about the changes as the reactants form new products during the course of the reaction,.

Atomic isotopes are closely related. Consider, for example, the isotopes of **radium**. They are atoms of the same element, so they have the same number of protons and electrons. The only way they differ is in the number of neutrons. One isotope has 202 neutrons, and the other isotope has 203 neutrons. The isotope with the extra neutron is a little heavier than its sister. Therefore, the isotopes behave differently. For example, they may have different reaction rates. Isotopes of heavier elements, like radium, however, are so close in weight (257 neutrons are not that much heavier than 256 neutrons) that this difference is insignificant.

The difference is much larger however, for the isotopes of the lightest elements because one neutron changes the **mass** of the **atom** by a larger percentage. So deuterium, which has one neutron, is heavier than hydrogen, which has none. And in a chemical reaction, deuterium reacts much more slowly than hydrogen. This chemical difference is put to use by scientists in the technique of **kinetic isotope effects**. Besides hydrogen and its isotope deuterium, researchers use the isotopes of **boron**, **oxygen**, **nitrogen** and **carbon**.

The reason isotopes H and D behave differently relates to their ''**dissociation energy**,'' or the amount of energy required to break their bonds with other atoms. Generally, the lighter the atom, the less energy is needed to excite its bond to its breaking point, and the lower the ''dissociation energy.'' Consider, for example, two bonds, a carbon-hydrogen (C-H) and a carbon-deuterium (C-D) bond. The atoms at each end of the bond can be viewed as two balls vibrating back and forth on two ends of a spring. The lighter hydrogen atoms are like ping-pong balls, and the heavier deuterium atoms like golf-balls. The golf-balls are harder to get moving than the ping pong balls just because they are heavier, and more energy is required vibrate them to the point where they will separate. In this example then, the C-D bond will take more energy to break than the C-H bond.

A reaction involving deuterium will proceed more slowly than the same reaction with hydrogen reactants provided the isotope is involved in the ''rate limiting step.'' In a chemical reaction, a rate limiting step is the slowest part which sets the pace for the whole reaction. The remainder of the reaction occurs extremely quickly in comparison. To find out which part of the reaction is the rate limiting step, chemists look for a kinetic isotope effect. For example, consider a two-step reaction process where the first step involves combining two molecules, and the second step involves breaking a C-H bond. One of these steps occurs more slowly than the other, and so the rate of this step ''limits'' the rate for the whole reaction. Scientists can run this reaction twice, once with deuterium and once with hydrogen. If the reaction is much slower with deuterium, then the ''rate-limiting step'' is probably the one in which the C-H bond is broken. This difference in reaction rates is called a primary KIE because the isotopes were part of the bond broken during the rate limiting step.

Scientists also evaluate secondary KIEs. A KIE is secondary when the heavier isotope is not part of a bond which breaks during the rate limiting step, but is close by in the same molecule. Such a secondary effect is much smaller than a primary KIE, but it can provide useful information about the ''transition state'' of a reaction. A ''transition state'' is the term for a temporary chemical structure which may be formed during the course of the reaction, but which is not present as either a reactant or product. Therefore, the ''transition state'' compound does not exist as a separate substance. Nevertheless, we can observe its effects on secondary KIEs. A larger secondary KIE indicates the transition state compound looks more like the products, a smaller one means it is more like the reactants.

So far, we have explained how KIEs are used to investigate the middle of a reaction. Sometimes however, these effects can be used differently, as a way to 'sort' isotopes.

For example, investigators at Oak Ridge National Laboratory in the United States use KIEs to remove an isotope which is a threat to our environment. **Tritium**, another isotope of hydrogen, contains two neutrons, and is in waste-water from **nuclear reactors**. It must be disposed of carefully and is easier to deal with if it is in a concentrated form. One way to concentrate this isotope is to use KIEs.

The primary KIE for this reaction is very strong; the reaction occurs 124 times faster for **water** with hydrogen than for water with tritium. So, when the water contains both isotopes, the lighter hydrogen-water reacts much more quickly than the heavier tritium- water. After a short while, most of the hydrogen resides in the product, while the reagents contain a concentrated form of tritium. With the isotopes in different chemical substances, the chemists at Oak Ridge find them much easier to separate from each other. Applications like this one are part of the reason why KIEs are a growing area of fruitful research.

KINETIC MOLECULAR THEORY

Several properties of a gaseous sample can be explained readily by the features of a model of gas behavior called the kinetic-molecular theory or simply kinetic theory. This theory views the molecules as very small, very hard spheres that travel constantly and randomly at high speed.

Although scientists now think of kinetic-molecular theory as a very simple model of the behavior of atoms and molecules in the gas phase, it took over one hundred years for it to be accepted by the scientific community. Daniel Bernoulli (1700-1782) published an article describing a theory on this subject in 1743. However, it wasn't until the latter half of the next century that a statistical view developed by **James Clerk Maxwell** and **Ludwig Boltzmann** convinced physicists and chemists that individual molecules in a gas could be viewed as independent, rapidly moving bodies.

The kinetic molecular theory is important to chemists and other scientists who need a molecular viewpoint in their work—biologists, material scientists, geologists, chemical engineers, medical scientists, environmental scientists and others. They do not use this theory to perform calculations or predict properties of **gases** but apply it constantly in their consideration of how gaseous substances behave because it provides the most fundamental picture of **matter** at the molecular level. If we think of molecules as particles in constant agitation—bumping into each other at high **velocity** and exchanging **energy**—we develop a view that enables us to correlate a vast array of observations: the expandibility and **compressibility** of gases; their ability to take the shape of and fill whatever container they occupy; their **miscibility**; their rates of **diffusion** through membranes. In addition, all the **gas laws** can be derived from mathematical considerations based on the model in the kinetic theory.

The following statements are the basis of a model of gases. As such, the statements do not necessarily describe what a gas really is but rather suggest how we can think of a gas to explain its behavior. An underlying premise is that an ideal gas obeys the gas laws perfectly, and its behavior may be adequately described by kinetic theory.

Real gases sometimes behave ideally. The model for **ideal gases** has been improved to account for properties of non-ideal gases by comparing the properties observed for real gases with those predicted from the kinetic-molecular model of an ideal gas.

The kinetic-molecular model of an ideal gas consists of five statements:

1. Molecules are very small and far apart. Most of a container of gas is empty space. This picture is reasonable for gases because they are highly compressible. A common example of gas compression occurs when a bicycle tire is inflated with a hand pump. In fact, an entire industry has been built around compressed gases. To transport gaseous substances such as propane or acetylene economically, they must be compressed greatly to fit into containers of a reasonable size.

In an ideal gas, the total **volume** occupied by the molecules themselves is considered to be so small compared to the volume of the gas that it is negligible.

2. Molecules are very hard spheres that bounce off each other without losing energy in encounters called elastic collisions.

Very hard balls, such as billiard balls or steel ball bearings, undergo nearly elastic collisions. In elastic collisions, one **molecule** hits another and energy may be transferred, but the total energy of motion of the molecules remains the same. If one could construct a frictionless table (like an air hockey table) with rigid, hard sides, a group of very hard balls set in motion would keep moving—bouncing off each other and the sides of the table—for a long time. When this idea is applied to gases, it means the gas molecules in an insulated container with hard walls will bounce around continuously without losing any energy.

3. The motion of molecules is random, and properties that depend on the motion of the molecules will be the same in all directions.

Collisions of gas molecules with the walls cause the pressure that a gas exerts on its container. If a box contains a gas sample, a pressure gauge placed on any of the sides gives the same reading. (This assumes we use a laboratory-scale box so that the effect of gravity is negligible.)

4. Because molecules in gases are far apart, they act independently and neither attract nor repel each other.

Several kinds of attractive forces are known: gravitational, electrical and magnetic. All of them decrease rapidly as two objects move farther apart. The gravitational force between objects as light as molecules is small and generally ignored. Because molecules in the gas phase are considered to be far apart in the kinetic-molecular theory, these forces are ignored and the molecules are viewed as totally independent entities.

5. Properties of gases we observe, such as **temperature** and pressure, are due to average motions of a very large number molecules.

A great triumph of Boltzmann, Maxwell and others was to convince physicists and chemists to view gases statistically. From this viewpoint, a sample of a gas is a collection of molecules with a distribution of speeds. At one given temperature, some molecules move much faster than others and some much slower. In the statistical model, a particular temperature is characterized by a certain average speed of the molecules. Ensembles of gas molecules that have the same average kinetic energy ($KE_{avg,1} = KE_{avg,2}$) have the same temperature ($T_1 = T_2$). Knowing the temperature or pressure of a gas tells us nothing about one particular molecule but does give us information about the entire sample.

If the simple view of the kinetic-molecular theory governed all gases under all conditions, it would be very easy to predict their properties. But reality is more varied and interesting than this simple model. Points 1 and 4 in kinetic theory may be modified when the behavior of real gases is considered.

If molecules were really infinitely small, gases would be infinitely compressible, but this is not the case. As real gases are compressed, all become liquid or solid, and the liquid or solid is very difficult to compress further. At low pressures, the relationship of pressure and volume is accurately described by Boyle's Law: the volume is inversely proportional to the pressure and (P)(V) is constant. The gases behave ideally. As the pressure increases and the volume becomes small, the straight line relationship no longer holds; (P)(V) is no longer constant, and real behavior deviates from the ideal.

At high pressure, gases occupy smaller volumes than are predicted by ideal theory. For any real gas there is an abrupt change in the slope of the line of a plot of pressure vs. volume that occurs as the gas condenses to form a liquid. This abrupt change in slope that marks the deviation from ideality occurs at different values of pressure for each gas.

If molecules in a gas acted independently, the pressure of a gas would depend directly on the number of molecules hitting the wall (the concentration) as long as the temperature is kept constant. This feature of the kinetic-molecular theory also corresponds to Boyle's Law. However, a direct proportionality is not observed over the entire range of pressure. If Boyle's Law were valid for all gases under all conditions, there would be only one line for a plot of pressure times volume (PV) vs. volume and that line would have a constant value of 22.414 L atm at all pressures. This is not the actual case.

The simple kinetic molecular theory outlined above and the Ideal Gas Law are only valid when the pressure is low and the temperature high. At low pressures, molecules in a gas are far apart. Under these conditions the space taken up by the molecules is negligible compared to the total volume of the gas. This corresponds to the first statement in the kinetic theory. As the pressure increases, however, the volume occupied by the gas becomes smaller and the space taken up by the molecules is no longer negligible in comparison. At high pressures the volume of a real gas differs from ideal because the actual space taken up by the molecules becomes significant. For real gases, the model must be modified to account for the actual volume the molecules of a gas occupy.

When pressure is low, molecules are too far apart for any attractive forces to influence their behavior. Because both the effect of molecular size and attractive forces are insignificant at low pressures, gases behave ideally. At high pressures, molecules are forced closer together and attractive forces become significant. Furthermore, when the temperature is high, gas molecules move so rapidly that any attractive forces between them are overcome. Correspondingly, as temperature decreases, molecules move more slowly and attractive forces become significant. At low temperatures, gas volumes are smaller than the ideal volumes predicted by theory because the molecules move more slowly and the effects of intermolecular attractive forces have a chance to manifest themselves. This view requires modification of the simple model to include intermolecular attractive forces which become significant at high pressures and low temperatures.

Note these two effects work in opposite directions, for one increases and the other decreases volume. Real gases tend to behave ideally under moderate conditions where neither of the two effects is significant. Under conditions of low temperature and/or high pressure, the observed deviation depends upon which effect is dominant.

A relatively early and quite successful attempt to modify kinetic molecular theory to provide a better approximation to the behavior of real gases was accomplished by **Johannes van der Waals**. He added terms to modify the pressure and volume in the ideal gas equation, adding a n2a/V2 term, where n is the number of mols, V is the observed volume, and a is a constant with a different value of different gases, to the observed pressure, increasing the value of the pressure over the ideally predicted quantity, to correct for the loss due to the influence of attractive forces between the molecules and an nb term, where n is the number of mols and b is a constant with different values for different gases, to decrease the total volume occupied by the gas, thereby accounting for the portion of the volume taken up by the molecules themselves.

KINETICS

Chemical kinetics is the study of the rates of reactions and of the factors that determine the speed of a reaction. This information is of used to help scientists understand reactions such as those involved in biochemical processes such as **metabolism** and **photosynthesis**; reactions that take place in the atmosphere, such as ozone formation and depletion, geochemical processes such as mineral formation and chemical changes that take place in the ocean and many other natural systems. Information from kinetics studies are also of great practical value. Industrial chemists use their knowledge of reaction rates and the factors that influence them to produce materials in a reasonable amount of time with the least loss of valuable reagents. Analytical chemists, including those who design medical diagnostic kits, use kinetics information to design detection reactions that can be carried out rapidly under moderate conditions and with a high degree of reproducibility. Besides the practical value of knowing how fast a reaction proceeds and what factors influence the rate of a particular reaction, studies of kinetics also provide chemists with a way to determine how reactions occur at the molecular level, often allowing a step-by-step analysis.

The study of chemical kinetics is concerned about what is going on during a reaction, not just about the identity or amounts of substances produced in a reaction. **Stoichiometry** and **thermodynamics** focus on reactants and products; what actually happens to the reacting species during a chemical change is not the focus of these approaches. Kinetics studies focus not only on how fast a reactions occurs under a particular set of conditions, but on trying to understand the steps that take place at the molecular level.

Some reactions, such as explosions and acid-base neutralizations, occur almost instantaneously. Others, such as the **corrosion** of the Statue of Liberty or the decomposition of marble statues by **acid rain**, are very slow. Some reactions occur when two molecules collide, while others are only complete after a complicated series of processes. As a result, the study of kinetics is complex and can involve many different techniques to allow the study of reactions that take place over a wide range of times and conditions.

Some reactions occur in a single step. For example, an **oxygen molecule** binds to hemoglobin in a single step when it collides with an **iron atom** inside the protein. However, many reactions are actually accomplished by a series of single-step reactions. For example, the formation of sugars from **carbon** dioxide and **water** by photosynthesis requires dozens of reactions. The sequence of steps by which a reaction proceeds is called the mechanism. An important objective of kinetics studies is to deduce reaction mechanisms. If a correct **reaction mechanism** can be found, we can understand what the most important factors are that determine the rate of a reaction and, in many cases, can learn how to adjust the rate of the reaction to suit our purposes. It may also be desirable to increase the rate of a reaction used or to decrease the rate of a reaction that causes a detrimental process like corrosion.

To alter the rate of a reaction, the kineticist typically determines the step of a reaction that is the slowest, the rate determining step. Once the rate determining step is identified, reaction conditions can often be modified to make it proceed faster and thereby speed up the entire reaction.

One method of speeding up reactions is to use a catalyst, a species that increases the rate of a chemical reaction but is not changed by the process. Ideally, the catalyst can be recovered in its original form when the reaction is over. The catalyst increases the rate of a step by providing an alternate route, allowing the overall reaction to be accomplished without proceeding through the slow, rate-determining step.

A scientist who studies reaction rates is called a kineticist. How does a kineticist evaluate reactions? The first step is to focus on one particular reaction. The kineticist may be interested in large problems like air **pollution** or the production of a pharmaceutical or may wonder how a particular **enzyme** works or how a new class of molecules that has just been prepared reacts. But the kineticist focuses on a particular reaction that seems to be most crucial to the process in question. This may be the one that limits the rate of the entire process or one that produces or destroys a particular substance of interest for example, the destruction of ozone by a chlorofluorocarbon in the study of atmospheric reactions. The kineticist then chooses conditions that are relevant to the reaction in question for example, conditions like those in the upper atmosphere where ozone destruction occurs.

Next the kineticist must choose appropriate experimental methods for studying the reaction. There must be a means of accurately determining when the concentrations of substances involved in the reaction change, and there must be a way to positively identify the products of the reaction. The sensitivity of the method chosen to study the reaction determines the quantities of reactants that can be used in the study. The experimental methods must also be capable of measuring the quantities of substances involved in the reaction under the actual conditions of the reaction, which may occur in the gas, liquid or solid phase and at very high, moderate or low **temperature**. The kineticist must also consider how reaction conditions can be varied to test the possible ways a reaction occurs at the molecular level. The methods must allow different concentrations of substances to be used and the temperature at which the reaction takes place to be varied. Once the data are collected, the kineticist must be prepared to change the original experimental plan if unexpected results occur.

An important reason to carefully study reaction rates is to be able to predict the rate of a reaction under specific conditions, such as reactant concentrations, **pH**, temperature and so forth. A rate law is a mathematical relationship that defines how a reaction depends on concentrations. For example, when a radioactive **isotope** decays, the rate at which the original isotope decreases and the new species is formed is directly proportional to the amount of the radioactive isotope originally present. This can also be stated: rate = k[isotope]. The isotope carbon-14 is used to date the time that ancient plants or animals died because this isotope is present at a very nearly constant **concentration** in the atmosphere but decreases in concentration when a plant or animal dies because it is longer exchanging its carbon compounds with the atmosphere and other living things. The decrease in the amount of carbon-14 is given by the equation:

rate= k [C-14].

This equation is the rate law for the reaction. It expresses the change in concentration of a particular species as a function time in terms of the concentration, or concentrations, of species involved in the reaction. The rate law is usually written in terms of concentrations that can be determined experimentally. The term "k" is the rate constant. The rate constant is typically represented by a small letter k, large K being used for equilibrium constants. The units of the rate constant are determined by the concentration terms of the rate law and the time units used. For carbon-14 dating, the units of the rate constant, k, can be expressed as per **mole** per liter per hour or 1/(mol/l hr).

To characterize the affect that changing a particular reactant has on the rate of a reaction, kineticists use the term "reaction order." When the rate of a reaction is directly related to the concentration of a substance, it is said to be "first order" in that substance. This is the case for radioactive decomposition.

The reaction of **chlorine** atoms and ozone, which has the rate law rate = $k[Cl][O_3]$ is first order in chlorine atoms and first order in ozone. The order of the entire rate law, called the reaction order, is the sum of all the exponents of the concentrations in the rate law. For radioactive decay, the order is 1, and the reaction is a first order reaction. For the reaction of chlorine atoms and ozone, the order is 2. This is a second order reaction. The order of the reaction is determined by the rate law, which is always determined experimentally. Only when a reaction is known to be an elementary reaction can we state the order of its rate law from its chemical equation.

Another important term used to characterize chemical reactions kinetics is "**half-life**." The term half-life is very commonly used in reference to radioactive decay reactions, but it is a useful concept for other reactions as well. The half-life of a reaction is the time required to convert half the initial concentration of a reagent to product. The half-life is given the symbol $t_{1/2}$. For a first order reaction the half- life is easily calculated: it is equal to the natural logarithm of 2 (ln2 = 0.693) divided by the rate constant. For second or higher order reactions the calculation is considerably more complex.

Even though kinetics is concerned with the rates of reactions, there is an important relationship between rates of a reaction and the equilibrium constant for the reaction. Chemical changes are dynamic. When a reaction reaches equilibrium, there is no longer a net change in the concentrations of the reactants and products, but reactions still occur. The key feature of a system at equilibrium is that the net rate of formation of each species is equal to the rate at which it is changed into another species. The equilibrium constant for the reaction is equal to the rate constant of the forward reaction divided by the rate constant for the reverse reaction. This is a general result that applies to all reactions under equilibrium conditions.

The rate of a reaction can be changed not only by changing the concentrations of reactants, but also by changing temperature. **Energy** is required to distort molecules in such a way that chemical bonds will be broken and different bonds formed. The energy that is required to activate molecules for a chemical reaction is the **activation energy** of the reaction. There is a "hill" between the reactants and the products regardless of whether the energy of the products is lower or higher than that of the reactants.

Activation energy theory states that molecules must have a certain amount of energy available to initiate a reaction. This available energy comes from the motion of the molecules and is directly related to the temperature; the higher the temperature, the faster the molecules move. Consequently, if the temperature of a **solution** in which a reaction takes place is increased, a greater fraction of the molecules will have an energy equal to or above the activation energy and the reaction proceeds faster. By measuring reaction rates at more than one temperature, kineticists obtain the information they need to be able to predict reaction rates at other temperatures.

See also Activated complex

King, Reatha Clark (1938-)

American chemist

Reatha Clark King is an African American chemist whose early research in fluoride **chemistry** aided NASA's space program. A large part of her career has been devoted to academic and scientific administration.

Reatha Belle Clark King was born in Pavo, Georgia, on April 11, 1938, the second of three daughters born to Willie and Ola Watts Campbell Clark. Her father, an illiterate farm worker, and her mother, a domestic servant, divorced when King was a young child. Shortly afterward, King moved with her mother and sisters to Moultree, Georgia. There, her life centered around activities at the Mt. Zion Baptist Church and her studies at school. During summers and spare time, King and her sisters earned money by gathering tobacco and picking cotton. Having no role models in scientific professions, they aspired to be hairdressers, teachers, or nurses. King graduated from high school in 1954, the valedictorian of her class.

Awarded a scholarship to Clark College in Atlanta, King originally set out to become a home economics teacher. This changed when she enrolled in an introductory chemistry class, a requirement for a home economics major. The course was taught by Alfred Spriggs, an African American chemist who had received his Ph.D. from Washington University, and the subject, coupled with the professor's dynamic personality, inspired King to change her career path. As King's new mentor, Spriggs encouraged her to continue beyond college and obtain a Ph.D. After completing her undergraduate work, she received a Woodrow Wilson Scholarship and enrolled in the chemistry graduate program of the University of Chicago. There she was drawn to the study of **physical chemistry** and developed a strong interest in the area of **thermochemistry**. In the spring of 1963, she received her Ph.D. in chemistry.

Six months after leaving the University of Chicago, she accepted a research position at the National Bureau of Standards, in Washington D.C., and began work on developing materials that could safely contain the highly corrosive compound **oxygen** difluoride. Her research on other fluoride and intermetallic compounds had important applications for the use of rockets in the NASA space program. Her reputation at the bureau was one of professionalism and perseverance; she often stayed in the lab overnight to supply quickly needed analyses. Her superiors cited King with an outstanding performance rating, and she won the Meritorious Publication Award for a 1969 **paper** on fluoride flame **calorimetry**.

In 1968 King left the Bureau of Standards to taking a teaching position at York College in New York City. Two years later she was appointed Assistant Dean of Natural Sciences and Mathematics. Shortly after her promotion to Associate Dean, she took a leave of absence from York College to obtain an M.B.A. from Columbia University. In 1977 she became the president of Metropolitan State University in St. Paul, Minnesota. Under her leadership the small college increased its number of graduates five-fold, added a graduate program in management, and expanded its general curriculum. In 1988, King left academia to become the President and Executive Director of the General Mills Foundation. She also serves on several corporate boards, including that of Exxon Corporation.

King has received a host of honorary doctorate degrees from institutions, including Alverno College, Carleton College, Empire State College, Marymount Manhattan College, Nazareth College of Rochester, Rhode Island College, Seattle University, Smith College, and the William Mitchell College of Law. In 1988 she was named the Twin Citian of Year for Minneapolis-St. Paul, Minnesota. She is married to N. Judge King, with whom she has two sons, N. Judge III and Scott.

Kirchhoff, Gustav Robert (1824-1887)

German physicist

Gustav Robert Kirchhoff was reported to have a fastidious and dull personality. However, examining the results of his research career shows constant evidence of imagination, sincerity, and wit. He was able to do what was proclaimed impossible; that is, identify elements making up the sun. Kirchhoff's collaboration with **Robert Bunsen** was due to a sincere desire to help a friend solve interesting puzzles. A story told about him shows that he did have wit. He told a banker friend that he was investing his money into his research identifying the elements of the sun. The banker countered with "Why do I care if there is **gold** on the sun? I cannot bring it to earth." Later, Kirchhoff won an award for his work and was paid in British gold sovereigns. Kirchhoff showed the gold coins to his banker friend and said they had come from the sun.

Kirchhoff was born in 1824 in the city of Königsberg, Prussia. The son of a law councilor, he grew up in a comfortably affluent family. Because he attended the University of Königsberg, he did not have to leave his hometown until he was twenty-three. While Kirchhoff was in college, he developed the laws of electrical circuits that bear his name. This discovery, published in 1845, was followed up two years later with a detailed application of these laws to circuit networks. He graduated in 1847. A third paper covering solid conductors was published the year after he graduated. These papers ensured that his reputation and influence in scientific research circles was beginning to rise.

After a short, unpaid period of teaching at the University of Berlin, Kirchhoff received an honored position of professor at Breslau in 1850. There, he began a long-term friendship with Robert Bunsen. When Bunsen received a position at Heidelberg, Kirchhoff followed him there. In 1854, he accepted a professorship that was not as high a position as the one that he held at Breslau. Kirchhoff remained at Heidelberg for twenty-one years. During that time, he achieved great success with both his collaborative projects with Bunsen and his independent research.

Kirchhoff and Bunsen were very busy in the years 1854 through 1860, working on developing and improving their spectroscope methodology. While they made many advancements, the major breakthrough came in 1857 with the invention of the **Bunsen burner**. The Bunsen burner could **heat** experimental substances with very little light of its own. Kirchhoff's major contribution to **spectroscopy** was recognizing that a prism could produce a measurable spectrum. During this time of initial spectroscope experimentation, they learned that they could identify the presence of an element in a substance through the spectrum of light given off when the substance was heated to the point of glowing.

Between 1859 and 1861, Kirchhoff and Bunsen began to publish their findings, including the "how-to" of spectroscopy. Kirchhoff presented his study of Fraunhofer spectral lines in 1859. In this study, he established the foundation for Kirchhoff's law of emissive to absorptive powers. This law stated that the ratio of emissive to absorptive powers was a function of **temperature** and wavelength. That is, all substances will have the same ratios at a specific temperature and wavelength. In addition, he announced that he had found a way to identify the elements present on the sun. The report of this study gave some hints about the effectiveness of the new methodology, spectroscopy. The next year, Kirchhoff and Bunsen presented their seminal work on spectroscopy announcing the discovery of two previously unknown elements, **cesium** and **rubidium**.

Kirchhoff progressed with research independent of Bunsen and modified the spectroscope by adding prisms. In 1861, he announced the identification of even more terrestrial elements found on the sun. At age 51, Kirchhoff left Heidelberg to accept the position Chair of Mathematical Physics in Berlin. His later work was in the area of theoretical physics. He died in 1887 in Berlin at the age of 73.

In discussing the products (discoveries and inventions) of the collaboration between Kirchhoff and Bunsen, it should be noted that the collaboration was mutually beneficial. What brought the two together was a mutual respect for the talent of the other. Bunsen had the background in **photochemistry**, an exciting new field; Kirchhoff had the mathematical and Newtonian interests. Bunsen presented the problem: how to measure the light produced and absorbed in chemical reactions. Kirchhoff was the one who recognized the challenge as a measurement of the absence or presence of components of light. Since light passing through a prism is divided into a spectrum, he suggested integrating a prism into the structure of the light-measuring instrument. In this way, a spectrum developed through the light emitted in a chemical reaction could be compared with a "standard" spectrum, thus identifying the components that were present or not.

This insight was the foundation for their invention of the spectroscope. The methodology was completed with the invention of the Bunsen burner. This invention was needed because it was difficult to use other burners in measuring light emitted from heated elements because those other burners gave off their own light. An extra source of light interfered with an accurate measurement of the target spectrum. The Bunsen burner produced so little light as to be insignificant to the measurements.

With this new methodology, Kirchhoff and Bunsen began recording the spectral lines of all known elements. One day, they found a mineral that emitted previously unrecognized spectral lines. This could only mean that it was a newly discovered element. They named it after its strongest spectral line, which was blue. The name cesium is derived from the Latin word for blue. Less than a year later, they found another element with a strong red line. They named it rubidium, derived from the Latin word for red.

Working on his own, Kirchhoff recognized that there was an inverse relationship between the spectrum from **sodium** and the spectrum from solar light. The dark line of the solar spectrum recorded by Fraunhofer was in the same position as a bright line of the sodium spectrum. Kirchhoff reasoned that, if he shined them together through the spectroscope, the sodium bright line would cancel out the sun's dark line. Instead, the dark line became darker. Investigating this phenomenon through many experiments, Kirchhoff discovered that, when light passes through a gas, the expected spectrum of that gas is filtered out. If light passes through sodium, it loses dominant components of the sodium spectrum. The only explanation, evident to Kirchhoff, for the sun's spectral phenomenon was that the sun's light must pass through sodium in the sun's atmosphere. Many replicated his findings and a new means to study what was previously thought impossible was made real.

Most of Kirchhoff's work with Bunsen was based on the spectroscope methodology. Although the invention of the spectroscope was a joint project, it was based on the original research need of Bunsen and on Bunsen's concept of how to resolve that need. Kirchhoff influenced Bunsen as a colleague and Bunsen influenced a later generation of researchers as mentor. Kirchhoff could have shown the banker the financial benefit of science through the discovery of cesium. Kirchhoff had helped Bunsen with the work that led to that discovery. A student of Bunsen discovered that cesium could increase the illumination of gas lamps while reducing costs. This recommendation saved the previously declining gas light industry.

Kirchhoff did offer direct influence on future researchers. His contributions to the collaboration with Bunsen were essential and the important discoveries they made together. Furthermore, Kirchhoff made many independent contributions to physics. The field of **chemistry** (as well as other fields of science) is always affected by developments in physics.

KISTIAKOWSKY, GEORGE B. (1900-1982)

American chemist

George B. Kistiakowsky's career exemplifies the melding of pure and applied sciences. Trained as a chemist and working in the areas of molecular structures and mechanisms of chemical reactions, Kistiakowsky later became a key member of the **Manhattan Project**, which developed the world's first atomic bomb, and eventually served as President Dwight D. Eisenhower's special assistant on science and technology. In the realm of pure sciences, Kistiakowsky's contributions include both teaching and carrying out highly acclaimed research in such fields as **spectroscopy**, **thermodynamics**, and chemical **kinetics**. Toward the applied end of the science spectrum, Kistiakowsky developed chemical **explosives** and rocket propellants and contributed to the development of nuclear weapons. Moreover, he played a role in national science policy, arguing for a broader and more balanced approach to scientific research in the late 1950s and 1960s and later working on problems of nuclear arms control and disarmament.

George Bogdan Kistiakowsky was born on November 18, 1900, in Kiev, Ukraine. His parents, Mary Berenstam and

Bogdan Kistiakowsky, were of Cossack background. His father was a professor of international law at the University of Kiev. Young George was sent to schools in both Kiev and Moscow. But this peaceful beginning was not to last. In 1918 Kistiakowsky joined the White Army to fight against the Bolsheviks. He served in the infantry and tank corps (even while suffering from typhus) until 1920, when he fled. After a brief internment in the Ottoman Empire, he made his way to Yugoslavia. There an uncle helped him pursue his education, enabling him to enroll in the University of Berlin in 1921.

At Berlin, Kistiakowsky studied **chemistry** with Max Bodenstein, receiving his doctorate in 1925. He emigrated to the United States in 1928 to study at Princeton University with a fellowship in **physical chemistry** from the International Education Board. In that year, he married Hildegard Moebius and later had a daughter, Vera, with her. In 1928 Kistiakowsky became a staff member in Princeton's chemistry department.

In 1930 Kistiakowsky left Princeton and moved to Harvard as an assistant professor of chemistry. There, his career moved ahead rapidly. In 1933 he became an associate professor. Also in this year he became a naturalized United States citizen. He took on the Abbott and James Lawrence Professorship in Chemistry in 1938 as a full professor. During these years he established himself as both a highly regarded teacher and an original researcher. At that time, his most important research was in the fundamental mechanisms of chemical reactions.

In 1940 the United States government recognized the need for a strong, central defense program. Kistiakowsky was asked to become a consultant for the National Defense Research Committee. He later became acting chair of the explosives section of this group. In 1942 he assumed the chair of the entire explosives division. In this role he organized the preparation, testing, and manufacture of both explosives and rocket propellants. His work covered the areas of performance and safety testing as well as explosive characteristics such as fragmentation and shock wave propagation.

One of Kistiakowsky's particular projects during this time was the development of the "Aunt Jemima" powder. This product was an explosive that looked like flour and could even be safely baked into breads and cookies without losing its efficacy as an explosive. The powder was sent to the Chinese in flour bags and was used against the Japanese occupying forces.

From 1941 to 1943 Kistiakowsky worked with the Manhattan Project. His role in this coalition of military strategists and scientists was to manufacture the conventional explosives for detonating atomic bombs. This task made good use of Kistiakowsky's understanding of chemical kinetics and thermodynamics, as it demanded very precise control of explosive characteristics in order to achieve the required uniform and rapid compression of the nuclear core, which would result in the now typical atomic explosion. In July of 1945 a test explosion was carried out with Kistiakowsky's detonating device, and a month later the first atomic bomb was dropped on Nagasaki, Japan. The war years were a turbulent period in Kistiakowsky's personal life. He was divorced from Moebius in 1942. In 1945 he married Irma E. Shuler. (He was to divorce again in 1962, when he married his third wife, Elaine Mahoney.)

Early in 1946, Kistiakowsky returned to Harvard, where he was to serve as chair of the chemistry department from 1947 to 1950. He continued his research on shock waves, molecular spectroscopy, and the kinetics of chemical reactions. He also served on several university committees. Kistiakowsky's involvement with the government resumed in the early 1950s. From 1953 to 1958 he served as a member of the ballistic missiles advisory committee for the Department of Defense. In that capacity, he urged the rapid development of intercontinental ballistic missiles. He was also a member of the United States delegation at the Conference for Prevention of Surprise Attack in Geneva in 1958.

Kistiakowsky joined the President's Science Advisory Council in 1957. Two years later, he succeeded James R. Killian, Jr. as special assistant for science and technology to President Eisenhower. Over the next two years he had broad oversight roles in national science policy, helping coordinate research and development efforts among the various branches of government. He also argued passionately for an increased emphasis on science education.

Kistiakowsky published about two hundred scientific articles during his career along with two books: *Photochemical Processes* (1928) and *A Scientist at the White House* (1976). This self-described "poor Russian immigrant" received numerous awards and honors over the years. He was given honorary doctor of science degrees by Harvard (1955), William and Mary (1958), and Oxford (1959). Recognition from branches of the United States military includes the Army Ordnance Award, Navy Ordnance Award, and Exceptional Services Award from the Air Force. The American Chemical Society bestowed several awards on him, and he was given the President's Medal for Merit in 1946.

Kistiakowsky was a member of the National Academy of Sciences, American Academy of Arts and Sciences, American Chemical Society, American Philosophical Society, American Physical Society, and Sigma Xi. He was also a foreign member of the Royal Society of London and an honorary fellow of the Chemical Society of London. Kistiakowsky died in Cambridge, Massachusetts, on December 7, 1982.

Klaproth, Martin Heinrich (1743-1817)

German chemist

Klaproth was born in Wernigerode, Germany, on December 1, 1743. The most significant event in his young life was a fire that destroyed the family home when he was eight years old. Suddenly impoverished by the disaster, the Klaproth family was unable to pay for Martin's formal education. Instead, he was apprenticed to an apothecary (pharmacist) in nearby Quedlinburg at the age of 15.

In later years, he moved on to other apothecaries in Hanover, Berlin, Danzig, and back to Berlin. Along the way, he became interested in **chemistry** and began educating himself. His last stop on this series of journeys was at the famous Schwanenapotheke pharmaceutical-chemistry laboratory owned by the Rose family in Berlin. There he studied briefly under Valentin Rose. When Rose died in 1771, Klaproth became director of the laboratory and took responsibility for raising Rose's two sons. In 1780, Klaproth set up his own chemistry laboratory and, seven years later, gave up his career in pharmacy to devote himself full time to chemical research.

Klaproth made three important contributions in the field of chemistry. First, he was instrumental in promoting **Antoine-Laurent Lavoisier**'s new anti-phlogistic theories of chemistry. Klaproth's acceptance of these ideas required some courage since they conflicted with the older theory of **phlogiston** that had been proposed by his own countryman, **Georg Stahl**, in the late seventeenth century. After repeating Lavoisier's experiments, however, Klaproth was convinced that Lavoisier was correct. His own activities on behalf of Lavoisier's ideas were instrumental in convincing his German colleagues to adopt the new ''French Chemistry.''

Klaproth's second achievement was his role in the creation of **analytical chemistry**. Along with **Joseph-Louis Proust**, **Nicholas-Louis Vauquelin** and William Hyde Wollaston, he developed techniques and established principles that are now standard in the field of analysis. He emphasized, for example, the importance of drying or igniting precipitates to constant weight in an analysis.

Finally, Klaproth was involved in the discovery of at least a half dozen new elements. In 1789, for example, he obtained from the ore pitchblende a yellow compound that he thought contained a new element. A colleague suggested that he name the element *klaprothium*. However, the discoverer preferred the name **uranium**, after the planet Uranus that had been discovered eight years earlier. In the same year, Klaproth also isolated from the gemstone zircon a new element which he called **zirconium**.

Over the next decade, Klaproth also confirmed the discoveries of **strontium**, **titanium**, **tellurium**, and **cerium**. In addition, he carried out important original research on the **rare earth elements**.

In 1810, at the age of 67, Klaproth was offered a position as the first professor of chemistry at the new University of Berlin. He accepted the appointment and remained at the university until his death on January 1, 1817.

Martin Heinrich Klaproth.

KLUG, AARON (1926-)

English molecular biologist

Aaron Klug made many breakthroughs that advanced the knowledge of the basic structures of molecular biology, but he is best known for his creation of the new technique of crystallographic **electron microscopy** which made possible not only his own scientific discoveries but those of many other scientists as well. For his development of this technique as well as for his contributions to scientific knowledge, he was awarded the Nobel Prize in **chemistry** in 1982. He was also knighted as Sir Aaron Klug by Queen Elizabeth II in 1988.

Klug was born on August 11, 1926, in Zelvas, Lithuania, the son of Lazar Klug, a cattle dealer, and Bella Silin Klug. When he was two years old, he and his parents emigrated to Durban, South Africa, where members of his mother's family were already established. He was educated in the Durban public schools and, while attending Durban High School from 1937 to 1941, he developed an interest in science. Klug became especially interested in microbiology through reading Paul De Kruif's well-known book, *Microbe Hunters,* first published in 1926. He entered the University of Witwatersrand in Johannesburg in 1942 to take the premedical curriculum but extended his courses to include additional chemistry, mathematics, and physics before graduating with a B.S. degree in 1945. He then attended the University of Cape Town on a scholarship to take a master's degree in physics. There he first learned the techniques of X-ray **crystallography**, a method of determining the arrangement of atoms within a crystal. This is accomplished by studying the patterns formed on a photographic plate after a beam of X rays is deflected by the crystal. This methodology was to be basic to much of his later research. He married Liebe Bobrow in 1948 and eventually became the father of two sons, Adam and David Klug.

In 1949 Klug and his wife moved to Cambridge, England, where he had received a fellowship to study at the Cavendish Laboratory of Cambridge University. He hoped to work in the research group examining biological materials under the direction of **Max Perutz** and **John Kendrew**. When he found that there were no positions available in that group, he decided to study the **molecular structure** of steel and wrote a thesis on the changes that occur when molten steel solidifies, for which he received his Ph.D. in 1952. Still wishing to use his training to study biological materials, Klug in 1953 obtained a fellowship to work at Birkbeck College in London. Here he came under the influence of **Rosalind Franklin**, a reticent scientist who had pioneered X-ray crystallographic analysis. This technique had made a vital contribution to the discovery of the double-helical structure of deoxyribonucleic acid (DNA), an accomplishment that had won a Nobel Prize for **Francis Crick** and **James Watson**. Franklin introduced Klug to the X-ray study of the structure of viruses, an undertaking that was to occupy him for several years. They worked together to determine the structural nature of the tobacco mosaic virus, which attacks tobacco plants. After Franklin's death in 1958, Klug became the director of the Virus Structure Research Group at Birkbeck College.

In 1962 Klug returned to Cambridge to accept a position at the Laboratory of Molecular Biology recently established by the British Medical Research Council. Here he found himself stimulated by a large group of distinguished scientists, including Francis Crick, Max Perutz, and John Kendrew, all three of whom won Nobel Prizes in 1962. Over the next thirty years Klug himself became an increasingly important part of the research team, becoming joint head of the division of structural studies in 1978 and director of the entire laboratory in 1986.

Klug's most important contribution to scientific research was the development over many years of a technique which came to be known as crystallographic electron microscopy. **X-ray crystallography** had proved adequate for many biological discoveries such as the double-helical structure of **DNA**. However, many complex biological molecular structures were simply not available in crystal form suitable for **X-ray diffraction**. The obvious alternative was the use of powerful electron microscopes which could magnify an object up to a million times and reproduce it on a photographic plate, or micrograph. But the problem with electron micrographs was that they presented an essentially two-dimensional picture of a three-dimensional object. The micrographs did, however, contain in a confused form much of the information necessary for a three-dimensional reconstruction of the object, especially if the object was examined from different angles in successive micrographs.

Klug's new idea was essentially to combine the techniques of the **electron microscope** and X-ray crystallography by doing X-ray diffractions of the electron micrographs themselves, just as one would do with a crystal. Then the researcher could gradually put together a three-dimensional reconstruction of the object under consideration, filtering out extraneous specks on the photographic plates that confused the picture. The process was a complex one that required, among other things, very sophisticated mathematical analysis in the course of the work. Armed with the new research technique, Klug and his colleagues were able to reveal the structures and modes of operation of many basic biological materials such as viruses, animal cell walls, subcellular particles, chromatin from the genetic material of the cell **nucleus**, and various **proteins**. His accomplishments were rewarded in 1982 when he received the Nobel Prize. He has also been awarded the H. P. Heineken Prize from the Royal Netherlands Academy of Arts and Sciences, Columbia University's Louisa Gross Horwitz Prize, and a host of honorary degrees.

KOLBE, ADOLF WILHELM HERMANN (1818-1884)

German chemist

Adolf Wilhelm Hermann Kolbe was born in Göttingen, Germany. In 1832 he entered the Göttingen Gymnasium, where he met Robert Bunsen. As a result of Bunsen's influence, Kolbe decided to study **chemistry**, and four years later he entered the University of Göttingen. There he studied under **Friedrich Wöhler** and met **Jöns Berzelius**, who impressed him greatly. Unfortunately, his friendship with Berzelius caused him to support the latter's theory of **vitalism** long after it had been discredited by most researchers. In 1842 Kolbe left the university to become an assistant to Bunsen in Marburg, Germany. He studied gas analysis while there and finished his doctorate by studying the effect of **chlorine** on **carbon** disulfide. While working on his doctorate he also succeeded in producing **acetic acid** from inorganic compounds, which was impossible, according to the doctrines of vitalism.

Upon Bunsen's recommendations in 1845 Kolbe acquired a post with the School of Mines in London, where he stayed until 1847. While there Kolbe and chemist **Edward Frankland** discovered how to convert nitriles into **fatty acids**. Kolbe returned to Marburg with Frankland and began publishing their research results. Kolbe also studied the effects of galvanic current on organic compounds and succeeded in producing ethane from fatty acid salts. Because of his adherence to Berzelius's theories, Kolbe incorrectly thought that he had succeeded in producing radicals instead of ethane. In 1859 he succeeded using phenol and **carbon dioxide** to produce salicylic acid, which led to the cheaper production of **acetylsalicylic acid**, or aspirin. The two reactions came to be called Kolbe's synthesis.

Eventually Kolbe returned to Marburg to teach at the university, taking over Bunsen's chair. He moved to Leipzig in 1865 and built the largest, best equipped laboratory of the time. He became the editor of the *Journal for Practical Chemistry* in 1870. He continued writing until his death in 1884 near Leipzig.

Kolthoff, Izaak Maurits (1894-1993)

American analytical chemist

Izaak Maurits Kolthoff, described by many as the father of modern **chemistry**, was a professor and department head at the University of Minnesota for thirty-five years. His research was at the forefront of major developments in the field of **analytical chemistry**, involving investigations of the reagents used in determining **pH** levels, crystalline precipitates, polarography and research on rubber and **plastics**. He also invented a method of synthesizing rubber, a discovery of immense value during World War II. The American Chemical Society's book *A History of Analytical Chemistry,* termed his contributions to the field of analytical chemistry as "monumental."

Kolthoff was born in Almelo, the Netherlands, on February 11, 1894, to Moses and Rosetta (Wysenbeek) Kolthoff. His chemistry career began informally in the kitchen of his home when he was fifteen. When his mother accidentally added **sodium** carbonate (soda) to a pot of chicken soup instead of **sodium chloride** (table salt), her son readily resolved the situation. He added **hydrochloric acid** to correct the pH level of the liquid until a strip of litmus **paper** turned pink. This treatment was effective, and the soup was saved.

Kolthoff received his formal training at the University of Utrecht in the Netherlands. Initially, he studied pharmacy, but later switched to analytical chemistry. After earning his Ph.D. in 1918, he taught at the University of Utrecht for nearly ten years. Kolthoff focused his research on the reagents used in determining pH, the acidity or alkalinity of a **solution**, and the significance of pH in bacteriology and industry, as well as in analytical chemistry. In 1927, after a lecture tour in the United States, he was offered a position at the University of Minnesota as professor and head of the analytical chemistry department. He later became a naturalized United States citizen.

While at the University of Minnesota, Kolthoff researched the properties of crystalline precipitates, and also studied polarography, which is an electrochemical method of analysis that involves passing an electric current through the solution being investigated. During World War II, the American government asked him to turn his attention to researching a method of making synthetic rubber. Kolthoff and his associates succeeded in inventing a low-temperature method of producing high-quality synthetic rubber. Even after the war ended, most synthetic rubber was made by the Kolthoff method.

When government funding for rubber research ended in 1955, Kolthoff began research on improving the chemistry of rubber and plastics. He and his research team also analyzed **acids and bases** in non-water based reactions, and conducted studies involving the analysis of metal ions that were later used to help reduce **water pollution**.

Kolthoff became professor emeritus in 1962. He wrote or coauthored about 900 papers and books, including the standard work *Textbook of Quantitative Inorganic Analysis.* The recipient of numerous honors and awards, including the Polarographic Medal from the Polarographic Society of England, the Nichols Medal from the American Chemical Society, and the first Kolthoff Gold Medal from the Academy of Pharmaceutical Science, he was also honored by the University of Minnesota, when a chemistry building there was named Kolthoff Hall in 1973. He received five honorary doctorates, was knighted by the government of the Netherlands, and was elected to the National Academy of Sciences in 1958.

Kolthoff's interests extended far beyond the laboratory and classroom. He enjoyed tennis and horseback riding in his spare time. He traveled to the Soviet Union on a least two occasions for scientific meetings, where, as he reported in an article for *Science,* he "was impressed by the large number of Chinese studying for advanced degrees at Moscow University." Remarking on the usefulness of educational and scientific links between countries, he regretted the United States policy of discouraging such contact. "This is too bad," he commented, "for scientific visitors like myself are in a unique position to establish much-needed relations with countries like China. From scientific contacts frequently come exchanges of opinion on other matters."

Domestically, Kolthoff devoted considerable time and effort in furthering his political beliefs. During World War II, he helped German scientists fleeing the Nazis to find jobs at the University of Minnesota. When he spoke out against McCarthyism during the 1950s, Senator Joseph McCarthy accused him of belonging to subversive groups, but never brought Kolthoff to testify before his Senate committee. In 1961 Kolthoff joined several colleagues in protesting nuclear weapons.

Kolthoff spent much of his time during his last years at the University of Minnesota library. He passed away on March 4, 1993, in St. Paul, Minnesota, at the age of 99; his *Star Tribune* obituary described him as an analytical chemist who helped to "revolutionize science in this century."

See also Rubber, synthetic; Rubber, vulcanized

Kornberg, Arthur (1918-)

American biochemist

Arthur Kornberg discovered deoxyribonucleic acid (DNA) polymerase, a natural, chemical tool which scientists could use to make copies of **DNA**, the giant **molecule** that carries the genetic information of every living organism. The achievement won him the 1959 Nobel Prize in medicine or physiology (which he shared with **Severo Ochoa**). Since his discovery, laboratories around the world have used the **enzyme** to build and study DNA. This has led to a clearer understanding of the biochemical basis of **genetics**, as well as new strategies for treating cancer and hereditary diseases.

Kornberg was born in Brooklyn, New York, on March 3, 1918, to Joseph Kornberg and Lena Katz. An exceptional student, he graduated at age fifteen from Abraham Lincoln High School. Supported by a scholarship, he enrolled in the premedical program at City College of New York, majoring

Arthur Kornberg.

in biology and **chemistry**. He received his B.S. in 1937 and entered the University of Rochester School of Medicine. It was here that his interest in medical research blossomed and he became intrigued with the study of enzymes—the protein catalysts of chemical reactions. During his medical studies, Kornberg contracted hepatitis, a disease of the liver that commonly causes jaundice, a yellowing of the skin. The incident prompted him to write his first scientific paper, "The Occurrence of Jaundice in an Otherwise Normal Medical Student."

Kornberg graduated from Rochester in 1941 and began his internship in the university's affiliated institution, Strong Memorial Hospital. At the outbreak of World War II in 1942, he was briefly commissioned a lieutenant junior grade in the United States Coast Guard and then transferred to the United States Public Health Service. From 1942 to 1945, Kornberg served in the **nutrition** section of the division of physiology at the National Institutes of Health (NIH) in Bethesda, Maryland. He then served as chief of the division's enzymes and **metabolism** section from 1947 to 1952.

During his years at NIH, Kornberg was able to take several leaves of absence. He honed his knowledge of enzyme production, as well as isolation and purification techniques, in the laboratories of Severo Ochoa at New York University School of Medicine in 1946, of **Carl Cori** and **Gerty Cori** at the Washington University School of Medicine in St. Louis in 1947, and of H. A. Barker at the University of California at Berkeley in 1951. Kornberg became an authority on the **biochemistry** of enzymes, including the production of **coenzymes**—the **proteins** that assist enzymes by transferring chemicals from one group of enzymes to another. While at NIH, he perfected techniques for synthesizing the coenzymes diphosphopyridine nucleotide (DPN) and flavin adenine dinucleotide—two enzymes involved in the production of the energy-rich molecules used by the body.

To synthesize coenzymes, Kornberg used a chemical reaction called a **condensation reaction**, in which **phosphate** is eliminated from the molecule used to form the enzymes. He later postulated that this reaction was similar to that by which the body synthesizes DNA. The topic of DNA synthesis was of intense interest among researchers at the time, and it closely paralleled his work with enzymes, since DNA controls the biosynthesis of enzymes in cells.

In 1953, Kornberg became professor of microbiology and chief of the department of microbiology at Washington University School of Medicine in St. Louis. That year was a time of great excitement among researchers studying DNA; **Francis Crick** and **James Watson** at Cambridge University had just discovered the chemical structure of the DNA molecule. At Washington University, Kornberg's group built on the work of Watson and Crick, as well as techniques Ochoa had developed for synthesizing RNA—the decoded form of DNA that directs the production of proteins in cells. Their goal was to produce a giant molecule of artificial DNA.

The first major discovery they made was the chemical **catalyst** responsible for the synthesis of DNA. They discovered the enzyme in the common intestinal bacterium *Escherichia coli,* and Kornberg called it DNA polymerase. In 1957, Kornberg's group used this enzyme to synthesize DNA molecules. Although the molecules were biologically inactive, this was an important achievement; it proved that this enzyme does catalyze the production of new strands of DNA, and it explained how a single strand of DNA acts as a pattern for the formation of a new strand of nucleotides—the building blocks of DNA.

In 1959, Kornberg and Ochoa shared the Nobel Prize for their "discovery of the mechanisms in the biological synthesis of ribonucleic acid and deoxyribonucleic acid." The *New York Times* quoted Nobel Prize recipient **Hugo Theorell** as saying that Kornberg's research had "clarified many of the problems of regeneration and the continuity of life."

In the same year he received the prize, Kornberg accepted an appointment as professor of biochemistry and chairman of the department of biochemistry at Stanford University. He continued his research on DNA biosynthesis there with Mehran Goulian. The two researchers were determined to synthesize an artificial DNA that was biologically active, and they were convinced they could overcome the problems which had obstructed previous efforts.

The major problems Kornberg had encountered in his original attempt to synthesize DNA were twofold: the complexity of the DNA template he was working with, and the presence of contaminating enzymes called nucleases which damaged the growing strand of DNA. At Stanford, Kornberg's group succeeded in purifying DNA polymerase of contaminating enzymes, but the complexity of their DNA template remained an obstacle, until Robert L. Sinsheimer of the California Institute of Technology was able to direct them to a simpler one. He had been working with the genetic core of Phi X174, a virus that infects *Escherichia coli.* The DNA of Phi X174 is a single strand of nitrogenous bases in the form of a **ring** which, when broken, leaves the DNA without the ability to infect its host.

But if the dilemma of DNA complexity was solved, the solution raised yet another problem. The DNA ring in Phi X174 had to be broken in order to serve as a template. But when the artificial copy of the DNA was synthesized in the test tube, it had to be reformed into a ring in order to acquire infectivity. This next hurdle was overcome by Kornberg's laboratory and other researchers in 1966; they discovered an enzyme called *ligase,* which closes the ring of DNA. With their new knowledge, Kornberg's group added together the Phi X174 template, four nucleotide subunits of DNA, DNA polymerase, and ligase. The DNA polymerase used the template to build a strand of viral DNA consisting of 6,000 building blocks, and the ligase closed the ring of DNA. The Stanford team then isolated the artificial viral DNA, which represented the infectious, inner core of the virus, and added it to a culture of *Escherichia coli* cells. The DNA infected the cells, commandeering the cellular machinery that uses genes to make proteins. In only minutes, the infected cells had ceased their normal synthetic activity and begun making copies of Phi X174 DNA.

Kornberg and Goulian announced their success during a press conference on December 14, 1967, pointing out that the achievement would help in future studies of genetics, as well as in the search for cures to hereditary diseases and the control of viral infections. In addition, Kornberg noted that the work might help disclose the most basic processes of life itself. The Stanford researcher has continued to study DNA polymerase to further understanding of the structure of that enzyme and how it works.

Kornberg has used his status as a Nobel Laureate on behalf of various causes. On April 21, 1975, he joined eleven speakers before the Health and Environment Subcommittee of the House Commerce Committee to testify against proposed budget cuts at NIH, including ceilings on salaries and the numbers of personnel. The witnesses also spoke out against the tendency of the federal government to direct NIH to pursue short-term projects at the expense of long-term, fundamental research. During his own testimony, Kornberg argued that NIH scientists and scientists trained or supported by NIH funding "had dominated the medical literature for twenty-five years." His efforts helped prevent NIH from being ravaged by budget cuts and overly influenced by politics.

Later that year, Kornberg also joined other Nobel Prize winners in support of **Andrei Sakharov**, the Soviet advocate of democratization and human rights who had been denied permission to travel to Sweden to accept the Nobel Prize in physics. Kornberg was among thirty-three Nobel Prize winners to send a cable to Soviet President Nikolai V. Podgorny, asking him to permit Sakharov to receive the prize.

Kornberg received the Paul-Lewis Laboratories Award in Enzyme Chemistry from the American Chemical Society, 1951, the Scientific Achievement Award of the American Medical Association, 1968, the Lucy Wortham James Award of the Society of Medical Oncology, 1968, the Borden Award in the Medical Sciences of the Association of American Medical Colleges, 1968, and the National Medal of Science, 1980. He is a member of the National Academy of Sciences, the American Academy of Arts and Sciences, the American Society of Biological Scientists, and a foreign member of the Royal Society of London. In addition, he is a member of the American Philosophical Society and, from 1965 to 1966, served as president of the American Society of Biological Chemists. Kornberg has been married to Sylvy Ruth Levy Kornberg since 1943. His wife, who is also a biochemist, has collaborated on much of his work. They have three sons.

KOSSEL, ALBRECHT (1853-1927)

German biochemist

Albrecht Kossel was one of the earliest scientists to apply the exact methods of **organic chemistry** to problems in the chemistry of living tissue. His investigations into the cell substance nuclein revealed that it contained both protein and nonprotein (nucleic acid) parts. His research into protein components led to the discovery of the amino acid histidine. For his work on cell chemistry and **proteins**, Kossel won the Nobel Prize for Physiology or Medicine in 1910.

Karl Martin Leonhard Albrecht Kossel was born on September 16, 1853, in Rostock, Germany. He was the eldest son of a merchant father, also named Albrecht Kossel, and Clara Jeppe Kossel. Botany was Kossel's first love, but his father saw no future in that, so in 1872 Kossel entered the University of Strasbourg to study medicine instead. While there, he came under the influence of Ernst Felix Immanuel Hoppe-Seyler, one of the forefathers of the then-emerging field of **biochemistry**. In 1877, Kossel passed the state medical exam and began working as an assistant at Hoppe-Seyler's institute of **physical chemistry**, also in Strasbourg. He received his doctor of medicine degree the following year.

Kossel remained an assistant in Hoppe-Seyler's laboratory until 1881, when he qualified as a lecturer in physiological chemistry. Two years later, he was appointed director of the chemical division at the Berlin physiological institute by another leading German scientist, Emil Heinrich Du Bois-Reymond. From 1887 to 1895, Kossel was a professor at the University of Berlin.

Beginning in 1879 and continuing for many years, Kossel undertook what proved to be trailblazing research on the makeup of the cell substance nuclein. This substance had been discovered a decade before by another of Hoppe-Seyler's star pupils, Johann Frederick Miescher. However, nuclein was still a vague entity when Kossel first set about defining its composition.

Kossel soon determined that nuclein was made up of two parts, one protein and one not. Thus, the word nuclein was eventually replaced by the more specific term nucleoprotein, and the nonprotein portion came to be called nucleic acid. **Nucleic acids** differed from any other **natural products** that were known up to that point. When broken down, they produced **carbohydrates** and nitrogen-bearing compounds called purines and pyrimidines. Kossel further isolated two purines (adenine and guanine), as well as three pyrimidines (thymine, cytosine, and uracil). In addition, Kossel correctly concluded that the function of nuclein was somehow tied in to the formation of flesh tissue. His writings foreshadowed many important later developments, including modern investigations of nucleic acids as the storers and transmitters of genetic data.

In 1895, Kossel left Berlin for the University of Marburg, where he was a professor and director of the physiological institute. That same year, he began work that lasted for more than three decades as editor of the *Zeitschrift für physiologische Chemie,* a noted journal founded by Hoppe-Seyler that was for a time the only periodical in the world devoted exclusively to biochemistry. It was primarily in this journal that Kossel's own papers appeared. Then in 1901, Kossel moved again—this time to the University of Heidelberg, where he remained a professor and administrator until his retirement in 1924. Thereafter, he held the post of director of that city's institute for protein investigation.

Starting in the 1890s, Kossel's attention turned more and more to research on proteins. In particular, he studied the proteins in fish sperm cells, which proved simpler than those in other cells. He developed an elaborate theory to explain how complex ordinary proteins could be built from the simple bases present in spermatozoa. Unfortunately, his elegant explanation proved wrong. It was decades before anyone realized that the crucial compounds for this purpose are not the proteins but the nucleic acids, which are present in spermatozoa in their full complexity.

Given the technical limitations of his time, Kossel was remarkably successful at elucidating the makeup of proteins. He discovered histidine, an amino acid that is the chief component of protein. He also devised a method for comparing the **amino acids** in the sperm of different fish species. In the laboratory, Kossel was never satisfied with purely chemical findings; he always strove to understand the biological meaning of his discoveries. In this regard, he was a true pathfinder.

Based on such impressive achievements, Kossel received many honors in addition to the 1910 Nobel Prize. Notable among these were honorary degrees from universities in Cambridge, England; Edinburgh, Scotland; Dublin, Ireland; Ghent, Belgium; and Greifswald, Germany. He was also a member of various societies, including the Royal Society of Sciences in Sweden. Among Kossel's students over the years were several who achieved later prominence, including **Phoebus Aaron Theodor Levene**, who in 1909 became the first chemist to show that nucleic acids contain a sugar (ribose). Twenty years later, Levene demonstrated that other nucleic acids contain a different sugar (deoxyribose), thus defining the two types of nucleic acid: ribonucleic acid (**RNA**) and deoxyribonucleic acid (**DNA**).

Kossel married Luise Holtzmann in 1886. They had two children: a daughter, Gertrude, and a son, Walther. The latter, born in 1888, went on to become a distinguished physicist. Kossel died on July 5, 1927, in Heidelberg, Germany, after a brief illness. Upon his death, obituary notices were carried by such respected journals as *Nature, Science,* and the *Journal of the American Medical Association.*

Krebs cycle

The Krebs cycle is a set of biochemical reactions that occur in the mitochondria. The Krebs cycle is the final common pathway for the **oxidation** of food molecules such as sugars and **fatty acids**. It is also the source of intermediates in biosynthetic pathways, providing **carbon** skeletons for the synthesis of **amino acids**, nucleotides, and other key molecules in the cell. The Krebs cycle is also known as the **citric acid** cycle, and the tricarboxylic acid cycle. The Krebs cycle is a cycle because, during its course, it regenerates one of its key reactants.

To enter the Krebs cycle, a food **molecule** must first be broken into two- carbon fragments known as acetyl groups, which are then joined to the carrier molecule **coenzyme** A (the A stands for acetylation). Coenzyme A is composed of the **RNA** nucleotide adenine diphosphate, linked to a pantothenate, linked to a mercaptoethylamine unit, with a terminal S-H. **Dehydration** of this linkage with the OH of an acetate group produces acetyl CoA. This reaction is catalyzed by pyruvate dehydrogenase complex, a large multi-enzyme complex.

The acetyl CoA linkage is weak, and it is easily and irreversibly hydrolyzed when Acetyl CoA reacts with the four-carbon compound oxaloacetate. Oxaloacetate plus the acetyl group form the six-carbon citric acid, or citrate. (Citric acid contains three carboxylic acid groups, hence the alternate names for the Krebs cycle.)

Following this initiating reaction, the citric acid undergoes a series of transformations. These result in the formation of three molecules of the high-energy **hydrogen** carrier NADH (nicotinamide adenine dinucleotide), one molecule of another hydrogen carrier $FADH_2$ (flavin adenine dinucleotide), one molecule of high-energy GTP (guanine triphosphate), and two molecules of **carbon dioxide**, a waste product. The oxaloacetate is regenerated, and the cycle is ready to begin again. NADH and $FADH_2$ are used in the final stages of **cellular respiration** to generate large amounts of ATP.

The details of the cycle are as follows:

1. Oxaloacetate condenses with acetyl coA to form citrate. This reaction is catalyzed by the **enzyme** citrate synthase.

2. Citrate undergoes isomerization to isocitrate, via a dehydration/rehydration step involving the temporary creation of cis-aconitate. This reaction, catalyzed by aconitase, prepares the molecule for decarboxylation in the next step.

3. Isocitrate undergoes an oxidative decarboxylation, in which it both loses a CO_2 molecule and becomes more oxidized. Isocitrate is first converted to oxalosuccinate by removal of a hydrogen, and formation of NADH. Oxalosuccinate is then decarboxylated, becoming the five-carbon compound alpha-ketoglutarate. This reaction is catalyzed by isocitrate dehydrogenase.

4. Alpha-ketoglutarate is itself decarboxylated, forming another NADH and CO_2. The four-carbon product, succinate,

is temporarily linked to coenzyme A, forming the weakly linked succinyl CoA. This reaction is catalyzed by the alpha-ketoglutarate dehydrogenase complex.

5. Succinyl CoA is cleaved by succinyl CoA synthetase to release succinate. In the process, guanosine diphosphate (GDP) is linked to inorganic **phosphate** to form guanosine triphosphate (GTP). GTP is used in the cell for **protein synthesis** and signal cascades, or the **energy** can be transferred to create ATP for most common energy-requiring tasks.

6. The four-carbon succinate undergoes a series of rearrangements to regenerate oxaloacetate. It is first oxidized to form fumarate (catalyzed by succinate dehydrogenase), with a hydrogen transfer to FAD to form $FADH_2$. Fumarate is hydrated by fumarase to form malate, which is then oxidized by malate dehydrogenase to form oxaloacetate, with the creation of a final NADH.

As a central metabolic pathway in the cell, the rate of the Krebs cycle must be tightly controlled to prevent too much, or too little, formation of products. This regulation occurs through inhibition or activation of several of the enzymes involved. Most notably, the activity of pyruvate dehydrogenase is inhibited by its products, acetyl CoA and NADH, as well as by GTP. This enzyme can also be inhibited by enzymatic addition of a phosphate group, which occurs more readily when ATP levels are high. Each of these actions serves to slow down the Krebs cycle when energy levels are high in the cell. it is important to note that the Krebs cycle is also halted when the cell is low on **oxygen**, even though no oxygen is consumed in it. Oxygen is needed further along in cell respiration though, to regenerate NAD^+ and FAD. Without these, the cycle cannot continue, and pyruvic acid is converted in the cytosol to lactic acid by the **fermentation** pathway.

The Krebs cycle is also a source for precursors for biosynthesis of a number of cell molecules. For instance, the synthetic pathway for amino acids can begin with either oxaloacetate or alpha-ketoglutarate, while the production of **porphyrins**, used in hemoglobin and other **proteins**, begins with succinyl CoA. Molecules withdrawn from the cycle for biosynthesis must be replenished. Oxaloacetate, for instance, can be formed from pyruvate, carbon dioxide, and **water**, with the use of one ATP molecule.

See also Bioenergetics

KREBS, EDWIN G. (1918-)

American biochemist

In the 1950s Edwin G. Krebs and his longtime associate **Edmond Fischer** discovered reversible protein phosphorylation, a fundamental biological mechanism. Together Krebs and Fisher's work illuminates the basic processes that regulate many vital aspects of cell activity, such as **protein synthesis**, cell **metabolism**, and hormonal responses to stress. Medical application of their discoveries has helped in research on Alzheimer's disease, organ transplants, and certain kinds of cancer, and in 1992 the two scientists shared the Nobel Prize for physiology or medicine. In addition to his contributions in the field of biochemical research, Krebs has also been recognized for his teaching and administrative abilities.

Edwin Gerhard Krebs was born to William Carl Krebs and Louisa Helena Stegeman Krebs in Lansing, Iowa, on June 6, 1918. He was the third of four children. His father, a Presbyterian minister, died while Krebs was in his first year of high school. In order to keep Krebs's two older brothers enrolled at the University of Illinois in Urbana, Louisa Krebs moved the family from Greenville, where Edwin Krebs grew up, to the university town. There she rented a house big enough for her family, with extra rooms to rent out to students, keeping the family together and helping the children continue their education. She had been a teacher herself before her marriage.

In 1940, after completing his high-school and undergraduate work in Urbana, Krebs entered medical school at Washington University School of Medicine in St. Louis, Missouri. To Krebs's way of thinking, medicine had the advantage of being directly related to people. His general interest in science he attributed to concerns about economic security. At Washington University he received classical medical training and was also introduced to medical research. He had the opportunity to work under Arda A. Green, who was associated with **Carl Ferdinand Cori** and **Gerty T. Cori**. The Coris were a husband-and-wife team who had won the Nobel Prize in 1947 for research on carbohydrate metabolism and the **enzyme** phosphorylase. Krebs's later collaboration with Fischer at the University of Washington in Seattle had its beginning in the research conducted by the Coris.

After receiving his medical degree in 1943 and completing an eight-month residency in internal medicine at Barnes Hospital in St. Louis, Krebs became a medical officer in the navy, serving in that capacity until 1946. This was the only period in his career during which Krebs was a practicing physician. Due to the unavailability of a resident position, and on the advice of one of his professors, Krebs now began studying science. Because of his background in **chemistry**, Krebs chose to work in **biochemistry** and was accepted by the Coris as a postdoctoral fellow in their laboratory. For two years, while working for the Coris, Krebs studied the interaction of protamine (a basic protein) with rabbit muscle phosphorylase. This work seemed so rewarding to him that he decided to continue his efforts in the field of research, and when in 1948 he was invited by Hans Neurath to join the faculty in the department of biochemistry at the University of Washington, he jumped at the opportunity to become assistant professor.

At this time Neurath's department greatly emphasized protein chemistry and enzymology (enzymes are **proteins** that act as catalysts in biochemical reactions). Work in the Coris' laboratory had established that the enzyme phosphorylase existed in active and inactive forms, but what controlled its activity was unknown. Combining his experience on mammalian skeletal muscle phosphorylase with Edmond Fischer's experience with potato phosphorylase after Fischer joined the department, Krebs and Fischer teamed up to uncover the molecular mechanism by which phosphorylase makes **energy** available to a contracting muscle. What they discovered was reversible protein phosphorylation. An enzyme called protein kinase takes **phosphate** from adenosine triphosphate (ATP), the supplier of energy to cells, and adds it to inactive phosphorylase, changing the shape of the phosphorylase and consequently switching it on. Another enzyme, called protein phosphatase, reverses this process by removing the phosphate from phosphorylase, thus deactivating it. Protein kinases are present in all cells.

Once it became evident that reversible protein phosphorylation was a general process, the impact of Krebs and Fischer's work was immeasurable. Their collaboration opened the field of biochemical research and paved the way to much of the work done in the area of biotechnology and **genetic engineering**. Protein phosphorylation has even been posited as the basis of learning and memory. Medical applications have included development of the drug cyclosporin, which blocks the body's immune response by interfering with phosphorylation to prevent rejection of transplants. As important as what happens when the process functions normally is what happens when it goes awry: protein kinases are involved in almost 50 percent of cancer-causing oncogenes.

Recognition for Krebs's work came through various awards besides the Nobel Prize. In 1988 Krebs and Fischer shared the Passano Award for their research, and Krebs was one of four scientists to share the Lasker Award for Basic Medical Research in 1989. He was co-recipient of the Robert A. Welch Award in Chemistry in 1991, followed by the Nobel Prize a year later. Besides concentrating his research on protein phosphorylation, Krebs has investigated signal transduction and carbohydrate metabolism.

In 1968 Krebs had left the University of Washington to accept the position of founding chairman of the department of biological chemistry at the University of California in Davis. When he returned to Washington in 1977 he became chairman of the department of pharmacology. In both positions he was able to assist in the recruitment of talented faculty, which Krebs considers critical to the continued development of the field of biochemistry. From 1977 until 1983, Krebs was associated with the Howard Hughes Medical Institute as well.

Krebs was married on March 10, 1945, to Virginia Deedy French, and they have three children, Sally, Robert, and Martha. As a young boy, Krebs loved to read historical novels about the Civil War and the settling of the West. He credits his wife for keeping him aware of the other aspects of living besides his work. The University of Washington dean of medicine, Philip Fialkow, commented in the *Daily,* a campus newspaper, that Krebs (along with Fischer) was "simply a joy to work with... [and] very considerate of students and the young faculty," giving credit to the contributions of others and accepting institutional responsibility when called on.

KREBS, HANS ADOLF (1900-1981)

English biochemist

Few students complete an introductory biology course without learning about the **Krebs cycle**, an indispensable step in the process our bodies perform to convert food into **energy** on which we subsist. Also known as the **citric acid** cycle or tricarboxylic acid cycle, the Krebs cycle derives its name from one of the most influential biochemists of our time. Born in the same year as the twentieth century, Hans Adolf Krebs spent the greater part of his eighty-one years engaged in research on intermediary **metabolism**. First rising to scientific prominence for his work on the ornithine cycle of **urea** synthesis, Krebs shared the Nobel Prize for physiology and medicine in 1953 for his discovery of the citric acid cycle. Over the course of his career, the German-born scientist published, oversaw, or supervised a total of more than 350 scientific publications. But the story of Krebs's life is more than a tally of scientific achievements; his biography can be seen as emblematic of biochemistry's path to recognition as its own discipline.

On August 25, 1900, Alma Davidson Krebs gave birth to her second child, a boy named Hans Adolf. The Krebs family—Hans, his parents, sister Elisabeth and brother Wolfgang—lived in Hildesheim, in Hanover, Germany. There his father Georg practiced medicine, specializing in surgery and diseases of the ear, nose, and throat. Hans developed a reputation as a loner at an early age. He enjoyed swimming, boating, and bicycling, but never excelled at athletic competitions. He also studied piano diligently, remaining close to his teacher throughout his university years. At the age of fifteen, the young Krebs decided he wanted to follow in his father's footsteps and become a physician. But World War I had broken out, and before he could begin his medical studies, he was drafted into the army upon turning eighteen in August of 1918. The following month he reported for service in a signal corps regiment in Hanover. He expected to serve for at least a year, but shortly after he started basic training the war ended. Krebs received a discharge from the army to commence his studies as soon as possible.

Krebs chose the University of Göttingen, located near his parents' home. There, he enrolled in the basic science curriculum necessary for a student planning a medical career and studied anatomy, histology, embryology and botanical science. After a year at Göttingen, Krebs transferred to the University of Freiburg. At Freiburg, Krebs encountered two faculty members who enticed him further into the world of academic research: Franz Knoop, who lectured on physiological **chemistry**, and Wilhelm von Möllendorff, who worked on histological staining. Möllendorff gave Krebs his first research project, a comparative study of the staining effects of different dyes on muscle tissues. Impressed with Krebs's insight that the efficacy of the different dyes stemmed from how dispersed and dense they were rather than from their chemical properties, Möllendorff helped Krebs write and publish his first scientific paper. In 1921, Krebs switched universities again, transferring to the University of Munich, where he started clinical work under the tutelage of two renowned surgeons. In 1923, he completed his medical examinations with an overall mark of "very good," the best score possible. Inspired by his university studies, Krebs decided against joining his father's practice as he had once planned; instead, he planned to **balance** a clinical career in medicine with experimental work. But before he could turn his attention to research, he had one more hurdle to complete, a required clinical year, which he served at the Third Medical Clinic of the University of Berlin.

Krebs spent his free time at the Third Medical Clinic engaged in scientific investigations connected to his clinical duties. At the hospital, Krebs met Annelise Wittgenstein, a more experienced clinician. The two began investigating physical and chemical factors that played substantial roles in the distribution of substances between blood, tissue, and cerebrospinal fluid, research that they hoped might shed some light on how pharmaceuticals such as those used in the treatment of syphilis penetrate the nervous system. Although Krebs published three articles on this work, later in life he belittled these early, independent efforts. His year in Berlin convinced Krebs that better knowledge of research chemistry was essential to medical practice.

Accordingly, the twenty-five-year-old Krebs enrolled in a course offered by Berlin's Charité Hospital for doctors who wanted additional training in laboratory chemistry. One year later, through a mutual acquaintance, he was offered a paid research assistantship by **Otto Warburg**, one of the leading biochemists of the time. Although many others who worked with Warburg called him autocratic, under his tutelage Krebs developed many habits that would stand him in good stead as his own research progressed. Six days a week work began at Warburg's laboratory at eight in the morning and concluded at six in the evening, with only a brief break for lunch. Warburg worked as hard as the students. Describing his mentor in his autobiography, *Hans Krebs: Reminiscences and Reflections,* Krebs noted that Warburg worked in his laboratory until eight days before he died from a pulmonary embolism. At the end of his career, Krebs wrote a biography of his teacher, the subtitle of which described his perception of Warburg: ''cell physiologist, biochemist, and eccentric.''

Krebs's first job in Warburg's laboratory entailed familiarizing himself with the tissue slice and manometric (pressure measurement) techniques the older scientist had developed. Until that time biochemists had attempted to track chemical processes in whole organs, invariably experiencing difficulties controlling experimental conditions. Warburg's new technique, affording greater control, employed single layers of tissue suspended in **solution** and manometers (pressure gauges) to measure chemical reactions. In Warburg's lab, the tissue slice/manometric method was primarily used to measure rates of respiration and glycolysis, processes by which an organism delivers **oxygen** to tissue and converts **carbohydrates** to energy. Just as he did with all his assistants, Warburg assigned Krebs a problem related to his own research—the role of heavy **metals** in the **oxidation** of sugar. Once Krebs completed that project, he began researching the metabolism of human cancer tissue, again at Warburg's suggestion. While Warburg was jealous of his researchers' laboratory time, he was not stingy with bylines; during Krebs's four years in Warburg's lab, he amassed sixteen published papers. But Warburg had no room in his lab for a scientist interested in pursuing his own research. When Krebs proposed undertaking studies of intermediary metabolism that had little relevance for Warburg's work, the supervisor suggested Krebs switch jobs.

Hans Adolf Krebs.

Unfortunately for Krebs, the year was 1930. Times were hard in Germany, and research opportunities were few. He accepted a mainly clinical position at the Altona Municipal Hospital, which supported him while he searched for a more research-oriented post. Within the year he moved back to Freiburg, where he worked as an assistant to an expert on metabolic diseases with both clinical and research duties. In the well-equipped Freiburg laboratory, Krebs began to test whether the tissue slice technique and manometry he had mastered in Warburg's lab could shed light on complex synthetic metabolic processes. Improving on the master's methods, he began using saline solutions in which the concentrations of various ions matched their concentrations within the body, a technique which eventually was adopted in almost all biochemical, physiological, and pharmacological studies.

Working with a medical student named Kurt Henseleit, Krebs systematically investigated which substances most influenced the rate at which urea—the main solid component of mammalian urine—forms in liver slices. Krebs noticed that the rate of urea synthesis increased dramatically in the presence of ornithine, an amino acid present during urine production. Inverting the reaction, he speculated that the same ornithine produced in this synthesis underwent a cycle of conversion and synthesis, eventually to yield more ornithine and urea. Scientific recognition of his work followed almost immediately, and at the end of 1932—less than a year and a half after he began his research—Krebs found himself appointed as a *Privatdozent* at the University of Freiburg. He immediately embarked on the more ambitious project of identifying the intermediate steps in the metabolic breakdown of carbohydrates and **fatty acids**.

But Krebs was not to enjoy his new position in Germany for long. In the spring of 1933, along with many other German scientists, he found himself dismissed from his job as a result of Nazi purging. Although Krebs had officially and legally renounced the Jewish faith twelve years earlier at the urging of his patriotic father, who believed wholeheartedly in the assimilation of all German Jews, this legal declaration proved insufficiently strong for the Nazis. In June of 1933, he sailed for England to work in the **biochemistry** lab of Sir **Frederick Gowland Hopkins** of the Cambridge School of Biochemistry. Supported by a fellowship from the Rockefeller Foundation, Krebs resumed his research in the British laboratory. The following year, he augmented his research duties with the position of demonstrator in biochemistry. Laboratory space in Cambridge was cramped, however, and in 1935 Krebs was lured to the post of lecturer in the University of Sheffield's Department of Pharmacology by the prospect of more lab space, a semi-permanent appointment, and a salary almost double the one Cambridge was paying him.

His Sheffield laboratory established, Krebs returned to a problem that had long preoccupied him: how the body produced the essential **amino acids** that play such an important role in the metabolic process. By 1936, Krebs had begun to suspect that citric acid played an essential role in the oxidative metabolism by which the carbohydrate pyruvic acid is broken down so as to release energy. Together with his first Sheffield graduate student, William Arthur Johnson, Krebs observed a process akin to that in urea formation. The two researchers showed that even a small amount of citric acid could increase the oxygen absorption rate of living tissue. Because the amount of oxygen absorbed was greater than that needed to completely oxidize the citric acid, Krebs concluded that citric acid has a catalytic effect on the process of pyruvic acid conversion. He was also able to establish that the process is cyclical, citric acid being regenerated and replenished in a subsequent step. Although Krebs spent many more years refining the understanding of intermediary metabolism, these early results provided the key to the chemistry that sustains life processes. In June of 1937, he sent a letter to *Nature* reporting these preliminary findings. Within a week, the editor notified him that his paper could not be published without a delay. Undaunted, Krebs revised and expanded the paper and sent it to the new Dutch journal *Enzymologia,* which he knew would rapidly publicize this significant finding.

In 1938, Krebs married Margaret Fieldhouse, a teacher of domestic science in Sheffield. One year later, Margaret, who was thirteen years younger than Hans, gave birth to the first of the couple's three children. In the winter of the same year, the university named him lecturer in biochemistry and asked him to head their new department in the field. Married to an Englishwoman, Krebs became a naturalized Englishman in September, 1939, three days after World War II began.

The war affected Krebs's work minimally. He conducted experiments on vitamin deficiencies in conscientious objectors, while maintaining his own research on metabolic cycles. In 1944, the Medical Research Council asked him to head a new department of biological chemistry. Krebs refined his earlier discoveries throughout the war, particularly trying to determine how universal the Krebs cycle is among living organisms. He was ultimately able to establish that all organisms, even microorganisms, are sustained by the same chemical processes. These findings later prompted Krebs to speculate on the role of the metabolic cycle in evolution.

In 1953, Krebs received one of the ultimate recognitions for the scientific significance of his work—the Nobel Prize in physiology and medicine, which he shared with **Fritz Lipmann**, the discoverer of co-enzyme A. The following year, Oxford University offered him the Whitley professorship in biochemistry and the chair of its substantial department in that field. Once Krebs had ascertained that he could transfer his metabolic research unit to Oxford, he consented to the appointment. Throughout the next two decades Krebs continued research into intermediary metabolism. He established how fatty acids are drawn into the metabolic cycle and studied the regulatory mechanism of intermediary metabolism. Research at the end of his life was focused on establishing that the metabolic cycle is the most efficient mechanism by which an organism can convert food to energy. When Krebs reached Oxford's mandatory retirement age of sixty-seven, he refused to stop researching and made arrangements to move his research team to a laboratory established for him at the Radcliffe Hospital. On November 22, 1981, Krebs died at the age of eighty-one.

KROTO, HAROLD WALTER (1939-)

English physical chemist

University of Sussex professor Harold Walter Kroto was awarded the 1996 Nobel Prize for chemistry along with Rice University professors **Robert F. Curl, Jr.**, and **Richard E. Smalley** for their discovery of a new form of the element **carbon**, called Carbon 60. The third molecular form of carbon (the other two forms are diamonds and graphite), C60 consists of 60 atoms of carbon arranged in hexagons and pentagons and is called a ''buckminsterfullerene,'' ''fullerene,'' or by its nickname ''Buckyball'' in honor of Buckminster Fuller, whose geodesic domes it resembles. Kroto made the discovery in 1985, and Curl and Smalley confirmed his findings.

Harold Kroto was born on October 7, 1939, in Wisbech, Cambridgeshire, England to Heinz and Edith Kroto. Raised in Bolton, Lancashire, England, he graduated with a degree in chemistry from the University of Sheffield in 1961 and received his Ph.D. there in 1964. In 1963 he married Margaret Henrietta Hunter, with whom he would have two children. His Ph.D. work involved high-resolution electronic spectra of **free radicals** produced by flash photolysis—chemical decomposition by the action of radiant **energy**. His postdoctoral work was conducted at the National Research Council in Ottawa, Canada. After completing this work he spent a year at Bell Laboratories in New Jersey, where he also studied **quantum chemistry**. He began teaching and researching at the University of Sussex in 1967. He was appointed full professor in 1985 and a Royal Research Professor in 1991.

In the 1970s Kroto launched a research program at the University of Sussex to look for **chains** of carbon in interstellar space. Kroto worked with scientists at the National Research Council, where he did his postdoctoral studies, to find the space molecules and did actually find several chains in the years between 1975-78. He speculated that the molecules might be products of red-giant stars, but did not know how the chains themselves were formed. His search to find an explanation led him to not only the carbon-60 **molecule**, but also resulted in new areas of carbon chemistry study. In 1992 Kroto won the Italgas Prize for Chemistry for this work.

In 1984 Kroto borrowed from his wife's bank account and flew to Houston find out exactly how the carbon chains were formed. He had heard of the work being done in laser **spectroscopy** by Richard Smalley and Robert Curl at Rice University in Texas and thought he could use their laser apparatus to simulate the temperatures in space needed to create the carbon chains. Smalley and Curl had been looking at **semiconductors** like **silicon** and **germanium** in the laser apparatus, and had no reason to look at simple carbon. But that September, the scientists turned the laser beam on a piece of **graphite** and found something they were not looking for, a molecule that had 60 carbon atoms. Carbon had previously been known to have only two molecular forms, **diamond** and graphite. They surmised correctly that this was a third form of carbon, and that it had a cage-like structure resembling a soccer ball, or a geodesic dome. They named the structure buckminsterfullerene, which became fullerene, and also by the nickname "buckyballs."

Two things made this find unusual. One reason is that fundamental research led to the discovery, and the second is that a corporation did not fund it. Evidence for the existence of large carbon **clusters** had existed before, but Smalley, Curl, and Kroto were the first scientists to fully identify and stabilize carbon-60. In October of 1996, all three were recognized for this remarkable discovery with the Nobel Prize in Chemistry. At the time of the announcement Kroto thought he might be winner, but decided to go to lunch instead of waiting. He had been depressed because just two hours before the announcement he had been turned down for research funding by the British government for the very same research that had won him the Nobel Prize. That same year Kroto was knighted for winning the Nobel Prize.

The Royal Swedish Academy of Sciences, which grants the Nobel, heralded the breakthrough. "From a theoretical viewpoint, the discovery of the **fullerenes** has influenced our conception of such widely separated scientific problems as the galactic **carbon cycle** and classical aromaticity, a keystone of **theoretical chemistry**," stated the Academy citation. Aromaticity refers to the chemical stability of organic compounds.

Kroto's research team in Sussex continues fundamental work on the fullerene, looking at its basic chemistry as well as the way it has changed how carbon-based materials are viewed. Interdisciplinary work is also being conducted on the interstellar applications of carbon microparticles. But Kroto is not one to speculate on the uses of carbon-60. In an interview in the newsletter *Science Watch*, Kroto explains that "fundamental scientists are not necessarily the best people to ask about applications. People like myself have, in a sense, spent a lifetime avoiding applications. We're puzzled about interesting things for their own sake, and we follow up on them." Kroto compares the discovery of the fullerene to the discovery of **lasers**; it was almost a decade before scientists discovered a practical use for lasers.

When Kroto began his research into carbon chains, his primary interest was finding carbon-60 in space. "It was discovered by simulating certain astrophysical conditions, after all," Kroto explained in the *Science Watch* interview. This desire was fulfilled when geochemist Jeffrey Bada of the Scripps Institution of Oceanography in La Jolla, California found fullerenes inside rocks extracted from a 1.8 billion-year-old meteor crash in Canada. Bada thought that the molecules had been formed by their proximity to a red-giant star. Kroto himself had suggested almost 20 years ago that this might be the place to find carbon chains.

KRYPTON

Krypton is the fourth member of Group 18 of the **periodic table**, a group of elements known as the noble **gases** or inert gases. Krypton's **atomic number** is 36, its atomic **mass** is 83.80, and its chemical symbol is Kr.

Properties

Krypton is a colorless, odorless, tasteless gas with a **density** of 3.64 grams per liter, nearly three times the density of air. Its **boiling point** is -243.2°F (-152.9°C) and its freezing point is -250.8°F (-157.1°C). Like other elements in Group 18, krypton is chemically inert. A few compounds of the element have been made for research purposes, but they have no commercial application.

Occurrence and Extraction

The abundance of krypton in the atmosphere is thought to be about 0.000108-0.000114%, making it the seventh most common gas in the atmosphere. The element is also thought to be present in the Earth's crust as the result of natural radioactive processes, but its abundance there is too small to estimate. Krypton can be prepared commercially by the fractional **distillation** of liquid air.

Discovery and Naming

Krypton was discovered in 1898 by the Scottish chemist and physicist Sir **William Ramsay** and the English chemist Morris William Travers (1872-1961). In the same year, Ramsay and Travers discovered two other noble gases, **xenon** and **neon**. The gases were discovered by careful analysis of the products that remained as liquid air evaporated over a period of time. The element was named after the Greek word *kryptos* meaning "hidden."

Richard Kuhn.

Uses

The only important commercial use of krypton is in making various kinds of lamps. When an electrical current is passed through an evacuated tube containing krypton gas, a very bright white light is produced. Tubes of this kind are used in specialized lamps, like those used along airport runways. The light produced is so bright that it can be seen for distances of up to 1,000 ft (300 m) even in foggy conditions. Lights of this kind do not produce a continuous beam but give off pulses that last no more than about 10 microseconds. Krypton gas is also used in making ''neon'' lights with a yellow **color**.

KUHN, RICHARD (1900-1967)

German chemist

Richard Kuhn was a Nobel Prize-winning organic chemist who devoted much of his life to studying the synthesis of **vitamins** and carotenoids, the fat-soluble yellow pigments that are found in plants. He researched the **chemistry** of algae sex cells and optical **stereochemistry**, and spent a great deal of time understanding **carbohydrates**. He was determined to succeed in his work by uncovering the practical applications of substances in the fields of medicine and agriculture. Later in his career, Kuhn concentrated on studying how the body fights disease using organic compounds.

Kuhn was born in Vienna, Austria, on December 3, 1900, to Hofrat Richard Clemens, a hydraulics engineer, and Angelika (Rodler) Kuhn, an elementary school teacher. After spending almost ten years of his life at home under the educational guidance of his mother, Kuhn entered the Döbling Gymnasium, where he attended classes with future Nobel Prize-winning physicist **Wolfgang Pauli**. After graduating from the Gymnasium in 1917, he was drafted into the German (Austro-Hungarian) army and served until World War I ended in November of 1918.

Once Kuhn was discharged from the military, he entered the University of Vienna where a professor of medical chemistry, Ernst Ludwig, turned his interests towards chemistry. Just three semesters after entering the university, Kuhn transferred to the University of Munich, where he studied chemistry under noted scientist **Richard Willstätter**. Kuhn received his Ph.D. in 1922 for his thesis, ''On the Specificity of Enzymes in Carbohydrate Metabolism.'' He worked briefly as Willstätter's assistant before leaving Munich in 1926 to join the Federal Institute of Technology at Zurich, a Swiss technological high school, where he spent three years as professor of general and **analytical chemistry**.

In 1929 Kuhn left Zurich and joined the University of Heidelberg's newly established Kaiser Wilhelm Institute for Medical Research (renamed the Max Planck Institute in 1950, with Kuhn's assistance) as both a professor of **organic chemistry** and director of the institute's chemistry department. Kuhn would turn down a number of other offers to spend the remainder of his career at the institute; he became its director in 1937.

Kuhn was particularly interested in how the chemistry of organic compounds was related to their function in biological systems. His early work concentrated on carotenoids. One such substance was carotene, the pigment found in carrots, whose chemical formula had been determined earlier by Willstätter at the University of Munich. After further research, Kuhn discovered that carotene was a precursor in the chemical production of vitamin A and that nature uses all kinds of chemical structures for biological actions. In addition, Kuhn and his colleagues discovered that carotenoids existed in numerous plants and animals and that vitamin A was an essential part of maintaining the body's mucous membranes.

At the time, Kuhn was just one of two scientists working with carotene; **Paul Karrer** at the University of Zurich was the other. The two men would remain fierce competitors throughout their careers. Through Kuhn's investigations, he found two distinct compounds in carotene: beta-carotene, which bends light, and alpha-carotene, which does not. Two years later Kuhn's work led to the discovery of a third form of carotene, called gamma-carotene. These compounds have exactly the same chemical formulas but different molecular structures; therefore, they are known as **isomers**.

In the 1930s, Kuhn's turned his attention to researching members of the water-soluble vitamin B group. Working with other scientists, he painstakingly isolated and crystallized a small amount of vitamin B2 (riboflavin) from skim milk. By determining the structure of riboflavin, Kuhn was able to clearly explain the chemical composition and to eventually synthesize this compound. He also demonstrated that B2 plays a primary role in respiratory **enzyme** action and provided the key to how vitamins function and what their applications are in living systems. For this work, Kuhn was offered the 1938 Nobel Prize. Then, in the late 1930s, Kuhn and three other coworkers determined both the chemical composition and **molecular structure** of adermin, now commonly referred to as vitamin B6, which acts against skin disease and helps to regulate the metabolism of the nervous system.

Although Kuhn was to be awarded the Nobel Prize in 1938, he did not actually receive it until the late 1940s. Due to the political climate in Germany and the fact that a Nazi concentration camp prisoner, Carl von Ossietzky, was honored with the Nobel Peace Prize in 1934, Hitler instituted a policy forbidding German citizens from accepting the award. As a result, Kuhn was forced to turn down the award and was not properly honored until after the war ended in 1945. He received his medal and certificate in 1949 at a special ceremony in Stockholm for his work with carotenoids and vitamins.

The 1940s saw Kuhn expand his research to include carbohydrates. Kuhn researched **alkaloid** glycosides, which appear in potatoes and tomatoes, and tried to unlock their pigments and biological structures. He also returned to researching milk, from which he extracted carbohydrates using **chromatography**. In doing so, he greatly improved the use of chromatography, which is the chemical separation of mixtures into their original form. After becoming a professor of **biochemistry** at the Max Planck Institute in 1950, Kuhn focused much of his effort on the study of organic substances that are instrumental in the body's resistance to infection. His investigations into a variety of "resistance" factors effective against cholera and influenza uncovered the molecular interaction between an organism and its attacker. He also went on to identify pantothenic acid, an important ingredient in hemoglobin formation and the release of **energy** from carbohydrates, and **para-aminobenzoic acid** (PABA), a compound that proved useful in the synthesis of **anesthetics**.

Known to have an upbeat and outgoing personality, Kuhn enjoyed such activities as billiards, chess and tennis. He was also a skilled violinist who frequently played with a chamber ensemble for public enjoyment. Kuhn met his wife, Daisy Hartmann, while he was a professor and she a student at the Federal Institute of Technology at Zurich. They would marry in 1928; the couple had four daughters and two sons together.

In addition to winning the Nobel Prize, Kuhn was awarded the Pasteur, Paterno and Goethe Prizes for his work. He was a member of numerous national and international scientific societies and received honorary degrees from a variety of institutions including the Munich Technical University, the University of Vienna, and the University of St. Maria, Brazil. A charter senate member of the Max Planck Society for the Advancement of Science, he later served as vice-president. Kuhn became the editor of the chemical journal, *Annalen der Chemie,* in 1948, and served as president of the German Chemical Society. He published more than seven hundred scientific papers and received over fifty distinctions. Shortly before his death, the University of Heidelberg gave its first commemorative medal struck in honor of a scientist. Kuhn died July 31, 1967, at age sixty-six, in Heidelberg, Germany.

KWOLEK, STEPHANIE LOUISE (1923-)

American chemist

Stephanie Louise Kwolek is best known for her discovery of liquid crystalline solutions of synthetic aromatic polyamides and their fibers. These fibers became the first of a new generation of high performance fibers, and the basis for the commercialization of Du Pont's Kevlar brand fibers.

Kwolek graduated with a degree in **chemistry** from the Carnegie Institute of Technology in 1946, and joined Du Pont in the same year. She moved to Du Pont's Pioneering Research Laboratory at the Experimental Station when it opened four years later. There she was encouraged and supported by Hale Church, director of the laboratory at that time, as she struggled to gain intellectual acceptance in a male-dominated research community.

Kwolek began her career scouting for new polymers. In 1964, Kwolek began experimenting with poly-p-Phenylene-terephthalate (PPD-T) and polybenzamide (PBA). She became the first person to prepare pure monomers that could be used to synthesize PBA.

The intermediates needed in this process proved extremely sensitive to moisture and **heat**, so were readily susceptible to **hydrolysis** and premature self-polymerization. Kwolek discovered a suitable solvent and identified the low-temperature polymerization conditions required to produce an unusual polymer **solution** that was fluid and cloudy, rather than clear and viscous. Upon spinning this solution, she found that tough fibers with startling properties had formed. These fibers were much stronger and stiffer than any previously made synthetic fiber. The discovery ushered in a stimulating period of research at Du Pont. The search for a candidate for commercialization finally came up with Kevlar.

Kwolek has been honored on many occasions for her pioneering studies. In 1980 she received an award for Creative Invention from the American Chemical Society. In 1995, she was inducted into the National Inventors Hall of Fame.

L

LAND, EDWIN H. (1909-1991)

American inventor

Edwin H. Land was the driving force behind the Polaroid Corporation's engineering and marketing successes. He was the first to figure out how to manufacture practical and useful polarized screens during the 1930s, and he produced revolutionary optics for the military during World War II. But it was the development of the instant camera that made his company famous, and he was able to dominate the instant-photography market with cameras that first produced pictures in sepia tones, then in black and white, and finally in **color**. One who routinely discarded conventional wisdom, Land believed that market research was not necessary; he claimed that any invention would sell if people believed it was something they could not live without.

An only child, Edwin Land was born to Martha F. and Harry Land on May 7, 1909, in Bridgeport, Connecticut. His father ran a salvage and scrap metal business; the family was well-off and Land had a comfortable upbringing. In his youth he dreamed of being an inventor and idolized **Michael Faraday**, **Thomas Alva Edison**, and Alexander Graham Bell. Even as a boy, Land was very interested in polarized light. He entered Harvard at age seventeen in 1926. While walking along Broadway in New York City that same year, he was overwhelmed by the glare from headlights and store signs that shone in his eyes. Land perceived safety hazards in all that glare, and he determined that polarized lights could reduce it. He left Harvard at the end of the school year to pursue this idea and did not return for three years.

Land's parents provided an allowance that enabled him to stay in New York and work on this idea. He studied at the New York Public Library and even found a laboratory at Columbia University whose window was habitually unlocked. He would climb in at night and conduct various experiments. During this period Land met Helen Maislen, a graduate of Smith College who began assisting him in his research. They were married in 1929, and Land returned to Harvard that same year. This time the university provided him with a laboratory to conduct his research.

It had been known since the eighteenth century that certain kinds of crystals could affect the direction of light waves. In his effort to develop a method for polarizing light, Land was searching for a crystal that could not only reduce glare but was stable and economical enough to be produced commercially. He conceived of the idea of two plates which would absorb the light waves that were not wanted and transmit those that were. He then succeeded in aligning millions of microscopic **iodine** crystals in one direction, thus creating the first polarizer. As Mark Olshaker wrote in the *Instant Image:* "Land's singular achievement was in discovering a way to synthesize a sheet material that could align light waves in the desirable planes of vibration. The invention was a combined achievement of chemistry and optics." Land presented a paper on his discovery at a physics colloquium at Harvard in February, 1932. In June he left the university, one semester short of a degree, and never returned.

With a Harvard graduate student named George Wheelwright, who had been one of his teachers, Land formed Land-Wheelwright Laboratories, Inc. in June of 1932. The two men worked on developing methods of manufacturing polarized sheets made of crystals trapped in nitrocellulose. On November 30, 1934, Eastman Kodak gave Land-Wheelwright an order for ten thousand dollars worth of polarizing filters. Kodak wanted a polarizer laminated between two sheets of optical **glass**, but neither Land nor Wheelwright had any idea how to manufacture such an item. Nonetheless, they accepted the order—a decision typical of the way Land would work in the future. Their persistence paid off, and Land-Wheelwright Laboratories invented what they dubbed "Polaroid," with which they fulfilled their contract with Eastman Kodak.

Land had a flair for the dramatic which he put to good use in marketing his inventions. For example, when he was trying to sell his polarizers for use as sunglasses, he rented a room at a hotel and invited executives from the American Op-

Edwin H. Land.

tical Company to meet him there. The late afternoon sun produced a glare on the windowsill; Land put a fishbowl there and the glare rendered the goldfish inside it invisible. When the executives arrived, Land handed them each a sheet of polarizer and they were able to see the fish instantly. Land told them that from now on their sunglasses should be made with polarized glass, and the company bought the idea.

Land gave his first press conference on **polarization** on January 30, 1936, at the Waldorf Astoria Hotel. He repeated it for the National Academy of Sciences and the New York Museum of Science soon after that. The press coverage Land expected from this last presentation was overshadowed by the abdication of King Edward VIII in England. But his sales ability once again came through. On August 10, 1937, a group of investors, impressed with Land-Wheelwright's accomplishments, put up $375,000 to fund the Polaroid Corporation. Furthermore, they gave Land controlling interest in the company.

With the money Land purchased some competing patents on polarization and decided that the 1939 New York World's Fair would be an excellent way to demonstrate to automakers and the American public the virtues of polarized headlights. Chrysler rented Polaroid space in one of its booths, and Land played a three-dimensional movie he had invented which graphically illustrated how much improved polarized headlights were. The twelve-minute film was well-received by the public; 150,000 people saw it in its first two months, but Detroit never bought Land's system. Polaroid nevertheless enjoyed good success with its Polaroid Day Glasses and the dual polarized windows they installed on Union Pacific's Copper King rail cars. By January of 1940 the company had about 240 employees, and it moved from Boston to Cambridge.

Early in 1941 the Navy invited companies to bid on a contract to develop an altitude finder. Before the other companies even replied, Polaroid invented the item and presented it to the Navy. During the war, Polaroid provided the Army and Navy with various types of goggles, as well as improved gunsights for Sherman tanks. Land drove the company hard, convinced that the Allies' only hope of winning lay in the superiority of the science behind the effort. To this end, he invented ''vectography,'' a system that took three-dimensional photographs. Such photos were invaluable in reconnaissance efforts, especially when camouflage was employed by the enemy. Polaroid also worked on the guided missile program known as the SX–70 or ''Project Dove.'' The company's work was deemed valuable enough for them to have won four Navy ''E'' pennants for excellence.

Two of Land's employees, **Robert B. Woodward** and William E. Doering, solved a critical problem for the Allies. With the Japanese controlling Java, and thus the world supply of cinchona trees, quinine—the only cure known for malaria—was unavailable. Woodward and Doering synthesized **quinine** and Polaroid waived the royalties from this synthetic compound and gave them to the government with no commercial limitations. Woodward later won the 1965 Nobel Prize in chemistry for synthesizing **cortisone**.

By 1943 Land had a three-year-old daughter named Jennifer. While vacationing in Santa Fe just before Christmas, Land's daughter asked him why they could not see the pictures they had taken during the day. Land recalled his reaction to this question, as quoted in the *New York Times:* ''As I walked around that charming town I undertook the task of solving the puzzle she had set me. Within the hour, the camera, the film, and the **physical chemistry** became so clear to me.'' Several years of intense work at Polaroid followed, and on February 21, 1947, Land demonstrated his instant camera at the winter meeting of the Optical Society of America. Although the images were in sepia, public reaction was so enthusiastic that Polaroid came under much pressure to manufacture the camera quickly even though they did not yet have the capability to do it. The camera also had another revolutionary feature: it linked aperture size and shutter speed, which eliminated much of the guesswork in using the camera.

In addition to his consumer products, Land also developed a new optical system for the Sloan-Kettering Cancer Institute in 1948 that enabled scientists to observe living human cells in their natural color. Near the end of 1953 he also invented a microscope that used light invisible to the human eye for illumination. Both of these inventions were a great aid in cancer research. In 1954 President Eisenhower appointed Land the head of an intelligence committee to study how to prevent a future attack like the one at Pearl Harbor. Land's recommendation to Eisenhower was to establish a system of aerial reconnaissance; this was the beginning of the U2 spy plane project, which carried several Polaroid developments on board.

In 1952 Polaroid introduced true black-and-white film for their instant camera. But unlike the sepia prints which held their images very well, the black-and-white ones faded over time. Land pressed his company to find a solution, which they did in only four weeks. Yet he was not happy with it, since it required swabbing the photo after pulling it from the camera; Land wanted his instant camera to require only one step. In addition, research continued simultaneously to develop an instant camera that produced color photos. Land even had his own private laboratory where he did much work on this.

It was not until April of 1972 that Land was finally able to reveal to the public a long-time dream brought to reality: the SX–70 instant color camera. The introduction of color had meant overcoming a host of technological obstacles. His achievement was recognized with the October 27, 1972, cover of *Life* magazine, which showed Land taking pictures of children from behind his SX–70.

Land was awarded the Hood Medal from the Royal Photographic Society in London in 1935. He was named one of America's Modern Pioneers by the National Association of Manufacturers on February 27, 1940. On December 6, 1963, Land received the Presidential Medal of Freedom, awarded by Lyndon Johnson less than three weeks after John F. Kennedy's assassination. In 1967, Land received the National Medal of Science, which was again presented by President Johnson. In February 1977 Land was inducted into the Inventor's Hall of Fame by the American patent office. All told, Land received 533 patents.

Land also started an inner city program for disadvantaged blacks during the 1960s, as well as the Rowland Foundation in 1965. Land later founded the Rowland Institute for Science. His financial gifts to various institutions, such as the Massachusetts Institute of Technology (MIT) and Harvard University, were always anonymous. He was a visiting professor at MIT, a member of Harvard's visiting committees for astronomy, chemistry, and physics, and a fellow at MIT's School for Advanced Study. Land received an honorary doctorate from Harvard in 1957, and others from such notable institutions as Yale, Tufts, Columbia University, Loyola, and Washington University. In 1951 Land was elected President of the American Academy of Arts and Sciences and served for two years. Land retired in August of 1982, and died March 1, 1991, of undisclosed causes.

LANDFILL

The term sanitary landfill was first used in the 1930s to refer to the *compacting* of solid waste materials. Initially adopted by New York City and Fresno, California, the sanitary landfill used heavy earth-moving equipment to compress waste materials and then cover them with soil. The practice of covering solid waste was evident in Greek civilization over 2,000 years ago, but the Greeks did it without compacting.

Today, the sanitary landfill is the major method of disposing of waste materials in North America and other developed countries, even though considerable efforts are being made to find alternative methods, such as **recycling** and composting. Among the reasons that landfills remain a popular alternative are their simplicity and versatility. For example, they are not sensitive to the shape, size, or weight of a particular waste material. Since they are constructed of soil, they are rarely affected by the chemical composition of a particular waste component or by any collective incompatibility of comingled wastes. By comparison, composting and incineration require uniformity in the form and chemical properties of the waste for efficient operation.

About 80% of the solid waste generated in the United States still is dumped in landfills. In a sanitary landfill, refuse is compacted each day and covered with a layer of dirt. This procedure minimizes odor and litter, and discourages insect and rodent populations that may spread disease. Although this method does help control some of the **pollution** generated by the landfill, the fill dirt also occupies up to 20% of the landfill space, reducing its waste-holding capacity. Sanitary landfills traditionally have not been enclosed in a waterproof lining to prevent leaching of chemicals into groundwater, and many cases of groundwater pollution have been traced to landfills.

Historically, landfills were placed in a particular location more for convenience of access than for any environmental or geological reason. Now more care is taken in the siting of new landfills. For example, sites located on faulted or highly permeable rock are passed over in favor of sites with a less-permeable foundation. Rivers, lakes, floodplains, and groundwater recharge zones are also avoided. It is believed that the care taken in the initial siting of a landfill will reduce the necessity for future clean-up and site rehabilitation. Due to these and other factors, it is becoming increasingly difficult to find suitable locations for new landfills. Easily accessible open space is becoming scarce and many communities are unwilling to accept the siting of a landfill within their boundaries. Many major cities have already exhausted their landfill capacity and must export their trash, at significant expense, to other communities or even to other states and countries.

A landfill has three stages of decomposition. The first one is an aerobic phase. The solid wastes that are biodegradable react with the **oxygen** in the landfill and begin to form **carbon** dioxide and **water. Temperature** during this stage of decomposition in the landfill rises about 30°F (16.7°C) higher than the surrounding air. A weak acid forms within the water and some of the minerals are then dissolved. The next stage is the anaerobic one in which microorganisms that do not need oxygen break down the wastes into **hydrogen**, **ammonia**, **carbon dioxide**, and inorganic acids.

In the third stage of decomposition in a landfill, **methane** gas is produced. Sufficient amounts of water and warm temperatures have to be present in the landfill for the microorganisms to form the gas. About half of the gas produced during this stage will be **carbon dioxide**, but the other half will be methane. Systems of controlling the production of methane gas are either passive or active. In a passive system, the gas is naturally vented into the atmosphere, and may include venting trenches, cutoff walls, or gas vents to direct the gas. An active system employs a mechanical method to remove the methane gas and can include recovery wells, gas collection lines, a gas burner, or a burner stack. Both active and passive systems have monitoring devices to prevent explosions or fires.

While landfills may outwardly appear simple, they need to operate carefully and follow specific guidelines that include where to start filling, wind direction, the type of equipment used, method of filling, roadways to and within the landfill, the angle of slope of each daily cell, controlling contact of the waste with groundwater, and the handling of equipment at the landfill site.

Considerations have to be made regarding the soil that is used as a daily cover, which is usually 6 in (15.2 cm) thick, an intermediate cover of 1 ft (30.5 cm), and a final cover of 2 ft (61 cm). The compacting of the solid waste and soil has to be considered as well, so that the biological processes of decomposition can take place properly.

Shredding of solid wastes is one method of saving space at landfills. Another method is baling of wastes. The advantages to shredding are twofold. The material can be compacted to a greater **density**, thereby extending the life of the landfill, and it can be compacted more quickly as well. Less cover is required and there is also less danger of spontaneous fire. Landfills using shredded materials produce more organic decomposition than those disposing of unshredded solid wastes. The advantages of baling are an increase in landfill life because of an increase in waste density. Hauling times are reduced, as are litter, dust, odor, fires, traffic, noise, earth moving, and land settling. Less heavy equipment is needed for the cover operation and the amount of time it takes for the land to stabilize is reduced.

When the secure landfill reaches capacity, it is capped by a cover of clay, plastic, and soil, much like the bottom layers. Vegetation is planted to stabilized the surface and make the site more attractive. Sump pumps collect any fluids that filter through the landfill either from rain water or from waste leakage. This liquid is purified before it is released. Monitoring wells around the site ensure that the groundwater does not become contaminated. In some areas where the water table is particularly high, above-ground storage may be constructed using similar techniques. Although such facilities are more conspicuous, they have the advantage of being easier to monitor for leakage.

Langmuir, Irving (1881-1957)

American chemist and physicist

Irving Langmuir was a renowned chemist, physicist, and industrial researcher who worked at the General Electric Company for more than forty years. In addition to developing important improvements to the light bulb and other devices, he made significant contributions to the understanding of chemical forces and reactions that occur at the boundaries between different substances. For his work in this field of research, known as surface **chemistry**, Langmuir received the 1932 Nobel Prize for chemistry. He was only the second American—and the first scientist employed as an industrial researcher—to attain such an honor.

Langmuir, the third of four sons, was born on January 31, 1881, in Brooklyn, New York, to Charles Langmuir, an insurance executive of Scottish descent, and Sadie Comings Langmuir, the daughter of an anatomy professor and a descendant, on her mother's side of the family, of settlers who came to America on the Mayflower. Langmuir received part of his secondary-school education in France, where his father had been posted for several years. In 1903, Langmuir completed his baccalaureate degree in metallurgical engineering at the Columbia University School of Mines, where he had studied chemistry, physics, and mathematics with equal interest. Thereafter, Langmuir went to the University of Göttingen in Germany to pursue his postgraduate degree. There he studied under noted physical chemist and future Nobel laureate **Walther Nernst**. Langmuir wrote his doctoral dissertation on the chemical reactions between a glowing **platinum** wire and hot **gases** at low pressure; this research formed the foundation for later work at the General Electric Company (GE). Upon returning to the United States in 1906, Langmuir taught chemistry at the Stevens Institute of Technology in Hoboken, New Jersey.

By 1909, Langmuir had become dissatisfied with the lack of research opportunities at the Stevens Institute and applied for a summer job at the GE research laboratory in Schenectady, New York. He began his career with GE at a time when the laboratory gave its professional scientists the freedom to pursue personal research interests whether or not they related to company goals. Langmuir, who valued professional and financial success as well as scientific inquiry and accomplishment, was well suited for work at GE, and his summer position stretched into a lifelong career. He became known among his colleagues for his analytical mind, creativity, ambition, and excellent research skills.

Langmuir's first assignment at the laboratory was to help perfect a new type of electric lamp (light bulb), one that used a **tungsten** metal wire as a light emitter instead of the more fragile **carbon** filament pioneered by **Thomas Edison** and others. Scientists at the time thought that the tungsten filament would work best if a vacuum were created in the **glass** bulb. Over time, however, the tungsten filament became brittle when set aglow by electric current; it also blackened the inside of the bulb. Langmuir began to study the problem, and within four years, he found that the lamp's lifetime and efficiency could be improved greatly by filling the bulb with a mixture of inert gases (nitrogen and **argon**, for example) and by using a coiled filament. The result was a more energy-efficient and longer-lasting light bulb, for which Langmuir received a patent in 1916. The new light bulb design was extremely profitable for GE and saved its customers millions of dollars on their electric bills.

Early in his tenure at GE, Langmuir also studied vacuum tubes, which were becoming increasingly important to developments in radio broadcasting and the control of electrical power—innovations that would revolutionize society in Langmuir's lifetime. His research led him to invent an enhanced vacuum pump that was one hundred-fold more powerful than any existing vacuum pump. The new pump greatly improved the manufacture of vacuum tubes and was widely used by industry. Langmuir's other technical contributions include a **hydrogen** welding torch, an improved electric stove, and a new kind of gauge for measuring gas concentrations.

In later years at GE, Langmuir continued his research in several other important areas. He developed theories explaining chemical reactions of gases at high **temperature** and low pressure. He also devised useful explanations about the structure of the **atom**. Langmuir's studies of electricity's effects on gases prompted him to coin the term ''**plasma**'' for the unstable mixture produced when gases are charged with large amounts of **electricity**. Langmuir's plasma research paved the way for scientific progress in physics, astrophysics, and thermonuclear fusion (atomic reactions used in nuclear weapons and energy).

Langmuir's most acclaimed work was perhaps in the field of surface chemistry. He studied and defined principles of **adsorption**, the phenomenon he observed in the tendency of gas molecules to cling in a single layer to surfaces of **liquids** or **solids**, and in the behavior of thin films of oil on the surfaces of liquids. Langmuir's body of experimental techniques and theories had applications in many fields, including biology, chemistry, and optics. For his contributions in surface chemistry, Langmuir won the Nobel Prize in 1932. In that same year he became assistant director of the GE laboratory, the position he held until he retired in 1950.

Beginning in 1938, Langmuir increasingly turned his attention to atmospheric science and meteorology, lifelong interests he pursued even after his retirement. During World War II, he investigated methods of aircraft de-icing and invented a machine to produce smoke screens that would shield troops from enemy observation. Later, as head of Cirrus, a joint program of the United States Army Signal Corps, the Navy, and the Air Force, Langmuir helped develop ways of seeding clouds with dry ice and **silver** iodide to make rain and snow.

Langmuir married Marion Mersereau in 1912 and adopted two children, Kenneth and Barbara. In his leisure time, he hiked, sailed, and piloted his own plane, once observing a solar eclipse from an altitude of 9,000 ft (2,730 m). His interest in the world extended well beyond the realm of atoms and molecules to mountain climbing and classical music. In 1935 he ran (unsuccessfully) for a seat on the city council of Schenectady, New York. He actively supported wilderness conservation and control of atomic **energy**. Achieving celebrity in his own time, Langmuir was also a popular public speaker. He received numerous awards and medals, fifteen honorary degrees, and sixty-three patents. He wrote more than two hundred scientific papers and reports from 1906 to 1956, and since 1985, his name has been honored as the title of the American Chemical Society's journal of surfaces and colloids. Upon his death from a heart attack in Massachusetts on August 16, 1957, the *New York Times* ran his obituary on the front page.

Irving Langmuir.

LANTHANUM

Lanthanum is the third element in Row 6 of the **periodic table**. It is sometimes classified as a **rare earth element**, a group of elements that follow it in Row 6 and are also known as the lanthanides. Lanthanum's **atomic number** is 57, its atomic **mass** is 138.9055, and its chemical symbol is La.

Properties

Lanthanum is a white metal that is both ductile and malleable. It is relatively soft and can be cut with a sharp knife. The element's melting point is 1,688°F (920°C), its **boiling point** is 6,249°F (3,454°C), and its **density** is about 6.18 grams per cubic centimeter. Lanthanum is chemically very active, reacting with both cold **water** and most acids. It also reacts with **oxygen** in moist air.

Occurrence and Extraction

Lanthanum is a relatively common element with an abundance of about 18 parts per million in the Earth's crust. Although it is about as abundant as **copper** and **zinc**, it is much more difficult to obtain than those two elements. Copper and zinc often occur in large deposits that are easily mined, while lanthanum occurs in a variety of minerals that are spread widely throughout the earth. Its most common minerals are monazite, bastnasite, and cerite. Lanthanum is separated from other **rare earth elements** by an extended series of processes based on the differential **solubility** of the elements' compounds.

A magnified view of lanthanum aluminate crystal. ***(Photograph by Michael W. Davidson, National Audubon Society Collection/Photo Researchers, Inc. Reproduced by permission.)***

Discovery and Naming

Credit for the discovery of lanthanum is usually given to the Swedish chemist **Carl Gustaf Mosander**. Mosander was very much interested in an unusual black rock found near the town of Bastnas, Sweden, in the 1830s. Over the next 60 years, chemists discovered a total of seven new elements in that rock, one of them being lanthanum, discovered in 1839. The element was named after the Greek word *lanthanein*, meaning "to hide."

Uses

One of the oldest use of lanthanum metal is in the production of misch metal, an **alloy** that produces sparks when struck. Misch metal is used to make the flint in cigarette lighters. Compounds of lanthanum are used to make phosphors, special kinds of **glass**, and optical fibers.

LASERS

A laser is a device that produces a very intense beam of light with properties that make it an essential piece of equipment across the whole spectrum of science.

Normal "white" light is a mixture of different wavelengths, each wavelength associated with a specific frequency and **energy**. This is why white light separates into a "rainbow" when it is beamed through a prism. Each **color** in the rainbow is a different wavelength of light, with short wavelength blue light at the energetic end of the spectrum and long wavelength red light at the low-energy end. Lasers are different because they produce light of only certain wavelengths, all of it traveling in the same direction. The light is also *coherent*, which means all the peaks and troughs of the different waveforms match up peak to peak and trough to trough. They are like soldiers that are not only marching shoulder to shoulder and at the same speed, but are in step too.

Lasers need three things to make them work. The first is a material called the *active medium*, which can be charged up with energy and then made to give up this energy in a controlled way. The second is a way of charging up the active medium with the necessary energy and the third is an optical cavity to amplify the radiation.

The word laser is an acronym for Light Amplification by Stimulated Emission of Radiation and suggests how lasers work. In a **molecule**, each **electron** can be assigned a specific energy. One way to understand this is to imagine the electrons to be whizzing round the nuclei in an "orbital" that is similar to rather like the orbits of the planets revolving around the Sun. Each orbital and therefore each electron within it has a certain energy associated with it. Unlike the planets however, an electron is given extra energy it can jump from one orbital to one with higher energy.

In most materials in their normal state, there will be far more electrons in the unexcited lower **energy level** orbitals then there are in the more energetic excited ones. If energy that matches an energy gap between these different levels is put in, the electrons are excited from low energy state orbitals to more energetic excited orbitals. If these excited oribitals are *metastable*, this means the electrons stay in these orbitals for a short period of time before decaying back to their original orbitals. Excitation of electrons to metastable orbitals therefore causes a *population inversion*, because there will be more molecules with electrons populating the higher energy levels than the lower.

The electrons in active mediums give up this extra energy and decay back to their ground states in two ways. The first is called *spontaneous emission*. This involves an electron spontaneously jumping down from an excited orbital, giving off the energy as a particle of light called a **photon**. This photon is equal in energy to the gap between the energy levels, but is given off in a completely random direction. The second way is called *stimulated*, which occurs when a photon with energy that matches the gap in energy between the two levels hits an excited molecule. The photon then effectively "knocks" the electron down from its excited state by causing it to emit a photon. The important thing here is that the emitted photon not only travels in the same direction and with the same energy as the one which caused it to be emitted, but that it is also coherent with it. These identical photons can now travel through the active medium, knocking down more electrons from their excited states. This causes a snowball effect, releasing a whole cascade of coherent photons.

If a mirror is placed at either end of the active medium, an optical cavity is formed, where the light radiation bounces back and forth between the mirrors. As long as the active medium is charged up and there is "room" in the lower energy levels for the excited molecules to decay to, the light will sweep back and forth through the active medium getting *amplified* with each pass. Any photons produced in the cavity that do not travel in line with the mirrors are not reflected back into the active medium and lose energy and fade away. If the mirror at one end of the optical cavity is a partial mirror, it reflects only part of the radiation back into the cavity letting the rest out. The radiation emitted through this mirror is called a laser beam.

One factor determining the output of a laser is the length of the optical cavity. It is only possible for certain wavelengths—the so-called *longitudinal modes*—to exist in the cavity as all others will interfere and cancel each other. A particular active medium will also be capable of supporting only some of the longitudinal modes, and it is the combination of these two factors that shapes the output.

There are several different types of active medium, which the type used gives the laser its name. Gas lasers use **gases** such as **argon**, **carbon** dioxide, **neon**, and **nitrogen**. Solid state lasers can use crystals (ruby for example), **glass** (Nd^{3+} in silicate glass, for example) or **semiconductors**. Another type uses a dye as the active medium and hence is called a dye laser. These are important in **chemistry** because it is very easy to change the frequency of the laser output and select a desired frequency for a particular experiment.

The nature of the active medium also determines how the energy to cause the population inversion is supplied. An electrical discharge supplies the energy to a gas laser as **electricity**. Optical pumping, where a light is shone on the medium, is used in crystal and glass lasers. Chemical lasers work by reacting two species (such as **hydrogen** and **chlorine** gases) together to form products with an inverted population. Hybrid electrical discharges work in the same way, but an electric discharge excites the products first, causing them to react.

Lasers operate in two modes. The first is called continuous wave and occurs because the molecules remain in the upper, metastable excited levels longer than in the lower. This means there is always room in the lower levels where the molecules can decay. These lasers can then operate continuously. The second mode of operation is pulsed operation. In this mode the upper level is pumped up before the lower level has had time to empty and, consequently, there is no room in the lower levels for the metastable states to decay to. The radiation from these lasers are therefore produced in short pulses. With both modes of operation, there are advantages and disadvantages that dictate where these different types of lasers are used.

In many applications pulses of an extremely short duration are required. There are several ways of achieving this. For a lower limit of nanoseconds (a one-thousand millionth or 10-9 of a second), Q switching is used. This uses an optical switch, which ''spoils'' the optical cavity by blocking one of the mirrors. This allows the active medium to charge up, but there is nowhere for the energy to go. When a very short electric pulse is applied to the switch, the cavity is momentarily restored and the radiation rushes out in one big pulse.

For even shorter pulses, mode-locked lasers are used. One type of mode locked laser contains a dye cell in the optical cavity, which absorbs light at the operational frequency of the laser. Normally the dye absorbs all of the energy of the laser and nothing happens. If by chance, however, the active medium releases a big pulse, it forces its way through the dye by saturating it with so much light that it cannot absorb any more and allows the rest of the pulse through. This pulse is then free to bounce back and forth in the cavity, soaking up all the energy from the active medium. When the big pulse is not traveling through it, the dye absorbs all other radiation, and so the only output of the laser is the big pulse. The frequency of the pulses in such a laser is fixed by the time it takes the pulse to complete the passage from one end of the cavity to the other. These pulses can be so short that quantum effects such as the uncertainty principle come into play and mean that instead of just one specific frequency, there are many. This can be solved to an extent by beaming in radiation from another laser which singles out just one of the possible frequencies present. These pulses can be even shorter than femtoseconds (a femtosecond is a thousand-million-millionth or 10-15 of a second).

Lasers are extremely useful in **spectroscopy**, because they combine a focusable beam with a precise frequency and high energy. In absorbtion spectroscopy, a beam is split in two, one part being sent through the sample, the other going straight to a detector. By comparing the intensity of the two beams it is possible to see whether the one that passed through the sample was absorbed by it to any extent. By doing this at a number of different frequencies it is possible to get a very accurate abortion spectrum of even the most dilute samples.

Lasers with these extremely short pulses have given rise to hyperfast spectroscopy, in which it is possible to study processes such as bonds breaking step by step. This has huge implications for the study of the incredibly short-lived transition states that are characteristic of so many chemical reactions.

Lasers have also revolutionized Raman spectroscopy. Raman spectroscopy relies on the distortion—called the change in polarisability—of the electron cloud in a molecule when it is exposed to light radiation. Low-intensity normal light produces Raman spectra, but they are very faint. The intense radiation from lasers produces very clear Raman spectra.

Coherent Anti-stokes Raman Spectroscopy (CARS) has been developed on the back of developments in lasers using the nonlinear effects on the polarisability of molecules when placed in an even more powerful field then that used for normal Raman spectroscopy. The fact that the probe in Raman spectroscopy is light means that CARS can be used to look at the chemistry that occurs in hostile and otherwise inaccessible environments such as jet engine exhausts.

Lasers have also revolutionized the separation of chemical isotopes, which had otherwise been extremely difficult, if not impossible. Laser radiation of a very specific frequency picks out a molecule with **isotope** specific features in the spectrum and can then excite just the isotope leaving the rest of the molecules unaffected. The selectively excited isotope can then be captured either by ionizing it with more laser radiation or using the enhanced **reactivity** of an excited molecule to make it undergo a chemical reaction. This technique has facilitated the separation of isotopes of **uranium**, which has had a big impact on both nuclear power and nuclear weapons.

The quantum nature of light, in which it can be described as both a wave and a particle, has made it possible to use lasers to trap and cool individual atoms. The momentum of laser beams can be used to repeatedly ''punch'' atoms to slow them down. The wave nature of light means that beams can interfere with each other, causing regions in space with high and low laser intensity. The combination of these two techniques have made it possible to trap individual atoms. First a laser is used to slow the atoms down, and then they are caught in an ''optical cage'' formed by the interference pattern that arises when many laser beams are crossed at a point in space.

LAURENT, AUGUSTE (1807-1853)

French chemist

Born in 1807, Auguste Laurent was the son of a wine merchant. When he became an adult, he took the political stance as a Republican in an authoritarian society. Therefore, although he conducted brilliant research, he could not gain a respectable position and lived most of his life in poverty. Furthermore, he had an abrasive personality and made enemies out of his colleagues. He is famous for his **nucleus** theory of chemical reaction, which was highly controversial at that time.

Even though Laurent was a student of **Jean Baptiste André Dumas**, Dumas refused to stand behind Laurent's theory. Only a short time later, Dumas came out with his own theory that essentially was the same as Laurent's theory except in terminology. Laurent worked closely with **Charles Gerhardt** and it is difficult to determine who offered what in the collaborations. When Gerhardt tried to resolve the inequalities in molecular weights between physical and organic science, it was Laurent who developed the order of the **solution**. Laurent established that **molecular weight** refers to the total of all atomic weights in a substance, while equivalent weight refers to the weight in any given reaction. At times, equivalent weight would equal molecular, but not necessarily.

After the Revolt of 1848, his politics were no longer an issue and he received a moderate government position. Laurent died of tuberculosis at age 46 in 1853. With his abrasiveness out of the debate, French scientists felt free to look again at his studies and were impressed. Some of those who blocked his advancement in life sought for government support for his widow and orphans.

LAUTERBUR, PAUL C. (1929-)

American chemist

Paul C. Lauterbur invented and developed the use of **nuclear magnetic resonance** (NMR) to create images of organs and other tissues in the human body. **Magnetic resonance imaging (MRI)**, as it is also called, has become an important tool in modern medicine as it offers a method for looking at soft tissues in the body without the use of X rays or surgery.

Paul Christian Lauterbur was born on May 6, 1929, in Sidney, Ohio, to Edward Joseph Lauterbur, a mechanical engineer, and Gertrude Wagner Lauterbur, a homemaker. The oldest of three children, Paul's brother, Edward Jospeph II, died at the age of sixteen. He also has a younger sister, Margaret McDonough.

In an interview with Lee Katterman, Lauterbur said that he spent his entire childhood in Sidney, Ohio. He was interested in all kinds of science, and he credits an aunt, Anna Pauline Lauterbur, a schoolteacher, as an important resource in helping him satisfy his curiosity. He attended the Case Institute of Technology (now part of Case Western Reserve University) in Cleveland, where he majored in **chemistry**. After graduating from Case Institute in 1951, Lauterbur moved to Pittsburgh where he joined a research group at the Mellon Institute. He pursued his interest in organosilicon chemistry, the study of organic compounds primarily composed of **carbon**, **hydrogen**, and **silicon**. The Mellon Institute was affiliated with the University of Pittsburgh, so in addition to his research Lauterbur began to take graduate classes on a part-time basis.

In 1953 Lauterbur was drafted into the U.S. Army and assigned to the service's Army Chemical Center Medical Laboratories in Edgewood, Maryland. For the next two years, he helped establish the army's first nuclear magnetic resonance **spectroscopy** laboratory. NMR machines, only recently available from commercial sources, were proving to be valuable tools for determining the structure of chemical compounds. In nuclear magnetic resonance spectroscopy, a **chemical compound** is exposed to a magnetic field and a radio signal. Certain atoms in the compound, such as hydrogen, carbon 13, and **fluorine** 19, then absorb **energy** in patterns that provide information about the arrangement of these atoms.

After his stint in the army, Lauterbur returned to the Mellon Institute in 1955 on the condition that it buy an NMR machine. He organized an NMR laboratory at the institute and continued his research on organosilicon chemistry and the refinement of NMR as an analytical tool. He also took more classes and in 1962 received a Ph.D. in chemistry from the University of Pittsburgh. In 1963 Lauterbur moved to New York to become an associate professor of chemistry at the State University of New York (SUNY) at Stony Brook. By now, he was concentrating his research on NMR studies, refining ways that the information obtained using this tool could be used to interpret the structure of chemical compounds. Lauterbur spent the 1969–70 school year on a sabbatical at Stanford University in California. The same year Lauterbur was promoted to professor of chemistry at SUNY, Stony Brook.

Lauterbur said in the interview with Katterman that the idea to use nuclear magnetic resonance to create images came to him during the summer of 1971. Until this time, NMR primarily had been applied to studying molecular structures of individual chemical compounds or simple mixtures. The typical output from an NMR spectrograph was graphical data that required interpretation to help deduce chemical structures. Lauterbur determined that NMR technology could be used to peer inside the human body and produce images that might be used to distinguish tissues or to spot abnormalities representing the early stages of illness. He coined the word *zeugmatography* to describe the use of NMR for making images. Lauterbur wanted to distinguish his method of image formation and the physics underlying it from other techniques, such as those using X rays, based on other physical properties and principles. Although he first thought of NMR imaging in 1971, his idea did not appear in scientific literature until 1973 after various patent applications could be filed.

Initially Lauterbur spent time confirming that NMR could be used to make images that were reliable and reproducible. He had to make sure that the magnets necessary for whole body studies could be made. Lauterbur also undertook many experiments testing NMR imaging of biological and nonbiological systems. Lauterbur said there were many skeptics of his proposal at first. Some scientists claimed his ideas violated established physical principles. Others said that while the physics might be right, the technique would never be practical. Still others suggested that the medical profession would never accept the new technology. And Lauterbur remembers many rejected grant applications. ''Fortunately a few were funded,'' he told Katterman, and through it all, he remained confident. ''Once I thought of it,'' he said of NMR imaging, ''it was clear it would work. I just had to work out the details.'' After spending some fourteen years developing the idea at the Stony Brook campus, Lauterbur joined the faculty of the University of Illinois at Urbana-Champaign where he continued his work.

At about the same time that Lauterbur was working on his new NMR techniques, other scientists were developing computerized tomography (CT) scanning, the use of multiple X-ray images to create pictures of two-dimensional ''slices'' of the body, and this process reached practical use in hospitals ahead of Lauterbur's NMR method. In an interview with the *Chicago Tribune* in 1990, Lauterbur said that the effect of CT technology had a mixed impact on his own research. ''The precedent with X rays had both a negative and positive effect on our work,'' said Lauterbur. ''It was negative in that people looked at CT and asked why we need another expensive imaging medium. A lot of analysts and companies said that. But it was positive in persuading people that a big, expensive medical technology could be worthwhile.''

Once it was introduced, magnetic resonance imaging—as Lauterbur's NMR technique is now called—quickly gained wide acceptance as a tool for medical diagnosis. MRI is especially good for contrasting different types of soft tissue clustered together, such as a tumor embedded in healthy brain tissue. It also has the potential for the study of biological function, such as following **metabolism** in the brain or other organs. Lauterbur and others have been working on a type of **microscopy** based on nuclear magnetic resonance that may be able to produce images of individual cells in tissue. With this technique, it is becoming possible to look inside thick, opaque tissues or other materials and see structural details that cannot be seen by light microscopy.

Lauterbur has received many awards recognizing his work with NMR. In 1984 he received the prestigious Albert Lasker Clinical Medical Research Award. The following year the General Motors Cancer Research Foundation presented Lauterbur with the Charles F. Kettering Prize, which comes with a $100,000 award and a gold medal. In 1988 he was awarded the National Medal of Science and in 1989 the National Medal of Technology. In 1990 Lauterbur became the first recipient of the Bower Award, a $290,000 prize given by the Franklin Institute of Philadelphia. Lauterbur has received honorary degrees from Carnegie-Mellon University in Pittsburgh, l'Université de L'Etat à Liège in Belgium, and Nicolaus Copernicus Medical School in Kraków, Poland.

Lauterbur married M. Joan Dawson in 1984 and has one child, Mary Elise, from the marriage. He also has two children, Daniel and Sharyn, from a previous marriage to Rose Mary Caputo.

Antoine-Laurent Lavoisier.

LAVOISIER, ANTOINE-LAURENT (1743-1794)

French chemist

Antoine-Laurent Lavoisier, the father of modern chemistry, was the first scientist to explain how things burn. He developed the first rational system for naming chemical compounds, which is still in use today, and established the practice of *accurate measurement*, which is the basis for all valid quantitative experiments.

Lavoisier grew up in Paris, France, the product of a sophisticated urban culture and a well-to-do bourgeois lifestyle. Lavoisier's mother adored her first-born, and after her premature death, he received equally doting care from his young aunt. The boy was treated to an excellent education at an exclusive school, where he proved to be a brilliant student.

Lavoisier's father was an influential attorney and the vocation of law was a family tradition. Although Lavoisier earned his law degree, he never became intrigued by the profession. Instead, he began studying with some of France's most distinguished scientists in the fields of astronomy, mathematics, botany, geology, and chemistry. By the time he embarked on his scientific career, he had gained entrance into the leading intellectual circles of the day and had been exposed to a great variety of scientific pursuits.

The breadth of Lavoisier's curiosity is reflected in his earliest research. Before he turned twenty-five, he had discovered the composition of the mineral gypsum (plaster of Paris), collaborated in producing a geological atlas of France, and explored the possibility of using street lights in France's large towns. Lavoisier was elected to France's Academy of Sciences in 1768.

Lavoisier's experiments with gypsum illustrate one of his strengths as a chemist—accurate measurement. By carefully measuring the amount of **water** given off when gypsum was heated, Lavoisier showed that the mineral is composed partly of water. Although a handful of earlier scientists had paid careful attention to measurement, most notably **Joseph Black** and **Henry Cavendish**, it was Lavoisier who convinced the majority of chemists that accurate measurements are essential to experimental success and scientific progress.

As a member of the scientific academy, Lavoisier served on many boards and committees that were appointed by the government to improve public welfare. He set up a model farm for applying scientific methods to agriculture; he helped standardize the national system of weights and measures, which laid the foundation for today's **metric system**; and he investigated prison reform and many other topics of public interest.

In 1768 Lavoisier dispelled the ancient notion that earth could be created from water. People thought this was possible because a solid sediment appears when water is heated for several days. Lavoisier conducted a long, tedious experiment to disprove this idea. By carefully weighing the **glass** container and the water it held before and after his test, Lavoisier proved that the sediment was made up of material that had been eaten away from the container during heating.

In 1771 Lavoisier married Marie Paulze, who was barely 14 years old. Madame Lavoisier was not only beautiful but also intelligent. She soon began to collaborate with her husband in his scientific work, translated English works into French for him, and illustrated his works. Although Lavoisier was fairly wealthy, scientists earned little money in those days, and he was forced to invest his funds in a profit-making venture in order to support his research. Unfortunately, he chose to invest in a private agency that collected taxes for the government. (Marie's father was an executive of the firm.) The French peasants viciously hated these tax collectors, who were notorious for gouging extra profits, but the money Lavoisier made from this position allowed him to continue with his scientific research. This job, however, would later contribute to his death.

During the early 1770s Lavoisier began heating substances in air to see whether they would burn. Using two huge magnifying lenses, he placed **diamond** in a ray of sunshine that was filtered through the lenses. The diamond slowly disappeared and **carbon dioxide** gas accumulated. This test proved that diamonds are made of carbon, or at least contain **carbon**. Lavoisier also showed that diamonds would not burn without air. Intrigued by the process of **combustion**, Lavoisier went on to burn other substances such as **phosphorus** and **sulfur**. Again, by carefully measuring the materials and containers, he showed that these elements gain weight when they are heated in air.

According to the accepted theory of the day, a substance called **phlogiston** is released during combustion. But Lavoisier realized that materials were not losing phlogiston, they were combining with a portion of the air, which increased their weight. Around the same time, **Joseph Priestley** had isolated a gas that greatly promoted combustion. He called it dephlogisticated air because it absorbed phlogiston so readily. Lavoisier realized that this gas was precisely the same as that part of the air which reacts with substances during combustion. After repeating and expanding Priestley's experiments, Lavoisier announced his new theory. Stating that phlogiston is required if an object is to burn, he re-named Priestley's gas **oxygen** and argued that air also contains a second gas, which does not support combustion. Although Lavoisier called it azote, it was soon given its modern name of **nitrogen**. Lavoisier's discovery additionally implied that the total weight of the substances taking part in a chemical reaction remains the same before and after the reaction. Today we call this fundamental concept the law of **conservation of mass**. A popular version of this law says that **matter** can be neither created nor destroyed.

In 1775 Lavoisier was put in charge of the government's **gunpowder** manufacturing operation. He and Marie moved to the arsenal, where they lived for many years. There, using his investment earnings, Lavoisier set up a magnificent private laboratory—the best in Europe at that time. One of his most valuable pieces of equipment was a chemical **balance** that could weigh objects with great precision. Leading scientists from France met regularly at Lavoisier's laboratory, and world-renowned figures such as Thomas Jefferson (1743-1826) and Benjamin Franklin also visited.

As an offshoot of his studies of combustion, Lavoisier began to explore the process of respiration. Working with French physicist Pierre Laplace, Lavoisier measured the **heat** given off when guinea pigs and sparrows digest food. These experiments, reported in 1789, showed that animals' **energy** and warmth depends on oxygen intake. Lavoisier and Laplace also demonstrated that nitrogen plays no part in respiration; it is only the oxygen in the air that is needed to support animal life.

As his reputation grew, Lavoisier was recruited to pursue new research avenues. Louis-Bernard Guyton de Morveau (1737-1816), a colleague who was trying to write a history of chemistry, turned to Lavoisier for help. It was Lavoisier who pinpointed the biggest problem—language. For centuries alchemists had deliberately tried to keep their discoveries secret from common people, so they gave new substances absurd names, such as butter of **arsenic** or sugar of **lead**, that were meaningless to the uninitiated. In collaboration with other chemists, including **Claude Berthollot**, Lavoisier developed a new, logical system for naming chemical substances in 1787. They decided that a chemical name should indicate the elements that make up the compound; for example, **hydrogen** sulfide contains hydrogen and sulfur.

In 1789, just two years after introducing this system, Lavoisier published the first truly modern chemical textbook, *Elementary Treatise on Chemistry*. In it he not only stated the *law of conservation of mass*, but also revived the definition of *element*, which had been suggested earlier by **Robert Boyle**, and listed all the substances thought to be elements at that time. Lavoisier emphasized that the list probably contained some compounds that could be decomposed only with advanced scientific techniques. Overall, the list was remarkably accurate.

Anxious to prove himself, Lavoisier often discredited the contributions of other chemists to his work. For example, even though Lavoisier deserves full credit for figuring out the combustion process, he neglected to mention the information he got from Priestley. Similarly, Lavoisier repeated research done by Cavendish, who discovered *inflammable air* and burned it to produce water. When Lavoisier named this gas hydrogen, he failed to point out that Cavendish had performed the original experiments. Possibly, these omissions were not intentional; Lavoisier simply wanted, very badly, to discover an element himself—something he would never accomplish.

With the outbreak of the French Revolution in 1789, Lavoisier's position as an administrator in the government's tax collection agency automatically made him a target of hatred. After being barred from his laboratory, Lavoisier fled his home but was caught and arrested a few months later. When he protested that he was a scientist, not a taxman, he was told: "The Republic has no need of scientists." Lavoisier was charged with ridiculous crimes. Testifying against him was an age-old enemy, Jean-Paul Marat (1743-1793), whom Lavoisier had prevented, with good reason, from joining the Academy of Sciences. Eager for revenge, Marat accused Lavoisier of diluting commercial tobacco and cutting off Paris's air supply by building a defensive wall around the city. On May 8, 1794, Lavoisier was sentenced to death and guillotined.

Désirée Le Beau.

Le Beau, Désirée (1907-1993)

American chemist

Désirée Le Beau was an early pioneer in what is known today as rubber **recycling**. A **colloid** chemist (one concerned with substances made up of tiny insoluble, nondiffusable particles suspended in a medium of different matter), she developed methods of reclaiming natural and synthetic rubbers, primarily from old tires, to be used to produce new products, and held several patents for these processes. Le Beau made significant strides in this field during World War II, when American rubber sources were unavailable due to the conflict.

Born to Phillip and Lucy Le Beau in Teschen, Austria-Hungary (now Poland), on February 14, 1907, Le Beau worked briefly in Paris before Hitler's aggressions in Europe prompted her to relocate to the United States. Le Beau's lifelong career began as a college fluke, when she inadvertently sat down in the **chemistry** laboratory instead of her intended pharmacy laboratory. By the time she realized the mistake, she knew she wanted to study chemistry. Le Beau first completed undergraduate studies at the University of Vienna, then earned her Ph.D. in chemistry in 1931 from the University of Graz in Austria, where she minored in physics and mathematics.

In 1932 Le Beau accepted a position with the Austro-American Rubber Works in Vienna. She then moved to Paris in 1935 to join the Société de Progrès Technique as a consultant. The following year, she came to the United States and filled a post as Research Chemist with Dewey & Almy Chemical Company in Massachusetts. From 1940 to 1945, Le Beau served as a Research Associate at Massachusetts Institute of Technology's Department of Chemical Engineering and Division of Industrial Cooperation. In 1945, she was appointed Director of Research at Midwest Rubber Reclaiming Company in Illinois, the world's largest independent reclaiming company at the time, where she studied the structures of natural and synthetic rubbers and clays. There, Le Beau developed a tie pad for railroads using reclaimed rubber. One of her patents was for this process, which she developed in 1958. She also held patents for a method of producing reclaimed rubber in particulate form (formed of tiny separate particles), and for methods of reclaiming using certain amines and acids. In 1950 she was named a Currie lecturer at Pennsylvania State College.

Le Beau married Henry W. Meyer on August 6, 1955. In addition to her contributions in science, Le Beau spoke English, French, German, Swedish, and Latin. She embraced as hobbies music, interior decorating, swimming and horseback riding. Her many honors included the 1959 Society of Women Engineers Achievement Award for "her significant contributions to the field of rubber reclamation." Le Beau was also the first woman to chair the American Chemical Society's Division of Colloid Chemistry, as well as the first woman to chair the Society's St. Louis section. She was elected a Fellow of the American Institute of Chemists, and she authored numerous publications on reclamation.

LE BEL, JOSEPH (1847-1930)

French chemist

Joseph Le Bel's most famous contribution to **chemistry** was shared with another chemistry. Le Bel and **Jacobus Henricus van't Hoff** were both working on the same problem of light rotating power of compounds. Although they were students of Wurtz, they had little interaction with each other; each worked out the **solution** and published their findings essentially at the same time. Working against Le Bel's receiving primary credit for the discovery was the fact that van't Hoff published his **paper** shortly before Le Bel published his. However, van't Hoff produced no more papers about **stereochemistry** and left the field for **physical chemistry** shortly after the publication. Le Bel, on the other hand, conducted research in stereochemistry for the remainder of his career.

Le Bel was born in 1847 into a family who made a fortune in petroleum. As a young man, he was involved in his family's business and attended college at the École Polytechnique. In 1873, he became a student of Wurtz, working in his laboratory. At that time, Wurtz was working on the problem of rotation in optical compounds.

Le Bel published his theory in 1874. His discovery stressed that a three dimensional model was more appropriate in explaining the makeup of a light rotating organic compounds. **Louis Pasteur** had shown that light sensitive crystals tended to have mirror images. **Friedrich Kekulé** showed the similarity between the structure of organic compounds and crystals. Le Bel had studied Louis Pasteur's findings and van't Hoff was a student of Kekulé. Starting from these already similar theoretical foundations, their discoveries coincided. The overlap of their research explains that if an optically active **carbon** compound is asymmetric, a mirror image of the compound should be expected. Such a compound is asymmetric when the substituents around the carbon **atom** are different. The coincidence of their similar research is not so mysterious when one considered that they concurrently had the same engrossed teacher. Furthermore, the field was ready for such a solution since the classic structural theory was so ineffective in explaining optical compounds.

Even while he was doing research, Le Bel still had responsibilities in his family business and divided his time between business and research. At age 42, he sold his business and retired on the capital. He was then freed to spend all his time on research. In 1891, he reported another discovery concerning optical activities, but he could not substantiate it. This discovery that pentavalent **nitrogen** might contain optical compounds was later experimentally verified by the English chemist William Pope. In his later years, Le Bel pursued other interests, including paleontology. He also conducted botany studies in his own garden.

Le Bel was not a typical researcher. Although he liked to do research, he never sought nor was he offered an academic position. Although many researchers followed his research, he did not have any students to carry on his legacy. Le Bel did not receive many accolades for his achievements. However, he was elected President of the French Chemical Society. Unfortunately, some members delayed his membership in the Academy of Sciences due to personal grievances against him. He finally was accepted in 1929, a year before his death.

LE CHÂTELIER, HENRY LOUIS (1850-1936)

French chemist

Le Châtelier's relatives and close family friends included many engineers and scientists. Members of this circle were involved in **lime** and cement production, railway construction, mining, and **aluminum** and steel manufacturing. France's leading chemists often visited the Le Châtelier home, and all of the Le Châtelier children pursued scientific or technological careers. In later life, Le Châtelier said that his family had strongly influenced his research pursuits.

After serving as an army lieutenant in the Franco-Prussian War of 1870-1871, Le Châtelier finished college at the Ècole Polytechnique, earning a degree in science and engineering in 1875. Two years later, he became a **chemistry** professor at the Ecole des Mines, where he began research on cement, **ceramics**, and **glass**. Some of his experiments required the measurement of very high temperatures, for which the equipment available at the time was inadequate. To measure these temperatures more accurately, Le Châtelier developed the thermocouple, an instrument consisting of a **platinum** wire and a platinum-rhodium **alloy** wire that measures **temperature** as a function of the difference in voltage between the two wires. He also introduced the use of known boiling and melting points as standards for calibrating thermocouples. Around the same time, Le Châtelier developed an optical pyrometer, another temperature measurement device. Although other methods have replaced the pyrometer, Le Châtelier's equipment was useful at the time, and scientists continue to employ thermocouples in high-temperature research.

In the early 1880s, the French government began investigating the cause and prevention of mining disasters. As an Ecole des Mines professor, Le Châtelier took part in research on gas explosions, which involved studying the ignition temperature, flame speed, and other conditions affecting explosions of **methane**, **carbon** monoxide, and **hydrogen** mixtures. Le Châtelier's results, applied to acetylene **combustion**, enabled other chemists to develop the oxyacetylene torch now used for cutting and welding steel. During this period, Le Châtelier also helped discover safer **explosives** and made several improvements to the miner's safety lamp, which had been invented in the early 1800s by **Humphry Davy**.

Le Châtelier applied his earlier work on gas explosions to the chemical reactions occurring in blast furnaces, which are used to manufacture steel, and determined why certain **gases** were unexpectedly present in the furnace exhaust. By explaining blast furnace chemistry, Le Châtelier enabled industrial engineers to develop furnaces that could reach higher temperatures by preheating the combustion air with hot exhaust gases.

Le Châtelier's scientific experience culminated in the discovery for which he is best known today—Le Châtelier's principle. Announced in 1884, the principal states that when a system is in equilibrium and one of the factors affecting it is changed, the system will respond by minimizing the effect of the change. Essentially, the principle predicts the direction that a chemical reaction will take when pressure, temperature, or any other condition is altered. By employing *Le Châtelier's principle*, scientists were able to maximize the efficiency of chemical processes. For example, **Fritz Haber** utilized the principle to develop a practical process for synthesizing **ammonia** from **nitrogen** and hydrogen. Le Châtelier himself had tried this but was unsuccessful.

For the rest of his career, Le Châtelier continued teaching. In addition to his position at the Ecole des Mines, he held posts at the College de France and at the Sorbonne. After working for the French government during World War I, Le Châtelier retired from the Ecole des Mines in 1919 at age 69.

Le Châtelier's principle

The modern interpretation of Le Châtelier's principle states that if a system at equilibrium, which occurs when the rate of the forward reaction equals the rate of the reverse reaction, is disturbed, the system will shift so as to relieve the disturbance. Thus, Le Châtelier's principle asserts that the equilibrium of a system can be changed by variations in **temperature**, pressure, **volume**, or in **concentration** of the reactants or products.

Henri Louis Le Châtelier was a French chemist who became interested in the conditions needed for equilibrium in chemical reactions after examining some unexpected results at a mine's furnace. The reaction of **iron** oxides with **carbon** monoxide in the furnace was believed to produce iron and **carbon dioxide**. However, the **gases** emerging from the furnace contained a considerable amount of **carbon monoxide**. Le Châtelier realized that the carbon monoxide formed carbon dioxide and carbon in a reversible reaction, for which the iron oxides acted as a catalyst. In 1888, he stated what became known as the Le Chatelier principle: every change in one of the factors of an equilibrium occasions a rearrangement of the system in such direction that the factor in question experiences a change in the sense opposite to the original change.

Suppose you remove a substance from, or add a substance to, a system in chemical equilibrium to alter the concentration of the substance. The system will shift in order to restore the initial concentration of the removed or added substance. For example, consider the following system in chemical equilibrium:

$2NO_{(g)} + O_{2(g)}\ 2NO_{2(g)}$

If more **oxygen** gas is added to the system, the concentration of oxygen will increase. Le Châtelier's principle predicts that the reaction will shift in the forward direction to produce more **nitrous oxide**. This will alleviate the stress put upon the system by consuming the added oxygen gas. In general, when more reactant is added to, or some product is removed from, an equilibrium mixture, the reaction shifts in the forward direction, and more products are produced. And when more product is added to, or some reactant is removed from, an equilibrium mixture, the reaction shifts in the reverse direction, and more reactants are produced.

Changing the pressure of a chemical system in equilibrium can be attained by changing the volume. **Boyle's law** states that at a fixed temperature, the pressure multiplied by the volume equals a constant so, for example, doubling the pressure is equivalent to halving the volume. Le Châtelier's principle states that altering the pressure (or the volume) of a system in equilibrium will cause the system to shift in the direction required to reduce the pressure of the system to the equilibrium pressure. The pressure is reduced by the equilibrium shifting in the direction that produces fewer moles of gas. Increasing the pressure (by decreasing the volume) of the above chemical system will cause the equilibrium to shift in the forward direction because there are two moles of product and three moles of reactant.

There are two important points to remember when changing the volume and pressure of a system in equilibrium. First, **liquids** and **solids** are nearly incompressible and are therefore not affected by pressure changes. Thus, they are ignored when considering the effects of pressure on an equilibrium system. Second, if in the balanced chemical equation there are the same number of moles on both sides of the reaction, a pressure or volume change will not affect the equilibrium.

Altering the temperature can also affect a system in equilibrium. For an endothermic reaction (heat is required), the amounts of products are increased at equilibrium by an increase in temperature. For an exothermic reaction (heat is produced), the amounts of products are increased at equilibrium by a decrease in temperature. Consider the following equation:

$N_{2(g)} + 3H_{2(g)}\ 2NH_{3(g)} + 91.8kJ$

An increase in temperature will cause this reaction to shift in the reverse direction, producing more **nitrogen** and **hydrogen**, because the reverse direction uses the excess **heat** supplied. An increase in the forward direction would produce even more heat since the forward reaction is exothermic. Therefore, the shift caused by a change in temperature depends on whether the reaction is exothermic or endothermic.

Adding a catalyst to a system in equilibrium has no effect on the equilibrium. A catalyst is a substance that increases the rate of a reaction but is not consumed by the reaction. Therefore, adding a catalyst merely speeds up the attainment of equilibrium.

Lead

Lead is the heaviest element in Group 14 of the **periodic table**, a group often known as the **carbon** family. Lead has an **atomic number** of 82, and atomic **mass** of 207.2, and a chemical symbol of Pb.

A dish constructed of lead. ***(Photograph by Rich Treplow, Photo Researchers, Inc. Reproduced by permission.)***

Properties

Lead is a heavy, soft, gray solid that is both ductile and malleable. The metal has a shiny surface when it is first cut, but it slowly tarnishes and becomes dull as it reacts with **oxygen** in the air. Lead is easily bent, cut, pulled, and otherwise worked to produce specific shapes. The melting point of lead is 621.3°F (327.4°C), its **boiling point** is about 3,180°F (about 1,750°C), and its **density** is 11.34 grams per cubic centimeter. Lead is not a good conductor of **electricity**, **heat**, sound, or vibrations. It is a moderately active metal that dissolves very slowly in **water** and in most cold acids.

Occurrence and Extraction

Lead is relatively common in the Earth's crust with an abundance of about 13-20 parts per million. It rarely occurs as a free element, and is found most commonly as a compound in the form of galena (lead sulfide; PbS), anglesite (lead **sulfate**; $PbSO_4$), cerussite (lead carbonate; $PbCO_3$), and mimetite (a mixed lead chloride arsenate; $PbCl_2 \cdot Pb_3(AsO_4)_2$).

The largest producers of lead in the world are Australia, China, the United States, Peru, Canada, Mexico, and Sweden. In the United States, more than 90% of all the lead produced comes from a single state, Missouri. Lead is extracted from its ores by first converting the ore to lead **oxide** and then heating the oxide with **charcoal** (pure carbon): $PbO_2 + C \text{ —heat→ } CO_2 + Pb$. The lead produced by this process is usually not very pure and can be further refined electrolytically. A significant amount of lead is now being recovered through **recycling** programs. For example, old car **batteries** were once just thrown away. Now they are sent to recycling plants where lead can be extracted and used over and over again.

Discovery and Naming

Lead is one of the oldest elements known to humans. One of the oldest lead objects known is a small statue found in Egypt from the First Dynasty, about 3400 B.C.. The metal is mentioned in many ancient manuscripts, including the Bible. No one is quite sure how lead got its modern name. At one time, it was known by the Latin term *plumbum*, from which it gets its chemical symbol.

Uses

Throughout history, lead has been used for many important applications, including water and sewer pipes, cable coverings, type metal, paints, food and tobacco wrappings, and **gasoline** additives. Over the past few decades, however, evidence of the toxic effects of lead on humans and other animals has begun to accumulate. Today, the element is recognized as a serious environmental hazard and, as a result, the lead industry has begun to undergo a dramatic change. Many of the products for which lead was used in the past are now being made of other elements.

On the other hand, some traditional uses of lead have not declined at all and, in many cases, have actually increased. The best example is the lead storage battery present in nearly every car and truck. Researchers have been looking for a substitute for the lead storage battery that would provide electrical current efficiently, but that would not pose an environmental hazard. So far, they have been largely unsuccessful.

A small percentage of lead is used to make compounds with many different uses, including insecticides, waterproofing materials, varnishes, stains for **glass** and **ceramics**, high **explosives**, specialized industrial paints, photography, cloud seeding, and **semiconductors** and photovoltaic devices.

Health Issues

Low levels of exposure to lead can produce symptoms such as nausea, vomiting, extreme tiredness, high blood pressure, and convulsions. Higher levels of exposure can cause brain damage that can result in mental retardation. Lead poisoning is usually not a serious problem for the general public, although certain groups are at special risk. For example, people who work in factories where lead is used may inhale lead fumes and accumulate increasingly large amounts of the metal in their bodies. Young children are also at risk for lead poisoning because they tend to eat materials such as dirt, **paper**, chalk, and paint chips that may contain lead compounds.

LEBLANC, NICOLAS (1742-1806)

French chemist

Nicolas Leblanc was born in Ivoy-le-Pré France. Emulating his guardian, a doctor, Leblanc studied medicine and, in 1780, became physician to the Duke of Orléans. Supported financially by the Orléans family, Leblanc was able to devote time to research. He studied crystallization and then developed a commercially important process to produce soda (sodium carbonate) out of **salt** (sodium chloride). The Leblanc process was crucial to industrial-chemical progress in the nineteenth century, as economical supplies of soda ash were necessary for the widespread manufacture of both soap and **glass**.

In 1790, Leblanc, the Duke of Orléans, a chemist named J.J. Dizé, and Henry Shée established a company to produce soda ash using the Leblanc process. Between 1791 and 1793 they built a factory at St. Denis, near Paris, France. By then, however, the French Revolution was well underway. When the Duke was guillotined in 1793, the factory was closed and later nationalized in 1794. The prize Leblanc had won for developing the process was never paid. The Committee of Public Safety forced Leblanc to publish details of his process without remuneration.

Leblanc finally regained control of his factory in 1800, but by then it needed extensive and expensive renovations. Leblanc was unable to raise the necessary capital, and the settlement of his claims against the French government in 1805 was much less than he had hoped for. Financially ruined, Leblanc committed suicide in 1806. His process, however, remained in use until it was finally replaced by **Ernest Solvay's** process in the 1860s.

LEE, YUAN T. (1936-)

American chemist

Yuan T. Lee, a professor of **chemistry** at the University of California, shared the 1986 Nobel Prize in chemistry with **Dudley Herschbach** and **John C. Polanyi** for their work in the field of chemical **kinetics**, also called reaction dynamics. Herschbach and Lee worked together on their study of what happens when individual particles of **matter** collide and chemically change. Lee's major contribution to the prize-winning effort was to improve the instruments. Before Lee worked on the apparatus, Herschbach's research was very limited in scope. With the changes Lee made, the equipment could be used to study almost any chemical change. The new apparatus is so useful it is described as a "universal" machine, and a laboratory "workhorse" by colleagues at the University of California.

Yuan Tseh Lee was born in Hsinchu, Taiwan, on November 29, 1936. Lee's father was an artist and art teacher; his mother taught elementary school. When the Japanese occupied Taiwan during World War II, the people of Hsinchu were sent to live in the mountains; there Lee could not go to school. After the war he finished elementary school. Lee graduated from Hsinchu High School in 1955. His academic standing was so high that he won admission to the National Taiwan University without taking an entrance examination. Lee majored in chemistry at the University. He was influenced by a biography of **Marie Curie**. Lee received his B.S. in 1959. Two years later he earned a M.S. degree from the National Tsinghua University. In 1962 Lee attended the University of California at Berkeley. He married Bernice Chinli Wu, a childhood acquaintance, in 1963. They have three children, a daughter and two sons. In 1968 Lee moved to the University of Chicago as an assistant professor. He became an associate professor in 1971, and a full professor in 1973. The next year he returned to the University of California at Berkeley as a professor of chemistry. He continued his research at the Lawrence Berkeley Laboratory. In 1974 Lee also became a United States citizen.

Yuan T. Lee. *(AP/Wide World. Reproduced by permission.)*

Lee received his Ph.D. in 1965 from the University of California at Berkeley, then stayed on as a postdoctoral student for a year and a half. He designed and constructed the apparatus needed to facilitate his research into the reaction that occurs when a stream of neutral **hydrogen** atoms collides with a stream of positively charged **nitrogen** ions. Since 1963 Herschbach had been doing similar research at Harvard, but he did not have the apparatus he needed for very sophisticated experiments. In 1967 Herschbach invited Lee to join him. In less than a year at Harvard, Lee succeeded in designing the intricate universal crossed-molecular-beam apparatus that would provide the means for Herschbach's research. Herschbach's experiments were not entirely new. Physicists had used streams of colliding **subatomic particles** to study the nature of matter for many years. But he was the first to explore chemical reactions in depth by this method. With Lee's contribution they could analyze the results of collisions between individual particles to determine the products and the **energy** of the products in detail never before possible. Herschbach had tried hot wire detectors but only a certain few reactions could be evaluated by this method. Lee adapted a very sensitive **mass** spectrometer as a detector to replace the hot wire. A mass spectrometer has a slit which allows particles of different mass to be sorted and detected electronically as the slit is moved across a field of bombarding product particles. The product particles are what result when the molecular beams cross paths and collide. The results of scanning the products are recorded. With the sophisticated, specially adapted spectrometer, not only were the results better from each test reaction, but many more reactions could now be studied.

Lee was awarded the Ernest Orlando Lawrence Memorial Award of the United States Energy Research and Development Agency in 1981. Two years later he received an award from the Atomic Energy Commission. In 1986 he received the **Peter Debye** Award from the American Chemical Society, the National Medal of Science from the National Science Foundation, and shared the Nobel Prize in chemistry with Herschbach and Polanyi "for their contributions concerning the dynamics of elementary chemical processes." The citation for the Nobel Prize describes the method used by Herschbach and Lee as "one of the most important advances within the field of reaction dynamics." Lee is described as a modest, quiet person, extremely talented and deeply committed to his research. A colleague at the University of California says Lee "combines two qualities rarely found in the same scientist: brilliance and attention to detail." He adds, Lee is "the kind of guy who comes into the lab at midnight to see how things are going." Herschbach describes Lee as "the Mozart of **physical chemistry**" because he has "a precise touch" when he is at work in the laboratory. In his presentation at the Nobel award, Lee said that "the experimental investigation of elementary chemical reactions is presently a very exciting period.... In the remaining years of the twentieth century, there is no doubt that the experimental investigation of the dynamics and mechanics of elementary chemical reactions will play a very important role."

LEHN, JEAN-MARIE (1939-)

French chemist

Jean-Marie Lehn shared the 1987 Nobel Prize in **chemistry** for his contributions to the field of supramolecular or host-guest chemistry, particularly for his development of the crown ethers known as a cryptands and his work on synthesizing artificial enzymes. His work in this area has numerous applications in medicine and industry and in chemical and biochemical research.

Jean-Marie Pierre Lehn was born to Pierre and Marie (Salomon) Lehn on September 30, 1939, in Rosheim, in the Alsace region of northeastern France. His father had an unusual career combination as both a baker and an organist. After completing a diverse curriculum of chemistry, classics, and philosophy at the Collège Freppel in 1957, Lehn continued his studies at the University of Strasbourg, where his attention was turned to **organic chemistry** by Guy Ourisson, one of his professors.

After earning his bachelor's degree in 1960, Lehn joined Ourisson as a researcher at the National Center for Scientific Research (CNRS) in France. During his six-year tenure there, he researched the chemical and physical properties of the enzymes used to synthesize vitamin A, and spent a year in the United States as a visiting professor at Harvard University, working with **Roald Hoffmann** on quantum mechanics and with **R. B. Woodward** on the composition of vitamin B_{12}. He received his Ph.D. from the University of Strasbourg in 1963.

Lehn returned to the University of Strasbourg as an assistant professor of chemistry in 1966. By this time, he had already begun studies of the human nervous system in order to determine biological and chemical relationships within the human body and was making great strides in the area of physical organic chemistry. It was during this period that he coined the term "supramolecular chemistry," later called "host-guest chemistry." Lehn defined supramolecular chemistry as a process whereby molecules recognize one another and selectively connect or bond, though the movement occurs quickly and the structural **bonding** is not permanent. In 1969, Lehn accepted an associate professorship at the Université **Louis Pasteur**, which became a full professorship in 1970.

Several years earlier, the American chemist **Charles John Pedersen** had published the results of his research on molecules known as ethers, in which **carbon** and **oxygen** atoms are strung together to form a crown-like shape. He referred to these molecules as crown ethers. Pedersen's work in the early 1960s was quite astounding because it enabled him to achieve so many practical applications in his laboratory. The shape of the crown **ether**, he found, allowed it easily to capture, or bind with, a metal **ion**. By altering the **ring** of atoms, he could create designer crown ether molecules, or hosts, to capture specific ions, or guests.

Lehn expanded upon Pedersen's work by showing that crown ethers could be made three-dimensional by adding layers. The **molecule** now formed a cavity, or crypt, that increased the number of contact points to which the metal ions could adhere, thus making the crown ether more selective with regard to the type of molecule it would capture. Lehn called these structures cryptands. They resembled, he said, chemical "locks" that only particular molecular "keys" would fit. Concurrently, another American chemist, **Donald J. Cram**, was reporting significant findings in molecular selectivity. Cram and Lehn both discovered techniques for synthesizing crown ether molecules into artificial enzymes. Lehn, Pedersen, and Cram shared the 1987 Nobel Prize in chemistry for their work in "elucidating mechanisms of molecular recognition, which are fundamental to enzymic catalysis, regulation, and transport." Through their combined efforts, they have made it possible to create synthetically molecules and enzymes with enormous pharmacological and research applications; one **enzyme** developed by Lehn, for instance, binds with the neurotransmitter acetylcholine. Because of this research, scientists have the capability of "caging" certain toxic materials, thus either rendering them harmless or making them easier to extract from soil or **water**. Host-guest chemistry is also useful in the purification of **metals**.

In 1965, Lehn married Sylvie Lederer, with whom he had two sons, David and Mathias. In addition to listening to music and traveling, Lehn enjoys playing the piano, carrying on the musical tradition of his father. In 1979, Lehn moved to Paris, where he became a professor of chemistry and chair of chemistry of molecular interactions at the Collège de France. Director of the laboratories in both Strasbourg and Paris, he has traveled internationally as a visiting professor and has received a number of awards, including the Paracelsus Prize in 1982, the von Humboldt Prize in 1983, and the Vermeil Medal of Paris in 1989, in addition to the Nobel Prize. He was made chevalier of the Ordre National du Mérite in 1976, chevalier of the Légion d'Honneur in 1983, and officier of the Légion d'Honneur in 1988. Lehn is recognized internationally for his work and has published numerous papers and chapters in books on the subject. He continues to build a distinguished career in chemistry.

LELOIR, LUIS F. (1906-1987)

Argentine biochemist

Luis F. Leloir began a career in medicine but found himself drawn to the relatively more tractable problems posed by **biochemistry**. His early research involved investigations of **fatty acids** in the liver, which led to the discovery of antihypertensives. In a subsquent search for the ''missing link'' in the conversion of **carbohydrates** in the body into **energy**, Leloir discovered a group of substances called sugar nucleotides, which allowed him and others to determine the precise mechanism of energy conversion. Leloir also discovered **glycogen**, which is synthesized along with nucleotides and is the major store of energy in animal cells. Leloir's work with sugar nucleotides won him the Nobel Prize in **chemistry** in 1970.

Luis Federico Leloir was born on September 6, 1906, in Paris, France. Leloir's grandparents on both sides were immigrants to Argentina from France and Spain. When they moved to Argentina they invested in land, which turned to considerable profit as cattle and crops took on great importance in Argentine industry. This money would serve Leloir well later, as it would allow him to follow a career solely in scientific research at a time when such opportunities were very scarce in Argentina.

Leloir's parents, Federico and Hortensia Aguirre Leloir, were in France in 1906 only for a visit, and returned to their home in Buenos Aires, Argentina, when Leloir was two years old. Federico Leloir was educated as a lawyer, though he never practiced in that field. The younger Leloir grew up in a house filled with books on a variety of topics. Later in his life Leloir maintained there was no specific reason for his foray into the field of science, as it was clearly not a family tradition. He described himself as ill-suited to a career in music, sports, politics, or law, but acknowledged that he had a tremendous capacity for teamwork. Leloir completed his primary and secondary education in Buenos Aires, and then enrolled at the University of Buenos Aires to study medicine. He graduated with a medical degree in 1932, followed by employment in the hospital of the university as an intern. He found medicine somewhat limited in terms of the treatment options available at the time, and had no confidence in his own ability to diagnose and treat his patients.

Luis F. Leloir. *(Corbis Corporation. Reproduced by permission.)*

He decided to try a position in research at the Institute of Physiology, still at the university, to help develop new options in treatment for physicians and to work on a Ph.D. degree. He worked under Bernardo Houssay, a Nobel winner in 1947 in the area of adrenal gland research, and consequently developed an interest in biochemistry. His doctoral thesis was written on the influence of the adrenal glands on carbohydrate **metabolism**, and his thesis won the annual prize of the faculty for best thesis. Leloir's relationship with Houssay would last the rest of Houssay's life, and Leloir described him as an intellectual inspiration.

In 1936 Leloir left Argentina to spend a year in Cambridge, England, conducting postdoctoral work in **enzyme** research at the Biochemical Laboratory of Cambridge University. He then returned to Buenos Aires and began research on breakdown of fatty acids in the liver. Eventually, his work led to a collaborative discovery of the peptide hypertensin, so named by this group because of its vasoconstrictive action. Vasoconstriction is the constriction of blood vessels, which causes high blood pressure or hypertension. Another group of scientists at Eli Lilly, a pharmaceutical company in Indianapolis, made a similar discovery around the same time, and named their peptide angiotensin. Both groups used these different names for the same peptide for several years, fighting over which it should be called, until finally a compromise was reached. Today the peptide is known as angiotensin. Later, in

1946, Leloir and his group issued a book based on their research findings in this area called *Renal Hypertension.*

Though averse to political involvement, Leloir was affected by the Argentine government's decision in 1943 to dismiss Houssay from his position at the university. Houssay had innocently signed a letter that was interpreted as antigovernment, and the country's politics at the time were in a state of upheaval. Many others at the university resigned their positions in support of Houssay, and Leloir decided it would be a good time for him to work abroad. He had married Amelie Zuherbuhler in 1943; together they left for the United States with no positions secured. After a short time in New York City, the Leloirs settled in St. Louis, Missouri, where Leloir worked as a research assistant in a biochemistry laboratory at Washington University. Later he moved to the Enzyme Research Laboratory of the College of Physicians and Surgeons of Columbia University in New York.

In 1945 Leloir returned to Argentina to work again under Houssay, who had been reinstated at the university. Leloir, though, had begun to hatch a plan to start a private research institute, and slowly began gathering the necessary team. Houssay was eventually removed from his post again, this time for being "over age," but Leloir had his team assembled, and finally received backing from Jaime Campomar. The owner of a textile firm, Campomar had expressed an interest to Houssay in sponsoring a research institute specifically in the area of biochemistry. Thus the Institute for Biochemical Investigations was begun.

The future of Leloir's institute was in question after Campomar died in 1957. In a bit of a last-ditch effort, Leloir applied to the National Institutes of Health in the United States for funding, and to his surprise obtained it. The institute continued to receive monies from the NIH for several years until rules for granting money to foreign applicants were changed. In 1958 the government of Argentina offered assistance as well, giving the institute a former girls' school for a new home. Further financial backing came a short while later after the formation of the Argentine National Research Council, and the institute became associated with the faculty of the University of Buenos Aires.

Scientists at this time were familiar with the idea that carbohydrates are broken down by the body into simpler sugars for energy. Beginning in the late 1940s, Leloir believed there was a "missing link" in the understanding of this process, and set out to find it. What he eventually discovered was a group of substances, now known as sugar nucleotides, that are responsible for the conversion into energy of sugars stored in the body. The discovery of these substances helped Leloir, among others, to determine specifically the process of carbohydrate conversion into energy. Leloir also found that a complex sugar called glycogen is synthesized with these sugar nucleotides, stored in the liver and muscles, and then broken down by the body into simpler glucose as energy is needed.

For his work with sugar nucleotides, Leloir was awarded the Nobel Prize in chemistry in 1970. He was only the third Argentine to receive the Nobel in any field, and the first in the area of chemistry. He instantly became a national hero, and was later honored as the subject on a postage stamp. Leloir was somewhat leery of the Nobel, telling *Newsweek* that his prize money would be spent on further research, "if I'm ever allowed to work again in the peace and quiet that I'm used to."

Leloir played a major role in the establishment of the Argentine Society for Biochemical Research as well as the Panamerican Association of Biochemical Sciences. Among his memberships were the National Academy of Sciences (United States) and the American Academy of Arts and Sciences. He was elected to membership in the Royal Society of London in 1972, and to the French Academy of Sciences in 1978. In addition to the Nobel, he received prizes and honors from universities all over the globe, including the Gairdner Foundation Award in 1966 and honorary degrees from the Universities of Paris, Granada (Spain), and Tucumán (Argentina). In 1971 he was keynote speaker at a biochemistry symposium held in his honor.

Leloir was known to be courteous and accessible. He has been given credit for performing major scientific research with limited funding. He often used homemade apparatus and gadgets, and encouraged inventions for use in his laboratory. In one instance Leloir constructed makeshift gutters out of waterproof cardboard to protect the library in his research laboratory from a leaky roof. Leloir died on December 2, 1987, in Buenos Aires, leaving his wife, one daughter, and several grandchildren.

LESTER, JR., WILLIAM ALEXANDER (1937-)

American chemist

William Alexander Lester, Jr. has had a distinguished career in research, teaching, and administration. As the associate dean of the University of California at Berkeley's College of **Chemistry** as well as faculty senior scientist at Lawrence Berkeley Laboratory, he has long been at the forefront of molecular **collision theory**—a field which seeks to explain and predict the essential properties of individual molecules and their effects on each other. Developing the computational methods to accomplish this has also involved him in the most current issues surrounding high-performance computing. Lester has also directed and organized the nation's first unified effort at conducting research in chemistry using computational methods.

Lester was born in Chicago, Illinois, on April 24, 1937, to William and Elizabeth Clark Lester. He received his bachelor's degree in chemistry from the University of Chicago in 1958 and his master's degree in chemistry from the university a year later. For a year he did postgraduate work at Washington University in St. Louis, Missouri, but received his Ph.D. in chemistry in 1964 from the Catholic University of America in Washington, DC. Lester began his professional career as a theoretical physical chemist at the National Bureau of Standards in 1961. In 1964 he became research associate and assistant director of the Theoretical Chemistry Institute of the University of Wisconsin, Madison. In 1968 Lester joined the staff of the IBM Research Laboratory at San Jose, California, and conducted research in **quantum chemistry** and molecular collisions. His studies on molecular collisions helped to explain such everyday phenomena as **combustion**, sound propagation, and atmospheric reactions. This research also provided the basis for understanding in microscopic detail the chemical reactions that take place during those processes. In 1975 Lester moved to Yorktown Heights, New York, to join the IBM T.J. Watson Research Center, where he became director of research on the planning staff of the vice president. A year later he returned to California to manage the molecular interactions group at the IBM laboratory in San Jose.

In 1978 Lester was chosen to be the director of the National Resource for Computation in Chemistry (NRCC). Located at the University of California's Lawrence Berkeley Laboratory, this new unit was dedicated to the development of new computational methods for chemists. He also assumed the position of associate director and senior staff scientist at the Lawrence Berkeley Laboratory that year. In 1981 he became professor of chemistry at the University of California, Berkeley, and faculty senior scientist at the Materials and Molecular Research Division at Lawrence. When he became associate dean of the College of Chemistry at Berkeley in 1991, he remained as faculty senior scientist at Lawrence.

In addition to his research, Lester has edited the proceedings of a major conference and served on the editorial board of the *Journal of Physical Chemistry* (1979–1981), *International Journal of Quantum Chemistry* (1979–1987), *Journal of Computational Chemistry* (1980–1987), and *Computer Physics Communications* (1981–1986). Besides extensive service on governmental advisory committees and national and international panels, he chaired the Gordon Conference on Atomic and Molecular Interactions in 1978. In 1993 he became a member of the National Science Foundation's Blue Ribbon Panel on High Performance Computing.

Among the many honors and awards Lester has received is the IBM Outstanding Contribution Award in 1974. He also received the Percy L. Julian Award in Pure or Applied Research in 1979 and the Outstanding Teacher Award in 1986, both given by the National Organization of Black Chemists and Chemical Engineers; he also won the Catholic University of America Alumni Achievement Award in Science in 1983. He is a fellow of the American Association for the Advancement of Science, the American Physical Society, and a member of the American Chemical Society. He married Rochelle Diane Reed in 1959; they have two children.

LEVENE, PHOEBUS AARON THEODOR (1869-1940)

American biochemist

Phoebus Aaron Theodor Levene made important contributions to the understanding of the **nucleic acids**. He identified the sugars in nucleic acid nucleotides and discovered the precise structure of individual nucleotides.

Levene was born in Sager, Russia, and soon moved with his family to St. Petersburg, where his father owned several custom shirtmaking shops. He interrupted his medical studies at the Imperial Military Medical Academy in St. Petersburg to come to the United States in 1891 with his parents and seven brothers and sisters. He returned to Russia that same year to complete his studies, then came back to New York in 1892. There he practiced medicine and at the same time studied organic **chemistry** at Columbia University and worked in the medical school's physiology laboratory.

Levene gave up his medical practice after contracting tuberculosis. He began performing research on the disease at the Saranac Lakes, NY, sanitorium where he recuperated. He continued his biochemical studies in Germany, working in Marburg with **Albrecht Kossel** and becoming interested in nucleic acids. He also studied protein composition with the organic chemist **Emil Fischer** in Berlin. In 1905, while working for the state of New York, he was asked to lead the biochemical section of the new Rockefeller Institute for Medical Research (now Rockefeller University).

Levene isolated and identified the carbohydrate portion of nucleic acid and defined the nucleotide as a phosphate—base—sugar. He showed that there are two nucleic acids by identifying (in 1909) ribose, the sugar in ribonucleic acid (**RNA**) and (in 1929) deoxyribose, the sugar in deoxyribonucleic acid (**DNA**). His work was the foundation for the later identification of nucleic acids' linear structure by the British organic chemist **Alexander Todd**.

Upon receiving a prestigious award from the American Chemical Society in 1931, Levene spoke of "the revolt of the biochemists" against any restrictions on knowledge, especially on the mysteries of life itself. He was certain that future biochemists would continue work to solve these mysteries, as indeed they have. Among Levene's numerous awards and honors was membership in the National Academy of Sciences.

LEWIS, GILBERT NEWTON (1875-1946)

American chemist

Gilbert Newton Lewis is best known for his theory on the natures of **acids and bases**, as well as for his explanation of chemical **bonding**. However, he also made successful contributions to **thermodynamics** (a branch of physics concerned with the properties of heat), **photochemistry**, and **isotope** separation, which is the sorting of a mixture of isotopes—or a species of an element with the identical **atomic number** but different atomic mass—into its respective isotopic constituents. In addition, his "**electron** dot" structures have provided a visual model for how molecules hold together.

Lewis's ideas about chemical bonding are still among the first theories taught to beginning **chemistry** students. When Lewis began his career, theories of **energy** and energy flow (thermodynamics) were almost never applied to practical chemical problems. His efforts made considerable progress toward finding ways of integrating thermodynamic principles into real chemical systems. Lewis helped to revolutionize chemical education in the United States, making it more investigative and analytical rather than simply descriptive.

Lewis was born on October 23, 1875, in Weymouth, Massachusetts, to Francis Wesley and Mary Burr (White)

Lewis. His early education was informal and private; he learned to read while still very young and was educated at home until age thirteen, when he enrolled in a prep school affiliated with the University of Nebraska. J. H. Hildebrand noted in the *Annual Review of Physical Chemistry* that Lewis regarded his unusual schooling as "an advantage that... occurred frequently in the careers of distinguished men, that of having 'escaped some of the ordinary processes of formal education.'" In 1894 Lewis transferred from the University of Nebraska to Harvard University, where he received a B.S. degree in chemistry two years later. He continued his studies at Harvard—pursuing his Ph.D. in chemistry under the direction of **Theodore William Richards**—while teaching part time at the Phillips Academy in Andover, Massachusetts. He received his doctorate in 1899, at the age of twenty-four. His Ph.D. thesis, "Some Electrochemical and Thermochemical Relations of **Zinc** and **Cadmium** Amalgams," formed the basis of his first scientific publication.

In the following years, Lewis taught at Harvard, did postdoctoral work in Germany with chemists **Wilhelm Ostwald** and **Walther Nernst**, and spent a year in the Philippines working for the Bureau of Science in Manila. It was during these years that Lewis began thinking about many of his most important contributions to chemical theory. While in Germany, Lewis made the acquaintance of **Albert Einstein**, and he later published several papers on Einstein's theories, although his work in this field met with indifferent success. Distressed by the rigidity of thought at Harvard, Lewis moved to the Massachusetts Institute of Technology in 1905 as an associate professor. In the research group of chemist Arthur Amos Noyes, he reached full professor status, becoming acting director of research less than five years later, a remarkable achievement in such a short period of time. In 1912, Lewis moved to California to take the position of dean in the College of Chemistry at the University of California, Berkeley, where he remained for the rest of his working life, except for a decorated tour of duty with the Army Chemical Warfare Service in France during World War I.

At Berkeley, Lewis established himself as a motivating force for both researchers and students. Although too nervous in front of large groups to teach much himself, he influenced the entire process of chemical education. He redesigned the chemistry curriculum to include the teaching of thermodynamic principles. His textbook *Thermodynamics and the Free Energy of Chemical Substances,* coauthored with Merle Randall and published in 1923, was widely studied by generations of pupils. In order to attract more students to the discipline, Lewis paid particular attention to the education of first-year students, involving them in research early in their studies. He gave his colleagues great freedom of choice in their research projects and insisted on a broad base of chemical knowledge rather than rigid specialization. He encouraged discussion and debate among all of his colleagues and students (an unusual practice at that time), demanded that the university provide adequate money for facilities and equipment, and gradually built one of the best chemistry programs in the world. Nearly three hundred Ph.D. degrees in chemistry were awarded at Berkeley during Lewis's tenure, and several of the department's graduate students, including **Glenn Seaborg, Willard Libby,** and **Melvin Calvin**, eventually won Nobel Prizes.

A cornerstone of thermodynamics is a quantity called **free energy**, a mathematical combination of several energy-related characteristics of a system. Free energy tells the chemist how likely a reaction is to occur, how complete an individual chemical reaction is likely to be, and how much work must go into or come out of the reacting system in order to accomplish the reaction. Until Lewis began working in the area, however, the theory was entirely mathematical; very little actual measured data existed to enable practical calculation of this important quantity for real reactions. While at the Massachusetts Institute of Technology, Lewis began meticulous work measuring the free energy values associated with several chemical processes, both organic and inorganic. Between 1913 and 1923, he and his frequent collaborator Merle Randall published several papers and a book on free energy, in which they summarized virtually all the known data on the subject. Lewis also showed how this free energy data could be applied to many different problems in chemistry. For example, he formulated the concept of *activity,* which, because it relied on thermodynamic principles in addition to chemical **concentration**, better explained the behavior of reactions occurring in **solution**.

By the time Lewis began thinking about chemical bonding, the electron had been discovered and was believed to participate in bonding. The **periodic table** was largely laid out, and repeating properties had been observed to occur every eight elements. Several different ideas had been advanced to explain how atoms hooked together to make molecules, but none of them were entirely satisfactory. Lewis had been thinking about the relationship between the eight membered rows on the periodic table in relation to the number of electrons for some time, but it was not until 1916 that he formally proposed his theory of bonding.

Atoms, according to the theory Lewis proposed, have eight electrons around them on the outside. These eight electrons occupy the corners of a cube. When bonding occurs, particularly between atoms which do not have a full complement of eight electrons, the pooled electrons from both atoms pair up, with a pair shared between the bonding atoms. Sharing pairs of electrons enables each **atom** in a **molecule** to obtain the greatest stability. He symbolized these pairs with dots, a convention still used today.

No one took this theory seriously when Lewis first proposed it, but in 1919 the noted chemist **Irving Langmuir** further developed Lewis's idea about an octet of electrons conferring the greatest stability. Langmuir was more persistent and articulate in convincing the chemical community that the theory was important, and so it was for a time called the Lewis-Langmuir electron dot theory. Lewis took issue with the compound name, feeling that he had originated the theory, but historians largely agree that Langmuir made important contributions and that the theory would have languished even longer had it not been for his activism on its behalf. In 1923, Lewis published a book on the theory, *Valence and the Structure of Atoms and Molecules.*

In 1933, after a decade of frustration in his areas of interest, Lewis abruptly started working on a completely new subject, isotope separation. The identity of an element depends not on its atomic weight, but rather on the number of protons in the atom's **nucleus**. Isotopes are differently weighted forms of a single element; most **separation science** depends on exploiting the differences in chemical reactions of different elements, but isotopes of the same element are very difficult to separate because their chemistry is identical. Lewis, working with the isotopes of **hydrogen**, managed to prepare nearly pure **water** containing only the **deuterium** isotope of hydrogen. This is the so-called ''heavy water'' important in nuclear reaction research. This was his major contribution to this area, however his subsequent investigations on isotopes were largely unsuccessful.

In his 1923 book on **valence**, Lewis had successfully applied his chemical bonding theory to acids and bases, but once again it had received little notice. At that time, chemists thought that acids must have a hydrogen **ion** to donate and that a base must be able to receive a hydrogen ion. They were at a loss, however, to explain acidic behavior in compounds that possessed no donatable hydrogen. In 1938, Lewis gave a lecture at Philadelphia's Franklin Institute in which he insisted that this obstacle could be removed by defining an acid as any electron-pair recipient and a base as any electron-pair donor. These definitions cover almost all chemicals in any reaction, including hydrogen ion transfers, and are still used today.

The last years of his life found Lewis engaged in the study of photochemistry—the interaction of light energy with chemical compounds to cause reactions. As with most of his other areas of interest, Lewis had begun to investigate light and **color** early in his career, but his most solid published work occurred later. He collaborated with the respected chemist Melvin Calvin on a review paper concerning photochemistry, and he then experimented in the areas of **fluorescence and phosphorescence**. Eventually, he began to connect these experiments with the rapidly developing field of quantum mechanics, once again combining theory and practice as he had done with thermodynamics and chemical reactions decades earlier. Lewis's interests outside of chemistry included economics and prehistoric glaciation in the Americas; his last paper, published posthumously, concerned the thermodynamics of ice ages. He officially retired in 1941, although he continued his scientific work until his death from a heart attack in his Berkeley laboratory on March 23, 1946. He left behind his wife, Mary Hinckley Sheldon Lewis, whom he had married in 1912, and three children. His two sons followed him to careers in chemistry.

LEWIS STRUCTURES

In 1913, the British scientist **Henry G. J. Moseley** determined the frequency and wavelength of x rays emitted by a large number of elements. By this time, the number of protons in the **nucleus** of some of the lighter elements had been determined. Moseley found the wavelength of the most energetic x ray of an element decreased systematically as the number of protons in the nucleus increased. Moseley then hypothesized the idea could be turned around: he could use the wavelengths of x rays emitted from heavier elements to determine how many protons they had in their nuclei. After Moseley's work, the idea that the periodic patterns in chemical **reactivity** might actually be due to the number of electrons and protons in atoms intrigued many chemists. Among the most notable was **Gilbert Newton Lewis** of the University of California at Berkeley.

Lewis explored the relationship between the number of electrons in an **atom** and its chemical properties and the kinds of substances formed when elements reacted together to form compounds, and the ratios of atoms in the formulas for these compounds. Lewis concluded that chemical properties change gradually from metallic to nonmetallic until a certain ''stable'' number of electrons is reached. An atom with this stable set of electrons is a very unreactive species. But if one more **electron** is added to this stable set of electrons, the properties and chemical reactivity of this new atom changes dramatically: the element is again metallic, with the properties like elements of Group 1. Properties of subsequent elements change gradually until the next stable set of electrons is reached and another very unreactive element completes the row. A stable number of electrons is defined as the number of electrons found in an unreactive or ''noble'' gas. Lewis suggested electrons occupied specific areas around the atom, called shells. The noble gas atoms have a complete octet of electrons in the outermost shell.

Dmitry Mendeleev put elements together in a family because they had similar reactivities and properties; Lewis proposed that elements have similar properties because they have the same number of electrons in their outer shells. Many observations of the chemical behavior of elements are consistent with this idea: the number of electrons in the outer shell of an atom determines the chemical properties of an element. These electrons give the element its chemical ''worth'' or value and are called its **valence** electrons.

G. N. Lewis extended his ideas about the importance of the number of valence electrons from the properties of elements to the **bonding** of atoms together to form compounds. He proposed that atoms bond with each other either by sharing electrons to form covalent bonds or by transferring electrons from one atom to another to form ionic bonds. Each atom forms stable compounds with other atoms when all atoms achieve complete shells. An atom can achieve a complete shell by sharing electrons, by giving them away completely to another atom, or by accepting electrons from another atom.

Many important compounds are formed from the elements in rows two and three in the **periodic table** (the elements **lithium** to chlorine). Lewis predicted these elements would form compounds in which the number of electrons about each atom would be a full shell, like the noble **gases**. The noble gases of rows two and three, **neon** or **argon**, each has eight elect rons in the outermost valence shell. Thus, Lewis's rule has become known as the **octet rule** and simply states that there should be eight electrons in the outer shell of an atom in a compound. An important exception to this is **hydrogen** for

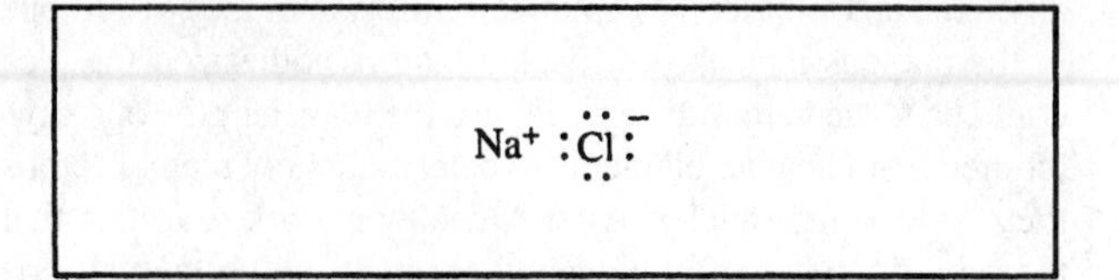

Figure 1. ***(Illustration by Electronic Illustrators Group.)***

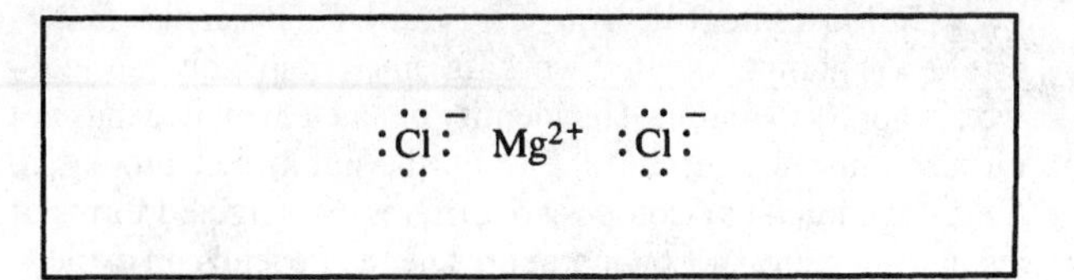

Figure 2. ***(Illustration by Electronic Illustrators Group.)***

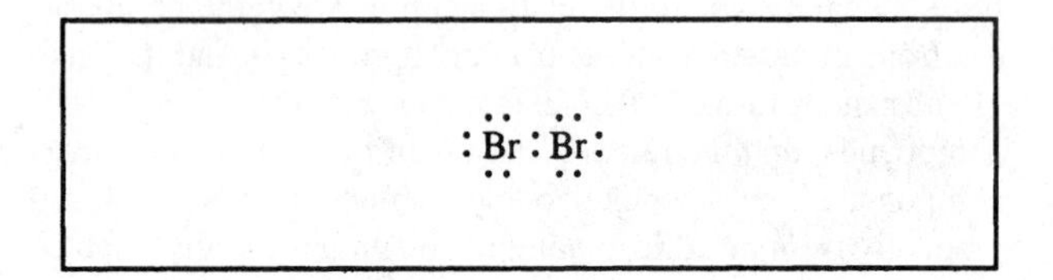

Figure 3. ***(Illustration by Electronic Illustrators Group.)***

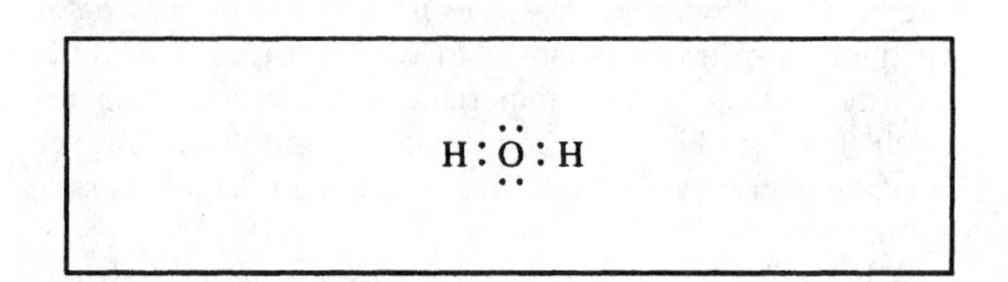

Figure 4. ***(Illustration by Electronic Illustrators Group.)***

which a full shell consists of only two electrons. The octet rule is followed in so many compounds it is a useful guide. It is not a fundamental law of chemistry and many exceptions are known, especially for compounds of elements below row three of the periodic table. However, he octet rule is a good starting point for learning how chemists view compounds and how the periodic table can be used to make predictions about the likely existence, formulas and reactivities of chemical substances.

The structure and bonding of a compound determine its chemical and physical properties. Lewis's idea of stable, filled electron shells can be used to predict which atom is bonded to which other atom in a **molecule**.

A "Lewis structure" of a compound consists of the symbols of each element surrounded by "dots" representing the valence electrons. In many cases, Lewis's octet rule is followed by taking one or more electrons from one atom to form a **cation** and donating the electron or electrons to another atom to form an **anion**. When electrons are transferred from one atom to another to form an ionic compound, the valence electron transferred is shown with the anion. Metallic elements on the far left of the periodic table can give electrons away and elements on the far right can readily accept electrons: the metallic elements having relatively low electronegativities and the non-metallic elements at the right of the periodic table having relatively high electronegativities. When these elements combine, ionic bonds result. An example of an ionic compound is **sodium** chloride. To construct the Lewis structure, we attempt to have completely filled or empty shells. If the sodium atom loses one electron and the **chlorine** atoms gains one, the ions will have stable octets. The electron transferred from the sodium atom to form the sodium cation (figure 1).

One of the electrons of the chloride **ion** in NaCl originates with the sodium atom, but once a compound is formed, we cannot determine the atom a particular electron came from. The sodium ion has the same electron configuration as neon, the noble gas immediately before it on the periodic table. The ion is described as being isoelectronic with the noble gas. Likewise, the chloride ion is isoelectronic with argon.

In the case of the ionic compound **magnesium** chloride, the magnesium atom has two electrons more than a full shell. If each of the two chlorine atom gains one electron, both the magnesium cation, Mg2+, and the two chloride anions, Cl–, have full shells. We symbolize the electronic structure of magnesium chloride as shown in figure 2.

The magnesium ion is isoelectronic with neon; the chloride ions, with argon.

Lewis structures are most commonly used in compounds in which atoms do not fully transfer electrons but share them: covalent compounds. Lewis defined a **covalent bond** as a union between two atoms resulting from the sharing of two electrons. Each covalent bond is a pair of electrons shared by two atoms. For covalent compounds, the shared electrons are symbolized by dots between the bonding atoms.

Elemental **bromine**, Br_2, is an example of a covalent compound. Each bromine has seven electrons in its outer shell and requires one electron to achieve a noble gas configuration. Each can pick up the needed electron by sharing one with the other bromine atom. The Lewis structure of bromine is shown in figure 3.

For covalent compounds, a simplified convention is used to show which atoms are connected: a line drawn between the bound atoms, giving what we will call a line-bond representation. For bromine, this representation is simply Br-Br, with all of the non-bonding valence electrons omitted and the line between the two atoms symbolizing the shared pair of electrons.

Water is the combination of atoms of two nonmetallic elements, hydrogen and **oxygen**. Each hydrogen atom requires just one electron to fill its shell because the first shell (the number of electrons of the noble gas helium) has only two electrons. Oxygen lacks two electrons compared with neon, the nearest noble gas. If each hydrogen can obtain one electron by sharing electrons from the oxygen atom and the oxygen atom can share one electron from each of the two hydrogen atoms, every atom will have a full shell of electrons, and two covalent

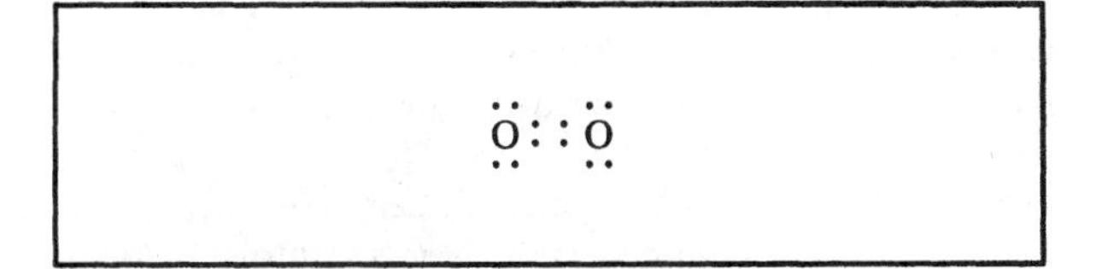

Figure 5. ***(Illustration by Electronic Illustrators Group.)***

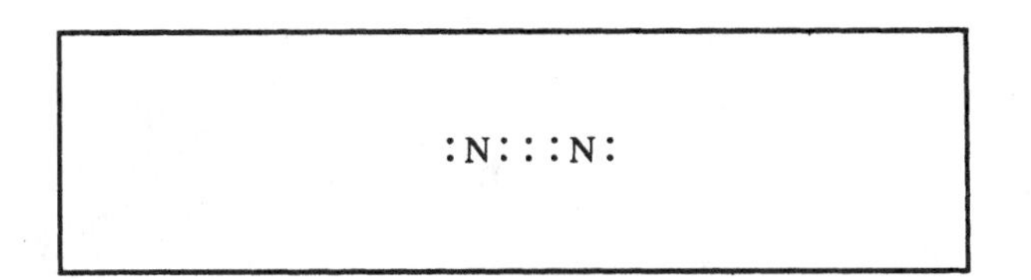

Figure 6. ***(Illustration by Electronic Illustrators Group.)***

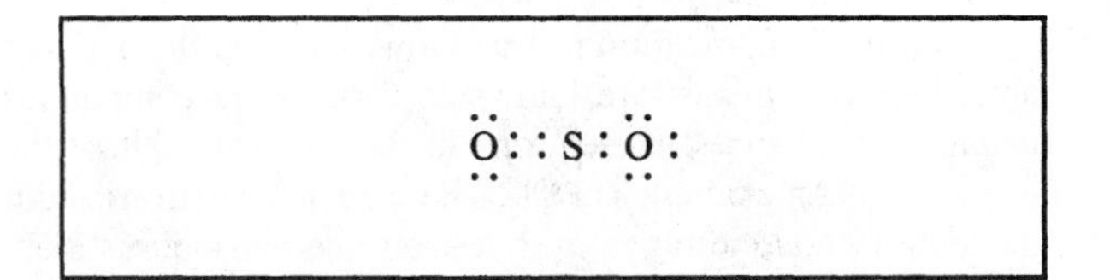

Figure 7. ***(Illustration by Electronic Illustrators Group.)***

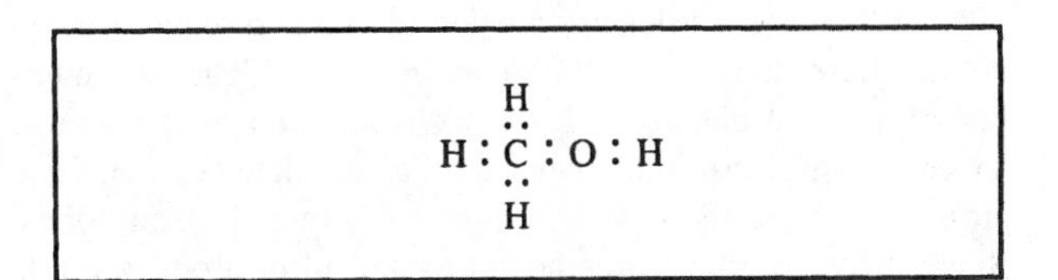

Figure 8. ***(Illustration by Electronic Illustrators Group.)***

bonds will be formed as a result of sharing two pairs of electrons. We can symbolize this result as shown in figure 4.

This example illustrates an important use for Lewis structures of covalent compounds: they yield information about connectivity, a consideration of what atom is bonded to what other atom. Before Lewis structures existed, chemists had no theory to predict what the formula for H_2O meant in terms of the structure of the compound: the structure could be H-H-O or H-O-H. The octet theory supplies a logical framework for making this decision and clearly predicts both hydrogen atoms are bound to centrally located oxygen, H-O-H, as opposed to H-H-O.

For any compound, the Lewis structure is constructed by first drawing all the valence electrons around each atom and then, if possible, arranging the valence electrons between atoms to obtain an octet of electrons around each atom other than hydrogen and a pair of electrons for each hydrogen atom. The octet can be obtained either by transferring electrons to form ions or by sharing electrons to form covalent bonds between atoms.

For many compounds Lewis's octet rule cannot be satisfied unless atoms share more than one pair of electrons. These compounds contain multiple bonds. An example is the **diatomic** molecule O_2. Each oxygen atom has six valence electrons. If a single bond is made by sharing one electron from each atom, each oxygen then has seven electrons, one short of an octet. To complete an octet, each oxygen must share two electrons in a covalent bond. The resultant bond is called a double bond, and it is symbolized by two lines connecting the atoms, where each line corresponds to an electron pair (figure 5).

Even more electrons can be shared. To obtain filled valence shells on the atoms of **nitrogen** molecules, it is necessary to share three pairs of electrons, forming a triple bond (figure 6).

Double bonds and triple bonds are referred to collectively as multiple bonds.

The elements hydrogen, nitrogen, oxygen, **fluorine**, chlorine, bromine and **iodine** consist of diatomic molecules. All have single bonds except oxygen (double) and nitrogen (triple). The noble gases are composed of monatomic molecules. They have no need to share electrons because each already has an octet. All other elements have multiatomic molecular structures that allow them to attain filled valence shells.

Certain atoms tend to form only one covalent bond. Thus, these atoms are only found at the ends of molecules or of polyatomic ions—they are terminal atoms. The most common of the terminal atoms are hydrogen (there are more compounds of hydrogen than of any other element) and fluorine. The other elements in the halogen family are terminal atoms in all but a few compounds they form with each other or oxygen and hydrogen can bond in some cases, such as borane compounds, with more than one atom. The elements that are typically terminal atoms are useful to remember for drawing Lewis structures of more complex molecules.

The number of bonds for Lewis structures of a great many compounds can be determined using a few rules. It is useful to first consider a method to handle molecules with only one central (non-terminal) atom and then generalize the method to more complex molecules and consider only compounds of elements in the first two rows of the periodic table (hydrogen to fluorine). For these molecules, the rules are:

1. Consider only valence electrons 2. Count the total number of valence electrons for all atoms; this is the number of electrons you have. 3. Add all of electrons needed to complete a shell for every atom; two for hydrogen, eight for others. This is the number of electrons you need. 4. Subtract the number of electrons you have from the number you need and divide by two. This gives you the number of two-electron bonds. 5. If the number of two electron bonds is equal to the number of atoms bound to the central atom, every bond is a single bond. If the number of two electron bonds is more than the number of atoms bound to the central atom, the excess will be multiple bonds. 6. Use hydrogen and **halogens** as terminal atoms (except for boranes and interhalogen compounds)and then use the least electronegative atom as the central atom.

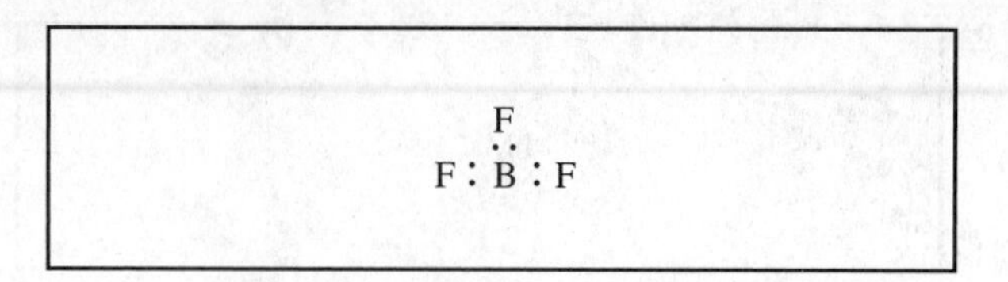

Figure 9. *(Illustration by Electronic Illustrators Group.)*

When these rules are applied to **sulfur** dioxide, for example, these steps lead to the following results. The total number of valence electrons if 18, six from each of the atoms. Twenty-four electrons are needed for a complete octet on three atoms, therefore there will be three (6/2) two electron bonds. Since there are two atoms bound to the sulfur atom (the less electronegative element), there must be one double bond. The Lewis structure is shown in figure 7.

A limitation of this Lewis structure is the perception that there are two different kinds of bonds in sulfur dioxide. In fact, the double and single bonds are averaged over both oxygen atoms. However, the structure is correct in identifying a total of three two-electron bonds between the oxygen atoms and the sulfur atom and the connectivity oxygen atom to sulfur atom to oxygen atom.

Lewis structures can also be written for more complex molecules that have a chain of bonded atoms instead of a single central atom. Often the atoms comprising a complex molecule can be arranged in several ways and still satisfy the octet rule for each atom. Hence, to write Lewis structures for complex molecules, the order in which atoms are connected must be known first: some structural information is required. With the structural information in hand, the process for constructing a Lewis structure is to begin at one end of the molecule and treat terminal atoms first. When the filled-valence-shell rule for the terminal atoms of the first atom of the chain, the number of bonds formed by one of the atoms along the chain is then determined. The process is followed atom by atom along the chain of the molecule.

The following example illustrates how to determine the number of bonds in the molecule **methanol**, which has the **structural formula** CH_3OH. Each of the three terminal hydrogen atoms bound to the **carbon** atom shares one its electron and one of the carbon atom's electrons in a single two-electron bond. The one remaining valence electron on the carbon atom is available to be shared with the oxygen atom and the carbon atom needs only a single electron to complete its octet, having three shared pairs in addition to the one unshared electron. Therefore, it shares one electron from the oxygen atom to form a single two-electron bond. The oxygen atom now has one electron shared from the carbon atom in addition to its six valence electrons. It needs only one electron to complete its octet, which it obtains by sharing the electron from its terminal hydrogen atom, forming a single two-electron single bond. All hydrogen atoms now have their two-electron valence shell satisfied and the carbon and oxygen atoms have their eight electron valence shells complete. The Lewis structure of methanol is shown in figure 8.

Two exceptions to the octet rule are common: molecules with atoms having fewer than eight electrons and those which have more than eight in the valence shell after sharing or electron transfer. Those with less than an octet are often called electron deficient and are typical of certain elements with an odd number of electrons, such as **boron** and nitrogen. A typical reaction of such compounds is to accept an additional electron from an anion or to share electrons with an anion to gain a stable octet. An example is boron trichloride, which has the Lewis structure shown in figure 9.

Elements in the third row and lower rows of the periodic table, heavy elements, are known to form many compounds having more than eight electrons in the valence shell of the heavy atom. An example is PCl_5. There are ten electrons about the central **phosphorus** atom: five from the phosphorus atom itself and one from each of the five chlorine atoms.

While Lewis structures provide some guides to chemical structure, they have significant limitations. Perhaps the biggest limitation is that Lewis structures do not provide a guide to **molecular geometry**. For example, the Lewis structure for **methane**, CH_4, is a flat, cross-shape. Yet, the methane molecule is a three-dimensional, tetrahedral molecule. Another limitation concerns the electrons not involved in chemical bonds. For example, the Lewis structure for oxygen, shows all non-bonding electrons paired, but experiments reveal the two non-bonding electrons on the oxygen molecule are not paired. The Lewis structure does predict correctly, however, that the bond in oxygen is a double bond. Many complex compounds of the **transition elements** cannot be accurately depicted by Lewis structures.

Even with such limitations, Lewis structures apply to a very broad range of covalent compounds and are tools professional chemists often use as their first approach to chemical bonding because they enable us to predict connectivity.

LI, CHOH HAO (1913-1987)

American biochemist

Choh Hao Li is credited with isolating and purifying eight of the ten different **hormones** manufactured in the front portion, or adenohypophysis, of the pituitary gland. The pituitary is the pea-sized governing gland which is located at the base of the brain, and it manufactures hormones responsible for **metabolism**, reproduction, growth, and maturation. One of these is adrenocorticotropin (ACTH), an adrenal booster used in the treatment of arthritis. Another is somatotropin, or human growth hormone (HGH), which has proven particularly effective on children whose growth has slowed or stopped because of hypopituitary problems. Li was director of the Hormone Research Laboratory of the University of California at Berkeley, and he received the Lasker Award for Basic Medical Research in 1962.

Li was born in Canton, China, on April 21, 1913, to Mewching Tsui and Kan-chi Li. His father was an industrialist with a high-school education, but Li's parents made sure their eleven children received college educations. Li attended Pui-

Choh Hao Li.

Ying Middle School in Canton, where he developed a curiosity about nature at an early age. He continued his studies at the University of Nanking, earning his B.S. degree in **chemistry** there in 1933, and he stayed for two years as a chemistry instructor.

Li came to the United States in 1935 to study chemistry at the University of California. Because the chemistry department there had never heard of the University of Nanking, he was initially admitted on a six-month probationary period. But they were impressed with his work, and he was allowed to stay. He earned his Ph.D. in chemistry in 1938, supporting himself by teaching Chinese to children in Chinatown, and he was immediately appointed research associate with experimental biologist Herbert M. Evans at the Institute of Experimental Biology. He would remain at the University of California for the rest of his career.

In 1938, Li first began research on the adenohypophysis of the pituitary gland. With Evans and Miriam E. Simpson, he made his first significant contribution to pituitary hormone research in 1940, when he isolated the interstitial cell-stimulating hormone (ICSH) from sheep glands and pig glands. ICSH is a fertility hormone which controls ovulation in women and the production of sperm in men. This hormone is the basis for modern oral contraceptives. In 1950, Li was appointed director of the Hormone Research Laboratory, part of the University of California's School of Medicine. He continued his investigations of pituitary hormones, isolating and identifying them and attempting to develop methods to synthesize them.

By 1953, Li and his colleagues isolated and purified the adrenocorticotropic hormone (ACTH). This hormone regulates the adrenal cortex, which produces **sex hormones** as well as hormones that govern metabolic functions; it also produces adrenaline. Li found that an injection of ACTH can stimulate the cortex to increase the secretion of **cortisone** and that it was therefore as effective as cortisone itself in treating arthritis. But it was not until many years later, in 1973, that he managed to synthesize it. In 1956, Li and his associates isolated and purified human growth hormone (HGH), though determining its structure took them until 1966. This accomplishment has important applications in medicine, as the hormone can stimulate human growth and has been used with cases of hyperpituitary dwarves to help them achieve a normal size. But HGH can also be beneficial to sex hormones, the production of **antibodies**, and the healing of wounds, and has been proven effective in reducing blood **cholesterol** levels, developing breast tissue, and increasing the milk of nursing mothers; in addition, it can affect diabetes, obesity, and giantism. In 1992, the National Institute on Aging initiated studies to determine whether HGH can actually reverse or slow the aging process.

In 1978 Li discovered beta-endorphin, a substance produced in the brain that acts as a painkiller. This discovery initiated a number of investigations into new treatments for pain. In addition to its morphine-like characteristics, beta-endorphin can exert behavioral effects on both animals and humans when it is administered intravenously. Li was also the first to synthesize growth factor 1, which has important resemblances to insulin.

Li was the recipient of many prestigious awards for his contributions to medical research. Among them were the Ciba Award from the Endocrine Society, which he received in 1947. He was also awarded the Emory Prize of the American Academy of Arts and Sciences, and the Lasker Award for Basic Medical Research, which Mrs. Lyndon B. Johnson presented to him in 1962. He received the Scientific Achievement Award from the American Medical Association in 1970, the National Award from the American Cancer Society in 1971, and the Nichols Medal in 1979.

Li married Annie Lu on October 1, 1938; they had a son and two daughters. Li died on November 28, 1987 at age seventy-four.

LIBAVIUS, ANDREAS (1560-1616)

German iatrochemist and author

Andreas Libavius, born around 1560, lived and worked in Germany at a time when the field of **chemistry** was still trying to grow out of its ancient association with the mystical **alchemy**. Libavius sought to resolve some of the philosophical issues stirred up with the death of **Paracelsus**. His career began well into a generation after Paracelsus tried to give life to chemistry by focusing its purpose on developing medicines. Paracelsus was ineffective and inaccurate in presenting his ideas. After his death, there was a great division among those studying chemistry. The division was between those who wished to study the effective application of generalizable principles and those who wished to learn and practice the ancient ways. The iatrochemists who sought to develop medicines out of minerals were, in a primitive way, working to grow the field into a true science. Galenists were those who still relied on the mystical solutions. The Galenist tradition was handed down from antiquity without the benefit of any scientific test. It involved cabalistic ceremonies for developing healing substances. Paracelsus recognized something about what was scientifically needed but because he could not accurately communicate his ideas, those associated with the mystical only gained more strength.

This was the situation that Libavius sought to moderate. Libavius was originally a professor of history and poetry before earning a degree in medicine. He was attracted to the thinking of the iatrochemists who followed Paracelsus' tradition. However, Libavius resented the disorganization, inconsistencies, and blatant untruths found in Paracelsus' teachings. In addition, Libavius noted that Paracelsus knew the potential of chemicals but emphasized the elixir of these chemicals and not the substance found in the residuals. The iatrochemists were beginning to recognize that the elixir usually contained little more than **water** and acid, but the philosophy that their study was based on did not allow the logical **solution** of studying the residuals. The philosophy stated that medicines worked because they contained mostly the pure substance. When the pure is extracted, what is left behind is the impure or unnatural. It took the new thinking of Libavius to give the framework necessary for appropriate experimentation with the residuals.

Libavius wrote a book in 1597 that collected the comprehensive knowledge of chemistry to that time. He described chemistry in terms that made the search for medicines only a part of a greater discipline. Other parts included metallurgy and industrial applications. He further introduced new classifications with logical connections as a means to bolster up the new discipline. In his book, he established standard methods of study and of describing known chemicals.

Although the book was a necessary key in drawing the scientific aspects of chemistry away the **transmutation** of ancient alchemy, it did not have a direct effect on the teaching of chemistry. The book was poorly distributed particularly because it was written in Latin. However, the sections on laboratory methods were written in German and found popular appeal.

As the chemists of the sixteenth century were a closely communicating society, it is reasonable to assume that the influence of Libavius was greater than the acceptance of his book. Twelve years after the book's publication, the University of Marburg became the first university to offer formal instruction in chemistry in line within the guidelines set down by Libavius. He died five years later in the year 1616.

LIBBY, WILLARD F. (1908-1980)

American chemist

Chemist Willard F. Libby developed the radiocarbon dating technique used to determine the age of organic materials. With applications in numerous branches of science, including archaeology, geology, and geophysics, radiocarbon dating has been used to ascertain the ages of both ancient artifacts and geological events, such as the end of the Ice Age. In 1960, Libby received the Nobel Prize for his radiocarbon dating work. During World War II, Libby worked on the **Manhattan Project** to develop an atomic bomb and was a member of the Atomic Energy Commission for several years in the 1950s. An outspoken scientist during the Cold War between the U.S. and the former Soviet Union, Libby advocated that every home have a fallout shelter in case of nuclear war. Opposed to bans against nuclear weapons testing, Libby was considered by some to be a pawn for a federal administration that wished to continue the arms race. Libby, however, was a strong proponent of the progress of science, which he believed resulted in more benefits than detriments for the human race.

Willard Frank Libby was born to Ora Edward and Eva May Libby on December 17, 1908, on a farm in Grand Valley, Colorado. In 1913, the family, which included Libby and his two brothers and two sisters, moved to an apple ranch north of San Francisco, California, near Sebastopol, where Libby received his grammar school education. A large boy who would eventually grow to be six-feet three-inches tall, Libby developed his legendary stamina while working on the farm. He played tackle for his high school football team and was called ''Wild Bill,'' a nickname used by some throughout Libby's life. After graduating from high school in 1926, Libby enrolled at the University of California, Berkeley. He made money for college by building apple boxes, earning one cent for each box and sometimes $100 in a week. ''I was the fastest box maker in Sonoma County,'' he told Theodore Berland, who interviewed Libby for his book *The Scientific Life.*

Although Libby was interested in English literature and history, he felt obligated to seek a more lucrative career and entered college to become a mining engineer. By his junior year, however, Libby became interested in **chemistry**, spurred on by the discussions of his boarding house roommates, who were graduate students in chemistry. Libby took on an heavy course load, focusing on mathematics, physics, and chemistry. After receiving his B.S. in chemistry in 1931, he entered graduate school at Berkeley and studied under the American physical chemist **Gilbert Newton Lewis** and Wendell Latimer, who were pioneering the **physical chemistry** field.

During graduate school, Libby built the United States' first Geiger-Muller tube for detecting **radioactivity**, which results from the spontaneous disintegration of an atom's **nucleus**. Libby refined the mechanism in order to detect minute amounts of radioactivity in elements not previously thought to be radioactive, including **samarium**. Libby continued to make his own Geiger counters throughout his life, claiming that they were much more sensitive than those manufactured for the open market.

Libby received his Ph.D. in 1933 and was appointed an instructor in chemistry at Berkeley. After the Japanese bombed

Pearl Harbor in 1941, Libby, who was on a year sabbatical as a Guggenheim Fellow at Princeton University, joined a group of scientists in Chicago, Illinois, to work on the Manhattan Project, a government-sponsored effort to develop an atomic bomb. During this time, he worked with American chemist and physicist **Harold Urey** at Columbia University on gaseous **diffusion** techniques for the separation of **uranium** isotopes (isotopes are different forms of the same element having the same **atomic number** but a different number of protons). After the war, he accepted an appointment as a professor of chemistry at the University of Chicago and began to conduct research at the Institute of Nuclear Studies.

In 1939, scientists at New York University had sent radiation counters attached to balloons into the earth's upper atmosphere and discovered that **neutron** showers were created by cosmic rays hitting atoms. Further evidence indicated that these neutrons were absorbed by **nitrogen**, which then decayed into radioactive carbon–14. In addition, two of Libby's former students, Samuel Ruben and Martin Kamen, made radioactive carbon–14 in the laboratory for the first time. They used a cyclotron (a circular device that accelerates charged particles by means of an alternating electric field in a constant magnetic field) to bombard normal carbon–12 with neutrons, causing it to decay into carbon–14.

Intrigued by these discoveries, Libby hypothesized that radioactive carbon–14 in the atmosphere was oxidized to **carbon** dioxide. He further theorized that, since plants absorb **carbon dioxide** through **photosynthesis**, all plants should contain minute, measurable amounts of carbon–14. Finally, since all living organisms digest plant life (either directly or indirectly), all animals should also contain measurable amounts of carbon–14. In effect, all plants, animals, or carbon-containing products of life should be slightly radioactive.

Working with Aristide von Grosse, who had built a complicated device that separated different carbons by weight, and graduate student Ernest C. Anderson, Libby was successful in isolating radiocarbon in nature, specifically in **methane** produced by the decomposition of organic **matter**. Working on the assumption that carbon–14 was created at a constant rate and remained in a **molecule** until an organism's death, Libby thought that he should be able to determine how much time had elapsed since the organism's death by measuring the **half-life** of the remaining radiocarbon isotopes. (Half-life is a measurement of how long it takes a substance to lose half its radioactivity.) In the case of radiocarbon, Libby's former student Kamen had determined that carbon–14's half-life was 5,370 years. So, in approximately 5,000 years, half of the radiocarbon is gone; in another 5,000 years, half of the remaining radiocarbon decays, and so on. Using this mathematical calculation, Libby proposed that he could determine the age of organisms that had died as many as 30,000 years ago.

Since a diffusion column such as von Grosse's was extremely expensive to operate, Libby and Anderson decided to use a relatively inexpensive Geiger counter to build a device that was extremely sensitive to the radiation of a chosen sample. First, they eliminated 99% of the **background radiation** that occurs naturally in the environment with eight-inch thick

Willard F. Libby.

iron walls to shield the counter. They then used a unique chemical process to burn the sample they were studying into pure carbon lampblack, which was then placed on the inner walls of a Geiger counter's sensing tube.

Libby first tested his device on tree samples, since their ages could be determined by counting their rings. Next, Libby gathered tree and plant specimens from around the world and discovered no significant differences in normal age-related radiocarbon distribution. When Libby first attempted to date historical artifacts, however, he found his device was several hundred years off. He soon realized that he needed to use at least several ounces of a material for accurate dating. From the Chicago Museum of Natural History, Libby and Anderson obtained a sample of a wooden funerary boat recovered from the tomb of the Egyptian King Sesostris III. The boat's age was 3,750 years; Libby's counter estimated it to be 3,261 years, only a 3.5% difference. Libby spent the next several years refining his technique and testing it on historically significant—and sometimes unusual—objects, such as prehistoric sloth dung from Chile, the parchment wrappings of the Dead Sea Scrolls, and **charcoal** from a campsite fire at Stonehenge, England. Libby saw his new dating technique as a way of combining the physical and historical sciences. For example, using **wood** samples from forests once buried by glaciers, Libby determined that the Ice Age had ended 10,000 to 11,000 years ago, 15,000 years later than geologists had previously believed. Moving on to man-made artifacts from North America and Europe (such as a primitive sandal from Oregon and charcoal specimens from various campsites), Libby dispelled the notion of an Old and New World, proving that the oldest dated human settlements around the world began in approximately the same era. For many years after Libby's discovery of radiocarbon dating, the journal *Science* published the results of dating studies by Libby and other scientists from around the world. In 1960, Libby was awarded the Nobel Prize in chemistry for his work in developing radiocarbon dating. In his acceptance speech, as quoted in *Nobel Prize Winners,* Libby noted that radiocarbon dating "may indeed help roll back the pages of history and reveal to mankind something more about his ancestors, and in this way, perhaps about his future." Further progress in radiocarbon dating techniques extended its range to approximately 70,000 years.

In related work, Libby had shown in 1946 that cosmic rays produced **tritium**, or hydrogen–3, which is also weakly radioactive and has a half-life of 12 years. This radioactive form of **hydrogen** combines with **oxygen** to produce radioactive **water**. As a result, when the U.S. tested the Castle hydrogen bomb in 1954, Libby used the doubled amount of tritium in the atmosphere to date various sources of water, deduce the water-circulation patterns in the U.S., and determine the mixing of oceanic waters. He also used the method to date the ages of wine, since grapes absorb rain water.

In 1954, U.S. President Dwight D. Eisenhower appointed Libby to the Atomic Energy Commission (AEC). Although he continued to teach graduate students at Chicago, Libby drastically reduced his research efforts and plunged vigorously into his new duties. Previously a member of the commission's General Advisory Committee, which developed commission policy, Libby was already acquainted with the inner workings of the commission. He soon found himself embroiled in the nuclear fallout problem. Upon a recommendation by the Rand Corporation in 1953, Libby formed and directed Project Sunshine and became the first person to measure nuclear fallout in everything from dust, soil, and rain to human bone.

As a member of the AEC, Libby testified before the U.S. Congress and wrote articles about nuclear fallout. He noted that all humans are exposed to a certain amount of natural radiation in sources such as drinking water. He went on to point out that the combination of the body's natural radioactivity, cosmic radiation, and the natural radioactivity of the earth's surface was more hazardous than fallout resulting from nuclear testing. Libby believed, and most scientists concurred, that the effects of nuclear fallout on human genetics were minimal.

Many scientists, however, thought that Libby was merely a "yes man" for the federal administration. In reply, Libby often responded to what he considered misguided thinking. He once wrote to the French physician and author Albert Schweitzer, who had publicly declared that future generations would probably suffer from fallout, that he doubted whether Schweitzer was aware of the most recent data on the subject and that nuclear testing was necessary for the defense effort and the free world's survival. Even after Libby resigned from the AEC in 1959 to resume his scientific studies, he continued to argue with zeal about the necessity for nuclear testing. He also urged the nation's industrial community to employ isotopes in factories and on farms and was a member of the international Atoms for Peace project that supported **nuclear energy** production for non-military purposes. Libby's experiences in Washington, DC, convinced him that more scientists needed to be in positions of political power and not just advisers. As a result, he was pleased when U.S. President John F. Kennedy appointed **Glenn T. Seaborg**, a nuclear chemist, chair of the AEC in 1961.

After retiring from the AEC, Libby took a position in the chemistry department at the University of California, Los Angeles (UCLA), largely due to his first wife's desire to live in California again. Libby had married Leonor Lucinda Hickey in 1940, and the couple had twin daughters. When the family moved to their new home in Bel-Air, California, Libby proceeded to build his own fallout shelter, using sandbags and railroad ties, for approximately $30. During the Cold War, Libby believed that every home should have a fallout shelter in case of nuclear war and wrote a series of articles for the Associated Press news service proposing this necessity. According to Berland, Libby once complained, after hearing a physician say that perhaps it would have been better if scientists had never discovered the power of the **atom**, that the only way to stop such inevitable discoveries was to "kill all the scientists." He went on to tell Berland that physicians would then have to "go back to witchcraft" for treating people.

In 1962, Libby received a joint appointment as director of the Institute of Geophysics and Planetary Physics. He believed that the new frontier in science was outer space and said that the U.S. must support a large space exploration program in order to prevent the Soviets from controlling outer space, which would probably enable them to rule the world.

Libby and his wife Leonor divorced in 1966. Libby later married Leona Woods Marshall, a professor of environmental engineering at UCLA. He retired in 1976 and died in Los Angeles on September 8, 1980, from complications suffered during a bout of pneumonia.

LIEBIG, JUSTUS VON (1803-1873)

German chemist

Justus von Liebig did not make his reputation with a single discovery or innovation, but rather with tremendous versatility. He conducted research in organic and inorganic **chemistry**, **agricultural chemistry**, physiology and **biochemistry**, making significant contributions to the study of **acids and bases**, the chemistry of **ether**, the systematization of **organic chemistry**, and the production of industrial dyes as well as synthetic **fertilizers**. Liebig is considered to be one of the most important chemists of the nineteenth century.

Born in Darmstadt, Germany, on May 12, 1803, Liebig was the son of a merchant who sold pharmaceuticals, dyes, and salts, so he developed a keen interest in chemistry early in his youth. By the time he was nineteen, he had earned his Ph.D.

at Erlangen, and at the recommendation of Alexander von Humboldt was hired to work in the laboratory of **Joseph Gay-Lussac**, where he remained for two years. In 1825, he was appointed chairman of chemistry at the obscure University of Giessen, where he proceeded to build an excellent chemistry program. Considered a great teacher, he was among the first to focus on laboratory instruction as a means of educating chemists. He remained at Giessen for 27 years before he moved to the University of Munich, staying there until his death on April 18, 1873.

Liebig is probably best known to chemistry students as the inventor of the Liebig condenser, a **distillation** apparatus found in most every chemical laboratory but he also played a part in numerous key discoveries. For example, in the early 1820s Liebig and **Friedrich Wöhler** had been conducting individual research on inorganic chemicals, and Gay-Lussac noticed that a compound Liebig called **silver** fulminate had the same chemical formula as a compound Wöhler called silver cyanate, though the compounds were different chemically. Informed of this peculiarity by Gay-Lussac, **Jöns Berzelius** was inspired to derive the theory of isomers. Liebig and Wöhler began working together in organic chemistry, a field that lacked systematic theory at the time, and introduced a degree of methodical analysis to the field, including techniques to determine the content of various elements such as **carbon**, **hydrogen**, and **halogens** in organic chemicals. Upon discovering the benzoyl radical (C_6H_5CO-), they also attempted to find a way to define *all* organic chemicals as combinations of radicals (groups of molecules that tend to act as a unit). Though they failed, their efforts to present organizing principles in organic chemistry stimulated more successful attempts later.

Liebig made other noteworthy scientific contributions including data on **fermentation** and the calorific content of foods. His studies of plant biochemistry were particularly important to the future of agriculture, for he was among the first to theorize that plants obtain carbon not from organic compounds in the soil, but from atmospheric **carbon dioxide**. Liebig also discovered that plants required simple mineral compounds from the soil to survive. This led him to develop the earliest artificial fertilizers.

Justus von Liebig.

LIGAND FIELD THEORY

Developed in the first half of the 20th century to describe the **electron** systems of transition-metal complexes (a phenomenon within the branch of inorganic chemistry), ligand field theory is an extension of crystal field theory. Thus, it is also known as ''adjusted crystal field theory.'' Ligand field theory and its predecessors represented huge advances in scientists' understanding of the structure, **bonding**, and behavior of transition metal complexes. Chemists use ligand field theory to interpret the spectra (colors) of transition **metals** and to understand how they react with other substances.

Somewhat similar to **Linus Pauling**'s **valence**-bond model, ligand field theory is a subset of molecular orbit theory. It is more accurate and more powerful than its predecessor (crystal field theory) for explaining the chemical properties of transition metal complexes. To understand ligand field theory, it is first necessary to know the basic precept of crystal field theory, since they are so closely related.

Physicist Hans Bethe introduced the idea in 1929 to explain the magnetic and spectral properties of inorganic materials. A way to analyze inorganic complexes' electron structures, crystal field theory assumes that an inorganic complex consists of a central metal **atom** surrounded by ionic ligands. It can be used to predict kinetic and chemical properties, spectral and magnetic characteristics, thermodynamic properties, and reaction mechanisms. It explains the properties of coordination complexes in terms of the electrostatic pull between the ligands and the metal **ion** and how this interaction changes the orbitals' energies (in this case the *d* orbitals).

The ligand field theory is different from crystal field theory because it takes into account the overlap of electrons as they orbit the central metal atom. In addition, ligand field theory is applied to transition-metal complexes because these employ covalent (electron sharing) bonds, rather than ionic (electron transferring) bonds to stick together. Crystal field theory cannot explain the covalent bonding that occurs in coordination complexes.

Limestone. ***(Photograph by Andrew J. Martinez, Photo Researchers, Inc. Reproduced by permission.)***

The ligand field theory is especially useful for explaining the spectroscopic, magnetic, and optical characteristics of ions in the transition metal and rare-earth groups. Many transition metals have an incomplete *d* orbital, meaning there is a gap in the area where their electron is usually found. Examples of transition metals include **vanadium**, **chromium**, **iron**, **cobalt**, **manganese**, **nickel**, **scandium**, **yttrium**, **zirconium**, **silver**, and **gold**.

The ligand field theory's main concern is the effects of ligands on the **energy** levels of the central atom or ion. Ligands are the ions, molecules, or atoms that surround a central ion in a complex. They produce an electrical field that influences the electron system of that central ion, and are usually the source of donated electrons within a coordination complex. The ligand field's strength and symmetry determine the optical, stability, and magnetic characteristics of a transition-metal complex.

Ligand field theory, like crystal field theory, concentrates on what happens when ligands split the central metal atom's inner orbitals. For instance, the size of the split in a complex's crystal field indicates what frequency of light it will absorb, and therefore, what **color** it will be.

See also Crystal field theory

Lime

Calcium oxide (CaO), more commonly known as lime or quicklime, has been studied by scholars as far back as the pre-Christian era. In his book *Historia Naturalis,* for example, Pliny the Elder discussed the preparation, properties, and uses of lime. Probably the first scientific paper on the substance was Dr. **Joseph Black**'s ''Experiments Upon Magnesia, Alba, Quick-lime, and Some Other Alkaline Substances,'' written in 1755.

Lime does not occur naturally since it reacts so readily with **water** (to form hydrated lime) and **carbon** dioxide (to form limestone). It is synthetically produced in very large quantities by the heating of limestone. For many years, calcium oxide has ranked among the top 10 chemicals in the United States in terms of production. Other common names by which the compound is known include burnt lime, unslaked lime, fluxing lime, and calx.

In its pure form, calcium oxide occurs as white crystals, white or gray lumps, or a white granular powder. It has a very high melting point of 4,662°F (2,572°C) and a **boiling point** of 5,162°F (2,850°C). It dissolves in and reacts with water to form calcium hydroxide and is soluble in acids and some organic **solvents**.

Like other calcium compounds, calcium oxide is used for many construction purposes, as in the manufacture of bricks, mortar, plaster, and stucco. Its high melting point makes it attractive as a refractory material, as in the lining of furnaces. The compound is also used in the manufacture of various types of **glass**. Common soda-lime glass, for example, contains about 12% calcium oxide, while high-melting aluminosilicate glass contains about 20% calcium oxide. One of the new forms of glass used to coat surgical implants contains an even higher ratio of calcium oxide, about 24% of the compound.

Lime is also used in the production of pulp and paper, the removal of hair from animal hides, clarifying cane and beet sugar, in poultry feeds, and as a drilling fluid.

In 1816, Thomas Drummond invented limelight, and in 1825 its first use was during a survey of Ireland. Drummond's early limelight was based on a torch that burned **hydrogen** and **oxygen**, which had been developed by **Robert Hare** (1781-1858), an American chemist. Although limelight was considered for lighthouse applications, the cost of production was too high. In 1837, limelight found its niche in the theater, where it was used not only to spotlight actors but also to create realistic special effects such as moonlight on a river or clouds moving through the sky. The light was produced by pointing a hot torch at a solid block of lime (calcium oxide). When heated, the lime gave off a bright, soft white light that was easy to focus on a small area by using a mirror as a reflector. However, limelight had a major disadvantage; it required constant attention by a stagehand to keep turning the block of lime and tending the gas torch. In the late 1800s, limelight began to be replaced by electric arc spotlights. However, its name has lived on; we still speak today of ''basking in the limelight'' of popularity or attention.

Lipids

Lipids are a group of organic compounds characterized by their presence in plants and animals, and their insolubility in **water**. They are extracted from cells using **solvents** such as **ether** or **benzene** that are non-polar. They are composed of mainly **hydrogen** and **carbon** and include materials such as fats, phospholipids, **steroids**, and **waxes**. Lipids have important functions in the body, being involved in such activities as **energy** storage and generation, the construction of cell membranes, and systemic chemical signaling. They also have a protective function in nature.

The most common type of lipids are fats. Fats are large molecules composed of two smaller types of molecules, **fatty acids** and **glycerol**. Glycerol is a polyhydroxy **alcohol** containing three carbon atoms and three hydroxyl groups. Fatty acids can be thought of as having two ends. One end is the ''head'' which contains a carboxylic acid group (COOH). The other end is the ''tail'' which is composed of a long carbon skeleton that can have anywhere from 6 to 20 or more atoms in length. It is this long tail group that makes fats insoluble in water. Nearly all naturally occurring fatty acids have an even number of carbon molecules.

A fat is an ester made when three fatty acids are joined on the glycerol **molecule** through an ester linkage. This is caused by a **dehydration** reaction between the hydroxyl group from the glycerol and the carboxyl groups of the fatty acids. Fats are also known as triglycerides reflecting the molecular composition. The fatty acids in a fat can each be the same or there can be two or three different types. The most common animal fat is tristearin, which is composed of glycerol and three **stearic acid** molecules.

Pure fats have no **color**, odor, or **taste**. However, impurities give fat a natural brownish color. Fats are slippery and less dense than water. When exposed to air for extended periods of time, fats become rancid and develop an unpleasant taste and odor.

Fatty acids are characterized by their carbon chain lengths and the number and location of their double bonds. Saturated fatty acids are those that contain no double bonds in their carbon **chains**. The term is used because all of the bonds that could possibly be occupied by hydrogen atoms are filled. Conversely, unsaturated fatty acids have one or more double bonds between carbon atoms in their chains. For this reason, the shape of an unsaturated fatty acid is slightly kinked wherever a double bond occurs. Some common saturated fatty acids include palmitic and stearic acid. Unsaturated fatty acids include palmitoleic, oleic, linoleic, and linolenic acid. Linoleic and linolenic acid are essential fatty acids that are needed to make critical compounds in the body. However, the body can not produce them so they must be included in a regular diet.

Animal fats such as bacon grease, butter, and lard tend to be saturated. This **molecular structure** tends to cause them to solidify at room **temperature**. In contrast, the fats from fish and plants are mostly unsaturated and liquid at room temperature. The kinks in the carbon chains do not allow the molecules to pack closely enough to form **solids** at these temperatures. Since these materials are liquid they are typically referred to as oils.

Unsaturated fats may be chemically modified by a process called **hydrogenation**. This involves the addition of hydrogen to the carbon chain thereby removing the double bonds and producing hydrogenated vegetable oils. Products such as peanut butter and margarine use hydrogenated vegetable oils to prevent lipids from separating out as **liquids**.

Saturated fats have been implicated as possible causes for atherosclerosis. A symptom of this cardiovascular disease is the formation of plaques on the internal linings of blood vessels. They impede the flow of blood and can lead to a heart attack.

Fats have an important role in biochemical **metabolism**. The primary function of fats is to store energy. One gram of fat contains almost twice as much energy as a gram of carbohydrate. Mammals store fats in specialized tissues composed of adipose cells. This tissue not only stores energy but also provides insulation and protection to vital organs in the body.

Fats can be reduced in a **solution** of **sodium** hydroxide to produce glycerol and another kind of lipid, soap. From a chemical standpoint, soap is a **salt** formed by the reaction of a base, such as **potassium** or **sodium hydroxide**, with fatty acids. In this reaction, the triglycerides are reduced to their component fatty acids. The base then neutralizes them into salts. A byproduct of this method of soap production is glycerin. This process, called saponification, is the basis for the modern day production of bar soap. Detergents are similar to soaps and are produced in an analogous reaction however, the starting material is derived from linear, alkyl compounds. The addition of a **sulfate** group helps prevent some of the drawbacks associated with soaps in hard water.

Since soaps and detergents are salts they separate into their component ions in a solution of water. One end is composed of an **ion** that has two ends with different **solubility** characteristics. The fatty, **hydrocarbon** portion or ''tail'' is hydrophobic (water hating) and associates with the oily particles in the solution. The carboxylate end or ''head'' is hydrophilic (water loving) and tends to associate with the aqueous phase. This unique structure is responsible for the cleansing ability of soaps and detergents. For this reason they are generally known as surface-acting agents, or surfactants.

Phospholipids are another type of lipid similar to fats in that they are composed of long chain fatty acids connected to a glycerol molecule. However, phospholipids have only two fatty acids attached to the **bonding** sites on the glycerol molecule. The third site is occupied by a **phosphate** group which has a negative electrical charge. This phosphate group makes phospholipids polar like surfactants. An additional smaller charged ion is typically linked with the phosphate group. A common phospholipid is phosphatidylcholine or lecithin.

The molecular structure of phospholipids alters their solubility in water. Since the phosphate group is water soluble, this part of the molecule tends to be hydrophilic. The fatty acid tails, however, remain hydrophobic. When put in a solution of water, phospholipids assemble into structures which minimize contact between the water and the hydrophobic tails. One structure, known as a micelle, is a small, spherical cluster of molecules that have the phosphate groups on the surface and the fatty acid tails on the interior. Lamellar sheets are another structure that phospholipids take on depending on the **concentration** of the phospholipid in the solution. Phospholipids are used by cells to produce the cell membrane. They are arranged in a double layer, or bilayer, with the phosphate groups on the outside and the fatty acids groups on the inside. This structure allows cells to be protected from the external environment.

Related to phospholipids are sphingolipids. They are composed of fatty acids bonded to the amino alcohol sphingosine. The most important sphingolipids are sphingomyelins. A common sphingomyelin is ceramide. It is present in most animal cell membranes and provides protection and insulation.

Another class of lipid are steroids. A steroid is characterized by a carbon skeleton made up of four connected rings. Steroids have various functions that are determined by the groups attached to their rings. They are found throughout the body.

Cholesterol is one of the most important types of steroids. It is a component of all membranes in animal cells. It is also a precursor from which nearly all other steroids are synthesized. Cholesterol is the starting material for bile acids, steroid **hormones**, and vitamin D. It is also the precursor for **sex hormones** such as estrogen, progesterone, and testosterone. Bile acids are produced from cholesterol in the liver. The most significant bile acid in humans is cholic acid. It aids in digestion by emulsifying fats in the small intestine. Vitamin D is a steroid produced in the skin through a reaction between cholesterol and sunlight. One problem with cholesterol is that high levels in the body have been found to contribute to heart disease so doctors suggest limiting dietary cholesterol.

Adrenocorticoid hormones are lipids formed in the adrenal gland. They are involved in the maintenance of water and electrolyte balance and in various metabolic reactions. An example of these compounds is cortisol.

Waxes are another type of lipid. They are similar to fats and phospholipids in that they are esters, however, they are based on monohydroxy alcohols instead of glycerol. Most waxes are solid at room temperature but have low melting points. In nature, they are produced by plants and animals as waterproof coatings. This waterproofing characteristic makes them important industrial compounds. Waxes are primarily used to make candles, **cosmetics**, and coatings. A common example is beeswax that is composed primarily of myricyl palmitate, formed by the reaction of palmitic acid with an alcohol. Another important wax is lanolin that is derived from wool. It is used for a variety of medical and cosmetic purposes. Other important waxes include carnauba wax and spermaceti.

See also Soaps and detergents

LIPMANN, FRITZ (1899-1986)

American biochemist

Fritz Lipmann was one of the leading architects of the golden age of **biochemistry**. His landmark paper, "Metabolic Generation and Utilization of Phosphate Bond Energy," published in 1941, laid the foundation for biochemical research over the next three decades, clearly defining such concepts as group potential and the role of group transfer in biosynthesis. Most biochemists clearly recognized that Lipmann had revealed the basis for the relationship between metabolic energy production and its use, providing the first coherent picture of how living organisms operate. His discovery in 1945 of **coenzyme** A (CoA), which occurs in all living cells and is a key element in the **metabolism** of **carbohydrates**, fats, and some **amino acids**, earned him the 1953 Nobel Prize in physiology or medicine. Lipmann also conducted groundbreaking research in **protein synthesis**. Lipmann was an instinctual researcher with a knack for seeing the broader picture. Lacking the talent or inclination for self-promotion, he struggled early in his career before establishing himself in the world of biochemistry.

Fritz Albert Lipmann was born on June 12, 1899, in Königsberg, the capital of East Prussia (now Kaliningrad, Russia). The son of Leopold, a lawyer, and Gertrud Lachmanski, Lipmann grew up in a happy and cultured surroundings and fondly remembered the peaceful years at the turn of the century. He counted his only brother Heinz, who would pursue the arts as opposed to science, as one of the two people who most influenced him in his formative years. The other was Siegfried (Friedel) Sebba, a painter who would remain his friend for life. From these two, he first learned to appreciate the arts, an avenue of interest that he used to escape the confines and pressures of his laboratory investigations.

Early on, Lipmann demonstrated a diffidence in academic pursuit that would belie his future success. He admitted that he was never very good at school, even when he reached the university. After graduating from the gymnasium, Lipmann decided to pursue a career in medicine, largely due to the influence of an uncle who was a pediatrician and one of his boyhood heroes. In 1917 he enrolled in the University of Königsberg but had his medical studies interrupted in 1918 as he was called to the medical service during World War I. Serving near the front during the last days of the war, he first learned to exert authority and never forgot the grim experience of severely wounded men receiving bad care.

In 1919 he was discharged from the army and went to study medicine in Munich and Berlin. Lipmann's brother was a literature student in Munich, and Lipmann became involved with his brother's circle of artistic friends while he lived in Schwabing, which Lipmann called the Greenwich Village of the city at that time. Throughout his life he maintained fond memories of Berlin. He eventually returned to Königsberg to complete his studies and obtained his medical degree from Berlin in 1922. Even though he cared about patients, Lipmann became more intrigued by what went on inside the human body. This interest was further cultivated when during his practical year of medical studies he worked in the pathology department in a Berlin hospital and took a three-month course in modern biochemistry taught by Peter Rona. At the same time, Lipmann was troubled by his concerns over the ethics of profiting from providing necessary medical services. The final turning point came when he went to the University of Amsterdam on a half-year stipend to study pharmacology. There, he first became versed in biochemical problems and the working of a biological laboratory. He left Amsterdam bent on a new career as a researcher.

Returning to Königsberg, Lipmann, who had no money, lived with his parents while he studied **chemistry** in the university for the next three years. Looking for a laboratory to do research in for his thesis, he chose to work with biochemist **Otto Meyerhof**, whose physiological investigations focused on the muscle. For the most part, Lipmann worked on inhibition of glycolysis (the breakdown of glucose by enzymes) by fluoride in muscle contraction and did his doctoral dissertation on metabolic fluoride effects. During this time in Berlin, Lipmann met many of the era's great biochemists, including Karl Lohmann, who discovered adenosine triphosphate (ATP—a compound that provides the chemical energy necessary for a host of chemical reactions in the cell) and taught Lipmann about phosphate ester chemistry, which was to play an important role in Lipmann's later research. Lipmann also met his eventual lifelong companion while attending one of the masked balls popular at that time. Freda Hall, an American-born German and an artist, would become his wife.

Over the next ten years, Lipmann continued with a varied but not very lucrative research career. In *The Roots of Modern Biochemistry,* Freda remembered her husband as a very "unusual young man" who "seemed to be certain of a goal" but "had no position, no prospects, and it did not seem to trouble him." Although he was interested in his work, Freda recalled that "at no time was Fritz the obsessed scientist without other interests. He always had time for fun," which included tennis matches, bicycle races, and the theater.

Lipmann spent a short time in Heidelberg when Meyerhof moved his laboratory there but then returned to Berlin and worked with Albert Fischer on tissue culturing and the study of metabolism as a method to measure cell growth. But soon uniformed followers of Hitler began to appear in the streets of Berlin; both Lipmann and Freda had unpleasant encounters, and once Lipmann was beaten up. Realizing that they would soon have to leave Germany, Lipmann, through Fischer's intervention, received an offer to work at the Rockefeller Foundation (now Rockefeller University). Before leaving for the United States, Lipmann and Freda Hall were married on June 21, 1931. As it turned out, Freda's birth in Ohio made her an American citizen, thus greatly reducing obstacles to immigration. At the Rockefeller Foundation, Lipmann worked in the laboratory of chemist **Phoebus Aaron Theodor Levene**, who had conducted research on the egg yolk protein, which he called vitellinic acid, and found that it contained 10 percent bound phosphate (that is, phosphate strongly attached to other substances). Lipmann's interest in this protein, which served as food for growing animal tissues, led him to isolate serine phosphate from an egg protein.

At the end of the summer of 1932, Lipmann and his wife returned to Europe to work with Fischer, who was now in the Biological Institute of the Carlsberg Foundation in Copenhagen, Denmark. Free to pursue his own scientific interests, Lipmann delved into the mechanism of **fermentation** and glycolysis and eventually cell energy transformation. In the course of these studies, Lipmann found that pyruvate **oxidation** (a reaction that involves the loss of electrons) yielded ATP. Lohmann, who first discovered ATP, had also found that creatine phosphate provides the muscle with energy through ATP. Further work led Lipmann to the discovery of acetyl phosphate and the recognition that this phosphate was the intermediate of pyruvate oxidation. A discovery that Lipmann said was his most impressive work and had motivated all his subsequent research.

Despite his belief in his work, Lipmann had still to make his mark in research. In his book, *Wanderings of a Biochemist,*

Fritz Lipmann.

Lipmann would remember his efforts at the institute and throughout that decade as a time of personal scientific development that set the stage for his later discoveries. "In the Freudian sense," said Lipmann, "all that I did later was subconsciously mapped out there; it started to mature between 1930 and 1940 and was more elaborately realized from then on."

But before Lipmann could piece together his formula for the foundation of how organisms produce energy, once again the rise of the Nazis forced him and his wife to flee to the United States; they were nearly penniless. Fortunately, Lipmann acquired a research fellowship in the biochemistry department of Cornell University Medical College. His work with pyruvate oxidation and ATP had germinated and set him on a series of investigations that led to his theories of phosphate bond energy and energy-rich phosphate bond energy. During a vacation on Lake Iroquois in Vermont, Lipmann began his essay "Metabolic Generation and Utilization of Phosphate Bond Energy," in which he introduced the squiggle to represent energy-rich phosphate, a symbol subsequently used by other researchers to denote energy-rich metabolic linkages. In this essay, Lipmann also first proposed the notion of group potential and the role of group transfer in biosynthesis.

This essay was the turning point in Lipmann's career. Prior to its publication, Lipmann had contributed disparate pieces to the puzzle of biosynthesis, but through his natural scientific instinct and the ability to see the broader picture, he had now laid the foundation for the basis of how living organisms function. Although his essay covered a wide range of topics, including carbamyl phosphate and the synthesis of **sulfate** esters, his elucidation of the role of ATP in group activation (such as amino acids in the synthesis of proteins), foretold the use of ATP in the biosynthesis of macromolecules (large molecules). In more general terms, he identified a link between generation of metabolic energy and its utilization. A prime example of ATP's role in energy transmission was the transfer of phosphor potential from ATP to provide the energy needed for muscles to contract.

Despite the growing acknowledgement that Lipmann had written a groundbreaking paper in biochemistry, he soon found himself without a solid job prospect when Dean Burk, whose lab Lipmann worked in, left for the National Institutes of Health. Burk was reluctant to take Lipmann with him because of Lipmann's lack of interest in Burk's cancer research. While Lipmann's renown had grown, he had also antagonized other researchers, especially by his insistence that the term "bond energy" had been misused and his replacement of the term with "group potential" to refer to the capacity of a biochemical bond to carry **potential energy** for synthesis. It also took many years for the squiggle to be fully accepted as a way to denote energy-carrying bonds. Fortunately, Lipmann gained an unusual appointment in the Department of Surgery at the Massachusetts General Hospital through the support of a Ciba Foundation fellowship. "This was really one of the lucky breaks in my life," Lipmann recalled in his autobiography. Soon he received growing support from the Commonwealth Fund as more and more people began to recognize the importance of his work. Building upon his group transfer concept, Lipmann delved into the nature of the metabolically active acetate, which had been postulated as an "active" intermediary in group activation. In 1945, working with a potent **enzyme** from pigeon liver extract as an assay system for acetyl transfer in animal tissue, Lipmann and colleagues at Massachusetts General Hospital discovered Coenzyme A (CoA), the "A" standing for the activation of acetate. (Coenzymes are organic substances that can attach themselves to and supplement specified **proteins** to form active enzyme systems.) Eventually, CoA would be shown to occur in all living cells as an essential component in the metabolism of carbohydrates, fats, and certain amino acids. In 1953 Lipmann received the Nobel Prize in physiology or medicine for his discovery specifically of the acetyl-carrying CoA, which is formed as an intermediate in metabolism and active as a coenzyme in biological acetylations. (Lipmann shared the prize with his old colleague and friend, **Hans Krebs**, from Berlin.) Although proud of the Nobel Prize, Lipmann often stated that he believed his earlier work on the theory of group transfer was more deserving.

In 1957 Lipmann once again found himself at the Rockefeller Institute, twenty-five years after his first appointment there. Lipmann was to spend the next thirty years at the institute, primarily working on the analysis of protein biosynthesis. He and his colleagues contributed greatly to our understanding of the mechanisms of the elongation step of protein synthesis (stepwise addition of single amino acids to the primary protein structure).

Lipmann's productive career included 516 publications between 1924 and 1985. His 1944 paper on acetyl phosphate is a citation classic, having been cited in other works more than seven hundred times. His work on high-energy phosphate bonds and group transfer discoveries propelled biochemistry to the forefront of physiological research for nearly three decades. In addition to the Nobel Prize, Lipmann received the National Medal of Science in 1966 and was elected a foreign member of the Royal Society in London.

In 1959 the Lipmanns, who had a son Stephen Hall, bought a country home with Fritz's Nobel Prize money. Although his wife described him essentially as a city person, Lipmann enjoyed the country and often strolled the twenty acres of woods that surrounded his home with his Australian terrier, Pogo, named after the satiric comic strip character popular in the 1960s and 1970s. A private man who avoided political and social issues, Lipmann did, however, sign the Nobel laureate public appeal letters seeking prohibition of the **hydrogen** bomb and asking for freedom for the Polish Worker's Union. Lipmann's talent for writing was evident in the easy-to-follow and informative format of his scientific essays and in his autobiography. Still, he was given to preoccupation, and a colleague fondly recalled Lipmann once combing an auditorium after a lecture in search of his shoes, which he had left behind in going to the podium.

Lipmann's unique ability to see the entire scientific picture set him apart from many of his contemporaries. Interestingly, this ability also translated into his noted penchant for spotting four-leaf clovers almost anywhere. He kept them in books, manuscripts, and wallets, perhaps reflecting his own estimation that he had been fortunate in a life and career that allowed him to follow his instincts so successfully.

Despite failing strength, Lipmann continued to work until he suffered a stroke on July 17, 1986, and died seven days later. "One evening I heard him say: I can't function anymore," recalled Freda Hall in *The Roots of Modern Biochemistry,* "and that was that." Lipmann's ashes were scattered along his walking path in the woods that surrounded his home.

LIPSCOMB, JR., WILLIAM NUNN (1919-)

American physical chemist

William Nunn Lipscomb, Jr. worked in several of the subdivisions of **chemistry** in his career, but was awarded the 1976 Nobel Prize in chemistry for his studies of chemical **bonding** in **boron** compounds, particularly the boron hydrides. These materials break some of the conventional rules of chemistry, and Lipscomb's theories have expanded chemists' understanding of how atoms can be bonded to one another.

Lipscomb was born on December 9, 1919, to William and Edna (Porter) Lipscomb. He grew up in Lexington, Kentucky, after his family had moved there from Cleveland, Ohio, when he was a year old. He graduated from the University of Kentucky with a degree in chemistry in 1941. After serving with the United States Office of Scientific Research and Development during World War II, he finished his doctorate in **physical chemistry** at the California Institute of Technology in 1946. His thesis advisor was the distinguished Nobel Laureate **Linus Pauling**. Lipscomb taught for several years at the University of Minnesota, and in 1959 moved to Harvard University, where he spent the rest of his working life.

The bonding in boranes—a compound consisting of boron and hydrogen—had puzzled chemists ever since the German chemist **Alfred Stock** had synthesized them. (Boranes do not normally occur in nature; they are purely man-made compounds.) Normally, chemists expect a bond to consist of two electrons located between two atoms; this is standard bonding theory, originally advanced by American chemist **Gilbert Newton Lewis**. Additionally, they expect most atoms to be stable if their outer **electron** shells contain eight electrons, in either bonded or non-bonded pairs. Boranes do not appear to possess sufficient outer-shell electrons to conform to either of these ideas, and are thus classified as "electron deficient" compounds. Several scientists, including Pauling himself, had tried to explain electron deficiency, but none of the theories proved satisfactory. H. C. Longuet-Higgins had advanced the theory that **hydrogen** could be linked to two boron atoms, forming a "three-center" bond with hydrogen bridging the two borons.

Lipscomb and his co-workers Bryce Crawford, Jr., and W. H. Eberhardt expanded on this idea in 1954 when they suggested in the *Journal of Chemical Physics* that, in addition to the three-center bond, a regular **covalent bond** could hold boron and hydrogen atoms together under certain circumstances. Each bond still consisted of two electrons, but the covalent link was made possible because more than two atoms regarded themselves as bonded by the two electrons. Depending on the boron compound, the three atoms linked by the single bond might be three borons, or two borons and a hydrogen. Lipscomb knew how radical this idea was; his 1954 paper with Crawford and Eberhardt stated, "We have even ventured a few predictions, knowing that if we must join the ranks of boron hydride predictors later proved wrong, we shall be in the best of company."

Lipscomb's research also expanded the understanding of the unusual shape of large boron molecules. Because of the three-center bonds, these molecules fold up into basket or cage-like structures that are chemically very stable. (Chemical stability does not occur in the smaller boranes, however, which are volatile, reactive, and occasionally explosive). Lipscomb's contributions to the structure determination and molecular orbital theories of these molecules allowed the planned synthesis of many more of them, including boron combined with other elements. **Carboranes**, for example, are boron cages with **carbon** embedded in the structure; they are used in the manufacture of highly stable polymers that resist both **heat** and chemicals. Many other elements are now known to make stable cage structures, but Lipscomb's elucidation of boron chemistry laid much of the groundwork for later discoveries.

Later in his career, Lipscomb worked on biochemical problems, particularly the structure determination of **proteins**. As in his boron work, he used the technique of **X-ray diffraction** in his investigations. A success in this area was his determination of the structure of carboxypeptidase A, a digestive **enzyme** and an enormous **molecule** with a **molecular weight** of 34,400. He next worked on the enzyme trans-carbamylase, which is involved in the synthesis of **nucleic acids** in the body.

In 1976 Lipscomb was awarded the Nobel Prize in chemistry for his work on the chemical bonding in boron compounds. Russell N. Grimes, writing in *Science*, noted that theories advanced by Lipscomb had been utilized in regard to many elements, "but it is fair to say that it all began with boron hydrides—and it is here that Lipscomb's powerful insight and creative imagination have been so effectively felt." In addition to the Nobel Prize, Lipscomb has garnered many other awards, among them the George Ledlie Prize of Harvard University and the Distinguished Service Award of the American Chemical Society.

Lipscomb married Mary Adele in 1944; the couple had a son and a daughter, but were divorced in 1983. Lipscomb married Jean Evans, a lettering artist, soon afterwards. A well-read and enthusiastic individual, affectionately called "Colonel" by his associates, Lipscomb is also an excellent musician. His particular interest is chamber music, and he has played clarinet in chamber groups with professionals. Russell Grimes wrote in *Science* that "the dominant personal characteristics of Professor Lipscomb have been an unfailing scientific imagination, a refusal to accept the limitations imposed by current dogma, an ability to perceive relationships often missed by others, and above all, a delight in the intellectual challenge of uncovering scientific truth."

LIQUEFACTION OF GASES

Since the 1600s, chemists have known that **temperature** can determine whether a substance exists as a gas, liquid, or solid. Flemish chemist **Johannes van Helmont** who coined the term gas to describe **carbon** dioxide, used the term vapors to describe substances that became gaseous only when heated, such as **water**. During the late 1700s, scientists learned that when a gas is cooled, its **volume** is reduced by a predictable amount. Cooling slows down the motion of the gas molecules, so they take up less space. Similarly, pressurizing a gas, or forcibly squeezing its molecules closer together, reduces its volume. Eventually, through cooling and compression, the volume of a gas can be reduced by so much that its molecules collapse upon each other and come into contact, changing into a liquid. Compression and cooling soon became the twin tools of scientists attempting to liquefy **gases**.

The first scientist to liquefy a substance that normally exists as a gas was **Gaspard Monge**, a French mathematician, who produced liquid **sulfur** dioxide in 1784. However, most gases were not liquefied until the mid-1800s, beginning in 1823 when English chemist **Michael Faraday** liquefied **chlorine**. Faraday pressurized chlorine gas inside a curved **glass** tube that was submerged at one end in a beaker of crushed ice. Under pressure, the gas changed into liquid chlorine when cooled by the ice near the end of the tube. Faraday also liquefied **carbon dioxide, hydrogen** sulfide, and hydrogen bromide in a similar manner. More than 20 years later, after pursuing other research, Faraday returned to gas liquefaction. By then, more effective cooling agents had been developed. Despite the combined effects of cooling and compression, Faraday was unable to liquefy several gases, such as **oxygen** and hydrogen, which he called "permanent" gases. Then in the late 1840s, Irish physical chemist Thomas Andrews (1813-1885) suggested that every gas has a precise temperature—called the critical temperature—above which the gas cannot be liquefied even under greater pressure. Andrews reached his conclusion by observing the behavior of pressurized liquid carbon dioxide.

Andrews' concept of critical temperature soon led to a breakthrough in the liquefaction of the so-called permanent gases. To reach temperatures low enough to liquefy these gases, two scientists independently came up with the idea of using a "cascade" process that reduces temperature step by step. In this method, one liquefied gas is used to cool a second gas that has a lower critical temperature; then the second gas, when liquefied, is used to cool another gas with an even lower critical temperature; and so on.

In 1877, French physicist Louis Paul Cailletet (1832-1913) succeeded in liquefying three permanent gases—oxygen, **nitrogen**, and **carbon monoxide** using the cascade process.

Around the same time, Swiss chemist Raoul Pierre Pictet (1846-1929) liquefied oxygen using methods very similar to Cailletet's, and propriety was vigorously debated, although Cailletet demonstrated priority. Because Pictet's equipment was more elaborate, he was able to produce greater quantities of liquid oxygen. Pictet's interest in gas liquefaction arose from his attempts to produce ice artificially for use as a refrigerant.

Although most gases had been liquefied by the late 1800s, commercial production of them was still unfeasible until German chemist Karl von Linde (1842- 1934) invented a continuous process for producing large quantities of liquid air (mostly nitrogen and oxygen) in 1895. British chemical engineer William Hampson (1859- 1926) invented a simila r liquefaction method around the same time. Von Linde became an engineering professor in Munich, where he became interested in low-temperature research. In 1876, he developed the first practical refrigerator.

Von Linde's commercial liquefaction process makes use of a phenomenon called the Joule-Thomson effect, named for its discoverers, English physicist **James Joule** and British physicist Lord Kelvin (**William Thomson**). These scientists had shown in 1853 that a compressed gas becomes cooler when it expands, assuming it does not absorb **heat** from its surroundings. In von Linde's process, which is still the basis for modern practice, liquid air is cooled and compressed, then allowed to expand, cooling it even more. The cold air is constantly recycled to cool more incoming compressed air. Because of the cumulative cooling effect, the air gradually becomes cold enough to liquefy. Von Linde's process immediately became a commercial success and laid the foundation for today's liquid air production industry.

Von Linde also developed more economical methods of separating the liquid oxygen and liquid nitrogen, which both found many practical uses in research and industry. Incandescent light bulbs filled with nitrogen, for example, lasted longer than earlier bulbs that used a vacuum instead of a filler gas. Liquid nitrogen is also quite useful for instantaneous freezing. In biological research, liquid nitrogen is used to freeze blood cells, sperm, tissues, and even whole small organisms. When frozen, the cells stop normal activities, allowing scientists to examine a "freeze-frame" of cellular life.

Meanwhile, hydrogen gas had stubbornly resisted all scientific attempts to liquefy it until 1898, when Scottish chemist and physicist **James Dewar** (1842-1923) applied the Linde process in larger, more efficient equipment. Dewar used liquid air to pre-cool compressed hydrogen, reducing its temperature enough to liquefy the gas by expansion. (Some gases such as hydrogen have a very low inversion temperature; if the gas is expanded above this temperature, it becomes warmer instead of cooler.) A year later, in 1899, Dewar succeeded in solidifying hydrogen.

Dewar, a professor at universities in London and Cambridge, had become interested in the field of extremely low temperatures during the 1870s, when permanent gases first began to be liquefied. In 1891, Dewar produced liquid oxygen in large quantities and studied its magnetic properties.

Near the end of the century, a new family of elements called the *inert*, or nonreactive, gases was discovered by British chemist Sir **William Ramsay** (1852-1916) and his coworkers. These gases—argon, **helium**, and neon—presented another challenge to scientists interested in liquefaction. Dewar came very close to liquefying helium, but his sample of the gas also contained **neon**, which froze and blocked the valves on his equipment. A few years later, in 1908, Dutch physicist Heike Kamerlingh Onnes (1853-1926) succeeded in producing liquid helium using techniques similar to Dewar's. The compressed helium gas was pre-cooled by liquid hydrogen before undergoing expansion cooling, as in the Linde process. Kamerlingh Onnes' equipment was rather elaborate; later, other scientists developed simpler helium liquefiers that could produce greater amounts of the liquid gas.

Other research related to gas liquefaction was concerned with temperature measurement. British chemist Morris William Travers (1872-1961) was the first scientist to measure the temperature of liquid gases accurately. After working in India at a newly-founded science institute, Travers returned to England and became director of a glass manufacturing plant, which led to his early interest in furnaces and fuel technologies, such as **coal** gasification. In 1894, Travers began working with Ramsay on isolating the new inert gases. He shares credit with Ramsay for discovering **krypton** in 1898. Independently of Dewar, Travers constructed equipment for liquefying hydrogen that he and Ramsay used to obtain neon from air. Late in life, Travers wrote a biography of Ramsay that recounted their exciting work on the inert gases.

Today, key commercial uses of liquefied gas include liquefied **refrigerants** for cryogenic application and liquefied petroleum gases for fuel use.

LIQUID CRYSTALS

A liquid crystal is a substance that apparently falls between the solid and liquid states. The three well-known states of **matter** are **solids**, **liquids**, and **gases**. These three **states of matter** exist because of the different degrees of order and **energy** in molecules. The molecules in each state have a different degree of order or randomness about them. For example, a solid is composed of molecules with a large amount of order and lower kinetic energy. The molecules in a crystalline solid occupy a specific place in the arrangement of the solid and are also oriented in a specific manner. The molecules in a solid vibrate a bit, but they maintain this highly ordered arrangement. The molecules in a solid can be thought of as balls connected with springs. The balls can vibrate due to the contractions and expansions of the springs between them, but overall they stay in the same place with the same orientation.

Another type of solid exists, an amorphous solid, that does not crystallize. An amorphous solid has molecules that do not move, but there is no orderly arrangement between them, i.e. the molecules are arranged randomly within the solid. An example of an amorphous solid is **glass**. An amorphous solid has a definite shape without the geometric regularity of crystalline solids. Glass can be molded into any shape. If glass is shattered, the resulting pieces are irregularly shaped. A crystalline solid exhibits regular geometrical shapes when shattered. Amorphous solids tend to hold their shape, but they also tend to flow very slowly. It is not easy to deform a solid because of the strong attractive forces within the structure. Solids tend to be hard, highly ordered, and not easily deformed.

The molecules in a liquid, in contrast, are not located in a specific place in the substance, are free to move, and are not oriented in a particular way. These molecules move randomly, colliding with other molecules or the walls of the container that holds the liquid. Thus, a liquid is not as orderly as a solid, and the molecules have a greater kinetic energy. Attractive forces between molecules in liquids may be strong, but they are not as strong as those in solids. Liquids, therefore, tend to deform much more easily than solids. They are somewhat more easily compressed than solids, but less easily than gases. A liquid does not have a shape, instead it takes the shape of the container in which it is placed.

Every substance exists in a specific state of matter at a given **temperature**. A change in temperature changes the stability of the molecules in a substance, resulting in a **change of state**. As the temperature of a substance is increased, so is the kinetic energy of the molecules. As the kinetic energy of the molecules of a solid increases, they begin to vibrate more and more rapidly. At a certain point, the attractive forces between the molecules are no longer sufficient to hold the molecules in place, and they begin to move. This is the melting of a solid, or the change of state from a solid to a liquid. The change of state occurs at a precise temperature which is called the melting point of the substance. At that temperature, a phase transition is said to have occurred.

The study of liquid crystals began in 1888 with the study of the phase transitions of a substance called chloesteryl myristate by the Austrian botanist Friedrich Reinitzer (1857-1927). A **molecule** of chloesteryl myristate consists of mostly **carbon** and **hydrogen** atoms. It is a common substance that is found in the human body in cell membranes as well as in the deposits that cause atherosclerosis, or hardening of the arteries. This substance is a solid at room temperature, 68°F (20°C). If chloesteryl myristate is slowly heated, it will melt at 160°F (71°C). At this point, the liquid is very cloudy and viscous. It does not resemble other liquids such as **alcohol** or **water**. If heating of this liquid continues, another change occurs at 185°F (85°C). At this point, the cloudy liquid becomes clear and less viscous, and now resembles other liquids. If the temperature is allowed to increase, no further changes occur, even at temperatures above 392°F (200°C).

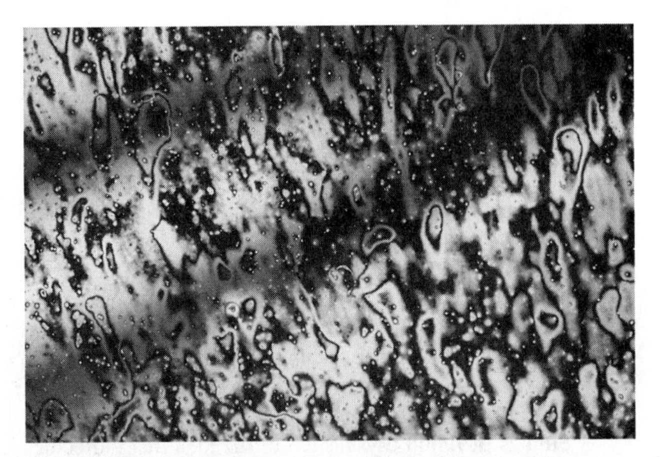

Liquid crystals. ***(Photograph by Michael W. Davidson, Photo Researchers, Inc. Reproduced by permission.)***

If a phase transition is a structural change that occurs at a constant temperature with the absorption of **heat**, then it would appear that chloesteryl myristate went through two phase transitions in the above experiment. The first change was from a solid to the cloudy liquid-like state, and the second change was from the cloudy state to the clear liquid. Actually, the first phase transition was not the transition from a solid to a liquid, rather, it was the transition from a solid to a liquid crystal state.

The liquid crystal state is a fluid state. The molecules in a liquid crystal do move. The liquid crystal flows and takes the shape of its container. It differs from the liquid phase because of its **viscosity** and cloudiness. When a solid melts to a liquid, the molecules in the solid lose two kinds of order—positional order (the idea that the molecules in a solid occupy specific places in the solid) and orientational order (the idea that the molecules in a solid are oriented in a particular way). When a solid changes into a liquid crystal, the positional order is lost, but some of the orientational order remains. The molecules in a liquid crystal are free to move, only they remain oriented in a particular direction as they move. This orientation is not as rigid as in a solid, but on average the molecules tend to stay orientated very close to the way in which they were oriented in the solid. Because the liquid crystal maintains a different order than either a liquid or a solid, it is considered to be a different state of matter altogether. A liquid crystal is also more responsive to electric and magnetic fields than either a solid or a gas, i.e., it responds to much weaker electric and magnetic fields.

A liquid crystal is considered to be more similar to a liquid than to a solid. The reason for this is the latent heat of the liquid crystal to liquid phase transition. The latent heat of a transition is the amount of energy required to cause a phase transition. For cholesteryl myristate, the latent heat of a solid to a liquid transition is 65 calories/gram, and the latent heat for the liquid crystal to liquid transition is only 7 calories/gram. The fact that the latent heat of this transition is so low means the liquid crystal phase is more similar to the liquid phase than to the solid phase.

Studies of liquid crystals led to the idea that molecules possess certain characteristics which make them likely or unlikely to be a liquid crystal at some temperature. Molecules that are likely to be liquid crystals at some temperature have an elongated shape. They also must have some rigidity in the center of the molecule and flexibility at the ends of the molecule. Molecules that do not possess these characteristics are unlikely to be liquid crystals.

There are three different types of liquid crystal phases. The first type is called the nematic liquid crystal phase. Molecules in the nematic liquid crystal phase stay parallel to one another. The second type is the cholesteric liquid crystal phase, also called the chiral nematic liquid crystal phase. Molecules in the cholesteric liquid crystal phase are aligned at a slight angle to one another. A substance can be one of the two, but not both. The third type of liquid crystal phase—the smectic phase—can either be the only phase a substance possesses, or it may be a phase that occurs at a temperature below one of the other two phases. Molecules in the smectic phase have orientational as well as a small amount of positional order. Molecules in this phase tend to arrange themselves in layers.

There are many uses for liquid crystals. One of the most familiar uses is in liquid crystal displays (LCDs). LCDs require less power than other types of displays, which makes them ideal for battery-operated products such as watches, radios, and calculators. Forehead **thermometers** incorporate liquid crystal temperature sensors, making use of the fact that the **color** reflected by a liquid crystal changes with temperature. Because of this property, liquid crystals are also used in fish tank thermometers. Liquid crystals are also important biologically, as mentioned earlier. Liquid crystal structures are present in the cell membrane, the nuclear membrane, the endoplasmic reticulum of a cell, and in the chloroplast membrane in plants. Further research into the properties of liquid crystals is certain to provide even more uses for substances in this state.

LIQUIDS

A liquid is a state of matter intermediate between a solid and a gas. A liquid, at constant **temperature** can readily change its shape (like a gas but unlike a solid) but it is incapable of changing its **volume** (like a solid but unlike a gas).

A substance can be said to be a liquid when its physical properties are that it has a fixed volume, a moderate to high **density**, an indefinite shape (it takes the shape of its container), and it generally flows easily.

The **change of state** from a solid to a liquid is melting, from a liquid to a solid is freezing. The change from a liquid to a gas is boiling and from a gas to a liquid is condensing. With a pure substance each of these changes of state occurs at a specific temperature. This is an indication of a pure substance, an impure substance or mixture would change states at an indefinite temperature, i.e., at a range of temperatures around the correct one. The melting and freezing point of a liquid are always the same temperature, the **boiling point** and the condensation point are also at a characteristic temperature. Some materials, such as **carbon** dioxide, do not exist in a liquid state, instead they can change directly form solid to gas, a process known as **sublimation**. A liquid can change to a gaseous state at a temperature below its boiling point, this process is know as evaporation.

All of these changes of state and the properties of a liquid can be explained by considering the molecules that make up the liquid. In a liquid the molecules are in constant motion, but they are always touching neighbouring molecules. Because of this the order achieved by the molecules is very local involving only adjacent molecules. There is no overall order within the liquid. The motion of the molecules is a function of the **energy** that is present in the system, in other words it is a measure of the temperature of the system. The motion of molecules in a liquid can be observed indirectly. If solid particles of a small size (such as pollen grains) are placed on the surface of a liquid they are hit by molecules of the liquid, this causes the solid to move in a random manner. This is known as **Brownian motion**. If the temperature is decreased the molecules lose energy and they start to become more ordered as their movement decreases. This is the change form a liquid to a solid. The change form a liquid to a gas is the reverse process, as the energy of the molecules increases they become less ordered and the distances between the molecules increases. Eventually the molecules are acting independently of each other, this is characteristic of a gas. During evaporation some molecules spontaneously achieve sufficient energy to release them from the body of the liquid and they transform into the gas or vapor phase. This extra energy is usually from molecular collisions—as one **molecule** hits another it transfers some of its energy across. When these collisions occur near the surface of the liquid the energy obtained may be enough to allow the molecule to escape from its neighbours. The closer the liquid is to its boiling point the more rapidly evaporation takes place. Examples of this can be seen after rain, a puddle of **water** on the ground will rapidly disappear due to evaporation. If a liquid is stored in a sealed container this change form liquid to vapor will still occur. In this situation however an equilibrium will be reached between the number of molecules evaporating and the number of molecules condensing back into the liquid phase. The pressure exerted by the vapor molecules, once equilibrium has been achieved, is the **vapor pressure**.

When a liquid boils it does so at a specific temperature. This temperature can be changed by altering the pressure.

When the pressure is increased the boiling point is increased and when the pressure is decreased the boiling point decreases. An autoclave or pressure cooker operates on this principle. By increasing the pressure inside these vessels the temperature of the water can be greatly increased allowing the killing of bacteria or the cooking of food as appropriate. Decreasing pressure can be seen to have an effect with mountaineers, at the top of Mount Everest water will boil at a specific temperature which is lower than the standard temperature of 212°F (100°C) at which water boils at a pressure of one atmosphere. The boiling point of a liquid can tell us the nature of the **intermolecular forces** within the liquid, those with a high boiling point have stronger attractive forces between the molecules than those with lower boiling points.

When liquids are added to each other one of three thing may occur, depending on the liquids used. Firstly there may be a chemical reaction, secondly there may be complete mixing of the liquids, and thirdly the liquids may not mix or react at all. Chemical reactions in liquids occur in the same manner as all other chemical reactions. Liquids that mix together completely are said to be miscible. The liquids in a mixture such as this can be separated by fractional **distillation**. When two liquids are added together and they do not mix they are said to be immiscible. Immiscible liquids can be separated by decantation or by use of a separating funnel. Many **solids** will dissolve into a liquid to produce a **solution**. The liquid is the solvent and the solid is the solute.

A liquid has a boundary layer with anything it touches, this boundary can act like a form of skin around the liquid. For example if a drop of liquid is placed on a sheet of **glass** it may produce a bead shape. This is due to surface tension. The molecules in a liquid are equally attracted to each other but those at the surface will have an overall inward attraction. Surface tension acts to reduce the surface of a liquid to the minimum. This effect can also be seen when a liquid is in a tube such as a measuring cylinder. When closely observed the liquid is not totally flat. Depending on the liquid the surface may be concave or convex, exactly which situation depends on the relative cohesive and adhesive forces of the materials involved. With water the surface is concave, with **mercury** it is convex. This is called the **meniscus**.

Some liquids can pass **electricity**, they are conductors. Liquid metals, liquid metal/non metal compounds, aqueous solutions of metal/non metal compounds, and aqueous solutions of acids. This characteristic is used in **electrolysis** and it occurs because the liquids will allow a flow of electrons. With the non metal liquids mentioned when electricity is passed through them they will decompose.

Gravity pulls a liquid down in its container, this exerts pressure on the container and on anything in the liquid. With an increase in depth of the liquid the pressure exerted is greater (this is why dams are thicker at the base than at the top). Unlike **gases** liquids cannot be compressed. This fact is used in hydraulics. Liquids under pressure are used to transmit forces. This pressure is transmitted through the whole body of the liquid. This allows us to increase the force applied at a particular point.

Liquids are a state of matter with properties intermediate between solids and gases. The fact that water is a liquid at standard temperature and pressure is one of the main reasons life has evolved in the manner it has on the Earth. Liquids have a wide range of uses, many of which are related to what the substance is. Because of the smaller volume of a liquid it is often more convenient to transfer substances which we normally think of as gases from one location to another as liquids. This change is often brought about by increasing the pressure or sometimes by decreasing the temperature.

Two bottles filled with clear liquid. ***(Photograph by Charles D. Winters, National Audubon Society Collection/Photo Researchers, Inc. Reproduced by permission.)***

See also States of matter

LITHIUM

Lithium is the first element in Group 1 of the **periodic table**, a group of elements generally known as the **alkali metals**. It is the lightest of all metals. Lithium's **atomic number** is 3, its atomic **mass** is 6.941, and its chemical symbol is Li.

Properties

Lithium is a very soft, silvery metal with a melting point of 356.97°F (180.54°C), a **boiling point** of 2,435°F (1,335°C), and a **density** of 0.534 grams per cubic centimeter. Lithium is extraordinarily soft for a metal with a rating of 0.6 on the Mohs scale, softer even than talc, whose Mohs rating is 1.

Lithium is an active metal, although, as one would predict from its place in the periodic table, not as active as **sodium**, **potassium**, and the other alkali metals. For example, it reacts slowly with cold **water**, releasing **hydrogen** gas. It does not react with **oxygen** at room **temperature**, although it does react at higher temperatures to form lithium **oxide** (LiO_2). Under the proper conditions, the element also reacts with **sulfur**, hydrogen, **nitrogen**, and the **halogens**.

Occurrence and Extraction

Lithium is about the 15th most abundance element in the Earth's crust with an abundance of about 0.005%. It never occurs as a free element, but is found in minerals such as spodumene, petalite, and lepidolite. Some lithium is also found in seawater, primarily as dissolved lithium chloride (LiCl). The world's largest producer of lithium is the United States, followed by Australia, Russia, Canada, Zimbabwe, Chile, and China. In the United States, lithium is produced at three large mines in Nevada and North Carolina.

The first step in extracting lithium from its ores is to convert the ore to lithium chloride, which is then electrolyzed to obtain the pure element: 2LiCl —electric current→ 2Li + Cl_2.

Discovery and Naming

Credit for the discovery of lithium is usually given to the Swedish chemist John August Arfwedson (or Arfvedson; 1792-1841) who isolated the metal from petalite in 1817. Arfwedson was not able to prepare pure lithium because it was too active. That step was accomplished about a year later by the Swedish chemist William Thomas Brande (1788-1866) and the English chemist Sir **Humphry Davy**, both of whom used a method similar to the one employed today for extracting lithium from lithium chloride by **electrolysis**. Arfwedson suggested the name lithium after the Greek word *lithos* for "stone."

Uses

Some lithium metal is used to make alloys, especially with **aluminum** or **magnesium**. These alloys tend to be strong, but very light. They are used for armor plates and in aerospace applications.

Most lithium is used in the form of a compound. For example, the addition of a small amount of lithium carbonate (Li_2CO_3) to a **glass** or ceramic adds strength to the final product. Examples of the use of lithium carbonate glass are in the production of shock-resistant cookware (such as the Pyrex brand) and black-and-white television tubes. About 40% of the lithium used in the United States now goes to these applications.

Another industrial application of lithium carbonate is in the production of aluminum metal. Adding lithium carbonate to the reaction mixture reduces the temperature at which aluminum is produced from its ores, thus providing a significant cost savings to the industry. Another compound of lithium, lithium stearate, is added to petroleum to make a thick lubricating grease that does not break down at high temperatures or become hard at low temperatures. The grease is used in military, industrial, automotive, aircraft, and marine applications.

Health Issues

An important new use for lithium carbonate was discovered in 1949 by Australian physician John Cade (1912-1980). Cade found that patients with bipolar disorder (manic-depressive syndrome) obtained relief from their symptoms when treated with lithium carbonate. Prior to Cade's discovery, there had been essentially no successful treatment for bipolar disorder. The drug can have uncomfortable side effects for some patients, but it can also provide a level of relief that is possible with no other treatment.

Lomonosov, Mikhail Vasilevich (1711-1765)

Russian chemist and physicist

The Russian chemist and physicist Mikhail Vasilevich Lomonosov proposed advanced scientific theories, but the diversity of his activities and interests hindered him from gaining widespread recognition.

Mikhail Lomonosov was born on November 8, 1711, in the village of Denisovka. There being few opportunities for education in his native village, he ran away at the age of 19 to Moscow, where he entered a theological seminary and began to study for the priesthood.

Having displayed outstanding abilities as a scholar, young Lomonosov was chosen in 1735 to attend lectures given at the Academy of Sciences in St. Petersburg. This experience changed the whole direction of his career. The St. Petersburg Academy was at this time promoting a series of studies on the material resources of Siberia for which it needed trained chemists and metallurgists. From 1736 to 1741 Lomonosov studied these subjects in Germany, first at the University of Marburg, where he gained a thorough grounding in the basic sciences, and later at the famous mining academy at Freiburg.

On his return to Russia, Lomonosov became a member of the St. Petersburg Academy, and the remainder of his life was devoted almost exclusively to its affairs. He soon emerged as the contentious leader of the group of native Russian scientists in the academy opposed to the clique of foreign members, largely German, who had been imported into its membership at its foundation to stimulate Russian science. In 1745 he was appointed professor of **chemistry** at the academy, where he built a chemical laboratory for instruction and research. Here he gave one of the earliest courses in practical instruction in chemistry.

Although Lomonosov published much on various aspects of physics and chemistry, his works were mainly in the form of dissertations with a limited circulation. His many activities seem to have prevented him from completing many of his projects, and much of his original work was never published.

Lomonosov's physical and chemical work was characterized by its emphasis on the use of atomic and molecular modes of explanation. In a century when most scientists regarded **heat** as material substance, he argued that heat was in fact a form of motion—the result of the motion of the molecules which constitute **matter**. His essentially physical approach to chemistry led him to place great emphasis on quantitative measurements. In this he was certainly ahead of his time, although the claims that he anticipated **Antoine Laurent Lavoisier** in stressing the **conservation of mass** in chemical reactions and recognizing the chemical role of air in **combustion** are certainly exaggerated.

Lomonosov's other scientific interests were **electricity**, light, **mineralogy**, meteorology, and astronomy. He observed the transit of Venus in 1761 and concluded that Venus had an atmosphere "similar to, or perhaps greater than that of the earth." He also made significant contributions to the philological study of the Russian language, including the development of a scientific vocabulary, and wrote a controversial history of Russia.

Although Lomonosov was a man of immense talent, his creative energies were somewhat thwarted by his domineering nature and quarrelsome disposition. He died from influenza in St. Petersburg on April 4, 1765.

Mikhail Vasilevich Lomonosov.

London, Fritz (1900-1954)

American physicist

Fritz London discovered practical applications of quantum mechanics and is known primarily for his work with the explanation of the **covalent bond** in **hydrogen**. His theories regarding **superconductivity** and superfluids were also important contributions to the development of these areas of science. Fritz Wolfgang London was born on March 7, 1900, in Breslau, Germany (now Wroclaw, Poland). He was the elder of two sons born to Luise (Hamburger) and Franz London, who taught mathematics at the University of Breslau as well as at the University of Bonn. Fritz London received a classical education in high school, and his university studies followed the same pattern. He attended the universities of Bonn, Frankfurt, and Munich, receiving his Ph.D. in philosophy from Munich in 1921; his dissertation involved a study of symbolic logic. In the four-year period following his graduation, London taught at various secondary schools in Germany and continued to study and write about philosophy. He then decided to pursue a new field of interest—theoretical physics—and returned to Munich to study under German theoretical physicist **Arnold Sommerfeld**.

One of the first topics on which London worked at Munich was the hydrogen **molecule**. He was interested in the question of whether or not modern quantum mechanics could provide an explanation for the covalent bond—the chemical bond involving shared electrons—that holds two hydrogen atoms together in a hydrogen molecule. That task became the model for much of London's subsequent work.

Until the 1930s, quantum mechanics, or wave mechanics, had been regarded as a totally sufficient system—in fact, the only correct system—for dealing with atomic-scale particles and their interactions with **energy**. The practical application was much less clear, and London began to explore the ways in which quantum principles could be used to explain visible phenomena. His research into the hydrogen molecule in the late 1920s, along with German physicist Walter Heitler, is recognized for advancing the existing knowledge of chemical **bonding**. In his 1954 introduction to London's second volume of *Superfluids,* Felix Bloch described the duo's success with the hydrogen molecule as an illustration of "the direct connection between pure quantum phenomena and some of the most striking facts of **chemistry**." In 1927 London and Heitler published a paper detailing the results of their analysis.

After this triumph, London continued to work on similar problems, looking for the applications of quantum theory to various types of chemical reactions and to the nature of forces between two molecules. Most of his work was brought together in a book on molecular theory scheduled to be published by the German firm of Springer. By the time the manuscript reached the publisher, however, London had left Germany, and Springer refused to honor its contract. Although London continued to work on the book in English translation, it never appeared in print.

In 1932 London turned his energies toward a new topic: superconductivity, or the complete disappearance of electrical resistance in a substance, seen particularly at low temperatures. That topic was one his younger brother Heinz London had selected for his doctoral research at the University of Breslau. Over the next few years, the London brothers worked together to develop a new theory of superconductivity. In the course of their research, they found that an extremely thin outer layer in the superconducting material contains the electrical current. A key element in their studies was a new perspective on the subject provided by the 1933 discovery by W. Meissner and R. Ochsenfeld that superconducting materials tend to expel a magnetic field within it prior to cooling. Bloch, in his introduction to volume two of London's *Superfluids,* called the Londons' 1934 paper on superconductivity "a decisive step forward by indicating the direction in which a solution [to the problem of superconductivity] has to be sought."

London's work on superconductivity seemed to lead naturally to a related topic: superfluidity. First described by the Russian physicist **Pyotr Kapitsa** in the late 1930s and early 1940s, superfluidity is the tendency of a fluid to flow without resistance, much as electrons flow without resistance in a superconductor. Liquid **helium** below the **temperature** of 2.19 K, for example, may flow upward along the sides of a container and pass through tiny cracks that would be impervious, or impenetrable, to other **liquids**. London spent a large fraction of the last two decades of his life trying to solve that challenge. The result of that effort was the monumental two-volume work *Superfluids,* the second part of which was published after his death.

London's research work was interrupted in the 1930s by the rise of Nazi dictator Adolf Hitler in Germany. Both London and his brother left Germany in 1933 for England, where they both accepted appointments at Oxford University's Clarendon Laboratory. Fritz then spent two years at the Institut Henri Poincaré in Paris before taking a job as professor of **theoretical chemistry** at Duke University in North Carolina. He remained at Duke until his death of a heart attack in Durham, North Carolina, on March 30, 1954. He was survived by his wife, the former Edith Caspary, an artist, and their two children.

Lone pair

Valence electrons of atoms in covalent compounds may be shared between atoms to form covalent bonds or they may be unshared. Pairs of electrons that are not shared in covalent bonds are lone pairs. Lone pairs of electrons play a critical role in determining the three-dimensional shape of molecules. A molecule's shape, in turn, determines many of its physical properties and its chemical **reactivity**.

A very useful method for understanding and predicting molecular shapes that relies upon the role of the lone pair is called the Valence Shell **Electron** Pair Repulsion (VSEPR) theory. This method has been developed extensively by Professor R. J. Gillespie of MacMaster University. It correctly accounts for the structures of most covalent compounds of elements other than transition metal complexes.

The basic tenet of the VSEPR theory is that atoms in a covalent **molecule** are arranged to minimize the repulsions of valence shell electrons of the bonded atoms. All valence electrons must be considered. As a result the shape of the **ammonia** molecule, NH_3 is not flat triangular as it would be from considering only the three **hydrogen atom** bonds. Instead, it is pyramidal, with the three hydrogen atom bonds pushed away from a plane, and towards each other, by the requirement for space of the lone pair of electrons on the **nitrogen** atom.

The role of the lone pair in VSEPR theory is even more important than a **bonding** pair of electrons. The repulsions of lone pair of electrons for other electrons are considered to be stronger than the repulsions of bonded electrons. Because bonded electrons are strongly attracted to the nuclei of the two bonded atoms, they are localized between the two atoms. Lone pair electrons are more diffuse and extend into a larger **volume**. Therefore, they can come closer to other electrons and repel them more strongly. The order of repulsive interactions is: lone pair-lone pair, lone pair-bonding pair, bonding pair-bonding pair. Not only do the lone pair electrons lead to a certain type of shape, pyramidal in the case of ammonia, but they also affect bond angles. If the repulsions of lone pair electrons and bonding electrons were the same the hydrogen-nitrogen-hydrogen bond angles would be those of a perfect tetrahedron, 109°. Instead, the lone pair electrons push the bonding electrons closer together, resulting in a bond angle of approximately 107° in ammonia. In **water**, Where there are two lone pairs of electrons on the **oxygen** atom to repel the bonding electrons of each O-H bond, the H-O-H angle is squeezed to 105°.

Lubricant

Whenever one body moves over the surface of another, it encounters a resistance to its movement from the other surface. This resistance is called friction. Although friction may exist between a solid and fluid surface, friction is primarily a surface phenomenon that arises between **solids**.

Lubricants are used for one or more of the following purposes: to reduce friction, to prevent wear, to prevent adhesion, to aid in distribution of a load, to cool moving parts, and/or to prevent **corrosion**. Besides oils and greases, many **plastics**, solids, and even **gases** are now used as lubricants. The chief limitations on many of these materials is their ability to replenish themselves, to dissipate frictional **heat**, to withstand high **temperature** environments, and to remain stable in working environments. Types of lubricants include petroleum fluids, synthetic fluids, greases, solid films, working fluids, gases, plastics, animal fat, metallic and mineral films, and vegetable oils.

When a small amount of lubricant is placed between two solid surfaces, the surfaces will remain in contact, but the friction between the two surfaces will be reduced. In this regime, lubricants act by reducing the shear strength at the surface through the formation of a surface film held by physical **adsorption** to, chemisorption to, or chemical reaction with the solid surface.

When larger amounts of a viscous lubricant are placed between two sliding surfaces, the separation between the surfaces increases as the sliding speed increases or as the vertical load on the surfaces decreases. As the distance between the surfaces increases, the amount of solid-solid contact between the surfaces decreases. In this case, the resistance to sliding is produced by harder asperities on the surfaces and by adhesion between points of solid-solid contact. At the point that the solid surfaces are completely separated by the lubricating film, the lubrication becomes hydrodynamic, i.e., governed only by the lubricant, and the frictional resistance becomes very small, being determined by the fluid shear of the lubricant. Hydrodynamic lubrication has been achieved in thrust bearings using mineral oil, synthetic lubricants, gas, **water**, and other lubricants.

The coefficient of friction is defined as the ratio of the frictional force between two bodies in contact parallel to the contacting surfaces to the force at right angles to contact that presses the bodies together. Clean **metals** in vacuum have a coefficient of friction between themselves of about unity. Lubrication by **liquids** can give a coefficient of friction for roller bearings of about $10^{(-3-4)}$, for ball bearings of about $10^{(-5)}$, and for hydrostatically pumped oil pads of about $10^{(-7)}$. Solids such as **graphite**, **molybdenum** disulfide, **Teflon**, **lead**, babbit, **silver**, and metal oxides, with coefficients of friction of about $10^{(-1-2)}$, are sometimes used as dry film lubricants, often in conjunction with fluid or grease lubricants. Lubricating greases are usually petroleum oils thickened with dispersions of soap; synthetic oils with soap or inorganic thickeners; or oils containing silicon-based dispersions.

See also Adhesives and glues; Petrochemicals

LULL, RAMON (1232-1316)

Spanish alchemist

Ramon Lull was born in Spain in 1232, the son of a nobleman. He married at age twenty-five. Six years later after asserting he had had a series of spiritual experiences, he set about to accomplish what he perceived to be his life mission. Lull would establish colleges where missionaries could train to convert the Arabic people. He wanted his students to clearly present the superiority of Christianity over any other way of thinking. They would need training in other languages and the best knowledge on a wide range of subjects. To this end, he developed what he called the "art of finding the truth." This method linked all knowledge together as a means to show that all could come to know the Creator by knowing the creation.

It is one of the ironies of history that a man such as Ramon Lull should be described as an alchemist. Studies of his work since 1950 show that Lull was clearly against the **alchemy**. It was one of the influences leading faithful away from Christianity. What he did promote was the teaching of the currently best knowledge of medicine. This was four hundred years before the practice of making medicine would begin to separate from the ancient ways of alchemy. Surely, the best knowledge of medicine at the time of Lull was intricately connected with alchemy. So, even though in his lifetime he tried to steer people away from the mystical influences of alchemists, because his books contained methods associated with alchemists, he has been believed to be one of them. Furthermore, although he stressed reason and study for determining truth, his faith was professedly based on mystical experience. The alchemists were attracted to the mysticism in his work and quoted him. Many authors wrote works of alchemy in his name, but these works have been dated to years after Lull's death.

A study of Lull's authentic work shows that he did have much of value in his thinking. He did not study at one of the Christian schools, because he had not mastered Latin and the schools wanted younger students. He was essentially self-taught and, therefore, not limited in what he chose to study. He was one of the few scholars of his time who was free to study classical works not directly associated with Christianity. Furthermore, because he was not learning under the church supervision, he did not have to seek approval for his opinions and challenged himself to search for the truth beyond popular opinions and to organize that search. This is similar to the modern role of the scientist. **Andreas Libavius**, whose writings influenced **chemistry** away from alchemy and toward becoming a science, surely knew of Lull. In the writings of Libavius, there is the same emphasis on understanding the best knowledge available.

Lull lived to be 84. Late in life, he did see his dream of missionary colleges become a reality. The Council of Vienna approved the development of these colleges four years before he died. It is another irony that he should be described as a martyr. Lull constantly stressed collecting the facts and making a rational assumption from those facts. The evidence suggesting he was martyred is very scant and does not hold up in light of other evidence. Lull died in early 1316, most likely at Mallorca.

LUTETIUM

Lutetium is the heaviest, rarest, and most expensive of the **rare earth elements**, the elements found in Row 6 of the **periodic table** between Groups 2 and 3. Lutetium's **atomic number** is 71, its atomic **mass** is 174.97, and its chemical symbol is Lu.

Properties

Lutetium is a silvery white metal that is quite soft and ductile. It has a melting point of 3,006°F (1,652°C), a **boiling point** of 6,021°F (3,327°C), and a **density** of 8.49 grams per cubic centimeter. The element reacts slowly with cold **water** and with most acids.

Occurrence and Extraction

Lutetium is thought to be very rare in the Earth's crust with an abundance of about 0.8-1.7 parts per million. The most common ore of lutetium is monazite, in which its **concentration** is about 0.003%. The metal is produced by first converting its ores to lutetium fluoride (LuF_3) and then treating that compound with **sodium** metal: $3Na + LuF_3 \rightarrow Lu + 3NaF$.

Feodor Lynen.

Discovery and Naming

Lutetium was discovered almost simultaneously in 1907 by three researchers, George Urbain (1872-1938) of France, Karl Auer (Baron von Welsbach; 1858-1929) of Germany, and Charles James (1880-1926) of the United States. The element was one of nine new elements discovered in an unusual black rock known as ytterite first discovered in 1787 by a Swedish army officer, Carl Axel Arrhenius (1757-1824). The name of the element, suggested by Urbain, comes from the ancient name for the city of Paris, Lutecia.

Uses

Because lutetium is so expensive (currently, about $75 a gram), it has almost no commercial use. The one exception is that it is sometimes used as a **catalyst** in the petroleum industry.

LYNEN, FEODOR (1911-1979)

German biochemist

Feodor Lynen was a well-respected biochemist whose work led to a better understanding of how cells make and use **cholesterol** and other materials necessary for life. His discovery of the structure of acetyl-coenzyme A led to a detailed description of the steps of several important life processes, including the **metabolism** of both cholesterol and **fatty acids**. Aside from influencing **biochemistry**, his work was also important to medicine because cholesterol was known to contribute to heart attacks, strokes, and other circulatory diseases. For his work on cholesterol and the fatty acid cycle, Lynen shared the 1964 Nobel Prize in medicine or physiology with German-American biochemist **Konrad Emil Bloch**.

Feodor Felix Konrad Lynen was born in Munich, Germany, on April 6, 1911, the seventh of eight children. His father, Wilhelm L. Lynen, was a professor of engineering at the Munich Technical University. His mother, Frieda (Prym) Lynen, cared for the family. Lynen showed an early interest in his older brother's **chemistry** experiments but remained undecided about his career throughout secondary school. He considered medicine and even thought of becoming a ski instructor. Ultimately, he enrolled in the Department of Chemistry at the University of Munich in 1930, where he studied with German chemist and Nobel laureate **Heinrich Wieland**. Wieland was Lynen's principal teacher both as an undergraduate and graduate student. On February 12, 1937, Lynen received his doctorate degree. Three months later, on May 17, he married Wieland's daughter, Eva, with whom he would have five children: Peter, Annemarie, Susanne, Eva-Marie, and Heinrich.

Upon his graduation, Lynen stayed at the University of Munich in a postdoctoral research position. In 1942, he was appointed a lecturer, and he became an associate professor in 1947. Throughout his years with the University, where he stayed until his death, Lynen supervised the research of nearly ninety students, many of whom reached leading positions in academia or industry.

In the early 1940s, Germany entered into war against much of Europe and, eventually, the United States. Lynen, however, was exempt from both military service and work in Nazi paramilitary organizations because of a permanently damaged knee that resulted from a ski accident in 1932. The onset of World War II made it difficult to continue working in Munich and, in an effort to maintain his research, Lynen moved his lab to a small village, Schondorf on the Ammersee, about eighteen miles from Munich. In 1945, the University's Department of Chemistry in Munich was completely destroyed. In the chaotic aftermath of Germany's surrender, scientific research halted altogether. Lynen eventually continued his work at various lab facilities, but did not return to the rebuilt Department of Chemistry until 1949.

In the first years after the war, German scientists were spurned by their European and American colleagues. Only four German biochemists were invited to attend the First International Congress of Biochemistry held in Cambridge, England, in July of 1949. Lynen, one of the four, made an ideal good-will ambassador for Germany because of his good sense of humor and the fondness he had for parties. "I believe the problem of mankind is its lack of simple joys," Lynen is quoted as saying in *Current Biography Yearbook.* "I think one should drink for fun occasionally." His cheery nature and solid research drew many foreign scientists to Munich. His magnetic personality was formally recognized years later when, in 1975, he was chosen to serve as president of the Alexander von Humboldt Foundation, an institution devoted to fostering relations between Germany and the international scientific community.

During the 1940s, Lynen began studying how the living cell changes simple chemical compounds into sterols and **lipids**, complex **molecules** that the body needs to sustain life. The long sequence of steps and the roles various enzymes and **vitamins** played in this complicated metabolic process were not well understood. After World War II, Lynen began to publish his early findings. At the same time, he became aware of similar work being conducted in the United States by Bloch. Eventually, Lynen and Bloch began to correspond, sharing their preliminary discoveries with each other. By working in this manner, the scientists determined the sequence of thirty-six steps by which animal cells produce cholesterol.

One of the breakthroughs in the cholesterol synthesis work came in 1951 when Lynen published a paper describing the first step in the chain of reactions that resulted in the production of cholesterol. He had discovered that a compound known as acetyl-coenzyme A, which is formed when an acetate radical reacts with **coenzyme** A, was needed to begin the chemical **chain reaction**. For the first time, the chemical structure of acetyl-coenzyme A was described in accurate detail. By solving this complex biochemical problem, Lynen established his international reputation and created a new set of challenging biochemical problems. Determining the structure of acetyl-coenzyme A supplied Lynen with the discovery he needed to advance his research.

Lynen, who had remained an enthusiastic skier even after his 1932 accident, suffered a second serious ski injury at the end of 1951. (Although the second accident left him with a pronounced limp, Lynen continued to hike, swim, mountain climb, and ski.) During his rehabilitation, he contemplated how the structure and action of acetyl-coenzyme A made it a likely participant in other biochemical processes. Upon his return to the lab, Lynen began investigating the role of acetyl-coenzyme A in the biosynthesis of fatty acids and discovered that, as with cholesterol, this substance was the necessary first step. Lynen also investigated the catabolism of fatty acids, the chemical reactions that produce **energy** when fatty acids in foods are burned up to form **carbon** dioxide and **water**.

In 1953, Lynen was made full professor at the University of Munich. A year later, he was named director of the newly established Max Planck institute for Cell Chemistry. At a time when other universities were attempting to coax Lynen away from Munich, this position ensured that he would stay.

In addition to elucidating the role of acetyl-coenzyme A, Lynen's research revealed the importance of many other chemicals in the body. One of the most significant of these was his work with the vitamin biotin. In the late 1950s, Lynen demonstrated that biotin was needed for the production of fat.

Lynen and Bloch shared the Nobel Prize in medicine or physiology in 1964, largely because the Nobel Committee recognized the medical importance of their work. Medical authorities knew that an accumulation of cholesterol in the walls of arteries and in blood contributed to diseases of the circulatory system, including arteriosclerosis, heart attacks, and strokes. In its tribute to Lynen and Bloch, the Nobel Committee noted that a more complete understanding of the metabolism of sterols and fatty acids promised to reveal the possible role of cholesterol in heart disease. Any future research into the link between cholesterol and heart disease, the Nobel committee observed, would have to be based on the findings of Lynen and Bloch.

In 1972, Lynen moved to the Max Planck Institute for Biochemistry, which had just recently been founded. Between 1974 and 1976, Lynen was acting director of the Institute. He continued to oversee a lab at the University of Munich, however.

In 1976, on the occasion of his sixty-fifth birthday, more than eighty of Lynen's friends, students, and colleagues contributed essays to a book, *Die aktivierte Essigsäure und ihre Folgen,* in which they described their relationship with Lynen. They celebrated Lynen as a renowned scientist and a proud Bavarian. The author of over three hundred scholarly pieces, Lynen was also praised as a hard-working man who expected much of himself and his students. Six weeks after an aneurism operation, Lynen died on August 6, 1979.

M

MACINTOSH, CHARLES (1766-1843)

Scottish chemist

Charles Macintosh was born in Glasgow, Scotland, in 1766. Although he was supposed to become a merchant, like his father, the young Macintosh had a passion for **chemistry** and science. By the time he was twenty, he had opened a plant in Glasgow for producing **ammonia** from coal-gas waste. About the same time, he introduced the manufacture of **lead** and **aluminum** acetates to Britain. He also made advances in cloth-dyeing processes, opened Scotland's first alum works, helped devise a method for making bleaching powder, and developed improved methods of **iron** production. Macintosh is best known, however, for the **water**proof garment that bears his name.

A waste product of the Glasgow gasworks was naphtha, a volatile liquid **hydrocarbon** mixture. In 1819 Macintosh began experimenting with the naphtha and discovered that it dissolved rubber. Applying the knowledge of textiles he had gained as a dye-maker, Macintosh had the idea of using the liquid rubber to waterproof fabrics. He painted one side of wool cloth with the rubber **solution**, then laid a second thickness of cloth over it. The rubber interior made the resulting sandwich of cloth waterproof.

Macintosh patented his invention in 1823. Within a year, the chemist had a flourishing factory in Manchester, England, producing rainproof cloth for the British military, the Franklin Arctic expedition, and the general public. Thomas Hancock joined Macintosh and his other partners around 1829. Charles Macintosh & Company became famous for its "Macintosh Coat," the world's first raincoat, later known popularly as the mackintosh. (Where the *k* came from is a mystery.)

Macintosh was honored for his contributions to chemistry by his election in 1823 as a fellow of the Royal Society. He died at Dunchattan, near Glasgow, in 1843.

MACROMOLECULE

Macromolecules, also called polymers, are compounds that are made up of multiple, smaller repeating units called monomers. These monomers are generally bonded together in a chain-like manner creating a molecular backbone. Depending on the type of **monomer**, side groups may extend out from this primary backbone. When a single type of monomer is present in the macromolecule, it is known as a homopolymer. Copolymers contain two or more kinds of monomers in the backbone. While many linear macromolecules have a single long-chain backbone, others are branched, meaning the side groups from one backbone connect with another.

While macromolecules were used by ancient human civilizations, their chemistry only began to be described in the 1800s. The earliest work related to the production of macromolecules involved the conversion of natural macromolecules into more useful derivatives. The first significant development was by American inventor Charles Goodyear (1800-1860) when he developed a process for the vulcanization (cross-linking) of rubber in 1839. Other scientists made various plaster derivatives, artificial silks, and thermoplastics. The first truly artificial polymer, called Bakelite, was produced by Leo Hendrik Baekeland in 1907.

The groundwork for modern polymer science was laid by **Hermann Staudinger** in the 1920s. He showed that these materials were long, chain-like molecules and not colloids as previously thought. In 1928, Staudinger's models were confirmed by Meyer and Mark who used x-ray techniques to show the dimensions of natural rubber. By the 1930s, the leading scientists agreed that polymers were macromolecules. The size and shape of these molecules were responsible for the thickness of solutions they were in.

Macromolecules are used for innumerable purposes. Some important examples include **polyethylene** that is used for making plastic piping, packaging, and toys. **Polypropylene** is used to produce carpeting and bottles. **Polyvinyl chloride** (**PVC**) is an important material for floor tile and clothes. Other

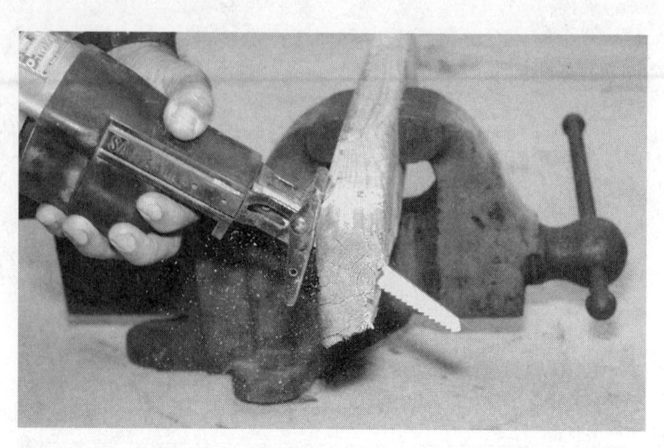

Magnesium alloys are often used in the manufacture of power tools, such as the reciprocating saw shown above. ***(Photograph by Robert J. Huffman. Reproduced by permission.)***

macromolecules include **polystyrene**, polybutadien, polyacrylinitrile, and **polyester**.

MAGNESIUM

Magnesium is the second element in Group 2 of the **periodic table**, a group of elements known as the **alkaline earth metals**. Magnesium has an **atomic number** of 12, an atomic **mass** of 24.305, and a chemical symbol of Mg.

Properties

Magnesium is a moderately hard, silvery white metal, the lightest of all structural metals. The element is easily molded, bent, cut, and otherwise fabricated. Magnesium's melting point is 1,204°F (651°C), its **boiling point** is 2,025°F (1,100°C), and its **density** is 1.738 grams per cubic centimeter.

Magnesium is a fairly active metal that reacts slowly with cold **water** and more rapidly with hot water. It reacts with **oxygen** at room **temperature** to form a thin skin of magnesium **oxide** (MgO) that protects the underlying metal from further **oxidation**. At higher temperatures, magnesium reacts vigorously with oxygen to produce a blinding white light. The metal also reacts with most acids and with some strong alkalis. It combines easily with many non-metals, including **nitrogen**, **sulfur**, **phosphorus**, and the **halogens**. It also reacts with a number of compounds, such as **carbon** monoxide (CO), **carbon dioxide** (CO_2), sulfur dioxide (SO_2), and **nitric oxide** (NO).

Occurrence and Extraction

The abundance of magnesium in the Earth's crust is estimated to be about 2.1%, making it the sixth most common element in the earth. It also occurs in seawater, usually as magnesium chloride ($MgCl_2$). According to some estimates a cubic mile of seawater may contain up to six million tons of magnesium metal.

Magnesium does not occur as a free element in nature, but it is found in many common minerals, such as dolomite [calcium magnesium carbonate; $CaMg(CO_3)_2$], magnesite [magnesium carbonate; $Mg(CO_3)$], carnallite (potassium magnesium chloride; $KMgCl_3$), and epsomite (magnesium **sulfate**; $MgSO_4$). The largest producer of magnesium ores is Turkey, followed by North Korea, China, Slovakia, Austria, and Russia. Magnesium produced in the United States comes from three sources: seawater, brine, and mines. The mines are large deposits of crystalline salts formed when ancient seas dried up and were buried underground.

Magnesium is extracted from its ores by one of two processes. In the first, the ore is converted to magnesium chloride ($MgCl_2$), which is then electrolyzed. In the second process, the ore is converted to magnesium oxide (MgO), which is then treated with the **alloy** ferrosilicon. The ferrosilicon reacts with magnesium oxide to yield pure magnesium metal.

Discovery and Naming

Magnesium compounds have been known and used by humans for many centuries. The first detailed scientific study of those compounds was conducted by the Scottish physician and chemist **Joseph Black** in the eighteenth century. Black is sometimes given credit for discovering the element because of his detailed report on the properties of its compounds.

Black was never able to produce pure magnesium metal, however, because compounds of the element tend to be very stable. It was not until 1808 the anyone was able to find a method for extracting the element from its compounds. In that year, the English chemist **Humphry Davy** found a way to extract pure magnesium metal from its compounds. Davy passed an electric current through molten magnesium oxide and obtained pure magnesium metal: 2MgO—electric current→ 2Mg + O_2. The procedure used by Davy was similar to one he used to isolate **sodium**, **potassium**, and a few other active metals. Magnesium was named after a region in Greece known as Magnesia from which large supplies of magnesium compounds have been extracted.

Uses

Metallic magnesium is sometimes used because of the brilliant white light it produces when it reacts with oxygen. For example, at one time, the flash bulbs used in photography contained a strip of magnesium metal which, when ignited, produced light by which the photograph could be taken. That application of magnesium has now become outdated because of the general availability of electronic flashes. Magnesium metal continues to be used in fireworks, however, when a flash of white light is desired.

Magnesium is also used widely in the manufacture of alloys. It improves the strength and durability of other metals with which it is alloyed. As an example, an alloy of magnesium and **aluminum** is often used in airplanes, automobiles, metal luggage, ladders, shovels and other gardening equipment, racing bikes, skis, race cars, cameras, and power tools because it is both strong and light.

The largest single used of magnesium compounds is in refractories used to line electric and high-temperature ovens and furnaces. Many magnesium compounds are also used in medicine. For example, magnesium **hydrogen** [$Mg(OH)_2$], magnesium **phosphate** [$Mg_3(PO_4)_2$], and magnesium silicate ($MgSiO_3$) are used as antacids; magnesium bromide (MgBr) is used as a sedative; a number of magnesium compounds are used as laxatives; and magnesium borate ($3MgO{\bullet}B_2O_3$) and magnesium sulfate ($MgSO_4$) are used as antiseptics.

Perhaps the best known compound of magnesium is magnesium sulfate, popularly known as Epsom salts. The compound is named after the town of Epsom, in Surrey, England. The story is told that a farmer named Henry Wicker discovered the curative effects of natural water containing magnesium sulfate in 1618 when he brought his cows to a spring to drink. The cows refused to drink the bitter-tasting waters, but neighbors soon learned that soaking in those same waters soothed their bodies and eased their aches and pains. Today, Epsom salts are used in bath water to relax sore muscles and remove rough skin.

Health Issues

Magnesium is essential for good health in both plants and animals. It forms part of the **chlorophyll molecule** that catalyzes the conversion of carbon dioxide and water to **carbohydrates** in green plants. Plants that do not get enough magnesium do not manufacture an adequate amount of chlorophyll, and their leaves tend to become yellow.

Magnesium also occurs in enzymes needed by animals. The amounts required are so small that magnesium deficiency diseases are rare. In some areas, however, people with poor diets may not get the magnesium they need, producing symptoms such as extreme agitation or aggressiveness.

Overexposure to magnesium is also rare, but may occur and, if it does, may produce health problems. For example, people who work around magnesium metal may inhale the fumes, experiencing symptoms such as irritation of the throat and eyes, damage to muscles and nerves, and loss of feeling and paralysis.

MAGNETIC MOMENT

Any current circulating in a planar loop produces a magnetic moment whose magnitude is equal to the product of the current and the area of the loop. When any charged particle is rotating, it behaves like a current loop with a magnetic moment. For a system of charges, the magnetic moment is determined by summing the individual contributions of each charge-mass-radius component.

It turns out that, both classically and quantum mechanically, there is a close connection between a particle's magnetic moment and its angular momentum. (The angular momentum of a particle moving in a circle is equal to the product of the particle's linear momentum and its perpendicular distance from the axis of revolution.) Classically, for the charged particle moving in a circle, the size of the magnetic moment is proportional to the magnitude of the particle's angular momentum. In quantum mechanics, angular momentum is quantized, i.e., it only assumes discrete values instead of the continuous range of possible values predicted by classical theory. A consequence of quantum theory and the quantization of angular momentum is that the magnetic moment also must be quantized.

The magnetic properties of **matter** can be thought of as arising from microscopic atomic currents that produce magnetic moments in that matter. In paramagnetic materials, permanent magnetic moments arise from the intrinsic angular momentum (spin) of individual electrons. In the absence of an applied magnetic field, these magnetic moments are randomly arranged, and there is no net magnetic moment. An applied magnetic field, however, aligns the moments in the field direction. In diamagnetic materials, all intrinsic magnetic moments are cancelled out by the pairing of electrons. Any remaining magnetic effects in diamagnetic materials are produced by the orbiting electrons.

See also Magnetism; Quantum chemistry

MAGNETIC RESONANCE IMAGING

Magnetic resonance imaging (MRI) is the application of **nuclear magnetic resonance** to produce images. MRI is now a common, though relatively expensive, diagnostic technique. MRI images can be very detailed and informative and the technique is non-invasive and it does not use highly energetic and potentially dangerous ionizing radiation. Because of the association many people make of the term nuclear with dangerous radiation, the term nuclear was eliminated from the original term for the technique (nuclear magnetic resonance imaging).

Nuclear magnetic resonance was developed in the 1950s by physicists as a means of probing the properties of the atomic **nucleus**. Theoreticians predicted that nuclei with particular ratios of protons and neutrons, giving rise to non-zero nuclear spin numbers, would show an asymmetry, or anisotropy, in a magnetic field that would result in a greater number of nuclei aligned in one direction relative to the north-south orientation of the magnetic field than in the opposite direction. The **energy** necessary to cause a nucleus aligned with the field to lose its orientation was found to be in the range of radio frequencies for the magnetic field developed by strong electromagnets.

Very soon it was found that the chemical environment of the **atom** in which an atomic nucleus resides affects the energy the absorbed by the nucleus to change its orientation in a particular magnetic field. For example, the nuclei of the **hydrogen** atoms bound to the **carbon** atom of methyl **alcohol**, CH_3OH, absorb a radio frequency different from the hydrogen atom bound to the **oxygen** atom. Chemists quickly adopted nuclear magnetic resonance as a means of determining **molecular structure**. Because the absolute energy absorbed by the nucleus and the relative energies of nuclei in different chemical environments increase in proportion to the size of the magnetic field, chemists have continually worked with physicists and engineers to develop instruments higher and higher magnetic fields. Modern instruments using superconducting magnets have magnetic fields over ten times as great as those of the permanent and electromagnet based instruments used in the 1960s when high resolution nuclear magnetic resonance (NMR) **spectroscopy** was developed as a common technique for organic structure determination. Much larger, more complex molecules, including important biomolecules, can now be studied by this method.

Along with the effect of chemical environments on producing differences in energies, scientists found that the time it takes for nuclei that are disoriented in the magnetic field by applied the radio frequency to recover their equilibrium distribution orientation (the relaxation time) depends on their environment. If the radio transmitter is turned off and then turned back again before the relaxation processes are the nuclei in the sample are complete, less energy will be absorbed from the second pulse of radio frequency energy. The relaxation times of nuclei are determined by using a sequence of carefully timed pulses and observing the amount of energy absorbed by each pulse.

Further experimentation revealed that the effect of the environment of a nucleus was not limited just to the atoms bonded to or even part of the same **molecule** that contains the particular nucleus, but to other molecules that are near it, such as the solvent, and the ability of the molecule in which the nucleus is contained to physically reorient itself in the magnetic field (its mobility). As instruments were developed that allowed scientists to use larger samples and developments in high resolution NMR spectroscopy encouraged scientists to investigate more complex materials, experiments on biological tissues indicated that the hydrogen nuclei in **water** contained within different tissues has different relaxation times—the basis of MRI.

A key to MRI is accurate location of the water whose relaxation time is measured by radio frequency pulse sequences. This is accomplished by the use of a magnetic field that varies in strength, referred to as a field gradient. The energy of radio frequency absorption by the hydrogen nuclei of the water molecules is great at the end of the field that has the greatest magnetic field strength and decreases proportionately as the field strength decreases, providing a means of locating the position of the water. The patient having an MRI scan performed is moved slowly through the magnet so that a series of images corresponding to field gradient "slices" can be obtained. The energy/time data are processed, enhanced and rendered into photograph-like images by powerful computer programs. Image enhancement for MRI and other diagnostic scanning techniques is an active area of scientific research.

The high resolution and high contrast of different tissues afforded by MRI, coupled with its non-invasive safety have made it the diagnostic method of choice in a many situations despite its high cost. MRI is a practical, life-saving tool that only have come about by the collaborative interplay of fundamental theory, technological advances, and experimental science.

MAGNETISM

Magnetism has intrigued humankind for millenia. Because magnetic and magnetically susceptible objects affect each other at a distance, without being in contact with each other, the effect appears to be magical. An understanding of magnetic effects that are much more subtle than the orientation of a compass needle by the Earth's magnetic field has contributed greatly to our current concepts of chemical **bonding** and **molecular structure**.

Naturally magnetized pieces of magnetite, Fe_3O_4 have been known for thousands of years and for the past several hundred years sailors and explorers have used the effect of Earth's magnetic field on an **iron** needle to help them maintain their direction. The great advance in the understanding of magnetism that has been exploited in **chemistry** only occurred in the last century when physicists **James Clerk Maxwell** and others determined relationships that link **electricity** with magnetism. When **Gilbert Newton Lewis** convincingly demonstrated the importance of electrons in chemical bonding and Irwin Schrödinger developed a mathematical model for the interactions of electrons in atoms that could be expanded to include a description of the magnetic properties of electrons in molecules, the stage was set for using experimentally measured magnetic properties to deduce information about the chemical structure of materials.

Moving electric charge creates a magnetic field. If the distribution of moving electric charge is not symmetric, the magnetic field will also be unsymmetric. In some materials, such as the natural magnet lodestone, the unsymmetric magnetic fields of individual atoms are aligned in the same direction within the bulk material. If this condition persists even in the absence of an external magnetic field, the material is a permanent magnet. Magnetite that does not act as a magnet may not have been exposed for a sufficient time or under other appropriate conditions to become magnetized by the Earth's magnetic field. Placing it close enough to a magnet or placing it in a strong electromagnetic field can magnetize non- magnetized magnetite. Similarly, metallic iron and many other **metals**, alloys, oxides, sulfides and other compounds containing iron and a number of other transition metals can be magnetized. Materials that can become magnetized are composed of atoms whose electrons can be predominately oriented asymmetrically and remain oriented asymmetrically once the external magnetic field is removed. The asymmetric **electron** distribution in such materials reinforces itself. Such materials are called ferro (like iron) magnetic.

Most materials are not ferromagnetic, but there are a variety of magnetic effects that other materials exhibit. In most materials, the electrons of the atoms are all paired with each other. The molecules of most substances in non-crystalline bulk materials are randomly oriented in **liquids** and **solids**. A bulk sample of a randomly oriented, spin-paired material will not be attracted into a magnetic field. In fact, such materials are very slightly (compared to the magnitude of ferromagnetic effects) repelled by a magnetic field. These materials are said to be diamagnetic. Although diamagnetic effects are usually very small, one class of materials shows strikingly large diamagnetism: superconductors. Superconducting materials are so strongly repelled by magnetic fields that they are levitated by them. There have been proposals to utilize this property of superconductors (the Meissner effect) to construct levitated trains that would experience nearly zero friction by traveling above magnetized tracks.

Some materials, particularly compounds of the transition metals but also some compounds of **nitrogen** and other elements with an odd number of **valence** electrons, have unpaired electrons. The magnetic moments resulting from the unpaired electrons often align themselves with an external magnetic field, drawing the materials into the magnetic field. This effect can be so great for compounds with atoms that have several unpaired electrons, such as those of **manganese** (II) with five unpaired electrons per manganese (II) **atom** that a sample can weigh 10% more in the field of a laboratory magnet than outside the field. Compounds that are attracted into a magnetic field are paramagnetic. The gain in weight at a particular magnetic field strength (the paramagnetic susceptibiliy) usually decreased inversely with the **temperature** on the absolute (Kelvin) scale, a phenomenon known as Curie's law.

By observing such factors as the degree to which a material becomes magnetized, the weight gained or lost in a magnetic field, the temperature dependence of such changes, and the dependence of such changes on the magnitude of the external magnetic field, chemists have determined that there are a number of different patterns that can be correlated with the interactions of electrons of one atom with neighboring atoms in a molecular or ionic structure or with the overall repeating structure of atoms in a crystalline structure. For example, in some structures, the magnetic moments of entire molecules within a crystal are aligned in opposite directions, a condition termed antiferromagnetism. If the magnetic moments of atoms within the molecules are aligned in opposite directions, the condition is antiferrimagnetism. Sometimes the **magnetic moment** in one direction is larger than that in the other and the material shows some degree of magnetization at high enough magnetic fields. These materials are called parasitic ferromagnets or parasitic ferrimagnets, respectively. The magnetic properties of some crystalline substances change drastically when they are subjected to pressure along one direction of the crystal. These substances are called piezomagnets, analogous to the terminology for piezoelectric crystals whose **electrical conductivity** changes when they are subjected to pressure.

The behavior of substances in a magnetic field may be used to deduce the number of unpaired electrons present in atoms of the substance, the ways that unpaired electrons on one atom interact with those on a nearby atom and to deduce short and long range structural effects in crystalline compounds. Our understanding of magnetism has progressed from the mysterious to fundamental science useful in solving modern chemical questions about the structure of **matter**.

Malleability

Malleability is one of several general physical properties of **metals** and metallic compounds. Chemically, elements can be classified as metals, metalloids, or non-metals based, in part, upon these physical properties. Malleability is the ease with

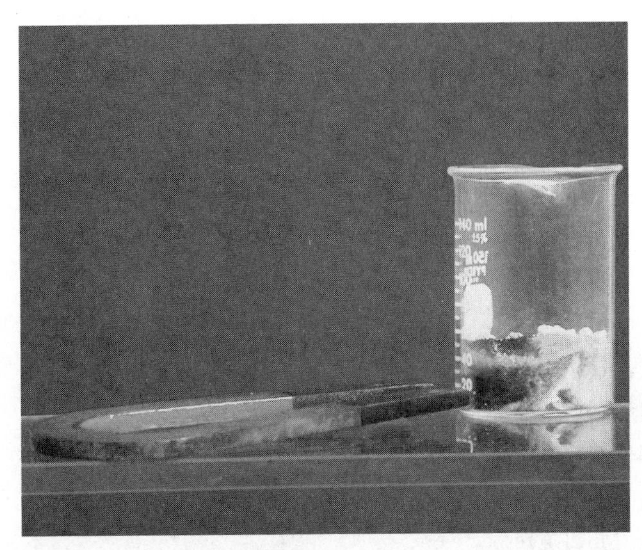

Magnet separating a mixture of sulfur and iron. ***(Photograph by Yoav Levy/Phototake NYC. Reproduced by permission.)***

which a metal can be hammered, forged, pressed, or rolled into thin sheets. Different metals vary in malleability. For example, **lead** is highly malleable and can be hammered flat easily. **Iron** requires considerably more effort to pound into a sheet and is therefore less malleable. Yet, both are metals. In contrast, nonmetallic elements, such as **carbon** or **sulfur**, shatter into pieces when hammered. Malleability is a valuable property because it allows metals to be shaped into useful forms. Pure **gold** is the most malleable metal. **Silver**, **aluminum**, lead, **tin**, and **copper** are also very malleable. Heating usually increases malleability. For instance, **zinc**, at standard temperatures is brittle, but becomes malleable at temperatures between 248°F (120°C) and 302°F (150°C). Also, impurities can adversely affect the malleability of metals, making them less pliable.

While some metals can have radically different properties overall (gold and **calcium**, for example), by definition all metals share some physical properties which help define their chemical identities. These include: high **electrical conductivity** (the ability to carry an electric charge), high thermal conductivity (the ability to transfer heat), varying degrees of luster (the ability to reflect light), **ductility** (the ability to be stretched into a wire), and malleability. Like other metallic properties, malleability is due to the loosely held electrons in a metal, which allow the metal atoms to slide past one another without experiencing the strongly repulsive forces that would shatter them.

See also Metallic bond

Manganese

Manganese as a transition metal, one of the elements that occur in the middle of the **periodic table** in Rows 4, 5, and 6 between Groups 2 and 13. Manganese has an **atomic number** of 25, and atomic **mass** of 54.9380, and a chemical symbol of Mn

Properties

Manganese is a steel-gray, hard, shiny metal that is so brittle it can not be machined in its pure form. The metal must always be alloyed if it is to be fabricated or worked to make some useful product. Manganese exists in four allotropic forms, the most common of which is stable to a **temperature** of about 1,300°F (700°C). The melting point of manganese **allotrope** is 2,273°F (1,245°C), its **boiling point** is about 3,800°F (2,100°C), and its **density** is 7.47 grams per cubic centimeter.

Manganese is a moderately active metal that combines slowly with **oxygen** in the air to form manganese dioxide (MnO_2). At higher temperatures, it reacts with oxygen more quickly and, in some cases, may actually burn with the release of a brilliant white light. Manganese reacts slowly with cold **water**, but more rapidly with hot water or steam. It dissolves in most acids and also reacts with **fluorine** and **chlorine** to produce manganese fluoride (MnF_2) and manganese chloride ($MnCl_2$), respectively.

Occurrence and Extraction

Manganese is the 12th most abundant element in the Earth's crust, with an estimated abundance of 0.085-0.10%. It never occurs as a free element, but only in ores such as pyrolusite, manganite, psilomelane, and rhodochrosite. Manganese also occurs commonly in conjunction with ores of **iron**. The largest producers of manganese ore in the world are China, South Africa, Ukraine, Brazil, Australia, Gabon, and Kazakstan.

Scientists believe that very large deposits of manganese exist on the ocean floor in the form of nodules that also contain **cobalt**, **nickel**, **copper**, and iron. So far, no economical method for mining these nodules has been developed.

Discovery and Naming

Certain compounds of manganese have been known for many centuries. For example, pyrolusite was used by artisans to give **glass** a beautiful purple **color**. Pyrolusite contains the compound manganese dioxide (MnO_2). By the mid-1700s, many chemists began to suspect that pyrolusite might contain a new element, but all efforts to isolate that element were unsuccessful. Finally, in 1774, the Swedish mineralogist Johnn Gottlieb Gahn (1745-1818) found a way to extract the element. He heated pyrolusite with **charcoal** to obtain nearly pure manganese: $MnO_2 + C \rightarrow CO_2 + Mn$. The origin of manganese's name is somewhat uncertain. Early chemists associated the element discovered by Gahn with the mineral called magnesia (now known as **magnesium**). Magnesia does not contain manganese, but, over time, the new element retained the name given to it.

Uses

By far the most important use of manganese is in the production of alloys, primarily steel alloys. Manganese is added to an **alloy** to increase hardness and improve resistance to **corrosion** and mechanical shock. The most common alloy of manganese is ferromanganese, which contains about 48% manganese, combined with iron and **carbon**. Ferromanganese steel is used to make tools, heavy-duty machinery, railroad tracks, bank vaults, construction components, and automotive parts. About 60% of all the manganese used in the United States goes to the production of ferromanganese.

Another 33% of the manganese produced in the United States is used in another alloy, silicomanganese. Silicomanganese contains manganese, **silicon**, carbon, and iron. It is used for structural components and heavy-duty springs.

A small amount of manganese is used to make a variety of manganese compounds, the most important of which is manganese dioxide. Manganese dioxide is used primarily in the manufacture of dry cells.

Health Issues

Manganese is one of the chemical elements that has both positive and negative effects on living organisms. A very small amount of the element is needed by plants and animals in order to remain healthy. It is used in the synthesis of certain enzymes. Very large doses of manganese can, however, lead to health problems, such as weakness, sleepiness, tiredness, emotional disturbance, and paralysis. The only individuals who need be concerned about these problems, however, are those who work with the metal directly, for example, in mines or smelters.

MANHATTAN PROJECT

In 1939, German scientists bombarded uranium-235 with neutrons. The **atom** split, releasing three new neutrons. The results were noted by other nuclear physicists all over the world, who recognized that if you put one **neutron** in and got three neutrons out (each of which could hit other **uranium** atoms), that the reaction could proceed geometrically—multiplying every time a neutron hit a uranium atom—with weapons potential. American scientists wrote a letter to President Roosevelt, signed by **Albert Einstein** as their spokesman, saying that "...the element uranium may be turned into a new and important source of **energy** in the immediate future...This new phenomenon would also lead to the construction of bombs..."

Roosevelt appointed Enrico Fermi, a Nobel Prize winner who came to the United States as fascism rose in Italy, as leader of the Manhattan Project. The name for the project was taken from the location of the office of Colonel James C. Marshall, who was selected by the U.S. Army Corps of Engineers to build and run the bomb's production facilities. (When the project was activated by the U.S. War Department in June 1942, it came under the direction of Colonel Leslie R. Groves.) The project's scientific research was conducted at the University of Chicago. In a squash court under Stagg Field, Fermi's team constructed a 20-foot cube of **graphite** embedded with uranium—the first nuclear pile. On December 2, 1942, ten **cadmium** rods (neutron absorbers) were removed from the pile. As the last rod was slowly removed, radiation detectors showed a **chain reaction**.

In Oak Ridge, Tennessee, gaseous **diffusion** of uranium hexafluoride was conducted in order to separate the fissionable

U-235 from the more abundant U-238. In Hanford, Washington, plutonium-239 was produced from uranium-239. In Los Alamos, New Mexico, the uranium bomb was designed and assembled. On July 16, 1945, a test bomb was exploded over the sand flats at Alamogordo, New Mexico. A second bomb was dropped on Hiroshima, Japan, on August 6, 1945. The third and last bomb produced during this time devastated Nagasaki, Japan, three days later and signaled the end of World War II.

MANOMETER

First invented in the seventeenth century, manometers are used to measure the pressure of **gases**. During the 1600s, scientists tried to explain natural phenomena in logical, rational ways, instead of relying on mystical or magical explanations. At this time, most people, including the great scientist Galileo, believed **water** pumps demonstrated the commonly held theory that nature abhors a vacuum, but because they were unaware that air and other gases exert pressure, they could not explain why a pump could not raise water more than 32 ft (9.7 m). In 1643, Italian physicist **Evangelista Torricelli** tried to explain this phenomenon. In his experiments, Torricelli used a vertical tube filled with **mercury** to measure air pressure, and in doing so, he created the first mercury barometer, proving that air has weight and exerts pressure on all objects and substances on Earth. Soon afterward, British chemist **Robert Boyle** used a mercury barometer to determine the relationship between the pressure of air and its **volume**. In these experiments both Torricelli and Boyle used rudimentary versions of the manometer.

The simplest manometers are open-ended, U-shaped tubes that are partially filled with mercury, oil, or some other liquid. When a container of gas is connected to one end of the tube, the pressure of the gas causes the liquid in the manometer to become displaced and rise up into the arms of the tube. If the pressure of the gas introduced into the manometer is greater than the atmospheric pressure, the liquid will rise up into the open-ended arm; if the atmospheric pressure is greater, liquid rises up into the closed portion of the tube. From the amount of liquid that is displaced, scientists can determine the pressure of the gas.

Since its invention, the manometer has been altered and used for several different purposes. In some laboratory manometers, one arm of the tube is inclined at an angle instead to provide more accurate measurements. Other manometers contain a sealed-off vacuum at one end, so that changes in atmospheric pressure don't have to be accounted for in calculations. In research and industry, a special type of manometer called a McLeod gauge is used to measure extremely low pressures. Other common types of manometers include the sphygmomanometer, which is used to measure blood pressure.

MARCUS, RUDOLPH A. (1923-)

American physical chemist

In recognition for his contributions to the theory of electron-transfer reactions in chemical systems, Rudolph A. Marcus was awarded the 1992 Nobel Prize in **chemistry**. Marcus received the news of his award while attending a meeting of the Electrochemical Society in Toronto, Canada; the award recognized work he had done in the 1950s and 1960s, but some controversy had surrounded his discoveries until they were validated in the 1980s. Marcus was able to understand the way a basic chemical reaction occurs, where **electron** transfers take place, how fast the electrons travel, and how they affect the results. His work was a breakthrough for chemists because it enabled them to make selections on the basis of bringing about specific outcomes. His work has been applied in a variety of areas, such as **photosynthesis**, electrically conducting polymers, **chemiluminescence** (cold light), and **corrosion**.

Rudolph Arthur Marcus, the only son of Myer and Esther Cohen Marcus, was born on July 21, 1923, in Montreal, Canada. He has described his parents as loving and noted that he admired his father's athletic abilities and his mother's musical talents. Marcus traced his interest in science to his high school years, when he explored mathematics and later chemistry. After graduating from high school, he attended McGill University, receiving his Bachelor of Science degree in 1943 and his doctorate in 1946. At McGill, he was supervised by Carl A. Winkler, who specialized in the rates of chemical reactions. Marcus's first work after McGill was in two research positions, one with the National Research Council (NRC) of Canada in Ottawa and the other at the University of North Carolina. At the NRC, he did experimental work under E. W. R. Steacie on free-radical reactions, which concern atoms or groups of atoms with unpaired **valence** electrons.

Marcus credited his move to the United States as the beginning of his theoretical work. For the first three months at North Carolina he read everything he could find on **reaction rate** theory. After concentrating on a particular problem, he was able within several months to consolidate several theories from early statistical ideas developed in the 1920s and 1930s. What had been called the Rice-Ramsperger-Kassel (RRK) theory became the Rice-Ramsperger-Kassel-Marcus (RRKM) theory, published in the early 1950s. Attempting to explain observed differences in reaction rates, Marcus identified simple mathematical expressions to explain how the **energy** of a molecular system is affected by structural changes. Counterintuitive and highly controversial, his contribution to the RRKM theory of unimolecular reactions—which related molecular properties and the lifespan of transition states to reaction rates—was validated by experimental findings announced in 1985 by Gerhard L. Closs of the University of Chicago and John R. Miller of Argonne National Laboratory.

After his postdoctoral grants, Marcus began the search for a faculty position; he received an offer from the Polytechnic Institute of Brooklyn, New York, and joined the staff as an assistant professor in 1951. Ten years later, he was the acting head of the division of **physical chemistry**. It was here that Marcus became an independent researcher. He experimented on gas phase and **solution** reaction rates at first, but a student brought his attention to a problem in polyelectrolytes. By 1960, Marcus felt he needed to commit his time completely to theoretical work.

While he was a faculty member at the University of Illinois at Champaign-Urbana from 1964 until 1978, Marcus concentrated his interest in electron transfer and reaction dynamics. During these years, he extended his knowledge into astronomy, including classical mechanics, celestial mechanics, quasiperiodic motion, and chaos. In a year spent as a visiting professor at Oxford and Munich from 1975 to 1976, Marcus explored electron transfer in photosynthesis. In 1978, he accepted an offer from the California Institute of Technology in Pasadena to become the Arthur Amos Noyes Professor of Chemistry. At Caltech, Marcus was influenced by the work that his colleagues were doing and he returned to the RRKM theory to treat more complicated problems. This was a fertile association for Marcus, and more than half of his articles were published after he went to Caltech.

The theories of electron-transfer reactions in chemical systems for which Marcus was recognized by the Nobel Committee have been applied extensively by other scientists in numerous areas. Several groups have used computer simulations in the study of electron transfer, employing Marcus's theory for the framework of their studies. One research group has been able to apply Marcus's theory to effects present in photosynthetic **proteins**. It was Marcus's ability to formulate a simple mathematical method for calculating the energy change that takes place in electron-transfer reactions that has made his work so valuable to other researchers. He was also able to find the driving force of the electron-transfer exchange.

It was while he was at North Carolina that Marcus met and married Laura Hearne, a graduate student in sociology, on August 27, 1949. They have three sons: Alan Rudolph, Kenneth Hearne, and Raymond Arthur. In 1958, Marcus became a naturalized U.S. citizen. He continues his father's interest in sports by skiing and playing tennis; his other leisure interests include music and history.

Marcus has received many awards and honors other than the Nobel Prize, notably the **Irving Langmuir** Award in Chemical Physics in 1978, the Wolf Prize in 1985, and the **Linus Pauling** Award in 1991. Marcus has been an active member of important scientific societies in his field and serves on the editorial boards of several scientific journals. He has lectured widely in the United States, Canada, Europe, the Middle East, and the Far East. He is highly respected and warmly regarded among his colleagues, who generally praise him for the warmth of his personality and his enthusiasm for his work.

MARIA THE JEWESS (c. first century A.D.)

Alexandrian alchemist

Maria the Jewess is credited with establishing the theoretical and practical foundations of **alchemy**, the forerunner of modern **chemistry** in the western world. She was one of the first chemists to combine the theories of alchemical science with the practical chemistry of the craft traditions. Although her theoretical contributions remained influential into the middle ages and beyond, Maria was more famous for her designs of laboratory apparatus.

Although nothing is known of her life, there are many references to Maria in ancient texts and she is believed to have lived in Alexandria, Egypt, in the first century A.D. Founded by Alexander the Great in 332 B.C., under the ruler Ptolemy, Alexandria became the center of Greek science, featuring an institute of higher learning called the Museum, the Great Library, a zoo, botanical gardens, and an observatory. However by the first century, the Greco-Roman world had entered an intellectual decline. Alchemy was the one science which continued to develop, at a time when most scientists believed there was nothing new to discover and that all important knowledge could be found in the works of the ancient Greeks.

The Egyptian goddess Isis was said to be the founder of alchemy; but the science probably originated with the women who used the chemical processes of **distillation**, extraction, and **sublimation** to formulate perfumes and cosmetics in ancient Mesopotamia. Likewise, Babylonian women chemists used recipes and equipment derived from the kitchen. Thus, ancient alchemy was identified with women, and the work of the early alchemists occasionally was referred to as *opus mulierum*, or "women's work." Artists working with dyes and theories of color were also important sources for the practical aspects of Egyptian alchemy; but alchemical theory was steeped in the Gnostic tradition, centered in Alexandria. Gnosticism was a mixture of Jewish, Chaldean, and Egyptian mysticism, neoplatonism and Christianity. In alchemy, as in Gnosticism, the male and female elements were considered to be of equal importance.

Alchemy was a secretive science—perhaps to protect its practitioners from persecution; however, both the mystical cults and the crafts also had traditions of secrecy. In any case, it was common for alchemists to write under the name of a deity or famous person. Thus, Maria wrote under the name of Miriam the Prophetess, sister of Moses. In addition, she is referred to in alchemical literature as Maria the Jewess, Mary, Maria Prophetissa, and Maria the Sage, as well as Miriam. Maria's many alchemical treatises have been expanded, corrupted, and confused with other writings over the ages. However, fragments of her work, including one called the *Maria Practica*, are extant in ancient alchemical collections. She was quoted often by other early alchemists, particularly the Egyptian encyclopediast Zosimus (c. 300). Maria the Jewess also may have been the author of "The Letter of the Crown and the Nature of the Creation by Mary the Copt of Egypt" which was found in a volume of Arabic alchemical manuscripts, translated from Greek. This work summarized the major theories of Alexandrian alchemy and described the manufacture of colored glass, as well as other chemical processes.

Although the ultimate goal of the alchemist was to transmute common **metals** into **silver** and **gold**, the ancient alchemists were scientists who were examining the nature of life and of chemical processes. Although their science was based in Aristotelian theory, they were the first true experimenters. Maria believed that metals were living males and females and that the products of her laboratory experiments were the result of sexual generation. The early alchemists believed that the base metals were evolving toward the perfect metal—gold—and they clearly distinguished between gilding or forming alloys of base metals to simulate gold and silver, and true transmutation. By transferring the ''spirit'' or vapor of gold to a base metal, as measured by the transfer of color, alchemists saw themselves as encouraging a natural process.

Maria invented, and improved on, techniques and tools that remain basic to laboratory science today and her writings described her designs for laboratory apparatus in great detail. Her water bath, the *balneum mariae* or ''Maria's bath,'' was similar to a double-boiler and was used to maintain a constant **temperature**, or to slowly heat a substance. Two thousand years later, the water bath remains an essential component of the laboratory. In modern French, the double-boiler is called a *bain-marie*.

Distillation was essential to experimental alchemy and Maria invented a still or *alembic* and a three-armed still called the *tribikos*. The liquid to be distilled was heated in an earthenware vessel on a furnace. The vapor condensed in the *ambix*, which was cooled with sponges, and a rim on the inside of the *ambix* collected the distillate and carried it to three copper delivery spouts fitted with receiving vessels. Maria described how to make the **copper** tubing from sheet metal that was the thickness of a pastry pan. Flour paste was used to seal the joints.

Maria studied the effects of arsenic, mercury, and sulfur vapors on metals, softening the metals and impregnating them with colors. For these experiments she invented the *kerotakis* process, her most important contribution to alchemical science. Her apparatus also came to be known as the *kerotakis*, a cylinder or sphere with a hemispherical cover, set on a fire. Suspended from the cover at the top of the cylinder was a triangular palette, used by artists to heat their mixtures of pigment and wax, and containing a copper-lead alloy or some other metal. Solutions of **sulfur**, **mercury**, or **arsenic** sulfide were heated in a pan near the bottom of the cylinder. The sulfur or mercury vapors condensed in the cover and the liquid condensate flowed back down, attacking the metal to yield a black sulfide called ''Mary's Black.'' This was believed to be the first step of transmutation. A sieve separated impurities from the black sulfide and continuous refluxing produced a gold-like alloy. Plant oils such as attar of roses also were extracted using the *kerotakis*.

Maria has been credited with inventing or improving upon the hot-ash bath and the dung-bed as laboratory heat sources and perfecting processes for producing phosphorescent gems. Maria's theoretical work included the concept of the macrocosm, or universe, and the microcosm, or individual body, and she applied this concept to the processes of distillation and reflux.

By the third century, the alchemists of Alexandria were being persecuted and their texts were destroyed. Much of this work was rescued by the Arabs, who venerated Maria and adopted her alchemical theories. However, when alchemy was rediscovered in medieval Europe, it was primarily in the form of charlatanism. Laboratory chemistry advanced very little from the time of Maria to the mid-seventeenth century.

MARK, HERMAN (1895-1992)

American polymer chemist and educator

Herman Francis Mark was born on May 3, 1895, in Vienna, Austria, the son of Herman Carl Mark, a Jewish surgeon who had converted to Lutheranism, and Lili Mueller Mark. In July 1913, he graduated from *Gymnasium* (high school). In the fall of that year, Mark enlisted in the Austrian army as an *einjahrig-Freiwilliger* (one-year volunteer) in order to discharge his military obligation so that he could start his university studies in the fall of 1914. World War I broke out, and Mark spent five years in a mountain infantry regiment on the Italian front. He was wounded three times, earned 15 medals, and became Austria's most decorated company-grade officer. In November 1918, Mark's division was captured by the Italians. He spent 11 months in a prison camp at a former convent, where he studied Italian, French, and English and organized a general **chemistry** course. In 1919, Mark returned to Vienna and continued his study of chemistry at the university, which he had begun in 1915 during his recuperation from a war wound. In July 1921 Mark received his doctorate *summa cum laude*.

In 1922 Mark married Marie Schramek. Although she was Roman Catholic, the couple became ardent Zionists and visited Israel several times. Also, in that same year Mark joined the Kaiser Wilhelm Institute in Berlin-Dahlem. Here, he used such newly developed experimental methods as **x-ray diffraction** to study the structures of natural textile fibers (e.g., **cellulose**, silk, and wool), which he showed to consist of long-chain molecules with molecular weights greater than 100,000. During the period 1927-32 at IG Farbenindustrie, Germany's largest chemical cormpany located in Ludwigshafen am Rhein, where Mark had become an Assistant Director, he worked on **electron** diffraction and the synthesis and practical application of his results. However, he continued his fundamental studies of macromolecules and published two books. With Carl Wulff, Mark developed a process for the catalytic production of styrene from ethyl **benzene**, which lowered the cost of styrene and made possible the manufacture from it of **polystyrene** and Buna S synthetic rubber.

Sensing the increasing Nazi threat, Mark left Germany to become Professor of Chemistry at his alma mater, the University of Vienna, where he served from 1932-38. Here, he taught **physical chemistry** and developed the world's first curriculum in polymer science and technology. At that time, only a few laboratories, mostly industrial ones, cultivated the subject, and no organized university courses were available. He continued his research, wrote two more books, and traveled extensively abroad and lectured repeatedly at international conferences to publicize his laboratory as well as the new discipline of polymer science.

After helping Jewish colleagues leave Austria, Mark was arrested on March 12, 1938, the day after the German annexation (*Anschluss*), and imprisoned and interrogated by the Gestapo in Vienna. In April, disguised as Alpine tourists, Mark, his wife, two young sons, and his Jewish niece fled to neutral Switzerland, taking their valuables in the form of **platinum** coat hangers. They eventually migrated to Hawkesbury,

Ontario, where Mark became Research Manager for the Canadian International Pulp and Paper Company. He worked there from 1938 to 1940, and founded the *Polymer Bulletin* as well as the monograph series *High Polymers and Related Substances.*

After successfully completing his assignment of modernizing equipment, training personnel, and applying recent fundamental results to practical production problems at Hawkesbury, in 1940 Mark became Adjunct Professor at the Polytechnic Institute of Brooklyn, New York, which provided a teacher's visa and a consultantship with the DuPont Company. In 1942 he became Professor. During World War II, Mark directed a number of research projects for the United States government. In 1944, he founded the Institute of Polymer Research, the first of its kind in the United States. He continued as its Director until 1964. From 1961 to 1974 he served as Dean of the Faculty at Brooklyn Polytechnic and continued his activity as a lecturer and writer long after his retirement in 1964. Mark was the author of more than 600 publications and the recipient of numerous honorary degrees and awards, including the United States National Medal of Science, which recognized his lifetime achievement in polymer science. Mark continued his extensive travels, visiting throughout his career more than a thousand scientists and engineers in more than a hundred countries.

Mark died in Austin, Texas, on April 6, 1992, less than a month before his 97th birthday. Although not the world's first polymer chemist, he was known as the father of polymer science because of his many contributions to polymer science research and education, first in Europe and then in the United States. Nobel chemistry laureate **Linus Pauling** regarded Mark "a pioneer in modern structural chemistry and an important early contributor to its development."

Marker, Russell (1903-1995)

American chemist and industrialist

Russell Earl Marker is credited with discovering the process of creating progesterone from a steroid found in Mexican yams. One of the functions of progesterone is inhibiting ovulation. He thus prepared the path for the development of the birth control pill. The process was named Marker degradation. He recognized immediately the potential in his discovery and tried to interest American investors but to no avail. He then moved to Mexico where he found eager business partners, an American and a German. He named his company Syntex, "Synt" for Synthesis and "ex" for Mexico. However, a short time after starting the company, he had a falling out with his partners and he returned to the United States.

Marker had an interesting personality. While working on his Ph. D., like **James Prescott Joule**, he refused to study the courses that did not interest him. He even finished his dissertation, but could not be awarded a Ph. D. because of the courses he had not fulfilled. Still, without his doctorate, he managed to find a research position at Pennsylvania State where he worked from 1934 until 1944 when he left for Mexico. When he returned from Mexico, he accepted a teaching position at Pennsylvania State where he stayed until his death in 1995, at age 92. He never did finish his doctorate.

Although he may have been gruff with his business partners, he did had a sense of humor. He named newly discovered compounds after his friends. Rockogen was named for a friend whose nickname was Rocky. Someone returned the favor and so a compound is named after Marker, marcogen.

Marker's discovery is important because of the impact that it has had on twentieth century society. The development of the birth control pill has been assumed to have initiated the "sexual revolution," which has greatly influenced modern western culture. Synthetic progesterone has also had a positive effect on helping women through menopause and other conditions of low **hormones**.

Martin, A. J. P. (1910-)

English biochemist

One of Great Britain's most noted biochemists, A.J.P. Martin developed techniques of **paper chromatography** in addition to pioneering the separation of **gases** by chromatography, which revolutionized basic research in organic **chemistry**. For the development of paper partition chemistry, Martin and his colleague **Richard Synge** were awarded the Nobel Prize in chemistry in 1952.

Archer John Porter Martin, the only son of four children, was born on March 1, 1910, in London, England, to William Archer Porter Martin, a physician, and Lilian Kate Brown, a nurse. Martin attended high school in Bedford and graduated in 1929. Intent on becoming a chemical engineer, he entered Peterhouse, Cambridge, on a merit scholarship and studied chemistry, physics, mathematics, and **mineralogy**. While at Cambridge, he met Professor John Burdon Sanderson Haldane, became interested in **biochemistry**, and changed his major to that subject. He received a B.S. degree in 1932 and became a researcher in the **physical chemistry** laboratory of the university. With a colleague, Nora Wooster, he published an article in *Nature* in 1932, describing the preparation and mounting of deliquescent materials, solid substances that become liquid as they absorb moisture from the air.

In their laboratories at Cambridge, Martin and Synge became interested in chromatography in the early 1930s. Chromatography is a technique that separates parts of a mixture as it moves over a porous solid. **Richard Willstätter**, a German scientist, had developed a technique for separating plant pigments but had not been able to separate more complex substances. Martin and Synge had been searching for a method that would isolate the constituents of carotene, a ruby-red pigment that is present in various plants and animals and is the precursor of vitamin A. They found the compounds would move in columns or zones in a tube packed with porous materials like starch, **cellulose**, or silica gel.

While continuing his work at Cambridge, Martin held the Grocers' Scholarship for original medical research from 1934 to 1936. He worked under Sir Charles Martin, and he later credited this distinguished scientist as the greatest influence on his work in biochemistry. He received his M.A. degree in 1935 and a Ph.D. in 1936 from the University of Cambridge. In 1938 Martin became a biochemist at the Wool Industries Research Association Laboratory in Leeds. He stayed there through 1946. His studies involved the composition of wool felting and amino acid analysis. While working at this laboratory, he conceived the idea of separating **amino acids** using porous paper.

Several previous researchers had been frustrated in their efforts to break down **proteins** into amino acids. Building on their earlier studies at Cambridge, Martin and Synge devised paper partition chromatography in 1944. In this new technique, substances move in columns on sheets of paper instead of in an absorbent in a **glass** tube. The technique proved an instant success in separating proteins into their base amino acids. Martin and Synge worked the paper chromatography technique in the following manner. A drop of amino acid mixture is put near the bottom of a strip of porous paper and allowed to dry. The edge of the paper is dipped in a solvent which spreads through the strip by capillary action. As the solvent goes through the dried amino acid mixture, different amino acids move with the solvent at varying rates. Hence, the amino acids are separated and can be studied. The news of this new technique spread rapidly throughout the scientific community. Because of chromatography, other scientists made great strides. For example, **Frederick Sanger** was able to identify the order of amino acids in the insulin **molecule**, and **Melvin Calvin** was able to work out the mechanism of **photosynthesis**. Paper chromatography also contributed to advances in the biochemical knowledge of the sterols, an enormous array of substances that play a role in the life processes. Martin's work in chromatography received the highest recognition in 1952, when he and Synge were awarded the Nobel Prize in chemistry.

Martin worked at Boots Pure Drug Research Company, Nottingham, from 1946 to 1948. During this time, he and colleagues R. Consden and A. H. Gordon identified lower **peptides** in complex mixtures using paper chromatography. He joined the Medical Research Council at Lister Institute, Chelsea, London, in 1948 and accepted a post as head of the physical chemistry division of the National Institute of Medical Research in Mars Hill, London, in 1952. Continuing to apply paper chromatography, Martin studied sugars in a variety of substances and the partition of **fatty acids**. In 1953 his studies led him to gas-liquid chromatography, a method of separating volatile substances by blowing them down a long tube filled with inert gas. Gas-liquid chromatography is an adaptation of the paper chromatography technique.

Martin has written many articles on chromatography for journals and worked at major international universities. From 1956 to 1959 he was a chemical consultant, and he acted as a director to Abbotsbury Laboratories from 1959 to 1970. He was consultant to the Wellcome Foundation from 1970 to 1973. He has held professorships at Eindoven Technological University, Holland; at the University of Sussex; at the University of Houston, Texas; and at the Ecole Polytechnique Fédérale de Lausanne in France. He was elected a fellow of the Royal Society in 1950 and in 1951 received the Berzelius **Gold** Medal from the Swedish Medical Society. He is also the recipient of honorary doctorates from the universities of Leeds, Glasgow, and Urbino.

Martin married Judith Bagenal, a teacher, on January 9, 1943. They have two sons and three daughters. While in college he developed a love for the self-defense art of Jiu-Jitsu and in the past has enjoyed gliding and mountaineering. The Martins reside in Cambridge, England.

MARVEL, CARL SHIPP (1894-1988)

American organic chemist and polymer chemist

Carl Shipp (''Speed'') Marvel was born on September 11, 1894, on a farm in Waynesville, Illinois, the son of John Thomas Marvel and Mary Lucy Wasson Marvel. Although he was expected to be a farmer, his mother insisted that all her children attend college. Marvel's uncle, a former high school teacher, urged him to study science because the next generation of farmers was going to need scientific knowledge. In 1911, Marvel entered Illinois Wesleyan University, in Bloomfield, Illinois. During his junior year he discovered his forte—synthesizing organic compounds. He developed a ''library of smells'' and used his nose as an ''infrared spectroscope'' for recording the odor of every organic compound available to him.

After receiving his B.A. and M.S. degrees in 1915, Marvel immediately entered the University of Illinois at Urbana on a scholarship to study **chemistry**, still expecting to return to the family farm. Because of a deficiency in courses taken, his first semester consisted, not of the customary four courses, but of five, of which four were laboratory courses. Marvel's classmates nicknamed him ''Speed,'' a sobriquet that he used throughout his career, even in official correspondence, because he could work late in the laboratory, sleep until the last moment, and still get to breakfast before the dining hall closed at 7:30 A.M.

Between 1917 and 1918 Marvel worked in the Organic Chemical Manufactures unit, producing organic chemicals previously imported from Germany before World War I, a ''spin- off'' of which was the journal in book format, *Organic Syntheses*. Nearly 20% of the 264 preparations in Collective Volume I, published in 1932, were submitted or checked by Marvel. In 1919, he returned to full-time graduate studies and received his Ph.D. degree in 1920. He had intended to work in industry, but because industrial jobs were then scarce, he remained at Illinois, becoming Instructor (1920-21), Associate (1921-23), Assistant Professor (1923-27), Associate Professor (1927- 30), Professor of Organic Chemistry (1930-53), and Research Professor of Organic Chemistry (1953-61) until his retirement in 1961. On December 26, 1933, he married Alberta Hughes, a librarian and former high school English teacher, with whom he had a son and daughter.

With **Roger Adams** and Reynold C. Fuson, Marvel made Illinois' **organic chemistry** program preeminent in the United States. After retiring, he became Professor of Chemistry at the

University of Arizona, serving there from 1961 to 1978. Following his second retirement, Marvel continued working almost daily in his laboratory with a small group of postdoctoral fellows, until the summer before his death. Marvel died of renal failure in Tucson on January 4, 1988, at the age of 93.

Marvel's first 60 or 70 articles involved mostly preparative organic chemistry, e.g., syntheses of amino acids, organometallic compounds, and acetylenes. However, he soon moved into the areas of rearrangements, free radical chemistry, magnetic susceptibility, **hydrogen bonding**, stereoisomerism, the structure of organomercury and organophosphorus compounds, and most important of all polymers. He was the first American academic chemist to study the synthesis, structure, and mechanism of formation of synthetic polymers. He also introduced several generations of research students, postdoctoral students, and colleagues to polymer science. Furthermore, he also influenced the increasing industrial research on polymers and made **polymer chemistry** a popular field of study in numerous academic laboratories. His experience was very valuable to the American government during the World War II research on synthetic rubber.

After Marvel became a Du Pont consultant in 1928, he became even more involved with polymers. During his 60 years with Du Pont he gave 19,000 consultations. Beginning in 1933, he determined the structure of copolymers of **sulfur** dioxide and alpha-olefins and developed initiators for their preparative polymerization reactions. In 1937, he began to study the polymerization mechanism and structure of **vinyl** polymers, which led to the preparation and polymerization of many new monomers. During World War II, he directed a group of as many as a hundred chemists on the government's synthetic rubber program. During the next decade this experience enabled him to synthesize a large series of new polymers. Marvel visited Germany on a technical intelligence team, and on returning he developed the German cold synthetic rubber process for American industry.

Marvel developed the new technique of cyclopolymerization while experimenting with the syntheses of heat-resistant polymers for NASA. He synthesized polymers with repeating rigid heterocyclic or benzenoid groups in the main chain and polymers with repeating benzimidazole units (polybenzimidazoles or PBIs), heat-resistant macromolecules of high **molecular weight**, one of the most significant advances in the chemistry of high **temperature** polymers of the 1960s. In 1980, PBI became the first new man-made fiber to be produced commercially in nearly a decade. As a substitute for potentially carcinogenic fiberglass and **asbestos**, it is used in suits for astronauts and fire fighters.

An accomplished ornithologist who identified almost all 650 species of American birds, Marvel wrote two ornithological books, *Unusual Feeding Habits of a Cape May Warbler* (1928) and *The Blue Grossbeck in Western Ontario* (1950). He was active in the American Chemical Society, where he held many offices including President in 1945. Marvel received numerous domestic and foreign honors and awards. During his long career he wrote more than 500 articles and four books, and held 52 patents. In spite of his voluminous research Marvel considered teaching his greatest contribution; 176 students earned doctorates under his supervision, and 150 postdoctoral fellows worked with him. A list of his students reads like a "Who's Who" of American chemistry, and it is difficult to find organic polymer chemists anywhere in the world who have not had some relationship to Marvel. According to former American Chemical Society president Charles C. Price, "Marvel exerted an extraordinary influence on American chemistry. He stood at the center of its transformation from a small and provincial enterprise to a world-renowned technologically rich, and continuously innovative science."

MASS

Mass is a measure of the amount of **matter** present in an object. Matter is anything that has mass and **volume**. Therefore, by definition, all matter has mass. The mass of a particular object is constant. The only way the mass of an object can change is if matter is added to or taken away from it. The more matter an object contains, the more massive the object is. Mass and volume are directly proportional to each other. If an object's mass is doubled, so is its volume. The *Système International d'Unités* (S.I.) unit for mass is the kilogram (kg). The kilogram is a rather large quantity, however, so for smaller objects the gram (one-thousandth of a kilogram) or the milligram (one-thousandth of a gram) is used.

People often confuse an object's mass with its weight. The weight of an object is often expressed, incorrectly, in grams. The mass of an object is determined by comparing that object with a standard mass, or a **balance**. The weight of an object is a measurement of how much gravitational pull is exerted on the object. A weight measurement is taken using a spring scale and is expressed in Newtons (N) or pounds (lb). Mass and weight are related, however. A more massive object will weigh more than a less massive object. The weight of an object is actually the product of its mass and the acceleration due to gravity.

Mass is a measurement of all matter. An elephant has mass, as does an **electron** spinning around the **nucleus** of an **atom**. As can be imagined, the masses of atoms are very small. If the mass of an atom were expressed in grams, the number would be extremely tiny. For example, an atom of oxygen-16 has a mass of 0.657×10^{-23} g. Chemical calculations and measurements of such small numbers are very cumbersome. It is much more convenient in these cases to use relative atomic masses. Scientists created a relative scale of atomic mass by arbitrarily choosing the carbon-12 atom as a standard and expressing the mass of all other atoms by comparing them to this standard. The mass of the carbon-12 nuclide was assigned a mass of 12 atomic mass units (amu). One amu is exactly one-twelfth the mass of a carbon-12 atom. One amu is equal to 1.660540×10^{-27} kg. Every other atom is expressed in terms of the carbon-12 atom. For example, a hydrogen-1 atom has the mass of 1/12 a carbon-12 atom, or 1 amu. An oxygen-16 atom has about 16/12 the mass of a carbon-12 atom, or 16 amu. The atomic masses on the **periodic table** are expressed in amu.

Looking at the periodic table, one would notice that the atomic masses are not whole numbers, as would be suggested above. This is because there are different isotopes for each atom, i.e., atoms with different atomic masses, and the atomic mass expressed on the periodic table is a weighted average of all of these isotopes. The masses of **subatomic particles** (electrons, protons, and neutrons) are also expressed using amu. An electron has a mass of.00005486 amu, a **proton** has a mass of 1.007276 amu, and the mass of a **neutron** is 1.008665 amu. These would be very small numbers if expressed in kilograms.

Two terms related to mass which are frequently encountered in **chemistry** are the **mass number** of an atom and the molar mass of a substance. The mass number of an atom is simply the sum of the protons and neutrons in the atom. Since the protons and neutrons have a mass of approximately 1 amu each, and the mass of an electron is so small, the mass number is a good approximation of the atomic mass of an element. The molar mass of a substance describes the relationship between moles and mass. The molar mass is equal to the mass that one **mole** of a substance has. It is expressed in units of grams per mole.

Chemists often refer to a theory called the law of **conservation of mass**, especially when balancing equations for chemical reactions. The law of conservation of mass states that matter can neither be created nor destroyed. This means that when two or more compounds or elements combine in a chemical reaction, the total mass of the reactants has to equal the total mass of the products. The law of conservation of mass applies to the individual atoms involved in a reaction as well. Because each atom has a particular mass, and mass is conserved, this means that the number of atoms of each element involved in a reaction must be the same before the reaction occurs and after the reaction is completed. A chemical reaction is the rearrangement, not the creation or destruction, of atoms. The law of conservation of mass is symbolized by a balanced equation for a chemical reaction, i.e., the same number of atoms of each element are found on the reactant side and the product side.

Another definition of mass is a measure of the inertia of an object. An object's inertia is its resistance to changes in its motion. It is more difficult to change the motion of a more massive object. For example, if you need to push a couch across the floor, you need to exert a force on the couch in order to make it move. As you move the couch, it resists your pushing or pulling on it. This resistance is the inertia of the couch. Once an object is moving, it requires a force to stop it. For example, a speeding car needs the force of friction applied from the brakes to slow it down to a stop. As you press on the brake pedal of the car, you should notice that the car is resisting. This is due to the car's inertia. The more mass an object has, the more inertia it has. The car has a greater inertia than the couch. Therefore, a larger force must be used on the car to overcome its inertia.

Another property dealing with the mass of moving objects is momentum. All moving objects have momentum. An object with a greater momentum will be harder to stop. The momentum of an object depends on two things, the mass of the object and its **velocity**. Therefore, two identical cars traveling at different speeds will have different momentum. The car that is traveling at a greater velocity will have the greater momentum. Likewise, a two-ton truck traveling at the same velocity as a sports car will have more momentum because it has more mass.

The mass of an object is also related to force. The force of an object is the product of its mass and its acceleration. If either the mass or the acceleration of an object is large, the object will have a large force. The force of gravity between two objects depends on mass. The size of the force of gravity depends on two things, the masses of the objects and the distance separating them. As the masses of the objects increase, so does the force of gravity. This is why we notice the force of gravity between us and Earth much more than the force of gravity between us and the person standing next to us. Gravitational forces of attraction always exist between two objects of any mass, but it takes an object as large as a planet for this force to become noticeable.

Mass is also directly related to **energy**. The kinetic energy of an object depends on the mass of the object and the velocity. A large mass or a high velocity will increase the kinetic energy of an object. **Albert Einstein** (1879-1955) developed the theory of relativity in 1905 which related mass and energy. According to this theory, even objects with very small masses can have a very large amount of energy. Einstein showed that mass and energy can be converted into each other. His famous equation, $E = mc^2$, further explained the law of **conservation of energy**. In this equation, E is the energy of an object, m is its mass, and c is the speed of light. Einstein showed that matter can, in fact, be destroyed. When matter is destroyed, it is converted into energy. The same holds true for energy. When energy is destroyed, it is converted into matter. Einstein's theory explained why **nuclear reactions** did not seem to abide by the law of conservation of mass or the law of conservation of energy. **Nuclear fusion** is the process that changes mass to energy. In fact, during this process, a very tiny mass produces a huge amount of energy.

The mass of an object is more than simply how much matter is present. It determines factors such as inertia, momentum, and force. The mass of an object is also directly related to energy. Mass is a property of an object that can account for many other behaviors of the object, whether the object is a jumbo jet or a proton.

MASS DEFECT

All **matter** in the universe is composed of atoms. Atoms, in turn, are comprised of differing combinations of **subatomic particles**. The subatomic composition of an **atom** determines which element it is, and therefore what physical and chemical properties it has. The subatomic construction of an atom is comprised of three particles. Protons are the positively charged particles that reside in the **nucleus** (the massive center of an atom). Neutrons carry no charge and also are found in the nucleus. Electrons are small, negatively charged particles, having almost no **mass** and moving with great **velocity** within shells that surround the nucleus of the atom. The individual masses of these three subatomic particles have been deduced by scientists and are standard constants. Together, the combination of protons, neutrons, and electrons constitute the entire mass of an atom, the bulk of which is found in the nucleus.

Because the total number of each of the three subatomic particles of a given atom of any element is known, and the individual mass of each particle is also known, the total mass of the atom can be calculated. For example, the element **helium** (He) has two protons and two neutrons in its nucleus, with two orbiting electrons. If the masses of all six particles are added together, the total should equal the entire mass of a single atom of helium (its atomic mass). However, when the mass of a single helium atom is actually measured, it is always *less than* the combined weights of its component particles. This loss of mass, called the mass defect, is observed for all atoms of all the elements. How is this possible?

One of the most startling scientific advances of the twentieth century was **Albert Einstein**'s now famous equation $E=mc^2$. This revolutionary formula demonstrates the equivalence of matter and **energy**. That is, energy can be transformed into matter, and matter into energy. Einstein's most memorable contribution to physics explains why an atom of helium weighs less than its component particles, i.e. it explains the origin of the mass defect.

The combination of protons and neutrons to form the nucleus of an atom requires energy to hold the nucleus together. The mass defect arises when protons and neutrons combine to form the nucleus and a bit of their mass is converted into this nuclear binding energy, which stabilizes the nucleus and holds it together. Therefore, the nuclear binding energy can also be defined as the amount of energy required to break apart a nucleus into its component subatomic particles. From the mass defect for the atom, Einstein's equation can be used to calculate the nuclear binding energy for that atom.

Mass defect also works in reverse. If a nucleus is cleaved (in a process called fission), the resulting subatomic particles have slightly less mass than they did while in the nucleus. The lost mass is converted into energy, which is released in the nuclear reaction. The **nuclear energy** that is released per atom is quite large, so when large numbers of atoms are involved in a fission reaction, the energy release is huge. This is the reaction that produces nuclear power and atomic bombs. In **nuclear reactors**, atoms are cleaved allowing the release of their subatomic protons and neutrons in controlled chain reactions. These controlled reactions produce an enormous quantity of **heat** that is used to generate **electricity**. The mass defect that occurs during the fission reactions of an atomic bomb explosion, however, is the product of uncontrolled chain reactions.

Even the Sun in our solar system experiences mass defect. The massive amounts of light energy and heat energy generated by the Sun are the result of the fusion of **hydrogen** atoms to form helium atoms. When hydrogen protons combine to form the nucleus of a helium atom, the resulting helium atom has less mass than the combined hydrogen protons had originally. The loss of mass in the Sun, or mass defect, occurs as matter is transformed into energy, and this energy warms the Earth with rays of sunshine.

MASS NUMBER

The **mass** number of an **atom** is the total number of protons plus neutrons in its **nucleus**.

Different isotopes of the same element have different mass numbers because their nuclei contain different numbers of neutrons. In the written symbol for a particular **isotope**, the mass number is written at the upper left of the symbol for the element, as in $^{238}{}_{92}U$, where 92 is the **atomic number** of **uranium** (U) and 238 is the mass number of this particular isotope. The symbol is read ''uranium-238.'' The mass number is always a whole number—just a count of the particles. It differs from the exact mass of the atom in atomic mass units (amu) that is often known and expressed to six decimal places. (One amu is exactly one-twelfth of the mass of an atom of carbon-12, ^{12}C, and is equal to approximately 1.66×10^{-24} g.) There are two reasons why the mass number of an atom is different from its exact mass. First, neutrons and protons don't happen to weigh exactly one amu apiece; the **proton** actually weighs 1.0072765 amu and the **neutron** weighs 1.0086650 amu. Second, when neutrons and protons are bound together as an atomic nucleus, the nucleus has less mass than the sum of the masses of the neutrons and protons. The difference in mass, when expressed in **energy** units according to Einstein's formula $E=mc^2$, is called the binding energy of the nucleus.

To understand this strange-sounding situation, we can think of the binding energy as the strength of the ''glue'' that holds the protons and neutrons together as a nucleus. It is, therefore, the amount of energy that would be required to break the ''glue'' and pull the nucleus apart into its individual neutrons and protons. But if energy must be added to an object in order to pull it apart, and if energy and mass are equivalent, then we could say that mass had to be added to pull it apart. The separated particles will therefore have more mass than when they were bound together as a nucleus.

MASS SPECTROMETRY

Mass spectrometry is an instrumental method of obtaining structure and mass information about either molecules or atoms by generating ionized particles and then accelerating them in a curved path through a magnetic field. Heavier particles are more difficult for the magnetic field to deflect around the curve, and thus travel in a straighter path than lighter particles. Consequently, by the time the particles reach the detector, a mixture of ions will have separated into groups by mass (or more specifically the mass-to-charge ratio of the individually weighted ions.) The ions are produced from neutral molecules and atoms by stressing them with some form of **energy** to knock off electrons. In the case of molecules, fragmentation as well as **ionization** usually occurs. Each type of **molecule**

breaks up in a characteristic manner, so a skilled observer can interpret a mass spectrum much like an archaeologist can reconstruct an entire skeleton from bone fragments. A mass spectrum can help establish values for ionization energy (the amount of energy it takes to remove an **electron** from a neutral **atom** or molecule) and molecular or atomic mass for unknown substances. The extremely high sensitivity of mass spectrometry makes it indispensable for analyzing trace quantities of substances, so it is widely used in environmental, pharmaceutical, forensic, flavor, and fragrance analysis. The petroleum industry has used mass spectrometry for decades to analyze hydrocarbons.

The basic principle underlying mass spectrometry was formulated by **Joseph J. Thomson** (the discoverer of the electron) early in the century. Working with cathode ray tubes, he was able to separate two types of particles, each with a slightly different mass, from a beam of **neon** ions, thereby proving the existence of isotopes. (Isotopes are atoms of the same element that have slightly different atomic masses due to the presence of differing numbers of neutrons in the **nucleus.**) The first mass spectrometers were built in 1919 by **Francis W. Aston** and A. J. Dempster.

There are five major parts to a mass spectrometer: the inlet, the ionization chamber, the mass analyzer, the detector, and the electronic readout device. The sample to be analyzed enters the instrument through the inlet, usually as a gas, although a solid can be analyzed if it is sufficiently volatile to give off at least some gaseous molecules. In the ionization chamber, the sample is ionized and fragmented. This can be accomplished in many ways—electron bombardment, chemical ionization, laser ionization, electric field ionization—and the choice is usually based on how much the analyst wants the molecule to fragment. A milder ionization (lower electric field strength, less vigorous chemical reaction) will leave many more molecules intact, whereas a stronger ionization will produce more fragments. In the mass analyzer, the particles are separated into groups by mass, and then the detector measures the mass-to-charge ratio for each group of fragments. Finally, a readout device—usually a computer—records the data.

Mass spectrometers are often used in combination with other instruments. Since a mass spectrometer is an identification instrument, it is often paired with a separation instrument like a chromatograph. Sometimes two mass spectrometers are paired, so that a mild ionization method can be followed by a more vigorous ionization of the individual fragments.

MASSIE, SAMUEL P. (1919-)

American chemist

Samuel P. Massie's outstanding educational career as a **chemistry** professor has led him to be recognized as a leader in the field of chemical education. For his contributions to both the scientific and academic community, Massie received the 1961 Chemical Manufacturers Association award for "excellence in chemistry teaching" and "in recognition of service to the scientific community in instructing and inspiring students." Massie was nominated for this award by his colleagues and students at Fisk University.

Samuel Proctor Massie was born in North Little Rock, Arkansas, on July 3, 1919. An excellent student, Massie graduated from high school at the age of thirteen. He then attended Dunbar Junior College. When Massie graduated from Dunbar he went to the Agricultural Mechanical and Normal College of Arkansas, where he received a bachelor of science degree in 1938. In 1939 he began working as a laboratory assistant in chemistry at Fisk University, where he also obtained his master of arts degree in 1940. Between 1940 and 1941 Massie was an associate professor of mathematics at the Agricultural Mechanical and Normal College of Arkansas. He was a research associate in chemistry at Iowa State University beginning in 1943 and received a doctorate in **organic chemistry** from that university in 1946. That same year he returned to Fisk University as an instructor. The year following, Massie married Gloria Tompkins. They have three sons.

During his long teaching career, Massie held several positions, including professor and chair of the chemistry department at Langston University from 1947 to 1953. From 1953 to 1960 he held a similar post at Fisk University. In addition to his regular teaching position at Fisk, Massie also served as Sigma Xi Lecturer at Swarthmore College. In 1960 Massie became associate program director at the National Science Foundation (NSF), an agency that works to support a national science policy by sponsoring research, science curriculum development, teacher training, and various other programs. In 1961 Massie went on to become chair of **pharmaceutical chemistry** at Howard University, a post he gave up in 1963 to become president of North Carolina College at Durham. In 1966 he joined the faculty at the United States Naval Academy as a professor of chemistry, also serving as chair of the chemistry department from 1977 to 1981.

In addition to the award from the Chemical Manufacturers Association in 1961, Massie received an honorary doctorate from the University of Arkansas in 1970. In 1980 he was named Outstanding Professor by the National Organization of Black Chemists, and in 1981 he received a Distinguished Achievement Citation from Iowa State University. He is a member of the American Chemical Society. Massie has been active outside academia as chair of the Maryland State Board for Community Colleges, and he has also served on the Governor's Science Advisory Council. He has contributed to his community in many capacities, including membership on the Board of Directors of the Red Cross, and distinguished service with United Fund.

MATERIALS CHEMISTRY

Materials **chemistry** is a branch of chemistry that is concerned with the discovery of new materials and the improvement of existing materials for use in the chemical industry. The field of materials chemistry is continually growing since new materials with unique and interesting properties are being discovered almost every day. Materials chemistry is concerned with several different classes of materials, including **metals**, **ceramics**, polymers, and **composites**.

Materials chemists spend a lot of time and money researching and developing metallic materials. A metallic material is not necessarily a pure metal. It can be any inorganic material containing one or more of the metallic elements in its structure. Metals conduct **electricity** and are often good thermal conductors as well. A metal has a well-ordered crystalline structure that gives the material strength as well as a high melting point. Metals can be classified into two basic categories—ferrous and nonferrous. Metals that contain **iron** are ferrous. Nonferrous metals include **zinc**, **aluminum**, and **copper**. Alloys of metals are metal mixtures created by melting different metals together and allowing them to solidify as a solid **solution**. Metallic materials with new, desirable properties can be created using alloys. Materials chemists study and develop new alloys that can make a metal product stronger, lighter, or more conductive. The metallic materials they develop can then be used to make automobiles, appliances, wires, or airplanes.

Materials chemists also study the development of ceramics. The ceramic materials they create could be traditional ceramics, advanced ceramics, or glasses. Ceramics are also inorganic materials. They consist of metals and nonmetals bound together in a crystalline or non-crystalline structure. Some ceramics are a mixture of crystalline and non-crystalline structures. Ceramics are generally quite resistant to **temperature** changes, are hard **solids**, and are resistant to **corrosion** by chemical agents. They are not very conductive, electrically or thermally. It is this property of ceramics that makes them useful for insulating materials. Even though ceramics tend to be quite hard, they are also very brittle.

Traditional ceramics are composed mostly of clay. The clay is the material that gives ceramics the ability to be molded and worked into various shapes and forms. The clay is mostly composed of hydrated aluminum **silicates** (Al_2O_3*SiO_2*H_2O). In addition to clay, ceramics often contain silica (SiO_2) which raises the melting point of the product and feldspars that produce the glassy quality of fired ceramics. Materials chemists study the effect of manipulating the relative amounts of each of these components on the quality of the ceramics produced.

Advanced ceramics are used for more technical applications where a very specific material is required. These materials are often used in electronics. Advanced ceramics have a pure chemical composition, in contrast with the mixed composition of the traditional ceramics. Compounds that are used for advanced ceramics include aluminum **oxide** (Al_2O_3) and zirconia (ZrO_2). Scientists who specialize in materials chemistry study different chemical compounds that can be used for advanced ceramics and that have desirable physical and chemical properties.

A **glass** is a substance which is non-crystalline yet almost completely undeformable. This state is called the vitreous state. Vitreous substances, when heated, will transform slowly through stages of decreasing **viscosity**. As a sample of glass is heated, it becomes more and more deformable, eventually reaching a point where it resembles a very viscous liquid. At lower temperatures glasses are only very slightly deformable. Glasses tend to bend and elongate under their own weight, especially when formed into rods, plates, or sheets. Glasses can be either organic or inorganic materials. Many definitions of glass include the idea that they do not crystallize as they solidify. Since the definition of solidification is the act of crystallization, this idea of glass as a non-crystalline solid may not be entirely correct. Crystallization occurs when the atoms or molecules of a substance arrange themselves in a systematic, periodic fashion. The atoms or molecules of glass do not exhibit this **periodicity**. This has led to the idea that glass is an extremely viscous, or ''supercooled'' liquid.

Glass is often referred to as an amorphous solid. An amorphous solid has a definite shape without the geometric regularity of crystalline solids. Glass can be molded into any shape. If glass is shattered, the resulting pieces are irregularly shaped. A crystalline solid exhibits regular geometrical shapes when shattered. Amorphous solids tend to hold their shape, but they also tend to flow very slowly. If left undisturbed for a long period of time, glass will very slowly crystallize. Once it crystallizes, it is no longer considered to be glass. At this point, it has devitrified. This crystallization process is extremely slow and in many cases may never occur.

Glasses are made differently than the other ceramics. They are heated and cooled relatively rapidly so the materials do not crystallize. Most glasses are based on silica (SiO_2) with other oxides added in small amounts. These additives give the glass characteristic properties. The materials chemist studies and develops new additives that will increase the strength, transparency, or the ability to mold a particular glass.

Materials chemistry also deals with materials made from polymers. Polymers are long **chains** of repeating smaller units, called monomers. These materials are usually noncrystalline, although some can be partly crystalline. Polymers include **plastics** and rubbers. They usually have low melting points, are good insulators, and can be molded and shaped easily. There are three classes of polymers—thermoplastics, thermosets, and **elastomers**. Thermoplastics consist of long **carbon** chains that are covalently bonded to chains of other atoms. They can be melted at a low temperature and shaped into a mold as they cool back into a solid. Thermoplastics are used to make plastic containers such as soda bottles and milk jugs. Thermoset polymers are made of a network of carbon atoms. They are formed into a shape and then set using a chemical reaction that involves **heat** and pressure. Once they are set, they cannot be reformed into a different shape. Rubbers are elastomers. These elastic polymers are made of long carbon chains without additional atoms bonded to the structure. Polymers are used in many applications, including electrical insulation, appliances, plastic containers, epoxies, and automobile tires. The monomers used to make a polymer give the polymer its properties. Materials chemists develop new monomers and combinations of monomers to create the most efficient and productive polymers.

Materials chemistry also involves the production of composite materials. These materials are formed from two or more different materials combined together. They take advantage of the unique properties of each component (for example, glass fibers in a polymer matrix) and are designed to optimize the strength, **density**, electrical properties, and cost of the materials. Some examples of composite materials include metal matrix compounds, polymer matrix compounds, and ceramic matrix compounds. These compounds can be used for products that need greater strength and lower weight, such as airplane coatings and sports equipment. The material chemist identifies combinations of materials that can be used to create composites with particular characteristics.

MATTER

In ancient Greece, some philosophers, most notably Heraclitus, believed that everything in the world was in a state of fluctuation. Others argued that there must be some permanence, otherwise it would not be possible to see anything as being real.

The fifth century Greeks were apparently the first to attribute structure to matter. They postulated that matter consisted of very small particles that were firmly bound together in the solid state, but which could change position to accommodate compression and deformation. At higher temperatures, these particles were thought to slide past each other and to eventually separate, as the material object underwent melting and evaporation. The Greeks called these particles atoms, Greek for indivisible. This philosophical position offered a reconciliation between the views of a world in fluctuation and one that is permanent.

In 1687, Isaac Newton published his *Principia*, in which he described his laws of motion. As had his Greek predecessors, Newton viewed matter as passive and inert, and as consisting of "solid, massy, impenetrable, movable particles." Thus, for Newton and his contemporaries, there was little distinction between the properties of matter in the material world and the building blocks of which it was composed.

In 1785, the French chemist, Antoine Lavoisier, proposed his Law of Conservation of Matter, which states that matter can neither be created nor destroyed, only changed into different forms. This law has since been superseded by the Law of Conservation of **Mass** and **Energy**, which takes into account Albert Einstein's observation that mass and energy are interchangeable under certain conditions.

Today, scientists believe that matter is made up of molecules; that molecules consist of atoms bound together; that each **atom** contains a **nucleus** surrounded by electrons; that the nucleus contains protons and neutrons, bound together; and that protons and neutrons consist of quarks bound together by gluons.

The modern viewpoint is consistent with the definition of matter as anything that occupies space and has mass. Mass is the material property that shows up as weight when the object is acted upon by gravity. Mass can be measured from an object's tendency to resist moving, i.e., its inertia. The ratio of an object's mass to the **volume** it takes up is known as its **density**. The basic unit of mass in **chemistry** is the gram (1 kilogram = 1000 grams).

Matter may exist in the solid, liquid, or gaseous state. **Solids** have definite sizes and shapes; **liquids** have definite volumes, but no fixed shapes; and **gases** possess neither shapes nor fixed volumes.

Matter that is made up of only one type of atom is called an element. Loosely, if a substance contains two or more kinds of atoms joined together and grouped in a definite way, that substance is called a compound; mixtures have a composition that is not uniform, but heterogeneous.

James Clerk Maxwell.

The physical properties of matter are those that can be observed when no change is taking place in the **atomic structure**. Examples include **color**, density, melting point, and hardness. Chemical properties have to do with a substance's tendency to react with other substances. An example is iron's tendency to rust in moist air.

In the case of ordinary chemical transformations, the mass of the products always equals the mass of the reactants.

See also Conservation of mass; Energy

MAXWELL, JAMES CLERK (1831-1879)

Scottish mathematician and physicist

James Clerk Maxwell was a researcher who opened a new paradigm with his electromagnetic theory, influencing generations of researchers. He was without a doubt a child prodigy. At an early age, he was solving geometric problems and writing explanations that intrigued academics. Just as he considered how charged particles interact with their surrounding area, one might consider the interaction of the conditions of his inherent nature and the environment of his early childhood. Maxwell's life could make a good case study for the strength of the influences of heredity vis à vis environment as he had strong influences from both sources.

Maxwell was born into a distinguished family with notable accomplishments among his ancestors. He had an almost ideal childhood, living in the country in close warm association with his parents and relatives. Like Lord Kelvin, he experienced the death of his mother during his childhood, an experience that drove him closer to his father. Nevertheless, Maxwell did not have the personality of a high achiever. In fact, many reported him as being "eccentric," mostly due to his shyness and his rustic ways. Fortunately, with his eccentric personality came many admirable personality traits. He constantly demonstrated a great imagination, planning experiments to study what others considered impossible. He was a clear communicator, explaining complex topics to learned audiences.

James Clerk Maxwell was a descendent of the Clerk and the Maxwell families both with a distinguished heritage. His father inherited a house in Edinburgh and land in the countryside. Maxwell was born in 1831 in Edinburgh while his parents were waiting for their country house to be built. They moved shortly after he was born. His father was a lawyer but was not very aggressive in pursuing new business. John Clerk Maxwell enjoyed studying science and building mechanical devices. As young as three years old, Jamesy was following his father insisting to know how everything worked. He was very close to his father all of his life. Maxwell's mother died suddenly when he was eight years old. For two years after his mother's death, he was educated by a series of tutors, but none were found suitable for Maxwell and his unique way of learning. His father and his aunt arranged for him to begin studies at the Edinburgh Academy. His first year at the Academy was difficult as the children teased him for the way he talked and dressed. They considered Maxwell stupid and nicknamed him "Dafty." From the second year at the Academy, he started to show his true capabilities and his classmates were less cruel.

About this time, Maxwell became interested in geometry. He did a study of ovals and double-foci ellipses. Just after he turned 15, he wrote a paper on what he had learned from his drawing and measuring ovals. His father showed the paper to Professor Forbes who read it to the Edinburgh Royal Society. Those who were impressed with this work recognized that most of the conclusions were already known. However, they were amazed that the eloquence and intricacy in the description of the issues came from so young a person with so limited formal education. In 1847, at age 16, Maxwell began his college studies at the University of Edinburgh. He spent three years there and during this time, he contributed two more papers to the Edinburgh Royal Society. When he finished his studies at Edinburgh, his father sent him to Peterhouse, but shortly after beginning there, he transferred to Trinity where he believed he had a better chance for a fellowship. Maxwell studied at Trinity from early 1851 until he graduated in 1854. After graduation, he was awarded the fellowship. Maxwell then applied for a position at Marischal College to be close to his ailing father. However, his father did not live much longer. After his father's death in April 1855, he accepted the position at Marischal.

In 1858, he married the well-educated Katherine Dewar. Two years later, he had to leave Marischal, the victim of an institutional merger. He was immediately invited to teach at King's College, London. It was in London that he did his most prominent work. He remained there until he resigned his post (probably due to exhaustion) in the spring of 1865. He spent most of the next five years at his country home writing a book on his theory. He considered himself retired.

To stay involved in academia, he did consulting work for Cambridge. His encouraging Cambridge to offer courses on **heat** and electromagnetism directly influenced the foundation of the Cavendish Laboratory. It was only natural that the first Cavendish professorship should be offered to him and he accepted. During his eight years as Cavendish professor, he worked to prepare for publication the experiment papers Cavendish had written. It is well accepted that this self-imposed responsibility was influential in bringing due respect to Cavendish's work. In May 1879, as the school year wound down, it was obvious to many that Maxwell's health was beginning to fail. He tried to return to Cambridge in the autumn, but he could scarcely walk. On November 5, 1879, Maxwell died of abdominal cancer at the age of 48.

Maxwell's work leading to his kinetic theory of **gases** and his theory of electromagnetic fields was a logical advance from **James Prescott Joule**'s work. Both researchers measured the **velocity** of gas molecules and both recognized that heat was not the fluid that it once was thought to be. The importance of Maxwell's work was the direction that it gave these new understanding. Joule only showed the scientific community what was possible to measure and what might be proven. Maxwell went forward with detailed mathematical models that left no holes unfilled, with one important exception. Maxwell used statistics to show the high probability that proposed laws would direct the behavior of **matter**. Discussing the probability of natural law took science away from determinism. This opened the door for the modern study of physics. Einstein's theory of relativity and the recently nurtured **Chaos theory** could not have been developed except for this new philosophical direction.

Maxwell began measuring the average velocity of a gas **molecule** with the objective to investigate whether the perceived random order of their movement could be predicted with some degree of accuracy. What he found was that the greater the velocity of the molecules, the greater the heat generated. There was a direct relationship between the amount of movement among the molecules and the amount of heat in a gas. In this experimental demonstration, heat was shown undeniably to be a property of particle movement and not a fluid flowing from one object to another. Furthermore, Maxwell's findings showed that the movement of particles could be controlled through increasing or reducing heat.

Maxwell understood **Michael Faraday**'s theory of electric and magnetic fields. He worked to demonstrate what Faraday could not explain himself through complex calculations. Assuming that the space surrounding a charged particle contained a field of force, Maxwell created a mathematical model demonstrating all the possible phenomena of electric and magnetic fields. Through this model, Maxwell demonstrated that the electric and magnetic fields worked together. He coined the term "electromagnetic" to name this new breakthrough.

This discovery is important for **chemistry** because it ultimately led to the discovery of the **electron. J. J. Thomson** discovered the electron when he was investigating the effects of the electromagnetic field on gases, applying the principles that Maxwell had established. Research on the effects of light on elements was furthered by Maxwell's work. His subsequent work on the velocity of the oscillation of electromagnetic fields demonstrated that light should be considered a form of electromagnetic radiation. This consideration effected light effect theories.

Maxwell was a brilliant man who appeared at times to live in another world. Considering that he recognized deeply the order and potential in some things that others take for granted (such as light and heat), perhaps he did live in another world. However, it was evident by his appreciation of a good laugh with friends that he was not so distant. Furthermore, he could not ignore another brilliant mind that was not yet recognized. When he felt that Willard Gibbs did not receive the respect that he deserved, he built a three-dimensional model of the phenomenon Gibbs was describing. Maxwell named the model after Gibbs. He did this in his dying days; two weeks after he finished it, James Clerk Maxwell died.

MAYOW, JOHN (1641-1679)

English chemist and physiologist

Mayow, a member of a well-established family in Cornwall, studied at Oxford from which he received a bachelor's degree in 1665 and a doctorate in civil law in 1670. He also studied medicine, and although he received no degree in the field, Mayow entered medical practice for a short time after leaving Oxford. He spent much of the 1670s in London where he became acquainted with Robert Hooke, who recommended Mayow's election as fellow of the Royal Society in 1678.

From the late seventeenth century to the early twentieth century scholars have debated Mayow's contributions to science. Oxford printed his first publication, *Tractatus duo* in 1668. These two short tracts addressed respiration and rickets, medical issues of the day. "De respiratione" cited his English contemporaries, **Robert Boyle**, Nathaniel Highmore, Thomas Willis, and the Italian Marcello Malpighi and included references to experiments on respiration by Robert Hooke and Richard Lower. Some scholars have questioned Mayow's originality, citing his familiarity with the scientific literature of the day as proof that his ideas were "borrowed."

Mayow proposed that the purpose of respiration was to convey a supply of fine nitrous particles from the air to the blood and that "nitrous air" was necessary for life and the beating of the heart. Not only did the nitrous particles react with the "sulphureous" particles in the blood to produce animal **heat** and change the blood from purple to scarlet, but also they violently interacted with the "volatile **salt**" of the blood to create tiny explosions which inflated and contracted the heart muscles. Failure to inhale enough nitrous particles would lead to the stoppage of the heartbeat, which would in turn bring about death because life required the distribution of animal spirits throughout the body. Mayow showed that if a mouse were kept in a closed container over **water**, the water would rise into the container until one-fourteenth of the original **volume** of air was diminished, and the mouse would die. He obtained similar results with burning candles; the flame would be extinguished when one-fourteenth of the original volume of air was diminished. These results demonstrated that air contained at least two different types of particles and that one of the particles was essential for both life and burning.

Unfortunately, Mayow died at an early age, and no one carried on his work. Shortly after Mayow's death, Georg Stahl's **phlogiston** theory displaced Mayow's ideas. Not until a nearly a century later would Antoine Lavoisier's work demonstrate that some of Mayow's theories about respiration and **combustion** were correct.

MCCOLLUM, ELMER VERNER (1879-1967)

American biochemist, organic chemist, and nutritionist

Elmer Verner McCollum was a distinguished biochemist, organic chemist, and nutritionist who studied **nutrition** and its effects on health and **metabolism**. The discoverer of **vitamins** A, B, and D, McCollum also found the nutritional cause of rickets, a bone disease. His twenty-six years of research at the School of Hygiene and Public Health at the Johns Hopkins Medical School in Baltimore continued his earlier research and made him an authority on how diets lacking certain vitamins or trace minerals can cause diseases. He developed a biological method to analyze foods and popularized the use of rats as an extremely valuable experimental model. By modifying the rats' diets, McCollum produced animals with rickets. Because of these studies, rickets, a disease in which bones become deformed, is virtually nonexistent in developed countries. In a field of research that had few findings of importance for decades, he was instrumental in ushering in a new era.

McCollum, the fourth of five children and the first son, was born on March 3, 1879, near Fort Scott, Kansas, to Cornelius Armstrong McCollum, a farmer believed to be of Scottish descent, and Martha Catherine Kidwell McCollum, the daughter of mountain people also originally from Scotland. McCollum grew up on a farm, and as a child he was very inquisitive, a trait he had all of his life. Oddly enough for a man who would devote his life to researching nutrition and the role vitamins play in having good health, as an infant he suffered from scurvy, a nutritional disease later determined to be caused by the lack of vitamin C. He was cured when his mother fed him apple peels, which contain this vitamin. His parents under-

Elmer Verner McCollum.

stood the value of advanced education, and his mother in particular was determined that her children should receive higher education. He attended the University of Kansas in Lawrence, where as a sophomore he became interested in organic **chemistry**. At the end of his junior year, he was elected to membership in Sigma Xi, the honorary scientific society. He graduated with a baccalaureate degree in 1903 and a master's of arts degree a year later. McCollum continued his education at Yale University in New Haven, Connecticut, and graduated in 1906 with a doctorate in **organic chemistry**.

After an unsuccessful search for a position in the field of organic chemistry, McCollum began to study **biochemistry**. In 1907 he became an instructor in **agricultural chemistry** at the Wisconsin Agricultural Experimental Station and quickly rose through the academic ranks, becoming a full professor in 1913. In his early work he performed chemical analyses on the diets of dairy cattle and studied how different diets affected the cows' health and reproductive capacities. However, he thought the procedures were long and tedious in such large animals. Fortuitously, references in scientific journals about using mice in other experiments led him to develop a rat colony for his nutritional research. This was the first rat colony used in the study of nutritional aspects of disease. The rats were fed various diets, each one lacking certain substances. Their short life span quickly enabled McCollum to determine what effect the diets had. This work helped to popularize the use of rats in other experimental situations.

In 1912 McCollum noticed that rats fed a diet deficient in certain fats grew normally again when the fats were added back in. This discovery was the first of a fat-soluble nutrient, which McCollum named vitamin A. Subsequent studies have shown that vitamin A helps makes teeth and bones strong and is necessary for normal vision and healthy skin. McCollum later demonstrated that a water-soluble substance, which he called vitamin B, was also necessary for normal health. He had created the alphabetical **nomenclature** for vitamins.

Although pleased with his rapid rise at the Wisconsin Agricultural Experimental Station, McCollum felt somewhat restricted in his research. At this time, funds from the Rockefeller Foundation established a School of Hygiene and Public Health at Johns Hopkins Medical School in Baltimore. Highly respected among his peers, McCollum was offered a position as the first faculty member—professor of biochemistry—which he assumed in 1917. He was made emeritus professor in 1945.

A coincidence led to the discovery for which McCollum is best known: producing rickets in rats. The rats had bending, fractures, and swellings of the ends of the long bones in their legs. But McCollum did not recognize it as rickets until John Howland, professor of pediatrics at Johns Hopkins Medical School, stopped by McCollum's office to inquire whether anyone had ever produced rickets in rats. After McCollum described his rats, he and Howland concluded the rats had rickets. The discovery threw new light on the disease. McCollum identified the missing factor in the rats' diets that resulted in rickets, which he named vitamin D. (Another researcher had discovered a different vitamin, so *C* was taken.) Vitamin D aids in the absorption of **calcium**, which is used in bones and teeth, and is found in fortified milk, fish such as tuna and sardines, dairy products, and egg yolks. It also is formed in the body when a person is exposed to sunlight. Because of this discovery, people changed their diets to include vitamin D, and now rickets is rare in most developed countries. Some of McCollum's research efforts focused on trace elements, simple substances essential to life in very small quantities. Through his research he started to isolate these substances, McCollum showed that a deficiency of calcium eventually would produce muscular spasms. Other trace elements he focused on include **fluorine**, **zinc**, and **manganese**.

In addition to his laboratory research, McCollum lectured widely about progress and problems in nutritional studies. In 1923 he started writing articles for the popular press in *McCall's* magazine. In honor of his outstanding contributions in the field of nutrition, he received many awards. He was a member of numerous national and international organizations devoted to public health, including the international committee on vitamin standards of the League of Nations in 1931, the food and nutrition board of the National Research Council starting in 1942, and the World Health Organization. In 1948 the McCollum-Pratt Institute was created at Johns Hopkins for the study of trace elements. Years after his retirement in 1946, he remained actively interested in nutrition and related health fields and published the comprehensive book *A History of Nutrition* in 1957. He died in Baltimore, Maryland, November 15, 1967.

McConnell, Harden Marsden (1927-)

American chemist

Harden McConnell is known for his innovative work dealing with **nuclear magnetic resonance** (NMR). He created his namesake equation, the McConnell Relation, in the late 1950s; the equation has proved beneficial in studying atomic and **molecular structure**. He also introduced the technique of spin labeling that has aided scientists in identifying the motion and location of molecules, and he has used the procedure to research several subjects including the relationship of **antibodies** and antigens in immune systems. His more recent work has used silicone technology to detect changes within human cells.

Harden Marsden McConnell was born to George and Frances (Coffee) McConnell in Richmond, Virginia, on July 18, 1927. He attended George Washington University, where he majored in **chemistry** and minored in mathematics. After receiving his B.S. degree in 1947, McConnell enrolled at the California Institute of Technology where he earned a Ph.D. in chemistry, with a minor in physics, in 1951. After completing a two-year National Research Council fellowship at the University of Chicago in 1952 with Robert S. Mulliken and John Platt, McConnell began his career at Shell Development Company in Emeryville, California, where he worked as a research chemist for the next four years.

In 1956, McConnell married Sofia Glogovac, with whom he would eventually have two sons and a daughter. The same year McConnell was married, Caltech delegated a team of faculty members, including **Linus Pauling**, who had won the Nobel Prize for chemistry two years earlier, to recruit him as an assistant professor of chemistry. McConnell accepted the position and was subsequently promoted to associate professor in 1958, professor in 1959, and professor of chemistry and physics in 1963.

First at Shell and then at Caltech, McConnell spent a great deal of time developing theoretical models for the purpose of explaining nuclear magnetic resonance (NMR) spectra in organic systems. NMR deals with the magnetic forces that exist in various atoms and result from the spin of the atom's **nucleus**. By subjecting atoms to both an alternating magnetic field and a radio signal of specific frequency, the nucleus of the atoms can be made to reverse their spin. The radio frequency at which this reverse occurs reveals much information about the atoms, and the spectra produced by the process is also important to scientists. During the 1950s, McConnell published a highly influential series of 12 papers on the subject. This work validated chemical shielding and the use of NMR in determining molecular electronic structure and chemical **kinetics**. Using his background in mathematics, McConnell applied **group theory** to the analysis of NMR spectra of complex molecules with several spin-spin couplings between different nuclei.

One of McConnell's papers, written with D. B. Chesnut in 1958, introduced what has become known as the McConnell Relation, an equation that deals with the relationship between spin **density** the **proton bonding** of a **carbon atom**. The March 25, 1993, issue of *The Journal of Physical Chemistry,* dedicated to McConnell's achievements, notes that "This relationship has made possible a means of testing theoretical predictions of **molecular orbital theory** and **valence** bond theory." The *Journal of Physical Chemistry* also points out that the McConnell Relation has been important in allowing scientists to fully understand the organization of free radicals—atoms or molecules that contain an **electron** that is not paired with another electron in the same orbital.

As a result of his research, McConnell was often able to theoretically predict scientific occurrences that, years later, turned out to be correct. His 1963 theoretical analysis of ferromagnetics (iron substances that possess magnetic attraction in the absence of magnetic fields) preceded experimental validation by twenty years. It also took research chemists a decade to empirically verify his 1961 conclusions (with Himan Sternlicht) about the existence and properties of paramagnetic excitons in organic crystals. In the words of the *Journal of Physical Chemistry*'s tribute, "Not only has McConnell been at the forefront of major theoretical developments in interpreting magnetic resonance spectra, but his theoretical contributions have always been balanced by significant experimental innovations and systematic verification in his laboratory."

In 1964, McConnell accepted a professorship at Stanford University and was elected the following year to the National Academy of Sciences. In 1979, he was named Robert Eckles Swain Professor of Chemistry at Stanford; he served as head of the Department of Chemistry from 1989 to 1992. Shortly after joining the Stanford faculty, he introduced a fruitful technique called spin labeling. In a free radical, the spin of the unpaired electron creates a **magnetic moment** that can be detected by electron spin resonance (ESR) **spectroscopy**. When a free radical is chemically bonded to another **molecule**, it acts as a "tag" or "marker" to make the molecule recognizable to scientists, so that its motion or orientation with respect to surrounding molecules can be studied. The technique has been used in many applications, including the dynamics of nonmembrane **proteins** (notably hemoglobin), **biological membranes**, and investigation of the antibody-antigen combining site—a subject of importance in studying immune systems and resistance to disease.

His invention of spin labeling brought McConnell the **Irving Langmuir** Award in Chemical Physics in 1971. McConnell used spin labeling to solve several biochemical problems that had defied previous techniques. He was the first to demonstrate the existence of conformational intermediates during oxygenation of hemoglobin, and he helped Syva Associates develop a fast, convenient, and sensitive test for the presence in the human body of hard drugs such as **morphine**. During the mid-1970s, McConnell applied spin labeling to several topics related to biological membranes, including membrane fusion and the interactions of **hormones** on cell membranes. He also used spin labels to investigate immune responses, developing new information about the interactions of antibodies and antigens.

In 1983, McConnell founded the Molecular Devices Corporation to develop hybrid instruments applying **silicon** technology to biology and **biochemistry**. An instrument called Threshold was produced in 1988 to perform rapid and very sensitive immunoassays (determinations of the amounts of an antigen or antibody present in a sample). Four years later, the company introduced a microphysiometer to detect substances (including toxins, drugs, and hormones) that affect biological cells. This instrument can be used to detect changes that might occur within the cell. For example, cellular metabolic changes caused by virus infection can be continuously monitored with the microphysiometer.

McConnell's contributions to science are not only reflected in research journals and in the instruments and techniques used in laboratories but also in the activities of the students and colleagues he has inspired. During his teaching career, he has mentored over 75 graduate students and over 60 postdoctoral fellows, as well as collaborating with numerous colleagues. In summarizing his career up to the age of 65, *The Journal of Physical Chemistry* concludes "McConnell is without question one of the most imaginative, stimulating, and productive chemists of this generation. His profound insights and his incisive theories and experiments have provided fertile ground for literally thousands of subsequent investigators." In addition to the Langmuir Award in 1971, McConnell has received the 1961 California Section Award of the American Chemical Society, the 1962 National American Chemical Society Award in Pure Chemistry, and the National Medal of Science in 1989.

McMillan, Edwin M. (1907-1991)

American physicist

Edwin M. McMillan's first important discovery was made in 1940 when he, Philip Abelson, and **Glenn T. Seaborg** produced and identified samples of transuranium elements, later named **neptunium** and **plutonium**. After World War II, McMillan became involved in the development of particle accelerators. Simultaneously with but independent of the Russian physicist V. I. Veksler, McMillan found a way to compensate for the relativistic **mass** increase that occurs in high **energy** accelerators. He won a share of the 1951 Nobel Prize for **chemistry** (with Seaborg) for his discovery of neptunium and a share of the 1963 Atoms for Peace Award (with Veksler) for his work on accelerators.

McMillan was born in Redondo Beach, California, on September 18, 1907. His father was Edwin Harbaugh McMillan, a physician, and his mother was Anna Marie Mattison. The McMillan family moved to Pasadena when Edwin was a year old. He attended local primary and secondary schools, graduating from Pasadena High School in 1924.

After completing his B.S. and M.S. degrees at the California Institute of Technology in 1928 and 1929, McMillan enrolled at Princeton University for his graduate study. In 1932, he was awarded a Ph.D. degree for his thesis, entitled "Electric Field Giving Uniform Deflecting Force on a Molecular Beam." McMillan then received a National Research fellowship that allowed him to begin his postdoctoral studies at the University of California at Berkeley.

In 1934, Ernest Orlando Lawrence, inventor of the cyclotron, established the Berkeley Radiation Laboratory on the University of California campus and invited McMillan to join its staff. McMillan maintained his relationship with the laboratory (later renamed the Lawrence Radiation Laboratory) for the next forty years. He was made associate director of the laboratory in 1954 and director in 1958, a post he held until his retirement in 1973.

In the late 1930s, McMillan turned his attention to one of the most exciting topics in scientific research at the time: the bombardment of atomic nuclei by neutrons. This type of research had been inspired by a series of experiments conducted by Enrico Fermi in the mid-1930s. Fermi had found that nuclei will often capture a **neutron** and undergo a nuclear transformation in which they are converted to a new element one place higher in the atomic table than the original element. Over a period of time, Fermi used this technique to transform more than 60 different elements.

The one element in which Fermi was most interested, however, was **uranium**. It was obvious that neutron capture by a uranium **nucleus** would result in the formation of the next heavier element, an element that does not exist naturally on the earth. When Fermi actually conducted this experiment, however, he obtained confusing results that could not be interpreted as the formation of a new element. A few years later, **Otto Hahn**, **Fritz Strassmann**, and **Lise Meitner** correctly interpreted Fermi's experiments, showing that the bombardment of uranium with neutrons had resulted in **nuclear fission**.

Working first with Abelson and later with Seaborg, McMillan repeated Fermi's original experiments. They bombarded uranium with neutrons and found that, while fission did occur, a small fraction of the uranium nuclei did undergo the kind of transformation to a heavier element that Fermi had anticipated. Later studies found that the uranium-235 **isotope** undergoes fission, while the uranium-238 isotope, under the proper conditions, is transformed to a heavier element. McMillan and Abelson suggested the name neptunium for the element after the planet Neptune, located one planet beyond Uranus, the namesake of uranium. A year later, McMillan and Seaborg found that the radioactive decay of neptunium produces yet another element, heavier than itself, an element they called plutonium. Again, the element's name was taken from that of a planet, Pluto, the furthest from the sun. The two researchers were later to share the 1951 Nobel Prize in chemistry for their research on the transuranium elements.

With the onset of World War II, McMillan left Berkeley to conduct military research at the Massachusetts Institute of Technology, at the United States Navy Radio and Sound Laboratory in San Diego, and finally at the Manhattan Project laboratories at Los Alamos. At the war's conclusion, McMillan returned to Berkeley and the radiation laboratory, and immediately became immersed in problems of accelerator design. The cyclotron, invented by Lawrence in the early 1930s, had served the research community well for more than a decade, but the limits of the machine's usefulness were now becoming apparent. The most serious problem with traditional cyclotrons was relativistic mass increase.

Relativistic mass increase refers to the fact that as particles gain **velocity** in an accelerator, they also gain mass; the mass gain subsequently causes them to lose velocity. Before long, high energy particles begin to fall out of phase with the electrical fields that are used to accelerate them, and they become lost inside the machine. McMillan's **solution** for this problem was simple and elegant. He modified the electrical field in the accelerator so that, as particles speed up and gain mass, the rate at which the electrical field changes direction slows down. In that way, the electrical field can be kept in phase with the particles even while they gain energy and mass.

For many years thereafter, this concept was employed in the development of more advanced circular accelerators, such as the synchrotron and synchrocyclotron. In 1963, McMillan was awarded a share of the Atoms for Peace award with Veksler, who had developed the same concept at about the same time.

McMillan married Elsie Walford Blumer on June 7, 1941. She was the daughter of the former dean of the Yale University school of medicine and the sister of Mrs. Ernest Lawrence. The McMillans had three children, Ann, David, and Stephen. McMillan died of complications from diabetes on September 7, 1991, in El Cerrito, California.

Edwin M. McMillan.

Measurements and calculations

Observations, which are an integral part of any science, can be classified into two main types: qualitative and quantitative. In general, qualitative observations are non-numeric. As a result they are not exact and are often not reproducible (two people generally do not describe a particular **color** in exactly the same way). Reproducibility is an important aspect of science and so it is preferable to use quantitative observations or measurements to describe an experiment. Measurements are made with instruments that allow us to exceed the limitations of our senses. Some instruments, such as a ruler that can be used to measure length, or a **balance** that can be used to measure **mass**, are simple to use. Other more complicated instruments, such as a spectrophotometer (used for measuring the relative intensities of light in different parts of the spectrum), or a **nuclear magnetic resonance** (NMR) machine are more difficult to operate. Measurements can be made by one person and duplicated by another, thus making them independent of the measurer.

Measurements consist of two parts: a number and a unit. The number part of the measurement specifies the size, while the unit part of the measurement allows for standards. It would not be useful to say that the distance from A to B is five. This particular statement does not allow us to determine if the distance in question is 5 in or 5 mi. In science, the standard measuring units are metric.

Some quantities are easier to measure than others. Mass, length, and **temperature**, for example, can be measured directly, with one reading from one instrument. **Volume**, however, is not as easy to measure. For example, the volume of a solid, such as a rock, can be measured by dropping it into **water** and measuring the volume of the water that is displaced. The volume of a liquid can be measured directly in a graduated cylinder. The volume of a gas, however, depends on the container in which it is, the pressure surrounding it, and its temperature.

Although it is not easy to measure the volume of a gas directly, it is quite easy to calculate the volume of a gas. Calculations are used when a quantity is determined by a mathematical combination of other measurements. For example, the volume of a regular solid such as a block can be calculated by multiplying its length by its width and its height ($V = l \times w \times h$), and the **density** of a solid can be calculated by dividing its mass by its volume ($d = m/V$). When performing calculations, it is vitally important to ensure that units are consistent. For example, the volume of an object cannot be expressed in standard units if the length, width, and height are all measured using different units. In this case, it would be necessary to convert two of the units to the same as the third one. In more complex calculations, use of incompatible units can result in a completely meaningless value for the quantity being calculated. Thus, when making a measurement or when performing a calculation using measured quantities, it is extremely impor-

Lise Meitner.

tant to ensure that the size of the measurement as well as its unit are noted.

See also International System of Units; Metric system

MEITNER, LISE (1878-1968)

Austrian physicist

The prototypical female scientist of the early twentieth century was a woman devoted to her work, sacrificing family and personal relationships in favor of science; modestly brilliant; generous; and underrecognized. In many ways Austrian-born physicist Lise Meitner embodies that image. In 1938, along with her nephew Otto Robert Frisch, Meitner developed the theory behind **nuclear fission** that would eventually make possible the creation of the atomic bomb. She and lifelong collaborator **Otto Hahn** made several other key contributions to the field of nuclear physics. Although Hahn received the Nobel in 1944, Meitner did not share the honor—one of the more frequently cited examples of the sexism rife in the scientific community in the first half of this century.

Elise Meitner was born November 7, 1878, to an affluent Vienna family. Her father Philipp was a lawyer and her mother Hedwig travelled in the same Vienna intellectual circles as Sigmund Freud. From the early years of her life, Meitner gained experience that would later be invaluable in combatting—or overlooking—the slights she received as a woman in a field dominated by men. The third of eight children, she expressed interest in pursuing a scientific career, but her practical father made her attend the Elevated High School for Girls in Vienna to earn a diploma that would enable her to teach French—a much more sensible career for a woman. After completing this program, Meitner's desire to become a scientist was greater than ever. In 1899, she began studying with a local tutor who prepped students for the difficult university entrance exam. She worked so hard that she successfully prepared for the test in two years rather than the average four. Shortly before she turned twenty three, Meitner became one of the few women students at the University of Vienna.

At the beginning of her university career in 1901, Meitner could not decide between physics or mathematics; later, inspired by her physics teacher **Ludwig Boltzmann**, she opted for the latter. In 1906, after becoming the second woman ever to earn a Ph.D. in physics from the University of Vienna, she decided to stay on in Boltzmann's laboratory as an assistant to his assistant. This was hardly a typical career path for a recent doctorate, but Meitner had no other offers, as universities at the time did not hire women faculty. Less than a year after Meitner entered the professor's lab, Boltzmann committed suicide, leaving the future of the research team uncertain. In an effort to recruit the noted physicist **Max Planck** to take Boltzmann's place, the university invited him to come visit the lab. Although Planck refused the offer, he met Meitner during the visit and talked with her about quantum physics and radiation research. Inspired by this conversation, Meitner left Vienna in the winter of 1907 to go to the Institute for Experimental Physics in Berlin to study with Planck.

Soon after her arrival in Berlin, Meitner met a young chemist named Otto Hahn at one of the weekly symposia. Hahn worked at Berlin's Chemical Institute under the supervision of **Emil Fischer**, surrounded by organic chemists—none of whom shared his research interests in radiochemistry. Four months older than Hahn, Meitner was not only intrigued by the same research problems but had the training in physics that Hahn lacked. Unfortunately, Hahn's supervisor balked at the idea of allowing a woman researcher to enter the all-male Chemical Institute. Finally, Fischer allowed Meitner and Hahn to set up a laboratory in a converted woodworking shop in the Institute's basement, as long as Meitner agreed never to enter the higher floors of the building.

This incident was neither the first nor the last experience of sexism that Meitner encountered in her career. According to one famous anecdote, she was solicited to write an article by an encyclopedia editor who had read an article she wrote on the physical aspects of **radioactivity**. When she answered the letter addressed to Herr Meitner and explained she was a woman, the editor wrote back to retract his request, saying he would never publish the work of a woman. Even in her collaboration with Hahn, Meitner at times conformed to gender roles. When British physicist Sir **Ernest Rutherford** visited their Berlin laboratory on his way back from the Nobel ceremonies in 1908, Meitner spent the day shopping with his wife Mary while the two men talked about their work.

Within her first year at the Institute, the school opened its classes to women, and Meitner was allowed to roam the building. For the most part, however, the early days of the collaboration between Hahn and Meitner were filled with their investigations into the behavior of beta rays as they passed through **aluminum**. By today's standards, the laboratory in which they worked would be appalling. Hahn and Meitner frequently suffered from headaches brought on by their adverse working conditions. In 1912 when the Kaiser-Wilhelm Institute was built in the nearby suburb of Dahlem, Hahn received an appointment in the small radioactivity department there and invited Meitner to join him in his laboratory. Soon thereafter, Planck asked Meitner to lecture as an assistant professor at the Institute for Theoretical Physics. The first woman in Germany to hold such a position, Meitner drew several members of the news media to her opening lecture.

When World War I started in 1914, Meitner interrupted her laboratory work to volunteer as an x-ray technician in the Austrian army. Hahn entered the German military. The two scientists arranged their leaves to coincide and throughout the war returned periodically to Dahlem where they continued trying to discover the precursor of the element **actinium**. By the end of the war, they announced that they had found this elusive element and named it **protactinium**, the missing link on the **periodic table** between **thorium** (previously number 90) and **uranium** (number 91). A few years later Meitner received the Leibniz Medal from the Berlin Academic of Science and the Leibniz Prize from the Austrian Academy of Science for this work. Shortly after she helped discover protactinium in 1917, Meitner accepted the job of establishing a radioactive physics department at the Kaiser Wilhelm Institute. Hahn remained in the **chemistry** department, and the two ceased working together to concentrate on research more suited to their individual training. For Meitner, this constituted a return to **beta radiation** studies.

Throughout the 1920s, Meitner continued her work in beta radiation, winning several prizes. In 1928, the Association to Aid Women in Science upgraded its Ellen Richards Prize—billing it as a Nobel Prize for women—and named Meitner and chemist Pauline Ramart-Lucas of the University of Paris its first recipients. In addition to the awards she received, Meitner acquired a reputation in physics circles for some of her personal quirks as well. Years later, her nephew Otto Frisch, also a physicist, would recall that she drank large quantities of strong coffee, embarked on ten mile walks whenever she had free time, and would sometimes indulge in piano duets with him. By middle age, Meitner had also adopted some of the mannerisms stereotypically associated with her male colleagues. Not the least of these, Hahn later recalled, was absent-mindedness. On one occasion, a student approached her at a lecture, saying they had met earlier. Knowing she had never met the student, Meitner responded earnestly, "You probably mistake me for Professor Hahn."

Meitner and Hahn resumed their collaboration in 1934, after Enrico Fermi published his seminal article on "transuranic" uranium. The Italian physicist announced that when he bombarded uranium with neutrons, he produced two new elements—number 93 and 94, in a mixture of lighter elements. Meitner and Hahn joined with a young German chemist named **Fritz Strassmann** to draw up a list of all the substances the heaviest natural elements produced when bombarded with neutrons. In three years, the three confirmed Fermi's result and expanded the list to include about ten additional substances that resulted from bombarding these elements with neutrons. Meanwhile, physicists **Irène Joliot-Curie** and Pavle Savitch announced that they had created a new radioactive substance by bombarding uranium by neutrons. The French team speculated that this new mysterious substance might be thorium, but Meitner, Hahn, and Strassmann could not confirm this finding. No matter how many times they bombarded uranium with neutrons, no thorium resulted. Hahn and Meitner sent a private letter to the French physicists suggesting that perhaps they had erred. Although Joliet-Curie did not reply directly, a few months later she published a paper retracting her earlier assertions and said the substance she had noted was not thorium.

Current events soon took Meitner's mind off these professional squabbles. Although her father, a proponent of cultural assimilation, had all his children baptized, Meitner was Jewish by birth. Because she continued to maintain her Austrian citizenship, she was at first relatively impervious to the political turmoil in Weimar Germany. In the mid-1930s she had been asked to stop lecturing at the university but she continued her research. When Germany annexed Austria in 1938, Meitner became a German citizen and began to look for a research position in an environment hospitable to Jews. Her tentative plans grew urgent in the spring of 1938, when Germany announced that academics could no longer leave the country. Colleagues devised an elaborate scheme to smuggle her out of Germany to Stockholm where she had made temporary arrangements to work at the Institute of the Academy of Sciences under the sponsorship of a Nobel grant. By late fall, however, Meitner's position in Sweden looked dubious: her grant provided no money for equipment and assistance, and the administration at the Stockholm Institute would offer her no help. Christmas found her depressed and vacationing in a town in the west of Sweden.

Back in Germany, Hahn and Strassmann had not let their colleague's departure slow their research efforts. The two read and reread the paper Joliet-Curie had published detailing her research techniques. Looking it over, they thought they had found an explanation for Joliet-Curie's confusion: perhaps instead of finding one new substance after bombarding uranium, as she had thought, she had actually found two new substances. They repeated her experiments and indeed found two substances in the final mixture, one of which was **barium**. This result seemed to suggest that bombarding uranium with neutrons led it to split up into a number of smaller elements. Hahn immediately wrote to Meitner to share this perplexing development with her. Meitner received his letter on her vacation in the village of Kungalv, as she awaited the arrival of her nephew, Frisch, who was currently working in Copenhagen under the direction of physicist **Niels Bohr**. Frisch hoped to discuss a problem in his own work with Meitner, but it was clear soon after they met that the only thing on her mind was Hahn and Strassmann's observation. Meitner and Frisch set off for a walk in the snowy woods—Frisch on skis, with his aunt trotting along—continuing to puzzle out how uranium could possibly yield barium. When they paused for a rest on a log, Meitner began to posit a theory, sketching diagrams in the snow.

If, as Bohr had previously suggested, the **nucleus** behaved like a liquid drop, Meitner reasoned that when this drop of a nucleus was bombarded by neutrons, it might elongate and divide itself into two smaller droplets. The forces of electrical repulsion would act to prevent it from maintaining its circular shape by forming the nucleus into a dumbbell shape that would—as the bombarding forces grew stronger—sever at the middle to yield two droplets—two completely different nuclei. But one problem still remained. When Meitner added together the weights of the resultant products, she found that the sum did not equal the weight of the original uranium. The only place the missing **mass** could be lost was in **energy** expended during the reaction.

Frisch rushed back to Copenhagen, eager to test the revelations from their walk in the woods on his mentor and boss, Bohr. He caught Bohr just as the scientist was leaving for an American tour, but as Bohr listened to what Frisch was urgently telling him, he responded: ''Oh, what idiots we have been. We could have foreseen it all! This is just as it must be!'' Buoyed by Bohr's obvious admiration, Frisch and Meitner spent hours on a long-distance telephone writing the paper that would publicize their theory. At the suggestion of a biologist friend, Frisch coined the word ''fission'' to describe the splitting of the nucleus in a process that seemed to him analogous to cell division.

The paper ''On the Products of the Fission of Uranium and Thorium'' appeared in *Nature* on February 11, 1939. Although it would be another five and a half years before the American military would successfully explode an atom bomb over Hiroshima, many physicists consider Meitner and Frisch's paper akin to opening a Pandora's box of atomic weapons. Physicists were not the only ones to view Meitner as an important participant in the harnessing of **nuclear energy**. After the bomb was dropped in 1944, a radio station asked First Lady Eleanor Roosevelt to conduct a transatlantic interview with Meitner. In this interview, the two women talked extensively about the implications and future of nuclear energy. After the war, Hahn found himself in one of the more enviable positions for a scientist—the winner of the 1944 Nobel prize in chemistry—although, because of the war, Hahn did not accept his prize until two years later. Although she attended the ceremony, Meitner did not share in the honor.

But Meitner's life after the war was not without its plaudits and pleasures. In the early part of 1946, she travelled to America to visit her sister—working in the U.S. as a chemist—for the first time in decades. While there, Meitner delivered a lecture series at Catholic University in Washington, DC. In the following years, she won the Max Planck Medal and was awarded numerous honorary degrees from both American and European universities. In 1966 she, Hahn, and Strassmann split the $50,000 Enrico Fermi Award given by the Atomic Energy Commission. Unfortunately, by this time Meitner had become too ill to travel, so the chairman of the AEC delivered it to her in Cambridge, England, where she had retired a few years earlier. Meitner died just a few weeks before her 90th birthday on October 27, 1968.

Mendeleev, Dmitry Ivanovich (1834-1907)

Russian chemist

One of the most unlikely success stories in the history of chemistry is that of Dmitry Ivanovich Mendeleev (also Dmitri Mendeléev, Mendeleef, and Mendeleeff). Mendeleev was born in Tobolsk in western Siberia on February 8, 1834. He was the youngest child in a family of either 14 or 17 children (records do not agree). His father, a teacher at the Tobolsk gymnasium (high school) lost his job after he became blind when Dmitry was still quite young. His mother tried to take over support of the family by building a glassworks in the nearby town of Axemziansk.

Mendeleev was only an average student. He learned science from a brother-in-law who had been exiled to Siberia because of revolutionary activities in Moscow. Dmitry completed high school at the age of 16, but only after the family had experienced more bad luck. His father died of tuberculosis and his mother's glassworks burned down.

In 1850, his mother decided to see that her two youngest children received a college education. She and the children traveled by horse first to Moscow, then on to St. Petersburg. Through the efforts of a family friend, she was able to enroll Dmitry at the Central Pedagogical Institute in St. Petersburg. A few months later, exhausted from her efforts, his mother died.

Mendeleev graduated from the Pedagogical Institute in 1855 and then traveled to France and Germany for graduate study. While at Heidelberg with **Robert Bunsen**, he discovered the phenomenon of critical **temperature**, the highest temperature at which a liquid and its vapor can exist in equilibrium. Credit for this discovery is usually given to Thomas Andrews (1813-1885) who made the same discovery independently two years later.

In 1861, Mendeleev returned to St. Petersburg, where he became professor of chemistry at the Technological Institute. Six years later, he was also appointed professor of general chemistry at the University of St. Petersburg, a post he held until 1890. In that year, he resigned his university appointment in a dispute with the Minister of Education. Three years later he was appointed Director of the Bureau of Weights and Measures, a post he held until his death on February 2, 1907. Mendeleev is remembered as a brilliant scholar, interesting teacher, and prolific writer. Besides his career in chemistry, he was interested in art, education, and economics. He was a man of strong opinions who was not afraid to express them, even when they might offend others. He was apparently bypassed for a few academic appointments and honors because of his irascible nature.

The achievement with which Mendeleev's name will forever be associated was his development of the periodic law.

In 1868, he set out to write a textbook in chemistry, *Principles in Chemistry*, that was later to become a classic in the field. Mendeleev wanted to find some organizing principle on which he could base his discussion of the 63 chemical elements then known. After attending the Karlsruhe Congress in 1860, he thought that the **atomic weight**s of the elements might provide that organizing principle. He began by making cards for each of the known elements. On each card, he recorded an element's atomic weight, valence, and other chemical and physical properties. Then he tried arranging the cards in various ways to see if any pattern emerged. Mendeleev was apparently unaware of similar efforts to arrange the elements on the basis of their weights made by J. A. R. Newlands (1838-1898) only a few years earlier.

Eventually he was successful. He saw that, when the elements were arranged in ascending order, according to their weights, their properties repeated in a predictable, orderly manner. That is, when the cards were laid out in sequence, from left to right, the properties of the tenth element (**sodium**) were similar to those of the second element (**lithium**), the properties of the eleventh element (**magnesium**) were similar to those of the third element (**beryllium**), and so on.

When Mendeleev laid out all 63 elements according to their weights, he found a few places in which the law appeared to break down. For example, **tellurium** and **iodine** were in the wrong positions when arranged according to their weights. Mendeleev solved this problem by inverting the two elements, that is, by placing them where they ought to be according to their properties, even if they were no longer in the correct sequence according to their weights.

Mendeleev hypothesized that the atomic weights for these two elements had been incorrectly determined. He happened to be incorrect in this assumption, and it was not until **Henry Moseley** discovered **atomic number**s in 1914 that the real explanation for inversion was found.

Mendeleev made one other critical hypothesis. He found three places in the periodic table where elements appeared to be missing. The blank spaces occurred when Mendeleev insisted on keeping elements with like properties underneath each other in the table, regardless of their weights. He predicted not only that the three missing elements would be found, but also what the properties of those elements would be.

Mendeleev's law was soon vindicated when the three missing elements were found in 1875 (**gallium**), 1879 (**scandium**), and 1885 (**germanium**).

Dmitry Ivanovich Mendeleev.

Mendelevium

Mendelevium is a transuranium element in the actinide series. It is denoted by the atomic symbol Md. It has an **atomic number** of 101 and an approximate atomic weight of 258. It is a man-made radioactive element which has no stable nuclides.

The longest-lived **isotope** of the element, mendelevium-258, has a **half-life** of 56 days. This period is long enough to allow the investigation of some chemical and physical properties. As expected, the element appears to be chemically similar to thulium, the element above it in the **periodic table**.

Mendelevium was produced in 1955 by the research team of **Albert Ghiorso**, Gregory Choppin (1927-), Stanley Thompson (1912-), and **Glenn T. Seaborg** at the University of California at Berkeley. A tiny sample of **einsteinium**-253 was bombarded with **alpha particles** in the university's 60-inch cyclotron. The product of the reaction was 17 atoms of an isotope with mass 256 and a half-life of 77 minutes.

The research team recommended naming the element *mendelevium* in honor of **Dmitry Mendeleev**, the developer of the periodic table.

Meniscus

The meniscus is the curved, upper surface of a liquid formed when in a tube. The meniscus is produced by surface tension. With some **liquids**, such as **water**, the meniscus is concave. This means that the liquid at the edge of the tube stretches up above the surface of the water. Other liquids, for example **mercury**, have a convex meniscus. With a convex meniscus the liquid stands up away from the bulk of the liquid and there is a gap between the liquid and the wall of the container. These differences are caused by the surface tension and cohesive and adhesive forces that are acting between the liquid and the container.

John Mercer.

With water in a **glass** tube the adhesive forces are greater than the cohesive forces. This means that the water is attracted to the glass and at the edges the water is attracted up the glass wall, rather than being attracted to the rest of the water. With mercury in a glass tube, cohesive forces are greater than the adhesive forces. This means that the mercury is attracted to itself more than it is attracted to the glass walls of the tube. This gives the effect of the mercury clumping together so the top of the liquid is a little bump standing upright with as little contact between the end of the mercury and the glass walls as possible.

When the **volume** of a liquid is being read from a scale, the reading should be taken from the bottom of the meniscus to ensure an accurate reading.

MERCER, JOHN (1791-1866)

English chemist

During the mid-to-late 1800s, there were a number of important advances in the textiles industry: the inventions of the flying shuttle, the spinning jenny, the **water** frame spinner, and the mechanical loom all helped to fuel the Industrial Revolution. However, none of these inventions would have had as great an impact as they did without similar advances in the textiles themselves—for example, the development of new materials and methods for processing these materials. One of the most important contributors to this field was **John Mercer**, the father of textile **chemistry**.

Mercer grew up in Lancashire, England—an area that would soon become the hub of the English textile industry. He first entered that industry as a boy, working as a bobbin-winder and, later, as a weaver. At the age of sixteen, he became drawn to the art of dyeing. He set up a small dye laboratory in the Mercer home, and there experimented with new mixtures and colors. He became quite skilled at the manufacture of dyes, and that year he entered into partnership with an investor to open a dyeing shop. Their business, though small, was quite successful, and Mercer was only drawn away by an offer to become an apprentice at a print shop in nearby Oakenshaw. His time there was, unfortunately, somewhat wasted: he was prevented by a spiteful foreman from gaining any real experience, and after a year he was relieved of his apprenticeship.

Mercer spent several years as a simple weaver before once again becoming a dyer. His return to the profession was sparked by an interest in chemistry. In his home laboratory, he experimented with a number of chemicals, eventually producing a new orange dye that (unlike previous dyes) was ideal for calico-printing. In 1818 he was once again employed by the Fort brothers (who had owned the Oakenshaw print shop) as a **color** chemist; there he invented a number of dyes of yellow, orange, and indigo. He was made a partner in 1825.

While his employ with the Fort brothers was profitable, it took away a great deal of Mercer's free time—time that he had previously spent in his laboratory developing new chemicals for textile processing. The partnership was dissolved in 1848 and, at the age of fifty-seven, Mercer finally had both the time and financial resources to pursue the research that had been put off.

His first experiment turned out to be his most important. For years he had wondered about the effect upon cotton fabric of certain sodas, acids, and chlorides. He soon found that, when treated with these caustic chemicals, the material would become thicker and shorter; this made the cotton stronger, shrink-resistant, and more easily dyed. It also imparted to the material a lustrous sheen that became highly valued by textile manufacturers. Mercer called his chemical process mercerization and patented it in 1850.

Mercer himself was most interested in how mercerization aided the dyeing process. When chemically treated, the cotton fibers would swell, becoming more absorbent; mercerized fabrics require about 30% less dye than untreated fabrics. It soon became apparent, however, that mercerization could be applied to many other materials, including parchment and woolen fabric. Today, mercerization is still an important part of the cotton finishing process.

Mercer was made a Fellow of the Royal Society. He died in 1866 of a prolonged illness, brought about by falling into a reservoir of cold water.

MERCURY

Mercury is a transition metal, one of the elements found in the middle of the **periodic table** in Rows 4, 5, and 6 between Groups 2 and 3. Its **atomic number** if 80, its atomic **mass** is 200.59, and its chemical symbol is Hg.

Properties

Mercury is the only liquid metal and the only liquid element except for **bromine**. Mercury has a freezing point of -37.93°F (-38.85°C), a **boiling point** of 673.9°F (356.6°C), and a **density** of 13.59 grams per cubic centimeter.

Two physical properties of special interest are mercury's high surface tension and its superior ability to conduct an electric current. The metal's surface tension is great enough to permit a steel needle to be floated on its surface. Its ability to conduct an electric current allows the metal to be used in a "mercury switch," a small **glass** capsule in which the metal can flow from one end of the capsule to the other. In this way, the capsule can open and close an electric circuit depending on the angle at which it is tipped.

Mercury is a moderately active metal that does not react with **oxygen** at room **temperature** very readily, although it does react with oxygen and most acids at elevated temperatures.

Occurrence and Extraction

Mercury is a relatively rare element in the Earth's crust with an abundance estimated to be about 0.5 parts per million. It usually does not occur as an element, and its most common compound is cinnabar (mercuric sulfide; HgS). Cinnabar usually occurs as a dark red powder more commonly known as vermilion or Chinese vermilion.

The largest producer of mercury outside the United States is Spain. U.S. production numbers are not announced in order to protect U.S. industries from revealing important company secrets. Other producers of the metal are Kyrgyzstan, Algeria, China, and Finland. In the United States, mercury is produced as a byproduct of **gold** mining in California, Nevada, and Utah.

Discovery and Naming

Mercury is rather easily prepared from cinnabar by a process that was known to ancient humans. The Greek philosopher Theophrastus (372-287 B.C.), for example, described a method for preparing mercury by rubbing cinnabar with vinegar in a clay dish. The oldest sample of mercury metal dates to about the fifteenth or sixteen century B.C. It was found in an Egyptian tomb at Kurna, stored in a small glass container.

The poisonous effects of mercury were also known to ancient civilizations. Slaves who worked in Roman mercury mines, for example, often died of exposure to the metal. Strangely enough, trees and plants around mercury mines were unaffected by the element.

Mercury amalgams have also been known and used for centuries. One of the most important uses in ancient societies was to remove **silver** from its ores. Silver dissolves in mercury, forming an amalgam. The amalgam can then be heated to drive off the mercury, leaving pure silver behind.

Droplets of mercury, the only liquid metal. ***(Photograph by Dr. Jeremy Burgess/Science Photo Library, National Audubon Society Collection/ Photo Researchers, Inc. Reproduced by permission.)***

Mercury has long been known also by the name quicksilver, because it is a silver liquid. The Latin term for "watery silver" is *hydrargyrum*, a word from which the element was given its chemical symbol of Hg.

Mercury is now extracted from its ores by a method that has been used for hundreds of years. Cinnabar is heated in air until the mercuric sulfide of which it is made breaks down to yield pure mercury metal: HgS—heat→ Hg + S.

Uses

The most important use of mercury metal today is in the manufacture of **chlorine**. Chlorine is produced by the **electrolysis** of aqueous **sodium** chloride: 2NaCl—electric current→ $2Na + Cl_2$. The problem with this reaction is that sodium reacts with **water** as soon as it is formed, sometimes violently. To avoid this problem a layer of mercury may be placed on the bottom of the container in which the electrolysis is carried out. In this case, the sodium dissolves in the mercury as soon as it forms, preventing an explosive reaction between sodium and water. For many years, the "mercury cell" was a very popular method for producing chlorine. Today, companies are looking for other methods to make chlorine because of the toxic hazards posed by mercury to workers who handle the material and to the environment.

The second most important use of mercury in the United States is in switches and other electrical applications. Again, increasing concerns about the environmental hazards of mercury are prompting companies to move to greater reliance on electronic switches. The toxic effects of mercury also account for the decreasing use of the element in fluorescent lamps and mercury **batteries**, two applications which once counted for a significant portion of the mercury consumed each year.

Some compounds of mercury that are still being used include mercuric arsenate ($HgHAsO_4$) for waterproofing paints; mercuric chloride (mercury bichloride; corrosive sublimate; $HgCl_2$) for tanning of leather, preservation of **wood**, textile printing, and engraving; mercuric cyanide [$Hg(CN)_2$] in germicidal soaps and photography; and mercuric **oxide** (HgO) as a red or yellow pigment in paint, a **fungicide**, and a disinfectant.

Health Issues

Scientists have become fully aware of the toxic effects of mercury only in the past few decades. One of the most important discoveries they have made is that the form in which mercury occurs determines how toxic it will be. Mercury fumes, for example, tend to enter the body easily and exert toxic effects much more easily than liquid mercury metal itself. Short term, limited contact with mercury can cause nausea, vomiting, diarrhea, and stomach pain. More extensive exposure can lead to kidney damage and death in a relatively short period of time.

MERRIFIELD, R. BRUCE (1921-)

American biochemist

American biochemist R. Bruce Merrifield is recognized as a leading figure in peptide research. He was awarded the 1984 Nobel Prize in **chemistry** for his development of an automated laboratory technique for rapidly synthesizing peptide **chains** in large quantities. His research into peptide-protein and nucleic acid chemistry has significantly advanced the fields of **biochemistry**, molecular biology, and pharmacology.

Merrifield was born July 15, 1921, in Fort Worth, Texas, the only son of George E. and Lorene Lucas Merrifield. Two years after his birth, the family moved to California, moving frequently during the Great Depression as his father sought work as a furniture salesman. By Merrifield's count he attended more than forty schools before the family finally settled in Montebello, California. It was at a high school there that he became interested in science, especially chemistry and astronomy. He joined the astronomy club, eventually building his own telescope, grinding its mirror himself, and won runner-up honors in the annual science contest. When he graduated high school in 1939, Merrifield initially entered Pasadena Junior College, but at the end of two years transferred to the University of California at Los Angeles (UCLA) to continue his studies in the field of chemistry. Working in the laboratory of Max S. Dunn, he assisted in the synthesis of a complex amino acid called dihydroxyphenylalanine (DOPA), essential in nerve transmission and used in the treatment of certain illnesses such as Parkinson's disease.

After receiving a Bachelor of Arts degree with a major in chemistry from UCLA in 1943, Merrifield went to work at the Philip R. Park Research Foundation as a chemist. During his stay at the lab, Merrifield assisted in growth experiments feeding test animals a synthetic amino acid diet. The experience lasted only a year, but it was enough to convince Merrifield that to pursue his goals he would need to return to school. An Anheuser-Busch Inc. fellowship allowed him to continue his studies at UCLA's graduate school in the chemistry department. He served as a chemistry instructor from 1944 until 1947, then returned to Professor Dunn's laboratory as a research assistant from 1948 through 1949. His work with Dunn included the development of microbiological methods for the study of yeast purines and the pyrimidines, organic bases that make up such compounds as the nucleotides and **nucleic acids**. In 1949, he received his Ph.D. in biochemistry. Upon graduating, Merrifield was offered an appointment as assistant chemist at the Rockefeller University, then known as the Rockefeller Institute for Medical Research, in New York City. He remained at the institute, becoming a John D. Rockefeller Jr. Professor—the institution's highest academic rank—in 1984.

At the Institute, Merrifield first worked as an assistant to Dr. Dilworth W. Woolley, whom he considered a profound influence on his later career. At this time, having recognized that **proteins** are the key components of all living organisms, he chose to focus on this aspect of their research. Their studies centered on peptide growth factors that Dr. Woolley had discovered and on the dinucleotide growth factors Merrifield had discovered during his graduate study. Merrifield's research required him to isolate biologically active **peptides**. For his experiments, he had to synthesize analogues of the materials. However, the research pointed up the need to improve the techniques of peptide synthesis. The methods pioneered by **Emil Fischer** at the turn of the century for peptide synthesis were laborious and time consuming. Preparing a single experiment's sample could take months of work. Fischer's process involved activating the link in a chain of peptides to which one would attach the next peptide. Before the bond could be activated, however, all the other bonds in the chain would have to be protected. Then, after the required bond had formed, all other protected bonds needed to be cleared of their chemical caps. The process would then be repeated however many times necessary to gradually build the desired **molecule**.

In May 1959, a note in his laboratory book records the moment Merrifield conceived the idea of a new approach to the synthesis. Understating the urgency, his note read: "There is a need for a rapid, quantitative, automatic method for synthesis of long chain peptides." Realizing that the critical step involved activating the peptide bond while protecting all others in a long chain, Merrifield hit on the notion of anchoring the chain to a solid base. The first amino acid in the sequence would be tightly bonded to a polymeric support, an insoluble foundation that would not react during the peptide bond additions but, following the reactions, could easily be removed by the proper **solvents**. This solid-phase method acted like the frame of a loom, holding the ends of the chain taunt while link after link was sewn in. Not only did this save one of the most cumbersome steps in the Fischer process, the need to purify the intermediate product prior to the addition of another bond, it also allowed much more reagent to be flushed in due to the secure hold at the polymeric end. The reactions could be driven longer and harder, producing a purer product. Purities of

99.5% were achieved, as Merrifield announced at a 1962 meeting of the Federation of American Societies of Experimental Biology.

Despite being simplified, however, the steps were still so repetitive Merrifield felt they could be automated. Working in the basement of his house, he devised the first prototype of an automated peptide synthesizer by 1965. The machine was a technological success. In 1969, Merrifield used his box of computer switches filled with jars and tubes to carry out the complete synthesis of one of the first enzymes he had begun his work on years earlier. It took 369 separate chemical reactions in 11,391 steps to make the ribonuclease molecule, but it took far less time than before. However, the practical advantage had a price. Because the step of isolating—and thereby purifying—the intermediates had been circumvented, trace side reactions were no longer washed out of the product during a protein's synthesis. While Merrifield could boast that each stage's yield had a high degree of purity, the overall synthesis after so many steps may have had an appreciable number of undesirable side products. Advances since then, such as in liquid **chromatography**, a method of identifying substances using light, and the use of stronger reagents, have dramatically improved product purity. But the key advance was in the speed and simplification of a once unimaginably complex task. The process—which has never been patented, either by Merrifield or Rockefeller University—has been used widely and is seen as one of the fundamental techniques of genetic and biochemical research.

Merrifield was invited to be a Nobel Guest Professor and traveled to Uppsala University in 1968. Since 1969 he has edited the *International Journal of Peptide and Protein Research.* In addition to his Nobel Prize, he has received many academic and professional awards, including: the Lasker Award for Basic Medical Research (1969); the Gairdner Award (1970); the Intra-Science Award (1970); American Chemical Society Award for Creative Work in Synthetic Organic Chemistry (1972); the Nichols Medal (1973); the Instrument Specialties Company Award of the University of Nebraska (1977); the Second Alan E. Pierce Award of the American Peptide Symposium (1979); the Ralph F. Hirschmann Award in Peptide Chemistry, American Chemical Society (1990); and the Josef Rudinger Award (1990). Merrifield and his wife, Elizabeth, live in Cresskill, New Jersey. They have six children: daughters Nancy, Betsy, Cathy, Laurie, and Sally, and their son James. In the 1980s, Elizabeth, a trained biologist, joined Dr. Merrifield in his lab at Rockefeller University.

Meselson, Matthew Stanley (1929-)
Stahl, Franklin William (1930-)

American molecular biologists

Matthew Meselson and Franklin Stahl are best known for their joint demonstration of semiconservative replication of deoxyribonucleic acid (**DNA**). This means that when a cell prepares to divide into two new cells and duplicates its double-stranded DNA, each new cell receives one original (or parent) strand and the newly-replicated (copied) "child" version of its complementary strand. The other possibility, conservative replication, would give both parent DNA strands to one of the new cells and both new strands to the other new cell.

Meselson was born in Denver and received his bachelor's degree in liberal arts from the University of Chicago in 1951. He earned his Ph.D. in physical **chemistry** from the California Institute of Technology in 1957, remaining on its faculty for three years. Franklin Stahl, who was born in Boston, earned a bachelor's degree from Harvard University in 1951 and a Ph.D. in biology from the University of Rochester in 1956. He worked with Meselson while doing postdoctoral research at Caltech during 1955-1957.

In 1957 the American botanist J. Herbert Taylor and colleagues showed that plant cell chromosomes were transmitted semiconservatively. Meselson and Stahl then investigated the replication of DNA, using a technique that they, along with molecular biologist Jerome R. Vinograd, had developed, in which labeled **solids** are separated in a **centrifuge** and their **density** is measured. Meselson and Stahl began by growing several generations of *E. coli* bacteria in an environment of nitrogen-15, a stable **nitrogen isotope** that is heavier than the common form, nitrogen-14. They then switched the bacteria to a nitrogen-14 environment for several more generations. After using a centrifuge to separate the DNA by density, they found that some contained only nitrogen-14, some only nitrogen-15, and some contained equal amounts of the two isotopes. Next, they used **heat** to separate the double strands of DNAs that contained both isotopes. The scientists discovered that one strand contained only nitrogen-14 and its partner only nitrogen-15. This confirmed that one strand came from a parent and one was more recently replicated—the method predicted by **James Watson** and **Francis Crick** when they determined the double-helix shape of DNA. Meselson, Vinograd, and another colleague extended this technique, which is widely used in nucleic acid research. Other scientists later confirmed that natural cell division follows the same pattern.

Since 1960, Meselson has been on the biology faculty at Harvard University. His other studies have included observing how DNA-based viruses produce their **proteins** within a cell, how DNA repairs itself, and gene evolution. He is a member of the U.S. National Academy of Sciences and received its 1963 award for molecular biology. Stahl's other work has included studying the genetics of bacteriophages (viruses that infect bacteria). He has been on the faculty at the University of Missouri and, since 1970, has been a member of the molecular biology faculty at the University of Oregon. He is a member of the U.S. National Academy of Sciences.

Metabolism

The term metabolism refers to all the chemical changes that take place in the body's tissues when the cells are producing both **energy** and essential new organic materials. While these metabolic activities are many and varied, most of them fall into two broad categories: anabolic processes and catabolic ones. These two quite different processes take place constantly and simultaneously.

Anabolism is the cell's synthesizing or building-up phase. Through anabolic reactions, simple substances—usually the molecules of glucose, **fatty acids**, and **amino acids** that have been derived from foodstuffs—are combined in various ways to form more complex substances, such as the new cellular material needed for growth and tissue maintenance. By combining amino acids, for instance, the cells can form structural **proteins** and use them to repair or replace worn-out tissues. The cells can also form functional proteins, such as **enzymes**, **antibodies**, and **hormones**.

Catabolism, the cell's degradative or breaking-down phase, is almost exactly the reverse. Through catabolic reactions, complex compounds—proteins, fats, and **carbohydrates**—are broken down primarily to produce the energy needed for all metabolic activities. Not all the energy is used up at once: as complex nutrient molecules are broken down and oxidized—or burned—as fuel, only about 60% of the energy released is used for immediate needs. The rest is stored in the chemical bonds that link certain **atoms**, most notably those of the phosphate adenosine triphosphate (ATP). When the body needs energy, enzymes break these chemical bonds and release the stored energy for use by the cells.

Enzymes help regulate most metabolic activities. For example, hormones secreted by special cells in the pancreas help determine whether metabolic reactions will be largely anabolic, as is usually the case soon after a meal is eaten, or catabolic, generally during periods when additional energy is needed. And thyroxine, a hormone secreted by the thyroid gland, is one of several hormones that help determine the rate at which these activities will occur—the metabolic rate or, roughly, the rate at which the body uses up energy in the course of its various metabolic reactions.

The metabolic rate is influenced by a number of factors, such as the individual's age, sex, level of activity, state of health, and, of course, the amount of the hormone thyroxine he or she secretes. Probably the most influential factors, though, are the **temperature** of the surrounding environment and the calorigenic effects of the foods most recently eaten. Because an individual's metabolic rate can give physicians a great deal of important information, it often needs to be measured. And since almost all the energy used by the body is eventually converted to heat, the metabolic rate is usually calculated by measuring the amount of heat loss an individual displays during resting (or basal) conditions. The person's basal metabolic rate (BMR) is then judged to be normal or abnormal by comparing it to standardized rates—that is, rates that reflect the average BMRs of healthy individuals of various ages taken under identical and standardized conditions.

Back in the 1830s, the German physiologist Theodor Schwann (1810-1882) coined the word metabolism for the chemical changes that take place in living tissues. Fittingly enough, it was during Schwann's lifetime that many important metabolic concepts were first formulated. In 1828, for instance, **Friedrich Wöhler** discovered that urea, an organic product present in urine, could be manufactured in the laboratory from inorganic chemicals, a discovery that stimulated interest in studying the body's own chemicals. Later, Wöhler also showed that a chemical taken by mouth somehow had combined with other chemicals when it appeared in the urine, one of the first studies that demonstrated chemical changes definitely could—and did—occur inside the body.

In 1842, **Justus von Liebig** showed that animal **heat** stemmed almost entirely from the **oxidation** of recently consumed foods. Liebig, who also believed that not all foods provided equal amounts of heat, pioneered a number of studies aimed at determining the caloric values of different foods. In the 1850s, Claude Bernard (1813-1878), the noted French physiologist, made an important contribution to the study of metabolism. For one thing, Bernard discovered a starch-like substance in the liver that he named glycogen. He went on to show that glycogen was composed of simple blood sugars stored in the liver and could be broken down into glucose when needed. This process, Bernard argued, indicated that the body's various mechanisms appeared to work together, in an integrated fashion, to maintain a constant and well-balanced inner environment.

In the 1890s, German physiologist Max Rubner (1854-1932) showed that the energy used and released by animals followed the same basic chemical principles that inanimate systems were already known to follow, and **Eduard Buchner's** experiments with yeast proved that metabolic processes in organisms resulted strictly from chemical **reactions** and were not energized by a ''life force.'' In the years that followed, numerous scientists—among them Hermann von Helmholtz (1821-1894), German physiologist Eduard Pflüger (1829-1910), and **Arthur Harden**—were able to establish most of the important metabolic parameters.

During the twentieth century, therefore, many metabolic studies began to center around the concept of intermediary metabolism. The concept, introduced in the 1860s by **Karl von Voit**, held that chemical reactions were often much longer and more complicated than had previously been realized, and that it was vital for chemists to search for the intermediate steps that helped a chemical reaction get from one stage to another. Among those scientists working on intermediary metabolism were two German physiologists, Gustav Embden (1874-1933) and **Otto Meyerhof**, who, in 1933, were able to work out the complicated sequences of chemical reactions that are involved in the breakdown of glycogen. Four years later, **Hans Krebs** introduced a model of what he called the citric acid cycle (also called the **Krebs cycle**), a central feature of one of the body's major catabolic pathways. And in the late 1940s and 1950s, German biochemist **Fritz Lipmann** made a number of valuable contributions to intermediary metabolism through his discovery of **coenzyme** A and by his emphasis on the vital role played by **phosphates** in various metabolic pathways.

One focus of recent research is the role of metabolism in the aging process. Studies have found that reducing calorie intake substantially, but not to the point of malnutrition, extends the life span of animals ranging from spiders and fleas to mice and monkeys. The increased longevity goes beyond the effects of simply avoiding diseases that are linked to obesity. Scientists are now trying to better understand this observation.

METALLIC BOND

The physical and chemical properties of **metals** are consistent with the idea that the electrons in these materials are highly mobile. They are, for example, good conductors of **electricity**, and they easily lose electrons to form positive ions. The high thermal conductivity of metals can be explained by postulating frequent collisions between electrons. And the shiny surfaces of metallic **solids** are consistent with the loosely held electrons being able to absorb and emit light over a wide range of wavelengths.

In the **electron** sea model of metallic **bonding**, the metallic lattice is represented as a regular array of positive ions, i.e., metal atoms minus their **valence** electrons, with the valence electrons wandering relatively undisturbed through the lattice. Although they are slowed down in the presence of a positive **ion**, the electrons are able to avoid becoming trapped by these positive centers. The strong attractive forces between the electrons and metal ions in the electron sea results in the metallic bond. These forces contribute to the high melting points, great strength, and good conducting properties of metals.

The electron sea model relates the strength of the metallic bond to the charge of the positive ions that occupy the lattice sites. The metallic bonds in **magnesium**, for example, are predicted to be stronger than those in **sodium**, since the charge of the positive ions in magnesium is twice that of the ions in sodium. The theory similarly explains why magnesium has a higher melting point (1,202°F [650°C]) than sodium (208.4°F [98°C]).

The metallic bond can be thought of as an unsaturated **covalent bond**. In the case of the **hydrogen molecule**, the bonding is *saturated* because no additional hydrogen atoms can share the covalent bond. Each hydrogen molecule contains two electrons, the maximum number of electrons that can be accommodated without violating the **Pauli exclusion principle**, i.e., the fundamental principle that governs the electronic configurations of atoms and molecules containing more than one electron.

If one considers the metal **lithium**, however, there is no such problem combining multiple atoms. This is because there are six unfilled electron states for each lithium **atom**. There is no limit in the number of lithium atoms that may be combined into a single molecule because lithium forms a crystal with each atom surrounded by eight nearest neighbors. And because each atom has one electron to contribute to bonding, each of the eight nearest neighbor bonds involves one-fourth of an electron on average, rather than two as in ordinary covalent bonds. Thus, the bonds are highly unsaturated, as are the bonds in other metals. There are so many unoccupied electron states in metals that an electron can wander freely without being limited by the Pauli exclusion principle.

The tendency of metals to easily deform is consistent with the unsaturated nature of the metallic bond. Because metal atoms have a large number of vacancies in their outermost electron shells, the bonds do not show directional preference, and the metal atoms can easily rearrange relative to each other.

See also Crystallography; Electrical conductivity

METALLOCENES

Metallocenes are a type of organometallic complex, i.e., a **chemical compound** in which a metal **atom** is bonded to an organic structure. With the metallocenes, one or more aromatic rings are bonded to the metal **ion** by their pi electrons (electrons that occur in a certain type of orbit around the atom or molecule).

Ferrocene was the first metallocene to be discovered. This reddish-orange crystalline solid melts at 343°F (173°C). It is also one of a subgroup called the ''sandwich compounds'' because its two aromatic rings are parallel to each other—one above and one below the plane containing the metal atom (in this case, iron).

Metallocenes are formed by the combination of ionic cyclopentadiene, a reactive but aromatic organic **anion**, with derivatives of the transition **metals** or metal halides. (Transition metals are divided into three main series and include **chromium**, **cobalt**, **hafnium**, **iron**, **titanium**, **vanadium**, **ruthenium**, **rhodium**, **zirconium**, **tungsten**, **molybdenum**, **osmium**, and **nickel**. Halides, in this case, are compounds of **halogens** with another element.) Ionic cyclopentadiene's aromatic rings coordinate with the positive ions in the transition metals, forming a metallocene.

Metallocenes are known for their vibrant colors. For instance, those based on titanium are green, those made of chromium (chromocene) are red, cobalt metallocenes (cobaltocene) are purple, and osmium metallocenes are yellow. Some of these metallocenes are more reactive than others in terms of stability. For instance, scientists must be extremely careful when dealing with chromocene, since, when finely powdered, it tends to spontaneously burst into flame after a few moments in the open air. The compounds also tend to be paramagnetic, meaning their magnetic moments can be aligned in the direction of the applied field. Only those metallocenes in the iron group (ferrocene, ruthenium, and osmium) are dimagnetic, which means that their magnetization is in the opposite direction of the applied field.

Another property of metallocenes is that they are easily oxidized to **cation** (positive ion) forms of transition metals. Cobalticinium and ferricinium are examples. In addition, when the central metal atom of a metallocene is in a stable oxidative state (i.e., its ion has lost or gained all the electrons it is going to), the metallocene is impervious to decomposition by air, high temperatures, bases, dilute acids, or **water**. Otherwise, however, many metallocenes are unstable when exposed to air, **heat**, or water.

Metallocenes are useful in industrial **chemistry** as reducing agents (deoxidizers), anti-knock agents for internal **combustion** engine fuels, absorbers of ultraviolet light, and free radical scavengers. However, perhaps their most important use is as catalysts in creating organic compounds. For example, they are central to the most common industrial method for making **acetic acid**, which is the main ingredient in vinegar and essential in the manufacture of some **solvents**, acetate fibers, and flavors. New metallocene catalysts are used in the preparation of new **plastics**, including new methods for the polymerization of alkenes such as ethylene and propylene.

METALLOIDS

The **periodic table** is divided into **metals** on the left and nonmetals on the far right. Separating them are metalloids or semimetals. Metalloids are on either side of a stair-like line and have properties of both metals and non-metals. There are seven metalloids: **boron**, **silicon**, **germanium**, **arsenic**, **selenium**, **antimony**, and **tellurium**. They conduct **electricity**, but not as well as metals and are called **semiconductors**. Their ability to conduct electricity increases with **temperature**. Semiconductors are used in electronic and for making transistors. A computer chip can hold more than a million transistors. Other devices made from semiconductors include hand-held calculators, digital watches, and desktop computers.

Germanium is a relatively rare element found in transistors. Selenium, although rarer than germanium, has several allotropic forms. Electric conductivity of selenium is increased when light shines on it. In addition to being used as a semiconductor in photoelectric cells and copy machines, selenium is also used in **ceramics**, **glass**, and medicine. Like selenium, tellurium is used in electronic devices. The most common form is as a **silver**, brittle metal-like substance. It is the only element that is found in combination with **gold**.

Silicon reacts chemically like **carbon** although it does not form multiple bonds. It is used in making alloys with **iron**, **copper**, and bronze. Silicones are **chains** of silicon and **oxygen** and have organic groups attached. These polymers, depending on the length of the chain and the pattern of the chain are used for **water** repellents, oils, lubricants, polishes, antacids, and in paints that resist **heat**. Pure silicon can be mixed with antimony, arsenic or boron and used to make integrated circuits. Arsenic has two allotropes, yellow arsenic and metallic arsenic, which is brittle. Its major use is as a poison in weed killers and insecticides.

Antimony is used in the manufacture of vases, **cosmetics**, and ornaments. It also has two allotropes, a metallic, brittle silver white form, and a soft, yellow unstable form. Antimony is used as an **alloy** with **lead** in **batteries**, making it more resistant to acid.

See also Allotrope; Computer chips

METALS

Within the **periodic table** there are some 90 elements that can be described as metals. They all have various characteristics in common ranging from **bonding** to chemical nature.

The metals are very different from the nonmetals and they are also a more unified group than the nonmetals, showing a greater similarity of properties within the group. Broadly speaking, the metals are elements that conduct **electricity**, are malleable, and are ductile. Another group of elements—the **metalloid** or semi-metal elements—share some properties with the metals and some with the nonmetals. There are eight of these elements and they are **semiconductors**.

Metals are usually **solids** at standard **temperature** and pressure (STP). One exception to this is **mercury**, which is a liquid at STP. As is to be expected from the fact that most metals are solids at STP, the majority of metals also have melting and boiling points that are high. Metals are usually hard and dense materials that are capable of conducting electricity—a test for **electrical conductivity** is the most common test used to determine whether or not an element is a metal. Electricity is not the only thing that metals conduct well. They are also good conductors of **heat**. Metals all form an **oxide** that is basic in nature. A basic metal oxide when dissolved in **water** gives an alkaline **solution**.

Metals have a shape that can be easily changed by hammering, i.e., they are malleable. Metals are also ductile—they can be drawn out into a long wire. With the exception of **gold** and **copper**, metals are silvery gray in **color** and all metals take a polish well.

Chemically, the atoms of a metallic element are bonded to their neighbors by metallic bonds, producing a giant metallic lattice structure. Metallic elements have relatively few electrons in their outermost shells. When metallic bonds are formed these outermost electrons are lost into a pool of free or mobile electrons. Thus, the metallic lattice structure is actually comprised of positive ions packed closely together and a pool of freely moving electrons surrounding them. These free electrons are referred to as delocalized because they are not restricted to orbiting one particular **ion** or **atom**. It is this pool of delocalized electrons that allows a metal to conduct an electrical charge—the electrons are free to move/to pass the charge along. Alloys also have this type of bonding, allowing them to conduct electricity as well. For this same reason, metals are also good conductors of heat.

The close packing of the metal ions (the ions are packed as close as they can possibly be) explains the high **density** of the majority of the metals. A metal with a high molecular **mass** will have a greater density than one with a lower molecular mass even though their atomic radii may be similar. **Lead** and **aluminum** have similar atomic sizes, but lead has a much larger molecular mass and consequently it has a much higher density. The materials at the left side of the periodic table are all metals and the molecular mass of these elements increases toward the bottom of the table. The lighter metals—the alkali metals—have a much lower density than the others and, in fact, some metals can float on the surface of water, e.g. **lithium** and **sodium**.

The **malleability** of metals is due to the regular arrangement of ions within the metallic lattice. The bonds holding the lattice in place are strong, but they are not inflexible. Layers of ions can slip over each other without the structure of the

Assorted metals. At the center of the image is copper; the other metals are (clockwise from the left) aluminum pellets, nickel-chrome ore, nickel bars, titanium bars, iron-nickel ore, niobium bars, and chromium granules. ***(Photographs by Klaus Guldbrandsen/Science Photo Library, Photo Researchers, Inc. Reproduced by permission.)***

molecule being destroyed. This also explains why metals are ductile. Both of these characteristics are more noticeable when the metal is hot. The metallic lattice is also responsible for the appearance of some metals. When a metal is examined under a microscope it is seen to have a crystalline structure that is made of regions called grains. The smaller the grain size the more closely packed are the ions of the metal and the stronger and harder it is. If hot metal is allowed to cool slowly, the resulting grains are large in size making a metal that is easy to shape. This process is known as annealing. When a hot metal is cooled quickly, the crystals produced are small. When this cooling is carried out in water, it is called quenching. Quenching will produce a metal that is strong, hard, and brittle.

The chemical properties of metals are also characteristic and can be used to distinguish elements (such as silicon) that may appear to be metals on physical grounds. Chemically speaking, one of the best ways to determine if an element is metallic is to look at its oxide. When a metal is burned in air to produce the oxide, the resulting compound has a **pH** of 7 or higher, i.e., it is alkaline. Another characteristic of most metals is that they liberate **hydrogen** gas when they are reacted with dilute **hydrochloric acid** or dilute **sulfuric acid**. The appropriate metal chloride or **sulfate** is also produced. **Platinum** will not undergo this reaction and many metals need a strong oxidizing reagent to react (mercury and **silver** require nitric acid).

Metals can be arranged into a series based on how reactive they are. For example, **potassium** and sodium are found at the beginning of the list because of their high **reactivity**. They burn in air to produce an oxide, they react violently with cold water, and they also react violently with dilute acids. At the end of the reactivity series are metals such as mercury, silver, and platinum. These metals essentially do not react with air, nor are they attacked by steam. They require either a strong oxidizing agent to react with an acid or they undergo no reaction at all. This reactivity series is used to predict how a metal will behave in a given reaction.

Many materials are referred to as metals when in fact they are not. The true metals are actually elements, whereas the false metals are alloys—composites of elements. For example, the element **iron** is a metal. Steel, however, is not a true metal, since it is an **alloy** containing a mixture of iron and carbon—the relative ratios of the two materials control the physical characteristics of the product. Alloys have different properties than the materials from which they are produced, so by careful blending the exact properties required can be manufactured.

Metals have a wide range of uses. For example, copper is used to conduct electricity in cables (an excellent conductor and very ductile). **Tin** is used to coat cans for food storage (non-poisonous and **corrosion** resistant). Aluminum is used as kitchen foil (high malleability). Iron is used as a fencing material (easily workable and relatively resistant to corrosion). Alloys made from metals have different uses. Steel—an alloy of iron and carbon—is used widely in the construction industry (high strength and easy to work). Solder—tin and lead—is used for joining metals together (low melting point). Brass—copper and zinc—is fashioned into ornaments, buttons, and screws (high strength and low weight).

Some metals are referred to as heavy metals or transition metals. These are the elements between groups II and III in the periodic table. The are often used as catalysts and they generally form more than one positive ion. Other metals are referred to as **precious metals**. Precious metals are relatively rare and as a result have a high value in the marketplace.

Metals are rarely encountered in their elemental state in nature. They must first be extracted from the ground as an ore, which is then treated to release the metal. Some metals may be extracted from their natural state by **electrolysis** (for example sodium), while others may need more drastic treatment (such as iron or zinc).

See also Alkaline earth metals; Heavy metals and heavy metal poisoning; Transition elements

METHANE

Methane (CH_4) is the simplest member of the alkanes. At standard **temperature** and pressure, it is a colorless, odorless, nontoxic gas. Methane melts at -296.5°F (-182.5°C) and boils at -259.06°F (-161.7°C). It is virtually insoluble in **water**, but it will dissolve in organic **solvents**. Complete **combustion** of methane occurs with a blue flame (very little soot is produced) and produces **carbon** dioxide and water.

Methane is produced when organic **matter** is digested bacteria in the absence of air, creating **natural gas**. This gas contains 50-90% methane. Most natural gas lies with **coal** and oil deposits buried deep underground and is a product of the decomposition of ancient swamps and bogs. Like coal and oil, methane is especially useful as a fuel for cooking, heating, and even the operation of some motor vehicles. Many factory furnaces burn methane gas, and utilities use it to generate **electricity**. Methane is also a key raw material for making solvents and other organic chemicals. It was first synthesized from carbon and **hydrogen** in 1904 by a Russian-American chemist, Vladimir Ipatiev (1867-1952). Today, methane is prepared industrially by passing **carbon monoxide** and hydrogen over a **nickel** catalyst at moderate temperature. (This reaction is reversible and temperature governs the overall direction.) In the laboratory the normal method of methane preparation is to combine one part **sodium** acetate and two parts **sodium hydroxide**. When heated this produces methane gas and a solid residue of **sodium carbonate**.

People have been aware of methane's existence for thousands of years. As early as 940 B.C., the Chinese piped the gas through hollow bamboo rods and burned it to evaporate seawater and produce **salt**. Though it was used as a lighting fuel as early as the second century A.D. until modern scientific methods were developed, methane was regarded mainly as a natural marvel rather than a useful fuel. Ancient people may have accidentally discovered methane seeping up from the ground when they noticed that breathing the gas made them light-headed and uncoordinated. Thinking they were in the presence of a supernatural power, they erected temples of worship near these sites.

By the late 1700s, scientists had begun to explore the nature of **gases** more systematically. The earliest gases to be identified were **carbon dioxide**, hydrogen, and **oxygen**, but scientists were aware that some sort of "flammable air" could be found near marshes, swamps, cesspools, and dung heaps, where organic matter was in the process of decay. In the 1770s, Italian physicist Alessandro Volta (1745-1827) read a paper by Benjamin Franklin (1706-1790) describing a natural source of flammable air. Volta's friend and fellow scientist, P. Carlo Giuseppe Campi, had discovered such a source of gas in Italy, and Volta, fascinated, began combing the countryside for the signs of gas in Italian marshes. His testing of these marsh gases began when he found a source of gas at Lake Maggiore in November 1776. Volta was able to isolate methane from the gas in 1778, and his careful experimentation with the combustion of gases also led directly to Antoine Lavoisier's discovery of the composition of water in 1783. Although Volta later became famous for his work on electricity, it was the discovery of methane that gave his scientific reputation its first big boost.

In certain proportions, mixtures of methane and air can explode when ignited. This happened frequently in coal mines in the 1800s, when methane gas seeping from coal seams was ignited by the candles and lamps that miners carried to light their way. In 1815, British chemist **Humphry Davy** tested samples of this explosive coal-gas, called "firedamp." He confirmed that it was mainly methane and that it would ignite only at high temperatures. Thus was born the Davy lamp, in which the flame is surrounded by wire gauze to dissipate **heat** and prevent ignition of flammable gases. This invention was the first giant step forward in the safety of coal mining.

After Ipatiev synthesized methane in 1904, the "manufactured gas" was first used in American cities as fuel for gas lights. As simple techniques were developed for extracting methane from coal or oil, and as more pipelines were built to bring natural gas from drilling fields to the cities, the value of methane as a convenient fuel and as a building block for the organic chemicals industry began to rise steadily. Today, scientists are using new sources, such as sewage and waste in **landfills**, to produce methane. Furthermore, researchers are finding new ways to use methane—as fuel for cars and other vehicles, for example.

However, methane has also been implicated as a major contributor to the **greenhouse effect**, meaning that it can accumulate in the atmosphere, trap the Earth's heat, and cause global warming. In addition to seeping from swamps, bogs, and rice paddies or leaking from pipelines, methane is emitted in large quantities by cows, termites, and other animals that digest plants. Although the **concentration** of methane in the atmosphere has increased rapidly in the past few decades, the latest scientific data show that this increase has slowed down. Research is focusing on identifying the natural sources of atmospheric methane via chemical "fingerprints."

Methane is not restricted to the planet Earth. In the early 1900s, astronomers used **spectroscopy** to photograph light waves being emitted from Jupiter and Saturn, the two largest planets in our solar system. The spectra of the light indicated that the atmospheres of these planets contain methane. Since then, astronomers have determined that all of the *gas giants*—Jupiter, Saturn, Uranus, and Neptune—have atmospheres containing significant amounts of methane.

See also Alkane functional group

METHANOL

A computer-generated model of methanol molecules. ***(Photograph by Ken Eward/Science Source, National Audubon Society Collection/ Photo Researchers, Inc. Reproduced by permission.)***

Methanol, which is also known as methyl **alcohol**, **wood** spirit, and wood naptha, is the simplest example of an alcohol. It has the chemical formula CH_3OH. At standard **temperature** and pressure, it is a colorless liquid with a sweet smell. It is completely miscible with **water**. Methanol is poisonous to humans and in small doses it causes blindness, paralysis, and death. It is mixed with **ethanol** to render it undrinkable. With a **boiling point** of 149°F (65°C) it is a volatile liquid. The boiling point is higher than would be anticipated from the **molecular weight** alone, due to the presence of **hydrogen** bonds within the liquid. Methanol was first discovered in 1661 by **Robert Boyle**.

Methanol is incorporated into various plant oils as a methyl ester. It was originally produced by the destructive **distillation** of wood. This produces a mixture of wood gas (hydrogen and methane), pyroligneous acid (ethanoic acid, methanol and propanone), wood tar (various decomposition products), and **charcoal**. Modern industrial production is carried out over a **zinc** and **chromium oxide** catalyst at a high pressure and moderate temperature. Hydrogen is reacted with **carbon** monoxide or **carbon dioxide** (carbon monoxide and hydrogen **gases** in a mixture are called synthesis gas). A less extreme manufacturing technique is the partial **oxidation** of hydrocarbons. Oxidation occurs readily in air in the presence of a **platinum** or **nickel** catalyst. This produces the aldehyde analogue, methanal (acetaldehyde). The methanal can then be used in the manufacture of thermosetting **plastics**. Methanol is a useful solvent for many inorganic salts and organic compounds and it is quite commonly encountered as a solvent for paints and varnishes. The majority of methanol is used in the manufacture of methanal, with smaller amounts of methanoic acid (formic acid) and chloromethane also being manufactured. Methanol can also be used as antifreeze in car engines. In the mid 1990s, the annual United States production of methanol was some 3.5 megatonnes.

Methanol is an example of a primary alcohol. As has already been stated oxidation yields the aldhyde methanal, but further oxidation of this will yield the carboxylic acid methanoic acid. Methanol can be reduced to give **methane**. This reaction is carried out under pressure by heating the methanol with red phosphorous and an aqueous **solution** of hydrogen iodide. This is one of the most efficient reducing methods for organic compounds. The methyl **halide** can be produced by reacting methanol with phosphorous halides, hydrogen halides, or, in the case of methyl chloride, thionyl chloride.

If excess methanol is heated with concentrated **sulfuric acid** one **molecule** of water is eliminated form two molecules of methanol to produce dimethyl **ether**.

Methanol is the simplest occurring alcohol. It has many uses but it is chiefly used in the manufacture of other organic compounds.

METHYL GROUP

The methyl group is the **alkyl group** with the chemical formula $-CH_3$. The methyl group is sometimes abbreviated to -Me. The methyl group is derived from **methane**, CH_4. The methyl group, along with other alkyl groups is a modifier of the behavior of the **functional group** of the **molecule** in which it occurs. Characteristics such as melting point and **boiling point** generally increase with an increase in molecular **mass**. A methyl group has a low molecular mass, and as such, it has a minimal effect on the properties of other molecules of which it is attached to. To illustrate this point, consider the boiling points of the first three members of the carboxylic acid homologous series. Methanoic, or formic, acid (HCOOH) has a boiling point of 212°F (100°C), ethanoic, or acetic, acid (CH_3COOH) has a boiling point of 244°F (118°C), and propanoic acid (CH_3CH_2COOH) has a boiling point of 286°F (141°C).

The methyl carbocation, CH_3^+ is the least stable carbocation.

The addition of a methyl group to a compound is known as methylation. In aliphatic **chemistry**, this is carried out as a substitution of a **hydrogen atom** in a group. In aromatic chemistry, it may involve the substitution of one of the hydrogen atoms in the **ring** structure. This can be carried out using the Friedel-Crafts reaction.

Methylation is an important reaction in biology. DNA methylation, for example, is an important method for the regulation of gene transcription. Disorders in these methylation pathways, especially relating to the two vitamins (folic acid and S-adenosylmethionine) are associated with human disease. Inadequate dietary folic acid during pregnancy, for example, may increase the risk of birth defects.

METHYLENE GROUP

The methylene group is the divalent group :CH_2. It is a modifier of the behavior of the **functional group** of a compound. This group of atoms is unsaturated and it is incapable of an independent existence due to the fact that other atoms will readily react with it. This is because the group has two potential chemical bonds unused and is, as a result, very reactive. The methylene group is often found in the central part of other organic molecules acting as a joining **molecule** for several other groups such as alkyl or aryl groups. It is often formed as an intermediate in organic reactions. For example, in the substitution reactions of alkanes, such as the chlorination of **methane**, methylene dichloride is formed (CH_2Cl_2).

METRIC SYSTEM

The metric system is a system of measurement of decimal units based on the meter. For scientific purposes the SI units of the metric system are adopted. SI (Système Internationale d'Unites) uses seven different fundamental units: the meter, kilogram, second, ampere, kelvin, candela, and **mole**. There are also two supplementary units: the radian, and the steradian. All other units are derived from these units by multiplication or division.

The basic unit of length is the meter (sometimes spelled metre). The original meter is a metal rod stored in Sèvres, France, and its length initially was defined as a fraction of the circumference of Earth. The meter is now defined more precisely as the distance that light can travel in a vacuum in 1/299,792,458 of a second. The internationally recognized abbreviation of the meter is m. Meters are divided up into smaller units so a decimeter is one tenth of a meter, a centimeter is one hundredth of a meter, and a millimeter is one thousandth of a meter. Larger units are measured in kilometers, which are one thousand meters. The word meter for a unit of measurement was coined in France in the eighteenth century. The word is taken from the Latin *metrum* and the Greek *metron*, both of which mean measure. One meter is equivalent to 1.09 yards and one inch is equivalent to 2.54 centimeters.

The kilogram (or kilogramme)is the basic unit of **mass**. The symbol for the kilogram is kg. A kilogram is one thousand grams. In common with many of the other SI standards, the standard kilogram is stored at the Bureau Internationale des Poids et Mesures in Sèvres near Paris, France. One thousand kilograms is a tonne (or metric ton). This is equivalent to 2204.6 pounds.

The second is the basic unit of time. The second is the only SI unit that is not truly decimal in nature. The modern internationally recognized definition of a second is that it is the duration of 9,192,631,770 periods of radiation corresponding to the transition between two hyperfine levels of the ground state of cesium-133. Alternatively, a second is one sixtieth of a minute or one six-hundreth of an hour. The term second comes from the fourteenth century old French *pars minuta secunda*. This translates as the second small part (of an hour), the first small part being a minute. Sixty seconds make a minute, 60 minutes comprise an hour, 24 hours make a day, 7 days a week, and 52 weeks a year. Smaller units of time are measured in milliseconds which are one thousandth of a second and microseconds, which are one millionth of a second. The recognized abbreviation for a second is s.

The ampere is the basic SI unit of electrical current. One ampere is defined as the constant current that, when maintained in two parallel conductors of infinite length and negligible cross-section placed 1 meter apart in a vacuum, produces a force of 2×10^{-7} newton per meter between them. Historically, an ampere was defined as the current that, when passed through a **solution** of **silver** nitrate, deposits silver at a rate of 0.001118 grams per second. The symbol for an ampere is A. The ampere is named for the French physicist André-Marie Ampère (1775-1836) who made major discoveries in the fields of **electricity** and **magnetism**.

The SI unit of thermodynamic **temperature** is the kelvin. The kelvin is described as 1/273.16 of the thermodynamic temperature of the triple point of **water**. For all practical purposes, the size of the degree is the same as the Celsius degree. The zero point on the Kelvin scale is known as absolute zero and it is theoretically the coldest temperature achievable. The triple point of water is at 273.16K and this is the same temperature as 32°F (0°C). The kelvin (symbol K) is named after **William Thomson**, the first Baron Kelvin, a British physicist who worked extensively on **thermodynamics**. He invented the Kelvin scale.

The candela is the metric or SI unit of luminous intensity, that is of light. One candela is defined as the intensity, in a perpendicular direction, of a surface of 1/600,000 square meter of a black body at the temperature of freezing **platinum** under a pressure of 101,325 newtons per square meter. The symbol for a candela is cd. The name is derived from the Latin *candle*, meaning to glitter.

The mole is the basic SI unit of the amount of substance present. One mole is defined as an amount containing the same number of particles as there are atoms in 0.12 kg of carbon-12. The symbol for a mole is mol. The name was coined in the twentieth century from the Greek *mol*, which is an abbreviated form of the word **molecule**.

The first of the two supplementary units is the radian. The radian (symbol rad) is a unit of plane angle. One radian is defined as the angle formed between two radii of a circle that cut off on the circumference an arc with a length the same as the radius. One radian is the same as 57.296 degrees. The name is derived from radius.

The steradian is also a unit of angle, but it is a unit of solid angle. A steradian is defined as the angle, having its vertex in the center of a sphere, that cuts off an area of the surface of the sphere equal to the square of the length of the radius. The symbol for steradian is sr.

Many of the SI units can be divided by the use of an appropriate modifier. For example a tenth of a unit is a deciunit, one hundredth is a centiunit, a thousandth is a milliunit, and a millionth is a microunit. Multiplication of units works in a similar manner such that a kilounit is one thousand units and a megaunit is one million units. For convenience in normal speech some of these units are not used, instead older, imperial measurements are used. For example, some countries use inches, feet, and miles instead of meters and the units derived from the meter.

Many of the units that we use in the metric system are many centuries old and were originally defined in a much more logical fashion. The defining units to which all could be compared are stored in Sèvres, France, at the Bureau Internationale des Poids et Mesures. Originally, one would have to travel there and make copies of the standards. This proved to be inconvenient, introducing inaccuracies that were compounded when copies of the copies were made. The modern definitions are such that a suitably equipped laboratory can recreate the standard identically without recourse to the original. This gives a true world standard for all of these quantities that can easily be manufactured anywhere.

The metric system is an internationally recognized and agreed system of measurement. It is used universally in scientific communications, although not all aspects of it are integrated into everyday life.

MEYER, JULIUS LOTHAR (1830-1895)

German chemist

Julius Meyer, most often known as Lothar Meyer, was the son of a prominent physician and a physician's daughter. Although Lothar intended to become a physician, physical frailties temporarily ended his education at age 14. His father, believing that an avoidance of all mental exertion would cure Lothar's severe headaches, secured for him a gardener's position at the grand duke of Oldenburg's summer palace. Within a year a vast improvement in his heath allowed Meyer to return to studies. After graduating the gymnasium (high school) at Oldenburg in 1851, he began medical studies at the University of Zurich, but after two years he transferred to Würzburg which granted him an M.D. in 1854. Fascination with the **chemistry** of **gases** as exhibited by his dissertation, "On the Gases of the Blood," led him to Heidelberg to study with **Robert Bunsen**, the era's authority in gasometric analysis. In 1858, the University of Breslau awarded Meyer the Ph.D. for research on the effect of **carbon** monoxide on the blood. The following year he became the director of the laboratory in the Physiological Institute at Breslau and remained there until 1868 when he became chemistry professor at Karlsruhe Polytechnic Institute. He moved to Tübingen in 1876 and taught there until his death.

Meyer's interest in the chemical teachings of Jöns Berzelius, the creator of the system of using initial letters for symbols of chemical elements, coupled with the his work with gases apparently prepared him to tackle the chaos in chemical terminology and **nomenclature** that existed in the mid-nineteenth century. In 1860, **Friedrich Kekulé** organized an international conference at Karlsruhe to remedy the confusion. At the meeting Meyer acquired a copy of Stanislao Cannizarro's presentation on the significance of Avogadro's hypothesis in establishing atomic and molecular weights. Deeply impressed by this pamphlet, Meyer wrote *Die modernen theorien der Chemie* (1864; "Modern Chemical Theory"), which included a table of 28 elements arranged in order of increasing atomic weight and grouped into six families, each member of which had the same **valence**. The table left a space for an unknown element with an atomic weight of 73.1 (now known to be germanium). Although other chemists had offered arrangements of elements before, Meyer believed that the key to understanding chemical properties of elements, even those as yet undiscovered, lay in an arrangement according to atomic weight. The book's first edition appeared later in English, French, and Russian. In 1868 notes prepared for a new edition included an expanded table of 52 elements arranged in fifteen families which included the B-subgroups. A modified version of this table appeared in "die Natur der Chemischen Elemente als Function ihrer Atomgewichte" (1870;"The nature of the chemical elements as a function of their atomic weights") months after **Dmitry Mendeleev** published a similar table. Both Meyer and Mendeleev received credit for developing the **periodic table**, and they shared the Royal Society's Davy Medal in 1882 for their accomplishment. For years thereafter, Meyer devoted himself tirelessly to recalculating the atomic weights on the basis of **hydrogen**, assigning it the value of one.

Meyer conducted a large **volume** of chemical work in other fields including molecular volumes, solubilities, **diffusion**, **boiling point**, **chlorine** and **oxygen** carriers, and **mass** action. In **organic chemistry** he proposed the first "centric" formula for the structure of **benzene** in which each carbon **atom** used three of its four valence bonds and the unused fourth bond pointed to the center of the **ring**. His study of the effects of time, **temperature**, **solvents**, and the **concentration** of reagents led him to advocate quantitative studies.

MEYERHOF, OTTO (1884-1951)

American biochemist

In the course of a long and distinguished career both in Germany and the United States, Otto Meyerhof helped lay the foundations for modern **bioenergetics**, the application of the principles of **thermodynamics** (the science of physics in relation to **heat** and mechanical action) to the analysis of chemical

Otto Meyerhof.

processes going on within the living cell. Meyerhof's was the first attempt at explaining the function of a cell in terms of physics and **chemistry**; his research into the chemical processes of the muscle cell paved the way for the full understanding of the breakdown of glucose to provide body **energy**. For his discovery of the fixed relationship between the consumption of **oxygen** and the **metabolism** of lactic acid in the muscle, Meyerhof shared the 1922 Nobel Prize for Physiology or Medicine.

Born Otto Fritz Meyerhof on April 12, 1884, in Hannover, he was the second child and first son of Felix Meyerhof, a Jewish merchant, and Bettina May Meyerhof. Brought up in a comfortable middle class home, Meyerhof attended secondary school at the Wilhelms Gymnasium in Berlin where the family had moved. As a 16-year-old, he developed kidney trouble, necessitating a long period of bed rest. It was during this time that Meyerhof began reading extensively, especially the works of Goethe, which were to influence him deeply in his later life. A trip to Egypt in 1900 also provided him with a lasting love of archaeology. Once his health had improved, Meyerhof began studying medicine, receiving his medical degree from the University of Heidelberg in 1909. As a doctoral dissertation, Meyerhof wrote on the psychological theory of mental disturbances. His interests in psychology and psychiatry were ongoing, though he soon changed the direction of his professional passions.

From 1909-1912 Meyerhof was an assistant in internal medicine at the Heidelberg Clinic, and it was there he came under the influence of physiologist **Otto Warburg** who was researching the causes of cancer. Meyerhof soon joined this young man in a study of cell respiration, for Warburg was examining the changes that occur when a normal living cell becomes cancerous. It was this early work that set Meyerhof on a course of biochemical research. He also spent some time at the Stazione Zoologica in Naples, and by 1913 had joined the physiology department at the University of Kiel where, as a young lecturer, he first introduced his theory of applying the principles of thermodynamics to the analysis of cell processes. It was one thing to lecture about such an application; quite another to apply such principles to his own research.

Meyerhof decided to focus on the chemical changes occurring during voluntary muscle contraction, applying thermodynamics to the study of cellular function. The chemical and mechanical changes occurring during muscle contraction were on a large enough scale that he could measure them with the primitive instruments he had at Kiel, and there had been some earlier research on which to build. Specifically, physiologist Archibald Vivian Hill in England had observed the heat production of muscles, and others, including physiologist Claude Bernard of France and English biochemist **Frederick Gowland Hopkins**, had shown both that **glycogen**, a carbohydrate made up of **chains** of glucose molecules, is stored in liver and muscle cells, and that working muscles accumulated lactic acid. But Meyerhof's research had to be put on hold during World War I.

In 1914, as Europe was becoming increasingly unsettled, Meyerhof married Hedwig Schallenberg, a painter. The couple had three children in the course of their marriage: George Geoffrey, Bettina Ida, and Walter Ernst. There was still time for some research during this period. In 1917 he was able to show that the carbohydrate **enzyme** systems in animal cells and yeast are similar, thus bolstering his philosophical conviction in the biochemical unity of all life. During the war, Meyerhof served briefly as a German medical officer on the French front, after the war resuming his research into the physicochemical mechanics of cell function and being appointed assistant professor at the University of Kiel. It was while examining the contraction of frog muscle in 1919 that Meyerhof proved in a series of careful experiments that there was a quantitative relationship between depletion of glycogen in the muscle cell and the amount of lactic acid that was produced. He also demonstrated that this process could occur without oxygen, in what is termed anaerobic glycolysis. Meyerhof went on to demonstrate further that when the muscle relaxed after work, then molecular oxygen would be consumed, oxidizing part of the lactic acid—actually only about one-fifth of it. He concluded that the energy created by this oxidative process was used to convert the remaining lactic acid back to glycogen, and thus the cycle could begin again in the muscle cell. Such an understanding, though not completely explaining all the steps in such a metabolic pathway, did pave the way for other research into the glucose cycle and energy production in the body. Meyerhof won the 1922 Nobel Prize for this pioneering work, sharing it with Hill whose work on heat in muscle contraction had inspired much of Meyerhof's own research. The prize could not have come at a better time. Just prior to the announcement, Meyerhof had been passed over for the position of chair of the physiology department at the University of Kiel, largely because of anti-Semitism. The Nobel Prize secured for him laboratory space at the Kaiser Wilhelm Institute for Biology in Berlin, where he was appointed a professor in 1924. During his stay there, from 1924-1929, he trained many notable biochemists, **Hans Adolf Krebs**, **Fritz Lipmann**, and **Severo Ochoa** among them. He brought his colleague Karl Lohmann into the group, and Meyerhof and his team continued with further research at Berlin-Dahlem where the institute was located. In 1925 Meyerhof managed to extract the glycolytic enzymes of muscle, thereby making it possible to isolate these enzymes and study the individual steps involved in the complex pathway in the muscle cell from glycogen to lactic acid. It took another two decades to fully delineate the pathway, and many other brilliant scientists worked on it, but it is today called the Embden-Meyerhof pathway in recognition of Meyerhof's groundbreaking work.

In 1926 researchers in the U.S. and in England discovered a new phosphorylated compound in muscle, phosphocreatine. Searching for another possible source of chemical energy in addition to lactic acid, Meyerhof tested this compound and found that its breakdown did produce heat. But more importantly, in 1929, Meyerhof's assistant, Lohmann, discovered adenosine triphosphate (ATP). Meyerhof and Lohmann went on to show that ATP is the most important **molecule** that powers the biochemical reactions of the cell, and to demonstrate its role in muscle contraction as well as in other energy-requiring processes of biological systems. That same year Meyerhof was appointed director of the new Kaiser Wilhelm Institute for Medical Research in Heidelberg, and for the first time had expansive and modern work space at his disposal.

Meyerhof remained at Heidelberg until 1938, continuing his work with ATP and providing striking evidence of the unity of biochemical processes amid the amazing diversity of life forms. But such philosophical proofs would not help him in the new Germany. Starting with Hitler's rise to power in 1933, Jewish scientists like Meyerhof had been steadily emigrating. Finally in 1938 he realized that he could no longer stay. His daughter and oldest son were already out of the country. Together with his wife and youngest child, Meyerhof, on the pretext of taking a few weeks vacation, escaped to Paris where he continued his research at the Institute of Physicochemical Biology. In 1940, when the Germans invaded France, Meyerhof and his family were forced to flee once again, this time to Spain and eventually, via an arduous trip over the Pyrenees, to neutral Portugal where they caught a ship to the United States. There he joined his friend Hill—with whom he had shared the 1922 Nobel Prize—at the University of Pennsylvania where the Rockefeller Foundation had created a professorship in physiological chemistry for him. He continued research into the bioenergetics of cell function, summering at Woods Hole Marine Laboratory on Cape Cod and meeting with other refugees from the European conflagration. He suffered a severe heart attack in 1944, but recovered and became a United States citizen in 1946. In 1949 he was elected to the National Academy of Sciences in recognition of his life's work that included not only original research, but also the publication of over 400 scientific articles. He died of a second heart attack on October 6, 1951.

MICHEL, HARTMUT (1948-)

German biochemist

A biochemist known for his wide range of technical skills in the laboratory, Hartmut Michel has spent his career studying the complex **proteins** involved in **photosynthesis**. He shared the 1988 Nobel Prize in **chemistry** with **Johann Deisenhofer** and **Robert Huber** for a project he initiated: a detailed analysis of the **atomic structure** of a cluster of proteins called the photosynthetic reaction center.

Hartmut Michel was born July 18, 1948, in Ludwigsburg, Germany, to Karl and Freda Kachler Michel. His father was a joiner and his mother was a dressmaker, though in previous generations, Michel's family on both sides had been farmers. Michel graduated in 1967 from the Friedrich Schiller Gymnasium, with a strong interest in molecular biology and **biochemistry**, but before pursuing his education, served for two years in the military.

In 1969, Michel enrolled in the University of Tubingen. During 1972, he did laboratory work at the University of Munich and at the Max Planck Institute for Biochemistry at Martinsred, and by the end of that year, he had decided to pursue a career in academic research. Michel passed his exams in 1974, and began work in the Friedrich Miescher Laboratory of the Max Planck Society in Tubingen under Dieter Oesterhelt, a biochemist. Oesterhelt moved to the University of Wurzburg in 1975, and Michel went with him. It was here he completed work for his Ph.D. in 1977.

A good portion of Michel's work under Oesterhelt was devoted to the study of photosynthesis. This is considered the most important chemical reaction in the biosphere. At its most basic level, photosynthesis is the conversion of **water** and **carbon** dioxide with the use of sunlight into **oxygen** and nutrients, but the process is actually extremely complicated and not well understood. At the time Michel began his work, it was surmised that an area of protein, known as the photosynthetic reaction center, was a major actor in the process of photosynthesis; scientists believed that the electrons somehow picked up the charge that drove the reaction here, but little else was known about it.

Scientists usually study proteins, as they study many substances, by crystallizing them. The crystallized form of a substance is a good subject because crystals are characterized by a very organized internal atomic structure—they are predictable **solids**. X-ray **crystallography** is a tool commonly used in analysis of crystals. This process uses radiation of known wavelength, x rays; the radiation is aimed at crystals, and crystallographers study the reaction of the rays after they come in contact with the crystal. **Electron** clouds present in the crystal cause interference with the x rays, so knowing how the direction of the rays is changed tells the scientist where the electrons are located and thus determines atomic structure.

X-ray crystallography was first used in 1912 by Max Laue, so it was not a new procedure. The difficulty for Michel was not in finding the technology to examine the crystal, but in finding a method to create a crystal out of the protein. It was considered almost impossible to crystallize membrane proteins, the type present in the photosynthetic reaction center. Water is generally used in crystallization, but membrane proteins interact with water as part of their function, so they are not water soluble. Michel decided that this did not mean membrane proteins could not be crystallized, but that they must be crystallized with a different **solution**. He originally worked on this problem with bacteriorhodopsin, one of the halobacteria, which means salt-loving. By using different detergents instead of water, he was able to form a two-dimensional crystal and a very small three-dimensional crystal, neither suitable for study with x-ray crystallography.

At this point in his research, Michel had planned to pursue a postdoctoral position. His limited success, however, excited him enough to move once again with Oesterhelt, this time to the Max Planck Institute for Biochemistry in Martinsred in 1979. Once settled, Michel decided to try a different membrane protein, choosing that of *rhodopseudomonas viridis,* a purple bacterium containing the simplest known photosynthetic reaction center. Success came two years later; in 1981 he formed a crystal that could be studied.

With a crystal well-formed enough to study, Michel turned to Robert Huber, a department head at the Max Planck Institute, to help find a suitable expert in x-ray crystallography to run the tests. Huber directed Michel to Johann Deisenhofer, and a four-year collaboration was begun. By the end of this period, they had identified and placed more than 10,000 atoms in the membrane protein. This work led to the award of the 1988 Nobel Prize in chemistry to Michel, Huber, and Deisenhofer. In receiving this award, the scientists were credited for work that had implications far beyond photosynthesis, including the understanding of respiration, nerve impulses, hormone action, and the process of nutrient introduction to the cells. The understanding of photosynthesis alone, however, is an important advance in scientific knowledge; their research may even make it possible to create artificial reaction centers, which would have implications for many aspects of technology.

Before the award of the Nobel Prize, Michel won two other awards jointly with Deisenhofer—the Biophysics Prize of the American Physical Society in 1986, and the Otto-Bayer Prize in 1988. Michel was also the recipient of the 1986 Otto Klung Prize for Chemistry and the Leibniz Prize of the German Research Association in 1986. Michel is a member of the European Molecular Biology Organization, the Max Planck Society, the Society for Biological Chemistry, the German Chemists' Society, and the Society for Physical Biology. In 1987, he was named department head and director at the Max Planck Institute for Biophysics in Frankfurt/Main. His work continues in the area of photosynthesis and crystallization, and he was the editor of *Crystallization of Membrane Proteins* in 1989.

Michel has been married to Ilona Leger-Michel since 1979. They have a son and a daughter together, and a daughter from his wife's previous marriage. In his spare time, Michel enjoys reading history, traveling, and growing orchids.

MICROSCOPY

Microscopy is the study of items using a microscope, a device designed to make small things appear larger.

There are several different forms of microscope available to the researcher in a modern laboratory but all owe their origin to the microscopes that used optical lenses to enlarge objects. The first record we have of an artificial lens being used for magnefication dates from 1267. Roger Bacon's work *Perspectiva* described viewing minute objects through a lesser segment of a sphere of **glass** or crystal to enlarge them. Spectacles were in use shortly after this period to correct vision and enlarge objects, but it was not until 1595 that the first device that could truly be considered a microscope was made. This microscope was prepared by Zacharias Jansen (1580-c.1638) in Holland. This was the first **compound microscope** in that it employed two separate lenses that could be moved relative to each other by the means of a sliding tube. This allowed the microscope to zoom (to change its magnification) from 3x to 9x. This system was eventually improved by Robert Hooke (1635-1703) who added a third lens attached to the viewing (eyepiece) lens. This was carried out using an eyepiece from a telescope, an optical instrument with a longer pedigree than the microscope. Once the first Jansen microscopes had been made, word spread rapidly throughout the world and the seventeenth century saw many microscope manufacturers and users appear. It was at this time that the word microscope was first used by an Italian scientific society which included Galileo (1564-1642) as a member.

Some of the early work that was carried out using these primitive microscopes is still highly regarded today. In 1660 Marcello Malpighi (1628-1694) was able to prove the blood circulation theories of William Harvey (1578-1657) by discovering the presence of capillaries connecting arteries and veins in the body as well as identifying many microscopic structures in the human body. Some five years later, Robert Hooke (1635-1703) published the first pictures of the cells making up living organisms. Prior to these works, it had been assumed that the microscope was nothing more than a toy. At this period some microscopes were 2 ft in length and illuminated by oil lamps.

A Dutch amateur scientist named Antoni van Leeuwenhoek (1632-1723) is often credited with the first microscopes. In fact Leeuwenhoek was merely capable of making better lenses than anyone else which made his instruments more effective, he was able to produce magnifications from 50x to 200x with excellent resolution. He made the first observations and descriptions of bacteria, protozoa, and spermatozoa. This is all the more incredible when it is considered that Leeuwenhoek had returned to using only a single lens. It was not until the 1800s that compound microscopes were capable of similar resolution. The period from then until the late twentieth century has seen much improvement both in the mechanics and optics of microscopes. Of particular problem has been aberration—distortions introduced by the lenses themselves. It is often necessary to introduce new lenses to correct faults in other lenses.

Modern light microscopes are capable of a maximum magnification of something in the order of 1,500x. This is achieved by viewing the subject through oil which allows better transmission of the light into the glass of the lens which it is in contact with. This technique is called oil immersion.

For higher magnifications other instruments and techniques must be used. To overcome the limit of the compound light microscope electrons were the first successful alternative. It was in 1932 that the **electron** microscope was first developed. In many ways it is similar to the light microscope, with the exception that electrons are used to view the image and the electrons are focused by electronic fields and magnets rather than glass lenses. This development led to magnifications in the order of 500,000x. Because electrons are not visible to the human eye, the eyepiece is replaced by a television screen.

The next development in the microscope field was towards the end of the twentieth century when the **atomic force microscope** was developed. In rapid succession the **field ion microscope** and the scanning tunnelling microscope soon joined these microscopes. This latter device is capable of such magnifications and resolutions that groups of atoms can be visualised. This group of instruments can magnify 1,000,000x and they generally work by measuring distortions in electric fields around atoms.

Microscopy is a fascinating subject which has quite literally given us a whole new view of the world. In the 400 years since the field started massive advances have been made, both in terms of what is technically possible and also in what can be discovered using a microscope. During the last 50 years of the twentieth century, the reliance on light and light microscopes has reduced. The twenty-first century will hopefully open up new ways of looking more closely at the world.

See also Atomic force microscope; Electron microscope; Field ion microscope

MIDGLEY, THOMAS (1889-1944)

American engineer and chemist

Though trained as an engineer, Thomas Midgley, Jr. is best known as an industrial chemist. He was primarily responsible for four important advances in the field of chemistry: discovering effective antiknock additives for **gasoline**; developing a practical process for the extraction of **bromine** from seawater; advancing knowledge of the vulcanization of rubber and the composition of both natural and synthetic rubbers; and developing nontoxic and nonflammable **gases** for use in refrigeration and air-conditioning. As a result of his endeavors, he was awarded all four of the major American medals for achievements in **chemistry**.

Midgley was born on May 18, 1889, in Beaver Falls, Pennsylvania, the son of Thomas Midgley and Hattie Lena Emerson Midgley. His father was a prolific inventor, especially of improvements in automobile tires; he was also an entrepreneur, whose business ventures usually proved unprofitable. When he was about six years old, Midgley moved with his family to Columbus, Ohio, where he attended elementary school and the early years of high school. In 1905 he went to Betts Academy, a preparatory school in Stamford, Connecticut. Midgley entered Cornell University in 1907 to study mechanical engineering, and he graduated in 1911.

On August 3, 1911, Midgely married Carrie M. Reynolds. They would eventually have two children. He went to work that same year as a draftsman and designer for the National Cash Register Company in Dayton, Ohio. He left after a year to work with his father in a small company the older man had established to manufacture his improved auto tires. This venture failed, and in 1916 Midgley joined the Dayton Engineering Laboratories Company, known as Delco. The company was headed by Charles Franklin Kettering, a noted engineer and inventor who became Midgley's boss, mentor, and friend; their close relationship was to last until the end of Midgley's life.

Kettering soon put him to work investigating the problem of knock in internal-combustion engines, ignoring Midgley's protest that he was an engineer, not a chemist. Midgley was to spend the next five years on this project, becoming in the process a largely self-taught chemist. ''Knock'' was an audible pinging sound that developed in internal-combustion engines when they were driven near their maximum load capacity. The knock became worse at high engine-compression ratios, and it could destroy an engine if it continued long enough. Since higher compression ratios were essential to improve engine power and fuel efficiency, the problem had to be solved. Midgley soon determined that knock occurred after ignition of the fuel and that it was the result of a sudden increase in pressure and **temperature** within the engine cylinders. He also determined that it was caused by the fuel rather than by the engine itself. The problem then became one of finding a substance to add to gasoline that would lower the temperature, and hence the pressure, within the cylinder and thus end the knock.

Searching for a gasoline additive, Midgley and his coworkers at first simply tested a large number of chemical compounds in a process of trial and error. Some substances were found which effectively ended the knock, but they all had some serious drawback. Some were expensive to produce and some had a foul exhaust odor. As time passed and the research grew more sophisticated, Midgley and his colleagues discovered that all of the substances that reduced knock contained chemical elements that occupied a certain part of the **periodic table**. It then became possible to try compounds of other elements in the same area of the table. On December 9, 1921, they tested tetraethyl **lead** in an engine and found that a minute amount of it completely suppressed knock.

There were some problems even with tetraethyl lead. In the first place, it tended to foul engine valves and spark plugs. It was eventually discovered that adding bromine to the additive would solve this problem, but bromine was a scarce chemical. Midgely solved this by inventing a method for extracting bromine from seawater, where it is present in very small quantities. A more serious problem was the fact that lead compounds are poisonous and are especially dangerous to workers producing them. At the time tests done by the United States Bureau of Mines concluded that tetraethyl lead could be manufactured safely if proper precautions were followed. It was not until the 1970s that growing concern about lead **pollution** led to a ban on lead compounds in gasoline and the substitution of less toxic substances as antiknock additives.

Delco was absorbed by General Motors in 1920, and in 1924 General Motors and Standard Oil of New Jersey jointly formed a new concern, the Ethyl Corporation, whose purpose was to manufacture tetraethyl lead. Midgley became a vice president of the new organization but continued his chemical research for General Motors. In 1926 he became interested in natural and synthetic rubber and persuaded Alfred Sloan, the president of General Motors, to fund a research project on these materials. His research resulted in a series of scientific papers which greatly advanced the knowledge of the exact composition of natural rubber; he also improved the understanding of the process of vulcanization of rubber, and he was able to outline possible methods for the production of synthetic rubber. In 1928 Sloan ended financial support of this work because it seemed unlikely to produce any practical commercial results, but Midgley continued the work on his own time and with his own money for many years. He considered his rubber research to be the most truly scientific work he had done, precisely because it produced no immediate practical results.

In 1928, Kettering asked Midgley to do research on a new refrigerant suited for home use. The Frigidaire division of General Motors was then in serious financial difficulties, due primarily to the deficiencies of the refrigeration equipment it produced. One of the most basic problems with all the refrigerators then being made was the refrigerants that were being used, such as **sulfur** dioxide, methyl chloride, **ammonia**, and butane. All were either toxic or inflammable. Kettering and Midgley agreed that a suitable refrigerant must be stable, noncorrosive, nontoxic, nonflammable; they also wanted a substance which had a **boiling point** between –0 and –40° centigrade, and they wanted it to be at least relatively cheap.

When Midgley examined his periodic table, he found that all elements of sufficient **volatility** for this purpose were clustered on the right-hand side. After rejecting all the elements which were either too unstable or too toxic, he was left with **carbon**, **nitrogen**, **oxygen**, **hydrogen**, **fluorine**, sulfur, **chlorine**, and bromine. Midgley and his assistants examined the physical properties of these elements for flammability and toxicity, and they decided that some compound of fluorine would be ideal. After experimentation, they synthesized dichlorodifluoromethane (soon called ''freon''). Subsequent testing that the compound was stable and that it met the other criteria. Midgley revealed his discovery in April 1930 at the annual meeting of the American Chemical Society in Atlanta. He breathed in some freon and then exhaled to extinguish a candle flame, thus dramatically demonstrating that the gas was both nontoxic and nonflammable. General Motors and the Du Pont Company joined together in August 1930 to form Kinetic Chemicals Inc. for the production of freon. Midgley became vice president of the new company. Freon soon became the standard refrigerant for home use.

Midgley devoted most of the remainder of his life to research at the laboratories of Ohio State University in Columbus. He had been awarded the Nichols Medal of the American Chemical Society in 1922; he subsequently received the Perkin Medal of the Society of Chemical Industry in 1937, the Priestley Medal of the American Chemical Society in 1941, and the Willard Gibbs Medal of the American Chemical Society in 1942. He was granted honorary degrees by the College of Wooster in 1936 and Ohio State University in 1944. He published 57 scientific papers and was awarded 117 patents in the course of his career. He was president of the American Chemical Society at the time of his death in 1944.

Midgely was stricken with polio in 1940. In the *National Academy of Sciences, Biographical Memoirs,* Kettering remembers Midgely computing the odds of a man his age catching polio as ''substantially equal to the chances of drawing a certain individual card from a stack of playing cards as high as the Empire State building.'' Although severely crippled, Midgely remained active, making the best he could of his infirmities. He rigged up a system of ropes and pulleys to assist him in rising from bed. On November 2, 1944, he somehow entangled himself in the apparatus and strangled to death at his home in Worthington, Ohio. He was fifty-five.

MILLER, ELIZABETH C. (1920-1987)
MILLER, JAMES A. (1915-)

American biochemists

Elizabeth C. Miller and James A. Miller are known for their ground-breaking research into the mechanism of chemical carcinogenesis. The Millers' discoveries laid the foundations for understanding the metabolic interactions with carcinogenic chemicals that produce cancer in experimental animals. Their work sparked intensive research into carcinogenesis and public interest in **carcinogens**.

James A. Miller was born in 1915 in Dormont, Pennsylvania, a small town just south of Pittsburgh. His father, John, was the manager of circulation for the *Pittsburgh Press,* and his mother, Emma Stenger, was a homemaker. Two brothers died in their youth. In 1929, his mother died and his father became seriously ill. ''All the children had been taught to earn their keep,'' Miller said in an interview. ''It fell to the four boys to stick together, absent mother and father.'' Despite the economic pressures, Miller completed high school in 1933, gaining high grades in science.

Miller credits the National Youth Administration, a New Deal youth employment program, with giving him his first job in chemistry—filling reagent bottles at the **chemistry** department at the University of Pittsburgh. Within two years, Miller began the day program and an honors chemistry course. At the University of Pittsburgh, Miller got a job in an animal room lab with Charles Glen King, a well-known biochemist who had crystallized the first vitamin, vitamin C, and Max Schultze, who had trained at the University of Wisconsin. ''I finally found myself,'' Miller said. He graduated with a B.S. in chemistry in 1939. Schultze urged Miller to apply for a Wisconsin Alumni Research Foundation (WARF) scholarship in **biochemistry** at the University of Wisconsin, which he was awarded in 1939. He began laboratory research on the **metabolism** of recently identified chemicals that could induce cancer in animals. Miller received an M.S. and Ph.D. degrees in biochemistry from the University of Wisconsin in 1941 and 1943 respectively. Miller met Elizabeth Cavert in his second year at Wisconsin when he became her teaching assistant in a biochemistry lab. Miller soon found her to be an outstanding student and the two shared research interests. In August 1942, the couple married.

Elizabeth Cavert Miller was born on May 2, 1920, in Minneapolis, the second daughter of Mary Elizabeth Mead and William Lane Cavert. Her father was the Director of Research in Agricultural Economics at the Federal Land Bank in Minneapolis, Minnesota. Her mother was a graduate of Vassar College. In 1941, Elizabeth Cavert received a bachelor's degree in biochemistry from the University of Minnesota and was elected to Phi Beta Kappa. She also received a Wisconsin Alumni Research Foundation (WARF) scholarship and began graduate work in a joint biochemistry and home economics program. Initially, she was denied entry to the biochemistry program, which Miller attributed to a sex bias and the lack of jobs for graduating male Ph.D.s.

Before Cavert met James A. Miller, her goal of pursuing biochemistry research seemed unattainable. Miller became an important advocate, however, and succeeded in convincing Dr. Carl Baumann to take her on as a biochemistry graduate student. Cavert obtained an M.S. degree in biochemistry in 1943 and a Ph.D. in 1945. She began to study the metabolism of the vitamin pyridoxine in mice in Baumann's lab.

In 1944, Miller joined the new McArdle Laboratory for Cancer Research at the University of Wisconsin and continued to study experimental chemical carcinogenesis. Elizabeth A. Miller joined McArdle as a postdoctoral fellow in 1945. There, the Millers began their productive partnership in research into the mechanisms of chemical carcinogenesis. ''When we started our work, little was known about chemical carcinogenesis,'' said Miller.

In 1947, the Millers became the first researchers to demonstrate that a foreign chemical, an aminoazo dye caused cancer in rats, by binding with **proteins** in the liver in a process referred to as covalent binding. In tissues that were not sensitive to the carcinogenic effect of the azo dye, there was no binding. The Millers' subsequent research described the molecular events leading to metabolic activation of a large number of carcinogens. In 1949, Miller further demonstrated that one chemical may alter the carcinogenicity of a second chemical by influencing its enzymatic metabolism. Allan Conney, chairman of cancer research, Rutgers University said in a commemorative interview published in the *Journal of NIH Research* in 1992: ''This study set the stage for many aspects of modern **toxicology** and led to an enhanced understanding of mechanisms of toxicity of drugs, environmental toxins, and carcinogens.''

After the structure of DNA was discovered in 1953, the Millers realized that DNA played a major role in the binding of chemical carcinogens. The Millers and their associates were the first to recognize that initiation of carcinogenesis is dependent on metabolic reactions of carcinogenic chemicals with DNA. They also demonstrated that mutagenicity depends upon alteration of genetic material. The Millers' work stimulated intensive research on the binding of carcinogens to DNA, the mechanisms of mutagenesis, the activation of proto-oncogenes, and the inactivation of tumor suppressor genes.

In 1960, the Millers reported that a metabolite proved to be much more carcinogenic than its parent compound and produced tumors in tissues including the site of administration. This research demonstrated that the initiation of carcinogenesis depended on metabolic activation to electrophiles, a major unifying concept of their research. These findings not only were significant for cancer research, they also opened up a new field of study of drug interactions in metabolic studies in toxicology and pharmacology.

Between 1968 and 1971, the Millers and their associates were the first to demonstrate that chemical carcinogens are potential **mutagens**, with mutagenicity dependent on metabolic conversion and access to the genetic material. This work set the stage for more rapid testing of potential mutagens and risk assessments of chemicals in humans. Subsequently, the Millers evaluated the carcinogenecity of **food additives**, contaminants, drugs, environmental pollutants, and industrial chemicals, stimulating a growing public awareness and concern about potential carcinogens.

The Millers' commitment to cancer research and public health policy spanned more than forty-five years. Elizabeth C. Miller was editor of *Cancer Research,* the journal of the American Association for Cancer Research (AACR), between 1954–64; James A. Miller was associate editor between 1978–81. Elizabeth served as president of the American Association for Cancer Research between 1976 and 1978 and was twice elected to its board of directors. She was appointed to President Carter's Cancer Panel of the National Cancer Institute from 1978 to 1980. The Millers were concurrently admitted to the National Academy of Sciences in 1978. They participated in grant review and policy committees for numerous groups including the National Cancer Institute, the American Cancer Society, and the National Academy of Science. In 1973, Elizabeth became associate director of the McArdle Laboratory and served in this capacity until her retirement in 1987. Both were appointed WARF Senior Distinguished Research Professor of Oncology and Emeritus Professor of Oncology. By the time they retired in the 1980s, the Millers had written more than 300 papers on chemical carcinogenesis and mentored 42 McArdle researchers.

The Millers' preeminent contributions to the study of carcinogenesis have been recognized with over 25 awards, including the Papanicolaou, 1975, First Founder's Award from Chemical Industry Institute of Toxicology, 1978, and Mott Award from General Motors Cancer Research Foundation, 1980.

The Millers had two children, Linda Ann, a fiber artist, and Helen Louise, an associate professor of botany. When her children were young, Elizabeth held a half-time appointment, but worked full-time in research and administration. The family enjoyed hiking, camping, and travel. Elizabeth died on October 14, 1987, of kidney cancer. In 1988, Miller remarried Barbara Butler, a teacher of religious studies. In 1992, the *Journal of NIH Research* commemorated the Millers' 45-year contributions to cancer research by reprinting the 1947 landmark study and interviewing Miller.

Miller, Stanley Lloyd (1930-)

American chemist

Stanley Lloyd Miller is most noted for his experiments that attempted to replicate the chemical conditions that may have first given rise to life on Earth. In the early 1950s, he demonstrated that **amino acids** could have been created under primordial conditions. Amino acids are the fundamental units of life; they join together to form **proteins**, and as they grow more complex they eventually become **nucleic acids**, which are capable of replicating. Miller has hypothesized that the oceans of primitive Earth were a **mass** of molecules, a prebiological "soup," which over the course of a billion years became a living system.

Miller was born in Oakland, California, on March 7, 1930, the youngest of two children. His father, Nathan Harry Miller, was an attorney and his mother, Edith Levy Miller, was a homemaker. Miller attended the University of California at Berkeley and received his B.S. degree in 1951. He began his graduate studies at the University of Chicago in 1951.

In an autobiographical sketch entitled "The First Laboratory Synthesis of Organic Compounds under Primitive Earth Conditions," Miller recalled the events that led to his famous experiment. Soon after arriving at the University of Chicago, he attended a seminar given by **Harold Urey** on the origin of the solar system. Urey postulated that the Earth was reducing when it was first formed—in other words, there was an excess of molecular **hydrogen**. Strong mixtures of **methane** and **ammonia** were also present, and the conditions in the atmosphere favored the synthesis of organic compounds. Miller wrote that when he heard Urey's explanation, he knew it made sense: "For the nonchemist the justification for this might be explained as follows: it is easier to synthesize an organic compound of biological interest from the reducing atmosphere constituents because less chemical bonds need to be broken and put together than is the case with the constituents of an oxidizing atmosphere."

After abandoning a different project for his doctoral thesis, Miller told Urey that he was willing to design an experiment to test his hypothesis. However, Urey expressed reluctance at the idea because he considered it too time consuming and risky for a doctoral candidate. But Miller persisted, and Urey gave him a year to get results; if he failed he would have to choose another thesis topic. With this strict deadline, Miller set to work on his attempt to synthesize organic compounds under conditions simulating those of primitive earth.

Miller and Urey decided that ultraviolet light and electrical discharges would have been the most available sources of **energy** on Earth billions of years ago. Having done some reading into amino acids, Miller hypothesized that if he applied an electrical discharge to his primordial environment, he would probably get a deposit of hydrocarbons, an organic compound containing **carbon** and hydrogen. As he remembered in "The First Laboratory Synthesis of Organic Compounds": "We decided that amino acids were the best group of compounds to look for first, since they were the building blocks of proteins and since the analytical methods were at that time relatively well developed." Miller designed an apparatus in which he could simulate the conditions of prebiotic Earth and then measure what happened. A **glass** unit was made to represent a model ocean, atmosphere, and rain. For the first experiment, he filled the unit with the requisite "primitive atmosphere"—methane, hydrogen, **water**, and ammonia—and then submitted the mixture to a low-voltage spark over night. There was a layer of hydrocarbons the next morning, but no amino acids.

Miller then repeated the experiment with a spark at a higher voltage for a period of two days. This time he found no visible hydrocarbons, but his examination indicated that glycine, an amino acid, was present. Next, he let the spark run for a week and found what looked to him like seven spots. Three of these spots were easily identified as glycine, alpha-alanine, and beta-alanine. Two more corresponded to a-amino-n-butyric acid and aspartic acid, and the remaining pair he labeled A and B.

At Urey's suggestion, Miller published "A Production of Amino Acids under Possible Primitive Earth Conditions" in May 1953 after only three-and-a-half months of research. Reactions to Miller's work were quick and startling. Articles evaluating his experiment appeared in major newspapers, and a Gallup poll even asked people whether they thought it was possible to create life in a test tube; 79% of the respondants said no.

After Miller finished his experiments at the University of Chicago, he continued his research as an F. B. Jewett Fellow at the California Institute of Technology from 1954-1955. Miller established the accuracy of his findings by performing further tests to identify specific amino acids. He also ruled out the possibility that bacteria might have produced the spots by heating the apparatus in an autoclave for 18 hours (15 minutes is usually long enough to kill any bacteria). Subsequent tests conclusively identified four spots that had previously puzzled him. Although he correctly identified the a-amino-n-butyric acid, what he had thought was aspartic acid (commonly found in plants) was really iminodiacetic acid. Furthermore, the compound he had called A turned out to be sarcosine (N-methyl glycine), and compound B was N-methyl alanine. Other amino acids were present but not in quantities large enough to be evaluated.

Although other scientists repeated Miller's experiment, one major question remained: was Miller's apparatus a true representation of the primitive atmosphere? This question was finally answered by a study conducted on a meteorite that landed in Murchison, Australia, in September 1969. The amino acids found in the meteorite were analyzed and the data compared to Miller's findings. Most of the amino acids Miller had found were also found in the meteorite. On the state of scientific knowledge about the origins of human life, Miller wrote in "The First Laboratory Synthesis of Organic Compounds" that "the synthesis of organic compounds under primitive earth conditions is not, of course, the synthesis of a living organism. We are just beginning to understand how the simple organic compounds were converted to polymers on the primitive earth.... Nevertheless we are confident that the basic process is correct."

Miller's later research has continued to build on his famous experiment. He is looking for precursors to ribonucleic acid (RNA). "It is a problem not much discussed because there is nothing to get your hands on." He is also examining the natural occurrence of **clathrate hydrates**, compounds of ice and **gases** that form under high pressures, on the Earth and other parts of the solar system.

Miller has spent most of his career in California. After finishing his doctoral work in Chicago, he spent five years in the department of **biochemistry** at the College of Physicians and Surgeons at Columbia University. He then returned to California as an assistant professor in 1960 at the University of California, San Diego. He became an associate professor in 1962 and eventually full professor in the department of **chemistry**.

Miller served as president of the International Society for the Study of the Origin of Life (ISSOL) from 1986-1989. The organization awarded him the Oparin Medal in 1983 for his work in the field. Outside of the United States, he was recognized as an Honorary Councillor of the Higher Council for Scientific Research of Spain in 1973. In addition, Miller was elected to the National Academy of Sciences and he belongs to Sigma Xi and Phi Beta Kappa. His other memberships include the American Chemical Society, the American Association for the Advancement of Science, and the American Society of Biological Chemists. Miller is unmarried.

MILLIKAN, ROBERT A. (1868-1953)

American physicist

Robert A. Millikan vaulted from obscurity to international fame on the strength of his classic experiment designed to measure the charge on the **electron**. His "oil-drop" method for determining the electron charge earned him the 1923 Nobel Prize for physics. He solidified his role as a leader in American science by presiding over the rise of the California Institute of Technology into a world-famous center of scientific research.

Millikan was born March 22, 1868, the second of six children, to Silas Franklin Millikan and Mary Jane Andrews Millikan in Morrison, Illinois. His mother, a graduate of Oberlin College in Ohio, served as dean of women at Olivet College in Michigan before moving to Illinois, and his father later earned a degree at the Oberlin Theological Seminary and became a Congregational preacher. The family moved to Maquoketa, Iowa, in 1875.

Millikan graduated with high marks from Maquoketa High School in 1886. Following in his parents' footsteps, he then entered the Oberlin college preparatory program, and then the college itself the next year. He followed their classical course of study, taking classes in higher mathematics and basic physics, along with Latin and Greek. In 1889, he was asked to teach an introductory course in physics; the physics program at that time contained a large number of Greek terms, and he seemed to have been asked simply because he had done so well in Greek. His interest in physics began with his efforts to prepare himself for teaching this course. Millikan earned a B.A. in 1891, but stayed two more years as a physics tutor, taking additional science courses. He earned an M.A. in 1893 for his analysis of Silvanus P. Thomson's 1884 book, *Dynamic Electric Machinery*. On the strength of this achievement, Millikan earned a fellowship to Columbia University in New York as its sole graduate student in physics.

At Columbia, Millikan gravitated to Michael I. Pupin of the electrical engineering department. Pupin schooled Millikan in mathematical precision in experimentation. In the summer of 1894, Millikan, with the aid of Columbia professor of physics Ogden Rood, enrolled at the Ryerson Laboratory at the University of Chicago. There, Millikan first met **Albert Michelson**, the noted scientist who had measured the speed of light in 1879–1880. Michelson's emphasis on precise measurement and rigorous attention to detail, as well as his faith in the evolutionary pace of scientific progress, appealed strongly to Millikan.

Millikan received his Ph.D. in 1895 for his research on the **polarization** of incandescent light. With financial assistance from Pupin, Millikan went to Europe for postgraduate study. This was a common practice for American scientists at the time, and he studied under such luminaries as **Jules Henri Poincaré**, **Max Planck**, and **Walther Nernst**. He was in Europe when the revolutionary discoveries of x rays and **radioactivity** were made. Millikan returned to the United States in 1896 to take a position as a physics instructor at the University of Chicago, where Michelson was still head of the department of physics and director of the Ryerson Laboratory. Millikan had a heavy teaching load at first; he was responsible for establishing the physics curriculum and preparing textbooks. In 1906, he published his *First Course in Physics,* which was widely used.

Despite his teaching success, Millikan remained anxious to pursue original research in physics. He was aware of the rapid progress in Europe on **atomic theory**, and he understood that advancement in his profession depended upon research results. Initial work in **thermodynamics** met with little success. In April 1902 he took time off from research to marry Greta Irvin Blanchard, the daughter of a wealthy local businessman, and to travel in Europe. The couple later had three sons. After his honeymoon, Millikan focused his attention on radioactivity, and then on the behavior of electrons in **metals**. In particular, he studied the **photoelectric effect**, the ability of certain electromagnetic waves to detach electrons from metal surfaces.

In 1908, Millikan, now an assistant professor, began to focus on a problem involving precise measurement: the charge on the electron. The electron was playing an increasingly important role in the atomic theories then being assembled, and the precise measurement of its charge had become a pressing problem. The standard technique for measuring the charge had been developed by British physicist H. A. Wilson. First, **water** droplets would be given an electric charge, and the rate at which they fell would be measured. Next, an electric field would be introduced to attract the charged droplets upward and oppose the force of gravity. By determining the strength of the electric field and the **mass** of the water droplets, one could calculate the charge on the droplets and on the electrons.

Millikan tried Wilson's technique but found that the rapid evaporation of the water droplets made measurement difficult and the results erratic. He decided to measure the charge on a single drop of water by balancing it between the electric field and gravity. He observed in 1909 that the charge on any drop was an integral multiple of a fundamental value. His findings, however, were still complicated by evaporation of the droplets. In 1909 Millikan hit upon the idea of using slow-evaporating oil drops instead of water. He could now observe charged drops for several hours, instead of only seconds. By balancing charged oil drops between an electric field and gravity, Millikan proved that the electron was a fundamental particle with a fundamental charge. In 1913, he published his value for the charge, which remained the accepted value for decades. The determination of this value, and the ingenuity of the experiment, earned Millikan international recognition and the 1923 Nobel Prize in physics.

After determining the charge on the electron, Millikan returned to his research on the photoelectric effect. In 1905, **Albert Einstein** had explained the phenomenon by suggesting that small packages or quanta of light—photons—were responsible for knocking the electrons off the metal surfaces. Millikan, like many other physicists of the day, resisted this particle view of light, preferring instead to consider light as waves. He set out to test the validity of Albert Einstein's hypothesis by observing the liberation of electrons from metals by ultraviolet light. By careful observation and elimination of the errors of previous methods, in 1916 Millikan was able to verify the validity of Einstein's calculations, although Millikan himself refused to abandon the wave theory of light.

Europe was engulfed in World War I by this time. As it became more likely that the United States would enter the war, George Ellery Hale, the astronomer and observatory builder, successfully lobbied for the National Academy of Sciences to establish a National Research Council, dedicated to research for the war effort. Millikan assisted Hale in his plans and in 1917 took a leave of absence from the University of Chicago to serve as head of a committee on submarine detection. Soon after the United States entered the war, Millikan joined the U.S. Army Signal Corps and became director of the Corps Division of Science and Research, rising to the rank of lieutenant colonel. His work for the National Research Council in antisubmarine warfare helped to establish the council as a permanent body.

Millikan returned to the University of Chicago after the war. But in 1921, Hale and chemist Arthur A. Noyes persuaded him to chair the Executive Council of the California Institute of Technology in Pasadena. Millikan was also named director of its Norman Bridge Laboratory. Despite his initial misgivings, Millikan quickly turned the Institute into an internationally respected center of research. He lured talented students to the school by inviting such prominent scientists as Einstein, Michelson, Arnold Sommerfeld, Paul Ehrenfest, and C. V. Raman to take faculty assignments. By the end of the 1920s, the California Institute of Technology was the leading research institution in the United States, boasting of such faculty giants as Thomas Hunt Morgan, Theodore Kármán, and **Linus Pauling**.

Many Americans of the 1920s considered Millikan to be the foremost American scientist of the day. He continued to be active in the National Research Council, and from 1922-1932, he was the American representative to the Committee on International Cooperation of the League of Nations. He served as president of the American Association for the Advancement of Science in 1929. It was also during this period that he earned the Hughes Medal of the Royal Society of London and the Faraday Medal of the British Chemical Society. He would eventually hold honorary doctorates from 25 universities and belong to 21 foreign scientific academies.

Millikan still found time outside his administrative duties to conduct original research. With graduate student Carl F. Eyring, Millikan studied the ability of strong electric fields to draw electrons out of cold metals. His classical explanation for the phenomenon was eventually replaced by a quantum mechanical approach. He also studied the ultraviolet spectra of light elements with graduate student Ira S. Bowen. Their research supported Sommerfeld's relativistic explanation of spectra, rather than the atomic model of **Niels Bohr**; the conflict was resolved in 1925 by the hypothesis of electron spin.

Most of Millikan's research at the Institute, however, focused on cosmic rays, the penetrating rays that had been discovered by Austrian physicist Victor Hess in 1912. Millikan's initial objective was to determine whether the rays came from space or from radioactive elements in the earth. He launched a large number of sounding balloons, even mounting an expedition to the top of Pike's Peak in 1923, to detect and measure **ionization** of atmospheric **gases** by the rays. These experiments failed to settle the issue, and in 1925 Millikan decided to approach the problem by measuring the variation in ionization from Lake Arrowhead (5,000 feet above sea level) to Muir Lake (12,000 feet above sea level) in California. The atmosphere between the two lakes had the absorbing power of six feet of water, and a cosmic source for the rays could be assumed if the intensity of ionization at Lake Arrowhead matched the intensity of ionization six feet lower at Muir Lake. Millikan discovered this to be the case. Moreover, he observed that the ionization effects continued day and night, and that the rays were about 18 times more energetic than the most energetic gamma rays then known. It was Millikan who dubbed these ionizing rays "cosmic."

Millikan became embroiled in controversy, however, when he attempted to build upon his cosmic ray work. He failed to convince other scientists that the cosmic rays were photons emitted by the spontaneous fusing of **hydrogen** atoms into heavier elements; he abandoned the theory himself by 1935. In 1932, Arthur Holly Compton detected that the intensity of the rays varied with latitude, suggesting that the rays were charged particles deflected by the earth's magnetic field. Millikan fiercely denied Compton's assertion, and he supported his argument by conducting his own experiments that failed to detect this latitude effect. In 1933, however, after determining that a local irregularity in cosmic ray intensity had interfered with his observations, Millikan ceded that some cosmic rays were charged particles.

Millikan remained a determined conservative throughout his career, both in politics and in science. True to his scientific training under Michelson and Pupin, he preferred to think of scientific progress as evolutionary and not revolutionary. Though he encouraged original research, he never quite accepted the revolutionary ideas of the new quantum mechanics. In his later years, he even fought back attempts to reexamine his determination of the charge on the electron. In politics he was a staunch Republican; he was strongly opposed to the New Deal and, after the war, to the formation of the National Science Foundation. Despite his conservative inclinations, however, he was a religious modernist, promoting the compatibility of science and religion even as the 1925 Scopes trial pitted science against fundamentalist Christianity.

Millikan retired from his professorship and presidency of the California Institute in 1946. He then wrote his autobiography, which was published in 1950. He died in Pasadena on December 19, 1953.

MILSTEIN, CÉSAR (1927-)

English biochemist

César Milstein conducted one of the most important late-twentieth-century studies on **antibodies**. In 1984 he was granted the Nobel Prize for physiology or medicine, shared with Niels K. Jerne and Georges Kohler, for his outstanding contributions to this field. Milstein's research on the structure of antibodies and their genes, through the investigation of deoxyribonucleic acid (DNA) and ribonucleic acid (RNA), has been fundamental for a better understanding of how the human immune system works.

Milstein was born on October 8, 1927, in the eastern Argentine city of Bahía Blanca, one of three sons of Lázaro and Máxima Milstein. He studied **biochemistry** at the National University of Buenos Aires from 1945-1952, graduating with a degree in **chemistry**. Heavily involved in opposing the policies of President Juan Peron and working part-time as a chemical analyst for a laboratory, Milstein barely managed to pass with poor grades. Nonetheless, he pursued graduate studies at

César Milstein. *(Photo Researchers, Inc. Reproduced by permission.)*

the Instituto de Biología Química of the University of Buenos Aires and completed his doctoral dissertation on the chemistry of aldehyde dehydrogenase, an **alcohol enzyme** used as a catalyst, in 1957.

With a British Council scholarship, he continued his studies at Cambridge University from 1958-1961 under the guidance of **Frederick Sanger**, a distinguished researcher in the field of enzymes. Sanger had determined that an enzyme's functions depend on the arrangement of **amino acids** inside it. In 1960 Milstein obtained a Ph.D. and joined the Department of Biochemistry at Cambridge, but in 1961, he decided to return to his native country to continue his investigations as head of a newly-created Department of Molecular Biology at the National Institute of Microbiology in Buenos Aires.

A military coup in 1962 had a profound impact on the state of research and on academic life in Argentina. Milstein resigned his position in protest of the government's dismissal of the Institute's director, Ignacio Pirosky. In 1963 he returned to work with Sanger in Great Britain. During the 1960s and much of the 1970s, Milstein concentrated on the study of antibodies, the protein organisms generated by the immune system to combat and deactivate antigens. Milstein's efforts were aimed at analyzing myeloma **proteins**, and then **DNA** and **RNA**. Myeloma, which are tumors in cells that produce antibodies, had been the subject of previous studies by Rodney R. Porter, MacFarlane Burnet, and **Gerald M. Edelman**, among others.

Milstein's investigations in this field were fundamental for understanding how antibodies work. He searched for mutations in laboratory cells of myeloma but faced innumerable difficulties trying to find antigens to combine with their antibodies. He and Köhler produced a hybrid myeloma called hybridoma in 1974. This cell had the capacity to produce antibodies but kept growing like the cancerous cell from which it had originated. The production of monoclonal antibodies from these cells was one of the most relevant conclusions from Milstein and his colleague's research. The Milstein-Köhler paper was first published in 1975 and indicated the possibility of using monoclonal antibodies for testing antigens. The two scientists predicted that since it was possible to hybridize antibody-producing cells from different origins, such cells could be produced in massive cultures. They were, and the technique consisted of a fusion of antibodies with cells of the myeloma to produce cells that could perpetuate themselves, generating uniform and pure antibodies.

In 1983 Milstein assumed leadership of the Protein and Nucleic Acid Chemistry Division at the Medical Research Council's laboratory. In 1984 he shared the Nobel Prize with Köhler and Jerne for developing the technique that had revolutionized many diagnostic procedures by producing exceptionally pure antibodies. Upon receiving the prize, Milstein heralded the beginning of what he called "a new era of immunobiochemistry," which included production of molecules based on antibodies. He stated that his method was a by-product of basic research and a clear example of how an investment in research that was not initially considered commercially viable had "an enormous practical impact." By 1984 a thriving business was being done with monoclonal antibodies for diagnosis, and works on vaccines and cancer based on Milstein's breakthrough research were being rapidly developed.

In the early 1980s Milstein received a number of other scientific awards, including the Wolf Prize in Medicine from the Karl Wolf Foundation of Israel in 1980, the Royal Medal from the Royal Society of London in 1982, and the Dale Medal from the Society for Endocrinology in London in 1984. He is a member of numerous international scientific organizations, among them the U.S. National Academy of Sciences and the Royal College of Physicians in London. His hobbies include walking, outdoor cooking, and attending the theater. Milstein is married to biochemist Celia Prilleltensky; they have no children.

MINERALOGY

Mineralogy, as the name suggests, is the study of minerals. There are several definitions of "mineral," depending on how they are studied. Chemists would define minerals as particular chemical elements or compounds. Dietitians would define a mineral as a substance that is required in the diet for good **nutrition**, for example, **calcium**, **potassium**, and **magnesium**. Sometimes all of **matter** on Earth is classified as either animal, vegetable, or mineral, which would suggest that all inorganic substances are minerals. This is incorrect, although all minerals are inorganic substances. Rocks in the Earth's crust are composed of one or more minerals. A mineral in the geologic sense is a naturally occurring, inorganic, crystalline solid. A particular mineral has a specific chemical composition. Each mineral has its own physical properties such as **color**, hardness, and **density**.

Most minerals are chemical compounds that are made of two or more different elements. The composition of a mineral is shown by its chemical formula, which states each of the chemical elements present in the mineral as well as the ratios of each element. For example, the mineral quartz has the chemical formula SiO_2. This means that quartz is made of the elements **silicon** (Si) and **oxygen** (O). The formula also shows that for every one silicon **atom**, two oxygen atoms are present. The mineral orthoclase has the chemical formula $KAlSi_3O_8$. A **molecule** of orthoclase contains one potassium (K), one **aluminum** (Al), three silicon, and eight oxygen atoms. Some minerals always have the same chemical formula. Quartz always is composed of SiO_2 and halite is always made of **sodium** (Na) and **chlorine** (Cl), with the chemical formula NaCl. Some minerals can have more than one chemical formula, depending on their composition. Sometimes an element can substitute for another in a mineral. This occurs when the atoms of two elements are the same charge and close to the same size. For example, an **iron** (Fe) atom and a magnesium (Mg) atom are both about the same size, so they can substitute for each other. The chemical formula for the mineral olivine is $(Mg,Fe)_2SiO_4$. The (Mg,Fe) indicates that either magnesium, iron, or a combination of the two may be present in an olivine sample.

Native elements are minerals that are composed of only one element. These are the substances that dietitians call minerals. Examples of native elements include **gold** (Au), **silver** (Ag), and **platinum** (Pt). Two other native elements are **graphite** and **diamond**, both of which are entirely made of **carbon** (C).

All minerals are crystalline **solids**. A crystalline solid is a solid consisting of atoms arranged in an orderly three-dimensional matrix. This matrix is called a crystal lattice. A crystalline solid is composed of molecules with a large amount of order. The molecules in a crystalline solid occupy a specific place in the arrangement of the solid and do not move. The molecules not only occupy a certain place in the solid, but they are also oriented in a specific manner. The molecules in a crystalline solid vibrate a bit, but they maintain this highly ordered arrangement. The molecules in a crystalline solid can be thought of as balls connected with springs. The balls can vibrate due to the contractions and expansions of the springs between them, but overall they stay in the same place with the same orientation. It is not easy to deform a crystalline solid because of the strong attractive forces at work within the structure. Crystalline solids tend to be hard, highly ordered, and very stable.

Under ideal conditions, mineral crystals will grow and form perfect crystals. An ideal condition would be in a place where the crystals are allowed to grow slowly without disturbances, such as in a cavity. A perfect crystal has crystal faces (planar surfaces), sharp corners, and straight edges. The external crystal form is controlled by the internal structure. When the atoms in a crystal are arranged in a perfectly orderly fashion, the crystal will also be formed in a perfect orderly fashion. Even if a perfect crystal is not formed, the internal crystalline structure can be shown. Many minerals exhibit a property called cleavage. A mineral that has cleavage will break or split along planes. If the internal structure is formed in an orderly crystal arrangement, then the breaks will occur along the planes of the internal crystal structure.

There have been over 3,500 minerals identified and described. Only about two dozen of these are actually common. There are many reasons why there are a limited number of minerals. For one, there are only a certain number of chemical elements that can combine to form chemical compounds. Some combinations of elements are unstable, such as a potassium-sodium or a silicon-iron compound. In addition, only eight elements are found abundantly in the Earth's crust, where minerals are formed. Oxygen and silicon alone account for more than 74% of the Earth's crust. These factors place a limit on the number of possible minerals.

The minerals that have been discovered and studied can be placed into one of five groups. These groups are the silicate minerals, carbonate minerals, oxides, sulfides, and halides. The silicate minerals are those that contain silica, a combination of silicon and oxygen. Examples of **silicates** include quartz, orthoclase, and olivine. The silicate minerals are the most common, making up approximately one-third of all known minerals. They are composed of building blocks called the silica tetrahedron. A silica tetrahedron is one silicon atom and four oxygen atoms. The atoms are arranged in a four-faced pyramidal structure (the tetrahedron) with the silicon atom in the center. The silicon atom has a 4+ charge, and each of the four oxygen atoms have a 2- charge. As a result, a silicon tetrahedron has a net charge of 4-. Because of this charge, it does not occur in isolation in nature. A silica tetrahedron is always bound to other atoms or molecules.

A silica tetrahedron can bind to positively charged ions. In these minerals, the silicon to oxygen ratio is 1:4. An example of a mineral with this structure is olivine. Silica tetrahedra also can form single **chains**. Minerals with this structure have a silicon to oxygen ratio of 1:3, and include enstatite, $MgSiO_3$. Silica tetrahedra within such chains have a net electrical charge of 2-, so they need to be balanced by positive ions such as Mg^{2+} which link parallel chains together. Silica tetrahedra can link together to form a double chain structure. Minerals with this structure have a silicon to oxygen ratio of 4:11. Each double tetrahedra has a net charge of 6-, so positive ions need to link the double chains together. The silica tetrahedra may also form a sheet structure where three oxygen atoms of each tetrahedron are shared by adjacent tetrahedra. In this case, the silicon to oxygen ratio is 2:5. These sheets also carry a negative charge which is balanced by positive ions bound between sheets.

There are two types of silicate minerals, the ferromagnesian and the nonferromagnesian silicates. Ferromagnesian silicates are those containing iron, magnesium, or both. These minerals tend to be dark colored and more dense than the nonferromagnesian silicates. An example of a ferromagnesian silicate is olivine. Nonferromagnesian silicates do not have iron and magnesium. These minerals are light colored and less dense. The most common nonferromagnesian silicates are the feldspars.

The carbonate minerals contain the carbonate **ion**, $(CO_3)^{2-}$. Calcite ($CaCO_3$), the main component of limestone, is an example of a carbonate mineral. The oxides are minerals that contain an element combined with oxygen. An example of an **oxide** is hematite, Fe_2O_3. The sulfides contain a **cation** combined with **sulfur** (S^2). An example of a sulfide is galena (PbS), which is **lead** (Pb) combined with sulfur. The halides all contain halogen elements, such as chlorine and **fluorine** (F). Examples of halite minerals include halite (NaCl) and fluorite (CaF_2).

All minerals posses specific physical properties such as color, luster, crystal form, cleavage, fracture, hardness, and specific gravity. The physical characteristics of a mineral depend on its internal structure and chemical composition. The physical properties of minerals can be used for identification purposes by mineralogists. Color is the least reliable of the physical properties. Many minerals display a variety of colors due to impurities. Some generalizations can be made, however. Ferromagnesian silicates, for example, are usually black, brown, or dark green. The luster of a mineral refers to the way in which light is reflected off of the mineral. Two types of luster can be displayed—metallic or nonmetallic.

The crystal form of a mineral is also a physical property specific for the type of mineral being observed. This property is most easily observed when the mineral has formed a perfect crystal. Cleavage is the tendency of a mineral to break or split along planes. There are different types of cleavage that correspond to the different internal crystal structures that make up individual minerals. Fracture is when a mineral does not break along smooth planes, rather along irregular surfaces. Some minerals display cleavage, others display fracture.

The hardness of a mineral is its resistance to being scratched. The Mohs hardness scale can be used to determine how hard a mineral is by determining what will scratch its surface. The specific gravity of a mineral is the ratio of its weight to the weight of an equal **volume** of **water**. For example, a mineral with a specific gravity of 4.0 is four times as heavy as water. The specific gravity of a mineral is determined by its composition and structure.

Mineralogy is an interesting science which studies the nature of minerals—naturally occurring, inorganic crystalline solids. There are many different minerals, each with its own properties determined by its chemical composition. Mineralogists continue to search for new, useful minerals in the Earth's crust.

MINERALS (DIETARY)

The minerals (inorganic nutrients) that are relevant to human **nutrition** include **water**, **sodium**, **potassium**, chloride, **calcium**,

phosphate, **sulfate**, **magnesium**, **iron**, **copper**, **zinc**, **manganese**, **iodine**, **selenium**, and **molybdenum**. **Cobalt** is a required mineral for human health, but it is supplied by vitamin B $_{12}$. Cobalt appears to have no other function, aside from being part of this vitamin. There is some evidence that **chromium**, **boron**, and other inorganic elements play some part in human nutrition, but the evidence is indirect and not yet convincing. Fluoride seems not to be required for human life, but its presence in the diet contributes to long term dental health. Some of the minerals do not occur as single atoms, but occur as molecules. These include water, phosphate, sulfate, and selenite (a form of selenium). Sulfate contains an **atom** of **sulfur**. We do not need to eat sulfate, since the body can acquire all the sulfate it needs from protein.

The statement that various minerals, or inorganic nutrients, are required for life means that their continued supply in the diet is needed for growth, maintenance of body weight in adulthood, and for reproduction. The amount of each mineral that is needed to support growth during infancy and childhood, to maintain body weight and health, and to facilitate pregnancy and lactation, are listed in a table called the Recommended Dietary Allowances (RDA). This table was compiled by the Food and Nutrition Board, a committee that serves the United States government. All of the values listed in the RDA indicate the daily amounts that are expected to maintain health throughout most of the general population. The actual levels of each inorganic nutrient required by any given individual is likely to be less than that stated by the RDA. The RDAs are all based on studies that provided the exact, minimal requirement of each mineral needed to maintain health. However, the RDA values are actually greater than the minimal requirement, as determined by studies on small groups of healthy human subjects, in order to accomodate the variability expected among the general population.

The RDAs for adult males are 800 mg of calcium, 800 mg of **phosphorus**, 350 mg of magnesium, 10 mg of iron, 15 mg of zinc, 0.15 mg of iodine, and 0.07 mg of selenium. The RDA for sodium is expressed as a range (0.5-2.4 g/day). The minimal requirement for chloride is about 0.75 g/day, and the minimal requirement for potassium is 1.6-2.0 g/day, though RDA values have not been set for these nutrients. The RDAs for several other minerals has not been determined, and here the estimated safe and adequate daily dietary intake has been listed by the Food and Nutrition Board. These values are listed for copper (1.5-3.0 mg), manganese (2-5 mg), fluoride (1.5-4.0 mg), molybdenum (0.075-0.25 mg), and chromium (0.05-0.2 mg). In noting the appearance of chromium in this list, one should note that the function of chromium is essentially unknown, and evidence for its necessity exists only for animals, and not for human beings. In considering the amount of any mineral used for treating mineral deficiency, one should compare the recommended level with the RDA for that mineral. Treatment at a level that is one tenth of the RDA might not be expected to be adequate, while treatment at levels ranging from 10-1,000 times the RDA might be expected to exert a toxic effect, depending on the mineral. In this way, one can judge whether any claim of action, for a specific mineral treatment, is likely to be adequate or appropriate.

People are treated with minerals for several reasons. The primary reason is to relieve a mineral deficiency, when a deficiency has been detected. Chemical tests suitable for the detection of all mineral deficiencies are available. The diagnosis of the deficiency is often aided by tests that do not involve chemical reactions, such as the hematocrit test for the red blood cell content in blood for iron deficiency, the visual examination of the neck for iodine deficiency, or the examination of bones by densitometry for calcium deficiency. Mineral treatment is conducted after a test and diagnosis for iron-deficiency anemia, in the case of iron, and after a test and diagnosis for hypomagnesemia, in the case of magnesium, to give two examples.

A second general reason for mineral treatment is to prevent the development of a possible or expected deficiency. Here, minerals are administered when tests for possible mineral deficiency are not given. Examples include the practice of giving young infants iron supplements, and of the food industry's practice of supplementing infant formulas with iron. The purpose here is to reduce the risk for iron deficiency anemia. Another example is the practice of many women of taking calcium supplements, with the hope of reducing the risk of osteoporosis.

Most minerals are commercially available at supermarkets, drug stores, and specialty stores. There is reason to believe that the purchase and consumption of most of these minerals is beneficial to health for some, but not all, of the minerals. Potassium supplements are useful for reducing blood pressure, in cases of persons with high blood pressure. The effect of potassium varies from person to person. The consumption of calcium supplements is likely to have some effect on reducing the risk for osteoporosis. The consumption of selenium supplements is expected to be of value only for residents of Keshan Province, China, because of the established association of selenium deficiency in this region with ''Keshan disease.''

MISCIBILITY

Miscibility refers to the ability of a gas or liquid to dissolve uniformly in another gas or liquid. **Gases** mix with each other in all proportions, but this may or may not apply to **liquids**, where miscibility depends on chemical affinity. Ethanol, an **alcohol**, and **water** are miscible because they are chemically similar, but **benzene** and water are only slightly miscible because of the very large differences in their chemical properties.

Some liquids are essentially insoluble in other liquids, such as **gasoline** in water (and water in gasoline). Such liquids are said to be immiscible. Other liquids are only soluble up to a point. In the case of water and ethyl **ether** ($CH_3CH_2OCH_2CH_3$), it is possible to dissolve up to about 4 g of ethyl ether in 100 g of water, but the addition of more ethyl ether results in the production of separate layers of the less dense diethyl ether above the denser water layer. And some liquids are completely soluble in other liquids, regardless of the amounts combined; such liquids are said to be *miscible* in each other in all proportions. (Traditionally, the term **solu-**

bility has been used interchangeably for miscibility in reference to liquids, even though it should strictly only be applied to **solids**.)

There are several fundamental rules governing the miscibility of liquids in other liquids. First, the solubility of liquids in liquids increases with increasing **temperature**. Second, the more similar two compounds are in terms of polarity, the more likely that one is soluble in the other. Polar compounds dissolve in polar compounds, and non-polar compounds dissolve in non-polar compounds. (Polar molecules dissolve in polar molecules because the dipole of one attracts the dipole end of the other.) Thus, benzene and **carbon** tetrachloride, being both non-polar, dissolve in each other, but neither will appreciably dissolve in water, which is polar.

Both alcohols and ethers with up to three or four carbons are miscible in water because the OH groups in these molecules form **hydrogen** bonds with the water molecules. Alcohols and ethers with higher molecular weights are not miscible in water, however, because the water molecules can not completely surround those molecules. The **molecule** 1-heptanol, for example, consists of an alkyl chain of seven carbons and an OH group. The OH group forms hydrogen bonds with water molecules, but the alkyl portion of the molecule exerts no attraction on the water molecules. This part of the molecule is called hydrophobic, meaning water-hating. Because this part of the molecule cannot be surrounded by water, 1-heptanol is immiscible in water.

In aqueous solutions, globular **proteins** usually turn their polar groups outward toward the aqueous solvent, and their non-polar groups inward, away from the polar molecules. The non-polar groups prefer to interact with each other, and exclude water from these regions, leading to immiscibility. This type of interaction is usually weaker than hydrogen **bonding**, and usually acts over large surface areas.

Many gases are miscible with liquids. The miscibility of gases in liquids almost always decreases with increasing temperature, and increases with increasing pressure. For example, more **oxygen** is dissolved in the blood at higher than normal pressures.

See also Polar bond, Polarization

MITCHELL, PETER D. (1920-1992)

English biochemist

Peter D. Mitchell was awarded the 1978 Nobel Prize in **chemistry** for his chemiosmotic theory, which explained how organisms use and synthesize **energy**. In his Nobel Prize address, Mitchell honored his long association with Professor David Keilin of Cambridge University, whose work provided the takeoff point for Mitchell's discoveries. Keilin had discovered cytochromes—electron-carrier **proteins** that assist in energy transfer via a respiratory chain. Mitchell's revolutionary chemiosmotic hypothesis changed the way scientists view energy transformation, and though it was initially viewed as controversial, it eventually won almost universal acceptance.

In 1961, when Mitchell's idea was first introduced, it was greeted by some in the scientific community with skepticism: what he was proposing was radically different than the prevailing thought on energy conversion at that time, and those opposing his conclusions questioned the validity of his research. Also, although Mitchell viewed his small research staff and unconventional laboratory at Glynn House mansion in Cornwall as positive elements conducive to productive research, others viewed his unorthodox working environment with suspicion. Mitchell's chemiosmotic theory generated intense debate, but the positive result for science as a whole was the creation of much additional scientific experimentation and productivity attempting to prove or disprove his theory, and advancing the discipline of bioenergetics—the study of energy exchanges and transformations between living things and their environments—in the process.

Peter Dennis Mitchell was born in Mitcham, Surrey, England, on September 29, 1920, the son of Christopher Gibbs Mitchell, a civil servant, and Kate Beatrice Dorothy Taplin Mitchell. He received his secondary education at Queens College in Taunton, England, and was admitted to Jesus College at Cambridge University in 1939. A graduate student of James F. Danielli in the department of **biochemistry** at Cambridge, Mitchell earned his doctorate degree in 1951. He taught biochemistry at Cambridge from 1951 until 1955, when he left to develop a chemical biology unit at Edinburgh University. He remained there until 1963, when poor health caused him to look for a calmer working atmosphere.

Mitchell found a peaceful environment in an eighteenth-century manor house in Cornwall. The manor house, called Glynn House, was in disrepair, and was restored by Mitchell and converted to family living quarters and a research laboratory. Glynn Research Laboratories was organized and directed in 1964 by Mitchell and his colleague, Jennifer Moyle, whose background work was instrumental to Mitchell's development of the chemiosmotic hypothesis. By the time Mitchell received the Nobel Prize, the laboratory had grown to require a staff of six.

The intriguing question of how organisms take energy from their surroundings and transform it for use in specialized functions, such as movement and respiration, was thought to have been answered by a theory called chemical coupling. This theory postulated that energy was carried down the respiratory chain by an unknown high-energy intermediate compound formed during **oxidation**. The energy derived from the intermediate compound was thought to form a "universal energy currency" known as adenosine triphosphate (ATP).

The search was on to identify the energy-rich intermediary when Mitchell upset prevailing thought by proposing that the process was an electrical, not a chemical, one. He coined the term "proticity" to explain the process by which protons flow across cell membranes to synthesize ATP. Mitchell likened this process to the way **electricity** moves from a high **concentration** to a concentration low enough to power an electric appliance. Laboratory experiments crucial for the support of his chemiosmotic theory were successfully carried out during the 1960s by Mitchell and Moyle at Glynn Research Laboratories, as well as in other research labs throughout the world. These experiments included identifying the membrane protons that provide a link to the movement of other molecules across the cell membrane and showing that the membrane also serves to halt the movement of other molecules.

Recognition for his work on cell energy transfer culminated in Mitchell's receipt of the Nobel Prize in 1978. Later, Mitchell and his staff at the Glynn laboratory studied the biochemical actions involved in energy transfer within cells, seeking precise details of this complex process. Those contributions advanced scientific knowledge of how cells use, transform, and generate energy. Mitchell maintained that he was just one more link in science's intellectual and historical chain. He believed that the practice of science was a continuing process, whereby one scientist builds on the discoveries and knowledge of another, and was quick to give credit to those whose past work had advanced and made possible his own. In the December 15, 1978, issue of *Science,* Frank Harold quoted Mitchell as saying, "Science is not a game like golf, played in solitude, but a game like tennis in which one sends the ball into the opposing court and expects its return."

Many awards other than the Nobel Prize were presented to Mitchell. Among them were the CIBA Medal and Prize, Biochemical Society, England; the Warren Triennial Prize, Massachusetts General Hospital; the Louis and Bert Freedman Award of the New York Academy of Sciences; the Lewis S. Rosenstiel Award for Distinguished Work in Basic Medical Research of Brandeis University; and the Copley Medal of the Royal Society. He held memberships in various professional societies and received honorary degrees from universities in Berlin and Chicago, as well as from numerous British institutions. Although immersed in his work, Mitchell found time to participate in local affairs, respond to environmental issues, and restore medieval farmhouses. He and his wife, Mary Helen French, were married in 1958; they had six children: Jeremy, Daniel, Jason, Gideon, Julia, and Vanessa. Peter Mitchell died at Glynn House, Bodmin, on April 10, 1992.

MIXTURE, CHEMICAL

Despite their presence in all fields of **chemistry**, chemical mixtures cannot be represented by any formula. This is because they contain two or more distinct chemical components that are either heterogeneous or homogeneous. The substances in a mixture, no matter how closely they have intermingled, have no firm chemical interaction or **bonding** between or among themselves.

A heterogeneous mixture combines substances that have distinct and different phases, such as ice cubes and **water**. In homogenous mixtures, the atoms or molecules of the ingredients are interspersed with each other to form a single phase, as in a mixture of **gases**.

There are four types of mixtures which are either heterogeneous or homogeneous. The first type of mixture is called a **solution**, in which chemical components (whether gas, solid, or liquid) combine to form a single-phase, homogeneous mixture whose components are evenly distributed. An azeotropic mixture is one in which at least two **liquids** combine to make a solution whose composition remains the same upon **distillation**. A **colloid** is a mixture whose components dissolve spontaneously into a solvent; they have the unusual property of not being able to pass through a membrane because they contain small, solid particles (the colloid, or dispersed phase) thoroughly mixed into a liquid (dispersion medium). The fourth type of mixture is a **suspension**. A suspension consists of two or more substances formed into a mixture that appears at first to be homogeneous. However, after a period of time, the mixture becomes heterogeneous. An example of a suspension is cough syrup.

A mixture's parts may or may not be dispersed evenly, but its components can usually be separated from each other, at least theoretically, through physical means (distillation, crystallization, etc.). They retain their own chemical characteristics despite being mixed together. This is in contrast to chemical compounds, whose components react with one other and thus cannot easily be taken back apart.

One minor exception to this rule can occur when the components of a mixture have attractive or repulsive forces that act upon each other. In this case, the reaction among the components may affect the properties of the overall mixture. For instance, the total **volume** of several liquids all mixed together might actually be greater than the sum of all the liquids' individual volumes before they were mixed together.

Chemical mixtures can be both natural and artificial. Some examples of natural mixtures, chemically speaking, are **wood**, blood, vegetable oils, milk, petroleum, air, marble, latex, and ocean water. Artificial mixtures include alloys, cement, **glass**, perfumes, paint, and some **plastics**.

When dealing with chemical mixtures in the laboratory, scientists and researchers will sometimes use a mathematical technique known as "alligation." This is a formula used to figure out the proper proportions of chemicals to put into the mixture. It can also calculate the value of a certain property of a mixture by determining the values of that property in its respective component. Alligation works because the ingredients in a mixture retain their unique chemical characteristics.

See also Azeotrope; Colloid; Solution; Suspension

MOISSAN, HENRI (1852-1907)

French chemist

Henri Moissan, who occupied the chair of Inorganic **Chemistry** at the Sorbonne from 1900 until his death, was the first person to isolate and characterize the element **fluorine**. He also invented the high-temperature electric arc furnace and prepared many new fluorine compounds, elemental transition **metals**, and carbides, borides, and silicides. For his work on fluorine and the electric furnace he was awarded the Nobel Prize in chemistry in 1906.

Ferdinand Frédéric Henri Moissan was born in Paris on September 28, 1852, the son of Francois Ferdinand and Josephine Mitel Moissan. The family lived modestly; his father was a clerk and his mother a seamstress. In 1864 they moved to Meaux, about twenty-five miles east of Paris. He was educated at the municipal school and was influenced by a mathematics and science teacher who gave him private lessons. His family, however, could not afford to pay for him to complete the courses in physics and classical languages that would have given him his baccalaureate, a necessity for university admission. In 1870 he was apprenticed briefly to a watchmaker before joining the army to defend Paris from the Prussians. He was finally apprenticed to a pharmacist in 1871. He planned to enroll in the three-year course at the Ecole Supérieure de Pharmacie to earn the only qualification open to him, pharmacist second-class. Instead he was attracted into the laboratories of Edmond Frémy's School of Experimental Chemistry at the Paris Museum of Natural History. There he engaged in research, supported himself by tutoring, and finished the courses needed for his degree. He received the baccalaureate in 1874, and thereafter a series of higher degrees: *license* (B.S.) in 1877, pharmacist first-class in 1879, and a doctorate in 1880.

Although his earlier research had been in the chemistry of plant respiration, his doctoral dissertation dealt with pyrophoric **iron** and various oxides of that metal. He was interested only in **inorganic chemistry**, and this was the direction he would take for the rest of his career. In the course of this research he also concerned himself with **chromium** salts, and he developed a process for the preparation of pure chromium by **reduction** in a stream of **hydrogen**. Moissan began teaching at the École Supérieure de Pharmacie, and in 1880 he was named associate professor. In 1882 he married Marie Leonie Lugan, the daughter of a Meaux pharmacist who took an interest in Moissan's scientific progress and provided financial support. In 1885 they had their only child, Louis Ferdinand Henri, who was killed in 1915 in one of the early battles of World War I.

In 1884 Moissan attempted to isolate the element fluorine. This was a long-standing problem that had eluded many notable scientists of the day. It was also a dangerous problem because of the high toxicity of fluoride compounds. Moissan's mentor Frémy had claimed production of fluorine by **electrolysis** (passing an electric current through it) of molten **potassium** fluoride, but at the **temperature** necessary for fusion the gas immediately attacked the **platinum** electrodes and thus could not be isolated. Frémy had also produced anhydrous hydrogen fluoride, but it could not be electrolyzed because it did not conduct **electricity**.

Moissan found that potassium acid fluoride which had been dissolved in hydrogen fluoride did conduct electricity. With an apparatus consisting of a platinum U-tube capped with plugs of fluorite (calcium fluoride) and fitted with iridium-platinum electrodes that resisted attack by fluorine, he produced the gaseous element at –50°C to keep its **reactivity** to a minimum. This was in 1886, and over the next several years he devoted his major research efforts to reactions and compounds of fluorine, producing thionyl fluoride and sulfuryl fluoride and the unreactive **carbon** tetrafluoride and **sulfur** hexafluoride, as well as a number of organic alkyl fluorides. Later, with James Dewar, he produced liquid fluorine and, in 1903, solid fluorine. The element proved reactive down to its

Three tubes of multicolored liquid. ***(Photograph by Phototake NYC. Reproduced by permission.)***

liquefaction temperature of –188°C. It is in fact the most reactive of all the elements, forming compounds even with the "inert" **gases krypton**, **xenon**, and **radon**; **water** itself "burns" in a fluorine atmosphere with a visible flame.

After an excursion into the chemistry of **boron**, during which time he produced the pure element and studied a number of its compounds, Moissan became interested in the laboratory production of diamonds. To force carbon from its **graphite** form into the **density** of a **diamond**, great **heat** and pressure was required. Moissan reasoned that this might be accomplished by dissolving carbon into molten iron at very high temperature, then quickly cooling the **solution** in such a way that the **mass** formed a solid "skin" that would generate pressure on the still-liquid interior. However, the crystals he produced in this way were considered by later researchers not to be true diamonds because the pressure required to create diamonds is about five times greater than Moissan's method produced. The most important result of these experiments proved to be an instrument he had developed to conduct them. Moissan devised an electric arc furnace capable of producing temperatures as great as 3,500°C.

The design of the furnace was simple. Two blocks of **lime** were laid on top of one another; the lower block was grooved to admit the electrodes and their leads, and the center had room for a crucible. A smaller hollow in the upper block formed a lid for the chamber. The electric furnace opened a new world of high-temperature chemistry. Within a few years of its development, Moissan prepared pure samples of metallic **vanadium**, chromium, **manganese**, **zirconium**, **niobium**, **molybdenum**, **tantalum**, **tungsten**, **thorium**, and **uranium**. He also synthesized previously unknown metallic borides, carbides, and silicides, including the extremely hard and refractory **silicon** carbide, which was later produced by a more practical method under American patent as the abrasive Carborundum.

Moissan was professor of **toxicology** at the Ecole Surérieure de Pharmacie from 1886-1899, when he became professor of inorganic chemistry. The following year, he accepted the chair of Inorganic Chemistry in the Faculty of Sciences of the University of Paris (Sorbonne), which he held until his death. In addition to receiving the Nobel Prize in 1906, Moissan's other awards and honors include the Prix Lacaze of the French Academy of Sciences in 1887, and he was elected to that body in 1891. He received the Davy Medal of the Royal Society of London in 1896 and was made a foreign member of that society in 1905. Moissan's health was adversely affected by his work with fluorine and its compounds. He was stricken with appendicitis and died on February 20, 1907, after an operation.

Molality

Molality is the **concentration** of a **solution** given in terms of the number of moles of solute in 1 kg of solvent.

The molality of a solution is given the symbol *m*. The molality of a solution can be found by dividing the number of moles of solute by the **mass** of solvent in kilograms.

It is important to note the distinction between molality and **molarity**. Molality is a concentration expressed in terms of the mass of solvent, whereas molarity is given in terms of the **volume** of the solution. In some instances the molality and molarity may in fact have the same numerical value, but it should not be assumed that this is always the case.

If 1kg of solvent has two moles of solute dissolved in it than it is a 2 molal solution. This is generally abbreviated to 2 *m*. A 1*m* solution has 1 **mole** of solute dissolved in 1 kg of solvent and from this it can be seen that the molal unit is an abbreviation for moles per kilogram. One major difference between the molality and molarity is that the molality does not vary with **temperature** as mass is a constant measure. Molarity can alter with temperature, as it is a function of the volume of the solution, and the volume of solution is affected by temperature (due to expansion and contraction). Molality is a temperature-independent term and molarity is a temperature-dependant term.

When **water** is the solvent, the molality and molarity of a dilute solution are essentially equal because 1 kg of solvent is nearly the same as 1 kg of solution and, because of the **density** of water, 1kg of a dilute solution will be very close to 1 L.

Molality used to be called weight molarity which was given the symbol m_w. This term may still be encountered in some older text books.

In dilute solutions where the number of moles of solvent is much greater than the number of moles of solute, the following calculation can be used. The molality of a solution is equivalent to 1,000 times the mole fraction concentration of the solute divided by the **molecular weight** of the solvent.

The **ionic strength** of an electrolytic solution (one that will conduct electricity) is one half of the sum of all the squares of the charge on the ionic species under consideration, multiplied by the molality of the solution. If discussing a mixture, the summation is carried out for all of the electrolytes in the solution.

Molarity

The molarity of a **solution** is the **concentration** of the solution expressed in terms of the number of moles of solute in one liter of solution.

The molarity of a solution is given the symbol *M*. It can be calculated by dividing the number of moles of solute by the **volume** of the solution in liters.

It is important to note the distinction between molarity and **molality**. Molarity is the concentration in terms of the volume of solution and molality is the concentration expressed in terms of the **mass** of the solvent.

The molarity of a solution is the most commonly used way of describing the concentration. An example of this would be describing a solution as a 1 molar solution, or saying it had a strength of 1 *M*. This would mean that for every liter of solution there is one **mole** of solute. It can be seen from this that the 1 *M* solution has one mole per liter, so the symbol *M* stands for moles per liter. If we are aware of the molarity of a solution we can calculate the number of moles of solute in a given volume of solution. It can then be seen that the molarity of a solution is a conversion factor between the volume of the solution and the number of moles of solute.

To illustrate the above, 4 L of a solvent and 2 moles of a solute would be able to produce a 0.5 *M* solution (2 divided by 4). The number of moles present in any amount of the solution can be calculated by multiplying the molarity by the volume. For example, in 3 L of the 0.5 *M* solution there are 1.5 moles (0.5 *M* x 3 L).

It should be noted that the volume of a solution is dependent upon the **temperature**. At higher temperatures the solution will expand and at lower temperatures the solution will contract. As a result, the molarity of a solution is a temperature-dependent quantity. This is different from molality which is a temperature-independent quantity. For maximum accuracy, the temperature the molarity was measured at should be stated.

If the solvent used is **water** the molarity and the molality are essentially the same. This is because 1 kg of solvent is nearly equal to 1 kg of solution and the nature of water's **density** means that 1 kg of a dilute solution will be very close to 1 L in volume.

For dilute solutions the volume of the solvent is essentially the same as the volume of the solution and most calculations are made with this assumption.

MOLE

In chemistry, a mole is a certain number of particles, usually of atoms or molecules. Just as a dozen particles (abbreviated doz.) would be 12 particles, a mole of particles (abbreviated mol) is 6.022137×10^{23} particles. This number, usually shortened to 6.02×10^{23}, is known as **Avogadro's number** in honor of Count **Amedeo Avogadro**, an Italian professor of chemistry and physics at the University of Turin who was the first person to distinguish in a useful way between atoms and molecules. It is such a huge number (more than 600 billion trillion) because atoms and molecules are so incredibly tiny that we must have huge numbers of them before we can do anything useful with them.

A standard unit for counting numbers of particles is needed in chemistry, because atoms and molecules react with one another particle by particle. The amount of a chemical reaction—how much of the chemicals are used up or produced—is determined by the numbers of particles that are reacting. Weighing the chemicals would not tell us anything very meaningful unless we knew how to translate those weights into actual numbers of atoms or molecules. For example, if one mole of substance A requires one mole of substance B to react with completely, we need to know how much of substance B to weigh out in order to have just the right amount, without any shortage or waste.

The mole is the translation factor between weights and numbers of particles. One mole of any substance weighs a number of grams that is equal to the atomic or **molecular weight** of that substance. Thus, if the atomic weights of **iron** and **silver** are 55.85 and 107.9, respectively, then 1.95 oz (55.85 g) of iron and 3.78 oz (107.9 g) of silver each contains 6.02×10^{23}atoms. Putting it the other way, a mole of iron (that is, 6.02×10^{23} atoms of iron) would weigh 1.95 oz (55.85 g), while a mole of silver would weigh 3.78 oz (107.9 g).

Iron and silver are elements, and are made up of atoms. **Sodium chloride** (table salt) and sucrose (cane sugar), on the other hand, are compounds, and are made up of molecules. Nevertheless, the mole still works: a mole of salt or sugar means 6.02×10^{23} *molecules* of them. The molecular weights of salt and sugar are 58.45 and 342.3, respectively. Thus, 2.05 oz (58.45 g) of salt and 11.98 oz (342.3 g) of sugar contain the same number of molecules: 6.02×10^{23}.

See also Atom; Atomic weight/mass; Molecule

MOLECULAR FORMULA

The molecular formula specifies the actual number of atoms of each element in a molecule.

The conventional form for writing a molecular formula is to write the symbol for each element followed by a subscript, indicating the actual number of those atoms present in a molecule. When only one **atom** of an element is present, the subscript is omitted. For example, the molecular formula for **water**, H_2O, specifies that there are two **hydrogen** atoms and one **oxygen** atom present in each **molecule** of water.

Mole quanties of elements and compounds. ***(Photograph by Yoav Levy/Phototake NYC. Reproduced by permission.)***

It is important to remember that the molecular formula—in contrast to the simpler empirical formula that specifies only the relative number of atoms or moles present in a compound—identifies the actual number of atoms present in a molecule. For example, the molecular formula for glucose (a sugar important in many biological reactions), $C_6H_{12}O_6$ specifies that in each molecule of glucose there are six **carbon** atoms, 12 hydrogen atoms, and six oxygen atoms. In contrast, the empirical formula for glucose, CH_2O, specifies only that there are two hydrogen atoms for every carbon atom and one oxygen atom for every carbon atom in one molecule of glucose. When dealing with moles of glucose, the empirical formula for glucose, CH_2O, specifies only that there are two moles of hydrogen atoms for every mole of carbon atoms and one mole of oxygen atoms for every mole of carbon atoms in a mole of glucose.

More information is required to construct a molecular formula than is required to obtain the empirical formula of a substance. The empirical formula can be obtained from the elemental analysis of a substance. To obtain the molecular formula, the total molecular mass must be determined experimentally. The molecular formula is then determined from both the **empirical formula** and the molecular mass of a substance.

The molecular formula of a compound is always an integer multiple (e.g., 1, 2, 3,...) of the empirical formula. If the empirical formula of a compound is known, the molecular formula can be determined by the experimental determination of the molecular weight of the compound.

There are two steps to determining the molecular formula once the molecular weight of a compound has been experimentally determined. The first step is to divide the experimentally determined molecular weight of the compound by the molecular weight of the empirical formula in order to determine the integer multiple that represents the number of empirical formula units in the molecular formula. In the second step, the molecular formula is obtained by multiplying the subscripts of the empirical formula by the integral multiple of empirical formula units.

For example, there are many **carbohydrates** or saccharides that have the empirical formula CH_2O and that have a molecular formula that is an integer multiple of CH_2O so that they can, as a group be generally described by the formula $(CH_2O)_n$, where *n* is an integer representing the number of empirical formula units in the molecular formula of the carbohydrate.

If the molecular weight of a carbohydrate (simple sugars) with an empirical formula of CH_2O is experimentally determined by combustion analysis to be 180 g/**mole**, an integer multiple of six is obtained by dividing the experimentally determined molecular weight of 180 g/mole by 30g/mole (the theoretical weight of the empirical formula unit). This means that there are six empirical formula units in the molecular formula. When the subscripts of the empirical formula are then multiplied by the integer multiple of six the result yields the molecular formula of glucose ($C_6H_{12}O_6$).

In the converse, a molecular formula may be reducible to a simple or empirical formula if all its subscripts are divisible by a common denominator. **Nicotine**, for example, has a molecular formula of $C_{10}H_{14}N_2$ that can be divided to obtain the empirical formula of $_5H_7N$.

Some experimental methods can be used to directly determine the molecular formula (i.e., the actual number of atoms that exist in a molecule). X-ray determination can, for example, can allow the actual determination of the positions of atoms in some solids and thus allow direct calculation of the molecular formula.

There are compounds that have the same molecular and empirical formula. For example, methane has both an empirical formula and a molecular formula of CH_4. The simplest whole number ratio of hydrogens to carbon atoms is four hydrogen atoms to every carbon atom. Because CH_4 is also the molecular formula this specifies that in a molecule of methane, four hydrogen atoms are bonded to a single carbon atom.

The empirical formula for water, H_2O, is also the correct molecular formula for water. The molecular formula for water specifies that two hydrogen atoms are bound to a single oxygen atom.

Compounds may also share an empirical formula but dramatically differ in their molecular formulae. For example, acetylene and benzene both have an empirical formula of CH. Acetylene has a molecular formula of C_2H_2 while benzene, has a molecular formula of $_6H_6$.

Some materials do not actually exist as isolated molecules so it is technically impossible to give a molecular formula for such substances. For example, table **salt** (NaCl) does not exist as a molecular substance but rather as an ionic bonded crystal lattice. Regardless, the terms molecular mass and molecular formula are often casually (but now incorrectly) applied to these ionic substances. The reason for this confusion in terminology is that the term molecule originally was defined to be any aggregate of atoms bonded (by either ionic or covalent bonds) together in close enough order to be considered as a discrete physical structure. In this regard a molecule was considered to be a unit of—or the smallest particle of—a compound that retained the chemical properties of the substance.

Ionic compounds, however, are composed of ions, not covalently bonded atoms. For ionic compounds formula mass should be used instead of molecular mass and empirical formula, simplest formula or formula unit) should be used instead molecular formula.

Substances can be molecular (linked together by covalent bonds) or ionic (associated by ionic electrical attraction). Molecular substances are described by their molecular formula (e.g., H_2O or CH_4. Ionic substances are described by the formula unit (e.g., NaCl or MgF_2). When dealing with ionic compounds, the smallest whole-number subscripts are always used.

The molecular formula does not provide information on how atoms are arranged in a molecule. The structural formula is needed to determine the actual arrangement of the atoms in a molecule. Methyl alcohol, for example, consists of a carbon atom that has three hydrogen atoms and one oxygen atom bonded to it. The oxygen atom is, in turn, bonded to a fourth hydrogen atom. The molecular formula is CH_4O does not convey this specificity of arrangement as does the structural formula, CH_3OH.

Like the molecular formula, the structural formula of a substance gives the exact number of atoms of each element per molecule. In addition, however the structural formula depicts how the atoms are bonded to each other. As a result, the structural formula is essential for the determination of molecular shape and the properties associated with particular molecular geometric arrangements.

Just as there are substances with the same empirical formula that differ in their molecular formula, there are substances that have the same empirical and molecular formula. The only way to differentiate these substances is by determining the structural formula for each substance. For example, *cis* dibromoethene and *trans* dibromoethene have the same empirical formula (CHBr) and the same molecular formula ($C_2H_2Br_2$) and must, therefore, be distinguished by each having a unique structural formula.

See also Chemical notation; Empirical formula; Formula, chemical; Molecular structure; Molecule; Structural formula

MOLECULAR GEOMETRY

Molecular geometry describes and predicts the arrangement of atoms in a **molecule**.

When dealing with very simple molecules such as **methane** (CH_4), **water** (H_20), or **Ammonia** (NH_3) the rules and principles of molecular geometry accurately predict the shapes of the molecules. As molecules become increasingly complex it becomes very difficult—but not impossible—to predict and describe complex geometric arrangements of atoms.

The number of bonds between atoms, the types of bonds, and the presence of lone **electron** pairs on the central **atom** in the molecule influence the arrangement of atoms about the central atom in a molecule.

Use of valance shell electron pair repulsion theory (VSEPR) allows chemists to predict the shape of a molecule. According to VESPR theory, molecules will be shaped so as to minimize repulsion between electrons. (Repulsion between **valence** electrons causes these pairs to be as far apart as possible.) Because they are all negatively charged, electrons repel one another. As a result of this repulsion the atoms of a covalently bonded molecule assume a shape around the central atom that maximizes the distance between the outermost or valence electrons. This means that repulsion between electrons in a molecule is at a minimum when the bond angles allow for the greatest separation of the valence electrons in three dimensional space. Bond angles are calculated using the central atom as the vertex of the bond angle.

To learn how to predict molecular shapes it is often useful to consider the oversimplified view of molecules with independent electron orbitals (e.g., *s* and *p* orbitals).

In the case of methane (CH_4), the farthest apart in space the four carbon-hydrogen bonds can be around the central **carbon** atom in three-dimensional space is when the bonds are pointed at the corners of a tetrahedron. When the bonds are pointed toward the corners of a tetrahedron the bond angles are 109.5° and the molecule is said to be a tetrahedral molecule.

When only two atoms are bonded together, as is the case with molecular **oxygen** (O_2), the resulting bond angles must be 180° and therefore the molecule is a linear molecule.

If three atoms are bonded to a central atom and all of the bonds lie in the same plane then the three bond angles must be 120° apart. Such a planar molecule is termed a trigonal planar molecule.

Identifying types of bonds and lone pairs of electrons is as critical to accurately predicting the shape of a molecule as is determining the number of bonds. Electron pairs not only assume the position of bonds in arranging themselves about the central atom but because of their charge **density** they can actually take up more space than the electrons in a **covalent bond**.

The molecular shape of ammonia (NH_3) provides an excellent example of the influence of electron pairs. Although in each molecule there are three **hydrogen** atoms bonded to the central **nitrogen** atom, the nitrogen atom also carries a lone electron pair. As with the case of methane, these four sets of electron pairs (three pairs participating in covalent nitrogen-hydrogen bonds and the one **lone pair** on the nitrogen atom) would find minimum repulsion by spreading themselves out around the nitrogen molecule so that they were pointed at the four corners of a tetrahedron. The lone electron pair, however, with a higher charge density (charge per unit of space) needs more space. In the presence of a lone pair of electrons the bonded electrons are forced to crowd together a bit to make additional space available to the lone electron pair. As a result, in ammonia the bond angles between the central nitrogen atom and the three hydrogen atoms are about 107° and molecule is termed a pyramidal molecule. The bonds between the central nitrogen and the three surrounding hydrogen atoms are pointed at the corners of the base of a triangular pyramid. The lone electron can be thought of a pointing toward the apex of the pyramid.

Water (H_2O) has, of course two oxygen-hydrogen covalent bonds. In addition, there are two electron pairs on the central oxygen molecule. In the same way that the lone electron pair on nitrogen distorts the bond angles in ammonia, the lone electron pairs on the oxygen atom in a water molecule force the two covalent bonds between oxygen and hydrogen to assume a bond angle of approximately 105°. Water is termed a bent molecule.

Bent molecules such as water produce polar molecules if, as in the case of water, the bonded atoms have different **electronegativity** values.

This oversimplified view of molecular geometry has its limitations. Whenever there is more than one electron in an atom (i.e., all atoms heavier than hydrogen) the electron orbitals interact or hybridize. For example, the presence of a *s* orbital electron makes a *p* orbital electron somewhat *s*-like and, in turn, presence of the *p* orbital electron makes an *s*-orbital electron more *p*-like. Hybrid orbitals are named by combining the names of the participating orbitals. An *s* orbital and three *p* orbitals will hybridize to form an sp^3 orbital that has the characteristics of both *s* and *p* orbitals.

The carbon atom at the center of a methane (CH_4) molecule spreads out it's four valence electrons into four sp^3 orbitals pointed at the corners of a tetrahedron with bond angles of 109.5°.

In the case of ammonia (NH_3), the five valence electrons surrounding nitrogen—two electrons occupying the outermost 2*s* orbital and three electrons in three 2*p* orbitals—hybridize to form four sp^3 orbitals that, if equal, would separate themselves in three dimensional space by pointing at the corners of a tetrahedron. One of these orbitals, however, contains the lone electron pair, and the three remaining sp^3 orbitals make additional space available by pointing at the corners of a pyramid with a triangular base.

See also See also Hybridization; Octet rule; Valence

MOLECULAR MODELING

For much of the history of **chemistry**, the proof of any chemical reaction was in the doing. The bench top was the proving ground. The blackboard or the back of the envelope might

Molecular graphics. ***(Photograph by Ken Eward. BioGrafx/Science Source, National Audubon Society Collection/Photo Researchers, Inc. Reproduced by permission.)***

serve as the site for working out a reaction but actually making the compound was the true test. This is still true today but in the past 50 years, the availability of computers and computation software, along with a deeper understanding of the theoretical underpinnings of chemistry, have made the job of the synthetic chemistry much easier. Testing out products, reagents, and reactions using minutes or hours of computer time instead of days or months at the bench has greatly increased productivity and design in research.

Computers have enabled the facile storage and retrieval of structures and other chemical information. The interesting mathematical problems posed by the new quantum mechanics required massive amounts of computational power. Simple calculations could require years when done with pad and pencil. Indeed, many dissertations arose from solving just one aspect of a much larger problem.

Solving quantum mechanical calculations wasn't the only use for computers. **Crystallography**, with its massive matrices of interdependent variables and large data sets of observations, was also begging for some form of automation of calculations. In essence, the computer freed the theoretical chemist and crystallographer to think about the process instead of doing the arithmatic. Of course, this is a sweeping statement that neglects the work required to set up and debug the computer programs but the chemistry, like most of the physical sciences, was definitely ready for computers to come along. To this day, some of the largest users of computational time and power on any campus in the country are chemists.

As computers have evolved, so to have the uses to which they have been applied. With the relatively slow and simple machines of the early 1960s, the level of sophistication for any program was severely limited. A measure of the effect of computers on chemistry can be observed in the number of crystal structures accomplished and submitted to the Cambridge Crystallographic database. In the early 1960s, this number was in the hundreds. By the 1980s, in the thousands. By the mid-1990s, the number of structures is in the hundreds of thousands and beyond the capacity of the database to track. That this growth in structures parallels the growth in computer power is an indication of the effect that computers have had on chemistry. And it is not just in shear numbers but the level of complexity of the structures that has increased. Where in the mid-1980s, protein crystal structures were still a rarity—the 1988 Nobel Prize in chemistry was awarded to **Johann Deisenhofer**, **Robert Huber**, and **Hartmut Michel** for the structural elucidation of a bacterial photosynthetic reaction center. By the mid-1990s, several hundred structures were registered with and submitted to the Brookhaven Protein Data Bank.

Crystallographic structures are, in one sense, a model of a **chemical compound**. They are **electron density** maps that provide a minimum in the least squares fit of the theoretically calculated structure factor with the experimentally observed data. That is, they are a representation of the **molecule** based on x-ray or **neutron** diffraction data. However, they are only one type of model. Expanding the number and accuracy of the models in chemistry has been an underlying driving force for **theoretical chemistry** over the last two decades of the 20th century.

In exploring the chemical world, chemists have always come up against a barrier. Atoms are tiny. They are far too tiny to see by ordinary methods (hence, the need for x-ray diffraction Crystallography). This presents a number of problems. For example, how does one visualize whether or not a particular neurotransmitter will fit into a particular receptor? How does one know which **conformation** of a simple alkane will be the most thermodynamically stable in a non-polar solvent? How about in a polar solvent? The best **solution** is difficult to come by through intuition or guess work. Computational chemistry has eliminated some of the guesswork.

In dealing with molecules, two basic approaches have been used. From a historical perspective, the first approach was to model compounds using pure quantum mechanical calculations. That is, the orbitals of the molecules were composed by a combination of atomic orbitals with increasingly sophisticated levels of complexity and mathematical accuracy. The size of the basis set often determined the accuracy of the results and the agreement with experimental values. The quantum or *ab initio* approach to calculating molecular properties and features is the ''best.'' Unfortunately, it is also the most time consuming. Historically, even simple molecules, with only a few atoms, would take days to calculate. The advent of high speed super computers and parallel processing has decreased the length of time involved in a calculation but this approach is still not ''fast.'' Certainly, computer programs running these algorithms are not interactive.

The difficulties with the mathematically rigorous approach of *ab initio* spawned a number of simpler solutions. In essence, these approaches all have in common the fact that they do not try to calculate the real parameters for the electrons in each **atom** of a molecule but use observable variables to model molecular behavior. Molecular mechanics, for example, models bond lengths and angles based upon a minimization of molecular **energy** as determined using the stretching force constants for bonds (available from **infrared spectroscopy**), torsional energies (which are also obtained from spectroscopy), and idealized bond lengths (from crystallographically-determined structures). The arrangement of atoms in such a molecule is the best arrangement for the parameters available but ultimately, the quality of the structure depends on the quality of the parameters. In the early history of molecular mechanics, there were some difficulties but as the techniques have been tried, tested, and refined, they now provide answers with a greater certainty or that have a higher degree of confidence. More importantly, they do this in an interactive fashion which allows the researcher to visualize the results and refine the model.

This ability to "visualize" chemical compounds and their interactions has been critical to the development of a number of areas of chemistry and **biochemistry** but probably the most important is in the area of drug design. The ability to "dock" molecules into active sites for enzymes allows researchers to, in the first instance, detect the active site by using the naturally active compound and, secondly, to design synthetic molecules with the same docking properties but that are not chemically active. That is, one approach to turning off an **enzyme** is to block its active site with a non-reactive compound. This is a bit like sticking the wrong key in a car's ignition and then getting the key stuck. It prevents the car from working.

Interactive docking programs are one of the ways that molecular modeling is helping to shape industrial processes. Other approaches, though, have included the modeling of molecules on surfaces. For example, a metal catalyst needs to bind a molecule prior to reaction. The mode of binding, the orientation with respect to the surface, and the energies involved all affect the catalytic properties of the metal and the rate of conversion. Modeling this form of interaction using a computer simulation is one of the ways that computational chemistry is bridging the gap with industry. Matching computer generated predictions with industrial results is also providing some assurance for the accuracy of the models which, in turn, allows for more confidence in their predictions. But it also facilitates an understanding of what the chemistry means. That is, why certain reactions proceed and others don't or why some molecules never achieve a 1:1 coverage of a surface.

An example is the model of **oxygen dissociation** on a **rhodium** surface developed by Andrew Rappe of the University of Pennsylvania. Intuitively, complete coverage might be expected but experimentally only a 50% coverage is observed. Modeling the process indicates that repulsive force between the dissociating molecule and neighbouring oxygen increases the energy of the transition state limiting the possible coverage.

Molecular modeling has advanced and continues to advance to a level where extremely large systems can now be studied. By adopting the appropriate model, even systems with millions of atoms can be rendered computationally tractable. For example, investigating pore creation in the surface of **silicates** or testing the lubricating properties of wear inhibitors are expanding both an understanding of molecules and the computational programs required. It is the synergy between the computer programs and the type of problems which need to be addressed that drives forward much of the work in computational chemistry and molecular modeling.

Also of interest is the synergy between the problems addressed by chemists and those in other disciplines. For example, the same algorithms that allow protein chemists to model complex protein structures provide the three dimensional computer simulated characters used in movies. The force field interactions provide the basis for "bumping" into objects in computer games. A visual form of hypertext mark-up language (HTML) is being developed for chemists but will see applications in home shopping. The progress and level of computer programming exceeds just the needs of computational chemistry. Arguably, this is one area of chemistry that is leaving its mark as it develops.

However, molecular modeling is developing and keeping pace with the latest developments in computers. Chemists have no end of problems that certainly can be visualized on a computer screen. The number of algorithms and the variety of programs available for solving molecular mechanics or *ab initio* calculations is growing. Sorting out drug interactions, catalyst design, thermodynamic properties, or biochemical pathways are all within the power of molecular modeling techniques. And, although they are computer generated and not real atoms, it does give the chemist a chance to examine chemistry at the level of the atom.

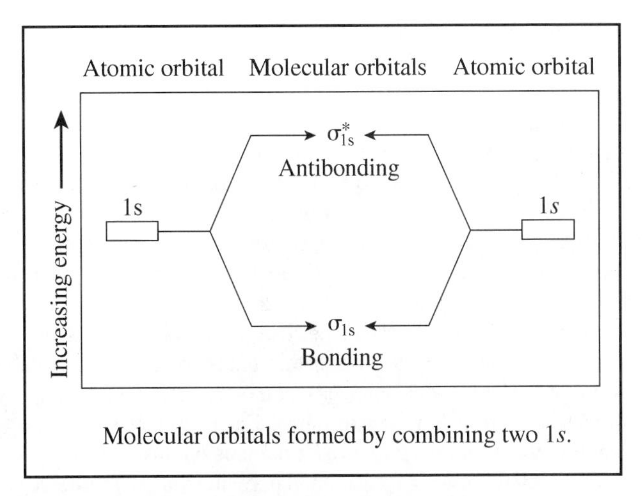

Figure 1. ***(Illustration by Electronic Illustrators Group.)***

MOLECULAR ORBITAL THEORY

Atoms are the fundamental building blocks of all **matter**. This is the basis of **chemistry**. But if they are the bricks, then what is the mortar? What holds the atoms together to form molecules? The answer is ''bonds'' which, of course, doesn't really explain anything because it raises the next question—what is a bond? It is the nature of science that the answer to one question invariably leads to another question.

A simple thought experiment can be used to determine the basic nature of bonds. Electrostatics tells us that oppositely charged particles attract and that similarly charged particles repel. This means that nuclei will repel each other, as will electrons. Simply trying to push two nuclei together will not allow for the formation of a **molecule**. But attractions do exist between electrons and nuclei and this tends to favor molecule formation. Here's what happens when two **hydrogen** atoms approach to within **bonding** distance. Each **atom** has a single **electron** that is attracted to its own **nucleus** but also is attracted to the nucleus of the other atom. There are four such interactions—each of the two electrons are attracted to two nuclei. These four interactions overcome the electron-electron and nuclear-nuclear repulsions as they develop, allowing a molecule to form. In this case, a bond is a pair of shared electrons.

While this is qualitatively the picture of bonding, it is by no means complete. For example, if pure electrostatic interaction was all that was necessary then the more electrons and protons an atom had, the stronger it would bond. **Metals**, such as **gold**, would have so many bonding interactions that they would never melt or be malleable. The simplistic view of bonding outlined above is represented by **Lewis structures**, for example, and a great deal of chemistry can be done without ever pursuing the exact nature of the chemical bond any further. Most synthetic chemists don't need to know the details.

Various models have been put forward to explain molecular interactions. The two most dominant are the ''**valence** bond approach'' and ''molecular orbital theory.'' They are very similar in the way that they describe molecules and come to roughly the same conclusion but from slightly different angles. In grossly simplified terms, the valence bond approach fills the atoms with electrons and then interacts the orbitals, whereas molecular orbital theory interacts the orbitals and then fills them with electrons.

Orbitals are a mathematical construct used to explain the location of electrons around atoms and their physical properties. No one has ever seen an orbital. But they are completely consistent with all that we know about atoms and the way that electrons behave in atoms. This may sound like metaphysics—and it is—but it is important to remember that orbitals are a consequence of the quantum mechanics and the equations that underlay it. This is important because representing orbitals by equations means that adding together equations should give new combinations or orbitals. This is exactly what molecular orbital theory does. By treating the molecule as a single unit and all of the atoms in the molecule as part of that unit, a new set of orbitals for the whole molecule are constructed out of linear combinations of atomic orbitals (LCAO).

Linear combinations of atomic orbitals have the property that orbitals are conserved. Thus, if two orbitals are combined, then the product is two orbitals. If three orbitals are combined, then three new orbitals are created. The two new orbitals created from the original orbitals, designated Ψ_1 and Ψ_2, are $(\Psi_1+\Psi_2)$ and $(\Psi_1- \Psi_2)$, i.e., the sum and difference of the original orbitals. This is what is meant by the term ''linear combination''.

Orbitals, though, are really electron **density** distributions and the symbol Ψ represents the **solution** to the wave function. The combination $(\Psi_1+\Psi_2)$ is a ''piling up'' of electron density. This is a simplistic view, but useful as this combination is bonding. It results in an increase in the electron density between the atomic nuclei. But if $(\Psi_1+\Psi_2)$ is an increase in electron density, then $(\Psi_1- \Psi_2)$ must represent the removal of electron density from between the nuclei. And if an increase in electron density is bonding, then what is a decrease? The answer is that $(\Psi_1- \Psi_2)$ represents an anti- bonding orbital—a combination that negates the bonding interaction. The **energy** of a bonding orbital is decreased relative to the atomic orbital and the anti-bonding orbital is increased. The energy differences are equal and opposite in sign so that energy is conserved in the system. The energy of the two atomic orbitals is the same as the combination of the bonding and anti-bonding orbitals.

Consider the interaction of two hydrogen atoms. The molecular combination has a bonding and anti-bonding pair of orbitals generated from the 1s atomic orbital on each atom (Figure 1). Each atom also has a single electron. In forming H_2, these electrons fill the molecular orbitals by the Aufbau principle—from the lowest energy orbital up. The result in H_2 is that both electrons reside in the bonding combination. The resulting molecule is stabilized relative to the individual atoms. On the other hand, consider trying to form the dihelium molecule, He_2. In this case, the 1s orbital also would form a bonding and anti-bonding pair. But with each atom contributing two electrons, the result would be the filling of both the bonding and anti-bonding combinations. The energy required to put the electrons into the anti-bonding orbital is slightly more (due the overlap integral) than the energy that is obtained from the electrons in the bonding orbital. This is the reason that He_2 doesn't form. It is energetically unfavorable relative to the separate atoms. Indeed, this is the reason that none of the noble **gases** occurs as a **diatomic** molecule. It would cost energy and there is no profit in it.

Now, consider the hydrogen molecule again. What are the consequences of combining a hydrogen atom with a hydrogen ion—H with H^+? In this case, the bonding orbital would only have one electron in it, but this enough to cause the molecule to form. The result is H_2^+, which is a real, experimentally observable molecule containing a half bond between the hydrogens. The energy of this bond is about one half that of the hydrogen-hydrogen interaction and the bond length is 106 pm versus the 74.2 pm of H_2, as would be expected for a weaker interaction. Similarly, it is possible to combine H^- with H to give H_2^-. Again, there is a half bond between the atoms, but in this case it is composed of two electrons in the bonding orbital and one in the anti-bonding. The bond order is defined as one half of the difference between the number of bonding and anti-bonding electrons.

Consider a slightly more complicated molecule, dilithium. In fact, it is a rather ordinary molecule. In this case, each

lithium atom contributes two orbitals to the molecule, a 1s and a 2s orbital. And much as you would expect, these result in bonding and anti- bonding combinations. The 1s orbitals overlap with each other and the 2s orbitals do likewise. The 1s orbital on one atom doesn't interact with the 2s on the other due to the relative energy differences. The 2s orbital is much higher in energy and inaccessible to the 1s orbital. Dilithium, therefore, has two bonding combinations and two anti-bonding combinations and with three electrons contributed from each atom, the net result is a single bond (i.e. one half of four minus two).

So far, as described, molecular orbital theory hasn't really differentiated itself from the Lewis model. Bonds are still the result of electron pairs residing in bonding molecular orbitals. It explains why simply adding more electrons and protons doesn't result in massive bonding. The anti-bonding orbitals insure that the bond order never gets too high. But this can be explained in a Lewis approach using lone pairs. What does molecular orbital theory tell us that makes chemists believe that it is a good model for bonds? Consider the p-orbitals. Unlike the s-orbitals, which are spherically symmetric, the p-orbitals have direction. They orient along the x, y, and z axes of a Cartesian coordinate system. If a line joining the two nuclei is defined as the z-axis, then it is fairly easy to see that the P_z orbitals will be pointing at each other. The result is a bonding and anti-bonding combination, usually designated σ_p and $\sigma_p{}^*$, respectively, to distinguish them from the σ_s and $\sigma_s{}^*$ orbitals that are generated from the s-orbitals. The σ means that the orbital is spherically symmetric around the bond axis, while the asterisk is used to designate anti-bonding orbitals.

But what about the interaction of the P_x and P_y orbitals? Their interaction is much weaker because of the distance separating them, but they do overlap with their opposite partner to give a bonding and anti-bonding combination. These combinations are designated π_p and $\pi_p{}^*$ where the symbol π means that they have a nodal plane that includes the bond axis. Viewed down the bond axis, σ-bonds look like s-orbitals and π-bonds look like p-orbitals.

Since P_x and P_y are equivalent, they have the same energy, but they don't interact with each other because of their spatial orientation. The result is that there are two sets of π_p and $\pi_p{}^*$ orbitals created with the same energy. Furthermore, since the p-orbital overlap isn't as strong as the s-orbital overlap, the resulting difference in energy between the bonding and anti-bonding orbital is not as large. The result is that π_p is filled after σ_p, but $\pi_p{}^*$ is filled before $\sigma_p{}^*$. That is, the filling order for a molecule such as **oxygen** is: σ_{1s}, $\sigma_{1s}{}^*$, σ_{2s}, $\sigma_{2s}{}^*$, σ_{2p}, $2(\pi_p)$, 2(pi;$_p{}^*$), $\sigma_{2p}{}^*$. With each oxygen contributing eight electrons, the highest occupied molecular orbital (HOMO) is the anti-bonding π interaction and since there are two degenerate orbitals, each orbital gets one electron. This results in oxygen being a "diradical" and paramagnetic, both of which properties can be confirmed experimentally. This result is explained only by molecular orbital theory.

The HOMOs of oxygen could also be designated as SOMOs or singly occupied molecular orbitals. The next orbital up in energy, which is left unoccupied, is the LUMO or lowest unoccupied molecular orbital. Collectively, these are termed the "frontier orbitals," because it is at the frontier of the molecular orbitals that chemistry happens. For example, it is these orbitals that are involved when oxygen binds to hemoglobin or when it reacts with combustible materials.

A computerized image of the molecular structure of iodine pentaflouride. ***(Photograph by Kenneth Eward/BioGrafx, Photo Researchers, Inc. Reproduced by permission.)***

Molecular orbital theory is the best explanation of molecular bonding that chemists have. In its simplest form, it provides qualitative explanations for the formation and **reactivity** of molecules. It can also be used with rigorous calculations to provide the **energy level** diagrams for molecules that qualitatively explain the results of photoelectron **spectroscopy**. In addition, it can also **lead** to daunting calculations as each atom contributes its orbitals to the whole molecule. While this explains such phenomena as the hyperfine and superhyperfine coupling observed for the magnetic **resonance** spectroscopies, it does destroy the intuitive picture of one bond/two electrons that was originally used to describe bonding. In the end, though, the advantages of molecular orbital theory vastly outweigh the disadvantages.

MOLECULAR STRUCTURE

The notion that molecules might have specific geometric structures arose in the second half of the nineteenth century. Prior to that time, much debate existed among chemists as to whether molecules had definite shapes that could be determined experimentally. For example, French chemist **Charles Gerhardt** emphasized his belief that they did not by writing different formulas for the same compound. He wrote BaO,SO_3; BaS,O_4; and BaO_2, SO_2, for **barium sulfate**.

Friedrick Kekulé was one of the first chemists to attack the problem of molecular structure. In 1857, he suggested that the **carbon atom** was tetravalent, that is, it could bond with four other atoms. He developed the tool of structural formulas to illustrate this concept, although the formulas created by **Ar-**

chibald Couper at about the same time were far simpler and more efficient to use than were those of Kekulé. Neither Kekulé nor Couper carried their analysis of molecular architecture much beyond the two-dimensional diagrams that could be drawn on **paper**. Couper, for example, showed the bonds on a carbon atom directed at the four corners of a square.

In 1874, **Jacobus Van't Hoff** and **Joseph Le Bel** independently developed a three-dimensional concept of the carbon atom. They proposed that the four carbon bonds were directed towards the corners of a tetrahedron. When they constructed models of organic compounds using tetrahedral atoms, they were able to explain a number of phenomena, including the existence of optical isomer s, first discovered by **Louis Pasteur** in 1848. Other chemists were unimpressed by the ideas of Van't Hoff and Le Bel. **Adolph Kolbe**, for example, doubted the reality of atoms, molecules, and chemical bonds, and warned that thinking of them in concrete, structural terms was carrying theory too far.

The great German chemist Hermann Helmholtz expressed similar concerns. The work of Van't Hoff and Le Bel was justified, however, because of its successes in explaining many physical phenomena. Other chemists began to explore other consequences of molecular structure. The technique soon had spectacular success in the field of organic **chemistry**, where many puzzling experimental results were eventually explained. Attempts to describe the molecular structure of inorganic compounds were less successful. In fact, there was limited progress in this field in the twenty years following the work of Van't Hoff and Le Bel.

At first, structures for simple compounds, such as binary salts, could be drawn. But a number of more complex compounds escaped explanation. Then, in 1893, German chemist **Alfred Werner** successfully applied structural theory to inorganic compounds. Waking early one morning, Werner had arrived at a fully formed **solution** in his sleep. He began writing what turned out to be his most important scientific paper on the spot and had finished it by the next afternoon. His theory of *coordination compounds* showed that the structure of molecules could be understood in terms of geometry, rather than simply in terms of chemical bonds between atoms. In working with organometallic compounds, for example, he placed the metal atom at the center of a geometric figure (such as a cube) and surrounded it with other atoms, ions, and groups of atom at the corners of the figure. He suggested that the metallic atom was attached to the surrounding groups by ''secondary valences.'' Werner's suggestion was enormously fruitful. He used the theory, for example, to predict the existence of geometric isomers, two forms of a compound that differ from each other only in the position of a single atom, **ion**, or group of atoms. In 1907, he synthesized a pair of geometric isomers that confirmed his predictions.

The theory of molecular structure reached its highest development in the work of **Linus Pauling** in the 1920s and 1930s. Pauling applied the theory of quantum mechanics to electrons and showed how the formation of chemical bonds resulted in more stable configurations than existed in free atoms. Pauling's work not only provided a clearer understanding of the chemical bond, but also explained the structures of molecules. He demonstrated that the most stable configuration for some molecules was some intermediary structure between two other structures. The bonds in a **benzene molecule**, for example, are neither pure single nor pure double bonds, but some kind of ''compromise'' that results from the shifting of electrons back and forth between the two possibilities. This theory of **resonance** explained a number of phenomena that had previously been a mystery.

In the 1940s, Pauling applied these ideas to the complex molecules that make up living systems, especially **proteins** and **nucleic acids**. One approach he found highly productive was model-building. Beginning with tinker-toy-type equipment, he constructed molecular structures that seemed reasonable based on available experimental evidence. Then he refined the models until he could produce a structure that accounted for all existing data. In the early 1950s, he used this technique to demonstrate how polypeptides could adopt a helical structure. He very nearly achieved similar success in unraveling the structure of nucleic acids. This work—using the physical structure of chemical molecules to explain biological phenomena—led to the development of the science now known as molecular biology.

In the 1980s, computer technology had improved to the point where it could be used to construct graphic models of molecular structures. This led to the development of a new science known as computational chemistry. In this area of chemistry, scientists input data about compounds such as molecular composition or chemical behavior, and the computer displays a picture of the molecular structure. This technology will be an important part of the development of synthetic drugs and new catalysts.

MOLECULAR WEIGHT

The molecular weight of a **molecule** indicates how heavy that molecule is relative to an **atom** of **carbon** (with six protons and six neutrons). Saying that the molecular weight of **water** is 18 means that the water molecule is 18/12 as heavy as a carbon-12 atom, or 18/16 as heavy as an oxygen-16 atom. In general the molecular weight of a molecule is the sum of the atomic weights of its constituent atoms.

The gram molecular weight (GMW) of a substance is defined as the weight in grams of one **mole** (6×10^{23} molecules, known asAvogadro's number). The molecular weight of water is 18.02; its gram molecular weight is 18.02 g. Given the gram molecular weight of a substance, the weight of an individual molecule can be calculated by dividing that weight by **Avogadro's number**.

The first direct approach to determining molecular weight was proposed by two French scientists in 1819, Pierre Louis Dulong (1785-1838) and Alexis-Thérèse Petit (1791-1820). They suggested that the amount of **heat** required to raise the **temperature** of an atom of a solid material by a given amount should be independent of the type of atom, with the result that the gram atomic weight should be inversely proportional to the material's specific heat. Although the law of Dulong and Petit proved a fair approximation for many elements, it was far from exact for many others, and it was not at all helpful in determining the atomic weights of gaseous elements.

In 1811, the Italian physicist Lorenzo Romano Amadeo Carlo Avogadro di Quaregua e di Cerreto, known as Avogadro, concluded that equal volumes of all **gases** at the same temperature and pressure contain the same number of molecules. Unfortunately, Avogadro's ideas had little influence on the work of his contemporaries, and it was not until 1860 that another Italian scientist, **Stanislao Cannizzaro**, pointed out that Avogadro's hypothesis could be used as a basis for determining atomic and molecular weights.

The classical method of determining the molecular weight of a gas is to use **density** data. Experimentally, a few milliliters of a volatile liquid are placed in a stoppered flask containing a small orifice. The flask is then heated to a temperature above the **boiling point** of the liquid. As the liquid evaporates, its vapor replaces the air in the flask. The flask is then allowed to cool, and the vapor condenses as air re-enters the flask. The **mass** of the vapor is then calculated, and used in conjunction with the ideal gas law to determine the molecular weight of the liquid. This method works quite well for many gases and volatile **liquids**, but it cannot be used for substances that decompose on heating, such as **urea**.

The molecular weights of such substances have been determined by measuring those physical properties of solutions that depend primarily on the concentrations of solute particles and their molecular weight. Such properties include the lowering of **vapor pressure**, the elevation of boiling point, and the lowering of freezing point and **osmotic pressure**. Freezing point depressions have often been used because they are comparatively easy effects to observe. Osmotic pressure measurements have been used for solutes of high molecular weight.

Modern methods for measuring molecular weight include chromatographic methods and **mass spectrometry**.

See also Avogadro's law; Ideal gases; Thermochemistry

MOLECULE

It was not until knowledge about atoms and elements was gained that the make-up of the millions of different substances around us was understood. All the scientific knowledge we have today indicates that these different substances are made from only 92 different kinds of atoms that make up the naturally occurring elements.

These atoms are able to join together in millions of different combinations and arrangements to form all the substances in the universe. A molecule is formed when two or more of the same or different kinds of atoms join together. The newly formed substance is called a compound. Although the identity of the elements stays the same because the number of protons is the same, the physical and chemical properties of the compound are different from the properties of the elements from which they formed because the arrangement of the outermost electrons is different.

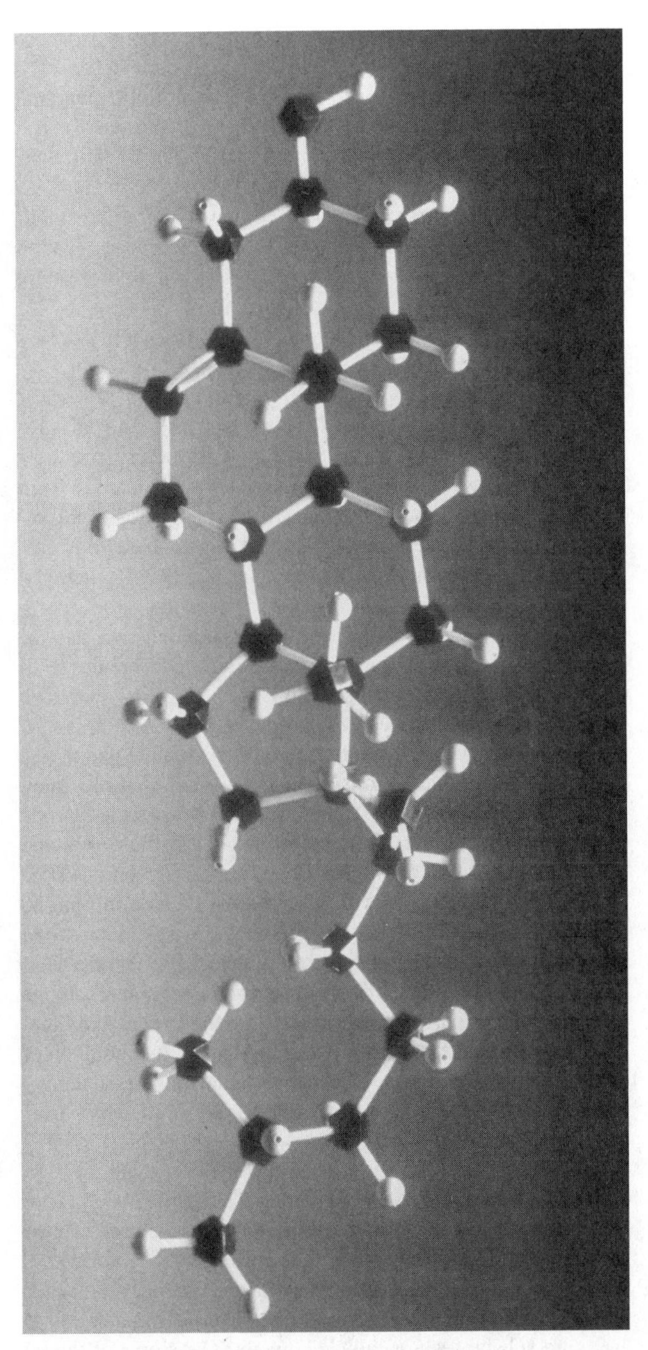

A molecule. *(Digtal Stock. Reproduced by permission.)*

For centuries chemists and physicists believed that it was possible to transmute one element, such as **lead**, into another, such as **gold**. When the corpuscular theory of **matter** was developed and accepted (which could explain but not predict chemical changes in terms of transmutations), this belief was strengthened. By the middle of the eighteenth century, however, virtually all chemists and physicists believed that transmutations of matter into other kinds of matter were not possible. But lack of knowledge about the elements, the basic building blocks of all matter, hindered any real understanding of the nature of matter and the formation of new substances.

During the period between 1789 and 1803, **Antoine-Laurent Lavoisier** defined the elements as substances that could not be separated by fire or some other chemical means and **John Dalton** defined atoms as small, indestructible and invisible particles. These ideas cleared the way for understanding the makeup of all the substances in the universe. Dalton assumed that each kind of element had its own kind of **atom**, which was different from the atoms of all other elements. He also assumed that chemical elements kept their identity during all chemical reactions.

During the early nineteenth century, chemical experiments centered mainly around taking measurements of substances involved in chemical reactions both before and after the reaction. It was found that elements always reacted to form a new substance in the same ratio. If different ratios of the reacting substances were used, different substances were produced. Dalton's **atomic theory** went on to say that chemical compounds are the new substances that form when atoms combine with each other; that a specific compound always has the same kinds of atoms in the same ratio; and that chemical reactions do not involve a change in the atoms themselves but in the way they are arranged.

In 1809, French chemist **Joseph-Louis Gay-Lussac** and others began doing numerous experiments with **gases** by measuring the amounts of the gases that actually reacted. They found that two volumes of **hydrogen** reacted with one **volume** of **oxygen** to form two volumes of **water**, and that one volume of hydrogen gas reacted with one volume of **chlorine** gas to form two volumes of hydrogen chloride gas. In 1811, **Amedeo Avogadro** hypothesized that equal volumes of different gases, when at the same **temperature** and **pressure**, contained the same number of particles. These experimental results and theories eventually led to the determination of the number of atoms in the substances. The name molecule was later assigned to particles made up of more than one atom, which may be of the same or different atoms.

Of all the naturally occurring substances around us every day, there are only 92 that cannot be chemically changed to simpler substances. Two or three of these substances are so rare in nature that their natural occurrence is questionable. There are also 17 other known substances that are man-made in sophisticated instruments like cyclotrons. Together, these 109 pure substances are called elements. The atoms of the 92 naturally occurring elements are the building blocks for all of the substances in the universe.

When atoms join together in various combinations of kind and number, they form molecules. When molecules are made from two or more different kinds of atoms, the substances are called compounds. If molecules are made from only one kind of atom, the substances are elements. New combinations of the atoms produce new molecules and therefore different substances.

Special kinds of chemical formulas, called molecular formulas, are used to represent the kinds of atoms and the number of each kind of atom in a molecule. The simplest molecules are composed of just two atoms (usually a single atom is not referred to as a molecule), which may be the same or different. Oxygen gas (O_2), hydrogen gas (H_2), and nitrogen gas (N_2), are made up of molecules composed of just two atoms of oxygen, hydrogen, or nitrogen respectively. Since these substances are composed of only one kind of atom, they are elements. **Carbon monoxide** (CO) is a gas with molecules composed of one atom of **carbon** and one atom of oxygen and **carbon dioxide** (CO_2) is a gas with molecules composed of one atom of carbon and two atoms of oxygen. Water molecules (H_2O) are composed of one atom of oxygen and two atoms of hydrogen. These substances are compounds because the molecules that make it up have two kinds of atoms. Many molecules, especially those in living things such as sugar, fat, or protein molecules and molecules of deoxyribonucleic acid (**DNA**) or ribonucleic acid (**RNA**), are much larger and more complex.

It is possible for all the different substances in the universe to be produced from only 92 naturally occurring elements because there are endless ways to combine these 92 kinds of atoms, which are the building blocks for all molecules. The 26 letters of the alphabet can be compared to the 92 different kinds of atoms. Different words can be formed in many different ways: by using different letters, such as dog and dig and dodge and dug and dugout; or by using the same letters with different arrangements, such as dog and God and good; or mate and tame and meat, or met and meet and teem. Because of all these possibilities, the 26 letters of the alphabet form all the millions of words of the English language. Similarly, new substances form when different kinds or different numbers of atoms join, or when the same kinds of atoms join in different arrangements. And, just like the letters of the alphabet, the 92 naturally occurring elements form all the millions of different substances in the entire universe including all the various **metals**, plastics, materials for building, fabrics, all parts of all living things, etc. It is the kind, number, and arrangement of atoms within the molecule that determines what the substance is.

All molecules have a definite **mass** and size that are dependent on the atoms from which the molecule is made. The mass is equal to the sum of the masses of all the individual atoms in the molecular structure. The size is not only dependent on the atomic components of the molecule, but also on the arrangement of the atoms within the molecule and how tightly they are joined together.

When atoms join other atoms to form molecules, the chemical and physical properties of the compounds are different from those of the elements from which they were formed. These include such things as **color**, hardness, conductivity, state (solid, liquid, gas), etc. When letters are used to form new words, such as dog, God and good from the letters g, o and d, the meanings of the new words cannot be discovered by observing or studying either the letters g, o and d, or the other words. The new words formed have new and different meanings. This is also true when new molecules form. The properties of the new substances cannot be found by studying the properties of the elements from which they formed or the properties of other similar molecules.

The **molecular formula** for sugar is $C_6H_{12}O_6$, which indicates that a sugar molecule is made up of six atoms of carbon, 12 atoms of hydrogen, and six atoms of oxygen. Carbon is an element that is black and often is found in powder form. It is well-known as the major component of coal or the black that appears on burnt toast. Pure hydrogen is a the lightest gas known and pure oxygen is a gas present in the air and needed for living things to breathe and for fires to burn. When these three substances chemically combine in the ratio of $C_6H_{12}O_6$ to form sugar, which is a white, crystalline solid that has a sweet taste and is soluble in water, the properties of the sugar are unlike those of the pure elements from which it was formed. However, if sugar is reacted under extreme conditions (a chemical change), the original substances, carbon, hydrogen, and oxygen could be recovered.

Hydrogen and oxygen atoms can join to form water molecules. Once again, the properties of water (often used to extinguish fires) are completely different from the properties of oxygen gas (needed to support burning). These same two elements, hydrogen and oxygen, also form another common substance, **hydrogen peroxide**, with a molecular formula of H_2O_2. Hydrogen peroxide, in its undiluted form, can cause serious burns. When diluted, it is often used for bleaching and as an antiseptic in cleansing wounds. These properties are completely different from the properties of the elements from which it is made, hydrogen and oxygen, as well as from the similar molecule, water. You could not boil potatoes in H_2O_2 instead of H_2O without deadly effects. The properties of hydrogen peroxide differ greatly from those of hydrogen, oxygen, and water because each of the substances has its own specific number, kind, and arrangement of atoms.

While the molecular formula gives the basic information about what atoms are joined together and how many of each kind of atom, this formula does not give the whole story. The arrangement of the atoms within the molecule must also be considered since different arrangements of the same atoms within a molecule produce different substances. The molecular formulas for ethyl **alcohol** (formed from the **fermentation** of grains and fruits and present in all wines and liquors) and for methyl ether (sometimes used in refrigeration but not the same ether used as an anesthetic) are identical, C_2H_6O. However, the chemical properties are very different. This is because the atoms are arranged differently within the molecule. Molecular formulas cannot convey this information. Different kinds of formulas, called structural formulas, are needed to show molecular arrangements. In the case of ethyl alcohol, the oxygen atom is joined to a carbon atom and to a hydrogen atom. But in the case of methyl **ether**, the oxygen atom is joined to two carbon atoms. This different arrangement of atoms within the molecule is responsible for imparting different chemical properties to the compound. Thus, the new chemical properties are not only dependent on how many of each kind of atom have joined together, but on how these atoms are arranged within the molecule.

Much of the research in the field of chemistry today involves the formation of new substances by trying to change atoms within a molecule or the arrangement of the same atoms. This latter is a particularly important area when the arrangement of the atoms is changed from what is called a right-handed molecule to a left-handed molecule or vice-versa. Molecules such as these behave as they do because of their shape, much like gloves fit either the right or the left hand because of their shape. Some of the current research on new fat-free commercial products, such as ice cream or cooking oil, involves right- or left-handed versions of the original fat molecules. These mirror-images of the original molecules cannot be absorbed or used by the body because of the different shape of the molecule. Yet because they are often so similar, enough of the original properties of the fat remain intact to make it useful as a substitute. So far, the taste and their inability to withstand high temperatures in cooking are problematic.

Similarly, many powerful drugs in use today have both right- and left-handed versions. Most drugs contain both forms of the molecules because this mixture is cheaper and easier to produce. However, often only one version gives the desired effect while frequently the other produces unwanted side effects. Although it is chemically much more difficult and more expensive to produce drugs of only one "handedness," patients needing these drugs are finding the purer, single-handed version much easier to tolerate. Drug companies are beginning to pay attention and research is being done to produce drugs of only one "handedness" at a reasonable cost.

When compounds are formed, the identity of the atoms, which is associated with the number of protons in the **nucleus**, does not change. For example, oxygen atoms are still oxygen atoms whether they are part of oxygen gas molecules, water molecules, carbon dioxide molecules, etc., because the number of protons is unchanged.

But unlike mixtures, where two or more substances are mixed together but there is little or no interaction among the atoms of the various substances, there is interaction among the atoms within molecules. New compounds are formed when the atoms within the molecule form a chemical bond. These bonds are a sort of "glue" that hold the atoms together within the molecule. Bonds involve only the outermost electrons of the atoms, that is, those in the highest shell or energy level. It is this change in **electron** arrangements that is responsible for the new properties observed when compounds are formed. There are two major types of bonds, ionic and covalent.

An **ionic bond** forms when the outermost electrons are transferred from one atom to another. One atom loses one or more electrons and another atom gains these electrons. **Sodium** metal is a soft, shiny, and very reactive metal that is stored under kerosene to keep it from reacting with the oxygen in the air. Sodium atoms have only one electron at the highest energy level and would be more stable if they got rid of this electron. Chlorine is a very poisonous green gas involved in the purifying process of swimming pools and responsible for the characteristic smell around them. Chlorine has seven electrons at the highest energy level and would be much more stable with eight electrons at this level. When sodium and chlorine come in contact with each other, there is an instantaneous reaction. Neutral atoms of sodium metal give up one electron with their negative charge and form particles, called ions, with net charges of +1.

Neutral atoms of chlorine gain negatively-charged electrons from sodium atoms and form particles, also called ions, with net charges of -1.

Throughout these changes, the number of protons and neutrons in the nucleus and all of the innermost electrons around the nucleus stay the same. Only an insignificant amount of the mass of the atom is associated with electrons, so the mass of the atom also stays essentially the same. However, a chemical change has occurred and the original properties of the atoms have changed because of the new arrangement of the electrons. The newly formed sodium **ion** and the chloride ion are electrically attracted or bonded to each other because opposite charges attract each other. The substance formed is ordinary table salt which is a white, salty, crystalline solid, properties that are very different from the original elements of sodium and chlorine. The bond that formed is called an ionic bond and sodium chloride is called an ionic compound. An ionic bond is formed when the ''glue'' between atoms is the force of attraction between opposite charges. In fact, salt crystals are formed by the very neat and orderly arrangement of alternating sodium and chloride ions. Ionic bonds are most often formed between atoms of metals and atoms of non-metals.

Often, the outermost electrons of an atom are shared with the outermost electrons of another atom. Electrons move around the nucleus of an atom at very high speeds and, when electrons are shared between two atoms, the nuclei move so close together that the shared electrons spend part of their time near both nuclei simultaneously. This sharing of electrons by the nuclei of two atoms simultaneously is the ''glue'' that is holding them together within the molecule. This type of bond is called a **covalent bond**. Electrons that are shared can be contributed by either or both atoms involved in the bond formation.

Carbon atoms have four outermost electrons and need eight to be more stable. If one carbon atom shares each of its outermost electrons with a hydrogen atom, which needs only two electrons to become more stable, a molecule of **methane** is formed. The formula for the new substance formed is CH_4. Methane is often called marsh gas because it forms in swamps and marshes from the underwater decomposition of plant and animal material. It is widely distributed in nature and about 85% of **natural gas** is methane.

When things are shared, like sharing the sofa with a friend, they are not always shared equally. This is also true of electrons involved incovalent bonds. At times the electrons involved in **bonding** are shared equally between the nuclei of two atoms and the bond is called a pure covalent bond. More often, however, the sharing is unequal and the electrons spend more time around the nucleus of one atom than of the other. The bond formed is called a polar covalent bond. Usually covalent bonds (both pure covalent and polar covalent) form when atoms of non-metallic elements bond to atoms of other non-metallic elements.

The ''glue'' or bonding that holds atoms of metals close to each other is usually referred to as a **metallic bond**. It is formed because the outermost electrons of the metal atoms form a sort of ''sea'' of electrons as they move freely around the nuclei of all the metal atoms in the crystal.These mobile electrons are responsible for the electrical and **heat** conductivity of the metals.

See also Avogadro's number; Change of state; Chemical compound; Element, chemical; Neutron; Proton; States of Matter; Structural formula

MOLINA, MARIO (1943-)

American chemist

Mario Molina is an important figure in the development of a scientific understanding of our atmosphere. Molina earned national prominence by theorizing, with fellow chemist **F. Sherwood Rowland**, that **chlorofluorocarbons** (CFCs) deplete the Earth's **ozone** layer. In his years as a researcher at the Jet Propulsion Lab at CalTech and a professor at the Massachusetts Institute of Technology (MIT), Molina has continued his investigations into the effects of chemicals on the atmosphere.

Mario José Molina was born in Mexico City on March 19, 1943. His father was Roberto Molina-Pasquel; his mother, Leonor Henriquez. Following his early schooling in Mexico, he graduated from the Universidad Nacional Autónoma de México in 1965 with a degree in **chemical engineering**. Immediately upon graduation, Molina went to West Germany to continue his studies at the University of Freiburg, acquiring the equivalent of his master's degree in polymerization **kinetics** in 1967. Molina then returned to Mexico to accept a position as assistant professor in the chemical engineering department at his alma mater, the Universidad Nacional Autónoma de México.

In 1968, Molina left Mexico to further his studies in physical **chemistry** at the University of California at Berkeley. He received his Ph.D. in 1972 and became a postdoctoral associate that same year. His primary area of postdoctoral work was the chemical laser measurements of vibrational **energy** distributions during certain chemical reactions. The following year, 1973, was a turning point in Molina's life. In addition to marrying a fellow chemist, the former Luisa Y. Tan (the couple have one son, Felipe), Molina left Berkeley to continue his postdoctoral work with physical chemist, Professor F. Sherwood Rowland, at the University of California at Irvine.

Both Molina and Rowland shared a common interest in the effects of chemicals on the atmosphere. And both were well aware that every year millions of tons of industrial pollutants were bilged into the atmosphere. Also, there were questions about emissions of **nitrogen** compounds from supersonic aircraft. What impact did these various chemical discharges have on the envelope of air that surrounds the Earth? Molina and Rowland decided to conduct experiments to determine what happens to chemical pollutants that reach both the atmosphere directly above us but also at stratospheric levels, some ten to twenty-five miles above the earth. Both men knew that within the stratosphere, a thin, diffuse layer of ozone gas encircles the planet which acts as a filter screening out much of the sun's most damaging ultraviolet radiation. Without this ozone shield, life could not survive in its present incarnation.

The two scientists concentrated their research on the impact of a specific group of chemicals called chlorofluorocarbons, which are widely used in such industrial and consumer products as aerosol spray cans, pressurized containers, etc. They found that when CFCs are subjected to massive ultraviolet radiation they break down into their constituent chemicals: **chlorine**, **fluorine**, and **carbon**. It was the impact of chlorine on ozone that alarmed them. They found that each chlorine **atom** could destroy as many as 100,000 ozone molecules before becoming inactive. With the rapid production of CFCs for commercial and industrial use—millions of tons annually—Molina and Rowland were alarmed that the impact of CFCs on the delicate ozone layer within the stratosphere could be life-threatening.

Mario Molina published the results of his and Rowland's research in *Nature* magazine in 1974. Their findings had startling results. Molina was invited to testify before the House of Representatives's Subcommittee on Public Health and Environment. Suddenly CFCs were a popular topic of conversation. Manufacturers began searching for alternative propellant **gases** for their products.

Over the next several years, Molina refined his work and, with Rowland, published additional data on CFCs and the destruction of the ozone layer in such publications as *Journal of Physical Chemistry, Geophysical Research Letter* and in a detailed piece entitled "The Ozone Question" in *Science.* In 1976, Mario Molina was named to the National Science Foundation's Oversight Committee on Fluorocarbon Technology Assessment.

In 1982, Molina became a member of the technical staff at the Jet Propulsion Laboratory at CalTech; two years later he was named senior research scientist, a position he held for an additional five years. In 1989, Mario Molina left the West coast to accept the dual position of professor of **atmospheric chemistry** at the MIT's department of Earth, atmosphere and planetary sciences, and professor in the department of chemistry. In 1990, he was one of ten environmental scientists awarded grants of $150,000 from the Pew Charitable Trusts Scholars Program in Conservation and the Environment. In 1993, he was selected to be the first holder of a chair at MIT established by the Martin Foundation, Inc., "to support research and education activities related to the studies of the environment."

Molina has published more than fifty scientific papers, the majority dealing with his work on the ozone layer and the chemistry of the atmosphere. In 1992 Molina and his wife, Luisa, wrote a monograph entitled "Stratospheric Ozone" published in the book *The Science of Global Change: The Impact of Human Activities on the Environment* published by the American Chemical Society.

His later work has also focused on the atmosphere-biosphere interface which Molina believes is "critical to understanding global climate change processes." He is the recipient of more than a dozen awards including the 1987 American Chemical Society Esselen Award, the 1988 American Association for the Advancement of Science Newcomb-Cleveland Prize, the 1989 NASA Medal for Exceptional Scientific Advancement, and the 1989 United Nations Environmental Programme Global 500 Award.

Mario Molina.

MOLYBDENUM

Molybdenum is a transition metal, one of the elements that occur in Rows 4, 5, and 6 between Groups 2 and 13 in the **periodic table**. Its **atomic number** is 42, its atomic **mass** is 95.94, and its chemical symbol is Mo.

Properties

Molybdenum is a silvery white solid that usually occurs as a dark gray or black powder with a metallic luster. Its melting point is about 4,730°F (2,610°C), its **boiling point** is between 8,670 and 10,040°F (between 4,800 and 5,560°C), and its **density** is 10.28 grams per cubic centimeter.

Molybdenum is not very active. It does not react with cold acids or with **oxygen** at room **temperature**, although it does become more active at higher temperatures. For example, it does react with hot, strong sulfuric or nitric acids. It also reacts with oxygen at high temperatures to produce molybdenum trioxide (MoO_3).

Occurrence and Extraction

Molybdenum never occurs as the free element in nature, but is most commonly found in the minerals molybdenite (primarily molybdenum sulfide; MoS_2) and wulfenite ($PbMoO_4$). It is moderately common in the Earth's crust with an abundance of about 1-1.5 parts per million. About two-thirds of all the molybdenum in the world comes from Canada, Chile, China, and the United States. The element is extracted from its ores by passing hot **hydrogen** over molybdenum trioxide: $MoO_3 + 3H_2 \rightarrow 3H_2O + Mo$.

Discovery and Naming

Molybdenum was discovered in 1781 by the Swedish chemist Peter Jacob Hjelm (1746-1813), although Hjelm's work did not become generally known for nearly a century. Hjelm discovered the element while working with molybdenite, an interesting mineral that is very similar to the **graphite** used to make "**lead**" pencils. Most chemists thought that molybdenite was a form of lead, a perception that was perpetuated when the name molybdenum was selected for the element. In Greek, the word *molybdos* actually means lead.

Uses

About 75% of the molybdenum used in the United States is made into alloys of steel and **iron**. Nearly half of these alloys, in turn, are used to make stainless and heat-resistant steels. Molybdenum alloys can be found in airplane, spacecraft, and missile parts; spark plugs; propeller shafts; rifle barrels; and boiler plate. The most widely used compound of molybdenum is molybdenum sulfide, which is still used as a **lubricant**, as it was more than 200 years ago. Other compounds of molybdenum are used as protective coatings in materials used at high temperatures; as solders; as catalysts; as additives to animals feeds; and as pigments and dyes in glasses, **ceramics**, and enamels.

Gaspard Monge.

Monge, Gaspard (1746-1818)

French mathematician

Monge was born into a merchant family and received a typical public school education. His instinctive aptitude for mathematics and science was so strong that he was placed in charge of the physics course at the Collège de la Trinité, where he was also a student. Though he never again received any formal science education, Monge managed to enjoy a spectacularly successful scientific career.

In 1764 he drew up a large-scale plan of his hometown, Beaune, France, that was so admired by military officers that he was given the position of draftsman and technician at the École Royale du Génie at Mezières. Monge's technique of using geometry to solve construction details, which normally required tedious calculations, was the foundation of what is called descriptive geometry. Monge designed fortresses for the army, which guarded Monge's unique methods as a military secret.

Monge's interests expanded to include the physical sciences and **chemistry**. In 1783 he synthesized **water** and in collaboration with Jean François Clouet (1751-1801) was the first to liquify **sulfur** dioxide, a substance that at room **temperature** is normally a gas but changes to a liquid at -98° F (-72° C).

After the French Revolution, Monge also became involved in public affairs and served on the committee to create the **metric system** of measurement. This marked the beginning of a long participation in French political life. By 1792 Monge was one of the most recognized and respected of French scientists and was instrumental in the creation of a number of important public institutions. Among these was the École Polytechnique, which Monge helped form and served as its first director. The École Polytechnique evolved into one of the premier training centers for French scientists and engineers. During this time, Monge developed a strong friendship with Napoleon Bonaparte (1769-1821), who appointed Monge to important committees involved in organizing the newly conquered regions in the expanding French sphere of influence, such as Italy and the Middle East. When Napoleon became Emperor he bestowed upon Monge the title of senator for life, which forever linked Monge to the political fortunes of Bonaparte.

When Napoleon fell in 1815, Monge was also discredited. His steady climb to the pinnacle of success quickly became a plummet to disgrace. His honors were stripped from him and he was expelled from the Institut de France. Exhausted physically and spiritually, Monge suffered various forms of political harrassment until his death in 1818. Despite the dim view of him held by the restored French monarchy, tribute was paid to Monge by his many friends, colleagues, and students at the École Polytechnique, where he was still respected as one of the towering figures in French scientific life.

MONOMER

A monomer is a an organic **molecule** that is involved in polymerization reactions. During these reactions, they are chemically linked forming macromolecules, or polymers, which have large molecular weights. Monomers are the starting materials for a wide range of materials. Numerous kinds are available from both natural and synthetic sources.

While polymers have been used by humans for centuries, it was not until the early twentieth century that their **chemistry** was understood. The groundwork for modern understanding of polymer science was laid by Nobel laureate **Hermann Staudinger** in the 1920s. He theorized that polymers were composed of simpler molecules, or monomer units, that were linked together in a chainlike fashion. In 1928, Meyer and Mark used x-ray techniques to confirm Staudinger's models. By the 1930s the monomer/polymer model was accepted by most polymer scientists.

The first type of polymers that were used by humans were naturally occurring. The primary categories for natural polymers include polysaccharides, **proteins**, and **nucleic acids**. These materials are made up of specific monomers including saccharides, **amino acids**, and nucleotides, respectively.

Amino acids are the basic monomer building blocks for proteins. They contain an amino group ($-NH_2$), a carboxylic acid group (-COOH), and a side chain which differs from one amino acid to the next. There are a total of twenty common amino acids that comprise, when bonded together, the protein structures. The simplest amino acid is glycine which contains a **hydrogen atom** as its side group. Other examples of amino acids are alanine, leucine, cysteine, arginine, and tryptophan.

Simple **carbohydrates**, or monosaccharides, are monomers that bond together to form polysaccharides. Monosaccharides are composed of a number of linked **carbon** atoms and a **functional group**. They are classified by these two characteristics. Aldoses have an aldehyde functional group while ketoses contain a ketone. The most important monosaccharides are hexoses such as glucose and galactose. They react to form polymers such as starch, **glycogen**, chitin, and **cellulose**.

Another type of natural monomer are nucleotides. These compounds create the polymers ribonucleic and deoxyribonucleic acids, the genetic information common to all living organisms. Nucleotides consist of three distinct sections: the cyclic five-carbon monosaccharide ribose, a **phosphate** diester (that links the riboses by connection to the hydroxy groups of the riboses), and a heterocyclic base. The bases (thus termed because they contain basic **nitrogen** atoms) are purines or pyrimidines, that provide the code which is translated into proteins in living systems. Nucleotides are linked together by the phosphate groups to form polynucleotide polymers.

Synthetic polymers are produced through two types of polymerization reactions. Condensation reactions depend on a repeated reaction between functional groups on the monomers. Addition polymerization reactions use free radical monomers to produce polymers. The chemistry of these monomers is slightly different.

Monomers involved in condensation reactions typically contain two reactive groups within the molecule. For example, polyesters are created by the reaction of a dihydric **alcohol**, such as ethylene glycol, which contains two hydroxyl (-OH) groups and a diacid, like adipic acid, containing two carboxyl groups (-COOH). During the reaction, these groups can react with either end of the monomer. This reaction continues until there are no reactive groups left. The result is a long chain, high **molecular weight** polymer. The production of polysulfide polymers is a result of a **condensation reaction** between dialkalide monomers with sulfide monomers. **Nylon** is produced from diamine and diacid monomers. These molecules have reactive amine (-NH) groups which bond with the carboxyl groups. Polyurethanes are created from dihydric alcohols and diisocyanate monomers. Other condensation monomers include epichlorohydrin, phenol, and **formaldehyde**.

Addition polymerization reactions are characterized by monomers that contain double bonds. When polymerization is initiated the double bond is broken and a free radical is formed. This free radical attacks the double bonds on other monomers creating a longer chain molecule and an additional free radical. The reaction continues until a termination step is reached. Important monomers which undergo these reactions include styrene, methylmethacrylate, **vinyl** acetate, vinyl chloride, and other acrylates.

A micrograph image of monosodium glutamate (MSG). ***(Photograph by Science Photo Library, National Audubon Society Collection/Photo Researchers, Inc. Reproduced by permission.)***

MONOSODIUM GLUTAMATE (MSG)

Monosodium glutamate (MSG) is the **sodium salt** of the naturally occurring amino acid, glutamic acid. It is denoted by the chemical formula $C_5H_8NNaO_4 \bullet H_2O$. The compound was first isolated in 1886 and became an important industrial material in 1908 when it was discovered to enhance the flavor of certain foods. The MSG **molecule** can exist in two different structural forms called isomers. Only monosodium L-glutamate has a flavor enhancing effect, while the R form does not. MSG is commonly believed to be responsible for a condition known as "Chinese Restaurant Syndrome," however current evidence has failed to establish a link between the two.

The development of MSG as an important industrial material began with the isolation of glutamic acid from a wheat protein called gluten in 1886. In 1890, the chemical structure of glutamic acid was identified. The flavor enhancing ability of MSG was discovered by the Japanese chemist, Ikeda Kibunae (1864-1936). From a kelp-like seaweed that was traditionally used to add flavor to Japanese food, he isolated MSG and patented a method for its production in 1908. Commercial production of this flavor-enhancing agent soon followed and Japan's first major chemical industry was born.

MSG typically exists as a white, odorless crystalline powder that is soluble in **water** and **alcohol**. It does not have a melting point per se, but it decomposes when it is heated. MSG can be produced in three ways. The oldest method is by the **hydrolysis** of vegetable **proteins**. In this method, waste protein such as wheat gluten, is combined with water and acid. It is then heated causing the proteins to breakdown into their component **amino acids**. The glutamic acid is then isolated by crystallization. It is made into MSG by reacting the glutamic acid crystals with **sodium hydroxide**. A second production method involves using a microorganism to produce MSG. The bacteria *Micrococcus glutamicus* converts some **carbohydrates** into glutamic acid. This glutamic acid is then converted to MSG by a partial **neutralization** process. The final method for producing MSG involves an organic reaction based on acrylate.

While MSG is tasteless by itself, it is used to enhance the flavor of meat, fish, and vegetables. It is reported that when MSG is used in foods at concentrations below 0.9% it adds a unique flavor that is neither bitter, salty, sweet, or sour. MSG is used by the food industry in canned, frozen, and dried foods and also in Oriental food. The pharmaceutical industry also has employed MSG, with sugar, to improve the palatability of bitter drugs.

In 1968, the safety of MSG was scrutinized when a report was published suggesting that MSG caused a disease known as "Chinese Restaurant Syndrome" (CRS). Symptoms of this syndrome include burning, numbness, fever, and a tightness in the upper body. Although many subsequent studies failed to show any link between MSG and these symptoms, the safety of MSG is still questioned by some people.

Moore, Stanford (1913-1982)

American biochemist

Stanford Moore's work in protein **chemistry** greatly advanced understanding of the composition of enzymes, the complex **proteins** that serve as catalysts for countless biochemical processes. Moore's research focused on the relationship between the chemical structure of proteins, which are made up of strings of **amino acids**, and their biological action. In 1972 he was awarded the Nobel Prize in chemistry with longtime collaborator **William Howard Stein** for providing the first complete decoding of the chemical composition of an **enzyme**, ribonuclease (RNase). This discovery provided scientists with insight into cell activity and function, which has had important implications for medical research.

Stanford Moore. *(Corbis Corporation. Reproduced by permission.)*

Moore was born on September 4, 1913, in Chicago, Illinois, but spent most of his childhood in Nashville, Tennessee, where his father, John Howard Moore, was a professor at Vanderbilt University's School of Law. His mother was the former Ruth Fowler. In 1935 Moore earned a B.A. in chemistry from Vanderbilt. Moore continued his education in **organic chemistry** at the graduate school of the University of Wisconsin and completed the Ph.D in 1938. Though Moore considered attending medical school, he accepted a position in 1939 as a research assistant in the laboratory of German chemist Max Bergmann at the Rockefeller Institute for Medical Research (RIMR), later renamed the Rockefeller University.

Bergmann's research group focused on the structural chemistry of proteins. During his early years at RIMR, Moore questioned whether proteins actually had specific structures. The direction of research in Bergmann's laboratory was greatly influenced by the arrival of William Howard Stein. At Bergmann's suggestion, Moore and Stein began a long-lived and successful collaboration. With the exception of his wartime service, the years abroad, and a year at Vanderbilt University, Moore remained at RIMR all of his professional life. In 1952 he became a member of the institute, and his association with

RIMR/Rockefeller University continued until his death in 1982.

From 1942-1945, Moore worked as a technical aide in the National Defense Research Committee of the Office of Scientific Research and Development (OSRD). As an administrative officer for university and industrial projects, Moore studied the action and effects of mustard gas and other chemical warfare agents. As a technical aide, Moore also coordinated academic and industrial studies on the actions of chemical agents. In 1944 he was appointed to the project coordination staff of the Chemical Warfare Service and continued to contribute to this research until the end of the war in 1945.

Though his initial investigation of chromatography—the process of separating the components of a solution—began in the late 1930s, the war interrupted this work. After the war and following the death of Max Bergmann in 1944, Moore returned to RIMR to resume work with William Howard Stein. This marked a productive period for the two men, leading to their work on chromatographic methods for separating amino acids, **peptides** (compounds of two or more amino acids), and proteins. Moore's work in **chromatography** was influenced by the methods of **paper** chromatography developed by **A. J. P. Martin** and **Richard L. M. Synge** of England. However, limitations in these earlier methods prohibited the study of protein chemistry; new techniques in chromatography had to be developed so that amino acids could be separated. Moore and Stein utilized column chromatography, in which a column or tube is filled with material that separates the components of a **solution**. In 1948 they successfully separated amino acids by passing the solution through a column filled with potato starch. The process was time consuming, however, and presented inadequate separations for amino acid analysis. To facilitate the procedure, Moore and Stein replaced the filler material with a synthetic **ion** exchange resin, which separated components of a solution by electrical charge and size. In 1949 they successfully separated amino acids from blood and urine.

Further interruptions in the work at RIMR occurred in 1950, when Moore held the Francqui Chair at the University of Brussels (where he established a laboratory for amino acid analysis), and in 1951, when he was a visiting professor at the University of Cambridge (England) working with **Frederick Sanger** on the amino acid sequence of insulin.

In 1958 Moore and Stein contributed to the development of the automated amino acid analyzer. This instrument facilitated the complete amino acid analysis of a protein in twenty-four hours; previously employed procedures, such as chromatography, had required up to one week. The automated technique afforded researchers a tool with which to separate and study the large chemical sequences in protein molecules. This instrument is used worldwide for the study of proteins, enzymes, and **hormones** as well as the analysis of food. While Moore and Stein endeavored to determine the chemical composition of proteins, scientists concurrently worked to determine their three-dimensional structure.

By 1959 Moore and Stein had determined the amino acid sequence of pancreatic ribonuclease (RNase), a digestive enzyme that breaks down ribonucleic acid (RNA) so that its components can be reused. They discovered that ribonuclease is made up of a chain of 124 amino acids, which they identified and sequenced. This marked the first complete description of the chemical structure of an enzyme, a discovery that earned Moore and Stein the 1972 Nobel Prize in chemistry (an award shared with **Christian Boehmer Anfinsen** of the National Institutes of Health). Understanding protein structure is essential to understanding biological function, which opens the door to the treatment of disease. Moore's findings influenced research in **neurochemistry** and the study of such diseases as sickle-cell anemia. Scientists later discovered that related ribonucleases are present in nearly all human cells, which prompted studies in the fields of cancer and malaria research.

In 1969, Moore returned to Rockefeller University following one year spent at Vanderbilt University School of Medicine as a visiting professor of health sciences. Upon his return, the team of Moore and Stein resumed their studies of protein chemistry with an investigation of deoxyribonuclease, the enzyme that breaks down deoxyribonucleic acid (DNA).

In addition to scientific work, Moore also served as an editor of *The Journal of Biological Chemistry,* treasurer and president of the American Society of Biological Chemistry, and president of the Federation of American Societies for Experimental Biology. Moore received many honors, including membership in the National Academy of Sciences and receipt of the Richards Medal of the American Chemical Society and the Linderstrøm-Lang Medal. He was also awarded honorary degrees by the University of Brussels, the University of Paris, and the University of Wisconsin.

Moore was diagnosed with amyotrophic lateral sclerosis (Lou Gehrig's disease) some time in the early 1980s. On August 23, 1982, he committed suicide in his home in New York City. Upon his death Moore, who never married, left his estate to Rockefeller University.

MORLEY, EDWARD WILLIAMS (1838-1923)

American chemist and physicist

Originally trained for the ministry, Edward Williams Morley decided instead in 1868 to pursue a career in science, the other great love of his life. At first he devoted himself primarily to teaching, but gradually became engaged in more original research. His work can be divided into three major categories: the first two involved the determination of the **oxygen** content of the atmosphere and efforts to evaluate Prout's hypothesis; his third field of research involving experiments on the **velocity** of light with Albert Michelson, are those for which his name will always be most famous.

Morley was born in Newark, New Jersey, on January 29, 1838. His mother was the former Anna Clarissa Treat, a schoolteacher, and his father was Sardis Brewster Morley, a Congregational minister. According to a biographical sketch of Morley in the December 1987 issue of the *Physics Teacher,* the Morley family had come to the United States in early Colonial days and "was noted for its deep patriotism and religious devotion."

Edward Williams Morley.

Morley's early education took place entirely at home, and he first entered a formal classroom at the age of nineteen when he was admitted to Williams College in Williamston, Massachusetts, as a sophomore. His plans were to study for the ministry and to follow his father in a religious vocation. But he also took a variety of courses in science and mathematics, including astronomy, **chemistry**, calculus, and optics. His courses at Williams were a continuation of an interest in science that he had developed at home as a young boy.

Morley graduated from Williams as valedictorian of his class with a bachelor of arts degree in 1860. He then stayed on for a year to do astronomical research with Albert Hopkins. Morley's biographers allude to the "careful" and "precise" calculations required in this work as typical of the kind of research Morley most enjoyed doing. Certainly the later research for which he is best known was also characterized by this high level of precision and accuracy.

After completing his work with Hopkins, Morley entered the Andover Theological Seminary to complete his preparation for the ministry, while concurrently earning his master's degree from Williams. Morley graduated from Andover in 1864, but rather than finding a church, he took a job at the Sanitary Commission at Fortress Monroe in Virginia. There he worked with Union soldiers wounded in the Civil War.

His work completed at Fortress Monroe, Morley returned to Andover for a year and then, failing to find a ministerial position, took a job teaching science at the South Berkshire Academy in Marlboro, Massachusetts. It was at Marlboro that Morley met his future wife, Isabella Ashley Birdsall. The couple was married on December 24, 1868. They had no children.

Morley had finally received an offer in September, 1868, to become minister at the Congregational church in Twinsburg, Ohio. He accepted the offer but, according to biographers David D. Skwire and Laurence J. Badar in the *Physics Teacher,* became disenchanted with "the low salary and rustic atmosphere" at Twinsburg and quickly made a crucial decision: he would leave the ministry and devote his life to science.

The opportunity to make such a change had presented itself shortly after Morley arrived in Twinsburg when he was offered a job teaching chemistry, botany, geology, and **mineralogy** at Western Reserve College in Hudson, Ohio. Morley accepted the job and when Western Reserve was moved to Cleveland in 1882, Morley went also. While still in Hudson, Morley was assigned to a full teaching load, but still managed to carry out his first major research project. That project involved a test of the so-called Loomis hypothesis that during periods of high atmospheric pressure air is carried from upper parts of the atmosphere to the Earth's surface. Morley made precise measurements of the oxygen content in air for 110 consecutive days, and his results appeared to confirm the theory.

In Cleveland, Morley became involved in two important research studies almost simultaneously. The first was an effort at obtaining a precise value for the atomic weight of oxygen, in order to evaluate a well-known hypothesis proposed by the English chemist **William Prout** in 1815. Prout had suggested that all atoms are constructed of various combinations of **hydrogen** atoms.

Morley (as well as many other scientists) reasoned that should this hypothesis by true, the atomic weight of oxygen (and other elements) must be some integral multiple of that of hydrogen. For more than a decade, Morley carried out very precise measurement of the ratios in which oxygen and hydrogen combine and of the densities of the two **gases**. He reported in 1895 that the atomic weight of oxygen was 15.897, a result that he believed invalidated Prout's hypothesis.

Even better known than his oxygen research, however, was a line of study carried out by Morley in collaboration with Albert A. Michelson, professor of physics at the Case School of Applied Science, adjacent to Western Reserve's new campus in Cleveland. Morley and Michelson designed and carried out a series of experiments on the velocity of light. The most famous of those experiments were designed to test the hypothesis that light travels with different velocities depending on the direction in which it moves, a hypothesis required by current theories regarding the way light is transmitted through space. A positive result for that experiment was expected and would have confirmed existing beliefs that the transmission of light is made possible by an invisible "ether" that permeates all of space.

In 1886 Michelson and Morley published their report of what has become known as the most famous of all negative experiments. They found no difference in the velocity with which light travels, no matter what direction the observation is made. That result caused a dramatic and fundamental rethinking of many basic concepts in physics and provided a critical piece of data for **Albert Einstein**'s theory of relativity.

The research on oxygen and the velocity of light were the high points of Morley's scientific career. After many years of intense research, Morley's health began to deteriorate. To recover, he took a leave of absence from Western Reserve for a year in 1895 and traveled to Europe with his wife. When he returned to Cleveland, he found that his laboratory had been dismantled and some of his equipment had been destroyed. Although he remained at Western Reserve for another decade, he never again regained the enthusiasm for research that he had had before his vacation.

Morley died on February 24, 1923, in West Hartford, Connecticut, where he and Isabella had moved after his retirement in 1906; she predeceased him by only three weeks. Morley had been nominated for the Nobel Prize in chemistry in 1902 and received a number of other honors including the Davy Medal of the Royal Society in 1907, the Elliot Cresson Medal of the Franklin Institute in 1912, and the Willard Gibbs Medal of the Chicago section of the American Chemical Society in 1917. He served as president of the American Association for the Advancement of Science in 1895 and of the American Chemical Society in 1899.

MORPHINE

Morphine is the most effective naturally occurring compound used for the relief of pain in medicine and surgery. It also induces sleep and produces euphoria. The active ingredient in opium, from which it is derived, morphine is highly addictive with repeated use. Its story is the story of the founding of **alkaloid chemistry**, which grew out of the study of plant bases and plays an essential role in medicine. In 1805, opium was widely used for its euphoric effects, and a German pharmacist named Friedrich Sertürner (1783-1841) decided to investigate the components of poppy juice, from which opium is derived. He found an unknown acid, converted it into a crystalline precipitate, and named it principium somniferum. Having determined that this substance was the active ingredient in opium, in 1809 he recommended the cultivation of the poppy on a large scale as a way to further the national economy since morphine was used in the production of the popular drug and poppy seed oil. In 1815, Sertürner and three young volunteers each took three 30 mg doses of principium somniferum over a period of 45 minutes, and were not fully themselves again until several days later. In 1817, he published a paper describing the drug, in which he changed its name to *morphium*, after Morpheus, the Greek god of dreams. The same year, the name was changed to morphine by the French chemist **Joseph Gay-Lussac**. During the 1800s, the French physiologist François Magendie advanced the use of morphine in medicine, administering it both orally and by injection. Morphine's greatest medical advantage is its depressant action that causes the threshold of pain to rise, relieving pain many other **analgesics** are unable to control. Its **narcotic** properties also produce a calming effect, protecting the body's system in traumatic shock. Its greatest disadvantage, however, is its addictiveness. Morphine is an alkaloid, meaning that it is an organic compound that contains **carbon**, **hydrogen** and **nitrogen**, and forms a water-soluble **salt**. The chemistry of alkaloids is crucial in medicine; the analgesic properties of the opium alkaloids are a case in point, but other examples of important alkaloids include **strychnine**, a respiratory stimulant; codeine, a painkiller; and conine, an active ingredient in hemlock. Sertürner's groundbreaking research provided the foundations for the field of alkaloid chemistry and these contributions to medicine. Morphine's popularity on the Civil War battlefields boosted its general use in the treatment of many kinds of discomfort, and a leading British doctor called morphine "God's own medicine." However, thousands of people worldwide were tragically addicted. Many chronicles of addiction have been written, including Eugene O'Neill's (1888-1953) semi-autobiographical play *Long Day's Journey Into Night*. Once addicted, a person is likely to experience severe symptoms of withdrawal, including pain, hyperventilation, restlessness, and confusion. In 1898, the Bayer corporation synthesized heroin from morphine and marketed it as an antidote to morphine addiction, but the concurrent moral reform movements were beginning to give rise to anti-opiate sentiments, and morphine's popularity and acceptance began to decline. Today, morphine is used most often for short-term, post-operative pain control, while methadone is used to treat opiate addiction and for long-term pain control.

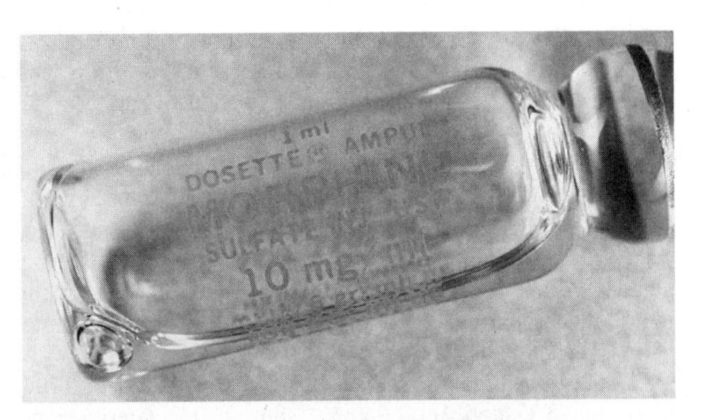

A sample of morphine, a narcotic drug. ***(Photograph by Scott Camazine, Photo Researchers, Inc. Reproduced by permission.)***

MOSANDER, CARL GUSTAF (1797-1858)

Swedish chemist

A large share of the credit for unraveling the complex nature of the **rare earth elements** goes to Carl Gustaf Mosander. Mosander was born in Kalmar, Sweden, on September 10, 1797. He was educated as a physician and pharmacist and served as an army surgeon for many years.

Perhaps his most important professional association was with the eminent Swedish chemist **J. J. Berzelius**. Mosander lived with Professor and Mrs. Berzelius for many years and worked as Berzelius' assistant at the Stockholm Academy of

Sciences. Eventually, Mosander became curator of minerals at the Academy and, in 1832, succeeded Berzelius as Permanent Secretary of the Academy. Mosander was also Professor of **Chemistry** and **Mineralogy** at the Caroline Institute for many years.

Mosander became interested in the rare earth elements in the late 1830s. Fifty years earlier, a Swedish army officer, Carl Axel Arrhenius, had discovered a new mineral that he named *ytterite* near the small town of Ytterby. Chemists spent much of the next century trying to separate the mineral into its many chemically-similar parts.

The first breakthrough in this effort occurred in 1794 when Johan Gadolin (1760-1852) showed that ytterite contained a large fraction of a totally new **oxide**, which he called *yttria*. A decade later, M. H. Klaproth, Berzelius, and Wilhelm Hisinger (1766-1852) showed that ytterite also contained a second oxide, which they called *ceria*.

Mosander first concentrated his efforts on the ceria part of ytterite. In 1839, he found that the ceria contained a new element which he named **lanthanum** (for hidden). Mosander did not publish his results immediately, however, because he was convinced that yet more discoveries were to be made. He was not disappointed in these hopes. In 1841, he identified a second new component of ceria. He named the component *didymium*, for ''twin,'' because it was so closely related to lanthanum. Later research showed that didymium was not itself an element, but a complex mixture of other rare earth elements.

In 1843, Mosander turned his attention to the yttria portion of ytterite. He was able to show that the yttria consisted of at least three components. He kept the name yttria for one and called the other two *erbia*, and *terbia*. The last two of these components are now known by their modern names of **erbium** and **terbium**. Mosander is acknowledged as the discoverer, then, of three elements: lanthanum, erbium, and terbium. Mosander died in Ångsholm, Sweden, on October 15, 1858.

MOSELEY, HENRY GWYN JEFFREYS (1887-1915)

English physicist

Henry Gwyn Jeffreys Moseley's great contribution to science was his ordering of the elements by examining their x-ray spectra, thus creating the **periodic table**. His work, which ranks among the most profound discoveries of the twentieth century, was cut short when the scientist was killed during World War I. Although he never received a major award for his research, Moseley's accomplishments laid the groundwork for a number of later scientific developments.

Henry Gwyn Jeffreys Moseley (widely known as ''Harry'') was born on November 23, 1887, at Weymouth, England. He was the only son of Henry Nottidge Moseley, a professor of human and comparative anatomy at the University of Oxford. He had also served as a naturalist on the famous voyages of the *Challenger* in that vessel's studies of the world's oceans. Both sides of the family included a number of eminent scientists, including a paternal grandfather who was the first professor of natural philosophy at King's College, London, and a great grandfather who was an expert on tropical diseases.

Moseley's father died when Harry was four years old. The family then moved to the small town of Chilworth, in Surrey, where Harry and his two sisters began their education. Moseley's interest in science surfaced at a young age. He was fortunate in having the constant encouragement of his mother and friends in this interest which was to lead not only to a professional career in science, but also to a lifelong fascination with natural history.

In 1896, at the age of nine, Moseley was enrolled at the Summer Fields school, an institution that specialized in preparing boys for Eton. Five years later, he won a King's Scholarship that allowed him to enroll at Eton. At Eton, Moseley studied with T. C. Porter, one of the first scientists in England to work with x rays. Although Moseley studied a number of other subjects, x rays became the topic in which he was interested during his career in science.

After leaving Eton in 1906, Moseley was awarded a scholarship to continue his education at Trinity College, Oxford. Though he earned only second class honors in science, Moseley was able to get letters of recommendation that allowed him to take a position at the University of Manchester, where he worked with **Ernest Rutherford**. His assignment at Manchester involved a full teaching load, but he still found time to carry out an ambitious program of original research.

At the conclusion of his first year at Manchester, Moseley was relieved of his teaching responsibilities and allowed to devote all his time to research. The topic he selected for that research was the diffraction of x rays, a phenomenon that had just been discovered by the German physicist Max Laue. For a period of some months, Moseley collaborated with C. G. Darwin, a mathematical physicist, on a study of the general characteristics of diffracted x rays. In the fall of 1913, however, the two men went their separate ways, and Moseley began to focus on the nature of the spectra produced by scattered x rays.

Moseley saw that when x rays are beamed at certain crystalline materials, they are diffracted by atoms within the crystals, forming a continuous spectrum on which is superimposed a series of bright lines. The number and location of these lines are characteristic of the element or elements being studied. Much of the basic research on x-ray **spectroscopy** had been done in England by **William Lawrence Bragg**, with whom Moseley studied for a short period of time at the University of Leeds.

In his own research, Moseley devised a system that allowed him to study the **x-ray diffraction** pattern produced by one element after another in an orderly and efficient arrangement. Very quickly, he found that the frequencies of one set of spectral lines, the ''K'' lines, differed from element to element in a very consistent and orderly way. That is, when the elements were arranged in ascending order according to their atomic masses, the frequency of their K spectral lines differed from each other by a factor of one. To Moseley, the meaning of these results was clear. Some property inherent in the structure of atoms was responsible for the regular, integral change he observed. That property, he decided, was the charge on the **nucleus**. When the elements are arranged in ascending order to their atomic masses, he pointed out, they are also arranged in ascending order according to their nuclear charge. The main difference is that the variation in atomic masses between adjacent elements is never consistent, whereas the variation in nuclear charge is always precisely one. This property is such a clear defining characteristic of atoms and elements that it was given the special name of **atomic number**. Of all the properties of an **atom**, atomic number has come to be the single most important characteristic by which an atom can be recognized.

The implications of Moseley's discovery were manifold and profound. In the first place, the concept of a unique, identifying and characteristic number of an element—its nuclear charge—provided a new basis for the periodic table. The Russian chemist **Dmitry Ivanovich Mendeleev**'s original proposal for arranging the elements on the basis of their atomic masses had worked extraordinarily well, but it did have its flaws. Moseley found, however, that the discrepancies found in Mendeleev's proposal disappeared when the elements were arranged according to their atomic numbers. It was obvious, therefore, that atomic number was an even more fundamental property of atoms and elements than was atomic **mass**.

Moseley's research also allowed him to predict the number and location of elements still missing from the periodic table. Since each element could be assigned its own unique atomic number (hydrogen=1, helium=2, lithium=3, beryllium=4, and so on), any missing numbers must mean that an element remains to be discovered for this particular position in the periodic table. Based on that logic, Moseley was able to predict with confidence the existence of an as-yet-undiscovered element #43, between **molybdenum** (#42) and **ruthenium** (#44) and another between **neodymium** (#60) and **samarium** (#62). Furthermore, he was able to predict the spectral pattern to be expected for each missing element and thereby provide a valuable tool in the search for those elements.

As his initial work on x-ray spectra was being concluded, Moseley decided to resign his position at Manchester and return to Oxford in the hope of securing a position as professor of experimental physics. At Oxford, Moseley continued to work on x-ray spectra, concentrating on the study of the relatively unknown group of elements known as the lanthanides.

In June of 1914, Moseley left for a meeting of the British Association for the Advancement of Science scheduled to be held in Australia. He received word while still aboard ship of the outbreak of World War I and decided to return to England immediately. He enlisted in the Royal Engineers and, after completing an eight-month training period, was sent to the Turkish battlefront at Gallipoli. There, during the battle of Sulva Bay on June 15, 1915, he was killed by a sniper's bullet. Moseley was never married and, in his short lifetime, received no great honors. His contributions to science, however, were enormous, in that many later advances in physics and **chemistry** were based on his work.

MÖSSBAUER, RUDOLF (1929-)

German physicist

Rudolf Mössbauer's study of the recoilless emission of gamma rays and nuclear **resonance** florescence led to the discovery of methods for making exact measurements in solid-state physics, archeology, biological sciences, and other fields. His measurement method was used to verify **Albert Einstein's** theory of relativity and is known as the Mössbauer effect. He was honored with a 1961 Nobel Prize in physics for his work.

Rudolf Ludwig Mössbauer was born on January 31, 1929, in Münich, Germany. He was the only son of Ludwig and Erna (Ernst) Mössbauer. Ludwig Mössbauer was a phototechnician who printed **color** post cards and reproduced photographic materials. Mössbauer grew up during a difficult time in Germany, during the disruptions accompanying the rise of Adolf Hitler's National Socialism (Nazi party) and the onset of World War II. Still, he was able to complete a relatively normal primary and secondary education, graduating from the Münich-Pasing Oberschule in 1948. His plans to attend a university were thwarted because, due to Germany's loss in the war, the number of new enrollments was greatly reduced.

Through the efforts of his father, Mössbauer was able to find a job working as an optical assistant first at the Rodenstock optical firm in Münich and later for the U.S. Army of Occupation. Eventually Mössbauer saved enough money from these jobs to enroll at the Münich Institute for Technical Physics. In 1952 he received his preliminary diploma (the equivalent of a B.S. degree) from the institute, and three years later he was awarded his diploma (the equivalent of an M.S. degree). For one year during this period, 1953–54, he was also an instructor of mathematics at the institute.

After receiving his diploma in 1955, Mössbauer began his doctoral studies at the Münich Institute for Technical Physics. His advisor there was Heinz Maier-Leibnitz, a physicist with a special interest in the field of nuclear resonance fluorescence. As a result of Maier-Leibnitz's influence, Mössbauer undertook his thesis research in that field. His research was not carried out primarily in Münich, however, but at the Max Planck Institute for Medical Research in Heidelberg, where Mössbauer received an appointment as a research assistant.

Nuclear resonance fluorescence is similar to widely-known phenomena such as, for example, the resonance in tuning forks. When one tuning fork is struck, it begins to vibrate with a certain frequency. When a second tuning fork is struck close to the first, it begins vibrating with the same frequency, and is said to be "resonating" with the first tuning fork. Fluorescence is a form of resonance involving visible light. When light is shined on certain materials, the atoms that make up those materials may absorb electromagnetic **energy** and then re-emit it. The emitted energy has the same frequency as the original light as a result of the resonance within atoms of the material. This principle explains the ability of some materials to glow in the dark after having been exposed to light.

The discovery of fluorescence by R. W. Wood in 1904 suggested some obvious extensions. If light, a form of electromagnetic radiation, can cause fluorescence, scientists asked, can other forms of electromagnetic radiation do the same? In 1929, W. Kuhn predicted that gamma rays, among the most penetrating of all forms of electromagnetic radiation, would also display resonance. Since gamma rays have very short wavelengths, however, that resonance would involve changes in the atomic **nucleus**. Hence came the term nuclear resonance fluorescence.

For two decades following Kuhn's prediction, relatively little progress was made in the search for nuclear resonance fluorescence. One reason for this delay was that such research requires the use of radioactive materials, which are difficult and dangerous to work with. A second and more important factor was the problem of atomic recoil that typically accompanies the emission of **gamma radiation**. Gamma rays are emitted by an atomic nucleus when changes take place among the protons and neutrons that make up the nucleus. When a gamma ray is ejected from the nucleus, it carries with it a large amount of energy resulting in a ''kick'' or recoil, not unlike the recoil experienced when firing a gun. Measurements of gamma ray energy and of nuclear properties become complicated by this recoil energy.

Researchers looked for ways of compensating for the recoil energy that complicated gamma ray emission from radioactive nuclei. Various methods that were developed in the early 1950s had been partially successful but were relatively cumbersome to use. Mössbauer found a **solution** to this problem. He discovered that a gamma emitter could be fixed within the crystal lattice of a material in such a way that it produced no recoil when it released a gamma ray. Instead, the recoil energy was absorbed by and distributed throughout the total crystal lattice in which the emitter was imbedded. The huge size of the crystal compared to the minute size of the emitter **atom** essentially ''washed out'' any recoil effect.

The material used by Mössbauer in these experiments was iridium–191, a radioactive **isotope** of a platinum-like metal. His original experiments were carried out at very low temperatures, close to those of liquid air, in order to reduce as much as possible the kinetic and thermal effects of the gamma emitter. Mössbauer's first report of these experiments appeared in issues of the German scientific journals *Naturwissenschaften* and *Zeitschrift fur Physic* in early 1958. He described the recoilless release of gamma rays whose wavelength varied by no more than one part in a billion. Later work raised the precision of this effect to one part in 100 trillion.

In the midst of his research on gamma ray emission, Mössbauer was married to Elizabeth Pritz, a fashion designer. The couple later had three children, two daughters and a son. In 1958, Mössbauer was also awarded his Ph.D. in physics by the Technical University in Münich for his study of gamma ray emission.

The initial reactions to the Mössbauer papers on gamma ray emission ranged from disinterest to doubt. According to one widely repeated story, two physicists at the Los Alamos Scientific Laboratory made a five-cent bet on whether or not the Mössbauer Effect really existed. When one scientist was in fact able to demonstrate the effect, the scientific community gained interest.

Physicists found a number of applications for the Mössbauer Effect using a system in which a gamma ray emitter fixed in a crystal lattice is used to send out a signal, a train of gamma rays. A second crystal containing the same gamma ray emitter set up as an absorber so that the gamma rays travelling from emitter to absorber would cause resonance in the absorber. Therefore, the emission of gamma rays stays resonating and constant until a change, or force such as gravity, **electricity**, or **magnetism** enters the field. By noting the changes in the gamma ray field, unprecedented measurements of these forces became available.

One of the first major applications of the Mössbauer Effect was to test Einstein's theory of relativity. In his 1905 theory, Einstein had predicted that photons are affected by a gravitational field, and therefore an electromagnetic wave should experience a change in frequency as it passes near a massive body. Astronomical tests had been previously devised to check this prediction, but these tests tended to be difficult in procedure and imprecise in their results.

In 1959, two Harvard physicists, Robert Pound and Glen A. Rebka, Jr., designed an experiment in which the Mössbauer Effect was used to test the Einstein theory. A gamma ray emitter was placed at the top of a sixty-five-foot tower, and an absorber was placed at the bottom of the tower. When gamma rays were sent from source to absorber, Pound and Rebka were able to detect a variation in wavelength that clearly confirmed Einstein's prediction. Today, similar experimental designs are used for dozens of applications in fields ranging from theoretical physics to the production of synthetic **plastics**.

Mössbauer's Nobel Prize citation in 1961 mentioned in particular his ''researches concerning the resonance absorption of gamma radiation and his discovery in this connection of the effect which bears his name.'' The Nobel citation went on to say how Mössbauer's work has made it possible ''to examine precisely numerous important phenomena formerly beyond or at the limit of attainable accuracy of measurement.'' The physics community acknowledged the enormous scientific and technical impact the discovery would eventually make.

After receiving his Ph.D. in 1958, Mössbauer took a position as research fellow at the Institute for Technical Physics in Münich. Two years later he was offered the post of full professor at the Institute, but, according to his entry in *Nobel Prize Winners,* declined the offer because he was ''frustrated by what he regarded as the bureaucratic and authoritarian organization of German universities.'' Instead, he accepted a job as research fellow at the California Institute of Technology and was promoted to full professor a year later, shortly after the announcement of his Nobel award.

In 1964, Mössbauer once more returned to Münich, this time as full professor and with authority to reorganize the physics department there. In 1972, he took an extended leave of absence from his post at Münich to become director of the Institute Max von Laue in Grenoble, France. After five years in France, he returned to his former appointment in Münich. In addition to the Nobel Prize, Mössbauer has received the Science Award of the Research Corporation of America (1960), the Elliott Cresson Medal of the Franklin Institute (1961), the Roentgen Prize of the University of Giessen (1961), the Bavarian Order of Merit (1962), the Guthrie Medal of London's Institute of Physics (1974), the Lomonosov **Gold** Medal of the Soviet Academy of Sciences (1984), and the Einstein Medal (1986).

MÜLLER, PAUL (1899-1965)

Swiss chemist

Paul Müller was an industrial chemist who discovered that **dichlorodiphenyltrichloroethane** (DDT) could be used as an insecticide. This was the first insecticide that could actually target insects; in small doses it was not toxic to humans and yet it was stable enough to remain effective over a period of months. When DDT was introduced in 1942, the effects it would have on the environment were not well understood. It was widely hailed, in particular for its ability to reduce the incidence of tropical diseases by reducing insect populations. For his work with DDT and the role his discovery played in the fight against diseases such as typhus and malaria, Müller was awarded the 1948 Nobel Prize in medicine or physiology.

Paul Hermann Müller was born in Olten, Switzerland, on January 12, 1899, to Gottlieb and Fanny Leypoldt Müller. His father was an official on the Swiss Federal Railway, and the family moved to Lenzburg and then to Basel, where Müller was educated until the age of seventeen. After finishing his secondary education, Müller worked for several years in a succession of jobs with local chemical companies. In 1919, he entered the University of Basel to study **chemistry**. He did his doctoral work under F. Fichter and H. Rupe, and his dissertation examined the chemical and electrochemical reactions of m-xylidine and some related compounds. Xylidines are used in the manufacture of dyes, and when Müller received his Ph.D. in 1925 he went to work in the dye division of the J. R. Geigy Corporation, a very large Swiss chemical company. Müller married Friedel Rügsegger in 1927; they had two sons and a daughter. Müller initially conducted research on the **natural products** that could be derived from green plants, and the compounds he synthesized were used as pigments and tanning agents for leather. In 1935, he was assigned to develop an insecticide. At that time the only available insecticides were either expensive natural products or synthetics ineffective against insects; the only compounds that were both effective and inexpensive were the **arsenic** compounds, which were just as poisonous to human beings and other mammals. Müller noticed that insects absorbed and processed chemicals much differently than the higher animals, and he postulated that for this reason there must be some material that was toxic to insects alone. After testing the biological effects of hundreds of different chemicals, in 1939 he discovered that the compound DDT met most of his design criteria. First synthesized in 1873 by German chemist Othmar Zeidler, who had not known of its insecticide potential, DDT could be sprayed as an **emulsion** with **water** or could be mixed with talcum or chalk powder and dusted on target areas. It was first used against the Colorado potato beetle in Switzerland in 1939; it was patented in 1940 and went on the market in 1942.

Müller had set out to find a specific compound that would be cheap, odorless, long-lasting, fast in killing insects, and safe for plants and animals. He almost managed it. DDT in short term application is so non-toxic to human beings that it can be applied directly on the skin without ill effect. It is cheap and easy to make, and it usually needs to be applied only once during a growing season, unlike biodegradable **pesticides** which must often be applied several times, in larger amounts and at much higher cost. Typhus and malaria are very severe, often fatal illnesses, which are carried by body lice and mosquitoes respectively; in the 1940s several potentially severe epidemics of these diseases were averted by dusting the area and the human population with DDT. The insecticide saved many lives during World War II and increased the effectiveness of Allied forces. Soldiers fighting in both the Mediterranean and the tropics were dusted with DDT to kill lice, and entire islands were sprayed by air before invasions.

Despite these successes, environmentalists were concerned from the time DDT was introduced about the dangers of its indiscriminate use. DDT was so effective that all the insects in a dusted area were killed, even beneficial ones, eradicating the food source from many birds and other small creatures. Müller and other scientists were actually aware of these concerns, and as early as 1945 they had attempted to find some way to reduce DDT's toxicity to beneficial insects, but they were unsuccessful. Müller also believed that insecticides must be biodegradable.

Hailed as a miracle compound, DDT came into wide use, and the impact on beneficial insects was not the only problem. Because it was such a stable compound, DDT built up in the environment; this was a particular problem once it began to be used for agricultural purposes and applied over wide areas year after year. Higher animals, unharmed by individual small doses, began to accumulate large amounts of DDT in their tissues (called bio-accumulation). This had serious effects and several bird species, most notably the bald eagle, were almost wiped out because frequent exposure to the chemical caused the shells of their eggs to be thin and fragile. Many insects also developed resistances to DDT, and so larger and larger amounts of the compound needed to be applied yearly, increasing the rate of bio-accumulation. The substance was eventually banned in many countries; in 1972 it was banned in the United States.

In addition to the 1948 Nobel Prize in physiology or medicine, Müller received an honorary doctorate from the University of Thessalonica in Greece in recognition of DDT's impact on the Mediterranean region. He retired from Geigy in 1961, continuing his research in a home laboratory. He died on October 13, 1965.

MULLIKEN, ROBERT S. (1896-1986)

American chemical physicist

Robert S. Mulliken's early interests and education were in the field of **chemistry**, especially in the structure of molecules. He eventually discovered that quantum mechanical theory offered an effective tool for the study of the topic and switched his allegiance to the field of physics, a subject that he taught for

Robert S. Mulliken.

more than three decades at various institutions. His work ultimately contributed to the establishment of a new interdisciplinary subject, combining these two great fields of science in what is known as chemical physics. Mulliken's research has led to a new understanding of the way in which atoms are held together in molecules and the ways in which molecules interact with each other. In recognition of these accomplishments, he was awarded the 1966 Nobel Prize in chemistry.

Robert Sanderson Mulliken was born in Newburyport, Massachusetts, on June 7, 1896. His father was Samuel Parsons Mulliken, a professor of **organic chemistry** at the Massachusetts Institute of Technology (MIT). His mother was Katherine W. Mulliken, which was her maiden name as well as her married name. Robert's father was the first Mulliken in many generations not to choose a seafaring career. In fact, his grandfather (Samuel Parson's father) had chosen to become a ship's captain rather than attend college, as his parents had wanted.

Mulliken's interest in chemistry developed early, largely as a result of his father's work. Through his father, Robert Mulliken made an important professional contact early in his life. Arthur A. Noyes, later to become one of the nation's and world's leading physical chemists, was also a resident of Newburyport and faculty member at MIT. While still in high school, the young Robert proofread galleys of his father's textbooks, books that were later to become standards in the field and make the elder Mulliken famous. Robert Mulliken wrote in ''Molecular Scientists and Molecular Science: Some Reminisces'' that the proofreading experience helped him to become ''well acquainted with the rather formidable names of organic compounds.''

In a 1975 interview with the *Journal of Chemical Education,* Mulliken praised the education he had at Newburyport High School. While there, he studied a full academic program, including Latin, French, German, physics, biology, and mathematics. The salutatory address he gave at his high school graduation made it clear that his scientific interests were fixed even at that early age. The topic of his address was ''The Electron: What It Is and What It Does.''

There seemed little doubt that Mulliken would follow his father to MIT and major in chemistry. In 1913, he was awarded a Wheelwright Scholarship, given to young men from Newburyport that allowed him to enroll at MIT. Once settled into MIT, Mulliken began to consider the possibility of concentrating in **chemical engineering** rather than chemistry. During one summer, he took part in a Chemical Engineering Practice School that required him to spend a week at each of a half dozen chemical plants in Massachusetts and Maine. He found the experience highly rewarding.

Mulliken eventually returned to chemistry, however, and graduated with a bachelor of science in that field in 1917, just as the United States was entering World War I. Instead of going on to graduate school, therefore, he took a job in war-related research at American University in Washington, DC. His job was to work on the development of poison **gases**, but he was not very good at the work. On one occasion, he spilled mustard gas on the floor, much to the horror of his superior, James B. Conant. He later came down with a severe case of the flu and was still recovering in the hospital when the war ended. After the war, Mulliken briefly held a job at the New Jersey **Zinc** Company, where he studied the effects of zinc **oxide** and **carbon** black on the compounding of rubber. He apparently realized rather quickly that this was not the kind of work he wanted to do, and he enrolled in a doctoral program in chemistry at the University of Chicago.

He chose Chicago because he wanted especially to work with Professor W. D. Harkins on the study of atomic nuclei. The work to which he was assigned involved finding ways to separate the isotopes of **mercury** from each other. Mulliken experienced extraordinary success in this line of research and, in 1921, he received his Ph.D. for this work. He went on to improve the system by which mercury isotopes can be separated from each other and later called the construction of this ''**isotope** factory'' one of his proudest accomplishments.

One reason for his sense of pride was that it confirmed that he had finally become a ''proper experimentalist.'' In his earlier years, he had not been very proficient in the laboratory (witness the mustard gas event), leading Noyes to express some doubt that he could become a successful researcher. Although Mulliken's practical skills obviously had improved at Chicago, it was ultimately as a theorist, and not an experimentalist, that he was to gain his greatest fame.

In 1923, Mulliken applied to have his National Research Council Fellowship extended for two more years. He was told, however, that he would have to go to a different institution and study a different field of chemistry. His request to work with British physicist **Ernest Rutherford** on **radioactivity** was denied by the board, so he chose instead to go to Harvard. At Harvard, he became interested in studying the spectral lines of **diatomic** molecules, a subject he began with the compound **boron** nitride (BN). His hypothesis was that when analyzed scientists would find isotopes (such as boron–10 and boron–11) present in the **molecule**. When he actually carried out this research, his hypothesis was confirmed. Faint bands reflecting the presence of less abundant isotopes—bands that no one had noticed to that point—were apparent.

A turning point in Mulliken's career was marked by two visits to Europe in 1925 and 1927. During these visits, he made an effort to visit all of the major researchers who were working on molecular spectra. But, inevitably, he also came into contact with physicists who were developing the quantum theory.

The 1920s were a decade of revolution in physics as researchers were developing, elaborating, and refining the new view of physics arising out of the work of **Max Planck**, Louis Victor Broglie, **Albert Einstein**, **Erwin Schrödinger**, and others. A very few chemists—**Linus Pauling**, perhaps most obvious—were also looking for ways in which quantum theory could be used to explain chemical phenomena.

Mulliken had already been introduced to quantum theory at MIT and Chicago. But it was his trips to Europe that really focused his attention on the problem that had now become foremost in his mind: how quantum theory could be used to explain the structure of molecules. He explained later in "Molecular Scientists and Molecular Science: Some Reminisces" that his earlier studies of spectra had "led naturally to attempts also to understand molecular electronic states as more or less like those of atoms."

Perhaps the most important single event on his visits to Europe was his meeting with the German chemist, Friedrich Hund. Hund had been working on many of the same problems as had Mulliken with one important exception. While Mulliken was using a deterministic form of quantum mechanics, stressing a more narrow range of likely outcomes for experiments in the field, Hund had employed the uncertainty principle developed by **Werner Karl Heisenberg** in his study of molecular electronic structure. The uncertainty principle states that it is impossible to specify precisely both the position and momentum of a particle at the same time. Applying such a concept to the study of quantum mechanics opened up possibilities for Hund's work. Mulliken began to adopt Hund's approach and soon achieved success. In 1928, he published a classic **paper** that was crucial in his being awarded the Nobel Prize in 1966. In that paper, Mulliken proposed an entirely new model for **molecular structure**. Previously, chemists had assumed that when two or more atoms combine to form a molecule, the atoms tend to retain their independent characteristics. The molecule was thought to be an assemblage of essentially independent, recognizable atoms "tied together" in some fashion.

Mulliken argued that individual atoms lose their identify when they combine to form a molecule. Electrons that once "belonged" to one or another **atom** in the molecule now became part of an overall molecular structure. The **bonding** electrons involved in the molecular structure now had **energy** levels whose quantum mechanical descriptions were determined by *molecular* characteristics, not *atomic* characteristics.

In succeeding decades, Mulliken derived a number of specific applications from this general theory. For example, he found ways to measure the relative ionic character of a chemical bond and the properties of conjugated double bonds using this model.

Mulliken's first academic appointment was as assistant professor of physics at the Washington Square College of New York University in 1926. The appointment was significant in that Mulliken, trained as a chemist, was now recognized as a physicist also. He held the New York position until 1928 when he was invited to join the faculty at the University of Chicago. Again, this appointment was in the department of physics, as associate professor. In 1931, he was promoted to full professor.

Mulliken continued to hold an appointment at Chicago for the rest of his academic career, a period of more than five decades. From 1956 to 1961, he was Ernest de Witt Burton Distinguished Service Professor of Physics and Chemistry and, at the end of that period, he was given a joint appointment as professor of physics and chemistry. Beginning in 1965, Mulliken also spent part of each year at Florida State University in Tallahassee, where he was a distinguished research professor of chemical physics at the Institute of Molecular Biophysics.

Mulliken was married on December 24, 1929, to the former Mary Helen von Noé, daughter of a colleague in the geology department at the University of Chicago. The Mullikens had two daughters, Lucia Maria and Valerie Noé. Mrs. Mulliken accompanied her husband on many of his overseas trips, one of which was particularly noteworthy. During their 1932 trip to Europe, Mrs. Mulliken came down with an infection of the appendix, a condition that grew increasingly more severe as they moved from city to city on the continent. Eventually an operation in Berlin solved the problem. Their stay in Berlin was further troubled, however, by the growing aggressiveness of Nazi supporters. In "Molecular Scientists and Molecular Science: Some Reminisces" Mulliken described how upset his German colleagues had become about "coming events" and how he himself could "see and hear the Nazi storm troopers marching up and down in the street" below his hotel room early every morning.

During World War II, Mulliken served as director of editorial work and information at the University of Chicago's **Plutonium** Project, a division of the **Manhattan Project**, set up to develop the atomic bomb. He also served briefly as Scientific Attaché at the American Embassy in London in 1955. As Mulliken approached formal retirement age in 1961, he was

Kary Mullis. *(Archive Photo. Reproduced by permission.)*

offered a number of prestigious teaching and lecturing assignments. He was the Baker Lecturer at Cornell University in 1960 and the Silliman Lecturer at Yale in 1965. In the latter year, he also served as John van Geuns Visiting Professor at the University of Amsterdam.

Mulliken continued working on problems of molecular structure, molecular spectra, and molecular interactions throughout his life. In 1952, for example, he applied quantum mechanical theory to an analysis of the interaction between Lewis acid and base molecules. In the 1960s, his studies of molecular structure and spectra ranged from the simplest of cases (hydrogen and diatomic **helium**, for example) to complex molecular aggregates. During this decade, Mulliken also received most of the major awards available to chemists in the United States, including the **Gilbert Newton Lewis**, **Theodore William Richards**, John G. Kirkwood, and Willard Gibbs Medals and the **Peter Debye** Award. Mulliken died in Arlington, Virginia, on October 31, 1986.

Mullis, Kary (1944-)

American biochemist

Kary Mullis is a biochemist who designed polymerase **chain reaction** (PCR), a fast and effective technique for reproducing specific genes or **DNA** fragments that is able to create billions of copies in a few hours. Mullis invented the technique in 1983 while working for Cetus, a California biotechnology firm. After convincing his colleagues of the importance of his idea, they eventually joined him in creating a method to apply it. They developed a machine which automated the process, controlling the chain reaction by varying the **temperature**. Widely available because it is now relatively inexpensive, PCR has revolutionized not only the biotechnology industry, but many other scientific fields, and it has important applications in law enforcement, as well as history. Mullis shared the 1993 Nobel Prize in **chemistry** with **Michael Smith** of the University of British Columbia, who also developed a method for manipulating genetic material.

Kary Banks Mullis was born in Lenoir, North Carolina, on December 28, 1944, the son of Cecil Banks Mullis and Bernice Alberta (Barker) Fredericks. He grew up in Columbia, North Carolina, a small city in the foothills of the Blue Ridge Mountains, where his temperament and his curiosity about the world set him apart from others. As a high school student, for example, Mullis designed a rocket that carried a frog some 7,000 feet in the air before splitting open and allowing the live cargo to parachute safely back to earth. Even at a young age, Mullis was considered a maverick and nonconformist. He entered Georgia Institute of Technology in 1962 and studied chemistry. As an undergraduate, he created a laboratory for manufacturing poisons and **explosives**. He also invented an electronic device stimulated by brain waves that could control a light switch.

Upon graduation from Georgia Tech in 1966 with a B.S. degree in chemistry, Mullis entered the doctoral program in **biochemistry** at the University of California, Berkeley. In Berkeley at that time there was growing interest in hallucinogenic drugs; Mullis taught a controversial **neurochemistry** class on the subject. His thesis adviser, Joe Nielands, told *Omni* that as a graduate student Mullis was ''very undisciplined and unruly; a free spirit.'' Yet at the age of twenty-four, he wrote a paper on the structure of the universe that was published by *Nature* magazine. He was awarded his Ph.D. in 1973, and he accepted a teaching position at the University of Kansas Medical School in Kansas City, where he stayed for four years. In 1977, he assumed a postdoctoral fellowship at the University of California, San Francisco. After two years there, discouraged by universities and uncertain what to do with his life, he left and took a job in a restaurant. One day his graduate advisor encountered him there and convinced him that he was wasting both his mind and his education waiting tables. In 1979, he accepted a position as a research scientist with a growing biotech firm, Cetus Corporation, in Emeryville, California, which was in the business of synthesizing chemicals used by other scientists in genetic cloning.

At Cetus, Mullis was bored by the routine demands of corporate life. He spent much of his time sunbathing on the roof and writing computer programs that would automatically respond to certain kinds of administrative requests. ''I'd no real responsibilities for about two years,'' he told *Omni* magazine. ''I was playing,'' he told *Parade Magazine.* ''I think really good science doesn't come from hard work. The striking advances come from people on the fringes, being playful.'' He conceived of polymerase chain reaction (PCR) while driving out to his ranch in Mendocino county and thinking, as he describes it, somewhat ''randomly'' about ways to look at individual sections of the genetic code.

Reproducing deoxyribonucleic acid or DNA had long been an obstacle to anyone working in molecular biology. The most effective way to reproduce DNA was by cloning, but however much of a scientific advance this process represented, it was still cumbersome in certain respects. DNA strands are long and complicated, composed of many different chromosomes; the problem was that most **genetic engineering** projects were tasks that involved tiny fragments of the DNA **molecule**, almost infinitesimal sections of a single strand. Cloning works by inserting the DNA into bacteria and waiting while the reproducing bacteria creates copies of it. The cloning process is not only time-consuming, it replicates the whole strand, increasing the complexity. The revolutionary advantage of PCR is its selectivity: It is a process that reproduces specific genes on the DNA strand millions or billions of times, effectively allowing scientists to amplify or enlarge parts of the DNA molecule for further study.

Mullis remembers that it took a long time to convince his colleagues at Cetus of the importance of this discovery. ''No one could see any reason why it wouldn't work,'' he told *Popular Science.* ''But no one seemed particularly enthusiastic about it either.'' Once they had become convinced of its importance, however, PCR became the focus of intensive research at Cetus. Scientists there developed a commercial version of the process and a machine called the Thermal Cycler; with the addition of the chemical building blocks of DNA, called nucleotides, and a biochemical catalyst called polymerase, the machine would perform the process automatically on a target piece of DNA. The machine is so economical that even a small laboratory can afford it, and the technique, as one microbiologist told *Time,* ''can reproduce genetic material even more efficiently than nature.''

The selectivity of the PCR process, as well as the fact that it is simple and economical, have profoundly changed the course of research in many fields. In an interview with *Omni,* Mullis remarked of PCR: ''It's so widely used by molecular biologists that its future direction is the future of molecular biology itself.'' In the field of genetics, the process has been particularly important to the Human Genome Project—the massive effort to map human DNA. Nucleotide sequences that have already been mapped can now be filed in a computer, and PCR enables scientists to use these codes to rebuild the sequences, reproducing them in a Thermal Cycler. The ability of this process to reproduce specific genes, thus effectively enlarging them for easier study, has made it possible for virologists to develop extremely sensitive tests for acquired immunodeficiency syndrome (AIDS), capable of detecting the virus at early stages of infection. There are many other medical applications for PCR, and it has been particularly useful for diagnosing genetic predispositions to diseases such as sickle cell anemia and cystic fibrosis.

PCR has also revolutionized evolutionary biology, making it possible to examine the DNA of woolly mammoths and the remains of ancient humans found in bogs. PCR can also answer questions about more recent history; it has been used to identify the bones of Czar Nicholas II of Russia who was executed during the Bolshevik revolution, and scientists at the National Museum of Health and Medicine in Washington, DC, are preparing to use PCR to amplify DNA from the hair of Abraham Lincoln, as well blood stains and bone fragments, in an effort to determine whether he suffered from a disease called Marfan's syndrome. In law enforcement, PCR has made genetic ''fingerprinting'' more accurate and effective; it has been used to identify murder victims, and to overturn the sentences of men wrongly convicted of rape. Some have suggested the PCR can be used to create tags or markers for industrial and biotechnological products, including oil and other hazardous chemicals, to insure that they are used and disposed of in a safe manner.

Cetus awarded Mullis only ten-thousand dollars for developing the PCR patent. Frustrated both by the size of this award and the restrictions the company continued to place on his scientific research, Mullis left Cetus in 1986. He became director of molecular biology at Xytronyx, a San Diego research firm, but two years later, again frustrated with the routine of corporate research, he left to become a private biochemical research consultant. The Du Pont Corporation challenged Cetus for Mullis's patent for PCR, filing suit in the late 1980s; they argued that while working for them in the early 1970s, Nobel laureate **Har Gobind Khorana** had written a paper which anticipated the process. Although he had already left Cetus, Mullis agreed to testify on their behalf, and in February 1991 a federal jury decided against Du Pont. That same year, Cetus sold the process to Hoffman-LaRouche, Inc., for 300 million dollars, the most money ever paid for a patent. When he invents something again, Mullis told *Omni,* ''I'm not going to hand it over to some company like Cetus without something saying it's mine. If anyone makes $300 million off it, I'm going to be part of that.''

In 1990, Mullis received both the Preis Biochemische Analytik Award from the German Society of Clinical Chemistry and the Allen Award from the American Society of Human Genetics; in 1991, he received the Gairdner Foundation Award and the National Biotech Award. In 1993, he was presented with the Japan Prize, in addition to the Nobel Prize. He is a member of the American Chemical Society. He has been married and divorced three times and is the father of three children. He works and lives in an apartment overlooking the Pacific Ocean in La Jolla, California.

Mutagen

Mutagens are chemicals or physical factors (such as radiation) that increase the rate of **mutation** in the cells of bacteria, plants, and animals (including humans). Most mutagens are of natural origin and are not just a modern phenomenon: even dinosaurs were susceptible to mutagens. Mutagens can be found in the food we eat, the air we breath, or the ground we walk on. Very small doses of a mutagen usually have little effect while large doses of a mutagen could be lethal. **DNA** in the nuclei of all cells encodes **proteins** that play important structural and functional (metabolic) roles in the cell. Mutagens typically disrupt the DNA of cells, causing changes in the proteins that the cell produces which can lead to abnormally fast growth (cancer), or even cell death. In rarer incidences mutagens can even cause protein changes that are beneficial to the cell.

The first mutagens to be identified were **carcinogens**, or cancer-causing substances. Early physicians detected tumors in patients more than 2,000 years before the discovery of chromosomes and DNA. In 500 B.C., the Greek Hippocrates named crab-shaped tumors cancer (meaning crab).

In England in 1775, Dr. Percivall Pott wrote a paper on the high incidence of scrotal cancer in chimney sweeps who were typically boys small enough to fit inside chimneys and clean out the soot. Pott suggested that chimney soot contained carcinogens that could cause the growth of the warts seen in scrotal cancer. Over a 150 years later, chimney soot was found to contain hydrocarbons capable of mutating DNA.

In France in the 1890s, Bordeaux wine workers showed an unusually high incidence of skin cancer on the back of the neck. These workers spend their days bending over in the fields picking grapes, so exposing the back of their necks to the sun. The ultraviolet (uv) radiation in natural sunlight was later identified as a mutagen.

Mutagens can be found in foods, beverages, and drugs. Sometimes a substance is mutagenic because it is converted in the body into something harmful. Regulatory agencies are responsible for testing food and drugs to insure that the public is not unknowingly exposed to mutagens. However, some mutagen-containing substances are not tightly controlled. One such substance is found in the tobacco of cigars and cigarettes.

Some mutagens occur naturally, and some are synthetic. Cosmic rays from space are natural, but they are mutagenic. Some naturally occurring viruses are considered mutagenic since they can insert themselves into host DNA. **Hydrogen** and atomic bombs are man-made, and they emit harmful radiation. Radiation from nuclear bombs and gaseous particles from **nitrogen** mustard and acridine orange have been used destructively in war. On the other hand, some mutagens are used constructively to kill bacteria that could grow in human foods, such as the small doses of nitrites used to preserve some meat. Even though nitrites can be mutagenic, without the nitrites these meats could cause botulism; hence, the benefit outweighs the risk involved.

Mutagens affect DNA in different ways. Some mutagens such as nitrogen mustard bind to a base and cause it to make a different amino acid. These mutagens cause point mutations, because they change the genetic code at one point, so changing a protein's amino acid sequence.

Mutagens such as acridine orange work by deleting or inserting one or more bases into the DNA **molecule**, so shifting the frame of the triplet code for an amino acid. Deletion and insertion mutations causing ''frame-shift'' mutations can change a long string of **amino acids**, which can severely alter the structure and function of a protein product.

Normal cells recognize cues from their environment and respond with specific reactions, but cells impaired by a mutation do not behave or appear normal, and are said to be transformed.

Mutations in somatic (body) cells are not transferred to offspring. Mutated DNA can only be passed to the next generation if it is present in a germ cell such as spermatozoa and ova (eggs), each of which contribute half of the DNA of the new organism.

Chemical mutagens are classified as alkylating agents, cross-linking agents, and polycyclic **aromatic hydrocarbons** (PAHs). Alkylating agents act by adding molecular components to DNA bases, so altering the protein product. Cross-linking agents create covalent bonds with DNA bases, while PAHs are metabolized by the human body into other potentially mutagenic molecules.

Radiation is a potent mutagen, classified as UV or ionizing radiation. UV radiation causes covalent bonds to form between neighboring thymine bases in the DNA, so altering the DNA product at that location. Ionizing radiation includes the x rays, gamma rays, and **subatomic particles**. Ionizing radiation alters the way two strands of DNA interact. This high **energy** radiation passes through cells and tissues, cutting up any DNA in its path. Ionizing radiation can rearrange entire sections of the chromosomes, altering relatively long stretches of DNA.

Mutagens are often associated with specific cancers in humans. Aromatic amines are mutagens that can cause bladder cancer. Tobacco taken in the form of snuff contains mutagens that can cause nose tumors. Tobacco smoke contains mutagens such as PAHs and nitrosamine (a type of alkylating agent), as well as toxins such as **carbon** monoxide, cyanide, **ammonia**, **arsenic**, and radioactive **polonium**. Although tobacco products are legally and widely available, many physicians and government agencies warn about the health risks linking smoking with several types of cancer and heart disease.

In 1973, Bruce Ames introduced the most widely-used test to identify potential mutagens. Suspected mutagens are mixed with a defective strain of the bacteria salmonella, which only grows if it is mutated. Substances which allow the salmonella to grow are considered mutagenic.

In addition to mutagen-induced DNA changes, spontaneous mutations occur in the dividing cells of the human body every day. The nuclei of the cells have repair enzymes that remove mutations and restore mutated DNA to its original form.

If these natural DNA repair mechanisms fail to keep up with the rate of mutation or the repair mechanisms themselves are defective, then disease can result. This latter case is seen in lung cancer due to cigarette smoking, where the **nicotine** in the smoke is thought to block an important repair process in the lungs.

MUTATION

In general, deoxyribonucleic acid, or **DNA**, is a very stable **molecule**. As it conveys genetic information from one generation to the next, errors rarely occur; however, they are not totally absent. Alterations in the DNA are called mutations. Mutations are not necessarily harmful to an organism. Some mutations go completely unnoticed, while others are the driving force behind evolution. However, some mutations are harmful; the root cause of many cancers, for example, is genetic mutation.

Mutations are divided into two categories: germline mutations and somatic mutations. Germline mutations can be passed from one generation to the next. The mutation occurs in the DNA of an egg or sperm cell. Any offspring resulting from the egg or sperm will inherit the mutation from the affected parent. Somatic mutations are mutations that occur in any cells other than egg or sperm cells. These mutations cannot be passed on to the following generation.

To understand mutations and how they can affect an organism, it is important to understand how DNA functions as genetic material. DNA is a double-stranded, spiral-shaped molecule. Each strand is formed by a chain of nucleotide molecules. The nucleotides are composed of deoxyribose (a type of sugar), a **phosphate** group, and one of four nitrogen-containing bases—adenine, thymine, cytosine, or guanine. The deoxyribose and phosphate groups form the backbone of DNA strands, and the bases face one another like the teeth of a zipper. The two strands of the double helix are joined by pairing between nucleotide bases occupying the same position on opposite strands. Adenine always pairs with thymine, and vice versa; cytosine always pairs with guanine, and vice versa.

One strand of DNA serves as the coding strand; the other strand is referred to as the matching or complementary strand. This system helps ensure that errors are kept to a minimum. The coding strand is read in three-nucleotide segments known as triplets. There are 61 triplet combinations, each of which codes for a specific amino acid. (Amino acids are the building blocks of **proteins**.) Since there are only 20 **amino acids**, most have more than one associated triplet code.

Mutations affect how the genetic code is read. In most cases, mutations hinge on the substitution, insertion, or deletion of one or more nucleotides. The substitution of one nucleotide for another may or may not have an effect on an organism. Because each amino acid has more than one triplet code, a nucleotide substitution may simply change one triplet to another that codes for the same amino acid. This type of mutation is a silent mutation. If the substitution results in a triplet that codes for a different amino acid, and the resultant protein is not noticeably different, then it is called a neutral mutation. Substitutions that cause noticeable effects are missense and nonsense mutation. Unlike a neutral mutation, in which a protein functions normally, a missense mutation produces an abnormal protein that functions incorrectly. Nonsense mutations prevent the entire gene from being read and yield only partial instructions for building a protein.

The insertion or deletion of one or more nucleotides can result in a more serious type of mutation—a frameshift mutation. Because the genetic code is read in three-nucleotide segments, an insertion or deletion of one or more nucleotides can throw off the reading frame. From the point of the mutation onward, the code is skewed and a very different, potentially very useless, protein results.

Mutations are either spontaneous or induced. Spontaneous mutations occur due to internal cellular mistakes. For example, during DNA replication, there is a one in a billion chance that the incorrect nucleotide will be added to a DNA strand under construction. A DNA proofreading mechanism usually, but not always, corrects these errors. Induced mutations are caused by environmental factors, specifically chemicals or radiation. The factors can physically damage the DNA, setting the stage for future mistakes in reading or copying the genetic code.

Biological repair mechanisms are the body's defense against potentially harmful mutations. The significance of DNA being a double-stranded molecule is very obvious during repair efforts. Because damage usually occurs to only one strand, the second strand can act as a template to direct reconstruction. First, the damaged or incorrect section is snipped out of the DNA. Next, polymerases, the enzymes that are responsible for constructing DNA, help fill in the gap. Finally, the patch is sealed into place by another set of enzymes, called ligases.

N

NAKANISHI, KOJI (1925-)

Japanese-born American chemist

Koji Nakanishi believes in conventional values, like devotion to family and respect for authority. Yet, his career and life have been anything but conventional. In his autobiography, *A Wandering Natural Products Chemist,* Nakanishi describes himself as a ''hybrid'' who grew up in Europe, Egypt, and Japan and whose philosophy and behavior in life have been influenced by both Eastern and Western thought. A dedicated, world-renowned organic chemist with wide-ranging interests, Nakanishi is also an amateur magician who delights in mystifying his audience and likes ''drinking in quiet bars with friends.'' Although he has spent the past 30 years primarily in the United States conducting research and teaching, Nakanishi has always maintained strong ties with his homeland.

In his autobiography, Nakanishi says that his wife accuses him of having too many professional interests instead of specializing. He also modestly asserts that he does not consider himself ''brilliant.'' However, with the dedicated help of his students and colleagues, Nakanishi has made important contributions in elucidating the structures of natural bioactive compounds and their modes of action. For example, his work in the isolation and structural studies of visual pigments has led to new insights into the mechanism of vision. As a result, he has gained international respect and numerous honors, including the Imperial Prize and the Japan Academy Prize, which represent the highest honors for a Japanese scholar.

Nakanishi was born in the hills of Hong Kong on May 11, 1925, the eldest of Yuzo and Yoshiko Nakanishi's four sons. In Japanese, the characters for KO in Nakanishi's first name are identical to Hong, which means perfume. (Hong Kong means perfume harbor.) Nakanishi's father, Yuzo, worked in international banking, and the family moved to Lyon, France, soon after Nakanishi's birth. His father was then transferred to England, where Nakanishi learned English at a young age and ''proper English manners.'' The family eventually relocated to Alexandria, Egypt, and then to Japan in 1935.

During the next 10 years, Nakanishi was to experience one of the most tumultuous times in Japanese history. A fanatic militarism was on the rise in Japan, culminating in World War II and ultimate defeat at the hands of the Allies. While he was interested in **chemistry** and biology, Nakanishi chose chemistry as a major without any specific plans or ambitions. After high school, he applied to Tokyo Imperial University and was the only applicant from his school not to be accepted. ''The failure was devastating to both my mother and myself,'' he writes. Because of the country's military build-up, Nakanishi and his family knew he would be drafted unless he went to college.

Nakanishi eventually was accepted by Nagoya Imperial University, located between Tokyo and Kyoto. He enjoyed the university's pastoral setting and the young and bright staff. It was also here that he met his future wife, Yasuko, who was a laboratory assistant. He eventually moved in with his future wife's family after his apartment was bombed during an air raid. In his autobiography, Nakanishi recounts the numbing effects of war, describing his engagement to Yasuko as ''strangely emotionless because I did not think we would live to be married.''

Following the war, Nakanishi and his colleagues quickly emersed themselves in chemistry, trying to catch up on lost time. In 1946, he joined the research group of Fujio Egami, who was carrying on Japan's tradition of excellence in **natural products** chemistry. After completing his graduate work in structural studies of the red crystalline antibiotic actinomycin, Nakanishi went on to become a Garioa Fellow at Harvard. Unfortunately, because of his financial situation, he was forced to leave his wife and newly born daughter Keiko in Japan for two years. (Nakanishi also has a son, Jun.) He returned to Nagoya University in 1952 as an assistant professor of chemistry. Shortly afterward he was diagnosed with tuberculosis and required to rest for several months. Nakanishi used this time to complete his translation of *Organic Chemistry* (the original work was coauthored by his mentor at Harvard, Louis Fieser) in a three-volume set. The book, which became a best-selling

Koji Nakanishi. ***(Camera One. Reproduced by permission.)***

chemistry book well into the 1960s, brought financial security to Nakanishi and his family.

In 1969, after stints as a professor of chemistry at Tokyo Kyoiku and Tohoku universities in Japan, Nakanishi accepted a professorship at Columbia University in New York. ''I had no pressing reason to leave Tohoku University except that, subconsciously, I was interested in trying out life abroad,'' writes Nakanishi. This interest would drastically change the Nakanishis' life. Although his plan was to conduct advanced research in chemistry for 10 years in the United States and then return to Japan to upgrade his home country's efforts in the field, Nakanishi has spent 30 years at Columbia. Yet, he still found time to direct the Suntory Institute for Bioorganic Research (SUNBOR) in Japan and fulfilled his dream of improving Japanese science by starting the first true international postdoctoral system in Japan in 1980.

Over his career, Nakanishi has been noted for his multidisciplinary and international collaborations. He has directed research that has characterized more than 180 natural products, including antimutagens from plants and metal-sequestering compounds from sea squirt blood. He has also worked with visual pigments and wasp toxins, an effort that has led Nakanishi and colleagues to synthesize a series of compounds structurally similar to the venom of a type of Egyptian wasp. These synthetic compounds are 33 times more powerful than the natural venom and will help in obtaining pure samples of glutamic acid **receptors** on the surface of nerve cells for further study.

Like wasp venom, many of the compounds Nakanishi has worked with can be found only in minuscule amounts. As a result, much of his research has focused on methods stressing isolation and purification of such compounds and on new approaches to structure elucidation. Beginning in the late 1980s, his worked has centered on investigations into the interaction of bioactive molecules with receptor molecules.

A man of many interests professionally, Nakanishi also enjoyed building miniature railroads for many years and collects paintings and sculptures of bulls and cows. ''Cows are pastoral, never appear to be rushing about, and give one a peaceful feeling; I am the opposite,'' he writes. But his favorite hobby has been magic, which he took up as a way to entertain people at graduations, weddings, and parties. When he received the prestigious Imperial Prize, Nakanishi performed magic rope tricks for Japan's Crown Prince during the reception and dinner.

As for the future of chemistry, Nakanishi is optimistic. ''Medicines and pharmacy are built completely around organic compounds,'' he writes in his autobiography. ''In this interdisciplinary era, if we want to solve the mode of action of bioactive compounds for the purpose of uncovering the mysteries of life and to develop more active compounds, chemistry simply has to play a central role.''

NANOCHEMISTRY

In nanochemistry, the rules of **bonding** are extended from atoms and molecules to larger objects having the dimension of from 1-100 nanometers, where one nanometer is one-billionth of a meter. In this regime, new chemical and physical properties appear that depend on the size and shape of the particles. Nanochemistry undertakes the synthesis of precisely defined nanoparticles to achieve novel materials for specific applications in such fields as advanced catalysis, ultrathin films and membranes (separation and adsorption), storage devices, and optical and electronic devices. The nanometer regime is of particular interest because it is this region where phenomena associated with atomic and molecular interactions begin to be strongly influenced by the macroscopic properties of materials.

In mechanochemistry, individual chemical reactions at the molecular or atomic level are achieved by pressing the reactants together to overcome the normal reaction barriers. This requires positioning of the reactants on a nanometer scale to precisely bring the reactants together.

The fabrication of thin metallic wires, or nanowires, is a critical technology for communications technology. Methods that produce wires 20 times thinner, longer, and better-defined than conventional ones have been based on nanotechnology.

In gas phase condensation, the vapor evolved from an appropriate precursor undergoes a reaction that produces nanometer-sized particles. **Indium tin oxide** particles made by this method are conductive, and can be processed to form transparent thin films that are antistatic. Other particles containing yttria and some lanthanides are phosphors that be processed to form light-emitting devices.

Nanostructured chip devices prepared from nanoscale materials (including complex oxides, carbides, borides and nitrides) have been successfully mass produced in kilogram quantities.

In the field of biology, biochemists have been studying vesicles, the small membrane sacs found within cells. Vesicles transport important molecules from cell to cell to keep cells functioning properly. Because vesicles are too small (from tens of nanometers to a few micrometers) to study by traditional chemical analysis, nanochemistry techniques have been employed to study these structures.

Surface scientists have been able to exploit atomic self-organization processes for the controlled fabrication of nanostructures at surfaces having well defined size and shape. This approach is based on a detailed understanding of the microscopic pathways of **diffusion**, nucleation and aggregation, processes governed by activation barriers for migrating atoms. The knowledge of the energetic hierarchy of the participating atomic diffusion events allows one to synthesize nanostructures of a desired size, dimension, and shape in large quantities by exploiting the laws of nature.

In nanocomposites, one of the constituent phases has dimensions—length, width, or thickness, in the nanometer range. Because the building blocks of the nanocomposite have the dimensions of nanometers, they have enormous surface areas and large interfaces with other phases. The properties of the nanocomposite result from interactions at these interfaces. In conventional **composites** where the building blocks have dimensions on the order of microns (one-millionth of a meter), there is less interfacial contact between phases, and consequently less effect by the interfaces on properties of the composite.

One naturally-occurring nanocomposite is bone. Bone consists to a large extent of nanoscale, platelike crystals of hydroxyapatite, $Ca_{10}(PO_4)_6(O\ H)_2$ dispersed in a matrix of collagen fibers. Conventional attempts to synthesize hydroxyapatite for artificial bone or bone implant materials have been fraught with difficulties, but scientists have found that these problems can be largely overcome if hydroxapatite is synthesized as a nanostructured material.

In the field of catalysis, chemists have been searching for catalysts that are stable at high temperatures. The complex oxide **barium** hexaaluminate (BHA), or $BaO{\cdot}_6Al_2\ O_3$, has been of interest it retains its catalytic activity at high temperatures. But conventional methods to synthesize BHA tend to reduce the material's surface area, and hence the activity. But when nanoparticles of BHA are prepared, a final grain size of 30 nanometers is achieved, and a large surface area and catalytic activity are achieved.

In organic/inorganic nanocomposites, a nanoscale inorganic filler is typically dispersed in an organic polymer matrix. The filler carries the load, and its large surface area in contact with the matrix reduces the mobility of the polymer **chains**. **Nylon** nanocomposites containing small amounts of clay that are capable of withstanding high **temperature** environments have been fashioned into automobile air intake covers.

Chemists have also been looking at **carbon** nanotubes (honeycomb **graphite** lattices rolled into cylinders having nanoscale thicknesses) as fillers in nanocomposites. The nanotubes, by virtual of their **electrical conductivity**, make the composite conductive as well. Nanotubes have been found by experiment to be stiffer than carbon fibers (which are thicker than 1 micron), but less brittle. Thus, nanocomposites containing carbon nanotubes are able to withstand much greater deformations before breaking than carbon fiber composites.

Nanochemistry, which is actually a branch of solid-state materials **chemistry**, is an emerging discipline with a promising future, and, as a consequence, many of the techniques employed are still evolving.

NAPHTHALENE

Naphthalene is a white, crystalline solid at room **temperature** (melting point, 176°F/80°C) with a strong coal-tar odor. Soluble in **benzene**, absolute **alcohol**, and **ether**, it is derived either from boiling coal-tar oils with subsequent crystallization and **distillation**, or from petroleum fractions following various catalytic processing operations. Naphthalene is toxic when inhaled. It is used as a moth repellent, insecticide, **fungicide**, **lubricant**, preservative, and antiseptic. Naphthalene has the chemical formula is $C_{10}H_8$. Structurally, the **molecule** consists of two benzene rings fused in such a way that they share two **carbon** atoms. (Figure 1).

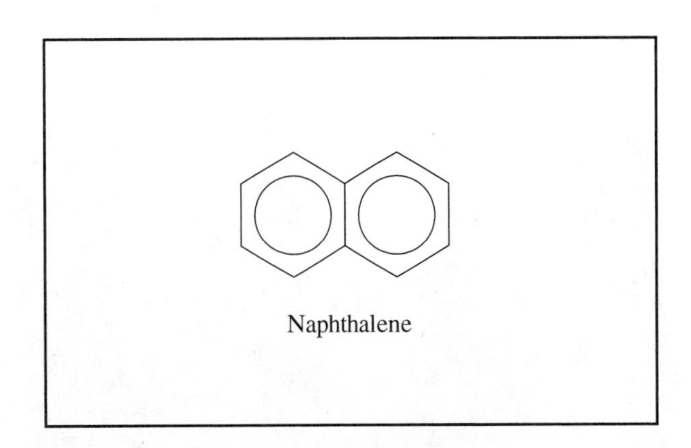

Figure 1. ***(Illustration by Electronic Illustrators Group.)***

Auguste Laurent (1807-1853), working with Jean-Baptiste Dumas (1800-1884) in France, succeeded in preparing pure naphthalene from **coal** tar and in studying its halogen derivatives in 1831. Out of Laurent's studies of naphthalene arose the modern notion that all compounds can be thought of as being derived from hydrocarbons by substitutions, with the number of carbon atoms the sole basis for classification into a series.

See also Petrochemicals

NARCOTICS

One of the greatest challenges facing science today is understanding the physical basis of thought. The brain is a physical entity filled with cells that interact chemically via biochemical

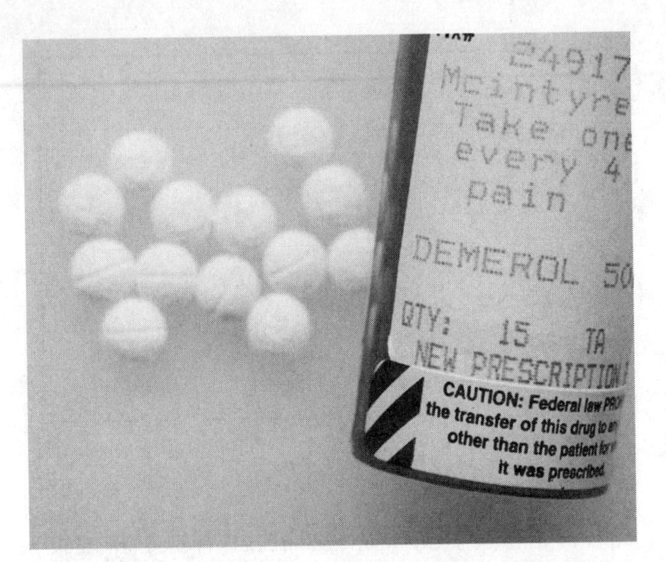

Demerol tablets, a pain-killing narcotic. ***(Photograph by Will and Deni McIntyre, Photo Researchers, Inc. Reproduced by permission.)***

pathways. Research during the past fifty years has made considerable progress in understanding the scientific basis for the brain activity—understanding the **biochemistry** of the brain. But why certain compounds produce feelings, thoughts, and memories is still an area of exploration and research.

Narcotics are a class of compounds that affect the physical brain but also alter the mind. Specifically, they are compounds that produce narcosis or general anesthesia and relief from pain or analgesia. They alter the minds perception of pain through the actions of the brain but they also induce a sense of euphoria and well-being that has not yet been tied to specific biochemical reactions. That is, the brain is the physical entity that is composed of the neurons in our skulls. The actions of narcotics on the brain are reasonably well understood with respect to their anesthetic and analgesic properties. The mind, on the other hand, is our self —our thoughts and feelings and such—and the effects of narcotics on the mind are less clear, particularly with respect to the sense of euphoria and tranquility that the compounds induce.

Many different pharmaceutical agents are narcotics. However, in legal terms in the United States, only those compounds that are also addictive are defined as narcotics. Their use is carefully regulated or prescribed and a significant amount of **energy** and money has been spent dealing with illegal narcotics. The use of legal narcotics is regulated through the medical community.

The resinous juices from the seed casing of the opium poppy (*Papaver somniferum*) are the source of naturally occurring narcotics. This material is a complex mixture of about 20 different alkaloids or **nitrogen**-containing organic bases. The principal **alkaloid**, **morphine**, on makes up about 10% of this mixture. Other compounds, such as codeine, can be found within this collection of alkaloids.

Historically, opium has been one of the most important medicines because of its **narcotic** effects. The Greek physician Galen (c.130-c.200), administered opium juice as a calming agent for asthmatics, to relieve the pain of headaches, gallbladder infections, colic, and kidney stones, and for the treatment of congestive heart failure. But the psychoactive effects of opium have probably been used for even longer and opium probably only takes a back seat to **alcohol** as the most common drug used by humanity. Certainly, writings from ancient Sumeria dated to 4000 B.C. seem to describe the euphoric effects of opium. The ancient Greeks appear to have used opium ''recreationally'' as well as a medicine. The plant-derived drug called nepenthe, described by Homer in the *Odyssey* was most likely opium as the symptoms are consistent. The importance of opium in Roman culture is evident by the fact that Somnus, the Roman god of sleep, is frequently depicted with opium poppies or a container filled with opium. Sleep, a sense of well being, and mild euphoria were the characteristics of opium.

This idyllic view of opium juice persisted well into the nineteenth century. Opium dens were an established part of society and many of the English romantic writers, for example, were frequent users. Samuel Taylor Coleridge's (1774-1834) well known poem *Kubla Khan* was a result of an vision he experienced under the influence of opium. Thomas De Quincey (1785-1859) and Elizabeth Barrett Browning (1806-1861) were also users, despite an increasing awareness of the potential for addiction.

Admittedly, though, because the extract from the opium poppy contains a mixture of compounds and is diluted by sugars, **water**, and other phytochemicals, it is not nearly as addictive as the pure compounds. In 1805, a young German chemist, Friedrich Sertürner (1783-1841) was able to extract the psychoactive substance from opium poppies and demonstrate its efficacy. He named the substance morphine after Morpheus, the Greek god of dreams and visions. The actual isolation of morphine had one intellectually important side effect. It induced other chemists to search for and extract the active components from other medicinal plants, such as salicylic acid from willow bark which, when acetylated, is aspirin.

As long as opium was in the form of a plant extract, it could only be administered orally and this hid much of its addictive properties. The pure compound, morphine, could be diluted in water and, with the invention of the hypodermic syringe in 1853, injection directly into the blood stream became the favored method of delivery. This both intensified the psychoactive properties of the compound and increased the speed with which they were realized. The analgesic and anesthetic properties aided in surgery and recovery by allowing for pain management. Both the American Civil War and the Franco-Prussian War saw the widespread use and overuse of morphine. Addiction to morphine was labeled ''the soldier's disease.'' The fact that morphine also induced constipation meant that it was applied as a treatment for dysentery, another common complaint amongst soldiers.

With the realization of the addictive potential of morphine, chemists began to investigate possible alternatives. Methylation of the two alcohol groups on the morphine **molecule** produced codeine, which was not as addictive but also not as effective. Acetylation, similar to the acetylation of salicylic acid, increased the potency of morphine by making the compound more soluble in brain fat and facilitating transfer across the blood brain barrier. This new compound, heroin, was a much better analgesic and was advertised as being non-addictive. Indeed, as late as 1900, it was thought to be a ''safe'' drug. Many cough syrups and other over-the-counter drugs contained heroin as their active principle. Of course, with the benefit of history, we now know of the damaging effects of heroin and other narcotics. The societal cost, in both monetary terms and human potential, is enormous. Still, researchers have yet to find a better drug for controlling pain and this has led to calls for its use in the treatment of terminal cancer patients and others suffering from severe pain.

The question of how narcotics act—what it is that they do in the brain that so profoundly affects the mind—has, for the most part, been solved. The story begins with the isolation of neuroreceptors specific for the opiates by Solomon Snyder and co-workers in 1973. They achieved this by using a combination of ''agonists,'' which are compounds that act at a receptor site to generate some measurable change in body function, and ''antagonists,'' which act to inhibit agonists. By selective **filtration** of neuronal membranes and radioisotope labeling, they were able to show that specific neuroreceptors are sensitive to the opiates. Further identification of the location of the neuroreceptors within the brain, via thin tissue slicing, showed that the opiates act to block pain in the brain stem. The sense of euphoria may be tied to the abundance of **receptors** within the limbic system, which is the part of the brain that is thought to be related to emotions.

Of course, finding very specific opiate receptors in the brain begs the question of why they are there. We did not evolve to eat poppies and yet one in every million receptor sites seems to be ideally suited to binding compounds from them. Scientists quickly realized that there must be naturally occurring compounds that acted like the opiates. They termed these compounds endorphins, short for endogenous morphines. In Scotland, John Hughes and Hans Kosterlitz took advantage of the ability of opiates to induce intestinal contractions to test for the naturally occurring analogues. By the end of 1975, they had succeeded in finding two short peptide **chains**, made up of five **amino acids** that differed in only the terminal amino acid, that appeared to fill the natural role of the opiates. These were the body's own narcotics, designed to intercept and reduce the sensation of pain. They named their new discovered compounds enkephalins, a term derived from Greek for in the head, and this has led to confusion every since. Strictly speaking, the enkephalins are the two amino acid sequences discovered by Hughes and Kosterlitz. The term endorphins is reserved for the larger class of natural opiates of which several other examples have been discovered.

The actual binding site of the narcotics, both natural and plant derived, has been deduced from the common morphology of the morphine family of compounds and the amino acid sequence. The tyrosine residue in the amino acid along with the peptide chain bind to a specific receptor site using a combination of Van der Waals interaction and **hydrogen bonding**. This understanding has aided in the development and synthesis of analogues for morphine. Unfortunately, potency with regard to analgesic and anesthetic properties and an addictive potential seem to be tied together. Still, this understanding has lead to new drugs such as Naloxone which is a potent opiate antagonist. Indeed, heroin patients suffering from massive overdoses can be ''cured'' in a matter of moments through the injection of Naloxone. But while it treats the immediate systems, it does nothing for the underlying psychological causes of the addiction.

Giulio Natta.

NATTA, GIULIO (1903-1979)

Italian chemist

Giulio Natta was a highly regarded Italian chemist who, during an active career spanning almost fifty years, worked closely with the Italian chemical industry to create many new processes and products. His studies of high polymers and his discovery of the hard plastic substance **polypropylene** ushered in the age of **plastics** with immense world-wide impact. For his work in this field he shared the 1963 Nobel Prize in **chemistry** with the German chemist **Karl Ziegler**.

Natta was born on February 26, 1903, in Imperia, Italy, a resort town located about sixty miles southwest of Genoa on the Ligurian Sea. His parents were Francesco Natta, a lawyer and judge, and Elena Crespi Natta. He received his primary and secondary education in Genoa. Having read his first chemistry book at age twelve, he quickly became fascinated by the topic. At age sixteen, Natta graduated from high school and entered the University of Genoa, intending to study mathematics. Finding this subject too abstract, he then transferred to the Milan Polytechnic Institute in 1921 and earned his doctorate in **chemical engineering** in 1924, at the early age of twenty-one.

Following graduation, Natta remained at the Polytechnic Institute as an instructor. He was promoted to assistant professor of general chemistry in 1925, and to full professor in 1927. Moving to the University of Pavia in 1933, Natta served as professor and director of the school's chemical institute. Next, in 1935, he became professor and chairman of the department of **physical chemistry** at the University of Rome, then professor and director of the institute of **industrial chemistry** at the Turin Polytechnic Institute in 1937. He returned to the Milan Polytechnic Institute as professor and director of the Industrial Chemical Research Center in 1938 and remained there for the rest of his career. He married Rosita Beati, a professor of literature at the University of Milan, in 1935. The couple had a daughter, Franca, and a son, Giuseppe. In his younger days, Natta was an enthusiastic skier, mountain climber, and hiker.

Natta's decision to seek a degree in chemical engineering rather than in chemistry, or some other pure science, was the earliest manifestation of the basic attitude which characterized his entire career. He was always concerned with the practical results of scientific research and believed that science should serve chiefly to meet the needs of business and industry. His research in the 1920s and early 1930s was largely devoted to x-ray and **electron** diffraction analyses to determine the structure of various inorganic substances. One of his early practical breakthroughs was the discovery of an effective catalyst for the synthetic production of the important chemical, **methanol**.

In the early 1930s Natta became interested in the chemistry of polymers, or large molecules, thus shifting his focus from inorganic to **organic chemistry**, the study of **carbon** compounds. In the late 1930s, the Italian government headed by Benito Mussolini actively promoted scientific research in order to increase Italy's self-sufficiency in the production of vital materials. Natta, using his recently acquired expertise in **polymer chemistry**, contributed to the national effort especially in the development of new methods to produce synthetic rubber. Following the conclusion of World War II, Natta continued his research in polymer chemistry, his work being subsidized by the large Milan chemical firm, the Montecatini Company.

In 1952 Natta attended a lecture at Frankfurt, Germany, given by the chemist, Karl Ziegler, then the director of the Max Planck Institute for Coal Research in Mulheim. Ziegler spoke about his recently discovered Aufbau (''growth'') reaction, which could be used to create large molecules from ethylene molecules, a gaseous product of the refining of petroleum. Though Ziegler had lectured frequently on his discovery, Natta was one of the first scientists other than Ziegler to grasp its potential importance for the production of high polymers (very large molecules), a venture which might have significant practical applications. Natta and representatives of the Montecatini Company soon invited Ziegler to Milan where they reached an agreement under which Montecatini would have commercial rights to exploit Ziegler's discoveries in Italy and Ziegler and Natta would exchange information on their respective research projects.

In the autumn of 1953, Ziegler and his research staff developed a catalytic process which allowed them to synthesize from ethylene a true high polymer, linear **polyethylene**, a plastic substance much harder and stronger than any plastic then known. Ziegler promptly patented the new substance but not the process which had produced it. Natta learned of the new discovery through representatives of the Montecatini Company who had been stationed in Ziegler's laboratory in Mulheim. Natta and his research group decided to try Ziegler's catalytic process on propylene, another gaseous product of petroleum which was much cheaper than ethylene. On March 11, 1954, they synthesized linear polypropylene, another high polymer with even more desirable chemical properties than polyethylene. The new plastic proved capable of being molded into objects stronger and more **heat** resistant than polyethylene. It could be spun into fiber stronger and lighter than **nylon**, spread into clear film, or molded into pipes as sturdy as metal ones.

Natta and his associates followed up on their discovery with a careful series of x-ray and **electron diffraction** experiments which demonstrated conclusively the exact nature of the polymer they had created. It was a molecular chain structure in which all of the subgroups were arranged on the same side of the chain. The substance had a high degree of crystallinity which was the cause of its strength. Natta and his colleagues soon discovered other high polymer plastics, including **polystyrene**. Natta did not inform Ziegler of his discovery of polypropylene until after he had filed for a patent on the new substance. Ziegler was greatly disturbed by what he considered to be Natta's failure to live up to their earlier agreement to share their research and the incident disrupted their previously close friendship to the extent that they were not on speaking terms for many years. The two scientists patched up their quarrel sufficiently to appear together at the 1963 ceremony in Stockholm at which they jointly received the Nobel Prize in chemistry.

In 1959 Natta contracted Parkinson's disease and was already seriously crippled by it at the time of the Nobel Prize award ceremony. He retired from active work in the early 1970s and died at Bergamo, Italy, on May 2, 1979, from complications following surgery for a broken femur bone. In the course of his career, Natta authored or coauthored over five hundred scientific papers and received nearly five hundred patents. He was the recipient of numerous gold medals and at least five honorary degrees for his scientific contributions.

Natural gas

Natural gas is an indispensable **energy** resource throughout most of the industrialized world. In American homes, natural gas is used in furnaces, stoves, **water** heaters, clothes dryers, and other appliances. The fuel also supplies energy for numerous industrial processes and provides raw materials for making many products that we use every day. The largest sources of natural gas in the United States are found in Alaska, Texas, Oklahoma, western Pennsylvania, and Ohio.

The Chinese were the first people known to have discovered and used natural gas. As early as 940 B.C., they found gas underground and piped it through hollow bamboo poles to the seashore, where they burned it to boil off ocean water and collect the leftover **salt**. By 615 B.C., the Japanese were producing gas from similar wells. Other ancient civilizations may have accidentally discovered natural gas seeping up from the ground. When they learned that the gas would burn continuously, they built temples to house mysterious "eternal fires." People traveled from faraway lands to visit these temples and marvel at this supernatural phenomenon.

In colonial North America, several natural gas seepages were discovered when they were accidentally set on fire. In the 1770s, French missionaries reported "pillars of fire" in the Ohio River valley, and George Washington described in wondering terms a "burning spring" on the banks of a river in West Virginia. Most scientists believe that natural gas was created by the same forces that formed petroleum, another fossil fuel.

In 1821, an American gunsmith named William Aaron Hart drilled the first natural gas well in the United States. It was covered with a large barrel, and the gas was directed through wooden pipes that were replaced a few years later with **lead** pipe. Although the well was only 27 ft (8.2 m) deep, it produced enough gas to illuminate nearby houses and stores in Fredonia, New York. Through the 1830s and 1840s, a few other gas wells were drilled in New York, Pennsylvania, and West Virginia. The first company to distribute and sell natural gas was established in Fredonia in 1865. But by then, oil had been discovered near Titusville, Pennsylvania, and in the oil rush that followed, natural gas was practically forgotten. Also, about 300 companies were already selling "manufactured" gas made from **coal** or oil. Gas lighting systems that burned manufactured gas had been established in many cities long before natural gas was discovered.

For many years, natural gas was used only in places that happened to be near gas wells, mainly because early pipes were unable to transport it much farther. In some towns that used natural gas for street lighting, the lights were left on during the day because it cost more to turn them off than it did to burn the gas. When new wells produced gas along with oil, the gas was usually just burned off, or flared.

In the late 1800s, the introduction of electric lighting nearly killed off the gas industry. However, customers of manufactured gas continued to use the fuel for cooking and heating, so gas companies never completely died out. Gradually, the natural gas industry began to recover as pipeline technology was improved and as larger quantities of natural gas were discovered. The first "long-distance" pipeline—only about 25 mi (40.2 km) long and less than a foot (30 m) in diameter—was built in the early 1870s to serve Rochester, New York. It was made of pine logs with holes bored through them. **Iron** pipe was also tested at that time in a 5.5 mi (8.8 km) pipeline serving 250 customers in Titusville, Pennsylvania, but for decades pipelines remained relatively short in length and small in diameter.

Then in the early 1900s, huge amounts of natural gas were found in Texas and Oklahoma, and in the 1920s modern seamless steel pipe was introduced. The strength of this new pipe, which could be electrically welded into long sections, allowed gas to be carried under higher pressures and thus in greater quantities. For the first time, natural gas transportation became profitable, and the American pipeline network grew by leaps and bounds through the 1930s and 1940s. By 1950, almost 300,000 mi (482,700 km) of gas pipelines had been laid—a length greater than that used to pipe oil. Soon it became routine for natural gas to be transported over distances of several hundred miles to the major centers of population and industry. As natural gas became available at prices lower than manufactured gas, customers switched to the cheaper fuel and consumption of natural gas increased phenomenally. Despite the loss of the lighting market, natural gas emerged as the most important heating and cooking fuel, and it gradually increased its share of the industrial market as well. Between 1940 and 1955, production of natural gas in the United States multiplied more than threefold. Gas now supplies more than one-fourth of all energy consumed in America.

In western Europe, however, the use of natural gas was virtually unknown until after World War II, when gas fields began to be developed in France and the Netherlands in the 1950s and in the North Sea, by England and Norway, in the 1960s. An American company came up with the idea of exporting natural gas to Europe by liquefying it at very cold temperatures and shipping it. In 1959, the first cargo of liquefied natural gas (LNG) crossed the Atlantic and was delivered to a specially built terminal in England.

Despite the huge amounts of natural gas that have been produced and consumed, new discoveries have continued to increase gas "reserves"—the amount of gas that is potentially recoverable. Exploring for natural gas is a complex process that demands the skills of geologists, physicists, chemists, and engineers. Once the right clues have been found above ground, surveyors map the area and samples of surface rocks are closely examined. Then underground structures are explored from the surface by means of instruments that identify rock layers from sound waves (seismographs) or from changes in gravity and **magnetism** (gravity meters and magnetometers). Still, no aboveground technique can prove the presence of gas, and an expensive well must be drilled for confirmation. On land, wells typically cost about $500,000. Unusually deep wells, and offshore wells drilled in water, cost much more.

Natural gas, like petroleum, is a mixture of many organic substances. It consists mainly of **methane**, the simplest **hydrocarbon**, as well as small amounts of heavier, more complex hydrocarbons such as ethane, butane and propane. Some natural gas also contains impurities such as **hydrogen** sulfide (''sour'' gas), **carbon** dioxide (''acid'' gas), and water (''wet'' gas). Before entering the transmission pipeline, natural gas is processed to remove impurities and extract valuable hydrocarbons. **Sulfur** and **carbon dioxide** are sometimes recovered and sold as byproducts. Propane and butane are usually liquefied under pressure and sold separately as LPG (liquefied petroleum gas). LPG is commonly used as a substitute for natural gas in rural areas that are not served by transmission pipelines.

Natural gas has its origins in decayed living **matter**, most likely as the result of the action of bacteria upon dead animal and plant material. In order for most bacteria to effectively break down organic matter to hydrocarbons, there must be low levels of **oxygen** present. This would mean that the decaying matter was buried (most likely under water) before it could be completely degraded to carbon dioxide and water. Conditions such as this are likely to have been met in coastal areas where sedimentary rocks and marine bacteria are common. The actions of **heat** and pressure along with bacteria produced a mixture of hydrocarbons. The smaller molecules that exist as **gases** were then either trapped in porous rocks or in underground reservoirs where they formed sources of hydrocarbon fuels.

Recently, in light of environmental concerns, natural gas has begun to be reconsidered as a fuel for generating **electricity**. Natural gas is the cleanest burning fossil fuel, producing mostly just water vapor and carbon dioxide. Several gas power generation technologies have been advanced over the years, including a process developed by Meredith Gourdine, an American physicist, that uses the principles of electrogasdynamics (EGD). Cars have also been produced that use natural gas as their primary fuel source. In the future, these cars may become more common because of their lower environmental impact.

NATURAL PRODUCTS

While there is single, no specific definition that applies to all natural products, they are generally defined as organic compounds that are formed as a product or byproduct of a living organism. Examples of natural products include fibers (e.g. cotton, silk, wool, hemp); fuels (e.g., oil and **natural gas**); construction materials such as **wood** and rubber; and animal byproducts like leather. These and many other natural products have been used throughout recorded history. One of the most important applications of natural products is in the area of health sciences.

The **chemistry** of natural products prior to 1945 was much different than the high-tech science of today. Before the middle of the century, chemists in this field were intent merely on isolating and determining the structure of natural products. However, over the last several decades, scientists have developed a host of analytical tools that greatly improved the study of these natural materials. For example, infrared and visible/ultraviolet **spectroscopy** was introduced in the early 1940s; **nuclear magnetic resonance** imaging (NMR) was developed by 1960, and the development of ligand/receptor interaction theory was introduced in 1990. These tools have supported major advances in the analysis of natural products.

Similar advances in chemistry have resulted in the development of naturally derived raw materials. These are compounds that originate in nature but require chemical processing before they can be used. This processing may include **distillation** or other methods or separation or purification that may be necessary to be applied before they are usable form. Naturally derived materials include chemicals used **fatty acids** and alcohols used in detergents and other industrial applications. Cetyl **alcohol**, for example, may be derived from coconut oil. With other breakthroughs in chemistry, it is now possible to synthesize some natural products, like rubber, from non-natural sources. However, these synthetic versions do not qualify as natural products despite their virtually identical chemical makeup. There is an important branch of chemistry dedicated to synthetic natural products that is crucial to industry and medicine. For instance, because the Pacific yew tree is endangered, the taxol we need to fight cancer may soon have to be entirely synthesized.

A large number of drugs are based on natural products and can be categorized into four major groups: terpenes (which include both taxoids and steroids); glycosides (sugar derivatives that include flavonoids, saponins, anthraglycosides and digitalis compounds); alkaloids (like belladonnas, camptothecins, opiates, rauwolfias and vincas) and miscellaneous substances (including plant-derived **vitamins**, psoralens, ephedrines and salicylates.) Global sales of these drugs based was estimated to be over $20 billion in 1997 according to one market research firm, Business Communication Company (BCC) of Norwalk, Connecticut. Nearly half of these sales come from prescription drugs with the remainder coming from herbal remedies. This trend is expected to continue to grow at an average annual growth rate 6.3% through 2002.

A large part of this growth is anticipated to come from research into herbal remedies. Historically, there has been little incentive for manufacturers to test and research these products, since these materials cannot be patented in the same way as synthetic drugs. But in since the late 1980s, the number of Americans using natural products to heal their ailments has steadily grown. Even though there is meager scientific data to support the claimed benefits of some products, they are tremendously popular. For this reason the federal government has allocated a $160 million annual budget for research in this field. Plans are ongoing to regulate this sector in order to protect consumers from fraudulent drug claims. As of the 1999, two federal agencies (the Food and Drug Administration [FDA] and the Federal Trade Commission [FTC]) are still closely reviewing the trying to take a more discriminating approach to regulating the natural products industry.

The rising interest in natural products chemistry is due to two reasons: (1) Our increasing awareness that the planet contains myriad natural substances potentially beneficial to humans and (2) Our realization that the producers of these substances are becoming extinct at an alarming rate. In the last two decades, scientists have discovered a wealth of new chemicals in marine and terrestrial organisms that seem to offer hope for finding cures for our most devastating illnesses.

Neodymium

Neodymium is one of the **rare earth elements**, the elements that are found in Row 6 between Groups 2 and 3 in the **periodic table**. Neodymium's **atomic number** is 60, its atomic **mass** is 144.24, and its chemical symbol is Nd.

Properties

Neodymium is a soft, malleable metal that can be cut and shaped fairly easily. It has a melting point of 1,875°F (1,024°C), a **boiling point** of 5,490°F (3,030°C), and a **density** of 7.0 grams per cubic centimeter. Neodymium is a somewhat active metal that combines with **oxygen** at room **temperature** to form a yellowish coating of neodymium **oxide** (Nd_2O_3). It also reacts with most acids, with the release of **hydrogen**.

Occurrence and Extraction

Neodymium is one of the elements that belie the name ''rare earth'' since its abundance is about 12-24 parts per million in the Earth's crust. That places it about 27th among the chemical elements in terms of abundance. The term *rare earth* more properly applies to the difficulty of separating the members of this family from each other than to their abundance in the earth. The most common ores of neodymium are monazite and bastnasite, as is the case with all rare earth elements.

Discovery and Naming

Neodymium was discovered in 1885 by the Austrian chemist Carl Auer (Baron von Welsbach; 1858-1929). Neodymium was one of seven new elements discovered in a mineral known as cerite, first found near the Swedish town of Bastnas in the late 1700s. The element was given the name neodymium for the Greek expression that means ''new twin.'' The name comes from the fact that the element was found in close proximity with a second new element, later given the name **praseodymium**, or ''green twin.''

Uses

Neodymium and its compounds have a number of important uses. One is in a type of laser known as a neodymium **yttrium aluminum** garnet (Nd:YAG) laser, used especially for treating bronchial cancer and certain types of eye disorders. Another important use of neodymium is in the manufacture of very strong magnets. The neodymium-iron-boron (NIB) magnet, for example, is one of the strongest magnets known. It is so strong that it must be handled with care. Two NIB magnets can attract each other so strongly that they can smash into each other and shatter. NIB magnets about an inch in diameter and a quarter inch thick are used in stereo audio speakers. Neodymium compounds are also used in special types of lamps that produce very pure, bright, white light.

Neon

Neon is the second element in Group 18 of the **periodic table**, a group of elements known as the inert or noble **gases**. Neon's **atomic number** is 10, its atomic **mass** is 20.179, and its chemical symbol is Ne.

Properties

Neon is a colorless, odorless, tasteless gas with a **boiling point** of -410.66°F (-245.92°C), a freezing point of -415.5°F (-248.6°C), and a **density** of 0.89994 grams per liter. Neon is chemically inert. No compound of the element has ever been made.

Occurrence and Extraction

Neon occurs in the Earth's atmosphere where it is the fifth most abundant gas, with an abundance of about 18.2 parts per million (0.0182%). It is produced for commercial use by the fractional **distillation** of liquid air.

Discovery and Naming

Neon was discovered in 1898 by two British chemists, **William Ramsay** and Morris Travers (1872-1961). Ramsay and Travers attempted to identify the very small amount of liquid left after **oxygen**, **nitrogen**, **carbon** dioxide, and **argon** had been extracted from liquid air by traditional methods of fractional distillation. They analyzed the remaining liquid spectroscopically and found three new sets of spectral lines. They attributed those lines to the presence of three new elements, which they named neon, **krypton**, and **xenon**. The name neon is taken from the Greek word *neos* meaning new.

Uses

The most important single use for neon is in the manufacture of ''neon'' lights. A ''neon'' light is a lamp filled with a gas (such as neon) at low pressure through which an electric current flows. The electric current causes the gas to fluoresce, giving off some distinctive **color**. A variety of gases are used in ''neon'' lights, each chosen for the distinctive color it produces. Neon is also used in the manufacture of **lasers** employed in industry and surgery.

Neptunium

Neptunium is a transuranium element, one of the elements found in Row 7 of the **periodic table** after **uranium** (atomic number 92). Neptunium's **atomic number** is 93, its atomic **mass** is 237.0482, and its chemical symbol is Np.

Properties

All isotopes of neptunium are radioactive, the longest lived being neptunium-237 with a half life of 2,140,000 years. The element exists as a silvery white metal with a melting point of 1,180°F (640°C) and a **density** of 20.45 grams per cubic centimeter. Neptunium is a fairly active element from which some unusual compounds, such as neptunium dialuminide ($NpAl_2$) and neptunium beryllide ($NpBe_3$), have been made. The element also forms more traditional compounds, such as neptunium dioxide (NpO_2) and neptunium trifluoride (NpF_3).

Occurrence and Extraction

Neptunium occurs in only very small amounts in nature, as the byproduct of the radioactive decay of uranium. The element is now produced routinely for commercial use in **nuclear reactors**.

Discovery and Naming

Neptunium was discovered in 1940 by a pair of physicists at the University of California at Berkeley, **Edwin M. McMillan** and Philip H. Abelson (1913-). They produced the element artificially by bombarding uranium atoms with neutrons. McMillan and Abelson suggested the name neptunium for the element in honor of the planet Neptune. Uranium, the element before neptunium in the periodic table, had earlier been named for the planet Uranus.

Uses

Neptunium and its compounds have been made for research purposes, but they have no commercial applications of any consequence.

NERNST, WALTHER (1864-1941)

German chemist

Walther Nernst made a significant breakthrough with his statement of the Third Law of **Thermodynamics**, which holds that it should be impossible to attain the **temperature** of absolute zero in any real experiment. For this accomplishment, he was awarded the 1920 Nobel Prize for **chemistry**. He also made contributions to the field of **physical chemistry**. While still in his twenties, he devised a mathematical expression showing how electromotive force is dependent upon temperature and **concentration** in a galvanic, or electricity-producing, cell. He later developed a theory to explain how ionic, or charged, compounds break down in **water**, a problem that had troubled chemists since the theory of **ionization** was proposed by **Svante A. Arrhenius**.

Born Hermann Walther Nernst in Briesen, West Prussia on June 25, 1864, he was the third child of Gustav Nernst, a judge, and Ottilie (Nerger) Nernst. He attended the gymnasium (high school) at Graudenz (now Grudziadz), Poland, where he developed an interest in poetry, literature, and drama. For a brief time, he considered becoming a poet. After graduation in 1883, Nernst attended the universities of Zurich, Berlin, Graz, and Würzburg, majoring in physics at each institution. He was awarded his Ph.D. summa cum laude in 1887 by Würzburg. His doctoral thesis dealt with the effects of **magnetism** and **heat** on **electrical conductivity**.

Nernst's first academic appointment came in 1887 when he was chosen as an assistant to professor **Friedrich Wilhelm Ostwald** at the University of Leipzig. Ostwald had been introduced to Nernst earlier in Graz by Svante Arrhenius. These three, Ostwald, Arrhenius, and Nernst, were to become among the most influential men involved in the founding of the new discipline of physical chemistry, the application of physical laws to chemical phenomena.

The first problem Nernst addressed at Leipzig was the **diffusion** of two kinds of ions across a semipermeable membrane. He wrote a mathematical equation describing the process, now known as the Nernst equation, which relates the electric potential of the ions to various properties of the cell.

In the early 1890s, Nernst accepted a teaching position appointment at the University of Göttingen in Leipzig, and soon after married Emma Lohmeyer, the daughter of a surgeon. The Nernsts had five children, three daughters and two sons. In 1894, Nernst was promoted to full professor at Göttingen. At the same time, he also received approval for the creation of a new Institute for Physical Chemistry and **Electrochemistry** at the university.

At Göttingen, Nernst wrote a textbook on physical chemistry, *Theoretische Chemie vom Standpunkte der Avogadroschen Regel und der Thermodynamik* (*Theoretical Chemistry from the Standpoint of Avogadro's Rule and Thermodynamics*). Published in 1893, it had an almost missionary objective: to lay out the principles and procedures of a new approach to the study of chemistry. The book became widely popular, going through a total of fifteen editions over the next thirty-three years.

During his tenure at Göttingen, Nernst investigated a wide variety of topics in the field of solution chemistry. In 1893, for example, he developed a theory for the breakdown of ionic compounds in water, a fundamental issue in the Arrhenius theory of ionization. According to Nernst, **dissociation**, or the dissolving of a compound into its elements, occurs because the presence of nonconducting water molecules causes positive and negative ions in a crystal to lose contact with each other. The ions become hydrated by water molecules, making it possible for them to move about freely and to conduct an electric current through the solution. In later work, Nernst developed techniques for measuring the degree of hydration of ions in solutions. By 1903, Nernst had also devised methods for determining the **pH** value of a solution, an expression relating the solution's hydrogen-ion concentration (acidity or alkalinity).

In 1889, Nernst addressed another fundamental problem in solution chemistry: **precipitation**. He constructed a mathematical expression showing how the concentration of ions in a slightly soluble compound could result in the formation of an insoluble product. That mathematical expression is now known as the **solubility** product, a special case of the ionization constant for slightly soluble substances. Four years later, Nernst also developed the concept of **buffer** solutions —solutions made of bases, rather than acids—and showed how they could be used in various theoretical and practical situations.

Around 1905, Nernst was offered a position as professor of physical chemistry at the University of Berlin. This move was significant for both the institution and the man. Chemists at Berlin had been resistant to many of the changes going on in their field, and theoretical physicist and eventual Nobel Prize winner **Max Planck** had recommended the selection of Nernst to revitalize the Berlin chemists. The move also proved to be a stimulus to Nernst's own work. Until he left Göttingen, he had concentrated on the reworking of older, existing problems developed by his predecessors in physical chemistry. At Berlin, he began to search out, define, and explore new questions. Certainly the most important of these questions involved the thermodynamics of chemical reactions at very low temperatures.

Attempting to extend the Gibbs-Helmholtz equation and the Thomsen-Berthelot principle of maximum work to temperatures close to absolute zero—the temperature at which there is no heat—Nernst eventually concluded that it would be possible to reach absolute zero only by a series of infinite steps. In the real world, that conclusion means that an experimenter can get closer and closer to absolute zero, but can never actually reach that point. Nernst first presented his "Heat Theorem," as he called it, to the Göttingen Academy of Sciences in December of 1905. It was published a year later in the *Nachrichten von der Gesellschaft der Wissenschaften zu Göttingen.* The theory is now more widely known as the Third Law of Thermodynamics. In 1920, Nernst was awarded the Nobel Prize in chemistry in recognition of his work on this law.

The statement of the Heat Theorem proved to be an enormous stimulus for Nernst's colleagues in Berlin's chemistry department. For at least a decade, the focus of nearly all research among physical chemists there was experimental confirmation of Nernst's hypothesis. In order to accomplish this objective, new equipment and new techniques had to be developed. Nernst's Heat Theorem was eventually integrated into the revolution taking place in physics, the development of quantum theory. At the time he first proposed the theory, Nernst had ignored any possible role of quantum mechanics. A few years later, however, that had all changed. In working on his own theory of specific heats, for example, **Albert Einstein** had quite independently come to the same conclusions. He later wrote that Nernst's experiments at Berlin had confirmed his own theory of specific heats. In turn, Nernst eventually realized that his Heat Theorem was consistent with the dramatic changes being brought about in physics by quantum theory. Even as his work on the Heat Theorem went forward, Nernst turned to new topics. One of these involved the formation of **hydrogen** chloride by photolysis, or chemical breakdown by light **energy**. Chemists had long known that a mixture of hydrogen and **chlorine gases** will explode when exposed to light. In 1918, Nernst developed an explanation for that reaction. When exposed to light, Nernst hypothesized, a **molecule** of chlorine (Cl_2) will absorb light energy and break down into two chlorine atoms (2Cl). A single chlorine **atom** will then react with a molecule of hydrogen (H_2), forming a molecule of hydrogen chloride and an atom of hydrogen (HCl + H). The atom of hydrogen will then react with a molecule of chlorine, forming a second molecule of hydrogen chloride and another atom of chlorine. The process is a **chain reaction** because the remaining atom of chlorine allows it to repeat.

In 1922, Nernst resigned his post at Berlin in order to become president of the Physikalisch-technische Reichsanstalt. He hoped to reorganize the institute and make it a leader in German science, but since the nation was suffering from severe inflation at the time, there were not enough funds to achieve this goal. As a result, Nernst returned to Berlin in 1924 to teach physics and direct the Institute of Experimental Physics there until he retired in 1934.

Walther Nerst.

In addition to his scientific research, Nernst was an avid inventor. Around the turn of the century, for example, he developed an incandescent lamp that used rare-earth **oxide** rather than a metal as the filament. Although he sold the lamp patent outright for a million marks, the device was never able to compete commercially with the conventional model invented by Thomas Alva Edison. Nernst also invented an electric piano that was never successfully marketed.

The rise of the Nazi party in 1933 brought an end to Nernst's professional career. He was personally opposed to the political and scientific policies promoted by Adolf Hitler and his followers and was not reluctant to express his views publicly. In addition, two of his daughters had married Jews, which contributed to his becoming an outcast in the severely anti-Semitic climate of Germany at that time.

Walther Nernst was one of the geniuses of early twentieth-century German chemistry, a man with a prodigious curiosity about every new development in the physical sciences. He was a close colleague of Einstein, and was a great contributor to the organization of German science—he was largely responsible for the first Solvay Conference in 1911, for example. In his free time, he was especially fond of travel, hunting, and fishing. Nernst also loved automobiles and owned one of the first to be seen in Göttingen. Little is known about his years after his retirement. Nernst died of a heart attack on November 18, 1941, at his home at Zibelle, Oberlausitz, near the German-Polish border.

NEUFELD, ELIZABETH F. (1928-)

American biochemist

Elizabeth F. Neufeld is best known as an authority on human genetic diseases. Her research at the National Institutes of Health (NIH) and at University of California, Los Angeles (UCLA), provided new insights into mucopolysaccharide storage disorders (the absence of certain enzymes preventing the body from properly storing certain substances). Neufeld's research opened the way for prenatal diagnosis of such life-threatening fetal disorders as Hurler syndrome. Because of this research, she was awarded the Lasker Award in 1982 and the Wolf Prize in Medicine in 1988.

Neufeld was born Elizabeth Fondal in Paris, on September 27, 1928. Her parents, Jacques and Elvire Fondal, were Russian refugees who had settled in France after the Russian revolution. The impending occupation of France by the Germans brought the Fondal family to New York in June 1940. Her parents' experience led them to instill in Neufeld a strong commitment to education "They believed that education was the one thing no one could take from you," she said in a 1993 interview.

Neufeld first became interested in science while a high school student, her interest sparked by her biology teacher. She attended Queens College in New York, receiving her bachelor of science degree in 1948. She worked briefly as a research assistant to Elizabeth Russell at the Jackson Memorial Laboratory in Bar Harbor, Maine. From 1949 to 1950, she studied at the University of Rochester's department of physiology. In 1951, she moved to Maryland, where she served as a research assistant to Nathan Kaplan and Sidney Colowick at the McCollum-Pratt Institute at Johns Hopkins University. In 1952, Neufeld moved again, this time to the West Coast. From 1952 to 1956, she studied under W. Z. Hassid at the University of California, Berkeley. She received her Ph.D. in comparative **biochemistry** from Berkeley in 1956 and remained there for her postdoctoral training. She first studied cell division in sea urchins. Later, as a junior research biochemist (working again with Hassid) she studied the biosynthesis of plant cell wall polymers—which would prove significant when she began studying Hurler syndrome and related diseases.

Neufeld began her scientific studies at a time when few women chose science as a career. The historical bias against women in science, compounded with an influx of men coming back from the World War II and going to college, made positions for women rare; few women could be found in the science faculties of colleges and universities. Despite the "overt discrimination" Neufeld often witnessed, she decided nonetheless to pursue her interests. "Some people looked at women who wanted a career in science as a little eccentric," she said, "but I enjoyed what I was doing and I decided I would persevere."

After spending several years at Berkeley, Neufeld moved on to NIH in 1963, where she began as research biochemist at the National Institute of Arthritis Metabolism and Digestive Diseases. It was during her time at NIH that Neufeld began her research on mucopolysaccharidoses (MPS), disorders in which a complex series of sugars known as mucopolysaccharides cannot be stored or metabolized properly. Hurler syndrome is a form of MPS. Other forms of MPS include Hunter's Syndrome, Scheie Syndrome, Sanfillipo, and Morquio. These are all inherited disorders. Defectively, metabolized sugars accumulate in fetal cells of victims. The disorders can cause stunted physical and mental growth, vision and hearing problems, and a short life span.

Because some plant cell wall polymers contain uronic acids (a component of mucopolysaccharides), Neufeld, from her work with plants, could surmise how the complex sugars worked in humans. When she first began working on Hurler syndrome in 1967, she initially thought the problem might stem from faulty regulation of the sugars, but experiments showed the problem was in fact the abnormally slow rate at which the sugars were broken down.

Working with fellow scientist Joseph Fratantoni, Neufeld attempted to isolate the problem by tagging mucopolysaccharides with radioactive **sulfate**, as well as mixing normal cells with MPS patient cells. Fratantoni inadvertently mixed cells from a Hurler patient and a Hunter patient—and the result was a nearly normal cell culture. The two cultures had essentially "cured" each other. Additional work showed that the cells could cross-correct by transferring a corrective factor through the culture medium. The goal now was to determine the makeup of the corrective factor or factors.

Through a combination of biological and molecular techniques, Neufeld was able to identify the corrective factors as a series of enzymes. Normally, the enzymes would serve as catalysts for the reactions needed for cells to metabolize the sugars. In Hurler and other MPS patients, **enzyme** deficiency makes this difficult. A further complication is that often the enzymes that do exist lack the proper chemical markers needed to enter cells and do their work. Neufeld's subsequent research with diseases similar to MPS, including I-Cell disease, showed how enzymes needed markers to match with cell **receptors** to team with the right cells.

This research paved the way for successful prenatal diagnosis of the MPS and related disorders, as well as genetic counseling. Although no cure has been found, researchers are experimenting with such techniques as gene replacement therapy and bone marrow transplants.

In 1973, Neufeld was named chief of NIH's Section of Human Biochemical Genetics, and in 1979 she was named chief of the Genetics and Biochemistry Branch of the National Institute of Arthritis, Diabetes, and Digestive and Kidney Diseases (NIADDK). She served as deputy director in NIADDK's Division of Intramural Research from 1981 to 1983.

In 1984, Neufeld went back to the University of California, this time the Los Angeles campus, as chair of the biological **chemistry** department, where she continues her research. In addition to MPS, she has done research on similar disorders such as Tay-Sachs disease. But her concerns go beyond research. She strongly believes that young scientists just starting out need support and encouragement from the scientific community, because these scientists can bring new and innovative perspectives to difficult questions and issues. At the same time, young scientists can learn much from the experience of established scientists. In her capacity as department chair, Neufeld encourages interaction among established scientists, young scientists, and students.

Neufeld has chaired the Scientific Advisory Board of the National MPS Society since 1988 and was president of the American Society for Biochemistry and Molecular Biology from 1992 to 1993. She was elected to both the National Academy of Sciences (USA) and the American Academy of Arts and Sciences in 1977, and was named a fellow of the American Association for Advancement in Science in 1988. In 1990, she was named California Scientist of the Year.

Married to Benjamin Neufeld (a former official with the U.S. Public Health Service) since 1951, she is the mother of two children. Although, her work takes up a great deal of her time, she enjoys hiking when she gets the chance, and travel ''when it's for pleasure and not business.''

NEUROCHEMISTRY

Neurochemistry is the area of science that focuses on the **chemistry** of the nervous system. The chemicals of the nervous system are called neurotransmitters. These chemicals are used to transmit signals between cells and carry information to and from the brain. As early as 1904, scientists suspected that chemicals might be involved in the transmission of nerve impulses. However, acetylcholine, the first neurotransmitter to be identified, was not isolated until the early 1920s.

To understand neurochemistry, it may be helpful to review the structure and function of the nervous system and its component parts. The nervous system contains cells called neurons. Neurons come in many different shapes and sizes, but they all have the same basic parts: dendrites and cell body, axons, and axon terminals. Dendrites are extensions of the cell body. The dendrites and cell body are where signals are received. The cell body also contains the organelles responsible for cell **metabolism**, growth, and repair. The axon is a special threadlike extension of the neuron. Some neurons may have axons more than 3 feet long. Axons are responsible for carrying signals away from the neuron. At the very ends of an axon are axon terminals. The axon terminals release neurotransmitters that travel from the neuron to the next cell. If the next cell is another neuron, the space between the two cells is called the synapse. In general, the signal travels across the synapse from the axon of one cell to the dendrite of the next. The signal is carried by means of neurotransmitters, which will be described in greater detail below.

Over the last 150 years, scientists have learned much about the chemistry of nerve functions. In 1848 E. Dubois-Reymond conducted experiments in which he applied current to the cut ends of muscles fibers that were bathed in saline **solution**. His work demonstrated the existence of a resting potential (E_{RP}), a potential voltage difference between the interior cytoplasm and the exterior aqueous portion phase of the cell. This resting potential occurs when the cell has not been stimulated. It is usually on the order of several tens of millivolts and is relatively constant. This value does, however, vary species to species. By 1939 researchers Curtis and Cole and Hodgkin and Huxley had inserted an electrode in the interior of squid giant axon and measured the potential difference across the axon membrane. These measurements supported the theory that resting potential is tied to **ionic strength**.

The resting potential changes when the cell is stimulated. The stimulus may be external or internal. External stimuli are received through the senses—touch, **taste**, smell, sight, and sound. Internal stimuli may come from other nerve cells or be self-generated, as in the case of pacemaker cells. Whatever the stimulus, it changes the cell membrane of the neuron so that positively charged particles called ions can flow into it. This state is called the action potential. The work of Hodgkins and Huxley showed the potential across the membrane approached zero at the peak of the nerve impulse. In fact, the action potential does not just go to zero, but it actually reverses so that the interior of the axon becomes more positive. This change in potential results from the flow of **sodium** and **potassium** ions across cell membranes. Membranes conduct current through ionic channels which switch from open to closed states and back again. In 1952, Hodgkin and Huxley proposed there were charged molecular entities responsible for the opening of these channels. However, it was not until in 1973 that the existence of a gating current across a channel was demonstrated. Now scientists know of at least 10 different classes of membrane channels through which charges can travel.

The charges builds as it passes through the membrane and travels down the axon. When it reaches the axon terminals, neurotransmitters are released and the neuron returns to its resting potential state. Once a neurotransmitter is released from a neuron, it attaches to specialized molecules, or **receptors**, on the next cell. The relationship between a neurotransmitter and its receptor is very specific, comparable to the relationship between a lock and a key. When the neurotransmitter binds with the receptor, the next cell responds accordingly.

A cell's response to these chemicals hinges on whether it is an excitatory or an inhibitory neurotransmitter. Excitatory neurotransmitters transfer an impulse to the next cell. For example, if the next cell is a muscle cell, the neurotransmitter causes it to contract. Or, if the next cell is a neuron, the neurotransmitter acts as a stimulus to generate an action potential. Inhibitory neurotransmitters are used to slow or block the transmission of a nerve impulse by another neuron. This inhibition is useful in situations such as blocking pain signals.

Acetylcholine is a neurotransmitter that has the interesting ability to act as both an excitatory or an inhibitory neurotransmitter. In skeletal muscle, acetylcholine is the main excitatory neurotransmitter; in heart muscle, it acts as an inhibitory neurotransmitter. The reason it can act in both ways is because the two muscle types have separate types of acetylcholine receptors. Skeletal muscle has nicotinic acetylcholine receptors; heart muscle has muscarinic acetylcholine receptors.

The two receptor types work in opposite ways. When acetylcholine binds to a nicotinic acetylcholine receptor, the cell membrane allows the entry of positively charged ions and an action potential is generated. Acetylcholine binding to a muscarinic acetylcholine receptor causes positively charged ions to flow out of the cell, reducing or eliminating any action potential.

Acetylcholine is one of several neurotransmitters that enable fast chemical transmission of a nerve impulse. Fast chemical transmissions require just a few milliseconds (a millisecond is one-thousandth of a second) to transmit a signal from one cell to another. In contrast, there are slow chemical transmissions which require hundreds of milliseconds to transmit an impulse. Not only is the transmission much slower—comparatively speaking—but the response in the receptor cell lasts much longer.

The reason for the difference in speed and duration of response is that neurotransmitters that cause a fast chemical transmission generate a direct response in the receptor cell. Neurotransmitters associated with slow chemical transmissions have to act through a secondary system. When the neurotransmitter binds to its receptor, it activates the second messenger system. The secondary system takes over from there and is responsible for generating the response in the receptor cell. There are several types of secondary systems found in the nervous system.

One neurotransmitter that uses slow chemical transmission is dopamine. Dopamine is the most abundant chemical in a class of neurotransmitters called catecholamines. This class also includes norepinephrine and epinephrine, for which dopamine is a precursor **molecule**. Dopamine is found in the brain and illustrates the important role that neurotransmitters have in controlling the proper function of the nervous system. There are several medical conditions in which the levels of dopamine in the brain are affected. For example, in Parkinson's disease, the neurons that produce dopamine are damaged and there is not enough dopamine for proper transmission of signals. As a result, a person with Parkinson's disease experiences slowed movements, rigidity, and tremors. On the opposite extreme, too much dopamine can cause some forms of schizophrenia, a mental illness in which a person suffers disruptions of perception and thought processes.

The field of neurochemistry encompasses two related fields, neuropharmacology and neurotoxicology. Both of these fields focus on the effects that chemicals from outside the body can have on the nervous system. Neuropharmacology is concerned with the effects of drugs on the nervous system; neurotoxicology centers on the effects of poisons on the nervous system. Both of these fields are based on the fact that chemicals other than neurotransmitters can react with receptors in the nervous system. The activity of these chemicals stems from their molecular resemblance to the body's neurotransmitters. It was through research in neuropharmacology that scientists discovered two new types of neurotransmitters, the enkephalins and the endorphins.

In a reversal of the usual order of things, the receptors for enkephalins and endorphins were demonstrated before the neurotransmitters themselves had been identified. The receptors for both enkephalins and endorphins are opiate receptors. As the name might suggest, cells with opiate receptors respond to opium and its derivatives, **morphine** and heroin. Juice from the opium poppy has been used since ancient times for pain relief. In 1803, its active ingredient—morphine—was isolated in the laboratory. Pure morphine was widely used, but its toxic and addictive properties made it unwise to use for long-term pain control. Chemical modifications of morphine were meant to lessen these properties but only produced drugs such as heroin.

In the early 1970s, several scientists demonstrated how morphine and its derivatives act in the nervous system. This research proved the long-suspected existence of opiate receptors, but it raised an interesting new question: Does the body itself produce a substance that attaches to these receptors? Opiate receptors were shown to be clustered in areas of the spinal cord and brain which were associated with pain transmission and perception. Scientists reasoned that the body itself must produce substances that act as pain killers. Within a few years, researchers had isolated two classes of peptide molecules, the enkephalins and the endorphins. These **peptides** are composed of **amino acids**, but are much smaller than **proteins**. Some of the enkephalins contain only five amino acids. Yet, like the other neurotransmitters, they exert a powerful effect on the nervous system.

NEUTRALIZATION

Neutralization is a chemical reaction in which a base reacts with an acid to create **water** and a **salt**. In chemical terms, neutralization results from the interaction of ions in the acid and base. This reaction can occur between organic and inorganic compounds, between two inorganic compounds, between two organic compounds, and by adding a base to an acid or vice versa.

A **chemical compound** that has been neutralized is no longer either a base or an acid but a salt, while a neutralized **solution** has equal numbers of protonated and deprotonated forms of the solvent (forms in which protons have been added and taken away, respectively). When strong **acids and bases** interact, **hydrogen** ions (H^+) combine with hydroxyl ions (HO^-) to form molecules of water.

A perfectly neutralized solution will have a **pH** of 7.0. In other words, the solution will be the salt dissolved in plain water. However, this is fairly difficult to accomplish because most chemical reactions begin with unbalanced amounts of acids and bases. Thus, a solution whose acids and bases have finished reacting with each other might end up with a pH of 7.2 or a value lower or higher than that. For instance, if a chemist poured a strong acid into a beaker with a strong base, those ingredients would neutralize each other, but would have a final pH less than 7.0. Likewise, combining a weak acid and a weak base would result in neutralization, but the product would have a pH higher than 7.0.

Neutralization plays an important part in the **chemistry** of everyday life. Without it, the carbonic acid produced during breathing would remain acidic and throw an organism's system out of **balance**. Fortunately, there is bicarbonate, which can act as either an acid or base, in the bloodstream at all times. This neutralizes carbonic acid and transports **carbon** dioxide in the blood. Bicarbonate is an example of a chemical **buffer**, without which humans (because of the huge number of acid- and base-producing processes going on within them at all times) could not maintain a virtually constant blood pH of 7.4. Likewise, the stomach produces **hydrochloric acid**, which requires neutralization in the duodenum by bile and pancreatic fluids that are bases (alkaline).

Another practical example of the chemical process of neutralization is **acid rain**. A potent threat to the environment of industrialized countries and their neighbors, acid rain occurs when **precipitation** picks up industrial chemicals as it falls to earth. In this manner, **sulfuric acid**, **nitric acid**, and other acidic substances change the usual pH of rain from its normal value of 5.6 to a value between 2.1 and 4.5. This acidic moisture gradually damages vegetation, soil, and many organisms. However, researchers have begun experimenting with neutralizing the acidity by introducing powdered limestone (calcium carbonate) into environments suffering from excessive acidity. Like the bicarbonate in blood, **calcium** carbonate helps return environmental pH to closer-to-normal levels by neutralizing the excessive acid in the water and soil.

See also Acids and bases

NEUTRINO

The neutrino, whose symbol is the Greek letter *nu*, is a neutral massless particle that is created during beta decay and travels at the speed of light. Its existence was suggested to explain missing **energy** and angular momentum when a beta particle was emitted from a radioactive element. Wolfgang Pauli suggested in 1930 that beta decay must involve a third particle that would be neutral, have negligible rest **mass** and a spin of one-half. Enrico Fermi named this particle the *neutrino*, Italian for little neutral one, and its existence was accepted without further evidence. It was not until 1953 that Frederick Reines and Clyde Cowan devised an experiment to detect the neutrinos that should be produced by the **nuclear reactions** taking place in a nuclear reactor. They succeeded, and Reines was awarded the 1995 Nobel Prize in physics for that work.

Discoveries in the field of physics during the second half of the twentieth century have led to the development of a new classification system for atomic and **subatomic particles** that is based on their interactions. Under this system, the neutrino is classified as a lepton, the category for light particles (*lepton* is derived from a Greek word meaning light) with a spin of one-half that interact via the weak nuclear force but do not respond to the strong force. Because they interact with **matter** only via the weak force, neutrinos are hard to detect. Nevertheless, they have been detected on earth and in space. A very important source of neutrinos turned out to be supernovae: studies on SN1987 show that most of the energy of the supernova was emitted as neutrinos. The core of the exploding star was converted to neutrons during the explosion when protons and electrons were forced together in reverse beta decay, producing the neutrinos.

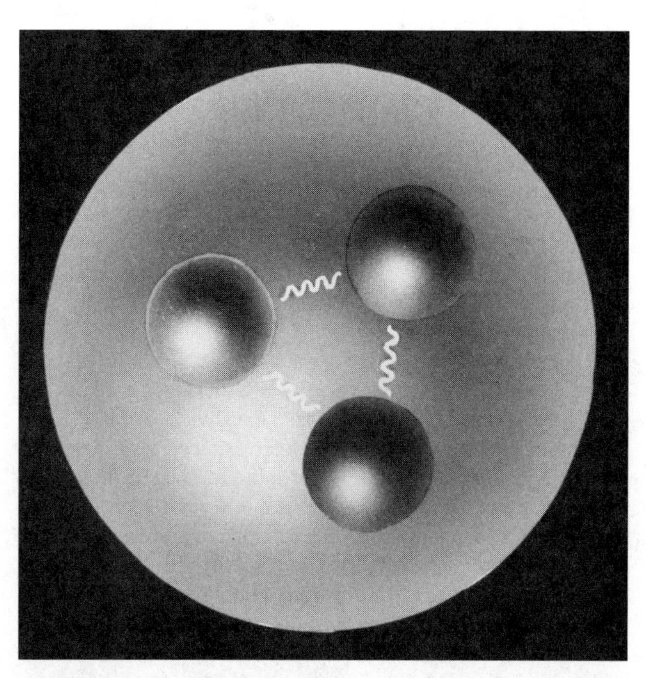

A diagram of the structure of the neutron. It consists of three smaller, fundamental particles called quarks, which are bound together by the strong nuclear force. *(Photograph by Michael Gilbert/Science Photo Library, Photo Researchers, Inc. Reproduced by permission.)*

NEUTRON

The discovery of the nuclear **atom** by **Ernest Rutherford** in 1911 was a critical step forward in improving our understanding of the fundamental nature of **matter**. However, it was immediately clear to scientists that Rutherford's atomic model was incomplete. The simplest and most obvious data showed without question that atoms must consist of more than nuclei, which contain protons, and electrons, located outside the **nucleus**.

Those data come from measurements of **atomic number** and atomic **mass**. An element's atomic number is equal to the total positive charge on—the number of protons in—the nuclei of atoms of which the element is made. The element's atomic mass is equal to the total mass of all particles that make up the atom. Since the mass of a single **proton** on the atomic mass scale is 1, and that of the **electron** is 0.0055, the mass of the atom must be very nearly equal to the mass of the protons present in the atom if protons and electrons are the only particles present in the atom. In such a case, an element's atomic number and atomic mass should be about equal. However, with the exception of **hydrogen**, this is never the case.

Scientists considered a number of ways in which this dilemma could be resolved. One popular notion was that a nucleus might contain both protons and electrons. The electrons in the nucleus could balance the electrical charge of some of the protons without adding any significant mass to the atom. Unfortunately, there were a number of rather clear reasons that electrons could not exist in the nucleus, and this theory was abandoned.

Rutherford himself suggested that some sort of neutral particle might exist in the nucleus and, during the 1920s, he and a graduate student, James Chadwick (1891-1974), searched for such a particle. They were unsuccessful, however, largely because neutrons (the particles for which they were seeking) do not readily ionize atoms and are not, therefore, detected by any standard tools such as cloud chambers or Geiger counters. The clue that led to the resolution of this problem showed up in the research of **Irène Joliot-Curie** and her husband, **Frédéric Joliot-Curie**, in France and of Walther Bothe (1891-1957) in Germany in the early 1930s. The Joliot-Curies and Bothe all observed that the bombardment of light elements, such as **beryllium** and **lithium**, with alpha particles resulted in the production of a strange, intense form of radiation. They were unable, however, to identify that radiation.

Chadwick, in England, repeated the experiments of the Joliot-Curies and Bothe with greater success. He directed the beam of "strange radiation" at a piece of paraffin and observed that protons were ejected from the paraffin. He concluded that the radiation must consist of particles with no charge and a mass about equal to that of the proton. That particle was the neutron.

Chadwick's discovery was a bonanza for nuclear physicists. Unlike the proton, electron, and other charged subatomic particles, the neutron is not repelled by either the nucleus or the orbital electrons in an atom. It is a much more efficient "bullet," then, in initiating nuclear reactions.

For the next three decades, the neutron was regarded as a fundamental particle that could not be broken down into anything simpler. In the early 1960s, however, the American physicist Robert Hofstadter was able to observe internal structure in both neutrons and protons. Hofstadter passed a beam of electrons close to each type of particle and studied the patterns that resulted from the electrons' deflection. He discovered that both protons and neutrons contain a central core of positively charged matter that is surrounded by two shells. In the neutron, one shell is negatively charged, just balancing the positive charge in the particle's core.

See also Atomic weight/mass

NEUTRON ACTIVATION ANALYSIS

Almost all radioactive isotopes emit characteristic gamma rays, and many chemical elements can be recognized by their gamma ray spectra. **Neutron** activation analysis is based on the detection and measurement of characteristic gamma rays emitted from radioactive isotopes produced in a sample upon irradiation with neutrons. Depending on the source of the neutrons, their energies and the treatment of the samples, the technique takes on several differing forms. Neutron activation analyses may employ thermal neutrons (with energies below 0.5 eV); epithermal neutrons (with energies between 0.5 eV and about 0.5 MeV); or fast neutrons (with energies above 0.5 MeV). Neutron activation analysis is used when the amount of an element in a specimen is so small that techniques of ordinary chemical analysis are impractical. Neutron activation analysis is capable of detecting the presence of as little as 10^{-9} grams of an element in an unknown specimen.

It is usually the case in neutron activation analysis that both unknown samples and standard materials containing known elemental concentrations are irradiated with thermal neutrons in a nuclear reactor. Following a decay period, a high resolution gamma ray spectrum is obtained to determine the intensity and energies of the gamma lines emitted. By comparing the activities induced in the standards and unknowns, it is possible to compute the abundance of elements in the unknown samples.

The collision of the neutron with a **nucleus** in the unknown material produces a compound nucleus in an excited state. This compound nucleus almost immediately begins to decay into a more stable configuration by emitting one or more characteristic gamma rays. (In many cases, this new configuration may also emit one or more characteristic delayed gamma rays, at a rate that depends on the **half-life** of the radioactive nucleus.)

As an example, when the stable **isotope** of **cobalt** (cobalt-59) is subjected to intense neutron irradiation, the radioactive isotope cobalt-60 is produced, which undergoes beta decay with a half life of 5.27 years, to give nickel-60. This **nickel** isotope emits two gamma rays with energies of 1.17 MeV and 1.33 MeV, both having the same intensity. If an unknown substance is irradiated with neutrons, giving rise to two gamma rays of equal intensity and having energies of 1.17 MeV and 1.33 MeV, one can be reasonably sure that the unknown sample contained cobalt. One could go a step farther and calculate the amount of cobalt present based on the known neutron flux and the neutron capture cross- section of cobalt-59.

Neutron activation analysis was discovered in 1936 when it was observed that samples containing certain **rare earth elements** became highly radioactive after exposure to a source of neutrons. This observation led to experiments in which measurements were made of the induced **radioactivity** in samples that had been exposed to **nuclear reactions** for purposes of qualitatively and quantitatively identifying the elements present in the samples.

Neutron activation analysis has found applications in archeology (to characterize archeological specimens e.g., pottery, obsidian, chert, basalt, and limestone, and to relate the artifacts to their origins through their chemical signatures); **biochemistry** (to study biochemical processes in animals); environmental studies (to characterize and monitor contaminated radioactive ore-processing sites; to study the fate of hazardous elements; and to study **selenium** distribution in aquatic species); pharmaceutical testing; medicine (to study fat absorption in cystic fibrosis; and to study thyroid action development); nutritional epidemiology (for example, in cancer studies); forensic investigations (to analyze evidence in criminal cases); geological science (to identify the processes involved in the formation of various rocks); studies of semiconductor materials and other high-purity materials (to measure trace- and ultra trace-element concentrations of impurities and/or dopants); and soil science (to quantify the distribution of agricultural chemicals).

NICKEL

Nickel is a transition metal, one of the elements in the middle of the **periodic table** in Rows 4, 5, and 6 between Groups 2 and 13. Nickel's **atomic number** is 28, its atomic **mass** is 58.69, and its chemical symbol is Ni

Properties

Nickel is a silvery white metal and is both ductile and malleable. Its melting point is 2,831°F (1,555°C), its **boiling point** is 5,135°F (2,835°C), and its **density** is 8.90 grams per cubic centimeter. Nickel is one of three naturally occurring elements that is strongly magnetic, the other two being **iron** and **cobalt**.

Nickel is a relatively unreactive element that does not react with **oxygen** or **water** at room **temperature**. At higher temperatures, it becomes more active reacting, for example, with oxygen to form nickel **oxide** (NiO) and with steam to give nickel oxide and **hydrogen** gas (H_2).

Occurrence and Extraction

Nickel ranks about 22nd among the chemical elements found in the Earth's crust with an abundance of about 0.01-0.02%. Nickel is thought to be much more abundant in the Earth's core, which may consist almost entirely of nickel and iron. One argument for this theory is the relatively abundant presence of nickel in meteorites.

The most common ores of nickel are pentlandite, pyrrhotite, and garnierite. The element is also found as an impurity in ores of iron, **copper**, cobalt, and other **metals**. The only nickel mine in the United States is located in Riddle, Oregon. The largest producer of nickel in the world is Russia, followed by Canada, New Caledonia, Australia, and Indonesia. The largest single deposit of nickel in the world is located at Sudbury Basin, Ontario, discovered in 1883. It covers an area 17 miles (27 kilometers) wide and 37 miles (59 kilometers) long. Some experts believe the Sudbury deposit was created when a meteorite struck the area in the distant past.

Discovery and Naming

Nickel was discovered by the Swedish mineralogist Axel Fredrik Cronstedt (1722-1765) in 1751. He was studying a new mineral from a cobalt mine in Sweden that had been given to him for analysis. He assumed that the ore would contain cobalt or copper, but he found a portion of the mineral that had properties of neither element. He decided that the unexplained part of the mineral must be a new element.

Cronstedt suggested the name nickel for the element as a shortened version of the German word *Kupfernickel*, or "Old Nick's copper." That term had been used for many years by German miners who frequently encountered a useless mass of earthy material (nickel) while they were searching for copper.

Uses

The most important use of nickel is in the manufacture of alloys, a use to which about 80% of the nickel consumed in the United States is put. About two-thirds of that amount goes into stainless steel, used to make many common household appliances, such as coffee makers, toasters, and pots and pans, as well as medical and industrial equipment.

Nickel is also used to make superalloys, alloys made primarily of iron, cobalt, or nickel. Superalloys are very resistant to **corrosion** and retain their properties at high temperatures. They are widely used in jet engine parts and gas turbines. Nickel is also popular in the manufacture of **batteries**. Nickel-cadmium (nicad) and nickel-metal hydride batteries are among the most popular of these batteries. They are used in a great variety of appliances, such as hand-held power tools, compact disc players, pocket recorders, camcorders, cordless and cellular telephones, scanner radios, and laptop computers.

Health Issues

Nickel can cause a number of health disorders, ranging from the relatively mild to the very serious. For example, some people are allergic to nickel and develop a rash if exposed to the metal. The rash becomes itchy and may form watery blisters. A common way in which a person discovers that he or she is allergic to nickel is by wearing jewelry that contains a nickel **alloy**. People who work with nickel, especially those who inhale nickel fumes, may develop more serious problems. These problems may include respiratory disorders and, on a long-term basis, cancer.

NICOTINE

In its pure state, nicotine is a colorless, oily, acrid liquid. It has the chemical formula $C_{10}H_{14}N_2$, and the systematic name 3-(1-methyl-2 pyrrolidyl) pyridine. It is toxic and it turns yellowish brown in air and light. Nicotine has a **boiling point** of 476°F (247°C). Nicotine is the principal **alkaloid** in tobacco (*Nicotiana tabacum*). While the tobacco plant is indigeous to North America, it is now commercially cultivated and naturalized in most sub tropical countries. The word nicotine comes from J. Nicot, the man who introduced tobacco to France in the sixteenth century.

Nicotine is present in cigarettes, cigars, pipes, chewing tobacco, and snuff. A stimulant, nicotine increases the pulse rate and blood pressure of the smoker. The initial side effects of nicotine are dizziness and nausea, but nicotine is also highly addictive. Users claim that nicotine aids in relaxation and helps combat anxiety, and it is also an appetite suppressant. Without regular exposure to nicotine, however, users exhibit classic withdrawal symptoms. Nicotine addiction is also very difficult to break; only 20% of those who attempt to quit smoking are successful on their first try.

Like most alkaloids, nicotine exerts its effects at **receptors** for chemicals that transmit nerve impulses. Specifically, nicotine acts at the nicotinic receptor class for the transmitter acetylcholine. Outside the brain, nicotinic receptors are found primarily in the sympathetic nervous system. Thus, nicotine use triggers sympathetic nervous system effects throughout the body. These effects largely account for nicotine's impact on the user's health. Among nicotine's effects on the human body is constricting small arteries. This raises the blood pressure and makes the heart pump harder, as does the faster heart rate that nicotine produces. Yet, because nicotine also constricts the arteries supplying the heart muscle, the organ receives less blood. When buildups of fatty plaque have already narrowed a person's heart arteries, this may be enough to trigger heart pain (angina) or an actual heart attack. At the same time, elevated blood pressure increases the risk of stroke, with possible disability or even death. Nicotine use may also worsen other circulatory problems, including poor circulation in the hands and feet and some men's difficulty in obtaining an erection. Nicotine in any form should not be used during pregnancy, as it may harm the fetus or cause miscarriage. Nicotine passes into breast milk and may cause problem in nursing babies whose mothers smoke.

Nicotine is also used as an insecticide to control aphids. In such applications, the nicotine is extracted from the stem and the leaf mid ribs of tobacco plants (those sections not used in the manufacture of smoking tobacco)and distilled. Nicotine is a strong, fast acting poison and it is usually applied as a 0.5% **solution** in **water**.

Niobrium samples. ***(Photograph by Russ Lappa/Science Source, National Audubon Society Collection/Photo Researchers, Inc. Reproduced by permission.)***

NIOBIUM

Niobium is a transition element. It is the fifth element in Row 5 of the **periodic table**. Niobium's **atomic number** is 41, its atomic **mass** is 92.9064, and its chemical symbol is Nb.

Properties

Niobium is a shiny gray metal with a melting point of 4,474°F (2,468°C), a **boiling point** of 8,901°F (4,927°C), and a **density** of 8.57 grams per cubic centimeter. Niobium is a relatively inert element, although it does react with **oxygen** and concentrated acids at high temperatures.

Occurrence and Extraction

Niobium occurs primarily in two minerals, columbite and pyrochylore. A second element, **tantalum**, is always present in these minerals along with niobium. Separating the two elements from each other is very difficult. The abundance of niobium in the Earth's crust is estimated to be about 20 parts per million.

Discovery and Naming

The story of niobium's discovery is a long and fascinating tale. The element was for many years confused with its "twin," tantalum. The two elements always occur together in nature and have very similar properties. Discovery of the element was first announced in 1801 by the English chemist Charles Hatchett (1765- 1847). However, it was more than 40 years before enough research had been completed that chemists generally recognized Hatchett's claim to discovery of the element.

Hatchett originally suggested the name *columbium* for the element after the mineral, columbite, in which it was found. A half century later, the German chemist Heinrich Rose (1795-1864), after confirming Hatchett's original discovery, proposed a different name—niobium. Rose made the suggestion because in Greek mythology, Niobe is the daughter of the god Tantalus, from whom the name tantalum comes. In 1949, the element was finally and officially named niobium. However, many metallurgists still use the older name of columbium for the element.

Marshall Warren Nirenberg.

Uses

Niobium is used primarily in the manufacture of alloys for use in products such as nuclear reactor components, airplanes and space vehicles, and skateboards. Niobium alloys are becoming more popular for the manufacture of jewelry because they are light weight and do not produce allergic skin reactions. Niobium alloys are also used in the construction of superconducting magnets. The most powerful magnet in the world is one at the Lawrence Berkeley Laboratory in Berkeley, California. It is made of niobium and **tin** and is three times as strong as the best magnet previously made.

Nirenberg, Marshall Warren (1927-)

American biochemist

Marshall Warren Nirenberg is best known for deciphering the portion of **DNA** (deoxyribonucleic acid) that is responsible for the synthesis of the numerous protein molecules which form the basis of living cells. His research has helped to unravel the DNA genetic code, aiding, for example, in the determination of which genes code for certain hereditary traits. For his contribution to the sciences of genetics and cell **biochemistry**, Nirenberg was awarded the 1968 Nobel Prize in physiology or medicine with **Robert W. Holley** and **Har Gobind Khorana**.

Nirenberg was born in New York City on April 10, 1927, and moved to Florida with his parents, Harry Edward and Minerva (Bykowsky) Nirenberg, when he was ten years old. He earned his B.S. in 1948 and his M.Sc. in biology in 1952 from the University of Florida. Nirenberg's interest in science extended beyond his formal studies. For two of his undergraduate years he worked as a teaching assistant in biology, and he also spent a brief period as a research assistant in the **nutrition** laboratory. In 1952, Nirenberg continued his graduate studies at the University of Michigan, this time in the field of biochemistry. Obtaining his Ph.D. in 1957, he wrote his dissertation on the uptake of hexose, a sugar **molecule**, by ascites tumor cells.

Shortly after earning his Ph.D., Nirenberg began his investigation into the inner workings of the genetic code as an American Cancer Society (ACS) fellow at the National Institutes of Health (NIH) in Bethesda, Maryland. Nirenberg continued his research at the NIH after the ACS fellowship ended in 1959, under another fellowship from the Public Health Service (PHS). In 1960, when the PHS fellowship ended, he joined the NIH staff permanently as a research scientist in biochemistry.

After only a brief time conducting research at the NIH, Nirenberg made his mark in genetic research with the most important scientific breakthrough since **James D. Watson** and **Francis Crick** discovered the structure of DNA in 1953. Specifically, he discovered the process for unraveling the code of DNA. This process allows scientists to determine the genetic basis of particular hereditary traits. In August of 1961, Nirenberg announced his discovery during a routine presentation of a research paper at a meeting of the International Congress of Biochemistry in Moscow.

Nirenberg's research involved the genetic code sequences for **amino acids**. Amino acids are the building blocks of protein. They link together to form the numerous protein molecules present in the human body. Nirenberg discovered how to determine which sequences patterns code for which amino acids (there are about 20 known amino acids).

Nirenberg's discovery has led to a better understanding of genetically determined diseases and, more controversially, to further research into the controlling of hereditary traits, or **genetic engineering**. For his research, Nirenberg was awarded the 1968 Nobel Prize for physiology or medicine. He shared the honor with scientists Har Gobind Khorana and Robert W. Holley. After receiving the Nobel Prize, Nirenberg switched his research focus to other areas of biochemistry, including cellular control mechanisms and the cell differentiation process.

Since first being hired by the NIH in 1960, Nirenberg has served in different capacities. From 1962-1966 he was Head of the Section for Biochemical Genetics, National Heart Institute. Since 1966 he has been serving as the Chief of the Laboratory of Biochemical Genetics, National Heart, Lung and Blood Institute. Other honors bestowed upon Nirenberg, in addition to the Nobel Prize, include honorary membership in the Harvey Society, the Molecular Biology Award from the National Academy of Sciences (1962), National Medal of Science presented by President Lyndon B. Johnson (1965), and the Louisa Gross Horwitz Prize for Biochemistry (1968). Nirenberg also received numerous honorary degrees from distinguished universities, including the University of Michigan (1965), University of Chicago (1965), Yale University (1965), University of Windsor (1966), George Washington University (1972), and the Weizmann Institute in Israel (1978). Nirenberg is a member of several professional societies, including the National Academy of Sciences, the Pontifical Academy of Sciences, the American Chemical Society, the Biophysical Society, and the Society for Developmental Biology.

Nirenberg married biochemist Perola Zaltzman in 1961. While described as being a reserved man who engages in little else besides scientific research, Nirenberg has been a strong advocate of government support for scientific research, believing this to be an important factor for the advancement of science.

NISHIZUKA, YASUTOMI (1932-)

Japanese biochemist

Yasutomi Nishizuka is a celebrated biochemist who discovered protein kinase C, an **enzyme** which controls the biology of cells. In further studies, he and his group found that tumor-promoting agents could trigger unregulated cell growth by activating protein kinase C. In 1989, Nishizuka won the Lasker Basic Medical Research Award "for his profound contributions to the understanding of signal transduction in cells and for his discovery that **carcinogens** trigger cell growth by activating protein kinase C." In 1988, he received The Order of Culture from the Emperor of Japan. Nishizuka is professor and chairman of the department of **biochemistry** at Kobe University School of Medicine and director of the Biosignal Research Center in Kobe.

Nishizuka was born on July 12, 1932. He received his medical degree in 1957 and his Ph.D. in 1962, both from Kyoto University. For the next two years, Nishizuka was a research associate in the laboratory of Osamu Hayaishi in the department of medical **chemistry** at Kyoto University. While still a research associate, Nishizuka was named an NIH International Postdoctoral Research Fellow in 1964; he went to Rockefeller University in New York City, where he worked in the laboratory of **Fritz Lipmann** for two years. Nishizuka remained on the faculty of Kyoto University until 1969, when he was appointed professor and chairman of the department of biochemistry at Kobe University Medical School.

It was at Kobe University, in 1977, that Nishizuka and his group announced the discovery of protein kinase C, with characteristics which resemble an enzyme. An enzyme is an organic catalyst produced by living cells but capable of acting independently; they are **proteins** that can cause chemical changes in other substances without being changed themselves. At first, the role of protein kinase C in intracellular signalling was not recognized. But later Nishizuka and his colleagues showed that it could be activated by tumor-promoting agents known as phorbol esters. These substances can remain in a cell membrane, causing it to continually produce protein kinase C—which can lead to uncontrollable cell growth, the basis of many types of carcinogenesis, or cancer.

The work done by Nishizuka and his team initiated many new lines of research; scientists began looking for substances that activate protein kinase C. Nishizuka's work also revealed the overwhelming importance of protein kinase C in the maintenance of normal health in all living things above the level of unicellular microorganisms. Exploration of the enzyme continues. It is now understood that it is part of a large family of proteins with multiple sub-species exhibiting individual enzymological characteristics and distinct patterns of tissue distribution.

Among the many prizes Nishizuka has received for his work are the Award of the Japan Academy in 1986; the Cultural Merit Prize from the Japanese Government in 1987; the Alfred P. Sloan Jr. Prize in 1988 from the General Motors Cancer Research Foundation; the Gairdner Foundation International Award in 1988; the Order of Culture in 1988 from the Emperor of Japan; the Albert Lasker Basic Medical Research Award in 1989, and the Kyoto Prize in 1992. In 1994, he received the Dale Medal from the British Endocrine Society. He was elected a foreign associate of the National Academy of Sciences in the United States in 1988 and a foreign member of the Royal Society of the United Kingdom in 1990. In addition to being a member of the Japan Academy, he is also a foreign associate of l'Academie des Sciences in France and a foreign honorary member of the American Academy of Arts and Sciences. The magazine *Science* noted that of the ten most cited Japanese papers of the 1980s, five were written by Nishizuka.

In June of 1992, Nishizuka was appointed director of the Biosignal Research Center at Kobe University and continues in the position. Nishizuka is married and has two daughters. The family lives in Ashiya.

NITRATES AND NITRITES

Nitrates are compounds containing the **anion** NO_3^-. Nitrites are compounds containing NO_2^-. Nitrates are salts derived from **nitric acid** and nitrites are salts or esters of nitrous acid.

All nitrates are readily soluble in **water**. Many nitrates are ionic in nature, but heavy metal nitrates and anhydrous nitrates have covalently bonded nitrate groups. Nitrates are formed from reacting nitric acid with metal oxides, hydroxides, or carbonates. Decomposition of nitrates on heating will give the metal **oxide**, **nitrogen** dioxide, and **oxygen**. The presence of nitrates in a compound can be shown by the following test. **Iron** (II) sulphate is added to the test **solution** and then concentrated **sulfuric acid** is carefully added down the side of the test tube. If a nitrate is present, a brown **ring** will form where the two **liquids** meet.

When **ammonia** is reacted with nitric acid, ammonium nitrate is produced. This is the most important nitrogenous fertilizer. Because of the high **solubility** of nitrates in water, it is very easy for nitrate fertilizer to be washed from the area of application into local rivers and water supplies. This is a problem for at least two reasons. First, the nitrate acts as a food supply for algae. This leads to a massive increase in numbers of algae called an algal bloom. As the algae die, their decomposition removes oxygen from the water, which in turn means that other organisms, such as fish, cannot survive in the polluted water. Second, when nitrates enter the drinking water supply, they can pose a threat to human health. Nitrates in drinking water have been linked to blue baby syndrome, a condition where the oxygen-carrying capacity of an infant's blood is impaired. Ammonium nitrate is also a major constituent of **explosives**.

Nitrites and nitrates are an important part of the **nitrogen cycle**. Nitrites are found in the soil where they are produced from ammonia by nitrifying bacteria such as *Nitrosomonas*. These nitrites are in turn transformed into nitrates that can be assimilated by plants through the action of other bacteria, such as *Nitrobacter*.

Nitrates and nitrites also are used as preservatives. They are added to cured meats such as bacon, hot dogs, and ham to retard the growth of bacteria (particularly the causative organism for botulism food poisoning) and to preserve the flavor and **color** of the meat. There is still some concern over the use these **food additives** because the nitrites react with the stomach acids to produce nitrous acid. The nitrous acid can then react with **amino acids** to produce nitrosamines which have been shown to cause cancer in laboratory animals.

In the 1970s the Food and Drug Administration (FDA) responded to public concerns by dramatically cutting back on the quantity of nitrates and nitrites that could be added to foods. By 1981, however, a thorough study of the issue by the National Academy of Sciences showed that nitrates and nitrites are only a minor source of nitrosamine compared to smoking, drinking water, **cosmetics**, and industrial chemicals. Based on this study, FDA finally decided in January 1983 that nitrates and nitrites are safe to use in foods.

NITRIC ACID

Nitric acid has the chemical formula HNO_3. At standard **temperature** and pressure it is a colorless, corrosive liquid and it decomposes at a temperature of 181°F (83°C). If nitric acid is not pure it is generally colored due to the presence of **nitrogen** dioxide dissolved in the liquid. The decomposition of nitric acid into nitrogen, **water**, and **oxygen** is a naturally occurring reaction which is catalyzed by light.

Nitric acid is produced industrially by the **oxidation** of **ammonia** over a **platinum** catalyst at a high temperature. The resultant gas is then dissolved in water to give nitric acid. Nitric acid is completely ionized in **solution** and as such it is a strong acid and an efficient oxidizing reagent. Nitric acid is sufficiently strong as an oxidizing agent to be able to oxidize **sulfur** and **phosphorus** directly to **sulfuric acid** and **phosphoric acid**. **Carbon** is oxidized to directly produce **carbon dioxide** and nitrogen dioxide. A number of **metals** are not affected by **hydrochloric acid** and other acids but nitric acid will dissolve them and release oxides of nitrogen in the process. **Copper** will react with nitric acid to produce a blue solution of copper (II) nitrate along with water and nitrogen dioxide. The only common metals not dissolved by nitric acid are **gold** and platinum.

When nitric acid reacts with a metal the products that result are dependant upon the **concentration** of the acid used. This process is complicated by the fact that nitric acid acts as a strong oxidizing agent as well as an acid. Reactive metals will react with nitric acid to give the metal nitrate and **hydrogen** gas. Less reactive metals do not produce hydrogen at all. With dilute acid one of the principal products is nitrogen monoxide, while with concentrated acid it is nitrogen dioxide. In practical terms both reactions give the brown fumes normally associated with nitrogen dioxide, because the nitrogen monoxide rapidly reacts with the oxygen in the air to produce nitrogen dioxide. Treating metals (iron, **chromium**, **aluminum**, and calcium) with concentrated nitric acid can produce a protective **oxide** layer over the metal that halts further reactions.

Nitric acid reacted with a base will give the nitrate of the **salt** and water. The reaction with carbonates gives the nitrate salt, water, and carbon dioxide.

Nitric acid can be thermally decomposed to give water, nitrogen dioxide, and oxygen. If the products are collected over water only oxygen is collected as the nitrogen dioxide dissolves in the water.

Nitric acid is used in the production of **fertilizers** (after the nitric acid is turned into ammonium nitrate), dyes, **explosives** (including **trinitrotoluene** [TNT]), various polymers (such as **nylon** and terylene), and some drugs. The old (alchemical) name of nitric acid is aqua fortis meaning strong water. In the 1990s production of nitric acid in the United States was in excess of 6 megatons.

NITRIC OXIDE

Nitric **oxide** (sometimes called **nitrogen** monoxide) has the chemical formula NO. It is a colorless gas at standard **temperature** and pressure and it is slightly toxic. Nitric oxide is generally prepared by reacting **copper** and **nitric acid** or industrially by **oxidation** of **ammonia** over a catalyst (this is the first step in the manufacture of nitric oxide by the Ostwald process). At high pressure decomposition occurs to give other oxides of nitrogen, chiefly N_2O and NO_2. This latter oxide can also be produced by oxidation with atmospheric **oxygen**. Nitric oxide re acts with **halogens** to produce a compound with the chemical formula of halogen NO. These are the nitrosyl halides that act as strong oxidizing agents. Loss of an **electron** will give NO^+, the nitrosonium **ion** which is also a strong oxidizing agent.

Nitric oxide is an important **hormone** in the human body. It causes the muscles surrounding blood vessels to relax, which allows an increased passage of blood. It can also help the immune system kill invading parasites.

Nitric oxide is also present in the atmosphere. Atmospheric nitric oxide is either naturally occurring (produced by electrical discharges) or is present as a pollutant from the **com-**

bustion of organic matter and also from the internal combustion engine. In non-polluted air nitric oxide is found at levels of 0.01 ppm, but in smog it can be present in concentrations as high as 0.2 ppm.

NITRO FUNCTIONAL GROUP

Nitration with **nitric acid** alone, or in combination with **acetic acid**, acetic anhydride, or **sulfuric acid**, are common examples of adding nitro ($-NO_2$) functional groups to **aromatic hydrocarbons**. Aliphatic (i.e, straight or branched chain hydrocarbon) nitration is less common than aromatic, but propane can be nitrated under pressure to yield nitropropanes.

The nitro group is a very strong deactivating group, i.e, when added to a **benzene ring**, the ring becomes less reactive. This deactivation results from the very strong electron-withdrawing property of the nitro group. Nitrobenzene, for example, undergoes nitration at a rate that is one-ten-thousandth the rate exhibited by benzene. In nitrated alkanes (nitroalkanes), depending on the location of the **hydrogen atom** in the **molecule**, a hydrogen **ion** may be produced by the reaction of the nitroalkane with a strong base, as a result of the powerful electron-withdrawal effects of an adjacent nitro group. The resulting anion may react with **aldehydes** and **ketones**, in a reaction known as a nitro aldol reaction. Nitro group are reduced easily to amine groups. The most widely used method of preparing aromatic amines, for example, involves a nitration of the aromatic ring and subsequent conversion of the nitro group to an amino group.

When **glycerol** is treated with a mixture of sulfuric and nitric acid, a yellow oily liquid **nitroglycerin** is formed. Nitroglycerin is capable of exploding without warning, even during its preparation. This liquid has the basic structure of glycerol, but with the glycerol molecule's three –OH groups substituted by nitro groups. The explosive properties of nitrog lycerine are due to the high exothermicity associated with that molecule's decomposition, and the large **volume** of gaseous products that are produced.

Another explosive material, nitrocellulose (also known as **cellulose** trinitrate or gun cotton), has three of the free hydroxyl groups of each glucose unit of the cellulose molecule substituted with nitro groups.

Most of the aromatic nitro compounds containing more than one nitro group that are produced have substituted groups in the meta positions; para and ortho dinitro compounds are comparatively rare. (In order to name aromatic molecules containing more than one nitro group, one must identify the positions of the substituted groups. An aromatic molecule containing two nitro groups will have three positional isomers. If the nitro groups are adjacent (or 1, 2–), they are termed ortho; if they are at opposite ends of the ring (or 1,4–), they are called para; if they assume the one remaining position, they are referred to as meta, or 1,3–). When nitrobenzene is nitrated with nitric and sulfuric acids, for example, 93% of the substituted groups end up at meta positions. This is because the electron-withdrawing nitro group from the first nitration directs the next incoming nitro group to preferential (meta) positioning on the benzene ring.

In the nitro group, the central **nitrogen** atom is involved in both a double bond and a single bond. The O–N=O bond angle is very nearly tetrahedral (i.e., approximately 120°).

Nitrated hydrocarbons have found use as **solvents**, **explosives**, dyes, and analytical reagents; and as intermediates for the synthesis of primary amines. Commerically important nitro compounds include nitrobenzene; nitrotoluenes; dinitrobenzenes; nitrochlorobenzenes; 2,4,6-trinitrotoluene (better known as the explosive TNT); nitroglycerin, and picric acid.

NITROGEN

Nitrogen is the first element in Group 15 of the **periodic table**. The elements that make up this group are sometimes known as the nitrogen family, after nitrogen itself. Nitrogen has an **atomic number** of 7, an atomic **mass** of 14.0067, and a chemical symbol of N.

Properties

Nitrogen is a colorless, odorless, tasteless gas with a **density** of 1.25046 grams per liter, slightly less than that of air (density = 1.29 grams per liter). Nitrogen has a **boiling point** of -320.42°F (-195.79°C) and a freezing point of -346.02°F (-210.01°C). When it freezes, nitrogen becomes a white solid that looks somewhat like snow. Nitrogen is slightly soluble in **water**, to the extent of about two liters of nitrogen to 100 liters of water.

Nitrogen is a relatively inactive element, evidenced by the fact that free nitrogen is abundant in the atmosphere. Nitrogen does combine with **oxygen** in the presence of lightning or a spark, the product of the reaction being nitric oxide: $N_2 + O_2$ —electrical energy→ 2NO. This reaction occurs commonly in the atmosphere and is the mechanism by which elemental nitrogen is converted to a compound. It is the first step in the process by which nitrogen is made available to plants and animals, where it is used primarily in the formation of **amino acids** and **proteins**. The conversion of elemental nitrogen to compounds of nitrogen is known as "nitrogen fixation."

Occurrence and Extraction

Nitrogen is the most abundant gas in the atmosphere, where it makes up 78.084% of the air we breathe. The element is also present in the Earth's crust, primarily in the form of nitrates and nitrites. Nitrogen is produced commercially by the fractional **distillation** of liquid air. Air is first cooled to the point at which all of the **gases** in it liquefy (close to -328°F/-200°C). The liquid air is then allowed to evaporate slowly. The first major gas to escape from liquid air as it warms is nitrogen. The nitrogen is captured and stored in steel cylinders for later transportation and use. Large amounts of nitrogen are made by this process. Typically, more than a trillion cubic feet of the element are produced in the United States alone each year, making it the second most important industrial chemical after **sulfuric acid**.

Discovery and Naming

Nitrogen was discovered in the early 1770s, probably by a number of different chemists at almost the same time. The reason for this multiple discovery was that chemists were just learning how to capture, store, and study gases during this period. Prior to that time, all gases were thought of as different forms of air. They were, in fact, not called gases, but ''airs.'' Credit for the discovery of nitrogen is often given to the Scottish physician and chemist **Daniel Rutherford** because he was the first person to announce his discovery and give a detailed description of nitrogen. It seems likely that the English chemist **Henry Cavendish** (1731-1810) discovered nitrogen at about the same time, if not earlier. However, Cavendish failed to publish his results until somewhat later than Rutherford.

The modern name for the element was suggested in 1790 by the French chemist Jean Antoine Claude Chaptal (1756-1832) based on the fact that it occurs in both **nitric acid** and nitrates. Thus, nitrogen means ''nitrate and nitric acid'' (*nitro-*) and ''origin of'' (*-gen*).

Uses

The most important use of nitrogen as an element is in situations where an inert gas is needed. For example, librarians often wish to store books and other precious documents where they will be protected from oxygen in the air (which would cause them to deteriorate). They may use sealed cases filled with nitrogen gas for such purposes. Nitrogen does not react with and damage the documents, as would oxygen. A similar application is the common incandescent light bulb, which may be filled with nitrogen, **argon**, or some other inert gas. The inert gas prevents the wire filament inside the bulb from reacting chemically with oxygen and burning out quickly.

Liquid nitrogen is also used very widely in situations where very cold temperatures are required. For example, it is used to quick dry foods, the first step in preserving them as frozen foods. Liquid nitrogen is also used by physicians to freeze and kill small tumors and other damaged tissue. After being frozen, the tumors or damaged tissue simply fall off, leaving healthy tissue behind.

Nitrogen is also the starting point for the manufacture of dozens of very important compounds. The most important of these compounds by far is **ammonia** (NH_3). The discovery of a method by which nitrogen can be artificially ''fixed'' by German chemist **Fritz Haber** (1868-1934) in 1905 was one of the great milestones in the history of **chemistry**. Haber found that the two elements could be made to combine under high pressure and **temperature** in the presence of an appropriate catalyst. Ammonia gas is annually one of the top five chemicals in terms of production in the United States. Its most important use is in the manufacture of synthetic **fertilizers**, although very large amounts are also used to make nitric acid (HNO_3), ammonium **sulfate** [$(NH_4)_2SO_4$], and ammonium nitrate (NH_4NO_3). These compounds are also used to make fertilizers as well as **explosives**, fireworks, insecticides and **herbicides**, and rocket fuels. Ammonium sulfate is also used in water and sewage treatment systems, as a food additive, in the tanning of leather, and in fireproofing materials.

The process of freezing food often involves the usage of liquid nitrogen. Frozen foods commonly found in grocery stores are produced this way. ***(Photograph by Robert J. Huffmann. Reproduced by permission.)***

Health Issues

Nitrogen is absolutely essential to all living organisms. It is an essential element from which all amino acids and all proteins are made. Nitrogen is also an important component of **nucleic acids**, the chemicals in cells that carry genetic information and that direct the biochemical operations of all cells.

NITROGEN CYCLE

All plants and animals need certain organic **nitrogen** compounds to live. During the 1700s, scientists knew that nitrogen was obtained by plants in order to make **proteins**; however the origin of the nitrogen remained unknown. One obvious possible source was air, which is nearly 80% nitrogen. But in the late 1700s, H. B. de Saussure discovered that most plants cannot extract, or assimilate, nitrogen from the air. Instead, he concluded, they must absorb it somehow from the soil through their roots. Most people thought that nitrogen had to be restored to the soil by the addition of humus, manure, or other decaying organic **matter**. Farmers routinely added these **fertilizers** to their soil to supply nitrogen compounds in a form that plants could use.

Then in the mid-1800s, Jean Boussingault proved that plants could flourish without organic fertilization as long as other sources of nitrogen, such as nitrates or ammonium salts, were supplied. Eventually, these chemical fertilizers took the place of the farmer's smelly, germ-breeding manure pile and compost heap. Boussingault also showed that beans, peas, clover, and other legumes can replenish the soil's nitrogen by assimilating it from the air. Most other plants, Boussingault found, depend entirely on fertilizers for their nitrogen, because they cannot obtain it from the air as legumes do. Still, no one had figured out how leguminous plants assimilate nitrogen.

Boussingault and his contemporaries believed that the process was strictly chemical.

Then in 1862, **Louis Pasteur** first suggested that microorganisms might be involved. During the 1880s, Pierre Berthelot grew plants in soil that had been sterilized in order to test Louis Pasteur's hypothesis. When he found that nitrogen is not assimilated in sterile soil, he concluded that microorganisms were involved in the process.

Soon chemists were studying how bacteria and other microorganisms break down and recreate nitrogen compounds. By the late 1880s, scientists had learned that certain bacteria live in a *symbiotic*, or mutually beneficial, relationship with leguminous plants. Today, we know that the process takes place through tiny hairs on the plant's roots, and it involves the production of nodules, or lumps, that contain colonies of bacteria. These ''nitrogen-fixing'' bacteria convert nitrogen from the air into **ammonia**. Once ammonia has been formed, it is converted by ''nitrifying'' bacteria into nitrates—compounds that can be used by plants to make proteins and other organic nutrients. Some bacteria that live free in the soil are also capable of directly assimilating nitrogen from the air, and in tropical regions, certain algae perform this function. Other types of nitrifying bacteria produce nitrate compounds from ammonium salts in the soil, rather than from atmospheric nitrogen. This process requires soil aeration and a neutral or alkaline soil environment. Still other types of bacteria ''denitrify'' soil by breaking down its nitrogen compounds and returning free nitrogen to the air.

During the **nitrogen cycle**, organic material decays and forms ammonia; bacteria use nitrogen from ammonia or from the air to make nitrate compounds; nitrates are absorbed by plants and used to make proteins; plants are eaten by people and other animals; and when plants and animals die and decay, their nitrogen compounds are recycled to the soil. More recently, scientists have also discovered that, when thunderstorms discharge **electricity** into the atmosphere, nitrogen oxides are formed. These compounds are literally rained onto the earth in the form of nitric and nitrous acids that replenish the earth's nitrogen content by forming nitrates and nitrites in the soil.

However, harmful amounts of **acid rain** can result when automobile exhaust, power plants, and other artificial sources overload the atmosphere with nitrous oxides. Acid rain falling back on Earth causes acidification of lakes and streams and contributes to damage of trees at high elevations. In addition, acid rain hastens the decay of building materials and paints. Nitrogen **oxide** emissions also contribute to smog and poor visibility.

Nitroglycerin

A highly volatile explosive, nitroglycerin was first produced by Italian chemist Ascanio Sobrero (1812-1888) in 1847. At about the same time German-Swiss professor **Christian Schönbein** discovered guncotton. Both inventions, given their high degree and speed of detonation, had the potential to immediately replace **gunpowder** as the chief military and commercial explosive. (Nitroglycerin's speed of explosion, for example, proved to be 25 times faster than that of gunpowder.)

Sobrero, guided by scientific curiosity rather than military interests, found that by combining a commonly used skin lotion, **glycerol**, with **nitric acid** and **sulfuric acid**, a colorless, oily liquid of enormous power resulted. However, the grim potential of his discovery so frightened him that he refused to publish his findings for nearly a year and then almost in secret. Consequently, nitroglycerin went largely unnoticed for many years, while the manufacture and use of guncotton, despite several notorious accidents, spread throughout Europe.

It was not until the mid-1860s, when Swedish manufacturer and inventor **Alfred Nobel** (1833-1896) made the important discoveries, successively, of the blasting cap and dynamite, that nitroglycerin became a widely used and respected explosive. Since that time—and even more since Nobel's development of nitroglycerin-based blasting **gelatin** in 1875—its impact on the mining and construction industries has been profound. Previously, these industries were hampered by gunpowder's inability to obliterate large sections of rock. Nitroglycerin makes up about 75% of the materials used in the manufacture of dynamite.

Sobrero's substance also found use as a key ingredient in two smokeless powders, ballistic and **cordite**, from which all modern bullets derive their construction. Lastly, and perhaps most reassuring to the memory of both Sobrero and Nobel, nitroglycerin, in minute quantities, is regularly relied upon by doctors for the treatment of heart disease. Its properties of expansion have proven ideal for dilating coronary arteries in patients suffering from hypertension, cardiac pain, and the threat of heart attack.

Nitroglycerin ($C_3H_5N_3O_9$, also known as nitroglycerol, glyceryl trinitrate, and trinitroglycerin) is a pale yellow, viscous liquid that is extremely sensitive to shock and **heat**. It is soluble in **ethanol** and **ether**, and slightly soluble in **water**. Nitroglycerin is derived from the glycerol **molecule**, a common biological molecule that is a building block for triglyceride fats and oils, where all the -OH groups have been replaced by $-NO_2$ groups.

Nitroglycerin is an explosive because no outside source of **oxygen** is needed for its complete **combustion**. But it is nitroglycerin's rate of decomposition reaction that makes it such a violent explosive. The molecule decomposes almost instantaneously when a supersonic shock wave passes through it. The instantaneous destruction of all molecules in a sample is known as detonation, and the rapid expansion of hot **gases** that results is what gives rise to the destructive blast. In the decomposition, four moles of nitroglycerin decompose into approximately 30 moles of **carbon** dioxide, water, **nitrogen**, and oxygen. Each **mole** of gas produced occupies approximately 1,000 times the **volume** of the nitroglycerin.

Dynamite, which contains nitroglycerin, is safe to handle because the nitroglycerin contained in it is diluted with about 25% of an inert substance, diatomaceous earth.

See also Explosives; Lipids

NITROUS OXIDE

Nitrous oxide, or "laughing gas" (dinitrogen monoxide, N_2O) is the only inorganic gas that is practical for clinical **anesthetics**. It was first synthesized in 1776 by **Joseph Priestley** and was named "diminished nitrous air." He noted its effects when inhaled, namely, lightheadedness, a mild hysteria, and a diminished sensation to pain, and soon became a novelty at carnivals and parties. In the 1790s, **Sir Humphry Davy** noted that nitrous oxide might be helpful in surgical procedures, but his observation went ignored by the medical profession.

By the mid 1800s, however, surgery had advanced to the point where anesthesia was desirable. In 1844, Gardner Q. Colton (1814-1898) an American showman, entrepreneur, and partially trained physician gave a public demonstration of the effects of nitrous oxide which was a financial success. He then began lecturing in other cities. In December 1844, at a Hartford, Connecticut demonstration, Horace Wells, a dentist, asked Colton to extract one of his teeth while he was under the effects of the gas. The procedure was successful and without complication. Wells immediately began using nitrous oxide in his dental practice, and later made a controversial claim that he was the first to use the gas in dentistry as an anesthetic, which Colton acknowledged. In 1863, Colton was again lecturing in Hartford when he met J.H. Smith, another dentist. Colton and Smith extracted over a thousand teeth in a month using nitrous oxide as an anesthesia.

The popularity of nitrous oxide as an anesthetic eventually waned. The reason why this was so was that it never produced complete general anesthesia. Finally in 1879, Paul Bert (1833-1886), a French physiologist experimenting with gases under pressure showed that 85% nitrous oxide (a rather high concentration) would cause complete general anesthesia only at the hyperbaric pressure of 1.2 atmospheres. Thus, "laughing gas" was considered best used as an adjunct to other anesthetics at normal pressure and temperature. Today, this adjunctive role of nitrous oxide to additional anesthetics remains its primary mode of use. By decreasing the amount of other agents required to produce general anesthesia nitrous oxide greatly decreases their undesirable side-effects.

Chemically, nitrous oxide is a colorless gas with a slightly sweet odor and no appreciable taste at normal ambient temperature and pressure. It is supplied in steel cylinders as a colorless liquid which is under pressure and in equilibrium with its gas phase. As it is released from the cylinder, some of the liquid nitrous oxide returns to the gas phase, the heat from the walls of the tank and from the surrounding air providing the heat required for vaporization. The tank becomes cold as a result, and the pressure inside the cylinder remains essentially constant until all the liquid has evaporated. Considering the number of electronic and electro-surgical devices in operating rooms and dentists' offices, it is fortunate that nitrous oxide is not flammable. It will, however, support combustion just as oxygen will and should be used with caution whenever electrocautary or laser instruments are being used in the head and neck areas.

The effect of nitrous oxide on vital functions circulation, blood pressure, central nervous system blood flow, and respiratory drive) is minimal. Hence, in its role as an adjunctive anesthetic agent it is nearly ideal. It has no muscle relaxant qualities however and in many procedures must also be accompanied by a paralyzing agent in order to allow the surgeon access without muscle tension resistance.

See also Anesthetics

Alfred Nobel.

NOBEL, ALFRED (1833-1896)

Swedish inventor, industrialist, and philanthropist

Owner of more than 350 patented inventions during his lifetime, Nobel is best known as the discoverer of dynamite and the man who upon his death bequeathed much of his large estate to support the annual Nobel Prizes for accomplishments in physics, **chemistry**, economics, physiology and medicine, literature, and the promotion of peace.

Born in Stockholm, Nobel received his education from private tutors and from various apprenticeships. Like his father, a manufacturer of mines and other **explosives**, Nobel displayed an avid interest in engineering and chemistry and as a young man worked for a time in the laboratory of French chemist Théophile Jules Pelouze (1807-1867), who is regarded by some as the inventor of guncotton (most accord the honor to **Christian Schönbein**). After extensive travels, through which he acquired the sharp skills of a businessman and the distinct advantages of a multilinguist, Nobel returned to Sweden in 1863 for the singular purpose of safely manufacturing **nitroglycerin**.

Almost two decades earlier, Ascanio Sobrero (1812-1888) had invented this oily liquid, but it proved so volatile as to preclude its widespread use. Instead, **gunpowder** and guncotton dominated the explosives industry, despite their own shortcomings. Through his own studies and experiments, begun as early as 1859, Nobel had familiarized himself with Sobrero's compound of glycerine treated with **nitric acid**, and had even exploded small quantities of it under **water**. Sharing his interest, his father during this same time designed a method for the large-scale production of the explosive. In Nobel's mind, all that remained was to devise a special blasting charge to ensure a predictable detonation of the nitroglycerine by shock rather than **heat**, which he already knew to be a dangerously imperfect firing method. The result was Nobel's first important invention, the **mercury** fulminate cap.

A fatal factory accident the following year, in which Nobel's brother Emil was killed, led the inventor to continue his research with nitroglycerine, this time in the hope of discovering a benign substance to absorb the liquid explosive, thereby making it safe for manipulation and transportation, without seriously diminishing its eruptive characteristics. After exhaustive experimentation, Nobel found a nearly perfect substance, kieselguhr. When saturated with nitroglycerine, this porous clay became a highly desirable explosive, which Nobel termed dynamite and patented in 1867.

Virtually overnight, dynamite revolutionized the mining industry, for it was five times as powerful as gunpowder, relatively easy to produce, and reasonably safe to use. Nobel acquired a vast fortune from this invention, which spawned an intricate network of factories, sales representatives, and distributors in several industrialized countries around the world. Despite the enormous demands of his business ventures, which required that he travel almost continuously and engage repeatedly in legal battles, Nobel persevered with his scientific research. Less than satisfied with the qualities of *kieselguhr*, which occasionally leaked nitroglycerine as well as somewhat reduced the liquid's power, he began experimenting with nitroglycerine and collodion, a low **nitrogen** form of guncotton. He found that these two substances formed a gelatinous **mass** which, with modifications, possessed a high resistance to water and an explosive force greater than that of dynamite. The invention, perfected in 1875, became known by a variety of names, including blasting **gelatin**, Nobel's Extra Dynamite, saxonite, and gelignite.

One of Nobel's last significant discoveries was closely related to his work with blasting gelatin. Like a number of other inventors, Nobel was in search of a smokeless powder to replace gunpowder. In 1888 he introduced ballistite, a mixture of nitroglycerine, guncotton, and camphor which could be cut into flakes and used as a propellant; the substance was particularly valuable for its ability to burn ferociously without exploding. A year later, two British scientists invented a smokeless powder based on ballistite called **cordite**. To Nobel, the invention represented an infringement on his patent; his suit to recover damages, however, was unsuccessful.

Nobel died in 1896, but despite his long and successful career developing and manufacturing explosives, he was a devoted humanitarian who wished to aid efforts that might bring about lasting peace as well as beneficial advancements in technology. To this end he composed a handwritten will which, though problematic and fiercely contested, led to the creation of the Nobel Foundation, which grants monetary prizes for contributions to scientific research and efforts toward world peace. The first prizes were awarded in 1901 and the tradition continues today.

NODDACK, IDA TACKE (1896-1979)

German chemist

Working with fellow chemist Walter Noddack (her future husband) and x-ray specialist Otto Berg, Ida Tacke discovered element 75, **rhenium**, in 1925, thus solving one of the mysteries of the **periodic table** of elements introduced by Russian chemist **Dmitry Ivanovich Mendeleev** in 1869. Ida Tacke Noddack's continuing study of the periodic table also led her to be the first to suggest in 1934 that physicist Enrico Fermi had not made a new element in an experiment with **uranium** as he thought, but instead had discovered **nuclear fission**. Her prediction was not verified until 1939.

Ida Tacke was born in Germany on February 25, 1896, and studied at the Technical University in Berlin, where she received the first prize for **chemistry** and metallurgy in 1919. In 1921, soon after receiving her doctorate, she set out to isolate two of the elements that Mendeleev had predicted when he proposed the Periodic System and displayed all known elements in a format now called the periodic table. Mendeleev had left blank spaces on his table for several elements that he expected to exist but that had not been identified. Two of these, elements 43 and 75, were located in Group VII under **manganese**.

Assuming that these elements would be similar in their properties to manganese, scientists had been searching for them in manganese ores. Tacke and Walter Noddack, who headed the chemical laboratory at the Physico-Technical Research Agency in Berlin, focused instead on the lateral neighbors of the missing elements, **molybdenum**, **tungsten**, **osmium**, and **ruthenium**. With the assistance of Otto Berg of the Werner-Siemens Laboratory, who provided expertise in analyzing the x-ray spectra of substances, Tacke and Noddack isolated element 75 in 1925 and named it rhenium, from *Rhenus,* Latin for the Rhine, an important river in their native Germany. It took them another year to isolate a single gram of the element from 660 kilograms of molybdenite ore. They also believed they had discovered traces of element 43, which they dubbed masurium. Later research, however, did not confirm their results. Now known as **technetium**, element 43 has never been found in nature, although it has been produced artificially.

In 1926, Ida Tacke married Walter Noddack. They would work together in their research until Walter Noddack's death in 1960, and together would publish some one hundred scientific papers. The Noddacks were awarded the Leibig Medal of the German Chemical Society in 1934 for their discovery of rhenium.

In 1934 Ida Noddack challenged the conclusions of Enrico Fermi and his group that they had produced transuranium elements, artificial elements heavier than uranium, when they bombarded uranium atoms with **subatomic particles** called neutrons. Although other scientists agreed with Fermi, Noddack suggested he had split uranium atoms into isotopes of known elements rather than added to uranium atoms to produce heavier, unknown elements. She had no research to support her theory, however, and for five years her hypothesis that atomic nuclei had been split was virtually ignored. ''Her suggestion was so out of line with the then-accepted ideas about the atomic **nucleus** that it was never seriously discussed,'' fellow chemist **Otto Hahn** would later comment in his autobiography. In 1939, after much research had been done by many scientists, Hahn, **Fritz Strassmann** and **Lise Meitner** discovered that Noddack had been right. They named the process nuclear fission.

The Noddacks moved from Berlin to the University of Freiburg in 1935, to the University of Strasbourg in 1943, and to the State Research Institute for Geochemistry in Bamberg in 1956. In 1960, Walter Noddack died. Ida Noddack received the High Service Cross of the German Federal Republic in 1966. During her life, she received honorary membership in the Spanish Society of Physics and Chemistry and the International Society of Nutrition Research, as well as an honorary doctorate of science from the University of Hamburg. Ida Noddack retired in 1968 and moved to Bad Neuenahr, a small town on the Rhine. She died in 1979.

NOMENCLATURE

Chemical nomenclature is the terminology used to describe compounds and processes in **chemistry**. Its name is dervived from the Greek *nomen clatura*, which literally means list of names.

Many chemical processes are named after their original discoverers. A compound cannot be assigned an arbitrary name, for there are certain rules that must be followed. These rules are administered by the **International Union of Pure and Applied Chemistry**, or IUPAC. The forerunner of this organization was founded in Switzerland in 1892. All chemists follow their rules and use a common system for the naming of chemicals.

The majority of nomenclatural work that is being carried out today is concerned with **organic chemistry,** because there are far more new organic compounds currently being made and discovered. For example, there are over six million compounds of **carbon** already known, and without a rigid system describing these compounds there would be much confusion. From an IUPAC name, it is possible to work out the constituents of a particular compound. In the majority of cases, it is also possible to get an indication of the relative placing of different components within the **molecule**. When an IUPAC name is employed, chemists worldwide has a very good idea about what is being discussed.

There are four basic steps in naming a new organic compound. We can take one of the simplest organic classes of compound as an example. The following are the steps used in working out a name for a new alkane. These steps are essentially the same for other organic compounds. First, the longest straight chain of carbon atoms is identified in the compound. The name of this chain is then used to provide the base name of the compound. It should be noted that when the compound is drawn, the longest straight chain might not be drawn in a straight line. It is the longest chain ignoring side branches that is important. With an alkane, once the longest chain has been identified, that name is us ed as the base name and the chemical is said to be a substituted form of that compound. For example, if the longest chain were found to be five carbon atoms, then the compound would be identified as a substituted pentane. Any groups attached to the main chain are called substituents, they are there in place of **hydrogen** atoms. The second s tep is to number the atoms in the longest chain, starting with the end that is closest to a substituent. If there were a **methyl group** attached to one of the carbons, the numbering would start so that it gave the carbon with the methyl group the smallest possible number. Next, each substituent group is named and its location is given by the number of the carbon **atom**. The final step deals with the situation where there are several substituent groups present. Each group is listed alphabetically and where there are more than one of a particular group, the number is given by a prefix. The prefixes used include mono- for one, di- for two, tri- for three, and so on. For instance, a compound named 2,3-dimethylpentane would have a chain of five carbon atoms. At the second and third carbon atoms, instead of a full complement of hydrogens, each carbon atom would have a methyl group substituting one of the carbons. If a substance were comprised of several groups of the same type, the approximate number of constituents could be indicated by the prefix oligo-, meaning that it has a low number of components, or poly- if there were many constituents. Thus, if we have a short chain of nucleotides, we could talk about an oligonucleotide. If we had a long chain of **peptides**, we would refer to a polypeptide.

NONMETAL

The known elements are classified as **metals**, nonmetals, or metalloids. Unlike the metals, which are shiny, the nonmetals vary greatly in appearance. Of the approximately 20 elements that are considered to be nonmetals at room temperature: **carbon**, phosphorous, **sulfur**, and **iodine** are **solids**; **bromine** is a liquid; **hydrogen**, **oxygen**, **nitrogen**, **fluorine**, and **chlorine** are **diatomic gases**; and **helium**, **neon**, **argon**, **krypton**, **xenon**, and **radon** are monotomic gases.

Nonmetals in the solid state are usually brittle and lack metallic luster. With the exception of **graphite**, they are poor conductors of **electricity**. There are wide variations in other physical properties of the nonmetals. For example, helium has a melting point of -452.2°F (-269°C) compared to carbon's melting point of over 6,332°F (3,500°C). Hardness ranges from that of **diamond** to the softness of white phosphorous. Nonmetals show an even greater variation in chemical properties than do metals. Some (argon, helium, neon) form almost no compounds; others (chlorine, fluorine) have high chemical activities. The nonmetals are neither malleable nor ductile; if drawn out or hammered, they shatter. In general, most nonmetals are brightly colored. Compounds formed between a metal and nonmetal are usually ionic; those formed between nonmetals are always covalent.

Nonmetals tend to gain electrons in chemical reactions. For example, the elements oxygen and fluorine tend to acquire electrons when they react with metals. The tendency of nonmetals to gain electrons is evident in their high electronegativities (ranging from 2.1 to 4.0), compared to less than 2.0 for metals. (The **electronegativity** of an **atom** is a measure of the tendency of that atom to attract electrons to itself when forming a **covalent bond**.)

Among the nonmetals, the noble gases show no tendency to combine with each other, and show little tendency to react with the atoms of other elements. In contrast most other nonmetals tend to form polyatomic molecules in the gaseous state. These two characteristics of the noble gases derive from the unusual stabilities of the **electron** configurations in these elements.

When a pair of nonmetals form only one compound, the name of the element whose symbol appears first in the compound's chemical formula (the element of lower electronegativity) is written first. The second portion of the name is formed by adding the suffix -ide to the stem of the name of the second nonmetal, an example being hydrogen chloride (HCl). (Many binary compounds of nonmetals have common names, e.g., **water**, **ammonia**, and **nitric oxide**.) If more than one binary compound is formed by two nonmetals, the Greek prefixes, di-, tri-, tetra-, penta-, hexa-, etc., are used to designate the number of atoms in each element. Examples among the oxides of nitrogen include dinitrogen pentoxide, nitrogen dioxide, and nitrogen oxide.

In general, nonmetals (with exceptions such as argon) combine readily with active metals, and slightly less readily with one another. For example, chlorine and fluorine will react violently with most active metals, but will not react with oxygen. Sulfur and phosphorous, on the other hand, burn brightly in oxygen.

Any given nonmetal may have a variety of **oxidation** numbers, with the maximum oxidation number—the number of electrons involved in bond formation—being equal to the **periodic table** group number. For example, sulfur, which is in Group VIA, has a maximum positive oxidation number of +6, but it can also exhibit oxidation numbers of +4, +2, and -2. In the case of binary covalent compounds, the positive oxidation number is assigned to the more electropositive element, and the negative oxidation number to the more electronegative element.

There is no sharp dividing line between the metals and nonmetals; the metalloids (B, Si, Ge, As, Se, Sb, and Te), also called semimetals, form a buffer between the two. These elements share some of the properties of metals and nonmetals.

The nonmetal hydrogen exhibits much of the physical properties and many of the chemical properties of metals; it behaves something like the metals in the way it undergoes chemical reactions. For example, hydrogen combines much more readily and violently with nonmetals than with metals.

The oxides, hydrides, and halides are volatile covalent compounds that have low melting points. Oxides of nonmetals, when soluble, combine with water to form acids.

The nonmetals do not react with weak acids. Some are attacked by bases, but the reactions do not exhibit a definite pattern.

NORMALITY

Normality is a measure of **concentration** of a chemical species in **solution**. It is defined as the number of **equivalents** of solute per **volume** of solution in liters.

This definition is dependent upon the type of chemical reaction under consideration. For acid-base reactions, one equivalent of acid is the quantity that supplies one **mole** of H^+ ions; whereas one equivalent of base is the amount that reacts with one mole of H^+ ions. In oxidation-reduction reactions, an equivalent is the quantity of substance that either gains (oxidant) or loses (reductant) one mole of electrons. Though not as commonly used, an equivalent in **precipitation** and/or complexation reactions is the amount of the substance that provides or reacts with one mole of reacting **cation** if it is univalent, one-half of a mole if it is divalent, one-third of a mole if it is trivalent, *etc.* At the equivalence point of a titrimetric analysis, irrespective of the reaction type, one equivalent of analyte will have reacted with one equivalent of titrant.

Products formed as the result of acid-base and/or oxidation-reduction reactions often depend upon the experimental conditions. For example, when H_3PO_4 reacts as an acid to form the HPO_4^{2-} **ion** it loses two H^+ ions. Thus, one mole of H_3PO_4 (98.0 grams) will be two equivalents. If one mole of H_3PO_4 is dissolved in sufficient **water** to form 1 liter of solution, its concentration is given as 1 Molar (abbreviated as 1 M) or 2 Normal (abbreviated as 2 N). If only a single H^+ is removed, the concentration is only 1 N. The normality of the H_3PO_4 depends upon the number of H^+ ions that are removed. Similarly, in oxidation-**reduction** reactions, the calculated normality depends upon the number of electrons gained or lost be one formula unit of the substance (*i.e.*, Normality = number of electrons transferred x Molarity). Concentrations expressed as normalities are meaningless unless accompanied by knowledge of how the solution is to be used and the chemical reaction(s) involved.

See also Concentration; Equivalent; Molarity

NORRISH, RONALD G. W. (1897-1978)

English physical chemist

The English chemist Ronald G. W. Norrish spent his academic life studying reaction **kinetics**, a discipline in **chemistry** concerned with rates of chemical reactions and factors influencing those rates. Norrish received the 1967 Nobel Prize for chemistry—which he shared with a former student, **George Porter**, and German scientist **Manfred Eigen**—for his work in this realm. A pioneer researcher in flash photolysis (chemical reactions induced by intense bursts of light), Norrish developed a process which allowed minute intermediate stages of a chemical reaction to be measured and described. He also contributed to chemistry an understanding of chain reactions, **combustion**, and polymerization (the formation of large molecules from numerous smaller ones). Over his career, Norrish was awarded the Liversidge Medal of the Chemical Society and the Davy Medal of the Royal Society, both in 1958, and the Bernard Lewis Gold Medal from the Combustion Institute in 1964. In addition, he was a member of scientific academies in eight foreign countries.

Ronald George Wreyford Norrish was born on November 9, 1897, in Cambridge, England. The son of Amy and Herbert Norrish, he attended the Perse Grammar School and won a scholarship to study natural sciences at Emmanuel College in Cambridge University. Although Norrish entered Cambridge in 1915, World War I intervened and he served in France as a lieutenant in the Royal Field Artillery. Captured by the Germans in 1918, he spent a year in a prisoner of war camp before being repatriated. Norrish then returned to his academic career at Cambridge and finished his bachelor of science degree in chemistry by 1921.

Norrish studied for his doctorate under the renowned physical chemist E. K. Rideal, who directed him to investigations of chemical kinetics and **photochemistry** (the effect of light upon solutions of **potassium** permanganate). By 1924 he had earned his Ph.D. in chemistry, staying on at the university to become a fellow of Emmanuel College and then, in 1925, a demonstrator in chemistry. The following year Norrish married Anne Smith who was a lecturer at the University of Wales. They would eventually have twin daughters together.

Norrish served for the rest of his academic and research life at Cambridge University. He became the Humphrey Owen Jones Lecturer in Physical Chemistry in 1930, then seven years later, he was named professor of **physical chemistry** as well as the director of the department of physical chemistry. He retained this position until 1965, when he retired.

Norrish's early work at Cambridge involved the photochemistry of rather simple compounds, such as **ketones**, **aldehydes**, and **nitrogen** peroxide. He discovered that light breaks down these compounds in one of two directions, creating either stable molecules or unstable "**free radicals**," which are molecules that have unpaired electrons. As a corollary to this work, Norrish and his laboratory also began studying chemical chain reactions. Working with M. Ritchie, Norrish was able to describe the process by which **hydrogen** and **chlorine** react when initiated by light. Studies of other chain reactions led Norrish and his fellow workers to a study of **hydrocarbon** combustion, building on **Nikolai N. Semenov**'s work in branching chain reactions to describe the means by which **methane** and ethylene are combusted. They discovered that **formaldehyde** formation is a necessary intermediate step in such a **chain reaction**.

Norrish also conducted an investigation into the mechanics of polymerization, primarily in **vinyl** compounds. It was Norrish who coined the term gel effect to describe the final slowing-down stages of polymerization as a **solution** undergoing the process becomes increasingly semi-fluid or viscous. With the advent of World War II, Norrish's laboratory work increasingly involved military projects, such as research into gun-flash suppression. Norrish became chairman of the Incendiary Projectiles Committee during this period and also assisted in the development of incendiary devices.

After the war, Norrish worked with Porter to pioneer the study of flash-photolysis. This involved the measurement of very fast chemical reactions while exposing the substance to extremely strong and short blasts of light. Unstable molecules turned into free radicals, thus resulting in a dissociative reaction. Intermediate stages and products of such fast chemical reactions were then gauged by use of spectrographic analysis—the illumination by weaker flashes of light following at millisecond intervals upon the initial flash. Such analysis went a long way toward proving intermediate stages of reactions which had been, until the Norrish-Porter work, only theoretical.

Norrish and Porter continued their research together from 1949-1965, perfecting their technique to allow analysis of short-lived intermediate compounds down to a thousandth of a millionth of a second. They published numerous articles and opened new vistas of research in fast reactions. For such work, Norrish and Porter shared the 1967 Nobel Prize for chemistry with Eigen, who was doing similar work (although he employed a "relaxation technique," whereby small disturbances of equilibrium were induced rather than the intense ones elicited by flash-photolysis).

After his retirement in 1965, Norrish remained a senior fellow at his old college, Emmanuel. Having lived in Cambridge most of his life, Norrish felt an abiding affection for all things dealing with the university. Famous for his hospitality, Norrish held at-homes with an eclectic blend of cultural personalities in attendance. He died on June 7, 1978, in Cambridge.

NORTHROP, JOHN HOWARD (1891-1987)

American biochemist

John Howard Northrop, a Nobel laureate in **chemistry**, is best known for his work on the purification and crystallization of enzymes, which regulate important body functions like digestion and respiration. Northrop's studies on the chemical composition of enzymes enabled him to confirm the hypothesis that enzymes are proteins—a discovery that spurred much ad-

John Howard Northrop.

ditional research on these critical catalysts of biochemical reactions. For this discovery, Northrop was awarded the Nobel Prize in chemistry in 1946. In addition to these studies, he also contributed to the development of techniques for isolating—and thus identifying—a variety of substances, including bacterial viruses and valuable antitoxins.

Northrop was born in Yonkers, New York, on July 5, 1891, to John Isaiah and Alice Belle (Rich) Northrop. The Northrops hailed from a long list of notable ancestors; well-known relations include the Reverend Jonathan Edwards, president of Princeton University in 1758. Prior to his son's birth, Isaiah Northrop was killed in a laboratory fire at Columbia University, where he taught zoology. Alice Northrop, a biology teacher at Normal (Hunter) College, influenced her son's interest in zoology and biology. John Northrop received a B.S. in 1912 from Columbia University, where he majored in **biochemistry**. He continued his studies in Columbia's chemistry department and was awarded an M.A. in 1913 and a Ph.D. in 1915.

In 1915 Northrop accepted a position in the laboratory of **Jacques Loeb** at the Rockefeller Institute for Medical Research (RIMR). Loeb, an experimental physiologist, headed RIMR's laboratory of general physiology. There Northrop studied the effect of environmental factors on heredity through experimentation with *Drosophila* spp. (fruit fly). He developed a method for producing *Drosophila* spp. free of microorganisms, which revolutionized studies that investigated factors affecting the flies' lifespan. Using these flies, Northrop and Loeb demonstrated that **heat** affected the life and health of the flies—not light or expenditure of **energy**, as previously believed.

Northrop's work in Loeb's laboratory was interrupted by the advent of World War I. At that time he became involved with research geared toward wartime concerns. Northrop developed a **fermentation** process for **acetone** that was used in the production of **explosives** and airplane wing coverings. As a result of these efforts, he was commissioned a captain in the U.S. Army Chemical Warfare Service. He was subsequently sent to the Commercial **Solvents** Corporation in Terre Haute, Indiana, to oversee the plant development of acetone production.

Although Loeb and Northrop remained close associates, Northrop was ready for independent work upon his return to the institute in 1919. At this time he studied the digestive enzymes pepsin and trypsin; these studies continued throughout the 1920s and 1930s, but Northrop was also interested in a myriad of other scientific investigations. He studied vision in the *Limulus* crab; with RIMR colleague Moses Kunitz he analyzed the chemical composition of **gelatin**; and he worked with Paul De Kruif on bacterial suspensions.

Shortly after Loeb's death in 1924, Northrop transferred to the institute's Princeton, New Jersey, department of animal pathology; at this time he was made a full member of the institute. The animal pathology department was opened in 1917 to study basic research in animal diseases and later expanded in 1931 to include a department of plant pathology. Inspired by the work of **James B. Sumner**, who had isolated and crystallized an **enzyme** called urease, Northrop continued his studies of the protein-splitting enzymes pepsin, trypsin, and chymotrypsin; he eventually isolated and crystallized all three substances. In 1929, Northrop and M. L. Anson developed the **diffusion** cell, a relatively simple means for isolating materials. In 1931, with Kunitz, Northrop validated the usefulness of the phase rule **solubility** method of studying the purity of substances, which tests for the homogeneity of dissolved **solids**. By applying this testing method to crystalline pepsin, chymotrypsin, and trypsin, he corroborated Sumner's controversial belief that enzymes were **proteins**. The research of this period was presented in *Crystalline Enzymes* (1939), written by Northrop, Kunitz, and Roger Herriott.

Northrop's investigation of bacteriophages (viruses that attack bacteria) began in the 1920s, but did not flower until the 1930s. He and associate **Wendell Stanley** applied their techniques for isolating enzymes to crystallizing the tobacco mosaic virus—isolation being the first step in determining the chemical composition of any substance. From 1936 to 1938 Northrop examined the chemical nature of bacteriophages and successfully isolated purified nucleoprotein (protein plus **DNA** and RNA) from cultures of *Staphylococcus aureus,* the bacteria that causes boils; this finding was one of the earliest indications that nucleoproteins are an essential part of a virus. Using his ability to isolate and crystallize substances, in 1941 Northrop produced the first crystalline antibody, for diphtheria. He would later, with W. F. Goebel, produce an antibody for pneumococcus.

With the start of World War II, Northrop was once again called on to become involved with government research undertaken at RIMR. In 1941, RIMR and the U.S. Office of Scientific Research and Development (OSRD) initiated the investigation of lethal **gases** used in battle. One of Northrop's biggest wartime achievements was developing the Northrop Titrator and the portable, battery-operated Northrop Field Titrator. These devices measure the **concentration** of mustard gas in the air at some distance from the gassed zone. The Northrop Titrator is considered an important prototype for subsequent development of the more sophisticated defensive instruments used in chemical warfare today.

In 1949, RIMR's Princeton facility closed, prompting Northrop's move to the University of California, Berkeley. During his tenure as visiting professor, Northrop maintained his association with RIMR and was named professor emeritus in 1961. While at Berkeley, Northrop continued his work with bacteriophages. He conducted research on the life cycle of *B. megatherium* cells from their normal stage to that of a phage. He also investigated the origin of bacterial viruses and found that they were mutations of normal cells.

In addition to the Nobel Prize, which he shared with Stanley and Sumner, Northrop was also awarded the W. B. Cutting Traveling Fellowship and the Stevens Prize of the College of Physicians and Surgeons of Columbia University, and he was elected to the National Academy of Sciences. Beginning in 1924, he also served on the editorial board of the *Journal of General Physiology,* an association that spanned a sixty-two year period.

Northrop retired from Berkeley in 1959 and moved to Wickenburge, Arizona, where he died on May 27, 1987. He left his wife, Louise Walker, whom he married in June, 1918, and two children, Alice Havemeyer and John.

NUCLEAR CHEMISTRY

In some ways, **chemistry** can be viewed as having nothing whatsoever to do with the **nucleus** of the atoms. **Bonding**, **spectroscopy**, chemical **reactivity**, and such are all a consequence of the electrons of which an **atom** is partially composed. The nucleus, in chemical terms, is merely there. It contains the **mass** of the atom but it does not participate in chemistry.

Of course, this is a very limited view of chemistry as a whole and, in particular, the role of the nuclear chemistry. Just saying that the nucleus contains all of an atom's mass is not saying nearly enough about its role. The concept of the "atom"—the smallest unique piece of **matter** for any substance—can be dated back to the Greek philosopher, Democritus, but it took another 2,500 years before the concept really captured the imagination of scientists. Even **John Dalton**, who put forward the theory of the atom at the beginning of the **chemical revolution**, did not think of them as anything more than simple, solid spheres. For most practicing scientists, atoms were a way of arranging the elements and talking about compounds, but not something to be taken too seriously.

However, as the 19th century progressed, a better understanding of the atom developed. Radiation was observed, implying that there must be some structure within the atom. The **electron** was discovered and determined to be smaller than the atom. If the atoms was the "smallest, most indivisible component of matter," then what could the electron be? What was this radiation that scientists were observing? These are the questions that ultimately lead to the discovery of the nucleus and the formation of the twin sciences, nuclear physics and nuclear chemistry. Indeed, they are such twins that it is often difficult to tell whether a scientist researching the nucleus is a chemist or a physicist.

Atom nuclei are composed of two particles, the **proton** and the **neutron**. These particles are not fundamental. They are made up of quarks, gluons, and such but for the purposes of constructing the elements in the **periodic table**, they are the basic building blocks. The proton is slightly lighter than the neutron (1.0073 amu and 1.0087 amu, respectively, where an "amu" is an atomic mass unit) but it is also positively charged. The magnitude of the charge is equal to that on the electron but of opposite sign. The electron is much smaller than either nuclear particle or **nucleon**, with a mass of only 0.0005 amu. This is why it is accurate to say that the mass of an atom is principally located in the nucleus as the electrons contribute only a very small amount.

The nucleus is also incredibly small in size. Relatively speaking, it is about 100,000 times as small as the atom itself—about the equivalent of a golf ball sitting center field in the Los Angeles Coliseum, with the electron being a flea hopping around the outer concourse. But it is this tiny lump of protons and neutrons that determines the elemental nature of the atom. Each element has a unique number of protons, called its atomic number. That is, all **hydrogen** atoms—whether here on Earth or in the farthest reaches of outer space—have only a single proton in their nucleus. All **oxygen** atoms have eight protons and all **uranium** have ninety-two. The one characteristic that determines an atoms identity is the number of protons in its nucleus, not the number of electrons nor neutrons. In a neutral atom, the number of electrons is the same as the number of protons but atoms can be ionized, either oxidized or reduced, and end up with a different electron count.

The number of neutrons, on the other hand, can actually vary within a single element. As an example, consider hydrogen. It occurs in three different forms. The vast majority of hydrogen occurs as the simplest of all elements. It has a single proton for its nucleus and a single electron occupying an orbital. The second form of hydrogen has a nucleus with a proton and a neutron and is called **deuterium**. The third form has a proton and two neutrons and is called **tritium**. This particular arrangement of nucleons is unstable and so tritium readily undergoes radioactive decay to yield a **helium** atom. This decay process involves the conversion of a neutron to a proton via the ejection of an electron, which is an example of nuclear math. Two particles can combine to make one or be made by the decay of a single particle.

Each of these different forms of hydrogen are called isotopes (Greek for equal in place) and hydrogen is unique in having named isotopes. All the other elements use numbers to designate their isotopes, such as carbon-14 and uranium-235.

Within every element, up to and including uranium, there is at least one arrangement of protons and neutrons that yields a natural **isotope**. However, not all isotopes are stable. Indeed, many isotopes spontaneously convert to other elements through **radioactivity**. It is the instability of the nuclear configuration that is the source of radioactivity.

As an example, consider tritium. The arrangement of two neutrons to one proton means that the nucleus is "neutron rich." Neutrons, themselves, are unstable and decompose if they are removed from the nucleus. It is only through binding to the proton that neutrons have any stability at all. It is easy to see then that while the first neutron and proton might be stable, adding a second would generate some instability. The neutron in tritium decays as a consequence. This is a very simplistic view of the reasoning behind the process. Nuclear decay is a function of the "weak force" that has been studied by nuclear physicists and is beyond the scope of this discussion. Suffice it to say, though, that a roughly one to one ratio of neutrons to protons is stable for the lighter elements. As the **atomic number** increases, more neutrons are required to mediate the strong repulsive forces between the positively charged protons. The ratio of neutrons to protons increases so that uranium, for example, has 92 protons in its nucleus but about 148 neutrons, depending upon the isotope.

The lifetime of a radioactive nucleus is randomly determined. That is, any single unstable atom could undergo a nuclear reaction now or five years from now or five million years from now. If the lifetime of any single atom is undetermined, what does a **half-life** for a radioactive species mean?

Chemists routinely deal in **mole** quantities of substances and since a mole contains 6.022 x 1023 atoms or molecules, chemists invariably deal in statistics. With that many atoms, it is possible to say that one of them will decay in such and such a period of time. Not which one but that one will. Further, using calculus, it is relatively straightforward to show that half of the atoms initially present will decay in a specified time. This is called the half-life. For example, the half-life of tritium is 12.26 years. That means that if we had a mole of tritium atoms right now, in 12.26 years, we would have half a mole. At that point, we would be starting with half a mole, and in another 12.26 years we would have half of that quantity or a quarter of our original mole of tritium. In the following 12.26 years, we would be down to one eighth, and so on. It would take approximately 79 half-lives or 969 years before the tritium atoms actually run out.

The half-life of an isotope can vary in length from nanoseconds to millions or billions of years. This means that for some isotopes of different elements, their radioactivity is very short-lived. For others, they will still be emitting radiation so far into the future, that it is hard to imagine. This, of course, has led to many concerns about the use of nuclear power as a source of **energy**. How to dispose of radioactive waste is a major difficulty, particularly when that waste will still be radioactive many, many generations from now.

Nuclear medicine employs radioactive isotopes in the detection and curing of disease. Everything from brain scans to determining bone **density** to imaging the lungs employs radioisotopes. In essence, the process involves binding a radioisotope to a compound that has a specific affinity for an organ or region of the body. Using a variety of radiation detectors, the medical technician is able to view the soft tissue. For example, Positron Emission Tomography (PET) uses the positron (a positive electron) emitted by carbon-11 that has been incorporated into glucose to scan brain tissue for metabolic activity.

Radioactivity has also seen use in the treatment of cancer. There are two basic approaches that have been adopted. Historically, the first was to expose the whole patient to radioactive rocks or powders in the hope that the damaging radiation would preferentially kill the cancer cells. This approach has been refined so that tightly-focused beams are now employed. The second approach relies on the chemical inclusion of the radioactive atom into a **molecule** that is targeted for the tumor cells. This takes the radiation into the body and allows it to be more effective. Unfortunately, it is hard to focus chemical compounds on single organs with the result that other parts of the body are affected. The latter approach is a significant area of research in nuclear chemistry, finding the right delivery compound for the right type of radioactive isotope.

Radioactivity also finds use in preserving foods in which the gamma ray emissions from cobalt-60 are used to eradicate any bacteria or other decay causing agents. This method of preservation has lead to a significant increase in shelf life for produce but also to controversy over the potential harmful side effects. These concerns arise from the fact that just as radiation can be emitted from an atom, it can also be absorbed which leads to the formation of "unnatural" compounds within the food. These, in turn, could be harmful and may lead to disease. The process of testing the proposed use of radiation for food is meant to safeguard against this possibility.

One of the more interesting uses of nuclear chemistry is the natural incorporation of carbon-14 into living cells. Carbon-14 is a naturally occurring radioactive isotope of the **carbon** that makes up every cell in our bodies. It is generated, via irradiation, in the upper atmosphere by the bombardment of **nitrogen** with neutrons from cosmic rays. In the form of **carbon dioxide**, it is incorporated into plant cells which pass it along to other organisms. As long as an organism is alive, it accumulates carbon-14. But when it dies, the amount of carbon-14 present is fixed and thereafter, it decreases. By detecting both carbon-14 and its decay product, nitrogen-14, scientists are able to determine how long ago something was alive. In the case of an ancient artifact made of, for example, leather, this provides a date for its death. Assuming that the leather object was made shortly thereafter, carbon-14 dating provides a good method for determining its age. This technique is one of the most important uses of nuclear chemistry for archeologists.

NUCLEAR FISSION

Nuclear fission is a process in which the **nucleus** of an **atom** splits, usually into two pieces. This reaction was discovered when a target of **uranium** was bombarded by neutrons. Fission fragments were shown to fly apart with a large release of **energy**. The fission reaction was the basis of the atomic bomb, which was developed by the United States during World War II. After the war, controlled energy release from fission was applied to the development of **nuclear reactors**. Reactors are utilized for production of **electricity** at nuclear power plants, for propulsion of ships and submarines, and for the creation of radioactive isotopes used in medicine and industry.

The fission reaction was discovered in 1938 by two German scientists, **Otto Hahn** and **Fritz Strassmann**. They had been doing a series of experiments in which they used neutrons to bombard various elements. If they bombarded **copper**, for example, a radioactive form of copper was produced. Other elements became radioactive in the same way. When uranium was bombarded with neutrons, however, an entirely different reaction seemed to occur. The uranium nucleus apparently underwent a major disruption.

The evidence for this supposed process came from chemical analysis. Hahn and Strassmann published a scientific paper showing that small amounts of **barium** (element 56) were produced when uranium (element 92) was bombarded with neutrons. It was very puzzling to them how a single **neutron** could transform element 92 into element 56.

Lise Meitner, a long-time colleague of Hahn who had left Germany due to Nazi persecution, suggested a helpful model for such a reaction. One can visualize the uranium nucleus to be like a liquid drop containing protons and neutrons. When an extra neutron enters, the drop begins to vibrate. If the vibration is violent enough, the drop can break into two pieces. Meitner named this process fission because it is similar to the process of cell division in biology. It takes only a small amount of hyperlink energy to start the vibration which leads to a major breakup.

Scientists in the United States and elsewhere quickly confirmed the idea of uranium fission, using other experimental procedures. For example, a cloud chamber is a device in which vapor trails of moving nuclear particles can be seen and photographed. In one experiment, a thin sheet of uranium was placed inside a cloud chamber. When it was irradiated by neutrons, photographs showed a pair of tracks going in opposite directions from a common starting point in the uranium. Clearly, a nucleus had been photographed in the act of fission.

Another experimental procedure used a Geiger counter, which is a small, cylindrical tube that produces electrical pulses when a radioactive particle passes through it. For this experiment, the inside of a modified Geiger tube was lined with a thin layer of uranium. When a neutron source was brought near it, large voltage pulses were observed, much larger than from ordinary **radioactivity**. When the neutron source was taken away, the large pulses stopped. A Geiger tube without the uranium lining did not generate large pulses. Evidently, the large pulses were due to uranium fission fragments. The size of the pulses showed that the fragments had a very large amount of energy.

To understand the high energy released in uranium fission, scientists made some theoretical calculations based on Albert Einstein's famous equation $E=mc^2$. The Einstein equation says that **mass** m can be converted ito energy E, and the conversion factor is a huge number, c, which is the **velocity** of light, squared. One can calculate that the total mass of the fission products remaining at the end of the reaction is slightly less than the mass of the uranium atom plus neutron at the start. This decrease of mass, multiplied by ^{2}, shows numerically why the fission fragments are so energetic.

Through fission, neutrons of low energy can trigger off a very large energy release. With the imminent threat of war in 1939, a number of scientists began to consider the possibility that a new and very powerful "atomic bomb" could be built from uranium. Also, they speculated that uranium perhaps could be harnessed to replace **coal** or oil as a fuel for industrial power plants.

In general, **nuclear reactions** are much more powerful than chemical reactions. A chemical change such as burning coal or even exploding TNT affects only the outer electrons of an atom. A nuclear process, on the other hand, causes changes among the protons and neutrons inside the nucleus. The energy of attraction between protons and neutrons is about a million times greater than the chemical binding energy between atoms. Therefore, a single fission bomb, using **nuclear energy**, might destroy a whole city. Alternatively, nuclear electric power plants theoretically could run for a whole year on just a few tons of fuel.

In order to release a substantial amount of energy, many millions of uranium nuclei must split apart. The fission process itself provides a mechanism for creating a so-called **chain reaction**. In addition to the two main fragments, each fission event produces two or three extra neutrons. Some of these can enter nearby uranium nuclei and cause them in turn to fission, releasing more neutrons, which cause more fissions, and so forth. In a bomb explosion, neutrons have to increase very rapidly, in a fraction of a second. In a controlled reactor, however, the neutron population has to be kept in a steady state. Excess neutrons must be removed by some type of absorber material.

In 1942, the first nuclear reactor with a self-sustaining chain reaction was built in the United States. The principal designer was Enrico Fermi (1901-1954), an Italian physicist and the 1938 Nobel Prize winner in physics. He had emigrated to the United States to escape from Benito Mussolini's fascism. Fermi's reactor design had three main components: lumps of uranium (the fuel), blocks of **carbon** (the moderator, which slows down the neutrons), and control rods made of **cadmium** (an excellent neutron absorber). The reactor was built at the University of Chicago. When the pile of uranium and carbon blocks was about 10 ft (3 m) high and the cadmium control rods were pulled out far enough, Geiger counters showed that a steady-state chain reaction had been successfully accomplished. The power output was only about 200 watts, but it was enough to verify the basic principle of reactor operation. The power level of the chain reaction could be varied by moving the control rods in or out.

General Leslie R. Groves was put in charge of the project to convert the chain reaction experiment into a usable military weapon. Three major laboratories were built under wartime conditions of urgency and secrecy. Oak Ridge, Tennessee, became the site for purifying and separating uranium into bomb-grade material. At Hanford, Washington, four large reactors were built to produce another possible bomb material, **plutonium**. At Los Alamos, New Mexico, the actual work of bomb design was started in 1943 under the leadership of the physicist J. Robert Oppenheimer (1904-1967).

The element uranium is a mixture of two isotopes, uranium-235 and uranium-238. Both isotopes have 92 protons in the nucleus, but uranium-238 has three additional neutrons. Both isotopes have 92 orbital electrons to **balance** the 92 protons, so their chemical properties are identical. When uranium is bombarded with neutrons, the two isotopes have differing nuclear reactions. A high percentage of the uranium-235 nuclei undergo fission, as described previously. The uranium-238, on the other hand, simply absorbs a neutron and is converted to the next heavier **isotope**, uranium-239. It is not possible to build a bomb out of natural uranium. The reason is that the chain reaction would be halted by uranium-238 because it removes neutrons without reproducing any new ones.

The fissionable isotope, uranium-235, constitutes only about 1% of natural uranium, while the non-fissionable neutron absorber, uranium-238, makes up the other 99%. To produce bomb-grade, fissionable uranium-235, it was necessary to build a large isotope separation facility. Since the plant would require much electricity, the site was chosen to be in the region of the Tennessee Valley Authority (TVA). The technology of large-scale isotope separation involved solving many difficult, unprecedented problems. By early 1945, the Oak Ridge Laboratory was able to produce kilogram amounts of uranium-235 purified to better than 95%.

An alternate possible fuel for a fission bomb is plutonium-239. Plutonium does not exist in nature but results from radioactive decay of uranium-239. Fermi's chain reaction experiment had shown that uranium-239 can be made in a reactor. However, to produce several hundred kilograms of plutonium required a large increase from the power level of Fermi's original experiment. Plutonium production reactors were constructed at Hanford, Washington, located near the Columbia river to provide needed cooling **water**. A difficult technical problem was how to separate plutonium from the highly radioactive fuel rods after irradiation. This was accomplished by means of remote handling apparatus that was manipulated by technicians working behind thick protective **glass** windows.

With uranium-235 separation started at Oak Ridge and plutonium-239 production under way at Hanford, a third laboratory was set up at Los Alamos, New Mexico, to work on bomb design. In order to create an explosion, many nuclei would have to fission almost simultaneously. The key concept was to bring together several pieces of fissionable material into a so-called critical mass. In one design, two pieces of uranium-235 were shot toward each other from opposite ends of a cylindrical tube. A second design used a spherical shell of plutonium-239, to be detonated by an ''implosion'' toward the center of the sphere.

The first atomic bomb was tested at an isolated desert location in New Mexico on July 16, 1945. President Truman then issued an ultimatum to Japan that a powerful new weapon could soon be used against them. On August 8, a single atomic bomb destroyed the city of Hiroshima with over 80,000 casualties. On August 11, a second bomb was dropped on Nagasaki with a similar result. The Japanese leaders surrendered three days later.

The decision to use the atomic bomb has been vigorously debated over the years. It brought a quick end to the war and avoided many casualties that a land invasion of Japan would have cost. However, the civilians who were killed by the bomb and the survivors who developed radiation sickness left an unforgettable legacy of fear. The horror of mass annihilation in a nuclear war is made vivid by the images of destruction at Hiroshima.

The first nuclear reactor designed for producing electricity was put into operation in 1957 at Shippingsport, Pennsylvania. From 1960-1990, more than 100 nuclear power plants were built in the United States. These plants now generate about 20% of the nation's electric power. Worldwide, there are over 400 nuclear power stations.

The most common reactor type is the pressurized water reactor (abbreviated PWR). The system operates like a coal-burning power plant, except that the firebox of the coal plant is replaced by a reactor. Nuclear energy from uranium is released in the two fission fragments. The fuel rod becomes very hot because of the cumulative energy of many fissioning nuclei. A typical reactor core contains hundreds of these fuel rods. Water is circulated through the core to remove the **heat**. The hot water is prevented from boiling by keeping the system under pressure.

The pressurized hot water goes to a heat exchanger where steam is produced. The steam then goes to a turbine, which has a series of fan blades that rotate rapidly when hit by the steam. The turbine is connected to the rotor of an electric generator. Its output goes to cross-country transmission lines that supply the electrical users in the region. The steam that made the turbine rotate is condensed back into water and is recycled to the heat exchanger. This method of generating electricity was developed for coal plants and is known to be very reliable.

Safety features at a nuclear power plant include automatic shutdown of the fission process by insertion of control rods, emergency water cooling for the core in case of pipeline breakage, and a concrete containment shell. It is impossible for a reactor to have a nuclear explosion because the fuel enrichment in a reactor is intentionally limited to about 3% uranium-235, while almost 100% pure uranium-235 is required for a bomb. The worst accident at a PWR would be a steam explosion, which could contaminate the inside of the containment shell. This scenario was played out at the Chernobyl nuclear power plant in the Ukrainian republic of the former USSR on April 26, 1986. Considered the world's worst nuclear power plant disaster, the accident occurred when human errors led to a dangerous heat and steam buildup, triggering two enormous explosions at the Chernobyl plant.

The fuel in the reactor core consists of several tons of uranium. As the reactor is operated, the uranium content gradually decreases because of fission, and the radioactive waste products (the fission fragments) build up. After about a year of operation, the reactor must be shut down for refueling. The old fuel rods are pulled out and replaced. These fuel rods, which are very radioactive, are stored under water at the power plant site. After five to ten years, much of their radioactivity has decayed. Only those materials with a long radioactive lifetime remain, and eventually they will be stored in a suitable underground depository.

There are vehement arguments for and against nuclear power. The various advantages and problems should be thoroughly aired so that the general public can evaluate for itself whether the benefits outweigh the risks. Additional electric power plants will be required in the future to supply a growing world population that desires a higher standard of living.

All methods of producing electricity have serious environmental impacts. The main objections to nuclear power plants are the fear of possible accidents, the unresolved problem of nuclear waste storage, and the possibility of plutonium diversion for weapons production by a terrorist group. The issue of waste storage becomes particularly emotional because leakage from a waste depository could contaminate ground water. Chemical dump sites have leaked in the past, so there is distrust of all hazardous wastes.

The main advantage of nuclear power plants is that they do not cause atmospheric **pollution**. No smokestacks are needed because nothing is being burned. France initiated a large-scale nuclear program after the Arab oil embargo in 1973 and has been able to reduce its **acid rain** and **carbon dioxide** emissions by more than 40%. Nuclear power plants do not contribute to the global warming problem. Shipments of fuel are minimal so the hazards of coal transportation and oil spills are avoided.

Environmentalists are divided in their opinions of nuclear power. It is widely viewed as a hazardous technology but there is growing concern about atmospheric pollution making nuclear power more acceptable.

NUCLEAR FUSION

Nuclear fusion is the process by which two light atomic nuclei combine to form one heavier atomic **nucleus**. As an example, a **proton** (the nucleus of a **hydrogen** atom) and a **neutron** will, under the proper circumstances, combine to form a deuteron (the nucleus of an **atom** of "heavy hydrogen"). In general, the **mass** of the heavier product nucleus is less than the total mass of the two lighter nuclei.

When a proton and neutron combine, for example, the mass of the resulting deuteron is 0.00239 atomic mass unit less than the total mass of the proton and neutron combined. This "loss" of mass is expressed in the form of 2.23 MeV (million **electron** volts) of kinetic **energy** of the deuteron and other particles and as other forms of energy produced during the reaction. Nuclear fusion reactions are like **nuclear fission** reactions, therefore, in the respect that some quantity of mass is transformed into energy.

Some typical fusion reactions

The particles most commonly involved in nuclear fusion reactions include the proton, neutron, deuteron, a triton (a proton combined with two neutrons), a **helium**-3 nucleus (two protons combined with a neutron), and a helium-4 nucleus (two protons combined with two neutrons). Except for the neutron, all of these particles carry at least one positive electrical charge. That means that fusion reactions always require very large amounts of energy in order to overcome the force of repulsion between two like charged particles. For example, in order to fuse two protons with each other, enough energy must be provided to overcome the force of repulsion between the two positively charged particles.

Naturally occurring fusion reactions

As early as the 1930s, a number of physicists had considered the possibility that nuclear fusion reactions might be the mechanism by which energy is generated in the stars. Certainly no familiar type of chemical reaction, such as **oxidation**, could possibly explain the vast amounts of energy released by even the smallest star. In 1939, the German-American physicist Hans Bethe worked out the mathematics of energy generation in which a proton first fuses with a **carbon** atom to form a **nitrogen** atom. The reaction then continues through a series of five more steps, the net result of which is that four protons disappear and are replaced by one helium atom.

Bethe chose this sequence of reactions because it requires less energy than does the direct fusion of four protons and, thus, is more likely to take place in a star. Bethe was able to show that the total amount of energy released by this sequence of reactions was comparable to that which is actually observed in stars.

The Bethe "**carbon cycle**" is by no means the only nuclear fusion reaction that one might conceive. A more direct approach, for example, would be one in which two protons fuse to form a deuteron. That deuteron could, then, fuse with a third proton to form a helium-3 nucleus. Finally, the helium-3 nucleus could fuse with a fourth proton to form a helium-4 nucleus. The net result of this sequence of reactions would be the combining of four protons (hydrogen nuclei) to form a single helium-4 nucleus. The only net difference between this reaction and Bethe's carbon cycle is the amount of energy involved in the overall set of reactions.

The term less energy used to describe Bethe's choice of **nuclear reactions** is relative, however, since huge amounts of energy must be provided in order to bring about any kind of fusion reaction. In fact, the reason that fusion reactions can occur in stars is that the temperatures in their interiors are great enough to provide the energy needed to bring about fusion. Since those temperatures generally amount to a few million degrees, fusion reactions are also known as thermonuclear (thermo = **heat**) reactions.

Fusion reactions on Earth

The understanding that fusion reactions might be responsible for energy production in stars brought the accompa-

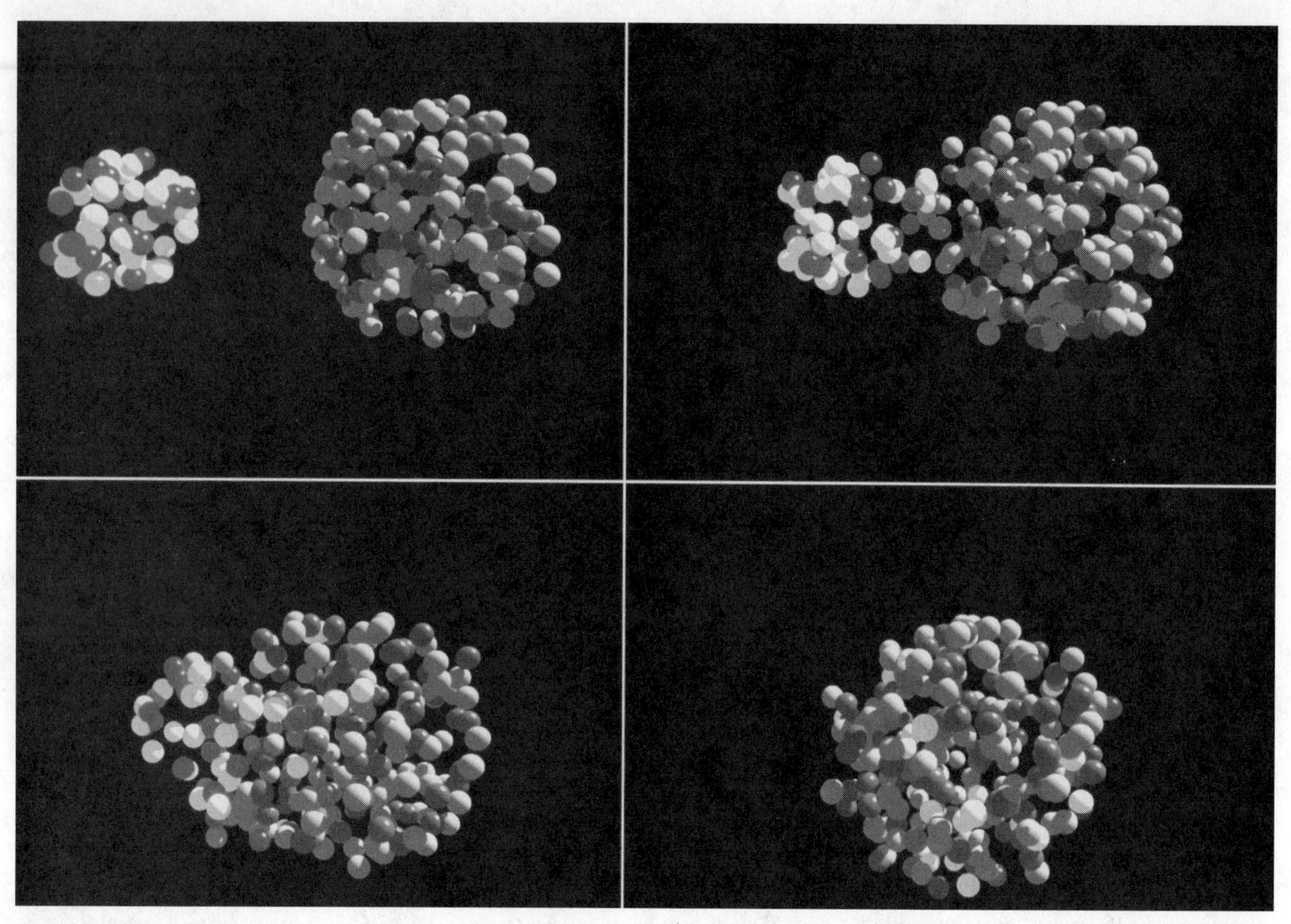

A computerized simulation sequence of gold and nickel nuclei joining, or fusing, to make a nielsbohrium heavy nucleus. Each nucleus type has a specific number of positively-charged protons. ***(Photograph by Jens Konopka and Henning Weber/Science Photo Library, Photo Researchers, Inc. Reproduced by permission.)***

nying realization that such reactions might be a very useful source of energy for human needs. Imagine that it would be possible to build and operate a small star on the outskirts of your community that operated on nuclear fusion. That power plant would be able to supply all of the community's energy needs as far into the future as anyone could see.

The practical problems of building a fusion power plant are incredible, however, and scientists are still a long way from achieving that goal. A much simpler challenge, however, is to construct a fusion power plant that does not need to be controlled, that is, a fusion bomb.

Scientists who worked on the first fission (''atomic'') bomb during World War II were aware of the potential for building an even more powerful bomb that operated on fusion principles. Here is how it would work.

The core of the fusion bomb would consist of an fission bomb, such as the one they were then developing. That core could then be surrounded by a casing filled with isotopes of hydrogen. Isotopes of hydrogen are various forms of hydrogen that all have a single proton in their nucleus, but may have zero, one, or two neutrons. The nuclei of the hydrogen isotopes are the proton, the deuteron, and the triton.

Imagine that a device such as the one described here could be exploded. In the first fraction of a second, the fission bomb would explode, releasing huge amounts of energy. In fact, the **temperature** at the heart of the fission bomb would reach a few millions degrees, the only way that humans know of for producing such high temperatures.

That temperature would not last very long, but in the microseconds that it did exist, it would provide the energy for fusion to begin to occur within the casing surrounding the fission bomb. Protons, deuterons, and tritons would begin fusing with each other, releasing more energy, and initiating other fusion reactions among other hydrogen isotopes. The original explosion of the fission bomb would have ignited a small star-like reaction in the casing surrounding it.

From a military standpoint, the fusion bomb had one powerful advantage over the fission bomb. For technical reasons, there is a limit to the size one can make a fission bomb. But there is no technical limit on the size of a fusion bomb. One simply makes the casing surrounding the fission bomb larger and larger, until there is no longer a way to lift the bomb into the air so that it can be dropped on an enemy.

On August 20, 1953, the Soviet Union announced the detonation of the world's first fusion bomb. It was about 1,000 times more powerful than was the fission bomb that had been dropped on Hiroshima less than a decade earlier. Since that date, both the Soviet Union (now Russia) and the United States have become proficient at manufacturing fusion bombs.

As research on fusion weapons was going on, attempts were also being made to develop peaceful uses for nuclear fusion. The concept of a star power plant just outside the city was never out of sight for a number of nuclear scientists.

The problems to be solved in controlling the nuclear fusion reaction have, however, been enormous. The most obvious challenge is simply to find a way to hold the nuclear fusion reaction in place as it occurs. One can't build a machine made out of metal, plastic, **glass**, or any other common kind of material. At the temperatures at which fusion occurs, any one of these materials would vaporize instantly. So how does one contain the nuclear fusion reaction?

Traditionally, two general approaches have been developed to solve this problem: magnetic and inertial containment. To understand the first technique, imagine that a mixture of hydrogen isotopes has been heated to a very high temperature. At a sufficiently high temperature, the nature of the mixture begins to change. Atoms totally lose their electrons, and the mixture consists of a swirling mass of positively charged nuclei and electrons. Such a mixture is known as a **plasma**.

One way to control that plasma is with a magnetic field. One can design such a field so that a swirling hot mass of plasma within it can be held in any kind of a shape one chooses. The best known example of this approach is a doughnut-shaped Russian machine known as a tokamak. In the tokamak, two powerful electromagnets create fields that are so powerful that they can hold a hot plasma in place as readily as a person can hold an orange in her hand.

The technique, then, is to **heat** the hydrogen isotopes to higher and higher temperatures while containing them within a confined space by means of the magnetic fields. At some critical temperatures, nuclear fusion will begin to occur. At that point, the tokamak is producing energy by means of fusion while the fuel is being held in **suspension** by the magnetic field.

A second method for creating controlled nuclear fusion makes use of a laser beam or a beam of electrons or atoms. In this approach, hydrogen isotopes are suspended at the middle of the machine in tiny hollow glass spheres known as microballoons. The microballoons are then bombarded by the laser, electron, or atomic beam and caused to implode. During implosion, enough energy is produced to initiate fusion among the hydrogen isotopes within the pellet. The plasma thus produced is then confined and controlled by means of the external beam.

The production of useful nuclear fusion energy by either of these methods depends on three factors: temperature, containment time, and energy release. That is, it is first necessary to raise the temperature of the fuel (the hydrogen isotopes) to a temperature of about 100 million degrees. Then, it is necessary to keep the fuel suspended at that temperature long enough for fusion to begin. Finally, some method must be found for tapping off the energy produced by fusion.

A measure of the success of a machine in producing useful fusion energy is known as the Lawson confinement parameter, the product of the **density** of particle in the plasma and the time the particles are confined. That is, in order for controlled fusion to occur, particles in the plasma must be brought close together and they must be kept together for some critical period of time. All of this must take place, of course, at a temperature at which fusion can occur.

The two nuclear reactions now most commonly used for power production purposes are designated as D-D and D-T reactions. The former stands for deuterium-deuterium and involves the combination of two **deuterium** nuclei to form a helium-3 nucleus and a free neutron. The second reaction stands for deuterium-tritium and involves the combination of a deuterium nucleus and a **tritium** nucleus to produce a helium-4 nucleus and a free neutron. The most common form of an inertial confinement machine, for example, uses a fuel that consists of equal parts of deuterium and tritium.

Hope for the future

Research on controlled fusion power has now been going on for a half century with somewhat disappointing results. Some experts believe that success may be ''just around the corner,'' but others argue that the problems of an economically feasible fusion power plant may never be solved.

In recent years, scientists have begun to explore approaches to fusion power that depart from the more traditional magnetic and inertial confinement techniques. One such approach is called the PBFA process. In this machine, electric charge is allowed to accumulate in capacitors and then discharged in 40-nanosecond micropulses. **Lithium** ions are accelerated by means of these pulses and forced to collide with deuterium and tritium targets. Fusion among the lithium and hydrogen nuclei takes place, and energy is released. Thus far, however, the PBFA approach to nuclear fusion has been no more successful than has that of more traditional methods.

Cold fusion

The scientific world was astonished in March of 1989 when two electrochemists, Stanley Pons and Martin Fleischmann, reported that they had obtained evidence for the occurrence of nuclear fusion at room temperatures. During the **electrolysis** of heavy **water** (deuterium oxide), it appeared that the fusion of deuterons was made possible by the presence of **palladium** electrodes used in the reaction. If such an observation could have been confirmed by other scientists, it would have been truly revolutionary. It would have meant that energy could be obtained from fusion reactions at moderate temperatures.

The Pons-Fleischmann discovery was the subject of immediate and intense scrutiny by other scientists around the world. It soon became apparent, however, that evidence for cold fusion could not consistently be obtained by other researchers. A number of alternative explanations were developed by scientists for the fusion results that Pons and Fleischmann believed they had obtained. Today, some scientists are still convinced that Pons and Fleischmann had made a real and important breakthrough in the area of fusion research. Most researchers, however, attribute the results they reported to other events that occurred during the electrolysis of the heavy water.

NUCLEAR MAGNETIC RESONANCE

Nuclear magnetic **resonance** (NMR) is the effect produced when a radiofrequency field is imposed at right angles to a (usually much larger) static magnetic field to perturb the orientation of nuclear magnetic moments generated by spinning electrically charged atomic nuclei. When the perturbed spinning nuclei interact with the very large (10,000-50,000 gauss) static magnetic field, characteristic spectral shifts and fine structure are produced that reflect the molecular or chemical environment seen by the **nucleus**. **Hydrogen** nuclei, **fluorine**, carbon-13, and oxygen-17 all have distinctive magnetic properties that make them suitable for NMR studies.

The sensitivity of nuclear magnetic resonance to **molecular structure** has made it a valuable research tool in organic **chemistry**, enabling chemists to determine hydrogen locations in crystals, something that cannot be done using **x-ray diffraction**. Nuclear magnetic resonance has also been used to study **electron** densities, chemical **bonding**, the compositions of mixtures, and to make purity determinations.

The basic requirements for NMR **spectroscopy** are that the magnetic field be homogeneous over the **volume** of the sample, that there be a radiofrequency field rotating in a plane perpendicular to the static field, and that there be a means of detecting the interaction of the radiofrequency field with the sample.

Techniques that have been developed for the observation of NMR signals fall into two categories: pulsed and continuous wave. In the case of pulsed methods, an applied rotating (or alternating) magnetic field with a frequency at or near the Larmor frequency (i.e., frequency of precession) of the nucleus to be studied is directed at a right angle to the static field. If the rotating field is applied at exact resonance, the nuclei precess about that field as though there was no static field. Continuous wave methods are either high resolution or broadline. Broad line widths are produced by most oriented molecules exhibiting strong magnetic dipolar interactions, so broadline spectroscopy does not permit measurements of chemical shifts and spin-spin coupling. High resolution spectroscopy, on the other hand, has been used to identify molecules, to measure subtle electronic effects, to determine structure, to study reaction intermediates, and to follow the motion of molecules or groups of atoms within molecules. For high resolution studies, the magnetic field must be uniform to 1 part in 10^8 for a 100MHz instrument if a resolution of 1 MHz is to be obtained. In the case of broadline studies, 5 parts in 10^6 may be adequate.

Nuclear magnetic resonance has been used to study the physics and chemistry of **solids**, including **metals**, **semiconductors**, magnetic solids, and organic materials. Physical phenomena studied include conduction-electron paramagnetism; spin waves and magnetic fluctuations in ordered magnetic materials; metal- insulator transitions; charge **density** wave phenomena; spin-freezing in spin glasses; and frequency shift and spin-lattice relaxation effects. At low temperatures, NMR has been used to make **temperature** measurements and to study the superfluid phases of ^{3}He.

In the fields of **organic chemistry** and materials science, NMR has been used to study polymers, amorphous systems, and complex molecular solids. In many of these systems, the NMR line widths of the nuclei are dominated by dipolar fields arising from neighboring magnetic moments. These systems exhibit complex NMR spectra due to shifts in nuclear magnetic resonance frequencies.

In the case of complex molecules in liquid environments, the molecules undergo a tumbling motion, producing very sharp NMR spectra. The technique used to study these systems is known as Fourier transform NMR spectroscopy.

Nuclear magnetic resonance has been adapted to medical studies in the form of magnetic resonance imaging (MRI), a technique that is capable of producing high quality images of the internal human body without requiring invasive surgery. Begun as a tomographic imaging technique because it produced an image corresponding to a thin slice of the human body, MRI later evolved into a three-dimensional imaging technique.

Like NMR, magnetic resonance imaging is based on the absorption and emission of radiofrequency **energy** under the influence of a magnetic field. On a molecular level, the human body consists largely of hydrogen-rich fat and **water** molecules, and approximately 63% of the human body contains hydrogen atoms. The hydrogen nuclei emit signals that can be observed by magnetic resonance imaging.

When exposed to a magnetic field, the spin of the protons in the hydrogen nuclei of water, which ordinarily assume random orientations, line up. Although short pulses of radio waves briefly disturb this spin alignment, the spins promptly realign in the direction of the magnetic field. In the process of realigning, small signals are produced that can be picked up by sensitive scanners. The signals are then compiled by computer and an image is formed.

Because water is the major component of soft tissue, MRI creates excellent images of soft tissue and organs, but poor images of dense structures like bone. MRI thus complements x-ray imaging, which by contrast is fine for dense structures but useless for soft tissue.

See also Magnetism

NUCLEAR POWER

Nuclear power is any method of doing work that makes use of **nuclear fission** or fusion reactions. In its broadest sense, the term refers to both the uncontrolled release of **energy**, as in fission or fusion weapons, and to the controlled release of energy, as in a nuclear power plant. Most commonly, however, the expression nuclear power is reserved for the latter of these two instances.

An aerial view of the nuclear explosion, code-named *Seminole*, at Enewetak Atoll in the Pacific Ocean on June 6, 1956. This atomic bomb was detonated at ground level and had the same explosive force of 13.7 thousand tonnes of TNT. It was detonated as part of *Operation Redwing*, an American military program which tested systems for atomic bombs. *(Photograph by U.S. Department of Energy/Science Photo Library, Photo Researchers, Inc. Reproduced by permission.)*

The world's first exposure to nuclear power came with the detonation of two fission (''atomic'') bombs over Hiroshima and Nagasaki, Japan, events that brought World War II to a conclusion. A number of scientists and laypersons perceived an optimistic aspect of these terrible events. They hoped that the power of **nuclear energy** could be harnessed to do much of the work that all human societies face. Those hopes have been realized to only a modest degree, however. Some serious problems associated with the use of nuclear power have never been satisfactorily solved and, after three decades of progress in the development of controlled nuclear power, interest in this energy source has leveled off and, in many nations, declined.

The nuclear power plant

A nuclear power plant is a system in which energy released by fission reactions is captured and used for the generation of **electricity**. Every such plant contains four fundamental elements: the reactor, coolant system, electrical power generating unit, and the safety system.

The source of energy in a nuclear reactor is a fission reaction in which neutrons collide with nuclei of uranium-235 or plutonium-239 (the fuel), causing them to split apart. The products of any fission reaction include not only huge amounts of energy, but also waste products, known as fission products, and additional neutrons. A constant and reliable flow of neutrons is insured in the reactor by means of a moderator, which slows down the speed of neutrons, and control rods, which control the number of neutrons available in the reactor and, hence, the rate at which fission can occur.

Energy produced in the reactor is carried away by means of a coolant, a fluid such as **water**, liquid **sodium**, or **carbon** dioxide gas. The fluid absorbs **heat** from the reactor and then begins to boil itself or to cause water in a secondary system to boil. Steam produced in either of these ways is then piped into the electrical generating unit where it turns the blades of a turbine. The turbine, in turn, powers a generator that produces electrical energy.

The high cost of constructing a modern nuclear power plant reflects in part the enormous range of safety features needed to protect against various possible mishaps. Some of those features are incorporated into the reactor core itself. For example, all of the fuel in a reactor is sealed in a protective coating made of a **zirconium alloy**. The protective coating, called a cladding, helps retain heat and **radioactivity** within the fuel, preventing it from escaping into the plant itself.

Every nuclear plant is also required to have an elaborate safety system to protect against the most serious potential problem of all, loss of coolant. If such an accident were to occur, the reactor core might well melt down, releasing radioactive materials to the rest of the plant and, perhaps, to the outside environment. To prevent such an accident from happening, the pipes carrying the coolant are required to be very thick and strong. In addition, back-up supplies of the coolant must be available to replace losses in case of a leak.

On another level, the whole plant itself is required to be encased within a dome-shaped containment structure. The containment structure is designed to prevent the release of radioactive materials in case of an accident within the reactor core.

Another safety feature is a system of high-efficiency filters through which all air leaving the building must pass. These filters are designed to trap microscopic particles of radioactive materials that might otherwise be vented to the atmosphere. Other specialized devices and systems have also been developed for dealing with other kinds of accidents in various parts of the power plant.

Types of nuclear power plants

Nuclear power plants differ from each other primarily in the methods they use for transferring heat produced in the reactor to the electricity generating unit. Perhaps the simplest design of all is the boiling water reactor plant (BWR) in which coolant water surrounding the reactor is allowed to boil and form steam. That steam is then piped directly to turbines, whose spinning drives the electrical generator. A very different type of plant is one that was popular in Great Britain for many years, one that used **carbon dioxide** as a coolant. In this type of plant, carbon dioxide gas passes through the reactor core, absorbs heat produced by fission reactions, and is piped into a secondary system. There the heated carbon dioxide gas gives up its energy to water, which begins to boil and change to steam. That steam is then used to power the turbine and generator.

In spite of all the systems developed by nuclear engineers, the general public has long had serious concerns about the use of such plants as sources of electrical power. Those concerns vary considerably from nation to nation. In France, for example, more than half of all that country's electrical power now comes from nuclear power plants. The initial enthusiasm for nuclear power in the United States in the 1960s and 1970s soon faded, and no nuclear power plants have been constructed in this country in more than a decade.

One concern about nuclear power plants, of course, is an echo of the world's first exposure to nuclear power, the atomic bomb blasts. Many people fear that a nuclear power plant may go out of control and explode like a nuclear weapon. And, in spite of experts' insistence that such an event is impossible, a few major disasters have instigated the fear of nuclear power plants exploding. By far the most serious of those disasters was the explosion that occurred at the Chernobyl Nuclear Power Plant near Kiev in the Ukraine in 1986.

On April 16, of that year, one of the four power-generating units in the Chernobyl complex exploded, blowing the top off the containment building. Hundreds of thousands of nearby residents were exposed to lethal or damaging levels of radiation and were removed from the area. Radioactive clouds released by the explosion were detected as far away as western Europe. More than a decade later, the remains of the Chernobyl reactor remain far too radioactive for anyone to spend more than a few minutes in the area.

Critics also worry about the amount of radioactivity released by nuclear power plants on a day-to-day basis. This concern is probably of less importance than is the possibility of a major disaster. Studies have shown that nuclear power plants are so well shielded that the amount of radiation to which nearby residents are exposed is no more than that of a person living many miles away.

In any case, safety concerns in the United States have been serious enough to essentially bring the construction of new plants to a halt in the last decade. Licensing procedures are now so complex and so expensive that few industries are interested in working their way through the bureaucratic maze to construct new plants.

Perhaps the single most troubling issue for the nuclear power industry is waste management. After a period of time, the fuel rods in a reactor are no longer able to sustain a **chain reaction** and must be removed. These rods are still highly radioactive, however, and present a serious threat to human life and the environment. Techniques must be developed for the destruction and/or storage of these wastes.

Nuclear wastes can be classified into two general categories, low-level wastes and high-level wastes. The former consist of materials that release a relatively modest level of radiation and/or that will soon decay to a level where they no longer present a threat to humans and the environment. Storing these materials in underground or underwater reservoirs for a few years or in some other system is usually a satisfactory way of handling these materials.

High-level wastes are a different matter. The materials that make up these wastes are intensely radioactive and are likely to remain so for thousands of years. Short-term methods of storage are unsatisfactory because containers leak and break open long before the wastes are safe.

For more than two decades, the United States government has been attempting to develop a plan for the storage of high-level nuclear wastes. At one time, the plan was to bury the wastes in a **salt** mine near Lyons, Kansas. Objections from residents of the area and other concerned citizens made that plan infeasible. More recently, the government decided to construct a huge crypt in the middle of Yucca Mountain in Nevada for the burial of high-level wastes. Again, complaints by residents of Nevada and other citizens have delayed placing that plan into operation. The government insists, however, that

Yucca Mountain will eventually become the long-term storage site for the nation's high-level radioactive wastes. Until that site is actually put into operation, however, those wastes are in "temporary" storage at nuclear power sites throughout the United States.

History

The first nuclear reactor was built during World War II as part of the **Manhattan Project** to build an atomic bomb. The reactor was constructed under the direction of Enrico Fermi in a large room beneath the squash courts at the University of Chicago. It was built as the first concrete test of existing theories of nuclear fission.

Until the day on December 2, 1942, when the reactor was first put into operation, scientists had relied entirely on mathematical calculations to determine the effectiveness of nuclear fission as an energy source. It goes without saying that the scientists who constructed the first reactor were taking an extraordinary chance.

That reactor consisted of alternating layers of **uranium** and uranium **oxide** with **graphite** as a moderator. **Cadmium** control rods were used to control the **concentration** of neutrons in the reactor. Since the various parts of the reactor were constructed by piling materials on top of each other, the unit was at first known as an atomic "pile." The moment at which Fermi directed the control rods to be withdrawn occurred at 3:45 p.m. on December 2, 1945, and that date can legitimately be regarded as the beginning of the age of controlled nuclear power in human history.

Nuclear fusion power

Many scientists believe that the ultimate **solution** to the world's energy problems may be in the harnessing of nuclear fusion. A fusion reaction is one in which two small nuclei combine with each other to form one larger **nucleus**. As an example, two **hydrogen** nuclei may combine with each other to form the nucleus of an **atom** known as **deuterium**, or heavy hydrogen.

Many scientists now believe that fusion reactions are responsible for the production of energy in stars. They hypothesize that four hydrogen atoms fuse with each other in a series of reactions to form a single **helium** atom. An important byproduct of these fusion reactions is the release of an enormous amount of energy. In fact, gram-for-gram, a fusion reaction releases many times more energy than does a fission reaction.

The world was introduced to the concept of fusion reactions in the 1950s when the Soviet Union and the United States exploded the first fusion ("hydrogen") bombs. The energy released in the explosion of each such bomb was more than 1,000 times greater than the energy released in the explosion of a single fission bomb.

As with fission, scientists and non-scientists alike expressed hope that fusion reactions could someday be harnessed as a source of energy for everyday needs. This line of research has been much less successful, however, than research on fission power plants. In essence, the problem has been to find a way of containing the very high temperatures produced (a few million degrees Celsius) when fusion occurs. Optimistic reports of progress on a fusion power plant appear in the press from time to time, but some authorities now doubt that fusion power will ever be an economic reality.

NUCLEAR REACTIONS

Nuclear decay, **neutron** capture, fission, and fusion are the four major nuclear reactions.

The understanding, use, and control of nuclear reactions is one of the most profound scientific accomplishments. Although chemical reactions of everyday experience generally take place between the electrons surrounding an **atom** there are important and fundamental processes that take place in the **nucleus** of an atom. Nuclear reactions are different than chemical reactions. Nuclear energies are several orders of magnitude larger than the energies involved in chemical reactions.

In 1903, French physicist, Antoine **Henri Becquerel** won the Nobel Prize in physics for his work with spontaneous **radioactivity**. In 1898, French scientists, **Marie Curie** and **Pierre Curie**, two-time Nobel Laureates, discovered the naturally occurring radioactive elements **polonium** and **radium**.

In the 1930s, French physicists **Frédéric Joliot-Curie** and **Irène Joliot-Curie** demonstrated artificial radioactivity by bombarding stable atoms with nuclear particles to create radioactive isotopes (radioisotopes). In 1932 British scientists, Sir John Cockcroft (1897-1967) and Ernest Walton (1903-1995), were the first disintegrate the nucleus by bombarding it with high **energy** projectile-particles.

Before World War II, German chemists **Otto Hahn** and **Fritz Strassmann** discovered **nuclear fission**. During the war there was a frantic race to develop atomic weapons. The United States assembled a team of leading physicists and chemists directed by J. Robert Oppenheimer (1904-1967) at Los Alamos, New Mexico, to take part in project code named Trinity. In 1945, they produced the world's first atomic explosion using a fission reaction. In August 1945, atomic bombs were dropped on the Japanese cities of Hiroshima and Nagasaki. As a result the Japanese surrendered and the World War II ended.

In elements heavier than **hydrogen** (hydrogen's nucleus has only a single proton), the nucleus is composed of protons and neutrons. The number of protons is described by the **atomic number** and is unique for each element. Along with the protons are neutrons, particles of **mass** similar to the **proton**, but without any electrical charge. The sum of the atomic mass of the protons and neutrons determines the atomic mass or atomic weight of the nucleus.

Elements can be found that do not have identical numbers of neutrons, these elements are isotopes of one another and they have nuclei with different atomic mass. Nuclei with differing weights, that is, unique combinations of protons and neutrons are called nuclides.

Not all atomic nuclides are stable. Radioactive nuclides are unstable and undergo various nuclear reactions that transform the particles within the nucleus and simultaneously release energy. It is important to remember that radioactivity is a process not a term to be applied to the energy or substance emitted as a result of the radioactive process.

Nuclear reactions include alpha decay (the emission of a **helium** nucleus, He^{+}), beta decay (a reaction where a neutron is transformed into a proton and a high energy **electron** is emitted), positron decay (a reaction where a nuclear proton converts into a neutron and a high-energy positron is emitted), and decay that produces **gamma radiation** (a high energy **photon** emission).

A common misconception is that x rays are a product of nuclear reactions. In fact, x rays are the result of high energy electron transitions (e.g., jumps between electron energy levels or orbits) in heavy elements.

Scientists study nuclear reactions by accelerating atomic particles (e.g. neutrons) that, upon collision with the target nucleus, transfer enough energy to stimulate nuclear reactions. When some elements are bombarded with protons, neutrons, or other accelerated (and therefore highly energetic) atomic particles, their nuclei may be transformed to create unstable isotopes that are radioactive. These radioactive isotopes are created by nuclear reactions termed neutron capture.

In neutron capture reactions an element's atomic number (Z) remains unchanged but its atomic mass increases by one because a neutron is added to the nucleus.

Two other important nuclear reactions, fusion and fission, also start with the capture of a neutron. In these reactions, however, the energy levels are so high that the transformed nucleus is very unstable. At the lower energy levels involved in simple neutron capture, the element simply becomes radioactive, that is, it undergoes decay reactions to form decay products.

The nuclear reactions of fission and fusion have profound scientific and controversial social consequences. Fission involves the splitting of the atomic nucleus. During fission reactions, a nucleus is split into two nuclei. When, under the right circumstances, **uranium** is bombarded with neutrons it can undergo nuclear fission to produce **barium**, **krypton**, and three neutrons (the basis of the chain reaction).

The use of fission reactions in **nuclear reactors** remains socially and politically controversial. Fission's most common use is as a trigger in fusion reaction bombs. Other debates usually concern the safe construction and operation of nuclear power facilities and the proper disposal of the dangerous radioactive products of termed nuclear waste.

Fusion is the nuclear reaction that fuels the Sun and stars, the reaction involves the combining of two nuclei into one nucleus. At stellar temperatures of 10-15 million degrees Celsius, hydrogen is converted to helium. The energy from these reactions provides the energy to sustain life on Earth. The basic reaction combines or fuses four hydrogen atoms into one helium atom with the emission of tremendous energy generated by converting mass into energy according to Einstein's equation $E = mc^2$.

The potential benefits to mankind for the proper development of fusion based technology are enormous. However, until methods are developed to control the reaction, its use will be limited to nuclear weapons.

As stars age the supply of hydrogen for fusion reactions is used up and stars must use increasingly heavier elements in their fusion reactions. The mass of the star determines what fuels it can use (up to iron). Except for the transformation of elements by nuclear reactions, all of the atoms in the universe heavier than helium are the products of the nuclear reactions that take place in dying stars.

See also Radiation chemistry; Radioactivity

NUCLEAR REACTORS

A nuclear reactor is a device by which **energy** is produced as the result of a nuclear reaction, either fission or fusion. At the present time, all commercially available nuclear reactors make use of fission reactions, in which the nuclei of large atoms such as **uranium** (the fuel) are broken apart into smaller nuclei, with the release of energy. It is theoretically possible to construct reactors that operate on the principle of **nuclear fusion**, in which small nuclei are combined with each other with the release of energy. But after a half century of research on fusion reactors, no practicable device has yet been developed.

When neutrons strike the **nucleus** of a large **atom**, they cause that nucleus to split apart into two roughly equal pieces known as fission products. In that process, additional neutrons and very large amounts of energy are also released. Only three isotopes are known to be fissionable—uranium-235, uranium-233, and plutonium-239. Of these, only the first, uranium-235, occurs naturally. Plutonium-239 is produced synthetically when nuclei of uranium-238 are struck by neutrons and transformed into **plutonium**. Since uranium-238 always occurs along with uranium-235 in a nuclear reactor, plutonium-239 is produced as a byproduct in all commercial reactors now in operation. As a result, it has become as important in the production of nuclear power as uranium-235. Uranium-233 can also be produced synthetically by the bombardment of **thorium** with neutrons. Thus far, however, this **isotope** has not been put to practical use in nuclear reactors.

Nuclear fission is a promising source of energy for two reasons. First, the amount of energy released during fission is very large compared to that obtained from conventional energy sources. For example, the fissioning of a single uranium-235 nucleus results in the release of about 200 million **electron** volts of energy. In comparison, the **oxidation** of a single **carbon** atom (as it occurs in the burning of **coal** or oil) releases about four electron volts of energy. When the different masses of carbon and uranium atoms are taken into consideration, the fission reaction still produces about 2.5 million times more energy than does the oxidation reaction.

Second, the release of neutrons during fission makes it possible for a rapid and continuous repetition of the reaction. Suppose that a single **neutron** strikes a one gram block of uranium-235. The fission of one uranium nucleus in that block releases, on an average, about two to three more neutrons. Each of those neutrons, then, is available for the fission of three more uranium nuclei. In the next stage, about nine neutrons (three from each of three fissioned uranium nuclei) are released. As long as more neutrons are being released, the fission of uranium nuclei can continue.

The interior of the containment building at the Trojan nuclear power station, Rainier, Oregon. In the center of the frame is the reactor core, under 35 feet (11 m) of water. On each side of the core are the four generators that make the superheated steam. This steam is them piped to the main electricity generating turbines in an adjacent building. ***(Photograph by U.S. Department of Energy/Mark Marten, Photo Researchers, Inc. Reproduced by permission.)***

A reaction of this type that continues on its own once under way is known as a **chain reaction**. During a nuclear chain reaction, many billions of uranium nuclei may fission in less than a second. Enormous amounts of energy are released in a very short time, a fact that becomes visible with the explosion of a nuclear weapon.

Arranging for the uncontrolled, large-scale release of energy produced during nuclear fission is a relatively simple task. Fission (atomic) bombs are essentially devices in which a chain reaction is initiated and then allowed to continue on its own. The problems of designing a system by which fission energy is released at a constant and useable rate, however, are much more difficult.

Reactor core

The heart of any nuclear reactor is the core, which contains the fuel, a moderator, and control rods. The fuel used in some reactors consists of uranium **oxide**, enriched with about 3-4% of uranium-235. In other reactors, the fuel consists of an **alloy** made of uranium and plutonium-239. In either case, the amount of fissionable material is actually only a small part of the entire fuel assembly.

The fuel elements in a reactor core consist of cylindrical pellets about 0.6 in (1.5 cm) thick and 0.4 in (1.0 cm) in diameter. These pellets are stacked one on top of another in a hollow cylindrical tube known as the fuel rod and then inserted into the reactor core. Fuel rods tend to be about 12 ft (3.7 m) long and about 0.5 in (1.3 cm) in diameter. They are arranged in a grid pattern containing more than 200 rods each at the center of the reactor. The materials that fuel these pellets are made of must be replaced on a regular basis as the proportion of fissionable nuclei within them decreases.

A nuclear reactor containing only fuel elements would be unusable because a chain reaction could probably not be sustained within it. The reason is that nuclear fission occurs best with neutrons that move at relatively modest speeds, called thermal neutrons. But the neutrons released from fission reactions tend to be moving very rapidly, at about 1/15 the speed of light. In order to maintain a chain reaction, therefore, it is necessary to introduce some material that will slow down the neutrons released during fission. Such a material is known as a moderator.

The most common moderators are substances of low atomic weight such as heavy **water** (deuterium oxide) or **graphite**. Hydrides (binary compounds containing hydrogen), hydrocarbons, and **beryllium** and beryllium oxide have also been used as moderators in certain specialized kinds of reactors.

A chain reaction could easily be sustained in a reactor containing fuel elements and a moderator. In fact, the reaction might occur so quickly that the reactor would explode. In order to prevent such a disaster, the reactor core also contains control rods. Control rods are solid cylinders of metal constructed of some material that has an ability to absorb neutrons. One of the **metals** most commonly used in the manufacture of control rods is **cadmium**.

The purpose of control rods is to maintain the ratio of neutrons used up in fission compared to neutrons produced during fission at about 1:1. In such a case, for every one new neutron that is used up in causing a fission reaction, one new neutron becomes available to bring about the next fission reaction.

The problem is that the actual ratio of neutrons produced to neutrons used up in a fission reaction is closer to 2:1 or 3:1. That is, neutrons are produced so rapidly that the chain reaction goes very quickly and is soon out of control. By correctly positioning control rods in the reactor core, however, many of the excess neutrons produced by fission can be removed from the core and the reaction can be kept under control.

The control rods are, in a sense, the dial by which the rate of fission is maintained within the core. When the rods are inserted completely into the core, most neutrons released during fission are absorbed, and no chain reaction occurs. As the rods are slowly removed from the core, the rate at which fission occurs increases. At some point, the position of the control rods is such that the 1:1 ratio of produced to used up neutrons is achieved. At that point, the chain reaction goes forward, releasing energy, but under precise control of human operators.

Reactor types

In most cases, the purpose of a nuclear reactor is to capture the energy released from fission reactions and put it to some useful service. For example, the **heat** generated by a nuclear reactor in a nuclear power plant is used to boil water and make steam, which can then be used to generate **electricity**. The way that heat is removed from a reactor core is the basis for defining a number of different reactor types.

For example, one of the earliest types of nuclear reactors is the boiling water reactor (BWR) in which the reactor core is surrounded by ordinary water. As the reactor operates, the water is heated, begins to boil, and changes to steam. The steam produced is piped out of the reactor vessel and delivered (usually) to a turbine and generator, where electrical power is produced.

Another type of reactor is the pressurized water reactor (PWR). In a PWR, coolant water surrounding the reactor core is kept under high pressure, preventing it from boiling. This water is piped out of the reactor vessel into a second building where it is used to heat a secondary set of pipes also containing ordinary water. The water in the secondary system is allowed to boil, and the steam formed is then transferred to a turbine and generator, as in the BWR.

Some efforts have been made to design nuclear reactors in which liquid metals are used as **heat transfer** agents. Liquid **sodium** is the metal most often suggested. Liquid sodium has many attractive properties as a heat transfer agent, but it has one serious drawback. It reacts violently with water and great care must be taken, therefore, to make sure that the two materials do not come into contact with each other.

At one time, there was also some enthusiasm for the use of **gases** as heat transfer agents. A group of reactors built in Great Britain, for example, were designed to use **carbon dioxide** to move heat from the reactor to the power generating station. Gas reactors have, however, not experienced much popularity in other nations.

Applications

At the end of World War II, great hopes were expressed for the use of nuclear reactors as a way of providing power for many human energy needs. For example, some optimists envisioned the use of small nuclear reactors as power sources in airplanes, ships, and automobiles. These hopes have been realized to only a limited extent. Nuclear-powered submarines, for example, have become a practical reality. But other forms of transportation seldom make use of this source of energy.

Instead, the vast majority of nuclear reactors in use today are employed in nuclear power plants where they supply the energy needed to manufacture electrical energy. In a power reactor, energy released within the reactor core is transferred by a coolant to an external building in which are housed a turbine and generator. Steam obtained from water boiled by reactor heat energy is used to drive the turbine and generator, thereby producing electrical energy.

Reactors with other functions are also in use. For example, a breeder reactor is one in which new reactor fuel is manufactured. By far the most common material in any kind of nuclear reactor is uranium-238. This isotope of uranium does not undergo fission and does not, therefore, make any direct contribution to the production of energy. But the vast numbers of neutrons produced in the reactor core do react with uranium-238 in a different way, producing plutonium-239 as a product. This plutonium-239 can then be removed from the reactor core and used as a fuel in other reactors. Reactors whose primary function it is to generate plutonium-239 are known as breeder reactors.

Research reactors may have one or both of two functions. First, such reactors are often built simply to test new design concepts for the nuclear reactor. When the test of the design element has been completed, the primary purpose of the reactor has been accomplished.

Second, research reactors can also be used to take advantage of the various forms of radiation released during fission reactions. These forms of radiation can be used to bombard a variety of materials to study the effects of the radiation on the materials.

Nuclear spectroscopy

Nuclear **spectroscopy** is a powerful tool in the arsenal of scientists studying the structure of **matter** based upon the reactions that take place in excited atomic nuclei. Nuclear spectroscopy is a widely used technique to determine the composition of substances because it is more sensitive than other spectroscopic methods and can detect the trace presence of elements that may only be present on the order of parts per billion in an unknown substance.

Nuclear spectroscopic technology allowing the identification of trace elements in soil and **water** samples has increasing use in ecological, agricultural, and geological research. Applications of nuclear spectroscopic principles are important to the development of non-invasive diagnostic tools used by physicians.

A number of methods can be used to excite atomic nuclei and then measure their decaying gamma ray emissions as the atoms return to normal **energy** levels (i.e., their ground state). The emissions are then analyzed and separated into an emission spectrum that is characteristic for each element. Excitation can be accomplished by colliding nuclei, heavy **ion** beams, and a number of other methods, but the fundamental purpose remains to measure the spectral properties of a sample as a tool to learn something about the quantum structure of the atoms in the sample.

Like other forms of spectroscopy, the fundamental measurements of nuclear spectroscopy involve recording the emissions or absorption of photons by atoms. The specific emissions or absorptions reflect the energy levels, spin states, parity, and other properties of an atom's structure (e.g., quantized energy levels).

A **qualitative analysis** identifies the components of a substance or mixture. **Quantitative analysis** measures the amounts or proportions of the components in a reaction or substance.

Because each element—and each nuclide (i.e., an atomic **nucleus** with a unique combination of protons and neutrons)—emits or absorbs only specific frequencies and wavelengths of electromagnetic radiation, nuclear spectroscopy is a qualitative test (i.e., a test designed to identify the components of a substance or mixture) to determine the presence of an element or **isotope** in an unknown sample.

In addition, the strength of emissions and absorption for each element and nuclide can allow for a quantitative measurement of the amount or proportion of the element in an unknown. To perform quantitative tests, that is, to measure amounts of an element present, the measured spectrum needs to be narrowed down to analysis of photons with specific energies (i.e., electromagnetic radiation of specific wavelength or frequency). Quantitative computation using Beer's Law is then applied to the measured intensities of **photon** emission or absorption. Many other spectroscopic methods use this technique (e.g., atomic **absorption spectroscopy** and UV-visible light spectroscopy) to determine the amount of a element present.

One of most widely used methods of nuclear spectroscopy used to determine the elemental composition of substances is Nuclear Activation Analysis (NAA).

In **neutron** activation analysis the goal is to determine the composition of an unknown substance by measuring the energies and intensities of the gamma rays emitted after excitation and the subsequent matching of those measurements to the emissions of gamma rays from standardized (known) samples. In this regard, **neutron activation analysis** is similar to other spectroscopic measurements that utilize other portions of the **electromagnetic spectrum**. Infrared photons, x-ray fluorescence, and spectral analysis of visible light are all used to identify elements and compounds. In each of these spectroscopic methods, a measurement of electromagnetic radiation is compared with some known quantum characteristic of an atomic nucleus, **atom**, or **molecule**. With NAA, of course, high energy gamma ray photons are measured.

Neutron activation analysis involves a comparison of measurements from an unknown sample with values obtained from tests with known samples. Depending on which elements are being tested for, the samples are irradiated with energetic neutrons. The process of **radioactivity** results in the emission of products of **nuclear reactions** (in this case, gamma rays) that are measurable by instruments designed for that purpose. After a time (dependent of the length of radiation), the gamma rays are counted by gamma ray sensitive spectrometers. Because the products of the nuclear reactions are characteristic of the elements present in the sample and a measure of amounts of the amounts present, neutron activation analysis is both a qualitative and quantitative tool.

Although NAA usually involves the measurement of gamma rays emitted from the radioactive sample, more complex techniques also measure beta and positron emissions.

Nuclear magnetic resonance (NMR) is another form of nuclear spectroscopy that is widely used in medicine.

NMR is based on the fact that a **proton** in a magnetic field had two quantized spin states. The actual magnetic field experienced by most protons is, however, slightly different from the external applied field because neighboring atoms alter the field. As a result, however, a picture of complex structures of molecules and compounds can be obtained by measuring differences between the expected and measured photons absorbed. NMR spectroscopy as an important tool used to determine the structure of organic molecules.

When a group of nuclei are brought into resonance—that is, when they are absorbing and emitting photons of similar energy (electromagnetic radiation, e.g., radio waves, of similar wavelengths)— and then small changes are made in the photon energy, the resonance must change. How quickly and to what form the resonance changes allows for the non-destructive (because of the use of low energy photons) determination of complex structures. This form of NMR is used by physicians as the physical and chemical basis of a powerful diagnostic technique termed magnetic resonance imaging (MRI).

See also Beer's law; Bohr theory; Electromagnetic spectrum; Nuclear reactions; Spectroscopy

NUCLEIC ACIDS

Nucleic acids are divided into two **molecule** types: deoxyribonucleic acid (DNA) and ribonucleic acid (RNA). **DNA** and **RNA** are chemically very similar, and both are involved in the transmission of genetic information within living organisms. However, each has its own particular role in that transmission.

Nucleic acids were discovered in 1869 by German scientist Johann Friedrich Miescher (1844-1895). Miescher called the newly discovered material nuclein—later known as nucleic acid—because it was found in the **nucleus** of cells. (The nucleus is a structure inside plant and animal cells.) Later it was found that it was not necessary for a cell to have a nucleus to have nucleic acid.

By the late nineteenth century, most scientists accepted the idea that inheritable characteristics were somehow linked to cellular **chemistry**. Nucleic acid was considered a potential candidate, but most scientists favored **proteins**. Proteins seemed to be a more logical choice because they are more abundant in cells than nucleic acid and their chemical structure is more diverse.

In the 1940s, scientists began to reconsider the possibility that nucleic acid, specifically DNA, might be the correct candidate for storing genetic information after all. Discoveries by scientists such as **Erwin Chargaff**, **Oswald Avery**, Colin MacLeod (1909-1972), Maclyn McCarty (1911-), Alfred Hershey (1908-1997), and Martha Chase confirmed that DNA was the carrier of the genetic code. (Later in the century, it was discovered that RNA is sometimes used to store genetic information. This situation occurs in a class of viruses called retroviruses. The human immunodeficiency virus, or HIV, is a type of retrovirus.)

By the early 1950s, research had shifted to uncovering the structure of DNA. In 1953, **Francis Crick** and **James Watson** published their findings on the structure of DNA. The so-called Watson-Crick model depicts a double-stranded, spiral-shaped molecule, the building blocks of which are nucleotides. Each nucleotide contains a **phosphate** group, a type of sugar molecule called deoxyribose, and one of four nitrogen-containing bases. The nitrogen-containing bases belong to two classes of molecules—purines and pyrimidines. Both classes have molecular rings composed of **carbon** and **nitrogen**. Pyrimidine molecules are single-ring molecules. Cytosine and thymine are included in this group. Purines are larger two-ring molecules. Guanine and adenine are included in this group.

The backbone of each DNA strand is a chain of linked phosphate and deoxyribose molecules. When the two strands are lined up, the nitrogen-containing bases face one another like the teeth of a zipper. The strands are joined by **hydrogen** bonds between bases. Adenine always bonds with thymine, and vice versa; guanine always pairs with cytosine, and vice versa. Virtually all living organisms contain DNA as a double-stranded molecule.

The genetic code holds information for constructing proteins that form and maintain living organisms. Proteins are composed of smaller molecules called **amino acids**. The information encoded in DNA indicates which amino acids to use, as well as the order in which they should appear. From the time that DNA was identified as the carrier of the genetic code, it was recognized that the transmission of information was not direct. RNA was quickly identified as forming the link between DNA and the proteins for which it coded.

The existence of RNA has been known since the early 1940s, based on experiments carried out by Jean Brachet and Torbjorn Caspersson (1910-). However, its role in the cell was not clarified until the late 1950s.

Unlike DNA, RNA is almost always a single-stranded molecule. It is constructed as a long chain of nucleotides. Another difference between DNA and RNA is that RNA contains ribose, rather than deoxyribose. Both RNA and DNA nucleotides contain a phosphate group, and three of the four nitrogen-containing bases are the same. However, where DNA would have thymine, RNA has a different base called uracil. This base is chemically very similar to thymine and is also a pyrimidine.

There are three types of RNA, each of which is named according to its function. Chemically, however, they are the same molecule. One strand of the DNA—the coding strand or the template—serves as a master copy of the genetic code. The cell uses RNA to construct a working copy of that strand.

This RNA molecule is called messenger RNA, or mRNA. The order of nucleotides in mRNA is complementary to the nucleotide order of the coding strand of DNA. This complementary order follows the same rules as base pairing between two strands of DNA, except that uracil is used in place of thymine. For example, if cytosine occupies a certain position in the DNA strand, guanine will be used in that position on the RNA strand.

The process of transferring the genetic code to mRNA is called transcription. Once the DNA has been transcribed, the RNA is modified and then released into the cell's interior. The next step is translation of the code, which relies on two types of RNA: ribosomal RNA (rRNA) and transfer RNA (tRNA). The rRNA is a component of cellular structures called ribosomes. The ribosomes serve as a site for translation, the process in which proteins are constructed based on the genetic code.

To start translation, a ribosome attaches itself to a strand of mRNA. The ribosome serves to hold the mRNA in place and direct the speed of translation. As the ribosome gradually moves along the length of the mRNA, tRNA carries the appropriate amino acids to a growing protein molecule. When the ribosome reaches the end of the mRNA, it detaches from the mRNA and a new protein molecule is released.

See also Protein synthesis

Nucleon

Nucleon is the term used to refer collectively to the particles that are found in the **nucleus** of an **atom**, the protons and the neutrons. The proton is a positively charged particle, while the **neutron** is neutral, but they have similar masses, the neutron being heavier than the **proton** by only one-thousandth of an atomic **mass** unit. An attractive force known as the strong nuclear force acts between the nucleons to hold the nucleus together. It acts between two neutrons, between two protons, or between a neutron and a proton, as long as they are no farther apart than a distance of one fermi (10^{-15} meters).

After developments in modern physics led to a proliferation of **subatomic particles**, by the 1960s physicists suggested a classification system that would group particles as either fermions or bosons. Fermions are the particles that exhibit half-integer spin and satisfy the **Pauli exclusion principle**, while bosons exhibit integer spin and are not limited by the exclusion principle. Therefore, nucleons are classified as fermions. A second classification divides particles into those that interact via the strong force, called hadrons, and those that do not, called leptons. Thus, nucleons are classified as hadrons. Hadrons that carry nucleon number one, as do both nucleons, are called baryons. In 1963 Murray Gell-Mann proposed that hadrons are made up of smaller charged particles which he called quarks, with each baryon containing three quarks. The only difference between the two nucleons is the number of each type of quark: protons have two ''up'' quarks and one ''down'' quark which give it a charge of +1, while neutrons have two down quarks and one up quark which give it a charge of 0. Gell-Mann was awarded the Nobel Prize in physics in 1969 for this work.

Nucleophile

A nucleophile is any negative **ion** or neutral **molecule** whose electronic configuration consists of at least one unpaired **electron** pair. In chemical reactions, the nucleophile attacks the positive center of some other molecule or positive ion, and then donates a pair of electrons to form a **covalent bond**. Nucleophiles are produced in the formation of **acids and bases**, and in covalent **carbon bonding** in organic compounds.

Thus, reagents that seek a **proton** or some other positive center are called nucleophiles (from **nucleus**, i.e., the positive part of the **atom**, plus *phile* from the Greek word for love). Electrophiles, on the other hand, are reagents that, in their reactions, seek extra electrons to achieve a stable **valence** shell of electrons. For example, in their reactions, nucleophiles consisting of ions with a negatively charged carbon atoms seek a proton or some other positive center to which they can donate their electron pair and thereby achieve electrical neutrality. In a similar way, both hydroxide ions and **water** molecules may act as nucleophiles by reacting with alkyl halides to produce alcohols.

In nucleophilic substitution reactions, the unshared pair of electrons on the nucleophile is used in a new carbon-nucleophile bond. In the case of nucleophilic substitution reactions involving the polar carbon-halogen bond of an alkyl **halide** and a negative nucleophile, the nucleophile seeks the carbon atom that bears the halogen atom. This carbon has a partial positive charge because the halogen atom pulls the electrons in the carbon-halogen bond toward itself. The halide retains the electron pair of the carbon-halogen bond when it leaves. Such leaving groups are usually less nucleophilic than the reacting nucleophile, so these reactions usually do not immediately reverse themselves.

Nucleus

The word nucleus is a derivative of the Latin word *nux*, meaning nut or kernel. Around 1909, **Ernest Rutherford** gave that name to the dense, central part of the **atom** that he and his colleagues Hans Geiger (1882-1945) and Ernest Marsden discovered by bombarding thin sheets of **gold** foil with alpha particles. The results of the experiment led Rutherford to propose that all the positive charge of an atom and almost all its **mass** are located in the nucleus. He developed this model based on the behavior of the positively charged alpha particles as they hit the gold foil. Although most of them passed right through, some were deflected at various angles, including angles that directed the alpha particles back toward their source. Since most of the alpha particles passed through undeflected, Rutherford concluded that the positive part of the atom was not evenly distributed throughout the atom. He explained the deflections as the result of the electric repulsion between like charges—a positively charged **alpha particle** and the positive part of the atom. He used measurements of the deflection angles and the number of particles deflected at that angle to calculate the size of the nucleus. Its diameter is about 10^{-15} meters (one femtometer, now known as a fermi) which is ten thousand times smaller than the diameter of the atom itself. About half a century later, Robert Hofstadter (1915-1990) won the 1961 Nobel Prize in physics for his work verifying that the radius of an atomic nucleus is about one fermi (10^{-15} meters). In his experiments, Hofstadter collided electrons with atoms at high energies.

How the positive charges are able to stay in such close proximity in the confines of the nucleus is another question that was raised. The force, labeled the strong nuclear force, would have to be attractive between the nuclear particles and act at very small distances. At that time only two forces were recognized. The first, gravity, is only significant when large masses are involved. The second, electric force, acts only between charged particles. It is an attractive force between particles of opposite charge, and a repulsive force between particles of same charge. The nuclear force is about one hundred times stronger than this repulsive force between the protons. Rutherford suggested that there is another particle in the nucleus that has mass but no charge. He called this particle the **neutron**, and postulated that the strong nuclear force acts between the protons and the neutrons to keep the nucleus together. This particle was not discovered until 1932 by English physicist James

Chadwick (1891-1974) Chadwick won the 1935 Nobel Prize in physics for this discovery. In 1934, Hideki Yukawa (1907-1981) was the first to attempt to explain the way in which this nuclear force operates. He suggested that the strong nuclear force results from the exchange of a particle between the neutrons and the protons; he named that exchange particle a meson. In 1947, English physicist Cecil Powell (1903-1969) observed Yukawa's mesons, now called pi-mesons or pions, in the upper atmosphere, where they were produced by cosmic ray collisions.

Maria Goeppert Mayer shared in the 1963 Nobel Prize in physics for developing the shell model theory of the structure of atomic nuclei. This theory suggests that just like electrons in an atom, protons and neutrons within the nucleus are arranged in quantum shells that fill according to specific magic numbers.

It has been proved that the mass of an atomic nucleus is less than the sum of the masses of the protons and neutrons that it contains. Using Einstein's mass-energy equivalency equation, this mass difference multiplied by the speed of light squared gives the binding **energy** of the nucleus. When the sum of the masses of the particles in a nucleus is less than the mass of the nucleus itself, the nucleus will not be stable and that atom will be radioactive (unstable).

NUTRITION

Nutrition is the biological process by which an organism, such as a plant or animal, takes in and utilizes food. This food is converted to **energy** which is used to keep the organism alive. While plants and animals have distinctly different nutritional requirements, there are some commonalities.

Theories about nutrition have been posed throughout history. During the time of the ancient Greeks, Anaxagoras (c.500-c.428 B.C.) suggested that nutrition was a result of the fact that everything contained a small amount of everything else. For example, he believed that when food was eaten, the part of the food that was hair would become part of the hair while the part of it that was muscle would become muscle. Theories about plant nutrition were also advanced. For example, Aristotle believed that plants got all their nutritional needs from the soil. This theory was widely accepted until the seventeenth century when a Belgian physician named **Johannes Baptista van Helmont** ran an experiment to see whether that was true. He thought that if plants grew by absorbing soil, then the amount of soil that he started with would become less overtime as the plant grew. After five years he found that almost no soil was lost. He concluded that the main compound responsible for the increase in the plant's **mass** was the **water** that he gave the plant. In the eighteenth century, Stephen Hales (1677-1761) postulated the notion that plants were also nourished by the air. Later, all of these scientists were found to be at least partially correct.

Plants grow mainly by accumulating water in their cells. In this case, water is a nutrient because it supplies the **hydrogen** and some **oxygen** that is incorporated into compounds for photosynthesis. Most of the water that enters a plant is lost by transpiration. The water that is left behind works as a solvent and helps maintain the form of soft tissue. The dry weight of a plant is about 95% organic material and 5% inorganic minerals. **Carbon**, oxygen, and hydrogen are the most abundant elements found in the dry weight of a plant. **Nitrogen**, **sulfur**, and **phosphorus** are also relatively abundant.

Since plants and other photosynthetic organisms can produce many of their own nutrition requirements they are known as autotrophs. This means they can readily transform inorganic compounds into biologically useful organic compounds. And while plants can produce much of their own nutritional needs such as starch, they still require at least seventeen essential nutrients.

To some extent the minerals in a plant reflect the mineral content of the soil they are growing in. Some nutrients are essential. Elements required in large amounts by plants are known as macronutrients. Nine have been identified. These include carbon, oxygen and hydrogen which are the major components of the plant's organic compounds. Nitrogen and phosphorus are important components of the plant's genetic material, the **nucleic acids**. Nitrogen is also a key component of **proteins**, **hormones**, and coenzymes. Phosphorus is found in the phospholipid bilayer of the cell membranes. Other macronutrients include **potassium** which helps maintain water balance and is involved in **protein synthesis**. **Calcium** is an important part of the formation of cell walls. It also helps activate some enzymes and is involved in the way plant cells respond to stimuli. **Magnesium** is a component of **chlorophyll**, the energy producing organelle in plant cells.

In addition to the macronutrients, plants also require smaller amounts of micronutrients to grow properly. Most of these function as cofactors in biological reactions. These include **chlorine** which is involved in **photosynthesis** and water balance control. **Iron** makes up part of the **cytochromes**, the organelles which control **electron** transport. Other micronutrients include **boron**, **manganese**, **zinc**, **copper**, **molybdenum**, and **nickel**. Mineral deficiency can weaken or even kill a plant. Typically, when a plant is deficient in a nutrient, a variety of symptoms might result. For example, if a plant is deficient in magnesium, a key part of chlorophyll, the leaves turn yellow.

Plants take in their nutrient requirements in different ways. Some nutrients are absorbed from the soil by the roots. Others are absorbed through the air. **Carbon dioxide**, the most important, is taken in through the air. Some plants get nutrition by being predators. For example, mistletoe grows on trees and supplements its nutrition by absorbing nutrients from the tree. Carnivorous plants have specialized leaves that can act as traps for insects.

Unlike plants, animals are heterotrophs, meaning they are unable to live on inorganic nutrients alone. They need a supply of organic compounds for energy and growth. There are different types of animals and they can be classified by their nutrition requirements. Herbivores eat plants, algae, and autotrophic bacteria. Carnivores eat other animals. Omnivores eat both plants and other animals. Animals are also categorized by the way in which they get there food. Suspension feeders sift food particles from the water. This includes animals such as clams and whales. Substrate feeders live in their food source and eat their way out. This is the strategy of many insect larvae. Fluid feeders suck nutrients from a host. Aphids or mosquitoes are examples. Bulk feeders eat large pieces of food.

Eating is called ingestion and it is the first step in food processing. Digestion is the second step and this involves the breaking down of food into molecules. Digestion breaks down chemical bonds in a process called enzymatic **hydrolysis**. Absorption is the third step. Here the appropriate compounds are taken in by the animal's cells. Elimination is the final stage in which unused and waste material are passed out of the digestive system.

Food has three primary functions for animals. First, it acts as fuel for various reactions. This energy runs all the chemical reactions and provides energy for locomotion. Food also provides the organic raw materials for the body to make new molecules. This is how animals get their source of carbon and nitrogen. And finally food provides the essential nutrients which animals can not make for themselves. Essential nutrients include some **amino acids**. There are eight amino acids that must be ingested because the human body can not produce them themselves. Certain **fatty acids** are also essential. These are unsaturated fatty acids, an example of which is linoleic acid. **Vitamins** are another required nutrient. There are 13 vitamins required in a typical human diet. They act as coenzymes and are required in small amounts. Minerals are inorganic materials that are essential to proper nutrition. For example, they are involved in such things as bone and tooth formation, iron **metabolism** and muscle activity. Minerals that are required include such things as calcium, phosphorus, and **sodium**. There are over 17 different minerals that are required in a human diet.

See also Minerals (dietary)

Nylon

Nylon is a synthetic fiber derived from **petrochemicals**. Generically known as a group of polyamides, nylon is chemically distinguished by its amide linkages to two aromatic rings. This chemical arrangement produces such properties in the fiber as strength, elasticity, durability, and its ability to withstand numerous washings. Commonly manufactured nylons include nylon 6 and nylon 6,6. Nylon 6 is formed from a petrochemical that contains six carbons, while nylon 6,6 is created from a reaction between two petrochemicals that contain six carbons each.

Nylon was first developed at Du Pont by American chemist **Wallace Hume Carothers** and his research team. Du Pont had made a commitment to find an artificial substitute for silk. They knew that the market for silk stockings was a $70 million business and an inexpensive material with the properties of silk was bound to be successful. After several years of research, Carothers had almost given up on finding a substitute, when he and the team discovered the secret of cold drawing.

In 1930, Carothers and Julian Hill, his assistant, developed equipment that led to the breakthrough. They wanted to

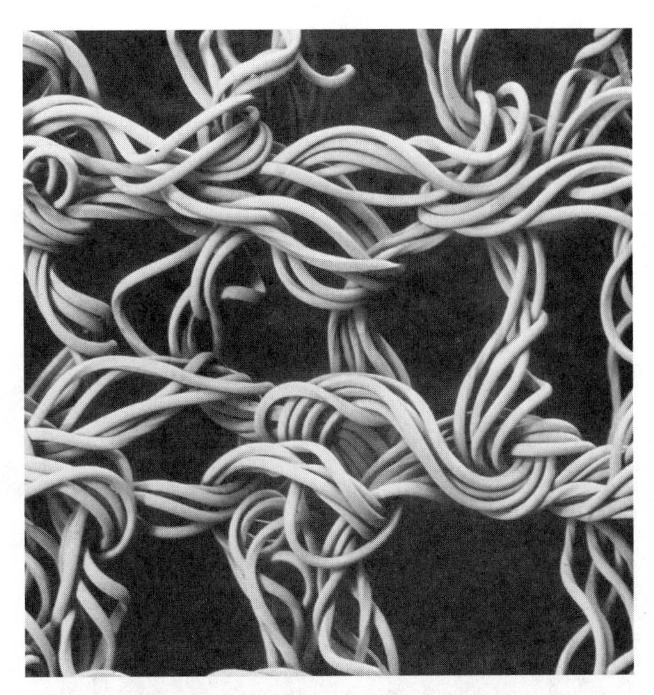

A color enhanced scanning electron microscope image of a nylon stocking, magnified 300 times. ***(Photograph by Meckes/Ottawa, Photo Researchers, Inc. Reproduced by permission.)***

form long chain **polyester**. The polyester was formed by reacting diacids and diols. This reaction also formed **water**, which caused the polyester to revert to the original reactants. The equipment they invented removed the water as it was formed, resulting in a high **molecular weight** polyester. Hill noticed that he could stretch the material into a fiber. He and the other research members stretched the polyester down the hall of the laboratory. They realized that the fiber grew silkier and stronger as it was drawn out. By doing this they had inadvertently discovered the process of cold drawing. Cold drawing orients the molecules into a long linear chain and fosters strong **bonding** between molecules. Unfortunately, the polyester Julian Hill used had a low melting **temperature** and was unsuitable for textile applications.

Eventually, Carothers was persuaded to try the experiment using polyamides. In 1934, Donald Coffman, another assistant, drew out the first nylon fiber and by 1935, a polyamide called nylon 6,6 was identified as the most suitable substitute for silk. The first nylon stockings went on sale in the United States in 1938. By 1941, 60 million pairs were sold. During World War II, however, stockings became rare as nylon production was diverted to the war effort for use in parachutes, mosquito netting, ropes, blood filters, and sutures.

Today, nylon is forced through small holes in a flat plate, called spinnerettes, to form fibers and then stretched by passing it through a pair of rollers rotating at different speeds. This process stretches the fiber several hundred percent and increases its strength by more than 90%. Nylon is stronger than steel by weight and is almost inflammable. It is used in clothing, laces, toothbrushes, strings on musical instruments, sails, fish nets, carpets, and other products requiring strong, lightweight fibers.

See also Synthetic fibers

O

OCHOA, SEVERO (1905-1993)

Spanish biochemist

Spanish-born biochemist Severo Ochoa spent his life engaged in research into the workings of the human body. In the 1950s, he was one of the first scientists to synthesize the newly discovered ribonucleic acid (RNA) in the laboratory. This was the first time that scientists managed to combine molecules together in a chain outside a living organism. For this work, Ochoa received the Nobel Prize for physiology or medicine. In addition to his laboratory work, Ochoa, who was trained as a physician in Spain, taught **biochemistry** and pharmacology to many generations of New York University medical students.

Severo Ochoa was born on September 24, 1905, in Luarca, a small town in the north of Spain. Named after his father, a lawyer, Ochoa was the youngest son in the family. When he was seven, his parents decided to move to Málaga. The move enabled Ochoa to attend a private school, which prepared him for admission to Málaga College. Although Ochoa now knew that he wanted to be a scientist, he was unsure about the field of science he should pursue. Because he found mathematics at his school quite taxing, he decided not to study engineering, a field where mathematical skills are essential. Instead, he chose biology. In 1921, after he graduated from his college with a degree in biology, Occhoa spent completing the prerequisite course for medical school (physics, **chemistry**, biology, and geology. In 1923, he matriculated at the University of Madrid's medical school.

At Madrid, Ochoa hoped to study under the noted neurohistologist Santiago Rámon y Cajal, but was disappointed when he learned that the professor had retired. Rámon y Cajal still ran a laboratory in Madrid, but Ochoa seemed hesitant to approach him. Nevertheless, Ochoa realized that it was essential to gain experience in the laboratory.

The Medical School had no research facilities; fortunately, Ochoa's physiology professor directed a small laboratory under the aegis of the Council for Scientific Research. Working with a classmate, Ochoa first mastered the rather routine laboratory procedure of isolating creatinine (a white crystalline compound) from urine. He then moved to the more demanding task of studying the function and **metabolism** of creatine, a nitrogenous substance, in muscle. The summer following his fourth year of medical school he spent in a Glasgow laboratory, continuing work on this problem. Ochoa received his medical degree in 1929.

Wishing to further his scientific education, Ochoa applied for a post-doctoral position under the direction of Otto Meyerhoff at the Kaiser-Wilhelm Institute in a suburb of Berlin. Because he did not need financial help, Ochoa turned down a fellowship offered by the Council for Scientific Research, explaining that the fellowship should go to a less fortunate young scientist. He remained in Germany for a year.

On July 8, 1931, Ochoa married Carmen García Cobian, daughter of a lawyer and business, and moved with his wife to England, where he had a University of Madrid fellowship to study at the National Institute for Medical Research in London. In England, Ochoa met Sir Henry Hallett Dale, would later, in 1936, win the Nobel Prize for physiology or medicine for his discovery of the chemical transmission of nerve impulses. During his first year at the Institute, Ochoa studied glyoxalase, an **enzyme**; the following year, he started to work directly under Dale, investigating how the adrenal glands affected the chemistry of muscular contraction. In 1933, he returned to his alma mater, the University of Madrid, where he became lecturer in physiology and biochemistry.

Relatively soon, Ochoa was offered a new position. One of the heads of the Department of Medicine decided to start an Institute for Medical Research, with section for biochemistry, physiology, microbiology, and experimental medicine. The institute would be partly supported by the University of Madrid, which agreed to house it; additional support would come from wealthy patrons, who planned to provide a substantial budget for equipment, salaries, and supplies. The institute's director offered Ochoa the directorship of the physiology section, a position which provided him with a staff

Severo Ochoa.

of three. However, a few months after Ochoa bagan work, civil war broke out in Spain. In order to continue his work, Ochoa had to leave country. In September, 1936, he and his wife went to Germany, hardly a stable country at that time.

When Ochoa arrived, he found that his mentor Meyerhoff, who was Jewish, was under considerable political and personal pressure. The German scientist had not allowed this to interfere with his work, though Ochoaa did find, to his surprise, that the type of research Meyerhoff conducted and changed dramatically in the six years since he had seen him last. As he wrote of the laboratory in a retrospective piece for the *Annual Review of Biochemistry:* ''When I left in 1930 it was basically a physiology laboratory. Glycolysis and **fermentation** in muscle or yeast extracts or partial reactions of these processes catalyzed by purified enzymes, were the main subjects of study.'' Meyerhof's change in research orientation influenced Ochoa's work, despite the fact that his work with the German scientist ended, less tha a year later, when Meyerhof fled to France.

Before Meyerhof left, however, he ensured that his younger colleague would not be stranded in Germany, arranging that Ochoa receive a six-month fellowship at the Marine Biological Laboratory in Plymouth, England. Although the fellowship was a short one, Ochoa enjoyed his time there, not the least because his wife started working with him in the laboratory. Their collaboration later led to the publication of a joint paper in *Nature*. At the end of the six months, though, Ochoa had to move on, and friends at the lab found him a post as a research assistant at Oxford University. Two years later, when England entered the war, Oxford's Biochemistry Department shifted all its efforts to war research, in which Ochoa, as an alien, could not participate. So, in 1940, the Ochoas picked up stakes again, this time to cross the Atlantic to work in the laboratory of **Carl Ferdinand Cori** and **Gerty T. Cori** in St. Louis. Part of the Washington University School of Medicine, the Cori lab was renowned for its cutting edge research on enzymes, as well as work with intermediary metabolism of **carbohydrates**. This work involved studying the biochemical reaction in which carbohydrates produce **energy** for cellular operations. Ochoa worked there for a year before New York University persuaded him to take a job as a research associate in medicine at he Bellevue Psychiatric Hospital, where he would, for the first time, have graduate and post-doctoral students working under his direction.

In 1945, Ochoa was promoted to assistant professor of biochemistry at the medical school. Two years later, when the pharmacology chair retired, Ochoa accepted the offer to succeed him. He remained chairperson for nine years, taking a sabbatical in 1949 to serve as visiting professor at the University of California. His administrative work did ot deter him from pursuing his research projects in biochemistry, however. In the early 1950s, he isolated one of the chemical compounds necessary for **photosynthesis** to occur, triphosphopyridine nucleotide, known as TPN. Ochoa continued his work in the area of intermediary metabolism, expanding the work of **Hans Adolf Krebs**, who posited the idea that of a cycle through which food is metabolized into adenosine triphosphate, or ATP, the **molecule** that provide the cell with energy. Ochoa discovered that one molecules of glucose, when burned with **oxygen** produced 36 ATP molecules. When the chairman of the department resigned in 1954, Ochoa took the opportunity to return to the department full time, as chair and full professor.

Once more fully devoted to biochemistry research, Ochoa turned his attention to a new field: the rapidly growing area of deoxyribonucleic acid (DNA) research. Earlier in his career, enzymes had been the hot new molecules for biochemists to study; now, after, after the critical work of **James Watson** and **Francis Crick** in 1953, **nucleic acids** were fascinating scientists in the field. Ochoa was no exception. Drawing on his earlier work with enzymes, Ochoa began investigating which enzymes played roles in the creation of nucleic acids in the body. Although most enzymes assist in breaking down materials, Ochoa knew that he was looking for an enzyme that helped combine nucleotides into the long **chains** that were nucleic acids. Once he isolated these molecules, he hoped he would be able to synthesize **RNA** and **DNA** in the lab. In 1955, he found a bacterial enzyme in sewage that appeared to play just such role. When he added this enzyme to a **solution** of nucleotides, he discovered that the solution became viscous, like jelly, indicating that RNA had indeed formed in the dish. The following year, **Arthur Kornberg**, who had studied with Ochoa in 1946, applied these methods to synthesize DNA.

In 1959, five years after he assumed the directorship of the biochemistry department, Ochoa shared the Nobel Prize for physiology or medicine with Kornberg, for the discovery of enzymes that help produce nucleic acids. Ochoa, who was delighted to share the prize with his old colleague, was, by this time, no stranger to academic honors. In addition to several honorary degrees from American and European universities, including Oxford, Ochoa had also received the Carl Neuberg Medal in biochemistry, in 1951, as well as the Charles Leopold Mayer Prize, in 1955. He served as chairperson of NYU's biochemistry department for twenty years, until the summer of 1974. When he retired from that position, he rejected the department's offer to make him an emeritus professor, preferring to remain on staff as a full professor. But even that could not keep Ochoa fully occupied. In 1974, he joined the Roche Institute of Molecular Biology in New Jersey.

In 1985, he returned to his native Spain as a professor of biology at the Autonomous University of Madrid to continue his lifelong fascination with biochemical research. At the age of 75, Ochoa wrote a retrospective of his life, which he titled *Pursuit of a Hobby*. In the introduction to this piece, he explained the choice of his title. Namely, at a party given in the forties in honor of two Nobel laureate chemists, Ochoa, who at the time taught pharmacology at New York University, listed biochemistry as his hobby in the guest register. Sir Henry Dale, one of the party's honorees, joked, "Now that he is a pharmacologist, he has biochemistry as a hobby." Ochoa concluded this tale by saying: "In my life biochemistry has been my only and real hobby." Ochoa died in 1993.

OCTET RULE

The octet rule is used to describe the attraction of elements towards having, whenever possible, eight valence-shell electrons (4 **electron** pairs) in their outer shell. Because a full outer shell with eight electrons is relatively stable, many atoms lose or gain electrons to obtain an electron configuration like that of the nearest noble gas. Except for **helium** (with a filled 1s shell), noble **gases** have eight electrons in their **valence** shells. The octet rule is used when drawing Lewis dot structures.

The components of table **salt**, **sodium**, and chloride ions, illustrate the octet rule. Sodium (Na) with an electron configuration of $1s^22s^22p^63s^1$ sheds its outermost $3s$ electron and, as a result, the Na^+ **ion** has an electron configuration of $1s^22s^22p^6$. This is the same electron configuration as **neon**, a noble gas (i.e., highly stable and relatively nonreactive).

Chlorine (Cl), on the other hand, has an electron configuration of $1s^22s^22p^63s^23p^5$. Chlorine needs one electron to fill its outermost third shell with eight electrons. When chlorine takes on the electron shed by sodium then the Cl^+ ion's electron configuration becomes $1s^22s^22p^63s^23p^6$. This is the same configuration as **Argon**, another noble gas.

The octet rule, however, does not accurately predict the electron configurations of all molecules and compounds. Not every **nonmetal**, nor metal, can form compounds that satisfies the octet rule. As a result, the octet rule must be used with caution when predicting the electron configurations of molecules and compounds. Some atoms violate the octet rule and surround themselves with more than four electron pairs.

In general, the octet rule works for representative **metals** (Groups IA, IIA) and nonmetals, but not for the transition, inner-transition or post-transition elements. These elements seek additional stability by having filled half-filled or filled orbitals *d* or *f* subshell orbitals.

ODORS AND OLFACTION

Olfaction, the act of smelling, is as primitive as it is sophisticated. The human sense of smell, although only a fraction as sensitive as most other animals', not only warns us of danger and helps bond babies and mothers, but also allows us to appreciate the finest of wines and perfumes. **Chemistry** is at the heart of all olfactory perception, a mechanism by which the brain processes a certain type of information.

The chemical theory of olfaction holds that substances emit particles that waft to the olfactory **receptors** via **diffusion** and/or convection. Once there, these particles cause chemical reactions that lead to perception of different odors. However, some materials, such as those with high **molecular weight**, have no odor, while others emit huge quantities of particles that can travel into the sensory organs.

Many organisms have the ability to perceive odor, which requires the presence of certain organs and other apparatus. In terms of the chemistry of smell, the key physical requirements in humans and other higher animals are the olfactory receptor cells, which become stimulated in the presence of odorants; olfactory bulbs, which receive the electrical stimuli from the receptor cells; and olfactive lobes, which distribute the perception to different areas of the brain. In addition, the vomeronasal organ in the nose is a part of a chemosensory system that is independent of the olfactory system. This organ may be associated with the reason that smells can affect human behavior so effectively.

Chemists measure odors according to their intensity, using what they call a "threshold value." On the lowest end of the scale is "detection threshold," which represents the minimum amount of intensity for a person to be able to detect a stimulus. This scale is key to the chemistry of odor, because a single stimulus can be perceived very differently as its intensity rises. For instance, the substance indole, which comes from **coal** tar and animal feces, actually smells like jasmine at its detection threshold. Although it smells strongly like its sources at higher intensities, its lowest threshold value makes indole a popular ingredient in the fragrance industry.

Understanding and being able to use the chemistry of odor is intrinsically important in numerous industries. Most of the time, manufacturers use this knowledge to make their products smell good—or at least better than the products do on their own. There are two main techniques for improving odor: masking and counteraction.

The process of chemical masking consists of reducing the olfactory perception of a particular odor stimulus by adding another odorant. Masking does not remove or chemically alter the original odor, but merely covers it up with another scent (often one of the aldehydes) so that the brain perceives less or none of the target odor. For this reason, masking is also called "reodorant," as opposed to "deodorant." A potential problem with reodorants is that they raise the overall odor level, which sometimes results in an overwhelming olfactory sensation. Any example of masking is spraying perfume into a diaper-changing room. The perfume will reduce, but not eliminate, the brain's olfactory perception of dirty diapers, but the combination of odors might be even more unpleasant and overwhelming than the original odor alone.

Odor counteraction is a more effective method of changing or eliminating an odor. Similar to the chemical concept of **neutralization**, odor counteraction consists of mixing two odorous materials at a precise ratio to produce a less-intense odor than those of the separate ingredients. Technically speaking, this process is known as "compensation." Counteraction rarely results in the absence of odor, because there is almost always some residual scent. However, the process of **oxidation** (i.e., increasing the **valence** of the odor **molecule** by removing some of its electrons) can eliminate the offending odor molecule and leave no trace odor at all. An example of counteraction is mixing citral, a pale-yellow liquid used in the perfume and flavoring industries, to neutralize the amines emitted from decaying meat in the dumpster behind a restaurant.

Sometimes, however, industrial chemists are called upon to give a product an unpleasant or sharp odor. This is less common, but an important factor in such areas as oil and gas manufacturing. For example, some of the fuel **gases** are completely odorless, so there is no indicator (to humans, at any rate) when a leak occurs. Because such an event could be life threatening, inserting an odorant into these gases is crucial to provide a warning of exposure.

There are a whole range of "aroma chemicals," as they are known, that come in a variety of dispersal media, including powder, aerosol, liquid, solid, and electric. These chemicals were originally all obtained from natural sources (e.g., jasmine scent from jasmine flowers, vanilla flavoring from vanilla beans, etc.), but as the amount of agricultural land has declined and the science of chemistry has grown more sophisticated, "synthetics" have become more common and much less expensive than "naturals."

Once chemists began to identify the chemical compounds responsible for a material's characteristic odor, they could make the odor right in the laboratory. Thus, perfume and fragrance makers could simply buy quantities of the synthesized odor they needed rather than search the world over for the exact plant or animal that carried the scent. Thus, perfume makers no longer need natural ambergris, the sperm whale intestinal secretion so good at "carrying" a fragrance, because synthetic versions of the rare and expensive natural substance are now abundant. Likewise, food manufacturers no longer have to use natural grape to flavor a juice or dessert—they need only to add some artificial methyl anthranilate to achieve the same effect at much less expense.

In the ever-expanding science of odor, there are precise categories that help to make odor perception as objective as possible. The main categories in the perfumery industry, for instance, are fruity, citrus, green, floral, and woody. The science has become so precise that chemists have been able to assign characteristic odors to entire functional groups. They know that the ketone group contains aroma chemicals that produce mainly musky scents, for example, while those in the **hydrocarbon** group produce warm, woody odors. For the increasing number of products that consumers buy based merely on odor, this knowledge represents very big business.

OIL RECOVERY

The process of oil recovery is essentially the process of getting oil from the places where oil exists in the ground (whether onshore or offshore) and into processing plants for refining so the oil is suitable for industrial and residential purposes. The ways to recover oil through a conventional well bore are known as primary, secondary, and tertiary (enhanced), but some unconventional methods are also becoming popular.

Oil, which is usually called crude oil in its most basic form, is a valuable fuel whose chemical makeup is a mixture of **hydrocarbon** fuels: **kerosene**, dissolved natural gas, naphtha, light and heavy heating oils, diesel fuel, tars, **benzene**, and **gasoline**. It is formed over millions of years by the action of **heat** and pressure on organic material buried deep within rock, and typically exists in combination with **salt water**, **natural gas**, and soil. Most of the world's oil comes from huge, seemingly inexhaustible subterranean patches of porous, oil-permeated rock. The oil is confined to a certain location, or "trap," by other layers of impermeable rock (usually types of shale) and/or faults.

Throughout recorded history, people have been using oil (in the form of kerosene) for lighting and other purposes. "Recovery" in those early days of oil use generally meant skimming it off the surfaces of seeps and standing bodies of water, although the Chinese seem to have drilled for fuel in the fourth century A.D. Beginning in about the 1850s, however, whale oil—another major source of fuel—began to become less abundant and people started looking for ways to get more crude oil. A man in Pennsylvania, Edwin Drake (1819-1880), dug what is considered to be the first modern oil well in 1859, and in 1861 the first oil refinery went into business to separate crude oil into all its "fractions," or useful parts.

As oil quickly became a crucial, if not indispensable, product for much of the industrialized world, demand increased annually. The methods people invented for recovering oil from its hidden recesses in the earth becoming more sophisticated as well. At first, oil explorers were content to drill down into the earth where they thought there might be a deposit of oil and then stand by as it exploded out of the ground in a "blowout" or "gusher" as the oil was released from the tremendous pressures that helped create it. Collecting oil forced out by natural pressure caused by water and gas "drives" is known as primary recovery.

As the first pressurized wells began to stop expelling their oil because they were constantly being drawn on, new ways of extracting the remaining oil came about. These artificial oil-recovery methods are known as secondary techniques. One of the tools that became popular for this eventuality was the legendary pumpjack or beam pump, which could raise the oil to the surface artificially. When these pumps stop bringing oil up, sometimes more powerful hydraulic pumps are installed.

The most reliable methods for improving oil extraction were water and gas injection through what came to be known as injection wells. These techniques (still in use today) consist of pumping water or gas in through the injection wells to force the oil up to the surface or into a well from which it can be extracted.

Tertiary, or enhanced, oil recovery (EOR) is perhaps the most common category of modern oil recovery methods. It comes into play after a well's natural drives have stopped working and secondary techniques have recovered as much oil as they can. Even then, there may be as much as 95% of a well's oil remaining.

The main EOR techniques are chemical flooding, water flooding, and thermal recovery. Chemical flooding consists of injecting water that contains surfactants, which lower the water's surface tension and so increase its wetting and spreading capabilities. This causes the remaining oil in the well to mix with the special water, as it normally would not, so that the oil can more easily move through the reservoir's porous rock and toward the well bore. Similarly, injecting alkaline (caustic) solutions into a well causes certain types of crude oil to form surfactants, which have the same effect. Chemical flooding using a micellar-polymer mixture along with surfactants makes the water more viscous (thicker) so it slows down as it passes through rock and picks up more oil on its way to the well bore.

Water flooding is when well operators inject huge volumes of water into "dead" wells near still-producing wells. The water, moving toward the area of lesser pressure, moves some of the oil in the dead wells through the rock and toward the producing wells in the same reservoir. From there, the oil and water are pumped out and separated. This is the most popular and least-expensive EOR technique.

In thermal recovery, heat is the most important agent. Some crude oil is so thick that it will not move toward the well bore on its own. In these cases, well operators sometime use a steam drive, or continuous steam injection, to force the oil out. Not only does the steam reduce the oil's **viscosity** to make it more moveable, it condenses inside the well to form hot water that transports the oil to the bore. In addition, the heat vaporizes the natural **gases** in the oil, which provides another drive to make the oil move.

Cyclic steam injection, or "huff and puff," also involves sending hot steam down the well, but then the injection stops and the well sits closed for perhaps 72 hours to "soak." Upon opening the well, the thinned oil and condensed water flow out. Some crews light a fire in a reservoir to create "in situ **combustion**" or "fire flooding." With sufficient air injected into the well for the fire to burn, the fire starts spreading toward the production wells, pushing the heat-thinned oil ahead of it.

Retorting is the main procedure used for recovering oil from oil shale, a plentiful but difficult-to-process and environmentally-hazardous substance. Oil shale is a sedimentary rock that has large amounts of organic, fossil-based material mixed in with it (kerogen). Retorting is the process of using heat to decompose the kerogen, converting it into a liquid called shale oil. Some experts believe that as the world's conventional sources of oil run out, we will be forced to turn to oil shale to fill our **energy** needs, should these remain unchanged.

Powdered catalyst being poured into a person's hand at an oil refinery. Aluminum-silica gel powders are used as catalysts to break up, or crack, the heavier components (fractions) of crude oil into lighter, more valuable fractions, such as petroleum. *(Photograph by Ed Young/Science Photo Library, Photo Researchers, Inc. Reproduced by permission.)*

OIL REFINING

Oil refining refers to the processes involved in converting crude oil into useful petroleum products. Crude oil is often called "black **gold**" because of its thick, black consistency and high value. Crude oil is a form of petroleum, a complicated mixture of different hydrocarbons. A **hydrocarbon** is a **chemical compound** composed solely of **hydrogen** (H) and **carbon** (C). Hydrocarbon molecules in petroleum can contain from one to 50 carbon atoms. Crude oil itself is practically useless. The different fractions of the crude oil, however, are considered to be some of the most useful compounds on Earth. Oil refining turns crude oil into fuels and materials which can be used to make thousands of products. Almost half of the world's **energy** sources originate from crude oil. Petroleum products make up the majority of the items used by people every day, including bathing suits, toothpaste, soda bottles, clothing, and pharmaceuticals.

Petroleum is a fossil fuel. The other two fossil fuels are **coal** and **natural gas**. The fossil fuels are nonrenewable resources, that is, they cannot be replaced once they are used up. There is only a certain amount of petroleum in the Earth, and once it is used, alternative forms of energy and products will have to be used. Petroleum was formed from ancient plants and animals which died and were buried under layers of sediments such as sand, silt, clay, or mud. As more sediment buried the organic material, intense pressure built up. This pressure along with the high temperatures which resulted from being pushed towards the center of the Earth converted the organic material into petroleum.

Petroleum is sometimes called "buried sunshine" because the energy stored in petroleum originated from the sun. Essentially, the Earth's core is heated by residual **radioactivity** from the sun. The sun's radiant energy is converted by plants into chemical energy (sugar and starches) through the process of **photosynthesis**. When these plants are eaten by animals, this chemical energy is either transformed into another form of chemical energy (fats or muscle) or converted to mechanical or thermal energy. Whenever an animal consumes another animal, a similar process occurs. Petroleum originated as ancient plants and animals which were buried under many layers of sediment, therefore, even the fuels we use originated from the sun. The energy we use every day is the result of the transformations of solar energy over millions of years.

Oil refining separates the various fractions of petroleum by a process called fractional **distillation** and takes place in a large plant called a refinery. Distillation is the heating up of a liquid above its **boiling point**. At this point, the liquid becomes a gas, in other words, it vaporizes. The vapor is then allowed to cool until it condenses, or changes back into a liquid. The different fractions of petroleum have different boiling points, depending on the **molecular weight** of the compound. The more carbon atoms a hydrocarbon has, the higher its boiling point will be. The boiling point of a substance is also the **temperature** at which it condenses, therefore, the different fractions of petroleum also condense at different temperatures. As the crude oil vapor is allowed to cool, the different fractions will condense at different temperatures. As the fractions condense, they can be drawn off, separating the petroleum into its useful components. Fractional distillation at a refinery takes place in a large structure called a fractionating tower. Fractionating towers can be as tall as 30 meters or more.

Asphalt is a hydrocarbon composed of many carbon atoms, therefore, it has an extremely high boiling point. Asphalt does not yet vaporize at 385°C, therefore, as the other fractions vaporize it is left behind as a thick liquid. Asphalt is drawn out of the bottom of the fractionating tower. The different fractions, in order of highest temperature to lowest temperature, condense as follows; lubricating oils, industrial fuel oils, diesel fuel, heating fuels, jet fuel, **kerosene**, and lastly, **gasoline**. The lower the temperature at which it condenses, the farther up the tower the fraction will travel. Therefore, gasoline is collected at the very top of the tower.

Gasoline ranges from four to 12 carbon atoms per **molecule** and its boiling point is approximately 200°C. Kerosene contains between 10 and 14 carbon atoms. The boiling point of kerosene is between 180 and 290°C. The middle distillates, those which condense in the middle of the fractionation tower, include heating oil, gas-turbine fuel, and diesel. These compounds contain 12-20 carbon atoms and have boiling points from 185-345°C. Wide-cut gas oil, which includes lubricating oil and **waxes**, boils from 345-540°C. These compounds contain 20-36 carbon atoms. Finally, asphalt, which contains more than 36 carbon atoms per molecule, boils above 540°C. The larger the molecule, in other words, the more carbon atoms a molecule contains, the higher the boiling point. The larger hydrocarbons have limited commercial use and need to be further refined. This process is called cracking. During cracking, the hydrocarbons are heated to high temperatures which break some of the carbon-carbon bonds, creating a mixture of smaller hydrocarbon molecules.

Crude oil can be converted into many useful products. One barrel of crude oil gives the following percent yield: gasolines, 46.7%; fuel oil, 28.6%; jet fuel, 9.1%; **petrochemicals**, 3.8%; coke, 3.5%; asphalt and road oil, 3.1%; liquefied **gases**, 2.9%; lubricants, 1.3%; kerosene, 0.9%; and waxes, 0.1%. Most of the products which are made from petroleum require processing beyond that of the refinery. Two products which come directly from the crude oil and do not require further processing are asphalt and waxes. Asphalt is the main material used for building roads, and waxes are used in such products as floor polish and to coat milk cartons. These two petroleum products are considered raw materials because they do not require extra processing.

Most of the products which come from crude oil require processing beyond the refinery. For example, lubricants are a group of petroleum products which require processing. These substances reduce friction between the moving parts of equipment. The largest group of petroleum products which require processing are fuels. The majority of the crude oil is used for various fuels. Fuels which come from petroleum are easily burned and create large amounts of thermal energy. Petroleum fuels are preferred over other combustible materials such as coal or **wood** because they are cleaner, easier to store, and easier to transport from one location to another. Almost all of the fuels used for transportation and the majority of the fuels used for **heat** and **electricity** come from petroleum products.

OLAH, GEORGE A. (1927-)

Hungarian-born American chemist

The recipient of the 1994 Nobel Prize for **chemistry**, Olah is primarily known for his crucial work on reactive intermediates in hydrocarbons. The complex chemistry of hydrocarbons, compounds of **carbon** and **hydrogen**, includes the study of numerous reactions, which are sometimes extremely difficult to record. Reactive intermediates, or substances acting as the intermediate steps of a chemical **reaction mechanism**, are so short-lived and elusive that chemists used to regard them as purely hypothetical entities.

Before there was empirical evidence for the existence of reaction intermediates, chemists believed that carbon ions, or positively charged atoms, played an intermediary role in **hy-**

George A. Olah. *(Archive Photo. Reproduced by permission.)*

drocarbon reactions; the action of these intermediaries, however, was imagined to be so rapid, maybe measurable in millionths of a second, that scientists only postulated their existence. Olah, however, did not doubt the existence of reactive intermediates were real, deciding, in fact, to empirically prove their existence. In order to identify a reactive intermediate, Olah needed a substance that would somehow arrest the reaction mechanism, thus enabling the observer to capture processes which cannot be seen under normal circumstances. The substances that worked, he found, were **superacids** (a superacid is an extremely powerful acid—for example, more than a trillions times the strength of sulfuric acid). Olah subsequently created a superacid which could extract individual atoms from hydrocarbon compounds. What remained when a hydrocarbon compound was exposed to a superacid, Olah noticed, was an alkyl (an alkyl is a univalent group created when a hydrogen **atom** is removed from an open-ended hydrocarbon compound) carbon **ion**, which, although unstable, was measurable. The carbon ion, or **cation**, was the reaction intermediary.

Born in Budapest, Hungary, on May 22, 1927, the son of Julius Olah and Magda Krasznai, Olah received his Ph.D. from the Technical University of Budapest in 1949. That year, he married Judith Lengyel; they have two sons. Olah taught at the Technical University from 1949 to 1954, subsequently joining the Hungarian Academy of Sciences, where he served as associate director of the Central Chemical Research Institute from 1954 to 1956. When the Soviet Union crushed the Hungarian revolt in 1956, Olah and his family fled to the West. In 1957, he joined Dow Chemical in Sarnia, Ontario, where he worked as a research scientist from 1957 to 1965. From 1965 to 1977, he was professor of chemistry at Case Western University. Since 1977, he has worked as professor of chemistry and director of the Hydrocarbon Research Institute at the University of Southern California.

Following his discovery, in 1962, that superacids could neutralize the extreme **reactivity** of cations, Olah has worked on developing new superacids to be used in both industry and fundamental research. As Olah's work has opened vast areas of research, many younger chemists have contributed to the search for new superacids. Significantly, the study of intermediate reactants has also resulted in numerous industrial applications, particularly in fuel synthesis. For example, the synthesis of high-octane **gasoline** is one of the notable industrial uses of Olah's original research. In essence, Olah has created the scientific instruments for creating cleaner and more efficient fuels.

Olah's awards include the 1966 Baekeland Award, the 1989 American Chemical Society Award, the 1993 Pioneer of Chemistry Award given by the American Institute of Chemists, and the Mendeleev Medal, which he received from the Russian Academy of Sciences in 1992.

ONSAGER, LARS (1903-1976)

American chemist

Born in Norway, Lars Onsager received his early education there before coming to the United States in 1928 to do graduate work at Yale University. After receiving his Ph.D. in theoretical **chemistry** he stayed on at Yale and ultimately spent nearly all of his academic career at that institution. Onsager's first important contribution to chemical theory came in 1926 when he showed how improvements could be made in the Debye-Hückel theory of electrolytic **dissociation**. His later (and probably more significant) work involved non-reversible systems—systems in which differences in pressure, **temperature**, or some other factor are an important consideration. For his contributions in this field, Onsager received a number of important awards including the Rumford Medal of the American Academy of Arts and Sciences, the Lorentz Medal of the Royal Netherlands Academy of Sciences, and the 1968 Nobel Prize in Chemistry.

Lars Onsager was born in Oslo (then known as Christiania), Norway, on November 27, 1903. His parents were Erling Onsager, a barrister before the Norwegian Supreme Court, and Ingrid Kirkeby Onsager. Onsager's early education was somewhat unorthodox as he was taught by private tutors, by his own mother, and at a somewhat unsatisfactory rural private school. Eventually he entered the Frogner School in Oslo and did so well that he skipped a grade and graduated a year early. Overall, his early schooling provided him with a broad liberal education in philosophy, literature, and the arts. He is said to have become particularly fond of Norwegian epics and continued to read and recite them to friends and family throughout his life.

In 1920, Onsager entered the Norges Tekniski Høgskole in Trondheim where he planned to major in **chemical engineering**. The fact that he enrolled in a technical high school suggests that he was originally interested in practical rather

Lars Onsager. *Archive Photo. Reproduced by permission.)*

than theoretical studies. Onsager had not pursued his schooling very long, however, before it became apparent that he wanted to go beyond the everyday applications of science to the theoretical background on which those applications are based. Even as a freshman in high school, he told of making a careful study of the chemical journals, in order to gain background knowledge of chemical theory.

One of the topics that caught his attention concerned the chemistry of solutions. In 1884, **Svante Arrhenius** had proposed a theory of ionic dissociation that explained a number of observations about the conductivity of solutions and, eventually, a number of other **solution** phenomena. Over the next half century, chemists worked on refining and extending the Arrhenius theory.

The next great step forward in that search occurred in 1923, when Onsager was still a student at the Tekniski Høgskole. The Dutch chemist **Peter Debye** and the German chemist Erich Hückel, working at Zurich's Eidgenössische Technische Hochschule, had proposed a revision of the Arrhenius theory that explained some problems not yet resolved—primarily, whether ionic compounds are or are not completely dissociated (''ionized'') in solution. After much experimentation, Arrhenius had observed that dissociation was not complete in all instances.

Debye and Hückel realized that ionic compounds, by their very nature, already existed in the ionic state *before* they ever enter a solution. They explained the apparent incomplete level of dissociation on the basis of the interactions among ions of opposite charges and **water** molecules in a solution. The Debye-Hückel mathematical formulation almost perfectly explained all the anomalies that remained in the Arrhenius theory.

Almost perfectly, but not quite, as Onsager soon observed. The value of the molar conductivity predicted by the Debye-Hückel theory was significantly different from that obtained from experiments. By 1925, Onsager had discovered the reason for this discrepancy. Debye and Hückel had assumed that most—but not all—of the ions in a solution move about randomly in ''Brownian'' movement. Onsager simply extended that principle to *all* of the ions in the solution. With this correction, he was able to write a new mathematical expression that improved upon the Debye-Hückel formulation.

Onsager had the opportunity in 1925 to present his views to Debye in person. Having arrived in Zurich after traveling through Denmark and Germany with one of his professors, Onsager is reported to have marched into Debye's office in Zurich and declared, ''Professor Debye, your theory of electrolytes is incorrect.'' Debye was sufficiently impressed with the young Norwegian to offer him a research post in Zurich, a position that Onsager accepted and held for the next two years.

In 1928, Onsager emigrated to the United States where he became an associate in chemistry at Johns Hopkins University. The appointment proved to be disastrous: he was assigned to teach the introductory chemistry classes, a task for which he was completely unsuited. One of his associates, Robert H. Cole, is quoted in the *Biographical Memoirs of Fellows of the Royal Society:* ''I won't say he was the world's worst lecturer, but he was certainly in contention.'' As a consequence, Onsager was not asked to return to Johns Hopkins after he had completed his first semester there.

Fortunately, a position was open at Brown University, and Onsager was asked by chemistry department chairman Charles A. Krauss to fill that position. During his 5-year tenure at Brown, Onsager was given a more appropriate teaching assignment, statistical mechanics. His pedagogical techniques apparently did not improve to any great extent, however; he still presented a challenge to students by speaking to the blackboard on topics that were well beyond the comprehension of many in the room.

A far more important feature of the Brown years was the theoretical research that Onsager carried out in the privacy of his own office. In this research, Onsager attempted to generalize his earlier research on the motion of ions in solution when exposed to an electrical field. In order to do so, he went back to some fundamental laws of **thermodynamics**, including Hermann Helmholtz's ''principle of least dissipation.'' He was eventually able to derive a very general mathematical expression about the behavior of substances in solution, an expression now known as the Law of Reciprocal Relations.

Onsager first published the law in 1929, but continued to work on it for a number of years. In 1931, he announced a more general form of the law that applied to other non-equilibrium situations in which differences in electrical or magnetic force, temperature, pressure, or some other factor exists. The Onsager formulation was so elegant and so general that some scientists now refer to it as the Fourth Law of Thermodynamics.

The Law of Reciprocal Relations was eventually recognized as an enormous advance in **theoretical chemistry**, earning Onsager the Nobel Prize in 1968. However, its initial announcement provoked almost no response from his colleagues. It is not that they disputed his findings, Onsager said many years later, but just that they totally ignored them. Indeed, Onsager's research had almost no impact on chemists until after World War II had ended, more than a decade after the research was originally published.

The year 1933 was a momentous one for Onsager. It began badly when Brown ended his appointment because of financial pressures brought about by the Great Depression. His situation improved later in the year, however, when he was offered an appointment as Sterling and Gibbs Fellow at Yale. The appointment marked the beginning of an affiliation with Yale that was to continue until 1972.

Prior to assuming his new job at Yale, Onsager spent the summer in Europe. While there, he met the future Mrs. Onsager, Margarethe Arledter, the sister of the Austrian electrochemist H. Falkenhagen. The two apparently fell instantly in love, became engaged a week after meeting, and were married on September 7, 1933. The Onsagers later had three sons, Erling Frederick, Hans Tanberg, and Christian Carl, and one daughter, Inger Marie.

Onsager had no sooner assumed his post at Yale when a small problem arose: the fellowship he had been awarded was for postdoctoral studies, but Onsager had not as yet been granted a Ph.D. He had submitted an outline of his research on reciprocal relations to his alma mater, the Norges Tekniski Høgskole, but the faculty there had decided that, being incomplete, it was not worthy of a doctorate. As a result, Onsager's first task at Yale was to complete a doctoral thesis. For this thesis, he submitted to the chemistry faculty a research **paper** on an esoteric mathematical topic. Since the thesis was outside the experience of anyone in the chemistry or physics departments, Onsager's degree was nearly awarded by the mathematics department, whose chair understood Onsager's findings quite clearly. Only at the last moment did the chemistry department relent and agree to accept the judgment of its colleagues, awarding Onsager his Ph.D. in 1935.

Onsager continued to teach statistical mechanics at Yale, although with as little success as ever. (Instead of being called ''Sadistical Mechanics,'' as it had been by Brown students, it was now referred to as ''Advanced Norwegian'' by their Yale counterparts.) As always, it was Onsager's theoretical—and usually independent—research that justified his Yale salary. In his nearly four decades there, he attacked one new problem after another, usually with astounding success. Though his output was by no means prodigious, the quality and thoroughness of his research was impeccable.

During the late 1930s, Onsager worked on another of Debye's ideas, the **dipole** theory of dielectrics. That theory had, in general, been very successful, but could not explain the special case of **liquids** with high dielectric constants. By 1936, Onsager had developed a new model of dipoles that could be used to modify Debye's theory and provide accurate predictions for all cases. Onsager was apparently deeply hurt when Debye rejected his paper explaining this model for publication in the *Physikalische Zeitschrift,* which Debye edited. It would be more than a decade before the great Dutch chemist, then an American citizen, could accept Onsager's modifications of his ideas.

In the 1940s, Onsager turned his attention to the very complex issue of phase transitions in **solids**. He wanted to find out if the mathematical techniques of statistical mechanics could be used to derive the thermodynamic properties of such events. Although some initial progress had been made in this area, resulting in a theory known as the Ising model, Onsager produced a spectacular breakthrough on the problem. He introduced a ''trick or two'' (to use his words) that had not yet occurred to (and were probably unknown to) his colleagues—the use of elegant mathematical techniques of elliptical functions and quaternion algebra. His solution to this problem was widely acclaimed.

Though his status as a non-U.S. citizen enabled him to devote his time and effort to his own research during World War II, Onsager was forbidden from contributing his significant talents to the top-secret Manhattan Project, the United State's research toward creating atomic weapons. Onsager and his wife finally did become citizens as the war drew to a close in 1945.

The postwar years saw no diminution of Onsager's **energy**. He continued his research on low-temperature physics and devised a theoretical explanation for the superfluidity of **helium** II (liquid helium). The idea, originally proposed in 1949, was arrived at independently two years later by Princeton University's **Richard Feymann**. Onsager also worked out original theories for the statistical properties of **liquid crystals** and for the electrical properties of ice. In 1951 he was given a Fulbright scholarship to work at the Cavendish Laboratory in Cambridge; there, he perfected his theory of diamagnetism in **metals**.

During his last years at Yale, Onsager continued to receive numerous accolades for his newly appreciated discoveries. He was awarded honorary doctorates by such noble universities as Harvard (1954), Brown (1962), Chicago (1968), Cambridge (1970), and Oxford (1971), among others. He was inducted to the National Academy of Sciences in 1947. In addition to his Nobel Prize, Onsager garnered the American Academy of Arts and Sciences' Rumford Medal in 1953 and the Lorentz Medal in 1958, as well as several medals from the American Chemical Society and the President's National Medal of Science. Upon reaching retirement age in 1972, Onsager was offered the title of emeritus professor, but without an office. Disappointed by this apparent slight, Onsager decided instead to accept an appointment as Distinguished University Professor at the University of Miami's Center for Theoretical Studies. At Miami, Onsager found two new subjects to interest him, biophysics and **radiation chemistry**. In neither field did he have an opportunity to make any significant contributions, however, as he died on October 5, 1976, apparently the victim of a heart attack.

Given his shortcomings as a teacher, Onsager still seems to have been universally admired and liked as a person. Though modest and self-effacing, he possessed a wry sense of humor. In *Biographical Memoirs*, he is quoted as saying of research, "There's a time to soar like an eagle, and a time to burrow like a worm. It takes a pretty sharp cookie to know when to shed the feathers and... to begin munching the humus." In a memorial some months after Onsager's death, Behram Kursunoglu, the director of the University of Miami's Center for Theoretical Studies, described him as a "very great man of science—with profound humanitarian and scientific qualities."

OPARIN, ALEKSANDR IVANOVICH (1894-1980)

Russian biochemist

Aleksandr Ivanovich Oparin was a prominent biochemist in the former Soviet Union whose achievements were recognized throughout the international scientific community. He is best known for his theory that life on earth originated from inorganic **matter**. Although a belief that life formed through spontaneous generation was prevalent up to the nineteenth century, that theory was disputed by the development of the microscope and the experiments of French scientist **Louis Pasteur**. Oparin's materialistic approach to the subject was responsible for a renewed interest in how life on earth originated. His book *The Origin of Life* outlined his basic theory, which was that life originated as a result of evolution acting on molecules created in the primordial atmosphere through **energy** discharges. In addition to his work on the **origin of life**, he played a major role in the development of technical botanical **biochemistry** in the Soviet Union.

Oparin was born near Moscow on March 2, 1894. He was the youngest child of Ivan Dmitrievich Oparin and Aleksandra Aleksandrovna. He had a sister, Aleksandra, and a brother, Dmitrii. His secondary education was marked by his achievements in science. He studied plant physiology at Moscow State University, graduating in 1917. He was a graduate student and teaching assistant there from 1921 to 1925. He also studied at other institutes of higher learning in Germany, Austria, Italy, and France, but it is thought that he never earned a graduate degree (he was awarded a doctorate in biological sciences in 1934 by the U.S.S.R. Academy of Sciences).

Aleksei N. Bakh, Oparin's mentor during his years of graduate study, was to have great influence on Oparin's later role in the development of Soviet biochemistry. Bakh was well known internationally for his research in medical and industrial **chemistry**, and played an important role in the organization of the chemical industry in Russia. After the Russian Revolution in 1917, Bakh helped develop the chemical section of the National Economic Planning Council (VSNKh) and founded its Central Chemical Laboratory. Oparin studied plant chemistry with Bakh in 1918, and from 1919 through 1925, he worked under Bakh at the VSNKh and the Central Chemical Laboratory. Bakh and Oparin cofounded the Institute of Biochemistry at the Academy of Sciences of the Soviet Union in Moscow in 1935. Oparin was appointed deputy director of the institute and held that position until 1946. After Bakh's death that same year, Oparin assumed the director's position, which he held until his death.

The practical aspects of Oparin's work during his association with Bakh in the early thirties involved biochemical research for increasing production in the food industry, work that was of extreme importance to the Soviet economy. Through his study of enzymatic activity in plants, he found that it was necessary for molecules and enzymes to combine in order to create starches, sugars, and other **carbohydrates** and **proteins**. He was able to show that this biocatalysis was the basis for producing many food products in nature. He held a post from 1927 through 1934 as assistant director and head of the laboratory at the Central Institute of the Sugar Industry in Moscow, where he conducted research on tea, sugar, flour, and grains. During this same period, he also taught technical biochemistry at the D. I. Mendeleev Institute of Chemical Technology. As professor at the Moscow Technical Institute of Food Production from 1937 to 1949, he continued his research of plant processes and began the study of **nutrition** and **vitamins**.

Oparin's biochemical research on plant enzymes and their role in plant **metabolism**, so important for its practical application, would also be important for what was to be the focus of his career, the question of how life first appeared on earth. His first **paper** on this subject was presented to a meeting of the Moscow Botanical Society in 1922. This paper, which was never published, was revised and published in 1924 by the *Moscow Worker*. In it, Oparin discussed the problem of spontaneous generation, arguing that any differences between living and nonliving material could be attributed to physicochemical laws. This work went largely unnoticed, and Oparin did not seriously consider the topic again until the mid-thirties. In 1936, he published *The Origin of Life,* which modified and enlarged his earlier ideas. His ideas at this time were influenced not only by contemporary international thinking on astronomy, **geochemistry**, **organic chemistry**, and plant enzymology, but also by the dialectic philosophy espoused by Friedrich Engels, and the work of H. G. Bungenburg de Jong on colloidal coacervation. Translated into English in a 1938 edition, *The Origin of Life* was also revised and updated in 1941 and 1957. Although the later versions amended the original, the concept that life arose through a natural evolution of matter remained central, and he often described this concept metaphorically by comparing life to a constant flow of liquid in which elements within are constantly changed and renewed.

Oparin's theory that the origin of life had a biochemical basis was based on his suppositions concerning the condition of the atmosphere surrounding the primeval earth and how those conditions interacted with primitive organisms. It was his idea that the primeval atmosphere (consisting of **ammonia**, **hydrogen**, **methane** and water) in conjunction with energy (probably in the form of sunlight, volcanic eruptions, and lightning) gave this primitive matter its metabolic ability to grow and increase. He speculated that the first organisms had appeared in ancient seas between 4.7 and 3.2 billion years ago. These living organisms would have evolved from a nonliving coagulate, or gel-like, **solution**. Oparin argued that a separation process called coacervation occurred within the **gel**, causing nonliving matter at the multimolecular level to be chemically transformed into living matter. He further theorized that this chemical transformation was dependent upon protoenzymatic catalysts and promoters contained in the coacervates. From there, a process of natural selection began, which resulted in the formation of increasingly complex organisms and, eventually, primitive systems of life. Although others, such as de Jong and T. H. Huxley, would postulate that life arose from a kind of "sea jelly," Oparin's theory that nonliving material was a catalyst for the formation of living organisms is considered by many to be his special contribution to the issue.

His suppositions on life's origins were not merely theoretical. In laboratory experiments, he showed how molecules might combine to produce the needed protein structure for transformation. Experiments of other scientists, such as **Stanley Lloyd Miller**, **Harold Urey**, and **Cyril Ponnamperuma**, confirmed his initial experiments on the chemical structure necessary to produce life. Ponnamperuna took the work a step further when he altered Oparin's original experiments and was able to easily produce nucleotides, dinucleotides, and adenosine triphosphate, which also contribute to the formation of life. Building on Ponnamperuna's research, Oparin was able to produce droplets of gel that he called protobionts. He believed these protobionts were living organisms because of their ability to metabolize and reproduce. Although later research of scientists in both the Soviet Union and the West would develop independently of Oparin's biochemical experiments, he must be given credit for putting the question of the origin of life into the realm of modern science. It has been said that his work in this area opened the door, and scientists in the West walked through.

Oparin was a man of his time, and his thinking was greatly influenced by Charles Darwin's theory of natural selection and the ideological climate of dialectical materialism which pervaded Soviet society during the 1930s. Although Oparin was never a Communist Party member, both his writings and his research methods reflect a bias toward dialectical materialism. However, it has been suggested that his denigration of the science of genetics and his support of Trofim Denisovich Lysenko and the Marxist-Leninist ideology which permeated and controlled Soviet genetics at that time may have resulted from political pressure and a desire to protect his career, as much as philosophical and scientific belief. Whatever the reasons, Oparin used his influence and prestige as chief administrator of the U.S.S.R. Academy of Sciences from 1948 through 1955 to implement policies that advanced Lysenko's views at the expense of the advancement of Soviet genetics. The influence of Lysenko and the Marxist-Leninist view of biology waned after Stalin's death in 1953. In 1956, as the result of a petition by 300 scientists calling for his resignation, Oparin was removed from his top position in the academy's biology division. He was replaced by Vladimir A. Engelhardt, a leading Soviet advocate of molecular biology. The 1950s saw an international explosion in the growth of molecular biology, but Oparin was severely critical of its principles. Although he considered the discoveries made by Watson and Crick concerning **DNA** to be important, he was skeptical of the idea of a genetic code, calling it "mechanistic reductionism." He did, however, support DNA research within his own Institute of Biochemistry during this time, and was a coauthor of papers discussing DNA and **RNA** in coacervate droplets.

Although Oparin's influence in Soviet science weakened in the early sixties, his international reputation, based on the origin of life theory, remained strong. This, coupled with his political reliability, led his government to send him abroad as a Soviet representative. Traveling by scientists in the Soviet Union was severely restricted in the 1950s, but Oparin was sent on official Soviet business not only to countries in the Eastern bloc and Asia, but to Europe and the United States as well. He also represented his country at international scientific and political conferences, such as the World Peace Council and the World Federation of Scientists.

His work brought him numerous honors. His awards from the Soviet Union include the A. N. Bakh Prize in 1950, the Elie Metchnikoff Gold Prize in 1960, the Lenin Prize in 1974, and the Lomonosov Gold Medal in 1979. The International Society for the Study of the Origin of Life elected him as its first president in 1970. He also was elected a member of scientific societies in Finland, Bulgaria, Czechoslovakia, East Germany, Cuba, Spain, and Italy.

Beginning in 1965 and continuing through 1980, the Soviet Union placed new emphasis on the science of genetics and molecular biology. However, Oparin's Institute of Biochemistry remained a stronghold of old-style biochemistry, and it eventually was bypassed by more progressive research institutions. Oparin died of heart disease in Moscow on April 21, 1980.

OPTICAL ACTIVITY

Certain molecules differ from one another only in the fact that one is the mirror image of the other. These molecules are known as **enantiomers** (from the Greek word for opposite). Although enantiomeric pairs have identical physical and chemical properties, the fact that the two molecules are mirror images of each other causes them to exhibit different optical behavior.

An ordinary light beam consists of a group of electromagnetic waves that vibrate in all directions. When a beam of such light passes through a polarizer, only those waves vibrating in a specific plane are transmitted. Almost all molecules are theoretically capable of producing a slight rotation of the plane of plane-polarized light, with the magnitude and direction of the rotation in part dependent on the molecule's orientation when it is struck by the beam. In an *optically inactive* **solution**, the beam is as likely to encounter a **molecule** in its mirror-image orientation as in its original orientation, with the result that all the rotations produced by individual molecules cancel out.

If a polarized beam is passed through a solution consisting of only one enantiomeric type, however, the beam does not encounter any of the molecule's mirror images and so emerges with its plane of **polarization** rotated. Separate enantiomers rotate the plane of polarized light by the same amount, but in opposite directions. Because these molecules show different behaviors under plane polarized light, they are said to be *optically active*.

See also Stereochemistry

ORGANIC CHEMISTRY

Organic **chemistry** is the study of the compounds of **carbon**. It is the single largest branch of chemistry, and the one with the most direct impact on the daily lives of most people in the world. The number of organic compounds is well over a million, and thousands of new compounds are created or discovered every year. In addition to the practical uses of organic compounds as drugs, fuels, and industrial chemicals, the study of organic compounds provides new information about chemical **bonding**, reactions, and other processes unavailable from the study of other types of compounds.

Why is carbon at the center of this vast field of study, rather than, say, sodium? The answer is that carbon's **electron** configuration allows it to form four bonds, and its size and **electronegativity** mean these bonds will be primarily covalent or polar covalent, rather than ionic. The ability to form multiple bonds means carbon can form **chains**, allowing the creation of large and complex carbon backbones, as it does in molecules such as **proteins**, **DNA**, **plastics**, and other polymers. In addition, carbon normally has no lone pairs of electrons in its compounds, preventing the kind of lone-pair repulsion that makes **nitrogen** or **oxygen** chains unstable.

The covalent character of carbon's bonds, even with such electronegative elements as oxygen, means that its bonds continue to be highly directional, unlike the ionic bonds of **sodium**, for instance, which are equally strong in all directions. This directionality gives the individual **molecule** its identity separate from other molecules, allowing it to participate as a unit in the highly complex series of reactions found in living organisms or in the modern chemistry laboratory.

The first organic compounds, and still the most complex, are those from living organisms. Most organisms share perhaps a thousand or more similar or identical compounds, such as **amino acids**, sugars, and nucleotides. Many organisms, however, make unique compounds, which function in signaling, metabolic pathways, or defense. Such compounds have proven to be a potent source of new drugs and enzymes. Examples include taxol from the Pacific yew, curare from a group of South American shrubs, and aspirin, from the bark of willow trees. While many useful compounds originally isolated from living organisms are now made synthetically, others are either too complex or too expensive to synthesize, and continue to be harvested from the natural source.

The other source of new organic compounds is from deliberate modification of existing ones in the laboratory. Changing the types of bonds or side chains can often have significant effects on the properties of the compound, and may give it important new characteristics.

While carbon's bonding characteristics form the theoretical basis of organic chemistry, the subject (and, indeed, the world) would be rather dull if carbon only bonded to itself. **Diamond**, **graphite**, and the **fullerenes** would be the only carbon compounds. The vast range of organic compounds arises from the addition of different atoms onto a carbon skeleton. Hydrocarbons are compounds in which **hydrogen** is the only other **atom**. The simplest is **methane**, CH_4. Larger hydrocarbons include all the compounds used as **gasoline** and heating fuels, as well as petrochemical-based **waxes**.

Substitution of a halogen such as **chlorine** for a hydrogen on a **hydrocarbon** creates a new class of compounds, the alkyl halides. The halogen is called the **functional group** of this class, meaning it is the group that gives the class its characteristic properties.

The study of functional groups provide the theoretical framework for understanding the reactions and behavior of the various classes of organic compounds. For instance, a **halide** is an electronegative atom, which withdraws electrons from the carbon it is bonded to. This leaves carbon with a partial positive charge, which will, in turn, serve to attract negative groups during reactions. The electron-withdrawing nature of the halides, then, strongly influences the chemical behavior of the parent compound. Compounds may have more than one functional group, which influence the behavior in different ways.

Other important functional groups include: OH, the hydroxyl group, which makes the parent compound an **alcohol**; O, which makes the compound an **ether**; C=O, the **carbonyl group**, which makes the compound an aldehyde if the group is on a terminal carbon, or a ketone if it is between two carbons; COOH, the carboxylic acid group, which makes the compound an acid; NH_2, the **amine group**, which makes the compound an amine; NH, the **amide group**, which makes the compound an amide; C_6H_5, the **phenyl group**; and SH, the thiol group.

In addition, the chemistry of a compound is strongly affected by whether it is a straight-chain molecule or a **ring** compound, and whether that ring is aliphatic or aromatic.

The study of organic compounds includes the determination of the identity and structure of existing compounds, the synthesis of new ones, and the determination of the step-by-step mechanisms and other parameters of reactions.

Analysis is concerned with determining the structure and identity of a compound. It requires the isolation and purification of a sample of the compound, followed by a series of physical and chemical tests. Simple tests include determination of melting and boiling point. **Molecular weight** was once commonly determined through colligative property analysis, but is now more likely to be done with a **mass** spectrometer. Determination of the atomic makeup of the compound can be approached through **combustion** analysis combined with a variety of qualitative tests for various functional groups.

Structure determination is most often performed with one or another type of **spectroscopy**. **Infrared spectroscopy** is a principal tool for functional group analysis, while Nuclear Magentic **Resonance** (NMR) spectroscopy gives detailed information on the position of hydrogens and other atoms. X-ray **crystallography** can solve the three-dimensional structure, especially important for larger compounds that could exist in any one of many different conformations. Proposed structures are often confirmed by creation of derivative compounds, whose successful formation from the parent compound depends on the predicted structure. Structure determination is a prerequisite for synthesizing the compound, as would be needed for manufacturing a compound isolated from a natural source.

Synthesis means creation of a compound from simpler or more readily available starting materials. For many syntheses, this means using such compounds as hydrocarbons and alcohols, plus inorganic acids, **metals**, **halogens**, or other compounds. The synthesis of **urea** from hydrogen cyanide and **ammonia** by **Friedrich Wöhler** in 1828 is often said to mark the beginning of organic chemistry as a separate discipline.

Working out the series of reactions for an organic synthesis is often like solving a logic puzzle or a maze, in that the starting point and ultimate goal are known, but the intermediate steps in between are unknown. Techniques for planning organic syntheses have been highly refined by Elias Corey, among others, who was awarded the 1990 Nobel Prize for his work.

Reactions in a multi-step synthesis must be chosen for their ability to create intermediates that can go on to react in later steps, and which will not engage in undesired reactions instead. ''Blocking groups'' may need to be added to functional groups to protect them from such undesired reactions, while preserving them for reaction later in the synthesis.

Intermediates must be formed in high yield, so that by the end of the synthesis, a meaningful amount of final product is available. One of the principal barriers to high yield is the creation of both **enantiomers** of a chiral compound, when only one is desired. Under most conditions, these mirror-image molecules will be formed in equal amounts, which must then be separated and half discarded before proceeding. The development of enantiomer-selective catalysts for ''asymmetric synthesis'' is one of the most intensely studied areas in organic chemical synthesis today.

To fully understand a reaction, the step-by-step mechanism by which products become reactants must be worked out. A mechanism may involve only a single step and a single molecular collision, but more often involves several steps, in which very short-lived intermediates are formed and consumed in the course of forming products. Reaction mechanisms can be used to determine the best conditions under which to run the reaction, or to devise reactants with slight structural changes that alter the **reactivity** in desired ways. In addition, the determination of a mechanism can be a purely theoretical challenge, whose results can shed light on the nature of **matter** and its interactions.

Reaction mechanisms are determined through a variety of means, most importantly spectroscopy, which is used to determine the structure of unstable intermediates. Computer modeling may also play a part, allowing the calculation of the quantum mechanical changes of proposed intermediates. Simple **kinetics** experiments play their part as well by varying the **concentration** of starting materials, it is possible to determine which is involved in the rate-determining step.

While organic chemistry is more than a century old, it remains one of the most prolific branches of chemistry, with cutting-edge research constantly pushing into new areas. The frontiers of organic chemistry include many different areas. Catalysis has been expanded greatly with the use of asymmetric catalysts for organic syntheses. Organometallic chemistry is the study of carbon-metal compounds, such as the **organolithiums**. These compounds are of both theoretical and practical interest, for they, too, may allow enantioselectivity in reactions. Polymers continue to be of great importance in chemistry, and new polymers are being synthesized for specific uses. Electrically conductive polymers are finding uses in the electronics industry, while thin films that change their transparency with **temperature** are being developed as coatings from windows.

The border between organic chemistry and **biochemistry** continues to be one of the most fruitful areas of research, especially as it concerns use of biomolecules as organic reagents or catalysts. Enzymes are especially important in this regard because of their strict enantioselectivity. Reactions that exploit these natural catalysts are increasingly being used in synthetic pathways. Finally, drug discovery and design is an important application of organic chemistry to medicine, and is one of the most profitable areas of research.

ORGANIC FUNCTIONAL GROUP

Organic functional groups are the section of the organic **molecule** which controls the reactivity of the molecule. For example the presence of multiple bonds between **carbon** atoms increases the molecule's reactivity. Functional groups undergo characteristic reactions, these occur regardless of the size of the molecule. The reactions of an organic compound are controlled largely by the **functional group**, or groups, it contains. For further information on each of these groups refer to their main entry in the book. Several functional groups may be contained in any one molecule, giving that molecule the properties of all of the functional groups present.

Molecules or atoms attached to functional groups act as modifiers of the functional group and they alter such characteristics as **boiling point** and melting point of the substance. The carbon-carbon single bonds and the carbon-hydrogen bonds are the unreactive parts of the molecule. Normally the designating letter R is given to the modifier, the **alkyl group**, including the carbon it is bonded to. So for example the **alcohol** has a general formula of ROH. Where two or more alkyl groups are present (e.g. an ether) the symbols R, R', R'', etc. are used. These may be the same or different molecules. An alkyl group is a straight chain molecule, but if the modifier is in the form of a **ring** structure it is more properly called an aryl group. Homologous series are produced for each functional group. These show slight changes in characteristics as increasing numbers of carbon atoms are added as modifier groups.

See also Alkane functional group; Alkene functional group; Alkyne functional group; Allene functional group; Allyl functional group; Ether functional group; Ester functional group; Ether fuctional group; Ketone functional group; Nitro functional group; Oxirane functional group; Thiol functional group

ORGANOBORANES

Organoboranes are organic compounds of **boron** having one or more alkyl or aryl functional groups attached directly to the boron **atom**. Alkylboranes can be conveniently prepared by the addition of diborane, B_2H_6, to a carbon-carbon double or triple bond:

$6\ RCH{=}CH_2 + B_2H_6 \rightarrow 2\ (RCH_2CH_2)_3B$
$4(CH_3)_2C{=}CHCH_3{+}B_2H_6{\rightarrow}2((CH_3)_2CHC(CH_3H)_2BH$
$2\ (CH_3)_2C{=}C(CH_3)_2 + B_2H_6 \rightarrow 2\ (CH_3)_2CHC(CH_3)_2BH_2$

Alkylboranes can also be prepared from the reaction of boron halides and metal alkyls, such as alkyllithium compounds or Grignard reagents (RMgX, where R denotes an alkyl chain). Boron halides undergo metathesis (double decomposition) reactions with alkyllithium reagents according to the hard-soft acid-base principle. Boron is a softer acid than **lithium**, and **carbon** in these molecules is a very soft base. Metathesis generates the lithium **halide** and organoborane.

Hydroboration of most alkenes proceeds directly to the trialkylborane; however, it is possible to prepare dialkylboranes, R_2BH, and monoalkylboranes, RBH_2, if there is sufficient steric hindrance about the carbon-carbon double and triple bond. The partially substituted borane derivatives, R_2BH and RBH_2, form dimers with three- centered two-electron bonds between the borons and two bridging **hydrogen** atoms. Most alkylboranes spontaneously ignite in air. Controlled **oxidation**, followed by dissolution in **water**, yields dihydroxy(alkyl)boranes ($RB(OH)_2$). The various aryl derivatives can be prepared from boron halides using the Grignard reaction.

The term *organoborane* is generally reserved for boron-containing compounds that have at least one carbon-boron bond. Boron is trivalent, and two of the chemical bonds can involve non-carbon atoms. Other important organoboranes include: alkoxydialkylboranes (R_2BOR); dialkoxyalkylboranes ($RB(OR)_2$); chlorodialkylboranes (R_2BCl); dichloro(alkyl)borane ($RBCl_2$); hydroxydialkylborane (R_2BOH); and dihydroxy(alkyl)borane. Trialkoxyborane ($B(OR)_3$) is at times listed as an organoborane even though it does not have the boron-carbon bond required by definition. Organoboranes undergo substitution reactions with complete retention of **stereochemistry**. This facilitates the manufacture of stereoisomers of pharmaceutically important organic compounds. Organoboranes are used as synthetic reagents in organic **chemistry** in the preparation of acylic and cylic saturated and unsaturated hydrocarbons, terpenes, **amino acids**, steriods, carbohydrates, and numerous other organic molecules.

ORGANOCUPRATES

Organocuprates are organic compounds comprised of metallic **copper** directly attached to a **carbon atom**. In these compounds, one or more **hydrogen** atoms in the parent organic **molecule** have been replaced by copper, usually with the establishment of a **valence** bond between the copper atom and the carbon atom. Organocuprate (organocopper) reagents are any one of several organic reagents containing copper and often other **metals**; they are important in conjugate-addition reactions and in the displacement of leaving groups.

Many organocuprates exist as metallobiomolecules, i.e., biomolecules that contain one or more metallic elements. Metallobiomolecules are complex coordination compounds whose metal-containing sites are frequently involved in **electron** transfer, the binding of exogenous molecules, and catalysis. Many metallobiomolecules, including the organocuprates tyrosinase, superoxide dimutase, and cytochrome oxidase, are present as enzymes in the human body; the **enzyme** cytochrome oxidase, for example, catalyzes a reaction in the electron transport system. Other metallobiomolecules, such as the organocuprate ceruloplasmin, are present as carrier **proteins**.

Some organocuprates have commercial importance. Copper pththalocyanine or $C_{32}H_{16}N_8Cu$ (also known as Pigment Blue 15), is used in paints, alkyd resin enamels, printing **inks**, lacquers, rubber, resins, papers, tinplate printing, and colored chalks. And the **chemical compound** Maculure is used to control flies in poultry and cattle operations.

See also Coordination chemistry; Free radicals

ORGANOHALOGENS

Organohalogens are organic compounds that contain one or more halogen atoms (fluorine, **chlorine**, **bromine** or iodine). The halogen atom(s) is covalently bonded to a **carbon atom**. Relatively few organohalogen compounds are found in the terrestrial organisms of nature, but organohalogen compounds are much more common within marine organisms. An organohalogen with an essential role in human biochemistry is thyroxin, the iodine-containing hormone that helps to regulate human metabolism. Simple organohalogens may be produced synthetically from the radical chlorination/bromination of saturated alkanes:

$CH_4 + Cl_2 \rightarrow CH_3Cl + HCl$
$CH_4 + Br_2 \rightarrow CH_3Br + HBr$

They may also be produced from the chlorination of alkylbenzenes:

$C_6H_5CH_3 + Cl_2 \rightarrow C_6H_5CH_2Cl + HCl$

The electrophilic substitution of **aromatic hydrocarbons** using ferric chloride catalyst also produces organohalogens:

$C_6H_6 + Cl_2 \rightarrow C_6H_5Cl + HCl$

And orgaohalogens can be produced from addition reactions involving chlorine/bromine gas (and/or **hydrogen** chloride/hydrogen bromide) and unsaturated alkene and alkyne hydrocarbons:

$CH_2{=}CH_2 + Cl_2 \rightarrow ClCH_2CH_2Cl$
$CH_2{=}CH_2 + Br_2 \rightarrow BrCH_2CH_2Br$

$CH_2=CH_2 + HCl \rightarrow CH_3CH_2Cl$

The corresponding fluoro- and iodo-derivatives are prepared by displacement reactions: $CH_3CH_2CH_2Br + KF \rightarrow CH_3CH_2CH_2F + KBr$, where the bromine (or chlorine) atom is replaced with a **fluorine** (or iodine) atom.

Saturated methanes with the general **molecular formula** of CF_xCl_{4-x} containing both fluorine and chlorine atoms are formally referred to as chlorofluorocarbons. These substances, also known as freons, were once used extensively as coolant fluids in refrigerators and air conditioners, and as propellants in aerosol cans. Freons are unreactive, and they remain in the atmosphere for long periods of time. Eventually freons reach altitudes where they are decomposed into chlorine atoms by strong ultraviolet radiation. The resulting chlorine atoms that are generated catalyze the depletion of the ozone layer in the stratosphere. Government regulations have required industries to replace freons with chemicals that do less harm to the ozone layer. The appliance, automotive and building industries have switched from freons to hydrofluorocarbons (i.e., CH_2FCH_3 and $CHCl_2CF_3$) for home refrigerators, car air conditioners and cooling systems in commercial buildings.

Organohalogens are synthesized by the chemical industry to be utilized as chemical intermediates, **solvents**, and specialty chemicals such as **refrigerants**, **pesticides** for crop protection, and **herbicides** for weed control. Several organochlorine and organobromine compounds are used as medicines in treating diseases and infections. Halothane, $CHClBrCF_3$, is a nonirritating and nonflammable anesthetic used in very low concentrations with **oxygen** gas or with a nitrous oxide-oxygen gas mixture. Highly substituted alkyl fluorides, perfluoroalkanes, are now used as artificial blood substitutes because of their oxygen transporting properties. The organochlorine polymer polyvinyl chloride (with a $-(CH_2CClH)_n-$ backbone) is used in the manufacture of records, packing materials, floor tiles, plumbing materials, and raincoats.

See also Ozone layer depletion

Organolithiums

Organolithiums are a type of organometallic compound in which the **carbon** of an organic group is directly bonded to a **lithium atom**. The term is also sometimes used for compounds in which a **nitrogen** or other non-metallic atom of an organic group is directly bonded to a lithium atom. Organolithiums are critical to the preparation of many important pharmaceuticals, agricultural compounds, and specialty organic chemicals.

Organolithiums were first identified in 1917 as products from the reaction of lithium metal and organomercury compounds. A simpler preparation discovered in 1930, reaction of an organic **halide** with lithium metal, made possible their current widespread use.

Organolithium-initiated polymerization of isoprene was discovered in 1957. When carried out in **hydrocarbon** solvent, this reaction produces a synthetic rubber similar to natural rubber. This discovery led to the first large scale application of organolithiums. They are now used in the preparation of polymers such as styrene-butadiene-styrene (SBS) rubber. SBS rubber is a common component of tire treads and rubber-soled shoes.

Organolithiums undergo similar chemical reactions to those of Grignard reagents, another important class of organometallic compounds. However, organolithiums are generally more reactive and hence more useful. Most organolithiums react vigorously with air and **water**. The more reactive burn spontaneously in air. To avoid such reactions, the compounds and their solutions must be handled under a protective atmosphere of nitrogen or **argon**.

Organolithiums have physical properties similar to organic compounds. Organolithiums made from nonconjugated hydrocarbons are colorless **liquids** or low-melting, white, crystalline **solids**. Compounds with conjugated hydrocarbon groups are often colored. Most organolithiums are soluble in hydrocarbon **solvents**, which are cheaper than the **ether** solvents required by Grignard reagents. Although commonly represented as monomers, organolithiums tend to form oligomers of two to six (or more) **monomer** units. The degree of association depends on the organic group, solvent, **concentration**, and **temperature**.

Organometallics

An organometallic compound is an organic compound in which a metal **atom** is attached directly to a **carbon**, with the exclusion of metallic salts of organic acids. Organometallic compounds have been prepared using practically all the **metals**. The first known organometallic compound was diethylzinc $Zn(C_2H_5)_2$. Other examples include Grignard compounds such as methyl **magnesium** iodide CH_3MgI; metallic alkyls such as butyllithium C_4H9Li, tetraethyl **lead** $Pb(C_2H_5)_4$, triethyl **aluminum** $(C_2H_5)_3Al$, tetrabutyl titanate $Ti(OC_4H_9)_4$, **sodium** methylate CH_3ONa, **copper** pththalocyanine $C_{32}H_{16}N_8Cu$; and **metallocenes**. Some organometallic compoun ds are highly toxic or flammable; others are coordination compounds (i.e., compounds formed between a metal **ion** and a nonmetallic ion or **molecule**, with the nonmetallic component referred to as a complexing agent). Many organometallic compounds are powerful catalysts.

The carbon-metal bonds in organometallic range in nature from essentially ionic to primarily covalent. The identity of the metal atom has much greater influence on the carbon-metal bond that the structure of the organic component. Carbon-sodium and carbon-potassium bonds are largely ionic; while carbon-lead, carbon-tin, carbon-thallium, and carbon-mercury bonds are mostly covalent. Carbon-lithium and carbon-magnesium bonds lie between the two extremes.

Organometallic bonds become more reactive as the percentage of ionic character of the metal-carbon bond increases. Thus, alkyl sodium and alkyl **potassium** compounds are highly reactive, and are very powerful bases. They react explosively with **water**, and burst into flame when exposed to air. Organomercury and organolead compounds are less reac tive, and remain stable in air. All organometallic compounds are poisonous. They are generally soluble in nonpolar **solvents**.

The metal-carbon bonds of organometallic compounds of **lithium** and magnesium have relatively large ionic characters, which make them strong bases and powerful nucleophiles.

Organomagnesium halides were discovered by the French chemist **Victor Grignard** in 1900; they are now known as Grignard reagents. They are usually prepared by reacting an organic **halide** with magnesium metal in an **ether** solvent. Grignard reagents are very strong bases, and so react with any compound that has a **hydrogen** atom attached to an electrone gative element such as **oxygen**, **nitrogen**, or **sulfur**. They react with water and alcohols in what are essentially acid-base reactions, as if they contained a carbanion; they will also remove protons that are much less acidic than those of water and alcohols. They carry out nucleophilic attack at saturated carbons when they react with ethylene **oxide** epoxy groups (oxiranes). A particularly important reaction undergone by Grignard reagents occurs with the unsaturated carbons of carbonyl groups. In the case of **aldehydes** and **ketones** containing carbonyl groups, the highly nucleophilic Grignard reagent contributes its **electron** pair to form a bond with the carbon atom. One electron pair of the carb onyl group shifts to the oxygen atom. This step results in the formation of an alkoxide ion that is subsequently protonated, leading to the formation of an **alcohol** and a compound MgX2 (where X is a halogen). The reaction of Grignard reagents with carbonyl compounds is used to prepare primary, secondary, and tertiary alcohols.

Organolithium reagents, which like Grignard reagents are very strong bases, react with compounds containing carbonyl groups in essentially the same way to produce alcohols. These reagents are often prepared by reducing an organic halide with lithium metal in an ether solvent. As an example, butyl bromide can be reacted with lithium metal in diethyl ether to create a **solution** of the metallic alkyl butyllithium.

ORIGIN OF LIFE

No one knows exactly how life originated on Earth. The planet is approximately 4.6 billion years old. Fossils of microorganisms similar to blue-green bacteria are present in rocks three billion years old. This suggests that life evolved within the first billion years after the Earth was formed, at a time when the planet's environments were very different from those of today. Experiments performed over the past 50 years suggest that many important ingredients of organisms, including amino acids and nucleic acids, may have formed under pre-life conditions existing on Earth; their presence must have facilitated the actual genesis of the first organisms.

Background of the origin of life

All organisms are made of chemicals rich in the same kinds of organic, or **carbon**-containing compounds. Moreover, the same 20 amino acids combine to make up the enormous diversity of proteins occurring in living things. In addition, all organisms have their genetic blueprint encoded in nucleic acids, either DNA or RNA. These **nucleic acids** are encoded with the information needed to synthesize specific **proteins** from **amino acids**. One class of proteins, known as **enzymes**, acts to regulate the activity of nucleic acids and other biochemical functions essential to life. Enzymes do this by greatly increasing the speed of specific chemical reactions (that is, they are catalysts). In addition to acting as enzymes, proteins provide structure for cells and assist in obtaining nutrients, functions which cells require for survival. These two types of molecules, nucleic acids and proteins, are so essential to life that many scientists assume that they, or closely related compounds, were present in the first life forms.

Theories of the origin of life

All cultures have developed stories to explain the origin of life. During Medieval times, for example, southern Europeans believed that small creatures such as insects, amphibians, and mice appeared by ''spontaneous generation'' in old clothes or piles of garbage. The Italian physician Francesco Redi challenged this belief in 1668, when he showed that maggots come from eggs laid by flies, and not spontaneously from the decaying **matter** in which they are found.

An series of experiments conducted in the 1860s by the French microbiologist **Louis Pasteur** also helped to disprove the idea that life originated by spontaneous generation. Pasteur sterilized two containers, both of which contained a broth rich in nutrients. He exposed both containers to the air, but one had a trap in the form of a loop in a connecting tube, which prevented dust and other particles from reaching the broth. Bacteria and mold quickly grew in the open container and made its broth cloudy and rank, but the container with the trap remained sterile. Pasteur interpreted this experiment to indicate that the microorganisms did not arise spontaneously in the open container, but were introduced by dust and other contaminants that could not reach the container with the trap.

Although Redi, Pasteur, and other scientists thoroughly disproved the theory of spontaneous generation as an explanation for the origin of life, they raised a new question: If organisms can arise only from other organisms, how then did the first organism arise?

Charles Darwin, the famous English naturalist, suggested that life might have first occurred in ''some warm little pond'' rich in minerals and chemicals, and exposed to electricity and light. Darwin believed that once the first living beings appeared, all other creatures that have ever lived could have evolved from them. Many of the laboratory experiments that have been conducted to shed light on the origin of life have been variations of the ''warm little pond'' that Darwin mused about.

Another early, influential explanation of the origin of life was provided by the Soviet scientist **Aleksandr Oparin**, and the British scientist J.B.S. Haldane. Oparin and Haldane suggested in the 1920s that the atmosphere of billions of years ago was very different from that which exists today. The modern atmosphere contains about 79% nitrogen and 20.9% oxygen, and only trace quantities of other gases. Because of the **oxygen** gas, which is a relatively reactive compound, this is an oxidizing atmosphere. Oparin noted that oxygen interferes with the formation of organic compounds necessary for life, by oxidizing their hydrogen atoms. Oparin therefore reasoned that the atmosphere present when life began was a reducing atmosphere, which contained little or no oxygen, but had high concentrations of **gases** that can react to provide **hydrogen** atoms to synthesize compounds needed to create life. Oparin and Haldane suggested that this primordial, chemically reducing atmosphere consisted of hydrogen, ammonia, methane, and additional simple hydrocarbons (molecules consisting only of carbon and hydrogen atoms).

According to their theory, energy for rearranging atoms and molecules into organic forms that promoted the genesis of life is thought to have came from sunlight, lightning, and geothermal heat. This model of the environment of genesis became popular among scientists after a graduate student named **Stanley Miller**, studying at the University of Chicago, designed an experiment to test it in 1953. Miller filled a closed **glass** container with a mixture of the gases that Oparin and Haldane suggested were in the ancient, pre-life atmosphere. In the bottom of the container there was a reservoir of **water**, and above it an apparatus caused electrical arcs to crackle. After one week of reaction, Miller found that dilute concentrations of amino acids and other organic chemicals had formed from the contained gases and water. In the years since Miller reported his results, other researchers have performed more sophisticated ''warm little pond'' experiments, and have been to synthesize additional amino acids and even the building blocks of nucleic acids, the molecules that organize into **RNA** and **DNA**, which encode the genetic information of organisms.

Subsequent research influenced by these experiments led many scientists to believe that the **concentration** of organic molecules in the primordial, nutrient-laden, warm ''ponds'' (which may have been tidal pools, puddles, or shallow lakes) increased progressively over time. Eventually, more complex molecules formed, such as **carbohydrates**, **lipids**, proteins, and nucleic acids. **Energy** driving these reactions was probably supplied by ultraviolet radiation or **electricity**. The assembly of more complex compounds from simpler ones may have occurred on the surface of oily drops floating on the water surface, or on mineral surfaces.

Some scientists believe that the young Earth was too inhospitable a place for life to have developed on its surface. They believe that a more likely environment for genesis was the vicinity of deep-sea vents, or holes in the crust under the ocean from which hot, mineral-laden water flows.

Many scientists today believe that the pre-life atmosphere may not have been as strongly reducing as the one proposed by Oparin and Haldane and used in Miller's experiment. They believe that volcanoes added **carbon monoxide**, **carbon dioxide**, and nitrogen to the early atmosphere, which may even have contained traces of oxygen. Nevertheless, more recent experiments of the Miller-type, run using a less reducing atmosphere, have also resulted in the synthesis of organic compounds. In fact, all 20 of the amino acids found in organisms have been created in the laboratory under experimental conditions designed to mimic what scientists believe the pre-life Earth was like billions of years ago.

Cellular **metabolism** links amino acids together using specific enzymes to form particular proteins. When this happens, a hydrogen molecule and a hydroxyl group (OH) are removed from the amino acids, which then link up into a protein chain, while the hydrogen and hydroxyl link up as a water **molecule**. Without enzymes, amino acids do not link up in this way, or as a biochemist might describe it, the polymerization reaction does not proceed. If this is true, then how, before life began and cells existed, could amino acids join to form proteins? One possibility is that amino acids may have joined together on hot sand, clay, or even rock. Laboratory experiments have shown that amino acids and other organic building blocks of larger molecules, called polymers, will join together if dilute solutions of them are dripped onto warm sand, clay, or rock. The larger molecules formed in this way have been named proteinoids. It is easy to imagine Darwin's ''warm little pond,'' complete with amino acids splashing onto hot volcanic rocks. Clay and **iron** pyrite have particularly favorable properties making them good ''platforms'' for the formation of larger molecules from smaller building blocks.

Proteinoids can cluster together into droplets that separate, and that may protect their components from degrading influences of the surrounding environment. In this way, the droplets are like extremely simple cells, although they can not reproduce. Such droplets are called *microspheres*. When fats (also known as lipids) are present, the microspheres that form are even more cell-like. If a mixture of linked amino acids called polypeptides, sugars called polysaccharides, and nucleic acids is shaken, droplets called coacervates will form. All of these kinds of prebiotic droplets are called protobionts, and they may represent a stage in the genesis of life.

The formation of amino acids and other organic compounds is presumed to have been a necessary step in the genesis of life. However, another step in the process must have happened soon after: self-replicating molecules would have to be capable of forming if life was to exist. Scientists presume that the first self-replicating molecules were similar to the nucleic acids of organisms.

Once molecules that could self-replicate were formed, the process of evolution would account for the subsequent development of life. As such, the particular molecules best adapted to the local environmental conditions would have duplicated themselves more efficiently and more often than competing molecules. Eventually, primitive cells appeared, and perhaps coacervates or other protobionts played a role at this stage in the genesis of life. Once cells became established, evolution by natural selection could have resulted in the development of all of the life forms that have ever existed on Earth.

Most living cells today store genetic information in DNA. The information of DNA is transferred to RNA by a process known as transcription, and the RNA then forms proteins, including enzymes, by translating the information dictated by the DNA. The enzymes facilitate the biochemical cellular functions necessary to maintain life and reproduce it. Many scientists believe it is unlikely that all of the components of this complex sequence of events, DNA to RNA to protein, evolved at the same time. Some scientists propose that, in fact, RNA appeared before DNA. This view has been strengthened by the discovery that some forms of RNA, called *ribozymes*, can act like non-protein enzymes to catalyze biological reactions. These scientists suggest that RNA was capable of ordering the sequence of amino acids, forming proteins, and replicating itself in a type of "RNA world," in which RNA was more important than DNA.

Scientists who favor the hypothesis of an "RNA world" suggest that RNA might have been able to self-replicate even before DNA and protein enzymes had evolved. As such, single-stranded RNA might have been able to assume a shape that allowed it to line up amino acids in specific sequences to create specific protein molecules. RNA molecules capable of causing amino acids to link up to form a protein could have had an advantage in replication and survival, compared with other RNA molecules. At that point, molecular evolution and natural selection could have taken over in furthering the development of life. The RNA that produced useful protein enzymes, for example, would have survived better than those that did not.

Critics of these ideas say that the evidence for self-replicating RNA is weak. Instead, they suggest that other organic molecules, rather than nucleic acids, were the first self-replicating chemicals capable of storing genetic information. According to this idea, these simple hereditary systems were later replaced by nucleic acids during the course of evolution.

Radio astronomers have found that organic molecules, which might have played an important role in the formation of life, are present in dust clouds in outer space. Organic molecules are also known to be present in meteors that have fallen to Earth's surface. These observations provide further evidence that chemicals important for the genesis of life may have been present on the pre-life Earth. The presence of complex organic compounds outside of our solar system suggests that the formation of compounds important for life is more likely than once thought.

The presence of organic compounds in outer space also suggests to a few scientists that life may not have actually originated on Earth. Instead, they suggest that genesis may have occurred somewhere in outer space, and organisms later arrived on Earth. Most researchers discount this hypothesis, because they feel that ionizing radiation and the great extremes of temperature in space would have killed any organisms before they could have reached the Earth. The suggestion of an extraterrestrial origin of life also suffers from the drawback that it merely shifts the compelling questions about genesis from Earth to another place in the universe.

A theory known as *panspermia* suggests that organic precursors to life arrived to Earth with meteors. Once here, the organic compounds arranged themselves into molecules that eventually led to the development of life. This theory simplifies the problem of explaining the origin of life, by suggesting that the formation of simple organic compounds did not have to take place on Earth.

Nevertheless, the genesis of organisms from simple organic compounds, however diverse and abundant they may have been, is not yet satisfactorily explained by this or any other theory of the origin of life.

OSBORN, MARY J. (1927-)

American biochemist

Mary J. Osborn is the first person to demonstrate the mode of action of methotrexate, a major cancer chemotherapeutic agent and folic acid antagonist (in other words, it opposes the physiological effects of folic acid). Best known for her research into the biosynthesis of a complex polysaccharide known as lipopolysaccharide—a **molecule** that is essential to bacterial cells—Osborn helped to identify a potential target for the development of new antibiotics and chemotherapeutic agents.

Mary Jane Osborn was born in Colorado Springs, Colorado, on September 24, 1927, and raised in west Los Angeles and Beverly Hills, California. Her father, Arthur Merten, had an eighth-grade education and was a machinist; her mother, Vivian, went to secretarial classes and also taught school. "Both parents were high achievers and their ambitions for me were considerable," Osborn told Laura Newman in an interview. Osborn noted that her background was somewhat atypical for girls growing up in the 1930s. She recalled reading a book for young girls about being a nurse when she was ten years of age. "I got very interested in being a nurse, but when I told my parents, they asked me, 'Why don't you want to be a doctor?'" Osborn credited her parents for their early support of her interest in science; from her mother and father she gained "a very naive and blind assumption that I could do whatever I wanted to do." In describing her academic progress as a girl Osborn noted, "The thing that amazes me about my primary and secondary education is that I remained interested in biology. What I remember of the teaching was pretty awful."

Osborn entered the University of California at Berkeley as a pre-med student. "By senior year I realized that there was no way in the world that I wanted to treat patients." She then pursued **biochemistry** courses. Osborn recalled, "I realized that I liked bench research and could do it well. I was good at planning experiments and thinking about the results and going on to the next step." She was awarded a B.A. in physiology from the University of California at Berkeley in 1948, then went on to the University of Washington, attaining a Ph.D. in biochemistry in 1958. Osborn's thesis examined the functions of the **vitamins** and enzymes whose action depended on folic acid. In 1957, Osborn reported the mode of action of methotrexate, which became a major cancer chemotherapeutic agent, especially for leukemia.

In 1959, Osborn moved into a new area, the study of the structure and building blocks—or biosynthesis—of a molecule complex polysaccharide named lipopolysaccharide. Lipopolysaccharide is unique to a certain class of bacteria that includes pathogens such as salmonella, shigella, and the cholera bacillus. Abundant on the surface of these bacteria, lipopolysaccharide is responsible for major immunological reactions and for the bacteria's characteristic toxicity. Osborn's work led to a new understanding of a previously unknown mechanism of polysaccharide formation.

For her contributions to biochemistry, Osborn was accepted as a fellow of the American Academy of Arts and Sciences in 1977 and was elected to the National Academy of Sciences in 1978. Other major distinctions include having served as president of the American Society of Biological Chemists from 1981 to 1982 and as president of the Federation of American Societies for Experimental Biology from 1982 to 1983. She has been appointed to numerous scientific advisory councils, including the National Institute for General Medical Sciences, National Institutes of Health Division of Research Grants, and the National Science Board. In addition, Osborn has served as editor of several journals, including *Biochemistry, Journal of Biological Chemistry*, and the *Annual Review of Biochemistry*.

Osborn became professor of microbiology at the University of Connecticut Health Center School of Medicine in 1968 and she has been head of the department since 1980. Her interest in the development of antibiotics and chemotherapeutic agents continues on into the 1990s. She is married to a painter, Ralph, and they have no children. In her leisure time, Osborn gardens.

Osborne, Thomas Burr (1859-1929)

American biochemist

Osborne, the son of a banker, was born in New Haven, Connecticut, and did both his undergraduate and graduate work at Yale University. After getting his Ph.D. in 1885, he joined the recently-established Connecticut Agricultural Experiment Station as an analytical chemist. Four years later, at the suggestion of the station's director, Osborne began an investigation into the **proteins** of plant seeds—an investigation which eventually became his lifelong work.

The young chemist began by studying oat kernels, in time managing to isolate an alcohol-soluble protein and a globulin from them. Intrigued, he turned to other seeds and, over the next three years, isolated the proteins of at least 32 different plant species, including nuts, legumes, and cereal grains. Subjecting them to an intensive chemical analysis, he found, to his surprise, that the proteins of different species were distinctly different from each other. The differences were especially marked, he noted, in the amino acid content of the various proteins. Although findings like these contradicted the well-known (and widely accepted) doctrine of Justus von Liebig—that only four kinds of protein existed in nature, albumin, casein, fibrin and **gelatin**, and that they were all pretty much alike—Osborne became increasingly convinced he was on the right track.

In 1909, Osborne invited another biochemist, Lafayette B. Mendel (1872-1935), then working at his alma mater, Yale, to join him in his ongoing investigations, now directed toward probing into nutritional properties of plant proteins. Mendel accepted and the two biochemists proceeded to work together for almost twenty more years. They co-wrote roughly a hundred papers and made a number of important discoveries. For example, they found that two **amino acids** in particular, lysine and tryptophan, were essential for the normal growth of animals. Furthermore, laboratory rats were unable to manufacture these substances within their bodies and thus had to rely on dietary lysine and tryptophan to survive.

The outstanding accomplishment of the Osborne and Mendel collaborations was made in 1913 with the discovery of the substance that later proved to be vitamin A. In that year, the two researchers determined that butter contains a fat-soluble essential nutrient (a nutrient they also found in cod-liver oil). Unfortunately, they published their results three weeks after Elmer McCollum had announced his discovery of the same substance. McCollum, therefore, received most of the credit for the discovery.

Oscillating reactions

The study of oscillating reactions is the study of periodic changes in chemical reactions. In an oscillating chemical reaction, the concentrations of the reactants and products change with time in a periodic or quasi-periodic manner. Chemical oscillators exhibit chaotic behavior, in which concentrations of products and the course of a reaction depend on the initial conditions of the reaction.

Scientists have a long-standing fascination with the complexities of oscillating systems. In the seventeenth century, the English-Irish chemist **Robert Boyle** reported the periodic "flaring up" of **phosphorus** in contact with the air.

The classic modern example of an oscillating reaction is the Belousov-Zhabotinsky reaction that yields a red **solution** that turns blue at varying intervals of time. In a stirred vessel, the Belousov-Zhabotinsky reaction mixture will change **color** from red to blue dozens or hundreds of times before equilibrium is established. If the mixture is poured into a shallow vessel, the oscillation will be triggered at randomly spaced points and give rise to outgoing waves of alternating red and blue color.

Another fundamental example is provided by the Bray reaction, the first identified homogeneous isothermal chemical oscillator, which is a complex reaction of iodate, **iodine** and **hydrogen** peroxide. As **hydrogen peroxide** decomposes to **oxygen** and **water**, the resulting rate of the evolution of oxygen and I_2 vary periodically.

A distinguishing feature of oscillating reactions is the phenomena of autocatalysis. In autocatalytic reactions the an increasing rate of reaction increases with the **concentration** of the reactants. Autocatalytic reactions eventually achieve a steady state (where the net production of products is zero) that can be determined by setting all the time derivatives equal to zero and solving the resulting algebraic equations for the concentrations of reactants and products. In oscillating reactions, small changes may result in a dramatic departure from the steady state.

Many skeptics of oscillating reactions once dismissed apparently oscillating reactions as aberrations due to contamination. Concerns that oscillating reactions could not exist because of apparent violations of thermodynamic laws have recently been refuted by careful studies that establish that oscillating reactions proceed in accord with thermodynamic laws.

Oscillating chemical reactions are unlike the oscillations of a pendulum. Oscillating chemical reactions do not have to pass through an equilibrium point during each oscillating cycle. Although this seems counter-intuitive (outside of experience with the natural world) it is in perfect accord with quantum theory.

Because a closed system must eventually reach equilibrium, closed systems can sustain oscillating chemical reactions for only a limited time. Sustained oscillating reactions require an open system with a constant influx of reactants, **energy** and removal of products.

Oscillating reactions, a common feature of biological systems, are best understood within the context of nonlinear chemical dynamics and **chaos theory** based models that are used to predict the overall behavior of complex systems. Chaotic and oscillating systems are unpredictable, but not random.

Deterministic equations describe chotic and oscillating systems, and a key feature is that such systems are so sensitive to their initial conditions that future behavior is inherently unpredictable beyond some relatively short period of time. The goal of scientists studying oscillating reactions is to determine mathematical patterns or repeatable features that establish relationships to observable phenomena related to the oscillating reaction.

Oscillating systems can interact with interesting results. Much like waves can cancel one another out, two oscillating systems can interact to produce a steady state of oscillation. On the other hand, the joining of two systems already at steady state can cause rhythmogenesis in which the two systems will depart from the steady state.

The exact mechanisms (a series intermediate reactions or steps) of oscillating reactions are elusive and difficult to obtain for all but the simplest reactions. The chemical equations and mechanisms commonplace to stoichiometric **chemistry** describe only the overall reactions, they do not specify the molecular transformations that place between colliding molecules.

Oscillating reactions are thought to play key roles in biological morphogenesis and geologic stratigraphy.

OSMIUM

Osmium is in Row 6 of the **periodic table**, one of the elements known as the transition **metals**. It has an **atomic number** of 76, an atomic **mass** of 190.2, and a chemical symbol of Os.

Properties

Osmium is a bluish white, shiny metal with a melting point of about 5,400°F (about 3,000°C), and a **density** of 22.5 grams per cubic centimeter. These values are all among the highest for the elements. Osmium is not very ductile, malleable, or, in general, workable, thereby reducing the number of practical applications it has. It is a relatively inactive element, combining with **oxygen** only at high temperatures and reacting only with acids after long exposure.

Occurrence and Extraction

Osmium is very rare with an abundance no greater than about one part per billion in the Earth's crust. Its most common ore is osmiridium. The element occurs in all ores of **platinum**. It is obtained commercially as a byproduct of the extraction of platinum from its ores.

Discovery and Naming

Osmium was discovered in 1804 by the English chemist **Smithson Tennant**. Tennant was analyzing the element platinum and found that he continually obtained an unusual black powder with properties different from that of platinum itself. Tennant was eventually able to determine that the black powder was a mixture of two new elements, osmium and **iridium**. He suggested the name osmium for one of the elements from the Greek word *osme*, meaning odor. The name came from one of the osmium compounds with which Tennant was working, osmium tetroxide (OsO_4).

Uses

Osmium has relatively few uses, its primary application being in alloys with platinum or iridium. The addition of small amounts of osmium to these metals greatly increases their hardness. Osmium alloys are sometimes used in pen tips and in specialized laboratory equipment.

Health Issues

Osmium tetroxide is a particularly interesting compound. It has considerable demand for use as a catalyst in research, but it is highly reactive and very toxic. It is shipped in small **glass** containers that carry no labels and are not marked in ink. The compound would react violently with either label or ink, so instructions provided separately explain how to use the material.

OSMOSIS

Osmosis is the passage of **water** from a weak **solution** to a strong solution through a semi permeable membrane.

A semi-permeable membrane is otherwise known as a selectively permeable membrane. It is a type of barrier that will allow the passage of some molecules but not others. In osmosis, water passes from one side of this membrane to the other. The water will move only in one direction—it will move from the side where the water is in highest **concentration** to the other side of the membrane. The liquid that the water moves into is a strong solution, so the water is acting to dilute the solution. If one side of the membrane has a solution of sugar and the other side has water then the water will flow into the sugar solution, diluting it. The pore size of the membrane is such that water molecules can readily move through but the larger solute molecules cannot. The water diffuses from one side to another until the concentration is the same on both sides of the membrane. Once the concentration of the solute and the water is the same on both sides the process of osmosis stops. Osmosis will occur only across a **diffusion** gradient.

Osmosis is a passive process as opposed to an active one. Osmosis requires no **energy** to drive it. Osmosis generally requires a living cell membrane but it will also happen with suitable non-living membranes such as Visking **dialysis** tubing.

All solutions have an osmotic potential. The osmotic potential is the amount of net movement that can occur when the solution is compared to pure water. The **osmotic pressure** of a solution is the pressure that has to be exerted to halt osmosis.

Osmosis occurs in natural systems in cells and in organs. In humans, osmosis occurs in the kidneys to recover the water form waste materials of the body. In plants, osmosis occurs for example at root hairs, allowing the uptake of water from the soil. Individual cells can prevent water rushing into them by osmoregulation. If water were to rush into cells unchecked then the cells would most likely burst. Plant cells are surrounded by a rigid **cellulose** cell wall. If there is a strong solution external to the cell, water will flow out from the cell into the strong solution. This will result in the cytoplasm and the cell membrane pulling away from the cell wall, a process known as plasmolysis. At this point, the cell is incapable of supporting weight; it is nearly empty of water and it is said to be flaccid. If this situation occurs for too long the cell will dry out and die, but this does not frequently occur in nature. When there is a weaker solution outside the cell water will flow into the cell, allowing the cell to become more supportive. There will be a greater pressure inside the cell and it will become turgid. If water continues to flow into the cell, the cell becomes increasingly turgid. Eventually, as the cell presses against the rigid cell wall, osmosis slows down and eventually stops. This is due to the pressure exerted by the wall of the plant cell. This is called wall pressure and it halts osmosis. A fully turgid cell gives the maximum support for the plant. Generally, in a healthy plant the cells alternate between being flaccid and fully turgid.

Similar reactions can be seen in animal cells although they lack the rigid cell wall. In some simple organisms, such as amoeba, water is actively excreted from the cell in vacuoles. Red blood cells will burst if placed in fresh water and if they are placed in too strong a solution they will shrink. The kidneys regulate the concentration of water in the blood **plasma**.

If a cell is surrounded by water, the water will flow into the cell. The osmotic pressure of the external solution is lower than that of the internal solution. In this situation, the external solution is said to be hypotonic to the cell. If water flows from the cell into the stronger solution outside, the external solution is said to be hypertonic. If both solutions are of the same concentration and there is no net flow of water then the solutions are said to be isotonic to each other. The flow of water into a cell is endomosis, the flow out is exomosis.

See also Solution; Solvents.

Osmotic pressure

Osmotic pressure is a measure of the extra pressure that has to be exerted to counteract **osmosis**. The osmotic pressure of a **solution** is the force that has to be exerted to halt osmosis. If a U-shaped **glass** tube were divided in two by a semipermeable membrane and filled with two solutions the osmotic pressure could be shown. If a solution were on the left-hand side of the membrane and pure solvent were on the right-hand side

Friedrich Wilhelm Ostwald.

then movement would occur from the right to the left. This movement could be halted by applying pressure to the left hand arm (the one containing the solution). The pressure exerted is the osmotic pressure. The osmotic pressure obeys a law similar in form to that of the ideal gas law, PiV = nRT, where Pi is the osmotic pressure, V is the **volume** of the solution, n is the number of moles of solute, R is the ideal gas constant, and T is the absolute **temperature**. Since the number of moles of solute divided by the volume of the solution is the **molarity** (M) of the solution this equation can be rewritten as Pi = MRT. If two solutions are separated by a semipermeable membrane and they have the same osmotic pressure they are said to be isotonic. If one solution has a lower osmotic pressure it is said to hypotonic to the more concentrated solution, which is termed the hypertonic solution.

Ostwald, Friedrich Wilhelm (1853-1932)

German physical chemist

Around the turn of the twentieth century, Friedrich Wilhelm Ostwald was responsible for organizing physical **chemistry** into a discipline distinct from **organic chemistry**. He wrote a basic textbook on the subject and co-founded a journal that provided physical chemists with a forum for their theories and experimental results. For his work in the measurement of chemical reactions, **electrochemistry**, and the acceleration of chemical reactions by the use of catalysts, Ostwald won the 1909 Nobel Prize for chemistry. He was also a prolific writer in both the philosophy and psychology of science, and he culminated a long academic career with valuable independent research into **color** theory.

Ostwald was born on September 2, 1853, in Riga, Latvia (now Estonia) into a family of master artisans. His parents, Gottfried and Elisabeth (Leuckel) Ostwald, were descendants of German immigrants; his father was a master cooper who had been a painter as a young man. The humanities were emphasized in the Ostwald home, and young Friedrich Wilhelm learned to paint and play the viola and piano; he was also an avid reader. These passions stayed with him throughout his life, but at an early age he also became enthralled with chemical experimentation, creating his own fireworks when he was eleven. He studied at the Riga Realgymnasium and in 1872 enrolled at the University of Dorpat (now the State University of Tartu in Estonia), where he studied both chemistry and physics. At this time, chemists were almost exclusively concerned with research on organic molecules, but Ostwald's natural inclinations led him to a study of **physical chemistry**. He received his bachelor's degree in 1875 from Dorpat and stayed on to lecture and complete his master's degree in 1876 and his Ph.D. in 1878.

Ostwald's early interests in the measurement of chemical reactions were spurred by the work of Julius Thomsen, who had measured the **heat** accompanying chemical reactions. Ostwald had realized that other properties could serve equally well for such measurements; his master's thesis had concerned the **density** of substances by **volume** in a watery **solution**, and his doctoral thesis dealt with optical refraction. The result of the laboriously repetitive experiments he performed were affinity tables for twelve acids. During this period, he became increasingly interested in the subject of chemical affinities, or the combinational reactions between various chemicals. In 1881 Ostwald was appointed professor of chemistry at the Riga Polytechnic University, where he expanded his research into chemical affinities by measuring the rate at which chemical changes take place. He confirmed his earlier measurements of volume and density with measurements of the **velocity** at which acids will split esters into **alcohol** and organic acid, and he was able to assign precise numerical values to chemical reactions and affinities.

Ostwald's name was becoming known for such discoveries, and he was soon joined in his work establishing physical chemistry by two younger scientists: **Svante Arrhenius** from Sweden and Jacobus Van't Hoff from Holland. In 1884, Arrhenius sent Ostwald his doctoral dissertation; hotly contested by many of the scientists at his university, it concerned affinity and **electrical conductivity**. Ostwald immediately saw the importance of Arrhenius's work, and he recognized that it included the beginning of the idea of electrolytic **dissociation** and thus of **ionization**, or the conversion of a neutral **atom** into a positive **ion** and a free **electron**. He did all he could to sponsor Arrhenius. In 1886, Ostwald became interested in van't Hoff's work on the similarities between solutions and **gases**; he was no longer working alone in physical chemistry.

In 1885 Ostwald had begun work on the *Lehrbuch der allgemeinen Chemie,* a textbook of general chemistry which he finished in 1887, the year he received an appointment at the University of Leipzig as the first professor of physical chemistry in Germany. There van't Hoff joined him as an assistant and the two soon created a center for the study of physical chemistry and founded the influential journal for the new discipline, *Zeitschrift für physikalische Chemie.* In 1889, Ostwald published a book on **analytical chemistry**, *Grundriss der allgemeinen Chemie,* which further distinguished physical chemistry from organic chemistry. By this time, Ostwald had begun to understand the world in terms of **energy**; he believed that everything could be reduced to that single concept, and this was a theme that would dominate the rest of his life and work. By 1898 Leipzig University had created a physical chemistry institute, a training and research center where much of Ostwald's later work was done.

Ostwald's research into the dynamics of chemical reactions in solutions led to the dilution law of 1888, which established a relationship between elecrolytic dissociation and conductivity. Arrhenius's theory of electrolytic dissociation states that atoms will come apart (dissociate) in **water**, creating charged ions that have gained or lost electrons. Excited by this theory, Ostwald and his researchers turned their attention to electrochemistry, seeing it as a model for chemical reactions that are accelerated or catalyzed by weak bases or acids. This led to his most important research, which was in the area of catalysis, whereby a substance is used to speed up a chemical reaction but remains unaffected by that reaction. Though the process of catalysis had been described some sixty years earlier, Ostwald made such processes measurable and also connected them with his own work on chemical affinity. Although his theory that catalysis operates simply by having catalysts present and that catalysts do not take part in the reaction is now known to be incorrect, his work on catalysis was otherwise productive. In 1901, Ostwald's work led to the process for converting **ammonia** to **nitric acid**, which was accomplished by burning ammonia in the presence of **platinum**. This process was patented in 1902 and allowed **mass** manufacturing of the basic component of **explosives**. Renewed interest in catalysis also led to great strides in the chemical industry: oil, for example, is transformed into fuel and **natural gas** by the catalytic process. In 1909, Ostwald was awarded the Nobel Prize in chemistry for this work.

In 1905, Ostwald spent a year as an exchange professor at Harvard University in the United States, spreading the word of physical chemistry across the Atlantic. The following year he retired from his chair of physical chemistry at Leipzig, tired of administrative duties and political infighting. His working life was far from over, however. He bought an estate in a Leipzig suburb, which he dubbed 'Energie,' and there he continued a number of experimental and writing projects. Along with editing professional journals, he also worked on a history and classification of people who were considered geniuses, reprints of significant papers in chemistry and physics, and a three-volume autobiography. Some of his many interests included the philosophy and history of science, pacifism, internationalism, and the creation of a world language—he was very interested in Esperanto while he was a visiting professor at Harvard and later he wrote his own language, Ido. Ostwald also made important contributions to color theory—standardization of colors and a theory of color harmony. This was an outgrowth of his own interest in painting.

In 1880 Ostwald had married the daughter of a medical doctor in Riga, Helene von Reyher. They had three sons and two daughters. One of his sons, Wilhelm Wolfgang, would grow up to be a well-known chemist himself. Until his death from uremia in early April of 1932, Ostwald continued to work tirelessly for the causes he espoused.

OXALIC ACID

Oxalic acid is the more common name of ethanedioic acid. The name ethanedioic acid communicates that the **molecule** has two **carbon** atoms (as in ethane) and two acid groups (COOH). It has the chemical formula $C_2H_2O_4$.

It is a white solid used in removal of certain kinds of stains, in removing **calcium** ions from solutions, and in tanning leather. It occurs naturally and is toxic. The **potassium** and calcium salts of oxalic acid are found naturally in cabbage, spinach, and rhubarb leaves, and are also found in the bark of some species of eucalyptus trees. The **metabolism** of sugar by many species of mold results in the production of oxalic acid. Ingestion of large amounts can cause kidney damage, convulsions, and death.

The most common uses of oxalic acid are in tanning leather and removing rust and ink stains. In stain removal, it acts as a reducing agent (a substance that donates electrons to other substances) and is relied on by most dry cleaners for this purpose. **Iron** rust stains contain iron in its oxidized form (Fe III); the oxalic acid reduces it to its colorless reduced form (Fe II). Oxalic acid is also used to clean **metals** in many industries and is also used in the purification of **glycerol** (glycerin).

Few people ingest toxic amounts of oxalic acid directly. However, if a child or pet swallows antifreeze (which typically contains ethylene glycol and has a sweet taste), enzymes in the body will metabolize the ethylene glycol to oxalic acid, which is the reason antifreeze is toxic.

In many industrial processes oxalic acid is used to remove calcium ions from solutions. The reaction of calcium ions with oxalic acid produces an insoluble solid, calcium oxalate.

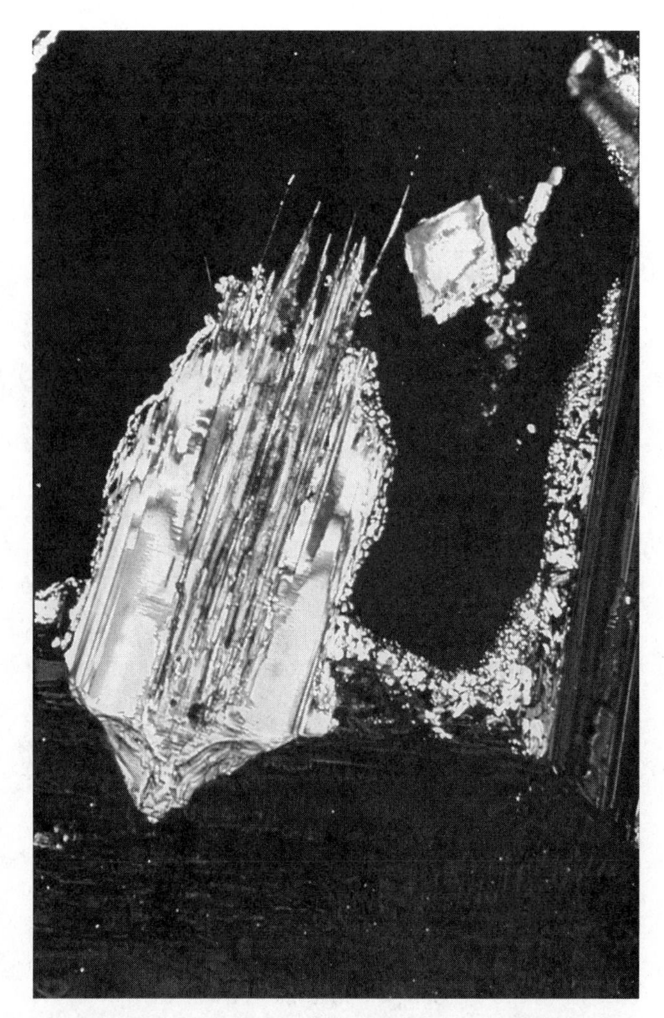

A polarized light micrograph of oxalic acid crystals. ***(Photograph by John Hendry/Science Photo Library, National Audubon Society Collection/Photo Researchers, Inc. Reproduced by permission.)***

OXIDATION

Oxidation and **reduction** reactions are essential to life. Oxidation of nutrients provides plants and animals with the **energy** they need to survive. When an organism dies, oxidation reactions are responsible for the decay of the organic **matter**. Oxidation reactions are also used by chemists to synthesize pharmaceuticals, textiles, dyes, paints and a multitude of other important products. **Oxygen** is third on the list of bulk chemicals produced per year in the United States.

One of the most challenging questions about the **origin of life** on Earth is how early organisms protected themselves from oxygen in the atmosphere. All organic matter exposed to oxygen tends to be converted into **carbon** dioxide and **water**, and most minerals are oxidized forms of elements. Avoiding oxidation of materials is of practical concern as well: **metals** are painted or lubricated to inhibit oxidative **corrosion** and **wood** products are coated to prevent the oxidative degradation known as decay.

The original definition of oxidation centered on the class of reaction in which one or more oxygen atoms from oxygen itself to another species was added to an element or compound. It became apparent, however, that other substances besides oxygen could add oxygen atoms to molecules. Species that cause oxidation are called oxidants or oxidizing agents. It is important to note that oxidation and reduction always occur in tandem: if a species in a reaction is oxidized, then some thing else must be reduced.

Most metals are readily oxidized when exposed to the atmosphere. Many metal oxides are powdery, non-malleable, non-ductile substances. Only as the free metal do they have significant tensile strength and can they be shaped and formed into useful tools or other objects. Engineers and scientists continually strive to develop better methods to prevent corrosion of metals by atmospheric oxygen.

The current approach to understanding oxidation reactions has been developed from the idea that electrons are crucial to chemical **bonding**. When a metal is oxidized, it is converted from an uncharged **atom** to a **cation** by losing electrons. The definition of oxidation of a chemical species, including a particular atom within a **molecule** as well as entire molecules or ions, has been broadened to include any process in which the species loses electrons.

Most compounds are covalent, not ionic. Covalent compounds can undergo oxidation: for example, the conversion of **carbon monoxide** to **carbon dioxide**. To encompass all oxidation and reduction reactions, it is necessary to determine the gain and loss of electrons for covalent as well as ionic and elemental species. Electrons are assigned to atoms in covalent and ionic compounds and the oxidation number of each atom is determined.

To determine if a change in the number of electrons of an atom has occurred, we must have some way of assigning the starting and ending states of atoms in terms of number of electrons. This requires determining the oxidation states of the atoms. Oxidation states are specified by oxidation numbers. A species is oxidized in a reaction if it undergoes a increase in oxidation number.

A few simple rules are used to assign oxidation numbers. These rules are the result of assigning all electrons in a **covalent bond** to the more electronegative atom in the bond. Since only **valence** electrons are shared or transferred, the maximum positive oxidation number of an atom is the number of valence electrons. The rules are given in order of preference: if two rules apply, the first one is used.

1. The oxidation number of an atom in an elemental substance is zero.

2. The oxidation number of a monatomic **ion** is the charge on the ion.

3. The oxidation number of **fluorine** in all of its compounds is -1, because fluorine is the most electronegative element.

4. Because the **hydrogen** atom has only one **electron** to share and can accommodate only one electron, the oxidation number of hydrogen is +1 except when it forms compounds with metals, in which it is -1.

5. Because oxygen is the second most electronegative element, its oxidation number is -2 except when it bonds to fluorine [OF_2: O = +2] or to itself, as in O_2 (zero), H_2O_2 (-1), or HO_2 (-1/2).

6. The oxidation number of the **halogens** other than fluorine is -1, unless they are bonded to oxygen or a more electronegative halogen.

7. The sum of oxidation numbers for all atoms in a polyatomic ion equals the charge on the ion, and the sum of all oxidation numbers for atoms in a neutral molecule equals zero.

Oxidation numbers are not ionic charges. They can be computed for atoms in any compound, ionic or covalent. When two elements may combine in several different ratios, the oxidation number of an atom is specified by a Roman numeral in parentheses within the name of the compound. For example, three oxides of **nitrogen** are N_2O (nitrogen(I) oxide), NO (nitrogen(II) oxide) and NO_2 (nitrogen(IV) oxide), and two chlorides of **iron** are $FeCl_2$ (iron(II) chloride) and $FeCl_3$ (iron(-III) chloride).

For some compounds specific names were used to indicate an **oxidation state** before the current Roman numeral method was adopted. It is still common to see the older **nomenclature**. For example, copper(I) chloride is also known as cuprous chloride; copper(II) chloride, cupric chloride. **Copper** has two oxidation states: +1 and +2. The -ous ending on the name of the metal specifies the lower of the two oxidation states. The -ic ending specifies the higher oxidation state. Iron also has two common oxidation states: +2 and +3. For the compounds of iron, then, ferrous designates iron(II) and ferric, iron(III). Some elements combine is several ratios with different oxidation states, leading to a complex series of names. For example, the older names for the oxoacids and the corresponding ions of **chlorine** are: perchloric acid ($HClO_4$), perchlorate ion (ClO_2^-), chloric acid ($HClO_3$), chlorate ion (ClO_2^-), chlorous acid ($HClO_2$), chlorite ion (ClO_2^-), hypochlorous acid ($HClO$), and hypochlorite ion (ClO^-). In such as extensive series of names, the prefix ''per-'' indicates an oxidation state higher than the one designated with the -ate or -ic suffix and an oxidation state lower than the one with the -ite or -ous suffix is designated by the prefix ''hypo-.'' The naming system using Roman numerals is preferred, because the oxidation state of the element is specified clearly.

In a chemical reaction involving oxidation of one or more atoms, other atoms must be reduced to the same extent: electrons are not gained or lost in an overall chemical reaction. Since changes in oxidation numbers indicate changes in the number of electrons assigned to atoms, the overall change in oxidation numbers of all the atoms involved in the reaction must also be zero. This fact is critical in balancing chemical equations for oxidation-reduction reactions.

Oxidation-reduction reactions cover a wide range of chemical changes, including reactions of atoms of the same element in different oxidation states. Three categories that encompass all these different oxidation-reduction reactions are: atom transfer reactions, electron transfer reactions, and **disproportionation** reactions. In an atom transfer oxidation-reduction reaction, the atom being oxidized or reduced is bonded with a different species in the product than in the reactant; it has been chemically transferred to form a new species. Reactions of non-metallic elements to form covalent compounds are atom transfer reactions. Because we exist in an oxygen rich atmosphere, the most prevalent class of atom transfer oxidation-reduction reactions involves oxygen. Such reactions include **combustion**, oxidative corrosion of metals, and metabolic oxidation of foods. **Metabolism** is just a form of controlled combustion, as is the rusting of iron. The complete metabolism of cane sugar and its complete combustion yield the same products: carbon dioxide and water.

Reactions in which oxidation states change without the transfer of either the reductant or oxidant atoms are defined as electron transfer reactions. Electron transfer reactions occur in the process of metabolism and **photosynthesis** and are also used in large scale industrial processes when appropriate oxidants and reductants are mixed together. Electrochemical processes are also electron transfer reactions. Unlike most other types of chemical reactions, the reactants in an electrochemical process can be spacially separated as long as there is a means for electrons to flow from the reductant to the oxidant. Electrochemical oxidation is used industrially to protect and beautify metal surfaces by anodizing them and to produce many other products and it is used in methods to analyze the metal contents of samples (anodic stripping) as well as in other electroanalytical techniques.

Reactions in which atoms of the same element are both oxidized and reduced are disproportionation reactions. An example is the reaction of two molecules of **hydrogen peroxide** to form two molecules of water and one molecule of oxygen. The oxidation number for both oxygen atoms in hydrogen peroxide is -1. In the products, the oxidation number of the oxygen atom in water is -2, while that in molecular oxygen is 0.

The **periodic table** can be used to predict ability of species to act as oxidizing agents. The **reactivity** of nonmetallic elements often parallels their electronegativity: the more electronegative the element, the more powerful an oxidant. Fluorine is the strongest oxidant of all the elements. Oxygen, the next most electronegative element, is also a powerful oxidant, as are chlorine and **bromine**. Elemental nitrogen, however, is only capable of oxidizing a few metals such as **lithium** and **magnesium**. Ozone, an unstable, allotropic form of oxygen, is a more powerful oxidizing agent than O_2. In acidic **solution**, it is one of the most powerful oxidizing agents known. When ozone functions as an oxidizing agent, one of its oxygens is reduced; the other two form a molecule of O_2, another example of a disproportionation reaction.

The reactivities of polyatomic ions as oxidants often follows a sequence in which the most highly oxidized central atom is each series of oxoanions of a particular element is the strongest oxidant. Thus, perchlorate ion is a stronger oxidant than chlorate ion, followed by chlorite ion and hypochlorite ion. There is a direct relationship within each series of oxoanions between oxidizing ability and oxidation state of the central atom. From one series of oxoanions to another, the reactivity depends both on the oxidation state of the central atom and its **electronegativity**. The more electronegative the central atom and the higher its electronegativity, the stronger the oxidant. Thus, chlorate ion is a stronger oxidant than bromate ion and so forth.

Liquid bleach used in the laundry is an alkaline solution of hypochlorite ion. The hypochlorite ion is an oxidizing agent that sterilizes water by ridding it of microorganisms. A convenient way to prepare solutions of hypochlorite ion is an **oxidation-reduction reaction**. When bubbled through alkaline water, chlorine gas disproportionates to form chloride ion and hypochlorite ion.

Oxidation and reduction reactions are essential to life, abound in geological systems, and are mainstays of the chemical manufacturing industry. They differ from other reactions in that electrons are transferred from one species to another, thereby changing the oxidation number of atoms involved in the chemical change. Oxidation and reduction always occur together, for as one species is oxidized and loses electrons, another must gain electrons and be reduced. Balancing oxidation-reduction reactions requires consideration of the numbers of electrons gained and lost by atoms in the process, as well as to the conservation of matter.

Rust, corrosion, spoilage of food, and combustion are all oxidations. Metabolism is a biochemically important oxidation process. The range of organic oxidation reactions is vast. Carbon in its most reduced form (oxidation number -4) is found in alkanes, which are typically fuels like **methane**, propane, and octane. Next are compounds such as **methanol** CH_3OH) or the alkenes—compounds with carbon-carbon double bonds (oxidation number -2). **Aldehydes** contain carbons at an oxidation state of 0 (**carboxylic acids**, +2; and finally carbon dioxide, at +4, used as a fire extinguisher).

See also Anodized surfaces; Electrochemistry

OXIDATION STATE

The **oxidation** state of an **atom**, also called the oxidation number, is a numerical value that represents the charge an **ion** has or an atom appears to have when its electrons are counted based on a set of commonly accepted rules. The value (+ or -) of the oxidation state symbolizes the apparent charge for each atom in a chemically bonded **molecule** based on the distribution of the electrons in the molecule. For ionically bonded compounds, the oxidation state of an atom is equal to the ionic charge of the atom. The charge of an atom in an ionic compound represents the **electron** distribution of the compound. The oxidation state of an atom bound in a covalent compound is the average charge for the atom based on the **electronegativity** values for each atom involved in the molecule. In this case, the oxidation state is equal to the charge the atom in a covalent compound would have if the electrons in the molecule were distributed around the atoms with the greatest electronegativities.

The oxidation state of an atom is not necessarily equal to the charge of the atom. This difference is expressed in the notation of oxidation state versus ionic charge. The oxidation state is written as a superscript with the charge followed by the number, for example, **aluminum** (Al^{+3}) or **fluorine** (F^{-1}). Ionic charges, on the other hand, are written as a superscript with the number followed by the charge. Examples of ionic charge notation include the **hydrogen** ion (H^{1+}) and the **oxygen** ion (O^{2-}). Oxidation states do not represent an actual physical property of an atom, as ionic charges do. An ionic charge of 1- means the atom has gained an electron, whereas an oxidation state of -1 means that atom has a greater attraction to a **bonding** electron. Oxidation states are simply a way of keeping track of the electrons involved in a molecule during a chemical reaction.

Oxidation and **reduction** are two processes that are quite common in chemical reactions. Oxidation occurs when an atom loses one or more electrons, becoming more positive. Reduction occurs when an atom gains one or more electrons, becoming more negative. Oxidation always takes place at the same time as reduction. When an atom loses an electron during oxidation, that electron needs to go somewhere. It is picked up by another atom during the process of reduction. Reactions which involve oxidation and reduction are called oxidation-reduction reactions, or simply redox reactions.

Redox reactions involve the transfer of electrons between atoms. The atom that undergoes reduction, i.e., the atom that receives the electron in a reaction, is called the oxidizing agent. The oxidizing agent causes the oxidation of an atom by receiving an electron from that atom and is itself reduced. The atom that undergoes oxidation, i.e., the atom that loses the electron in a reaction, is called the reducing agent. The reducing agent causes the reduction of an atom by donating an electron and is itself oxidized.

When an atom is oxidized, it loses an electron and becomes more positive. When this occurs, its oxidation state increases. Likewise, when an atom is reduced, it gains an electron and becomes more negative. When this occurs, its oxidation state decreases. In every redox reaction, an atom must increase its oxidation state and an atom must decrease its oxidation state. A change in oxidation state of an atom indicates that a redox reaction has taken place. Assigning oxidation states to the chemical compounds involved in a redox reaction is important because it allows every electron to be accounted for, which in turn allows for the formation of balanced equations for the reaction.

As mentioned earlier, the oxidation state of an atom is not necessarily the same as the ionic charge of an atom. The oxidation state of an atom can be determined by studying the bonding characteristics of that atom, but this requires time and facilities that many people do not have. As a general rule, shared electrons in a compound are assumed to belong to the atom that has a higher electronegativity, but this does not help much in assigning oxidation states, either. To make the process simpler, a set of rules has been developed to determine the oxidation state of an atom involved in a reaction. First, the oxidation state of any atom in an uncombined element is equal to zero. For example, the oxidation state of an elemental **carbon** (C) atom is zero, as is the oxidation state of an elemental **sulfur** (S) atom. The second rule specifies that the oxidation state of any monoatomic ion is equal to its ionic charge. A **chlorine** ion (Cl^{1-}) has an oxidation state of -1, for example, and a **magnesium** ion (Mg^{2+}) has an oxidation state of +2. The third rule gives the oxidation states of many elements in a compound according to the element's position in the **periodic table**. Elements in group 1A on the periodic table always have an oxidation state of +1. Elements in group 2A on the periodic table always have an oxidation state of +2. Aluminum (Al) is always +3 and fluorine is always -1. The oxidation state of hydrogen is always +1 if it is combined with a **nonmetal**, and the oxidation state of oxygen is usually -2 in most ions and compounds. The fourth rule says that the oxidation states of all the atoms in a compound must total zero. The last rule specifies that the oxidation states of all the atoms in a polyatomic ion must total the charge of the ion.

Not every atom follows the rules for assigning oxidation states. Many nonmetals, for example, can have more than one oxidation number. For example, sulfur can have an oxidation state of +4 or +6. The Stock system of **nomenclature** eliminates some of the resulting confusion by placing the oxidation state of a substance in Roman numerals within parentheses as part of the name of a compound. For example, sulfur with an oxidation state of +4 forms the compound SO_2, named sulfur (IV) **oxide**. The form of sulfur with an oxidation state of +6 forms the compound SO_3, named sulfur (VI) oxide. In general, however, the rules for assigning oxidation states can be followed and used to name compounds, write formulas, or **balance** chemical equations.

With the assistance of the above rules, oxidation states can be determined for the atoms in almost any compound. The oxidation state of an atom can give valuable information about that atom. For example, the oxidation number describes how well an atom is going to bond with other atoms. A positive oxidation state indicates that the atom can undergo oxidation, i.e., lose electrons. A negative oxidation state indicates that an atom can undergo reduction, i.e., gain electrons. A negative oxidation state indicates a stronger attraction for electrons than does a positive oxidation state. The larger the negative value, the larger the attraction for the electrons. For example, an atom with an oxidation state of -3 will have a stronger attraction for electrons than one with an oxidation state of -1. Similarly, the larger the positive value of an atom's oxidation state, the smaller the attraction for the electrons. An atom with an oxidation state of +3 will have a weaker attraction for electrons than one with an oxidation state of +1. The oxidation number of an atom can also be used to predict how it will combine with other atoms as well as what the resulting chemical formula and balanced chemical equation will be.

OXIDATION-REDUCTION REACTION

A type of chemical reaction that is significant to our daily lives is oxidation-reduction reactions. The term **oxidation** was originally used to describe reactions in which an element combines with **oxygen**. In contrast, **reduction** meant the removal of oxygen. By the turn of this century, it became apparent that oxidation always seemed to involve the loss of electrons and did not always involve oxygen. In general oxidation-reduction reactions involve the exchange of electrons between two species.

An oxidation reaction is defined as the loss of electrons, while a reduction reaction is defined as the gain of electrons. The two reactions always occur together and in chemically equivalent quantities. Thus, the number of electrons lost by one species is always equal to the number of electrons gain by another species. The combination of the two reactions is known as a redox reaction. Species that participate in redox reactions are described as either reducing or oxidizing agents. An oxidizing agent is a species that causes the oxidation of another species. The oxidizing agent accomplishes this by accepting electrons in a reaction. A reducing agent causes the reduction of another species by donating electrons to the reaction.

Many common redox reactions involve metal species in **solution**, for example: $Zn(s) + Cu^{2+}(aq) \rightarrow Zn^{2+}(aq) + Cu(s)$. The metal, **zinc** (Zn), is converted (oxidized) to zinc ions (Zn^{2+}) by the loss of two electrons, while the **copper ion** (Cu^{2+}) is reduced to copper metal (Cu) by the gain of two electrons. The oxidizing agent in this reaction is the copper ions that accept two electrons from zinc, causing zinc to be oxidized. The reducing agent in the reaction is zinc that is oxidized by donating two electrons to the copper ions.

The above reaction can be separated into two separate reactions called half reactions, one for oxidation and one for reduction. $Zn(s) \rightarrow Zn^{2+}(aq) + 2e^-$ (oxidation half reaction) $Cu^{2+}(aq) + 2e^- \rightarrow Cu(s)$ (reduction half reaction). When the two reactions are combined into a single redox equation, the electrons canceled each other out. So while electrons are shown in the individual half reactions, the **electron** transfer is implied in the redox reaction.

In general, a strong oxidizing agent is a species that has an attraction for electrons and can oxidize another species. The standard voltage reduction of an oxidizing agent is a measure of the strength of the oxidizing agent. The more positive the species' standard reduction potential, the stronger the species is as an oxidizing agent.

In reactions where the reactants and products are not ionic, there is still a transfer of electrons between species. Chemists have devised a way to keep track of electrons during chemical reactions where the charge on the atoms is not readily apparent. Charges on atoms within compounds are assigned oxidation states (or oxidation numbers). An oxidation number is defined by a set of rules that describes how to divide up electrons shared within compounds. Oxidation is defined as an increase in **oxidation state**, while reduction is defined as a decrease in oxidation state. Since an oxidizing agent accepts electrons from another species, a component **atom** of the oxidizing agent will decrease in oxidation number during the redox reaction.

There are many examples of oxidation-reduction reactions in the world. Important processes that involve oxidation-reduction reactions include **combustion** reactions that convert **energy** stored in fuels into thermal energy, the **corrosion** of **metals**, and **metabolism** of food by our bodies. The burning of **natural gas** is an oxidation-reduction reaction that releases energy [$CH_4(g) + 2O_2(g) \rightarrow CO_2(g) + 2H_2O(g)$ + energy]. During the consumption of food, the human body uses a sequence of redox reactions to burn carbohydrates that provide energy [$C_6H_{12}O_6(aq) + 6O_2(g) \rightarrow 6CO_2(g) + 6H_2O(l)$]. In both examples, the carbon-containing compound is oxidized, and the oxygen is reduced. **Batteries**, which supply electrical energy, use spontaneous oxidation-reduction reactions to start a car or power a calculator.

OXIDE

An oxide is a compound in which one or more **oxygen** atoms are bonded to another type of **atom**, often a metal. In oxidative reactions, oxygen combines chemically with another substance.

In **nonmetal** oxides, the **bonding** is primarily covalent (characterized by the sharing of electrons)rather than ionic (accomplished as a result of opposite charges). Many familiar nonmetallic oxides, including **carbon** monoxide, **carbon dioxide**, and **sulfur** dioxide, exist as molecules. Most nonmetals can form more than one oxide. Normally the more highly oxidized oxide forms when the nonmetal burns at ordinary temperatures in an abundance of oxygen, allowing the maximum number of oxygens to bond, e.g., CO_2. High temperatures and/or little oxygen favor formation of the less oxidized oxide, e.g., CO. Many nonmetal oxides react with **water** to form acids; these substances are called acid anhydrides.

When oxygen reacts with metal, the reaction product is usually an ionic solid containing an oxide **ion**. Many transition **metals** form more than one oxide. In these cases, it is usually the oxide containing the positive ion of highest charge that is formed at ordinary temperatures. Many ionic oxides react with water to form the corresponding metal hydroxide in water. Compounds that react with water to form hydroxide ions are called basic anhydrides.

A few transition metals form oxides in which the bonding is more covalent than ionic. **Titanium** oxide and **manganese** oxide have macromolecular structures more like that of SiO_2. Neither of these compounds react with water to form hydroxide ions.

Most metal oxides can be reduced to the pure metal by heating them in the presence of an element that has a strong affinity for oxygen. Carbon, in the form of coke, is most frequently used for this purpose. In this reaction, the decomposition products of coke react with the metal oxide to produce carbon dioxide and the pure metal.

In the case of hematite ore (Fe_2O_3), a mixture of the ore, coke, and limestone is first placed in a blast furnace, then pure or compressed air is introduced at the base of the furnace to burn the coke, whose decomposition products remove the oxygen atoms from the hematite.

Oxides of highly active metals cannot be reduced in the presence of coke, but their pure metals can sometimes be obtained by **electrolysis**. An example is the **reduction** of bauxite to yield pure **aluminum** by means of electrolysis.

In the case of certain ores containing relatively inactive metals such as **mercury**, separation can be achieved by heating the ore in air, i.e., by oxidative **calcination** (also known as roasting). In the case of more chemically active metals, e.g., **zinc**, **roasting** produces the metal oxide instead of the free metal.

Although the most common **anion** formed by oxygen is monatomic, oxygen can also form **diatomic** ions known as peroxides (bivalent) and superoxides (monovalent). Metal peroxides undergo a violent reaction with water to form a **solution** containing **hydrogen** peroxide. Superoxides react with water to produce molecular oxygen and **hydrogen peroxide**.

See also Acids and bases; Oxidation-reduction reaction

Oxirane functional group

The oxirane functional group is a subset of organic cyclic ether compounds. "Cyclic" refers to the fact that the oxirane atoms—two carbons and an **oxygen** in a three-membered ring—and "ether" that indicates that there is a **carbon**-oxygen-carbon bonding arrangement. More commonly known as epoxides, the oxirane group is an essential component of many industrial products and processes and is a common functional group introduced during the **metabolism** of many compounds.

What makes epoxides so useful in synthetic chemistry is the high strain of the **ring** structure. In this functional group, there is an oxygen **atom** tenuously connected to two neighboring carbon atoms by single bonds, making a saturated, three-sided ring. The resulting strain on the bonds weakens the carbon-oxygen bond, allowing the epoxide to readily break open the three-membered ring.

Epoxidation, the conversion of an olefin (an unsaturated alkene **hydrocarbon**) to a cyclic ether through reaction with **hydrogen peroxide**, a peracid, or oxygen, is an important chemical technique. The process produces such key industrial chemicals as ethylene oxide, the simplest epoxide. Ethylene oxide is a disinfectant, fumigant, and sterilizer, and its derivatives have a wide range of industrial and commercial uses. For instance, ethylene glycol (formed by adding **water** to, or hydrolyzing, ethylene oxide) is used for antifreeze formulations and making polyester fibers; **polyethylene** glycol goes into **cosmetics**, **lubricants**, **paints**, and drugs; ethylene glycol ethers are useful for brake fluids and solvents; and ethanolamine (formed by adding **ammonia**, or ammonolyzing, ethylene oxide) is often an ingredient in detergents and **soaps**.

One of the biggest uses of epoxides is in the manufacture of epoxy resins. These high-strength polymers are created by the reaction of less-rigid, one-dimensional polymers with epoxides to produce a much more rigid three-dimensional polymer. These resins all contain the characteristic three-sided ring of the oxirane functional group.

Epoxy resins, or epoxies, are ubiquitous in such industries as electronics, where they are often used to encapsulate or embed components; boating and automotive manufacture, where they serve as protective coatings and sealants; and construction, where they form many kinds of tough, durable structures. Epoxy resins are popular in general because of their low shrinkage rate, strength, excellent adhesion, and resistance to chemicals.

Epoxides are so useful and sought after commercially that chemists have invented a number of processes to create them. For many years, chlorohydrins treated with alkalies were the standard source, but demand for epoxides has increased to keep pace with the development of many new industrial applications. This demand has led to significant developments in **oxidation** techniques, although direct oxidation using oxygen is commercially viable only for forming ethylene oxide from ethylene. Epichlorohydrin is an important epoxide whose creation comes mainly from hydrochlorination of allyl chloride. Another major epoxide, propylene oxide, comes from propene reacting with organic hydroperoxides (more reactive equivalents of hydrogen peroxides). Many epoxides still come from hydrogen peroxide because it is inexpensive and a good donor of oxygen atoms, but most often epoxides are made by adding an oxygen atom to the double bond of an olefin.

Oxygen

Oxygen is the first element in Group 16 of the **periodic table**, a group of elements sometimes known as the oxygen family. Oxygen has an **atomic number** of 8, an atomic **mass** of 15.9994, and a chemical symbol of O.

Properties

Oxygen is a colorless, odorless, tasteless gas with a **density** of 1.429 grams per liter, slightly greater than that of air (1.29 grams per liter). Its **boiling point** is -297.33°F (-182.96°C) and its freezing point is -361.2°F (-218.4°C). Liquid oxygen has a slightly bluish **color** to it and is slightly magnetic.

Oxygen exists in three allotropic forms: **diatomic** oxygen, or dioxygen (O_2), atomic or nascent oxygen (O), and ozone (O_3). Atomic oxygen is a highly reactive species that forms diatomic oxygen readily and reacts with other elements and compounds easily. **Ozone** also tends to be unstable and break down into dioxygen and nascent oxygen and to react readily with other substances.

Ozone has a slightly bluish color as both a gas and a liquid. It has a boiling point of -169.4°F (-111.9°C), a freezing point of -315°F (-193°C), and a density of 2.144 grams per liter.

Oxygen is a relatively inactive substance at room **temperature**, as evidenced by its abundance in the atmosphere. It does react slowly with a number of elements and compounds, however. For example, it reacts slowly with many **metals** to form the metal **oxide** in a process sometimes known as rusting, and it reacts with many organic compounds in the process known as decay.

At elevated temperatures, however, oxygen becomes much more active. It reacts with a variety of substances in the process known as **combustion**. The combustion of **wood** and fossil fuels (coal, oil, and **natural gas**) is arguably one of the most important chemical processes in human society. The **oxidation** of any organic material, whether by decay or combustion, results in the formation of **carbon** dioxide (CO_2) and **water** (H_2O), as well as **carbon monoxide** (CO) and other products in many instances.

Occurrence and Extraction

Oxygen occurs in the Earth's atmosphere primarily as an element. It makes up 20.948% of the atmosphere, the second most abundant gas after **nitrogen**. Oxygen also occurs in the hydrosphere in the form of water, of which it makes up nearly 89% by weight, and in the Earth's crust. In the earth, it has an abundance of about 45%, making it by far the most abundant element in the crust. It occurs in all kinds of minerals, such as oxides, carbonates, nitrates, sulfates, and phosphates. Oxygen is produced commercially by the fractional **distillation** of liquid air.

Discovery and Naming

Oxygen was discovered almost simultaneously in about 1774 by the Swedish chemist **Carl Wilhelm Scheele** and the English chemist **Joseph Priestley**. Both chemists followed a similar approach in their research, heating compounds of oxygen until they broke down. In his classic experiment, for example, Priestley heated red oxide of **mercury** (mercuric oxide; HgO) and found that he obtained liquid mercury and a new gas: $2HgO \xrightarrow{heat} 2Hg + O_2$. Priestley carried out a number of tests on the new gas, including breathing it himself. He described the sensation by saying that the new gas, ''was not sensibly different from that of common air, but I fancied that my breast felt peculiarly light and easy for some time afterwards. Who can tell but that, in time, this pure air may become a fashionable article in luxury? Hitherto, only two mice and myself have had the privilege of breathing it.'' Priestley's prediction came to pass at the end of the twentieth century with the introduction of ''oxygen bars'' at which patrons could, for a price, spend a few minutes breathing pure oxygen.

Oxygen was named by the French chemist **Antoine-Laurent Lavoisier** a few years after its discovery. Lavoisier thought that oxygen was present in all acids, so he suggested the name from two Greek words, *oxy-*, for ''acidic,'' and *-gen*, for ''forming.'' Lavoisier was wrong about the presence of oxygen in all acids, but the name was retained for the element.

Uses

The most important single application of oxygen is in metallurgy where it is used to extract metals from their ores. For example, oxygen is used to burn off carbon, **silicon**, and other impurities present in **iron** ore during the process of making steel. The **carbon dioxide** thus formed escapes into the air, while the silicon dioxide formed becomes part of a ''slag'' that is scraped off the molten steel produced in the reaction.

Oxygen is also used in the production of other metals, such as **copper**, **lead**, and **zinc**. These metals occurs in the earth in the form of sulfides, such as copper sulfide (CuS), lead sulfide (PbS), and zinc sulfide (ZnS). The first step in recovering these metals is to convert them to oxides:

$$2CuS + 3O_2 \rightarrow 2CuO + 2SO_2$$
$$2PbS + 3O_2 \rightarrow 2PbO + 2SO_2$$
$$2ZnS + 3O_2 \rightarrow 2ZnO + 2SO_2$$

The oxides thus formed are then heated with carbon to make the pure metals:

$$2CuO + C \rightarrow 2Cu + CO_2$$
$$2PbO + C \rightarrow 2Pb + CO_2$$
$$2ZnO + C \rightarrow 2Zn + CO_2$$

Another use of oxygen is in high-temperature torches. The oxyacetylene torch, for example, produces **heat** by burning acetylene gas (C_2H_2) in pure oxygen. The torch can produce temperatures of 5,400°F (3,000°C), hot enough to cut through steel and other tough alloys.

One of the best known applications of oxygen is in the health care field. Individuals sometimes develop respiratory conditions, such as bronchitis or emphysema, which make it very difficult for them to breathe. In such conditions, they may be provided with oxygen masks that make it easier for their lungs to obtain the oxygen needed by the body.

Oxygen is also used extensively in the chemical industry as the starting point in making a variety of compounds. Sometimes, the steps in getting from oxygen to the final compound are lengthy and complex. For example, ethylene gas (C_2H_4) can be treated with oxygen to form ethylene oxide (CH_2H_4O). Ethylene oxide, in turn, is used to produce ethylene glycol [$CH_2CH_2(OH)_2$], which is an antifreeze as well as the starting point in the manufacture of **polyester** fibers, film, plastic containers, bags, packaging materials, and many other consumer products. Thousands of oxygen-containing compounds have important commercial uses. Many of these compounds are discussed under other elements in this book.

Health Issues

Nearly all organisms require oxygen for their survival. Most animals, as an example, can survive for weeks and even months without food and for many days without water. But they can not survive more than a few minutes without oxygen.

Ozone

Ozone is **oxygen** in a **triatomic** state. Ozone is an **allotrope** of oxygen with the chemical formula 0_3.

Ozone is depicted as a **resonance** structure. Oxygen can form a double bond with itself to give **diatomic** oxygen. The triatomic **molecule** ozone contains one single bond and one such double bond. The location of these bonds constantly switches (i.e., resonates) between the two possible bond configurations.

The name ozone comes from the Greek *ozon*, meaning smell. At atmospheric temperatures, ozone is a colorless gas with an odor similar to **chlorine** that can usually be detected at a level of about 0.01 parts per million.

High in the atmosphere, ozone plays an important protective role by diminishing the amount of potentially damaging ultraviolet radiation reaching the Earth. In sufficient **concentration**, however, ozone is a poison that at lower atmospheric levels is a pollutant that can be damaging to health. Ozone is also a strong oxidizing agent used in many industrial processes for bleaching and sterilization. Although ozone is often used in **water** treatment, the largest commercial application of ozone is in the production of pharmaceuticals, synthetic lubricants, and other commercially useful organic compounds.

In the atmosphere, ozone is formed predominantly by electric discharges (e.g., lightning). In the laboratory, ozone can be extracted form a mixture of oxygen and ozone by fractionation.

Ozone can also be formed by ultraviolet light. Ultraviolet light is very energetic and when it strikes the atmosphere it can break down some oxygen molecules producing highly energized oxygen atoms (free radicals). These **free radicals** can then react with molecular oxygen to produce ozone. The absorption of energetic light radiation also triggers the decomposition of ozone. As a result, ozone is an unstable molecule that exists in a dynamic equilibrium of formation and destruction. Consequently, the protective ozone layer is also in dynamic equilibrium.

The area where ozone is formed at the fastest rate is in the atmosphere at a height of approximately 50km. At this height there is a **balance** between the number of free radicals made and the concentration of diatomic oxygen is sufficiently high to ensure that reactive collisions occur.

The protective ozone layer is found in the upper reaches of the atmosphere (between 30 km and 90 km) where it absorbs ultraviolet radiation that, in excess, can be harmful to biological organisms. The potential detrimental effects of increased exposure to ultraviolet light due to a lessening of atmospheric ozone are of great concern. Holes in the ozone layer, or a global breakdown of stratospheric ozone would **lead** to increasing doses of ultraviolet radiation at the Earth's surface. Scientists fear that significant increases exposure to ultraviolet light will increase risks of cancer in animal skin, eyes, and immune systems. Studies have shown that high ultraviolet radiation doses can supply the needed **energy** for chemical reactions that produce highly reactive radicals that have the potential to damage **DNA** and other cell regulating chemicals and structures.

Ozone is one of the atmospheric trace elements that are important in the regulation of the global climate. Although the atmosphere consists of mainly of **nitrogen** and oxygen, approximately, one percent of the Earth's atmosphere is made of small amounts of other **gases**. Trace gases include water vapor, **carbon** dioxide, **nitrous oxide**, **methane**, **chlorofluorocarbons** (CFCs), and ozone. Because the amount of trace gases in the atmosphere is small, human activities can significantly affect the proportions of atmospheric trace gases.

Evidence indicating that the ozone equilibrium was being tipped toward overall **ozone layer depletion** as a consequence of human use of CFCs in refrigeration and aerosol propellants (a pressurized gas used to propel substances out of a container) resulted in restrictions in the use of CFCs in many industrialized countries. Consumer aerosol products in the United States have not used ozone-depleting substances such as CFCs since the late 1970s. Federal regulations, including the Clean Air Act and Environmental Protection Agency (EPA) regulations restrict the use of ozone-depleting substances.

Ozone played a critical role in the development of life on Earth. Once primitive plants evolved, oxygen started to accumulate in the atmosphere. Some of this oxygen was converted into ozone and the developing ozone layer gave needed protection from disruptively energetic ultraviolet radiation. As a consequence, complex organic molecules which would otherwise have been destroyed began to accumulate.

As well as being found high in the atmosphere ozone can be found at ground level. At these locations it is regarded as a pollutant. Ozone at ground level can be manufactured as part of photochemical smog. This is brought about by the dissassociation of oxides of nitrogen which produce oxygen free radicals. These free radicals can react with diatomic oxygen to produce ozone. Pollutant ozone can also be a by-product of the action of photo copiers and computer printers. Low level ozone is usually found at a concentration of less than 0.01 parts per million whereas in photochemical smog it can be encountered at levels as high as 0.5 parts per million. Levels of ozone exposure between 0.1 and 1 part per million cause headaches, burning eyes, and irritation to the respiratory passages in humans. Elderly people, asthma sufferers, and those exercising in photochemical smog suffer the greatest adverse effects.

Some plant species (e.g., the tobacco plant) are particularly sensitive to low lying ozone. The presence of excessive ozone causes a characteristic spotting of the leaves. High ozone levels are also known to damage structural material such as rubber.

Replacing more dangerous chlorine gas, ozone is used in many **waste treatment** facilities to purify water. Ozone is responsible for disinfecting the water and the efficient removal of trace elements such as **pesticides**. Ozone kills bacteria and other small life forms and it reacts with organic compounds. During the process the ozone is transformed to molecular oxygen.

Ozone is a strong electrophilic reagent which can attack the double bond found in alkenes. Ozone is so efficient at this process that the reaction will continue beyond the simple addition stage and the alkene will actually be split apart at the location of the double bond. It is possible that the two carbons which were part of the double bond structure will both accept an oxygen **atom** during this reaction. This will then give two carbonyl groups. This ozone mediated splitting of an alkene is called ozonolysis.

Ozonolysis occurs in a two stage reaction. The first stage is the production of a compound called an ozonide. Most ozonides are thick oily **liquids** that are foul smelling. They are also explosive. They readily decompose during the second stage of the ozonolysis reaction. This is a method of preparation for both **aldehydes** and **ketones**. Inorganic ozonides also exist.

Ozone is a much stronger oxidizing agent than diatomic oxygen and it will form oxides under conditions where diatomic oxygen would not. For example ozone will oxidize all of the common **metals** with the exception of **gold** and **platinum**.

Because ozone rapidly decomposes it cannot be stored for long and when it is stored it is done so at low temperatures to reduce the rate of decomposition. Ozone decomposition is catalyzed by metals such as **silver**, platinum, and **palladium**. Transition metal oxides can also catalyze this decomposition.

The melting point of ozone is -324°F (193°C) and the **boiling point** is -169°F (112°C). Ozone is faintly blue in **color**.

See also Ozone layer depletion

OZONE LAYER DEPLETION

The ozone layer is a part of the atmosphere between 18.6 mi and 55.8 mi (30 and 90 km) above the ground. The ozone present is responsible for blocking potentially harmful ultraviolet radiation reaching the surface of the Earth. The ozone layer is gradually being destroyed by human activity.

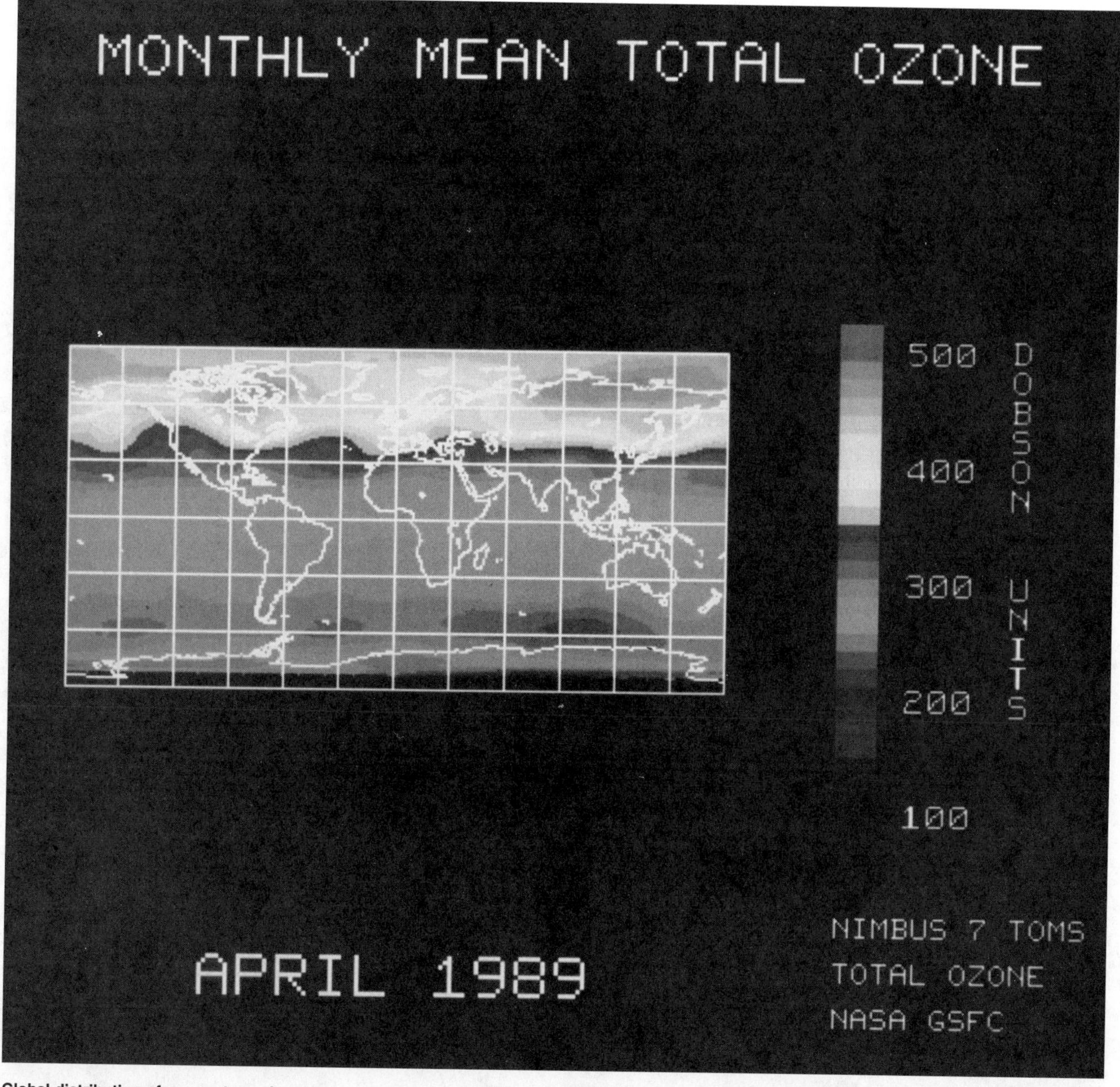

Global distribution of upper atmosphere ozone, mapped over a Mercator projection, in April 1989. The shading represent Dobson units, a measure of ozone concentration, as shown on the scale to the right of image. The data for this map were gathered by the Total Ozone Mapping Spectrometer (TOMS) instrument on the Nimbus-7 satellite. ***(Photograph by NASA/Science Source, Photo Researchers, Inc. Reproduced by permission.)***

Ozone is constantly created and destroyed in natural processes (manufactured by the action of lightning on **oxygen** and destroyed by the action of ultraviolet radiation) but the amounts **balance** each other out so there is no net increase or decrease due to natural processes. In 1970 **Paul Crutzen** showed that naturally occurring oxides of **nitrogen** can catalytically destroy ozone. In 1974 **F. Sherwood Rowland** and **Mario Molina** demonstrated that **chlorofluorocarbons** (CFCs) could also destroy ozone. In 1995 all three were jointly awarded the Nobel prize for **chemistry**.

The CFCs that were observed as being damaging included Freon 11 ($CFCl_3$) and Freon 12 (CF_2Cl_2). These chemicals are widely used in industry and the home. They have uses as propellants in aerosol spray cans, refrigerant **gases**, and foaming agents for blown **plastics**. One problem associated with these gases is their relative lack of **reactivity**. When released there is very little that will break them down and as they are not soluble in **water** they are not removed from the atmosphere by rain. As a consequence of these two characteristics once released they tend to concentrate in the upper regions of the atmosphere. It is estimated that some several million tonnes of CFCs are present in the atmosphere.

Once in the upper atmosphere the CFCs are exposed to high **energy** radiation which can cause disassociation of the **molecule**, producing free **chlorine** atoms. This atomic chlorine reacts readily with ozone to produce chlorine monoxide and molecular oxygen. The chlorine monoxide can further react to produce molecular oxygen and more atomic chlorine. This all accelerates the destruction of ozone beyond its natural ability to regenerate. Overall there is a net **reduction** in the amount of ozone present in the upper atmosphere. This has led to a thinning of the ozone layer that in extreme cases has led to the production of a hole where there is an absence of ozone completely. Ozone holes have so far only been detected in the Southern Hemisphere principally over Antarctica between September and October. The majority of this loss is at an altitude between 7.44 mi and 18.6 mi (12 and 30 km) and in the late 1990s evidence was seen that suggested losses were also occurring at other altitudes. In total over 70% of the ozone above Antarctica is lost during these periods. These holes are increasing in size and in the late 1990s holes were detected over Australia and dramatic thinning of the ozone layer in the Northern Hemisphere was also recorded during the winter months. In the Northern Hemisphere losses of some 30% have been recorded at an altitude of 12.4 mi (20 km).

In 1987 the Montreal Protocol was signed with the appropriate countries agreeing to reduce CFC production. A more stringent set of guidelines was introduced in 1992 whereby 100 countries agreed to stop CFC production by 1996. There will be a lag effect from these treaties because of the long life that CFCs experience in the atmosphere, so any reduction in depletion from these protocols will not be seen for several years to come, the exact length of time is not known.

Refrigerants and propellants are not the only damaging items to the ozone layer. High flying aircraft, ranging from commercial craft such as Concorde to military spy planes and the space shuttle, contribute to ozone depletion. The engines of these vehicles, particularly those traveling at supersonic speeds, produce oxides of nitrogen which will also attack the ozone layer.

In the absence of the ozone layer harmful ultraviolet radiation is able to reach the surface of the Earth in higher doses. This can lead to increases in skin cancers.

The ozone layer is being depleted by human activity at a rate faster than it can replenish itself. This has lead to holes appearing in the ozone layer. Despite agreements to limit production and use of damaging CFCs, the damage seen will not peak for several years due to the long-lived nature of CFCs.

P

PAINTS AND COATINGS

Paints and coatings are materials applied to surfaces for the purpose of coloring, decorating, or protecting them. In general, they are viscous fluids with characteristic colors, drying times and flow properties which are determined by formulation. They include products such as lacquer, varnish, and specialty coatings designed for architectural, commercial and industrial applications. Used by humans for centuries, paints and coatings technology has steadily improved over the years. Today, it represents a multibillion dollar industry.

The earliest evidence of the use of paint is cave paintings in Europe which demonstrate that people have been using paint for at least 30,000 years. Paint and coating use in the Middle East dates back over 10,000 years ago. They used various materials such as **lime**, **copper**, silica, **iron** oxides, and chalk to produce numerous colors. They also used resins from milk casein and plant sap. Crude varnishes and lacquers were first made over 2,000 years ago in Asia using materials from insect secretions.

The paint and coatings industry got its start during the early 1700s. At this time, the first American paint mill was opened and operated by Thomas Child in Boston. An important advancement was announced in 1867 by D. R. Averill when he was granted the first patent on ''ready mixed'' paints. By the mid 1880s, numerous paint factories opened all over the United States. This industrial growth was aided by a mechanized manufacturing process. Another factor that increased the market size of the paint and coatings industry was the rapid introduction of new products that required painting. For example, the automobile industry made extensive use of paints and coatings to beautify and protect car parts. In the twentieth century various polymeric materials were invented which led to more effective and easier to use paints and coatings. Latex **emulsion** paints were introduced during the 1940s and have become the preferred material.

In general, paints are mixtures of one or more pigments in a liquid formulation. A coating is a liquid mixture of some polymeric material. They are formulated differently depending on the application. The ingredients in paints can be grouped into four categories including pigments, binders or resins, **solvents** and additives.

The pigments that are incorporated into paint formulations are responsible for a paint's appearance and performance qualities. The appearance qualities refer to the **color** and opacity. Performance qualities refer to such things as chemical resistance, weatherability, fade resistance and particle size and shape. Pigments can be classified by their functions as either prime or inert. Prime pigments are organic or inorganic materials that create the paint color. The inorganic compounds include such things as **titanium** dioxide which produces white colors. Iron oxides produce browns, yellows, and reds. Organic compounds such as phthalocyanine produce greens and blues. Inert pigments are added as fillers or extenders to make paint last longer. They also add to the protective coating nature of paint. They include materials such as clay, talc, **calcium** carbonate, and silica.

The binder is often the most important ingredient in a paint or coating formulation. It is the material responsible for locking all the other ingredients in place on the surface. Binders must have specific properties such as good adhesion, abrasion resistance, the ability to cure at room **temperature**, **water** resistance and ultraviolet light resistance. They are typically incorporated into the formulation through a latex emulsion. Latex emulsions are made up of polymeric materials suspended in an aqueous solvent by a surfactant.

The type of binder in a paint or coating is dependent on the end use of the product. The most durable exterior house paints are based on acrylic co-polymers of materials like methyl methacrylate and butyl acrylate. These types of binders are also used for interior semigloss or gloss paints. **Vinyl** acetate is another common binder. When copolymerized with butyl acrylate it makes an excellent interior flat paint. Other binders include styrene-acrylic copolymers and alkyd resins. Alkyd resins are produced by the curing reaction of a material like phthalic anhydride with **glycerol** in the presence of a fatty acid.

The solvent typically makes up the bulk of a paint formulation. It plays a variety of roles including aiding in spreading, keeping the binder and pigments dissolved or dispersed, and determining drying time. A popular solvent is mineral spirits, a petroleum distillate composed of aliphatic hydrocarbons. It has many desirable characteristics such as low cost, low odor, and good drying time. Water is another solvent however, it is typically used in combination with another solvent. Organic solvents such as glycol ethers are also commonly employed.

Additives represent compounds that are put into paint formulations at minor levels. They can however, have a significant effect on the final characteristics and performance of the product. Additives are chosen based on the type of product being made. They include such materials as thickeners, surfactants, defoamers, biocides, and pigment dispersants. For solvent-based paints, mildewcides, driers, and antiskinning compounds are added.

The fundamental purpose of paint is decorating. The purpose for coatings is protection. In a practical sense, most paints also provide protection. Similarly, most coatings alter the appearance of a surface. For this reason the terms paints and coatings are usually used interchangeably. Paints and coatings work by creating a thin film on the surface to which they are applied. When a paint is applied the solvents evaporate and the binder, pigments and additives are left behind as a hard, colored film on the surface.

Paints and coatings are grouped by their performance characteristics in three primary categories; architectural paints, industrial coatings, and commercial finishes. Architectural paints are colored formulations that are used to decorate and protect structural surfaces. They are typically air drying materials that are applied with brushes and rollers or spayed from aerosol cans. These paints can be further classified by their base formulations. They include solvent based and water based formulas.

Solvent thinned paints work through an evaporation mechanism. When they are put on a surface, they help spread the pigment. As the solvent evaporates, a film is formed and the pigment gets locked onto the surface. The most common solvent based paints are latex. They have a relatively quick drying time, less odor than other coatings, and are easy to clean. They are composed of latex polymers which have pigments included. After the film is applied to a surface, the emulsion particles coagulate a form a stable film.

Oil paints are made up of a **suspension** of pigments in an oil such as linseed oil that dries. As it dries, a film is formed by a reaction between **oxygen** in the air and the evaporating oil. Catalysts are sometimes added to improve the cross-linking reaction. Once an oil paint is dried, the film is cured and insoluble. It can only be removed using a paint stripper which degrades the polymer.

Oil varnishes consist of a polymer (natural or synthetic) dissolved in an oil. Various additives are also added such as catalysts to improve the cross-linking reaction with oxygen. When these materials dry they take on the form of a clear, hard film.

Other important coatings include enamel, lacquers, solvent free products. Enamel is a pigmented oil varnish. A polymer is included to make it harder and glossier coating. Lacquers are composed of pigmented polymer solutions. A thin film is formed as the solvent evaporates. Since there is no subsequent cross linking reaction, these coated surfaces have poor resistance to certain solvents. Environmental concerns have led to the development of reduced solvent paints and coatings. Some products, called plastisols, contain no solvent and consist of polymers dispersed in a liquid plasticizer. They are coated on a surface and then **heat** cured.

Paints and coatings play an important role in delivering high quality durable goods, housings, furniture and thousands of other products. Commercially, they are used to coat products like display cases, beverage cans, appliances and automobiles. For home use they are used to protect and decorate the interior and exterior of the house. In the United States, about 900 million gallons of paint are sold annually. In 1997, this represented a market of about $16.4 billion.

PALLADIUM

Palladium is in Row 5, Group 10 of the **periodic table**. Along with **ruthenium**, **rhodium**, **osmium**, **iridium**, and **platinum**, it belongs to the platinum group of **metals**. These metals are sometimes also called the noble metals because they are not very reactive. Palladium has an **atomic number** of 46, an atomic **mass** of 106.42, and a chemical symbol of Pd.

Properties

Palladium is a relatively soft, silver-white metal that is both malleable and ductile. It has a melting point of 2,829°F (1,554°C), a **boiling point** of 5,100°F (2,800°C), and a **density** of 12.0 grams per cubic centimeter. A physical property of particular interest is palladium's ability to absorb **hydrogen** gas, somewhat like a sponge absorbs **water**. When a surface is coated with finely divided palladium metal, hydrogen gas passes into the space between palladium atoms, allowing the metal to absorb up to 900 times its own weight in hydrogen gas.

Palladium is the most reactive member of the platinum group. It combines poorly with **oxygen** at room temperatures, but will catch fire if ground into a powder. Palladium does not react with most acids at room **temperature**, but will do so with hot acids. The metal also reacts with **fluorine** and **chlorine** when very hot.

Occurrence and Extraction

Palladium generally occurs in the Earth's crust in conjunction with other members of the platinum family. Its abundance is no more than about 1-10 parts per trillion, making it one of the rarest of the chemical elements. More than 90% of the palladium mined in the world comes from Russia and South Africa. Very small amounts are also produced in the United States and Canada.

Discovery and Naming

Palladium was discovered in 1803 by the English chemist William Hyde Wollaston (1766-1828). The ores with which

Wollaston was working had come originally from Brazil, where they had been called by names such as *prata* ("**silver**"), *ouro podre* ("worthless, or spoiled, **gold**"), and *ouro branco* ("white gold"). The ores were actually combinations of many **precious metals**, including gold and members of the platinum group of elements. Wollaston discovered two new elements in the ores, rhodium and palladium. He suggested the name palladium for one of the elements in honor of the asteroid Pallas, which had been discovered at about the same time.

Uses

Palladium has two primary uses: as a catalyst and in making jewelry and specialized alloys. Palladium catalysts are used in refining and cracking petroleum in the process of making high quality **gasoline** and other products. It is also used in the production of essential chemicals, such as **sulfuric acid** (H_2SO_4), and **paper** and fabrics. The catalytic converters used in automobiles may also contain a palladium catalyst.

Palladium is generally alloyed with other precious metals, such as gold and silver, as well as with **copper**. The alloys are used in a variety of products, such as ball bearings, springs, **balance** wheels in watches, surgical instruments, electrical contacts, and astronomical mirrors. Palladium alloys are also used quite widely in dentistry. Few compounds of palladium have any commercial value.

PAPER

Derived from **cellulose** (plant) fibers that are made into pulp and felted (pressed) together, paper is one of the most common of man-made materials. There are approximately 7,000 types of paper, used not for writing or printing, but for products like paper bags and cardboard boxes.

The oldest extant writing surfaces include Babylonian clay tablets and Indian palm leaves. Around 3000 B.C., the Egyptians developed a writing material using papyrus, the plant for which paper is named. This substance was composed of strips cut from stalks of the papyrus reed, which were dried, laid across each other crosswise, and glued together to form a somewhat nubby writing surface. Other early materials were parchment, made from the untanned skins of sheep or goats and vellum, or fine parchment, made from calf-, kid-, and lambskin. Used in Europe from the second century B.C., they were expensive and impractical—the equivalent of a 200-page book, for example, would require twelve animal skins.

The invention of paper similar to the kind used today is credited to Cai Lun, a Minister of Public Works under Emperor He Di during China's Han Dynasty (202 B.C. to A.D. 202). The Chinese had earlier written on bamboo, bone, and silk, but these surfaces were either too heavy and cumbersome or too expensive. In the year A.D. 105, Cai Lun devised a way to make the bark of the mulberry tree into paper. The process he used contained the basic elements still found in paper mills today: the raw material was chopped and mixed with **water** to form a pulp, which was spread thin on porous screens, drained, and dried in sheets. The Chinese also used hemp, rag, and fish nets to make paper. At first, paper was used for clothing, household articles, decorative arts, and for wrapping. By the fourth century, papermaking was firmly established in China, although the technology would not reach Europe for another 1,000 years.

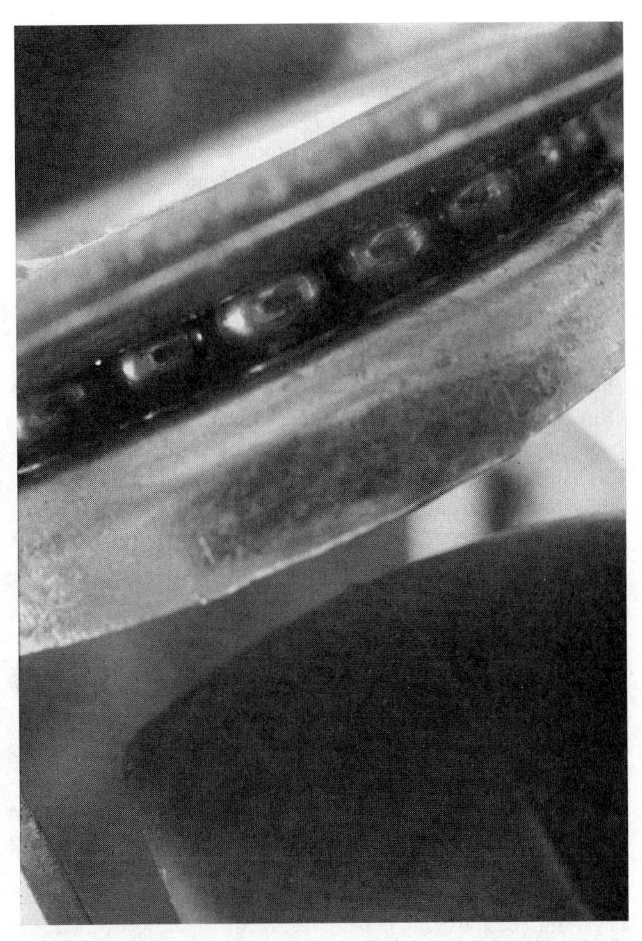

Palladium alloys are used to make ball bearings. They are shown here in a castor, the part of the wheel that enable a chair to swivel and roll. *(Photograph by Robert J. Huffman. Reproduced by permission.)*

The Chinese managed to keep papermaking a secret for 500 years, and it was not until the seventh century that is spread to Korea and Japan. Chinese prisoners (from eighth-century battles fought in what is now Turkistan) taught their Arab captors to make paper, which led to the establishment of papermaking workshops at Baghdad, Damascus, and Samarkind. The Spanish Moors finally brought paper to Europe in the twelfth century; they used essentially the same process as the Chinese, producing a very absorbent product similar to modern plotting paper. Europeans initially referred to paper as *cloth parchment*, and used linen rags (sometimes with cotton or straw added) to produce it. By the year 1276, a paper mill had been established in Italy's Appenine region. The invention of the printing press in 1450 created an increased demand for paper. During the sixteenth century, sizing (treating paper with starch to stiffen it and prevent smudging) was introduced, and papers began to be coated with opaque mineral powders (such as lime) to improve quality.

The first American paper mill was established near Philadelphia in 1690; by 1810 there were 185 paper mills in the United States. The papermaking process remained quite slow and tedious until 1750, when a Dutch inventor designed a machine that reduced the time it took to break rags down to fiber. In 1798, Nicholas Louis Robert, a Frenchman, invented a machine that made paper in continuous rolls. The English Foudrinier brothers, Henry and Sealy financed his idea and produced a successful model in 1805. The device, which bears their name, is still used in modern papermaking. The Foudrinier machine was first used in the United States in 1816, when Thomas Gilpin and his brother Joshua built America's first mechanized paper mill. They "borrowed" a state-of-the-art design from English inventor John Dickinson and even persuaded Dickinson's foreman to defect to their operation. The Gilpins' monopoly on American papermaking was broken only when improved technology became widely available.

By the early nineteenth century, **wood** and other plant pulps began to replace the rags that had been used earlier, particularly in Europe. In 1840, a method for reducing logs to chips was invented, thus facilitating pulp production. In 1867 an American inventor named Tilghman discovered that wood fibers could be easily separated if the chips were soaked in a sulfurous acid **solution**. In 1883, German inventor Carl Dahl incorporated **sodium sulfate** into his papermaking process, producing a very strong pulp that became known as kraft (which means strength in German). Kraft pulp is still used, in bleached and unbleached forms, for such products as paper bags, milk cartons, and paper cups. Most paper now produced is made from wood, either harvested particularly for that purpose or from lumber or woodworking waste. Wood pulp is created by both *mechanical* and *chemical* processes designed to break down the wood fibers. Once the pulp has been produced, it is washed to remove any chemicals; foreign **matter** is filtered out, and the pulp is drained, bleached, washed, and formed into sheets that are pressed by heavy rollers and dried by air suction. Cylindrical colander presses put a smooth finish on the paper, which is finally wound into large rolls. Some paper, however, is made from such plants as cotton, rice, wheat, cornstalks, hemp, and jute; very high quality "rag" paper is still derived from cotton rags.

By 1991, Americans were producing nearly 70 million tons (64 million metric tons) of paper per year and consuming 660 lb (300 kg) per person each year. With increasing concern about the environmental implications of forest depletion and waste disposal, many communities have begun paper **recycling** programs, chopping newspaper, magazines, and other waste papers into a pulp that is then treated to remove ink. The end product, called secondary fiber, is made into paperboard, some forms of printed paper, napkins, and towels. Despite the continued promise of a "paperless office" because of computer technology, paper consumption continues to grow. The future of paper might be affected by such technology, however. In 1994, Xerox corporation announced its development of "smart paper." Such paper will have dataglyphs, which can microencode digital information, embedded within it. Smart paper can count copies made from it, assure authenticity, and become part of the inter-office routing system.

PARA-AMINOBENZOIC ACID

Para-aminobenzoic acid, or PABA, is most frequently encountered as an ingredient in sunscreen. Alternate names for this chemical are p-aminobenzoic acid and 4-aminobenzoic acid. Its chemical structure features a **benzene ring**. The benzene **molecule** is a six-carbon ring with alternating single and double bonds between the **carbon** atoms. These double bonds allow PABA to absorb the ultraviolet radiation carried in sunlight.

Ultraviolet radiation cannot be seen, but it is responsible for much of the damage that sunlight inflicts on people's skin. Short-term exposure to ultraviolet radiation leads to skin tanning—or burning—if a person is fair-skinned or stays in the sun too long. Long-term exposure, however, can damage the skin. The chances of premature skin aging increase due to the total amount of sun exposure during a person's life.

A more serious problem associated with long-term sun exposure is skin cancer. Ultraviolet radiation triggers genetic changes in skin cells. These changes are usually—but not always—repaired by gene-repair mechanisms. As a result of unrepaired damage, several types of skin cancer are possible, some of which can be disfiguring or even life-threatening. When PABA is used on the skin's surface, ultraviolet rays are absorbed before they enter the skin. This absorption prevents most sun-linked damage and lessens the risk of sunburn.

Some sunscreens use chemicals that are closely related to PABA. These closely related chemicals, or derivatives, include glyceryl-p-aminobenzoate, menthyl anthranilate, padimate A, and padimate O. PABA and its derivatives may be absorbed into the top layers of the skin, but there is no evidence that they can reach the deeper layers. They are considered safe when used as directed, but certain individuals experience skin irritation when using PABA-containing sunscreens.

PARKES, ALEXANDER (1813-1890)

English metallurgist and chemist

Born on December 29, 1813, in Birmingham, England, Parkes began his career as an apprentice in the art metal trade and then moved on to an **electroplating** firm, where he silver-plated diverse objects such as spider webs and plants. His work with **silver** solutions and the chemicals used to produce them—namely **phosphorus** and **carbon** disulfide—led him to investigate solutions of rubber and **cellulose** nitrate. In 1841 he patented a method of waterproofing fabrics by coating them with rubber. He received a second patent in 1843 for an electroplating process.

In 1855 Parkes patented the first plastic. By dissolving cellulose nitrate in **alcohol** and camphor-containing **ether**, he produced a hard solid that could be molded when heated, which he called Parkesine (later known as celluloid). Unfortunately, Parkes could find no market for the material. John Wesley Hyatt, an American chemist, would rediscover celluloid and market it successfully as a replacement for ivory in the 1860s. Parkes died in London on June 29, 1890.

PARTICULATE NATURE OF MATTER

Since the times of the ancient Greeks, scientists and philosophers have generally agreed that everything that takes up space, or **matter**, is made up of tiny bits of material that have come to be known as elementary particles. The only thing that has changed over the years is the scale of what we mean by the term particle.

The Greek philosopher **Democritus**, who conceived of the particulate nature of matter, believed that what he called atoms (from the Greek word atomos for indivisible) were the smallest constituents of matter. However, scientific advances have shown us that there are at least several smaller kinds of particles that make up each **atom**. Nevertheless, the atom remains the most stable and convenient way to understand and discuss the nature of matter—especially since it is the smallest particle that can enter into chemical combination.

The **atomic theory** of Democritus and his school fell by the wayside for many years as the science of Aristotle came into vogue. Aristotle preferred to think of the nature of matter in more concrete terms, disdaining such intangible abstractions as atoms. Therefore, he concluded that all matter consisted of four elements: air, fire, wind, and **water**. He added **ether** to the list to account for the heavens.(This last concept would persist into the 1800s.) Although there was much value in Aristotle's theories, some of which has been absorbed into modern-day science, he turned out to be wrong about the nature of matter. However, it would not be until about the start of the nineteenth century that Democritus's view would return to the forefront of science.

In the field of **chemistry**, there were several important events that solidified the concept of particulate matter in the early days of quantitative science. In the late 1700s, French chemist **Antoine Lavoisier** discovered that although the different components in a chemical reaction might change radically in terms of appearance and form, their total original **mass** remained the same. This prompted him to conclude that some basic part of the components did not change, although he could not see this part. In about 1800, scientists studying chemical reactions began wondering whether the behavior of matter could be ruled by something that even their microscopes could not perceive.

The next step came when another French chemist, **Joseph-Louis Proust**, extended Lavoisier's work by demonstrating conclusively that every pure **chemical compound** maintains the same constant proportions in weight of the elements that make it up. Soon afterward, in 1808, **John Dalton** proposed a scenario in which every element is composed of a huge number of identical atoms and that chemical compounds come about through the combination of a smaller number of atoms from different elements.

In 1811, Italian chemist and physicist **Amadeo Avogadro** cleared up some of the problems with Dalton's theories by showing that the basic components of a pure element were not necessarily individual atoms, but rather compound atoms existing as **diatomic** ("two-atom") molecules (e.g., HCl or H_2). A scientific conference in 1860 officially adopted a combined form of Dalton's and Avogadro's theories, and soon afterward chemists began creating an accurate compilation of relative atomic masses. Yet what came to be known as the atomic-molecular theory of Dalton and Avogadro was developed without any real, direct proof of the particulate nature of matter.

The first hard evidence of the existence of fundamental particles came in 1897, when English physicist **Joseph Thomson** discovered electrons within an atomic **nucleus**. This was the end of the belief that the atom is the smallest particle that makes up matter. British physicist **Ernest Rutherford** confirmed and advanced Thomson's findings in the early 1900s, when he and his colleagues showed that every atom consists of a positively charged nucleus in the center that is surrounded by what they termed a "cloud" of negatively charged electrons. They also found that the atomic nucleus itself (except for **hydrogen**) consists of positive **proton**s and electrically neutral **neutron**s. Collectively, these components are known as **subatomic particles**.

Rutherford's work gradually led to the conclusion that the different properties of chemicals arise from the different numbers of electrons in their respective clouds. However, it became clear that an atom's weight is determined by the contents of its nucleus (i.e., its protons and neutrons). This weight is the number assigned to each of the chemical elements in the **periodic table**.

Since the days of Rutherford, physicists and chemists have theorized that atoms of matter contain even smaller particles than electrons, protons, and neutrons. Since 1945, researchers have predicted and/or confirmed the existence of several hundred smaller particles, many of which only come into existence under extreme conditions and for very short times. These infinitesimal particles are usually grouped into four main categories: the mesons, the baryons, the leptons, and the photons (the most basic unit of electromagnetic radiation). Other categories include the hadrons and the bosons. In addition, as British physicist Paul Dirac predicted in 1930, every kind of elementary particle has an **antiparticle** (the basis of antimatter).

In 1963, physicists Murray Gell-Mann and George Zweig proposed the existence of an even smaller particle that they named the quark, which they postulated is the basis for the baryons and mesons. Subsequent research has experimentally confirmed the presence of six kinds of quarks, and in 1995, physicists actually detected one of them (the "top" quark).

As science becomes ever more sophisticated, it seems possible that even the unimaginably small quark will be found to contain yet tinier components. However, these discoveries only strengthen the fundamental truth of the particulate matter of nature.

Louis Pasteur.

PASTEUR, LOUIS (1822-1895)

French chemist and microbiologist

Louis Pasteur was one of the most extraordinary scientists in history, leaving a legacy of scientific contributions which include an understanding of how microorganisms carry on the biochemical process of **fermentation**, the establishment of the causal relationship between microorganisms and disease, and the concept of destroying microorganisms to halt the transmission of communicable disease. These achievements led him to be called the founder of microbiology.

After his early education Pasteur went to Paris, studied at the Sorbonne, then began teaching **chemistry** while still a student. After being appointed chemistry professor at a new university in Lille, France, Pasteur began work on yeast cells and showed how they produce **alcohol** and **carbon** dioxide from sugar during the process of fermentation. Fermentation is a form of **cellular respiration** carried on by yeast cells, a way of getting **energy** for cells when there is no **oxygen** present. He found that fermentation would take place only when living yeast cells were present.

Establishing himself as a serious, hard-working chemist, Pasteur was called upon to tackle some of the problems plaguing the French beverage industry at the time. Of special concern was the spoiling of wine and beer, which caused great economic loss and tarnished France's reputation for fine vintage wines. Vintners wanted to know the cause of l'amer, a condition that was destroying the best burgundies. Pasteur looked at wine under the microscope and noticed that when aged properly the liquid contained little spherical yeast cells. But when the wine turned sour, there was a proliferation of bacterial cells which were producing lactic acid. Pasteur suggested that heating the wine gently at about 120°F would kill the bacteria that produced lactic acid and let the wine age properly. Pasteur's book *Etudes sur le Vin*, published in 1866 was a testament to two of his great passions—the scientific method and his love of wine. It caused another French Revolution—one in wine-making, as Pasteur suggested that greater cleanliness was need to eliminate bacteria and that this could be done with **heat**. Some wine-makers were aghast at the thought but doing so solved the industry's problem.

The idea of heating to kill microorganisms was applied to other perishable fluids like milk and the idea of pasteurization was born. Several decades later in the United States the pasteurization of milk was championed by American bacteriologist Alice Catherine Evans who linked bacteria in milk with the disease brucellosis, a type of fever found in different variations in many countries.

In his work with yeast, Pasteur also found that air should be kept from fermenting wine, but was necessary for the production of vinegar. In the presence of oxygen, yeasts and bacteria break down alcohol into acetic acid—vinegar. Pasteur also informed the vinegar industry that vinegar production could be increased by adding more microorganisms to the fermenting mixture. Pasteur carried on many experiments with yeast. He showed that fermentation can take place without oxygen (*anaerobic* conditions), but that the process still involved living things such as yeast. He did several experiments to show (as Lazzaro Spallanzani had a century earlier) that living things do not arise spontaneously but rather come from other living things. To disprove the idea of spontaneous generation, Pasteur boiled meat extract and left it exposed to air in a flask with a long S-shaped neck. There was no decay observed because microorganisms from the air did not reach the extract. On the way to performing his experiment Pasteur had also invented what has come to be known as sterile technique, boiling or heating of instruments and food to prevent the proliferation of microorganisms.

In 1862 Pasteur was called upon to help solve a crisis in another ailing French industry. The silkworms that produced silk fabric were dying of an unknown disease. So armed with his microscope, Pasteur went to the south of France in 1865. He found the tiny parasites that were killing the silkworms and affecting their food, mulberry leaves. His solution seemed drastic at the time. He suggested destroying all the unhealthy worms and starting with new cultures. The solution worked and French silk scarves were back in the marketplace.

Pasteur then turned his attention to human and animal diseases. He had believed for some time that microscopic organisms cause disease and that these tiny microorganisms could travel from person to person spreading the disease.

Other scientists had expressed this thought before, but Pasteur had more experience using the microscope and identifying different kinds of microorganisms such as bacteria and fungi.

In 1868, Pasteur suffered a stroke and much of his work thereafter was carried out by his wife Marie Laurent Pasteur. After seeing what military hospitals were like during the Franco-Prussian War, Pasteur impressed upon physicians that they should boil and sterilize their instruments. This was still not common practice in the nineteenth century.

Pasteur developed techniques for culturing and examining several disease-causing bacteria. He identified *Staphylococcus pyogenes* bacteria in boils and *Streptococcus pyogenes* in puerperal fever. He also cultured the bacteria that cause cholera. Once when injecting healthy chickens with cholera bacteria, he expected the chickens to get sick. Unknown to Pasteur, the bacteria were old and no longer virulent. The chickens failed to get the disease, but instead they received immunity against cholera. Thus Pasteur discovered that weakened microbes make a good vaccine by imparting immunity without actually producing the disease.

Pasteur then began work on a vaccine for anthrax, a disease that killed many animals and infected people who contracted it from their sheep and thus was known as "woolsorters' disease." Anthrax causes sudden chills, high fever, pain, and can affect the brain. Pasteur experimented with weakening or attenuating the bacteria that cause anthrax, and in 1881 produced a vaccine that successfully prevented the deadly disease.

Pasteur's last great scientific achievement was developing a successful treatment for rabies, a deadly disease contracted from bites of an infected, rabid dog. Rabies, or hydrophobia, first causes terrible pain in the throat that prevents swallowing, then brings on spasms, fever, and finally death. Pasteur knew that rabies took weeks or even months to become active. He hypothesized that if people were given an injection after being bitten, it could prevent the disease from manifesting. After methodically producing a rabies vaccine from the spinal fluid of infected rabbits, Pasteur sought to test it. In 1885 nine-year-old Joseph Meister, who had been mauled and bitten by a rabid dog, was brought to Pasteur, and after a series of shots of the new rabies vaccine, the boy did not develop any of the deadly symptoms of rabies. Pasteur's triumphant success was a great relief to many worldwide.

To treat cases of rabies, the Pasteur Institute was established in 1888 with monetary donations coming from all over the world. It later became one of the most prestigious biological research institutions in the world. When Pasteur died in 1895 he was well-recognized for his outstanding achievements in science.

PATTERSON, CLAIRE (1922-)

American geochemist

Claire Patterson's contributions to science ranged from developing dating techniques using radioactive decay to alerting the public about the dangers of **lead**, although he began his career as an emission and **mass** spectroscopist with the **Manhattan Project** in 1944. Much of Patterson's professional career was spent as a research associate of environmental science at California Institute of Technology (Caltech) until his retirement in 1992. His early studies focused on using isotopes to date rocks—a process that enabled him to estimate the age of the earth. Later research on lead levels in the environment and their dangers to public health proved fundamental to the enactment of the Clean Air Act in 1970. He was elected to the National Academy of Sciences in 1987; his other honors include having a mountain and an asteroid named after him.

Claire Cameron Patterson was born in Des Moines, Iowa, on June 2, 1922, to Claire Cameron Patterson, a rural letter carrier, and his wife Vivian Ruth (Henny) Patterson. He obtained an A.B. in **chemistry** in 1943 from Grinnell College in Iowa, an M.S. in 1944 from the University of Iowa, and a Ph.D. in 1951 from the University of Chicago. Patterson met his wife Lorna (McCleary) Patterson while at Grinnell; the two married in 1944 and had two sons and two daughters. Lorna worked as a high school chemistry and physics teacher until her retirement.

Early in his career at Caltech, Patterson conducted studies to determine the age of various rocks. He did this by using the radioactive elements **uranium** and **thorium** to measure the rate of decay of an element's isotopes (forms of atoms) into lead. Decay is the process whereby rocks and minerals are transformed into new, more stable chemical combinations. While studying meteorites using his method, Patterson solved a problem that had mystified scientists for over 300 years—he calculated the age of the earth. Patterson determined the earth to be about 4.6 billion years old. Estimates prior to his 1953 calculations put the earth at approximately three billion years old. Subsequent research by others has not overturned Patterson's findings.

In 1973, twenty years after his historic finding, Patterson was elected to the National Academy of Science (NAS), one of the highest honors to be bestowed on a U.S. scientist or engineer. Caltech vice president and provost, geophysicist Barclay Kamb, as quoted in a California Institute of Technology press release, commented on the occasion by heralding Patterson's "thinking and imagination [as] so far ahead of the times that he has often gone misunderstood and unappreciated for years, until his colleagues finally caught up and realized he was right."

Patterson was often recognized as a scientist who did not follow the status quo. This reputation was gained in part by his inclination to challenge the findings of his peers and his disdain for what he calls "ivory-tower scientists." In a 1990 interview with the *Los Angeles Times,* Patterson asserted that science as it is practiced in United States has lost its imagination and creativity. It was this view of progress that prompted novelist Saul Bellow to pattern his disillusioned scientist in *The Dean's December* after Patterson.

Patterson's renegade stature may have originated during World War II when, as a member of the Manhattan Project, he contributed to the development of the atomic bomb by analyzing the uranium isotopes that went into the bombs. He told the *Los Angeles Times* that developing such a destructive force was "the greatest crime that science has committed yet," and that science, which "remains an abstract, beautiful refuge within the mind," is often misguided.

Continuing his work with isotopes, Patterson began studying the evolution of the earth's crust from the mantle (an area between the earth's outer crust and its inner core). During this research, he discovered that millions of years ago the amount of lead stored in plankton (microscopic plant and animal life) and in ocean sediments was only 1/10th to 1/100th the amount of lead now being introduced into the ocean by rivers. Lead, the end product of the decay of radioactive minerals, is normally one of the rarest elements in ocean **water**.

Interested in this dramatic increase in the ocean's lead levels, Patterson set out to measure the amount of lead present in the atmosphere, the polar ice caps, and other oceans to see if they showed similar shifts. In December of 1963, Patterson, Tsaihwa J. Chow of University of California's Scripps Institution of Oceanography, and M. Tatsumoto of the U.S. Geological Survey in Denver reported the results of a study in which they had sampled waters of the Pacific and Atlantic oceans and the Mediterranean sea. The scientists determined that nearly 500,000 tons of lead were annually entering the oceans in the Northern Hemisphere. But more startling than the amount was the fact that quantities of lead this high were apparently a new phenomenon; for centuries the rate had remained at 10,000 tons per year.

To arrive at these estimates, small amounts of a particular form of lead (an **isotope** called lead 208), called a tracer, were added to lead-free containers. When the container was submerged in the ocean, the water mixed with the tracer lead. Using chemicals, the scientists removed the tracer lead, and from the residue they were able to calculate the proportions of tracer lead to the lead already present in the sea water.

The three geochemists proposed that the increase in lead was the result of automobile exhaust emissions, which introduced small particles of lead **oxide** into the atmosphere. From the 1930s to the 1960s, the lead content of **gasoline** increased from near zero to about 175,000 tons a year in the United States alone. Winds carried vehicles' exhaust particles through the atmosphere to the world's rivers and oceans either as dry particles or in the form of rain. Once thought a natural, relatively benign part of the environment, lead was being consumed by humans in hundreds of times the quantities taken in by pre-industrial people, Patterson's studies revealed. These high levels of lead were contributing to damage to bones, kidneys, blood circulation, and brain cells, especially in children. Armed with Patterson's research, environmentalists and scientists successfully lobbied for passage of the Clean Air Act of 1970. Enactment of the legislation reduced lead emissions from cars and trucks by about 96%, but Patterson remained concerned about the quantity of lead remaining in the atmosphere, which may continue to affect public health for years to come.

Patterson also determined that biological laboratories are highly contaminated with industrial lead. This finding had wide ramifications because it implied that previous biochemical knowledge was based on studies of biological systems that were grossly contaminated by lead. Patterson believed that knowledge of the biochemistries of natural systems unaffected by lead toxicity probably do not exist.

As a result, Patterson formulated strict and sterile laboratory procedures which since have become standard in the field. To protect his experiments from lead drifting through the air, Patterson pressurized and lined his lab with plastic. Before entering his lab, he washed his hands with distilled water and put on a lab coat and surgical cap.

Patterson, with the help of graduate students, post doctoral fellows, and research associates from all over the United States and the world, has focused on delineating the extent to which industrial lead during the past two centuries has altered natural biogeochemical cycles of lead. They have measured lead in the earth's atmosphere, hydrosphere (the water on Earth and in the atmosphere), and ecosystems on both land and water. Patterson's research has demonstrated that the magnitude of global **pollution** can be determined only after pre-industrial chemical levels are established. To do this, Patterson analyzed lead concentrations in buried skeletal remains of pre-Colombian Southwest Native Americans. The natural level of lead in the remains was found to be about 1/1000th of the average amount of lead in twentieth-century adult Americans.

Patterson's most recent work has involved investigating the factors that encouraged humans to poison the earth's biosphere with lead. To do this, he studied the development of lead technologies and production over the past 10,000 years. In the process he utilized isotopes to determine methods of medieval metallurgy (the science of making and manipulating metals). Another alarming statistic he uncovered was that tuna packed in cans sealed with lead contains 10,000 times more lead than tuna from the pre-industrial era. This study from 1980 also found that lead pollution in United States and its effects on public health have been vastly underestimated.

Based on his study, Patterson proposed that there is a feedback link (a self-perpetuating mechanism) between the development of engineering technologies and the emergence of social institutions that define a culture. He reports that the geometries of the brains of those who lived in cultures that used **metals** and alloys for utilitarian purposes appear to be different from those who used lead for artistic purposes.

Despite Patterson's criticism of some scientific institutions' methods and focus, he has received many honors. These include the J. Lawrence Smith medal given by the National Academy of Science in 1975, the V.M. Goldschmidt medal given by the Geochemical Society in 1980, and the Professional Achievement award given by the University of Chicago in 1980.

PAULI EXCLUSION PRINCIPLE

In 1913, **Niels Bohr** proposed a model of the **hydrogen atom** that introduced the concept of quantized **energy** levels. Bohr assumed that the energy which an **electron** in an atom may have is quantized, i.e., it can possess only certain amounts of energy. Bohr assumed that electrons rotate around the **nucleus** in orbits with distances from the nucleus related to a quantum number n. The lowest quantized **energy level** has n=1; the next highest has n=2, etc. The properties that are predicted by Bohr's model match the experimentally observed values.

Efforts were begun immediately to extend Bohr's model of the hydrogen atom, which has only one electron, to multi-electron atoms.

The electrons of an atom will assume a configuration which has the lowest possible energy, called the ground state. In the hydrogen atom, the single electron is located in the orbit for which the quantum number n=1. **Helium** has two electrons. In its ground state, both electrons are located in the lowest energy orbit with n=1.

The **lithium** atom has three electrons. Again it might be assumed that in the ground state, the lowest energy configuration would require all three electrons to be in the n=1 orbit. Experiment indicates, however, that while two electrons are in the n=1 orbit, the third is in the orbit corresponding to n=2. Atoms of the element **beryllium** have four electrons, two with n=1 and two with n=2.

For **boron**, with 5 electrons, two have n=1, two have n=2, and we might suppose that the fifth will have n=3. But for some reason, unexplained by the Bohr model, the energy of this electron is much closer to the n=2 level than to that of n=3. A more sophisticated theory of **atomic structure** was needed.

In 1923, Louis de Broglie (1892-1987) proposed that particles possess wave properties. In 1925, to explain why all electrons in an atom will not be in the lowest n=1 energy state, Wolfgang Pauli (1900-1958) proposed that no two electrons can simultaneously have the same wave properties. This proposal evolved into the Pauli exclusion principle.

In 1926, **Erwin Schrödinger** (1887-1961) proposed a theory, known as wave mechanics, to explain the properties of particles such as electrons. The solutions of the theory's wave equation for hydrogen match exactly the predictions of the **Bohr theory**. Moreover, Schrödinger's wave mechanics has been successfully applied to systems with more than one electron.

Each **solution** of Schrödinger's wave mechanical equation for atomic systems, called an orbital, has a set of four quantum numbers associated with it. The principal quantum number n may have any positive, non-zero integer value. It is the primary determinant of the allowed energy levels of electrons. The angular momentum quantum number l may have only integer values from 0 to (n-1). The energy of an orbital depends slightly on the value of l. The magnetic quantum number m(sub)l may have any of the integer values -l to +l (including 0); and the spin quantum number s may have only the values +1/2 and -1/2. Each orbital has a different set of these four quantum numbers.

The Pauli exclusion principle may now be stated as requiring that no two electrons in an atom may occupy the same orbital, i.e. may simultaneously possess the same values of the four quantum numbers. Since the spin quantum number may have only the values +1/2 and -1/2, and since the value of spin affects neither the energy nor the rest of the wave function, we often speak of each orbital containing two electrons, one with spin +1/2 and the other with spin -1/2.

The quantum numbers corresponding to the possible energy levels of an atom are as follows (with energy increasing as the value of n and l increase):

There is only one orbital with n=1, because when n=1, l=0 and m=0. This is called the 1s orbital and two electrons may occupy it (one with spin +1/2 and one with spin -1/2).

There are five orbitals with n=2: a 2s orbital with n=2, l=0, m=0, containing two electrons; three 2p orbitals, of slightly higher energy, for which n=2, l=1, and m has the values -1, 0, and +1; each of these p orbitals may contain two electrons; six electrons may thus be accommodated in these p orbitals.

There are nine orbitals with n=3: a 3s orbital with two electrons; three 3p orbitals of slightly higher energy that may contain a total of six electrons; and five 3d orbitals with n=3, l=2, and m has values of -2, -1, 0, +1, and +2. The 3d orbitals have slightly higher energy than the 3p orbitals, and may contain a total of ten electrons.

Similar orbital schemes may be constructed for larger values of the principal quantum number.

Recall that the ground state of an atom is the state of lowest energy, with the electrons of the atom occupying the orbitals of lowest energy.

The lone electron of the hydrogen atom will occupy the 1s orbital in the ground state. The two electrons of the helium atom are both in the 1s orbital, one with spin +1/2, the other with spin -1/2. The third electron of lithium must have n=2 since there are no more unfilled orbitals with n=1. Both the third and fourth electrons of beryllium reside in the 2s orbital, each with a different spin quantum number. The fifth electron of boron cannot be accommodated in the filled 2s orbital; the next lowest orbital is 2p.

Continuing this process, using the Pauli exclusion principle, the ground state electron configuration of the elements of the **periodic table** can be explained. For instance, a **chlorine** atom has seventeen electrons. In the ground state, they will be located in the orbitals of lowest possible energy: two electrons in the 1s orbital, two in 2s, six in 2p, two in 3s, and five in 3p. A **bromine** atom has thirty-five electrons, located as follows: two in 1s, two in 2s, six in 2p, two in 3s, six in 3p, ten in 3d, two in 4s, and five in 4p.

The electrons of molecular bonds can be understood in a similar way. Instead of orbitals located on a single atom, the wave theory treatment of the electrons of molecular bonds results in a set of molecular orbitals which spread over two or more atomic centers. The electrons that are shared by the atoms to form the bond belong to one of these molecular orbitals. The ground state of the **molecule** is that in which the **bonding** electrons are in the molecular orbitals of lowest energy. Again, the Pauli exclusion principle holds: only two electrons may be located in each molecular orbital, each with a different spin.

In addition to electrons, the Pauli exclusion principle applies to all sub-atomic particles with half-integral spins, known as fermions, such as neutrons and protons. The Pauli exclusion principle does not apply, however, to particles with integral spin, known as bosons, such as photons.

Linus Pauling. *(AP/Wide World. Reproduced by permission.)*

PAULING, LINUS (1901-1994)

American chemist

Linus Pauling is the only person ever to win two unshared Nobel Prizes. His 1954 prize for **chemistry** was given in recognition of his work on the nature of the chemical bond, while the 1963 Nobel Peace Prize was awarded for his efforts to bring about an end to the atmospheric testing of nuclear weapons. Pauling has made important contributions to a number of fields, including the structure of **proteins** and other biologically important molecules, **mineralogy**, the nature of mental disorders, nuclear structure, and **nutrition**.

Linus Carl Pauling was born in Portland, Oregon, on February 28, 1901. He was the first of three children born to Herman Henry William Pauling and Lucy Isabelle (Darling) Pauling, usually called Belle. Herman Pauling was a druggist who struggled continuously to make a decent living for his family. With his business in Portland failing, Herman moved his family to Oswego, seven miles south of Portland, in 1903. He was no more successful in Oswego, however, and moved on to Salem in 1904, to Condon (in northern Oregon) in 1905, and finally back to Portland in 1909. A year later, Herman died of a perforated ulcer, leaving Belle to care for the three young children.

Linus was a precocious child who read every book he could get his hands on. At one point, his father wrote the local paper asking for readers to suggest additional books that would keep his son occupied. Pauling's interest in science was apparently stimulated by his friend Lloyd Jeffress during grammar school. Jeffress kept a small chemistry laboratory in a corner of his bedroom where he performed simple experiments. Pauling was intrigued by these experiments and decided to become a chemical engineer.

During his high school years, Pauling continued to pursue his interest in chemistry. He was able to obtain much of the equipment and materials he needed for his experiments from the abandoned Oregon Iron and Steel Company in Oswego. Since his grandfather was a night watchman at a nearby plant, Linus was able to "borrow" the items he needed for his own chemical studies. Pauling would have graduated from Portland's Washington High School in 1917 except for an unexpected turn of events. He had failed to take the necessary courses in American history required for graduation and so did not receive his diploma. The school corrected this error 45 years later when it awarded Pauling his high school diploma—after he had been awarded two Nobel Prizes.

In the fall of 1917, Pauling entered Oregon Agricultural College (OAC), now Oregon State University, in Corvallis. He was eager to pursue his study of science and signed up for a full load of classes. Finances presented a serious problem, however. His mother was unable to send him any money for his expenses and, in fact, Pauling was soon forced to help her pay family bills at home. As a result, Pauling regularly had to work 40 hours or more in addition to studying and attending classes. By the end of his sophomore year, it became apparent that Pauling could not afford to stay in school. He decided to take a year off and help his mother by working in Portland. At the last minute, however, those plans charged. OAC offered him a job teaching **quantitative analysis**, a course he had completed as a student only a few months earlier. The $100-a-month assignment allowed him to return to OAC and continue his education.

During his last two years at OAC, Pauling learned about the work of **Gilbert Newton Lewis** and **Irving Langmuir** on the electronic structure of atoms and the way atoms combine with each other to form molecules. He became interested in the question of how the physical and chemical properties of substances are related to the structure of the atoms and molecules of which they are composed and decided to make this topic the focus of his own research. An important event during Pauling's senior year occurred when he met Ava Helen Miller, a student in one of his classes. Ava Helen and Linus were married on June 17, 1923, and later had four children: Linus, Jr., born in 1925, Peter Jeffress, born in 1931, Linda Helen, born in 1932, and Edward Crellin, born in 1937.

After graduation from OAC in 1922, Pauling entered the California Institute of Technology (Caltech) in Pasadena. At the time, Caltech was a vigorous, growing institution that served as home to some of the leading researchers in chemistry and physics in the nation. Pauling quickly became immersed in a heavy load of classes, seminars, lectures, and original research. He was assigned to work with Roscoe Gilley Dickinson on the x-ray analysis of crystal structure. As a result of this research, his first paper, "The Crystal Structure of Molybdenite,'' was published in the *Journal of the American Chemical Society* (*JACS*) in 1923. During his remaining years at Caltech, Pauling was to publish six more papers on the structure of other minerals.

Pauling was awarded his Ph.D in chemistry *summa cum laude* in 1925 and decided to continue his studies in Europe. He planned to spend time with three leading researchers of the time, **Arnold Sommerfeld** in Munich, **Niels Bohr** in Copenhagen, and **Erwin Schrödinger** in Zurich. Sommerfeld, Bohr, and Schrödinger were all working in the new field of quantum mechanics. The science of quantum mechanics, less than a decade old at the time, was based on the revolutionary concept that particles can sometimes have wave-like properties, and waves can sometimes best be described as if they consisted of massless particles. Pauling had been introduced to quantum mechanics at OAC and was eager to see how this new way of looking at **matter** and **energy** could be applied to his own area of interest, the electronic structure of atoms and molecules. After two years in Europe, Pauling was determined to make this subject the focus of his future research. He and Ava Helen left Zurich in the summer of 1927 and returned to Caltech, where Pauling took up his new post as Assistant Professor of Theoretical Chemistry.

Pauling's first few years as a professor at the college were a time of transition. He continued to work on the x-ray analysis of crystals, but also began to spend more time on the quantum mechanical study of atoms and molecules. He wrote prolifically, turning out an average of ten papers a year during his first five years at Caltech. His reputation grew apace, and he was promoted quickly to associate professor in 1929 and then to full professor a year later. By 1931, the American Chemical Society had awarded Pauling its Langmuir Prize for ''the most noteworthy work in pure science done by a man 30 years of age or less.''

In the meanwhile, Pauling had spent another summer traveling through Europe in 1930, visiting the laboratories of Laurence Bragg in Manchester, **Herman Mark** in Ludwigshafen, and Sommerfeld in Munich. In Ludwigshafen, Pauling learned about the use of **electron** diffraction techniques to analyze crystalline materials. Upon his return to Cal Tech, Pauling showed one of his students, L. O. Brockway, how the technique worked and had him build an **electron diffraction** instrument. Over the next 25 years, Pauling, Brockway, and their colleagues used the diffraction technique to determine the **molecular structure** of more than 225 substances.

In some ways, the 1930s mark the pinnacle of Pauling's career as a chemist. During that decade he was able to apply the principles of quantum mechanics to solve a number of important problems in chemical theory. The first major paper on this topic, ''The Nature of the Chemical Bond: Applications of Results Obtained from the Quantum Mechanics and from a Theory of Paramagnetic Susceptibility to the Structure of Molecules,'' appeared in the April 6, 1931 issue of *JACS*. That paper was to be followed over the next two years by six more on the same topic and, in 1939, by Pauling's magnum opus, *The Nature of the Chemical Bond, and the Structure of Molecules and Crystals*. The book has been considered one of most important works in the history of chemistry, and the ideas presented in the book and the related papers are the primary basis upon which Pauling was awarded the Nobel Prize for chemistry in 1954.

Pauling's work on the chemical bond proved crucial in understanding three important problems in chemical theory: bond **hybridization**, bond character, and **resonance**. Hybridization refers to the process by which electrons undergo a change in character when they form bonds with other atoms. For example, the **carbon atom** is known to have two distinct kinds of **bonding** electrons known as 2s and 2p electrons. Traditional theory had assumed, therefore, that carbon would form two types of bonds, one type using 2s electrons and a second type using 2p electrons. Studies had shown, however, that all four bonds formed by a carbon atom are identical to each other. Pauling explained this phenomenon by illustrating that carbon's electrons change their character during bonding. All four assume a new energy configuration that is a hybrid of 2s and 2p energy levels. Pauling called this hybrid an sp^3 **energy level**.

In a second area, Pauling examined the relationship between two kinds of chemical bonding: ionic bonding, in which electrons are totally gained and lost by atoms, and covalent bonding, in which pairs of electrons are shared equally between atoms. Pauling was able to show that ionic and covalent bonding are only extreme states that exist in relatively few instances. More commonly, atoms bond by sharing electrons in some way that is intermediate between ionic and covalent bonding.

Pauling's third accomplishment involved the problem of the **benzene molecule**. Until the 1930s, the molecular structures that were being written for benzene did not adequately correspond to the properties of the substance. The German chemist **Friedrich Kekulé** had proposed a somewhat satisfactory model in 1865 by assuming that benzene could exist in two states and that it shifted back and forth between the two continuously. In his work on the chemical bond, Pauling suggested another answer. Quantum mechanics showed, he said, that the most stable form of the benzene molecule was neither one of Kekulé's structures, but some intermediate form. This intermediate form could be described as the superposition of the two Kekulé structures formed by the rapid interconversion of one to the other. This ''rapid interconversion'' was later given the name resonance.

By the mid-1930s, Pauling was looking for new fields to explore. The questions he soon addressed concerned the structure of biological molecules. These molecules are complex substances that are found in living organisms and can contain thousands of atoms in each molecule rather than the relatively simple molecules that Pauling had studied previously that contained only twenty or thirty atoms. This was a surprising choice for Pauling, because earlier in his career he had mentioned that he wasn't interested in studying biological molecules. One reason for his change of heart may have been the changes taking place at Caltech itself. In an effort to expand the institution's mission beyond chemistry and physics, its administration had begun to build a new department of biology. Among those recruited for the department were such great names as Thomas Hunt Morgan, Theodosius Dobzhansky, Calvin Bridges, and Alfred Sterdevant. Pauling's almost daily interaction with these men opened his eyes to a potentially fascinating new field of research.

The first substance that attracted his attention was the hemoglobin molecule. Hemoglobin is the substance that transports **oxygen** through the bloodstream. Pauling's initial work with hemoglobin, carried out with a graduate student Charles Coryell, produced some fascinating results. Their research showed that the hemoglobin molecule undergoes significant structural change when it picks up or loses an oxygen atom. In order to continue his studies, Pauling decided he needed to know much more about the structure of hemoglobin, in particular, and proteins, in general.

Fortunately, he was already familiar with the primary technique by which this research could be done: **x-ray diffraction** analysis. The problem was that x-ray diffraction analysis of protein is far more difficult than it is for the crystalline minerals Pauling had earlier worked with. In fact, the only reasonably good x-ray pictures of protein available in the 1930s were those of the British crystallographer, **William Astbury**. Pauling decided, therefore, to see if the principles of quantum mechanics could be applied to Astbury's photographs to obtain the molecular structures of proteins.

The earliest efforts along these lines by Pauling and Coryell in 1937 were unsuccessful. None of the molecular structures they drew based on quantum mechanical principles could account for patterns like those in Astbury's photographs. It was not until eleven years later that Pauling finally realized what the problem was. The mathematical analysis and the models it produced were correct. What was wrong was Astbury's patterns. In the pictures he had taken, protein molecules were tilted slightly from the position they would be expected to have. By the time Pauling had recognized this problem, he had already developed a molecular model for hemoglobin with which he was satisfied. The model was that of a helix, or spiral-staircase-like structure in which a chain of atoms is wrapped around a central axis. Pauling had developed the model by using a research technique on which he frequently depended: model building. He constructed atoms and groups of atoms out of pieces of paper and cardboard and then tried to fit them together in ways that would conform to quantum mechanical principles. Not surprisingly, Pauling's technique was also adopted by two contemporaries, **Francis Crick** and **James Watson**, in their **solution** of the **DNA** molecule puzzle, a problem that Pauling himself very nearly solved.

Pauling also turned his attention to other problems of biological molecules. In 1939, for example, he developed the theory of complementarity and applied it the subject of **enzyme** reactions. He later used the same theory to explain how genes might act as templates for the formation of enzymes. In 1945, Pauling attacked and solved an important medical problem by using chemical theory. He demonstrated that the genetic disorder known as sickle-cell anemia is caused by the change of a single amino acid in the hemoglobin molecule.

The 1940s were a decade of significant change in Pauling's life. He had never been especially political and, in fact, had voted in only one presidential election prior to World War II. But he rather quickly began to immerse himself in political issues. One important factor in this change was the influence of his wife, Ava Helen, who had long been active in a number of social and political causes. Another factor was probably the war itself. As a result of his own wartime research on explosions, Pauling became more concerned about the potential destructiveness of future wars. As a result, he decided while on a 1947 boat trip to Europe that he would raise the issue of world peace in every speech he made in the future, no matter what topic.

From that point on, Pauling's interests gradually shifted from scientific to political topics. He devoted more time to speaking out on political issues, and the majority of his published papers dealt with political rather than scientific topics. In 1958 he published his views on the military threat facing the world in his book *No More War!* Pauling's views annoyed many of his scientific colleagues, fellow citizens, and many legislators. In 1952 he was denied a passport to attend an important scientific meeting in London, and in 1960 he was called before the Internal Security Committee of the United States Senate to explain his antiwar activities. Neither professional nor popular disapproval could sway Pauling's commitment to the peace movement, however, and he and Ava Helen continued to write, speak, circulate petitions, and organize conferences against the world's continuing militarism. In recognition of these efforts, Pauling was awarded the 1963 Nobel Prize for Peace.

At the age of 65, when many men and women look forward to retirement, Linus Pauling had found a new field of interest: the possible therapeutic effects of vitamin C. Pauling was introduced to the potential value of vitamin C in preventing colds by biochemist Irwin Stone in 1966. He soon became intensely interested in the topic and summarized his views in a 1970 book, *Vitamin C and The Common Cold.* Before long, he became convinced that the vitamin was also helpful in preventing cancer.

Pauling's views on vitamin C have received relatively modest support in the scientific community. Many colleagues tend to feel that the evidence supporting the therapeutic effects of vitamin C is weak or nonexistent, though research on the topic continues. Other scientists are more convinced by Pauling's argument, and he is regarded by some as the founder of the science of orthomolecular medicine, a field based on the concept that substances normally present in the body (such as vitamin C) can be used to prevent disease and illness.

Pauling's long association with Caltech ended in 1964, at least partly because of his active work in the peace movement. He accepted an appointment at the Study of Democratic Institutions in Santa Barbara for four years and then moved on to the University of California at San Diego for two more. In 1969 he moved to Stanford University where he remained until his compulsory retirement in 1974. In that year, he and some colleagues and friends founded the Institute of Orthomolecular Medicine, later to be renamed the Linus Pauling Institute of Science and Medicine, in Palo Alto.

Pauling died of cancer at his ranch in the Big Sur area of California on August 19, 1994. He was 93 years old.

PAYEN, ANSELME (1795-1871)

French chemist

Payen was the son of an entrepreneur who had started several chemical production factories. Payen studied **chemistry** first with his father and later with **Nicolas Louis Vauquelin** and **Michel-Eugène Chevreul**.

In 1815 Payen was promoted to head of his father's borax production plant. While there Payen discovered a method of preparing borax from boric acid which was readily and cheaply available from Italy. Because production costs for this method were very low, Payen was able to undersell his competitors, the Dutch, who until that time had a monopoly on borax.

Five years later Payan began investigating the production of sugar from sugar beets. He developed a process in which **charcoal** was used to decolor the sugar. Charcoal filters were later used in gas masks since **carbon** filters absorbed dangerous organic **gases**. Payan continued his research and in 1833 developed a chemical from malt extract that catalyzed the starch-to-sugar conversion. He named the organic catalyst diastase, which was the first **enzyme** produced in concentrated form. Enzymes discovered since then are named with the *ase* suffix, the pattern started by Payen.

Payen went on to study woods and during his research began to separate out a substance resembling starch. He named this substance, which he found in abundance in the cell walls of all the plants he studied, **cellulose**. Again, Payen established a naming system: **carbohydrates** end with *ose*. Although Payen was the first to isolate cellulose, it was not until the late 1930s through the work of **Wanda Farr** that it was discovered how plants produce this carbohydrate.

Cellulose, a natural polymer, was the building block of many other inventions. Treated with various acids and additives it became the main ingredient in nitrocellulose, guncotton, collodion, rayon, celluloid, cellophane, and many other products.

In 1835 Payen accepted a professorship at the Ecole Centrale des Arts et Manufactures and concentrated exclusively on research. He died in Paris during the Franco-Prussian War in 1871 after refusing to leave the city as the Prussian army advanced.

PEDERSEN, CHARLES (1904-1990)

American chemist

Charles Pedersen is credited with the discovery of crown ethers. He won a Nobel Prize in **chemistry** for this discovery but not for eighteen years after he first published his research. Pedersen was born in Korea in 1904 to a Norwegian engineer and a Japanese woman. His father was working in a **gold** mine and his mother's family had a business near the mine. Pedersen attended college in the United States. After he earned his master's degree, he found a position as industrial chemist at the Du Pont Company. He chose to continue with research, although it was not essential to his position. Pedersen was working on the problem of removing contaminants from **gasoline** when he recognized the **carbon** to **oxygen** to carbon bond of an **ether** in an unusual compound. What was unusual was the model of the ether was the zigzag of a crown and it tended to surround metal ions. This discovery was soon applied in the synthetic research to manipulate reactions, particularly in organic compounds. Pedersen died in 1990.

PELLETIER, PIERRE-JOSEPH (1788-1842)

French chemist

A pharmacist from a family of pharmacists, Pierre-Joseph Pelletier is known for his research into vegetable bases and the resulting contributions of **alkaloid chemistry** to the field of medicine. Working with Joseph-Bienaimé Caventou (1795-1877), he introduced a rational basis into alkaloid chemistry.

Pelletier began his research in 1809, became a pharmacist in 1810, received a doctor of science degree in 1812, became a professor at the École de Pharmacie in Paris in 1825, and became the school's assistant director in 1832. He experimented with gum resins and coloring **matter** in plants. In place of the harsh **distillation** methods of extraction common in laboratories at the time, he used the milder method of solvent extraction. Inspired by the German pharmacist Friedrich Sertürner's (1783-1841) investigations into poppy extracts (from which the analgesic drug **morphine** is derived), he began to examine the newly discovered class of vegetable bases, or alkaloids.

Alkaloids are organic compounds which contain **carbon**, **hydrogen** and **nitrogen** and which form water-soluble salts. They perform a variety of functions in medicine; for example, some are **analgesics** (pain-killers), and others are respiratory stimulants. The study of their chemistry provided the groundwork for further chemical discoveries.

In 1817, Pelletier and the French physiologist François Magendie isolated the alkaloid emetine. Pelletier began to work with Caventou, isolating **chlorophyll** in 1817, **strychnine** in 1818, brucine in 1819, **quinine** in 1820 and **caffeine** in 1821. They demonstrated that alkaloids contained **oxygen**, hydrogen and carbon, but did not believe nitrogen was present until 1823. Then, using elementary closed-tube analyses in which the alkaloids were combusted, they discovered that nitrogen was present in the compounds.

As medicine quickly found a place for alkaloids, a shift occurred in the field. Instead of crude plant extracts, chemists began to isolate and produce natural and synthetic alkaloid compounds for medicinal uses. For example, the bark of the cinchona tree had long been used in the treatment of malaria, but Pelletier and Caventou's isolation of the alkaloid quinine from this bark made the pure drug available, and eventually led to the synthesis of the drugs used to prevent malaria today. Their isolation of quinine earned Pelletier and Caventou the Montyon Prize from the Academy of Sciences.

Pelletier was admitted into admitted into the Academy of Sciences in 1840 and retired the same year due to illness. He died in 1842.

A micrograph of *Penicillium notatum* growing inside a petri dish. This is the species of fungus which led Alexander Fleming to discover the antibiotic penicillin. *(Photograph by Andrew McClenaghan/Science Photo Library, Photo Researchers, Inc. Reproduced by permission.)*

PENICILLIN

One of the major advances of twentieth-century medicine is the discovery of penicillin. Penicillin is a member of the class of drugs known as antibiotics. These drugs either kill (bacteriocidal) or arrest the growth of (bacteriostatic) bacteria, fungi (yeast), as well as several other classes of infectious organisms. Antibiotics are ineffective against viruses. Prior to the advent of penicillin bacterial infections such as pneumonia and sepsis (blood poisoning—the infecting organism invades the bloodstream) were usually fatal. Once the use of penicillin became widespread, fatality rates from pneumonia dropped precipitously.

The discovery of penicillin marked the beginning of a new era in the fight against disease. Scientists had known since the mid-nineteenth century that bacteria were responsible for some infectious diseases, but they were virtually helpless to stop them. Then, in 1928, Alexander Fleming (1881-1955), a Scottish bacteriologist working at St. Mary's Hospital in London, stumbled onto a powerful new weapon.

Fleming's research centered on the bacterium *Staphylococcus*, a class of bacteria that caused infections such as pneumonia, abscesses, post-operative wound infections, and sepsis. To study these bacteria he grew them in glass Petri dishes on a substance called agar, in his laboratory. In August 1928, he noticed that some of the Petri dishes in which the bacteria were growing had become contaminated with mold, which he later identified as belonging to the *Penicillum* family.

Fleming noted that bacteria in the vicinity of the mold had died. Exploring further, Fleming found that the mold killed several, but not all, types of bacteria. He also found that an extract from the mold did not damage healthy tisssue in animals. However, growing the mold and collecting even tiny amounts of the active ingredient—penicillin—was extremely difficult. Fleming did, however, publish his results in the medical literature in 1928.

Ten years later, other researchers picked up where Fleming had left off. Working in Oxford, England, a team led by Howard Florey (1898-1968), an Australian, and **Ernst Chain**, a refugee from Nazi Germany, came across Fleming's study and confirmed his findings in their laboratory. They too had problems growing the mold and found it very difficult to isolate the active ingredient

Another researcher on their team, Norman Heatley, developed better production techniques, and the team was able to produce enough penicillin to conduct tests in humans. In 1941 the team announced that penicillin could combat disease in humans. Unfortunately, producing penicillin was still a very difficult process and supplies of the new drug were extremely limited. Working in the United States, Heatley and other scientists improved production and began making large quantities of the drug. Owing to this success, penicillin was available to treat wounded soldiers by the latter part of World War II. Fleming, Florey, and Chain were awarded the Noble Prize in medicine or physiology. Heatley received an honorary M.D. from Oxford University in 1990.

Penicillin's mode of action is blocking the construction of cell walls in certain bacteria. The bacteria must be reproducing for penicillin to work, thus there is always some lag time between dosage and response.

Bacteria are divided into gram-positive and gram-negative classes based on a laboratory test called the gram-staining procedure. The gram-staining procedure uses special dyes that cause gram-negative bacteria to turn pink and gram-positive bacteria to become purplish-black. Classifying bacteria as gram-negative or gram-positive aids in identifying the bacterial species.

The mechanism of action of penicillin at the molecular level is still not completely understood. It is known that the initial step is the binding of penicillin to penicillin-binding **proteins** (PBPs) which are located in the cell wall. Some PBPs are inhibitors of cell autolytic enzymes which literally eat the cell wall and are most likely necessary during cell division. Other PBPs are enzymes which are involved in the final step of cell wall synthesis called transpeptidation. These latter enzymes are outside the cell membrane and link cell wall components together by joining glycopeptide polymers together to form peptidoglycan. The bacterial cell wall owes its strength to layers composed of peptidoglycan (also known as murein or mucopeptide). Peptidoglycan is a complex polymer composed of alternating N-acetylglucosamine and N-acetylmuramic acid as a backbone off of which a set of identical tetrapeptide side chains branch from the N-acetylmuramic acids, and a set of identical peptide cross-bridges also branch. The tetrapeptide side chains and the cross-bridges vary from species to species, but the backbone is the same in all bacterial species.

Each peptidoglycan layer of the cell wall is actually a giant polymer molecule because all peptidoglycan chains are cross-linked. In gram-positive bacteria there may be as many as 40 sheets of peptidoglycan, making up to 50% of the cell wall material. In gram-negative bacteria, there are only one or two sheets (about 5-10% of the cell wall material). In general, penicillin G, that is, the penicillin that Fleming discovered, has high activity against gram-positive bacteria and low activity against gram-negative bacteria (with some exceptions).

Penicillin acts by inhibiting peptidoglycan synthesis by blocking the final transpeptidation step in the synthesis of peptidoglycan. It also removes the inactivator of the inhibitor of autolytic enzymes, and the autolytic enzymes then lyse the cell wall, and the bacterium ruptures. This latter is the final bacteriocidal event.

Since the 1940s, many other antibiotics have been developed. Some of these are based on the **molecular structure** of penicillin, but others are completely unrelated. At one time it was believed that science had conquered bacterial infections. However, in the late twentieth century, bacterial resistance to antibiotics—including penicillin—was recognized as a potential threat to this success. A classic example is the *Staphylococcus* bacterium, the very species Fleming had found killed by penicillin on his petri dishes. In 1999 essentially all *Staphylococcus* bacteria are resistant to penicillin G. Continuing research so far has been able to keep pace with emerging resistant strains of bacteria. We must be judicious about our use of antibiotics, however, so that we continue to be able to combat bacterial disease.

See also Antibiotic drugs; Biological membranes; Enzyme

Pennington, Mary Engle (1872-1952)

American chemist

Mary Engle Pennington was a bacteriological chemist who revolutionized methods of storing and transporting perishable foods. Denied a B.S. degree in 1895 because she was a woman, Pennington went on to head the U.S. Department of Agriculture's food research lab. As persuasive as she was resourceful, Pennington was able to convince farmers, manufacturers, and vendors to adopt her techniques. She developed methods of slaughtering poultry that kept them fresh longer, discovered ways to keep milk products from spoiling, and determined how best to freeze fruits and vegetables. Pennington was the first female member of the American Society of Refrigerating Engineers. She eventually went into business for herself as a consultant and investigator in the area of perishable foods.

Pennington was born October 8, 1872, in Nashville, Tennessee. She was the first of two daughters born to Henry and Sarah B. Molony Pennington. Pennington spent most of her early life in Philadelphia, where her family moved to be closer to their Quaker relatives. With her father, a successful label manufacturer, she shared a love of gardening.

Pennington found her way to the field of **chemistry** through a library book on that subject. Her interest prompted her to enter the Towne Scientific School of the University of Pennsylvania, an uncommon occurrence for a woman at that time. In 1895 she received a certificate of proficiency, having been denied a B.S. because of her gender. Not to be deterred, Pennington continued academic work, earning her Ph.D. at age twenty-two from the University of Pennsylvania with a major in chemistry and minors in zoology and botany. This degree was conferred under an old statute that made exceptions for female students in "extraordinary cases." Pennington then accepted a two-year fellowship at the university in chemical botany, followed by a one-year fellowship in **physical chemistry** at Yale.

From 1898 to 1906 Pennington served as instructor in physiological chemistry at Women's Medical College. During this same period, she started and operated a clinical laboratory performing analyses for physicians, and was a consultant to Philadelphia regarding the storage of perishable foods during the marketing process. Her reputation for quality work led to an appointment as head of the Philadelphia Department of Health and Charities Bacteriological Laboratory. One of her first goals here was the improvement of the quality of milk and milk products. Her natural gift of persuasion aided her in convincing ice-cream manufacturers and vendors to adopt simple steps to help avoid bacterial contamination of their foods.

The Pure Food and Drug Act was passed in 1906, and the U.S. Department of Agriculture planned to establish a research laboratory to help provide scientific information for prosecutions under the act. Specifically, this lab would be concerned with the quality of eggs, dressed poultry, and fish. With the encouragement of Harvey W. Wiley, chief of the chemistry section of the USDA and a longtime family friend, Pennington took and passed the civil service exam in 1907 under the name M. E. Pennington. Unaware that Pennington was a woman, the government gave her a post as bacteriological chemist. Wiley promoted her to head the food research lab in 1908. That same year she delivered an address for Wiley to a startled all-male audience at the First International Congress of Refrigeration.

During Pennington's tenure the laboratory effected alterations in the warehousing of food, its packaging, and use of refrigeration in transport. Pennington eventually developed techniques that were commonly used for the slaughter of poultry, ensuring safe transport and high quality long after the butchering occurred. In the area of eggs, a highly perishable item especially in warm weather, she again used her powers of persuasion. She worked to convince farmers to collect and transport eggs more frequently during warmer weather. She is also credited with developing the egg cartons that prevent excessive breakage during transport.

During World War I, Pennington consulted with the War Shipping Administration. The United States had forty thousand refrigerated cars available for food transport at the start of the war. Pennington determined only three thousand of these were truly fit for use, with proper air circulation. Following the war she was recognized for her efforts with a Notable Service Award given by Herbert Hoover.

Pennington made another career change in 1919 when she accepted a position as manager of research and development for New York's American Balsa Company, a manufacturer of insulating material. In 1922 she made her final career move, starting her own business in New York as a consultant and investigator in the area of perishable foods. She was particularly interested in frozen foods, helping to determine the best strains of fruits and vegetables for freezing, and the best method for freezing them.

Pennington was the author of books, articles, pamphlets, and several government bulletins. She gave many addresses and was the recipient of several awards, including the American Chemical Society's 1940 Garvan Medal to honor a woman chemist of distinction. Pennington, in fact, was one of the first dozen female members of the society. She was the first female member of the American Society of Refrigerating Engineers, and the first woman elected to the Poultry Historical Society's Hall of Fame. She served as director of the Household Refrigeration Bureau of the National Association of Ice Industries from 1923 to 1931.

Pennington earned herself the reputation for always producing quality work. She was accepted in industry even while she was working for the government in enforcing the Pure Food and Drug Act. She maintained her interest in gardening and botany, growing flowers in her apartment. She was a lifelong member of the Quaker Society of Friends. Pennington, who never married, was still working as a consultant and as vice president of the American Institute of Refrigeration when she died on December 27, 1952, in New York City.

PENTYL GROUP

Unbranched alkanes are compounds containing only **carbon** and **hydrogen** atoms in continuous **chains**. Branched-chain alkanes, on the other hand, contain short carbon chains (known as alkyl groups) attached to longer carbon skeletons. The convention is to name the **alkyl group** using the Greek word for the number of carbon atoms present. Thus, the alkyl group with five carbons is referred to as a pentyl group, which is the name obtained by dropping the *-ane* suffix from the five-carbon unbranched chain (pentane, having the **structural formula** $CH_3CH_2CH_2CH_2CH_3$), and substituting a *-yl* ending.

The pentyl group (also known as an amyl group), therefore, corresponds to the five-carbon aliphatic group C_5H_{11}. This group has eight possible constitutional as isomeric arrangements (excluding optical isomers); the structure of the simplest of these, i.e., the n-pentyl group, is $CH_3CH_2CH_2CH_2CH_2$ (where the n- prefix is an abbreviation for "normal." Compounds containing pentyl groups usually occur as as mixtures of several isomers. Because the as boiling points of these compounds are close and their other properties are similar, the expense and difficulty of separating them is seldom warranted. Amyl compounds, i.e., those containing pentyl groups, exist in fusel oils; they can also be formed from petroleum pentanes.

The amyl alcohols ($C_5H_{11}OH$), which are among the simplest of compounds containing pentyl groups, have the following eight isomeric structures (excluding optical isomers): n-amyl **alcohol**, primary or $CH_3(CH_2)_4OH$; 2-methyl-1-butanol or $CH_3CH_2CH(CH_3)CH_2OH$; isoamyl alcohol, primary or $(CH_3)_2CHCH_2CH_2OH$; 2-pentanol or $CH_3CH_2CH_2CH_2OCH_3$; 3-pentanol or $CH_3CH_2CHOHCH$ $_2CH_3$; tert-amyl alcohol or $(CH_3)_2C(OH)CH_2CH_3$; sec-isoamyl alcohol or $CH_3CHOHCH(CH_3)_2$; and 2,2-dimethyl-1-propanol or $(CH_3)_3CCH_2OH$. The isomers n-amyl alcohol, primary; 2-pentanol; and 3-pentanol exist as linear, unbranched chains of carbon and hydrogen atoms, whereas the isomers 2 methyl-1-butanol;isoamyl alcohol, primary; tert-amyl alcohol; sec-isoamyl alcohol;and 2,2 dimethyl-1 propanol are branched compounds.The isomers 2-methyl-1-butanol; 2-pentanol; and sec-isoamyl alcohol are chiral and exhibit **optical activity**.

See also Petrochemicals

PEPTIDES

A peptide is an organic **molecule** consisting of two or more **amino acids** linked through an amide linkage called a peptide bond. Peptides are formed when the amino group ($-NH_2$) of one amino acid, links with the **carboxyl group** (-COOH) of an adjacent amino acid. An amide linkage (-CO-NH-) results. If two amino acids are linked, the compound is called a dipeptide, if three, a tripeptide, and so on. Longer **chains** containing a few amino acids are often called oligopeptides or polypeptides if they contain as many as 50. Still longer chains are called **proteins**. Peptides may have many different properties depending on the nature of the amino acids they contain, and the order in which they are linked. Some peptides function as **hormones**—a few as antibiotics and others as important participants in **metabolism**.

For example, an octapeptide with a hormone-like function, called angiotensin, is formed in the blood and targets the circulatory system, causing an increase blood **pressure**. Oxytocin, useful in obstetrics because it promotes uterine contraction, is an unusual form because it contains nine amino acids arranged in a ring structure. Two peptides, made by bacteria, with antibiotic activity, include gramicidin, which inhibits oxidative phosphorylation (the synthesis of adenosine triphosphate [ATP] that obtains **energy** from **electron** transport occurring as a result of aerobic respiration), and bacitracin, used for treatment of bacterial infections of the eye.

Several peptides have neuroactive (stimulating neural tissues) properties and function similarly to other neurotransmitters but with some important differences. Classic transmitters like acetylcholine act on synaptic receptors (a cellular entity that is the mechanism for genetic cross over) for only milliseconds, while neuropeptides may act for seconds, hours, or even days. Neuropeptides are released in lower concentrations than other transmitters, and they have higher potency. Examples of neuropeptides include neurotensin, somatostatin, cholecystokinin, and the opioid peptides. Opioid peptides, including endorphins, enkephalins, and dynorphins, have been widely studied because opiate drugs like **morphine** bind (latch on) to the same receptor sites and mimic their pain-killing action.

Peptides are formed routinely in the digestive tract of animals when protein is broken down. As peptide bonds in the protein are hydrolyzed, the complex molecule is first converted to a number of smaller peptides. The peptides are subsequently converted to individual amino acids. Two enzymes involved in this digestive process are pepsin, released in the acid environment of the stomach, and trypsin, secreted by the pancreas and active in the duodenum (part of the small intestine) and upper jejunum (part of the small intestine just beyond the duodenum).

See also Cellular respiration; Enzyme; Neurochemistry

Perey, Marguerite (1909-1975)

French physicist

Marguerite Perey is best known for her discovery of **francium**, the 87th element in the **periodic table**. Francium, a rare, highly unstable, radioactive element, is the heaviest chemical of the alkali metal group. Perey's work on francium and on such scientific occurrences as the **actinium** radioactive decay series led to her admission to the French Academy of Sciences. Perey was the first woman to be admitted to the two-hundred-year-old Academy—even **Marie Curie** had been unable to break the sex barrier.

Marguerite Catherine Perey was born in Villemomble, France, in 1909. As a child, she showed an interest in science and wanted to become a doctor. Her father's early death, however, left her family without the resources for such an education. Nonetheless, Perey was able to study physics and showed a talent for scientific endeavors. Because of her technical prowess, she was able to secure a position as a lab assistant (initially for a three-month stint) in Marie Curie's laboratory at the Radium Institute in Paris. Curie, for all her influence, made an unpretentious first impression, so much so that Perey, upon first meeting her at the Institute, thought she was the lab's secretary. This incident, combined with Curie's tendency to be aloof with strangers, might have portended a short career at the Curie lab for Perey. In fact, after the initial meeting, Perey thought she would only stay at the Radium Institute for her three months and leave. But Curie saw that Perey was both talented and dedicated, and she encouraged the younger woman, thus building a working relationship that extended beyond Perey's initial intentions.

Perey worked with Curie until the latter's death in 1934; thereafter she continued her mentor's research. Perey discovered the sequence of events that lead to the process known as the actinium radioactive decay series. This research inadvertently led to her most important discovery. She was aware of the existence of actinouranium, actinium-B, actinium-C, and actinium-D as part of the decay series she was trying to interpret. During this time, scientists were still trying to discover what they then believed to be the only three elements missing in the periodic table (which at the time contained 92 elements). One of these was Element 87. As Perey attempted to confirm her results of actinium radioactive decay, she found that other elements kept cropping up, disrupting the procedure. One of the elements was Element 87, with an atomic weight of 223. The element was highly charged—in fact, the most electropositive of all the elements. Because of this property, she considered naming it catium (from **cation**, which is a term for positively charged ions). But the word sounded too much like ''cat'' to her colleagues. As a result, she decided on francium, in honor of her homeland (and the place where the element had been discovered).

The following year, Perey took a position at France's National Center for Scientific Research. She remained there until 1949, when she became a professor of nuclear physics at the University of Strassbourg. She later became director of Strassbourg's Nuclear Research Center, holding that post for the rest of her life. By the time of her admission into the French Academy, Perey had already been diagnosed with the cancer that would slowly kill her. (She was undergoing treatment at the time of her appointment and was unable to attend the ceremonies.) She remained at the Nuclear Research Center and continued to conduct research. Eventually, the battle against the cancer grew more fierce, and, after a fifteen-year struggle, she succumbed in Louveciennes, France, on May 14, 1975.

Pericyclic reactions

Pericyclic reactions are reactions that involve bond changes in a circle of atoms. In pericyclic reactions, bonds are made or broken in a concerted cyclic transition state. This means that there are no intermediates formed in the course of the reaction (i.e., the reaction is concerted), and the transition state (i.e., the unstable arrangement that the atoms form during the reaction) involves six orbitals and six electrons.

Pericyclic rearrangements are classified based on the migration of a sigma bond, i.e., a **covalent bond** directed along the line joining the centers of two atoms. In this classification scheme, the reactions fall into four major categories: 1) electrocyclic reactions in which there is a concerted formation of a sigma bond between the two ends of certain linear conjugated systems, or the reverse reaction in which the sigma bond is broken (an example is the **ring** opening of cyclobutenes); 2) cycloaddition reactions involving the concerted formation of two or more sigma bonds between the ends of two or more specific conjugated systems (e.g., **Diels-Alder reaction**s used to synthesize 6-membered rings) and the reverse reactions involving the concerted cleavage of two or more sigma bonds; 3) sigmatropic reactions involving the concerted migration of an **atom** or group of atoms from one point of attachment to another point of attachment, during which one sigma bond is broken and another sigma bond is made (e.g., **alkyl group** rearrangements or shifts); and 4) ene reactions involving the formation and cleavage of unequal numbers of sigma bonds in a concerted cyclic transition state.

Pericyclic reactions share a number of characteristics. First, their rates of reaction are scarcely influenced by the solvent. Second, they ordinarily do not require catalysts to proceed. Third, the reactions are ordinarily very stereospecific. Fourth, light or **heat** can frequently promote the reactions. And fifth, very few enzymes capable of catalyzing such reactions are known.

See also Stereochemistry

PERIODIC TABLE

The periodic table is an essential part of the language of **chemistry**. It has much in common with a thesaurus, providing a guide to similarities and differences among the elements. From the way elements are organized in the periodic table, we can predict their behavior and write chemical formulas of compounds using just a few general guidelines. Using such rules is not the same as understanding why elements in certain areas the periodic table behave as they do, but the trends that arise from the arrangement of elements in the periodic table allows a chemist to remember useful facts about the types of compounds formed from specific elements and their chemical reactions.

In the second half of the nineteenth century, data from laboratories in France, England, Germany and Italy were assembled into a pamphlet by **Stanislao Cannizzaro**, a teacher in what is now northern Italy. In this pamphlet, Cannizzaro demonstrated a way to determine a consistent set of atomic weights, one weight for each of the elements then known. Cannizzaro distributed his pamphlet and explained his ideas at an international meeting held in Karlsruhe, Germany, in 1860 organized to discuss new ideas about the theory of atoms. When **Dmitry Ivanovich Mendeleev** returned from the meeting to St. Petersburg, Russia, he pondered Cannizzaro's list of atomic weights along with an immense amount of information he had gathered about the properties of elements. Mendeleev found that when he arranged the elements in order of increasing atomic weight, similar properties were repeated at regular intervals; they displayed **periodicity**. He used the periodic repetition of chemical and physical properties to construct a chart much like the Periodic Table we currently use.

Early in the twentieth century work initiated by **Joseph John Thomson** in England led to the discovery of the **electron** and, later, the **proton**. In 1932, James Chadwick, also in England, proved the existence of the **neutron** in the atomic **nucleus**. The discovery of these elementary particles and the experimental determination of their actual weights led scientists to conclude that different atoms have different weights because they contain different numbers of protons and neutrons. However, it was not yet clear how many **subatomic particles** were present in any but the simplest atoms, such as **hydrogen**, **helium**, and **lithium**.

In 1913, a third British scientist, **Henry G. J. Moseley**, determined the frequency and wavelength of x rays emitted by a large number of elements. By this time, the number of protons in the nucleus of some of the lighter elements had been determined. Moseley found the wavelength of the most energetic x ray of an element decreased systematically as the number of protons in the nucleus increased. Moseley then hypothesized the idea could be turned around: he could use the wavelengths of x rays emitted from heavier elements to determine how many protons they had in their nuclei. His work set the stage for a new interpretation of the periodic table.

Moseley's results led to the conclusion that the order of elements in the periodic chart was based on some fundamental principle of **atomic structure**. As a result of Moseley's work, scientists were convinced that the periodic nature of the properties of elements is due to differences in the numbers of subatomic particles. As each succeeding element is added across a row on the periodic chart, one proton and one electron are added. The number of neutrons added is unpredictable but can be determined from the total weight of the **atom**.

When Mendeleev placed elements in his periodic table, he had all elements arranged in order of increasing relative atomic weight. However, in the modern periodic table, the elements are placed in order of the number of protons in the nucleus. As atomic weight determinations became more precise, discrepancies were found. The first case of a heavier element preceding a lighter one in the modern periodic table occurs for **cobalt** and **nickel** (58.93 and 58.69, respectively). In Mendeleev's time, both atomic weights had been determined to be 59. Mendeleev grouped both elements together, along with **iron** and **copper**. From the work of Moseley and others, the number of protons in the nuclei of the elements cobalt and nickel had been found to be 27 and 28. Therefore, the order of these elements, and the reason for their similar behavior with respect to other members of their chemical families, arises because nickel has one more proton and one more electron than cobalt.

Whereas Mendeleev based his order of elements on **mass** and chemical and physical properties, the arrangement of the table now arises from the numbers of subatomic particles in the atoms of each element. The stage was now set for examining how the number of subatomic particles affects the chemistry of the elements. The role of the electrons in determining chemical and physical properties was obscure early in the 20th century, but that would soon change with the pivotal work of **Gilbert N. Lewis**.

After Moseley's work, the idea that the periodic patterns in chemical **reactivity** might actually be due to the number of electrons and protons in atoms intrigued many chemists. Among the most notable was G. N. Lewis of the University of California at Berkeley. Lewis explored the relationship between the number of electrons in an atom and its chemical properties, the kinds of substances formed when elements reacted together to form compounds, and the ratios of atoms in the formulas for these compounds. Lewis concluded that chemical properties change gradually from metallic to nonmetallic until a certain ''stable'' number of electrons is reached.

An atom with this stable set of electrons is a very unreactive species. But if one more electron is added to this stable set of electrons, the properties and chemical reactivities of this new atom change dramatically: the element is again metallic, with the properties like elements of Group 1. Properties of subsequent elements change gradually until the next stable set of electrons is reached and another very unreactive element completes the row.

A stable number of electrons is defined as the number of electrons found in an unreactive or ''noble'' gas. Lewis suggested electrons occupied specific areas around the atom, called shells. The noble gas atoms have a complete octet of electrons in the outermost shell.

The observation that each element starting a new row has just one electron in a new shell opens the door to relating

Main–Group Elements | Transition Metals | Main–Group Elements

Atomic number 86 (222) Atomic weight
Symbol Rn
Name radon

Period	1 I A	2 II A		3 III B	4 IV B	5 V B	6 VI B	7 VII B	8	9 VIII B	10	11 I B	12 II B	13 III A	14 IV A	15 V A	16 VI A	17 VII A	18 VIII A
1	1 1.00794 H hydrogen																		2 4.002602 He helium
2	3 6.941 Li lithium	4 9.012182 Be beryllium												5 10.811 B boron	6 12.011 C carbon	7 14.00674 N nitrogen	8 15.9994 O oxygen	9 18.9984032 F fluorine	10 20.1797 Ne neon
3	11 22.989768 Na sodium	12 24.3050 Mg magnesium												13 26.981539 Al aluminum	14 28.0855 Si silicon	15 30.973762 P phosphorus	16 32.066 S sulfur	17 35.4527 Cl chlorine	18 39.948 Ar argon
4	19 39.0983 K potassium	20 40.078 Ca calcium		21 44.955910 Sc scandium	22 47.88 Ti titanium	23 50.9415 V vanadium	24 51.9961 Cr chromium	25 54.9305 Mn manganese	26 55.847 Fe iron	27 58.93320 Co cobalt	28 58.69 Ni nickel	29 63.546 Cu copper	30 65.39 Zn zinc	31 69.723 Ga gallium	32 72.61 Ge germanium	33 74.92159 As arsenic	34 78.96 Se selenium	35 79.904 Br bromine	36 83.80 Kr krypton
5	37 85.4678 Rb rubidium	38 87.62 Sr strontium		39 88.90585 Y yttrium	40 91.224 Zr zirconium	41 92.90638 Nb niobium	42 95.94 Mo molybdenum	43 (98) Tc technetium	44 101.07 Ru ruthenium	45 102.90550 Rh rhodium	46 106.42 Pd palladium	47 107.8682 Ag silver	48 112.411 Cd cadmium	49 114.82 In indium	50 118.710 Sn tin	51 121.75 Sb antimony	52 127.60 Te tellurium	53 126.90447 I iodine	54 131.29 Xe xenon
6	55 132.90543 Cs cesium	56 137.327 Ba barium	67-70 *	71 174.967 Lu lutetium	72 178.49 Hf hafnium	73 180.9479 Ta tantalum	74 183.85 W tungsten	75 186.207 Re rhenium	76 190.2 Os osmium	77 192.22 Ir iridium	78 195.08 Pt platinum	79 196.96654 Au gold	80 200.59 Hg mercury	81 204.3833 Tl thallium	82 207.2 Pb lead	83 208.98037 Bi bismuth	84 (209) Po polonium	85 (210) At astatine	86 (222) Rn radon
7	87 (223) Fr francium	88 (226) Ra radium	89-102 †	103 (262) Lr lawrencium	104 (261) Rf rutherfordium	105 (262) Db dubnium	106 (263) Sg seaborgium	107 (264) Bh bohrium	108 (265) Hs hassium	109 (268) Mt meitnerium	110 (269) Uun ununnilium	111 (272) Uuu unununium	112 (277) Uub ununbiium		114 (289) Uuq ununquadium		116 (289) Uuh ununhexium		118 (293) Uuo ununoctium

Inner–Transition Metals

*Lanthanides	57 138.9055 La lanthanum	58 140.115 Ce cerium	59 140.90765 Pr praseodymium	60 144.24 Nd neodymium	61 (145) Pm promethium	62 150.36 Sm samarium	63 151.965 Eu europium	64 157.25 Gd gadolinium	65 158.92534 Tb terbium	66 162.50 Dy dysprosium	67 164.93032 Ho holmium	68 167.26 Er erbium	69 168.93421 Tm thulium	70 173.04 Yb ytterbium
† Actinides	89 (227) Ac actinium	90 232.0381 Th thorium	91 (231) Pa protactinium	92 238.0289 U uranium	93 (237) Np neptunium	94 (244) Pu plutonium	95 (243) Am americium	96 (247) Cm curium	97 (247) Bk berkelium	98 (251) Cf californium	99 (252) Es einsteinium	100 (257) Fm fermium	101 (258) Md mendelevium	102 (259) No nobelium

chemical properties to the number of electrons in a shell. Mendeleyev put elements together in a family because they had similar reactivities and properties; Lewis proposed that elements have similar properties because they have the same number of electrons in their outer shells.

Many observations of the chemical behavior of elements are consistent with this idea: the number of electrons in the outer shell of an atom (the **valence** electrons) determines the chemical properties of an element. G. N. Lewis extended his ideas about the importance of the number of valence electrons from the properties of elements to the **bonding** of atoms together to form compounds. He proposed that atoms bond with each other either by sharing electrons to form covalent bonds or by transferring electrons from one atom to another to form ionic bonds. Each atom forms stable compounds with other atoms when all atoms achieve complete shells. An atom can achieve a complete shell by sharing electrons, by giving them away completely to another atom, or by accepting electrons from another atom.

Many important compounds are formed from the elements in rows two and three in the Periodic Table. Lewis predicted these elements would form compounds in which the number of electrons about each atom would be a full shell, like the noble **gases**. The noble gases of rows two and three, **neon** or **argon**, each has eight electrons in the outermost valence shell. Thus, Lewis's rule has become known as the **octet rule** and simply states that there should be eight electrons in the outer shell of an atom in a compound. An important exception to this is hydrogen for which a full shell consists of only two electrons.

The octet rule is followed in so many compounds it is a useful guide. However, it is not a fundamental law of chemistry. Many exceptions are known, but the octet rule is a good starting point for learning how chemists view compounds and how the periodic chart can be used to make predictions about the likely existence, formulas and reactivities of chemical substances.

Elements in a vertical column of the periodic table typically have many properties in common. After all, Mendeleev used similarities in properties to construct a periodic table in the first place. Because they show common characteristics, elements in a column are known as a family. Sometimes a family had one very important characteristic many chemists knew about: that characteristic became the family name. Four important chemical family names of elements still widely used are the **alkali metals**, the alkaline earths, the **halogens**, and the noble gases. The alkali metals are the elements in Group 1, excluding hydrogen, which is a special case. These elements—lithium, **sodium**, **potassium**, **rubidium**, **cesium** and **francium**—all react with **water** to give solutions that change the **color** of a vegetable dye from red to blue. These solutions were said to be highly alkaline or basic; hence the name alkali metals was given to these elements.

The elements of Group 2 are also metals. They combine with **oxygen** to form oxides, formerly called "earths," and these oxides produce alkaline solutions when they are dissolved in water. Hence, the elements are called alkaline earths.

The name for Group 17, the halogens, means **salt** former because these elements all react with metals to form salts.

The name of group 18, the noble gases, has changed several times. These elements have been known as the **rare gases**, but some of them are not especially rare. In fact, argon is the third most prevalent gas in the atmosphere, making up nearly 1% of it. Helium is the second most abundant element in the universe—only hydrogen is more abundant. Another name used for the Group 18 family is the inert gases. But **Neil Bartlett**, while at the University of British Columbia in Vancouver, Canada, showed over 30 years ago that several of these gases can form well-defined compounds. The members of Group 18 are now known as noble gases—they do not generally react with the common elements but do on occasion, especially if the common element is as reactive as **fluorine**.

Knowing the chemistry of four families of the periodic table— groups 1, 2, 17, and 18, the alkali metals, the alkaline earths, the halogens and the noble gases— enables us to divide the elements in the periodic chart into other general categories: metals and nonmetals. We all automatically think of metals as shiny, hard but ductile substances that conduct **electricity**. Groups 1 (excluding hydrogen) and 2 are families of metallic elements. Groups 17 and 18 contain elements with very different properties perhaps best described by what they are not—they are not metals, and hence are called nonmetals. Between Groups 1 and 2 and Groups 17 and 18 is a dividing line between these two types of elements. Most periodic charts have a heavy line cutting between **aluminum** and **silicon** and descending downward and to the right in a stair-step fashion. Elements to the left of the line are metallic; those to the right, nonmetallic. The boundary is somewhat fuzzy, however, because the properties of elements change gradually as one moves across and down the chart, and some of the elements touching that border have a blend of characteristics of metals and nonmetals; they are frequently called semi-metals or metalloids.

The elements in the center region of the table, consisting of dozens of metallic elements in Groups 3-12, including the lanthanide and actinide elements, are called the **transition elements** or transition metals. The other elements, Groups 1,2, and 13-18, are called the representative elements.

There is a correlation among the representative elements between the number of valence electrons in an atom and the tendency of the element to act as a metal, **nonmetal** or **metalloid**. Among the representative elements, the metals are located at the left and have few valence electrons. The nonmetals are at the right and have nearly a full shell of electrons. The metalloids have an intermediate number of valence electrons.

The structure and bonding of a compound determine its chemical and physical properties. Lewis's idea of stable, filled electron shells can be used to predict what atom is bonded to what other atom in a **molecule**. In many cases, Lewis's octet rule is followed by taking one or more electrons from one atom to form a **cation** and donating the electron or electrons to another atom to form an **anion**. Metallic elements on the far left of the periodic table can give electrons away and elements on the far right can readily accept electrons. When these elements combine, ionic bonds result. An example of an ionic compound is **sodium chloride**. The electron transferred from the sodium atom to form the sodium cation, Na^+, is shown with the chloride anion, Cl^-. In covalent bonds, electrons are shared between atoms. Lewis defined a **covalent bond** as a union between two atoms resulting from the sharing of two electrons. Thus, a covalent bond must be considered a pair of electrons shared by two atoms. Elemental **bromine**, Br_2, is an example of a covalent compound. Each bromine atom has seven electrons in its outer shell and requires one electron to achieve a noble gas configuration. Each can pick up the needed electron by sharing one with the other bromine atom. Water is a compound formed by the combination of atoms of two nonmetallic elements, hydrogen and oxygen. Each hydrogen atom requires just one electron to fill its shell because the first shell (the number of electrons of the noble gas helium) has only two electrons. Oxygen lacks two electrons compared with neon, the nearest noble gas. If each hydrogen can obtain one electron by sharing electrons from the oxygen atom and the oxygen atom can share one electron from each of the two hydrogen atoms, every atom will have a full shell of electrons, and two covalent bonds will be formed as a result of sharing two pairs of electrons.

The tendency for particular elements to form ionic or covalent compounds was known before Mendeleev had constructed his version of the periodic table, but there was no simple way to remember which elements formed which types of compounds until the periodic table was formulated. We can now use the table instead of memorizing each element's properties separately. One of the most important properties of an element that can be used to predict bonding characteristics is whether the element is metallic or nonmetallic.

Pure metals are typically shiny and malleable. These properties were evident to people thousands of years ago when elements such as **gold**, **silver**, copper, **tin**, **zinc**, and **lead** were recognized as materials of a particular class—the metals. Other properties might require some equipment to determine, for example, **electrical conductivity**. Chemists have found metals also have common chemical properties. Metals combine in similar ways with other elements and form compounds with common characteristics. Metals combine with nonmetals to form salts. In salts, the metals tend to be cations. Salts conduct electricity well when melted or when dissolved in water or some other **solvents** but not when they are solid.

Most pure metals, when freshly cut to expose a new surface, are lustrous, but most lose this luster quickly by combining with oxygen, **carbon** dioxide or hydrogen sulfide to form oxides, carbonates or sulfides. Only a few metals such as gold, silver and copper are found pure in nature, uncombined with other elements.

Nonmetals in their elemental form are usually gases or **solids**. A few are shiny solids, but instead of being metallic grey they are typically black (boron, carbon as graphite), colorless (carbon as diamond) or highly colored (violet **iodine**, yellow sulfur). At room **temperature**, only one of them is a liquid (bromine).

Nonmetallic elements combine with metallic elements to form salts. In salts, the nonmetallic elements tend to be anions. Nonmetals accept electrons in forming anions while metals donate electrons to form cations. This reflects a periodic property of elements: as you move from left to right across a row on the periodic chart, you move from atoms of metals which tend to give up electrons relatively easily to atoms of nonmetals which do not readily give up electrons in forming chemical bonds. At the start of the next row, the trend is repeated. This periodic property is referred to as **electronegativity**. The more readily atoms accept electrons in forming a bond, the higher their electronegativity. Metals are characterized by low electronegativities; nonmetals, by high electronegativity. Electronegativity increases across a row on the periodic chart.

Nonmetallic elements combine with each other to form compounds. Although some nonmetallic elements form solutions when mixed with other nonmetallic elements, most react with other nonmetals to form new substances. For example, at the high temperatures and pressures of an internal **combustion** engine, **nitrogen** and oxygen gases from the atmosphere react to form nitrogen oxides such as **nitric oxide**, NO, and nitrogen dioxide, NO_2. Nonmetallic elements form covalent bonds with each other by sharing electron pairs. This tendency to bond by sharing electrons reflects the periodic trend described above: elements on the right side of the periodic chart do not give up electrons easily when forming bonds; their electronegativity is high. They tend to either accept electrons from metals to form salts or share electrons with other nonmetals to form covalent compounds.

Metalloids typically show physical characteristics intermediate between the metals and nonmetals. For example, the electrical conductivity of the metalloid **germanium**, which is used in semiconductor devices, is 2.2×10^4 ohm^{-1} cm^{-1} and its thermal conductivity is 0.60 watts cm^{-1} K^{-1}. By comparison, the electrical conductivity of a metal such as copper is over 10^6 ohm^{-1} cm^{-1} and its thermal conductivity is 4.01 watts cm^{-1} K^{-1} thermal conductivity is and the electrical and thermal conductivities of non-metals are orders of magnitude lower than those of germanium. Metalloids typically act more like nonmetals than metals in their chemistry. They more often combine with nonmetals to form covalent compounds rather than salts, but they can do both. This reflects their intermediate position on the periodic table. They can, however, also form alloys with metals and with the other metalloids. **Semiconductors** are typically made from combinations of two metalloids. The minor constituent, for example germanium, is said to be "doped" into the major constituent, which is often silicon.

The boundaries between metals, nonmetals and metalloids are arbitrary. The changes in properties as one moves from element to element on the chart are gradual.

Mendeleev used several properties when he decided how to arrange the elements in the periodic table. He considered metallic and nonmetallic properties, deciding potassium should be placed in the column with sodium, for example. He also considered the ratio of the number of atoms of the metallic element to the number of atoms of the nonmetallic element in salts. On this basis, he put **calcium**, which forms $CaCl_2$, in the column under **magnesium**, which forms $MgCl_2$. Another important property he used was the acidic or basic character of oxides formed from the elements.

Because we exist in an atmosphere containing about 20% oxygen and oxygen is quite reactive, most elements can be found in nature as oxides. The alkali metals (Group 1) and alkaline earths (Group 2) were so named because the metallic oxides formed when the metals reacted with oxygen produced basic solutions when dissolved in water. Metallic oxides are known as basic anhydrides (anhydrous, meaning without water), because basic solutions are formed when they are added to water.

Nonmetallic elements combine with oxygen to form oxides, many of which, such as **carbon dioxide**, **sulfur** dioxide and nitrogen dioxide, are gases. When oxides of nonmetallic elements are dissolved in water they tend to form acidic solutions or neutral solutions. Nonmetal oxides that form acidic solutions when dissolved in water are called acid anhydrides.

Transition metals react with oxygen to form a wide variety of oxides, some of which are basic and some acidic. A few transition metals are relatively unreactive and may be found in nature as pure elements.

A great deal of chemical information is contained in the periodic table. The organization of the modern table gives us insight into the importance of composition of the nucleus and number of electrons in the outer shell of an atom. Several useful theories enable us to use the number of valence electrons to write formulas of compounds and to predict physical properties and chemical reactivities of elements. An element is defined by the number of protons in the nucleus of its atoms, but its chemical reactivity is determined by the number of electrons in its outer shell. Elements in a family of the periodic table have the same number of electrons in the outer shell. Because they have the same number of valence electrons, elements in a family often exhibit similar chemical properties.

Metals and nonmetals differ in their tendencies to lose valence electrons, accept valence electrons, or share valence electrons in forming compounds. The octet rule, proposed by G. N. Lewis, is a useful guideline to predict formulas and the likely existence of compounds. Ionic compounds, formed when electrons are transferred from one atom to another to make ions, have vastly different properties compared to covalent compounds, which are formed when pairs of electrons are shared between atoms.

Chemical behavior, such as the acidic or basic properties of oxides, is also periodic. Physical properties, such as the ability to conduct electricity or **heat** and **malleability** or brittleness, may also be predicted from the position of an element on the periodic table.

Periodicity

Elements on the **periodic table** are arranged in groups according to their **electron** configuration. All elements in a vertical column have the same number of electrons in their outer **energy** level. Their distance from the **nucleus** increases as you go down a column. Electrons in the outer **energy level** are the electrons involved in chemical reactions. Since all the elements in a column have the same number of electrons available for chemical change, their properties are similar and are often referred to as families. The horizontal rows or periods also have predictable trends in characteristics because as you move left to right in a row only one electron is added changing the **atomic number** by one. Variations of properties are based on electron configuration. Predictable trends that occur repeatedly in a definite pattern are periodic properties or periodicity. **Dmitry Ivanovich Mendeleev** arranged the known elements according to **mass** in the development of the first table and found elements with similar properties were grouped together. This was prior to the knowledge of electron configuration. Properties and position of an element on the periodic table are based on their electron configuration. The most predictable and regular patterns are found in the s and p block elements (groups 1,2, 13 through 17). The d block elements do not show the same regularity in properties as the main group elements because the electrons in both the s and d sublevels can be involved in chemical reactions. The periodicity of the following properties can be predicted; **valence** electrons, atomic and ionic radii, **ionization** energy, electron affinity, and **electronegativity**.

Valence electrons are the number of electrons available that can be lost, gained or shared in the formation of a compound. Since elements are grouped in vertical columns according to the number of outer shell electrons their group number represents the valence electrons. Group one and two have one and two outer shell electrons. Groups 13-18 have ten less electrons than the group number. For example, Group 14 elements have four valence electrons or four electrons in their outer shell available to react.

The **atomic radius** is half the distance between the nuclei of the same **atom** in an element or a compound. Atomic size, generally increases down a group because electrons are being added to higher energy levels at a greater distance from the nucleus increasing the size of the electron cloud. Atomic size decreases from left to right in a row or period because electrons are being added to the same energy level. As a result, the attraction between the positively charged nucleus and the number of electrons increases, pulling the electrons toward the center of the atom.

The ionic radius refers to the radius of an **ion** in an ionic compound. Atoms gain or lose electrons to form ions. Group one **metals** lose one electron and group two metals lose two electrons to become cations. Group 15-17 elements form anions more frequently by gaining electrons. Anions, therefore have more electrons than their neutral atom and are larger. Positively charged cations are smaller than their neutral atoms because they lose electrons. Going left to right across a period cations decrease in size as more electrons are lost. Beginning with group 15, anions decrease in size because fewer electrons are being added. Going down in a group electrons are being added to higher levels, increasing the size of the ions.

Ionization energy or ionization potential can also be predicted. The ionization energy is the energy needed to remove an electron from a gaseous atom. The first ionization energy refers to the energy needed to remove the most loosely held outer shell electron. Ionization energy decreases as you go down a column or group because the outer shell electrons are farther and farther away from the nucleus and are less influenced by its positive charge. The electrons are shielded (shielding effect) by inner shell electrons. Therefore, less energy is required to remove an electron. The element at the top of the column is held more tightly because the outer shell electron is closer to the positively charged nucleus. Ionization energies generally tend to increase across a row or period because electrons are being added to the same energy level and the attraction between the nucleus and the increasing number of negatively charged electrons in the shell holds them more tightly. Second, third and higher ionization energies progressively increase because each successive electron is attracted to a larger positive charge. A large jump in ionization energy generally indicates removal of the electron from a filled level or sublevel. Group 18, the noble **gases** have the highest first ionization energies because their energy level is full.

Electron affinity is the energy released when an electron is added to a neutral atom. Electron affinity can have positive or negative values. The addition of one electron is an exothermic reaction and the electron affinity value will be negative. Positive values indicate energy is needed (endothermic) to add another electron. An alternate definition of electron affinity is the attraction an atom has for an electron. Electron affinity is affected by the same factors as ionization energy. An element's valence shell that loses electrons easily will have little desire to attract electrons. Electron affinity increases from left to right in a row indicating a greater ease in acquiring electrons and decreases from top to bottom in a group. Metals give up electrons easily and have low electron affinities, nonmetals gain electrons more easily and have high electron affinities.

Electronegativity refers to the ability of an atom in a compound to attract electrons. This differs from ionization energy and electron affinity which describes the properties of single atoms. Electronegativity increases from left to right since metals tend to give up electrons and non metals tend to gain electrons. In the formation of a compound such as **magnesium** chloride, **chlorine** attracts electrons in the compound. **Fluorine** is the most electronegative element because it has a high ionization energy and a small radius. **Cesium** and **francium** have the lowest electronegative values. Electronegativity decreases from top to bottom because larger atoms ionize easily and do not attract electrons. Smaller atoms which do not ionize readily attract electrons. The electronegativity scale ranges from 0.0 to 4.0 with 4.0 being the highest value. **Linus Pauling** developed this scale and in 1954 received a Nobel prize in **chemistry** for his work on chemical **bonding** which included electronegativity.

Trends in atomic and ionic radii are influenced most by the number of electrons in the element. Ionization energy, electron affinity, and electronegativity are directly related to nuclear charge and the distance of the outer shell electrons from the nucleus. Metallic and nonmetallic characteristics show a pattern also. Metallic characteristics increase down a column and decrease left to right. Periodicity is also observed in atomic numbers. Going down a column on the periodic table the atomic numbers differ by eight between rows one and two and two and three. Between rows three and four and four and five the difference in atomic numbers are eighteen. The difference in atomic numbers between rows five and six is thirty two. For example, **calcium** is in row four and has an atomic number of 20. The element below it is **strontium** in row five with an atomic number of 36. Periodicity continues to be an important scientific tool in understanding the behavior and interaction of elements.

See also Electron; Electronegativity; Periodic table

PERKIN, WILLIAM HENRY (1838-1907)

English chemist

William Perkin is considered to be the father of the synthetic dye and perfume industries.

Perkin was born in London, England, and as a child attended the City of London School. There he came into contact with **Michael Faraday** who fostered his fledgling fascination with **chemistry**. In 1853, Perkin entered the Royal College of Chemistry where, at seventeen, he was named an assistant to the school's director, a renowned German chemist named **August Wilhelm von Hofmann**. Although Hofmann was a brilliant chemist, he was awkward with laboratory work and depended on talented assistants to help him in his research on **coal** tar and its derivatives.

It was under Hofmann's tutelage in 1856 that Perkin experienced his first major success. That year, Perkin spent his Easter vacation attempting to synthesize **quinine** from **aniline**, a coal-tar derivative. Although he failed to produce artificial quinine, the results of his experiment determined the course of his career. As part of his process, Perkin mixed aniline with **potassium** dichromate and **alcohol**, which yielded a purple liquid. Thinking it might be useful as a dye, Perkin named the liquid *aniline purple* and sent a sample to a silk dyeing firm. When the company sent back for more dye, it became clear that this was a lucrative business opportunity so Perkin convinced his father and brother to invest in a company to produce the new dye. Soon the company began marketing aniline purple, which became known as mauve (from the French word for the plant previously used to make violet).

While his family tended to the practical aspect of the business, Perkin headed up the company's research department. His experiments led to the development of more dyes, including violets and rosanilines. Over the next few years, he introduced several other colors into his company line: aniline red (1859), aniline black (1863), and alkalate magenta (1864). In 1868, Perkin used the work of two German chemists, Carl Graebe (1841-1927) and Carl Liebermann (1842-1914), as a basis for synthesizing alizarin, the chemical component of the madder plant essential in dye making. While Graebe and Liebermann had developed a workable synthesis process, it was too expensive to be of practical use. Perkin came up with a cost-effective production version of his fellow chemists' process, and by 1871, his company was producing two hundred twenty tons of alizarin annually. Within a short time, Perkin's curiosity and drive paid off as his synthetic dyes replaced natural dyes all over the world.

Perkin's further experimentation led to his discovery of a method for changing the structure of organic compounds on

Max Perutz.

a molecular level. Using this process, known as the ''Perkin synthesis,'' he produced a coumarin, a synthetic perfume which has been described as smelling like fresh hay or vanilla. Although he technically retired at age thirty-six, he launched a second career in the synthetic perfume business. He later teamed up with B.F. Duppa to research and develop other aspects of the synthetic fragrance field. Their accomplishments include the development of a process for producing glycine, racemic acid, and **tartaric acid**, as well as significant research into the similarities between tartaric acid and maleic acids. In 1889, Perkin received the Davy medal of the Royal Society, and the British government recognized Perkin's contribution to science, industry, and his country by knighting him in 1906. He died one year later, on July 14, in Sudbury, England.

Perutz, Max (1914-)

English crystallographer and biochemist

Max Perutz pioneered the use of x-ray **crystallography** to determine the **atomic structure** of **proteins** by combining two lines of scientific investigation—the physiology of hemoglobin and the physics of **x-ray crystallography**. His efforts resulted in his sharing the 1962 Nobel Prize in **chemistry** with his colleague, biochemist **John Kendrew**. Perutz's work in deciphering the diffraction patterns of protein crystals opened the door for molecular biologists to study the structure and function of enzymes—specific proteins that are the catalysts for biochemical reactions in cells. Known for his impeccable laboratory skills, Perutz produced the best early pictures of protein crystals and used this ability to determine the structure of hemoglobin and the molecular mechanism by which it transports **oxygen** from the lungs to tissue. A passionate mountaineer and skier, Perutz also applied his expertise in x-ray crystallography to the study of glacier structure and flow.

Perutz was born in Vienna, Austria, on May 19, 1914. His parents were Hugo Perutz, a textile manufacturer, and Adele Goldschmidt Perutz. In 1932, Perutz entered the University of Vienna, where he studied **organic chemistry**. However, he found the university's adherence to classical organic chemistry outdated and backward. By 1926 scientists had determined that enzymes were proteins and had begun to focus on the catalytic effects of enzymes on the chemistry of cells, but Perutz's professors paid scant attention to this new realm of research. In 1934, while searching for a subject for his dissertation, Perutz attended a lecture on organic compounds, including **vitamins**, under investigation at Cambridge University in England. Anxious to continue his studies in an environment more attuned to recent advances in biochemical research, Perutz decided he wanted to study at Cambridge. His wish to leave Austria and study elsewhere was relatively unique in that day and age, when graduate students seldom had the financial means to study abroad. But Hugo Perutz's textile business provided his son with the initial funds he would need to survive in England on a meager student stipend.

In 1936, Perutz landed a position as research student in the Cambridge laboratory of **John Desmond Bernal**, who was pioneering the use of x-ray crystallography in the field of biology. Perutz, however, was disappointed again when he was assigned to research minerals while Bernal closely guarded his crystallography work, discussing it only with a few colleagues and never with students. Despite Perutz's disenchantment with his research assignments and the old, ill-lit, and dingy laboratories he worked in, he received excellent training in the promising field of x-ray crystallography, albeit in the classical mode of mineral crystallography. ''Within a few weeks of arriving,'' Perutz states in Horace Freeland Judson's *Eighth Day of Creation: Makers of the Revolution in Biology,* ''I realized that Cambridge was where I wanted to spend the rest of my life.''

During his summer vacation in Vienna in 1937, Perutz met with Felix Haurowitz, a protein specialist married to Perutz's cousin, to seek advice on the future direction of his studies. Haurowitz, who had been studying hemoglobin since the 1920s, convinced Perutz that this was an important protein whose structure needed to be solved because of the integral role it played in physiology. In addition to making blood red, hemoglobin red corpuscles greatly increase the amount of oxygen that blood can transport through the body. Hemoglobin also transports **carbon** dioxide back to the lungs for disposal.

Although new to the **physical chemistry** and crystallography of hemoglobin, Perutz returned to Cambridge and soon obtained crystals of horse hemoglobin from Gilbert Adair, a leading authority on hemoglobin. Since the main goal of x-ray crystallography at that time was to determine the structure of any protein, regardless of its relative importance in biological activity, Perutz also began to study crystals of the digestive **enzyme** chymotrypsin. But chymotrypsin crystals proved to be unsuitable for study by x ray, and Perutz turned his full attention to hemoglobin, which has large crystal structures uniquely suited to x-ray crystallography. At that time, microscopic protein crystal structures were ''grown'' primarily through placing the proteins in a **solution** which was then evaporated or cooled below the saturation point. The crystal structures, in effect, are repetitive groups of cells that fit together to fill each space, with the cells representing characteristic groups of the molecules and atoms of the compound crystallized.

In the early 1930s, crystallography had been successfully used only in determining the structures of simple crystals of **metals**, minerals, and salts. However, proteins such as hemoglobin are thousands of times more complex in atomic structure. Physicists William Henry Bragg and **William Lawrence Bragg**, the only father and son to share a Nobel Prize, were pioneers of x-ray crystallography. Focusing on minerals, the Braggs found that as x rays pass through crystals, they are buffeted by atoms and emerge as groups of weaker beams which, when photographed, produce a discernible pattern of spots. The Braggs discovered that these spots were a manifestation of Fourier synthesis, a method developed in the nineteenth century by French physicist Jean Baptiste Fourier to represent regular signals as a series of sine waves. These waves reflect the distribution of atoms in the crystal.

The Braggs successfully determined the amplitude of the waves but were unable to determine their phases, which would provide more detailed information about crystal structure. Although amplitude was sufficient to guide scientists through a series of trial and error experiments for studying simple crystals, proteins were much too complex to be studied with such a haphazard and time consuming approach.

Initial attempts at applying x-ray crystallography to the study of proteins failed, and scientists soon began to wonder whether proteins in fact produce **x-ray diffraction** patterns. However, in 1934, Desmond Bernal and chemist **Dorothy Crowfoot Hodgkin** at the Cavendish laboratory in Cambridge discovered that by keeping protein crystals wet, specifically with the liquid from which they precipitated, they could be made to give sharply defined X-ray diffraction patterns. Still, it would take twenty-three years before scientists could construct the first model of a protein **molecule**.

Perutz and his family, like many other Europeans in the 1930s, tended to underestimate the seriousness of the growing Nazi regime in Germany. While Perutz himself was safe in England as Germany began to invade its neighboring countries, his parents fled from Vienna to Prague in 1938. That same summer, they again fled to Switzerland from Czechoslovakia, which would soon face the onslaught of the approaching German army. Perutz was shaken by his new classification as a refugee and the clear indication by some people that he might not be welcome in England any longer. He also realized that his father's financial support would certainly dwindle and die out.

As a result, in order to vacation in Switzerland in the summer of 1938, Perutz sought a travel grant to apply his expertise in crystallography to the study of glacier structures and flow. His research on glaciers involved crystallographic studies of snow transforming into ice, and he eventually became the first to measure the **velocity** distributions of a glacier, proving that glaciers flow faster at the surface and slower at the glacier's bed.

Finally, in 1940, the same year Perutz received his Ph.D., his work was put to an abrupt halt by the German invasions of Holland and Belgium. Growing increasingly wary of foreigners, the British government arrested all ''enemy'' aliens, including Perutz. ''It was a very nice, very sunny day—a nasty day to be arrested,'' Perutz recalls in *The Eighth Day of Creation.* Transported from camp to camp, Perutz ended up near Quebec, Canada, where many other scientists and intellectuals were imprisoned, including physicists Herman Bondi and Tom Gold. Always active, Perutz began a camp university, employing the resident academicians to teach courses in their specialties. It didn't take the British government long, however, to realize that they were wasting valuable intellectual resources and, by 1941, Perutz followed many of his colleagues back to his home in England and resumed his work with crystals.

Perutz, however, wanted to contribute to the war effort. After repeated requests, he was assigned to work on the mysterious and improbable task of developing an aircraft carrier made of ice. The goal of this project was to tow the carrier to the middle of the Atlantic Ocean, where it would serve as a stopping post for aircrafts flying from the United States to Great Britain. Although supported both by then British Prime Minister Winston Churchill and the chief of the British Royal Navy, Lord Louis Mountbatten, the ill-fated project was terminated upon the discovery that the amount of steel needed to construct and support the ice carrier would cost more than constructing it entirely of steel.

Perutz married Gisela Clara Peiser on March 28, 1942; the couple later had a son, Robin, and a daughter, Vivian. After the war, in 1945, Perutz was finally able to devote himself entirely to pondering the smeared spots that appeared on the x-ray film of hemoglobin crystals. He returned to Cambridge, and was soon joined by John Kendrew, then a doctoral student, who began to study myoglobin, an enzyme which stores oxygen in muscles. In 1946 Perutz and Kendrew founded the Medical Research Council Unit for Molecular Biology, and Perutz became its director. Many advances in molecular biology would take place there, including the discovery of the structure of deoxyribonucleic acid (**DNA**).

Over the next years, Perutz refined the x-ray crystallography technology and, in 1953, finally solved the difficult phase dilemma with a method known as isomorphous replacement. By adding atoms of mercury—which, like any heavy metal, is an excellent x-ray reflector—to each individual protein molecule, Perutz was able to change the light diffraction pattern. By comparing hemoglobin proteins with **mercury** attached at different places to hemoglobin without mercury, he found that he had reference points to measure phases of other hemoglobin spots. Although this discovery still required long and assiduous mathematical calculations, the development of computers hastened the process tremendously.

By 1957, Kendrew had delineated the first protein structure through crystallography, again working with myoglobin. Perutz followed two years later with a model of hemoglobin. Continuing to work on the model, Perutz and Hilary Muirhead showed that hemoglobin's reaction with oxygen involves a structural change among four subunits of the hemoglobin molecule. Specifically, the four polypeptide **chains** that form a tetrahedral structure of hemoglobin are rearranged in oxygenated hemoglobin. In addition to its importance to later research on the molecular mechanisms of respiratory transport by hemoglobin, this discovery led scientists to begin research on the structural changes enzymes may undergo in their interactions with various biological processes. In 1962, Perutz and Kendrew were awarded the Nobel Prize in chemistry for their codiscoveries in x-ray crystallography and the structures of hemoglobin and myoglobin, respectively. The same year, Perutz left his post as director of the Unit for Molecular Biology and became chair of its laboratory.

The work of Perutz and Kendrew was the basis for growing understanding over the following decades of the mechanism of action of enzymes and other proteins. Specifically, Perutz's discovery of hemoglobin's structure led to a better understanding of hemoglobin's vital attribute of absorbing oxygen where it is plentiful and releasing it where it is scarce. Perutz also conducted research on hemoglobin from the blood of people with sickle-cell anemia and found that a change in the molecule's shape initiates the distortion of venous red cells into a sickle shape that reduces the cells' oxygen-carrying capacity.

In *The Eighth Day of Creation,* Judson remarks that Perutz was known to have a ''glass thumb'' for the difficult task of growing good crystals, and it was widely acknowledged that for many years Perutz produced the best images of crystal structures. In the book, published in 1979, Perutz's long-time colleague Kendrew remarks that little changed over the years, explaining, ''If I had come into the lab thirty years ago, on a Saturday evening, Max would have been in a white coat mounting a crystal—just the same.'' Perutz retired in 1979.

PESTICIDES

Pesticides are chemicals that are used to kill insects, weeds, and other organisms to protect humans, crops, and livestock. There have been many substantial benefits of the use of pesticides. The most important of these have been: (1) an increased production of food and fiber because of the protection of crop plants from pathogens, competition from weeds, defoliation by insects, and parasitism by nematodes; (2) the prevention of spoilage of harvested, stored foods; and (3) the prevention of debilitating illnesses and the saving of human lives by the control of certain diseases.

Unfortunately, the considerable benefits of the use of pesticides are partly offset by some serious environmental damages. There have been rare but spectacular incidents of toxicity to humans, as occurred in 1984 at Bhopal, India, where more than 2,800 people were killed and more than 20,000 seriously injured by a large emission of poisonous methyl isocyanate vapor, a chemical used in the production of an agricultural insecticide.

A more pervasive problem is the widespread environmental contamination by persistent pesticides, including the presence of chemical residues in wildlife, in well water, in produce, and even in humans. Ecological damages have included the poisoning of wildlife and the disruption of ecological processes such as productivity and nutrient cycling. Many of the worst cases of environmental damage were associated with the use of relatively persistent chemicals such as DDT (**dichlorodiphenyltrichloroacetic acid**). Most modern pesticide use involves less persistent chemicals.

One way that pesticides can be classified is according to their intended pest target. The major categories of pesticides classified by target are **fungicides**, **herbicides**, insecticides, acaricides, molluscicides, nematicides, rodenticides, avicides, and antibiotics.

Fungicides prevent the germination of most fungal spores and protect crop plants and animals from fungal pathogens. Examples of fungicides include captan, maneb, zeneb, dinocap, folpet, pentachlorphenol, methyl bromide, carbon bisulfide, and chlorothalonil (Bravo). Fungicides are among the most widely used pesticides in the United States.

Herbicides kill weedy plants, thus decreasing the competition for desired crop plants. They function by blocking photosynthesis, acting as hormones to disrupt plant growth and development, or killing soil microorganisms essential for plant growth. Examples of herbicides include 2,4-D, 2,4,5-T, paraquat, dinoseb, diaquat, atrazine, Silvex, and linuron.

Insecticides kill insect defoliators and vectors of deadly human diseases such as malaria, yellow fever, plague, and typhus. The various types of insecticides are described in more detail in below.

Acaricides kill mites, which are pests in agriculture, and ticks, which can carry encephalitis of humans and domestic animals. Molluscicides destroy snails and slugs, which can be pests of agriculture or, in waterbodies, the vector of human diseases such as schistosomiasis. Nematicides kill nematodes, which can be parasites of the roots of crop plants. Rodenticides control rats, mice, gophers, and other rodent pests of human habitation and agriculture. Avicides kill birds, which can depredate agricultural fields. Antibiotics treat bacterial infections of humans and domestic animals.

Pesticides can also be categorized by the use to which they are applied. The most important use-categories of pesticides are in human health, agriculture, and forestry.

Pesticides are used in various parts of the world to control species of insects and ticks that play a critical role as vectors in the transmission of human disease-causing pathogens. The most important of these diseases and their vectors are: (1) malaria, caused by the protozoan *Plasmodium* and spread to humans by an *Anopheles*-mosquito vector; (2) yellow fever and related viral diseases such as encephalitis, also spread by mosquitoes; (3) trypanosomiasis or sleeping sickness, caused by the protozoans *Trypanosoma* spp. and spread by the tsetse fly *Glossina* spp.; (4) plague or black death, caused by the bacterium *Pasteurella pestis* and transmitted to people by the flea *Xenopsylla cheops*, a parasite of rats; and (5) typhoid fever, caused by the bacterium *Rickettsia prowazeki* and transmitted to humans by the body louse *Pediculus humanus*.

The incidence of all of these diseases can be reduced by the judicious use of pesticides to control the abundance of their vectors. For example, there are many cases where the local abundance of mosquito vectors has been reduced by the application of an insecticide to their aquatic breeding habitat, or by the application of a persistent insecticide to walls and ceilings of houses, which serve as a resting place for these insects. The use of insecticides to reduce the abundance of the mosquito vectors of malaria has been especially successful, although in many areas this disease is now re-emerging because of the evolution of insecticide-tolerance by mosquitoes.

Modern, technological agriculture employs pesticides for the control of weeds, insects, and plant diseases, all of which cause large losses of crops. In agriculture, insect pests are regarded as competitors with humans for a common food resource. Sometimes, defoliation can result in a total loss of the crop, as in the case of acute infestations of locusts. More commonly, defoliation causes a reduction in crop yields. In some cases, insects may cause only trivial damage in terms of the quantity of biomass that they consume, but by causing cosmetic damage they can greatly reduce the economic value of the crop. For example, codling moth (*Carpocapsa pomonella*) larvae do not consume much of the apple that they infest, but they cause great esthetic damage by their presence and can render produce unmarketable.

In agriculture, weeds are considered to be any plants that interfere with the productivity of crop plants by competing for light, water, and nutrients. To reduce the effects of weeds on agricultural productivity, fields may be sprayed with a herbicide that is toxic to the weeds but not to the crop plant. Because there are several herbicides that are toxic to dicotyledonous weeds but not to members of the grass family, fields of maize, wheat, barley, rice, and other grass-crops are often treated with those herbicides to reduce weed populations.

There are also many diseases of agricultural plants that can be controlled by the use of pesticides. Examples of important fungal diseases of crop plants that can be managed with appropriate fungicides include: late blight of potato, apple scab, and *Pythium*-caused seed- rot, dampingoff, and root-rot of many agricultural species.

In forestry, the most important uses of pesticides are for the control of defoliation by epidemic insects and the reduction of weeds. If left uncontrolled, these pest problems could result in large decreases in the yield of marketable timber. In the case of some insect infestations, particularly spruce budworm (*Choristoneura fumiferana*) and gypsy moth (*Lymantria dispar*), repeated defoliation can cause the death of trees over a large area of forest. Most herbicide use in forestry is for the release of desired conifer species from the effects of competition with angiosperm herbs and shrubs. In most places, the quantity of pesticide used in forestry is much smaller than that used in agriculture.

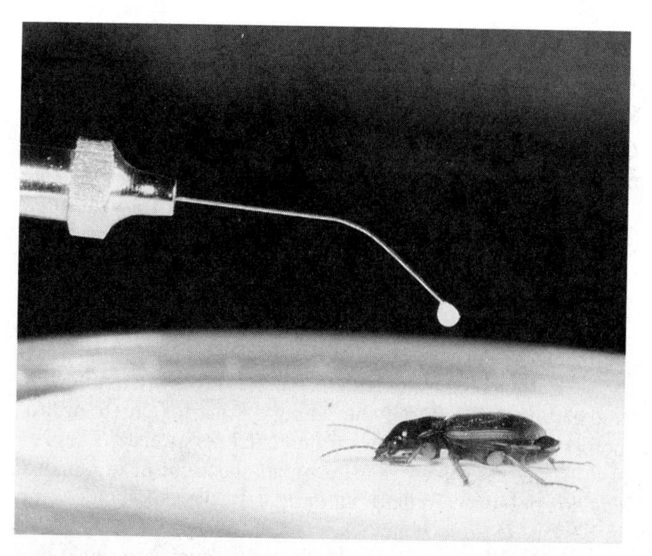

A micro-applicator for testing pesticides on individual insects. *(Photograph by Nigel Cattlin/Holt Studios, Photo Researchers, Inc. Reproduced by permission.)*

Pesticides can also be classified according to their similarity of chemical structure. The following paragraphs describe pesticides grouped by chemical similarity.

Inorganic pesticides include compounds of **arsenic**, **copper**, **lead**, and **mercury**. Some prominent inorganic pesticides include Bordeaux mixture, a complex fungicide with several copper-based active ingredients, used for fruit and vegetable crops. Various arsenic compounds are used as nonselective herbicides, as soil sterilizers, and sometimes as insecticides. These inorganic pesticides tend to be highly toxic to a broad spectrum of organisms, are persistent in the environment, and also tend to bioaccumulate.

Organic pesticides are a diverse group of chemicals. Some are produced naturally by certain plants, and analogs of these natural organic pesticides have also been synthesized by chemists. Among the natural organic pesticides extracted from plants are important insecticides such as the alkaloid nicotine and other nicotinoids [largely extracted from tobacco (*Nicotiana tabacum*)]. These nicotine insecticides are often applied as the salt nicotine **sulfate**. Another natural organic insecticide is pyrethrum, a complex of chemicals extracted from the daisy- like plant *Chrysanthemum cinerariaefolium*. Natural organic pesticides tend to be fast-acting and effective against a broad range of insect pests.

The great majority of organic pesticides are synthesized by chemists. Included in this category of synthetic organic pesticides are **organometallic** fungicides (including methylmercury) and phenols (e.g., pentachlorophenol) used as a fungicide in the preservation of wood.

Organochlorine insecticides include chemicals such as DDT, methoxychlor, heptachlor, chlordane, toxaphene, aldrin, endrin, dieldrin, and lindane. Most of these insecticides are neurotoxins. They have the advantage of being cheap, fast-acting, easy to apply, and effective against a broad range of insects. On the negative side, they are persistent in the environment and tend to bioaccumulate. The environmental problems caused by DDT have been well documented and the use of this insecticide has been banned in the United States since 1972.

Organophosphorus pesticides are a diverse group of chemicals, most of which are used as insecticides, acaricides, or nematicides. The organophosphates generally have a high acute toxicity to arthropods but a short persistence in the environment. Some of the insecticides are highly toxic to non-target organisms such as fish, birds, and mammals. Some prominent examples are the insecticides parathion, fenitrothion, malathion, and phosphamidon. Glyphosate, an important herbicide, is not very toxic to animals.

Carbamate pesticides (e.g., carbaryl (Sevin), aminocarb, and carbofuran) generally have a high acute toxicity to arthropods but only a moderate environmental persistence. They are fast-acting, effective against a broad spectrum of organisms, and do not bioaccumulate. Unfortunately, these pesticides tend to be very toxic to honey bees.

Biological pesticides are bacteria, fungi, or viruses that are toxic to pests. One of the most widely used biological insecticides is a preparation manufactured from spores of the bacterium *Bacillus thuringiensis*, or B.t. Because this insecticide has a relatively specific activity against leaf-eating lepidopteran pests and a few other insects such as blackflies and mosquitoes, its non-target effects are small.

The intended ecological effect of a pesticide application is to control a pest species, usually by reducing its abundance to an economically acceptable level. In a few situations, this objective can be attained without important non-target damage. However, whenever a pesticide is broadcast- sprayed over a field or forest, a wide variety of on-site, non-target organisms are affected. In addition, some of the sprayed pesticide invariably drifts away from the intended site of deposition, and it deposits onto non-target organisms and ecosystems. The ecological importance of any damage caused to non-target, pesticide-sensitive organisms partly depends on their role in maintaining the integrity of their ecosystem. From human perspective, however, the importance of a non-target pesticide effect is also influenced by specific economic and esthetic considerations.

PETERMANN, MARY LOCKE (1908-1975)

American biochemist

Mary Locke Petermann isolated and worked out the structure of animal ribosomes, organelles that are now known as the sites of **protein synthesis** in cells. She began her original investigation of the particles (for a time they were known as "Petermann's particles") because they were interfering with her studies of **DNA** and **RNA**. Her work was fundamental and pioneering; her continued work established the importance of ions in stabilizing ribosomes and elucidated ribosomal transformations.

Peterman was born in Laurium, Michigan, on February 25, 1908, one of three children and the only daughter of Albert Edward and Anna Mae Grierson Petermann. Her mother was a graduate of Ypsilanti State Teachers' College. Her father, a graduate of Cornell University, became a lawyer for Calumet and Hecla Consolidated Copper Company in Calumet, Michigan, after World War I; he later was president and general manager. The Petermann family lived in a large company house and enjoyed high status in the community.

After graduating from Calumet High School in 1924, Petermann spent a year at a Massachusetts preparatory school before entering Smith College. In 1929, she graduated from Smith with high honors in **chemistry** and membership in Phi Beta Kappa. After a year at Yale University as a technician, she spent four years working at the Boston Psychopathic Hospital, investigating the acid-base **balance** of mental patients. In 1936 she entered the University of Wisconsin; she received a Ph.D. degree in physiological chemistry in 1939, with a thesis project on the role of the adrenal cortex in **ion** regulation.

In 1939 Petermann became the first woman chemist on the staff of the Department of Physical Chemistry at the University of Wisconsin. She remained at Wisconsin as a postdoctoral researcher until 1945. During these six years she and Alwin M. Pappenheimer began to investigate the physical chemisty of **proteins**. Petermann discovered what were at first called "Petermann's particles" but were named ribosomes at a meeting of the Biophysical Society in 1958. (It was at this meeting that George Palade, a research scientist who had independently played a pivotal role in discovering ribosomes, called Petermann "the mother of the particles.") Ribosomes are where protein synthesis occurs in a cell. Petermann's research isolated several types of ribosomes and clarified their properties. She also pioneered the study of **antibodies**. This research later led to **Rodney Porter** winning a Nobel Prize in 1972 for his work on the structure of immunoglobulins.

After leaving the University of Wisconsin in 1945, Petermann accepted the position of research chemist at Memorial Hospital in New York City to explore the role of **plasma** proteins in cancer. (According to Mary L. Moller, Petermann had been recommended to the director, Cornelius Rhoads, as "the girl out in Wisconsin.") In 1946 she was appointed Finney-Howell Foundation fellow at the newly founded Sloan-Kettering Institute, where she explored the role of nucleoproteins in cancer. She became an associate member of the institute in 1960, the first woman member in 1963, and member emeritus in 1973 when she retired. Concurrent with her work at Sloan-Kettering, she also taught **biochemistry** in the Sloan-Kettering Division, Graduate School of Medical Sciences, Cornell University. In 1966, she became the first woman appointed a full professor at Cornell. She authored or co-authored almost 100 scientific papers.

As the Sloan Award recipient in 1963, Petermann was honored for what the accompanying citation explained was her "many basic and distinguished contributions to the knowledge of the relevance of proteins and nucleoproteins in abnormal growth. An even greater contribution has been her fundamental work on the nature of the cell ribosome." Petermann used her award money to work for a year in the Swedish laboratory of Nobel laureate **Arne Tiselius**. She also lectured in several

European countries, including England and France. In 1966 she received the Garvan Medal of the American Chemical Society, which honors contributions made by women scientists, an honorary doctorate from Smith College, and the Distinguished Service Award from the American Academy of Achievement.

Petermann never married. In 1974, the year before her death, she organized the Memorial Sloan-Kettering Cancer Center Association for Professional Women and served as its first president. She died in Philadelphia on December 13, 1975, of intestinal cancer, which had been misdiagnosed as a "nervous stomach" earlier that year. In 1976 the Educational Foundation of the Association for Women in Science named one of its graduate scholarships in her honor.

PETROCHEMICAL

Oil, or petroleum, is a complicated and diverse mixture of many **hydrocarbon** compounds, not easily characterized. Different deposits yield significantly different chemical compositions ranging from light crudes to heavy viscous tar sands. In addition, within an oil reservoir and during the life of an oil well, the chemical composition can, and most often does, change. However, in broad terms, petroleum can be divided into six fractions obtained from fractional **distillation** of the crude oil. With the understanding that the exact percentage and chemical of the fractions is neither fixed nor constant, the following outlines the general composition of each.

The "gas" fraction is the lightest, consisting of the organic **gases** from **methane** to the butanes. These compounds, at normal pressure and **temperature**, have a **boiling point** below 100° F (40° C). They usually constitute a minor component of an oil deposit although most reservoirs have a gas cap. While methane generally serves as a fuel and is often simply burned off at a refinery, the other light gases can serve as the starting materials for **plastics** and polymers or as feedstock for a number of other petrochemicals.

The **gasoline** fraction extends from compounds with 5-12 **carbon** atoms, with boiling points in the range of 100-400°F (40-200° C). These compounds generally constitute the largest fraction of crude oil but demand for gasoline easily outstrips the **concentration** available in petroleum, necessitating the complicated **chemistry** of an oil refinery. The gasoline fraction can be further divided into the "straight run gasoline" and "naphtha." While it is possible to run an automobile on "straight run," generally, both fractions are combined in some form of catalytic reformer with a view to enhancing the properties of the resulting gasoline. In particular, the formation of branched alkanes increases the octane number and combustibility of the resulting hydrocarbon.

The **kerosene** fraction extends from compounds with 12-16 carbon atoms, with boiling points in the 350-525°F (175-275° C) range. These compounds generally constitute the second largest fraction of petroleum and are used for a number of different purposes including, after chemical processing or refinement, enhancing the gasoline production from a barrel of crude oil. This is achieved through hydrocracking that uses **hydrogen** to break up the larger molecules into fragments suitable in size. Kerosene, itself, is used in Diesel fuel, jet fuel, and home heating. When the original oil deposits were tapped in Pennsylvania in 1859, kerosene was the valuable fraction as it was used in home heating and lamps. The more dangerous and volatile gasoline fraction was burnt.

The gas oil fraction extends from compounds with 15-18 carbon atoms, with boiling points in the 475-750°F (250-400° C) range. Although catalytic cracking allows some of these compounds to be used in gasoline, the majority is used as furnace oil, diesel oil, and as industrial fuels.

The lubricating oils are compounds consisting of more than 17 carbon atoms, with boiling points in excess of 575°F (300° C), while the residues are **chains** longer than 20 carbon atoms and boiling points in excess of 650° F (350° C). The lubricants find use in automotive and industrial applications while the residues are used for making **waxes**, asphalt, and coke.

Of course, there is some overlap of the boiling points and molecular size for these fractions. The distillation process does produce mixtures. In general, though, the purpose behind oil refineries is to shift the heavy and light fractions towards gasoline.

About 90% of the crude oil going into a refinery comes out in much the same form. It is oil for use as fuel, whether it is gasoline, jet fuel, or heating oil. A small fraction, typically about 10%, is used for the synthesis of several thousand petrochemicals. This fraction is most typically drawn from the lighter fractions such as ethylene which is used for making **polyethylene** plastic while the preformed aromatic rings in the heavier fractions find use in such products as phthalic anhydride. Indeed, some portion of every fraction of crude oil finds its way into petrochemical production. This lowly 10% of crude oil accounts for over 80% of all of the organic compounds that we use. Everything from pharmaceuticals to plastics to foods are made from this small protion of the petroleum industry.

Historically, the first petrochemical was isopropanol, produced in bulk by Standard Oil of New Jersey in 1920. From this modest beginning, though, over 3,000 products can now be defined as organic petrochemicals. And some of the most important inorganic compounds, such as sulphuric acid and **ammonia**, rely upon the petrochemical industry for their feedstocks. In addition, the use of petroleum has changed the way in which some traditional compounds are obtained, such as **acetic acid** (or vinegar) and **ethanol**.

Ethylene is probably the simplest and most widely used of all petrochemicals. It is a simple compound consisting of two carbons joined by a double bond. But it is the double bond that makes ethylene such a desirable starting material as it affords the opportunity to functionalize the **molecule**. Indeed, a significant portion of the hydrogen generated by a refinery is obtained from the catalytic conversion of ethane (two carbons joined by a single bond) to ethylene and a hydrogen molecule. The hydrogen is an incidental byproduct of the more important ethylene production.

Ethylene is used in the synthesis of polyethylene, which is probably the most common form of plastic presently used.

Indeed, the production of polymers is probably the dominate use of petrochemicals. For example, halogenation of ethylene provides dichloroethane which can lose an HCl to give a chlorinated ethylene used in the production of poly(vinyl choride) or PVC. The structural modification to the polymer backbone makes this a much tougher and more resilient plastic used in **water** pipes and garden furniture. **Oxidation** of ethylene to ethylene **oxide** leads to the formation of detergents and ethanolamine (an important industrial chemical). Addition of water to ethylene oxide generates ethylene glycol used in antifreeze and the formation of **polyester** fibers. Addition of ethylene, through a Friedel-Crafts alkylation, to **benzene** affords ethylbenzene, which can be dehydrogenated to give styrene. Styrene is subsequently used to make the ubiquitous **polystyrene** and synthetic rubbers.

Ethylene is also used in the synthesis of two small molecules of significant industrial and commercial importance. The first is acetic acid and the second is ethanol. Both are compounds that were heavily used prior to the development of the petrochemical industry and were available from the **fermentation** of sugar. Indeed, the synthesis of ethanol predates recorded history and acetic acid is oxidized ethanol or wine gone bad. However, the industrial demands for these two compounds far exceeds their capacity for production from traditional routes. Greater than 95% of all the industrial ethanol used in North America is synthesized from the catalytic **hydrolysis** of the ethylene double bond. This is usually achieved via the addition of an inorganic acid, such as sulphuric acid across the double bond, followed by cleavage of the sulphate group, which illustrates the integration of the chemical industry. Petrochemicals are used to make **sulfuric acid** which is used to make other petrochemicals.

Acetic acid is made directly using the Wacker or Monsanto Process. The addition of **oxygen** to the ethylene group is achieved using a **palladium** chloride based organometallic intermediate. This intermediate is, in turn, restored using **copper** chloride that is recharged with oxygen. The net result is a complicated industrial process that hides behind the straightforward addition of O_2 to ethylene. Since its introduction, this method of making acetic acid has dominated the market. More to the point, it has facilitated many reactions that depend upon acetic acid as a reagent or solvent. Even common aspirin requires the use of acetic acid.

Ethylene is only one of the many petrochemicals used for the synthesis of other compounds. Acetylene or **ethyne** is used to also make **vinyl** chloride and vinyl acetate, both of which are converted to polymers. Acrylonitrile is another product of acetylene, formed in a reaction with hydrogen cyanide. It is used in the synthesis of acrylic fibers used in the manufacture of clothing. Even dry-cleaning fluid, which is generally trichloroethylene, can be made from acetylene.

Petrochemicals are not just confined to small molecules and polymers. Propene, in reaction with **carbon monoxide** and hydrogen via the OXO process results in n-butyl and iso-butyl **alcohol**. With benzene, it is used to synthesize first cumene and then **acetone** and phenol. Butene is used in the synthesis of methyl ethyl ketone via a process that involves the addition of sulfuric acid across the double bond, followed by hydrolysis and oxidation to the ketone. It doesn't matter whether the starting material is 1-butene or 2-butene as the final product is the same. The methyl ethyl ketone is used as a solvent for paints and lacquers.

Benzene is reacted via a number of different synthetic routes to yield everything from the cumene previously mentioned (and ultimately phenol and acetone) to **aniline**. The latter is the basis for a number of polymerization reactions yielding rubber-like chemicals and polyurethanes. But aniline is also used to make synthetic dyes. **Hydrogenation** of benzene to a cyclohexene and cleavage affords the six membered precursors necessary for 6,6-nylon. Toluene is converted to benzoic acid and through a series of conversions to **polyurethane**. In combination with **nitric acid**, toluene gives **trinitrotoluene** (TNT).

The list of products that come directly from petroleum sources or use petrochemicals in their synthesis is prodigious. It is said of the chemical industry that it is its own biggest consumer since a significant amount of production is sold by one company to another for further processing. Without petrochemicals, society as we know it would be radically different.

PH

The pH of a **solution** is a measure of the **hydronium ion** (H_3O^+) **concentration** in that solution. The pH is measured on a scale from 0-14. The hydronium ion in a solution results from the selfionization of **water**.

A drop of pure water is not composed entirely of H_2O molecules. Water always contains small quantities of two ions, hydronium ions and hydroxyl ions (OH^-). The presence of both of these ions makes water an amphoteric substance. An amphoteric substance can act either as an acid or as a base. An acid is a substance which can donate H^+ ions. When a water **molecule** acts as an acid, it becomes an OH^- ion. A base accepts H^+ ions. When a water molecule acts as a base, it becomes an H_3O^+ ion. The self-ionization of water is a reaction that occurs between two water molecules. One water molecule donates an H^+ ion to the second water molecule, forming the hydronium and hydroxyl ions. During this reaction, water acts both as an acid and as a base. In a sample of pure water at 25°C, the concentrations of both ions are 1.0×10^{-7} moles per liter (M).

For most solutions, the concentration of hydronium ion is quite small. Even in the strongest acid solutions, such as the **hydrochloric acid** (HCl) in the human stomach, the concentration of hydronium ions is only about 1×10^{-2} M. In a strong basic solution, the hydronium ion concentration is even less, as low as 1×10^{-12} or 1×10^{13} M.

Expressing the hydronium ion concentration of solutions in terms of **molarity** (moles per liter) can be cumbersome. In 1909, **Soren Sorensen**, a Danish biochemist, proposed what is now known as the pH scale. Sorensen developed a simple equation to express the hydronium ion concentrations logarithmically. The pH of a solution is -1 times the logarithm (of the base 10) of the hydronium ion concentration (expressed in moles per liter). The equation for the pH of a solution is: pH = - log[H_3O^+]. This equation is made even simpler when one realizes that a number's logarithm, to the base 10, is equal to the exponent of the number when 10 is the base. For example, for the number 10^{-3}, the log (10^3) equals 3. Likewise, the log (10^{-5}) equals 5 and the log (10^{-8}) equals 8. Therefore, the pH for a solution with a hydronium ion concentration of 10^{-1} M is 1. The pH of a solution with a hydronium ion concentration of 10^9 M is 9, and so on.

The pH scale ranges from 0-14. The smaller the pH of a solution, the more acidic the solution. A neutral solution has a pH of 7. Any pH value below 7 indicates an acidic solution, and any pH value above 7 indicates a basic solution. The pH for several common substances can give an idea of what the somewhat arbitrary pH values represent. The pH of lemon juice, which is quite acidic, is 2.5. The pH of a banana is 4.7. Coffee has a pH near 5.2. Saliva, which is somewhat neutral, has a pH of approximately 6.5. Pure water has a pH of 7, it is completely neutral. Human blood has a pH of 7.3, also quite neutral. Borax, a base, has a pH of approximately 9.2. Limewater has a pH of 10.8, and the pH of bleach, a relatively strong base, is about 12.5. Since the pH scale is based on a logarithmic scale, for every one-unit change in pH there is actually a tenfold change in the hydronium ion concentration. For example, if the pH of a solution drops from 9 to 8, the hydronium ion concentration has increased by a factor of 10, from 10^{-9} M to 10^{-8} M.

The pH of a solution can be measured in one of two ways. The simplest method is to use a product called an acid-base indicator. An indicator is a weak acid or base that undergoes a change in **color** whenever it gains or loses a hydronium ion. Litmus is an example of an indicator. In the presence of an acidic solution, litmus will gain hydronium ions and turn red. In the presence of a basic solution, litmus will lose hydronium ions and turn blue.

Each indicator has a different equilibrium constant, meaning that each has a different pH range over which it changes color. There are many indicators commercially available that turn specific colors in specific pH ranges. For example, thymol blue changes from a red color at pH 0 to yellow above pH 2. Methyl red changes from red at pH 4 to yellow around pH 6. Bromthymol blue changes from yellow at pH 4 to blue around pH 7, and phenolphthalein changes from colorless at pH 7 to pink at pH 9. Common household substances can also be used as crude pH indicators. Red cabbage juice covers the entire pH range. Grape juice is bright pink when added to an acidic solution and yellow when added to a basic solution.

A good way to pinpoint the exact pH of a solution is to test it with several indicators which operate over a broad range of pH values. Strips of **paper** are available to chemists which are coated with a combination of indicators. Placing a drop of solution on the paper will result in the paper changing color. The color of the paper can then be compared with a color chart to pinpoint the solution's exact pH. The color change in acid-base indicators is due to a chemical reaction between the indicator and the hydronium ions in the solution. An acid-base indicator is much like water in that it can have both an acid and a base form. Adding an acid to an indicator adds hydronium ions to the solution, increasing the concentration of the acid form of the indicator. Adding a basic solution to an indicator removes hydronium ions from the indicator, increasing the concentration of the basic form of the indicator.

The second method of measuring the pH of a solution is much more accurate than using acid-base indicators. This method involves using a device called a pH meter. A special electrode which measures the concentrations of hydronium ions is simply placed in a solution, and the pH of the solution is displayed on the pH meter. Although this method is much more accurate than reading color changes on indicators, it is also much more expensive. The acid-base indicators serve as a good measurement of the pH of solutions whenever exact precision is not necessary.

Sometimes it is necessary to maintain a solution at a constant pH. This is especially true in bodily fluids such as blood, which needs to be kept neutral (between a pH of 7.35 and 7.45). If the pH of blood is allowed to vary outside of this range, serious illness or death may occur. Buffers are substances which control the pH of a solution. A **buffer** is usually a mixture of **acids and bases**. This mix of acids and bases allows the buffer to release or absorb hydronium ions, which keeps the pH of a solution constant. The most common buffers are mixtures of weak acids and their conjugate bases. A buffer cannot keep the pH of a solution under absolute control under all conditions. There is a limit to the ability of a buffer to maintain a constant pH, called the buffer capacity. In general, the greater the concentration of buffer in a solution, the greater the buffer capacity.

PHARMACEUTICAL CHEMISTRY

Pharmaceutical **chemistry** is one of the most active branches of applied chemistry, although much of the research in general chemistry and other biological and physical sciences are useful to it. Pharmaceutical chemistry concentrates on the chemical composition and properties of drugs and other medicinal substances. All of the substances that appear in the *National Formulary* and the *U.S. Pharmacopeia* are the result, at least in part, of efforts in the field of pharmaceutical chemistry.

It is impossible to underestimate the importance of this branch of chemistry to the welfare of all humans and animals. The discipline has produced innumerable substances crucial to preventative, therapeutic, and diagnostic purposes. For instance, pharmaceutical chemistry was behind the discovery that **alcohol** can be converted to **ether**, a highly effective anesthetic. Likewise, the science helped to give us lifesaving treatments from such natural substances as insulin (diabetes), **penicillin** (infections), thyroid (hormone-related illnesses), and digitalis (heart problems).

Pharmaceutical chemistry has historically focused on finding and isolating the active chemicals in plants (known in the industry as "raw materials" or "components"), and it continues to do so today. This trend began in 1805, when an apothecary (a latter- day pharmacist) isolated **morphine**, another powerful anesthetic, from opium, a derivative of the poppy plant. From that point until fairly recently, pharmaceutical chemistry was also known as plant chemistry. The field continues to be a major center of research—particularly because experts believe that more than 97% of the world's plants remain unidentified and unexamined. This factor is what lends a sense of urgency to pharmaceutical chemistry and helps to make it such an active field of research: many species of plants are becoming extinct faster than scientists even learn of their existence, much less identify and analyze them for any possible beneficial qualities.

There are certain goals, activities, and practices that help to distinguish this branch of chemistry from the others. Although these are not exclusive to the field, they tend to occur especially in pharmaceutical chemistry. There is the most potential for exciting discovery in the isolation and identification of agents in natural sources that could, for instance, prevent cancer or cure the common cold. However, in the day-to-day world of medicinal chemistry, as it is also called, scientists frequently spend a great deal of time making artificial copies of natural medicinal agents, whether because the natural source is now unavailable or because the artificial version is more pure. They also investigate which form of an agent will be the most effective, keeping in mind how easy the formula will be to make and dispense, and ascertain whether there will be any biological or chemical incompatibilities among a certain prescription's various ingredients.

Pharmaceutical chemists also ''semisynthesize'' drugs, which involves changing a natural medicinal substance so that its therapeutic properties are intensified or modified in some more favorable way. In addition, since a 1906 law went into effect that made analysis of drug quality and potency a legitimate research area, pharmaceutical chemists are concerned with making sure all medications are safe, uniform, and practical in terms of their quality and dosage. In fact, the subdiscipline of pharmaceutical testing has arisen to manage and regulate the products of pharmaceutical chemistry.

PHASE CHANGES

A phase is defined as a physically distinct region. Under certain conditions, every substance may exist in one or more phases. It may be a solid, a liquid, or a gas (or vapor). Also, some substances have allotropic solid forms. For instance, **carbon** may exist as either **graphite** or **diamond** in its solid phase.

It has long been known that most substances can change from one phase to another. Aristotle (384-322 B.C.) noted that **water** could be changed into ''air'' (gas) and could come out of ''air'' as well.

The phase change from liquid to gas is called vaporization or boiling (depending on the circumstances), and the opposite change, from gas to liquid, condensation; from solid to liquid, melting, and the opposite, from liquid to solid, freezing; and from solid to gas, **sublimation**.

Energy must be added or removed from substances during phase changes; e.g., **heat** must be added as water boils; the evaporation of sweat removes heat from the skin; and a refrigerator removes heat from liquid water to form ice.

In general, phase changes are accompanied by changes in **volume**. At a particular pressure, the same weight of a substance occupies a larger volume as a gas than as a liquid or solid. Most substances expand when they melt. An important exception is water: ice occupies a larger volume than liquid water.

Phases and phase changes can be understood on the basis of molecular behavior. The molecules of the gas phase are moving randomly within their container, and there is great disorder. The motion is rapid, and the attractive forces between molecules are not great enough to significantly affect their relative motion. As the **temperature** is lowered, the molecular translational energy decreases until a point is reached where the attractive forces between molecules are able to cause a coalescence into the liquid phase. The liquid molecules possess translational energy and are still moving about, but they are now much closer to each other. Thus the volume of the liquid phase is smaller than that of the vapor phase. Moreover the now-effective **intermolecular forces** cause the molecules to arrange themselves into ordered groups. This grouping is transitory and is constantly changing, but at any given instant there is considerably more order in the liquid phase than there was in the vapor. When the temperature is lowered further, the translational energy continues to decrease until a second point is reached at which the attractive forces between molecules succeed in stopping the translational motion completely, and the molecules arrange themselves in fixed, almost totally ordered patterns relative to each other. A decrease in volume usually (but not always) accompanies this solidification as the molecules pack more tightly together.

Vaporization of a liquid may occur over a wide range of temperature and pressure. Wet clothes will dry, and a pan of water will slowly evaporate to dryness. Under virtually all conditions, some of the molecules near the surface of a liquid attain enough translational energy to pull away from the attraction of their neighbors, overcome the opposing external pressure, and escape into the gas or vapor phase. Similarly, molecules in the vapor phase occasionally strike the surface and are captured by the attraction of molecules in the liquid phase. When the liquid is not confined, most gas phase molecules will move away from the liquid, and few will reunite with the liquid. In this case, the liquid will eventually evaporate completely. If, however, the liquid and gas are confined within a container, a point is reached when the number of molecules returning to the liquid phase from the vapor is equal to the number escaping. This circumstance is called equilibrium.

Molecules in the vapor phase run into the walls of the container and thus exert a pressure on the container. When the vapor is in equilibrium with the liquid, this pressure is called the **vapor pressure**. Its value depends on the temperature. For a given external pressure, there is a temperature at which the vapor pressure of the liquid is equal to the external pressure. Under this condition, the liquid can vaporize quickly and completely. The term boiling is reserved for vaporization which occurs under these conditions. A liquid may boil over a range of temperatures, depending on the external pressure. As the pressure increases, the temperature of boiling must also increase for the molecules to be energetic enough to escape en masse into the vapor phase.

The normal **boiling point** of a substance is a characteristic of the substance and is defined as the temperature at which its liquid to gas phase transition occurs when the external pressure is equal to that exerted by the atmosphere at sea level. This pressure is equal to 14.7 lbs/in^2 and is equivalent to the pressure exerted by a column of **mercury** 29.9 in (76 cm) tall. The normal boiling point of water is 212°F (100°C), and its normal melting point is 32°F (0°C). At an external pressure of 27.3 in (70 cm), water boils at 207.9°F (97.7°C); at 31.2 in (80 cm), the boiling point is 214.5°F (101.4°C).

In a manner similar to the vaporization of a liquid, sublimation of a solid may occur over a wide range of temperature. Wet clothes hung out in sub-freezing temperatures will dry, and ice will disappear without melting. Individual molecules at the surface of the solid become energetic enough to break away into the gas phase and sublimation of the solid occurs slowly. A solid confined in a container will reach a characteristic equilibrium vapor pressure at a given temperature.

There is one unique combination of values of temperature and pressure at which a substance may exist in equilibrium simultaneously as a solid, liquid, and gas, and transitions may occur between the phases. This is called the triple point. The triple point of water occurs at an external pressure of.18 in (.46 cm) and a temperature of 32.02°F (0.01°C).

When the temperature is increased sufficiently, a point is reached beyond which a substance can only exist as a gas: a phase transition to its liquid or solid form is not possible, regardless of the external pressure exerted on it. This unique point is known as the critical temperature. The critical temperature of water is 705.3°F (374°C).

Phenol functional group

The phenol **functional group** is an -OH attached to a **benzene ring**. There may be one or more hydroxyl molecules attached to the ring structure.

The phenol functional group is acidic. Phenol-containing compounds are soluble in alkalis, producing the appropriate metallic **salt** in a compound called a phenate. Phenol-containing compounds are unusual amongst organic compounds because of their **solubility** in **water**. This solubility increases if more hydroxyl groups are added to the ring structure. The solubility is due to **hydrogen bonding**. The simplest compound containing the phenol group is actually called phenol, it has the formula C_6H_5OH. The phenolic group is C_6H_4OH. The phenol functional group is phenol with the loss of one hydrogen producing a negatively charged **molecule**.

Phenol was originally isolated from **coal** tar in 1834 by Friedlieb Ferdinand Runge (1795-1867), who named it carbolic acid. The name phenol was first used in 1841 by **Charles-Frédéric Gerhardt**.

All simple phenol-containing compounds are **liquids** at standard **temperature** and pressure. The phenols all have a characteristic sweet smell and they are colorless when pure. As well as being soluble in water, phenol-containing compounds are also soluble in organic **solvents**. The benzene ring of the phenol group is an unsaturated compound (i.e., there are double bonds present). The benzene ring is responsible for some of the characteristic reactions of phenol-containing materials. The remainder of the characteristic reactions are due to the presence of the hydroxyl group. The unsaturated benzene ring readily reacts with **halogens**, rapidly discoloring **bromine** water. Nitration will occur in a similar manner with dilute **nitric acid**, producing nitrophenols. **Sulfuric acid** will react in similarly to give phenosulfonic acids. The reactions of the -OH group are similar to those of the alcohols, but with modification by the ring structure. The reaction with any electropositive metal produces the appropriate salt and hydrogen. In general, alkalis do not react with alcohols, but they do react with phenolic compounds. The phenate is produced along with water (a phenate is the phenol group with the hydrogen from the hydroxyl replaced by the metal). Phenol-containing compounds can be reduced to benzene by passing the vapor over a heated **zinc** catalyst. A particularly important reaction is with the aldehyde methanal. This produces a category of compounds called phenol methanal resins, these are extensively used in the **plastics** industry.

Compounds containing the phenol group are widely found in nature and they include many plant pigments, tannins, and naturally occurring alkaloids.

Compounds containing the phenol functional group are extensively used in industry, and in the late 1990s, the annual production of such compounds in the United States was in excess of 2 megatonnes. Phenol-containing compounds are widely used as disinfectants, dyes, **explosives**, perfumes, **plastics**, and pharmaceuticals.

Phenyl group

The phenyl group is a monovalent organic radical, derived from **benzene**. It has the chemical formula C_6H_5 and it is sometimes abbreviated to -Ph. The phenyl group is a modifier of organic functional groups. The phenyl group is basically a benzene **ring** with a **hydrogen** removed. The phenyl group is a ring structure, and as such, it is an aryl group (an organic group derived from an aromatic hydrocarbon with the absence of one hydrogen).

Similar to other organic modifier groups, the phenyl group alters the physical and chemical properties of the substance it bonds to. For example, the **boiling point** of a chemical can be increased by replacing a hydrogen **atom** with the phenyl group. In general as the **molecular weight** of a compound increases, the boiling point increases. The phenyl group is unsaturated, that is there are double bonds present that do not have the maximum number of hydrogens attached to them. As such, compounds containing the phenyl group are able to participate in all of the reactions of an unsaturated compound. They are able to undergo **hydrogenation** with hydrogen adding across the double bonds. **Bromine water** added to a compound with the phenyl group will decolorize. This is a standard test for unsaturated compounds.

A view of an insect trap baited with phermones in an orchard. Insect phermones are odors produced by insects to attract each other for reproduction. Traps baited with artificial phermones are used to biologically control insects. *(Photographs by Astrid and Hanns Frieder Michler/Science Photo Library, Photo Researchers, Inc. Reproduced by permission.)*

The polymer **polystyrene** contains the phenyl group as part of its repeat unit. The organic chemical phenol could also be called phenyl hydroxide, it has the chemical formula C_6H_5OH.

Phenylalanine contains the phenyl group and is one of only three **amino acids** containing an aromatic group (a ring structure). As part of the **metabolism** of phenylalanine, **oxygen** is used to break the ring structure, and eventually after several intermediate steps, the aliphatic (straight chain) molecules fumarate and acetoacetate are produced. The absence of the functioning **enzyme** that catalyzes this reaction produces a disease called phenylketonuria, which is genetically controlled and leads to mental disorders. **Aniline** itself contains the phenyl group and its formula can be written as $PhNH_2$. Aniline is used in the manufacture of a number of products including dyes, pharmaceuticals, **antioxidants**, and vulcanization accelerators in rubber compounds.

The phenyl group is often encountered as an intermediate in many organic reactions. This is particularly common in reactions occurring in **solution**. The phenyl group by itself is unstable and, as such, will rapidly react with some other compound.

Pheromones

Pheromones are volatile chemical compounds secreted by insects and animals. They act as chemical signals between individuals influencing physiology and behavior in a manner similar to **hormones**. Pheromones are important to a variety of behaviors including mate attraction, territorality, trail marking, danger alarms, and social recognition and regulation.

The term pheromone is derived from the Greek words *pherein* (to transfer) and *hormainein* (to excite). In animals, they are produced in special glands and are released through body fluids, including saliva and perspiration. Most pheromones are biogenetically derived blends of two or more chemicals that must be emitted in exact proportions to be biologically active.

There is a remarkable diversity in the **stereochemistry** of pheromones. Insects are sensitive to and utilize chirality to sharpen the perception of pheromone messages. The configurations of pheromones are critical. Stereoisomers of pheromones, for example, can also be inhibitors of the pheromone action.

Pheromones are found throughout the insect world. They are active in minute amounts. In fact, the pheromones released by some female insects (e.g., Silkworm Moth) are recognized by the male of the species as far as a mile away. The pheromone secreted by the female gypsy moth can be detected by the male in concentrations as low as one **molecule** of pheromone in 1xIO^{17} molecules of air. Insects detect pheromones with specialized chemosensory organs.

At close range, pheromones continue to be released dictating specific behaviors. Another common example of pheromones in action is the trailing behavior of ants. Scout ants release pheromones that guide other ants to the location of food. In boars, pheromones found in boar saliva are known to cause the female to assume a mating position.

An increasingly important use of pheromones involves the control of insects. Because insects rely on phermomones, these compounds have been used by farmers as a method to control harmful insects. Using insect sex attractant pheromones, scientists have been able to produce highly specific traps and insecticides.

Pheromone traps are used to control the insects such as the European corn borer that damages millions of dollars of crops each year. The European corn borer larvae feed on and bore into the corn plant. Cavities produced by borers reduce the strength of the corn and interfere with plant physiology, including the translocation of **water** and nutrients. European corn borer pheromone traps contain a substance that mimics (i.e., acts like) a part of the chemical communication system used by female moths when they are are receptive to mating. Male moths are attracted to and captured by the pheromone trap that is coated with a sticky substance that retains attracted insects.

Research continues on insect pheromones. It is believed that these compounds hold the key to developing insecticides that can kill only harmful insects while being harmless to humans and beneficial insects.

The search for human aphrodisiacs (stimulants to sexual response) is as old as human history. Although the scientific evidence with regard to human pheromones is contradictory and highly debatable, pheromones are often used as an olfactory aphrodisiac in fragrances and perfumes.

The first discovery related to human pheromones was reported the early 1970s. At this time low **molecular weight** aliphatic acids, called copulins, were found in the vaginal secretion of women. At the time, it was believed that these compounds could stimulate male sexual response. They were thought to work as did their chemical counterparts excreted by monkeys, baboons and chimpanzees. In the late 1970s more alleged human pheromones were discovered in human perspiration and urine. Some studies suggest a role for pheromones in the regulation and synchronization of the human female menstrual cycle.

The organ responsible for detecting pheromones in animals is a chemosensory structure in the nose called the vomeronasal organ (VNO). In lower animals, this organ detects substances that mediate sexual and territorial behaviors in species. It was once generally believed that humans did not have a VNO. Embryological texts asserted that this organ disappeared during embryonic development. In the 1980s, however, investigations refuted this alleged disappearance. Subsequent research suggested that a functioning VNO was present in near two small holes on the hard divider in the nose. A group of cells similar to nerve cells are located behind these holes. These cells, which make up the VNO, transmit a signal to the hypothalamus in the brain. The stimulating effect on the hypothalamus results in the production of **proteins** that may influence behavior.

PHLOGISTON

During the seventeenth century, chemists needed a broad conceptual system to explain what caused the chemical reactions they observed. The phlogiston theory of **combustion** was the first comprehensive theory of **chemistry** and it served to satisfy the need for a broad system of explanation. This theory was based on phlogiston, a substance that chemists thought was contained in all combustible substances. The phlogiston theory was the first theory of chemistry that was able to explain a wide range of phenomena.

An example of a reaction that the phlogiston theory could explain is the preparation of **metals** from their ores by heating with **charcoal** (smelting). The transformation of an earthy substance, the ore, into a metal by this process appeared to be much the same whether the metal involved was **iron**, **tin**, or **copper**. The fact that metals became ores when heated and could be changed back into metals in the presence of charcoal was hard to reconcile without imagining the addition or subtraction of some substance. Therefore, it seemed plausible to assume that in each instance the ore, when heated with charcoal, took up a metallizing principle that conferred upon the earth the properties of a metal. This metallizing principle was first named *terra pinguis* (fertile earth) by **Johann Becher** in 1669. When metals were calcined, the *terra pinguis* escaped, leaving behind a metallic calx (what we today call an oxide). **Georg Ernst Stahl**, in 1737, renamed Becher's *terra pinguis* phlogiston (from the Greek word for burning). It was almost exactly similar to *terra pinguis*, in that when a metal burned, it released phlogiston leaving a calx behind.

The phlogiston theory of combustion had strong explanatory power. During the **smelting** process, burning charcoal with the ore resulted in the metal. Since charcoal burns in air leaving only a small quantity of ash, it was clear that charcoal was rich in phlogiston. The reason a metal formed when its calx was heated with charcoal was therefore because the phlogiston left the charcoal and united with the calx.

It was easily observable that a burning substance would cease burning if the supply of air was cut off. **Robert Boyle** had already demonstrated in the mid-seventeenth century that combustion would not occur in a vacuum. The phlogiston theory of combustion was able to explain this by assigning to air the ability to absorb phlogiston. When the air was fully saturated with phlogiston, combustion ceased. Combustion, therefore, according to the phlogsiton theory, was completely impossible in a vacuum because there was no air to absorb the phlogiston.

Combustion of antimony in chlorine. ***(Photograph by Yoav Levy/ Phototake NYC. Reproduced by permission.)***

Phlogiston was accepted by the 1770s despite some of its anomalous properties. Chemists could not satisfactorily explain why metals gained weight during combustion. If metals gave off their phlogiston during their combustion, should they not lose weight? Most chemists were not concerned with this detail because during the first half of the eighteenth century chemistry was not concerned with **quantitative analysis**. It was enough for most chemists that the phlogiston theory gave a good qualitative explanation of combustion. But towards the end of the eighteenth century, some chemists reconciled themselves to this gain in weight upon combustion by supposing that phlogiston had a negative weight, and therefore upon its expulsion from a metal on combustion would result in the metal getting heavier.

In 1772, **Antoine-Laurent Lavoisier** began working on the problem of combustion. He was not satisfied with the explanation of the weight gain of metals on combustion. Guyton de Morveau (1737-1816), a colleague of Lavoisier, had performed experiments showing that metals did without question increase in weight on combustion. Lavoisier used these results to propose his first theory of combustion—in all cases of combustion where an increase in weight is observed, air is absorbed, and that when a calx is burned with charcoal, air is liberated. This air he later called **oxygen**. This oxygen was what combined with metals during combustion and was what caused of the increase in a metal's weight on combustion.

Lavoisier's oxygen theory of combustion was almost the exact opposite of the phlogiston theory of combustion. In the phlogiston theory, phlogiston is released during combustion, and in the oxygen theory, oxygen is absorbed during combustion.

PHOSPHATE

A phosphate is a **salt** that is prepared from orthophosphoric acid. Phosphates are usually **sodium** salts but can also include **potassium**, **calcium**, **magnesium**, and ammonium salts as well. The phosphates have a wide range of applications in the chemical industry. The mono-, di-, and tri-metallic salts of **phosphoric acid**, for example, are used extensively in many industrial applications. Polyphosphates are a group of phosphates that are created from polymeric chemical compounds. These compounds are produced from the condensation reactions of several phosphoric acid molecules. The chemical formula for the phosphate **ion** is PO_4^{3-}. The phosphate ion is a polyatomic ion, in other words, a group of covalently bonded atoms that together carry an electrical charge. Polyatomic ions can combine with oppositely charged ions, through ionic **bonding**. When they do so, they form ionic compounds.

Because the phosphates have many industrial applications, they are produced in great quantities by the chemical industry. Phosphoric acid (HPO_3) is required in the production of phosphates, as well as sodium, potassium, and calcium hydroxides [NaOH, KOH, and $Ca(OH)_2$]. Phosphates can exist in a variety of hydrated forms, each with a different stability. The two most stable species of phosphates are the two extremes of **hydrates**, the anhydrous salt and the fully hydrated salt. These are the two phosphate species that are produced in the greatest quantities, although phosphates of every degree of hydration are available. Many grades of phosphates are also commercially available in terms of purity, from low-grade to "technical" and "for use in foods."

The production of phosphates is highly automated. The **pH** of the reaction vessel must be kept within certain specific ranges, depending on the properties and chemical nature of the phosphate being produced. The pH values correspond to the various stages of the **neutralization** of phosphoric acid involved in the production of phosphates. The pH is adjusted by adding alkali or **ammonia** (to raise the pH) as necessary and then crystallizing out the salt at the optimal **temperature** for crystallization. For example, both disodium and dipotassium **hydrogen** phosphates are prepared in a **solution** with a pH value of 8.5, attained by the addition of alkali to phosphoric acid. If the anhydrous form is being prepared, the solution is then concentrated to remove the **water** and crystallized. The crystals are separated from any remaining solution by centrifugation. In general, the production of sodium phosphates is much less expensive than the potassium phosphates. As a result, more sodium phosphates are produced by industry. Potassium phosphates are produced in quantities when they are specifically needed, for example, when working with bacterial cultures the potassium ion is much more biologically compatible than the sodium ion.

Phosphates are used in a wide range of applications. Phosphates can act as buffers to control the pH values of various solutions. They are also used in water treatment and the production of sugar solutions, the etching of **metals**, in polymerization processes, and in the textile industry. Phosphates can aid in the emulsification and protection of colloidal suspensions for the food industry as well. Certain phosphates, monoammonium dihydrogen phosphate and diammonium hydrogen phosphate, in particular, evolve non-flammable **gases** when heated and are used as flame retardants. These same two phosphates are also used as nutrients for cultures of microorganisms. Sodium, potassium, and calcium phosphates are all used in the chemical and pharmaceutical industries. Condensed phosphates, or polyphosphates, are the choice for water treatment and detergents because of their ability to form strong complexes with the unwanted ions in hard water.

Polyphosphates are prepared differently from other phosphates. Orthophosphoric acid is sometimes heated to **dehydration** to form polyphosphoric acids, which are then neutralized to form polyphosphates. More commonly, the salts of orthophosphoric acid are simply heated to form the polyphosphates. There are several specific polyphosphates which are produced in the chemical industry. For example, the tetra-alkali metal ion dimers sodium pyrophosphate ($Na_4P_2O_7$) and potassium pyrophosphate ($K_4P_2O_7$), which are used to make liquid detergents, to etch metal, to stabilize emulsions, and as **food additives**. Another polyphosphate that is produced commercially is the dimeric disodium salt, disodium dihydrogen pyrophosphate ($Na_2H_2P_2O_7$), which is used in both the food and mining industries. Another group of polyphosphates which are useful industrially are the ultraphosphates, which have the general chemical formula $(MePO_3)_n$. Sodium hexametaphosphate (n=6) is one of the most important of the ultraphosphates, used in both detergents and dispersing agents. Another polyphosphate which is used in detergents is sodium tripolyphosphate, $Na_5P_3O_{10}$. This polyphosphate acts as a support for **silicates** and bleaching agents in the detergents.

Phosphates are just as important biologically as they are industrially. The phosphate **buffer** system in the cytoplasm of our cells is extremely important in maintaining the pH of the cytoplasm between 6.4 and 7.4. The phosphate buffer system consists of the hydrogen phosphate ion (HPO_4^{2-}) and the dihydrogen phosphate ion ($H_2PO_4^-$). The **DNA** in our cells also relies on phosphates. Phosphate groups make up a significant part of the DNA backbone and provide the sites where H^+ ion transfer can occur.

One of the most important applications of phosphates, biologically and economically, is their use in **fertilizers**. Until the middle of the nineteenth century, when the use of phosphates in fertilizers began, the amount of **phosphorus** compounds that could be used by plants was very low. The phosphorus content of soils was so low, in fact, that the crop yields were declining. Two reasons the phosphorus levels in the soil were so low is the fact that phosphorus only makes up about 0.1% of the Earth's crust, and naturally occurring phosphorus compounds, such as those found in manure, are generally insoluble in aqueous solutions. The development of phosphate fertilizers increased the yield of crops. The extra phosphorus available in the soil gave rise to stronger, healthier plants.

With these benefits of phosphates have come problems. Phosphates that leach out of the soil and drain into lakes, streams and ponds are causing major ecological disturbances. A build-up of phosphates in stagnant waters causes the eutrophication of the waters. Phosphates from fertilizers are not the only cause, either. Recent improvements in detergents have called for the addition of more phosphates which also find themselves in water systems. Legislation has been passed which requires the amounts of phosphates in effluents to be maintained below a maximum level. Research efforts have been established to find alternatives to phosphates, especially in detergents. Some possible alternatives include trisodium citrates, zeolites, or nitrilotriacetic acid (NTA), all of which could also cause hazardous health effects in large amounts. Despite these problems associated with phosphates, however, they are still considered some of the most important compounds in the chemical industry and the benefits of their use far outweigh their hazards.

Phosphoric acid

Phosphoric acid, also termed orthophosphoric acid, is a substance with the chemical formula H_3PO_4. At high concentrations in aqueous **solution**, it is a relatively weak acid. The **pH** value at which a one molar solution exists as half orthophosphoric acid and half dihydrogen **phosphate ion**, $H_2PO_4^-$, the pK_1, is 2.15. At a pH of 7.20, the pK_2 a one molar solution of phosphoric acid exists as equal concentrations of dihydrogen phosphate ion and monohydrogen phosphate ion, $H_2PO_4^{2-}$ and at pH 12.37, the pK_3 the solution consists of equal concentrations of monohydrogen phosphate and phosphate ion, PO_4^{3-}. At lower concentrations, a greater proportion of deprotonation occurs. Salts of all three anions can be prepared by using solutions of appropriate pH. Phosphoric acid is not an oxidizing acid: the **phosphorus atom** in the phosphate ion is fully oxidized.

Orthophosphoric acid is produced in quantities of millions of tons each year, principally from a process in which rocks containing phosphate minerals such as **calcium** fluoride orthophosphate, $Ca_5(PO_4)_3F$, are crushed and dissolved in aqueous **sulfuric acid** solutions to produce solutions of phosphoric acid and, in this case, hydrofluoric acid, leaving behind insoluble portions of the rock and gypsum (hydrated calcium **sulfate**, $CaSO_4.2H_20$). Most of the phosphoric acid that is commercially produced is converted to salts used in **fertilizers**. Much purer solutions of orthophosphoric acid are produced by **roasting** phosphorpentoxide or phosphorus (IV) **oxide**, P_4O_{10}, in the presence of **water**. This process is used to make the orthophosphoric acid used in soft drinks (the major use), detergents, and other consumer items. Phosphoric acid gives soft drinks a slightly tart **taste**.

Phosphoric acid and its phosphate derivatives are essential in human **nutrition**. The so-called ''backbone'' of **DNA** as well as the linkages in **RNA** are phosphate diesters. The ''energy-storage'' **molecule** prevalent in all living systems is adenosine triphosphate (ATP). Adenosine triphosphate is the nucleotide adenine found in DNA and RNA with an extended phosphate chain. When the terminal unit of the triphosphate chain is hydrolyzed in a cell to form adenosine diphosphate (ADP) and phosphate ion **energy** is released due to hydration of the phosphate ion. Energy is stored when ADP and hydrated phosphate ion are converted by a coupled biochemical reaction to form ATP. A common misconception is that there is a weak phosphate bond in ATP. Not so, the energy is released and absorbed in hydrating and dehydrating the highly charged phosphate ion.

Industrially, phosphoric acid and its derivatives are used in metal cleaning and treatment. The phosphate ion is strongly attracted to metal ions, particularly 2+ and 3+ ions, due to its high negative charge, so it is good at removing ferric and ferrous ions in rust, calcium and **magnesium** salts that are impurities and contaminants, and so forth. It is also used as a catalyst in the production of ethylene and in the manufacture of numerous commercial products.

Phosphorus

Phosphorus is the second element in Group 15 of the **periodic table**. The elements in this group are sometimes referred to as the **nitrogen** family of elements. Phosphorus has an **atomic number** of 15, an atomic **mass** of 30.97376, and a chemical symbol of P.

Properties

Phosphorus exists in three allotropic forms, named after their colors: white (or yellow) phosphorus; red phosphorus; and black (or violet) phosphorus. White phosphorus is a waxy, transparent solid with a melting point of 111°F (44.1°C), a **boiling point** of 536°F (280°C), and a **density** of 1.88 grams per cubic centimeter. If kept in a vacuum, white phosphorus sublimes if exposed to light. White phosphorus is also phosphorescent, giving off a beautiful greenish white glow. It does not dissolve in **water**, although it does dissolve in many organic **liquids**, such as **benzene**, chloroform, and **carbon** disulfide. White phosphorus sometimes appears slightly yellowish because it contains traces of red phosphorus.

Red phosphorus is a red powder produced by heating white phosphorus in the presence of a catalyst. Red phospho-

Two phosphorus compounds are used to make the coating found on the tips of safety matches. ***(Photograph by Robert J. Huffman. Reproduced by permission.)***

rus does not melt when heated, but sublimes at a **temperature** of 416°C (781°F). Its density is 2.34 grams per cubic centimeter. It does not dissolve in most liquids.

Black phosphorus looks like **graphite**. It can be made by exposing the white **allotrope** of phosphorus to high pressures. The black allotrope has a density of 3.56-3.83 grams per cubic centimeter. One of its unusual properties is that it conducts an electric current although it is a **nonmetal** and not a metal.

White phosphorus is the most active form of the element, with a tendency to catch fire spontaneously at room temperature. For this reason, white phosphorus is usually stored under water in chemical laboratories as a safety precaution. All allotropes of phosphorus also combine with the **halogens** and with **metals** to form compounds known as phosphides: $3Mg + 2P \rightarrow Mg_3P_2$ (magnesium phosphide).

Occurrence and Extraction

The abundance of phosphorus in the Earth's crust is estimated to be 0.12%, making it the 11th most common element. It usually occurs as a **phosphate**, such as **calcium** phosphate [$Ca_3(PO_4)_2$], the major component of phosphate rock. The United States is the largest producer of phosphate rock in the world, with about 13,000,000 metric tons being mined each year. That amount is about a third of the world's total phosphate rock. Nearly 90% of phosphate rock comes from two states: North Carolina and Florida. Other producers of phosphate rock include Morocco, China, Russia, Tunisia, Jordan, and Israel.

Discovery and Naming

Phosphorus was discovered in 1669 by the German physician Hennig Brand (ca. 1630-1692). Brand is famous in the history of **chemistry** as being the last of the alchemists. Brand was convinced that the key to changing base metals into **gold** could be found in human urine. In the process of heating and purifying urine for his experiments, he obtained a white wax substance that glowed in the dark. The element was named for this phenomenon, the process of phosphorescence.

The dangerous properties of elemental phosphorus were discovered early in Brand's research. The story is told that one of his servants left some phosphorus on top of Brand's bed. Later that night, the bed covers burst into flame when the phosphorus caught fire spontaneously.

Uses

Elemental phosphorus has relatively few uses because of its tendency to ignite spontaneously. However, many of its compounds are widely used. More than 90% of all phosphate rock mined in the United States is converted to synthetic fertilizer. Phosphorus is one of three macronutrients needed by growing plants, the other two being nitrogen and **potassium**. The production of synthetic **fertilizers** to meet the needs of farmers in the United States and around the world is now by far the greatest application for compounds of phosphorus.

Another familiar, although much less important, use of phosphorus is the manufacture of **wood** and **paper** safety matches. Phosphorus pentasulfide (P_2S_5) and phosphorus sesquisulfide (P_4S_3) are used to coat the tip of the match, providing a material that ignites easily when scratched. Another compound of phosphorus with many uses is phosphorus oxychloride ($POCl_3$), used in the manufacture of **gasoline** additives, in the production of certain kinds of **plastics**, as a fire retardant agent, and in the manufacture of transistors for electronic devices.

Health Issues

Phosphorus is essential to the health of plants and animals. One of the most important compounds in living cells is adenosine triphosphate (ATP), an energy-carrying **molecule** that makes possible many of the chemical reactions that take place in cells. Phosphorus is also critical to the development of strong bones and teeth in the form of the compound fluorapatite. Finally, phosphorus is a critical component of **nucleic acids** that carry genetic information in cells and that provide the instructions which direct the activities of those cells.

On the other hand, elemental phosphorus is highly toxic. Swallowing even a speck of white phosphorus can cause diarrhea with loss of blood; damage to the liver, stomach, intestines, and circulatory system; and coma. A piece of white phosphorus no larger than 50-100 milligrams (about 0.0035 ounce) can even cause death. Interestingly enough, red phosphorus seems to be much less toxic than is white phosphorus.

PHOSPHORUS CYCLE

The **phosphorus** cycle is one of many mineral nutrient cycles demonstrating the constant cycling of elements through organisms and the environment. The **nitrogen** cycle and **carbon** cycle are the two most well known examples of these natural cycles. The phosphorus cycle differs from these other cycles in that it does not have an atmospheric component to the cycle.

Plants are the first link in the phosphorus cycle. The plants obtain their phosphorus directly from the soil. In what form this phosphorus is found in the soil is dependant upon the **pH**. In alkaline conditions, above pH 7, the phosphorus is encountered as hydrogenphosphate (HPO_4^{2-}). At more acidic pHs, the phosphorus is encountered as dihydrogenphosphate ($H_2PO_4^-$). The dihydrogenphosphate is taken up more quickly by plants than the hydrogenphosphate. Once the **molecule** is in the plant it remains in the form PO_4^{3-}. In this form it can be encountered in a number of molecules of importance to living organisms. These molecules include **nucleic acids**, phosphorylated **carbohydrates**, and fats.

Animals can obtain their phosphorus directly form the plants if the animal is a herbivore. If the animal is not an herbivore the phosphorus is obtained from eating other animals which are herbivores. Once an organism has died, be it plant or animal, the phosphorus is returned directly to the soil by the action of decomposers such as bacteria and fungi. It is returned as a **phosphate ion**.

In most regions phosphorus is in short supply. Because of this, most phosphorus is encountered when incorporated into living organisms. The exception to this is phosphorus deep in the Earth. This phosphorus has no chance of being assimilated into a plant and consequently it does not take place in the phosphorus cycle, unless there is some major upheaval which exposes it.

The phosphorus found in soil originally came from rocks. The release of phosphorus from rocks is a very slow process. Phosphorus, once in the soil, is relatively immobile, leading to localized areas of differing concentrations. Because of this phosphorus is often the limiting factor on plant growth. This is especially true for aquatic plants. Phosphorus uptake is greatly aided by the presence of fungi. Certain species of fungi are able to form a symbiotic relationship with plants called a mycorrhiza. These mycorrhizal relationships benefit plants by allowing increased uptake of limiting nutrients, such as phosphorus. This is achieved partly by the biological action of the fungus but it is mostly a function of the increased surface area. The fungus grows over the root of the plant and covers the root in a sheath. From this sheath, individual strands of fungal mycelium spread out into the soil. This ensures that absorption of material takes place from a much larger **volume** than would previously have occurred. This makes more phosphorus available to the plant. In return the fungus receives food in the form of leached photosynthate from the plant.

As can be seen the phosphorus cycle is a relatively simple cycle with immediate **recycling** and assimilation of any phosphorus found in a dead organism. Because phosphorus is often a limiting factor for the growth of plants the application of phosphate **fertilizers** is a very common practice. Excess phosphorus that finds its way into **water** supplies can be problematic. Phosphorus also is a limiting factor for aquatic plants. When it is encountered as agricultural runoff this can **lead** to massive growth of the aquatic plants. This can in turn lead to blocked waterways and deoxygenation of the water, leading to the death of aquatic animals.

The phosphorus cycle is a simple example of a mineral nutrient cycle.

PHOTOCHEMISTRY

The term photochemistry is interpreted broadly to include a range of both chemical and physical changes that result from interaction of **matter** with electromagnetic radiation from the infrared range far into the ultraviolet. A photochemical process is characterized by the participation of an excited electronic state in the chemical or physical transformation. An excited state is a configuration of a **molecule** or **ion** with an **energy** higher than the most stable configuration (the ground state) of the molecule or ion. An electronic excited state involves the pattern of **electron** distribution in the **atom** or molecules that is different from the ground state electron distribution. In order for a chemical transformation to occur, the excited state must have an energy greater that is than the ground state by an amount similar to that of the energy of a chemical bond (on the order of one hundred to several hundred kilojoules/mol). Electromagnetic energy of this order of magnitude occurs in the near infrared to ultraviolet regions of the spectrum.

A physical photochemical change can take place when the energy difference between an excited state and a ground state is lower than that needed to break chemical bonds. An example of a physical photochemical process is fluorescence, whereby a molecule or ion absorbs electromagnetic radiation and a very short time later (typically in the millisecond to nanosecond **half-life** range) emits lower energy light at longer wavelength (so-called red-shifted, that is, shifted toward the infrared region of the spectrum from the visible region). Another process closely related to fluorescence is phosphorescence, which is characterized by a longer lifetime of the excited state: phosphorescent objects continue to emit light after the exciting light source is extinguished whereas fluorescent objects cease emitting light almost immediately after the source is turned off. Chemiluminesce also involves light emission but differs from **fluorescence and phosphorescence** in that the excitation energy originates in a chemical reaction rather than illumination with a light source. Natural sources of **chemiluminescence** include fireflies, glow-worms and a variety of bacteria and fungi. Chemiluminescent products include "cold-light" glow sticks that can be used where there may be a danger from **heat** sources or sparks as well as many novelty items.

Chemical processes that are characterized as photochemical reactions can either occur as a result of rearrangements of chemical bonds within an excited molecule, intramolecular reactions, or by the collision of an excited molecule with another molecule and subsequent alteration of chemical bonds, an intermolecular reaction. Photochemical reactions may produce different products than thermal reactions because of the difference in the pattern of electron distribution in the excited state relative to the ground state and/or to the relatively high energy of the excited state. Heating a molecule or ion, even to temperatures several hundred degrees Celsius above room **temperature** cause typically cause changes in the translational, rotational and vibrational states but does not cause electronic excitation.

A type of reaction that is utilized in nature and by synthetic chemists that exploits the difference in reaction paths of thermal and photochemical reactions is isomerization about a carbon-carbon double bond. The four atoms bound directly to the two **carbon** atoms of a carbon-carbon double tend to lie in a plane with the two carbon atoms. If we symbolize the four atoms as R_{1a}, R_{1b}, R_{2a}, and R_{2b}, where the first two are bound to one carbon atom and the latter two to the other carbon atom with both "a" substituents oriented in one direction, up, and both "b" substituents in the other direction, down, the energy required to rotate one carbon atom relative to the other to achieve a configuration such that the "1a" and "2b" substituent would be oriented in the same direction and "1b" and "2a" in the other direction would be high—an energy similar to that needed to break a chemical bond. Depending on the chemical nature of the different substituents on the carbon atoms of a carbon-carbon double bond, the electronic distribution of an excited state may favor the rotated configuration of the molecule relative to the starting configuration. Thus, electronic excitation may lead to a product that is rotated relative to the original structure whereas heating (thermal excitation) might not cause any change. An important photochemically induced carbon-carbon bond rotation, a photoisomerization, occurs in the visual response of the eye. As light reaches the retina, the compound 11-*cis*-retinal undergoes photoisomerization to form 11-*trans*-retinal. The shape of the *trans* is much different from the *cis* compound, leading to a series of subsequent changes in structures of the visual **proteins**. Organic chemists take advantage of the differences between photoisomerization products and thermally favored structures to synthesize useful compounds and synthetic intermediates that are used to make useful compounds.

The most important natural photochemical process is certainly **photosynthesis**. Through the absorption of light by **chlorophyll** molecules in green plants and certain photosynthetic bacteria, energy is transferred through a series of chemical reactions that typically convert adenosine diphosphate (ADP) and **phosphate** ion to adenosine triphosphate (ATP) and that convert triphosphopyridine nucleotide (TPN) into its reduced form TPNH. The energy that is stored in ATP and TPNH is then used in a series of reactions to convert **carbon dioxide** and **water** into **carbohydrates** and **oxygen** molecules. It appears very likely that photosynthesis was essential in the development of the oxygen present in the earth's atmosphere. In our geological time, it maintains the oxygen/carbon dioxide **balance** in the atmosphere and produces organic materials in plants that are essential for maintaining both plant and animal life. The **resonance** or the chlorophyll molecule allows it to absorb light in the visible region of the spectrum very efficiently. The great stability of the porphyrin-like **ring** system ensures that the excited chlorophyll molecule will not decompose but will maintain its structure long enough to transfer its energy to other molecules so that it can be used to cause the appropriate chemical reactions to occur. The actual mechanism of photosynthesis is very complex with much still to be learned. A great deal of evidence suggests that chlorophyll molecules as well as other molecules that can accept energy to form electronically excited states and transfer the energy efficiently are arranged in the chloroplast in a specific orientation with one another so that they can act cooperatively.

Many important reactions that occur in the atmosphere are photochemical. In the upper atmosphere, species such as **chlorine**, **chlorofluorocarbons** (freons), **nitrogen** oxides, **ozone** and molecular oxygen absorb ultraviolet and far ultraviolet radiation that causes the molecules to decompose into atoms. Individual atoms also absorb ultraviolet and far ultraviolet solar radiation to form highly reactive excited states. For example, oxygen atoms absorb energy to form an excited state that react with ozone to form molecular oxygen 10,000 times faster than ground state oxygen atoms does. Photochemical experiments carried out in a laboratory in the 1970s by Sherwood Rowland and his coworkers led them to conclude that chlorofluorocarbons could lead to **ozone layer depletion**, which protects us from harmful ultraviolet radiation by absorbing it in the upper atmosphere. Their research teams subsequently carried out experiments in the atmosphere that supported their hypothesis, leading to a worldwide effort to reduce the release of chlorofluorocarbons into the atmosphere.

The most familiar use of photochemistry is in photography. Both black and white and **color** film photochemistry relies predominantly on the photosensitivity of microcystalline grains of **silver** halides. Darkened images form as the result of photoreduction of Ag (I) ions to Ag (0) metal atoms. Initial exposure of the film to light typically leads to a very small proportion of reduced Ag (I) atoms. Developers contain reducing agents that cause further **reduction** at the sites where some reduction initially occurred. Other **metals** are sometimes used by photographers and artists to produce special effects.

Production of a reversible photo-induced color change is known as photochromism. When a photochromic material is exposed to light, the light absorption properties of the material change. For example, when a certain photochromic material that is colorless under normal circumstances is exposed to a particular energy range of ultraviolet light its absorption characteristics could be altered so that it would now also absorb red visible light, leaving the blue end of the spectrum to be reflected. The material would then appear to be blue. However, as soon as the ultraviolet light is turned off, the material would again appear to be colorless even though it is still exposed to white light. Eyeglasses and aircraft windows that darken on exposure to bright outdoor light that has a higher amount of ultraviolet radiation and regain their clarity in less intense indoor light contain photochromic compounds.

Optical brighteners in laundry products represent the most widespread commercial utilization of photochemically active substances. Optical **bleaches** absorb in the ultraviolet region and emit light in the visible region, typically in the higher energy blue-violet end of the visible spectrum. As a result, the materials that contain optical brighteners appear to reflect more light than is incident upon them and the blue-violet emission tends to offset yellowish reflected light making the material appear whiter.

Photochemistry involves physical or chemical processes involving electronically excited molecules or ions. Important natural photochemical processes include photosynthesis, in which the light energy absorbed by chlorophyll molecules is converted to chemical energy necessary for sustaining life and vision, in which a structural change that is photo-induced leads to structural changes in proteins which signal the brain, registering a visual image. Commercial applications of photochemistry include photography, optical-brighteners and luminescent products.

See also Kinetics

PHOTOELECTRIC EFFECT

The process in which visible light, x rays, or gamma rays strike **matter** and cause an **electron** to be ejected. The ejected electron is called a photoelectron.

The photoelectric effect was discovered by Heinrich Hertz in 1897 while performing experiments that led to the discovery of electromagnetic waves. Since this was just about the time that the electron itself was first identified, the phenomenon was not really understood. It soon became clear in the next few years that the particles emitted in the photoelectric effect were indeed electrons. The number of electrons emitted depended on the intensity of the light but the **energy** of the photoelectrons did not. No matter how weak the light source, the maximum kinetic energy of these electrons stayed the same. The energy however was found to be directly proportional to the frequency of the light. The other perplexing fact was that the photoelectrons seemed to be emitted instantaneously when the light was turned on. These facts were impossible to explain with the then current wave theory of light. If the light were bright enough it seemed reasonable, given enough time, that an electron in an **atom** might acquire enough energy to escape regardless of the frequency. The answer was finally provided in 1905 by **Albert Einstein** who suggested that light, at least sometimes, should be considered to be composed of small bundles of energy or particles called photons. This approach had been used a few years earlier by **Max Planck** in his successful explanation of black body radiation. In 1907, Einstein was awarded the Nobel Prize in physics for his explanation of the photoelectric effect.

Einstein's explanation of the photoelectric effect was very simple. He assumed that the kinetic energy of the ejected electron was equal to the energy of the incident **photon** minus the energy required to remove the electron from the material, which is called the work function. Thus the photon hits a surface, gives nearly all its energy to an electron and the electron is ejected with that energy less whatever energy is required to get it out of the atom and away from the surface. The energy of a photon is given by $E = h\gamma = hc/\lambda$ where γ is the frequency of the photon, λ is the wavelength, and c is the **velocity** of light. This applies not only to light but also to x rays and gamma rays. Thus the shorter the wavelength the more energetic the photon.

Many of the properties of light such as interference and diffraction can be explained most naturally by a wave theory while others, like the photoelectric effect, can only be explained by a particle theory. This peculiar fact is often referred to as wave-particle duality and can only be understood using quantum theory which must be used to explain what happens on an atomic scale and which provides a unified description of both processes.

The photoelectric effect has many practical applications which include the photocell, photoconductive devices and solar cells. A photocell is usually a vacuum tube with two electrodes. One is a photosensitive cathode which emits electrons when exposed to light and the other is an **anode** which is maintained at a positive voltage with respect to the cathode. Thus when light shines on the cathode, electrons are attracted to the anode and an electron current flows in the tube from cathode to anode. The current can be used to operate a relay which might turn a motor on to open a door or **ring** a bell in an alarm system. The system can be made to be responsive to light, as described above, or sensitive to the removal of light as when a beam of light incident on the cathode is interrupted, causing the current to stop. Photocells are also useful as exposure meters for cameras in which case the current in the tube would be measured directly on a sensitive meter.

Closely related to the photoelectric effect is the photoconductive effect which is the increase in **electrical conductivity** of certain non metallic materials such as **cadmium** sulfide when exposed to light. This effect can be quite large so that a very small current in a device suddenly becomes quite large when exposed to light. Thus photoconductive devices have many of the same uses as photocells.

Solar cells, usually made from specially prepared **silicon**, act like a battery when exposed to light. Individual solar cells produce voltages of about 0.6 volts but higher voltages and large currents can be obtained by appropriately connecting many solar cells together. **Electricity** from solar cells is still quite expensive but they are very useful for providing small amounts of electricity in remote locations where other sources are not available. It is likely however that as the cost of producing solar cells is reduced they will begin to be used to produce large amounts of electricity for commercial use.

Iodine compounds are use in the production of photographic film. ***(Photograph by Robert J. Huffman. Reproduced by permission.)***

PHOTOGRAPHY, CHEMISTRY OF

Without **chemistry**, photography, as it is known, would not exist. Photography is a complex array of chemical reactions. Each step, from the manufacture of the film through the final prints or slides being viewed, has its own particular chemical processes.

The film that is used in most cameras is a thin base of **gelatin** (peptides of collagen) which is bonded to a polymeric plastic sheet cut into strips. The gelatin is hardened during manufacture by cross-linkage of the polypeptides. This is so the gelatin matrix is not damaged by mechanical movement through the camera. The gelatin acts as a support for the light-sensitive material that is used to capture an image. The gelatin protects the light sensitive material and it also enhances its sensitivity. The light-sensitive material used is a **silver halide**. With modern camera film silver bromide generally is used. Other photographic emulsions may use other silver halides; for example silver chloride is used for any photography using infrared light. All film is a sandwich with a scratch-resistant coating, an **emulsion** layer (containing the silver halide), a gelatin support layer, and an anti-halation layer. When light strikes the film the silver halide decomposes to give silver. This is indistinguishable from the rest of the film and is called a latent image.

In order to see the latent image in the film, the image first must be treated or developed. The chemical that does this is called a developer. When the developer is added to the film it converts the silver ions into black, colloidal silver. This is a **reduction** process. The silver halide not exposed to light remains unaffected at this point. Next, an acidic chemical bath, called a stop bath, is added to halt any further change or development in the silver. The stop bath's acidic nature rapidly neutralizes any alkaline developer which has not been removed. A **solution** of ethanoic acid is commonly used for this purpose. The next stage of the process is the addition of chemical called a fixer, because it fixes the image permanently in place. The fixer reacts with the unconverted silver halide to produce a **water** soluble compound. One fixative used commonly is **sodium** thiosulfate (or hypo). A complex of soluble silver thiosulfate is produced. When the film is washed the last remains of the unexposed silver halides are washed away. This whole process produces a black-and-white negative image of the subject photographed.

Producing color negatives requires a similar process but there are three layers of light sensitive emulsion. Each layer is sensitive to a different wavelength range of light. There is a blue-sensitive layer, a yellow-sensitive layer, and a green-sensitive layer. Within each of these layers there are dyes which are released when the film is developed.

The third type of film, transparency film, produces slides (a positive image on the film). A similar process of development occurs here but there is a two-stage development. First, the black-and-white latent image is developed and then the rest of the unexposed material is chemically fogged. The color dyes then are developed. Afterwards the silver is bleached out using the fixer as before.

These processes provide an image on a piece of film. With the slide film this is the final product. With the negative types of film a positive print is required. The negative is projected onto light-sensitive **paper** and the paper is then developed using similar processes to those previously described. This gives a positive image of the original subject on a permanently fixed sheet of silver halide-impregnated paper.

There are many variations on the basic pattern above but all photography employing film uses some form of the basic processes outlined. The photographic process was first shown by Louis-Jacgues-Mandé Daguerre (1789-1851) in 1838 and many others have adapted it over the years to give the process that is now used.

PHOTON

One of the fundamental dichotomies in classical physics was that between **energy** and **matter**. By the late nineteenth century, most scientists agreed that matter is composed of tiny, discrete particles with measurable **mass**. At the time, the **atom** was considered to be the ultimate particle of which all matter consists. Energy, on the other hand, was thought to have no material basis, but corresponded to anything that was able to cause a change of position in some form of matter. Most forms of energy were thought to travel through space as waves. The duality between particles and waves was, therefore, a fundamental tenet of physical theory. By the turn of the century, however, the wave-particle duality began to come apart fairly rapidly. A critical factor in this change was Albert Einstein's analysis of the **photoelectric effect**. Einstein showed that the emission of electrons from a metal that has been exposed to light can best be explained by assuming that light consists of tiny "packages" of energy. The energy of each package and its relationship with the wavelength of the light is translated in the equation $E = c \div \lambda$, where E is **Planck's constant** (6.62617×10^{-34} J-sec), c is the speed of light (2.99792×10^{8} m sec^{-1}), and λ is the wavelength in meters. The energy is directly proportional to the frequency of the photon that is equal to c/λ and expressed in sec^{-1}. The concept of an energy "package'' was not originally conceived by Einstein. In 1903, Joseph J. Thomson had found it useful to talk about energy ''specks'' in describing his work on **ionization** of **gases**. There is no question, however, that the clearest expression of this concept came from Einstein's work. In the 1920s, Arthur Holly Compton proposed the name *photon* (from the Greek *phos*, for ''light'') for this ''package'' of energy. Compton's own work involved the scattering of x rays by atoms in a crystalline lattice. Compton adopted Einstein's approach to explain the fact that x rays reflected from the crystals have longer wavelengths than the incident x rays. By assigning x rays the property of momentum, a property traditionally reserved for particles, Compton was able to calculate exactly the observed properties of the scattered x rays. He later wrote that his experiments probably ''were the first to give, at least to physicists in the United States, a conviction of the fundamental validity of the quantum theory.'' The photon concept is important historically because it marks the first time that language normally used for particles was applied to energy phenomena. Consequently, scientists began to study both matter and energy as two manifestations of a common, or at least related, phenomenon. For example, Werner Heisenberg adopted the photon concept in his development of quantum theory. In fact, the photon is usually described today as the quantum of light, a particle with zero charge and zero mass that mediates the transfer of the electromagnetic force. In modern particle physics theory, the photon is described as a boson, a particle with integral spin that acts as carrier of the electromagnetic force.

PHOTOSYNTHESIS

The ultimate source of **energy** for life on Earth is the sun. Plants are able to transform the light energy from the sun into chemical energy through a process called photosynthesis. Through absorption of light energy, plants, algae, and a few types of bacteria transform **carbon** dioxide and **water** into **carbohydrates**. Carbohydrates serve as fuel for their growth and **metabolism**. Other organisms benefit indirectly from photosynthesis. **Oxygen** is a waste product of photosynthesis. By releasing oxygen into the environment, plants make the air breathable for many other life forms.

The presence of oxygen in relation to plants was first demonstrated by the English chemist, **Joseph Priestley** in 1772. He noted that air that had been depleted through burning candles could be made breathable again by plants. Seven years later, a Dutch doctor, Jan Ingen-Housz, showed that plants required sunlight in order to make air breathable. Ingen-Housz also demonstrated that it was only the green parts of plants that had this ability. However, neither scientist knew anything of oxygen's existence. In 1782, Swiss scientist Jean Senebier proved that the plants also needed ''fixed air''—what is now called carbon dioxide—in addition to sunlight. His discovery was followed in 1804 by another Swiss scientist, Nicholas Théodore de Saussure, proving that water was also a necessary factor.

By the early 19th century, scientists had at least a general understanding of photosynthesis. They knew that green plants, if given **carbon dioxide**, water, and sunlight, would produce organic material and oxygen. Building on this foundation, scientists began to tease apart the details of how the plants accomplished this process.

Giant kelp plants growing in turquoise water, Monterey Bay Aquarium. ***(Photograph by Robert J. Huffman.)***

A big step forward was the discovery that chloroplasts were the site of oxygen generation in plant cells. Chloroplasts are cell organelles comparable to the mitochondria of animal cells. These organelles contain a cell's energy-generating mechanisms. Discovering the function of chloroplasts prompted scientists to scrutinize the organelle's structure and chemical makeup.

Not all cells capable of photosynthesis have chloroplasts, however. Cells are generally divided into two categories: prokaryotes and eukaryotes. Eukaryotes have chloroplasts, but prokaryotes do not. The difference between eukaryotes and prokaryotes is that eukaryotes have a **nucleus**, while prokaryotes do not. The nucleus is a cellular structure that stores genetic material separately from the rest of the cell interior. Plants and certain types of algae are eukaryotes. Blue-green algae and photosynthetic bacteria are prokaryotes. The following discussion of photosynthesis will concentrate on photosynthetic eukaryotes—i.e., plants and algae.

Chloroplasts contain **chlorophyll**, a green-colored pigment. Chlorophyll is the key **molecule** in photosynthesis. There are several types of chlorophyll, but the predominant form is chlorophyll *a*. Chlorophyll molecules are composed of a central porphyrin **ring** and side **chains**. These side chains differ slightly between types of chlorophyll, but the general appearance of the molecule is very similar. The porphyrin ring is actually a complex multi-ring structure composed of carbon, **hydrogen**, **nitrogen**, and oxygen atoms.

Light is broken down into units called photons which move in waves. The distance between waves is called the wavelength. Visible light contains wavelengths ranging from 350 nm to 800 nm. Longer wavelengths contain less energy; short wavelengths contain more. Wavelengths correspond with the colors of the rainbow—called the visible light spectrum—and each **color** represents a different **energy level**.

The colors that an object appears to be are those that are reflected; colors that are absorbed are not seen. Chlorophyll absorbs energy from the red portion of the spectrum, approximately 680-700 nm, and reflects the green portion. For that reason, the leaves and stems of plants appear green. When a **photon** is absorbed by a chlorophyll molecule, its energy causes the chlorophyll to enter an energy-rich excited state.

Within a chloroplast, chlorophyll molecules are densely packed together. Because the chlorophyll molecules are so close to each other, the energy they absorb circulates from molecule to molecule. This energy is eventually transmitted to a reaction center, thought to be a special arrangement of chlorophyll and protein molecules. There are two types of reaction centers in chloroplasts, each of which corresponds with a photosystem. The P700 reaction center is associated with photosystem I; P680 is linked with photosystem II. The names of the photosystems arise from the wavelength of light which they most efficiently absorb.

Both photosystems serve as trigger points for **electron** transport chains. Electron transport chains operate through the **reduction** and **oxidation** of chemicals or chemical complexes that make up its links. Although reduction usually means that something is lost, it has the opposite meaning in **chemistry**. As a chemical term, reduction means that an electron is gained. Oxidation means that an electron is lost.

When P680 absorbs the energy of a photon, it goes from a ground state to an excited state. This change is seen at the atomic level. In the ground state, the electrons of an **atom** are at a low-energy configuration. In an excited state, an electron jumps from a normal low-energy position, or orbital, to a high-energy orbital. The high-energy position is unstable, and the electron is quickly transferred to another chemical, pheophytin. Pheophytin is the first link in the photosystem II electron transport chain.

As P680 returns to its ground state, it picks up an electron to replace the one transferred to pheophytin. This electron is taken from a water molecule. In chemical terms, P680 is reduced and water is oxidized. The water molecule is split into hydrogen ions (positively charged atoms) and oxygen. For every two water molecules broken down in this manner, four hydrogen ions and one oxygen molecule are produced. The hydrogen ions can be used for other reactions, but the molecular oxygen is released as a useless waste product.

The electron transport chain that picks up an electron from P680, leads to P700. The chain itself comprises several chemical links. The extra electron is passed from one link to the next in a series of reductions and oxidations. As the electron proceeds to P700, it encounters a chemical complex called cytochrome *bf*. The final link in the photosystem II electron transport chain is plastocyanin.

The P700 reacts to a photon in the same manner as P680. When the photon's energy is absorbed, P700 is transformed from its ground state to an excited state. In its excited state, P700 transfers an electron to the photosystem I electron transport chain. In returning to its ground state, P700 is reduced by plastocyanin, courtesy of photosystem II.

The photosystem I electron transport chain leads to an **enzyme** called ferredoxin-NADP reductase. (Enzymes are protein molecules that trigger speedy chemical reactions.) Once this enzyme has the extra electron available, it is able to transform NADP+ to NADPH and a hydrogen **ion**. NADPH is the abbreviation for nicotinamide adenine dinucleotide **phosphate**. This molecule is a high-energy fuel used for certain chemical reactions in the cell.

Under certain circumstances, P700 is raised to its excited state and the electron passes through only the first few links of the photosystem I electron transport chain. Instead of proceeding to the last link, the electron is shunted over to the cytochrome *bf* complex of photosystem II. The end result of this process—called cyclic photophosphorylation—is the production of a different high-energy fuel. This fuel is adenosine triphosphate, or ATP. Both ATP and NADPH are use to drive the chemical reactions that produce carbohydrates.

Both photosystem I and photosystem II require sunlight. As a result, this part of photosynthesis falls under the heading of light reactions. The actual generation of carbohydrates is a dark reaction, because it can occur in the absence of light. Carbohydrate construction, or synthesis, is accomplished through the Calvin cycle. The Calvin cycle is a series of chemical reactions through which carbon dioxide is used to build glucose. Glucose, in turn, is used as an eventual building block for sucrose, starch, and other carbohydrates.

The first step of the Calvin cycle is the addition of carbon dioxide to an acceptor molecule. This step was first described by **Melvin Calvin** in the 1940s, and the entire cycle is named in his honor. The cycle begins with the addition of carbon dioxide to another molecule, ribulose-1,5-bisphosphate. This reaction is triggered, or catalyzed, the enzyme ribulose 1,5-bisphosphate carboxylase/oxygenase—often called by its short name, rubisco. Rubisco is able to add either carbon dioxide or molecular oxygen to a ribulose-1,5-bisphosphate molecule. The addition of carbon dioxide leads into the Calvin cycle; the addition of oxygen leads to a different process called photorespiration.

When rubisco binds carbon dioxide to ribulose-1,5-bisphosphate, a six-carbon molecule is temporarily formed. This molecule quickly splits into two molecules of 3-phosphoglycerate. At this stage, ATP enters the cycle by transferring a phosphate group to the molecule. The resulting molecule is reduced by NADPH to form glyceraldehyde-3-

phosphate. At this point, glyceraldehyde-3-phosphate can go in one of two directions. A certain quantity of the molecule is dedicated to completing the cycle by regenerating the ribulose-1,5-bisphosphate stores. The remaining glyceraldehyde-3-phosphate continues on toward glucose, sucrose, starch, and other carbohydrate production.

Some plants have the ability to concentrate carbon dioxide levels within their cells before it enters the Calvin cycle. This ability helps circumvent photorespiration. As mentioned above, photorespiration occurs when rubisco adds oxygen, rather than carbon dioxide, to ribulose-1,5-bisphosphate. This is a wasteful reaction for the plant because it uses up ATP and reduces carbohydrate synthesis. Generally rubisco will use carbon dioxide in preference to oxygen, but in hot, sunny climates, the carbon dioxide near plants is rapidly used up. With lower levels of carbon dioxide in the area, oxygen uptake would increase and photorespiration would occur.

Some plants have evolved a mechanism to reduce photorespiration. These plants are called C4 plants, because they create a four-carbon molecule as an intermediary between carbon dioxide and the start of the Calvin cycle. The C4 plants concentrate carbon dioxide levels prior to the Calvin cycle through the Hatch-Slack pathway. The Hatch-Slack pathway is named for its discovers, Australian biochemists, Marshall Hatch and Roger Slack. Examples of C4 plants are corn, sorghum, and sugar cane, both of which do well in hot, sunny conditions. Plants which add carbon dioxide directly to the Calvin cycle are called C3 plants, because they create the three-carbon molecule, 3-phosphoglycerate. The C3 plants do best in temperate climates. Wheat is an example of a C3 plant. Regardless of how carbon dioxide enters the Calvin cycle, the products are the same.

PHYSICAL CHEMISTRY

Physical chemistry is the branch of **chemistry** which deals with the physical foundations of chemical substances and processes. Thus, for example, physicial chemists strive to understand the physical mechanisms underlying chemical reactions. It is important to note that research in physical chemistry depends on the development of new techniques and instruments.

Physical chemistry is generally divided into a variety of branches, including chemical physics (the study of the chemical propereties of **atoms** and **molecules**) and **theoretical chemistry** (the mathematical interpretation of quantum and thermodynamic processes).

Essentially, the physical chemist studies **matter**—its changes, structure, and chemical equilibrium.

Matter can undergo two types of changes, chemical and physical. A chemical change occurs when a substance reacts chemically to form a new substance. An example of this kind of change is **combustion**. Thus, when **wood**, a compound composed of mainly **carbon** (C), **hydrogen** (H), and **oxygen** (O), burns, it changes to ashes, consisting mostly of carbon, **water**, and **carbon dioxide**.

An example of a physical changes would be chopping a piece wood into smaller pieces or melting ice. The physical chemist is concerned concerned with the rate of change in a substance, which is the study of reaction **kinetics**. **Diffusion** and **ionization** are both examples of physical changes which would be studied by a physical chemist. The diffusion of **gases**, for example, is usually described in terms of kinetic energy of the gas particles. The physical chemist would study the molecular forces between the particles (a transport process), not the kinetic **energy** of the particles. Ionization concerns the movement of **ions** in a solution, which is also a transport process. The study of transport processes in physical chemistry is done by scientists who specialize in non-equilibrium dynamics.

Physical chemists explore many types of chemical change. The rate of chemical reactions often depends on the concentration of the reactants involved. Reaction rate can also be affected by conditions such as **temperature**, agitation of the reaction mixture, or addition of **catalysts**. Scientists working in physical chemistry have formulated several laws which can be used to predict the rate of chemical reaction, including the rate coefficient (which explains how the cocentrations of the reactants affect the rate) and the Arrhenius rate law (which describes the role of temperature in reaction kinetics).

The study of chemical kinetics is either based on the statistical model of **thermodynamics** or the calculated trajectories of the molecules involved in a reaction. The statistical model of thermodynamics is called the activated complex theory, and is the main tenet of the field of molecular reaction dynamics. The calculation of trajectories is called the particle dynamics approach. Physical chemists involved in particle dynamics calculate the paths of a molecule and converts these into rate coefficients. Historically, these methods of determining the chemical kinetics of a reaction has been primarily theoretical. Experimental methods are now being developed which allow scientists to actually observe the molecules involved in a chemical reaction—methods such as "molecular beams." During a chemical reaction, a beam of one reactant is directed at a beam of the other reactant. Scattering of the products is observed to determine the trajectories of the molecules involved. This technique enables scientists to get an accurate description of the chemical kinetics for many types of reactions.

Physical chemists study the structure of matter by using several innovative techniques. The study of molecular and atomic structure is called **quantum chemistry**, or quantum mechanics. For example, the structure of many atoms and molecules can be predicted by solving the Schrödinger's equation. The numerical solution of this equation allows scientists to determine the wave functions and geometries of molecules. Other equations have been developed to predict the shape and electron distribution of molecules. These equations are so advanced that they can be used to determine the charges in structure of a molecule during each phase of a reaction

The shape, size, and electron distribution of molecules can also be determined by using **spectroscopy**. For categories of spectroscopic techniques used by physical chemists are absorption spectroscopy, emission spectroscopy, Raman spectroscopy, and resonance techniques. The size and shape of a molecule can be determined by using absorption spectroscopy.

In this this procedure, the amount of radiation emitted by a molecule is observed at different wavelengths. The frequency of light emitted from atoms in an excited state is measured in emission spectroscopy. These data allow scientists to identify molecules, as well as the processes which occur during a reaction. Light emitted from molecules is measured in Raman spectrometry, which is used, in combination with other forms of spectrometry, to provide additional information. Resonance techniques are used to measure the frequencies of excited molecules in resonance with surrounding radiation. These techniques, which include **nuclear magnetic resonance** (NMR), electron spin resonance (ESR), and Mossbauer spectroscopy, provide valuable information about the structure of a molecule.

Physical chemists use other techniques to determine the structure of atoms and molecules as well. For example, diffraction techniques involve directing radiation toward a sample and observing the direction of the resulting scatter. An example of a diffraction technique is x-ray diffraction, which provides detailed information about the atoms in crystals or large biological molecules. Electrical properties such as **dipole** moments, magnetic properties, optical properties, and the Faraday effect all give important information about the structure of molecules as well.

Matter in the state of equilibrium is studied by physical chemists specializing in chemical thermodynamics and equilibrium **electrochemistry**. Chemical thermodynamics is concerned with the response of a reaction system to changes in external variables such as **pressure**, **temperature**, or composition of the reaction mixture. **Thermochemistry** is a branch of chemical thermodynamics, which is the study of the energy involved in a chemical reaction. The energy involved in a reaction is directly related to its efficiency. Physicial chemists specializing in thermodynamics can apply their research findings to improve the efficiency of such devices as as **batteries**, engines, and refrigerators. Scientists can also use the thermodynamics of a reaction to determine how to increase the yield of a chemical reaction for industrial purposes.

The laws of thermodynamics are valuable tools for the physical chemist. For example, the change in enthalpy can be used to predict the **heat** involved in a particular reaction. The heat involved in a reaction is an indication of the **energy** involved, and therfore an indication of the efficiency of a reaction. Gibbs free energy is another important thermodynamic property. The Gibbs free energy of as reaction describes the amount of work done in the reaction, and this information can be used to explain complicated processes such as those involved in biological growth.

Equilibrium electrochemistry studies the behavior of chemical reactions which release energy in the form of electron flow. This branch of physical chemistry determines the equilibria of ionic reactions and their response to changes in external variables. Applications of equilibrium electrochemistry include the analysis of electrochemical cells, fuel cells, and the power of batteries.

Physical chemistry is an important branch of chemistry which studies the processes involved in chemical reactions. Chemical kinetics, the structure of atoms and molecules, and the equilibria of chemical reactions are studied by the physical chemist. The methods involved in physical chemistry are extremely complicated and quite advanced. Physical chemists use a wide variety of techniques and procedures in their research. Physical chemistry provides the basis for all other branches of chemistry by explaining the fundamental processes involved in the formation of chemical substances. A growing, vitally important field, this branch of chemistry is also important for determining the efficiency and productivity of technologies developed by other chemists.

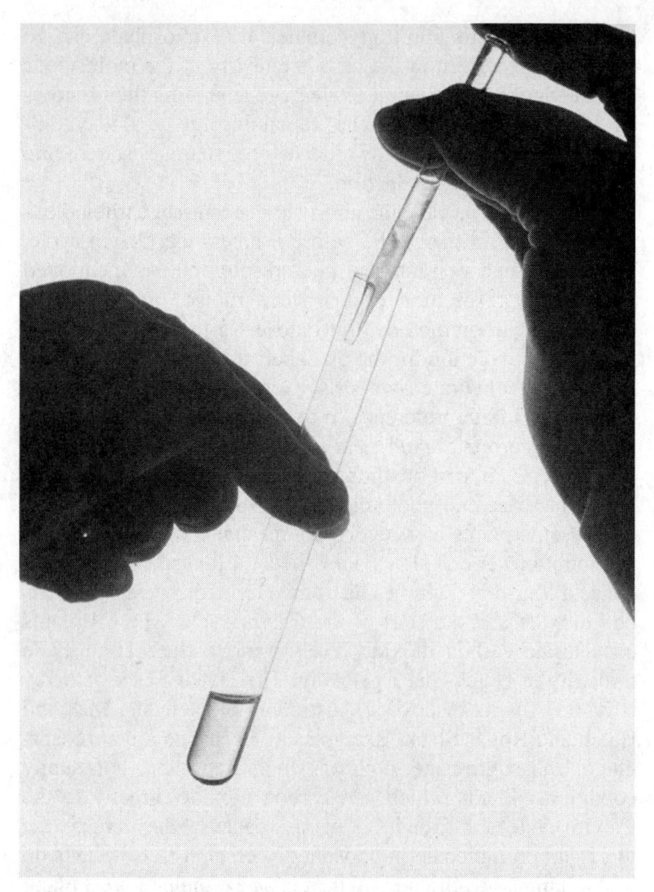

Pipettes are used in the laboratory to transfer liquids to or from a test tube. ***(Photograph by Oscar Buriel/Latin Stock/Science Photo Library, Photo Researchers, Inc. Reproduced by permission.)***

PIPETTE

A pipette is a piece of volumetric glassware used to transfer quantitatively a desired **volume** of **solution** from one container to another. Pipettes are calibrated at a specified **temperature** (usually 68°F [20°C] or 77°F [25°C]) either to contain (TC) or to deliver (TD) the stated volume indicated by the etched/painted markings on the pipette side. Pipettes that are marked TD generally deliver the desired volume with free drainage; whereas in the case of pipettes marked TC the last drop must be blown out or washed out with an appropriate solvent.

For high-accuracy chemical analysis and research work, a volumetric transfer pipette is preferred. Volumetric transfer pipettes are calibrated to deliver a fixed liquid volume with free drainage, and are available in sizes ranging from 0.5-200 mL. Class A pipettes with volumes greater than 5 mL have a tolerance of +/- 0.2 % or better. The accuracy and precision of the smaller Class A pipettes and of the Class B pipettes are less. The Ostwald-Folin pipette is similar to the volumetric transfer pipette, except that the last drop should be blown out. Mohr and serological pipettes have graduated volumetric markings, and are designed to deliver various volumes with an accuracy of +/- 0.5-1.0 %. The volume of liquid transferred is the difference between the volumes indicated before and after delivery. Serological pipettes are calibrated all the way to the tip, and the last drop should be blown out. The calibration markings on Mohr pipettes, on the other hand, begins well above the tip. Lambda pipettes are used to transfer very small liquid volumes down to 1 microliter. Dropping pipettes (i.e., medicine droppers) and Pasteur pipettes are usually uncalibrated, and are used to transfer **liquids** only when accurate quantification is not necessary.

Automatic dispensing pipettes and micropipettes are also available commercially. Automatic dispensing pipettes, in sizes ranging from 1-2,000 mL, permit fast, repetitive delivery of a given volume of solution from a dispensing bottle. Micropipettes consist of a cylinder with a thumb-operated airtight plunger. A disposable plastic tip attaches to the end of the cylinder, the plunger is depressed, and the plastic tip is immersed in the sample solution. The liquid enters the tip when the plunger is released. The solution never touches the plunger. Micropipettes generally have fixed volumes, however, some models have provisions for adjustable volume settings. Micropipettes are extremely useful in clinical and biochemical applications where errors of +/- 1 % are acceptable, and where problems of contamination make disposable tips desirable.

PLANCK, MAX (1858-1947)

German physicist

Max Planck is best known as one of the founders of the quantum theory of physics. As a result of his research on **heat** radiation he was led to conclude that **energy** can sometimes be described as consisting of discrete units, later given the name *quanta*. This discovery was important because it made possible, for the first time, the use of matter-related concepts in an analysis of phenomena involving energy. Planck also made important contributions in the fields of **thermodynamics**, relativity, and the philosophy of science. He was awarded the 1918 Nobel Prize in physics for his discovery of the quantum effect.

Max Karl Ernst Ludwig Planck was born on April 23, 1858, in Kiel, Germany. His parents were Johann Julius Wilhelm von Planck, originally of Göttingen, and Emma Patzig, of Griefswald. Johann had previously been married to Mathilde Voigt, of Jena, with whom he had two children. Max was the fourth child of his father's marriage to Emma.

Johann von Planck was descended from a long line of lawyers, clergyman, and public servants and was himself Professor of Civil Law at the University of Kiel. Young Max began school in Kiel, but moved at the age of nine with his family to Munich. There he attended the Königliche Maximillian Gymnasium until his graduation in 1874.

Max Planck.

As a child, Planck demonstrated both talent in and enthusiasm for a variety of fields, ranging from mathematics and science to music. He was accomplished at both piano and organ and gave some thought to a career in music. He apparently abandoned that idea, however, when a professional musician told him that he did not seem to have the commitment needed for that field. Planck did, however, maintain a life-long interest in music and its mathematical foundations. Later in life, he held private concerts in his home which featured eminent musicians, such as Joseph Joachim and Maria Scherer, as well as fellow scientists, including **Albert Einstein**, often with Planck at the piano.

Planck entered the University of Munich in 1874 with plans to major in mathematics. He soon changed his mind, however, when he realized that he was more interested in practical problems of the natural world than in the abstract concepts of pure mathematics. Although his course work at Munich emphasized the practical and experimental aspects of physics, Planck eventually found himself drawn to the investigation of theoretical problems. It was, biographer Hans Kango points out in *Dictionary of Scientific Biography,* "the only time in [his] life when he carried out experiments."

Planck's tenure at Munich was interrupted by illness in 1875. After a long period of recovery, he transferred to the University of Berlin for two semesters in 1877 and 1878. At Berlin, he studied under a number of notable physicists, including Hermann Helmholtz and **Gustav Kirchhoff**. By the fall of 1878, Planck was healthy enough to return to Munich and his studies. In October of that year, he passed the state examination for higher level teaching in math and physics. He taught briefly at his alma mater, the Maximillian Gymnasium, before devoting his efforts full time to preparing for his doctoral dissertation. He presented that dissertation on the second law of thermodynamics in early 1879 and was granted a Ph.D. by the University of Munich in July of that year.

Planck's earliest field of research involved thermodynamics, an area of physics dealing with heat energy. He was very much influenced by the work of Rudolf Clausius, whose work he studied by himself while in Berlin. He discussed and analyzed some of Clausius's concepts in his own doctoral dissertation. Between 1880 and 1892, Planck carried out a systematic study of thermodynamic principles, especially as they related to chemical phenomena such as **osmotic pressure**, boiling and freezing points of solutions, and the **dissociation** of **gases**. He brought together the papers published during this period in his first major book, *Vorlesungen über Thermodynamik,* published in 1897.

During the early part of this period, Planck held the position of Privat-Dozent at the University of Munich. In 1885, he received his first university appointment as extraordinary professor at the University of Kiel. His annual salary of 2,000 marks was enough to allow him to live comfortably and to marry a childhood sweetheart from Munich, Marie Merck. Marie was eventually to bear Planck three children.

Planck's personal life was beset with tragedy. Both of his twin daughters died while giving birth: Margarete in 1917, and Emma in 1919. His son, Karl, also met an untimely death when he was killed during World War I. Marie had predeceased all her children when she died on October 17, 1909. Planck later married Marga von Hoessli, with whom he had one son.

Planck's research on thermodynamics at Kiel soon earned him recognition within the scientific field. Thus, when Kirchhoff died in 1887, Planck was considered a worthy successor to his former teacher at the University of Berlin. Planck was appointed to the position of assistant professor at Berlin in 1888 and assumed his new post the following spring. In addition to his regular appointment at the university, Planck was also chosen to head the Institute for Theoretical Physics, a facility that had been created especially for him. In 1892, Planck was promoted to the highest professorial rank, ordinary professor, a post he held until 1926.

Once installed at Berlin, Planck turned his attention to an issue that had long interested his predecessor, the problem of black body radiation. A black body is defined as any object that absorbs all frequencies of radiation when heated and then gives off all frequencies as it cools. For more than a decade, physicists had been trying to find a mathematical law that would describe the way in which a black body radiates heat.

The problem was unusually challenging because black bodies do not give off heat in the way that scientists had predicted that they would. Among the many theories that had been proposed to explain this inconsistency was one by the German physicist Wilhelm Wien and one by the English physicist John Rayleigh. Wien's explanation worked reasonably well for high frequency black body radiation, and Rayleigh's appeared to be satisfactory for low frequency radiation. But no one theory was able to describe black body radiation across the whole spectrum of frequencies. Planck began working on the problem of black body radiation in 1896 and by 1900 had found a solution to the problem. That solution depended on a revolutionary assumption, namely that the energy radiated by a black body is carried away in discrete "packages" that were later given the name *quanta* (from the Latin, *quantum,* for "how much"). The concept was revolutionary because physicists had long believed that energy is always transmitted in some continuous form, such as a wave. The wave, like a line in geometry, was thought to be infinitely divisible.

Planck's suggestion was that the heat energy radiated by a black body be thought of as a stream of "energy bundles," the magnitude of which is a function of the wavelength of the radiation. His mathematical expression of that concept is relatively simple: $E = h\nu$, where E is the energy of the quantum, ν is the wavelength of the radiation, and h is a constant of proportionality, now known as **Planck's constant**. Planck found that by making this assumption about the nature of radiated energy, he could accurately describe the experimentally observed relationship between wavelength and energy radiated from a black body. The problem had been solved.

The numerical value of Planck's constant, h, can be expressed as 6.626×10^{-34} J/second, an expression that is engraved on Planck's headstone in his final resting place at the Stadtfriedhof Cemetery in Göttingen. Today, Planck's constant is considered to be a fundamental constant of nature, much like the speed of light and the gravitational constant. Although Planck was himself a modest man, he recognized the significance of his discovery. Robert L. Weber in *Pioneers of Science: Nobel Prize Winners in Physics* writes that Planck remarked to his son Erwin during a walk shortly after the discovery of the quantum concept, "Today I have made a discovery which is as important as Newton's discovery." That boast has surely been confirmed. The science of physics today can be subdivided into two great eras, classical physics, involving concepts worked out before Planck's discovery of the quantum, and modern physics, ideas that have been developed since 1900, often as a result of that discovery. In recognition of this accomplishment, Planck was awarded the 1918 Nobel Prize physics.

After completing his study of black body radiation, Planck turned his attention to another new and important field of physics: relativity. Albert Einstein's famous paper on the theory of general relativity, published in 1905, stimulated Planck to look for ways on incorporating his quantum concept into the new concepts proposed by Einstein. He was somewhat successful, especially in extending Einstein's arguments from the field of electromagnetism to that of mechanics. Planck's work in this respect is somewhat ironic in that it had been Einstein who, in another 1905 paper, had made the first productive use of the quantum concept in his solution of the photoelectric problem.

Throughout his life, Planck was interested in general philosophical issues that extended beyond specific research questions. As early as 1891, he had written about the importance of finding large, general themes in physics that could be used to integrate specific phenomena. His book *Philosophy of Physics,* published in 1959, addressed some of these issues. He also looked beyond science itself to ask how his own discipline might relate to philosophy, religion, and society as a whole. Some of his thoughts on the correlation of science, art, and religion are presented in his 1935 book, *Die Physik im Kampf um die Weltanschauung.*

Planck remained a devout Christian throughout his life, often attempting to integrate his scientific and religious views. Like Einstein, he was never able to accept some of the fundamental concepts of the modern physics that he had helped to create. For example, he clung to the notion of causality in physical phenomena, rejecting the principles of uncertainty proposed by Heisenberg and others. He maintained his belief in God, although his descriptions of the Deity were not anthropomorphic but more akin to natural law itself.

By the time Planck retired from his position at Berlin in 1926, he had become probably the most highly respected scientific figure in Europe, if not the world, except for Einstein. Four years after retirement, he was invited to become president of the Kaiser Wilhelm Society in Berlin, an institution that was then renamed the Max Planck Society in his honor. Planck's own prestige allowed him to speak out against the rise of Nazism in Germany in the 1930s, but his enemies eventually managed to have him removed from his position at the Max Planck Society in 1937. The last years of his life were filled with additional personal tragedies. His son by his second marriage, Erwin, was found guilty of plotting against Hitler and executed in 1944. During an air raid on Berlin in 1945, Planck's home was destroyed with all of his books and papers. During the last two and a half years of his life, Planck lived with his grandniece in Göttingen, where he died on October 4, 1947.

Planck's constant

Light emitted by a heated solid object is a well-known phenomenon: even before the age of modern science, alchemists noted the correlation between an object's **temperature** and the kind of light it emitted. For example, if a metal object was being heated, a red glow gradually turned to white as the temperature increased.

Max Planck (1858-1947) was the first to notice, in 1900, a characteristic regularity in the behavior of heated objects. Assuming that the atoms of a particular heated object will vibrate at a particular frequency v, Planck noticed that only certain **energy** levels were possible. These energy levels could only be captured if the value of v was multiplied by a whole number (n) and a constant (h) whose value was 6.626×10^{-34} J/second. The upshot of Planck's discovery was that atoms, unlike objects in the macroscopic world, obeyed energy constraints that could only be explained in terms of a physical constant multiplied by a whole number. As Darrell D. Ebbing and Steven D. Gammon explained, under similar constraints, a car would, for example, increase its speed only by increments of five miles per hour: from zero to five, from five to ten, and so on. In other words, a speed of, say, seven miles per hour would be impossible. The quantum rule may seem unreasonable in our world; in the atomic world, however, energy levels are determined by Planck's constant.

While Planck himself was somewhat uncomfortable with his discovery, younger scientists quickly grasped the universality of Planck's constant. Thus, **Albert Einstein** (1879-1955) posited that Planck's formula $E = hv$ could be applied to light. According to Einstein, Planck's formula accurately expresses the wave-particle duality of light: E represents the energy of a light particle (photon), and v represents the frequency of light as a wave. By introducing the concept of light particle, or **photon**, in 1905, Einstein explained the **photoelectric effect**, or the process whereby an object emits electrons when exposed to light. Einstein theorized that electrons are ejected by photons. According to Einstein, the photoelectric effect clearly manifests the dual nature of light: light approaches an object as a particle (photon) but is absorbed as a wave by the **electron** that is about to be ejected. Following Einstein's insight, scientists later realized that the wave-particle duality applied to **subatomic particles** in general.

See also Bohr theory; Electromagnetic spectrum; Quantum chemistry

Plasma

The term plasma refers to a condition of **matter** sufficiently different from **solids**, **liquids**, and **gases** to have earned the description, "the fourth state of matter." The state develops when a gas is heated to such a high **temperature** that all atoms in the gas are ionized. In this state, matter consists of positively charged ions and electrons in apparently random motion. The name plasma was given to this state by the American chemist **Irving Langmuir** in 1920.

The study of plasma-like materials actually goes as far back as the 1830s when the English physicist, **Michael Faraday**, passed electrical discharges through gases at low pressures. Faraday's research was extended and expanded by **William Crookes** in the 1870s. Crookes was apparently the first scientist to suggest that the ionized gas within his **glass** tubes might be a fourth state of matter.

Plasma research during the 1920s and 1930s was largely a matter of interest to astronomers. It was apparent that, at temperatures present in stars, matter almost certainly exists as plasma. Understanding the composition and properties of stars, therefore, required some understanding of the nature of plasma.

Some of the earliest breakthroughs in plasma research were accomplished by Hannes Olof Göst Alfvén, a Swedish astrophysicist. Alfvén developed a theory to explain the behavior of plasma in the presence of magnetic fields. His work forms the basis of the modern science known as magnetohydrodynamics (MHD). For this research, Alfvén was awarded a share of the 1970 Nobel Prize for physics.

Research on **nuclear fusion** in the 1940s shifted the focus of plasma research from the stars to laboratories on Earth. Nuclear fusion reactions—the combination of two small nuclei to produce one larger nucleus—occur only at very high temperatures, greater than 10 million° C. At such temperatures, similar to those present in the core of a star, matter exists as a plasma.

Scientists attempting to find ways to use fusion reactions as a practical source of **energy** discovered that trapping plasma inside magnetic fields was the best way to control such reactions. Finding the most practical method of achieving such controlled reactions, however, is an enormously difficult task, one that has still not been perfected after a half century of research.

One of the earliest suggestions for solving the problem of plasma confinement was offered by the Russian physicist Igor Evgenevich Tamm in 1950. Tamm outlined a model by which the magnetic field surrounds the plasma and ''pinches'' it together. Another approach was proposed by the American astrophysicist, Lyman Spitzer, Jr. Spitzer's interest in controlled fusion grew out of his earlier research on fusion reactions in the stars. His model calls for a twisting magnetic field to be wrapped around the hot plasma in an arrangement that came to be known as a stellerator.

The most common mechanism for controlling plasma reactions today is called a tokamak, originally designed by the Russian physicist Lee Artsimovich in the late 1950s. A tokamak consists of a *toroidal* (hollow, doughnut-shaped) tube in which the hot plasma is contained by a strong magnetic field.

PLASTICS

The word ''plastic'' was originally an adjective describing a material that could flow. Its first use as a noun came with the introduction of Bakelite and Celluloid in the early 1900s. The word is now used to mean a substance which can be pressed into definite shape and retains its shape when the pressure is removed.

Plastics are long-chain molecules known as polymers. The starting materials are called monomers. **Proteins**, starches, and resins such as frankincense and myrrh are natural polymers. Synthetic polymers are used in **fibers** and in molded and extruded **solids** which are commonly called plastics or resins. **Hermann Staudinger**, who was awarded the Nobel Prize in **chemistry** in 1953, introduced the term **macromolecule** in the 1920s. One of his first compounds was **polystyrene**, with a **molecular weight** of approximately 600,000, formed by a chain of 5,700 styrene units.

In terms of the chemistry involved, the two main types of polymers are formed by (1) addition or (2) condensation. Addition polymerization uses a catalyst to open the double bond of an alkene **monomer** forming fragments with single electrons (free radicals). Thousands or hundreds of thousands or even millions of these monomer fragments then add to each other in a **chain reaction** to form polymers. More than one compound can be used as the monomer, producing copolymers or terpolymers. Depending on the catalyst, it is possible to have two monomers link alternately or randomly.

Condensation polymers are usually formed between dibasic organic acids and dihydroxyalcohols or diamines. **Water** is eliminated from the two functional groups allowing the chain to grow.

Crosslinking is possible with either type of polymer. For a condensation polymer, the presence of more than two functional groups on one or both monomers allows the chain to grow laterally as well as linearly. With addition polymers, a free radical chain fragment can attack the growing chain to produce side-chains.

Choice of monomer(s), chain length, extent of crosslinking, molding conditions, and additives results in plastics that can be tailored to an incredible number of types of material.

In terms of manufacturing, plastics are categorized as thermosetting or thermoplastic. The chemical structure of thermosetting plastics is altered by **heat**, which causes crosslinking to form one large **molecule**. Objects are formed using compression molding and cannot be resoftened. In addition to molded objects, thermosetting resins find use in laminates and foams. Thermoplastic resins remain stable when heated, and can be injection molded—squirted under pressure into cooler molds, or extruded—pushed out of a shaped orifice to form sheets, rods, or tubes.

Plastics based on **cellulose** (primarily **wood** pulp) include cellulose esters such as cellulose acetate, diacetate, and triacetate, and mixed esters such as cellulose acetate butyrate. The first cellulose resin, cellulose nitrate or celluloid, was invented by John Wesley Hyatt, who was trying to find a substitute for ivory in billiard balls in order to win a contest. He didn't win, but he received a patent for his invention in 1870. Cellulose nitrate was commonly used in World War I to seal the fabric wings of fighter planes, but cellulose nitrate is extremely flammable, and the planes were soon nicknamed ''flaming coffins.'' This problem spurred the production of the less flammable cellulose diacetate. At the end of the war, Camille and Henri Dreyfus found themselves with a large factory and a product that now had no use. They developed a method for spinning their cellulose diacetate into fiber; they also sponsored the research which resulted in new textile dyes. Cellulose acetate is used as a thermoplastic for photographic and movie film and for moldings. It has good impact strength, is easy to **color**, and has fair outdoor stability. The mixed ester, cellulose acetate butyrate, has better weatherabilitly and finds use as pipe and telephone housings. Cellulose xanthate is produced from wood cellulose in reaction with **sodium** hydroxide and **carbon** disulfide. It is usually spun into fibers as rayon, but it is also cast in films as cellophane. Cellulose nitrate, the original plastic in its class, is cheap and durable, takes a high polish and good color, and can be rolled or molded into specific shapes. As celluloid, it is used today in combs and brushes. It is also used as artificial leather for auto seat covers and furniture.

The first completely synthetic plastic, Bakelite, is a phenol-formaldehyde resin. The **condensation reaction** between phenol and **formaldehyde**, resulting in resinous products, had been known since 1870. Dr. **Leo Baekeland** discovered in 1909 that moldings could be made if the reaction was carried out under heat and pressure. Bakelite is still used today because of its good electrical properties, high heat and acid resistance, and slow burning rate. It does not color well, so is used for moldings for electrical parts and as a backing material for laminates, where it is usually colored black. Its primary use today is as an adhesive in the production of plywood and particle board.

Nylon was the first synthetic fiber not based on natural materials such as cellulose. **W. H. Carothers** of E. I. DuPont de Nemours first produced nylon in 1931 from hexamethylenediamine and adipic acid, a six-carbon amine and a six-carbon acid. The resulting product is known as nylon 66. Other amines and acids result in nylons with different properties. Although we think of nylon as a material that is spun into fibers, the plastic has other uses as well. Nylon gears and other moving parts have low coefficients of friction and excellent wear-resistance and impact and tensile strengths. They are self-extinguishing in case of fire. Nylons are thermoplastic, and so can be extruded into various shapes.

Polyesters were first produced in England in 1941 by condensing ethylene glycol with terephthalic acid. This first **polyester** was called Terylene in England and Dacron (by DuPont) in the United States. We think of polyesters as fibers, but they can be used as thermosetting plastics. They find use in laminates in aircraft, boats, and furniture. Dacron is used as tubing to replace diseased blood vessels. Polyesters impregnated into **glass** cloth or fibers forms a composite with strength approaching that of steel. Polyesters can also be cast into clear sheets. **Polyethylene** terephthalate is used to make soft drink bottles and can also be made into thin films known as Mylar, used for packaging frozen foods, as a skin substitute for burn victims, and for recording tape. A new type of polyester, tradenamed Olestra, is marketed to snack-food producers as a fat substitute that is not absorbed by the body.

Melamine-formaldehyde resins exhibit excellent resistance to heat, flames, abrasion, and stains. These thermosetting resins are easy to color and find use in laminates. Urea-formaldehyde resins are similar but with somewhat inferior properties. They are used in adhesives. Both of these formaldehyde-based resins are thermosetting. Epoxy resins have outstanding adhesion to **metals** and are widely used in surface coatings and adhesives. Polyurethanes, another of the thermosetting plastics, are widely used as flexible foams.

Two thermoplastic condensation polymers are Mylar (ethylene glycol/terephthalic acid) and the polycarbonates. Mylar is known for its toughness and tear-strength in films. Polycarbonates have excellent impact strength at high temperatures and are self-extinguishing. The polycarbonate trademarked as Lexan is used in "bulletproof" glass and in astronaut's helmets. A sheet with a thickness of one inch can stop a 0.38-caliber bullet fired from twelve feet away.

Polyethylene, a thermoplastic, is the most important plastic on the market. The monomer, **ethene** or ethylene, is obtained in large quantities from the cracking of petroleum. Polyethylene was developed shortly before the start of World War II and was used for insulating cables in radar. Today high **density** polyethylene, with its linear **chains**, is used for radio and television cabinets, toys, and large-diameter pipes. Low density polyethylene, a crosslinked polymer, is used for plastic bags, refrigerator dishes, electrical insulation, films and sheeting, coatings, pond liners, and moisture barriers.

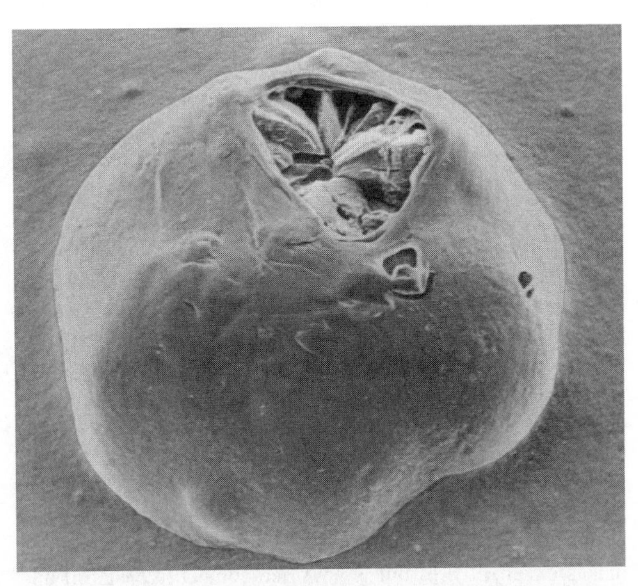

A scanning electron micrograph (SEM) of the surface of a sheet of biodegradable plastic. The spherical object that dominates the image is one of the many granules of starch embedded in the surface of the plastic. When the plastic is buried, the starch takes up water and expands, thereby breaking up the plastic so that more of it exposed to the bacteria that feeds on it. ***(Photograph by National Audubon Society Collection/Photo Researchers, Inc. Reproduced by permission.)***

Vinyl monomers can be represented as CH2=CHX and vinylidene monomers as CH2=CY2. The monomers can be thought of as substituted ethenes, and the polymerization process proceeds in a similar way.

Polystyrene is a thermoplastic which is easily colored and fabricated. It is molded into appliance housing and parts. Polystyrene is most familiar in its foamed form, trade-named Styrofoam, used as insulation and cushioning. Styrofoam is formed by expansion molding. Beads of polystyrene containing four to seven percent by weight of a low-boiling liquid are heated with steam or hot air. The pellets puff and are molded into the shape of the mold cavity. Expanded polystyrene is also used for egg cartons, meat trays, coffee cups, and packaging material. A terpolymer of styrene with acrylonitrile and butadiene (ABS resin) is a synthetic rubber.

Polymethylmethacrylate, with trade names of Lucite and Plexiglas, is well known as a glass substitute, especially in uses such as skylights where its impact strength and resistance to shatter is important. It is also used in signs, lenses, automotive parts such as taillights, and bowling balls.

Polyvinyl chloride, or PVC, is familiar as pipe. Waldo Simon, who died in 1999 at age 100, discovered vinyl in the

1920s while attempting to turn polyvinyl chloride—then considered worthless—into glue. At his wife's request, he first used it to waterproof a shower curtain.

Polyvinylidene chloride is known as Saran wrap. Polytetrafluoroethylene is **Teflon**, used for electrical insulation as well as on pots and pans as a nonstick coating. Polyvinyl acetate is important in latex paints and adhesives. Polyacrylonitrile is known as Orlon.

Natural rubber is a polymer of isoprene. As World War II approached and the supply of natural rubber was cut off, the search began for synthetic rubbers. By the beginning of the war, Germany was producing butadiene/styrene copolymers which were acceptable as rubber. The procedure was duplicated by the United States. Butadiene/acrylonitrile copolymers and acrylonitrile/butadiene/styrene terpolymers are also used as synthetic rubbers.

Polychloroprene is trademarked as Neoprene, and is a polymer or copolymer of 2-chloro-1,3-butadiene. Neoprene makes excellent tires but is not price-competitive. Its oil and gasoline-resistance makes it particularly suitable to hoses. It is heat-and flame-resistant, so finds uses in wire insulation.

Hypalon is a chlorosulfonated polyethylene. Polyethylene is treated with **chlorine** and **sulfur** dioxide; sulfonyl chloride radicals are formed, which become part of the crosslinking of the polymer. Hypalon is so chemical-resistant that it can be used for handling concentrated **sulfuric acid**. It is also flame-resistant.

Silicones, with their backbones of **silicon** and **oxygen** atoms, do not decompose at high temperatures and so are used as automobile brake fluids, lubricants, and electrical insulators. They are also stable at low temperatures, so are used as sealants in space vehicles. Because they are water-repellent, they are used as insulating varnish on electric motors and for high **temperature** greases and lubricants. Silicone rubber is easily molded, serving as a replacement for body parts. Liquid silicone in urethane bags was used extensively for breast implants until reports of leakage in 1992.

Recycling of plastics lags behind that for **aluminum** cans, **paper**, and glass, but should increase as new uses are developed. PET (polyethylene terephthalate) from soft drink bottles finds new life as fiberfill and carpet fibers. HDPE (high density polyethylene) becomes one-gallon milk jugs and pipe. Recycled polystyrene is used to make plastic "lumber." Polyvinyl chloride can be converted into sewer drainpipes.

See also Polymer chemistry

PLATINUM

Platinum is a transition metal in Group 10 of the **periodic table**. It is also classified with a group of similar **metals**, including **ruthenium**, **rhodium**, **palladium**, **osmium**, and **iridium**, with similar properties. This group of metals is known as the platinum group. Platinum has an **atomic number** of 78, an atomic **mass** of 195.08, and a chemical symbol of Pt.

Properties

Platinum is a silver-gray, shiny metal that is both ductile and malleable. It is so malleable that it can be hammered into a thin sheet no more than about 100 atoms thick. Such a sheet is thinner than **aluminum** foil. Platinum has a melting point of about 3,223°F (1,773°C), a **boiling point** of about 6,921°F (3,827°C), and a **density** of 21.45 grams per cubic centimeter. Its density is one of the highest of all elements.

Platinum is a relatively inactive metal that does not corrode or tarnish in air. It is not attacked by most acids, although it does dissolve in **aqua regia** (a mixture of concentrated **nitric acid** and 3-4 parts of hydrochloric acid). Platinum also dissolves in hot alkalis such as hot **sodium** hydroxide or **potassium** hydroxide. An unusual property of platinum is that, in powdered form, it tends to absorb large quantities of **hydrogen** gas.

Occurrence and Extraction

The platinum metals tend to occur together in nature and are relatively difficult to separate from each other. Platinum is one of the rarest of elements with an abundance estimated to be about 10 parts per billion in the Earth's crust. The world's largest supplier of platinum by far is South Africa, which produces about 110,000 kilograms of the metal annually. The next largest source of the metal, Canada, produces less than 10,000 kilograms a year. The only other large producer of platinum is the United States, which gets most of the metal from the Stillwater Mine in Montana.

Discovery and Naming

The first known reference to platinum can be found in the writing of Italian physician, scholar, and poet Julius Caesar Scaliger (1484-1558). Scaliger reported seeing the metal during a visit to Central America in 1557. Natives referred to the metal as *platina*, or "little **silver**." They knew of no use for the metal and regarded it as a nuisance in their search for **gold** and silver. The first complete description of platinum was given by the Spanish military leader Don Antonio de Ulloa (1716-1795). While serving in South America from 1735 to 1746, de Ulloa collected samples of platinum and later wrote a report about his studies on the metal. The name by which the element was known to Central and South Americans was retained as its official chemical name.

Uses

The most important use of platinum is as a catalyst in a variety of industrial operations. For example, platinum is used as a catalyst in the refining and fractionating of petroleum. It is also employed in the production of **fertilizers**, **plastics**, synthetic fibers, drugs and pharmaceuticals, and dozens of other everyday products. The application with which most people are likely to be familiar is in catalytic converters in automobile exhaust systems. The metal is also used in other parts of motor vehicle systems, such as spark plugs.

A well-known, but commercially less important, use of platinum is in the manufacture of jewelry. The metal's hardness, resistance to **corrosion**, and beauty make it ideal for objects such as bracelets, necklaces, earrings, pins, and watch bands.

Platinum also finds applications in systems where chemical inactivity is an important property. For example, artificial pacemakers are often made of platinum alloys because the platinum will not interact with body tissue and cause an allergic reaction. Small amounts of platinum are also used in the manufacture of alloys used in electromagnets. The platinum-**cobalt** magnet is one of the strongest magnets known.

See also Precious metals

PLUTONIUM

Plutonium is a transuranium element, one of the elements that follows **uranium** in Row 7 of the **periodic table**. Its **atomic number** is 94, its atomic **mass** is 244.0642, and its chemical symbol is Pu.

Properties

All isotopes of plutonium are radioactive. The most stable isotopes are plutonium-242 and plutonium-244 with half lives of 376,300 years and 82,600,000 years respectively. If these isotopes were present when the Earth was created, they would long ago have decayed and disappeared from the Earth's crust. Plutonium is a silvery-white metal with a melting point of 1,183°F (639.5°C) and a **density** of 19.816 grams per cubic centimeter. The element is very active and forms a number of compounds.

Occurrence and Extraction

Microscopic amounts of plutonium may exist in the Earth's crust as a result of naturally-occurring fusion reactions, but natural sources are of interest only as a **matter** of curiosity to scientists. For all practical purposes, the element is prepared synthetically in particle accelerators and in **nuclear reactors**.

Discovery and Naming

Plutonium was the first transuranium element to be discovered. It was prepared artificially in 1941 by researchers at the University of California at Berkeley (UCB). Because of wartime concerns, the discovery was not announced until 1946. The element was found when **Edwin McMillan**, Philip Abelson (1913-), and other members of the UCB team bombarded uranium with deuterons. The discoverers suggested the name plutonium in honor of the planet Pluto (uranium was named after the planet Uranus).

Uses

Plutonium has two important applications. First, one of its isotopes, plutonium-239, is one of the few commercially available isotopes to undergo **nuclear fission**. It can be used, therefore, as a fuel in nuclear reactors and in nuclear weapons. One of plutonium's great advantages in this regard is that it is produced as a "waste product" in nuclear reactors fueled with isotopes of uranium. The "waste product" can, thus, be re-used in other nuclear reactors and nuclear weapons.

A second use of plutonium is based on the very large amounts of **heat** it produces as it undergoes radioactive decay.

A contraband plutonium sample. Plutonium is a radioactive element occurring naturally in uranium ore and is also manufactured as a by-product in nuclear reactors. This sample was seized by German custom officers in 1994. *(Photograph by Jurgen Scriba/Science Photo Library, Photo Researchers, Inc. Reproduced by permission.)*

This heat can be used to generate **electricity** in certain specialized situations in which efficiency is more important than cost. For example, plutonium generators have been used to provide electrical power on space probes and space vehicles. They have also found some application in artificial pacemakers for people with heart conditions. The **isotope** used for this purpose is plutonium-238 because the radiation it emits poses no health threat to the wearer's body.

Health Issues

Plutonium is one of the most toxic substances known to humans. In the body, it tends to concentrate in bones leading to the development of bone cancer. Men and women who work with the element are required to wear protective clothing and handle the element with remote-control devices.

POCKELS, AGNES (1862-1935)

German physicist

Agnes Pockels was a self-taught physicist who conducted her experiments at home, publishing her findings in numerous papers over a 40-year period. Her pioneering work on surface films laid the foundation for future work in this field.

Pockels was born on February 14, 1862 in Brunswick, Germany to Captain Theodor Pockels and his wife. Her father's service in the Royal Austrian army required the family to move many times during her childhood, including an unfortunate post in a part of northern Italy with a high malaria infestation. This stay caused serious health problems for the whole family, which persisted for the rest of their lives. When, at age 41, Captain Pockels' illness forced an early discharge from the army, he returned with family to Brunswick in 1881.

Pockels excelled in all her classes at Municipal High School for Girls. She shared a love of the natural sciences with her younger brother Friedrich who went on to become a professor of physics. After graduation, Pockels remained at home since, women were not accepted for higher education at that time. When women were later admitted to higher education, her parents refused her request to apply. Struggling with her own frail health, she managed the household, assumed the care of her younger brother, Friedrich, until his marriage in 1900, and nursed both her father and mother until their deaths in 1906 and 1914. Despite these responsibilities, she remained dedicated to her role as caretaker writing in the diary she kept for most of her life, ''Like a soldier, I stand firm at my post caring for my aged parents.'' This tenacious nature was evident as she pursued her education independently. She read whatever textbooks she could find, and launched a series of experiments that engaged her for half a century.

At 18, Pockels became fascinated with the effect films had on water. She performed her experiments mainly in the kitchen where she did household chores. In fact, some of her first observations made while washing dishes. ''This is really true and no joke or poetic license,'' Pockels sister-in-law later wrote, ''what millions of women see every day without pleasure and are anxious to clean away, i.e. the greasy washing-up water, encouraged this girl to make observations and eventually to... scientific investigation.''

For the next 10 years, Pockels researched surfactants, substances that reduce surface tension when dissolved in an aqueous solution. She added **salts** to a **solution**, noted the resulting stream of currents, and measured the changes in the surface tension with a float attached to a balance placed on the liquid surface. In 1891 she sent a letter describing her work to Lord Rayleigh, an established researcher in the same field. Lord Rayleigh was so impressed with the quality and findings of her work, he had the translation from German to English published in the journal *Nature* in 1891.

Pockels received wide recognition for her contributions to the understanding of surface layers and surface films. The techniques she developed were used by physical chemists to define the physical properties of organic molecules before the advent of X-ray diffraction. Her methods of certifying a clean surface, essential in this type of work, were universally adopted as standard practice in this branch of physics. In 1931 she was awarded the Laura Leonard Prize for Quantitative Investigation of the Properties of Surface Layers an Surface Films and in 1932, she received an honorary doctorate from Carolina-Wilhelmina University in Brunswick, Germany. That same year, W. Ostwald published a tributary review of her remarkable work, stating that ''every colleague who is now engaged on surface layer or film research will recognize that the foundations for the quantitative method in this field,...had been laid by (her) observations over 50 years ago.''

POISONS AND TOXINS

Poisons and toxins are molecules that are harmful to living organisms. it is important to note that virtually any substance can be harmful at high enough concentrations—as Paracelsus (1493-1541) said in the sixteenth century, ''the dose makes the poison.'' In common usage, the word *poison* designates a substance that is harmful in relatively small quantities. There is no absolute limit for this designation, although it generally is reserved for substances that are lethal in adult humans at doses below 100 grams. Toxin is a synonym for poison.

Poisons include both naturally produced compounds and chemicals manufactured by humans. Natural poisons are produced by species of bacteria, fungi, protists, plants, and animals. Modern society produces and uses thousands of chemicals, including more than 65,000 manufactured in the United States. Many of these are poisonous, and their regulation is an important part of the work of the Occupational Safety and Health Agency, or OSHA. Cleanup of toxic wastes and spills is overseen by the Environmental Protection Agency (EPA).

Poisons are toxic because they disrupt metabolic processes or destroy tissue through chemical reactions with cells. While the exact mechanisms of action of poisons are almost as varied as the number of poisons themselves, there are several broad classes of mechanisms.

The brain requires large amounts of **oxygen** to function, and cannot survive if deprived of it for more than ten minutes. Asphyxiation occurs when the lungs cannot provide enough oxygen for the brain to continue functioning. Because of this, otherwise harmless **gases** such as **nitrogen** or **carbon** dioxide can be poisonous in high concentrations. **Carbon monoxide** binds to hemoglobin, the **molecule** that carries oxygen in the blood stream, and in this way leads to asphyxia.

Neural transmission relies on the rapid and precise delivery of chemical signals along neurons and across the spaces between them. Curare, botulinum toxin, and organophosphorus insecticides each interfere with neural transmission.

Curare, a poison derived from a South American shrub, blocks the action of an **enzyme** vital for intracellular signaling in neurons, the Na^+/K^+ ATPase. This protein creates the electrochemical gradient whose breakdown carries information from the dendrite to the axon. Death by curare is usually due to paralysis of the respiratory muscles, which cannot contract without neural stimulation, or by fibrillation of the heart, whose cells become desynchronized in their contractions.

Botulinum toxin, from the bacterium *Clostridium botulinum*, prevents release of the neurotransmitter acetylcholine. Acetylcholine is released by neurons to provoke muscle contraction. Botulinum toxin inactivates a key protein in the axon that allows vesicles of acetylcholine to fuse with the axon membrane, which releases the acetylcholine onto the muscle surface. Death by botulinum toxin is usually due to respiratory paralysis.

Organophosphorus insecticides bind to and inactivate acetylcholinesterase. This enzyme breaks down acetylcholine after its release, to prevent continued muscle contraction in the absence of a new signal. Without this enzyme, muscle continues to contract uncontrollably, leading to tremor, convulsions, and death. Organophosphates are also used in the most common types of **chemical weapons**.

The heart can be affected by a variety of substances that change its excitability or the force of its contractions. Digitalis, derived from foxglove, is one of the oldest herbal medicines for the heart. In small doses, it is prescribed for congestive heart failure to increase heart output. In slightly larger doses, it is deadly. Digitalis is a type of steroid, called a cardiac glycoside. Other cardiac glycosides are produced by various South American frogs of the genus *Bufo*. Like curare, this poison isused on arrowtips for hunting.

Cell **energy** production is dependent on mitochondria, small organelles that transfer the energy from sugar and other foods to create adenosine triphosphate, or ATP. in one step, high-energy electrons are transferred along a chain of **proteins** called **cytochromes**. Cyanide binds permanently to these cytochromes, inactivating them. At another step, the energy from these electrons is used to pump H^+ ions across the mitochondrial membrane, creating a gradient that is later harnessed to actually produce ATP. Dinitrophenol, a chemical used in dye manufacture, is a membrane-soluble H^+ carrier. By shuttling H^+ ions back across the mitochondrial membrane, ATP production is prevented. As with many other poisons, neurons and heart muscle cells are most gravely affected.

Chronic exposure to some agents leads to long-term deterioration of health. **Lead** inhibits production of hemoglobin, causing anemia, as well as interfering with a variety of other biologic pathways. Lead builds up in the tissues, and chronic lead poisoning can cause mental retardation and sterility.

Some chemicals directly affect the cell's **DNA** (deoxyribonucleic acid). Chemicals that cause changes in DNA sequence, or mutations, are called **mutagens**. If these sequence changes eliminate the control mechanisms that keep a mature cell from dividing, the cell may become cancerous. Substances that cause cancer are called **carcinogens**. **Asbestos** can cause lung cancer when its fibers are inhaled over a long period of time. Chemicals in cigarette smoke are also carcinogens.

The three major routes of exposure for poisons in the human body are the skin, the lungs, and the gastrointestinal tract. Intact skin provides a barrier against many poisons, but because of its large surface area and direct contact with the environment, it is a likely route for accidental exposure through spills, for instance. The lungs provide direct and rapid entry into the bloodstream for most small molecules, although larger particles such as asbestos are trapped in the mucus lining the respiratory passages. Ingestion into the gastrointestinal tract will inactivate many poisons, especially proteins, though digestion. Nonetheless, ingestion is the principal means of entry of botulinum toxin, for instance, which is a protein.

The dose of a poison a person is exposed to is not equivalent to the amount that enters the body, but rather the amount that reaches the target organ. Biochemical transformation in the liver is the principal means by which potential poisons are detoxified. Liver cells contain a variety of enzyme systems designed to modify foreign compounds to make them more soluble and less reactive. Excretion through the kidneys is the major means of ridding the body of a toxin.

In some cases, biochemical transformation creates the poison from an otherwise harmless substance. For instance, ethylene glycol, a component of antifreeze, is harmless until it is transformed by the body. Acted on by the enzyme that normally metabolizes **ethanol**, it is transformed into **oxalic acid**, which can crystalize in the kidneys, causing kidney failure.

Two poisons can interact synergistically, causing increased toxicity at doses that individually would be lower than the toxic threshold. For instance, asbestos exposure in a smoker increases the risk of lung cancer 40-fold over the risk from wither alone.

One way to measure the toxicity of a compound is by its "LD_{50}" the dose that kills 50% of experimental animals receiving it. However, the LD_{50} varies among different animals. For instance, in rats, the LD_{50} for dinitrophenol is 30 milligrams per kilogram of body weight, while in cats it is 75 mg/kg. The over-the-counter analgesic acetaminophen has an LD_{50} of 338 mg/kg in the mouse, but almost eight times that—2404 mg/kg—in the rat. In humans, the minimum lethal dose is approximately 10 grams, or 150 mg/kg.

Relatively few poisons have specific antidotes. Instead, the toxic effects are reduced as the compound is metabolized and excreted over time. Administration of ethanol (drinking alcohol) is an antidote for ethylene glycol poisoning, because it occupies the metabolizing enzyme until the ethylene glycol is excreted. An antibody against botulinum toxin is available, which binds to the toxin and inactivates it. However, it must be administered very soon after ingestion, and provokes an anaphylactic immune reaction in approximately 10% of people taking it, which can be as life-threatening as the toxin itself.

See also Bioenergetics; Enzymes; Neurochemistry

POLANYI, JOHN C. (1929-)

Canadian chemist

John C. Polanyi, a pioneer in the field of reaction dynamics, made major contributions toward scientists' knowledge of the molecular mechanisms of chemical reactions. His work on the use of infrared **chemiluminescence** paved the way for the development of powerful chemical **lasers**. In recognition of his achievement, he was awarded the Nobel Prize in **chemistry** in 1986.

Polanyi was born on January 23, 1929, in Berlin, Germany, to Michael Polanyi, a chemistry professor, and Magda Elizabeth Kemeny Polanyi, both of Hungarian descent. Polanyi's family moved to Manchester, England, when he was four years old. There, his father took a position as professor of chemistry at Manchester University. Polanyi attended Manchester Grammar School as a child, and enrolled at Manchester University in 1946. That same year, his father stopped teaching chemistry and joined the university's philosophy department.

Polanyi's father had focused his research on the molecular basis of chemical reactions. Polanyi, who had taken his father's last chemistry classes, began to conduct his own chemistry research under the supervision of Ernest Warhurst, one of his father's former students. Where Warhurst, the senior Polanyi and their colleagues investigated the probability that a chemical reaction would result from a collision between molecules, the young Polanyi began to investigate the motions of the newly created reaction products.

Initially, Polanyi had been only marginally interested in science. As a student, he was more enthusiastic about politics,

John C. Polanyi. *(Corbis Corporation. Reproduced by permission.)*

writing poetry, and newspaper editing. Eventually, however, he developed an interest in chemistry, especially ''reaction dynamics,'' as the study of molecular motions in chemical reactions would eventually be called. He went on to earn a Ph.D. in chemistry in 1952, and then moved to Ottawa to conduct his postdoctoral work at Canada's National Research Council. There he attempted to determine whether the **transition state theory** of reaction rates, which his father had helped to develop, could predict the rates at which reactions would occur. He concluded that scientists had insufficient understanding of the forces in the transition state to accomplish this.

After two years in Ottawa, Polanyi worked for several months in the laboratory of **Gerhard Herzberg**, studying vibrational and rotational motions in molecular **iodine**. In 1954, he was invited by the chemist Hugh Stott Taylor to a postdoctoral fellowship at Princeton University. There, the research of Taylor's colleagues, Michael Boudart and David Garvin, impressed Polanyi. In their study of the vibrations produced when atomic **hydrogen** chemically reacted with **ozone**, the reaction emitted a visible glow. From this, Polanyi concluded that it should be possible to determine the vibrational and rotational excitation in newly formed reaction products from the wavelengths of the infrared radiation arising from chemical reactions.

In 1956, Polanyi returned to Canada to take a position as lecturer in the chemistry department at the University of Toronto. He advanced to assistant professor in 1957, and to full professor in 1974. At the University of Toronto, Polanyi and graduate student Kenneth Cashion conducted experiments on the reaction of atomic hydrogen and molecular **chlorine**. The reaction emitted a faint infrared light ''chemiluminescence.'' The study was significant because it suggested a way to obtain quantitative information, for the first time, concerning the vibrational and rotational **energy** released in chemical reactions. Polanyi's subsequent report, ''An Infrared Maser Dependent on Vibrational Excitation,'' followed up on Arthur L. Schawlow's and Charles Townes' 1958 proposal that light could be amplified by passing it through a medium containing highly excited atoms and molecules, a proposal that led to the development of the laser (*l*ight *a*mplification by *s*timulated *e*mission of *r*adiation). Polanyi realized that products of the hydrogen-chlorine reaction—and similar chemical reactions—would provide a medium suitable for a laser—a chemical laser. His report, published in the *Journal of Chemical Physics* in September 1960, after its initial rejection by *Physical Review Letters,* paved the way for the University of California, Berkeley's George Pimentel to develop the chemical laser, one of the most powerful lasers that exist.

In 1986, Polanyi and two other scientists, **Dudley R. Herschbach** and **Yuan T. Lee**, shared the Nobel Prize in chemistry for their contributions to ''the development of a new field of research in chemistry—reaction dynamics.'' Polanyi was cited for his work on ''the method of infrared chemiluminescence, in which the extremely weak infrared emission from a newly formed **molecule** is measured and analyzed.'' He was also recognized for his use of ''this method to elucidate the detailed energy disposal during chemical reactions.''

Polanyi's interests extended far beyond the laboratory. Beginning in the late 1950s, he became active in the arms control debate. In an article he'd written for the *Bulletin of the Atomic Scientists* after attending an arms control meeting in Moscow, he was struck by the ''symmetry of fears'' between the Soviets and Western powers that prompted the arms build-up as a precaution against surprise attacks. His concern as a scientist over ''the mounting spiral of precaution, fear, increased precaution, increasing fear'' led him to become the founding chairman of the Canadian Pugwash Group. He was also an active member of the American National Academy of Sciences' Committee on International Security Studies and the Canadian Center for Arms Control and Disarmament. In addition, he has given many lectures on the subject of arms control and has written approximately sixty articles on this topic.

Polanyi's contributions have been officially recognized by various quarters. His many honors and awards, in addition to the Nobel Prize, include the Marlow Medal of the Faraday Society, the Steacie Prize for the Natural Sciences, the Centennial Medal of the Chemical Society, the Remsen Award, and the Royal Medal of the Royal Society of London. He has been awarded more than two dozen honorary degrees from universities in Canada and the United States, including Harvard University and Rensselaer Polytechnic Institute. In recognition of his accomplishments, the Canadian government appointed him an officer and, later, a companion of the Order of Canada and a member of the Privy Council. A fellow of the Royal Society of Canada and the Royal Society of London, he is a foreign member of the American Academy of Arts and Sciences, the American National Academy of Sciences, and belongs to the Pontifical Academy of Rome.

Polanyi married musician Anne Ferrar Davidson in 1958. The couple has two children. Although Polanyi is more knowledgeable about art, literature and poetry than music, he and his wife have collaborated in writing professionally performed skits, for which she wrote the music and he wrote the words. For relaxation, he enjoys skiing and walking; he no longer engages in the white **water** canoeing and aerobatics he enjoyed when he was younger.

POLAR BOND

The combination of forces that holds two atoms in a fixed relationship within a **molecule** is called a chemical or molecular bond.

In 1916, **Gilbert Newton Lewis** proposed that covalent bonds are formed by the sharing of electrons between the bonded atoms. Quantum mechanics later affirmed the shared paired **covalent bond**, providing a theoretical basis for Lewis' proposal.

In the quantum mechanical interpretation of the covalent bond, atomic orbitals on each **atom** overlap, forming a molecular orbital that stretches over both nuclei and is occupied by two electrons, one from each of the two bonded atoms. When these two atoms have the same **electronegativity** (i.e., the ability to attract electrons from other atoms), the resulting bond is purely covalent: the shared electrons are equally attracted to and shared by both nuclei. When, however, the bonded atoms have different electronegativities, the atom with the greater attraction for electrons pulls the shared electrons toward itself, away from the less electronegative atom. The shared electrons, therefore, spend more time in the vicinity of the more electronegative atom, and as a consequence, it possesses a partial negative charge. The less electronegative atom has a partial positive charge since the shared electrons spend less time near it. There is, therefore, a separation of charges in the bond, with a partial positive charge at one end of the bond and a partial negative charge at the other. Such a bond is called a polar covalent bond.

In the extreme case, when the electronegativity differences of the bonded atoms are quite large, the less electronegative atom loses control of its **electron** to the more electronegative atom and becomes a positive **ion**. The more electronegative atom, now in complete possession of the electrons, becomes a negative ion, and purely electrostatic forces hold the two together in an **ionic bond**.

POLARIZATION OF LIGHT

Light, like radio transmissions and nuclear radiation, is emitted in the form of electromagnetic waves. Specifically, light travels in transverse waves—that is, while traveling along a generally straight path, light waves will vibrate in many directions perpendicular to that path. However, there are several methods available to filter light, causing it to vibrate in only one plane perpendicular to the path of travel. Such filtered light is called polarized light.

When an **electron** is excited it will vibrate; if excited further, it will sometimes emit a **photon** of light. Since the excited electron vibrates in only one direction, that photon will also vibrate along a single plane and, therefore, be polarized. If all electrons in a material vibrated in the same direction, all the light emitted would be plane-polarized. In reality, though, the material's electrons vibrate in an almost infinite number of directions, producing unpolarized light. In order for ordinary light to be polarized it must either pass through or bounce off a polarizing substance. Depending upon its individual properties, this substance will either align the light along a single plane or will remove the light not vibrating along that plane.

The first person to observe the polarization of light was Etienne-Louis Malus (1775-1812), who found that light passing through a piece of Iceland spar crystal was split into two beams. Thinking that each beam was aligned with some mystical ''pole of light'' (similar, in theory, to the poles of a magnet) Malus described the two beams as being ''polarized.'' More precise work on the subject was conducted by a classmate of Malus', Jean-Baptiste Biot (1774-1862), during the final years of the eighteenth century.

Biot's research was further advanced by Augustin Jean-Fresnel (1788-1827), who used the phenomenon to redesign lenses for lighthouses and, more importantly, to support his theory of the transverse nature of light. At this time, debate raging between Europe scientists over whether to subscribe to Isaac Newton's (1642-1727) particle theory of light or, like Fresnel, to the much-doubted theory that light acted as a wave.

In 1828 the Scottish physicist William Nicol (1768-1851) used two pieces of Iceland spar to construct a polarizing prism. As light entered the prism it was split into two beams (just as Malus had observed); upon reaching the second crystal one beam would exit the prism while the second, now polarized, would pass through. This was the first reliable instrument for obtaining polarized light, and, through its use, a generation of scientists began to truly understand the nature of light.

Today, there are four basic methods for polarizing light: absorption, reflection, double refraction, and scattering. The most popular method, absorption, requires a material whose molecules allow only one plane of light to pass through, absorbing all other planes. Such a material is called a *dichroic*. A common man-made dichroic is *polaroid*, discovered in 1938 by the American entrepreneur **E. H. Land**. Land figured that, if a large crystal could be used to polarize light, then thousands of tiny crystals should be capable as well. He mixed the crystals into a clear plastic **solution** which was cooled and stretched into thin sheets; this had the effect of aligning all the crystals and keeping them from drifting, which is a problem in single-crystal systems. Within a few years polaroid replaced nearly all polarizing crystals in scientific and industrial use, and Land built his Polaroid Corporation from its revenues.

Polarization of light by reflection is found more in nature than in industry. When light strikes a flat surface it is polarized to some extent, depending on the angle at which it strikes the surface. If the angle is zero or 90 degrees, the light will not be polarized; if it is less than 90 degrees, some polarization will occur. At one particular angle light will be completely polarized. This phenomenon is called Brewster's law (after its discoverer, Sir David Brewster [1781-1868]) and the polarizing angle is often referred to as Brewster's angle. Many different materials, such as **water**, snow, and **glass**, can polarize light through reflection, though the angle at which this occurs is different for each material.

When Nicol perfected his polarizing prism, he was using the method called double refraction. Certain crystals will split incoming light into two separate rays; Iceland spar is one such crystal, as are calcite and quartz. Each ray produced is polarized, although they are aligned perpendicular to each other—if one ray vibrated up-and-down, the other would vibrate side-to-side. Such polarizing crystals, called birefringent crystals, are seldom used for polarization today, for they are far less versatile than man-made dichroics.

The last method for polarizing light is called scattering. It occurs when light enters a system of many particles that can absorb energy and re-emit it. Many gases, including air, scatter light. As a photon of light strikes a gas molecule it will excite the molecule, causing it to vibrate. When the molecule rereleases the photon it, too, vibrates along the same plane. If many gas molecules are caused to vibrate along one plane, the scattered light will be, to some degree, polarized. Polarization by scattering is primarily a natural phenomenon and has found almost no use in industry.

While instrumental in the study of light, polarization has many practical uses as well. For example, polarized sunglasses have been developed; in addition to blocking direct sunlight, they also absorb the polarized light that is reflected from ground, water, or snow.

Certain plastic and crystalline materials have the unusual effect of twisting polarized light. With such materials, polarized light enters vibrating along one plane; upon exiting, however, it vibrates along a completely different plane, while still remaining polarized. These materials are said to display **optical activity**, and are rapidly gaining popularity among engineers—for example, liquid crystal displays (LCDs) rely upon polarization by optically active **liquid crystals**.

See also Crystallography; Radioactivity

POLLUTION

Pollution can be defined as unwanted or detrimental changes in a natural system. Usually, pollution is associated with the presence of toxic substances in some large quantity, but pollution can also be caused by the presence of excess quantities of **heat** or by excessive fertilization with nutrients.

Because pollution is judged on the basis of degradative changes, there is a strongly anthropocentric bias to its determination. In other words, humans decide whether pollution is occurring and how bad it is. Of course, this bias favors species, communities, and ecological processes that are especially desired or appreciated by humans. In fact, however, some other, *less-desirable* species, communities, and ecological processes may benefit from what we consider pollution.

An important aspect of the notion of pollution is that ecological change must actually be demonstrated. If some potentially polluting substance is present at a **concentration** or intensity that is less than the threshold required to cause a demonstrable ecological change, then the situation would be referred to as contamination, rather than pollution.

This aspect of pollution can be illustrated by reference to the stable elements, for example, **cadmium**, **copper**, **lead**, **mercury**, **nickel**, **selenium**, **uranium**, etc. All of these are consistently present in at least trace concentrations in the environment. Moreover, all of these elements are potentially toxic. However, they generally affect biota and therefore only cause pollution when they are present at water-soluble concentrations of more than about 0.01-1 parts per million (ppm).

Some other elements can be present in very large concentrations, for example, **aluminum** and **iron**, which are important constituents of rock and soil. Aluminum constitutes 8-10% of the earth's crust and iron 3-4%. However, almost all of the aluminum and iron present in minerals are insoluble in **water** and are therefore not readily assimilated by biotic community and cannot cause toxicity. In acidic environments, however, ionic forms of aluminum are solubilized, and these can cause toxicity in concentrations of less than one part per million. Therefore, the bio-availability of a chemical is an important determinant of whether its presence in some concentration will cause pollution.

Most instances of pollution result from the activities of humans. For example, anthropogenic pollution can be caused by:

(1) the emission of **sulfur** dioxide and **metals** from a smelter, causing toxicity to vegetation and acidifying surface waters and soil,

(2) the emission of waste heat from an **electricity** generating station into a river or lake, causing community change through thermal stress, or

(3) the discharge of nutrient-containing sewage wastes into a water body, causing eutrophication.

Most instances of anthropogenic pollution have natural analogues, that is, cases where pollution is not the result of human activities. For example, pollution can be caused by the emission of sulfur dioxide from volcanoes, by the presence of toxic elements in certain types of soil, by thermal springs or vents, and by other natural phenomena. In many cases, natural pollution can cause an intensity of ecological damage that is as severe as anything caused by anthropogenic pollution.

An interesting case of natural air pollution is the Smoking Hills, located in a remote and pristine wilderness in the Canadian Arctic, virtually uninfluenced by humans. However, at a number of places along the 18.63 miles (30 km) of seacoast, bituminous shales in sea cliffs have spontaneously ignited, causing a fumigation of the tundra with sulfur dioxide and other pollutants. The largest concentrations of sulfur dioxide (more than two parts per million) occur closest to the combustions. Further away from the sea cliffs the concentrations of sulfur dioxide decrease rapidly. The most important chemical effects of the air pollution are acidification of soil and fresh water, which in turn causes a solubilization of toxic metals. Surface soils and pond waters commonly have pHs less than 3, compared with about pH 7 at non-fumigated places. The only reports of similarly acidic water are for volcanic lakes in Japan, in which natural pHs as acidic as 1 occur, and **pH** less than 2 in waters affected by drainage from **coal** mines.

At the Smoking Hills, toxicity by sulfur dioxide, acidity, and water-soluble metals has caused great damage to ecological communities. The most intensively fumigated terrestrial sites have no vegetation, but further away a few pollution-tolerant species are present. About one kilometer away the toxic stresses are low enough that reference tundra is present. There are a few pollution-tolerant algae in the acidic ponds, with a depauperate community of six species occurring in the most-acidic (pH 1.8) pond in the area.

Other cases of natural pollution concern places where certain elements are present in toxic amounts. Surface mineralizations can have toxic metals present in large concentrations, for example copper at 10% in peat at a copper-rich spring in New Brunswick, or surface soil with 3% lead plus **zinc** on Baffin Island. Soils influenced by nickel-rich serpentine minerals have been well-studied by ecologists. The stress-adapted plants of serpentine habitats form distinct communities, and some plants can have nickel concentrations larger than 10% in their tissues. Similarly, natural soils with large concentrations of selenium support plants that can hyperaccumulate this element to concentrations greater than 1%. These plants are poisonous to livestock, causing a toxic syndrome known as *blind staggers*.

Of course, there are many well-known cases where pollution is caused by anthropogenic emissions of chemicals. Some examples include:

(1) Emissions of sulfur dioxide and metals from smelters can cause damage to surrounding terrestrial and aquatic ecosystems. The sulfur dioxide and metals are directly toxic. In addition, the deposition of sulfur dioxide can cause an extreme acidification of soil and water, which causes metals to be more bio-available, resulting in important, secondary toxicity. Because smelters are point sources of emission, the spatial pattern of chemical pollution and ecological damage displays an exponentially decreasing intensity with increasing distance from the source.

(2) The use of **pesticides** in agriculture, forestry, and around homes can result in a non-target exposure of birds and other wildlife to these chemicals. If the non-target biota are vulnerable to the pesticide, then ecological damage will result. For example, during the 1960s urban elm trees in the eastern United States were sprayed with large quantities of the insecticide DDT, in order to kill beetles that were responsible for the transmission of Dutch elm disease, an important pathogen. Because of the very large spray rates, many birds were killed, leading to reduced populations in some areas. (This was the "silent spring" that was referred to by Rachel Carson in her famous book by that title.) Birds and other non-target biota have also been killed by modern insecticide-spray programs in agriculture and in forestry.

(3) The deposition of acidifying substances from the atmosphere, mostly as acidic **precipitation** and the dry deposition of sulfur dioxide, can cause an acidification of surface waters. The acidity solubilizes metals, most notably aluminum, making them bio-available. The acidity in combination with the metals causes toxicity to the biota, resulting in large changes in ecological communities and processes. Fish, for example, are highly intolerant of acidic waters.

(4) Oil spills from tankers and pipelines can cause great ecological damage. When oil spilled at sea washes up onto coastlines, it destroys seaweeds, invertebrates, and fish, and their communities are changed for many years. Seabirds are very intolerant of oil and can die of hypothermia if even a small area of their feathers is coated by petroleum.

(5) Most of the lead shot fired by hunters and skeet shooters misses its target and are dispersed into the environment. Waterfowl and other avian wildlife actively ingest lead shot because it is similar in size and hardness to the grit that they ingest to aid in the mechanical abrasion of hard seeds in their gizzard. However, the lead shot is toxic to these birds, and each year millions of birds are killed by this source in North America.

Humans can also cause pollution by excessively fertilizing natural ecosystems with nutrients. For example, freshwaters can be made eutrophic by fertilization with **phosphorus** in the form of **phosphate**. The most conspicuous symptoms of eutrophication are changes in species composition of the phytoplankton community and, especially, a large increase in algal biomass, known as a *bloom*. In shallow waterbodies there may also be a vigorous growth of vascular plants. These primary responses are usually accompanied by secondary changes at higher trophic levels, including arthropods, fish, and waterfowl, in response to greater food availability and other habitat changes. However, in the extreme cases of very eutrophic waters, the blooms of algae and other microorganisms can be noxious, producing toxic chemicals and causing periods of **oxygen** depletion that kill fish and other biota. Extremely eutrophic waterbodies are polluted because they often cannot support a fishery, cannot be used for drinking water, and have few recreational opportunities and poor aesthetics.

Pollution, therefore, is associated with ecological degradation, caused by environmental stresses originating with natural phenomena or with human activities. The prevention and management of anthropogenic pollution is one of the greatest challenges facing modern society.

POLONIUM

Polonium is the heaviest element in Group 16 of the **periodic table**, a group of elements often known as the **oxygen** family. Polonium has an **atomic number** of 84, an atomic **mass** of 208.9824, and a chemical symbol of Po.

Properties

All isotopes of polonium are radioactive, the most stable being polonium-209 with a half life of 102 years. The element has a melting point of 489°F (254°C), a **boiling point** of 1,764°F (962°C), and a **density** of 9.4 grams per cubic centimeter. Polonium has chemical problems similar to the other members of the oxygen family and is the most metallic member of that family. In general, scientists are much less interested in the element's chemical properties than its radioactive properties.

Occurrence and Extraction

Polonium is an extremely rare element, with an abundance estimated at no more than about 3×10^{-10} parts per million. For all practical purposes, the only polonium used today is produced synthetically in particle accelerators.

Discovery and Naming

Polonium was the first element to be discovered by **Marie Curie** and her husband Pierre as they analyzed pitchblende, an ore of **uranium**, in 1898. The Curies found that for each ton of pitchblende they processed, they were able to extract about 100 micrograms of a new element, which they name in honor of Madame Curie's native land, Poland.

Uses

Polonium has two minor uses. First, it can be used as a compact and efficient source of **heat** production in specialized situations where cost is of little consequence. An example of such a situation is a space probe. Polonium is also used to remove static **electricity** from photographic film where the electrical charge would otherwise damage the quality of a photograph.

Health Issues

Polonium is an extremely toxic material. Perhaps its greatest practical threat is in cigarette smoking. The products found in tobacco have been found to contain minute amounts of polonium. The radiation given off by the element is so intense, however, that it must be considered as one of the health risks posed by cigarette smoking.

POLYBROMINATED BIPHENYLS (PBBS)

A mixture of compounds having from one to ten bromine atoms attached to a biphenyl ring, analogous to **polychlorinated biphenyls (PCBs)**. Manufactured as fire retardants, PBBs were banned after a 1973 Michigan incident when pure product was accidently mixed with cattle feed and distributed throughout the state. PBBs were identified as the cause of weight loss, decreased milk production, and mortality in many dairy herds. Approximately 30,000 cattle, 1.5 million chickens, 1,500 sheep, 6,000 hogs, 18,000 pounds (8,172 kg) of cheese, 34,000 pounds (15,436 kg) of dried milk products, 5 million eggs, and 2,700 pounds (1,225 kg) of butter were eventually destroyed at an estimated cost of $1 million. Although human exposures have been well-documented, long term epidemiological studies have not shown widespread health effects.

POLYCHLORINATED BIPHENYLS (PCBS)

A mixture of compounds having from one to 10 **chlorine** atoms attached to a biphenyl **ring** structure. There are 209 possible structures theoretically; the manufacturing process results in approximately 120 different structures. PCBs resist biological and **heat** degradation and were used in numerous applications, including dielectric fluids in capacitors and transformers, **heat transfer** fluids, hydraulic fluids, plasticizers, dedusting agents, adhesives, dye carriers in carbonless copy **paper**, and pesticide extenders. The United States manufactured PCBs from 1929-1977, when they were banned due to adverse environmental effects and ubiquitous occurrence. They bioaccumulate in organisms and can cause skin disorders, liver dysfunction, reproductive disorders, and tumor formation. They are one of the most abundant organochlorine contaminants found throughout the world.

POLYCYCLIC AROMATIC HYDROCARBONS

Polycyclic **aromatic hydrocarbons** are a class of organic compounds having two or more fused aromatic **ring** structures each based on the structure of **benzene**. While usually referring to compounds made of **carbon** and **hydrogen**, PAH also may include fused aromatic compounds containing **nitrogen**, **sulfur**, or cyclopentene rings. Some of the more common PAH include **naphthalene** (2 rings), anthracene (3 rings), phenanthrene (3 rings), pyrene (4 rings), chrysene (4 rings), fluoranthene (4 rings), benzo(a)pyrene (5 rings), benzo(e)pyrene (5 rings), perylene (5 rings), benzo(g,h,i)perylene (6 rings), and coronene (7 rings).

PAH are formed by a variety of human activities including incomplete **combustion** of fossil fuels, **wood**, and tobacco; the incineration of garbage; **coal** gasification and liquefaction processes; **smelting** operations; and coke, asphalt, and petroleum cracking; they are also formed naturally during forest fires and volcanic eruptions. Low molecular-weight PAH (those with four or fewer rings) are generally vapors, while heavier molecules condense on submicron, breathable particles. It is estimated that more than 800 tons of PAH are emitted annually in the United States. PAH are found worldwide and are present in elevated concentrations in urban **aerosols**, and in lake sediments in industrialized countries. They also are found in developing countries due to coal and wood heating, open burning, coke production, and vehicle exhaust.

The association of PAH with small particles gives them atmospheric residence times of days to weeks, and allows them to be transported long distances. They are removed from the atmosphere by gravitational settling and are washed out during **precipitation** to the Earth's surface, where they accumulate in soils and surface waters. They also directly enter **water** in discharges, runoff, and oil spills. They associate with water particulates due to their low water **solubility**, and eventually accumulate in sediments. They do not bioaccumulate in biota to any appreciable extent, as they are largely metabolized.

Many but not all PAH have carcinogenic and mutagenic activity; the most notorious is benzo(a)pyrene, which has been shown to be a potent carcinogen. Coal tar and soot were implicated in the elevated skin cancer incidence found in the refining, shale oil, and coal tar industries in the late nineteenth century. Subsequent research led to the isolation and identification of several **carcinogens** in the early part of this century, including benzo(a)pyrene. More recent research into the carcinogenicity of PAH has revealed that there is significant additional biological activity in urban aerosols and soot beyond that explained by known carcinogens such as benzo(a)pyrene. While benzo(a)pyrene must be activated metabolically, these other components have direct biological activity as demonstrated by the Ames test. They are polar compounds, thought to be mixtures of mono- and dinitro-PAH and hydroxy-nitro derivatives. Tobacco smoking exposes more humans to PAH than any other source.

POLYESTER

Polyesters are a class of linear, thermoplastic polymers, characterized by an ester group contained in their primary molecular backbone. They are the result of a reaction between a diacid (such as terephthalic acid) and a diol (such as ethylene glycol). First produced commercially in the late 1920s, polyesters have become important compounds used in a wide variety of industries. The most economically important types of polyesters include poly(ethylene terephthalate) (PET) and poly(butylene terephthalate) (PBT).

The chemical reactions for making polyesters were investigated in 1901 and resulted in the production of glyptal polyesters. These reactions involved the combination of a diacid with a diol. The reaction was called a **condensation reaction** because the two initial types of monomers combined to produce a longer chain polymer and **water** as a byproduct. Linear polyesters were not produced until the 1930s, when **W. H. Carothers** systematically investigated reactions of diols with diacids. Carothers was not successful in producing a polyester fiber and switched the focus of his research. In 1942, **John Whinfield** and W. Dickson made the first high **molecular weight** PET. After these fibers were produced other polyesters were discovered and have since become important compounds.

Polyesters have many characteristics that make them useful. They have high melting points and maintain good mechanical properties up to 283°F (175°C). They are generally resistant to **solvents** and other chemicals. When made into a fiber they demonstrate low moisture absorption and good resistance to abrasion. As a film they have a high tensile strength, are stiff, and have a high impact strength.

The versatility of polyesters can be demonstrated by the wide variety of products in which they are used. When spun into a fiber, polyesters are used to produce textiles, yarns, ropes and tire cord. When drawn out as a thin film, polyesters are used for making magnetic tape for audio and video recording, packaging materials and photographic film. Polyester can also be molded into many forms and used for making such things as bottles and containers that package consumer goods. They are also used to make parts for the automotive and electronics industries.

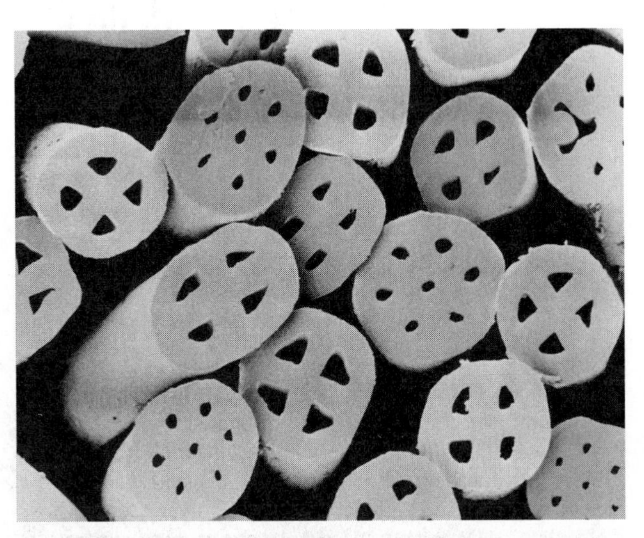

A scanning electron micrograph of polyester fibers. ***(Photograph by Science Photo Library/National Audubon Society Collection/Photo Researchers, Inc. Reproduced by permission.)***

POLYETHYLENE

Polyethylene is a long chain polymer produced by the polymerization reaction of liquid ethylene at high temperatures and pressures. It is a solid plastic that has a milky transparent appearance.

The discovery of polyethylene in the 1930s greatly aided the Allied war effort during World War II because it was used as insulation for cables vital to the Allied information network. Polyethylene was accidently discovered by J. Swallow and M. Perrin at Imperial Chemical Industries in Britain. In 1933, while researching the effects of high pressure on chemical reactants, a fellow scientist, R. Gibson, managed to produce a waxy solid from ethylene and benzaldehyde. He repeated the experiment with ethylene alone with no success. The experiment had taken place at 280°F (138 °C) with ethylene at a pressure of 1,400 atmospheres. Theorizing that a higher pressure was needed, the three colleagues set to work designing and building improved laboratory equipment.

Nearly three years later, Swallow and his companions tried the experiment at a higher **temperature**. In the course of the polymerization process the pressure dropped and they added more ethylene to compensate. The procedure yielded approximately 0.28 oz (8 g) of polyethylene. After inspecting the equipment they discovered that there had been a leak. The polyethylene they had added contained **oxygen**. Further investigation revealed that the oxygen was vital to the polymerization process. The polyethylene, like polyesters, benefitted from the cold drawing technique developed by **Wallace Carothers** at DuPont.

Polyethylene, or polythene as it was marketed in Britain, began as a polymer with little practical use. It would have remained so if an employee had not noticed that the mechanical properties of polyethylene were comparable to those of *gutta percha*, a natural product used to insulate telegraph cables. Polyethylene was used to insulate the cables laid between France and Britain, providing a crucial line of communication towards the end of World War II. After the war, polyethylene film was used for packaging, liners, tank and pool covers, and drop cloths.

The early polyethylene was a low **density** branched polymer. This means that there was a good deal of empty space in each **molecule**, and the molecules were formed in a branching pattern. These characteristics accounted for the fact that this polyethylene was not a very strong material. However, in 1953, German scientist **Karl Ziegler** and his staff discovered a method of producing high density, linear polyethylene (whose molecules had less empty space and were arranged in rope-like strands, both of which give a substance greater tensile strength). Ziegler was researching organometallic compounds (carbon compounds that contain metals) and their effects on polymerization reaction. He attempted to polymerize ethylene using catalysts at a much lower pressure than that used by the British process.

At first, Ziegler was puzzled when an experiment yielded a dimer (a compound of two radicals, rather than the several radicals which compose a polymer) of polyethylene rather than the expected low **molecular weight** polymer. He discovered trace amounts of **nickel** on the laboratory equipment had inhibited the reaction. As a result of his investigation of this occurrence, Ziegler eventually succeeded in producing a very high molecular weight polyethylene with a very high melting point using metal chloride and organoaluminum compounds as catalysts. Polyethylene could now be carried out at low temperatures and normal pressure, which greatly simplified the industrial production process.

High density polyethylene is currently used in dishes, squeezable bottles, and other soft **plastics** products. It is recyclable. Linear low density polyethylene (LLPE) now substitutes for the older branched low density polyethylene (LDPE). LLPE is formed using the Ziegler process using **hydrogen** to regulate molecular weight. All of the various polyethylenes may be processed by injection molding or extrusion.

POLYMER CHEMISTRY

Polymer chemistry is the field of study concerned with the production, classification, and modification of macromolecules or polymers. Polymers are produced by a polymerization reaction of simpler units called monomers. While polymers have been utilized for centuries, it was not until the twentieth century that their chemistry was understood. Many of the organic materials on Earth are polymers. These include natural materials like **nucleic acids, protein, rubber,** and cotton, as well as synthetic materials like **nylon, polyester** and **polyvinyl chloride.**

Polymers have been used by humans for centuries. However, the field of polymer chemistry has been around only since the 1800s. The first experiments in the development of this branch of **chemistry** were performed by technologists who attempted to convert natural polymers into more useful derivatives. In 1839, Charles Goodyear produced the first significant development when he created a process for making cross-linked, or **vulcanized rubber.** This material was harder and more useful than the native material. Other scientists developed processes to make plaster derivatives, thermoplastics and artificial silks. The first synthetic polymer, called Bakelite (a polymer made from phenol and **formaldehyde**) was invented by **Leo Baekeland** in 1907.

In the early part of the 1900s advances in polymer technology were achieved largely without the advantage of understanding the chemistry behind them. The groundwork for modern understanding of polymer science was laid by Nobel laureate **Hermann Staudinger** in the 1920s. He theorized that the structure of these compounds were long, chainlike **molecules** and not aggregates or cyclic compounds as previously thought. In 1928, Staudinger's models were confirmed by Meyer and Mark who used x-ray techniques to show the dimensions of natural rubber. By the 1930s most polymer scientists accepted Staudinger's model of polymeric molecules. The development and of polymer chemistry since the 1940s has been extensive. Today, polymers are used in nearly every facet of life.

For an area of science to develop, a method of naming compounds is required to facilitate communication between researchers in the field. The materials studied in polymer chemistry however, are more complex than those of traditional areas of chemistry and standard naming systems were inadequate. In the early years of polymer chemistry, names given to compounds were not consistent. Some were named after the origin of their starting materials, others were named after their inventor. Eventually, a source-based naming system was adopted. In this system, polymeric compounds are named by adding the prefix Apoly- to the name of their starting monomers. For example, polystyrene is the material that results from the polymerization of styrene monomers. This nomenclature system is slowly being replaced by a structure-based system that was adopted by the **International Union of Pure and Applied Chemistry (IUPAC).**

Polymer science is concerned with the classification, creation, and modification of macromolecules. An important aspect of polymer classification is the structure of polymers. In general, polymers can be grouped into three structural groups, linear, branched, and cross-linked. Linear polymers are long, chainlike molecules covalently bonded in a rigid manner. Side groups are also chemically bonded to atoms in the chain to give the polymer its unique characteristics. Examples of linear polymers include polyethylene, polyvinyl chloride and polypropylene. Depending on the location of the side groups and the nature of the starting monomers, linear polymers can have various structural isomers denoted as isotactic, syndiotactic, or atactic. These isomers exhibit different characteristics such as variable melting temperatures and viscosities.

Branched polymers are similar to linear polymers however, they have irregularly spaced extensions of the polymer

chain that are attached to the primary polymer backbone. These polymers tend to occupy more volume per unit and have reduced densities. An example of this type of polymer is low density polyethylene. Linear and branched polymers are classified as thermoplastics which means they will soften when heated and harden when cooled. Crosslinked polymers, which are the third type of structural group of polymers, are thermoset polymers which do not soften with heat but decompose. Cross-linked polymers consist of two or more chains that are covalently bonded. An example of this type of polymer is vulcanized rubber.

Although chemical composition determines some characteristics of polymers, the bulk properties of polymers are dependent on the organization of their chains. While the chains of crosslinked polymers are covalently bonded, linear and branched polymers interact with weaker bonds such as **van der Waals forces** and hydrogen bonding. These bonds are responsible for characteristics such as melting and boiling points. The strength of these bonds is determined by the organization of the polymer chains. Amorphous polymers have chains that are random and disordered. As more of the chains are caused to line up in an ordered manner, the polymer becomes more crystalline.

Polymers can be further characterized by specific bulk properties such as rheology, solubility, viscosity and molecular weight. The rheology of a polymer is indicative of its physical characteristics. A polymer may be glassy, leathery, rubbery or liquid. Its rheological characteristics help determine for what applications a polymer can be used. Similarly, the solubility of a polymer in a solvent is another important characteristic. This is particularly critical for polymers that are the basis for paints and coatings. The viscosity of a polymer is a measure of its thickness. It is the most widely used measurement for characterizing polymers and determining their molecular weight. The molecular weight of a polymer is indicative of the average number of monomers that make-up a polymer chain. Other methods of molecular weight determination include gel permeation chromatography, end-group analysis, and osmometry.

To ensure the quality of polymeric materials, specific tests have been developed by the American Society for Testing and Materials (ASTM). Stress tests can measure the amount of force that can be applied to a material before it breaks. This test gives an indication of how well the material will perform for certain applications. Other physical tests measure such parameters as tensile strength, impact strength, compressive strength, hardness, elasticity and electrical properties.

Polymer chemistry is such an important area of study because nearly every material in the world is polymeric. This includes such diverse materials as the nucleic acids that make-up the **DNA** of living things, the paint used to coat walls, and plastics used everyday. Many of these materials contain one type of monomer and are considered homopolymers. Some materials are known as copolymers because they contain two or more different monomeric units. In general, polymers can be classified as either natural or synthetic depending on how they are produced.

The first polymers to be used by humans were naturally occurring polymers. There is a wide variety of naturally occurring polymers and their science and production is still largely not understood. The primary categories for natural polymers include polysaccharides, proteins, nucleic acids, and polyisoprenes.

Polysaccharides are the most abundant organic compounds on Earth. They are linear or branched polymers that are composed of repeating carbohydrate units. They are used by living organisms as food, and also provide structure. Important types of polysaccharides include cellulose, starch, and chitin. Proteins are more complex polymers made up of repeating amino acid units. These polymers are critical to all living systems providing structure, and catalyzing biochemical reactions. Nucleic acids are also polymers, composed of repeating units called nucleotides. A nucleotide is a combination of a ribose or deoxyribose sugar, a purine or pyrimidine base, and a phosphate group. They are the building blocks for genetic information in all living organisms. The final class of important natural polymers is polyisoprenes. These are hard plastics such as rubber produced by various plants.

Numerous synthetic polymers have been developed over the years. Two general methods for their production including condensation polymerization and addition polymerization. These methods generally apply for both homopolymers and copolymers. Condensation polymerization is a method in which two types of monomers are joined together with a characteristic loss of atoms. This can involve a reaction between a di-alcohol and a di-acid. Examples of polymers produced in this way include polyesters, polyamides, polyurethanes, polyethers, and polysulfones.

Addition polymerization involves the joining of monomers without the loss of atoms. In this reaction, there are three key steps including initiation, propagation, and termination. In the initiation phase, some monomers are converted to free radicals. These free radicals react with other monomers during the propagation step, growing the polymeric chains and creating more free radicals. The polymerization continues through the termination step in which free radical production stops. Various materials are produced in this method including polyacrylates, polyvinyl chloride, polytetrafluoroethylene (Teflon), polyethylene, and polypropylene.

While polymers themselves have useful characteristics they are typically modified to improve their characteristics. Some of the first useful polymers were chemically modified natural polymers such as vulcanized rubber and cellulose acetate. Synthetic polymers are also modified by having various materials added. Fillers are added to polymers to make them stronger, more workable and reduce costs. They include materials such as wood, chopped paper, cotton, glass flakes, chalk, or sand. Reinforcement materials such as fiber glass are also added to polymeric materials. Other property modifiers include plasticizers which improve polymeric films, stabilizers that make polymers more temperature and UV resistant, colorants, biocides, and flame retardants which help prevent burning.

Polymers are used in many forms and for numerous applications. Polymeric fibers have become important in the textile industry. Particularly notable have been materials like

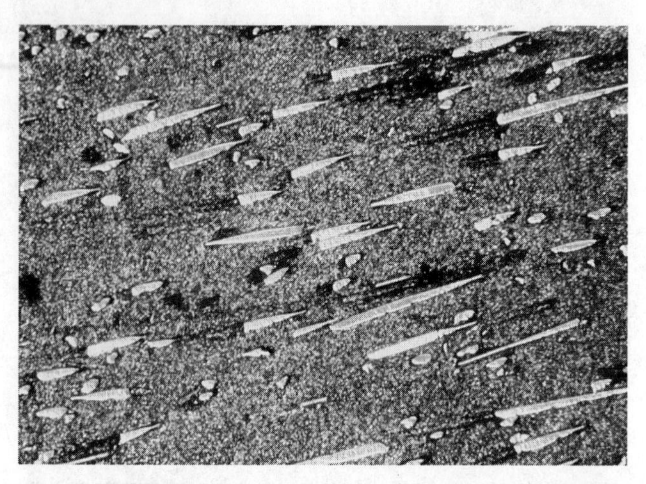

A light micrographic image of glass fibers in a polypropylene matrix. Polypropylene refers to any of a number of tough, flexible, synthetic plastics, which become soft when heated and harden on recooling without appreciable changes to their properties. *(Photograph by Astrid and Hanns Frieder Michler/Science Photo Library, Photo Researchers, Inc. Reproduced by permission.)*

polyester, nylon, cotton and polyurethane. Elastomers are another form of polymeric materials. These are rubbery materials used for everything from tires to carpets to gloves. Polybutadiene, polyacrylates and polyisoprene are important examples. Films and sheets of polymers have been used for packaging materials, movie films, and construction. Most notable are regenerated cellulose films. Polymeric foams have been used for such things as building insulation, flotation devices and furniture cushions. Polystyrene, polyvinyl chloride, and polyethylene are all examples. Molded plastics are another area where polymers find application. Additionally, the paints and coatings industry is based on polymers.

See also Plastics

Polypropylene

Polypropylene is a linear **macromolecule**, or polymer, composed of repeating units of isopropane. It is a thermoplastic material which gets softer with heating and hardens when cooled. First produced in the late 1950s, polypropylene has become an important plastic used in a variety of industries.

The **chain reaction** polymerization process for producing polypropylene was first developed in 1954 by **Giulio Natta**. He used a catalyst discovered by **Karl Ziegler** to produce polypropylene from propylene. This reaction was important because it could be run at room **temperature** and low pressures. The first commercial production of polypropylene occurred in 1957 by Montecatini in Italy. It rapidly became a significant industrial compound because of its good physical and economic properties. For their work, Ziegler and Natta were awarded the Nobel Prize in **Chemistry** in 1963. Over the years, improvements have been made in the types of catalysts used for producing polypropylene. Improvements have also been found for controlling the structure of the resulting product. In 1993, worldwide production of polypropylene was greater than 14 million tons.

Polypropylene is a lightweight, low **density** material that has a high degree of crystallinity compared to other polymers. It also has high tensile strength, stiffness, and hardness. It has good gloss and a high softening temperature range which makes it suitable for sterilization. Additionally, it has good electrical properties, is resistant to moisture and is chemically inert. It also has excellent barrier properties making it resistant to permeation of **water** vapors and other **gases**.

Polypropylene is used for a variety of purposes. It can be spun into a filament that is useful for making rope, webbing and cordage. It is also used for making carpeting. It can be molded into different forms and used for parts in home appliances and automobiles. Additionally, polypropylene is used for household product packaging.

Polystyrene

Polystyrene is a high **molecular weight** polymer made up of repeating monomeric units of styrene. It is a clear, solid plastic that softens at about 185°F (85°C). It is produced through a **free radical** polymerization reaction of styrene monomers. First produced during the 1930s, polystyrene has become a valuable material with numerous applications for consumer products, packaging, and insulation.

The development of polystyrene began with the discovery in 1839 of styrene. By 1925, German scientists had included styrene in a synthetic rubber. The polymerization reactions which produce polystyrene were developed in 1937 by scientists at the Bakelite Corporation. Today, polystyrene is produced on an industrial scale using a bulk, free radical polymerization process.

Various characteristics make polystyrene a unique and useful material. Polystyrene is clear, easily colored and fabricated. This makes it an excellent material for producing consumer products like toys, housewares, light fixtures, lenses and appliances. It has fair mechanical properties but good resistance to **acids and bases**. It is also a good electric insulator. Polystyrene can be physically modified into expanded polystyrene which is an ideal packaging and insulation material. Polystyrene can also be chemically modified to produce copolymer resins that have better impact resistance. These materials are useful for making dishes, tumblers, and telephone parts.

Polyurethane

Polyurethanes are a class of linear polymers characterized by a molecular backbone that contains carbamate groups (-NHCOO). These groups are also called urethane groups and are the result of a chemical reaction between a diisocyanate with a polyol. First investigated in the late 1930s, polyurethanes have become important materials used for a wide variety of applications from building insulation to athletic apparel.

Research in the **chemistry** of polyurethane materials was initiated by the German chemist, Friedrich Bayer in 1937. This

Polyurethane foam. ***(Photograph by Charles D. Winters, National Audubon Society Collection/Photo Researchers, Inc. Reproduced by permission.)***

early work was based on toluene diisocyanate reacted with dihydric alcohols. One of the first crystalline polyurethane fibers was Perlon U. In 1953, commercial production of a flexible polyurethane foam was begun in the United States. This material was used as foam insulation. In 1956, more flexible, less expensive foams were made. In the late 1950s moldable polyurethanes were produced. Also, spandex fibers, polyurethane coatings, and thermoplastic **elastomers** were introduced.

Polyurethanes have many characteristics that make them useful for a variety of applications. They are produced in four different forms including fibers, elastomers, coatings, and cross-linked foams. Polyurethane fibers are lightweight and have a high degree of stretchability which makes them good for applications like swimsuits and other athletic apparel. The elastomers can be stretched but will eventually return to their original shape. They are resistant to abrasion and grease, and have good hardness. This makes them useful as base materials for small wheels. Polyurethane coatings demonstrate a resistance to solvent degradation and have good impact resistance. These coatings are used on surfaces where these characteristics are important such as bowling alleys and dance floors. Foam polyurethanes have high-impact strength. They are used for the production of pillows and cushions.

Polyvinyl chloride (PVC)

Polyvinyl chloride (PVC) is a **macromolecule** that results from the free-radical polymerization of **vinyl** chloride. First discovered in the early 1870s, PVC has become widely distributed second only to **polyethylene** in annual production volume. It has many desirable characteristics finding use as a construction material, packaging resin, and textiles coating.

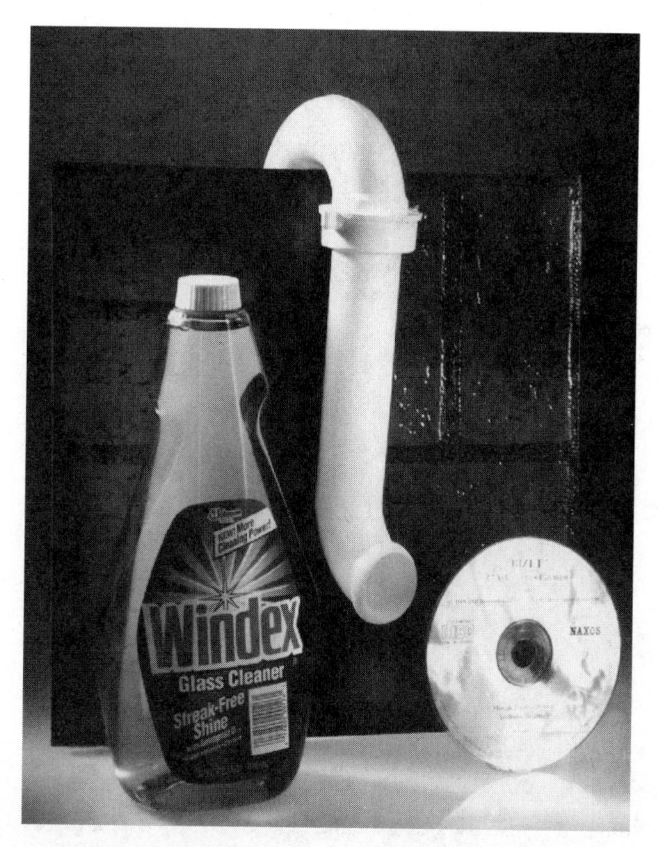

Products made of polyvinyl chloride. ***(Photograph by Charles D. Winters, National Audubon Society Collection/Photo Researchers, Inc. Reproduced by permission.)***

The discovery of PVC was first described in 1872. It was found when a container of vinyl chloride was exposed to sunlight and a white solid was produced. This material was of interest because it was resistant to degradation by **potassium** hydroxide or **water**. From 1912 to 1926 German scientists attempted to process PVC into more usable products. In 1926, it was found that when PVC was boiled in tricresyl **phosphate** it became elastic. This new reaction made PVC an important material and produced the first thermoplastic elastomer.

Depending on the grade required, PVC can be produced by a variety of techniques including **suspension**, **emulsion**, **solution**, microsuspension, and **mass** polymerization. Suspension polymerization is the most often used procedure. It involves the suspension of **monomer** particles in water. Stabilizers prevent coalescence while the polymerization reaction proceeds. After 80-90% of the reaction is complete, excess monomer and water is removed and the polymer is dried with hot air.

The bulk properties of PVC depend on the production method. It can be produced as droplets suspended in water, a thin membrane, a powder or crystals. To improve the workability of PVC resin various ingredients are added to the resin including stabilizers, impact modifiers, lubricants, plasticizers, biocides and flame retardants.

PVC is employed in a variety of applications. The principle market for PVC is as a base material for pipes and fit-

Cyril Ponnamperuma. *(AP/Wide World. Reproduced by permission.)*

tings. It is also used as a weathering material because it is less damaged by exposure to sunlight. PVC makes an excellent resin for clear bottles that hold consumer products. In addition, PVC is used as an insulating material for cables and wires.

Ponnamperuma, Cyril (1923-1994)

Sri Lankan-born American chemist

Cyril Ponnamperuma, an eminent researcher in the field of **chemical evolution**, rose through several National Aeronautics and Space Administration (NASA) divisions as a research chemist to head the Laboratory of Chemical Evolution at the University of Maryland, College Park. His career focused on explorations into the **origin of life** and the ''primordial soup'' that contained the precursors of life. In this search, Ponnamperuma took advantage of discoveries in such diverse fields as molecular biology and astrophysics.

Born in Galle, Ceylon (now Sri Lanka) on October 16, 1923, Cyril Andres Ponnamperuma was educated at the University of Madras (where he received a B.A. in Philosophy, 1948), the University of London (B.Sc., 1959), and the University of California at Berkeley (Ph.D., 1962). His interest in the origin of life began to take clear shape at the Birkbeck College of the University of London, where he studied with J. D. Bernal, a well-known crystallographer. In addition to his studies, Ponnamperuma also worked in London as a research chemist and radiochemist. He became a research associate at the Lawrence Radiation Laboratory at Berkeley, where he studied with **Melvin Calvin**, a Nobel laureate and experimenter in chemical evolution.

After receiving his Ph.D. in 1962, Ponnamperuma was awarded a fellowship from the National Academy of Sciences, and he spent one year in residence at NASA's Ames Research Center in Moffet Field, California. After the end of his associate year, he was hired as a research scientist at the center and became head of the chemical evolution branch in 1965.

During these years, Ponnamperuma began to develop his ideas about chemical evolution, which he explained in an article published in *Nature.* Chemical evolution, he explained, is a logical outgrowth of centuries of studies both in **chemistry** and biology, culminating in the groundbreaking 1953 discovery of the structure of deoxyribonucleic acid (DNA) by **James Watson** and **Francis Crick**. Evolutionist Charles Darwin's studies affirming the idea of the ''unity of all life'' for biology could be extended, logically, to a similar notion for chemistry: protein and nucleic acid, the essential elements of biological life, were, after all, chemical.

In the same year that Watson and Crick discovered **DNA**, two researchers from the University of Chicago, **Stanley Lloyd Miller** and **Harold Urey**, experimented with a primordial soup concocted of the elements thought to have made up earth's early atmosphere—methane, **ammonia**, **hydrogen**, and **water**. They sent electrical sparks through the mixture, simulating a lightening storm, and discovered trace amounts of **amino acids**.

During the early 1960s, Ponnamperuma began to delve into this primordial soup and set up variations of Miller and Urey's original experiment. Having changed the proportions of the elements from the original Miller-Urey specifications slightly, Ponnamperuma and his team sent first high-energy electrons, then ultraviolet light through the mixture, attempting to recreate the original conditions of the earth before life. They succeeded in creating large amounts of adenosine triphosphate (ATP), an amino acid that fuels cells. In later experiments with the same concoction of primordial soup, the team was able to create the nucleotides that make up nucleic acid—the building blocks of DNA and ribonucleic acid (**RNA**).

In addition to his work in prebiotic chemistry, Ponnamperuma became active in another growing field: exobiology, or the study of extraterrestrial life. Supported in this effort by NASA's interest in all matters related to outer space, he was able to conduct research on the possiblity of the evolution of life on other planets. Theorizing that life evolved from the interactions of chemicals present elsewhere in the universe, he saw the research possibilities of spaceflight. He experimented with lunar soil taken by the *Apollo 12* space mission in 1969. As a NASA investigator, he also studied information sent back from Mars by the unmanned Viking, Pioneer, and Voyager probes in the 1970s. These studies suggested to Ponnamperuma, as he stated in an 1985 interview with *Spaceworld,* that ''Earth is the only place in the solar system where there is life.''

In 1969, a meteorite fell to earth in Muchison, Australia. It was retrieved still warm, providing scientists with fresh, uncontaminated material from space for study. Ponnamperuma and other scientists examined pieces of the meteorite for its chemical make-up, discovering numerous amino acids. Most important, among those discovered were the five chemical bases that make up the nucleic acid found in living organisms. Further interesting findings provided tantalizing but puzzling clues about chemical evolution, including the observation that light reflects both to the left and to the right when beamed through a **solution** of the meteorite's amino acids, whereas light reflects only to the left when beamed through the amino acids of living **matter** on Earth. ''Who knows? God may be left-handed,'' Ponnamperuma speculated in a 1982 *New York Times* interview.

Ponnamperuma's association with NASA continued as he entered academia. In 1979, he became a professor of chemistry at the University of Maryland and director of the Laboratory of Chemical Evolution—established and supported in part by the National Science Foundation and by NASA. He continued active research and experimentation on meteorite material. In 1983, an article in the science section of the *New York Times* explained Ponnamperuma's chemical evolution theory and his findings from the Muchison meteorite experiments. He reported the creation of all five chemical bases of living matter in a single experiment that consisted of bombarding a primordial soup mixture with **electricity**.

Ponnamperuma's contributions to scholarship include hundreds of articles. He wrote or edited numerous books, some in collaboration with other chemists or exobiologists, including annual collections of papers delivered at the College Park Colloquium on Chemical Evolution. He edited two journals, *Molecular Evolution* (from 1970 to 1972) and *Origins of Life* (from 1973 to 1983). In addition to traditional texts in the field of chemical evolution, he also co-authored a software program entitled ''Origin of Life,'' a simulation model intended to introduce biology students to basic concepts of chemical evolution.

Although Ponnamperuma became an American citizen in 1967, he maintained close ties to his native Sri Lanka, even becoming an official governmental science advisor. His professional life has included several international appointments. He was a visiting professor of the Indian Atomic Energy Commission (1967); a member of the science faculty at the Sorbonne (1969); and director of the UNESCO Institute for Early Evolution in Ceylon (1970). His international work included the directorship of the Arthur C. Clarke center, founded by the science fiction writer, a Sri Lankan resident. The center has as one of its goals a **Manhattan Project** for food synthesis.

Ponnamperuma was a member of the Indian National Science Academy, the American Association for the Advancement of Science, the American Chemical Society, the Royal Society of Chemists, and the International Society for the Study of the Origin of Life, which awarded him the A. I. Oparin Gold Medal in 1980. In 1991, Ponnamperuma received a high French honor—he was made a Chevalier des Arts et des Lettres. Two years later, the Russian Academy of Creative Arts awarded him the first Harold Urey Prize. In October 1994, he was appointed to the Pontifical Academy of Sciences in Rome. He married Valli Pal in 1955; they had one child. Ponnamperuma died on December 20, 1994, at Washington Adventist Hospital.

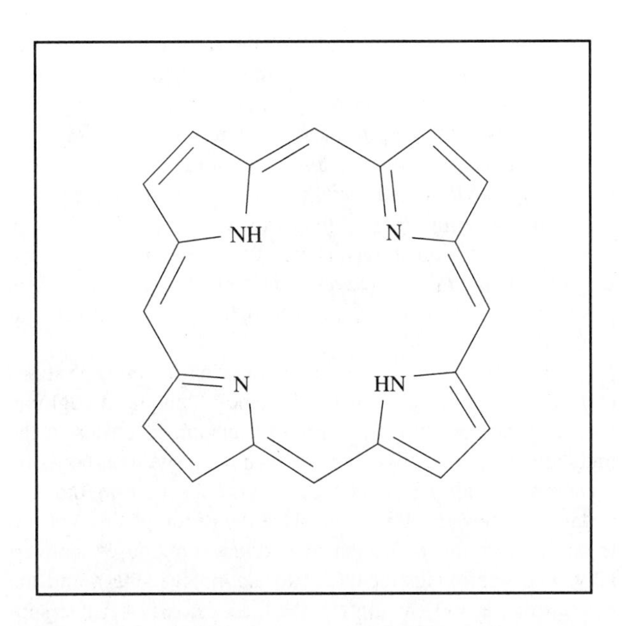

A porphyrin ring. ***(Illustration by Electronic Illustrators Group.)***

Porphyrins

The porphyrins and their closely related molecules are essential to all plants and animals. An **iron** porphyrin, called heme, is the portion of hemoglobin that carries **oxygen** from the lungs to our body's cells. The heme of another protein, myoglobin, carries oxygen within muscle cells. Other iron porphyrins in the **cytochromes** transport electrons in metabolic processes. A closely related type of **molecule** is **chlorophyll**, a key component of the photosynthetic process.

A porphyrin is a macrocyclic (large cycle) molecule that is made up of four smaller pyrrolenine rings. Each pyrrolenine **ring** is made up of a **nitrogen atom** and four **carbon** atoms. In the porphyrin ring, they are linked to each other by a carbon atom that bridges from the carbon atom nearest the nitrogen atom of one ring to the carbon atom nearest the nitrogen atom of the next ring (called the alpha carbon). The porphyrin ring is unsaturated, having alternating carbon-carbon single bonds and double bonds throughout the ring. The class of porphyrin molecules is very large because the ring can be substituted at many positions. In biochemically derived porphyrins, the **hydrogen** atoms on the carbon atoms of each nitrogen-containing ring that are farthest from the nitrogen atom (called the beta carbon) are replaced by carbon-containing groups such as a **methyl group** ($-CH_3$), and **ethyl group** ($-CH_2CH_3$), a carboxymethyl group ($-CH_2COOH$), or one of a large variety of other groups. The substituents of porphyrins profoundly affect their solubilities and their interactions with other molecules. They also change the ease with which the porphyrin ring can be oxidized or reduced. Substituents can greatly affect the oxidation-reduction properties of the metalloporphyrin complex.

The most common types of synthetic porphyrins have substituents on the carbon atoms bridging the five-membered nitrogen-containing rings. These synthetic porphyrins are often much less susceptible to photodegradation and other decomposition reactions than the biochemically derived porphyrins and have proven to be very useful in studying properties and reactions of the porphyrin ring system. They are also being studied intensely as candidates for building molecular scale computing devices and switches among other applications. The phthalocyanine ring system, which is used industrially in dyes and other applications, is closely related to the porphyrin ring system, having a **benzene** ring attached to the beta carbon positions of each of a porphyrin ring's four nitrogen-containing rings.

The metal complexes of porphyrins in living systems carry out many critical functions: hemoglobin and myoglobin transport oxygen, the cytochromes transport electrons in the **metabolism** of foods by which oxygen reacts with carbon- and hydrogen-containing molecules to produce **carbon dioxide**, **water** and **energy**, catalases catalyze the decomposition of potentially harmful **hydrogen peroxide**, peroxidases convert alkyl hydroperoxides (R-OOH) to alcohols (R-OH), and the cytochrome-P 450 enzymes in the liver detoxify organic substances by the conversion of -CH groups to more water soluble -COH groups. In these processes, the unusually extensive unsaturated network of the porphyrin ring plays an important role. The highly unsaturated ring tightly binds the iron atom, preventing it from being removed by oxygen to form highly stable iron oxides. It also allows electronic charge that develops at the iron atom to be spread throughout the ring. This allows high formal charges (Fe(IV), for example, that is postulated to develop in cytochrome P-450 reactions) to be stabilized.

Another consequence of the extensive unsaturated ring of porphyrins is its very great chemical stability. When carbon-carbon double and single bonds alternate appropriately, they average. The energy of the molecule is lower, that is the molecule is thereby more stable, because electrons are spread over a greater area, decreasing repulsions. The amount that the energy is lowered by this mechanism is called **resonance** energy. The resonance energy of the porphyrin ring is about 1,000 kilojoules/mol, many times that of a benzene ring and equivalent to the energy needed to break several chemical bonds. This energy lends so much stability to the porphyrin structure that porphyrin ring compounds are the only complex molecular structures other than hydrocarbons of all the compounds originally present in plants and animals to remain intact in oil and oil shale.

Porphyrins, metalloporphyrins, and related macrocyclic species such as the chlorophylls are noted for their very intense coloration (chlorophyll comes from the words for **color** and plant and cytochrome comes from the words color and cell). A 10 micromolar **solution** of any of these molecules typically absorbs over 90% of the light at some part of the visible spectrum but allows approximately 90% of light in other regions in the visible spectrum to be transmitted or reflected. Differences in the regions of light absorption for the different species leads to the color differences. The metalloporphyrin protoporphyrin IX iron(II) in hemoglobin gives blood its intense red color (that turns to brown as the Fe(II) atom is oxidized to Fe(-III) when hemoglobin decomposes in air) and the **magnesium** complex of a related macrocylic ring system in chlorophyll gives leaves their bright green coloration. The intense coloration of the porphyrin ring system and the related ring system of the chlorophylls (there are plant and bacterial chlorophylls that differ structurally) is a property of the very extensive **electron** delocalization, resulting from the extensive single bond-double bond alternation. This property also leads to the intense luminescence and electron transfer capability of porphyrins and many metalloporphyins.

Not only are the natural porphyrins essential to life, but synthetic porphyrins and related ring systems have found uses as cancer treatment agents, diagnostic reagents, analytical reagents and dyes. There is a great deal of research underway worldwide to synthesize and understand the **chemistry** of new porphyrins with unusual and useful properties.

PORTER, GEORGE (1920-)

English chemist

Sir George Porter shared the Nobel Prize in **chemistry** in 1967 with his former teacher, **Ronald G. W. Norrish**, and **Manfred Eigen** for their contributions to the study of rapid chemical reactions. Porter's efforts included research on flash photolysis, which has been used widely in the fields of **organic chemistry**, **biochemistry**, and photobiology. Porter, who is praised for having an outgoing personality and being a great promotor of science education, has also contributed to the scientific education of nonspecialists and children, especially through his role in helping prepare television programs in Great Britain.

Porter was born on December 6, 1920, to John Smith Porter and Alice Ann (Roebuck) Porter in Stainforth, West Yorkshire, where he received his early education at Thorne Grammar School. With the award of an Ackroyd Scholarship, he entered Leeds University in 1938 to study chemistry and received his bachelor of science degree in 1941. While at Leeds he also studied radio physics and electronics, and he drew upon this background while serving in the Royal Navy Volunteer Reserve as a radar specialist during World War II. At the end of the war, Porter entered Emmanuel College at Cambridge University to do graduate work. There he met and studied under Norrish, who had pioneered research in the area of photochemical reactions in molecules. Porter received his doctorate degree from Cambridge in 1949.

Using very short pulses of **energy** that disturbed the equilibrium of molecules, Porter and Norrish developed a method to study extremely fast chemical reactions lasting for only one-billionth of a second. The technique is known as flash photolysis. First, a flash of short-wavelength light breaks a chemical that is photosensitive into reactive parts. Next, a weaker light flash illuminates the reaction zone, making it possible to measure short-lived **free radicals**, which are especially reactive atoms that have at least one unpaired **electron**. Flash photolysis made it possible to observe and measure free radicals for the first time and also to study the sequence of the processes of reactants as they are converted into products. When

Porter won the Nobel Prize in 1967, he was praised, along with Norrish and Eigen, for making it possible for scientists around the world to use their techniques in a wide range of applications, opening many passageways to scientific investigation in **physical chemistry**. In his own work, Porter was able to apply his methods from his early work with **gases** to later work with solutions. He also developed a method to stabilize free radicals, which is called matrix isolation. It can trap free radicals in a structure of a supercooled liquid (a glass). Porter also made important contributions in the application of laser beams to photochemical studies for the purpose of investigating biochemical problems. Some practical applications of photochemical techniques include the production of fuel and chemical feedstocks.

In 1949, Porter became a demonstrator in chemistry at Cambridge University and an assistant director of research in the Department of Physical Chemistry in 1952. While he was at the British Rayon Research Association as assistant director of research in 1954, Porter used his method of flash photolysis to record organic free radicals with a lifetime as short as one millisecond. Also at the Rayon Association, he worked on problems of light and the fading of dye on fabric.

Porter was appointed professor of physical chemistry at Sheffield University in 1955, and in 1963 he became the head of the chemistry department and was honored as Firth Professor. During his years at Sheffield, Porter used his flash photolysis techniques to study the complex chemical interactions of **oxygen** with hemoglobin in animals. He also investigated the properties of **chlorophyll** in plants with the use of his high-speed flash techniques. He was able to improve his techniques to the degree that he could examine chemical reactions that were more than a thousand times faster than with the use of flash tubes. Porter also studied chloroplasts and the primary processes of **photosynthesis**.

In 1966, Porter also became Fullerian Professor of Chemistry at the Royal Institution in London and the Director of the Davy Faraday Research Laboratory. He left there to take the position of chair for the Center for Photomolecular Sciences at Imperial College in London in 1990. During his career, Porter received many other honors and awards in addition to the Nobel Prize. He was knighted in 1972, and he has been granted numerous honorary doctorate degrees and awarded prizes from British and American scientific societies, including the Robertson Prize of the American National Academy of Sciences and the Rumford Medal of the Royal Society, both in 1978.

Porter has been active outside scientific circles in the promotion of science to the general public. His concern about communication between scientists and the rest of society induced him to participate as an adviser on film and television productions. He has been praised for his activities in educating young people and people in nonscientific fields about the value of science. He was an active participant during his service with the Royal Institution in a science program series for British Broadcasting Company television (BBC-TV) called *Young Scientist of the Year.* Another BBC-TV program in which he participated was called *The Laws of Disorder* and *Time Machines.* Porter has also served on many policy and institutional committees that are involved in promoting science and education in Europe, England, and America.

Porter married Stella Brooke in 1949 and they have two sons, John Brooke Porter and Christopher Porter. His outgoing personality is considered an asset in promoting scientific knowledge. He has been an active contributor to scientific journals and has also played the role of advisor to industry. Besides sailing, Porter spends some leisure time vacationing on the coast of Kent with his family.

PORTER, RODNEY (1917-1985)

English biochemist

Rodney Porter was a biochemist who spent most of his professional life investigating the chemical structure and functioning of **antibodies**, a class of **proteins** which are also called immunoglobulins. Since 1890 scientists had known that antibodies are found in the blood serum and provide immunity to certain illnesses. However, when Porter began his research in the 1940s, little was known about their chemical structure, or how antigens (substances that cause the body to produce antibodies) interacted with them. Using the results of his own research as well as that of **Gerald M. Edelman**, Porter proposed the first satisfactory model of the immunoglobulin **molecule** in 1962. The model allowed the development of more detailed biochemical studies by Porter and others that led to a better understanding of the way in which antibodies worked chemically. Such understanding was key to research on the prevention and cure of a number of diseases and the solution to problems related to organ transplant rejection. For his work, Porter shared the 1972 Nobel Prize in physiology or medicine with Edelman.

Rodney Robert Porter was born October 8, 1917, in Newton-le-Willows, near Liverpool in Lancashire, England. His mother was Isobel Reese Porter and his father, Joseph L. Porter, was a railroad clerk. "I don't know why I became interested in [science]," Porter once told the *New York Times.* "It didn't run in my family." He attended Liverpool University, where he earned a B.S. in **biochemistry** in 1939. During World War II he served in the Royal Artillery, the Royal Engineers, and the Royal Army Service Corps, and participated in the invasions of Algeria, Sicily, and Italy. After his discharge in 1946, he resumed his biochemistry studies at Cambridge University under the direction of **Frederick Sanger**.

Porter's doctoral research at Cambridge was influenced by Nobel laureate Karl Landsteiner's book, *The Specificity of Serological Reactions,* which described the nature of antibodies and techniques for preparing some of them. Antibodies, at the time, were thought to be proteins that belonged to a class of blood-serum proteins called gamma globulins. From Sanger, who had succeeded in determining the chemical structure of insulin (a protein that metabolizes **carbohydrates**), Porter learned the techniques of protein **chemistry**. Sanger had also demonstrated tenacity in studying problems in protein chemistry involving amino acid sequencing that most believed impossible to solve, and he was a model for the persistence Porter would show in his later work on antibodies.

Fortunately, Porter chose rabbits to experiment on for his research. Although this was not known at the time, the antibody system is not as complex in this animal as it is in some. The most important antibody, or immunoglobulin, in the blood is called IgG, which contains more than 1,300 **amino acids**. The problem of discovering the active site of the antibody—the part that combines with the antigen—could be solved only by working with smaller pieces of the molecule. Porter discovered that an **enzyme** from papaya juice, called papain, could break up IgG into fragments that still contained the active sites but were small enough to work with. He received his Ph.D. for this work in 1948.

Porter remained at Cambridge for another year, then in 1949 he moved to the National Institute for Medical Research at Mill Hill, London. There, he improved methods for purifying protein mixtures and used some of these methods to show that there are variations in IgG molecules. He obtained a purer form of papaya enzyme than had been available at Cambridge and repeated his earlier experiments. This time the IgG molecules broke into thirds, and one of these thirds was obtained in a crystalline form which Porter called fragment crystallizable (Fc).

Obtaining the Fc crystal was a breakthrough; Porter now was able to show that this part of the antibody was the same in all IgG molecules, since a mixture of the different molecules would not have formed a crystal. He also discovered that the active site of the molecule (the part that binds the antigen) was in the other two-thirds of the antibody. These he called fragment antigen-binding (or FAB) pieces. After Porter's research was published in 1959, another research group, led by Gerald M. Edelman at Rockefeller University in New York, split the IgG in another way—by separating amino acid **chains** rather than breaking the proteins at right angles between the amino acids as Porter's papain had done.

In 1960 Porter was appointed professor of immunology at St. Mary's Hospital Medical School in London. There he repeated Edelman's experiments under different conditions. After two years, having combined his own results with those of Edelman, he proposed the first satisfactory structure of the IgG molecule. The model, which predicted that the FAB fragment consisted of two different amino acid chains, provided the basis for far-ranging biochemical research. Porter's continuing work contributed numerous studies of the structures of individual IgG molecules. In 1967 Porter was appointed Whitley Professor of Biochemistry and chairman of the biochemistry department at Oxford University. In his new position, Porter continued his work on the immune response, but his interest shifted from the structure of antibodies to their role as **receptors** on the surface of cells. To further this research, he developed ways of tagging and tracing receptors. He also became an authority on the structure and **genetics** of a group of blood proteins called the complement system, which binds the Fc region of the immunoglobulin and is involved in many important immunological reactions.

Porter married Julia Frances New in 1948. They had five children and lived in a farmhouse in a small town just outside of Oxford. Porter was killed in an automobile accident a few weeks before he was to retire from the Whitley Chair of Biochemistry. He had been planning to continue as director of the Medical Research Council's Immunochemistry Unit for another four years; he had also intended to continue his laboratory work, attempting to crystallize one of the proteins of the complement system. Porter's awards in addition to the Nobel Prize include the Gairdner Foundation Award of Merit in 1966 and the Ciba Medal of the Biochemical Society in 1967.

POTASSIUM

Potassium is the third element in Group 1 of the **periodic table**. The elements in this group are commonly referred to as the **alkali metals**. Potassium has an **atomic number** of 19, an atomic **mass** of 39.0983, and a chemical symbol of K.

Properties

Potassium is a soft, silvery-white metal with a melting point of 145°F (63°C), a **boiling point** of 1,420°F (770°C), and a **density** of 0.862 grams per cubic centimeter. Its density is low enough that it will float if placed into **water**. Its melting point is low enough that the **heat** of a candle is sufficient to melt the metal.

Potassium is a very active metal that reacts violently with water to produce **hydrogen** gas: $2K + 2H_2O \rightarrow 2KOH + H_2$. So much heat is produced in this reaction that the hydrogen gas tends to ignite and burn. Potassium also reacts readily with all acids and with many nonmetals, such as **sulfur**, **fluorine**, **chlorine**, **phosphorus**, and **nitrogen**.

Occurrence and Extraction

Potassium is the eighth most abundant element in the Earth's crust with an abundance of about 2.0-2.5%. It is just slightly less abundant than its alkali cousin, **sodium**. Potassium is much too active to occur as a free element. It is found most commonly in the minerals sylvite (potassium chloride; KCl), sylvinite (sodium potassium chloride; $NaCl \bullet KCl$), carnallite (potassium **magnesium** chloride; ($KCl \bullet MgCl_2$), langbeinite (potassium magnesium **sulfate**; $K_2SO_4 \bullet 2MgSO_4$), and polyhalite (calcium magnesium potassium sulfate; $2CaSO_4 \bullet MgSO_4K_2SO_4$).

The term potash is commonly used to describe any ore of potassium. The term originally referred to a specific compound of the element, potassium carbonate (K_2CO_3), but has now become generalized to include almost any naturally occurring mineral of the element or any compound derived from a mineral.

The most important source of potash in the United States is a mine near Carlsbad, New Mexico, that produces sylvinite. About 85% of the country's potash comes from that source. Potash is also obtained from huge long-buried **salt** blocks formed when ancient seas evaporated.

Discovery and Naming

Potash has been used by humans for many centuries. It was generally known as *vegetable alkali* because it was obtained from ashes produced when plants were burned. The term was used to distinguish the material from *mineral alkali*, which we now know as soda ash or **sodium hydroxide**.

Chemists long suspected that both potash and soda ash contained previously undiscovered elements, but they were unable to extract those elements from the native materials. Finally, in the early 1800s, the great English chemist and physicist **Humphry Davy** found a way to break down these compounds to their elements. He first melted the potash and then passed an electric current through the molten material. He discovered that a tiny liquid droplet of metal formed at one electrode of the apparatus. That droplet of metal was the first sample of potassium ever seen by humans. Davy suggested that the name potassium be given to the new element, based on its historical source, potash.

Uses

Potassium is rarely used as an element because it is so reactive. It has been used occasionally as a heat exchange medium in **nuclear reactors**. However, sodium is usually preferred over potassium for this application.

By far the most important compound of potassium is potassium chloride, used primarily to make synthetic fertilizer. In an average year, about 1.5 billion kilograms (3.4 billion pounds) is produced in the United States, 85% of which goes to the fertilizer industry.

Many other compounds of potassium are commercially important, although no use begins to compare with the amount of potash used for **fertilizers**. Some examples of other important potassium compounds include: 1) potassium bicarbonate (baking soda; $KHCO_3$) used in baking powders, antacids, **food additives**, and fire extinguishers; 2) potassium bisulfate ($KHSO_3$) used as a food preservative, in the bleaching of textiles and straw, in wine- and beer-making, and in leather tanning; 3) potassium bitartrate (cream of tartar; $KHC_4H_4O_6$) used in baking powders, ''tinning'' of metals, and food additives; 4) potassium bromide (KBr) used in photographic film and engraving; 5) potassium chromate (K_2CrO_4) used in dyes and stains, **explosives** and fireworks, leather tanning, safety matches, and fly **paper**; 6) potassium hydroxide (caustic potash; KOH) used in paint remover, the manufacture of specialized soaps, fuel cells and **batteries**, bleach, food additives, and **herbicides**; 7) potassium nitrate (nitre or saltpeter; KNO_3) used in explosives and fireworks, matches, rocket fuel, the manufacture of **glass**, and food curing.

Health Issues

Potassium is essential to both plant and animal life. It is one of the three macronutrients (in addition to nitrogen and phosphorus) required by plants. In humans, potassium plays a number of essential roles, such as transmitting chemical messages between nerve cells, causing the contraction of muscle cells, aiding in the digestion of foods, and maintaining proper function of the eyes.

Potential Energy

Matter may store **energy** even when at rest. Energy that is stored and held for future use is referred to as potential energy. If this is confusing, recall that energy is the capacity to do work, regardless of whether that energy involves motion (kinetic energy) or not (potential energy).

The most familiar form of potential energy involves the pull of Earth's gravity. The law of **conservation of energy** tells us that a boulder at rest on the edge of a cliff has potential energy in an amount equal to the amount of work it took to raise the boulder to that height from ground level. (Mathematically, the work done in lifting an object is equal to the weight of the object times the distance it is raised.)

But Earth's gravity is not required for the existence of potential energy. An example of an object containing non-gravitational potential energy is a stretched bow. When one draws the bowstring, the energy required to stretch it is transferred into the bow. Neptune has potential energy because it can expend energy by falling into the Sun due to the Sun's gravitational pull. Moreover, a nail within the magnetic field of a magnet has potential energy because it can expend energy by moving toward the magnet.

The chemical energy in fossil fuel, for example, is a form of potential energy. This energy is converted to kinetic energy when a chemical change takes place in the material. There is also potential energy in electric **batteries**, wound alarmclocks, and in the food we eat.

Praseodymium

Praseodymium is one of the **rare earth elements** found in Row 6 of the **periodic table**. The elements in this group are not especially rare, although they are quite difficult to separate from each other. Praseodymium has an **atomic number** of 59, an atomic **mass** of 140.9077, and a chemical symbol of Pr.

Properties

Praseodymium is a soft, malleable, ductile metal with a yellowish, metallic shine. It has a melting point of 1,710°F (930°C), a **boiling point** of about 5,800°F (about 3,200°C), and a **density** about 6.78-6.81 grams per cubic centimeter. Two allotropes of the element exist, accounting for the slight variation in densities.

Praseodymium is a moderately active element that reacts with **oxygen** in moist air to form a greenish yellow scale of praseodymium **oxide** (Pr_2O_3) that acts to protect the metal from further **corrosion**. Praseodymium also reacts with **water** and with acids with the production of **hydrogen** gas.

Occurrence and Extraction

Praseodymium is one of the more common rare earth elements with an abundance of about 3.5-5.5 parts per million in the Earth's crust. It occurs primarily with the other rare earth elements in two minerals, monazite and bastnasite. The element is extracted from its ores by converting the ore first to praseodymium fluoride (PrF_3) and then electrolyzing the fluoride: $2PrF_3$—electric current→ $2Pr + 3F_2$.

Discovery and Naming

Praseodymium was one of seven new elements discovered in a mineral known as cerite first discovered in 1751 near the Swedish town of Bastnas. The discovery of praseodymium is credited to the Austrian chemist Carl Auer (Baron von Welsbach; 1858-1929) who found that a previously discovered element, didymium, was actually a mixture of two other new elements. Auer named the two new elements **neodymium** (for "new twin") and praseodymium (for "green twin"), the latter name because of the **color** of many of praseodymium's compounds.

Uses

One of the oldest uses for praseodymium is in the manufacture of misch metal, a pyrophoric metal (a metal that gives sparks when struck) used to make lighter flints and tracer bullets. Compounds of praseodymium are also used to give color to **glass**, **ceramics**, enamels, and other materials. The characteristic color provided by compounds of praseodymium in such cases is a bright yellow. Praseodymium is also added to the **carbon** in carbon arc lamps to produce a more brilliant white light.

Precious metals

Gold, **silver**, and **platinum** have historically been valued for their beauty and rarity. They are the precious **metals**. Platinum usually costs slightly more than gold, and both metals are about 80 times more costly than silver. Precious metal weights are given in Troy ounces (named for Troyes, France, known for its fairs during the Middle Ages) a unit approximately 10% larger than 1 oz (28.35 g).

The ancients considered gold and silver to be of noble birth compared to the more abundant metals. Chemists have retained the term noble to indicate the resistance these metals have to **corrosion**, and their natural reluctance to combine with other elements.

The legends of King Midas and Jason's search for the golden fleece hint at prehistoric mankind's early fascination with precious metals. The proof comes in the gold and silver treasure found in ancient Egyptian tombs and even older Mesopotamian burial sites.

The course of recorded history also shows twists and turns influenced to a large degree by precious metals. It was Greek silver that gave Athens its Golden Age, Spanish gold and silver that powered the Roman empire's expansion, and the desire for gold that motivated Columbus to sail west across the Atlantic. The exploration of Latin America was driven in large part by the search for gold, and the Jamestown settlers in North America had barely gotten their "land legs" before they began searching for gold. Small amounts of gold found in North Carolina, Georgia, and Alabama played a role in the 1838 decision to remove the Cherokee Indians to Oklahoma. The California gold rush of 1849 made California a state in 1850, and California gold fueled northern industry and backed up union currency, two major factors in the outcome of the Civil War.

Gold

Since ancient times, gold has been associated with the sun. Its name is believed derived from a Sanskrit word meaning "to shine," and its chemical symbol (Au) comes from *aurum*, Latin for gold. Pure gold has an exceedingly attractive, deep yellow **color** and a specific gravity of 19.3. Gold is soft enough to scratch with a fingernail, and the most malleable of metals. A block of gold about the size of a sugar cube can be beaten into a translucent film some 27 ft (8 m) on a side. Gold's purity is expressed either as fineness (parts per 1,000) or in karats (parts per 24). An **alloy** containing 50% gold is 500 fine or 12 karat gold. Gold resists corrosion by air and most chemicals but can be dissolved in a mixture of nitric and hydrochloric acids, a **solution** called **aqua regia** because it dissolves the "king of metals."

Gold is so rare that one ton of average rock contains only about eight pennies worth of gold. Gold ore occurs where geologic processes have concentrated gold to at least 250 times the value found in average rock. At that **concentration** there is still one million times more rock than gold and the gold is rarely seen. Ore with visible gold is fabulously rich.

Gold most commonly occurs as a pure metal called native gold or as a natural alloy with silver called electrum. Gold and silver combined with **tellurium** are of local importance. Gold and silver tellurides are found, for example, in the mountains around the old mining boom-town of Telluride, Colorado. Gold is found in a wide variety of geologic settings, but placer gold and gold veins are the most economically important.

Placer gold is derived from gold-bearing rock from which the metal has been freed by weathering. Gravity and running **water** then combine to separate the dense grains of golds from the much lighter rock fragments. Rich concentrations of gold can develop above deeply weathered gold veins as the lighter rock is washed away. The "Welcome Stranger" from the gold fields of Victoria, Australia, is a spectacular 2,516 oz (71.5 kg) example of this type of occurrence.

Gold washed into mountain streams also forms placer deposits where the stream's **velocity** diminishes enough to deposit gold. Stream placers form behind boulders and other obstructions in the stream bed and where a tributary stream merges with a more slowly moving river. Placer gold is also found in gravel bars where it is deposited along with much larger rocky fragments.

The discovery of place gold set off the California gold rush of 1849 and the rush to the Klondike in 1897. The largest river placers known are in Siberia, Russia. Gold-rich sands there are removed with jets of water, a process known as hydraulic mining. An fascinating byproduct of Russia's hydraulic mining is the unearthing of thousands of woolly mammoths, many with flesh intact, locked since the Ice Age in frozen tundra gravel.

Stream placer deposits have their giant ancient counterparts in paleoplacers, and the Witwatersrand district in South Africa outproduces all others combined. Gold was reported from the Witwatersrand (White Waters Ridge) as early as 1834, but it was not until 1886 that the main deposit was discovered. From that time until today, it has occupied the paramount position in gold mining history. Witwatersrand gold was deposited between 2.9 and 2.6 billion years ago in six major fields, each produced by an ancient river system.

Placer and paleoplacers are actually secondary gold deposits, their gold having been derived from older deposits in the mountains above. The California 49ers looked upstream hoping to find the mother lode, and that's exactly what they called the system of gold veins they discovered.

Gold veins

Vein gold is deposited by hot subterranean water known as a hydrothermal fluid. Hydrothermal fluids circulate through rock to leach small amounts of gold from large volumes of rock and then deposit it in fractures to form veins. Major U.S. gold vein deposits have been discovered at Lead in the Black Hills of South Dakota and at Cripple Creek on the slopes of Pike's Peak, Colorado. Important vein deposit are also found in Canada and Australia. All these important deposits where located following the discovery of placer gold in nearby streams.

Gold's virtual indestructibility means that almost all the gold ever mined is still in use today. It is entirely possible that some gold atoms that once graced the head of Cleopatra now reside in your jewelry, stereo, or teeth. Today, gold is being mined in ever increasing amounts from increasingly lower-grade deposits. It is estimated that 70% of all gold recovered has been mined in this century. Each year nearly 2,000 tons are added to the total. Nevada currently leads the nation in gold production, and the Republic of South Africa is the world's leading gold-producing nation.

Gold has traditionally been used for coinage, bullion, jewelry and other decorative uses. Gold's chemical inertness means that gold jewelry is nonallergenic and remains tarnish-free indefinitely. For much the same reasons gold has long been used in dentistry. Modern industry is consuming increasing quantities of gold, mostly as electrical contacts in micro circuitry.

Silver

Silver is a brilliant white metal and the best metal in terms of thermal and **electrical conductivity**. Its chemical symbol, Ag, is derived from its Latin name, *argentum*, meaning "white and shining." Silver is not nearly as precious, dense, or noble as gold or platinum. The ease with which old silverware tarnishes is an example of its chemical **reactivity**. Although native silver is found in nature, it most commonly occurs as compounds with other elements, especially **sulfur**.

Hydrothermal veins constitute the most important source of silver. The Comstock Lode, a silver bonanza 15 mi (24 km) southeast of Reno, Nevada, is a well known example. Hydrothermal silver veins are formed in the same manner as gold veins, and the two metals commonly occur together. Silver, however, being more reactive than gold, can be leached from surface rocks and carried downward in solution. This process, called supergene enrichment, can concentrate silver into exceedingly rich deposits at depth.

Mexico has traditionally been the world's leading silver producing country, but the United States, Canada, and Peru each contribute significant amounts. Although silver has historically been considered a precious metal, industrial uses now predominate. Significant quantities are still used in jewelry, silver ware, and coinage; but even larger amounts are consumed by the photographic and electronics industries.

Platinum

Platinum, like silver, is a beautiful silver-white metal. Its chemical symbol is Pt and its name comes from the Spanish world for silver (*plata*), with which it was originally confused. Its specific gravity of 21.45 exceeds that of gold, and, like gold, it is found in pure metallic chunks in stream placers. The average crustal abundance of platinum is comparable to that of gold. The melting point of platinum is 3,219°F (1,769°C), unusually high for a metal, and platinum is chemically inert even at high **temperature**. In addition, platinum is a catalyst for chemical reactions that produce a wide range of important commodities.

Platinum commonly occurs with five similar metals known as the platinum group metals. The group includes **osmium**, **iridium**, **rhodium**, **palladium**, and **ruthenium**. All were discovered in the residue left when platinum ore was dissolved in aqua regia. All are rare, expensive, and classified chemically as noble metals.

Platinum is found as native metal, natural alloys, and as compounds with sulfur and **arsenic**. Platinum ore deposits are rare, highly scattered, and one deposit dominates all others much as South Africa's Witwatersrand dominates world gold production. That platinum deposit is also in the Republic of South Africa.

Placer platinum was discovered in South Africa in 1924 and subsequently traced to a distinctively layered igneous rock known as the Bushveld Complex. Although the complex is enormous, the bulk of the platinum is found in a thin layer scarcely more than three feet thick. Nearly half of the world's historic production of platinum has come from this remarkable layer.

The Stillwater complex in the Beartooth mountains of southwestern Montana also contains a layer rich in platinum group metals. Palladium is the layer's dominant metal, but platinum is also found. The layer was discovered during the 1970s, and production commenced in 1987.

Platinum is used mostly in catalytic converters for vehicular **pollution** control. Low-voltage electrical contracts form the second most common use for platinum, followed closely by dental and medical applications, including dental crowns, and a variety of pins and plates used internally to secure human bones. Platinum is also used as a catalyst in the manufacture of **explosives**, fertilizer, **gasoline**, insecticides, paint, plastic, and pharmaceuticals. Platinum crucibles are used to melt high-quality optical **glass** and to grow crystals for **computer chips** and **lasers**. Hot glass fibers for insulation and **nylon** fibers for textiles are extruded through platinum sieves.

Because of their rarity and unique properties, the demand for gold and platinum are expected to continue to increase. Silver is more closely tied to industry, and the demand for silver is expected to rise and fall with economic conditions.

PRECIPITATION

Precipitation is a process that causes dissolved substances to separate from a **solution** as a solid. The resulting solid is referred to as a precipitate. In terms of physical processes, precipitation is the opposite of dissolution. Precipitation also describes a different process that takes place in the atmosphere, the condensation of **water** vapor to form rain droplets, snow or hail.

In chemical reactions ionic compounds that dissociate (break apart) in solution are termed ionic salts. Any resulting dissociated components (*e.g.*, ions) that do not contribute to the formation of the precipitate are termed spectators (*e.g.*, spectator ions). The components that react to form the precipitate are termed the precipitate's constituent or contributing components.

Whether a precipitate will form depends on both the solute **concentration** and its saturation **solubility** at the specified **temperature**, that is, the maximum amount of a substance that can be dissolved in the given solvent. When an ionic solute is dissolved in water equilibrium is established between the solid phase and the various hydrated ionic species. At equilibrium the rate of formation of the dissociated ions exactly equals the rate of their recombination to form the solid solute. This equilibrium is governed mathematically by the solubility product, K_{sp}, which in the case of **silver** chloride is given by $AgCl_{(solid)} \rightleftharpoons Ag^+_{(aq)} + Cl^-_{(aq)}$ $K_{sp}=[Ag^+][Cl^-]$ the product of the molar concentrations of the silver and chloride ions. Precipitation occurs whenever the numerical value of $[Ag^+][Cl^-]$ is larger than the solubility product. This happens at supersaturation and the net reaction favors recombination rather than **dissociation** of components in solution. For precipitation to occur there must be at least a slight supersaturation of the substance in solution.

Pollutants or impurities in solution also contribute to components in solution and, therefore, can affect the rates and type of precipitation that occur. Selective precipitation is a process that separates ions out of a solution by deliberate creation of new precipitates introduced to solution. Selective precipitation was the basis behind the **qualitative analysis** scheme that was used years ago to identify which ions were present in unknown aqueous solutions.

A familiar example of a precipitation reaction is that observed in the limewater test for **carbon** dioxide; the formation of the characteristic milky-white precipitate. Before the wastewater is disinfected and discharged, phosphorous is removed by chemical precipitation through reactions with **aluminum**, **iron**, **calcium**, and clay minerals in soils.

Covalent compounds also precipitate when the solution becomes supersaturated. In the case of organic compounds this may occur when two species react to form a less soluble species. A simple compound in solution can precipitate if the solvent evaporates sufficiently, if the solution is cooled, or if a nonsolvent is added to the solution to reduce the solute's solubility. For example, the organic compound phenanthrene can be precipitated from an ethanolic solution by the addition of water. Phenanthrene is considerably more soluble in **ethanol** than in water.

Precipitation is also used to describe the phase transitions that take place between the water molecules in the Earth's atmosphere (composed primarily of **nitrogen** and oxygen). Water molecules suspended in the atmosphere are constantly moving between phase states. This means that water is constantly changing from ice (solid), to liquid, to gas and back again as it moves among its phase states. When the temperature drops below what is termed the dew-point, there is a net condensation and a cloud formation can occur. Fog is simply condensation at ground level, that is a cloud on the ground. When the rate of condensation in the atmosphere becomes great enough larger and larger drops of water are formed until water precipitation occurs in the form of rain (liquid) or snow (solid).

Pollutants such as **sulfur** and nitrogen oxides in the atmosphere can travel hundreds of miles from the source of the **pollution** during which time they are converted into sulfuric and nitric acids. The deposition of these acids by atmospheric precipitation, (*e.g.*, Acid Rain) is an important ecological problem.

See also Condensation reaction; Evaporation; Freezing and melting; Ionization; Phase changes; Reaction rate; Solvents; Sublimation; Suspension; Vapor pressure; Waste treatment; Water purification

PREGL, FRITZ (1869-1930)

Austrian analytical chemist

The work of Fritz Pregl is an example of the maxim that every difficulty is an opportunity. It was the problems inherent in analyzing organic **matter** that motivated Pregl to take microanalysis into new realms of exactitude, developing new instrumentation for the precise measurement of such substances. Such microanalytic tools paved the way for later biochemical research on pigments, **hormones**, and **vitamins**. Pregl's innovations in the field earned him the 1923 Nobel Prize in **chemistry**.

Pregl was born on September 3, 1869, in Laibach, Austria (now Ljubljana, Republic of Slovenia), the only son of Friderike Schlacker and Raimund Pregl, the treasurer of a bank in nearby Krain (now Kranj). Though his father died when he was quite young, Pregl finished Gymnasium or high school in Laibach before he and his mother moved in 1887 to Graz, where he studied medicine at the University of Graz. Early in his academic career, Pregl demonstrated the intelligence and skill that would become more evident in his subsequent work as an analytical chemist. While still a student, his physiology professor, Alexander Rollett, made him an assistant in his laboratory. Upon gaining his medical degree in 1893, Pregl began to practice medicine with a specialty in ophthalmology but also stayed on part-time at Rollett's laboratory.

Becoming an assistant lecturer in physiology and histology at the University of Graz, and working in Rollett's laboratory, Pregl increasingly turned his attention to biological and physiological chemistry, focusing on organic matter. His early research centered on human physiology and, in particular, the properties of bile and urine. His research on the reaction of cholic acid, which is found in bile, and the causality of the high ratio of **carbon** to **nitrogen** in human urine, won him a university lectureship at Graz in 1899. In 1904, Pregl went to Germany to study chemistry with **Friedrich Wilhelm Ostwald** in Leipzig and **Emil Fischer** in Berlin. Fischer was a 1902 Nobel laureate in **organic chemistry** for his sugar and purine research, and Ostwald, a physical chemist, would win the Nobel in 1909 for his work in catalysis.

Returning to Graz in 1905, Pregl renewed his bile research and began protein investigations, having been intrigued by Fischer's recent work on the structure of **proteins**. He also became an assistant at the medical and chemical laboratory of the University of Graz, a position which provided him with valuable research space. In 1907 he was appointed as the forensic chemist for central Styria, the province of which Graz is the capital. In the course of his chemical investigations, Pregl continually came up against one problem: the methods of analysis employed by organic chemistry were much too cumbersome, lengthy, and overly complex for the new discipline of **biochemistry** in which he was becoming increasingly involved. In particular, Pregl found that he would have to prepare large amounts of test samples if he used traditional analytical methods in his studies on bile acids. Because these acids are complicated proteins, only small quantities can be isolated from liver bile, a process that is both time-consuming and costly: Pregl's research in bile acid alone would require processing several tons of raw bile in order to refine enough of the acid for traditional analysis. It was to overcome such difficulties that he set to improve the methods of microanalysis, thereby altering the direction of his research from biochemistry to **analytical chemistry**.

By the time Pregl entered the field, microanalysis was already over seventy years old, pioneered by **Justus von Liebig**, who had developed the **combustion** method. In Liebig's technique, proportionate amounts of elements in an organic substance could be determined by burning the substance in a **glass** tube under conditions that would convert the carbon to **carbon dioxide** (CO_2) and all the **hydrogen** into **water**. The water and CO_2 would in turn be absorbed by other materials such as **potassium** hydroxide or a **lime** and soda mixture, and the change of weight in the respective absorbing materials would thus give the relative amounts of carbon and hydrogen in the combusted substance. Additionally, a contemporary of Pregl's, Friedrich Emich, at the Technical University of Graz had shown the reliability of working with small quantities of substances in an inorganic framework. Pregl set out to achieve Emich's measurement techniques with organic material.

It was Pregl's achievement to build upon Liebig and Emich's developments, and to refine and improve them to the point where substantially less of the organic substances were required for analysis. In 1910 he left Graz for Innsbruck, where he took the position of professor of medical chemistry at the University of Innsbruck. With this position, Pregl could devote more time to his research. His first priority was to find or create a **balance** that would accurately weigh much smaller amounts of substances than those currently available. He turned to W. H. Kuhlman, a German chemist who had recently developed a microbalance accurate to between 0.01 and 0.02 milligrams; Pregl found that with careful adjustments he could accurately utilize Kuhlman's balance to within 0.001 milligrams.

Fritz Pregl. *(Corbis Corporation. Reproduced by permission.)*

Pregl also took on the combustion analysis of carbon and hydrogen, improving that process by scaling down the size of the analytic equipment and adding a universal filling for the combustion tube that consisted of a mixture of lead chromate and **copper oxide** set in between two pieces of **silver**. This adaptation improved the absorption of the carbon dioxide and water. With such refinements, Pregl was able to obtain accurate analyses with between 2–4 milligrams of an organic substance—and fairly accurate readings with only 1 milligram—a significant **reduction** compared to the 0.2 to 1 grams needed for Liebig's method. With the new materials employed, Pregl was also able to reduce the time needed for such analysis from three hours to an hour. Pregl and his team also went on to devise new microanalytic techniques for boiling substances to determine their **molecular weight** by creating apparatus that impeded the substances' contamination with air. This allowed determinations to be made with greatly reduced amounts of such substances. Pregl made these advances known in two public presentations: in 1911 at the German Chemical Society in Berlin, and in 1913 at a scientific congress in Vienna.

Although improved techniques since Pregl's time now allow scientists to work with organic samples of only a few tenths of a milligram, his microanalytic improvements were revolutionary in their day and opened the way to new vistas of biochemical research in both science and industry. World renowned, Pregl returned to the University of Graz in 1913 as a full professor at the Medicochemical Institute, and here he perfected the methods he had pioneered, remaining in Graz—despite other tantalizing offers—until his death. In 1916, in the midst of the World War I, he was made dean of the medical school, and in 1920 became vice chancellor of the university. His major publication on his findings, *Die quantitative organische Mikroanalyse,* was published in 1917 and has since gone through numerous editions and translations. He subsequently won the Lieben Prize and membership in the Vienna Academy of Sciences. Pregl continued his research into a wide range of organic substances, employing his own methods of microanalysis on bile acids, enzymes, and sera. He also employed microanalysis in forensics, determining poisonous alkaloids from minute amounts of substance.

In 1923 Pregl was awarded the Nobel Prize in chemistry for his advances in microanalysis of organic substances. Though his work was an improvement rather than an invention, it was a well deserved honor for a man who tirelessly devoted his life to the cause of science. A life-long bachelor, Pregl's only pleasures aside from his research were mountain climbing and bicycling. He was also devoted to his students, lending both money and support when needed. In 1929 he endowed an award for chemistry through the Vienna Academy of Sciences, the Fritz Pregl Prize, which continues to provide yearly stipends to promising students. Pregl died following an illness in 1930 at the age of sixty-one.

PRELOG, VLADIMIR (1906-1998)

Croatian-Swiss organic chemist, stereochemist, and educator

Vladimir Prelog, the son of Milan Prelog, a Croatian high school teacher, and his wife Mara (*née* Cettolo), was born on July 23, 1906 in Sarajevo, Bosnia-Herzogovina, a province of the Austro-Hungarian Empire. He served as a ''flower boy'' at the ill-fated visit of Franz Ferdinand, Archduke of Austria, and his wife, whose assassination on July 28, 1914 triggered World War I.

Prelog's parents separated in 1915, and Prelog went to live with his paternal aunt and grandmother in Zagreb, where he spent his first three years in high school (*Realgymnasium*). At the age of 12 he began to experiment in his home **chemistry** laboratory. In 1918, Prelog joined his father in Osijek, where he attended a science-based high school. Encouraged in his love of chemistry by his chemistry teacher, he wrote his first article at age 15.

In 1921, Prelog moved to Zagreb, where he graduated from high school in 1924. He then entered the Chemical Engineering School of the Czech Institute of Technology in Prague, where he spent his spare time assisting his mentor and lifelong friend, Docent Rudolf Luke, with his research in **alkaloid** chemistry and receiving his Ing.-Chem. degree in 1928. His father was forced to retire early for political reasons and could no longer support him financially, forcing Prelog to complete his graduate studies as quickly as possible. His professor assigned him a **natural products** project, the constitution of the aglycone of a new glycoside, earning him his doctorate in 1929.

Because unemployment was severe in both Czechoslovakia and Yugoslavia, Prelog was unable to find an institutional research position. The Docent's former classmate hired him to design and direct a laboratory to prepare commercially unavailable chemicals. Prelog married Kamila Vítek on October 31, 1933. The couple had one son, Jan, born in 1949. In 1932 Prelog served in the Royal Yugoslav Navy, attaining the rank of sublieutenant. He contracted tropical malaria, providing the impetus for continuing his previous research on antimalarial drugs for another two decades.

Prelog returned to Zagreb in 1935 as a poorly paid lecturer (Docent, 1935-40) and subsequently Associate Professor (1940-41) at the inadequately equipped university. He was simultaneously appointed a consulting chemist for a company where the royalties from his commercial production of sulfanilamide enabled him to spend several months in 1937 working with fellow Croat **Leopold Ružička** at the Eidgenössische Technische Hochschule (Federal Institute of Technology, ETH) in Zürich, Switzerland.

Aided by Nobel laureate Ružička, in 1941 Prelog migrated from German-occupied Zagreb to neutral Switzerland, where he remained for the rest of his life at the ETH. He served in various capacities: as a *Privat-Dozent* (1942-47); Associate Professor, (1947-50); and Professor (1950-76). In 1957, Prelog succeeded Ružička as Laboratory Director. Because he disliked administration and the use of power (He was totally devoid of autocratic tendencies), in 1965 he persuaded the ETH President to establish a collegial leadership for the laboratory in which all the professors participated, an unusual arrangement at that time. The freedom from administrative duties allowed Prelog to pursue his research more fully and to devote himself to other activities, such as consulting, traveling, lecturing, and board membership work. He served as Director of the Swiss pharmaceutical giant CIBA, Ltd. of Basel (later Ciba-Geigy, now Novartis) from 1963 to 1978, which had financially supported his work since 1943. He became a Swiss citizen in 1959 but considered himself a citizen of the world. In 1976, at the age of 70, he retired from the ETH but remained active in research.

The author of some 400 papers with about a hundred doctoral candidates and about a hundred co-workers, Prelog worked on a wide variety of research problems, mostly stereochemical projects. He abandoned his work on alkaloids, **steroids**, and terpenes of plant and animal origin and until 1970 devoted himself to compounds with novel types of structures and intriguing biochemical properties of interest to his industrial sponsors, which he isolated from microbial cultures. The first result of this research was a lactone carboxylic acid (Prelog-Djerassi lactone) that plays an important role in the synthesis of macrolides. Prelog's name is also associated with the Cahn-Ingold-Prelog (CIP) **nomenclature** system now universally used for the unambiguous specification of stereoisomers.

Prelog's research involved the determination of the structures for many natural products, including antiobiotics, nonactin, boromycin (the first discovered boron-containing natural product), ferrioxamins, and rifamycins; heterocyclic compounds; dyes; medicinal products; organ extracts; **stereochemistry** of medium-sized **ring** compounds from dicarboxylic acid esters by acyloin condensation; transannular reactions; asymmetric syntheses; stereoselectivity of microbial and enzymatic reactions; alicyclic chemistry; chirality in large and complicated molecules; and chemical topology. Together with Nobel chemistry laureate Sir Derek Harold Richard Barton, Prelog is recognized as a founder of conformational analysis.

Prelog was elected to membership in many scientific societies: foreign member, U.S. National Academy of Sciences, Accademia dei Lincei (Rome), USSR Academy of Sciences; honorary member, American Academy of Arts and Sciences, Leopoldina (Halle/Saale), Royal Irish Academy, Royal Danish Academy of Sciences, and Academy of Pharmaceutical Sciences; and fellow, Royal Society of Great Britain (1962). He held honorary doctorates from Zagreb (1954), Liverpool (1963), Paris (1963), Cambridge (1969), Brussels (1969), and Manchester (1971) universities.

Prelog's honors include the Werner Gold Medal, Schweizerische Chemische Gesellschaft (1945); Stas Medal (1962); Medal of Honor, Rice University (1962); Marcel Benoist Award (1965); Hofmann Medal (1968); Humphry Davy Medal (1968); Roger Adams Award in Organic Chemistry, American Chemical Society (1969); and Paracelsus Medal (1976). In 1975, along with Sir **John Warcup Cornforth**, he was awarded the Nobel prize in chemistry ''for his work on the stereochemistry of organic molecules and reactions.'' A warm, compassionate man, modest in behavior and reticent in personal matters, as well as a captivating raconteur with a seemingly infinite repertoire of humorous jokes and anecdotes, Prelog died in Zürich on January 7, 1998, at the age of 91 after a brief illness.

PRESSURE

Pressure is a physical force exerted over a surface. Specifically, it is defined as a force (F) divided by unit area (A). When expressed relative to a pure vacuum, this quotient is referred to as absolute pressure. When referred to relative to atmospheric pressure it is called gauge pressure. Pressure is an important concept in many chemicaland physical processes. For example, many types of reactor vessels and boilers must be monitored to ensure they do not exceed specific pressure limits, lest they rupture. In atmospheric science, an understanding of air pressure, also known as barometric pressure, is critical for making weather predictions.

Pressure is expressed in various units depending on the application. For example, air pressure is usually given in terms of atmospheres (atm), torr, bars, or millimeters of **mercury** (mm Hg at 0°C). (Although air pressure varies with altitude and weather conditions, on average the air pressure at sea level can support a column of mercury 760 mm in height.) Industrially, pressure is described in terms of pounds per square inch (psi) or kilograms per square meter. In the **International System of Units**, pressure is given as pascals (Pa), which are defined as 1.0 newtons per square meter. The centimeter-gram-second system quantifies pressure in dynes per square centimeter. In terms of equivalent units, 1 atm = 760 mm Hg = 760 torr = 1.01325 bars = 14.6960 psi = 101,325 Pa = 1,013,250 dynes/cm^2.

Historically, pressure has been measured with gauges that monitor the displacement of a mechanical element. One of the earliest measuring tools was a liquid-filled tube known as a **manometer**. Manometers were commonly used by scientists in the 1700s and 1800s and are still used today to calibrate other pressure gauges. A manometer consists of a cylindrical **glass** U-shaped tube that is partially filled with liquid. One end of the tube is exposed to the process that is generating pressure; the other end may be sealed or left open depending on the design of the instrument. The pressure exerts a force on the surface of the liquid and causes it to move a specific distance that is proportional to the magnitude of the force. The pressure differential can be measured by comparing the height of the liquid to calibrated marks on the side of the tube. Another type of manometer is the inverted bell-type that consists of two invert ed U-shaped tubes, or bells, mounted on a **balance** beam. Each of these bells is located over a pressure inlet tube in such a way that the bell that is subjected to the highest pressure will rise. This type of apparatus is used to measure very small pressure changes such as those found in drying appliances like conveyor dryers and kilns.

Another type of pressure-sensing element is the Bourdon tube, which is a flexible tube made from steel, **beryllium**, **copper**, and other special alloys. The tube, which is flattened slightly and sealed on one end, is formed into a coil. When the open end is exposed to pressure, the tube uncoils to a degree that corresponds to the magnitude of the pressure. Positive pressure causes the tube to expand and a vacuum (negative pressure as compared to air pressure) causes it to contract. As the coil moves, it displaces a needle or other indicator that points along a pre-calibrated scale to give the **pressure** measurement. Bourdon tubes have the advantage of being low cost, simply designed, and useful for measuring both high and low pressure. They also have several disadvantages: they are not extremely accurate at low pressures, the mechanical linkage to the indicator needle may require amplification to give a meaningful reading, and process materials may accumulate inside the tube since one end is sealed.

Other important mechanical pressure sensors include the bellows and diaphragm type elements. The bellows employs a coiled spring instead of a coiled tube as the responsive element. When exposed to pressure, the spring stretches and moves a needle on a measuring gauge. The diaphragm is a flexible disk made of sheet metal with ridges carved into it. When the disk is exposed to pressure it deforms to one side or the other, depending on which side is exposed to the higher pressure. The disks are very sensitive to small changes in pressure and therefore are useful in low pressure measurements as low as 13 Pa (0.1 torr).

Today, mechanical pressure devices have been supplemented with electrical sensors and transmitters. Certain devices, like piezoelectric crystals, may directly register pressure changes. These crystals operate on the principle that certain quartz compounds can be cut and aligned in such a way that they will trigger a small electric current when pressure is applied to them. Other electronic sensing devices depend on a mechanical sensor like a Bourdon tube or a diaphragm to register a change in pressure, and then they use electronic components to amplify and transmit the signal. The mechanical element transmits the pressure change to the magnetic coil of a transformer, which in turn translates the pressure to an electronic signal that can be relayed to a variety of output devices. Examples of this type of gauge include strain gauges, which measure the change in electrical resistance of a metal wire exposed to pressure via a mechanical sensor such as a bourdon tube or a bellows, and capacitive pressure transducers that register changes in capacitance when an elastic element is displaced by pressure.

PRICHARD, DIANA GARCÍA (1949-)

American chemical physicist

Diana García Prichard is a research scientist who conducts fundamental photographic materials research for the Eastman Kodak company. Her graduate work on the behavior of gas phases that she completed at the University of Rochester was lauded for its inventiveness and received unusual attention and recognition by the scientific community. She is also an active leader in the Hispanic community and has garnered numerous awards for her work.

Prichard was born in San Francisco, California, on October 27, 1949. Her mother, Matilde (Robleto) Dominguez García, was originally from Granada, Nicaragua. Her father, Juan García, was from Aransas Pass, Texas, and was of Mexican and Native American descent. He worked as a warehouse foreman at Ray-O-Vac. Although both of her parents received little education, they knew well the value of schooling and saw that Prichard appreciated the worth and the joys of learning. After graduating from El Camino High School in South San Francisco, Prichard entered the College of San Mateo and received her LVN degree (nursing) in 1969.

After taking some years to care for her two children, Erik and Andrea, Prichard chose a dramatic career shift and reentered academia in 1979. Interested in things scientific ever since she was young, and always intrigued and attracted by the thinking process and creativity required to do real scientific research, she enrolled at California State University at Hayward and earned her B.S. degree in chemistry/physics in 1983. She then continued her post graduate education at the University of Rochester in New York, obtaining her M.S. degree in **physical chemistry** in 1985. Continuing at Rochester, she entered the doctoral program and earned her Ph.D. in chemical physics in 1988.

Her graduate studies at Rochester emphasized optics, electronics, automation, vacuum technology, and signal processing with data acquisition and analysis. During this graduate work on the high resolution infrared absorption spectrum (which basically involves telling how much or what type of atoms or molecules are present), she was able to construct the first instrument ever to be able to measure van der Waals **clusters**. Named after Dutch Nobel prize-winning physicist, **Johannes Diderick Waals**, the van der Waals equation accounts for the non-ideal behavior of **gases** at the molecular level. An ideal or perfect gas is one which always obeys the known **gas laws**. The van der Waals equation allows scientists to predict the behavior of gases that do not strictly follow these laws by factoring in specific corrections. Van der Waals clusters are weakly bound complexes that exist in a natural state but are low in number. Prichard's work allowed other scientists to produce these rare clusters by experimental methods and thus be able to study them. Her graduate publications on this subject were themselves cited in more than one hundred subsequent publications.

Upon graduation, Prichard accepted a position with Eastman Kodak of Rochester, New York. A research scientist in the firm's PhotoScience Research Division, she conducts basic studies in **silver halide** materials for photographic systems. A member of Sigma Xi and Sigma Pi Sigma honor societies as well as a national board member of the Society for Hispanic Professional Engineers (SHPE) and a charter member of the Hispanic Democratic Women's Network of Washington, D.C., she also served on the Clinton/Gore Transition Cluster for Space, Science and Technology in 1992.

Prichard founded a program in Rochester called "Partnership in Education" that provides Hispanic role models in the classroom to teach science and math to limited English proficient students. She has also co-founded, within Eastman Kodak, the Hispanic Organization for Leadership and Advancement (HOLA). She is married to Mark S. Prichard, also a research scientist at Eastman Kodak. As to what she is most proud of in her career, she says that it is the fact that although her parents had little schooling, she was nevertheless able to come to love learning, obtain an advanced degree, and work in a professional field that she truly loves.

PRIESTLEY, JOSEPH (1733-1804)

English chemist

Born into a poor family in a village near Yorkshire, England, Priestley lost his mother at an early age and was sent to live with his aunt, a devout Protestant. He was educated at religious schools that endorsed nonconformist beliefs; he never formally studied science, but he did excel as a scholar of languages, logic, and philosophy. Priestley became a country preacher, but eventually turned to teaching. While employed at the Warrington Academy in the 1760s, he argued that school curriculums should reflect contemporary discoveries, rather than following outdated classical models. In 1762 Priestley married Mary Wilkinson, the sister of one of his schoolmates.

In 1766 Priestley visited London, England, and met Benjamin Franklin, who was trying to settle a dispute between the American colonies and the British government.

Soon after meeting Franklin, Priestley took over a pastorate in Leeds, England. Next door to his home was a brewery, and Priestley's scientific curiosity was aroused by the layer of heavy gas hovering over the huge **fermentation** vats. Priestley began experimenting with this gas, which we know today as **carbon dioxide**. Finding that the gas was heavier than air and that it could extinguish flames, Priestley realized he had isolated the same gas **Joseph Black** had designated as *fixed air*. Conducting various experiments with this gas, he found that when dissolved in **water**, a bubbly drink was produced. Priestley had invented *soda water*, or *seltzer*.

During this period Priestley also wrote a history of optics and an immensely successful history of electrical research. He discovered that **carbon** conducts **electricity**, for example, and he learned that an *electrostatic* charge collects on the outer surface of a charged object. As the first scientist to predict a relationship between electricity and **chemistry**, Priestley anticipated the new field of **electrochemistry**. At this time, Priestley also gave the modern name rubber to a Brazilian tree-sap product that had just been introduced to Europe. It was Priestley who told draftsmen that the material could be used to erase, or "rub out," pencil marks on their drawings.

Soon Priestley turned to chemistry, particularly the study of **gases**. During the early 1770s he developed new methods of collecting gases in the laboratory and prepared several gases unknown to chemists at the time. Priestley adapted a device called the pneumatic trough, filling it with liquid **mercury** instead of water to obtain samples of gases. In this way, he was able to isolate gases such as **sulfur** dioxide, **ammonia**, and **hydrogen** chloride. (Ammonia and **hydrochloric acid** were known earlier, but only as **liquids**.)

Priestley also discovered **nitrous oxide** (N_2O) years before Sir **Humphry Davy** popularized the gas's properties. Other gases isolated and identified by Priestley include **nitrogen** dioxide and **silicon** fluoride. As a result of these accomplishments, Priestley was elected to the French Academy of Sciences.

In 1773 Priestley won a lucrative post as librarian and companion to Lord Shelburne (1737-1805), a liberal politician. Priestley and his employer both sympathized with the colonial American rebels, who were then ready to begin the American Revolution, and an essay on government published by Priestley in 1768 provided Thomas Jefferson (1743-1826) with ideas for writing the Declaration of Independence.

Priestley's most famous scientific research was done during his eight years with Lord Shelburne. Because most of the gases he had studied were created by heating various substances, Priestley obtained a large magnifying lens. In 1774 Priestley used this lens to discover **oxygen**. Although Swedish chemist **Carl Wilhelm Scheele** had discovered the gas just a few years earlier, Priestley's results were reported first, and he usually gets the credit for the discovery. Priestley found that mercuric oxide, when heated, breaks down to form shiny globules of elemental mercury, while giving off a gas with unusual properties. A smoldering ember of **wood**, for example, burst into flames when exposed to the gas. Also, a mouse trapped in a container of the gas became frisky and survived for a longer time than it would when trapped in ordinary air. And when Priestley inhaled the gas, he reported feeling "light and easy." Priestley realized that this same gas was produced by plants, enabling them to restore "used-up" air to its original freshness.

Joseph Priestley.

In keeping with the scientific theory accepted at that time, Priestley named the gas dephlogisticated air because it absorbed **phlogiston** so readily. Phlogiston was thought to be the substance that gives materials their ability to burn. Supposedly, during **combustion**, phlogiston is released from burning material and absorbed by the surrounding air or gas. When Priestley reported his findings, he unknowingly gave **Antoine-Laurent Lavoisier** the key to a new theory of combustion that contradicted the phlogiston theory. It was Lavoisier who later expanded on Priestley's work, re-named the gas oxygen, and explained how substances burn by combining chemically with oxygen.

While conducting his experiments, Priestley continued to speak out aggressively on political and religious issues. He not only supported the American colonists' war with England, but also sympathized with supporters of the French Revolution. Priestley's religious allegiance had shifted toward the Unitarian Church, which was also unpopular in England at the

Ilya Prigogine. *(AP/Wide World Photo. Reproduced by permission.)*

time. Priestley eventually settled in Birmingham, England, where he served as a chapel minister and joined the Lunar Society, a club of respected scientists and inventors, but in 1791 an angry ''Church and King'' mob retaliated against France's supporters and burned down Priestley's home and laboratory, destroying much of his research.

Priestley escaped to London, but even there, his beliefs were barely tolerated. After the French people beheaded their king, declared war on England, and offered to make Priestley a citizen, Priestley gave in to public outrage and left for America. There he became a personal friend of Jefferson and of other politicians.

Prigogine, Ilya (1917-)

Belgian chemist

Ilya Prigogine was awarded the 1977 Nobel Prize in **chemistry** for his pioneering work on nonequilibrium **thermodynamics**. He revolutionized chemistry by introducing the concepts of irreversible time and probability in his approach to unstable chemical states and providing mathematical models of dissipative structures and their self-organization, the processes by which disorder progresses to order. Taking a highly philosophical approach to science, Prigogine has redefined the framework of the laws of nature, insisting on a less deterministic understanding of the natural world.

Prigogine was born in Moscow, Russia, on January 25, 1917, two months before the collapse of the czarist regime and nine months before the Bolshevik revolution. His father, Roman, was a chemical engineer and a factory owner, and his mother, Julia (Wichman) Prigogine, had studied music at the Moscow conservatory of music. Prigogine's brother, Alexandre, who was four years older, also became a chemist.

Four years after he was born, Prigogine's father decided to leave Russia because of the restrictions placed on private ownership and enterprise by the new Soviet government. After a year in Lithuania, the family moved to Berlin where they remained until 1929. Their stay in Berlin coincided with the terrible inflation of the early twenties and the beginnings of Nazism in the latter half of the decade. After two of his business attempts failed, and with growing Nazi strength boding ill for Jewish émigrés, Prigogine's father took the family to Brussels, Belgium, where Prigogine, then twelve, was enrolled in the Latin-Greek section of the Athénée d'Ixelles, a secondary school with a strict classical curriculum.

At this stage in his life, Prigogine was interested not in science but in history, archeology, art, and music; his mother claimed he could understand musical notes before he could understand words. Taught to play the piano by his mother, he even considered a career as a concert pianist, later regretting that he did not have as much time to devote to his music as he would have liked. He also read widely in the classics and philosophy during his teen years, and was particularly impressed by Henri Bergson's *L'évolution créatrice* and the Bergsonian view of the nature of time. He became a chemist as the result of a chance occurrence after his family had decided, and Prigogine concurred, that he should become a lawyer. Feeling that a first step to that end would be to learn about the criminal mind, he sought information about criminal psychology. In so doing, he came across a book dealing with the chemical composition of the brain, and he was so intrigued that he abandoned the law and took up chemistry.

Prigogine enrolled in the Free University of Brussels in 1935 to study chemistry, as had his brother before him. In 1939 he received his master's degree; he also won a prize for his performance of some Schumann pieces in a piano competition. Under the direction of Théophile De Donder, Prigogine received his Ph.D. in 1941 with the thesis, ''The Thermodynamic Study of Irreversible Phenomena.'' De Donder was the founder of the Brussels school of thermodynamics (the branch of physics that deals with the behavior of **heat** and related phenomena) and one of the first scientists to attempt to deal with the thermodynamics of systems not at equilibrium, that is, not in **balance**. Prigogine has credited De Donder with developing the mathematical apparatus needed for studying nonequilibrium states, and these tools would prove important to Prigogine's own work. Another professor who had an influence on the course of Prigogine's career was Jean Timmermans, a chemist and experimentalist interested in applying classical thermodynamics to the study of solutions and other complex systems. In 1957, Prigogine co-wrote *The Molecular Theory of Solutions,* which explained that at a low **temperature**, liquid **helium** would spontaneously separate into two phases, one of helium–3 and the other of helium–4. This was later confirmed experimentally. The book also develops methods for dealing with polymer solutions, some of which are still in use.

With his early interest in the nature of time, it was natural that Prigogine should be attracted to a study of the second law of thermodynamics that states that any spontaneous change in a closed system (one where neither **matter** nor **energy** flows into or out of the system) occurs in the direction that increases entropy, the measure of unavailable energy in a system or the measure of its disorder. This law indicates that as time passes in a closed system, disorder always increases, leading Sir **Arthur Stanley Eddington** to refer to the second law as supplying "the arrow of time." The move toward entropy described in the second law is irreversible, which contrasts with all other physical laws in which processes are reversible in time. This contrast begged the question of how the reversible, random workings of molecular and atomic motions could **lead** to processes that have a preferred direction in time. Furthermore, the second law, when extended to the largest known systems, suggests that the universe is moving toward eventual decay, a point when all energy and matter will reach a uniform state of **equilibrium** known as heat death.

Intrigued by these issues, Prigogine moved his focus away from the ideal "closed" system described in the second law and instead studied open systems that exchange matter and energy with an outside environment. Prigogine's first success in dealing with irreversible processes and open systems not at equilibrium came in 1945. In his doctoral research, he showed that for systems not too far from equilibrium, changes take place so as to achieve a steady state in which the production of entropy is at a minimum. This is true near equilibrium where the flux (or flow) of energy or matter through the system is directly proportional to the force creating that flux; that is, the flux and the force are linearly related. But such a steady state, once established, is stable and continues unchanged; it cannot evolve into a new state. Prigogine's work in this area led to his book *Thermodynamics of Irreversible Processes.*

In 1947 Prigogine succeeded De Donder to become full professor at the Free University of Brussels, where he assembled an interdisciplinary group to study irreversible processes. He went on to show that far from equilibrium, where fluxes and forces are no longer linearly related, a system can become unstable and evolve new, organized structures spontaneously. Prigogine called these organizations dissipative structures and developed the mathematical means of describing them. Prigogine theorized that such structures can be maintained as long as the energy and material fluxes are kept up. The process by which a new order evolves is labeled self-organization. In a nonlinear system there exist points—Prigogine referred to these as moments of choice or bifurcation points—at which the system is unstable, and small fluctuations can grow to a macroscopic or large size, creating a new structure. Randomness enters at the bifurcation points, so that predictions with respect to outcomes can only be expressed as probabilities. Common examples, although complicated and difficult to analyze, include the development of conduction cells in **liquids** heated from below (known as the Bénard instability), or the abrupt change from smooth flow to turbulent flow as the **velocity** of a liquid passing through a pipe is increased.

Prigogine's findings, while important, remained largely theoretical into the 1960s. Attempting to confirm his ideas, Prigogine worked with G. Nicolis and Réné Lefever to devise a simple mathematical model now called the Brusselator to better test his theories. Then in 1965 the Belousov-Zhabotinskii reaction, discovered in 1951 in the Soviet Union, became widely known abroad. One version of the reaction, in which the dissipative structures can be seen and do not have to be revealed by elaborate measurements, is a **solution** of malonic acid and bromate **ion** in **sulfuric acid** and ferrous phenanthroline (ferroin). Depending on the temperature and concentrations of the various species, the **color** of the solution may change back and forth from red to blue, or a pattern of red and blue may be formed that is either stationary or moves through the solution in a regular manner. These patterns gave striking visual proof of the existence of Prigogine's dissipative structures. In 1968 Richard Noyes at the University of Oregon was able to establish the mechanism of the reaction and, using Prigogine's work, to describe the phenomenon exactly.

The various processes that take place in cells involve complicated cycles of reactions catalyzed by special **proteins** called **enzymes**. Many of these enzymatic cycles meet the requirements for the formation of dissipative structures. For example, the breakdown of sugar in a cell has been shown to occur on a regular, periodic basis. Consequently, Prigogine's work is of great interest to biologists and biochemists. In fact, it was suggested by Alan Mathison Turing in 1952 that instabilities in chemical reaction systems could explain the patterns of stripes on a zebra or spots on a leopard. On a still larger scale, the thermodynamics of irreversible systems may explain how evolution, a process that gives rise to ever more specialized forms, is compatible with a physical picture of the world in which systems inevitably move from an ordered to a disordered state.

Prigogine and others have also applied the principles of irreversible thermodynamics to such disparate systems as the development of traffic patterns on a highway in response to driving conditions, the aggregation of slime molds in response to the depletion of nutrients in their environment, and the buildup of giant termite mounds in which a large number of independent termites behave in an orderly, seemingly purposeful, and intelligent fashion. On a larger scale, Prigogine's research allows a somewhat different and brighter view of the universe's ultimate fate. As explained in *Omni,* the theory of dissipative structures "offers a guardedly optimistic alternative to the pessimistic view of mankind's future—that winding down of nature toward a kind of heat death."

In 1949 Prigogine became a Belgian citizen, so that his Nobel Prize in 1977 was the first given to a Belgian. He was named director of the Instituts Internationaux de Physique et de Chimie (the Solvay Institute) in 1959, a post in which he continued after his retirement from the Free University in 1985. From 1961 to 1966 he spent time at the University of Chicago, and since 1967 spends three months of each year as director of the Ilya Prigogine Center for Statistical Mechanics and Thermodynamics at the University of Texas. He and his wife Marina Prokopowicz, an engineer whom he married in 1961, live in Brussels.

Prigogine has attempted to explain the implications of his work for the general public in two books: *From Being to Becoming: Time and Complexity in the Physical Sciences* in 1980 and, with Isabelle Stengers, *Order out of Chaos: Man's New Dialog with Nature* in 1984. For the latter work, Prigogine was made *commandeur* of the Ordre des Arts et des Lettres by the French government, an honor he especially prized because it is usually given to recognize achievement in the arts. Among the many honors conferred on him are the honorary foreign memberships in the U.S. National Academy of Sciences and the Academy of Sciences of the U.S.S.R. (now Russia).

PROMETHEUM

Prometheum is one of the **rare earth elements** found in Row 6 of the **periodic table.** Elements in this group are also known as the lanthanides after the first element in that group, **lanthanum.** Prometheum has an **atomic number** of 61, an atomic **mass** of 144.9128, and a chemical symbol of Pm.

Properties

All isotopes of prometheum are radioactive, the most stable being prometheum-145 with a half life of 17.7 years. The element is a silver-white metal with a melting point of 2,120°F (1,160°C) and a **density** of 7.2 grams per cubic centimeter. Other properties of the element have not been thoroughly studied since researchers are more interested in its radioactive properties than its chemical or physical properties.

Occurrence and Extraction

Prometheum has never been found in the Earth's crust although it has been observed in the spectra of some stars in the galaxy of Andromeda. The element can be prepared artificially in particle accelerators and is a byproduct of **nuclear fission** reactions.

Discovery and Naming

The existence of an element with atomic number of 61 was fairly well accepted by the beginning of the nineteenth century. The periodic chart developed by Russian chemist **Dmitry Mendeleev** showed that an element should be found between **neodymium** (atomic number 60) and **samarium** (atomic number 62). However, searches for the missing element produced no positive results until 1945. In that year, scientists working at the Oak Ridge National Laboratory in Oak Ridge, Tennessee, discovered the new element among fission products formed during reactions occurring the laboratory's fission reactor. The element was named after the Greek god Prometheus who, according to legend, stole fire from the gods and brought it to Earth for human use.

Uses

Prometheum has a limited number of commercial applications. For example, it is used to provide **energy** in a battery-like device where more traditional, bulkier **batteries** would be an inconvenience. Such devices have been used to provide energy in space probes and space vehicles. The element has also been used to measure the thickness of objects, a measurement that can be made on the basis of the amount of the element's radiation that passes through the object.

PROPYL GROUP

The propyl group is a **functional group** derived from propane (C_3H_8) by the removal of a **hydrogen atom**. It has the chemical formula C_3H_7-. The propyl group exists in two distinct isomeric forms. The first is normal propyl (n propyl or 1 propyl) which has the chemical formula $CH_3CH_2CH_2$-. The second is isopropyl (2 propyl) which has the formula $(CH_3)_2CH$-. Like all isomers the properties of the two propyl forms are slightly different. For example, propan-1-ol has a **boiling point** of 206°F (96.6°C), whereas propan-2-ol has a boiling point of 180°F (82°C). In this example, the first **molecule** is a primary **alcohol** and the second is a secondary alcohol. Each has slightly different chemical as well as physical properties.

In common with other alkyl groups, both forms of the propyl group modify the behavior of a functional group. As molecular **mass** increases, for example, the boiling point of a liquid will also increase. As can be seen in the previous example, the arrangement of the propyl group can also have an effect. This is due to the fact that with propan-2-ol the hydroxyl group of the alcohol is protected and consequently unavailable for hydrogen **bonding**. As a result, there are less **intermolecular forces** to overcome so less **energy** has to be added to the system to overcome these internal forces.

PROSTAGLANDINS

Prostaglandins are a type of chemical which can be found in almost all tissues of mammals. Prostaglandins exist in cyclic form, and are composed of fatty acids. They exert their functions both locally (in the basic vicinity of their production), as well as traveling through the bloodstream to act on distant target organs (in an endocrine manner). There are many types of prostaglandins, including those labeled PGA1, PGE1, PGE2, and PGI1.

Prostaglandins work on a number of different target tissues, including the uterus, ovaries, and fallopian tubes. Prostaglandins are known to be involved in the initiation and maintenance of uterine contractions during labor, as well as in the production of the uncomfortable cramps occurring during menstrual periods. Prostaglandins are also involved in regulating the size of the blood vessels of the kidney. Prostaglandins affect the actions of the blood cells responsible for clotting (platelets). Prostaglandins are involved in the production of fever. Prostaglandins are also involved in the process of inflammation, and anti-inflammatory medications such as aspirin and ibuprofen work by interfering with the production of prostaglandins.

PROTACTINIUM

Protactinium is the next-to-last heaviest element, just preceding **uranium** (atomic number 92) in the **periodic table**. Protactinium's **atomic number** is 91, its atomic **mass** is 231.03588, and its chemical symbol is Pa.

Properties

All isotopes of protactinium are radioactive. Protactinium-231 is the most stable **isotope** with a half life of 3.276 x 10^4 years. The element is a bright shiny metal with a melting point of about 2,840°F (1,560°C) and a **density** of 15.37 grams per cubic centimeter. The element is fairly active and reacts with **oxygen**, the **halogens**, and **hydrogen**. Detailed properties of the element and its compounds have not been well studied, however.

Occurrence and Extraction

The amount of protactinium in the Earth's crust is too small to estimate accurately. The best estimate of its abundance is based on the fact that a ton of its most common ore, pitchblende, yields about 0.1 part per million of the element.

Discovery and Naming

Protactinium was discovered in 1913 by the German-American physicist Kasimir Fajans (1887-1975) and his colleague, O. H. Göhring. Fajans and Göhring were analyzing the mixture of substances found when isotopes of uranium undergo radioactive decay. They originally suggested the name of *brevium* for the element because of its short half life (1.175 minutes for protactinium-231), although the element's name was later changed to protoactinium. This name was suggested because the element produces **actinium** (atomic number 89) when it decays, making it "first actinium" or "protoactinium." The spelling of the name was later changed slightly to its current form

Uses

Neither protactinium nor its compounds have any commercial uses. The element is still very rare and sells for about $300 per gram for research purposes.

PROTEIN SYNTHESIS

Proteins are synthesized according to instructions carried in the genetic material, **DNA** (deoxyribonucleic acid). Protein synthesis can be divided into three phases: transcription, translation, and post-translation modification. The first two phases rely on the **nucleic acids**, DNA and **RNA** (ribonucleic acid), and on associated enzymes. Enzymes are proteins that speed up chemical reactions in living **matter**. The final phase involves other enzymes, but no nucleic acids. The finer details of protein synthesis differ between cells that have no **nucleus** (prokaryotes) and cells that do have a nucleus (eukaryotes). However, the overall process is basically the same.

Proteins are a very common type of **molecule** in living matter. One thing that all proteins have in common is that they are composed of **chains** of smaller molecules, called **amino acids**. A protein's ability to function properly is very closely related to having the correct sequence of amino acids. In some cases, a single misplaced amino acid can make a protein useless.

The first phase of protein synthesis is transcription of DNA. Like proteins, DNA is made up of a chain of smaller molecules. These smaller molecules are called nucleotides. There are four types of nucleotides in DNA, each containing one of four bases—adenine, guanine, cytosine, or thymine. DNA is a double-stranded, spiral-shaped molecule, often referred to as a double helix. When the two strands of DNA are lined up, their nucleotide bases face one another like the teeth of a zipper. The strands are joined by **hydrogen** bonds between the bases. Pairing between bases follows a strict pattern. Adenine will only pair with thymine, and vice-versa; cytosine will only pair with guanine, and vice-versa. Although there are only four types of nucleotides, the order in which they appear along the DNA strand can convey a lot of information.

However, the information flow from DNA to protein is not direct. One strand of the double helix—the coding strand or template strand—serves as a master copy of the genetic code. The cell uses RNA to construct a working copy from that strand. RNA is chemically similar to DNA. Key differences are that it is single-stranded, and uracil is used as one of the four bases rather than thymine.

Transcription begins when the **enzyme**, RNA polymerase, binds to the coding strand of DNA just before the start of a gene. A gene is the stretch of DNA that codes for a particular protein or polypeptide. (Polypeptides are composed of amino acids, but are smaller than proteins.) The RNA polymerase binds to DNA with the help of other proteins called transcription factors. Transcription factors can help a cell control the level of gene expression, i.e., the amount of protein synthesized.

Once RNA polymerase binds to the DNA, it begins to construct a strand of RNA. The two strands of DNA separate and, for a short while, the coding strand is paired with RNA as a matching strand. The base pairing between the DNA strand and the RNA strand follows the same rules as base pairing between two strands of DNA. The key exception is that uracil is used in place of thymine.

Once the DNA has been transcribed, the mRNA (messenger RNA) detaches from the DNA. It may be modified before moving on to the next phase of protein synthesis. Modification of mRNA is only used by eukaryotes. The mRNA is made up of coding and noncoding portions. Coding portions are called exons; noncoding portions are known as introns. Modification stabilizes the mRNA and snips out introns. Why introns exist is unknown. They may represent changes that have been added over the course of time, or they may be remnants of old genes that are no longer needed. In any case, they are removed before the mRNA is released from the nucleus into the interior of the cell.

The next phase of protein synthesis is translation of the code, which relies on two other types of RNA: ribosomal RNA (rRNA) and transfer RNA (tRNA). Although there are three types of RNA, they are all composed of adenine, cytosine, guanine, and uracil. The differences between the RNA types are based more on their function and their associated proteins, rather than the nucleic acid itself.

The rRNA is a component of a cellular structure called a ribosome. Ribosomes serve as sites for translation. To start translation, a ribosome attaches itself to a strand of mRNA. The ribosome serves to hold the mRNA in place and direct the speed of translation. The transcribed code carried on the mRNA is read three nucleotides at a time. There are 64 possible mRNA triplets, or codons. Sixty-one of them code for a specific amino acid; the remaining three serve as stop signals to end translation. Most amino acids have more than one associated codon.

As the ribosome gradually moves along the length of the mRNA, tRNA transports amino acids to the growing protein molecule. The tRNA contains a nucleotide sequence called the anticodon. The anticodon contained in the tRNA determines which amino acid it will carry. When the anticodon pairs up with a corresponding mRNA codon, it releases the amino acid to the growing protein chain. When the ribosome reaches one of the three stop codons, it detaches from the mRNA and a new protein molecule is released. Once released, the new protein may or may not undergo modification.

PROTEINS

Proteins are large, organic, nitrogen-containing molecules that are essential both to the structure and to the function of all living cells. Exceedingly complex themselves, the protein molecules have as their structural units much simpler compounds called **amino acids**. Typically, the amino acids in each **molecule** are strung together in chain-like fashion, with the longer **chains** folded into ribbons, spirals, and other three-dimensional forms.

The types of amino acids on the chain (over 22 varieties have so far been identified in animal protein), the number of them (a chain can contain a few amino acids or several thousand), and the particular configuration all help differentiate one protein from another. Because proteins have such a wide range of specific functions and characteristics, they can be classified in a number of ways. Probably the simplest way is by function. One of protein's main functions is to serve as "raw material" for the growth, maintenance, and repair of all living tissues. Proteins make up at least 50% of most animal cells and they constitute much of the solid **matter** in muscles, organs, and endocrine glands. Specialized proteins (keratins, for instance) are used to form skin, hair, and nails; other proteins (collagen) help form connective tissue; and still others (globulins and albumins) make up the soluble or semi-soluble molecules of all cells.

A number of proteins also have highly specialized functions in the regulation of body processes. Some of these regulators include the nucleoproteins (protein plus **DNA** and **RNA**) that contain the blueprints for the synthesis of all proteins; catalytic proteins (or enzymes) used to activate biological processes; hormonal proteins that start and stop cellular **metabolism**; contractile proteins (myosin and actin) that regulate muscular contraction; blood proteins (hemoglobin and lipoprotein) that **transport proteins** throughout the system; and the serum albumin and serum protein that regulate **osmotic pressure** during **osmosis** and maintain **water balance**.

When the amount of fats and **carbohydrates** in the diet is insufficient, protein can also serve as a source of **energy**. However, the protein "burned" to provide energy can no longer be used to synthesize new protein, resulting over time in a serious deficiency problem. While protein deficiency is rare in developed countries, it still occurs in young children in many poorer countries, usually in the form of kwashiorkor, in which the child's body is swollen and growth-retarded, or marasmus, in which the child suffers from emaciation.

Proteins were first recognized as natural organic molecules in the 1830s when Dutch chemist Gerardus Mulder began systematically analyzing a number of plant and animal products. In several apparently unrelated test materials—silk, egg white, blood, and gelatin—Mulder discovered the same nitrogen-containing organic substance, a substance that appeared much more complicated in constitution than either fats or carbohydrates. After further research, Mulder proclaimed it "unquestionably the most important of all known substances in the organic kingdom. Without it, no life appears possible on the planet." Mulder (perhaps at the suggestion of a contemporary, **Jöns Berzelius**) decided to name the substance protein, from the Greek word proteius, roughly meaning to come first.

Even before protein had a name, a few researchers were aware that nitrogen-containing foods were somehow important to life. In 1816, French physiologist François Magendie demonstrated in dogs that **nitrogen** was essential to life and that the nitrogen could only be supplied in the diet through certain foods. In 1827, English biochemist **William Prout** classified foods into four main categories: *water*, *saccharinous* (carbohydrates), *oleaginous* (fats), and *albuminous* (proteins). In 1842, **Justus von Liebig** found that the protein acquired through nutritional sources was necessary for building body tissues, and protein, along with fats and carbohydrates, quickly became recognized as a dietary essential. Liebig also suggested that the urinary level of nitrogen could be used to measure the amount of protein eaten by an individual.

In the 1870s, **Karl von Voit** established the principles of nitrogen equilibrium in the body that are used as the criteria for human protein requirements. By this time, protein **chemistry** was being studied by a number of investigators, including

German biochemist Ernst Hoppe-Seyler, who devised a classification system in 1875 that is still employed today (because of the large number of proteins, several classification systems are possible; chemical composition, physical properties, and **solubility** are among the key criteria used for grouping proteins).

By the turn of the century, more than half of protein's structural units—the amino acids—had been identified. The exact nature of these compounds was still unknown, however. The first clues were provided by **Frederick Gowland Hopkins**, who discovered tryptophan in 1900 and conducted countless nutritional experiments thereafter. Through his feeding tests, Hopkins suggested that certain amino acids were more nutritionally important than others and that, while some could be synthesized in the body, others had to be supplied in the diet. By 1907, **Emil Fischer** greatly furthered structural research by demonstrating precisely how amino acids were linked within proteins. In 1911, Thomas B. Osborne, after investigating with Lafayette B. Mendel the protein in seed plants, provided definite proof that certain amino acids, such as lysine and tryptophan, were not only nutritionally essential but could not be synthesized using laboratory rats. The puzzle was finally finished by William Cummings Rose (once Mendel's student), who in 1935 isolated the last nutritionally important amino acid to be discovered, threonine. Rose was able to prove that while roughly half the amino acids in protein could be synthesized in the body, the other half—the so-called essential amino acids—had to be supplied through various ''essential'' foods.

Current research efforts have been focused on *protein folding*, the three-dimensional structure of a protein molecule. Scientists are trying to discover a way to predict a protein's folded structure from the amino acid sequence. Knowing the three-dimensional structure of a protein gives insight into its interaction with other molecules, reaction **kinetics** and dynamics, and evolutionary relationships with similar proteins in other organisms. Because many factors and intermediate structures influence the folding of a particular protein, this has become a laborious task which will require further study.

PROTON

A subatomic particle, a proton (*proton* means ''first'' in Greek) is the fundamental unit of positive **electricity** in the **atom**, with a charge equal to an **electron**'s negative charge. However, a proton's **mass** is 1,836 times the mass of an electron. Protons, along with neutrons, constitute an atom's nucleus. In **chemistry**, the number of protons in a the atom of a particular element defines that **atomic number** (*Z*); an element's atomic number is its identity tag. With one exception (**hydrogen**), chemical elements in their ideal form (excluding isotopes) have an equal number of protons and neutrons. The sum of protons and neutrons constitutes an element's atomic mass. The atomic number and mass of hydrogen, because it has one proton and no **neutron**, both have the value of one.

Early atomic theory assumed that atoms were indivisible. However, when **Michael Faraday** demonstrated, in 1834, that chemical elements were electrical in nature, scientists embarked on a series of experiments which challenged, and eventually demolished, the fundamental assumptions of traditional **atomic theory**. For example, in 1878, **William Crookes**, who predicted the discovery of isotopes, started experimenting with cathode rays, which occurred when electricity was discharged between two metal plates in a tube (later named a Crookes tube) with gas at an extremely low pressure. Crookes believed these rays were negatively charged electrons. In 1897, however, **J. J. Thomson** identified them as subatomic particles, eventually named electrons. Further experiments led to the discovery of protons. For example, when scientists drilled holes in the positively charged (**anode**) plate of the Crookes tube, they detected rays moving in the opposite direction (from the anode to the negatively charged **cathode**) of the cathode rays. Named ''canal rays,'' these rays were studied by Wilhelm Wien (1864-1928). In 1905, Wien identified some of these rays are hydrogen ions. Researchers later established that rays with the least mass were protons.

A visualization of how a proton is produced from three quarks. The proton is a constituent of all atomic nuclei, containing three 'valence' quarks, two 'up' quarks. The quarks are held together by the strong nuclear force, seen as the exchange of gluons (straight lines). ***(Photograph by ArSciMed/Science Photo Library, Photo Researchers, Inc. Reproduced by permission.)***

In 1909, **Ernest Rutherford** instructed his younger colleague Hans Geiger (1882-1945) and Ernst Marsden (1889-1970), who was still an undergraduate, to perform the **gold** foil experiment. The experimenters bombarded a thin gold foil with **alpha particles** (helium atoms without electrons). Since the current model of the atom was a positively charged sphere electrons inserted throughout (the plum pudding model), they expected the alpha particles to penetrate the foil without resistance. Indeed, most of the alpha particles did, but a small num-

Joseph-Louis Proust.

ber were strongly repelled. These results indicated that the gold atom had a positively charged nucleus. In 1919, Rutherford identified the proton as the fundamental unit of positive electrical charge in the atom. The neutron, the major particle constituting the atom's nucleus was discovered in 1932 by James Chadwick (1891-1974). Approximately equal to a proton in mass, the neutron has no charge.

PROUST, JOSEPH-LOUIS (1754-1826)

French chemist

Until he was twenty, Joseph Proust followed dutifully in his father's footsteps, learning **chemistry** from him and eventually becoming his apprentice in a pharmacy in Angers, France. But in 1774 Proust left for Paris, against his family's wishes and apprenticed himself to another pharmacist. By 1776 he had won a position at a Paris hospital, where he worked as a chemist and pharmacist while lecturing at the Royal Palace.

In 1778 Proust went to Spain, having obtained the post of chemistry professor. In 1780 he returned to France and stayed there for five years; during this time he taught chemistry and experimented with the new scientific sport of ballooning. In 1784 Proust participated in a ballooning event at Versailles that was witnessed by French and Swedish royalty.

In 1785 Proust accepted a lucrative teaching position offered by the Spanish government. He spent the next twenty years in Spain at various posts in Madrid and Segovia, thus missing the French Revolution and the rise to power of Napoleon Bonaparte (1769-1821). In addition to teaching chemistry, Proust worked for the Spanish government, frequently conducting geological surveys and analyzing the nation's mineral resources.

While in Spain, Proust also began studying different types of sugar. He was the first to identify the sugar that comes from grapes as glucose. In 1799, when the chemical laboratories of Segovia and Madrid were merged, Proust became director of the new, lavishly equipped facility. While there, Proust published his law of constant composition, which later evolved into the law of definite proportions. At the time, most chemists agreed with Claude Berthollet, who believed the composition of a compound would vary according to the amounts of reactants used to produce it. In contrast, Proust proposed that pure reactants always combine in the same proportions to produce exactly the same compound.

Proust based his theory on careful analysis of **copper** carbonate, which he prepared in various ways. He also compared his laboratory results with analyses of copper carbonate taken from mineral rocks. Proust found that all pure copper carbonate samples had the same composition—that is, the same proportion of copper, **carbon**, and **oxygen**—no matter how they were produced or where they came from. In analyses of other compounds, Proust obtained similar results.

For about eight years, Proust and Berthollet engaged in a friendly controversy over this issue, but, in the end, Proust was proved to be right. Berthollet had used impure reactants in his experiments, and thus he had analyzed the products inaccurately. Meanwhile, **John Dalton** had been formulating his **atomic theory** which was published in 1808. In this theory, Dalton rephrased Proust's law, calling it the *law of multiple proportions*. Although it is unclear whether Dalton was directly influenced by Proust, the *law of constant composition* provided evidence for Dalton's atomic theory, which in turn provides an explanation for Proust's observations.

PROUT, WILLIAM (1785-1850)

English chemist

Prout was born in Gloucestershire, England. He was trained as a physician, but very early became interested in the **chemistry** of living organisms. Most of Prout's original research and thought involved the chemistry of **nutrition**. In 1827, he suggested dividing foods into the three large classifications— **carbohydrates**, fats, and proteins—that are still used by nutritionists today.

His most important discovery in the field of nutrition was his recognition of the existence of **hydrochloric acid** in the stomach. Scientists were—and, to some extent, still are—amazed that the presence of such a potent acid in living organisms does not cause them serious damage. Prout's findings were published in an 1824 edition of the *Quarterly Journal of Science and the Arts*.

Prout made other important contributions to **organic chemistry**, such as the development of methods for urinalysis and the analysis of organic compounds by means of **combustion**.

Prout's name will be forever famous not for any specific experimental work, but for a hypothesis he developed about the composition of **matter**. Prout examined the atomic weights of the elements as they were known in the early 1800s. He observed that most of these weights were whole number multiples—or close to whole number multiples—of the weight of **hydrogen**. Prout concluded from this observation that hydrogen is the fundamental element from which all other elements are made. He called hydrogen the protyle, or "first matter," from which all other matter is formed by condensation.

Prout published his hypothesis anonymously in 1815 in Thomas Thomson's *Annals of Philosophy*. A year later, Thomson identified Prout as the author of the anonymous article. Debate raged over Prout's hypothesis for a century. When he first proposed the idea, atomic weights were so inaccurately known that variations from whole numbers could easily be blamed on faulty analyses. As experimental techniques improved, however, it became more obvious that atomic weights like those of **chlorine** (35.5) and **magnesium** (24.3) were accurate. As a consequence, Prout's hypothesis fell in disfavor.

The discovery of isotopes in 1913, however, changed all that. Chemists finally realized that the atomic weights of *isotopes* were nearly whole numbers, even if those of elements were not. Thus, Prout's hypothesis might indeed have some validity. Today, we realize that the most massive particle in a hydrogen **atom**, the **proton**, is indeed the building block of which all other atoms are made and which determines exactly what element those atoms will be. We can, therefore, accept Prout's hypothesis in a revised and modernized form.

PSYCHOTROPIC DRUGS

Psychotropic drugs have revolutionized the treatment of mental illness, perhaps most profoundly in schizophrenia and depression. Psychotropic is a word derived from the Greek *psyche*, meaning the mind, and *tropos*, to turn or change. Psychotropic drugs are used to treat people with: 1) Clearly diagnosed primary psychiatric illness as defined by the *Diagnostic and Statistics Manual of Mental Disorders*, (*DSM- IV*); 2) certain medical conditions—such as specific types of epilepsy; 3) emotionally distressing and extreme behavior which drastically interfere with a person's ability to function; 4) severe dysfunctions resistant to other types of treatment; 5) withdrawal difficulties associated with other psychotropic medications, **alcohol**, **nicotine**, **caffeine**, and opiates; and 6) sedation during dental or medical procedures not requiring **anesthetics**. Psychotropic drugs fall into four primary categories: antipsychotics, mood stabilizers, anti-anxiety agents, and antidepressants. Each drug is specifically designed and/or prescribed to alter abnormal thought (such as hallucinations, delusions, distortions, and paranoias), abnormal moods (such as extremes of euphoria and depression), and disruptive behaviors (especially those caused by delusions of grandeur or paranoia).

Mental illnesses have been recorded throughout history: Babylonians and Egyptians believed demonic possession to be the cause and used magic, religious rites, or plants and herbs as "cures." Hippocrates attributed "hysteria" to a woman's uterus, and blamed "melancholia" (depression) on black bile, which he attempted to treat with purgatives. Roman writer Cicero (106-43 B.C.), defied these theories by suggesting melancholia was psychological and that individuals were responsible for the way they thought or felt. Thus began the conflict about the cause of mental illness—was it psychological and treatable with "talk therapy" (psychotherapy); or was it biological requiring medication? While this discussion still rages today and psychotherapy plays an important role in the treatment of many mental illnesses, psychotropic drugs have positively affected the lives of millions of mentally ill people previously untreatable.

Although synthetic sedatives appeared around the late 1800s and barbiturate use began in the early 1900s, 1949 saw the first truly revolutionary treatment of mental illness when Australian psychiatrist, John F. J. Cade, accidentally discovered the amazing benefit of **lithium** in manic-depression (bipolar disorder). This psychotropic drug is classified as a mood stabilizer. The antipsychotic drug, chlorpromazine (Thorazine), initially developed as an antihistamine, revolutionized the treatment of schizophrenia with its introduction in 1954. Within eight months of its release, more than two million patients were taking it, many of whom left mental institutions to lead relatively normal lives. Haloperidol (Haldol), clozapine (Clozaril), risperidone (Risperdal), and olanzapine (Zyprexa), are also effective in treating psychoses. In the late 1950s, the antidepressant imipramine (Tofranil) became the first of several tricyclic antidepressants to meet with considerable success in treating depression. In 1980s, monoamine oxidase inhibitors gained increased popularity in treating certain types of depression not responsive to the tricyclics, and the selective serotonin reuptake inhibitors—the first being Prozac introduced in the late 1980s—gained immense popularity for their effectiveness, even for people suffering from life-long depression. **Barbiturates**, highly addictive and sedative antianxiety agents widely prescribed until the 1960s, were replaced by **benzodiazepines** such as chlordiazepoxide (Librium). In the early 1990s, six of the highest-selling prescription drugs in the United States were benzodiazepines.

Undoubtedly, ever-increasing knowledge about the function of the brain will allow more specific development and targeted prescription of psychotropic medications, while research into adjunct therapies—such as light therapy for seasonal affective disorder—and genetics will all play an important role in controlling and alleviating mental illness.

PYROTECHNICS

The science and art of pyrotechnics is much more than just a way to make fireworks. It is also used in all kinds of **explosives**, fuses and flares, and dispersal and powering mechanisms. Pyrotechnics has been used in celebrations and military

A fireworks display over the United States Capitol Building, Washington, D.C. ***(Photograph George Chan, Photo Researchers, Inc. Reproduced by permission.)***

campaigns for as long as 2,500 years, beginning with the Greeks. In general, they produce some combination of **heat**, light, noise, motion, and smoke when they are ignited. For this reason, people use pyrotechnics for everything from Fourth of July celebrations to spacecraft.

Black powder is perhaps the earliest known pyrotechnic. The Chinese discovered it before 1000 A.D. during the Sung Dynasty, using a compound of **charcoal**, **potassium** nitrate, and **sulfur** to produce powerful rockets and fireworks. Pyrotechnics came to Europe in the thirteenth century, where they quickly revolutionized warfare with their use as propellants in cannons and rockets, and in 1677, miners applied the science to blasting rock for the first time. Its use continues today as both an effective explosive, whether for blasting roads through mountains or for making noise with fire crackers and model rockets.

A pyrotechnic mixture's primary ingredients are an oxidizer (usually perchlorate, nitrate, chlorate, or peroxide of **strontium**, **barium**, or potassium), a binder to hold everything together (dextrin, red gum, paraffin, or shellac), and a fuel (charcoal, **antimony** sulfide, sulfur, **boron**, **titanium**, **magnesium**, aluminum). There might also be a delay element (pyrotechnic devices in their own right), as in hand grenades, and there is always an igniter. The igniter takes many forms, including black powder-impregnated fuses, **batteries** that create hot spots, or friction generators, but some pyrotechnics employ impact to ignite a primer material.

Unlike the techniques of ancient times, when the ingredients were often separate until the moment of reaction, today's formulas are self-contained (i.e., manufacturers put the oxidizer, fuel, binder, and igniter all together in a single package). This works because pyrotechnic chemical reactions are heat-releasing (exothermic) and self-sustaining once ignition occurs. The most reactive fuels are those that are very exothermic and tend to combine with **oxygen** at low temperatures.

Sometimes manufacturers will add another chemical designed to produce a special effect. For instance, for a fireworks show the engineers might add magnesium to the fuel- oxidizer-binder mix to make a bright-white light, strontium salts to give a scarlet color, or **aluminum** and **iron** shavings to make white and gold sparks. Adding an unstable burner such potassium benzoate can produce the familiar whistling sound at fireworks shows, while rapid burners such as powdered titanium makes huge booming sounds. Making fireworks explode into stars and other shapes relies upon such materials as **sodium** metal, which tends to burn or suddenly ignite when flying through the air. For an airshow, the planes might be loaded with **sodium bicarbonate** and different organic dyes so they trail colored smoke as the burning mixture volatizes the materials and the dyes condense into smoke.

Military use of pyrotechnics has always been extensive. Modern armies frequently rely upon flares to illuminate a large area using flares. These pyrotechnics often use powdered magnesium or aluminum as the fuel, and when ignited produce a white glare that (when exploded overhead) leaves no shadow in which the enemy can hide. Pyrotechnics are also used to make signaling smoke that can alert others of the enemy's presence or communicate to allied troops, or a thick haze of obscuring smoke in which to retreat or advance.

Far from the exciting world of warfare and fireworks, pyrotechnics remain important tools in many industries. In automobiles, for example, the thermal decomposition of sodium azide produces a large amount of **nitrogen** gas that acts as a propellant that causes air bags to open on impact. To keep traffic away from the scene of an accident, highly visible pyrotechnics known as flares often come in handy. For the U.S. Space Shuttle, pyrotechnics play an important part in getting the craft into orbit. Ammonium perchlorate (an oxidizer) accounts for 70% of the solid propellant in the shuttle's booster rockets because it generates so much gas when burning.

Q

QUALITATIVE ANALYSIS

Qualitative analysis involves determining the composition of an unknown compound or mixture. Qualitative analysis can also analyze a mixture to determine the precise percentage composition of the sample in terms of elements, radicals, and compounds. Whereas **quantitative analysis** involves determining how much of a material is present in a sample (quantity), qualitative analysis involves identifying what materials are present in a sample (quality).

Qualitative analysis was the primary method of analysis until the seventeenth century. The alchemists used this method of analyzing substances in their attempts to transmute **lead** into **gold**. **Robert Boyle** was one of the first chemists to perform chemical reactions where one individual substance was differentiated from others by the use of specific reagents. In the eighteenth and nineteenth centuries, analyses based upon the different solubilities of metal ions became a useful way to determine the composition of mixtures. The use of this qualitative method became more widespread when the ionic theory and the theory of solutions were developed in the late nineteenth century.

One of the primary methods of qualitative analysis involves **precipitation** reactions. These reactions allow you to identify what chemical entities are present in a substance. One example is using the relative solubilities of metal ions to determine what substances compose a sample. In this quantitative method, metal cations (positive ions) are detected by the presence of a characteristic precipitate. For instance, **silver** cations give a white precipitate when they are in a solution of chloride ions. This method was developed in the eighteenth and nineteenth centuries, and is still in use late in the twentieth century.

Another modern qualitative analytical method is **chromatography**. Chromatography is a technique for analyzing mixtures of **gases**, **liquids**, or dissolved substances. It was invented in 1906 by **Mikhail Tswett**, a Russian botanist. He originally devised column chromatography to separate and then identify the substances that make up plant pigments. A vertical glass tube is packed with an absorbing material (called the stationary phase) and the sample being analyzed is poured into the column. Different components of the sample move through the column at different rates and are absorbed by the stationary phase at different speeds, separating the mixture into layers.

There are several forms of chromatography. Gas chromatography is a technique for separating mixtures of gases. This apparatus consists of a very long tube that contains a stationary phase. This phase can be either a solid, or a nonvolatile liquid. The sample being analyzed is usually a volatile liquid that is carried through the stationary phase of the column by a gas. The components of the mixture pass through the column at different rates and are detected as they leave, either by measuring their conductivity or by a flame detector. The components are then identified by the time they take to pass through the column.

Most analyses today are complete analyses which are a combination of both quantitative and qualitative analytical methods. These kinds of analyses determine the kind and the amount of each component in a mixture, an important part of **chemistry** for the last 400 years.

QUANTITATIVE ANALYSIS

Quantitative analysis involves measuring the proportions of known components in a mixture. Whereas **qualitative analysis** involves identifying what materials are present in a sample (quality), quantitative analysis involves determining how much of a material is present in a sample (quantity).

Chemists did not begin to use quantitative methods of analysis in earnest until the mid-eighteenth century. The development of **chemistry** into a science unto its own with robust theories and laws mirrored the introduction of quantitative analytical methods. Chemists could argue more strongly their views about the composition of substances when they could provide numbers and exact amounts. The two chief classes of quantitative analysis, gravimetric and volumetric, were developed in the eighteenth and nineteenth centuries.

The first quantitative analyses in the mid-eighteenth century were gravimetric. The experiments conducted with gravimetric techniques helped develop the theory of **combustion**, the law of chemical combination, and the concept of equivalent weight. Volumetric analysis was also developed in the eighteenth century, but its use in determining quantities of a substance in a mixture was limited until chemists reached a better understanding of solutions in the nineteenth century.

Gravimetric analysis involves weighing substances both before and after chemical reactions. **Antoine-Laurent Lavoisier** ushered in the new system of chemistry by dismantling the **phlogiston** theory of combustion using gravimetric analysis to determine that the weight gained by **metals** upon combustion was equal to the loss of weight of **oxygen** in the reaction vessel. Gravimetric analyses today frequently involve **precipitation** reactions. In these reactions, the amount of a species in a sample is determined by precipitating it from **solution**. The precipitate is then filtered from the solution, dried, and weighed. Gravimetric analyses tend to be very simple and accurate, but because they involve much time-consuming work, chemists today tend to use modern instrumental methods like gas chromatography- **mass** spectrometry.

The other main branch of quantitative methods is volumetric analysis. These analyses are based on **titration**. Titration is a procedure for determining the **concentration** of a sample by carefully adding a just enough of a solution with known concentration to react with the sample. If you know the **volume** and concentration of substance with known concentration that just reacts with sample, you can determine the concentration of the sample in solution. This method of analysis is called titration.

To find when a reaction is just complete, an indicator is added to the solution. An indicator is a substance that changes colour when a reaction approaches completion. An example of volumetric quantitative analysis is the following: a flask contains a solution with an unknown amount of HCl. This solution is titrated with 0.207M NaOH. It takes 4.47 mL of NaOH to complete the reaction. What is the concentration of the HCl?

Most quantitative techniques, such as gravimetric and volumetric techniques, were replaced in the 1960s with instrumental methods. These instrumental techniques usually include both qualitative and quantitative analytical methods. Gas chromatography-mass **spectroscopy**, for instance, can give both quantitative and qualitative information. This is a technique for separating mixtures of **gases**, **liquids**, or dissolved substances into their constituents and then identifying what substances make up the mixtures and their relative amounts. Other techniques that will give both qualitative and quantitative analyses are spectrophotometry.

QUANTUM CHEMISTRY

Quantum **chemistry** involves the application of the principles of quantum theory to chemistry.

Quantum chemistry helps predict the way atoms combine to form molecules, and the way molecules interact with each other. Applications of quantum chemistry include the calculation of molecular structures, prediction of properties, analysis of spectroscopic data, and the description of chemical reactions in terms of individual molecular events.

Quantum theory states that **energy** is transferred between systems in discrete amounts. The theory arose from the inability of classical physics to adequately explain certain physical phenomena, such as blackbody radiation and the **photoelectric effect**. To account for these phenomena, the concept of quantization was introduced in 1901 by the German physicist **Max Planck**.

Planck devised his quantum hypothesis while studying blackbody radiation. The term *perfect blackbody* applies to bodies that act as perfect radiators or absorbers of **heat**. Although there are no known perfect blackbodies, by definition, if a blackbody acted as a perfect absorber, all of the light that hit the blackbody would be absorbed. The blackbody would then return to its normal energy level by emitting thermal radiation, or heat. The spectrum of this emission is called a *blackbody emission*.

The properties of blackbody radiation depend solely on the **temperature** used to excite the radiation. Early attempts to account for the distribution of the radiation using classical physics did not match the observed phenomena. Classical physics yielded a distribution that predicted that at a certain temperature, the frequency of the blackbody radiation shoots up to infinity. In 1900, Planck tackled the problem of blackbody radiation with **thermodynamics**. He assumed that radiation of a certain frequency, v, could be excited only in discrete steps of energy of magnitude hv, where h is **Planck's constant** (introduced to satisfy observable phenomena). Planck termed the discrete amounts of energy *quanta*.

Although Planck's formula for the distribution of the blackbody radiation matched the observed distribution of blackbody radiation, the introduction of the quantization of energy and of the constant h was originally thought by Planck to be nothing more than a calculation device. In 1905, however, **Albert Einstein** showed that Plank's constant was a fundamental constant of nature. Einstein used both the notion of quantization and Planck's constant, h, to explain the photoelectric effect.

Expanding on Planck's theory of blackbody radiation, Einstein assumed that light was transmitted in as a stream of particles termed photons. By treating light as a stream of photons, Einstein was able to explain the photoelectric effect.

The photoelectric effect occurs when photons of light at a certain frequency excite and cause the ejection of electrons from a metal. Regardless of intensity, light with a frequency below the photoelectric threshold—characteristic of the individual metal—will not cause electrons (also, in these circumstances termed *photoelectrons*) to be emitted from the metal. Above the photoelectric threshold, photoelectrons are emitted in proportion to the intensity of incident light.

When a **photon** strikes the surface of a metal, it raises the energy level of the metal and causes the ejection of electrons. This ejection of an **electron** only occurs if the energy overcomes the energy binding the electron to the metal. The energy required to remove an electron from the metal is called the *work function*, which is written as hv (Planck's constant multiplied by the required frequency of light). If hv is less than the work function, an electron will not be ejected; but if the frequency is large enough to overcome the work function, an electron is ejected, and the photoelectric effect is observed.

As long as the frequency of the light is greater than the work function of the metal, any photon can eject an electron. The intensity of light depends on the number of photons in the light. Accordingly, the more photons, the more intense, or brighter the light. At low intensities of light only a few collisions occur, at higher intensities more collisions and ejections occur. Each photon in a low intensity beam of light, however, carries the same energy as photons in a high intensity beam. Therefore the ejection of electrons again depends only on the frequency of light.

The work of Planck and Einstein regarding the quantization of energy, revolutionized physics and chemistry. The concept of quantization and its application to physical phenomena at the beginning of the twentieth century formed what is now called the old quantum theory.

The old quantum theory was refined by the work of **Louis-Victor de Broglie**, **Erwin Schrödinger**, and **Werner Heisenberg** in the 1920s. De Broglie proposed in his doctoral thesis (published in 1924) that particles, such as electrons, could also be described as waves. This proposal led Schrödinger to publish his wave equation in 1926. The Schrödinger wave equation, applied to a system, gives the *wave function* of the system. The wave function of a system contains information regarding the properties of the system. In quantum chemistry, the Schrödinger's wave equation is applied to chemical systems, like atoms and molecules, and his system of quantum mechanics is called wave mechanics.

Working at about the same time, Heisenberg formulated matrix mechanics, which was the first complete and self-consistent theory of quantum mechanics. Matrix mathematics was well-established by the 1920s, and Heisenberg applied this tool to quantum mechanics. In 1926, Heisenberg put forward his uncertainty principle that states that two complementary properties of a system, such as position and momentum, can never both be known exactly. This proposition helped cement the dual nature of particles (e.g., light can be described as having both wave and a particle characteristics).

Schrödinger's wave mechanics and Heisenberg's matrix mechanics initially looked quite different from each other, but they are actually mathematically equivalent. When these two systems were put together they forced a complete revision of classical physics. De Broglie, Schrödinger, and Heisenberg's development of Planck's quantum theory form what is now called the new quantum theory.

Quantum theory was firmly applied to chemistry in the late 1920s and early 1930s by **Linus Pauling**. In 1931, Pauling published a paper that used quantum mechanics to explain how two electrons, from two different atoms, are shared to make a **covalent bond** between the two atoms. Pauling's work provided the connection needed in order to fully apply the new quantum theory to chemistry.

One of the main points in quantum chemistry, proposed by Heisenberg, is that, because an electron is not a classical particle located at a definite point in space (as classical physics dictated), even a single electron can surround the **nucleus** of an **atom**. The region the electron occupies is called an orbital. Each atom can contain many orbitals, which are also called shells. These shells are all interpenetrating like ripples on a puddle and are spread around the nucleus of the atom. Some of the shells are concentrated further from the nucleus than others. Because the binding energy between the negative electrons and the positive nucleus decreases with distance, the further electrons are from the nucleus, the less energy is required to remove them from the atom.

This first orbital forms a shell around the nucleus and is assigned a principal quantum number (n) of n. Additional orbital shells are assigned values n, n, n, etc. The orbital shells are not spaced at equal distances from the nucleus and the radius of each shell increases rapidly as the square of n. Increasing numbers of electrons can fit into these orbital shells according to the formula $2n^2$. Accordingly, the first shell can hold up to two electrons, the second shell (n) up to eight electrons, the third shell (n) up to 18 electrons, Subshells or suborbitals (designated *s,p,d,* and *f*) with differing shapes and orientations allow each element a unique electron configuration.

Atoms tend to be more stable with either empty or full outer shells (some also seek half-filled shells) and will lose or gain electrons to gain those configurations. For example, **helium** has two electrons (first orbital full) and neon has 10 electrons (first and second orbitals full). Both of these elements are stable and relatively unreactive. **Hydrogen**, in contrast, has one electron; so its first orbital is not full. Because it needs another electron to fill its first orbital, hydrogen is highly reactive.

This is the basis of quantum chemistry. When Pauling used quantum mechanics to explain how two electrons from two different atoms shared their electrons to form a bond, quantum chemistry became a field unto its own. Quantum chemistry can detail the structure of all the electrons, and the way that electrons are arranged around the nucleus. Using quantum chemistry, each electron can be characterized by a set of four quantum numbers. Each quantum number describes the value of a property of the electron.

The four quantum numbers that characterize an electron are the principal quantum number, the orbital quantum number, the magnetic quantum number, and the spin quantum number. The principal quantum number, n, gives the main energy level, or the distance of the electron from the nucleus. The orbital quantum number, l, gives the angular momentum of the electron. The possible values of l range from $(n - 1)$ to 0. Therefore electrons in the first shell ($n = 1$) can only have angular momentum of zero. The magnetic quantum number, m, gives the energies of electrons in an external magnetic fields. Finally, the spin quantum number, m_s, gives the spin of the individual electrons and has values of +1/2 or -1/2. All of these quantum numbers define the quantum state of the electron.

Quantum chemistry is an exciting field of research. There are two types of quantum chemistry, computational and non-computational. Computational quantum chemistry is concerned with the numerical computation of molecular structures. Typical activities in computational quantum chemistry include calculating the electron distributions in molecules, predicting molecular geometries, and screening molecules for pharmacological activity. Noncomputational quantum chemistry is concerned with formulating expressions for the properties of molecules and their reactions. Typical problems in noncomputational quantum chemistry include analyzing features of potential energy surfaces for rate constants and other aspects of molecular collisions, analyzing the symmetry properties of molecules, and manipulating the functions that represent interactions in molecules.

See also Bohr theory; Molecule; Octet rule; Schrödinger's equation

Quarterman, Lloyd Albert (1918-1982)

American chemist

Lloyd Albert Quarterman was one of only a handful of African Americans to work on the "**Manhattan Project**," the team that developed the first **atom** bomb in the 1940s. He was also noted as a research chemist who specialized in fluoride **chemistry**, producing some of the first compounds using inert **gases** and developing the "**diamond** window" for the study of compounds using corrosive **hydrogen** fluoride gas. In addition, later in his career, Quarterman initiated work on synthetic blood.

Quarterman was born May 31, 1918, in Philadelphia. He attended St. Augustine's College in Raleigh, North Carolina, where he continued the interest in chemistry he had demonstrated from an early age. Just after he completed his bachelor's degree in 1943 he was hired by the U.S. War Department to work on the production of the atomic bomb, an assignment code-named the Manhattan Project. Originally hired as a junior chemist, he worked at both the secret underground facility at the University of Chicago and at the Columbia University laboratory in New York City; the project was spread across the country in various locations. It was the team of scientists at Columbia which first split the atom. To do this, scientists participated in trying to isolate an **isotope** of **uranium** necessary for **nuclear fission**; this was Quarterman's main task during his time in New York.

Quarterman was one of only six African American scientists who worked on the development of the atomic bomb. At the secret Chicago facility, where the unused football stadium had been converted into an enormous, hidden laboratory for the "**plutonium** program," Quarterman studied quantum mechanics under renowned Italian physicist Enrico Fermi. When the Manhattan Project ended in 1946, the Chicago facilities were converted to become Argonne National Laboratories, and Quarterman was one of the scientists who stayed on. Although his contributions included work on the first nuclear power plant, he was predominantly a fluoride and nuclear chemist, creating new chemical compounds and new molecules from fluoride solutions. Dr. Larry Stein, who worked at Argonne at the same time as Quarterman, told interviewer Marianne Fedunkiw that Quarterman was very good at purifying hydrogen fluoride. "He helped build a still to purify it, which he ran." This was part of the research which led to the production of the compound **xenon** tetrafluoride at Argonne. Xenon is one of the "inert" gases and was thought to be unable to react with other molecules, so Quarterman's work in producing a xenon compound was a pioneering effort.

After a number of years at Argonne National Laboratories, Quarterman returned to school and received his master's of science from Northwestern University in 1952. In addition to his fluoride chemistry work, Quarterman was a spectroscopist researching interactions between radiation and **matter**. He developed a **corrosion** resistant "window" of diamonds with which to view hydrogen fluoride. He described this to Ivan Van Sertima, who interviewed him in 1979: "It was a very small window—one-eighth of an inch. The reason why they were one-eighth of an inch was because I couldn't get the money to buy bigger windows. These small diamonds cost one thousand dollars apiece and I needed two for a window." Diamonds were necessary because hydrogen fluoride was so corrosive it would eat up **glass** or any other known container material. Quarterman was able to study the x-ray, ultraviolet, and Raman spectra of a given compound by dissolving it in hydrogen fluoride, making a cell, and shining an electromagnetic beam through the **solution** to see the vibrations of the molecules. His first successful trial was run in 1967.

Quarterman also began research into "synthetic blood" late in his career but he was thwarted by what he described as "socio-political problems" and later fell ill and died before he could complete it. Besides holding memberships in the American Chemical Society, American Association for the Advancement of Science, and Scientific Research Society of America, Quarterman was an officer of the Society of Applied **Spectroscopy**. He also encouraged African American students interested in science by visiting public schools in Chicago, and was a member of the National Association for the Advancement of Colored People. In recognition for his contributions to science, Quarterman's alma mater, St. Augustine's College, departed from 102 years of tradition to award him an honorary Ph.D. in chemistry in 1971 for a lifetime of achievement. He was also cited for his research on the Manhattan project in a certificate, dated August 6, 1945, by the Secretary of War for "work essential to the production of the Atomic Bomb thereby contributing to the successful conclusion of World War II."

Quarterman was also a renowned athlete. During his university days at St. Augustine's College he was an avid football player. Van Sertima, who interviewed Quarterman three years before his death, later wrote, "As he spoke, the shock of his voice and his occasional laughter seemed to contradict his illness and I began to see before me, not an aging scientist, but the champion footballer." Quarterman died at the Billings Hospital in Chicago in the late summer of 1982. He donated his body to science.

Quinine

Quinine is an **alkaloid** obtained from the bark of several species of the cinchona tree. Until the development of synthetic drugs, quinine was used as the primary treatment of malaria, a disease that kills over 100 million people a year. The cinchona tree is native to the eastern slopes of the Andes Mountains in South America. Today, the tree is cultivated throughout Central and South America, Indonesia, India, and some areas in Africa. The cinchona tree contains more than 20 alkaloids of which quinine and quinidine are the most important. Quinidine is used to treat cardiac arrhythmias.

South American Indians have been using cinchona bark to treat fevers for many centuries. Spanish conquerors learned of quinine's medicinal uses in Peru, at the beginning of the 17th century. Use of the powdered ''Peruvian bark'' was first recorded in religious writings by the Jesuits in 1633. The Jesuit fathers were the primary exporters and importers of quinine during this time and the bark became known as ''Jesuit bark.'' The cinchona tree was named for the wife of the Spanish viceroy to Peru, Countess Anna del Chinchón. A popular story is that the Countess was cured of the *ague* (a name for malaria the time) in 1638. The use of quinine for fevers was included in medical literature in 1643. Quinine did not gain wide acceptance in the medical community until Charles II was cured of the *ague* by a London apothecary at the end of the 17th century. Quinine was officially recognized in an edition of the London Pharmacopoeia as ''Cortex Peruanus'' in 1677. Thus began the quest for quinine. In 1735, Joseph de Jussieu, a French botanist, accompanied the first non-Spanish expedition to South America and collected detailed information about the cinchona trees. Unfortunately, as Jussieu was preparing to return to France, after 30 years of research, someone stole all his work. Charles Marie de la Condamine, leader of Jussieu's expedition, tried unsuccessfully to transfer seedlings to Europe.

Information about the cinchona tree and its medicinal bark was slow to reach Europe. Scientific studies about quinine were first published by Alexander von Humboldt and Aimé Bonpland in the first part of the 18th century. The quinine alkaloid was separated from the powdered bark and named ''quinine'' in 1820 by two French doctors. The name quinine comes from the Amerindian word for the cinchona tree, quinaquina, which means ''bark of barks.'' As European countries continued extensive colonization in Africa, India and South America, the need for quinine was great, because of malaria. The Dutch and British cultivated cinchona trees in their East Indian colonies but the quinine content was very low in those species. A British collector, Charles Ledger, obtained some seeds of a relatively potent Bolivian species, *Cinchona ledgeriana*. England, reluctant to purchase more trees that were possibly low in quinine content, refused to buy the seeds. The Dutch bought the seeds from Ledger, planted them in Java, and came to monopolize the world's supply of quinine for close to 100 years.

During World War II, the Japanese took control of Java. The Dutch took seeds out of Java but had no time to grow new trees to supply troops stationed in the tropics with quinine. The United States sent a group of botanists to Columbia to obtain enough quinine to use throughout the war. In 1944, synthetic quinine was developed by American scientists. Synthetic quinine proved to be very effective against malaria and had fewer side effects, and the need for natural quinine subsided. Over the years, the causative malarial parasite became resistant to synthetic quinine preparations. Interestingly, the parasites have not developed a full resistance to natural quinine.

The chemical composition of quinine is $C_2OH_24N_2O_2 \cdot H_2O$. Quinine is derived from cinchona bark, and mixed with **lime**. The bark and lime mixture is extracted with hot paraffin oil, filtered, and shaken with **sulfuric acid**. This **solution** is neutralized with **sodium** carbonate. As the solution cools, quinine **sulfate** crystallizes out. To obtain pure quinine, the quinine sulfate is treated with **ammonia**. Crystalline quinine is a white, extremely bitter powder. The powdered bark can also be treated with **solvents**, such as toluene, or amyl **alcohol** to extract the quinine. Current biotechnology has developed a method to produce quinine by culturing plant cells. Grown in test tubes that contain a special medium that contains absorbent resins, the cells can be manipulated to release quinine, which is absorbed by the resin and then extracted. This method has high yields but is extremely expensive and fragile.

Medicinally, quinine is best known for its treatment of malaria. Quinine does not cure the disease, but treats the fever and other related symptoms. Pharmacologically, quinine is toxic to many bacteria and one-celled organisms, such as yeast and plasmodia. It also has antipyretic (fever-reducing), analgesic (pain-relieving), and local anesthetic properties. Quinine concentrates in the red blood cells and is thought to interfere with the protein and glucose synthesis of the malaria parasite. With treatment, the parasites disappear from the blood stream. Many malarial victims have a recurrence of the disease because quinine does not kill the parasites living outside the red blood cells. Eventually, the parasites make their way into the blood stream, and the victim has a relapse. Quinine is also used to treat myotonic dystrophy (muscle weakness, usually facial) and muscle cramps associated with early kidney failure. The toxic side effects of quinine, called Cinchonism, include dizziness, tinnitus (ringing in ears), vision disturbances, nausea, and vomiting. Extreme effects of excessive quinine use include blindness and deafness.

Quinine also has non-medicinal uses, such as in preparations for the treatment of sunburn. It is also used in liqueurs, bitters, and condiments. The best known nonmedicinal use is its addition to tonic **water** and soft drinks. The addition of quinine to water dates from the days of British rule in India-quinine was added to water as a prevention against malaria. About 40% of the quinine produced is used by the food and drug industry, the rest is used medicinally. In the United States, beverages made with quinine may contain not more than 83 parts per million cinchona alkaloids.

QUINONE FUNCTIONAL GROUP

The **chemical compound** quinone, or 1,4-benzoquinone, is a unicyclic, or **ring** phenolic compound. Quinone is a derivative of **benzene**, with two hydrogens replaced by two oxygens, each double bonded to a ring **carbon** (Figure 1). Quinone exists as yellow crystals, with a sharp odor similar to **chlorine**

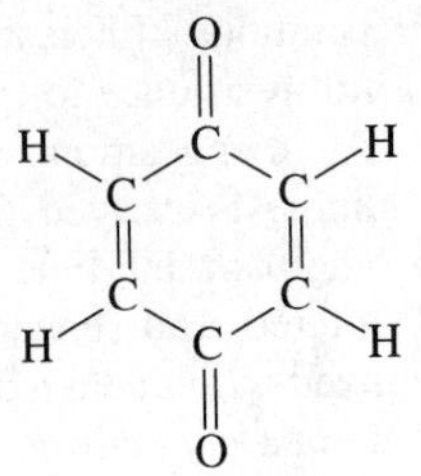

Figure 1. 1, 4-benzoquinone

Figure 1. 1,4-benzoquinone. ***(Illustration by Electronic Illustrators Group.)***

bleach. The chemical formula for quinone is $C_6H_4O_2$, and its **molecular weight** is 108.1 grams per **mole**. Quinone is used as a chemical intermediate of complex reactions, an inhibitor of polymerization, a photographic chemical, an oxidizing agent, and a tanning agent. Benzoquinone and napthoquinone dyes also exist.

When a quinone cyclic **molecule** is the active portion of a larger molecule, it is referred to as being a *quinone* **functional group**. Quinone functional groups have powerful biological activity and are found in the most basic life-sustaining chemical reactions. For example, quinone functional groups are part of molecules participating in both **cellular respiration** and **photosynthesis**. Ubiquinone, or **coenzyme** Q, is a ubiquitous electron-shuttling molecule found in nearly all cells. Coenzyme Q is an essential **electron** and **proton** carrier that functions in the production of biochemical **energy** during respiration in aerobic organisms. It also has antioxidant and membrane stabilizing properties that serve to prevent the damage to cells by **free radicals** that results from normal metabolic processes. The structure of coenzyme Q consists of a quinone ring attached to a larger side chain molecule. Similarly, in photosynthesis, **chlorophyll** and the quinones are joined together by a protein and other molecules to form the photosynthetic reaction center, where energy from sunlight is converted into chemical energy.

R

Radiation Chemistry

Radiation **chemistry** is the study of how radioactive elements interact with other materials and how these radioactive materials can be used for different processes. Radiation exists as three different types: alpha rays, beta rays, and gamma rays. Alpha rays consist of **helium**-4 nuclei (two protons and two neutrons) and have a positive charge. Beta rays consist of electrons and have a negative charge. Gamma rays are electromagnetic waves.

When a radioactive element decays it emits particles of one or more of the radiation types. As this happens the element changes its nature. As protons are lost from the **nucleus** in the form of alpha particles the element changes from one element to another. Because of the different number of elementary particles in the **atom** the element is an **isotope**. There is a gradual change of one element into an isotope of another until a stable state is reached. This is radioactive decay, where the change is brought about by loss of material from the atom. The path from element to element is known as a radioactive decay series. Each radioactive element decays at its own rate independent of the form it is in (i.e., whether it is elemental or part of a compound). If the element has an **atomic number** less than 83, then when decay takes place beta particles usually are emitted. If the atomic number is higher than 83 then **alpha particles** usually are emitted. **Gamma radiation** is only emitted in conjunction with alpha or beta particles. These types of radiation all have the ability to knock electrons out of atoms that they hit. This produces positively charged ions and so these forms of radiation are known as ionizing radiation. Ionizing radiation is capable of affecting the physical and chemical properties of the molecules with which it interacts.

When radioactive material decays it does so spontaneously and it is impossible to state when a given atom will decay. How quickly this decay occurs is measured by the half-life of the material. The half-life is the time taken for half of the material to decay to another element. The half-life can range from 10^{7} seconds to 10^{9} years. As a material decays it can change from one radioactive material to another before eventually reaching a non radioactive and stable element. An example of this can be seen in the decay of **uranium**-238. Uranium-238 emits an alpha particle and changes to **thorium**-234, which then emits a beta particle to allow it to change to **palladium**-234. Loss of another beta particle changes the palladium-234 into uranium-234. This is an example of part of a radioactive decay series. By knowing the steps in a radioactive decay series and the half-life of the materials involved the amount of **radioactivity** in a sample of material can be used to calculate the age of that material. This is known as radioactive dating and the most well known example of this is **carbon** dating, because it relies on the decay of radioactive carbon-14.

Many materials also have a radioactive form which can be used in place of the normal, non-radioactive element. Many chemical and biological pathways have been worked out using this fact. If a radioactive form of an element is placed in a compound it is possible to follow where this element is and work out the pathway followed. The radioactive isotopes used for this purpose are known as tracers. This has been employed extensively in tracing metabolic pathways through living organisms. One example of how this has been used is with **photosynthesis**. Plants have been placed into an atmosphere containing **carbon dioxide** manufactured with radioactive carbon-14. The radioactive carbon is the tracer **molecule** and the carbon dioxide has been labelled or tagged. The plant is allowed to take up the carbon dioxide as normal and then the tissues are examined to see where the radioactive carbon has been assimilated. This can be done simply by laying part of the plant onto a photographic plate. The radioactivity will affect the plate and when developed it can be seen where the radioactivity has been incorporated. This is known as an autoradiograph.

A more detailed analysis can be carried out by grinding up the tissue and then performing **chromatography**. The chromatogram produced can then be used to give an autoradiograph so the compounds that the radioactive material has been incorporated into can be readily identified. This can be carried out using several individual organisms where each one is examined at different times. This allows the changes that occur with time in a metabolic pathway to be worked out. Photosynthesis and cell respiration pathways were both worked out using this technique.

Radioisotope tracers also have many medical applications. When an organ or tissue is under investigation a radioactive tracer can be injected into the patient and allowed to get to the target organ. The absorption rate of the material is then examined. The rates are known for healthy organs and if an organ is not functioning correctly then a different rate will be detected. **Iodine**-131 can be used to study the thyroid, **iron**-59 for red blood cells, phosphorous-32 for eyes, liver and tumors, **technetium**-99 for heart, bones, liver and lungs, and **sodium**-23 for the circulatory system. All of these elements have short half lives and they are administered in doses that should not be problematic to the patient. Their subsequent detection is carried out using appropriate equipment. A Geiger counter held to the neck will give a rough indication of thyroid activity but normally more sensitive imaging is used which can enable a picture of the whole area to be built. One application of this is in positron emission tomography or PET analysis. This can be used to study blood flow, metabolic rates, and other biological processes.

Gamma radiation is used to irradiate food, this is high-**energy** radiation which kills microorganisms that can be responsible for food spoilage.

A particle accelerator can be used to bombard a target with charged particles. The addition of these particles can **lead** to the creation of a new element. Many elements at the end of the **periodic table** are very radioactive and highly unstable. They have only been created for tiny fractions of a second using particle accelerators. These short-lived particles do provide us with insights into various chemical processes, however. The stability of atoms and nuclei is being studied using this sort of research. Radioactive materials are all potential fuel sources, (gamma radiation produces **heat** to heat **water** which is used to drive turbines to generate electricity). Some of the artificially created materials may prove useful in this area.

As already stated, radiation can produce charged ions which can affect the environment around them. These can be used to accelerate chemical reactions by radioactive production of ions that are free to react with other materials. The chemical reactions that radioactive materials take part in are the same as the ones that non radioactive materials will take part in although energy supplied to the system may make reactions proceed at a faster rate. The degradation of a radioactive material by decay is known as fission, the splitting of material. The opposite process, **nuclear fusion**, where nuclei are joined together, also releases energy. For example, the power of the sun is produced by the fusion of radioactive nuclei of **hydrogen** known as **deuterium** to provide a nucleus of helium. Much energy has been expended to try to recreate this process on Earth under safe conditions to provide energy. As of 1999, this has not been successful.

See also Carbon dating

RADIOACTIVITY

Radioactivity is the process in which unstable atomic nuclei become more stable by spontaneously emitting highly energetic particles and/or **energy**. A sample of material is said to be radioactive if some of its atomic nuclei are emitting such radiation. The radiations emitted by unstable nuclei are capable of ionizing **matter** and disrupting molecules, including **DNA**; they are therefore a biological hazard in prolonged or intense exposures.

Radioactivity is important to society for two reasons. First, it is produced in large amounts by **nuclear fission** in nuclear power plants, and the safe disposal of radioactive waste is a problem. Second, radioactivity is widely used as a diagnostic and therapeutic tool in many important medical applications. It is therefore both a burden and a blessing to society.

Stable and unstable nuclei

Every atomic **nucleus** consists of a certain number of protons, strongly bound to a certain number of neutrons. Among the almost limitless number of kinds of nuclei that can be concocted by combining various numbers of protons with various numbers of neutrons, only certain combinations will be stable. The rest will be unstable or radioactive: they will spontaneously change their proton-neutron composition. Atoms that are radioactive are radionuclides, often called radioisotopes. (Nuclide is a generic term meaning ''kind of nucleus,'' just as species is used to denote a specific kind of plant or animal. A radionuclide is a radioactive nuclide.)

Making different nuclides out of protons and neutrons is similar to making different kinds of molecules out of atoms. You can't just make a **molecule** out of any old combination of atoms and expect it to hold together indefinitely, if at all, because atoms bind together according to certain rules of chemical **bonding**. For example, two **hydrogen** atoms and one **oxygen atom** form a perfectly stable molecule, H_2O. But two hydrogen atoms and two oxygen atoms make the unstable molecule H_2O_2, **hydrogen peroxide**. This compound will slowly break down all by itself into **water** and oxygen. If you try to make a molecule out of two hydrogen atoms and three oxygen atoms, it won't hold together long enough for you to name it.

Similarly, the nucleus that consists of two protons and two neutrons is absolutely stable; it is a helium-4, the **helium isotope** of **mass** number 4, symbolized ^4He. But try to use three neutrons, to make a nucleus of helium-5 (^5He), and it'll blow itself apart after only 10^{-21} seconds.

There are certain natural ''rules'' that govern how many protons and neutrons can bind together to form a stable nucleus. While these rules of nuclear binding aren't understood as deeply as are the rules for molecular bonding, their effects are well known and predictable. Nuclides that are constructed of ''rule-breaking'' numbers of protons and neutrons will be unstable to varying degrees. They will spontaneously break up, emitting particles and energy in order to change themselves into more stable nuclei. That is, they will be radioactive. This is an example of the general principle that nature always tries to increase the stability of a system, and in order to accomplish that it will use any avenue that is open to it—that is, any process that is energetically possible. In the case of unstable nuclei, there appear to be three options: the nucleus can split apart, it can emit particles, or it can emit pure energetic radiation. The general term for all of these changes is nuclear disintegration. Radioactivity, then, is any process by which unstable nuclei disintegrate in order to become more stable.

When a radioactive nucleus disintegrates by emitting an alpha or beta particle, it changes into a different nucleus—that is, one with a different number of protons and/or neutrons. In those cases in which the number of protons changes, the new nucleus has a different **atomic number**, and it therefore belongs to a different element. In other words, the radioactive atom has undergone a **transmutation** from one element to another. When this fact was first proposed at the beginning of the twentieth century, it was very difficult for the scientific world to accept because until that time the notion of changing one element into another had existed solely in the realm of **alchemy** and magic.

Any observable sample of a radionuclide will contain a huge number of atoms. As the unstable nuclei of these atoms continue to disintegrate one after the other, changing themselves into other kinds of atoms, there will be fewer and fewer of the original kind left. The more unstable that particular radionuclide is, the faster its atoms will be disappearing. Radionuclides are known that are so unstable that their half-lives are only the tiniest fractions of a second that can be measured. Others are known that are so very slightly unstable that their half-lives are trillions and quadrillions of years.

Many nuclides appear to be absolutely stable, and presumably their atoms will last forever. There are no avenues open to them, no spontaneous, energetically possible processes that could change their numbers of protons and neutrons to a more stable combination. Of the roughly 2,000 known nuclides, only 264 are stable. The rest are all radioactive in one way or another. Some of these radionuclides (such as the isotopes of **uranium** and radium) occur naturally on earth, but the vast majority of them have been made artificially in **nuclear reactors** and in particle accelerators, commonly known as ''atom smashing'' machines.

History of radioactivity

In December 1895, the German physicist Wilhelm Roentgen (1845-1923) announced his discovery of mysterious penetrating rays—he called them x rays, and we still do—that could go right through an object such as a human hand, making an image of the bones on a photographic plate placed behind the hand. Because these rays came from fluorescent (glowing) spots in Roentgen's **glass** vacuum tubes, scientists immediately began to test a wide variety of fluorescent materials—materials that glow after being exposed to light—to see if they also emitted x rays.

In early 1896, the French physicist **Henri Becquerel**, working in Paris, was examining many chemical substances that were known to be fluorescent. He first exposed them to bright sunlight to make them fluoresce and then observed whether they emitted any rays that could go through light-proof **paper** and expose a photographic plate that was wrapped inside. On one gray February day, he put some samples away in a drawer until he could expose them on the next sunny day. To his surprise, he found that these samples left an image on a photographic plate that was also stored in the drawer. Apparently, the samples were emitting some kind of penetrating radiation without even having to be exposed to light. He soon found that only compounds that contained the element uranium had this ability to emit radiation, and it didn't matter what chemical compounds the uranium atoms were in. Becquerel had discovered that uranium is radioactive.

What was startling about this discovery was that the uranium atoms were apparently a source of energy all by themselves; they didn't have to be ''activated'' by absorbing light energy or anything else. Where the uranium atoms were getting their energy was the big For more than 50 years, scientists had believed in the law of conservation of energy: that energy simply cannot come from nowhere. But they also believed strongly that atoms were unchangeable. How could uranium atoms be giving off radiation without changing?

In December 1897, **Marie Curie**, a Polish chemist working on her doctoral thesis at the Sorbonne in Paris, began to try to find out where this mysterious uranium energy was coming from. In the process, she discovered two new elements that are millions and billions of times more radioactive than uranium: **radium** and **polonium**. By inventing new chemical techniques for separating radioactive elements, Marie Curie laid the foundation for all of today's applications of radioisotopes in industry and medicine. She was the world's first radiochemist.

Marie Curie's work was an important step toward the understanding of radioactivity. However, it remained for **Ernest Rutherford** and **Frederick Soddy** to suggest in 1902 that radioactivity represents an actual disintegration of atoms. Later, Rutherford also discovered the atomic nucleus and the **proton**. In 1931 he was named Baron Rutherford of Nelson (the town in New Zealand where he was born) in recognition of his scientific achievements.

Lord Rutherford found two kinds of rays coming from the radioactive uranium atoms: a slightly penetrating kind that wouldn't even go through paper and a somewhat more penetrating kind that would go through thin sheets of metal. He called them alpha rays and beta rays, respectively. Later, a very penetrating, third kind of radiation, called gamma rays, was discovered. These three types of radiation, symbolized by the Greek letters α, β, and γ, still represent the three major types of radioactivity. We now know that alpha and beta ''rays'' are actually high-speed **subatomic particles**, while gamma rays are electromagnetic energy waves of pure energy.

Types of radioactivity

Nuclear chemists and physicists have found that certain combinations of neutrons and protons seem to make the most stable nuclei. In general, the most stable nuclei will be those that (a) contain nearly equal numbers of protons and neutrons, but with more neutrons than protons, and that (b) have an even (rather than odd) total number of protons plus neutrons. Nuclei that deviate too much from these rules will be unstable to various degrees.

The three kinds of radioactivity, alpha, beta, and gamma, come from nuclei that deviate in three different ways from the stability rules: alpha particles come from nuclei that have too many protons plus neutrons (that is, they are simply too big and heavy); beta particles come from nuclei that have too many protons or too many neutrons; and **gamma radiation** comes from nuclei that simply have too much energy.

Nuclei are known that contain up to 266 neutrons and protons, a very large number. That's a lot of particles to be packed into a **volume** that has a radius of only 10^{-12} centimeter, especially since almost half of the particles are protons—positively charged particles that, being all of the same charge, are trying hard to repel each other. So when a nucleus is too "big and fat," it tries to reduce by shooting off some of its particles. The most energetically favorable combination of particles that it can shoot off is a tight little package of two protons and two neutrons. Two protons and two neutrons, bound together, constitute a nucleus of helium-4. This nucleus, when shot out at high speed by an unstable nucleus, is called an alpha (α) particle.

Alpha particles, containing two protons, have a charge of +2. So when an **alpha particle** is shot off into the surrounding matter by a radioactive nucleus, it will interact very strongly with the negatively charged electrons in the matter. This has two effects: (1) the alpha particle tears electrons off many atoms as it passes through, that is, it ionizes many atoms, and (2) it slows down and stops very quickly because the **ionization** process uses up its energy. Alpha particles, therefore, cannot penetrate very far through matter before they slow down completely and stop. Even a sheet of paper will stop most alpha particles.

All nuclides that are heavier than **bismuth** (atomic number 83, **mass number** 209) are too heavy to be stable; they are radioactive and emit alpha particles. Among the commonly known alpha emitters are various isotopes of polonium, radium, **thorium**, uranium and all of the transuranium elements.

When a nucleus emits an alpha particle, it decreases its mass number—the total number of protons plus neutrons—by four units: it loses two protons and two neutrons. However, losing the two protons also decreases its nuclear charge or atomic number by two units, transforming it into the nucleus of a different element, two spaces to the left in the **periodic table**. For example, when a radium nucleus (atomic number 88) emits an alpha particle, it becomes a nucleus of **radon** (atomic number 86):

$^{226}Ra \rightarrow {}^{4}He + {}^{222}Rn + \text{Energy}$

(In these symbols, the superscript is the mass number, the total number of protons and neutrons in the nucleus.) The released energy is mostly in the form of kinetic (movement) energy of the alpha particles, which are emitted at speeds of around one-tenth the speed of light, depending on the particular radionuclide that is emitting them.

Alpha particles are both highly energetic and relatively highly charged, so they disrupt many atoms and molecules along their paths before they come to a stop. But they are not much of a hazard to living things because they stop so soon. They can only penetrate about a thousandth of an inch (0.03 mm) of **aluminum**, for example. Human skin will stop them, so they can't penetrate far enough to disrupt the cells in any vital organs. If alpha-emitting radionuclides are inhaled into the lungs or ingested into the stomach, however, they can do their damage locally to highly susceptible kinds of tissues, and can therefore be very dangerous. Inhaling radon gas has been blamed for many lung cancers in uranium miners, for example. There is radon gas in uranium mines because the disintegration of uranium leads to radium, which then forms radon as shown in the equation above. The radon is not yet stable, and it emits alpha particles itself.

A second means of disintegration that is open to too-heavy nuclides is spontaneous fissionspontaneous fission. Most of the transuranium elements have isotopes that disintegrate by fissioning (splitting) in addition to emitting alpha particles.

As previously stated, a nucleus must have roughly equal numbers of protons and neutrons in order to be stable. If a nucleus has too many protons for its number of neutrons, it will be radioactive. Simply shooting out an unwanted proton turns out to require more energy than the nucleus has to give, except in a few very rare cases. (Lutetium-151, which has an extremely large number of protons compared with its number of neutrons, has actually been observed to emit protons from some of its nuclei.)

If a nucleus with too many protons could transform one of them into a **neutron**, however, it would be improving its proportion of protons to neutrons by simultaneously losing a proton and gaining a neutron. And that is what it does, but because the proton has a positive charge and the neutron is neutral, the nucleus somehow has to get rid of a positive charge. It does that by creating and emitting a positronpositron, also known as a positive beta particle. For example, potassium-40, an isotope of **potassium** that constitutes 1.2% of all potassium atoms in nature (including those in our own bodies), is radioactive and emits positrons:

$^{40}K \rightarrow \beta^{+} + {}^{40}Ar + \text{Energy}$

Positrons are emitted at perhaps nine-tenths the speed of light, depending on the radionuclide that is emitting them. The positron is a light particle, identical to an ordinary **electron** except that its charge is +1 instead of -1. Theory predicts that every particle has an opposite, called an antiparticleantiparticle, and the positron is the **antiparticle** of the electronelectron. Antiparticles don't last long in our world of ordinary particles, however. As soon as a positron (or any antiparticle) meets an ordinary electron (or its ordinary counterpart), the two particles annihilate each other: they both disappear in a puff of energy.

Positrons emitted from radioactive nuclei are lightweight and have only a single unit of charge, so they don't interact as strongly with matter as alpha particles do. They can penetrate farther into matter—about a tenth of an inch of aluminum—before slowing down and annihilating. Therefore, they are also a greater hazard to humans than alpha particles are.

Another kind of radioactivity that accomplishes the same thing as positron emission is electron capture. This is the process in which a proton is converted into a neutron by the nucleus capturing a negative electron from one of the inner orbits of its atom. No particles are emitted, but some x rays are, due to the now-missing atomic electron. Radionuclides that have too many protons often disintegrate by both methods: some of the nuclei by positron emission and some by electron capture.

If a nucleus has too many neutrons in relation to its number of protons, it will try to become more stable by decreasing its number of neutrons. Ejecting a neutron is energetically unfavorable, however, except in one or two rare cases. (Lithium-11 has been observed to emit neutrons.) However, a nucleus with too many neutrons can convert one of them into a proton. To do this, it would have to find an extra positive charge, because the neutron is neutral and the proton is positively charged. But an object can also increase its positive charge by throwing out a negative charge, and that is what the nucleus does: it creates and emits a negative beta particle, which is identical to an ordinary electron except that it comes from a nucleus instead of from the outer parts of an atom.

A common emitter of negative beta particles is carbon-14, the radioactive isotope of **carbon** that is found in all living plants and animals:

$^{14}C \rightarrow \beta^- + {}^{14}N + \text{Energy}$

Negative beta particles penetrate matter in the same way as positive beta particles do, except that they don't annihilate.

A nucleus can have an unusually large amount of internal energy, just as the electrons in a whole atom can. In both cases, we say that the nucleus or the atom is excited, or in an excited state. A nucleus can find itself in an excited state when, for example, it has just been created through the disintegration of another radioactive nucleus. Just as an excited atom can dispose of its excess energy by emitting x rays, an excited nucleus can emit gamma rays. Gamma rays are electromagnetic radiation just like x rays except that they are generally of higher energy. When a nucleus emits gamma rays, the composition of its protons and neutrons does not change.

Most radionuclides that emit alpha and beta particles also emit gamma rays. This is because the nuclei into which they are converted are often created in excited states; these excited nuclei immediately get rid of their excess energy by emitting gamma rays. This is an important safety consideration because gamma rays are extremely penetrating and can cause biological damage all the way through the body. Almost any radioactive substance must be assumed to be emitting highly penetrating gamma rays, even if the substance is known to be ''only'' an alpha or beta emitter.

Radioactivity in nature

All of the elements heavier than bismuth (atomic number 83) are completely radioactive—that is, they have no stable isotopes. We still find quite a few of the elements in nature, either because they have such long half-lives that they haven't completely died out since the earth was formed some 4.5 billion years ago, or because they are constantly being produced by the disintegration of uranium or thorium, whose half-lives are indeed comparable to the age of the earth. (By an interesting coincidence, uranium-238, the principal isotope of uranium, has a **half-life** of 4.47×10^9 years, just about equal to the age of the earth. Thus, there is almost exactly half as much uranium left on earth today as there was when the earth was formed.)

There are three series of naturally occurring heavy radionuclides. Each one begins with a radionuclide that is long-lived enough to have survived since the earth was formed. By a sequence of alpha and beta disintegrations, it transforms itself into a series of other radionuclides, until it becomes a stable isotope.

One natural radioactive series begins with uranium-238 (atomic number 92), which undergoes a long sequence of disintegrations, producing radioactive isotopes of several elements until it reaches the stable isotope, lead-206. A second series of disintegrations begins with uranium-235 (half-life 7.04×10^8 years) and winds up as stable lead-207. The third series begins as thorium-232 (half-life 1.41×10^{10} years) and winds up at stable lead-208.

Along the way, these disintegration series produce radioactive isotopes of **protactinium**, thorium, **actinium**, radium, **francium**, radon, **astatine**, polonium, bismuth, **lead**, **thallium** and **mercury**. The first eight of these are those heavier-than-bismuth elements that we find in nature. Depending on the relationship between the half-lives of the radionuclides and the half-lives of their predecessors and successors in the sequences that produce them, various amounts of these elements exist on earth at the present time.

In addition to the heavy radioactive elements, 18 lighter elements have radioisotopes that are found in nature because their half-lives are long compared with the age of the earth. Two others, hydrogen-3 (tritium) and carbon-14, are constantly being produced by special processes. All of the naturally occurring radionuclides, both heavy and light, contribute to a certain amount of radioactivity to which everyone on earth is always being exposed, regardless of humanity's activities in nuclear technology.

Synthetic radioactivity

Of the roughly 1,700 radioactive nuclides that are known, only about 70 occur in nature. The rest have all been made synthetically: either they were found in the nuclear debris of man- made nuclear fission, or they have been produced in particle accelerators. During **nuclear reactions** carried out in accelerators, the numbers of protons and neutrons in atomic nuclei can be changed, thereby transforming them into different, and sometimes new, radionuclides. In fact, the elements with atomic numbers 43, 61, and 85 (technetium, **promethium**, and astatine, respectively) were unknown on earth until some of their radioactive isotopes had been produced synthetically. In addition, all of the elements with atomic numbers higher than uranium's (92) were discovered by making them synthetically in particle accelerators.

Applications of radioactivity

Both natural and synthetic radionuclides, generally referred to as radioisotopes, are of enormous value in science and industry, and in medical research, diagnosis, and therapy. Applications of radioisotopes are based primarily on two facts: radioactivity can be detected with such astounding sensitivity—the disintegration of single atoms can actually be detected—that extremely tiny amounts of radioactive material can be followed through complex biological and industrial processes by keeping track of where the radiation goes; and the radiations from radioactive materials can be used to destroy living cells, such as harmful microorganisms and human cancer cells.

RADIUM

Radium is the heaviest element in Group 2 of the **periodic table**. The elements in this group are sometimes referred to as the alkaline earth elements. Radium's **atomic number** is 88, its atomic **mass** is 226.0254, and its chemical symbol is Ra.

Properties

All isotopes of radium are radioactive with radium-226 being the most stable. It has a half life of 1,600 years. Radium is a brilliant white metal with a melting point of 1,292°F (700°C), a **boiling point** of 3,159°F (1,737°C), and a **density** of 5.5 grams per cubic centimeter. The element's chemical properties are similar to those of the other alkaline earth elements, except that it is the most reactive member of the family. For most scientists, the most interesting features of radium are not its chemical properties, but its **radioactivity**.

Occurrence and Extraction

Isotopes of radium are formed during the radioactive decay of **uranium** and are, therefore, found in all ores of uranium. However, the isotopes of radium themselves decay rather quickly, so that relatively small amounts of the element are present in the Earth's crust. By some estimates, its abundance may be no greater than 1×10^{-7} parts per million. The element can be extracted from pitchblende and other ores of uranium by a lengthy process of chemical separations. When the Curies first discovered the element, they purified less than a gram of the metal from more than seven tons of pitchblende ore.

Discovery and Naming

Radium was discovered in 1898 by **Marie Curie** and her husband Pierre. Like many other scientists of the time, the Curies had become fascinated by the discovery of radioactivity by the French physicist **Antoine-Henri Becquerel** in 1896. The Curies resolved to learn as much as they could about the source of radioactivity in pitchblende, the ore with which Becquerel originally worked. The story of the Curie's four-year struggle to purify, purify, and purify over again the components of pitchblende has now been told in popular novels, plays, and motion pictures. Eventually they discovered two new elements is the pitchblende, **polonium**, named after Madame Curie's homeland of Poland, and radium, named in honor of the newly discovered phenomenon of radioactivity.

Uses

When first discovered, radium was put to use in a number of ways, probably most importantly in the field of medicine. The radiation emitted by radium passes through the soft tissue in human bodies, but is blocked by bones and other hard material in the body. This radiation provides a way to study the body, therefore, without the need for surgery.

The initial enthusiasm for working with radium was soon tempered, however. The element itself is extremely dangerous to work with, and many investigators themselves developed cancer as a result of their exposure to its radiation. Madame Curie herself was one of the scientists who died from cancer as a result of her years of working with the element. Today, radium has virtually no important commercial applications. In fact, no more than about two kilograms (five pounds) of the element are produced each year. The greatest portion of that production is used for research purposes, although small amounts are still employed by the medical sciences.

RADON

Radon is the last member of the noble gas family, the group of elements that make up Group 18 in the **periodic table**. Radon has an **atomic number** of 86, an atomic **mass** of 222.0176, and a chemical symbol of Rn.

Properties

Radon is a colorless, odorless, tasteless gas with a **boiling point** of -79.2°F (-61.8°C) and a **density** of 9.72 grams per liter, making it about seven times as dense as air. Radon condenses to a clear, colorless liquid at its boiling point and then freezes to form a yellow, then orangish red solid. These changes provide a dramatic sight for the observer since radiation emitted by the element causes these colors to glow brilliantly.

Like other members of the noble gas family, radon is almost completely inert. In the early 1960s, a number of chemists found ways of making compounds of some noble **gases**, including radon, **xenon**, and **krypton**. In most cases, these compounds contain a noble gas and a halogen. The first compound of radon to be produced was radon fluoride (RnF). These compounds have no commercial value and are of research interest only.

Occurrence and Extraction

Radon is present in the atmosphere because it is constantly being formed during the radioactive decay of **uranium** and **radium**. However, it too decays, and its **concentration** is too small to be estimated. When needed, radon can be produced by the fractional **distillation** of liquid air.

Discovery and Naming

Radon was discovered by the German physicist **Friedrich Ernst Dorn** in 1900. Dorn's discovery came during the busy period following the discovery of **radioactivity** in 1896 by the

French physicist **Antoine-Henri Becquerel**. It was the third radioactive element to be discovered, after radium and **polonium**, first isolated by **Marie Curie** and her husband Pierre. At first, Dorn called the new element **emanation**, because of the radiation it emits. Eventually, however, the name radon became more popular because of the element's connection with *rad*ium.

Uses

Radon has relatively few commercial applications, all of them based on the radiation that it emits. For example, it is sometimes used in leak detection systems. Pipes and other kinds of containers can be filled with radon gas and then checked with a Geiger counter or other detection device. If there are any leaks in the system, radon gas will escape and be detected because of the radiation it emits.

Health Issues

Radon has become a topic of considerable environmental interest in recent years. The gas is constantly being formed by the decay of uranium and other radioactive elements in the Earth's crust. Under most circumstances, the gas escapes into the atmosphere and decays rather quickly. The radiation it emits poses no problem to any living organism.

An exception to that scenario may occur when houses, office buildings, and other structures are built on rock that contains radioactive materials. In such cases, the radon emitted during the decay of these materials escapes not into the atmosphere, but into the basement of the structure and, eventually, the structure itself.

In some respects, radon is a serious health hazard because of the radiation it emits. But radon tends to have a short half life and breaks down quickly. The problem is that it breaks down into elements that are solid, such as polonium-214, polonium-218, and lead-214. These elements are also radioactive and pose a greater threat to human health.

Scientists now believe that radon may cause as many as 20,000 cases of lung cancer each year. If so, that would make radon the second leading cause of this disease after smoking. The people most in danger from radon are those who smoke and are also exposed to radon.

Ramsay, William (1852-1916)

English chemist

The first two decades of William Ramsay's career were spent on a variety of comparatively insignificant studies, including work on the alkaloids, **water** loss in salts, the **solubility** of **gases** in **solids**, and a class of organic compounds known as the diketones. It was not until 1892 that he became engaged in the line of research for which he was eventually to win a Nobel Prize, the study of the inert gases. Those studies were to occupy Ramsay for most of the next decade and to win him worldwide fame for his participation in the discovery of five new chemical elements.

Ramsay was born on October 2, 1852, at Queen's Crescent, Glasgow, Scotland. He was the only child of **William Ramsay**, a civil engineer, and the former Catharine Robertson, who came from a family of physicians. In spite of this scientific background, young William showed no particular interest in the sciences and had a classical liberal education at Glasgow Academy.

William Ramsay.

When he entered the University of Glasgow at the age of fourteen in 1866, Ramsay chose to remain in a classical curriculum that included literature, logic, and mathematics, thinking that he might join the clergy. Over a period of time, however, his interests shifted toward the sciences and, from 1869 to 1870, he worked as an apprentice to a local chemist, Robert Tatlock. At the end of this period, Ramsay was ready to make a commitment to a career in **chemistry**, and in 1871 he enrolled in a doctoral program at the University of Tübingen under the noted organic chemist Rudolf Fittig. Ramsay received his Ph.D. from Tübingen only a year later at the early age of nineteen for a study of toluic and nitrotoluic acids.

After receiving his degree, Ramsay returned to Glasgow and became a research assistant at Anderson's College (later the Royal Technical College). At Anderson's, Ramsay's work dealt primarily with **organic chemistry**, especially with compounds related to **quinine** and cinchonine. Six years later, in 1880, he was appointed professor of chemistry at University

College, Bristol (later, Bristol University). During his seven years at Bristol, Ramsay worked with an assistant, Sydney Young, on the relationships between the physical properties of a liquid and the liquid's **molecular weight**.

Ramsay's appointment in 1887 as professor of chemistry at University College, London, marked a turning point in his career. For a few years he continued to work on a variety of problems, such as surface tension, the metallic compounds of ethylene, and the atomic weight of **boron**. But then, in late 1892, Ramsay was confronted with a puzzle that was to captivate him. That puzzle went back to a discovery made by **Henry Cavendish** in 1785. Cavendish had found that the compete removal of **oxygen** and **nitrogen** from a sample of air still left a small bubble of some additional unknown gas. The puzzle was confounded by the work of **Robert Strutt** (Lord Rayleigh) in the late 1880s that showed the **density** of nitrogen to be slightly different depending on whether the gas came from air or from a compound of nitrogen.

Ramsay decided to resolve this dilemma. He began by removing all of the nitrogen and oxygen from a sample of air by burning **magnesium** metal (which reacts with both) in the air. He found a small bubble of gas, similar to that reported by Cavendish a century earlier. But then Ramsay took an additional step that Cavendish could not have taken: he did a spectroscopic analysis of the gas bubble. The result of that analysis was a set of spectral lines that had never been seen before—the gas bubble was clearly a new element. Because of the inertness of the element, Ramsay suggested the name **argon** for the element, from the Greek *argos,* for "lazy."

The discovery of argon immediately posed new research possibilities. Determination of the element's atomic weight placed it between **chlorine** and **potassium** in the **periodic table**. Clearly the element was located in a new column in the table, a column that **Dmitry Ivanovich Mendeleev** could never have imagined when he proposed the periodic law in 1869. The challenge that Ramsay recognized was to locate other members of this new chemical family, those that made up column "0" (or column 18) in the periodic table.

Shortly after announcing the discovery of argon, Ramsay heard about another inert gas that had been discovered by the American chemist William Hillebrand. To see if Hillebrand's gas might also be argon, Ramsay heated a sample of the mineral clevite in **sulfuric acid** and had the gas produced tested by spectroscopic analysis. The results of that analysis showed that the gas was *not* argon, but it did have the same spectral lines as those of an element discovered in the sun in 1868 by Pierre Janssen and Joseph Lockyer, an element they had named **helium**. Ramsay's research showed that helium also existed on the Earth.

Over the next few years, Ramsay looked for the remaining missing inert gases in various minerals, always without success. Then in 1898 he decided on another approach. He and a colleague, Morris Travers, prepared fifteen liters of liquid argon, which they then allowed to evaporate very slowly. Eventually they identified three more new gases, **krypton**, **neon**, and **xenon**, which they announced to the world on June 6, June 16, and September 8, 1898, respectively.

Ramsay remained at London until his retirement in 1912. During the last decade of his tenure there, he became increasingly interested in **radioactivity**. Among his discoveries in this field was one made with **Frederick Soddy** in 1903, namely that helium is always a product of the radioactive decay of **radium**. This discovery was later explained when it was found that the alpha particles emitted by a radioactive substance are actually positively charged helium ions. In conjunction with Robert Whytlaw-Gray, Ramsay also determined the atomic weight of the one inert gas in whose discovery he was not involved, **radon**.

Ramsay was married to Margaret Buchanan in August, 1881; they had two children. After the outbreak of World War I, Ramsay attempted to carry on chemical research for military applications, but his health failed rapidly and he died on July 23, 1916, at his home in Hazlemere, Buckinghamshire, England. In addition to the 1904 Nobel Prize in chemistry for his discovery of the **rare gases**, Ramsay was awarded the 1895 Davy Medal of the Royal Society and the 1903 **August Wilhelm von Hofmann** Medal of the German Chemical Society. He was made a fellow of the Royal Society in 1888 and was knighted in 1902.

RAO, C. N. R. (1934-)

Indian chemist

An Indian professor of **chemistry**, C. N. R. Rao has been instrumental in the worldwide research into **superconductivity**. Superconductivity occurs when certain **metals** experience a total loss of electrical resistance, turning them into superconductors capable of carrying currents without any loss of **energy**. Electrical transmission through wires normally involves a substantial loss of energy; with superconductivity, this transmission could be vastly improved, saving costs. So far, superconductivity has occurred only at extremely cold temperatures, barring its use in commercial applications. Scientists have for years been working to create the phenomenon at normal temperatures.

Chintamani Nagesa Ramachandra Rao was born on June 30, 1934, in Bangalore, India, the son of Hanumantha Nagesa and Nagamma Nagesa Rao. In 1953 he earned a master's degree from Banares Hundu University; in 1958, a doctor of philosophy degree from Purdue University. In 1958 he became a research chemist at the University of California at Berkeley, returning to India in 1959 to work as a lecturer at the Indian Institute of Science in Bangalore. In 1960 he married Indumati. They have two children, Suchitra and Sanjay.

From 1963–76, Rao was a professor of chemistry at the Indian Institute of Technology in Kanpur. He served as head of the chemistry department from 1964 to 1968, and was dean of research for three years. He was chairman of the Solid State and Structural Chemistry Unit and Materials Research Laboratory at the Indian Institute of Science in Bangalore between 1976–84. Since 1984, Rao has been the director of the Institute of Science. Concurrent with his academic positions in India, Rao was a visiting professor at Purdue University in 1967–68, at Oxford University in 1974–75, and he held a fellowship at King's College of Cambridge University in 1983.

Since its discovery in 1911 by Heike Kamerlingh-Onnes, scientists had considered superconductivity to be a laboratory curiosity, able to be produced only at temperatures approaching absolute zero. In 1986, however, physicists J. George Bednorz and K. Alex Müller discovered an **alloy** that was superconductive at 30°K, a much higher **temperature** than previously known. This discovery led a number of scientists to examine the question. In 1987, Paul Ching-Wu Chu found an alloy that was superconductive at an even higher temperature, 95°K.

In 1989 three researchers at Purdue University—Jurgen Honig, Zbigniew Kąkol and Józef Spałek —discovered a superconductive material that did not contain **copper** as part of the alloy. All previous materials were copper-oxide based; the Princeton researchers used **nickel oxide**, the first time such a compound had been successfully utilized as a superconductor. While initial results of the experiments were still being analyzed, and the crystalline structure of the compound was still a mystery, Rao conducted similar tests with nickel oxide compounds at the Indian Institute of Science and confirmed their superconductivity. His research on the chemical properties of superconductive materials resulted in the publication of three books, *Chemical and Structural Aspects of High Temperature Superconductors,* 1988, *Bismuth and Thalium Cuprate Superconductors,* 1989, and *Chemistry of High Temperature Superconductors,* 1991. During his career Rao has published more than 25 books and 700 research papers.

Rao has received numerous awards for his contributions to chemistry, including the Marlow Medal of the Faraday Society of London in 1967, the Jawaharlal Nehru fellowship in 1973, the American Chemical Society Centennial foreign fellow in 1976, the Indian Chamber of Commerce and Industry Award for Physical Sciences in 1977, the Royal Society of Chemistry (London) Medal in 1981, the Padma Vibhushan Award from the President of India in 1985, the General Motors Modi Award in 1989, and the Heyrovsky gold medal from the Czechoslovak Academy of Sciences in 1989.

In addition, Rao has received honorary degrees from many universities, including Purdue University in 1982, the University of Bordeaux in 1983, and the University of Wroclaw (Poland) in 1989. For two years, 1985–87, he served as president of the **International Union of Pure and Applied Chemistry (IUPAC)**, a board of chemists who decide such issues as rules for naming new chemical compounds. Rao is an elected foreign member of the Slovenian Academy of Sciences, the Serbian Academy of Sciences, the American Academy of Arts and Sciences, the Russian Academy of Sciences, the Czechoslovak Academy of Sciences, and the Polish Academy of Sciences. He was a founding member of the Third World Academy of Sciences.

RARE EARTH ELEMENTS

Most elements of the **periodic table**, and in particular the metallic elements, are divided into groups based primarily on such chemical properties as **reactivity**, **oxidation** number, or the formulae of binary compounds. Between 1751, when the Swedish mineralogist A. F. Cronstedt discovered a heavy mineral that was later found to contain the element **cerium**, and about 1920, the designation *rare earth* was given to any unfamiliar, but naturally occurring metal **oxide**. However, the name used to describe this group of elements is now both inappropriate and a source of confusion: inappropriate because the elements are no longer rare, with the exception of **promethium** (^{147}Pm, Z=61) which has no stable isotopes. Part of the confusion lies in determining which elements should belong to the group. The number of elements that are included in the rare earth group varies from 14 to 32 depending on the literature citation. It is generally accepted that the elements in which the $4f$ sublevel is filling be identified as rare earth elements. This block of elements (Ce, Z=58 to Lu, Z=71) are also known as the Lanthanides. Interestingly, this list does not include **lanthanum** (La, Z=57) for which the $4f$ block of elements is named. Because of the similar chemical and physical properties among the 3A group elements (Sc, Y, La, but not Ac) and the Lanthanides, these 17 elements are collectively referred to as the rare earth elements.

Separation and identification of the rare earth elements is complicated by the remarkable similar physical properties of the Lanthanide elements and **yttrium**. From minerals discovered near Ytterby, Sweden by J. Gadolin and by Cronstedt nearly all of the rare earth elements were isolated. The difficulty in separating this group of **metals** stems from the both the common+3 **oxidation state** and the metal **ion** radius. Thus, minerals containing these elements are mixtures rather than simple stoichiometric compounds. Classical means of separation were based on the small differences in basicity and **solubility**. In order for these methods to be effective, the separation was repeated many times (usually a thousand times or more). The American chemist C. James performed 15,000 recrystalizations before pure **thulium** bromate ($Tm(BrO_3)_3$) was obtained. Only cerium, which can by oxidized to Ce^{4+} and **europium**, which can be reduced to Eu^{2+} may be separated based on chemical properties. Modern means of separation, including ion-exchange **chromatography** and liquid-liquid extraction techniques, were developed during World War II when it became necessary to find continuous flow methods to remove the rare earth elements from **uranium** and **thorium** ores. The existence of promethium was predicted by Branner in 1902 and confirmed by H. Moseley in 1914 with his work on atomic numbers. Finally in 1945, J. Marisnsky, L. Glendenin, and C. Coryell isolated promethium from uranium fission products purified by ion-exchange chromatography.

Chemical properties of the rare earth elements arise from the **electron** configuration. Without exception, the 3+ oxidation state is the most common and it arises from loss of the $6s$ and $5d$ electrons. For the Group 3A metals ions, the electron configuration is the same as the previous noble gas, that is, Sc^{3+} and **argon** are isoelectronic. The Lanthanide metals in general have the $[Xe]6s^25d^04f^n$ (where n=1 to 14). There are three exceptions to this electronic configuration: (a) cerium, for which the sudden contraction and decrease in **energy** of the $4f$ orbitals immediately after lanthanum is not yet sufficient so the $5d$ orbital is occupied with one electron; (b) **gadolinium**, having a half-filled $4f$ orbital; and (c) **lutetium**, having a completely filled 4*f* orbital. The gadolinium and lutetium exceptions result in a marked increase in radius compared to the slight decrease in metal **atom** radius for the other elements. Upon loss of three electrons, one finds a regular decrease in ionic radius (La^{3+}, [Xe]$4f^0$, 103.2 pm to 86.1 pm for Lu^{3+}, [Xe]$4f^{14}$) known as the "lanthanide contraction". The regular decrease is a result of an imperfect balancing of the increased nuclear charge with the directional shielding of the seven *f* orbitals.

Throughout the nineteenth century, known sources of minerals that contained the rare earth elements were scarce. This changed soon after the Austrian chemist, C. von Welsbach, perfected the thoria gas mantle that improves the light output of coal-gas flame and to this day has not been improved significantly. The cotton or silk fibers of the gas mantle are soaked in an aqueous **solution** that is 99% ThO_2 and 1% CeO_2, present as a catalyst. Additional sources of thorium were necessary due to the commercial success of the thoria gas mantle. A worldwide search for the mineral monazite, the principle source of thorium oxide, found that it was more plentiful than first thought. This ore usually contains less than 12% thorium oxide, however, oxides of the various Lanthanide elements account for about 45% of the **mass**. With the increased availability of the rare earth elements, though not yet pure metals, applications for their use were sought. A pyrophoric material known as "misch metal" contains 50% Ce, 25% La, and 25% of the other Lanthanide metals and is used in flints for lighters and other applications where an open flame is not desirable. A new generation of permanent magnets containing the rare earth elements **samarium** (Sm, Z=62) and **neodymium** (Nd, Z=60) are being constructed because these magnets display the biggest magnetic energy stored in a unit **volume**. This allows significant miniaturization for use in loud speakers, motors, and toys. Because *f-f* electronic transitions are forbidden by the LaPorte selection rule and long luminescence lifetimes (in the micro to millisecond range) applications of Lanthanide luminescence are found in the medical, telecommunications fields and in laser technology.

RARE GASES

The rare **gases**, also known as the noble gases, are a group of six gaseous elements found in small amounts in the atmosphere: **helium** (He), **neon** (Ne), **argon** (Ar), **krypton** (Kr), **xenon** (Xe), and **radon** (Rn). Collectively, they make up about 1% of the earth's atmosphere. They were discovered by scientists around the turn of the century, and because they were so unreactive, they were initially called the inert gases.

Helium was the first of the rare gases to be discovered. In fact, its discovery is unique among the elements since it is the only element to be first identified in another part of the solar system before being discovered on earth. In 1868, Pierre Janssen (1824-1907), a French astronomer, was observing a total solar eclipse from India. Janssen used an instrument called a spectroscope to analyze the sunlight. The spectroscope broke the sunlight into lines that were characteristic of the elements emitting the light. He saw a previously unobserved line in the solar spectrum, indicating the presence of a new element that Janssen named helium after the word *helios* (Greek for sun). A quarter of a century later, **William Ramsay** studied gases emitted from radioactive **uranium** ores. With help from two British experts on **spectroscopy**, **William Crookes** and Norman Lockyer (1836-1920), the presence of helium in earth-bound minerals was confirmed. Shortly thereafter, helium was also detected as a minor component in the earth's atmosphere.

The discovery of the remaining rare gases is credited to two men, Ramsay and Lord Rayleigh (1842-1919). Beginning in 1893, Rayleigh observed discrepancies in the **density** of **nitrogen** obtained from different sources. Nitrogen obtained from the air (after removal of **oxygen**, **carbon** dioxide, and **water** vapor) always had a slightly higher density than when prepared from a chemical reaction (such as heating certain nitrogen-containing compounds). Ramsay eventually concluded that the nitrogen obtained from chemical reactions was pure, but nitrogen extracted from the air contained small amounts of an unknown gas that accounted for the density discrepancy. Eventually, it was realized that there were several new gases in the air. The method used to isolate these new gaseous elements involved liquefying air (by subjecting it to high pressure and low temperature) and allowing the various gases to boil off at different temperatures. The names given to the new elements were derived from Greek words that reflected the difficultly in isolating them: Ne, *neos* (new); Ar, *argos* (inactive); Kr, *kryptos* (hidden); Xe, *xenon* (stranger). Radon, which is radioactive, was first detected as a gas released from **radium**, and subsequently identified in air. Ramsay and Rayleigh received Nobel Prizes in 1904 for their scientific contributions in discovering and characterizing the rare gases.

The rare gases form group 18 of the **Periodic table** of elements. This is the vertical column of elements on the extreme right of the Periodic table. As with other groups of elements, the placement of all the rare gases in the same group reflects their similar properties. The rare gases are all colorless, odorless, and tasteless. They are also monatomic gases, meaning that they exist as individual atoms.

The most noticeable feature of the rare gases is their lack of chemical **reactivity**. Helium, neon, and argon do not combine with any other atoms to form compounds, and it has been only in the last few decades that compounds of the other rare gases have been prepared. In 1962 **Neil Bartlett**, then at the University of British Columbia, succeeded in the historic preparation of the first compound of xenon. Since then, many xenon compounds containing mostly **fluorine** or oxygen atoms have also been prepared. Krypton and radon have also been combined with fluorine to form simple compounds. Because some rare gas compounds have powerful oxidizing properties (they can remove electrons from other substances) they have been used to synthesize other compounds.

The low reactivity of the rare gases is due to the arrangement of electrons in the rare gas atoms. The configuration of electrons in these elements makes them very stable and therefore unreactive. The reactivity of any element is due, in part, to how easily it gains or loses electrons, which is necessary for an **atom** to react with other atoms. The rare gases do not readily do either. Prior to Bartlett's preparation of the first xenon compound, the rare gases were widely referred to as the inert gases. Because the rare gases are so unreactive, they are harmless to living organisms. Radon, however, is hazardous because it is radioactive.

Most of the rare gases have been detected in small amounts in earth minerals and in meteorites, but are found in greater abundance in the earth's atmosphere. They are thought to have been released into the atmosphere long ago as by-products of the decay of radioactive elements in the earth's crust. Of all the rare gases, argon is present in the greatest amount, about 0.9 percent by **volume**. This means there are 0.2 gal (0.9 l) of argon in every 26.4 gal (100 l) of air. By contrast, there are 78 liters of nitrogen and 21 liters of oxygen gas in every 26.4 gal (100 l) of air. The other rare gases are present in such small amounts that it is usually more convenient to express their concentrations in terms of parts per million (ppm). The concentrations of neon, helium, krypton, and xenon are, respectively, 18, 5, 1, and 0.09 ppm. For example, there are only 1.32 gal (1.5 l) of helium in every million liters of air. By contrast, helium is much more abundant in the sun and stars and consequently, next to **hydrogen**, is the most abundant element in the universe. Radon is present in the atmosphere in only trace amounts. However, higher levels of radon have been measured in homes around the United States. Radon can be released from soils containing high concentrations of uranium, and can be trapped in homes that have been weather sealed to make heating and cooling systems more efficient. Radon testing kits are commercially available for testing the radon content of household air.

Most of the rare gases are commercially obtained from liquid air. As the **temperature** of liquid air is raised, the rare gases boil off from the mixture at specific temperatures and can be separated and purified. Although present in air, helium is commercially obtained from **natural gas** wells where it occurs in concentrations of between one and seven percent of the natural gas. Most of the world's helium supplies come from wells located in Texas, Oklahoma, and Kansas. Radon is isolated as a product of the radioactive decay of radium compounds.

The properties of each rare gas dictate its specific commercial applications. Because they are the most abundant, and therefore the least expensive to produce, helium and argon find the most commercial applications. Helium's low density and inertness make it ideal for use in lighter-than-air craft, such as balloons and blimps. Although helium has nearly twice the density of hydrogen, it has about 98 percent of hydrogen's lifting power. A little over 324.7 gal (1,230 l) of helium lifts 2.2 lbs (one kg). Helium is also nonflammable and therefore considerably safer than hydrogen, which was once widely used in gas-filled aircraft. Liquid helium has the lowest **boiling point** of any known substance (about -452°F; -269°C) and therefore has many low-temperature applications in research and industry. Divers breathe an artificial oxygen-helium mixture to prevent gas bubbles forming in the blood as they swim to the surface from great depths. Other uses for helium have been in supersonic wind tunnels, as a protective gas in growing **silicon** and **germanium** crystals and, together with neon, to make gas **lasers**.

Neon is well known for its use in neon signs. **Glass** tubes of any shape can be filled with neon and when an electrical charge is passed through the tube, an orange-red glow is emitted. By contrast, ordinary incandescent light bulbs are filled with argon. Because argon is so inert, it does not react with the hot metal filament and prolongs the bulb's life. Argon is also used to provide an inert atmosphere in welding and high-temperature metallurgical processes. By surrounding hot **metals** with inert argon, the metals are protected from potential **oxidation** by oxygen in the air. Krypton and xenon also find commercial lighting applications. Krypton can be used in incandescentlight bulbs and in fluorescent lamps. Both are also employed in flashing stroboscopic lights that outline commercial airport runways. Because they emit a brilliant white light when electrified, they are also used in photographic flash equipment. Due to the radioactive nature of radon, it has found medical applications in radiotherapy.

RATNER, SARAH (1903-)

American biochemist

Sarah Ratner is a biochemist whose research has focused on **amino acids**, the subunits of protein molecules. Her use of **nitrogen** isotopes to study **metabolism**—the chemical processes by which **energy** is provided for the body—resulted in the discovery of argininosuccinic acid, a substance formed by a sequence of reactions that take place in the liver. Ratner's awards for her work include the Carl Neuberg Medal from the American Society of European Chemists in 1959.

Ratner was born in New York City on June 9, 1903, the daughter of Aaron and Hannah (Selzer) Ratner. She received her bachelor of arts degree from Cornell University before proceeding to Columbia University for graduate studies, where she received an M.A. in 1927. Ratner worked as an assistant in **biochemistry** in the College of Physicians and Surgeons of Columbia University until she received her Ph.D. in biochemistry from the university in 1937. Following her graduation she was appointed a resident fellow at the College of Physicians and Surgeons and rose to the position of assistant professor. In 1946 she became an assistant professor of pharmacology at the New York University College of Medicine in New York City. Later, she became associated with the New York City Public Health Research Institute as an associate member of the division of **nutrition** and physiology and became a member of the department of biochemistry in 1957.

In her research Ratner used an **isotope** of nitrogen to study chemical reactions involving amino acids, particularly arginine. Isotopes are atoms of an element that have a different atomic **mass** than other atoms of the same element. Through her studies she discovered an intermediate **molecule**, called argininosuccinic acid, which forms when the amino acid citrulline is converted to arginine. Ratner determined that argininosuccinic acid plays an important role in the series of chemical reactions that occurs in the liver and leads to the formation of urine. This sequence of reactions is known as the **urea** cycle. Urea, a product of protein metabolism, has a high nitrogen content and is excreted by mammals.

The American Chemical Society honored Ratner with the Garvan Medal in 1961, and in 1974 she was elected to the National Academy of Sciences. In addition, she received research grants from the National Institutes of Health (NIH) for over twenty years, and from 1978 to 1979 she was the institutes' Fogarty Scholar-in-Residence and served as a member of the advisory council. She has received honorary doctorates from the University of North Carolina-Chapel Hill, Northwestern University, and State University of New York at Stony Brook.

REACTION MECHANISM

The term mechanism describes the series of steps by which an overall reaction occurs. The study of mechanisms of chemical reactions is one of the most obvious cases in science of the formation of an hypothesis, gathering of evidence, testing of tentative conclusions, positing of a theory and revision when new data are brought to light, an important illustration of the way science works. Chemists who attempt to deduce mechanisms, just as those who study chemical structures or equilibria, must be willing to give up their cherished conclusions when new, reliable data are presented.

From many studies, some important classes of mechanisms have been deduced. These serve as a guide for kineticists and provide us with insight about how reactions occur. Many "simple" reactions actually occur by one of these multi-step mechanisms.

Rate Determining Step

If an overall reaction consists of several steps, it is unlikely that every step proceeds at the same rate. In fact, in most cases, one step is much slower than all of the others. This slow step is the rate determining step, because the amount of product that can be formed by an overall process in a given period of time is limited by the slowest step in the series. The rate determining step is the target of study of the kineticist, who changes parameters such as concentrations of reactants and **temperature** to see the effects such changes have. All other steps proceed so rapidly that they do not affect the overall rate of the reaction. Only speeding up or slowing down the rate determining step changes the rate. The rate law for an overall reaction contains the concentrations of all species that participate in the rate determining step.

As we shall see, some reactions have more than one pathway by which they proceed. In this case, there will undoubtedly be rate determining steps along each pathway.

Consecutive Reactions

A consecutive reaction mechanism involves a series of steps that directly follow each other without any "branches." The product from one step is used in the following step. The chemical species that are formed between the initial reactants and the product are intermediates. For an overall reaction: $A + B \Rightarrow$product. There can be an intermediate stage that is reversible. For example, in reactions involving enzymes the formation of an enzyme-substrate complex is a rapidly reversible equilibrium, but the second step in which the newly formed product separates from the enzyme-substrate complex is essentially irreversible. This situation is symbolized by the equation: $E + S \Leftrightarrow ES \Rightarrow E + P$.

Parallel Reactions

Parallel mechanisms are a common type of mechanism involving two series of reactions that occur at the same time. Often the conditions under which the reaction is run, such as temperature or **pH**, determines which of the paths is most important. The observed rate law is the sum of the rate laws for each path. A rate law with separate terms is typical of reactions with parallel paths.

An interesting class of reactions is that in which a substance can be decomposed by more than one means. Many materials can be decomposed both by exposure to ultraviolet radiation from sunlight and by **oxygen** in the air. The rate expression describing the decomposition in this case has terms involving ultraviolet light intensity and oxygen **concentration**.
Chain Reactions

A **chain reaction** mechanism is a combination of consecutive series and parallel mechanisms. As the reaction proceeds, "branches" occur where the reaction can proceed by several paths, and there is also usually more than one way the reaction can be stopped. Chain mechanisms are involved in reactions in the atmosphere, including those responsible for the production and destruction of ozone, in the industrial production of commercial polymers, and in **nuclear reactors**.

The reaction that starts the reaction is an initiation reaction. Reactions that produce products that cause further reaction to occur are called chain propagating reactions. Reactions that consume species that are required to allow the reaction sequence to proceed are called termination steps.

A chain reaction that has received a lot of attention in recent years is the reaction by which Freons (chlorofluorocarbons) deplete the ozone layer that protects us from ultraviolet radiation. The culprit in this reaction is the **chlorine atom**, which acts as a catalyst in the propagation steps. The chlorine atom is produced by the decomposition of a Freon, such as Freon-12, dichlorodifluoromethane, by the very energetic light in the upper atmosphere: CCl_2F_2 + light$\Rightarrow$ $CClF_2$ + Cl Freon-12. The chlorine atoms react with ozone to produce oxychloride: $Cl + O_3 \Rightarrow ClO + O_2$. The oxychloride then decomposes and the chlorine atom is reformed: ClO + light$\Rightarrow$O + Cl. Both oxygen atoms and chlorine atoms can react with ozone: $O + O_3 \Rightarrow 2O_2$ and $Cl + O_3 \Rightarrow ClO + O_2$. The reaction continues until the chlorine atom meets another chlorine atom instead of an **ozone molecule**. This is a rare event because there are fewer chlorine atoms **ion** the upper atmosphere than ozone molecules, but the chlorine atom may also react with some other scavenger. termination: $Cl + Cl \Rightarrow Cl_2$.

Efforts to decrease the amount of freon **gases** released in the atmosphere have led to international agreements about

Half-lives of Some Radioactive Isotopes.

Radionuclide	Half-life (Days)	Radionuclide	Half-life (Days)
3H	4.50×10^3	90Sr	1.00×10^4
14C	2.09×10^6	99Mo	2.79
32P	14.3	99mTc	0.250
35S	87.1	99Tc	7.70×10^6
42K	0.52	109Pd	0.570
45Ca	16.4	111In	2.81
47Ca	4.90	129I	6.30×10^9
59Fe	45.1	131I	8.00
57Co	270	135I	0.280
72Ga	0.59	207T1	3.33×10^{-3}
58mCo	0.38	207Bi	1.53×10^{-3}
58Co	72.0	226Ra	5.84×10^5
60Co	1.9×10^3	235U	2.60×10^{11}
64Cu	0.538	236U	8.72×10^{-5}
67Cu	2.58		

(Illustration by Electronic Illustrators Group.)

the use and proliferation of these materials. Loss of even a few percent of the ozone layer could lead to thousands of additional cases of skin cancer per year.

The mechanisms of overall reactions typically consist of one of a few types: consecutive reactions occur as a linear series of reactions; parallel mechanisms feature reactants that are consumed or products that are formed by independent reactions occurring simultaneously; chain reactions typically involve combinations of consecutive and parallel reactions. The mechanistic description of chain reactions includes initiation, propagation, inhibition and termination steps.

REACTION RATE

The rate of a chemical reaction is typically specified as the rate of formation of a particular product of the reaction or the rate of disappearance of a particular reactant. These two rates are the same for some reactions, such as those in which the reaction of every **molecule** of a particular reagent leads to the production of one molecule of a particular product. Many reactions are more complex and are characterized by different rates for the different species involved in the reaction.

In order to be able to predict the rate of a reaction, the affect of changing the concentrations of each species involved in the reaction is determined experimentally. The experimental information is summarized in an equation called the rate law for the reaction. It expresses the change in **concentration** of a particular species as a function time in terms of the concentration, or concentrations, of species involved in the reaction. The rate law is usually written in terms of concentrations that can be determined experimentally. The rate law equation also contains one or more rate constants that are multiplied by the concentration terms. The rate constants account for the probability that the specific reaction takes place under the conditions the reaction rate are measured, the more probable and faster the reaction, the higher the value of the rate constant. The units of the rate constant are determined by the concentration terms of the rate law and the time units used.

A term that is closely related to reaction rate is the ''half- life'' of a reaction. The term **half-life** is very commonly used in reference to radioactive decay reactions, but it is a useful concept for other reactions as well. The half-life of a reaction is the time required to convert half the initial concentration of a reagent to product. The shorter the half-life, the faster the reaction rate. The half-life is given the symbol $t_{1/2}$. For a first order reaction the half- life is easily calculated: it is equal to the natural logarithm of 2 ($\ln_2 = 0.693$) divided by the rate constant. For second or higher order reactions the calculation is considerably more complex.

To alter the rate of a reaction, the kineticist typically determines the step of a reaction that is the slowest, the rate determining step. Once the rate determining step is identified, reaction conditions can often be modified to make it proceed faster and thereby speed up the entire reaction.

One method of speeding up reactions is to use a catalyst, a species that increases the rate of a chemical reaction but is not changed by the process. Ideally, the catalyst can be recovered in its original form when the reaction is over. The catalyst increases the rate of a step by providing an alternate route, allowing the overall reaction to be accomplished without proceeding through the slow, rate-determining step.

Radioactive isotopes are typically characterized by their half-lives. Isotopes with short half-lives (hours) are often useful for medical diagnosis, while isotopes with long half-lives are useful for archaeological and geological dating and as time standards. The half-life is used to determine how many nuclear transformations occur per **mole** and, hence, per gram of the **isotope**. The Table gives the half-lives for several important radionuclides. Extensive tables of half-lives and the corresponding nuclear processes are available in reference sources such as the *Handbook of Chemistry and Physics* published by the Chemical Rubber Company each year.

Rates of reactions are affected by factors other than concentration: they are also often highly dependent on **temperature** and many reactions can be affected by catalysts. Chemical reactions require a minimum or threshold **energy**, called the **activation energy**. At higher temperature more molecules have an energy at or above this activation energy. Hence the probability that a reaction will occur, and the reaction rate, increases with an increase in temperature.

The rate of a reaction can also be increased by use of an appropriate catalyst. A catalyst provides a different pathway for a reaction that has a lower activation energy. Hence, at a particular temperature more molecules will have the lower activation energy available through the catalytic pathway to products and the reaction via the catalytic pathway will be faster. Although the catalyst must be intimately involved in the reaction as part of the activation process, it is not consumed in the process.

In complex systems in which there are many possible reactants, a catalyst for one reaction may become a reactant for another and will, therefore, be consumed by the other reaction. This situation occurs in catalytic converters in automobiles when contaminants in **gasoline** react with the catalyst and ''poison'' it, rendering the converter inefficient or useless.

REACTIONS, TYPES OF

The number of possible chemical reactions is so vast that chemists have divided them into categories. Although there is no single, universally recognized classification schemes, there are relatively few classification schemes and some very important names of types of reaction are used generally. Most chemists divide reactions into groups based on structural features of molecules and ions that allow them to predict the products that would result if a particular **molecule** or **ion** undergoes a class of reaction. Often a chemist can use the classification of reactions and information about a molecule or ion's structure to predict how readily it would react with another molecule or ion. Five general types of reactions cover most situations: acid-base, oxidation-reduction (redox), addition, substitution, elimination, and rearrangement (also called metathesis) reactions. The first two are commonly used categories, but chemists divide other reactions in several different ways. Some of the other ways of classifying of reactions that area common include complexation, **hydrolysis** and polymerization reactions.

Any chemist's list of important types of reactions would include acid-base reactions. There have been several different definitions of **acids and bases**, but the most useful definitions of acids and bases comes from Brønsted-Lowry theory, which defines acid-base reactions as **proton** transfer reactions and views acids as proton donors and bases as proton acceptors. The Brønsted-Lowry theory was proposed in 1923 by two chemists working independently, **Johannes Nicolaus Brønsted** in Denmark and T. M. Lowry (1874-1936) in England.

Materials have been classified as acids and bases from the time of the great expansion of the Moslem Empire 1,300 years ago. The familiar term alkaline, which refers to basic substances, is derived from Arabic, as is the term alchemist, which originally referred to someone skilled in the art of producing useful substances such as **sulfuric acid**. Despite the recognition and use of acids and bases for many centuries, a theory to explain and predict acid-base reactions did not exist until the **Periodic Table** provided the framework to explain chemical behavior. Acids were first described as substances that tasted sour, tart, or biting. Alkaline substances did not have a sour **taste** but were caustic and felt slippery. Since it is not a good idea to use taste or touch to identify an unknown substance, it was a great improvement for the health and well being of experimentalists when the affect of acids and bases on the **color** of vegetable extracts became the operational definition. Litmus, which turns red in the presence of acid and blue in base, is the most commonly used of these dyes. Another characteristic of acids and bases has been recognized since their early identification: mixing one with the other can neutralize or destroy the characteristics of both.

Many important acid-base reactions take place in aqueous media. Because the free proton, H, does not exist in **water**, the proton in aqueous **solution** is depicted as H_3O^+ (aq), the hydrated proton or **hydronium ion**. The species donated by a Brønsted acid and accepted by a Brønsted base is the proton, H^+. The term proton in acid/base reactions refers to H^+, the species transferred. The symbol H_3O^+ (aq) refers to the proton as it exists in water, as a hydrated ion.

Any substance that dissociates in solution to produce hydroxide ions or that reacts in solution to produce hydroxide ions is a Brønsted base. This definition encompasses salts that contain the OH- ion, other anions that react with water to form the OH- ion, and some neutral molecules that we will learn about during our study of acids and bases. The hydroxide ion accepts protons to form water and is a very important base, although not the only base recognized in the Brønsted-Lowry theory.

In the nineteenth century, scientists observed that even the purest water conducted **electricity**. This means that ions must be present to carry the charge. If the water is pure, the water molecules themselves must be the source of the ions. Autoionization, self-ionization or autoprotolysis are three terms that describe the process by which one water molecule interacts with another to produce hydroxide ion and the hydronium ion in solution. In pure water, the number of hydronium ions produced is exactly equal to the number of hydroxide ions produced. The **concentration** of ions in pure water is very small, at 77°F (25°C), $[H_3O^+(aq)] = [OH^-(aq)] = 1.0 \times 10^{-7}$ M. When the concentration of hydronium ions in a solution equals the concentration of hydroxide ions, $[H_3O^+(aq)] = [OH^-(aq)]$, as in the case of pure water, the result is a neutral solution, neither acidic or basic.

The concentration of hydronium ions and hydroxide ions changes when solutions become either acidic or basic. To focus on the relationship between the concentrations of those ions, chemists use the relationship between the concentrations of ions as defined by the K_w of water. The K_w defines a quantitative characteristic of all aqueous solutions at 77°F (25°C): no **matter** what the concentrations of hydronium and hydroxide ions are, their product must always equal 1.0×10^{-14}. This fact is the basis of the **pH** scale, the negative of the exponent of ten corresponding to the hydronium ion concentration, a series of numbers derived from the concentration of hydronium ion and describing the acidity or basicity of a solution. The pH of a solutions is the negative of the exponent of 10, corresponding to the hydronium ion concentration. If $[H_3O^+(aq)] = [OH^-(aq)]$, the solution is neutral and the pH is 7.0. If an acid is added to water, the hydronium concentration will be greater than it is in pure water and the pH is below 7.0. If a base containing the hydroxide ion, such as **sodium** hydroxide, is added to water, the hydroxide ion concentration will be greater than it is in pure water and the pH will be above 7.0.

To understand the relationship how acid-base reactions affect the pH of a solution, it is useful to distinguish strong acids and strong bases from weak acids and bases. **Svante Arrhenius** and others used the observation that solutions of some substances conduct electric current to establish the existence of ions. Very early quantitative work on measuring the amount of electric current passing through solutions led to an improved understanding of the behavior of acids and bases. **Carboxylic acids** like **acetic acid** and many other species that do not ionize completely in aqueous medium are called weak acids. Weak acids vary in strength from those which exist primarily as hydrated molecules and only ionize to a very small extent to those which ionize substantially and are in equilibrium with only a small amount of undissociated molecules. The extent of their **ionization** is defined by the value of the equilibrium constant for their reaction. The K_{eq} values for the ionization of weak acids are designated as K_a, or acid **dissociation** constants. Strong acids, such as HCl, HBr, HI, and acids formed from certain polyatomic ions that have several **oxygen** atoms, such as perchloric acid, $HClO_4$, chloric acid, $HClO_3$, sulfuric acid, $_4$ and **nitric acid**, HNO_3 essentially ionize completely in water to form the hydronium ion. Although the other **hydrogen** halides are strong acids in water, HF is weak. Weak acids have dissociable protons like strong acids, but they simply do not dissociate completely. Most of the common acids formed from oxoanions other than those listed above are weak. All common carboxylic acids are weak acids. Another weakly acidic group commonly found in biologically important molecules is the **phosphoric acid** group, $-PO_3H$. Some of the molecules with this acidic group are the polynucleic acids, **DNA** and **RNA**, and molecules called phospholipids that compose membranes.

Reaction of sodium with chlorine. ***(Photograph by Yoav Levy/ Phototake NYC. Reproduced by permission.)***

Metallic oxides are bases because the **oxide** ions accept protons from water molecules, thereby generating hydroxide ions in solution. Because of this behavior upon hydrolysis, they are known as basic anhydrides. Although most metallic oxides are insoluble or only very slightly soluble in water, they readily react with acidic solutions to form hydroxide ions that are then neutralized. Common bases in water are the hydroxides of Group I **metals** and of **calcium**, **strontium**, and **barium** in Group II.

Metal oxides are not the only substances that hydrolyze to produce hydroxide ions and basic solutions. The compound **lithium** hydride, LiH, is a polar covalent solid that reacts with water to liberate hydrogen gas and form basic solutions of the metal hydroxide. Other metal hydrides react the same way with water. In fact, all ionic metal hydrides are strong bases. Another important non-ionic hydride is also basic in water. If **ammonia**, $NH_3(g)$, is dissolved in pure water, the solution is basic because the concentration of hydroxide ion in the medium increases because ammonia competes the hydroxide ion formed by the autoionization of water, forming the ammonium ion (NH_4) and leaving more hydroxide than hydronium ion in the solution. Ammonia and the organic amines, molecules in which one or more of the hydrogen atoms of ammonia are replaced by organic groups, are weak bases.

Although acid-base reactions are often called **neutralization** reactions, they do not all yield neutral solutions. If a strong acid is added to a solution of a strong base, each hydronium ion will react with a hydroxide ion to form a water and ions of a neutral **salt**. When the total amount of hydronium ion added equals the amount of hydroxide originally present (the equivalence point), the solution will be neutral, with a pH of 7.0. If more acid is added, the pH will be lower by the same amount as if the additional acid were being added to pure water. Addition of a strong base to a solution of a strong acid is just the reverse process. However, if a weak acid is added to a strong base to the equivalence, the salt that results will not be neutral. The **anion** formed by removal of the dissociable hydrogen **atom** of a weak acid will compete with the hydronium ion formed from the autoionization of water, leaving an excess of hydroxide ion. The pH of the solution will be greater than 7.0. It will be the same pH as one would observe if the salt itself were dissolved at the same concentration. If a strong acid is added to a weak base to the equivalence point, the resulting solution would be acidic. When a weak acid and weak base react, the resulting salt can be either basic or acidic depending on the relative weakness of the acid and base.

Oxidation-reduction (redox) reactions are processes in which the number of electrons assigned to atoms involved in the reaction change during the course of the reactions. The original definition of **oxidation** centered on the class of reaction in which one or more oxygen atoms from oxygen itself to another species was added to an element or compound. It became apparent, however, that other substances besides oxygen could add oxygen atoms to molecules. Species that cause oxidation are called oxidants or oxidizing agents. They remove electrons from the species being oxidized and are themselves reduced. Species that cause **reduction** are reductants or reducing agents. They contribute electrons to the species being reduced and are themselves oxidized. Oxidation and reduction always occur simultaneously and to the same extent (the same number of electrons is taken from the reductant as are added to the oxidant).

Oxidizing and reducing agents are identified by changes in the oxidation number that occur as a result of a reaction. Atoms and ions are assigned oxidation numbers based on the total number of electrons available in the entire molecule or ion and the atom's **electronegativity**. For example, electrons in a **covalent bond** are assigned to the more electronegative atom. Oxidation numbers are not ionic charges. They can be computed for atoms in any compound, ionic or covalent.

Oxidation-reduction reactions cover a wide range of chemical changes, including reactions of atoms of the same element in different oxidation states. Three categories that encompass all these different oxidation-reduction reactions are: atom transfer reactions, **electron** transfer reactions, and **disproportionation** reactions. In an atom transfer oxidation-reduction reaction, the atom being oxidized or reduced is bonded with a different species in the product than in the reactant; it has been chemically ''transferred'' to form a new species. Reactions of non-metallic elements to form covalent compounds are atom transfer reactions. Because we exist in an oxygen rich atmosphere, the most prevalent class of atom transfer oxidation-reduction reactions involves oxygen. Such reactions include **combustion**, oxidative **corrosion** of metals, and metabolic oxidation of foods.

Reactions in which oxidation states change without the transfer of either the reductant or oxidant atoms are defined as electron transfer reactions. Electron transfer reactions occur in the process of **metabolism** and **photosynthesis** and are also used in large scale industrial processes when appropriate oxidants and reductants are mixed together. Electrochemical processes are also electron transfer reactions. Unlike most other types of chemical reactions, the reactants in an electrochemical process can be spatially separated as long as there is a means for electrons to flow from the reductant to the oxidant.

Reactions in which atoms of the same element are both oxidized and reduced are disproportionation reactions. An example is the reaction of two molecules of **hydrogen peroxide** to form two molecules of water and one molecule of oxygen. The oxidation number for both oxygen atoms in hydrogen peroxide is -1. In the products the oxidation number of the oxygen atom in water is -2, while that in molecular oxygen is 0.

Addition (combination) reactions are processes in which two molecules or ions combine to form a single new species with no atoms left over. Organic molecules with double or triple bonds (alkenes and alkynes) undergo addition reactions. Examples include the addition of hydrogen to an unsaturated fat to make a saturated fat and addition of molecules such as $_2$, HCl or HBr to a double or triple carbon-carbon bond. Addition reactions of inorganic molecules occur when an atom has more than one **valence**. For example, **phosphorus** atoms can bind to three or five halogen atoms so PCl_3 can react with Cl_2 to form PCl_5. **Sulfur** can bond to two, three, or four oxygen atoms so sulfur dioxide can undergo an addition reaction to form sulfur trioxide and, in turn, sulfur trioxide can react with water to form a molecule of sulfuric acid (more properly designated dihydrogen **sulfate** if undissociated). From these examples, you may see that many inorganic addition reactions can also be classified as redox reactions.

In substitution reactions, one or more atoms or groups of atoms attached to a particular atom of one reactant is transferred to an atom of another reactant. One of the most important types of substitution reactions of metallic compounds is **precipitation**. These reactions have tremendous utility in the preparation of commercial products and in the testing and analysis of ions of clinical and environmental significance. They are also behind many day to day irritations: the formation of stains on porcelain, the encrusting of heating elements in coffee makers, and the formation of bathtub rings. In natural water systems, many common minerals are formed by anion substitution-precipitation reactions, among them carbonates, phosphates and the sulfate containing rocks. Both qualitative **solubility** rules and quantitative expressions of equilibrium constants for solubility, K_{sp}, are useful in discussing the **chemistry** behind these occurrences and in controlling them. They also enable us to make reliable predictions about what will happen when systems containing ions are perturbed by the addition of other ions, by the presence of dissolved **gases**, or by changing pH.

The dissolution of varying amounts of a substance in water results in what are qualitatively described as unsaturated, saturated, and supersaturated solutions. For systems not at equilibrium, determining the solubility reaction quotient, Q_{sp}, is useful for predicting what happens as the system returns to equilibrium. The value of Q_{sp} is compared to the K_{sp} for salts in solution. If Q_{sp} K, no precipitate will form in the solution; if Q_{sp} K_{sp}, one predicts a precipitate will form. When the calculated Q_{sp} is less than the K_{sp}, the solution is unsaturated. An unsaturated solution can dissolve more solute. When the concentrations of the ions in solution are such that $Q_{sp} = K_{sp}$, the solution is saturated. No more solute can dissolve in the solvent; it is ''holding'' the maximum amount. When the concentrations are such that Q_{sp} is greater than K_{sp}, either of two results are observed. In the simplest cases, a precipitate forms until $Q_{sp} = K_{sp}$: until the concentration of ions in solution as defined by the equilibrium constant is reached. At this point, the solution is saturated. A solution is saturated with respect to a solid if some of the solid is visible in the bottom of the container. The maximum concentration of ions in the solution is in equilibrium with the precipitated solid.

Sometimes, however, no precipitate forms until Q_{sp} greatly exceeds K_{sp} or until the solution is disturbed. When Q_{sp} K_{sp} for a solution in such cases, the solution is supersaturated. The ions in the solution are at a higher concentration than would be predicted by the K_{sp} for the dissolved material.

Ion substitution reactions are put to great practical use in a process known as **ion exchange**. Ions causing hard water can be removed from solution and replaced by other ions through the intermediacy of inorganic zeolites or organic ion-exchange resins.

Substitution reactions of non-metallic elements are of two major types: nucleophilic and electrophilic substitution. In a large class of reactions called nucleophilic substitutions, reactions occur when a species with available electrons (a nucleophile), either in the form of an anionic charges or an unshared pair of electrons, is attracted to an electropositive site in a molecule. During the reaction, a new bond is formed between the **nucleophile** and another bond in the molecule is broken, thereby releasing a new species into the reaction medium. Nucleophilic substitution reactions can be envisioned by using two different mechanisms—one unimolecular and one bimolecular—and the type of **substitution reaction** that a given molecule undergoes depends on its structure and the conditions under which the reaction is carried out.

Another type of substitution reaction involves **free radicals**. These reactions are industrially important for the production of some useful small molecules and also play a major role in **atmospheric chemistry**.

Aromatic compounds like **benzene** that contain delocalized electrons undergo electrophilic substitution reactions. Electrophiles are electron-poor species that are attracted to the electron cloud of aromatic compounds and other species with loosely held electrons.

Inorganic compounds held together by covalent bonds also undergo substitution reactions. Compounds and ions containing **nitrogen** and sulfur are frequently nucleophilic. Phosphorus behaves similarly to nitrogen but has a more extensive **reactivity** because of its ability to expand its octet to form compounds with five or even six bonds. Substitution reactions of **boron**-containing compounds are directed by the ability of boron to form electron deficient molecules that can readily bond with electron rich species, especially those containing a pair of non-bonding electrons. The chemistry of **silicon** compounds is similar to that of **carbon**, but silicon forms extremely stable silicon-oxygen bonds that results in compounds like sand and quartz that are resistant to further chemical modification.

In elimination reactions, larger molecules expel smaller molecules like water, hydrogen halides, **carbon dioxide** and other **diatomic** gases to form new substances. Both organic and inorganic compounds experience elimination reactions.

Rearrangement reactions are isomerization. They include both intramolecular changes in covalent bond connectivity or orientation that requires bond breaking and reformation (including ''rotation'' about a carbon-carbon double bond) and structural changes caused by an intermolecular reaction, often catalytic. Many organic molecules undergo structural changes as a result of catalytic reactions with acid or base. For example, 1-butene ($CH_3CH_2CH{=}CH_2$) is converted to the more stable isomer, 2-butene ($CH_3CH{=}CHCH_3$), in the presence of acid.

Complexation reactions are processes in which molecules of ion (called legends) bond to a metal atom (usually a transition metal) by forming coordinate (dative) covalent bonds. In a coordinate covalent bond, both electrons are donated by the Logan, a Lewis base, to the metal atom, a Lewis acid. Complexation is often used to make a metal atom or ion more soluble. Metals in dietary supplements, for example, are often chelated (a **chelate** is a ligand that has more than one atom that bonds to the metal atom) to increase solubility. Iron(III) citrate, formed by adding a solution of **citric acid** or a citrate salt to a solid iron(III) salt, is very soluble, whereas uncomplexed Fe(III) is highly insoluble in water. Insoluble **silver** halides on photographic film can be dissolved by adding ammonia, which forms a silver diammine complex ion, $Ag(NH_3)$.

Hydrolysis is the reaction of a substance in solution with the solvent water. It is a specific case of a more general reaction called solvolysis, which simply refers to reaction of a dissolved species with solvent. Many other fluids can serve as **solvents** for chemical reactions, and if the solvents interact chemically with materials dissolved in them, the reactions are named in a corresponding fashion. If ammonia is the solvent, for example, the reaction with the solvent is called ammonolysis.

It is important to distinguish hydration, a physical process whereby water molecules surround a species in solution, and hydrolysis, a chemical reaction occurring between solvent and dissolved species. The species to whose symbols we have appended ''(aq)'' are hydrated; they are physically surrounded by water. When hydrolysis occurs, we will write a chemical equation to describe it, and new species will be formed in solution.

Polymerization reactions are addition reactions, principally of two types: condensation reactions and free-radical reactions. In a **condensation reaction**, atoms dissociate from the ends of two molecules, forming a new small molecule and allowing the remaining fragment of the two molecules to form a covalent linkage and, hence, a new larger molecule. The process can continue until very large polymers are formed. The **monomer** units can be the same or different. The largest class of condensation reactions is **dehydration**, in which water is the small molecule that is dissociated. Geopolymers such as silicate minerals and metal oxides and biopolymers such as **proteins**, **nucleic acids**, and polysaccharides are formed by dehydration condensation reactions. **Ceramics**, synthetic fibers and many other products are also produced by condensations reactions.

In free radical polymerization, monomers can add to each other without formation of a small molecule. A free radical has an atom with unpaired electron. Free radicals can react with molecules to form a new, larger free radical or can react with other free radicals to form a covalent bond. The process of free radical polymerization is initiated by a species that can remove an atom from a monomer and produce a free radical. The free radical adds to a monomer, producing another free radical that, in turn, adds to another monomer, increasing the length of the polymer chain. The growth is terminated by a reaction of free radicals that produces a molecule that is not a free radical. Synthetic polymers such as **polyethylene** and **polypropylene** are produced by free radical addition reactions.

The essence of chemistry is understanding and applying chemical reactions. Chemists have divided chemical reactions into several categories based upon the type of changes that occur to atoms and molecules during each type of process. Acid-base reactions involve the transfer of protons, typically resulting in some extent of salt formation and neutralization. Oxidation-reduction reactions result in a change in the number of electrons assigned to some of the atoms involved in the process. Addition reactions result in the formation of one molecule or polyatomic ion from more than one. In substitution reactions, atoms or groups of atoms originally bound to one particular atom are replaced by another atom or group of atoms. In elimination reactions, one group of atoms is removed from a molecule or polyatomic ion. Rearrangement processes result in isomerization. Complexation is the formation of a Lewis acid-Lewis base combination where a metal atom is the Lewis acid and ligands that form coordinate covalent bonds to the metal act as Lewis bases. Hydrolysis is the addition of a hydrogen atom and a hydroxyl group to separate atoms of a molecule or ion. Polymerization is an addition reaction, typically either a condensation reaction involving dehydration or a free radical process, in which many units of a molecule or molecular fragment are covalently combined.

See also Electrochemistry

Reactivity

Reactivity is the degree to which a particular **atom**, radical, or **molecule** is active. It has to do with how able a molecule is to combine chemically with another molecule, resulting in a chemical or physical change. Reactivity varies according to the substance under scrutiny and the materials with which that material comes into contact. It is a variable in a chemical reaction, which brings about the chemical mingling of two or more substances, and is a measurement of how fast one chemical reacts compared with another (i.e., its kinetic properties).

Reactivity is a relative term, and as such only has meaning when used in reference to a certain set of conditions. Thus, it is not a measurement that chemists use when dealing with reactions or general compound reactions. Instead, a chemical species (a group of substances with identical molecular entities) might be called reactive if it had a larger rate constant (reaction rate in terms of concentrations) than another species for a certain elementary reaction. However, chemists also use the term ''reactivity'' outside of elementary reactions to describe a chemical species' constants and equilibrium, as well as rate.

The reactivity-selectivity principle (RSP) is an important, although controversial, way for scientists to use reactivity as a measurement. Selectivity is the tendency of molecules to have preferences of substance with which to react. As a measurement, it reveals the extent to which a chemical species' reactivity rates differ in various substrates. According to the RSP, the more reactive a chemical is, the less selective it will be. In other words, a chemical that can react with many other substances (reactive) will not be particularly choosy (selective) about the chemicals with which it bonds. The reactivity-selectivity principle is controversial because some scientists think of it as a universal law and some see it as practically worthless.

There are certain patterns of reactivity that allow chemists to predict the outcome of a reaction, thus saving time and effort. In this sense, reactivity is a highly useful measurement. For instance, if a chemist knows that **sodium** reacts with **water** to make **hydrogen** gas and **sodium hydroxide**, he or she would be able to predict with confidence that **potassium** (in the same column of elements in the periodic table) would do the same thing when exposed to water. In other words, he or she could predict that the entire family of alkali metallic elements in that column has similar patterns of reactivity.

See also Reaction rate; Reactions (types of)

Receptors

The **plasma** membranes of cells contain specialized **proteins** called receptors. Receptors allow cells to receive chemical signals that direct cellular function and activity. These chemical signals, also called signaling molecules, include **hormones**, neurotransmitters, and local mediators. Signaling molecules have the ability to bind to receptors and trigger a series of biochemical events in a cell's interior.

There are hundreds of different signaling molecules, each of which will only bind to a certain receptor. Neurotransmitters are the chemicals used by the nervous system to transmit nerve impulses to and from the brain. Hormones are released into the bloodstream by specialized tissues called glands. They can influence cellular events throughout the body. On a more local level, local mediators are chemicals released by many cells types to influence other cells that are nearby.

The relationship between signaling molecules and their receptors is very specific. Each receptor on a cell's surface is made to accept only one signaling **molecule**. However, one cell may have many different receptors on its surface. As a result, the function and activity of that cell can be influenced by several different signaling molecules.

Although there are many different receptors, they can be divided into three main categories based on their underlying molecular machinery. The three classes of receptors are **ion** channel receptors, receptors linked to G protein, and catalytic receptors. In general, all three types feature receptors that are embedded in a cell's plasma membrane. Each of these receptors are in contact with the cell's interior as well as its external environment.

Ion channel receptors are typically found in connection with the nervous system and provide sites for neurotransmitter binding. The action of a common neurotransmitter, such as acetylcholine, illustrates how ion channel receptors work. Acetylcholine receptors are found on neurons (nerve cells) and on muscle cells. When acetylcholine binds to its receptor, it causes the cell membrane to depolarize. Depolarization means that the membrane allows charged particles, or ions, to flow into or out of the cell. This ion flow triggers a cellular response, such as continuation of a nerve impulse to another neuron or the contraction of a muscle cell.

G protein receptors are the most diverse category of receptors. These receptors are found in cells throughout the body and are used by neurotransmitters, hormones, and local mediators. The key feature of G protein receptors is that once the receptor binds a signaling molecule, it triggers the formation of a second messenger. Common second messengers include cyclic adenosine monophosphate (cAMP), inositol triphosphate, and diacylglycerol. In effect, these second messengers convey the signaling molecule's message to the rest of the cell. Second messengers may activate or deactivate enzymes within a cell or may help to control the cellular responses to other signaling molecules. The eventual response to the signaling molecule depends on the cell type and the function it is expected to perform.

The third category of receptor is the catalytic receptor. These receptors function as enzymes. When a catalytic receptor is occupied by a signaling molecule, a chemical reaction is triggered within the cell. A common reaction is the addition of a **phosphate** group to another **enzyme** that is inside the cell. The addition of a phosphate group—phosphorylation—is one method in which an enzyme can be switched from its inactive form to its active form. An inactive enzyme is nonfunctional; an active enzyme is able to trigger specific chemical reactions. The reactions unleashed within a cell correspond with its response to the signaling molecule.

Once a receptor has been activated there are several ways in which the cellular response comes to an end. The simplest method relies on the signaling molecule detaching from the receptor. Once the receptor is no longer occupied, the response mechanism is essentially switched off. Another method revolves around the receptor becoming less sensitive to the signaling molecule. In this case, once the cellular response has been achieved, further signals have little or no effect. More drastic desensitization can be accomplished by reducing the number of receptors in the plasma membrane. Receptor down-regulation is accomplished by drawing receptors that have bound signaling molecules into the cell's interior. Once inside, the receptors are snipped from the plasma membrane and destroyed.

See also Neurochemistry

RECYCLING

The **chemistry** of recycling involves the treatment of commercial products to remove and isolate components for reuse in other products. The materials most commonly recycled are **paper** products, **metals**, **glass**, and **plastics**.

Nearly one third of all newspapers, as well as very large quantities of magazines and corrugated cardboard, are recycled. Newsprint consists principally of **cellulose** fibers, binders, and other additives to improve the appearance and feel of the paper and **inks**. The objective of paper recycling processes is to recover the cellulose fiber and to eliminate or greatly disperse the ink. This is accomplished either by a series of washing and **filtration** steps or by pressure-steaming the paper.

Nearly half of the metal produced in the United States is recycled metal scrap, the majority of which is from automobiles. Recycling **iron** scrap saves nearly 0.75 of the **energy** required for steel production with a concomitant **reduction** in air and **water pollution**. Recycling **aluminum** requires nearly 20 times less energy than producing aluminum from bauxite. Recycling also eliminates the need to mine the bauxite, reducing environmental consequences.

Glass is usually separated by **color**, cleaned, and crushed into particles suitable for melting (cullet). The amount of energy and overall cost reduction from glass recycling is not as great as for metals or paper, but recycling of glass does reduce the amount of landfill space communities require.

Plastics are sorted by their **molecular structure** and reused accordingly. **Polyethylene** terephthalate (PET) products, such as soft drink bottles, are either cut into chips and used directly, for example to make carpets, or depolymerized to the **monomer** and reused as starting material for making products such as soda bottles. High **density** polyethylene products, such as milk containers, are cut into small pieces and remelted to make detergent bottles. Polycarbonate water bottles are used to make automobile bumpers. **Polystyrene** products are melted and converted to foam insulation. Mixed plastics products as made into products such as **wood** substitutes. In any application where the plastic need only be cut and remelted, there is considerable savings of process costs and environmental benefits. Plastics are organic polymers derived from petroleum. The monomers of which most plastics are made must be synthe-

Workers recycling construction debris, Red Hook Recycling Center, Brooklyn, New York. ***(Photograph by Vanessa Vick, Photo Researchers, Inc. Reproduced by permission.)***

sized and, subsequently, polymerized. There is likely to be an increased emphasis on plastics recycling as more products are made of plastic and space for **landfills** becomes scarce.

Recycling is a major industry, involving the reuse of substantial quantities of paper, plastics, metals, and glass. The benefits of recycling include conservation of natural resources, energy savings, and reduction of the need for waste disposal.

Reduction

Reduction reactions are essential to life. In **photosynthesis**, **carbon** dioxide and **water** are reduced to form the **carbohydrates** required as a source of **energy** for plant and animal life. Photosynthesis ultimately provides all energy, with the exception of nuclear power and the relatively small contributions from solar, wind, water, and geothermal power, for human and animal life and most of the fuel used by society and industry. If research into artificial means of using sunlight to convert water and **carbon dioxide** into carbohydrates without the intervention of green plants succeeds, the impact on our global energy problems will be enormous.

Reduction reactions are also used by chemists to synthesize pharmaceuticals, textiles, dyes, paints, and a multitude of other important products. A major industrial use of **hydrogen** is the reduction of unsaturated liquid vegetable oils to make edible solid fats like margarine.

The original definition of reduction reactions centered on the removal of **oxygen** from a species, most typically by the action of hydrogen. When oxygen is removed from a metal **oxide** to derive the metal itself, the **mass** of the metal oxide is reduced. Metal-containing ores have been reduced with carbon-containing materials to obtain **metals** for thousands of years. However, it was not until the eighteenth century that experimenters such as French chemist **Antoine Lavoisier** carefully weighed metal ores and metals produced from them to quantitatively establish that weight reduction occurs in such reactions. Species that cause molecules to lose oxygen are called reductants or reducing agents. **Oxidation** always occurs in tandem with reduction: if a species in a reaction is oxidized, then something else must be reduced.

Industrial processes to recover a metal from its oxidized forms are examples of reduction reactions. Many metal oxides are powdery, non-malleable, non-ductile substances. Only as the free metal do they have significant tensile strength and can they be shaped and formed into useful tools or other objects. Carbon is a reducing agent in metallurgical processes. In the case of making **iron** in a blast furnace, the carbon is added as coke, which is converted into carbon monoxide—the actual reducing agent.

The current approach to oxidation-reduction reactions has been developed from the idea that electrons are crucial to chemical **bonding**. When a metal oxide or **salt** is reduced to form the metal, a **cation** is converted to an uncharged **atom**. In reduction, electrons are gained by the cation. Reduction is most generally defined as a gain of electrons.

Most compounds are covalent, not ionic. To encompass all oxidation-reduction reactions, it is necessary to determine the gain and loss of electrons for covalent as well as ionic and elemental species. Electrons are assigned to atoms in covalent and ionic compounds and the oxidation number of each atom is determined. Reduction and oxidation involve a change in the number of electrons assigned to atoms that undergo the reaction. To determine if a change has occurred, the starting and ending states of atoms in terms of the number of electrons must be assigned. This requires determining the oxidation states of atoms, which are specified by oxidation numbers. A species is reduced in a reaction if it undergoes a decrease in oxidation number; oxidation occurs when the oxidation number of a species increases.

Oxidation-reduction (redox) reactions are classified into categories to facilitate our ability to understand trends and predict the course of possible reactions. The first redox reactions to be classified involved addition and removal of oxygen atoms from chemical species. The next definition developed focused on **electron** transfer. It is consistent with the earlier definition but broadens it to include reactions not involving oxygen atoms. We now recognize that redox reactions cover a wide range of chemical changes, including reactions of atoms of the same element in different oxidation states. Three categories that encompass all these different redox reactions are atom transfer redox reactions, electron transfer reactions, and **disproportionation** reactions.

Although the most common atom transfer oxidation-reduction reactions involve oxygen, several critical biochemical reactions as well as reactions used to synthesize organic compounds involve hydrogen atoms. Hydrogen atoms have three possible oxidation states: +1 when they are bound to more electronegative elements like other non-metals, 0 for elemental hydrogen, and -1 when they are bound to less electronegative elements like the metals. In acid-base reactions, the hydrogen atoms remain in the +1 **oxidation state**. In oxidation-reduction reactions of hydrogen, its oxidation state changes.

In biochemical systems, the hydrogen atoms are typically produced and used one at a time and are not present as **diatomic** hydrogen gas. For example, the **enzyme** nitrogenase converts atmospheric **nitrogen** into **ammonia** in a series of steps involving hydrogen atoms.

One example of the use of hydrogen atom transfer in organic synthesis is **hydrogenation**, i.e., the addition of hydrogen to a carbon-carbon double bond. Hydrogenation converts an unsaturated oil to a saturated oil, which is an essential process in making solid margarine from liquid vegetable oil.

Electron transfer reactions are essential steps in many biological processes, including photosynthesis, in which light energy induces the transfer of electrons from **chlorophyll** molecules to chemical intermediates in a very complex reaction sequence. The energetic electrons released from chlorophyll eventually cascade through a series of redox reactions, resulting in the production of organic compounds such as carbohydrates and the liberation of oxygen. The carbon in the carbohydrates is reduced relative to the starting material, carbon dioxide.

We can use the **Periodic Table** to predict the **reactivity** of atoms in each of category of redox reaction. The most straightforward reactions to predict are those that occur between elements themselves. Metals act as reductants, providing electrons to nonmetallic elements. The relative reactivity of metals in electron transfer reactions in aqueous medium is commonly displayed as an activity series. The activity series is a list of elements in order of the ease with which they lose electrons in redox reactions. Metals high on the activity series react more vigorously with oxidants including the **halogens** and the **hydronium ion** than those lower on the scale. Metals higher on the series are also capable of reducing ions of metals lower on the series. For example, **zinc** is higher on the activity series than **copper** and a **solution** of a copper salt will corrode metallic zinc because zinc metal reduces copper ion. Copper metal does not react with a solution of a zinc salt to produce copper ion and metallic zinc.

The reaction of metals with acid to form hydrogen gas was one of the earliest oxidation-reduction reactions of the metals to be characterized. The conversion of hydronium ion to hydrogen or the reverse reaction is, in fact, used as the reference reaction for oxidation-reduction reactions. The relative activity of metals to act as reductants is judged by their ability to reduce hydronium ion to hydrogen, and the relative reactivity of oxidants is judged by their ability to oxidize hydrogen to hydronium ion.

Hydrogen transfer reactions play an important role in transformations in organic synthesis. The borohydride ion (BH_4^-) in salts like **sodium** borohydride and the **aluminum** hydride ion (AlH_4^-) in **lithium** aluminum hydride are often used as sources of hydride ions that react with and reduce many organic compounds. Aluminum hydrides are powerful reducing agents, while borohydrides are more selectively reactive. However, both BH_4^-) and AlH_4^-) react with **ketones** to reduce them efficiently to alcohols. Literally dozens of hydride sources have been developed for use in organic synthesis, and many hydride transfer agents are highly specific for one type of substrate or for one particular reduction. Some agents may reduce an acid to an aldehyde and stop there, while others may reduce an acid all the way to an **alcohol**.

Reduction reactions (e.g., photosynthesis) are essential to life and are important in the chemical manufacturing industry. Reductions, especially those carried out in the metal processing industry, often require the use of carbon or **carbon monoxide** to remove oxygen from ores. In other types of reductions, especially those involving organic compounds, hydrogen is added to molecules. Other species can serve as sources of hydrogen or electrons and likewise participate in redox reactions as reductants.

Reese, Charles L. (1862-1940)

American chemist

The developer of nonfreezing dynamite (safe for use in cold-weather situations), permissible **explosives** (for use in underground **coal** mines), and the industrial explosive sodatol (created from TNT and picric acid), Reese is an important American chemist of the early twentieth century.

Born in Baltimore, Maryland, Reese received his doctorate in Heidelberg, Germany, while studying under Robert Bunsen, inventor of the spectroscope and **Bunsen burner**. In 1902, after a brief but distinguished career in academics, he assumed the position of chemical director at the Du Pont Company, which he held until his retirement in 1924.

Virtually all of the achievements for which Reese is known came through his work at Du Pont, where he oversaw the rapid expansion of one of the largest and most advanced research departments in the world. He is less well known for his involvement with efforts during World War I to secretly escort two German scientists, who possessed valuable knowledge regarding dyestuff and synthetic-nitrates, to the United States. This successful mission may well have been one of the important turning points for both the war and the field of **chemistry**, which had long been dominated by the Germans. He died in 1940.

Refrigerants

Refrigeration is the cooling of substances below ambient temperatures by extracting **heat** from them. The medium that absorbs the heat from the substance to be cooled is called the refrigerant. **Refrigerants** can be classified into two main groups. The first absorb heat by undergoing a **change of state** (for example from liquid to vapor), the second absorb heat without undergoing a change of state (for example they may absorb heat as a liquid and remain a liquid, albeit a hotter one). If the refrigerant is only used to transport ''cold'' from one place to another it is called a secondary refrigerant.

The most widespread refrigeration method is the vapor compression cycle, which relies on the evaporation of a liquid. This method is used in industry as well as in domestic refrigerators. The vapor compression cycle relies predominantly on two physical principles to work; the tendancy for **liquids** to form vapors and the latent heat of evaporation.

If we partially fill a closed empty container with a volatile liquid, some of it will evaporate to fill the empty part of the container with vapor. Because the vapor is trapped by the container walls it exerts a pressure on them, called the saturated vapor **pressure** (or just vapor pressure). This pressure depends only on the type of liquid and the **temperature** of the container. If some of the vapor is removed by pumping it away then more of the liquid will evaporate to take its place and maintain the **vapor pressure**.

Energy is required to release a **molecule** from a liquid into the gaseous state (that is, of a liquid evaporating). This is because the molecules in a liquid are all very close together and so can interact with each other through **Van der Waals forces**, polar interactions, and/or **hydrogen bonding**. These forces are about ten times weaker than covalent molecular bonds, but are still strong enough so that it takes a relatively large amount of energy to break them and allow the molecule to escape into the vapor. This energy can be supplied to the molecules as heat, in which case it is called the latent heat of evaporation. The process by which evaporating molecules remove heat from their surroundings, can be used to cool substances.

The vapor compression cycle works in three stages:

1)The liquid refrigerant is placed in an evaporator and the pressure is lowered until the **boiling point** of the refrigerant is just below the temperature to which we want to cool our substance. The boiling point of a liquid is determined by the equilibrium point for molecules escaping from the liquid to the vapor and molecules returning to the liquid from the vapor. The lower the total pressure of gas over the liquid, the more readily molecules can escape from the liquid phase.

2) The vapor is pumped to a compressor, where the pressure is raised until the condensation temperature (the temperature at which the vapor turns into a liquid) is just above that of the cooling medium. As the temperature is now lower than that needed for the vapor to exist, the refrigerant condenses, giving up the latent **heat of vaporization** it absorbed to the cooling medium. The cooling medium can either be **water** or, in the case of a home refrigerator, the air in the room (this is why the air feels hot behind a refrigerator).

3) The liquid is now pumped back to the evaporator via an expansion valve, which lowers the pressure of the refrigerant so that the cycle can be repeated.

The early vapor compression machines that began to be built in the 19th century used many different types of refrigerants. **Ether**, **ammonia**, **sulfur** dioxide, and even water were used in various systems. Many of these were quite unsuitable, as they were either toxic, flammable, unstable, or required impractical pressures to work.

Research was undertaken in the late 1920s to find a compound that would be stable, inflammable, non-toxic and have a boiling point that would allow it to be used at reasonable pressures. The research revealed flurocarbon compounds as being the most suitable, and dichlorodifluoromethane ($CHC1_2F$) was the first to be used as a refrigerant. Fluorocarbons are compounds that usually contain less than four **carbon** atoms and one or more **fluorine** atoms. They may also contain hydrogen, **chlorine**, or **bromine** atoms. If they contain only fluorine, chlorine and carbon they are called **chlorofluorocarbons**, or CFCs for short.

In the 1970s, the chemists **F. Sherwood Rowland** and Marino proposed from results of their laboratory experiments that CFCs were playing a significant part in the destruction of the ozone layer. Atmospheric experiments provided very strong support for their hypothesis. The high stablility of CFCs means that, if they are released into the atmosphere, they are not broken down until they have been carried high up into the part of the earth's atmosphere called the stratosphere. Once in the stratosphere, they are broken apart by the powerful ultraviolet radiation from the sun, releasing chlorine. This chlorine then catalyzes the conversion of ozone (O_3) into dioxygen (O_2), reducing the ozone layer and allowing the suns harmful ultraviolet radiation to penetrate the earths lower atmosphere.

Replacement refrigerants have now been developed which are less harmful to the ozone layer. The first generation were hydroflurochlorocarbons (HCFCs), which still contain chlorine, but also contain hydrogen, which makes them much more reactive. These are broken down before they can be carried up to the ozone layer and so are less damaging to the ozone layer than CFCs. However, as they still contain chlorine there is still the possibility that they could cause some damage and so hyrdofluorocarbons (HFCs) were developed as the second generation. HFCs contain only carbon, hydrogen, and fluorine, and so are completely harmless to the ozone layer. This makes them the most viable replacement for CFCs.

REICHMANIS, ELSA (1953-)

Australian-American chemist

Elsa Reichmanis is a chemist and engineer who has worked to develop sophisticated chemical processes and materials that are used in the manufacture of integrated circuits, or **computer chips**. She has served as supervisor of the Radiation Sensitive and Applications Group at AT&T Bell Laboratories in Murray Hill, New Jersey, since 1984. As of 1994 she holds eleven patents, and she received the R & D 100 Award for one of the one hundred most significant inventions of 1992. She received the 1993 Society of Women Engineers (SWE) Annual Achievement Award for her contributions in the field of integrated circuitry.

Several of Reichmanis's patents are for the design and development of organic polymers—known as resists—that are used in microlithography. Microlithography is the principal process by which circuits, or electrical pathways, are imprinted upon the tiny **silicon** chips that drive computers. During the multi-stage process of chip manufacture, layers of resist material are applied to a silicon base and exposed to patterns of ultraviolet light. Portions of the resists harden, becoming templates for the application of subsequent layers of positively and negatively charged **semiconductors** that serve as the channels through which electric current travels. Reichmanis received the 1992 award for the development of a resist material called Camp–6, which will be used in the late 1990s to make the next generation of integrated circuits smaller and more powerful than ever before.

Reichmanis was born December 9, 1953, in Melbourne, Australia. She completed her undergraduate studies in **chemistry** at Syracuse University in 1972. She then performed her doctoral studies in **organic chemistry**, also at Syracuse, as a university research fellow. She earned her Ph.D. in 1975 at age twenty-two with a perfect grade point average. She was a postdoctoral fellow for scientific research at Syracuse from 1976 to 1978, when she left academia for the private sector. In 1978, Reichmanis accepted a position as a member of the technical staff of the organic chemistry research and development department at AT&T Bell Laboratories in New Jersey. In 1984, she was promoted to her current position as supervisor of the Radiation Sensitive and Applications Group at AT&T.

Reichmanis's awards and appointments are numerous. In 1986, she was a member of a National Science Foundation panel to survey Japanese technology in advanced materials. She also served as a member of a National Research Council committee to survey materials research opportunities and needs for the electronics industry. Reichmanis served on the American Chemical Society (ACS) advisory board from 1987 to 1990, and was chair-elect of the ACS Division of Polymeric Materials in 1994. She has authored nearly one hundred publications, and co-edited three books that were presented at American Chemical Society symposia. She was plenary lecturer at the 1989 International Symposium of Polymers for Microelectronics.

An American citizen, Reichmanis is a member of the American Association for the Advancement of Science, and the Society of Photographic Instrumentation Engineering. As the mother of four children, Reichmanis encourages women to embrace both career and family. ''If something interests you and you like doing it, then go for it,'' she told contributor Karen Withem in an interview. ''If you ask yourself, 'How can I manage having both a career and children?'—you'll never do it. If you just do it, things will fall into place.''

REICHSTEIN, TADEUS (1897-1996)

Polish-Swiss organic chemist

It is now known that the **hormones** of the adrenal gland are essential to controlling many challenges to the human body, from maintaining a proper **balance** between **water** and **salt** to responding to stress. Tadeus Reichstein is one of those responsible for this knowledge; **Edward Kendall** and **Philip Hench** also played an important role in these efforts, and the three men shared the 1950 Nobel Prize in physiology or medicine. Reichstein's work has had effects throughout medicine—in the treatments of Addison's disease and rheumatoid arthritis, for example, and in the understanding of the fundamental biochemical processes of steroid hormone **metabolism**.

The eldest son of Gustava and Isidor Reichstein, Tadeus was born on July 20, 1897, near Warsaw in Poland. After moving first to Kiev in the Ukraine and then to Berlin, the family settled in Zürich and became Swiss citizens. Tadeus attended the Eidgenössiche Technische Hochshule and graduated in 1920 with a **chemical engineering** degree. He worked briefly in a factory, then returned to the Eidgenössiche Technische Hochshule where he earned his doctorate in organic **chemistry** in 1922.

For several years thereafter Reichstein continued to work with his doctoral advisor, **Hermann Staudinger**, who would later win the 1953 Nobel Prize in chemistry. Reichstein's early work focused on identifying and isolating the chemical species in coffee that give it its flavor and aroma. This interest in plant products was to remain with Reichstein throughout his career. He had an early success when he discov-

Tadeus Reichstein. ***(Corbis Corporation. Reproduced by permisission.)***

ered how to synthesize the newly discovered compound ascorbic acid (vitamin C). He published this method in 1933, and later that year Reichstein developed a second method of synthesis which is still widely used in the commercial production of this dietary supplement.

In 1934, Reichstein began work on what he originally believed to be a single hormone produced by the cortex or outer layers of the adrenal glands. He soon realized, however, that the adrenals were producing a milieu of active substances. His work began with 1,000 kg (more than a ton) of adrenal glands that had been surgically removed from cattle. His first stage of purification resulted in one kilogram (about 2.2 lb) of biologically active extract. He established that the extract was biologically active by injecting it into animals whose adrenal cortices had been removed; if the compound was active it replaced what was missing as a result of the operation and allowed the animal to survive. The next stage of purification reduced the kilogram of extract to 25 g (less than one ounce), only about one-third of which proved to be the critical hormone mixture. Instead of one hormone, this sample contained no fewer than twenty-nine distinct chemical species.

Reichstein isolated the twenty-nine species and then individually examined them. He identified the first four which were found to be biologically active, and later synthesized one of them. It was also Reichstein who demonstrated that these compounds were all **steroids**. Steroids are a group of chemicals which share a particular structure of four linked carbon-based rings; other important compounds having steroid structure include the **sex hormones**, **cholesterol**, and vitamin D.

Reichstein built on his earlier work with plant extracts to synthesize the steroid hormones. He and his colleagues developed several different methods to this end, though a process that used an animal waste product (ox bile) proved to be the most economical. One of the most important syntheses that Reichstein accomplished was that of aldosterone, which controls both water balance and sodium-potassium balance in the body. Aldosterone has been widely used in medical practice. Reichstein's work was also critical to the eventual syntheses of desoxycorticosterone, which for many years was the preferred treatment for Addison's disease, and **cortisone**, which is used for treating rheumatoid arthritis. It was principally for this latter accomplishment that Reichstein shared the 1950 Nobel Prize in chemistry.

Reichstein moved to the University of Basel in 1938 where he was appointed director of the Pharmaceutical Institute; in 1946 he became head of the **organic chemistry** division. Here he turned his attention to plant glycosides, a group of compounds with wide-ranging biological effects. They are the basis for a number of widely used drugs, and one of these, digitalis, has proven useful in controlling the heart rate. Reichstein was able to identify both the plants and the parts of the plants that contained glycosides, and his contributions were critical for initiating many botanical studies. He was one of the first researchers to realize the value of the tropical rain forests to the pharmaceutical industry. His work has also been pivotal in the field of chemical taxonomy, where the identities of plants are determined through their chemical composition—a method which has a higher degree of certainty than identification through visible characteristics. This technique has had broad applications in the development of both natural insecticides and drugs.

Reichstein was presented with an honorary doctorate from the Sorbonne in 1947. He received the Marcel Benoist Award in 1947, the Cameron Award in 1951, and a medal from the Royal Society of London in 1968. He is a foreign member of both the Royal Society and the National Academy of Sciences.

Reichstein married Henriette Louise Quarles van Ufford in 1927, while still at the Eidgenössiche Technische Hochschule. They had one daughter. He retired from his academic posts in 1967, but continued to work in the laboratory until 1987. He died on August 1, 1996, in Basel, Switzerland, at the age of 99.

Resonance

Some molecules and ions cannot be described using a single Lewis structure. This is because of resonance, the **bonding** in molecules and ions that result in more than one Lewis structure. Resonance is a feature of the valence-bond theory. In the

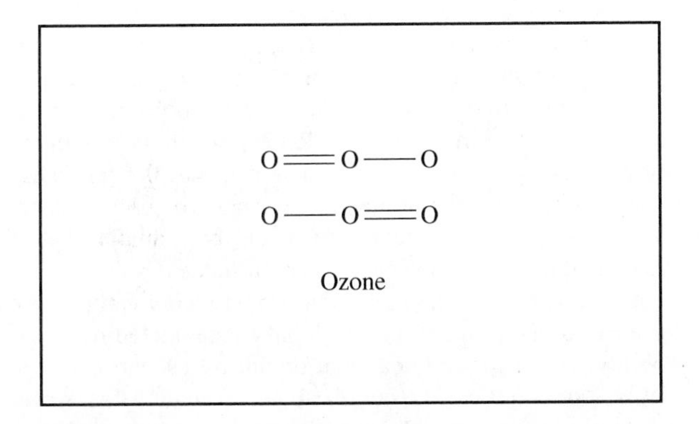

Figure 1. ***(Illustration by Electronic Illustrators Group.)***

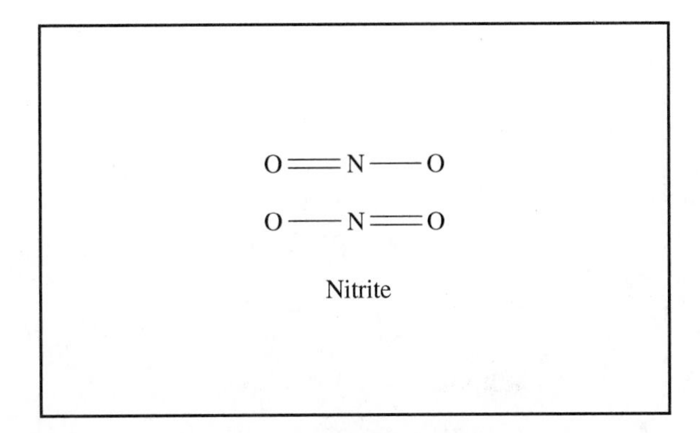

Figure 2. ***(Illustration by Electronic Illustrators Group.)***

valence-bond theory, Schrödinger's equation is used to determine the structure of molecules. If more than one **solution** to Schrödinger's equation is possible, then more than one **molecular structure** must be possible. According to the **valence** bond theory, various combinations of the solutions to Schrödinger's equations depict possible molecular structures. These different structures are called resonance structures or contributing structures. The actual structure of the **molecule** is called the resonance hybrid, a structure which represents the average of each of the resonance structures. It was originally believed that a molecule resonated between the different possible structures, much like a plucked guitar string vibrates back and forth. These different structures were thought to transfer **energy** to and from each other, thus the term "resonance."

An example of a molecule which demonstrates resonance is ozone, O_3 (Figure 1). Ozone can be represented by two different **Lewis structures**, O=O-O and O-O=O. Each structure represents two different bonds between the **oxygen** atoms, a single bond or a double bond. The difference between the two structures is the location of the double bond relative to the single bond. When resonance was first being investigated, scientists thought that ozone split its time between the two structures, spending half of its time with one structure and the other half with another structure. Later experiments showed, however, that both of the bonds involved in an ozone molecule are identical to each other. This resulted in a paradigm shift and scientists today theorize that ozone actually has one structure that is an average between the two possible Lewis structures. This average structure is the resonance hybrid for ozone. If drawing the Lewis structure for ozone, the two possible structures are drawn with a double headed arrow between them, showing that the **ozone** molecule exhibits resonance and exists in nature as a resonance hybrid.

Another example of a molecule which exists as a resonance hybrid in nature is NO_2. This molecule has 17 valence electrons. The three atoms are arranged in a linear fashion, and the placement of the first four valence electrons can be determined quite easily. The problem lies in the placement of the remaining 13 electrons. Even with the use of double bonds, the last 13 valence electrons cannot be placed such that each **atom** has eight valence electrons, satisfying the **octet rule**. Instead of one structure, two are possible: O=N-O and O-N=O. Again, notice that the difference between the two structures is in the placement of the double bond relative to the single bond (Figure 2). Neither of these structures satisfies the octet rule. In each of the structures, one **electron** is left unpaired on the **nitrogen** atom. In nature, atoms are only stable if they contain paired electrons. If any unpaired electrons are present on an atom, the atom is said to be unstable. In this example, both structures are equally likely to occur and no single Lewis structure can accurately describe this molecule. Lewis structures are simply models for representing the structure of molecules and are not completely accurate. Experiments show that both of the bonds in an NO_2 molecule are identical. The bond lengths between the atoms are between those found in single bonds and those found in double bonds. The resulting resonance hybrid is more stable than either Lewis structure.

In order to determine the Lewis structure of a molecule, one must know the number of valence electrons for the atoms involved and also understand the octet rule. The octet rule is the principle that describes the bonding in atoms. Individual atoms are unstable unless they have an octet of electrons in their highest **energy level**. The electrons in this level are called valence electrons. When atoms gain, lose, or share electrons with other atoms, they satisfy the octet rule and form chemical compounds. An example of this is the formation of molecular **fluorine** (F_2) from two individual fluorine atoms. A fluorine atom has seven valence electrons. According to the octet rule, eight valence electrons is the most stable configuration, so each fluorine atom needs one more electron. If each fluorine atom shares one electron with the other, the octet rule will be satisfied and a covalent bond is formed. Another example of a molecule formed by covalent bonds in order to satisfy the octet rule is **ammonia** (NH_3). A nitrogen atom has five valence electrons, so it needs three more to satisfy the octet rule. **Hydrogen** has one valence electron, so it needs one more to fill its outer energy level (hydrogen has electrons in only the first energy level, which fills when two electrons are present, unlike the other energy levels which require eight). If one nitrogen atom shares an electron with each of three hydrogen atoms, and the three hydrogen atoms each share an electron with the nitrogen atom, three covalent bonds are formed and the octet rule is satisfied for all four atoms.

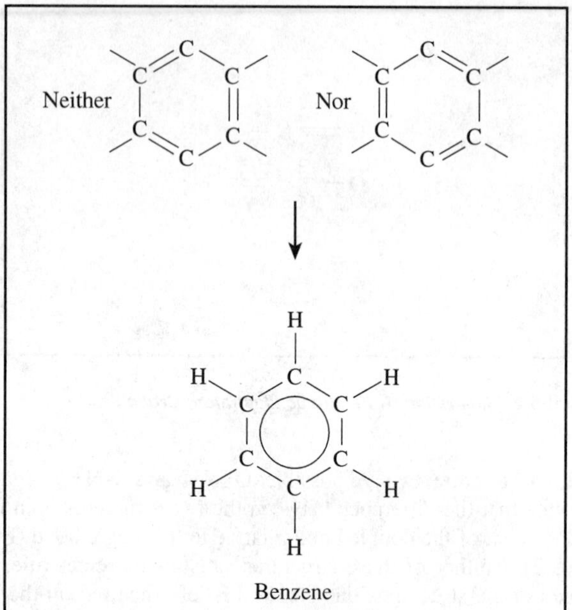

Figure 3. *(Illustration by Electronic Illustrators Group.)*

The octet rule is more or less a guideline for the purpose of determining Lewis structures for molecules. It is not always strictly adhered to, as demonstrated by the hydrogen atom in the previous example. The octet rule is a useful way to determine the structure of common molecules, especially those important biologically. Even after the development of the octet rule, however, scientists still had difficulty determining the structure of **benzene**, C_6H_6. Chemists were baffled as to why the structure of a molecule with such a simple chemical formula could be so difficult to determine. In the early 1870s, **F. A. Kekulé** theorized that benzene could be represented using two structures and that the benzene molecule in nature actually resonated, or vibrated back and forth, between these two possibilites. Kekulé's theory was not entirely correct but was accepted for many years. Once the Schrödinger equation was developed and used to explain resonance structures, the true nature of benzene was discovered. Benzene exists in nature not as a molecule oscillating between two structures, but as a resonance hybrid representing the average of each of the possible structures (Figure 3).

The above examples demonstrated the problem of finding structures to represent molecules which contain multielectron atoms. Once the Schrödinger equation was developed to show the exact three-dimensional structure of the hydrogen atom, the theory of resonance was developed to explain the multiple solutions sometimes obtained using this equation. Another theory developed to explain molecular structure is the **molecular orbital theory** of **hybridization**. This theory is difficult to understand and requires complicated mathematics. The valence-bond theory, using Schrödinger's equation, is a very useful theory to scientists because it is easily understood and quite applicable. The resonance structures obtained by the Schrödinger equation are the Lewis structures which scientists are familiar with and are comfortable using. Originally, chemists often preferred resonance theory over molecular orbital theory because of these reasons. Recent developments in the study of molecular structure have shown that the resonance theory does have some limitations. It cannot be used to accurately describe all molecular structures, for example, that of cyclobutadiene. This molecule is very unstable despite its similarities to benzene. The molecular orbital theory can account for many of the properties of cyclobutadiene, including its instability. This theory is not as comfortable as resonance theory, that is, it does not use the familiar Lewis structures that scientists have used for years. Despite this deviation from the tenets of classical **chemistry**, the molecular orbital theory is used extensively. The molecular orbital theory has gained popularity with chemists, but the theory of resonance hybrids is still in use by many scientists today.

RHENIUM

Rhenium is a transition metal, one of the elements found in the middle of the **periodic table**. It has an **atomic number** of 75, an atomic **mass** of 186.207, and a chemical symbol of Re.

Properties

Rhenium is a silvery metal that is both ductile and malleable. It has a melting point of 5,756°F (3,180°C), a **boiling point** of 10,166°F (5,630°C), and a **density** of 21.02 grams per cubic centimeter. All three of these values are among the highest for any chemical element. Rhenium is a moderately stable metal that does not react with **oxygen** or most acids very readily. It does react with concentrated **nitric acid** (HNO_3) and concentrated **sulfuric acid** (H_2SO_4).

Occurrence and Extraction

Rhenium is one of the rarest elements in the world with an abundance estimated at about one part per million in the Earth's crust. The major rhenium-producing countries of the world are Chile, Germany, the United Kingdom, and the United States. The metal is usually found as an impurity in **copper** and **molybdenum** ores.

Discovery and Naming

Rhenium was discovered in 1925 by a German research team that included Walter Noddack (1893-1960), **Ida Tacke Noddack**, and Otto Berg. The team named the element in honor of the Rhineland in western Germany.

Uses

About three-quarters of all rhenium consumed in the United States is used in the manufacture of superalloys—alloys containing **iron**, **cobalt**, or **nickel**. Such alloys have the ability to withstand very high temperatures and attack by oxygen and are used in making jet engine parts and gas turbine engines. Other rhenium alloys are used in making **temperature** control devices, such as thermostats; vacuum tubes, like those in a television set; and electromagnets, electrical contacts, and thermocouples. Finally, a relatively small amount of rhenium is used to make catalysts for the petroleum industry. The compounds of rhenium have virtually no commercial applications.

Rhodium

Rhodium can be classified as a transition metal, a member of the **platinum** group of **metals**, and as a precious metal. It is located in Group 9 of the **periodic table**, below **cobalt** (atomic number 27) and above **iridium** (atomic number 77). Rhodium's **atomic number** is 45, its atomic **mass** is 102.9055, and its chemical symbol is Rh.

Properties

Rhodium is a silver-white metal with a melting point of 3,571°F (1,966°C), a **boiling point** of about 8,100°F (about 4,500°C), and a **density** of 12.41 grams per cubic centimeter. It is a very good conductor of both **electricity** and **heat**. Chemically, rhodium is relatively inactive. It combines slowly with **oxygen** at high temperatures, but does not react with strong acids. It reacts with **chlorine** and **bromine**, but not, interestingly enough, with **fluorine**. It is one of a handful of elements that does not react with fluorine.

Occurrence and Extraction

Rhodium is one of the rarest elements on Earth with an abundance estimated at about 0.1 parts per billion. The element is found in combination with other members of the platinum group of metals in ores such as rhodite, sperrylite, and iridosimine. It is extracted as a byproduct from the recovery of platinum from its ores.

Discovery and Naming

Rhodium was discovered in about 1804 by the English chemist and physicist William Hyde Wollaston (1766-1828). Wollaston found two new elements, rhodium and **palladium**, in an ore found originally in South America. Wollaston suggested the name rhodium for one of the elements because the first compound of the element he prepared had a beautiful pinkish rose **color**. The Greek word for "rose" is *rhodon.*

Uses

Rhodium is used almost exclusively in the manufacture of alloys, especially alloys of platinum. Rhodium is harder than platinum and it has a higher melting point than platinum. It improves these two properties in an **alloy** containing the two elements. Rhodium alloys are used primarily for research and industrial applications to an limited extent. They are used to make thermocouples and to coat mirrors. Some compounds of rhodium have limited use as catalysts.

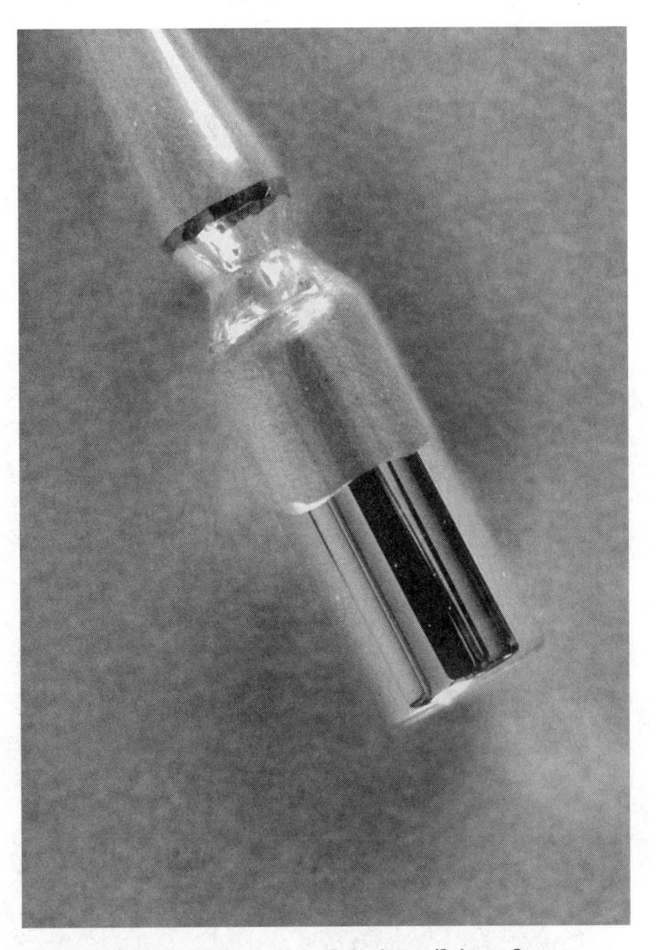

Rhenium samples. *(Photograph by Russ Lappa/Science Source, National Audubon Society Collection/Photo Researchers, Inc. Reproduced by permission.)*

Richards, Ellen Swallow (1842-1911)

American chemist

Ellen Swallow Richards was an applied scientist, sanitary chemist, and the founder of home economics. For twenty-seven years she was employed by the Massachusetts Institute of Technology (MIT), where she taught **chemistry** and developed methods for the analysis of air, **water**, and consumer products. Her work as a scientist and educator led to improvements in the home and opened the door to scientific professions for women.

Swallow was born on December 3, 1842, in Dunstable, Massachusetts. She was the only child of Peter Swallow, a teacher, farmer, and store keeper, and Fanny Gould Taylor, a teacher. She was educated at home by her parents until the family moved to Westford, Massachusetts, in 1859. There she attended Westford Massachusetts Academy, where she enrolled in mathematics, French, and Latin. In 1863 she graduated from the academy, and the family relocated to Littleton, Massachusetts.

Ellen Swallow Richards.

Swallow worked at an assortment of jobs—storekeeping, tutoring, housecleaning, cooking, and nursing—to earn enough money to continue her education. Because of her mother's ill health, she struggled with exhaustion and mental depression for a period of several years.

By 1868 Swallow had saved enough money to attend Vassar college, where she excelled in astronomy and chemistry. Her chemistry professor, convinced that science should be applied to practical problems, contributed to Swallow's developing interest in consumer and environmental science. Receiving a bachelor of arts degree in 1870, Swallow decided to apply to MIT to further her study of chemistry and became one of the first women students at that institution. She received a bachelor of science degree from MIT in 1873. In that same year, after submitting a thesis on the estimation of **vanadium** in **iron** ore, she received her masters of arts degree from Vassar. Although she continued her studies at MIT an additional two years, she was never awarded a doctorate.

Swallow married Robert Hallowell Richards, a professor of mining engineering, on June 4, 1875. The couple had no children and were able to devote their full support to each other's professional career. In her leisure time Richards enjoyed gardening, entertaining, and traveling. She also took an active interest in improving her own home. At one time she boasted of having year-round hot water and a telephone.

Richards helped establish a laboratory at MIT for women. While still an undergraduate, she had taught chemistry at the girls high school in Boston through a project funded by the Woman's Education Association. With the help of this association, Richards convinced MIT of the need for a women's lab, and in 1876, armed with the title of assistant, she began teaching chemical analysis, **industrial chemistry**, **mineralogy**, and biology to a handful of women students. In addition to their traditional studies, the students assisted in testing a variety of consumer products for composition and adulterations. After seven years, in which four students graduated and the rest were accepted as regular MIT students, the laboratory closed.

In 1884 MIT opened a new laboratory for the study of sanitation, and Richards was appointed assistant and instructor in sanitary chemistry. Her teaching duties included instruction in air, water, and sewage analysis. In addition, she was responsible for completing a two-year survey of Massachusetts inland waters (begun in 1887 for the state board of health). Her success in analyzing nearly forty thousand water samples was attributed to her knowledge of methodology, apparatus, and her excellent supervisory and record-keeping skills. The water survey work and her involvement with environmental chemistry were significant contributions to the new science of ecology.

Richards was a pioneer in the effort to increase educational opportunities for women. She was one of the founders of the Association of Collegiate Alumnae, which later changed its title to the American Association of University Women. She organized the science section for the Society to Encourage Studies at Home, a correspondence school founded in 1887 by Anna Tickenor. Her correspondence with students provided insight into the daily life and problems faced by women in the home. Richards learned that women were seeking help with a wide range of problems, not all of which were scientific in nature, including manners of dress, food preparation, and exercise.

In 1890 Richards opened the New England Kitchen in Boston as a means of demonstrating how wholesome foods could be selected and prepared. In 1899 she organized and chaired a summer conference at Lake Placid, New York, that established the profession of home economics. Conference participants explored new ways of applying sociology and economics to the home and developed courses of study for schools and colleges. Later she helped found the American Home Economics Association and provided financial support for its publication, the *Journal of Home Economics.*

In addition to her work at the sanitation laboratory, Richards consulted, lectured, authored ten books, and published numerous papers, including bulletins on **nutrition** for the United States Department of Agriculture. In 1910, in recognition of her commitment to education, she was appointed to supervise the teaching of home economics in public schools by the council of the National Education Association. In that same year she was awarded an honorary doctorate from Smith College. Richards died of heart disease in 1911 at the age of 68.

RICHARDS, THEODORE WILLIAM (1868-1928)

American chemist

Theodore William Richards, a professor of **chemistry** at Harvard University, was the first American chemist to receive a Nobel Prize. The prize was awarded to Richards in 1914 in chemistry for his accurate determination of the atomic weights of twenty-five chemical elements. He was renowned for his unsurpassed laboratory skill in chemical analysis. His work provided essential fundamental data for practical and theoretical chemists and physicists, and his graduate program at Harvard produced many eminent educators and research scientists.

Richards, the fifth of six children, was born on January 31, 1868, in Germantown, Pennsylvania, to William Trist Richards, a painter of seascapes, and Anna Matlack Richards, a writer and poet. Until he was fourteen, Richards received all his education from his mother, who had little regard for the public schools of Germantown. His interest in chemical experiments began when he was ten; when he was thirteen, he attended lectures in chemistry at the University of Pennsylvania. At the age of fourteen, Richards enrolled in Haverford College as a sophomore and graduated at the head of his class with a specialty in chemistry in 1885. Upon graduation, Richards enrolled at Harvard to study chemistry with Josiah Cooke, a professor whom he had met on summer vacation when he was six years old. Richards received his second baccalaureate, with *summa cum laude* distinction, at Harvard in 1886, and remained to study for the Ph.D. under Cooke's supervision. He received his doctorate in 1888, when he was twenty.

Richards's dissertation research marked the beginning of his study of atomic weights, which formed the major field of investigation in his long career. Richards published over 150 papers on atomic weights, beginning with his doctoral research on the atomic weight of **hydrogen**. At the time, chemists theorized that all elements were made from hydrogen and that there should be an integral ratio between the atomic weight of hydrogen and other elements. Although the theory called for the ratio of **oxygen** to hydrogen in **water** to be exactly 16, Richards's careful laboratory work, which involved difficult manipulations of **gases**, showed the ratio was actually 15.869 and strongly suggested that the theory was erroneous.

In the 1888–89 academic year, Richards received a fellowship and visited analytical laboratories in Europe. He returned to Harvard to become an assistant in the **analytical chemistry** course. He was promoted to instructor in 1891, assistant professor in 1894, and full professor in 1901 (after rejecting an offer from the University of Göttingen). He taught analytical chemistry until 1902 and **physical chemistry** from 1895, after the death of Josiah Cooke. He was chairman of the chemistry department from 1903 to 1911 and director of the Wolcott Gibbs Laboratory from 1912 until his death in 1928.

The measurement of atomic weights coincides with the beginning of modern chemistry, with **Antoine-Laurent Lavoisier**'s conception of chemical elements and with **John Dalton**'s **atomic theory**, which established that atoms are the building blocks of **matter**. The chemists of the nineteenth century had determined the atomic weights of all the known elements, and **Dmitry Mendeleev** based his **periodic table** of the elements on these values. However, all values of the atomic weights at the time were relative values, where ratios of atomic weights were actually determined in chemical compounds, and an error in a crucial ratio meant that the atomic weights of several elements would be inaccurate. For example, if the **silver** to **chlorine** ratio in silver chloride was inaccurate, then the atomic weight of **sodium** from **sodium chloride** or **potassium** from potassium chloride would consequently be erroneous.

Theodore William Richards.

Richards found that the long-accepted atomic weight values of the French chemist Jean-Servais Stas were incorrect because of several experimental errors which had been previously overlooked. The crucial part of the analysis of atomic weight involves the complete collection of a pure precipitate; Richards showed that Stas's compounds were impure and that Stas had not accounted for all of the chemical product. Richards and his students were able to redetermine accurately the atomic weights of twenty-five elements, and other chemists who had studied with Richards added thirty more elements. In his studies, Richards showed that an element's physical origin does not affect its atomic weight (he determined that terrestrial and meteoric **iron** have identical values). Richards's determination of physical constants, which were used by all chemists, won the admiration of the chemistry community and led to his Nobel Prize.

Richards's work was also important to the dramatic new discoveries of early twentieth-century scientists in **radioactivity** and the structure of the atomic **nucleus**. Richards determined that the atomic weight of ordinary **lead** was different from that of lead which came from the radioactive decay of **uranium**. This critical evidence corroborated theories of radioactivity, and because of Richards's reputation for accuracy, the experimental result was readily accepted by many scientists.

In addition to his work on atomic weights, Richards also directed research in physical chemistry. Before he began to teach physical chemistry at Harvard, he was sent to Germany to study with two leaders in the field, Nobel Prize winners **Friedrich Wilhelm Ostwald** and **Walther Nernst**. Richards invented an improved calorimeter for measuring **heat** in chemical reactions and published sixty papers in **thermochemistry**. He also contributed to the field of **electrochemistry**. His sixty graduate students became distinguished professors at other universities and continued the research studies they began at Harvard. Richards's influence on American chemical research in analytical and physical chemistry was exceptional.

In addition to the Nobel Prize, Richards won scientific awards from many nations. He also served as president of the American Chemical Society (1914), the American Association for the Advancement of Science (1917), and the American Academy of Arts and Sciences (1920–1921). Outside of his scientific interests, he enjoyed sketching, sailing, and golf.

In 1896, Richards married Miriam Stuart Thayer, daughter of Professor Joseph H. Thayer of the Harvard divinity school. Their son William studied chemistry with his father at Harvard and went on to teach at Princeton. Another son, Greenough, became an architect, and their daughter, Grace, married James B. Conant, one of Richards's graduate students. Conant became professor of **organic chemistry** and president of Harvard University. Richards died in Cambridge, Massachusetts, on April 2, 1928.

RING

A *ring* compound, also called a cyclic compound, is one whose molecules contain atoms chemically bonded together to create a closed chain or circle. Generally, such molecules are organic, and so the ring configuration is composed of **carbon** atoms. Ring compounds most often contain five or six atoms to form pentagonal or hexagonal structures. Ring compounds may also be assembled to form double-ring structures. For example, 10 carbon atoms can be arranged into two conjoined hexagons, each hexagon sharing two carbon atoms with the other. Similarly, nine atoms can form a pentagon bonded to a hexagon, each ring sharing two atoms with the other ring. Many cyclic, or ring compounds exhibit the special properties of aromatic compounds. Aromatic compounds are volatile and have strong, characteristic odors. They are any of a large class of ring compounds that have alternating single and double bonds, including **benzene** and compounds that resemble benzene. Common aromatic compounds other than benzene include toluene, **naphthalene**, and anthracene, all of which are present in **coal** tar or creosote. Ring structures are generally very stable. Exceptions include three member ring compounds, like ozone. Ozone, consisting of three **oxygen** atoms bonded into a strained triangular ring, is highly unstable, and therefore very reactive.

RISI, JOSEPH (1899-1993)

Canadian chemist

Joseph Risi, the principal force behind the rise of university research laboratories in French Canada, was born March 13, 1899, in Ennetbürgen, near Lucerne, Switzerland, to Alois and Marie Rothenfluh Risi. His father was a cabinetmaker. The exacting and slow nature of his father's trade persuaded Joseph Risi to find another path in life. He decided on teaching. After completing his secondary education at the Collège St.-Michel in Zug, in 1918, he enrolled in the Catholic University of Fribourg, which had received university status only in 1909. He first finished the four diplomas (*licences*) then required of Swiss science teachers—in mathematics, physics, **chemistry**, and biology. He stayed on for a doctorate in **organic chemistry**, which he completed in 1925 under the direction of A. Bistrzycki.

For the preceding five years, the ecclesiastical hierarchy of Laval University, in Quebec City, had been trying to establish an advanced school of chemistry. They had called in chemists (Paul Cardinaux, Julian J. Gutensperger, and Carl Faessler) and a physicist (Alphonse Cristen) from the University of Fribourg. But in 1925 Cardinaux and Cristen were fired, the former for financial improprieties and the latter for a sexual escapade. Other chemists came and went from both Switzerland and France. Offered a position as lecturer in organic chemistry, Risi immigrated to Canada in 1925. He became a full professor at the university in 1931.

Risi plunged immediately into organizing Laval's chemistry program. He supervised the installation of new laboratories, the funding for which had taken five years to obtain. In addition to teaching organic chemistry, he taught **mineralogy** and botany, served as librarian of his school, and from 1931 to 1936 was the scientific overseer of the first marine biological station on the St. Lawrence River, at Trois-Pistoles. Most importantly, he began original research with local students.

The research was inspired by the environment of French Canada. From the time of his arrival at Laval, Risi had tirelessly promoted the economic development of the region through **industrial chemistry**. By 1930, he was focusing on the aromatics of maple sugar. His interest was stimulated by the discovery of a vegetable product in South Africa that could mimic the distinctive **taste** of maple sugar—a development that threatened an important local industry. Beginning with the results of an experiment reported in a German publication, Risi and a student discovered that the aromatic could be produced artificially from oak resin, and in the process they identified a way to distinguish the true maple-syrup taste. This work was followed by a study of the chemical properties of Canadian rhubarb. At the same time, Risi explored the new domain of **polymer chemistry** (polymers are natural or synthetic materials, like rubber and **plastics**, that have a high **molecular weight** and are composed of repeating units). He and a student presented an early description of the polymerization of styrene. In all, Risi would publish more than fifty scientific papers during his career. The chemistry of aromatics from the natural world captured the attention of a number of chemists in the 1920s. Musk was the starting point for the Croatian-Swiss chemist **Leopold Ružička**, for example, who then moved on to explore the composition of **hormones** and who garnered a Nobel prize in 1939 for his work in this field. A measure of Risi's eminence is found in his being invited, in 1937, to propose a candidate for the Nobel Prize in chemistry. Although he was not the first to do so, Risi recommended Ružička to the Nobel authorities. By this act, Risi established Laval as the premier research university of French Canada.

Beginning about 1940, Risi became interested in the chemistry of **wood** products; in 1948 he joined the faculty of Forestry Engineering at Laval, and in 1950 he became director of the Canadian Institute of Forestry Products. Elected a fellow of the Royal Society of Canada in 1954, he directed Laval's graduate school from 1960 until his retirement in 1971.

Risi had married Alice Neuhaus in 1926; they had four children. A Roman Catholic, Risi enjoyed playing chess and fishing in his spare time. He died on July 21, 1993.

RITTER, JOHANN WILHELM (1776-1810)

German physicist and chemist

Ritter, the inventor of the dry-cell battery and discoverer of ultraviolet light, was an important contributor to the annals of science. However, during his lifetime he earned little respect from his peers in the European scientific community—chiefly due to his questionable philosophical opinions, which he incorporated into his experimental findings. Today he is considered an influential, if eccentric, researcher.

Ritter was born in Samitz (now a part of Chojnow, Poland) in 1776. At the age of fourteen he was sent to Liegnitz to serve as apprentice to an apothecary. While his apprenticeship allowed him to learn a trade, it also fostered in Ritter a profound interest in **chemistry**. During his nineteenth year he inherited a sum of money, enough to leave the apothecary and devote himself to academia. He enrolled at the University of Jena in April, 1796; he remained there as an instructor after receiving his degree.

Ritter's early research was with **electricity**, particularly in the way electrical charges could contract animal muscles. He began to form a general theory of nature, in which the organic and inorganic realms were intricately related. His experiments seemed to validate this philosophy, which shaped the direction of much of his future research.

The most fruitful period of Ritter's career began in 1800. In that year he succeeded in separating **hydrogen** and **oxygen** from **water** using **electrolysis**, collecting samples of each element. He continued to experiment with electricity, eventually using a current to separate molecules of **copper** in order to plate them to other **metals**; this introduced the art of **electroplating**. This process is now used for plating **gold**, **silver**, and other metals.

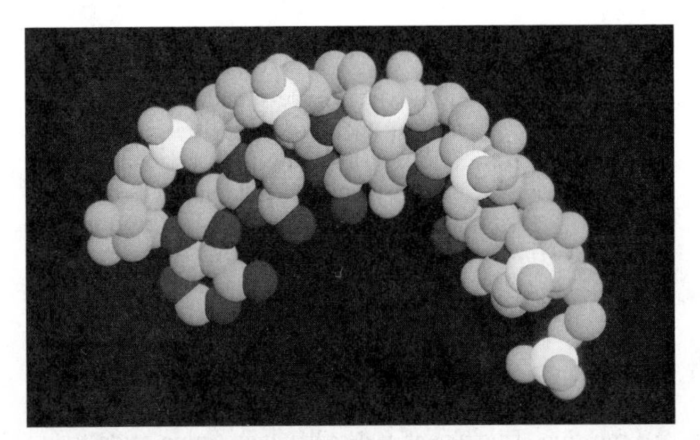

A computerized graphic of ribonucleic acid (RNA). ***(Photograph by Ken Eward. National Audubon Society Collection/Photo Researchers, Inc. Reproduced by permission.)***

In 1802 Ritter invented the first dry-cell battery, the electrolytes of which were in the form of a paste, rather than an easily spilled liquid. A year later he constructed an electrical storage battery.

The year 1800 also marked Ritter's first experiments with light as a chemical catalyst. It was known that silver chloride would break down rapidly when exposed to sunlight, and that blue light would initiate the process faster than red light. However, Ritter was surprised to find frequencies higher than the blue light of the visible spectrum; he therefore concluded that some invisible light existed beyond the blue end of the spectrum. (This was not a particularly bold conclusion, since William Herschel had announced the presence of infrared light just a year earlier.) Ritter named the new light *ultraviolet*, and sometimes referred to it as "chemical rays" because of its ability to break down certain compounds.

In all, Ritter published thirteen volumes describing his research and a score of journal articles; however, he lived in relative anonymity. He often announced his discoveries in short, mysterious letters, and more prominent scientists usually rediscovered and publicized phenomena studied by Ritter. In his later years, Ritter devoted himself to such dubious activities as water divining and metal witching. He died in 1810 of heart failure, after five years of increasing financial burden and declining professional consideration.

RNA

The letters RNA stand for ribonucleic acid, a type of nucleic acid. There are two types of nucleic acid—DNA (deoxyribonucleic acid) and RNA. **DNA** carries a cell's genetic information, but this information cannot be accessed without RNA. In some organisms, such as in certain viruses, RNA is the carrier of genetic information. It has also been discovered that RNA can sometimes act as a catalyst of chemical reactions.

RNA is almost always a single-stranded **molecule** constructed as a long chain of nucleotides. Each nucleotide contains a ribose molecule (a type of sugar), a **phosphate** group, and one of four nitrogen-containing cyclic units. The nitrogen-containing units are the bases: adenine, cytosine, guanine, and uracil. The nucleotides of DNA and RNA differ in two respects: an -OH group of the sugar present in RNA is -H in DNA and one of the bases of DNA is thymine rather than uracil.

DNA is a double-stranded molecule that contains the blueprint for constructing **proteins** necessary for cell structure and function. This blueprint details which **amino acids** to use and in what order they should appear. However, the information flow from DNA to protein is not direct. One strand of the DNA—the coding strand—serves as a master copy of the genetic code. The cell uses RNA to construct a working copy from that strand. Construction of this working copy is called transcription.

Transcription begins when an **enzyme**, RNA polymerase, binds to the coding strand of DNA just before the start of a gene. A gene is the stretch of DNA that codes for a particular protein or polypeptide. The RNA polymerase binds to DNA with the help of other proteins called transcription factors.

Once RNA polymerase binds to the DNA, RNA construction begins. The two strands of DNA separate and, for a short while, the coding strand is paired with RNA as a matching strand. The base pairing between the DNA strand and the RNA strand follows the same rules as base pairing between two strands of DNA. The key exception is that uracil is used in place of thymine. The mRNA (messenger RNA) detaches from DNA when transcription is complete. It may be modified before the next step, release into the cell's interior.

The next phase is translation of the code contained in mRNA. Translation relies on two other types of RNA: ribosomal RNA (rRNA) and transfer RNA (tRNA). Although there are three types of RNA, their differences are based more on their function and their associated proteins, rather than the nucleic acid itself.

The rRNA is found within a cellular structure called a ribosome. To start translation, a ribosome attaches itself to a strand of mRNA. The ribosome holds the mRNA in place and directs the speed of translation. The transcribed code carried on the mRNA is read one triplet—three nucleotides—at a time. There are 64 possible mRNA triplets, or codons. Sixty-one of them code for a specific amino acid; the remaining three serve as stop signals to end translation. Most amino acids have more than one associated codon.

As the ribosome gradually moves along the length of the mRNA, tRNA transports amino acids to the growing protein molecule. The tRNA contains a nucleotide sequence called the anticodon. The anticodon determines which amino acid the tRNA carries. When the anticodon pairs up with a corresponding mRNA codon, it releases the amino acid to the growing protein chain. Once the mRNA is translated, it is broken down by a special enzyme called ribonuclease. The component nucleotides can then be recycled by the cell.

In some cases, RNA serves as the carrier of genetic information. This situation appears with several viruses such as the tobacco mosaic virus and a class of viruses called retroviruses. Retroviruses include the human immunodeficiency virus (HIV), which causes acquired immunodeficiency syndrome (AIDS).

In the case of HIV, the virus contains RNA and an enzyme called reverse transcriptase. When the virus encounters a particular type of immune system cell, it injects its contents into it. The reverse transcriptase goes to work to construct a DNA molecule from the RNA. (This situation is the opposite of the usual mode of transcription, i.e., from DNA to mRNA.) The newly constructed DNA then attaches to the rest of the cell's DNA where it can hide for months or years before it is used.

The discovery that RNA can carry genetic information has led to other discoveries as well. One such discovery has been that an RNA molecule can sometimes act as an enzyme by catalyzing the modification of itself or other RNA molecules. This finding may provide insight into the origins of life. An outstanding scientific question regarding the origins of life has been whether enzymes or genes came first. Discovering that RNA can act as both a carrier of genetic information and as an enzyme may provide an answer to this question.

RNA is also taking a place in the areas of gene therapy and **genetic engineering**. Antisense technology makes use of a type of RNA called antisense RNA. Typically, a cell only contains mRNA that corresponds to the coding strand of DNA. However, by inserting an inverted copy of that gene into a cell's genetic material, it is possible to make the cell produce mRNA for both the coding strand and the matching strand. The two mRNA strands are complementary to one another and form a double-stranded molecule much like DNA. This pairing prevents the mRNA from being translated and the abnormal double-stranded RNA is often destroyed by ribonuclease.

Gene therapy is also making use of catalytic RNA through the construction of ribozymes. Ribozymes are catalytic RNA molecules that destroy RNA that has a specific sequence of nucleotides. A potential application of ribozymes might be treating certain diseases that are characterized by the presence of certain proteins. For example, if the mRNA for a viral protein were destroyed prior to being translated, new virus particles could not be constructed and the viral infection would be halted.

See also Protein synthesis

Roasting

There are three basic steps in the process of converting metal ores to free metals: preliminary treatment in which the metal-containing mineral is separated from less desirable parts of the ore or in which it is transformed to a compound which is more easily reduced; the **reduction** process itself, and, finally, purification or refining.

The reduction process which produces the free metal involves heating a metal **oxide** at high temperatures with **carbon** in the form of coke. This step is known as **smelting**, A flux is sometimes used that combines chemically with the infusible materials (gangue) in the ore to form slag, which floats on top of the metal and can be drawn off.

Many metal ores such as iron(II) sulfide (pyrites), mercury(II) sulfide (cinnabar), and lead(II) sulfide (galena), or **zinc** carbonate (smithsonite) must be converted to the oxide before reduction. This preliminary step is known as roasting. For carbonate ores, this involves a simple loss of **carbon dioxide** leaving the desired metal oxide. For sulfide ores, the roasting process uses **oxygen** in a blast furnace, hearth furnace, or Bessemer converter to convert the metal sulfide into the metal or the metal oxide plus **sulfur** dioxide. In the past, the evolution of sulfur dioxide from roasting facilities was a major cause of **acid rain**, but most sulfur dioxide is now captured and used to make **sulfuric acid**. Little **heat** is needed in the roasting of sulfide ores because the **oxidation** of the sulfide **ion** to sulfur dioxide is exothermic.

ROBERTS, RICHARD J. (1943-)

English biochemist

For decades scientists assumed that genes are continuous segments within deoxyribonucleic acid (DNA), the chemical template of heredity. In 1977, however, Richard. J. Roberts, a thirty-four year old British scientist working with adenovirus, the same virus that causes the common cold and pink eye, discovered that genes (the functional units of heredity) can be composed of several separate segments rather than of a single chain along the **DNA** strand. For his discovery of "split genes," Roberts was awarded the Nobel Prize in 1993.

Richard John Roberts was born on September 6, 1943, in Derby, England, a mid-sized industrial city about forty miles northeast of Birmingham. His father, John Roberts, was a motor mechanic, while his mother, Edna (Allsop) Roberts, took care of the family and served as Richard's first tutor. In 1947, the Roberts family moved to Bath, where Richard spent his formative years. At St. Stephen's junior school, Roberts encountered his first real mentor, the school's headmaster known only to the students as Mr. Broakes. Here he was exposed to a variety of mentally-stimulating games, ranging from crossword to logical puzzles. "Most importantly, I learned that logic and mathematics are fun!," Roberts wrote in a brief autobiography for the Nobel Foundation.

At the City of Bath Boys School (now Beechen Cliff School), Roberts became enamored with the life and literature of detectives, as they represented the ultimate puzzle solvers. His young career path changed abruptly, however, when he received a **chemistry** set from his parents. His ever supportive father had a large chemistry cabinet constructed and, with the aid of a local chemist who supplied the myriad chemicals he needed, Roberts soon discovered how to assemble fireworks and other concoctions not found in a beginner's chemistry manual. "Luckily I survived those years with no serious injuries or burns. I knew I had to be a chemist," he wrote in the Nobel Foundation autobiography.

At the age of seventeen, Roberts entered Sheffield University, where he concentrated in chemistry. His initial introduction to **biochemistry** was totally negative, he recalled in his autobiography: "I loathed it. The lectures merely required rote learning and the laboratory consisted of the most dull experiments imaginable." After graduating with honors in 1965, Roberts remained at Sheffield to study for his doctoral degree under David Ollis, his undergraduate professor of **organic chemistry**. But the direction of Roberts' scientific interests were profoundly altered after reading a book by **John Kendrew** on **crystallography** and molecular biology. Roberts became hooked on molecular biology and was later invited to conduct his postdoctoral work as part of a research team assembled by his colleague, Jack Strominger, a professor of biochemistry and molecular biology at Harvard University.

In 1969, Roberts left the English countryside and moved to Cambridge, Massachusetts, where he spent the next four years deciphering the sequence of nucleotides in a form of ribonucleic acid known as tRNA. Using a new method devised by English biochemist **Frederick Sanger** at Cambridge, he was able to sequence the **RNA molecule**, while teaching other scientists Sanger's technique. His creative work with tRNA led to the publication of two papers in *Nature* and an invitation by genetic pioneer and Nobel laureate, **James Watson**, to join his laboratory in Cold Spring Harbor, Long Island, New York.

In 1972, Roberts moved to Long Island to research ways to sequence DNA. American microbiologists Daniel Nathans and Hamilton Smith had shown that a restriction **enzyme**, Endonuclease R, could split DNA into specific segments. Roberts thought that such small segments could be used for DNA sequencing and began looking for other new restriction enzymes to expand the repertoire. (Enzymes are complex **proteins** that catalyze specific biochemical reactions.) He noted in his autobiography that his laboratory was responsible for discovering or characterizing three-quarters of the world's first restriction enzymes. In 1977, he developed a series of biological experiments to "map" the location of various genes in adenovirus and found that one end of a messenger ribonucleic acid (mRNA) did not react as expected. With the use of an **electron** microscope, Roberts and his colleagues observed that genes could be present in several, well-separated DNA segments. As he told the *New York Times,* "Everybody thought that genes were laid out in exactly the same way, and so it came as a tremendous surprise that they were different in higher organisms, such as humans."

In 1986, Roberts married his second wife, Jean. He is the father of four children, Alison, Andrew, Christopher and Amanda. He moved back to Massachusetts in 1992 to join New England Biolabs, a small, private company in Beverly, Massachusetts, involved in making research reagents, particularly restriction enzymes. He serves as joint research director. In 1993, Roberts was awarded the Nobel Prize for his discovery of "split genes." The Nobel Committee stated that, "The discovery of split genes has been of fundamental importance for today's basic research in biology, as well as for more medically oriented research concerning the development of cancer and other diseases."

Robert Robinson.

Robinson, Robert (1886-1975)

English chemist

Robert Robinson worked on many types of chemical problems, but he received the 1947 Nobel Prize for his work with the alkaloids, complex nitrogen-containing natural compounds that often exhibit high biological activity. His work in synthesis, identification, and reaction theory make him one of the founders of modern organic **chemistry**. Robinson summed up his philosophy about basic research when he said in his Nobel address, ''In both [chemistry and physics] it is in the course of attack of the most difficult problems, without consideration of eventual applications, that new fundamental knowledge is most certainly garnered.... Such contributions as I have been able to make are to the science itself and do not derive their interest from the economic or biological importance of the substances studied.''

Robinson was born to the inventor William Bradbury Robinson and Jane (Davenport) Robinson on September 13, 1886 near Chesterfield, England. His very large family included eight half-siblings from his father's first marriage, as well as four younger children. The family moved to New Brampton when Robinson was three years old. He received an excellent private education from the Fulneck School, run by the Moravian Church, and entered Manchester University in 1902. Robinson's family had manufactured bandages and other medical products for nearly a century and he was expected to enter the family business, so his father insisted that he study chemistry instead of mathematics. While at Manchester, Robinson studied under William H. Perkin, Jr., and after graduating with high honors in 1905, he worked in Perkin's private laboratory for five years before finishing his Ph.D. in 1910. In 1912, Robinson moved to Australia to take a teaching position at the University of Sydney. He returned to England in 1915 and held university appointments at Liverpool, St. Andrews, and Manchester, before finally landing at Oxford as Waynflete Professor of Chemistry, succeeding his mentor Perkin. Robinson remained at Oxford until his retirement in 1955. He also spent time as a consultant to the dye and petroleum industries.

Robinson's interests spanned all of **organic chemistry** (molecular structure elucidation, theoretical considerations, and synthesis), and most of them originated in Perkin's laboratory. He first studied such plant pigments as brazilin, a dyestuff obtainable from brazilwood, and the group of red/blue flower pigments called anthocyanins. He also worked on some of the steroid **hormones**, and synthesized several artificial estrogens. As did many scientists of the time, during World War II Robinson worked on war-related research efforts—from **explosives** to anti-malarial drugs to **penicillin**. Later in his life, Robinson became interested in **geochemistry**, particularly the origin and composition of petroleum. His work convinced him that plants must synthesize chemicals in certain ways, and he proposed a biosynthesis pathway (later confirmed by radioactive tracers) for some of the plant alkaloids. His contributions to chemical theory also include ideas on the **electron** distribution (and therefore the chemical reactivity) of aromatic compounds like **benzene**.

Alkaloids, although not the largest natural chemical compounds, are arguably the most complex, since they always contain **nitrogen** and usually some combination of **carbon** rings. Alkaloids as a group have profound biochemical effects on living things; cocaine, **morphine** and opium all belong to this class of **natural products**, as do many natural poisons. Robinson elucidated the structure of morphine and **strychnine**, and synthesized the alkaloids papaverine, hydrastine, narcotine, and tropinone.

In addition to receiving the 1947 Nobel Prize in chemistry for his work with the alkaloids, Robinson was knighted in 1939; he was also awarded the Order of Merit in 1949, and the Longstaff, Faraday, Davy, Royal, and Copley medals. In addition, he was an active member of numerous professional organizations around the world: at different times during his career, he served as president of the Royal Society, the British Association for the Advancement of Science, and the Society of the Chemical Industry. With the help of Nobel Laureate **Robert B. Woodward**, Robinson established the organic chemistry journal *Tetrahedron.*

Robinson married Gertrude Maude Walsh in 1912; they had a son and a daughter. Robinson's hobbies included music, literature, gardening, and photography, but his most enthusiastic pursuits outside of science were mountain-climbing (he and his wife explored ranges all over the world) and chess. He won several chess championships, served as president of the British Chess Federation, and collaborated on a book entitled *The Art and Science of Chess*. Three years after his wife's death in 1954, Robinson married Stearn Hillstrom. He retired from Oxford in 1955, and died on February 8, 1975.

ROELOFS, WENDELL L. (1938-)

American biochemist

Wendell L. Roelofs was instrumental in developing insect sex attractants—substances used to attract insects—for pest control in crops. An organic chemist by training, Roelofs has identified more than 100 attractants of different insect species, using a technique that was hailed as a major breakthrough. Roelofs conducted tests in fields to determine how to use the attractants to lure male insects to traps or to confuse them, thus preventing them from mating.

Wendell Lee Roelofs was born on July 26, 1938, in Orange City, Iowa, to Edward and Edith Beyers Roelofs. His father was a life insurance salesman and former superintendent of schools. Roelofs was the youngest of three boys; his two brothers also became scientists, one of them a chemist and the other an electrical engineer. As an undergraduate at Central College in Pella, Iowa, Roelofs majored in **chemistry**. He earned his bachelor's degree in 1960 and subsequently married Marilyn Joyce Kuiken. The couple raised four children: Brenda Jo, Caryn Jean, Jeffrey Lee, and Kevin Jon.

Roelofs attended graduate school at Indiana University in Bloomington and studied **organic chemistry**. He wrote his doctorate thesis on biologically active compounds with potential use in medicine. For his post-doctoral work, Roelofs moved on to the Massachusetts Institute of Technology (MIT). While looking for a job in 1965, Roelofs heard of an opening in the entomology department at Cornell University's New York State Agricultural Experiment Station. The department chair, Paul Chapman, was looking for an organic chemist to explore insect sex attractants called **pheromones**.

Rachel Carson's book, *Silent Spring*, published in 1962, had raised consciousness about the overuse of toxic **pesticides** to control insects and the need for alternatives. Female insects relied on pheromones to attract mates. Instead of poisons, pheromones could be used to prevent insects from mating and multiplying. Roelofs had never even taken a college course in biology, but the job piqued his interest in the subject. He was hired as an assistant professor, and his new research soon led him to the interface of the disciplines of chemistry and biology.

Every insect species used a unique blend of chemicals as a sex attractant; when Roelofs began his work in 1965, no more than a few had been identified. After nearly 30 years of research, one team of German researchers had discovered the composition of one sex attractant. The work required removing the glands from thousands of female insects, extracting the pheromone, and running it through tests to determine its chemical composition. The first task Roelofs faced at the agricultural station was to begin a mass breeding program to raise insects. He decided to study the voracious pest of apple crops, the redbanded leafroller moth. After extracting the pheromone from approximately 50,000 female moths, Roelofs used a new instrument called a gas chromatograph. He identified the pheromone's composition after about two years.

Roelofs then developed an even greater shortcut to identifying the pheromones. German researchers had been studying the response of silkworm antennae to pheromones by using an instrument called an electroantennogram. The antenna of a male moth was hooked up to a machine that recorded each time the moth responded to a pheromone. Roelofs realized that he could use the technique on male moths to identify sex attractants. Using the electroantennogram reduced identification time to a matter of days and, in some cases, hours. Roelofs isolated the sex attractants of more than 100 species, including the grape berry moth, the tobacco budworm moth, and the potato tuberworm.

As an anecdote about how the work became virtually routine, Roelofs recalled that he stopped identifying pheromones after those of the major pests of interest had been described. But upon retiring, a professor from the University of Michigan requested that Roelofs pinpoint one last pheromone. Roelofs offered to spend two days on it and pledged to quit if he could not get results. Within two days, he had found a blend that worked.

Once Roelofs knew the composition of the redbanded leafroller pheromone, he made an artificial blend in the lab. He tested it in the field, confirming that it did indeed attract male moths. From 1969 to 1972 Roelofs and his colleagues explored how to use the redbanded leafroller pheromones in pest control. They laced traps with pheromones to lure males. Moth populations could be suppressed, they found, using as few as one trap per tree. Since males detected pheromones in extremely minute amounts, the researchers also tried releasing enough pheromone to completely confuse and disorient them. In addition to using attractants to disrupt mating, Roelofs used them as a tool to lure insects to traps where their numbers could be monitored. Pesticide applications could then be reduced to times when they were strictly needed and most effective.

Throughout his career, Roelofs's work took him to many parts of the world. He joined delegations to the People's Republic of China in 1976, Japan in 1977, and the Soviet Union in 1978. Roelofs went to New Zealand in 1983 to help research the pheromones of pests attacking kiwifruit crops. Researchers there had successfully identified the pheromones, yet when the pheromones were used in the field, the insects failed to respond. Roelofs and his colleagues discovered the underlying reason: the populations were actually composed of different species that looked very similar but used different pheromones.

Continuing their work in the United States, Roelofs and his colleague Timothy J. Dennehy, an associate professor of entomology, found a way to use pheromones against grape berry moths, the most serious insect pest for grape crops in North America east of the Rocky Mountains. At the time, more than 100 tons of pesticides were required to control the insects each year in New York state alone. In 1984 Roelofs and Dennehy found a way to seal grape berry moth pheromones inside the hollow plastic and wire ties used to keep the grape plants on their trellises. The ties leaked the pheromones slowly over 100 days, distracting the males from finding the females. Experiments showed that vineyards that had been treated with the pheromone generally had less than one percent damage, compared to approximately 23% in untreated areas.

As Roelofs advanced in his career, he gained pleasure from exploring different facets of pheromone research, from chemical analysis to the design of traps. He investigated how insects made pheromones in their bodies. Once pheromones were in the air, he studied how males honed in on the source. In the basement of a campus building, Roelofs and his colleagues built a wind tunnel for flying insects. A pheromone was released into the tunnel, where they could watch an insect as it navigated toward the source.

Although Roelofs had not foreseen that his work would lead into insect **biochemistry**, he was pleased with the outcome and timing of his career. "I got in at the ground floor of pheromone research. The field was wide open," Roelofs told Miyoko Chu in an interview. One of the joys of his work, Roelofs said, was the privilege of being able to work in different subfields, including molecular biology, endocrinology, and behavior.

In his spare time, Roelofs coaches a youth league football team of kids aged eleven and twelve. Roelofs likened a cooperative effort in the laboratory to teamwork in football. With a coach's natural ability, he fostered an atmosphere where people could contribute their academic strengths and interests. "With our wide range of interests, we can always follow the most interesting lead whether it's my area of expertise or not," he told Chu. "That's how we stay at the forefront. It's synergistic. There's more creativity among us all."

In 1978 Roelofs was named the Liberty Hyde Bailey Professor of Insect Biochemistry at the New York State Agricultural Experiment Station. He was awarded the 1982 Wolf Prize in Agriculture, considered the most prestigious international award in that field. The following year, former U.S. President Ronald Reagan awarded him the National Medal of Science.

Roentgen Equivalent Man (REM)

Roentgen equivalent man (rem) is a unit of radiation dose equivalent that describes not only the amount of radiation that a human is exposed to, but also the biological damage that can result from exposure to the various types of radiation. The actual numerical value is computed by multiplying the exposure in rads (*i.e.,* absorption of 1 x 10^{-2} Joules of **energy** per kilogram of matter) by a normalizing factor that takes into account the fact that alpha rays cause more damage to human tissue than either x-ray or **gamma radiation**. The normalizing factor for x-ray and gamma radiation is unity. The dosage equivalent in rems for x-ray and gamma radiation is simply the dosage in rads. The normalizing factor for exposure to alpha particles is 20, for thermal neutrons is two, for fast neutrons is 10, for protons is 10, and for beta particles is one.

The National Council on Radiation Protection (NCRP) and Measurements recommends that the annual dosage for nonoccupational exposure not exceed 0.17 rem above the background radiation. For occupational exposure, the current federal safety standard is set at 5 rems per person per year. The average American is exposed to approximately 0.1-0.2 rem per year from natural sources such as cosmic rays, **radon** in the air and transuranic elements in various building materials, and from medically-related procedures such as diagnostic chest and dental x rays. Research and medical personnel who are exposed to radiation on a daily basis are required to wear a radiation badge, which contains unexposed pieces of photographic film. The darkening of the film is used as an index to measure the total amount of radiation that a person was exposed to during the time period that the badge was worn. Exposure to large radiation dosages can be lethal, or can lead to cancer and birth defects.

See also Radiation chemistry

Rose, William Cumming (1887-1985)

American biochemist

Born in South Carolina, William Rose graduated from North Carolina's Davidson College and received his Ph.D. from Yale in 1911. After post-doctoral work in Germany, he taught at the University of Texas from 1913 to 1922, then joined the staff of the University of Illinois where he remained until his retirement in 1955.

Throughout his career, Rose centered most of his attention on the **amino acids**, the nitrogen-containing organic compounds found in all protein molecules. It was already well known, of course, that **proteins** were essential to life. Still, since the early 1900s, researchers were aware that not all proteins were equally valuable. In 1912, for instance, **Frederick Gowland Hopkins** had shown that the protein in corn, called zein, could not sustain life in laboratory rats if it was the sole protein fed to them. But if casein, the protein in milk, was added to their diet, the rats once again thrived. What was the basic difference between the two proteins? Increasingly, biochemists—Rose, among them—suspected that the amino acid content of the proteins might be involved.

To discover the nutritional importance (if any) of the amino acids, Rose decided to feed his laboratory rats a diet composed not of intact proteins, but of various free amino acids. He began with a protein known to be nutritionally important: casein. Very quickly, he discovered that his rats did well on a rodent diet consisting of casein broken down into its various amino acids. However, if he reversed the process, and started with the amino acids widely believed to make up casein and formed together into the supposedly correct proportions, the lab rats suddenly began losing weight. The obvious conclusion: casein must contain another component—a nutritionally essential one—that had so far gone unnoticed.

Determined to find that unknown component, Rose began extracting and then testing various fragments of casein in the rodent diet. Finally, in 1935, he managed to isolate the right fragment. It proved to be threonine, the most important amino acid to be discovered.

Since threonine obviously couldn't be manufactured by the rats themselves, then its presence in the diet was essential. If one essential amino acid existed, Rose reasoned, there must be others. For the next few years, Rose manipulated first the rodent diet and, in the 1940s, the diets of human volunteers. By the mid-1940s he was able to show that, of the roughly 20 amino acids present in most protein molecules, about half could be synthesized by the body, the other half had to be supplied by the diet. In humans, the essential dietary amino acids are: isoleucine, leucine, lysine, methionine, phenylalanine, threonine, tryptophan, valine and, in infants and children, histidine.

Rose's findings not only helped solve some of the puzzling questions about protein **nutrition** but also clarified the concept of essential and nonessential amino acids. For his work on proteins and amino acids, Rose was honored with the National Medal of Science in 1967. He died in 1985 at the age of 98.

Rowland, F. Sherwood (1927-)

American atmospheric chemist

In 1974 **F. Sherwood Rowland** and his research associate, **Mario Molina**, first sounded the alarm about the harmful effects of **chlorofluorocarbons**, or CFCs, on the earth's ozone layer. CFCs, which have been used in air conditioners, refrigerators, and aerosol sprays, release **chlorine** atoms into the upper atmosphere when the Sun's ultraviolet light hits them; chlorine then breaks down atmospheric **ozone** molecules, destroying a shield that protects the earth from damaging ultraviolet rays. In the mid–1980s a National Aeronautics and Space Administration (NASA) satellite actually confirmed the existence of a continent-sized hole in the ozone layer over Antarctica. By the early 1990s NASA and National Oceanographic and Atmospheric Administration scientists were warning that yet another ozone hole, this one over the Arctic, could imperil Canada, Russia, Europe, and, in the United States, New England. This news might have been gratifying affirmation for Rowland, a professor of **chemistry** at the University of California at Irvine, but rather than rest on his laurels he continued to steadfastly—and soberly—warn the world of the ozone danger. His efforts have won him worldwide renown and prestigious awards, including the Charles A. Dana Award for Pioneering Achievement in Health in 1987, the **Peter Debye** Award of the American Chemical Society in 1993, the Roger Revelle Medal from the American Geophysical Union for 1994, and the Japan Prize in Environmental Science and Technology, presented to Rowland by the Japanese emperor in 1989.

Frank Sherwood Rowland always seemed destined to do something in science. Born June 28, 1927, in Delaware, Ohio, the son of a math professor, Sidney A. Rowland, and his wife, Latin teacher Margaret (Drake), Rowland said in an interview with Joan Oleck that math always came easy for him. "I always liked solving puzzles and problems," he said. "I think the rule we had in our family that applied even to my own children was you had your choice in school as to *what order* you took biology, chemistry, and physics, but not *whether.*"

Sidetracked by World War II, Rowland was still in boot camp when peace arrived. In 1948 he received his bachelor of arts degree from Ohio Wesleyan University; after three years—the summers of which he spent playing semiprofessional baseball—he obtained his master's from the University of Chicago. His Ph.D. came a year later, in 1952. That same year he married the former Joan E. Lundberg; the couple would eventually have a son and daughter.

The year 1952 was a banner one in Rowland's life; along with marriage and his doctorate he got his first academic job, an instructorship in chemistry at Princeton University, where he would remain four years. In 1956 Rowland moved his family west to the University of Kansas, where he was a professor for eight years, and then farther west still, to Irvine, California, where he took over as chemistry department chairman at the University of California in Irvine in 1964. He has stayed at Irvine ever since, enjoying stints as Daniel G. Aldrich, Jr., Professor of Chemistry from 1985 to 1989 and as Donald Bren Professor of Chemistry since 1989.

At Chicago, Rowland's mentor had been **Willard F. Libby**, winner of the Nobel Prize for his invention of carbon–14 dating, a way to determine the age of an object by measuring how much of a radioactive form of **carbon** it contains. The **radioactivity** research Rowland conducted with Libby led the young scientist eventually to **atmospheric chemistry**. Realizing, as he told Oleck, that "if you're going to be a professional scientist one of the things you're going to do is stay out ahead of the pack," Rowland looked for new avenues to explore. In the 1970s Rowland was inspired by his daughter's dedication to the then-fledgling environmental movement and by the tenor of the times: 1970 was the year of the first Earth Day. In 1971 the chemist helped allay fears about high levels of **mercury** in swordfish and tuna by showing that preserved museum fish from a hundred years earlier contained about the same amount of the element as modern fish.

Later events pushed him further in the direction of environmental concerns. At a meeting in Salzburg, Austria, Rowland met an Atomic Energy Commission (AEC) staffer who was trying to get chemists and meteorologists into closer partnerships. Sharing a train compartment with the AEC man to Vienna, Rowland was invited to another professional meeting. And it was there, in 1972, that he first began to think about chlorofluorocarbons in the atmosphere.

In those days, production of CFCs for household and industrial propellants was doubling every five years. A million tons of CFCs were being produced each year alone, but scientists were not particularly alarmed; it was believed they were inert in the atmosphere. Rowland, however, wanted to know more about their ultimate fate. Ozone, a form of **oxygen**, helps make up the stratosphere, the atmospheric layer located between eight and thirty miles above the earth. Ozone screens out dangerous ultraviolet rays, which have been linked to skin cancer, malfunctions in the immune system, and eye problems such as cataracts. Performing lab experiments with Molina, Rowland reported in 1974 that the same chemical stability that makes CFCs useful also allows them to drift up to the stratosphere intact. Once they rise thorough the ozone shield, Rowland and Molina warned, they pose a significant threat to ozone: each chlorine **atom** released when CFCs meet ultraviolet light can destroy up to one hundred thousand ozone molecules.

Sounding the alarm in the journal *Nature* in June of 1974 and in a subsequent presentation to the American Chemical Society that September, Rowland attracted attention: A federal task force found reason for serious concern; the National Academy of Sciences (NAS) confirmed the scientists' findings; and by 1978 the Environmental Protection Agency (EPA) had banned nearly all uses of CFC propellants. There were setbacks: In the 1980s President Ronald Reagan's EPA administrator, Anne Gorsuch, dismissed ozone depletion as a scare tactic. And Rowland himself discovered that the whole matter was more complex than originally thought, that another chemical reaction in the air was affecting calculations of ozone loss. The NAS's assessment of the problem was similarly vague, generalizing future global ozone losses as somewhere between 2 and 20 percent.

Then came a startling revelation. In the mid–1980s a hole in the ozone shield over Antarctica the size of a continent was discovered; NASA satellite photos confirmed it in 1985. The fall in ozone levels in the area was drastic—as high as 67 percent. These events led to increased concern by the international community. In 1987 the United States and other CFC producers signed the Montreal Protocol, pledging to cut production by 50 percent by the end of the millennium. Later, in the United States, President George Bush announced a U.S. plan to speed up the timetable to 1995.

There were more accelerations to come: Du Pont, a major producer, announced plans to end its CFC production by late 1994, and the European Community set a 1996 deadline. And producers of automobile air conditioning and seat cushions—two industries still using CFCs—began looking for alternatives. These goals only became more urgent in the face of the 1992 discovery of another potential ozone hole, this one over the Arctic. Scientists have attributed the extreme depletion of ozone over the poles to weather patterns and seasonal sun that promote an unusually rapid cycle of chlorine-ozone chain reactions.

In addition to the development of holes in the ozone layer, the atmosphere is further threatened because of the time delay before CFCs reach the stratosphere. Even after a complete ban on CFC production is achieved, CFCs will continue to rise through the atmosphere, reaching peak concentrations in the late 1990s. Some remained skeptical of the dangers, however. In the early 1990s a kind of "ozone backlash" occurred, with a scientist as prominent as Nobel Prize-winning chemist **Derek Barton** joining those who called for a repeal of the CFC phaseout pact.

Meanwhile Rowland continued his examination of the atmosphere. Every three months, his assistants have fanned out around the Pacific Ocean to collect air samples from New Zealand to Alaska. The news from his research has been sobering, turning up airborne compounds that originated from the burning of rain forests in Brazil and the aerial **pollution** of oil fields in the Caucasus mountains. "The major atmospheric problems readily cross all national boundaries and therefore can affect everyone's security," Rowland said in his President's Lecture before the American Association for the Advancement of Science (AAAS) in 1993. "You can no longer depend upon the 12-mile offshore limit when the problem is being carried by the winds." An instructive reminder of the international nature of such insecurity was given by the arrival only 2 weeks later in Irvine, California, of trace amounts of the radioactive fission products released by the 1986 Chernobyl nuclear reactor accident in the former Soviet Union.

Rowland has said in interviews that he's pleased with the progress he's helped set in motion. "One of the messages is that it is possible for mankind to influence his environment negatively," Rowland told Oleck. "On the other side there's the recognition on an international basis that we can act in unison. We have the [Montreal] agreement, people are following it, and it's not only that they have said they will do these things but they *are* doing them because the measurements in the atmosphere show it. People have worked together to solve the problem."

RUBBER, SYNTHETIC

Natural rubbers, before vulcanization, tend to be sticky and soft at high temperatures, while at low temperatures they are brittle and stiff, making them difficult to process. Because of these properties as well as the difficulties associated with obtaining adequate and affordable supplies of natural rubber, the search for natural rubber substitutes began.

In 1906 Farbenfabriken of Elderfeld began to search for a viable production process. Fritz Hofmann was appointed head of the research group. He attempted to polymerize isoprene, which was known to be a component of rubber. Convinced that the purity of the isoprene was paramount, he spent two years researching methods of producing pure isoprene. Finally he developed a six-step process. He next attempted to polymerize the pure isoprene. None of the techniques discussed in the scientific journals of the time were successful. Ultimately Hofmann simply heated the isoprene in an autoclave. Various viscous liquids were used in the heating process to emulsify the rubber. The product was comparable to natural rubber in toughness and elasticity and was much less sticky. But the mixture had to be shaken at 140°F (60°C) for several weeks. Because of the long production time and high cost, the process was abandoned as impractical.

Once World War I started and German access to natural rubber supplies was cut off, Farbenfabriken developed two grades of synthetic rubber, both made from 2,3-dimethyl butadiene. The softer grade was formed by pretreating the *monomers* (single units) with oxygen and polymerizing them at 149°F (65°C). The firmer grade used an initiator (usually preformed rubber) and was polymerized at 86°F (30°C). Despite the initiators both reactions took several weeks to complete.

After the war, the price of natural rubber fell below the cost of production for synthetic rubber. It was not until natural rubber producers raised the price from $.17/lb. to $1.21/lb. that interest in synthetic rubbers resumed. Buna rubber was made of butadiene (Bu) by sodium polymerization. Buna S was a copolymer of butadiene and styrene; Buna N used butadiene and acrylonitrile as copolymers. Both were developed by **Hermann Staudinger** and his research team, and patented in 1929. Both were more resistant to oil, gasoline, and aging than natural rubber.

Neoprene is another synthetic rubber discovered in the pre-World War II years. Father Julius Nieuwland (1878-1936) of Notre Dame University shared his research with **Wallace Carothers** of Du Pont. Father Nieuwland was researching the polymerization of acetylene, and his research showed strong similarities between the structure of polymerized acetylene and natural rubber.

Arnold Collins, a research chemist on Carothers' team, purified a sample of Nieuwland's acetylene and produced a small amount of a liquid which, when left out over a weekend, formed an *elastomeric* (rubber-like) solid. After additional research, a reaction converting acetylene to vinyl acetylene, using copper chloride as a catalyst and hydrogen chloride as an additive, was patented. The product of the reaction was 2-chlorobutadiene, called chloroprene because of its similarity to isoprene. Du Pont renamed it neoprene and began to market it in 1930. It was more expensive than natural rubber but had an even greater resistance to oil, **gasoline**, and **ozone**.

Thiokol Chemical Corporation produced a synthetic rubber from ethylene dichloride and sodium polysulfide. It was developed by J. C. Patrick, Thiokol's president, and marketed in 1929. Although it emitted an unpleasant odor, it found use in fuel tank linings for aircraft and for windshield seals for cars.

GR-S (government rubber, styrene type) came on the market during World War II, when there were again restricted supplies of natural rubber.

Studies by the German scientist **Karl Ziegler** and by the Italian **Giulio Natta** showed that organometallic compounds greatly increased the speed of some reactions. By using organometallic catalysts, scientists were able to prepare copolymers of styrene and butadiene very quickly, with reactions completing in less than twenty-four hours. GR-S rubbers are now commonly called SBRs (styrene-butadiene rubbers). Ziegler's catalyst system led to the development of ethylene-propylene rubber in 1960. Du Pont succeeded in crosslinking the ethylene-propylene rubber by adding small amounts of a third monomer (termonomer) in 1961; by 1967, Uniroyal, Copolymer Corporation, and Jersey Standard were producing ethylene-propylene rubber using a termonomer patented by Du Pont. Ethylene-propylene polymers (EPMs) are most often used in abrasion-resistant applications.

Today there are many synthetic rubbers on the market. Silicone rubbers are linear polymers (polymers whose molecules are arranged in long chain-like structures) derived from dimethyl silicone. They are difficult to process but are stable over a wide range of temperatures -130–601°F (-90–316°C). They are used in wire and cable insulation and gasket applications.

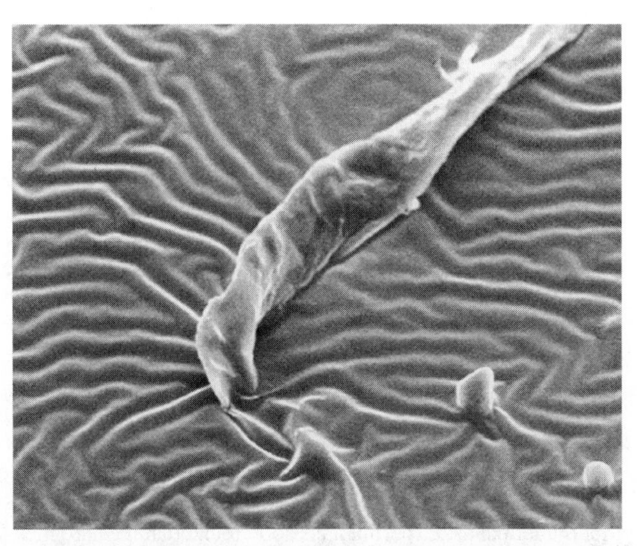

A scanning electron micrograph (SEM) of latex rubber used in the manufacture of condoms. ***(Photograph Andrew Syred/Science Photo Library, Photo Researchers, Inc. Reproduced by permission.)***

Besides these synthetic rubbers, many styrene-based polymers have been used in sporting goods. Thermoplastic elastomers are currently the fastest reacting and most economical synthetic rubbers available.

Most developments in the rubber industry since 1955 have been more characterized by technology transfer rather than innovation. The major West German synthetic rubber producer Chemische Werk Hüls decided to abandon its own obsolete technology and purchase a GR-S plant from Firestone in 1954. Synthetic rubber production began in Great Britain and Italy in 1957, in France and Japan two years later, and in Brazil in 1962. By the early 1970s, Japan had become a major producer of SBR as well as of all-cis polybutadiene and polyisoprene.

See also Neoprene; Synthetic rubber

Rubber, vulcanized

When rubber was first introduced to Europe from the New World, it was considered a marvelous novelty. Early explorers were amazed to find Caribbean natives playing games with bouncing balls made from the milky white juice (latex) of certain trees. A French explorer, Charles Marie de la Condamine, brought back samples of Indian-made rubber from the Amazon Valley in 1738 and set about promoting interest in the new substance. It was variously called *caoutchouc* (from the South American Indian word for it), gum elastic, and India rubber—"India" because the substance came from the West Indies, which Columbus thought were India, and "rubber" from the observation of British chemist **Joseph Priestley** that the substance rubbed out **lead** pencil marks.

Europeans were fascinated with rubber's attributes: it was elastic, waterproof, strong, springy, and moldable. Since

CH_3 H CH_3 H
C=C C=C
CH_2 CH_2 CH_2 CH_2

Structural formula of natural rubber. ***(Illustration by Electronic Illustrators Group.)***

latex coagulates quickly, rubber always arrived in Europe in solid form, usually as bottles. Manufacturers like Thomas Hancock (1786-1865) in England sliced up the bottles to make rubber novelties like shoe lasts, tobacco pouches, and rings that were used as garters and wristbands—the Western world's first rubber bands. Experimenters soon found that the hardened gum could be dissolved in turpentine and then reshaped. Cloth soaked in the liquid became waterproof, but it also smelled like turpentine. Hancock and **Charles Macintosh** solved that problem in the early 1820s by dissolving rubber in naphtha instead. Hancock also designed commercially successful rubber manufacturing machines.

A great craze for India-rubber products ensued, both in England and the United States. Five hundred pairs of India rubber boots were imported to Boston, Massachusetts, in 1823. In England, Macintosh began producing rubber-lined waterproof coats. Manufacturers vied with one another to produce rubber overshoes, coats, caps, wagon covers, and suspenders. Hancock developed rubber tubing from which he cut rubber bands and hoses. But natural rubber's most serious flaw soon showed itself: it is unstable over varying temperatures. People soon discovered that their overshoes became stiff and brittle in cold weather, and that in the **heat** their raincoats dissolved into a stinking, gummy mess that had to be disposed of by burial.

In 1839, Charles Goodyear (1800-1860), an American inventor who had devoted himself to improving the usefulness of rubber, discovered the answer. Goodyear, with absolutely no knowledge of **chemistry**, had spent five years mixing crude gum elastic with every possible substance, with the idea that sooner or later something would work. He had no success with **salt**, sugar, castor oil, ink, soap, or even cottage cheese. Magnesia, quicklime, and **nitric acid** all seemed promising for a time, but failed. Goodyear then experimented with Nathaniel Hayward's process of mixing rubber with sulphur. While doing so, Goodyear had a famous accident: he spilled some of his rubber-sulphur compound on a hot stove and was amazed to find that instead of melting, as natural rubber would have, it ''charred like leather'' and lost its stickiness. Goodyear noticed a tiny line of perfectly cured rubber on the edge of the piece. Further tests revealed that the cured rubber remained flexible even when left outdoors overnight in intense cold.

Goodyear spent the next five years perfecting his rubber-sulphur curing-by-heat process, securing a patent for it in 1844. The process became known as vulcanization—named after Vulcan, the Roman god of fire. A sample of Goodyear's vulcanized rubber found its way across the Atlantic to Hancock in England. Hancock studied and experimented with the sample and soon took out a British patent for his version of rubber-sulphur vulcanizing. Other variations of the vulcanizing process were developed, one by **Alexander Parkes** in 1846 which used sulphur monochloride vapor and another by S. J. Peachy using sulphur dioxide gas.

Vulcanizing made rubber a practical, eminently usable product. The rubber industry flourished, spewing out hundreds of everyday items (including waterproof clothing and footwear, fire hoses, rubber bands, mattresses, combs, and balloons), and contributed greatly to the process of industrialization. Rubber provided the electrical and communications industries with the effective insulation they badly needed. Rubber seals perfected industrial machinery. Perhaps most importantly, the use of rubber for pneumatic tires helped to expand both the automotive and rubber industries.

Today, natural rubber, which is essentially cis-1,4-polyisoprene, remains the strongest of all rubbers, with its excellent dynamic properties such as resistance to fatigue. In some products the rubber of choice is determined solely by properties (e.g., aircraft tires which require 100% natural rubber) but in many products the choice between between natural and synthetic rubbers hinges on price and properties. World consumption of natural rubber in 1996 was 6.13 million metric tons, corresponding to 39% of world consumption of all **elastomers** (i.e., natural plus synthetic rubber).

RUBIDIUM

Rubidium is the fourth member of the alkali metal family, consisting of elements in Group 1 of the **periodic table**. Its **atomic number** is 37, its atomic **mass** is 85.4678, and its chemical symbol is Rb.

Properties

Rubidium is a soft, silvery metal with a melting point of 102°F (39°C), a **boiling point** of 1,270°F (688°C), and a **density** of 1.532 grams per cubic centimeter. It is one of the most active elements, reacting vigorously with **oxygen**, **water**, acids, and the **halogens**. It is so active, in fact, that it is usually stored under some organic liquid in order to prevent it from combining explosively with oxygen in the air.

Occurrence and Extraction

Rubidium is a relatively common element with an abundance of about 35-75 parts per million. Its most common ores are lepidolite, carnallite, and pollucite. It is also found in seawater and in mineral springs. The pure metal is obtained by the **electrolysis** of molten rubidium chloride (RbCl): 2RbCl —electric current→ $2Rb + Cl_2$.

Discovery and Naming

Rubidium was discovered in 1861 by **Robert Bunsen** and **Gustav Kirchoff**. Bunsen and Kirchoff contributed to the development of the science of **spectroscopy** and used that science to analyze a number of well- known substances. During this analysis, they observed spectral lines that could not be attributed to any known element and assumed that they had found a new element. They proposed the name of rubidium for the element because of the dark red **color** of the most prominent of its spectral lines. In Greek, the word for "deep red" is *rubidus.*

Uses

Rubidium and its compounds have relatively few uses. It has seen some application in the manufacture of atomic clocks for specialized purposes in which very precise timekeeping is essential. It is also used occasionally in the manufacture of photocells, although other **alkali metals** are usually preferred for this purpose.

Ružička, Leopold (1887-1976)

Croatian-Swiss chemist

Leopold Ružička worked in what he referred to as the "borderland" between bio-organic **chemistry** and **biochemistry**. His studies of odorous **natural products** led to his discovery of **carbon** rings with many more carbon atoms than had been originally thought possible. His research also contributed important information on how living things biosynthesize some **steroids** and sex **hormones**. For this work he shared the 1939 Nobel Prize in chemistry.

Leopold Stephen Ružička was born on September 13, 1887, to Stjepan and Amalija (Sever) Ružička. He was the first of two boys. They lived at first in Vukovar in Eastern Croatia (later part of Yugoslavia). His father, a cooper, died when Ružička was about four years old, and the family then moved to Osijek to live with relatives. Ružička attended elementary and high school in Osijek, where he received a classical education (Latin and Greek), and was initially determined to enter the Catholic priesthood. As a teenager, he changed his interests to chemistry, and upon graduation began to look for graduate schools in Germany and Switzerland. He eventually settled on the *Technische Hochschule* in Karlsruhe, Germany, choosing it over the Swiss Federal Institute of Technology (Eidgenossische Technische Hochschule, or ETH) in Zurich, because it provided more flexibility in courses and did not require an entrance examination in descriptive geometry.

Ružička obtained his doctorate in only four years, under the direction of **Hermann Staudinger** at Karlsruhe. He then assisted Staudinger in research on the natural products in the *Chrysanthemum* species; these chemicals, called pyrethrins, were of particular interest as insecticides. In September of 1912, they both moved to ETH, where Ružička had originally considered studying, when Staudinger replaced **Richard Willstätter** as professor of organic and **inorganic chemistry**. Reflecting on these events later in his life, Ružička wrote, in *Annual Review of Biochemistry,* "The fact that I went to Karlsruhe for my training was a very important factor in my life. If I had taken my doctorate degree with Willstätter I should have gone to Germany with him in 1912, and two years later Germany was at war.... That war ended in 1918 with the destruction of the Habsburg empire and the beginning of bad times in Germany." Instead, he had established sufficient residency in Zurich by 1917 to obtain Swiss citizenship, and avoided the devastation caused during World War I.

In 1916, Ružička started his own research program, supported financially by a Geneva perfume company. (His position at ETH carried no salary until 1925, two years after he was named a professor.) The University of Utrecht, in the Netherlands, offered him a job as an **organic chemistry** professor in 1926. After three years there, he went back to Zurich to take on the job of directing ETH. During much of his career he was also supported financially by the Rockefeller Foundation.

Ružička studied various organic compounds early in his career, but in 1921 his most fruitful work began—on the structure and synthesis of several natural compounds important to the fragrance industry. (His collaborations with the Swiss pharmaceutical and perfume industries was to continue throughout his working life.) Before Ružička's discoveries, chemists thought that **ring** structures containing more than eight carbons would be unstable, because no one had been able to synthesize large rings. Ružička's research on muscone (obtained from the male musk deer) and civetone (from both male and female civet cats), however, indicated rings with as many as seventeen carbons—a huge number. He was able to synthesize some of these very large rings with new procedures developed by his research group.

Another line of research dealt with isoprene. Biochemists are interested in how living things bio-synthesize large molecules; they had known for some time that isoprene is one of nature's favorite building blocks. Ružička found many more large biochemicals that were constructed from isoprene units, and he formulated a rule of thumb called the "isoprene rule" for predicting biosynthesis based on this starting material. Ružička also synthesized testosterone and androsterone, the male **sex hormones**. In recognition of these successes he was awarded the Nobel Prize in chemistry in 1939, which he shared with **Adolf Butenandt**.

Ružička conducted research in an era when instrumentation was primitive by contemporary standards. The elucidation of **molecular structure**, therefore, depended entirely upon the observation of chemical reactions and the purification of reaction products. In this process an unknown compound would be exposed to various well-characterized reagents; if it reacted to give certain products, the chemist knew that the original **molecule** contained particular arrangements of atoms. (Ružička, for example, frequently used dehydrogenation—the removal of **hydrogen** atoms—to gain information about molecular structure.) Once these arrangements had been identified, the chemist would attempt to synthesize the original compound, and then compare the original and the synthetic. If they matched, that was taken as good evidence that the perceived structure was at least partly correct. This time-consuming, "wet" chemistry often gave ambiguous results, and polite arguments frequently occurred in scientific literature as the chemistry community debated the structure of a complicated new molecule. Often old rules had to give way when new discoveries were made. In his Nobel lecture, Ružička said, ''Experience has shown that there is no rule governing the architecture of natural compounds which is valid without exception, and which would enable us to dispense with the need to test its validity accurately for every new compound to be examined.''

Ružička married Anna Housmann in 1912; they were divorced in 1950. In 1951, he married Gertrud Acklin. He was an avid gardener and collector of paintings, so much so that he once said his chemistry had suffered as a result of his hobbies. He established an important collection of Dutch and Flemish Masters of the seventeenth century, as well as an art library on that general period, which he later gave to the Zurich art museum. During World War II, he worked to secure the escape of several Jewish scientists from the Nazis, and founded the Swiss-Yugoslav Relief Society. He was instrumental in providing refuge in Switzerland to the future Nobel Laureate **Vladimir Prelog**, who succeeded Ružička as the director of ETH when the latter retired in 1957. Ružička died on September 26, 1976.

Ruthenium

Ruthenium belongs to the **platinum** group of **metals** found in the middle of Rows 5 and 6 of the **periodic table**. Its **atomic number** is 44, its atomic **mass** is 101.07, and its chemical symbol is Ru.

Properties

Ruthenium is a hard, silvery white metal with a shiny surface and a melting point of 4,172-4,442°F (2,300-2,450°C), a **boiling point** of about 7,052-7,500°F (3,900-4,150°C), and a **density** of 12.41 grams per cubic centimeter. It is a relatively inert element that does not react with **oxygen**, most acids, or **aqua regia** (a mixture of concentrated **nitric acid** and 3-4 parts of hydrochloric acid).

Occurrence and Extraction

Ruthenium is one of the rarest elements on Earth with an abundance estimated at about 0.4 parts per billion. It usually occurs in combination with other members of the platinum family in ores of platinum. It is obtained as a byproduct of the purification of those ores and during the refining of **nickel** metal.

Discovery and Naming

Ruthenium was ''discovered'' at least three times in the first half of the nineteenth century. The first discovery was made by the Polish chemist Jedrzej Sniadecki (1768-1838) in 1808. Sniadecki suggested the name *vestium* for the element in honor of the asteroid Vesta. Sniadecki's work could not be confirmed by other chemists, however, and he eventually dropped his claim to having found the element. About 20 years later, this scenario was repeated when the Russian chemist Gottfried W. Osann again found element 44. Once again, Osann's discovery could not be confirmed and his claim was rejected. Finally, in 1844, the Russian chemist Carl Ernst Claus (1796-1864) was able to show conclusively that element 44 existed and it was the same element that both Sniadecki and Osann had found. Claus suggested the name of ruthenium for the element after the ancient name of Russia, *Ruthenia*.

Uses

The most important single use of ruthenium is in the manufacture of alloys with platinum or **palladium**. Ruthenium adds hardness and resistance to **corrosion** to such alloys. They are used to make electrical contacts, medical instruments, and some types of very expensive jewelry. Ruthenium is also used to make catalysts for research and industrial applications.

Rutherford, Daniel (1749-1819)

Scottish chemist

An uncle of the great novelist Sir Walter Scott (1771-1832), Rutherford was born in Edinburgh, Scotland. Following in the footsteps of his father, who had been a professor of medicine at the University of Edinburgh for nearly 40 years, Rutherford studied medicine at the university.

One of Rutherford's teachers was the chemist **Joseph Black**, who was renowned for his work on **carbon** dioxide, which he called fixed air. At the time, scientists knew that **carbon dioxide** does not support **combustion** or respiration. Black himself had shown that in an atmosphere of carbon dioxide, small animals die and candles do not burn.

Black directed Rutherford to study the component of air that does not support combustion which Rutherford called mephitic, meaning noxious or poisonous air. After repeating Black's experiments with candles and mice, Rutherford removed carbon dioxide from mephitic air by passing the air through limewater, a strong alkali that was known to absorb carbon dioxide.

Rutherford's results were surprising. The gas that was left over contained no carbon dioxide, yet it was still mephitic—it would not support life or combustion. Rutherford called the gas phlogisticated air. According to the theory of combustion accepted at the time, a substance called **phlogiston** is released from burning materials and absorbed by the surrounding air. Thus Rutherford believed that if a gas could no longer support combustion, it had absorbed all of the phlogiston it could hold.

Rutherford reported his experiments on phlogisticated air in 1772 in his doctoral dissertation. Although such chemists as **Henry Cavendish** and **Joseph Priestley** may have been aware of the gas, Rutherford's report was the first published record describing a gas in the air, other than carbon dioxide, that extinguishes life. A few years later, **Antoine-Laurent Lavoisier** developed the modern theory of combustion which states that **oxygen** is required for combustion to occur and explained the behavior of Rutherford's gas, which we now call **nitrogen**.

RUTHERFORD, ERNEST (1871-1937)
New Zealand-English physicist

Ernest Rutherford's explanation of **radioactivity** earned him the 1908 Nobel Prize in **chemistry**, but his most renowned achievement was his classic demonstration that the **atom** consists of a small, dense **nucleus** surrounded by orbiting electrons. He also demonstrated the **transmutation** of one element into another by splitting the atom. His direction of laboratories in Canada and Great Britain led to such triumphs as the discovery of the **neutron** and helped to launch high-energy, or particle, physics, which concentrates on the constitution, properties, and interactions of elementary particles of **matter**.

Rutherford was born the fourth of twelve children on August 30, 1871, to James and Martha Thompson Rutherford on the South Island of New Zealand near the village of Spring Grove. Both parents had arrived in New Zealand as children, not long after Great Britain annexed the territory into the Commonwealth in 1840. Rutherford's father, of Scottish descent, logged, cultivated flax, worked in construction, and pursued other endeavors with a mechanical inventiveness inherited from his wheelwright father, George. Martha Rutherford was a schoolteacher of English descent.

Rutherford's early success in school earned him a scholarship to Nelson College, a secondary school in a village on the north end of New Zealand's South Island. He then received a scholarship to Canterbury College at Christchurch, New Zealand, where he earned his bachelor of arts degree in 1892. He continued studying at Canterbury, earning a master of arts degree with honors in mathematics and mathematical physics in 1893 and a bachelor of science degree in 1894. In New Zealand, Rutherford met Mary Newton, the woman who would become his wife in 1900. She was the daughter of the woman who provided Rutherford with lodging while he studied at Canterbury. Rutherford and his wife had one daughter, Eileen (1901–1930), who married Ralph Fowler, a laboratory assistant of Rutherford's in the 1920s and 1930s.

While working toward his bachelor of science degree, Rutherford researched the effects of electromagnetic waves, produced by rapidly alternating electrical currents, on the magnetization of **iron**. He observed that, contrary to contemporary expectations, iron did magnetize in high-frequency electromagnetic fields. Conversely, he also showed that electromagnetic waves could demagnetize magnetized iron needles. On the basis of this observation, Rutherford devised a device for picking up electromagnetic waves produced at a distance. Italian physicist Guglielmo Marconi would later parlay the same principles into the development of wireless telegraphy, or radio.

These experiments earned Rutherford a scholarship in 1895 derived from profits from London's Great Exhibition of 1851. Rutherford attended Trinity College at Cambridge University to work under the direction of English physicist **J. J. Thomson** at the Cavendish Laboratory as the university's first research student. The laboratory had been established in 1871 for research in experimental physics and was first led by electromagnetism pioneer **James Clerk Maxwell**. Rutherford's

Ernest Rutherford.

demonstration of his electromagnetic detector greatly impressed Thomson and other scientists at the Cavendish almost immediately.

Thomson invited Rutherford in 1896 to assist him in studies of the effects of X rays, which had been discovered in 1895 by Wilhelm Conrad Röntgen, on the electrical properties of **gases**. Thomson and Rutherford demonstrated that x rays broke gas molecules into electrically and positively charged ions, making the gas electrically conductive. This work brought Rutherford widespread recognition in the British scientific community for the first time.

In 1897, Rutherford took up the study of radioactivity, the phenomenon discovered almost accidentally by French physicist **Henri Becquerel** in 1896. He began by studying the radioactive emissions of **uranium**, systematically wrapping uranium in successive layers of **aluminum** foil to observe the penetrating ability of these emissions. He concluded that uranium emitted two distinct types of radiation: a less penetrating type, which he called "alpha," and a more penetrating type, "beta." He also later observed what was described by French physicist Paul Villard as "gamma" radiation, the most penetrating type of all.

Not assured of a professorship at Cambridge, Rutherford applied for and was appointed Second MacDonald Professor of Physics at McGill University in Montreal, Canada, in

1898. McGill University appealed to Rutherford especially because it had perhaps the best-equipped laboratory in North America, if not the world, at the time. At McGill, Rutherford turned from studying uranium to **thorium**, another radioactive element. Although thorium emits alpha and **beta radiation** as does uranium, emission patterns for thorium substances seemed erratic. Rutherford determined in 1899 that an **emanation**, or new radioactive substance, was being produced. He also observed that the radioactivity of the emanation gradually decreased geometrically with time, an occurrence now known as the **half-life** of a radioactive substance, which is a measurement of the time it takes for half of a substance to decay. In 1901, Rutherford forged a partnership with **Frederick Soddy**, an Oxford chemistry demonstrator based at McGill who first encountered Rutherford in a debate on the existence of **subatomic particles**. Rutherford wanted Soddy to help him study the thorium emanations and to explain curious observations of radioactive substances noticed in Europe by Becquerel and by Sir **William Crookes**, who discovered **thallium**. Both had isolated the active parts of uranium from an apparently inert part. However, Becquerel also observed that the active part soon lost its activity, while the inert remainder regained its activity.

Rutherford and Soddy isolated the active part of radioactive thorium, which they named thorium-X, from the apparently-inert parent thorium. They charted how thorium-X gradually lost its radioactivity while the original thorium regained its activity, illustrating that thorium-X had its own distinctive half-life, which was much shorter than the half-life of thorium. Soddy tried to get thorium-X to interact chemically with other reagents without success. From these observations, Rutherford and Soddy put together the modern understanding of radioactivity in 1903. Thorium-X was a product of the disintegration or decay of thorium. In nature, radioactive elements and their products decay simultaneously. However, when the product is separated out, it continues to decay but is not replenished by decaying thorium atoms, so its radioactivity falls off. Meanwhile, the inert parent thorium eventually regains its radioactivity as it generates new radioactive products. Rutherford's explanation of radioactivity at the atomic level is what caused a sensation in scientific circles. He explained that radioactivity—alpha, beta, and gamma radiation—was the physical manifestation of this disintegration, the pieces of the thorium atom that were released as it decayed. In other words, thorium was steadily being transformed, or transmuted, into a new element that was lower in **atomic number**. It was this work that earned Rutherford the 1908 Nobel Prize in chemistry.

Rutherford received, and turned down, offers to teach at Yale and Columbia Universities in the United States. He became a Fellow of the Royal Society in 1903 and received the Rumford Medal in 1904. His books on radioactivity became standard textbooks on the subject for years and he was a popular speaker. He attracted a number of talented associates at McGill, the most famous of whom was **Otto Hahn**, a German physicist who would, with Austrian physicist **Lise Meitner**, demonstrate the fissioning of uranium in 1939.

In 1904, Rutherford was the first to suggest that radioactive elements with extremely long half-lives might provide a source of **energy** for sustaining the **heat** of the Earth's interior. This would supply a means for estimating the age of the Earth in the billions of years, allowing plenty of time for evolution by natural selection to proceed along lines outlined by the naturalist Charles Darwin in 1859.

In 1906, Sir Arthur Schuster offered Rutherford his chair as professor of physics at Manchester University in Great Britain. Eager to return to what was then the center of the scientific world, Rutherford accepted the position in 1907. Rutherford was again blessed at Manchester with a well-equipped laboratory and talented associates from around the world, such as Hans Geiger, who would develop the radioactivity counter; Charles Darwin, grandson of the famous naturalist; and physicists **Niels Bohr**, Ernest Marsden, and **H. G. J. Moseley**.

Rutherford proceeded with his study of radioactive emissions, particularly the high-energy alpha particles. His research was slowed at first when a sample of **radium**, his favorite alpha source, was sent by the Radium Institute of the Austrian Academy of Sciences in Vienna to a rival, **William Ramsay**, discoverer of the noble gases. Rutherford had to request and await another sample from the Institute before he could proceed in earnest with his work.

Rutherford wanted to determine precisely the nature of the alpha particles. In 1903, at McGill, he had succeeded in deflecting alpha particles in electric and magnetic fields, proving they had a positive charge. He was certain that the relatively massive particle must be equivalent to **helium** nuclei, which consist of two protons. At Manchester in 1908, Rutherford and his colleagues proved experimentally through spectroscopic means that the alpha particles were indeed nuclei of helium atoms.

In 1908, Rutherford and Geiger devised a method for counting alpha particles precisely. Alpha particles were fired into a nearly evacuated tube with a strong electric field. The resulting ionizing effect in the gas could be detected by an electrometer, a device that measures electric charge in a gas. The alpha particles could then be detected visually as well as they struck a **zinc** sulfide screen to cause an identifiable flash or scintillation. Geiger would build upon this technique in developing the electric radiation counter that bears his name.

In 1909, Rutherford had instructed Marsden to study the scattering of alpha particles at large angles. Marsden observed that when alpha particles were fired at **gold** foil, a significant number of particles were deflected at unusually large angles; some particles were even reflected backward. **Metals** with a larger atomic number (such as lead) reflected back even more particles. It was not until late in 1910 that Rutherford postulated from this evidence the modern concept of the atom, which he announced early in 1911. He surmised that the atom did not resemble a ''plum pudding'' of positively charged nuclear particles with electrons embedded within like raisins, as suggested by Thomson. Instead, Rutherford suggested that the atom consisted of a very small, dense nucleus surrounded by orbiting electrons. Geiger and Marsden provided the mathematics to support the theory and Bohr linked this concept with quantum theory to produce the model of the atom employed today. After World War I broke out in 1914, Rutherford was called upon to serve in the British Navy's Board of Invention and Research. His main area of research for the Board was in devising a method for detecting German U-boats at sea. His work established principles applied later in the development of sonar.

The work on **alpha particle** scattering continued during the war. Marsden observed in 1914 that alpha particles fired into **hydrogen** gas produced anomalous numbers of scintillations. Rutherford first concluded that the scintillations were being caused by hydrogen nuclei. However, Rutherford later observed the same effect when alpha particles were fired into **nitrogen**. After a long series of experiments to exclude all possible explanations, in 1919 Rutherford determined that the alpha particles were splitting the nitrogen atoms and that the extraneous hydrogen atoms were remnants of that split. Nitrogen was thus transmuted into another element. In 1925, English physicist Paul Maynard Stewart Blackett used the cloud chamber apparatus devised by Scottish physicist C. T. R. Wilson to verify Rutherford's observation and to show that the atom split after it had absorbed the alpha particle.

In 1919, Rutherford was persuaded to succeed Thomson as director of the Cavendish Laboratory at Cambridge, a post he would hold until his death. Rutherford directed the Cavendish during its most fruitful research period in its history. English atomic physicist John D. Cockcroft and Irish experimental physicist Ernest Walton constructed the world's first particle accelerator in 1932 and demonstrated the transmutation of elements by artificial means. Also in the early 1930s, English physicist James Chadwick confirmed the existence of the neutron, which Rutherford had predicted at least a decade earlier. Rutherford, with Chadwick, continued to bombard and split light elements with alpha particles. With Marcus Oliphant and Paul Harteck, Rutherford, in 1934, bombarded **deuterium** with deuterons (deuterium nuclei), achieving the first fusion reaction and production of **tritium**.

Rutherford was involved significantly in national and international politics during this period, albeit not for himself but for the sake of science. He worked with the civilian Department of Scientific and Industrial Research (DSIR) to obtain grants for his scientific team and served as president of the British Association for the Advancement of Science from 1925 to 1930. Beginning in 1933, Rutherford served as president of the Academic Assistance Council, established to assist refugee Jewish scientists fleeing the advance of Nazi Germany. When the Soviet Union prevented Russian physicist Pyotr Kapitsa, a promising Cavendish scientist, from returning to Great Britain from the Soviet Union in 1934, Rutherford launched an ultimately futile effort to convince the Soviets to release him. Rutherford maintained close relations through correspondence with leading scientists in Europe, North America, Australia, and New Zealand. Despite his preference for experiment, Rutherford corresponded with Bohr, German physicist **Max Planck**, American physicist **Albert Einstein** and other theoretical physicists transforming physics with relativity and quantum mechanics. He also remained close to his mother in New Zealand, exchanging letters with her frequently until her death in 1935.

In his long and distinguished career, Rutherford's most prestigious award, aside from his Nobel Prize, may have been the Order of Merit that he received from King George V in 1925. The Order of Merit is Britain's highest civilian honor. Rutherford was knighted in 1914 and made a peer (Baron Rutherford of Nelson or Lord Rutherford) in 1931. He died from complications after surgery on a strangulated hernia on October 19, 1937, in Cambridge. His cremated remains were buried near the graves of Isaac Newton and Charles Darwin at Westminster Abbey in London.

Rydberg constant

All **matter** radiates electromagnetic **energy** when it is heated. Because each element is different, the radiation it emits under conditions of heating is unique and characteristic of that element. Line emission spectra, produced by heating a very hot gas under pressure, consists of a series of bright lines separated by dark regions. Each of the lines in a spectral series corresponds to a characteristic frequency or wavelength.

Ordinary atomic optical spectra suggest that the motion of the electrons in the outer part of an **atom** depend as much on the interactions between the electrons as on the attractive forces between the electrons and the atomic **nucleus**, both forces being of about the same magnitude. In 1890, the Swedish spectroscopist Johannes Rydberg (1854-1919) identified a general relationship between the spectra of **hydrogen** and those of many other atoms. This relationship demonstrated the similarity of the spectral lines of elements with larger atomic numbers to the hydrogen spectral series.

Mathematically, Rydberg showed that the spectral terms in different spectral series can be represented by a formula. This formula contains a term which also appears in the mathematical representation predicting the characteristic wavelengths of some lines in the hydrogen spectrum. This common term is known as the Rydberg constant; its value (for wavelength measured in meters) is 1.097×10^7/m.

Because Rydberg's formula includes a term characteristic of the hydrogen spectra, the formula has been interpreted to mean that the spectra that obey it are representative of the last step of a process by which a neutral atom is built up by capturing and binding electrons to a nucleus, one at a time. That is to say, the force that is required to bind an **electron** to a nucleus shielded by previously bound electrons is very similar to the force experienced by an electron from the nucleus of a hydrogen atom when the electron is the same distance from the nucleus as in the shielded case.

See also Electromagnetic spectrum; Heat; Spectroscopy

S

SABATIER, PAUL (1854-1941)

French chemist

Paul Sabatier, who shared the 1912 Nobel Prize in **chemistry** with his countryman **Victor Grignard**, spent thirty-two years of a fifty-year career studying heterogeneous catalysis, especially the catalytic **hydrogenation** of organic compounds over finely divided **metals**.

Born on November 5, 1854, in Carcassone, France, Sabatier attended school in Carcassone, where his uncle was a teacher. An older sister helped tutor him, taking Latin and mathematics for that purpose. When his uncle transferred to the Toulouse Lycée, Sabatier followed. While at Toulouse, he used his free time to attend a public course in physics and chemistry that gave him a taste for science.

Accepted at both the École Polytechnique and the École Normale Supérieure in 1874, he entered the latter and graduated at the head of his class in 1877. He worked as an instructor in Nîmes for a year, but teaching secondary school physics was not what he wanted, and he returned to Paris as an assistant to Marcellin Berthelot at the Collège de France. There, in 1880, he earned his doctoral degree with a thesis on metallic sulfides.

After a year of teaching physics at the Faculté des Sciences at Bordeaux, he returned to Toulouse in 1882 to teach physics at the Faculté des Sciences there. In 1883, his duties expanded to include chemistry, and in 1884, at the age of thirty, the earliest allowable, he was appointed Professor of Chemistry. He remained in that post for the rest of his career, refusing offers from the Sorbonne to succeed **Henri Moissan** and from the Collège de France to succeed Berthelot. He was chosen Dean of the Faculty of Science in 1905, an office which he held for over twenty-five years. In addition to his research and teaching during this period, he was instrumental in the creation of schools of chemistry, agriculture, and electrical engineering at Toulouse. Even after his official retirement in 1929, he continued, by special permission, to lecture until failing health forced him to stop in 1939. Sabatier died on August 14, 1941.

Sabatier was a man of great reserve, and there is little information available about his private life. His marriage to Mlle. Herail was ended by her death in 1898. He never remarried, and their four daughters were raised with the help of his older sisters.

After receiving the Nobel Prize in 1912, Sabatier was elected a year later the first member of the Academy of Sciences, who did not reside in Paris. He had been a corresponding member since 1901, but the residency requirement kept him from full membership until a special class of six nonresident members was created, in part so that he could become a full member without having to move to Paris. He was made a Chevalier of the Légion d'Honneur in 1907 and named Commander in 1922. Among the many other honors bestowed on him by various organizations in different countries were the Davy Medal from the Royal Society in 1915 and the Franklin Medal from the Franklin Institute in Philadelphia in 1933. He received honorary doctoral degrees from the universities of Pennsylvania (in 1926, in conjunction with the Philadelphia Sesquicentennial celebration), Louvain, and Saragossa.

For his doctoral research and during his first fifteen years at Toulouse, Sabatier worked in the area of **inorganic chemistry**. His early work on the sulfides, **hydrogen** sulfides, and polysulfides of alkali and **alkaline earth metals** helped to clarify a complicated area of chemistry. He prepared the first pure sample of dihydrogen disulfide and was the first to make **silicon** monosulfide and tetraboron monosulfide as well as **boron** and silicon selenides. He carried out a number of thermochemical studies of the hydration (addition of H_2O) of metal chlorides and chromates and various **copper** compounds and was a pioneer in the use of absorption **spectroscopy** to study chemical reactions. **Absorption spectroscopy** exploits the unique patterns of light absorption characteristic of chemical substances to identify them. Spectroscopes scatter the light with a prism so that the dark absorption lines become visible in the spectrum.

Paul Sabatier.

In the 1890s it occurred to Sabatier to see if **nitric oxide** would produce a compound with **nickel** analogous to the recently discovered compound of nickel and **carbon** monoxide. These experiments, conducted with the chemist Jean-Baptiste Senderens, were not very fruitful, though some **nitrogen** compounds of copper, **cobalt**, nickel, and **iron** were obtained by the reaction of nitrogen dioxide with the metal. Sabatier then thought to use acetylene, an organic compound, but learned that Moissan and François Moreau had passed acetylene over powdered nickel made by heating nickel oxide with hydrogen and reported the formation of only carbon, some liquid hydrocarbons, and a gas they thought to be hydrogen.

In 1897, after being assured that Moissan and Moreau had no plans to continue their acetylene studies further, Sabatier and Senderens tried the reaction using the gas ethylene, another **hydrocarbon**. The experiment was successful and thus solidified Sabatier's switch to organic (carbon-based) chemistry. The result was again the formation of carbon, liquid hydrocarbons, and a gas, but on analyzing the gas, they found it to be mostly ethane and only a little hydrogen. Appreciating that the ethane could only have arisen through the addition of hydrogen to the ethylene (hydrogenation), they tried passing a mixture of ethylene and hydrogen over finely divided nickel and found that the smooth hydrogenation of ethylene took place at only a little above room **temperature** (30-40°C). For the next thirty-two years, Sabatier and his students investigated the heterogeneous catalysis (a process in which a third substance, or catalyst, influences the rate of a chemical reaction) of a variety of organic reactions by metals and metal oxides.

On the basis of his studies, Sabatier explained the catalytic action by the formation of unstable intermediate compounds between the catalyst and the reactant(s). This view, opposed to an earlier theory that the effect was due only to local extremes of pressure and temperature in small pores of the catalyst, proved to be correct and revolutionized **organic chemistry**.

In 1912, Sabatier's work was recognized by the shared award of the Nobel Prize in chemistry. The following year, he summed up his fifteen years of work on catalysis and reviewed the accumulating literature in the field in the book *La Catalyse en chimie organique.* Although his pioneer work was basic to the development of important industrial processes such as the catalytic cracking of petroleum to increase the yield of **gasoline** and the hydrogenation of vegetable oils to make shortening, Sabatier did not interest himself in such practical applications, nor did he profit from them.

SALT

When referring to a salt, chemists usually mean something other than the seasoning, or **sodium** chloride (NaCl). In chemical terms, a salt is a compound that is the result of a reaction between a base and an acid, or **neutralization**. For example, NaCl can be formed by the reaction of **hydrochloric acid** (HCl) with **sodium hydroxide** (NaOH). Salts are composed of ions rather than molecules, so the chemical symbol for a salt indicates the proportion of the elements that compose it (e.g., the symbol NaCl shows that table salt is made up of equal numbers of sodium and chloride ions). All salts are ionic compounds which contain the **cation** of a base, other than OH^- or O_2^-, and the **anion** of an acid, other than H^+.

One of the identifying characteristics of most salts is that they have an ionic lattice (a regular arrangement of ions) when in a solid crystal state and completely dissociate (break down into smaller components) when in **solution**. However, there are many different categories of salts, including:

(1) simple salts, such as NaCl, which have only one kind of positive **ion**.

(2) basic salts, such as **aluminum** hydroxide dichloride ($Al(OH)Cl_2$), which contain at least one hydroxyl group that can act as a base.

(3) acidic salts, such as sodium dihydrogen **phosphate** (NaH_2PO_4), which still has an acidic **proton** that can neutralize a base.

(4) amphoteric salts, such as aluminum hydroxide ($Al(OH)_3$), which can react with either bases or acids depending on the conditions.

(5) mixed salts, in which there are two or more different cations or anions. An example is sodium **zinc** uranyl acetate, $NaZn(UO_2)_3(CH_3COO)_9 \cdot 6H_2O$.

(6) double salts, in which two simple salts crystallize together such as iron(II) ammonium **sulfate** hexahydrate

($Fe(NH_4)_2)(SO_4)\bullet 6H_2O$) which can be crystallized from an aqueous solution of iron(II)sulfate ($FeSO_4$) and ammonium sulfate ($[NH_4]_2SO_4$).

A salt is classified as acidic if it is capable of neutralizing base, basic if it contains a hydroxyl (OH) group, or normal. A normal salt does not have any acidic protons or hydroxyl groups. There are also **hydrates**, such as the above mentioned iron(II) ammonium sulfate hexahydrate, which are salts that contain **water** even in their solid crystalline form. Epsom salt (magnesium sulfate heptahydrate $MgSO_4\bullet 7H_2O$) and Glauber's salt (sodium sulfate decahydrate $Na_2SO_4\bullet 10H_2O$) are examples of hydrates.

The **solubility** of a salt is a complex phenomenon and so only generalizations regarding salt solubility can be given. Generally the rate of dissolution increases with increasing surface area; thus, large crystals dissolve faster when ground into powder. Furthermore, the rate decreases with time since as the solid is dissolved the surface area becomes smaller as there is less and less crystallite.

The identity of the individual ionic components of a salt normally has a large effect on the salt's solubility. All common inorganic acids are soluble in water. All common salts of the **alkali metals**, such as sodium and **potassium**, and of the ammonium ion, NH_4^+, are soluble in water. All common chlorides are soluble in water with the exceptions of AgCl, Hg_2Cl_2, and $PbCl_2$. Bromides and iodides have similar solubilities to chlorides with some exceptions. Many common metal hydroxides (with the exceptions of the salts of the alkali metals and some of the alkaline earth metals) are insoluble in water.

Upon dissolution, the resulting solution will show the properties of each individual ionic species. For instance, dissolving cupric (copper(II)) salts in water will cause the water to turn bright blue, since this is the natural **color** of the cupric cation. Some salts can actually react with water in a process called **hydrolysis** to reverse neutralization by forming an acid and a base. Usually this is only possible, though, when the salt's "parents" were a weak acid and a weak base.

From an energetic standpoint, the ease of dissolution of a salt depends to a large extent on the crystal lattice **energy**. The crystal lattice energy is the strength of the interactions between the particles in the lattice. It is defined as the energy evolved in the formation of one **mole** of formula units in the crystalline state from the constituents in the gaseous state. For an ionic solid: $M^+(gas) + X^-(gas) \rightarrow MX(solid)$ + crystal lattice energy. The reverse of this reaction can be considered as an initial step in dissolving a salt. Additional factors that influence dissolution include solvation of the formed ions and energetic requirements to change the bulk properties of the solvent that accompany dissolution. Solvation is the generally energetically favorable interaction of individual solvent molecules with the ions released upon dissolution. The **balance** of these energetic factors determines whether the salt dissolves and if so, how much energy is evolved.

There are many reactions besides neutralization that produce salts. An acid and a salt can react to make an entirely different salt and acid, while two salts can react with each other to yield two new salts. Also, a metal can combine with a **nonmetal** to make a salt. For example, **chlorine** gas and sodium metal yield **sodium chloride** (NaCl). Another salt yielding reaction is that of metal oxides with acids; thus **calcium oxide** reacts with **nitric acid** to form calcium nitrate ($Ca[NO_3]_2$) and water.

A scanning electron micrographic (SEM) image of sodium chloride, or pure salt, recrystallized from distilled water. The crystal is built up from a cubic lattice of sodium chloride ions. ***(Photograph by Dr. Jeremy Burgess/Science Photo Library, Photo Researchers, Inc. Reproduced by permission.)***

See also Anion; Cation; Ion

SAMARIUM

Samarium is a rare earth element, one of the elements that occupy the space in Row 6 of the **periodic table** between **lanthanum** (atomic number 57) and **hafnium** (atomic number 72). Samarium's **atomic number** is 62, its atomic **mass** is 150.4, and its chemical symbol is Sm.

Properties

Samarium is a yellowish metal with a melting point of 1,962°F (1,072°C), a **boiling point** of 3,450°F (1,900°C), and a **density** of 7.53 grams per cubic centimeter. It is the hardest and most brittle of the **rare earth elements**.

Samarium is a fairly reactive metal that reacts with **water**, **oxygen**, acids, the **halogens**, and some other elements.

It ignites in pure oxygen at a **temperature** of about 300°F (150°C).

Occurrence and Extraction

Samarium is a moderately abundant element. It is estimated to occur at about 4.5-7 parts per million in the Earth's crust. It is found in the minerals samarskite, cerite, orthite, ytterbite, and fluorspar. It can be produced by heating samarium **oxide** (Sm_2O_3) with **barium** or lanthanum metal: $2Ba + Sm_2O_3 \rightarrow Ba_2O_3 + 2Sm$.

Discovery and Naming

Samarium was one of seven new elements found in an unusual mineral originally discovered near the town of Bastnas, Sweden, in the 1830s. The new element was ''discovered'' at least three times in the period 1879-80 by the French chemists Jean-Charles Galissard de Marignac (1817-1894), **Paul-Émile Lecoq de Boisbaudran**, and Eugène-Anatole Demarçay (1852-1904), all working independently. Credit for the discovery of samarium is often given to all three men. The element's name was taken from a mineral in which it occurs, samarskite which, in turn, was named after a Russian mine official, Colonel Samarski.

Uses

Like other rare earth elements, samarium is used to add **color** or special optical properties to **glass**. It is also used to make **lasers** and very strong electromagnets. One of the strongest magnets available is made from an **alloy** of samarium and **cobalt**. Samarium-cobalt (SmCo) magnets retain their magnetic properties at high temperatures and are not very reactive. They are used in certain types of airplane motors.

SAMUELSSON, BENGT (1934-)

Swedish biochemist

Bengt Samuelsson shared the 1982 Nobel Prize for physiology or medicine with his compatriot **Sune K. Bergström** and British biochemist John R. Vane ''for their discoveries concerning **prostaglandins** and related biologically active substances.'' Because prostaglandins are involved in a diverse range of biochemical functions and processes, the research of Bergström, Samuelsson, and Vane opened up a new arena of medical research and pharmaceutical applications.

Bengt Ingemar Samuelsson was born on May 21, 1934, in Halmstad, Sweden, to Anders and Kristina Nilsson Samuelsson. Samuelsson entered medical school at the University of Lund, where he came under the mentorship of Sune K. Bergström. Called ''the father of prostaglandin **chemistry**,'' Bergström was on the university faculty as professor of physiological chemistry. In 1958, Samuelsson followed Bergström to the prestigious Karolinska Institute in Stockholm, which is associated with the Nobel Prize awards. There, Samuelsson received his doctorate in medical science in 1960 and his medical degree in 1961, and he was subsequently appointed as an assistant professor of medical chemistry. In 1961, he served as a research fellow at Harvard University, and then in 1962 he rejoined Bergström at the Karolinska Institute, where he remained until 1966.

At the Karolinska Institute, Samuelsson worked with a group of researchers who were trying to characterize the structures of prostaglandins. Prostaglandins are hormone-like substances found throughout the body, which were so named in the 1930s on the erroneous assumption that they originated in the prostate. They play an important role in the circulatory system, and they help protect the body against sickness, infection, pain, and stress. Expanding on their earlier research, Bergström, Samuelsson, and other researchers discovered the role that arachidonic acid, an unsaturated fatty acid found in meats and vegetable oils, plays in the formation of prostaglandins. By developing synthetic methods of producing prostaglandins in the laboratory, this group made prostaglandins accessible for scientific research worldwide. It was Samuelsson who discovered the process through which arachidonic acid is converted into compounds he named endoperoxides, which are in turn converted into prostaglandins.

Prostaglandins have many veterinary and livestock breeding applications, and Samuelsson joined the faculty of the Royal Veterinary College in Stockholm in 1967. He returned to the Karolinska Institute as professor of medicine and physiological chemistry in 1972. Samuelsson served as the chair of the department of physiological chemistry from 1973 to 1983, and as dean of the medical faculty from 1978 to 1983, combining administrative duties with a rigorous research schedule. During 1976 and 1977, Samuelsson also served as a visiting professor at Harvard University and the Massachusetts Institute of Technology.

During these years, Samuelsson continued his investigation of prostaglandins and related compounds. In 1973, he discovered the prostaglandins which are involved in the clotting of the blood; he called these thromboxanes. Samuelsson subsequently discovered the compounds he called leukotrienes, which are found in white blood cells (or leukocytes). Leukotrienes are involved in asthma and in anaphylaxis, the shock or hypersensitivity that follows exposure to certain foreign substances, such as the toxins in an insect sting.

In the wake of such research, prostaglandins have been used to treat fertility problems, circulatory problems, asthma, arthritis, menstrual cramps, and ulcers. Prostaglandins have also been used medically to induce abortions. As noted by *New Scientist* magazine, the 1982 Nobel Prize shared by Bergström, Samuelsson, and Vane acknowledged that they had ''carried prostaglandins from the backwaters of biochemical research to the frontier of medical applications.'' In 1983, succeeding Bergström, Samuelsson was appointed as president of the Karolinska Institute.

The importance of Samuelsson's research has been recognized by numerous awards and honors in addition to the Nobel Prize. Such acknowledgments include the A. Jahres Award in medicine from Oslo University in 1970; the Louisa Gross Horwitz Prize from Columbia University in 1975; the Albert Lasker Medical Research Award in 1977; the Ciba-

Geigy Drew Award for biomedical research in 1980; the Gairdner Foundation Award in 1981; the Bror Holberg Medal of the Swedish Chemical Society in 1982; and the Abraham White Distinguished Scientist Award in 1991. Samuelsson has published widely on the **biochemistry** of prostaglandins, thromboxanes, and leukotrienes.

Samuelsson married Inga Karin Bergstein on August 19, 1958; they have three children.

SANGER, FREDERICK (1918-)

English biochemist

Frederick Sanger's important work in **biochemistry** has been recognized by two Nobel Prizes for **chemistry**. In 1958 he received the award for determining the arrangement of the **amino acids** that make up insulin, becoming the first person to thusly identify a protein **molecule**. In 1980 Sanger shared the award with two other scientists, being cited for his work in determining the sequences of **nucleic acids** in deoxyribonucleic acid (**DNA**) molecules. This research has had important implications for genetic research, and taken in conjunction with Sanger's earlier work on the structure of insulin, represent considerable contributions to combatting a number of diseases.

Frederick Sanger was born in Rendcombe, Gloucestershire, England, on August 3, 1918. His father, also named Frederick, was a medical doctor, and his mother, Cicely Crewsdon Sanger, was the daughter of a prosperous cotton manufacturer. Young Frederick attended the Bryanston School in Blandford, Dorset, from 1932 to 1936 and was then accepted at St. John's College, Cambridge. By his own admission, Sanger was not a particularly apt student. Later in life he wrote in *Annual Review of Biochemistry* that "I was not academically brilliant. I never won scholarships and would probably not have been able to attend Cambridge University if my parents had not been fairly rich."

Upon arriving at Cambridge and laying out his schedule of courses, Sanger found that he needed one more half-course in science. In looking through the choices available, Sanger came across a subject of which he had never heard—biochemistry—but that sounded appealing to him. "The idea that biology could be explained in terms of chemistry," he later wrote in *Annual Review of Biochemistry,* "seemed an exciting one." He followed the introductory course with an advanced one and eventually earned a first–class degree in the subject.

Sanger rapidly discovered his strengths and weaknesses in science. Although he was not particularly interested in or skilled at theoretical analysis, he was a superb experimentalist. He found that, as he later observed in *Annual Review of Biochemistry,* he could "hold my own even with the most academically outstanding" in the laboratory. This observation was to be confirmed in the ingenious experiments that he was to complete in the next four decades of his career.

After receiving his bachelor's degree from St. John's in 1939, Sanger decided to continue his work in biochemistry. Though World War II had just begun, Sanger avoided service in the English army because his strong Quaker pacifist beliefs qualified him as a conscientious objector. Instead, he began looking for a biochemistry laboratory where he could serve as an apprentice and begin work on his Ph.D. The first position he found was in the laboratory of protein specialist, N. W. Pirie. Pirie assigned Sanger a project involving the extraction of edible protein from grass. That project did not last long as Pirie left Cambridge, and Sanger was reassigned to Albert Neuberger. Neuberger changed Sanger's assignment to the study of lysine, an amino acid. By 1943, Sanger had completed his research and was awarded his Ph.D. for his study on the **metabolism** of lysine.

Frederick Sanger.

After receiving his degree, Sanger decided to stay on at Cambridge, where he was offered an opportunity to work in the laboratory of A. C. Chibnall, the new Professor of Biochemistry. Chibnall's special field of interest was the analysis of amino acids in protein, a subject in which Sanger also became involved. The structure of **proteins** had been a topic of considerable dispute among chemists for many years. On the one hand, some chemists were convinced that proteins consisted of some complex, amorphous material that could never be determined chemically. Conversely, other chemists believed that, while protein molecules might be complex, they did have a structure that could eventually be unraveled and understood.

Probably the most influential theory of protein structure at the time of Sanger's research was that of the German chemist **Emil Fischer**. In 1902, Fischer had suggested that proteins consist of long **chains** of amino acids, joined to each other head to tail. Since it was known that each amino acid has two reactive groups, an amino group and a **carboxyl group**, it made sense that amino acids might join to each other in a continuous chain. The task facing researchers like Sanger was to first determine what amino acids were present in any particular protein, and then to learn in what sequence those amino acids were arranged. The first of these steps was fairly simple and straight-forward, achievable by conventional chemical means. The second was not.

The protein on which Sanger did his research was insulin. The reason for this choice was that insulin—used in the treatment of diabetes—was one of the most readily available of all proteins, and one that could be obtained in very high purity. Sanger's choice of insulin for study was a fortuitous one. As proteins go, insulin has a relatively simple structure. Had he, by chance, started with a more complex protein, his research would almost certainly have stretched far beyond the ten years it required.

In 1945, Sanger made an important technological breakthrough that made possible his later sequencing work on amino acids. He discovered that the compound dinitrophenol (DNP) will bond tightly to one end of an amino acid and that this bond is stronger than the one formed by two amino acids **bonding** with one another. This fact made it possible for Sanger to use DNP to take apart the insulin molecule one amino acid at a time. Each amino acid could then be identified by the newly discovered process of **paper chromatography**. This was a slow process, requiring Sanger to examine the stains left by the amino acids after they were strained through paper filters, but the technique resulted in the eventual identification of all amino acid groups in the insulin molecule.

Sanger's next objective was to determine the sequence of the amino acids present in insulin, but this work was made more difficult by the fact that the insulin molecule actually consists of two separate chains of amino acids joined to each other at two points by sulfur-sulfur bonds. In addition, a third sulfur-sulfur bond occurs within the shorter of the two strands. Despite these difficulties, Sanger, in 1955, announced the results of his work: he had determined the total structure of insulin molecule, the first protein to be analyzed in this way. Sanger's work in this area was considered important because it involved proteins—''the most important substances in the human body,'' as Sanger described them in a *New York Times* report on his work. Proteins are integral elements in both the viruses and toxins that cause diseases and in the **antibodies** that prevent them. Sanger's research, in laying the groundwork for future work on proteins, greatly increased scientists' ability to combat diseases. For his important work on proteins, Sanger was awarded the Nobel Prize in chemistry in 1958.

Writing in *The Annual Review of Biochemistry,* Sanger referred to the decade after completion of the insulin work as the ''lean years' when there were no major successes.'' Part of this time was taken up with various research projects aimed at learning more about protein structure. In one series of experiments, for example, he explored the use of radioactive isotopes for sequencing. The work was not particularly productive, however, and Sanger soon undertook a new position and a new area of research.

In 1962 he joined the newly established Medical Research Council (MRC) Laboratory of Molecular Biology at Cambridge, a center for research that included such scientists as **Max Perutz**, **Francis Crick**, and Sydney Brenner. This move marked an important turning point in Sanger's career. The presence of his new colleagues—and Crick, in particular—sparked Sanger's interest in the subject of nucleic acids. Prior to joining the MRC lab, Sanger had had little interest in this subject, but he now became convinced of their importance. His work soon concentrated on the ways in which his protein-sequencing experiences might be used to determine the sequencing of nucleic acids.

The latter task was to be far more difficult than the former, however. While proteins may consist of as few as 50 amino acids, nucleic acids contain hundreds or thousands of basic units, called nucleotides. The first successful sequencing of a nucleic acid, a transfer **RNA** molecule known as alanine, was announced by **Robert William Holley** in 1965. Sanger had followed Holley's work and decided to try a somewhat different approach. In his method, Sanger broke apart a nucleic acid molecule in smaller parts, sequenced each part, and then determined the way in which the parts were attached to each other. In 1967, Sanger and his colleagues reported on the structure of an RNA molecule known as 5S using this technique.

When Sanger went on to the even more challenging structures of DNA molecules, he invented yet another new sequencing technique. In this method, a single-stranded DNA molecule is allowed to replicate itself but stopped at various stages of replication. Depending on the chemical used to stop replication, the researcher can then determine the nucleotide present at the end of the molecule. Repeated applications of this process allowed Sanger to reconstruct the sequence of nucleotides present in a DNA molecule.

Successful application of the technique made it possible for Sanger and his colleagues to report on a 12 nucleotide sequence of DNA from bacteriophage λ in 1968. Ten years later, a similar approach was used to sequence a 5,386 nucleotide sequence of another form of bacteriophage. In recognition of his sequencing work on nucleic acids, Sanger was awarded his second Nobel Prize in chemistry in 1980, shares of which also went to **Walter Gilbert** and **Paul Berg**. Their work has been lauded for its application to the research of congenital defects and hereditary diseases and has proved vitally important in producing the artificial genes that go into the manufacture of insulin and interferon, two substances used to treat diseases.

In 1983, at the age of 65, Sanger retired from research. He was beginning to be concerned, he said in the *Annual Review of Biochemistry,* about ''occupying space that could have been available to a younger person.'' He soon found that he very much enjoyed retirement, which allowed him to do many things for which he had never had time before. Among these were gardening and sailing. He also had more time to spend with his wife, the former Margaret Joan Howe, whom he had married in 1940, and his three children, Robin, Peter Frederick, and Sally Joan.

During his career, Sanger received many honors in addition to his two Nobel Prizes. In 1954, he was elected to the Royal Society and in 1963 he was made a Commander of the Order of the British Empire. He has been given the Corday-Morgan Medal and Prize of the British Chemical Society, the Alfred Benzons Prize, the Copley Medal of the Royal Society, and the Albert Lasker Basic Medical Research Award, among other honors.

SCANDIUM

Scandium is the first element in Group 3 of the **periodic table**. It has an **atomic number** of 21, an atomic **mass** of 44.9559, and a chemical symbol of Sc.

Properties

Scandium is a silvery white metal that develops a slight pink or yellow tint when exposed to air. It has a melting point of 2,800°F (1,538°C), a **boiling point** of about 4,900°F (2,700°C), and a **density** of 2.99 grams per cubic centimeter. Scandium is chemically similar to other **rare earth elements**. It reacts readily with acids, but not with **oxygen**.

Occurrence and Extraction

Scandium has an abundance of about 5-6 parts per million. It appears to be more abundant in the Sun and in some stars than it is on Earth. It is found in more than 800 different minerals, the most important of which are thortveitite and wolframite. Scandium is produced as a byproduct of other industrial operations, such as the mining of fluorite and **tantalum** ores.

Discovery and Naming

The existence of an element with an atomic number of 21 was predicted in 1869 by **Dmitry Mendeleev**. Mendeleev developed the periodic table and found that no known element was available to fill the space between **calcium** (atomic number 20) and **titanium** (atomic number 22). Mendeleev predicted that element 21 would be discovered and announced the properties he anticipated for the element. About ten years later, the Swedish chemist Lars Nilson (1840-1899) found the element. He suggested the name scandium for the element in honor of the region known as Scandinavia.

Uses

Scandium has relatively few commercial uses, the most important being in alloys. Scandium adds low density, resistance to **corrosion**, and a high melting point to alloys. Such alloys have been found to be especially desirable in sporting equipment, such as baseball bats, lacrosse sticks, and bicycle frames. No compound of scandium has any important commercial use.

SCANNING TUNNELING MICROSCOPE

The latter half of the twentieth century has opened to scientists an entirely new world: the world of atomic and **subatomic particles**. With the inventions of the **electron** microscope and the field **ion microscope**, scientists have been able to observe the microcosm as never before.

Until the early 1980s, however, one mystery that eluded researches was the nature of the surfaces of substances. Since the arrangement of atoms in the surface of a substance differs greatly from that of its bulk, it requires other methods of analysis. Scientists had lacked a mechanism for studying the intricacies of surfaces until 1981, when German physicists Gerd Binnig and Heinrich Rohrer invented the scanning tunneling microscope (STM).

The word *tunneling* used here describes an effect of quantum mechanics theorized upon for years and first verified in the laboratory in 1960. It was known that an electron orbits about the **nucleus** of an **atom**, its motion random and diffuse. When an atom is placed very close to another, the cloud-orbits of the surface atoms will overlap slightly. The resulting **diffusion** of electrons is called tunneling.

When Binnig and Rohrer met in 1978 they were both working at IBM research laboratories in different cities, each studying the atomic structures of surfaces. They decided to combine their efforts toward using tunneling to explore these structures. By 1980, they had constructed a prototype STM, and in the spring of 1981, they succeeded in obtaining microscopic images using electron tunneling.

The STM that Binnig and Rohrer had built was actually based upon the field ion microscope invented by Erwin Wilhelm Müller. The field ion microscope uses a tiny sharpened needle placed within a cathode-ray tube; as an electrical field is applied, metal ions are emitted from the tip of the needle, creating an image of the metal's **atomic structure** upon the cathode screen.

In Binnig and Rohrer's device, a similar needle is placed in a vacuum, above a specimen to be scanned at a height of less than one nanometer. The sharper the needle, the more precise is the STM reading; the best needle tips are only one or two atoms wide. A very low voltage is applied, causing the overlapping clouds of electrons to tunnel from the needle to the specimen. This electron flow is called a tunneling current, and by measuring this flow irregularities in the surface of the specimen can be determined. The scanning tip is swept over the sample, so that the entire surface may be mapped.

Even the very first tests of the STM showed it to be extremely powerful. When scanning crystals of calcium-iridium, Binnig and Rohrer resolved surface hills only one atom high. Maintaining the very small distance between the needle tip and the specimen proved to be difficult, however, since noises as unobtrusive as a footstep would jar the instrument. Binnig and Rohrer used magnets to suspend the microscope over a table equipped with shock absorbers to solve this problem. Other improvements increased the magnification of the STM, and today's tunneling microscopes can resolve features as small as one hundred-billionth of a meter, or about one-tenth the width of a **hydrogen** atom. It has also been discovered that STMs are equally useful in air, **water**, and cryogenic fluid media.

The incredibly precise three-dimensional images provided by STMs have found varied applications in a number of industries. They are used for quality control in manufacturing digital recording heads as well as in the construction of compact audio disk stampers. In 1991, the STM was used to move and place 35 atoms of **xenon** in a predetermined pattern; this ability to manipulate **matter** at the level of a single atom may allow scientists to customize molecules, possibly creating ultramicroscopic data storage chips.

Because it is effective in many media and uses a very low voltage, the STM can be used to study the atomic structure of living and biologic matter if that matter readily conducts electrons. Most biologic matter does not, however, and must be coated with a thin layer of a conducting substance. The STM is standard equipment in most atomic research laboratories. It is undeniably the most powerful optical tool yet invented, and for its invention Gerd Binnig and Heinrich Rohrer shared the 1986 Nobel Prize for Physics, along with Ernst Ruska, the inventor of the **electron microscope**.

SCHALLY, ANDREW V. (1926-)

Polish-American biochemist

Andrew V. Schally helped conduct pioneering research concerning **hormones**, identifying three brain hormones and greatly advancing scientists' understanding of the function and interaction of the brain with the rest of the body. His findings have proved useful in the treatment of diabetes and peptic ulcers, and in the diagnosis and treatment of hormone-deficiency diseases. Schally shared the 1977 Nobel Prize with French-born American endocrinologist **Roger Guillemin** and **Rosalyn Yalow** (an American scientist whose work in the discovery and development of radioimmunoassay, the use of radioactive substances to find and measure minute substances—especially hormones—in blood and tissue, helped Schally and Guillemin isolate and analyze peptide hormones).

Andrew Victor Schally was born on November 30, 1926, in Wilno, Poland, to Casimir Peter Schally and Maria Lacka Schally. His father served in the military on the side of the Allies during World War II, and Schally grew up during Nazi occupation of his homeland. The family later left Poland and immigrated to Scotland, where Schally entered the Bridge Allen School in Scotland. He studied **chemistry** at the University of London and obtained his first research position at London's highly regarded National Institute for Medical Research. Leaving London for Montreal, Canada, in 1952, Schally entered McGill University, where he studied endocrinology and conducted research on the adrenal and pituitary glands. He obtained his doctorate in **biochemistry** from McGill in 1957. Also in 1957, Schally became an assistant professor of physiology at Baylor University School of Medicine in Houston, Texas. There he was able to pursue his interest in the hormones produced by the hypothalamus.

Scientists had long thought that the hypothalamus, a part of the brain located just above the pituitary gland, regulated the endocrine system, which includes the pituitary, thyroid and adrenal glands, the pancreas, and the ovaries and testicles. They were, however, unsure of the way in which hypothalamic hormonal regulation occurred. In the 1930s British anatomist Geoffrey W. Harris theorized that hypothalamic regulation occurred by means of hormones, chemical substances secreted by glands and transported by the blood. Harris was able to support his hypothesis by conducting experiments that demonstrated altered pituitary function when the blood vessels between the hypothalamus and the pituitary were cut. Harris and others were unable to isolate or identify the hormones from the hypothalamus.

Schally devoted his work to identifying these hormones. He and Roger Guillemin, who also worked at Baylor University's School of Medicine, were engaged in research to unmask the chemical structure of corticotropin-releasing hormone (CRH). Their efforts, however, were unsuccessful—the structure was not determined until 1981. The two then focused their work, independently, on other hormones of the hypothalamus. Schally left Baylor in 1962, when he became director of the Endocrine and Polypeptide Laboratory at the Veterans Administration (VA) Hospital in New Orleans, Louisiana. Also that year, Schally became a U.S. citizen and took on the post of assistant professor of medicine at Tulane University Medical School.

Schally's first breakthrough came in 1966 when he and his research group isolated TRH, or thyrotropin-releasing hormone. In 1969 Schally and his VA team demonstrated that TRH is a peptide containing three **amino acids**. It was Guillemin, though, who first determined TRH's chemical structure. The success of this research made it possible to decipher the function of a second hormone, called luteinizing-hormone releasing factor (LHRH). Identified in 1971, LHRH is a decapeptide and controls reproductive functions in both males and females. The chemical makeup of the growth-releasing hormone (GRH) was also discovered by Schally's team in 1971. Schally was able to show that GRH, a peptide consisting of ten amino acids, causes the release of gonadotropins from the pituitary gland. These gonadotropins, in turn, cause male and female **sex hormones** to be released from the testicles and ovaries. In conjunction with this, Schally was able to identify a factor that inhibits the release of GRH in 1976. Guillemin, however, had determined its structure earlier and named it somatostatin. Subsequent studies by Schally showed that somatostatin serves multiple roles, some of which relate to insulin production and growth disorders. This led to speculation that the hormone could be useful for treating diabetes and acromegaly, a growth-disorder disease.

The hormone research done by Schally and his colleagues was tedious and expensive. Thousands of sheep and pig hypothalami were required to extract the smallest amount of hormone. These organs were solicited from many area slaughterhouses and required immediate dissection to prevent the hormones from degrading. Their accomplishment of isolating the first milligram of pure thyrotropin-releasing hormone, Guillemin stated, cost many times more than the NASA space mission that brought a kilogram of moon rock back to earth.

Schally's intense years of hard work and accomplishment were capped by the Nobel Prize, but he has also received many other awards and honors. In 1974 he was given the Charles Mickle Award of the University of Toronto, and the Gairdner Foundation International Award. He received the Borden Award in the Medical Sciences of the Association of American Medical Colleges in 1975 and, that same year, the Lasker Award and the Laude Award. He has held memberships in the National Academy of Sciences, the American Society of Biological Chemists, the American Physiology Society, the American Association for the Advancement of Science, and the Endocrine Society. In the years prior to receiving the Nobel Prize, Schally and his colleagues published more than 850 papers. Married to Brazilian endocrinologist, Ana Maria de Medeiros-Comaru, Schally often lectures in Latin America and Spain. He and his first wife, Margaret Rachel White, have two children.

Scheele, Carl Wilhelm (1742-1786)

Swedish chemist

Scheele's fame has been eclipsed by other chemists who made the same discoveries just a little earlier or who followed through on their experiments more thoroughly. Still, Scheele, who began training at age fourteen to be an apothecary has been recognized by Isaac Asimov as the greatest pharmacist in history.

Scheele was born and raised in the town of Stralsund in Swedish-controlled Pomerania. He eventually moved to Sweden, where he met and befriended many famous scientists, including chemist Johann Gottlieb Gahn (1745-1818) and mineralogist Torbern Olof Bergman (1735-1784). He worked at a series of drugstores, finally establishing his own pharmacy in 1776. Although Scheele could easily have chosen a more secure life as a professor at one of Europe's prominent universities, he preferred to practice pharmacy and to concentrate on his own experiments. He even turned down an opportunity to become the royal druggist to Prussia's Frederick II (1712-1786).

Like many of his contemporaries, Scheele tested many chemicals on himself. In his laboratory he prepared and tasted some of the most poisonous **gases** known to man, including **hydrogen** cyanide. Scheele's habit of being his own guinea pig may have contributed to his death. When he died, his symptoms resembled those of **mercury** poisoning.

Scheele is best remembered for discovering **oxygen** in 1771 and 1772. He prepared the gas by heating several different oxygen-containing compounds, such as mercuric **oxide**. Scheele found that many substances require oxygen in order to burn, so he called the oxygen *fire air*. Like most chemists at that time, Scheele used the **phlogiston** theory of **combustion** to describe chemical reactions. According to this theory, which has since been proved wrong, a substance called phlogiston is released from burning materials, and, in order for combustion to take place, air is required to absorb the phlogiston. Although Scheele modified this theory, stating that heat—and not phlogiston—was the necessary ingredient, it was left to French chemist **Antoine-Laurent Lavoisier** to develop the modern theory of combustion, which states that substances burn by combining chemically with oxygen. (In fact, in corresponding with Lavoisier, Scheele provided him with valuable information about oxygen that enabled the French scientist to complete his theory.)

Carl Wilhelm Scheele.

In studying how oxygen reacts with different compounds, Scheele weighed the gas and found that another gas, which he termed *foul air*, was also present in the atmosphere. Scheele conducted tests in which all of the oxygen and **carbon** dioxide in a container of air were consumed, leaving only foul air behind; today we know this gas is **nitrogen**.

Unfortunately, the publication of Scheele's discoveries was sidetracked by a run of bad luck. He wrote about his experiments in *Chemische Abhandlung von der Luft und dem Feuer* (*Chemical Observation and Experiments on Air and Fire*), which he finished in 1775. Publication, however, was first delayed by Scheele's colleague Bergman, who took a long time to write an introduction for the book, and then by printing problems; the book did not appear in print until 1777. Because British chemist **Joseph Priestley**, who was doing similar experiments with oxygen, was the first to report his findings, most people credit him with discovering the element.

Scheele's pattern of anticipating or independently duplicating others' discoveries is also illustrated by his research on **chlorine**. In 1774 Scheele produced chlorine by combining **hydrochloric acid** with a **manganese** compound. However, he failed to recognize chlorine as an element in its own right, believing instead that it was a compound of oxygen. More than thirty years later, British chemist Sir **Humphry Davy** isolated chlorine and declared its elemental nature, thus earning the credit for its discovery.

Although Scheele also had a hand in isolating and studying many other elements, he was far more successful in discovering new compounds, especially such organic acids found in plants as **tartaric acid**. Working with plant juices, Scheele went on to discover **citric acid**, which he crystallized from lemon juice, and malic acid, which he found in apples. Scheele investigated more than twenty kinds of fruits and berries as well as other vegetable materials, and he obtained **oxalic acid** from sugar. Turning to the animal kingdom, Scheele studied eggs, blood and glue, and he discovered uric acid which is normally excreted in bodily waste but can cause kidney stones. He also identified lactic acid.

Two additional compounds he studied were named for him: scheelite, a **calcium** tungstate mineral used in high-speed tools, and **copper** *arsenite*, which is called Scheele's green. Scheele also developed new methods for preparing **arsenic** acid, and he demonstrated that **graphite** is a form of carbon.

More than fifty years before the development of photography, Scheele discovered the effect of light on **silver** compounds. Salts of silver modified by the action of light were later used in the first photographic emulsions.

SCHÖNBEIN, CHRISTIAN (1799-1868)

German chemist

Schönbein was a chemist who is best known for his discovery and research on **ozone** (O_3).

Schönbein's family was financially unable to continue his formal education past the age of fourteen, so he was apprenticed to a chemical and pharmaceutical company in Boeblingen. For the next seven years, Schöenbein worked at the plant and studied **chemistry**, philosophy, mathematics and several languages in his spare time. In 1820 he finished his apprenticeship and moved to Augsburg where, for a short time, he translated scientific papers from French to German. He soon found a position with another chemical company where he remained for three years.

In 1823, Schönbein took a post under Friedrich Froebel (1782-1852) teaching chemistry, physics, and **mineralogy**. After three years with Froebel, Schönbein moved to England to teach mathematics and natural history. He stayed in England only one year, then moved again in 1827, this time to France. There he was awarded an honorary Ph.D. and appointed a professor of physics and chemistry at the University of Basel. It was during these years that Schönbein conducted his most significant research.

Schönbein began his studies by exploring the passivity of **iron**. Although he reached several interesting conclusions, the methods he used to conduct his studies were qualitative and relied heavily on analogies rather than empirical data. Thus, most of his research into passive iron has since been disproved. Schönbein's major accomplishment came almost coincidentally. Because his laboratory at Basel was poorly ventilated, he was able to notice an unusual odor that was produced during the electrolytic decomposition of **water**. He observed that this odor was similar to the smell produced around large electrical apparatus machines. He theorized that in both cases this smell was due to a new type of gas. This same gas, he discovered, was also produced when **phosphorus** oxidized. He also discovered the gas was positively charged and that it resembled **chlorine** and **bromine** in its chemical properties. Schönbein named the gas ozone (from the Greek *ozon*, meaning "odor") because of its peculiar smell.

Another of Schönbein's major discoveries was also a fortunate accident. He discovered guncotton while experimenting with nitric and sulfuric acids in his kitchen. He spilled some acid and, not wanting his wife to know that he had been using the kitchen as a laboratory, quickly cleaned the liquid up with an apron and hung it on the stove to dry. Much to Schönbein's surprise, once the apron dried, it burst into flames. He then realized that the **nitric acid** had reacted with the **cellulose** in the cotton apron to form nitrocellulose. Because the guncotton, as Schönbein named it, would burn without creating a great deal of smoke, it briefly flourished as a replacement for black **gunpowder**. Schönbein died in 1868.

SCHRÖDINGER EQUATION

In 1924, Louis-Victor de Broglie showed that any particle, such as an **electron**, has wave-like properties. Electrons up to that point were treated as hard particles, much like billiard balls. De Broglie's theory changed this view though, and in 1926, Erwin Schröodinger formulated a wave equation that treated an electron as a wave, rather than as a particle.

A wave is an any oscillating disturbance that moves through space, or a medium like air or **water**. To a physicist, a wave of a quantum entity is essentially the same as a wave on a lake. Common waves, like those on lakes, are described by mathematical equations, called wave equations. If common everyday waves can be described by equations, then an electron's wave should also be able to be described by a mathematical wave equation. Schrödinger was the first to mathematically describe an electron's wave.

A wave equation mathematically describes the way that a wave moves. In quantum mechanics, the **solution** of the wave equation gives a wavefunction. The wavefunction describes all of the properties of the nature of an electron, or any other quantum entity. Solving Schrödinger's wave equation then yields the wavefunction for an electron. When calcualted, Schrödinger's wave equation yields the probability of where the electron is in a certain space around the **nucleus**; it does not say anything about where the electron actually is in the **atom**.

Schrödinger's wave equation is a mathematical beast that involves second-order differential equations and can be either time-dependent or time-independent. Though difficult to solve, it is not impossible. Indeed there are many possible solutions to Schrödinger's wave equation. But only some of these solutions make sense physically. For the wave equation to work for electrons, the equation must be continuous, that is there should be no sudden breaks or holes if the wavefunctions were plotted, and the solutions must also be single valued everywhere, which means there should be only one probablility for finding an quantum entity at one point. Solutions that satisfy the boundary conditions can only be found for certain values of **energy**. Thus, Schrödinger's wave equation provides support for Max Planck's assertion that energy is quantized. Only certain values of energy are possible; there is no continuous spectrum of energies possible for electrons and other quantum entities to have.

Schrödinger's wave equation can be applied to entities other than electrons. For example, the application of the Schrodinger wave equation to the **hydrogen** atom results in a series of possible wavefunctions, called orbitals, which describe the areas of space where an electron can be. There are four quantum numbers, n, l, m, and m_s, that describe the energy levels, angular momentum, magnetic properties, and spin of the atom. Schrödinger's wave equation has even been applied to the Universe in an attempt to combine quantum mechanics with cosmology!

At about the same time as Schrödinger was developing his wave equation for quantum entities, Werner Heisenberg was developing his matrix mechanics to understand the nature of quantum entities. Instead of yielding probabilities of an electron's location in an atom, Heisenberg's matrix mechanics yields the position and the momentum of an electron in an atom. Heisenberg's matrix mechanics represents the particle interpretation of electrons, whereas Schrödinger's wave mechanics represents the wave interpretation of electrons.

The standard explanation of what went on in the quantum world, which held sway from the 1930s to the 1980s, is the Copenhagen interpretation. This was offered as a way to picture both the wave and particle nature of quantum entities. According to the Copenhagen interpretation of quantum mechanics, it is meaningless to ask what electrons and other quantum entities are doing when we are not looking at them. All we can do is calculate the probability that a particular experiment will come up with a particular result. If we measure the wavelength of an electron, for example, the electron is a wave. If we instead measure the momentum of an electron, the electron is a particle. The Copenhagen interpretation was replaced in the 1980s by an interpretation that has quantum particles such as electrons acting as waves and particles at the same time.

Schrödinger, Erwin (1887-1961)

Austrian physicist

Erwin Schrödinger shared the 1933 Nobel Prize for physics with English physicist Paul Dirac in recognition of his development of a wave equation describing the behavior of an **electron** in an **atom**. His theory was a consequence of French theoretical physicist **Louis Victor de Broglie**'s hypothesis that particles of **matter** might have properties that can be described by using wave functions. Schrödinger's wave equation provided a sound theoretical basis for the existence of electron orbitals (energy levels), which had been postulated on empirical grounds by Danish physicist **Niels Bohr** in 1913.

Erwin Schrödinger.

Schrödinger was born in Vienna, Austria, on August 12, 1887. His father, Rudolf Schrödinger, enjoyed a wide range of interests, including painting and botany, and owned a successful oil cloth factory. Schrödinger's mother was the daughter of Alexander Bauer, a professor at the Technische Hochschule. For the first eleven years of his life, Schrödinger was taught at home. Though a tutor came on a regular basis, Schrödinger's most important instructor was his father, whom he described as a "friend, teacher, and tireless partner in conversation," as Armin Hermann quoted in *Dictionary of Scientific Biography.* From his father he also developed a wide range of academic interests, including not only mathematics and science but also grammar and poetry. In 1898, he entered the Akademische Gymnasium in Vienna to complete his precollege studies.

Having graduated from the Gymnasium in 1906, Schrödinger entered the University of Vienna. By all accounts, the most powerful influence on him there was Friedrich Hasenöhrl, a brilliant young physicist who was killed in World War I a decade later. Schrödinger was an avid student of Hasenöhrl's for the full five years he was enrolled at Vienna. He held his teacher in such high esteem that he was later to remark at the 1933 Nobel Prize ceremonies that, if Hasenöhrl had not been killed in the war, it would have been Hasenöhrl, not Schrödinger, being honored in Stockholm.

Schrödinger was awarded his Ph.D. in physics in 1910 and was immediately offered a position at the University's Second Physics Institute, where he carried out research on a number of problems involving, among other topics, **magnetism** and dielectrics. He held this post until the outbreak of World War I, at which time he became an artillery officer assigned to the Italian front. As the War drew to a close, Schrödinger looked forward to an appointment as professor of theoretical physics at the University of Czernowitz, located in modern-day Ukraine. However, those plans were foiled with the disintegration of the Austro-Hungarian Empire, and Schrödinger was forced to return to the Second Physics Institute.

During his second tenure at the Institute, on April 6, 1920, Schrödinger married Annemarie Bertel, whom he had met prior to the War. Not long after his marriage, Schrödinger accepted an appointment as assistant to Max Wien in Jena, but remained there only four months. He then moved on to the Technische Hochschule in Stuttgart. Once again, he stayed only briefly—a single semester—before resigning his post and going on to the University of Breslau. He received yet another opportunity to move after being at the University for only a short time: he was offered the chair in theoretical physics at the University of Zürich in late 1921.

The six years that Schrödinger spend at Zürich were probably the most productive of his scientific career. At first, his work dealt with fairly traditional topics; one **paper** of particular practical interest reported his studies on the relationship between red-green and blue-yellow **color** blindness. Schrödinger's first interest in the problem of wave mechanics did not arise until 1925. A year earlier, de Broglie had announced his hypothesis of the existence of matter waves, a concept that few physicists were ready to accept. Schrödinger read about de Broglie's hypothesis in a footnote to a paper by American physicist **Albert Einstein**, one of the few scientists who did believe in de Broglie's ideas.

Schrödinger began to consider the possibility of expressing the movement of an electron in an atom in terms of a wave. He adopted the premise that an electron can travel around the **nucleus** only in a standing wave (that is, in a pattern described by a whole number of wavelengths). He looked for a mathematical equation that would describe the position of such ''permitted'' orbits. By January of 1926, he was ready to publish the first of four papers describing the results of this research. He had found a second order partial differential equation that met the conditions of his initial assumptions. The equation specified certain orbitals (energy levels) outside the nucleus where an electron wave with a whole number of wavelengths could be found. These orbitals corresponded precisely to the orbitals that Bohr had proposed on purely empirical grounds thirteen years earlier. The wave equation provided a sound theoretical basis for an atomic model that had originally been derived purely on the basis of experimental observations. In addition, the wave equation allowed the theoretical calculation of **energy** changes that occur when an electron moves from one permitted orbital to a higher or lower one. These energy changes conformed to those actually observed in spectroscopic measurements. The equation also explained why electrons cannot exist in regions between Bohr orbitals since only non-whole number wavelengths (and, therefore, non-permitted waves) can exist there.

After producing unsatisfactory results using relativistic corrections in his computations, Schrödinger decided to work with non-relativistic electron waves in his derivations. The results he obtained in this way agreed with experimental observations and he announced them in his early 1926 papers. The equation he published in these papers became known as ''the Schrödinger wave equation'' or simply ''the wave equation.'' The wave equation was the second theoretical mechanism proposed for describing electrons in an atom, the first being German physicist **Werner Karl Heisenberg**'s matrix mechanics. For most physicists, Schrödinger's approach was preferable since it lent itself to a physical, rather than strictly mathematical, interpretation. As it turned out, Schrödinger was soon able to show that wave mechanics and matrix mechanics are mathematically identical.

In 1927, Schrödinger was presented with a difficult career choice. He was offered the prestigious chair of theoretical physics at the University of Berlin left open by German physicist **Max Planck**'s retirement. The position was arguably the most desirable in all of theoretical physics, at least in the German-speaking world; Berlin was the center of the newest and most exciting research in the field. Though Schrödinger disliked the hurried environment of a large city, preferring the peacefulness of his native Austrian Alps, he did accept the position.

Hermann quoted Schrödinger as calling the next six years a ''very beautiful teaching and learning period.'' That period came to an ugly conclusion, however, with the rise of National Socialism in Germany. Having witnessed the dismissal of outstanding colleagues by the new regime, Schrödinger decided to leave Germany and accept an appointment at Magdalene College, Oxford, in England. In the same week he took up his new post he was notified that he had been awarded the 1933 Nobel Prize for physics with Dirac.

Schrödinger's stay at Oxford lasted only three years; then, he decided to take an opportunity to return to his native Austria and accept a position at the University of Graz. Unfortunately, he was dismissed from the University shortly after German leader Adolf Hitler's invasion of Austria in 1938, but Eamon de Valera, the Prime Minister of Eire and a mathematician, was able to have the University of Dublin establish a new Institute for Advanced Studies and secure an appointment for Schrödinger there.

In September, 1939, Schrödinger left Austria with few belongings and no money and immigrated to Ireland. He remained in Dublin for the next seventeen years, during which time he turned to philosophical questions such as the theoretical foundations of physics and the relationship between the physical and biological sciences. During this period, he wrote one of the most influential books in twentieth-century science, *What Is Life?* In this book, Schrödinger argued that the fundamental nature of living organisms could probably be studied and understood in terms of physical principles, particularly those of quantum mechanics. The book was later to be read by and become a powerful influence on the thought of the founders of modern molecular biology.

After World War II, Austria attempted to lure Schrödinger home. As long as the nation was under Soviet occupation, however, he resisted offers to return. Finally, in 1956, he accepted a special chair position at the University of Vienna and returned to the city of his birth. He became ill about a year after he settled in Vienna, however, and never fully recovered his health. He died on January 4, 1961, in the Alpine town of Alpbach, Austria, where he is buried.

Schrödinger received a number of honors and awards during his lifetime, including election into the Royal Society, the Prussian Academy of Sciences, the Austrian Academy of Sciences, and the Pontifical Academy of Sciences. He also retained his love for the arts throughout his life, becoming proficient in four modern languages in addition to Greek and Latin. He published a book of poetry and became skilled as a sculptor.

Seaborg, Glenn T. (1912-1999)

American nuclear chemist

Glenn T. Seaborg.

Glenn T. Seaborg was a pioneering nuclear chemist whose work with isotopes and transuranium elements contributed to the development of nuclear technology in medicine, **energy**, and weapons. His early research on the identification of radioisotopes advanced radiological imaging techniques and radiotherapy; he was the co-discoverer of technetium–99m, one of the most widely used radioisotopes in nuclear medicine. Seaborg's most significant accomplishments resulted from his discovery of a number of transuranic elements, including **plutonium**, which is used to build the atomic bomb. His contributions to the ''atomic age'' began during World War II with his work on the **Manhattan Project** to develop nuclear weapons. His impeccable reputation as a scientist and administrator earned him an appointment as chairman of the Atomic Energy Commission, where he was influential in the development and testing of **nuclear energy** and weapons and in the establishment of nuclear arms control agreements.

Glenn Theodore Seaborg was born in Ishpeming, Michigan, on April 19, 1912, to Swedish immigrants Herman Theodore Seaborg, a machinist, and Selma O. (Erickson). Both had come to the United States in 1904, entering through the historic landmark Ellis Island in New York City, and then traveling to Ishpeming. Swedish was Seaborg s first language, and his his early upbringing included the cultural traditions of his parents homeland. His parents moved to southern California when he was ten, in part to seek better educational opportunities for him and his sister. He was educated in the multi-cultural Watts district of Los Angeles, where he gained the ability to deal effectively with people of different backgrounds, a skill that was useful to him in later life. Although his parents urged the young Seaborg to focus on commercial studies in high school, a dynamic **chemistry** teacher sparked his interest in science. After graduating as valedictorian of his class in 1929, Seaborg enrolled at the University of California, Los Angeles (UCLA), earning his tuition through a series of odd jobs, including stevedore and farm laborer. After graduating from UCLA in 1934 with a degree in chemistry, Seaborg pursued his graduate work at the University of California, Berkeley, which had renowned chemistry and physics departments. There, he studied under chemist **Gilbert Newton Lewis** and physicist Ernest Orlando Lawrence, who had a twenty-seven-inch cyclotron (a circular device that serves as an accelerator in which charged particles are propelled by an alternating electric field in a constant magnetic field). Seaborg often worked all night, conducting research with the cyclotron when it was not being used by Lawrence and his colleagues. He earned a Ph.D. in 1937 for his thesis concerning the interaction of ''fast'' neutrons with **lead** and was appointed to the faculty at Berkeley, where he worked as an assistant in Lewis's laboratory. Shortly thereafter, Seaborg met Lawrence's secretary, Helen L. Griggs, whom he married in 1942.

While still working on his Ph.D., Seaborg began to collaborate with physicist Jack Livingood on the chemical separations that occurred in the cyclotron to produce radioactive isotopes (different forms of the same element having the same **atomic number** but a different number of protons). Radioisotopes have vital applications in the field of radiology, both in radiotherapy and diagnosing disease with radiological imaging techniques. Seaborg, Livingood, and colleagues discovered iodine–131, iron–59, cobalt–60, and technetium–99m, one of the most widely used radioisotopes in nuclear medicine.

In 1934, physicist Enrico Fermi attempted to create new elements by irradiating **uranium** with neutrons. Uranium was the element with the ''heaviest'' nuclei on the **periodic table**, a chart that systematically groups together the elements with similar properties in order of increasing atomic number. In 1939, the German scientists **Otto Hahn** and **Fritz Strassmann** were able to split a uranium **nucleus** by bombarding it with neutrons, and this fission reaction produced a powerful release of energy. Edwin M. McMillan and Philip Hauge Abelson, working with the cyclotron at the Lawrence Radiation Laboratory at Berkeley, discovered the first transuranium element (elements with atomic number higher than 92), element 93. Although they also created **nuclear fission** with their experiments, McMillan and Abelson noticed that some of the nuclei bombarded by neutrons did not undergo fission but had decayed through **electron** emission, creating the new element. Since uranium was named after Uranus, the seventh planet in our solar system, they named their new element **neptunium**, after Neptune, the eighth planet.

Seaborg soon took over the work of searching for element 94 when McMillan left Berkeley to conduct war research on the East Coast. Working with Joseph W. Kennedy and graduate student Arthur C. Wahl, Seaborg discovered the chemically unstable element 94 in 1941 by bombarding neptunium with deuterons, the nuclei of the **hydrogen isotope deuterium**, and named the new element plutonium–238, after the ninth planet, Pluto. During the course of these experiments, Seaborg and colleagues isolated the isotope plutonium–239, which was a fissionable isotope with potential for use in the development of nuclear weapons and nuclear power. The following year, Seaborg, John W. Gofman, and Raymond W. Stoughton identified the isotope uranium–233, which would lead to the use of **thorium**, an abundant element, as a source of nuclear fuel.

On April 19, 1942, Seaborg left Berkeley to work on the Manhattan Project in Chicago, a large–scale scientific effort supported by the government with the goal of creating an atomic bomb. He led a group of scientists, including B. B. Cunningham and L. B. Werner, in the difficult task of developing methods for chemically extracting plutonium–239 from uranium in amounts that could be used to produce nuclear energy. Seaborg later reflected on this work as being the most exciting efforts of his career as he and his colleagues worked feverishly to develop an atomic weapon before Germany.

Because of their nearly identical chemical makeup, separating plutonium from uranium was a difficult task. In the course of their work, Seaborg and colleagues pioneered ultramicrochemical analysis, a technique used in working with minute amounts of radioactive material, and discovered that minuscule amounts of plutonium existed in pitchblende and carnotite ores. Seaborg was a primary influence in the decision to use plutonium instead of uranium for the first atomic-bomb experiments. By 1944, Seaborg's group had achieved success in isolating large amounts of plutonium, which enabled the Manhattan Project scientists to construct two nuclear weapons.

After Seaborg met his primary responsibility of developing enough plutonium to construct atomic weapons, he returned his attention to the transuranium elements. In 1944, Seaborg postulated that the element **actinium** and the 14 elements heavier than it were similar in nature and should be placed in a separate group on the periodic table. Known as the actinide concept, this theory helped scientists to accurately predict the chemical properties of heavier elements in the periodic table, and was the most significant alteration to the table since it was devised by **Dmitry Mendeleev** in 1869. Interestingly, **Niels Bohr** had predicted this alteration to the table many years earlier. Using his new actinide concept, Seaborg and colleagues began to predict the chemical makeup of other possible transuranics and, as a result, discovered element 95, **americium**, and element 96, **curium**. Seaborg applied for and received patents for these elements, the only person ever to do so for a chemical element.

When Seaborg returned to Berkeley in 1946, he began to assemble a premier group of scientists (many of whom were colleagues on the Manhattan Project) to search for transuranium elements. From 1948 to 1959, under Seaborg's guidance as associate director of the university's Lawrence Radiation Laboratory, this group isolated and identified the transuranic elements **berkelium**, (element 97), **californium** (element 98), **einsteinium** (element 99), **fermium** (element 100), mendelevium (element 101), and nobelium (element 102). In 1951, Seaborg shared the Nobel Prize for chemistry with McMillan for their groundbreaking work in discovering transuranic elements. In 1974, scientists working under Seaborg discovered element 106 (unnilhexium).

Patriotism required Seaborg and his colleagues on the Manhattan Project to contribute to the war effort, but once they achieved success in creating atomic weapons, they urged the government not to use this new-found means of **mass** destruction on the Japanese, who refused to surrender despite the fall of Germany. Seaborg and six others signed the Franck Report, which recommended that the government merely demonstrate the bomb's terrible power by inviting the Japanese to watch a detonation. Yet the bomb was used twice, and the enormous loss of human life that resulted from it inspired Seaborg and other scientists to crusade actively for the control of nuclear weapons. They realized that the creators of such weapons were morally obligated to contribute to the outcome of a debate that might determine the survival of life on earth.

Seaborg's accomplishments as an administrator at the University of California, Berkeley, were nearly as impressive as his scientific discoveries. In 1958, he became the second vice-chancellor at the university and helped guide the institution during a period of dynamic growth. There was an ambitious building program at Berkeley, and the College of Environmental Design and the Space Sciences Laboratory were established at this time. He was also the faculty representative from Berkeley to the panel that oversaw the development of the Pac Ten athletic conference, which aimed at resolving abuses and corruption in California's collegiate athletic system. During this period, Seaborg also contributed to national educational reform by chairing the steering committee that created a new chemistry curriculum to help improve science education in schools. This curriculum was, in part, a response to the Sivoet Union s launching of *Sputnik*, the first artificial satelite, in 1957, which seemed to demonstrate Soviet superiority in science. Seaborg also served on the National Commission on Excellence in Education, which published *A Nation at Risk* (1983), a book that delineated the failings of the U.S. educational system.

Seaborg's knack for handling difficult issues was tested during his tenure as chairman of the Atomic Energy Commission (AEC) in the turbulent decade that stretched from 1961 to 1971. He entered the debate raging over society's concerns about nuclear power and its potential to harm the environment and produce mass destruction. Seaborg received this appointment during a conversation with President Kennedy, which he described in his book *Kennedy, Khrushchev and the Test Ban* as "the telephone call that changed my life." He continued, "Within a few days I accepted, and soon I was plunged into a new kind of chemistry, that of national and international events."

In 1961, Seaborg was a member the American delegation that signed the Limited Nuclear Test Ban Treaty, outlawing nuclear testing in the atmosphere, **water**, and space. He also contributed to the ratification of the Non-Proliferation Treaty of 1970 by developing safeguards for handling nuclear materials, and by ensuring that those products meant for industrial and medical technology would not be illegally diverted to weapons' manufacture. In spite of these victories, Seaborg was disappointed by the failure to develop a "*comprehensive* treaty to prohibit *any* nuclear weapons testing....[D]espite some near misses," he wrote, "this glittering prize, which carried with it the opportunity to arrest the viciously spiraling arms race, eluded our grasp."

In 1971, he returned to the University of California at Berkeley as University Professor of Chemistry. He continued to foster international cooperation in science and arms control. He was one of the founders and in 1981 president of the International Organization for Chemical Sciences in Development (IOCD), an organization that seeks solutions to Third World problems through scientific collaboration. In the 1980s, he also published two influential books, the first of which was *Kennedy, Khrushchev and the Test Ban* (1981). Based on his scrupulous memoirs, this book was an attempt "to contribute some facts and insights not previously published" about nuclear testing." Seaborg hoped that it might aid future negotiations by providing "a wider understanding of what is involved in the achievement of an important arms control agreement." He also wrote *Stemming the Tide: Arms Control in the Johnson Years,* which was published in 1987. His last book was an autobiography, titled *A Chemist in the White House: From the Manhattan Project to the End of the Cold War*, which he published in 1998. In addition to his books, Seaborg wrote more than 500 scientific papers.

Florence B. Seibert.

During the course of his career, Seaborg acquired many honors and appointments, including election to the American Association for the Advancement of Science and ten foreign national academies of science. Many of the younger scientists, however, disapproved of Seaborg, and, suspicious of nuclear power, they opposed his appointment to the National Academy of Sciences.

A passionate conservationist, Seaborg helped to establish the "Golden State Trail" in California and served as vice-president of the American Hiking Society. He died in 1999.

SEIBERT, FLORENCE B. (1897-1991)

American biochemist

A biochemist who received her Ph.D. from Yale University in 1923, Florence B. Seibert is best known for her research in the **biochemistry** of tuberculosis. She developed the protein substance used for the tuberculosis skin test. The substance was adopted as the standard in 1941 by the United States and a year later by the World Health Organization. In addition, in the early 1920s, Seibert discovered that the sudden fevers that sometimes occurred during intravenous injections were caused by bacteria in the distilled **water** that was used to make the protein solutions. She invented a **distillation** apparatus that prevented contamination. This research had great practical significance later when intravenous blood transfusions became widely used in surgery. Seibert authored or coauthored more than a hundred scientific papers. Her later research involved the study of bacteria associated with certain cancers. Her many honors include five honorary degrees, induction into the National Women's Hall of Fame in Seneca Falls, New York (1990), the Garvan Gold Medal of the American Chemical Society (1942), and the John Elliot Memorial Award of the American Association of Blood Banks (1962).

Florence Barbara Seibert was born on October 6, 1897, in Easton, Pennsylvania, the second of three children. She was the daughter of George Peter Seibert, a rug manufacturer and merchant, and Barbara (Memmert) Seibert. At the age of three she contracted polio. Despite her resultant handicaps, she completed high school, with the help of her highly supportive parents, and entered Goucher College in Baltimore, where she studied **chemistry** and zoology. She graduated in 1918, then worked under the direction of one of her chemistry teachers, Jessie E. Minor, at the Chemistry Laboratory of the Hammersley Paper Mill in Garfield, New Jersey. She and her professor, having responded to the call for women to fill positions vacated by men fighting in World War I, coauthored scientific papers on the chemistry of **cellulose** and **wood** pulps.

Although Seibert initially wanted to pursue a career in medicine, she was advised against it as it was ''too rigorous'' in view of her physical disabilities. She decided on biochemistry instead and began graduate studies at Yale University under Lafayette B. Mendel, one of the discoverers of Vitamin A. Her Ph.D. research involved an inquiry into the causes of ''protein fevers''—fevers that developed in patients after they had been injected with protein solutions that contained distilled water. Seibert's assignment was to discover which **proteins** caused the fevers and why. What she discovered, however, was that the distilled water itself was contaminated—with bacteria. Consequently, Seibert invented a distilling apparatus that prevented the bacterial contamination.

Seibert earned her Ph.D. in 1923, then moved to Chicago to work as a post-graduate fellow under H. Gideon Wells at the University of Chicago. She continued her research on pyrogenic (fever causing) distilled water, and her work in this area acquired practical significance when intravenous blood transfusions became a standard part of many surgical procedures.

After her fellowship ended, she was employed part-time at the Otho S. A. Sprague Memorial Institute in Chicago, where Wells was the director. At the same time, she worked with Esmond R. Long, whom she had met through Wells's seminars at the University of Chicago. Supported by a grant from the National Tuberculosis Association, Long and Seibert would eventually spend thirty-one years collaborating on tuberculosis research. Another of Seibert's long-time associates was her younger sister, Mabel Seibert, who moved to Chicago to be with her in 1927. For the rest of their lives, with the exception of a year in Sweden, the sisters resided together, with Mabel providing assistance both in the research institutes (where she found employment as secretary and later research assistant) and at home. In 1932, when Long moved to the Henry Phipps Institute—a tuberculosis clinic and research facility associated with the University of Pennsylvania in Philadelphia—Seibert (and her sister) transferred as well. There, Seibert rose from assistant professor (1932–1937), to associate professor (1937–1955) to full professor of biochemistry (1955–1959). In 1959 she retired with emeritus status. Between 1937 and 1938 she was a Guggenheim fellow in the laboratory of **Theodor Svedberg** at the University of Upsala in Sweden. In 1926 Svedberg had received the Nobel prize for his protein research.

Seibert's tuberculosis research involved questions that had emerged from the late-nineteenth-century work of German bacteriologist Robert Koch. In 1882 Koch had discovered that the tubercle bacillus was the primary cause of tuberculosis. He also discovered that if the liquid on which the bacilli grew was injected under the skin, a small bite-like reaction would occur in people who had been infected with the disease. (Calling the liquid ''old tuberculin,'' Kock produced it by cooking a culture and draining off the dead bacilli.) Although he had believed the active substance in the liquid was protein, it had not been proven.

Using **precipitation** and other methods of separation and testing, Seibert discovered that the active ingredient of the liquid was indeed protein. The next task was to isolate it, so that it could be used in pure form as a diagnostic tool for tuberculosis. Because proteins are highly complex organic molecules that are difficult to purify, this was a daunting task. Seibert finally succeeded by means of crystallization. The tiny amounts of crystal that she obtained, however, made them impractical for use in widespread skin tests. Thus, she changed the direction of her research and began working on larger amounts of active, but less pure protein. Her methods included precipitation through ultrafiltration (a method of filtering molecules). The result, after further purification procedures, was a dry powder called TPT (Tuberculin Protein Trichloracetic acid precipitated). This was the first substance that was able to be produced in sufficient quantities for widespread use as a tuberculosis skin test. For her work, Seibert received the 1938 Trudeau Medal from the National Tuberculosis Association.

At the Henry Phipps Institute in Philadelphia, Seibert continued her study of tuberculin protein molecules and their use in the diagnosis of tuberculosis. Seibert began working on the ''old tuberculin'' that had been created by Koch and used by doctors for skin testing. As Seibert described it in her autobiography *Pebbles on the Hill of a Scientist,* old tuberculin ''was really like a soup made by cooking up the live tubercle bacilli and extracting the protein substance from their bodies while they were being killed.'' Further purification of the substance led to the creation of PPD (Purified Protein Derivative). Soon large quantities of this substance were being made for tuberculosis testing. Seibert continued to study ways of further purifying and understanding the nature of the protein. Her study in Sweden with Svedberg aided this research. There she learned new techniques for the separation and identification of proteins in **solution**.

Upon her return from Sweden, Seibert brought the new techniques with her. She began work on the creation of a large batch of PPD to serve as the basis for a standard dosage. The creation of such a standard was critical for measuring the degree of sensitivity of individuals to the skin test. Degree of sensitivity constituted significant diagnostic information if it was based upon individual reaction, rather than upon differences in the testing substance itself. A large amount of substance was necessary to develop a standard that ideally would be used world-wide, so that the tuberculosis test would be comparable wherever it was given. Developing new methods of purification as she proceeded, Seibert and her colleagues created 107 grams of material, known as PPD-S (the S signifying "standard"). A portion was used in 1941 as the government standard for purified tuberculins. Eventually it was used as the standard all over the world.

In 1958 the Phipps Institute was moved to a new building at the University of Pennsylvania. In her memoirs, Seibert wrote that she did not believe that the conditions necessary for her continued work would be available. Consequently, she and Mabel, her long-time assistant and companion, retired to St. Petersburg, Florida. Florence Seibert continued her research, however, using for a time a small laboratory in the nearby Mound Park Hospital and another in her own home. In her retirement years she devoted herself to the study of bacteria that were associated with certain types of cancers. Her declining health in her last two years was attributed to complications from childhood polio. She died in St. Petersburg on August 23, 1991.

In 1968 Seibert published her memoirs, which reveal her many friendships, especially among others engaged in scientific research. She particularly enjoyed international travel as well as driving her car, which was especially equipped to compensate for her handicaps. She loved music and played the violin (privately, she was careful to note).

SELENIUM

Selenium is the third element in Group 16 of the periodic family. The members of this family are sometimes referred to as the chalcogens, meaning "ore-forming." Selenium has an **atomic number** of 34, an atomic **mass** of 78.96, and a chemical symbol of Se.

Properties

Selenium exists in a number of allotropic forms. One **allotrope** is an amorphous red powder, while a second allotrope has a bluish, metallic appearance. Other allotropes have properties intermediate between these two allotropes. The crystalline form of selenium has a melting point of 423°F (217°C), a **boiling point** of 1,265°F (685°C), and a **density** of 4.5 grams per cubic centimeter.

Chemically, selenium is a **metalloid**. It is a fairly reactive element that combines readily with **hydrogen**, **fluorine**, **chlorine**, and **bromine**. It also combines with a number of **metals** to form compounds known as selenides as, for example, **magnesium** selenide (MgSe). Selenium burns in **oxygen** with a bright blue flame to produce selenium dioxide (SeO_2), which has a characteristic odor of rotten horseradish.

Occurrence and Extraction

Selenium is a rare element with an abundance in the Earth's crust of about 0.05-0.09 parts per million. It is widely distributed in the Earth's crust and occurs in no one mineral from which it can be extracted with profit. It is most commonly obtained as a byproduct of the extraction of **copper**, **iron**, and **lead** from their ores. The major producers of selenium are Japan, Canada, Belgium, the United States, and Germany.

Discovery and Naming

Selenium was discovered in 1818 by the Swedish chemists **Jöns Jakob Berzelius** and J. G. Gahn (1745-1818). The two men were studying the chemicals used in making **sulfuric acid** at a plant where they had just become part owners. Among the chemicals they found a material they thought at first to be the element **tellurium**. Upon further analysis, they found that the material was not tellurium, but a new element with similar chemical properties. They suggested the name selenium for the element after the Greek word *selene*, meaning "moon." The name seemed appropriate because of the new element's close association with tellurium, which had been named from the Latin word *tellus*, meaning "Earth."

Uses

The two most important uses of selenium are in glassmaking and in electronics. The addition of selenium to **glass** can either add a beautiful ruby red **color** to the final product or, if iron is present, it can **balance** the green color caused by the iron and produce a clear, colorless product. Selenium is of growing importance in the field of electronics, where it is used to make photovoltaic cells, laser printers, and plain-paper photocopiers.

Health Issues

Selenium has a dual biological effect on plants and animals. It is required in very small amounts to maintain good health of both plants and animals, but in larger amounts it can produce health problems. For example, birds visiting the Kesterson Reservoir in California are exposed to very high levels of selenium in the **water** and, over the past decades, have shown an increasing number of genetic deformities caused by the element.

SELF-ORGANIZING SYSTEMS

Self-organizing systems are systems in which lasting structures spontaneously appear as time goes on. Self-organization is a phenomenon common to biology, economics, sociology and many other fields in addition to chemistry. Within chemistry it is particularly associated with the so-called Brussels school, which has developed around Nobel laureate chemist **Ilya Prigogine**. Self-organization is also called autopoiesis, from the corresponding expression in Greek.

Self-organization is amply apparent in the biological world. Highly complex multi-cellular organisms develop from single cells. New species evolve to fill new ecological niches, and organisms living in the same ecosystem exchange matter and chemical energy in complex patterns. In contrast, isolated chemical systems tend towards equilibrium states which are homogeneous and unchanging. This observation has led some philosophers to claim that the organizing tendency apparent in living systems is evidence for a vital force, operating outside the realm of physics and chemistry. The development of thermodynamic theory for non-equilibrium systems over the past half-century has provided a framework for understanding self-organization without the need for a mysterious organizing force.

Possibly the simplest example of self-organization is the so-called Bénard instability (named after the physicist Henri Bénard, who did poneering work in thies field in the early 1900s) which can be observed by heating a layer of oil in a flat-bottomed pan. Because there is a temperature difference between the bottom and top of the oil, heat energy must flow through the layer from the warmer side to the cooler side. As long as the temperature difference is small, this can be accomplished by the migration of individual water molecules from the warmer bottom layer, where their average energy is a greater than average for the fluid as a whole to the top layer, where the energy per molecule is a bit less. As the temperature difference increases, the difference in density between top and bottom lead to the formation of organized convective cells, with a net flow of warm oil upward in some regions areas balanced by a net downward flow of cooler oil in others. The Bénard instability illustrates two important features of self-organization. There is a *reduction of symmetry*, from the initial homogeneous structure to one that is spatially organized, and the process results from an *instability to fluctuations* that develops as the homogeneous system is displaced further from thermal equilibrium.

An important early contribution to the theory of self-organizing systems was made by British mathematician, logician, and computer Alan Matheson Turing (1912-1954), who also wrote on chemistry and biology. In a paper entitled "The Chemical Basis of Morphogenesis," originally published by the Royal Society in 1952, and also included in a volume of Turing's collected works entitled *Morphogenesis*, Turing examined a situation in which the chemical reacting compounds in an initially homogeneous system could give rise to standing concentration waves which could then guide the process of cellular differentiation. In Turing's model, the reactants may be called the initiator and the inhibitor. The assumed overall chemical reaction conditions are such that an increase in the concentration of one the initiator, results in further synthesis of both itself *(autocatalysis)* and the inhibitor (cross-catalysis), while an increase of the concentration of the inhibitor results in decreased production of the initiator. Further, the inhibitor must diffuse more rapidly than the initiator, so that an initial increase in the concentration of the initiator in a small region of the mixture, a spontaneous fluctuation) sets up a net flow of the inhibitor into the surrounding region, establishing a concentration gradient and suppressing further fluctuations. This concentration gradient, in effect, allows the cells to sense their position in the developing organism

Initially unaware of Turing's contribution, Prigogine and his collaborators approached the study of self-organization from the standpoint of chemical **thermodynamics**. According to the second law of thermodynamics, any isolated system will eventually attain a state of thermodynamic equilibrium, characterized by maximum **entropy**, a condition of maximum randomness and homogeneity. In open systems, by contrast, although the combined entropy of the system and its environment must increase, steady states can be attained in which the entropy generated within the system is transported into the environment by the flow of **heat** energy or **matter**, so that the entropy of the system remains constant. Prigogine and his collaborators were then able to show that if the steady state were far enough from equilibrium and the possible chemical reactions among the components included the possibility of autocatalysis, the homogeneous state can become unstable with respect to a spatially organized state. These states, in which the concentrations of the reactants vary over macroscopic distances are termed *dissipative structures*, since they require the continued dissipation of **energy** to exist.

Simple examples of dissipative structures include the flames of candles and **Bunsen burners**, for which the flows of reactants and energy out of the system (the visible flame) are quite apparent. A more striking example is afforded by the *Belousov-Zhabotinsky reaction*, discovered by the Russian chemist Boris Belousov in 1950. This is a complex autocatalytic **oxidation-reduction reaction** in which one of the active species changes color on oxidation. When the mixture is poured into a shallow vessel, concentration fluctuations give rise to oscillations randomly spaced points and give rise to outgoing waves of alternating color, leading to complex structures as time goes on. In a closed vessel the structure disappears as equilibrium is achieved. In a flow system, with continued addition of reactants, the dissipative structures can exist indefinitely. Numerous other self-organizing chemical systems are, of course, possible and many arecurrently being actively studied by researchers

Chaos theory is the study of systems that behave in a complex and apparently unpredictable manner. The field is also known as *deterministic chaos*, as it is mainly devoted to the study of systems the behavior of which can in principle be calculated exactly from equations of motion. Chaos does not occur in chemical systems at or near equilibrium, but may be apparent in explosions, chemical reactors and biochemical systems.

SEMENOV, NIKOLAI N. (1896-1986)

Russian physical chemist and physicist

Nikolai N. Semenov was a physical chemist and physicist who was the first Soviet citizen living in Russia to win the Nobel

Prize. His scientific work focused on chain reactions and their characteristic "explosiveness" during chemical transformations. This influenced the development of greater efficiency in automobile engines and other industrial applications where controlled **combustion** was involved, such as jet and rocket engines. Enjoying important academic success, he also played a significant role as a spokesperson for the Soviet scientific community. He was instrumental in establishing institutions where physical **chemistry** could be studied, and he collaborated in creating a journal dedicated to the field. In addition, he actively participated in scientific conferences dealing with **physical chemistry**.

Nikolai Nikolaevich Semenov was born on April 16, 1896, in Saratov, Russia, to Nikolai Alex and Elena (Dmitrieva) Semenov. He graduated from Petrograd University (later renamed Leningrad; now called St. Petersburg, its original name) in 1917, the year of revolution that led to the establishment of Communism in Russia. Semenov had shown an interest in science from the time he entered Petrograd University at age sixteen in 1913 to study physics and mathematics. He published his first paper at the age of twenty on the subject of the collision of molecules and electrons. After graduation from Petrograd, Semenov accepted a post in physics at the Siberian University of Tomsk, but in 1920, he returned to Petrograd where he was associated with the Leningrad Institute of Physics and Technology for eleven years. In 1928, Semenov organized the mathematics and physics departments at the Leningrad Polytechnical Institute. He became the head of the Institute of Physical Chemistry of the Soviet Academy of Sciences in 1931, where he remained for more than thirty years. In 1944, Semenov became the head of the department of chemical kinematics at the University of Moscow.

The branch of physical chemistry concerned with the rates and conditions of chemical processes, called chemical kinetics, dominated Semenov's research from his earliest studies. His work led to the understanding of the sequence of chemical reactions and provided insight into the conversion of substances into products. Along with some of his colleagues, Semenov felt that physics held the key to understanding chemical transformations. The branch of science referred to as chemical balances was a consequence of their work.

Semenov was awarded the Nobel Prize in chemistry in 1956 with English chemist **Cyril Hinshelwood** for their researches into the mechanism of chemical reactions. Both scientists had worked independently for twenty-five years on chemical chain reactions and their importance in explosions. There is wide agreement in the scientific community that Semenov and Hinshelwood were responsible for the development of **plastics** and the improvement of the automobile engine. There remains some controversy over whether their work on chain reactions contributed to atomic research.

Other experiments by Semenov had culminated in his theory of thermal explosions of mixtures of **gases**. As a result of this research, he increased the understanding of **free radicals**—highly unstable atoms that contain a single, unpaired **electron**. Semenov demonstrated that when molecules disintegrate, energy-rich free radicals are formed. His extensive works on this subject were published first in Russian in the 1930s and later in English.

Nikolai N. Semenov.

In subsequent research, Semenov found that the walls of an exploding chamber can influence a **chain reaction** as well as the substances within the chamber. This concept was particularly beneficial in the development of the combustion engine in automobiles. Semenov's chemical chain reaction theory and his observations on the inflammable nature of gases informed the study of how flames spread, and had practical applications in the oil and chemical industries, in the process of combustion in jet and diesel engines, and in controlling explosions in mines. This work was based on Semenov's earlier investigations of condensation of steam on hard surfaces and its reaction under electric shock.

Semenov made substantial contributions to the development of Soviet scientific institutions and journals. He was active in the training of Soviet scientists and the organizing of important institutions for scientific research in physical chemistry. His long association with the Academy of Sciences of the U.S.S.R. earned him an appointment as a full member in 1932. When the Academy moved to Moscow in 1944, Semenov began teaching at Moscow University. Semenov's theories of combustion, explosion, and problems of chemical kinetics, along with a bibliography of his work by the Academy, were published during the 1940s and 1950s and helped secure his role in his field.

Semenov was not immune to the politics of his country. He became a member of the Communist Party in 1947, and he was the person who answered criticism of the Soviet Union from the *Bulletin of the Atomic Scientists,* a publication of the United States. The *Bulletin* challenged Soviet scientists to protest against Soviet restrictions on release of scientific publications from the country. Semenov replied that there were no such restrictions and accused the American scientists of ignoring their own government restrictions. It was discovered later that some Soviet publications had been arriving regularly at the Library of Congress in Washington, D.C.

In his own country, Semenov was highly regarded. He had received the Stalin Prize, the Order of the Red Banner of Labor, and the Order of Lenin, the latter seven times. He served his country in the political capacity of deputy in the Supreme Soviet in the years 1958, 1962, and 1966, and he was made an alternate to the Central Committee of the Communist Party in 1961. While he was a loyal Soviet citizen, he did work diligently for freedom in scientific experimentation.

On September 15, 1924, he married Natalia Nikolaevna Burtseva, who taught voice, and they had a son, Yurii Nikolaevich, and a daughter, Ludmilla Nikolaevna. Semenov enjoyed hunting, gardening, and architecture in his leisure time. He died in 1986.

SEMICONDUCTORS

Electric current is a flow of electrons through an object. In a conductor, electric current can flow in either direction. **Metals** are the best conductors of **electricity**, and even some **liquids** and **gases** can conduct electrical current. The study of certain substances, such as **gallium**, **silicon**, and **germanium**, revealed that they too had the ability to conduct electricity, but they had special properties. These substances came to be called semiconductors.

In the 1870s, German-born physicist Karl Ferdinand Braun (1850-1918) observed that a crystal called galeria (an ore of **lead** sulfide) was able to conduct electricity in one direction only. This property soon made galeria very useful in electrical devices. Its one-directional conduction gave it the ability to convert alternating current into direct current. Similarly, it could also change radio waves, with their positive and negative alternation, into electrical pulses that could easily be made into audible sounds. Thus the crystals were important as an early electronic component, especially in radios.

In 1928 Swiss-born American physicist Felix Bloch (1905-1983) studied the conditions under which some crystals carry electricity and developed a theory of "bands" to explain how they worked. According to Bloch, normally the electrons of a crystal's atoms are frozen in **valence** bands and cannot conduct electricity. But when certain crystals receive the right amount of **energy** (in the form of light or heat), electrons can "jump" from the valence bands into a normally "forbidden band" and begin to conduct electricity.

Physicists dubbed these crystals semiconductors and in subsequent decades, leaned to better manipulate them. In 1945 a team of American physicists at Bell Telephone Laboratories—John Bardeen (1908-1991), Walter Houser Brattain (1902-1987), and William Bradford Shockley (1910-1989)—created the first transistor using semiconductors. Like Braun's crystals, transistors could conduct electricity directionally, and they could do it with tremendous precision. Vacuum tubes had long since replaced crystals in electronic devices, but transistors were compact, simple, versatile, and functioned at room **temperature**. They soon took the place of vacuum tubes in a multitude of electronic devices.

Shockley also discovered that a technique called doping, or adding small amounts of impurities to a semiconductor, can increase the amount of electricity that it can carry. Incorporating a small number of dopant atoms or ions in a crystal lattice can have a very dramatic effect on the **electrical conductivity** of the crystal. The effect on the nonmetallic (semiconducting) elements silicon and germanium is especially striking. In extremely pure specimens, these materials have practically zero electrical conductivity because their valence electrons are confined by the four covalent bonds that each **atom** of the crystal makes with its neighbors. When traces of **arsenic** or **boron** are added, however, there is a sharp rise in electrical conductivity.

This effect can be understood from valence considerations. An atom of arsenic, which has five valence electrons and an **atomic radius** close to those of silicon and germanium, can easily fit into the germanium or silicon lattice structure, but to do so, it must give up a valence **electron**. This electron is then free to move through the crystal under an applied electrical field. This type of electrical conduction is typical of n-type semiconductors.

If an atom of boron, with three valence electrons, is inserted into the germanium or silicon lattice structure, an electron deficiency arises at the site of substitution, as there are now seven surrounding electrons rather than the usual eight. In an electric field, an electron from a neighboring site can fill in this "hole," resulting in the propagation of the electron deficiency. This type of conduction is known as p-type conduction.

Swift progress in semiconductor technology has made even transistors seem unwieldy. Miniaturized circuits made of silicon dioxide and gallium arsenide semiconductors can now carry the equivalent of up to a million transistors on a "chip" the size of a postage stamp. Besides being essential to computers, these integrated circuits are widely used in televisions, radios, and even automobiles.

See also Computer chips; Crystallography; Electrical conductivity; Periodic Table

SEPARATION SCIENCE

In **chemistry**, separation science refers to the collection of techniques chemists use to isolate one material from another. These are roughly divided into two categories: mechanical and chemical. Some of the most common separation techniques are leaching, flotation, **filtration**, **chromatography**, and centrifugal force.

Leaching is something like making tea, where the separation consists of removing the tea essence from the tea leaves.

In this example, the tea leaves would be put in a container and then be flooded with a solvent (hot water). The solvent would separate, or ''leach out,'' the target substance by percolating through the original substance. A chemical separation method, leaching works mainly to isolate soluble materials from a solid mixture.

Flotation is a separation technique that allows the isolation of powdered mineral ore particles. It is especially useful and economical for separating **metals** from low-grade ores. With flotation, particles are literally floated by various means to the surface of an aqueous media, rather than letting them settle or remain in **solution** according to their specific gravities. The key is in determining the affinity, or preference, of a certain type of particle for attaching to **water**, gas bubbles, or oil, so that the particles come to the surface to be collected.

There are several kinds of flotation separation techniques, but the most popular is ''froth'' flotation. This method relies on the ability of gas bubbles, which are rising through an aqueous solution, to trap certain minerals. The bubbles form a particle- containing froth at the surface that can be skimmed off. The ''skin'' or ''film'' process uses the aqueous media's surface tension to hold up certain kinds of difficult-to-wet particles as they are deposited on the surface, while the remainder of the mixture becomes wet and sinks. The ''bulk-oil'' process uses the surface-wetting properties of the target particle. When a container of oil and water is shaken, the target material will either pass into the oil and the remainder of the material will stay in the aqueous portion or vice versa. Thus, when the oil and water naturally separate, the desired material remains with either the oil or the water and can be removed.

Filtration is one of the most economical and simplest ways to separate substances from each other. Most filtration methods work on the same principle—they use a porous filter to remove **solids** from **liquids**. Sand is a good and inexpensive filter through which a solution of solids and liquids can be passed to separate out the liquid. Usually the sand will be in a container with holes in the bottom to allow the liquid to pass through into a container. Some laboratories use pressure from the top or a vacuum from the bottom to speed the separation process, and others use diatomaceous earth or coagulants such as **aluminum sulfate** to stop or pick up even the smallest particles of solids as they travel through the filter.

Another effective filtration method is the filter press. This comes in three different configurations: the chamber press, the rotary filter, and the plate-and-frame filter. The chamber press and plate-and-frame filter use a series of vertical filters through which a solution is horizontally forced. The rotary filter is more complex because it uses several individual filters. It has a drum that rolls in place in a special trench. As it rotates into the trench, the filtering part of the drum passes a vacuum so that a **mass** of filtered solids accumulates there. This is removed when the filter section rolls back out of the trench. As particles are removed from the mixture, the liquid goes to holding tanks inside the device.

Chromatography, which has many variations, is one of the most important separation methods and it finds wide use both in industrial and academic settings. It can be performed on large scale for chemical separation or on microscale for analytical purposes. All chromatographic methods rely on differences in the affinities of the various members of a group of dissolved or gaseous chemicals for a certain adsorbent. Typically all chromatographic methods have a mobile and a stationary phase. The mixture is placed in the mobile phase that is then passed through the adsorbent-containing stationary one. The different components of the mixture have different affinities for the adsorbent in the stationary phase; these differences in affinities result in different rates of passage through the stationary phase and result in the separation.

Different methods of chromatography include column chromatography methods and planar chromatography methods. In column chromatography, the mixture is dissolved in a liquid phase or vaporized and then passed through a **glass** or metal tube containing the stationary phase. Gas chromatography (GC) is an example of a column chromatography method. The mixture is vaporized and injected into a column that contains the adsorbent on a stationary phase. GC can be performed on preparative or analytical scale. GC itself has several different variants including gas-liquid chromatography (GLC) in which the stationary phase is made up of a liquid. High- performance liquid chromatography (HPLC) is another example of a column method. The mixture is dissolved in a liquid phase and passed through a stationary phase that has several different types. HPLC has both preparative and analytical variants. Thin-layer chromatography (TLC) is a typically form of planar chromatography. The mixture is dissolved in a liquid and allowed to move by capillary action through a thin layer of material that is typically supported on a plate. This thin layer contains the adsorbent. TLC also has both preparative and analytical variants.

Using centrifugal force to separate solids from liquids and unmixable liquids from each other is a staple of **industrial chemistry**. It is based on the **centrifuge**, a machine that spins its contents either vertically or horizontally to increase the normal effect of gravity. In a rotating vertical centrifuge, the denser particles will generally move to the outside of the cylinder, while the lighter particles remain near the center. In a rotating horizontal centrifuge, the denser particles tend to move to one end, displacing the lighter materials to the other end.

See also Analytical chemistry

SEWAGE TREATMENT

Sewage is wastewater discharged from a home, business, or industry. Sewage is treated to remove or alter contaminants in order to minimize the impact of discharging wastewater into the environment. The operations and processes used in sewage treatment consist of physicochemical and biological systems.

The concerns of those involved in designing sewage treatment systems have changed over the years. Originally, the biochemical **oxygen** demand (BOD) and total suspended **solids** (TSS) received most of the attention. This was primarily because excessive BOD and TSS levels could cause severe and readily apparent problems, such as oxygen deficits that led to odors and fish kills, and sludge deposits that suffocated benthic organisms. By removing BOD and TSS, other contaminants were also removed and other benefits were realized; so even today, some discharge permits contain only limits for BOD and TSS. However, many permits now contain limits on other contaminants as well, and these limits, as well as other requirements, are constantly changing.

Among the first contaminants to be added to the requirements for discharge permits were nutrients. The most commonly regulated nutrients are phosphorus and nitrogen. Originally removing **phosphorus** and **nitrogen** could only be done through expensive, advanced methods. But scientists have recently discovered ways to accomplish enhanced removals of nutrients in conventional biological treatment plants with relatively minor operational and structural adjustments.

The most recently regulated pollutants are toxicants. There are regulations for specific toxic agents, and there are the generic-type regulations, which specify that the toxicity to certain test organisms should not exceed a certain level. For example, the wastewater discharged from a particular municipality may be restricted from killing more than 50% of the *Ceriodaphnia* in an aquatic toxicity test. The municipality would not need to determine what is causing the toxicity, just how to minimize its effects. Efforts to understand the causes of toxicity are referred to as toxicity reduction evaluations. The generic limit can therefore sometimes turn into a more specific standard, in the view of the municipality or industry, when the identity of a toxicant is determined; the general regulatory limit might remain, but treatment personnel are more cognizant of the role that a certain pollutant plays in overall, effluent toxicity.

The systems used to treat sewage can be divided into stages. The first stage is known as preliminary treatment. Preliminary treatment includes such operations as flow equalization, screening, comminution (or grinding), grease removal, flow measurement, and grit removal. Screenings and grit are taken to a landfill. Grease is directed to sludge handling facilities at the plant.

The next stage is primary treatment, which consists of gravity settling to remove suspended solids. Approximately 60% of the TSS in a domestic wastewater is removed during primary settling. Grease that floats to the surface of the sedimentation tank is skimmed off and handled along with the sludge (known as primary sludge) collected from the bottom of the tank.

The next stage is secondary treatment, which is designed to remove soluble organics from the wastewater. Secondary treatment consists of a biological process and secondary settling. There are a number of biological processes. The most common is activated sludge, a process in which microbes, also known as biomass, are allowed to feed on organic matter in the wastewater. The make-up and dynamics of the biomass population is a function of how the activated sludge system is operated. There are many types of activated sludge systems that differ based on the time wastewater remains in the biological reactor and the time microbes remain there. They also differ depending on whether air or oxygen is introduced, how gas is introduced, and where wastewater enters the biological reactor, as well as the number of tanks and the mixing conditions.

There are also biological treatment systems in which the biomass is attached. Trickling filters and biological towers are examples of systems that contain biomass adsorbed to rocks plastic. Wastewater is sprayed over the top of the rocks or plastic and allowed to trickle down and over the attached biomass, which removes materials from the waste through sorption and biodegradation. A related type of attached-growth system is the rotating biological contactor, where biomass is attached to a series of thin, plastic wheels that rotate the biomass in and out of the wastewater.

It is important to note that each of the above biological systems is aerobic, meaning that oxygen is present for the microbes. Anaerobic biological systems are also available in both attached and suspended growth configurations. Examples of the attached and suspended growth systems are, respectively, anaerobic filters and upflow anaerobic sludge blanket units.

The end-products of aerobic and anaerobic processes are different. Under aerobic conditions, if completely oxidized, organic matter is transformed into products that are not hazardous. But an anaerobic process can produce **methane** (CH_4), which is explosive, and **ammonia** (NH_3) and **hydrogen** sulfide (H_2S), which are toxic. There are thus special design considerations associated with anaerobic systems, though methane can be recovered and used as a source of **energy**. Some materials are better degraded under anaerobic conditions than under aerobic conditions. In some cases, the combination of anaerobic and aerobic systems in a series provides better and more economical treatment than either system could alone. Many substances are not completely mineralized to the end-products mentioned above, and other types of intermediate metabolites can be considered in selecting a biological process.

Biomass generated during biological treatment is settled in secondary clarifiers. This settled biomass or secondary sludge is then piped to sludge-management systems or returned to the biological reactor in amounts needed to maintain the appropriate biomass level. The hydraulic detention time of secondary clarifiers is generally in the area of two hours.

As mentioned above, biological systems are designed on the basis of hydraulic residence time and sludge age. In a conventional activated-sludge system, sewage is retained in the reactor for about five to seven hours. Biomass, due to the recycling of sludge from the secondary clarifier, remains in the reactor, on average, for about ten days.

Disinfection follows secondary clarification in most treatment plants. Disinfection is normally accomplished with **chlorine**. Due to the potential environmental impact of chlorine, most plants now dechlorinate wastewater effluents before discharge.

Some facilities use another stage of treatment before disinfection. This stage is referred to as tertiary treatment or advanced treatment. Included among the more commonly used advanced systems are adsorption to activated **carbon**, filtration through sand and other media, **ion exchange**, various membrane processes, nitrification-denitrification, coagulation-flocculation, and fine screening.

The treatment systems used for municipal sewage can be different from the systems used by industry, for industrial wastes can pose special problems which require innovative applications of the technologies available. Additionally, industrial wastes are sometimes pretreated before being discharged to a sewer, as opposed to being totally treated for direct discharge to the environment.

See also Landfills; Waste treatment

Sex hormones

The sex **hormones** are **steroids** that control sexual maturity and reproduction—androgens for male functions and estrogen and progesterone for female functions. They are produced mainly by the female ovaries and male testes, which are endocrine glands. Sex hormones are secreted throughout a person's life, but production greatly increases at puberty and normally decreases in old age.

Men's and women's bodies produce both male and female hormones, synthesizing them from **cholesterol** and intermediate compounds. For example, females produce androgens in the ovary and adrenal gland (also an endocrine gland), but almost all ovarian androgens are converted to estrogens. After menopause, a female's fat and skin cells convert androgens to estrogens. Males produce estrogen in three ways: converting testosterone in the testes, converting androgens in fat and skin cells, and synthesizing it in the adrenal glands. At puberty, the hypothalamus in the brain produces increased amounts of gonadotropin-releasing hormone. It, in turn, stimulates the nearby pituitary gland to secrete two glycoprotein hormones (those with both carbohydrate and protein structures): follicle-stimulating hormone (FSH) and luteinizing hormone (LH). Finally, these two hormones signal the sex glands (gonads) to produce the sex hormones.

Females produce three estrogens—estradiol, estriol, and estrone—which stimulate growth of the ovaries and begin preparing the uterus for pregnancy. Progesterone maintains uterine conditions during pregnancy; it also acts on the central nervous system in a way that is not yet understood. During the monthly reproductive cycle, FSH stimulates growth of an ovarian body called the graafian follicle, which encloses the egg. LH aids in the rupturing of the follicle, sending the egg to the fallopian tubes. It also promotes growth of the corpus luteum when the ovary prepares to release the egg into the uterus. The corpus luteum, in turn, secretes large amounts of estrogens and progesterone. If no pregnancy occurs within 10-12 days, the corpus luteum withers, the uterus sheds the blood supply formed to nourish a fetus (menstrual flow), production of estrogens and progesterone drops dramatically, and the cycle begins again. Without pregnancy, progesterone produced by the corpus luteum controls changes in the uterine lining during the monthly cycle. In pregnancy, the corpus luteum continues making progesterone, but a much larger amount is produced by the placenta. Estrogens also control the body's secondary sex characteristics, including breast and pelvic development and the distribution of fat and muscle.

In males, LH stimulates the development of the testes, which produce two androgens—testosterone and androsterone. When FSH activates the testes' sperm-forming cells, testosterone maintains spermatogenesis, which is the ten-week process resulting in sperm ready for release by ejaculation from the penis. The androgens also promote the secondary sex characteristics of muscle growth, lowered voice range, the Adam's apple, and increased body hair.

As with other hormones, sex hormones were intensively studied during the 1920s. Discovery of their steroid structure and relationship to other steroids was the key to isolation and synthesis. The first breakthroughs came with the female hormones. In 1929, the American biochemist **Edward Doisy** isolated a crystalline form of oestrone, as did several other teams of scientists. Five years later, the German biochemist **Adolf Butenandt** and his colleagues, as well as three other groups, isolated progesterone. Meanwhile, in 1931, the American scientists H. L. Fevold, F. L. Hisaw, and S. L. Leonard discovered luteinizing hormone and follicle-stimulating hormone. That same year, Butenandt and Kurt Tscherning isolated the male hormones. The Swiss biochemist **Leopold Ružička** soon determined the structure of testosterone. In 1934, Ružička partly synthesized androsterone from cholesterol, after proposing its structure. This was the first synthesis of a sex hormone and the first proof of the relationship between cholesterol and sex hormones. Butenandt's group showed that the hormones were related to cholesterol and bile acids, and in 1939 converted cholesterol into progesterone. For their work in demonstrating the structure of steroids, including the sex hormones, Ružička and Butenandt shared the 1939 Nobel Prize in **chemistry**.

Commercial synthesis came next. In the 1930s, Austrian chemists were synthesizing male and female hormones from soybean sterols (solid alcohols), but the process was expensive because it was hard to separate the sterols. The American chemist **Percy Julian** discovered a much easier way to separate sterols, permitting inexpensive synthesis of both progesterone and testosterone. The American chemist **Carl Djerassi** is also noted for synthesizing estrone and estradiol from plant materials. Medical uses of the hormones soon followed. In 1941, the American surgeon Charles Huggins (1901-1997) was the first to use **chemotherapy** (chemical treatment of disease) when he treated prostate cancer with female sex hormones. For his work he received the 1966 Nobel Prize in physiology or medicine.

Today both male and female hormones are used to treat many kinds of cancer. Estrogen is also administered to treat menopause-related conditions and osteoporosis (the loss of bone calcium). In addition to its use in the treatment of cancer, testosterone is administered by injection to treat men's sexual dysfunction, and in 1993 transdermal (skin) patches were developed.

The first female oral contraceptive (birth control pill)—developed by the Americans Gregory Pincus (1903-1967), Min-Chueh Chang (1908-), and John Rock (1890-1984)—contained a synthetic progesterone, called progestin, developed by Djerassi. It was approved by the U.S. Food and Drug Administration in 1960. Today a variety of pills containing varying amounts of progesterone and estrogen are available by prescription. Progesterone-containing implants and injections were also approved in the early 1990s.

Estrogen therapy has also been linked with decreased heart risk in women who have had a heart attack. A study at the University of Washington, Seattle showed that women (who have no history of a coronary disease) given estrogen therapy after a heart attack were a third less likely to have another. Further research is necessary to determine if this therapy would be beneficial to all women who have had a heart attack and to identify any long-term effects of such treatment.

SHELDRAKE, RUPERT (1942-)

English biochemist

Rupert Sheldrake, a British biochemist, is best known for his controversial hypothesis of "formative causation," or the idea that nature itself has memory. According to Sheldrake's theory, every system in the universe—molecules, cells, crystals, organisms, societies—reacts in similar or established patterns in response to invisible fields of influence. This is known as "morphic **resonance**." Sheldrake purports that the invisible field, known as a "morphic field," is where established patterns collect to influence a like activity that may be taking place contemporaneously. An example that he often uses to convey this idea more readily is that of crystallization; in his book *The Rebirth of Nature*, Sheldrake explained morphic resonance thus: "The development of crystals is shaped by morphogenetic fields with an inherent memory of previous crystals of the same kind. From this point of view, substances such as **penicillin** crystallize the way they do not because they are governed by timeless mathematical laws but because they have crystallized that way before; they are following habits established through repetition." Sheldrake further claims that morphic resonance transcends time and space.

Born Alfred Rupert Sheldrake on June 28, 1942, in Newark Notts, England, Sheldrake received his Ph.D. in **biochemistry** from Cambridge University. He was a research fellow of the Royal Society and a fellow of and director of studies in cell biology and biochemistry at Clare College at Cambridge. He studied philosophy at Harvard from 1963 to 1964 as Frank Knox Fellow in the special studies program. Beginning in 1974, Sheldrake conducted research on tropical plants at the International Research Institute in India, as well as in Malaysia. He is married to Jill Purce, has two sons, and lives in London. Sheldrake's father was an herbalist and pharmacist. Sheldrake credits his strong interest in plants and animals to both of his parents, who encouraged him in his studies.

Sheldrake developed the necessary emotional detachment required of one pursuing scientific study during his early years at Cambridge. He came to believe—as he was taught—that nature was, in fact, a lifeless mechanistic system without purpose. But a tension persisted between his scientific studies and his personal experiences. He felt the two bore little relationship to each other and were often irreconcilable. He later came to see this conflict as rooted in the mechanistic view of nature—nature as lifeless as opposed to nature as alive and evolving. Sheldrake's hypothesis of formative causation, with its morphic fields creating morphic resonance, subscribes to the latter view.

Sheldrake's books, *A New Science of Life* (1981), *The Presence of the Past* (1988), and *The Rebirth of Nature* (1991) address his theory of formative causation—one not openly embraced by the scientific community at large. Sheldrake's theory that nature has memory and is, therefore, alive challenges the basic foundations of modern science. According to Sheldrake, the conventional scientific approach has been unable to answer the questions relative to morphogenesis—how things come into being or take form—because of their mechanistic outlook.

Sheldrake's hypothesis has elicited much criticism from his contemporaries. Joseph Hannibal, writing for *Library Journal* on *The Rebirth of Nature,* stated, "This new work is even more unorthodox—some might say outrageous—as Sheldrake attempts to combine scientific, religious, and even mystical views." Critic Patrick H. Samway wrote in *America,* "Sheldrake's methodology parallels in many ways that of the Jesuit paleontologist Teilhard de Chardin, who formulated his view that the world in its entirety is developing toward the Omega Point."

Some of Sheldrake's contemporaries who do not subscribe to the theory of formative causation do, nonetheless, believe that science must be open to new possibilities. "Science is not threatened by the imaginative ideas of the Sheldrakes of the world," wrote fellow scientist James Lovelock in *Nature,* "but those who would censor them." Lovelock went on to say, "Sheldrake is a threat, but only to the established positions of those who teach and practice an authoritarian science. A healthy scientific community would accept or reject formative causation as the evidence appeared." Critic Theodore Roszak allowed in *New Science,* "If for no better reason than to exercise their wits against a first class polemic, his critics should value this work. Finding answers to his questions will fortify their ideology." And though terms such as "unrealistic," "fanciful," and "off-the-wall" have been used to describe Sheldrake's hypothesis of formative causation, his theory has indeed received significant attention from the scientific community.

SILICATES

Silicates are compounds of **silicon**, **oxygen** and other elements. Silica (silicon dioxide), silicate, and aluminosilicate minerals make up over 90% of the Earth's crust. Naturally occurring silicates are important building materials and make up the clays upon which **ceramics** and brickmaking are based. Silica is the basis of **glass** technology, and synthetic silicates are important in detergent and adhesive applications.

Like **carbon**, the silicon **atom** has a marked tendency to form four covalent bonds. Unlike carbon, however, it has little or no tendency to form multiple bonds. In silica, each silicon atom is bonded to four oxygen atoms and each oxygen atom is bonded to two silicons (Figure 1). Energetic considerations prohibit two silicon atoms from sharing more than one **bonding** oxygen. One can visualize the structures of silica by thinking of the SiO_4 units as tetrahedra sharing corners. A little experimentation with models reveals that numerous different silica structures are possible.

Silica exists in several crystalline forms, in a large number of colloidal forms, and as an amorphous solid. At atmospheric pressure silica exists in three basic crystalline forms. Quartz exists from low temperatures up to 1,598°F (870°C), undergoing a phase transition from alpha quartz to beta quartz at 1,063°F (573°C). Tridymite is stable up to 2,678°F (1,470°C) and crystoballite up to the melting point at 3,110°F (1,710°C). Because there is a significant difference in crystal structure among the three forms, cooling tridymite or crystoballite below 572°F (300°C) results in metastable forms that retain the high **temperature** structure. Cooling molten silica below the melting point results in fused silica, a rigid, transparent substance, chemically unreactive to the vast majority of substances. By melting silica mixed with **sodium** hydroxide or **calcium oxide** one is able to interrupt the long **chains** of silicon-oxygen bonds to form more easily melted glasses.

Soluble silicates can be obtained by heating alkali metal carbonates and silica. Aqueous solutions of sodium silicate are called "**water** glass." As the **pH** of water glass is lowered, colloidal particles of amorphous silica are formed. At still lower pH these particles join together to form silica **gel**, a rigid material with water trapped between chains of small particles. Partial **dehydration** to about 4% water by weight produces the commercial silica gel product, a highly porous material often used as a drying agent or adsorbent for **chromatography**. Replacing the water with **alcohol** and evaporating the alcohol above its critical temperature results in a silica **aerogel**, in which the original arrangement of silica particles is retained. Aerogels can have densities as low as 0.02 grams per cubic centimeter.

There are over 600 known silicate minerals. The minerals can be grouped on the basis of the ways in which silica tetrahedra can be connected together. In a few materials, called orthosilicates, the tetrahedra share no oxygen atoms but exist as quadruply charged negative ions, with the compensating charge provided by doubly or triply charged cations. Orthosilicate minerals include willemite, in which the cations are **zinc** ions, phenacite, with **beryllium** cations, and forsterite, with **magnesium** cations. Zircon is a **zirconium** orthosilicate, sometimes used as a substitute for **diamond** in jewelry.

When the silica tetrahedra share one corner a disilicate **ion** is formed and the compounds are called pyrosilicates or sorosilicates. Pyrosilicates are relatively rare, although they are formed by a number of the lanthanide elements. By sharing two oxygen atoms, silicates can form either cyclic or chain structures. The most common **ring** structure involves six silicon atoms and eighteen oxygen atoms. This structure occurs in the mineral beryl, a beryllium **aluminum** silicate. Chain silicates are called pyroxenes or inosilicates. When parallel single chains are bound together by shared oxygen to form a double chain, the structure is called an amphibole. This structure is found in some of the **asbestos** minerals.

Silicates in which the tetrahedra share three corners are called phyllosilicates or sheet silicates, because they tend to cleave into thin sheets. A sheet of silica tetrahedra will have three oxygen atoms in the sheet and the fourth oxygen protruding to one side. The spacing between the protruding oxygen atoms is comparable to the spacing between certain of the hydroxyl groups in layers of magnesium hydroxide (brucite) or aluminum hydroxide (gibbsite). The silica oxygens can thus replace these hydroxyl groups to form a bilayer or trilayer structure. A large number of sheet silicates are known. These include clay minerals such as kaolinite and talc, micas, montmorillonites, and the remainder of the asbestos minerals including chrysotile (white asbestos).

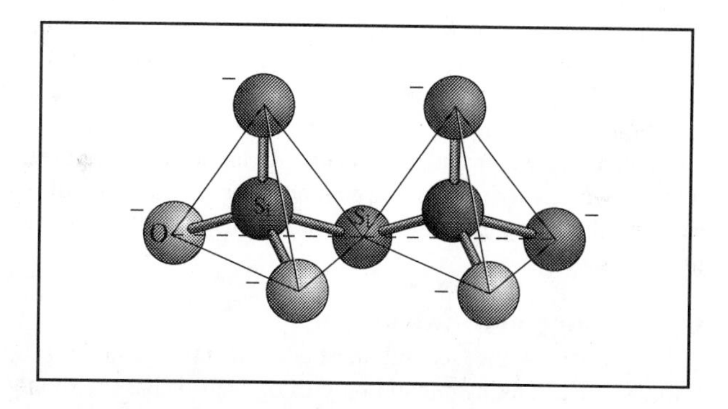

Figure 1. Silicates. ***(Illustration by Electronic Illustrators Group.)***

When all four corners of the tetrahedra are shared, the mineral is called a tectosilicate. Silicate minerals can be derived from the quartz, tridymite, and crystoballite structures by replacing a subset of the silicon ions with ions of a lower charge, together with placing other cations in places between tetrahedra to achieve overall neutrality. This group of minerals includes feldspars, which are aluminosilicates with additional mono- or divalent cations, and the ultramarines, which incorporate bimolecular anions. Perhaps the most useful group of tectosilicate minerals, at least from a chemical standpoint, are the zeolites, aluminosilicates with alkali metal or alkaline earth ions and usually including tightly bound water molecules. These structures incorporate large channels, large enough to accommodate ions and small molecules. Zeolites are used for **ion exchange**, for example removing calcium ions from "hard" water, replacing them with sodium ions, and as molecular sieves, to separate smaller molecules, which fit into the channels, from larger ones that do not. Zeolite-based catalysts have also come to play a major role in petroleum refining.

SILICON

Silicon is the second element in Group 14 of the **periodic table**. It has an **atomic number** of 14, an atomic **mass** of 28.0855, and a chemical symbol of Si.

Properties

Silicon exists in two allotropic forms, one of which consists of shiny, grayish black needle-like or crystal plates. The other **allotrope** is an amorphous brown powder. The melting point of the crystalline allotrope is 2,570°F (1,410°C), its **boiling point** is 4,270°F (2,355°C), and its **density** is 2.33 grams per cubic centimeter. Silicon is a relatively hard element with a hardness of 7 on the Mohs scale. Silicon is a semiconductor, a property which determines some of its most important uses.

As its two allotropic forms might suggest, silicon is a **metalloid**. It is relatively inactive at room **temperature**, and resists attack by **water** and most acids. At higher temperatures, it reacts with many **metals**, **oxygen**, **nitrogen**, **sulfur**, **phosphorus**, and the **halogens**. It also forms a number of alloys in the molten state.

Occurrence and Extraction

Silicon is the second most abundant element in the Earth's crust, ranking second only to oxygen. It has an abundance estimated to be about 27.6%. According to some experts, as much as 97% of the Earth's crust may be made of rocks that contain silicon, oxygen, and one or more other elements. Silicon has also been detected in the Sun and stars and is found in certain types of meteorites known as aerolites, or ''stony meteorites.'' Silicon never occurs free in nature, but is usually found as a compound with oxygen, **magnesium**, **calcium**, phosphorus, and/or other elements. The most common minerals are those that contain silicon dioxide (SiO_2) in one form or another. Compounds that contain silicon and oxygen with one or more other elements are known as **silicates**.

Silicon is extracted from its ores by heating silicon dioxide with carbon: $SiO_2 + C \text{ —heat→ } CO_2 + Si$. Many situations in which silicon is used require that the element be very pure. Silicon with a purity of more than 99.97% is called hyperpure silicon.

Discovery and Naming

The discovery of silicon as an element evaded chemists for many years because of the stability of most silicon compounds. Chemists had little reason to believe that a new element existed in sand, silicates, and other earthy materials, nor did they have the technology to extract the element from its compounds. One researcher with perhaps the greatest reason to hope for success in producing silicon was the English chemist and physicist **Humphry Davy**. Davy had developed a technique by which unusually stable compounds could be decomposed into their constituent elements. He used this method to prepare **sodium**, **potassium**, calcium, and other elements for the first time. He was unsuccessful, however, in producing silicon by the same method.

The first successful effort in the search for silicon was achieved by the Swedish chemist **Jöns Jakob Berzelius**. In 1823, Berzelius electrolyzed a molten mixture of potassium metal and potassium silicon fluoride (K_2SiF_6) and obtained a small sample of pure silicon: $4K + K_2SiF_6 \text{ —heat and electricity→ } 6KF + Si$. The new element was named by the Scottish chemist Thomas Thomson (1773-1852) because of the element's presence in the mineral flint (*silex* or *silicis* in Latin). He added the ending *-on* because of the element's similarity to carb*on*.

Uses

Probably the best known use of silicon is in transistors, photovoltaic cells, rectifiers, and other electronic devices. The largest single use of the element, however, is in the production of alloys, especially various forms of steel. The process by which **iron** is extracted from its ores always results in a product in which iron, silicon, and other elements are present in the form of an **alloy**. The manufacture of various types of steel involves primarily the enhancement or removal of one or more of these components.

One of the most common silicon-iron alloys, for example, is ferrosilicon. This alloy is used for two major purposes. First, it can be added to other forms of steel to improve their strength and toughness. Second, it can be added during the steel-making process to remove impurities from the steel that is being made. The **aluminum** industry also produces a number of silicon alloys. These alloys are used primarily to make molds and in the welding process. Alloys of silicon, aluminum, and magnesium are very resistant to **corrosion** and are used in the construction of large buildings, bridges, and transportation vehicles, such as ships and trains.

Hundreds of silicon compounds are used for a variety of practical applications. The most common silicon compound, silicon dioxide, is used in the manufacture of **glass**, **ceramics**, abrasives, as a food additive, in water **filtration** systems, as insulating material, in **cosmetics** and pharmaceuticals, and in the manufacture of **paper**, **rubber**, and insecticides. A compound known as silicon carbide (SiC), also known as carborundum, is one of the hardest substances known. It is used as a refractory material and as an abrasive. Another category of silicon compounds are the silicones—compounds containing silicon, oxygen, and one or more organic groups. Silicones are used to make toys (such as Silly Putty and Superballs), lubricants, weatherproofing materials, adhesives, foaming agents, brake fluids, cosmetics, polishing agents, electrical insulation, surgical implants, and parts for automobile engines.

Silliman, Benjamin (1779-1864)

American scientist

The most prominent and influential man of science in America during the early 19th century, Benjamin Silliman was a chemist, naturalist, and editor.

Benjamin Silliman was born on Aug. 8, 1779, in what is now Trumbull, Conn., and brought up in nearby Fairfield. He entered Yale in 1792 at the age of 13, graduating in 1796. He spent 2 years partly at home and partly teaching in a private school in Connecticut, then returned to Yale to begin studying law and to tutor. He was admitted to the bar in 1802.

That same year, with no background for the position, Silliman was appointed to the newly established professorship of **chemistry** and natural history at Yale, with permission to qualify himself for the job before beginning his duties. His preparation included attending lectures at the Philadelphia Medical School; work with the chemist **Robert Hare**; occasional visits to John Maclean, professor of chemistry at Princeton; and 2 years at Edinburgh, Scotland. In 1808 he assumed full professorial duties at Yale, lecturing in chemistry, geology, and **mineralogy**.

Although Silliman was a competent researcher, he was not an original scientist. However, he was without a peer in his contributions to the institutional development of science. During nearly 50 years as a professor, he was instrumental in establishing the sciences at Yale, arranging for the college to receive the finest mineral collection in America, aiding in establishing the Yale Medical School in 1813, and persuading the Yale Corporation to establish the ''department of philosophy and the arts,'' where science could be studied intensively. Within a few years the department had grown into the Yale Scientific School, which subsequently became the Sheffield Scientific School, Yale's most distinctive contribution to American education in the 19th century.

In July 1818 Silliman issued the first number of the *American Journal of Science and Arts,* of which he was founder, proprietor, and sole editor for almost 20 years. Devoted to the publication of original papers, notices, and reviews in the broad field of the natural and physical sciences, it won international acclaim. Even more important, he, and later his junior editors, used the journal to introduce the latest in European science to American readers. In its pages Americans first learned of such advances as the natural system of classification, the classification of rocks in terms of the fossils they contained, the chemical approach to mineralogy, and Darwin's theory of evolution.

A brilliant lecturer, Silliman was much in demand by popular audiences for lectures on chemistry, geology, and the bearing of science on religion throughout the 1830s and 1840s. He spent his last years compiling memoirs and conducting his voluminous correspondence. He died on Nov. 24, 1864.

Benjamin Silliman.

SILVER

Silver is classified as a transition metal, one of the elements in the middle of the **periodic table**, and also as a precious metal. Precious **metals** are elements that are rare in the Earth's crust, are attractive to look at, and are chemically quite inert. Silver has an **atomic number** of 47, an atomic **mass** of 107.868, and a chemical symbol of Ag.

Properties

Silver is a soft, white metal with a shiny, attractive surface. It is the most ductile and malleable of all metals. It has two other important physical properties: high electrical and thermal conductivity. In addition, it reflects light very well. Silver's melting point is 1,762°F (961.5°C), its **boiling point** is about 3,600-4,000°F (2,000-2,200°C), and its **density** is 10.49 grams per cubic centimeter.

Silver is a relatively inert element that does not react with **oxygen** or most acids at room **temperature**. It does tend to react slowly with **sulfur** and sulfur compounds, forming a black coating on the metal of silver sulfide (Ag_2S). It is the silver sulfide coating that causes the tarnishing of silver utensils, jewelry, and other objects over time.

Occurrence and Extraction

Silver is relatively rare in the Earth's crust with an abundance estimated at about 0.1 parts per million. It is also found in seawater with an abundance of about 0.01 parts per million. The element usually occurs in association with other metal ores, especially those of **lead**. The most common silver ores are argentite (Ag_2S); ceragyrite (''horn silver''; AgCl); proustite ($3Ag_2SAs_2S_3$); and pyragyrite ($Ag_2Sb_2S_3$). The largest producers of silver in the world are Mexico, Peru, the United States, Canada, Poland, Chile, and Australia. In the United States, more than two-thirds of all silver comes from mines in Nevada, Idaho, and Arizona.

Discovery and Naming

Silver is one of a handful of elements known to ancient peoples. It was probably discovered after **gold** and **copper**. These two elements often occur in the native state and have distinctive colors. Silver may also occur in the native state, but less often than gold and silver. Archaeologists have found silver objects dating to 3400 B.C., and the metal is mentioned a number of times in the Bible.

The element's name goes back to at least the twelfth century A.D. and seems to have come from an old English word for the metal, *seolfor*. The element's chemical symbol is taken from the Latin name for the metal, *argentum*, which may have been derived from the Greek word *argos*, meaning ''shiny'' or ''white.''

Uses

The primary use for silver is in the manufacture of photographic film. The metal is first converted to a compound: silver chloride (AgCl), silver bromide (AgBr), or silver iodide (AgI). These compounds are then used to make photographic film. The compounds are all decomposed easily by light, producing the black spots (individual silver crystals) found on exposed film. The second most important use for silver is in electrical and electronic equipment. Silver is actually the preferred metal for use in electrical equipment because it conducts **electricity** so well. However, it is far too expensive to use for most situations and is reserved for specialized devices. For example, electrical devices on spacecrafts, satellites, and aircraft must work reliably and efficiently in situations where they can not be easily repaired. In such cases, the cost of using silver is not as important as it would be in a home appliance, for example.

Silver is often used, of course, in coins, art objects, and jewelry, usually in the form of an **alloy**. One metal with which silver is often alloyed is gold. Silver adds hardness to gold, which is generally too soft to be used in a pure form. Another use of silver alloys is in dental amalgams. Silver amalgams are highly desirable because they do not break down or react with bodily fluids. Silver is also used in specialized **batteries**, such as silver-zinc and silver-cadmium batteries.

Health Issues

Silver is a mildly toxic element. Its health effects are generally of concern only to workers who come into contact with its vapors. When deposited on the skin, it can produce a bluish appearance known as argyria or argyrosis.

SIMON, DOROTHY MARTIN (1919-)

American chemist

Dorothy Martin Simon has been responsible for several significant advances in space engineering, particularly in the area of **combustion**. By relating the fundamental properties of flame to each other through the principles of **heat** and **mass** transfer and chemical reaction, she helped establish the present-day theory of flame propagation and quenching. She also contributed to the development of ablative coatings, which protect missiles from heat damage upon reentering the Earth's atmosphere. In recognition of these accomplishments, as well as her success in executive management and public speaking, the Society of Women Engineers presented Simon with their Achievement Award in 1966.

Simon was born Dorothy Martin on September 18, 1919, in Harwood, Missouri. Her parents were Robert William Martin, head of the **chemistry** department at Southwest Missouri State College, and Laudell Flynn Martin. Simon attended high school at Greenwood Laboratory School in Springfield, where she won the highest sports honor while also earning the highest grade-point average in the school's history. After graduation, she attended the college at which her father taught, where she received a bachelor's degree with honors in 1940. Once again, she was class valedictorian. From there, she went on to the University of Illinois, where her thesis research on active deposits from **radon** and thoron gas was among the earliest work on radioactive fallout. She obtained a Ph.D. in chemistry in 1945.

Upon completing college, Simon first spent a year as a chemist at the Du Pont Company in Buffalo, New York. During this time, she studied the chemical reactions involved in producing the synthetic fiber now known as Orlon. In 1946, she began working for the Atomic **Energy** Commission (AEC) at Oak Ridge Laboratory in Tennessee and the Argonne Laboratory in Illinois. Among her accomplishments while with the AEC was the isolation of a new **isotope** of **calcium**.

In 1949, Simon began six years with the National Advisory Committee for Aeronautics, the agency that evolved into NASA. These proved to be her most fruitful years as a researcher. During this period, her work elucidating the fundamental nature of flames was recognized with a Rockefeller Public Service Award. Simon used the stipend of $10,000 to visit university and technical laboratories in England, France, and the Netherlands. She studied at Cambridge University with **Ronald G. W. Norrish**, who later won the Nobel Prize in chemistry.

In 1955 she spent a year as group leader at the Magnolia Petroleum Company in Dallas. Then in 1956, Simon began a lengthy association with Avco Corporation, where her early work addressed the design problems of reentry vehicles for intercontinental ballistic missiles. Her research dealt with ablation cooling —a method of protecting the missile body from extreme heat while reentering the Earth's atmosphere by absorbing the heat in a shielding material that is changing phase. This was the topic of a **Marie Curie** Lecture that Simon delivered at Pennsylvania State University in 1962.

Soon Simon's interests turned toward management within the giant conglomerate. She was appointed the first female corporate officer at Avco in 1968. In her capacity as vice president of research, she was responsible for guiding the company's various high-tech divisions. At that time, she was one of the few women to have scaled such heights on the corporate ladder, a fact that was recognized by Worcester Polytechnic Institute when conferring an honorary doctorate upon Simon in 1971. The institute cited her position as "perhaps the most important woman executive in American industry today." Simon later received a second honorary doctorate, this one from Lehigh University. She is a fellow of the American Institute of Chemists, as well as a member of the American Chemical Society and the American Institute of Aeronautics and Astronautics.

Simon is known as an outstanding speaker, who has frequently lectured and written on the challenges of space, research management, and women in science. She served on President Jimmy Carter's Committee for the National Medal of Science, the National Research Council's National Materials Advisory Board, the Department of Defense's Defense Policy Advisory Committee, and the Department of Commerce's Statutory Committee for the National Bureau of Standards. In her free time, she enjoys traveling, cooking, and gardening. Simon was married on December 6, 1946, to Sidney L. Simon—a leading scientist in his own right who became vice president at Sperry Rand. He died in 1975. Simon currently makes her home in Pittsboro, North Carolina.

SINGER, MAXINE (1931-)

American biochemist and geneticist

Maxine Singer, a leading scientist in the field of human genetics, is also a staunch advocate of responsible use of biochemical genetics research. During the height of the controversy over the use of recombinant deoxyribonucleic acid (DNA) techniques to alter genetic characteristics, she advocated a cautious approach. She helped develop guidelines to **balance** calls for unfettered genetics research as a means of making medically valuable discoveries with demands for restrictions on research to protect the public from possible harm. After the **DNA** controversy waned, Singer continued to contribute to the field of genetics, researching cures for cancer, hemophilia, and other diseases related to genetics.

Singer was born on February 15, 1931, in New York City, to Hyman Frank, an attorney, and Henrietta (Perlowitz) Frank, a hospital admissions officer, children's camp director, and model. Singer received her B.A. from Swarthmore College in Pennsylvania in 1952, and earned her Ph.D. in **biochemistry** from Yale in 1957. From 1956 to 1958 she worked as a U.S. Public Health Service postdoctoral fellow at National Institute for Arthritis, **Metabolism** and Digestive Diseases (NIAMD), National Institutes of Health (NIH), in Bethesda, Maryland. She then became a research chemist on the staff of the section on enzymes and cellular biochemistry from 1958 to 1974. There she conducted DNA research on tumor-causing viruses as well as on ribonucleic acid (RNA). In the early 1970s, Singer also served as a visiting scientist with the Department of Genetics of the Weizman Institute of Science in Rehovot, Israel.

While Singer was working at NIH, scientists learned how to take DNA fragments from one organism and insert them into the living cells of another. This "recombinant DNA" could direct the production of **proteins** in the foreign organism as if the DNA was still in its original home. This technique had the potential of creating completely new types of organisms. On one hand, the new research brought unprecedented opportunities to discover cures for serious diseases, to develop new crops, and otherwise to benefit humanity. Yet the prospect of creating as-yet-unknown life forms, some possibly hazardous, was frightening to many.

In 1972, one of Singer's colleagues and personal friends, **Paul Berg** of Stanford University, was the first to create recombinant DNA molecules. He later voluntarily stopped conducting related experiments involving DNA manipulation in the genes of tumor-causing viruses because of some scientists' fears that a virus of unknown properties might escape from the laboratory and spread into the general population.

Although Berg's self-restraint was significant, the catalyst for the debate over gene-splicing was the 1973 Gordon Conference, an annual high-level research meeting. Singer, who was co-chair of the event, was approached by several nucleic acid scientists with the suggestion that the conference include consideration of safety issues. Singer agreed. She opened the discussion with an acknowledgment that DNA manipulation could assist in combatting health problems, yet such experimentation brought to bear a number of moral and ethical concerns.

Maxine Singer.

The scientists present decided, by ballot, to send a public letter about the safety risks of recombinant DNA research to the president of the National Academy of Sciences, and asked *Science* magazine to publish it. Singer and her co-chair, Dieter Söll of Yale University, wrote the letter warning that organisms of an unpredictable nature could result from the new technique, and suggested that the National Academy of Sciences study the problem and recommend guidelines. Concern generated by this letter led to another meeting at the Asilomer Conference Center in Pacific Grove, California, where a debate ensued. Such proceedings—to consider the ethical issues arising from the new DNA research—were unprecedented in the scientific community. Immediately after the Asilomer Conference concluded, a NIH committee began formulating guidelines for recombinant DNA research.

In helping develop the guidelines, Singer advocated a careful analytic approach. In 1976, she presented four principles to the committee to be used in drafting the guidelines. She advised that certain experiments posed such serious hazards that they should be banned altogether; that experiments with lesser or no potential hazards should be permitted if their benefits are unobtainable through conventional methods and if they are properly safeguarded; that the more risk in an experiment, the stricter the safeguards should be; and that the guidelines should be reviewed annually.

Singer provided a calm voice of reason throughout the public debate over gene-splicing that followed. Committees of lay people, such as the Coalition for Responsible Genetic Research, held demonstrations calling for a complete ban on recombinant DNA research. Some members of the media made analogies to the nightmarish vision contained in Aldous Huxley's book *Brave New World,* which described a genetically altered society. When sent to address a public forum on the issue in 1977, for example, Singer responded to accusations that scientists ignore public concerns. As Clifford Grobstein recounted in his book, *A Double Image of the Double Helix: The Recombinant-DNA Debate,* Singer maintained that "scientists recognize their responsibility to the public... (but) dispute over the best way to exercise responsibility must not be confused with the negation of it." According to Grobstein, Singer explained that "while freedom of inquiry is a democratic right, it is clearly unacceptable to cause harm in the name of research. But [Singer] warned that levels of anxiety are not necessarily directly related to levels of real risk."

During her career, Singer has also served on the editorial Board of *Science* magazine and has contributed numerous articles. In her writing for that publication about recombinant DNA research, she stressed the benefits to humanity that recombinant DNA techniques could bring, especially in increasing the understanding of serious and incurable disease. After the NIH guidelines were implemented, she told *Science* readers that "under the Guidelines work has proceeded safely and research accomplishments have been spectacular." By 1980, when public near-hysteria had waned, Singer called for a "celebration" of the progress in molecular genetics. In *Science* she wrote: "The manufacture of important biological agents like insulin and interferon by recombinant DNA procedures," as well as the failure of any "novel hazards" to emerge, was evidence of the value of the cautious continuation of DNA research.

In 1974, Singer accepted a new position at NIH as chief of the Section of Nucleic Acid Enzymology, Division of Cancer Biology and Diagnosis (DCBD) at the National Cancer Institute in Bethesda, Maryland. In 1980 she became chief of the DCBD's Laboratory of Biochemistry. She held this post until 1988, when she became president of the Carnegie Institution, a highly regarded research organization in Washington, D.C. Singer remains affiliated with the National Cancer Institute, however, as scientist emeritus, where she continues her research in human genetics.

In addition to her laboratory research, Singer has devoted considerable time and **energy** to other scientific and professional pursuits. In 1981, she taught in the biochemistry department at the University of California at Berkeley. A skilled and prolific writer, she has issued more than one hundred books, articles, and papers. Most are highly technical, including numerous articles published in scientific journals. Singer also compiled a graduate-level textbook with Paul Berg on molecular genetics called *Genes and Genomes: A Changing Perspective.* Reviewers gave the work high praise for its clear presentation of difficult concepts. Marcelo Bento Soares in *Bioscience* also commented that the book was "superbly written" and "magnificently captures the sense of discovery, understanding, and anticipation that has followed the so-called recombinant DNA breakthrough."

Singer has also written extensively on less technical aspects of science. She and Berg authored a book for laypeople on **genetic engineering**, and she continued to promote the benefits of recombinant DNA techniques and battle public suspicion and fear long after the controversy peaked in the 1970s. In the early 1990s, for example, Singer issued an article encouraging the public to try the first genetically engineered food to reach American supermarket shelves. In describing the harmlessness of the "Flavr Savr" tomato, she decried public objections that eating it was dangerous, unnatural, or immoral to readers of the *Asbury Park Press.* Pointing out that "almost all the foods we eat are the product of previous genetic engineering by cross-breeding," Singer said that the small amount of extra DNA in the tomato would be destroyed in the digestive tract, and that people already consume the DNA present in the other foods in their diets. Moreover, she said the decision to eat a genetically altered tomato did not reduce her admiration for nature's creations.

In addition to her writing and lecturing, Singer has served on numerous advisory boards in the United States and abroad, including science institutes in Naples, Italy, Bangkok, Thailand, and Rehovot, Israel. She also has served on an advisory board to the Pope and as a consultant to the Committee on Human Values of the National Conference of Catholic Bishops. She worked on a Yale committee that investigated the university's South African investments, and serves on Johnson and Johnson's Board of Directors. Concerned about the quality of science education in the United States, she started First Light, a science program for inner-city children.

Singer travels extensively and maintains long work weeks to accommodate all her activities. She married Daniel Singer in 1952; the couple have four children: Amy Elizabeth, Ellen Ruth, David Byrd, and Stephanie Frank. Singer is the recipient of more than forty honors and awards, including some ten honorary doctor of science degrees and numerous commendations from NIH.

Skou, Jens C. (1918-)

Dutch biochemist

Jens C. Skou was one of three men who shared the 1997 Nobel Prize in **Chemistry**. Skou received the award in recognition of discovering the first "molecular pump," Na^+, K^+ ATPase, an **enzyme** that promotes movement through the membrane surrounding a cell and maintains the **balance** of **sodium** ions (Na^+) and **potassium** ions (K^+) in a living cell.

Jens Christian Skou was born October 8, 1918, in Lemvig, Denmark, to Magnus Martinus Skou, a timber merchant, and Ane-Margrethe (Jensen Knak) Skou. He received his M.D. degree (cand.med.) from the University of Copenhagen in 1944. Ten years later, he received his Doctor of Medical Sciences degree (dr.med.) from Aarhus University. In 1948 he married Ellen-Margrethe (Nielsen); they have two children, Hanne and Karen.

After receiving his M.D., Skou went for clinical training at the Hospital at Hjørring and Orthopaedic Clinic at Aarhus, Denmark. He remained there until 1947, when he became an assistant professor in the University of Aarhus's Institute of Physiology. In 1954 the same year he received his Doctor of Medical Sciences degree, he became associate professor at the institute. In 1963 Skou became a full professor and was named chairman of the Institute of Physiology. From 1978-1988, he was professor of biophysics at the University of Aarhus.

Skou has devoted his career to both education and research. He has served as an advisor for many Ph.D. and doctor of medical science students and as an examiner at doctoral dissertation presentations. He has published more than 90 papers on his research, which has investigated the actions of local **anesthetics** and what mechanisms made them work, as well as the work that earned him the 1997 Nobel Prize, the transport of sodium and potassium ions through the cell membranes.

A cell's health depends on maintaining a balance between its inner chemistry and that of the cell's surroundings. This balance is controlled by the presence of the cell membrane, the wall between the cell's inner workings and its environment.

For more than 70 years, scientists have known that one of the delicate balances that are maintained involves ions (electrically charged particles) of the elements sodium (Na) and potassium (K). A cell maintains its inner **concentration** of sodium ions (Na^+) at a level lower than that of its surroundings. Similarly, it maintains its inner concentration of potassium ions (K^+) at a level higher than its surroundings.

This balance is not static, however. In the 1950s, English researchers Alan Hodgkin and Richard Keynes found that sodium ions rush into a nerve cell when it is stimulated. After the stimulation, the cell restores its original sodium/potassium levels by transporting the extra sodium out through its membrane. Scientists suspected that this transport involved the compound adenosine triphosphate (ATP). ATP was discovered in 1929 by German chemist Karl Lohmann. Further research by Franz Lippman between 1939 and 1941 showed that ATP carries chemical **energy** in the cell. It has been called the cell's ''energy currency.'' Scientists noticed that, when ATP's presence was inhibited, cells did not rid themselves of the extra sodium that they absorbed during stimulation.

In the 1950s, Skou began his investigations into the workings of ATP. For his experimental material he chose finely found nerve membranes from crabs. He wanted to find out if there was an enzyme in the nerve membranes that degraded ATP, and that could be involved with the transport of ions through the membrane.

He did find such an ATP-degrading enzyme, which needed ions of **magnesium**. In his experiments, Skou found that he could stimulate the enzyme by adding sodium ions—but there was a limit to the stimulation he could achieve. Adding small amounts of potassium ions, however, stimulated the enzyme even more. In fact, Skou noted that the enzyme—called ATPase—reached its maximum point of stimulation when he added quantities of sodium and potassium ions that were the same as those normally found in nerve cells. This evidence made Skou hypothesize that the enzyme worked with an **ion** ''pump'' in the cell membrane.

Skou published his first paper on ATPase in 1957. Years of further experimentation followed. In them, Skou learned more about this remarkable enzyme. He learned that different places on the enzyme attracted and bound ions of sodium and potassium.

When ATP breaks down and releases its energy, it become adenosine diphosphate (ADP) and releases a **phosphate** compound. Skou's work discovered that this freed phosphate bound to the ATPase as well, a process known as phosphorylation. The presence or absences of this phosphate changed the enzyme's interaction with sodium and potassium ions, Skou discovered. When the ATPase lacked a phosphate group, it became dependent on potassium. Similarly, when it has a phosphate, it became dependent on sodium.

This latter discovery was key to learning just how ATPase moved sodium out of the cell. ATPase molecules are set into the cell membrane, and they consist of two parts, one which stabilizes the enzyme and the other which carries out activity.

Part of the enzyme pokes inside the cell. There, one ATP **molecule** and three sodium ions can bind to it at a time. A **phosphorus** group is taken from the ATP to bind to the enzyme, and the remaining ADP is released. The enzyme then changes shape, carrying the attached sodium ions with it to the outside of the cell membrane. There, they are released into the cell's surroundings, as is the attached phosphorus. In place of the three sodium ions, two potassium ions attach themselves to the enzyme, which again changes shape and carries the K^+ into the cell's interior.

This activity uses up about one-third of the ATP that the body produces each day, which can range from about half of a resting person's body weight to almost one ton in a person who is doing strenuous activity.

Thanks to this molecular pump, the cell is able to maintain its balance of potassium ions on the inside and sodium ions on the outside, maintaining the electrical charges that allow cells to pass along or to react to stimulation from nerve cells.

This enzyme is important for other reason as well. For example, the pump's action on the balance of sodium and potassium makes it possible for the cell to take in nutrients and to expel waste products. If the molecular pump were to stop—as it can when a lack of nourishment or **oxygen** shuts down ATP formation—the cell would swell up, and it would be unable to pass along nerve impulse. If this were to happen in the brain, unconsciousness would rapidly follow.

Since Skou discovered ATPase, scientists have found other molecular pumps hard at work in the cell. They include

H^+, K^+-ATPase, which produces stomach acid, and $Ca2^+$-ATPase, which helps control the contraction of muscle cells.

In addition to the Nobel Prize, Skou has received much recognition for his work. He is a regular participant and organizer of symposia on transmembrane transport. In addition, he has received the Leo Prize, the Novo Prize, the Swedish Medical Association's Anders Retzius gold medal, and the Dr. Eric K. Ferntroms Big Nordic Prize.

He is a member of the Danish Royal Academy, and has served on a number of its committees and science foundation board. He is also a member of the Danish Royal Society, the Deutsche Akademie der Naturfoscher, Leopoldins, and the European Molecular Biology Organization (EMBO). He is a foreign associate of the American National Academy of Sciences. In addition, Skou is an honorary member of the Japanese Biochemical Society and the American Physiological Society. He received and honorary doctorate from the University of Copenhagen. Skou lives in Denmark.

SMALLEY, RICHARD ERRETT (1943-)

American chemist and physicist

American scientist Richard E. Smalley is best known as one of the winners of the 1996 Nobel Prize for **Chemistry**, along with fellow Rice University professor **Robert F. Curl, Jr.**, and Briton **Harold W. Kroto** from the University of Sussex, for the discovery of a new **carbon molecule**, the buckminsterfullerene. It was given that name, or more simply "fullerene," in honor of architect Buckminster Fuller, whose geodesic dome the carbon molecule resembles. A pioneer of supersonic beam laser **spectroscopy**, Smalley is also renowned for his elaborate supersonic beam experiments, which use **lasers** to produce and study **clusters**, aggregates of atoms that occur for a short time under specific conditions. The discovery of **fullerenes** promises to be the basis for not only a new area of carbon chemistry, but also a way to produce remarkably strong and lightweight materials, new drug delivery systems, computer **semiconductors**, solar cells, and superconductors.

Smalley was born in Akron, Ohio, on June 6, 1943. Smalley's mother, Esther Virginia Rhoads, was from a furniture manufacturing family. Smalley credits his mother with sparking his interest in science. He spent many hours with her collecting samples from a local pond and looking at them under the microscope. His mother taught him to love literature and nature and the practical skill of mechanical drawing. His father, Frank Dudley Smalley, Jr., was the CEO of a trade journal for farm implements, *Implement and Tractor*. His father taught him machinery repair as well as woodworking. In his autobiogrpahy published on-line through Rice University's web site, Smalley believes that these childhood activities were the perfect preparation for a scientific career.

Several more events inspired Smalley to become a scientist. One was the launching of *Sputnik* in 1957. Another was his aunt, Dr. Sara Jane Rhoads, who was one of the first women in the United States to achieve a full professorship in chemistry. Smalley used to refer to this bright, active woman as "the Colossus of Rhoads." She encouraged Smalley to study chemistry and one of Smalley's best memories is of working in her **organic chemistry** laboratory at the University of Wyoming.

Smalley's aunt also encouraged him to attend Hope College in Holland, Michigan, which was known for its undergraduate programs in chemistry. Smalley spent two years at Hope College but decided to transfer to the University of Michigan after one favorite professor died and another retired. When Smalley graduated in 1965, he decided to take a job rather than go directly to graduate school. He worked for three years at Shell Chemical Company's **polypropylene** manufacturing plant and at their **Plastics** Technical Center in Woodbury, New Jersey. There Smalley learned what he called in his autobiography "real-world applications of chemistry." It was also there that Smalley met his wife, Judith Grace Sampieri, a secretary for Shell. They were married on May 4, 1968.

Smalley enjoyed his work at Shell but he knew it was time to begin graduate school. His graduate school prospects became entangled with several near misses in the Vietnam War draft. He was close to accepting an offer from the University of Wisconsin when he discovered that graduate students were no longer automatically deferred from the draft. His industrial deferment was still valid, so he decided to stay at Shell. However, that deferment eventually expired, so he decided to reapply to graduate school anyway and take his chances. He applied to Princeton University because his wife's family lived there. In the fall of 1968 he was in fact drafted, but within a week of that event, his wife became pregnant and he was reclassified. The Smalley's son Chad Richard was born on June 9, 1969.

That fall the Smalleys moved to Princeton, New Jersey, and Smalley began his Ph.D. work. Here Smalley learned a concentrated style of research as well as chemical physics and molecular systems. In 1973 Smalley began his postdoctoral research with Professor Don Levy at the University of Chicago. Part of his oral exam was three original research proposals; in researching topics, Smalley became interested in the work of Nobel Prize winner Yuan Lee and Stuart Rice. Yuan Lee had built a universal molecular beam apparatus and had used it to slow down molecules. This was the germ of Smalley's future Nobel Prize-winning work. His collaboration with Don Levy on supercooled molecules led to supersonic beam laser spectroscopy. This technology allowed scientists to examine molecules with the kind of detail only achieved before on atoms.

Smalley became an assistant professor in the chemistry department of Rice University in Houston, Texas, in 1976. He was aware of Rice University professor Robert Curl's work with laser spectroscopy and had wanted to collaborate with him. Smalley's first work was building a supersonic beam apparatus similar to one he had used at the University of Chicago. His first proposal to the National Science Foundation was for a larger apparatus that would allow him to increase the beam's intensity and be able to study a larger variety of molecules.

At the same time Smalley was using his laser apparatus to examine molecules, a professor at Sussex University in England, Harold Kroto, was researching **chains** of carbons in space. Kroto thought these chains might be the products of red-giant stars, but was not sure how the chains actually formed. In 1984, Kroto traveled to the United States to use Smalley's beam apparatus. He thought that he could use the machine to simulate the temperatures in space needed to form the carbon chains. Smalley and Curl had had no reason to look at simple carbon in their complex laser apparatus. It was something of a favor as well as a break in their research when Kroto asked them to look at carbon in order to verify his research. So that September, the scientists turned the laser beam on a piece of **graphite** and found something they were not looking for, a molecule that had 60 carbon atoms. Carbon had previously been known to have only two molecular forms, **diamond** and graphite. They surmised correctly that this was a third form of carbon and that it had a cage-like structure resembling a soccer ball, or a geodesic dome. They named the structure buckminsterfullerene, which later became known as fullerene, and also by the nickname "buckyball."

Evidence for the existence of large carbon clusters had existed before, but Smalley, Curl, and Kroto were the first scientists to fully identify and stabilize carbon-60. In October of 1996, all three were recognized for this remarkable discovery with the Nobel Prize in Chemistry.

Fullerene research took off quickly, and today scientists can manufacture pounds of buckyballs in a day. Extraordinarily stable because of their **molecular structure** and resistant to radiation and chemical destruction, fullerenes have many potential uses.

Smalley's research group is now looking at the tubular versions of fullerenes. In his autobiography Smalley writes that he is "convinced that major new technologies will be developed over the coming decades from fullerene tubes, fibers, and cables, and we are moving as fast as possible to bring this all to life."

SMELTING

Reductive **calcination**, or smelting, is a process of heating ores to a high temperatures in the presence of a reducing agent such as **carbon** (coke), and a fluxing agent to remove the accompanying clay and sand (gangue). Smelting, which involves the **reduction** of metal oxides in ores to pure metal, is followed by separation of the gangue of the liquid metal and slag (i.e., the fused mixture of impurities and flux).

Iron ore is the ore most frequently smelted. The ore, typically containing 20% clay and sand, is first heated in an air blast furnace with coke and limestone (the fluxing agent) to a **temperature** above the melting point of iron and the slag. The coke is oxidized to **carbon dioxide**, which changes to **carbon monoxide** at high temperatures. The carbon monoxide reduces the ores, i.e., converts the metal oxides to metal, by taking on **oxygen**, which changes the carbon monoxide back to carbon dioxide. The difference in densities of the molten iron and molten slag allow each material to be removed separately from the furnace.

Secondary smelting is a related process used to recover nonferrous **metals** and alloys from new and used scrap and

Lead smelting. ***(Photograph by Chris Jones, The Stock Market. Reproduced by permission.)***

dross. Examples of materials that may be subjected to secondary smelting include scrap **aluminum**, babbitt, brass, bronze, **copper**, **iridium**, **lead**, **magnesium**, **nickel**, platinum-group metals, **precious metals**, **tin**, and **zinc**.

See also Oxidation-reduction reaction

SMITH, MICHAEL (1932-)

English-Canadian biochemist

Michael Smith began his professional research career in salmon physiology and endocrinology, but returned to the chemical synthesis that had been his first interest, including the chemical synthesis of deoxyribonucleic acid (DNA). Smith experimented with isolating genes and invented site-directed mutagenesis, a technique for deliberately altering gene sequences. Smith's work was hailed as having tremendous implications for genetic studies and the understanding of how individual genes function, and already has been applied in the study of disease-producing viruses. In 1993 Smith shared the Nobel Prize in **Chemistry** independently with **Kary Mullis**. The Royal Swedish Academy of Sciences credited Smith and Mullis with having revolutionized basic research and saluted the possibilities offered by their research toward the cure of hereditary diseases.

Smith was born in Blackpool, England, on April 26, 1932. His parents were Rowland Smith, a market gardener, and Mary Agnes Armstead Smith, a bookkeeper who also helped with the market gardening. Smith was admitted to Arnold School, the local private secondary school, with a scholarship he earned based on his examination results (this examination was taken, at the time, by all English children when they finished their primary education). Without this scholarship, Smith would have had little opportunity for advanced education, as his parents did not have the money to pay for it. While at Arnold School, Smith became involved in scouting, which eventually led to a life-long interest in camping and other outdoor activities.

After graduating from Arnold School in 1950, Smith enrolled at the University of Manchester in order to study chemistry, realizing a natural inclination toward the ''hard'' sciences. He moved rapidly through school, receiving a B.Sc. in 1953, and a Ph.D. in chemistry in 1956, both sponsored by scholarship. Smith's desire following completion of his Ph.D. was to earn a fellowship on the West Coast of the United States. This did not work out, but he was accepted into biochemist **Har Gobind Khorana**'s laboratory in Vancouver, Canada. Smith's original plan in migrating to Canada was to work for a year, then return to England and work for a chemical company. However, his experience working with Khorana, who would win the Nobel Prize in 1968 for his contributions to genetics, changed his plans. Smith decided university research was the path he wanted to take and that British Columbia, with its natural beauty, would be his home. Smith is now a Canadian citizen.

Smith stayed with the Khorana group and moved with it in 1960 to the Institute for **Enzyme** Research at the University of Wisconsin. (Smith had recently married Helen Christie. The couple later separated, but they had three children, Tom, Ian, and Wendy.) Until then, Smith's work in Canada had been in several different areas of chemical synthesis. In 1961 he decided it was time for a change and decided to re-locate to the West Coast. Smith accepted a position as head of the chemistry section of the Vancouver Laboratory of the Fisheries Research Board of Canada. His work there was mainly in salmon physiology and endocrinology, but he also continued to work in chemical synthesis.

In 1966 Smith entered the academic field, taking an appointment as associate professor of **biochemistry** and molecular biology at the University of British Columbia (UBC), and bringing with him an interest in chemically synthesized **DNA** (the **molecule** of heredity). Also beginning in 1966 Smith held a concurrent position as medical research associate of the Medical Research Council of Canada. He was made full professor in 1970, and has continued his teaching duties ever since. In 1986 he was asked to establish a biotechnology laboratory on the campus of UBC, which he has headed since that time.

Smith has taken three sabbaticals from his duties at the University of British Columbia, spending three months in 1971 at Rockefeller University in New York, one year during 1975 and 1976 at the Medical Research Council laboratory in Cambridge, and eight months in 1982 at Yale University. The middle excursion was spent in English biochemist **Frederick Sanger**'s laboratory learning about DNA sequence determination, essential to Smith's later research.

Smith was first able to isolate genes using chemical synthesis in 1974. Slowly he developed what became known as site-directed mutagenesis, a technique that allows gene sequences to be altered deliberately. More specifically, it involves separating one strand of a piece of DNA and producing a mirror image of it. This mirror image can then be used as a probe into a gene. It can also be used with chemical enzymes—proteins that act as catalysts in biochemical reactions—that are able to cut and splice DNA in living cells. Jeffrey Fox, editor of *Bioscience,* called this process the ''intellectual bombshell that triggered protein engineering ,'' as quoted in the Toronto *Globe and Mail.* Smith's findings were published in 1978 in *Journal of Biological Chemistry.* This **paper** lays the foundation of the research Smith has done since. The paper concludes, ''This new method of mutagenesis has considerable potential in genetic studies. Thus, it will be possible to change and define the role of regions of DNA sequence whose function is as yet incompletely understood.''

Smith, in demonstrating that biological systems are chemical, has allowed scientists to tinker systematically with genes, altering properties one at a time to see what effect each alteration may have on the gene's functioning. Genes are the building blocks for countless **proteins** that make up skin, muscles, bone, and **hormones**. Changes in the expression of these proteins reveal to the scientist how his or her tinkering has altered the gene function. This process has been used specifically to study disease-producing viruses, such as those that cause cancer. The eventual goal is to uncover the functioning of the genes, so drugs to combat the viruses can be developed.

After being several times a nominee, Smith was awarded the Nobel Prize in Chemistry in 1993 jointly with Kary Mullis from California. Their work was not collaborative, though both dealt with biotechnology. Announcing the award, the Royal Swedish Academy of Sciences credited Smith for having ''revolutionized basic research and entirely changed researchers' way of performing their experiments,'' as quoted in the Toronto *Globe and Mail.* The academy further said Smith's work holds great promise for the future with the ''possibilities of gene therapy, curing hereditary diseases by specifically correcting mutated code words in the genetic material.''

The award money from the Nobel Prize amounted to close to $500,000 for Smith. With it he established an endowment fund, half of which will be earmarked to aid research on molecular genetics of the central nervous system, specifically in relation to schizophrenia research. The other half is to be divided between general science awareness projects and the Society for Canadian Women in Science and Technology in an effort to induce more women to pursue careers in science. He also convinced both the provincial and federal governments to contribute to his funds.

In addition to his receipt of the Nobel Prize, Smith has garnered numerous other honors in the course of his career, including the Gairdner Foundation International Award in 1986, and the Genetics Society of Canada's Award of Excellence in 1988. He has assumed several administrative responsibilities, including becoming acting director of the Biomedical Research Center, a privately funded research institute, in 1991, and is a member of the Canadian Biochemical Society, the Genetics Society of America, and the American Association for the Advancement of Science. He is a fellow of the Chemical Society of London, the Royal Society of Canada, and the Royal Society of London, and has served on several medical committees, such as the advisory committee on research for the National Cancer Institute of Canada. He is a popular speaker, and has delivered over 150 addresses throughout the world during the course of his career. His scientific research articles number more than two hundred.

SOAPS AND DETERGENTS

Soaps and detergents are cleaning ingredients that are able to remove oil particles from surfaces because of their unique chemical properties. Soaps are created by the chemical reaction of a **fatty acid** with an **alkali metal** hydroxide. In a chemical sense, soap is a **salt** made up of a **carboxylic acid** and an alkali like sodium of potassium. Soaps are a specific type of the more general category of compounds called detergents. The cleaning action of soaps and detergents is a result of their ability to surround oil particles on a surface and disperse it in water. Bar soap has been used for centuries and continues to be an important product for bathing and cleaning. It is also a mild antiseptic and ingestible antidote for certain poisons.

The exact origin of soap is not known, but records suggest that it was known as early as 600 B.C. by the Phoenicians. It was also used by the ancient Romans, as is evidenced by the writings of Pliny, who described a method for making soap by boiling goat tallow with alkali wood ashes. During the eighth century, soap-making was common in the southern countries of Europe. However, the production methods were costly, and there was a general negative social attitude toward cleanliness. This made soap a luxury item used primarily as a cosmetic, and available only to the rich.

It was not until the late eighteenth century that soap became widely available and affordable. One important development was made in 1790 by the French chemist **Nicholas Leblanc**. He invented a method for producing sodium hydroxide (caustic soda) from chalk, salt, sulfuric acid, and coal. Since these ingredients were relatively inexpensive, it significantly reduced the cost of a soap-making process that involved the reaction of natural fats and oils with caustic soda. The method was further refined in 1823, when **Michel-Eugéne Chevreul** the process by which fates are hydrolized by water to fatty acids and glycerols to produce soap. As the cost of soap production fell and attitudes about cleanliness changed, soap-making was poised to become an important industry.

By the early nineteenth industry, soap-making was an established industry. Important companies included Colgate-Palmolive, and Proctor and Gamble. Soap was used for most cleaning needs, including personal hygiene, laundry, and dishwashing. During World War II, fats were in short supply, which prompted companies to develop synthetic detergents. They were first introduced as laundry detergent for automatic washing machines in 1946. Steadily, the synthetic detergents replaced soap. By 1953, their production exceeded that of soap for the first time. Because these early detergents were not biodegradable like soap, they became a public nuisance, causing sewage problems. Detergent manufacturers responded by developing biodegradable, linear alkyl sulfonates. By 1965, production of non-biodegradable synthetic detergents was halted.

Chemically speaking, soap is the salt formed by the reaction of an alkali metal, such as **potassium** or **sodium**, with carboxylic acids. It is produced through a chemical reaction, known as saponification, between triglycerides and a base, such as **sodium hydroxide**. In this reaction, the triglycerides are reduced to their component fatty acids. The base then neutralizes them into salts. A byproduct of this method of soap production or glycerin. Detergents are formed similarly. However, the staring material is derived from linear, alkyl compounds. These materials are reacted with sulfuric acid, then neutralized, and converted to a salt.

Soaps and detergents have the general chemical formula RCOOX. The R represents a **hydrocarbon** chain made up of anywhere from eigth to 22 **carbon** atoms bonded to each other and to **hydrogen atom**. The X represents an alkali metal—any of the elements found in the first column on the **periodic table**. An example of a soap molecule is sodium stearate ($NaC_{18}H_{35}O_2$), which is made from steric acid. The detergent sodium sulfate (Na_2SO_4) is made from lauric acid.

Since soap and detergents are salts, they separate into their component ions in a solution of water. The portion of the molecule that has a cleansing effect is $RCOO^-$. The two ends of the **ion** have different solubility characteristics. The carboxylate end ($-COO^-$), or ''head,'' is hydrophilic, and tends to associate with the aqueous phase. The hydrocarbon portion (R), or ''tail,'' is lipophilic, and associated with the oily particles in the solution. The unusual molecular structure is responsible for the surface action and solubility of soaps and detergents. For this reason they are generally known as surfactants.

In a system composed of soap and water, the surfactant molecules tend to be uniformly dispersed. However, thus mixture is not a true solution, because the hydrocarbon portions of the surfactant ions are attracted to each other and form structures called micelles. Micelles are spherical aggregates of surfactant molecules that have the molecular tails in their interior and the hydrophilic heads on their exterior. This structure reduces the surface tension between the incompatible species. When oil is present in the system, itgets incorporated into these micelles, and can be rinsed away.

Detergents and soaps can be classified by their ioic nature. Soaps and sulfate detergents have a negatively charged ion, and are called anionic. Cationic detergents have a positively charged ion. There are also non-ionic detergents which have no charge when placed in a water solution. Finally, amphoteric detergents have either a positively or negatively charged ion, depending on the **pH** of the system they are incorporated into.

Soap manufacture before World War II was done by a ''full-boiled'' process. In this method, fats and oils were mixed in large, open kettles, and caustic soda was added. The system was heated, and tons of salt were added to make the soap precipitate out and float to the top. The soap was then skimmed off and processed into flakes or bars. The disadvantage to this system that it took an excessive amount of energy and time. In fact, it required a full six days to complete a single batch.

After World War II, a more continuous process was developed. In this method of production, fats and oils are reacted directly with caustic soda. By using higher temperatures—250° F (120°C) and pressures (2 atmospheres) the reaction is accelerated. The glycerin (another name of for **glycerol**) is removed from the system, and soap is isolated by using centrifugation and neutralization processes. This production method proved to be more desirable than the ''full-boiled'' process. It was energy- and time-efficient, allowing greater control of the composition and concentration of th soap, and it allowed the recovery of glycerin.

Detergent manufacture is similar to that of soap. The starting material is typically a vegetable oil or petroleum product. Vegetable oils, such as coconut, palm kernel, or canola contain an appropriate fatty acid distribution. The oils are first reacted with **sulfuric acid**, which convertts them into **sulfates**. These materials are then neutralized and converted into salts. To improve the characteristics of the soap before selling it, various ingredients can be added. For example, the foam of pure soap can be imroved by the addition of fatty acids or alkanolamides. Glycerin can also be added to reduce the harshness of the soap on the skin. Antibacterial compounds, such as triclosan, can be incorporated into the final soap or detergent product. Additionally, fragrances, **dyes**, and preservatives are used to modify esthetic characteristics.

While soaps are excellent, biodegradable cleansing ingredients, they suffer from the drawback of forming hard water deposits. Hard water contains amounts of **calcium** and **magnesium** ions. The carboxylate ion in soap reacts with these ions, forming a water-insoluble salt that remains deposited on fabric and other surfaces. These hard water plaques can dull fabric colors and cause undesirable rings on bath-tubs and sinks. An additional drawback of soaos is that they do not function properly at acidic pHs. Under these conditions, soap ions do not dissociate into their component ions. As a result, they lose their surcae active characterustics. Since detergents do not form hard water deposits and can are effective under a wide range of pH conditions, they are often preferred to soap.

Soaps and detergents are primarily used for their cleansing ability. However, soap has also proved its effectiveness as a mild antiseptic. It is also an ingestible antidote for mineral acid or heavy metal poisoning. Furthermore, specialized metallic soaps are employed as additive in **paints**, **inks**, and lubricating oils.

There is considerable environmental concern about the accumulation of detergents in the ecosystem. In 1997, Union Carbide devloped a non-ionic detergent that can reduce the pollution caused by waste water at institutional cleaners. Under normal conditions, this detergent functions just like a regular detergent. However, when acid is added to the system, the detergent splits into two non-toxic fragments. This was the first chemical to gain approval as environmentally safe under the United States Environmental Protection Agency's Environmental Technology Initiative.

SODDY, FREDERICK (1877-1956)

English chemist

Frederick Soddy's major contribution to science was his discovery of the existence of isotopes in 1913, an accomplishment for which he was awarded the 1921 Nobel Prize in **chemistry**. That discovery came as the result of extensive research on the radioactive elements carried out first with British physicist **Ernest Rutherford** at McGill University and later with British chemist Sir **William Ramsay** at London University. Among Soddy's contributions during this period was his recognition of the relationship between **helium** gas and **alpha particle** emanations—the latter being the ejection of a type of nuclear particle during a radioactive transformation—as well as his enunciation of the disintegration law of radioactive elements (which states that when a substance decays, it emits a particle and is transformed into a totally new substance). Soddy's most important work was carried out while he was lecturer in **physical chemistry** at the University of Glasgow between 1904 and 1919. Later in life Soddy's interests shifted to politics and economics, although he was able to make relatively little lasting impact in these fields.

Soddy was born on September 2, 1877, in Eastbourne, England. He was the seventh and last child of Benjamin Soddy and Hannah (Green) Soddy. His mother died eighteen months after his birth and his father was a successful and prosperous corn merchant in London who was already fifty-five years of age when Frederick came into the world. Soddy's interest in science, evident from an early age, was further developed at Eastbourne College by its science master, R. E. Hughes. Hughes and Soddy coauthored a paper on the reaction between dry **ammonia** and dry **carbon** dioxide in 1894, when Soddy was only seventeen years old. Hughes encouraged Soddy to continue his education in chemistry at Oxford. Teacher and student agreed, however, that an additional year of preparation would be desirable before going on to the university, so Soddy spent a year at University College, Aberystwyth, in 1895. In that year, he won the Open Science Postmastership Scholarship, offered by Merton College, Oxford, and in 1896, he enrolled at that institution.

During his years at Merton College, Soddy published his first independent paper, on the life and work of German chemist Victor Meyer, which was received as an accomplished paper for a young undergraduate. He stood for his chemistry examination in 1898 and was awarded a First Class in the Honors School of Natural Science. One of his examiners was Sir William Ramsay, with whom he was later to collaborate in London.

Soddy stayed on at Oxford for two years following his graduation. The chemical research he pursued during this period led to no substantial results. By 1900, however, he felt he was ready to move on and applied for a position that opened in the chemistry department at the University of Toronto. Deciding to pursue the post aggressively, Soddy traveled to Canada to make his case in person. When he failed to receive the Toronto appointment, he traveled on to Montreal, where he accepted a position as a junior demonstrator at McGill University. The McGill appointment may have been attractive both because of the superb physical facilities provided by the young institution and because of the presence of a rising young star at the university, Ernest Rutherford. In any case, Soddy's family fortune made it possible for him to accept the modest annual salary without hardship.

By the fall of 1900, Soddy and Rutherford had begun to collaborate on studies of the disintegration of radioactive elements. These studies led to a revolutionary theory of nuclear disintegration. Prior to the Soddy-Rutherford research, scientists were unclear as to what happens during nuclear decay. The most common notion was that radioactive materials give off some form of **energy**, such as X rays, without undergoing any fundamental change themselves. Rutherford and Soddy were able to demonstrate that the process is more substantial than previously believed and that, in the process of decaying, the composition of a radioactive substance is altered.

In 1903, Soddy returned to London. He wanted to work with Ramsay on a study of the gaseous products of radioactive decay. In his brief stay at London, the two were able to demonstrate that helium is always produced during the disintegration of **radium**. Five years later, Rutherford was to confirm that connection when he showed that alpha particles are doubly-charged helium nuclei.

In the spring of 1904, Soddy accepted an appointment as lecturer in physical chemistry at Glasgow University. Before moving to Scotland, however, he also accepted another commission, that from the extension service of London University. In this assignment, Soddy gave a series of lectures on physical chemistry and **radioactivity** at venues in Western Australia. At the conclusion of the tour in the fall of 1904, Soddy returned to Great Britain by way of New Zealand and the United States, to begin what was to be a ten-year tenure at Glasgow.

Soddy's work at Glasgow was primarily concerned with the chemical identification of the elements involved in the radioactive decay of **uranium** and radium, which was the subject of intense investigation by a number of scientists. The problem was that the disintegration of uranium and radium appeared to result in the formation of about a dozen new elements, elements that were tentatively given names such as uranium X, radium A, radium B, radium C, radium D, radium E, radium F, ionium, and mesothorium. How all these elements could be fitted into the few remaining spaces in the **periodic table** was entirely unclear.

By 1907, some clues to the answer to this problem had begun to appear; H. N. McCoy and W. H. Ross at the University of Chicago showed that two of the elements produced during radioactive decay, **thorium** and radiothorium, were chemically identical to each other. Soon, similar results were being announced for other pairs, such as ionium and radium, mesothorium and thorium K, and radium D and **lead**. These results were similar to those being obtained by Soddy in his own laboratory. By 1910, he began to formulate a possible explanation for the research findings. In a paper published that year, he first raised the possibility that many of the products of radioactive decay are not different from each other, but are variations of the same element. He began to develop the concept of different forms of a chemical element with identical chemical properties, but different atomic weights.

That idea came to fruition in 1913 when Soddy first proposed the term **isotope** for these forms of an element. Soddy's paper published in *Chemical News* summarized his views on isotopes. He wrote that "it would not be surprising if the elements... were mixtures of several homogeneous elements of similar but not completely identical atomic weights." It was for this hypothesis that Soddy would be awarded the 1921 Nobel Prize for chemistry.

Frederick Soddy.

At this time Soddy was also working on an explanation of the patterns observed during radioactive decay. In 1911, he pointed out that each time an element loses an alpha particle, it changes into a new element whose **atomic number** is two less than that of the original element. This generalization became known as the Displacement Law. Shortly thereafter, A. S. Russell and Kasimir Fajans independently extended that law to include beta decay, in which an element's atomic number increases by one after the loss of a beta particle. In 1914 Soddy left Glasgow to take a chair in chemistry at the University of Aberdeen. His major work there involved the determination of the atomic weight of lead extracted from the radioactive ore Ceylon thorite. He showed for the first time that the atomic weight of an element (lead, in this case) can differ significantly and consistently from its normally accepted value as published in the periodic table.

During World War I, Soddy was involved in military research for the marine sub-committee of the Board of Inventions and Research. The major part of this work involved the development of methods for extracting ethylene from **coal** gas.

At the conclusion of the war, Soddy was appointed to the Lees Chair in Chemistry at Oxford University. He remained in this post until 1936. Soddy's interest in scientific research had largely dissipated by the time he reached Oxford, and he published no original research in chemistry during his seventeen years there. Instead, he showed interest in social, political, and economic issues, motivated to some extent by a feeling that progress in science had not produced or had not been accompanied by a comparable development of human civilization. He became—and remained—actively involved in a number of social and political causes, including the women's suffrage movement and the controversy over the status of Ireland.

Soddy's academic career ended in 1936 when he took early retirement from Oxford. The occasion for this decision was the unexpected death of his wife Winifred Moller (Beilby) Soddy. The couple had been married in 1908 and, although childless, had been happy together. They enjoyed traveling and spent some of their most pleasant moments in mountain climbing. Winifred's death from a coronary thrombosis was so distressing that Soddy almost immediately left Oxford.

Even after his retirement, Soddy continued to think, write, and speak about current events. He was particularly concerned with his fellow scientists who, he believed, had not demonstrated sufficient social conscience about the difficult issues their own research had brought about. Soddy died in Brighton, England, on September 22, 1956. According to one provision of his will, a trust was to be established to study social problems in various regions of the country.

Sir Alexander Fleck, Soddy's former student, colleague, and biographer, has described Soddy's personality in *Biographical Memoirs of Fellows of the Royal Society* as "complex." On the one hand, he was often kind and generous to friends and fellow workers, and could be "a live and inspiring leader" to those students he worked with in small groups. On the other hand, he seems to have been, more generally, a failure as a teacher. "His mental processes were different from those of the ordinary run of students so that the latter could not easily follow the words with which he clothed his thoughts," Fleck wrote. Soddy held very strong moralistic views on a number of issues and was not hesitant to make those views known and to defend them with vigor and little tact. Fleck observes that "he very frequently found himself in acrimonious discussions" during his tenure at Oxford, although his personal life appears to have been filled with personal happiness and many enjoyable social events.

SODIUM

Sodium is the second member of the alkali family, the group of elements that make up Group 1 of the **periodic table**. Sodium has an **atomic number** of 11, an atomic **mass** of 22.98977, and a chemical symbol of Na.

Properties

Sodium is a silvery white metal with a waxy appearance. It is soft enough to be cut with a knife. When first cut, the surface is bright and shiny, but it quickly becomes dull as sodium reacts with **oxygen** in the air to form sodium **oxide** (Na_2O). Sodium's melting point is 208.1°F (97.82°C), its **boiling point** is 1,618°F (881.4°C), and its **density** is 0.968 grams per cubic centimeter. Sodium is a good conductor of **electricity**.

Sodium is a very active element that combines with oxygen at room **temperature** and burns with a brilliant golden-yellow flame. It reacts violently with **water** to produce **hydrogen** gas, which may be ignited by the **heat** of the reaction. To avoid having sodium react with oxygen or water vapor in the air, it is usually stored under **kerosene**, naphtha, or some other organic liquid with which it does not react. Sodium also reacts readily with most nonmetals and dissolves in **mercury** to form a sodium amalgam.

Occurrence and Extraction

Sodium is too reactive to occur free in nature, although it is present in the Earth's crust in a great many different compounds. It is the seventh most abundant element in the crust, with an estimated abundance of about 2.27%. The most common compound of sodium is halite, also known as rock **salt**. Halite is nearly pure **sodium chloride** ($NaCl$). Halite is found in underground deposits known as salt domes, formed when ancient seas dried up and were buried under the earth. Sodium chloride is also abundant in seawater and brine. It can be obtained very easily from these sources simply by trapping the seawater or brine and allowing the water to evaporate. The technology for obtaining sodium chloride by this process is many centuries old.

The usual method for obtaining pure sodium metal is by the **electrolysis** of molten sodium chloride: $2NaCl$ —electric current→ $2Na + Cl_2$. Since the demand for pure sodium metal is quite limited, another method is often used to extract the element in a different form. When aqueous sodium chloride is electrolyzed, the products are **sodium hydroxide** ($NaOH$) and **chlorine** gas: $2NaCl + 2H_2O$ —electric current→ $2NaOH + H_2 + Cl_2$. The demand for sodium hydroxide is very great, so this method is the one most commonly used.

Discovery and Naming

Probably the best known compound among ancient civilizations was **sodium carbonate** (Na_2CO_3), commonly known as soda. Soda is one of the most common ores of sodium found in nature and it was used very early in human history to make **glass**. The Egyptians called soda *natron*, from which the Romans later derived the term *natrium*. It is from these names that sodium's modern chemical symbol, Na, is derived.

Most compounds of sodium are very stable, so it is not surprising that the element itself was not discovered until fairly recent times. The great English chemist and physicist Sir **Humphry Davy** first prepared a sample of pure sodium metal by electrolyzing molten sodium chloride. In essence, the system developed by Davy is still used in some cases to prepare pure sodium. Davy went on to apply his method for preparing sodium to the extraction of **potassium**, **calcium**, and other active **metals**.

Uses

Sodium metal has a relatively small, but important, number of uses. For example, it is sometimes used as a heat exchange medium in nuclear power plants. It is much more difficult to work with than water, but more efficient, as a heat exchange fluid. Sodium is also used to make certain types of fluorescent bulbs, known as sodium vapor lamps. Elemental sodium is also used to extract certain metals from their ores, **titanium** being one example: $4Na + TiCl_4 \rightarrow 4NaCl + Ti$ Dozens of sodium compounds have important applications in research, industrial operations, and everyday life. Probably the three most important of these compounds are sodium chloride ($NaCl$), sodium carbonate (Na_2CO_3) and **sodium bicarbonate** ($NaHCO_3$). Sodium chloride is probably best known as a flavor enhancer in foods, but its addition to foods also helps to preserve the food from decay. Salting of foods as a method of food preservation goes back many centuries. The primary use of sodium chloride in terms of amounts produced, however, is in the preparation of other sodium compounds.

Sodium carbonate is also used primarily as a raw material in the manufacture of dozens of important sodium compounds. In addition, it is used in **water purification** systems and in the production of many commercial products, such as **glass**, pulp and **paper**, **soaps and detergents**, and textiles. Sodium bicarbonate may be best known to many people as an additive to food that causes baked goods to rise during baking. The compound is also used in mouthwashes, cleaning solutions, wool and silk cleaning systems, fire extinguishers, and mold preventatives in the timber industry.

Health Issues

Sodium is an essential nutrient for both plants and animals. It humans, for example, it helps control the amount of fluid present in cells, the transmission of nerve impulses between cells, and the movement of muscles.

Sodium Benzoate

Sodium benzoate is the sodium **salt** of benzoic acid. It is an aromatic compound denoted by the chemical formula $C_7H_5NaO_2$ with a **molecular weight** of 144.11. In its refined form, sodium benzoate is a white, odorless compound that has a sweetish, astringent **taste**. It is soluble in **water**. Sodium benzoate has antimicrobial characteristics it is typically used as a preservative in food products.

Sodium benzoate is supplied as a white powder or flake. During use it is mixed dry in bulk **liquids** where it promptly dissolves. Approximately 1.75 oz (50 g) will readily dissolve in 3 fl oz (100 ml) of water. In contrast, benzoic acid has a significantly lower water **solubility** profile. When placed in water, sodium benzoate dissociates to form sodium ions and benzoic acid ions. It is a weak organic acid that contains a **carboxyl group**. Benzoic acid occurs naturally in some foods including cranberries, prunes, cinnamon, and cloves. It is also formed by most vertebrates during **metabolism**.

Sodium benzoate is an antimicrobial active against most yeast and bacterial strains. It works by dissociating in the system and producing an amount of benzoic acid. Benzoic acid is highly toxic to microbes however, it is less effective against molds. Overall, it has more effect as the **pH** of a system is reduced with the optimal functional range between pH 2.5-4.0. The antimicrobial effect is also enhanced by the presence of **sodium chloride**.

There are three methods for the commercial preparation of sodium benzoate. In one method, **naphthalene** is oxidized with **vanadium** pentoxide to give phthalic anhydride. This is decarboxylated to yield benzoic acid. In a second method, toluene is mixed with **nitric acid** and oxidized to produce benzoic acid. In a third method, benzotrichloride is hydrolyzed and then treated with a mineral acid to give benzoic acid. Benzotrichloride is formed by the reaction of **chlorine** and toluene. In all cases, the benzoic acid is further refined to produce sodium benzoate. One way this is done is by dissolving the acid in a **sodium hydroxide solution**. The resulting chemical reaction produces sodium benzoate plus water. The crystals are isolated by evaporating off the water.

Some toxicity testing has shown sodium benzoate to be poisonous at certain concentrations. However, research conducted by the U.S. Department of Agriculture (USDA) has found that in small doses and mixed with food, sodium benzoate is not deleterious to health. Similar conclusions were drawn about larger doses taken with food, although certain physiological changes were noted. Based on this research and subsequent years of safety data, the United States government has determined sodium benzoate to be generally recognized as safe (GRAS). It is allowed to be used in food products at all levels below 0.1%. Other countries allow higher levels, up to 1.25%.

Studies investigating the accumulation of sodium benzoate in the body have also been done. This led to the discovery of a natural metabolic process that combines sodium benzoate with glycine to produce hippuric acid, a material that is then excreted. This excretion mechanism accounts for nearly 95% of all the ingested sodium benzoate. The remainder is thought to be detoxified by **conjugation** with glycuronic acid.

Sodium benzoate has been used in a wide variety of products because of its antimicrobial and flavor characteristics. It is the most widely used food preservative in the world being incorporated into both food and soft drink products. It is used in margarines, salsas, maple syrups, pickles, preserves, jams, and jellies. Almost every diet soft drink contains sodium benzoate as do some wine coolers and fruit juices. It is also used in personal care products like toothpaste, dentifrice cleaners, and mouthwashes. As a preservative, sodium benzoate has the advantage of low cost. A drawback is its astringent taste which can be avoided by using lower levels with another preservative like **potassium** sorbate.

In addition to its use in food, it is used as an intermediate during the manufacture of dyes. It is an antiseptic medicine and a rust and mildew inhibitor. It is also used in tobacco and pharmaceutical preparations. In the free-acid form, it is used as a **fungicide**. A relatively recent use for sodium benzoate is as a **corrosion** inhibitor in engine coolant systems. Sodium benzoate has recently been incorporated into **plastics**, like **polypropylene**, where it has been found to improve clarity and strength.

A common compound of sodium, sodium bicarbonate, produces a fizzing reaction. It is an ingredient in such medications as Alka-Seltzer. *(Photograph by Robert J. Huffman. Reproduced by permission.)*

SODIUM BICARBONATE

Sodium bicarbonate ($NaHCO_3$) is a white crystalline powder commonly known to as baking soda. It is classified as an acid **salt**, formed by combining an acid (carbonic) and abase (sodium hydroxide), and it reacts with other chemicals as a mild alkali. At temperatures above 300°F (149°C), sodium bicarbonate decomposes into **sodium carbonate** (a more stable substance), **water**, and **carbon** dioxide.

The native chemical and physical properties of sodium bicarbonate account for its wide range of applications, including cleaning, deodorizing, buffering, and fire extinguishing. Sodium bicarbonate neutralizes odors chemically, rather than masking or absorbing them. Consequently, it is used in bath salts and deodorant body powders. Sodium bicarbonate tends to maintain a **pH** of 8.1 (7 is neutral) even when acids, which lower pH, or bases, which raise pH, are added to the **solution**. Its ability to tabletize makes it a good effervescent ingredient in antacids and denture cleaning products. Sodium bicarbonate is also found in some anti-plaque mouthwash products and toothpaste. When sodium bicarbonate is used as a cleaner in paste form or dry on a damp sponge, its crystalline structure provides a gentle abrasion that helps to remove dirt without scratching sensitive surfaces. Its mild alkalinity works to turn up **fatty acids** contained in dirt and grease into a form of soap that can be dissolved in water and rinsed easily. Sodium bicarbonate is also used as a leavening agent in making baked goods such as bread or pancakes. When combined with an acidic agent (such as lemon juice), **carbon dioxide** gas is released and is absorbed by the product's cells. As the gas expands during baking, the cell walls expand as well, creating a leavened product.

In addition to its many home uses, sodium bicarbonate also has many industrial applications. For instance, sodium bicarbonate releases carbon dioxide when heated. Since carbon dioxide is heavier than air, it can smother flames by keeping **oxygen** out, making sodium bicarbonate a useful agent in fire extinguishers. Other applications include air **pollution** control (because it absorbs **sulfur** dioxide and other acid gas emissions), abrasive blastings for removal of surface coatings, chemical manufacturing, leather tanning, oil well drilling fluids (because it precipitates **calcium** and acts as a lubricant), rubber and plastic manufacturing, **paper** manufacturing, textile processing, and water treatment (because it reduces the level of **lead** and other heavy metals).

Imported from England, sodium bicarbonate was first used in America during colonial times, but it was not produced in the United States until 1839. In 1846, Austin Church, a Connecticut physician, and John Dwight, a farmer from Massachusetts, established a factory in New York to manufacture sodium bicarbonate. Dr. Church's son, John, owned a mill called the Vulcan Spice Mills. Vulcan, the Roman god of forge and fire, was represented by an arm and hammer, and the new sodium bicarbonate company adopted the arm and hammer logo as its own. Today, the Arm & Hammer brand of baking soda is among the most widely recognized brand names.

Named after **Nicolas Leblanc**, the French chemist who invented it, the Leblanc process was the earliest means of manufacturing soda ash (Na_2CO_3), from which sodium bicarbonate is made. **Sodium chloride** (table salt) was heated with **sulfuric acid**, producing sodium **sulfate** and **hydrochloric acid**. The sodium sulfate was then heated with **coal** and limestone to form sodium carbonate, or soda ash.

In the late 1800s, another method of producing soda ash was devised by **Ernest Solvay**, a Belgian chemical engineer. The Solvay method was soon adapted in the United States, where it replaced the Leblanc process. In the Solvay process, carbon dioxide and **ammonia** are passed into a concentrated solution of sodium chloride. Crude sodium bicarbonate precipitates out and is heated to form soda ash, which is then further treated and refined to form sodium bicarbonate of *United States Pharmacopoeia* (U.S.P.) purity.

Although this method of producing sodium bicarbonate ash is widely used, it is also problematic because the chemicals used in the process are pollutants and cause disposal problems. An alternative is to refine soda ash from tronaore, a natural deposit.

Sodium bicarbonate, comes from soda ash obtained either through the Solvay process or from trona ore, a hard, crystalline material. Trona dates back 50 million years, to when the land surrounding Green River, Wyoming, was covered by a 600 sq mi (1,554 sq km) lake. As it evaporated over time, this lake left a 200-billion-ton deposit of puretrona between layers of sandstone and shale. The deposit at the Green River Basin is large enough to meet the entire world's needs for soda ash and sodium bicarbonate for thousands of years.

Because the synthetic process used in the Solvay method presented some pollution problems, Church & Dwight Co. Inc. is basing more and more of its manufacturing on trona mining. Another large producer of soda ash, the FMC Corporation, also relies on trona to manufacture soda ash and sodium bicarbonate. Trona is mined at 1,500 ft (457.2 m) below the surface. FMC's mine shafts contain nearly 2,500 mi (4,022.5 km) of tunnels and cover 24 sq mi (62 sq km). Fifteen feet (4.57) wide and 9 ft (2.74 m) tall, these tunnels allow the necessary equipment and vehicles to travel through them.

The quality of sodium bicarbonate is controlled at every stage of the manufacturing process. Materials, equipment, and the process itself are selected to yield sodium bicarbonate of the highest possible quality. According to FMC sources, when the company constructed plants, it chose materials and equipment that would be compatible with the stringent quality requirements for making pharmaceutical grade sodium bicarbonate. FMC also uses *Statistical Process Control* (SPC) to maintain unvarying daily quality, and key operating parameters are charted to maintain process control. Product quality parameters are recorded by lot number, and samples are kept for two to three years.

All U.S.P. grades meet the *United States Pharmacopoeia and Food Chemicals Codex* specifications for use in pharmaceutical and food applications. In addition, food grade sodium bicarbonate meets the requirements specified by the U.S. Food and Drug Administration as a substance that is *Generally Recognized as Safe* (GRAS).

At the turn of the twentieth century, 53,000 tons (48,071 metric tons) of sodium bicarbonate were sold annually. While the population increased dramatically, sales by 1990 were down to about 32,000 tons (29,024 metric tons) per year. Self-rising flour and cake and biscuit mixes have decreased the demand for baking soda as an important baking ingredient. Nevertheless, demand for the product is still significant. Commercial bakers (particularly cookie manufacturers) are major users of this product. One of the most important attributes of sodium bicarbonate is that, when exposed to **heat**, it releases carbon dioxide gas (CO_2) which makes the baking goods rise. Sodium bicarbonate is also used in the pharmaceutical and health industries, and it has other industrial applications as well. It therefore continues to be an important product for today and for the future.

SODIUM CARBONATE

Sodium carbonate is a **chemical compound** which conforms to the general formula Na_2CO_3.

It is commonly referred to as soda ash because it was originally obtained from the ashes of burnt sea weeds. Now, soda ash is primarily manufactured by a method known as the Solvay process. Currently, it is one of the top industrial chemicals, in terms of volume, produced in the United States. It is mostly used in the manufacture of **glass**, but is also used in the manufacture of other products and is an important precursor to many of the sodium compounds used throughout industry.

The process for obtaining sodium carbonate has changed significantly over time. It was originally produced by burning seaweeds that were rich in sodium. When the weeds were burned, sodium would be left in the ashes in the form of sodium carbonate ($Na_2SO_4 + CaCO_3 \rightarrow CaSO_4 + Na_2CO_3$). Although this process was effective, it could not be used to produce large amounts.

The first process that allowed production of significant amounts of sodium carbonate was a synthetic process known as the LeBlanc process, developed by the French chemist **Nicolas LeBlanc**. In this process, **salt** reacts with **sulfuric acid** to produce sodium **sulfate** and **hydrochloric acid** ($NaCl + H_2SO_4 \rightarrow Na_2SO_4 + HCl$). The sodium sulfate was heated in the presence of limestone and **coal** and the resulting mixture contained **calcium** sulfate and sodium carbonate, which was then extracted.

Two significant problems with the LeBlanc process, including high expense and significant **pollution**, inspired a Belgian chemical engineer named **Ernest Solvay** (1838-1922) to develop a better process for creating sodium carbonate. In the Solvay process, **ammonia** and **carbon** dioxide are used to produce sodium carbonate from salt and limestone. Initially, the ammonia and **carbon dioxide** reacts with **water** to form the weak electrolytes, ammonium hydroxide and carbonic acid. These ions react further and form **sodium bicarbonate**. Since the bicarbonate barely dissolves in water, it separates out from the **solution**. At this point, the sodium bicarbonate is filtered and converted into sodium carbonate by heating.

Synthetic production is not the only method of obtaining sodium carbonate. A significant amount is mined directly from naturally occurring sources. The largest natural sources for sodium carbonate in the United States, are found around Green River, Wyoming and in the dried-up desert lake Searles in California.

At room **temperature**, sodium carbonate (Na_2CO_3) is an odorless, grayish-white powder which is hygroscopic. This means when it is exposed to air, it can spontaneously absorb water molecules. Another familiar compound that has this hygroscopic quality is sugar. Sodium carbonate has a melting point of 1,564° F (851° C), a **density** of 2.53 g/cm³, and is soluble in water. A water solution of soda ash has a basic **pH** and a strong alkaline **taste**. When it is placed in a slightly acidic solution, it decomposes and forms bubbles. This effect, called effervescence, is found in many commercial **antacid** products which use sodium carbonate as an active ingredient.

Anhydrous (without water) sodium carbonate can absorb various amounts of water and form **hydrates** that have slightly different characteristics. When one water **molecule** per molecule of sodium carbonate is absorbed, the resulting substance, sodium carbonate monohydrate, is represented by the chemical formula Na_2CO_3•HOH. This compound has a slightly lower density than the anhydrous version. Another common hydrate is formed by the absorption of 10 water molecules per molecule of sodium carbonate. This compound, Na_2CO_3•10HOH, known as sodium carbonate decahydrate, exists as transparent crystals that readily effervesce when exposed to air.

Sodium carbonate is utilized by many industries during the manufacture of different products. The most significant user is the glass industry, which uses sodium carbonate to decompose **silicates** for glass making. The cosmetic industry uses it while manufacturing soap. The chemical industry uses it as a precursor to numerous sodium containing reagents. It is also important in photography, the textile industry, and water treatment. In addition to these industrial applications, sodium carbonate is used in medicine as an antacid.

Sodium chloride. ***(JLM Visuals. Reproduced by permission.)***

SODIUM CHLORIDE

Sodium chloride (chemical formula NaCl), known as table **salt**, rock salt, sea salt, and the mineral halite, is an ionic compound consisting of cube-shaped crystals composed of the elements sodium and **chlorine**. This salt has been of importance since ancient times and has a large and diverse range of uses. It can be prepared chemically and is obtained by mining and evaporating **water** from seawater and brines.

Sodium chloride is colorless in its pure form. It is somewhat hygroscopic or absorbs water from the atmosphere. The salt easily dissolves in water. Its dissolution in water is endothermic, meaning it takes some **heat** energy away from the water. Sodium chloride melts at 1,474° F (801° C) and it conducts **electricity** when dissolved or in the molten state.

An ionic compound such as sodium chloride that contains the elements sodium and chlorine is held together by an **ionic bond**. This type of bond is formed when oppositely charged ions attract. This attraction is similar to that of two opposite poles of a magnet. An **ion** or charged **atom** is formed when the atom gains or loses one or more electrons. It is called a **cation** if a positive charge exists and an **anion** if a negative charge exists.

Sodium (Na) is an alkali metal and tends to lose an **electron** to form the positive sodium ion (Na^+). Chlorine (Cl) is a **nonmetal** and tends to gain an electron to form the negative chloride ion (Cl^-).

The oppositely charged ions Na^+ and Cl^- attract to form an ionic bond. Many sodium and chloride ions are held together this way, resulting in a salt with a distinctive crystal shape. The three-dimensional arrangement or crystal lattice of ions in sodium chloride is such that each Na^+ is surrounded by six anions (Cl^-) and each Cl^- is surrounded by six cations (Na^+). Thus, the ionic compound has a **balance** of oppositely charged ions and the total positive and negative charges are equal.

Sodium chloride, found abundantly in nature, occurs in seawater, other saline waters or brines and in dry rock salt deposits. It can be obtained by mining and evaporating water from brines and seawater. This salt can also be prepared chemically by reacting **hydrochloric acid** (HCl) with **sodium hydroxide** (NaOH) to form sodium chloride and water. Countries leading in the production of salt include the United States, China, Mexico, and Canada.

Two ways of removing salt from the ground are by room and pillar mining room and pillar mining and **solution** mining. In the room and pillar method, shafts are sunk into the ground and miners use techniques such as drilling and blasting to break up the rock salt. The salt is then removed in such a way that empty rooms remain that are supported by pillars of salt.

In mining, solution water is added to the salt deposit to form brine. Brine is a solution of sodium chloride and water that may or may not contain other salts. In one technique, a well is drilled in the ground and two pipes (a smaller pipe placed inside a larger one) are placed in it. Fresh water is pumped through the inner pipe to the salt. The dissolved salt forms brine which is pumped through the outer pipe to the surface and then removed.

A common way to produce salt from brine is by evaporating the water using vacuum pans. In this method brine is boiled and agitated in huge tanks called vacuum pans. High quality salt cubes form and settle to the bottom of the pans. The cubes are then collected, dried, and processed.

Solar evaporation of seawater to obtain salt is an old method that is widely used today. It uses the sun as a source of **energy**. This method is successful in places that have abundant sources of salt water, land for evaporating ponds, and hot, dry climates to enhance evaporation. Seawater is passed through a series of evaporating ponds. Minerals contained in the seawater precipitate or drop out of solution at different rates. Most of them precipitate before sodium chloride and therefore are left behind as the seawater is moved from one evaporating pond to another.

Since ancient times, the salt sodium chloride has been of importance. food preservative. It has been used in numerous ways including the salting of food flavoring and preserving of food and even as a form of money. This salt improves the flavor of food items such as breads and cheeses, and it is an important preservative in meat, dairy products, margarine and other items, because it retards the growth of microorganisms. Salt promotes the natural development of **color** in ham and hot dogs and enhances the tenderness of cured meats like ham by causing them to absorb water. In the form of iodized salt, it is a carrier of **iodine**. (Iodine is necessary for the synthesis of our thyroid **hormones** which influence growth, development and metabolic rates).

The chemical industry uses large amounts of sodium chloride salt to produce other chemicals. Chlorine and sodium hydroxide are electrolytically produced from brine. Chlorine products are used in metal cleaners, **paper** bleach, **plastics** and water treatment. The chemical soda ash, which contains sodium, is used to manufacture **glass**, **soaps**, paper and water softeners. Other chemicals, produced as a result of sodium chloride reactions, are used in ceramic glazes, metallurgy, curing of hides and photography.

Sodium chloride has a large and diverse range of uses. It is spread over roads to melt ice by lowering the melting point of the ice. The salt has an important role in the regulation of body fluids. It is used in medicines and livestock feed. In addition, salt caverns are used to store chemicals such as petroleum and **natural gas**.

SODIUM HYDROXIDE

Sodium hydroxide (NaOH), also known as lye, caustic soda, or sodium hydrate, is an extremely caustic (corrosive and damaging to human tissue) white solid that readily dissolves in **water**, **alcohol**, and glycerin. It absorbs **carbon** dioxide and moisture from the air. Sodium hydroxide is used in the manufacturing of soaps, rayon, and **paper**, in petroleum refining and finds uses in homes as drain cleaners and oven cleaners. Sodium hydroxide is one of the strongest bases commonly used in industry. Solutions of sodium hydroxide in water are at the upper limit (most basic) of the **pH** scale. Sodium hydroxide is made by the **electrolysis** (passing an electric current through a solution) of solutions of **sodium chloride** (table salt) to produce sodium hydroxide and **chlorine** gas.

Sodium hydroxide crystals. ***(Photograph by Charles D. Winter, National Audubon Society Collection/Photo Researchers, Inc. Reproduced with permission.)***

Two of the more common household products containing sodium hydroxide are drain cleaners such as Drano, and oven cleaners such as Easy-Off. When most pipes are clogged, it is with a combination of fats and grease. Cleaners that contain sodium hydroxide (either as a solid or already dissolved in water) convert the fats to soap that easily dissolves in water. In addition when sodium hydroxide dissolves in water, a great deal of **heat** is given off. This heat helps to melt the clog. Sodium hydroxide is very damaging to human tissue (especially eyes). If a large amount of solid drain cleaner is added to a clogged drain, the heat produced can actually boil the water, leading to a splash in the eyes of a **solution** caustic enough to cause blindness. Some drain cleaners also contain small pieces of **aluminum** metal. Aluminum reacts with sodium hydroxide in water to produce **hydrogen** gas. The bubbles of hydrogen gas help to agitate the mixture, helping to dislodge the clog. Oven cleaners work by converting built up grease (fats and oils) into soap that can then be dissolved and wiped off with a wet sponge.

Sodium hydroxide is used to neutralize acids and as a source of sodium ions for reactions that produce other sodium compounds. In petroleum refining, it is used to neutralize and remove acids. The reaction of **cellulose** with sodium hydroxide is a key step in the manufacturing of rayon and cellophane.

SOLIDS

Solids represent one of the three **states of matter**, **liquids** and **gases** representing, respectively, the other two. A solid has a fixed **volume**, high molecular **density**, a definite shape, and it foes not flow. These characteristics are exactly opposite to those of a gas. A liquid could be described as an intermediate form. A solid can change into a liquid by melting, a liquid changes into a solid by freezing, and some solids can change directly into a gas by **sublimation**. The **temperature** at which each of these events occurs is a fixed property of the solid. Chemists know these temperatures for pure substances; however, if the solid is not a pure substance, these temperature are harder to determine. In fact, an unexpected melting point is an indication that a substance is impure. The melting point and the freezing point are identical. A substance with a high melting point has strong **intermolecular forces**.

In a solid, the molecules are in close contact, and there is little intermolecular movement. In addition, a solid has a high degree of regularity in molecular arrangement.

When a solid is heated there is very little expansion due to the effects of **heat**. Similarly, when a solid is placed under **pressure** there is very little change to the solid. The particles constituting a solid are held rigidly in place and can only vibrate. As the temperature increases, they vibrate more vigorously, remaining rigidly in place until the liquid phase is reached.

While a solid usually has a much higher density than a liquid, this is not true for **water**. Solid water, or ice, is less dense than water. This is is a crucial feature, because it enables ice to float on water. This, in combination with other features of water, has been instrumental in the development of life on this planet.

The inner structure of a solid depends on the type of **bonding** that exists between the solid's constitutive molecules. As a general rule, regardless of the bonding, the particles tend to form the closest packed arrangement of the largest particles present. Solids with a highly ordered structure are said to be crystalline in nature. For example, particles of metal will form a lattice in regular arrangement.

A solid can be dissolved in a liquid. When this occurs, the solid is called called a solute, the liquid is a solvent, and the result of the process is a **solution**. Any solid will not dissolve in any liquid. If a solid is an organic compound, then it will generally only dissolve in an organic solvent. Similarly, if the compound is inorganic, an organic solvent, such as water, is needed. Sometimes a solid will not dissolve even if the appropriate solvent is used; in that case, the solid is said to be insoluble.

Kinetic theory explains the distribution of molecules and molecular **energy** within a solid. The intermolecular forces holding the particles together in a solid are strong enough not just to hold them in together but to virtually hold them in place. The intermolecular forces present within a solid may be powerful enough to fold the molecules in position but they are much weaker than the ionic bonds that are present. Less energy is required to melt a solid than is needed to break the bonds in the molecules. Because of the closeness of the particles in a solid, a solid (and a liquid) is sometimes referred as the condensed phase.

Solids can occur in a crystalline or non-crystalline (amorphous) form. Crystalline solids have their particles in well-defined and ordered arrangements. Such solids usually have flat surfaces or faces, adjoining surfaces making a definite angle. A **diamond** is a perfect example of a crystalline solid.

Amorphous solids are fundamentally different: their particles are not structured in an orderly manner. These solids do not have well-defined shapes or angles. Most amorphous solids are mixtures of particles that do not stack together well, and this may be because they are composed of large, complex molecules, or due to a considerable inbalance between the sizes of the molecules. Rubber is an example of an amorphous solid. If a crystalline solid with a regular arrangement is heated to a sufficiently high temperature, some of the molecular bonds will break. This temperature is usually higher than the melting point. If the liquid is then cooled rapidly, there is no opportunity for these bonds to re-form. When the solid is reformed, the molecules are no longer able to align in an ordered manner. Consequently, the formerly crystalline solid becomes amorphous. Due to its lack of an orderly structure, an amorphous solid does not have a definite melting point.

Because a crystalline solid is regular, we can see the inner form of the entire solid by looking at a fragment. The smallest repeating unit of the solid is the unit cell. A larger unit is a crystal lattice. The crystal lattice is a three-dimensional array of atoms. The principal lattice structures are simple cubic, body-centered, and face-centered. The simple cubic consists of eight atoms forming a cube. The body-centered form is a cube with a ninth **atom** in the center. In the face-centered cubic, there are six atoms added to the eight of the basic cube, one for each face. Regardless of the lattice composing a solid, the particles are always placed in the closest possibly proximity, maximizing the attractive force between them, so they hold the solid's shape as rigidly as possible. This is known as close packing.

The arrangement of particles in a crystal can be seen by using a technique called **x-ray diffraction**. This technique projects x rays through a crystal and records the beam on photographic film. If the crystal is sufficiently thin, the researcher can infer the particle arrangement from the picture produced.

It is the bonding in a solid that is responsible for the solid's specific characteristics. For example, molecular crystals are held together by a variety of weaker forces; as aresult, these solids are soft and have a lower melting point. Strong solids are formed with a structure called a covalent network. Diamond, **graphite**, and sand (silicone dioxide) represent this class. Their molecules are bonded to each other in layers. Another type of bonding is found in ionic solids, exemplified by

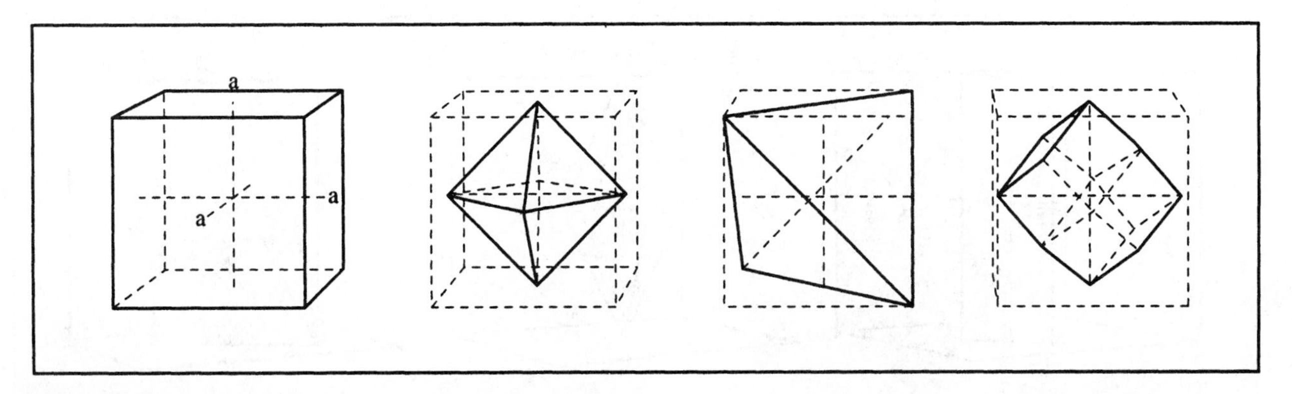

Cubic crystals. *(Illustration by Electronic Illustrators Group.)*

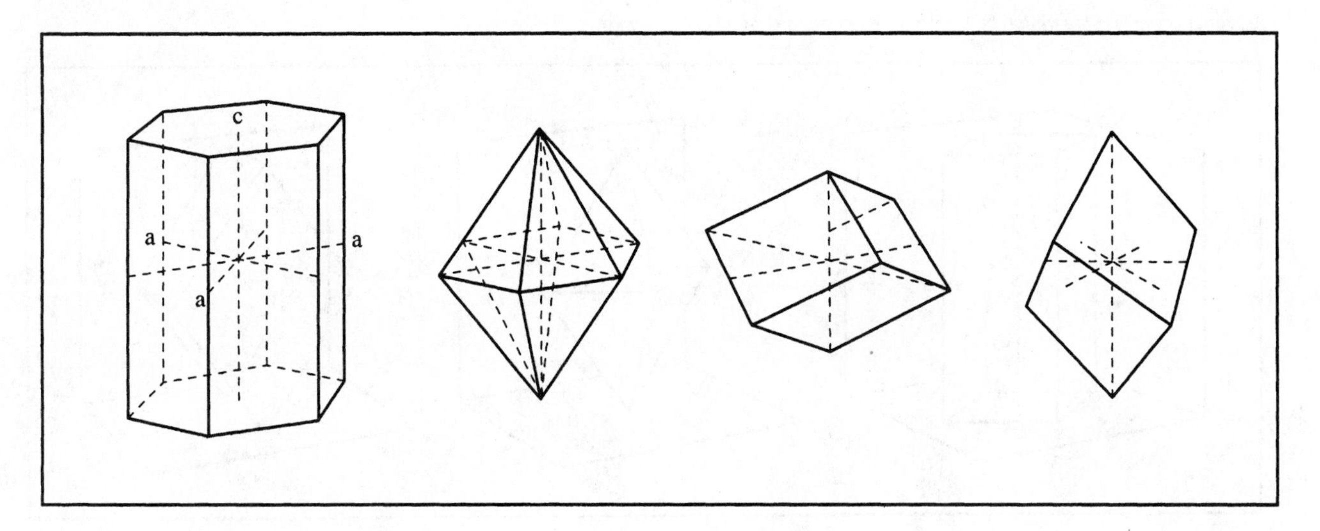

Hexagonal crystals *(Illustration by Electronic Illustrators Group.)*

sodium chloride (NaCl), or table salt, crystals. Here, the particles are held together by ionic bonds, each **ion** being surrounded by several oppositely charged ions within the inner structure of the solid.

SOLID-STATE CHEMISTRY

Chemists often make use of the observed shapes of crystals to help in their identification. The description of the shapes of crystals is the subject of the science of **crystallography**. Every crystal can be classified in one of the following six crystal systems:

Cubic. Three axes are at right angles and a re of equal length.

Hexagonal. Three equal lateral axes are intersecting at angles of 60° and vertical axis of variable length are at right angles (as in the hexagonal prism).

Tetragonal. All three axes are at right angles and the two lateral axes are equal.

Orthorhombic. Three unequal axes are at right angles to each other.

Monoclinic. Three unequal axes that have only one oblique intersection.

Triclinic. Three unequal axes that intersect at oblique angles.

Five-fold symmetry was ignored for a long period of time in science, especially in crystallography. This was because of a proof of the impossibility of a five-fold symmetry axis in a crystal medium. Only one-, two-, three-, four-, and six-fold symmetry axes are possible. Three-dimensional structures with five-fold symmetry cannot be packed to fill all available space and cover a surface without gaps. This was an official understanding until 1985 when the Kroto/Smalley (1996 Nobel Prize in Chemistry) research team discovered the carbon-60 molecule—a truncated icosahedron with 12 pentagons, 20 hexagons, 60 vertices, and 90 edges. Its discovery opened a new branch in crystallography.

The basic physical properties of C-60 are: **density** 1.67 g/cm^3, diameter 0.71 nm, lattice constant 1.42 nm, **ionization**

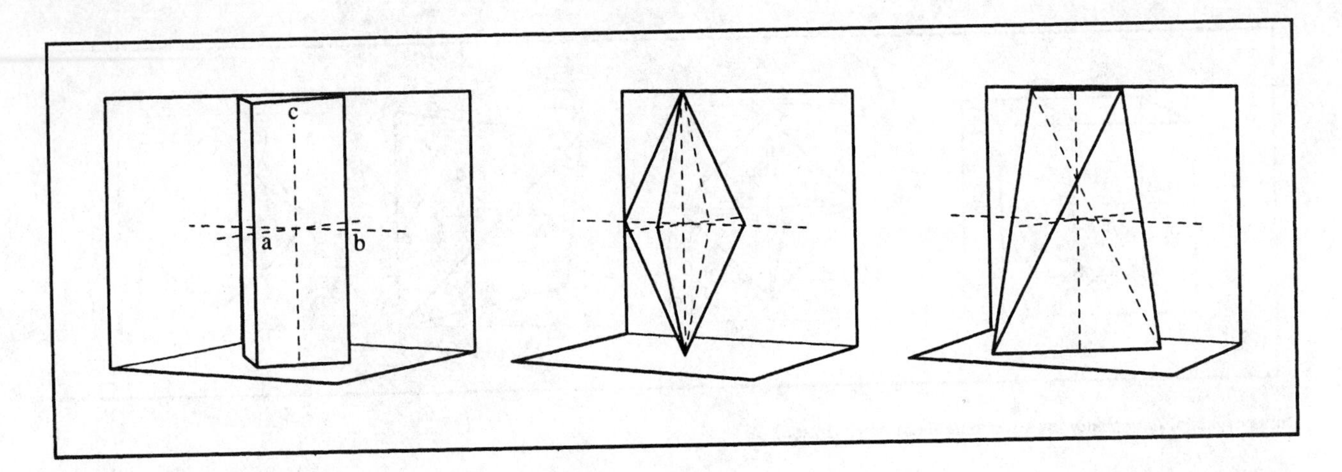

Tetragonal crystals. *(Illustration by Electronic Illustrators Group.)*

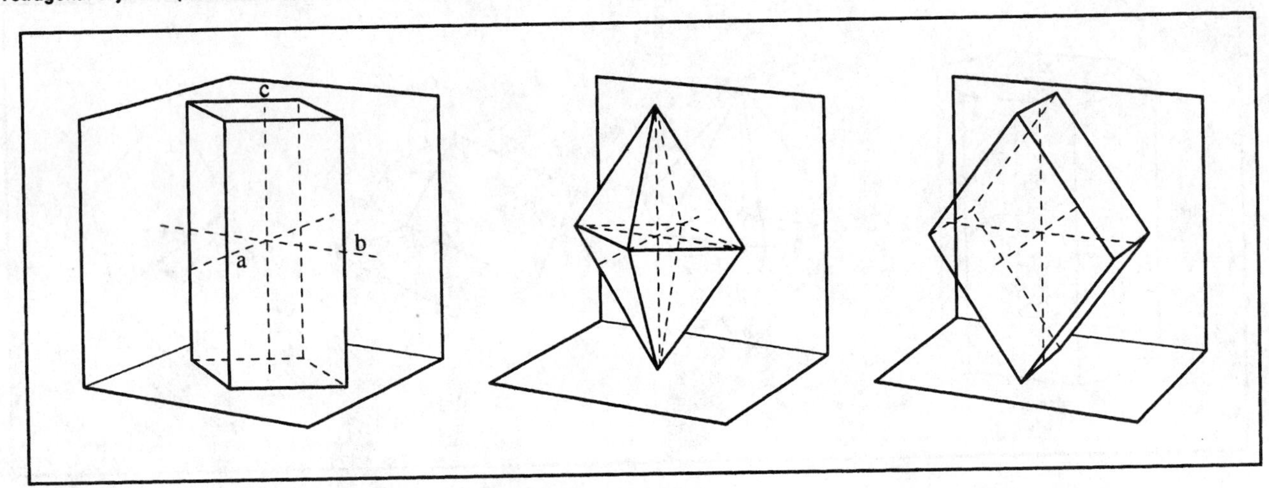

Orthorhombic crystals. *(Illustration by Electronic Illustrators Group.)*

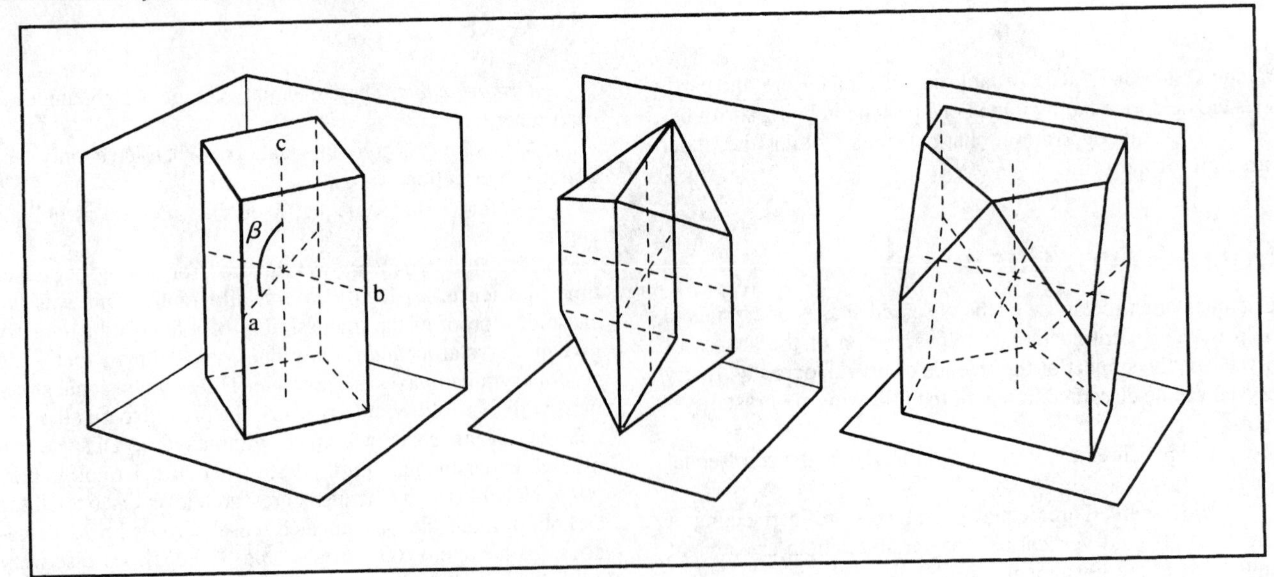

Monoclinic crystals. *(Illustration by Electronic Illustrators Group.)*

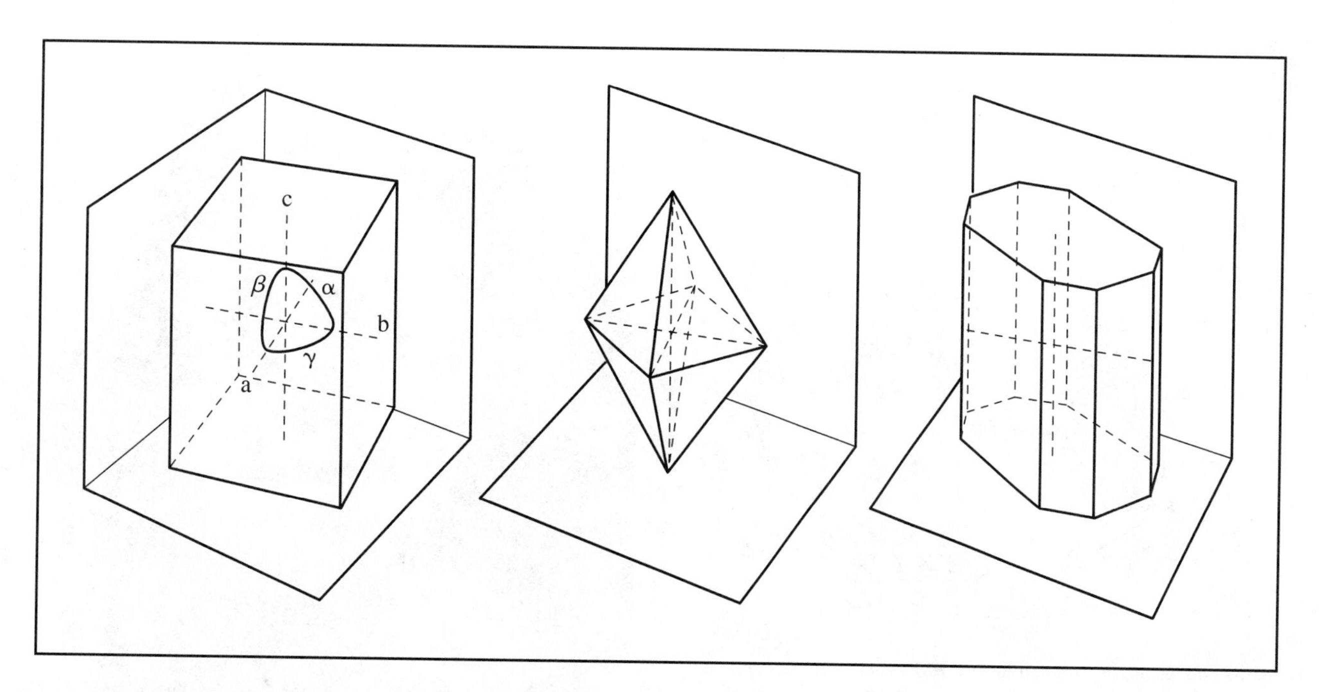

Triclinic crystals. ***(Illustration by Electronic Illustrators Group.)***

potential 7.6 eV (12.16x10^{-19}J). In the crystal state (central-cubic), it shows only rotation. Experimental results strongly indicate that rotation in crystal is faster than in **solution** (rotational **diffusion** constant is 3x10^{10} vs. 1.8x10^{10} per second).

Solomon, Susan (1956-)

American atmospheric chemist

Susan Solomon played a key role in discovering the cause of a major threat to the earth—the loss of the protective **ozone** layer in the upper atmosphere. Ozone protects all life on earth from large amounts of damaging ultraviolet radiation from the sun. Solomon, an atmospheric chemist, was first to propose the theory explaining how **chlorofluorocarbons**, **gases** used in refrigerators and to power aerosol spray cans, could in some places on the globe lead to ozone destruction in the presence of stratospheric clouds.

Solomon said in an interview with Lee Katterman that she recalls "exactly what got me first interested in science. It was the airing of Jacques Cousteau on American TV when I was nine or ten years old." Solomon said that as a child she was very interested in watching natural history programming on television. This sparked an interest in science, particularly biology. "But I learned that biology was not very quantitative," said Solomon in the interview. By the time she entered the Illinois Institute of Technology, Solomon met her need for quantitative study by choosing **chemistry** as her major at the Illinois Institute of Technology. A project during Solomon's senior year turned her attention toward **atmospheric chemistry**. The project called for measuring the reaction of ethylene and hydroxyl radical, a process that occurs in the atmosphere of Jupiter. As a result of this work, Solomon did some extra reading about planetary atmospheres, which led her to focus on atmospheric chemistry.

During the summer of 1977, just before entering graduate school at University of California at Berkeley, Solomon worked at the National Center for Atmospheric Research (NCAR) in Boulder, Colorado. She met research scientist Paul Crutzen at NCAR, who introduced her to the study of ozone in the upper atmosphere. In the fall at Berkeley, Solomon sought out Harold Johnston, a chemistry professor who did pioneering work on the effects of the supersonic transport (SST) on the atmosphere. Solomon credits Crutzen and Johnston for encouraging her interest in atmospheric chemistry. After completing her course work toward a Ph.D. in chemistry at Berkeley, Solomon moved to NCAR to do her thesis research with Crutzen.

She received a Ph.D. in chemistry in 1981 and then accepted a research position at the National Oceanic and Atmospheric Administration (NOAA) Aeronomy Laboratory in Boulder, Colorado. Initially, Solomon's research focused on developing computer models of **ozone** in the upper atmosphere. Ozone is a highly reactive **molecule** composed of three atoms of **oxygen**. By comparison, the oxygen that is essential to the **metabolism** of living things is a relatively stable combination of two oxygen atoms. In the upper atmosphere between about 32,000 and 100,000 feet altitude, a layer of ozone exists that absorbs much of the sun's deadly ultraviolet radiation, thereby protecting all life on earth.

In 1985 scientists first reported that, during the months of spring in the Southern Hemisphere (September and October), the **density** of the ozone layer over Antarctica had been

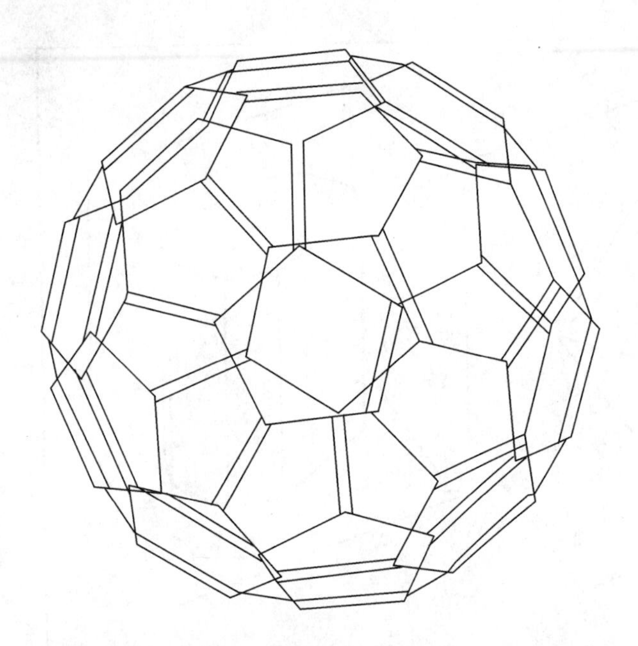

Carbon-60 molecule. *(Illustration by Electronic Illustrators Group.)*

decreasing rapidly in recent years. The cause of this ''hole'' in the ozone layer was unknown and many scientists began to look for its cause. In 1986 the scientific community wanted to send some equipment to Antarctica to measure atmospheric levels of ozone and **nitrogen** dioxide. Much to the surprise of her scientific colleagues, Solomon volunteered to travel to Antarctica to get the needed measurements; until then, she had concentrated on theoretical studies, but the chance to understand the cause of the ozone hole prompted Solomon to take up experimental work. Solomon led an expedition to Antarctica during August, September, and October of 1986, where she and co-workers measured the amounts of several atmospheric components, including the amount of **chlorine** dioxide in the upper atmosphere. The level of this atmospheric chemical was much higher than anyone expected and provided an important clue in determining why the ozone hole had appeared. Back at her NOAA lab in Boulder, Solomon wrote a research article that provided a theoretical explanation for the ozone hole. Solomon showed how the high level of chlorine dioxide was consistent with fast chemical destruction of ozone triggered by reactions occurring on stratospheric clouds. The extra chlorine dioxide was derived from chlorofluorocarbons released into the atmosphere from sources such as foams and leaking refrigeration equipment. Solomon returned to Antarctica for more measurements in August of 1987. Her explanation for the cause of the ozone hole is now generally accepted by scientists, and has led many countries of the world to curtail the production and use of chlorofluorocarbons.

Solomon's scientific studies to uncover the likely cause of the ozone hole have led to public recognition and many awards. In 1989, Solomon received the **gold** medal for exceptional service from the U.S. Department of Commerce (the agency that oversees the NOAA) She has testified several times before congressional committees about ozone depletion and is increasingly sought out as an expert on ozone science and policy (although the latter role is one she does not welcome, Solomon admitted in her interview, since she considers herself a scientist and not a policy expert).

Susan Solomon.

Solomon was born on January 19, 1956, in Chicago, Illinois. Her father, Leonard Solomon, was an insurance agent. Susan's mother, Alice Rutman Solomon, was a fourth-grade teacher in the Chicago public schools. She has one brother, Joel. She married Barry Lane Sidwell on September 20, 1988, and has a stepson by the marriage. She continues to study the atmospheric chemistry of ozone and has added Arctic ozone levels to her research subjects.

Solubility

Solubility is the ability of one substance to fully dissolve in another substance which is often present in a different state of **matter**. The word soluble comes from the fourteenth century, from the Latin word *solvere* meaning to dissolve.

The solubility of a substance is the amount of that substance that will dissolve in another substance under specified conditions. When solutions of **solids** or **liquids** in liquids are considered, the solubility is given in terms of the weight dissolved in a given weight or **volume** of the solvent at a specified **temperature**.

Substances are generally soluble in similar **solvents**. For example inorganic salts are generally soluble in **water** but not in organic solvents. Organic compounds are generally soluble in organic solvents but not in water. Organic solvents include such compounds as **ethanol** and propanone. Organic solvents are commonly encountered in many household items such as glues, **paints**, **aerosols**, and varnishes.

The most commonly encountered solutions are of solids dissolved in liquids. The solid that dissolves in a liquid is the solute and the liquid in which it dissolves is the solvent. If a solid can dissolve in a liquid, it is said to be soluble in that liquid; if not, it is said to be insoluble. As we add more solid to a liquid the **solution** becomes more concentrated. The greater the solubility of a substance the more concentrated it is possible to make the solution. The **concentration** of a solution is usually quoted in terms of **mass** of solute dissolved in a particular volume of solvent. This is generally expressed in grams per liter.

When we add solute to solvent a point is reached where no more will dissolve, under the specified conditions. The solution is saturated. The concentration of the solute in a saturated solution is the solubility of the solute in that solvent at that temperature. Solubility is the mass of solute that will saturate 100 g of water at the specified temperature. Saturation of a solution is also defined as the point where the solution is in equilibrium with the undissolved solute. If less solute is added to the solvent then the solution is said to be unsaturated. It is possible to dissolve more solute into a solvent than is required to give a saturated solution. This yields a supersaturated solution. Such solutions can often be prepared by utilizing the greater solubility possible at higher temperatures. The solute is dissolved into the solvent at a high temperature and then the solution is slowly cooled. Such solutions are unstable and the addition of a tiny amount of the solute will cause all of the excess solute to crystallize out of solution.

Solubility can be affected by such processes as **hydrogen bonding** between the solute and solvent. The greater the amount of hydrogen bonding that occurs the greater the solubility of the solute in that liquid.

A graph showing how the solubility of a solid changes with temperature is known as a solubility curve. In general the solubility of most solids will increase with an increase in temperature. At 68°F (20°C) **potassium** nitrate has a solubility of 30 g in 100 g of water. If the temperature is raised to 104°F (40°C) the solubility of potassium nitrate in 100 g of water will increase to 65 g. This pattern is not always the same. For example, if **copper sulfate** is considered under the same conditions the solubility is 21 g and 27 g. Other substances show no change. **Sodium** chloride has a solubility of 30 g at both temperatures.

For most purposes a substance which has a solubility of less than 0.01 moles per liter is generally regarded as insoluble. With ionic compounds certain salts are generally soluble or insoluble. For example nitrates are soluble, as are most chlorides, bromides, and iodides (those with **silver**, **mercury** or **lead** as the **cation** are exceptions to this). Carbonates and hydroxides are generally insoluble. The solubility of a substance whose **anion** is basic will be affected by the **pH** of the solution. As the acidity of the solution increases the solubility of the basic anion also increases. Also with regard to ions the solubility of a slightly soluble **salt** is decreased by the presence of a second solute, if the second solute provides a common **ion**.

Gases can also dissolve in liquids. However as the temperature increases the solubility of the gas generally decreases. To illustrate the solubility of gases in water we can consider the gases found in the air. The solubility of these gases in water is quite small but the amount of **oxygen** that dissolves in water is sufficient to support aquatic life. Normally the composition of gases in air is 79% **nitrogen** and 20% oxygen. When air is dissolved in water the composition is 61% nitrogen and 37% oxygen. This is due to the greater solubility of oxygen in water than nitrogen in water. This gives an enrichment of oxygen in the water that is of great importance for living organisms. In 100 g of water at 68°F (20°C) the solubility of oxygen is 0.004 g, if the temperature is increased to 104°F (40°C) then the solubility decreases to 0.003 g. Under the same conditions nitrogen shows solubilities of 0.002 g and 0.0015 g. **Carbon** dioxide is even more soluble in water, but this is because there is a chemical reaction occurring to produce carbonic acid. With the temperature conditions previously described **carbon dioxide** has a solubility of 0.17 g and 0.05 g.

The solubility of gases decreases with an increase in temperature but the solubility increases with an increase in pressure. Soda drinks contain carbon dioxide gas that has been dissolved under pressure. When the pressure is released, by opening the can or bottle, the carbon dioxide comes out of solution. It is this escape of carbon dioxide that gives these drinks their fizz. If the can or bottle is left standing then the drink will go flat, i.e., the fizz will be lost as all of the carbon dioxide comes out of solution. This will happen more quickly if the liquid is warm.

The disease suffered by divers known as the bends (decompression sickness) is an example of the same phenomenon. As the diver descends in the water the pressure increases. If the diver is breathing an air mixture more nitrogen will dissolve in the divers blood than at the surface. This is due to the greater solubility of gases under pressure. If the diver comes back to the surface too quickly there will be a rapid decrease in the pressure he is experiencing. The nitrogen will come out of solution too quickly, producing bubbles of gas. These bubbles may stop blood flow and damage nerves. Decompression sickness can be a fatal condition. To reduce the likelihood that this problem will occur, divers may use mixtures of **helium** and oxygen. Helium has a much lower solubility in blood than nitrogen.

Henry's law describes the solubility of gases in liquids. It states that the mass of gas that dissolves in a given volume of liquid at constant temperature is directly proportional to the pressure of the gas. This law only holds true if there is no chemical reaction between the liquid and the gas.

SOLUTION

A solution is a homogenous mixture of two or more substances in which the particles are completely dispersed. A colloidal solution is one in which some of the components are present as small particles that have not dissolved. The term solution comes from the Latin *solvere*, to dissolve or to release.

A solution may contain a **liquid** or a **solid** evenly dispersed through it. This may occur by the material being totally dispersed throughout the solvent or it may be that the solid is held in suspension. For a solution to be truly regarded as a solution the solute must be evenly dispersed (the mixture must be homogenous) and there must be no tendency for the solute to settle out of solution. A colloidal solution has the same properties as described. The physical properties of a solution vary depending on the proportions of the components of the system. Solutions can be formed from a solid in a liquid, a gas in a liquid, a solid in a solid, a liquid in a solid, and a gas in solid. The air we breathe is a solution of several **gases**. Brass is a solid solution of **zinc** in **copper**. Sea **water** is a solution of several solids dissolved in water. Each of the substances in a solution is a component of the solution. The solvent is the component that is present in the greatest quantity. The other components are solutes.

The amount of substance dissolved in another substance is the concentration, or strength, of the solution. This is generally given in terms of grams per liter, or as a molar fraction. A solution with only a small amount of material dissolved in it is said to be weak or dilute. A solution with a high proportion of material dissolved in it is said to be strong or concentrated. If a solution has the maximum amount of material dissolved in it, it is said to be saturated. The maximum amount that can be dissolved into a solution is termed solubility. As the **temperature** increases the solubility of a solid in a liquid generally increases. The solubility of a gas in a liquid decreases as the temperature increases. It is possible to dissolve more material into a liquid than the saturated solution should hold. The solution is produced by heating the liquid and adding excess solute. This will then dissolve and by gradual cooling a supersaturated solution will be produced. A supersaturated solution is unstable and the addition of a small piece of the solute to a supersaturated solution will cause the excess dissolved solute to precipitate out.

A solution will form when the various intermolecular attractive forces between the particles of solute and solvent are similar in size to the forces between either the solute **molecules** themselves or between the solvent molecules. With sodium chloride and water, the saturated condition is rapidly reached because the attraction between the sodium chloride ions and the polar water molecules is so strong. The attractive force experienced is easily sufficient to break the lattice structure that **sodium chloride** is normally encountered in. **Hydrogen** bonding plays a strong part in this particular reaction. The separate ions are actually surrounded by water molecules. This is a process known as hydration. The more general term is solvation, hydration refers only to the situation where water is the solvent. To overcome the bonds that are present between the sodium and chloride ions and the bonds that are present between the water molecules, energy is needed. Consequently energy is drawn into the system, however energy is also given up by the system as new bonds form. The overall net effect may be of an endothermic or **exothermic reaction**. **Sodium hydroxide** dissolved into water is an exothermic process, ammonium nitrate dissolved in water is an endothermic process.

When ionic compounds form a solution they are capable of conducting electricity. How well they are able to do this depends on how complete the **disassociation** into ions is in a solution. For example, **methanol** in solution in water does not form ions so electricity cannot be passed. Mercury chloride partially disassociates so it can pass **electricity** poorly. Sodium chloride in water disassociates completely and consequently it is a good conductor of electricity. A substance that passes electricity by this manner is known as an electrolyte.

The solute can be reclaimed from a solution by a process of evaporation. Sea water, which is a solution containing water and sodium chloride as its main components, can undergo this process. If sea water is allowed to evaporate a white solid is left behind in the container. This is sodium chloride (common table salt). If the evaporation is carried out slowly then the crystals of sodium chloride produced are large in size. A quick evaporation will yield small crystals. The water is driven off as a vapor, this can be collected and condensed to give pure solute.

A solution has different properties when compared to any of the components of the system. Salt when dissolved in water lowers the freezing point and raises the **boiling point**. Exactly how much effect this has depends on the concentration of solute. The increase in boiling point and the decrease in freezing point are directly proportional to the number of solute molecules per mole of solvent molecules. This characteristic is commonly encountered in everyday life. **Salt** is added to water to make it boil at a higher temperature that is used in cooking. Calcium chloride can be added to ice on roads to reduce the freezing point, the practical effect being that the ice is no longer at a sufficient temperature to remain as cold so the ice melts.

If a solution and solvent or two solutions of different strength are separated by a semi permeable membrane, osmosis can occur. **Osmosis** is the passage of water from a weak solution to a strong solution through a semi permeable membrane. The net effect of osmosis is to try to average out the concentration between the two solutions. A solution can have the solute removed, it can be purified, by reverse osmosis. Reverse osmosis is used to prepare drinking water from sea water.

If large particles of material are placed into a liquid they may not dissolve. For example fine clay in water will merely settle to the bottom of the container due to the action of gravity. The dispersed particles of a solution are of molecular size. Between these two extremes there are particles which are larger than molecules but not so large that they will settle out due to the action of gravity. The solution containing particles of this size is termed a colloidal solution or often just a colloid. Many examples of colloids using different phases are commonly encountered and many of their properties are similar to those of normal, molecular solutions.

A solution of two gases generally has each gas acting as if the other component were not present. This is due to the large intermolecular distances encountered between gas molecules. This is most obviously encountered when the pressure of the gas solution is looked at. This is covered by **Dalton's law of partial pressure**.

See also Colloid; Solubility; Solvents

Solvay, Ernest (1838-1922)

Belgian chemist and industrialist

Ernest Solvay was born in Rebecq-Rognon, near Brussels, Belgium, to a family of industrial chemists. Although his only formal education was at local schools, young Solvay read and experimented widely in **chemistry** and **electricity**. He joined his father's salt-making business, and then, at age twenty-one, went to work with his uncle at a gasworks factory near Brussels.

Solvay began experimenting with the ammonia-soda reaction as a means of producing soda ash, which was then in great demand by the **glass** and soap industries. Most soda ash at the time was produced by the Leblanc process of converting common **salt**, a method which was expensive and created unusable byproducts. Solvay solved the practical problems of conducting the ammonia-soda process on a large scale, unaware that many chemists had tried and failed to do this over the past fifty years. He devised carbonating towers, which allowed large amounts of **ammonia**, salt **solution**, and **carbon** dioxide to be mixed; his process also allowed the recovery of expensive ammonia, which could then be reused.

Solvay patented his process in 1801 and founded a company with his brother, Alfred. They built a factory in 1863 and began commercial production of soda ash, using the Solvay process in 1865. Solvay protected his process with many patents, but offered other manufacturers licensing rights to use his technology. By 1890 he had established plants in many foreign countries.

By the end of the 1800s the Solvay process produced most of the world's soda ash. Solvay became very wealthy as a result, and he used his money to endow schools and, in particular, to found he Solvay International Institutes of Chemistry, of Physics, and of Sociology. These institutes held well respected international conferences; those on physics were particularly noted for their inquiries into **atomic structure** and quantum mechanics. Solvay died in Brussels in 1922.

Solvents

In a **solution**, a solvent is the substance which comprises the bulk of the solution. The substance that is dissolved into the solvent is the solute. A solvent is a substance capable of dissolving another substance. If some of the components of this system chemically react with each other then the inert substance dispersing the reactants is the inert solvent.

One of the most commonly encountered solvents is **water**. Water is an excellent solvent for many ionic substances but not for the majority of organic substances. Because water is a polar **molecule**, the water molecules can easily break down ionic crystals and separate the component ions from each other. The water molecules then form a shell around the ions keeping them further apart. Organic compounds and other covalently bonded molecules do not dissolve well in water. One exception to this is some of the alcohols. Because of the **hydrogen bonding** that is possible, some alcohols will dissolve in water.

A solvent other than water is required for other organic chemicals. Organic solvents (non-aqueous solvents) are capable of dissolving covalent compounds that are not ionic. An example of an organic solvent is cyclohexane. **Iodine** is virtually insoluble in water but it will readily dissolve in cyclohexane. Organic solvents are used in a variety of commonly encountered substances including correction fluid, glue, paint, varnish, and **aerosols**. Some of these organic solvents are flammable and can potentially form explosive mixtures with air. It is for these reasons that aerosols and other devices containing organic solvents need to be used carefully and in locations with good ventilation.

It is important to realize that the solvent does not need be a liquid and the solute does not need to be a solid. Solutions can be formed in which a gas is dissolved in another gas, or in a liquid or a solid. Solid-solid and liquid-liquid solutions are also common.

The **temperature** of the solvent will control how much solute can dissolve in it. In most cases, the higher the temperature the more solute will dissolve in the solvent. One exception to this is when a gas dissolves in water. The **solubility** of a gas in liquid typically decreases with temperature.

With some solvents a chemical reaction occurs with the solute. This means that more of the solute will apparently dissolve in the solvent, but the original solute will not be recoverable by evaporation.

The physical properties of a solvent are altered once a solute has been added. Exactly how these properties are affected depends upon the solvent-solute combination in use. For example, when solutes are added to water the freezing point is lowered and the **boiling point** is raised.

Sonochemistry

Sonochemistry involves the study of chemical changes that occur as the direct result of sound or ultrasound waves traveling through either a homogeneous or heterogeneous liquid medium. The sound wave, as it passes through the **solution**, consists of expansion (negative-pressure) and compression (positive-pressure) waves. Expansion waves **lead** to formation of tiny bubbles, which may be filled with a dissolved gas, vapor of the liquid solvent or may be almost empty depending upon the pressure and the forces holding the liquid molecules together. The bubbles grow in size until they are collapsed rapidly by the compression region of the sound wave. The rapid collapse generates intense localized heating and associated high-energy **chemistry**. Effective temperatures within the gas-phase reaction zone may reach as high as 5000 K, and effective pressures of several hundreds of atmospheres have been reported. Localized heating produces significantly enhanced chemical reaction rates and increased product yields, and in some instances leads to chemicals that are unobtainable under ordinary reaction conditions.

Solid-liquid interfaces have also been subjected to ultrasound irradiation. Cavitation still occurs; however, the bubbles collapse asymmetrically directing a jet of liquid at the solid surface at velocities exceeding 100 meters/second. The liquid jet as it strikes the solid surface can cause localized erosion (and even melting) and surface pitting. The surface defects and deformations that are created give a highly reactive solid surface. The effects of ultrasound on liquid-solid interfaces has been a **matter** of intense investigation, particularly in applications involving organometallic reactions, development of heterogeneous catalytic systems, and activation of the lesser reactive **metals**. Also, ultrasound waves have been found to enhance **mass** transport near metal electrode surfaces. This latter observation has proved to be extremely useful in the synthesis of both organic and inorganic chalcogenides.

SORENSEN, SOREN PETER LAURITZ (1868-1939)

Danish chemist

The **pH** scale, invented by Soren Peter Lauritz Sorensen, ''has become so much a part of scientific literature and its influence so important a factor in considering biological problems that one wonders how theories of acidity and alkalinity were ever formulated without a knowledge of Sorensen's fundamental conceptions,'' A. J. Curtin Cosbie commented in *Nature*. Potential of **hydrogen**, or pH, is a simplified measure of the acidity of any given mixture. Though of immense value to scientist and layman alike, the pH scale is only one of Soren Sorensen's many achievements in a career devoted to the application of classical physico-chemical methods to the new realms of biochemistry and specifically to fermentations problems. His research on **enzymes** and **proteins** in particular—for which the invention of the pH scale was merely a methodological improvement—were invaluable, laying the groundwork for precise and thorough studies of these nitrogenous compounds.

Sorensen was born on January 9, 1868, at Havrebjerg, Slagelse, Denmark, the son of a farmer, Hans Sorensen. Sorensen was a high-strung, nervous youth, suffering from epileptic-like attacks and a pronounced stammer. His schoolwork became something of a refuge for him and upon graduation from high school in Soro in 1886, he entered the University of Copenhagen. Initially Sorensen intended to study medicine, but was diverted from this course after studying **chemistry** under S. M. Jorgensen, who fired him with an interest in the structure of inorganic compounds. In 1889 Sorensen was already proving his academic worth, winning a gold medal for an essay on chemical radicals. A second gold medal award came in 1896 for his research into **strontium** compounds. It was during this period of study that Sorensen's interest was turning toward research in **analytical chemistry**, a development that would later become very important for the progress of his work, as he would be able to blend his flair for experimentation with a precise attention to detail. After receiving his Master of Science degree in 1891, he worked as an assistant at the chemistry laboratory of the Danish Polytechnic Institute, consulted for the Royal Naval Dock Yard, and also found time to assist on a geological survey of Denmark. In 1899 he received his doctorate in chemistry, writing his dissertation on cobaltic oxalates.

Throughout the period of study for his doctorate, Sorensen focused on inorganic chemistry and related questions. In 1901, however, this focus changed with an appointment to the directorship of the prestigious Carlsberg Laboratory in Copenhagen. Sorensen, age thirty-three, took over from Johann Kjeldahl, whose work had been primarily in **biochemistry**, and it was with similar investigations that Sorensen would spend the rest of his scientific career. He became interested in **proteins** and especially in **amino acids**, successfully synthesizing ornithine, proline, arginine, and several others. His interest in analytical methods led him to further research into the measurement of **nitrogen** concentration and improving methods of **titration**, the process by which the **concentration** of ingredients in a **solution** is determined by adding carefully measured amounts of a reagent until a desired chemical reaction occurs.

Much of Sorensen's fame rests in his papers on enzyme study, the first of which was a study of enzyme action utilizing the titration method. But it was in the second paper, published in 1909, where he examined the electromotive force or EMF method for determining hydrogen **ion** concentration, that Sorensen addressed the topic that would lead to his pH scale. Other investigators had already suggested that hydrogen ion concentrations could be used as valid indicators of the acidity or alkalinity of a solution, and the hydrogen electrode also had become the standard for such measurements. Sorensen sought to simplify the cumbersome analytical apparatus then used to measure acidity and alkalinity, and devised the familiar scale numbered 1-14 in which 7 represents a neutral solution, the acids are represented by numbers lower than that, and alkalines or bases by numbers above 7. Other scientists, such as **Leonor Michaelis** with his book on hydrogen ion concentration, and Arnold Beckman with a simplified pH meter, helped to popularize the pH method, which is simple enough to be grasped by laymen, yet accurate enough for scientific use.

From methodological work and studies on enzymes, Sorensen and his laboratory turned to the investigation of proteins, applying many of the classical principles of chemistry to their description and characterization, and by 1917 the laboratory had succeeded in crystallizing egg albumen, a pioneering step in the characterization of proteins. Sorensen was able to determine the molecular weight of the albumen by osmotic measurement and determine its acid and base capacity by pH. He and his assistants—later including his wife, Magrethe Hoyrup Sorensen—went on to study serum proteins, lipoproteins, and the complexes of hemoglobin and **carbon monoxide**.

In addition to his life-long work at the Carlsberg Laboratory, Sorensen was also involved in the role of science in industry, working with spirits, yeast, and **explosives**. He also contributed knowledge in the medical field, his first love at university, researching epilepsy and diabetes. Honors and awards attested to his contributions to science: he was president of the Danish Scientific Society and an honorary member of societies in both Europe and the United States. Sorensen died in Copenhagen on February 12, 1939, following a yearlong illness. In a memorial address delivered to the Chemical Society, E. K. Rideal remarked, "By his death the world loses a perfect example of a man whose devotion to scientific accuracy and consistency should serve as an example to many who, claiming to be scientific in this ever-accelerating age of speed, serve their science badly by neglecting the solid in their search for the superficial and spectacular."

See also Acids and Bases

SPECTROSCOPY

The absorption, emission, or scattering of electromagnetic radiation by atoms or molecules is referred to as spectroscopy. A transition from a lower **energy** level to a higher level with transfer of electromagnetic energy to the **atom** or **molecule** is called absorption; a transition from a higher **energy level** to a lower level is called emission (if energy is transferred to the electromagnetic field); and the redirection of light as a result of its interaction with **matter** is called scattering.

When atoms or molecules absorb electromagnetic energy, the incoming energy transfers the quantized atomic or molecular system to a higher energy level. Electrons are promoted to higher orbitals by ultraviolet or visible light; vibrations are excited by infrared light, and rotations are excited by microwaves. Atomic-absorption spectroscopy measures the **concentration** of an element in a sample, whereas atomic-emission spectroscopy aims at measuring the concentration of elements in samples. UV-VIS **absorption spectroscopy** is used to obtain qualitative information from the electronic absorption spectrum, or to measure the concentration of an analyte molecule in **solution**. Molecular fluorescence spectroscopy is a technique for obtaining qualitative information from the electronic fluorescence spectrum, or, again, for measuring the concentration of an analyte in solution.

Infrared spectroscopy has been widely used in the study of surfaces. The most frequently used portion of the infrared spectrum is the region where molecular vibrational frequencies occur. This technique was first applied around the turn of the twentieth century in an attempt to distinguish **water** of crystallization from water of constitution in **solids**. Forty years later, the technique was being used to study surface hydroxyl groups on oxides and interactions between adsorbed molecules and hydroxyl groups.

Ultraviolet spectroscopy takes advantage of the selective absorbance of ultraviolet radiation by various substances. The technique is especially useful in investigating biologically active substances such as compounds in body fluids, and drugs and narcotics either in the living body (*in vivo*) or outside it (*in vitro*). Ultraviolet instruments have also been used to monitor air and water **pollution**, to analyze dyestuffs, to study **carcinogens**, to identify **food additives**, to analyze petroleum fractions, and to analyze pesticide residues. Ultraviolet photoelectron spectroscopy, a technique that is analogous to x-ray photoelectron spectroscopy, has been used to study **valence** electrons in **gases**.

Electron spectroscopy for chemical analysis. ***(Photograph by Joseph Nettis. Photo Researchers, Inc. Reproduced by permission.)***

Microwave spectroscopy, or molecular rotational **resonance** spectroscopy, addresses the microwave region and the absorption of energy by molecules as they undergo transitions between rotational energy levels. From these spectra it is possible to obtain information about **molecular structure**, including bond distances and bond angles. One example of the application of this technique is in the distinction of trans and gauche rotational isomers. It is also possible to determine **dipole** moments and molecular collision rates from these spectra.

In **nuclear magnetic resonance** (NMR), resonant energy is transferred between a radio-frequency alternating magnetic field and a **nucleus** placed in a field sufficiently strong to decouple the nuclear spin from the influence of atomic electrons. Transitions induced between substates correspond to different quantized orientations of the nuclear spin relative to the direction of the magnetic field. Nuclear magnetic resonance spectroscopy has two subfields: broadline NMR and high resolution NMR. High resolution NMR has been used in inorganic and organic **chemistry** to measure subtle electronic effects, to determine structure, to study chemical reactions, and to follow the motion of molecules or groups of atoms within molecules.

Electron paramagnetic resonance is a spectroscopic technique similar to nuclear magnetic resonance except that microwave radiation is employed instead of radio frequencies. Electron paramagnetic resonance has been used extensively to study paramagnetic species present on various solid surfaces. These species may be metal ions, surface defects, or adsorbed molecules or ions with one or more unpaired electrons. This technique also provides a basis for determining the **bonding** characteristics and orientation of a surface complex. Because the technique can be used with low concentrations of active sites, it has proven valuable in studies of **oxidation** states.

Atoms or molecules that have been excited to high energy levels can decay to lower levels by emitting radiation. For atoms excited by light energy, the emission is referred to as atomic fluorescence; for atoms excited by higher energies, the emission is called atomic or optical emission. In the case of molecules, the emission is called fluorescence if the transition occurs between states of the same spin, and phosphorescence if the transition takes place between states of different spin.

In x-ray fluorescence, the term refers to the characteristic x rays emitted as a result of absorption of x rays of higher frequency. In electron fluorescence, the emission of electromagnetic radiation occurs as a consequence of the absorption of energy from radiation (either electromagnetic or particulate), provided the emission continues only as long as the stimulus producing it is maintained.

The effects governing x-ray photoelectron spectroscopy were first explained by **Albert Einstein** in 1905, who showed that the energy of an electron ejected in photoemission was equal to the difference between the **photon** and the binding energy of the electron in the target. In the 1950s, researchers began measuring binding energies of core electrons by x-ray photoemission. The discovery that these binding energies could vary as much as 6 eV, depending on the chemical state of the atom, led to rapid development of x-ray photoelectron spectroscopy, also known as Electron Spectroscopy for Chemical Analysis (ESCA). This technique has provided valuable information about chemical effects at surfaces. Unlike other spectroscopies in which the absorption, emission, or scattering of radiation is interpreted as a function of energy, photoelectron spectroscopy measures the kinetic energy of the electrons(s) ejected by x-ray radiation.

Mössbauer spectroscopy was invented in the late 1950s by Rudolf Mössbauer, who discovered that when solids emit and absorb gamma rays, the **nuclear energy** levels can be separated to one part in 10^{14}, which is sufficient to reflect the weak interaction of the nucleus with surrounding electrons. The Mössbauer effect probes the binding, charge distribution and symmetry, and magnetic ordering around an atom in a solid matrix. An example of the Mössbsauer effect involves the Fe-57 nuclei (the absorber) in a sample to be studied. From the ground state, the Fe-57 nuclei can be promoted to their first excited state by absorbing a 14.4-keV gamma-ray photon produced by a radiocactive parent, in this case Co-57. The excited Fe-57 nucleus then decays to the ground state via electron or gamma-ray emission. Classically, one would expect the Fe-57 nuclei to undergo recoil when emitting or absorbing a gamma-ray photon (somewhat like what a person leaping from a boat to a dock observes when his boat recoils into the lake); but according to quantum mechanics, there is also a reasonable possibility that there will be no recoil (as if the boat were embedded in ice when the leap occurred).

When electromagnetic radiation passes through matter, most of the radiation continues along its original path, but a tiny amount is scattered in other directions. Light that is scattered without a change in energy is called Rayleigh scattering; light that is scattered in transparent solids with a transfer of energy to the solid is called Brillouin scattering. Light scattering accompanied by vibrations in molecules or in the optical region in solids is called Raman scattering.

In vibrational spectroscopy, also known as Raman spectroscopy, the light scattered from a gas, liquid, or solid is accompanied by a shift in wavelength from that of the incident radiation. The effect was discovered by the Indian physicist C. V. Raman in 1928. The Raman effect arises from the inelastic scattering of radiation in the visible region by molecules. Raman spectroscopy is similar to infrared spectroscopy in its ability to provide detailed information about molecular structures. Before the 1940s, Raman spectroscopy was the method of choice in molecular structure determinations, but since that time infrared measurements have largely supplemented it. Infrared absorption requires that a vibration change the dipole moment of a molecule, but Raman spectroscopy is associated with the change in polarizability that accompanies a vibration. As a consequence, Raman spectroscopy provides information about molecular vibrations that is particularly well suited to the structural analysis of covalently bonded molecules, and to a lesser extent, of ionic crystals. Raman spectroscopy is also particularly useful in studying the structure of polyatomic molecules. By comparing spectra of a large number of compounds, chemists have been able to identify characteristic frequencies of molecular groups, e.g., methyl, carbonyl, and hydroxyl groups.

See also Chemical compound; Fluorescence and phosphorescence; Magnetism; Quantum chemistry

SPEDDING, FRANK HAROLD (1902-1984)

Canadian American chemist

Frank Harold Spedding played an important role in the **Manhattan Project**—the U.S. government's effort to develop an atomic bomb during World War II. With his colleagues, he also devised chemical techniques for isolating rare-earth elements, thus making them available for industry at affordable cost.

Spedding was born in Hamilton, Ontario, Canada, on October 22, 1902, but his photographer father, Howard Leslie Spedding, and mother, the former Mary Ann Elizabeth Marshall, were American citizens. While at the University of Michigan, he majored in metallurgy, receiving his B.S. degree in 1925. He earned his M.S. degree in **analytical chemistry** there in 1926, and was awarded his Ph.D. in **physical chemistry** from the University of California, Berkeley, in 1929. While working on his doctorate at UC, Spedding studied the mathematics underlying the chemical relationships among properties of rare-earth **metals**, the elements consisting of **scandium** and **yttrium**, and the fifteen elements from **lanthanum** to **lutetium**. These elements exist in nature only as oxides, and are never found in the pure mineral state.

While at UC he collaborated with noted chemist **Gilbert Newton Lewis** in developing methods for concentrating heavy **water**, which contains **deuterium**, the **hydrogen isotope** with a **mass** double that of ordinary hydrogen (i.e., 2 instead of 1). The young chemist was a National Research Council fellow at UC from 1930 to 1932 and a chemistry instructor there from 1932 to 1934.

In 1931 he married Ethel Annie MacFarlane; they would eventually have one daughter, Mary Ann Elizabeth. Spedding moved to Cambridge, England, in 1934 to study **theoretical chemistry** and physics on a Guggenheim Fellowship. The following year, he moved to Cornell University, where he taught from 1935 to 1937. During this time, he collaborated at Cornell with **Hans Bethe**, the noted physicist and one of the developers of the atomic bomb, on studies of the atomic structures of rare-earth elements. He continued his studies of these elements at Iowa State University, where he began to develop methods for separating individual rare-earth elements and their isotopes. He was to stay at Iowa State University for much of the remainder of his career, becoming professor of physical chemistry from 1937 to 1941, professor of chemistry from 1941 to 1973, professor of physics from 1950 to 1973, and professor of metallurgy from 1962 to 1973. He was named distinguished professor of science and humanities in 1957, and emeritus professor of chemistry, physics, and metallurgy in 1973.

During World War II, Spedding began dividing his time between the University of Chicago, where scientists were trying to evoke a self-sustaining nuclear **chain reaction**, and Iowa. As director of the **Plutonium** Project, he directed chemical and metallurgical research to support the chain reaction program from 1942 to 1943. During 1942, Spedding and his colleagues developed processes for producing high-purity **uranium** using a column of resinous material that attracted metallic ions, a process called **ion** exchange. Using this technique, the team produced a third of the pure uranium that was used for the first self-sustaining nuclear chain reaction, which occurred at Stagg Field in Chicago on December 2, 1942.

Spedding's lab turned the new uranium purification process over to industry, which scaled up the process. The lab then concentrated its attention on producing pure **thorium** and **cerium**. Iowa State University received the Chemical Engineering Achievement Award for its wartime efforts, and Spedding himself received an honorary LL.D. degree from Drake University. Because the team of experts in pure materials, high-temperature metallurgy, and rare elements that Spedding had assembled was too valuable to disband after the war, the government asked Iowa State to operate a national laboratory, called the Institute for Atomic Research (later renamed the Ames Laboratory), with Spedding as director. He also served as principal scientist of the Ames Laboratory in 1968.

Spedding continued his research on the purification of rare-earth elements and the production of rare earth alloys and compounds. In the 1950s, Spedding's team developed processes for the large-scale production of yttrium, which the Atomic Energy Commission needed for research.

During his career he was awarded the William H. Nichols Medal of the American Chemical Society's New York Section for his "outstanding contributions in the constitution, properties and chemistry of the rare earth and actinide elements." He also received the James Douglas Medal of the American Institute of Mining, Metallurgical, and Petroleum Engineers in 1961 and the Francis J. Clamer Medal of the Franklin Institute in 1969. The National Academy of Sciences elected him to membership in 1952. In addition to his other honors, Spedding was technical representative for the Atomic Energy Commission at the Geneva Conference on Peaceful Uses of Atomic Energy in 1955, and a U.S. Department of State representative at the Fifth World Power Conference in Vienna in 1956. He was a member of the American Physical Society, the American Association for the Advancement of Science, the Faraday Society, and the American Association of University Professors.

After his retirement, Spedding devoted much of his time to researching and writing papers on the properties of the rare earths. He died on December 15, 1984, in Ames, Iowa.

STAHL, FRANKLIN W. (1929-)

American molecular biologist

Franklin W. Stahl, in collaboration with **Matthew Meselson**, discovered direct evidence for the semiconservative nature of deoxyribonucleic acid (**DNA**) replication in bacteria. In experiments, Stahl and Meselson showed that when a double stranded DNA **molecule** is duplicated, the double strands are separated and a new strand is copied from each "parent" strand forming two new double stranded DNA molecules. The new double stranded DNA molecules contain one conserved "parent" strand and one new "daughter" strand. Therefore, the replication of a DNA molecule is semiconservative: it retains some of the original material while creating some new material. The understanding of the semiconservative nature of DNA in replication was a major advancement in the field of molecular biology.

Franklin William Stahl, the youngest of three children, was born on October 6, 1929, in Boston, Massachusetts, to Oscar Stahl, an equipment specialist with New England Telephone and Telegraph, and Eleanor Condon Stahl, a homemaker. He received a baccalaureate degree from Harvard University in 1951 in the area of biological sciences; he continued with graduate studies in the field of biology, earning a Ph.D. degree at the University of Rochester in New York, in 1956. From 1955 to 1958, Stahl was a research fellow at the California Institute of Technology, where he collaborated with Matthew Meselson on the semiconservative replication experiment.

In 1952, having just graduated from Harvard, Stahl attended a course at Cold Spring Harbor Laboratories in New York given by A. H. Doermann. Doermann was well known for research on bacteriophages, microscopic agents that destroy disease-producing bacteria in a living organism. This course gave Stahl his first exposure to the genetics of bacteriophages. The subject so fascinated him that he spent his summers in the laboratory of Dr. Doermann while working on his doctorate at the University of Rochester during the school year. Bacteriophage genetics would later become the major focus of his own laboratory's scientific research. Stahl would also come to teach the same course at Cold Spring Harbor.

After receiving his doctorate in biology, Stahl moved to California to work in the laboratory of Max Delbrück at the California Institute of Technology as a postdoctoral fellow. While at Cal Tech, he began a collaboration with graduate student Meselson to design an experiment to describe the nature of DNA replication from parent to offspring using bacteriophages. The idea was to add the substance 5-bromouracil, which would become incorporated into the DNA of a T4 bacteriophage upon its replication during a few rounds of reproduction. Phage samples could then be isolated by a **density** gradient centrifugation procedure which was originally designed by Stahl, Meselson, and Jerome Vinograd. It was thought that the phage samples containing the incorporated 5-bromouracil would separate in the density gradient centrifugation to a measurable degree based on the length of the new strands of DNA acquired during replication. Several attempts to obtain measurable results were unsuccessful. Despite these first setbacks, Stahl had confidence in the theory of the experiment. After further contemplation, Stahl and Meselson decided to abandon the use of the T4 bacteriophage and the labelling substance 5-bromouracil and turned to the use of a bacteria, *Escherichia coli*, with the heavy **nitrogen isotope** 15N as the labelling substance. This time, when the same experimental steps were performed using the new substitutions, the analysis of the density gradient centrifugation samples showed three distinct types of bacterial DNA, two from the original parent strands of DNA and one from the new offspring. Analysis of the new offspring showed each strand of DNA came from a different parent. Thus the theory of semiconservative replication of DNA had been proven.

After spending 1958 at the University of Missouri as an associate professor of zoology, Stahl took a position as associate professor of biology and research associate at the Institute of Molecular Biology, located at the University of Oregon in Eugene, Oregon. In 1963, he was awarded status of professor; he was appointed acting director of the Institute from 1973 to 1975. Stahl has held a concurrent position as resident research professor of molecular genetics at the American Cancer Society.

Stahl set up his own laboratory at the Institute, contributing further to the scientific research and understanding in the area of bacteriophage genetics, as well as the genetic recombination of bacteriophages and fungi. In the early years at Eugene, he continued to focus his research in the area of genetic recombination and replication in bacteriophages using the techniques of density gradient and equilibrium centrifugations. Through the years, he was able to map the DNA structure of the T4 bacteriophage. The experiments involved T4 bacteriophages inactivated by decay of DNA incorporated radionucleotides or by X-irradiation of the DNA that would cause breaks in the DNA sequence. By performing reactivation-crosses of these bacteriophage, Stahl studied the patterns in which markers on the DNA were ''knocked out'' or lost. Although the inactivated phages are unable to produce offspring themselves, they can contribute particular markers to their offspring when they are grown in the presence of rescuing phages that supply the functions necessary for phage development. By the pattern of markers seen in the offspring of these reactivation crosses, a map can be constructed. From the map constructed, the correlated knockout markers reflected a linkage relationship in the form of a circle. With this map in hand, particular DNA sequences could be shown to be important for various functions of the bacteriophage.

Much of Stahl's later work focused on the bacteriophage Lambda, which has a more complex structure than bacteriophage T4, and its replication inside of a bacterial cell. He determined particular ''hot spots'' in the DNA sequence that were susceptible to various mutations or recombinations during the process of replication. These ''hot spots'' were particular sites in the DNA sequence of the phage that tended to show crossing over between two DNA strands of the **chromosome**. The resulting mutations (Chi mutations), which occurred at four or five particular sites in the Lambda phage, conferred a particular large plaque forming character by accelerating the rate of crossing over at these sites. These mutations affected the overall function of the bacteriophage, sometimes causing complete inactivation. Through further studies of genetic recombination in bacteriophages, Stahl became known to the scientific world as an expert on their structure and life cycle.

From 1964–85, Stahl held several year-long positions as visiting professor or volunteer scientist in various universities throughout the world. He was the volunteer scientist in the Division of Molecular Genetics for the Medical Research Council in Cambridge, England. He took a sabbatical leave from Oregon and conducted research in the Medical Research Council Unit of Molecular Genetics at the University of Edinburgh, Scotland, as well as at the Laboratory of International Genetics and Biophysics in Naples, Italy. He held the position of Lady Davis Visiting Professor in the Genetics Department at Hebrew University in Jerusalem, Israel. Stahl also taught courses on bacterial viruses at Cold Spring Harbor Laboratories and in Naples. He is a member of the National Academy of Sciences, the American Academy of Arts and Sciences, and the Viral Study Section of the National Institutes of Health.

During a personal interview, when asked what was a leading factor in choosing scientific research in biology as a career, Dr. Stahl responded, ''The currency of science is truth and understanding opposed to power and money, leading to contact with people that are fun and exciting.'' This difference between the world of scientific research and the business world has been a primary factor in influencing the path of Stahl's scientific career. His main goal as a scientist was never fame or fortune, but the opportunity to interact with exciting people and share innovative ideas. According to Dr. Stahl, his greatest contribution to the scientific world has been "the ability to act as mentor to graduate students and scientists by giving encouragement and direction to their research." He has gained great pleasure in passing on his knowledge and experiences to future generations of scientists.

Throughout his career as a scientist, Stahl has written over one hundred articles published in major scientific journals, received numerous honors including two honorary degrees, and been awarded the prestigious Guggenheim Fellowship award three times. Franklin Stahl married Mary Morgan, also a scientist, in 1955; the couple had three children, Emily, Joshua and Andy. Mary worked alongside Frank in the laboratory, frequently carrying on experiments while Frank took care of administrative responsibilities. He enjoys watching "Monday Night Football" and getting together with family members.

STAHL, GEORG ERNST (1660-1734)

German chemist

The German chemist and medical theorist Georg Ernst Stahl was the founder of the **phlogiston** theory of **combustion** and the author of a theory of medicine based upon vitalistic ideas.

Georg Stahl was born on Oct. 21, 1660, at Anspach in Bavaria, the son of a Lutheran pastor. Although brought up in an extremely pious and religious household, he early showed an enthusiasm for **chemistry**. By the age of 15 he had mastered a set of university lecture notes on the subject as well as a difficult treatise by Johann Kunckel.

Stahl studied medicine at the University of Jena, where he graduated in 1683. Here he came under the influence of iatrochemical theories, which gave an interpretation of physiological processes in terms of chemistry. He was later to become a strong opponent of this school of medical theory. Following graduation he taught at the University of Jena for 10 years.

In 1694 Stahl was invited to fill the second chair of medicine at the newly founded University of Halle. He owed his appointment to the recommendation of the holder of the first chair of medicine, Friedrich Hoffmann. They made Halle one of the most important medical schools of the early 18th century, although their careers there were punctuated by frequent quarrels. For 22 years Stahl lectured at Halle and wrote an impressive list of works on chemistry and medicine. His lectures were said to have been dry and intentionally difficult; he is alleged to have had a low opinion of the intellectual capacity of his students at Halle.

Stahl's most notable contribution to chemistry was his famous phlogiston theory of combustion, which became one of the main unifying theories of 18th-century chemistry. He maintained that all substances which burned contained a combustible principle called phlogiston (from the Greek *phlogos,* a flame) which was liberated during the combustion process. This principle of phlogiston was present not only in such obvi-

Georg Ernst Stahl. *(Corbis Corporation. Reproduced by permission.)*

ously combustible substances as **wood**, wax, oils, and other organic materials but also in inorganic substances such as **sulfur** and **phosphorus** and even in **metals**. Thus when a metal was calcined by heating (a process now known as oxidation), the metal was said to lose its phlogiston. Conversely, when the metallic calx was reduced again to the metal, phlogiston was taken up.

This theory also offered the first explanation of why **charcoal** was used in the **smelting** of metallic ores. Charcoal was a substance rich in phlogiston (since on burning it left no residue), and in the smelting process the phlogiston passed from the charcoal to the ore to give the pure metal. One of the major achievements of this theory was that it offered a comprehensive explanation of so many seemingly disparate chemical phenomena. In developing his theory, Stahl drew from the earlier ideas on combustion of the late-17th-century German chemist J.J. Becher.

As a medical theorist, Stahl opposed the purely chemical and mechanistic explanations of living phenomena current in his time. He emphasized the gulf between living and nonliving materials, stating that the distinctive feature of the former was that they possessed a soul which prevented their decomposition. His reintroduction of animistic or vitalistic ideas into physiology had great influence on 18th-century medical theory.

Stahl retired from academic life in 1716 to take up appointment as physician to King Frederick I of Prussia. He held this post until his death on May 14, 1734.

Wendell Meredith Stanley.

STANLEY, WENDELL MEREDITH (1904-1971)

American biochemist

Wendell Meredith Stanley was a biochemist who was the first to isolate, purify, and characterize the crystalline form of a virus. During World War II, he led a team of scientists in developing a vaccine for viral influenza. His efforts have paved the way for understanding the molecular basis of heredity and formed the foundation for the new scientific field of molecular biology. For his work in crystallizing the tobacco mosaic virus, Stanley shared the 1946 Nobel Prize in **chemistry** with **John Howard Northrop** and **James B. Sumner**.

Stanley was born in the small community of Ridgeville, Indiana, on August 16, 1904. His parents, James and Claire Plessinger Stanley, were publishers of a local newspaper. As a boy, Stanley helped the business by collecting news, setting type, and delivering papers. After graduating from high school he enrolled in Earlham College, a liberal arts school in Richmond, Indiana, where he majored in chemistry and mathematics. He played football as an undergraduate, and in his senior year he became team captain and was chosen to play end on the Indiana All-State team. In June of 1926 Stanley graduated with a bachelor of science degree. His ambition was to become a football coach, but the course of his life was changed forever when an Earlham chemistry professor invited him on a trip to Illinois State University. Here, he was introduced to **Roger Adams**, an organic chemist, who inspired him to seek a career in chemical research. Stanley applied and was accepted as a graduate assistant in the fall of 1926.

In graduate school, Stanley worked under Adams, and his first project involved finding the stereochemical characteristics of biphenyl, a **molecule** containing **carbon** and **hydrogen** atoms. His second assignment was more practical; Adams was interested in finding chemicals to treat leprosy, and Stanley set out to prepare and purify compounds that would destroy the disease-causing pathogen. Stanley received his master's degree in 1927 and two years later was awarded his Ph.D. In the summer of 1930, he was awarded a National Research Council Fellowship to do postdoctoral studies with **Heinrich Wieland** at the University of Munich in Germany. Under Wieland's tutelage, Stanley extended his knowledge of experimental **biochemistry** by characterizing the properties of some yeast compounds.

Stanley returned to the United States in 1931 to accept the post of research assistant at the Rockefeller Institute in New York City. Stanley was assigned to work with W. J. V. Osterhout, who was studying how living cells absorb **potassium** ions from seawater. Stanley was asked to find a suitable chemical model that would simulate how a marine plant called *Valonia* functions. He discovered a way of using a water-insoluble **solution** sandwiched between two layers of **water** to model the way the plant exchanged ions with its environment. The work on *Valonia* served to extend Stanley's knowledge of biophysical systems, and it introduced him to current problems in biological chemistry.

In 1932, Stanley moved to the Rockefeller Institute's Division of Plant Pathology in Princeton, New Jersey. He was primarily interested in studying viruses. Viruses were known to cause diseases in plants and animals, but little was known about how they functioned. His assignment was to characterize viruses and determine their composition and structure.

He began work on a virus that had long been associated with the field of virology. In 1892, D. Ivanovsky, a Russian scientist, had studied tobacco mosaic disease, in which infected tobacco plants develop a characteristic mosaic pattern of dark and light spots. He found that the tobacco plant juice retained its ability to cause infection even after it was passed through a filter. Six years later M. Beijerinck, a Dutch scientist, realized the significance of Ivanovsky's discovery: the **filtration** technique used by Ivanovsky would have filtered out all known bacteria, and the fact that the filtered juice remained infectious must have meant that something smaller than a bacterium and invisible to the ordinary light microscope was responsible for the disease. Beijerinck concluded that tobacco mosaic disease was caused by a previously undiscovered type of infective agent, a virus.

Stanley was aware of recent techniques used to precipitate the tobacco mosaic virus (TMV) with common chemicals. These results led him to believe that the virus might be a protein susceptible to the reagents used in protein chemistry. He set out to isolate, purify, and concentrate the tobacco mosaic virus. He planted Turkish tobacco plants, and when the plants were about six inches tall, he rubbed the leaves with a swab of linen dipped in TMV solution. After a few days the heavily infected plants were chopped and frozen. Later, he ground and mashed the frozen plants to obtain a thick, dark liquid. He then subjected the TMV liquid to various enzymes and found that some would inactivate the virus and concluded that TMV must be a protein or something similar. After exposing the liquid to more than 100 different chemicals, Stanley determined that the virus was inactivated by the same chemicals that typically inactivated **proteins**, and this suggested to him, as well as others, that TMV was protein-like in nature.

Stanley then turned his attention to obtaining a pure sample of the virus. He decanted, filtered, precipitated, and evaporated the tobacco juice many times. With each chemical operation, the juice became more clear and the solution more infectious. The end result of two-and-one-half years of work was a clear concentrated solution of TMV which began to form into crystals when stirred. Stanley filtered and collected the tiny, white crystals and discovered that they retained their ability to produce the characteristic lesions of tobacco mosaic disease.

After successfully crystallizing TMV, Stanley's work turned toward characterizing its properties. In 1936, two English scientists at Cambridge University confirmed Stanley's work by isolating TMV crystals. They discovered that the virus consisted of ninety-four percent protein and six percent nucleic acid, and they concluded that TMV was a nucleoprotein. Stanley was skeptical at first. Later studies, however, showed that the virus became inactivated upon removal of the nucleic acid, and this work convinced him that TMV was indeed a nucleoprotein. In addition to chemical evidence, the first **electron** microscope pictures of TMV were produced by researchers in Germany. The pictures showed the crystals to have a distinct rod-like shape. For his work in crystallizing the tobacco mosaic virus, Stanley shared the 1946 Nobel prize in chemistry with John Howard Northrop and James Sumner.

During World War II, Stanley was asked to participate in efforts to prevent viral diseases, and he joined the Office of Scientific Research and Development in Washington D.C. Here, he worked on the problem of finding a vaccine effective against viral influenza. Such a substance would change the virus so that the body's immune system could build up defenses without causing the disease. Using fertilized hen eggs as a source, he proceeded to grow, isolate, and purify the virus. After many attempts, he discovered that **formaldehyde**, the chemical used as a biological preservative, would inactivate the virus but still induce the body to produce **antibodies**. The first flu vaccine was tested and found to be remarkably effective against viral influenza. For his work in developing large-scale methods of preparing vaccines, he was awarded the Presidential Certificate of Merit in 1948.

In 1948, Stanley moved to the University of California in Berkeley, where he became director of a new virology laboratory and chair of the department of biochemistry. In five years Stanley assembled an impressive team of scientists and technicians who reopened the study of plant viruses and began an intensive effort to characterize large, biologically important molecules. In 1955 **Heinz Fraenkel-Conrat**, a protein chemist, and R. C. Williams, an electron microscopist, took TMV apart and reassembled the viral **RNA**, thus proving that RNA was the infectious component. In addition, their work indicated that the protein component of TMV served only as a protective cover. Other workers in the virus laboratory succeeded in isolating and crystallizing the virus responsible for polio, and in 1960 Stanley led a group that determined the complete amino acid sequence of TMV protein. In the early 1960s, Stanley became interested in a possible link between viruses and cancer.

Stanley was an advocate of academic freedom. In the 1950s, when his university was embroiled in the politics of McCarthyism, members of the faculty were asked to sign oaths of loyalty to the United States. Although Stanley signed the oath of loyalty, he publicly defended those who chose not to, and his actions led to court decisions which eventually invalidated the requirement.

Stanley received many awards, including the Alder Prize from Harvard University in 1938, the Nichols Medal of the American Chemical Society in 1946, and the Scientific Achievement Award of the American Medical Association in 1966. He held honorary doctorates from many colleges and universities. He was a prolific author of more than 150 publications and he co-edited a three **volume** compendium entitled *The Viruses.* By lecturing, writing, and appearing on television he helped bring important scientific issues before the public. He served on many boards and commissions, including the National Institute of Health, the World Health Organization, and the National Cancer Institute.

Stanley married Marian Staples Jay on June 25, 1929. They had met at the University of Illinois, where they both were graduate students in chemistry. They coauthored a scientific paper together with Adams, which was published the same year they were married. The Stanleys had three daughters and one son. On June 15, 1971, while attending a conference on biochemistry in Spain, Stanley died from a heart attack.

STATES OF MATTER

All material must exist in one of the three forms of **matter**—a solid, a liquid, or a gas. These are different physical states of being and each form has implications for the substance in question.

When we consider chemical substances most can exist in any of the three states. Which state of matter is encountered depends upon the physical conditions that they are being studied under. If the conditions are not specified then standard **temperature** and pressure is assumed. As a result of this it can be said that **sodium** chloride (table salt) is a solid. What in fact should be said is that at a pressure of one atmosphere (1.013 x 10^5Nm ^{2}) and a temperature of 273.15 K (32°F or 0°C) **sodium chloride** is a solid.

The three states of matter have different ways of responding to changes in temperature and pressure. All three will show an increase in **volume** (expansion) when the temperature is increased and a decrease in volume (contraction) when the temperature is lowered. This effect is most noticeable with a gas and least noticeable with a solid, with a liquid being intermediate between the two extremes.

The difference between the states of matter are due to the differences between the amounts of **energy** their molecules have. A solid has molecules that are relatively immobile. All the molecules comprising a solid are in close contact with their neighboring molecules. The molecules are not free to move away from each other. This means that a solid has a definite shape and a definite volume, neither of which change much as the conditions of the environment alter. As the temperature increases the molecules are able to vibrate more vigorously. As the temperature decreases the molecules move more slowly and they become more aligned, this makes it easier for the transmission of **electricity**, if the solid is a conductor.

A liquid is very similar to a solid in many respects. The molecules of a liquid are also in close proximity to their neighbors. The liquid molecules are vibrating faster than those of a solid. A liquid has a fixed volume although its shape is not fixed; it will flow to take on the shape of its container. The layers of molecules in a liquid are more capable of moving over each other.

When a gas is considered the situation is very different. Within a gas the molecules have very high energy. The molecules of a gas are not touching any of their neighbors and they are free to act independently. This allows a gas to have neither a fixed volume nor shape. A gas will expand to fill the container into which it is placed. The properties of a gas are described by a series of equations known as the **gas laws**, these are **Boyle's law**, **Charles's law**, and the constant volume law.

The theory by which the physical properties of the three states of matter is explained by reference to the motion of the molecules making up the material is know as the kinetic theory of matter.

When a solid changes into a liquid it is by a process called melting. When a liquid changes into a gas it is by boiling. A gas changing to a liquid is condensation and a liquid changing to a solid is freezing. Some substances are capable of going directly from a solid to a gas, this is a process known as **sublimation**.

The **bonding** that is present has a strong influence on the state of matter of a material. Strong **intermolecular forces**, **van der Waals forces**, are characteristic of a solid. It is these forces that strongly hold together the molecules of a solid. With a liquid the molecules are also held closely together by intermolecular forces, although not as strongly as in a solid. The intermolecular forces in a liquid are not strong enough to keep the molecules from slipping past one another. It is this characteristic that makes the pouring of **liquids** a practicality. With **solids** the intermolecular forces virtually lock the molecules in place. The molecules in a solid can take up and retain a regular structure, a lattice. Kinetic energy has the tendency to speed up the movement of particles and force them apart whereas intermolecular forces tend to draw molecules together and stop them from moving. The particles of a solid and a liquid are fairly close together compared to those of a gas, and solids and liquids are called the condensed states. By altering the kinetic energies a solid can change to a liquid and then to a gas. The kinetic energy that is applied in a situation such as this has to be sufficient to overcome the various van der Waals forces that are in operation within the **molecule**.

If pressure is increased on a substance it forces the molecules closer together which in turn strengthens the intermolecular forces. If the pressure applied is sufficient then a state change to a more ordered form can be brought about. Hydrogen chloride in solution provides good indication of the relative strengths of intermolecular bonds compared to molecular bonds. To turn the hydrogen chloride from a liquid into a gas or vapor, only 16 kJ /mol is needed to overcome the intermolecular bonds that are present. Breaking the **covalent bond** to disassociate the molecule would require 431 kJ/mol of energy. It is due to this difference in energies required to break the different bonds that the majority of substances change from one state to another upon heating rather than disassociating. The majority of these intermolecular forces are due to the polarity of the molecules under consideration, the greater the polarity the stronger the resultant intermolecular forces.

It is possible for all three states of matter to exist in the same place and at the same time. This is known as the triple point and it only occurs at one set of temperature and pressure conditions. For example the triple point of **water** is at a pressure of 4.58 torr and a temperature of 32.0176°F (0.0098°C). At these conditions ice, water and water vapor can all exist together. This relationship can be shown in a phase diagram. A phase diagram is a plot of temperature against pressure for a particular substance and it shows the equilibria which exist between different states of matter (phases) under different physical conditions. The equilibria between the two phases are indicated by a line. There is only one point in a phase diagram where all three phases meet, this is the triple point. A phase change is a change from one state of matter to another.

The state of matter of the substances in a chemical equation can be indicated by adding the initial letter of the state—s for solid, l for liquid, and g for gas—within parentheses following the formulas.

See also Gases; Gases, Behavior and properties of

STAUDINGER, HERMANN (1881-1965)

German organic and polymer chemist

Hermann Staudinger's interest in organic **chemistry** was wide-ranging and he made many important contributions in that field. He is principally known, however, for his concept of the "**macromolecule**," or polymer, a long chain of repeating chemical units. Although initially this idea was greeted with incredulity and scorn in the chemistry community, Staudinger eventually overcame his critics' objections with patient explanation, careful research, and dogged insistence. Polymers are now known to be extraordinarily useful substances, ubiquitous in natural systems as well as human society. The entire **plas-**

tics and materials science industry bases itself on polymers, and the science of molecular biology was immeasurably aided by the concept of macromolecules. Staudinger was awarded a Nobel Prize in 1953, three years after he had retired from active research.

Staudinger was born March 23, 1881, in Worms, Germany, to Dr. Franz and Auguste (Wenck) Staudinger. His father was a philosopher and professor at various German institutions of secondary education, and was interested in social reform. Staudinger graduated from the Gymnasium at Worms in 1899 (a German *gymnasium* is roughly the equivalent of an American prep school) and began his university studies at Halle under the guidance of the botanist Professor Klebs. While there, Staudinger began a lifelong detour from his original interest in botany. His family encouraged him to study chemistry to provide a strong background for his biological investigations, and he not only took their advice, but actually stayed in chemistry for most of the rest of his career. He studied in Darmstadt under the direction of professors Kolb and Stadel, and in Munich under the direction of Professor Piloty. In 1901 he returned to the University of Halle and in 1903 finished his doctoral work under the direction of Professor Vorlander. Although Staudinger had begun his studies in the analytical subdivision of chemistry, Vorlander's influence and ideas caused him to develop an intense interest in theoretical **organic chemistry**. His doctoral thesis was on the malonic esters of unsaturated compounds.

Initially, Staudinger's organic chemical investigations were relatively routine, although they resulted in the synthesis of some interesting new classes of organic molecules. In 1905, while in his first teaching position (as instructor) with Professor Thiele in Strassburg, he discovered the first ketene (colorless, poisonous gasses). Ketenes as a class are extremely reactive—they even react with traces of **water** and the **oxygen** in the air—and Staudinger and other researchers investigated their properties and chemistry for several years. In 1907, he prepared a special dissertation on the ketenes and was awarded the title of assistant professor. Shortly thereafter, he accepted an associate professor position at the Technische Hoshschule in Karlsruhe. Staudinger began many basic organic synthesis projects in Karlsruhe, including a new synthesis of isoprene (a constituent part of rubber), although some of the newer projects slowed when he moved to the Swiss Federal Institute of Technology (Eidgenossische Technische Hochschule) in Zurich in 1912. The Zurich position was a prestigious one, but the teaching load was many hours per week, so he curtailed his research in some areas. He proved to be a dedicated teacher and instilled in his students an appreciation not only for chemistry, but also for the power of technology in society.

During his fourteen years in Switzerland, Staudinger continued investigations on the chemistry of the ketenes, oxalyl chloride, and several materials that are shock-sensitive, that is, they explode when bumped or dropped. He also continued work on pyrethrin insecticides, and when they could not be easily synthesized, drew on his botanical interests and suggested that new strains of chrysanthemum (from which natural pyrethrines are extracted) might yield better quantities than any

Hermann Staudinger.

laboratory. During World War I, much of Staudinger's work was driven by wartime shortages. He investigated the aromas of pepper and coffee to see if synthetic substitutes could be produced for those foods, and attempted to synthesize some important pharmaceuticals. He was successful enough to patent some of the artificial flavors and fragrances, although generally they proved uneconomical if the natural material was available. In 1926, Staudinger accepted a position as director of the chemical laboratories at Freiburg University, where he remained until his retirement in 1951. He continued to be internationally respected in his field, winning the LeBlanc medal given by the French Chemical Society in 1931 and the Cannizzarro Prize in Rome in 1933.

In 1924, Staudinger wrote in *Berichte der Deutschen Chemischen Gesellschaft,* "The molecules of rubber...have entirely different sizes...and these can be changed, by **temperature** for example.... It is very important here to use the idea of the **molecule**...; [the] particles are held together by normal chemical bonds and in the structural sense we are dealing with very long **carbon chains**. The polymerization of isoprene to these long chains...goes on until a sufficiently large, little reactive, and thus strongly saturated molecule...has been formed. For those...particles in which the molecule is identical with the primary particle and in which the individual atoms of the... molecule are linked by normal valences, we propose to differentiate the type by the term *macromolecules.*"

The move to Freiberg University signaled Staudinger's break with traditional organic chemistry. He had first proposed the idea of macromolecules in 1920 while still in Zurich, but at Freiberg he gave up most of his other chemical research to pursue the study of rubber and synthetic polymers. It was a decision which caused his colleagues some consternation because he was well respected in his field, and they felt he would do damage to his reputation by working on such unpromising materials from such an unorthodox point of view.

When Staudinger began work on his theory of very high **molecular weight** compounds, several entrenched ideas about molecular weight existed. Chemists believed, for example, that the size of a molecule was governed by its "unit cell," or smallest nonrepeating piece, and that its molecular weight was a fixed number. For low molecular weight compounds, those principles remained true. Many chemical methods had been devised for determining the structure of molecules, and nonrepeating structures with unit cells of up to 5000 atomic **mass** units had been elucidated. The relatively new science of X-ray **crystallography** also helped in structure determination. Additionally, researchers knew of compounds, soaps for example, which aggregated (clumped together) in water and other **solvents**, instead of dissolving in the normal way. These aggregates possessed some of the same unusual physical characteristics as rubber and other known polymers. Many chemists therefore insisted that Staudinger had mistaken aggregates of smaller molecules for single large molecules.

With carefully designed experiments, Staudinger gradually accumulated evidence that a group of extremely large molecules indeed existed. They did not comprise oddly associated clumps of smaller molecules, and were not themselves single unit cells. Instead, these molecules resembled chains of repeating units, strung together and bonded to each other—like pearls on a wire. Additionally, because any number of units might be bonded together during synthesis, these large molecules had differing molecular weights, depending on the length of the chain.

While working in this area, Staudinger developed some new analytical methods and discovered a relationship between the **viscosity** of a polymer **solution** and its molecular weight. He had developed the method in desperation when he could not get funding for more sophisticated equipment, but it was soon widely used in industry because it was inexpensive, fast, and accurate. The equation, now called Staudinger's Law, allows a fairly simple estimation of molecular weight by measuring the "drag" or "stickiness" of a liquid flowing through a small tube (viscosity).

During World War II, the Freiburg University chemistry facilities were virtually destroyed in an Allied air bombardment in November 1944, and it was several years before they were fully operational again. Staudinger's work slowed after the enormous stresses of the war, although he still found the **energy** to start and edit two new journals, one on macromolecular chemistry. He gave many talks and wrote prolifically on the subject of macromolecules until the end of his career.

In spite of the evidence, however, chemists had difficulty in accepting his conclusions. Spirited discussions bordering on uproar often greeted him when he gave scientific lectures; his persistence and patience in the face of such hostility became legendary. Historians have noted such resistance before on the part of the scientific community whenever a truly revolutionary idea is advanced. Some have speculated that chemists simply found the idea of huge molecules too messy; they preferred the neat, tidy unit cell with its fixed structure and weight, for which they had developed many good analytical procedures.

As time went on, however, evidence from other areas of scientific study built unequivocal support for macromolecules. X-ray crystallographers had refined their techniques to the point where they could obtain structural information on polymeric materials. Various microscope and optical techniques also lent important evidence, particularly for biological molecules. Staudinger had speculated for years that living systems must require macromolecules to function, and the new science of molecular biology began to lend vigorous support to that idea.

So controversial were Staudinger's macromolecular theories, that it was not until 1953, when he was seventy-two years of age, that he was finally rewarded with the Nobel Prize for his efforts. The presenter of the prize, Professor Fredga, noted in his speech, "In the world of high polymers, almost everything was new and untested. Long standing, established concepts had to be revised or new ones created. The development of macromolecular science does not present a picture of peaceful idylls."

After his retirement, Staudinger once again took up his original interest in botany, although his biological bent had always been apparent even in his chemical work. His wife, the former Magda Woit, was a Latvian plant physiologist who often participated in his research, collaborated in writing many of his papers, and made some important connections of her own between his macromolecular theories and the molecules of biology. Staudinger listed some of her considerable contributions in his Nobel Prize address, and dedicated his autobiography to her. They married in 1927, and she survived him upon his death on September 8, 1965.

STEARIC ACID

Stearic acid (or n-octadecanoic acid) has a chemical formula of $CH_3(CH_2)_{16}COOH$ and is a straight chain **molecule**. It is one of the most commonly found saturated **fatty acids** and it occurs as glycerides in the majority of animal and vegetable fats. It can account for 25% of the composition of animal fat. Stearic acid was isolated and named by French chemist **Michel-Eugène Chevreul**, who also isolated palmitic acid and oleic acid.

Stearic acid is particularly abundant in the harder fats with higher melting points. Pure stearic acid has a melting point of 158°F (70°C) and a **boiling point** of 709°F (376°C) at which **temperature** it shows thermal decomposition. Stearic acid, along with palmitic acid, can be used to make candles. Stearic acid is also used in the manufacture of suppositories. Like all fatty acids stearic acid is insoluble in **water**, but is soluble in **ether** and hot **alcohol**.

The **sodium** and **potassium** salts of stearic acid are soaps, sodium or potassium stearate. Stearic acid can be utilized in the Krebs tricarboxylic acid cycle where it can yield some 180 molecules of ATP (compared to sugar which will yield only 38 molecules).

See also Lipids

Stein, William Howard (1911-1980)

American biochemist

William Howard Stein, in partnership with **Stanford Moore**, was a pioneer in the field of protein **chemistry**. Although other scientists had previously established that **proteins** could play such roles as that of enzymes, **antibodies**, **hormones**, and **oxygen** carriers, almost nothing was known of their chemical makeup. Stein and Moore, during some forty years of collaboration, were not only able to provide information about the inner workings of protein molecules, but also invented the mechanical means by which that information could be extracted. Their discovery of how protein **amino acids** function was accomplished through a study of ribonuclease (RNase), a pancreatic **enzyme** that assists in the digestion of food by catalyzing the breakdown of **nucleic acids**. But their work could not have been accomplished without the development of a technology to assist them in collecting and separating the amino acids contained in ribonuclease. Their invention of the fraction collector and an automated system for analyzing amino acids was of great importance in furthering protein research, and these devices have become standard laboratory equipment.

Stein and Moore began their collective work in the late 1930s under Max Bergmann at the Rockefeller Institute (now Rockefeller University). After Bergmann's death in 1944, the pair developed the protein chemistry program at the Institute and began their research into enzyme analysis. Except for a brief period during World War II when Moore served with the Office of Scientific Research and Development in Washington D.C., and the two years when Stein taught at the University of Chicago and Harvard University, the partnership continued uninterrupted until Stein's death in 1980. Their joint inventions and co-authorship of most of their scientific papers were said to make it impossible to separate their individual accomplishments. Their combined efforts were acknowledged in 1972 with the Nobel Prize in chemistry. According to Moore, writing about Stein in the *Journal of Biological Chemistry* in 1980, they received the award ''for contributions to the knowledge of the chemical structure and catalytic function of bovine pancreatic ribonuclease.'' **Christian Anfinsen** shared the Nobel Prize with Stein and Moore for related research.

The son of community-minded parents, Stein was born in New York City on June 25, 1911. He was the second of three children. His father, Fred M. Stein, was involved in business and retired at an early age to lend his services to various health care associations in the community. The scientist's mother, Beatrice Borg Stein, worked to improve recreational and educational conditions for underprivileged children. From an early age, Stein was encouraged by his parents to develop an interest in science. He received a progressive education from grade school on, attending the Lincoln School of the Teacher's College of Columbia University, transferring at sixteen years of age to Phillips Exeter Academy for his college preparatory studies. He graduated from Harvard University in 1933, then took a year of graduate study in **organic chemistry** there. Finding that his real interest was **biochemistry**, he completed his graduate studies at the College of Physicians and Surgeons of Columbia University, receiving his Ph.D. in 1938. His dissertation concerned the amino acid composition of elastin, a protein found in the walls of veins and arteries. This work marks the beginning of his long search to understand the chemical function of proteins.

The successful research being done at the Rockefeller Institute under the direction of Max Bergmann caught Stein's attention. He pursued post-graduate studies there in 1938, spending his time improving analytical techniques for purifying amino acids. Moore joined Bergmann's group in 1939. There, he and Stein began work in developing the methodology for analyzing the amino acids glycine and leucine. Their work was interrupted when the United States entered World War II. Then, Bergmann's laboratory was given over to the study of the physiological effects of mustard **gases**, in the hope of finding a counteractant.

The group's efforts to find accurate tools and methods for the study of amino acid structure increased in importance when they assumed the responsibility of establishing the Institute's first program in protein chemistry. Looking for ways to improve the separation process of amino acids, they turned to partition **chromatography**, a filtering technique developed during the war by the English biochemists **A. J. P. Martin** and **Richard Synge**. Building on this technology, as well as that of English biochemist **Frederick Sanger**'s column chromatography and the ion-exchange technique of Werner Hirs, Stein and Moore went on to invent the automatic fraction collector and develop the automated system by which amino acids could be quickly analyzed. This automated system replaced the tedious two-week sequence that was previously required to differentiate and separate each amino acid.

From then on, the isolation and study of amino acid structure was advanced through these new analytical tools. Ribonuclease was the first enzyme for which the biochemical function was determined. The discovery that the amino acid sequence was a three-dimensional, chain-like structure that folds and bends to cause a catalytic reaction was a beginning for understanding the complex nature of enzyme catalysis. Stein and Moore were certain that this understanding would result in crucial medical advances. By 1972, the year Stein and Moore shared the Nobel Prize, other enzymes had been analyzed using their methods.

Because he was extremely eager to see that research done in laboratories all over the country be disseminated as widely and as quickly as possible, Stein devoted many years in various editorial positions to the *Journal of Biological Chemistry*. Under his leadership, the journal became a leading biochemistry publication. He had joined the editorial board in 1962 and became editor in 1968. He only held the latter post for one year, however. While attending an international meeting in Denmark, he contracted Guillain-Barré Syndrome, a rare disease often causing temporary paralysis. In grave danger of dying, he managed to recover somewhat. The illness left him a quadriplegic, confined to a wheelchair for the rest of his life. Although he remained involved with the work of his colleagues both in the laboratory and at the *Journal,* he was unable to participate actively.

In addition to the Nobel Prize, Stein shared with Moore the 1964 Award in Chromatography and Electrophoresis and the 1972 Theodore Richard Williams Medal of the American Chemical Society. He served as chairperson of the U.S. National Committee for Biochemistry from 1968 to 1969, as trustee of Montefiore Hospital, and as board member of the Hebrew University medical school. He married Phoebe L. Hockstader on June 22, 1936. They had three sons: William Howard, Jr., David, and Robert. Stein died in Manhattan on February 2, 1980.

STEREOCHEMISTRY

Stereochemistry is the study of the three dimensional shape of molecules and the effects of shape upon the properties of molecules.

Dutch chemist **Jacobus Hendricus van't Hoff**, the winner of the first Nobel Prize in Chemistry in 1901, pioneered the study of **molecular structure** and stereochemistry. Van't Hoff's ideas were not readily accepted by the scientific community in the late 1800s. His original paper was only 13 pages long, including one page of diagrams. This brief paper, however, gave rise to a powerful explanation of how molecules are structured and how they react with other molecules. Van't Hoff proposed that the concept of an asymmetrical carbon **atom** explained the existence of numerous isomers that had baffled the chemists of the day. Van't Hoff's work gave eventual rise to stereochemistry when he correctly described the existence of a relationship between a molecule's optical properties and the presence of an asymmetrical **carbon** atom.

The stereochemistry of carbon is important in all biological processes. Stereochemistry is also important in geology, especially **mineralogy**, with dealing with **silicon** based **geochemistry**.

Assuming that the all reactants are present, inorganic reactions are chiefly governed by **temperature**, that is, temperature is critical to determining whether or not a particular reaction will proceed. In biological reactions, however, the shape of the molecules is also a critical factor. Small changes in the shape or alignment of molecules can determine whether or not a reaction will proceed. In fact, one of the critical roles of enzymes in **biochemistry** is to lower the temperature requirements for chemical reactions. With the proper enzymes present, biological temperatures suffice to allow reactions to proceed. This leaves the stereochemistry of molecules as a controlling factor in biological and organic (molecules and compounds with carbon) reactions (assuming all the reactants are present) is the shape and alignment of the reacting molecules.

The **molecular geometry** around an atom depends upon the number of bonds to other atoms and the presence or absence of lone pairs of electrons associated with the atom.

Some compounds differ only in their shape or orientation in space. Compounds that have the same **molecular formula** are called isomers (Figure 1). Stereoisomers are isomers (i.e., they have the same **molecular weight** and formula) but that differ in their orientation in space. No **matter** how a stereoisomer is rotated it presents a different picture than its stereoisomer counterpart. Most importantly, stereoisomers are not spatially superimposable.

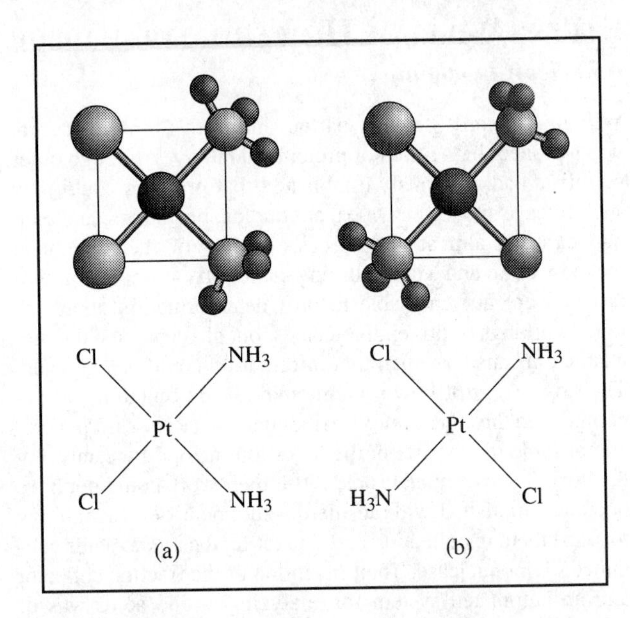

Figure 1. A planar complex, diamminedichloroplatinum [$Pt(NH_3)_2Cl_2$], presenting two geometric isomers: the *cis* isomer (A) and the *trans* (B) isomer. *(Illustration by Electronic Illustrators Group.)*

Enantiomers are stereoisomers that are mirror images, that is, they cannot map onto one another. If the molecules were two dimensional we would say that the molecules, just like human hands, could not be laid on top or superimposed upon each other.

Stereoisomers that rotate polarized light are called optical isomers. With the help of an instrument called a polarimeter, molecules are assigned a sign or rotation, either (+) for dextrorotatory molecules that rotate a plane of polarized light to the right, or (–) for levorotatory molecules that rotate a plane of polarized light to the left. Enantiomers differ in the direction that they rotate a plane of polarized light and in the rate that they react with other chiral molecules. Racemic mixtures contain equal amounts of the two enantiomers.

Symmetry is a term used to describes molecules made of equivalent parts. When a **molecule** is symmetrical it has portions that correspond in shape, size, and structure so that they could be mapped or transposed on one another. Bilateral symmetry means that a molecule can be divided into two corresponding parts. Radial symmetry means that if a molecule is rotated about an axis that a certain number of degrees rotation (always less than 360°) it looks identical to the molecule prior to rotation.

A molecule is said to be symmetrical if it can be divided into equal mirror image parts by a line or a plane. Humans are roughly bilaterally symmetrical. Draw a line down the middle of the human body and the line divides the body into two mirror image halves. If a blob of ink were placed on a piece of paper, and then the paper was folded over and then unfolded again, you would find two ink spots—the original and the image—symmetrical about the fold in the paper. Molecules and complexes can have more than just two planes of symmetry.

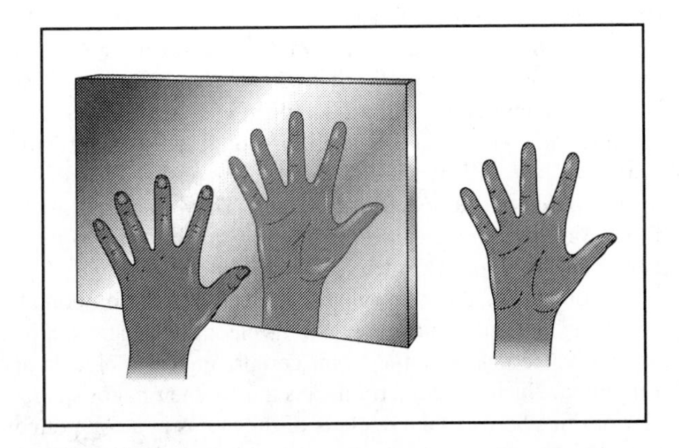

Figure 2. Chirality: the reflection of the left hand looks like the right hand, but it is impossible to superimpose the left hand on the right hand. The term is derived from the Greek word *cheir*, which means hand. *(Illustration by Electronic Illustrators Group.)*

Human hands are excellent examples of the concept of handedness. The right and left hands are normally mirror images of each other. The single major difference between them is the direction one takes to go from the thumb to the fingers. This sense of direction is termed handedness, that is, whether a molecule or complex has a left and right orientation. Two molecules that are mirror images of each other, alike in every way except for their handedness, are called enantiomers.

Handedness can have profound implications. Some medicines are vastly more effective in their left-handed configuration than in their right-handed configuration. In most cases biological systems make only one of the forms. Usually, only one of the forms is effective in cellular chemical reactions.

A molecule that is not symmetric—that is, a molecule without a plane of symmetry—is termed dissymmetric (or asymmetric), or chiral. A molecule is said to be chiral if it lacks symmetry and its mirror images are not superimposable (Figure 2). To be chiral, a molecule must lack symmetry, that is, a chiral molecule can not have any type or symmetry.

Carbon atoms with four sp^3 hybridized orbitals form four bonds about the central carbon atom. When the central carbon bonds with differing atoms or groups of atoms the carbon is termed a stereogenic carbon atom. Bromochlorofluoromethane is an example of such a molecule. The central carbon, with four sp^3 bonds oriented (pointing) to the corners of a tetrahedron, is bonded to a **bromine, chlorine, fluorine** and **methane** atoms. There is no symmetry to this molecule, and it exists in the two enantiomeric forms.

Stereogenic carbon atoms may be described as corresponding to either a R and S designation. Although the rules for determining this designation can be complex, for simple molecules and compounds the determination is easily accomplished with the help of a model of the molecule. The four different bonded groups are assigned a priority. When assigning priority to groups, atoms that are directly bonded to the carbon atom have their priority based upon their **atomic number**. The atom with the highest atomic number has highest priority and atom with the lowest atomic number the lowest priority. As a result, **hydrogen** atoms bonded to the stereogenic carbon have the lowest priority. If isotopes are bonded then the **isotope** with the largest **mass** has the higher priority. The molecule is then turned so that the lowest priority group is farthest away from view. If one must take a counterclockwise path from the highest to lowest priority group the carbon configuration is assigned as sinister (S). If the path from highest to lowest priority groups is clockwise, then the carbon is assigned as rectus (R).

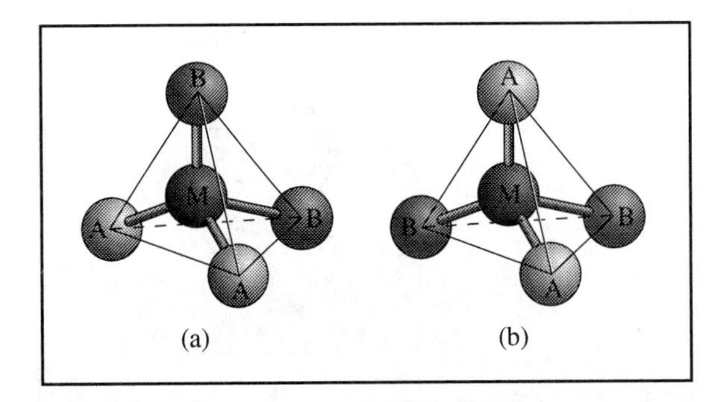

Figure 3. A tetrahedral complex (consisting of a metal atom M and ligands A and B) with no geometric isomers. Molecule A can be rotated to look exactly like molecule B. *(Illustration by Electronic Illustrators Group.)*

The rectus (R) and sinister (S) descriptor relates to the structure of an individual carbon. In contrast, dextro (+) and levo (–) properties are based on the collective property of a large ensemble of the molecules.

A molecule can have more than one stereogenic carbon. The number of stereoisomers can be determined by the $2n$ rule, where n equals the number of stereogenic carbons. Thus, if one stereogenic carbon is present there are two possible stereoisomers, with two stereogenic carbons there are four possible stereoisomers. Any chemical reaction that yields predominantly one stereoisomer out of several stereoisomer possibilities is said to be a stereoselective reaction.

Sometimes it is difficult to tell whether or not two molecules or complexes will exhibit stereochemical properties. If two molecules or complexes have the same molecular formula they are candidates for stereochemical analysis.

The first step is to determine if the two molecules or complexes are superimposable. If they are, they are identical structures and will not exhibit stereochemical properties.

The second step is to determine if the atoms are connected to each other in the same order. If the atoms are not connected in the same order then the molecules or complexes are constitutional **isomers.** If the atoms are connected in the same order, but do not spatially superimpose, then they are stereoisomers.

The next step is to see if the stereoisomers can be made identical by rotating them around a single bond in the molecule or complex then they are called conformational isomers. Stereoisomers that can not be so rotated are called configurational isomers.

The last step is to analyze the configurational isomers to determine whether they are enantiomers, or **diastereomers.**

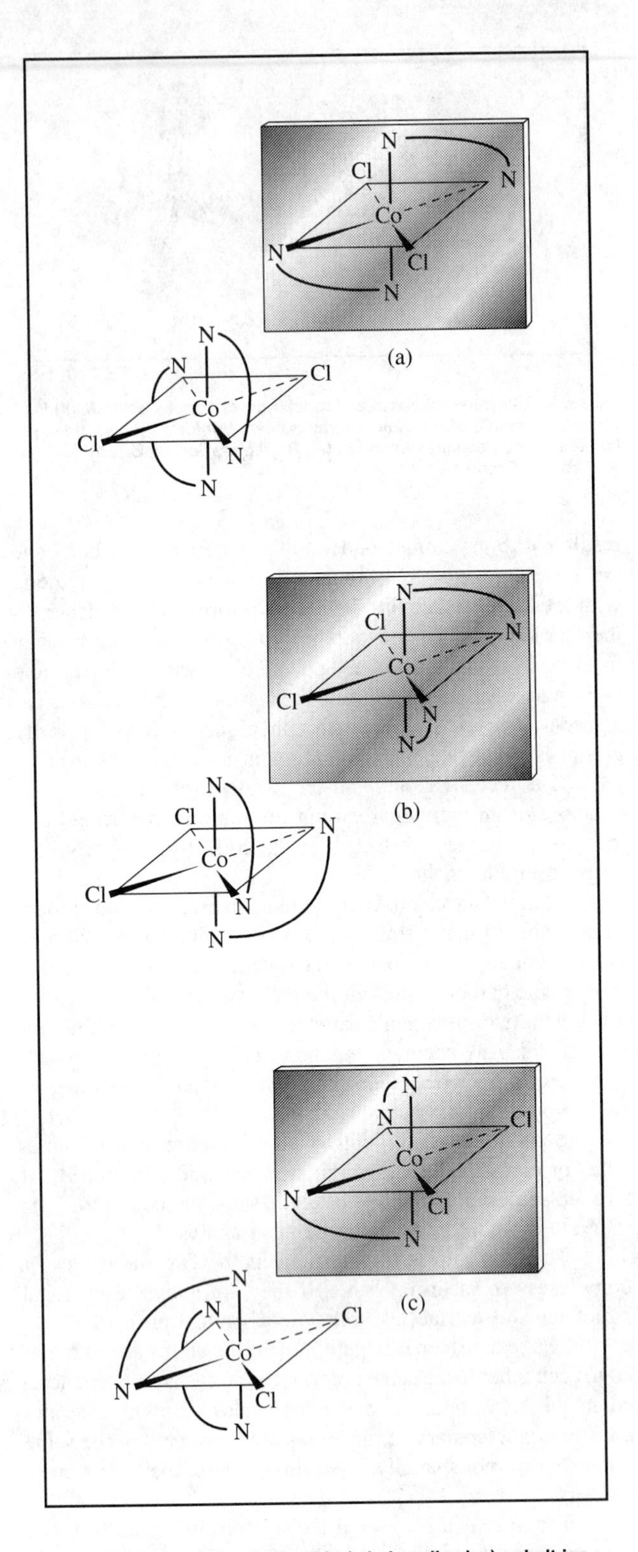

Figure 4. Isomers of the dichlorobis (ethylenediamine) cobalt ion $[CoCl_2(en)_2]$. A is the *trans* isomer. B and C, while both *cis* isomers, are not identical. They are enantiomers (optical isomers), which means they are nonsuperimposable mirror images of each other. *(Illustration by Electronic Illustrators Group.)*

Those that are mirror images are enantiomers (Figure 4). Those stereoisomers that are not mirror images of each other are diastereomers (the prefix dia indicated opposite or across from as in diagonal) or cis-trans isomers. Diastereoisomers can also be characterized as *cis* (Latin for ''on this side'') or *trans* (Latin for ''across'') when they differ in the positions of atoms or groups relative to a reference plane. They are cis-isomers if the atoms are on the same side of the plane or trans-isomers if they are on opposite sides of the reference plane.

Sometimes the **energy** of a molecule or a compound, that is, the particular **energy level** of its electrons depends upon the relative geometry of the atoms comprising the molecule or compound. Nuclear geometry means the geometrical or spatial relationships between the **nucleus** of the atoms in a compound or molecule (e.g., the balls in a ball and stick model). When a molecule or compound's energy is related to its shape this is termed a stereoelectronic property.

Stereoelectronic effects arise from the different alignment of electronic orbitals with different arrangements of nuclear geometry. It is possible to control the rate or products of some chemical reactions by controlling the stereoelectronic properties of the reactants.

See also Bonding; Geochemistry; Isomorphism; Mineralogy; Molecular geometry; Optical activity

STEROIDS

Steroids are any one of a large number of compounds that are related to sterols. Steroids include bile acids, D **vitamins**, certain **hormones**, and certain carcinogenic substances. Sterols are alcohols with a cyclic **nucleus**. An example of a steroid is **cholesterol**. In the human organism, cholesterol is the parent compound of all steroid hormones.

The accompanying illustration shows some of the most important steroid hormones. Their similarity to cholesterol is obvious. In the adrenal glands located above the kidneys, the synthesis of hydrocortisone, aldosterone, and a portion of the body's testosterone are produced. All human females produce a small amount of testosterone in this manner. It is the testes of males, however, that synthesize the greatest quantity of testosterone and account for male secondary sexual characteristics (e.g., facial hair, deep voice, etc.) In the ovaries of the female, estrogens and other female hormones are produced. Estradiol is an example of an estrogen, and it is also the basis for most oral contraceptives. Also illustrated are some synthetic steroids. Ethynyl estradiol combined with norethindrone form the mixture of the majority of birth control pills. These molecules fit receptor sites for naturally produced hormones and trick the body into a false state of pregnancy, so no egg is ever released from the ovaries. Conversely, RU-486 binds strongly to progesterone **receptors** and prevents a fertilized egg from implanting in the uterus. This has been commonly known as the ''morning after'' pill.

Important in all these steroidal compounds is their **stereochemistry** or 3-dimensional configuration. These molecules are not flat, but rather, they take on a shape that complements

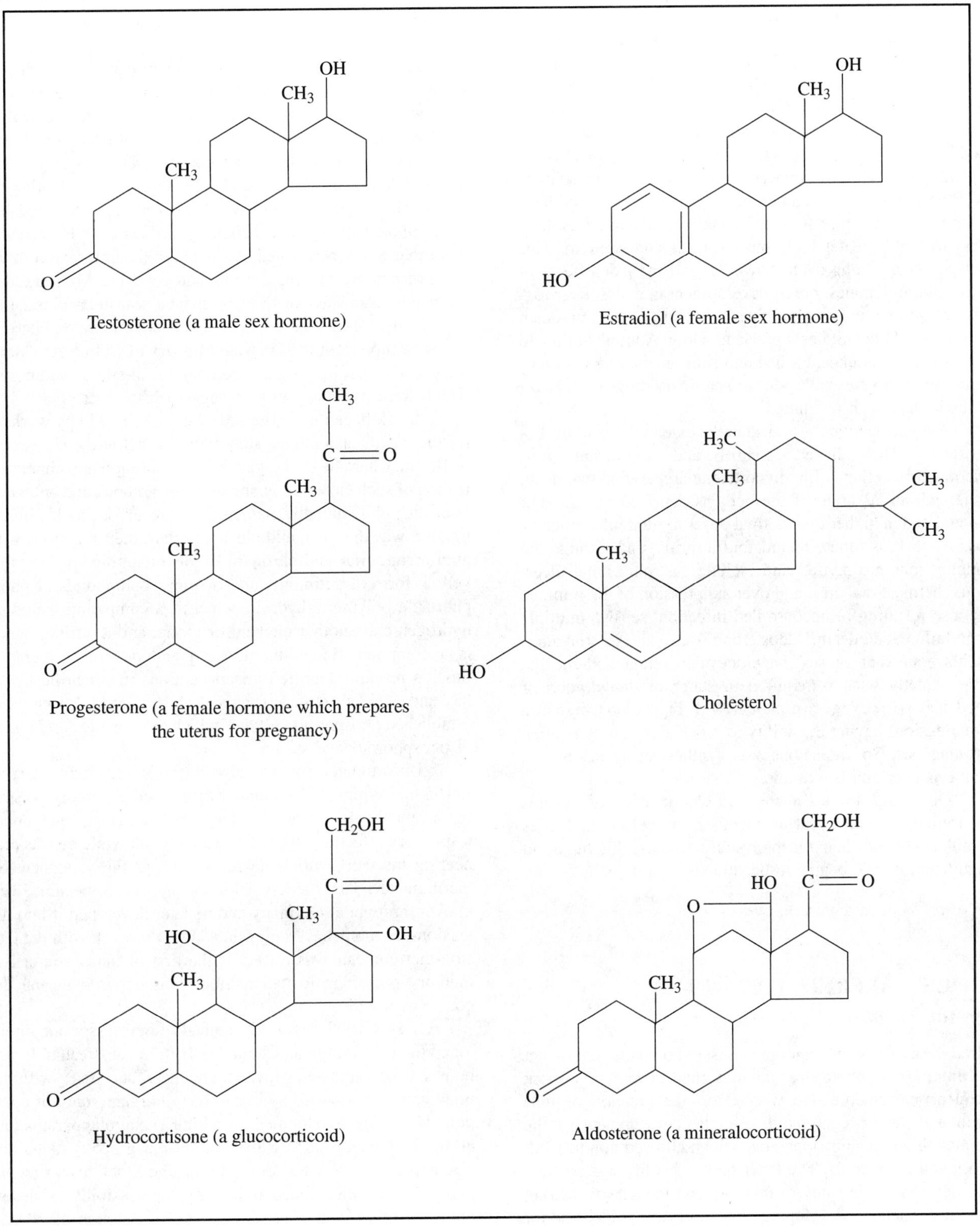
OH
CH3
CH3
O
Testosterone (a male sex hormone)
OH
CH3
HO
Estradiol (a female sex hormone)
CH3
C=O
CH3
CH3
O
Progesterone (a female hormone which prepares the uterus for pregnancy)
H3C
CH3
CH3
CH3
CH3
HO
Cholesterol
CH2OH
C=O
CH3
HO
OH
CH3
O
Hydrocortisone (a glucocorticoid)
CH2OH
HO
C=O
O
CH3
O
Aldosterone (a mineralocorticoid)

Common steroid hormones. ***(Illustration by Electronic Illustrators Group.)***

their receptor site. Thus, their synthetic counterparts must also have the proper stereochemistry.

The glucocorticoids and mineralocorticoids synthesized in the adrenal gland are vital for homeostasis and response to stress and physical trauma. Androgens such as testosterone are builders of muscle **mass** in males and females. A number of synthetic compounds based on testosterone have been developed in the hope of keeping a severely debilitated patient (such as a major burn victim) from catabolizing their own muscle tissue. Nandrolone is such a compound; unfortunately, athletes began to abuse such androgens by taking massive doses in an effort to build greater body muscle mass and strength. This strategy remains unproven, however, and the side effects of virilization in females, breast development in males, liver toxicity, inappropriately aggressive behavior, and even death forced the United States Food and Drug Administration to make such compounds a scheduled substance along with narcotics. Some athletes have gone to veterinary drugs in an effort to circumvent such regulation.

Adrenal steroids are absolutely essential for body homeostasis. The glucose control and significant anti-inflammatory actions of hydrocortisone are so vital that death ensues within 72 hours if the body is deprived of this substance. Used in higher than normal doses to treat inflammatory diseases such as lupus, rheumatoid arthritis, and asthma, the complications are myriad and include gastrointestinal ulceration, chemical diabetes, and over suppression of the immune response resulting in uncontrolled infection, sepsis,, multiple organ failure, and death. Aldosterone is also an essential adrenal hormone that causes the kidneys to retain **sodium** and **water**, exactly what is required in states of **dehydration** or blood loss. However, in a patient with high blood **pressure** (hypertension) having the ability to block aldosterone is often advantageous. Spirinolactone was synthesized to act in just such a manner and is extremely successful.

Obviously, the **chemistry** and **biochemistry** of steroids is extremely complex, but as chemists unravel the mysteries of molecules and their receptor site activities, this huge but magnificent puzzle begins to be solved.

See also Chemical contraceptives

Stock, Alfred (1876-1946)

German chemist

Alfred Stock was an experimentalist who made significant contributions to **chemistry** and designed several important **chemical instruments**. He worked on the creation of new **boron** and **silicon** compounds and the development of the chemical high-vacuum apparatus, which allowed him to work with volatile materials. The latter part of his life was particularly devoted to the study of **mercury** and mercury poisoning and, in particular, developing precautionary guidelines for other scientists to follow in order to avoid suffering from it. Stock contracted mercury poisoning while working with the substance in laboratories since his time in school; he was afflicted with the disease the rest of his life.

Stock was born in Danzig, West Prussia (now Gdansk, Poland) on July 16, 1876. His father was a banking executive. As a schoolboy, Stock developed an early interest in science, and he earned scholarships that allowed him to pursue a degree in chemistry at the University of Berlin in 1894. He chose to work at the chemical institute at the university, directed by **Emil Fischer**, but had to wait a year for space in the lab, which was overcrowded. He finally began his doctoral research in 1895 under the auspices of organic chemist Oscar Piloty. During his summer breaks from school, Stock worked in the private laboratory of the Dutch physical chemist Jacobus van't Hoff. It was there that Stock performed his first significant research in the areas of **magnesium** and oceanic **salt** deposits. After graduating magna cum laude in 1899, he spent a year in Paris assisting the chemist **Henri Moissan** at the Ecole superieure de Pharmacie, with support of the Prussian Ministry of Culture. At Moissan's lab he first investigated compounds of silicon and boron, which were to occupy him throughout his career.

Stock began his professional career in 1900 by working for nine years as a lab assistant with Fischer at the University of Berlin. There he investigated the preparation and characterization of such elements as **phosphorus**, **arsenic**, and **antimony** (a brittle, white metallic element). One result of Stock's investigation was that he could clearly explain their reactions with **hydrogen**, **sulfur** and **nitrogen**. He also identified an unstable yellow form of antimony, and two new compounds of phosphorus: a polymeric hydride, which is a compound including hydrogen and another element or group, and a nitride, which is a compound including nitrogen and one other element. Stock's research clarified misconceptions in scientific literature and established the existence of three of today's four well-established phosphorus sulfides, which are organic compounds of phosphorous and sulfur.

Unfortunately, it was also during these early years in Berlin that symptoms associated with Stock's mercury poisoning would begin surfacing. Headaches, dizziness and upper respiratory infections started plaguing Stock while he was pioneering his work with a device known as the vapor-tension thermometer. His success with the apparatus became well known throughout Germany and he later developed it into his tension-thermometer. Many years later, the work with the tension-thermometer was traced as the first of many sources of mercury poisoning to which Stock was exposed during the course of his life.

In July 1909, Stock was named full professor and director of the new Inorganic Chemistry Institute at Breslau. It was here that he surpassed previous chemists' successes with his imaginative work with hydrocarbons, inorganic **carbon** compounds and the development of a high-vacuum apparatus that allowed Stock to work with volatile and gaseous materials. The apparatus was later referred to as the Stock high-vacuum pump. He also envisioned at this time the possibility of developing the equivalent of organic chemistry's carbon-based system around boron, an element whose unanticipated potential he was just beginning to discover. His work with borohydrides, however, was interrupted with the outbreak of World War I. Stock was then charged with studying carbon subsulfide, an irritant, to determine its effectiveness as a war gas. The gas was never used, however, due to problems with polymerization, which is a chemical reaction in which molecules combine to form larger molecules that contain repeating structural units.

Stock left Breslau in April 1916 to continue his research and take charge of **Richard Willstätter**'s laboratory at the Kaiser Wilhelm Institute near Berlin. It was not long, however, before the military moved in and took over the institute. Since his still undiagnosed physical problems (including, by that time, an acute loss of hearing) kept him from serving in the military, he and his staff moved their equipment to the University of Berlin so that they could continue their work. When the war ended, he returned to the Kaiser Wilhelm Institute and continued to study silicon and boron hydrides. His work yielded a number of halogen and alkyl derivatives which, in turn, led to the discovery of new compounds such as silyl amines and silicones. In the process, Stock developed a chemistry based on silicon; this was similar to his work with boron at Breslau. At the suggestion of Hans Goldschmidt, Stock also collaborated on the production of metallic **beryllium**. This substance had become a worthwhile element to pursue because it is a metal which had possible applications in industry, so a beryllium study group was formed. By 1940, their new technique for making the material yielded enough beryllium to significantly reduce the market value per kilogram, thus increasing the element's cost-effectiveness in scientific experiments.

Stock's unexplained medical problems kept growing worse. Besides headaches, vertigo, respiratory infections, and deafness, he now also suffered frequent numbness. None of these symptoms was alleviated by medical treatment. In 1923, he suffered virtually total hearing and memory loss, and he almost didn't make it through the winter of 1924. At that time, many scientists in addition to Stock were unknowingly being exposed to mercury poisoning. It wasn't until after he saw similar symptoms in a colleague that Stock finally realized the **volatility** of this odorless substance. He began researching mercury poisoning, often experimenting on himself. As a result of his investigations, Stock published several articles outlining the dangers of mercury and offered up numerous precautionary guidelines for working with the substance.

It was a difficult decision, but the opportunity to establish a new mercury-free laboratory convinced Stock to leave the Kaiser Wilhelm Institute in 1926 to become the director of the Chemical Institute at Karlsruhe. The next ten years of his life were devoted exclusively to the study of mercury poisoning and borohydrides. His concepts and working models of laboratory rooms equipped with extensive safety precautions were sought after by scientists from around the world. Further experiments proved that inhaled mercury vapor was much more dangerous than ingested mercury because the vapor, entering through the nose, moved more quickly into the pituitary gland, where it wreaked havoc on the body. Stock also pioneered a teaching method during his tenure at Karlsruhe which used reflected light to project chemical objects on a large screen. Stock worked with his lecture assistant, Hans Ramser, and with Carl Zeiss-Jena to create this apparatus, which was called an epidiascope.

Stock married Clara Venzky in August of 1906 and later had two daughters. He was the president of the Verein Deutscher Chemiker (Association of German Chemists) in Paris in 1927 and later the Deutsche Chemische Gesellschaft (German Chemical Society) from 1936 to 1938. He was a guest professor at Cornell University for several months in 1932 under the George Fisher Baker Nonresident Lectureship in Chemistry. The last ten years of Stock's life were nearly unbearable, both physically and professionally. His mercury poisoning became debilitating and interfered with his work. His political differences with the Nazi government were increasing. In 1936, at the age of 60, he asked for his retirement. He returned with his family to Berlin where he continued to trace and validate the chemical path of mercury poisoning. By 1940, movement became difficult as he developed hardening of the muscles. Stock relinquished his laboratories in 1943 because they were needed for the war effort, and he and his wife moved to Bad Warmbrunn in Silesia to live with his brother-in-law. As the Russians were approaching in 1945, Stock and his wife again moved and sought shelter with an old friend, Ernst Kuss, in Dessau. The Stocks finally found refuge in a barracks in Aken, a small city on the Elbe. Stock died in the early morning of August 12, 1946.

Stoichiometric laws

Stoichiometry is the study of quantitative relationships between substances involved in chemical changes. That the Universe is a constantly changing place is readily apparent from any walk down the street. People and cars pass by, the sky is filled with moving clouds, the businesses and storefronts keep changing, and everything seems to be in motion. It is, in some ways, surprising that ancient philosophers often argued for the permanence of the Universe and unchanging nature. Some even argued against motion itself, saying it was but an illusion. In many ways, these thought experiments inhibited the development of science. But in understanding the constant nature of the Universe, they were right; just not in the way that they thought.

Underlying **chemistry** are a number of principles or laws that provide the basis for the relationships between the reactants and products in a chemical reaction. This relationship, called the stoichiometry of a reaction, can be used to predict how much reactant is needed to create a certain amount of product or to predict how much of the product will be formed from a certain amount of reactant. It also defines the proportions in which reactants will combine and the proportion in which products will be formed. It should be noted, emphatically, that while the expression of these stoichiometric laws is a human trait, they are not of human invention.

The "Law of Conservation of **Matter**" was first expounded by the Russian chemist, **Mikhail Lomonosov**, who investigated the reactions of compounds with air in closed containers. The weight, before and after the reaction, remained constant despite the observable change. Matter was neither being created nor destroyed, merely converted from one form to another. Unfortunately, working in Russia in the 1750s, his results did not get the attention nor he the recognition that he most likely deserved. The great French chemist, **Antoine Lavoisier**, is most often credited with the formulation of this law. Prior to the development of **atomic theory**, it was simply the absolute statement that there is no measureable change in the total **mass** of the chemical compounds upon reaction.

Methods of verifying this law required the careful observation of chemical reactions in sealed vessels. For example, the **calcination** of **metals** in sealed containers demonstrated that no change in mass occurred. Further, opening the vessels demonstrated the presence of a vacuum within and increase in weight only upon exposure to air. That is, when air is not the limiting reagent within the vessel, the reaction can proceed to a much greater extent. Arguably, this simple experiment - which Lavoisier repeated and refined - provides much of the basis for an understanding of stoichiometric relationships and laws, from the concept of stoichiometric quantities and limiting reagents to the conservation of matter.

The Law of Conservation of Mass can be seen using the simple decomposition of **water**, according to the equation:

$$2H_2O \rightarrow 2H_2 + O_2$$

The data for this experiment demonstrate that 11.2 g of **hydrogen** and 88.8 g of **oxygen** are produced by 100 g of water, regardless of the water's source. Furthermore, these elements can be converted back into 100 g of water if the experiment is done carefully (particularly as hydrogen and oxygen form an explosive mixture). Similarly, the burning of a candle, the rusting of a metal, or the act of living all obey the conservation law and this can be demonstrated, provided the reactions are carried out in a closed system to prevent the loss of gaseous products.

With the advent of atomic theory, it was soon realized that it is not just matter or mass that is conserved but atoms as well. The more modern formulation of the law states that atoms can be neither created nor destroyed during a chemical reaction, only converted or rearranged into new substances. But this was pre-dated by the Laws of Constant Composition and Definite Proportion. The first says that the elemental composition of a compound is the same, regardless of the source of the compound. That is, water is water and is always "H_2O", whether it is obtained from the snow pack of a Himalayan mountain or the bathroom tap. The second law is an alternative way of looking at the same thing: "When elements combine to form a **chemical compound**, they do so in a definite proportion by mass." That is, all molecules of **calcium** carbonate are the same, consisting of a calcium **ion** and carbonate ion, and the carbonate ion is further composed of a **carbon atom** and three oxygen atoms. The composition is invariant; the mass proportions are fixed.

The realization of these laws facilitated chemical analysis. Instead of looking for a variety of different compounds in, for example, each sample of **calcium carbonate**, chemists now knew the composition that was to be expected. This realization also posed the question of why this should be so. Atomic theory explains both propositions if it is assumed that atoms are indivisible and form complexes in fixed ratios. That is, one oxygen atom is capable of picking up two hydrogen atoms in forming water. This, of course, **lead** to questions on the nature of molecules and the interaction of atoms which, in turn, lead to the question of **bonding**. As with most of science, the answer to one question is invariably the starting point for the next one.

One other subtle point that the Law of Constant Composition makes but that is often lost on the general public is that chemicals know no providence. Given a **molecule** of, for example, **acetic acid**, it is impossible to determine whether that specific molecule was made via the Wacker Process in a chemical plant, as the by-product of a biochemical pathway, or through the gradual **oxidation** of **ethanol** in wine into vinegar (acetic acid by a different name). The distinction between natural and artificial chemical is a false one, used by advertisers to market a product and usually at a higher cost!

The Law of Multiple Proportions states that "in different compounds containing the same elements, the masses of one of the elements compared with the fixed mass of one of the other elements will always be in a ratio of small whole numbers." This principle was first expounded by **John Dalton** and the subsequent verification by experiment provided the strongest support, at the time, for his atomic theory.

In order to understand the Law of Multiple Proportions, consider the compounds **carbon monoxide** and **carbon dioxide**. If the first is composed of 12 g of carbon and 16 g of oxygen, then for the second, if 12 g of carbon are present, there must be some small number multiple of 16 g of oxygen. In this case, the small number is "2" and leads to the chemical formula of CO_2. With our advanced understanding of chemistry and atomic theory, this result may seem a little redundant but when it was first proposed, an understanding of chemical composition and molecular formulae did not exist. Rather, this law provided a method for determining the chemical formula from observation.

As an example, consider the problem of three unknown compounds. Each is composed of **chlorine** and oxygen. Chemical analysis tells us that if we have 17.8 g of chlorine, then the first has 8 g of oxygen, the second 16 g of oxygen, and the third 24 g of oxygen. Clearly, then, the simplest whole number ratio for these three compounds is 1:2:3, implying the presence of one oxygen atom in the first, two in the second, and three in the third, or at least some multiple of these numbers as the analysis will only give us the **empirical formula**. Furthermore, if we know that chlorine has a molar mass of 35.5 g (give or take), we can easily come to the conclusion that these three compounds are: ClO, ClO_2, and ClO_3. By studying multiple proportions for a variety of compounds, it is possible to actually come up with a consistent set of atomic weights for all of the elements involved.

The Stoichiometric Laws were fundamental observations for chemical reactions. They allowed for the systematic classification of reactions and the development of the atom as the building block of molecules. They transformed chemistry from a collection of observations to a consistent and theoretically sound discipline. They provided the basis for understanding the atomic nature of chemistry and, in many ways, are the pillars upon which modern chemistry has been built.

STOICHIOMETRY

That chemical elements are made of atoms (the smallest piece of **matter** that still retained the chemical features of the element) was an idea first put forth by Greek philosophers. Revitalization of the **atomic theory** was, in part, credited to **John Dalton**, an English school tutor and practicing chemist, during the early 1800s. The reason that his work on chemical atomism, if not physical atomism, was so widely accepted was that it worked to explain the relationships that had been noted and tested by the chemists of the latter part of the 1700s. In particular, it gave a rationale for the work of an obscure German chemist, Jeremias Richter.

Richter ''invented'' stoichiometry, which can be described as the systematic study of the quantitative changes between substances undergoing a chemical transformation or reaction. Just as Johannes Kepler had provided a mathematical harmony to the stars, Richter was determined to provide a mathematical basis for **chemistry**. Without an understanding of the atomic basis of matter, it was a task that was doomed for failure but Richter did generate considerable insight into the definite proportions of chemical reagents for reactions. His analysis led him to the conclusion that in, for example, a double decomposition of a **salt**, the proportions of the products was provided by the proportions of the reagents. That is, he recognized that a chemical reaction such as the combination of **sodium sulfate** with **barium** chloride resulted in precipitated barium sulfate and soluble **sodium chloride**, and that the proportions or ratios of the constituent groups was fixed by the starting materials.

Richter's work laid the foundation for other chemists to establish the laws of proportionality. For example, Ernst Gottfried Fisher published a table demonstrating the amount of base required to neutralize 1,000 parts of **sulfuric acid**. Further, he expanded the table to then show much of each acid was required to neutralize the corresponding bases. From this table, it is easy to see that 672 parts of **ammonia** require 712 parts of muriatic acid or 979 parts of phosphoric to be neutralized. This was the start of the systematization of **analytical chemistry** and the beginning of rationalization of the atomic **mass** of the elements. The implication is that each ''**atom**'' of muriatic acid weighs more than each atom of ammonia but less than each atom of **phosphoric acid**. Of course, it was not until much later that Dalton proposed an Atomic Theory of Matter.

The Atomic Theory provided a method for determining and systematizing the changes between reactants and products. Chemical equations began to emerge. This required symbols for the elements but that was not a critical issue. Rather, simply being able to begin to express reactions as mathematical entities with a view to balancing the various reagents impelled chemistry forward. Stoichiometry enabled chemists to begin to make sense of myriad equations and compounds.

Consider, for example, cooking food. Within any cookbook, there are invariably recipes which instructs the used to ''add two cups of flour'' or ''mix in three tablespoons of vanilla.'' These are empirically derived instructions that originate from the trial and error process that is consider a good recipe. But once a recipe has been established, the ratios are fixed and the product is defined. The question remains, however, why is it two cups of flour and not three cups? There is no underlying atomic theory of cooking equivalent to the atomic theory of matter, although there is an understanding of the chemistry that occurs during cooking. Stoichiometry provides this understanding by showing that each and every time, specific ratios are required for proper reaction.

This, of course, is fundamentally important to the practice of chemistry from the laboratory bench top to the multiton producing chemical manufacturing sites. If a **solution** needs to be neutralized, excess acid or base is not only wasteful of the **chemical compound** but can actually change the results of the reaction. Too much acid in neutralizing a basic solution can result in acid catalyzed cleavage of the product, and with too little acid the solution remains alkaline and unusable. Understanding stoichiometry allows chemical engineers to carefully **balance** all of the demands of a chemical process.

Most chemists encounter stoichiometry as high school or undergraduate students studying chemistry. It is presented in the form of balancing equations. ''Why do equations balance? Because they are stoichiometric.'' is a fairly common question and response. Ultimately, equations balance because of the law of conservation of matter. Atoms can be neither created nor destroyed, so if they occur on one side of a reaction, they must be found, somewhere, on the other side. Learning how to properly balance equations is one of the most important skills that chemists acquire.

Furthermore, equations are scale invariant. This is important. Consider the following reaction:

$$2H_2 + O_2 \rightarrow 2H_2O$$

Taken literally, this reaction says that two molecules of **hydrogen** will react with one **molecule** of **oxygen** and yield two molecules of **water**. But it also tells us that if we react four molecules of hydrogen with two molecules of oxygen, we will end up with four molecules of water. All that we have done is multiple the coefficients of the equation by two. And we can actually use any number. One of the most convenient is **Avogadro's number**, 6.022×10^{23} molecules per **mole**. This number has one unusual property. It is the number of atoms or molecules that make up the gram **molecular weight**. That is, 12.011 grams of **carbon** contains 6.022×10^{23} atoms of carbon. Similarly, 15.9994 grams of oxygen is composed of 6.022×10^{23} atoms of oxygen but, because each molecule of oxygen is composed of two atoms, it is only 3.011×10^{23} molecules of oxygen. This is an important distinction to remember in dealing with stoichiometry. There is an equivalence of the atomic and molecular mass as determined by the formula with the mole quantity of a substance. Thus, the above **combustion** of hydrogen requires two moles of the hydrogen molecule, each of which weighs 2.0159 grams per mole. Thus, the reaction is the combination of 4.0318 grams of hydrogen with 31.9988 grams of oxygen, which produces exactly 36.0306 grams of water.

Further, that the ideal ratio for hydrogen to oxygen is almost 1 to 8. But what happens to a chemical reaction if the reagents are not in an ideal ratio? For example, if there is more than 31.9988 grams of oxygen present? The answer is that the remainder does not react. It remains as oxygen, usable in some future reaction but not at the present. This is an important concept as it leads to the idea of a limiting reagent. One of the more important considerations in any industrial process is to use up enough chemicals to cause the reaction to proceed with out using up more than necessary. Any more than necessary wastes money and effort. The limiting reagent in any reaction is the species that will run out first. It is the one that is in short supply.

It is also the species that any discussion of yield is based on. The percentage yield is the number of molecules or moles of the limiting reagent that are turned into product(s). In our above reaction, if we have exactly two moles of hydrogen and it is the limiting reagent, then producing two moles of water mean that the reaction had a 100% yield, regardless of how many moles of oxygen are present. The limiting reagent always limits the yield of product and defines the yield for the reaction. This is also of importance to industry as the yield of a reaction can vary under different circumstances and any industrial process should be optimized to give the best yield of the product, thus minimizing waste.

The concept of stoichiometry emerged out of the analytical techniques of the chemists of the 18th century. It provided the necessary mathematical relations for the development of the atomic theory and ultimately provided the basis for the development of the balance chemical equation to describe a reaction. Stoichiometry provides the chemist with an understanding of the quantities of reactants required for a reaction to occur and the amount of product that will be formed. But, in a subtle way, it has also provided us with the element masses of the **periodic table** and is one of the fundamental practical tools of chemistry.

Strain energy

The strain **energy** is the **potential energy** stored in a body when that body is *elastically* deformed. It is equal to the amount of work required to produce this deformation. A body deforms elastically until the applied stress reaches some maximum value, beyond which the body exhibits permanent d eformation. The body need not break at this maximum stress; it will simply not return to its original dimensions if stressed any more. As long as the elastic limit is not exceeded, the strain is directly proportional to the stress. Familiar examples of bodies that behave elastically include rubber bands, diving boards, and springs.

If a uniform rod is stretched slowly and elastically, energy will be stored in the rod as elastic energy. The elastic energy stored per unit **volume**, or strain energy **density**, is equal to one-half the stress raised to the second power divided, in general, by the Young's modulus. The Young's modulus is defined as the longitudinal stress divided by the longitudinal strain in the elastic region of deformation; strain is defined as the relative change in dimensions or shape in a body as the result of an applied stress, and is a dimensionless quantity; and stress is the magnitude of the applied force per unit area, usually measured in units of pounds per square inch or pascals.

See also Ductility

Strassmann, Fritz (1902-1980)

German chemist

Fritz Strassmann's experiments with the bombardment of **neutrons** on **uranium** atoms resulted in the discovery of **nuclear fission**, which has been employed in making the world's first atomic bomb and nuclear **energy**. He also developed a widely-used geological dating method employing radiation techniques, and served as a well-respected teacher of nuclear **chemistry**.

Friedrich Wilhelm Strassmann, the youngest of nine children, was born on February 22, 1902, in Boppard, Germany, to Richard Strassmann, a court clerk, and Julie Bernsmann. Strassmann attended primary schools in Cologne and Düsseldorf and secondary school at the Düsseldorf Oberrealschule. During this time, he became interested in chemistry. However, the death of his father in 1920 and the desperate economic conditions of post-World War I Germany made it impossible to enroll at a university. Instead, he entered the Technical Institute at Hanover, where he supported himself as a private tutor.

In 1924 Strassmann was awarded his diplomate (comparable to a master's degree) in chemistry at Hanover. He then stayed on to pursue a doctoral program in **physical chemistry** under Hermann Braune. Strassmann completed that program in 1929 and was awarded his Ph.D. for a thesis on the **solubility** of **iodine** in **carbon dioxide**. In the *Dictionary of Scientific Biography* it is noted that Strassmann's choice of physical chemistry as a major was largely due to the increased likelihood of finding employment in that field during a tight labor market.

Shortly after receiving his Ph.D., Strassmann was offered a scholarship to the Kaiser Wilhelm Institute for Chemistry in Berlin by its director, **Otto Hahn**. That scholarship was renewed twice and then, in 1932, Strassmann was invited to continue his work at the institute with Hahn, first at no salary and later at a minimal wage.

The 1930s were a difficult time for Strassmann, partly because of his restrictive financial situation and partly because of his views regarding the policies of Adolf Hitler's new regime. Although he was not in any immediate personal or professional danger, Strassmann decided to refuse all offers of employment that would have required his joining the Nazi party. He continued his work, therefore, with Hahn and later, **Lise Meitner** at the Kaiser Wilhelm Institute.

The subject on which Strassmann, Hahn, and Meitner were working during the mid–1930s was the bombardment of uranium by neutrons. The great Italian physicist **Enrico Fermi** had found that the bombardment of an element by neutrons often results in the formation of a new element one place higher in the **periodic table**. Fermi had successfully applied this technique to a number of elements during the early 1930s.

The one example that most intrigued Fermi—and other scientists familiar with his work—was the bombardment of uranium. Should earlier patterns hold, they realized, such a reaction would result in the formation of an element one place higher than uranium (number 92) in the periodic table, that is, element 93. Since no such element was known to exist naturally, a successful experiment of this design would result in the first artificially produced element.

Unfortunately, the uranium bombardment yielded results that were not easily interpreted. Original indications seemed to be that the products of the reaction were largely elements close to uranium in the periodic table, **thorium**, **radium**, and **actinium**, for example. Such results would not have been surprising since all **nuclear reactions** known up to that time involved changes in **atomic number** between reactants and products of no more than one or two places.

More detailed studies by Strassmann yielded troubling results, however. His chemical analysis consistently showed the presence of **barium**, whose atomic number is 36 less than that of uranium. The formation of barium seemed so unlikely to Strassmann and Hahn that their reports of their work were hedged with doubts and qualifications. In fact, it was not until Lise Meitner, then a refugee in Sweden, received word of the Hahn-Strassmann experiment that the puzzle was solved. With a fearlessness that Strassmann and Hahn had lacked, Meitner stated forthrightly that barium *had* been formed and that this change had occurred as a result of the splitting apart of the uranium **nucleus**.

The significance of this discovery can hardly be overestimated. Within a decade, it had been put to use in the development of the world's first atomic bombs and, shortly thereafter, was being touted as one of humankind's great new sources of power for peacetime applications. For his part in this historic event, Strassmann was awarded the 1966 Enrico Fermi Prize of the U.S. Atomic Energy Commission.

A second line of research in which Strassmann's analytical skills resulted in an important discovery was that of geological dating. Beginning in 1934, he and Ernst Walling studied the radioactive decay of **rubidium**, a process that results in the formation of **strontium**. Strassmann's careful analysis of the characteristics of this process eventually led to its use in the dating of geological strata, a process that has become a standard tool in geology.

Strassmann remained at the Kaiser Wilhelm Institute during the war, first in Berlin and later in Tailfingen, where the institute was moved to avoid destruction by bombing. After the war, he was appointed professor of inorganic and **nuclear chemistry** at the University of Mainz. At Mainz he became involved in complex negotiations over the construction of two new institutes. One was a new physical facility for the **Max Planck** Institute of Chemistry, successor to the former Kaiser Wilhelm Institute in Berlin. The second was a new Institute for Nuclear Chemistry, designed as part of the newly reestablished University of Mainz. Although beset by all manner of political, technical, and economic problems, Strassmann made major contributions to the establishment of both institutions. In addition, he saw through to completion the construction of the TRIGA Mark II nuclear reactor in Mainz.

Strassmann died in Mainz on April 22, 1980. He was remembered not only for his discoveries in nuclear fission and geological dating and for his skills as an administrator of scientific research, but also for his qualities as a teacher. His achievements were recognized in 1969 when the Society of German Chemists established the Fritz Strassmann award, to be given to an outstanding young nuclear chemist each year. Strassmann was married twice, the first time on July 20, 1937, to Maria Heckter, a former pupil. The couple had one son, Martin. Three years after Maria's death in 1956, Strassmann was married a second time, to Irmgard Hartmann, a good friend of his first wife.

STRONTIUM

Strontium is the fourth element in Group 2 of the **periodic table**, a group of elements known as the **alkaline** earth **metals**. It has an **atomic number** of 38, an atomic **mass** of 87.62, and a chemical symbol of Sr.

Properties

Strontium is a silvery white metal that combines with **oxygen** of the air to form a thin film of strontium **oxide** (SrO). The film gives the metal a yellowish hue and protects it from further **oxidation**. Strontium has a melting point of 1,395°F (757°C), a **boiling point** of 2,491°F (1,366°C), and a **density** of 2.6 grams per cubic centimeter.

Strontium is a very active element that is normally kept under **kerosene**, mineral oil, or some other organic liquid to keep it from reacting with oxygen in the air. In a finely divided form, the metal may catch fire spontaneously and burn vigorously. Strontium is active enough to combine even with **hydrogen** and **nitrogen** when heated, behaviors unusual for most metals. Strontium also reacts with cold **water** to produce hydrogen gas as one product: $Sr\ 2H_2O \rightarrow Sr(OH)_2 + H_2$

Occurrence and Extraction

Strontium is relatively abundant in the Earth's crust, ranking about 15th among the elements found there. Its most common minerals are celestine (primarily strontium **sulfate**; $SrSO_4$) and strontianite (primarily strontium carbonate; $SrCO_3$). The most important sources of strontium worldwide are Mexico, Spain, Turkey, and Iran. A small amount of the element is also obtained from mines in California and Texas.

Strontium metal is still extracted by a method originally developed by the English chemist and physicist **Humphry Davy**, the **electrolysis** of molten stronium chloride ($SrCl_2$): $SrCl_2$ —electric current→ $Sr + Cl_2$

Discovery and Naming

Credit for the discovery of strontium usually goes to the Irish physician Adair Crawford (1748-1795). Crawford was interested in the study of minerals and other chemicals and spent some time analyzing the mineral known as baryte, a primary source of the element **barium**. Crawford was surprised to find that a small portion of baryte was always left over after all other known elements had been identified. He decided that the remaining material must be a new element. He suggested naming the element *strontia* after a region in Scotland from which his sample of baryte had come. Nearly two decades later, Davy showed that strontia was a compound of a metal and oxygen, a metal to which the name strontium was then given.

Uses

Strontium and its compounds have relatively few commercial uses. Some compounds are added to **glass** and **ceramics** to give them a beautiful red **color**. Compounds of strontium are also used to provide the red colors seen in a fireworks display.

Health Issues

One radioactive **isotope** of strontium, strontium-90, is of some interest in the field of health. The isotope is produced in relatively large amounts during the explosion of nuclear weapons. When such an explosion takes place in the atmosphere, the isotope eventually falls back to earth and coats grass, leaves, and other plant material. Cattle, sheep and other domestic animals then eat that plant material and ingest the strontium-90. It becomes part of the milk that may be passed on to humans as a food.

This is significant because the human body treats strontium in much the same way it treats **calcium**, the element above it in the periodic table. The body uses the strontium to build bones and teeth. But strontium-90 is radioactive, giving off radiation that can cause harm to the body, most seriously, bone cancer.

The health risks posed by strontium-90 became an important issues during the 1950s and 1960s when the United States, the then-Soviet Union, China, and a few other nations tested nuclear weapons in the atmosphere. Concerns about the potential health risks of strontium-90 were one of the major factors that convinced those nations to discontinued atmospheric testing of nuclear weapons. Such tests are now almost entirely conducted underground where strontium-90 can not be released to the atmosphere.

STRUCTURAL FORMULAS

The term formula in **chemistry** can have a variety of meanings. The **empirical formula** of a compound is that which is obtained through laboratory research. By analyzing the chemical composition of a compound in the laboratory and knowing the atomic weights of the elements in the compound, one can calculate the simplest, or empirical, formula for that compound. For example, the analysis of ethane shows that the ratio of **carbon** to **hydrogen** atoms in the **molecule** is 1:3. Therefore, the empirical formula for ethane, is CH_3.

If one also knows the **molecular weight** of a compound, the **molecular formula** for that compound can be determined. Knowing the molecular weight of ethane to be 30, for example, allows one to calculate the molecular formula for the compound to be C_2H_6.

Finally, the **structural formula** of a compound shows not only the kind and actual number of atoms, but also their arrangement in space. The structural formula for ethane, for example, is H H I I H - C - C - H I I H H.

Structural formulas are exceedingly important to chemists, especially organic chemists, since they make it possible to design schemes of synthesis and to predict the properties of compounds. The writing of structural formulas was not possible until at least some chemists were convinced that chemical formulas were more than a convenience in representing compounds, and that they could actually show the structure of atoms in a molecule. That situation developed in the late 1850s and early 1860s. The term chemical structure was first coined in 1861 by the Russian chemist Alexander Butlerov (1828-1892). The first true structural formulas were written by **Friedrich Kekulé** and **Archibald Couper**.

In 1857, Kekulé suggested that every carbon **atom** was capable of combining with (''**bonding**'' to) four other atoms. The number four was specific and definite. A carbon atom never joined with one, two, three, five, or some other number of atoms. Further, Kekulé said, in many instances, carbon atoms joined with each other in long **chains**. Kekulé represented the molecules formed in this way with small circles and ellipses that some colleagues jokingly referred to as ''Kekulé's sausages.'' Couper's suggestion of using dotted lines and short dashed lines for the bonds between atoms was a much better system and was soon widely adopted. Other chemists struggled to find ways of representing the structure of molecules in three dimensions. Kekulé and Couper had shown how to depict the arrangement of atoms in a molecule in two dimensions, on a piece of **paper**. But it was obvious to others that molecules really exist in three dimensional space and had to be represented as such.

Among the first efforts to move in this direction was that of **August Wilhelm von Hofmann**. In 1865, Hofmann made models of molecules using sticks and croquet balls. However, those models showed only planar arrangements of atoms. The first real breakthrough in this field came with the work of **Joseph Le Bel** and **Jacobus van't Hoff**. Working independently, the two chemists developed the concept of a tetrahedral carbon atom. They pointed out that the four bonds in a carbon atom were directed in three dimensions, toward the four corners of a tetrahedron. A major success of this model was its ability to explain an unknown phenomenon of optical isomers that had been known since the time of **Louis Pasteur**. With the work of Le Bel and van't Hoff, the theory of structural formulas essentially reached the form in which it is used today.

STRUCTURAL PROTEINS

Structural **proteins** fulfill a variety of roles within and outside of cells. Structural proteins within the cell give it shape, and allow it to withstand stress from the environment or from its own motions. Structural proteins are also used during mitosis and cytokinesis, the processes by which a cell reproduces itself. Structural proteins in bone and cartilage provide strength and support for muscles to work against, and structural proteins in skin, hair, and nails provide a tough outer covering to the body.

Within cells, several proteins help create and maintain the cell's dynamic structure. Together, these proteins form the cytoskeleton. Individual molecules of actin link together to form filaments. These actin filaments are often used to form cables or meshwork to counteract stresses. In the amoeba, actin filaments are rapidly disassembled and reassembled, allowing this single-celled organism to rapidly change its shape as it moves. Alpha and beta tubulin link together to form microtubules, the longest structural fibers within the cell. Microtubules are used to transport cell organelles, and during cell division, pull chromosomes apart into the newly formed daughter cells. Microtubules also form the structural scaffolding for cilia and the tails of sperm.

Collagen is secreted by cells of cartilage and bone to form a dense, resilient matrix. Collagen is formed from three helices of protein wound around one another. The triple helix is extensively cross-linked with others, allowing it to be both flexible and tough. The cells of skin, hair, and nails are packed with keratin. Keratin is naturally quite flexible, but can be strengthened and stiffened by increasing the number of disulfide cross-bridges that link different **chains**. Elastin is even more flexible due to its very loose network of cross-links, and the weak **hydrogen** bonds holding different strands together. Elastin gives arteries the ability to expand in response to stretch, and then contract again, an action aided by surrounding muscle tissue.

STRUTT, JOHN WILLIAM (1842-1919)

English physicist

In 1873 John William Strutt's father, the second Baron Rayleigh, died and Strutt succeeded to that title. He is, therefore, almost universally referred to in the scientific literature as Lord Rayleigh. While the majority of his work dealt with sound and optics, Rayleigh may be most familiar to the layperson as the discover of the rare gas **argon**. For this accomplishment he was awarded the 1904 Nobel Prize in physics. Rayleigh served for a period of five years as director of the Cavendish Laboratory at Cambridge University. With that exception, he spent nearly all of his adult life at his home in Terling Place where he constructed a well-equipped scientific laboratory. There he carried out experiments on a remarkable variety of subjects that led to the publication of some 450 papers.

John William Strutt was born at Langford Grove, near Maldon, in Essex, on November 12, 1842. He was the eldest son of John James Strutt, second Baron Rayleigh, and the former Clara Elizabeth Vicars. Strutt's health as a child was not very good, and he was unable to remain at school for very long. He attended Eton and Harrow for about one term each and spent three years at a private school in Wimbledon. Finally, in 1857, his education was entrusted to a private tutor, the Reverend George Townsend Warner, with whom he stayed for four years.

In 1861 Strutt entered Trinity College, Cambridge, where he studied mathematics under the famous teacher E. J. Routh. During his four years at Trinity, Strutt went from being a student of only adequate skills to one who captured major prizes at graduation. One examiner is reported to have said that Strutt's answers were better than those found in books.

Following graduation, Strutt was elected a fellow of Trinity College, a position he held until 1871. In 1868 he took off on the extended "Grand Tour" vacation traditional among upper class Englishmen, except that he chose to visit the post-Civil War United States rather than the continent of Europe. In 1871, at the conclusion of his tenure at Trinity, Strutt was married to Evelyn Balfour, sister of Arthur James Balfour, later to be prime minister of Great Britain in 1902. The Strutts had three sons, Robert John, Arthur Charles, and Julian.

Within a few months of his marriage, Strutt became seriously ill with a bout of rheumatic fever. As his health returned, he decided to make a recuperative visit to Egypt and Greece with his young bride. It was on a trip down the Nile during this vacation that he began the scientific work that was to occupy his attention for most of the rest of his life, a massive work on *The Theory of Sound.*

Shortly after the Strutts returned to England in the spring of 1873, his father died and Strutt succeeded to the hereditary title of Baron Rayleigh. He also took up residence in the family mansion at Terling Place, Witham, where he was to live for most of the rest of his life. He soon constructed a modest, but well-equipped, laboratory in which he was to carry out experiments for the next forty years. At first he divided his time between the laboratory and the many chores associated with the maintenance of the Rayleigh estate. Gradually he spent less time on the latter, and after 1876 he left the management of his properties to his younger brother Edward.

Rayleigh always seemed to be a man of unlimited interests. The two fields to which he devoted the greatest amount of time, however, were sound and optics. In 1871, for example, he derived a formula expressing the relationship between the wavelength of light and the scattering of that light produced by small particles, a relationship now known as Rayleigh scattering. One of his first projects in his new Terling laboratory was a study of diffraction gratings and their use in spectroscopes (instruments with which scientists may study the electromagnetic spectrum). Rayleigh appeared to be totally satisfied with his life and work at Terling Place. Then, in 1879, **James Clerk Maxwell**, the first Cavendish Professor of Experimental Physics at Cambridge, died. The post was offered first to Sir **William Thomson**, who declined, and then to Lord Rayleigh. With considerable reluctance, he accepted the appointment with the understanding that he would remain for only a limited period of time. During his tenure at Cambridge, Rayleigh made a number of changes that placed the young Cavendish Laboratory on a firm footing and prepared it for the period of unmatched excellence that was to follow in succeeding decades. The most important experimental work carried out under his auspices was a reevaluation of three electrical standards, the volt, ampere, and ohm. This work was so carefully done that its results remained valid until relatively recently.

In 1884 Rayleigh resigned his post at Cambridge and returned to his work at Terling Place. Over the next decade he took on an even more diverse set of topics, including studies of electromagnetism, mechanics, capillarity, and **thermodynamics**. One of his major accomplishments during this period was the development of a law describing radiation from a black body, a law later known as the Rayleigh-Jeans law.

At the end of the 1880s, Rayleigh began work on the problem for which he is perhaps best known, his discovery of the inert gas argon. That work originated as a by-product of Rayleigh's interest in Prout's hypothesis. In 1815 the English chemist **William Prout** had argued that all elements are made of some combination of **hydrogen** atoms. An obvious test of this hypothesis is to find out if the atomic weights of the elements are exact multiples of the atomic weight of hydrogen.

By 1890 most scientists were convinced that Prout's hypothesis was not valid. Still, Rayleigh was interested in examining the problem one more time. In so doing, he made an unexpected discovery, namely that the atomic weight of **nitrogen** varied significantly depending on the source from which it was obtained. The clue that Rayleigh needed to solve this puzzle was a report that had been written by the English chemist **Henry Cavendish** in 1795. Cavendish had found that whenever he removed **oxygen** and nitrogen from a sample of air, there always remained a small bubble of some unknown gas.

To Rayleigh, Cavendish's results suggested a reason for his own discovery that the atomic weight of nitrogen depends on the source from which it comes. Nitrogen taken from air, he said, may include a small amount of the unknown gas that Cavendish had described, while nitrogen obtained from **ammonia** would not include that gas. Still, Rayleigh was not entirely sure how to resolve this issue. As a result, he wrote a short note to *Nature* in 1892 asking for ideas about how to solve the nitrogen puzzle.

The answer to that note came from **William Ramsay**, who was working on the same problem at about the same time. Eventually, the two scientists, working independently, obtained an answer to the problem of the mysterious gas. They discovered that Cavendish's "tiny bubble" was actually a previously unknown element, an inert gas to which they gave the name argon, from the Greek *argos*, for "inert." On January 31, 1895, Rayleigh and Ramsay published a joint paper in the *Philosophical Transactions of the Royal Society* announcing their discovery of argon. A decade later, in 1904, Rayleigh was given the Nobel Prize in physics and Ramsay the Nobel Prize in **chemistry** for this discovery.

Even after returning to Terling Place in 1884, Rayleigh remained active in a number of professional positions. He was appointed professor of natural philosophy at the Royal Institute in 1887 and gave more than a hundred popular lectures there over the next fifteen years. In 1885 he became secretary of the Royal Society, a post he held until his election as president of the organization in 1905. When he left that post in 1908, he became chancellor of Cambridge University, a position he held until his death at Terling Place on June 30, 1919.

In addition to the 1904 Nobel Prize, Rayleigh received the Royal Medal of the Royal Society in 1882, Italy's Bressa Prize in 1891, the Smithsonian Institution's Hodgkins Prize in 1895, Italy's Matteuci Medal in 1895, the Faraday Medal of the Chemical Society in 1895, the Copley Medal of the Royal Society in 1899, the Rumford Medal of the Royal Society in 1914, and the Elliott Cresson Medal of the Franklin Institute in 1914.

Strychnine

Strychnine is a poison with several medicinal effects. It is most commonly used as a poison for small mammals. Strychnine has the chemical formula $C_{21}H_{22}N_2O_2$ and it is the principal **alkaloid** found in the seeds from plants in the Strychnos family, including *Strychnos nux-vomica* (Quaker buttons or poison nut). Strychnine can account for 5% of the weight of the nuts of this tree, which is a native of Indian and Malaysia. When concentrated it can be crystalized out into colorless prisms with a melting point of 520°F (270°C). Strychnine has a stimulatory effect on all parts of the nervous system and in large doses it can produce convulsions and death. The senses are made more acute and blood pressure increases whilst the heart rate decreases, it can be effectively used for heart failure. Strychnine is used as a poison for rats and other vermin, it is particularly effective in this role due to the rapidity with which it is absorbed.

Strychnine is odorless but it has a very bitter **taste**. Medicinally it has been used to cure constipation, to stimulate the appetite, and in cases of heart failure.

Stubbe, JoAnne (1946-)

American chemist

JoAnne Stubbe's research has helped scientists understand the ways in which enzymes catalyze, or cause, chemical reactions. Her major research efforts have focused on the mechanism of nucleotide reductases, the enzymes involved in the biosynthesis of deoxyribonucleic acid (**DNA**), the **molecule** of heredity. Her work has led to the design and synthesis of nucleotide analogs—structural derivatives of nucleotides—that have potential antitumor, antivirus, and antiparasite activity. In 1986 the American Chemical Society honored Stubbe with the Pfizer Award which is given annually to scientists under forty for outstanding achievement in enzyme **chemistry**.

Stubbe was born June 11, 1946. She earned a B.S. in chemistry with high honors from the University of Pennsylvania in 1968 and her Ph.D. in chemistry at University of California at Berkeley under the direction of George Kenyon in 1971. Stubbe's first two publications in scientific journals outlined the mechanism of reactions involving the enzymes enolase, which metabolizes **carbohydrates**, and pyruvate kinase.

Following completion of her doctorate, Stubbe spent a year at the University of California at Los Angeles doing postdoctoral research in the department of chemistry with Julius Rebek. In 1972 she accepted a post as assistant professor of chemistry at Williams College in Massachusetts where she stayed until 1977. In late 1975 she accepted a second postdoctoral fellowship, took a leave of absence from her teaching duties, and spent a year and a half at Brandeis University on a grant from the National Institutes of Health (NIH).

From 1977 to 1980 Stubbe was assistant professor in the department of pharmacology at the Yale University School of Medicine. She then began a seven-year association with the University of Wisconsin at Madison, beginning as assistant professor and rising to full professor of **biochemistry** in 1985. In 1987 the Massachusetts Institute of Technology (MIT) beckoned, and Stubbe accepted a position as professor in the department of chemistry. In 1992 she was named John C. Sheehan Professor of Chemistry and Biology.

Stubbe's research focused on the mechanism of enzymes called ribonucleotide reductases, which catalyze the rate-determining step in DNA biosynthesis. This mechanism involves radical (that is, with at least one unpaired electron) intermediates and requires protein-based radicals for catalysis. Ribonucleotide reductases are major targets for the design of antitumor and antiviral agents, because inhibiting these enzymes interferes with the biosynthesis of DNA and cell growth. In collaboration with colleague John Kozarich, Stubbe has also explained the mechanism by which the antitumor antibiotic bleomycin degrades DNA. Bleomycin is used to kill cancer cells, a function that is thought to be related to its ability to bind to and degrade DNA. Other research interests of Stubbe's include the design of so-called suicide inhibitors and mechanisms of DNA repair enzymes.

Stubbe has published over eighty scientific papers and has been recognized frequently for her research achievements. She was the recipient of a NIH career development award, the Pfizer Award in enzyme chemistry in 1986, and the ICI-Stuart Pharmaceutical Award for excellence in chemistry in 1989. She received a teaching award from MIT in 1990 and the Arthur C. Cope Scholar Award in 1993. Stubbe was elected to the American Academy of Arts and Sciences in 1991 and to the National Academy of Sciences in 1992. Stubbe is a member of the American Chemical Society, the American Society for Biological Chemists, and the Protein Society. She has been active on several committees, including review boards for the NIH grants committee and the editorial boards for various scientific journals.

SUBATOMIC PARTICLES

For nearly a century after the English scientist **John Dalton** announced his **atomic theory** in 1803, the concept that **matter** consisted of atoms and that atoms were tiny, indivisible particles was widely accepted among scientists. The concept of the **atom** and the origin of the word atom actually dated back to ancient Greece and the scientist/philosopher **Democritus** who defined atoms as matter ''unable to be cut.''

The atomic theory was profoundly shaken in the 1890s with English physicist **J. J. Thomson**'s discovery of the **electron**. Thomson's work made it obvious that atoms are not indivisible, but, in fact, consist of smaller, subatomic particles.

In addition, the existence of the electron implied the existence of at least one other subatomic particle. Because electrons are negatively charged and atoms are electrically neutral, some positively charged subatomic particle must exist to **balance** the electron's negative charge. That particle, the **proton**, was discovered by English physicist **Ernest Rutherford** in 1919. Rutherford had already won the 1908 Nobel prize for his work regarding the disintegration of the elements, and the **chemistry** of radioactive substances.

Strong evidence also existed for the presence of yet a third subatomic particle. Experiments demonstrating a discrepancy between **atomic number** (number of protons in an atom) and atomic weight suggested that atomic nuclei might contain a particle approximately equal in **mass** to the proton, but electrically neutral. This particle, the **neutron**, was discovered by James Chadwick in 1932. Chadwick's discovery merited the 1935 Nobel Prize in Physics.

As a result of these discoveries, a new atomic model emerged that included protons, neutrons, and electrons. The model was attractive to scientists because of its simplicity and because it explained most of the known physical and chemical phenomena. Yet, even before Chadwick's discovery of the neutron, a suggestion had been made that still another subatomic particle existed.

In 1928, Paul Dirac (1902-1984) predicted the existence of a positively charged electron, the positron. Four years later Dirac's positron was also found in a cosmic ray shower and in 1933 Dirac was awarded the Nobel Prize in Physics for his extension of atomic theory. The discover of the positron, Carl David Anderson (1905-1991), was awarded 1936 Nobel Prize in Physics.

Dirac's theory suggested that, like the electron, all particles should have an **antiparticle** from which they differed only in their electrical charge. The search for other antiparticles—including the antiproton—involved far more difficult technical problems than did the search for the positron. In fact, nearly 30 years elapsed before the existence of the antiproton was confirmed. Afterwards the family of subatomic particles rapidly expanded.

The Japanese physicist, Hideki Yukawa (1907-1981), developed a theory to explain the force of attraction that holds protons and neutrons together in the **nucleus**. Yukawa called the force the ''strong'' force. His calculations showed that such a force would be carried by a particle whose mass would measure between the mass of an electron and the mass of a proton.

The search for Yukawa's medium mass particle began almost immediately, In 1938 scientists thought they had found the particle and suggested the name mesotron, later shortened to meson, for it. Unfortunately, studies showed that the new particle did not interact with protons and neutrons and could not, therefore, be the carrier of the strong force. What scientists had discovered, however, was another subatomic particle, one that was eventually renamed the *mu* meson, or muon.

Yukawa's prediction was confirmed in 1947, when the English physicist, Cecil Powell (1903-1969), discovered, discovered another type of meson in cosmic ray showers. This type of meson did interact with protons and neutrons and did meet Yukawa'a criteria. The particle was given the name *pi* meson, or pion. Yukawa was then awarded the 1949 Nobel Prize in Physics for his nuclear force theories that predicted the existence of mesons and Powell received the 1950 Nobel Prize in Physics for his work developing the photographic methods used in the study of nuclear processes.

At the very time that Yukawa was developing his meson theory, a technological breakthrough was taking place that was to revolutionize the study of subatomic particles. That breakthrough began with the invention of machines that can accelerate particles to very high energies. Experiments conducted with these accelerators soon revealed the existence of well over 100 new subatomic particles.

Before long, physicists were overwhelmed with the increasing numbers of newly discovered subatomic particles—and with the terms used to describe the seemingly bewildering variety of properties these particles exhibited.

Some for example, had half-integer spins (1/2, 3/2, 5/2, or 7/2), while others had only integral spins (0,1,2). Some exist only in neutral states while others (e.g., pion and delta particles) existed in one or more charged (or neutral) states. Some had very short lives (as short as 10^{-23} second), while others had lifetimes much longer than predicted. Particles with unusually long lifetimes were said to have a property labeled ''strangeness.''

Finally, many of the new particles decayed in ways that suggested the existence of certain basic properties that were (usually) conserved during decay. For example, certain types of heavy particles (baryons) always decayed in such a way that a property described as a baryon number seemed to be conserved.

To add even more confusion to the study of subatomic particles, evidence increasingly pointed to the existence of particles without mass. The concept was, and is, difficult to comprehend and can only be fully understood in terms of the interconvertability of mass and **energy**. Albert Einstein's use of the concept of photons in 1905 was one of the first references to a particle without mass. In 1931, Wolfgang Pauli (1900-1958) predicted the existence of the massless **neutrino** to explain away mathematical problems with beta decay.

Scientists are traditionally uncomfortable with disorganization and uncertain **nomenclature**. Efforts were made to organize the mounting data regarding subatomic particles. Some proposals were relatively simple, merely grouping and organizing particles according to common properties. For example, particles were classified according to their spin. Those with half-integral spins were called fermions, while those with integral spins were classified as bosons. All particles acted upon by the strong force were grouped together as hadrons. Particles were also grouped according to their mass. The lightest particles were named leptons, those with medium mass, mesons, heavier particles were called baryons, and the heaviest particles of all, hyperons. These classification systems and this terminology have changed somewhat with time. As scientists have learned more about the composition of various particles, the basis on which they are classified the particles has also changed.

Particles also began to take on whimsical names. Hadrons were determined to be composed of various combinations of three even smaller, more elementary particles named quarks. The name quark was derived from Irish writer James Joyce's novel *Finnegan's Wake* and a character's cry, ''Three quarks for Muster Mark.'' According to the physicist Murray Gell-Mann (1929-) who named the new elementary particle, the name quark was appropriate because of its association with the number three which is the number of quarks that always occur together in nature.

In order to account for all known properties of the hadrons, three types of quarks were hypothesized. These quarks were eventually named the up (u), down (d) and strange (s) quarks. The terms up, down and strange are not descriptive. There is nothing up about the ''up'' quark, nor is there anything strange in the conventional sense about the ''strange'' quark.

In 1961, a scheme known as the ''eightfold way'' was proposed for organizing all known subatomic particles. According to this strategy, particles are grouped according to various properties, such as spin and strangeness. Much like the validity of the **periodic table** was enhanced by its successful prediction of undiscovered elements, the validity of the new system seemed to be established in 1964 with the discovery of the W^- particle. The W^- particle had been predicted in order to fill a gap in one of the eightfold way categories. The general theory that hadrons are bound states of smaller particles held together by the strong force is also known as the bootstrap model.

Over the past two decades, physicists have refined their methods of classifying subatomic particles to produce what is commonly known as the Standard Model. According to the Standard Model, two basic types of particles exist: quarks and leptons. Quarks and leptons are thought to be fundamental particles. Because they have no space dimensions they can not be subdivided. In this regard they are similar to geometric points. Six kinds of each particle are arranged in pairs according to mass and **energy level**.

Energy level one refers to the normal conditions of everyday experience. At this energy level, only up and down quarks, the electron and the electron neutrino exist. Although Energy level two can be produced by most particle accelerators, it is characteristic of comic ray events. At this energy level, two more quarks, the strange and charm quarks, exist. Leptons, the muon, and the muon neutrino also inhabit level two.

Finally, energy level three—attainable in only the most powerful particle accelerators—is characteristic of the energy level that existed at the creation of the universe. At this level, the ''bottom'' and ''top'' quark exist.

According to the Standard Model, all other subatomic particles consist of some combination of quarks and their antiparticles. A proton, for example, is thought to consist of two up quarks and one down quark. Also the neutral *pi* meson is made up of an up quark and its antiparticle or of a down quark and its antiparticle. In general, baryons consist of three quarks and mesons of one quark and its corresponding antiquark.

The discovery in 1994 of the top quark added validity to the Standard Model proposition that all matter consists of six types of quarks named, in order of increasing mass, up, down, strange, charm, bottom, and top. In addition, there are six types of particles—including the electron—named leptons.

The Standard Model accounts for the strong, weak, and electromagnetic fields and forces as interactions of quarks and leptons, and provides explanations for the strong force and **nuclear reactions** (decay).

In addition to quarks and leptons, there are groups of other fundamental particles These mediating particles include the graviton (proposed carrier of the graviational force), photons (carriers of electromagnetic or light energy), gluons (mediators of the strong force), and, as carriers of the weak force, the W^+, W^-, and the Z bosons.

The weak force allows a quark to change to another type of quark, or a lepton to another lepton. When these transformations occur, quarks and leptons are said to have changed "flavor." Weak force interactions allow quarks and leptons decay to lighter quarks and leptons. When a particle decays, it is replaced by two or more daughter particles. Energy is used in these transformations, however, and therefore the sum of the masses of the daughter particles is less than the mass of the parent particle.

Because of these energy considerations, the stable matter of everyday experience contains only electrons and the lightest two quarks (up and down) that combine to form protons and neutrons.

More powerful accelerators will be needed to determine if there is a more elemental composition to quarks and leptons. There are still other important questions left unanswered by the Standard Model. At the extremely high temperatures (energy levels) modern cosmologists predict would have existed at and immediately after the Big Bang, the model breaks down. In addition, the graviton, proposed mediator of the gravitational force, remains unobserved.

See also Nuclear reactions; Radiation chemistry; Radioactivity

Sublimation

The transition of molecules from the solid phase into the vapor (or gas) phase is called sublimation.

At temperatures below the freezing point of a substance, the transition between solid and liquid phases (melting) does not generally occur. However, the direct transition from the solid to vapor phase does take place. For instance, wet clothes will freeze solid at temperatures below 32°F, but the clothes will become dry by sublimation without the ice melting. Another well known example is the evaporation of dry ice (solid **carbon** dioxide).

Sublimation can be understood on the basis of molecular behavior. In the solid phase, the molecules do not possess translational **energy**, e.g., they do not move around. The attractive forces between molecules have succeeded in causing the molecules to arrange themselves into fixed, ordered patterns relative to each other. They do, however, possess vibrational energy, with the molecules vibrating about their fixed centers of gravity. From time to time, individual molecules at the surface of the solid become energetic enough in their vibration to break the attractive hold their neighbors have on them and escape into the gas phase. Thus, sublimation occurs. A solid confined in a container will reach a characteristic equilibrium **vapor pressure** at a given **temperature**. Since the vibrational energy which the molecules of the solid phase possess depends on the temperature, they have more energy at higher temperatures and thus may escape more readily. The equilibrium vapor pressure will thus increase as the temperature increases.

Virtually all **solids** sublime to some extent. The aroma of solids (e.g., soap, chocolate) results from the interaction of our nasal sensors with molecules of the substance that have escaped into the vapor phase. Substances with very low vapor pressures, such as **metals**, generally do not have detectable aromas.

Substitution reaction

In substitution reactions, one atom or radical replaces another. The most common type of substitution reactions of compounds of **metals** is **ion** exchange accompanied by **precipitation**. The most common substitution reactions of compounds of nonmetallic elements are nucleophilic and electrophilic substitution.

Ion substitution reactions abound in geology in processes like sedimentation, mineral formation, and the development of underground caverns with exotically shaped stalactites and stalagmites. In the human body, the products of ion substitution reactions form bones and teeth. Chemists have applied ion substitution reactions to numerous tasks, ranging from the preparation of commercial products such as baking soda to the removal of undesirable metal ions from drinking **water** and the **recycling** of metal ions dissolved in waste water.

A quantitative description of the substitution **chemistry** of minerals and salts, requires a more precise definition of the term **solubility**. When a material dissolves in a liquid, it is soluble in that fluid. In that sense, solubility is a qualitative concept. However, solubility is quantitatively defined, too, as the amount of a solute that dissolves in a given quantity of solvent at a specific **temperature**. Quantitative information determined from the concentrations of ions in **solution** in equilibria with precipitates formed from them is used to assess which reactions will actually occur based on the concentrations of ions available.

The key to **quantitative analysis** of the composition of **salt** solutions is the solubility product of the salt, K_{sp}. The expression for K_{sp} is similar to an equilibrium constant for the **dissociation** of a weak acid or base. The concentrations of the products of the reaction—in this case, the ions formed upon dissolution of the salt—appear in the numerator of the mathematical expression raised to a power equivalent to their exponent in the balanced chemical equation. However, the **concentration** of the solid does not appear in the solubility product expression. As long as any solid is in contact with the solution, the concentration of the substance within the solid phase is constant. The symbolic expression of the K_{sp} for **magnesium** hydroxide in aqueous solution in equilibrium with solid magnesium hydroxide is: $Mg(OH)_2 \rightleftharpoons Mg^{2+}(aq) + 2 OH^-(aq)$ $K_{sp} = [Mg\ Mg^{2+}(aq)][OH^-(aq)]^2$

Precipitation reactions resulting from **anion** substitution are widespread in nature, ranging from mineral and sediment deposition to bone formation. They are also important industrially: magnesium is extracted from seawater by precipitating $Mg(OH)_2$ as described previously, **silver** is precipitated as **halide** salts to make photographic film, and **sodium** bicarbonate (baking soda) is precipitated by mixing solutions of **sodium chloride** and ammonium carbonate produced from limestone.

The concentrations used in the solubility product are equilibrium concentrations in the presence of solid. If the system is not at equilibrium, an expression with the same form as the solubility product, the solubility reaction quotient, Q_{sp}, is useful for predicting how the system will behave if concentrations are changed. Similar to K_{sp}, the expression for Q_{sp} shows the product of concentrations of ions in solution raised to powers equivalent to their coefficients in balanced equations. If a value calculated for $Q_{sp} < K_{sp}$ for a given salt, then that salt will not precipitate; if $Q_{sp} = K_{sp}$, then the solution is at equilibrium and the maximum amount of salt is dissolved and no precipitate forms; if $Q_{sp} > K_{sp}$, then precipitation will occur until the concentrations adjust such that $Q_{sp} = K_{sp}$. Such situations are frequently encountered when two solutions of known concentration are mixed. In calculations involving solutions that have been mixed, remember that mixing two solutions dilutes them both. The concentrations of the ions after mixing must be used in the expression, and they must be calculated on the basis of the total **volume** of the solution.

When the calculated Q_{sp} is less than the K_{sp}, the solution is unsaturated. An unsaturated solution can dissolve more solute. When the concentrations of the ions in solution are such that $Q_{sp} = K_{sp}$, the solution is saturated. No more solute can dissolve in the solvent; it is "holding" the maximum amount. When the concentrations are such that Q_{sp} is greater than K_{sp}, either of two results are observed. In the simplest cases, a precipitate forms until $Q_{sp} = K_{sp}$: until the concentration of ions in solution as defined by the equilibrium constant is reached. At this point, the solution is saturated. A solution is saturated with respect to a solid if some of the solid is visible in the bottom of the container. The maximum concentration of ions in the solution is in equilibrium with the precipitated solid. Sometimes, however, no precipitate forms until Q_{sp} greatly exceeds K_{sp} or until the solution is disturbed. When $Q_{sp} > K_{sp}$ for a solution in such cases, the solution is supersaturated. The ions in the solution are at a higher concentration than would be predicted by the K_{sp} for the dissolved material. A saturated solution can be identified by visual inspection if it contains undissolved solute.

In natural water systems, many common minerals are formed by anion substitution-precipitation reactions, among them carbonates, phosphates, and the **sulfate**-containing rocks. Some of the most spectacular examples are the stalagmites and stalactites that form from the precipitation of **calcium** carbonate as water and CO_2 partially evaporate as the solution drips from the ceiling to the floor of a cave. Other substances that are slightly soluble in water also precipitate on the surface of these geological formations as the water evaporates, including magnesium carbonate and phosphates. The composition of stalactites and stalagmites depends on the salts present in the water that flows over them which, in turn, depends on the types of rocks over which the water previously passed. To completely explain the formation of stalactites and stalagmites and many other reactions that take place among ions in solution, the effects of **pH** and dissolved **gases** must also be taken into account.

Ion substitution reactions are put to great practical use in a process known as **ion exchange**. Ions causing hard water can be removed from solution and replaced by other ions through the intermediacy of inorganic zeolites or organic ion-exchange resins.

In substitution reactions involving covalently bound species, a free radical, ion, or **molecule** takes the place of an another that is removed from the starting material during the course of the reaction. Except for the inclusion of **free radicals**, that is exactly how we described substitution reactions involving ions. The major difference is that, in the substitution reactions of covalent molecules, covalent bonds are broken and formed instead of ionic ones. This difference is responsible for many of the features of organic reactions that are the subject of extended study in **organic chemistry** courses. The interaction of two ions can be thought of as the attraction of two charged spheres. The interaction is the same no matter where the two spheres touch. The attraction is omnidirectional and isotropic, extending in all directions and the same in all directions.

For ions dissolved in water, the rapid exchange of water molecules around the **cation** usually provides the space for an anion to approach the cation. The exchange of water molecules, which controls the access of the anions to the cation, is typically very fast: about one-billionth of a second for 1+ ions, one millionth of a second for 2+ ions, and a millisecond or less for 3+ ions at room temperature. This ready access coupled with the ability of the anion to interact effectively at any site on the cation makes ionic reactions very fast. Any orientation in which the cation and anion collide can lead to a reaction.

In contrast, covalent bonds are highly directed. For example, in the compound bromomethane, the four atoms bound to the **carbon** maintain the same relative geometry as the molecule tumbles about in solution or in the gas phase. For a substitution reaction to occur, one or more bonds in specific regions about the carbon **atom** must be broken. Correspondingly, new bonds form only in specific locations about the central carbon atom. Consequently, the species involved in the reaction must be specifically aligned to react. Because a specific alignment is required, all collisions will not result in product formation. This makes reactions of covalently bound species very sensitive to the structure of the molecules involved and also tends to make reactions requiring the breaking and forming of covalent bonds slower than ionic reactions. The reaction in organic chemistry that is most closely analogous to the anion substitution reactions of salts is nucleophilic substitution. **Nucleophile** is the name given to the species to which the new bond is formed as a result of the reaction; the species replaced is called the leaving group. The nucleophile substitutes for the leaving group on the reactant.

Substitution reactions can be carried out using many different organic molecules as starting materials. The common characteristic shared by all the molecules is an electropositive carbon. The reactions can involve either anionic nucleophiles such as iodide ion or neutral molecules such as amines. The common characteristic of all nucleophiles is the availability of electrons, either in the form of a negative charge or an unshared pair of electrons. Some examples of nucleophiles are the hydrosulfide ion, HS^-, the cyanide ion, CN^-, iodide ion $^-$, hydroxide ion, OH^-, and **ammonia**, NH_3. Such **electron** rich species react most readily at the most positively polarized portion of the organic molecule. Because the **nucleus** of an atom is positively charged, the electron rich species are called nucleophiles, from nucleus-loving or lovers of positive charge. Nucleophilic substitution always refers to the replacement of a group on an electropositive atom by an electron-rich species.

Nucleophilic substitution reactions can be either unimolecular or bimolecular. The key step in a unimolecular reaction is the formation of the carbon cation, or carbonium ion, by the departure of the leaving group. After the leaving group departs, any nucleophile present in the solution can take its place by forming a bond to the carbonium ion. In a bimolecular nucleophilic substitution reaction, the crucial step requires the involvement of two separate species: the nucleophile and the original intact molecule. The intermediate stage in this reaction involves a carbon atom with five species bound to it. The type of nucleophilic substitution reaction that a molecule undergoes depends on its structure and the conditions of the reaction.

Substitution reactions of organic compounds can also involve free radicals. Free radicals (uncharged atoms or groups of atoms with unpaired electrons) of **chlorine** can be obtained simply by exposing the gases to ultraviolet light or by heating chlorine to high temperature. Chlorine atoms readily replace **hydrogen** atoms bound to carbon. The industrial production of chlorocarbons like chloroform, $CHCl_3$, and methylene chloride, CH_2Cl_2, involves chlorine free radicals in substitution reactions. In practice, the chlorination of a molecule like **methane** (CH_4) results in the formation of all possible chloromethanes (CH_3Cl, CH_2Cl_2, $CHCl_3$, CCl_4 and HCl. This is a common result of free radical reactions: they tend to be unspecific. The yield of a specific product can often be optimized by using different ratios of methane to chlorine in the starting mixture or by using precursors other than methane. **Chlorofluorocarbons (CFCs)**, the chemicals now being phased out of production and use because of their link to ozone depletion in the atmosphere, are produced by similar substitution reactions Compounds with extensive electron delocalization due to **resonance**, such as **benzene**, undergo a different type of reaction: electrophilic substitution. In 1877, two chemists, Charles Friedel and James Mason Crafts, reported the reaction of methyl chloride (CH_3Cl) with benzene (C_6H_6) in the presence of **aluminum** chloride. The Electrophiles, a species with strong affinity for negative charge, attacks the electron rich aromatic **ring**. In this reaction the **electrophile** is produced when the methyl chloride reacts with the aluminum chloride to form an electropositive CH_3 group. The aluminum in $AlCl_3$ is an electron deficient atom and accepts a pair of electrons from the chlorine atom in $CHCl_3$ to form the electropositive CH_3 as a reactive electrophile. This species substitutes for a **proton** on an electron rich carbon in the benzene ring. This reaction is still called the Friedel-Crafts reaction.

Compounds containing double bonds, like ethylene, and even some alcohols like isopropanol, $(CH_3)_2CHOH$, can participate in Friedel-Crafts reactions as electrophiles. In Friedel-Crafts reactions, as in all electrophilic aromatic substitutions, the attacking group is an electropositive species whose attraction to a site of electron **density** in a molecule determines the course of the reaction. Other electrophilic aromatic substitution reactions include brominaion, using the reagents **bromine** and iron(III) bromide to produce electropositive bromine Br^+, sulfonation, which uses **sulfur** trioxide and **sulfuric acid** to produce electropositive hydrogen sulfate, HSO_4^+ and nitration, which uses nitric and sulfuric acids to produce electropositive nitrosonium ion, NO^+. These reactions all require the presence of a positively charged electrophile that is attracted to an electron-rich site. In many cases, the electrophile is produced in the reaction medium by the interaction of a neutral compound (like methyl chloride) with another substance ($AlCl_3$) that is not directly involved in the reaction.

Boron, **silicon**, **nitrogen**, **phosphorus** and sulfur are among the most prevalent of the elements other than carbon that form covalent compounds. The substitution reactions of covalent inorganic compounds of these elements are central to their industrial applications. In addition, the biochemical reactions of nitrogen and sulfur are often nucleophilic substitution reactions. Phosphorus-containing compounds can act as either electrophiles, when the phosphorus is bound to highly electronegative elements such as **fluorine** or **oxygen**, or as nucleophiles, when the phosphorus is bound to other elements. The chemistry of boron is quite diverse. As far as substitution reactions are concerned, boron is frequently electron deficient in its simple compounds, and substitution reactions on boron compete with addition reactions. Silicon behaves somewhat similarly to carbon, but compounds of silicon and hydrogen or silicon and the **halogens** (the silanes) are much more reactive than their carbon counterparts.

Substitution reactions are useful means by which chemists convert available compounds into new materials. The most common type of substitution reaction for compounds of metallic elements is precipiation. In a large class of reactions called nucleophilic substitutions, reactions occur when a covalently bound species with available electrons, either in the form of an anionic charges or an unshared pair of electrons, is attracted to an electropositive site in a molecule. During the reaction, a new bond is formed between the nucleophile and another bond in the molecule is broken, thereby releasing a new species into the reaction medium. Nucleophilic substitution reactions can be envisioned by using two different mechanisms—one unimolecular and one bimolecular—and the type of substitution reaction that a given molecule undergoes depends on its structure and the conditions under which the reaction is carried out.

Another type of substitution reaction involves free radicals. These reactions are industrially important for the production of some useful small molecules and also play a major role in **atmospheric chemistry**.

Aromatic compounds like benzene that contain delocalized electrons undergo electrophilic substitution reactions. Electrophiles are electron-poor species that are attracted to the electron cloud of aromatic compounds and other species with loosely held electrons.

Compounds and ions containing nitrogen and sulfur are frequently nucleophilic, and much of the chemistry of these compounds is presented in the section on substitution reactions on carbon. The chemistry of nitrogen and sulfur compounds is important in **biochemistry** in the synthesis of **amino acids**. Nitrogen is also a major component of the nucleotide bases. Phosphorus behaves similarly to nitrogen but has a more extensive **reactivity** because of its ability to expand its octet to form compounds with five or even six bonds. The inorganic compound phosphine is very similar to ammonia in its substitution reactions. Phosphorus has a major structural role in biochemistry because its **phosphate** esters form the backbone of polynucleotides. In addition, the **hydrolysis** of phosphate esters provides a significant source the use of precipitation reactions to prepare ionic compounds by anion substitution. In this, substitution reactions of boron-containing compounds are directed by the ability of boron to form electron deficient molecules that can readily bond with electron rich species, especially those containing a pair of non-bonding electrons. The chemistry of silicon compounds is similar to that of carbon, but silicon forms extremely stable silicon-oxygen bonds which results in compounds like sand and quartz that are resistant to further chemical modification.

Living systems use substitution reactions for processes in vivo as diverse as producing bone and teeth to producing active forms of **hormones** to detoxifying poisons. Industry applies substitution reactions analogous to many prevalent in biological systems to the production of compounds as simple as chloroform, as widely used as aspirin, and as structurally complicated as dyes and pharmaceuticals.

SULFATE

A sulfate is a **chemical compound** containing the **ion** SO_4^{2-}, or its corresponding radical. Many sulfates are found in nature as mineral deposits, including **barium** sulfate ($BaSO_4$, barite), **calcium** sulfate (gypsum, alabaster, and selenite), and **strontium** sulfate (celestite). **Potassium** sulfate (K_2SO_4) and **magnesium** sulfate ($MgSO_4 \cdot 7H_2O$ or Epsom salts) also occur naturally. Some sulfates are produced as byproducts in other manufacturing operations. For instance, **sodium** sulfate is produced during battery **recycling** and rayon manufacturing.

Sulfates are salts or esters of **sulfuric acid**, H_2SO_4. One or both of the hydrogens are replaced with a metal or a radical (ammonium or ethyl). Sulfates that have both hydrogens replaced are known as normal sulfates. Those with only one **hydrogen** replaced are known as hydrogen sulfates, acid sulfates, or bisulfates. Double sulfates contain two different **metals** and two sulfate radicals. Organic sulfates are esters and are formed by reacting an **alcohol** with cold sulfuric acid or by the reaction of sulfuric acid with a double bond in an alkene. The latter reaction produces alkyl hydrogen sulfate, which can further be broken down into an alcohol by heating with **water**.

Most metal sulfates are soluble in water, except for barium, **lead**, and strontium. Calcium sulfate is only slightly soluble. Calcium sulfate exists in two forms, as anhydrite ($CaSO_4$) and as gypsum ($CaSO_4 \cdot 2H_2O$). Heating gypsum to 120°C forms the hemihydrate, $(CaSO_4)_2 \cdot H_2O$, otherwise known as plaster of paris. Plaster of paris is widely used as a mold or model material in the metal and ceramic industries. It has also been used as a raw material in **glass** manufacturing.

Other sulfates are used in a variety of agricultural and industrial applications. Over half a million tons of potassium sulfate is used in the United States as fertilizer for non-grain and specialty crops like tobacco. Ammonium sulfate ($(NH_4)_2 \cdot SO_4$) is also used as a fertilizer, in addition as an ingredient to prepare other ammonium compounds and for fireproofing. Sodium sulfate or **salt** cake (Na_2SO_4) is used to make detergents, textiles and **paper** (for the bleaching process). Magnesium sulfate is also used in the paper and fertilizer industries.

Many sulfates are used in the ceramic or glass industries. Strontium sulfate is sometimes used to produce iridescence in glass and pottery glazes, and can also be used as a fining agent (to remove bubbles in the molten glass) in crystal glass. Sodium sulfate is also used as a fining agent. Barium sulfate can be used in the manufacture of porcelain enamels, to reduce dimpling defects or to improve workability. Magnesium sulfate is also used in the manufacture of enamels. **Cobalt** sulfate ($CoSO_4 \cdot 7H_2O$) is used as a blue colorant in **ceramics**, ferrous sulfate ($FeSO_4 7H_2 O$) as a red colorant, and **copper** sulfate ($CuSO_4 \cdot 5H_2O$ is sometimes used to make ruby glass.

SULFUR

Sulfur is the second element in Group 16 of the periodic family, a group of elements known as the **oxygen** family or the chalcogens. The term chalcogen means ''ore-forming'' and is used for members of this family because its first two members, oxygen and sulfur, are present in most important metallic ores. Sulfur has an **atomic number** of 16, an atomic **mass** of 32.064, and a chemical symbol of S.

Properties

Sulfur exists in three allotropic forms, two of which are yellow crystalline **solids**, and the third of which is an amorphous thick, brown, plastic liquid. The two crystalline allotropes are known as the α (alpha) and β (beta) forms of the element. The α-form is a bright yellow solid that usually changes to the β-form at a **temperature** of 202°F (94.5°C), although it can be melted if heated quickly to 235°F (112.8°C). The β-form has a melting point of 246°F (119°C). The two allotropes have densities of 2.06 grams per cubic centimeter (α-form) and 1.96 grams per cubic centimeter (β-form). Neither **allotrope** dissolves in **water**, but both dissolve in organic **liquids**, such as **benzene** (C_6H_6), **carbon** tetrachloride (CCl_4), and carbon disulfide (CS_2).

Sulfur is relatively active, burning in air with a characteristic pale blue flame and giving off a very obvious strong, choking odor. The odor is due to the formation of sulfur dioxide (SO_2) during **combustion**. Sulfur also combines with a number of other elements, sometimes quite easily. For example, when sulfur is heated with **magnesium**, magnesium sulfide (MgS) is formed. Sulfur also reacts with **hydrogen** gas to form hydrogen sulfide, another sulfur compounds with a very distinctive odor of rotten eggs: H_2 + S —heated→ H_2S.

Occurrence and Extraction

At one time, extensive layers of pure sulfur existed throughout the Earth's crust. The element could be mined easily, much the way **coal** near the Earth's surface is mined today. Now, that source of sulfur has been depleted, although deeper beds of sulfur still exist that can also be mined. The method used to extract the sulfur is to pass steam and compressed air through pipes down into the sulfur bed. The sulfur is melted by the steam and then forced back up to the surface by the compressed air.

Sulfur is also present in many important minerals, such as barite (barium **sulfate**; $BaSO_4$), celestite (strontium sulfate; $SrSO_4$), cinnabar (mercury sulfide; HgS), galena (lead sulfide; PbS), pyrites (iron sulfide; FeS_2), sphalerite (zinc sulfide; ZnS), and stibnite (antimony sulfide; Sb_2S_3). In all of these cases, the sulfur-containing mineral is a major ore for the metallic component of the ore.

The abundance of sulfur is the Earth's crust is estimated to be about 0.05%, making it the 16th most abundant element. The largest producers of sulfur in the world are the United States, Canada, China, Russia, Mexico, and Japan.

Discovery and Naming

Sulfur is one of the elements known to ancient peoples. Its abundance, distinctive **color**, and characteristic form of combustion probably made it well known thousands of years ago. Many religious traditions, in fact, make reference to burning sulfur as a particularly terrible form of suffering that comes to unbelievers. The Bible, for example, mentions the torments of being exposed to "fire and brimstone," thought to refer to burning sulfur. The origin and meaning of the element's name are now lost in history.

Uses

Elemental sulfur is used as an insecticide and in the vulcanization of rubber. By far the most important use for the element, however, is in the manufacture of **sulfuric acid** (H_2SO_4), the single most important industrial chemical. Each year, there is more sulfuric acid produced in the United States than any other chemical. In fact, nearly twice as much sulfuric acid is produced as the next most important chemical, **nitrogen**. In the United States alone, the annual production of sulfuric acid amounts to nearly 50 million metric tons.

Nearly three quarters of all the sulfuric acid produced in the United States goes to the production of synthetic **fertilizers**. The compound is also used in the treatment of **copper** ores; the production of **paper** and paper products; the manufacture of other agricultural chemicals; and the production of **plastics**, synthetic rubber, and other industrial and commercial chemical products.

Health Issues

Sulfur is a macronutrient for both plants and animals. It is used primarily to make **proteins** and **nucleic acids**. Sulfur-deficiency disorders are very rare, but not unheard of. For example, abnormally low levels of sulfur can cause itchy and flaking skin and improper development of hair and nails. In plants, a deficiency of sulfur can cause the young leaves on a plant to begin turning yellow.

SULFUR CYCLE

Sulfur is an abundant element in Earth's crust, occurring mostly in minerals, but also a variety of substances in **water**, the atmosphere, and organisms. The storage of sulfur in the various compartments of Earth and its biosphere, and the many transfers occurring among them, is referred to as the sulfur cycle.

Sulfur (S) occurs in the environment in many chemical forms, including organic and mineral compounds, either of which can be chemically transformed by both biological and inorganic processes. Sulfur dioxide (SO_2) is a gas that is toxic to plants at concentrations less than one part per million (ppm) in the atmosphere, and to animals at larger concentrations. Natural sources of emissions of sulfur dioxide include volcanic eruptions and forest fires. Large emissions are also associated with human activities, especially the burning of **coal** and oil to generate **electricity** and the processing of sulfide metal ores. Gaseous **hydrogen** sulfide (H_2S), with a strong smell of rotten eggs, is also emitted to the atmosphere, usually in situations where organic sulfur compounds are being decomposed in the absence of **oxygen**. In the atmosphere, both sulfur dioxide and hydrogen sulfide become oxidized to **sulfate** (SO_4^{2-}), a negatively charged **ion** (or anion) that attracts positively charged cations, such as ammonium (NH_4^+), **calcium** (Ca^{2+}), and hydrogen ion (H^+). The resulting fine particulates (i.e., $CaSO_4$), can serve as condensation nuclei for the formation of ice crystals, which may settle from the atmosphere as rain or snow, delivering the sulfate to terrestrial and aquatic ecosystems. If the sulfate is mostly balanced by hydrogen ion (as H_2SO_4), the **precipitation** will be acidic, and high rates of input may damage some freshwater ecosystems. This environmental problem is sometimes called "**acid rain.**"

Enormous amounts of sulfur occur in association with **metals**, as chemically reduced minerals (or sulfides). The most common of these are **iron** sulfides (such as FeS_2), but all heavy metals can occur as sulfide minerals. When metal sulfides are exposed to an oxygen-rich environment, certain bacteria known as *Thiobacillus thiooxidans* begin to oxidize the sulfide, generating sulfate as a product, and using **energy** liberated by the reaction to sustain their growth and reproduction (this autotrophic process is called chemosynthesis). In situations where large amounts of sulfides are being oxidized in this way, an enormous amount of acidity is associated with the sulfate product. This environmental problem is sometimes called "acid-mine drainage."

Sulfur is an important nutrient for organisms, as it is a key constituent of certain **amino acids**, **proteins**, and other biochemicals. Plants satisfy their nutritional need for sulfur by assimilating simple mineral compounds from the environment. This occurs mostly as sulfate dissolved in water taken up by roots, or as gaseous sulfur dioxide absorbed by foliage from the atmosphere. Animals obtain sulfur by eating the biomass of plants or other animals. Sulfur is rarely a deficient nutrient for animals, but in some kinds of intensively managed agriculture its availability can be a limiting factor for plant productivity, and application of a sulfate-containing fertilizer may be beneficial.

SULFURIC ACID

Sulfuric acid (H_2SO_4) is a clear, oily mineral acid that has a **boiling point** of approximately 554°F (290°C). It is made from the mineral **sulfur**. More than 80% of all sulfur produced is used to make sulfuric acid. Because of the acid's strength and high boiling point, it is useful in making other acids. Sulfuric acid alone dissolves many **metals**; in combination with **hydrochloric acid** (a mixture called **aqua regia**), it can even dissolve **gold** and **platinum**. Diluting sulfuric acid with **water** can be dangerous because the reaction releases a lot of **heat**. To prevent explosive spattering, the acid must be added to the water, rather than vice versa.

Scientists probably learned to make sulfuric acid some time after the year 1000. Around 1300, sulfuric acid was first described by alchemists. The alchemists called sulfuric acid *oil of vitriol*, a reference to its corrosive nature. In 1595, a German alchemist named **Andreas Libavius** wrote clear instructions on how to prepare sulfuric acid as well as other chemical substances. When sulfuric acid and other strong acids became widely available during the Middle Ages, they launched an experimental revolution, enabling alchemists to quickly decompose substances without high temperatures.

Methods of producing sulfuric acid have evolved over the centuries. The first small factory for manufacturing acids and other chemicals was set up by German chemist Johann Rudolf Glauber (1604-1670) in the mid-1600s. In 1746, British chemist John Roebuck (1718-1794) developed an inexpensive process for producing larger quantities of sulfuric acid by using sturdy **lead** containers instead of **glass** jars. By the late 1700s, sulfuric acid had become an important industrial chemical. French chemist Jean Antoine Claude Chaptal (1756-1832) established a sulfuric acid manufacturing plant at Montpellier in 1781, which introduced commercial production of the acid in France. A newer method of making sulfuric acid, called the *contact process*, was invented in 1831 by a British vinegar manufacturer. In this method, sulfur dioxide gas (SO_2) and air are converted catalytically into sulfur trioxide (SO_3), a gas that combines explosively with water to form sulfuric acid. The reaction is controlled by first dissolving the gas in concentrated sulfuric acid. Because of the purity of the acid produced, the contact process replaced the lead-chamber process during the late 1800s.

The sulfuric acid industry has grown with the demand for organic chemicals. At first, America depended on Europe for its supply of sulfur and other raw industrial chemicals. But around 1900 German-American chemical engineer **Herman Frasch** invented a way to produce sulfur from deep underground deposits in the Gulf Coast, giving America its own supply of raw material for making sulfuric acid.

Nearly all manufactured products that we use today, like **gasoline**, detergents, and **batteries**, depend in some way on sulfuric acid for their production. Industry uses it in greater quantities than any other chemical. That is why industrial production of sulfuric acid is an excellent indicator of a nation's general economic prosperity.

Nearly 70% of today's sulfuric acid production goes into **fertilizers**. **Oil refining**, particularly gasoline purification, uses the next greatest amount of the acid. The steel industry also uses large quantities. Other products made with sulfuric acid include textiles, paint, **explosives**, dyes, fabrics, **pesticides**, medicines, and synthetic rubber. In 1998, annual consumption of sulfuric acid in the United States was close to 50 million tons.

See also Acids and bases

SUMNER, JAMES B. (1887-1955)

American biochemist

Biochemist James B. Sumner's natural perseverance was strengthened by his efforts to overcome a handicap suffered in an accident as a youth. He set out to isolate an **enzyme** in 1917—a task believed to be impossible at the time. By 1926, he had crystallized an enzyme and proven it was a protein, but spent many years defending the veracity of discovery. Sumner's achievement was finally recognized in 1946, when he shared the Nobel Prize in **chemistry** for proving that enzymes can be crystallized.

James Batcheller Sumner was born just south of Boston in Canton, Massachusetts, on November 19, 1887, the son of Charles and Elizabeth Kelly Sumner. They were an old New England family, whose ancestors had arrived in 1636 from Bicester, England, and Sumner's relatives included industrialists as well as artists. His own family was wealthy, and his father owned a large country estate. As a boy Sumner was interested in firearms and enjoyed hunting, a hobby that led to tragedy when he lost his left forearm and elbow to an accidental shooting. The handicap was doubly traumatic since Sumner had been left-handed, but he trained himself to use his right arm instead. He continued to participate in sports, including tennis, canoeing, and clay pigeon shooting, and he would learn to perform intricate laboratory procedures with only one arm.

Sumner received his early education at the Eliot Grammar School, and he graduated from the Roxbury Latin School in 1906. He enrolled at Harvard to study electrical engineering, but discovered he was more interested in chemistry and graduated with a degree in that discipline in 1910. He then joined the Sumner Knitted Padding Company, which was managed by an uncle, but stayed only a few months before he was offered the chance to teach chemistry for one term at a college in New Brunswick. This appointment was followed by a position at the Worcester Polytechnic Institute in Worcester, Massachusetts. He remained there for just one term as well before enrolling in the doctoral program at Harvard in 1912.

Sumner conducted his doctoral work under the supervision of Otto Folin, who had originally told him that a one-armed man could not possibly succeed in chemistry. Sumner completed his master's degree in 1913 and his doctorate in 1914. Part of his doctoral thesis was published in 1914 as "The Importance of the Liver in **Urea** Formation from Amino Acids" in the *Journal of Biological Chemistry.* While on a trip to Europe after completing his graduate studies, Sumner was offered an appointment as assistant professor of **biochemistry** at the Ithaca Division of the Cornell University Medical College. Initially detained by the beginning of World War I, he finally made it to Ithaca and discovered that he would also be teaching in the College of Arts and Sciences.

Sumner would spend his entire academic career at Cornell. He began as an assistant professor, a position he held for fifteen years, and then spent nine years as professor in the department of physiology and biochemistry at the Medical College. In 1938, Sumner was appointed professor of biochemistry in the department of zoology of the College of Arts and Sciences. In 1947 he became the founding director of the Laboratory of Enzyme Chemistry within the department of biochemistry.

Sumner's teaching load at Cornell was always very heavy. He enjoyed teaching and was regarded as an excellent professor, but his schedule did not leave much time for research. His research was also limited by the minimal equipment and laboratory help available at Cornell. In his Nobel lecture Sumner recounted why had chosen to work on enzymes. "At that time I had little time for research, not much apparatus, research money or assistance," he recalled. "I desired to accomplish something of real importance. In other words, I decided to take a 'long shot.' A number of persons advised me that my attempt to isolate an enzyme was foolish, but this advice made me feel all the more certain that if successful the quest would be worthwhile."

He chose to isolate urease, an enzyme that catalyzed the breakdown of urea into **ammonia** and **carbon** dioxide. He found that relatively large amounts of urease were present in the jack bean (*Canavalia ensiformis*). Sumner disagreed with the belief, then commonly held, that enzymes were low-molecular-weight substances which were easily adsorbed on **proteins** but were not in fact proteins themselves. He concentrated on fractionating the proteins of the jack bean, and this effort took him nine years. In 1926, he published a paper in the *Journal of Biological Chemistry* announcing that he had isolated a new crystalline globulin, which he believed to be urease, from the jack bean. Urease was the first enzyme prepared in crystalline form and the first that was proven to be a protein.

His results, and his interpretation of them, were not immediately accepted. On the contrary, he spent years engaged in a controversy with those who believed that enzymes contained no protein. One of his strongest opponents was **Richard Willstätter**, a German chemist who had won the Nobel Prize in 1915 for his work on **chlorophyll**. Willstätter had tried to produce pure enzymes and failed, and he argued that what Sumner had isolated was merely the carrier of the enzyme and not the enzyme itself. Although Sumner continued to publish additional evidence over the next few years, it was not until 1930 that he received support for his discovery, when **John Howard Northrop** of the Rockefeller Institute announced he had crystallized pepsin. Sumner was jointly awarded the 1946 Nobel Prize in chemistry; his co-recipients were Northrop and **Wendell Meredith Stanley**, also at the Rockefeller Institute, honored for their preparation of enzymes and virus proteins in pure form.

James B. Sumner.

After the crystallization of urease and the debates with Willstätter and others, Sumner continued his research on enzymes, among them peroxidases and lipoxidase, doing most of his own laboratory work. In 1937, he and his student Alexander L. Dounce crystallized the enzyme catalase and helped prove it was a protein. Sumner was the first to crystallize haemagglutinin concanavalin A, and he noted that this protein required the presence of a divalent metal to act. He also continued his original research work on new and improved laboratory methods. He was a prolific author, writing or contributing to about 125 research papers and a number of books.

In 1929 Sumner, who spoke Swedish, French, and German, went to the University of Stockholm to work on urease with **Hans von Euler-Chelpin** and **Theodor Svedberg**. He returned to Sweden in 1937 to work at the University of Uppsala on a Guggenheim fellowship, and while there he was awarded the Scheele Medal from the Swedish Chemical Society for his work on enzymes. Other professional honors include his election to both the Polish Institute of Arts and Sciences and the American Academy of Arts and Sciences. Sumner was also a member of many associations including the National Academy of Sciences, the American Association for the Advancement of Sciences, the Society for Experimental Biology and Medicine, and the American Society of Biological Chemists.

Sumner was married three times and divorced twice. He married his first wife, Bertha Louise Ricketts on July 20, 1915, the year after he completed his Ph.D. They had five children, and were divorced in 1930. Sumner married Agnes Paulina Lundquist of Sweden in 1931; they had no children. After his second divorce, Sumner married Mary Morrison Beyer in 1943; they had two sons, one of whom died as a child.

In 1955, while preparing for his retirement from Cornell, Sumner was diagnosed with cancer. He had intended to work at the Medical School of the University of Minas Gerais, in Belo Horizonte, Brazil, organizing an enzyme research program and laboratory. However, the day after attending a symposium held partly in his honor, he was hospitalized, and died on August 12, 1955, at the Roswell Park Memorial Institute in Buffalo, New York.

SUPERACIDS

The term *superacids* was introduced into the chemical literature in 1927 by James Conant and Norris Hall to describe solutions of sulfuric or perchloric acid in glacial **acetic acid**. They found that these nonaqueous-superacid solutions reacted with weak bases which did not react with either sulfuric or perchloric acid in **water**. Conant and Hall's initial report received little attention; however, interest in superacids increased dramatically in the 1960s. Since then, the reactivity of superacids with very weak bases has been studied extensively, particularly with regard to the protonation of saturated hydrocarbons in acid-catalyzed organic reactions. In 1972, Ronald Gillespie defined superacids as acids which are stronger than 100% **sulfuric acid**. Gillespie's definition has become the accepted definition for Brönsted superacids. Lewis superacids are generally considered to be those which are stronger than anhydrous **aluminum** trichloride ($AlCl_3$). In 1987, **George Olah** received the Nobel Prize in **chemistry** for his pioneering research in the activation of **carbon**-carbon and carbon-**hydrogen** bonds by superacids.

There are several classes of superacids: Brönsted, Lewis, conjugate Brönsted, conjugate Brönsted-Lewis, and solid superacids. The Brönsted superacids include perchloric acid ($HClO_4$), halosulfuric acids (e.g. fluorosulfuric acid (FSO_3H)), and perfluoroalkane sulfonic acids (e.g. trifluoromethanesulfonic acid (CF_3SO_3H)). Brönsted superacids behave no differently than typical Brönsted acids, except that they are stronger than the arbitrarily-chosen standard of sulfuric acid, and are thereby categorized as superacids. The Lewis superacids include antimony pentafluoride (SbF_5), arsenic pentafluoride (AsF_5), tantalum pentafluoride (TaF_5), and niobium pentafluoride (NbF_5). Like the Brönsted superacids, Lewis superacids are simply very strong Lewis acids.

Conjugate superacids involve the mixture of two strong acids to create an exceptionally acidic medium. Conant and Hall's sulfuric acid:acetic acid mixture provides an example of how conjugate Brönsted superacids function. Sulfuric acid is a strong acid. In aqueous solution, the acid-base reaction between sulfuric acid and water goes to completion:

$$H_2SO_4 + H_2O \rightarrow HSO_4^- + H_3O^+$$

In glacial acetic acid, on the other hand, no water is present, so the strongest base available is acetic acid. Therefore, the acid-base equilibrium between sulfuric acid and acetic acid lies far to the right:

$$H_2SO_4 + CH_3CO_2H \rightleftharpoons HSO_4^- + CH_3C(OH)_2^+$$

$CH_3C(OH)_2^+$ is a much stronger acid than H_3O^+, so some weak bases which do not react with H_3O^+ will react with $CH_3C(OH)_2^+$. Thus, the key to conjugate Brönsted superacidity is that no water is present to react as a base, so it is possible to generate stronger acids than just H_3O^+.

Conjugate Brönsted-Lewis superacids behave in a similar manner. The most common conjugate Brönsted-Lewis superacid is a mixture of fluorosulfuric acid and antimony pentafluoride, known as "Magic Acid." Antimony pentafluoride, a strong Lewis acid, stabilizes fluorosulfate, the deprotonated form of fluorosulfuric acid. Therefore, the presence of antimony pentafluoride shifts the fluorosulfuric acid autoprotolysis equilibrium to the right:

$$2HSO_3F + SbF_5 \rightleftharpoons H_2SO_3F^+ + SbF_5(SO_3F)^-$$

$H_2SO_3F^+$ is an extremely strong acid, and will readily protonate even exceedingly weak bases which are added to the Magic Acid. Thus, conjugate Brönsted-Lewis superacids create very strong Brönsted acids in nonaqueous media by promoting the autoprotolysis of the Brönsted acid through stabilization of the dissociated form of the acid.

Solid superacids are composed of solid media treated with either Brönsted or Lewis acids. The **solids** used include natural clays and minerals, metal oxides and sulfides, metal salts, and mixed metal oxides. Typical solid Brönsted superacids include titanium dioxide:sulfuric acid (TiO_2:H_2SO_4) and zirconium dioxide:sulfuric acid (ZrO_2:H_2SO_4) mixtures. The most common solid Lewis superacids involve the incorporation of antimony pentafluoride into metal oxides, such as silicon dioxide (SbF_5:SiO_2), aluminum oxide (SbF_5:Al_2O_3), or titanium dioxide (SbF_5:TiO_2). Solid superacids are typically not as strong as liquid phase-conjugate superacids. However, they are advantageous for organic catalysis, due to their ease of separation from liquid phase reaction mixtures. Therefore, a great deal of research has been devoted to the discovery and characterization of solid superacids since the early 1970s.

To date, the most significant application of superacids has been as catalysts for organic reactions such as carbon-carbon and carbon-hydrogen bond protolysis, alkane isomerization, and alkane alkylation. These reactions, some of which are run on a large scale in the petroleum industry, are initiated by the protonation of saturated hydrocarbons. Saturated hydrocarbons are very weak bases, so high temperatures are required to initiate these reactions with conventional acid or noble metal catalysts. Due to their extreme acidity, however, superacids are capable of protonating saturated hydrocarbons at much lower temperatures. Thus, the discovery of superacids has opened up a new realm of **organic chemistry** and may also significantly impact the petroleum industry.

See also Acids and bases; Alkane functional group; Hydrocarbon

SUPERCONDUCTIVITY

Superconductors are materials that exhibit zero electrical resistivity and become diamagnetic when they are cooled to a sufficiently low **temperature**. In the superconducting state, persistent electrical currents can flow without attenuation in superconducting rings for many years, and permanent magnets can be levitated by a superconductor. Because of these unique electrical and magnetic properties, superconductors have found wide applications in power transmission, magnetic **energy** storage, magnetometry, magnetic shielding, high-speed digital signal and data processing.

Superconductivity was discovered by the Dutch physicist H. Kammerling Onnes at the University of Leiden in 1911. After success in the liquification of **helium** (He), Onnes observed that the electrical resistivity of a **mercury** filament drops abruptly to an experimentally undetectable value at a temperature near 4.2 K, the **boiling point** of liquid He.

The temperature below which the resistance of the material reaches zero is referred to as the superconducting transition temperature, or the critical temperature T_c. Another unique characteristic of superconductors is their diamagnetic susceptibility, which was discovered by W. Meissner and R. Ochsenfeld in 1933. In a weak magnetic field H and below T_c, a persistent supercurrent is set up on the superconductor's surface. This persistent current induces a magnetic field which exactly cancels the external field. The interior of the superconductor remains field-free. This phenomenon is called the Meissner effect.

In general, superconductors are categorized as Type I and Type II. For Type I superconductors, e.g., most pure superconducting elements such as **lead** or **tin**, the superconductivity is completely destroyed when the superconductor is subjected to a magnetic field greater than the thermodynamic critical field H_c. In the case of Type II superconductors, e.g., some superconducting alloys and compounds such as Nb_3Sn, the superconducting state can persist beyond the Meissner region. Between a lower critical field H_{c1} and an upper critical field H_{c2}, the superconducting state coexists with tiny filaments of magnetic flux penetrating the superconductor. Above H_{c2} the flux bundles completely overlap, and the superconducting state is destroyed.

The underlying microscopic mechanism for superconductivity has been proposed to be an electron-lattice interaction. John Bardeen, Leon Cooper, and Robert Schrieffer derived a theory at the University of Illinois in 1957 that proposed that such an interaction, leading to a pairing of electrons, occur between electrons with opposite momentum and spin. These **electron** pairs are called Cooper pairs, and as described by Schrieffer, they condense into a single state and flow as a totally directionless fluid. The theory also predicted an energy gap exists in superconductors. This was further verified by the experiment of electron tunneling, which lead Brian Josephson to propose in 1962 that Cooper pairs would also tunnel from one superconductor to another through a thin insulating layer. Such a structure, called a Josephson junction, has for years been widely fabricated for superconductive electronic devices and hybrid circuits.

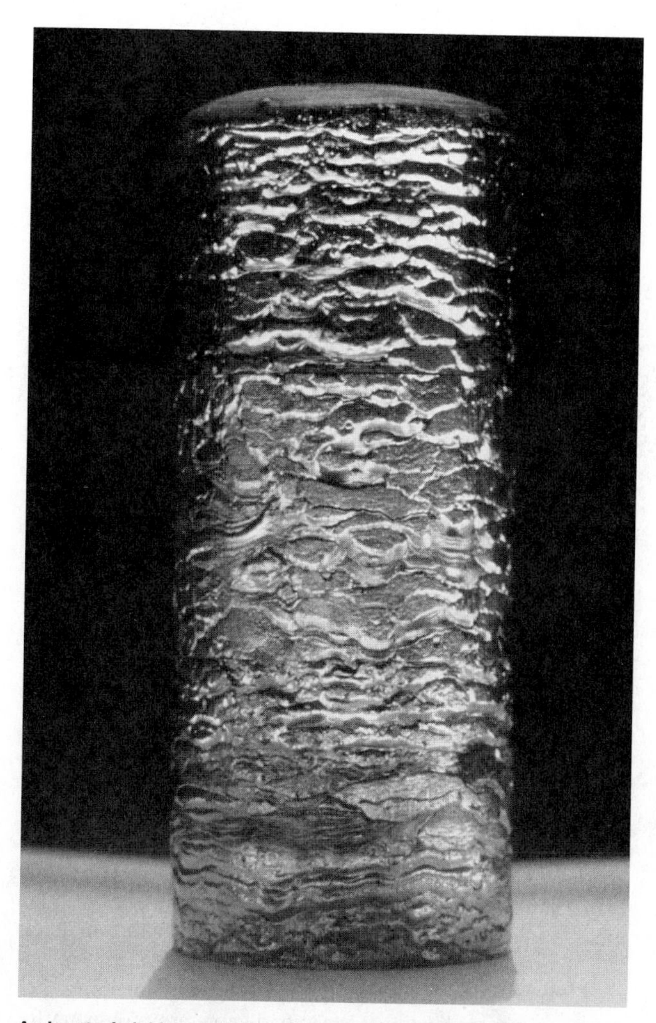

An ingot of niobium tantalum alloy, a superconducting metal material. Superconductivity occurs in certain materials at very low temperatures. Superconductors carry electrical current with little or no resistance and exclude magnetic fields. *(Photograph by Klaus Guldbrandsen/Science Photo Library, Photo Researchers, Inc. Reproduced by permission.)*

Although a variety of superconductors had been discovered and synthesized before 1987, they had to be cooled with liquid helium, which is an expensive coolant. The first superconductor with a critical temperature above 77 K (the boiling point of the less expensive liquid nitrogen) was the Type II superconductor $YBa_2Cu_3O_7$, which becomes superconducting at about 92 K. This Y-Ba-Cu-O compound was discovered in 1987; other superconducting Y-Ba-Cu-O compounds, with slight variations in composition, have since been discovered. The formula $YBaCuO_x$ is very convenient for describing this family of Type II superconductor, where x is the relative amount of **oxygen** present.

These high-temperature $YBaCuO_x$ superconductors are a mixture of metal oxides, which have the mechanical and physical properties of insulating **ceramics**. These materials contain planes in which **copper** and oxygen atoms are chemically bonded to each other. The special nature of the copper–oxygen chemical bond permits these materials to conduct **electricity** very well in preferential directions.

The $YBaCuO_x$ compounds are very sensitive to oxygen content. The high sensitivity of these superconductors to oxygen is due to the apparent ease with which that **atom** can move in and out of the crystal lattice. These superconductors usually contain more oxygen atoms than predicted by **valence** theory. There is a very subtle electronic charge **balance** between the one–dimensional copper-oxygen **chains** that exist in one direction, and the two–dimensional copper–oxygen pyramidal planes, where the superconductivity originates. With reference to the formula $YBaCuO_x$, a 90 K superconductor is obtained for $0<x<0.2$; a 60 K superconductor for $0.3<x<0.55$; and an antiferromagnetic semiconductor for $0.55<x<1.0$. In oxygen deficient $YBaCuO_x$, oxygen is removed from the copper–oxygen chains.

Superconductivity applications fall into two main areas—magnetics and electronics. Superconducting magnets can be used in magnetic **resonance** imaging, magnetic separation, magnetic levitation trains, and magnetic shielding. For power utility applications, superconductors are promising for magnetic energy storage, electrical power transmission, motors and generators. They are also used for the coatings for radio–frequency cavities. In electronic applications, thin–film interconnections and Josephson junctions are two key elements. Superconductors offer fast switching speeds and reduced wiring delays so that they are applicable for logic devices and memory cells. Superconducting field–effect transistors and Josephson–junction integrated circuits have been demonstrated. At the liquid **nitrogen** temperature, 77 K, superconductors can be further integrated with **semiconductors** to form hybrid devices. For sensor operation, superconducting quantum interference devices (SQIDs), again, based on the Josephson junction technology, are the most sensitive detector for measuring changes in magnetic field. They can detect very faint signals produced by the human brain (on the order of 10^{15} Tesla) and heart. SQID–based gradiometry is a very powerful instrument for non–destructive evaluation of materials. The increased energy gap in high temperature semiconductors allows the fabrication of superconducting electromagnetic radiation detectors used for over the spectrum from x–ray to the far infrared.

As time goes by, superconductors will find more and more applications in multidisciplinary technical fields. Recently, $YBaCuO_x$ has been shown to be a good material for the top and bottom electrodes of **oxide** ferroelectric thin–film capacitors that exhibit fatigue resistance superior conventional ones with Pt electrodes for dynamic random access memories (DRAMs). This suggests that when the microstructures and the properties of materials can be well-controlled and tailored, oxide superconductors may fulfill promises for many hybrid designs. We can also expect new hybrid fabrication technologies. New processes for thin films, thick films, wires and tapes may all be needed for the integration of a single superconductor–based instrument. Future growth in superconductor applications in the fields of electronic components, health, geology, transportation, and utilities can be expected.

Suspension

Suspensions are heterogeneous mixtures consisting of insoluble particulate **matter** distributed throughout a continuous medium. Dispersed material may range in size from particles of atomic and molecular dimensions to particles whose size is measured in millimeters. Generally, dispersed systems are classified on the basis of the mean particle diameter of the dispersed material. The dispersion is called a suspension if the particle is greater than 0.5 micrometers. Molecular dispersions have particle sizes less than 1.0 nanometer. Colloidal dispersions have particle sizes in the 1.0 nanometer to 0.50 micrometer range. Colloidal dispersions are sometimes referred to as colloidal suspensions.

Suspensions are encountered in the pharmaceutical industry, and are the end product of many industrial preparations involving paints, printing **inks**, **cosmetics**, agricultural and food products, and dyestuffs. Suspensions often experience gravitational settling and flocculation. The dispersed particles are generally visible under an ordinary light microscope. If the dispersion is of sufficiently low **viscosity** the particles can be observed to exhibit Brownian movement. The particles do not pass through normal filter **paper** nor do they dialyze through a semipermeable membrane. An example of a suspension would be the temporary dispersion that would result by mixing grains of loose sand or finely divided clay particles with **water**. The dispersed particles do not remain suspended indefinitely but eventually settle to the bottom of the container because of the gravitational pull.

Sutherland, Earl (1915-1974)

American biochemist

Earl Sutherland was a biochemist who expounded upon the manner in which **hormones** regulate body functions. His early work showed how the hormone adrenaline regulates the breakdown of sugar in the liver to release a surge of **energy** when the body is under stress. Later, Sutherland discovered a chemical within cells called cyclic adenosine 3'5'-monophosphate, or cyclic AMP. This chemical provided a universal link between hormones and the regulation of **metabolism** within cells. For this work, Sutherland was awarded the Nobel Prize in physiology and medicine in 1971.

Earl Wilbur Sutherland, Jr., the fifth of six children in his family, was born on November 19, 1915, in Burlingame, Kansas, a small farming community. His father, Earl Wilbur Sutherland, a Wisconsin native, had attended Grinnell College for two years and farmed in New Mexico and Oklahoma before settling in Burlingame to run a dry-goods business, where Earl Wilbur, Jr., and his siblings worked. Sutherland's mother, Edith M. Hartshorn, came from Missouri. She had been educated at a ''ladies college,'' and had received some nursing training. She taught Sutherland to swim at the age of five and then allowed him to go fishing by himself, a pastime that became a lifelong passion. While in high school, Sutherland also excelled in sports such as football, basketball, and tennis. In 1933 he entered Washburn College in Topeka, Kansas. Supporting his studies by working as an orderly in a hospital, Sutherland graduated with a B.S. in 1937. He married Mildred Rice the same year. Sutherland then entered Washington University Medical School in St. Louis, Missouri. There he enrolled in a pharmacology class taught by **Carl Ferdinand Cori**, who would share the 1947 Nobel Prize in medicine and physiology with his wife Gerty Cori. Impressed by Sutherland's abilities, Cori offered him a job as a student assistant. This was Sutherland's first experience with research. The research on the sugar glucose that Sutherland undertook in Cori's laboratory started him on a line of inquiry that led to his later groundbreaking studies.

Sutherland received his M.D. in 1942. He then worked for one year as an intern at Barnes Hospital while continuing to do research in Cori's laboratory. Sutherland was called into service during World War II as a battalion surgeon under General George S. Patton. Later in the war he served in Germany as a staff physician in a military hospital.

In 1945, Sutherland returned to Washington University in St. Louis. He was unsure whether to continue practicing medicine or to commit himself to a career in research. Sutherland later attributed his decision to stay in the laboratory to the example of his mentor Carl F. Cori. By 1953, Sutherland had advanced to the rank of associate professor at Washington University. During these years he came into contact with many leading figures in **biochemistry**, including **Arthur Kornberg**, **Edwin G. Krebs**, T. Z. Posternak, and others now recognized as among the founders of modern molecular biology. But Sutherland preferred, for the most part, to do his research independently. While at Washington University, Sutherland began a project to understand how an **enzyme** known as phosphorylase breaks down **glycogen**, a form of the sugar stored in the liver. He also studied the roles of the hormone adrenaline, also known as epinephrine, and glucagon, secreted by the pancreas, in stimulating the release of energy-producing glucose from glycogen.

Sutherland was offered the chairmanship of the Department of Pharmacology at Western Reserve (now Case Western) University in Cleveland in 1953. It was during the ten years he spent in Cleveland that Sutherland clarified an important mechanism by which hormones produce their effects. Scientists had previously thought that hormones acted on whole organs. Sutherland, however, showed that hormones stimulate individual cells in a process that takes place in two steps. First, a hormone attaches to specific **receptors** on the outside of the cell membrane. Sutherland called the hormone a ''first messenger.'' The binding of the hormone to the membrane triggers release of a molecule known as cyclic AMP within the cell. Cyclic AMP then goes on to play many roles in the cell's metabolism, and Sutherland referred to the molecule as the ''second messenger'' in the mechanism of hormone action. In particular, Sutherland studied the effects of the hormone adrenaline, also called epinephrine, on liver cells. When adrenaline binds to liver cells, cyclic AMP is released and directs the conversion of sugar from a stored form into a form the cell can use.

Sutherland made two more important discoveries while at Western Reserve. He found that other hormones also spur the release of cyclic AMP when they bind to cells, in particular, the adrenocorticotropic hormone and the thyroid-stimulating hormone. This implied that cyclic AMP was a sort of universal intermediary in this process, and it explained why different hormones might induce similar effects. In addition, cyclic AMP was found to play an important role in the metabolism of one-celled organisms, such as the amoeba and the bacterium *Escherichia coli,* which do not have hormones. That cyclic AMP is found in both simple and complex organisms implies that it is a very basic and important biological molecule and that it arose early in evolution and has been conserved throughout millennia.

In 1963 Sutherland became professor of physiology at Vanderbilt University in Nashville, Tennessee, a move which relieved him of his teaching duties and enabled him to devote more of his time to research. The previous year he and his first wife had divorced, and in 1963 Sutherland married Dr. Claudia Sebeste Smith, who shared with him, among other interests, a love of fishing. The couple later had two girls and two boys.

At Vanderbilt Sutherland continued his work on cyclic AMP, supported by a Career Investigatorship awarded by the American Heart Association. Sutherland studied the role of cyclic AMP in the contraction of heart muscle. He and other researchers continued to discover physiological processes in different tissues and various animal species that are influenced by cyclic AMP, for example in brain cells and cancer cells. Sutherland also did research on a similar molecule known as cyclic GMP (guanosine 3',5'-cyclic monophosphate). In the meantime, his pioneering studies had opened up a new field of research. By 1971, as many as two thousand scientists were studying cyclic AMP.

For most of his career Sutherland was well-known mainly to his scientific colleagues. In the early 1970s, however, a rush of awards gained him more widespread public recognition. In 1970 he received the prestigious Albert Lasker Basic Medical Research Award. In 1971 he was awarded the Nobel Prize for ''his long study of hormones, the chemical substances that regulate virtually every body function,'' as well as the American Heart Association Research Achievement Award. In 1973 he was bestowed with the National Medal of Science of the United States. During his career Sutherland was also elected to membership in the National Academy of Sciences, and he belonged to the American Society of Biological

Theodor Svenberg.

Chemists, the American Chemical Society, the American Society for Pharmacology and Experimental Therapeutics, and the American Association for the Advancement of Science. He received honorary degrees from Yale University and Washington University. In 1973 Sutherland moved to the University of Miami. Shortly thereafter, he suffered a massive esophageal hemorrhage, and he died on March 9, 1974, after surgery for internal bleeding, at the age of 58.

SVEDBERG, THEODOR (1884-1971)

Swedish chemist

Theodor Svedberg, helped to turn the arcane field of **colloid** chemistry into a vigorous and productive field of study. In so doing, he developed the ultracentrifuge, one of the most basic and useful tools in the modern biomedical laboratory, and an achievement for which he won the 1926 Nobel Prize in **chemistry**. Svedberg's work was not only innovative but cross-disciplinary, having valuable applications in a variety of fields, beginning with colloid chemistry. Colloids, of which milk fat and smoke are examples, are substances dispersed (as opposed to being dissolved) in a medium; colloids cannot be observed directly under the microscope, nor do they settle out under the force of gravity. Svedberg's development of the ultracentrifuge to study solutions was of enormous importance to biologists, who believed that gaining an understanding of colloids would help them to create models of biological systems.

Theodor Svedberg—called ''The Svedberg'' by his colleagues—was born on August 30, 1884, in Fleräng, Sweden, a small town near Gävle on the eastern coast. The only child of Elias Svedberg, a civil engineer employed at the local ironworks, and Augusta Alstermark Svedberg, the young Theodor often accompanied his father on long trips through the countryside, and performed simple experiments in a small laboratory at the ironworks under Elias's guidance.

Theodor attended the Karolinksa School in Örebro and showed a special aptitude for the natural sciences. Botany in particular peaked his interest, but he chose chemistry because of his interest in biological processes. His education progressed rapidly; he entered the University of Uppsala in January 1904, and received his B.S. in September of 1905 and his doctorate in 1907. He wrote his dissertation on colloids.

Until Svedberg's thesis describing his new method for producing colloidal solutions of **metals**, chemists made these mixtures by passing an electric arc between metal electrodes submerged in a liquid. Svedberg used an alternating current with an induction coil whose spark gap was submerged in a liquid to produce relatively pure colloidal mixtures of metals. The level of purity of these colloids, and the fact that the results were reproducible, permitted researchers to perform quantitative analyses during physicochemical studies. Svedberg's work propelled him quickly in the educational hierarchy at Uppsala, beginning with a lectureship in **physical chemistry** from 1907 to 1912. In 1912 he was awarded Sweden's first academic chair of physical chemistry, created by the University of Uppsala specifically for Svedberg and retained by him for thirty-six years.

Svedberg continued his work with colloids, using an ultramicroscope (a microscope that uses refracted light for visualizing specimens too small to be seen with direct light) to study the Brownian movement of particles. **Brownian motion**, the continuous random movement of minute particles suspended in liquid medium caused by collision of the particles with molecules of the medium, was named for the British botanist **Robert Brown**, who observed the phenomenon among pollen grains in **water**. Brownian movement was of great interest to a number of other researchers, including two future Nobel Prize winners, **Albert Einstein** and Jean Perrin. Perrin's work had provided verification of the theoretical work of Einstein and Marian Smoluchowski, and established definitively the existence of molecules. Perrin determined the size of large colloidal particles by measuring their rate of settling, a time-consuming process.

Using the ultramicroscope, Svedberg showed that the behavior of colloidal solutions obeys classical laws of physics and chemistry. But his method failed to distinguish the smallest particle sizes or determine the distribution of colloidal particles—the constant collisions of particles with water molecules kept the particles from settling out. In 1923 Svedberg and his colleague Herman Rinde began determining particle size distribution by measuring sediment accumulation in colloidal systems suspended on a **balance**. Although the technique itself was not new, Svedberg and Rinde increased its resolution by controlling air currents and other factors that disturbed the balance scale. While further refining this technique, Svedberg was also contemplating other approaches, especially **electrophoresis** (separation of particles in an electric field based on size and charge), and centrifugation.

Centrifugation—spinning solutions around a fixed circumference at high speed—mimics the force of gravity. Centrifuges were already being used to separate milk from cream and red blood cells from **plasma**. But fat globules and red cells are relatively large and heavy, and thus relatively easy to force out of **solution**. In order to force the much tinier and lighter colloidal particles out of solution, Svedberg needed a stronger **centrifuge** than was currently available.

In 1923 Svedberg accepted the offer of an eight-month guest professorship at the University of Wisconsin, where he taught and continued his research into centrifugation, electrophoresis, and **diffusion** of colloidal solutions. Working with J. Burton Nichols, Svedberg constructed the first ultracentrifuge, which could spin at up to thirty thousand revolutions per minute, generating gravitational forces thousands of times greater than earth's. This early ultracentrifuge was elaborate, equipped with both a camera and illumination for photographing samples during centrifugation. Using this device, Svedberg and Nichols determined particle size distributions and radii for **gold**, clay, **barium sulfate**, and arsenious sulfide.

Following his sabbatical in Wisconsin, Svedberg and his students continually increased the speed of successively higher-speed ultracentrifuges, pushing the limit from 100,000 g (gravitational force equivalent to that of earth's) at forty-five thousand revolutions per minute during the 1920s to 750,000 g by 1935. He used the machine to study **proteins**, which although huge molecules, retain their colloidal properties when in solution. Among the proteins whose weight and structure he studied using the ultracentrifuge were hemoglobin, pepsin, insulin, catalase, and albumin. The technique caught on, and Svedberg's invention became an invaluable tool used by most protein chemists. He extended his ultracentrifuge studies of large carbohydrate molecules, combining his interest in biomolecules with his interest in botany by undertaking a pioneering study of the complex sugars of the Lillifloreae family, which includes lilies and irises. His work contributed to the understanding of carbohydrate structure and provided a useful tool for later studies of evolution by biologists.

In 1926 Sweden, like much of Europe, was still recovering from the devastating effects of World War I, and research seemed destined to languish in an era of reduced government support and hopelessly outmoded facilities. Svedberg's Nobel Prize, however, gave Swedish science a boost, and led directly to the establishment in 1930 of Svedberg's proposed Institute of Physical Chemistry at Uppsala. Announced the same year that Perrin received the Nobel Prize for physics for his work with colloids, Svedberg's award greatly enhanced the recognition by science and society of the importance of the field of colloid chemistry to biological and physical processes. Svedberg became director of the new Institute for Physical Chemistry in 1931, allowing him to continue his research for the remainder of his career. During World War II, however, he was forced to switch his laboratory's research efforts to the development of polychloroprene (synthetic rubber), as well as other synthetic polymers. Despite this distraction from his main work, he was still able to devise ways to incorporate the use of the **electron** microscope and **X-ray diffraction** to study the properties of **cellulose** biomolecules. And he developed the so-called osmotic balance, which weighed colloid particles by separating particles through a permeable membrane.

On reaching mandatory retirement age in 1949, the Swedish government honored Svedberg with a promotion to emeritus professor and made a special exception to the retirement rule by appointing him lifelong director of the Gustav Werner Institute for Nuclear Chemistry; there he studied radiochemotherapy and the effects of radiation on macromolecules. Physical chemists also honored Svedberg by naming the so-called centrifugation coefficient unit after him: the svedberg unit, *s,* is equal to 1×10^{-13} seconds and represents the speed at which a particle settles out of solution divided by the force generated by the centrifuge. The coefficient depends on the **density** and shape of the particles, with specific values of *s* corresponding to specific masses measured in daltons, a unit that expresses relative atomic masses.

Svedberg was married four times, first to Andrea Andreen (1909), then to Jan Frodi Dahlquist (1916), Ingrid Blomquist Tauson (1938), and Margit Hallen Norback (1948); he had six daughters and six sons. He held memberships in the Royal Society, the American National Academy of Sciences, the Academy of Sciences of the USSR, among many other organizations. Svedberg received honorary doctorates from the universities of Delaware, Groningen, Oxford, Paris, Uppsala, Wisconsin, and Harvard. In addition, he was active in the Swedish Research Council for Technology and the Swedish Atomic Research Council. He died in Örebro, Sweden, on February 25, 1971.

SYMBOLS, CHEMICAL

The earliest chemical symbols were devised by alchemists, who often used and adapted the rich symbolism of astrology. For example, in Etienne Geoffroy's (1672-1731) table of chemical affinities, published in 1718, the astrological symbol for Mars is used to indicate **iron** (the symbol for the sun, a circle with a dot in the center, is used for **gold**). The symbolism is not accidental: the importance of the sun in astrology clearly parallels the exalted position of gold in **alchemy**. Geometrical figures, such as triangles, and circled letters were also used, N for **nickel**, M for **manganese**, etc.

John Dalton, famous for his atomic theory, also used circles surrounding capital letters: I for iron, Z for **zinc**, C for **copper**, L for **lead**, S for **silver**, and P for **platinum**.

It was **J. J. Berzelius**, who, in 1811, introduced our familiar one- or two-letter symbols for the chemical elements, with subscripts indicating how they are combined in compounds. One **atom** of an element is represented by a one- or two-letter symbol. Elements that were known first, or were

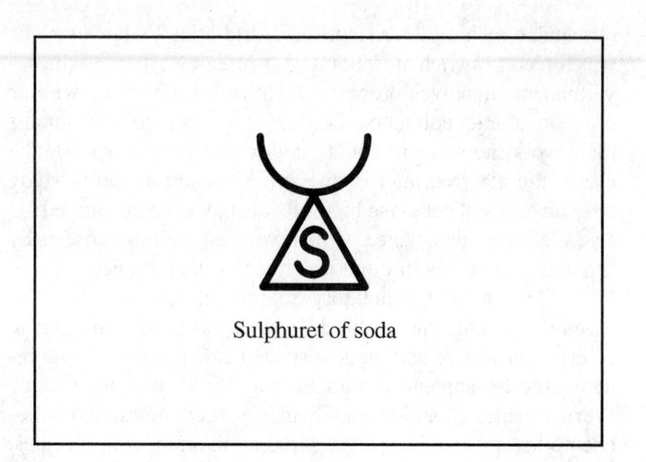

Chemical Symbol. *(Illustration by Electronic Illustrators Group.)*

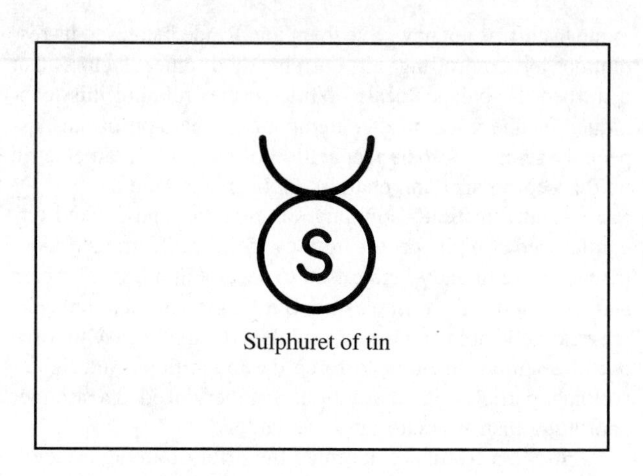

Chemical Symbol. *(Illustration by Electronic Illustrators Group.)*

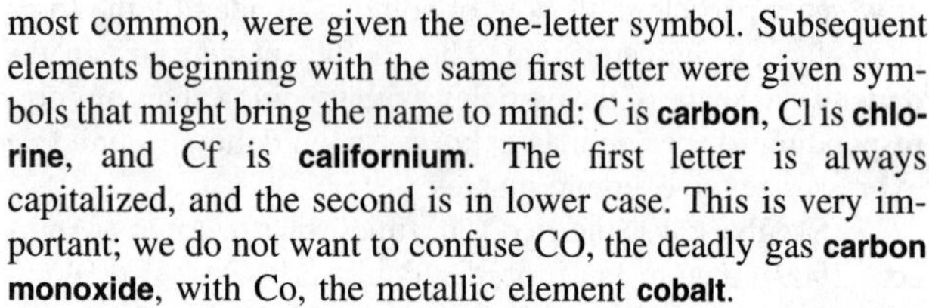

most common, were given the one-letter symbol. Subsequent elements beginning with the same first letter were given symbols that might bring the name to mind: C is **carbon**, Cl is **chlorine**, and Cf is **californium**. The first letter is always capitalized, and the second is in lower case. This is very important; we do not want to confuse CO, the deadly gas **carbon monoxide**, with Co, the metallic element **cobalt**.

Subscripts following the symbol for each atom are used to indicate how many are present in a compound. For example, a molecule of table sugar consists of 12 carbon atoms, 22 **hydrogen** atoms, and 11 oxygen atoms. The formula is $C_{12}H_{22}O_{11}$.

See also Chemical notation

SYNGE, RICHARD (1914-1994)

English biochemist and physical chemist

Richard Synge made important contributions in the fields of physical **chemistry** and **biochemistry**. He is best known for the development of partition **chromatography**, a collaborative effort undertaken with **A. J. P. Martin** in the late 1930s and early 1940s. As a result of their work, Synge and Martin received the 1952 Nobel Prize in chemistry.

Richard Laurence Millington Synge was born on October 28, 1914, in Liverpool, England, to Laurence Millington Synge, a stockbroker, and Katherine Charlotte (Swan) Synge. He was the oldest of three children and the only son. After growing up in the Cheshire area of England, he attended Winchester College, a private preparatory school, where he won a classics scholarship to attend Trinity College at Cambridge University. After listening to a speech given by the noted biochemist **Frederick Gowland Hopkins**, however, he decided to forego his education in the classics and instead pursue a degree in biochemistry at Trinity.

Synge undertook graduate studies at the Cambridge Biochemical Laboratory in 1936, receiving his Ph.D. in 1941. His doctoral research concerned the separation of acetyl **amino acids**. It was at this time that Synge first met Archer Martin, who was engaged in building a mechanism for extracting vitamin E. They began to work together on a separation process, which was delayed when Martin left for a position at the Wool Industries Research Laboratories in Leeds, England. Synge was able to join Martin there in 1939, when he received a scholarship from the International Wool Secretariat for his work on amino acids in wool.

Synge and Martin's work built on the **adsorption** chromatography techniques first developed by **Mikhail Tswett**, a Russian botanist, who evolved the procedure in his work on plant colors. Like Tswett, Synge and Martin's goal was to separate the various molecules that make up a complex substance so that the constituent molecules could be further studied. In order to achieve this goal, Tswett had filled a **glass** tube with powder, then placed a sample of the complex material to be studied at the top of the tube. When a solvent was trickled into the tube, it carried the complex material down into the powder. As the **solution** moved through the tube, the molecules of the different substances would separate and move at different speeds depending on their chemical attraction to the powder. While Tswett's technique was useful, it did not have universal application; there were a limited number of materials that could be used for the powder filling, and therefore only a limited number of substances could be identified in this manner.

In addition to Tswett's adsorption chromatography, there also existed the process of countercurrent solvent extraction. This technique involves a solution of two **liquids** that do not mix, such as **alcohol** and **water**. When a complex substance is applied to this solution, the molecules separate depending on whether they are more attracted to the water or the alcohol. Synge and Martin's breakthrough involved the combination of adsorption chromatography and countercurrent solvent extraction. This was achieved by using a solid substance adsorbent such as fine **cellulose** paper in place of Tswett's powder. In one application of the procedure, a complex mixture of molecules is spotted on one end of the **paper**, then that end is placed in a solution that might contain alcohol and water or chloroform and water. As the liquids flow through the paper, transporting the complex substance, the molecules in the substance separate depending both on their rate of adsorption by the paper and also by their affinity for either of the two liquids. When the process is completed, a series of spots is visible on the strip of paper. Each spot depicts one type of **molecule** present in the complex substance.

Synge and Martin had made early progress on partition chromatography during their time at Cambridge, but the need by industry and medicine for a more reliable technique spurred further research. At Leeds, they built a forty-unit extraction machine and experimented with various **solvents** and filtering materials. Their collaboration continued after Synge left to become a biochemist at the Lister Institute of Preventive Medicine in London in 1943, and by 1944 the improved cellulose filter method resulted. Later, they developed a two-dimensional chromatography process wherein two solvents flow at right angles to one another. This technique yielded an even sharper degree of molecular separation.

Partition chromatography was readily adopted by researchers for a variety of biochemical separations, especially those involving amino acids and **proteins**. Using the process in his doctoral research, Synge was able to separate and analyze the twenty amino acids found in protein. The technique was used in studies of **enzyme** action as well as in analyses of **carbohydrates**, **lipids**, and **nucleic acids**. Partition chromatography also became a useful tool for the food, drug, and chemical industries. Further experimentation with the process allowed proteins to be identified through the use of radioactive markers. The result of this marking was the ability to produce a photograph of the biochemical separation. The marking technique was used extensively by other biochemists, notably **Melvin Calvin** for his work in plant **photosynthesis**, and **Walter Gilbert** and **Frederick Sanger** for their research into **DNA** sequencing. All three would later receive the Nobel Prize for discoveries made using partition chromatography.

Continuing his research of amino acids and **peptides**, Synge traveled to the Institute of Biochemistry at the University of Uppsala, Sweden, in 1946. There, he and **Arne Tiselius**, the Swedish biochemist, studied other separation methods, especially **electrophoresis** and adsorption. Back home, Synge applied this knowledge toward the isolation of amino acids in rye grass in order to study their structure, a subject he collaborated on with J. C.Wood. He also used the new techniques to study the molecular makeup of plant juices, examining the juices' role as a stimulator in bacteria growth. Partition chromatography was an important factor in other research carried on by Synge at that time. With D. L. Mould, he separated sugars through electrokinetic ultrafiltration in order to study the metabolic process. With Mary Youngson, he studied rye grass proteins. He and E. P. White were able to isolate a toxin called sporidesmin, which produces eczema in sheep and other cud-chewing animals. Synge's findings in all of these areas benefited efforts by agriculture, industry, and medicine to improve human health and well-being.

In 1948 Synge accepted a position as director of the Department of Protein Chemistry at Rowett Research Institute in Scotland. From 1967 until his retirement in 1976, he was a biochemist with the Food Research Institute in Norwich. He closed out his academic career as an honorary professor in the School of Biological Sciences at the University of East Anglia from 1968 until 1984. In addition to the Nobel Prize, Synge received the John Price Wetherill Medal of the Franklin Institute in 1959. He held memberships in the Royal Society, the Royal Irish Academy, the American Society of Biological Chemists, and the Phytochemical Society of Europe. He married Ann Stephen, a physician and the niece of writer Virginia Woolf, in 1943. They had seven children. Synge died on August 18, 1994, of myelodysplastic syndrome.

Richard Synge.

Synthesis, Chemical

Chemical synthesis is the process of preparing inorganic and organic compounds by union of simpler, readily available starting materials through a series of planned reaction and isolation steps. Stoichiometric relationships derived from the balanced chemical reactions of each of the individual steps: a A + b B → c C + d D determine the maximum amount of interemediate(s) and/or product(s) (*i.e.*, chemicals C and D) that can be produced from the given starting materials (*i.e.*, chemicals A and B). For example, if one starts with a moles of chemical A plus b moles of chemical B, then one should theoretically make d moles of chemical D, assuming that the above reaction goes to completion. Chemical reactions rarely go to completion, and often some of the starting materials and products are consumed by side reactions.

The efficiency of a chemical reaction is expressed in terms of the percent yield, which is defined as:

Percent yield = 100 x (Product Synthesized/Theoretical Amount)

the ratio of the amount of product actually synthesized divided by the theoretical amount of product that should have been made based upon the balanced chemical reaction and limiting reagent. The ratio is multiplied then by 100 to convert to percent. In the case of a multi-step synthesis the overall percent yield equals the product of the percent yields of the individual steps. One- and two-step processes are generally more efficient than multi-step processes. Considerable effort is devoted in the chemical and pharmaceutical industries in designing more efficient synthetic methods. An increase in the percent yield by only a few tenths of a percent can save a company several millions of dollars in manufacturing costs.

Chemical synthesis is generally used whenever: (a) the desired compound cannot be easily obtained by a few-step reaction of readily available compounds; (b) it is not economically feasible to isolate the desired compound from a naturally occurring source; (c) there is an insufficient amount of the desired chemical in naturally occurring sources to meet consumer demand; or (d) the desired compound does not exist naturally. There is no difference between synthetic and natural compounds. Pure synthetic **ethanol** is identical to ethanol obtained from natural sources. Chemical synthesis can be performed on both the laboratory and industrial scale. Synthesis thus plays an important role in developing new chemicals.

Sulfuric acid, one of the world's most important industrial chemicals, is synthesized by a four-step contact process. **Sulfur** burns in dry air to form gaseous sulfur dioxide:

$S_{(solid)} + O_{2(gas)} \rightarrow SO_{2(gas)}$

which is then oxidized to sulfur trioxide in the presence of a **vanadium** (V) **oxide** catalyst:

$2\ SO_{2(gas)} + O_{2(gas)} +$ **heat** $\rightarrow 2\ SO_{3(gas)}$

The sulfur trioxide is absorbed in sulfuric acid to form disulfuric acid, $H_2S_2O_7$

$SO_{3(gas)} + H_2SO_4 \rightarrow H_2S_2O_7$

Dilution with **water** gives an aqueous sulfuric acid solution:

$H_2S_2O_7 + H_2O \rightarrow 2\ H_2SO_{4(aqueous)}$

Sulfuric acid is used mostly in manufacturing soluble **phosphate** and ammonium **sulfate** fertilizers; in petroleum refining; in **iron** and steel production; in making paints, pigments, dyes and rayon; and for battery acid. Two other very important inorganic compounds that are synthesized commercially are **ammonia** and **nitric acid**.

Synthesis of organic reactions requires a thorough knowledge of the reactions that expand **carbon atom** skeleton systems and that convert one organic **functional group** into another. Formation of carbon-carbon atom single bonds is the key construction step in many organic reactions. Such reactions require a carbon **nucleophile** to provide the two electrons for the chemical bond and a carbon **electrophile** to accept them. Typically, nucleophiles are either carbonanions, e.g., $RCH_2:^-$ as found in an alkyllithium compound or Grignard reagent) or the pi bonds in alkenes or aromatic **ring** systems. Electrophiles, on the other hand, are electron-deficient carbon atoms like those of carbonyls, e.g., $R_2C{=}O$) or carbon atoms which become **electron** deficient by departure of a negatively-charged leaving group. The carbon nucleophile and carbon electrophile combine to form the desired carbon-carbon bond as illustrated in the following representative unbalanced chemical reactions:

$R'_3CLi + RCH_2Cl \rightarrow R'_3CCH_2R$

$R'_2C{=}O + R_3CLi \rightarrow R_3CC(OH)R'_2$

$CN^- + RCH_2Cl \rightarrow NCCH_2R$

where R and R' are alkyl **chains** having molecular formulas of C_nH_{2n+1} and C_mH_{2m+1}, respectively. In each case the carbon atom skeleton is enlarged by at least one carbon atom. Once the desired skeleton system is obtained, is is usually easy to add functional groups or to alter (or remove or replace) existing functional groups using standard **oxidation**, **reduction**, **dehydrogenation**, dehalohydrogenation, and so on methods. The experimental conditions and reagents needed for the different functional group interconversions are wellestablished.

Polymerization is another type of chemical reaction employed in organic synthesis to couple small monomeric units into a much larger chain. The polymerization of ethylene to **polyethylene** can be represented as:

$n\ H_2C{=}CH_2 \rightarrow -[-CH_2CH_2-]_n-$

The reaction is started by an initiator, usually a free radical R•, that bonds to one of the carbon atoms of ethylene. The carbon- carbon double bond is converted into a single bond:

$R\bullet + CH_2{=}CH_2 \rightarrow RCH_2CH_2\bullet$

leaving one unpaired electron on the terminal carbon atom. The free radical species generated during the initiation step then joins with another ethylene **molecule** to form an even larger radical,

$RCH_2CH_2\bullet + H_2C{=}C_2 \rightarrow RC\ H_2CH_2CH_2CH_2\bullet$

which in turn reacts with an adjacent ethylene molecule. The process continues for several hundreds of steps. The propagation finally terminates whenever a molecule is produced that no longer has an unpaired electron. The combination of two radicals:

$R(-CH_2)_nCH_2\bullet + CH_2(CH_2-)_nR' \rightarrow R(-CH_2)_n CH_2CH_2(CH_2-)_nR'$

would be a possible termination step. **Polypropylene** and **polystyrene** are prepared in similar fashion from 1propene and styrene, respectively.

Polymerization can also be achieved through the condensation reactions that couple dissimilar ends of an organic molecule. For example, the condensation of 6-aminohexanoic acid,

$HOOC(CH_2)_5\ NH_2, nHOOC(C\ H_2)_nNH_2 + nHOOC-(CH_2)_nNH_2 \rightarrow -[-CO(CH_2)_5-NHC({=}O)-(CH_2)_5\ NH-]_n- + 2n\ H_2O$

produces a type of **nylon**. The OH portion of the carboxylic acid end of one molecule reacts with the **hydrogen** atom of the $-NH_2$ (amine) group of an adjacent 6-aminohexanoic acid molecule. A water molecule is released during the formation of the amide bond. A second type of nylon (nylon 6,6) can be prepared by reacting adipic acid with 1,6-hexanediamine, or alternatively by reacting adipoyl chloride with 1,6-hexanediamine. The latter chemical reaction generates HCl as a by-product, rather than water. There is often more than one synthetic route for preparing a desired chemical. Nylon is used as a clothing fiber in the textile industry. Organic synthesis is responsible for many of the coatings, adhesives, films, resins, polymers, dyes, medicines, **pesticides**, perfumes, soaps, preservatives, deodorants and the numerous other organic products that a person encounters in his/her everyday life.

Electrolytic cells are used in the commercial production of several important inorganic and organic compounds. **Sodium** hydroxide is made by the **electrolysis** of aqueous **sodium chloride** brine according to the following two half-reactions:

Anode reaction: $2\ Cl^- \rightarrow 2\ e^- + Cl_{2(gas)}$

Cathode reaction: $2\ H_2O + 2\ e^- \rightarrow H_{2(gas)} + 2\ OH^-$

Overall reaction: $2\ Cl^- + 2\ H_2O \rightarrow Cl_{2(gas)} + H_{2(gas)} + 2\ OH^-$

Chlorine and hydrogen **gases** are by-products of the reaction. The anode and cathode must be separated to prevent the chlorine gas from reacting with hydrogen to form **hydrochloric acid**, and from reacting with the hydroxide **ion** to form the hypochlorite ion, OCl^-. In commercial processes the anode and cathode compartments are separated by a **cation** exchange resin that allows only Na^+ ions pass between the two compartments. A concentrated sodium chloride brine is used to favor oxidation of the chloride ion over oxidation of water. The resulting **sodium hydroxide solution** is drained from the cathode compartment. Solid sodium hydroxide is obtained by evaporating the water. Sodium hydroxide is used as a base in a large number of industrial processes to neutralize acidic solutions, and is used in the manufacture of soaps, **paper** and textiles. The chlorine gas that is generated during the electrolysis is recovered, and is used both in waste water treatment and **water purification** plants, and in the production of specialty chemicals such as pesticides for crop protection, **herbicides** for weed control, and **vinyl** chloride (the starting material for polyvinyl chloride production). High-purity **copper** for making electrical wiring, **aluminum** for constructing light-weight containers, and adiponitrile for making nylon are also prepared by electrolysis methods.

Synthesis of the heavier transuranic elements, as well as of radioisotopes of the natural elements, is sometimes included under the broader umbrella of *chemical synthesis*. In 1940 the first of the transuranium elements, Np-239, was synthesized by bombarding a U-238 nuclei with neutrons.

Uncharged neutrons are effective projectiles for nuclear bombardment because they are not repelled when they approach a positively- charged nuclei. **Neutron** bombardment generally produces only small changes in **atomic number**. Larger changes in atomic number can be achieved during **nuclear fusion** reactions, which involves the formation of a heavier nuclei from two lighter nuclei. Researchers at the Lawrence Berkeley National Laboratory reportedly produced the unnamed element of atomic number 118 (mass number of 293) by bombardment of a Pb-208 target with an intense beam of high-energy Kr-86 ions. The two nuclei fused to form the new element with emission of a neutron. Element-118 subsequently decayed by **alpha particle** emission to give a second new element, Element-116 (mass number of 289). A joint Russian-American team of scientists reported the discovery of Element-114 (mass number of 289). The researchers produced the element through fusion of ^{244}Pu and ^{48}Ca nuclei.

Albert Szent-Györgyi.

Szent-Györgyi, Albert (1893-1986)

American biochemist, molecular biologist, and physiologist

Albert Szent-Györgyi was a controversial, charismatic, and intuitive scientist whose career took many paths in the course of his life: physiologist, pharmacologist, bacteriologist, biochemist, molecular biologist. In 1937 he was awarded the Nobel Prize in physiology or medicine for his work in isolating vitamin C and his advances in the study of intercellular respiration; in 1954 he received the Albert and Mary Lasker Award from the American Heart Association for his contribution to the understanding of heart disease through his research in muscle physiology. In later years, Szent-Györgyi moved into the **electron** sphere, where he studied **matter** smaller than molecules, seeking the substances that would define the basic building blocks of life. In his late seventies, he founded the National Foundation for Cancer Research.

Albert Szent-Györgyi von Nagyrapolt was born in Budapest, Hungary, on September 16, 1893, to Miklos and Josephine Szent-Györgyi von Nagyrapolt. He was the second of three sons. His father, whose family claimed a title and was said to have traced their ancestry back to the seventeenth century, was a prosperous businessman who owned a two-thousand-acre farm located outside Budapest. His mother came from a long line of notable Hungarian scientists.

As a student, Szent-Györgyi did not begin to develop his potential until his last two years in high school, when he decided to become a medical researcher. In 1911 he entered Budapest Medical School. His education was interrupted by World War I, when he was drafted into the Hungarian Army. He was decorated for bravery; but in 1916, disillusioned with the country's leadership and the progress of the war, he deliberately wounded himself in his upper arm. He was released from the army and sent back home, where he resumed his medical studies. He received his medical degree in 1917 and that same year married Cornelia (Nelly) Demeny. Their daughter, Cornelia (Little Nelly), was born in 1918.

The political situation in Hungary after the Austrian defeat caused many families to lose all they had. Szent-Györgyi's family was no exception. With Budapest under Communist rule, Szent-Györgyi decided to leave and accepted a research position at Pozony, Hungary, one hundred miles away. It was there, at the Pharmacological Institute of the Hungarian Elizabeth University, that Szent-Györgyi gained experience as a pharmacologist. In 1919 war broke out between Hungary and the Republic of Czechoslovakia. The Czechs seized Pozony, renaming it Bratislava. In order to continue his scientific training, Szent-Györgyi joined the millions of intellectuals who left Hungary during this time.

In 1921 he accepted a position at the Pharmaco-Therapeutical Institute of the University of Leiden in The Netherlands. This began a period of intense productivity for Szent-Györgyi: by the time he was twenty-nine years old, he had written nineteen research papers, and his research spanned the disciplines of physiology, pharmacology, bacteriology, and **biochemistry**. Szent-Györgyi is quoted by Ralph W. Moss, author of his biography, *Free Radical: Albert Szent-Györgyi and the Battle over Vitamin C,* as saying: "My problem was: was the hypothetical Creator an anatomist, physiologist, chemist or mathematician? My conclusion was that he had to be all of these, and so if I wanted to follow his trail, I had to have a grasp on all sides of nature." The scientist added that he "had a rather individual method. I did not try to acquire a theoretical knowledge before starting to work. I went straight to the laboratory, cooked up some senseless theory, and started to disprove it."

It was while in The Netherlands, as assistant to the professor of physiology at Groningen, that he presented the first of a series of papers on **cellular respiration** (the process by which organic molecules in the cell are converted to **carbon** dioxide and **water**, releasing energy), a question whose answer was considered central to biochemistry. Competing theories put forth on this question (one citing the priority of oxygen's role in the process; the other championing **hydrogen** as having the primary role) had caused biochemists to take one side or the other. Szent-Györgyi's contribution was that both theories were correct: active **oxygen** oxidized active hydrogen. Szent-Györgyi's research into cellular respiration laid the groundwork for the entire concept of the respiratory cycle. The paper discussing his theory is considered to be a milestone in biochemistry. Here, also, was the beginning of the work for which he was eventually given the Nobel Prize.

While still at Groningen, Szent-Györgyi began studying the role of the adrenal glands (responsible for secreting adrenaline and other important hormones), hoping to isolate a reducing agent (electron donor) and explain its role in the onset of Addison's disease. This work was to occupy him for almost a decade, produce unexpected results, and bring him worldwide attention as a scientist. He was sure he had made a breakthrough when **silver** nitrate added to a preparation of minced adrenal glands turned black. That indicated a reducing agent was present, and he set out to explain its function in oxidative **metabolism**. He thought the reducing agent might be a hormone equivalent to adrenalin. Frustrated because scientists in Groningen seemed unconvinced of the importance of his discovery, he wrote to Henry Hallett Dale, a prominent British physiologist. As a result of their correspondence, Szent-Györgyi was invited to England for three months to continue his work.

Unfortunately, his testing proved a failure—the **color** change of the silver nitrate turned out to be a reaction of adrenaline with the **iron** in the mincer in which he ground the adrenal glands. Szent-Györgyi returned to The Netherlands, where he continued his work on cellular respiration in plants, writing a paper on respiration in the potato. But increasing friction with the head of the laboratory caused him to resign his position. Unable to support his wife and daughter, he sent them home to Budapest. In August 1926 he attended a congress of the International Physiological Society in Stockholm, Sweden. It was there that his luck turned. The chairman of the event was Sir **Frederick Gowland Hopkins**, considered to be the greatest living biochemist of his day. Much to Szent-Györgyi's surprise, Hopkins referred to Szent-Györgyi's paper on potato respiration in his address to the congress. After the address, Szent-Györgyi introduced himself to Hopkins, who invited him to Cambridge, where he was to remain until he returned to Hungary in 1932, eventually becoming president of the University of Szeged.

With the assurance of a fellowship from the Rockefeller Foundation (the foundation was to be a source of much financial support throughout his career), Szent-Györgyi sent for his family, rented a house, and set to work. Hopkins became his mentor—and the man Szent-Györgyi regarded as having the most influence on him as a scientist. While at Cambridge, he was awarded a Ph.D. for the isolation of hexuronic acid, the name given to the substance he had isolated from adrenal glands. One of the puzzling things about this substance was its similarity to one also found in citrus fruits and cabbage. Szent-Györgyi set out to analyze the substance, but the main obstacle to doing this was obtaining a sufficient supply of fresh adrenal glands. He finally was able to isolate a small quantity of a similar substance from orange juice and cabbage, learning that it was a carbohydrate and a sugar acid.

In 1929 Szent-Györgyi made his first visit to the United States. It was at this time that he visited the scientific community at Woods Hole, Massachusetts. He then went on to the Mayo Clinic in Rochester, Minnesota, where he had been invited to use the research facilities to continue his work isolating the adrenal substance. He managed to purify an ounce of the substance, and sent ten grams of it back to England for analysis. Nothing came of this, however, as the amount sent was too small. After almost ten years, the research appeared to be at a dead end. Szent-Györgyi took what remained of the purified crystals and returned to Cambridge.

In 1928 Szent-Györgyi had been offered a top academic post at the University of Szeged in Hungary. He accepted, but did not take up his duties there until 1931 because of delays in completing the Szeged laboratory. At Szeged, in addition to his duties as teacher, Szent-Györgyi continued his research, still trying to solve the puzzle of the adrenal substance, hexuronic acid.

It had been known since the sixteenth century that certain foods, especially citrus fruits, prevented scurvy, a disease characterized by swollen gums and loosened teeth. Although scurvy could be prevented by including citrus fruit in the diet, isolation of the antiscurvy element from citrus eluded researchers. It was not until after World War I that drug companies began a concentrated search for the antiscorbutic element (now called vitamin C). Scientists in Europe and the United States began competing to be the first to isolate this element. Vitamin C was not unfamiliar to Szent-Györgyi, and he had written of its possible connection with hexuronic acid. Now he was able to positively identify hexuronic acid as vitamin C and not an adrenal hormone, as he had previously thought. He suggested the compound be called ascorbic acid, and continued his study of its function in the body, using vitamin C-rich Hungarian paprika as the source material.

Although Charles Glen King had also isolated Vitamin C and made the connection between it and hexuronic acid—and announced his findings just two weeks before Szent-Györgyi made his report, in 1937 Szent-Györgyi was given the Nobel Prize. His acceptance speech, ''Oxidation, Energy Transfer, and Vitamins,'' gave details of the extraordinary circumstances under which his discoveries were made.

In 1941 Szent-Györgyi and his wife were divorced. He married Marta Borbiro Miskolczy that same year. Bitterly opposed to Nazi rule in Hungary, he became an active member of the Hungarian underground. It was during the war years that he made some of his most important discoveries. His work during this time still concentrated on cellular respiration. His research in this area proved to be the basis for one of the fundamental breakthroughs in biology: the **citric acid** cycle. This cycle explains how almost all cells extract energy from food. It was during the war years that he also studied the chemical mechanisms of muscle contraction. His discoveries about how muscles move and function were fundamental to twentieth-century physiology, and made him a pioneer in molecular biology.

By 1944 Szent-Györgyi's outspoken opposition to Hitler's regime had put his life in danger. He and his wife went into hiding for the remainder of the war, surfacing in Budapest when the Russians liberated Hungary from the Nazis in 1945. Disillusioned with Soviet rule, he emigrated to the United States in 1947, and became an American citizen in 1954.

Szent-Györgyi and his wife settled in Woods Hole, Massachusetts. Research facilities were provided for him at the Marine Biological Laboratories. He struggled to find backing to continue his work. With the help of five wealthy businessmen, he set up the Szent-Györgyi Foundation (later called the Institute for Muscle Research), whose purpose was to raise money for muscle research and bring a group of Hungarian scientists to America to assist him. This endeavor met with partial success, but its full potential was not realized because of concerns about the legitimacy of the financial backing (and suspicion about the political loyalties of the Hungarians). As a result, Szent-Györgyi had a research team, but was unable to support them. To remedy this, he took a position in 1948 with the National Institutes of Health (NIH). He left there in 1950 for a short assignment at Princeton University's Institute for Advanced Studies. Then grants began to come in for his muscle research. Major funding came from Armour and Company (the Chicago meatpacking company), the American Heart Association, the Association for the Aid of Crippled Children, the Muscular Dystrophy Association, and NIH.

During these years, Szent-Györgyi and his team of researchers continued to make strides in the analysis of muscle protein. He also published three books: *Chemistry of Muscular Contraction, The Nature of Life,* and *Chemical Physiology of Contraction in Body and Heart,* and 120 scientific papers. These writings brought him to the attention of the American scientific community and had great influence on scientists worldwide.

Szent-Györgyi's wife Marta died of cancer in 1963. He married twice after her death. During the sixties and seventies, his opposition to the Vietnam War made him a hero to those connected to the peace movement. He wrote two books during this period that characterized his personal philosophy: *Science, Ethics, and Politics* and *The Crazy Ape* (which included his poem series, ''Psalmus Humanus and Six Prayers''). He spoke out against the war on numerous occasions, both in public lectures and through letters to newspapers and periodicals.

Szent-Györgyi was almost eighty years old when he founded the National Foundation for Cancer Research. Funding from the NFCR supported his research until the end of his life. For more than forty years, his research had been concerned with the development of a basic theory about the nature of life. Szent-Györgyi called this new field of endeavor ''submolecular biology.'' It was not just a cure for cancer that he was looking for, but a new way of looking at biology. He was convinced that his study of the structure of life at the level of electrons would not only make possible a cure for cancer but would also provide the knowledge to ensure the human body's optimum health.

Ralph Moss, the author of *Free Radical,* asked Szent-Györgyi for his philosophy of life shortly before the scientist's death of kidney failure on October 22, 1986. He scrawled on a piece of paper: ''Think boldly. Don't be afraid of making mistakes. Don't miss small details, keep your eyes open and be modest in everything except your aims.''

T

Takemine, Jokichi (1854-1922)

Japanese American chemist

Jokichi Takemine is best known for his isolation of the hormone epinephrine. He was also principally responsible for the Japanese government's gift of the cherry trees that bloom every spring in Washington, D.C.

Takemine was born into a Samurai family in Kanazawa, Japan. His father was a physician trained at the Dutch medical school in Nagasaki. At the age of eighteen Takemine began to study medicine in Osaka, but shortly transferred to the new school of science and engineering in Tokyo. After graduation he studied in Great Britain. Intrigued by the South Carolina displays at an exhibition in London, in the 1880s he came to the United States and learned the commercial fertilizer business. He met his future wife in New Orleans.

Returning to Japan, Takemine initiated Japan's own fertilizer industry along with various other manufacturing and chemical industries. He discovered a form of the **enzyme** diastase that converts starch to dextrin and sugar, making the commercial-scale **distillation** of grain **alcohol** possible. Returning to the United States in 1890, he and his family settled in Peoria, Illinois, where his Taka-diastase was used in liquor production.

He also conducted his own independent research and moved to New York to start his own laboratory. In 1897 he saw **John Jacob Abel**'s progress in isolating a form of epinephrine. Takemine developed a more successful isolation method of adding **ammonia** and allowing pure crystalline epinephrine to precipitate. The laboratory work was principally performed by a young assistant, Heizo Wooyenaka. Success came in 1901, when Wooyenaka left the laboratory one night without cleaning his equipment. When he returned the next morning, he found the crystals. Takemine promptly patented the process. One of the first people whose life was saved by the new product was Takemine's sister-in-law, who was hemorrhaging and in shock during childbirth.

A true internationalist, Takemine alternated long periods in the United States and Japan, and traveled worldwide to oversee his business interests. These included agreements with the Parke-Davis Pharmaceutical Co. in the West and the Sankyo Pharmaceutical Co. in Japan. In 1899, Takemine received a doctor of science degree from the University of Tokyo. His other honors included memberships in the Royal Chemical Society of England and the Japanese Academy of Science.

Tantalum

Tantalum is a transition metal in Group 5 of the **periodic table**. It has an **atomic number** of 73, an atomic **mass** of 180.9479, and a chemical symbol of Ta.

Properties

Tantalum is a very hard, malleable, ductile metal with a silvery bluish **color** when unpolished, but a bright silvery color when polished. It has a melting point of 5,425°F (2,996°C), a **boiling point** of 9,804°F (5,429°C), and a **density** of 16.69 grams per cubic centimeter. Its melting and boiling points are third highest among all elements after **tungsten** and **rhenium**.

Tantalum is one of the least reactive of all **metals**. At room **temperature** it reacts only with **fluorine** and certain fluorine compounds, although it does become more active above a temperature of about 300°F (150°C), at which point it begins to react with acids and alkalis.

Occurrence and Extraction

Tantalum ranks about 50 among the elements found in the Earth's crust with an abundance of about 1.7 parts per million. It occurs most commonly in the minerals columbite, tantalite, and microlite. The element always occurs with another chemical element, **niobium**.

Discovery and Naming

The discovery of tantalum is inextricably interwoven with the discovery of a second element, niobium. The two elements always occur together in the earth and are very difficult to separate from each other chemically. The distinction between the two elements was made in 1802 by the Swedish chemist and mineralogist Anders Gustaf Ekeberg (1767-1813). Ekeberg suggested the name tantalum for the element in honor of the Greek god Tantalus, a son of Zeus.

Uses

The commercial uses of tantalum are rather limited, the most important being in the production of microcapacitors. These capacitors are very small, but very efficient, and find use primarily in military weapons systems, aircraft and space vehicles, space communication systems, computers, and medical applications. Tantalum alloys are also used in laboratory equipment, weights for precise laboratory balances, fountain and ball point pen points, and a variety of medical and dental applications.

TARTARIC ACID

Tartaric acid, or 2,3- dihydroxybutanedioic acid, is denoted by the chemical formula $C_4H_6O_6$ and has a **molecular weight** of 150.09. It is a compound of particular importance to the historical development of organic **chemistry**. It has been known to humans for centuries. Ancient Greeks and Romans observed it in its **potassium** salt form as a by-product in the production of wine. The name tartaric acid is a derivation of the medieval, alchemical term *tartarus*. The acid was first isolated and studied in 1769 by **Carl Wilhelm Scheele**. He obtained the compound by boiling cream of tartar with chalk and then treating it with **sulfuric acid**.

As tartaric acid possesses two stereogenic **carbon** atoms, it exists in three stereoisomeric forms. The form of most importance is the (2R, 3R)-(+)- tartaric acid form, that is found in many fruits as both the free acid (tartaric) or as its salt (tartrate). In wine production (+)- tartaric acid partially separates as the mono potassium **salt**, potassium **hydrogen** tartrate (cream of tartar). In its solid form (+)- tartaric crystallizes as monoclinic sphenoidal prisms that melt between 334°F (168°C) and 338°F (170°C). It has a strong acid **taste** which is described as refreshing in a dilute aqueous **solution**. It is used as a flavoring agent in food products. Additionally, it is employed in the production of photography, tanning and **ceramics**, textile printing, and pharmaceuticals.

The remaining two stereoisomers of tartaric acid are the (-)- tartaric acid, the enantiomer of (+)- tartaric acid, and meso-tartaric acid (an achiral stereoisomer). (Stereoisomers are compounds that have identical **bonding** arrangements but with different spatial orientations of the bonds). (-)- tartaric acid is uncommon in nature, being found so far only in the leaves of an African plant. Meso-tartaric acid os not found in nature.

A pivotal experiment to the development of **organic chemistry** was the demonstration by **Louis Pasteur** that the mixed **sodium** and common salt of racemic tartaric acid (a racemate is an equal mixture of the two enantiomers) crystallizes from a cool solution (below 82.4°F [28°C]) as a conglomerate or mixture of the (+)- and (-)- salts. The two crystalline salts are physically separable — literally, with tweezers — by examination of the facial pattern of the crystal faces. Thus Pasteur established the principle of chirality as a fundamental property of organic chemical structure.

The commercial production of tartaric acid begins with potassium hydrogen tartrate obtained as a waste product from the wine industry. This material is first purified, then converted to **calcium** tartrate. The calcium tartrate is neutralized by the addition of sulfuric acid to produce calcium **sulfate** and tartaric acid. These materials are separated, and the tartaric acid is purified.

See also Stereochemistry

TASTE

In humans, taste is one of the five senses (along with sight, touch, smell, and hearing). Taste is chemically very similar to the sense of smell in that similar chemical reactions occur. The taste of something is the quality and flavor as perceived by an individual. As such it is a subjective experience. However, when something is tasted there are various chemical processes that take place.

The sense organ that provides us with the taste sensation is the tongue, or more specifically the taste buds on the tongue. The taste buds are small lumps on the surface of the tongue which are packed with chemical **receptors**. The tongue has many small projections making the surface very markedly ridged. The physical receptors for the different tastes are located in the pit at the bottom of adjacent projections. There are four different kinds of receptor, those that can detect bitter, sweet, **salt**, and sour tastes. All tastes are a combination of these four receptor types. Other flavors that can be detected are a combination of taste and smell (for example fruity and minty flavors). Because the cells are stimulated by chemicals they are called chemoreceptors.

The chemoreceptors of the tongue are distributed in a fixed pattern. At the tip of the tongue there are receptors for sweet and salt. Immediately behind these there are receptors for sweet and finally at the very back of the tongue there are receptors for bitter tastes. The receptors for sour tastes are at the sides of the tongue. It is possible to eat food and not get the full experience of the taste by missing part of the tongue. It is for this reason that wine tasters swill the wine around their mouth, they are ensuring the wine is catching all of the different taste receptors so they can detect all of the flavors in the wine.

For the taste buds to work effectively the food has to be in **solution**. When the tongue comes into contact with the food the tongue is wet. This allows some of the chemicals from the food to be taken into solution from where they are absorbed by the taste buds and the various chemoreceptors therein.

Foods may elicit a specific response without necessarily any reason. Completely unrelated food types may trigger the same reaction from the chemoreceptors on the tongue. However, a sour taste usually indicates that the substance is an acid or acidic in nature. The sense of taste is actually very limited and probably serves only as an indication of what is good to eat and what is not. As has already been stated the flavor of a substance is from a combination of the taste and the smell. If the sense of smell is not functioning, such as during a cold, it is difficult to distinguish between many kinds of food which would normally have distinct flavors.

Little is known about other organisms and taste, but it has been shown experimentally that flies can distinguish between chemicals that we would regard as sweet, sour, salt, and bitter. Most mammals probably experience taste in a similar way to humans. Part of taste in humans is genetically controlled and it is a well-known phenomena to find individuals who are unable to taste phenylthiocarbamide. This chemical normally tastes very bitter but some individuals are unable to detect any taste whatsoever.

Taste is a chemical process which takes place on the tongue using specific receptors to detect chemicals dissolved in saliva. The flavor of a food is a combination of the taste and the smell of the food.

TATUM, EDWARD LAWRIE (1909-1975)

American biochemist

Edward Lawrie Tatum's experiments with simple organisms demonstrated that cell processes can be studied as chemical reactions and that such reactions are governed by genes. With George Beadle, he offered conclusive proof in 1941 that each biochemical reaction in the cell is controlled via a catalyzing **enzyme** by a specific gene. The ''one gene-one enzyme'' theory changed the face of biology and gave it a new chemical expression. For the first time, the nature of life seemed within the grasp of science's quantitative methods. Tatum, collaborating with Joshua Lederberg, demonstrated in 1947 that bacteria reproduce sexually, thus introducing a new experimental organism into the study of molecular genetics. Spurred by Tatum's discoveries, other scientists worked to understand the precise chemical nature of the unit of heredity called the gene. This study culminated in 1953 with the description by **James Watson** and **Francis Crick** of the structure of **DNA**. Tatum's use of microorganisms and laboratory mutations for the study of biochemical genetics led directly to the biotechnology revolution of the 1980s. Tatum and Beadle shared the 1958 Nobel Prize in physiology or medicine with Joshua Lederberg for ushering in the new era of modern biology.

Tatum was born on December 14, 1909, in Boulder, Colorado, to Arthur Lawrie Tatum and Mabel Webb Tatum. He was the first of three children; a younger brother and sister would follow. Both of Edward's parents excelled academically. His father held two degrees, an M.D. and a Ph.D. in pharmacology. Edward's mother was one of the first women to graduate from the University of Colorado. Presumably an interest in science and medicine ran in the Tatum family: Edward would become a research scientist, his brother a physician, and his sister a nurse. As a boy, Edward played the French horn and trumpet; his interest in music lasted his whole life. He also enjoyed swimming and ice-skating.

Edward Lawrie Tatum. *(Archive Photo. Reproduced by permission.)*

In 1925, when Tatum was fifteen years old, his father accepted a position as a pharmacology professor at the University of Wisconsin. Tatum studied at the University of Chicago Experimental School and for two years at the University of Chicago before transferring and completing his undergraduate work at the University of Wisconsin. He almost became a geologist before deciding in his senior year to major in **chemistry**.

Tatum earned his A.B. degree in chemistry from the University of Wisconsin in 1931. In 1932 he earned his master's degree in microbiology. Two years later, in 1934, he received a Ph.D. in **biochemistry** for a dissertation on the cellular biochemistry and nutritional needs of a bacterium. Understanding the biochemistry of microorganisms such as bacteria, yeast, and molds would persist at the heart of Tatum's career.

After receiving his doctorate, Tatum remained at the University of Wisconsin for one year as a research assistant in biochemistry. He married the same year he completed his Ph.D. In Livingston, Wisconsin, Tatum wed June Alton, the daughter of a lumber dealer, on July 28, 1934. They eventually had two daughters, Margaret Carol and Barbara Ann.

From 1936 to 1937, Tatum studied bacteriological chemistry at the University of Utrecht in the Netherlands while on a General Education Board fellowship for postgraduate study. In Utrecht he worked in the laboratory of F. Kogl, who had identified the vitamin biotin. In Kogl's lab Tatum investigated the nutritional needs of bacteria and fungi. While Tatum was in Holland, he was contacted by geneticist George Beadle. Beadle, seven years older than Tatum, had done genetic studies with the fruit fly *Drosophila melanogaster* while in the laboratory of Thomas Hunt Morgan at the California Institute of Technology. Beadle, newly arrived at Stanford University, was now looking for a biochemist who could collaborate with him as he continued his work in genetics. He hoped to identify the enzymes responsible for the inherited eye pigments of *Drosophila*.

Upon his return to the United States in the fall of 1937, Tatum was appointed a research associate at Stanford University in the department of biological sciences. There he embarked on the *Drosophila* project with Beadle for four years. The two men successfully determined that kynurenine was the enzyme responsible for the fly's eye **color** and that it was controlled by one of the eye-pigment genes. This and other observations led them to postulate several theories about the relationship between genes and biochemical reactions. Yet they realized that *Drosophila* was not an ideal experimental organism on which to continue their work.

Tatum and Beadle began searching for a suitable organism. After some discussion and a review of the literature, they settled on a pink mold that commonly grows on bread known as *Neurospora crassa.* The advantages to working with *Neurospora* were many: it reproduced very quickly, its nutritional needs and biochemical pathways were already well known, and it had the useful capability of being able to reproduce both sexually and asexually. This last characteristic made it possible to grow cultures that were genetically identical and also to grow cultures that were the result of a cross between two different parent strains. With *Neurospora*, Tatum and Beadle were ready to demonstrate the effect of genes on cellular biochemistry.

The two scientists began their *Neurospora* experiments in March 1941. At that time, scientists spoke of ''genes'' as the units of heredity without fully understanding what a gene might look like or how it might act. Although they realized that genes were located on the chromosomes, they didn't know what the chemical nature of such a substance might be. An understanding of DNA (deoxyribonucleic acid, the **molecule** of heredity) was still twelve years in the future. Nevertheless, geneticists in the 1940s had accepted Gregor Mendel's work with inheritance patterns in pea plants. Mendel's theory, rediscovered by three independent investigators in 1900, states that an inherited characteristic is determined by the combination of two hereditary units (genes), one each contributed by the parental cells. A dominant gene is expressed even when it is carried by only one of a pair of chromosomes, while a recessive gene must be carried by both chromosomes to be expressed. With *Drosophila*, Tatum and Beadle had taken genetic mutants—flies that inherited a variant form of eye color—and tried to work out the biochemical steps that led to the abnormal eye color. Their goal was to identify the variant enzyme, presumably governed by a single gene, that controlled the variant eye color. This proved technically very difficult, and as luck would have it, another lab announced the discovery of kynurenine's role before theirs did. With the neurospora experiments, they set out to prove their one gene-one enzyme theory another way.

The two investigators began with biochemical processes they understood well: the nutritional needs of *Neurospora*. By exposing cultures of *Neurospora* to X rays, they would cause genetic damage to some bread mold genes. If their theory was right, and genes did indeed control biochemical reactions, the genetically damaged strains of mold would show changes in their ability to produce nutrients. If supplied with some basic salts and sugars, normal *Neurospora* can make all the **amino acids** and **vitamins** it needs to live except for one (biotin).

This is exactly what happened. In the course of their research, the men created, with X-ray bombardment, a number of mutated strains that each lacked the ability to produce a particular amino acid or vitamin. The first strain they identified, after 299 attempts to determine its **mutation**, lacked the ability to make vitamin B_6. By crossing this strain with a normal strain, the offspring inherited the defect as a recessive gene according to the inheritance patterns described by Mendel. This proved that the mutation was a genetic defect, capable of being passed to successive generations and causing the same nutritional mutation in those offspring. The X-ray bombardment had altered the gene governing the enzyme needed to promote the production of vitamin B_6.

This simple experiment heralded the dawn of a new age in biology, one in which molecular genetics would soon dominate. Nearly forty years later, on Tatum's death, Joshua Lederberg told the *New York Times* that this experiment ''gave impetus and morale'' to scientists who strived to understand how genes directed the processes of life. For the first time, biologists believed that it might be possible to understand and quantify the living cell's processes.

Tatum and Beadle were not the first, as it turned out, to postulate the one gene-one enzyme theory. By 1942 the work of English physician Archibald Garrod, long ignored, had been rediscovered. In his study of people suffering from a particular inherited enzyme deficiency, Garrod had noticed the disease seemed to be inherited as a Mendelian recessive. This suggested a link between one gene and one enzyme. Yet Tatum and Beadle were the first to offer extensive experimental evidence for the theory. Their use of laboratory methods, like X rays, to create genetic mutations also introduced a powerful tool for future experiments in biochemical genetics.

During World War II, the methods Tatum and Beadle had developed in their work with pink bread mold were used to produce large amounts of penicillin, another mold. Their basic research, unwittingly, thus had a very important practical effect as well. In 1944 Tatum served as a civilian staff member of the U.S. Office of Scientific Research and Development at Stanford. Industry, too, used the methods the men developed to measure vitamins and amino acids in foods and tissues.

In 1945, at the end of the war, Tatum accepted an appointment at Yale University as an associate professor of botany with the promise of establishing a program of biochemical microbiology within that department. Apparently the move was due to Stanford's lack of encouragement of Tatum, who failed to fit into the tidy category of biochemist or biologist or geneticist but instead mastered all three fields. In 1946 Tatum did indeed create a new program at Yale and became a professor of microbiology. In work begun at Stanford and continued at Yale, he demonstrated that the one gene-one enzyme theory applied to yeast and bacteria as well as molds.

In a second extremely fruitful collaboration, Tatum began working with Joshua Lederberg in March 1946. Lederberg, a Columbia University medical student fifteen years younger than Tatum, was at Yale during a break in the medical school curriculum. Tatum and Lederberg began studying the bacterium *Escherichia coli*. At that time, it was believed that *E. coli* reproduced asexually. The two scientists proved otherwise. When cultures of two different mutant bacteria were mixed, a third strain, one showing characteristics taken from each parent, resulted. This discovery of biparental inheritance in bacteria, which Tatum called genetic recombination, provided geneticists with a new experimental organism. Again, Tatum's methods had altered the practices of experimental biology. Lederberg never returned to medical school, earning instead a Ph.D. from Yale.

In 1948 Tatum returned to Stanford as professor of biology. A new administration at Stanford and its department of biology had invited him to return in a position suited to his expertise and ability. While in this second residence at Stanford, Tatum helped establish the department of biochemistry. In 1956 he became a professor of biochemistry and head of the department. Increasingly, Tatum's talents were devoted to promoting science at an administrative level. He was instrumental in relocating the Stanford Medical School from San Francisco to the university campus in Palo Alto. In that year Tatum also was divorced from his wife June. On December 16, 1956, he married Viola Kantor in New York City. Kantor was the daughter of a dentist in Brooklyn. Owing in part to these complications in his personal affairs, Tatum left the West Coast and took a position at the Rockefeller Institute for Medical Research (now Rockefeller University) in January 1957. There he continued to work through institutional channels to support young scientists, and served on various national committees. Unlike some other administrators, he emphasized nurturing individual investigators rather than specific kinds of projects. His own research continued in efforts to understand the genetics of neurospora and the nucleic acid **metabolism** of mammalian cells in culture.

In 1958, together with Beadle and Lederberg, Tatum received the Nobel Prize in physiology or medicine. The Nobel Committee awarded the prize to the three investigators for their work demonstrating that genes regulate the chemical processes of the cell. Tatum and Beadle shared one-half the prize and Lederberg received the other half for work done separately from Tatum. Lederberg later paid tribute to Tatum for his role in Lederberg's decision to study the effects of X-ray-induced mutation. In his Nobel lecture, Tatum predicted that "with real understanding of the roles of heredity and environment, together with the consequent improvement in man's physical capacities and greater freedom from physical disease, will come an improvement in his approach to, and understanding of, sociological and economic problems."

Tatum had a marked interest in social issues, including population control. In 1965 and 1966 Tatum organized other Nobel laureates in science to make public endorsements of family planning and birth control. These included statements to Pope Paul VI, whose encyclical against birth control for Catholics was issued at this time.

Tatum's second wife, Viola, died on April 21, 1974. Tatum married Elsie Bergland later in 1974 and she survived his death the following year, on November 5, 1975. Tatum died at his home on East Sixty-third Street in New York City after an extended illness. In a memoir written for the *Annual Review of Genetics,* Lederberg recalled that Tatum's last years were "marred by ill health, substantially self-inflicted by a notorious smoking habit." Lederberg noted, too, that Tatum's "mental outlook" was scarred by the painful death of his second wife.

In addition to the Nobel Prize, Tatum received the Remsen Award of the American Chemical Society in 1953 for his work in biparental inheritance and sexual reproduction in bacteria. In 1952 he was elected to the National Academy of Sciences. He was a founding member of the *Annual Review of Genetics* and joined the editorial board of *Science* in 1957. Tatum's collected papers occupy twenty-five feet of space in the Rockefeller University Archives and span the years from 1930 to 1975.

TAUBE, HENRY (1915-)

Canadian American chemist

A Canadian-born American chemist and professor at Stanford, Henry Taube has dedicated over fifty years to research and teaching. In addition to conducting important research in the mechanics of **electron** transfer reactions in complex **metals,** Taube has contributed greatly to increase our understanding major concepts in inorganic **chemistry** with applications in the chemical industry. For his work in this field, Taube received the Nobel Prize in chemistry in 1983.

Taube, the youngest of four boys, was born on November 30, 1915, in Neudorf, Saskatchewan, Canada, to Samuel Taube and Albertina Tiledetzski. His parents were Ukrainian peasants, who moved from their home near Kiev to Saskatchewan in 1911 to escape the tyranny of the tsar. Although his parents were uneducated and poor, they were very astute and worked hard to develop a farm. Taube himself considered farm work and its lessons of perseverance a valuable part of his education. At the age of thirteen, Taube was sent to a Lutheran boarding school, from where he moved to the University of Saskatchewan. The depression brought difficult times to the Taube family, and as a result of some failed investments, Taube's father was no longer able to support the young boy at school. However, a chemistry teacher took an interest in Taube and arranged for him to help in the laboratory in order

Henry Taube.

to continue his education. Although Taube received his best grades in chemistry, he loved English literature and wanted to become a writer.

Taube received the Bachelor of Science degree in 1935 and the Master of Science degree in **photochemistry** in 1937. Regardless, it was not until he received an opportunity to attend the University of California, Berkeley, that Taube became deeply interested in chemistry. While at Berkeley, he won the Rosenberg prize and was considered one of the most promising students. He received the Ph.D. in 1940, following which the university employed him as an instructor. Wanting to return to Canada, Taube applied for jobs at the major universities, but the opportunities never came. Instead he received a job offer from Cornell University, which he accepted in 1941.

At Cornell Taube did not find much interest in the kind of research he wanted to pursue, and by 1946 he was ready for a change. When an opportunity arose to work at the University of Chicago, a hub of scientific activity and innovative research at the time, Taube accepted. He considered his experience at Chicago a highly productive part of his career. Chosen to develop a course in advanced **inorganic chemistry**, Taube found little in published textbooks. He researched widely and became interested in complex metal or **coordination chemistry**. In the course of these investigations, he realized that work he had done on substitution of **carbon** in organic reactions could be related to inorganic complexes. In 1952 he wrote a seminal **paper** published in *Chemical Reviews* on the rates of chemical substitution, relating them to electronic structure. Although his research is outdated in parts, Taube's work was tremendously useful to chemists in planning experiments that depend on the differential in substitution rates.

Taube chaired the department of chemistry at the University of Chicago form 1956 to 1959. However, he did not enjoy administrative work, and in 1961 he took a position at Stanford University that would allow him to concentrate on research. Seeking to extend his work on substitution rates, he launched into research on **ruthenium** and **osmium**, elements with an electronic structure that fascinated him. Ruthenium is a rare metallic chemical element of the **platinum** group first found in ores from the Ural Mountains in Russia. It has remarkable back-bonding properties with chemicals such as **carbon monoxide**. The element osmium has an even greater back-bonding capacity than ruthenium. As Taube investigated the way electrons were transferred between molecules in chemical reactions, he noted unexpected changes in the electrical charge and shape of the molecules. Taube's important discovery was that before the transfer of electrons occurs, molecules build a ''chemical bridge.'' Previously, scientists had thought molecules simply exchanged electrons. Taube's work, identifying the intermediate step in electron exchanges, explained why some reactions among similar kinds of metals and ions occur at different rates.

Although inorganic reactions are important in developing principles of chemistry, interest in inorganic reactions had lagged behind the development of organic (carbon-based) chemistry. Taube's insights formed an important impulse for the further development of inorganic chemistry. For his work on the mechanisms of electron transfer, Taube won the 1983 Nobel Prize in chemistry, reversing the long-standing tradition of awarding achievements in **organic chemistry**.

In later work, Taube and several of his associates extended the scope of coordination chemistry with the study of ruthenium ammines and osmium, illuminating, for instance, the role of metals in catalysis. He was also able to show how the structure of the chemical bridge affects the electron transfer process in metals. In an interview with Richard Stevenson published in *Chemistry in Britain,* Taube described his work with chemistry as a love affair. ''I'm a chemist and I wanted to know what happens when you mix things.''

Taube is a member of fifteen societies, has received thirty-nine honors and awards, and became professor emeritus at Stanford in 1986. He has written over 330 scientific papers and articles. Taube married Mary Alice Wesche in 1952. They have two daughters and two sons. Until the mid–1960s he and his family continued to return to Saskatchewan during holidays to help with the family farm. Taube is semi-retired and spends only part of his time at Stanford tutoring small groups. He also works part-time as a consultant to various firms, including Catalytica, Inc. and Hercules, a chemical manufacturer. In his leisure time, Taube is a collector of classical vocal records from vintage 1897 to the most recent recordings, and he loves gardening.

TAUTOMERIZATION

Also known as dynamic isomerization, tautomerization is the process of one tautomer converting itself into another. Tautomers are structural isomers that exist in equilibrium with one another. Isomers are nothing more than two or more compounds that have identical molecular formulas (''iso'' means ''same''), but different chemical structures and physical properties.

Tautomerization usually involves the movement of a **hydrogen atom** between a different location on the **molecule**, resulting in two or more molecular structures. These structures are called tautomers, which exist in dynamic equilibrium with each other. In fact, unlike isomers, they are readily and directly interconvertible, meaning that with minor adjustments they can easily change from one isomeric form to the other.

Tautomers are interconvertible because of how easily their hydrogen atoms move around. These atoms are highly mobile, and can shift from place to place within the molecule. Often, this migration brings about rearrangement of a double bond as well.

Perhaps the most common tautomers are those between the keto (CH_3-CO-CH_2-$COOC_2H_5$) and enol (CH_3-C(OH)=CH-$COOC_2H_5$) forms of ethyl acetoacetate. Because acetoacetate contains these tautomers, it has the characteristics of both an unsaturated **alcohol** (from the enol) and a ketone (from the keto). Other kinds of tautomers are the three-carbon type and the amido-amidol type. There are also certain tautomers called desmotropes that can be separated from each other, unlike regular tautomers.

Just as there are different sorts of tautomers, there are different kinds of tautomerization. However, only two of these occur in laboratories with any regularity. The first is keto-enol tautomerization, after the enol-keto pair. This kind of tautomerization involves a shift of electrons and a hydrogen atom at the same time. The other type is known as ring-chain tautomerization. This phenomenon occurs with glucose when the aldehyde group in a sugar chain molecule reacts with a hydroxy group in the same molecule. The reaction gives the tautomer a cyclic shape. Lactam-lactim tautomerization, which chemists long suspected but which required the advent of sophisticated **spectroscopy** to prove, also occurs in a cyclic system.

TAYLOR, MODDIE (1912-1976)

African American chemist

Moddie Taylor gained distinction early in his career as an associate chemist on the U.S. **Manhattan Project**, which led to the development of the atomic bomb during World War II. A **chemistry** professor at Lincoln and later Howard universities, Taylor published a chemistry textbook in 1960 and served as head of the chemistry department at Howard from 1969 to 1976.

Moddie Daniel Taylor was born in Nymph, Alabama, on March 3, 1912, the son of Herbert L. Taylor and Celeste (Oliver) Taylor. His father worked as a postal clerk in St. Louis, Missouri, and it was there that Taylor went to school, graduating from the Charles H. Sumner High School in 1931. He then attended Lincoln University in Jefferson City, Missouri, and graduated with a B.S. in chemistry in 1935 as valedictorian and as a summa cum laude student. He began his teaching career in 1935, working as an instructor until 1939 and then as an assistant professor from 1939 to 1941 at Lincoln University, while also enrolled in the University of Chicago's graduate program in chemistry. He received his M.S. in 1939 and his Ph.D. in 1943. Taylor married Vivian Ellis on September 8, 1937, and they had one son, Herbert Moddie Taylor.

It was during 1945 that Taylor began his two years as an associate chemist for the top-secret Manhattan Project based at the University of Chicago. Taylor's research interest was in rare earth **metals** (elements which are the products of oxidized metals and which have special properties and several important industrial uses); his chemical contributions to the nation's atomic **energy** research earned him a Certificate of Merit from the Secretary of War. After the war, he returned to Lincoln University where he stayed until 1948, when he joined Howard University as an associate professor of chemistry, becoming a full professor in 1959 and head of the chemistry department in 1969.

In 1960, Taylor's *First Principles of Chemistry* was published; also in that year he was selected by the Manufacturing Chemists Association as one of the nation's six top college chemistry teachers. In 1972, Taylor was also awarded an Honor Scroll from the Washington Institute of Chemists for his contributions to research and teaching. Taylor was a member of the American Chemical Society, the American Association for the Advancement of Science, the National Institute of Science, the American Society for Testing Materials, the New York Academy of Sciences, Sigma Xi, and Beta Kappa Chi, and was a fellow of the American Institute of Chemists and the Washington Academy for the Advancement of Science. Taylor retired as a professor emeritus of chemistry from Howard University on April 1, 1976, and died of cancer in Washington, D.C., on September 15, 1976.

TECHNETIUM

Technetium is a transition element in Group 7 of the **periodic table**. It has an **atomic number** of 43, an atomic **mass** of 97.9072, and a chemical symbol of Tc.

Properties

All isotopes of technetium are radioactive, with the most stable being technetium-98, whose half life is about 4.2×10^6 years. The element is a silver-gray metal with a melting point of 4,000°F (2,200°C) and a **density** of 11.5 grams per cubic centimeter. The chemical properties are generally not of as much interest to scientists as are its **radioactivity**, although it is known that its properties are similar to those of **manganese** and **rhenium**, in the same group of the periodic table.

Occurrence and Extraction

Some scientists believe that minute amounts of technetium may be present in the Earth's crust, in conjunction with other radioactive materials, such as **uranium** and **radium**. However, it has never been found on the Earth, although it has been discovered in certain young stars. The element can be made artificially in particle accelerators.

Discovery and Naming

By the 1920s, scientists were relatively certain that an element with atomic number 43 must exist. But no element with that number had as yet been found to fill the designated location in the periodic table. Finally, in 1937, the element was found by Italian physicist Emilio Segrè (1905-89) and his colleague Carlo Perrier in a sample of **molybdenum** that had been bombarded with deuterons in a cyclotron at the University of California at Berkeley. Segrè and Perrier suggested the name technetium for the new element after the Greek word *technetos*, meaning ''artificial.''

Uses

Technetium is used in very limited amounts to make an **alloy** known as technetium-steel, which is very resistant to **corrosion**. Its applications are very limited, however, because of the radioactivity it contains. One **isotope** of technetium is of special value in medical diagnosis, technetium-99m. This isotope decays with a half life of about six hours, an ideal half life for a diagnostic radioisotope. The half life is ideal because it lasts long enough to produce the radiation needed to diagnose the condition of an organ, but it is short enough to present no serious long-term threat to the body.

TEFLON

Teflon is a fluorocarbon polymer known by the chemical name polytetrafluoroethylene. It is a slippery material that is chiefly known for its use in nonstick cookware and heart valves. It was first used in the gaskets and valves needed to concentrate **Uranium** 235 and resist the highly corrosive uranium hexaflouride gas.

Roy J. Plunkett of New Carlisle, Ohio, accidentally discovered Teflon. Plunkett graduated from Manchester College in 1932 with a B.A., and received a Ph.D from Ohio State University in 1936. There he became friends with future Nobel Prize winner **Paul Flory**. He then went to work at Du Pont's laboratories to research **refrigerants**, namely flourochlorohydrocarbons. To facilitate that research, he needed about 100 lb (45 kg) of tetraflouroethylene.

He developed a small pilot plant to produce the chemical, which was then stored in cylinders in a cold box. During a subsequent experiment on April 6, 1938, Plunkett and his assistant Jack Rebok realized that a full cylinder apparently contained no gaseous tetraflouroethylene. Instead of simply getting another tank, Plunkett began to investigate. Plunkett checked the valve on the tank and found a white powder. When he sawed the tank open it was full of a slippery, white, waxy substance. He theorized that the gas had polymerized.

Maria Telkes.

Teflon, as it came to be known, is resistant to strong acids, bases, **heat**, and **solvents**. Initially, it was extremely expensive to produce and was unavailable to the general public until 1960. It played an important role in the war effort, and in peacetime has come to be used in cookware, electrical insulators, space suits, and as nose cones, heat shields, and fuel tanks for space vehicles. Because Teflon is essentially inert, the body does not reject it, so it is used in artificial limb and joint replacements, sutures, heart valves, and pacemakers.

TELKES, MARIA (1900-1995)

Hungarian-born American physical chemist

Maria Telkes devoted most of her life to solar **energy** research, investigating and designing solar ovens, solar stills, and solar electric generators. She was responsible for the heating system installed in the first solar-heated home, located in Dover, Massachusetts. The importance of Telkes's work has been recognized by numerous awards and honors, including the Society of Women Engineers Achievement Award in 1952 (Telkes was the first recipient) and the Charles Greely Abbot Award from the American Section of the International Solar Energy Society.

Maria de Telkes, the daughter of Aladar and Maria Laban de Telkes, was born in Budapest, Hungary, on December 12, 1900. She grew up in Budapest and remained there to complete her high school and college education. Studying physical **chemistry** at Budapest University, she obtained a B.A. degree in 1920, then a Ph.D. in 1924. The following year, on a visit to her uncle in the United States, Telkes was hired as a biophysicist at the Cleveland Clinic Foundation investigating the energy associated with living things. Her studies looked at the sources of this energy, what occurs when a cell dies, and the energy changes which occur when a normal cell is transformed into a cancer cell. In 1937, the year she became an American citizen, Telkes concluded her research at the clinic and joined Westinghouse Electric as a research engineer. She remained at Westinghouse for two years, performing research and receiving patents on new types of thermoelectric devices, which converted **heat** energy into electrical energy.

In 1939, Telkes began working on solar energy, one of her greatest interests since her high school days. Joining the Massachusetts Institute of Technology Solar Energy Conversion Project, she continued her research into thermoelectric devices, with the heat energy now being supplied by the sun. She also researched and designed a new type of solar heating system which was installed in a prototype house built in Dover, Massachusetts, in 1948. Earlier solar heating systems stored the solar energy by heating **water** or rocks. This system differed in that the solar energy was stored as chemical energy through the crystallization of a **sodium** sulphate **solution**.

Telkes's expertise was also recruited by the United States government to study the production of drinking water from sea water. To remove **salt** from sea water, the water is vaporized to steam, then the steam is condensed to give pure water. Utilizing solar energy for vaporization of the water, she designed a solar still which could be installed on life rafts to provide water. This design was enlarged for use in the Virgin Islands, where the supply of fresh water was often a problem.

In 1953, Telkes moved to New York University and organized a solar energy laboratory in the college of engineering where she continued her work on solar stills, heating systems, and solar ovens. Transferring to the Curtiss-Wright company in 1958, she looked into the development of solar dryers and water heaters as well as the application of solar thermoelectric generators in space. Her position there, as director of research for the solar energy lab, also required her to design a heating and energy storage system for a laboratory building built by Curtiss-Wright in Princeton, New Jersey.

Working at Cryo-Therm from 1961 to 1963, Telkes developed materials for use in the protection of **temperature** sensitive instruments. Shipping and storage containers made of these materials were used for space and undersea applications in the Apollo and Polaris projects. In 1963, she returned to her efforts of applying solar energy to provide fresh water, moving to the MELPAR company as head of the solar energy application lab.

Telkes joined the Institute of Energy Conversion at the University of Delaware in 1969, where her work involved the development of materials for storing solar energy and the design of heat exchangers for efficient transfer of the energy. Her advancements resulted in a number of patents—both domestic and foreign—for the storage of solar heat. Her results were put into practical use in Solar One, an experimental solar heated building at the University of Delaware.

In 1977, the National Academy of Science Building Research Advisory Board honored Telkes for her contributions to solar heated building technology; previous honorees included Frank Lloyd Wright and Buckminster Fuller. In 1978, she was named professor emeritus at the University of Delaware, and retired from active research. She was, however, active as a consultant until about three years before her death. Telkes died on December 2, 1995, on a visit to Budapest in her native Hungary.

Tellurium sample. ***(Photograph by Russ Lappa/Science Source, National Audubon Society Collection/Photo Researchers, Inc. Reproduced by permission.)***

TELLURIUM

Tellurium is the fourth element in Group 16 of the **periodic table**, a group of elements sometimes known as the chalcogens. Tellurium has an **atomic number** of 52, an atomic **mass** of 127.60, and a chemical symbol of Te.

Properties

Tellurium is a grayish white solid with a shiny surface. Its melting point is 841.6°F (449.8°C), its **boiling point** is 1,814°F (989.9°C), and its **density** is 6.24 grams per cubic centimeter. Although the element has many metal-like properties, it is rather brittle and does not conduct an electric current very well.

Tellurium reacts with both acids and some alkalis. Its most interesting chemical property is that it combines with **gold** to form gold telluride (Au_2Te_3), a form in which much of the world's gold occurs.

Occurrence and Extraction

Tellurium is one of the rarest elements in the Earth's crust with an abundance estimated at about one part per billion. Its most common ore is sylvanite, a complex combination of gold, **silver**, and tellurium. Tellurium is obtained commercially as a byproduct in the refining of **copper** and **lead**.

Discovery and Naming

Tellurium was discovered in 1782 by the Austrian mineralogist Baron Franz Joseph Müller von Reichenstein (1740-1825). Von Reichenstein made his discovery while analyzing a sample of gold from a colleague that contained an unknown impurity. His tests indicated that the impurity was a new element, for which he suggested the name tellurium. The name was based on the Latin word *tellus*, meaning ''Earth.''

Uses

Tellurium is used primarily in alloys, to which it gives an improved quality of machinability. It is most commonly alloyed with steel, but may also be combined with copper and lead to improve their workability and to make them more resistant to vibration and fatigue. Small amounts of tellurium are used in the rubber and textile industries, primarily as catalysts. A growing application of tellurium is in electrical and electronic devices, such as photocopiers, printers, and infrared detection systems.

TEMIN, HOWARD MARTIN (1934-1994)

American Virologist and biochemist

Howard Temin was an American virologist who revolutionized molecular biology in 1965 when he found that genetic information in the form of ribonucleic acid (RNA) can be copied into deoxyribonucleic acid (DNA). This process, called reverse transcriptase, contradicted accepted beliefs of molecular biologists at that time, according to which **DNA** always passed on genetic information through **RNA**. Temin's research also contributed to a better understanding of the role viruses play in the onset of cancer. For this, he was featured on the cover of *Newsweek* in 1971, which hailed his discovery as the most important advance in cancer research in sixty years. In addition, Temin shared the 1975 Nobel Prize for physiology or medicine for his work on the Rous sarcoma virus. His discovery of the reverse transcriptase process contributed greatly to the eventual identification of the human immunodeficiency virus (HIV). Temin's later research focused on **genetic engineering** techniques. A vehement anti- smoker, he took every opportunity to warn against the dangers of tobacco, even in his acceptance speech for the Nobel Prize.

Howard Martin Temin was born in Philadelphia on December 10, 1934, to Henry Temin, a lawyer, and Annette (Lehman) Temin. The second of three sons, Temin showed an early aptitude for science and first set foot in a laboratory when he was only fourteen years old. As a student at Central High School in Philadelphia, he was drawn to biological research and attended special student summer sessions at the Jackson Laboratory in Bar Harbor, Maine. After graduation from high school, Temin enrolled at Swathmore College in Pennsylvania, where he majored and minored in biology in the school's honors program. He published his first scientific paper at the age of eighteen, and was described in his college yearbook as ''one of the future giants in experimental biology.''

After graduating from Swathmore in 1955, Temin spent the summer at the Jackson Laboratory and enrolled for the fall term at the California Institute of Technology in Pasadena. For the first year and a half, he majored in experimental embryology but then changed his major to animal virology. He studied under Renato Dulbecco, a renowned biologist, who worked on perfecting techniques for studying virus growth in tissue and developed the first plaque assay (a chemical test to determine the composition of a substance) for an animal virus. Temin received his Ph.D. in biology in 1959, and worked for another year in Dulbecco's laboratory. In 1960, he joined the McArdle Laboratory for Cancer Research at the University of Wisconsin—Madison, where he spent the remainder of his career as the Harold P. Rusch Professor of Cancer Research and the Steenbock Professor of Biological Sciences.

Temin began studying the Rous sarcoma virus (RSV) while still a graduate student in California. First identified in the early twentieth century by Peyton Rous, RSV is found in some species of hens, and was one of the first viruses known to cause tumors. In 1958, Temin and Harry Rubin, a postdoctoral fellow, developed the first reproducible assay *in vitro* (outside of an organism) for the quantitative measuring of virus growth. Accepting an appointment as assistant professor of oncology at Wisconsin in 1960, Temin continued his research with RSV. Using the assay method he and Rubin developed, Temin focused on delineating the differences between normal and tumor cells. In 1965, he announced his theory that some viruses cause cancer through a startling method of information transfer.

Scientists at the time thought that genetic information could only be passed from DNA to RNA. DNA is a long **molecule** comprised of two **chains** of nucleic units containing the sugar deoxyribose. RNA is a molecule composed of a chain of nucleic units containing the sugar ribose. For years, many of Temin's colleagues rejected his theory that some viruses actually reverse this mode of transmitting genetic information, and they cited a lack of direct evidence to support it. Temin, however, was convinced that RNA sometimes played the role of DNA and passed on the genetic codes that made a normal cell a tumor cell.

It took Temin several years, however, to prove his theory. Despite his progress in gathering evidence implicating DNA synthesis in RSV infection, many of his colleagues remained skeptical. Finally, in 1970, Temin, working with Satoshi Mitzutani, discovered a viral **enzyme** able to copy RNA into DNA. Dubbed ''a reverse transcriptase virus,'' this enzyme passed on hereditary information by seizing control of the cell and making a reverse transcript of the host DNA; in other words, the enzyme synthesized a DNA virus that contained all the genetic information of the RNA virus. This discovery was made simultaneously by biologist **David Baltimore** at his laboratory at the Salk Institute in La Jolla, California.

The work of Temin and Baltimore led to a number of impressive developments in molecular biology and recombinant DNA experimentation over the next twenty years, including characterizing retroviruses, a family of viruses that cause tumors in vertebrates by adding a specific gene for cancer cells. In 1975, Temin shared the Nobel Prize for physiology or medicine with his former mentor, Renato Dulbecco, and David Baltimore. These three scientists' research illustrated how separate avenues of scientific research could converge to produce significant advances in biology and medicine. Eventually, interdisciplinary research was to become a mainstay of modern science.

In 1987, Temin reflected on his discovery of viruses' roles in causing cancer. "I measure [my discovery's importance] by comparing what I taught in the experimental oncology course 25 years ago," said Temin in a University of Wisconsin press release, pointing out that the topic of viral carcinogenesis (the viral link to cancer) was rarely the focus of any lectures at that time. "Now, in the course we're teaching, between a third and half of the lectures are related directly or indirectly to viral carcinogenesis."

Temin's continuing efforts to understand the role viruses play in carcinogenesis had an important impact on acquired immunodeficiency syndrome (AIDS) research. Temin's discovery of reverse transcriptase provided scientists with the means to find and identify the AIDS virus His interest in genetic engineering and the causes of cancer eventually led him to another exciting discovery. He found a way to measure the **mutation** rate in retroviruses (viruses that engage in reverse transcriptase), which led to insights on the variation of cancer genes and viruses, such as AIDS. Determining the speed at which genes and viruses change provided vital information for devising attempts to vaccinate or treat viral diseases. His discovery of reverse transcriptase also led to the development of standard tools used by biologists to prepare radioactive DNA probes to study the genetic makeup of viral and malignant cells. Another genetic engineering technique that arose from this research was the ability to make DNA copies of messenger RNA, which could be isolated and purified for later study.

Temin was also interested in such areas as gene therapy, which uses gene splicing techniques to "genetically improve" the host organism. As he began to apply genetic engineering techniques to his research, he recognized legitimate concerns about producing pathogens (microorganisms that carry disease) that could escape into the environment. He also served on a committee that drew up federal guidelines in human gene therapy trials.

Temin's research convinced him that science was making progress in the fight against cancer. Temin said in a 1984 United Press International release: "We know the names of some of the genes which are apparently involved in cancer. If past history is a guide, this understanding will lead to improvement in diagnosis, therapy, and perhaps prevention."

Throughout his career, Temin continued to teach general virology courses for graduates and undergraduates. He also worked with students in his laboratory. "I get satisfaction from a number of things—from discovering new phenomena, from understanding old phenomena, from designing clever experiments—and from seeing students and post-doctoral fellows develop into independent and outstanding scientists," he stated in a University of Wisconsin press release.

A scientist and family man who shunned the spotlight after winning the Nobel Prize (which he kept in the bottom drawer of a file cabinet), Temin was committed to quietly searching for clues into the mysteries of cancer-causing viruses. Temin married Rayla Greenberg, also a geneticist, in 1962, and the couple had two daughters, Miriam and Sarah. A familiar site on the Wisconsin-Madison campus, Temin bicycled to work every day on his mountain bike. Although he preferred not to attract attention so he could better concentrate on his work, Temin did not hesitate to speak out about his beliefs. For example, Temin said in an *On Wisconsin* article, "I enjoy teaching and believe I have gained a lot from doing it. As a researcher, I'm able to present to students the newest work in certain areas. I see that as a benefit." Because of this dedication to academics, he became upset when researchers started to leave the University of Wisconsin-Madison in 1984 because of a state employee wage-freeze. Although his own salary was ensured through private and foundation support, Temin wrote the governor letters criticizing his lack of support for education and faculty researchers. Eventually, he reluctantly agreed to help the governor in developing salary proposals.

Temin also spoke out against cigarette smoking. During the award ceremonies for the Nobel Prize, he told the audience that he was "outraged" that people continued to smoke even though cigarettes were proven to contain **carcinogens**. He explained that eighty percent of all cancers were preventable because they resulted from environmental factors, such as smoking. "It was the most important general statement I could make about human cancer," he said later in a *People* magazine interview. "And I realized the Nobel Prize would give me an opportunity to speak out that a person does not ordinarily have." Temin went on to testify before the Wisconsin legislature and congress in support of anti-smoking bills. His research efforts in AIDS led him to urge the federal government to increase funding for further research into the AIDS epidemic. Despite living a lifestyle designed to minimize the risk of cancer, Temin, who never smoked, developed lung cancer in 1992. His illness was a rare form of cancer called adenocarcinoma of the lung, which is not usually associated with cigarette smoking. He died of this disease on February 9, 1994. In addition to the Nobel Prize, Temin received many other awards for his research, including the prestigious Albert Lasker Award in Basic Medical Research in 1974 and the National Medal of Science in 1992.

TEMPERATURE

Temperature is usually defined as the degree of an object's **heat** energy. The higher the **energy level**, the hotter the object.

When two objects are placed next to each other, heat **energy** will always travel from the hotter to the colder object. Temperature plays an important role in chemistry, as chemical reactions depend on the temperature of the environment. The rate of reaction generally increases with a temperature increase.

Temperature can have a significant impact on the behavior of substances. For example, at extremely low temperatures, some materials become superfluids, which "escape" out of containers. Other substances, such as **mercury**, become superconductors. **Superconductivity** was discovered in 1911, when Heike Kamerlingh Onnes (1853-1926), using liquid **helium**, cooled mercury to 4K, and discovered that the metal offered no resistance to **electricity**.

Since our perception of temperature is subjective (terms such as "hot" and "cold" are relative), scientists use **thermometers** to measure temperature. The accepted thermometer scale for scientific use is the Celsius (formerly called centigrade) scale, developed by the Swedish astronomer **Anders Celsius**. Celsius divided his scale into 100 degrees, 100 indicating the melting point of ice and 0 indicating the boiling point of water. After Celsius's death, his colleagues at the University of Uppsala reversed the scale, creating the modern form.

While the Celsius scale is widely used by scientists, the SI uses the Kelvin scale the measure temperature. Suggested by **William Thomson**, Lord Kelvin in 1848, this is an absolute scale, based on the concept of absolute zero, the lowest possible temperature, when all molecular movement stops. Although absolute zero still remains a purely theoretical concept, scientists working in the field of low-temperature physics have managed to create temperatures low enough to be measured in nanodegrees K. The Kelvin scale uses Celsius units, the main difference being that zero on the Kelvin scale is absolute. On the Kelvin scale, ice melts at 273.15K.

In the United States, temperature is also measured by the Fahrenheit scale, according to which ice melts as 32 degrees and water boils at 212 degrees. Since the Fahrenheit scale uses 80 units (from 32 to 212) for a temperature range covered by 100 Celsius units, it follows that nine Fahrenheit degrees equal five Celsius degreees.

TENNANT, SMITHSON (1761-1815)

English chemist

Tennant was largely self-educated and developed an early interest in **chemistry** as a hobby. He is best known for his discovery of two elements, **iridium** and **osmium**.

In 1781 Tennant spent a year studying at Edinburgh University, where he attended lectures given by **Joseph Black**, a chemist known for his work on **carbon** dioxide. Tennant then traveled to Sweden, where he met **Carl Wilhelm Scheele**, another chemist interested in **gases**. Although Tennant earned his medical doctorate from Cambridge University in 1796, he never practiced medicine, preferring to pursue his interest in chemistry.

In his first notable experiment, Tennant proved that diamonds are composed of pure carbon. French chemist **Antoine-Laurent Lavoisier** had previously shown that **charcoal** (carbon) and diamonds were in the same class of combustible materials, but Lavoisier thought that the true nature of diamonds might never be known. When Tennant burned a **diamond** inside a **gold** tube, it yielded the same amount of **carbon dioxide** as that produced from charcoal. Tennant insisted that this meant the two substances were chemically identical.

Tennant is best known, however, for discovering two new elements. For years scientists had tried without success to extract pure **platinum** from its ore. During his travels in Sweden, Tennant had discussed this problem with Scheele and others. Although other chemists suspected the presence of new **metals** in the black powder that was left over when platinum ore was chemically treated, it was Tennant who isolated and characterized them in 1803 and 1804. He named one iridium, from the Greek word for rainbow, because of the variety of colors produced by its compounds. The other he called osmium due to its distinctive smell; the Greek word for odor is osme.

Tennant's interest in platinum also led him and his fellow chemist William Hyde Wollaston to set up a business selling platinum boilers for making concentrated **sulfuric acid** and other products.

Tennant briefly taught chemistry at Cambridge University before he was killed in a horse riding accident in 1815.

TERBIUM

Terbium is a rare earth element, one of the elements found in Row 6 of the **periodic table**. Its **atomic number** is 65, its atomic **mass** is 158.9254, and its chemical symbol is Tb.

Properties

Terbium is a silvery gray metal soft enough to be cut with a knife. Its melting point is 2,473°F (1,356°C), its **boiling point** is about 5,070°F (2,800°C), and its **density** is 8.332 grams per cubic centimeter. The element is not very reactive chemically, although it does dissolve in most acids.

Occurrence and Extraction

Terbium is one of the rarest of the **rare earth elements**, ranking about 55th among the elements found in the Earth's crust. It is found with other rare earth elements in minerals such as monazite, cerite, gadolinite, xenotime, and euxenite. It can be produced by converting its ores first to terbium fluoride (TbF_3), and then electrolyzing the molten terbium fluoride: $2TbF_3 \rightarrow 2Tb + 3F_2$ or by reacting the terbium fluoride with **calcium** metal: $3Ca + 2TbF_3 \rightarrow 2Tb + 3CaF_2$

Discovery and Naming

Credit for the discovery of terbium is usually given to the Swedish chemist Carl Gustav Mosander. Mosander spent many years analyzing an unusual black rock found in 1787 near the town of Ytterby, Sweden, by a lieutenant in the Swedish army named Carl Axel Arrhenius (1757-1824). The rock proved to be a fruitful source of new elements, nine being discovered in it over the next century. Mosander identified one of those new elements in 1843 and suggested the name terbium for it in honor of the town of Ytterby.

Uses

Terbium has relatively few commercial uses, the most important of which is as a phosphor in cathode ray tubes, such as those used in television screens. Terbium compounds produce a green light when bombarded by electrons in such tubes. Terbium is also beginning to find some small use in fuel cells designed to operate at very high temperatures.

TESORO, GIULIANA CAVAGLIERI (1921-)

Italian-American chemist

Giuliana Cavaglieri Tesoro has built an international reputation as an expert on polymers, compounds consisting of large molecules formed by repeating units of smaller molecules. In a productive career during which she has been granted about 120 patents, Tesoro has made several important contributions to the field of textile **chemistry**. Among her accomplishments have been the development of the first antistatic chemical for synthetic fibers, the improvement of the permanent press property of textiles, and the development of flame-resistant fabrics. In honor of this research, Tesoro received the Society of Women Engineers' Achievement Award in 1978.

Tesoro was born in Venice, Italy, on June 1, 1921, one of three children born to Gino and Margherita Maroni Cavaglieri. Although her father had trained as a civil engineer, he worked as the manager of a large insurance company. He died when Tesoro was only twelve. By the time she was ready to begin her higher education in 1938, the rise of fascism in her native land meant that she could not enroll in a university there because of her Jewish ancestry. To escape such oppression, Tesoro went to Switzerland, where she briefly pursued training in X-ray technology. She immigrated to the United States in 1939, just before Italy officially entered World War II.

Tesoro, still in her teens and new to America, nevertheless set her sights high: She wanted to enter the graduate program at Yale, despite having little more than the equivalent of a high school education. As she recalled in an interview with Linda Wasmer Smith: ''I went to talk to the head of the chemistry department, and he said that I could enroll in the program if I could pass certain examinations. I studied, essentially on my own, for a number of months. Then the department head and a couple of chemistry professors gave me an oral exam, on the basis of which they decided that if I took some senior courses, I could enter the graduate school.'' This program was accelerated due to the war. In 1943, at the age of twenty-one, Tesoro completed her Ph.D.

Tesoro wasted no time establishing a solid track record in the chemical and textile industries. She worked first as a research chemist at American Cyanamid in Boundbrook, New Jersey. In 1944 she moved on to Onyx Chemical Company in Jersey City, New Jersey, where she served as chemical research director until 1955. From there, she moved again to a similar position at J. P. Stevens in Garfield, New Jersey, a job in which she remained from 1958 through 1968. After that came a year spent as a senior scientist at the Textile Research Institute. In 1969 Tesoro was named director of chemical research at Burlington Industries in Greensboro, North Carolina, a position she held for the next three years. During this period Tesoro became known as a prolific inventor of products and processes, and she was granted more than two dozen U.S. patents in 1970 alone. Her papers on applied topics ranging from antistatic finishes to flame retardants appeared in dozens of journals. In 1963 she was awarded the Olney Medal of the American Association of Textile Chemists and Colorists.

In 1973 Tesoro took a post as visiting professor at the Massachusetts Institute of Technology, and so embarked on a new phase of her career. She has since maintained ties with MIT, serving at various times in the roles of adjunct professor, senior research scientist, and senior lecturer. In 1982 she accepted a new appointment as research professor at Polytechnic University in Brooklyn, New York. As an academician, she has been able to pursue less pragmatic fields of study, and she revels in the change. ''I enjoy basic science-not data gathering, but rather concepts and things that remain important over a period of time. That's an attitude I've tried to impart to my students at the university as well,'' Tesoro explained in an interview with Linda Wasmer Smith.

Tesoro was a member of three National Research Council committees-on fire safety of polymeric materials, chemical protective clothing systems, and toxicity hazards of materials used in railway vehicles-between 1979 and 1985. She also served a term as president of the Fiber Society, and she has been a columnist for *Polymer News.* Tesoro enjoys travel; she has delivered invited papers and lectures around the United States, as well as in Western Europe, Israel, and China. She is a member of such organizations as the American Association for the Advancement of Science and the American Chemical Society, as well as a fellow of the Textile Institute in Great Britain. She was married to Victor Tesoro on April 17, 1943, in New York City. The couple have two children, Claudia and Andrew. They make their home in Dobbs Ferry, New York.

TETRATOMIC

Molecules of tetratomic elements contain four atoms. Elements are substances that cannot be chemically broken into two or more pure substances. Examples of elements include **carbon**, **hydrogen**, **potassium**, and **fluorine**. The smallest unit of a chemical element that has the properties of that element is called an **atom**. Atoms, in turn, combine with one another in infinite variation to produce the molecules of our universe. Many pure elements occur naturally as single atoms. For example, **aluminum**, **helium**, and **sodium** are *monatomic* elements, or elements whose atoms can exist alone, without forming molecules. Other elements only occur naturally as molecules made up of more than one atom of that same element chemically bonded together. For example, hydrogen is always found as a **molecule** in its elemental form (pure hydrogen). In other words, pure hydrogen gas never exists as single hydrogen atoms. Many elements ordinarily occur as **diatomic** molecules, or molecules consisting of two atoms chemically bonded. In addition to hydrogen, the elements **nitrogen**, **oxygen**, **chlorine**, **iodine** and **bromine**, all of which are **gases** at room **temperature**, also are diatomic. Oxygen, however, in addition to being diatomic can be found in a **triatomic** form called ozone. Fewer elements exist naturally as molecules consisting of four or more atoms. An example of a common and biologically important element that consists as molecules of four atoms chemically bonded together is the element **phosphorus**. Solid phosphorus usually occurs as a *tetratomic* molecule (tetra=four). Similarly, other elements are polyatomic. Crystalline **sulfur**, for example, occurs as molecules containing eight atoms.

THALLIUM

Thallium is the last element in Group 13 of the **periodic table**, a group of elements sometimes known as the **aluminum** family. Its **atomic number** is 81, its atomic **mass** is 204.37, and its chemical symbol is Tl.

Properties

Thallium is a heavy, bluish white metal that resembles **lead** (atomic number 82) in its physical properties. It is very soft and melts easily. Its melting point is 576°F (302°C), its **boiling point** is 2,655°F (1,457°C), and its **density** is 11.85 grams per cubic centimeter. Thallium is a fairly reactive element that combines with **oxygen** in the air to form a coating of thallium **oxide** (Tl_2O).

Occurrence and Extraction

Thallium is a rather uncommon element with an abundance of about 0.7 parts per million in the Earth's crust. Its most common minerals are crookesite, lorandite, and hutchinsonite. The element is usually obtained as a byproduct of the recovery of lead and **zinc** from their ores.

Discovery and Naming

Thallium was one of four elements first discovered by spectroscopic analysis. British physicist Sir **William Crookes** first observed the distinctive green spectral lines of the element in 1862. He suggested the name thallium for the element after the Greek word *thallos*, meaning ''budding twig.'' The element was discovered almost simultaneously through traditional chemical analysis by the French chemist Claude-Auguste Lamy (1820-1878).

Uses

Thallium is too rare and too expensive to have many practical uses. For many years, one of its compounds, thallium **sulfate** (Tl_2SO_4) was used as a rodenticide. Unfortunately, the mechanism by which it causes death in rats is also operative in humans. The danger to humans, especially children, posed by thallium sulfate became great enough for the compound to be banned for that purpose. Small amounts of the element are now used to make thallium sulfide (Tl_2S) for use in specialized kinds of photocells. The element has also been used experimentally to produce superconducting materials.

THÉNARD, LOUIS-JACQUES (1777-1857)

French chemist

Although Louis Thénard's farming family was large and poor, they gave education a high priority, scrimping and saving to send Thénard to school in a larger town. By age 17, he had moved on to study in Paris, where he became a professor and met his lifelong friend and co-worker, **Joseph Gay-Lussac**. Perhaps the contrasting personalities of the two chemists—Thenard's forceful, Gay-Lussac's reserved—made their collaboration more fruitful. For several years, they competed with British chemist **Humphry Davy** to discover and characterize new elements. To some degree, this competition reflected the uneasy political relations between France and England at the time.

In one case, the French team clearly won: Thénard and Gay-Lussac discovered the element **boron** just nine days before Davy. Also, they used a different technique to produce other elements in greater quantities than Davy had. The two chemists pioneered **photochemistry** by exploring the effect of light on various chemical mixtures, and they classified vegetable substances into three groups, one of which is now known as **carbohydrates**.

Thénard made another great discovery on his own. In 1818 he discovered **hydrogen** peroxide, a valuable compound used today as an antiseptic and bleaching agent. For several years he studied the chemical, defining its properties and using it to prepare new peroxide compounds.

Early in his career, Thénard became interested in how certain **metals** could promote chemical reactions—speed them up or prompt them at lower temperatures—without the metals themselves being affected. Then in 1823, he heard of an experiment in which **platinum** made hydrogen and **oxygen** combine at room **temperature**, and he decided to pursue his interest further. Working with French chemist and physicist Pierre-Louis Dulong (1785-1838), Thénard found that other metals such as (palladium, **rhodium**, and iridium) had the same effect as platinum. The two chemists studied numerous reaction conditions and contributed significantly to what is now known as the science of catalysis.

At the French government's request in 1804, Thénard created a blue pigment which today bears his name. Thénard's blue, made of **cobalt** compounds, was used to **color** fine porcelain. It replaced earlier colors that could not withstand the **heat** of porcelain-manufacturing furnaces. The compensation he received from France finally allowed Thénard to achieve financial security.

THEORELL, AXEL HUGO TEODOR (1903-1982)

Swedish biochemist

Axel Hugo Teodor Theorell (also known as Hugo Theorell) spent the majority of his career studying the action of **oxidation** enzymes, **proteins** essential for the metabolic process in plants and animals. His isolation of the yellow **enzyme** in the mid–1930s was a breakthrough toward a clearer understanding of the transformation in the cell of food into **energy**, called **cellular respiration**. Theorell's discoveries provided basic knowledge for the eventual creation of artificial life in the laboratory, and were essential to the study of such diseases as cancer and tuberculosis. In a related area of study, his work on the alcohol-burning enzymes led to a new method for testing the **alcohol** content in blood. He was the first to isolate myoglobin, a substance that gives certain muscles their red **color**. He also studied cytochrome c, a catalytic enzyme responsible for causing energy reactions in mitochondria, the cell's "powerhouse." Theorell was awarded the 1955 Nobel Prize in physiology or medicine for "his discoveries concerning the nature and mode of action of oxidation enzymes."

Theorell was born in Linköping, Sweden, on July 6, 1903, to Thure and Armida Bell Theorell. His father was a medical officer in the local militia and enjoyed singing; his mother was a gifted pianist. Young Axel absorbed their love of music, and developed an interest in his father's profession that led him to decide on a career in medicine. He received his bachelor of medicine degree (1924) and his doctor of medicine (1930) from the Karolinska Institute in Stockholm. He also studied at the Pasteur Institute in Paris. When a crippling attack of poliomyelitis made a career as a physician impractical, he decided instead to pursue research and teaching. His academic work while at Stockholm was an inquiry into the **chemistry** of **plasma lipids** (fatty acids) and their effect on red blood cells. A technique he developed at this time to separate the plasma proteins albumin and globulin was later to prove useful in his work on isolating enzymes (globular proteins) and coenzymes, which help to activate specific enzymes.

As professor of chemistry at Uppsala University from 1930 to 1936, Theorell expanded his research on plasma lipids to concentrate on myoglobin, a muscle protein whose oxygen-carrying capacities he compared to that of hemoglobin in the blood. By isolating (purifying) myoglobin, he was able to show its absorption and storage capacities, and to measure, using centrifugal force, its **molecular weight**. This determination of its physical properties showed that myoglobin was a separate protein from hemoglobin.

In 1933 Theorell received a grant from the Rockefeller Foundation that enabled him to further his study of enzymes with **Otto Warburg** at the Kaiser Wilhelm Institute (now the Max Planck Institute) in Berlin. Warburg had attempted without success to isolate the yellow enzyme. Using his own methods, Theorell accomplished the isolation. He further separated the yellow enzyme into two parts: the catalytic **coenzyme** and the pure protein apoenzyme. He also found that the main ingredient of the yellow enzyme is the plasma protein albumin. An important corollary to the research was Theorell's discovery of the chemical **chain reaction** necessary for cellular oxidation or respiration. These contributions brought a test-tube creation of life closer to reality, and advanced the study of the chemical differences between normal and cancerous cells.

Axel Hugo Teodor Theorell.

Returning to Stockholm, Theorell became head of the **biochemistry** department at the Karolinska Institute, part of a Nobel Institute established for the purpose of providing Theorell with further research opportunities. Under his direction, the department acquired a reputation for excellence that attracted biochemists from all over the world. It was here that Theorell continued his research on cytochrome c, succeeding in his attempts to purify it by 1939. He furthered this study that same year in the United States with his colleague, **Linus Pauling**, who discovered the alpha spiral (protein molecules arranged in a twisted-atom chain).

After World War II, a collaboration with **Britton Chance** of the University of Pennsylvania elucidated steps in the oxidation (breakdown) of alcohol and gave the process a name—the Theorell-Chance mechanism. Theorell's study of the enzymes that catalyze the oxidation, alcohol dehydrogenases, provided a new method for determining the level of alcohol in the bloodstream—a technique that came to be used by Sweden and West Germany to test the sobriety of their citizens. From a different perspective, Theorell's alcohol enzyme research pinpointed several bacterial strains, knowledge of which was thought to be useful in the treatment of tuberculosis.

Theorell published accounts of his findings in many scientific journals throughout Europe and the United States. His professional affiliations included membership in the Swedish Chemical Association, the Swedish Society of Physicians and Surgeons, the Royal Swedish Academy of Sciences, the International Union of Biochemistry, and the American Academy of Arts and Sciences. In addition to the 1955 Nobel Prize, he was awarded the **Paul Karrer** Medal in Chemistry of the University of Zurich, the Ciba Medal of the Biochemical Society in London, the Legion of Honor (France), and the Karolinska Institute 150th Jubilee Medal. Honorary degrees were bestowed upon him from Belgium, Brazil, the United States, and France.

His love for music continued throughout his life and played an important part in his social and community life. He played the violin and was active in Stockholm musical societies. In 1931 he married Elin Margit Alenius, a professional musician. They became parents of three sons. Theorell retired from the Nobel Institute in 1970. Afflicted with a stroke in 1974, his health deteriorated over the following years. He died on August 15, 1982, while vacationing on an island off the coast of Sweden.

THEORETICAL CHEMISTRY

Theoretical **chemistry** is the study of chemical phenomena with the help of mathematical models and computer calculations. It is different from bench chemistry done in laboratories in that theoretical chemistry is done outside of the laboratory, and is done using supercomputers.

Most of theoretical chemistry is based on quantum mechanics. In fact, another name for theoretical chemistry is **quantum chemistry**. Theoretical chemists devise mathematical models to determine the properties of **atom**s and **molecule**s, and to determine how atoms and molecules react with each other. These models are centered largely on quantum mechanical principles. Chemists perform numerically intense computations in order to make predictions from these models.

Though theoretical chemistry can be called chemistry out of the laboratory, there is a strong relationship between bench chemistry and theoretical chemistry. Theoretical chemistry allows chemists to study molecules that have not yet been isolated in a laboratory and molecules that are difficult to attain in the laboratory. It also allows chemists to study molecules without the fear of laboratory accidents and environmental hazards. The enormous advances made recently in computer technology allow computer modeling to be carried out for less cost than similar experimental work. Most importantly, theoretical work helps us better understand the behavior and properties of molecules. Theoretical chemistry and bench chemistry then have a strong, cooperative relationship.

There are limitations to what theoretical chemistry can do though. The accuracy of theoretical predictions is limited by the quality of the models. Although there is no reason to doubt the validity of quantum mechanics, the enormous complexity of **Erwin Schrödinger**'s wave equation for molecules compels chemists to use many mathematical approximations. Chemists must always check these models against experimental results, and these models must be refined and redefined based on these experimental results.

Theoretical chemistry is not only concerned with the properties of atoms and molecules. It is also involved in examining reaction pathways, reaction mechanisms, and reaction rates. Additionally, theoretical chemists also work in the **biochemistry** field. Protein molecules are large biological polymers of small molecules called **amino acids**, and each protein has a unique amino-acid sequence. These molecules spontaneously fold and coil into a characteristic three-dimensional shape in **solution**. The shapes of these **proteins** is very important because the folding and coiling produces unique surface features, such as grooves and indentations. It is these features that determine how the protein functions and interacts with other molecules. But it is nearly impossible to learn the folding patterns of proteins in the laboratory or using pen and **paper**. Instead, computational chemistry is very useful in determining the folding patterns of proteins.

Computational chemistry is a branch of theoretical chemistry, which applies the basic equations of quantum mechanics to chemical systems. It can predict bond lengths in small molecules, vibrational frequencies, and binding energies to levels of accuracy comparable to experimental methods. Computational chemistry has a very practical application in drug design. Computer-aided drug design uses computational chemistry to determine which molecules may react with the active sites of bacteria and viruses. This process cuts down the cost of conducting many laboratory experiments to determine the same information. Drugs for Alzheimer's disease, hypertension, and AIDS have already been designed using computer-aided drug design.

THERMOCHEMISTRY

Thermochemistry is the study of the **energy** changes that take place during chemical reactions. These energy changes may assume various forms. When air and **natural gas** are ignited, enough **heat** to supply a **Bunsen burner** is produced, but when glucose is formed during **photosynthesis**, light energy is absorbed in the reaction. Reactions that evolve heat are called exothermic; those that absorb heat are called endothermic.

The energy involved in a chemical reaction can be related to the differences in energy between the reactants and products. This energy is related to the heat content, or **enthalpy**, of the substances involved. "ΔH" is the change in the system's enthalpy, which is also known as the **heat of reaction**. For example, consider hydrogen peroxide, which breaks down to form liquid water and oxygen gas:

$$H_2O_2(aq) \rightarrow H_2O(l) + 1/2O_2(g) \quad \Delta H_{298} = -22.64\text{kcal}$$
(4–10)

In this case the enthalpy of the system drops and ΔH is negative. Therefore, the reaction is said to be exothermic. However, a change in the physical state of the reactants or products will cause a change in the heat of reaction. Consider what would happen if part of the energy liberated during the reaction went into vaporizing the water. In this case, less heat would be evolved:

$H_2O_2(aq) \rightarrow H_2O(g) + 1/2O_2(g)$ $\Delta H_{298} = -12.12.$ kcal

Finally, it is important to note that heats of reaction are additive. In other words, the energy change from each step of the reaction can be added together to derive the total energy change of the system. Therefore, enthalpy is considered to be a characteristic, or state property, of a system:

$H_2O_2(aq) \rightarrow H_2O(l) + 1/2O_2(g)$ $\Delta H_{298} = -22.64$ kcal (4–10)

$H_2O(l) \rightarrow H_2O(g)$ $\Delta H_{298} = +10.52$ kcal

$H_2O_2(aq) \rightarrow H_2O(g) + 1/2O_2(g)$ $\Delta H_{298} = -12.12$ kcal (4–12)

The basic laws of thermochemistry can be expressed in terms of the change in enthalpy. First, because enthalpy is directly proportional to **mass**, the change in enthalpy is proportional to the amount of substance that reacts or is produced in a reaction. Second, the change in enthalpy in a reaction is equal in magnitude to but opposite in sign to the change in the reverse reaction. Third, because enthalpy is a state property, the change in enthalpy must be independent of the path taken between the initial and final reaction states.

There are many types of energy besides heat that can be absorbed or evolved in chemical reactions. **Gasoline combustion** in an automobile engine produces mechanical as well as heat energy. An automobile battery produces enough electrical energy to drive the car's headlights.

The field of **thermodynamics** is concerned with all types of energy changes in physical systems. Thermodynamics arbitrarily divides energy changes into two categories: those involving heat and those involving work. By convention, the change in heat is positive when the system absorbs energy and negative when the system evolves heat. Similarly, the amount of work is taken to be positive when the system performs work, and negative when work is done on it. Given these conventions, one can define the total energy change of a system as the difference between the change in heat and the amount of work done. This energy change is frequently referred to as internal energy.

According to the first law of thermodynamics, which restates the law of **conservation of energy**, for reactions carried out at constant **volume**, the heat flow is exactly equal to the difference in internal energy between the reactants and the products. In the case of constant pressure conditions, the heat flow is equal to the difference in internal energy plus a contribution from the change in volume accompanying the reaction.

The internal energy of a system, like the enthalpy, is a state property, so it is fixed when the state of the system is specified. A system's change in enthalpy is related to its change in internal energy plus the amount of work (accompanied by a change in volume) that occurs when the reaction takes place against a constant pressure.

According to the second law of thermodynamics, in every energy transformation, some of the original energy is always changed into heat energy not available for later transformations. This law is just a formal statement of the everyday observation that other forms of energy often become heat energy, and that this heat eventually dissipates in to the surroundings. This law seems to imply that at some time in the future, when all energy has been converted to heat energy, no part of the universe will be warmer than any other part.

Today, thermal measurements usually involve heat introduced electrically or measured by electrical calibration, the convention has become to report heat measurements in joules. Since 1948, the thermochemical calorie has been defined as equivalent to 4.1840 joules.

When one adds heat to a system and raises its temperature, the average heat capacity between the initial and final temperatures is defined as the heat absorbed by the system divided by the change in temperature. In the limit that the amount of heat absorbed by the system is very small, and the final temperature is brought very close to the initial one, this quotient is equal to the heat capacity at that temperature.

If this process takes places at constant volume, no work is done and the heat capacity is determined from the ratio of the change in internal energy of the system divided by the change in temperature, again in the limit that these changes are very small. If the heating takes place at constant **pressure**, the heat capacity must include the work contribution. In this case, it is determined from the ratio of the change in enthalpy divided by the change in temperature, in the same limit of small changes. Heat capacity is ordinarily expressed as calories per gram per degree. The heat capacity is numerically, but not dimensionally, equivalent to another quantity, the specific heat.

Specific heat is defined as the ratio of substance's thermal capacity to that of water at 15°C, where the thermal capacity of a substance is the amount of heat needed to produce one unit change of temperature in a unit mass. Unlike heat capacity, specific heat is a dimensionless quantity.

Until about 1920, most measurements of heat were performed in such a way that the change in temperature of water or its equivalent was measured. These results were reported in calories, where one calorie was defined as the amount of heat required to raise the temperature of one gram of water by one degree Celsius. These experiments were performed at room temperature, which could range from 12 to 25°C. (The heat capacity of water over this temperature range, though not exactly constant, does not vary by much.)

The device used to measure the heat produced or consumed by a chemical reaction is called a **calorimeter**. One of the simplest calorimeters consists of an insulated cup of water with a **thermometer** inside. To measure the heat flow from a hot piece of metal to the water in the cup, one would first weigh the water in the cup, and measure its temperature. One could then calculate the heat flow from the piece of the metal to the water in the cup by measuring the difference in temperature of the water after the hot metal is dropped in the cup, followed by multiplying this difference by the weight of water in the cup and the specific heat of water. Because the heat flow from the metal to the water must be equal but opposite in sign to the heat flow for the water, one knows exactly how much heat was lost by the metal.

Although this simple cup calorimeter would be adequate to measure the heat flow in many chemical reactions, more complex reactions such as those involving gases or high temperatures require more sophisticated designs. For very precise measurements of heat flow, it is customary to use a bomb calorimeter. In this type of calorimeter, the heat liberated in a reaction is absorbed by the bomb (a heavy steel-walled container) and by the surrounding water, instead of by water alone.

Two other important chemical concepts are defined in terms of enthalpy. The molar **heat of formation** of a compound is equal to the change in enthalpy when one **mole** of the compound is formed from the constituent elements in their stable forms at 25°C and atmospheric pressure. Bond energy is defined as the change in enthalpy when one mole of bonds is broken in the gaseous state.

See also Bonding

THERMODYNAMICS

Thermodynamics is the study of the transformation of **energy**. In chemistry, thermodynamics refers to the transformations of energy associated with **chemical reactions**.

Thermodynamics deals with quantities (e.g., energy, entropy) known as state functions. A state function is a property of a system that does not depend on pathways, only on the initial and final states.

The First Law of Thermodynamics states that energy is conserved; it can neither be created nor destroyed. The Second Law of Thermodynamics states that, in a isolated system, entropy—a measure of amount of energy in a system unavailable to do work—must increase as time passes. The Third Law of Thermodynamics states that the entropy of a perfect crystal is zero when the **temperature** of the crystal is equal to absolute zero (0 K). There is a fourth law called the ''Zeroth'' law that states if two objects are each in thermal equilibrium with a third object, then all three bodies are in thermal equilibrium with each other.

Although energy is one of the most fundamental concepts in physics or chemistry— and is used in the fundamental definitions relating to thermodynamics— the term energy is difficult to define. Energy is usually defined as ''the ability to do work.'' The unit of energy in the **International System of Units** (SI), the joule (J), is replacing many older units (e.g., calorie).

Energy is a property of a system—not something that is exchanged between systems. Of course, the energy of a system can increase or decrease, but various mechanisms are required to accomplish these changes.. The mechanisms that accomplish changes in energy are forms of work. Heating is one such form of work.

Nobel prize winning physicist and renowned teacher of physics, **Richard Feynman**, once constructed an analogy between the law of conservation of energy and a mother's accounting for a child's indestructible blocks. Feynman, with his superb wit and intellect, ended his analogy by stating that accounting for energy was like accounting for the child's blocks except, ''there are no blocks.'' This was Feynman's way of saying that we do not have a precise definition for energy. We can describe what it does and we can account for it (as we do in the laws of thermodynamics) but precisely defining exactly what energy is remains difficult and elusive.

Whatever energy is described to be, the mechanisms of work (e.g., **heat**) are the ways that systems change their state of energy. Energy itself does not flow or move between objects, or between systems and surroundings.

The total internal energy in a system is best described as the sum of the forms of energy of its components. Chemistry is usually concerned with the internal energy (E) of a system where the internal energy is the sum of energy (e.g., potential, kinetic, etc.) of all the particles (**atoms, molecules**, compounds, or complexes) in the system.

Most chemical reactions are performed under constant pressure conditions (at atmospheric pressure). As a result, except for reactions involving **gases** most of the changes of energy in a system are accomplished through the transfer of heat. **Enthalpy** (H) is the thermodynamic property that measures of the heat or heat changes in a system. Enthalpy of formation is defined as the enthalpy change in a reaction that forms a compound from its constituent elements in their naturally-occurring forms.

When a chemical reaction requires heat in order for it to proceed spontaneously (i.e., proceed without further outside assistance), the reaction is said to be endothermic. When reactions yield heat, they are considered to be exothermic.

Most thermodynamic calculations are carried out at what are termed standard thermodynamic conditions. Standard thermodynamic conditions exist when all gases are at a pressure of one atmosphere, all **solids** and **liquids** are pure and aqueous solutions are concentrations of one mole solute per liter (1 M). Standard enthalpies of formation are almost always reported at a temperature of 298 K.

Just as matter is conserved, energy is always conserved in chemical reactions. That is, if the masses of reactants used in a reaction are constant then the masses of the products formed and the changes in energy between the reactants and products is also constant.

It is easy to recognize that the First Law of Thermodynamics is a law of conservation similar to the law of conservation of matter. Considered together, the two laws demonstrate the equivalence of **mass** and energy. Like **matter**, energy is conserved in all chemical reactions. Further, just as the masses of reactants and products are constant, the amount of energy emitted or absorbed in a specific chemical reaction is a constant. The First Law of Thermodynamics also demands conservation of energy. As a result, the sum of the energy or a system and its surroundings must remain constant. As the **energy level** of a system drops, the energy level of the system's surroundings must increase by the same amount.

Systems and surroundings are described by variables such as **pressure**, **volume**, temperature, specific heat, density, compressibility, and the thermal expansion coefficient. Regardless, the first law of thermodynamics demands that all of the energy must be accounted for because none is lost or destroyed.

Although, according to the First Law of Thermodynamics, energy can neither be created or destroyed, it can change its form. There are not, however, an unlimited mechanism to accomplish these transformations. The forms include nuclear energy, kinetic energy (i.e., the energy possessed by moving bodies), **potential energy**, heat energy, electrical energy, mechanical energy, light energy and chemical energy (i.e., energy stored in chemical bonds).

The Second law of Thermodynamics implies that the entropy of the universe is increasing over time. With regard to isolated system— those that can not do work upon their surroundings nor have work done upon them by the surroundings —entropy must also always increase over time.

When dealing with systems, entropy is a measure of the organization of a system. With regard to **chemistry**, a change in entropy is reflected in a change in the order or disorder of molecules. Because entropy is actually a measurement of disorder, an increase in entropy means an increase in the disorder of any system. Changes in entropy are measured in joules/Kelvin and is always related to absolute temperature (Kelvin scale).

As a consequence of the Second Law of Thermodynamics, natural process always move toward increasingly disorder or the lowest energy state. The increase in entropy within a system is spontaneous. During many processes— the organization of molecules in living things—entropy in the system may decreases, that is, the system becomes more ordered. It is important to note, however, that this decrease in entropy is never accomplished without work being done on a system. The earth is not an isolated system—liberalizing our definition of energy—the sun is constantly supplying energy to Earth's systems.

According to the Second Law of Thermodynamics atoms seek the lowest energy level (i.e., ground state). Energy in an excited atom is often reduced by conversion into light energy as **photons** are emitted by the atom. As the energy levels in the excited **hydrogen** atom decrease, the electron returns to lower energy orbitals closer to the **nucleus**.

One of the earliest statements of the first two laws of thermodynamics was put forth by German physicist Rudolf Clausius (1822-1888) in a paper published in 1865. Clausius stated that the energy of the universe is constant. and that the entropy of the universe tends to a maximum. Much of Clausius's work was based on earlier writings of French mathematician **Sadi Nicolas Léonard Carnot**. Carnot had meticulously studied and articulated the physical and chemical principles dealing with the operation of heat engines (steam engines) designed to convert heat into mechanical work.

Carnot's observations laid the intellectual and mathematical groundwork for the formal statement of the first two laws of thermodynamics. An ideal cycle would be performed by a perfectly efficient heat engine, that is, all the heat would be converted to mechanical work. Carnot's calculations dealing with thermodynamic cycles in steam engines proved that an ideal engine— one that was 100% efficient in the transformation of heat energy into mechanical energy could never exist. Because entropy increases some energy in every system is always made useless or unavailable.

There have been many attempts to build perpetual motion machines and energy contraptions that would have to violate the laws of thermodynamics if they were to operate as advertised. All have failed or been exposed as frauds either because they were purposefully designed to deceive (somewhat akin to magic tricks) or their proponents were careless in their analysis of the thermodynamics of the system and surroundings. Most of the time, wild claims to have discovered an exception to either of the first two laws of thermodynamics is simply sloppy ''accounting for the blocks.''

There are no known exceptions to the first two laws of thermodynamics.

The concept of entropy increasing, that is, of a isolated system becoming more random, with increasing amounts of energy unavailable to do work is what gives an arrow of time to reactions. We would find it most distressing to have water run uphill or shriveled fruit become ripe and then regress to an embryonic seed. The order of reactions in the natural world is based on the Second Law of Thermodynamics.

Third Law of Thermodynamics (the entropy of a perfect crystal is zero when the temperature of the crystal is equal to absolute zero (0 K)) implies that all molecular motion stops at absolute zero. The atoms in a perfectly pure crystal would be perfectly aligned and would not move. It is important to understand that this means that there is no movement between atoms. Atoms do not ''freeze'' and there is no cessation of movement of atomic particles within the atom.

The zeroth law of thermodynamics is commonly expressed as heat flowing from hot to cold objects. It would be unnatural to put a cube of ice in a warm drink and not have the ice melt (the increasing temperature allowing the constituent water molecule to move into the liquid phase) or for the drink not to get at least a bit cooler.

Much of the study of thermodynamics involves the use of sophisticated statistical analysis. Indeed, the application of statistical methods to the science is revolutionizing our views of natural processes in much the manner as did the concepts of relativity and quantum mechanics. In fact, an accurate depiction of the universe depends on the understanding and use of all three concepts.

It is a fundamental hypothesis of science, including chemistry, that natural phenomena are governed by laws that can be formulated in such a manner that they can be used to predict the outcome of events and reactions. Nonlinear thermodynamics is part of **chaos theory** and deals with efforts to find structure in systems that are so complex that they are usually termed ''unpredictable.''

In linear thermodynamics small changes produce small and predictable changes in systems. A certain measure of heat will always raise the energy level of a system by a certain amount. In nonlinear systems, the result of an action is highly dependent upon the conditions of the system. As a result, the change in energy in a nonlinear system would not only depend on the measure of heat but also on the state of the system.

Although nonlinear theory is more than half a century old, the application of high-speed computing rapidly advancing the application of the fundamental theories. A common feature of nonlinear systems is that their properties can be unexpected, counterintuitive and surprising. Non-linear thermodynamics deals with concepts such as self-organization.

See also Calorimetry; Endothermic reactions; Exothermic reactions; Energy transformations; Free energy; Heat and heat changes; Heat of combustion; Heat of formation; Heat of fusion; Heat of reaction; Heat of solution; Heat of vaporization; Heat transfer; Thermochemistry

THERMOLUMINESCENCE

Thermoluminescence is phosphorescence of a material due to heating.

Thermoluminescence occurs as a result of high-energy electrons trapped within the material being studied. The electrons are trapped at metastable sites within the mineral lattice of the material. When the material is heated the electrons are freed and this produces light very similar in quantity and quality to that of fluorescence. The amount of light produced is proportional to the **energy** that the material has absorbed as a result of exposure to ionizing radiation. While the mineral is at standard **temperature** and pressure the electrons are trapped inside the mineral and there is no thermoluminescence. The thermoluminescence does not manifest itself until a temperature of 932°F (500°C) is reached. The electrons are released at this temperature and they are able to return to their normal atomic orbits. As they do this photons of light are emitted in the visible spectrum.

A graph plotting light against temperature is called a glow curve and it is characteristic of a given mineral. The amount of light emitted is proportional to the sensitivity of the material and the amount of radiation absorbed. Comparison between the sample under test and a sample with a known amount of added radiation can be made to give an indication of when the sample was last exposed to ionizing radiation (a clock resetting event). A clock resetting event can be heating, so it can be calculated when a meteorite heated up in the atmosphere, or when a piece of pottery was fired, or when some material was burned.

Thermoluminescence is a technique which can be used to date material that has once been heated.

THERMOMETERS

A thermometer is an instrument used to measure **temperature**. Most modern thermometers tend to be a hollow tube of **glass** marked with a scale and filled with a liquid, either **mercury** or **alcohol**. The thermometer works because there is a constant ratio between the amount the liquid expands and the increase in temperature. The standard mercury thermometer was invented by German physicist **Gabriel Fahrenheit** in the early 1700s. (He devised the temperature scale that bears his name.) The liquid is stored in a reservoir at the base of the thermometer and the column is marked off with an appropriate scale. Strictly speaking, the linear relationship between the temperature and expansion of these **liquids** does not hold true. For greater accuracy and reliability, the liquid should be replaced with gas. A gas thermometer has the column filled with gas at a low pressure and there is a weight at the top of the column to maintain a constant pressure.

A gas thermometer has an advantage in that it can be used over all temperatures encountered. A mercury thermometer cannot be used below –38.2°F (–39°C) as this is the temperature at which mercury freezes. An alcohol thermometer will remain liquid down to a temperature of –175°F (–115°C) although it boils at 158°F (70°C).

Other types of thermometers includes the thermocouple, which is a pair of wires or **semiconductors** joined at both ends. One junction of the two materials is at a fixed temperature and the other is at the temperature to be measured. This results in the flow of an electric current and for a known pair of materials, the current flowing is directly proportional to the temperature difference. By connecting the system to a suitably calibrated ammeter, the temperature can be read. A similar system operates with a bimetallic strip. Two different types of metal are bonded together and the temperature can be calculated using the difference in the expansion shown by each metal, this forces the strip to bend one way or the other, the degree of bending being proportional to the temperature.

Another type of thermometer now used is known as the Galileo thermometer. This is named after the Italian scientist Galileo Galilei. This type of thermometer utilizes the principle that different liquids and mixtures of liquids have different densities at different temperatures. Sealed glass balls of mixtures of organic liquids are placed in a column of alcohol. The balls rise or fall depending upon the temperature and the appropriate temperature is engraved on the glass ball. It must be stressed this type of thermometer works on principles first worked out by Galileo, it was not invented by him nor was it in use during his time. Of the types of thermometer discussed this is the least accurate. The most accurate thermometer is the constant pressure gas thermometer.

Some thermometers are designed for specific purposes and consequently they operate over a very small range of temperatures. For example, a clinical thermometer is made to work only a few degrees either side of body temperature. It has a very fine capillary tube internally so that small changes in temperature can readily be seen. Another adaptation the clinical thermometer has is a small constriction in the tube so the mercury or alcohol cannot return to the bulb until it is reset. This allows the temperature to be viewed once the thermometer has been removed.

See also Temperature

THIOL FUNCTIONAL GROUP

Thiols are molecules which contain an SH group. They are structurally similar to alcohols which contain an OH group.

Chemically, thiols and alchohols react similarly. **Oxygen** and **sulfur** both appear on the same column of the **periodic table** with oxygen in the second row and sulfur in the third. In 1834, W. C. Zeise discovered thiols. Zeise named thiols as mercaptans because they readily react with **mercury** to form an insoluble **salt**. Thiols tend to be a clear liquid or white crystalline form. Characteristic of other sulfur-containing compounds, thiols have a stench that smells similar to rotten eggs.

Chemically, thiols occur readily in petroleum processes such as **distillation**. Biologically, thiols are found in the human body in protein linkages. The amino acid cysteine contains a thiol group, which bands with another cysteine to form disulfide bridges. Thiols are also found naturally in garlic, onions, coffee, and skunk secretions. Because of thiol's distinct odor, it is often added to odorless **gases** such as **methane** in order to detect leaks. It can also be used for **corrosion** protection of antifreeze, regulation of polymerization, and manufacture of insecticide and pharmaceuticals. In recent technology, thiols are used to create self-assembled monolayers (SAMs) on **gold** and other noble **metals**, which are protected from **oxidation** and other chemical effects.

Thomas, Martha Jane Bergin (1926-)

American chemist and engineer

Martha Jane Bergin Thomas made significant contributions to the development of phosphors, solid materials that emit visible light when activated by an outside **energy** source. In a productive career, she achieved many firsts, becoming the first female director at GTE Electrical Products and the first woman to receive the New England Award for engineering excellence from the Engineering Societies of New England. She is the holder of twenty-three patents, ranging from innovations in electric light technology to improvements in lamp manufacturing methods.

Thomas was born on March 13, 1926, in Boston. Her parents, both teachers, were John A. and Augusta Harris Bergin. "Even as a girl, I had an intense interest in science," Thomas recalled in an interview with Linda Wasmer Smith. After high school, she pursued that interest at Radcliffe, where she graduated with honors in 1945 at the age of nineteen. Her bachelor's degree in **chemistry** was supposed to be the initial step toward a medical degree. But then she was offered a job at Sylvania—later GTE Electrical Products—in Danvers, Massachusetts. It was the first nonteaching position to come her way, and Thomas accepted. So began her forty-five-year association with the company, during which she rose from junior technician to director of the technical services labs.

Thomas did not abandon her educational goals, however. She attended graduate school at Boston University, where she received an A.M. degree in 1950 and a Ph.D. two years later. In 1980 she became a part-time student once again; motivated by her new responsibilities as a manager, she obtained a master's degree in business administration from Northeastern University. Thomas has since been honored as a distinguished graduate of every institution from which she earned a degree: Boston Girl's Latin School, Radcliffe, Boston University, and Northeastern.

Thomas's first patent was for a method of etching fine **tungsten** coils that was designed to improve telephone switchboard lights. She went on to establish two pilot plants for the preparation of phosphors—the powdery substances used to coat the inside of fluorescent lighting tubes. Among her accomplishments was the development of a natural white phosphor that allowed fluorescent lamps to impart daylight hues. She also developed a phosphor that raised **mercury** lamp brightness by 10 percent. These contributions were noted by the Society of Women Engineers in 1965, when it named Thomas Woman Engineer of the Year. Thomas also was named New England Inventor of the Year at a 1991 event sponsored by Boston's Museum of Science, the Inventors Association of New England, and the Boston Patent Law Association.

In addition to her applied research, Thomas taught evening chemistry classes at Boston University from 1952 through 1970. She also served as an adjunct professor at the University of Rhode Island. She is a member of the American Chemical Society and the Electrochemical Society, a fellow of the American Institute of Chemists, and the author of numerous technical papers. In her free time, Thomas enjoys traveling, spending time with her family, and dabbling in arts and crafts.

While still in graduate school, Thomas met her future husband, a fellow chemist. She married George R. Thomas in Millbury, Massachusetts, in 1955. The couple have four daughters: Augusta, Abigail, Anne, and Susan. Thomas accorded family the highest priority in life. Yet despite the cultural norms at the time she raised her children, Thomas was never tempted to give up her scientific career for the role of full-time homemaker. As she explained to Linda Wasmer Smith in an interview, "My career was very intense; it had to be. If you were a woman in science then, you had to stay with it unequivocally. And that's what I did." Commented her husband, "If her career as a scientist was intense, then her career as a wife and mother was absolutely ferocious."

Thompson, Benjamin (1753-1814)

British-American chemist

Benjamin Thompson, the son of a New England farmer, had little formal schooling. After being injured in a fireworks accident while serving as a merchant's apprentice, he became a teacher in Rumford (now Concord), New Hampshire. There at age 19 he married a wealthy widow, Sarah Walker Rolfe. During the American Revolution, Thompson was an active Tory who served the British king by spying on his countrymen. After the fall of Boston in 1775, he fled to England, leaving behind his wife and daughter. During the war, Thompson served as undersecretary of state for the colonies and briefly as a British lieutenant colonel. In 1783 he joined the court of the elector of Bavaria where he instituted numerous social re-

Benjamin Thompson.

forms including workhouses for Munich beggars. Thompson also introduced Watt's steam engine and the potato to the continent. For his numerous services to Bavaria the elector made Thompson a count in the Holy Roman Empire in 1790, and he took the name Rumford in honor of his former home.

Thompson's greatest scientific discovery was an outgrowth of his interest in **gunpowder** and weapons. In Bavaria, where his job was to oversee the boring of cannons at the Munich Arsenal, he noticed that the metal got so hot it had to be constantly cooled with **water**. The prevalent theory of the day was that **heat** was a fluid called caloric which could be transferred from one substance to another, and during the process of boring the metal released caloric. Thompson, believing that the boring was removing far more caloric from the metal than it could possibly have contained, sought an explanation. Further experiments demonstrated that heat did not have weight, a characteristic of a fluid. In a paper to the Royal Society, ''An Experimental Enquiry concerning the Source of Heat excited by Friction'' (1798), he proposed that the mechanical work of the borer was generating the heat and that heat was a form of motion. The paper also included calculations on the amount of heat a quantity of mechanical **energy** produced, a value that, though too high, stood for a half century. During the early nineteenth century his theory of heat began challenging the caloric theory, but many physicists remained unconvinced that Thompson was correct until James Maxwell firmly established the kinetic theory of heat in 1871.

Thompson's interest in heat led to a study of convection and insulation. He demonstrated that convection was the principal means of heat loss that the best insulation inhibited convection currents, and that heat traveled through a vacuum only with great difficulty. Employing these principles, he improved fireplaces with the addition of an insulated box, invented a kitchen stove, a double boiler, and a drip coffee pot, and developed a calorimeter to measure heats of **combustion** of various fuels. His scientific interests included light, and among his inventions in this field were the shadow photometer and the Rumford lamp. For his achievements he was admitted to the Royal Society.

Thompson went to Paris in 1804 where he married the widow of **Antoine Lavoisier**. The marriage lasted only four years, and Thompson's daughter came to France to look after him until his death. In his later years, he attempted to reconcile with the United States, and though Thompson died in exile, he left most of his estate to his native land. Thompson's legacy reached beyond his theories and inventions. He started the Royal Institution in London in 1799 as a center for technological innovations and hired **Humphry Davy** and Thomas Young as research scientists and lecturers. To encourage continued research in heat and light, he established the Rumford medals of the Royal Society and the American Academy of Arts and Sciences and the Rumford professorship in physics at Harvard.

THOMSON, JOSEPH JOHN (1856-1940)

English physicist

Joseph John Thomson, who discovered the **electron** in 1897, won the 1906 Nobel Prize in Physics ''in recognition of the great merits of his theoretical and experimental investigations on the conduction of **electricity** by **gases**.'' The British scientist discovered that cathode rays consisted of negatively charged particles of subatomic size, which he called corpuscles, and which ultimately became known as electrons. He carried this theory to all **matter**, hypothesizing that matter consisted of negatively charged particles that were surrounded by positively charged particles, and that these charges neutralized each other. He became Sir Joseph in 1908 when he was knighted for his work. As evidence of his amazing mind and great teaching abilities, seven of his research assistants and his son, George, went on to win Nobel Prizes in physics.

Thompson, always known as ''J. J.,'' was born near Manchester, England, the son of a bookseller, who sent him to Owens College until an apprenticeship with a leading engineer opened up. However, his father died before the apprenticeship became available and, by the time it did, his family could not afford his fees. Encouraged by the excellent scientific professors at Owens, Thomson remained in the engineering program after receiving a small scholarship, and worked toward a scholarship in mathematics at Trinity College, Cambridge. He won the small scholarship, and entered Trinity in

1876. In 1880, he placed second in the final honors examinations in mathematics, and the college subsequently awarded him another scholarship in 1881. He studied three different lines of mathematical theory while continuing his own research interests. He was also strongly influenced by physicist **James Clerk Maxwell**, a fellow Englishman who proposed the theory of electromagnetics.

In 1882, as a fellow at Trinity, Thomson began designing his first mathematical models while contesting for an academic award called the Adams Prize. The subject matter was "a general investigation of the action upon each other of two closed vortices in a perfect incompressible fluid." He carried his research well beyond the subject matter to the theory of the vortex **atom**. In this theory, atoms of gas in a frictionless fluid are likened to smoke rings floating in the air, except that the atoms are eternal (never degrade). Thompson attempted to analyze the phenomenon mathematically and published his theories in *Treatise*. This work is described in the *Dictionary of Scientific Biography* Vol. III, as a "quantitative, mechanistic, and ultimate account of the physical world...perhaps the most glorious episode in this hopeless struggle."

Much of Thomson's work during this period focused on **atomic structure**, theories of chemical action, and the nature of light through "precise calculation and ingenious analogy applied with great virtuosity". He also wrote a dissertation for his fellowship that focused on an idea he had at Owens College-that "potential **energy** in a given system might be replaced by the kinetic energy of imaginary masses connected to it in an appropriate way."

In 1884, at the age of just 27, Thomson succeeded Lord Rayleigh as Cavendish Professor of Experimental Physics at Trinity. He remained in that role until 1919. On the advice of Maxwell, he decided to investigate the phenomenon of gas discharge in relation to cathode rays. Thomson suspected these rays to be streams of charged particles, "...matter highly charged with electricity and moving with great velocities, "because the path of the rays could be deflected or curved by a magnetic force field. He adhered to his line of thinking, even though German physicists were convinced cathode rays were an "aethere disturbance," or **ether** waves, similar to ultraviolet light, because of their ability to make **glass** fluoresce (or glow).

To conduct his experiments, Thomson used a glass tube with one end connected to a negative electrode (cathode), a positively charged electrode (anode) at the other end, and a tiny paddle wheel in the center. Upon running an electric current through a variety of gases inside the tube, he discovered that: 1) The current created different colors in different gases; 2) The **anode** (positive) end glowed; the negative (cathode) did not; and 3) The paddle wheel spun in the direction of the anode. From this, Thomson deduced that atoms of different gases produced different energies, as shown by the different colors; that, in order to make the paddle spin, the gases must contain particles; and that the current flowed from negative to positive. This latter finding was one of his most important discoveries.

Because Heinrich Hertz (1857-1894) had discovered that cathode rays could pass through thin **gold** or **aluminum**

Joseph John Thomson.

leafing inside a discharge tube, Thomson assumed the rays must consist of very minute components and so pursued the negatively charged particle theory. He began applying the theory of the vortex atom to the gas discharge and discovered an extremely important phenomenon-that gas discharge, like **electrolysis**, occurs through the disruption of chemical bonds. He found that, by placing positive and negative electrodes at the middle of the tube as well, he could "bend" or deflect the cathode ray. This ultimately led him to discover the existence of ions, charge carriers much smaller than atoms, and to theorize the structure of the atom. Because he found that negative particles produced in the discharge tube were the same in all residual gases and with different electrodes; and because negative particles could also be produced by heated **metals**, the action of ultraviolet light, or x rays on metals, he predicted that these tiny bodies must be the common component in all atoms.

In 1897, he gave a lecture to the Royal Institution in which he described in detail his apparatus for deflecting cathode rays in both magnetic and electric fields. Thompson called the tiny particles in the cathode rays "corpuscles." It was George Johnstone Stoney (1826-1911) who first proposed they be called "electrines" and eventually "electrons."

Thomson then turned his focus to determining the nature of positively charged ions, or positive electricity. This was to become his final important experimental work. Many years of painstaking research, in collaboration and comparison with other physicists, ultimately allowed him to show that **neon** gas was composed of two types of ions. Each one had a different charge, different **mass**, or both. He determined this by deflecting the stream of neon gas containing positive ions with both magnetic and electric fields. This produced two invisible streams (or traces), images of which he captured on a photographic plate. This led him to suspect different types of atoms of the same element, with the same **atomic number**, but with a different mass. Thompson devised many innovative experiments and types of equipment to help analyze these traces. An associate, **Francis Aston**, perfected the mass spectroscope, for which he also won a Nobel Prize.

Thomson was not only an ingenious scientist but also a gifted teacher. Twenty-seven of his students became fellows of the Royal Society and dozens became physics professors. He personally helped students get papers published and to gain placement in appropriate research institutions. He aimed at improving the level of scientific studies in secondary schools as well as universities; lectured at the Royal Institution where he became Professor of Natural Philosophy in 1905; established the Cavendish Physical Society in 1893, a seminar held every two weeks which reviewed recent works to help his scientists keep abreast of developments; and expanded the size of the Cavendish laboratory twice. His most important papers were published in *Philosophical Magazine*, which he helped become the premier journal on physics in all of England.

Other awards recognizing his tremendous contribution to science included the Order of Merit in 1912; presidency of the Royal Society beginning in 1915 (which placed on him the responsibilities of that Society's contribution to the efforts of World War I); and, in 1918, became master of Trinity, his old college at Cambridge. He continued to fulfill his duties in this capacity until just several months before his death. He was granted the tremendous honor of burial in London's Westminster Abbey.

THOMSON, WILLIAM (1824-1907)

Scottish mathematician and physicist

William Thomson, known to history as Lord Kelvin, was granted the first scientific peerage by Queen Victoria in 1892 for his unique consulting work that made possible the installation of the transatlantic cable linking the telegraph systems of America and England. The peerage was created especially for him, taking the name from the Kelvin River near Glasgow, Scotland. Thus, when his proposal for an absolute scale measuring heat was widely accepted, it was given the name Kelvin.

Thomson's life and personality was quite unique. He was a child prodigy who grew up in the environment of academia and became a professor at a young age. Thomson was enthusiastic and dramatic in his teaching style. Known as an expert on the dynamics of heat, he was also noted for having a wide range of interests in the sciences, particularly electricity and magnetism. As a science and technology authority, Thomson was willing to take on the unpopular side of a controversy. This characteristic almost ruined his career when he attempted to establish the age of the Earth, accepting the least popular premise concerning the origins of the planet. However, his gamble in supporting **James Prescott Joule** rewarded Thomson with a lifelong friendship and a competent research collaboration.

Thomson was born in Belfast, Northern Ireland in 1807. While he was still young, the family moved to Glasgow where his father received a position as a mathematics professor. His mother died when he was six years old, and young William became accustomed to attending his father's lectures. After attending his father's classes for many years, Thomson surprised many by actively participating. By age ten, he was ready for college. He was able to keep up with his much older classmates and he even surpassed them by writing his first scholarly treatise at the age of fourteen. Many were impressed with this work and a respected professor, representing Thomson, presented it in lecture to the Royal Society of Edinburgh. Thomson and his professors decided he should not present the paper himself because his young age would have undermined the respect his paper deserved.

After enrolling in Peterhouse College at age seventeen, Thomson achieved honors in a difficult mathematics program by age 21. Although his father had hoped that William would follow him into mathematics, Thomson was strongly interested in natural philosophy (as science was called in his time). He was very much interested in Fourier and other newly emerging theories of heat, having already achieved valuable experience at Regnault's Laboratory in Paris. Thus, when offered a position as professor of natural philosophy at the University of Glasgow in 1846, Thomson accepted. He remained a professor at Glasgow for the next 58 years.

Thomson's first achievement as a professor was an attempt to establish the age of the Earth based on mathematical models representing the difference in temperature between the Earth and the sun. This research was almost a fiasco, upsetting well- respected geologists. Although his basic geological premise was faulty, his proficient models and accurate calculations were impressive. More importantly, the principles of **thermodynamics** he employed were well thought out. Fortunately, because many paid attention to the strengths of his study, Thomson advanced his expertise on the properties and dynamics of heat.

At Glasgow, Thomson's students were fascinated with his youthfulness and his energetic lecturing style. He realized that his students would need a laboratory to fully comprehend the concepts he was presenting them. The university did not have a laboratory at that time. He built his own by converting an old wine cellar belonging to a more established member of the faculty.

In 1847, at a conference of the British association, Thomson listened as Joule presented findings on the effects of **heat** on **gases**. Joule's presentation was not convincing to most of the assembled scientists and he would have been ignored if Thomson had not come to his defense. Thomson not only supported Joule in this conference, but also agreed to collaborate with him on new research. From this work, came the

Joule-Thomson effect of heat conservation, presented in a paper in 1851. This concept —that gas allowed to expand in a vacuum will reduce its heat— became the foundation of an early refrigeration industry.

While studying the effects of heat on gases, Thomson recognized that the linear relationship between the heat and mass in gases was awkwardly graphed using the traditional Celsius Scale. In 1848, he proposed an absolute scale using the same range as the Celsius scale but with 0° set at the point with virtually no movement among the molecules (–273.18°C). This can be considered the point of absolutely no heat. In 1851, Thomson further elaborated how this new scale would make the principles of thermodynamics more clear in experiments. Many people from other fields also recognized how the absolute (later called Kelvin) scale could be very useful.

Thomson admitted that he learned much from Joule. However, for many years he tried to resolve the theoretical differences between the dynamic theories of heat that Joule was relying on and the principles he found to be true in the theories of **Sadi Carnot**. In his dissertation of 1851, he thoroughly presented the strengths of Carnot's theories reconciled to the dynamic theory. From this paper, the second law of thermodynamics was established. This law stated that heat transferred from hotter matter to colder matter has to release mechanical work. Furthermore, heat taken from colder matter to hotter matter requires the input of mechanical work. However, the second law of thermodynamics was not fully credited to Thomson because it was contemporaneously developed (separate from Thomson's work) by other researchers.

Thomson also excelled in other areas of science. His improvements in the conductivity of cable and in galvanometers were patented inventions without which the practicality of the transatlantic cable would not have been possible. Among Thomson's many honors, he was elected president of the Royal Society and held the post from 1890 to 1894. He continued to invent and study in his later life. After a short period of retirement from the university, he returned as a graduate student. Although Thomson maintained a sharp mind until his death, he was adamant against the changes in old paradigms that new discoveries brought at the end of the nineteenth century. His title, Baron Kelvin, died with him in 1907, as he had no children.

It is debatable whether Thomson contributed anything revolutionary to the field of chemistry. In his collaboration with Joule, Thomson certainly helped widen the fissure in Joule's findings. The Kelvin scale was progressive but was based on already known principles. Perhaps it could even be argued that his resistance to discoveries in atomic studies might have discouraged young researchers at the turn of the twentieth century. What is not debatable is that Thomson was unique in advancing scientific knowledge across a broad expanse of interests, and assisted in the progress of many glorious projects.

THORIUM

Thorium is a member of the actinide family, a group of elements in Row 7 of the **periodic table**. It has an **atomic number** of 90, an atomic **mass** of 232.0381, and a chemical symbol of Th.

William Thomson.

Properties

Thorium is a silvery white, soft, metal with physical properties similar to those of **lead**. It can be hammered, rolled, bent, cut, shaped, and welded rather easily. Its melting point is about 3,270°F (1,800°C), its **boiling point** is about 8,130°F (4,500°C), and its **density** is 11.7 grams per cubic centimeter.

All isotopes of thorium are radioactive. The most stable (thorium-232) has a half life of 1.405 x 10^{10} years. The element is relatively active, combining with **oxygen** to form thorium dioxide (ThO_2) and reacting with most acids.

Occurrence and Extraction

Thorium is a relatively abundant element with an abundance of about 15 parts per million in the Earth's crust. Its primary ores are thorite and monazite. The latter mineral is found somewhat commonly in beach sand, which may contain up to 10% of the element.

Discovery and Naming

Thorium was ''discovered'' twice by the Swedish chemist **Jöns Jakob Berzelius**. His first announcement was made in 1815, a report he later retracted when he found the substance he had discovered was actually a compound, **yttrium phosphate** (YPO_4). Ten years later, Berzelius once again announced the discovery of a new element, for which he suggested the name thorium. This time he was correct. The name he suggested was in honor of the Scandinavian god Thor.

Uses

Traditionally, thorium has had relatively few commercial uses. The one with which people are most likely to be familiar is in portable gas lanterns where it is used to make the mantle. Since gas mantles made with thorium are radioactive, their use has been phased out. There is some hope that thorium may eventually be used to replace **uranium** in **nuclear reactors** and nuclear weapons. Thorium itself is not fissionable, but it can be converted to uranium-233, an **isotope** which does undergo fission. At this point, the technology for converting thorium to uranium-233 has not been developed on a commercial scale.

THULIUM

Thulium is a rare earth element, one of the elements found in row 6 of the **periodic table.** It has an **atomic number** of 69, an atomic **mass** of 168.9342, and a chemical symbol of Tm.

Properties

Thulium is a silvery metal so soft it can be cut with a knife. It is both malleable and ductile and can be worked rather easily. Its melting point is 2,822°F (1,550°C), its **boiling point** is 3,141°F (1,727°C), and its **density** is 9.318 grams per cubic centimeter. Thulium is not a very active metal, although it does react slowly with **water** and more rapidly with most acids.

Occurrence and Extraction

Thulium is one of the rarest of the **rare earth elements** with an abundance in the Earth's crust of about 0.2-1 part per million. Its most common ores are monazite, euxenite, and gadolinite. The pure metal can be produced by treating thulium fluoride (TmF_3) with **calcium** metal: $2TmF_3 + 3Ca \rightarrow 3CaF_2 + 2Tm$

Discovery and Naming

Thulium is one of the elements discovered in yttria, a mineral first found by a Swedish army officer named Carl Axel Arrhenius (1757-1824) in 1787 near the town of Ytterby, Sweden. Eventually, yttria was to yield nine new elements. Thulium was found by the Swedish chemist Per Teodor Cleve (1840-1905) in 1879. He named the element for the ancient name of Scandinavia, Thule.

Uses

Thulium is too expensive to have many commercial applications. It is sometimes used in special types of **lasers** that are used for photographic purposes on satellites circling the Earth.

TILDON, J. TYSON (1931-)

African American biochemist

Discoverer of **Coenzyme** A Tranferase Deficiency, a disease of infants, J. Tyson Tildon and has also made major contributions to the establishment of the Sudden Infant Death Syndrome (SIDS) Institute at the University of Maryland School of Medicine. His research interests include developmental **neurochemistry** and the processes that control **metabolism**.

James Tyson Tildon was born April 7, 1931, in Baltimore, Maryland. He received his B.S. degree in **chemistry** from Morgan State College in 1954 and then worked for five years as a research assistant at Sinai Hospital, where he developed and used biochemical techniques to study vitamin deficiencies in humans and animal models. Subsequently, he spent a year as a Fulbright Scholar at the Institut de Biologie Physico-Chimique in Paris and, upon his return, matriculated to the doctoral program in **biochemistry** at Johns Hopkins University. After receiving his Ph.D. in 1965, Tildon accepted a two-year postdoctoral fellowship at Brandeis University, where his studies included an examination of how cells assume specialized functions during development. Tildon returned to Baltimore in 1967 to assume the post of assistant professor in the department of chemistry at Goucher College. The following year, he became research assistant professor in the department of pediatrics at the University of Maryland School of Medicine, and, in 1969, assistant professor in the department of biological chemistry. He has been a full professor of pediatrics since 1974 and a professor of biological chemistry since 1982. Tildon has also served as director of the Carter Clinical Laboratories for six years and director of pediatric research in the medical school's Department of Pediatrics for nine. In addition, he was a visiting scientist in the Laboratory of Developmental Biochemistry at the University of Groningen in the Netherlands, where he did research in the developmental neurobiology.

Among Tildon's contributions is the discovery of Coenzyme A (or CoA) Transferase Deficiency in infants. In his research, he demonstrated that the brains of infants use organic molecules called ketone bodies as an **energy** source during their first several weeks of life, disproving the previously held theory that glucose was the major energy source of the human brain at all ages of life. Tildon was also instrumental in establishing the SIDS Institute at the University of Maryland, one of the largest research programs dedicated to the study of SIDS, a disorder that causes an infant to abruptly stop breathing. Included in the discoveries made at the Institute are those of researcher Robert G. Meny, who found that babies suffered bradycardia, or abnormally slow heartbeat, before they stopped breathing. This finding has stimulated new research into the role of the heart in SIDS.

Tildon's interests extend beyond medicine and biochemistry and into other realms of research. In his book *The Anglo-Saxon Agony,* he points out that members of western civilization rely predominately on sight and hearing to gather information while ignoring the more personal senses of **taste**, smell, and touch. Rather than integrating all of the brain's responses to stimuli, Tildon suggests, the Anglo-Saxon approach is to separate thinking and feeling in an effort to be dispassionate. Yet, by excluding the emotional component, he writes, westerners fail to understand fully the human condition.

Among the many societies Tildon belongs to are Sigma Xi, the American Chemical Society, the Association for the Advancement of Science, the American Society for Biochemistry and Molecular Biology, and the Society for Experimental Biology and Medicine. He received the Maryland State Senate Citation for his work with SIDS in 1983, the City of Baltimore Citizen Citation in 1986, the Baltimore Chapter of the National Technical Association's Joseph S. Tyler, Jr. Award for Achievement in Science in 1986, the National Association of Negro Business and Professional Women's Club's Community Service Award in 1987, and the Humanitarian Award from the Associated Black Charities in 1991. In addition, he has served on several boards of directors, including those of the Mental Health Association of Metropolitan Baltimore, the Maryland Academy of Sciences, and the Associated Black Charities.

TIN

Tin is the fourth element in Group 14 of the **periodic table**. Its **atomic number** is 50, its atomic **mass** is 118.69, and its chemical symbol is Sn.

Properties

Two allotropic forms of tin exist, the α- and β-forms. The β-form of tin is also known as white tin and has a melting point of 450°F (232°C), a **boiling point** of 4,100°F (2,260°C), and a **density** of 7.31 grams per cubic centimeter. One of white tin's most interesting properties is its tendency to give off a strange screeching sound when it is bent. This sound is known as "tin cry."

The α-form is also known as gray tin and is formed when the β-form is cooled to temperatures of less than about 55°F (13°C). Gray tin is a gray amorphous powder. The change from white to gray tin explains an interesting phenomenon that has been observed in certain very old objects made of the metal. In the late nineteenth century, organ pipes in many cathedrals of Northern Europe began to crumble in very cold weather. The crumbling, known as "tin disease" at the time, is now known to have been caused by the conversion of α-tin to β- tin.

Tin is a relatively inactive metal at room **temperature**. It does not react with **oxygen** or with most cold acids, making it excellent as a protective coating for other **metals**. At higher temperatures, the metal does combine with many acids, the **halogens**, **sulfur**, **selenium**, and **tellurium**.

Occurrence and Extraction

Tin is moderately abundant in the Earth's crust, ranking about number 50 among elements found in the crust. Its abundance is estimated to be about 1-2 parts per million. The most common ore of tin is cassiterite, a form of tin **oxide** (SnO_2). The major producers of tin in the world are China, Indonesia, Peru, Brazil, and Bolivia. The metal can be extracted by heating tin oxide with carbon: $SnO_2 + C \rightarrow Sn + CO_2$.

Discovery and Naming

Tin and its alloys were well known to ancient peoples. Bronze, an **alloy** of tin and **copper**, was first produced at least as far back as 4000 B.C.. It was probably discovered accidentally when copper and tin compounds were heated together. Bronze rapidly became a very popular metal since it was harder and more durable than either copper or tin by itself. The impact of bronze on tool- and weapon-making was so profound that a whole period in human history between 5000 and 4000 B.C. became known as the Bronze Age.

The origin of the name tin is lost in history. Some scholars believe it was named for the Etruscan god Tinia. During the Middle Ages, the metal became known by its Latin name, *stannum*, from which its modern-day chemical symbol is derived.

Uses

The most important use of tin in the United States is the manufacture of solder, an alloy made of tin and **lead**. Bronze also continues to be a major end-product of the production of tin. The alloy is used in spark-resistant tools, springs, wire, electrical devices, **water** gauges, and valves. Another important application of tin is tinplating, the process by which a thin coat of tin is laid down over the surface of steel, **iron**, or some other metal. Tin is less reactive than the metal it covers, protecting the underlying metal from **corrosion**.

About a sixth of all tin consumed in the United States goes to the production of tin compounds. Tin chloride ($SnCl_2$), for example, is used in the manufacture of dyes, polymers, and textiles; in the silvering of mirrors; as a food preservative; as an additive in perfumes used in soaps; and as an anti-gumming agent in lubricating oils. Tin oxide (SnO_2) is used in the manufacture of special kinds of **glass**, ceramic glazes and colors, perfumes and **cosmetics**, and textiles; and as a polishing material for steel, glass, and other materials. Tin fluoride (stannous fluoride; SnF_2) and tin pyrophosphate ($Sn_2P_2O_7$) is used as a toothpaste additive to help protect against cavities.

Health Issues

Most compounds of tin are toxic. They tend to pose the most serious hazard when they get into the air and are inhaled. They may then cause health problems, such as nausea, diarrhea, vomiting, and cramps. The amount of tin absorbed by the human body from food stored in tin cans is too small to be of concern to consumers.

TISELIUS, ARNE (1902-1971)

Swedish chemist

Arne Tiselius was awarded the 1948 Nobel Prize in **chemistry** for his research in **electrophoresis** (the movement of mole-

Arne Tiselius.

cules based on their electric charge and their size) and for his investigations into **adsorption**, the inclination of certain molecules to cling to particular substances. Although the phenomenon of electrophoresis had been identified decades earlier, it did not become a useful technique for analyzing chemical compounds until Tiselius developed methods which delivered accurate results.

Arne Wilhelm Kaurin Tiselius was born in Stockholm, Sweden, on August 10, 1902, to Hans Abraham J. Tiselius, who was employed by an insurance company, and Rosa Kaurin Tiselius, the daughter of a Norwegian clergyman. Upon the death of Tiselius's father in 1906, Rosa relocated the family to Göteborg, Sweden, where Hans's family lived. Entering the gymnasium at Göteborg, Tiselius came under the tutelage of a chemistry and biology teacher who actively supported his student's interest in science. In 1921 Tiselius matriculated to the University of Uppsala—where his father had earned his degree in mathematics—and studied under the renowned physical chemist **Theodor Svedberg**. Earning his master's degree in chemistry, physics, and mathematics in 1924, Tiselius continued to work as Svedberg's research assistant in **physical chemistry**. Although Svedberg was interested in the electrophoretic properties of **proteins**, he turned the study of this over to his new assistant, and three years later Tiselius published his first paper jointly with Svedberg on the subject.

Tiselius would remain at Uppsala until his retirement in 1968, rising from researcher to full professor. His 1930 doctoral dissertation, which earned him a post as docent in the chemistry department, long stood as a standard in the field of electrophoresis. Sweden's first professorship in **biochemistry** was established for Tiselius at Uppsala in 1938. Besides his work in biochemistry, Tiselius had a strong interest in botany and ornithology and made frequent excursions into the Swedish countryside on photographic expeditions. On November 26, 1930, the year of his doctoral thesis, he married Ingrid Margareta Dalén, with whom he would have one son, Per, and a daughter, Eva.

Following his dissertation, Tiselius concentrated his attention in areas outside of chemistry. He expanded his research to include biochemical studies—not a typical element of the chemistry curriculum in those days—and became aware of the potential for exploiting the extremely specific electrical ''signature'' of proteins, as well as other substances. He became concerned, however, with the impurities in the substances under study, even those that had been carefully centrifuged, and turned to chromatography as a possible answer. In chromatographic analysis, light of a specific frequency is passed through a substance, and by using tables assembled over the course of many experiments, the ''chromatic signature'' of the particular sample can be detected. Tiselius applied this technique by looking into the properties of light **diffusion** through zeolite, a translucent mineral. While studying under Hugh S. Taylor from 1934 to 1935 at Princeton University's Frick Chemical Laboratory, Tiselius conceived of an accurate method to quantify the diffusion of **water** molecules through crystals of zeolite.

While at Princeton, Tiselius came to realize that a wealth of potential discoveries in the biochemical sciences awaited only the development of a method accurate enough to help separate and identify compounds. Returning to his original line of research, he completed a prototype of a new electrophoretic apparatus.

When Tiselius returned to Uppsala, he continued making improvements on his electrophoretic instrumentation. In one innovation, he filled a U-shaped tube with chemical **solvents**, added a **solution** containing the sample to be analyzed, then applied a charge to one end. As the elements migrated, they reached the solvents at different lengths along the tube. Tiselius constructed the tube so that test samples could be taken at various points along the path of migration and be analyzed to determine which of the original species had made it to that point. It was by using this technique that Tiselius was able to demonstrate that blood **plasma** contained a complex mix of different elements.

Tracking the movement of boundaries optically by a technique invented by August Toepler—the *Schlieren* method—Tiselius resolved the plasma into four distinct elements that showed up as separate bands in the tube. He was the first to isolate three of the blood proteins known as globulins, which he named *alpha*, *beta*, and *gamma*. These are important in many of the body's functions; the immunoglobulins, for example, are a critical factor in infection control. In the fourth band, located between those of *beta* and *gamma,* Tiselius discovered **antibodies**.

The method was a radical improvement but still dissatisfied Tiselius. At the time he was more interested in the breakdown products of polypeptides than in blood compounds. **Peptides** represent some of the most important proteins in the body and for a clear understanding of their function, it is essential to know their types. However, when the long chain of a polypeptide is broken down, the individual peptides are so similar in nature that even Tiselius's improved electrophoretic technique could not distinguish between them. Faced with this problem, he turned to adsorption methods of analysis, using the then-common column method. In this procedure, a mixture which contains a substance with a specific affinity for absorbing one peptide or another is flushed through a column (a tube or cylinder). The peptides which had been in the original mix can then be determined by analyzing the eluate (the wash which passed through the column).

In 1943 Tiselius introduced a critical improvement in the process. Research to that point had been carried out using a "frontal analysis method," which revealed the **concentration** of the components in a mixture but was unable to separate them for further study. Elution could accomplish separation, but had a major setback, "tailing," which is the corruption of one part of a solution by molecules from the other. Tiselius demonstrated that a simple modification to the old technique could reduce tailing, and this new method became known as "displacement analysis."

Other important work came out of Tiselius's laboratory throughout the 1940s, such as research on paper electrophoresis and zone electrophoresis. However, increasing demands from other sources took over his time and, in the summer of 1944, Tiselius became an advisor to the Swedish government. His responsibilities included sitting on a committee established to help improve conditions for advancing scientific research, with a focus on basic research. This was the beginning of a long and distinguished relationship with the Swedish Parliament, an association that ended only when Tiselius suffered a heart attack following an important meeting in Stockholm. He died the next morning on October 29, 1971.

Up to his last day, Tiselius followed an active schedule. Having accepted the four-year chairmanship of the Swedish Natural Science Research Council in 1946, he was instrumental in the creation of the Science Advisory Council to the Swedish government. Tiselius was elected vice president of the Nobel Foundation with membership on the Nobel Committee for Chemistry in 1947, one year before he was awarded his own Nobel Prize. That same year, at the International Congress of Chemistry held in London, he was elected vice president in charge of the section for biological chemistry of the International Union of Pure and Applied Chemistry—a body which he led as president four years later.

Among other honors Tiselius received were the Bergstedt Prize of the Royal Swedish Scientific Society in 1926, the Franklin Medal of the Franklin Institute in 1956, and the **Paul Karrer** Medal in Chemistry from the University of Zurich in 1961. He was also presented with numerous honorary degrees from universities, including those of Stockholm, Paris, Glasgow, Madrid, California at Berkeley, Prague, Cambridge, and Oxford. Tiselius was always interested in fields beyond his own and was concerned with the environmental, social, and ethical implications of science and technology. As president of the Nobel Foundation in 1960, he established the Nobel Symposium, perceiving the foundation as the perfect vehicle for raising awareness of the need to promote science as a solution to mankind's problems. This organization gathered a mix of Nobel laureates to discuss the implications of their work during symposia in each of the five prize fields.

TISHLER, MAX (1906-1989)

American chemist

Max Tishler is noted for taking the formulation of pharmaceutical chemicals out of the laboratory and onto the production floor. During his long career as an industrial research chemist, he received patents relating to more than 100 medicinal chemicals, **vitamins**, antibiotics, and **hormones**. In doing so he significantly improved human health and **nutrition** and laid the foundation for modern, large-scale process **chemistry** of complex compounds.

Tishler was born on October 30, 1906, in Boston, Massachusetts. His father's name was Samuel, and his mother's maiden name was Anna Gray. He attended Tufts College in Medford, Massachusetts, where he earned a B.S. in chemistry in 1928. During his high school and college years, he worked part time in a pharmacy, where he first became interested in the use of pharmaceutical chemicals to treat health problems.

After graduation from Tufts, Tishler studied **organic chemistry** at Harvard University, where he was a teaching fellow from 1930 to 1934. He received an M.A. degree from Harvard in 1933 and a Ph.D. in 1934, both in organic chemistry. Shortly after he graduated, he married Elizabeth M. Verveer. They had two sons during their marriage, Peter Verveer Tishler, who went on to become a noted physician specializing in the genetics associated with diseases, and Carl Lewis Tishler.

Tishler stayed on at Harvard, first as a research associate from 1934 to 1936 and then as an instructor in chemistry from 1936 to 1937. In 1937 George Merck, president of the pharmaceutical firm Merck and Company, persuaded Tishler to leave his teaching position at Harvard and become a research chemist at the company's laboratories in Rahway, New Jersey. One of Tishler's first assignments was to develop a method for making riboflavin, which would, in turn, allow economical production of vitamin B_2. His success in solving this problem led to processes for making other vitamins, such as vitamin A, vitamin K_1, and pantothenic acid.

In 1941 Tishler was named head of process development at Merck, and in 1944 he was promoted again to director of developmental research. During this period, Tishler continued to study complex chemical reactions and reduce them to definable processes for production in quantity. One of his investigations led to the development of a new drug, sulfaquinoxaline, which was used as a feed-additive to effectively combat parasite infections in poultry. This allowed a significant expansion of the poultry industry and opened up an entirely new field of pharmaceutical research specifically aimed at developing drugs to promote animal health.

In the late 1940s Tishler and his colleagues took on the difficult project of making the new therapeutic chemical **cortisone** in large quantities. In the laboratory the conversion of desoxcholic acid to cortisone was a complex process involving 42 separate steps which yielded less than one percent of the final product. By altering the chemical reactions, Tishler and his team at Merck were able to simplify the operations to work in a production environment, while at the same time increasing the product yield to 30 percent. Tishler's success proved that even the most sophisticated chemical process could be modified to work in a large-scale production facility, and his work made him one of the leaders in modern process chemistry.

In 1951 Tishler was awarded the Board of Directors' Scientific Award of Merck and Company, Inc. for his achievements. Tishler used the money from this award to establish the Max Tishler Visiting Lectureship at Harvard University and the Max Tishler Scholarship at Tufts University. In 1953 he was elected to the National Academy of Sciences.

Tishler was named vice president of scientific activities at Merck in 1954, and in 1956 he became president of the Merck laboratories. He held that position until 1969, when he took over as senior vice president of research and development until his retirement in 1970. During his 33 years at Merck, Tishler was responsible for the development of a wide range of new drugs for the treatment of infections, growth disorders, heart disease, hypertension, mental depression, and several inflammatory diseases, such as arthritis.

Tishler was honored for his contributions to industrial research when he was awarded the Industrial Research Institute Medal in 1961, and the Chemical Industry Award of the American Section of the Society of Chemical Industry in 1963. He received the Chemical Pioneer Award and the Gold Medal Award from the American Institute of Chemistry in 1968. In 1970 he was awarded the Priestley Medal, the highest award of the American Chemical Society.

After he retired from Merck in 1970, Tishler accepted a position as a professor of chemistry at Wesleyan University in Middletown, Connecticut, where he taught and conducted research into the chemistry of various natural substances. Some of the chemical compounds he investigated were leukomycin and prumycin, which are natural antibiotics, and cerulenin, a microbe which inhibits the production of fats. Tishler took over as chairman of the chemistry department during 1973-1974, and was named professor of the sciences emeritus in 1975. Concurrent with his work at Wesleyan, Tishler held advisory and directory positions with the Weizmann Institute of Science, the American Cancer Society, and the Sloan Kettering Institute.

Among his many honors, Tishler received nine honorary doctorate degrees and numerous lecture awards. In 1987, he received the National Medal of Science "for his profound contributions to the nation's health and for the impact of his research on the practice of chemistry." Max Tishler died on March 18, 1989, at the age of 82.

TITANIUM

Titanium is a transition metal, one of the elements found in Rows 4, 5, and 6 of the **periodic table**. It has an **atomic number** of 22, an atomic **mass** of 47.88, and a chemical symbol of Ti.

Properties

Titanium exists in two allotropic forms, one of which is a dark gray, shiny metal. The other **allotrope** is a dark gray amorphous powder. The metal has a melting point of 3,051°F (1,677°C), a **boiling point** of 5,931°F (3,277°C), and a **density** of 4.6 grams per cubic centimeter. At room **temperature**, titanium tends to be brittle, although it becomes malleable and ductile at higher temperatures. Chemically, titanium is relatively inactive. At moderate temperatures, it resists attack by **oxygen**, most acids, **chlorine**, and other corrosive agents.

Occurrence and Extraction

Titanium is the ninth most abundant element in the Earth's crust with an abundance estimated at about 0.63%. The most common sources of titanium are ilmenite, rutile, and titanite. The metal is often obtained commercially as a byproduct of the refining of **iron** ore. It can be produced from its ores by electrolyzing molten titanium chloride ($TiCl_4$): $TiCl_4$ —electric current→ $Ti + 2Cl_2$, or by treating hot titanium chloride with **magnesium** metal: $2Mg + TiCl_4 \rightarrow Ti + 2MgCl_2$.

Discovery and Naming

Titanium was discovered in 1791 by the English clergyman William Gregor (1761-1817). Gregor was not a professional scientist, but studied minerals as a hobby. On one occasion, he attempted a chemical analysis of the mineral ilmenite and found a portion that he was unable to classify as one of the existing elements. He wrote a report on his work, suggesting that the unidentified material was a new element. But he went no further with his own research. It was not until four years later that German chemist **Martin Heinrich Klaproth** returned to an investigation of ilmenite and isolated the new element. He suggested the name of titanium for the element in honor of the Titans, mythical giants who ruled the Earth until they were overthrown by the Greek gods.

Uses

By far the most important use of titanium is in making alloys. It is the element most commonly added to steel because it increases the strength and resistance to **corrosion** of steel. Titanium provides another desirable property to alloys: lightness. Its density is less than half that of steel, so a titanium-steel **alloy** weighs less than pure steel and is more durable and stronger.

These properties make titanium-steel alloys particularly useful in spacecraft and aircraft applications, which account for about 65% of all titanium sold. These alloys are used in airframes and engines and in a host of other applications, including armored vehicles, armored vests and helmets; in jewelry and eyeglasses; in bicycles, golf clubs, and other sports equipment; in specialized dental implants; in power-generating plants and other types of factories; and in roofs, faces, columns, walls, ceilings and other parts of buildings. Titanium alloys have also become popular in body implants, such as artificial hips and knees, because they are light, strong, long-lasting, and compatible with body tissues and fluids.

The most important compound of titanium commercially is titanium dioxide (TiO_2), whose primary application is in the manufacture of white paint. About half the titanium dioxide made in the United States annually goes to this application. Another 40% of all titanium dioxide produced is used in the manufacture of various types of **paper** and plastic materials. The compound gives "body" to paper and makes it opaque. Other uses for the compound are in floor coverings, fabrics and textiles, **ceramics**, ink, roofing materials, and catalysts used in industrial operations.

Yet another titanium compound of interest is titanium tetrachloride ($TiCl_4$), a clear colorless liquid when kept in a sealed container. When the compound is exposed to air, it combines with **water** vapor to form a dense white cloud. This property make it useful for skywriting, in the production of smokescreens, and in motion picture and television programs where smoke effects are needed.

TITRATION

In a titration experiment, the goal is to measure the amount of a **solution** of known **concentration** required to react completely with a given amount of a solution of unknown concentration. The solution with the amount is to be measured is called the titrant; the one with the unknown concentration is called the analyte. In general, the titrant is placed in a volumetric glassware called a burette and added slowly to a known **volume** of analyte until the reaction is complete. The point at which the reaction is complete is termed equivalence (or stoichiometric) point. Indicators usually detect this point. These are chemical substances that when added to the analyte, change **color** at the equivalence point.

The relationship between the analyte and the titrant solutions at the equivalence point of a titration is given by: CaVa = CtVt. Ca is the concentration of the analyte solution, Va is the volume of analyte solution required to reach the equivalence point, Ct is the concentration of the titrant solution, and Vt is the volume of the titrant solution required to reach the equivalence point. Therefore, knowing the concentration and volume of the titrant used to reach the equivalence point as well as the initial volume of the analyte allows calculation of the concentration of the analyte (provided that the coefficients of the analyte and titrant in the balanced reaction equation are the same).

A redox (reduction-oxidation) titration is one in which an oxidizing agent (or oxidant) is used to titrate a reducing

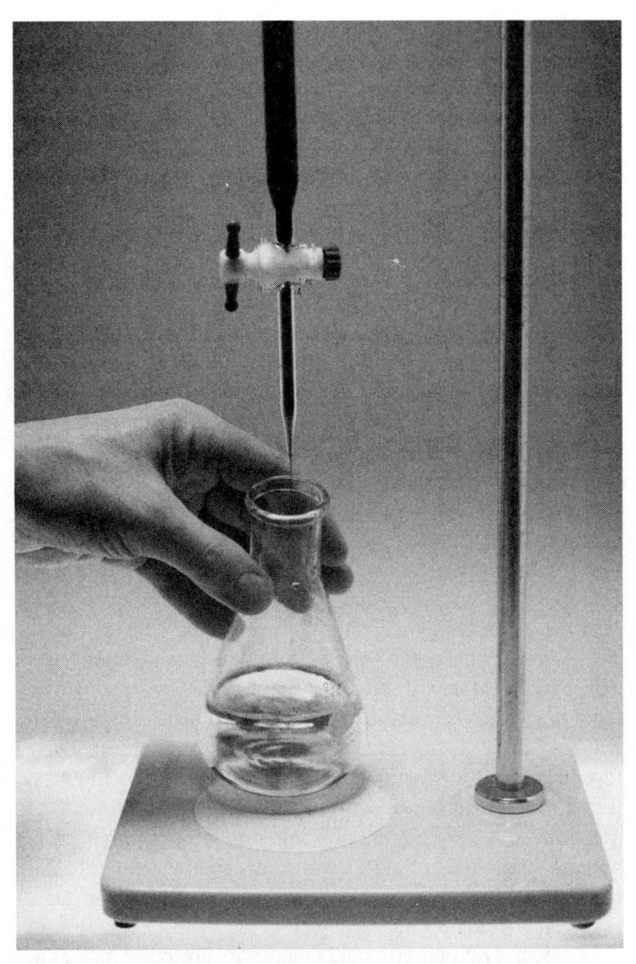

Acid-base titration. ***(Photograph by Charles D. Winters, Photo Researchers, Inc. Reproduced by permission.)***

agent (or reductant) or vice versa. In such a reaction, electrons are transferred between reactants. The reductant donates electrons and becomes oxidized while the oxidant gains electrons and is in turn reduced. A way to illustrate this is to look at the reaction between **potassium** and **iodine**, which can be described by two half-reactions:

$K(s) \rightarrow K^{+}(s) + e^{-}$

$I_2(g) + 2e^{-} \rightarrow 2I^{-}(s)$

The first half-reaction is an **electron** loss process in which the reductant, potassium metal (K), is oxidized into potassium **ion** (K^+). The second half-reaction is an electron gain process where the oxidant, iodine gas (I_2), is reduced into iodide ion (I^-). The overall redox process can be summed up in the following reaction in which the **oxidation** half-reaction has been multiplied by two to even up the electron count:

$2K(s) + I_2(g) \rightarrow 2K^{+}(s) + 2I^{-}(s) \rightarrow 2KI$

This can also be viewed in terms of oxidation states. An increase in **oxidation state** corresponds to an oxidation and a decrease corresponds to a **reduction**. In going from reactants to products, the oxidation state of potassium goes from 0 to 1, indicating oxidation, while that of iodine goes from 0 to 1, indicating reduction. Some ions change color depending on their oxidation state and this property can be used to determine the equivalence point, thus eliminating the need for an indicator. For example, when a **sodium hydrogen** sulfite ($NaHSO_3$) solution is used to titrate a potassium permanganate (KMnO4) solution, a significant color change is observed at the stoichiometric point. The initial $KMnO_4$ solution is purple due to the presence of MnO_4^- ions in which **manganese** is in the +7 oxidation state. At the equivalence point, the mixture turns pink indicating the presence of Mn^{2+} ions. Common reductants are the active **metals**, hydrogen, hydrogen sulfide, **carbon**, **carbon monoxide**, and sulfurous acid. Common oxidants include the **halogens** (especially **fluorine** and chlorine), **oxygen**, **ozone**, potassium permanganate, potassium dichromate, **nitric acid**, and concentrated **sulfuric acid**. Some substances are capable of acting either as reductants or as oxidants, e.g., **hydrogen peroxide** and nitrous acid. The **corrosion** of metals is a naturally occurring redox reaction. Industrially, many redox reactions are of great importance: **combustion** of fuels; **electrolysis** (oxidation occurs at the **anode** and reduction at the cathode); and metallurgical processes in which free metals are obtained from their ores.

An acid-base titration is one in which one solution is an acid and the other a base. According to the Bronsted-Lowry theory, an acid is a substance which can donate a **proton** (H^+) while a base is a substance which can accept a proton. Therefore, in an acid-base titration, instead of electrons being transferred (as is the case for redox titrations), protons are transferred. A very common example is the reaction between **sodium hydroxide** (NaOH) and **hydrochloric acid** (HCl) in aqueous solution:

$$NaOH(aq) + HCl(aq) \rightarrow NaCl(aq) + H_2O(l)$$

Here, hydrochloric acid donates its proton, which is accepted by the base, sodium hydroxide, resulting in the formation of two substances, a **salt** (sodium chloride), and **water**.

In an acid-base titration, the equivalence point is reached when exactly enough acid is present to convert all the base to a salt and water and can be pinpointed with the help of an indicator. Acid-base indicators have one color in acidic solution and a different color in basic solution. The exact point at which an indicator changes color is called the endpoint and depends on proton concentration (C_{H+}) or **pH** (where pH = $-\log C_{H+}$). The notation pH stands for ''power of hydrogen''. The endpoint is not to be mistaken for the equivalence point of the reaction being monitored. Since it is crucial that the endpoint of the indicator and the equivalence point of the reaction occur almost simultaneously, the choice of the indicator is a very important consideration and necessitates a thorough knowledge of the nature of the reaction in question. The indicator commonly used to locate the equivalence point of the sodium hydroxide-hydrochloric acid reaction is phenolphthalein which has an end point occurring at a pH of 8.2-10.

Instead of adding an indicator to observe the equivalence point, one can construct a graph on which the pH at regular intervals is plotted against the number of moles of added acid or base at these intervals; such a plot is called a titration curve or a pH curve and is usually sigmoid (S-shaped), with the inflection point, where the curve changes direction, corresponding to the equivalence point. The pH of a solution is usually monitored on a scale ranging from 0-14 (occasionally lower or higher than these limits). As the acidity increases, the pH number decreases. The relationship between the pH of the solution containing both an acid and a base and the equivalence point of the reaction between them depends on the nature of the acid and the base. In our example above, the acid in question (hydrochloric acid) is a strong acid and the base (sodium hydroxide) is also a strong base. Strong **acids and bases** are substances which are fully ionized in water solution according to the Bronsted-Lowry theory. Therefore, no molecular HCl or NaOH remain in water solutions and all acid-base **neutralization** reactions in water solution can be summarized by the equation:

$$H_3O^+(aq) + OH^-(aq) \rightarrow 2H_2O(l)$$

The equivalence point for the reaction of a strong acid with a strong base occurs when enough OH^- from the base has been added to react with all of the H_3O^+ from the acid. If HCl is used as the analyte, the analyte solution will initially have a pH < 7. As NaOH is added, the pH of the mixture will increase as OH^- reacts with H_3O^+. When a sufficient amount of base has been added to reach the equivalence point, a very large change in pH will occur as it reaches a value of 7 which indicates neutrality. If more NaOH is added, all its OH^- will remain unreactive as the pH continues to rise.

Titrations can also be used to determine the number of acidic or basic groups in an unknown compound. A specific weight of the compound is titrated with a known concentration of acid or base until the equivalence point has been reached. From the volume and concentration of added acid or base and the initial weight of the compound, the equivalent weight and thus the number of acidic or basic groups, can be computed. If a compound contains several different acidic or basic groups, the titration curve will show several sigmoid-shaped curves resembling steps, with a corresponding number of equivalence points. One industrial application of acid-base titration is in the attempt to overcome the damage from **acid rain** and related air **pollution**. The key is to limit the release of **sulfur** (SO_x) and **nitrogen** (NO_x) oxides which are both capable of acidifying the atmosphere and the **precipitation** that falls from it. SO_x and NO_x from coal-fired furnaces and **gasoline**-fueled vehicles account for the majority of human-made acidic emissions. A method called scrubbing is used to chemically remove the acidic sulfur dioxide by putting it into contact with **calcium** carbonate (limestone) which is a base. This method has the effect of neutralizing the acid:

$$2SO_2(g) + O_2(g) + 2CaCO_3(s) \rightarrow 2CaSO_4(s) + 2CO_2(g)$$

Scrubbing can be about 90% efficient and is so far the best alternative to using cleaner **coal**.

TIZARD, HENRY (1885-1959)

English physical chemist

Henry Tizard played a pivotal role in British military policy during World War II. He advised the government on a wide variety of military applications of scientific and technological innovations, including radar and the jet propulsion engine. His scientific education and military experience allowed him to communicate effectively with people in both areas.

Sir Henry Thomas Tizard was born on August 23, 1885, in Gillingham, Kent, to Captain Thomas Henry and Mary Elizabeth (Churchward) Tizard. The family was financially solid but not wealthy. As a navy hydrographer, his father had participated in extensive naval survey work around the world. He encouraged his son's early interest in science. The young Tizard looked forward to a naval career, but just before he was to enter naval school at 13, his left eye was damaged when a fly flew into it. His sight was impaired enough to bar him from naval service. He then enrolled at Westminster preparatory school, and later at Magdalene College, Oxford, where he studied physical **chemistry**. The day of his final examinations he was seriously ill with the influenza that would recur throughout his life, but he still managed to take first honors.

The center of **physical chemistry** research at the time was with **Walther Nernst** at the University of Berlin. Tizard enrolled there in 1908 to work toward a Ph.D. Although he stayed in Berlin only a year, Tizard had two experiences there that were to prove significant for Britain: he noted the powerful changes chemistry was bringing to Germany's technological (and therefore military) status, and he met Frederick Lindemann. Lindemann, a fellow chemistry student, was the son of an Alsatian who had become a naturalized Briton. In Berlin, Tizard and Lindemann studied together, practiced boxing at a gymnasium, and ice skated. This early friendship was to sour in the midst of the anxiety and political intrigue of World War II.

In 1911 Tizard became a fellow of Oriel College, Oxford, where he taught physical chemistry. In 1914 he embarked on a British Association tour of Australia, during which he met Cambridge's **Ernest Rutherford** (a noted physicist) and other eminent scientists. The onset of World War I cut the trip short, and Tizard returned to England to enlist in the Royal Garrison Artillery. He was soon transferred to the Royal Flying Corps, where he began a lifelong commitment to aviation technology. His training as a pilot during this period enabled Tizard to understand the practical problems of flying in a way that was rare among scientific advisers and much appreciated by the other aviators. It was also during this time (1915) that he married Kathleen Eleanor Wilson. They eventually had three sons, John, Richard, and David.

At the end of the war, Tizard became assistant deputy of the newly-formed Royal Air Force and encountered Winston Churchill, then Minister of Munitions, for the first time. He never got along well with Churchill, who like Lindemann, enjoyed the intricacies of political gamesmanship which Tizard disliked intensely. In 1919, partly at Tizard's urging, the professorship of experimental philosophy at Oxford went to Lindemann. The two men continued to see each other from time to time, but hostility was already flaring up between them. One colleague remembered them getting into a shouting match over the relatively trivial question of whether oranges should best be packed in symmetrical rows or in off-center layers.

Between the wars, Tizard consulted with the British petroleum industry. With Randall Pye, he studied adiabatic (heat neither lost nor gained) compression of **gases**, identifying the chemicals that were to prove most effective as fuel in internal **combustion** engines. Tizard also expanded his role as government adviser, working in the Department of Scientific and Industrial Research, where he hoped to encourage the application of scientific and technological discoveries. He found government and military leaders slow to assimilate new scientific knowledge, and this was to become a long-term frustration.

By the early 1930s Hitler's aggression posed a threat to Europe, and the British began to worry about air attacks. They had no way to detect incoming planes which, in any effective defense strategy, would have to be intercepted at the coastlines. Various "death ray" ideas were discussed, but no technology presented itself clearly. Tizard became chair of a committee (the Tizard Committee) to investigate the possibilities, one of which was the reflection of radio waves off the atmosphere. Lindemann soon joined the committee, urging study of aerial mines and balloon intercepts, which Tizard believed were unlikely to be of much use. Lindemann also wanted to find a protection against night attacks, whereas Tizard believed the daylight threat was greater. Tension between the two men increased, as Lindemann was close to Churchill and Tizard's power was advisory rather than executive. As it turned out, Lindemann was right about the danger of night attacks. Tizard's mistake in this regard tended to obscure his foresight in pushing radar research. He was largely responsible for establishing the country's chain of radar stations that enabled Britain to survive the Battle of Britain in 1940. Credit for his role was slow in coming.

In 1939 Tizard was asked to evaluate the feasibility of an atomic bomb. He tried unsuccessfully to obtain the option for all the **uranium** available from the Belgian Congo, but he did manage to obtain some uranium to send to the United States, where it became the first nuclear fuel. Working as an unofficial adviser to Lord Beaverbrook (William Maxwell Aitken), Tizard served as a conduit for classified information passing between Britain and the United States. In 1940 he headed a mission to America for this purpose, taking along the cavity magnetron (called the "heart of radar"), which the British had developed but the Americans were to make practicable.

Political rivalry with Lindemann and frustration with the government's inability to set clear priorities and lines of authority led Tizard to curtail his active government service in the last years of the war. He opposed the government's policy of random bombing of German cities because he believed it would be less effective than bombing U-boats. He became president of Magdalene College, Oxford, in 1942 and retired from the Air Ministry in 1943. After the war, Tizard continued to support increased rigor in Britain's technological development, advising the government on science and military policy through the early 1950s. He received many honors, including the Order of the Bath, several honorary degrees, and membership in the Royal Society. Tizard died in Fareham, Hampshire, on October 9, 1959.

Alexander Todd.

TODD, ALEXANDER (1907-1997)

English chemist

Alexander Todd was awarded the 1957 Nobel Prize in **chemistry** for his work on the chemistry of nucleotides. He was also influential in synthesizing **vitamins** for commercial application. In addition, he invesitgated active ingredients in cannabis and hashish and helped develop efficient means of producing **chemical weapons**.

Alexander Robertus Todd was born in Glasgow, Scotland, on 2 October 1907, to Alexander and Jane Lowrie Todd. The family, consisting of Todd, his parents, his older sister, and his younger brother, was not well-to-do. Todd's autobiography, *A Time to Remember,* recalls how through hard work his parents rose to the lower middle class despite having no more than an elementary education, and how determined they were that their children should have an education at any cost.

In 1918, Todd gained admission to the Allan Glen's School in Glasgow, a science high school; his interest in chemistry, which first arose when he was given a chemistry set at the age of eight or nine, developed rapidly. On graduation, six years later, he at once entered the University of Glasgow instead of taking a recommended additional year at Allan Glen's. His father refused to sign an application for scholastic aid, saying it would be accepting charity; because of superior academic performance during the first year, though, Todd received a scholarship for the rest of the course. In his final year at university, Todd did a thesis on the reaction of **phosphorus** pentachloride with ethyl tartrate and its diacetyl derivative under the direction of T. E. Patterson, resulting in his first publication.

After receiving his B.Sc. degree in chemistry with first-class honors in 1928, Todd was awarded a Carnegie research scholarship and stayed on for another year working for Patterson on optical rotatory dispersion. Deciding that this line of research was neither to his **taste** nor likely to be fruitful, he went to Germany to do graduate work at the University of Frankfurt am Main under Walther Borsche, studying **natural products**. Todd says that he preferred **Jöns Berzelius**'s definition of **organic chemistry** as the chemistry of substances found in living organisms to Gmelin's definition of it as the chemistry of **carbon** compounds.

At Frankfurt he studied the chemistry of apocholic acid, one of the bile acids (compounds produced in the liver and having a structure related to that of **cholesterol** and the steroids). In 1931, he returned to England with his doctorate. He applied for and received an 1851 Exhibition Senior Studentship which allowed him to enter Oxford University to work under **Robert Robinson**, who would receive the Nobel Prize in chemistry in 1947. In order to ease some administrative difficulties, Todd enrolled in the doctoral program, which had only a research requirement; he received his D.Phil. from Oxford in 1934. His research at Oxford dealt first with the synthesis of several anthocyanins, the coloring **matter** of flowers, and then with a study of the red pigments from some molds.

After leaving Oxford, Todd went to the University of Edinburgh on a Medical Research Council grant to study the structure of vitamin B_1 (thiamine, or the anti-beriberi vitamin). The appointment came about when George Barger, professor of medical chemistry at Edinburgh, sought Robinson's advice about working with B_1. At that time, only a few milligrams of the substance were available, and Robinson suggested Todd because of his interest in natural products and his knowledge of microchemical techniques acquired in Germany. Although Todd and his team were beaten in the race to synthesize B_1 by competing German and American groups, their synthesis was more elegant and better suited for industrial application. It was at Edinburgh that Todd met and became engaged to Alison Dale—daughter of Nobel Prize laureate Henry Hallett Dale—who was doing postgraduate research in the pharmacology department; they were married in January of 1937, shortly after Todd had moved to the Lister Institute where he was reader (or lecturer) in **biochemistry**. For the first time in his career, Todd was salaried and not dependent on grants or scholarships. In 1939 the Todds' son, Alexander, was born. Their first daughter, Helen, was born in 1941, and the second, Hilary, in 1945.

Toward the end of his stay at Edinburgh, Todd began to investigate the chemistry of vitamin E (a group of related compounds called tocopherols), which is an antioxidant—that is, it inhibits loss of electrons. He continued this line of research at the Lister Institute and also started an investigation of the active ingredients of the *Cannabis sativa* plant (marijuana) that showed that cannabinol, the major product isolated from the plant resin, was pharmacologically inactive.

In March of 1938, Todd and his wife made a long visit to the United States to investigate the offer of a position at California Institute of Technology. On returning to England with the idea that he would move to California, Todd was offered a professorship at Manchester which he accepted, becoming Sir Samuel Hall Professor of Chemistry and director of the chemical laboratories of the University of Manchester. At Manchester, Todd was able to continue his research with little interruption. During his first year there, he finished the work on vitamin E with the total synthesis of alpha-tocopherol and its analogs. Attempts to isolate and identify the active ingredients in cannabis resin failed because the separation procedures available at the time were inadequate; however, Todd's synthesis of cannabinol involved an intermediate, tetrahydrocannabinol (THC), that had an effect much like that of hashish on rabbits and suggested to him that the effects of hashish were due to one of the isomeric tetrahydrocannabinols. This view was later proven correct, but by others, because the outbreak of World War II forced Todd to abandon this line of research for work more directly related to the war.

As a member, and then chair, of the Chemical Committee, which was responsible for developing and producing chemical warfare agents, Todd developed an efficient method of producing diphenylamine chloroarsine (a sneeze gas), and designed a pilot plant for producing **nitrogen** mustards (blistering agents). He also had a group working on **penicillin** research and another trying to isolate and identify the "hatching factor" of the potato eelworm, a parasite that attacks potatoes.

Late in 1943 Todd was offered the chair in biochemistry at Cambridge University, which he refused. Shortly thereafter he was offered the chair in organic chemistry, which he accepted, choosing to affiliate with Christ's College. From 1963 to 1978, he served as master of the college. As professor of organic chemistry at Cambridge, Todd reorganized and revitalized the department and oversaw the modernization of the laboratories (they were still lighted by gas in 1944) and, eventually, the construction of a new laboratory building.

Before the war, his interest in vitamins and their mode of action had led Todd to start work on nucleosides and nucleotides. Nucleosides are compounds made up of a sugar (ribose or deoxyribose) linked to one of four heterocyclic (that is, containing rings with more than one kind of atom) nitrogen compounds derived either from purine (adenine and guanine) or pyrimidine (uracil and cytosine). When a **phosphate** group is attached to the sugar portion of the **molecule**, a nucleoside becomes a nucleotide. The **nucleic acids** (DNA and RNA), found in cell nuclei as constituents of the chromosomes, are **chains** of nucleotides. While still at Manchester, Todd had worked out techniques for synthesizing nucleosides and then attaching the phosphate group to them (a process called phosphorylating) to form nucleotides; later, at Cambridge, he worked out the structures of the nucleotides obtained by the degradation of nucleic acid and synthesized them. This information was a necessary prerequisite to **James Watson** and **Francis Crick**'s formulation of the double-helix structure of **DNA** two years later.

Todd had found the nucleoside adenosine in some coenzymes, relatively small molecules that combine with a protein to form an **enzyme**, which can act as a catalyst for a particular biochemical process. He knew from his work with the B vitamins that B_1 (thiamine), B_2 (riboflavin) and B_3 (niacin) were essential components of coenzymes involved in respiration and **oxygen** utilization. By 1949 he had succeeded in synthesizing adenosine—a triumph in itself—and had gone on to synthesize adenosine di- and triphosphate (ADP and ATP). These compounds are nucleotides responsible for **energy** production and energy storage in muscles and in plants. In 1952, he established the structure of flavin adenine dinucleotide (FAD), a **coenzyme** involved in breaking down **carbohydrates** so that they can be oxidized, releasing energy for an organism to use. For his pioneering work on nucleotides and nucleotide enzymes, Todd was awarded the 1957 Nobel Prize in chemistry.

Todd collaborated with **Dorothy Crowfoot Hodgkin** in determining the structure of vitamin B_{12}, the antipernicious anemia factor, which is necessary for the formation of red blood cells. Todd's chemical studies of the degradation products of B_{12} were crucial to Hodgkin's x-ray determination of the structure in 1955.

Another major field of research at Cambridge was the chemistry of the pigments in aphids. While at Oxford and working on the coloring matter from some fungi, Todd observed that although the pigments from fungi and from higher plants were all anthraquinone derivatives, the pattern of substitution around the anthraquinone **ring** differed in the two cases. Pigment from two different insects seemed to be of the fungal pattern and Todd wondered if these were derived from the insect or from symbiotic fungi they contained. At Cambridge he isolated several pigments from different kinds of aphids and found that they were complex quinones unrelated to anthraquinone. It was found, however, that they are probably the products of symbiotic fungi in the aphid.

In 1952 Todd became chairman of the advisory council on scientific policy to the British government, a post he held until 1964. He was knighted in 1954 by Queen Elizabeth for distinguished service to the government. Named Baron Todd of Trumpington in 1962, he was made a member of the Order of Merit in 1977. In 1955 he became a foreign associate of the United States' National Academy of Sciences. He traveled extensively and been a visiting professor at the University of Sydney (Australia), the California Institute of Technology, the Massachusetts Institute of Technology, the University of Chicago, and Notre Dame University.

A Fellow of the Royal Society since 1942, Todd served as its president from 1975 to 1980. He increased the role of the society in advising the government on the scientific aspects of policy and strengthened its international relations. Extracts from his five anniversary addresses to the society dealing with these concerns are given as appendices to his autobiography. In the forward to his autobiography, Todd reports that in preparing biographical sketches of a number of members of the Royal Society he was struck by the lack of information available about their lives and careers and that this, in part, led him to write *A Time to Remember.* Todd died on 10 January 1997, in his home city of Cambridge, England. He was 89.

TONAGAWA, SUSUMU (1939-)
Japanese American immunologist

Susumu Tonagawa made a major contribution to the understanding of the immune system by showing how gene fragments are rearranged in somatic cells to make functional immune system genes. For his work, he received the 1987 Nobel Prize in physiology or medicine.

Tonagawa was born in Nagoya, Japan, where his father was an engineer. Developing an interest in **chemistry** while in high school, Tonagawa took his undergraduate degree in that subject at the University of Kyoto in 1963. However, in his senior year he read papers on the operon theory by the French biochemists François Jacob and Jacques Monod, and subsequently switched to molecular biology for graduate studies, earning his Ph.D. in 1978 at the University of California, San Diego.

He began specializing in immunology while working at the Basel (Switzerland) Institute for Immunology, and in 1981 he joined the faculty of the Massachusetts Institute of Technology's Center for Cancer Research.

In the 1970s, some scientists believed that an individual inherited a separate gene for each of the millions of individual antibody molecules. Other scientists thought that the individual inherits a small number of genes that somehow diversify in specialized somatic (body) cells. Tonagawa made major contributions to resolving the debate.

Using purified messenger **RNA** (mRNA) for producing **antibodies** and observing the genes it hybridized with, Tonagawa was able to count the number of genes and show that there were far fewer of them than the number of antibodies that were produced. He next used the then-new restriction enzymes and **genetic engineering** techniques to demonstrate a theory by other scientists that the somatic cell uses a flexible, multi-step process to rearrange fragments from different genes in different ways, producing many different antibodies. Tonagawa's findings overcame older beliefs that one gene codes one polypeptide chain and that genes are unchanged during development and cell differentiation.

Other scientists have shown that **mutation**s further increase the genetic diversity of the immune system.

Tonagawa's work included investigating the genetic origins of antigen **receptors** of immune system T cells and, most recently, gene recombination in the central nervous system.

TORRICELLI, EVANGELISTA (1608-1647)
Italian physicist

Born near Ravenna, Torricelli was first educated in local Jesuit schools and showed such brilliance that he was sent to Rome to study with Galileo's former student Benedetto Castelli (1578-1643). Through Castelli he first corresponded with and met Galileo, finally becoming his secretary and assistant. A few months after Galileo's death in 1642, Torricelli accepted Galileo's old position as court mathematician and philosopher to the Grand Duke of Tuscany, a position he held until his own death, before his fortieth birthday. As a scientist Torricelli became well known for his study of the motion of fluids and was declared the father of hydrodynamics by Ernst Mach. Torricelli also conducted experiments on what we now call **gases**, though the term was not then in use. Most notably, Torricelli settled an argument about the nature of gases and the existence of the vacuum. Aristotle believed that a vacuum could not exist. Though Galileo disagreed, he contended that the action of suction (in a **water** pump, for example) was produced by a vacuum itself and not by the pressure of the air pushing on the liquid being pumped. Despite his argument, Torricelli noticed that water could be pumped only a finite distance through a vertical tube before it ceased to move any further and set out to examine this paradox, inventing the first barometer in the process. Torricelli first filled a one-ended **glass** tube with **mercury**, then immersed this open end in a dish of more mercury, placing the tube in a upright position. He found that about thirty inches of mercury remained in the tube, deducing that a vacuum had been created above the mercury in the tube, and that the mercury was held in place not by the vacuum, but by the pressure of air pushing down the mercury in the dish. Thus he demonstrated the existence of a vacuum, showed why pumps then in use could only move **liquids** vertically a certain distance (the distance determined by the pressure of the surrounding air), and created an instrument capable of measuring air

Evangelista Torricelli.

pressure. Torricelli's invention of the barometer led to a burst of both theoretical and experimental work in physics and meteorology. Torricelli also made a contribution to meteorology with his suggestion that wind was not caused by the "exhalations" of vapors from a damp earth, but by differences in the **density** of air which in turn were caused by differences in the air **temperature**. Torricelli's investigations in mathematics played an important role in scientific history as well. Based on Francesco Cavalieri's "geometry of indivisibles," Torricelli worked out equations upon curves, **solids**, and their rotations, helping to bridge the gap between Greek geometry and calculus. Along with the work of Rene Descartes, Pierre de Fermat, Gilles Personne de Roberval, and others, these works enabled Isaac Newton and Gottfried Wilhelm Leibniz to give calculus its first complete formulation. Though not as great a scientist as his older contemporary, Galileo, Torricelli continued the tradition of Italian scientific pioneering. This tradition was not to last long after his own death in the middle of the seventeenth century, however; by the beginning of the next century the center of scientific progress had shifted to northern Europe.

Toxicology

Toxicology is the study of poisons, or toxins, and how they affect living organisms. It also includes investigations of toxins in the environment, how they are distributed, and the risks they present to plants, animals, and people. Toxicology draws on several areas of science including biology, **chemistry**, mathematics, and physics.

Human awareness of poisons predates recorded history. The earliest humans used toxic plant and animal extracts for both hunting and warfare. In ancient times and during the Middle Ages, knowledge of toxins was well-developed. However, a parallel gain in knowledge also took place with regard to medicines. In the late Middle Ages, Paracelsus, a physician-alchemist wrote: "All substances are poisons; there is none which is not a poison. The right dose differentiates a poison from a remedy." At the time, his views were seen as revolutionary, but they mark the beginnings of modern toxicology.

By the late nineteenth and early twentieth centuries, many scientists were devoting their research efforts to finding out why certain chemicals were poisonous and identifying their effects in the body. Such efforts continue today as new toxins are identified and others are re-evaluated.

Toxins can be classified according to many different systems. For example, they can be grouped according to what organ or organ system they affect. They can also be grouped according to their purpose, such as a **pesticide** or a **food additive**. A more general classification scheme groups toxins according to their source—plant, animal, or created by humans. Toxins can also be classified by the type of injury, such as cancer or liver damage, that they cause. Some classification schemes are very specific and zero in on a toxin's molecular mode of action. Examples of such specific modes of action include toxins that inhibit a particular enzyme or toxins that damage genetic material.

Regardless of the classification scheme, toxins are assessed by a standard set of criteria. One of the first sets of questions in evaluating a toxin center on exposure, how it occurs, and at what frequency. How a toxin enters the body can have a great impact on what its effects might be. Toxins can enter the body through the skin, by the mouth, or by being inhaled. Exposure can occur due to a person's job or where that person lives. Exposure can also occur because of an accident or through a deliberate act such as a suicide attempt. Toxicologists divide the frequency of exposure into acute or chronic durations. Acute exposure corresponds to a short time frame; chronic exposure lasts for a long time frame.

The effects of exposure are evaluated by the reactions. For example, exposure to a toxin might cause a person to suffer an allergic reaction. Exposure to a different toxin might cause more serious effects such as trouble breathing, irregular heartbeat, or even death. Effects are not always immediately apparent. The effects of some exposures are not seen for years as in the case of cancer or damage to the nervous system.

Exposure to a toxin and the effects it causes are usually expressed by the dose-response relationship. This relationship describes the effects that certain amounts of toxin will have on a living organism. The concept of the dose-response relationship is seen in Paracelsus's writings; he makes it clear that the dose makes the poison. Any substance in a large enough quantity will have a negative effect. How large a quantity depends on the substance.

A toxin is also evaluated according to its **kinetics**. The kinetics of a toxin, or toxicokinetics, describes how the toxin is absorbed in the body, where it is distributed, and how the body handles the toxin. For example, if a toxin enters the body through the mouth, it may be absorbed from the stomach or the intestine. Once in the bloodstream, it is carried to the liver.

The liver contains many enzymes—proteins that trigger speedy chemical reactions—that can alter the chemical structure of toxins. Many toxins that enter the body eventually go through the liver, although there are some that will bind to other tissues and remain for a long time. In the liver, enzymatic alterations may lead to the toxin being quickly excreted or they may result in slowing the toxin's exit from the body. Other alterations may either make the toxin harmless or cause it to become harmful.

Determining how the body handles the toxin provides information on its effects and how to stop or reverse them. This information is very useful in making risk assessments and developing risk management plans. There are many natural and human-made chemicals in the environment. Some of these chemicals are toxins, but others are not. Using risk assessment, the ones that are dangerous can be identified. By identifying these chemicals, management efforts can be clearly focused on limiting or preventing exposure to them.

Transfermium elements

The transfermium elements are those elements with atomic numbers greater than 100, the **atomic number** for **fermium**. The transfermium elements are grouped together for a number of reasons. First, they are all prepared artificially. None occur in

Name	Atomic Symbol	Atomic Number	Atomic Mass
Mendelevium	Md	101	258
Nobelium	No	102	259
Lawrencium	Lr	103	262
Rutherfordium	Rf	104	261
Dubnium	Db	105	262
Seaborgium	Sg	106	263
Bohrium	Bh	107	264
Massium	Hs	108	265
Metinerium	Mt	109	268
Ununnilium	Uun	110	269
Unununium	Uuu	111	272
Ununbiium	Uub	112	277
Ununquadium	Uuq	114	289
Ununhexium	Uuh	116	289
Ununoctium	Uuo	118	293

Transfermium elements. ***(Illustration by Electronic Illustrators Group.)***

the Earth's crust (none, at least, have ever been discovered). Second, they are very difficult to create. In fact, no more than a few atoms of some transfermium elements have been created so far. Third, very little is known about the transfermium elements because so few of their atoms are available for study.

Still, the transfermium elements are of great interest to chemists and physicists because they help answer some fundamental questions about the **periodic table**. Scientists want to know if there is a limit to how heavy a chemical element can be and whether some of the very heavy elements are stable. They also want to know what the properties of these very heavy elements are.

Properties

No one know very much about the properties of the transfermium elements. It is impossible to measure their bulk properties, such as **color**, **malleability**, **ductility**, melting and boiling points, and densities. It has been possible to determine some fundamental chemical properties for a few of the elements. In 1997, for example, a German research team determined some chemical properties of element 106, seaborgium, working with only six atoms. One of their findings was that the element's properties are similar to those of **tungsten**, which is above it in the periodic table.

All of the transfermium elements are radioactive. They have anywhere from 13 isotopes (mendelevium) to one **isotope** (e.g., **hassium** and meitnerium). The half- lives of the transfermium elements are typically very short, often only a few seconds or less. In most cases, the elements radioactively decay so rapidly that scientists have very little opportunity to observe them and study their properties.

Discovery

All transfermium elements are made in particle accelerators (devices that make tiny particles like protons and small atoms move very fast). They are made by a process in which a heavy **nucleus** is bombarded with a relatively small nucleus. As an example, the transuranium element **americium** (atomic number 95) might be used as the target in an experiment and the nuclei of **neon** atoms (atomic number 10) might be used as the bullets. If a neon **atom** hits and americium atom just right it ''sticks'' making the americium atom heavier. A new element, atomic number 105, is produced by this reaction. That element is now known as dubnium. Dubnium and all other transfermium elements formed by this process are radioactive and decay very quickly.

This kind of experiment is easy to describe, but very difficult to carry out. In fact, research of this kind is carried on at only three laboratories in the world: the Joint Institute for Nuclear Research, in Dubna, Russia; the Lawrence Berkeley Laboratory at the University of California; and the Institute for Heavy **Ion** Research in Darmstadt, Germany. All three laboratories use large particle accelerators costing millions of dollars and operated by dozens of scientists from many different countries.

Assigning credit for the discovery of a transfermium element is very complicated. In most cases, no more than a handful of atoms is produced in a given reaction. For example, the Dubna group first claimed to have produced element 104 in 1964, but this report was questioned by many scientists. Five years later, the American scientific team also reported making element 104 and supported their report with somewhat stronger evidence.

Naming

One reason that scientists argue over the discovery of an element is related to the naming of the element. The custom in **chemistry** is that the scientist or scientists who discover an element earn the right to suggest a name for the element. For example, researchers at the Berkeley laboratory first discovered elements 97 and 98 and then suggested naming them **berkelium** and **californium** in honor of Berkeley, California, where the research was done.

The final decision as to a new element's name is made by the **International Union of Pure and Applied Chemistry (IUPAC)**. IUPAC appoints a committee to mediate disagreements as to which team actually discovered a new element and the element's name. In the case of the transfermium elements, two, and sometimes all three, laboratories have laid claim to the discovery of an element and have suggested their own names for the element. For example, the Russian team suggested the name of *kurchatovium* for element 104, in honor of the great Russian nuclear chemist Igor Vasilevich Kurchatov (1903-1960). The American team, on the other hand, proposed the name *rutherfordium* for the new element, in honor of the great British scientist Sir **Ernest Rutherford**.

IUPAC pondered the question of naming the transfermium elements for a very long time, more than 30 years in some cases. The organization made a ''final'' decision on these names in 1994, but reaction from the scientific community was so heated that it decided to reconsider its decision. Finally, in 1997, IUPAC really made a ''final'' decision, assigning the names listed in Figure 1. Elements 101 through 104, 106, 107, and 109 were all named for famous scientists, while elements 105 and 108 were named for the locations in which the Russian and German research teams, respectively, were located.

Future Research

Scientists have long expected that the list of transfermium elements would continue to increase. The task of making new elements with increasingly higher atomic numbers becomes more difficult, but there is no reason to expect that discoveries will not continue. A system has been developed for naming elements with atomic numbers greater than those for which official names have been developed. For example, element number 110 is unofficially known as ununnilium. In this system of naming, the grouping *un* represents ''one,'' while the grouping *nilium* repesents ''nil,'' or ''zero.'' Thus, un-un-nilium means one-one-zero, or 110. Similarly, element number 111 is known as unununium, and element number 112 is called ununbiium.

In January 1999, scientists at the Joint Institute for Nuclear Research in Russia reported producing one atom of element 114 in a nuclear reaction involving fusing a **calcium** atom with a **plutonium** atom. This claim has not yet been ratified.

Samples of transition metals: paint, titanium, wiring, copper plumbing, structural steel, iron, and gold. ***(Photograph by Charles D. Winters, Photo Researchers, Inc. Reproduced by permission.)***

Then, in June 1999, the scientific world was jolted with the announcement that two new transfermium elements had been found. They were not numbers 110, 111, or 112, as many people had expected, but numbers 118 and 116. The announcement was made by the Berkeley research team, who reported that the elements had been formed by the collision of **krypton** (atomic number 36) nuclei with **lead** (atomic number 82) nuclei. It is actually element number 118 that is formed in this reaction, but that element decays rapidly to form element number 116. It appears that elements number 114, 112, and 110 may also have been formed by the decay of 116, although evidence for these elements was not yet available.

TRANSITION ELEMENTS

Transition elements, or metals, are a category of materials in the **periodic table** that are set apart from the ''major group'' or ''A series'' elements by their ability to use either the penultimate and outermost **electron** shells to bond with other substances. In **industrial chemistry**, transition metals are especially valuable when in their metallic state. Most of the economically important **metals** (i.e., **silver, gold, platinum, nickel,** and **copper**) belong to the transition elements family, as do most of the elements used when high-performance metals are necessary.

One characteristic all the transition elements have in common, besides being metals, is their incomplete d electron orbitals, which fill as **atomic number** increases. This is partly why transition elements are so useful—the empty orbitals help them accommodate a wide variety of **bonding** interactions, which makes them good catalysts.

Chemists sometimes differ on which elements should be put in the transition category. Part of this confusion stems from whether the definition of "transition" can be taken to mean that an element's orbital was just completely filled or not. In general, based on this definition, the first three series (as viewed on the periodic table) include elements 21 though 30 (**scandium** through **zinc**), 39 through 48 (**yttrium** though **cadmium**), and 57 and 72-80 (**lanthanum** through **mercury**). The fourth group of transition metals comprises elements 89, 104-109, and the undiscovered elements 110 and 111. There is another group of transition metals called the "inner transition" elements, which comprise the lanthanides (58-71, **cerium** to **lutetium**), and the actinides (90-103, **thorium** to lawrencium). The inner transition elements differ only in the number of f-electrons in their penultimate shell. All have very similar characteristics, their outer-shell orbitals having the same number of electrons.

Like typical metals, the transition metals are mostly hard and strong. They conduct **electricity** and **heat**, have high densities, and boil and melt at high **temperatures.** However, unlike regular metals, the transition elements can easily form extremely stable coordination complexes. Some of these are widely used to recover metal from low-grade ores and to facilitate high-quality electroplating. Transition metals also produce complex **ions**, which are often intensely colored, and frequently have unpaired electrons in their d subshells. The latter property makes such transition elements paramagnetic, meaning their magnetic moments are roughly parallel. These metals will be attracted to a magnetic field.

One of the most important reasons that many industries require transition metals is because the elements produce metal oxide compounds. **Oxidation** usually causes corrosion, but the transition metal oxides have many useful characteristics. For instance, they have key roles in electronics, and their use as pigments makes paint much more weather resistant. The transition metal oxides' chemical stability makes them able to stand up to intense sun and years of exposure to the open air, so they are popular in the high-end automotive and house paint industries. The oxides also impart their **colors** to such pigments. Those with completely empty or full 3d subshells, for example, are white (zinc oxide and titanium oxide), but they come in a wide range of vibrant colors as well.

TRANSITION STATE THEORY

Chemical reaction rates are influenced by a variety of factors, including **temperature** and the presence of catalysts. For example, an increase in temperature will increase reaction rates. According to **collision theory**, the **reaction rate** is equal to the frequency of successful collisions. What is mainly required for a successful collision of reactants (molecules entering a reaction) is a minimal quantity of **energy** (activation energy) as well as a specific spatial orientation of the reacting molecules. In other words, collisions not satisfying these prerequisite do not lead to a reaction. Collision theory explains why an increase in temperature accelerates a reaction: the kinetic energy of molecules rises with temperature. However, another theory is needed to explain the phenomenon of activation energy: transition state theory.

Transition state theory does not define colliding molecules as compact objects; in fact, the term "collision," although current in chemical vocabulary, does not accurately describe the interaction of molecules during a chemical reaction. According to transition state theory, as molecules approach one another prior to a reaction, their orbitals connect. As a result, the orbitals become deformed, weakening the existing bond between molecules. Weakened, bonds can break, and new bonds can be formed. For example, when **hydrogen** iodide (HI) **molecule** interact with **chlorine** (Cl) molecules, chlorine, which is negatively charged, will attract the electrons shared by the hydrogen and **iodine** atoms in the HI molecule, weakening the H-I bond. In order for the reaction to proceed, the H-I bond must be broken, which enables the creation of a new bond: H- Cl. Indeed, HCl, along with a molecule of iodine, is the outcome of the reaction. There is a point, however, during the reaction, as the old bond is breaking and the new one is forming, when an intermediary grouping of atoms appears: the **activated complex**. For example, the activated complex for a HI + Cl reaction would be the unstable I-H-Cl group of atoms.

Transition state theory also describes the changes in energy during a reaction. During the reaction process, as molecules approach each other and connect, their kinetic energy turns into **potential energy**. As the process proceeds, potential energy rises, reaching a maximum at the point of contact. This energy rise is also known as a reaction profile. Each reaction has its characteristic reaction profile. For example, for the HI + Cl reaction, there needs to minimum of kinetic energy converted into potential energy if the reaction is to succeed. This minimum of energy is called **activation energy**, the level of energy needed to activate a reaction.

Transition state theory stems from the pioneering research by Michael Polanyi (1891-1976) and **Henry Eyring** in the 1920s and 1930s. Assuming that the current theory of chemical change was incomplete, Polanyi and Eyring proceeded to study the changes in potential energy during chemical reactions, eventually formulating a theory which was in accord with quantum mechanics.

TRANSMUTATION

Transmutation is the transformation of one element into another. This notion originated in **alchemy** but continues today as an important research area of physics and **chemistry**. The principal goal of alchemists was the conversion or transmutation of base **metals** like **lead** into **gold**. The alchemists searched from the Middle Ages to the sixteenth century, in one manner or another, for the "philosopher's stone," which was thought to be the vital ingredient needed to transmute lead to gold. The search for the means to transmute was considered a noble occupation; even Isaac Newton (1642-1727) devoted much of his time to alchemy. This hope of converting one element into another using alchemical means was almost wholly abandoned by the late eighteenth century because it was regarded as foolish trickery.

The goal of transmutation was finally realized in the early twentieth century in **radioactivity**. In 1903, **Ernest Rutherford** and **Frederick Soddy** made the astonishing discovery that natural radioactivity involves transmutation. Radioactivity involves the change of atoms of one chemical element into atoms of another element. The change, known as radioactive decay, occurs when an unstable **nucleus** spits out one or more particles and transforms to a new nucleus. This new nucleus—that of a different element—may be stable, or also unstable and so capable of undergoing another transmutation. There are two types of radioactive decay, alpha decay (ejection of a nucleus of **helium**, two protons and two neutrons tightly bound together) and beta decay (ejection of an electron). Both of these transmute an original radioactive nucleus into the nucleus of another element. Through radioactivity, it is indeed possible to transmute **uranium** to lead. The alchemists of today are those nuclear chemists who routinely transmute uranium to **plutonium** by bombarding uranium with neutrons.

TRANSPORT PROTEINS

Transport **proteins** are proteins that aid the movement of materials, either across **plasma** membranes in cells or through the circulatory system. Some transport proteins form channels through the membrane to allow passive flow of a substance down its **concentration** gradient. Others act as carriers, binding to a substance on one side of the membrane or in one region of the body, and releasing it on the other side or in another region. Some carrier proteins simply allow a material to move down its concentration gradient, equalizing concentrations of the material across the membrane. Others use **energy** to function, and act to create concentration gradients.

The chloride channel is an example of a channel protein. This protein allows the movement of chloride ions from one side of the membrane to the other. Movement of chloride ions often occurs when the cell deliberately transports positive ions, such as **sodium** or **potassium**. The chloride ions move in response to changes in charge within the cell, and allows the cell to minimize charge build- up.

The Na^+/K^+ ATPase uses cell energy, in the form of adenosine triphosphate (ATP), to move sodium ions into the cell, and potassium ions out. This allows the cell to create concentration gradients across the membrane for these two ions. These gradients can be used for a variety of purposes. Neurons (nerve cells) use these gradients for transmitting nerve signals down their length. In the resting state, the gradients are built up and maintained. When the nerve cell is stimulated, channel proteins open to allow the gradients to decay. This wave of gradient breakdown is called depolarization, and is the chemical change that underlies the functioning of the nervous system.

The sodium-glucose antiport is a passive co-transporter, meaning it moves sodium and glucose in opposite directions across the membrane without directly using ATP. It employs the sodium gradient built up by the Na^+/K^+ ATPase to drive glucose transport into the cell, against its concentration gradient.

The circulatory system also employs transport proteins. Hemoglobin is an **oxygen** transporter. Packed into red blood cells, it picks up oxygen in the lungs, and deposits it in the tissues. Hemoglobin allows the blood to carry much more oxygen than would dissolve in the blood plasma. HDL and LDL (high and low-density lipoprotein) are **cholesterol** transporters within the blood plasma. These transporters allow the otherwise insoluble cholesterol to be dissolved in the plasma.

TRIATOMIC

A **molecule** is triatomic when it consists of three atoms.Water is an example of a triatomic molecule consisting of two **hydrogen** atoms and one **oxygen atom** combined to form a single molecule. As these atoms combine, there is shifting of **valence** electrons, i.e., the atoms in the outside **electron** shell, of each atom. When triatomic molecules are formed, the outer **energy** level of each atom is usually completed, either as the result of electron pair sharing or of the loss or gain of electrons.

TRIBOLUMINESCENCE

Triboluminescence is a special case of luminescence, where luminescence is the emission of visible or invisible radiation unaccompanied by high **temperature** in any substance by the absorption of excited **energy**. Excitation sources include photons, charged particles, and chemical changes. An inorganic luminescent material, or phosphor, usually consists of a crystalline host material with a trace impurity added. In the case of triboluminescence, the excitation source is mechanical disruption. ZnS • Mn is an example of a material exhibiting triboluminescent behavior.

The word triboluminescence is derived from *tribo*, from the Greek word *tribein*, meaning to rub, and the Latin words *lumen*, meaning light, and *escens*, meaning characterized by. Thus, luminescence is the creation of light by means other than **heat**, as distinguished from incandescence, which describes light emission due solely to the temperature of a source.

Wint-O-Green Lifesavers™ produce an example of triboluminescence, which is essentially a two-step process. In the first step, the Lifesaver sugar crystals break, usually along planes with positive charges on one side and negative on the other. As the pieces of candy separate, the positive and negative charges attempt to recombine by jumping through the air like tiny lightning bolts. In the second step, the lightning bolts produce invisible ultraviolet light that is absorbed by the wintergreen molecules, causing them to fluoresce, or glow.

Some scientists believe that crystal structure and impurities are central to whether a material becomes triboluminescent. For many years, it was believed that only materials with an asymmetrical crystal structure would glow when crushed

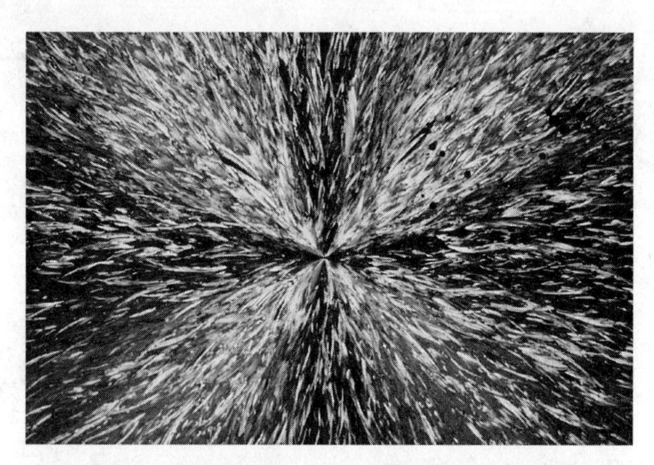

A transmitted light microscopic image of trinitrotoluene, or TNT, magnified 4 times. ***(Photograph by Michael W. Davidson, Photo Researchers, Inc. Reproduced by permission.)***

because the splitting of such a crystal places positive charges on one face and negative ones on the other. As expected, when these charges recombine, they crackle through the air like tiny lightning bolts.

However, several materials with symmetrical crystal structures also exhibit triboluminescent behavior. But these symmetrical crystals are found to lose their triboluminescence when any impurities present in the crystals are eliminated. These impurities are believed to produce local structural asymmetries that allow the symmetrical crystals to exhibit triboluminescence. But scientists are still very far from understanding the effect, with some evidence indicating that impurities may play a role in asymmetrical crystals as well. It is hoped that a better understanding of this effect will eventually provide explanations for other mysterious lights observed in nature, such as deep-sea luminescence.

Besides triboluminescence, other types of luminescence include photoluminescence (with **photon** excitation), electroluminescence (with electric field excitation), **chemiluminescence** (with chemical excitation), and **bioluminescence** (with biochemical excitation). The terms **fluorescence and phosphorescence** refer to specific characteristics of luminescent transitions. In fluorescence, the glow accompanying the emission of energy is very short (10^{-8}- 10^{-3} seconds), while in phosphorescence, the glow may last for several hours. This is because fluorescence involves an electronic transition from a higher to a lower electronic state, e.g., a triplet to singlet emission; whereas the electronic transition in phosphoresence is a same-state transition, e.g., a singlet to singlet emission. The light from the firefly is an example of phosphorescence.

See also Crystallography; Fluorescence and phosphorescence; Spectroscopy

Trinitrotoluene (TNT)

Trinitrotoluene (TNT), $CH_3C_6H_2(NO_2)_3$, is a crystalline, aromatic compound that can be explosive with the aid of a detonator, but is not as unstable as **nitroglycerin**. TNT is the product of toluene (C_7H_8), a colorless, liquid, aromatic **hydrocarbon**, after its been treated with a combination of nitric and sulfuric acids a process known as nitration. **Aromatic hydrocarbons** all contain benzene(C_6H_6) a **chemical compound** that is made up of six **carbon** atoms structurally arranged in a **ring**. Aromatic hydrocarbons can be isolated from **coal** tar.

TNT is a substance that traces its roots to the dye industry and later research by German chemist **Adolf von Baeyer**. It is the most powerful nonatomic military explosive of the twentieth century. Sources credit J. Wilbrand with its initial discovery in 1863. Although probably first employed in the Russo-Japanese War of 1904-05, TNT was not mass-produced or regularly used until its refinement by the Germans during World War I. Fired by long-range guns, TNT shells encased in steel exploded with a force of 2,250,000 pounds per square inch.

In both world wars, new forms of TNT were introduced. These included TNT in combination with such similarly volatile compounds as TNX, PETN, and RDX. One mixture, RDX-TNT, or cyclonite, with a detonation pressure of 4,000,000 pounds per square inch, is regarded as the most powerful of this new class of weaponry. It is especially forceful when combined with **aluminum** in the form of torpex. TNT itself is composed of **nitrogen**, **hydrogen**, carbon, and **oxygen**. Despite its violent potential when detonated, it is extremely safe to cast into shells and handle and is thus a preferred high explosive.

Tritium

Tritium is an **isotope** of the chemical element **hydrogen**. It has not only a single **proton** but also two **neutrons** in the **nucleus** of its **atoms**. Although technically it is still the element hydrogen, it has its own chemical symbol, T. Chemically, tritium reacts in exactly the same manner as hydrogen, although slightly slower because of its greater atomic weight. A tritium atom has almost three times the mass of a regular hydrogen atom: the atomic weight of tritium is 3.016 whereas the atomic weight of hydrogen is 1.008. Tritium is radioactive, with a **half-life** of 12.26 years. Its nucleus emits a low-energy beta particle, leaving behind an isotope of **helium**, helium-3, that has a single neutron in its atomic nucleus. (The common isotope of helium, helium-4, contains two neutrons in its atomic nucleus.) No gamma rays, which are high-energy electromagnetic radiation, are emitted in the decay of tritium, so the radioactive decay of tritium is of little hazard to humans.

The heavier atomic weight of tritium has an effect on the physical properties of this hydrogen isotope. For example, tritium has a **boiling point** of 25K (-415°F; -248°C), compared with ordinary hydrogen's boiling point of 20.4K (-423°F; -252.8°C). **Molecules** containing tritium show similar variances. For example, **water** made with tritium and having the formula T_2O has a melting point of 40°F (4.5°C), compared with 32°F (0°C) for normal water.

Tritium was present in nature at very low levels, about 1 atom every 10^{18} atoms of hydrogen, before atmospheric nuclear bomb testing. It is produced in the upper atmosphere, as highly-energetic neutrons in cosmic rays bombard **nitrogen** atoms, making a tritium atom and an atom of carbon-12.

Industrially, tritium is prepared by bombarding **deuterium** with other deuterium atoms to make a tritium atom and a regular hydrogen atom. The resulting two types of hydrogen can be separated by **distillation**. Another way to make tritium is to bombard lithium-6 atoms (the less-abundant isotope of **lithium**) with neutrons, which produces a helium atom and a tritium atom.

Due to the testing of nuclear weapons in the atmosphere (before such testing was banned), the tritium content of the atmosphere rose to approximately 500 atoms per 10^{18}, declining steadily ever since the ban due to radioactive decay.

Tritium is used in **nuclear fusion** processes because it is easier to fuse tritium nuclei than either of the other isotopes of hydrogen. However, because of its scarcity, it is commonly used with deuterium in fusion reactions:

$T + D \rightarrow He + neutron + energy$

This is the nuclear reaction that occurs in fusion bombs, or hydrogen bombs. Such weapons must be recharged periodically due to the radioactive decay of the tritium. Fusion reactions are also being used in experimental fusion reactors as scientists and engineers try to develop controllable nuclear fusion for peaceful power.

Tritium is used as a tracer because it is relatively easy to detect due to its **radioactivity**. In groundwater studies, tritium-labeled water can be released into the ground at one point, and the amount of tritium-labeled water that appears at other points can be monitored. In this way, the flow of water through the ground can be mapped. Such information is important when drilling oil fields, for example. Tritium can also be substituted for ordinary hydrogen in organic compounds and used to study biological reactions. Because of its radioactivity, it is easy to follow the tritium as it participates in biochemical reactions. In this way, specific metabolic processes at the cellular level can be monitored. Tritium is also used to make ''glow-in-the-dark'' objects by mixing tritium-containing compounds with compounds like **zinc** sulfide, which emit light when struck by alpha or beta particles from nuclear decay.

TSWETT, MIKHAIL (1872-1919)

Russian chemist and botanist

Although recognized only belatedly, Mikhail Tswett (sometimes spelled Tsvet) was the first to lay out in detail the methods of the separation technique called **chromatography**. Tswett himself regarded chromatography only as a tool in his chemical and biological studies; his purpose was to separate and identify the many different pigments in leaves and other plant parts, and he considered it merely an improvement on existing techniques such acid-extraction, base-extraction, and fractional crystallization. Since he first described this process, many kinds of chromatography have been developed, and no laboratory is considered complete without a number of chromatographic instruments.

Mikhail Semyonovich Tswett was born May 14, 1872, in Asti, in the northwest part of Italy about seventy miles from the Swiss border. His parents were Semyon Nikolaevich and Maria de Dorozza Tswett. His father was a Russian civil servant and his mother, who was very young, died soon after his birth. His father returned to Russia after her death, and left his son with a nurse in Lausanne. Tswett was educated in Lausanne and Geneva, becoming multilingual in the process. He received his secondary education at the Collège Gaillard in Lausanne and the Collège de St. Antoine in Geneva; he entered the University of Geneva in 1891, studying **chemistry**, botany, and physics. His baccalaureate in both physical and natural sciences was awarded in 1892. He began plant research during his undergraduate years, earning the Davy Prize while a doctoral student with a paper on plant physiology that was subsequently published. In 1896 he defended his thesis, ''Études de physiologie cellulaire,'' and received his doctoral degree.

Thereafter he moved to Russia, and in 1897 he began working at the laboratory of plant anatomy and physiology at the Academy of Sciences and the St. Petersburg Biological Laboratory. His academic horizon was limited by the fact that foreign degrees were not recognized in tsarist Russia, and he set to work earning another master's degree in botany at Kazan University. He finished in 1901, with a thesis in Russian whose title is translated ''The Physicochemical Structure of the **Chlorophyll** Grain.'' In 1902 Tswett became an assistant in the laboratory of plant anatomy and physiology at the University of Warsaw, which was under Russian control at that time, where he became a full professor in 1903. In 1907 he took on the additional task of teaching botany and microbiology at the Warsaw Veterinary Institute; a year later he was also teaching at the Warsaw Technical University. He resigned his teaching post at the University of Warsaw but took a second doctorate there in 1910 with a dissertation on plant and animal chromophils. This apparently led to his only book, published in the same year, whose title is translated as ''The Chromophils in the Animal and Vegetable Kingdoms.'' The book itself has never been translated. By 1914 Tswett's brief, brilliant research career was essentially at an end. The German invasion of Poland in 1915 forced the Technical University to move to Moscow, and then to Nizhni Novgorod in 1916. Tswett's time was largely consumed with organizing the work of the botanical laboratories after each of these moves. In 1917 he accepted a position at the University at Yuryev in Estonia, but that too was overrun by the German army a year later. The university moved to Voronezh in 1918, but Tswett's health, never robust, failed quickly, and he died of a heart ailment at age forty-seven, on June 26, 1919.

Tswett's strength as a scientist lay in how well he understood both chemistry and botany. He had always been interested in the internal molecular structures of plants, often inquiring what their purpose might be, and the work he did on chlorophyll was one of his most important research efforts. He had long doubted the contention, which was widely accepted at the time, that chlorophyll was a compound that actually existed in plants. He decided this belief was the result of a misunderstanding; he hypothesized that chemists had been confused either because chlorophyll was combined nearly inseparably with other molecules within the leaf or because a compound recovered by a particular separation technique might in fact be an artifact of the technique. He was able to demonstrate all of these misunderstandings in the work of others, both by his deployment of the chemical separation methods of the time (fractional **solution** and **precipitation**, **diffusion**, differential solution) and by the **adsorption** methods he developed, culminating in chromatography.

''Adsorbent'' means holding molecules on the surface of the material, not in the body, and chromatography is a process which employs substances which have this property. It is a separation technique in which a very finely powdered adsorbent material is held in a vertical tube or ''column.'' The mixture to be separated is placed on the top of the column, dissolved in as small an amount of solvent as possible, so that it forms a narrow band of adsorbed mixture; then more solvent is allowed to flow through the column, top to bottom. The molecules in the mixture are more or less strongly held by the adsorbent; those weakly held are washed down the column most rapidly, and those strongly held move less rapidly. After a suitable development time, the components of the mixture separate into a series of bands spaced along the column. The plug of wet adsorbent is blown out of the column onto a plate, where the bands can be cut apart and the components recovered separately. As the mixtures separated in these early experiments were colored, and the bands absorbed light in the visible spectrum, Tswett named the process *chromatography* (''color-writing''), and the developed separation he called a *chromatogram.* Even though most mixtures are not colored, this terminology is retained; the components must be detected by some means other than the eye. Many sophisticated varieties of chromatography are in use today: paper, thin-layer, gas-liquid, and **ion** exchange, to name but a few. Still, Tswett's column method has not been totally displaced.

Tswett used this technique to demonstrate that chlorophyll indeed does not exist in the plant as a free **molecule** but is complexed with albumin. He named this complex ''chloroglobin ,'' by analogy with the heme complex of the blood, hemoglobin. There was, however, widespread skepticism of his research methods, and this finding was sharply criticized. Tswett next analyzed the plant pigments themselves, which were understood at the time to be only two: green chlorophyll and yellow xanthophyll. Using not chromatography but the standard chemical methods of the time, he demonstrated that there are two chlorophylls: xanthophyll and carotene. This finding was hotly disputed, partly because chlorophyll passed the test of a single pure compound: it could be crystallized. Tswett was able to show that the ''crystallizable chlorophyll'' formed by lengthy extraction with hot **ethanol** was in fact another compound; it is known today as an ethyl ester formed by transesterification of one of chlorophyll's ester groups.

During the course of his pigment work Tswett had found that when he ground the plant leaves with powdered **calcium** carbonate to neutralize acids, all but carotene were adsorbed on the solid carbonate. He used this as a method to separate carotene. It is not clear that this led to his devising column chromatography, but once he had developed this technique he found that in addition to two chlorophylls there were four xanthophylls and, of course, carotene. These findings came to be accepted later, but mainly through the work of the German chemist **Richard Willstätter**.

The technique of column chromatography was not widely used in Tswett's lifetime, being regarded by his most vocal opponent, L. Marchlewski, as no more than a ''**filtration** experiment.'' It was only later in the century that his work was re-evaluated and his status as one of the originators, though probably not the sole inventor, of chromatography, was confirmed. This is his legacy today, although some would consider the plant pigment work to be at least as important.

TUNGSTEN

Tungsten is a transition metal, one of the elements that occupy the middle of the **periodic table**. It has an **atomic number** of 74, an atomic **mass** of 183.85, and a chemical symbol of W.

Properties

Tungsten is a hard, brittle solid whose **color** ranges from steel-gray to nearly white. It has the highest melting point of any element, 6,170°F (3,410°C), a **boiling point** of about 10,650°F (5,900°C), and a **density** of 19.3 grams per cubic centimeter. The metal is a good conductor of **heat** and **electricity**.

Tungsten is a relatively inactive metal that does not react with **oxygen** at temperatures of less than 750°F (400°C), nor does it react with most acids, although it does dissolve in **aqua regia** (a mixture of concentrated **nitric acid** and 3-4 parts of hydrochloric acid).

Occurrence and Extraction

Tungsten is a moderately rare element with an abundance in the Earth's crust estimated at about 1.5 parts per million. It occurs most commonly in the minerals scheelite (calcium tungstate ($CaWO_4$) and wolframite (iron **manganese** tungstate $(FeMn)WO_4$). The world's largest producers of tungsten are China, Russia, and Portugal. The pure metal is prepared by heating tungsten **oxide** (WO_3) with **aluminum** metal: $2Al + WO_3 \rightarrow W + Al_2O_3$ or by passing **hydrogen** gas over hot tungstic acid: $H_2WO_4 + 3H_2 \rightarrow W + 4H_2O$.

Discovery and Naming

Credit for the discovery of tungsten is often divided among three men: the Swedish chemist **Carl Wilhelm Scheele** and two Spanish scientists Don Fausto D'Elhuyard (1755-1833) and his brother, Don Juan José D'Elhuyard (1754-96). Scheele appears to have discovered tungstic acid (H_2WO_4) in about 1781. Although he recognized that he had produced a new substance, he was unable to extract elemental tungsten from the acid. Two years earlier, the D'Elhuyard brothers also managed to produce tungstic acid from a sample of wolframite, but this time they succeeded in isolating the new element itself.

The element's name comes from a Swedish phrase that means ''heavy stone.'' In some parts of the world, tungsten is still called by another name, wolfram. That name is derived from the German expression *Wolf rahm*, or ''wolf froth.'' The element's chemical symbol is taken from this German name for the element rather than its Swedish name.

Uses

The primary use of tungsten metal is the manufacture of alloys, to which the element gives hardness, strength, elasticity, and tensile strength. About 90% of all tungsten alloys are used in mining, construction, and electrical and metal-working machine. These alloys are used to make high- speed tools; heating elements in furnaces; parts for aircraft and spacecraft; equipment used in radio, television and radar; rock drills; metal-cutting tools; and similar equipment. A small but very important amount of tungsten is also used to make the filament in incandescent light bulbs.

Probably the most important compound of tungsten commercially is tungsten carbide (WC), which is very strong and has a very high melting point (5,036°F/2,780°C). It is the strongest structural material and is used to make parts for electrical circuits, cutting tools, cermets, and cemented carbide. Cermets are substances consisting of a ceramic combined with a metal, while cemented carbide is made by **bonding** tungsten carbide to a second metal, forming very strong products that do not weaken at very high temperatures.

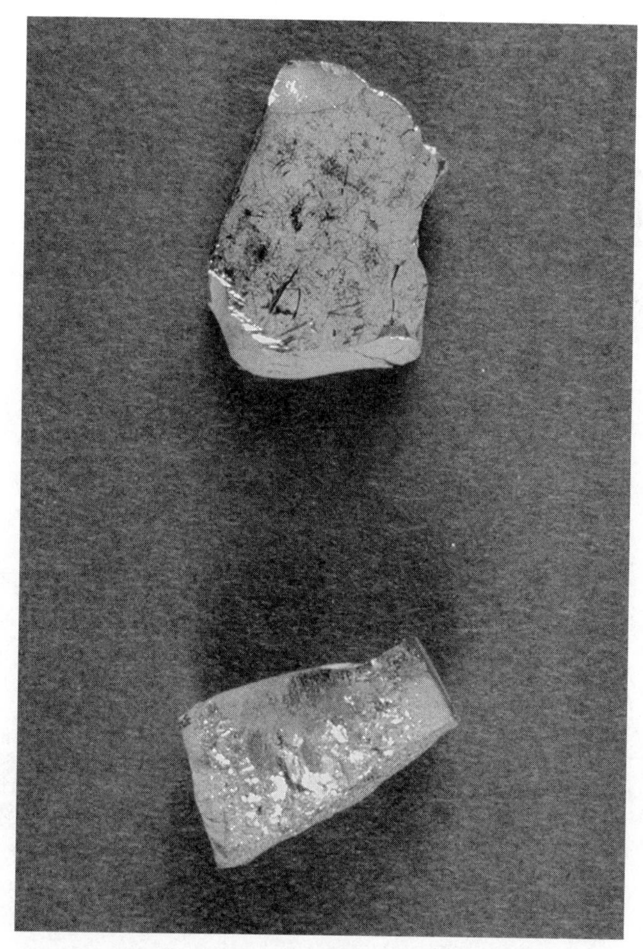

Tungsten samples. ***(Photograph by Russ Lappa/Science Source, National Audubon Society Collection/Photo Researchers, Inc. Reproduced by permission.)***

U

URANIUM

Uranium is the heaviest naturally occurring element. It is found in Row 7 of the **periodic table**. Uranium's **atomic number** is 92, its atomic **mass** is 238.0289, and its chemical symbol is U.

Properties

All isotopes of uranium are radioactive. The most common, uranium-238, is also the most stable with a half life of 4.468 x 10^9 years. Uranium is a silvery metal that is both ductile and malleable. It has a melting point of 2,070.1°F (1,132.3°C), a **boiling point** of about 6,904°F (3,818°C), and a **density** of 19.05 grams per cubic centimeter. The element is relatively reactive, combining with **oxygen**, **sulfur**, **fluorine**, **chlorine**, **bromine**, and **phosphorus**. It also dissolves in most acids and reacts slowly with **water** to form compounds that tend to be yellowish or green in **color**.

Occurrence and Extraction

Uranium is a relatively uncommon element in the Earth's crust with an abundance estimated at about 1-2 parts per million. The most common ore of the element is pitchblende, although it is also present in other minerals, such as uraninite, carnotite, uranophane, and coffinite. Uranium is extracted from its ores by converting them to uranium dioxide (UO_2), which is then converted to pure uranium metal with **hydrogen** gas: $UO_2 + 2H_2 \rightarrow 2H_2O + U$.

Discovery and Naming

Uranium was discovered in 1789 by the German chemist **Martin Klaproth**. Klaproth was analyzing the composition of pitchblende, which was thought at the time to be an ore of **iron** and **zinc**. Klaproth found, however, a small portion of pitchblende whose properties did not correspond to those of either of these elements. He decided that this portion was a new element. He proposed the name uranium for the element in honor of the planet Uranus, that had been discovered only a few years earlier in 1781.

Interestingly enough, Klaproth did not realize that uranium was radioactive. Indeed, the phenomenon of **radioactivity** was not to be discovered until nearly a century later. The material used widely by researchers interested in studying radioactivity at that time was pitchblende.

Uses

Compounds of uranium have had some modest use for many centuries, primarily as a coloring agent for **glass** and **ceramics**. Scientists have found glass made in Italy as early as 79 A.D. that was colored with uranium **oxide**. Uranium compounds continue to have very limited application as mordants and as filaments in light bulbs.

By far the most important application of uranium is in nuclear power plants and nuclear weapons. One of the less common isotopes of uranium, uranium-235, has the ability to undergo **nuclear fission**, the process by which **energy** is released for energy production in both weapons and power plants. Today, a very large fraction of nuclear weapons produced throughout the world contain uranium-235. Most of the more than 400 nuclear power plants that exist worldwide also used uranium-235 as their fuel.

UREA

Urea is a nitrogenous compound excreted in the urine by humans and most other mammals. The chemical structure of urea is $H_2N\text{-}C(O)\text{-}NH_2$. Urea is a colorless solid with a melting point of 270.8°F (132.7°C), possessing good **solubility** in both **water** and **ethanol**. A product of protein **metabolism**, urea is formed by the liver primarily from the **ammonia** that results when excess **amino acids** are deaminated (or broken down). In this process—basically, splitting off the amino acid's amine group—the resting ammonia is either used to make other nitrogen-containing compounds or transported to the liver, where it is converted to urea and then excreted by way of the urine.

Urea was first discovered in urine in 1773 by the French chemist Jean Rouelle. Over 50 years later, in 1828, it was syn-

Uranium disks. *(Photograph, U.S. Department of Energy.)*

thesized by the German chemist, Friedrich Wöhler—a feat more remarkable than it sounds. In the early 1800s, it was widely believed that an organic compound like urea—a product formed by the human body itself—could not be manufactured in the laboratory as though it were a common chemical. When Wöhler accidentally found he had synthesized crystals of urea while evaporating a **solution** of ammonium cyanate, he was surprised that it was possible. In his excitement, he dashed off a letter to his friend, the famous Swedish chemist **Jöns Berzelius**, and declared, "I can make urea without needing a kidney, whether of man or dog!" Wohler's historic preparation of "artificial" urea demonstrated to the scientific world that an organic compound could be synthesized, not only by a living organism, but by the working chemist. For many, then, Wöhler is considered the true father of organic **chemistry**.

UREY, HAROLD (1893-1981)

American chemist and physicist

In 1934 Harold Urey was awarded the Nobel Prize in **chemistry** for his discovery of **deuterium**, an **isotope**, or species, of **hydrogen** in which the atoms weigh twice as much as those in ordinary hydrogen. Also known as heavy hydrogen, deuterium became profoundly important to future studies in many scientific fields, including chemistry, physics, and medicine. Urey continued his research on isotopes over the next three decades, and during World War II his experience with deuterium proved invaluable in efforts to separate isotopes of **uranium** from each other in the development of the first atomic bombs. Later, Urey's research on isotopes also led to a method for determining the earth's atmospheric **temperature** at various periods in past history. This experimentation has become especially relevant because of concerns about the possibility of global climate change.

Harold Clayton Urey was born in Walkerton, Indiana, on April 29, 1893. His father, Samuel Clayton Urey, was a schoolteacher and lay minister in the Church of the Brethren. His mother was Cora Reinoehl Urey. Urey's father died when Harold was only six years old, and his mother later married another Brethren minister. Urey had a sister, Martha, a brother, Clarence, and two half-sisters, Florence and Ina.

After graduating from high school, Urey hoped to attend college but lacked the financial resources to do so. Instead, he accepted teaching jobs in country schools, first in Indiana (1911–1912) and then in Montana (1912–1914) before finally entering Montana State University in September of 1914 at the age of 21. Urey was initially interested in a career in biology, and the first original research he ever conducted involved a study of microorganisms in the Missoula River. In 1917 he was awarded his bachelor of science degree in zoology by Montana State.

The year Urey graduated also marked the entry of the United States into World War I. Although he had strong pacifist beliefs as a result of his early religious training, Urey acknowledged his obligation to participate in the nation's war effort. As a result, he accepted a job at the Barrett Chemical Company in Philadelphia and worked to develop high **explosives**. In his Nobel Prize acceptance speech, Urey said that this experience was instrumental in his move from **industrial chemistry** to academic life.

At the end of the war, Urey returned to Montana State University where he began teaching chemistry. In 1921 he decided to resume his college education and enrolled in the doctoral program in **physical chemistry** at the University of California at Berkeley. His faculty advisor at Berkeley was the great physical chemist **Gilbert Newton Lewis**. Urey received his doctorate in 1923 for research on the calculation of **heat** capacities and entropies (the degree of randomness in a system) of **gases**, based on information obtained through the use of a spectroscope. He then left for a year of postdoctoral study at the Institute for Theoretical Physics at the University of Copenhagen where **Niels Bohr**, a Danish physicist, was researching the structure of the **atom**. Urey's interest in Bohr's research had been cultivated while studying with Lewis, who had proposed many early theories on the nature of chemical **bonding**.

Upon his return to the United States in 1925, Urey accepted an appointment as an associate in chemistry at the Johns Hopkins University in Baltimore, a post he held until 1929. He interrupted his work at Johns Hopkins briefly to marry Frieda Daum in Lawrence, Kansas, on June 12, 1926. Daum was a bacteriologist and daughter of a prominent Lawrence educator. The Ureys later had four children, Gertrude Elizabeth, Frieda Rebecca, Mary Alice, and John Clayton.

In 1929, Urey left Johns Hopkins to become associate professor of chemistry at Columbia University, and in 1930 he published his first book, *Atoms, Molecules, and Quanta,* written with A. E. Ruark. Writing in the *Dictionary of Scientific Biography,* Joseph N. Tatarewicz called this work "the first comprehensive English language textbook on **atomic structure** and a major bridge between the new quantum physics and the field of chemistry." At this time he also began his search for an isotope of hydrogen. Since **Frederick Soddy**, an English chemist, discovered isotopes in 1913, scientists had been looking for isotopes of a number of elements. Urey believed that if an isotope of heavy hydrogen existed, one way to separate it from the ordinary hydrogen isotope would be through the vaporization of liquid hydrogen. Since heavy hydrogen would be more dense than ordinary hydrogen, Urey theorized that the lighter hydrogen atoms would vaporize first, leaving behind a mixture rich in heavy hydrogen. Urey believed that if he could obtain enough of the heavy mixture through a process of slow evaporation, spectroscopic readings would show spectral lines that differed from that of ordinary hydrogen.

With the help of two colleagues, Ferdinand Brickwedde and George M. Murphy, Urey carried out his experiment in 1931. The three researchers began with four liters of liquid hydrogen which they allowed to evaporate very slowly. Eventually, only a single milliliter of liquid hydrogen remained. This sample was then subjected to spectroscopic analysis which showed the presence of lines in exactly the positions predicted for a heavier isotope of hydrogen. This was deuterium.

The discovery of deuterium made Urey famous in the scientific world, and only three years later he was awarded the Nobel Prize in chemistry for his discovery. Since his wife was pregnant at the time, he declined to travel to Stockholm and was allowed to participate in the award ceremonies the following year. Urey's accomplishments were also recognized by Columbia University, and in 1933 he was appointed the Ernest Kempton Adams Fellow. A year later he was promoted to full professor of chemistry. Urey retained his appointment at Columbia until the end of World War II. During this time he also became the first editor of the new *Journal of Chemical Physics,* which became one of the principal periodicals in the field.

During the latter part of the 1930s, Urey extended his work on isotopes to other elements besides hydrogen. Eventually his research team was able to separate isotopes of **carbon**, **nitrogen**, **oxygen**, and **sulfur**. One of the intriguing discoveries made during this period was that isotopes may differ from each other chemically in very small ways. Initially, it was assumed that since all isotopes of an element have the same electronic configuration, they would also have identical chemical properties. Urey found, however, that the **mass** differences in isotopes can result in modest differences in the *rate* at which they react.

The practical consequences of this discovery became apparent all too soon. In 1939, word reached the United States about the discovery of **nuclear fission** by the German scientists **Otto Hahn** and **Fritz Strassmann**. The military consequences of the Hahn-Strassmann discovery were apparent to many scientists, including Urey. He was one of the first, therefore, to become involved in the U.S. effort to build a nuclear weapon, recognizing the threat posed by such a weapon in the hands of Nazi Germany. However, Urey was deeply concerned about the potential destructiveness of a fission weapon. Actively involved in political topics during the 1930s, Urey was a member of the Committee to Defend America by Aiding the Allies and worked vigorously against the fascist regimes in Germany, Italy, and Spain. He explained the importance of his political activism by saying that "no dictator knows enough to tell scientists what to do. Only in democratic nations can science flourish."

Harold Urey.

As World War II drew closer, Urey became involved in the Manhattan Project to build the nation's first atomic bomb. In 1940, he became a member of the Uranium Committee of the project, and two years later he was appointed director of the Substitute Alloys Materials Laboratory (SAML) at Columbia. SAML was one of three locations in the United States where research was being conducted on methods to separate two isotopes of uranium. As a leading expert on the separation of isotopes, Urey made critical contributions to the **solution** of the Manhattan Project's single most difficult problem, the isolation of uranium–235 from its heavier twin.

At the conclusion of World War II, Urey left Columbia to join the Enrico Fermi Institute of Nuclear Studies at the University of Chicago. In 1952 he was named Martin A. Ryerson Distinguished Service Professor there. The postwar period saw the beginning of a flood of awards and honorary degrees that was to continue for more than three decades. He received honorary degrees from more than two dozen universities, including doctorates from Columbia (1946), Oxford (1946), Washington and Lee (1948), the University of Athens (1951), McMaster University (1951), Yale (1951), and Indiana (1953).

The end of the war did not end Urey's concern about nuclear weapons. He now shifted his attention to work for the control of the terrible power he had helped to make a reality. Deeply conscious of a sense of scientific responsibility, Urey was opposed to the dropping of an atomic bomb on Japan. He was also aggressively involved in defeating a bill that would have placed control of nuclear power in the United States in the hands of the Department of Defense. Instead, he helped pass a bill creating a civilian board to control future nuclear development. In later years Urey explored peaceful uses of nuclear **energy**, and in 1975 he petitioned the White House to reduce production in nuclear power plants. He was also a member of the Union of Concerned Scientists.

Urey continued to work on new applications of his isotope research. In the late 1940s and early 1950s, he explored the relationship between the isotopes of oxygen and past planetary climates. Since isotopes differ in the rate of chemical reactions, Urey said that the amount of each oxygen isotope in an organism is a result of atmospheric temperatures. During periods when the earth was warmer than normal, organisms would take in more of a lighter isotope of oxygen and less of a heavier isotope. During cool periods, the differences among isotopic concentrations would not be as great. Over a period of time, Urey was able to develop a scale, or an ''oxygen thermometer,'' that related the relative concentrations of oxygen isotopes in the shells of sea animals with atmospheric temperatures. Some of those studies continue to be highly relevant in current research on the possibilities of global climate change.

In the early 1950s, Urey became interested in yet another subject: the chemistry of the universe and of the formation of the planets, including the earth. One of his first papers on this topic attempted to provide an estimate of the relative abundance of the elements in the universe. Although these estimates have now been improved, they were remarkably close to the values modern chemists now accept.

Urey also became involved in a study of the origin of the solar system. For well over 200 years, scientists had been debating the mechanism by which the planets and their satellites were formed. From his own studies, Urey concluded that the creation of the solar system took place at temperatures considerably less than those suggested by most experts at the time. He also proposed a new theory about the origin of the Earth's moon, claiming that it was formed not as a result of being torn from the Earth, but through an independent process of a gradual accumulation of materials.

Urey's last great period of research brought together his interests and experiences in a number of fields of research to which he had devoted his life. The subject of that research was the **origin of life** on Earth. Urey hypothesized that the Earth's primordial atmosphere consisted of reducing gases such as hydrogen, **ammonia**, and **methane**. The energy provided by electrical discharges in the atmosphere, he suggested, was sufficient to initiate chemical reactions among these gases, converting them to the simplest compounds of which living organisms are made, **amino acids**. In 1951, Urey's graduate student **Stanley Lloyd Miller** carried out a series of experiments to test this hypothesis. In these experiments, an electrical discharge passed through a **glass** tube containing only reducing gases resulted in the formation of amino acids.

In 1958 Urey left the University of Chicago to become Professor at Large at the University of California in San Diego at La Jolla. At La Jolla, his interests shifted from original scientific research to national scientific policy. He became extremely involved in the U.S. space program, serving as the first chairman of the Committee on Chemistry of Space and Exploration of the Moon and Planets of the National Academy of Science's Space Sciences Board. Even late in life, Urey continued to receive honors and awards from a grateful nation and admiring colleagues. He was awarded the Johann Kepler Medal of the American Association for the Advancement of Science (1971), the Priestley Medal of the American Chemical Society (1973), National Aeronautics and Space Administration (NASA) Exceptional Scientific Achievement Award (1973), and the 200th Anniversary Plaque of the American Chemical Society (1976). Urey died of a heart attack in La Jolla on January 5, 1981, at the age of 87.

URINE CHEMISTRY

In humans, urine is a mixture of 2-4% **urea**, some **salt**s, bile pigments (coloring), poisons, drugs, and **hormones**. The remainder of the **solution** is excreted **water**. The exact composition of urine is dependent upon diet, level of activity, and general health. If excess water is consumed, the urine is dilute, and after exercise it is concentrated. A normal adult will produce approximately 1.5l of urine per day.

The **chemistry** of urine is basically that of urea and as such it has been used in the production of dyes, as a fertilizer, as a food supplement for sheep, and in the production of urea **formaldehyde** resins. Urine has historically been used in dyeing materials until around 1828, when the commercial production of urea was first introduced.

In alternative medicine, urine is considered a curative for a variety of medical conditions. It is believed that the first urine of the day should be consumed for health purposes. Some also advocate bathing in urine to cure skin ailments. There is no hard scientific evidence to support such practices, although when it first leaves the body urine is a sterile liquid.

UV-VISIBLE SPECTROSCOPY

Ultraviolet and visible **spectroscopy** (UV-vis) is a reliable and accurate analytical laboratory assessment procedure that allows for the analysis of a substance. Specifically, ultraviolet and visible spectroscopy measures the absorption, transmission and emission of ultraviolet and visible light wavelengths by **matter**.

In the **chemistry** laboratory, ultraviolet and visible spectroscopy (UV-vis spectroscopy) is used to study molecules and inorganic ions in **solution**. Although they are distinct regions of the **electromagnetic spectrum**, the ultraviolet and visible regions of the electromagnetic spectrum are linked in UV-vis spectroscopy because similarities between the two regions allow many of the same research techniques and tools to be used for both regions.

Ultraviolet and visible light comprise only a small portion of the wide ranging electromagnetic radiation spectrum.

Although lower in frequency—and therefore lower in energy—than cosmic, gamma or x rays, ultraviolet and visible light are of a higher frequency and, therefore of higher **energy**, than infrared, microwave and radio waves.

The ultraviolet band of the electromagnetic spectrum is further separated into three regions termed UV-A, UV-B and UV-C. Although not all scientists agree on the exact division of these wavelengths, UV-A is generally considered to be light with wavelengths between 320-400 nm; UV-B wavelengths are generally considered to be those between 290-320 nm; and UV-C wavelengths usually fall between 200-290 nm.

In the practical sense, spectroscopy measures the absorption, emission, or scattering of electromagnetic radiation by atoms or molecules. By such measurements, the type of the atoms or molecules present in a sample, as well as a measure of their **concentration** or abundance, can be made to an astonishing degree of accuracy.

When ultraviolet or visible light strike atoms or molecules they can either bounce off or cause electrons to jump between energy levels.

Absorption of ultraviolet or visible light electromagnetic radiation causes **electron** to moves from lower energy levels to a higher energy levels. Ultraviolet-visible **absorption spectroscopy** measures the absorption of ultraviolet or visible light. Because the spectrum of an **atom** or **molecule** depends on its electron energy levels, UV-vis absorption spectra are useful for identifying unknown substances.

Emission of electromagnetic radiation—(e.g., ultraviolet and visible light) occurs when electrons move from higher energy levels to lower energy levels.

Scattering of electromagnetic radiation occurs when light is deflected or scattered in other directions. Rayleigh scattering describes light that is redirected at the same wavelength. Some interactions, however, may occur that result in a partial transfer of energy. As a result, the scattered electromagnetic radiation is of a different wavelength. Brillouin scattering and Raman scattering are examples of this type of scattering.

The way electromagnetic radiation affects atoms and molecules depends on the energy of the light. The energy is, in turn, dependent upon the frequency of the light. Ultraviolet and visible light promotes electrons into higher energy orbitals. Infrared light excites atomic and molecular vibrations. Microwaves excite atomic and molecular rotation.

Special instrumentation is used in UV-vis spectroscopy. **Hydrogen** or **deuterium** lights provide the source of light for ultraviolet measurements. **Tungsten** lamps provide the light for visible measurements. These light sources generate light at specific wavelengths. Deuterium lamps generate light in the UV range (190-380 nm). Tungsten-halogen lamps generate light in the visible spectrum (380-800 nm). **Xenon** lamps which can produce light in the UV and visible portions of the spectrum are used to measure both UV and visible spectra.

Spectrophotometry is the determination of a molecule or compound's identity. It also allows scientists to measure the amount of substance present in a sample. Spectrophotometry in the ultraviolet and visible spectrums has many applications in numerous scientific fields.

Ultraviolet and visible spectrophotometric methods were pioneered by American chemist Arnold Beckman, a member of the National Inventors Hall of Fame. Beckman had an early love for chemistry and completed college chemistry courses before he began his undergraduate studies at the University of Illinois. After completing his Ph.D. at the California Institute of Technology in 1928, Beckman went on to invent many measurement devices of scientific importance. In the 1940s, he invented a spectrophotometer that had the ability to measure the transmission of visible and ultraviolet and visible light through a target.

In UV spectroscopy, a beam of light is split into a sample and reference beams. As its name implies, the sample beam is allowed to pass through the target sample. Alternately, the reference beam passes through the control solvent or a portion of the solvent that does not contain the actual target. Once the light passes through the target sample of interest it is measured by a special meter termed a spectrometer designed to compare the difference in the transmissions of the sample and reference beams. Double beam UV spectroscopy instruments allow scientists to simultaneously measure transmissions through the target sample and solvent.

The wavelengths of ultraviolet or visible light absorbed by a substance result in a unique ultraviolet-visible spectroscopic signature for each substance.

The amount or percentage of a substance present is often of great importance. Chemists doing UV-vis spectroscopy use **Beer's Law** (Beer-Lambert law) to determine the relationship between absorbence and concentration of a substance in a solution. In general there is a linear relationship between the an increasing amount of substance and a decreasing percentage of light transmitted through the target sample. If there is more of a substance to absorb the UV light then less light will pass through to the detector. Correspondingly, less absorption by the target sample allows increased transmission light through the sample.

Ultraviolet and visible spectroscopy are increasingly important methods to detect atmospheric chemical reactions termed photochemical reactions and the products of those reactions. In photochemical reactions driven by ultraviolet and visible light, the light energy supplies energy sufficient to break bonds. Scientists use ultraviolet-visible spectroscopy to monitor the presence and abundance of man-made pollutants in the atmosphere. Atmospheric samples are taken and subjected to the laboratory analysis described above.

The potential detrimental effects of increased exposure to UV-B light are of great concern to scientists around the world. Many spectroscopic studies of the atmosphere are designed to measure the levels of stratospheric ozone that protect the Earth from damaging doses of UV-B radiation. Scientists fear that depletion of stratospheric **ozone** could lead to significant increases in the amount of UV-B reaching the Earth's surface. Studies have shown that excessive UV-B radiation may injure or increase risks of cancer in animal skin, eyes, and immune systems.

Physicians increasingly stress that people should protect themselves from the potential harmful effects of ultraviolet radiation when they go out in the sun for extended periods. Studies have shown that UV-radiation can drive the chemical reactions that produce highly reactive radicals that have the potential to damage **DNA** and other cell regulating chemicals and structures. The known harmful effects of ultraviolet radiation have caused a rapid cultural turn-around from time when a tanned body was considered a healthy body. Physicians now recommend the use of sunscreen products to lessen exposure to ultraviolet radiation. UV visible spectroscopy allows for the testing of UV transmission through potential sun blocking substances.

Although less energy is transmitted by a **photon** of visible light, light that we can see can also transmit substantial energy. In fact, all life on Earth depends on this fact. **Color** is based on the interaction of visible light with matter. There is, however, danger in overexposure to light. Mountain climbers, for example, must take special precautions that they are not blinded by the brilliant sky or glare off snow. According to legend, the ancient Greek mathematician Archimedes defended his native Syracuse during the Second Punic War by harnessing the power of visible light. The legend relates how the inventive Archimedes constructed a giant mirror that was used to reflect and focus the sun's rays upon approaching Roman ships until vessels burst into flames.

Astronomers study multiple wavelengths—including the ultraviolet and visible spectrums—in order to learn more about the objects of the universe.

See also Planck's constant

V

VALENCE

Valence refers to a number assigned to elements that reflects their ability to react with other elements and the type of reactions the element will undergo. The term valence is derived from the Latin word for strength and can reflect an element's strength or affinity for certain types of reactions.

The electrons in an **atom** are located in different **energy** levels. The electrons in the highest **energy level** are called valence electrons. In accord with the octet rule—and to become more energetically stable—atoms gain, lose, or share valence electrons in an effort to obtain a noble gas configuration in their outer shell. The configuration of electrons in an atom's outer shell determines its ability and affinity to enter into chemical reactions.

The valence number of an element can be determined by using a few simple rules relating to an element's location on the **periodic table**. In ionic compounds (formed between charged atoms or groups of atoms called ions) the valence of an atom is the number of electrons that atom will gain or lose to obtain a full outer shell. In group one of the periodic table, elements are assigned a valence number of 1. A valence number of 1 means that a group one element will generally react to lose one **electron** to obtain a full outer shell. Group two elements are assigned a valence number of 2. A valence number of 2 means that a group two element will generally react to lose two electrons to obtain a full outer shell. Group 17 elements are assigned a valence number of negative one (–1). A valence number of 17 means that a group two element will generally react to gain one electron to obtain a noble gas electron configuration. Reflecting an inability to react with other elements, Noble **gases**, already maintaining a stable arrangement of electrons, are assigned a valence of zero (O).

The term valence can also refer to the charge or **oxidation** number on an atom. In **Magnesium** atoms (Mg+2) the valence is +2. An atom or **ion** with a charge of +2 is said to be divalent.

In covalent compounds the valence of an atom may be less obvious. In this case it is the number of bonds formed, that is, whether the bonds are single, double, or triple bonds. A **carbon** atom with two single bonds and one double bond carries a valence of four (4). In **water** (H_2O), the valence of **oxygen** is 2 and the valence of **hydrogen** is 1. In both cases the valence number gives an indication of the number of bonds each atom forms.

Valence bond theory is similar to **molecular orbital theory** in that it is concerned with the formation of covalent bonds. Valence bond theory describes bonds in term of interactions between outer orbitals and hybridized orbitals to explain the formation of compounds.

Valence Shell Electron Pair Repulsion (VSEPR) Theory is one of the favored models to explain covalent bonds. This theory states that molecules will be shaped so as to minimize the repulsion that takes place between valence electrons. Because they are all negatively charged, valence shell electrons repel one another. VSEPR theory states that the atoms of a **molecule** will arrange themselves and assume a shape around a central atom so as to minimize repulsion between valence electrons.

See also Bonding; Molecular geometry; Octet rule

VAN DER WAALS FORCE

A Van der Waals force between molecules is a relatively weak intermolecular attraction. All neighboring molecules in **liquids** and **solids** attract each other. The nature and strength of these interactions depends on the types atom groups or functional groups that comprise the molecules. Some molecules are polar and some have **hydrogen** bonds. These relatively strong intermolecular interactions require specific structural features. Polar interactions require a nonsymmetric arrangement of bonds with atoms of different **electronegativity**—polar bonds. Hydrogen **bonding** requires that one species have a hydrogen

atom bonded to a highly electronegative atom such as **fluorine**, **oxygen**, or **nitrogen**. The other species must have a highly electronegative atom without a hydrogen atom bonded to it. However, all molecules interact with other molecules through Van der Waals interactions.

Van der Waals forces are the attractive forces of one transient **dipole** for another. A transient dipole is a temporary imbalance of positive and negative charge. At particular instants, even atoms that are spherical on average, such as those of the noble **gases**, will have greater **electron density** on one side of the atom than another. At that instant, the atom will possess a temporary dipole with a negative charge **concentration** on the side of the atom with greater electron density. If this happens in the case of an **argon** atom in liquid argon, for example, the argon atoms next to the one with temporary dipole would feel the effect of the dipole. An atom near the negative end of the dipole would have its own electrons slightly repelled from the negative concentration of charge, developing a dipole with its positive end near the negative charge of the original atom. An argon atom on the other side of the original temporary dipole would feel its electrons attracted to the positive end of the dipole, developing a dipole with the opposite orientation. In this way, temporary dipoles are propagated through a liquid or solid. The motion of the molecules in the liquid or solid soon disrupts the pattern, but similar events take place continually. The larger the size of atoms and the more electrons they possess, the greater the probability of forming substantial transient dipole interactions. Molecules which are non-polar and non-polar functional groups of molecules only experience Van der Waals interactions with other molecules or functional groups.

To understand the differences in properties of larger molecules, the additivity of intermolecular interactions becomes important. In effect, the interaction of each group of atoms of a **molecule** with a group of atoms of a neighboring molecule can be considered to be independent of the interactions of other groups of atoms of the molecules. The total **energy** required to move two molecules apart is the sum of all the energies of the individual interactions. The more groups and the stronger each individual interaction, the greater the sum of energy of interactions. Among the non-polar linear alkanes, the **boiling point** for a molecule with many -CH_2 groups, such as liquid octane— $CH_3(CH_2)_6 CH_3$—is higher than that of gaseous propane—$CH_3 CH_2 CH_3$—because of the greater number of Van der Waals interactions between the octane molecules. For large molecules such as the higher alkanes (heavy oils and waxes) and polymers such as **polyethylene**, the total attractive energy due to Van der Waals forces can be greater than the polar interactions or hydrogen bonding interactions of other, smaller molecules. Hence, molecules that have only Van der Waals interactions may still melt or boil at high temperatures.

VAN DER WAALS, JOHANNES DIDERIK (1837-1923)

Dutch physicist

Johannes Diderik van der Waals received his doctorate in physics from the University of Leiden at the relatively late age of thirty-six. His doctoral dissertation, "On the Continuity of Gaseous and Liquid States," quickly became known among his colleagues and made his reputation almost immediately. The Nobel Prize in physics, awarded him in 1910, recognized the line of work begun in his dissertation, eventually resulting in a famous **equation of state** relating the pressure, **volume**, and **temperature** of a gas. He also demonstrated why a gas cannot be liquified above its critical temperature. Van der Waals also investigated the weak nonchemical bond forces between molecules that now carry the name **van der Waals forces**.

Van der Waals was born in Leiden in the Netherlands on November 23, 1837. His parents were Jacobus van der Waals, a carpenter, and the former Elisabeth van den Burg. Van der Waals attended local primary and secondary schools and then took a job teaching elementary school in his hometown. In 1862 he began taking courses at the University of Leiden and, two years later, received the credentials necessary to teach high school physics and mathematics. He then accepted a job teaching physics in the town of Deventer and, a year later in 1866, became headmaster of a secondary school in The Hague.

During his year at Deventer, van der Waals married Anna Magdalena Smit, who bore him three daughters, Anne Madeleine, Jacqueline Elisabeth, and Johanna Diderica, and one son, Johannes Diderik. Biographers note that Anna Magdalena died while the children were still very young; Van der Waals never remarried.

While in The Hague, van der Waals continued to attend the University of Leiden on an informal basis. Since he had never studied Greek and Latin in high school, he was not allowed by federal law to enroll in a doctoral program. When that regulation was abolished in the late 1860s, van der Waals was admitted as a regular graduate student at Leiden. For his dissertation he chose to study the nature and behavior of the particles that make up **gases** and **liquids**.

Van der Waals's choice of topics, he later said, was strongly influenced by a paper written by the German physicist Rudolf Clausius in 1857. In that paper Clausius had argued that the molecules of a gas can be considered tiny points of **matter** in constant motion. From this initial premise, Clausius was able to derive theoretically a law relating gas pressure and volume originally stated empirically by **Robert Boyle** in 1662. It occurred to van der Waals that the molecules of both gases and liquids might be considered in the same way, as tiny points of matter. In such a case, according to van der Waals, there might be no fundamental difference between gases and liquids, the latter being only compressed gas at a low temperature.

It was this concept that van der Waals explored in detail in his doctoral thesis, presented to the faculty at Leiden in 1873. He pointed out that two fundamental assumptions of earlier **gas laws** were not valid. In the first place, such laws had assumed that the particles of which a material is made had no effective size. Van der Waals argued that they did have measurable volume and that such volume affected the behavior of a gas. A second assumption of gas laws was that gas particles do not interact with each other. Van der Waals argued instead that particles do indeed exert forces on each other.

Given these modifications in starting assumptions, van der Waals was able to develop an equation that more closely matches the actual behavior of gases. Laws such as those of Robert Boyle and J. A. C. Charles had been regarded as correct for "ideal" gases, but always failed to some extent when applied to any real gas. Under van der Waals's formulation, the revised gas law applied with remarkable precision to any real gas. Van der Waals's work earned him almost instantaneous fame among his colleagues. His thesis was translated into German, English, and French, and gained him notice in the science world. Van der Waals was elected to the Royal Dutch Academy of Sciences in 1877, and two years later he was appointed professor of physics at the newly created University of Amsterdam. He remained in that post for three decades, retiring in 1907, to be succeeded by his son.

Van der Waals continued to work on the relationship between gases and liquids for the rest of his career. In 1890 he suggested the notion of binary solutions—states in which a substance exists as both a gas and a liquid at the same time. The calculations that van der Waals made on binary solutions later proved crucial in the fledgling field of cryogenics, specifying the conditions under which a gas can be converted to a liquid. One of the pioneers of this field, Heike Kamerlingh Onnes, acknowledged his debt to van der Waals in an article in Eduard Farber's book *Great Chemists,* in which he said, "How much I was under the influence of its great importance as much as forty years ago may be best judged by my taking it then as a guide for my own researches."

For many students of science, van der Waals may be best known for the weak **intermolecular forces** that now carry his name. Originally called by him "pseudoassociation," these forces were hypothesized by van der Waals to explain the aggregation of particles in liquid solutions that occurred, for example, during the formation of binary solutions. Today, van der Waals forces are invoked to describe a host of situations in which rapidly shifting **electron** distributions in a **molecule** result in the formation of weak, but nonzero, transient attractions between molecules.

During the last ten years of his life, van der Waals gradually grew frail; he died in Amsterdam on March 8, 1923. Van der Waals had been elected to membership in the French Academy of Sciences, the British Chemical Society, the U.S. National Academy of Sciences, the Royal Academy of Sciences of Berlin, and the Russian Imperial Society of Naturalists.

Vanadium

Vanadium is a transition metal in Group 5 and Row 4 of the **periodic table**. It has an **atomic number** of 23, an atomic **mass** of 50.9415, and a chemical symbol of V.

Properties

Vanadium is a silvery white, ductile, malleable metal with a melting point of about 3,450°F (1,900°C), a **boiling point** of 5,400°F (3,000°C), and a **density** of 6.11 grams per cubic centimeter. It is fairly inactive at room **temperature**, not reacting with **oxygen**, **water**, or most cold acids. It becomes more reactive at higher temperatures.

Vanadium samples. ***(Photograph by Russ Lappa, National Audubon Society Collection/Photo Researchers, Inc. Reproduced by permission.)***

Occurrence and Extraction

Vanadium is relatively abundant in the Earth's crust, with an abundance estimated at about 100 parts per million. It is found in a number of minerals, including vanadinite, carnotite, roscoelite, and patronite. It is obtained commercially as a byproduct of the manufacture of **iron** since it is also present in most iron ores.

Discovery and Naming

Vanadium was "discovered" twice, the first time in 1801 by Spanish-Mexican metallurgist Andrés Manuel del Río (1764-1849). Del Río found the element for which he suggested the name *panchromium,* meaning "all colors." His colleagues in Europe were unable to confirm his discovery, however, and thought that del Río had mistaken **chromium** for a new element. About 30 years later, however, del Río's original findings were confirmed when Swedish chemist Nils Gabriel Sefström (1787-1845) rediscovered a new element in an ore of iron. He found that his "new element" was identical to the one del Río had reported three decades earlier. Sefström proposed the name vanadium for the element in honor of the Scandinavian goddess of love, Vanadis.

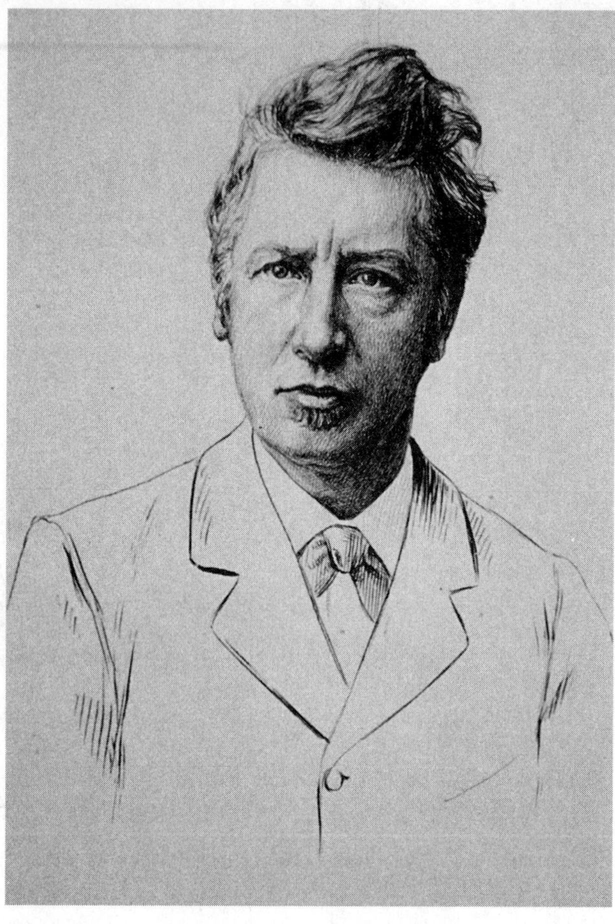

Jacobus Hernicus Van't Hoff.

Uses

About 90% of the vanadium produced is used in steel alloys, to which it adds strength, toughness, and resistance to **corrosion**. Such alloys are often used in space vehicles and aircraft; in building and heavy construction equipment; and in transportation, machinery, and tools. Relatively small amounts of vanadium are used to make compounds with commercial applications, the most important of which is vanadium pentoxide (V_2O_5). Vanadium pentoxide is used as a catalyst in many industrial operations, as a coloring material for **glass** and **ceramics**, in the dyeing of textiles, and in **batteries**.

VAN'T HOFF, JACOBUS HENRICUS (1852-1911)

Dutch chemist

Van't Hoff was born in Rotterdam, Holland, on August 30, 1852, the son of a doctor. As a young boy, Van't Hoff decided that he wanted to become a chemist. Doubting that he could earn a living as a chemist, his parents objected to this decision. Eventually, they allowed him to enroll at the Delft Polytechnic Institute, after which, in 1871, he entered the University of Leyden.

After further study with **Friedrich Kekulé** in Bonn, Germany, and C. A. Wurtz (1817-1884) in Paris, France, Van't Hoff returned to Holland (1876) as a lecturer in physics at the Veterinary College at Utrecht. Two years later, he was appointed chairman of the **chemistry** department at the University of Amsterdam, a post he held until 1896. His final position was professor of chemistry at the University of Berlin, where he remained until his death in Steglitz, Germany,on March 1, 1911.

Van't Hoff is best known for his contributions in the fields of **stereochemistry** (the three-dimensional structure of molecules) and dilute solutions. While still a young man, he attacked a problem that had been troubling chemists since the late 1840s. That problem had been posed by the great **Louis Pasteur**. Completing research previously begun by both biologists and chemists, Pasteur showed that some chemicals can exist in two forms that are identical with each other in all respects except the way in which they affect polarized light; these different forms are called optical isomers. One of the two optical isomers rotates polarized light to the left (and is, therefore, called the levorotary, or l-form, of the compound). The other rotates polarized light to the right (and is, therefore, called the dextrorotary, or d-form, of the compound). Pasteur's discovery set off considerable debate as to how two such forms of a compound could exist.

In 1874, Van't Hoff put forward a theory to explain the phenomenon of **optical activity**. He showed that a **carbon atom** that has four different groups attached to it can exist in two—and only two—different forms. These forms are mirror images of each other. Van't Hoff suggested that optically active compounds consist of molecules that contain such asymmetrical carbon atoms. One of the mirror images is levorotary and the other, dextrorotary. A similar theory was proposed independently at almost the same time by the French chemist, **Joseph Le Bel**. Therefore, Van't Hoff and Le Bel are given joint credit for the discovery of the asymmetric carbon atom.

Van't Hoff's theory met with vigorous criticisms from some major figures in the chemical world. **Adolph Wilhelm Hermann Kolbe**, for example, described Van't Hoff's ideas as "fanciful nonsense" and "supernatural explanations." As with any theory, however, the idea of asymmetric carbon atoms was ultimately evaluated on the basis of its success in explaining experimental observations. And, in this arena, it rapidly proved itself and was soon adopted by the vast majority of chemists.

Following his work on stereochemistry, Van't Hoff turned his energies to **thermodynamics**. He made some interesting discoveries in this field, but was unaware that others had made similar discoveries a decade earlier. The one notable success Van't Hoff experienced in chemical thermodynamics involved **solution** theory. In 1886, he demonstrated the fact that solute particles in a dilute solution act very much like the particles of which a gas is composed. Therefore, the laws that apply to **gases** could be applied, with modifications, to those dilute solutions. For this research he was awarded the first Nobel Prize for Chemistry in 1901. Van't Hoff died ten years later in 1911.

Vapor pressure

Vaporization of a liquid or **sublimation** of a solid may occur over a wide range of **temperature** and pressure. Wet clothes will dry, and a pan of **water** will slowly evaporate to dryness. Below the freezing point, frozen clothes will dry and a pan of ice cubes will slowly evaporate without first melting. Under virtually all conditions, some of the molecules near the surface of a liquid or solid attain enough **energy** to pull away from the attraction of their neighbors and escape into the gas or vapor phase. Similarly, molecules in the gas phase occasionally strike the surface and are captured by the attraction of molecules in the liquid or solid phase. When the liquid or solid is not confined, most gas phase molecules will move away from the liquid or solid, and few will reunite with the liquid or solid phase. In this case, the liquid or solid will eventually evaporate or sublimate completely. If, however, a liquid or solid is confined in a closed container, a point is reached when the number of molecules returning to the liquid or solid phase from the vapor is equal to the number escaping. This circumstance is called equilibrium.

Molecules in the vapor phase collide with the walls of the container and thus exert a force and a pressure (force per unit area). This pressure depends on the number of gas molecules in the container: with more molecules moving randomly about in the container, the number likely to run into the container increases; thus, the pressure exerted increases. When the vapor is in equilibrium with a solid or liquid, this pressure is called the vapor pressure. The vapor pressure is characteristic of the particular liquid or solid, and its value depends on the temperature. This temperature dependence may be understood by appealing to the kinetic molecular model for **solids** and **liquids**.

In the solid phase, the molecules do not possess translational energy, i.e., they do not move around. The attractive forces between molecules have succeeded in causing the molecules to arrange themselves into fixed, ordered patterns relative to each other. They do, however, still possess vibrational energy, with the molecules vibrating about their fixed centers of gravity. From time to time, individual molecules at the surface of the solid become energetic enough in their vibration to break the attractive hold of their neighbors, overcome the opposing external pressure, and escape into the gas phase.

In the liquid phase, the molecules possess translational energy and move about within the container. But, unlike the gas phase, the **intermolecular forces** cause the molecules to arrange themselves into ordered groups. This grouping is transitory and is constantly changing, but at any given instant there is considerably more order in the liquid phase than there is in the vapor. From time to time, individual molecules have enough translational energy when they collide with the surface of the liquid to escape into the gas phase.

Since the vibrational energy of the molecules in the solid phase and the translational energy of molecules in the liquid phase are both dependent on the temperature, more molecules are able to escape into the vapor phase at higher temperature, and the equilibrium vapor pressure will, therefore, increase as the temperature increases.

Because the strength of intermolecular forces differs significantly between different chemical substances, it is not surprising that the vapor pressure at a given temperature is a characteristic of the particular substance. The relative magnitudes of vapor pressure may be used as an indication of the relative strengths of intermolecular forces operating in various substances. Also, since the intermolecular forces are greater in the solid phase of a particular substance than in its liquid phase, the vapor pressures of liquids are, in general, higher than that for solids. In fact, many solids have vapor pressures which are so low that, for all practical purposes, sublimation does not occur. The vapor pressure of **tungsten** is such that it has been estimated that there is only one gaseous **molecule** of tungsten in the universe at room temperature.

For a given external pressure, there is a temperature at which the vapor pressure of a liquid is equal to the external pressure. Under this condition, the liquid vaporizes quickly and completely. Vaporization which occurs under these conditions is called boiling.

One of the most important applications of vapor pressure is the purification of liquids by **distillation**. In distillation, the liquid is changed completely into the vapor by boiling and then condensed back into the liquid phase. If there is an impurity in the liquid which has a significantly lower vapor pressure (or essentially no vapor pressure, as is the case with most solids) at the liquid's **boiling point**, the impurity will not boil and will not be significantly converted into its vapor. The impurity will be left behind while the liquid we wish to purify is completely converted into vapor. The vapor is isolated and then cooled below the boiling point. It will condense back into its liquid phase, now without the impurity. This process is used to remove **salt** from sea water to produce fresh drinking water.

Vauquelin, Louis Nicolas (1763-1829)

French chemist

The son of Nicolas Vauquelin, Louis Nicolas began his life as a field laborer on the estate that his father managed. Having been reprimanded at age fourteen by the estate master for taking scientific notes on his master's lectures, Vauquelin was rescued by the village priest and sent off to Paris to work in an apothecary shop.

Once there, he was introduced to chemist Antoine François de Fourcroy, and was made his laboratory assistant in 1783. In 1792 Vauquelin became manager of a pharmacy, but was forced to live outside France from 1793 to 1794 to escape the Reign of Terror. Upon his return, he became assistant professor of **chemistry** at the Ecole Centrale des Travaux Publiques. In 1795 he gained the title of master pharmacist.

While in this position Vauquelin made his two major discoveries. In 1797 he developed a **chromium** compound and succeeded in isolating chromium the following year. In 1798, he identified the element **beryllium** in the gems beryl and emerald. Beryllium was actually isolated by **Friedrich Wöhler** about thirty years later.

Vauquelin took professorships at the College de France in 1801 and at the Musee d'Histoire Naturelle in 1804. While at the museum, he reestablished his relationship with Fourcroy. When the latter died in 1809, Vauquelin was appointed to fill his position.

In 1806 Vauquelin discovered the amino acid asparagine, which is the identifying compound present in asparagus. He and Fourcroy set up a chemical factory in Paris in 1804 which Vauquelin continued to manage until 1822.

In 1827, he was elected to the French legislature; after serving in that capacity for two years, Vauquelin died at the age of sixty-six.

VELOCITY

The term *velocity* is often confused with speed; to understand the concept of velocity, it is helpful to first understand what is meant by speed. *Average speed* is determined by dividing the distance traveled by the time required to move that distance. If a car travels 50 miles in one hour, its average speed is 50 miles/one hour, or 50 miles per hour. Given the average speed that a body is traveling, the distance it travels is the product of its average speed and the time traveled. Thus, if the aforementioned car travels at 50 miles per hour travels for two hours, the distance traveled is 100 miles. It is more often the case, however, when a car is traveling between two points, that it will slow down or speed up to accommodate road conditions. For this reason, it is sometimes more helpful to speak of the car's instantaneous speed, which is the speed that the car is traveling at a given point on the trip.

The difference between velocity and speed is that speed only takes into account the distance that a body travels in a given time (i.e., rate of travel), while velocity takes into account the *direction traveled* as well as the travel speed. This is an important distinction. To the driver of an automobile traveling from San Francisco to New York, the average speed of the trip, which can easily be computed by dividing the mileage shown on the odometer by the travel time, is likely to be of more significance than the velocity. But to a weather forecaster, a storm front traveling 100 mph from the northeast is significantly different from one traveling at the same speed from the southwest; thus it is the velocity that is important in this case.

VINYL

"Vinyl" is a term commonly used in organic **chemistry** to describe a monosubstituted alkene, and to describe the polyalkane **plastics** derived from polymerization of monosubstituted ethylene (vinyl) monomers. It is not used in systematic organic **nomenclature**. The etymological origin of vinyl is uncertain, but it quite likely derived from the nineteenth century term for ethylene (ethene), "weingas" (vinegas, a reference to the role of **ethene** as a plant hormone). At one time the term vinyl referred primarily to a substituted ethylene, but vinyl now is used frequently to describe any substituted alkene. The use of vinyl is exemplified by the **molecule** vinyl chloride (its proper systematic name is chloroethylene), CH_2=CHCl. The CH_2=CH- segment is the vinyl **functional group**, covalently substituted by a chloro group (not ionically bonded to a chloride **anion**, as the non-systematic name "vinyl chloride" might be taken to imply). Among the other commonly encountered vinyl compounds are vinyl bromide, vinyl ester, vinyl acetate, and vinylbenzene (also called styrene).

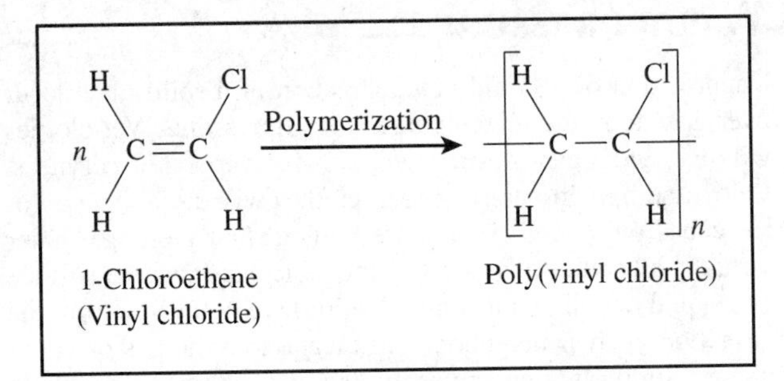

Figure 1. Polymerization of *n* 1-chloroethene molecules gives the chloro-substituted polyalkane polymer poly(vinyl chloride). ***(Illustration by Electronic Illustrators Group.)***

Many of these vinyl compounds are made by the reaction of acetylene with electrophiles. For example, HCl adds to acetylene (under pressure at 475K; and in the presence of $HgCl_2$ as a catalyst) to give vinyl chloride. Vinyl ethers are made by addition of alcohols to acetylene. In contrast, vinyl acetate is made by the metal-catalyzed oxidative acetylation of ethylene by **acetic acid**, in the presence of O_2 as the oxidant. Radical polymerization of vinyl chloride creates the plastic poly(vinyl chloride), where the vinyl chloride monomers are interconnected (with loss of the alkene double bond) forming the chloro-substituted polyalkane polymer (Figure 1). The vinyl polymers include poly(vinyl chloride) or PVC, poly(vinyl acetate), poly(vinylidene chloride), poly(vinyl alcohol), poly(vinyl acetal), poly(vinyl fluoride), poly(vinyl pyrrolidine, poly(vinyl carbazole), and poly(vinyl ether). Many of these are used as thermoplastics that are shaped and processed by means of injection molding and extrusion.

See also Alkene functional group; Polymer chemistry; Polyvinyl chloride

VIRTANEN, ARTTURI ILMARI (1895-1973)

Finnish biochemist

Artturi Ilmari Virtanen was a Finnish biochemist who discovered many of the nutritionally important components of plants, including **vitamins** and **amino acids**. His most important discovery, a method of preserving green fodder and silage, led to an improved understanding of the mechanism of plant decay. For his biochemical investigations in agriculture and **nutrition** he received the 1945 Nobel Prize in **chemistry**.

Virtanen was born January 15, 1895, in Helsinki, Finland, to Serafina (Isotalo) and Kaarlo Virtanen. He began his education at the Classical Lyceum in Viipuri (now Vyborg, Russia). Upon graduation, he entered the University of Helsinki to study biology, chemistry, and physics. Virtanen received his master of science degree in 1916 and worked briefly as an assistant chemist in the Central Industrial Laboratory of Helsinki. That same year he returned to the University to continue his studies and in 1919 he received his doctorate. Interested in a broad range of scientific subjects, Virtanen traveled extensively to pursue his studies. In 1920 he left for Zurich, Switzerland, to undertake postgraduate work in **physical chemistry**. The next year he had moved to Stockholm, Sweden, to study bacteriology, and in 1923 he worked on enzymology with **Hans von Euler-Chelpin**. At this time Virtanen discovered the subject that would become his life work, **biochemistry**.

In 1921 Virtanen was appointed Laboratory Director of the Finnish Cooperative Dairies Association in Valio, where he was responsible for controlling the manufacture of dairy products such as cheese and butter. A decade later, in 1931, he was appointed the Director of the Biochemical Research Institute in Helsinki. In that same year, he began teaching as professor of biochemistry at the Finland Institute of Technology. Late in 1939 he became professor of biochemistry at the University of Helsinki.

Virtanen was interested in both the theoretical and practical aspects of biochemistry. His first important study was an investigation of the **fermentation** reactions of some biologically important acids. His research showed how enzymes were necessary for bacterial fermentation. Enzymes, the complex organic catalysts of living cells, were thought to speed up chemical reactions without being chemically changed. Virtanen, who believed that enzymes were composed of **proteins**, began an exhaustive study of the protein content and **enzyme** activity of plants.

Proteins are rich in **nitrogen**, an essential element in both human and animal nutrition. Virtanen realized that nitrogen was plentiful in the atmosphere but almost completely unavailable to most plants and animals. He began to study legumes (peas, clover, and soybeans), which are able to convert atmospheric nitrogen into nitrogen compounds suitable for plant growth. Virtanen became interested in what happened to plants after they were cut and stored for cattle fodder. It was found that fodder stored as silage could lose one quarter to one half of its nutrients to bacterial decay. Animals consuming this silage produced poor quality butter, milk, and cheese. After study and experimentation, Virtanen discovered that the addition of a simple mixture of dilute hydrochloric and sulfuric acids to stored silage could slow down bacterial decay and prevent the destruction of nutritionally important vitamins and proteins.

Virtanen fed his acid-treated silage to dairy cows and tested the milk to determine the safety and effectiveness of his method. He discovered that milk remained rich in protein, carotene, and vitamin C, and that the milk cows remained healthy and strong. The method Virtanen devised for treating silage—called the Artturi Ilmari Virtanen (AIV) method—was first used in Finland in 1929.

Although most of Virtanen's scientific investigations had practical agricultural applications, his later work was more theoretical. He studied how bacteria in the root nodules of legumes synthesize nitrogen compounds, and how plant cells assimilate simple molecules into large complex vitamins. He discovered a red pigment in plant cells similar in structure and function to hemoglobin, the **molecule** that transports life-giving **oxygen** in human blood. He also studied the composition of plants, discovering many important amino acids—the building blocks of proteins. For his work on the AIV method of silage preservation and for his research in **agricultural chemistry** Virtanen received the 1945 Nobel Prize in chemistry.

After winning the Nobel Prize, Virtanen remained actively engaged in research. He published a book on animal and human nutrition and served on the editorial boards of several leading biochemical journals. He represented Finland on a United Nations Commission on Nutrition, and in 1948 he was elected president of Finland's State Academy of Science and Arts. Virtanen, remembered as one of Finland's leading scientists, died November 11, 1973, in the city of his birth, Helsinki.

Viscosity

Viscosity is a physical property of **liquids** which is a measure of a liquid's resistance to flow. In practical terms, it provides an indication of the thickness of a liquid. Its value is dependent on **intermolecular forces** in the liquid and the molecular complexity of the compound. When large intermolecular forces are present, a liquid will tend to be thick or highly viscous. For example, glycerin is thick because it has a high capacity to form **hydrogen** bonds with itself. Liquids that have complex **molecular structure** viscosity will also tend to be higher because the molecules can become entangled.

Any material that flows exhibits viscosity. The unit of measure typically used for viscosity is poise or centipoise. The viscosity measurement for **water** is defined as 1 centipoise. Oils are thicker, having values between 100 and 100,000 centipoise. High **molecular weight** polymers can have viscosity values in the millions of centipoise range.

Liquids can be classified by the type of flow behavior they exhibit. In Newtonian fluids, the shear stress is in direct proportion to the shear rate. This means that the viscosity measurement will remain constant even after several viscosity measurements. Examples of Newtonian fluids include water, glycerin, and light oils. In non-Newtonian fluids, the viscosity measurements change depending on shear stress. Shear thinning fluids exhibit a reduced viscosity value as the shear rate is increased. Shear thickening fluids have higher viscosity values as shear rate is increased.

The viscosity of certain systems is time-dependent. These systems are called thixotropic and are characterized by a difference in viscosity, depending on the speed at which the shear force is applied. An example of this type of system is ketchup. When left in a bottle undisturbed, ketchup remains thick. When enough force is put on the system it immediately thins.

Viscosity is an important quality control characteristic for a variety of products. It is used to ensure that such products as shampoos, hand creams, **paints**, and **inks** are stable. Additionally, it is used to measure the molecular weight of polymers and determine when a polymerization reaction is complete.

Vision, chemistry of

Vision is the ability to be able to see. In mammals the organ that is responsible for vision is the eye. Light enters the eye and it is then focused on the retina by the lens. The retina is a layer of cells that cover the interior surface of the eye. The retina contains two different types of cells—the rods and the cones. Both of these are known as photoreceptors because they respond to light. There are three million cones and 100 million rods in the human eye. Each of these types of cell performs a different function. The cones are responsible for **color** vision and they require high levels of light to operate. The rods are responsible for low light level vision and they operate in black and white. The cones are concentrated in the center of the back of the eye, at its most sensitive region, called the fovea. The rods are spread throughout the retina. The cones are sensitive to different wavelengths of light, there are yellow, blue, and green sensitive cones.

Rods and cones contain similar chemicals to allow them to operate. In the top of each cell, there is a chemical called rhodopsin. Rhodopsin consists of a protein called opsin bonded to a pigment called retinal. Retinal is also called neoretinene b or 11 *cis* retinene. It is the reddish purple color of retinal that gives rhodopsin its former name visual purple. When light hits the retinal, the **energy** is used to break a double bond contained in the **molecule**. Once the bond has been broken, the molecule can then rotate. The rotation changes the shape of the molecule from the *cis* to the *trans* arrangement and the retinal physically separates from the opsin. This separation triggers other mechanisms to send a nerve impulse to the brain. The process is known as the bleaching of rhodopsin because of the change in color that occurs. Once the brain receives this impulse, it interprets it as an indication of what is happening external to the body. This is the nerve impulse that causes vision. Sometimes the change in the rhodopsin molecule can occur spontaneously, creating random flashes of light or color in our vision. To overcome this spontaneous change, five adjacent rods or cones must all transmit a signal to the brain at the same time before it is interpreted as vision. Each cell can be stimulated by one **photon**. As a direct result, the smallest amount of light that will effectively stimulate our vision is five photons.

Once a cell has been stimulated, the molecules have to return to their original form. The opsin reattaches itself to the retinal and reforms the double bond to remake the rhodopsin molecule. The temporary blindness that occurs when looking at bright light is caused by cell overload and rhodopsin breakdown. As the molecules slowly return to their natural form and some are then disassociated, vision returns.

The pigment in the cones is less easy to break down and it is harder and slower for it to regenerate than the pigment in the rods. Color vision needs brighter conditions to operate in than the black and white vision controlled by the rods.

Vitalism

Vitalism is a school of thought which postulates that life cannot be fully explained in physical material terms. According to vitalists, life, which in the material world is manifested as a physical process, emerges as a result of an immaterial impulse. Aristotle, who is regarded as the founder of scientific vitalism, believed that the soul, as a modality of life-energy, kept the organism alive. According to Aristotle, the soul affects the organism without being connected to it in a physical sense.

Historians of science often identify René Descartes as the great intellectual force that facilitated the switch from Aristotelian metaphysics to the more sober mechanistic-materialistic paradigm of modern mainstream science. A fervent Catholic, Descartes altered Aristotle's terminology, retaining, however, the fundamental idea that an organism, being a physical thing, receives direction from a spiritual entity.

Although the powerful mechanistic paradigm created by Isaac Newton (1642-1727) dominated the physical sciences, many natural scientists revolted against what they saw as a lifeless, cold, and rigid conception of the universe. Often branded as purely speculative thinkers, the greatest representatives of vitalism in biology were nevertheless brilliant researchers and practical scientists. For example, Newton's younger contemporary, **Georg Ernst Stahl**, built a comprehensive medical theory and practice on vitalistic foundations. One of the greatest scientists of the eighteenth century, Marie-François-Xavier Bichat (1771-1802), the founder of histology, was a vitalist. Furthermore it was Karl Ernst von Baer, the eminent representative of nineteenth-century vitalism, who in 1827 made history with his discovery of the mammalian ovum.

Twentieth-century scientists and historians of science have often dismissed vitalism as basically obsolete, even unscientific, perhaps because it could not be proved. This insistence on empirical proof shows a profound misunderstanding of the essence of vitalism. Vitalism is an intellectual orientation, and not a mere hypothesis in need of material proof. During the first half of the twentieth century, vitalism's greatest exponents were Henri Bergson (1859-1941), who developed the concept of the *elan vital*, and Hans Driesch (1867-1941). While Bergson, who was primarily a philosopher, relied on secondary sources in biology, Driesch was a practicing biologist, who, in an experiment with sea urchins, showed that of a half of the egg, following the first division after fertilization, is destroyed, the remaining half will produce a complete, albeit smaller, embryo. In Driesch's view, this kind of regeneration clearly demonstrates that life follows a logic that is not determined by physical circumstances. Finally, the theory of morphogenetic fields, developed by **Rupert Sheldrake**, affirms the profoundly vitalistic idea that nature develops in harmony with invisible, immaterial, but powerful forces. According to Sheldrake, morphogenetic fields, like life itself, may not be detectable in a traditional sense, but biologists cannot afford to ignore them.

VITAMINS

Vitamins are organic substances found in food; people and animals need them in order to maintain life functions or prevent disease. About 15 different vitamin groups are necessary for the nutritional needs of humans. Only minute amounts are required to achieve their purpose, yet without them, life cannot be maintained. Some vitamins, including vitamins A, D, E, and K, are fat soluble, and are found in the fatty parts of food and body tissue. As such they can be stored in the body. Others, the most notable of which are vitamin C and all the B-complex vitamins, are water soluble. These are found in the watery parts of food and body tissue and cannot be stored by the body. They are excreted in urine and must be consumed on a daily basis.

Although vitamins were discovered early in the twentieth century, it was common knowledge long before that time that substances in certain foods were necessary for good health. Information about which foods were necessary developed by trial and error, with no understanding of why they promoted health. Scurvy, for example, had long been a dreaded disease of sailors. They often spent months at sea and, due to limited ways of preserving food without refrigeration, their diet consisted of dried foods and salted meats. In 1746, the English naval captain and surgeon James Lind (1716-1794) observed that 80 out of his 350 seamen came down with scurvy during a 10 week cruise. He demonstrated that this disease could be prevented by eating fresh fruits and vegetables during these long periods at sea. Because they lasted a long time, limes became the fruit of choice. In 1795, lemon juice was officially ordered as part of the seaman's diet. The vitamin C found in both limes and lemons prevented scurvy in sailors.

In 1897, Dutch physician Christian Eijkman (1858-1930) found that something in the hulls of rice not present in the polished grains prevented the disease beriberi. Soon after, British biochemist Sir **Frederick G. Hopkins** fed a synthetic diet of fats, **carbohydrates**, **proteins**, and minerals (but no vitamins) to experimental rats. The rats showed poor growth and became ill, leading Hopkins to conclude that there were some ''accessory food factors'' necessary in the diet. Eijkman and Hopkins shared the 1929 Nobel Prize in physiology and medicine for their important work on vitamins.

Finally, in 1912 Polish-American biochemist Casimir Funk (1884- 1967) published a paper on vitamin-deficiency diseases. He coined the word *vitamine* from the Latin *vita,* for life, and *amine,* because he thought that all these substances belonged to a group of chemicals known as amines. The *e* was later dropped when it was found that not all vitamins contained an amine group. Funk identified four vitamins (B_1 or thiamine, B_2, C, and D) as substances necessary for good health and the prevention of disease.

Since that time additional vitamins have been isolated from foods, and their relationship to specific diseases have been identified. All of these ''accessory food factors'' have been successfully synthesized in the laboratory. There is no difference in the chemical nature of the natural vitamins and those that are man-made even though advertisements sometimes try to promote natural sources (such as vitamin C from rose hips) as having special properties not present in the synthesized form.

Vitamins belong to a group of organic compounds required in the diets of man and animals in order to maintain good health-normal growth, sustenance, reproduction, and disease prevention. In spite of their importance to life, they are only necessary in very small quantities (the total vitamins required for one day weigh about one-fifth of a gram). Vitamins have no caloric value and are not a source of energy.

Vitamins cannot be synthesized by the cells of an organism but are vital for normal cell function. Certain plants manufacture these substances, and they are passed on when the plants are eaten as food. Not all vitamins are required in the diets of all animals. For example, vitamin C is necessary for man, monkeys, and guinea pigs, but not for animals able to produce it in their cells from other chemical substances. Nevertheless, all higher animals require vitamin C, and its function within organisms is always the same.

Vitamins were originally classified into two broad categories according to their solubilities in **water** or in fat. As more were discovered, they were named after letters of the alphabet. Some substances once thought to be vitamins were later removed from the category when it was found that they were unnecessary or that they could be produced by the organism. Four of the better known vitamins, A, D, E, and K, are fat soluble. Other vitamins such as vitamin C and the B complex vitamins are water soluble.

This difference in solubility is extremely important to the way the vitamins function within an organism and in the way they are consumed. Fat-soluble vitamins lodge in the fatty tissues of the body and can be stored there. It is, therefore, not necessary to include them in the diet every day. Because these vitamins can be stored, it is also possible, when consumed in excess, for them to build up to dangerous levels in the tissues and cause poisoning.

The Food and Drug Administration publishes a set of nutritional recommendations called the United States Recommended Dietary Allowances (USRDA) based on the needs of the average adult. These are based on the best information available, but are less than perfect because most of the experimental work is done on animals. Because the amounts of vitamins required are so small, precise work is very difficult.

A person who eats a balanced diet with plenty of fresh fruits and vegetables should receive adequate amounts of all the vitamins. Many vitamins, however, are very sensitive to heat, pressure cooking, cold, and other aspects of food preparation and storage and can be inactivated or destroyed.

There is controversy about vitamins among experts, some of whom believe higher doses are necessary to fight off some common diseases. According to these experts, the USRDA are really minimum requirements, and higher doses will keep a person healthier. Thus, many people worldwide take vitamin supplements as insurance that they are getting all they need. Overdosage on vitamins, especially the fat-soluble ones, can cause serious side effects, and in some cases they even interfere with the proper function of other nutrients.

Two very similar chemical forms of vitamin A exist: retinol$_1$ (the predominant form) and retinol$_2$. Vitamin A is present in animal tissue, mainly in liver, fish oil, egg yolks, butter, and cheese. Plants do not contain vitamin A, but they do contain beta carotene (in dark green leafy vegetables and in yellow fruits and vegetables, such as carrots, sweet potatoes, cantaloupe, corn, and peaches). Beta carotene is converted to vitamin A in the intestine and then absorbed by the body. The bodies of healthy adults who have an adequate diet can store several years' supply of this vitamin, but young children, who have not had time to build up such a large reserve, suffer from deprivation much quicker if they do not consume enough.

Vitamin A is necessary for proper growth of bones and teeth, for the maintenance and functioning of skin and mucous membranes, and for the ability to see in dim light. There is some evidence that it can help prevent cataracts and cardiovascular disease and, when taken at the onset of a cold, can ward it off and fight its symptoms. One of the first signs of a deficiency of this vitamin is "night blindness," in which the rods of the eye (necessary for night vision) fail to function normally. Extreme cases of vitamin A deficiency can lead to total blindness. Other symptoms include dry and scaly skin, problems with the mucous linings of the digestive tract and urinary system, and abnormal growth of teeth and bones.

Vitamin A is stored in the fatty tissues of the body, and is toxic in high doses. As early as 1596, Arctic explorers experienced vitamin A poisoning. In this region of extreme conditions, the polar bear was a major source of their food supply, and a quarter pound of polar bear liver contains about 450 times the recommended daily dose of vitamin A. Excessive amounts of vitamin A cause chronic liver disease, peeling of the skin of the entire body, bone thickening, and painful joints. It is nearly impossible to ingest beta carotene in toxic amounts, since the body will not convert excess amounts to toxic levels of vitamin A.

Although several chemical compounds provide vitamin D activity, only two of these substances are commonly found in the diet of animals: vitamin D_2, or calciferol, and vitamin D_3. (There is no vitamin D_1 since the substance originally given this name turned out to be a mixture of compounds, one of which was calciferol.) Vitamin D_2 is produced in animal tissues when a substance called ergosterol (present in yeast and other molds) is activated by sunlight. Vitamin D_3 is produced from a similar substance called 7-dehydrocholesterol which can be obtained from the liver of various fish. It too must be activated by sunlight in order to be effective. Since 7-dehydrocholesterol is present in the skin in large quantities and only needs sunlight to be activated, vitamin D is often called the "sunshine vitamin." It is hard to suffer a deficiency if one gets enough sunshine. Irradiated ergosterol is often added to milk as an additive. Storage and food preparation do not seem to affect this vitamin.

Vitamin D lets the body utilize **calcium** and **phosphorus** in bone and tooth formation. Deficiency of this vitamin causes a bone disease called rickets. This disease is characterized by bone deformities (such as bow legs, pigeon breast, and knobby bone growths on the ribs where they join the breastbone) and tooth abnormalities. In adults, bones become soft and porous as calcium is lost.

Excessive amounts of vitamin D cause nausea, diarrhea, weight loss, and pain in the bones and joints. Damage to the kidneys and blood vessels can occur as calcium deposits build up in these tissues.

Vitamin E is composed of a group of at least seven similar chemicals called the tocopherols. It is present in green leafy vegetables, wheat germ and other plant oils, egg yolks, and meat. The main function of this vitamin is to act as an antioxidant, particularly for fats. When oxidized, fats form a very reactive substance called peroxide, often very damaging to cells. Vitamin E is more reactive than the fatty acid **molecule** and, therefore, reacts instead of it. Because cell membranes are partially composed of fat molecules, vitamin E is vitally important in maintaining the nervous, circulatory, and reproductive systems and in protecting the kidneys, lungs, and liver.

All of the symptoms of vitamin E deficiency are believed to be due to the loss of the antioxidant protection it offers to cells. This protective effect also keeps vitamin A from oxidizing to an inactive form, and when vitamin E is lacking, vitamin A deficiency also frequently occurs. However, because vitamin E is so prevalent in foods, it is very difficult to suffer from a deficiency of this vitamin unless no fats are consumed in the diet. When it does occur, the symptoms include cramping in the legs, muscular dystrophy, and fibrocystic breast disease.

According to some current theories, many of the effects of aging are caused by the **oxidation** of fat molecules in cells. If this is true, then consuming extra vitamin E might counteract these effects because of its antioxidant properties.

Vitamin K was identified in 1934 by Danish biochemist **Henrik Dam**. Later, in 1943, the Nobel Prize in physiology and medicine was awarded jointly to Henrik Dam and American biochemist **Edward Doisy** for their work on this vitamin.

Vitamin K is composed of two groups of compounds, vitamins K_1 and K_2. Both can be derived from a synthetic compound called menadione (sometimes referred to as vitamin K_3), which has all the properties of vitamin K. The K vitamins are found in many plants (especially green leafy ones like spinach), in liver, and in the bacteria of the intestine. Nearly all higher animals must obtain these through their diets. Although the exact method by which vitamin K works in the body is not understood, it is known that vitamin K is vital to the formation in the liver of prothrombin-one of the chemicals necessary for blood clotting.

When vitamin K deficiency develops, it is rarely due to an incomplete diet; instead, it most often develops in people whose ability to absorb fats is impaired. The deficiency is characterized by the inability of the blood to clot, and manifests in unusual bleeding or large bruises under the skin or in the muscles. Drugs such as warfarin, used to treat blood clots, act by blocking the action of vitamin K and as such are double-edged swords. Newborn infants sometimes suffer from brain hemorrhage due to a deficiency of vitamin K.

What was once thought to be vitamin B was later found to be only one of many B vitamins. Today, more than a dozen are known, and they are frequently referred to as vitamin B complex. Thiamine was the first of these vitamins to be identified, and it was called. All of the B vitamins are water soluble.

Each of the B vitamins acts by combining with another molecule to form coenzymes. These coenzymes then work with enzymes to perform vital activities within the cell. Their functions within the cell vary, but all are somehow related to the release of energy. The most common members of this group of vitamins include vitamin B_1 (thiamine), vitamin B_2 (riboflavin), vitamin B_6 (pyridoxine), vitamin B_{12} (cobalamin), biotin, folate (also folacin or folic acid), niacin, and pantothenic acid.

Vitamin B_1 is present in whole grains, nuts, legumes, pork, and liver. It helps the body release energy from carbohydrates. Over 4,000 years ago the Chinese described a disease we know today as beriberi, which is caused by a deficiency of thiamine. It affects the nervous and gastrointestinal system and causes nausea, fatigue, and mental confusion. Thiamine is found in the husks or bran of rice and grains. Once the grain is milled and the husks removed, rice is no longer a source of this vitamin. Manufacturers today produce ''enriched'' rice, flour, etc. by adding thiamine back into the milled products.

Vitamin B_2 helps the body release energy from fats, proteins, and carbohydrates. It can be obtained from whole grains, organ meats, and green leafy vegetables. Lack of this vitamin causes severe skin problems. Vitamin B_6 is important in the building of body tissue as well as in protein metabolism and the synthesis of hemoglobin. A deficiency can cause depression, nausea, and vomiting. Vitamin B_{12} is necessary for the proper functioning of the nervous system and in the formation of red blood cells. It can be obtained from meat, fish, and dairy products. Anemia, nervousness, fatigue, and even brain degeneration can result from vitamin B_{12} deficiency.

Niacin is required to release energy from glucose. It is present in whole grains (but not corn), meat, fish, and dairy products. Inadequate amounts of this vitamin cause a disease called pellagra, which is characterized by skin disorders, weak muscles, diarrhea, and loss of appetite. Pellagra was once common in Spain, Mexico, and the southeastern United States, where a large component of the diet consisted of corn and corn products, since niacin exists in corn as part of a large, fibrous molecule that cannot be absorbed by the blood or used by the body. However, after it was discovered that treating corn with an alkaline solution (such as lime water) releases the niacin from the larger molecule and makes it available for the body to use, pellagra became much less common.

Pantothenic acid helps release energy from fats and carbohydrates and is found in large quantities in egg yolk, liver, nuts, and whole grains. Deficiency of this vitamin causes anemia. Biotin is involved in the release of energy from carbohydrates and in the formation of **fatty acids**. It is widely available from grains, legumes, and liver. A lack of biotin causes dermatitis.

Because of its association with the common cold, vitamin C, also known as ascorbic acid, is probably the best known of all the vitamins. Most animals can synthesize this vitamin in the liver, where glucose is converted to ascorbic acid. Humans, monkeys, guinea pigs, and the Indian fruit bat are exceptions and must obtain it from the diet. It is easily oxidized, and food storage or food processing and preparation frequently destroy its activity. Soaking fruits and vegetables in water for long periods also removes most of the vitamin C. Citrus fruits, berries, and some vegetables like tomatoes and peppers are good sources of vitamin C. The exact function of it in the body is still not well understood, but it is believed to be necessary for the formation of collagen, an important protein in skin, cartilage, ligaments, tendons, and bone. It also plays a role in the body's absorption, use, and storage of **iron**. It is an antioxidant and is therefore believed to offer protection to cells much as vitamin E does. There is increasing evidence that large amounts of vitamin C in the diet lessens the risk of heart disease and cancer.

A deficiency of vitamin C causes a disease called scurvy. Scurvy weakens the connective tissue in bones and muscles, causing bones to become very porous and brittle and muscles to weaken. As the walls of the circulatory system become affected, sore and bleeding gums and bruises result from internal bleeding. Anemia can occur because iron, which is critical to the transport of oxygen in the blood, cannot be utilized. Vitamin C is metabolized very slowly by the body, and deficiency diseases do not usually manifest themselves for several months.

Linus Pauling, the winner of two Nobel Prizes (one for chemistry and one for peace), believed that massive doses of vitamin C can ward off the common cold and offer protection against some forms of cancer. While scientific studies have been unable to prove this theory, they do suggest that vitamin C can at least reduce the severity of the symptoms of a cold. Some studies also suggest that vitamin C can lessen the incidence of heart disease and cancer. If this is true, it could be that the antioxidant properties of the vitamin help protect cells from weakening and breaking down much as vitamin E does. In fact, vitamins A, C, and E all play similar roles in the body, and it is difficult to distinguish among their effects.

See also Antioxidants

VOIT, KARL VON (1831-1908)

German physiologist

The son of a well-known architect, Voit originally planned to be a physician and obtained his medical degree in 1854. Voit had been the pupil of two of Germany's greatest chemists, **Friedrich Wöhler** and **Justus von Liebig**. Impressed by their work, he decided to switch careers, took more courses in **biochemistry** and eventually became almost as widely known as his mentors because of his basic research in **nutrition** and **metabolism**. In 1863, Voit joined the faculty of the University of Munich, where he did most of his research and where he remained for the rest of his life.

When Voit began his biochemical career, a major goal was to determine, as precisely as possible, what happens to **proteins**, fats and **carbohydrates** after they are absorbed into the body. He chose this research path partly because he liked to work with clear and accurately-measured figures and partly because he wanted to provide the sound, scientific evidence that would back up some of his friend Liebig's more sweeping nutritional theories. Liebig, for instance, theorized that *nitrog-*

Karl von Voit.

enous foods (proteins) were used by the body specifically to supply **energy** for muscular work, and one of Voit's early studies was designed to prove this theory.

Working with dogs trained to use a treadmill, Voit and another chemist, Theodor Bischoff (1807-1882), measured the protein in the animals' daily diets, then measured the amount of **urea** they excreted each day. (Urea, one of protein's end-products, was already being used as a measure of the amount of protein in the system.) After making his measurements under varying conditions, Voit was surprised to learn that his dogs excreted roughly the same amount of urea after intense activity as they did during periods of rest—suggesting that additional protein had not been ''burned up'' to fuel the muscular activity.

Voit persevered, however. Beginning in 1862, be worked with a chemist, Max von Pettenkofer (1818-1901), who designed an ingenious ''respiratory chamber'' that allowed them to perform increasingly more sophisticated metabolic studies. The chamber, large enough to hold a human, could measure (among other things) the consumption of **oxygen** and the production of **heat** before and after meals. It could also determine the amount of **nitrogen** and other **gases** being breathed out by a particular subject. (Even today, measuring the output of nitrogen, the major component of protein, is an important way to study protein metabolism). In the 1870s, Voit developed and refined the early tests in this area. By measuring the nitrogen contained in the diet, he explained, and comparing it to the nitrogen content of the patient's exhaled breath, an observer could determine the subject's nitrogen balance—could tell, in other words, whether the individual was storing too much nitrogen, excreting too much nitrogen or—like most healthy people—excreting roughly the same amount of nitrogen that he was taking in.

After a decade of studying a variety of subjects, both animal and human, Voit was able to publish monographs and articles that described, more clearly than ever, some of the intermediate steps in animal metabolism that until then had only been guessed at. His writings also disproved several more of Liebig's metabolic theories. Voit's articles appeared in both textbooks and medical journals and, in time, described the effects on metabolism of a great number of factors, including diet, fasting, and even illnesses such as diabetes and leukemia. Although some of his conclusions were controversial, Voit's emphasis on careful, detailed, and accurate investigative methods attracted many students to his courses. Today, he is considered by many to have laid much of the foundation for modern nutritional science.

VOLATILITY

A volatile substance is one which evaporates readily. Well-known examples are **gasoline** and dry ice. Although a qualitative concept, it has found considerable application in the development of **chemistry** as a science.

In the early years of chemistry, as scientists sorted out basic concepts, the term volatile was applied to materials which would yield **gases** upon various chemical treatments. A common substance which was widely studied was *sal volatile*, also known as volatile alkali. This substance is known today as ammonium carbonate. Ammonium carbonate is a solid, but chemical treatment can result in its breakdown, releasing **ammonia** and **carbon** dioxide, both gases under normal conditions. Such a disappearance of a solid into the gas phase was an intriguing phenomenon. Two prominent early chemists, **Robert Boyle** and **Joseph Black** were among those who studied the reactions of volatile alkali. Boyle also wrote of volatile nitre (saltpeter) which is either **potassium** or **sodium** nitrate. Under certain treatments, saltpeter yields **nitrogen** gas. **Antoine-Laurent Lavoisier** experimented with volatile sulphurous acid which readily yields the gas **sulfur** dioxide.

In present usage, the volatility of a substance refers to the relative ease with which it can be vaporized. Quantitative measures of volatility include **vapor pressure** and **boiling point**. Substances with high vapor pressures are highly volatile. Since the boiling point of a substance is the **temperature** at which its vapor pressure is equal to the atmospheric pressure, a substance whose boiling point is lower than another's is said to be more volatile. For instance, **bromine** and **mercury** both exist as **liquids** at room temperature, but bromine boils at 138°F (59°C) while mercury's boiling point is 693°F (367°C). Bromine is considerably more volatile than mercury. The boiling points of some common materials regarded as having high volatility are: chloroform, 142°F (61°C); **ether**, 95°F (35°C); and gasoline, 158-194°F (70-90°C).

Flasks and beakers with colored liquids, denoting the volume of each. ***(Photograph by Peggy & Ron Barnett, The Stock Market. Reproduced by permission.)***

VOLUME

Volume is the amount of space an object occupies.

This seems a simple concept, and it is for basic, solid objects. For a solid cube, for example, multiply length by width by height to obtain the volume. But what is the volume of a gas, which can easily expand and contract? Except for idealized mathematical models, the volume of an object depends on the composition of that object, and the environment around it. Scientists have spent significant effort to formulate how such quantities depend on their environmental quantities like pressure and **temperature**. The volume of a solid, for example, typically expands slightly as the temperature increases, but in general they are mostly resistant to changes in their shape or volume. A liquid like **water** will take the shape of whatever container it is poured into, but its volume stays the same. The gas in a balloon will take whatever shape the balloon has, and the volume will shrink if external pressure is applied to the balloon. (Ideal **gases**, in fact, obey **Boyle's Law**, which states that the volume is inversely proportional to the pressure, at constant temperature.)

Greek mathematicians, especially Archimedes, were among the first to derive formulas for the volumes of common shapes. It was not until the invention of calculus by Isaac Newton and Gottfried Wilhelm Leibniz in the seventeenth century that the volume for an arbitrary shape could be calculated. One can show, for example, that for an object of a certain volume, a sphere has the smallest ratio of surface area to volume. This explains why a cat curls up into a ball when it sleeps—it is trying to make itself like a sphere, so its surface area is as small as possible in order to minimize **heat** loss through its skin.

For a solid of an unusual or arbitrary shape, one could determine its volume by, for example, placing it in a full tub of water and measuring how much water spills over the edge. Lowering himself into a full tub, Archimedes is said to have been inspired to formulate his law of buoyancy.

Why does **matter** have volume? That is, since matter is composed of atoms which have a great deal of empty space inside them and between themselves, why doesn't the atoms just collapse into a tiny, solid clump that has an extremely small volume? Because the elements of the atoms—the electrons, protons, and neutrons—have quantities like electrons, electric charge that makes them repel one another if they get too close.

The volumes of some typical objects can vary enormously. For example, the human body has a volume of roughly 26 gal (0.1 cu m) or 100 l. A grown elephant's volume is about 1,500 gal (6 cu m), and the volume of the Earth is about 10^{21} cu m. By contrast, the volume of a **hydrogen atom** is far, far smaller, about 10^{-30} cu m.

W

WAAGE, PETER (1833-1900)

Norwegian chemist and mineralogist

Peter Waage was born on June 29, 1833, the son of shipmaster and ship owner Peter Pedersen Waage and Regine Lovise Wattne Waage, on the island of Hitterø (now Hidra) near Flekkefjord, Norway, about 200 miles southeast of Christinia (now Oslo). He was raised on this island, where his forefathers had lived as seamen for centuries. Because his father was usually at sea, Waage grew up mostly under the care of his mother, who was his first teacher. He was able to read by the age of about four. When Waage's precocity became known, it was decided that he should receive further education rather than follow the traditional family occupation of seafaring. As a youth, he had a large collection of minerals, plants, and insects, and some of his first publications dealt with **mineralogy** and **crystallography**.

Waage's first regular schooling began at Flekkefjord when he was 11. The school principal persuaded him to prepare to attend the University of Christiania by entering the fourth year of the Bergen Grammar School in 1849. He passed his matriculation examination *cum laudabilis* for the University of Christiania in 1854, the same year as **Cato Maximilian Guldberg**, with whom he began a lifetime friendship. Together with several other students, they established a small, informal club whose members met on Saturday afternoons to discuss physical and chemical problems. Waage studied medicine during his first three years at the university but switched to mineralogy and **chemistry** in 1857. He was awarded the Crown Prince's Gold Medal for his paper, ''Development of the Theory of the Oxygen-Containing Acid Radicals,'' which appeared in 1859, the same year as his book, *Outline of Crystallography*, coauthored with H. Mohn.

After graduating in 1859, Waage was awarded a scholarship in chemistry, enabling him to make a year's study tour of France and Germany (where most of his time was spent with Robert Wilhelm Bunsen at Heidelberg) the following spring. He was appointed Lecturer in Chemistry at the University of Christiania in 1861, and in 1866 he was promoted to Professor of the only chair of chemistry at the university.

Waage's name is intimately linked with that of his friend Guldberg primarily for their joint discovery of the law of **mass** action. This fundamental law of chemistry, which today is known to every beginning chemistry student, had several forerunners, but the combined efforts of the empiricist Waage and the theorist Guldberg were needed to produce the first general, exact, mathematical formulation of the role of the amounts of reactants in chemical equilibrium systems.

But Waage and Guldberg were also related through two marriages; Guldberg married his cousin Bodil Mathea Riddervold, daughter of cabinet minister Hans Riddervold, and the couple had three daughters. Waage married Bodil's sister, Johanne Christiane Tandberg Riddervold by whom he had five children, and after her death in 1869, he became Guldberg's brother-in-law a second time, in 1870, by marrying one of Guldberg's sisters, Mathilde Sofie Guldberg, by whom he had six children.

Guldberg and Waage's collaboration on the studies of chemical affinity that led to the law of mass action began immediately after Guldberg's return from abroad in 1862. Waage presented their first report to the Division of Science of the Norwegian Academy of Science and Letters on March 14, 1864, where it elicited little response. Even after its publication the following year in Norwegian, a language not read by many chemists, in the academy's journal, which was not accessible to many scientists, it failed to attract attention. Moreover, their work remained almost completely unknown to scientists as did the more detailed description of their theory published in 1867 in French. The theory did not become generally known until 1877 when German chemist Wilhelm Ostwald published an article that adopted the law of mass action and proved its validity by experiments of his own. The following year, Dutch chemist Jacobus Henricus van't Hoff derived the law from reaction **kinetics**, apparently without any awareness of Guldberg and Waage's previous work. Because their work had still not become universally known and van't Hoff had not recognized their priority, Guldberg and Waage published their previous work for a third time, this time in the German journal *Annalen der Chemie* and in German, the *lingua franca* of 19th-century chemistry. In 1884, in his *Études de Dynamique Chimique*, van't Hoff finally mentioned their work, thus assuring their priority.

After completing his collaboration with Guldberg, Waage concentrated more and more on practical problems and on social and religious work, dealing largely with **nutrition** and public health, such as his discovery of methods for producing unsweetened condensed milk and sterilized canned milk. He also developed an excellent, highly concentrated fish meal (*Professor Waages Fiskemel*) used on Norwegian ships and expeditions and exported to Sweden, Finland, Denmark, and Germany. Beer was then taxed according to the amount of malt used in its brewing, but Waage proposed that it be taxed according to its alcoholic content, and he developed a new method for determining this **concentration** by measuring the **boiling point**.

Among Waage's religious work with young people was his activity in the establishment and management of the Christiania Ynglingeforening (later the Oslo YMCA) and the Norwegian Christian Youth Association. He served as co- editor of the *Polyteknisk Tidsskrift* between 1859-60 and again in 1872- 1880, a journal devoted largely to practical applications of science, and was an active member and officer of various scientific societies and the recipient of many honors. He died in Christiania on January 13, 1900.

Wald, George (1906-1997)

American biochemist

George Wald first won a place in the spotlight as the recipient of a Nobel Prize for his discovery of the way in which hidden biochemical processes in the retinal pigments of the eye turn light **energy** into sight. Among Wald's important experiments were the effects of vitamin A on sight and the roles played by rod and cone cells in black and white and **color** vision. Outside the laboratory, his splendid lectures at Harvard to packed audiences of students generated great intellectual excitement. It was as a political activist during the turbulent 1960s, however, that Wald garnered further public recognition. Wald's personal belief in the unity of nature and the kinship among all living things was evidenced by the substantial roles he played in the scientific world as well as the political and cultural arena of the 1960s.

Wald's father, Isaac Wald, a tailor and later a foreman in a clothing factory, immigrated from Austrian Poland, while his mother, Ernestine Rosenmann Wald, immigrated from Bavaria. Most of Wald's youth was spent in Brooklyn, New York, where his parents moved after his birth on the Lower East Side of Manhattan on November 18, 1906. He attended high school at Brooklyn Tech, where he intended to study to become an electrical engineer. College changed his mind, however, as he explained for the *New York Times Magazine* in 1969, "I learned I could talk, and I thought I'd become a lawyer. But the law was man-made; I soon discovered I wanted something more real."

Wald's bachelor of science degree in zoology, which he received from New York University in 1927, was his ticket into the reality of biological research. He began his research career at Columbia University, where he was awarded a master's degree in 1928, working under Selig Hecht, one of the founders of the field of biophysics and an authority on the physiology of vision. Hecht exerted an enormous influence on Wald, both as an educator and a humanist. The elder scientist's belief in the social obligation of science, coupled with the conviction that science should be explained so the general public could understand it, made a great impression on the young Wald. Following Hecht's sudden death in 1947 at the age of 55, Wald wrote a memorial as a tribute to his colleague.

In 1932 Wald earned his doctorate at Columbia, after which he was awarded a National Research Council Fellowship in Biology. The two-year fellowship helped to support his research career, which first took him to the laboratory of **Otto Warburg** in Berlin. It was there, in 1932, that he discovered that vitamin A is one of the major constituents of retinal pigments, the light sensitive chemicals that set off the cascade of biological events that turns light into sight.

Warburg sent the young Wald to Switzerland, where he studied **vitamins** with chemist **Paul Karrer** at the University of Zurich. From there Wald went to **Otto Meyerhof**'s laboratory of cell **metabolism** at the Kaiser Wilhelm Institute in Heidelberg, Germany, finishing his fellowship in the department of physiology at the University of Chicago in 1934. His fellowship completed, Wald went to Harvard University, first as a tutor in **biochemistry** and subsequently as an instructor, faculty instructor, and associate professor, finally becoming a full professor in 1948. In 1968, he became Higgins Professor of Biology, a post he retained until he became an emeritus professor in 1977.

Wald did most of his work in eye physiology at Harvard, where he discovered in the late 1930s that the light-sensitive chemical in the rods—those cells in the retina responsible for night vision—is a single pigment called rhodopsin (visual purple), a substance derived from opsin, a protein, and retinene, a chemically modified form of vitamin A. In the ensuing years, Wald discovered that the vitamin A in rhodopsin is "bent" relative to its natural state, and light causes it to "straighten out," dislodging it from opsin. This simple reaction initiates all the subsequent activity that eventually generates the sense of vision.

Wald's research moved from rods to cones, the retinal cells responsible for color vision, discovering with his co-worker Paul K. Brown, that the pigments sensitive to red and yellow-green are two different forms of vitamin A that co-exist in the same cone, while the blue-sensitive pigments are located in separate cones. They also showed that color blindness is caused by the absence of one of these pigments.

For much of his early professional life, Wald concentrated his energy on work, both research and teaching. His assistant, Brown, stayed with him for over 20 years and became a full-fledged collaborator. A former student, Ruth Hubbard, became his second wife in 1958, and they had two children, Elijah and Deborah. (His previous marriage to Frances Kingsley in 1931 ended in divorce; he has two sons by that marriage, Michael and David.) Wald, his wife, and Brown together became an extremely productive research team.

By the late 1950s Wald began to be showered with honors, and during his career he received numerous honorary degrees and awards. After Wald was awarded (with Haldan K. Hartline of the United States and Ragnar Granit of Sweden) the Nobel Prize in physiology or medicine in 1967 for his work with vision, John E. Dowling wrote in *Science* that Wald and his team formed "the **nucleus** of a laboratory that has been extraordinarily fruitful as the world's foremost center of visual-pigment biochemistry."

As Wald's reputation flourished, his fame as an inspiring professor grew as well. He lectured to packed classrooms, inspiring an intense curiosity in his students. The energetic professor was portrayed in a 1966 *Time* article that summarized the enthusiasm he brought to teaching his natural science course: "With crystal clarity and obvious joy at a neat explanation, Wald carries his students from protons in the fall to living organisms in the spring, [and] ends most lectures with some philosophical peroration on the wonder of it all." That same year, the *New York Post* said of his lectures, "His beginnings are slow, sometimes witty.... The talk gathers momentum and suddenly an idea *pings* into the atmosphere—fresh, crisp, thought-provoking."

Six days after he received the Nobel Prize, Wald wielded the status of his new prestige in support of a widely popular resolution before the city council of Cambridge, Massachusetts—placing a referendum on the Vietnam War on the city's ballot of 7 November 1967. Echoing the sentiments of his mentor Hecht, he asserted that scientists should be involved in public issues.

The Cambridge appearance introduced him to the sometimes stormy arena of public politics, a forum from which he has never retired. The escalating war in Vietnam aroused Wald to speak out against America's military policy. In 1965, during the escalation of that war, Wald's impromptu denunciation of the Vietnam war stunned an audience at New York University, where he was receiving an honorary degree. Shortly afterward, he threw his support and prestige behind the presidential campaign of Eugene McCarthy. His offer to speak publicly on behalf of McCarthy was ignored, however, and he became a disillusioned supporter, remaining on the fringe of political activism.

Then on March 4, 1969, he gave an address at the Massachusetts Institute of Technology (MIT) that, "upended his life and pitched him abruptly into the political world," according to the *New York Times Magazine.* Wald gave "The Speech," as the talk came to be known in his family, before an audience of radical students at MIT. The students had helped to organize a scientists' day-long "strike" to protest the influence of the military on their work, a topic of much heated debate at the time.

Although much of the MIT audience was already bored and restless by the time Wald began, even many of those students who were about to leave the room stopped to listen as the Nobel laureate began to deliver his oration, entitled, "A Generation in Search of a Future." "I think this whole generation of students is beset with a profound sense of uneasiness, and I don't think they have quite defined its source," Wald asserted as quoted in the *New York Times Magazine.* "I think I understand the reasons for their uneasiness even better than they do. What is more, I *share* their uneasiness."

George Wald.

Wald's discourse evoked applause from the audience as he offered his opinion that student unease arose from a variety of troublesome matters. He pointed to the Vietnam War, the military establishment, and finally, the threat of nuclear warfare. "We must get rid of those atomic weapons," he declared. "We cannot live with them." Speaking to the students as fellow scientists, he sympathized with the their unease at the influence of the military establishment on the work of scientists, intoning, "Our business is with life, not death...."

The speech was reprinted and distributed around the country by the media. Through these reprints, Wald told readers that some of their elected leaders were "insane," and he referred to the American "war crimes" enacted in Vietnam. In the furor that followed, Wald was castigated by critics, many of whom were fellow academics, and celebrated by sympathizers. A letter writer from Piney Flats, Tennessee was quoted in the *New York Times Magazine* as saying, "So good to know there are still some intellects around who can talk downright horsesense." Wald summed up his role as scientist-political activist in that same article by saying, "I'm a scientist, and my concerns are eternal. But even eternal things are acted out in the present.'' He described his role as gadfly as putting certain controversial positions into words in order to make it, ''easier for others to inch toward it.''

His role as a Vietnam War gadfly expanded into activism in other arenas of foreign affairs. He served for a time as president of international tribunals on El Salvador, the Philippines, Afghanistan, Zaire, and Guatemala. In 1984, he joined four other Nobel Prize laureates who went with the ''peace ship'' sent by the Norwegian government to Nicaragua during that country's turmoil.

In addition to his interests in science and politics, Wald's passion included collecting Rembrandt etchings and primitive art, especially pre-Columbian pottery. This complex mixture of science, art, and political philosophy was reflected in his musings about religion and nature in the *New York Times Magazine:* ''There's nothing supernatural in my mind. Nature is my religion, and it's enough for me. I stack it up against any man's. For its awesomeness, and for the sense of the sanctity of man that it provides.''

In addition to the Nobel Prize, Wald received numerous awards and honors, including the Albert Lasker Award of the American Public Health Association in 1953, the Proctor Award in 1955 from the Association for Research in Ophthalmology, the Rumford Premium of the American Academy of Arts and Sciences in 1959, the 1969 Max Berg Award, and the **Joseph Priestley** Award the following year. In addition, he was elected to the National Academy of Science in 1950 and the American Philosophical Society in 1958. He is also a member of the Optical Society of America, which awarded him the Ives Medal in 1966. In the mid–1960s Wald spent a year as a Guggenheim fellow at England's Cambridge University, where he was elected an Overseas fellow of Churchill College for 1963–64. Wald also held honorary degrees from the University of Berne, Yale University, Wesleyan University, New York University, and McGill University.

Wald died on 12 April 1997, at his home in Cambridge, Massachusetts, at the age of 90.

WALKER, JOHN E. (1941-)

English biochemist

John E. Walker was awarded a share of the 1997 Nobel Prize in **Chemistry** for his research on the **enzyme** ATP synthase. That enzyme is responsible for the biologically critical **molecule** known as adenosine triphosphate (ATP) that provides the **energy** needed to drive a host of biochemical reactions in cells. Walker's research dovetailed with similar work carried out by the second 1997 Nobel Laureate, **Paul Boyer**, who devised a theory that explained the process by which the ATP synthase enzyme operates. Walker has been employed at the Laboratory of Molecular Biology of the Medical Research Council in Cambridge for more than two decades.

Walker was born on January 7, 1941, in Halifax, England. He attended the Rastrick Grammar School in Brighouse, Yorkshire, and then enrolled at St. Catherine's College,

John E. Walker. *(Archive Photo. Reproduced by permission.)*

Oxford, in 1960. He was awarded his B.A. Degree in Chemistry by St. Catherine's in 1964. Walker then spent four years as a research student at the Sir William Dunn School of Pathology at Oxford before earning his M.A. and D. Phil. degrees from Oxford in 1969.

Upon completion of his doctoral studies, Walker spent two years as a Postdoctoral Fellow at the School of Pharmacy at the University of Wisconsin. He then spent three additional years as a NATO Fellow at CNRS, in Gif-sur-Yvette, France, and as an EMBO Fellow at the Institut Pasteur in Paris. In 1974, Walker accepted an appointment as a member of the scientific staff at the Cambridge Laboratory of Molecular Biology of the Medical Research Council. He was assigned to the Division of Protein and Nucleic Acid Chemistry of the laboratory. In 1982, Walker was promoted to Senior Scientist at the laboratory, and in 1987, he was given a Special Appointment (Professorial Grade) at the laboratory.

Adenosine triphosphate (ATP) is one of the most important molecules in the cells of living organisms. It has been described as an ''energy-carrying'' molecule because it provides the energy needed to drive many essential biochemical reactions.

The energy carried by an ATP molecule is stored in its **phosphate** bonds. A molecule of ATP is produced through a series of steps in which a phosphate group is first attached to a molecule of adenosine monophosphate (AMP) to form adenosine diphosphate (ADP). ADP then adds a second phosphate group to form adenosine triphosphate (ATP). As each phosphate group is added to the growing molecule, it brings with it energy stored in the form of the chemical bonds by which the phosphate is attached to the core molecule.

ATP acts as an energy-provider because it tends to break down to form first ADP plus a phosphate group, and then AMP and a second phosphate group. Each time one of these steps occurs, energy is released. That energy is transferred to some other set of chemical reactants in a cell, making it possible for those reactants to form new compounds.

Scientists have long been very interested in learning precisely how ATP is formed and how it carries out its biochemical functions. Discovered in 1929 by the German chemist Karl Lohmann, ATP was first synthesized two decades later by the Scottish chemist **Alexander Todd**. The role of ATP in providing energy to cell reactions was first elucidated by the German-American biochemist **Fritz Lipmann** in the period 1939-1941.

One important line of ATP research has focused on the mechanism by which the molecule is formed. In 1960, the American biochemist Efraim Racker found a substance in the mitochondria of cells that appeared to be responsible for the synthesis of ATP. They called the enzyme F_0 F_1 ATPase, although it is now better known as ATP synthase. The original name for the enzyme comes from the fact that it consists of two parts, an F_0 domain that is attached to a cell membrane, and an F_1 domain that protrudes from the membrane.

During the 1950s, the American biochemist **Paul D. Boyer** developed a theory to explain how ATP synthase is able to produce ATP. Essentially, he argued that a flow of **hydrogen** ions within the cell membrane causes the F_0 domain of the enzyme to rotate in much the same way that the wind causes the blades on a windmill to turn. Boyer hypothesized that the turning of the F_0 membrane sequentially exposed structurally different regions of the F_1 domain in such as way as to make possible the ADP + phosphate reaction to occur. Walker's research completely identified the amino acid sequence of ATP synthase and **proteins** attached to it and determined much of the three-dimensional structure of the molecule. It was for this research that he was awarded the 1997 Nobel Prize in Chemistry.

Walker has received a host of honors and awards in addition to the Nobel Prize. Among these have been the A. T. Clay Gold Medal for academic distinction in 1959, the Johnson Foundation Prize of the University of Pennsylvania Medical School in 1994, the CIBA Medal and Prize of The Biochemical Society in 1995, the Peter Mitchell Medal of the European **Bioenergetics** Congress in 1996, and the Gaetana Quagliariello Prize for Mitochondrial Research from the University of Bari, Italy, in 1997. Walker has also been elected a Fellow of The Royal Society (1995), a Fellow of Sidney Sussex College at Cambridge (1997) and an Honorary Fellow of St. Catherine's College at Oxford (1997). He has also been asked to give named lectures at the University of California at San Diego (the Nathan Kaplan Memorial Lecture), Copenhagen (the Novo-Nordic Lecture), Sheffield University (the Krebs Lecture), the Biochemical Society (the CIBA Lecture), the EBEC Conference at Louvain, Belgium (the Peter Mitchell Lecture), and the Netherlands Biochemical Society (Lecture of the Year).

For two decades, Walker has also been active in a variety of professional responsibilities. He has served on the editorial board of *The Biochemical Journal*, **Biochemistry** *International*, *Molecular Biology*, *Journal of Bioenergetics*, and *Structure*. He is the author or co-author of more than 175 papers and book chapters. Walker is married and has two children.

Wallach, Otto (1847-1931)

German chemist

Otto Wallach was a highly regarded professor of **chemistry** whose curiosity about essential oils led to research that benefited both **organic chemistry** and an important industry. For his meticulous procedures and initiative in the study of terpenes, a class of compounds he identified and named as essential oils, Wallach received the Nobel Prize in chemistry in 1910. Essential oils, called ethereal oils in Wallach's time, are fragrant extracts from plant materials that are used in perfumes, flavorings, and medicines. Terpenes are responsible for much of the pleasant odor associated with essential oils. Wallach's work provided a scientific foundation for the fragrance industry.

Otto Wallach was born on March 27, 1847, in Königsberg, in East Prussia (Königsberg is now called Kaliningrad and is part of the reorganized former Soviet Union). Wallach's mother was Otillia Thoma Wallach; his father, Gerhard Wallach, was an official in the Prussian government whose post necessitated moves from Königsberg to Stettin (now in Poland and called Szczecin) and then to Potsdam, near Berlin, Germany. Wallach graduated from the Potsdam *Gymnasium* (high school). In school he became fascinated with chemistry and the history of art, and pursued both interests throughout his life. Early in 1867 he entered the University of Göttingen and received his doctorate in chemistry in 1869. Wallach never married.

Wallach's doctoral dissertation was on isomers of toluene. Isomers (*iso* means same; *mer* means parts) are substances with identical composition, but different arrangements of the parts, giving the substances different physical and chemical properties. Toluene is one of the products of **distillation** of **coal**, and among its many uses is as a solvent in the preparation of fragrances. Wallach's doctoral studies were to provide the background for his most important research.

After graduation Wallach worked briefly in Berlin, then went to the University of Bonn to assist **August Kekulé**, a renowned German professor of chemistry whose most noted contribution was the discovery of the **structural formula** of **benzene**, another important coal product and a substance similar to toluene. Kekulé was interested in turning much of his laboratory work over to a young assistant. Wallach remained at the University of Bonn until 1889, except for a period of two years, from 1871 to 1873, when he tried research at Aktiengesellschaft für Anilinfabrikation (Agfa).

In 1879, as a professor at the University of Bonn, Wallach was assigned to teach pharmacy. He had little background in the chemistry of essential (ethereal) oils, used in medicines. In Kekulé's laboratory he found abandoned samples of essential oils that Kekulé thought were too complex even to attempt to analyze. A very patient researcher, Wallach distilled and redistilled each oil sample until he could identify a pure substance. By 1881 he had repeated his procedure with different oils and identified eight pure, very similar, fragrant substances that he named terpenes, from the Greek *terebinthos* meaning turpentine.

Wallach did not work alone. While some colleagues worked on synthesizing new and similar compounds, he devoted most of his effort to studying how the terpenes he separated from essential oils were related. By 1887 Wallach discovered that all the terpenes he identified in essential oils are derived from a multiple of a particular arrangement of five **carbon** atoms, now called isoprene units. Some examples of terpenes found naturally are bayberry, rose oil, peppermint, menthol, camphor, and turpentine.

In 1889 Wallach was appointed director of the Chemical Institute at the University of Göttingen, where he continued his work on terpenes. In 1910 he received the Nobel Prize in chemistry "for his initiative work in the field of alicyclic substances," the terpenes, discovered to have carbon atoms arranged in rings, or cycles. Kekulé was the first person to identify a compound (benzene) as shaped like a **ring**.

One of the most important outcomes of Wallach's work was the research it spawned, which, combined with his own work, forever changed the fragrance industry. Before Wallach, there was little scientific background for production of fragrances. Processes used were much like the ones brought to Europe by the Crusaders. Wallach described the change his work inspired in his Nobel address. "The fragrant components of plants were merely distilled and the distillate brought to market. In this process, the products obtained were not always handled rationally, in the absence of all knowledge of their chemical nature, and the doors were wide open for every kind of falsification. When this was carried out with only some skill, the consumer was helpless against it.

"This has now been changed thoroughly. Thanks to the possibility of distinctly characterizing single components of the ethereal oils, we now possess a significant analytical system to detect falsifications and to guard against them." Wallach's work, combined with the work of others he inspired, was credited with putting science into the production of perfumes, flavorings, and medicines that use terpenes.

Many universities and scientific societies honored Wallach. In 1912 he received the Davy Medal, awarded to outstanding scientists in memory of the British chemist, Sir **Humphry Davy**, who identified a number of chemical elements through his pioneering work with **electrochemistry**. Wallach retired from Göttingen in 1915, but continued with research until he was eighty years old. In his lifetime Wallach published 126 papers on his research of essential oils. He died in Göttingen, Germany, on February 26, 1931, a month short of his eighty-fourth birthday.

WANG, JAMES C. (1936-)

Chinese American biochemist

James C. Wang is a biochemist who trained as a chemical engineer before turning to biophysical **chemistry** and molecular biology. Wang discovered deoxyribonucleic acid (DNA) topoisomerases (or local enzymes) and proposed a mechanism for their operation in the 1970s. He also studied the configuration (or topology) of **DNA**, an approach that proved fruitful in helping to explain how the structure of the double helix coils and relaxes.

Wang was born in mainland China on November 18, 1936. Less than a year later the Sino-Japanese War began. Wang lost his mother during the conflict, and shortly after it ended, his older sister also died. His father remarried, moving the family to Taiwan in 1949. Because of the war, Wang received only about two years of elementary education before starting junior high school in Taiwan. As a child, he wanted to study medicine, but his father encouraged him to become an engineer. A high school teacher inspired him to follow his interest in chemistry. **Chemical engineering** became Wang's course of study, and he earned a B.S. in 1959 from National Taiwan University. Continuing his studies in the field of chemistry, he earned a masters from the University of South Dakota in 1961, and a doctorate from the University of Missouri in 1964.

In the same year that Wang received his doctorate, he became a research fellow at the California Institute of Technology, remaining there until 1966. He then taught at the University of California in Berkeley from 1966 until 1977, when he joined the faculty at Harvard University. He was named the Mallinckrodt Professor of **Biochemistry** and Molecular Biology at Harvard in 1988.

Wang once noted that his interest in DNA topology came about by chance. His training in engineering and chemistry had led him to study the physical basis of chemical processes. He began to think about the questions raised by the double helix structure of DNA soon after its discovery by molecular biologists **James Watson** and **Francis Crick**. His own study of DNA confirmed the unique structure. But it was not clear how the two tightly intertwined strands could unravel at the speed at which the biochemical processes were thought to occur. Topologically, it did not seem possible for the strands to unravel at all. Wang studied supercoiling in *E. coli* bacteria and found that the rotation speed of an unraveling DNA strand is 10,000 revolutions per minute. He also found that the same **enzyme**, a DNA topoisomerase, is responsible for both breaking and rejoining the DNA strands.

Wang believed that the topological characteristics of the double helix affected all its chemical transformations, including transcription, replication, and recombination. In a 1991 interview in *Cell Science,* Wang remarked that he relied on intuition to further his scientific understanding, particularly when trying to make sense of "bits and pieces of information" that didn't obviously fit together. "There are times," he told the interviewer, "when results that seem to make no sense are key to new advances."

Wang has served on the editorial boards of several scholarly journals, including *Journal of Molecular Biology,*

Annual Review of Biochemistry, and *Quarterly Review of Biophysics.* He was a Guggenheim fellow in 1986, and was elected to the U.S. National Academy of Sciences in 1984. In 1961 Wang married a former classmate; they have two daughters.

WARBURG, OTTO (1883-1970)

German biochemist

Otto Warburg is considered one of the world's foremost biochemists. His achievements include discovering the mechanism of cell **oxidation** and identifying the iron-enzyme complex, which catalyzes this process. He also made great strides in developing new experimental techniques, such as a method for studying the respiration of intact cells using a device he invented. His work was recognized with a Nobel Prize for medicine and physiology in 1931.

Otto Heinrich Warburg was born on October 8, 1883, in Freiburg, Germany, to Emil Gabriel Warburg and Elizabeth Gaertner. Warburg was one of four children and the only boy. His father was a physicist of note and held the prestigious Chair in Physics at University of Berlin. The Warburg household often hosted prominent guests from the German scientific community, such as physicists **Albert Einstein, Max Planck, Emil Fischer**—the leading organic chemist of the late-nineteenth century, and **Walther Nernst**—the period's leading physical chemist.

Warburg studied **chemistry** at the University of Freiburg beginning in 1901. After two years, he left for the University of Berlin to study under Emil Fischer, and in 1906 received a doctorate in chemistry. His interest turned to medicine, particularly to cancer, so he continued his studies at the University of Heidelberg where he earned an M.D. degree in 1911. He remained at Heidelberg, conducting research for several more years and also making several research trips to the Naples Zoological Station.

Warburg's career goal was to make great scientific discoveries, particularly in the field of cancer research, according to the biography written by **Hans Adolf Krebs**, one of Warburg's students and winner of the 1953 Nobel Prize in medicine and physiology. Although he did not take up problems specifically related to cancer until the 1920s, his early projects provided a foundation for future cancer studies. For example, his first major research project, published in 1908, examined **oxygen** consumption during growth. In a study using sea urchin eggs, Warburg showed that after fertilization, oxygen consumption in the specimens increased 600 percent. This finding helped clarify earlier work that had been inconclusive on associating growth with increased consumption of oxygen and **energy**. A number of years later, Warburg did some similar tests of oxygen consumption by cancer cells.

Warburg was elected in 1913 to the Kaiser Wilhelm Gesellschaft, a prestigious scientific institute whose members had the freedom to pursue whatever studies they wished. He had just begun his work at the institute when World War I started. He volunteered for the army and joined the Prussian Horse Guards, a cavalry unit that fought on the Russian front. Warburg survived the war and returned to the Kaiser Wilhelm Institute for Biology in Berlin in 1918. Now 35 years old, he would devote the rest of his life to biological research, concentrating on studies of energy transfer in cells (cancerous or otherwise) and **photosynthesis**.

Otto Warburg.

One of Warburg's significant contributions to biology was the development of a **manometer** for monitoring cell respiration. He adapted a device originally designed to measure **gases** dissolved in blood so it would make measurements of the rate of oxygen production in living cells. In related work, Warburg devised a technique for preparing thin slices of intact, living tissue and keeping the samples alive in a nutrient medium. As the tissue slices consumed oxygen for respiration, Warburg's manometer monitored the changes.

During Warburg's youth, he had become familiar with Einstein's work on photochemical reactions as well as the experimental work done by his own father, Emil Warburg, to verify parts of Einstein's theory. With this background, Warburg was especially interested in the method by which plants converted light energy to chemical energy. Warburg used his manometric techniques for the studies of photosynthesis he conducted on algae. His measurements showed that photosynthetic plants used light energy at a highly efficient sixty-five percent. Some of Warburg's other theories about photosynthesis were not upheld by later research, but he was nevertheless considered a pioneer for the many experimental methods he developed in this field. In the late 1920s, Warburg began to develop techniques that used light to measure reaction rates and detect the presence of chemical compounds in cells. His "spectrophotometric" techniques formed the basis for some of the first commercial spectrophotometers built in the 1940s.

His work on cell respiration was another example of his interest in how living things generated and used energy. Prior to World War I, Warburg discovered that small amounts of cyanide can inhibit cell oxidation. Since cyanide forms stable complexes with heavy **metals** such as **iron**, he inferred from his experiment that one or more catalysts important to oxidation must contain a heavy metal. He conducted other experiments with **carbon** monoxide, showing that this compound inhibits respiration in a fashion similar to cyanide. Next he found that light of specific frequencies could counteract the inhibitory effects of **carbon monoxide**, at the same time demonstrating that the "oxygen transferring **enzyme**," as Warburg called it, was different from other enzymes containing iron. He went on to discover the mechanism by which iron was involved in the cell's use of oxygen. It was Warburg's work in characterizing the cellular catalysts and their role in respiration that earned him a Nobel Prize in 1931.

Nobel Foundation records indicate that Warburg was considered for Nobel Prizes on two additional occasions: in 1927 for his work on **metabolism** of cancer cells, then in 1944 for his identification of the role of flavins and nicotinamide in biological oxidation. Warburg did not receive the 1944 award, however, because a decree from Hitler forbade German citizens from accepting Nobel Prizes. Two of Warburg's students also won Nobel Prizes in medicine and physiology: Hans Krebs (1953) and **Axel Theorell** (1955).

In 1931 Warburg established the Kaiser Wilhelm Institute for Cell Physiology with funding from the Rockefeller Foundation in the United States. During the 1930s, Warburg spent much of his time studying dehydrogenases, enzymes that remove **hydrogen** from substrates. He also identified some of the cofactors, such as nicotinamide derived from vitamin B_3 (niacin), that play a role in a number of cell biochemical reactions.

Warburg conducted research at the Kaiser Wilhelm Institute for Cell Physiology until 1943 when the Second World War interrupted his investigations. Air attacks targeted at Berlin forced him to move his laboratory about 30 miles away to an estate in the countryside. For the next two years, he and his staff continued their work outside the city and out of the reach of the war. Then in 1945, Russian soldiers advancing to Berlin occupied the estate and confiscated Warburg's equipment. Although the Russian commander admitted that the soldiers acted in error, Warburg never recovered his equipment. Without a laboratory, he spent the next several years writing, publishing two books that provided an overview of much of his research. He also traveled to the United States during 1948 and 1949 to visit fellow scientists.

Even though Warburg was of Jewish ancestry, he was able to remain in Germany and pursue his studies unhampered by the Nazis. One explanation is that Warburg's mother was not Jewish and high German officials "reviewed" Warburg's ancestry, declaring him only one-quarter Jewish. As such he was forbidden from holding a university post, but allowed to continue his research. There is speculation that the Nazis believed Warburg might find a cure for cancer and so did not disturb his laboratory. Scientists in other countries were unhappy that Warburg was willing to remain in Nazi Germany. His biographer Hans Krebs noted, however, that Warburg was not afraid to criticize the Nazis. At one point during the war when Warburg was planning to travel to Zurich for a scientific meeting, the Nazis told him to cancel the trip and to not say why. "With some measure of courage," wrote Krebs, "he sent a telegram [to a conference participant from England]: 'Instructed to cancel participation without giving reasons.'" Although the message was not made public officially, the text was leaked and spread through the scientific community. Krebs believed Warburg did not leave Germany because he did not want to have to rebuild the research team he had assembled. The scientist feared that starting over would destroy his research potential, Krebs speculated.

In 1950 Warburg moved into a remodeled building in Berlin which had been occupied by U.S. armed forces following World War II. This new site was given the name of Warburg's previous scientific home—the Kaiser Wilhelm Institute for Cell Physiology—and three years later renamed the Max Planck Institute for Cell Physiology. Warburg continued to conduct research and write there, publishing 178 scientific papers from 1950 until his death in 1970.

For all of his interest in cancer, Warburg's studies did not reveal any deep insights into the disease. When he wrote about the "primary" causes of cancer later in his life, Warburg's proposals failed to address the mechanisms by which cancer cells undergo unchecked growth. Instead, he focused on metabolism, suggesting that in cancer cells "**fermentation**" replaces normal oxygen respiration. Warburg's cancer studies led him to fear that exposure to **food additives** increased one's chances of contracting the disease. In 1966 he delivered a lecture in which he stated that cancer prevention and treatment should focus on the administration of respiratory enzymes and cofactors, such as iron and the B **vitamins**. The recommendation elicited much controversy in Germany and elsewhere in the Western world.

Warburg's devotion to science led him to forego marriage, since he thought it was incompatible with his work. According to Karlfried Gawehn, Warburg's colleague from 1950 to 1964, "For him [Warburg] there were no reasonable grounds, apart from death, for not working." Warburg's productivity and stature as a researcher earned him an exemption from the Institute's mandatory retirement rules, allowing him to continue working until very near to the end of his life. He died at the Berlin home he shared with Jakob Heiss on August 1, 1970.

Waste Treatment

Many human processes produce waste. Before this waste is disposed of, it generally needs to be treated in some way. The exact method of treatment is dependant upon the nature of the waste. One example of waste treatment is the burning of fossil fuels. Fossil fuels are the fossilized remains of prehistoric plants and animals in the form of **coal**, oil, and **natural gas**. All fossil fuels contain **sulfur** and sulfur oxides as an impurity. High-grade fossil fuels have low levels of impurities and has a wide range of uses, from heating homes to powering industrial factories. This oxides the sulfur and releases the sulfur oxides into the atmosphere as a gas. These sulfur oxides can then react with **water** vapor in the clouds to produce **acid rain**. It is possible to remove this waste from exhaust **gases** before it reaches the atmosphere, though a process called scrubbing. With scrubbing, the waste gases are washed with water. This water is sprayed into the gases as they rise up the chimney. The impurities dissolve into the water and the water is then collected. This produces acidified water that can then be chemically neutralized. The sulfur oxides (along with oxides of **nitrogen** and other gases) are removed from the atmosphere so they can no longer be instrumental in producing acid rain.

Another waste product from industry is **heat**. Excess heat in an environment can alter the natural **balance** of organisms in the area. This can occur in terms of balance of species and balance of numbers. Heat is quite often put into the environment in the form of hot water. To remove the heat from the water, cooling pipes are placed next to the discharge pipes, or in some cases running inside them. The direction of flow in these cooling pipes is in the opposite direction to the flow of water in the discharge pipes. The cold water in the cooling pipes accepts some of the heat from the waste water by conduction through the walls of the pipe. This water can then be used to heat the factory rather than discharge the heat into the environment.

Automobiles produce a wide range of waste gases. These are usually treated in the exhaust before discharging into the atmosphere. Nitrogen oxides, **carbon** monoxide, and unburned hydrocarbons are the main exhaust pollutants and they are involved in the production of damaging photochemical smog. These are removed from the exhaust fumes by a device called a catalytic converter. The catalytic converter transforms **carbon monoxide** and unburned fuel into water and **carbon dioxide** and also reduces oxides of nitrogen to molecular nitrogen gas. This is done by passing the waste gases through a matrix of **platinum** and other noble **metals**. These work by adding or removing **oxygen** as appropriate. Between 76-96% of the waste products are removed from the exhaust gases by this process.

Many different waste materials are found in water that has passed through a city which must first be treated before it is discharged. Some of the material is organic in nature and is easily dealt with by microbial action. Bacteria are employed to feed on the organic waste and decompose it. **Solids** are removed in settling tanks where the solid material is allowed to settle to the bottom of a tank and is then carefully removed.

Some waste materials that are highly poisonous are considered toxic wastes. Toxic wastes can include such material as heavy metals or radioactive material. With some of this material, there is very little treatment that can be carried out. As a result, the material is placed inside storage containers and then placed in a storage facility. With radioactive waste, the material will eventually decay to non-radioactive materials, but this process may take thousands of years.

Some waste material is merely disposed of. Landfill sites have household waste deposited into them. In many cases, microbial degradation will break the compounds down so that the waste will eventually rot away. As this occurs, **methane** is often produced and released into the atmosphere. The methane gas may cause explosions, particularly where homes have been built on old landfill sites.

Some waste is treated so that it can be used again in a process called **recycling**. Newspapers are generally recycled **paper** (there is a limit to how often this can occur, as the fibres of the paper become shorter with each treatment). Many automobile manufacturers use recycled parts in the production of cars.

Virtually all processes that humans indulge in produce waste. Before the waste is discharged into the environment, it must be treated in some way to render it as harmless as possible. Many different waste products require many different treatments, some of which have been outlined here. The purpose of waste treatment is to make the waste less harmful to the environment in the broadest sense.

See also Sewage treatment; Water purification

WATER

Water is a **chemical compound** composed of a single **oxygen atom** bonded to two **hydrogen** atoms (H_2O) which are separated by an angle of 105°. Because of this asymmetrical arrangement, water molecules have a tendency to orient themselves in an electric field, with the positively charged hydrogen toward the negative pole and the negatively charged oxygen toward the positive pole. This tendency results in water having a large dielectric constant, which is responsible for making water an excellent solvent. Water is therefore referred to as the "universal solvent." Since mineral salts and organic materials can dissolve in water, it is the ideal medium for transporting life-sustaining minerals and nutrients into and through animal and plant bodies. Brackish and ocean waters may contain large quantities of **sodium** chloride as well as many other soluble compounds leached from the crust of the earth. For example, the **concentration** of mineral salts in ocean water is about 35,000 parts per million. Water is considered to be potable is it contains less than 500 parts per million of salts. Water can be reused indefinitely as a solvent because it undergoes almost no modification in the process.

Hydrogen **bonding**, which joins water **molecule** to water molecule, is responsible for other properties that make water a unique substance. These properties include its large **heat** capacity, which causes water to act as a moderator of **temperature** fluctuations; its high surface tension (due to cohesion among water molecules); and its adherence to other substances, such as the walls of a vessel (due to adhesion between water molecules and the molecules of a second substance).

The high surface tension makes it possible for surface-gliding insects and broad, flat objects to be supported on the surface of water. Adhesion of water molecules to soil particles is the primary mechanism by which water moves through unsaturated soils.

Hydrogen bonding is also responsible for ice being less dense than water. If ice did not float, all bodies of water would freeze from the bottom up, becoming solid masses of ice and destroying all life in them. In addition, from season to season, frozen water bodies would remain frozen, resulting in large changes in climate and weather, such as decreased **precipitation** due to reduced evaporation. Ice floats because as the temperature of water is lowered to -24.8°F (4°C), the tendency of water to contract as its molecular motion decreases is overcome by the strength of hydrogen bonding between molecules. At 4°C, water molecules start to structure themselves directionally along the lines of the hydrogen bonds, at angles of 105° As the temperature drops toward -32°F (0°C), spaces develop between the lines until the open, crystalline form characteristic of ice develops. Its openness produces a **density** slightly less than that of liquid water, and ice floats on the surface, with approximately nine-tenths submerged.

Water is the only common substance that occurs naturally on earth in three different physical states. The solid state, ice, is characterized by a rigid crystalline structure occurring at or below -32°F (0°C) and occupying a definite **volume** (found as glaciers and ice caps, as snow, hail, and frost, and as clouds formed of ice crystals). The liquid state exists over a definite temperature range -32 to 148°F (0°C to 100°C), but is not rigid nor does it have a particular shape. In other words, it has a definite volume but assumes the shape of its container. Liquid water covers three-fourths of the earth's surface in the form of swamps, lakes, rivers, and oceans as well as found as rain clouds, dew, and ground water. The gaseous state forms at temperatures above 148°F (100°C), and neither occupies a definite volume nor is rigid. In other words, it takes on the exact shape and volume of its container. It occurs naturally as fog, steam, and clouds. One phase does not suddenly replace its predecessor as the temperature changes, but for a time at the melting or **boiling point**, two phases will coexist. As water changes from the gaseous form to the liquid form, it gives off heat at about 540 calories per gram, and as it changes from the liquid form to the solid form, it gives off about 80 calories per gram. The turbulence of thunderstorms is in large part due to the release of large amounts of **energy** into the atmosphere as water condenses into water droplets or into crystals of ice (i.e., hail). Pressure affects the transition temperature between phases. For example, at pressures below atmospheric, water boils at temperatures under 100°C, so food will take longer to cook at higher elevations.

Water can pass directly from the solid phase to the gaseous phase without going through the liquid phase. This process occurs at low temperatures and greatly reduced pressures through a process called **sublimation**. Dehydrated foods are produced by sublimation, in which foods are quick-frozen and then placed in evacuation chambers. **Dehydration** by sublimation requires less energy than other methods, reduces physical deterioration that accompanies prolonged or excessive heating, and decreases the loss of volatile aromatic compounds responsible for flavor.

Liquid water is critical to sustain life. Without water to drink, a human will die in less than a week. Water in the form of perspiration is important in maintaining thermal stability in the body by dispersing large quantities of heat from the surface of the skin into the atmosphere. As the principal constituent of blood, water is the medium by which red blood cells transport oxygen (O_2) through the body, and by which **carbon** dioxide (CO_2) and other wastes are removed from the body. As a solvent, water in blood carries sugars and **proteins**, mineral salts, metabolites such as **urea**, **hormones**, and compounds that cause blood to clot.

The principal use of water in agriculture is for irrigation of crops, while lesser amounts are used for watering of livestock and cleaning of produce. Most industrial and manufacturing processes use large quantities of water to provide energy, remove unwanted heat, serve as a solvent, wash away impurities, function as a transport mechanism, and serve as a raw material. Water is used to keep individuals, homes, and communities clean, thus improving public health. Water is used to flush, wash, and dilute wastes from both households and industries. Because water is an excellent solvent and a convenient repository for wastes, it may contain many impurities that, if present in sufficient concentrations, constitute **pollution** and may require reclamation and treatment.

Water is a major geologic agent of change for modifying the earth's surface through erosion by water and ice. Water is also an important recreational medium, supporting fishing, swimming, and boating, and is a major factor in the tourism industry.

WATER PURIFICATION

Water purification is the process of changing undrinkable water to drinkable (potable) water.

Water is a very good solvent, it is this one fact which makes it difficult to obtain pure water without treatment. One solute commonly found in water is **sodium** chloride, or common table **salt**. Because of the high salt content of seawater, it is generally unfit for human consumption. In the United States, the legal limit for the salt content of municipal water supplies is set at 500 ppm. This is much lower than the 3.5% of salt in seawater or the 0.5% found underground in brackish water in some regions. The removal of salt from these waters is a particular type of water purification called **desalination**. Water and dissolved salts can be separated by **distillation** (water is volatile, the salts are not).

On a large scale, distillation is problematical and expensive. A more efficient desalination technique is reverse **osmosis** where, by applying pressure to the seawater, the water moves across a semi-permeable membrane, leaving the dissolved salt behind. The largest desalination plant in the world is located in Saudi Arabia, and it uses reverse osmosis to produce half of its country's drinking water. In 1992, a similar plant was opened in California that can produce 8 million gallons of drinking water a day.

Water purification occurs naturally when water evaporates from the oceans, leaving the salts and impurities behind. The water vapor then travels through the atmosphere and it eventually returns to the ground in the form of **precipitation**. When the water returns to the ground, it eventually finds its way into lakes and rivers which may be used to obtain fresh drinking water. In such a process, the water dissolves a number of particles, including sodium, **potassium**, **magnesium**, **calcium**, **iron**, chloride, **sulfate**, carbonate, **oxygen**, **nitrogen**, **carbon** dioxide, sewage and other human waste. Before this water can be used, it has to be purified, and in the United States this is usually a five-stage process. The first stage is a coarse **filtration** to remove large foreign bodies. This is carried out by allowing the water to flow through a mesh screen. The next stage is one of sedimentation. The water is allowed to stand to allow small particles to settle out. This process is aided by calcium **oxide** to the water followed by **aluminum** sulfate. A gelatinous **mass** of aluminum hydroxide is formed by reaction with some of the water. This gelatinous precipitate settles slowly and pulls down particles with it. The water is then filtered through a fine bed of sand and aerated to oxidize dissolved organic compounds. The final stage is sterlization of the water. This is commonly carried out by bubbling **chlorine** gas through the water, which produces a weak acid responsible for destroying all remaining bacteria. **Ozone** can also be used in this process in place of chlorine. After this the water is considered drinkable.

Spring water and other bottled waters are purified by porous rocks through which filters out all particles; the water is collected immediately at the source before other particles can dissolve into it. Other natural water purification systems include reed beds, which can concentrate many of the undesirable elements of water into the stems of the plants given sufficient time. Other water purification techniques include the usage of tablets that release chlorine into water, destroying harmful living organisms. Hand-operated reverse osmosis devices are also available and are of particular use to sailors.

See also Waste treatment

WATSON, JAMES D. (1928-)

American molecular biologist

James D. Watson won the 1962 Nobel Prize in physiology and medicine along with **Francis Crick** and Maurice Wilkins for discovering the structure of **DNA**, or deoxyribonucleic acid, which is the carrier of genetic information at the molecular level. Watson and Crick had worked as a team since meeting in the early 1950s, and their research ranks as a fundamental advance in molecular biology. More than thirty years later, Watson became the director of the **Human Genome Project**, an enterprise devoted to a difficult goal: the description of every human gene, the total of which may number up to one hundred thousand. This is a project that would not be possible without Watson's groundbreaking work on DNA.

James Dewey Watson was born in Chicago, Illinois, on April 6, 1928, to James Dewey and Jean (Mitchell) Watson.

James D. Watson.

He was educated in the Chicago public schools, and during his adolescence became one of the original Quiz Kids on the radio show of the same name. Shortly after this experience in 1943, Watson entered the University of Chicago at the age of fifteen.

Watson graduated in 1946, but stayed on at Chicago for a bachelor's degree in zoology, which he attained in 1947. During his undergraduate years Watson studied neither genetics nor biochemistry—his primary interest was in the field of ornithology; in 1946 he spent a summer working on advanced ornithology at the University of Michigan's summer research station at Douglas Lake. During his undergraduate career at Chicago, Watson had been instructed by the well-known population geneticist Sewall Wright, but he did not become interested in the field of genetics until he read **Erwin Schrödinger**'s influential book *What is Life?;* it was then, Horace Judson reports in *The Eighth Day of Creation: Makers of the Revolution in Biology,* that Watson became interested in "finding out the secret of the gene."

Watson enrolled at Indiana University to perform graduate work in 1947. Indiana had several remarkable geneticists who could have been important to Watson's intellectual development, but he was drawn to the university by the presence of the Nobel laureate Hermann Joseph Muller, who had demonstrated twenty years earlier that X rays cause **mutation**. Nonetheless, Watson chose to work under the direction of the Italian biologist Salvador Edward Luria, and it was under Luria that he began his doctoral research in 1948.

Watson's thesis was on the effect of X rays on the rate of phage lysis (a phage, or bacteriophage, is a bacterial virus). The biologist Max Delbrück and Luria—as well as a number of others who formed what was to be known as "the phage group"—had demonstrated that phages could exist in a number of mutant forms. A year earlier Luria and Delbruck had published one of the landmark papers in phage genetics, in which they established that one of the characteristics of phages is that they can exist in different genetic states so that the lysis (or bursting) of bacterial host cells can take place at different rates. Watson's Ph.D. degree was received in 1950, shortly after his twenty-second birthday.

Watson was next awarded a National Research Council fellowship grant to investigate the **molecular structure** of **proteins** in Copenhagen, Denmark. While Watson was studying **enzyme** structure in Europe, where techniques crucial to the study of macromolecules were being developed, he was also attending conferences and meeting colleagues.

From 1951 to 1953, Watson held a research fellowship under the support of the National Foundation for Infantile Paralysis at the Cavendish Laboratory in Cambridge, England. Those two years are described in detail in Watson's 1965 book, *The Double Helix: A Personal Account of the Discovery of the Structure of DNA.* (An autobiographical work, *The Double Helix* describes the events—both personal and professional—that led to the discovery of DNA.) Watson was to work at Cavendish under the direction of **Max Perutz**, who was engaged in the X-ray **crystallography** of proteins. However, he soon found himself engaged in discussions with Crick on the structure of DNA. Crick was twelve years older than Watson and, at the time, a graduate student studying protein structure.

Intermittently over the next two years, Watson and Crick theorized about DNA and worked on their model of DNA structure, eventually arriving at the correct structure by recognizing the importance of **X-ray diffraction** photographs produced by Rosalind Franklin at King's College, London. Both were certain that the answer lay in model-building, and Watson was particularly impressed by Nobel laureate **Linus Pauling**'s use of model-building in determining the alpha-helix structure of protein. Using data published by Austrian-born American biochemist **Erwin Chargaff** on the symmetry between the four constituent nucleotides (or "bases") of DNA molecules, they concluded that the building blocks had to be arranged in pairs. After a great deal of experimentation with their models, they found that the double helix structure corresponded to the empirical data produced by Wilkins, Franklin, and their colleagues. Watson and Crick published their theoretical **paper** in the journal *Nature* in 1953 (with Watson's name appearing first due to a coin toss), and their conclusions were supported by the experimental evidence simultaneously published by Wilkins, Franklin, and Raymond Goss. Wilkins shared the Nobel Prize with Watson and Crick in 1962.

After the completion of his research fellowship at Cambridge, Watson spent the summer of 1953 at Cold Spring Harbor, New York, where Delbruck had gathered an active group of investigators working in the new area of molecular biology. Watson then became a research fellow in biology at the California Institute of Technology, working with Delbruck and his colleagues on problems in phage genetics. In 1955, he joined the biology department at Harvard and remained on the faculty until 1976. While at Harvard, Watson wrote *The Molecular Biology of the Gene* (1965), the first widely used university textbook on molecular biology. This text has gone through seven editions, and now exists in two large volumes as a comprehensive treatise of the field. In 1968, Watson became director of Cold Spring Harbor, carrying out his duties there while maintaining his position at Harvard. He gave up his faculty appointment at the university in 1976, however, and assumed full-time leadership of Cold Spring Harbor. With John Tooze and David Kurtz, Watson wrote *The Molecular Biology of the Cell,* originally published in 1983 and now in its third edition.

In 1989, Watson was appointed the director of the Human Genome Project of the National Institutes of Health, but after less than two years he resigned in protest over policy differences in the operation of this massive project. He continues to speak out on various issues concerning scientific research and is a strong presence concerning federal policies in supporting research. In addition to sharing the Nobel Prize, Watson has received numerous honorary degrees from institutions, including one from the University of Chicago, which was awarded in 1961, when Watson was still in his early thirties. He was also awarded the Presidential Medal of Freedom in 1977 by President Jimmy Carter. In 1968, Watson married Elizabeth Lewis. They have two children, Rufus Robert and Duncan James.

Watson, as his book *The Double Helix* confirms, has never avoided controversy. His candor about his colleagues and his combativeness in public forums have been noted by critics. On the other hand, his scientific brilliance is attested to by Crick, Delbruck, Luria, and others. The importance of his role in the DNA discovery has been well supported by Gunther Stent—a member of the Delbruck phage group—in an essay that discounts many of Watson's critics through well-reasoned arguments.

Most of Watson's professional life has been spent as a professor, research administrator, and public policy spokesman for research. More than any other location in Watson's professional life, Cold Spring Harbor (where he is still director) has been the most congenial in developing his abilities as a scientific catalyst for others. His work there has primarily been to facilitate and encourage the research of other scientists.

WAXES

Waxes are thermoplastic materials (i.e., they soften when heated, and return to their original condition when cooled), but because they are not high polymers (i.e., macromolecules consisting of a large number of monomers), they are not considered to be **plastics**. Most waxes are **water** repellent, have smooth textures, low toxicities, and are free of objectionable odors.

Many waxes consist of simple esters, and therefore can be thought of as containing an acid part and an **alcohol** part.

The acid portions of the molecules are made up of a mixture of **fatty acids**, and the alcohol portions are simple long-chain alcohols. These waxes tend to be solid because they have relatively high molecular weights.

Ester waxes usually have higher melting points than fats (60 to 100°C), and they tend to be harder. In animals and plants, they provide protective coatings. The leaves of plants are coated with wax, which enables them to resist attack by micro-organisms and also to conserve water. The feathers of ducks are coated with wax, that prevents the feathers from absorbing water and the duck from becoming too heavy to swim or to fly.

Chemically, ester waxes are **lipids**, i.e., they are biological organic molecules that can be extracted from living organism using nonpolar **solvents** such as **ether**, chloroform, **carbon** tetrachloride, and **benzene**. They are also subject to **hydrolysis** in a base.

Important ester waxes include carnauba wax, lanolin, beeswax, and spermaceti. *Carnauba* wax is obtained from the leaves of the Brazilian palm tree. It is used in automobile and floor waxes, and in deodorant sticks. The wax contains large polyesters, which contribute to its hardness and durability. *Lanolin* is obtained from the solvent treatment of wool. It consists of **cholesterol** esters of higher fatty acids, and finds use in ointments, soaps, face creams, and suntan preparations. *Beeswax* is taken from honeycomb. It is a mixture of esters of alcohols and acids, and some high **molecular weight** hydrocarbons. It is used in shoe polish, candles, wax **paper**, and in the manufacture of artificial flowers. *Spermaceti* is a soft wax obtained from the head of the sperm whale that consists mainly of cetyl palmitate. Because this oil is extremely soft, it is used in the manufacture of **cosmetics**, and ointment medications. Like paraffin wax, it is also used in candles.

Paraffin waxes are mixtures of high molecular weight alkanes, and therefore are not esters. Whereas ester waxes are produced directly by living organisms, paraffin waxes consist of mixtures of high molecular weight hydrocarbons separated during the fractionation of petroleum.

The chief difference between a long chain alkane polymer such as **polyethylene** and a paraffin wax is their melting points, with the melting point of the polymer much higher than that of the wax. Whereas the paraffin wax forms a brittle solid, polyethylene is tough. For example, polyetylene is used to coat the surface of paper cups to prevent the paper from becoming wet. This toughness results from polyethylene's chain-folded morphology that promotes the formation of strong covalent bonds between various regions in the solid. The molecules in paraffin wax, on the other hand, are only held together by weak **van der Waals forces**. Portions of the polyethylene solid are rubbery, and these regions impart flexibility to the polymer; paraffin waxes, however, are 100% crystalline.

See also Paints and coatings; Petrochemicals

Chaim Weizmann.

WEIZMANN, CHAIM (1874-1952)

Russian-born British- Israeli organic chemist, biochemist, and biotechnologist

Chaim Weizmann was born on November 17, 1874, in Motol, Russia. In 1885, he migrated to Pinsk to attend high school, where he spent his spare time in Zionist activities. He later became president of the World Zionist Organization in 1921, president of the Hebrew University in 1932, and served as first president of the new State of Israel from its establishment in 1948 to his death in 1952.

After obtaining his higher education in Germany (Darmstadt Polytechnic Institute, 1893-94; Charlottenburg Polytechnic Institute, 1893-97) and Switzerland (University of Fribourg, 1897-99; Ph.D. 1899), he taught as a *Privat-Dozent* (unsalaried lecturer) at the University of Geneva and carried out basic and applied research at Manchester University, supplemented by industrial research. Weizmann became a British citizen in 1910.

A quest for synthetic rubber led to Weizmann's classic work on **fermentation** as a source of **acetone** in 1915, which was urgently needed by the British government for the manufacture of **cordite** (smokeless powder) during World War I. At the behest of the First Lord of the Admiralty, Winston Churchill, Weizmann's discovery of the acid- resistant microorganism *Clostridium acetobutylicum* used in the Weizmann process was utilized on an enormous scale in England, Canada, and the United States. This rapid wartime expansion from research laboratories to facilities which focused on industry was not only unique among the use of microbiological processes. It was also the forerunner of the production of **penicillin** during World War II and biotechnological processes of the present.

The first diplomatic act of international recognition of Zionism, the Balfour Declaration (1917) came about largely as a result of Weizmann's scientific and political efforts. Weizmann pursued his scientific research, along with his political activities, until the end of his life. In his later years he worked at the Weizmann Institute of Science in Rehovot, Israel, where he died on November 9, 1952.

WERNER, ALFRED (1866-1919)

Swiss chemist

Alfred Werner was a chemist and educator whose accomplishments included the development of the coordination theory of **chemistry**. This theory, in which Werner proposed revolutionary ideas about how atoms and molecules are linked together, was formulated in a span of only three years, from 1890 to 1893. The remainder of his career was spent gathering the experimental support required to validate his new ideas. For his work on the linkage of atoms and his coordination theory, Werner became the first Swiss chemist to win the Nobel Prize.

Werner was born December 12, 1866, in Mulhouse, a small community in the French province of Alsace. He was the last of four children born to Jean-Adam Werner, a factory foreman and farmer, and Salome Jeanette Tesche, daughter of a wealthy German family. Alsace was French when Werner was born but was annexed into Germany during the Franco-Prussian war. Although the Werner family maintained strong patriotic ties with France and continued to speak French in their home, young Werner began his education in German schools.

At age six he was enrolled at the Ecole Libre des Freres, partly because of his mother's recent conversion to Catholicism. In 1878 he entered the Ecole Professionelle, a technical school, and began studying chemistry. The family had moved from the city to take up residence on a nearby farm, where Werner's father was engaged in dairying. The farm provided an ideal place for young Werner to begin his experiments. During this time, an unpleasant explosion in his home lab almost ended his career in chemistry and forced him to move his vials and chemicals into the barn. Werner's earliest known work was a paper on **urea** that he submitted in 1885 to the director of the Mulhouse Chemie Schule. He was 18. Although the paper was scientifically unsound and showed youthful inexperience, it did reveal a talent for classification and systematization that would prove invaluable in later years.

In late 1885 Werner began serving a one year term of compulsory military duty. Stationed in the town of Karlsruhe, Werner enrolled in two **organic chemistry** courses taught at the Technical University there. After his tour of duty, he relocated to Zurich, Switzerland, to continue his education in chemistry at the Federal Institute of Technology. Werner excelled in chemistry but performed poorly in mathematical courses, especially descriptive geometry. After six semesters of work and completion of a paper describing the successful preparation of five compounds, he received a diploma in technical chemistry. A year later, in 1890, he was awarded a Ph.D.

Alfred Werner.

Werner's doctoral thesis in 1890 was his first publication and his most important work in organic chemistry. Along with his graduate advisor, Werner showed that the shape of **nitrogen** compounds are similar to **carbon** compounds. His second paper, ''Contribution to the Theory of Affinity and **Valence**,'' concerned the forces of attraction that hold carbon atoms together. Werner attacked the traditional theory that pictured atoms of carbon held together in rigid formations. He suggested that attractive forces emanate in all directions from the center of a central **atom**. Using this novel idea, Werner was able to derive kekulé formulas—notations for chemical structures in which valence bonds are illustrated with short lines—for organic carbon compounds.

His most important paper, ''Contribution to the Construction of Inorganic Compounds,'' was written in 1893.

Werner awoke at 2 a.m. one morning with the **solution** to the riddle of **molecular structure**. He began writing furiously and by 5 p.m. his monumental paper on coordination theory was finished. In his paper Werner proposed that single atoms or molecules could be grouped around a central atom according to simple geometrical principles. These coordination bonds were immensely successful in explaining the properties of observed compounds and in predicting the existence of unknown compounds.

During this time, Werner had been developing other dimensions of his career as well. In 1891 he went to Paris as a post-doctoral student and worked with the French chemist Pierre Berthelot on thermochemical problems. Werner began his teaching career during the summer semester of 1892, as a lecturer in **atomic theory** at the Federal Institute of Technology. In the fall of 1893, as a result of his almost overnight success with the publication of his theory, he was appointed professor of organic chemistry at the University of Zurich. In his first course, the chemistry of aromatic compounds, Werner proved to be a demanding professor whose exuberance and contagious enthusiasm for atoms and molecules inspired and enthralled students. Although Werner's theoretical and experimental work was primarily in the field of **inorganic chemistry**, it was not until 1902 that he was allowed to teach inorganic chemistry.

After writing his ground-breaking papers, Werner had set about immediately to prove his theory. In a span of some 25 years he painstakingly prepared over 8000 compounds and published his findings in more than 150 publications. In 1907 he succeeded in preparing a beautiful ammonia-violeo **salt**, a compound predicted by his theory. With this preparation his opponents finally conceded defeat. Werner's greatest experimental success came in 1911 with the successful resolution of optically active coordination compounds —substances able to deflect polarized light. A few years later he resolved carbon-free coordination compounds and ended forever carbon's dominance in **stereochemistry**. For his theoretical and experimental work on coordination theory, Werner was awarded the Nobel Prize in chemistry in 1913.

Werner married Emma Wilhelmine Giesker in 1894, the same year he became a Swiss citizen. They had two children, a boy and a girl. Werner was a robust man with a jovial sense of humor. He was a connoisseur of good foods and wine and enjoyed billiards and chess with friends and family. Werner's hobbies included photography, stamp collecting, mountain climbing, and ice skating.

Werner published prolifically in both organic and inorganic chemistry. He wrote two textbooks on inorganic and stereochemical topics. In addition to the Nobel Prize, he was the recipient of many awards and honorary degrees, including the prestigious Leblanc Medal of the French Chemical Society. Werner died on November 15, 1919, at a Zurich psychiatric institution, from arteriosclerosis of the brain. At his funeral he was remembered for his numerous contributions to science and teaching.

Harold Dadford West.

West, Harold Dadford (1904-1974)

African American biochemist

For forty-seven years Harold Dadford West was involved in biochemical research and education at Meharry Medical College. For thirteen of those years, he was president of the institution. He was selected to be the first honorary member of the National Medical Association, and the Science Center at Meharry was named for him.

Born in Flemington, New Jersey, on July 16, 1904, West was the son of George H. West and the former Mary Ann Toney. He attended the University of Illinois, where he received a bachelor of arts degree in 1925. He was an associate professor and head of the science department at Morris Brown College in Atlanta from 1925 to 1927. On December 27, 1927, West married Jessie Juanita Penn. They eventually had one daughter and one son.

In 1927 West joined the faculty of Meharry Medical College in Nashville, Tennessee, as an associate professor of physiological **chemistry**. Meharry Medical College had become an independent institution in 1915. Prior to that it was part of Central Tennessee College, established by the Freedmen's Aid Society of the Methodist Episcopal Church after the American Civil War in 1866. During his early years on the faculty of Meharry Medical College, West completed a master of arts degree and a doctorate. He was a recipient of a fellowship from the Julius Rosenwald Fund at the University of Illinois while he earned a master of arts degree in 1930. Following that he was a Rockefeller Foundation Fellow, receiving a doctorate degree from the same university in 1937. The title of his dissertation was "The Chemistry and Nutritive Value of Essential Amino Acids." In 1938 West became professor of **biochemistry** and chairperson of the department.

West's work in biochemical research was vast, including studies of tuberculosis and other bacilli, the antibiotic biocerin, and **aromatic hydrocarbons**. He worked with **amino acids**, becoming the first to synthesize threonine. As noted in the *Journal of the National Medical Association,* among his other investigations were "the role of **sulfur** in biological detoxification mechanisms; blood serum **calcium** levels in the Negro in relation to possible significance in tuberculosis; relation of B-vitamins, especially pantothenic acid, to detoxification of sulfa-drugs and susceptibility to bacillary disease."

West's studies were supported by the John and Mary R. Markle Foundation, the Nutrition Foundation, the National Institutes of Health, and the American Medical Association. His research papers were published in a number of professional journals, including the *American Journal of Physiology, Southern Medical Journal,* and *Journal of Biological Chemistry.*

In 1952 West was named the fifth president of Meharry Medical College, its first African American president. In 1963 he was the first black American to serve on the State Board of Education. West retired as president in 1965, returning to the position of professor of biochemistry. When he retired from Meharry in 1973 he became a trustee of the college. In his final years he worked on a complete history of the college. West died on March 5, 1974.

During his career, West was awarded two honorary degrees. In 1955 he received a doctor of laws from Morris Brown College, and in 1970 a doctor of science from Meharry Medical College. He was a member of many honorary and professional societies, including the American Chemical Society, the Society of Experimental Biology and Medicine, and the American Society of Biological Chemists. He was also elected to Sigma Xi, the scientific research society, which describes itself "as an honor society for scientists and engineers.... Its goals are to foster interaction among science, technology and society."

WHINFIELD, JOHN R. (1901-1966)

English textile chemist

John Rex Whinfield invented terylene, a synthetic **polyester** fiber that is equal to or surpasses **nylon** in toughness and resilience, and has become used universally as a textile fiber. The invention of terylene, also known as Dacron, was the culmination of many years of study and reasoning about the **molecular structure** and physical and chemical properties of polymers. Whinfield's major inventive work on terylene was carried out aside from his primary research in the small laboratory of a company that had little or no interest in research on fibers. He spent his life working as an industrial research chemist and eventually became director of the fibers division of Imperial Chemical Industries. Recognition for his work came in the later years of his life.

Whinfield was born February 16, 1901, in Sutton, Surrey, England, to John Henry Richard Whinfield, a mechanical engineer, and Edith Matthews Whinfield. As a boy, Whinfield showed an early interest in science and **chemistry**. He was educated at Merchants Taylors' School and Caius College of Cambridge, reading in natural sciences (1921) and chemistry (1922). In 1922 he married Mayo Walker, the daughter of the Rev. Frederick William Walker. She died in 1946, and in 1947 he married Nora Hodder of Worthing.

Whinfield was interested in the molecular makeup and properties of synthetic fibers, and to gain experience in fibers after graduating, he worked for a year without pay in the London laboratory of C. F. Cross and E. J. Bevan, who in 1892 had invented the "viscose reaction" for the production of rayon. In 1924, Whinfield was employed by the Calico Printers' Association as a research chemist, where he worked primarily on the chemistry of fabric dyeing and finishing. He continued his studies of the physical and chemical properties of synthetic fibers, however, and followed with interest the work of **Wallace Hume Carothers** in the United States, who in 1928 published the first of a long series of papers on condensation polymerization reactions. Carothers' work led to the invention of nylon, a polyamide; he had worked on but rejected the polyester group as a source of synthetic textile fibers because he thought the melting points were too low.

Whinfield's studies and rough analogies led him to believe that a polyester might work, specifically a polyester made from terephthalic acid and ethylene glycol. The latter chemical was available commercially, but terephthalic acid had been produced only in small quantities. Whinfield pressed his company to try some fiber work; in 1940 he was finally able to devote some time to the fiber research he had been thinking about, and in March 1941 he and his assistant, James T. Dickson, discovered a method of condensing terephthalic acid and ethylene glycol to produce a compound that could be drawn into fibers. Empirical work demonstrated—happily, because this had not been predictable from theoretical word—that the fibers had a high melting point and were resistant to hydrolytic breakdown. Whinfield and Dickson filed their patent on terylene in July 1941. Britain was engaged in World War II at the time, and terylene's potential utility for the war industry was examined briefly by the Ministry of Supply, for whom Whinfield had come to work during the war. It was known to be an important invention, but production was not thought to be practicable for the war effort, and registration of the patent was delayed until 1946, after the war.

The Calico Printers' Association decided not to develop terylene, and consequently sold their rights to the product to Imperial Chemical Industries (ICI), who obtained world manufacturing rights. Whinfield went to work for ICI in 1947. Du Pont in the United States independently prepared terylene and purchased the U.S. patent application filed by the British in 1946. Although Du Pont had been working on terylene, there was no question of the priority of its invention. Du Pont first called it "Fiber V," then "Dacron," and began full-scale production in the United States in 1953. ICI, after operating two pilot plants for several years, began commercial production of terylene fibers in 1955.

In the production of terylene, dimethyl terephthalate and ethylene glycol, derived from **coal**, air, **water**, and petroleum, are polymerized. Then the substance is "melt spun" into filaments. The filaments are stable, but springy; Whinfield found that the fibers would stretch to 10-25% of their original length before rupturing. Terylene was shown to be equal to nylon in its potential usefulness, and it contributed greatly to the popularity of "wash and wear" clothing.

At ICI, Whinfield worked first in the Fibers Development Department of the **plastics** division with W. F. Osborne, then in the Fibers Division, where he eventually became director. In 1954, he received a Commander of the Order of the British Empire (C.B.E.) for his work on terylene. The same year, he was engaged to advise on *Point of Departure*, an educational film on manmade fibers made by the Film Producers Guild. He was a clear explicator, but somewhat unexpectedly, he also proved to be an accomplished actor, and as a result played a leading role in the film. In 1955, he was elected an honorary fellow of the Textile Institute, and in 1956 he received the Perkin medal of the Society of Dyers and Colourists. During his tenure at ICI, Whinfield traveled widely, including to the former Union of Soviet Socialist Republics as a guest of the Russian government. He retired in 1963. In 1965, the University of York named its chemical library and a number of traveling fellowships after Whinfield.

Whinfield died on July 6, 1966, at Dorking, at age 65. In an obituary published in *Chemistry in Britain* (1967), P. C. Allen wrote, "[He] remained until the end of his life an essentially modest and simple man. He had a host of friends and no wonder for no one could be more charming companion, or when he was in the mood, a better talker. He wrote very clearly also; his publications are a model of clarity."

WIELAND, HEINRICH (1877-1957)

German chemist

Heinrich Wieland was one of the greatest organic chemists of the century, admired for the breadth of his knowledge and his devotion to arduous, painstaking research. Wieland is known for his studies on the structures of important complex **natural products**, from toad poisons to butterfly pigments. He also made major contributions to **biochemistry**, especially in the study of the mechanism of biological **oxidation**. His most famous work, for which he was awarded the Nobel Prize in **chemistry** in 1927, was the determination of the **molecular structure** of the bile acids. This research combined superb experimental skill with precise deductive reasoning and remains a model of organic chemical investigation.

Heinrich Otto Wieland was born on June 4, 1877, in Pforzheim, Germany, to Theodor and Elise Blom Wieland. Theodor Wieland was a pharmaceutical chemist, and Heinrich studied the subject in school in Pforzheim. At that time, instead of studying at a single university to obtain a degree, a student enrolled at several universities, listening to the lectures of the best professors. Wieland spent 1896 at the University of Munich, 1897 at the University of Berlin, and 1898 at the Technische Hochschule at Stuttgart. In 1899 he returned to Munich to work toward his Ph.D. under the direction of Johannes Thiele, in the laboratory of Adolf von Baeyer. After he received his Ph.D. in 1901, Wieland remained at Munich to do research, eventually becoming a lecturer in 1904 and a senior lecturer in 1913. In 1917 he was appointed professor at the Technische Hochschule in Munich, but was granted leave to work for **Fritz Haber**'s chemical warfare research organization at the Kaiser Wilhelm Institute in Berlin. At the end of World War I he returned to Munich, but left in 1921 to accept a professorship at the University of Freiburg. In 1925, Wieland returned to the University of Munich as professor and director of the Baeyer Laboratory, succeeding **Richard Willstätter**, who personally recommended Wieland for the position. By this time, Wieland was recognized as a world leader in **organic chemistry**, and he remained at Munich until his retirement in 1950.

Heinrich Wieland.

Wieland's early research was concerned with the chemistry of organic **nitrogen** compounds. He explored the addition of dinitrogen trioxide and nitrogen dioxide to carbon-carbon double bonds. A large series of papers described the reactions of aromatic amines (a type of organic compound derived from ammonia), especially their oxidations. One line of experiments led to the discovery of nitrogen **free radicals**, unusually reactive short-lived species in which nitrogen is bonded to two atoms, instead of the usual three atoms. Wieland published almost one hundred papers on organic nitrogen chemistry, which in itself was a notable achievement.

Another series of experiments led to Wieland's 1912 theory of biological oxidation, a process by which biologic substances are changed by combining with **oxygen** or losing electrons. For years, the accepted theory involved some kind of change to molecular oxygen inside the cell in which the oxygen becomes "activated" and reacts with the oxidizable substance. Wieland proposed that the oxidizable substance itself becomes "activated" and loses **hydrogen** atoms in the oxidation process. Wieland published more than fifty papers from 1912 to 1943 on biological oxidation and was able to demonstrate that many reactions proceed through **dehydrogenation** and could proceed in the absence of oxygen. He was challenged, however, by the German physiologist **Otto Warburg**, who showed that respiratory enzymes which contain **iron** (sometimes copper) do activate oxygen, and both types of oxidation mechanism are found in nature. Warburg received the Nobel Prize in 1931 for his contribution to understanding oxidation, but Wieland's work has been recognized as equally significant by biochemists.

In 1912, the year Wieland proposed his theory of biological oxidation, he published his first paper on the structure of the bile acids. This topic would occupy his interest for twenty years and earn him the Nobel Prize. Bile is a golden yellow liquid which is produced in the liver, stored in the gall bladder, and secreted in small amounts into the intestines. The **sodium** salts of bile acids, the principal constituent of bile, are essential to the digestion of fats. Although bile acids had been isolated early in the nineteenth century, their structural formulas were unknown when Wieland began his work. As the work progressed, it was shown by **Adolf Windaus**, a chemist at the University of Göttingen, that **cholesterol** and the bile acids share a common basic structure, allowing Windaus's research results on the structure of cholesterol (for which he won the Nobel Prize in 1928) to be used by Wieland, and vice versa. Later it was shown that the common basic structure, the steroid **nucleus**, is found in many naturally occurring sources, such as the sex **hormones**, adrenal hormones (cortisone), digitalis (a plant cardiac poison, used medicinally as a stimulant), and toad poison. Steroid chemistry became essential to the development of many powerful medicines, as well as oral contraceptives. The pioneering work of Wieland and his students on the bile acids became a foundation of modern pharmaceutical chemical research.

The work on bile acids was an enormous challenge for organic chemistry in the first quarter of the century. First, a procedure for isolation and purification of the various acids, obtained from ox bile, was required. Then, each acid had to be characterized and chemically related to the others. The acids each contain 24 **carbon** atoms and differ in the number of hydroxyl (alcohol) groups. Wieland used the method of selective degradation to break the acids into simpler compounds, thus allowing him to identify the smaller molecules. Although his work was somewhat simplified because he could use the results of Windaus, Wieland admitted in his Nobel Lecture that "the task would appear to be a long and unspeakably wearisome trek through an arid desert of structure." In this lecture he outlined the course of his research, showing the failures as well as successes. Although the structures of the bile acids and cholesterol appeared to be solved when Wieland and Windaus received their Nobel prizes, in fact a conclusion which they had made based on analogous reactions was not correct, and the final, unequivocal structures were proposed by Wieland and others in 1932.

In addition to the bile acids, Wieland also investigated other natural products. He contributed to the determination of the structures of **morphine**, lobeline, and **strychnine** alkaloids, as well as butterfly wing pigments and mushroom and toad poisons. He had a wide range of interests, encompassing all areas of organic chemistry, and for twenty years he was editor of the major chemical journal *Justus Liebigs Annalen der Chemie.* His work was recognized throughout his career, and he was honored by scientific societies and universities in many countries. In 1955 he was named the first recipient of the German Chemical Society's **Otto Hahn** Prize for Physics and Chemistry.

Wieland remained at the University of Munich during World War II. He had little regard for the Nazi government in Germany and made no secret of it. He protected Jews in his laboratory and in 1944 testified on behalf of students who had been accused of treason.

Wieland married Josephine Bartmann in 1908. All three of their sons became scientists: Wolfgang, a pharmaceutical chemist; Theodor, a professor of chemistry; and Otto, a professor of medicine. Their daughter, Eva, married **Feodor Lynen**, a professor of biochemistry who won the Nobel Prize in physiology or medicine in 1964. In addition to his love of family and his work, Wieland also enjoyed painting and music. He died in Starnberg, Germany, on August 5, 1957, two months after his eightieth birthday.

Wiley, Harvey Washington (1844-1930)

American chemist

The American chemist Harvey Washington Wiley established the methods and philosophy of food analysis. His writings and influence made him the "father of the Food and Drug Administration."

Harvey Wiley was born in Kent, Ind., on Oct. 18, 1844, the son of a farmer. His oldest sister Elizabeth Jane Wiley Corbett, became an early woman physician. A sturdy boy with a fine and receptive mind, Wiley advanced from a log-cabin schoolhouse to Hanover College in Indiana, where he majored in the humanities. He interrupted his studies to serve in the Union Army and then returned, graduating from Hanover in 1867.

Harvey Washington Wiley.

Wiley became an instructor in Latin and Greek (1868-1871) at Butler University while continuing his studies at Indiana Medical College, from which he received his medical degree in 1871; subsequent studies took him to Harvard and the University of Berlin. Meanwhile he became a professor of **chemistry** at Butler, then at Purdue University. Having served as Indiana's state chemist, he became chief chemist of the U.S. Department of Agriculture in 1883.

Wiley did a series of studies of food products and published several papers which established him among agricultural chemists. His achievements while at the Department of Agriculture were of both a technical and, uniquely, social character: he devised instruments and methods for processing glucose, grape sugar, and sorghum sugar and practically established the beet-sugar industry in the United States. Wiley also supervised the preparation of his landmark *Bulletin No. 13: Foods and Food Adulterants* (1887-1889), which covered all classes of food products and described methods of analysis. However, Wiley's dynamic personal qualities, expressed on the public platform and, informally, in such a private publication as *Songs of Agricultural Chemists* (1892) carried the subject beyond the arguments of technicians.

In 1902 Wiley established his famous "poison squad," a group of volunteers who became "human guinea pigs" to help determine the effect on digestion and health of preservatives, coloring **matter**, and other substances. His work was the base from which a variety of exposés and sensations, including patent medicines and processed beef, roused the nation, resulting in the passage of the Pure Food and Drug Act of 1906.

Subsequently Wiley found himself under fire by interests dissatisfied with his rigid application of standards. Controversy over administration of the act and its specific effect on industries continued through the presidencies of Theodore Roosevelt and William Howard Taft. Wiley, persuaded that the act had been betrayed, resigned his government post in 1912.

Wiley then became director of the bureau of foods for *Good Housekeeping,* published books on health and adulteration, and lectured widely and effectively. A man of excellent presence, magnetic and witty, he stirred general and professional audiences and was accorded national and international honors. In 1929 his retrospective *History of a Crime against the Food Law* provided inspiration for later crusaders. Active to the end, he died in Washington, D.C., on June 30, 1930.

WILKINSON, GEOFFREY (1921-)

English chemist

Geoffrey Wilkinson is best known for establishing the structure of a "sandwich **molecule**" he called ferrocene. In sandwich compounds a metal **atom** is the "filling" between two "slices" which are flat, typically carbon-based, rings. Since Wilkinson's original discovery of an **iron** filling between two cyclopentadienyl (five **carbon** atoms linked in a circle) slices, many different sandwiches have been built. Various metal fillings have been used, as have numerous other slices. Sandwich compounds have found widespread use as catalysts in industrial processes, and their previously unknown chemical **bonding** structure has proven to be of great theoretical interest to various branches of **chemistry**. Wilkinson's discovery revolutionized how chemists thought about chemical structure and opened up new avenues of chemical exploration.

Wilkinson was born on July 14, 1921, in Yorkshire, England, to Henry and Ruth Crowther Wilkinson. It was an uncle, the owner of a small chemical company in the town of Todmorden, who encouraged Wilkinson's interest in chemistry and had the most influence on his career choice. Wilkinson attended the Todmorden Secondary School and then the Imperial College of Science and Technology at the University of London. Supported by a scholarship, he obtained his B.S. degree in 1941 and his doctorate in 1946.

In 1942, while still a doctoral student, Wilkinson worked with the National Research Council on a joint atomic **energy** project. His work involved separating the various products of atomic fission reactions from one another so they could be studied and the fission process better understood. In the course of this work Wilkinson developed a new technique, ion-exchange **chromatography**, which has since proven useful in many chemical analyses. Using this technique, Wilkinson identified a number of new isotopes, atomic species which vary only in the number of neutrons within their nuclei. After the Second World War, Wilkinson continued his research on **nuclear chemistry** at the University of California at Berkeley. His work there focused on identifying neutron-deficient isotopes, products of atomic fission that are unstable because they have too few neutrons in their nuclei.

In 1950, Wilkinson moved to the Massachusetts Institute of Technology as a research associate. He had come to the end of what he considered to be his effectiveness in nuclear synthesis research, and began to study the chemical nature of the **transition elements**. These are the elements found in the center of the **periodic table**; they often have more than a single stable electrical charge state and unusual magnetic properties. One of Wilkinson's first major breakthroughs was to synthesize a compound in which the transition element **nickel** was chemically bound to **phosphorus** within a larger molecule.

Despite his decision to change the focus of his research, it was Wilkinson's expertise as a nuclear chemist that earned him a position as an assistant professor at Harvard University in 1951. He would only remain at Harvard for four years, but it was a very productive time in his life. It was here that he first deduced the sandwich-type **molecular structure**.

Early in 1952, while preparing to teach an **inorganic chemistry** course, Wilkinson read about a newly synthesized compound, bicyclopentadienyl iron (a chemical made from an iron atom bonded to two five-carbon rings). The structure proposed seemed unlikely to Wilkinson. Based on theories developed by **Linus Pauling**, he thought that the key to the structure must lie in the distribution of the so-called pi electrons in the cyclopentadienyl **ring**. Pauling's work indicated that this **ion** would have its pi electrons very evenly distributed in rings parallel to the plane of the carbon atoms. Wilkinson realized that a stable structure would result if the iron atom bonded through the pi electrons and was thus held equidistant from all of the carbon atoms in the cyclopentadienyl ring. In bicyclopentadienyl iron this would be possible only by having the iron atom ''sandwiched'' between the two flat cyclopentadienyl groups.

Along with a colleague, **Robert B. Woodward**, Wilkinson experimentally proved this novel structure in a few days. Woodward coined the term ferrocene, due to the similarity of the structure to the well-known compound **benzene**. Wilkinson rapidly adapted the synthesis methods to create a number of other sandwich compounds. For discovering this previously unknown class of chemical structure, Wilkinson shared the 1973 Nobel Prize in chemistry with **Ernst Otto Fischer**. Fischer had also worked on these compounds, and in presenting the award the committee congratulated both men for creating a new field of chemistry.

While at Harvard in the 1950s, Wilkinson also pioneered the use of the **nuclear magnetic resonance** (NMR) technique in chemical analysis. In NMR **spectroscopy**, chemists study the movement of atoms (most often hydrogen) within a magnetic field; each atom emits a distinct spectral line according to its bond. This technique helped to explain the concept of fluctionality, a theory which states that some chemical species may fluctuate back and forth from one bonding structure to another. This technique has since also found considerable use in medicine, where it is called magnetic resonance imaging or MRI. But despite the widespread recognition of the importance of his work, Harvard University did not offer Wilkinson tenure.

In 1955 Wilkinson returned to the Imperial College of Science and Technology to assume the chair of the inorganic chemistry department. Here he continued his work on transition elements and how they form complexes with organic species through pi electrons. In particular, Wilkinson concentrated on several ways in which transition metal complexes serve as catalysts. (A catalyst speeds up a chemical reaction and is then converted back into its original form, enabling it to serve for multiple cycles of the reaction.)

Wilkinson wrote widely, publishing more than four hundred articles on the transition **metals** and their complexes with organic compounds. He also coauthored *Advanced Inorganic Chemistry* (1962), which has remained a classic text in the field, and *Basic Inorganic Chemistry* (1976). Wilkinson received numerous honors in addition to the Nobel Prize. The French Chemical Society honored him in 1968 with its Lavoisier Medal, the Royal Society presented him with the Transition Metal Chemistry Award in 1972, and he won the Gallileo Medal from the University of Pisa, Italy, in 1973.

Wilkinson married Lise Solver Schau in 1951, soon after arriving at Harvard. They have two daughters.

WILLSTÄTTER, RICHARD (1872-1942)

German chemist

A gifted experimentalist, Richard Willstätter's pioneering work on **natural products**, especially chlorophylls and anthocyanins (plant pigments), was honored with the 1915 Nobel Prize in **chemistry**. In 1924 Willstätter, who was Jewish, resigned from his position at the University of Munich in protest against the anti-Semitism of some of the faculty. This act of conscience seriously hampered his research activity. In 1939 the anti-Semitic policies of the Third Reich forced him to emigrate to Switzerland, where he spent the remaining few years of his life.

Richard Martin Willstätter was born in Karlsruhe, Germany on August 13, 1872, the second of two sons of Max and Sophie Ulmann Willstätter. Willstätter's father was a textile merchant and his mother's family was in the textile business. Willstätter's education began in the classical Gymnasium in Karlsruhe. When he was eleven years old, his father moved to New York in search of better economic opportunities and to escape the circumscribed life in Karlsruhe; although this separation was meant to be short, it lasted seventeen years. Willstätter's mother took him and his brother to live near her family home in Nürnberg, a change to which Willstätter had difficulty adjusting, in part because of the more overt anti-Semitism he experienced there.

One effect of the move to a new school was that, although receiving good grades in his other subjects, he did poorly in Latin, the most important subject in the gymnasia of the time. A family council decided he should switch to the Realgymnasium and be educated for business instead of a profession. Ironically, it was at this time, stimulated by some home experiments and good teachers, that he decided to become a chemist. In his autobiography, Willstätter observed that excellence in academic subjects caused one to be disliked, while athletic excellence resulted in popularity. He was also attracted to medicine and might have become a physician instead of a chemist, but because of the longer schooling required his mother would not permit him to change. An interest in biological processes remained with him, though, and is evident in the kinds of chemical problems he attacked. Much later, while teaching at Zurich, he still thought of studying physiology and internal medicine, but the death of his wife ended the idea.

In 1890 the eighteen-year-old Willstätter entered the University of Munich and also attended lectures at the Technische Hochschule. In 1893 he began his doctoral studies and was assigned to do his research under Alfred Einhorn on some aspects of the chemistry of cocaine. It was at this time that **Adolf Baeyer**, the leading organic chemist in Germany, began to take Willstätter under his wing. Although Willstätter never worked directly for Baeyer, he thought of himself as Baeyer's disciple. Willstätter completed his doctoral work in a year and stayed on doing independent research, becoming a *privatdocent*, or unsalaried lecturer, in 1896.

In his work with Einhorn, Willstätter had come to suspect that the structure assigned to cocaine by Einhorn and others was incorrect. When he started his independent research, Einhorn forbade him to work on the cocaine problem. Willstätter, with Baeyer's approval, decided to work instead on the closely related chemical tropine, whose structure was suspected to be similar to that of cocaine; once the structure of tropine was known, the structure of cocaine could be easily derived. Willstätter showed that, indeed, the cocaine structure was not what it had been thought to be; for the remainder of his stay at Munich, Einhorn refused to speak to him. In 1902 Willstätter was appointed professor extraordinarius (roughly equivalent to associate professor), although Baeyer thought he should have accepted an industrial position. Baeyer, himself partly Jewish, also recommended that Willstätter be baptized, an act that would have removed the legal barriers he faced as a Jew. This Willstätter refused to consider. During Easter vacation in 1903 Willstätter met the Leser family from Heidelberg, and that summer he and Sophie Leser were married. Their son Ludwig was born in 1904 and their daughter, Margarete, in 1906.

In 1905 Willstätter accepted a call to the Eidgenössische Technische Hochscuhle in Zurich as professor of chemistry, beginning the most productive phase of his career. While at Munich he had begun an investigation into the chemical nature of **chlorophyll**, the green pigment in plants that converts light into **energy** through **photosynthesis**; at Zurich, he and his students made great strides in understanding this important material. They developed methods for isolating chlorophyll from plant materials without changing it or introducing impurities. Willstätter was then able to prove that the chlorophyll from different plants (he examined over two hundred different kinds) was substantially the same—a mixture of two slightly different compounds, blue-green chlorophyll a and yellow-green chlorophyll b, in a 3 to 1 ratio.

Richard Willstätter.

He also showed that **magnesium**, which had been found in chlorophyll by earlier workers, was not an accidental impurity but an essential component of these chlorophyll molecules, bonded in a way very similar to that in which **iron** is bonded in hemoglobin, the oxygen-carrying constituent of blood. The later work of others, especially **Hans Fischer**, in elucidating the detailed structures of the chlorophylls and hemoglobin would not have been possible without the pioneering work of Willstätter and his students. In 1913, Willstätter, in collaboration with his former student and good friend, Arthur Stoll, reviewed the work on chlorophyll in a book, *Untersuchungen über Chlorophyll.* In all, between 1913 and 1919, Willstätter published twenty-five papers in a series on chlorophyll. A preliminary step in the isolation of chlorophyll from plant materials yielded a yellow **solution** that on further study proved to contain carotenoid pigments. These had been described before, but Willstätter's work marked the beginning of our understanding of these materials that produce the **color** of tomatoes, carrots, and egg yolk.

In 1908, Willstätter suffered a devastating blow in the death of his wife after an operation for appendicitis had been delayed for thirty-six hours after the appendix had ruptured. He consoled himself with the care of his two children and with his work; in his autobiography he wrote that he took no vacations for the next ten years. During his stay at Zurich, Willstätter also did work on quinones and the mechanism of the **oxidation** of **aniline** to aniline black—a process of importance to the dye industry. He also completed a project begun eight years earlier, by synthesizing the chemical cyclooctatetraene and showing that it did not behave as an aromatic compound despite its structural similarities to **benzene**.

The Kaiser Wilhelm Institutes were founded in 1910 to afford outstanding scientists the chance to do research on problems of their own choosing, free of any teaching obligations. In 1911 Willstätter accepted the position of director of the Kaiser Wilhelm Institute of Chemistry and in 1912 moved into the new building at Berlin-Dahlem. The institute was situated next to the Institute for Physical Chemistry and **Electrochemistry**, headed by **Fritz Haber**, and a deep and lasting friendship developed between the two directors.

At Zurich, Willstätter had initiated a study of the pigments of various red and blue flowers, a class of compounds now known as anthocyanins. He began with dried cornflowers, or bachelor's button, because it was winter and they were commercially available. This choice, as it turned out, was not a good one; cornflowers only contained a percent or less of the pigment. In Berlin, Willstätter planted fields of double cornflowers, asters, chrysanthemums, pansies, and dahlias around the Institute and his residence. In these fresh flowers he found a much higher pigment content, up to 33 percent in blue-black pansies. Before World War I brought an end to this line of research, Willstätter published eighteen papers in an anthocyanin series between 1913 and 1916. He showed that the various shades of red and blue in these flowers as well as in cherries, cranberries, roses, plums, elderberries, and poppies all arose mainly from three closely related compounds, cyanidin, pelargonidin, and delphinidin chlorides, and were very dependent on the acidity or alkalinity of the flower. During the first year of the war, most of Willstätter's co-workers went into military service, and the flowers were taken to military hospitals instead of to the laboratory. Willstätter was bitterly disappointed by this interruption and could not bring himself to return to the problem after the war.

In 1915 Haber, who was in charge of Germany's chemical warfare work, asked Willstätter's assistance in developing the chemical absorption unit for a gas mask that would protect against **chlorine** and phosgene (a severe respiratory irritant). In five weeks, Willstätter came up with a canister containing activated **charcoal** and hexamethylenetetramine (also called urotropin). The use of charcoal was not new, but the use of hexamethylenetetramine was. When asked after the war how he had come to try so unusual a compound, he said that the idea had just popped into his head. For this work he received an Iron Cross, Second Class. He was also involved in an industrial research project with **Friedrich Bergius** on the **hydrolysis** of **cellulose** with **hydrochloric acid** to give dextrose, which could then be fermented to produce **alcohol**. The process, which was only perfected later, is now known as the Bergius-Willstätter process.

In the spring of 1915 Willstätter's ten-year-old son, Ludwig, died suddenly, apparently from diabetes. Willstätter wrote that his memory of the months following was blurred. Ironically, in November, while engaged in the work on gas masks, Willstätter learned that he had been awarded the 1915 Nobel Prize in chemistry in recognition of his work on chlorophylls and anthocyanins. Because of wartime conditions he did not travel to Stockholm to receive the prize until 1920, when a ceremony was held for a group of those who had been honored during the war. Willstätter made the trip in the company of fellow German awardees **Max Planck**, Fritz Haber, Max Laue, and Johannes Stark.

An offer of a full professorship to succeed Baeyer at Munich also came in 1915. This offer, recommended by Baeyer, was precipitated by an offer to succeed **Otto Wallach**, a pioneer in natural product chemistry, at Göttingen. Willstätter maintained that left to his own inclinations, he would have preferred Göttingen, because a medium-sized university would provide more contact with colleagues and greater interaction with different disciplines than was possible at large institutions. However, he accepted the appointment as professor and director of the state chemical laboratory in Munich and moved there in the spring of 1916.

He made two major demands before accepting the offer: that the old institute building be remodeled and a large addition to the chemical institute be built housing laboratories and a large lecture hall, and that a full professorship in **physical chemistry** be established. The first of these was contrary to the advice that the physical chemist **Walther Nernst** gave him before he left Berlin, "Don't ever build!" In fact, the construction, delayed by the war and post-armistice turmoil in Munich, was not completed until the spring of 1920.

At Munich, as before, Willstätter experienced the anti-Semitism that had troubled him during his earlier residence, and that finally brought about his resignation in 1924. The final straw was the refusal of the faculty to appoint the noted geochemist **Victor Goldschmidt** of Oslo, Norway, to succeed the mineralogist Paul von Groth, who had himself named Goldschmidt as the only one who could take his place. The sole reason for the refusal was that Goldschmidt was Jewish. When Willstätter's resignation became known, students and faculty joined in expressions of respect and confidence, urging him to reconsider. Nonetheless, he remained only for the time needed to see his students finish their research and to install **Heinrich Wieland** in his place. He received offers of positions at universities and in industry in Germany and abroad, but he declined all of them, finally leaving the university in September 1925 never to return.

Some of Willstätter's assistants continued work at the University, and in 1928 Wieland made room in what had been Willstätter's private laboratory for Willstätter's private assistant, Margarete Rohdewald, one of his former students. From 1929 until 1938 she collaborated with him in a series of eighteen papers on various aspects of **enzyme** research. It was an odd collaboration, conducted almost entirely over the telephone; Willstätter never saw her at work in the laboratory.

During the few years at Munich before his resignation, Willstätter began to concentrate his research on the study of enzymes. He had first encountered these biological catalysts in his early work on chlorophyll. Now he worked to develop methods for their separation and purification. His method for separation was to adsorb the materials on alumina or silica **gel** and then to wash them off using solutions of varying acidity, among other **solvents**. In this connection, Willstätter carried out a systematic study (comprised of nine papers) of **hydrates** and hydrogels during which he, with his assistants Heinrich Kraut and K. Lobinger, was able to show that **aluminum** hydroxide, silicic acid, ferric hydroxide, and stannic hydroxide do actually exist in solution and are not colloidal sols (dispersions of small solid particles in solution) of the corresponding oxides. Willstätter reported that this foray of an organic chemist into **inorganic chemistry** was not well received by inorganic chemists.

The enzyme studies were not as successful, in part because Willstätter thought that enzymes were relatively small molecules adsorbed on a protein or some other giant (polymer) **molecule**. The modern view, of course, is that enzymes are themselves **proteins**. Though Willstätter's chemical intuition failed him, there were positive results—for example, the enzymatic **reduction** of chloral and bromal resulted in the formation of trichloroethanol, a sedative (Voluntal), and tribromoethanol, an anesthetic (Avertin).

In 1938 the situation for Jews in Germany was becoming impossible. On a visit to Switzerland, Stoll tried to persuade Willstätter to stay, but he insisted on returning to Munich. There, after some trouble with the Gestapo, he was ordered to leave the country. After much red tape, which entailed the confiscation of much of his property, papers, and art collection, and an abortive attempt to leave unofficially, he entered Switzerland in March 1939 to stay for a while with Stoll and then to settle in the Villa Eremitaggio in Muralto. There he wrote his autobiography to pass the time. On August 3, 1942, Willstätter died of cardiac failure in his sleep. Among the honors received by Willstätter in addition to the Nobel prize were honorary membership in the American Chemical Society (1927), honorary fellowship in the Chemical Society (1927), the Willard Gibbs Medal for distinguished achievement in science from the Chicago Section of the American Chemical Society (1933), and election as foreign member of the Royal Society (1933). Willstätter's obituary by Sir **Robert Robinson** in *Obituary Notices of Fellows of the Royal Society,* has an eleven page bibliography, probably incomplete, listing over three hundred papers between 1893 and 1940.

WINDAUS, ADOLF (1876-1959)

German organic chemist

Adolf Windaus devoted his professional life to the investigation of the **chemistry** of **natural products**. He was awarded the Nobel Prize for chemistry in 1928 for his work on sterols, which led to his clarifying the chemical structure of **cholesterol**, and he is also noted for his discoveries of the structure of vitamin D, some of the B **vitamins**, and histamine. The impact of his work made it possible for many other scientists to study the structures of other natural products; for example, his work on cholesterol helped to establish the study of sex **hormones**. Windaus's research on digitalis was used in the treatment of heart disease, and his studies of vitamin D led to the development of irradiation, a process of exposing foods, such as milk and bread, to ultraviolet light in order to prevent nutritional deficiencies that could lead to disease.

Adolf Windaus.

Adolf Otto Reinhold Windaus came from a family of artisans and craftspeople on his mother's side and from weavers and clothing manufacturers on his father's side. He was born in Berlin to Adolf and Margarete (Elster) Windaus on December 25, 1876. In his youth, he attended the French Gymnasium in Berlin, where literature, not science, was the primary area of study. Young Windaus decided to become a physician after reading about the work of French chemist and microbiologist **Louis Pasteur** and German physician and microbiologist Robert Koch. His mother, who was a widow at the time of his decision, was disappointed, since she had hoped he would continue the long tradition of the family business.

Windaus's career in science began at the University of Berlin in 1895. The chemistry lectures given by **Emil Fischer** there were to be major influences which would shape his future. The physiological applications of Fischer's approach became the foundation of Windaus's investigations. After receiving a bachelor's degree in 1897 from the University of Berlin and abandoning any ideas of pursuing a career in medicine, he continued his studies at the University of Freiburg, where he was influenced by Heinrich Kiliani. Under Kiliani's direction he researched digitalis, which later was found to be a powerful stimulant to the heart and became widely used in the treatment of heart failure. Windaus wrote his dissertation on the chemistry of this substance and received his doctorate in 1899 from Freiburg.

After a year in military service, Windaus returned to Freiburg to work with Kiliani, turning now to the study of cholesterol. A seroid **alcohol** present in animal cells and body fluids, cholesterol regulates membrane fluidity and is involved in the process of **metabolism**. Because it was so widely found in animal cells, Windaus speculated that it must be closely connected with other important compounds. By 1906, he was appointed assistant professor at Freiburg. In 1913, Windaus moved to the University of Innsbruck in Austria to become a professor of applied medical chemistry. Two years later, he was at the University of Göttingen, where he was appointed director of the Laboratory for General Chemistry, succeeding chemist **Otto Wallach**. He remained at Göttingen for twenty-nine years, retiring in 1944.

While Windaus pursued his studies of natural products, a number of other chemists were working in related areas. During his investigation of cholesterol (the best known sterol) and associated substances, **Heinrich Wieland**, a colleague in Munich, was researching the structure of bile acids. By 1919, Windaus was able to show an affinity between sterines, a group of sterols he had established earlier, and bile acids. After this, the work between Wieland and Windaus in both of their laboratories proceeded in close collaboration and led to the clarification of the chemical structure of the sterol **ring** in 1932.

It was known that rickets could be cured with cod liver oil, which contained vitamin D. Some scientists, such as physiologist Alfred Hess in New York, felt that cholesterol was somehow involved with vitamin activity, which led him to ask Windaus to collaborate in efforts to find the chemical nature of vitamin D. Windaus's cooperation with scientists in New York and London resulted in the findings of other D vitamins and made Göttingen a center for vitamin research.

The results of the research taking place during the 1920s and 1930s on vitamins made it possible for Windaus to identify and characterize many other compounds formed in the process of the photochemical reactions under study. In 1927, Wieland was given the Nobel Prize in chemistry for his study of bile acids, and Windaus received the same award in 1928 for his discovery of the structure of sterols and their connection with vitamins. Windaus was granted numerous honorary degrees and other awards as well, including the Louis Pasteur Medal of the French Academy of Sciences in 1938 and the Goethe Medal of the Goethe Institute in 1941.

During his early work on cholesterol, Windaus also had collaborated with biochemist Franz Knoop. They studied the reaction of sugar with **ammonia**, hoping that they could convert sugar into **amino acids**, and possibly do the same for **carbohydrates** into **proteins**. This work led to the discovery of histamine, a compound that is significant in allergies and inflammation. Consequently, Windaus became involved with pharmaceutical companies that began to suggest problems for him to solve, and supplied him with much of the materials he needed for his work.

Windaus's work on a B vitamin, thiamine, helped to establish its correct structure and synthesis, while other work involved clarifying the structure of colchicine, a substance used in cancer therapy. Although Windaus abandoned the idea of becoming a physician early in his academic career, his contributions to **organic chemistry** paved the way for new medical treatments of disease.

Windaus's studies on cholesterol opened new research areas for many other investigators and led to an important branch of organic chemistry and **biochemistry**. He was considered a valuable collaborator because of the close work he did with chemists in Germany and other countries on natural products. He was generous with his students, giving them both freedom to pursue their research interests and full credit for contributions they made. His influence on other research was considerable. For instance, one of his students, Adolf Friedrich Johann Butenandt, presented the structure of sex **hormones** shortly after Windaus presented the structure of the sterol ring.

Windaus married Elisabeth Resau in 1915 and they had two sons, Gunter and Gustav, and a daughter, Margarete. While he was not sympathetic to the Nazi government during World War II, his reputation made it possible for him to continue his work without interference. After his retirement in 1944, he did not publish any further research, but a journal on which he had served editorially, the *Justus Liebigs Annalen der Chemie,* dedicated several volumes to him in 1957 in celebration of his eightieth birthday. He died at the age of eighty-two on June 9, 1959, at Göttingen.

WITTIG, GEORG (1897-1987)

German chemist

Organic chemist Georg Wittig's investigations led him to discover in 1953 a chemical process for synthesizing complex compounds such as vitamin A, vitamin D derivatives, **steroids**, and biological **pesticides**. Because of this process, known as the Wittig reaction, such compounds can now routinely be synthesized. For his work in organic synthesis, and especially for the Wittig reaction, he shared the 1979 Nobel Prize in **chemistry** with **Herbert C. Brown**.

Georg Friedrich Karl Wittig was born on June 16, 1897, in Berlin, Germany, to Gustav Wittig, a professor of fine arts at the University of Berlin, and Martha (Dombrowski) Wittig. He went to grade school at the Wilhelms-Gymnasium in Kassel. In 1916 he enrolled at the University of Tübingen, but interrupted his college years to serve in World War I. After moving to the University of Marburg in 1920, he began postgraduate studies in chemistry under the guidance of Karl von Auwers. After receiving his doctorate in 1923, Wittig stayed on at Marburg to teach and do research for many years. In 1932, he became associate professor at the technical university in Brunswick. He went to the University of Freiburg five years later in the capacity of associate professor. In 1944 he was appointed full professor and director of the University of Tübingen's Chemical Institute. After twelve years, he transferred to the University of Heidelberg, where he became emeritus professor in 1967. After retirement, he continued to work and publish with various students at the University of Heidelberg.

Among his peers, Wittig won renown as an original thinker and gifted deviser of experiments. During Wittig's tenure at Tübingen, he and his research team started working with a family of organic compounds called ylides. These compounds formed the basis of the Wittig reaction, which easily and predictably joins two **carbon** atoms from different molecules to form a double bond. The Wittig reaction's reliability enabled other chemists to pursue and publish findings on thousands of applications for linking large carbon molecules.

Prior to the Nobel Prize, Wittig received many accolades, including the Adolf von Baeyer Medal in 1953, the 1967 **Otto Hahn** Prize of the German Chemical Society, the 1972 **Paul Karrer** Medal in Chemistry from the University of Zurich and the 1975 **Roger Adams** Award from the American Chemical Society. He had also been granted honorary degrees from the universities of Hamburg, Tübingen, and Paris. Wittig married Waltraut Ernst in 1930. Together, they had two daughters. Wittig loved the out-of-doors and was an avid mountaineer. While young, he had shown considerable musical ability. Those who knew him often remarked that he could have had a career in music had his early inclinations led him away from chemistry. Wittig died on August 26, 1987, in Heidelberg at the age of ninety.

Wöhler, Friedrich (1800-1882)

German chemist

Friedrich Wöhler.

Wöhler was born in Eschersheim, Prussia. He began his medical studies at the Marburg University in 1820, but soon transferred to the University of Heidelberg. He received his M.D. in 1823 and began to study **chemistry**. He eventually studied for a year with **Jöns Berzelius**, a highly regarded chemist in Stockholm, Sweden. There he developed an interest in **inorganic chemistry**. Building upon the work of Hans Christian Oersted, Wöhler was able by 1828 to extract **aluminum** by heating a mixture of **potassium** and aluminum chloride in a **platinum** crucible. Wöhler used a similar technique to produce **beryllium** and went on to produce a variety of aluminum salts. Soon after, he created **calcium** carbide and was a close second in discovering **vanadium**.

Wöhler also disproved a major theory, **vitalism**, of his friend and mentor Berzelius. Vitalism stated that compounds were absolutely divided between the organic and the inorganic. Organic compounds supposedly could only be formed in the tissues of living organisms, where a postulated vital force was responsible for changing them. Based on this theory, it would be impossible to synthesize an organic compound in the laboratory from inorganic reactants. Berzelius believed that the rules governing inorganic compounds did not hold true for organic compounds. Another teacher of Wöhler, Leopold Gmelin, also adhered to Berzelius's theory.

While experimenting with ammonium cyanate in 1828, Wöhler heated **lead** cyanate with an **ammonia solution** and formed crystals that appeared to be **urea**. Further experiments proved that the crystals were urea crystals. He also determined that urea and ammonium cyanate have the same elements in the same proportion. They are isomers. Wöhler had succeeded in producing an organic compound from inorganic reactants. It was soon argued that ammonium cyanate was actually an organic substance and thus Wöhler's discovery was irrelevant to the debate over vitalism. However, his success encouraged other chemists to try to produce organics from inorganics. In 1845 **Adolf Kolbe** conclusively disproved vitalism when he produced **acetic acid** from the elements **carbon**, **hydrogen**, and **oxygen**. Finally Wöhler's most ardent critics agreed that Berzelius's theory had been soundly discredited.

Wöhler continued with his experiments and began combining his medical training with his chemical knowledge to study the body's **metabolism**. In 1832, after the death of his wife, he went to work with **Justus von Liebig** at Liebig's laboratory in Giessen, Germany. They collaborated in a study of bitter almonds, the source of cyanide, and proved that the pure oil of bitter almonds did not contain hydrocyanic acid and so was not poisonous. They continued to study the oil, today called benzaldehyde, and its reactions. They discovered that when subjected to different experiments, a group of atoms, later named benzoyl, did not change. They called this unchanging group "radicals." The theory was important to understanding the chemical behavior of organic compounds. In 1836 Wöhler accepted a position as a professor at the University of Göttingen. He continued his research into cyanides and aluminum, becoming the first to synthesize **boron**, **silicon**, silicon nitride and hydride, and **titanium**.

Wöhler was busy in his later years. He served as a professor of chemistry and pharmacy, director of laboratories, and inspector general for all of the pharmacies in Hanover, Germany. He also translated Berzelius's books and papers into German, and began studies in geology and on meteorites. Acquaintances and former students from all over the world sent him samples, and Wöhler eventually published 50 papers on these subjects. He published many papers and textbooks over his career and taught some 8,000 students, among them Rudolph Fittig (1835-1910) and Jewett. Jewett's student, Charles Hall (1863-1914) found the commercially feasible way to produce aluminum that had eluded Wöhler. Wöhler died in Gottingen.

Wood

Wood is the hard, fibrous substance found beneath the bark in trees, shrubs, and other similar plants. It is a result of the secondary growth processes of these plants. The cambium layer (the region between the xylem and phloem in vascular plants) divides to produce various new tissues in a process that results in secondary thickening. This increases the girth of the tree by the production of the new wood. Wood is actually a very complex organic material made of a variety of **carbohydrates**, lignin, inorganic materials, and other organic materials. Wood has and still continues to be used in the construction industry and in the manufacture of pulp and **paper**. **Charcoal** is produced from wood by slow **combustion** in a low supply of **oxygen**. Charcoal can be used as a fuel by burning it in excess oxygen, as a reducing agent in **industrial chemistry**, and it can be used to absorb **gases** and other particles. Charcoal is often one of the active ingredients in simple gas masks.

One of the main constituents of wood is **cellulose**, a carbohydrate polymer. Cellulose is used to manufacture rayon (viscose). Firstly, the cellulose is treated with an alkali and then **carbon** disulfide, to yield xanthated cellulose, that upon extrusion into an acid bath, gives rayon. When wood is pulped (as is done during paper manufacture) a liquid fraction is collected that is called tall oil. Tall oil contains mainly **fatty acids** and it can be used as a protective coating (a varnish), in the production of **soaps** (saponification), and as a plasticizer (a compound added to plastic or paint to improve the flexibility). Annual production of tall oil in the 1990s in the United States is in excess of 1 megatonne. Also from tall oil a solid resin called rosin (or colophony) can be obtained. Rosin is a translucent, amber solid that can also be found in the oil from pine trees. Rosin is mainly resin acids (monocarboxylic acids). Rosin can be used as a plasticizer, in the manufacture of varnishes and printing **inks**, and also to treat bows for stringed instruments. Steam **distillation** or **chromatography** of wood and wood residues can release terpenes. Terpenes are a large class of organic compounds that are widely employed in the pharmaceutical industry. Different terpenes can be obtained from different wood types and have characteristic physical properties often including distinctive odors. Terpenes are particularly common in the resins of coniferous trees. It is the terpenes that give the resins of conifers their characteristic smells.

Wood (along with other plant material), fossilised during the carboniferous period, is responsible for our **coal** and oil reserves.

Some wood can be produced very quickly and consequently it is usually a cheap and easily renewed resource. As a result of this it is still used extensively as a fuel and a building material. Wood can be divided into two types, hard and soft wood. Each has different physical properties. Hard wood is very dense and strong, and it is much slower growing and consequently is more expensive than soft wood. Hard woods are used where strength is needed. For example, the large wooden ships of the past were built of hard wood such as oak. Soft woods such as that obtained from conifers can be used where less physical strength is needed. They can be used in the manufacture of smaller structures or as is more common in the production of paper and pulp. Soft wood such as that obtained from conifers is very quick to grow and as such it is relatively cheap and easily renewable.

Wood is a versatile, natural product. It can be used directly as a building material or fuel. With minor treatment paper and pulp can be manufactured. With greater treatment a number of commercially important compounds can be obtained. Wood is obtained from trees as a result of secondary growth and different trees produce wood with different physical and chemical characteristics.

Woodward, Robert B. (1917-1979)

American organic chemist

Robert B. Woodward was arguably the greatest organic synthesis chemist of the twentieth century. He accomplished the total synthesis of several important **natural products** and pharmaceuticals. Total synthesis means that the **molecule** of interest—no **matter** how complex—is built directly from the smallest, most common compounds and is not just a derivation of a related larger molecule. In order to accomplish his work, Woodward combined physical **chemistry** principles, including quantum mechanics, with traditional reaction methods to design elaborate synthetic schemes. With Nobel Laureate **Roald Hoffmann**, he designed a set of rules for predicting reaction outcomes based on **stereochemistry**, the study of the spatial arrangements of molecules.

When Woodward won the Nobel Prize in chemistry in 1965, the committee cited his contributions to the "art" of organic synthesis. Upon Woodward's acceptance of the award, Bartlett, Westheimer, and Buchi wrote in *Science,* "Woodward's style is polished, showing an insight and sense of proportion that afford him strong convictions and a well-developed dramatic sense. In the laboratory, identifications and structural assignments must be complete, spectra exact, compounds not merely pure but beautifully crystallized, or he will not accept them. His lectures, given without notes or slides, are elegantly organized and illustrated with artistic blackboard formulas, with the key atoms shown in **color**.... Most of the polish comes naturally to a man with such intellectual vitality.''

Robert Burns Woodward was born in Boston on April 10, 1917, to Arthur and Margaret (Burns) Woodward. His father died when he was very young. Woodward obtained his first chemistry set while still a child and taught himself most of the basic principles of the science by doing experiments at home. By the time he graduated at the age of sixteen from Quincy High School in Quincy, Massachusetts, in 1933, his knowledge of chemistry exceeded that of many of his instructors. He entered the Massachusetts Institute of Technology (MIT) the same year but nearly flunked out a few months later, apparently impatient with the rules and required courses.

The MIT chemistry faculty, however, recognized Woodward's unusual talent and rescued him. They obtained funding and a laboratory for his work and allowed him complete freedom to design his own curriculum, which he made far more rigorous than the required one. Woodward obtained his doctorate degree from MIT only four years later, at the age of 20, and then joined the faculty of Harvard University after a year of postdoctoral work there.

Woodward spent virtually all of his career at Harvard but also did a significant amount of consulting work with various corporations and institutes around the world. As is true in most modern scientific endeavors, Woodward's working style was characterized by collaboration with many other researchers. He also insisted on utilizing the most up-to-date instrumentation, theories, and other available tools, which were sometimes looked upon with suspicion by more traditional organic chemists. He was known as an intense thinker, personally reserved and imperiously confident of his intellectual skills. His graduate students, however, still found ways to joke with him; one Halloween, noticing that he virtually always had the same color tie, office, and car, they painted his parking space ''Woodward blue.''

The design of a synthesis, the crux of Woodward's work, involves much more than a simple list of chemicals or procedures. Biochemical molecules exhibit not only a particular **bonding** pattern of atoms, but also a certain arrangement of those atoms in space. The study of the spatial arrangements of molecules is called stereochemistry, and the individual configurations of a molecule are called its stereoisomers. Sometimes the same molecule may have many different stereoisomers; only one of those, however, will be biologically relevant. Consequently, a synthesis scheme must consider the basic reaction conditions that will bond two atoms together as well as determine how to ensure that the reaction orients the atoms properly to obtain the correct stereoisomer.

Physical chemists postulate that certain areas around an **atom** or molecule are more likely to contain electrons than other areas. These areas of probability, called orbitals, are described mathematically but are usually visualized as having specific shapes and orientations relative to the rest of the atom or molecule. Chemists visualize bonding as an overlap of two partially full orbitals to make one completely full molecular orbital with two electrons. Woodward and Roald Hoffmann of Cornell University established the Woodward-Hoffmann rules based on quantum mechanics, which explain whether a particular overlap is likely or even possible for the orbitals of two reacting species. By carefully choosing the shape of the reactant species and reaction conditions, the chemist can make certain that the atoms are oriented to obtain exactly the correct stereochemical configuration. In 1970 Woodward and Hoffmann published their classic work on the subject, *The Conservation of Orbital Symmetry;* Woodward by that time had demonstrated repeatedly by his own startling successes at synthesis that the rules worked.

Robert B. Woodward.

Woodward and his colleagues synthesized a lengthy list of difficult molecules over the years. In 1944 their research, motivated by wartime shortages of the material and funded by the Polaroid Corporation, prompted Woodward—only twenty-seven years old at the time—and William E. Doering to announce the first total synthesis of **quinine**, important in the treatment of malaria. Chemists had been trying unsuccessfully to synthesize quinine for more than a century.

In 1947 Woodward and C. H. Schramm, another organic chemist, reported that they had created an artificial protein by bonding **amino acids** into a long chain molecule, knowledge that proved useful to both researchers and workers in the **plastics** industry. In 1951 Woodward and his colleagues (funded partly by Merck and the Monsanto Corporation) announced the first total synthesis of **cholesterol** and **cortisone**, both biochemical **steroids**. Cortisone had only recently been identified as an effective drug in the treatment of rheumatoid arthritis, so its synthesis was of great importance.

Woodward's other accomplishments in synthesis include **strychnine** (1954), a poison isolated from *Strychnos* species and often used to kill rats; colchicine (1963), a toxic natural product found in autumn crocus; and lysergic acid (1954) and reserpine (1956), both psychoactive substances. Reserpine, a tranquilizer found naturally in the Indian snake root plant *Rauwolfia,* was widely used to treat mental illness and was one of the first genuinely effective psychiatric medicines. In 1960, after four years of work, Woodward synthesized **chlorophyll**, the light **energy** capturing pigment in green plants, and in 1962 he accomplished the total synthesis of a tetracycline antibiotic.

Total synthesis requires the design and then precise implementation of elaborate procedures composed of many steps. Each step in a synthetic procedure either adds or subtracts chemical groups from a starting molecule or rearranges the orientation or order of the atoms in the molecule. Since it is impossible, even with the utmost care, to achieve one hundred percent conversion of starting compound to product at any given step, the greater the number of steps, the less product is obtained.

Woodward and Doering produced approximately a half a gram of quinine from about five pounds of starting materials; they began with benzaldehyde, a simple, cheap chemical obtained from **coal** tar, and designed a seventeen-step synthetic procedure. The twenty-step synthesis that led to the first steroid **nucleus** required twenty-two pounds of starting material and yielded less than a twentieth of an ounce of product. The best synthesis schemes thus have the fewest number of steps, although for some very complicated molecules, ''few'' may mean several dozen. When Woodward successfully synthesized chlorophyll (which has an elaborate interconnected **ring** structure), for example, he required fifty-five steps for the synthesis.

Woodward's close friend, Nobel Laureate **Vladimir Prelog**, helped establish the CIBA-Geigy Corporation-funded Woodward Institute in Zurich, Switzerland, in the early 1960s. There Woodward could work on whatever project he chose, without the intrusion of teaching or administrative duties. Initially, the Swiss Federal Institute of Technology had tried to hire Woodward away from Harvard; when it failed, the Woodward Institute provided an alternative way of ensuring that Woodward visited and worked frequently in Switzerland. In 1965 Woodward and his Swiss collaborators synthesized Cephalosporin C, an important antibiotic. In 1971 he succeeded in synthesizing vitamin B_{12}, a molecule bearing some chemical similarity to chlorophyll, but with **cobalt** instead of **magnesium** as the central metal atom. Until the end of his life, Woodward worked on the synthesis of the antibiotic erythromycin.

Woodward, who received a Nobel Prize in 1965, helped start two **organic chemistry** journals, *Tetrahedron Letters* and *Tetrahedron,* served on the boards of several science organizations, and received awards and honorary degrees from many countries. Some of his many honors include the Davy Medal (1959) and the Copley Medal (1978), both from the Royal Society of Britain, and the United States' National Medal of Science (1964). He reached full professor status at Harvard in 1950 and in 1960 became the Donner Professor of Science. Woodward supervised more than three hundred graduate students and postdoctoral students throughout his career.

Woodward married Irji Pullman in 1938 and had two daughters, Siiri and Jean. He was married for the second time in 1946 to the former Eudoxia Muller, who had also been a consultant at the Polaroid Corporation. The couple had two children, Crystal and Eric. An inveterate smoker and coffee-drinker, his only exercise was an occasional game of softball. Woodward died at his home of a heart attack on July 8, 1979, at the age of 62.

X

XENON

Xenon is the fifth element in Group 18 of the **periodic table**, a group of elements known as the noble **gases** or inert gases. Xenon has an **atomic number** of 54, an atomic **mass** of 131.29, and a chemical symbol of Xe.

Properties

Xenon is a colorless, odorless, tasteless gas with a **boiling point** of -162.6°F (-108.13°C), a melting point of -169.2°F (-111.8°C), and a **density** of 5.8971 grams per liter. It is almost entirely chemically inert. A small number of compounds have been made under research conditions, but no such compounds exist in the natural world.

Occurrence and Extraction

Xenon does not exist to any measurable extent in the Earth's crust, although it does occur to the extent of about 0.1 part per million in the Earth's atmosphere. The element has also been discovered in the atmosphere of Mars with about the same **concentration**. When needed, xenon can be produced by the fractional **distillation** of liquid air.

Discovery and Naming

Xenon was discovered in 1898 by the Scottish chemist and physicist Sir **William Ramsay** and the English chemist Morris William Travers (1872-1961). Ramsay and Travers found the new element by spectroscopic analysis of the residue left after **nitrogen**, **oxygen**, and **argon** had been removed from liquid air. They suggested the name xenon after the Greek word for ''stranger.''

Uses

The primary use of xenon is in fluorescent and ''**neon**'' lamps. The presence of xenon in such lamps results in a very bright, sun-like light used in photographic flash units, strobe lights, and airport runway lights.

X-RAY CRYSTALLOGRAPHY

X-ray **crystallography** is an experimental technique for determining the arrangement of atoms in a crystalline material using highly energetic electromagnetic radiation. X-ray crystallography provides the most direct and accurate means of establishing detailed molecular structure—the spatial relationships of atoms with each other including bond lengths and angles.

X-ray crystallography is based upon the idea that atoms regularly arranged in crystals scatter x rays in a manner analogous to the way that the regularly spaced grooves of a diffraction grating scatter light. The relationship between the observed points of x rays scattered from a crystal can be calculated using the same mathematics as used to interpret a diffraction pattern. Working backward from the positions of the observed ''diffracted'' x rays, the crystallographer calculates the positions of atoms in the crystal.

For diffraction phenomena, the spacing between diffraction points can be most accurately calculated from the diffraction pattern when the wavelength of radiation is about the same as the spacing. The x-radiation typically used for crystallography has a wavelength that is about the same as the distance between atoms linked by a **covalent bond** or the radius of an **ion** in an ionic crystal. For comparison, the x rays of a **molybdenum** source is 71 picometers (1/1,000,000,000,000 of a meter) and that of a **copper** source is 154 picometers (pm) while a typical C-C bond length is about 155 pm, the radius of the Na^+ ion is 102 pm and the radius of a Cl^- is 181 pm.

Not long after **John Dalton** published his **atomic theory** in 1808, Ludwig Seeber postulated in 1824 that the properties of **solids** could be explained by imagining that they possess an internal structure of atoms held at specific distances from each other together by a **balance** of attractive and repulsive forces. In 1850, Moritz Frankenheim and Auguste Bravais described 14 fundamental types of repeating lattice structures that could be used to describe the internal structures of crystalline solids. These structural types, now referred to as Bravais lattices, are

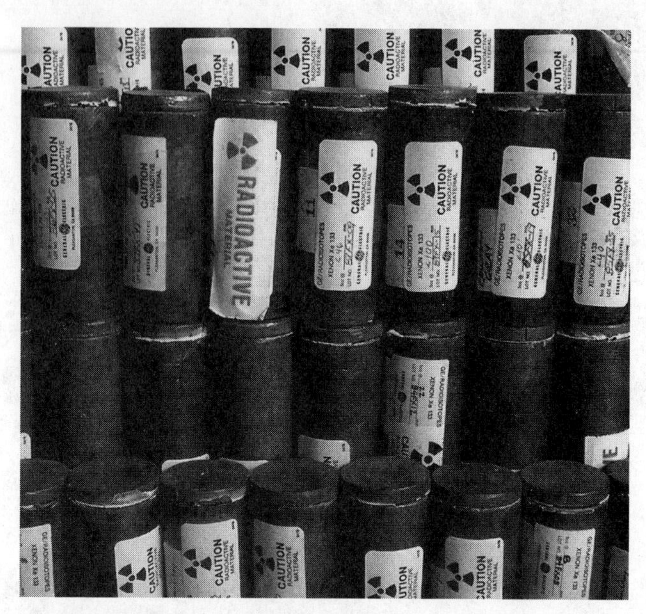

Lead containers used for radiactive xenon. ***(Photograph by Will & Deni McIntyre. National Audubon Society Collection/Photo Researchers, Inc. Reproduced by permission.)***

the types of repeating units or unit-cells that make up a regular or perfect crystal. The unit cells can be repeated in three dimensions with cells translated, inverted and/or rotated with respect to neighboring cells in three dimensions in a variety of ways. In 1890, Evgraph Federov and Artur Schoenflies determined 230 differences caused by unit cells that can be repeated in three dimensions. These are called space groups, crystallographic space groups, or Shoenflies groups. During this period, other scientists were investigating diffraction phenomena using **glass** and metal gratings. In 1910, Max von Laue published an equation that can be used to calculate diffraction maxima observed from two-dimensional cross gratings. The work of Bravais, Schoenflies, and von Laue was theoretical and was done before **x-ray diffraction** experiments had been performed.

X rays were discovered by Röntgen in 1895. Very soon after the discovery, Röntgen and others found that x rays penetrated **matter**—passing through flesh to show the outline of bones of the hand as shadows on film, for example. Scientists soon decided to see what would happen if they directed a beam of x rays through a crystal and they observed a pattern of spots of film that resembled a diffraction pattern observed. In 1912, W. Friedrich and P. Knipping published their interpretation of the diffraction of x rays by crystals. They extended the two-dimensional diffraction grating theory of Max von Laue to three dimensions. They reasoned that crystals act as three-dimensional diffraction gratings because the atoms inside crystals extend in regularly spaced repeating patterns in three dimensions. Further interpretation of the x-ray diffraction patterns of crystals by a father and son team, William Henry and **William Lawrence Bragg**, led to an equation for calculating the distance between repeated planes of ions in a crystalline **salt** using the observed separation of diffracted spots. They were able to ''solve'' the first structure from an x-ray diffraction pattern.

The Bragg equation, $2d \sin \theta = n\lambda$ is interpreted as follows. If an x-ray is incident on a plane of regularly spaced atoms in a crystal at an angle of incidence (θ) and if a detector is set to detect reflected (or scattered) x rays with a reflection angle of (θ) a diffracted beam of x rays will be detected if the angle satisfies the equation for an integer value of n (n = 1,2,3, etc.), where d is the distance between identical planes of atoms and (λ) is the wavelength of the x-radiation. The theory for x-ray diffraction that was developed by the Braggs and others also shows that there is a relationship between the internal structure of the crystal and the relative intensities of the diffracted beams.

Early in the study of x-ray crystallography the diffracted beams were detected using film. Many diffracted beams were detected at the same time and the positions of the spots were measured with a ruler. The relative intensities were either estimated or measured using a device that determined the darkness of the spots called a densitometer. Profiles of the spots (peaks) were integrated by hand using a tracing device (a polar planimeter). The next development was the diffractometer—a mechanical device that rotates the crystal and a detector such as a Geiger counter or scintillation counter relative to one another. Until the 1960s, the crystallographer would move the detector or crystal, which was mounted on a small platform called a goniometer that can be quite precisely rotated in several directions, by hand. When a diffracted beam was detected, the crystallographer would scan slowly across the beam and record its profile. From the 1960s on, diffractometers have been computer-controlled.

Crystallographers have continually improved their diffraction instruments. Electronic area detectors allow crystallographers to accurately record the intensities of many diffraction beams at once—analogous to a digital x-ray camera. This has the advantage not only of reducing the data acquisition time for any crystal but also reduce the time a crystal is exposed to x-radiation. Many biologically important molecules are destroyed by exposure to x-radiation and can only be investigated by a rapid exposure method. Most crystallographers use diffractometers in their own laboratories equipped with x-ray sources consisting of evacuated tubes that generate the x rays by acceleration of a high **energy** (tens of thousands of volts) against a metal target, usually copper or molybdenum. For special purposes, however, crystallographers can now use high-energy particle storage rings, synchotrons, at the national laboratories as an x-ray source.

The power of modern computers has greatly reduced the time necessary to perform calculations, from literally years for moderately complex molecules to a matter of hours. Bond distances and angles can now be determined much more precisely because of increased computing power. Approximations that were formerly used to simplify calculations can be dispensed with and calculations can be performed multiple times to obtain the best fit between the observed data and the structural model the crystallographer has deduced from the data.

The current theory of x-ray diffraction is built on the idea that the electrons of atoms are responsible for scattering x rays in a specific manner. In the x-ray diffraction experiment, a single crystal is subjected to a carefully collimated (well-defined) beam of x-radiation of a relatively narrow wavelength bandwidth. As the wavefront of the x-ray beam passes through the crystal, a very small amount is absorbed or scattered randomly but most of the beam interacts with electrons of the atoms in the crystal by causing the **electron** to vibrate. The vibrating electron then re-emits x rays of the same frequency (resonant scattering). Each electron that re-emits an x ray acts as an x-ray source. The entire set of re-emitting electrons in the crystal produce a set of waves that can interfere with each other. When waves interact that are of opposite phase (say one at maximum positive amplitude, the wave crest, and the other an maximum depth, the wave trough) they cancel each other out, but when they are of the same phase they add to each other, creating a more intense wave. The greater the number of electrons of an **atom**, the more likely the atom is to interact with and re-emit x rays. The pattern of diffracted rays and their intensity are determined from the arrangement of atoms and number of electrons on each atom in the crystal. The crystallographer uses the pattern and intensities of diffracted beams to construct a model of the electron **density** distribution in the crystal. Knowing the elemental composition of the compound, likely structures, the number of electrons of each element and likely bond distances and angles from the structures of similar compounds, the crystallographer constructs a trial structure, calculates what the diffraction pattern would be and compares the calculation with the observed data. If the match looks promising, the crystallographer then adjusts the structural model, recalculates the pattern and intensities and again makes the comparison with experimental data. This process is called refinement. When the crystallographer finds that the agreement between calculated and experimental data is good enough that she is confident that the major structural features are correct, she accounts for absorption by the crystal and vibration of the atoms within the crystal to arrive at precise bond distance and bond angles. To define the position and vibration characteristics of each atom, the solution of the equations to match diffraction data require at least one independent data point for every degree of freedom. For very large molecules with several hundred or thousands of atoms, it may not be possible to obtain enough data points to solve the equations for every atom. Therefore, for molecules such as **proteins** it may only be possible to determine (resolve) relative positions of groups of atoms.

X-ray crystallography is the experimental method that has provided the most direct and precise information about molecular structures in crystals. The method has been extended from simple compounds consisting of a few atoms to precise measurements of molecules with a few hundred atoms and identification of relative positions of groups of atoms for molecules with thousands of atoms because of improvements in computers. The structural information provided by x-ray crystallography has been the basis for a great expansion of our understanding in many areas of **chemistry**, **geochemistry**, and **biochemistry**. **James Watson** and **Francis Crick** used **Rosalind Franklin**'s x-ray diffraction results to construct the structure of **DNA**. X-ray diffraction results have shown the structural features in common for "high-temperature" **semiconductors** that have allowed researchers to synthesize new ones with predicted behavior. Structures of enzymes, drug-receptor pairs, heme proteins, and many other biochemically important molecules determined by x-ray crystallography have elucidated their mechanism of action.

X-RAY DIFFRACTION

Max von Laue (1879-1960) recognized that the regularly spaced planes of atoms and molecules in crystals resemble a diffraction grating. He and his coworkers, W. Friedrich and P. Knipping, showed in 1912 that a beam of x rays, with a wavelength comparable to the magnitude of crystal spacings, is diffracted when passed through crystalline substances. The observed diffraction patterns can be explained by considering the beam of x rays to be scattered by the atoms of the **molecule** in the crystal.

The solutions of the Schrödinger wave equation for atoms and molecules are sets of wave functions. Each of these wave functions corresponds to a **stationary state** in which the **atom** or molecule is allowed to be, and each of these allowed states has **energy** associated with it. There are two major types of interactions that molecules may have with electromagnetic radiation. In the first type, called **resonance** absorption, energy is absorbed from the radiation and the molecule changes from one allowed stationary state to another. The wavelength of the radiation is such that its energy is exactly equal to the difference in energy between two of the allowed stationary states. This is the type of absorption observed in **spectroscopy**.

The second type of interaction between a molecule and electromagnetic radiation is non-resonance absorption. In this instance, the wavelength of the radiation is such that its energy does not correspond to a difference in energy between any of the stationary states of the molecule. Nevertheless, molecules absorb energy from the radiation but change from stationary states to non-stationary states. The latter is an energy state that is not one of the solutions of the wave equation for the molecule. It is, therefore, not a stable state, and the molecule instantaneously re-emits the energy it has absorbed. This emission, which is of the same energy and wavelength as that absorbed, is given off in all directions, not just in the direction traveled by the incoming beam. The net result is that a part of the incoming radiation is scattered by the atoms in the molecules of the crystal. X rays are a part of the **electromagnetic spectrum**, with wavelengths from roughly 10^{-6}-10^{-8} cm. Although these wavelengths do not correspond to resonance absorption electrons in molecules, there is significant non-resonance absorption and scattering.

When the molecules of the system are arranged randomly relative to each other, as in a gas, the scattered radiation from one molecule bears no relation to that from the other molecules. If, however, the molecules are arranged in an orderly, three-dimensional array, as in a crystalline solid, and are separated by constant repeat distances in every direction, the radiation scattered from the various molecules will interact constructively in some directions, destructively in others. The net effect is the diffraction of the incident radiation by the array of molecules. The directions in which the radiation is diffracted are well defined so that if a photographic film is placed around the diffracting crystal, a large number of exposure spots, called Laue diffraction maxima, are obtained when the film is developed. Each of the spots is the result of one of the rays of diffracted radiation, and the placement of the spots on the film can be used to determine the spacing between molecules in the crystal.

For a molecular crystal, the various Laue diffraction maxima differ in intensity. Laue, William Henry Bragg (1862-1942), and **William Lawrence Bragg** (father and son) showed that the intensity of the x radiation diffracted in a particular direction is related to the three-dimensional structure of the molecules that make up the crystal. They and subsequent x-ray crystallographers developed methods that used the observed diffraction intensities to determine the structure of molecules. These methods have proved very productive, providing accurate molecular bond distances and angles and the structures of many complex molecules, including **proteins**.

Y

Ytterbium

Ytterbium is a rare earth metal, one of the elements found in Row 6 of the **periodic table**. It has an **atomic number** of 70, an atomic **mass** of 173.04, and a chemical symbol of Yb.

Properties

Ytterbium is a typical metal that is both ductile and malleable. It has a melting point of 1,515°F (824°C), a **boiling point** of 2,600°F (1,427°C), and a **density** of 7.01 grams per cubic centimeter. Ytterbium is a relatively reactive element that is usually stored in sealed containers to prevent its reacting with **oxygen** of the air.

Occurrence and Extraction

Ytterbium is one of the more common rare earth **metals** with an abundance in the Earth's crust of about 2.7-8 parts per million. Its most common ore is monazite, which is found in beach sands in Brazil, India, and Florida. It is extracted from its ores by heating **lanthanum** metal with ytterbium **oxide** (Yb_2O_3): $2La + Yb_2O_3 \rightarrow 2Yb + La_2O_3$.

Discovery and Naming

Ytterbium was ''discovered'' over a period of more than 20 years by three chemists: Jean-Charles-Galissard de Marignac (1817-94) of France, Lars Fredrik Nilson (1840-99) of Sweden, and Georges Urbain (1872-1938) of France. The three chemists all worked on a mineral known as yttria originally discovered near the town of Ytterby, Sweden, in 1787. The mineral eventually yielded nine new chemical elements. The element was given the name of ytterbium in honor of the town near which it was discovered.

Uses

Very small amounts of ytterbium are used in the manufacture of special steel alloys and in **lasers**. Generally speaking, however, it has almost no commercial application of significance.

Yttrium

Yttrium is the second element in Group 3 of the **periodic table**, one of the transition **metals**. It has an **atomic number** of 39, an atomic **mass** of 88.9059, and a chemical symbol of Y.

Properties

Yttrium has typical metallic properties with a melting point of 2,748°F (1,509°C), a **boiling point** of about 5,400°F (3,000°C), and a **density** of 4.47 grams per cubic centimeter. Yttrium is a moderately active element that reacts with cold **water** slowly and hot water more rapidly. It dissolves in both acids and alkalis. Yttrium does not react with **oxygen** at room **temperature**, but does so at higher temperatures. As a powder, it may react explosively with hot oxygen.

Occurrence and Extraction

Yttrium is a moderately abundant element in the Earth's crust with an abundance estimated at about 28-70 parts per million. As with many other elements, the abundance of yttrium varies in other parts of the solar system. Rocks brought back from the Moon, for example, tend to have a higher **concentration** of yttrium than those in the Earth's crust. The primary ore of yttrium is monazite, which occurs in beach sand in Brazil, India, Florida, and other parts of the world.

Discovery and Naming

Yttrium was the first new element to be discovered in a rock found in 1787 by a Swedish army lieutenant outside the town of Ytterby, Sweden. The rock was eventually to yield nine new elements. Yttrium was discovered by the Swedish chemist Carl Gustav Mosander, who named the element in honor of the town near which the rock was found.

Uses

About two-thirds of all yttrium produced goes to the manufacture of phosphors used in television picture tubes, computer monitors, and specialized fluorescent lights. An increasingly important use of the element is in the production of special **lasers** made of yttrium, **aluminum**, and synthetic garnet, the YAG laser. One use of the YAG laser is in making very precise measurements at long distances. As an example, the National Aeronautics and Space Administration used a YAG laser in 1996 to measure the dimensions of the asteroid Eros and to map its surface features.

Z

ZEEMAN EFFECT

According to classical electromagnetic theory, a charge rotating with a simple harmonic frequency should emit electromagnetic radiation of the same frequency. And when an external magnetic field is applied to this system, the frequency of oscillation should change. Depending on the orientation of the **atom** with respect to the magnetic field, the frequency should either increase, decrease, or remain the same. The observed splitting of the spectral lines into three components, which is now called the normal Zeeman effect (named after the Dutch physicist Pieter Zeeman who discovered it), provided one of the earliest estimates of the charge-to- **mass** ratio of the **electron**.

It is now known that this so-called normal Zeeman effect arises from electronic transitions between singlet states (i.e., states in which the contribution to angular momentum from the electron's intrinsic angular momentum, or spin, is zero and in which the total angular momentum is equal to the orbital angular momentum of the electron). Quantum mechanically, the change in angular frequency should be equal to the one-half the electron charge times the magnetic field divided by the mass of the electron, which is exactly the result predicted by classical electromagnetic theory.

In many cases, the frequency of the electron is split into more than three components under the application of an external magnetic field. This behavior is called the anomalous Zeeman effect. It arises when the contribution of electron spin to the total angular momentum is not zero, i.e., when the total magnetic moment is not parallel to the total angular momentum. A spin-orbit splitting of the **energy** levels arises from the interaction of the electron's spin magnetic moment with the internal magnetic field seen by the electron due to its orbital motion (even in the absence of an applied field). When an external magnetic field is applied, the Zeeman effect depends on the quantization of the electron orbital angular momentum, the total angular momentum, and the spin angular momentum. It is not possible to account for the anomalous Zeeman effect classically, as the complications that it introduces arise solely from quantum effects.

See also Electricity

ZIEGLER, KARL (1898-1973)

German chemist

Karl Ziegler had a long and distinguished career in diverse areas of **chemistry**. Although he is considered an organic chemist, he applied the methods and principles of inorganic, physical, and **analytical chemistry** to his research problems. He thought of himself as a chemist who carried out "pure" research, but his greatest contribution was a discovery that led to a revolution in "applied" research and was of great benefit to industry. This breakthrough related to catalysts—substances that provoke a chemical reaction—and Ziegler's work became the foundation of the modern **plastics** industry. The discovery and application of the "Ziegler catalysts" were rewarded with lucrative licensing fees and the 1963 Nobel Prize in chemistry, which Ziegler shared with **Giulio Natta**, an Italian chemist who significantly extended Ziegler's work.

Ziegler was born on November 26, 1898, in Helsa, Germany, to Luise (Rall) and Karl Ziegler, a Lutheran minister. As a youth he showed an early interest in chemistry, and had a laboratory at home. In 1916 he matriculated at the University of Marburg, from which his father had graduated, and was so advanced in his studies that he was able to complete his Ph.D. in 1920, when he was only twenty-one. His thesis adviser was Karl von Auwers, a noted organic chemist of his time. Ziegler remained at Marburg as a lecturer until 1925, then spent a year as a visiting lecturer at the University of Frankfurt. In 1926 Ziegler moved to the University of Heidelberg, receiving a promotion to professor the following year. He remained at Heidelberg until 1936, when he was appointed professor and director of the Chemical Institute at the University of Halle.

Karl Ziegler.

In 1943 Ziegler accepted the directorship of the Kaiser Wilhelm Institute (later known as the Max Planck Institute) for Coal Research in Müllheim, located in Germany's Ruhr valley. Ziegler agreed to the appointment on condition that he could work on any research project of his choice and not be limited to the chemistry of coal. It was at Müllheim that he discovered the catalyst that brought him great renown, but the discovery was a natural consequence of research which he had begun as a graduate student and developed over his long career.

From 1923 to 1943 Ziegler concentrated his research on **free radicals** (atoms or groups of atoms having one or more unpaired electrons), organometallic compounds and their reactions with double bonds, and the synthesis of large rings, which are cyclic compounds of molecules. He was primarily interested in the fundamental aspects of structural chemistry, such as the strength of the carbon-carbon bond. He studied the nature of free radicals while looking for compounds whose bonds could be broken easily to form a trivalent species, or one containing a **carbon atom** bound to three other substances. Usually free radicals exist only briefly and rapidly react to form normal tetravalent compounds—carbon most often has a **bonding** capacity of four—but Ziegler found many examples of complex free radicals that could survive and be manipulated like ordinary compounds, as long as reactive species such as **oxygen** were excluded.

In the course of his work on free radicals, Ziegler investigated the organic derivatives of reactive **metals**, such as **sodium** and **potassium**, and later, **lithium**. With a chemical composition similar to that of the organomagnesium compounds explored earlier by French chemist and Nobel Prize–winner **Victor Grignard**, lithium proved to be extremely useful in organic synthesis. Unlike Grignard's reagent, however, Ziegler found that certain organopotassium compounds could add to a carbon-carbon double bond to make a more complex organopotassium compound. Ziegler applied the reaction to butadiene, a compound that contains two double bonds, and found that the butadiene molecules could form long **chains**.

As this research progressed, Ziegler considered the problem of joining the ends of long chain molecules to form large rings. This problem had considerable practical importance, for example, in the synthesis of the natural perfume base, muscone. Ziegler eventually used a strong base—a material that accepts protons in solution—with a long-chain compound in very high dilution in order to prepare a large-ring ketone, a compound with fourteen to thirty-three carbons.

When Ziegler moved to the Institute for Coal Research in 1943, he continued the lines of research he had earlier developed. Ziegler tried many experiments to add lithium hydride to the carbon-carbon double bond, but the reaction was slow and unsuccessful. The compound lithium **aluminum** hydride was reported by Schlesinger at the University of Chicago in 1947, and Ziegler tried this as a substitute for lithium hydride in 1949. This was successful, and led Ziegler to the conclusion that the aluminum was the vital component. Ziegler found that organoaluminum compounds reacted with double-bond compounds at one hundred degrees centigrade to produce long chains of carbons attached to the aluminum atom. The organoaluminum compounds could be converted into long-chain alcohols (alcohols are characterized by a hydroxyl, or oxygen-hydrogen, group attached to a **hydrocarbon** chain) by allowing air into the reaction, and these alcohols were useful in the formulation of detergents.

In the course of the investigation of organoaluminum reactions in 1953, one experiment delivered a product that did not contain the expected long chains. The reaction had been carried out in an autoclave—an apparatus suited to special conditions such as high or low pressure or temperature—and careful analysis showed that the autoclave had been previously used in a reaction that contained **nickel**, with small traces of nickel salts remaining. Ziegler and his colleagues investigated the addition of other metal salts and found that in contrast with nickel, which caused the reaction to fail, certain salts dramatically improved the reaction. When ethylene, the simplest compound containing a carbon-carbon double bond, is bubbled into a hydrocarbon solvent containing a very small amount of an organoaluminum compound and **titanium** tetrachloride (a volatile liquid compound used now chiefly in skywriting and smoke screens because it fumes in moist air), there is formation of **polyethylene** (a long, straight hydrocarbon chain). The reaction conditions are very mild, consisting of atmospheric pressure and room **temperature**.

Polyethylene had been previously produced by the British company Imperial Chemicals Industries, but their method required temperatures up to two hundred degrees centigrade and pressures up to two thousand atmospheres. The ICI polyethylene had shorter chains, and the chains were branched; the substance was waxy and products made from it were soft and easily deformed. On the other hand, Ziegler's polyethylene was hard and rigid, and could be drawn into fibers. Many useful products could be made from Ziegler's low-pressure polyethylene, starting from inexpensive, abundant starting materials. Ziegler refined the process and investigated other catalyst systems. Chemical companies worldwide showed immediate interest in Ziegler's discovery and paid for the right to use it. Among those who extended Ziegler's work was Natta, a chemist at the University of Milan, who showed how the geometry of the polymer could be controlled by the catalyst, and made different polypropylenes, whose physical properties were determined by their molecular geometries.

The discovery of the Ziegler catalysts had a profound effect on the course of chemical research and development. Industrial and academic chemists turned their attention to the wide area of organometallic chemistry in order to understand the fundamental chemistry and to discover useful catalysts for polymerization and other commercial reactions. In Ziegler's Nobel lecture, he showed a world map that indicated large chemical plants that were producing polyethylene and other products based on his research only ten years after his initial discovery. Ziegler became wealthy as a result of his research, and when he was seventy years old, he established the Ziegler Fund for Research with ten million dollars. Ziegler had no political connections with the Nazi government, and he was welcomed at the Institute for Coal Research because his work could continue in the postwar period without interference from the Allies. After resurrecting the German Chemical Society in 1949, Ziegler served as its first president. He retired from the institute in 1969 after bringing great prestige and funding to it.

Ziegler had a long and happy marriage to Maria Kurz, whom he married in 1922. His daughter Marianne was a physician and his son Erhart was a physicist. Ziegler was able to enjoy himself outside the laboratory; among his hobbies were collecting paintings and hiking in the mountains. Ziegler died on August 12, 1973, after a short illness.

Ziegler-Natta Reaction

The revolutionary Ziegler-Natta reaction involves the catalytic polymerization of unsaturated hydrocarbons (alkenes) to yield non-branching, linear polymers. These polymers are known as *high-density* on account of their superior strength and high crystallinity. Indeed, high-density **polyethylene** (HDPE) is invaluable as a household plastic and high-density **polypropylene** (HDPP) as carpet fibers, amongst other innumerable applications. Professor **Karl Ziegler** in Germany accidentally discovered the Ziegler-Natta catalyst in 1953. Equipment from a previous experiment had not been cleaned thoroughly before a polymerization of ethylene and unexpected results were observed. After careful investigation a **nickel** residue was found and sparked the painstaking search for an ideal catalyst. The optimum catalyst was found to be a **solution** of **titanium** tetrachloride and triethylaluminium. The Italian chemist **Giulio Natta** successfully applied the catalysts to polymerize propylene. Furthermore, Natta investigated the spatial orientation of side-groups, or branches, in the new polymers. Traditional polymer synthesis results in the random (atactic) location of the side-groups either to the left or right of the polymer chain. However, the Ziegler-Natta catalysts yield polymers with all the side-groups positioned on the same side (isotactic) and thus, permitting the industrial synthesis of materials with properties identical to natural rubber. This extremely desirable structure owes its elasticity to the isotactic nature of the polymer, and for this reason these polymers are used for the manufacturing of car tires. The catalysts had revolutionized the polymer industry within ten short years and in 1963 Ziegler and Natta shared the Nobel Prize. The catalytic mechanism itself is chemically complex and while a number of theories have been proposed, it is still not entirely understood in 1999. The original catalyst was improved in the early 1970s by employing a **magnesium** dichloride support that dramatically increased catalytic activity and further minor modifications have been made since then. Homogenous metallocene catalyst of ethylene and propylene polymerization were discovered in the 1980s, and their application to the synthesis of new **plastics** remains an intense area of research.

A one ton specimen of zinc ingot. ***(Photo Researchers, Inc. Reproduced by permission.)***

See also Chirality

Zinc

Zinc is the first element in Group 12 of the **periodic table**, one of the transition **metals**. It has an **atomic number** of 30, an atomic **mass** of 65.38, and a chemical symbol of Zn.

Properties

Zinc is a bluish white metal that is neither ductile nor malleable. It has a melting point of 787.1°F (419.5°C), a **boiling**

point of 1,666°F (908°C), and a **density** of 7.14 grams per cubic centimeter. Zinc is a fairly active metal that dissolves in both acids and strong alkalis. It does not react with **oxygen** in dry air very readily, but it does react in moist air to form zinc carbonate ($ZnCO_3$). Zinc burns in oxygen with a blue flame.

Occurrence and Extraction

Zinc is about the 24th most abundant element in the Earth's crust with an abundance of about 0.02%. It never occurs free as an element, but is most commonly found in minerals such as smithsonite (zinc spar or zinc carbonate; $ZnCO_3$); spahlerite (zinc blende or zinc sulfide; ZnS); zincite (zinc **oxide**; ZnO); willemite (zinc silicate; $ZnSiO_3$); and franklanite (a complex of zinc, **manganese**, **iron**, and oxygen). Zinc is usually produced by first **roasting** the native ore to produce zinc oxide and then reducing the zinc oxide with **charcoal**.

Discovery and Naming

Zinc and its compounds were known to ancient peoples, but not well understood. The problem was that zinc, when heated, tends to boil away rather easily, so efforts to make tools, weapons, and other materials from zinc were generally unsuccessful. The element was probably most widely used and best known in the form of brass, an **alloy** of zinc and **copper**. The alloy occurs naturally in the earth and has been made by humans for untold centuries.

The first formal description of zinc's properties was probably that of the Swiss physician Paracelsus. Paracelsus was an alchemist as well as a physician, and he wrote about his experiences in working with zinc during the early 1500s. The name zinc was first used in 1651 and came from the German name for the element, *zink*. The original meaning of that word is not known.

Uses

The primary use of zinc is in galvanizing other metals, especially iron and steel. Galvanization is the process by which a thin layer of zinc is laid down on the surface of the base metal. The purpose of galvanization is to protect the underlying metal from **corrosion**. Zinc tends to be much less reactive with oxygen than are the metals that are galvanized.

The second largest use of zinc is in making alloys, the two most important of which are brass and bronze. Bronze is made primarily from copper and **tin**, but may also include a small amount of zinc. Among the uses of zinc alloys are automobile parts, roofing, gutters, **batteries**, organ pipes, electrical fuses, type metal, household utensils, and building materials.

Many zinc compounds have commercial applications. Some examples include zinc acetate [$Zn(C_2H_3O_2)$], used as a **wood** preservative, a dye for textiles, an additive for animal feeds, and glazing for **ceramics**; zinc chloride ($ZnCl_2$), used as a solder, for fireproofing materials, as a food preservative, as an additive in antiseptics and deodorants, as a dental cement, in petroleum refining, and in embalming and taxidermy products; zinc fluorosilicate ($ZnSiF_6$), as a mothproofing agent and hardener for concrete; zinc hydrosulfite (ZnS_2O_4), as a bleaching agent for textiles, straw, vegetable oils and other products and as a brightening agent for **paper** and beet and cane sugar juice; zinc oxide (ZnO), used in rubber production, as a white pigment in paint, to prevent the growth of mold on paints, and in the manufacture of **glass**, ceramics, tiles, and **plastics**, and zinc **sulfate** ($ZnSO_4$), used in the manufacture of rayon, as a supplement in animal feeds, and in the dyeing of textiles.

Health Issues

Zinc is an essential micronutrient for plants, humans, and other animals. In plants, a zinc deficiency interferes with plant reproduction. Plants deficient in the element may form flowers, but the flowers do not produce seeds.

In humans, zinc deficiencies can produce serious consequences. Zinc is used to build **DNA** molecules, so an insufficient supply of zinc can interfere with both growth and reproduction. For example, young children who do not get enough zinc in their diet may experience loss of hair, skin lesions, and stunted growth.

By contrast, an excess of zinc can also cause health problem. People who work with zinc metal may breathe in zinc dust and experience health problems, such as dryness in the throat, coughing, general weakness and aching, chills, fevers, nausea, and vomiting. One sign of zinc poisoning is a sweet **taste** in the mouth that cannot be associated with eating sweet foods.

ZIRCONIUM

Zirconium is the second element in Group 4 of the **periodic table**, one of the transition **metals**. It has an **atomic number** of 40, an atomic **mass** of 91.22, and a chemical symbol of Zr.

Properties

Zirconium usually occurs as a hard, grayish white metal, whose surface has a flaky appearance. It may also be prepared as a black or bluish black powder. Zirconium's melting point is 3,375°F (1,857°C), its **boiling point** is 6,471°F (3,577°C), and its **density** is 6.5 grams per cubic centimeter. Zirconium is a fairly inactive element that combines slowly with **oxygen** in the air, but does not react with most cold acids or with **water**. One of its most important and useful properties is its transparency to neutrons. Most metals tend to absorb neutrons that pass through them, but zirconium is largely transparent to them.

Occurrence and Extraction

Zirconium is a relatively common element in the Earth's crust with an abundance estimated at about 150-230 parts per million. Its most common ores are zircon (zirconium silicate; $ZrSiO_4$) and baddeleyite (zirconia or zirconium **oxide**; ZrO_2). The largest suppliers of zirconium in the world are South Africa and Australia.

Discovery and Naming

Zirconium was discovered in 1789 by the German chemist **Martin Heinrich Klaproth**. Klaproth found the element in a stone brought to him from the island of Ceylon (now Sri Lanka). The stone was commonly known as a jacinth (or hyacinth) stone and had been known for many years. In fact, Saint John talks about the jacinth stone as one of the jewels found in the walls surrounding Jerusalem. Klaproth chose zirconium as a name for the new element from the Persian name for the same stone, *zargun*, meaning ''gold-like.''

Uses

A very small amount of zirconium is used in the manufacture of alloys for products such as flash bulbs, rayon spinnerets, lamp filaments, precision tools, and surgical instruments. About 95% of all the element produced, however, is converted to compounds, the two most important of which are zircon (zirconium silicate) and zirconia (zirconium oxide). Synthetic zircon is used to make gemstones that resemble fine diamonds and as a refractory material in foundry molds and furnace linings. Zircon and zirconium are both used extensively as abrasives in industrial operations.

ZOSIMOS OF PANOPOLIS (circa 300 A.D.)

Egyptian alchemist

Zosimos was an Egyptian who wrote a book on **alchemy** in the fourth century A.D. There is not much more that can be observed about his life with any degree of certainty. He is mentioned in ancient books as a wise man and a mystical figure. These authors, writing centuries after he died, reported that Zosimos was born in Panopolis, Upper Egypt. One source refers to his home as Alexandria, where he is believed to have died. A famous alchemist associated with the holy place of Serapeum gave a book of Zosimos to the high priest there. However, the Serapeum was destroyed in 389 A.D. Thus, the time of Zosimos is established close to 300 A.D.

Many books attributed to Zosimos have been translated into Greek, Arabic, and French. However, the authenticity of these books is difficult to establish because of inaccurate and anachronistic references. The historically accepted figure of Zosimos is that of the author of a series of books on alchemy described in an ancient Arabic source. The books are addressed to Zosimos' sister, where the author attempts to warn her of her materialistic love for **gold**. Although Zosimos eventually capitulates to describing the apparatus and methods of alchemy, most of the books involve spiritual teachings. He believed that the **transmutation** of **metals** cannot come about without spiritual ceremony.

It is important for the student of **chemistry** to understand Zosimos because, in his writings, are the evidence and the origin of the mysticism inherent in ancient alchemy. This mysticism is what Paracelsus and Libavius warned against while attempting to transform chemistry into a branch of science. Interestingly, just as Zosimos entreated his sister to be more altruistic, Paracelsus and Libavius encouraged their students to use chemistry to serve others.

Zirconium alloys are used to make concrete drill bits. ***(Photograph by Robert J. Huffman. Reproduced by permission.)***

ZSIGMONDY, RICHARD (1865-1929)

German colloidal chemist

Although trained as an organic chemist, Richard Zsigmondy earned fame in the field of colloidal **chemistry**, the study of fine dispersions of a material in a **solution** of another substance. Colloids had been well known and widely used for centuries, but at the dawn of the twentieth century very little was known about their physical and chemical nature. To learn more about this class of materials, Zsigmondy invented a number of tools, including the ultramicroscope, with which he was able to study colloids more closely. Such equipment allowed Zsigmondy to make a number of fundamental discoveries about the composition and properties of colloids. For this work he was awarded the 1925 Nobel Prize in chemistry, the first and one of the few times this award has been given for research on colloids.

Richard Adolf Zsigmondy was born in Vienna on April 1, 1865. His father, Adolf Zsigmondy, a dentist and an inventor of surgical instruments, and his mother, the former Irma von Szakmáry, oversaw their four sons' home experiments in chemistry and physics. After Zsigmondy graduated from high

Richard Zsigmondy.

school in 1883, he enrolled at the Vienna Technische Hochschule, where he majored in chemistry. At the time, the Hochschule emphasized **organic chemistry**, which eventually became Zsigmondy's major field of study. However, he also became interested in the colorization of glasses, and he collaborated with a Prague chemist in some original research at a nearby **glass** factory. In 1887 Zsigmondy completed his studies at the Hochschule and began a graduate program in organic chemistry.

Some disagreement among scholars surrounds the conditions of Zsigmondy's doctoral work. Most authorities say that he attended the University of Munich and was granted his Ph.D. in organic chemistry in 1885. In one of the most complete biographies available, however, George Fleck in the American Chemical Society's *Nobel Laureates in Chemistry, 1901–1992,* claims that Zsigmondy did his research on **chlorine** derivatives at the Munich Technische Hochschule, which "was the basis for the doctor of philosophy degree awarded to Zsigmondy by the University of Erlangen in December 1889."

After receiving his doctorate, Zsigmondy became an assistant to A. A. Kundt at the University of Berlin's Institute of Physics. Kundt, an authority on the colorization of glass by inorganic materials, further encouraged Zsigmondy's interest in this field. In 1893 Zsigmondy became qualified to teach and accepted a job as *privatdozent* at the Technische Hochschule in Graz. He joined the Schott Glass Manufacturing Company in Jena, Germany, in 1897; there he continued his work on the colorization of glass and invented a product known as *Jena milk glass* that was later to attain wide commercial popularity.

Zsigmondy's work with colored glass led to an increased interest in colloids, which are often responsible for the colorization of glassy materials. Colloids are mixtures of two substances that do not form a solution, but that do not separate even after standing for long periods of time. For example, if one were to add powdered **iron** to **water**, the iron would not dissolve, but would remain suspended for some period of time. Eventually, however, the iron would settle at the bottom of the container.

Under certain conditions, however, the iron can be made into such fine particles that, although still not dissolved, remain in **suspension** essentially forever. The science of colloidal chemistry is devoted to a study of the ways in which such mixtures can be made, of their properties, and of the ways the particles can be made to settle out.

A fundamental problem with colloidal research is that, although not of atomic size, colloidal particles are too small to be seen with ordinary light microscopes. Direct observation of such particles was therefore impossible before Zsigmondy's time. The one method that was (and is) commonly available for the study of colloids relies upon the so-called Tyndall effect. The Tyndall effect occurs when colloidal particles scatter light shined through a mixture. A common example of this effect is the scattering of light that occurs when a beam of light shines through a smoky room.

Zsigmondy concluded that an instrument could be developed that makes use of the Tyndall effect. In this instrument, called the ultramicroscope, light is reflected off particles not in the same direction as the incident light, as in a conventional microscope, but at right angles to the incident beam.

From 1900 to 1907 Zsigmondy's work was supported by his family's own fortune, and he pursued his research without any official professional affiliation. During this time he was invited, however, to make use of the superb facilities at the Zeiss Optical Company in Jena for his research on the ultramicroscope. There he collaborated with Zeiss physicist H. F. W. Siedentopf ; together, the two men produced the first microscope of Zsigmondy's design, with which he soon made a number of discoveries about colloidal materials. For example, in 1898 Zsigmondy discovered that the valuable dye known as Cassius purple is actually a suspension of colloidal **gold** and stannic acid particles. In 1907 Zsigmondy returned to academia as assistant professor of **inorganic chemistry** and director of the Institute for Inorganic Chemistry at the University of Göttingen. Twelve years later he was promoted to full professor, a post he held for the rest of his life. Zsigmondy died from arteriosclerosis at his home in Göttingen on September 24, 1929.

Zsigmondy was married to Laura Luise Müller in 1903; the couple had two daughters. Zsigmondy's work was recognized by his election to the scientific academies of Göttingen, Vienna, Uppsala, Zaragoza, Valencia, and Haarlem. He was awarded honorary doctorates from the University of Königsberg and the Technische Hochschules at Vienna and Graz. Throughout his life, Zsigmondy's leisure-time passions were hiking and mountain climbing.

Sources Consulted

Books

Adams, Charles K. *Nature's Electricity.* Blue Ridge Summit, PA: Books, Inc., 1987.

American Men and Women of Science: A Biographical Directory of Today's Leaders in Physical, Biological, and Related Sciences, 1998-99, 20th edition. New Providence, NJ: R.R. Bowker, 1998.

Asimov, Isaac. *Asimov's Chronology of Science and Discovery.* New York: Harper & Row, Publishers, 1989.

Atherly, A.G., J.R. Girton, and J.F. McDonald. *The Science of Genetics.* Fort Worth, TX: Saunders College Publishing, 1999.

Atkins, P.W. *Molecular Quantum Mechanics,* 2nd edition. Oxford: Oxford University Press, 1983.

——. *Quanta: A Handbook of Concepts,* 2nd edition. Oxford: Oxford University Press, 1991.

——. *Physical Chemistry,* 6th edition. Oxford: Oxford University Press, 1997.

Azaroff, Leonid V. *Elements of X-Ray Crystallography.* New York: McGraw-Hill Book Company, 1968.

Bailey, Philip S., Jr., and Christina A. Bailey. *Organic Chemistry: A Brief Summary of Concepts and Applications,* 4th edition. Englewood Cliffs, NJ: Prentice Hall, 1989.

Baker, Arthur, and Konrad Krauskopf. *Introduction to Physics and Chemistry.* New York: McGraw-Hill Book Company, 1964.

Basolo, Fred, and Ronald C. Johnson. *Coordination Chemistry,* 2nd revised and enlarged edition. Science Reviews, 1989.

Bettelheim, Frederick A., and Jerry March. *Introduction to General, Organic, & Biochemistry.* Philadelphia: Saunders College Publishing, 1983.

Bockris, John O'M., and Amulya K. N. Reddy. *Modern Electrochemistry* New York: Plenum Press, 1973.

Boorse, Henry A., Lloyd Motz, and Jefferson Hane Weaver. *The Atomic Scientists: A Biographical History.* New York: John Wiley & Sons, Inc., 1989.

Boyer, Rodney. *Concepts in Biochemistry.* Pacific Grove, CA: Brooks/Cole Publishing Company, 1999.

Brock, William H. *The Norton History of Chemistry.* New York: W.W. Norton & Company, 1992.

Bruice, Paula Y. *Organic Chemistry.* Englewood Cliffs, NJ: Prentice-Hall, Inc., 1995.

Bynum, W.F., E.J. Browne, and Roy Porter, eds. *Dictionary of the History of Science.* Princeton, NJ: Princeton University Press, 1984.

Campbell, A. M. "Monoclonal Antibodies." In *Immunochemistry,* edited by Carol J. van Oss and Marc H. V. van Regenmortel. New York: Marcel Dekker, Inc., 1994.

Carroll, Felix A. *Perspectives on Structure and Mechanism in Organic Chemistry.* Pacific Grove, CA: Brooks/Cole Publishing Company, 1998.

The Chambers Dictionary of Science and Technology. Scotland: W & R Chambers, 1974.

Clifford, Martin. *Basic Electricity & Beginning Electronics.* Blue Ridge Summit, PA: TAB Books, 1973.

Cobb, Cathy, and Harold Goldwhite. *Creations of Fire: Chemistry's Lively History from Alchemy to the Atomic Age.* New York: Plenum Press, 1995.

Collings, Peter J. *Liquid Crystals: Nature's Delicate Phase of Matter.* Princeton, NJ: Princeton University Press, 1990.

The Concise Encyclopedia of Biochemistry and Molecular Biology, 3rd edition. Berlin: Walter de Gruyter, 1997.

Conant, James B. *The Overthrow of the Phlogiston Theory: The Chemical Revolution of 1775-1789*. Cambridge, MA: Harvard University Press, 1966.

Cotton, F. Albert. *Advanced Inorganic Chemistry: A Comprehensive Text,* 2nd edition. New York: Interscience Publishers, 1966.

Curtis, Helena. *Biology.* New York: Worth Publishers, Inc., 1983.

Dainith, John, Sarah Mitchell, Elizabeth Tootill, and Derek Gjertsen, eds. *Biographical Encyclopedia of Scientists,* 2nd edition. Bristol, UK: Institute of Physics Publishing, 1994.

Dainith, John, ed. *A Dictionary of Chemistry,* 3rd edition. Oxford: Oxford University Press, 1996.

Davies, Paul, and John Gribbin. *The Matter Myth: Dramatic Discoveries that Challenge Our Understanding of Physical Reality.* New York: Touchstone Books, 1992.

Davis, Raymond E., H. Clark Metcalfe, John E. Williams, and Joseph F. Castka, eds. *Modern Chemistry,* Teacher's Edition. Austin, TX: Holt, Rinehart and Winston, Inc. 1999.

Delgass, W. Nicholas, Gary L. Haller, Richard Kellerman, Jack H. Lunsford, eds. *Spectroscopy in Heterogeneous Catalysis*. New York: Academic Press, 1979.

Ebbing, Darrell D., and Steven D. Gammon. *General Chemistry,* 6th edition. Boston: Houghton Mifflin Company, 1999.

Elliott, W.H., and D.C. Elliott. *Biochemistry and Molecular Biology.* New York: Oxford University Press, 1997.

Elschenbroich, Christoph, and Albrecht Salzer. *Organometallics: A Concise Introduction,* 2nd revised edition. New York: VCH, 1992.

Ewing, Galen W. *Instrumental Methods of Chemical Analysis,* 4th edition. New York: McGraw-Hill Book Company, 1975.

Feldman, Anthony, and Peter Ford. *Scientists & Inventors.* New York: Facts on File, 1979.

Fieser, Louis F., and Mary Fieser. *Textbook of Organic Chemistry.* Boston: D.C. Heath and Company, 1950.

Fike, D.J. "Immunoglobulin Structure and Function." In *Clinical Immunology: Principles and Laboratory Diagnosis,* 2nd edition. Catherine Sheehan, editor. Philadelphia: Lippincott-Raven Publishers, 1997.

Flaste, Richard, ed. *What Everyone Needs to Know from Newton to the Knuckleball.* New York: Times Books/Random House, 1991.

Friend, J. Newton. *Man and the Chemical Elements: An Authentic Account of the Successive Discovery and Utilization of the Elements From the Earliest Times to the Nuclear Age,* 2nd revised edition. New York: Charles Scribner's Sons, 1961.

Frisch, Otto R. *The Nature of Matter*. New York: E.P. Dutton, 1972.

Garrity, John A., and Mark C. Carnes, eds. *American National Biography,* 7 volumes. New York: Oxford University Press, 1999.

Gillispie, Charles Coulston, editor-in-chief. *Dictionary of Scientific Biography,* 15 volumes. New York: Charles Scribner's Sons, 1985.

Gittwitt, Paul G. *Conceptual Physics,* 6th edition. New York: HarperCollins Publishers, 1989.

Goodman, H. Maurice. *Basic Medical Endocrinology,* 2nd edition. New York: Raven Press, 1994.

Goyer, R.A. "Toxic Effects of Metals." In *Casarett and Doull's Toxicology: The Basic Science of Poisons,* 5th edition. New York: McGraw-Hill Companies, Inc., 1996.

Gribben, John. *Q Is for Quantum: Particle Physics from A to Z.* London: Phoenix Giant Paperback, 1999.

Harre, Rom. *Great Scientific Experiments: Twenty Experiments that Changed Our View of the World.* Oxford: Phaidon Press Limited, 1981.

Heiserman, David L. *Exploring Chemical Elements and Their Compounds*. Blue Ridge Summit, PA: TAB Books, 1992.

Henisch, Heinz K. *Crystal Growth in Gels*. University Park, PA: The Pennsylvania State University Press, 1970.

Hoffman, Robert V. *Organic Chemistry: An Intermediate Text.* New York: Oxford University Press, 1997.

Hudson, John. *The History of Chemistry.* New York: Chapman & Hall, 1992.

Huizenga, John R. *Cold Fusion: The Scientific Fiasco of the Century.* Oxford: Oxford University Press, 1993.

Hunter, Robert J. *Introduction to Modern Colloid Science.* Oxford: Oxford University Press, 1993.

Iler, Ralph K. *The Chemistry of Silica.* New York: John Wiley & Sons, Inc., 1979.

Interrante, Leonard V., Lawrence A. Caspar, and Arthur B. Ellis, eds. *Materials Chemistry: An Emerging Discipline.* Washington, DC: American Chemical Society, 1995.

James, Laylin K, ed. *Nobel Laureates in Chemistry: 1901-1992.* American Chemical Society and the Chemical Heritage Foundation, 1993.

Jolly, William L. *Modern Inorganic Chemistry.* New York: McGraw-Hill Book Company, 1984.

Kane, Gordon. *The Particle Garden: Our Universe as Understood by Particle Physicists*. Reading, MA: Helix Books, 1995.

The Kirk-Othmer Encyclopedia of Chemical Technology, 4th edition. New York: John Wiley & Sons, Inc., 1993.

Kittel, Charles. *Thermal Physics.* New York: John Wiley & Sons, Inc., 1969.

Kleinsmith, L.J., and V.M. Kish. *Principles of Cell and Molecular Biology,* 2nd edition. New York: HarperCollins College Publishers, 1995.

Klug, William S., and Michael R. Cummings. *Concepts of Genetics,* 5th edition. Upper Saddle River, NJ: Prentice-Hall, Inc., 1997.

Krane, Kenneth S. *Modern Physics.* New York: John Wiley & Sons, Inc., 1983.

LeMay, H. Eugene, Herbert Beall, Karen M. Robblee, and Douglas C. Brower, eds. *Chemistry: Connections to Our Changing World,* Teacher's Edition. Upper Saddle River, NJ: Prentice-Hall, Inc., 1996.

Lerner, Rita G., and George L. Trigg, eds. *Encyclopedia of Physics,* 2nd edition. New York: VCH Publishers, Inc., 1991.

Lewis, Richard J., ed. *Hawley's Condensed Chemical Dictionary,* 13th edition. New York: Van Nostrand Reinhold, 1997.

Ley, Willy. *The Discovery of the Elements.* New York: Delacorte Press, 1968.

Lin, E.C.C., Richard Goldstein, Michael Syvanen. *Bacteria, Plasmids, and Phages: An Introduction to Molecular Biology.* Cambridge: Harvard University Press, 1984.

Lovelock, James. *The Ages of Gaia: A Biography of Our Living Earth,* revised and expanded edition. New York: W.W. Norton & Company, 1988.

——. *Gaia: A New Look at Life on Earth.* Oxford: Oxford University Press, 1995.

Lowry, Thomas H., and Kathleen Schueller Richardson. *Mechanism and Theory in Organic Chemistry,* 3rd edition. New York: HarperCollins Publishers, Inc., 1987.

Luciano, D.S., A.J. Vander, and J.H. Sherman, eds. *Human Anatomy and Physiology: Structure and Function,* 2nd edition. New York: McGraw-Hill Book Company, 1983.

Margulis, Lynn, and Dorion Sagan. *Slanted Truths: Essays on Gaia, Symbiosis, and Evolution.* New York: Copernicus, Springer-Verlag New York, Inc., 1997.

Mascetta, Joseph A. *Chemistry the Easy Way,* 3rd edition. Barron's, 1996.

Mason, Stephen F. *Chemical Evolution: Origin of the Elements, Molecules, and Living Systems.* Oxford: Claredon Press, 1991.

Masterson, William L., and Emil J. Slowinski, eds. *Chemical Principles,* 4th edition. Philadelphia: W.B. Saunders Company, 1977.

Maton, Anthea, Jean Hopkins, Susan Johnson, David LaHart, Maryanna Quon Warner, and Jill D. Wright, eds. *Exploring Physical Science,* Teacher's Edition. Upper Saddle River, NJ: Prentice-Hall, Inc., 1999.

Matthews, Christopher K., and K. E. Van Holde, eds. *Biochemistry,* 2nd edition. New York: Benjamin/Cummings Publishing Company, 1966.

McEvoy, G.K., ed. *AHFS Drug Information 1999.* Bethesda, MD: American Society of Health-System Pharmacists, Inc., 1999.

The McGraw-Hill Concise Encyclopedia of Science and Technology. New York: McGraw-Hill Book Company, 1986.

McGraw-Hill Modern Scientists and Engineers. New York: McGraw-Hill, Inc. 1980.

McQuarrie, Donald A., and Peter A. Rock. *Descriptive Chemistry.* New York: W.H. Freeman and Company, 1985.

Morrison, Robert Thornton, and Robert Neilson Boyd. *Organic Chemistry,* 5th edition. Boston: Allyn and Bacon, Inc., 1987.

Muir, Hazel, ed. *Larousse Dictionary of Scientists.* New York: Larousse Kingfisher Chambers, Inc., 1994.

Multhauf, Robert P. *The Origins of Chemistry.* London: Oldburne Book Co. LTD, 1966.

Otoxby, David W., H.P. Gillis, and Norman H. Nachtrieb, eds. *The Principles of Modern Chemistry.* Fort Worth, TX: Harcourt Brace College Publishers, 1999.

The Oxford Dictionary of Biochemistry and Molecular Biology. Oxford: Oxford University Press, 1997.

Pauling, Linus. *The Chemical Bond: A Brief Introduction to Modern Structural Chemistry.* Ithaca, NY: Cornell University Press, 1967.

Pitzer, Kenneth S. *Thermodynamics,* 3rd edition. New York: McGraw-Hill, Inc., 1995.

Porter, Roy. *The Greatest Benefit to Mankind: A Medical History of Humanity.* New York: W.W. Norton & Company, Inc., 1997.

Puddephatt, R. J., and P.K. Monaghan. *The Periodic Table of the Elements,* 2nd edition. Oxford: Clarendon Press, 1986.

Rayner-Canham, Geoffrey, Arthur Last, Robert Perkins, Mark van Roode. *Foundations of Chemistry.* Reading, PA: Addison-Wesley Publishing Company, 1983.

Russell, C.A., ed. *Recent Developments in the History of Chemistry.* London: The Royal Society of Chemistry, 1985.

Ryan, Charles W. *Basic Electricity: A Self-Teaching Guide,* 2nd edition. New York: John Wiley & Sons, Inc., 1986.

Salzberg, Hugh W. *From Caveman to Chemist: Circumstances and Achievements.* Washington, DC: American Chemical Society, 1991.

Schneider, Stephen H., and Penelope J. *Scientists on Gaia*. Cambridge, MA: The MIT Press, 1991.

Scott, Stephen K. *Oscillations, Waves, and Chaos in Chemical Kinetics*. New York: Oxford University Press, 1994.

Siegel, George J., Bernard W. Agranoff, R. Wayne Albers, and Perry B. Molinoff, eds. *Basic Neurochemistry: Molecular, Cellular, and Medical Aspects,* 5th edition. New York: Raven Press, 1994.

Simon, George P. *Ion Exchange Training Manual*. New York: Van Nostrand Reinhold, 1991.

Slater, J.C. *Introduction to Chemical Physics*. New York: Dover Publications, Inc., 1939.

Snell, R.S. *Clinical Anatomy for Medical Students,* 5th edition. Boston: Little, Brown and Company, 1995.

Solomons, T.W. Graham. *Fundamentals of Organic Chemistry,* 5th edition. New York: John Wiley & Sons, Inc., 1997.

Sperling, L. *Introduction to Polymer Science*. New York: John Wiley & Sons, Inc., 1992.

Spraycar, M., ed. *Stedman's Medical Dictionary,* 26th edition. Baltimore: Williams and Wilkens, 1995.

Stocchi, E. *Industrial Chemistry,* Volume 1. Translated by K.A.K. Lott and E.L. Short. Chichester, West Sussex, UK: Ellis Horwood Limited, 1990.

Swalin, Richard A. *Thermodynamics of Solids,* 2nd edition. New York: John Wiley & Sons, Inc., 1972.

Taber's Cyclopedic Medical Dictionary, 17th edition. Philadelphia: F.A. Davis Company, 1993.

Taylor, Hugh S. *A Treatise on Physical Chemistry: A Co-operative Effort by a Group of Physical Chemists*. 2 volumes, 2nd edition. New York: D. Van Nostrand Company, Inc., 1931.

Tipler, Paul. *Foundations of Modern Physics*. New York: Worth Publishers, Inc., 1969.

Tippens, Paul E. *Applied Physics,* 3rd edition. New York: Gregg Division, McGraw-Hill Book Company, 1984.

Tocci, Salvatore, and Claudia Viehland. *Chemistry: Visualizing Matter,* Annotated Teacher's Edition. Austin, TX: Holt, Rinehart and Winston, Inc., 1996.

Ullman's Encyclopedia of Industrial Chemistry. Weinheim, Germany: UCH, 1987.

Van Nostrand's Scientific Encyclopedia, 5th edition. New York: Van Nostrand Reinhold Company, 1976.

Weast, Robert C. *Handbook of Chemistry and Physics*. Cleveland, OH: CRC Press, 1975.

Weaver, Jefferson Hane. *The World of Physics: A Small Library of the Literature of Physics from Antiquity to the Present*. New York: Simon and Schuster, 1987.

Weisburger, E.K. "General Principles of Chemical Carcinogenesis." In *Carcinogenesis*. Edited by Michael P. Waalkes and Jerrold M. Ward. New York: Raven Press, Ltd., 1994.

Wenninger, J.A., and G.N. McEwen, Jr., eds. *International Cosmetic Ingredient Dictionary and Handbook,* 7th edition. Washington, DC: The Cosmetic, Toiletry, and Fragrance Association, 1997.

Who Was Who in America: Historical Volume, 1607-1896, revised edition. Chicago: Marquis Who's Who, Incorporated, 1967.

Willard, Hobart H., Lynn L. Merritt, Jr., John A. Dean, Frank A. Settle, Jr. *Instrumental Methods of Analysis,* 7th edition. Belmont, CA: Wadsworth Publishing Company, 1988.

Williams, E.T., and C.S. Nicholls, eds. *The Dictionary of National Biography: 1961- 1970.* New York: Oxford University Press, 1981.

Williams, Trevor I., ed. *A Biographical Dictionary of Scientists,* 3rd edition. New York: John Wiley & Sons, Inc., 1982.

Wolfe, Drew H. *General, Organic, and Biological Chemistry.* New York: McGraw-Hill Book Company, Inc. 1986.

Woolfson, M.M. *An Introduction to X-Ray Crystallography.* Cambridge: Cambridge University Press, 1970.

Journal Articles

Benin, A.L., J.D. Sargent, M.Dalton, and S.Roda. "High Concentrations of Heavy Metals in Neighborhoods near Ore Smelters in Northern Mexico." *Environmental Health Perspectives* 107, 4 (1999): 279-84.

Bertazzi, P.A., I. Bernucci, G. Brambilia, et al. "The Seveso Studies on Early and Long-term effects of Dioxin Exposure: A Review." *Environmental Health Perspectives* 106, 2 (1998): 625-33.

Bollen, M., S. Keppens, and W. Stalmans. "Specific Features of Glycogen Metabolism in the Liver." *Biochemistry Journal* 336, (1998): 19-31.

Bowles, R.L., R.P. Davie, and W.D. Todd. "A Method for the Interpretation of Brookfield Viscosities." *Modern Plastics* (November 15, 1955): 140-48.

Brandt, Laura A. "The Volume Baker's Sugar-Free Challenge." *Baking Management* (March 1999): 34-8.

Breslin, K. "Safer Sips: Removing Arsenic from Drinking Water." *Environmental Health Perspectives* 106, 11, (1998): A548-50.

Brown, S.D. " Has the Chemometrics Revolution Ended? Some Views on the Past, Present, and Future of Chemometrics." *Chemometrics and Intelligent Laboratories Systems* 30, 1 (November 1995): 49-58.

Chisholm, Michael. "Plastic Fantastic: From Aircraft Windows to Domestic Baths, Chemists Have Long Exploited the Unique Properties of Polymethyl Methacrylate." *Chemistry in Britain* (April 1998): 33-36.

Collins, F.S., A. Patrinos, E. Jordon, et al. "New Goals for the U.S. Human Genome Project." *Science* 282 (October 23, 1998): 682-89.

Doyle, Ellin. "Trans Fatty Acids." *Journal of Chemical Education* 74, 9 (September 1997): 1030-32.

Drozd, Joseph C. "Soaps from Genetically Engineered Laurate Canola." *Soap/Cosmetic/Chemical Specialties* (October 1996): 62-5.

Eubanks, M.W. "Hormones and Health." *Environmental Health Perspectives* 105, 5 (1997):482-87.

Fishman, Harvey M. "Fatty Acids: Part Two." *Happi* (December 1994): 30.

Frank, Ildiko E., and Bruce R. Kowalski. "Chemometrics." *Analytical Chemistry* 54, 5 (April 1982): 232R-43R.

Kiefer, David M. "The Tide Turns for Soaps." *Today's Chemist at Work* (October 1996): 70-4.

Kirschner, Elizabeth M. "Soaps and Detergents." *C & EN* (January 27, 1997): 30-44.

Lenton, Timothy M. "Gaia and Natural Selection." *Nature* 394 (July 30, 1998): 439-47.

Manuel, J. "NIEHS and CDC Track Human Exposure to Endocrine Disruptors." *Environmental Health Perspectives* 107, 1 (1999):A16.

McCarthy, Daniel C. "Genetic Engineering Puts a New Spin on Silk." *Biophotonics International* (August 1999).

Milius, S. "Octopus Suckers Glow in the Deep, Dark Sea." *Science News* 155 (March 13, 1999): 167.

Miller, R.D., and J. Michl. "Polysilane High Polymers." *Chemical Reviews* 89 (1989): 1359-1410.

Minamoto, T., M. Mai, and Z. Ronai. "Environmental Factors as Regulators and Effectors of Multistep Carcinogenesis." *Carcinogenesis* 20, 4 (1999): 519-27.

Nabors, Lyn O'Brien, and Eric Allen. "Gaining Weight in the Marketplace: Reduced-Fat Foods." *Chemtech* (December 1996): 50-2.

Ohr, Linda M. "A Sampling of Sweeteners." *Prepared Foods* (March 1998): 57.

Prigogine, Ilya, Gregorie Nicolis, and Agnes Babloyantz. "Thermodynamics of Evolution." *Physics Today* (November 1972): 23-8.

Scott, Stephen K. "Oscillations in Simple Models of Chemical Systems." *Accounts of Chemical Research* 20 (1987): 186-91.

Sita, L.R. "Heavy-Metal Organic Chemistry: Building with Tin." *Accounts of Chemical Research* 27 (1994): 191-97.

Steenland, K., L. Piacitelli, J. Deddens, et al. "Cancer, Heart Disease, and Diabetes in Workers Exposed to 2,3,7,8-tetrachlorodibenzo-*p*-dioxin." *Journal of the National Cancer Institute* 91, 9 (1999): 779-86.

Stein, Paul J. "The Sweetness of Aspartame." *Journal of Chemical Education* 74 9, (September 1997): 1112-13.

Thornton, Tricia. "The Twinkling Octopus." *Biophotonics International* (May/June 1999): 86.

U.S. Department of Health and Human Services, Public Health Services. National Toxicology Program. *The Sixth Report on Carcinogens*. 1998 Summary.

U.S. Department of Health and Human Services, Public Health Services. National Toxicology Program. 2,3,7, 8-Tetrachlorodibenzo-*p*-dioxin (TCDD). *The Eighth Report on Carcinogens*. 1998 Summary.

Wilkes, Ann Przybyla. "Gearing Up to Produce Snacks." *SNACKWorld* (March 1998): 16-18.

Winfree, A.T. "The Prehistory of the Belousov-Zhabotinsky Oscillator." *Journal of Chemical Education* 61, 8 (August 1984): 661-63.

Wold, S. "Chemometrics: What Do We Mean with It, and What Do We Want from It?" *Chemometrics and Intelligent Laboratory Systems* 30, 1 (November 1995): 109-15.

Zhao, Wenyan, Andrzej Kloczkowski, James E. Mark, Burak Erman, and Ivet Bahar. "Make Tough Plastic Films from Gelatin." *Chemtech* (March 1996): 32-8.

Websites

(Editor's Note: As the World Wide Web is constantly expanding, the URLs listed below may be altered and/or non-existant as of December 1999.)

"4-Aminobenzoic acid." http://www.chemfinder.com/cgi-win/cfserver.exe/ (3 Aug.1999).

Alexander, Steve. "The Pharmacology and Biochemistry of Cannabinoid Receptors." 1997. http://www.ccc.nottingham.ac.uk/~mqzwww/cannabinoid.html (14 July 1999).

American Dietetic Association. "Position of The American Dietetic Association: Vegetarian Diets." 1996-1999. http://www.eatright.org/adap1197.html (8 July 1999).

American Dietetic Association. "Position of The American Dietetic Association and the Canadian Dietetic Association: Nutrition for Physical Fitness and Athletic Performance for Adults." 1996-1999. http://www.eatright.org/afitperform.html (9 July 1999).

Barton, D.H.R., K. Nakanishi and O. Meth-Cohn. "Letter from the Editors." *Comprehensive Natural Products Chemistry* 1998. http://www.elsevier.co.jp/homepage/saa/conap/letter.htm (1999).

Beninger, Clifford. W, George L. Hosfield, M.G. Nair, Y. C.

Chang, and G.M. Strasburg. "Natural Products Chemistry: A New Approach to Understanding the Value of the Common Bean (*Phaseolus vulgaris l.*)." 1998. http://www.nalusda.gov/ttic/tektran/data/000008/94/0000089487.html (1999).

Berger, Dan. "Re: What Is the Difference between the Medical Nitroglycerin and the Explosive." 1997. http://128.252.223.112/posts/archives/may07/864073803.Ch.r.html (1999).

Brain, Marshall. "How Smoke Detectors Work." 1998. http://www.howstuffworks.com/smoke.htm (6 Aug. 1999).

Brown, Gilbert M., C.H. Ho, Leon Maya, Thomas J. Meyer, Bruce A. Moyer, Poonam M. Narula, Frederick V. Sloop, and Scott A. Trammel. "Utilization of Kinetic Isotope Effects for the Concentration of Tritium." 1998. http://www.doe.gov/em52/1998posters/id55103/sld001.htm (1999).

"Current Lab Highlights." *U.S. Department of Energy: Superconductivity for Electric Systems.* 1999. http://www.eren.doe.gov/superconductivity/labhighs.html (1999).

Dull, Robert W., and H. Richard Kerchner. "The Chemistry of Superconductors." *A Teacher's Guide to Superconductivity for High School Students.* 1996. http://www.ornl.gov/reports/m/ornlm3063r1/pt5.html (1999).

"Electrorefining: A Promising Technology for Treating DOE's Spent Nuclear Fuels." 1996. http://www.anl.gov/OPA/frontiers96/electroref.html (1999).

"Ethylene Oxide: Introduction." 1999. http://www.ethyleneoxide.com (1999).

Franks, Steve. "Extreme Hardness." *Properties of Diamonds.* 1997. http://www.mse.arizona.edu/classes/mse222/1997_diamond/hardness.htm (1999).

Friedman, Kenneth A., and Sharon M. Friedman. "Toxic Metals Backgrounder." *Reporting on the Environment: A Handbook for Journalists* 1994. http://www.lehigh.edu/~kaf3/books/reporting/hvymtl.html (23 July 1999).

"Gilbert N. Lewis, Chemist Who Isolated Heavy Water." http://www-itg.lbl.gov/ImgLib/COLLECTIONS/BERKELEY-LAB/PEOPLE/INDIVIDUALS/index/96602530.html (1999).

Graham, David. "Chocolate Cravings." 1996. http://www.popsci.com/content/science/news/961203.s.o.html (13 July 1999).

Hardcastle, Wayne L. "Information Paper for Management of Discarded Nitroglycerin Patches." 1998. http://chppm-www.apgea.army.mil/hmwp/Factsheets/NITROFAC.html (1999).

"Heavy Water." 1996. http://snodaq.phy.queensu.ca/SNO/D2O.html (1999).

"Heavy Water Production." *Bombs for Beginners.* 1998. http://www.fas.org/nuke/intro/nuke/heavy.htm (1999).

Hengge, Alvan. "Re: Biology and Heavy Water." 1998. http://madsci.wustl.edu/posts/archives/oct98/904868255.Bc.r.html (1999).

Iverson, Leslie. "Notes on the International Cannabinoid Research Society 1998 Symposium on Cannabinoids, La Grande Motte, France, 23-25 July 1998." 1998. http://www.erowid.org/entheogens/cannabis/uk_lords_report/Appendix4.shtml (14 July 1999).

Liljestrand, G. "Nobel Prize in Physiology or Medicine 1945: Presentation Speech." 1998. http://www.nobel.se/lareates/medicine-1945-press.html (3 Aug. 1999).

Los Angeles Cannabis Resource Center. "Cannabinoids in the Immune System." http://www.lacbc.org/immune.html (12 July 1999).

Los Angeles Cannabis Resource Center. "The Cannabinoid Revolution." http://www.lacbc.org/refsrev.html (12 July 1999).

Los Angeles Cannabis Resource Center. "Cannabinoids in the Brain." http://www.lacbc.org/brain.html (12 July 1999).

May, Paul. "Nitroglycerin." *Simple Molecules and Molecules in the Environment.* http://www.ch.ic.ac.uk/mim/environmental/html/nitroglyc_text.htm (1999).

Mihaly, Laszlo. "Interatomic Forces, Lattice Vibrations." 1996. http://solidstate.physics.sunysb.edu/book/prob/node21.html (1999).

Nakanishi, Koji. "Changing Trends in Structural Natural Products Chemistry." http://www.wspc.com.sg/books/chemistry/2416.html (1999).

National Safety Products, Inc. "Radon: An Intruder in Your Home." http://www.testproducts.com/radon-home.html (6 Aug. 1999).

National Safety Products, Inc. "Radon: How to Test Your Home." http://www.testproducts.com/radon-test.html (6 Aug. 1999).

National Safety Products, Inc. "Radon: You Can't Ignore the Facts." http://www.testproducts.com/radon1.html (6 Aug. 1999).

National Institute on Drug Abuse. "Researchers Discover Function for Brain's Marijuana-Like Compound." 1999. http://www.nida.nih.gov/MedAdv/99/NR-322.html (13 July 1999).

National Institute of Environmental Health Sciences. "Dioxin Research at the National Institute of Environmental Health Sciences (NIEHS)." 1994. http://www.niehs.nih.gov/oc/factsheets/dioxin.htm (1 July 1999).

National Human Genome Research Institute. "From Maps to Medicine: About the Human Genome Research Project." http://www.nhgri.nih.gov/Policy_and_public_affairs/Communications/Publications/Maps_to_me dicine/about.html (22 July 1999).

Oak Ridge National Laboratory. "Ethical, Legal, and Social Issues (ELSI) of the Human Genome Project." 1999. http://www.ornl.gov/hgmis/resource/elsi.html (26 July 1999).

Oak Ridge National Laboratory. "Potential Benefits of Human Genome Project Research." 1999. http://www.ornl.gov/hgmis/project/benefits.html (24 June 1999).

Oak Ridge National Laboratory. "Introduction." *Human Genome Program Report.* 1997. http://www.ornl.gov/hgmis/publicat/97pr/02intro.html (26 July 1999).

Phillips, Helen. "Of Pain and Pot Plants." 1998. http://helix.nature.com/nsu/981001-2.html (14 July 1999).

"Physics of Molecular Modeling." 1999. http://chemcca10.ucsd.edu/~chem215/lectures/lecture5/lecture5/node1.html (1999).

Press, Marina. "HUGO's Mission." *The Human Genome Organisation.* 1999 http://www.gene.ucl.ac.uk/hugo (26 July 1999).

"Questions and Answers about Tritium." *Community Relations Plan for Lawrence Berkeley Laboratory: Environmental Restoration Program.* 1993. http://www.lbl.gov/Community/FactSheet.Tritium.html (1999).

Rattray, Marcus. "A Brief Review of the History and Effects of Cannabis." http://www.umds.ac.uk/neupharm/can.htm (12 July 1999).

Society for Neuroscience. "Cannabinoids and Pain." 1999. http://gopher.sfn.org/briefings/cannabinoids.html (12 July 1999).

Taylor, Peter. "A Computational Chemistry Primer." 1996. http://www.sdsc.edu/GatherScatter/GSwinter96/taylor1.html (1999).

Thakkar, Ajit J. "Research Program." 1999. http://www.unb.ca/chem/ajit/research.htm (1999).

Travis, J. "Brain Doubles Up on Marijuana-like Agents." 1997. http://www.sciencenews.org/sn_arc97/8_23_97/fob2.htm (13 July 1999).

Trichopoulos, D., Frederick P. Li, and David J. Hunter. "What Causes Cancer?" 1996. http://www.sciam.com/0996issue/0996trichopoulos.html (13 July 1999).

Uranium Information Centre Ltd. "Smoke Detectors and Americium: Nuclear Issues Briefing Paper 35." 1997. http://www.uic.com.au//nip35.htm (6 Aug. 1999).

"Use of Marijuana in Neurological and Movement Disorders." *Medical Marijuana.* http://www.hivpositive.com/f-Nutrition/MedicalMarijuana/MM- Neurological.html (12 July 1999).

"What Is Heavy Water (D_2O)?" 1998. http://www.sci-ctr.edu.sg/ScienceNet/cat_physical/cat_che00244.html (1999).

"What Is Important about Hardness?" *The Physical Characteristics of Minerals* 1998. http://mineral.galleries.com/minerals/hardness.htm (1999).

Wiley, Robert. "Chapter Eleven - States of Matter." 1998. http://amug.org/~rwiley/chapter_eleven-intermolecu.htm (1999).

Historical Chronology

c.30,000 B.C.
Stone Age cultures use pigments to color various artifacts.

c. 6000 B.C.
Neolithic artisans fashion ornaments from gold.

c. 4000 B.C.
Early applied chemistry begins in Egypt with the extraction and working of metals, including copper, tin, and bronze. Egyptians are also familiar with eye paint and plaster of Paris.

c.3500 B.C.
Sumerians manufacture soap.

3400 B.C.
Bronze, an alloy of copper and tin, first appears in abundance in Sumeria. The Sumerians become expert in working gold, silver, copper, lead, and antimony.

c. 3000 B.C.
Iron is forged.

c. 2500 B.C.
The earliest known wholly glass objects are beads made in Egypt at this time. Early peoples may have discovered natural gas, which is created when lightening strikes sand. The Egyptians make glass beads by sand (silica), soda, lime, and other ingredients.

c.1500 B.C.
Use of iron becomes prevalent in the Mediterranean. It appears to have come there from the northeast, possibly beginning with the Hittites, and its use revolutionizes society.

c. 1000 B.C.
Phoenicians create several dyes, including the famous Tyrian purple, made from sea mollusks.

c. 1000 B.C.
Beginning of Greek metallurgy; the Greeks use iron.

c. 900 B.C.
Steel is manufactured in India.

c.450 B.C.
Empedocles (c.492-c.432 B.C.), Greek philosopher, first offers his concept of the composition of matter, postulating that it is made of four elements—earth, air, fire, and water. This notion is adopted by Aristotle and becomes the basis of chemical theory for over two millennia.

c. 450 B.C.
The Greek philosopher Leucippus first states the notion of atomism. He argues that upon continuous division of a substance, eventually a point would be reached beyond which further division was impossible. His disciple, the Greek philosopher Democritus ultimately names these small particles *atomos*, meaning indivisible.

c.350 B.C.
Aristotle (384-322 B.C.), Greek philosopher, offers his doctrine of the elements, stating that all things are composed of a basic material in combination with four qualities—hotness, dryness, coldness, and wetness. This theory eventually suggests the idea of transmutation (the changing of ordinary metals into gold or silvers) and gives rise to alchemy.

c.300 B.C.
Arthasastra, an ancient Indian manual on politics, discusses mining, metallurgy, medicine, pyrotechnics, poisons, and fermented liquors.

c.300 B.C.
Glass blowing first appears.

105 A.D. Modern papermaking from fiber is first traced to Ts'ai Lun (c. 50-118), an official in the imperial court of China.

c.300 Zosimus of Panopolis, Greek alchemist, founds a school in Alexandria where pupils are taught the basic chemical operations (filtration, fusion, sublimation, and distillation).

c.900 Rhazes (c.845-c.930), Persian physician and alchemist, uses his knowledge of chemistry in the practice of medicine.

c.900 Gunpowder is discovered in China. In the following century, it is used to make explosive bombs.

c.950 Members of Ism'iliya ("Brethren of Purity"), an Islamic sect, compile a corpus of alchemical works which are later attributed to Jabir ibn Hayyan, known in the West as Geber, a mysterious alchemist who may have lived in the eighth century.

c.1000 Al-Biruni (973-1048), Arab scholar, determines specific weights using a pycnometer similar to a modern instrument that measures and compares densities of solids and liquids. The instrument looks like a short, wide bottle with a thermometer-like device placed in its neck.

c.1100 First description of Greek fire (an incendiary device used as a weapon) and the use of saltpeter are given by Marcus Graecis in his *Liber ignium ad comburendos hostes*. Although the exact composition of Greek fire remains unknown, it appears to have been a petroleum-based mixture.

c.1100 Updated version of *Mappae clavicula* contains the first description of a liquid, distilled from wine, that will catch fire. It is named "alcohol" in the sixteenth century.

c. 1144 First translations of Arabic alchemical manuscripts in Spain, introducing European scholars to alchemy.

c.1150 Theophilus the Monk (who may have been Roger of Helmarshausen) writes an original work called *Diversarum artium schedula*. It provides information on painting, glass making, and metalwork, and contains the first direct Western reference to paper, or "Byzantine parchment."

1260 Albertus Magnus (1193-1280), German scholar also known as Labert of Bollstadt, writes his *De rebus metallicus et mineralibus*, in which he describes arsenic. Because of the accuracy of his description, he is sometimes credited with discovering arsenic. He is also considered the first important European alchemist.

c.1247 Roger Bacon (c.1220-1292), English scientist and philosopher, is among the first in the West to mention gunpowder and provide recipes for its manufacture. There is no doubt, however, that gunpowder originated in China, perhaps in the eleventh century.

c.1275 Arnold of Villanova (c.1235-1311), Spanish alchemist, first prepares pure alcohol. He also discovers carbon monoxide when he describes how wood burning under conditions of poor ventilation can give off poisonous fumes.

c.1300 Giles of Rome (1274-1316) proposes an atomic theory of matter based on the earlier ideas of the Spanish Hebrew scholar Ibn Gabirol (c. 1022-c. 1070), known in the West as Avicebron.

c.1310 An unknown Spaniard, called the "False Geber" because he publishes under Geber's name, writes four books on alchemy and is the first to describe sulfuric acid. The discovery of sulfuric acid is considered the greatest of chemical achievement of the Middle Ages, since it makes possible all manner of chemicals changes that vinegar, the strongest acid known to the ancients, cannot achieve.

c.1350 John of Rupescissa writes his *Liber lucis* in which he extolls the therapeutic value of the quintessence of wine (alcohol). He also describes how to build an alchemical furnace.

1500 Hieronymus Brunschwig (c.1450-c.1512) of Germany writes his *Kleines Distillierbuch* (c.1450-c.1512). This copiously illustrated work remains the most important book on distilling for decades.

1510 Two German books lay the foundation for industrial chemistry. *Bergwerkb'chlein* is dedicated to mineralogy and *Probierb'chlein* focuses on chemical tests and introduces quantitative concepts.

1520 Paracelsus (1493-1541), the Swiss physician and alchemist whose real name is Theophrastes Bombastus von Hohenheim, is appointed to the chair of medicine at Basel. He advocates the use of chemicals in medicine and recommends opium as a cure (naming it laudanum), as well as mercury and antimony. Among his books, *De natura rerum* and *Archidoxis* are of particular interest to chemistry.

1525 Philipp Ulstadt of Germany publishes *Coelum philosophorum*, a distillation book that adapts the hands-on techniques of the alchemists to medicine. It foreshadows the iatrochemical movements.

1540 Vanuccio Biringuccio (1480-c.1539) of Italy publishes his *De la pirotechnia* in Venice. This original work is the first of its kind in that it focuses solely on metallurgy. It also offers the first description of the amalgamation process used to extract silver from its ores.

1540 Valerius Cordus (1515-1544), German physician,

provides the first written description of how to prepare ether.

1540 Giovanni Ventura Rosetti writes his *Plictho dell'arte tentori*, which is the first book on dyeing fibers and fabric.

1555 Geronimo Ruscelli of Italy publishes *Secreti* in Venice under a different name. This compilation of chemical recipes has phenomenal success and sees more than 90 editions in virtually every European language.

1563 Bernard Palissy (c.1510-1589), French potter, publishes his *Recette veritable*, in which he discusses agriculture, geology, mining, and forestry. He discovers the Italian secret of producing majolica (pottery decorated with an opaque tin glaze) and is considered one of the most eminent chemists of France.

1574 Lazarus Ercker (c.1530-1594) of Germany publishes his *Beschreibung aller Furnemisten Mineralischen Ertzt und Bergwercks Arten*, which is the first manual of analytical, metallurgical chemistry. His text is especially valuable to the practicing assayers.

1589 Giambattista della Porta (1535-1615), Italian physicist, publishes the enlarged edition of his *Magiae naturalis*. He tells, among other things, how to prepare sulfuric acid and gives details on various laboratory apparatuses.

1597 *Alchemia*, the first textbook on chemistry, is written by the German alchemist Andreas Libavius (c. 1540-1616). He is also the first to describe the preparation of hydrochloric acid and is credited with planning the first true chemical laboratory.

1603 Nicholas Guibert (c. 1547-1620), French alchemist, publishes his *Alchymia*, which, despite its title, denies the possibility of transmuting metals. At this time, the words "alchemy" and "chemistry" are almost interchangeable.

1609 Johann Hartmann (1568-1631), Bavarian mathematician and iatrochemist, is nominated professor of "chymatria" at Marburg and is thus considered to be the first professor of chemistry in Europe.

1610 First chemistry textbook in French is *Tyrocinium chymicum*, written by French chemist, Jean Béguin (c.1550-c.1620) and published in Paris. He is also the first to mention obtaining acetone by distilling lead acetate.

1617 Angelo Sala (1576-1637), Italian physician, conducts careful experiments, in which he combines and decomposes substances while weighing them precisely. His experiments are influential, as he finds the weights to be in the same proportion.

1619 Daniel Sennert (1572-1637), German physician, publishes his *De chymicorum*, which is the first application of Greek atomic theory to chemistry. Sennert speaks of atoms and even of "second-level atoms," or molecules.

1630 Jean Rey (1582-1645), French physician, writes on the nature of air and its role in combustion, and lays the foundation for future chemical discoveries. He suggests a possible experiment for weighing air that Galileo actually performs.

1646 Johann Rudolf Glauber (1604-1670), German chemist, publishes the first of his five-volume *Furni novi philosophici* (1646-1649). This work gives his recipes for mineral acids and salts, including "sal mirabile"—the sodium sulfate residue that formed by the action of sulfuric acid on ordinary salt—which becomes known as "Glauber's salt."

1648 *Ortus medicinae* by Johann Baptista van Helmont (1580-1644), Flemish physician and alchemist, is published posthumously. He is the first to use quantitative methods in connection with a biological problem, and is therefore called the father of biochemistry. He is also the first to recognize that one air-like substance exists, and he names this vapor, or non-solid, "chaos," which in Flemish sounds like "gas."

1658 Franciscus Sylvius de la Boë (1614-1672), German-Dutch chemist and physiologist, builds the first university chemistry laboratory in Leyden, Holland, where he teaches from 1658 to 1672. He is also one of the first to appreciate the idea of chemical affinity (attractions).

1660 Robert Boyle (1627-1691), English physicist and chemist, publishes his *New Experiments Physico-Mechanicall*, in which he suggests that combustion and respiration are similar processes.

1661 Robert Boyle (1627-1691), English physicist and chemist, publishes his book, *The Sceptical Chymist*. Many consider this work as marking the beginnings of scientific chemistry, and call Boyle the father of chemistry. In it, he espouses the experimental method and breaks from the Greek notion of elements.

1662 Robert Boyle (1627-1691), English physicist and chemist, announces what becomes known as Boyle's Law, stating that when an ideal gas is under constant pressure, its volume and pressure vary inversely.

1665 Nicolaus Steno (1638-1686), Danish anatomist and geologist, briefly states what is now called the first law of crystallography. Also called the law of the constancy of crystalline angles, it states that crystals of a specific substance have fixed characteristic angles at which the faces, however distorted, always meet.

1665 Robert Hooke (1635-1703), English physicist, offers a thorough treatment of combustion in his book on microscopy called *Micrographia*. He also suggests that air is composed of two parts.

1666 Robert Boyle (1627-1691), English physicist and chemist, publishes *The Origine of Formes and Qualities*, in which he begins to explain all chemical reactions and physical properties through the existence of small, indivisible particles, or atoms.

1669 Johann Joachim Becher (1635-1682), German chemist, publishes his *Physica subterranea*, in which he is the first to attempt the formulation of a general theory chemistry. His concept of "terra pinguis" as the substance in air that burns forms the basis of the later phlogiston theory.

1671 Robert Boyle (1627-1691), English physicist and chemist, produces hydrogen by dissolving iron in hydrochloric or sulfuric acid, but he is unaware of his achievement.

1674 John Mayow (1641-1679), English physiologist, publishes his *Tractatus quinque medico-physici*, in which he compares respiration (breathing) to combusion. He is the first to state that the volume of air is reduced in respiration. He is also ahead of his time when he correctly asserts that the blood carries air from the lungs to all part of the body.

1674 Hennig Brand (c.1630-c.1692), German chemist, discovers phosphorus, which he finds in urine. This is the first discovery of an element that was not known in any earlier form.

1675 Nicolas Lémery (1645-1715), French chemist and physician, publishes his *Cours de chymie*, which becomes the authoritative textbook on chemistry for the next 50 years. He is an adherent of Boyle's, and advocates the experimental method.

1681 Johann Joachim Becher (1635-1682), German chemist, obtains tar from the distillation of coal. He also suggests that sugar is necessary for fermentation.

1687 Isaac Newton (1642-1727), English mathematician and physicist, offers a mathematical version of Boyle's Law on gases, showing how the pressure and volume of a gas are inversely proportional.

1688 John Clayton (1657-1725), English cleric, obtains methane and recognizes its flammable nature.

1699 Guillaume Amontons (1663-1705), French physicist, publishes his findings on gases, stating that each gas tested changes in volume by the same amount for a given change in temperature.

1702 Wilhelm Homberg (1652-1715), German physician, discovers boric acid, which he calls "sedative salt."

1709 John Freind (1675-1728), English physician and chemist, publishes *Praelectiones chemicae*, one of the earliest attempts to use Newtonian principles to explain chemical phenomena.

1718 Etienne-François Geoffroy (1672-1731), French apothecary, publishes his *Tables des différens rapports*, which offers a table of affinities between various acids and alkalis or metals. Chemistry eventually accepts his prophetic concept of affinity.

1723 Georg Ernst Stahl (1660-1723), German chemist, publishes his major work, *Fundamenta chymiae dogmaticae et experimentalis*, in which he develops his phlogiston theory. Although incorrect, it is the first comprehensive explanation of the phenomenon of combustion, and its popularity dominates for the next century. To Stahl, air is only the carrier of phlogiston, which he says is what is burning during combustion.

1728 Jacopo Bartolomeo Beccaria (1682-1766), Italian physician, discovers gluten in wheat flour. This is the first protein substance of plant origin to be found.

1732 Herman Boerhaave (1668-1738), Dutch physician, publishes his *Elementa chemicae*, whose comprehensiveness makes it the most popular chemical textbook for many decades. It serves chemistry as a great teaching book and presents a concise outline of all chemical knowledge.

1733 Georg Brandt (1694-1768), Swedish chemist, publishes the first accurate and complete study of arsenic and its compounds.

1735 Georg Brandt (1694-1768), Swedish chemist, discovers cobalt, which he later isolates in 1743. He becomes the first person to discover a metal that was entirely unknown to the ancients.

1736 Henri-Louis Duhamel du Monceau (1700-1782), French chemist and botanist, first uses the term "base" to indicate those substances that combine with acids to form salts. He is also the first to make a distinction between soda and potash (potassium).

1736 Antonio de Ulloa (1716-1795), Spanish mathematician, introduces platinum from South America into Europe and later writes an accurate description of it.

1737 First chair of chemistry is established at the faculty of medicine in Bologna, Italy, and is assigned to the German chemist Johann B. Becher.

1738 Daniel Bernoulli (1700-1782), Swiss mathematician, publishes his *Hydrodynamica* containing his kinetic theory of gases. This treatise becomes a work of major importance in both physics and chemistry. This is the first attempt at an explanation of the behavior of gases, which, he assumes, are composed of a vast number of tiny particles.

1739 Johann Andreas Cramer (1710-1777), German metallurgist, publishes his *Elementa artis docimasticae*, which is the first textbook on assaying and chemical analysis.

1740 First large chemical industry is founded by Joshua Ward (1685-1761) at Richmond, near London, to produce sulfuric acid. The acid is obtained by heating sulphur with saltpeter in iron capsules and condensing the vapor byproducts into large balls. Chemistry will prove to be the science most directly tied to industry.

1742 Anton Svab (1703-1768) of Sweden, also known as Swab, distills zinc from the alloy callamine.

1747 Andreas Sigismund Marggraf (1709-1782), German chemist, discovers that beets produce the same sugar as sugar cane. This leads to the development of the sugar industry.

1748 Jean-Antoine Nollet (1700-1770), French physicist, discovers osmosis, or osmotic pressure, while experimenting with water diffusing into sugar through a membrane.

1751 Axel Fredrik Cronstedt (1722-1765), Swedish chemist, discovers nickel.

1752 Henrik Theophilus Scheffer (1710-1759), Swedish chemist, publishes a detailed description of platinum and its properties, and shows that it is a distinct metal.

1754 Joseph Black (1728-1799), Scottish chemist, proves by quantitative experiments the existence of carbon dioxide, which he calls "fixed air." His proof is contained in his thesis done for his medical degree, and two years later he openly publishes the results.

1756 Joseph Black (1728-1799), Scottish chemist, publishes his paper "Experiments upon Magnesia Alba, Quicklime, and Some Other Alcaline Substances," in which he describes his discovery of carbon dioxide. This is regarded as the first work on quantitative chemistry.

1758 Axel Fredrick Cronstedt (1722-1756), Swedish chemist, initiates the classification of minerals by their chemical structure as well as by their appearance. He notes four kinds of minerals: earths, metals, salts, and bitumens.

1758 Andreas Sigismund Margraf (1709-1782), German chemist, introduces the flame test to chemistry.

1762 Joseph Black (1729-1799), Scottish chemist, first introduces the term as well as the notion of "latent heat." He is the first to realize that there is a distinction between the quantity of heat and the intensity of heat, and that it is only the latter that is measured as temperature. He shows that when ice is heated, it slowly melts, but its temperature remains the same. The ice, therefore, absorbs a quantity of "latent heat" in melting. Thus increasing the amount of heat it contains but not increasing its intensity.

1765 Karl Wilhelm Scheele (1742-1786), Swedish chemist, discovers prussic acid. Known also as hydrogen cyanide, this highly volatile, colorless liquid is extremely poisonous and is eventually used in industrial chemical processing.

1766 Henry Cavendish (1731-1810), English chemist and physicist, publishes a paper on "Factitious Airs" in the Royal Society's Philosophical Transactions, which relates his discovery of hydrogen, or what he calls "inflammable" air.

1768 Antoine Baume (1728-1804), French chemist, introduces the aerometer, an instrument for determining the amount of alcohol in a hydro-alcoholic mixture on the basis of density.

1770 Johan Gottlieb Gahn (1745-1818) and Karl Wilhelm Scheele, both Swedish chemists, discover phosphorus (phosphoric acid) in bones, noting that it is an essential component.

1770 Karl Wilhelm Scheele (1742-1786), Swedish chemist, discovers tartaric acid. One of the most widely distributed of the plant acids, it eventually assumes a wide variety of food and industrial uses.

1771 Peter Woulfe (c.1727-1803), English chemist, obtains picric acid by treating indigo with nitric acid. This substance dyes wool yellow, and is considered to be the first artificial organic dye.

1772 Daniel Rutherford (1749-1819), Scottish chemist, discovers nitrogen. Shortly thereafter Priestley, Scheele, and Cavendish also discover it independently.

1772 Louis-Bernard Guyton de Morveau (1737-1816), French lawyer and chemist, demonstrates for the first time that metals gain weight on calcination (the chemical change that occurs when metals are heated to just below their fusion or melting point).

1772 Joseph Priestley (1733-1804), English chemist, experiments with "fixed air" and writes his "Directions for Impregnating Water with Fixed Air," in which he details the production of seltzer water by using carbon and water. The distinctive taste of the seltzer, or soda, water brings Priestley much fame.

1774 Joseph Priestley (1733-1804), English chemist, discovers oxygen, or what he calls "dephlogisticated air." It had been isolated sometime between 1770 and 1773 by the Swedish chemist Karl Scheele (1742-1786), who called it "fire air," but Priestley publishes his results first. It is named by Antoine-Laurent Lavoisier (1743-1794), French chemist, on September 5, 1779.

1774 Manganese is discovered by Swedish mineralogist Johann Gottlieb Gahn (1745-1818) and Karl Wilhelm Scheele (1742-1786), Swedish chemist.

1774 Felice Fontana (1730-1805), Italian chemist, invents the eudiometer, a finely graded and calibrated tube for the volumetric measurement and analysis of gases.

1775 Torbern Olaf Bergman (1735-1784), Swedish mineralogist, publishes a new table of affinities (the attraction of one substance to another).

1775 Antoine-Laurent Lavoisier (1743-1794), French chemist, publishes his "Memoire," which contains his first major disavowal of the phlogiston theory, as well as a revision of his combustion theory.

1777 Carl Wenzel (1740-1793), German chemist, discovers the reaction rates of various chemicals, concluding that they are roughly proportional to the substance concentration.

1779 Antoine-Laurent Lavoisier (1743-1794), French chemist, first proposes the name "oxygen" for that part of the air that is breathable, and that is responsible for combustion.

1780 Claude-Louis Berthollet (1748-1822), French chemist, establishes the composition of ammonia and conducts the first thorough study of chlorine, introducing it as a bleaching agent.

c. 1780 Karl Wilhelm Scheele (1742-1786), Swedish chemist, discovers lactic acid. Found in the soil, in the blood and muscles of animals, and in fermented milk products, it is eventually used in food processing and for tanning leather and dyeing wool.

1781 Peter Jacob Hjelm (1746-1813), Swedish chemist, discovers molybdenum, a tough, malleable, silver-white metal used in alloys. He reduces molybdic acid to a metal state using carbon as a reducing agent.

1782 Franz Müller (1740-1825), Austrian mineralogist, discovers tellurium.

1782 Antoine-Laurent Lavoisier (1743-1794), French chemist, studies metals that have calcinated or rusted, and observes that their total weight is not changed but merely shifted around. This establishes an early version of the first law of conservation of matter.

1782 Torbern Olaf Bergman (1735-1824), Swedish mineralogist, publishes his *Skiagraphia regni mineralis*, in which he classifies minerals into four main groups: salts, earths, metals, and inflammable bodies.

1783 Nicolas Leblanc (1742-1806), French chemist, invents a process for the manufacture of soda (sodium hydroxide and sodium carbonate) from salt (sodium chloride). He soon opens a factory to manufacture soda that is much-needed in France for a variety of chemical purposes. This is the first chemical discovery that has an immediate commercial use. It is also the first major example of a technology being born out of an advance in theoretical science.

1783 Antoine-Laurent Lavoisier (1743-1794), French chemist, and Pierre-Simone de Laplace (1749-1827), French astronomer and mathematician, jointly publish a paper, "Mémoire sur la chaleur," which lays the foundations of thermochemistry. They demonstrate that the quantity of heat required to decompose a compound into its elements is equal to the heat evolved when that compound was formed from its elements.

1783 Juan Jose d'Elhuyar (1754-1796) and his younger brother, Don Fausto d'Elhuyar (1755-1833), both Spanish mineralogists, analyze a mineral called wolframite and discover a new metal, tungsten.

1783 Antoine-Laurent Lavoisier (1743-1794), French chemist, repeats the experiment conducted by Cavendish in 1766 and realizes that he is dealing with a separate gas. He calls this flammable gas "hydrogen," from the Greek phrase meaning "giving rise to water."

1783 Karl Wilhelm Scheele (1742-1786), Swedish chemist, discovers glycerine. Also called glycerol, it is a thick, clear, sweet-tasting liquid that comes to be used in making resins and gums for paints and as a softener in other products.

1784 Henry Cavendish (1731-1810), English chemist, establishes that water is formed when "inflammable air," or hydrogen, is burned in ordinary, or "dephlogisticated," air. He publishes his findings on the synthesis of water from two gases in *Philosophical Transactions*.

1784 Karl Wilhelm Scheele (1742-1786), Swedish chemist, discovers citric acid. This organic compound belongs to the family of carboxylic acids and is commonly found in nearly all plants and in many animal tissues and fluids. It comes to be used mainly as a flavoring agent in drinks.

1785 Martinus van Marum (1750-1837), Dutch botanist and physicist, obtains ozone by striking an electric spark in oxygen. He describes a very strong odor as "clearly the smell of electric matter," but does not recognize his discovery.

1786 Claude-Louis Berthollet (1748-1822), French chemist, and French mathematicians Gaspard Monge (1746-1818) and Alexandre-Theophile Vandermonde (1735-1796), publish a paper, "Mémoire sur le fer," which first establishes that the difference between iron and steel is due to carbon.

1787 Antoine-Laurent Lavoisier (1743-1794), French chemist, publishes *Methode de la nomenclature chimique* in collaboration with French chemist, Louis-Bernard Guyton de Morveau (1737-1816). This book gives the new chemistry a modern terminology and changes chemical nomenclature to correspond to the new antiphlogiston theory.

1787 Jacques-Alexandre Charles (1746-1823), French physicist, demonstrates that different gases all expand by the same amount with a given rise in temperature if the pressure is held constant. This becomes known as Charles's law, and also Gay-Lussac's law.

1788 Charles Blagden (1748-1820), English physician and secretary of the Royal Society, discovers that the lowering of the freezing point of a solution is proportional to the concentration of the solute. This becomes known as Blagden's law.

1789 Antoine-Laurent Lavoisier (1743-1794), French chemist, publishes *Traité élémentaire de chimie*, which defines the new science and becomes the first modern chemical textbook. In it, he unifies the subject and states his law of the indestructibility of matter (conservation of mass). He also popularizes chemistry and puts an end to the old phlogiston theories.

1789 Martin Heinrich Klaproth (1743-1817), German chemist, discovers uranium in pitchblende. This is the same substance that the Curies will refine in 1898. Later the same year, Klaproth discovers the element zirconium in the mineral zircon.

1789 William Higgins (c.1762-1825), Irish chemist, publishes his *Comparative View of the Phlogistic and Antiphlogistic Theories*, in which he anticipates both John Dalton's atomic theory and the chemical symbolism of Berzelius. His work is little known in his lifetime.

1790 Adair Crawford (1748-1795), Irish physician and chemist, and William Cumberland Cruikshank (1745-1800), Scottish surgeon and chemist, first suggest that a mineral found in Scottish lead mines contains a new element they call strontianite. In 1808, it is first isolated by English chemist Humphry Davy (1778-1829) who calls it strontium.

1791 William Gregor (1761-1817), English mineralogist, discovers titanium.

1792 Jeremias Benjamin Richter (1762-1807), German chemist, founds stoichiometry with the publication of his book *Anfangsgrunde der Stochymetrie*. His law of equivalent proportions, or law of equivalents, forms the basis of this form of quantitative chemistry, which states that substances react with each other in fixed proportions.

1794 Johan Gadolin (1760-1852), Finnish chemist, discovers a new mineral at a quarry in Ytterby, Sweden. It is called ytterbite, and is later found to contain over a dozen different elements which come to be called "rare earth" elements.

1795 Martin Heinrich Klaproth (1743-1817), German chemist, isolates a new metal and names it titanium, after the Titans of Greek mythology. He gives full credit to English mineralogist William Gregor (1761-1817), who first discovered it in 1791.

1796 Smithson Tennant (1761-1815), English chemist, conducts a series of quantitative combustion experiments and establishes for the first time that a diamond is chemically identical with carbon.

1798 Louis-Nicolas Vauquelin (1763-1829), French chemist, discovers the element chromium and prepares several of its compounds. He also discovers beryllium.

1798 Louis-Bernard Guyton de Morveau (1737-1816), French chemist, liquefies ammonia by using a mixture of ice and calcium chloride to cool the gas to -47.2°F (-44°C).

1799 Antoine-François de Fourcroy (1775-1809), French physician and chemist, isolates urea.

1799 Joseph-Louis Proust (1754-1826), French chemist, demonstrates that all compounds contain elements in certain, definite proportions. This comes to be known as Proust's law.

1800 William Nicholson (1753-1815), English chemist, and Anthony Carlisle (1768-1840), English anatomist, discover that water can be decomposed (hydrogen and oxygen separated) by an electric current. Using the new voltaic cell, they use electricity to produce a chemical reaction. Their discovery of electrolysis opens up the new field of electrochemistry.

1800 Johann Wilhelm Ritter (1776-1810), German physicist, repeats the Nicholson/Carlisle water electrolysis experiment but arranges to collect the two gases (hydrogen and oxygen) separately for the first time. In further experiments this year, he discovers electroplating when he passes an electric current through a solution of copper sulfate and produces metallic copper. The electric current separates the molecules of copper and allows it to be plated to other metals.

1801 John Dalton (1766-1844), English chemist, formulates the law of partial pressures for the components of a gaseous mixture. This states that each component of a mixture of gases exerts the same pressure that it would if it alone occupied the entire volume of the mixture, at the same temperature.

1801 Charles Hatchett (1765-1847), English chemist, discovers columbium, which is later called niobium.

1801 Franz Carl Achard (1753-1821), German chemist, develops Marggraf's sugar-from-beets discovery and opens the first European sugar refinery in Silesia.

1801 Sigismund Friedrich Hermdstaedt (1760-1833), German chemist, discovers sodium bicarbonate. Valentine Rose (1762-1807), German chemist, independently makes the same discovery.

1801 René Just Haúy (1743-1822), French mineralogist, founds the science of crystallography with the publication of *Traité de Minéralogie*. In it he explains that if crystalline forms are identical (or different), then that identity of difference is found in its chemical composition.

1801 Robert Hare (1781-1858), American chemist, invents the oxyhydrogen blowpipe, which is the ancestor of the modern welding torch. With it, Hare is able to melt sizeable quantities of platinum.

1801 Andreas Manuel del Rio (1764-1849), Spanish-Mexican mineralogist, discovers vanadium, but, unable to convince others that it is not chromium, allows himself to be persuaded and drops his claim to a new metal.

1802 Joseph-Louis Gay-Lussac (1778-1850), French chemist, shows for the first time that all gases expand equally, or by equal amounts with the same increase in temperature. It becomes known as Gay-Lussac's law. It is also known as Charles's law.

1802 Anders Gustaf Ekeberg (1767-1813), Swedish chemist, discovers tantalum.

1802 Domenico Pini Morichini (1773-1836), Italian chemist, discovers calcium fluoride in tooth enamel.

1802 Thomas Thomson (1773-1852), Scottish chemist, publishes his *System of Chemistry* and introduces a system of symbols for individual minerals using the first letters of their names.

1802 William Hyde Wollaston (1766-1828), English chemist and physicist, examines a candle flame through a prism and is able to distinguish dark lines that cross the spectrum. He dismisses them as natural boundaries between the colors, and does not know that he is one of the first to observe the different wave lengths of light.

1802 Thomas Wedgewood (1771-1805), English physicist, publishes *An Account of a Method of Copying Painting upon Glass, and of Making Profiles by the Agency of Light upon Nitrate Silver*. In this work, he describes his experiments with English chemist Humphry Davy (1778-1829), in which paintings on glass coated with silver nitrate are used as negatives while an image is made on another plate upon exposure to light. Since they do not know how to "fix" this image, it gets progressively darker.

c. 1803 William Hyde Wollaston (1766-1828), English chemist and physicist, discovers a method of making platinum malleable, allowing it to be hammered and molded into shapes mainly for laboratory apparatuses. He keeps this process secret, and amasses great wealth. He arranges to have the secret released after his death, the secret is released, and it forms the basis of modern powder metallurgy (the fabrication of metal forms out of a powder rather than molten metal).

1803 Claude-Louis Berthollet (1748-1822), French chemist, discovers that manner and rate of chemical reactions depended on more than affinities or the attraction of one substance for another. He shows that reactions are affected by relative quantities and temperature, and that many reactions are reversible. This is the first attempt to consider the physics of chemistry.

1803 William Henry (1774-1836), English physician and chemist, determines the effects of pressure on the solubility of gases in liquids.

1803 John Dalton (1766-1844), English chemist, states the law of multiple proportions which applies to two elements that could combine in more than one way. He also states his atomic theory which says that all elements are composed of extremely tiny, indivisible, indestructible atoms, and that all the known substances are composed of some combination of these atoms. He finally states that these atoms differ from each other only in mass, and that this difference can be measured.

1803 Jöns Jakob Berzelius (1779-1848), Swedish chemist, and Wilhelm Hisinger (1766-1852), Swedish mineralogist, discover cerium (which Klaproth also independently discovers).

1803 Jöns Jakob Berzelius (1779-1848), Swedish chemist, first suggests that acids and bases have opposite electrical charges. He discovers this while experimenting with a new voltaic pile and links this discovery to the concept of electrical polarity, which he then extends to the elements.

1803 William Henry (1774-1836), English physician and chemist, proposes what becomes Henry's law, stating that the amount of gas absorbed by a liquid is in proportion to the pressure of the gas above the liquid, provided no chemical reaction occurs.

1804 Smithson Tennant (1761-1815), English chemist, discovers iridium (from the Greek word for "rainbow") and osmium (from the Greek word for "smell").

1805 John Dalton (1766-1844), English chemist, publishes the first table of atomic weights and invents a new system of chemical symbols.

1805 Joseph-Louis Proust (1754-1826), French chemist, defends his law of constant, or definite, proportions over the views of French chemist Claude-Louis Berthollet (1748-1822), who argues wrongly that the constituents of a compound can unite in a continuous range of proportions.

1805 Joseph-Louis Gay-Lussac (1778-1850), French chemist, establishes that precisely two volumes of hydrogen combine with one volume of oxygen to form water.

1805 Antoine-François de Fourcroy (1775-1809), French chemist, identifies calcium and magnesium phosphate in bones.

1806 Humphry Davy (1778-1829), English chemist, performs experiments on electrolysis that lead him to state the electrochemical theory. He concludes that the production of electricity in simple electrolytic cells results from chemical action, and that chemical combination occurs between substances of opposite charge. He also discovers potassium when he passes a current through molten potash and isolates sodium from soda in 1807.

1806 Louis-Nicolas Vauquelin (1763-1829), French chemist, publishes his *Chimie appliqée aux arts*, which is the first book devoted specifically to industrial chemistry.

1808 John Dalton (1766-1844), English chemist, publishes his *New System of Chemical Philosophy*, in which he elaborates his atomic theory and makes his revolutionary ideas well-known.

1808 Joseph-Louis Proust (1754-1826), distinguishes among the different sugars in plants, and is the first to study glucose, the sugar in grapes.

1809 Joseph-Louis Gay-Lussac (1778-1850), French chemist, collaborates with Alexander von Humboldt (1769-1859), German naturalist, and states the law of gaseous volume. They discover that, in forming compounds, gases combine in proportions by volume that can be expressed in small whole numbers.

1809 Amedeo Avogadro (1776-1856), Italian physicist, introduces the concept of the relativity of acids. He states that a substance that is acid relative to another may be alkaline in respect to a third.

1810 Jöns Jakob Berzelius (1779-1848), Swedish chemist, first shows that the laws of combination and the atomic theory apply in both organic and inorganic theory.

1811 Amedeo Avogadro (1776-1856), Italian physicist, first proposes his theory of molecules, which is confirmed much later by modern chemistry. He states that equal volumes of all gases contain the same number of molecules if they are under the same pressure and temperature.

1811 Bernard Courtois (1777-1838), French chemist, discovers iodine.

1811 Joseph-Louis Gay-Lussac (1778-1850) and Louis-Jacques Thénard (1777-1857), both French chemists, determine the elementary composition of sugar for the first time.

1813 Humphry Davy (1778-1829), English chemist, publishes his *Elements of Agricultural Chemistry*, a pioneering work and the first textbook dealing with the application of chemistry to agriculture.

1815 Humphry Davy (1778-1829), English chemist, invents the safety lamp for coal miners by understanding the properties and nature of methane gas and oxygen.

1815 William Prout (1785-1850), English chemist and physiologist, publishes an anonymous article in which he states that the atomic weights of the elements are all integral multiples of that of hydrogen. This later becomes known as Prout's hypothesis.

1815 Joseph-Louis Gay-Lussac (1778-1850), French chemist, discovers cyanogen and describes the properties, preparation, and compounds of this organic radical in a paper titled "Recherche sur l'acide prussique."

1817 Johan August Arfvedson (1792-1841), Swedish chemist, discovers lithium.

1817 Friedrich Strohmeyer (1776-1835), German chemist, discovers the element cadmium.

1818 Jöns Jakob Berzelius (1779-1848), Swedish chemist, discovers selenium, publishes a table of atomic weights, and offers a system of chemical symbols. His weight table is based on a standard 100 for oxygen and becomes accepted in the twentieth century. His symbols use one or two letters of the Latin name and are essentially also retained in the twentieth century.

1818 Pierre-Louis Dulong (1785-1838), French chemist, and Alexis-Thérèse Petit (1791-1820), French physicist, show that the specific heat of an element is inversely related to its atomic weight (specific heat is the amount of heat required to raise the temperature of a substance by 1°C). This new law becomes useful in determining atomic weights.

1819 Jöns Jakob Berzelius (1779-1848), Swedish chemist, publishes his *Essai sur la thérie des proportions chemiques et sur l'influence chimique de l'electricité*, in which he offers his theory of electrochemical dualism. This states that chemical compounds are made up of two components—one charged negatively and the other positively.

1819 Eilhardt Mitscherlich (1794-1863), German chemist, discovers isomorphism. This chemical theory states that compounds of similar composition tend to crystallize together, or, conversely, that compounds with the same crystal form are analogous in chemical composition. This law becomes useful in establishing atomic weights.

1819 John Kidd (1775-1851), English chemist and physician, obtains naphthalene from coal tar, pointing the way toward the use of coal as a source of many important chemicals.

1821 Friedlieb Ferdinand Runge (1795-1867), German chemist, discovers caffeine.

1822 Anselme Payen (1795-1871), French chemist, discovers the filtering properties of charcoal in removing impurities from the sugar made from sugar beets. The adsorptive properties of charcoal will be put to use in the gas masks of World War I.

1823 Justus von Liebig (1803-1873) and Friedrich Wöhler (1800-1882), both German chemists, make a simultaneous discovery of isomerism. In this intriguing concept, two or more compounds can have the same chemical formula and yet have different structures and properties.

1823 Michel-Eugène Chevreul (1786-1889), French chemist, publishes his classic work which deals with oils, fats, and vegetable colors. He shows that fat is a compound of glycerol with an organic acid. This is one of the first works addressing the issue of the fundamental structure of a large class of compounds, and it has a revolutionary effect on the soap and candle industries.

1823 Michael Faraday (1791-1867), English physicist and chemist, devises methods for liquefying gases under pressure and is the first to produce laboratory temperatures below 0°F (-17.8°C).

1823 Jöns Jakob Berzelius (1779-1848), Swedish chemist, isolates silicon and describes it as a new element.

1823 Charles Macintosh (1766-1842), Scottish chemist and inventor, invents waterproofing by spreading dissolved rubber between two sheets of wool. He dissolves rubber in low-boiling naphtha and creates the "Macintosh," or raincoat.

1825 Michael Faraday (1791-1867), English physicist and chemist, discovers benzene, which later proves to be the starting point for understanding organic chemistry.

1825 Hans Christian Örsted (1777-1851), Danish physicist, isolates aluminum through a four-step process that involves a vacuum.

1825 Thomas Drummond (1797-1840), English engineer, and Goldsworthy Gurney (1793-1875), English inventor, independently produce limelight. They discover that heating lime in an alcohol flame burned in oxygen-rich air makes a stunningly bright artificial light. Drummond uses a parabolic mirror to reflect this light and turns it into spotlight. Impractical for street lighting, it is used to illuminate stages in the theater.

1825 Michel-Eugéne Chevreul (1786-1889), French chemist and Joseph-Louis Gay-Lussac (1778-1850), French chemist, employ Chevreul's earlier (1823) understanding of fats and take out a patent for the manufacture of candles from fatty acids. These new candles are harder than the old tallow candles, and give a brighter light.

1826 Jean-Baptiste-André Dumas (1800-1884), French chemist, develops a method for determining the molecular weight of a vapor from its density. It is still used today.

1826 Antoine-Jérôme Ballard (1802-1876). French chemist, discovers bromine and demonstrates that its nature and properties are analogous to iodine and chlorine.

1826 René-Joachim-Henri Dutrochet (1776-1847), French physiologist, conducts the first quantitative experiments on osmosis. He determines that the pressures involved during the diffusion of solutions are proportional to the solution concentrations. He is also the first to observe the motion of particles suspended in a liquid, later called Brownian motion.

1826 Otto Unverdorben (1806-1873), German chemist, first prepares aniline by distilling the indigo plant.

1827 Friedrich Wöhler (1800-1882), German chemist, first isolates metallic aluminum and describes its properties. He isolates beryllium in 1828.

1827 William Prout (1785-1850), English chemist and physiologist, first classifies the components of foods into carbohydrates, fats, and proteins. He uses the word saccharinous, oleaginous, and albuminous for these three respective groups.

1829 Johann Wolfgang Döbereiner (1780-1849), German chemist, proposes his "law of triads," which is based on the relation between an element's properties and its atomic weight number. His idea is ahead of its time and foreshadows Mendeleev's Periodic Table (1869).

1829 Jöns Jakob Berzelius (1779-1848), Swedish chemist, discovers thorium, which he names after Nordic god Thor.

1830 Nils Gabriel Sefström (1787-1845), Swedish chemist, discovers vanadium. It is isolated by Henry Enfield Roscoe (1833-1915) in 1869.

1830 Jöns Jakob Berzelius (1779-1848), Swedish chemist, introduces the scientific terms isomer, polymer, empirical formula, and rational formula.

1830 Karl von Reichenbach (1788-1869), German naturalist, discovers paraffin.

1831 Thomas Graham (1805-1869), Scottish physical chemist, discovers that the rate of diffusion of a gas is inversely proportional to the square root of its molecular weight. This is called Graham's law and is one of the founding principles of physical chemistry.

1831 Pierre-Jean Robiquet (1780-1840), French apothecary, and Jean-Jacques Collin (1784-1865), French chemist, discover the red dye alizarin, which they isolate from the root of the common madder plant.

1832 Pierre-Jean Robiquet (1780-1840), French apothecary, discovers codeine. Codeine is an alkaloid found in opium that is now used in prescription pain-relievers and cough medicines.

1832 Michael Faraday (1791-1867), English physicist and chemist, announces what are now called Faraday's laws of electrolysis. These two principles establish the intimate connection between electricity and chemistry and place the discipline of electrochemistry on a modern basis.

1832 Heinrich Wilhelm Ferdinand Wackenroder (1798-1854), German chemist, discovers carotene (carotin) in carrots. This organic compound is usually found as pigment in plants, giving them a yellow, red, or orange color, and is converted in the liver into vitamin A.

1832 Joseph-Louis Gay-Lussac (1778-1850), French chemist, introduces the concept and practice of titration into chemical practice. This method can determine the concentration of a dissolved compound by determining how much of it will react with another compound of a known concentration.

1833 Jean-Baptiste-André Dumas (1800-1884), French chemist, uses the combustion method to determine organic nitrogen. The analytical method helps to make organic analysis more quantitative.

1834 Eilhardt Mitscherlich (1794-1863), German chemist, first prepares nitrobenzene. He treats benzene with fuming nitric acid to produce this compound which is later used in perfumes.

1834 Friedlieb Ferdinand Runge (1795-1867), German chemist, isolates a substance he calls kyanol from coal tar. Kyanol was the primary industrial source of aromatic compounds until the end of World War I.

1834 Friedlieb Ferdinand Runge (1795-1867), German chemist, isolates phenol, or carbolic acid, by distilling coal tar.

1834 Emile Clapeyron (1799-1864), French engineer, arrives at an equation that involves the relationship of the heat of vaporization of a fluid, its temperature, and the increase in volume involved in its vaporization. His equation eventually contributes to the second law of thermodynamics.

1834 Jean-Baptiste-André Dumas (1800-1884), French chemist, offers his "type" theory of organic structure in which he suggests that the chemical properties of an organic compound might not be totally dependent on electrical properties (its positive or negative charge).

1834 Justus von Liebig (1803-1873), German chemist, revises the chemical symbolism of Berzelius and replaces his superscripts with subscripts. Now the number after the letter comes below it, and not above it.

1834 Anselme Payen (1795-1871), French chemist, separates a starch-like substance from wood. Because he obtains it from cell walls, he names it cellulose.

1836 John Frederic Daniell (1790-1845), English chemist, invents the electrochemical cell. Called the Daniell cell, it is made of copper and zinc, and is the first reliable source of electric current.

1836 Edmund William Davy (1785-1857), Irish chemist and cousin of English chemist, Humphry Davy (1778-1829), discovers and prepares acetylene. This gaseous hydrocarbon is later pressurized in cylinders for use in welding torches.

1838 Justus von Liebig (1803-1873), German chemist, gives a general definition of a radical. In a paper published on common radicals, he points out the usefulness to organic chemistry of many methods developed by inorganic chemists.

1838 Gerardus Johannes Mulder (1802-1880), Dutch chemist, coins the word protein.

1839 William Robert Grove (1811-1896), English physicist, constructs the first fuel cell using hydrogen and oxygen.

1839 Carl Gustav Mosander (1797-1858), Swedish chemist, discovers a new element, lanthanum, which he names after a Greek word meaning "hidden." Mosander's work reveals the complexity of the rare earth minerals.

1840 William Draper (1811-1882), English-American chemist, pioneers photochemistry with his recognition that light brings about chemical reactions through absorption of light energy by the molecules.

1840 Germain-Henri Hess (1802-1850), Swiss-Russian chemist, formulates the law that states that the quantity of heat evolved in a chemical change is the same no matter what chemical route the reaction takes (through a single stage or through many stages. This becomes known as Hess's law and makes Hess the founder of thermochemistry.

1840 Justus von Liebig (1803-1873), German chemist, publishes his *Die organische Chemie in ihrer Anwendung auf Agrikultur and Physiologie*, in which he introduces the use of mineral fertilizers.

1840 Christian Friedrich Schönbein (1799-1868), German-Swiss chemist, discovers ozone. Having studied the peculiar odor that surrounds electrical equipment, he finds that he can reproduce it by electrolyzing water or by allowing phosphorus to oxidize. He believes that it is a new element and names it after the Greek word for "smell." Other researchers later determine that ozone is a form of oxygen.

1840 Jöns Jakob Berzelius (1779-1848), Swedish chemist, first introduces the term "allotropy" to describe the existence of different varieties of an element. He converts charcoal into graphite and declares that the same element may have in different forms.

1841 Chemical Society is founded in London. This is the first such society to be composed of professional (non-student) chemists.

1842 Justus von Liebig (1803-1873), German chemist, publishes his *Die organische Chemie in ihrer Anwendung auf Physiologie und Pathologie*, which is the first formal treatise on organic chemistry as applied to physiology and pathology.

1842 John Bennett Lawes (1814-1900), English agricultural scientist, experiments with artificial fertilizer and develops superphosphate, the first commercial artificial fertilizer. He patents a method for its manufacture (first from animal bones and then from minerals) and sets up a factory.

1842 Eugène-Melchior Peligot (1811-1890), French chemist, first isolates the metal uranium.

1843 Carl Gustav Mosander (1797-1858), Swedish chemist, discovers the rare earth element erbium.

1843 Charles-Frédéric Gerhardt (1816-1856), French chemist simplifies chemical formula-writing, so that water becomes H_20 instead of the previous H_40_2.

1844 Karl Karlovich Klaus (1796-1864), Estonian chemist, discovers ruthenium.

1845 Christian Friedrich Schönbein (1799-1868), German-Swiss chemist, prepares guncotton. He discovers that a certain acid mixture combines with the cellulose in cotton to produce an explosive that burns without smoke or residue.

1845 Justus von Liebig (1803-1873), German chemist, begins four years of work on chemical fertilizers that eventually establish the foundation for what becomes the fertilizer industry.

1845 Adolph Wilhelm Hermann Kolbe (1818-1884), German chemist, first synthesizes acetic acid from inorganic materials.

1846 William Robert Grove (1811-1896), English physicist, offers the first experimental evidence for thermal dissociation. He shows that water (steam) in contact with a strongly heated wire will absorb energy and break up (dissociate) into hydrogen and oxygen.

1846 Ascanio Sobrero (1812-1888), Italian chemist, slowly adds glycerine to a mixture of nitric and sulfuric acids and first produces nitroglycerine. He is so impressed by the explosive potential of a single drop in a heated test tube and so fearful of its use in war that he makes no attempt to exploit it. It is another 20 years before Alfred Nobel learns the proper formula and puts it to use.

1846 Louis Pasteur (1822-1895), French chemist, discovers molecular asymmetry and demonstrates the existence of isomers, becoming one of the earliest scientists to deal with the three-dimensional structure of molecules.

1849 Charles-Adolphe Wurtz (1817-1884), French chemist, discovers amines as he synthesizes the first organic derivative of ammonia (called ethylamine).

1849 Edward Frankland (1825-1899), English chemist, isolates amyl alcohol, which is eventually used as a solvent for resins and other oily materials.

1850 Ludwig Wilhelmy (1812-1864) of Germany constructs one of the first mathematical equations for describing a rate of progress of a chemical reaction when he offers an algebraic formula to describe the laws according to which the hydrolysis of sugar takes place.

1850 Thomas Graham (1805-1869), Scottish physical chemist, studies the diffusion of a substance through a membrane (osmosis) and first distinguishes between crystalloids and colloids. He becomes the founder of colloidal chemistry.

1852 Alexander William Williamson (1824-1904), English chemist, publishes his study which shows for the first time that catalytic action clearly involves and is explained by the formation of an intermediate compound.

1852 James Joule (1818-1889) and William Thomson (1824-1907), both English physicists, show that when a gas is allowed to expand freely, its temperature drops slightly. This becomes known as the Joule-Thomson effect and is evidence that molecules of gases have a slight attraction for other molecules. Overcoming this attraction uses energy and causes a drop in temperature.

1852 Edward Frankland (1825-1899), English chemist, announces the theory of valence, in which he states that each type of atom has a fixed capacity for combining with other atoms. This concept will lead eventually to Mendeleev's Periodic Table.

1852 Abraham Gesner (1797-1864), Canadian geologist, prepares the first kerosene from petroleum. He obtains the liquid kerosene by the dry distillation of asphalt rock, treats it further, and calls the product kerosene after the Greek word *keros*, meaning oil.

1853 Henry Enfield Roscoe (1833-1915), English chemist, begins research that lays the foundation for quantitative photochemistry.

1853 Stanislao Cannizzaro (1826-1910), Italian chemist, discovers a method of converting a type of organic compound called an aldehyde into a mixture of an organic acid and an alcohol. It comes to be called the Cannizzaro reaction.

1853 Anders Jonas Ångström (1814-1874), Swedish physicist, demonstrates that the rays emitted by an incandescent gas have the same refrangibility (ability to be refracted) as the rays absorbed by the same gas.

1853 Hans Peter Jorgen Julius Thomsen (1826-1901), Danish chemist, works out a method of manufacturing sodium carbonate from the mineral cryolite. This mineral will soon become important to the production of aluminum.

1855 Henri-Etienne Sainte-Claire Deville (1818-1881), French chemist, first produces aluminum in a pure state. He produces the metal in quantity using his original method of heating aluminum chloride with metallic sodium.

1855 Robert Wilhelm Bunsen (1811-1899), German chemist, first uses a burner that is perforated at the bottom so that air is drawn in by the gas flow. It burns steadily with little light and no smoke. The Bunsen burner becomes standard equipment in chemistry laboratories.

1855 Justus von Liebig (1803-1873), German chemist, publishes his *Die Grundsatze der Agrikulturchemie*. This detailed study of chemical fertilizers lays the foundation for agricultural chemistry.

1855 Charles-Adolphe Wurtz (1817-1884), French chemist, develops a method of synthesizing long-chain hydrocarbons by reactions between alkyl halides and metallic sodium. This method is called the Wurtz reaction.

1856 William Henry Perkin (1838-1907), English chemist, produces the first synthetic dye. While trying to synthesize quinine, he accidentally discovers how to produce mauve from the impure aniline in cola tar. This creates a dyestuffs industry, while stimulating the development of synthetic organic chemistry.

1857 Rudolf Julius Emmanuel Clausius (1822-1888), German physicist, offers a new explanation of evaporation in terms of molecules and their velocities. He shows that evaporation produces a loss of energy in the liquid and a decrease in temperature.

1858 Friedrich August Kekulé von Stradonitz (1829-1896), German chemist, and Archibald Scott Couper (1831-1892), Scottish chemist, first develop symbols to represent the atom is always tetravalent (meaning always combines with four other atoms).

1858 Archibald Scott Couper (1831-1892), Scottish chemist, introduces the concept of bonds into chemistry, suggesting a dash, or a dotted line, to represent the chemical bond.

1859 James Clerk Maxwell (1831-1879), Scottish mathematician and physicist, studies the rings of Saturn and produces the first extensive mathematical development of the kinetic theory of gases. He shares this discovery of the distribution of molecular speeds in a gas with Ludwig E. Boltzmann (1844-1906), Austrian physicist, who accomplishes the same independently. It has come to be known as the Maxwell-Boltzmann theory of gases.

1859 Adolph Wilhelm Hermann Kolbe (1818-1884), German chemist, discovers the reaction named after him (Kolbe reaction) when he succeeds in using phenol and carbon dioxide to produce salicylic acid. This new method makes large-scale production of salicylic acid (for aspirin) possible, leading to cheaper production costs for this wonder drug.

1860 Robert Wilhelm Bunsen (1811-1899), German chemist, collaborates with Gustav Robert Kirchhoff (1824-1887), German physicist, and they develop the first spectroscope. Called the Kirchoff-Bunsen spectroscope, their device refracts light differently and opens up the entire new field of spectrum analysis.

1860 Cesium is the first element discovered using the newly-developed spectroscope. Robert Wilhelm Bunsen (1811-1899), German chemist, and Gustav Robert Kirchhoff (1824-1887), German physicist, name their new element cesium after its "sky blue" color in the spectrum.

1860 Jean-Servais Stas (1813-1891), Belgian chemist, begins work that leads to an accurate method of determining atomic weights. By 1865, he produces the first modern table of atomic weights using oxygen as a standard.

1860 First International Congress of Chemistry is held in Karlsruhe, Germany, and attracts 140 delegates from around the world.

1860 Stanislao Cannizzaro (1826-1910), Italian chemist, publishes the forgotten ideas of Italian physicist Amedeo Avogadro (1776-1856)—about the distinction between molecules and atoms—in an attempt to bring some order and agreement on determining atomic weights.

1861 William Crookes (1832-1919), English physicist, discovers the element thallium by using the newly-invented spectrum analysis. The following year, it is isolated by French chemist Claude-August Lamy (1820-1878).

1861 Alexander Mikhailovich Butlerov (1828-1886), Russian chemist, introduces the term "chemical structure" to mean that the chemical nature of a molecule is determined not only by the number and type of a atoms but also by their arrangement.

1861 Ernest Solvay (1838-1922), Belgian chemist, discovers a practical and inexpensive way of producing sodium bicarbonate. His new method requires much less heat and less fuel. He founds a company in 1863 to begin manufacture, and by 1913 he is producing virtually the entire world supply of sodium bicarbonate.

1861 Friedrich August Kekulé von Stadonitz (1829-1896), German chemist, publishes the first volume of *Lehrbuch der organischen Chemie*, in which he is the first to define organic chemistry as the study of carbon compounds.

1861 Robert Wilhelm Bunsen (1811-1899), German chemist, and Gustav Robert Kirchhoff (1824-1887), German physicist, discover the metal rubidium, using their new spectroscope.

1862 Charles Friedel (1832-1899), French chemist, prepares isopropyl, the first secondary alcohol and one which is later used as a solvent and as rubbing alcohol.

1863 Cato Maximilian Guldberg (1836-1902), Norwegian chemist and mathematician, and his brother-in-law Peter Waage (1833-1900), also a Norwegian chemist, first publish a paper which formulates the law of mass action. They state that the direction taken by a reaction is dependent not merely on the mass of the various components of the reaction, but upon the concentration or the mass present in a given volume. It is not acknowledged until much later.

1863 Adolf Johann Friedrich Wilhelm von Baeyer (1835-1917), German chemist, discovers barbituric acid, which becomes the basis for modern sleeping pills.

1863 Ferdinand Reich (1799-1882), German mineralogist, and his assistant Hieronymus Theodor Richter (1824-1898), examine zinc ore spectroscopically and discover the new, indigo-colored element iridium. It is used in the next century in the making of transistors.

1863 John Alexander Reina Newlands (1837-1898), English chemist, announces his arrangement of the elements in the order of their atomic weights. After discovering that their properties seem to repeat themselves in each group of seven elements, he calls this the Law of Octaves (after the musical scale). He is the first to do this and is a precursor of Dmitry Mendeleev (1869).

1864 Alexander Mikhailovich Butlerov (1828-1886), Russian chemist, prepares butyl alcohol, the first tertiary alcohol. It is later used as a solvent and in synthetic rubber.

1865 Friedrich August Kekulé von Stradonitz (1826-1896), German chemist, discovers the six-carbon ring structure of the benzene molecule.

1865 Alexander Parkes (1813-1890), English chemist, produces celluloid, the first synthetic plastic material. After working since the 1850s with nitrocellulose, alcohol, camphor, and castor oil, he obtains a material that can be molded under pressure while still warm. Parkes is unsuccessful at marketing his product, however, and it is left to the American inventor, John Wesley Hyatt (1837-1920), to make it a success.

1865 Johann Joseph Loschmidt (1821-1895), Austrian chemist, is the first to attempt to determine the actual size of atoms and molecules. He uses Avogadro's hypothesis to calculate the number of molecules in 22.4 liters of gas, and calls the resulting number Avogadro's number ($6.02x10^{23}$).

1866 Pierre-Eugène-Marcelin Berthelot (1827-1907), French chemist, synthesizes acetylene and uses it as a starting point for the synthesis of styrene, benzene, methyl alcohol, ethyl alcohol, and methane.

1867 Alfred Bernhard Nobel (1833-1896), Swedish inventor, invents dynamite, a safer and more controllable version of nitroglycerine. He combines nitroglycerine with "kieselguhr," or earth containing silica, and discovers that it could not be exploded without a detonating cap.

1868 William Henry Perkin (1838-1907), English chemist, synthesizes coumarin from a coal-tar derivative. The production of this pleasant-smelling white, crystalline substance marks the beginning of the synthetic perfume industry.

1868 August Friedrich Horstmann (1842-1929), German chemist, founds chemical thermodynamics.

1868 John Wesley Hyatt (1837-1920), American inventor, searching for an ivory substitute, improves on Parkes's celluloid (1865) and patents a method of making billiard balls out of what he names Celluloid. He produces this new plastic material by dissolving pyroxyline and camphor in alcohol and heating the mixture under pressure. He markets it successfully, and his plastics (although very flammable) are used for photographic film, combs, and baby rattles.

1868 Carl Hermann Wichelhaus (1842-1927), German chemist, introduces the term valency.

1868 Pierre-Jules-César Janssen (1824-1907), French astronomer, studies a total eclipse of the Sun and observes an unknown spectral line. He forwards the data to the English astronomer Joseph Norman Lockyear (1836-1920), who concludes it is an unknown element that he names helium, after the Sun.

1869 Dmitry Ivanovich Mendeleev (1834-1907), Russian chemist, and Julius Lothar Meyer (1830-1895), German chemist, independently put forth the Periodic Table of Elements, which arranges the elements in order of atomic weights. However, Meyer does not publish until 1870, nor does he predict the existence of undiscovered elements as Mendeleev does.

1869 Thomas Andrews (1813-1885), Irish physical chemist, first proposes the chemical concept of critical temperature. He suggests that every gas has a precise temperature, which he calls its critical temperature, above which it cannot be liquefied even under greater pressure. This leads to a breakthrough in the liquefaction of the so-called permanent gases (whose temperature must be lowered before pressure is exerted).

1871 Ernest Solvay (1838-1922), Belgian chemist, develops a new process for the manufacture of soda (from ammonia).

1873 Bernhard Christian Gottfried Tollens (1841-1918), German agricultural chemist, makes artificial methane, proving that Mendeleev's table is correct.

1873 Othmar Zeidler of Germany first synthesizes DDT (dichlorodiphenyltrichloroethane) by the reaction of chloral with chlorobenzene in the presence of sulfuric acid. He is unaware of its insecticidal properties.

1873 Johannes Diderck van der Waals (1837-1923), Dutch physicist, offers an equation that provides a molecular explanation of why there is a critical temperature for gas, above which it can only exist as a gas, and below which it can be condensed to a liquid,

1874 Jacobus Henricus Van't Hoff (1852-1911), Dutch physical chemist, and Joseph Achille Le Bel (1847-1930), French chemist, independently discover the theory of the relationship of optical activity to molecular structure. The asymmetric, three-dimensional molecular structure of carbon suggests to each scientist a theory of stereochemistry (the three-dimensional structure of molecules).

1875 Paul-Émile Lecoq de Boisbaudran (1838-1912), French chemist, discovers gallium, an element whose properties were predicted by Mendeleev.

1876 Josiah Willard Gibbs (1839-1903), American physicist, applies the laws of thermodynamics to chemical reactions and develops the modern concepts of "free energy" and "chemical potential" as the driving force behind chemical reactions.

1876 Josiah Willard Gibbs (1829-1903), American physicist, discovers the phase rule. He arrives at an equation that relates the variables (like temperature and pressure) to different phases (solid, liquid, gas). His work helps lay the foundation for chemical thermodynamics and generally for modern physical chemistry. The rule is later put into practical application by Dutch physical chemist Hendrik Willem Bakhuis Roozeboom (1854-1907).

1877 Louis-Paul Cailletet (1832-1913), French physicist, and Raoul-Pierre Pictet (1846-1929), Swiss chemist, independently liquefy the permanent gases. They both produce small quantities of liquid oxygen, nitrogen, and carbon monoxide.

1877 James Mason Crafts (1839-1917), American chemist, and Charles Friedel (1832-1899), French chemist, discover that aluminum chloride is a versatile catalyst for reactions tying together a chain of carbon atoms to a ring of carbon atoms. This becomes known as the Friedel-Crafts reaction.

1878 Adolf Johann Friedrich Wilhelm von Baeyer (1835-1917), German chemist, produces the first synthesized indigo blue dye. His distillation process has a great impact on the dye industry and becomes widely used to produce dyes from other materials.

1878 Jean-Charles Marignac (1817-1894), Swiss chemist, discovers the rare earth element ytterbium. In 1880, he discovers another rare earth element, gadolinium.

1878 Louis-Marie-Hillaire Bernigaud de Chardonnet (1839-1924), French chemist, invents rayon. He begins by producing fibers, which are made by forcing solutions of nitrocellulose through tiny holes and allowing the solvent to evaporate. The end product causes a sensation because it resembles silk. Rayon is the first artificial fiber to come into common use.

1879 Lars Fredrik Nilsson (1840-1899), Swedish chemist, discovers scandium, an element whose properties were predicted by Mendeleev.

1879 Constantin Fahlberg (1850-1910), German chemist, and Ira Remsen (1846-1927), American chemist, first prepare saccharin. Also called thibenzoyl sulfimide, it has a distinctly sweet taste.

1879 Paul-Émile Lecoq de Boisbaudran (1838-1912), French chemist, discovers samarium, a pale gray, lustrous, metallic element that is later used in alloys that form permanent magnets.

1879 Vladimir Markovnikov (1837-1904), Russian chemist, disproves the notion that carbon-based molecules can have only six-atom rings as he prepares molecules with rings of four carbon atoms. In 1889, he produces seven carbon atoms.

1879 Per Theodore Cleve (1840-1905), Swedish chemist, discovers two new rare earth elements, thulium and holmium.

c. 1880 Carl Oswald Viktor Engler (1842-1925), German chemist, begins his studies on petroleum. He is the first to states that it is organic in origin, and he can be considered the founder of the study of petroleum.

1882 François-Marie Raoult (1830-1901), French physical chemist, demonstrates that the lowering of the freezing point of a solution of a particular substance is proportional to the concentration of the dissolved substance in molecular terms. Known as Raoult's law, this leads to a valuable method for calculating molecular weights.

1882 Jules Violle (1841-1923), French physicist, design a method of keeping liquefied gas in that state. He is able to sustain low enough temperatures by the insulating properties of a vacuum.

1883 Frank Wigglesworth Clarke (1847-1931), American chemist and geophysicist, is appointed chief chemist to the U.S. Geological Survey. In this position, he begins an extensive program of rock analysis and is one of the founders of geochemistry.

1883 Johann Gustav Kjeldahl (1849-1900), Danish chemist, devises a method for the analysis of the nitrogen content of organic material. His method uses concentrated sulfuric acid and is simple and fast.

1884 Svante August Arrhenius (1859-1927), Swedish chemist, first proposes the concept of ions being atoms bearing electrical charges. He offers this theory as part of his dissertation, in which he argues that molecules of some compounds break up into charged particles when put in a liquid. His ideas are eventually proved correct, and he becomes one of the founders of modern physical chemistry.

1884 Jacobus Henricus Van't Hoff (1852-1911), Dutch physical chemist, publishes his *Etudes de dynamique chimique*, which lays the foundations for chemical kinetics and chemical thermodynamics.

1884 Henry-Louis Le Châtelise (1850-1936), French chemist, announces the principle that bears his name. It states that when one of the factors affecting a system is changed, the system will respond by minimizing the effect of the change. By applying this principle scientists are able to maximize the efficiency of chemical processes.

1884 Otto Wallach (1847-1931), German organic chemist, begins a life-long project to separate one terpene from another (terpenes are isomeric hydrocarbons found in essential oils) and to establish the structure of each. After 25 years of work, his research forms the basis for the modern perfume industry.

1885 Karl Auer, Baron von Welsbach (1858-1929), Austrian chemist, invents the incandescent gas mantle. His use of chemicals in the mantle produces a brilliant white glow, but is ultimately outdone by Edison's electric light.

1885 Karl Auer, Baron von Welsbach (1858-1929), Austrian chemist, discovers that what had been thought of as one element named didymium is in fact made out of two rare earth elements. He names them praseodymium and neodymium.

1886 Ferdinand-Frédéric-Henri Moissan (1852-1907), French chemist, devises a special electric furnace, in which he succeeds in preparing a pure sample of fluorine.

1886 William Crookes (1832-1919), English physicist, states his theory of cathode ray tubes which had been discovered in 1885 by German mathematical physicist Julius Plücker (1801-1868).

1886 Paul-Louis-Toussint Héroult (1863-1914), French metallurgist, and Charles Martin Hall (1863-1914), American chemist, independently invent an electrochemical process for extracting aluminum from its ore. This process makes aluminum cheaper and forms the basis of the huge aluminum industry. Hall makes the discovery in February of this year, and Héroult achieves his in April.

1886 Clemens Winkler (1838-1904), German chemist, discovers a new metallic element he names germanium. This new element confirms the validity of Mendeleev's Periodic Table.

1887 Wilhelm Ostwald (1853-1932), Russian-German physical chemist, founds the first learned journal devoted exclusively to physical chemistry. *Zeitschrift für physikalische Chemie* becomes an influential journal, and it is by Ostwald's efforts that physical chemistry becomes an organized, independent branch of chemistry.

1887 Viktor Meyer (1848-1897), German organic chemist, introduces the term "stereochemistry" for the study of molecular shapes.

1887 Emil Hermann Fischer (1852-1919), German chemist, studies the structure of sugars and shows the exact nature of the many varieties.

1887 Herman Frasch (1851-1914), German-American chemist, patents a method for removing sulfur compounds from oil. Once the foul sulfur smell is removed through the use of metallic compounds, petroleum becomes a marketable product.

1889 Frederick Augustus Abel (1827-1902), English chemist, and James Dewar (1842-1923), Scottish chemist and physicist, invent cordite and pioneer the production of smokeless powder. Their new mixture borrows form previous discoveries but proves safe to handle.

1890 Franz Hofmeister (1850-1922) of Germany obtains the first crystallization of a protein, albumin.

1890 Stephen Moulton Babcock (1843-1931), American agricultural chemist, perfects a test for determining the butterfat content of milk, and offers a standard method of grading milk.

1891 Charles Frederick Cross (1855-1935), English chemist, invents viscose rayon. He improves early rayon by dissolving cellulose in carbon disulfide and squirting the viscous solution out of fine holes. The fine threads of viscose rayon are formed as the solvent evaporates. By 1908, this viscous solution is forced through a narrow slit, forming thin, transparent sheets of what comes to be known as "cellophane."

1891 Edward Goodrich Acheson (1856-1931), American inventor, discovers that carbon heated with clay yields an extremely hard substance. He names it carborundum, and eventually finds it to be a compound of silicon and carbon. For half a century it remains second only to a diamond in hardness, and becomes very useful as an abrasive.

1891 Alfred Werner (1866-1919), German-Swiss chemist, develops his coordination theory which suggests that atoms or groups of atoms could be distributed around a central atom in accordance with fixed geometric principles regardless of valence.

1892 James Dewar (1842-1923), Scottish chemist and physicist, improves the Violle vacuum insulator by constructing a double-walled flask with a vacuum between the walls. He then coats all sides with silver so heat will be reflected and not absorbed. His "Dewar flask" becomes the first Thermos bottle.

1892 Standardization of organic chemistry nomenclature is achieved at the Geneva Conference which agrees that for name endings, all alcohols will be "ol," saturated hydrocarbons will end in "ane," aldehydes in "al," and ketones in "one."

1893 Ferdinand-Frédéric-Henri Moissan (1852-1907), French chemist, produces artificial diamonds in his electric furnace.

1894 John William Strutt Rayleigh (1842-1919), English physicist, and William Ramsay (1852-1916), Scottish chemist, succeed in isolating a new gas in the atmosphere that is denser than nitrogen and combines with no other element. They name it "argon," which is Greek for inert. It is the first of a series of rare gases with unusual properties whose existence had not been predicted.

1894 Wilhelm Ostwald (1853-1932), Russian-German physical chemist, determines that a catalyst can only speed up a reaction and not create one.

1895 William Ramsay (1852-1916), Scottish chemist, discovers helium in a mineral named cleveite. It had been speculated earlier that helium existed only in the Sun, but Ramsay proves it also exists on Earth. It is discovered independently this year by Swedish chemist and geologist Per Theodore Cleve (1840-1905). Helium is an odorless, colorless, tasteless gas that is also insoluble and incombustible.

1895 Paul Walden (1863-1957), Russian-German chemist, discovers he can alter the effect a substance produces without changing the substance itself, producing what comes to be known as the "Walden inversion." Studies of this phenomenon have since led to greater understanding of all stages of reaction mechanisms.

1896 Antoine-Henri Becquerel (1852-1908), French physicist, discovers radioactivity in uranium ore.

1897 Joseph John Thomson (1856-1940), English physicist, discovers the electron. He conducts cathode ray experiments and concludes that the rays consist of negatively charged "electrons" that are smaller in mass than atoms.

1897 Hamilton Young Castner (1859-1899), American chemist, invents the electrolytic method of isolating sodium and chlorine from brine, and forms, with his colleague Kellner, the Castner-Kellner electrochemical process for making caustic soda.

1897 Eduard Buchner (1860-1917), German chemist, publishes *Alkoholische Gärung ohne Hefezellen*, in which he proves that a cell-free extract of yeast can ferment sugar. This establishes that living cells are not essential for fermentation, and marks the beginning of enzyme chemistry.

1897 Richard Wilhelm H. Abegg (1869-1910), German chemist, begins to pursue the new electronic view of the atom and determines that it is its outermost electron shell that determines its properties.

1897 Paul Sabatier (1854-1941), French chemist, discovers via a failed experiment that nickel has properties as a catalyst. He develops nickel catalysis, which eventually makes possible the formation of edible fats like margarine and shortening from inedible plant oils.

1898 William Ramsay (1852-1916), Scottish chemist, discovers the new element krypton. He obtains it only after months of boiling down liquid air. He also obtains the inert gases xenon and neon with the same process.

1898 Hans Goldschmidt (1861-1923), German chemist, develops a method of producing an oxide-aluminum mixture called thermite that becomes ideal for welding.

1898 Marie Sklodowska Curie (1867-1934), Polish-French chemist, discovers thorium, which she proves is radioactive.

1898 Marie Sklodowska Curie (1867-1834) and her husband French chemist Pierre Curie (1859-1906) isolate a new, radioactive element they name polonium.

1898 Marie Sklodowska Curie (1867-1934) and Pierre Curie (1859-1906) discover the radioactive element radium. They spend the next four years refining eight tons of pitchblende to obtain a full gram of radium. Their discoveries in radioactivity inaugurate the investigation of this new field.

1899 André-Louis Debierne (1874-1949), French chemist, discovers the radioactive element actinium.

1899 James Dewar (1842-1923), Scottish chemist and physicist, builds a large device with which he first solidifies hydrogen. He is able to lower the temperature to 14 degrees above absolute zero.

1899 Julius Elster (1854-1920) and Hans Freidrich Geitel (1855-1923), German physicists, demonstrate that external effects do not influence the intensity of radiation. They are also the first to characterize radiation as being caused by changes that take place within the atom.

1899 William Jackson Pope (1870-1939), English chemist, discovers the first optically active (able to polarize light) compound that contains no carbon atoms. This proves that the Van't Hoff theory applies to atoms other than carbon.

1900 Moses Gomberg (1866-1947), Russian-American chemist, discovers triphenylmethyl, the first example of a free carbon radical. Carbon has four valences, but triphenylmethyl uses only three valences of the central carbon atom, leaving one free or unfilled. Free radicals eventually prove very important to chemical reactions.

1900 Paul Karl Ludwig Drude (1863-1906), German physicist, proposes the first model for the structure of metals. His model explains the constant relationship between electrical conductivity and the heat conductivity in all metals.

1900 Friedrich Ernst Dorn (1848-1916), German physicist, analyzes the gas given off by (radioactive) radium and discovers the inert gas he names radon. This is the first clear demonstration that the process of giving off radiation transmutes one element into another during the radioactive decay process.

1900 Thomas Alva Edison (1847-1931), American inventor, builds the first nickel alkaline battery.

1900 Vladimir Nikolaevich Ipatieff (1867-1952), Russian-American chemist, discovers the role catalysts can play when he finds that organic reactions taking place at high temperatures can be influenced in their course by varying the nature of the substance with which they are in contact.

1900 First direct-arc electric furnace for producing steel is put into operation at La Paz, France.

1901 Eugène-Anatole Demarcy (1852-1904), French chemist, discovers the rare-earth element europium.

1901 François-Auguste-Victor Grignard (1871-1935), French chemist, discovers the highly useful catalytic role that organic manganese halides can play in preparing organic compounds. His work produces an entire series of what come to be known as "Grignard reagents."

1901 Jacobus Henricus Van't Hoff (1852-1911), Dutch physical chemist, receives the Nobel Prize in Chemistry for the discovery of the laws of chemical dynamics and osmotic pressure in solution.

1902 Georges Claude (1870-1960), French technologist, liquefies air.

1902 Hermann Emil Fischer (1852-1919), German chemist, is awarded the Nobel Prize in Chemistry for his work on sugar and purine synthesis.

1903 Svante August Arrhenius (1859-1927), Swedish chemical is awarded the Nobel Prize in Chemistry for his electrolytic theory of dissociation.

1903 Ernest Rutherford (1871-1937), British physicist, and Frederick Soddy (1877-1956), English chemist, explain radioactivity by their theory of atomic disintegration. They discover that uranium breaks down and forms a new series of substances as it gives off radiation.

1904 Theodore William Richards (1868-1928), American chemist, improves Jean-Servais Stas's method of determining atomic weights and revises some of his figures.

1904 Joseph John Thomson (1856-1940), English physicist, formulates an atomic model and is one of the first to suggest a theory of structure of the atom (as a sphere of positive electricity in which negatively charged electrons are embedded). This model becomes known as the "plum pudding atom," and is soon replaced by the more useful model proposed by Ernest Rutherford (1871-1937), English physicist.

1904 Frederic Stanley Kipping (1863-1949), English chemist, discovers silicones. It is his pioneering work on these complicated substances that leads to the post-war boom in the use of silicones as greases, water repellants, and synthetic rubbers.

1904 William Ramsay (1852-1916), Scottish chemist, receives the Nobel Prize in Chemistry for the discovery of the inert gaseous elements in air, and for his determination of their place in the periodic system.

1905 Alfred Werner (1866-1919), German-Swiss chemist, publishes *Neure Anschauungen*, which prepares the way for modern developments in inorganic chemistry and the electronic theory of valency.

1905 Otto Hahn (1879-1968), German physical chemist, discovers radiothorium which is later determined to be an isotope of thorium.

1905 Richard Willstätter (1872-1942), German chemist, discovers the structure of chlorophyll.

1905 Johann Friedrich Wilhelm von Baeyer (1835-1917), German chemist, is awarded the Nobel Prize in Chemistry for his services in the advancement organic and industrial chemistry, through his work on organic dyes and hydroaromatic compounds.

1906 Mikhail Semenovich Tswett (1872-1919), Russian botanist, invents chromatography. It becomes a universally used method for separating and identifying organic substances.

1906 Hermann Walther Nernst (1864-1941), German physical chemist, discovers what becomes known as the third law of thermodynamics, which states that entropy change approaches zero at a temperature of absolute zero. From this he deduces the impossibility of attaining absolute zero.

1906 Ferdinand-Frédéric-Henri Moisson (1852-1907), French chemist, is awarded the Nobel Prize in Chemistry for his investigation and isolation of fluorine, and for the adoption in the service of science of the electric furnace named after him.

1906 Japan begins the production of monosodium glutamate as a flavor-enhancer for foods. By 1926, production reaches industrial proportions.

1907 Leo Hendrik Baekeland (1863-1944), Belgian-American chemist, produces synthetic resins on an industrial scale and creates Bakelite. This breakthrough is the first of the "thermosetting plastics," or plastics that, once set, will not soften under heat.

1907 Bertram Borden Boltwood (1870-1927), American chemist and physicist, discovers what he believes is a new element which he calls ionium. It is later determined to be a radioactive isotope of thorium.

1907 Georges Urbain (1872-1938), French chemist, discovers that last of the stable rare earth elements, and names it lutetium after the Latin name of Paris.

1907 Adolf Windaus (1876-1959), German chemist, first synthesizes histamine. This important compound is found in animal tissues and has major physiological effects.

1907 Eduard Buchner (1860-1917), German chemist, receives the Nobel Prize in Chemistry for his biochemical researches and his discovery of cell-free fermentation.

1908 William Ramsay (1852-1916), Scottish chemist, puts forward the electronic theory of valency.

1908 Heike Kamerlingh Onnes (1853-1926), Dutch physicist, first liquefies helium.

1908 Fritz Haber (1868-1934), German chemist, first synthesizes ammonia. Using iron as a catalyst, he combines nitrogen and hydrogen under pressure and forms ammonia. Using this method, which becomes known as the Haber process, he is able to fix nitrogen and create the potential for plentiful supplies of it for fertilizer or explosives.

1908 Ernest Rutherford (1871-1937), English physicist, is awarded the Nobel Prize in Chemistry for his investigations into disintegration of the elements and the chemistry of radioactive substances.

1909 Sören Peer Lauritz Sörensen (1868-1939), Danish chemist, first introduces the expression pH to denote the negative logarithm of the concentration of the hydrogen ion present. It becomes a standard measure of alkalinity and acidity.

1909 Alfred Stock (1876-1946), German chemist, first synthesizes boron hydrides (compounds of boron and hydrogen). Forty year later, boron hydrides prove useful to space exploration as additives to rocket fuel.

1909 Friedrich Wilhelm Ostwald (1853-1932), Russian-German physical chemist, is awarded investigations into the fundamental principles governing chemical equilibria and reaction rates.

1910 Otto Wallach (1847-1931), German organic chemist, is awarded the Nobel Prize in Chemistry for his services to organic chemistry and the chemical industry by his pioneer work in the field of alicyclic compounds.

1911 Fritz Pregl (1869-1930), Austrian chemist, first introduces organic microanalysis. He invents analytic methods that make it possible to determine the empirical formula of an organic compound from just a few milligrams of the substance.

1911 Ernst Rutherford (1871-1937), British physicist, discovers that atoms are made up of a positive nucleus surrounded by electrons. This modern concept of the atom replaces the notion of featureless, indivisible spheres that dominated atomistic thinking for 23 centuries—since Democritus (c. 470-c.380).

1911 Heike Kamerlingh Onnes (1853-1926), Dutch physicist, discovers superconductivity. While studying the properties of metals at very low temperatures, he finds that certain metals undergo a total loss of electrical resistance (and become "superconductive").

1911 Chaim Weizmann (1874-1952), Russian-English-Israeli-chemist, discovers how to use the bacteria from fermenting grain to synthesize acetone. Acetone is essential to the production of the explosive cordite.

1911 Marie Sklodowska Curie, Polish-French chemist, receives the Nobel Prize in Chemistry for the discovery of the elements radium and polonium, for the isolation of radium, and for investigating its compounds.

1912 Friedrich Karl Rudolf Bergius (1884-1949), German chemist, discovers how to treat coal and oil with hydrogen to produce gasoline.

1912 François-Auguste-Victor Grignard (1871-1935), French chemist, receives the Nobel Prize in Chemistry for the discovery of the so-called Grignard reagents, which greatly advance the progress of organic chemistry. Paul Sabatier (1854-1941), French chemist, also receives the Nobel Prize in Chemistry for his method of hydrogenating organic compounds in the presence of finely disintegrated metals.

1913 Henry Gwyn Jeffreys Moseley (1887-1915), English physicist, studies the x-ray frequency characteristics of the various elements and discovers a technique to determine the atomic number of the nucleus. He finds that the positive charge of the nucleus indicates the number of electrons present in every neutral atom.

1913 Niels Henrik David Bohr (1885-1962), Danish physicist, proposes the first dynamic model of the atom. It is seen as a very dense nucleus surrounded by electrons rotating in stationary elliptical orbits.

1913 Using the x-ray spectroscope which they invented, William Lawrence Bragg (1890-1971), Australian-English physicist, and his father, William Henry Bragg (1862-1942), make the first determinations of the structures of simple crystals and demonstrate the tetrahedral distribution of carbon atoms in diamonds. Their perfection of x-ray crystallography leads to the later examination of the molecule structure of thousands of crystalline substances.

1913 Chlorine is first used in the United States to purify water.

1913 Harry Brearly (1871-1948), English metallurgist, accidentally discovers a nickel-chromium alloy that is corrosion resistant. It becomes stainless steel.

1913 Jean-Baptiste Perrin (1870-1942), French physicist, publishes *Les Atomes*, in which he offers systematically obtained evidence of the size and number of atoms and molecules in a given volume of gum resin. For the first time, science has real evidence through observation for the existence of these tiny entities.

1913 Max Bodenstein (1871-1942), German physical chemist, develops the concept of a chain reaction in which one molecular change triggers the next, and so on.

1913 Leonor Michaelis (1875-1949), German-American scientist, works out an equation that describes how the rate of an enzyme-catalyzed reaction varies with the concentration of the substance taking part in the reaction. It is called the Michaelis-Menten equation, after him and his assistant Maud Lenora Menten.

1913 Alfred Werner (1866-1919), German-Swiss scientist, receives the Nobel Prize in Chemistry for his work on the linkage of atoms in molecules by which he opened up new fields of research, especially in inorganic chemistry.

1913 Frederick Soddy (1877-1956), English chemist, introduces the term and the concept of isotope. He suggests that different elements produced in radioactive transformations are capable of occupying the same place in the Periodic Table. He chooses the Greek words meaning "same place" to describe them.

1914 Theodore William Richards (1868-1928), American chemist, is awarded the Nobel Prize in Chemistry for his accurate determination of the atomic weight of a large number of chemical elements.

1915 Richard Martin Willstätter (1872-1942), German chemist, is awarded the Nobel Prize in Chemistry for his researchers on plant pigments, especially chlorophyll.

1916 Gilbert Newton Lewis (1875-1946), American chemist, suggests the phenomenon of shared electrons in nonionic compounds. He states that a bond between two elements can be formed by the sharing of a pair of electrons, and that the final result is that all atoms achieve the stable electron configuration of the inert gas atom.

1917 Otto Hahn (1879-1968), German physical chemist, and Lise Meitner (1878-1968), Austrian-Swedish physicist, discover the new element protactinium (atomic number 91). It is independently discovered by Polish chemist Kasimir Fajans (1887-1975), as well as by the English team of Frederick Soddy (1877-1956), John Cranston, and Alexander Fleck.

1918 Jaroslav Heyrovski (1890-1967), Czech physical chemist, invents "polarography," an extremely delicate analytical system for determining the concentration of ions in a solution of unknown composition.

1918 Fritz Haber (1868-1934), German chemist, is awarded the Nobel Prize in Chemistry for the synthesis of ammonia from its elements.

1919 Ernest Rutherford (1871-1937), English physicist, produces the first man-made transmutation. He is the first person ever to change one element into another when he shows that nitrogen, under alpha particle bombardment, ejects a hydrogen nucleus. In a very tiny way, this is also the first man-made nuclear reaction.

1919 Francis William Aston (1877-1945), English chemist and physicist, builds the mass spectrograph, which he uses to separate the isotopes of elements and determine their mass and relative abundance.

1920 Ernest Rutherford (1871-1937), English physicist, names the positively charged part of the atom's nucleus a "proton."

1920 Hermann Walther Nernst (1864-1941), German physical chemist, receives the Nobel Prize in Chemistry for his work in thermochemistry.

1921 Thomas Midgley, Jr. (1889-1944), American chemist, discovers the anti-knock properties of tetraethyl lead and develops the mixture of it and ethylene dibromide and dichlorine that is soon adopted throughout the world to increase the octane rating of gasoline.

1921 Frederick Soddy (1865-1956), English chemist, is awarded the Nobel Prize in Chemistry for his contributions to our knowledge of the chemistry of radioactive substances and his investigations into the origin and nature of isotopes.

1922 Francis William Aston (1877-1945), English chemist and physicist, receives the Nobel Prize in Chemistry for his discovery, by means of his mass spectrograph, of isotopes in a large number of non-radioactive elements, and for his enumeration of the whole-number rule.

1923 Theodor H. E. Svedburg (1884-1971), Swedish chemist, invents the ultracentrifuge, which makes it possible to obtain proteins in the pure state and to evaluate their molecular weight. For his work, Svedburg received the Nobel Prize in Chemistry in 1926.

1923 Louis-Victor-Pierre-Raymond de Broglie (1892-1987), French physicist, demonstrates that with any elementary moving particle there is an associated wave. This contributes to the new wave concept of the atom.

1923 Georg von Hevesy (1885-1966), Hungarian-Swedish chemist, and colleagues discover the new element hafnium.

1923 Johannes Nicolaus Brönsted (1879-1947), Danish chemist, studies hydrogen ions and their role in acid-base relationships, and clarifies what becomes the modern notion of acids and bases.

1923 Fritz Pregl (1869-1930), Austrian chemist, is awarded the Nobel Prize in Chemistry for his invention of the method of microanalysis of organic substances.

1925 Walter Karl Friedrich Noddack (1893-1960) and Ida Eva Tacke (1896-1979), both German chemists, discover the new element rhenium (atomic number 75). They marry in 1926 and continue to work together.

1925 John Masson Gulland (1898-1947), Scottish chemist, and Robert Robinson (1886-1975), English chemist, propose the definitive structure of morphine, solving a problem that had perplexed chemists for nearly four decades.

1925 Richard Adolf Zsigmondy (1865-1929), Austrian-German chemist, receives the Nobel Prize in Chemistry for his demonstration of the heterogeneous nature of colloid solutions, and for the methods he used, which have since become fundamental in modern colloid chemistry.

1926 Erwin Schrödinger (1887-1961), Austrian physicist, works out the fundamental mathematical equation of wave mechanics, which places Planck's quantum theory on a firm mathematical basis.

1926 James Batcheller Sumner (1887-1955), American biochemist, conducts the first crystallization of an enzyme. In extracting urease from the jack bean, he realizes that the crystals he isolates are both enzymes and proteins.

1926 Hermann Staudinger (1881-1965), German chemist, begins work on the nature of polymers (large molecules) and discovers that all of the various plastics being produced are similar polymers with simple units arranged in a straight line.

1927 Albert Szent-Györgyi (1893-1986), Hungarian-American physicist, discovers ascorbic acid, or vitamin C, while studying oxidation in plants.

1927 Fritz Wolfgang London (1900-1954), German-American physicist, works out a quantum mechanical treatment of the hydrogen molecule, which provides the theoretical basis for the study of molecules in terms of the new (quantum) physics.

1927 Heinrich Otto Wierland (1877-1957), German chemist, is awarded the Nobel Prize in Chemistry for his investigations of the constitution of the bile acids and related substances.

1928 Otto Paul Hermann Diels (1876-1954) and Kurt Alder (1902-1958), both German chemists, discover a technique of atomic combination that can be put to use in the formation of many synthetic compounds. It comes to be known as the Diels-Alder reaction (also called diene synthesis).

1928 Cyril Norman Hinshelwood (1897-1967), English physical chemist, studies chain-reaction mechanisms and shows that temperature determines whether a mixture of hydrogen and oxygen gases will explode or not. The Russian physical chemist Nikolay Nikolayevitch Semenov (1896-1986) independently discovers the same this year.

1928 Adolf Windaus (1867-1959, German chemist, receives the Nobel Prize in Chemistry for his research into the constitution of the sterols and their connection with vitamins.

1929 William Francis Giauque (1895-1982), American chemist, discovers that oxygen is a mixture of three isotopes. This leads to a debate between chemists and physicists concerning an atomic weight standard, which is not resolved until 1961.

1929 Arthur Harden (1865-1940), English biochemist and Hans Karl August Simon von Euler-Chelpin, German-Swedish chemist, are awarded the Nobel Prize in Chemistry for their investigations on sugar fermentation and fermentive enzymes.

1929 Julius Arthur Nieuwland (1878-1936), Belgian-American chemist, develops neoprene, the first successful synthetic rubber.

1930 Hans Fischer (1881-1945), German chemist, is awarded the Nobel Prize in Chemistry for his researches into the constitution of haemin and chlorophyll and especially for his synthesis of haemin.

1930 Nils Edlefsen (1893-1971) of the United States constructs the first cyclotron under the direction of the American physicist Ernest Orlando Lawrence (1901-1958). This first instrument is a small machine that is used to produce directed beams of charged particles. Over the next few years, Lawrence continues to build larger instruments, which eventually contribute to the discovery of new elements.

1930 Thomas Midgley, Jr. (1889-1944), American chemist, first prepares freon (difluorodichloromethane) as a refrigerant gas. It comes to be used in all refrigerators, freezers, and air conditioning systems. Although neither poisonous nor inflammable, it is later believed to play a major role in the depletion of the Earth's ozone layer.

1931 First synthetic fiber (nylon) is made by a group headed by American chemist Wallace Hume Carothers (1896-1937), while searching for a synthetic replacement for silk.

1931 Harold Clayton Urey (1893-1981), American chemist, discovers deuterium, one of the heavy isotopes of hydrogen.

1931 Linus Carl Pauling (1901-1994), American chemist, develops his theory of "resonance," which explains many phenomena in organic chemistry.

1931 Karl Bosch (1874-1940), German chemist, and

Friedrich Karl Rudolph Bergius (1884-1949), German chemist, receive the Nobel Prize in Chemistry for their contributions to the invention and development of chemical high pressure methods.

1932 James Chadwick (1891-1974), English physicist, proves the existence of the neutral particle of the atom's nucleus, called the neutron. It proves to be by far the most useful particle for initiating nuclear reactions.

1932 Irving Langmuir (1881-1957), American chemist, is awarded the Nobel Prize in Chemistry for his discoveries and investigations in surface chemistry.

1933 Gilbert Newton Lewis (1875-1946), American chemist, is the first to prepare a sample of water in which all the hydrogen atoms consist of deuterium (the heavy hydrogen isotope). Called "heavy water," this will later play an important role in the production of the atomic bomb.

1934 Leo Szilard (1898-1964), Hungarian-American physicist, and T. A. Chalmers discover that in producing atoms with high energy levels, nuclear transformations are followed by chemical effects (the breaking of bonds). This gives birth to the chemistry of excited states, or hot-atom chemistry.

1934 Arnold O. Beckman (1900-), American chemist and inventor, invents the pH meter, which uses electricity to accurately measure a solution's acidity or alkalinity.

1934 Harold Clayton Urey (1893-1981), American chemist, receives the Nobel Prize in Chemistry for his discovery of heavy hydrogen.

1934 Marcus Oliphant, Australian physicist, Paul Harteck, Austrian chemist, and Ernest Rutherford (1871-1937), English physicist, discover tritium, one of the heavy isotopes of hydrogen.

1935 Frédéric Joliot-Curie (1900-1958) and his wife Iréne Joliot-Curie (1897-1956), both French physicists, are awarded the Nobel Prize in Chemistry for their synthesis of new radioactive elements.

1935 John Howard Northrop (1891-1987), American biochemist, crystallizes chymotrypsin, a protein-splitting digestive enzyme of the pancreatic secretions.

1936 Peter Joseph William Debye (1884-1966), Dutch-American physical chemist, receives the Nobel Prize in Chemistry for his contributions to our knowledge of molecular structure through his investigations on dipole moments and on the diffraction of x rays and electrons in gases.

1937 Emilio Segre (1905-1989), Italian-American physicist, and Carlo Perrier bombard molybdenum with deuterons and neutrons to produce element 43, or technetium. This is the first element that does not exist in nature.

1937 William Thomas Astbury (1898-1961), English physicist, first obtains information about the structure of nucleic acids by means of x-ray diffraction.

1937 Michael Sveda, American chemist, accidentally discovers a sweet-tasting chemical that is eventually called cyclamate.

1937 Walter Norman Haworth (1883-1950), English chemist, receives the Nobel Prize in Chemistry for his investigations on carbohydrates and vitamin C. Paul Karrer (1889-1971), Swiss chemist, also receives the Nobel Prize in Chemistry for his investigations on carotenoids, flavins, and vitamins A and B_2.

1938 Nylon is first manufactured by DuPont. It is marketed in 1939.

1938 Richard Kuhn (1900-1967), Austrian-German chemist, first isolates vitamin B_6 (pyrodoxine) from skim milk.

1938 Richard Kuhn (1900-1967), Austrian-German chemist, receives the Nobel Prize in Chemistry for his work on carotenoids and vitamins.

1938 Teflon (tetrofluoroethylene) is accidentally discovered in the residue of refrigerant gases by Roy J. Plunkett, American chemist at the labs of the DuPont Company. It proves insoluble in everything, completely water-resistant, a good electrical insulator, heat-resistant, and, most importantly, fat will not stick to it.

1939 Linus Carl Pauling (1901-1994), American chemist, publishes *The Nature of the Chemical Bond*, a classic work that becomes one of the most influential chemical texts of the twentieth century.

1939 Marguerite Perey (1909-1975), French chemist, first isolates element number 87 from among the breakdown products of uranium. She names it francium, after her country.

1939 Adolf Friedrich Johann Butenandt (1903-1994), German chemist, receives the Nobel Prize in Chemistry for his work on sex hormones. Leopold Stephen Ružička (1887-1976), Croatian-Swiss chemist, also receives the Nobel Prize in Chemistry for his work on polymethylenes and higher terpenes.

1939 Paul Hermann Müller (1899-1965), Swiss chemist, discovers the insect-killing properties of DDT (dichlorodiphenyltrichloroethane). It is used during WW II to kill disease-carrying lice, fleas, and mosquitos and after the war, to kill agricultural pests. It is later proved to be a harmful environmental pollutant and its use in the U.S. is banned in 1972.

1940 Dale Raymond Corson, American physicist, K. R. McKenzie, and Emilio Segre (1905-1989), Italian-American physicist, artificially prepare element 85, astatine.

1940 Edwin Mattison McMillan (1907-1991), American physicist, and Philip Hauge Abelson (1913-), American physical chemist, prepare the first of the transuranium (a higher atomic number than uranium) elements, neptunium, element 93.

1940 Vincent du Vigneaud (1901-1978), American biochemist, identifies a compound called biotin as being what previously had been known as vitamin H.

1940 Martin David Kamen (1913-), Canadian-American biochemist, discovers the radioactive isotope carbon-14. With a half-life of 5,700 years, it becomes highly useful for biochemical and archaeological applications, such as dating ancient artifacts and sites.

1941 R. Sherr, Kenneth Thompson Bainbridge, American physicist, and H. H. Anderson produce artificial gold from mercury.

1941 Glenn Theodore Seaborg (1912-1999), American physicist, and his colleagues prepare the transuranium element 94, plutonium.

1941 Arnold O. Beckman (1900-), American physicist and inventor, invents the spectrophotometer. This instrument measures light at the electron level and can be used for many kinds of chemical analysis.

1942 Frank Harold Spedding (1902-1984), American physicist, develops the necessary methods to produce pure uranium in very large quantities for the U.S. atomic bomb effort. Spedding's laboratory produces two tons in November, to be used for the first "atomic pile."

1943 Lars Onsager (1903-1976), Norwegian-American chemist, works out the theoretical basis for the gaseous-diffusion method of separating uranium-235 from the more common uranium-238. This is essential for producing a nuclear bomb or nuclear power.

1943 Georg von Hevesy (1885-1966), Hungarian-Swedish chemist, receives the Nobel Prize in Chemistry for his work on the use of isotopes as tracers in the study of chemical processes.

1944 Glenn Theodore Seaborg (1912-1999), American physicist, and colleagues prepare the transuranium elements americum (number 95) and curium (number 96).

1944 Robert Burns Woodward (1917-1979), American chemist, first synthesizes quinine.

1944 Archer John Porter Martin (1910-) and Richard Laurence Millington Synge (1914-1994), both English biochemists, first develop paper chromatography. Their method of using porous filter paper to separate and identify the nearly identical but different types of amino acids proves an instant success.

1944 Otto Hahn (1879-1968), German physical chemist, receives the Nobel Prize in Chemistry for his discovery of nuclear fission.

1948 Germanium crystals are used by the Bell Telephone Company in the United States to build the first transistors.

1948 Arne Wilhelm Kaurin Tiselius (1902-1971), Swedish chemist, receives the Nobel Prize in Chemistry for his research on electrophoresis and absorption analysis, especially for his discoveries concerning the complex nature of the serum proteins.

1949 Glenn Theodore Seaborg (1912-1999), American physicist, and colleagues prepare the transuranium element berkelium (number 97).

1949 Dorothy Crowfoot Hodgkin (1910-1994), English biochemist, is the first to use an electronic computer in direct application to a biochemical problem. She uses a computer to work out x-ray data in her studies of the structure of penicillin.

1949 William Francis Giauque (1895-1982), American chemist, receives the Nobel Prize in Chemistry for his contributions in the field of chemical thermodynamics, particularly concerning the behavior of substances at extremely low temperatures.

1950 Glenn Theodore Seaborg (1912-1999), American physicist, and colleagues artificially prepare element 98, californium.

1950 Paul Karrer (1889-1971), Swiss chemist, and H. H. Inhoffen achieve the first total synthesis of carotenids.

1950 DuPont Company markets Orlon, the first acrylic fiber.

1950 Otto Paul Hermann Diels (1876-1954) and Kurt Adler (1902-1958), both German chemists, receive the Nobel Prize in Chemistry for their discovery and development of the diene synthesis (a technique of atomic combination that can be put to use in many different kinds of syntheses).

1950 The artificial sweetener cyclamate is first introduced to the market by Abbot Laboratories under the name Sucaryl. It does not have saccharin's aftertaste and becomes essential to the low-calorie soft drink industry.

1951 Edwin Mattison McMillan (1907-1991) and Glenn Theodore Seaborg (1912-1999), both American physicists, receive the Nobel Prize in Chemistry for their discoveries in the chemistry of the transuranium elements.

1952 Archer John Porter Martin (1910-), English biochemist, and A. T. James perfect gas chromatography, or vapor-phase chromatography. It is a variant of partition chromatography, and permits the separation and identification of mixture component.

1952 Archer John Porter Martin (1910-) and Richard Laurence Millington Synge (1914-1994), both English chemists, receive the Nobel Prize in Chemistry for their invention of partition chromatography. Martin protests that he should share his honor with his colleague A. T. James, rather than with Synge, and he gives James half his prize money.

1953 Karl Ziegler (1898-1973), German chemist, discovers a resin catalyst for the production of polyethylene. This produces a new, tougher product that has a much higher melting point than its old counterpart.

1953 Giulio Natta (1903-1979), Italian chemist, borrows Ziegler's 1953 idea of a new catalyst for polymers, and develops the first "isotactic" polymers.

1953 Hermann Staudinger (1881-1965), German chemist, receives the Nobel Prize in Chemistry for his discoveries in the field of macromolecular chemistry.

1954 Glenn Theodore Seaborg (1912-1999), American physicist, and colleagues artificially prepare einsteinium (element 99) and fermium (element 100).

1954 Linus Carl Pauling (1901-1994), American chemist, receives the Nobel Prize in Chemistry for his research into the nature of the chemical bond and its applications to the elucidation of the structure of complex substances.

1955 Vincent Du Vigneaud (1901-1978), American biochemist, receives the Nobel Prize in Chemistry for his work on biochemically important sulfur compounds, especially for the first synthesis of a polypeptide hormone.

1955 First synthetic diamonds are produces in the General Electric Laboratories.

1956 Cyril Norman Hinshelwood (1897-1967), English physical chemist, and Nikolay Nikolayevich Semenov (1896-1986), Russian physical chemist, receive the Nobel Prize in Chemistry for their research into the mechanism of chemical reaction.

1957 Alexander Robertus Todd (1907-1997), Scottish chemist, receives the Nobel Prize for his work on nucleotides and nucleotide coenzymes.

1958 Element 101 is artificially produced by Glenn Theodore Seaborg (1912-1999), American physicist, and colleagues, and is named mendelevium, after the originator of the Periodic Table. They also produce nobelium (element 102).

1958 Frederick Sanger (1918-), English biochemist, receives the Nobel Prize in Chemistry for his work on the structure of proteins, especially that of insulin.

1959 Jaroslav Heyrovský (1890-1967), Czech physical chemist, receives the Nobel Prize in Chemistry for his discovery and development of the polarographic methods of analysis.

1960 Chlorophyll is synthesized from man-made materials.

1960 Lyman Creighton Craig (1906-), American chemist, isolates and purifies "parathormone," the active principle of the parathyroid gland.

1960 Willard Frank Libby (1908-1980), American chemist, receives the Nobel Prize in Chemistry for his method to use carbon-14 for age determination in archeology and other branches of science.

1961 Albert Ghiorso (1915-), American physicist, and colleagues produce element 103 and lawrencium is added to the Periodic Table.

1961 Melvin Calvin (1911-1997), American biochemist, receives the Nobel Prize in Chemistry for his research on carbon dioxide assimilation in plants.

1962 Isotope carbon-12 is adopted as the official standard for atomic weights.

1962 Neil Bartlett (1932-), English chemist, successfully demonstrates that the noble gases are not incapable of reacting chemically when he forms a new noble gas compound called xenon platinofluoride.

1962 Max Ferdinand Perutz (1914-), Austrian-English biochemist, and John Cowdery Kendrew (1917-1997), English biochemist, receive the Nobel Prize in Chemistry for their studies of the structures of globular proteins.

1963 Karl Ziegler (1898-1973), German chemist, and Giulio Natta (1903-1979), receive the Nobel Prize in Chemistry for their discoveries in the field of the chemistry and technology of high polymers.

1964 Researchers at the Joint Nuclear Research Institute in Dubna, Russia, report they produced an isotope of element 104 by bombarding plutonium-242 with neon-22 nuclei. They propose the name kurchatovium, in honor of the Soviet nuclear physicist Igor Kurchatov (1903-1960). Element 104 is the first transactinide element, or the first to begin a new, rare-earth-type row in the Periodic Table. It is ultimately named rutherfordium in honor of Ernest Rutherford (1871-1937).

1964 Robert Burns Woodward (1917-1979), American

chemist, receives the Nobel Prize in Chemistry for his outstanding achievements in the art of organic synthesis.

1965 James M. Schlatter, American chemist, combines two amino acids and obtains a sweet-tasting substance. This chemical is about 200 times sweeter than sugar and is named aspartame. In 1983, it is approved for use in carbonated beverages. It becomes the most widely used artificial sweetener.

1966 Robert Sanderson Mulliken (1896-1986), American chemist, receives the Nobel Prize in Chemistry for his fundamental work concerning chemical bonds and the electronic structure of molecules by the molecular orbital method.

1967 Discovery of element 105 is first reported in the U.S. and it is named hahnium. This artificially produced tranferium element becomes a matter of dispute with Russian scientists who claim to have created it first. The Russian claim is verified and the element is ultimately named dubnium.

1967 Manfred Eigen (1927-), German physicist, Ronald George Wreyford Norrish (1897-1978), English chemist, and George Porter (1920-), English chemist, receive the Nobel Prize in Chemistry for their studies of extremely fast chemical reactions, effected by disturbing the equilibrium by means of very short pulses of energy.

1968 Lars Onsager (1903-1976), Norwegian-American chemist, receives the Nobel Prize in Chemistry for the discovery of the reciprocal relations bearing his name, which are fundamental for the thermodynamics of irreversible processes.

1969 Derek Harold Richard Barton (1918-1998), English chemist, and Odd Hassel (1897-1981), Norwegian physical chemist, receive the Nobel Prize in Chemistry for their contribution to the development of the concept of conformation and its application in chemistry.

1970 Luis Frederico Leloir (1906-1987), Argentine chemist, receives the Nobel Prize in Chemistry for his discovery of sugar nucleotides and their role in the biosynthesis of carbohydrates.

1971 Gerhard Herzberg (1904-1999), German-Canadian physicist, receives the Nobel Prize in Chemistry for his contributions to the knowledge of electronic structure and geometry of molecules, particularly free radicals.

1972 Vitamin B_{12} is first synthesized by American chemist Robert Burns Woodward (1917-1979). Although his process does not prove to be a practical source of this important compound, the research adds greatly to an understanding of complicated compounds.

1972 Gerhard Boehmer Anfinsen (1916-), American biochemist, receives the Nobel Prize in Chemistry for his work on ribonuclease, especially the amino acid sequence and the biologically active confirmation. Stanford Moore (1913-1982) and William Howard Stein (1911-1980), both American biochemists, receive the Nobel Prize in Chemistry for their contribution to the understanding of the connection between chemical structure and catalytic activity of the active center of the ribonuclease molecule.

1973 Ernst Otto Fischer (1918-), German chemist, and Geoffrey Wilkinson (1921-1996), English chemist, receive the Nobel Prize in Chemistry for their pioneering work, performed independently, on the chemistry of the organometallic, or "sandwich," compounds.

1974 Ozone layer damage from chlorofluorocarbons (CFCs) is first pointed out by F. Sherwood Rowland (1927-) and Mario Molina (1943-) of the United States.

1974 Paul J. Flory (1910-1985), American chemist, receives the Nobel Prize in Chemistry for his fundamental achievements, both theoretical and experimental, in the physical chemistry of macromolecules.

1975 John Warcup Cornforth (1917-), Australian-English chemist, receives the Nobel Prize for his work on the stereochemistry of enzyme-catalyzed reaction. Vladimir Prelog (1906-), Croatian-Swiss chemist, also receives the Nobel Prize in Chemistry for his research into the stereochemistry of organic molecules and reactions.

1976 William Nunn Lipscomb, Jr. (1919-), American chemist, receives the Nobel Prize in Chemistry for his studies of the structure of boranes, illuminating problems of chemical bonding.

1977 Ilya Prigogine (1917-), Russian-Belgian physical chemist, receives the Nobel Prize in Chemistry for his contributions to non-equilibrium thermodynamics, particularly the theory of dissipative structures.

1978 Peter Dennis Mitchell (1920-1992), English chemist, receives the Nobel Prize in Chemistry for his contribution to the understanding of biological energy transfer through the formulation of the chemiosmotic theory.

1979 Herbert Charles Brown (1912-), English-American chemist, and George Friedrich Wittig (1897-1987), German chemist, receive the Nobel Prize in Chemistry for their development of the use of boron- and phosphorous-containing compounds, respectively, into important reagents in organic synthesis.

1980 Paul Berg (1926-), American biochemist, receives

the Nobel Prize in Chemistry for his fundamental studies of the biochemistry of nucleic acids, with particular regard to recombinant DNA. Walter Gilbert (1932-), American microbiologist, and Frederick Sanger (1918-), English biochemist, also receive the Nobel Prize in Chemistry for their contributions concerning the determination of base sequences in nucleic acids.

1981 Kenichi Fukui (1918-1997), Japanese chemist, and Roald Hoffmann (1937-), Polish-American chemist, receive the Nobel Prize in Chemistry for their theories, developed independently, concerning the course of chemical reactions.

1982 Aaron Kluge (1926-) receives the Nobel Prize in Chemistry for his development of crystallographic electron microscopy and his structural elucidation of biologically important nucleic acid-protein complexes.

1985 The first fullerene (carbon 60) is discovered by American chemist Robert F. Curl, Jr. (1933-), English chemist Harold W. Kroto (1939-), and American physicist Richard E. Smalley (1943-). This molecule of carbon, popularly known as a "buckyball," is a cage-like molecule made up of 60 carbon atoms arranged in the shape of a soccer ball.

1985 Herbert Aaron Hauptman (1917-), American mathematician and crystallographer, and Jerome Karle (1918-), American crystallographer, receive the Nobel Prize in Chemistry for their outstanding achievements in the development of direct methods for the determination of crystal structures.

1986 Dudley Robert Herschbach (1932-), American chemist, Yuan Tseh Lee 1936-), Chinese-American chemist, and John Charles Polanyi (1929-), German-Canadian chemist, receive the Nobel Prize in Chemistry for their contributions concerning the dynamics of chemical elementary processes.

1987 Donald James Cram (1919-), American chemist, Jean-Marie Lehn (1939-), French chemist, and Charles John Pedersen (1904-1989), American chemist, receive the Nobel Prize in Chemistry for their development and use of molecules with structure-specific interactions of high selectivity.

1988 Johann Deisenhofer (1943-), Robert Huber (1937-), and Hartmut Michel (1948-), German biochemists, receive the Nobel Prize in Chemistry for the determination of the three-dimensional structure of a photosynthetic reaction center.

1989 Sidney Altman (1939-), Canadian-American molecular biologist, and Thomas Robert Cech (1947-), American biochemist and molecular biologist, receive the Nobel Prize in Chemistry for their discovery of catalytic properties of DNA.

1990 Elias James Corey (1928-), American chemist, receives the Nobel Prize in Chemistry for his development of the theory and methodology of organic synthesis.

1991 Researchers obtain the first evidence of medium-range structural order in glass. Using a sensitive neutron-scattering technique, they study the atomic structure of calcium silicate glass and discover calcium ions distributed in a manner similar to crystals. This discovery may lead to ways for optimizing the mechanical, optical, and other properties of glass by modifying this newly discovered structure.

1991 Arthur J. Epstein and Joel S. Miller, American chemists, develop the first molecular magnet that retains its magnetic properties at room temperatures and above.

1991 Andrew A. Griffith, American chemist, uses an atomic force microscope to obtain extraordinarily detailed images of the electrochemical reactions involved in corrosion.

1991 John V. Budding, American chemist, leads a group that compresses iron and hydrogen in a diamond cell, making the first observations of the chemical combination of iron and hydrogen under ultrahigh pressures to form iron hydride.

1991 Numerous papers on fullerenes are published, reflecting a significant increase of scientific interest in this substance. This unusual family of ball-shaped carbon molecules are hollow clusters of carbon atoms bonded into geodesic structures that resembles the geodesic domes popularized by the American engineer and architect R. Buckminster Fuller (1895-1983). Initial research indicates that fullerenes may contribute to the advance of high-temperature superconductors.

1991 Richard Robert Ernst (1933-), Swiss chemist, receives the Nobel Prize in Chemistry for his contributions in refining the technology of nuclear magnetic resonance spectroscopy.

1992 Researchers in the U.S. report the production of the first solid compound of the noble gas helium. A solid is formed when helium and nitrogen are mixed together and subjected to about 77,000 times normal pressure.

1992 Rudolph A. Marcus (1923-), Canadian American chemist, receives the Nobel Prize in Chemistry for his contributions to the theory of electron transfer reactions in chemical systems.

1993 Swiss researchers report making a mercury-containing ceramic material that starts to become superconducting (losing all resistance to the flow of electricity) when cooled to about 133 K.

1993 Joseph B. Lambert, American chemist, heads a team that reports success in making stable silicon (Si) cation, a positively charged form of the element that is attached to other atoms by three bonds. This may have important applications as a catalyst in speeding up polymerization reactions used in making adhesives, lubricants, and other silicon products.

1993 Jacob Israelachvili, American chemist, and co-workers report the first experimental observation of a state of ultra-low friction. They observe this phenomenon during studies of the co-called stick-slip motion of specially treated mica surfaces.

1993 Kary B. Mullis (1944-), American chemist, recives the Nobel Prize in Chemistry for his invention of the polymerase chain reaction method. Michael Smith (1932-), English Canadian biochemist, also receives the Nobel Prize for his discovery of site-directed mutagenesis (how to make a genetic mutation precisely at any spot in a DNA molecule).

1994 German researchers create an isotope of element 110 by bombarding a lead isotope with nickel atoms. With a mass of 270, this is the heaviest known element.

1994 Chemist Kyriacou C. Nicolau of the Scripps Research Institute in La Jolla, California, and colleagues, synthesize taxol, the powerful anti-cancer compound derived from the yew tree.

1994 Stephen B. H. Kent, biochemist at the Scripps Research Institute in La Jolla, California, and co-workers, successfully synthesize a protein. By mimicking natural chemical linking, Kent connects two peptides (one with 33 amino acids, the other with 39) and a copy of interleukin-8, a human immune system protein consisting of 72 amino acids.

1994 George A. Olah (1927-), Hungarian-American chemist, receives the Nobel Prize in Chemistry for his contributions to carbocation chemistry.

1995 Combinatorial chemistry, a technique which quickly surveys huge numbers of chemical combinations in order to select the most desirable molecular configurations, attract the attentions of chemical companies. Scientists predict the possibility of creating numerous new chemicals to serve the needs of industrial and pharmaceutical development.

1995 Paul Crutzen (1933-), Dutch meteorologist, Mario Molina (1943-), Mexican American chemist, and R. Sherwood Rowland (1927-), American atmospheric chemist, receive the Nobel Prize in Chemistry for their work in atmospheric chemistry, particularly concerning the formation and decomposition of ozone.

1996 Harold W. Kroto (1939-), of the University of Sussex, and Rice University's Robert F. Curl, Jr. (1933-) and Richard E. Smalley (1943-), receive the Nobel Prize in Chemistry for their discovery of a new molecular form of carbon. Although initially viewed with scepticism by the scientific community, the 60-atom molecules named fullerenes have found numerous applications in industry.

1996 Jagarlapudi Sarma, of the Indian Institute of Chemical Technology, in Hyderabad, India, determines that organic molecules tend to contain even numbers of atoms. Chemists hope that Sarma's discovery may lead to new insights on the nature of chemical bonding.

1997 After years of experimenting, scientists determine that 106 is the correct atomic number of seaborgium, the heavy element named after chemist Glenn Seaborg (1912-1999). Matthias Schadel, of the heavy ion research center in Darmstadt, Germany, and co-workers, affirms that seaborgium fits neatly in the group of elements which includes chromium, molybdenum, and tungsten.

1997 Researcher Fred Regnier, of Purdue University, develops a computer chip containing miniature chemical laboratories. Capable of performing complex analyses and experiments, these multiple laboratories herald an era of inexpensive and swift chemical research and experimentation.

1997 Paul Boyer (1918-), American biochemist, John E. Walker (1941-), English molecular biologist, and Jens C. Skou (1918-), Danish physiologist, share the Nobel Prize in Chemistry for their ground-breaking work on the enzymes that catalyze ATP (adenosine triphosphate), a cellular substance that fuels a variety of crucial physiological processes.

1998 Researchers determine that one half of every dose of Ritalin, the widely prescribed drug for ADHD (attention-deficit-hyperactivity disorder), has no therapeutic effect. Ritalin, as do many drugs, consists of two molecular forms of a compound, each structurally mirroring the other. Yu-Shin Ding, Joanna S. Fowler, and Nora Volkow, of the Brookhaven National Laboratory in Upton, New York, find that only one form of the pair is therapeutically significant, suggesting the desirability of using the drug in a purer form.

1998 Gali Steinberg-Yfrach, Edgardo N. Durantini, Anna L. More, Devens Gust, and Thomas A. More, researchers at Arizona State University in Tempe, synthesize a membrane capable of mimicking the entire process of bacterial photosynthesis.

1998 Theoretical chemist John A. Pople (1925-) and physicist Walter Kohn (1923-) receive the Nobel Prize in Chemistry for pioneering work in computational chemistry. Dr. Pople developed computational methods in quantum chemistry and Dr. Kohn developed density-functional theory.

1998 Researcher Katsuya Shimizu, of Osaka University, and co-workers, discover that oxygen acts as a superconductor if chilled to a temperature close to absolute zero.

1999 Ron M. A. Heeren, of the FOM Institute for Atomic and Molecular Physics, and co-workers determine the precise chemical composition of the paint, glaze, and varnish used by Rembrandt in painting his 1632 masterpiece, "The Anatomy Lesson of Dr. Nicolaes Tulp." This chemical analysis furnishes a restoration team with priceless information about the painting.

1999 Scientists at the Lawrence Berkeley Laboratory at the University of California announce the discovery of two new transfermium elements, numbers 118 and 116. The elements were formed by the collision of krypton nuclei with lead nuclei. It is actually element 118 that is formed in this reaction, but it rapidly decays to form element 116.

1999 Ahmed H. Zewail of the California Institute of Technology receives the Nobel Prize in Chemistry for demonstrating that a rapid laser technique can observe the motions of atoms in a molecule as they occur during a chemical reation.

General Index

A

B

C

D

E

F

G

H

I

J

K

M

N

O

P

Q

R

S

T

U

V

U

V

W

X

Y

Z